App. Alt.	OCT.—MAR. SUN Lower Limb	Upper Limb	APR.—SEPT. Lower Limb	Upper Limb	STARS PLANETS
° ′	′	′	′	′	′
0 00	−17·5	−49·8	−17·8	−49·6	−33·8
0 03	16·9	49·2	17·2	49·0	33·2
0 06	16·3	48·6	16·6	48·4	32·6
0 09	15·7	48·0	16·0	47·8	32·0
0 12	15·2	47·5	15·4	47·2	31·5
0 15	14·6	46·9	14·8	46·6	30·9
0 18	−14·1	−46·4	−14·3	−46·1	−30·4
0 21	13·5	45·8	13·8	45·6	29·8
0 24	13·0	45·3	13·3	45·1	29·3
0 27	12·5	44·8	12·8	44·6	28·8
0 30	12·0	44·3	12·3	44·1	28·3
0 33	11·6	43·9	11·8	43·6	27·9
0 36	−11·1	−43·4	−11·3	−43·1	−27·4
0 39	10·6	42·9	10·9	42·7	26·9
0 42	10·2	42·5	10·5	42·3	26·5
0 45	9·8	42·1	10·0	41·8	26·1
0 48	9·4	41·7	9·6	41·4	25·7
0 51	9·0	41·3	9·2	41·0	25·3
0 54	− 8·6	−40·9	− 8·8	−40·6	−24·9
0 57	8·2	40·5	8·4	40·2	24·5
1 00	7·8	40·1	8·0	39·8	24·1
1 03	7·4	39·7	7·7	39·5	23·7
1 06	7·1	39·4	7·3	39·1	23·4
1 09	6·7	39·0	7·0	38·8	23·0
1 12	− 6·4	−38·7	− 6·6	−38·4	−22·7
1 15	6·0	38·3	6·3	38·1	22·3
1 18	5·7	38·0	6·0	37·8	22·0
1 21	5·4	37·7	5·7	37·5	21·7
1 24	5·1	37·4	5·3	37·1	21·4
1 27	4·8	37·1	5·0	36·8	21·1
1 30	− 4·5	−36·8	− 4·7	−36·5	−20·8
1 35	4·0	36·3	4·3	36·1	20·3
1 40	3·6	35·9	3·8	35·6	19·9
1 45	3·1	35·4	3·4	35·2	19·4
1 50	2·7	35·0	2·9	34·7	19·0
1 55	2·3	34·6	2·5	34·3	18·6
2 00	− 1·9	−34·2	− 2·1	−33·9	−18·2
2 05	1·5	33·8	1·7	33·5	17·8
2 10	1·1	33·4	1·4	33·2	17·4
2 15	0·8	33·1	1·0	32·8	17·1
2 20	0·4	32·7	0·7	32·5	16·7
2 25	− 0·1	32·4	− 0·3	32·1	16·4
2 30	+ 0·2	−32·1	0·0	−31·8	−16·1
2 35	0·5	31·8	+ 0·3	31·5	15·8
2 40	0·8	31·5	0·6	31·2	15·4
2 45	1·1	31·2	0·9	30·9	15·2
2 50	1·4	30·9	1·2	30·6	14·9
2 55	1·7	30·6	1·4	30·4	14·6
3 00	+ 2·0	−30·3	+ 1·7	−30·1	−14·3
3 05	2·2	30·1	2·0	29·8	14·1
3 10	2·5	29·8	2·2	29·6	13·8
3 15	2·7	29·6	2·5	29·3	13·6
3 20	2·9	29·4	2·7	29·1	13·4
3 25	3·2	29·1	2·9	28·9	13·1
3 30	+ 3·4	−28·9	+ 3·1	−28·7	−12·9

					S
	′	′	′	′	′
3 30	+ 3·4	−28·9	+ 3·1	−28·7	−12·9
3 35	3·6	28·7	3·3	28·5	12·7
3 40	3·8	28·5	3·6	28·2	12·5
3 45	4·0	28·3	3·8	28·0	12·3
3 50	4·2	28·1	4·0	27·8	12·1
3 55	4·4	27·9	4·1	27·7	11·9
4 00	+ 4·6	−27·7	+ 4·3	−27·5	−11·7
4 05	4·8	27·5	4·5	27·3	11·5
4 10	4·9	27·4	4·7	27·1	11·4
4 15	5·1	27·2	4·9	26·9	11·2
4 20	5·3	27·0	5·0	26·8	11·0
4 25	5·4	26·9	5·2	26·6	10·9
4 30	+ 5·6	−26·7	+ 5·3	−26·5	−10·7
4 35	5·7	26·6	5·5	26·3	10·6
4 40	5·9	26·4	5·6	26·2	10·4
4 45	6·0	26·3	5·8	26·0	10·3
4 50	6·2	26·1	5·9	25·9	10·1
4 55	6·3	26·0	6·1	25·7	10·0
5 00	+ 6·4	−25·9	+ 6·2	−25·6	− 9·8
5 05	6·6	25·7	6·3	25·5	9·7
5 10	6·7	25·6	6·5	25·3	9·6
5 15	6·8	25·5	6·6	25·2	9·5
5 20	7·0	25·3	6·7	25·1	9·3
5 25	7·1	25·2	6·8	25·0	9·2
5 30	+ 7·2	−25·1	+ 6·9	−24·9	− 9·1
5 35	7·3	25·0	7·1	24·7	9·0
5 40	7·4	24·9	7·2	24·6	8·9
5 45	7·5	24·8	7·3	24·5	8·8
5 50	7·6	24·7	7·4	24·4	8·7
5 55	7·7	24·6	7·5	24·3	8·6
6 00	+ 7·8	−24·5	+ 7·6	−24·2	− 8·5
6 10	8·0	24·3	7·8	24·0	8·3
6 20	8·2	24·1	8·0	23·8	8·1
6 30	8·4	23·9	8·2	23·6	7·9
6 40	8·6	23·7	8·3	23·5	7·7
6 50	8·7	23·6	8·5	23·3	7·6
7 00	+ 8·9	−23·4	+ 8·7	−23·1	− 7·4
7 10	9·1	23·2	8·8	23·0	7·2
7 20	9·2	23·1	9·0	22·8	7·1
7 30	9·3	23·0	9·1	22·7	6·9
7 40	9·5	22·8	9·2	22·6	6·8
7 50	9·6	22·7	9·4	22·4	6·7
8 00	+ 9·7	−22·6	+ 9·5	−22·3	− 6·6
8 10	9·9	22·4	9·6	22·2	6·4
8 20	10·0	22·3	9·7	22·1	6·3
8 30	10·1	22·2	9·9	21·9	6·2
8 40	10·2	22·1	10·0	21·8	6·1
8 50	10·3	22·0	10·1	21·7	6·0
9 00	+10·4	−21·9	+10·2	−21·6	− 5·9
9 10	10·5	21·8	10·3	21·5	5·8
9 20	10·6	21·7	10·4	21·4	5·7
9 30	10·7	21·6	10·5	21·3	5·6
9 40	10·8	21·5	10·6	21·2	5·5
9 50	10·9	21·4	10·6	21·2	5·4
10 00	+11·0	−21·3	+10·7	−21·1	− 5·3

Additional corrections for temperature and pressure are given on the following page.

For bubble sextant observations ignore dip and use the star corrections for Sun, planets and stars.

A4 ALTITUDE CORRECTION TABLES—ADDITIONAL CORRECTIONS

ADDITIONAL REFRACTION CORRECTIONS FOR NON-STANDARD CONDITIONS

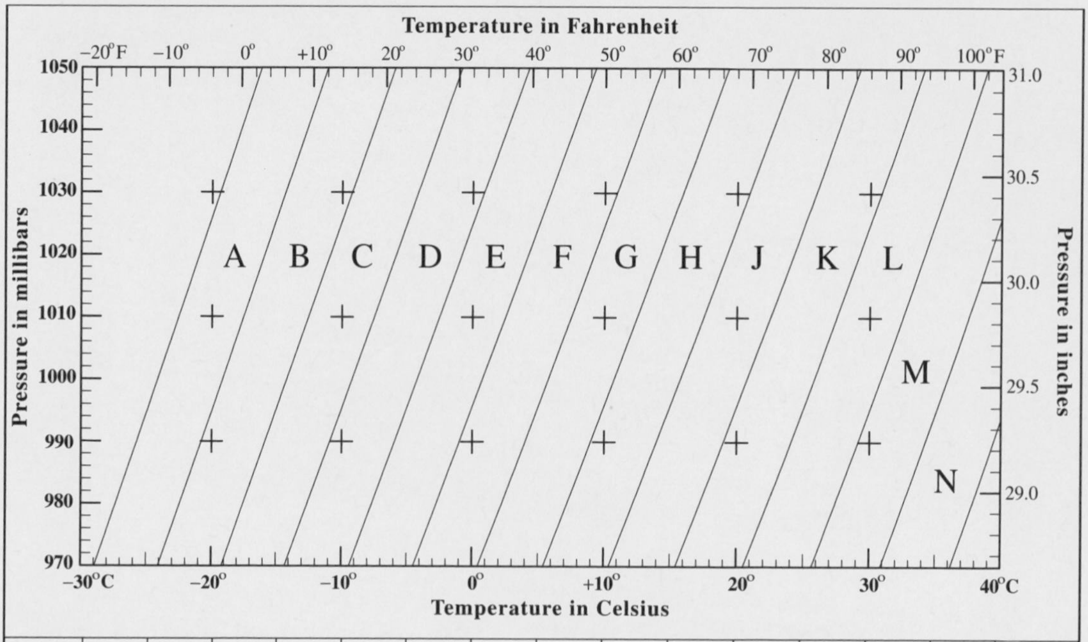

Temperature in Fahrenheit

−20°F −10° 0° +10° 20° 30° 40° 50° 60° 70° 80° 90° 100°F

Pressure in millibars: 1050 1040 1030 1020 1010 1000 990 980 970

Pressure in inches: 31.0 30.5 30.0 29.5 29.0

A B C D E F G H J K L M N

−30°C −20° −10° 0° +10° 20° 30° 40°C

Temperature in Celsius

App. Alt.	A	B	C	D	E	F	G	H	J	K	L	M	N	P	App. Alt.
° ′	′	′	′	′	′	′	′	′	′	′	′	′	′	′	° ′
00 00	−7·3	−5·9	−4·6	−3·4	−2·2	−1·1	0·0	+1·0	+2·0	+3·0	+4·0	+4·9	+5·9	+6·9	**00 00**
00 30	5·5	4·5	3·5	2·6	1·7	0·8	0·0	0·8	1·6	2·3	3·1	3·8	4·5	5·3	**00 30**
01 00	4·4	3·5	2·8	2·0	1·3	0·7	0·0	0·6	1·2	1·8	2·4	3·0	3·6	4·2	**01 00**
01 30	3·5	2·9	2·2	1·7	1·1	0·5	0·0	0·5	1·0	1·5	2·0	2·5	2·9	3·4	**01 30**
02 00	2·9	2·4	1·9	1·4	0·9	0·4	0·0	0·4	0·8	1·3	1·7	2·0	2·4	2·8	**02 00**
02 30	−2·5	−2·0	−1·6	−1·2	−0·8	−0·4	0·0	+0·4	+0·7	+1·1	+1·4	+1·7	+2·1	+2·4	**02 30**
03 00	2·1	1·7	1·4	1·0	0·7	0·3	0·0	0·3	0·6	0·9	1·2	1·5	1·8	2·1	**03 00**
03 30	1·9	1·5	1·2	0·9	0·6	0·3	0·0	0·3	0·5	0·8	1·1	1·3	1·6	1·8	**03 30**
04 00	1·6	1·3	1·1	0·8	0·5	0·3	0·0	0·2	0·5	0·7	0·9	1·2	1·4	1·6	**04 00**
04 30	1·5	1·2	0·9	0·7	0·5	0·2	0·0	0·2	0·4	0·6	0·8	1·0	1·3	1·5	**04 30**
05 00	−1·3	−1·1	−0·9	−0·6	−0·4	−0·2	0·0	+0·2	+0·4	+0·6	+0·8	+0·9	+1·1	+1·3	**05 00**
06	1·1	0·9	0·7	0·5	0·3	0·2	0·0	0·2	0·3	0·5	0·6	0·8	0·9	1·1	**06**
07	1·0	0·8	0·6	0·5	0·3	0·1	0·0	0·1	0·3	0·4	0·5	0·7	0·8	0·9	**07**
08	0·8	0·7	0·5	0·4	0·3	0·1	0·0	0·1	0·2	0·4	0·5	0·6	0·7	0·8	**08**
09	0·7	0·6	0·5	0·4	0·2	0·1	0·0	0·1	0·2	0·3	0·4	0·5	0·6	0·7	**09**
10 00	−0·7	−0·5	−0·4	−0·3	−0·2	−0·1	0·0	+0·1	+0·2	+0·3	+0·4	+0·5	+0·6	+0·7	**10 00**
12	0·6	0·5	0·4	0·3	0·2	0·1	0·0	0·1	0·2	0·2	0·3	0·4	0·5	0·5	**12**
14	0·5	0·4	0·3	0·2	0·1	0·1	0·0	0·1	0·1	0·2	0·3	0·3	0·4	0·5	**14**
16	0·4	0·3	0·3	0·2	0·1	0·1	0·0	0·1	0·1	0·2	0·2	0·3	0·3	0·4	**16**
18	0·4	0·3	0·2	0·2	0·1	−0·1	0·0	+0·1	0·1	0·2	0·2	0·3	0·3	0·4	**18**
20 00	−0·3	−0·3	−0·2	−0·2	−0·1	0·0	0·0	0·0	+0·1	+0·1	+0·2	+0·2	+0·3	+0·3	**20 00**
25	0·3	0·2	0·2	0·1	0·1	0·0	0·0	0·0	0·1	0·1	0·1	0·2	0·2	0·2	**25**
30	0·2	0·2	0·1	0·1	0·1	0·0	0·0	0·0	+0·1	0·1	0·1	0·1	0·2	0·2	**30**
35	0·2	0·1	0·1	0·1	−0·1	0·0	0·0	0·0	0·0	0·1	0·1	0·1	0·1	0·2	**35**
40	0·1	0·1	0·1	−0·1	0·0	0·0	0·0	0·0	0·0	+0·1	0·1	0·1	0·1	0·1	**40**
50 00	−0·1	−0·1	−0·1	0·0	0·0	0·0	0·0	0·0	0·0	0·0	+0·1	+0·1	+0·1	+0·1	**50 00**

The graph is entered with arguments temperature and pressure to find a zone letter; using as arguments this zone letter and apparent altitude (sextant altitude corrected for index error and dip), a correction is taken from the table. This correction is to be applied to the sextant altitude in addition to the corrections for standard conditions (for the Sun, stars and planets from page A2-A3 and for the Moon from pages xxxiv and xxxv).

THE NAUTICAL ALMANAC

FOR THE YEAR

2009

WASHINGTON:
Issued by the
Nautical Almanac Office
United States Naval Observatory
under the authority of the
Secretary of the Navy

TAUNTON:
Issued by
Her Majesty's
Nautical Almanac Office
United Kingdom
Hydrographic Office

U.S. GOVERNMENT PRINTING OFFICE
WASHINGTON: 2008

UNITED STATES

Washington: 2008

This is an official U.S. Government edition of this publication and is herein identified to certify its authenticity. Use of the 0-16 ISBN prefix is for U.S. Government Printing Office Official Editions only. The Superintendent of Documents of the U.S. Government Printing Office requests that any reprinted edition clearly be labeled as a copy of the authentic work with a new ISBN. See below for additional copyright information prior to reprinting.

For sale by the Superintendent of Documents, U.S. Government Printing Office Internet: bookstore.gpo.gov Phone: toll free (866) 512-1800; DC area (202) 512-1800 Fax: (202) 512-2250 Mail: Stop IDCC, Washington, DC 20402-0001

PREFACE

The British and American editions of *The Nautical Almanac*, which are identical in content, are produced jointly by H. M. Nautical Almanac Office, part of the UK Hydrographic Office's Maritime Safety Publications Branch, under the supervision of S. A. Bell and C. Y. Hohenkerk, and by the Nautical Almanac Office, United States Naval Observatory, under the supervision of S. E. Urban and M. T. Stollberg, to the general requirements of the Royal Navy and of the United States Navy. The Almanac is printed separately in the United Kingdom and in the United States of America.

The data in this almanac, on written application, may be made available for reproduction. It can be made available in a form suitable for direct photographic reproduction, to the appropriate almanac-producing agency in any country; language changes in the headings of the ephemeral pages can be introduced, if desired, during reproduction. Under this arrangement, this almanac, with minor modifications and changes of language, has been adopted for the Brazilian, Chilean, Greek, Indian, Indonesian, Italian and Mexican almanacs.

MICHAEL S. ROBINSON
Chief Executive
UK Hydrographic Office
Admiralty Way, Taunton
Somerset, TA1 2DN
United Kingdom

STEVEN W. WARREN
Captain, U.S. Navy
Superintendent, U.S. Naval Observatory
3450 Massachusetts Avenue NW
Washington, D.C. 20392-5420
U.S.A.

December 2007

For sale by the Superintendent of Documents, U.S. Government Printing Office
Internet: bookstore.gpo.gov Phone: toll free (866) 512-1800; DC area (202) 512-1800
Fax: (202) 512-2104 Mail: Stop IDCC, Washington, DC 20402-0001

ISBN 978-0-16-080195-2

RELIGIOUS CALENDARS

Epiphany	Jan. 6	Low Sunday	Apr. 19
Septuagesima Sunday	Feb. 8	Rogation Sunday	May 17
Quinquagesima Sunday	Feb. 22	Ascension Day—Holy Thursday	May 21
Ash Wednesday	Feb. 25	Whit Sunday—Pentecost	May 31
Quadragesima Sunday	Mar. 1	Trinity Sunday	June 7
Palm Sunday	Apr. 5	Corpus Christi	June 11
Good Friday	Apr. 10	First Sunday in Advent	Nov. 29
Easter Day	Apr. 12	Christmas Day (Friday)	Dec. 25

First Day of Passover (Pesach)	Apr. 9	Day of Atonement (Yom Kippur)	Sept. 28
Feast of Weeks (Shavuot)	May 29	First day of Tabernacles (Succoth)	Oct. 3
Jewish New Year 5770 (Rosh Hashanah)	Sept. 19		

Ramadân, First day of (tabular)	Aug. 22	Islamic New Year (1431)	Dec. 18

The Jewish and Islamic dates above are tabular dates, which begin at sunset on the previous evening and end at sunset on the date tabulated. In practice, the dates of Islamic fasts and festivals are determined by an actual sighting of the appropriate new moon.

CIVIL CALENDAR—UNITED KINGDOM

Accession of Queen Elizabeth II	Feb. 6	Birthday of Prince Philip, Duke of Edinburgh	June 10
St David (Wales)	Mar. 1	The Queen's Official Birthday†	June 13
Commonwealth Day	Mar. 9	Remembrance Sunday	Nov. 8
St Patrick (Ireland)	Mar. 17	Birthday of the Prince of Wales	Nov. 14
Birthday of Queen Elizabeth II	Apr. 21	St Andrew (Scotland)	Nov. 30
St George (England)	Apr. 23		
Coronation Day	June 2		

PUBLIC HOLIDAYS

England and Wales—Jan. 1†, Apr. 10, Apr. 13, May 4†, May 25, Aug. 31, Dec. 25, Dec. 28†
Northern Ireland—Jan. 1†, Mar. 17, Apr. 10, Apr. 13, May 4†, May 25, July 13, Aug. 31, Dec. 25, Dec. 28†
Scotland—Jan. 1, Jan. 2, Apr. 10, May 4, May 25†, Aug. 3, Dec. 25, Dec. 28†

CIVIL CALENDAR—UNITED STATES OF AMERICA

New Year's Day	Jan. 1	Labor Day	Sept. 7
Martin Luther King's Birthday	Jan. 19	Columbus Day	Oct. 12
Washington's Birthday	Feb. 16	Election Day (in certain States)	Nov. 3
Memorial Day	May 25	Veterans Day	Nov. 11
Independence Day	July 4	Thanksgiving Day	Nov. 26

†Dates subject to confirmation

PHASES OF THE MOON

New Moon	d h m	First Quarter	d h m	Full Moon	d h m	Last Quarter	d h m
		Jan.	4 11 56	Jan.	11 03 27	Jan.	18 02 46
Jan.	26 07 55	Feb.	2 23 13	Feb.	9 14 49	Feb.	16 21 37
Feb.	25 01 35	Mar.	4 07 46	Mar.	11 02 38	Mar.	18 17 47
Mar.	26 16 06	Apr.	2 14 34	Apr.	9 14 56	Apr.	17 13 36
Apr.	25 03 23	May	1 20 44	May	9 04 01	May	17 07 26
May	24 12 11	May	31 03 22	June	7 18 12	June	15 22 15
June	22 19 35	June	29 11 28	July	7 09 21	July	15 09 53
July	22 02 35	July	28 22 00	Aug.	6 00 55	Aug.	13 18 55
Aug.	20 10 02	Aug.	27 11 42	Sept.	4 16 03	Sept.	12 02 16
Sept.	18 18 44	Sept.	26 04 50	Oct.	4 06 10	Oct.	11 08 56
Oct.	18 05 33	Oct.	26 00 42	Nov.	2 19 14	Nov.	9 15 56
Nov.	16 19 14	Nov.	24 21 39	Dec.	2 07 30	Dec.	9 00 13
Dec.	16 12 02	Dec.	24 17 36	Dec.	31 19 13		

DAYS OF THE WEEK AND DAYS OF THE YEAR

	JAN.		FEB.		MAR.		APR.		MAY		JUNE		JULY		AUG.		SEPT.		OCT.		NOV.		DEC.	
Day	Wk	Yr	Wk	Yr	Wk	Yr	Wk	Yr	Wk	Yr	Wk	Yr	Wk	Yr	Wk	Yr	Wk	Yr	Wk	Yr	Wk	Yr	Wk	Yr
1	Th.	1	Su.	32	Su.	60	W.	91	F.	121	M.	152	W.	182	Sa.	213	Tu.	244	Th.	274	Su.	305	Tu.	335
2	F.	2	M.	33	M.	61	Th.	92	Sa.	122	Tu.	153	Th.	183	Su.	214	W.	245	F.	275	M.	306	W.	336
3	Sa.	3	Tu.	34	Tu.	62	F.	93	Su.	123	W.	154	F.	184	M.	215	Th.	246	Sa.	276	Tu.	307	Th.	337
4	Su.	4	W.	35	W.	63	Sa.	94	M.	124	Th.	155	Sa.	185	Tu.	216	F.	247	Su.	277	W.	308	F.	338
5	M.	5	Th.	36	Th.	64	Su.	95	Tu.	125	F.	156	Su.	186	W.	217	Sa.	248	M.	278	Th.	309	Sa.	339
6	Tu.	6	F.	37	F.	65	M.	96	W.	126	Sa.	157	M.	187	Th.	218	Su.	249	Tu.	279	F.	310	Su.	340
7	W.	7	Sa.	38	Sa.	66	Tu.	97	Th.	127	Su.	158	Tu.	188	F.	219	M.	250	W.	280	Sa.	311	M.	341
8	Th.	8	Su.	39	Su.	67	W.	98	F.	128	M.	159	W.	189	Sa.	220	Tu.	251	Th.	281	Su.	312	Tu.	342
9	F.	9	M.	40	M.	68	Th.	99	Sa.	129	Tu.	160	Th.	190	Su.	221	W.	252	F.	282	M.	313	W.	343
10	Sa.	10	Tu.	41	Tu.	69	F.	100	Su.	130	W.	161	F.	191	M.	222	Th.	253	Sa.	283	Tu.	314	Th.	344
11	Su.	11	W.	42	W.	70	Sa.	101	M.	131	Th.	162	Sa.	192	Tu.	223	F.	254	Su.	284	W.	315	F.	345
12	M.	12	Th.	43	Th.	71	Su.	102	Tu.	132	F.	163	Su.	193	W.	224	Sa.	255	M.	285	Th.	316	Sa.	346
13	Tu.	13	F.	44	F.	72	M.	103	W.	133	Sa.	164	M.	194	Th.	225	Su.	256	Tu.	286	F.	317	Su.	347
14	W.	14	Sa.	45	Sa.	73	Tu.	104	Th.	134	Su.	165	Tu.	195	F.	226	M.	257	W.	287	Sa.	318	M.	348
15	Th.	15	Su.	46	Su.	74	W.	105	F.	135	M.	166	W.	196	Sa.	227	Tu.	258	Th.	288	Su.	319	Tu.	349
16	F.	16	M.	47	M.	75	Th.	106	Sa.	136	Tu.	167	Th.	197	Su.	228	W.	259	F.	289	M.	320	W.	350
17	Sa.	17	Tu.	48	Tu.	76	F.	107	Su.	137	W.	168	F.	198	M.	229	Th.	260	Sa.	290	Tu.	321	Th.	351
18	Su.	18	W.	49	W.	77	Sa.	108	M.	138	Th.	169	Sa.	199	Tu.	230	F.	261	Su.	291	W.	322	F.	352
19	M.	19	Th.	50	Th.	78	Su.	109	Tu.	139	F.	170	Su.	200	W.	231	Sa.	262	M.	292	Th.	323	Sa.	353
20	Tu.	20	F.	51	F.	79	M.	110	W.	140	Sa.	171	M.	201	Th.	232	Su.	263	Tu.	293	F.	324	Su.	354
21	W.	21	Sa.	52	Sa.	80	Tu.	111	Th.	141	Su.	172	Tu.	202	F.	233	M.	264	W.	294	Sa.	325	M.	355
22	Th.	22	Su.	53	Su.	81	W.	112	F.	142	M.	173	W.	203	Sa.	234	Tu.	265	Th.	295	Su.	326	Tu.	356
23	F.	23	M.	54	M.	82	Th.	113	Sa.	143	Tu.	174	Th.	204	Su.	235	W.	266	F.	296	M.	327	W.	357
24	Sa.	24	Tu.	55	Tu.	83	F.	114	Su.	144	W.	175	F.	205	M.	236	Th.	267	Sa.	297	Tu.	328	Th.	358
25	Su.	25	W.	56	W.	84	Sa.	115	M.	145	Th.	176	Sa.	206	Tu.	237	F.	268	Su.	298	W.	329	F.	359
26	M.	26	Th.	57	Th.	85	Su.	116	Tu.	146	F.	177	Su.	207	W.	238	Sa.	269	M.	299	Th.	330	Sa.	360
27	Tu.	27	F.	58	F.	86	M.	117	W.	147	Sa.	178	M.	208	Th.	239	Su.	270	Tu.	300	F.	331	Su.	361
28	W.	28	Sa.	59	Sa.	87	Tu.	118	Th.	148	Su.	179	Tu.	209	F.	240	M.	271	W.	301	Sa.	332	M.	362
29	Th.	29			Su.	88	W.	119	F.	149	M.	180	W.	210	Sa.	241	Tu.	272	Th.	302	Su.	333	Tu.	363
30	F.	30			M.	89	Th.	120	Sa.	150	Tu.	181	Th.	211	Su.	242	W.	273	F.	303	M.	334	W.	364
31	Sa.	31			Tu.	90			Su.	151			F.	212	M.	243			Sa.	304			Th.	365

ECLIPSES

There are two eclipses of the Sun and one of the Moon.

1. *An annular eclipse of the Sun*, January 26. See map on page 6. The eclipse begins at $04^h\ 57^m$ and ends at $11^h\ 01^m$; the annular phase begins at $06^h\ 05^m$ and ends at $09^h\ 53^m$. The maximum duration of annularity is $7^m\ 51^s$.

2. *A total eclipse of the Sun*, July 21-22. See map on page 7. The eclipse begins at $23^h\ 58^m$ on July 21 and ends at $05^h\ 12^m$ on July 22; the total phase begins at $00^h\ 53^m$ on July 22 and ends at $04^h\ 18^m$ on July 22. The maximum duration of totality is $6^m\ 44^s$.

3. *A partial eclipse of the Moon*, on December 31. The umbral eclipse begins at $18^h\ 52^m$ and ends at $19^h\ 54^m$. The time of maximum eclipse is $19^h\ 23^m$ when 0·08 of the Moon's diameter is obscured. It is visible from northern Canada, Alaska, Australia, Indonesia, Asia, Africa, Europe and the Arctic regions.

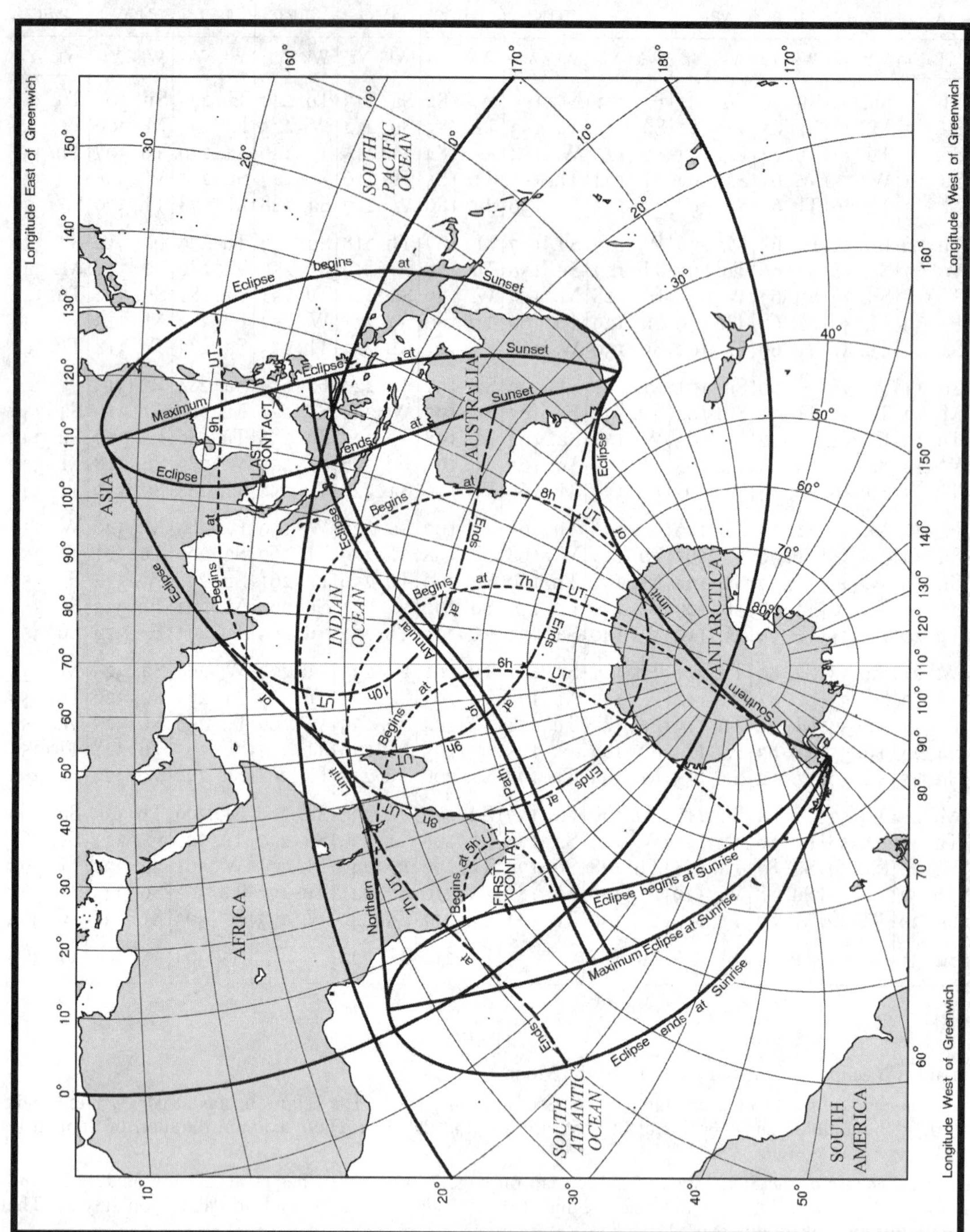

SOLAR ECLIPSE DIAGRAMS

The principal features shown on the above diagrams are: the paths of total and annular eclipses; the northern and southern limits of partial eclipse; the sunrise and sunset curves; dashed lines which show the times of beginning and end of partial eclipse at hourly intervals.

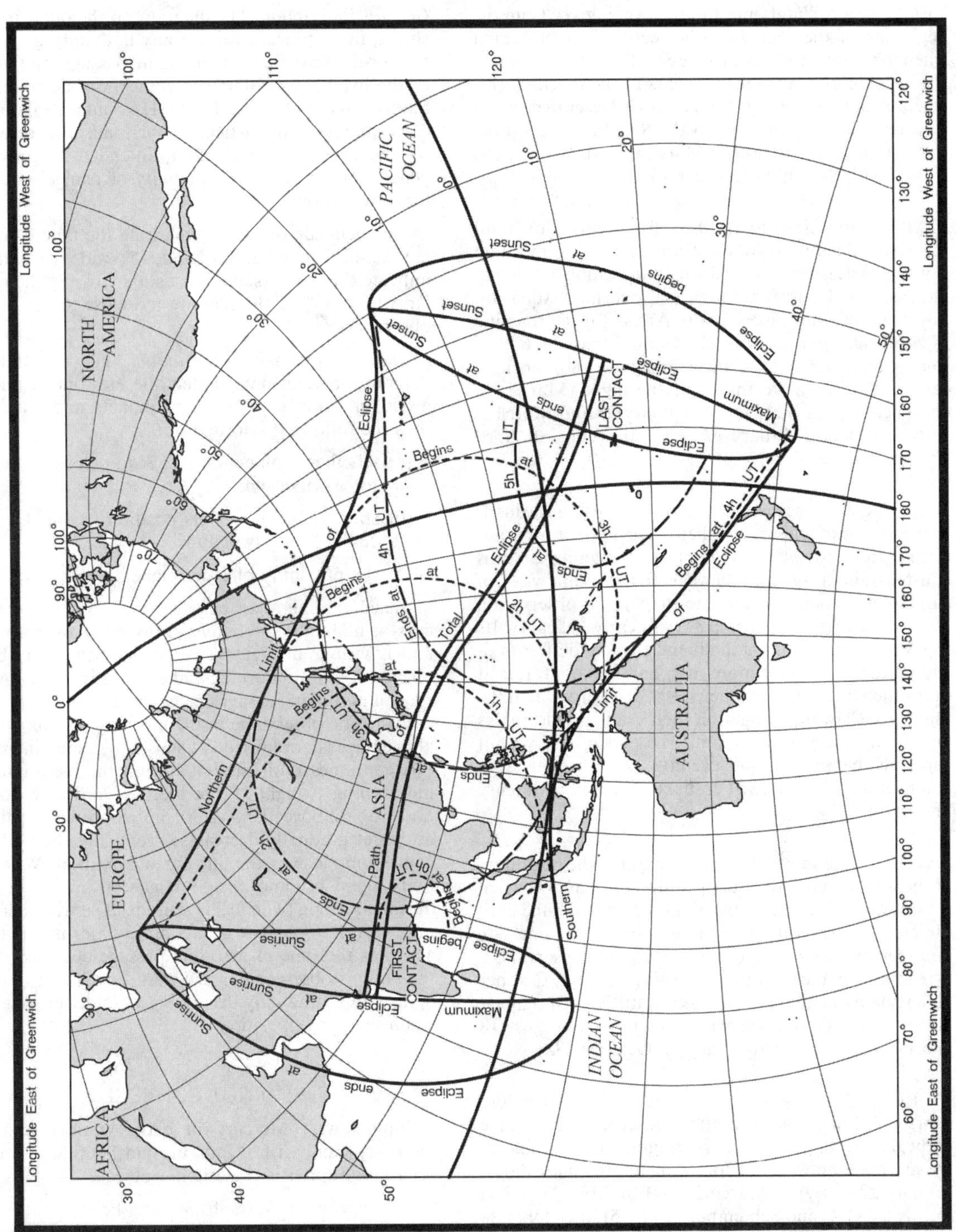

SOLAR ECLIPSE DIAGRAMS

Further details of the paths and times of central eclipse are given in *The Astronomical Almanac.*

VISIBILITY OF PLANETS

VENUS is a brilliant object in the evening sky from the beginning of the year until the second half of March when it becomes too close to the Sun for observation. At the beginning of April it reappears in the morning sky, where it can be seen until the start of December, when it again becomes too close to the Sun for observation. Venus is in conjunction with Mars on April 18 and June 19 and with Saturn on October 13.

MARS is too close to the Sun for observation until the start of February when it appears in the morning sky in Sagittarius. Its westward elongation gradually increases as it passes through Capricornus, Aquarius, Pisces, Cetus, into Pisces again, Aries, Taurus (passing 5° N of *Aldebaran* on July 27), Gemini (passing 6° S of *Pollux* on October 5), Cancer and into Leo, where it can be seen for more than half the night. Mars is in conjunction with Mercury on January 26 and March 1, with Jupiter on February 17 and with Venus on April 18 and June 19.

JUPITER can be seen in the evening sky in Sagittarius at the beginning of January and then passes into Capricornus during the first week of January, remaining in this constellation throughout the year. In the second week of January it becomes too close to the Sun for observation and reappears in the morning sky in early February. Its westward elongation gradually increases and after mid-May it can be seen for more than half the night. It is at opposition on August 14 when it is visible throughout the night. Its eastward elongation then gradually decreases and from mid-November until the end of the year it can only be seen in the evening sky. Jupiter is in conjunction with Mars on February 17 and with Mercury on February 24.

SATURN rises shortly before midnight at the beginning of the year in Leo and is at opposition on March 8, when it can be seen throughout the night. From mid-June until the end of August it is visible only in the evening sky, and then becomes too close to the Sun for observation. It reappears in the morning sky in Virgo in early October and remains in the morning sky until late December. Saturn is in conjunction with Mercury on August 18 and October 8 and with Venus on October 13.

MERCURY can only be seen low in the east before sunrise, or low in the west after sunset (about the time of beginning or end of civil twilight). It is visible in the mornings between the following approximate dates: January 27 (+1·9) to March 22 (−1·0), May 28 (+3·1) to July 6 (−1·4) and September 28 (+1·5) to October 23 (−1·1); the planet is brighter at the end of each period. It is visible in the evenings between the following approximate dates: January 1 (−0·7) to January 15 (+1·8), April 9 (−1·5) to May 9 (+3·0), July 22 (−1·2) to September 14 (+2·6) and November 22 (−0·7) to December 30 (+1·6); the planet is brighter at the beginning of each period. The figures in parentheses are the magnitudes.

PLANET DIAGRAM

General Description. The diagram on the opposite pag shows, in graphical form for any date during the yea the local mean time of meridian passage of the Su of the five planets Mercury, Venus, Mars, Jupiter, an Saturn, and of each 30° of SHA; intermediate line corresponding to particular stars, may be drawn i by the user if desired. It is intended to provide general picture of the availability of planets and star for observation.

On each side of the line marking the time of merid ian passage of the Sun a band, 45^{m} wide, is shaded t indicate that planets and most stars crossing the merid ian within 45^{m} of the Sun are too close to the Sun fo observation.

Method of use and interpretation. For any date th diagram provides immediately the local mean times o meridian passage of the Sun, planets and stars, an thus the following information:

(a) whether a planet or star is too close to the Su for observation;

(b) some indication of its position in the sky, espe cially during twilight;

(c) the proximity of other planets.

When the meridian passage of an outer planet oc curs at midnight the body is in opposition to the Su and is visible all night; a planet may then be observ able during both morning and evening twilights. As th time of meridian passage decreases, the body eventuall ceases to be observable in the morning, but its altitud above the eastern horizon at sunset gradually increases this continues until the body is on the meridian dur ing evening twilight. From then onwards the body i observable above the western horizon and its altitud at sunset gradually decreases; eventually the body be comes too close to the Sun for observation. When th body again becomes visible it is seen low in the eas during morning twilight; its altitude at sunrise increase until meridian passage occurs during morning twilight Then, as the time of meridian passage decreases to 0^{h} the body is observable in the west during morning twi light with a gradually decreasing altitude, until it onc again reaches opposition.

DO NOT CONFUSE

Jupiter with Mercury in early January and lat February and with Mars in mid-February; on al occasions Jupiter is the brighter object.

Mercury with Mars from late February to earl March and with Saturn in mid-August and in th second week of October; on all occasions Mercur is the brighter object.

Venus with Mars from mid-April to the start of Ma and from early June to early July and with Saturn i mid-October; on all occasions Venus is the brighte object.

LOCAL MEAN TIME OF MERIDIAN PASSAGE

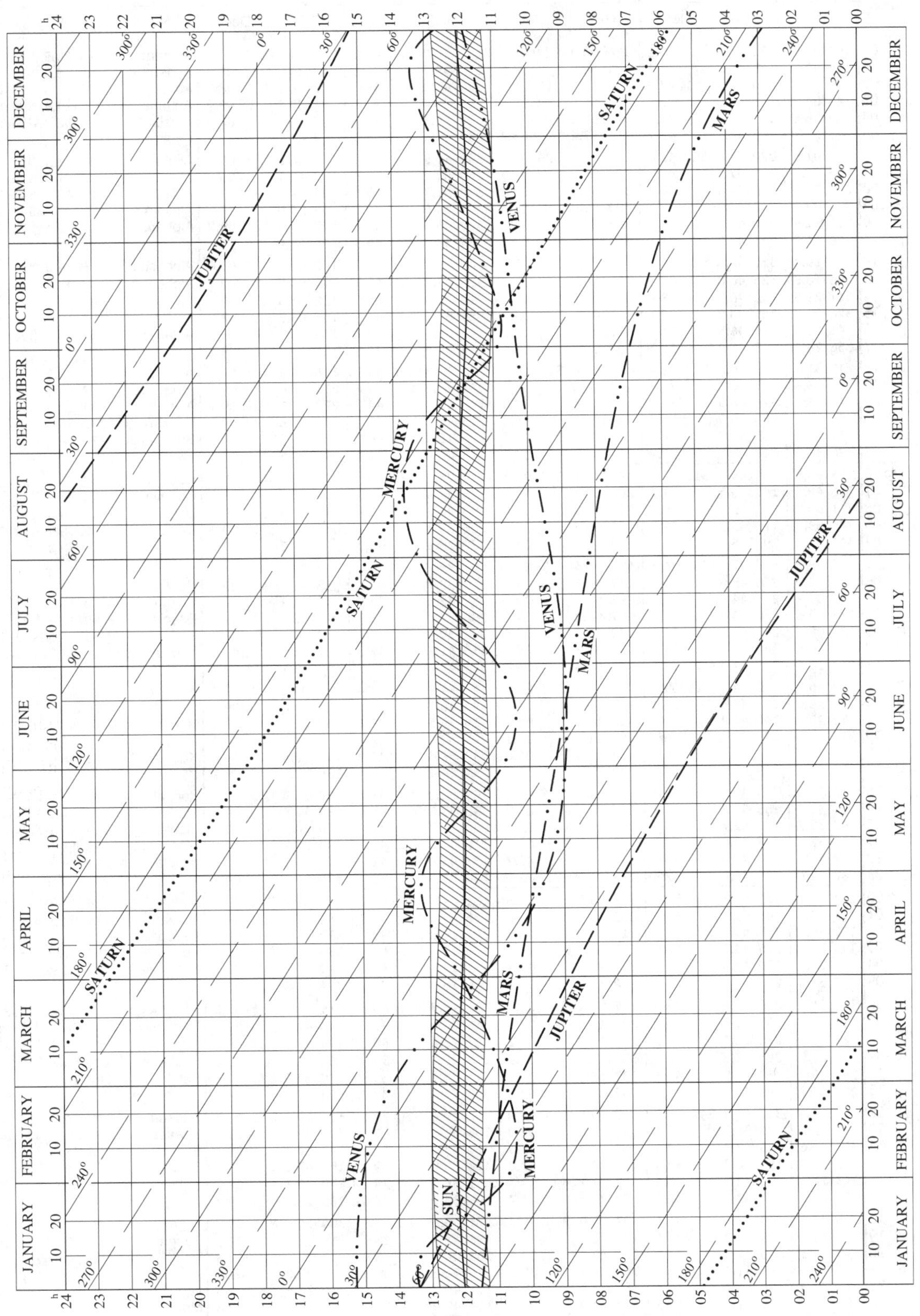

LOCAL MEAN TIME OF MERIDIAN PASSAGE

UT d	h		ARIES GHA	VENUS −4.4 GHA	VENUS Dec	MARS +1.3 GHA	MARS Dec	JUPITER −1.9 GHA	JUPITER Dec	SATURN +1.0 GHA	SATURN Dec
			° ′	° ′	° ′	° ′	° ′	° ′	° ′	° ′	° ′
1	00		100 46.8	130 48.4	S13 49.1	186 55.4	S24 05.6	159 37.8	S20 46.6	287 31.8	N 5 08.9
	01		115 49.2	145 48.2	48.1	201 55.8	05.5	174 39.7	46.5	302 34.3	08.9
	02		130 51.7	160 48.1	47.0	216 56.2	05.5	189 41.6	46.3	317 36.8	08.9
	03		145 54.2	175 48.0	. . 45.9	231 56.6	. . 05.5	204 43.4	. . 46.2	332 39.2	. . 08.9
	04		160 56.6	190 47.9	44.8	246 57.0	05.4	219 45.3	46.1	347 41.7	08.9
	05		175 59.1	205 47.7	43.7	261 57.4	05.4	234 47.1	46.0	2 44.1	08.9
	06		191 01.6	220 47.6	S13 42.6	276 57.8	S24 05.4	249 49.0	S20 45.9	17 46.6	N 5 08.9
	07		206 04.0	235 47.5	41.5	291 58.2	05.3	264 50.9	45.8	32 49.1	08.9
	08	T	221 06.5	250 47.4	40.4	306 58.6	05.3	279 52.7	45.6	47 51.5	09.0
	09	H	236 09.0	265 47.3	. . 39.3	321 59.0	. . 05.3	294 54.6	. . 45.5	62 54.0	. . 09.0
	10	U	251 11.4	280 47.1	38.2	336 59.4	05.2	309 56.4	45.4	77 56.5	09.0
	11	R	266 13.9	295 47.0	37.1	351 59.8	05.2	324 58.3	45.3	92 58.9	09.0
	12	S	281 16.4	310 46.9	S13 36.0	7 00.2	S24 05.2	340 00.2	S20 45.2	108 01.4	N 5 09.0
	13	D	296 18.8	325 46.8	34.9	22 00.6	05.1	355 02.0	45.1	123 03.8	09.0
	14	A	311 21.3	340 46.7	33.8	37 01.0	05.1	10 03.9	45.0	138 06.3	09.0
	15	Y	326 23.7	355 46.6	. . 32.7	52 01.4	. . 05.1	25 05.7	. . 44.8	153 08.8	. . 09.0
	16		341 26.2	10 46.5	31.6	67 01.8	05.0	40 07.6	44.7	168 11.2	09.1
	17		356 28.7	25 46.4	30.5	82 02.2	05.0	55 09.5	44.6	183 13.7	09.1
	18		11 31.1	40 46.2	S13 29.4	97 02.6	S24 05.0	70 11.3	S20 44.5	198 16.2	N 5 09.1
	19		26 33.6	55 46.1	28.3	112 03.0	04.9	85 13.2	44.4	213 18.6	09.1
	20		41 36.1	70 46.0	27.2	127 03.4	04.9	100 15.1	44.3	228 21.1	09.1
	21		56 38.5	85 45.9	. . 26.1	142 03.8	. . 04.8	115 16.9	. . 44.1	243 23.6	. . 09.1
	22		71 41.0	100 45.8	25.0	157 04.2	04.8	130 18.8	44.0	258 26.0	09.1
	23		86 43.5	115 45.7	23.9	172 04.6	04.8	145 20.6	43.9	273 28.5	09.1
2	00		101 45.9	130 45.6	S13 22.8	187 05.1	S24 04.7	160 22.5	S20 43.8	288 30.9	N 5 09.2
	01		116 48.4	145 45.5	21.7	202 05.5	04.7	175 24.4	43.7	303 33.4	09.2
	02		131 50.9	160 45.4	20.6	217 05.9	04.6	190 26.2	43.6	318 35.9	09.2
	03		146 53.3	175 45.3	. . 19.5	232 06.3	. . 04.6	205 28.1	. . 43.4	333 38.3	. . 09.2
	04		161 55.8	190 45.2	18.4	247 06.7	04.6	220 29.9	43.3	348 40.8	09.2
	05		176 58.2	205 45.1	17.3	262 07.1	04.5	235 31.8	43.2	3 43.3	09.2
	06		192 00.7	220 45.0	S13 16.2	277 07.5	S24 04.5	250 33.7	S20 43.1	18 45.7	N 5 09.2
	07		207 03.2	235 44.9	15.1	292 07.9	04.4	265 35.5	43.0	33 48.2	09.2
	08		222 05.6	250 44.8	14.0	307 08.3	04.4	280 37.4	42.9	48 50.7	09.3
	09	F	237 08.1	265 44.7	. . 12.9	322 08.7	. . 04.4	295 39.2	. . 42.7	63 53.1	. . 09.3
	10	R	252 10.6	280 44.6	11.8	337 09.1	04.3	310 41.1	42.6	78 55.6	09.3
	11	I	267 13.0	295 44.5	10.7	352 09.5	04.3	325 43.0	42.5	93 58.1	09.3
	12	D	282 15.5	310 44.4	S13 09.5	7 09.9	S24 04.2	340 44.8	S20 42.4	109 00.5	N 5 09.3
	13	A	297 18.0	325 44.3	08.4	22 10.3	04.2	355 46.7	42.3	124 03.0	09.3
	14	Y	312 20.4	340 44.2	07.3	37 10.7	04.1	10 48.5	42.2	139 05.5	09.3
	15		327 22.9	355 44.1	. . 06.2	52 11.1	. . 04.1	25 50.4	. . 42.0	154 07.9	. . 09.4
	16		342 25.3	10 44.0	05.1	67 11.5	04.0	40 52.3	41.9	169 10.4	09.4
	17		357 27.8	25 43.9	04.0	82 11.9	04.0	55 54.1	41.8	184 12.9	09.4
	18		12 30.3	40 43.9	S13 02.9	97 12.3	S24 03.9	70 56.0	S20 41.7	199 15.3	N 5 09.4
	19		27 32.7	55 43.8	01.8	112 12.7	03.9	85 57.8	41.6	214 17.8	09.4
	20		42 35.2	70 43.7	13 00.7	127 13.1	03.8	100 59.7	41.4	229 20.3	09.4
	21		57 37.7	85 43.6	12 59.6	142 13.5	. . 03.8	116 01.5	. . 41.3	244 22.7	. . 09.4
	22		72 40.1	100 43.5	58.4	157 13.9	03.7	131 03.4	41.2	259 25.2	09.5
	23		87 42.6	115 43.4	57.3	172 14.3	03.7	146 05.3	41.1	274 27.7	09.5
3	00		102 45.1	130 43.3	S12 56.2	187 14.7	S24 03.6	161 07.1	S20 41.0	289 30.2	N 5 09.5
	01		117 47.5	145 43.3	55.1	202 15.1	03.6	176 09.0	40.9	304 32.6	09.5
	02		132 50.0	160 43.2	54.0	217 15.5	03.5	191 10.8	40.7	319 35.1	09.5
	03		147 52.5	175 43.1	. . 52.9	232 15.9	. . 03.5	206 12.7	. . 40.6	334 37.6	. . 09.5
	04		162 54.9	190 43.0	51.8	247 16.3	03.4	221 14.6	40.5	349 40.0	09.5
	05		177 57.4	205 42.9	50.7	262 16.7	03.4	236 16.4	40.4	4 42.5	09.6
	06		192 59.8	220 42.9	S12 49.5	277 17.1	S24 03.3	251 18.3	S20 40.3	19 45.0	N 5 09.6
	07		208 02.3	235 42.8	48.4	292 17.5	03.3	266 20.1	40.2	34 47.4	09.6
	08	S	223 04.8	250 42.7	47.3	307 17.9	03.2	281 22.0	40.0	49 49.9	09.6
	09	A	238 07.2	265 42.6	. . 46.2	322 18.3	. . 03.2	296 23.9	. . 39.9	64 52.4	. . 09.6
	10	T	253 09.7	280 42.5	45.1	337 18.7	03.1	311 25.7	39.8	79 54.9	09.6
	11	U	268 12.2	295 42.5	44.0	352 19.1	03.1	326 27.6	39.7	94 57.3	09.7
	12	R	283 14.6	310 42.4	S12 42.8	7 19.5	S24 03.0	341 29.4	S20 39.6	109 59.8	N 5 09.7
	13	D	298 17.1	325 42.3	41.7	22 19.9	02.9	356 31.3	39.4	125 02.3	09.7
	14	A	313 19.6	340 42.3	40.6	37 20.3	02.9	11 33.2	39.3	140 04.7	09.7
	15	Y	328 22.0	355 42.2	. . 39.5	52 20.7	. . 02.8	26 35.0	. . 39.2	155 07.2	. . 09.7
	16		343 24.5	10 42.1	38.4	67 21.1	02.8	41 36.9	39.1	170 09.7	09.7
	17		358 27.0	25 42.1	37.2	82 21.5	02.7	56 38.7	39.0	185 12.2	09.7
	18		13 29.4	40 42.0	S12 36.1	97 21.9	S24 02.7	71 40.6	S20 38.9	200 14.6	N 5 09.8
	19		28 31.9	55 41.9	35.0	112 22.3	02.6	86 42.4	38.7	215 17.1	09.8
	20		43 34.3	70 41.9	33.9	127 22.7	02.5	101 44.3	38.6	230 19.6	09.8
	21		58 36.8	85 41.8	. . 32.8	142 23.1	. . 02.5	116 46.2	. . 38.5	245 22.0	. . 09.8
	22		73 39.3	100 41.7	31.6	157 23.5	02.4	131 48.0	38.4	260 24.5	09.8
	23		88 41.7	115 41.7	30.5	172 23.9	02.4	146 49.9	38.3	275 27.0	09.8
Mer. Pass.			h m 17 10.1	*v* −0.1	*d* 1.1	*v* 0.4	*d* 0.0	*v* 1.9	*d* 0.1	*v* 2.5	*d* 0.0

STARS Name	SHA	Dec
	° ′	° ′
Acamar	315 20.5	S40 16.2
Achernar	335 28.9	S57 11.7
Acrux	173 13.3	S63 08.7
Adhara	255 14.8	S28 59.0
Aldebaran	290 52.9	N16 31.7
Alioth	166 23.4	N55 54.3
Alkaid	153 01.5	N49 15.7
Al Na'ir	27 48.0	S46 55.2
Alnilam	275 49.4	S 1 11.7
Alphard	217 59.1	S 8 41.9
Alphecca	126 14.0	N26 40.8
Alpheratz	357 47.1	N29 08.6
Altair	62 11.8	N 8 53.5
Ankaa	353 18.8	S42 15.6
Antares	112 30.7	S26 27.1
Arcturus	145 58.8	N19 07.9
Atria	107 36.1	S69 02.6
Avior	234 18.9	S59 32.2
Bellatrix	278 35.2	N 6 21.5
Betelgeuse	271 04.6	N 7 24.6
Canopus	263 57.1	S52 42.0
Capella	280 38.9	N46 00.6
Deneb	49 34.3	N45 18.9
Denebola	182 36.9	N14 31.1
Diphda	348 59.2	S17 56.3
Dubhe	193 55.1	N61 41.8
Elnath	278 16.5	N28 37.0
Eltanin	90 48.2	N51 29.1
Enif	33 50.6	N 9 55.0
Fomalhaut	15 27.7	S29 34.6
Gacrux	172 04.8	S57 09.6
Gienah	175 55.7	S17 35.5
Hadar	148 53.1	S60 24.8
Hamal	328 04.4	N23 30.5
Kaus Aust.	83 48.6	S34 22.9
Kochab	137 20.1	N74 06.7
Markab	13 41.8	N15 15.3
Menkar	314 18.3	N 4 07.6
Menkent	148 11.7	S36 24.8
Miaplacidus	221 39.9	S69 45.1
Mirfak	308 44.9	N49 53.9
Nunki	76 02.8	S26 17.2
Peacock	53 24.9	S56 42.5
Pollux	243 31.4	N28 00.2
Procyon	245 02.9	N 5 12.1
Rasalhague	96 09.9	N12 33.1
Regulus	207 46.8	N11 55.3
Rigel	281 14.9	S 8 11.4
Rigil Kent.	139 56.8	S60 52.2
Sabik	102 16.7	S15 44.2
Schedar	349 44.6	N56 35.6
Shaula	96 26.8	S37 06.6
Sirius	258 36.3	S16 43.7
Spica	158 34.8	S11 12.6
Suhail	222 54.6	S43 28.0
Vega	80 41.7	N38 47.4
Zuben'ubi	137 09.3	S16 04.8

	SHA	Mer. Pass.
	° ′	h m
Venus	28 59.7	15 17
Mars	85 19.1	11 31
Jupiter	58 36.6	13 17
Saturn	186 45.0	4 45

UT d	h		SUN GHA	SUN Dec	MOON GHA	v	MOON Dec	d	HP
			° ′	° ′	° ′	′	° ′	′	′
1	00		179 08.4	S23 00.5	129 33.6	15.3	S 9 50.7	13.2	55.1
	01		194 08.1	00.3	144 07.9	15.3	9 37.5	13.2	55.2
	02		209 07.8	23 00.1	158 42.2	15.3	9 24.3	13.3	55.2
	03		224 07.5	22 59.9	173 16.5	15.4	9 11.0	13.3	55.2
	04		239 07.2	59.7	187 50.9	15.4	8 57.7	13.3	55.2
	05		254 06.9	59.5	202 25.3	15.4	8 44.4	13.4	55.2
	06		269 06.6	S22 59.3	216 59.7	15.4	S 8 31.0	13.4	55.3
	07		284 06.3	59.1	231 34.1	15.4	8 17.6	13.5	55.3
	08	T	299 06.0	58.8	246 08.5	15.4	8 04.1	13.5	55.3
	09	H	314 05.7	. . 58.6	260 42.9	15.4	7 50.6	13.5	55.3
	10	U	329 05.5	58.4	275 17.3	15.5	7 37.1	13.6	55.3
	11	R	344 05.2	58.2	289 51.8	15.5	7 23.5	13.7	55.4
	12	S	359 04.9	S22 58.0	304 26.3	15.4	S 7 09.8	13.6	55.4
	13	D	14 04.6	57.8	319 00.7	15.5	6 56.2	13.7	55.4
	14	A	29 04.3	57.6	333 35.2	15.5	6 42.5	13.8	55.4
	15	Y	44 04.0	. . 57.4	348 09.7	15.4	6 28.7	13.8	55.5
	16		59 03.7	57.1	2 44.1	15.5	6 14.9	13.8	55.5
	17		74 03.4	56.9	17 18.6	15.5	6 01.1	13.8	55.5
	18		89 03.1	S22 56.7	31 53.1	15.5	S 5 47.3	13.9	55.5
	19		104 02.8	56.5	46 27.6	15.5	5 33.4	13.9	55.5
	20		119 02.5	56.3	61 02.1	15.5	5 19.5	14.0	55.6
	21		134 02.2	. . 56.0	75 36.6	15.5	5 05.5	13.9	55.6
	22		149 01.9	55.8	90 11.1	15.5	4 51.6	14.1	55.6
	23		164 01.6	55.6	104 45.6	15.4	4 37.5	14.0	55.6
2	00		179 01.3	S22 55.4	119 20.0	15.5	S 4 23.5	14.1	55.7
	01		194 01.1	55.1	133 54.5	15.5	4 09.4	14.1	55.7
	02		209 00.8	54.9	148 29.0	15.4	3 55.3	14.1	55.7
	03		224 00.5	. . 54.7	163 03.4	15.5	3 41.2	14.1	55.7
	04		239 00.2	54.5	177 37.9	15.4	3 27.1	14.2	55.8
	05		253 59.9	54.2	192 12.3	15.5	3 12.9	14.2	55.8
	06		268 59.6	S22 54.0	206 46.8	15.4	S 2 58.7	14.2	55.8
	07		283 59.3	53.8	221 21.2	15.4	2 44.5	14.3	55.8
	08		298 59.0	53.6	235 55.6	15.4	2 30.2	14.2	55.9
	09	F	313 58.7	. . 53.3	250 30.0	15.4	2 16.0	14.3	55.9
	10	R	328 58.4	53.1	265 04.4	15.3	2 01.7	14.4	55.9
	11	I	343 58.1	52.9	279 38.7	15.4	1 47.3	14.3	56.0
	12	D	358 57.8	S22 52.6	294 13.1	15.3	S 1 33.0	14.3	56.0
	13	A	13 57.6	52.4	308 47.4	15.3	1 18.7	14.4	56.0
	14	Y	28 57.3	52.2	323 21.7	15.3	1 04.3	14.4	56.0
	15		43 57.0	. . 51.9	337 56.0	15.2	0 49.9	14.4	56.1
	16		58 56.7	51.7	352 30.2	15.3	0 35.5	14.4	56.1
	17		73 56.4	51.5	7 04.5	15.2	0 21.1	14.5	56.1
	18		88 56.1	S22 51.2	21 38.7	15.2	S 0 06.6	14.4	56.1
	19		103 55.8	51.0	36 12.9	15.1	N 0 07.8	14.5	56.2
	20		118 55.5	50.7	50 47.0	15.2	0 22.3	14.4	56.2
	21		133 55.2	. . 50.5	65 21.2	15.1	0 36.7	14.5	56.2
	22		148 54.9	50.3	79 55.3	15.0	0 51.2	14.5	56.3
	23		163 54.7	50.0	94 29.3	15.1	1 05.7	14.5	56.3
3	00		178 54.4	S22 49.8	109 03.4	15.0	N 1 20.2	14.5	56.3
	01		193 54.1	49.5	123 37.4	14.9	1 34.7	14.6	56.3
	02		208 53.8	49.3	138 11.3	15.0	1 49.3	14.5	56.4
	03		223 53.5	. . 49.0	152 45.3	14.9	2 03.8	14.5	56.4
	04		238 53.2	48.8	167 19.2	14.8	2 18.3	14.6	56.4
	05		253 52.9	48.6	181 53.0	14.8	2 32.9	14.5	56.5
	06		268 52.6	S22 48.3	196 26.8	14.8	N 2 47.4	14.6	56.5
	07	S	283 52.4	48.1	211 00.6	14.7	3 02.0	14.5	56.5
	08	A	298 52.1	47.8	225 34.3	14.7	3 16.5	14.6	56.6
	09	T	313 51.8	. . 47.6	240 08.0	14.6	3 31.1	14.5	56.6
	10	U	328 51.5	47.3	254 41.6	14.6	3 45.6	14.6	56.6
	11	R	343 51.2	47.1	269 15.2	14.6	4 00.2	14.6	56.7
	12	D	358 50.9	S22 46.8	283 48.8	14.5	N 4 14.8	14.5	56.7
	13	A	13 50.6	46.6	298 22.3	14.4	4 29.3	14.6	56.7
	14	Y	28 50.4	46.3	312 55.7	14.4	4 43.9	14.5	56.7
	15		43 50.1	. . 46.0	327 29.1	14.4	4 58.4	14.5	56.8
	16		58 49.8	45.8	342 02.5	14.3	5 12.9	14.6	56.8
	17		73 49.5	45.5	356 35.8	14.2	5 27.5	14.5	56.8
	18		88 49.2	S22 45.3	11 09.0	14.2	N 5 42.0	14.5	56.9
	19		103 48.9	45.0	25 42.2	14.1	5 56.5	14.5	56.9
	20		118 48.6	44.8	40 15.3	14.0	6 11.0	14.5	56.9
	21		133 48.4	. . 44.5	54 48.3	14.0	6 25.5	14.5	57.0
	22		148 48.1	44.2	69 21.3	13.9	6 40.0	14.5	57.0
	23		163 47.8	44.0	83 54.2	13.9	N 6 54.5	14.4	57.0
			SD 16.3	d 0.2	SD	15.1		15.3	15.4

Lat.	Twilight Naut.	Twilight Civil	Sunrise	Moonrise 1	Moonrise 2	Moonrise 3	Moonrise 4
°	h m	h m	h m	h m	h m	h m	h m
N 72	08 23	10 39	■	11 16	10 45	10 16	09 42
N 70	08 04	09 48	■	11 05	10 43	10 21	09 56
68	07 49	09 16	■	10 57	10 41	10 25	10 08
66	07 37	08 52	10 26	10 49	10 39	10 28	10 17
64	07 26	08 34	09 48	10 43	10 37	10 31	10 25
62	07 17	08 18	09 22	10 38	10 36	10 34	10 32
60	07 09	08 05	09 02	10 33	10 35	10 36	10 38
N 58	07 02	07 54	08 45	10 29	10 34	10 38	10 43
56	06 55	07 44	08 31	10 25	10 33	10 40	10 48
54	06 50	07 35	08 19	10 22	10 32	10 42	10 52
52	06 44	07 28	08 08	10 19	10 31	10 43	10 56
50	06 39	07 20	07 58	10 16	10 30	10 44	11 00
45	06 28	07 05	07 38	10 10	10 29	10 47	11 08
N 40	06 18	06 52	07 22	10 05	10 28	10 50	11 14
35	06 09	06 40	07 08	10 01	10 26	10 52	11 20
30	06 00	06 30	06 56	09 57	10 25	10 54	11 25
20	05 44	06 12	06 36	09 50	10 24	10 58	11 33
N 10	05 28	05 55	06 17	09 44	10 22	11 01	11 41
0	05 12	05 38	06 00	09 39	10 21	11 04	11 48
S 10	04 53	05 20	05 43	09 33	10 19	11 07	11 56
20	04 31	05 00	05 25	09 27	10 18	11 10	12 04
30	04 03	04 36	05 03	09 20	10 16	11 13	12 13
35	03 45	04 21	04 51	09 16	10 15	11 15	12 18
40	03 22	04 03	04 36	09 12	10 14	11 18	12 24
45	02 52	03 41	04 18	09 06	10 13	11 21	12 31
S 50	02 09	03 13	03 56	09 00	10 11	11 24	12 40
52	01 43	02 58	03 46	08 57	10 11	11 26	12 44
54	01 04	02 41	03 34	08 54	10 10	11 27	12 48
56	////	02 20	03 20	08 50	10 09	11 29	12 53
58	////	01 52	03 04	08 46	10 08	11 31	12 58
S 60	////	01 10	02 45	08 42	10 07	11 34	13 05

Lat.	Sunset	Twilight Civil	Twilight Naut.	Moonset 1	Moonset 2	Moonset 3	Moonset 4
°	h m	h m	h m	h m	h m	h m	h m
N 72	■	13 30	15 46	20 52	22 54	25 00	01 00
N 70	■	14 21	16 05	21 00	22 52	24 49	00 49
68	■	14 53	16 20	21 06	22 51	24 40	00 40
66	13 43	15 16	16 32	21 11	22 50	24 33	00 33
64	14 20	15 35	16 42	21 15	22 49	24 26	00 26
62	14 46	15 50	16 51	21 19	22 49	24 21	00 21
60	15 07	16 03	16 59	21 22	22 48	24 16	00 16
N 58	15 24	16 15	17 07	21 25	22 48	24 12	00 12
56	15 38	16 24	17 13	21 28	22 47	24 09	00 09
54	15 50	16 33	17 19	21 30	22 47	24 05	00 05
52	16 00	16 41	17 24	21 32	22 46	24 02	00 02
50	16 10	16 48	17 29	21 34	22 46	24 00	00 00
45	16 30	17 04	17 41	21 38	22 45	23 54	25 06
N 40	16 46	17 17	17 51	21 42	22 44	23 49	24 57
35	17 00	17 28	18 00	21 44	22 44	23 45	24 49
30	17 12	17 38	18 08	21 47	22 43	23 41	24 42
20	17 33	17 57	18 24	21 51	22 43	23 35	24 31
N 10	17 51	18 14	18 40	21 55	22 42	23 30	24 20
0	18 08	18 30	18 56	21 59	22 41	23 25	24 11
S 10	18 25	18 48	19 15	22 02	22 40	23 20	24 01
20	18 43	19 08	19 37	22 06	22 39	23 14	23 51
30	19 05	19 32	20 05	22 10	22 39	23 08	23 40
35	19 18	19 47	20 23	22 13	22 38	23 04	23 33
40	19 32	20 05	20 46	22 15	22 37	23 00	23 26
45	19 50	20 27	21 15	22 18	22 37	22 56	23 17
S 50	20 11	20 55	21 58	22 22	22 36	22 50	23 07
52	20 22	21 10	22 24	22 24	22 35	22 48	23 02
54	20 34	21 27	23 02	22 26	22 35	22 45	22 57
56	20 47	21 48	////	22 28	22 35	22 42	22 51
58	21 03	22 15	////	22 30	22 34	22 39	22 44
S 60	21 23	22 56	////	22 32	22 34	22 35	22 37

Day	SUN Eqn. of Time 00h	SUN Eqn. of Time 12h	SUN Mer. Pass.	MOON Mer. Pass. Upper	MOON Mer. Pass. Lower	MOON Age	MOON Phase
d	m s	m s	h m	h m	h m	d %	
1	03 26	03 40	12 04	15 49	03 28	05 21	
2	03 54	04 08	12 04	16 31	04 10	06 30	
3	04 22	04 36	12 05	17 14	04 52	07 40	

	UT	ARIES	VENUS −4.5		MARS +1.3		JUPITER −1.9		SATURN +0.9	
		GHA	GHA	Dec	GHA	Dec	GHA	Dec	GHA	Dec
	d h	° ′	° ′	° ′	° ′	° ′	° ′	° ′	° ′	° ′
SUNDAY	4 00	103 44.2	130 41.6	S12 29.4	187 24.3	S24 02.3	161 51.7	S20 38.1	290 29.5	N 5 09.9
	01	118 46.7	145 41.5	28.3	202 24.7	02.2	176 53.6	38.0	305 31.9	09.9
	02	133 49.1	160 41.5	27.1	217 25.1	02.2	191 55.5	37.9	320 34.4	09.9
	03	148 51.6	175 41.4	. . 26.0	232 25.5	. . 02.1	206 57.3	. . 37.8	335 36.9	. . 09.9
	04	163 54.1	190 41.4	24.9	247 25.9	02.0	221 59.2	37.7	350 39.4	09.9
	05	178 56.5	205 41.3	23.8	262 26.3	02.0	237 01.0	37.6	5 41.8	09.9
	06	193 59.0	220 41.2	S12 22.6	277 26.7	S24 01.9	252 02.9	S20 37.4	20 44.3	N 5 10.0
	07	209 01.4	235 41.2	21.5	292 27.1	01.9	267 04.7	37.3	35 46.8	10.0
	08	224 03.9	250 41.1	20.4	307 27.5	01.8	282 06.6	37.2	50 49.3	10.0
	09	239 06.4	265 41.1	. . 19.3	322 27.9	. . 01.7	297 08.5	. . 37.1	65 51.7	. . 10.0
	10	254 08.8	280 41.0	18.1	337 28.3	01.7	312 10.3	37.0	80 54.2	10.0
	11	269 11.3	295 41.0	17.0	352 28.7	01.6	327 12.2	36.8	95 56.7	10.0
	12	284 13.8	310 40.9	S12 15.9	7 29.1	S24 01.5	342 14.0	S20 36.7	110 59.2	N 5 10.1
	13	299 16.2	325 40.9	14.8	22 29.4	01.5	357 15.9	36.6	126 01.6	10.1
	14	314 18.7	340 40.8	13.6	37 29.8	01.4	12 17.8	36.5	141 04.1	10.1
	15	329 21.2	355 40.8	. . 12.5	52 30.2	. . 01.3	27 19.6	. . 36.4	156 06.6	. . 10.1
	16	344 23.6	10 40.7	11.4	67 30.6	01.3	42 21.5	36.2	171 09.1	10.1
	17	359 26.1	25 40.7	10.2	82 31.0	01.2	57 23.3	36.1	186 11.5	10.1
	18	14 28.6	40 40.6	S12 09.1	97 31.4	S24 01.1	72 25.2	S20 36.0	201 14.0	N 5 10.2
	19	29 31.0	55 40.6	08.0	112 31.8	01.0	87 27.0	35.9	216 16.5	10.2
	20	44 33.5	70 40.5	06.8	127 32.2	01.0	102 28.9	35.8	231 19.0	10.2
	21	59 35.9	85 40.5	. . 05.7	142 32.6	. . 00.9	117 30.8	. . 35.6	246 21.4	. . 10.2
	22	74 38.4	100 40.5	04.6	157 33.0	00.8	132 32.6	35.5	261 23.9	10.2
	23	89 40.9	115 40.4	03.5	172 33.4	00.8	147 34.5	35.4	276 26.4	10.3
MONDAY	5 00	104 43.3	130 40.4	S12 02.3	187 33.8	S24 00.7	162 36.3	S20 35.3	291 28.9	N 5 10.3
	01	119 45.8	145 40.3	01.2	202 34.2	00.6	177 38.2	35.2	306 31.4	10.3
	02	134 48.3	160 40.3	12 00.1	217 34.6	00.5	192 40.0	35.0	321 33.8	10.3
	03	149 50.7	175 40.3	11 58.9	232 35.0	. . 00.5	207 41.9	. . 34.9	336 36.3	. . 10.3
	04	164 53.2	190 40.2	57.8	247 35.4	00.4	222 43.8	34.8	351 38.8	10.3
	05	179 55.7	205 40.2	56.7	262 35.8	00.3	237 45.6	34.7	6 41.3	10.4
	06	194 58.1	220 40.2	S11 55.5	277 36.2	S24 00.2	252 47.5	S20 34.6	21 43.7	N 5 10.4
	07	210 00.6	235 40.1	54.4	292 36.6	00.2	267 49.3	34.4	36 46.2	10.4
	08	225 03.1	250 40.1	53.3	307 37.0	00.1	282 51.2	34.3	51 48.7	10.4
	09	240 05.5	265 40.1	. . 52.1	322 37.4	24 00.0	297 53.1	. . 34.2	66 51.2	. . 10.4
	10	255 08.0	280 40.0	51.0	337 37.8	23 59.9	312 54.9	34.1	81 53.7	10.5
	11	270 10.4	295 40.0	49.8	352 38.2	59.9	327 56.8	34.0	96 56.1	10.5
	12	285 12.9	310 40.0	S11 48.7	7 38.6	S23 59.8	342 58.6	S20 33.8	111 58.6	N 5 10.5
	13	300 15.4	325 39.9	47.6	22 39.0	59.7	358 00.5	33.7	127 01.1	10.5
	14	315 17.8	340 39.9	46.4	37 39.4	59.6	13 02.3	33.6	142 03.6	10.5
	15	330 20.3	355 39.9	. . 45.3	52 39.8	. . 59.5	28 04.2	. . 33.5	157 06.1	. . 10.6
	16	345 22.8	10 39.9	44.2	67 40.2	59.5	43 06.1	33.4	172 08.5	10.6
	17	0 25.2	25 39.8	43.0	82 40.6	59.4	58 07.9	33.2	187 11.0	10.6
	18	15 27.7	40 39.8	S11 41.9	97 41.0	S23 59.3	73 09.8	S20 33.1	202 13.5	N 5 10.6
	19	30 30.2	55 39.8	40.7	112 41.4	59.2	88 11.6	33.0	217 16.0	10.6
	20	45 32.6	70 39.8	39.6	127 41.8	59.1	103 13.5	32.9	232 18.5	10.7
	21	60 35.1	85 39.7	. . 38.5	142 42.2	. . 59.1	118 15.3	. . 32.8	247 20.9	. . 10.7
	22	75 37.5	100 39.7	37.3	157 42.6	59.0	133 17.2	32.6	262 23.4	10.7
	23	90 40.0	115 39.7	36.2	172 43.0	58.9	148 19.1	32.5	277 25.9	10.7
TUESDAY	6 00	105 42.5	130 39.7	S11 35.0	187 43.4	S23 58.8	163 20.9	S20 32.4	292 28.4	N 5 10.7
	01	120 44.9	145 39.7	33.9	202 43.8	58.7	178 22.8	32.3	307 30.9	10.7
	02	135 47.4	160 39.7	32.8	217 44.2	58.6	193 24.6	32.2	322 33.4	10.8
	03	150 49.9	175 39.6	. . 31.6	232 44.6	. . 58.6	208 26.5	. . 32.0	337 35.8	. . 10.8
	04	165 52.3	190 39.6	30.5	247 45.0	58.5	223 28.3	31.9	352 38.3	10.8
	05	180 54.8	205 39.6	29.3	262 45.4	58.4	238 30.2	31.8	7 40.8	10.8
	06	195 57.3	220 39.6	S11 28.2	277 45.8	S23 58.3	253 32.0	S20 31.7	22 43.3	N 5 10.9
	07	210 59.7	235 39.6	27.0	292 46.2	58.2	268 33.9	31.6	37 45.8	10.9
	08	226 02.2	250 39.6	25.9	307 46.6	58.1	283 35.8	31.4	52 48.2	10.9
	09	241 04.7	265 39.6	. . 24.8	322 46.9	. . 58.0	298 37.6	. . 31.3	67 50.7	. . 10.9
	10	256 07.1	280 39.6	23.6	337 47.3	58.0	313 39.5	31.2	82 53.2	10.9
	11	271 09.6	295 39.5	22.5	352 47.7	57.9	328 41.3	31.1	97 55.7	11.0
	12	286 12.0	310 39.5	S11 21.3	7 48.1	S23 57.8	343 43.2	S20 30.9	112 58.2	N 5 11.0
	13	301 14.5	325 39.5	20.2	22 48.5	57.7	358 45.0	30.8	128 00.7	11.0
	14	316 17.0	340 39.5	19.0	37 48.9	57.6	13 46.9	30.7	143 03.2	11.0
	15	331 19.4	355 39.5	. . 17.9	52 49.3	. . 57.5	28 48.8	. . 30.6	158 05.6	. . 11.0
	16	346 21.9	10 39.5	16.7	67 49.7	57.4	43 50.6	30.5	173 08.1	11.1
	17	1 24.4	25 39.5	15.6	82 50.1	57.3	58 52.5	30.3	188 10.6	11.1
	18	16 26.8	40 39.5	S11 14.4	97 50.5	S23 57.2	73 54.3	S20 30.2	203 13.1	N 5 11.1
	19	31 29.3	55 39.5	13.3	112 50.9	57.1	88 56.2	30.1	218 15.6	11.1
	20	46 31.8	70 39.5	12.2	127 51.3	57.0	103 58.0	30.0	233 18.1	11.1
	21	61 34.2	85 39.5	. . 11.0	142 51.7	. . 57.0	118 59.9	. . 29.9	248 20.5	. . 11.2
	22	76 36.7	100 39.5	09.9	157 52.1	56.9	134 01.8	29.7	263 23.0	11.2
	23	91 39.2	115 39.5	08.7	172 52.5	56.8	149 03.6	29.6	278 25.5	11.2
	Mer. Pass.	h m 16 58.3	*v* 0.0	*d* 1.1	*v* 0.4	*d* 0.1	*v* 1.9	*d* 0.1	*v* 2.5	*d* 0.0

STARS

Name	SHA	Dec
	° ′	° ′
Acamar	315 20.5	S40 16.2
Achernar	335 28.9	S57 11.7
Acrux	173 13.2	S63 08.8
Adhara	255 14.8	S28 59.1
Aldebaran	290 52.9	N16 31.7
Alioth	166 23.3	N55 54.2
Alkaid	153 01.4	N49 15.7
Al Na'ir	27 48.0	S46 55.2
Alnilam	275 49.4	S 1 11.7
Alphard	217 59.1	S 8 41.9
Alphecca	126 14.0	N26 40.8
Alpheratz	357 47.1	N29 08.6
Altair	62 11.8	N 8 53.5
Ankaa	353 18.9	S42 15.6
Antares	112 30.7	S26 27.1
Arcturus	145 58.8	N19 07.9
Atria	107 36.1	S69 02.5
Avior	234 18.9	S59 32.2
Bellatrix	278 35.2	N 6 21.5
Betelgeuse	271 04.6	N 7 24.6
Canopus	263 57.1	S52 42.0
Capella	280 38.9	N46 00.6
Deneb	49 34.3	N45 18.8
Denebola	182 36.9	N14 31.1
Diphda	348 59.2	S17 56.3
Dubhe	193 55.1	N61 41.8
Elnath	278 16.5	N28 37.0
Eltanin	90 48.2	N51 29.1
Enif	33 50.7	N 9 55.0
Fomalhaut	15 27.7	S29 34.6
Gacrux	172 04.8	S57 09.6
Gienah	175 55.7	S17 35.5
Hadar	148 53.1	S60 24.8
Hamal	328 04.4	N23 30.5
Kaus Aust.	83 48.6	S34 22.9
Kochab	137 20.1	N74 06.7
Markab	13 41.8	N15 15.3
Menkar	314 18.3	N 4 07.6
Menkent	148 11.7	S36 24.8
Miaplacidus	221 39.9	S69 45.1
Mirfak	308 44.9	N49 53.9
Nunki	76 02.8	S26 17.2
Peacock	53 24.9	S56 42.5
Pollux	243 31.4	N28 00.2
Procyon	245 02.8	N 5 12.1
Rasalhague	96 09.9	N12 33.1
Regulus	207 46.8	N11 55.2
Rigel	281 14.9	S 8 11.5
Rigil Kent.	139 56.8	S60 52.2
Sabik	102 16.7	S15 44.2
Schedar	349 44.6	N56 35.6
Shaula	96 26.8	S37 06.6
Sirius	258 36.3	S16 43.7
Spica	158 34.8	S11 12.6
Suhail	222 54.6	S43 28.1
Vega	80 41.7	N38 47.4
Zuben'ubi	137 09.3	S16 04.8

	SHA	Mer. Pass.
	° ′	h m
Venus	25 57.0	15 17
Mars	82 50.5	11 29
Jupiter	57 53.0	13 08
Saturn	186 45.5	4 33

UT		SUN GHA	SUN Dec	MOON GHA	v	MOON Dec	d	HP
d	h	° ′	° ′	° ′	′	° ′	′	′
4	00	178 47.5	S22 43.7	98 27.1	13.8	N 7 08.9	14.5	57.1
	01	193 47.2	43.5	112 59.9	13.7	7 23.4	14.4	57.1
	02	208 46.9	43.2	127 32.6	13.7	7 37.8	14.4	57.1
	03	223 46.6	. . 42.9	142 05.3	13.6	7 52.2	14.4	57.2
	04	238 46.4	42.7	156 37.9	13.5	8 06.6	14.4	57.2
	05	253 46.1	42.4	171 10.4	13.5	8 21.0	14.3	57.2
	06	268 45.8	S22 42.1	185 42.9	13.4	N 8 35.3	14.4	57.3
	07	283 45.5	41.9	200 15.3	13.3	8 49.7	14.3	57.3
	08	298 45.2	41.6	214 47.6	13.2	9 04.0	14.3	57.3
S	09	313 45.0	. . 41.3	229 19.8	13.2	9 18.3	14.2	57.4
U	10	328 44.7	41.1	243 52.0	13.0	9 32.5	14.3	57.4
N	11	343 44.4	40.8	258 24.0	13.0	9 46.8	14.2	57.4
D	12	358 44.1	S22 40.5	272 56.0	12.9	N10 01.0	14.1	57.5
A	13	13 43.8	40.3	287 27.9	12.9	10 15.1	14.2	57.5
Y	14	28 43.5	40.0	301 59.8	12.7	10 29.3	14.1	57.6
	15	43 43.3	. . 39.7	316 31.5	12.7	10 43.4	14.1	57.6
	16	58 43.0	39.4	331 03.2	12.5	10 57.5	14.0	57.6
	17	73 42.7	39.2	345 34.7	12.5	11 11.5	14.0	57.7
	18	88 42.4	S22 38.9	0 06.2	12.4	N11 25.5	14.0	57.7
	19	103 42.1	38.6	14 37.6	12.3	11 39.5	13.9	57.7
	20	118 41.9	38.3	29 08.9	12.2	11 53.4	13.9	57.8
	21	133 41.6	. . 38.1	43 40.1	12.1	12 07.3	13.9	57.8
	22	148 41.3	37.8	58 11.2	12.1	12 21.2	13.8	57.8
	23	163 41.0	37.5	72 42.3	11.9	12 35.0	13.8	57.9
5	00	178 40.7	S22 37.2	87 13.2	11.8	N12 48.8	13.7	57.9
	01	193 40.5	36.9	101 44.0	11.7	13 02.5	13.7	57.9
	02	208 40.2	36.7	116 14.7	11.7	13 16.2	13.6	58.0
	03	223 39.9	. . 36.4	130 45.4	11.5	13 29.8	13.6	58.0
	04	238 39.6	36.1	145 15.9	11.4	13 43.4	13.5	58.1
	05	253 39.3	35.8	159 46.3	11.3	13 56.9	13.5	58.1
	06	268 39.1	S22 35.5	174 16.6	11.2	N14 10.4	13.4	58.1
	07	283 38.8	35.2	188 46.8	11.2	14 23.8	13.3	58.2
	08	298 38.5	35.0	203 17.0	11.0	14 37.1	13.3	58.2
M	09	313 38.2	. . 34.7	217 47.0	10.9	14 50.4	13.3	58.2
O	10	328 38.0	34.4	232 16.9	10.7	15 03.7	13.2	58.3
N	11	343 37.7	34.1	246 46.6	10.7	15 16.9	13.1	58.3
D	12	358 37.4	S22 33.8	261 16.3	10.6	N15 30.0	13.1	58.3
A	13	13 37.1	33.5	275 45.9	10.4	15 43.1	13.0	58.4
Y	14	28 36.8	33.2	290 15.3	10.4	15 56.1	12.9	58.4
	15	43 36.6	. . 32.9	304 44.7	10.2	16 09.0	12.8	58.5
	16	58 36.3	32.6	319 13.9	10.1	16 21.8	12.8	58.5
	17	73 36.0	32.4	333 43.0	10.0	16 34.6	12.7	58.5
	18	88 35.7	S22 32.1	348 12.0	9.8	N16 47.3	12.7	58.6
	19	103 35.5	31.8	2 40.8	9.8	17 00.0	12.5	58.6
	20	118 35.2	31.5	17 09.6	9.6	17 12.5	12.5	58.6
	21	133 34.9	. . 31.2	31 38.2	9.5	17 25.0	12.4	58.7
	22	148 34.6	30.9	46 06.7	9.4	17 37.4	12.4	58.7
	23	163 34.4	30.6	60 35.1	9.2	17 49.8	12.2	58.8
6	00	178 34.1	S22 30.3	75 03.3	9.2	N18 02.0	12.2	58.8
	01	193 33.8	30.0	89 31.5	9.0	18 14.2	12.0	58.8
	02	208 33.5	29.7	103 59.5	8.9	18 26.2	12.0	58.9
	03	223 33.3	. . 29.4	118 27.4	8.7	18 38.2	11.9	58.9
	04	238 33.0	29.1	132 55.1	8.7	18 50.1	11.8	58.9
	05	253 32.7	28.8	147 22.8	8.5	19 01.9	11.7	59.0
	06	268 32.4	S22 28.5	161 50.3	8.4	N19 13.6	11.6	59.0
	07	283 32.2	28.2	176 17.7	8.2	19 25.2	11.6	59.0
T	08	298 31.9	27.9	190 44.9	8.1	19 36.8	11.4	59.1
U	09	313 31.6	. . 27.6	205 12.0	8.0	19 48.2	11.3	59.1
E	10	328 31.3	27.3	219 39.0	7.9	19 59.5	11.2	59.2
S	11	343 31.1	26.9	234 05.9	7.7	20 10.7	11.1	59.2
D	12	358 30.8	S22 26.6	248 32.6	7.6	N20 21.8	11.0	59.2
A	13	13 30.5	26.3	262 59.2	7.5	20 32.8	10.9	59.3
Y	14	28 30.3	26.0	277 25.7	7.3	20 43.7	10.8	59.3
	15	43 30.0	. . 25.7	291 52.0	7.2	20 54.5	10.6	59.3
	16	58 29.7	25.4	306 18.2	7.1	21 05.1	10.6	59.4
	17	73 29.4	25.1	320 44.3	6.9	21 15.7	10.4	59.4
	18	88 29.2	S22 24.8	335 10.2	6.8	N21 26.1	10.3	59.4
	19	103 28.9	24.5	349 36.0	6.7	21 36.4	10.2	59.5
	20	118 28.6	24.2	4 01.7	6.5	21 46.6	10.1	59.5
	21	133 28.4	. . 23.8	18 27.2	6.4	21 56.7	9.9	59.6
	22	148 28.1	23.5	32 52.6	6.3	22 06.6	9.8	59.6
	23	163 27.8	23.2	47 17.9	6.1	N22 16.4	9.7	59.6
		SD 16.3	d 0.3	SD 15.7		15.9		16.1

Lat.	Twilight Naut.	Twilight Civil	Sunrise	Moonrise 4	Moonrise 5	Moonrise 6	Moonrise 7
°	h m	h m	h m	h m	h m	h m	h m
N 72	08 19	10 29	■	09 42	08 54	□	□
N 70	08 01	09 42	■	09 56	09 24	08 11	□
68	07 47	09 12	11 28	10 08	09 46	09 10	□
66	07 35	08 49	10 19	10 17	10 04	09 45	08 58
64	07 25	08 31	09 44	10 25	10 18	10 11	09 59
62	07 16	08 16	09 19	10 32	10 31	10 31	10 33
60	07 08	08 04	09 00	10 38	10 41	10 47	10 59
N 58	07 01	07 53	08 43	10 43	10 50	11 01	11 19
56	06 55	07 43	08 30	10 48	10 58	11 13	11 36
54	06 49	07 35	08 18	10 52	11 06	11 24	11 50
52	06 44	07 27	08 07	10 56	11 12	11 33	12 03
50	06 39	07 20	07 58	11 00	11 18	11 42	12 14
45	06 28	07 05	07 38	11 08	11 31	12 00	12 37
N 40	06 18	06 52	07 22	11 14	11 41	12 14	12 56
35	06 09	06 41	07 09	11 20	11 51	12 27	13 12
30	06 01	06 31	06 57	11 25	11 59	12 38	13 25
20	05 45	06 12	06 36	11 33	12 13	12 57	13 49
N 10	05 30	05 56	06 19	11 41	12 25	13 14	14 09
0	05 13	05 39	06 02	11 48	12 37	13 30	14 28
S 10	04 55	05 22	05 45	11 56	12 48	13 46	14 48
20	04 33	05 02	05 27	12 04	13 01	14 03	15 08
30	04 05	04 38	05 05	12 13	13 16	14 23	15 33
35	03 47	04 24	04 53	12 18	13 24	14 34	15 47
40	03 25	04 06	04 39	12 24	13 34	14 48	16 03
45	02 56	03 45	04 21	12 31	13 45	15 04	16 23
S 50	02 14	03 16	04 00	12 40	13 59	15 23	16 48
52	01 49	03 02	03 50	12 44	14 06	15 33	17 01
54	01 13	02 45	03 38	12 48	14 13	15 43	17 14
56	////	02 25	03 25	12 53	14 22	15 55	17 31
58	////	01 59	03 09	12 58	14 31	16 09	17 50
S 60	////	01 20	02 50	13 05	14 42	16 26	18 14

Lat.	Sunset	Twilight Civil	Twilight Naut.	Moonset 4	Moonset 5	Moonset 6	Moonset 7
°	h m	h m	h m	h m	h m	h m	h m
N 72	■	13 42	15 52	01 00	03 28	□	□
N 70	■	14 29	16 10	00 49	03 00	06 00	□
68	12 44	14 59	16 25	00 40	02 39	05 02	□
66	13 52	15 22	16 37	00 33	02 23	04 29	07 15
64	14 27	15 40	16 47	00 26	02 10	04 04	06 14
62	14 52	15 55	16 56	00 21	01 59	03 45	05 41
60	15 12	16 08	17 03	00 16	01 50	03 30	05 16
N 58	15 28	16 18	17 10	00 12	01 41	03 17	04 56
56	15 42	16 28	17 16	00 09	01 34	03 05	04 40
54	15 54	16 37	17 22	00 05	01 28	02 55	04 26
52	16 04	16 44	17 27	00 02	01 22	02 47	04 14
50	16 13	16 51	17 32	00 00	01 17	02 39	04 03
45	16 33	17 07	17 43	25 06	01 06	02 22	03 41
N 40	16 49	17 19	17 53	24 57	00 57	02 09	03 23
35	17 03	17 31	18 02	24 49	00 49	01 57	03 09
30	17 14	17 41	18 10	24 42	00 42	01 47	02 56
20	17 35	17 59	18 26	24 31	00 31	01 30	02 34
N 10	17 52	18 15	18 41	24 20	00 20	01 15	02 15
0	18 09	18 32	18 58	24 11	00 11	01 01	01 57
S 10	18 26	18 49	19 16	24 01	00 01	00 48	01 39
20	18 44	19 09	19 38	23 51	24 33	00 33	01 21
30	19 05	19 33	20 05	23 40	24 16	00 16	00 59
35	19 18	19 47	20 23	23 33	24 06	00 06	00 46
40	19 32	20 05	20 45	23 26	23 55	24 32	00 32
45	19 49	20 26	21 14	23 17	23 43	24 15	00 15
S 50	20 11	20 54	21 56	23 07	23 27	23 54	24 32
52	20 21	21 08	22 20	23 02	23 20	23 44	24 19
54	20 32	21 25	22 55	22 57	23 12	23 33	24 05
56	20 46	21 45	////	22 51	23 03	23 20	23 49
58	21 01	22 11	////	22 44	22 52	23 05	23 29
S 60	21 20	22 49	////	22 37	22 41	22 48	23 04

Day	SUN Eqn. of Time 00^h	SUN Eqn. of Time 12^h	SUN Mer. Pass.	MOON Mer. Pass. Upper	MOON Mer. Pass. Lower	MOON Age	MOON Phase
d	m s	m s	h m	h m	h m	d %	
4	04 49	05 03	12 05	18 00	05 36	08 50	
5	05 16	05 30	12 05	18 49	06 24	09 61	
6	05 43	05 56	12 06	19 43	07 15	10 72	

UT		ARIES	VENUS −4.5		MARS +1.3		JUPITER −1.9		SATURN +0.9	
		GHA	GHA	Dec	GHA	Dec	GHA	Dec	GHA	Dec
d h		° ′	° ′	° ′	° ′	° ′	° ′	° ′	° ′	° ′
7 00		106 41.6	130 39.5	S11 07.6	187 52.9	S23 56.7	164 05.5	S20 29.5	293 28.0	N 5 11.2
01		121 44.1	145 39.5	06.4	202 53.3	56.6	179 07.3	29.4	308 30.5	11.3
02		136 46.5	160 39.5	05.3	217 53.7	56.5	194 09.2	29.2	323 33.0	11.3
03		151 49.0	175 39.5	. . 04.1	232 54.1	. . 56.4	209 11.0	. . 29.1	338 35.5	. . 11.3
04		166 51.5	190 39.5	03.0	247 54.5	56.3	224 12.9	29.0	353 37.9	11.3
05		181 53.9	205 39.5	01.8	262 54.9	56.2	239 14.7	28.9	8 40.4	11.3
06		196 56.4	220 39.6	S11 00.7	277 55.3	S23 56.1	254 16.6	S20 28.8	23 42.9	N 5 11.4
07	WEDNESDAY	211 58.9	235 39.6	10 59.5	292 55.7	56.0	269 18.5	28.6	38 45.4	11.4
08		227 01.3	250 39.6	58.4	307 56.1	55.9	284 20.3	28.5	53 47.9	11.4
09		242 03.8	265 39.6	. . 57.2	322 56.5	. . 55.8	299 22.2	. . 28.4	68 50.4	. . 11.4
10		257 06.3	280 39.6	56.0	337 56.9	55.7	314 24.0	28.3	83 52.9	11.4
11		272 08.7	295 39.6	54.9	352 57.3	55.6	329 25.9	28.1	98 55.4	11.5
12		287 11.2	310 39.6	S10 53.7	7 57.7	S23 55.5	344 27.7	S20 28.0	113 57.8	N 5 11.5
13		302 13.7	325 39.6	52.6	22 58.1	55.4	359 29.6	27.9	129 00.3	11.5
14		317 16.1	340 39.7	51.4	37 58.5	55.3	14 31.4	27.8	144 02.8	11.5
15		332 18.6	355 39.7	. . 50.3	52 58.8	. . 55.2	29 33.3	. . 27.7	159 05.3	. . 11.6
16		347 21.0	10 39.7	49.1	67 59.2	55.1	44 35.2	27.5	174 07.8	11.6
17		2 23.5	25 39.7	48.0	82 59.6	55.0	59 37.0	27.4	189 10.3	11.6
18		17 26.0	40 39.7	S10 46.8	98 00.0	S23 54.9	74 38.9	S20 27.3	204 12.8	N 5 11.6
19		32 28.4	55 39.8	45.7	113 00.4	54.8	89 40.7	27.2	219 15.3	11.7
20		47 30.9	70 39.8	44.5	128 00.8	54.7	104 42.6	27.0	234 17.8	11.7
21		62 33.4	85 39.8	. . 43.4	143 01.2	. . 54.6	119 44.4	. . 26.9	249 20.2	. . 11.7
22		77 35.8	100 39.8	42.2	158 01.6	54.5	134 46.3	26.8	264 22.7	11.7
23		92 38.3	115 39.9	41.0	173 02.0	54.4	149 48.2	26.7	279 25.2	11.7
8 00		107 40.8	130 39.9	S10 39.9	188 02.4	S23 54.3	164 50.0	S20 26.6	294 27.7	N 5 11.8
01		122 43.2	145 39.9	38.7	203 02.8	54.2	179 51.9	26.4	309 30.2	11.8
02		137 45.7	160 39.9	37.6	218 03.2	54.1	194 53.7	26.3	324 32.7	11.8
03		152 48.1	175 40.0	. . 36.4	233 03.6	. . 54.0	209 55.6	. . 26.2	339 35.2	. . 11.8
04		167 50.6	190 40.0	35.3	248 04.0	53.8	224 57.4	26.1	354 37.7	11.9
05		182 53.1	205 40.0	34.1	263 04.4	53.7	239 59.3	25.9	9 40.2	11.9
06		197 55.5	220 40.1	S10 32.9	278 04.8	S23 53.6	255 01.1	S20 25.8	24 42.7	N 5 11.9
07	THURSDAY	212 58.0	235 40.1	31.8	293 05.2	53.5	270 03.0	25.7	39 45.2	11.9
08		228 00.5	250 40.1	30.6	308 05.6	53.4	285 04.8	25.6	54 47.6	12.0
09		243 02.9	265 40.2	. . 29.5	323 06.0	. . 53.3	300 06.7	. . 25.4	69 50.1	. . 12.0
10		258 05.4	280 40.2	28.3	338 06.4	53.2	315 08.6	25.3	84 52.6	12.0
11		273 07.9	295 40.2	27.1	353 06.8	53.1	330 10.4	25.2	99 55.1	12.0
12		288 10.3	310 40.3	S10 26.0	8 07.2	S23 53.0	345 12.3	S20 25.1	114 57.6	N 5 12.1
13		303 12.8	325 40.3	24.8	23 07.6	52.9	0 14.1	25.0	130 00.1	12.1
14		318 15.3	340 40.3	23.7	38 08.0	52.8	15 16.0	24.8	145 02.6	12.1
15		333 17.7	355 40.4	. . 22.5	53 08.4	. . 52.6	30 17.8	. . 24.7	160 05.1	. . 12.1
16		348 20.2	10 40.4	21.3	68 08.7	52.5	45 19.7	24.6	175 07.6	12.2
17		3 22.6	25 40.5	20.2	83 09.1	52.4	60 21.5	24.5	190 10.1	12.2
18		18 25.1	40 40.5	S10 19.0	98 09.5	S23 52.3	75 23.4	S20 24.3	205 12.6	N 5 12.2
19		33 27.6	55 40.5	17.9	113 09.9	52.2	90 25.3	24.2	220 15.1	12.2
20		48 30.0	70 40.6	16.7	128 10.3	52.1	105 27.1	24.1	235 17.6	12.3
21		63 32.5	85 40.6	. . 15.5	143 10.7	. . 52.0	120 29.0	. . 24.0	250 20.0	. . 12.3
22		78 35.0	100 40.7	14.4	158 11.1	51.8	135 30.8	23.8	265 22.5	12.3
23		93 37.4	115 40.7	13.2	173 11.5	51.7	150 32.7	23.7	280 25.0	12.3
9 00		108 39.9	130 40.8	S10 12.0	188 11.9	S23 51.6	165 34.5	S20 23.6	295 27.5	N 5 12.4
01		123 42.4	145 40.8	10.9	203 12.3	51.5	180 36.4	23.5	310 30.0	12.4
02		138 44.8	160 40.9	09.7	218 12.7	51.4	195 38.2	23.3	325 32.5	12.4
03		153 47.3	175 40.9	. . 08.5	233 13.1	. . 51.3	210 40.1	. . 23.2	340 35.0	. . 12.4
04		168 49.8	190 41.0	07.4	248 13.5	51.1	225 42.0	23.1	355 37.5	12.5
05		183 52.2	205 41.0	06.2	263 13.9	51.0	240 43.8	23.0	10 40.0	12.5
06		198 54.7	220 41.1	S10 05.1	278 14.3	S23 50.9	255 45.7	S20 22.9	25 42.5	N 5 12.5
07		213 57.1	235 41.1	03.9	293 14.7	50.8	270 47.5	22.7	40 45.0	12.5
08		228 59.6	250 41.2	02.7	308 15.1	50.7	285 49.4	22.6	55 47.5	12.6
09	FRIDAY	244 02.1	265 41.2	. . 01.6	323 15.5	. . 50.5	300 51.2	. . 22.5	70 50.0	. . 12.6
10		259 04.5	280 41.3	10 00.4	338 15.9	50.4	315 53.1	22.4	85 52.5	12.6
11		274 07.0	295 41.4	9 59.2	353 16.3	50.3	330 54.9	22.2	100 55.0	12.6
12		289 09.5	310 41.4	S 9 58.1	8 16.7	S23 50.2	345 56.8	S20 22.1	115 57.5	N 5 12.7
13		304 11.9	325 41.5	56.9	23 17.1	50.1	0 58.6	22.0	131 00.0	12.7
14		319 14.4	340 41.5	55.7	38 17.5	49.9	16 00.5	21.9	146 02.5	12.7
15		334 16.9	355 41.6	. . 54.6	53 17.8	. . 49.8	31 02.4	. . 21.7	161 05.0	. . 12.7
16		349 19.3	10 41.7	53.4	68 18.2	49.7	46 04.2	21.6	176 07.5	12.8
17		4 21.8	25 41.7	52.2	83 18.6	49.6	61 06.1	21.5	191 10.0	12.8
18		19 24.3	40 41.8	S 9 51.0	98 19.0	S23 49.4	76 07.9	S20 21.4	206 12.5	N 5 12.8
19		34 26.7	55 41.9	49.9	113 19.4	49.3	91 09.8	21.2	221 15.0	12.8
20		49 29.2	70 41.9	48.7	128 19.8	49.2	106 11.6	21.1	236 17.5	12.9
21		64 31.6	85 42.0	. . 47.5	143 20.2	. . 49.1	121 13.5	. . 21.0	251 19.9	. . 12.9
22		79 34.1	100 42.1	46.4	158 20.6	48.9	136 15.3	20.9	266 22.4	12.9
23		94 36.6	115 42.1	45.2	173 21.0	48.8	151 17.2	20.7	281 24.9	12.9
Mer. Pass.		h m 16 46.5	v 0.0	d 1.2	v 0.4	d 0.1	v 1.9	d 0.1	v 2.5	d 0.0

STARS

Name	SHA	Dec
	° ′	° ′
Acamar	315 20.5	S40 16.2
Achernar	335 28.9	S57 11.7
Acrux	173 13.2	S63 08.8
Adhara	255 14.8	S28 59.1
Aldebaran	290 52.9	N16 31.7
Alioth	166 23.3	N55 54.2
Alkaid	153 01.4	N49 15.7
Al Na'ir	27 48.1	S46 55.2
Alnilam	275 49.4	S 1 11.8
Alphard	217 59.1	S 8 41.9
Alphecca	126 14.0	N26 40.8
Alpheratz	357 47.1	N29 08.6
Altair	62 11.8	N 8 53.5
Ankaa	353 18.9	S42 15.6
Antares	112 30.6	S26 27.1
Arcturus	145 58.8	N19 07.9
Atria	107 36.0	S69 02.5
Avior	234 18.9	S59 32.2
Bellatrix	278 35.2	N 6 21.5
Betelgeuse	271 04.5	N 7 24.5
Canopus	263 57.1	S52 42.0
Capella	280 38.9	N46 00.6
Deneb	49 34.3	N45 18.8
Denebola	182 36.9	N14 31.1
Diphda	348 59.2	S17 56.3
Dubhe	193 55.1	N61 41.8
Elnath	278 16.5	N28 37.0
Eltanin	90 48.2	N51 29.1
Enif	33 50.7	N 9 55.0
Fomalhaut	15 27.7	S29 34.6
Gacrux	172 04.7	S57 09.7
Gienah	175 55.7	S17 35.6
Hadar	148 53.0	S60 24.8
Hamal	328 04.5	N23 30.5
Kaus Aust.	83 48.6	S34 22.9
Kochab	137 20.0	N74 06.7
Markab	13 41.8	N15 15.3
Menkar	314 18.3	N 4 07.6
Menkent	148 11.7	S36 24.8
Miaplacidus	221 39.9	S69 45.1
Mirfak	308 44.9	N49 53.9
Nunki	76 02.8	S26 17.2
Peacock	53 24.9	S56 42.5
Pollux	243 31.3	N28 00.2
Procyon	245 02.8	N 5 12.1
Rasalhague	96 09.8	N12 33.1
Regulus	207 46.7	N11 55.2
Rigel	281 14.9	S 8 11.5
Rigil Kent.	139 56.7	S60 52.2
Sabik	102 16.6	S15 44.2
Schedar	349 44.6	N56 35.6
Shaula	96 26.8	S37 06.6
Sirius	258 36.3	S16 43.7
Spica	158 34.8	S11 12.6
Suhail	222 54.6	S43 28.1
Vega	80 41.6	N38 47.4
Zuben'ubi	137 09.3	S16 04.8

	SHA	Mer. Pass.
	° ′	h m
Venus	22 59.1	15 17
Mars	80 21.7	11 28
Jupiter	57 09.2	12 59
Saturn	186 47.0	4 21

Day	UT	SUN GHA	SUN Dec	MOON GHA	v	MOON Dec	d	HP
	d h	° ′	° ′	° ′	′	° ′	′	′
WEDNESDAY	7 00	178 27.6	S22 22.9	61 43.0	6.0	N22 26.1	9.5	59.7
	01	193 27.3	22.6	76 08.0	5.9	22 35.6	9.5	59.7
	02	208 27.0	22.3	90 32.9	5.7	22 45.1	9.2	59.7
	03	223 26.7	. . 21.9	104 57.6	5.7	22 54.3	9.2	59.8
	04	238 26.5	21.6	119 22.3	5.4	23 03.5	9.0	59.8
	05	253 26.2	21.3	133 46.7	5.4	23 12.5	8.8	59.8
	06	268 25.9	S22 21.0	148 11.1	5.2	N23 21.3	8.8	59.9
	07	283 25.7	20.7	162 35.3	5.1	23 30.1	8.5	59.9
	08	298 25.4	20.3	176 59.4	5.0	23 38.6	8.5	59.9
	09	313 25.1	. . 20.0	191 23.4	4.8	23 47.1	8.2	60.0
	10	328 24.9	19.7	205 47.2	4.7	23 55.3	8.1	60.0
	11	343 24.6	19.4	220 10.9	4.6	24 03.4	8.0	60.0
	12	358 24.3	S22 19.0	234 34.5	4.5	N24 11.4	7.8	60.1
	13	13 24.1	18.7	248 58.0	4.3	24 19.2	7.7	60.1
	14	28 23.8	18.4	263 21.3	4.2	24 26.9	7.5	60.1
	15	43 23.5	. . 18.1	277 44.5	4.1	24 34.4	7.3	60.1
	16	58 23.3	17.7	292 07.6	4.0	24 41.7	7.2	60.2
	17	73 23.0	17.4	306 30.6	3.8	24 48.9	7.0	60.2
	18	88 22.7	S22 17.1	320 53.4	3.8	N24 55.9	6.8	60.2
	19	103 22.5	16.7	335 16.2	3.6	25 02.7	6.7	60.3
	20	118 22.2	16.4	349 38.8	3.5	25 09.4	6.4	60.3
	21	133 21.9	. . 16.1	4 01.3	3.4	25 15.8	6.4	60.3
	22	148 21.7	15.7	18 23.7	3.3	25 22.2	6.1	60.4
	23	163 21.4	15.4	32 46.0	3.1	25 28.3	6.0	60.4
THURSDAY	8 00	178 21.2	S22 15.1	47 08.1	3.1	N25 34.3	5.8	60.4
	01	193 20.9	14.7	61 30.2	3.0	25 40.1	5.6	60.4
	02	208 20.6	14.4	75 52.2	2.8	25 45.7	5.4	60.5
	03	223 20.4	. . 14.1	90 14.0	2.8	25 51.1	5.2	60.5
	04	238 20.1	13.7	104 35.8	2.6	25 56.3	5.1	60.5
	05	253 19.8	13.4	118 57.4	2.6	26 01.4	4.9	60.6
	06	268 19.6	S22 13.0	133 19.0	2.4	N26 06.3	4.7	60.6
	07	283 19.3	12.7	147 40.4	2.4	26 11.0	4.4	60.6
	08	298 19.0	12.4	162 01.8	2.3	26 15.4	4.4	60.6
	09	313 18.8	. . 12.0	176 23.1	2.2	26 19.8	4.1	60.7
	10	328 18.5	11.7	190 44.3	2.1	26 23.9	3.9	60.7
	11	343 18.3	11.3	205 05.4	2.0	26 27.8	3.7	60.7
	12	358 18.0	S22 11.0	219 26.4	2.0	N26 31.5	3.5	60.7
	13	13 17.7	10.6	233 47.4	1.9	26 35.0	3.3	60.8
	14	28 17.5	10.3	248 08.3	1.8	26 38.3	3.2	60.8
	15	43 17.2	. . 09.9	262 29.1	1.7	26 41.5	2.9	60.8
	16	58 17.0	09.6	276 49.8	1.6	26 44.4	2.7	60.8
	17	73 16.7	09.3	291 10.4	1.6	26 47.1	2.5	60.9
	18	88 16.4	S22 08.9	305 31.0	1.6	N26 49.6	2.3	60.9
	19	103 16.2	08.6	319 51.6	1.4	26 51.9	2.2	60.9
	20	118 15.9	08.2	334 12.0	1.4	26 54.1	1.9	60.9
	21	133 15.7	. . 07.9	348 32.4	1.4	26 56.0	1.7	60.9
	22	148 15.4	07.5	2 52.8	1.3	26 57.7	1.4	61.0
	23	163 15.1	07.2	17 13.1	1.3	26 59.1	1.3	61.0
FRIDAY	9 00	178 14.9	S22 06.8	31 33.4	1.2	N27 00.4	1.1	61.0
	01	193 14.6	06.4	45 53.6	1.2	27 01.5	0.9	61.0
	02	208 14.4	06.1	60 13.8	1.1	27 02.4	0.6	61.0
	03	223 14.1	. . 05.7	74 33.9	1.1	27 03.0	0.4	61.1
	04	238 13.8	05.4	88 54.0	1.1	27 03.4	0.3	61.1
	05	253 13.6	05.0	103 14.1	1.1	27 03.7	0.0	61.1
	06	268 13.3	S22 04.7	117 34.2	1.0	N27 03.7	0.2	61.1
	07	283 13.1	04.3	131 54.2	1.0	27 03.5	0.4	61.1
	08	298 12.8	03.9	146 14.2	1.0	27 03.1	0.6	61.1
	09	313 12.6	. . 03.6	160 34.2	1.0	27 02.5	0.9	61.1
	10	328 12.3	03.2	174 54.2	0.9	27 01.6	1.0	61.2
	11	343 12.1	02.9	189 14.1	1.0	27 00.6	1.3	61.2
	12	358 11.8	S22 02.5	203 34.1	1.0	N26 59.3	1.4	61.2
	13	13 11.5	02.1	217 54.1	0.9	26 57.9	1.7	61.2
	14	28 11.3	01.8	232 14.0	1.0	26 56.2	1.9	61.2
	15	43 11.0	. . 01.4	246 34.0	1.0	26 54.3	2.1	61.2
	16	58 10.8	01.0	260 54.0	0.9	26 52.2	2.4	61.2
	17	73 10.5	00.7	275 13.9	1.0	26 49.8	2.5	61.2
	18	88 10.3	S22 00.3	289 33.9	1.1	N26 47.3	2.8	61.3
	19	103 10.0	21 59.9	303 54.0	1.0	26 44.5	2.9	61.3
	20	118 09.8	59.6	318 14.0	1.1	26 41.6	3.2	61.3
	21	133 09.5	. . 59.2	332 34.1	1.0	26 38.4	3.4	61.3
	22	148 09.3	58.8	346 54.1	1.2	26 35.0	3.6	61.3
	23	163 09.0	58.5	1 14.3	1.1	N26 31.4	3.8	61.3
		SD 16.3	*d* 0.3	SD 16.4		16.5		16.7

Lat.	Twilight Naut.	Twilight Civil	Sunrise	Moonrise 7	Moonrise 8	Moonrise 9	Moonrise 10
°	h m	h m	h m	h m	h m	h m	h m
N 72	08 15	10 18	■	□	□	□	□
N 70	07 58	09 36	■	□	□	□	□
68	07 44	09 07	11 08	□	□	□	□
66	07 32	08 46	10 12	08 58	□	□	□
64	07 23	08 28	09 40	09 59	□	□	12 32
62	07 14	08 14	09 16	10 33	10 46	11 37	13 26
60	07 07	08 02	08 57	10 59	11 25	12 23	13 59
N 58	07 00	07 51	08 41	11 19	11 52	12 53	14 23
56	06 54	07 42	08 28	11 36	12 14	13 15	14 42
54	06 48	07 34	08 16	11 50	12 32	13 34	14 58
52	06 43	07 26	08 06	12 03	12 47	13 50	15 12
50	06 38	07 19	07 57	12 14	13 00	14 04	15 24
45	06 28	07 04	07 38	12 37	13 27	14 32	15 49
N 40	06 18	06 52	07 22	12 56	13 49	14 54	16 09
35	06 09	06 41	07 09	13 12	14 06	15 12	16 26
30	06 01	06 31	06 57	13 25	14 22	15 28	16 40
20	05 46	06 13	06 37	13 49	14 48	15 54	17 05
N 10	05 31	05 57	06 20	14 09	15 11	16 17	17 25
0	05 15	05 41	06 03	14 28	15 32	16 39	17 45
S 10	04 57	05 24	05 46	14 48	15 54	17 00	18 05
20	04 35	05 04	05 29	15 08	16 16	17 23	18 25
30	04 08	04 41	05 08	15 33	16 43	17 50	18 49
35	03 50	04 26	04 56	15 47	16 59	18 06	19 03
40	03 29	04 09	04 41	16 03	17 17	18 24	19 20
45	03 00	03 48	04 25	16 23	17 40	18 46	19 39
S 50	02 19	03 21	04 04	16 48	18 08	19 14	20 03
52	01 56	03 07	03 54	17 01	18 22	19 28	20 15
54	01 23	02 51	03 42	17 14	18 38	19 43	20 28
56	////	02 31	03 29	17 31	18 57	20 02	20 43
58	////	02 06	03 14	17 50	19 21	20 25	21 00
S 60	////	01 30	02 56	18 14	19 52	20 54	21 22

Lat.	Sunset	Twilight Civil	Twilight Naut.	Moonset 7	Moonset 8	Moonset 9	Moonset 10
°	h m	h m	h m	h m	h m	h m	h m
N 72	■	13 56	16 00	□	□	□	□
N 70	■	14 38	16 17	□	□	□	□
68	13 06	15 07	16 30	□	□	□	□
66	14 02	15 28	16 42	07 15	□	□	□
64	14 34	15 46	16 52	06 14	□	□	10 33
62	14 58	16 00	17 00	05 41	07 39	09 07	09 38
60	15 17	16 12	17 07	05 16	07 00	08 21	09 05
N 58	15 33	16 23	17 14	04 56	06 33	07 51	08 41
56	15 46	16 32	17 20	04 40	06 12	07 28	08 21
54	15 58	16 40	17 26	04 26	05 54	07 10	08 04
52	16 08	16 48	17 31	04 14	05 39	06 54	07 50
50	16 17	16 55	17 36	04 03	05 27	06 40	07 38
45	16 36	17 10	17 46	03 41	05 00	06 12	07 12
N 40	16 52	17 22	17 56	03 23	04 39	05 50	06 51
35	17 05	17 33	18 04	03 09	04 21	05 31	06 34
30	17 17	17 43	18 12	02 56	04 06	05 15	06 19
20	17 37	18 00	18 28	02 34	03 41	04 49	05 53
N 10	17 54	18 17	18 43	02 15	03 19	04 25	05 31
0	18 10	18 33	18 59	01 57	02 58	04 04	05 11
S 10	18 27	18 50	19 17	01 39	02 38	03 42	04 50
20	18 45	19 09	19 38	01 21	02 16	03 19	04 28
30	19 06	19 33	20 05	00 59	01 51	02 52	04 02
35	19 18	19 47	20 23	00 46	01 36	02 36	03 46
40	19 32	20 04	20 44	00 32	01 18	02 17	03 28
45	19 49	20 25	21 13	00 15	00 58	01 55	03 07
S 50	20 09	20 52	21 53	24 32	00 32	01 26	02 39
52	20 19	21 06	22 16	24 19	00 19	01 12	02 26
54	20 31	21 22	22 48	24 05	00 05	00 56	02 10
56	20 43	21 41	////	23 49	24 37	00 37	01 52
58	20 58	22 06	////	23 29	24 13	00 13	01 29
S 60	21 16	22 41	////	23 04	23 42	25 00	01 00

Day	SUN Eqn. of Time 00^h	SUN Eqn. of Time 12^h	SUN Mer. Pass.	MOON Mer. Pass. Upper	MOON Mer. Pass. Lower	Age	Phase
d	m s	m s	h m	h m	h m	d	%
7	06 09	06 22	12 06	20 43	08 13	11	81
8	06 35	06 47	12 07	21 48	09 15	12	90
9	07 00	07 12	12 07	22 55	10 21	13	96

UT		ARIES	VENUS −4.5		MARS +1.3		JUPITER −1.9		SATURN +0.9	
	d h	GHA	GHA	Dec	GHA	Dec	GHA	Dec	GHA	Dec
		° ′	° ′	° ′	° ′	° ′	° ′	° ′	° ′	° ′
	10 00	109 39.0	130 42.2	S 9 44.0	188 21.4	S23 48.7	166 19.0	S20 20.6	296 27.4	N 5 13.0
	01	124 41.5	145 42.3	42.9	203 21.8	48.6	181 20.9	20.5	311 29.9	13.0
	02	139 44.0	160 42.3	41.7	218 22.2	48.4	196 22.8	20.4	326 32.4	13.0
	03	154 46.4	175 42.4	40.5	233 22.6	48.3	211 24.6	20.2	341 34.9	13.1
	04	169 48.9	190 42.5	39.3	248 23.0	48.2	226 26.5	20.1	356 37.4	13.1
	05	184 51.4	205 42.6	38.2	263 23.4	48.0	241 28.3	20.0	11 39.9	13.1
	06	199 53.8	220 42.6	S 9 37.0	278 23.8	S23 47.9	256 30.2	S20 19.9	26 42.4	N 5 13.1
	07	214 56.3	235 42.7	35.8	293 24.2	47.8	271 32.0	19.7	41 44.9	13.2
S	08	229 58.8	250 42.8	34.7	308 24.6	47.7	286 33.9	19.6	56 47.4	13.2
A	09	245 01.2	265 42.9	33.5	323 25.0	47.5	301 35.7	19.5	71 49.9	13.2
T	10	260 03.7	280 43.0	32.3	338 25.4	47.4	316 37.6	19.4	86 52.4	13.2
U	11	275 06.1	295 43.0	31.1	353 25.8	47.3	331 39.4	19.2	101 54.9	13.3
R	12	290 08.6	310 43.1	S 9 30.0	8 26.1	S23 47.1	346 41.3	S20 19.1	116 57.4	N 5 13.3
D	13	305 11.1	325 43.2	28.8	23 26.5	47.0	1 43.2	19.0	131 59.9	13.3
A	14	320 13.5	340 43.3	27.6	38 26.9	46.9	16 45.0	18.9	147 02.4	13.4
Y	15	335 16.0	355 43.4	26.5	53 27.3	46.7	31 46.9	18.7	162 04.9	13.4
	16	350 18.5	10 43.5	25.3	68 27.7	46.6	46 48.7	18.6	177 07.4	13.4
	17	5 20.9	25 43.5	24.1	83 28.1	46.5	61 50.6	18.5	192 09.9	13.4
	18	20 23.4	40 43.6	S 9 22.9	98 28.5	S23 46.3	76 52.4	S20 18.4	207 12.4	N 5 13.5
	19	35 25.9	55 43.7	21.8	113 28.9	46.2	91 54.3	18.2	222 14.9	13.5
	20	50 28.3	70 43.8	20.6	128 29.3	46.0	106 56.1	18.1	237 17.4	13.5
	21	65 30.8	85 43.9	19.4	143 29.7	45.9	121 58.0	18.0	252 20.0	13.6
	22	80 33.3	100 44.0	18.2	158 30.1	45.8	136 59.8	17.9	267 22.5	13.6
	23	95 35.7	115 44.1	17.1	173 30.5	45.6	152 01.7	17.7	282 25.0	13.6
	11 00	110 38.2	130 44.2	S 9 15.9	188 30.9	S23 45.5	167 03.5	S20 17.6	297 27.5	N 5 13.6
	01	125 40.6	145 44.3	14.7	203 31.3	45.4	182 05.4	17.5	312 30.0	13.7
	02	140 43.1	160 44.4	13.5	218 31.7	45.2	197 07.3	17.4	327 32.5	13.7
	03	155 45.6	175 44.5	12.4	233 32.1	45.1	212 09.1	17.2	342 35.0	13.7
	04	170 48.0	190 44.6	11.2	248 32.5	44.9	227 11.0	17.1	357 37.5	13.8
	05	185 50.5	205 44.7	10.0	263 32.9	44.8	242 12.8	17.0	12 40.0	13.8
	06	200 53.0	220 44.7	S 9 08.8	278 33.3	S23 44.7	257 14.7	S20 16.9	27 42.5	N 5 13.8
	07	215 55.4	235 44.8	07.6	293 33.7	44.5	272 16.5	16.7	42 45.0	13.8
	08	230 57.9	250 44.9	06.5	308 34.1	44.4	287 18.4	16.6	57 47.5	13.9
S	09	246 00.4	265 45.1	05.3	323 34.5	44.2	302 20.2	16.5	72 50.0	13.9
U	10	261 02.8	280 45.2	04.1	338 34.8	44.1	317 22.1	16.4	87 52.5	13.9
N	11	276 05.3	295 45.3	02.9	353 35.2	44.0	332 23.9	16.2	102 55.0	14.0
D	12	291 07.8	310 45.4	S 9 01.8	8 35.6	S23 43.8	347 25.8	S20 16.1	117 57.5	N 5 14.0
A	13	306 10.2	325 45.5	9 00.6	23 36.0	43.7	2 27.6	16.0	133 00.0	14.0
Y	14	321 12.7	340 45.6	8 59.4	38 36.4	43.5	17 29.5	15.8	148 02.5	14.0
	15	336 15.1	355 45.7	58.2	53 36.8	43.4	32 31.4	15.7	163 05.0	14.1
	16	351 17.6	10 45.8	57.0	68 37.2	43.2	47 33.2	15.6	178 07.5	14.1
	17	6 20.1	25 45.9	55.9	83 37.6	43.1	62 35.1	15.5	193 10.0	14.1
	18	21 22.5	40 46.0	S 8 54.7	98 38.0	S23 42.9	77 36.9	S20 15.3	208 12.5	N 5 14.2
	19	36 25.0	55 46.1	53.5	113 38.4	42.8	92 38.8	15.2	223 15.0	14.2
	20	51 27.5	70 46.2	52.3	128 38.8	42.6	107 40.6	15.1	238 17.5	14.2
	21	66 29.9	85 46.3	51.1	143 39.2	42.5	122 42.5	15.0	253 20.1	14.3
	22	81 32.4	100 46.5	50.0	158 39.6	42.3	137 44.3	14.8	268 22.6	14.3
	23	96 34.9	115 46.6	48.8	173 40.0	42.2	152 46.2	14.7	283 25.1	14.3
	12 00	111 37.3	130 46.7	S 8 47.6	188 40.4	S23 42.1	167 48.0	S20 14.6	298 27.6	N 5 14.3
	01	126 39.8	145 46.8	46.4	203 40.8	41.9	182 49.9	14.5	313 30.1	14.4
	02	141 42.3	160 46.9	45.2	218 41.2	41.8	197 51.7	14.3	328 32.6	14.4
	03	156 44.7	175 47.0	44.1	233 41.6	41.6	212 53.6	14.2	343 35.1	14.4
	04	171 47.2	190 47.2	42.9	248 42.0	41.4	227 55.5	14.1	358 37.6	14.5
	05	186 49.6	205 47.3	41.7	263 42.4	41.3	242 57.3	13.9	13 40.1	14.5
	06	201 52.1	220 47.4	S 8 40.5	278 42.7	S23 41.1	257 59.2	S20 13.8	28 42.6	N 5 14.5
	07	216 54.6	235 47.5	39.3	293 43.1	41.0	273 01.0	13.7	43 45.1	14.6
	08	231 57.0	250 47.6	38.1	308 43.5	40.8	288 02.9	13.6	58 47.6	14.6
M	09	246 59.5	265 47.8	37.0	323 43.9	40.7	303 04.7	13.4	73 50.1	14.6
O	10	262 02.0	280 47.9	35.8	338 44.3	40.5	318 06.6	13.3	88 52.6	14.7
N	11	277 04.4	295 48.0	34.6	353 44.7	40.4	333 08.4	13.2	103 55.2	14.7
D	12	292 06.9	310 48.1	S 8 33.4	8 45.1	S23 40.2	348 10.3	S20 13.1	118 57.7	N 5 14.7
A	13	307 09.4	325 48.3	32.2	23 45.5	40.1	3 12.1	12.9	134 00.2	14.7
Y	14	322 11.8	340 48.4	31.0	38 45.9	39.9	18 14.0	12.8	149 02.7	14.8
	15	337 14.3	355 48.5	29.9	53 46.3	39.8	33 15.8	12.7	164 05.2	14.8
	16	352 16.8	10 48.7	28.7	68 46.7	39.6	48 17.7	12.6	179 07.7	14.8
	17	7 19.2	25 48.8	27.5	83 47.1	39.4	63 19.5	12.4	194 10.2	14.9
	18	22 21.7	40 48.9	S 8 26.3	98 47.5	S23 39.3	78 21.4	S20 12.3	209 12.7	N 5 14.9
	19	37 24.1	55 49.1	25.1	113 47.9	39.1	93 23.3	12.2	224 15.2	14.9
	20	52 26.6	70 49.2	23.9	128 48.3	39.0	108 25.1	12.0	239 17.7	15.0
	21	67 29.1	85 49.3	22.8	143 48.7	38.8	123 27.0	11.9	254 20.2	15.0
	22	82 31.5	100 49.5	21.6	158 49.1	38.7	138 28.8	11.8	269 22.8	15.0
	23	97 34.0	115 49.6	20.4	173 49.5	38.5	153 30.7	11.7	284 25.3	15.1
	Mer. Pass.	h m 16 34.7	v 0.1	d 1.2	v 0.4	d 0.1	v 1.9	d 0.1	v 2.5	d 0.0

STARS

Name	SHA	Dec
	° ′	° ′
Acamar	315 20.5	S40 16.[illegible]
Achernar	335 28.9	S57 11.[illegible]
Acrux	173 13.1	S63 08.8
Adhara	255 14.7	S28 59.[illegible]
Aldebaran	290 52.9	N16 31.[illegible]
Alioth	166 23.3	N55 54.[illegible]
Alkaid	153 01.4	N49 15.[illegible]
Al Na'ir	27 48.0	S46 55.[illegible]
Alnilam	275 49.4	S 1 11.8
Alphard	217 59.0	S 8 41.9
Alphecca	126 14.0	N26 40.8
Alpheratz	357 47.1	N29 08.6
Altair	62 11.8	N 8 53.5
Ankaa	353 18.9	S42 15.6
Antares	112 30.6	S26 27.1
Arcturus	145 58.8	N19 07.9
Atria	107 36.0	S69 02.5
Avior	234 18.9	S59 32.2
Bellatrix	278 35.2	N 6 21.5
Betelgeuse	271 04.5	N 7 24.5
Canopus	263 57.1	S52 42.1
Capella	280 38.9	N46 00.6
Deneb	49 34.3	N45 18.8
Denebola	182 36.8	N14 31.1
Diphda	348 59.2	S17 56.3
Dubhe	193 55.0	N61 41.8
Elnath	278 16.4	N28 37.0
Eltanin	90 48.2	N51 29.1
Enif	33 50.6	N 9 55.0
Fomalhaut	15 27.7	S29 34.6
Gacrux	172 04.7	S57 09.7
Gienah	175 55.6	S17 35.6
Hadar	148 53.0	S60 24.8
Hamal	328 04.5	N23 30.5
Kaus Aust.	83 48.6	S34 22.9
Kochab	137 19.9	N74 06.7
Markab	13 41.8	N15 15.3
Menkar	314 18.3	N 4 07.6
Menkent	148 11.6	S36 24.8
Miaplacidus	221 39.8	S69 45.1
Mirfak	308 44.9	N49 53.9
Nunki	76 02.7	S26 17.2
Peacock	53 24.8	S56 42.5
Pollux	243 31.3	N28 00.2
Procyon	245 02.8	N 5 12.1
Rasalhague	96 09.8	N12 33.1
Regulus	207 46.7	N11 55.2
Rigel	281 14.9	S 8 11.5
Rigil Kent.	139 56.7	S60 52.2
Sabik	102 16.6	S15 44.2
Schedar	349 44.6	N56 35.6
Shaula	96 26.8	S37 06.6
Sirius	258 36.3	S16 43.7
Spica	158 34.8	S11 12.6
Suhail	222 54.6	S43 28.1
Vega	80 41.6	N38 47.4
Zuben'ubi	137 09.2	S16 04.8

	SHA	Mer. Pass.
	° ′	h m
Venus	20 06.0	15 17
Mars	77 52.7	11 26
Jupiter	56 25.4	12 50
Saturn	186 49.3	4 09

	UT	SUN GHA	SUN Dec	MOON GHA	v	MOON Dec	d	HP
	d h	° ′	° ′	° ′	′	° ′	′	′
	10 00	178 08.7	S21 58.1	15 34.4	1.2	N26 27.6	4.0	61.3
	01	193 08.5	57.7	29 54.6	1.2	26 23.6	4.2	61.3
	02	208 08.2	57.4	44 14.8	1.3	26 19.4	4.4	61.3
	03	223 08.0	. . 57.0	58 35.1	1.3	26 15.0	4.7	61.3
	04	238 07.7	56.6	72 55.4	1.4	26 10.3	4.8	61.3
	05	253 07.5	56.2	87 15.8	1.4	26 05.5	5.0	61.3
	06	268 07.2	S21 55.9	101 36.2	1.5	N26 00.5	5.3	61.3
	07	283 07.0	55.5	115 56.7	1.6	25 55.2	5.4	61.3
S	08	298 06.7	55.1	130 17.3	1.6	25 49.8	5.7	61.3
A	09	313 06.5	. . 54.7	144 37.9	1.6	25 44.1	5.8	61.3
T	10	328 06.2	54.4	158 58.5	1.8	25 38.3	6.1	61.3
U	11	343 06.0	54.0	173 19.3	1.8	25 32.2	6.2	61.3
R	12	358 05.7	S21 53.6	187 40.1	1.8	N25 26.0	6.5	61.3
D	13	13 05.5	53.2	202 00.9	2.0	25 19.5	6.6	61.3
A	14	28 05.2	52.8	216 21.9	2.0	25 12.9	6.8	61.3
Y	15	43 05.0	. . 52.5	230 42.9	2.1	25 06.1	7.0	61.3
	16	58 04.7	52.1	245 04.0	2.2	24 59.1	7.2	61.3
	17	73 04.5	51.7	259 25.2	2.3	24 51.9	7.4	61.3
	18	88 04.2	S21 51.3	273 46.5	2.4	N24 44.5	7.6	61.3
	19	103 04.0	50.9	288 07.9	2.5	24 36.9	7.8	61.3
	20	118 03.7	50.5	302 29.4	2.5	24 29.1	7.9	61.3
	21	133 03.5	. . 50.1	316 50.9	2.7	24 21.2	8.2	61.3
	22	148 03.3	49.8	331 12.6	2.7	24 13.0	8.3	61.3
	23	163 03.0	49.4	345 34.3	2.8	24 04.7	8.5	61.3
	11 00	178 02.8	S21 49.0	359 56.1	3.0	N23 56.2	8.6	61.3
	01	193 02.5	48.6	14 18.1	3.0	23 47.6	8.9	61.3
	02	208 02.3	48.2	28 40.1	3.2	23 38.7	9.0	61.3
	03	223 02.0	. . 47.8	43 02.3	3.2	23 29.7	9.1	61.3
	04	238 01.8	47.4	57 24.5	3.4	23 20.6	9.4	61.3
	05	253 01.5	47.0	71 46.9	3.4	23 11.2	9.5	61.2
	06	268 01.3	S21 46.6	86 09.3	3.6	N23 01.7	9.7	61.2
	07	283 01.0	46.2	100 31.9	3.7	22 52.0	9.8	61.2
	08	298 00.8	45.8	114 54.6	3.8	22 42.2	10.0	61.2
S	09	313 00.5	. . 45.4	129 17.4	3.9	22 32.2	10.2	61.2
U	10	328 00.3	45.1	143 40.3	4.0	22 22.0	10.3	61.2
N	11	343 00.1	44.7	158 03.3	4.1	22 11.7	10.4	61.2
D	12	357 59.8	S21 44.3	172 26.4	4.3	N22 01.3	10.6	61.2
A	13	12 59.6	43.9	186 49.7	4.4	21 50.7	10.8	61.1
Y	14	27 59.3	43.5	201 13.1	4.5	21 39.9	10.9	61.1
	15	42 59.1	. . 43.1	215 36.6	4.6	21 29.0	11.1	61.1
	16	57 58.8	42.7	230 00.2	4.7	21 17.9	11.2	61.1
	17	72 58.6	42.3	244 23.9	4.9	21 06.7	11.3	61.1
	18	87 58.4	S21 41.9	258 47.8	4.9	N20 55.4	11.5	61.1
	19	102 58.1	41.5	273 11.7	5.1	20 43.9	11.6	61.0
	20	117 57.9	41.1	287 35.8	5.2	20 32.3	11.7	61.0
	21	132 57.6	. . 40.6	302 00.0	5.4	20 20.6	11.9	61.0
	22	147 57.4	40.2	316 24.4	5.4	20 08.7	12.0	61.0
	23	162 57.2	39.8	330 48.8	5.6	19 56.7	12.1	61.0
	12 00	177 56.9	S21 39.4	345 13.4	5.7	N19 44.6	12.2	60.9
	01	192 56.7	39.0	359 38.1	5.9	19 32.4	12.4	60.9
	02	207 56.4	38.6	14 03.0	5.9	19 20.0	12.5	60.9
	03	222 56.2	. . 38.2	28 27.9	6.1	19 07.5	12.6	60.9
	04	237 56.0	37.8	42 53.0	6.2	18 54.9	12.7	60.9
	05	252 55.7	37.4	57 18.2	6.3	18 42.2	12.9	60.8
	06	267 55.5	S21 37.0	71 43.5	6.5	N18 29.3	12.9	60.8
	07	282 55.2	36.6	86 09.0	6.5	18 16.4	13.0	60.8
	08	297 55.0	36.2	100 34.5	6.7	18 03.4	13.2	60.8
M	09	312 54.8	. . 35.7	115 00.2	6.9	17 50.2	13.3	60.7
O	10	327 54.5	35.3	129 26.1	6.9	17 36.9	13.3	60.7
N	11	342 54.3	34.9	143 52.0	7.1	17 23.6	13.5	60.7
D	12	357 54.0	S21 34.5	158 18.1	7.2	N17 10.1	13.5	60.7
A	13	12 53.8	34.1	172 44.3	7.3	16 56.6	13.7	60.6
Y	14	27 53.6	33.7	187 10.6	7.4	16 42.9	13.7	60.6
	15	42 53.3	. . 33.3	201 37.0	7.5	16 29.2	13.8	60.6
	16	57 53.1	32.8	216 03.5	7.7	16 15.4	14.0	60.5
	17	72 52.9	32.4	230 30.2	7.8	16 01.4	13.9	60.5
	18	87 52.6	S21 32.0	244 57.0	7.9	N15 47.5	14.1	60.5
	19	102 52.4	31.6	259 23.9	8.0	15 33.4	14.2	60.5
	20	117 52.2	31.2	273 50.9	8.2	15 19.2	14.2	60.4
	21	132 51.9	. . 30.7	288 18.1	8.2	15 05.0	14.3	60.4
	22	147 51.7	30.3	302 45.3	8.4	14 50.7	14.4	60.4
	23	162 51.5	29.9	317 12.7	8.5	N14 36.3	14.5	60.3
		SD 16.3	*d* 0.4	SD 16.7		16.7		16.5

Lat.	Twilight Naut.	Twilight Civil	Sunrise	Moonrise 10	Moonrise 11	Moonrise 12	Moonrise 13
°	h m	h m	h m	h m	h m	h m	h m
N 72	08 09	10 07	■	☐	☐	15 19	18 27
N 70	07 53	09 28	■	☐	☐	16 07	18 43
68	07 40	09 02	10 52	☐	13 18	16 37	18 56
66	07 29	08 41	10 04	☐	14 30	16 59	19 06
64	07 20	08 25	09 34	12 32	15 06	17 17	19 15
62	07 12	08 11	09 11	13 26	15 31	17 31	19 22
60	07 04	07 59	08 53	13 59	15 51	17 43	19 28
N 58	06 58	07 49	08 38	14 23	16 08	17 54	19 34
56	06 52	07 40	08 25	14 42	16 22	18 03	19 39
54	06 47	07 32	08 14	14 58	16 34	18 10	19 43
52	06 42	07 25	08 04	15 12	16 44	18 18	19 47
50	06 37	07 18	07 55	15 24	16 54	18 24	19 50
45	06 27	07 04	07 37	15 49	17 13	18 38	19 58
N 40	06 18	06 51	07 21	16 09	17 29	18 49	20 04
35	06 09	06 41	07 08	16 26	17 43	18 58	20 10
30	06 02	06 31	06 57	16 40	17 54	19 07	20 15
20	05 47	06 14	06 37	17 05	18 14	19 21	20 23
N 10	05 32	05 58	06 20	17 25	18 31	19 33	20 30
0	05 16	05 42	06 04	17 45	18 47	19 45	20 37
S 10	04 59	05 25	05 48	18 05	19 03	19 56	20 44
20	04 38	05 06	05 31	18 25	19 20	20 08	20 51
30	04 11	04 43	05 10	18 49	19 40	20 22	20 59
35	03 54	04 29	04 58	19 03	19 51	20 30	21 04
40	03 32	04 13	04 45	19 20	20 04	20 40	21 09
45	03 05	03 52	04 28	19 39	20 19	20 50	21 15
S 50	02 25	03 25	04 08	20 03	20 38	21 03	21 22
52	02 03	03 12	03 58	20 15	20 46	21 09	21 26
54	01 33	02 56	03 47	20 28	20 56	21 15	21 29
56	00 33	02 38	03 34	20 43	21 07	21 23	21 33
58	////	02 14	03 20	21 00	21 20	21 31	21 38
S 60	////	01 41	03 02	21 22	21 34	21 40	21 43

Lat.	Sunset	Twilight Civil	Twilight Naut.	Moonset 10	Moonset 11	Moonset 12	Moonset 13
°	h m	h m	h m	h m	h m	h m	h m
N 72	■	14 10	16 08	☐	☐	12 07	10 52
N 70	■	14 48	16 24	☐	☐	11 17	10 33
68	13 25	15 15	16 37	☐	12 01	10 45	10 18
66	14 12	15 35	16 48	☐	10 49	10 21	10 06
64	14 43	15 52	16 57	10 33	10 12	10 03	09 55
62	15 05	16 06	17 05	09 38	09 46	09 47	09 46
60	15 23	16 17	17 12	09 05	09 25	09 34	09 39
N 58	15 38	16 28	17 19	08 41	09 07	09 23	09 32
56	15 51	16 37	17 24	08 21	08 53	09 13	09 26
54	16 02	16 45	17 30	08 04	08 40	09 04	09 20
52	16 12	16 52	17 35	07 50	08 29	08 56	09 16
50	16 21	16 58	17 39	07 38	08 19	08 49	09 11
45	16 40	17 13	17 49	07 12	07 58	08 33	09 01
N 40	16 55	17 25	17 59	06 51	07 41	08 21	08 53
35	17 08	17 36	18 07	06 34	07 26	08 10	08 46
30	17 19	17 45	18 15	06 19	07 14	08 00	08 40
20	17 39	18 02	18 30	05 53	06 52	07 44	08 29
N 10	17 56	18 18	18 44	05 31	06 33	07 29	08 19
0	18 12	18 34	19 00	05 11	06 15	07 15	08 10
S 10	18 28	18 51	19 17	04 50	05 57	07 01	08 01
20	18 45	19 09	19 38	04 28	05 38	06 46	07 51
30	19 05	19 32	20 05	04 02	05 15	06 29	07 40
35	19 17	19 46	20 22	03 46	05 02	06 19	07 33
40	19 31	20 03	20 43	03 28	04 47	06 07	07 25
45	19 47	20 23	21 10	03 07	04 29	05 54	07 16
S 50	20 08	20 50	21 49	02 39	04 06	05 37	07 06
52	20 17	21 03	22 11	02 26	03 54	05 29	07 00
54	20 28	21 19	22 40	02 10	03 42	05 20	06 55
56	20 41	21 37	23 34	01 52	03 27	05 09	06 49
58	20 55	22 00	////	01 29	03 10	04 58	06 42
S 60	21 12	22 32	////	01 00	02 49	04 44	06 34

Day	SUN Eqn. of Time 00^h	SUN Eqn. of Time 12^h	SUN Mer. Pass.	MOON Mer. Pass. Upper	MOON Mer. Pass. Lower	MOON Age	MOON Phase
d	m s	m s	h m	h m	h m	d %	
10	07 25	07 37	12 08	24 00	11 28	14 99	○
11	07 48	08 00	12 08	00 00	12 32	15 100	
12	08 12	08 23	12 08	01 02	13 30	16 97	

	UT d h	ARIES GHA ° ′	VENUS −4.5 GHA ° ′	VENUS Dec ° ′	MARS +1.3 GHA ° ′	MARS Dec ° ′	JUPITER −1.9 GHA ° ′	JUPITER Dec ° ′	SATURN +0.9 GHA ° ′	SATURN Dec ° ′
	13 00	112 36.5	130 49.7	S 8 19.2	188 49.9	S23 38.3	168 32.5	S20 11.5	299 27.8	N 5 15.1
	01	127 38.9	145 49.9	18.0	203 50.3	38.2	183 34.4	11.4	314 30.3	15.1
	02	142 41.4	160 50.0	16.8	218 50.7	38.0	198 36.2	11.3	329 32.8	15.2
	03	157 43.9	175 50.2	. . 15.6	233 51.1	. . 37.9	213 38.1	. . 11.1	344 35.3	. . 15.2
	04	172 46.3	190 50.3	14.4	248 51.4	37.7	228 39.9	11.0	359 37.8	15.2
	05	187 48.8	205 50.4	13.3	263 51.8	37.5	243 41.8	10.9	14 40.3	15.3
	06	202 51.2	220 50.6	S 8 12.1	278 52.2	S23 37.4	258 43.6	S20 10.8	29 42.8	N 5 15.3
	07	217 53.7	235 50.7	10.9	293 52.6	37.2	273 45.5	10.6	44 45.4	15.3
T	08	232 56.2	250 50.9	09.7	308 53.0	37.0	288 47.3	10.5	59 47.9	15.4
U	09	247 58.6	265 51.0	. . 08.5	323 53.4	. . 36.9	303 49.2	. . 10.4	74 50.4	. . 15.4
	10	263 01.1	280 51.2	07.3	338 53.8	36.7	318 51.1	10.3	89 52.9	15.4
E	11	278 03.6	295 51.3	06.1	353 54.2	36.5	333 52.9	10.1	104 55.4	15.4
S	12	293 06.0	310 51.5	S 8 05.0	8 54.6	S23 36.4	348 54.8	S20 10.0	119 57.9	N 5 15.5
D	13	308 08.5	325 51.6	03.8	23 55.0	36.2	3 56.6	09.9	135 00.4	15.5
A	14	323 11.0	340 51.8	02.6	38 55.4	36.1	18 58.5	09.7	150 02.9	15.5
Y	15	338 13.4	355 51.9	. . 01.4	53 55.8	. . 35.9	34 00.3	. . 09.6	165 05.5	. . 15.6
	16	353 15.9	10 52.1	8 00.2	68 56.2	35.7	49 02.2	09.5	180 08.0	15.6
	17	8 18.4	25 52.2	7 59.0	83 56.6	35.5	64 04.0	09.4	195 10.5	15.6
	18	23 20.8	40 52.4	S 7 57.8	98 57.0	S23 35.4	79 05.9	S20 09.2	210 13.0	N 5 15.7
	19	38 23.3	55 52.5	56.6	113 57.4	35.2	94 07.7	09.1	225 15.5	15.7
	20	53 25.7	70 52.7	55.4	128 57.8	35.0	109 09.6	09.0	240 18.0	15.7
	21	68 28.2	85 52.9	. . 54.3	143 58.2	. . 34.9	124 11.4	. . 08.8	255 20.5	. . 15.8
	22	83 30.7	100 53.0	53.1	158 58.6	34.7	139 13.3	08.7	270 23.1	15.8
	23	98 33.1	115 53.2	51.9	173 59.0	34.5	154 15.1	08.6	285 25.6	15.8
	14 00	113 35.6	130 53.3	S 7 50.7	188 59.4	S23 34.4	169 17.0	S20 08.5	300 28.1	N 5 15.9
	01	128 38.1	145 53.5	49.5	203 59.8	34.2	184 18.8	08.3	315 30.6	15.9
	02	143 40.5	160 53.7	48.3	219 00.1	34.0	199 20.7	08.2	330 33.1	15.9
	03	158 43.0	175 53.8	. . 47.1	234 00.5	. . 33.8	214 22.6	. . 08.1	345 35.6	. . 16.0
	04	173 45.5	190 54.0	45.9	249 00.9	33.7	229 24.4	07.9	0 38.2	16.0
	05	188 47.9	205 54.1	44.7	264 01.3	33.5	244 26.3	07.8	15 40.7	16.1
	06	203 50.4	220 54.3	S 7 43.5	279 01.7	S23 33.3	259 28.1	S20 07.7	30 43.2	N 5 16.1
W	07	218 52.9	235 54.5	42.4	294 02.1	33.2	274 30.0	07.6	45 45.7	16.1
E	08	233 55.3	250 54.7	41.2	309 02.5	33.0	289 31.8	07.4	60 48.2	16.2
D	09	248 57.8	265 54.8	. . 40.0	324 02.9	. . 32.8	304 33.7	. . 07.3	75 50.7	. . 16.2
N	10	264 00.2	280 55.0	38.8	339 03.3	32.6	319 35.5	07.2	90 53.2	16.2
	11	279 02.7	295 55.2	37.6	354 03.7	32.5	334 37.4	07.0	105 55.8	16.3
E	12	294 05.2	310 55.3	S 7 36.4	9 04.1	S23 32.3	349 39.2	S20 06.9	120 58.3	N 5 16.3
S	13	309 07.6	325 55.5	35.2	24 04.5	32.1	4 41.1	06.8	136 00.8	16.3
D	14	324 10.1	340 55.7	34.0	39 04.9	31.9	19 42.9	06.7	151 03.3	16.4
A	15	339 12.6	355 55.9	. . 32.8	54 05.3	. . 31.7	34 44.8	. . 06.5	166 05.8	. . 16.4
Y	16	354 15.0	10 56.0	31.6	69 05.7	31.6	49 46.6	06.4	181 08.3	16.4
	17	9 17.5	25 56.2	30.4	84 06.1	31.4	64 48.5	06.3	196 10.9	16.5
	18	24 20.0	40 56.4	S 7 29.2	99 06.5	S23 31.2	79 50.3	S20 06.1	211 13.4	N 5 16.5
	19	39 22.4	55 56.6	28.1	114 06.9	31.0	94 52.2	06.0	226 15.9	16.5
	20	54 24.9	70 56.7	26.9	129 07.3	30.9	109 54.1	05.9	241 18.4	16.6
	21	69 27.3	85 56.9	. . 25.7	144 07.7	. . 30.7	124 55.9	. . 05.8	256 20.9	. . 16.6
	22	84 29.8	100 57.1	24.5	159 08.1	30.5	139 57.8	05.6	271 23.5	16.6
	23	99 32.3	115 57.3	23.3	174 08.5	30.3	154 59.6	05.5	286 26.0	16.7
	15 00	114 34.7	130 57.5	S 7 22.1	189 08.9	S23 30.1	170 01.5	S20 05.4	301 28.5	N 5 16.7
	01	129 37.2	145 57.7	20.9	204 09.3	29.9	185 03.3	05.2	316 31.0	16.7
	02	144 39.7	160 57.8	19.7	219 09.6	29.8	200 05.2	05.1	331 33.5	16.8
	03	159 42.1	175 58.0	. . 18.5	234 10.0	. . 29.6	215 07.0	. . 05.0	346 36.1	. . 16.8
	04	174 44.6	190 58.2	17.3	249 10.4	29.4	230 08.9	04.8	1 38.6	16.9
	05	189 47.1	205 58.4	16.1	264 10.8	29.2	245 10.7	04.7	16 41.1	16.9
	06	204 49.5	220 58.6	S 7 14.9	279 11.2	S23 29.0	260 12.6	S20 04.6	31 43.6	N 5 16.9
	07	219 52.0	235 58.8	13.7	294 11.6	28.8	275 14.4	04.5	46 46.1	17.0
T	08	234 54.5	250 59.0	12.5	309 12.0	28.7	290 16.3	04.3	61 48.7	17.0
H	09	249 56.9	265 59.2	. . 11.3	324 12.4	. . 28.5	305 18.1	. . 04.2	76 51.2	. . 17.0
U	10	264 59.4	280 59.4	10.1	339 12.8	28.3	320 20.0	04.1	91 53.7	17.1
R	11	280 01.8	295 59.6	08.9	354 13.2	28.1	335 21.8	03.9	106 56.2	17.1
S	12	295 04.3	310 59.8	S 7 07.8	9 13.6	S23 27.9	350 23.7	S20 03.8	121 58.7	N 5 17.1
D	13	310 06.8	326 00.0	06.6	24 14.0	27.7	5 25.5	03.7	137 01.3	17.2
A	14	325 09.2	341 00.2	05.4	39 14.4	27.5	20 27.4	03.5	152 03.8	17.2
Y	15	340 11.7	356 00.3	. . 04.2	54 14.8	. . 27.3	35 29.3	. . 03.4	167 06.3	. . 17.2
	16	355 14.2	11 00.5	03.0	69 15.2	27.2	50 31.1	03.3	182 08.8	17.3
	17	10 16.6	26 00.7	01.8	84 15.6	27.0	65 33.0	03.2	197 11.3	17.3
	18	25 19.1	41 00.9	S 7 00.6	99 16.0	S23 26.8	80 34.8	S20 03.0	212 13.9	N 5 17.4
	19	40 21.6	56 01.2	6 59.4	114 16.4	26.6	95 36.7	02.9	227 16.4	17.4
	20	55 24.0	71 01.4	58.2	129 16.8	26.4	110 38.5	02.8	242 18.9	17.4
	21	70 26.5	86 01.6	. . 57.0	144 17.2	. . 26.2	125 40.4	. . 02.6	257 21.4	. . 17.5
	22	85 29.0	101 01.8	55.8	159 17.6	26.0	140 42.2	02.5	272 24.0	17.5
	23	100 31.4	116 02.0	54.6	174 18.0	25.8	155 44.1	02.4	287 26.5	17.5
	Mer. Pass.	h m 16 22.9	v 0.2	d 1.2	v 0.4	d 0.2	v 1.9	d 0.1	v 2.5	d 0.0

STARS

Name	SHA ° ′	Dec ° ′
Acamar	315 20.6	S40 16.2
Achernar	335 28.9	S57 11.7
Acrux	173 13.1	S63 08.8
Adhara	255 14.7	S28 59.1
Aldebaran	290 52.9	N16 31.7
Alioth	166 23.2	N55 54.2
Alkaid	153 01.3	N49 15.7
Al Na'ir	27 48.1	S46 55.2
Alnilam	275 49.4	S 1 11.8
Alphard	217 59.0	S 8 41.9
Alphecca	126 13.9	N26 40.8
Alpheratz	357 47.1	N29 08.6
Altair	62 11.8	N 8 53.5
Ankaa	353 18.9	S42 15.6
Antares	112 30.6	S26 27.1
Arcturus	145 58.7	N19 07.8
Atria	107 35.9	S69 02.5
Avior	234 18.9	S59 32.3
Bellatrix	278 35.2	N 6 21.5
Betelgeuse	271 04.5	N 7 24.5
Canopus	263 57.1	S52 42.1
Capella	280 38.9	N46 00.6
Deneb	49 34.3	N45 18.8
Denebola	182 36.8	N14 31.1
Diphda	348 59.2	S17 56.3
Dubhe	193 55.0	N61 41.8
Elnath	278 16.4	N28 37.0
Eltanin	90 48.2	N51 29.0
Enif	33 50.6	N 9 55.0
Fomalhaut	15 27.7	S29 34.6
Gacrux	172 04.7	S57 09.7
Gienah	175 55.6	S17 35.6
Hadar	148 52.9	S60 24.8
Hamal	328 04.5	N23 30.5
Kaus Aust.	83 48.5	S34 22.9
Kochab	137 19.9	N74 06.7
Markab	13 41.8	N15 15.3
Menkar	314 18.4	N 4 07.6
Menkent	148 11.6	S36 24.8
Miaplacidus	221 39.8	S69 45.2
Mirfak	308 44.9	N49 53.9
Nunki	76 02.7	S26 17.2
Peacock	53 24.8	S56 42.4
Pollux	243 31.3	N28 00.2
Procyon	245 02.8	N 5 12.1
Rasalhague	96 09.8	N12 33.0
Regulus	207 46.7	N11 55.2
Rigel	281 14.9	S 8 11.5
Rigil Kent.	139 56.6	S60 52.2
Sabik	102 16.6	S15 44.2
Schedar	349 44.6	N56 35.6
Shaula	96 26.8	S37 06.6
Sirius	258 36.3	S16 43.7
Spica	158 34.7	S11 12.6
Suhail	222 54.5	S43 28.1
Vega	80 41.6	N38 47.4
Zuben'ubi	137 09.2	S16 04.8

	SHA ° ′	Mer. Pass. h m
Venus	17 17.7	15 16
Mars	75 23.8	11 24
Jupiter	55 41.4	12 41
Saturn	186 52.5	3 57

	UT	SUN GHA	SUN Dec	MOON GHA	v	MOON Dec	d	HP
	d h	° ′	° ′	° ′	′	° ′	′	′
	13 00	177 51.2	S21 29.5	331 40.2	8.6	N14 21.8	14.5	60.3
	01	192 51.0	29.0	346 07.8	8.7	14 07.3	14.6	60.3
	02	207 50.8	28.6	0 35.5	8.8	13 52.7	14.7	60.2
	03	222 50.5	. . 28.2	15 03.3	8.9	13 38.0	14.7	60.2
	04	237 50.3	27.8	29 31.2	9.1	13 23.3	14.8	60.2
	05	252 50.1	27.3	43 59.3	9.1	13 08.5	14.8	60.1
	06	267 49.8	S21 26.9	58 27.4	9.3	N12 53.7	14.9	60.1
	07	282 49.6	26.5	72 55.7	9.4	12 38.8	15.0	60.1
T	08	297 49.4	26.1	87 24.1	9.5	12 23.8	15.0	60.0
U	09	312 49.1	. . 25.6	101 52.6	9.5	12 08.8	15.1	60.0
E	10	327 48.9	25.2	116 21.1	9.7	11 53.7	15.1	60.0
S	11	342 48.7	24.8	130 49.8	9.8	11 38.6	15.1	59.9
D	12	357 48.4	S21 24.3	145 18.6	9.9	N11 23.5	15.2	59.9
A	13	12 48.2	23.9	159 47.5	10.0	11 08.3	15.3	59.9
Y	14	27 48.0	23.5	174 16.5	10.1	10 53.0	15.3	59.8
	15	42 47.7	. . 23.0	188 45.6	10.2	10 37.7	15.3	59.8
	16	57 47.5	22.6	203 14.8	10.3	10 22.4	15.3	59.8
	17	72 47.3	22.2	217 44.1	10.4	10 07.1	15.4	59.7
	18	87 47.1	S21 21.7	232 13.5	10.4	N 9 51.7	15.5	59.7
	19	102 46.8	21.3	246 42.9	10.6	9 36.2	15.4	59.6
	20	117 46.6	20.8	261 12.5	10.7	9 20.8	15.5	59.6
	21	132 46.4	. . 20.4	275 42.2	10.7	9 05.3	15.5	59.6
	22	147 46.1	20.0	290 11.9	10.9	8 49.8	15.6	59.5
	23	162 45.9	19.5	304 41.8	10.9	8 34.2	15.6	59.5
	14 00	177 45.7	S21 19.1	319 11.7	11.0	N 8 18.6	15.5	59.5
	01	192 45.5	18.6	333 41.7	11.1	8 03.1	15.7	59.4
	02	207 45.2	18.2	348 11.8	11.2	7 47.4	15.6	59.4
	03	222 45.0	. . 17.8	2 42.0	11.3	7 31.8	15.6	59.3
	04	237 44.8	17.3	17 12.3	11.4	7 16.2	15.7	59.3
	05	252 44.5	16.9	31 42.7	11.4	7 00.5	15.7	59.3
	06	267 44.3	S21 16.4	46 13.1	11.5	N 6 44.8	15.7	59.2
W	07	282 44.1	16.0	60 43.6	11.6	6 29.1	15.7	59.2
E	08	297 43.9	15.5	75 14.2	11.7	6 13.4	15.7	59.1
D	09	312 43.6	. . 15.1	89 44.9	11.7	5 57.7	15.7	59.1
N	10	327 43.4	14.6	104 15.6	11.9	5 42.0	15.8	59.1
E	11	342 43.2	14.2	118 46.5	11.9	5 26.2	15.7	59.0
S	12	357 43.0	S21 13.7	133 17.4	11.9	N 5 10.5	15.7	59.0
D	13	12 42.8	13.3	147 48.3	12.1	4 54.8	15.8	58.9
A	14	27 42.5	12.8	162 19.4	12.1	4 39.0	15.7	58.9
Y	15	42 42.3	. . 12.4	176 50.5	12.1	4 23.3	15.7	58.9
	16	57 42.1	11.9	191 21.6	12.3	4 07.6	15.8	58.8
	17	72 41.9	11.5	205 52.9	12.3	3 51.8	15.7	58.8
	18	87 41.6	S21 11.0	220 24.2	12.4	N 3 36.1	15.7	58.7
	19	102 41.4	10.6	234 55.6	12.4	3 20.4	15.8	58.7
	20	117 41.2	10.1	249 27.0	12.5	3 04.6	15.7	58.7
	21	132 41.0	. . 09.7	263 58.5	12.5	2 48.9	15.7	58.6
	22	147 40.7	09.2	278 30.0	12.6	2 33.2	15.7	58.6
	23	162 40.5	08.8	293 01.6	12.7	2 17.5	15.7	58.5
	15 00	177 40.3	S21 08.3	307 33.3	12.7	N 2 01.8	15.6	58.5
	01	192 40.1	07.8	322 05.0	12.8	1 46.2	15.7	58.5
	02	207 39.9	07.4	336 36.8	12.8	1 30.5	15.6	58.4
	03	222 39.6	. . 06.9	351 08.6	12.9	1 14.9	15.6	58.4
	04	237 39.4	06.5	5 40.5	12.9	0 59.3	15.6	58.3
	05	252 39.2	06.0	20 12.4	13.0	0 43.7	15.6	58.3
	06	267 39.0	S21 05.5	34 44.4	13.0	N 0 28.1	15.6	58.3
	07	282 38.8	05.1	49 16.4	13.1	N 0 12.5	15.5	58.2
T	08	297 38.5	04.6	63 48.5	13.1	S 0 03.0	15.5	58.2
H	09	312 38.3	. . 04.1	78 20.6	13.2	0 18.5	15.5	58.1
U	10	327 38.1	03.7	92 52.8	13.2	0 34.0	15.5	58.1
R	11	342 37.9	03.2	107 25.0	13.2	0 49.5	15.4	58.1
S	12	357 37.7	S21 02.7	121 57.2	13.3	S 1 04.9	15.4	58.0
D	13	12 37.5	02.3	136 29.5	13.3	1 20.3	15.4	58.0
A	14	27 37.2	01.8	151 01.8	13.4	1 35.7	15.4	57.9
Y	15	42 37.0	. . 01.3	165 34.2	13.4	1 51.1	15.3	57.9
	16	57 36.8	00.9	180 06.6	13.4	2 06.4	15.3	57.8
	17	72 36.6	21 00.4	194 39.0	13.5	2 21.7	15.2	57.8
	18	87 36.4	S20 59.9	209 11.5	13.5	S 2 36.9	15.3	57.8
	19	102 36.2	59.5	223 44.0	13.5	2 52.2	15.1	57.7
	20	117 35.9	59.0	238 16.5	13.6	3 07.3	15.2	57.7
	21	132 35.7	. . 58.5	252 49.1	13.5	3 22.5	15.1	57.6
	22	147 35.5	58.0	267 21.6	13.6	3 37.6	15.1	57.6
	23	162 35.3	57.6	281 54.2	13.7	S 3 52.7	15.0	57.6
		SD 16.3	*d* 0.4	SD 16.3		16.1		15.8

Lat.	Twilight Naut.	Twilight Civil	Sunrise	Moonrise 13	14	15	16
°	h m	h m	h m	h m	h m	h m	h m
N 72	08 03	09 55	■	18 27	20 50	23 02	25 16
N 70	07 48	09 20	■	18 43	20 54	22 57	24 59
68	07 35	08 55	10 37	18 56	20 57	22 52	24 45
66	07 25	08 36	09 55	19 06	21 00	22 48	24 34
64	07 16	08 20	09 28	19 15	21 02	22 45	24 24
62	07 09	08 07	09 06	19 22	21 04	22 42	24 16
60	07 02	07 56	08 49	19 28	21 06	22 39	24 10
N 58	06 56	07 46	08 35	19 34	21 08	22 37	24 04
56	06 50	07 37	08 22	19 39	21 09	22 35	23 59
54	06 45	07 30	08 11	19 43	21 10	22 33	23 54
52	06 40	07 23	08 02	19 47	21 11	22 32	23 50
50	06 36	07 16	07 53	19 50	21 12	22 30	23 46
45	06 26	07 02	07 35	19 58	21 14	22 27	23 38
N 40	06 17	06 51	07 20	20 04	21 16	22 25	23 31
35	06 09	06 40	07 08	20 10	21 18	22 22	23 25
30	06 02	06 31	06 57	20 15	21 19	22 20	23 20
20	05 47	06 14	06 38	20 23	21 21	22 17	23 11
N 10	05 33	05 59	06 21	20 30	21 24	22 14	23 03
0	05 17	05 43	06 05	20 37	21 25	22 11	22 56
S 10	05 00	05 27	05 50	20 44	21 27	22 09	22 49
20	04 40	05 08	05 33	20 51	21 30	22 06	22 42
30	04 14	04 46	05 13	20 59	21 32	22 03	22 33
35	03 57	04 32	05 01	21 04	21 33	22 01	22 28
40	03 36	04 16	04 48	21 09	21 35	21 59	22 23
45	03 10	03 56	04 32	21 15	21 37	21 57	22 17
S 50	02 32	03 30	04 12	21 22	21 39	21 54	22 09
52	02 11	03 17	04 03	21 26	21 40	21 53	22 06
54	01 43	03 02	03 52	21 29	21 41	21 51	22 02
56	00 58	02 44	03 40	21 33	21 42	21 50	21 58
58	////	02 22	03 26	21 38	21 43	21 48	21 53
S 60	////	01 52	03 09	21 43	21 45	21 46	21 48

Lat.	Sunset	Twilight Civil	Twilight Naut.	Moonset 13	14	15	16
°	h m	h m	h m	h m	h m	h m	h m
N 72	■	14 24	16 17	10 52	10 12	09 40	09 08
N 70	■	14 59	16 32	10 33	10 05	09 41	09 17
68	13 42	15 24	16 44	10 18	09 59	09 42	09 25
66	14 23	15 43	16 54	10 06	09 53	09 42	09 31
64	14 51	15 59	17 03	09 55	09 49	09 43	09 36
62	15 13	16 12	17 10	09 46	09 45	09 43	09 41
60	15 30	16 23	17 17	09 39	09 42	09 43	09 45
N 58	15 44	16 33	17 23	09 32	09 38	09 44	09 49
56	15 57	16 41	17 29	09 26	09 36	09 44	09 52
54	16 07	16 49	17 34	09 20	09 33	09 44	09 55
52	16 17	16 56	17 38	09 16	09 31	09 45	09 57
50	16 25	17 02	17 43	09 11	09 29	09 45	10 00
45	16 43	17 16	17 53	09 01	09 24	09 45	10 05
N 40	16 58	17 28	18 01	08 53	09 21	09 46	10 10
35	17 11	17 38	18 09	08 46	09 17	09 46	10 13
30	17 22	17 48	18 17	08 40	09 14	09 46	10 17
20	17 41	18 04	18 31	08 29	09 09	09 47	10 23
N 10	17 57	18 20	18 46	08 19	09 05	09 47	10 28
0	18 13	18 35	19 01	08 10	09 00	09 48	10 33
S 10	18 29	18 51	19 18	08 01	08 56	09 48	10 38
20	18 46	19 10	19 38	07 51	08 51	09 48	10 43
30	19 05	19 32	20 04	07 40	08 46	09 49	10 49
35	19 17	19 46	20 21	07 33	08 43	09 49	10 53
40	19 30	20 02	20 41	07 25	08 39	09 49	10 56
45	19 46	20 22	21 08	07 16	08 35	09 49	11 01
S 50	20 05	20 47	21 45	07 06	08 30	09 50	11 07
52	20 15	21 00	22 05	07 00	08 27	09 50	11 09
54	20 25	21 15	22 32	06 55	08 25	09 50	11 12
56	20 37	21 32	23 15	06 49	08 22	09 50	11 15
58	20 51	21 54	////	06 42	08 19	09 51	11 19
S 60	21 07	22 23	////	06 34	08 15	09 51	11 22

Day	SUN Eqn. of Time 00^h	SUN Eqn. of Time 12^h	SUN Mer. Pass.	MOON Mer. Pass. Upper	MOON Mer. Pass. Lower	Age	Phase
d	m s	m s	h m	h m	h m	d	%
13	08 35	08 46	12 09	01 58	14 24	17	92
14	08 57	09 08	12 09	02 49	15 13	18	85
15	09 18	09 29	12 09	03 37	16 00	19	76

	UT	ARIES	VENUS −4.6		MARS +1.3		JUPITER −1.9		SATURN +0.9	
		GHA	GHA	Dec	GHA	Dec	GHA	Dec	GHA	Dec
	d h	° ′	° ′	° ′	° ′	° ′	° ′	° ′	° ′	° ′
	16 00	115 33.9	131 02.2	S 6 53.4	189 18.4	S23 25.6	170 45.9	S20 02.2	302 29.0	N 5 17.6
	01	130 36.3	146 02.4	52.2	204 18.8	25.4	185 47.8	02.1	317 31.5	17.6
	02	145 38.8	161 02.6	51.0	219 19.2	25.2	200 49.6	02.0	332 34.0	17.7
	03	160 41.3	176 02.8	. . 49.8	234 19.6	. . 25.0	215 51.5	. . 01.9	347 36.6	. . 17.7
	04	175 43.7	191 03.0	48.6	249 20.0	24.9	230 53.3	01.7	2 39.1	17.7
	05	190 46.2	206 03.2	47.4	264 20.4	24.7	245 55.2	01.6	17 41.6	17.8
	06	205 48.7	221 03.4	S 6 46.2	279 20.7	S23 24.5	260 57.0	S20 01.5	32 44.1	N 5 17.8
	07	220 51.1	236 03.7	45.0	294 21.1	24.3	275 58.9	01.3	47 46.7	17.8
	08	235 53.6	251 03.9	43.8	309 21.5	24.1	291 00.7	01.2	62 49.2	17.9
F	09	250 56.1	266 04.1	. . 42.6	324 21.9	. . 23.9	306 02.6	. . 01.1	77 51.7	. . 17.9
R	10	265 58.5	281 04.3	41.4	339 22.3	23.7	321 04.5	00.9	92 54.2	18.0
I	11	281 01.0	296 04.5	40.2	354 22.7	23.5	336 06.3	00.8	107 56.8	18.0
D	12	296 03.4	311 04.7	S 6 39.0	9 23.1	S23 23.3	351 08.2	S20 00.7	122 59.3	N 5 18.0
A	13	311 05.9	326 05.0	37.8	24 23.5	23.1	6 10.0	00.5	138 01.8	18.1
Y	14	326 08.4	341 05.2	36.6	39 23.9	22.9	21 11.9	00.4	153 04.3	18.1
	15	341 10.8	356 05.4	. . 35.4	54 24.3	. . 22.7	36 13.7	. . 00.3	168 06.9	. . 18.1
	16	356 13.3	11 05.6	34.2	69 24.7	22.5	51 15.6	00.2	183 09.4	18.2
	17	11 15.8	26 05.8	33.0	84 25.1	22.3	66 17.4	20 00.0	198 11.9	18.2
	18	26 18.2	41 06.1	S 6 31.9	99 25.5	S23 22.1	81 19.3	S19 59.9	213 14.4	N 5 18.3
	19	41 20.7	56 06.3	30.7	114 25.9	21.9	96 21.1	59.8	228 17.0	18.3
	20	56 23.2	71 06.5	29.5	129 26.3	21.7	111 23.0	59.6	243 19.5	18.3
	21	71 25.6	86 06.7	. . 28.3	144 26.7	. . 21.5	126 24.8	. . 59.5	258 22.0	. . 18.4
	22	86 28.1	101 07.0	27.1	159 27.1	21.3	141 26.7	59.4	273 24.5	18.4
	23	101 30.6	116 07.2	25.9	174 27.5	21.1	156 28.5	59.2	288 27.1	18.4
	17 00	116 33.0	131 07.4	S 6 24.7	189 27.9	S23 20.9	171 30.4	S19 59.1	303 29.6	N 5 18.5
	01	131 35.5	146 07.7	23.5	204 28.3	20.7	186 32.2	59.0	318 32.1	18.5
	02	146 37.9	161 07.9	22.3	219 28.7	20.5	201 34.1	58.8	333 34.7	18.6
	03	161 40.4	176 08.1	. . 21.1	234 29.1	. . 20.3	216 35.9	. . 58.7	348 37.2	. . 18.6
	04	176 42.9	191 08.4	19.9	249 29.5	20.1	231 37.8	58.6	3 39.7	18.6
	05	191 45.3	206 08.6	18.7	264 29.9	19.8	246 39.6	58.4	18 42.2	18.7
	06	206 47.8	221 08.8	S 6 17.5	279 30.3	S23 19.6	261 41.5	S19 58.3	33 44.8	N 5 18.7
S	07	221 50.3	236 09.1	16.3	294 30.7	19.4	276 43.4	58.2	48 47.3	18.8
	08	236 52.7	251 09.3	15.1	309 31.1	19.2	291 45.2	58.1	63 49.8	18.8
A	09	251 55.2	266 09.5	. . 13.9	324 31.5	. . 19.0	306 47.1	. . 57.9	78 52.4	. . 18.8
T	10	266 57.7	281 09.8	12.7	339 31.9	18.8	321 48.9	57.8	93 54.9	18.9
U	11	282 00.1	296 10.0	11.5	354 32.3	18.6	336 50.8	57.7	108 57.4	18.9
R	12	297 02.6	311 10.3	S 6 10.3	9 32.7	S23 18.4	351 52.6	S19 57.5	123 59.9	N 5 19.0
D	13	312 05.0	326 10.5	09.1	24 33.1	18.2	6 54.5	57.4	139 02.5	19.0
A	14	327 07.5	341 10.8	07.9	39 33.5	18.0	21 56.3	57.3	154 05.0	19.0
Y	15	342 10.0	356 11.0	. . 06.7	54 33.9	. . 17.8	36 58.2	. . 57.1	169 07.5	. . 19.1
	16	357 12.4	11 11.2	05.5	69 34.3	17.6	52 00.0	57.0	184 10.1	19.1
	17	12 14.9	26 11.5	04.3	84 34.6	17.3	67 01.9	56.9	199 12.6	19.2
	18	27 17.4	41 11.7	S 6 03.1	99 35.0	S23 17.1	82 03.7	S19 56.7	214 15.1	N 5 19.2
	19	42 19.8	56 12.0	01.9	114 35.4	16.9	97 05.6	56.6	229 17.6	19.2
	20	57 22.3	71 12.2	6 00.7	129 35.8	16.7	112 07.4	56.5	244 20.2	19.3
	21	72 24.8	86 12.5	5 59.5	144 36.2	. . 16.5	127 09.3	. . 56.3	259 22.7	. . 19.3
	22	87 27.2	101 12.7	58.3	159 36.6	16.3	142 11.1	56.2	274 25.2	19.4
	23	102 29.7	116 13.0	57.1	174 37.0	16.1	157 13.0	56.1	289 27.8	19.4
	18 00	117 32.2	131 13.2	S 5 55.9	189 37.4	S23 15.9	172 14.8	S19 55.9	304 30.3	N 5 19.4
	01	132 34.6	146 13.5	54.7	204 37.8	15.6	187 16.7	55.8	319 32.8	19.5
	02	147 37.1	161 13.8	53.5	219 38.2	15.4	202 18.5	55.7	334 35.4	19.5
	03	162 39.5	176 14.0	. . 52.3	234 38.6	. . 15.2	217 20.4	. . 55.5	349 37.9	. . 19.6
	04	177 42.0	191 14.3	51.1	249 39.0	15.0	232 22.2	55.4	4 40.4	19.6
	05	192 44.5	206 14.5	49.9	264 39.4	14.8	247 24.1	55.3	19 43.0	19.6
	06	207 46.9	221 14.8	S 5 48.7	279 39.8	S23 14.6	262 26.0	S19 55.1	34 45.5	N 5 19.7
	07	222 49.4	236 15.0	47.5	294 40.2	14.3	277 27.8	55.0	49 48.0	19.7
	08	237 51.9	251 15.3	46.2	309 40.6	14.1	292 29.7	54.9	64 50.5	19.8
S	09	252 54.3	266 15.6	. . 45.0	324 41.0	. . 13.9	307 31.5	. . 54.8	79 53.1	. . 19.8
U	10	267 56.8	281 15.8	43.8	339 41.4	13.7	322 33.4	54.6	94 55.6	19.8
N	11	282 59.3	296 16.1	42.6	354 41.8	13.5	337 35.2	54.5	109 58.1	19.9
D	12	298 01.7	311 16.4	S 5 41.4	9 42.2	S23 13.2	352 37.1	S19 54.4	125 00.7	N 5 19.9
A	13	313 04.2	326 16.6	40.2	24 42.6	13.0	7 38.9	54.2	140 03.2	20.0
Y	14	328 06.7	341 16.9	39.0	39 43.0	12.8	22 40.8	54.1	155 05.7	20.0
	15	343 09.1	356 17.2	. . 37.8	54 43.4	. . 12.6	37 42.6	. . 54.0	170 08.3	. . 20.0
	16	358 11.6	11 17.4	36.6	69 43.8	12.4	52 44.5	53.8	185 10.8	20.1
	17	13 14.0	26 17.7	35.4	84 44.2	12.1	67 46.3	53.7	200 13.3	20.1
	18	28 16.5	41 18.0	S 5 34.2	99 44.6	S23 11.9	82 48.2	S19 53.6	215 15.9	N 5 20.2
	19	43 19.0	56 18.3	33.0	114 45.0	11.7	97 50.0	53.4	230 18.4	20.2
	20	58 21.4	71 18.5	31.8	129 45.4	11.5	112 51.9	53.3	245 20.9	20.3
	21	73 23.9	86 18.8	. . 30.6	144 45.8	. . 11.2	127 53.7	. . 53.2	260 23.5	. . 20.3
	22	88 26.4	101 19.1	29.4	159 46.2	11.0	142 55.6	53.0	275 26.0	20.3
	23	103 28.8	116 19.4	28.2	174 46.6	10.8	157 57.4	52.9	290 28.6	20.4
	Mer. Pass.	h m 16 11.1	v 0.2	d 1.2	v 0.4	d 0.2	v 1.9	d 0.1	v 2.5	d 0.0

STARS

Name	SHA	Dec
	° ′	° ′
Acamar	315 20.6	S40 16.2
Achernar	335 29.0	S57 11.7
Acrux	173 13.0	S63 08.8
Adhara	255 14.7	S28 59.1
Aldebaran	290 52.9	N16 31.7
Alioth	166 23.2	N55 54.2
Alkaid	153 01.3	N49 15.7
Al Na'ir	27 48.1	S46 55.2
Alnilam	275 49.4	S 1 11.8
Alphard	217 59.0	S 8 41.9
Alphecca	126 13.9	N26 40.8
Alpheratz	357 47.1	N29 08.6
Altair	62 11.8	N 8 53.5
Ankaa	353 18.9	S42 15.6
Antares	112 30.6	S26 27.2
Arcturus	145 58.7	N19 07.8
Atria	107 35.9	S69 02.5
Avior	234 18.9	S59 32.3
Bellatrix	278 35.2	N 6 21.5
Betelgeuse	271 04.5	N 7 24.5
Canopus	263 57.1	S52 42.1
Capella	280 38.9	N46 00.6
Deneb	49 34.3	N45 18.8
Denebola	182 36.8	N14 31.1
Diphda	348 59.2	S17 56.3
Dubhe	193 54.9	N61 41.8
Elnath	278 16.5	N28 37.0
Eltanin	90 48.1	N51 29.0
Enif	33 50.7	N 9 55.0
Fomalhaut	15 27.7	S29 34.6
Gacrux	172 04.6	S57 09.7
Gienah	175 55.6	S17 35.6
Hadar	148 52.9	S60 24.8
Hamal	328 04.5	N23 30.5
Kaus Aust.	83 48.5	S34 22.8
Kochab	137 19.8	N74 06.6
Markab	13 41.9	N15 15.3
Menkar	314 18.4	N 4 07.6
Menkent	148 11.6	S36 24.8
Miaplacidus	221 39.8	S69 45.2
Mirfak	308 44.9	N49 53.9
Nunki	76 02.7	S26 17.2
Peacock	53 24.8	S56 42.4
Pollux	243 31.3	N28 00.2
Procyon	245 02.8	N 5 12.1
Rasalhague	96 09.8	N12 33.0
Regulus	207 46.7	N11 55.2
Rigel	281 14.9	S 8 11.5
Rigil Kent.	139 56.6	S60 52.2
Sabik	102 16.6	S15 44.2
Schedar	349 44.7	N56 35.6
Shaula	96 26.7	S37 06.6
Sirius	258 36.3	S16 43.8
Spica	158 34.7	S11 12.6
Suhail	222 54.5	S43 28.1
Vega	80 41.6	N38 47.4
Zuben'ubi	137 09.2	S16 04.8

	SHA	Mer. Pass.
	° ′	h m
Venus	14 34.4	15 15
Mars	72 54.9	11 22
Jupiter	54 57.4	12 32
Saturn	186 56.6	3 45

UT d h	SUN GHA ° ′	SUN Dec ° ′	MOON GHA ° ′	v ′	MOON Dec ° ′	d ′	HP ′
16 00 FRIDAY	177 35.1	S20 57.1	296 26.9	13.6	S 4 07.7	15.0	57.5
01	192 34.9	56.6	310 59.5	13.7	4 22.7	15.0	57.5
02	207 34.7	56.1	325 32.2	13.7	4 37.7	14.9	57.4
03	222 34.4	. . 55.7	340 04.9	13.8	4 52.6	14.8	57.4
04	237 34.2	55.2	354 37.7	13.7	5 07.4	14.9	57.4
05	252 34.0	54.7	9 10.4	13.8	5 22.3	14.7	57.3
06	267 33.8	S20 54.2	23 43.2	13.8	S 5 37.0	14.8	57.3
07	282 33.6	53.8	38 16.0	13.8	5 51.8	14.7	57.2
08	297 33.4	53.3	52 48.8	13.8	6 06.5	14.6	57.2
09	312 33.2	. . 52.8	67 21.6	13.8	6 21.1	14.6	57.2
10	327 33.0	52.3	81 54.4	13.8	6 35.7	14.5	57.1
11	342 32.8	51.8	96 27.2	13.9	6 50.2	14.5	57.1
12	357 32.5	S20 51.3	111 00.1	13.9	S 7 04.7	14.4	57.1
13	12 32.3	50.9	125 33.0	13.8	7 19.1	14.4	57.0
14	27 32.1	50.4	140 05.8	13.9	7 33.5	14.4	57.0
15	42 31.9	. . 49.9	154 38.7	13.9	7 47.9	14.2	56.9
16	57 31.7	49.4	169 11.6	13.9	8 02.1	14.3	56.9
17	72 31.5	48.9	183 44.5	13.9	8 16.4	14.1	56.9
18	87 31.3	S20 48.4	198 17.4	13.9	S 8 30.5	14.1	56.8
19	102 31.1	47.9	212 50.3	13.9	8 44.6	14.1	56.8
20	117 30.9	47.5	227 23.2	13.9	8 58.7	14.0	56.7
21	132 30.7	. . 47.0	241 56.1	13.9	9 12.7	13.9	56.7
22	147 30.5	46.5	256 29.0	14.0	9 26.6	13.9	56.7
23	162 30.2	46.0	271 02.0	13.9	9 40.5	13.8	56.6
17 00 SATURDAY	177 30.0	S20 45.5	285 34.9	13.9	S 9 54.3	13.8	56.6
01	192 29.8	45.0	300 07.8	13.9	10 08.1	13.6	56.6
02	207 29.6	44.5	314 40.7	13.9	10 21.7	13.7	56.5
03	222 29.4	. . 44.0	329 13.6	13.9	10 35.4	13.5	56.5
04	237 29.2	43.5	343 46.5	13.9	10 48.9	13.5	56.5
05	252 29.0	43.0	358 19.4	13.8	11 02.4	13.5	56.4
06	267 28.8	S20 42.5	12 52.2	13.9	S11 15.9	13.3	56.4
07	282 28.6	42.0	27 25.1	13.9	11 29.2	13.3	56.4
08	297 28.4	41.5	41 58.0	13.8	11 42.5	13.3	56.3
09	312 28.2	. . 41.0	56 30.8	13.9	11 55.8	13.2	56.3
10	327 28.0	40.5	71 03.7	13.8	12 09.0	13.1	56.2
11	342 27.8	40.0	85 36.5	13.8	12 22.1	13.0	56.2
12	357 27.6	S20 39.5	100 09.3	13.8	S12 35.1	12.9	56.2
13	12 27.4	39.0	114 42.1	13.8	12 48.0	12.9	56.1
14	27 27.2	38.5	129 14.9	13.8	13 00.9	12.8	56.1
15	42 27.0	. . 38.0	143 47.7	13.8	13 13.7	12.8	56.1
16	57 26.8	37.5	158 20.5	13.7	13 26.5	12.7	56.0
17	72 26.6	37.0	172 53.2	13.7	13 39.2	12.5	56.0
18	87 26.4	S20 36.5	187 25.9	13.7	S13 51.7	12.6	56.0
19	102 26.2	36.0	201 58.6	13.7	14 04.3	12.4	55.9
20	117 26.0	35.5	216 31.3	13.7	14 16.7	12.4	55.9
21	132 25.8	. . 35.0	231 04.0	13.6	14 29.1	12.3	55.9
22	147 25.6	34.5	245 36.6	13.7	14 41.4	12.2	55.9
23	162 25.4	34.0	260 09.3	13.6	14 53.6	12.1	55.8
18 00 SUNDAY	177 25.2	S20 33.5	274 41.9	13.5	S15 05.7	12.1	55.8
01	192 25.0	33.0	289 14.4	13.6	15 17.8	12.0	55.8
02	207 24.8	32.5	303 47.0	13.5	15 29.8	11.9	55.7
03	222 24.6	. . 32.0	318 19.5	13.5	15 41.7	11.8	55.7
04	237 24.4	31.5	332 52.0	13.5	15 53.5	11.7	55.7
05	252 24.2	31.0	347 24.5	13.5	16 05.2	11.7	55.6
06	267 24.0	S20 30.4	1 57.0	13.4	S16 16.9	11.6	55.6
07	282 23.8	29.9	16 29.4	13.4	16 28.5	11.4	55.6
08	297 23.6	29.4	31 01.8	13.3	16 39.9	11.4	55.5
09	312 23.4	. . 28.9	45 34.1	13.4	16 51.3	11.4	55.5
10	327 23.2	28.4	60 06.5	13.3	17 02.7	11.2	55.5
11	342 23.0	27.9	74 38.8	13.3	17 13.9	11.1	55.5
12	357 22.8	S20 27.4	89 11.1	13.2	S17 25.0	11.1	55.4
13	12 22.6	26.8	103 43.3	13.2	17 36.1	11.0	55.4
14	27 22.4	26.3	118 15.5	13.2	17 47.1	10.9	55.4
15	42 22.2	. . 25.8	132 47.7	13.1	17 58.0	10.8	55.4
16	57 22.0	25.3	147 19.8	13.2	18 08.8	10.7	55.3
17	72 21.8	24.8	161 52.0	13.0	18 19.5	10.6	55.3
18	87 21.6	S20 24.2	176 24.0	13.1	S18 30.1	10.5	55.3
19	102 21.4	23.7	190 56.1	13.0	18 40.6	10.4	55.2
20	117 21.2	23.2	205 28.1	13.0	18 51.0	10.4	55.2
21	132 21.0	. . 22.7	220 00.1	12.9	19 01.4	10.2	55.2
22	147 20.8	22.2	234 32.0	12.9	19 11.6	10.2	55.2
23	162 20.7	21.6	249 03.9	12.9	S19 21.8	10.1	55.1
	SD 16.3	*d* 0.5	SD 15.5		15.3		15.1

Lat. °	Twilight Naut. h m	Twilight Civil h m	Sunrise h m	Moonrise 16 h m	Moonrise 17 h m	Moonrise 18 h m	Moonrise 19 h m
N 72	07 55	09 43	■	25 16	01 16	03 56	■
N 70	07 41	09 11	11 39	24 59	00 59	03 12	■
68	07 30	08 48	10 23	24 45	00 45	02 43	05 05
66	07 20	08 30	09 46	24 34	00 34	02 22	04 19
64	07 12	08 15	09 21	24 24	00 24	02 05	03 50
62	07 05	08 03	09 01	24 16	00 16	01 51	03 27
60	06 59	07 52	08 44	24 10	00 10	01 39	03 10
N 58	06 53	07 43	08 31	24 04	00 04	01 29	02 55
56	06 48	07 35	08 19	23 59	25 21	01 21	02 42
54	06 43	07 27	08 08	23 54	25 13	01 13	02 31
52	06 38	07 20	07 59	23 50	25 06	01 06	02 21
50	06 34	07 14	07 51	23 46	25 00	01 00	02 13
45	06 25	07 01	07 33	23 38	24 47	00 47	01 54
N 40	06 16	06 50	07 19	23 31	24 36	00 36	01 40
35	06 09	06 40	07 07	23 25	24 26	00 26	01 27
30	06 01	06 31	06 56	23 20	24 18	00 18	01 16
20	05 47	06 14	06 38	23 11	24 04	00 04	00 58
N 10	05 33	05 59	06 22	23 03	23 52	24 42	00 42
0	05 19	05 44	06 07	22 56	23 41	24 27	00 27
S 10	05 02	05 29	05 51	22 49	23 30	24 12	00 12
20	04 42	05 11	05 35	22 42	23 18	23 56	24 37
30	04 17	04 49	05 15	22 33	23 05	23 38	24 15
35	04 00	04 35	05 04	22 28	22 57	23 28	24 03
40	03 40	04 20	04 51	22 23	22 48	23 16	23 48
45	03 15	04 00	04 36	22 17	22 38	23 02	23 31
S 50	02 39	03 35	04 17	22 09	22 26	22 45	23 09
52	02 19	03 23	04 08	22 06	22 20	22 37	22 59
54	01 54	03 09	03 57	22 02	22 14	22 29	22 48
56	01 16	02 52	03 46	21 58	22 07	22 19	22 35
58	////	02 31	03 32	21 53	21 59	22 07	22 20
S 60	////	02 04	03 17	21 48	21 51	21 55	22 02

Lat. °	Sunset h m	Twilight Civil h m	Twilight Naut. h m	Moonset 16 h m	Moonset 17 h m	Moonset 18 h m	Moonset 19 h m
N 72	■	14 39	16 26	09 08	08 31	07 26	■
N 70	12 42	15 10	16 40	09 17	08 50	08 12	■
68	13 58	15 33	16 51	09 25	09 06	08 42	07 57
66	14 35	15 51	17 01	09 31	09 19	09 05	08 43
64	15 00	16 06	17 09	09 36	09 30	09 23	09 14
62	15 20	16 18	17 16	09 41	09 39	09 38	09 37
60	15 37	16 29	17 22	09 45	09 47	09 50	09 55
N 58	15 50	16 38	17 28	09 49	09 54	10 01	10 11
56	16 02	16 46	17 33	09 52	10 00	10 11	10 24
54	16 12	16 54	17 38	09 55	10 06	10 19	10 36
52	16 22	17 00	17 43	09 57	10 11	10 27	10 46
50	16 30	17 06	17 47	10 00	10 16	10 34	10 55
45	16 47	17 20	17 56	10 05	10 26	10 48	11 14
N 40	17 01	17 31	18 04	10 10	10 34	11 00	11 30
35	17 14	17 41	18 12	10 13	10 41	11 11	11 44
30	17 24	17 50	18 19	10 17	10 48	11 20	11 55
20	17 43	18 06	18 33	10 23	10 59	11 36	12 15
N 10	17 59	18 21	18 47	10 28	11 08	11 50	12 33
0	18 14	18 36	19 02	10 33	11 18	12 03	12 49
S 10	18 29	18 52	19 18	10 38	11 27	12 16	13 06
20	18 46	19 09	19 38	10 43	11 37	12 30	13 24
30	19 05	19 31	20 03	10 49	11 48	12 46	13 44
35	19 16	19 44	20 19	10 53	11 55	12 56	13 56
40	19 29	20 00	20 39	10 56	12 02	13 07	14 10
45	19 44	20 19	21 05	11 01	12 11	13 19	14 26
S 50	20 03	20 44	21 40	11 07	12 21	13 35	14 47
52	20 12	20 56	21 59	11 09	12 26	13 42	14 57
54	20 22	21 10	22 24	11 12	12 32	13 50	15 08
56	20 33	21 27	23 00	11 15	12 38	14 00	15 20
58	20 47	21 47	////	11 19	12 45	14 10	15 35
S 60	21 02	22 14	////	11 22	12 52	14 22	15 52

Day	SUN Eqn. of Time 00^h	SUN Eqn. of Time 12^h	SUN Mer. Pass.	MOON Mer. Pass. Upper	MOON Mer. Pass. Lower	MOON Age	MOON Phase
d	m s	m s	h m	h m	h m	d %	
16	09 39	09 49	12 10	04 22	16 45	20 66	
17	09 59	10 09	12 10	05 07	17 29	21 56	
18	10 19	10 28	12 10	05 52	18 15	22 46	

UT d h	ARIES GHA ° ′	VENUS −4.6 GHA ° ′	VENUS Dec ° ′	MARS +1.3 GHA ° ′	MARS Dec ° ′	JUPITER −1.9 GHA ° ′	JUPITER Dec ° ′	SATURN +0.8 GHA ° ′	SATURN Dec ° ′	STARS Name	SHA ° ′	Dec ° ′
MONDAY 19 00	118 31.3	131 19.6	S 5 27.0	189 47.0	S23 10.6	172 59.3	S19 52.8	305 31.1	N 5 20.4	Acamar	315 20.6	S40 16.3
01	133 33.8	146 19.9	25.8	204 47.4	10.3	188 01.1	52.6	320 33.6	20.5	Achernar	335 29.0	S57 11.7
02	148 36.2	161 20.2	24.6	219 47.8	10.1	203 03.0	52.5	335 36.2	20.5	Acrux	173 13.0	S63 08.8
03	163 38.7	176 20.5	. . 23.4	234 48.2	. . 09.9	218 04.9	. . 52.4	350 38.7	. . 20.5	Adhara	255 14.7	S28 59.1
04	178 41.1	191 20.7	22.2	249 48.6	09.7	233 06.7	52.2	5 41.2	20.6	Aldebaran	290 52.9	N16 31.7
05	193 43.6	206 21.0	21.0	264 49.0	09.4	248 08.6	52.1	20 43.8	20.6			
06	208 46.1	221 21.3	S 5 19.8	279 49.4	S23 09.2	263 10.4	S19 52.0	35 46.3	N 5 20.7	Alioth	166 23.2	N55 54.2
07	223 48.5	236 21.6	18.6	294 49.8	09.0	278 12.3	51.8	50 48.8	20.7	Alkaid	153 01.3	N49 15.7
08	238 51.0	251 21.9	17.4	309 50.2	08.8	293 14.1	51.7	65 51.4	20.8	Al Na'ir	27 48.1	S46 55.2
09	253 53.5	266 22.2	. . 16.2	324 50.6	. . 08.5	308 16.0	. . 51.6	80 53.9	. . 20.8	Alnilam	275 49.4	S 1 11.8
10	268 55.9	281 22.5	15.0	339 51.0	08.3	323 17.8	51.4	95 56.4	20.8	Alphard	217 59.0	S 8 42.0
11	283 58.4	296 22.7	13.8	354 51.4	08.1	338 19.7	51.3	110 59.0	20.9			
12	299 00.9	311 23.0	S 5 12.6	9 51.8	S23 07.8	353 21.5	S19 51.2	126 01.5	N 5 20.9	Alphecca	126 13.9	N26 40.7
13	314 03.3	326 23.3	11.4	24 52.2	07.6	8 23.4	51.0	141 04.1	21.0	Alpheratz	357 47.1	N29 08.6
14	329 05.8	341 23.6	10.2	39 52.6	07.4	23 25.2	50.9	156 06.6	21.0	Altair	62 11.8	N 8 53.5
15	344 08.3	356 23.9	. . 09.0	54 53.0	. . 07.1	38 27.1	. . 50.8	171 09.1	. . 21.1	Ankaa	353 18.9	S42 15.6
16	359 10.7	11 24.2	07.8	69 53.4	06.9	53 28.9	50.6	186 11.7	21.1	Antares	112 30.5	S26 27.2
17	14 13.2	26 24.5	06.6	84 53.8	06.7	68 30.8	50.5	201 14.2	21.1			
18	29 15.6	41 24.8	S 5 05.4	99 54.2	S23 06.4	83 32.6	S19 50.4	216 16.7	N 5 21.2	Arcturus	145 58.7	N19 07.8
19	44 18.1	56 25.1	04.2	114 54.6	06.2	98 34.5	50.2	231 19.3	21.2	Atria	107 35.8	S69 02.5
20	59 20.6	71 25.4	03.0	129 55.0	06.0	113 36.3	50.1	246 21.8	21.3	Avior	234 18.9	S59 32.3
21	74 23.0	86 25.7	. . 01.8	144 55.4	. . 05.7	128 38.2	. . 50.0	261 24.4	. . 21.3	Bellatrix	278 35.2	N 6 21.5
22	89 25.5	101 26.0	5 00.6	159 55.8	05.5	143 40.0	49.8	276 26.9	21.4	Betelgeuse	271 04.5	N 7 24.5
23	104 28.0	116 26.3	4 59.4	174 56.2	05.3	158 41.9	49.7	291 29.4	21.4			
TUESDAY 20 00	119 30.4	131 26.6	S 4 58.2	189 56.6	S23 05.0	173 43.7	S19 49.6	306 32.0	N 5 21.4	Canopus	263 57.1	S52 42.1
01	134 32.9	146 26.9	57.0	204 57.0	04.8	188 45.6	49.4	321 34.5	21.5	Capella	280 38.9	N46 00.6
02	149 35.4	161 27.2	55.7	219 57.4	04.6	203 47.5	49.3	336 37.1	21.5	Deneb	49 34.3	N45 18.8
03	164 37.8	176 27.5	. . 54.5	234 57.8	. . 04.3	218 49.3	. . 49.2	351 39.6	. . 21.6	Denebola	182 36.8	N14 31.1
04	179 40.3	191 27.8	53.3	249 58.2	04.1	233 51.2	49.0	6 42.1	21.6	Diphda	348 59.2	S17 56.3
05	194 42.8	206 28.1	52.1	264 58.6	03.8	248 53.0	48.9	21 44.7	21.7			
06	209 45.2	221 28.4	S 4 50.9	279 59.0	S23 03.6	263 54.9	S19 48.7	36 47.2	N 5 21.7	Dubhe	193 54.9	N61 41.8
07	224 47.7	236 28.7	49.7	294 59.4	03.4	278 56.7	48.6	51 49.8	21.7	Elnath	278 16.5	N28 37.0
08	239 50.1	251 29.0	48.5	309 59.8	03.1	293 58.6	48.5	66 52.3	21.8	Eltanin	90 48.1	N51 29.0
09	254 52.6	266 29.3	. . 47.3	325 00.2	. . 02.9	309 00.4	. . 48.3	81 54.8	. . 21.8	Enif	33 50.7	N 9 55.0
10	269 55.1	281 29.7	46.1	340 00.6	02.7	324 02.3	48.2	96 57.4	21.9	Fomalhaut	15 27.8	S29 34.6
11	284 57.5	296 30.0	44.9	355 01.0	02.4	339 04.1	48.1	111 59.9	21.9			
12	300 00.0	311 30.3	S 4 43.7	10 01.4	S23 02.2	354 06.0	S19 47.9	127 02.5	N 5 22.0	Gacrux	172 04.6	S57 09.7
13	315 02.5	326 30.6	42.5	25 01.8	01.9	9 07.8	47.8	142 05.0	22.0	Gienah	175 55.6	S17 35.6
14	330 04.9	341 30.9	41.3	40 02.2	01.7	24 09.7	47.7	157 07.5	22.1	Hadar	148 52.8	S60 24.8
15	345 07.4	356 31.2	. . 40.1	55 02.6	. . 01.4	39 11.5	. . 47.5	172 10.1	. . 22.1	Hamal	328 04.5	N23 30.5
16	0 09.9	11 31.5	38.9	70 03.0	01.2	54 13.4	47.4	187 12.6	22.1	Kaus Aust.	83 48.5	S34 22.8
17	15 12.3	26 31.9	37.7	85 03.4	01.0	69 15.2	47.3	202 15.2	22.2			
18	30 14.8	41 32.2	S 4 36.5	100 03.8	S23 00.7	84 17.1	S19 47.1	217 17.7	N 5 22.2	Kochab	137 19.8	N74 06.6
19	45 17.3	56 32.5	35.3	115 04.2	00.5	99 18.9	47.0	232 20.2	22.3	Markab	13 41.9	N15 15.3
20	60 19.7	71 32.8	34.1	130 04.6	00.2	114 20.8	46.9	247 22.8	22.3	Menkar	314 18.4	N 4 07.6
21	75 22.2	86 33.2	. . 32.9	145 05.0	23 00.0	129 22.6	. . 46.7	262 25.3	. . 22.4	Menkent	148 11.6	S36 24.8
22	90 24.6	101 33.5	31.7	160 05.4	22 59.7	144 24.5	46.6	277 27.9	22.4	Miaplacidus	221 39.8	S69 45.2
23	105 27.1	116 33.8	30.5	175 05.8	59.5	159 26.4	46.5	292 30.4	22.5			
WEDNESDAY 21 00	120 29.6	131 34.1	S 4 29.3	190 06.2	S22 59.2	174 28.2	S19 46.3	307 33.0	N 5 22.5	Mirfak	308 45.0	N49 53.9
01	135 32.0	146 34.5	28.1	205 06.6	59.0	189 30.1	46.2	322 35.5	22.5	Nunki	76 02.7	S26 17.2
02	150 34.5	161 34.8	26.9	220 07.0	58.7	204 31.9	46.1	337 38.0	22.6	Peacock	53 24.8	S56 42.4
03	165 37.0	176 35.1	. . 25.7	235 07.4	. . 58.5	219 33.8	. . 45.9	352 40.6	. . 22.6	Pollux	243 31.3	N28 00.2
04	180 39.4	191 35.4	24.5	250 07.8	58.2	234 35.6	45.8	7 43.1	22.7	Procyon	245 02.8	N 5 12.0
05	195 41.9	206 35.8	23.3	265 08.2	58.0	249 37.5	45.7	22 45.7	22.7			
06	210 44.4	221 36.1	S 4 22.1	280 08.6	S22 57.8	264 39.3	S19 45.5	37 48.2	N 5 22.8	Rasalhague	96 09.8	N12 33.0
07	225 46.8	236 36.4	20.9	295 09.0	57.5	279 41.2	45.4	52 50.8	22.8	Regulus	207 46.7	N11 55.2
08	240 49.3	251 36.8	19.7	310 09.4	57.3	294 43.0	45.2	67 53.3	22.9	Rigel	281 14.9	S 8 11.5
09	255 51.8	266 37.1	. . 18.4	325 09.8	. . 57.0	309 44.9	. . 45.1	82 55.8	. . 22.9	Rigil Kent.	139 56.5	S60 52.2
10	270 54.2	281 37.4	17.2	340 10.2	56.7	324 46.7	45.0	97 58.4	23.0	Sabik	102 16.6	S15 44.2
11	285 56.7	296 37.8	16.0	355 10.6	56.5	339 48.6	44.8	113 00.9	23.0			
12	300 59.1	311 38.1	S 4 14.8	10 11.0	S22 56.2	354 50.4	S19 44.7	128 03.5	N 5 23.0	Schedar	349 44.7	N56 35.6
13	316 01.6	326 38.5	13.6	25 11.4	56.0	9 52.3	44.6	143 06.0	23.1	Shaula	96 26.7	S37 06.6
14	331 04.1	341 38.8	12.4	40 11.8	55.7	24 54.1	44.4	158 08.6	23.1	Sirius	258 36.3	S16 43.8
15	346 06.5	356 39.1	. . 11.2	55 12.2	. . 55.5	39 56.0	. . 44.3	173 11.1	. . 23.2	Spica	158 34.7	S11 12.6
16	1 09.0	11 39.5	10.0	70 12.6	55.2	54 57.8	44.2	188 13.7	23.2	Suhail	222 54.5	S43 28.1
17	16 11.5	26 39.8	08.8	85 13.0	55.0	69 59.7	44.0	203 16.2	23.3			
18	31 13.9	41 40.2	S 4 07.6	100 13.4	S22 54.7	85 01.5	S19 43.9	218 18.8	N 5 23.3	Vega	80 41.6	N38 47.4
19	46 16.4	56 40.5	06.4	115 13.8	54.5	100 03.4	43.8	233 21.3	23.4	Zuben'ubi	137 09.2	S16 04.8
20	61 18.9	71 40.9	05.2	130 14.2	54.2	115 05.3	43.6	248 23.8	23.4			
21	76 21.3	86 41.2	. . 04.0	145 14.6	. . 54.0	130 07.1	. . 43.5	263 26.4	. . 23.5			
22	91 23.8	101 41.6	02.8	160 15.1	53.7	145 09.0	43.4	278 28.9	23.5			
23	106 26.2	116 41.9	01.6	175 15.5	53.4	160 10.8	43.2	293 31.5	23.6			
Mer. Pass. (h m)	15 59.3	v 0.3	d 1.2	v 0.4	d 0.2	v 1.9	d 0.1	v 2.5	d 0.0			

	SHA ° ′	Mer. Pass. h m
Venus	11 56.2	15 14
Mars	70 26.2	11 20
Jupiter	54 13.3	12 24
Saturn	187 01.5	3 33

	UT	SUN GHA	SUN Dec	MOON GHA	v	MOON Dec	d	HP
	d h	° ′	° ′	° ′	′	° ′	′	′
	19 00	177 20.5	S20 21.1	263 35.8	12.8	S19 31.9	9.9	55.1
	01	192 20.3	20.6	278 07.6	12.8	19 41.8	9.9	55.1
	02	207 20.1	20.1	292 39.4	12.8	19 51.7	9.8	55.1
	03	222 19.9	. . 19.5	307 11.2	12.7	20 01.5	9.7	55.0
	04	237 19.7	19.0	321 42.9	12.7	20 11.2	9.5	55.0
	05	252 19.5	18.5	336 14.6	12.6	20 20.7	9.5	55.0
	06	267 19.3	S20 18.0	350 46.2	12.6	S20 30.2	9.4	55.0
	07	282 19.1	17.4	5 17.8	12.6	20 39.6	9.3	55.0
M	08	297 18.9	16.9	19 49.4	12.5	20 48.9	9.2	54.9
O	09	312 18.7	. . 16.4	34 20.9	12.5	20 58.1	9.1	54.9
N	10	327 18.6	15.8	48 52.4	12.4	21 07.2	9.0	54.9
D	11	342 18.4	15.3	63 23.8	12.4	21 16.2	8.8	54.9
A	12	357 18.2	S20 14.8	77 55.2	12.4	S21 25.0	8.8	54.8
Y	13	12 18.0	14.2	92 26.6	12.3	21 33.8	8.7	54.8
	14	27 17.8	13.7	106 57.9	12.3	21 42.5	8.6	54.8
	15	42 17.6	. . 13.2	121 29.2	12.3	21 51.1	8.5	54.8
	16	57 17.4	12.6	136 00.5	12.2	21 59.6	8.3	54.8
	17	72 17.2	12.1	150 31.7	12.1	22 07.9	8.3	54.7
	18	87 17.1	S20 11.6	165 02.8	12.2	S22 16.2	8.2	54.7
	19	102 16.9	11.0	179 34.0	12.0	22 24.4	8.0	54.7
	20	117 16.7	10.5	194 05.0	12.1	22 32.4	8.0	54.7
	21	132 16.5	. . 10.0	208 36.1	12.0	22 40.4	7.8	54.7
	22	147 16.3	09.4	223 07.1	11.9	22 48.2	7.7	54.6
	23	162 16.1	08.9	237 38.0	12.0	22 55.9	7.7	54.6
	20 00	177 15.9	S20 08.3	252 09.0	11.9	S23 03.6	7.5	54.6
	01	192 15.8	07.8	266 39.9	11.8	23 11.1	7.4	54.6
	02	207 15.6	07.3	281 10.7	11.8	23 18.5	7.3	54.6
	03	222 15.4	. . 06.7	295 41.5	11.8	23 25.8	7.2	54.6
	04	237 15.2	06.2	310 12.3	11.7	23 33.0	7.0	54.5
	05	252 15.0	05.6	324 43.0	11.7	23 40.0	7.0	54.5
	06	267 14.8	S20 05.1	339 13.7	11.6	S23 47.0	6.9	54.5
	07	282 14.7	04.5	353 44.3	11.6	23 53.9	6.7	54.5
T	08	297 14.5	04.0	8 14.9	11.6	24 00.6	6.6	54.5
U	09	312 14.3	. . 03.5	22 45.5	11.5	24 07.2	6.5	54.5
E	10	327 14.1	02.9	37 16.0	11.5	24 13.7	6.4	54.4
S	11	342 13.9	02.4	51 46.5	11.4	24 20.1	6.3	54.4
D	12	357 13.7	S20 01.8	66 16.9	11.4	S24 26.4	6.2	54.4
A	13	12 13.6	01.3	80 47.3	11.4	24 32.6	6.0	54.4
Y	14	27 13.4	00.7	95 17.7	11.3	24 38.6	5.9	54.4
	15	42 13.2	20 00.2	109 48.0	11.3	24 44.5	5.9	54.4
	16	57 13.0	19 59.6	124 18.3	11.3	24 50.4	5.7	54.4
	17	72 12.8	59.1	138 48.6	11.2	24 56.1	5.5	54.3
	18	87 12.7	S19 58.5	153 18.8	11.2	S25 01.6	5.5	54.3
	19	102 12.5	58.0	167 49.0	11.2	25 07.1	5.3	54.3
	20	117 12.3	57.4	182 19.2	11.1	25 12.4	5.3	54.3
	21	132 12.1	. . 56.9	196 49.3	11.1	25 17.7	5.1	54.3
	22	147 12.0	56.3	211 19.4	11.0	25 22.8	5.0	54.3
	23	162 11.8	55.7	225 49.4	11.0	25 27.8	4.8	54.3
	21 00	177 11.6	S19 55.2	240 19.4	11.0	S25 32.6	4.8	54.3
	01	192 11.4	54.6	254 49.4	11.0	25 37.4	4.6	54.2
	02	207 11.2	54.1	269 19.4	10.9	25 42.0	4.5	54.2
	03	222 11.1	. . 53.5	283 49.3	10.9	25 46.5	4.4	54.2
	04	237 10.9	53.0	298 19.2	10.9	25 50.9	4.2	54.2
	05	252 10.7	52.4	312 49.1	10.8	25 55.1	4.2	54.2
	06	267 10.5	S19 51.8	327 18.9	10.8	S25 59.3	4.0	54.2
W	07	282 10.4	51.3	341 48.7	10.8	26 03.3	3.9	54.2
E	08	297 10.2	50.7	356 18.5	10.7	26 07.2	3.7	54.2
D	09	312 10.0	. . 50.2	10 48.2	10.8	26 10.9	3.7	54.2
N	10	327 09.8	49.6	25 18.0	10.7	26 14.6	3.5	54.2
E	11	342 09.7	49.0	39 47.7	10.6	26 18.1	3.4	54.1
S	12	357 09.5	S19 48.5	54 17.3	10.7	S26 21.5	3.2	54.1
D	13	12 09.3	47.9	68 47.0	10.6	26 24.7	3.2	54.1
A	14	27 09.2	47.3	83 16.6	10.6	26 27.9	3.0	54.1
Y	15	42 09.0	. . 46.8	97 46.2	10.6	26 30.9	2.9	54.1
	16	57 08.8	46.2	112 15.8	10.5	26 33.8	2.7	54.1
	17	72 08.6	45.6	126 45.3	10.6	26 36.5	2.7	54.1
	18	87 08.5	S19 45.1	141 14.9	10.5	S26 39.2	2.5	54.1
	19	102 08.3	44.5	155 44.4	10.5	26 41.7	2.3	54.1
	20	117 08.1	43.9	170 13.9	10.5	26 44.0	2.3	54.1
	21	132 08.0	. . 43.4	184 43.4	10.5	26 46.3	2.1	54.1
	22	147 07.8	42.8	199 12.9	10.4	26 48.4	2.0	54.1
	23	162 07.6	42.2	213 42.3	10.4	S26 50.4	1.9	54.1
		SD 16.3	*d* 0.5	SD 14.9		14.8		14.8

Lat.	Twilight Naut.	Twilight Civil	Sunrise	Moonrise 19	20	21	22
°	h m	h m	h m	h m	h m	h m	h m
N 72	07 47	09 30	■	■	■	■	■
N 70	07 35	09 02	11 03	■	☰	☰	☰
68	07 24	08 40	10 09	05 05	☰	☰	☰
66	07 15	08 24	09 37	04 19	☰	☰	☰
64	07 08	08 10	09 13	03 50	05 43	☰	☰
62	07 01	07 58	08 54	03 27	05 05	06 41	08 02
60	06 55	07 48	08 39	03 10	04 39	06 03	07 12
N 58	06 50	07 39	08 26	02 55	04 18	05 36	06 41
56	06 45	07 31	08 15	02 42	04 01	05 15	06 18
54	06 40	07 24	08 05	02 31	03 47	04 57	05 59
52	06 36	07 18	07 56	02 21	03 34	04 43	05 43
50	06 32	07 12	07 48	02 13	03 23	04 30	05 29
45	06 23	06 59	07 31	01 54	03 01	04 03	05 01
N 40	06 15	06 48	07 18	01 40	02 42	03 43	04 38
35	06 08	06 39	07 06	01 27	02 27	03 25	04 20
30	06 01	06 30	06 56	01 16	02 14	03 10	04 04
20	05 47	06 14	06 38	00 58	01 52	02 45	03 38
N 10	05 34	06 00	06 22	00 42	01 32	02 23	03 15
0	05 20	05 45	06 07	00 27	01 14	02 03	02 54
S 10	05 04	05 30	05 53	00 12	00 56	01 43	02 32
20	04 44	05 13	05 37	24 37	00 37	01 22	02 09
30	04 20	04 52	05 18	24 15	00 15	00 57	01 43
35	04 04	04 39	05 07	24 03	00 03	00 42	01 27
40	03 45	04 23	04 55	23 48	24 25	00 25	01 09
45	03 20	04 05	04 40	23 31	24 05	00 05	00 48
S 50	02 46	03 41	04 21	23 09	23 40	24 20	00 20
52	02 27	03 29	04 13	22 59	23 28	24 07	00 07
54	02 04	03 15	04 03	22 48	23 14	23 51	24 41
56	01 31	02 59	03 52	22 35	22 58	23 33	24 22
58	////	02 40	03 39	22 20	22 39	23 10	23 59
S 60	////	02 15	03 24	22 02	22 15	22 42	23 29

Lat.	Sunset	Twilight Civil	Twilight Naut.	Moonset 19	20	21	22
°	h m	h m	h m	h m	h m	h m	h m
N 72	■	14 53	16 36	■	☰	☰	☰
N 70	13 20	15 21	16 49	■	☰	☰	☰
68	14 14	15 43	16 59	07 57	☰	☰	☰
66	14 46	15 59	17 08	08 43	☰	☰	☰
64	15 10	16 13	17 15	09 14	09 00	☰	☰
62	15 28	16 25	17 22	09 37	09 38	09 44	10 09
60	15 44	16 35	17 28	09 55	10 05	10 23	10 59
N 58	15 57	16 44	17 33	10 11	10 26	10 50	11 30
56	16 08	16 52	17 38	10 24	10 43	11 12	11 53
54	16 18	16 59	17 43	10 36	10 58	11 29	12 12
52	16 27	17 05	17 47	10 46	11 11	11 44	12 28
50	16 34	17 11	17 51	10 55	11 22	11 57	12 42
45	16 51	17 24	18 00	11 14	11 46	12 24	13 10
N 40	17 05	17 34	18 07	11 30	12 05	12 45	13 33
35	17 17	17 44	18 15	11 44	12 21	13 03	13 51
30	17 27	17 52	18 22	11 55	12 34	13 18	14 07
20	17 45	18 08	18 35	12 15	12 58	13 44	14 34
N 10	18 00	18 22	18 48	12 33	13 19	14 07	14 57
0	18 15	18 37	19 02	12 49	13 38	14 27	15 18
S 10	18 29	18 52	19 18	13 06	13 57	14 48	15 39
20	18 45	19 09	19 37	13 24	14 17	15 11	16 02
30	19 04	19 30	20 02	13 44	14 41	15 37	16 29
35	19 15	19 43	20 18	13 56	14 55	15 52	16 45
40	19 27	19 58	20 37	14 10	15 12	16 10	17 03
45	19 42	20 17	21 01	14 26	15 31	16 31	17 25
S 50	20 00	20 40	21 35	14 47	15 56	16 59	17 53
52	20 09	20 52	21 53	14 57	16 08	17 12	18 07
54	20 18	21 05	22 16	15 08	16 21	17 28	18 23
56	20 29	21 21	22 47	15 20	16 37	17 46	18 41
58	20 42	21 40	23 54	15 35	16 56	18 08	19 04
S 60	20 56	22 04	////	15 52	17 19	18 37	19 34

Day	SUN Eqn. of Time 00^h	12^h	Mer. Pass.	MOON Mer. Pass. Upper	Lower	Age	Phase
d	m s	m s	h m	h m	h m	d %	
19	10 38	10 47	12 11	06 38	19 02	23 37	
20	10 56	11 05	12 11	07 26	19 50	24 28	
21	11 13	11 22	12 11	08 15	20 40	25 20	

	UT	ARIES	VENUS −4.6		MARS +1.3		JUPITER −1.9		SATURN +0.8	
		GHA	GHA	Dec	GHA	Dec	GHA	Dec	GHA	Dec
	d h	° ′	° ′	° ′	° ′	° ′	° ′	° ′	° ′	° ′
	22 00	121 28.7	131 42.3	S 4 00.4	190 15.9	S22 53.2	175 12.7	S19 43.1	308 34.0	N 5 23.6
	01	136 31.2	146 42.6	3 59.2	205 16.3	52.9	190 14.5	42.9	323 36.6	23.6
	02	151 33.6	161 43.0	58.0	220 16.7	52.7	205 16.4	42.8	338 39.1	23.7
	03	166 36.1	176 43.3	. . 56.8	235 17.1	. . 52.4	220 18.2	. . 42.7	353 41.7	. . 23.7
	04	181 38.6	191 43.7	55.6	250 17.5	52.2	235 20.1	42.5	8 44.2	23.8
	05	196 41.0	206 44.0	54.4	265 17.9	51.9	250 21.9	42.4	23 46.8	23.8
	06	211 43.5	221 44.4	S 3 53.2	280 18.3	S22 51.6	265 23.8	S19 42.3	38 49.3	N 5 23.9
	07	226 46.0	236 44.7	52.0	295 18.7	51.4	280 25.6	42.1	53 51.9	23.9
T	08	241 48.4	251 45.1	50.8	310 19.1	51.1	295 27.5	42.0	68 54.4	24.0
H	09	256 50.9	266 45.5	. . 49.6	325 19.5	. . 50.9	310 29.3	. . 41.9	83 57.0	. . 24.0
U	10	271 53.4	281 45.8	48.4	340 19.9	50.6	325 31.2	41.7	98 59.5	24.1
R	11	286 55.8	296 46.2	47.2	355 20.3	50.3	340 33.0	41.6	114 02.0	24.1
S	12	301 58.3	311 46.5	S 3 46.0	10 20.7	S22 50.1	355 34.9	S19 41.5	129 04.6	N 5 24.2
D	13	317 00.7	326 46.9	44.8	25 21.1	49.8	10 36.7	41.3	144 07.1	24.2
A	14	332 03.2	341 47.3	43.6	40 21.5	49.5	25 38.6	41.2	159 09.7	24.3
Y	15	347 05.7	356 47.6	. . 42.4	55 21.9	. . 49.3	40 40.5	. . 41.0	174 12.2	. . 24.3
	16	2 08.1	11 48.0	41.2	70 22.3	49.0	55 42.3	40.9	189 14.8	24.4
	17	17 10.6	26 48.4	40.0	85 22.7	48.7	70 44.2	40.8	204 17.3	24.4
	18	32 13.1	41 48.7	S 3 38.8	100 23.1	S22 48.5	85 46.0	S19 40.6	219 19.9	N 5 24.5
	19	47 15.5	56 49.1	37.6	115 23.5	48.2	100 47.9	40.5	234 22.4	24.5
	20	62 18.0	71 49.5	36.4	130 23.9	48.0	115 49.7	40.4	249 25.0	24.5
	21	77 20.5	86 49.9	. . 35.1	145 24.3	. . 47.7	130 51.6	. . 40.2	264 27.5	. . 24.6
	22	92 22.9	101 50.2	33.9	160 24.7	47.4	145 53.4	40.1	279 30.1	24.6
	23	107 25.4	116 50.6	32.7	175 25.1	47.2	160 55.3	40.0	294 32.6	24.7
	23 00	122 27.9	131 51.0	S 3 31.5	190 25.5	S22 46.9	175 57.1	S19 39.8	309 35.2	N 5 24.7
	01	137 30.3	146 51.4	30.3	205 25.9	46.6	190 59.0	39.7	324 37.7	24.8
	02	152 32.8	161 51.7	29.1	220 26.4	46.3	206 00.8	39.5	339 40.3	24.8
	03	167 35.2	176 52.1	. . 27.9	235 26.8	. . 46.1	221 02.7	. . 39.4	354 42.8	. . 24.9
	04	182 37.7	191 52.5	26.7	250 27.2	45.8	236 04.5	39.3	9 45.4	24.9
	05	197 40.2	206 52.9	25.5	265 27.6	45.5	251 06.4	39.1	24 47.9	25.0
	06	212 42.6	221 53.3	S 3 24.3	280 28.0	S22 45.3	266 08.2	S19 39.0	39 50.5	N 5 25.0
	07	227 45.1	236 53.6	23.1	295 28.4	45.0	281 10.1	38.9	54 53.0	25.1
	08	242 47.6	251 54.0	21.9	310 28.8	44.7	296 11.9	38.7	69 55.6	25.1
F	09	257 50.0	266 54.4	. . 20.7	325 29.2	. . 44.5	311 13.8	. . 38.6	84 58.2	. . 25.2
R	10	272 52.5	281 54.8	19.5	340 29.6	44.2	326 15.7	38.5	100 00.7	25.2
I	11	287 55.0	296 55.2	18.3	355 30.0	43.9	341 17.5	38.3	115 03.3	25.3
D	12	302 57.4	311 55.6	S 3 17.1	10 30.4	S22 43.6	356 19.4	S19 38.2	130 05.8	N 5 25.3
A	13	317 59.9	326 56.0	15.9	25 30.8	43.4	11 21.2	38.0	145 08.4	25.4
Y	14	333 02.4	341 56.4	14.7	40 31.2	43.1	26 23.1	37.9	160 10.9	25.4
	15	348 04.8	356 56.7	. . 13.5	55 31.6	. . 42.8	41 24.9	. . 37.8	175 13.5	. . 25.5
	16	3 07.3	11 57.1	12.3	70 32.0	42.5	56 26.8	37.6	190 16.0	25.5
	17	18 09.7	26 57.5	11.1	85 32.4	42.3	71 28.6	37.5	205 18.6	25.6
	18	33 12.2	41 57.9	S 3 09.9	100 32.8	S22 42.0	86 30.5	S19 37.4	220 21.1	N 5 25.6
	19	48 14.7	56 58.3	08.7	115 33.2	41.7	101 32.3	37.2	235 23.7	25.7
	20	63 17.1	71 58.7	07.5	130 33.6	41.4	116 34.2	37.1	250 26.2	25.7
	21	78 19.6	86 59.1	. . 06.3	145 34.1	. . 41.2	131 36.0	. . 37.0	265 28.8	. . 25.8
	22	93 22.1	101 59.5	05.1	160 34.5	40.9	146 37.9	36.8	280 31.3	25.8
	23	108 24.5	116 59.9	03.9	175 34.9	40.6	161 39.7	36.7	295 33.9	25.9
	24 00	123 27.0	132 00.3	S 3 02.7	190 35.3	S22 40.3	176 41.6	S19 36.5	310 36.4	N 5 25.9
	01	138 29.5	147 00.7	01.5	205 35.7	40.0	191 43.4	36.4	325 39.0	26.0
	02	153 31.9	162 01.1	3 00.3	220 36.1	39.8	206 45.3	36.3	340 41.6	26.0
	03	168 34.4	177 01.5	2 59.1	235 36.5	. . 39.5	221 47.2	. . 36.1	355 44.1	. . 26.1
	04	183 36.9	192 01.9	57.9	250 36.9	39.2	236 49.0	36.0	10 46.7	26.1
	05	198 39.3	207 02.3	56.7	265 37.3	38.9	251 50.9	35.9	25 49.2	26.2
	06	213 41.8	222 02.8	S 2 55.5	280 37.7	S22 38.6	266 52.7	S19 35.7	40 51.8	N 5 26.2
	07	228 44.2	237 03.2	54.3	295 38.1	38.4	281 54.6	35.6	55 54.3	26.3
S	08	243 46.7	252 03.6	53.1	310 38.5	38.1	296 56.4	35.4	70 56.9	26.3
A	09	258 49.2	267 04.0	. . 51.9	325 38.9	. . 37.8	311 58.3	. . 35.3	85 59.4	. . 26.4
T	10	273 51.6	282 04.4	50.7	340 39.3	37.5	327 00.1	35.2	101 02.0	26.4
U	11	288 54.1	297 04.8	49.5	355 39.7	37.2	342 02.0	35.0	116 04.6	26.5
R	12	303 56.6	312 05.2	S 2 48.3	10 40.1	S22 37.0	357 03.8	S19 34.9	131 07.1	N 5 26.5
D	13	318 59.0	327 05.6	47.1	25 40.6	36.7	12 05.7	34.8	146 09.7	26.6
A	14	334 01.5	342 06.1	45.9	40 41.0	36.4	27 07.5	34.6	161 12.2	26.6
Y	15	349 04.0	357 06.5	. . 44.7	55 41.4	. . 36.1	42 09.4	. . 34.5	176 14.8	. . 26.7
	16	4 06.4	12 06.9	43.5	70 41.8	35.8	57 11.3	34.3	191 17.3	26.7
	17	19 08.9	27 07.3	42.3	85 42.2	35.5	72 13.1	34.2	206 19.9	26.8
	18	34 11.3	42 07.7	S 2 41.1	100 42.6	S22 35.2	87 15.0	S19 34.1	221 22.4	N 5 26.8
	19	49 13.8	57 08.2	39.9	115 43.0	35.0	102 16.8	33.9	236 25.0	26.9
	20	64 16.3	72 08.6	38.7	130 43.4	34.7	117 18.7	33.8	251 27.6	26.9
	21	79 18.7	87 09.0	. . 37.5	145 43.8	. . 34.4	132 20.5	. . 33.6	266 30.1	. . 27.0
	22	94 21.2	102 09.4	36.3	160 44.2	34.1	147 22.4	33.5	281 32.7	27.0
	23	109 23.7	117 09.9	35.1	175 44.6	33.8	162 24.2	33.4	296 35.2	27.1
	Mer. Pass.	h m 15 47.5	v 0.4	d 1.2	v 0.4	d 0.3	v 1.9	d 0.1	v 2.6	d 0.0

STARS

Name	SHA	Dec
	° ′	° ′
Acamar	315 20.6	S40 16.
Achernar	335 29.0	S57 11.
Acrux	173 13.0	S63 08.
Adhara	255 14.7	S28 59.
Aldebaran	290 52.9	N16 31.
Alioth	166 23.1	N55 54.
Alkaid	153 01.2	N49 15.
Al Na'ir	27 48.1	S46 55.
Alnilam	275 49.4	S 1 11.
Alphard	217 59.0	S 8 42.
Alphecca	126 13.9	N26 40.
Alpheratz	357 47.1	N29 08.
Altair	62 11.8	N 8 53.
Ankaa	353 18.9	S42 15.
Antares	112 30.5	S26 27.
Arcturus	145 58.7	N19 07.
Atria	107 35.8	S69 02.
Avior	234 18.8	S59 32.
Bellatrix	278 35.2	N 6 21.
Betelgeuse	271 04.5	N 7 24.
Canopus	263 57.1	S52 42.
Capella	280 38.9	N46 00.
Deneb	49 34.3	N45 18.
Denebola	182 36.7	N14 31.
Diphda	348 59.2	S17 56.
Dubhe	193 54.9	N61 41.
Elnath	278 16.5	N28 37.
Eltanin	90 48.1	N51 29.
Enif	33 50.6	N 9 55.
Fomalhaut	15 27.8	S29 34.6
Gacrux	172 04.6	S57 09.7
Gienah	175 55.5	S17 35.6
Hadar	148 52.8	S60 24.9
Hamal	328 04.5	N23 30.5
Kaus Aust.	83 48.5	S34 22.8
Kochab	137 19.7	N74 06.6
Markab	13 41.9	N15 15.3
Menkar	314 18.4	N 4 07.6
Menkent	148 11.5	S36 24.8
Miaplacidus	221 39.8	S69 45.2
Mirfak	308 45.0	N49 53.9
Nunki	76 02.7	S26 17.2
Peacock	53 24.8	S56 42.4
Pollux	243 31.3	N28 00.2
Procyon	245 02.8	N 5 12.0
Rasalhague	96 09.7	N12 33.0
Regulus	207 46.7	N11 55.2
Rigel	281 14.9	S 8 11.5
Rigil Kent.	139 56.5	S60 52.2
Sabik	102 16.5	S15 44.2
Schedar	349 44.7	N56 35.5
Shaula	96 26.7	S37 06.6
Sirius	258 36.3	S16 43.8
Spica	158 34.7	S11 12.6
Suhail	222 54.5	S43 28.2
Vega	80 41.6	N38 47.3
Zuben'ubi	137 09.1	S16 04.9

	SHA	Mer. Pass
	° ′	h m
Venus	9 23.1	15 12
Mars	67 57.7	11 18
Jupiter	53 29.3	12 15
Saturn	187 07.3	3 21

Day	UT d h	SUN GHA ° ′	SUN Dec ° ′	MOON GHA ° ′	v ′	MOON Dec ° ′	d ′	HP ′
THURSDAY	22 00	177 07.4	S19 41.7	228 11.7	10.5	S26 52.3	1.7	54.1
	01	192 07.3	41.1	242 41.2	10.4	26 54.0	1.6	54.1
	02	207 07.1	40.5	257 10.6	10.4	26 55.6	1.5	54.0
	03	222 06.9	. . 39.9	271 40.0	10.3	26 57.1	1.4	54.0
	04	237 06.8	39.4	286 09.3	10.4	26 58.5	1.2	54.0
	05	252 06.6	38.8	300 38.7	10.4	26 59.7	1.1	54.0
	06	267 06.4	S19 38.2	315 08.1	10.3	S27 00.8	0.9	54.0
	07	282 06.3	37.7	329 37.4	10.4	27 01.7	0.9	54.0
	08	297 06.1	37.1	344 06.8	10.3	27 02.6	0.7	54.0
	09	312 05.9	. . 36.5	358 36.1	10.4	27 03.3	0.6	54.0
	10	327 05.8	35.9	13 05.5	10.3	27 03.9	0.4	54.0
	11	342 05.6	35.3	27 34.8	10.3	27 04.3	0.3	54.0
	12	357 05.4	S19 34.8	42 04.1	10.4	S27 04.6	0.2	54.0
	13	12 05.3	34.2	56 33.5	10.3	27 04.8	0.1	54.0
	14	27 05.1	33.6	71 02.8	10.3	27 04.9	0.1	54.0
	15	42 04.9	. . 33.0	85 32.1	10.3	27 04.8	0.2	54.0
	16	57 04.8	32.4	100 01.4	10.4	27 04.6	0.3	54.0
	17	72 04.6	31.9	114 30.8	10.3	27 04.3	0.5	54.0
	18	87 04.4	S19 31.3	129 00.1	10.3	S27 03.8	0.6	54.0
	19	102 04.3	30.7	143 29.4	10.4	27 03.2	0.7	54.0
	20	117 04.1	30.1	157 58.8	10.3	27 02.5	0.8	54.0
	21	132 04.0	. . 29.5	172 28.1	10.4	27 01.7	1.0	54.0
	22	147 03.8	28.9	186 57.5	10.3	27 00.7	1.1	54.0
	23	162 03.6	28.4	201 26.8	10.4	26 59.6	1.2	54.0
FRIDAY	23 00	177 03.5	S19 27.8	215 56.2	10.4	S26 58.4	1.4	54.0
	01	192 03.3	27.2	230 25.6	10.3	26 57.0	1.5	54.0
	02	207 03.2	26.6	244 54.9	10.4	26 55.5	1.6	54.0
	03	222 03.0	. . 26.0	259 24.3	10.5	26 53.9	1.7	54.0
	04	237 02.8	25.4	273 53.8	10.4	26 52.2	1.9	54.0
	05	252 02.7	24.8	288 23.2	10.4	26 50.3	2.0	54.0
	06	267 02.5	S19 24.2	302 52.6	10.5	S26 48.3	2.2	54.0
	07	282 02.4	23.7	317 22.1	10.5	26 46.1	2.2	54.0
	08	297 02.2	23.1	331 51.6	10.4	26 43.9	2.4	54.0
	09	312 02.0	. . 22.5	346 21.0	10.6	26 41.5	2.5	54.0
	10	327 01.9	21.9	0 50.6	10.5	26 39.0	2.7	54.0
	11	342 01.7	21.3	15 20.1	10.5	26 36.3	2.7	54.0
	12	357 01.6	S19 20.7	29 49.6	10.6	S26 33.6	2.9	54.0
	13	12 01.4	20.1	44 19.2	10.6	26 30.7	3.0	54.0
	14	27 01.3	19.5	58 48.8	10.6	26 27.7	3.2	54.0
	15	42 01.1	. . 18.9	73 18.4	10.7	26 24.5	3.3	54.0
	16	57 00.9	18.3	87 48.1	10.6	26 21.2	3.4	54.0
	17	72 00.8	17.7	102 17.7	10.7	26 17.8	3.5	54.0
	18	87 00.6	S19 17.1	116 47.4	10.7	S26 14.3	3.6	54.0
	19	102 00.5	16.5	131 17.1	10.8	26 10.7	3.8	54.0
	20	117 00.3	15.9	145 46.9	10.8	26 06.9	3.9	54.0
	21	132 00.2	. . 15.3	160 16.7	10.8	26 03.0	4.0	54.0
	22	147 00.0	14.7	174 46.5	10.8	25 59.0	4.2	54.0
	23	161 59.9	14.1	189 16.3	10.9	25 54.8	4.2	54.0
SATURDAY	24 00	176 59.7	S19 13.5	203 46.2	10.9	S25 50.6	4.4	54.0
	01	191 59.5	12.9	218 16.1	10.9	25 46.2	4.5	54.1
	02	206 59.4	12.3	232 46.0	11.0	25 41.7	4.6	54.1
	03	221 59.2	. . 11.7	247 16.0	11.0	25 37.1	4.8	54.1
	04	236 59.1	11.1	261 46.0	11.0	25 32.3	4.9	54.1
	05	251 58.9	10.5	276 16.0	11.1	25 27.4	5.0	54.1
	06	266 58.8	S19 09.9	290 46.1	11.1	S25 22.4	5.1	54.1
	07	281 58.6	09.3	305 16.2	11.1	25 17.3	5.2	54.1
	08	296 58.5	08.7	319 46.3	11.2	25 12.1	5.3	54.1
	09	311 58.3	. . 08.1	334 16.5	11.2	25 06.8	5.5	54.1
	10	326 58.2	07.5	348 46.7	11.3	25 01.3	5.6	54.1
	11	341 58.0	06.9	3 17.0	11.3	24 55.7	5.7	54.1
	12	356 57.9	S19 06.3	17 47.3	11.3	S24 50.0	5.8	54.1
	13	11 57.7	05.7	32 17.6	11.4	24 44.2	5.9	54.1
	14	26 57.6	05.0	46 48.0	11.4	24 38.3	6.1	54.1
	15	41 57.4	. . 04.4	61 18.4	11.5	24 32.2	6.2	54.1
	16	56 57.3	03.8	75 48.9	11.5	24 26.0	6.2	54.1
	17	71 57.1	03.2	90 19.4	11.5	24 19.8	6.4	54.1
	18	86 57.0	S19 02.6	104 49.9	11.6	S24 13.4	6.5	54.2
	19	101 56.9	02.0	119 20.5	11.6	24 06.9	6.6	54.2
	20	116 56.7	01.4	133 51.1	11.7	24 00.3	6.8	54.2
	21	131 56.6	. . 00.8	148 21.8	11.7	23 53.5	6.8	54.2
	22	146 56.4	19 00.1	162 52.5	11.8	23 46.7	7.0	54.2
	23	161 56.3	S18 59.5	177 23.3	11.8	S23 39.7	7.0	54.2
		SD 16.3	d 0.6	SD 14.7		14.7		14.7

Lat. °	Twilight Naut. h m	Twilight Civil h m	Sunrise h m	Moonrise 22 h m	Moonrise 23 h m	Moonrise 24 h m	Moonrise 25 h m
N 72	07 39	09 17	■	■	■	■	■
N 70	07 27	08 52	10 40	■	■	■	■
68	07 18	08 32	09 56	■	■	■	■
66	07 09	08 17	09 27	■	■	■	10 02
64	07 02	08 04	09 05	■	■	09 41	09 22
62	06 56	07 53	08 48	08 02	08 41	08 52	08 54
60	06 51	07 43	08 33	07 12	07 57	08 21	08 33
N 58	06 46	07 35	08 21	06 41	07 28	07 57	08 15
56	06 41	07 27	08 10	06 18	07 05	07 38	08 01
54	06 37	07 21	08 01	05 59	06 47	07 23	07 48
52	06 33	07 15	07 53	05 43	06 31	07 09	07 37
50	06 29	07 09	07 45	05 29	06 18	06 57	07 27
45	06 21	06 57	07 29	05 01	05 50	06 32	07 05
N 40	06 14	06 47	07 16	04 38	05 28	06 12	06 48
35	06 07	06 37	07 04	04 20	05 10	05 55	06 34
30	06 00	06 29	06 55	04 04	04 55	05 40	06 21
20	05 47	06 14	06 37	03 38	04 28	05 16	06 00
N 10	05 35	06 00	06 22	03 15	04 06	04 54	05 41
0	05 21	05 46	06 08	02 54	03 44	04 34	05 23
S 10	05 05	05 32	05 54	02 32	03 23	04 15	05 06
20	04 47	05 15	05 39	02 09	03 00	03 53	04 47
30	04 23	04 54	05 21	01 43	02 34	03 28	04 25
35	04 08	04 42	05 10	01 27	02 18	03 13	04 12
40	03 49	04 27	04 58	01 09	02 00	02 56	03 57
45	03 25	04 09	04 44	00 48	01 38	02 36	03 39
S 50	02 53	03 46	04 26	00 20	01 10	02 10	03 17
52	02 35	03 35	04 18	00 07	00 57	01 58	03 06
54	02 14	03 22	04 09	24 41	00 41	01 43	02 54
56	01 45	03 07	03 58	24 22	00 22	01 26	02 41
58	00 56	02 49	03 46	23 59	25 06	01 06	02 24
S 60	////	02 26	03 32	23 29	24 40	00 40	02 05

Lat. °	Sunset h m	Twilight Civil h m	Twilight Naut. h m	Moonset 22 h m	Moonset 23 h m	Moonset 24 h m	Moonset 25 h m
N 72	■	15 08	16 46	■	■	■	■
N 70	13 45	15 33	16 58	■	■	■	■
68	14 29	15 53	17 07	■	■	■	■
66	14 58	16 08	17 15	■	■	■	13 20
64	15 20	16 21	17 22	■	■	12 00	13 59
62	15 37	16 32	17 28	10 09	11 15	12 48	14 26
60	15 51	16 41	17 34	10 59	11 59	13 19	14 46
N 58	16 04	16 50	17 39	11 30	12 28	13 42	15 03
56	16 14	16 57	17 43	11 53	12 50	14 00	15 18
54	16 23	17 04	17 47	12 12	13 09	14 16	15 30
52	16 32	17 10	17 51	12 28	13 24	14 30	15 41
50	16 39	17 15	17 55	12 42	13 38	14 41	15 50
45	16 55	17 27	18 03	13 10	14 05	15 06	16 11
N 40	17 08	17 38	18 11	13 33	14 26	15 25	16 27
35	17 20	17 47	18 17	13 51	14 44	15 41	16 40
30	17 29	17 55	18 24	14 07	15 00	15 55	16 52
20	17 46	18 10	18 37	14 34	15 26	16 19	17 12
N 10	18 01	18 24	18 49	14 57	15 48	16 39	17 30
0	18 15	18 37	19 03	15 18	16 09	16 58	17 46
S 10	18 30	18 52	19 18	15 39	16 29	17 17	18 02
20	18 45	19 09	19 37	16 02	16 52	17 37	18 19
30	19 03	19 29	20 01	16 29	17 17	18 00	18 39
35	19 13	19 41	20 16	16 45	17 32	18 14	18 50
40	19 25	19 56	20 34	17 03	17 50	18 30	19 03
45	19 39	20 14	20 57	17 25	18 11	18 48	19 19
S 50	19 57	20 36	21 29	17 53	18 37	19 11	19 37
52	20 05	20 47	21 46	18 07	18 50	19 22	19 46
54	20 14	21 00	22 07	18 23	19 04	19 35	19 56
56	20 24	21 15	22 35	18 41	19 21	19 49	20 07
58	20 36	21 33	23 19	19 04	19 42	20 05	20 20
S 60	20 50	21 55	////	19 34	20 08	20 25	20 35

Day	SUN Eqn. of Time 00^h	SUN Eqn. of Time 12^h	SUN Mer. Pass.	MOON Mer. Pass. Upper	MOON Mer. Pass. Lower	MOON Age	MOON Phase
d	m s	m s	h m	h m	h m	d	%
22	11 30	11 38	12 12	09 06	21 31	26	13
23	11 46	11 53	12 12	09 57	22 22	27	7
24	12 01	12 08	12 12	10 46	23 11	28	3

UT		ARIES	VENUS −4.7		MARS +1.3		JUPITER −1.9		SATURN +0.8		STARS		
		GHA	GHA	Dec	GHA	Dec	GHA	Dec	GHA	Dec	Name	SHA	Dec
d	h	° ′	° ′	° ′	° ′	° ′	° ′	° ′	° ′	° ′		° ′	° ′
25	00	124 26.1	132 10.3	S 2 33.9	190 45.0	S22 33.5	177 26.1	S19 33.2	311 37.8	N 5 27.1	Acamar	315 20.6	S40 16.3
	01	139 28.6	147 10.7	32.7	205 45.4	33.2	192 27.9	33.1	326 40.4	27.2	Achernar	335 29.0	S57 11.7
	02	154 31.1	162 11.1	31.5	220 45.9	32.9	207 29.8	33.0	341 42.9	27.2	Acrux	173 12.9	S63 08.8
	03	169 33.5	177 11.6	. . 30.3	235 46.3	. . 32.6	222 31.6	. . 32.8	356 45.5	. . 27.3	Adhara	255 14.7	S28 59.2
	04	184 36.0	192 12.0	29.1	250 46.7	32.4	237 33.5	32.7	11 48.0	27.3	Aldebaran	290 52.9	N16 31.7
	05	199 38.5	207 12.4	27.9	265 47.1	32.1	252 35.4	32.5	26 50.6	27.4			
	06	214 40.9	222 12.9	S 2 26.7	280 47.5	S22 31.8	267 37.2	S19 32.4	41 53.1	N 5 27.4	Alioth	166 23.1	N55 54.2
	07	229 43.4	237 13.3	25.6	295 47.9	31.5	282 39.1	32.3	56 55.7	27.5	Alkaid	153 01.2	N49 15.7
	08	244 45.8	252 13.7	24.4	310 48.3	31.2	297 40.9	32.1	71 58.3	27.5	Al Na'ir	27 48.1	S46 55.1
S	09	259 48.3	267 14.2	. . 23.2	325 48.7	. . 30.9	312 42.8	. . 32.0	87 00.8	. . 27.6	Alnilam	275 49.4	S 1 11.8
U	10	274 50.8	282 14.6	22.0	340 49.1	30.6	327 44.6	31.8	102 03.4	27.6	Alphard	217 59.0	S 8 42.0
N	11	289 53.2	297 15.1	20.8	355 49.5	30.3	342 46.5	31.7	117 05.9	27.7			
D	12	304 55.7	312 15.5	S 2 19.6	10 49.9	S22 30.0	357 48.3	S19 31.6	132 08.5	N 5 27.7	Alphecca	126 13.8	N26 40.7
A	13	319 58.2	327 15.9	18.4	25 50.3	29.7	12 50.2	31.4	147 11.1	27.8	Alpheratz	357 47.2	N29 08.6
Y	14	335 00.6	342 16.4	17.2	40 50.8	29.4	27 52.0	31.3	162 13.6	27.8	Altair	62 11.8	N 8 53.4
	15	350 03.1	357 16.8	. . 16.0	55 51.2	. . 29.1	42 53.9	. . 31.2	177 16.2	. . 27.9	Ankaa	353 18.9	S42 15.6
	16	5 05.6	12 17.3	14.8	70 51.6	28.8	57 55.7	31.0	192 18.7	27.9	Antares	112 30.5	S26 27.2
	17	20 08.0	27 17.7	13.6	85 52.0	28.5	72 57.6	30.9	207 21.3	28.0			
	18	35 10.5	42 18.2	S 2 12.4	100 52.4	S22 28.2	87 59.4	S19 30.7	222 23.9	N 5 28.0	Arcturus	145 58.6	N19 07.8
	19	50 13.0	57 18.6	11.2	115 52.8	27.9	103 01.3	30.6	237 26.4	28.1	Atria	107 35.7	S69 02.5
	20	65 15.4	72 19.1	10.0	130 53.2	27.6	118 03.2	30.5	252 29.0	28.2	Avior	234 18.8	S59 32.3
	21	80 17.9	87 19.5	. . 08.8	145 53.6	. . 27.3	133 05.0	. . 30.3	267 31.5	. . 28.2	Bellatrix	278 35.2	N 6 21.5
	22	95 20.3	102 20.0	07.6	160 54.0	27.0	148 06.9	30.2	282 34.1	28.3	Betelgeuse	271 04.5	N 7 24.5
	23	110 22.8	117 20.4	06.4	175 54.4	26.8	163 08.7	30.0	297 36.7	28.3			
26	00	125 25.3	132 20.9	S 2 05.2	190 54.9	S22 26.5	178 10.6	S19 29.9	312 39.2	N 5 28.4	Canopus	263 57.1	S52 42.1
	01	140 27.7	147 21.3	04.0	205 55.3	26.2	193 12.4	29.8	327 41.8	28.4	Capella	280 38.9	N46 00.6
	02	155 30.2	162 21.8	02.8	220 55.7	25.9	208 14.3	29.6	342 44.3	28.5	Deneb	49 34.3	N45 18.7
	03	170 32.7	177 22.2	. . 01.6	235 56.1	. . 25.6	223 16.1	. . 29.5	357 46.9	. . 28.5	Denebola	182 36.7	N14 31.0
	04	185 35.1	192 22.7	2 00.4	250 56.5	25.3	238 18.0	29.4	12 49.5	28.6	Diphda	348 59.2	S17 56.3
	05	200 37.6	207 23.2	1 59.2	265 56.9	24.9	253 19.8	29.2	27 52.0	28.6			
	06	215 40.1	222 23.6	S 1 58.0	280 57.3	S22 24.6	268 21.7	S19 29.1	42 54.6	N 5 28.7	Dubhe	193 54.8	N61 41.8
	07	230 42.5	237 24.1	56.9	295 57.7	24.3	283 23.5	28.9	57 57.2	28.7	Elnath	278 16.5	N28 37.0
	08	245 45.0	252 24.6	55.7	310 58.1	24.0	298 25.4	28.8	72 59.7	28.8	Eltanin	90 48.1	N51 29.0
M	09	260 47.5	267 25.0	. . 54.5	325 58.5	. . 23.7	313 27.3	. . 28.7	88 02.3	. . 28.8	Enif	33 50.6	N 9 55.0
O	10	275 49.9	282 25.5	53.3	340 59.0	23.4	328 29.1	28.5	103 04.8	28.9	Fomalhaut	15 27.8	S29 34.6
N	11	290 52.4	297 25.9	52.1	355 59.4	23.1	343 31.0	28.4	118 07.4	28.9			
D	12	305 54.8	312 26.4	S 1 50.9	10 59.8	S22 22.8	358 32.8	S19 28.2	133 10.0	N 5 29.0	Gacrux	172 04.5	S57 09.7
A	13	320 57.3	327 26.9	49.7	26 00.2	22.5	13 34.7	28.1	148 12.5	29.0	Gienah	175 55.5	S17 35.6
Y	14	335 59.8	342 27.4	48.5	41 00.6	22.2	28 36.5	28.0	163 15.1	29.1	Hadar	148 52.7	S60 24.9
	15	351 02.2	357 27.8	. . 47.3	56 01.0	. . 21.9	43 38.4	. . 27.8	178 17.7	. . 29.2	Hamal	328 04.5	N23 30.5
	16	6 04.7	12 28.3	46.1	71 01.4	21.6	58 40.2	27.7	193 20.2	29.2	Kaus Aust.	83 48.5	S34 22.8
	17	21 07.2	27 28.8	44.9	86 01.8	21.3	73 42.1	27.5	208 22.8	29.3			
	18	36 09.6	42 29.2	S 1 43.7	101 02.2	S22 21.0	88 43.9	S19 27.4	223 25.4	N 5 29.3	Kochab	137 19.6	N74 06.6
	19	51 12.1	57 29.7	42.5	116 02.7	20.7	103 45.8	27.3	238 27.9	29.4	Markab	13 41.9	N15 15.3
	20	66 14.6	72 30.2	41.3	131 03.1	20.4	118 47.6	27.1	253 30.5	29.4	Menkar	314 18.4	N 4 07.6
	21	81 17.0	87 30.7	. . 40.1	146 03.5	. . 20.1	133 49.5	. . 27.0	268 33.1	. . 29.5	Menkent	148 11.5	S36 24.9
	22	96 19.5	102 31.2	39.0	161 03.9	19.8	148 51.4	26.8	283 35.6	29.5	Miaplacidus	221 39.8	S69 45.2
	23	111 21.9	117 31.6	37.8	176 04.3	19.5	163 53.2	26.7	298 38.2	29.6			
27	00	126 24.4	132 32.1	S 1 36.6	191 04.7	S22 19.1	178 55.1	S19 26.6	313 40.7	N 5 29.6	Mirfak	308 45.0	N49 53.9
	01	141 26.9	147 32.6	35.4	206 05.1	18.8	193 56.9	26.4	328 43.3	29.7	Nunki	76 02.7	S26 17.2
	02	156 29.3	162 33.1	34.2	221 05.5	18.5	208 58.8	26.3	343 45.9	29.7	Peacock	53 24.8	S56 42.4
	03	171 31.8	177 33.6	. . 33.0	236 06.0	. . 18.2	224 00.6	. . 26.1	358 48.4	. . 29.8	Pollux	243 31.3	N28 00.2
	04	186 34.3	192 34.1	31.8	251 06.4	17.9	239 02.5	26.0	13 51.0	29.9	Procyon	245 02.8	N 5 12.0
	05	201 36.7	207 34.5	30.6	266 06.8	17.6	254 04.3	25.9	28 53.6	29.9			
	06	216 39.2	222 35.0	S 1 29.4	281 07.2	S22 17.3	269 06.2	S19 25.7	43 56.1	N 5 30.0	Rasalhague	96 09.7	N12 33.0
	07	231 41.7	237 35.5	28.2	296 07.6	17.0	284 08.0	25.6	58 58.7	30.0	Regulus	207 46.6	N11 55.2
T	08	246 44.1	252 36.0	27.0	311 08.0	16.6	299 09.9	25.5	74 01.3	30.1	Rigel	281 14.9	S 8 11.5
U	09	261 46.6	267 36.5	. . 25.9	326 08.4	. . 16.3	314 11.8	. . 25.3	89 03.8	. . 30.1	Rigil Kent.	139 56.4	S60 52.2
E	10	276 49.1	282 37.0	24.7	341 08.8	16.0	329 13.6	25.2	104 06.4	30.2	Sabik	102 16.5	S15 44.3
S	11	291 51.5	297 37.5	23.5	356 09.3	15.7	344 15.5	25.0	119 09.0	30.2			
D	12	306 54.0	312 38.0	S 1 22.3	11 09.7	S22 15.4	359 17.3	S19 24.9	134 11.5	N 5 30.3	Schedar	349 44.7	N56 35.5
A	13	321 56.4	327 38.5	21.1	26 10.1	15.1	14 19.2	24.8	149 14.1	30.3	Shaula	96 26.7	S37 06.6
Y	14	336 58.9	342 39.0	19.9	41 10.5	14.8	29 21.0	24.6	164 16.7	30.4	Sirius	258 36.3	S16 43.8
	15	352 01.4	357 39.5	. . 18.7	56 10.9	. . 14.4	44 22.9	. . 24.5	179 19.2	. . 30.5	Spica	158 34.6	S11 12.6
	16	7 03.8	12 40.0	17.5	71 11.3	14.1	59 24.7	24.3	194 21.8	30.5	Suhail	222 54.5	S43 28.2
	17	22 06.3	27 40.5	16.3	86 11.7	13.8	74 26.6	24.2	209 24.4	30.6			
	18	37 08.8	42 41.0	S 1 15.1	101 12.1	S22 13.5	89 28.4	S19 24.1	224 26.9	N 5 30.6	Vega	80 41.6	N38 47.3
	19	52 11.2	57 41.5	14.0	116 12.6	13.2	104 30.3	23.9	239 29.5	30.7	Zuben'ubi	137 09.1	S16 04.9
	20	67 13.7	72 42.0	12.8	131 13.0	12.9	119 32.2	23.8	254 32.1	30.7		SHA	Mer. Pass.
	21	82 16.2	87 42.5	. . 11.6	146 13.4	. . 12.5	134 34.0	. . 23.6	269 34.6	. . 30.8		° ′	h m
	22	97 18.6	102 43.0	10.4	161 13.8	12.2	149 35.9	23.5	284 37.2	30.8	Venus	6 55.6	15 10
	23	112 21.1	117 43.5	09.2	176 14.2	11.9	164 37.7	23.4	299 39.8	30.9	Mars	65 29.6	11 16
		h m									Jupiter	52 45.3	12 06
Mer. Pass.		15 35.8	v 0.5	d 1.2	v 0.4	d 0.3	v 1.9	d 0.1	v 2.6	d 0.1	Saturn	187 14.0	3 09

Day	UT d h	SUN GHA ° ′	SUN Dec ° ′	MOON GHA ° ′	MOON v ′	MOON Dec ° ′	MOON d ′	MOON HP ′
SUNDAY	25 00	176 56.1	S18 58.9	191 54.1	11.9	S23 32.7	7.2	54.2
	01	191 56.0	58.3	206 25.0	11.9	23 25.5	7.3	54.2
	02	206 55.8	57.7	220 55.9	11.9	23 18.2	7.3	54.2
	03	221 55.7	. . 57.1	235 26.8	12.0	23 10.9	7.5	54.2
	04	236 55.5	56.4	249 57.8	12.1	23 03.4	7.6	54.2
	05	251 55.4	55.8	264 28.9	12.1	22 55.8	7.7	54.2
	06	266 55.3	S18 55.2	279 00.0	12.1	S22 48.1	7.8	54.3
	07	281 55.1	54.6	293 31.1	12.2	22 40.3	7.9	54.3
	08	296 55.0	54.0	308 02.3	12.2	22 32.4	8.1	54.3
	09	311 54.8	. . 53.3	322 33.5	12.3	22 24.3	8.1	54.3
	10	326 54.7	52.7	337 04.8	12.4	22 16.2	8.2	54.3
	11	341 54.5	52.1	351 36.2	12.3	22 08.0	8.3	54.3
	12	356 54.4	S18 51.5	6 07.5	12.5	S21 59.7	8.4	54.3
	13	11 54.3	50.9	20 39.0	12.5	21 51.3	8.6	54.3
	14	26 54.1	50.2	35 10.5	12.5	21 42.7	8.6	54.3
	15	41 54.0	. . 49.6	49 42.0	12.6	21 34.1	8.7	54.3
	16	56 53.8	49.0	64 13.6	12.6	21 25.4	8.8	54.4
	17	71 53.7	48.4	78 45.2	12.7	21 16.6	8.9	54.4
	18	86 53.6	S18 47.7	93 16.9	12.7	S21 07.7	9.1	54.4
	19	101 53.4	47.1	107 48.6	12.8	20 58.6	9.1	54.4
	20	116 53.3	46.5	122 20.4	12.8	20 49.5	9.2	54.4
	21	131 53.2	. . 45.9	136 52.2	12.9	20 40.3	9.3	54.4
	22	146 53.0	45.2	151 24.1	12.9	20 31.0	9.4	54.4
	23	161 52.9	44.6	165 56.0	13.0	20 21.6	9.4	54.4
MONDAY	26 00	176 52.7	S18 44.0	180 28.0	13.0	S20 12.2	9.6	54.4
	01	191 52.6	43.3	195 00.0	13.1	20 02.6	9.7	54.5
	02	206 52.5	42.7	209 32.1	13.1	19 52.9	9.7	54.5
	03	221 52.3	. . 42.1	224 04.2	13.1	19 43.2	9.9	54.5
	04	236 52.2	41.4	238 36.3	13.3	19 33.3	9.9	54.5
	05	251 52.1	40.8	253 08.6	13.2	S19 23.4	10.0	54.5
	06	266 51.9	S18 40.2	An annular eclipse of the Sun occurs on this date. See page 5.				
	07	281 51.8	39.5					
	08	296 51.7	38.9					
	09	311 51.5	. . 38.3					
	10	326 51.4	37.6					
	11	341 51.3	37.0					
	12	356 51.1	S18 36.4	354 55.4	13.6	S18 11.5	10.7	54.6
	13	11 51.0	35.7	9 28.0	13.7	18 00.8	10.6	54.6
	14	26 50.9	35.1	24 00.7	13.6	17 50.2	10.8	54.6
	15	41 50.7	. . 34.4	38 33.3	13.8	17 39.4	10.9	54.6
	16	56 50.6	33.8	53 06.1	13.8	17 28.5	10.9	54.6
	17	71 50.5	33.2	67 38.9	13.8	17 17.6	11.0	54.7
	18	86 50.3	S18 32.5	82 11.7	13.8	S17 06.6	11.1	54.7
	19	101 50.2	31.9	96 44.5	14.0	16 55.5	11.2	54.7
	20	116 50.1	31.2	111 17.5	13.9	16 44.3	11.2	54.7
	21	131 49.9	. . 30.6	125 50.4	14.0	16 33.1	11.3	54.7
	22	146 49.8	30.0	140 23.4	14.1	16 21.8	11.4	54.7
	23	161 49.7	29.3	154 56.5	14.0	16 10.4	11.5	54.7
TUESDAY	27 00	176 49.6	S18 28.7	169 29.5	14.2	S15 58.9	11.5	54.7
	01	191 49.4	28.0	184 02.7	14.1	15 47.4	11.6	54.8
	02	206 49.3	27.4	198 35.8	14.2	15 35.8	11.7	54.8
	03	221 49.2	. . 26.7	213 09.0	14.3	15 24.1	11.7	54.8
	04	236 49.1	26.1	227 42.3	14.3	15 12.4	11.8	54.8
	05	251 48.9	25.4	242 15.6	14.3	15 00.6	11.9	54.8
	06	266 48.8	S18 24.8	256 48.9	14.4	S14 48.7	11.9	54.8
	07	281 48.7	24.1	271 22.3	14.4	14 36.8	12.0	54.8
	08	296 48.5	23.5	285 55.7	14.4	14 24.8	12.1	54.9
	09	311 48.4	. . 22.8	300 29.1	14.5	14 12.7	12.2	54.9
	10	326 48.3	22.2	315 02.6	14.5	14 00.5	12.2	54.9
	11	341 48.2	21.5	329 36.1	14.6	13 48.3	12.2	54.9
	12	356 48.0	S18 20.9	344 09.7	14.5	S13 36.1	12.4	54.9
	13	11 47.9	20.2	358 43.2	14.7	13 23.7	12.4	54.9
	14	26 47.8	19.6	13 16.9	14.6	13 11.3	12.4	55.0
	15	41 47.7	. . 18.9	27 50.5	14.7	12 58.9	12.5	55.0
	16	56 47.6	18.3	42 24.2	14.7	12 46.4	12.6	55.0
	17	71 47.4	17.6	56 57.9	14.8	12 33.8	12.6	55.0
	18	86 47.3	S18 17.0	71 31.7	14.8	S12 21.2	12.7	55.0
	19	101 47.2	16.3	86 05.5	14.8	12 08.5	12.7	55.0
	20	116 47.1	15.7	100 39.3	14.8	11 55.8	12.8	55.1
	21	131 46.9	. . 15.0	115 13.1	14.9	11 43.0	12.9	55.1
	22	146 46.8	14.4	129 47.0	14.9	11 30.1	12.9	55.1
	23	161 46.7	13.7	144 20.9	14.9	S11 17.2	12.9	55.1
		SD 16.3	d 0.6	SD	14.8	14.9		15.0

Lat. °	Twilight Naut. h m	Twilight Civil h m	Sunrise h m	Moonrise 25 h m	Moonrise 26 h m	Moonrise 27 h m	Moonrise 28 h m
N 72	07 29	09 04	11 35	▬	▬	10 22	09 39
N 70	07 19	08 41	10 20	▬	10 45	09 55	09 25
68	07 10	08 23	09 43	▬	10 03	09 34	09 14
66	07 03	08 09	09 17	10 02	09 34	09 18	09 05
64	06 57	07 57	08 57	09 22	09 12	09 04	08 58
62	06 51	07 47	08 41	08 54	08 54	08 53	08 51
60	06 46	07 38	08 27	08 33	08 39	08 43	08 45
N 58	06 42	07 30	08 15	08 15	08 27	08 34	08 40
56	06 37	07 23	08 05	08 01	08 16	08 27	08 35
54	06 34	07 17	07 57	07 48	08 06	08 20	08 31
52	06 30	07 11	07 49	07 37	07 58	08 14	08 28
50	06 27	07 06	07 41	07 27	07 50	08 08	08 24
45	06 19	06 54	07 26	07 05	07 33	07 56	08 17
N 40	06 12	06 45	07 14	06 48	07 19	07 46	08 11
35	06 05	06 36	07 03	06 34	07 08	07 38	08 05
30	05 59	06 28	06 53	06 21	06 58	07 30	08 00
20	05 47	06 14	06 37	06 00	06 40	07 17	07 52
N 10	05 35	06 01	06 23	05 41	06 24	07 06	07 45
0	05 22	05 47	06 09	05 23	06 10	06 55	07 38
S 10	05 07	05 33	05 55	05 06	05 55	06 44	07 31
20	04 49	05 17	05 40	04 47	05 40	06 32	07 24
30	04 26	04 57	05 23	04 25	05 22	06 19	07 15
35	04 11	04 45	05 13	04 12	05 11	06 11	07 10
40	03 53	04 31	05 02	03 57	04 59	06 02	07 05
45	03 31	04 14	04 48	03 39	04 45	05 52	06 58
S 50	03 00	03 52	04 31	03 17	04 27	05 39	06 50
52	02 44	03 41	04 24	03 06	04 19	05 33	06 47
54	02 24	03 29	04 15	02 54	04 10	05 26	06 43
56	01 59	03 15	04 05	02 41	03 59	05 19	06 38
58	01 20	02 58	03 53	02 24	03 47	05 11	06 33
S 60	////	02 37	03 40	02 05	03 33	05 01	06 28

Lat. °	Sunset h m	Twilight Civil h m	Twilight Naut. h m	Moonset 25 h m	Moonset 26 h m	Moonset 27 h m	Moonset 28 h m
N 72	12 51	15 22	16 57	▬	▬	16 13	18 26
N 70	14 07	15 45	17 07	▬	14 14	16 38	18 37
68	14 44	16 03	17 16	▬	14 56	16 57	18 46
66	15 09	16 17	17 23	13 20	15 24	17 12	18 53
64	15 29	16 29	17 29	13 59	15 45	17 24	18 59
62	15 45	16 39	17 35	14 26	16 01	17 34	19 04
60	15 59	16 48	17 40	14 46	16 15	17 43	19 09
N 58	16 10	16 56	17 44	15 03	16 27	17 50	19 13
56	16 20	17 03	17 48	15 18	16 37	17 57	19 16
54	16 29	17 09	17 52	15 30	16 46	18 03	19 19
52	16 37	17 15	17 56	15 41	16 54	18 08	19 22
50	16 44	17 20	17 59	15 50	17 02	18 13	19 25
45	16 59	17 31	18 07	16 11	17 17	18 24	19 30
N 40	17 12	17 41	18 14	16 27	17 30	18 32	19 35
35	17 23	17 50	18 20	16 40	17 40	18 40	19 39
30	17 32	17 57	18 26	16 52	17 49	18 46	19 42
20	17 48	18 12	18 38	17 12	18 05	18 57	19 48
N 10	18 03	18 25	18 50	17 30	18 19	19 07	19 54
0	18 16	18 38	19 03	17 46	18 32	19 16	19 58
S 10	18 30	18 52	19 18	18 02	18 44	19 25	20 03
20	18 44	19 08	19 36	18 19	18 58	19 34	20 08
30	19 01	19 27	19 59	18 39	19 13	19 45	20 14
35	19 11	19 39	20 13	18 50	19 22	19 51	20 17
40	19 23	19 53	20 31	19 03	19 32	19 58	20 21
45	19 36	20 10	20 53	19 19	19 44	20 06	20 25
S 50	19 53	20 32	21 24	19 37	19 58	20 15	20 30
52	20 01	20 42	21 39	19 46	20 05	20 20	20 32
54	20 09	20 54	21 59	19 56	20 12	20 24	20 35
56	20 19	21 08	22 23	20 07	20 20	20 30	20 38
58	20 30	21 25	22 59	20 20	20 29	20 36	20 41
S 60	20 43	21 45	////	20 35	20 40	20 42	20 44

Day	SUN Eqn. of Time 00^h	SUN Eqn. of Time 12^h	SUN Mer. Pass.	MOON Mer. Pass. Upper	MOON Mer. Pass. Lower	MOON Age	MOON Phase
d	m s	m s	h m	h m	h m	d	%
25	12 15	12 22	12 12	11 35	23 58	29	1
26	12 29	12 35	12 13	12 21	24 43	00	0 ●
27	12 41	12 48	12 13	13 05	00 43	01	1

UT d h	ARIES GHA ° ′	VENUS −4.7 GHA ° ′	VENUS Dec ° ′	MARS +1.3 GHA ° ′	MARS Dec ° ′	JUPITER −1.9 GHA ° ′	JUPITER Dec ° ′	SATURN +0.8 GHA ° ′	SATURN Dec ° ′
28 00 (WEDNESDAY)	127 23.6	132 44.0	S 1 08.0	191 14.6	S22 11.6	179 39.6	S19 23.2	314 42.4	N 5 30.9
01	142 26.0	147 44.5	06.8	206 15.0	11.3	194 41.4	23.1	329 44.9	31.0
02	157 28.5	162 45.0	05.6	221 15.5	10.9	209 43.3	22.9	344 47.5	31.1
03	172 30.9	177 45.6	. . 04.5	236 15.9	. . 10.6	224 45.1	. . 22.8	359 50.1	. . 31.1
04	187 33.4	192 46.1	03.3	251 16.3	10.3	239 47.0	22.6	14 52.6	31.2
05	202 35.9	207 46.6	02.1	266 16.7	10.0	254 48.9	22.5	29 55.2	31.2
06	217 38.3	222 47.1	S 1 00.9	281 17.1	S22 09.7	269 50.7	S19 22.4	44 57.8	N 5 31.3
07	232 40.8	237 47.6	0 59.7	296 17.5	09.3	284 52.6	22.2	60 00.3	31.3
08	247 43.3	252 48.1	58.5	311 18.0	09.0	299 54.4	22.1	75 02.9	31.4
09	262 45.7	267 48.7	. . 57.3	326 18.4	. . 08.7	314 56.3	. . 21.9	90 05.5	. . 31.4
10	277 48.2	282 49.2	56.1	341 18.8	08.4	329 58.1	21.8	105 08.0	31.5
11	292 50.7	297 49.7	55.0	356 19.2	08.0	345 00.0	21.7	120 10.6	31.6
12	307 53.1	312 50.2	S 0 53.8	11 19.6	S22 07.7	0 01.8	S19 21.5	135 13.2	N 5 31.6
13	322 55.6	327 50.8	52.6	26 20.0	07.4	15 03.7	21.4	150 15.8	31.7
14	337 58.0	342 51.3	51.4	41 20.4	07.1	30 05.5	21.2	165 18.3	31.7
15	353 00.5	357 51.8	. . 50.2	56 20.9	. . 06.7	45 07.4	. . 21.1	180 20.9	. . 31.8
16	8 03.0	12 52.3	49.0	71 21.3	06.4	60 09.3	21.0	195 23.5	31.8
17	23 05.4	27 52.9	47.9	86 21.7	06.1	75 11.1	20.8	210 26.0	31.9
18	38 07.9	42 53.4	S 0 46.7	101 22.1	S22 05.8	90 13.0	S19 20.7	225 28.6	N 5 32.0
19	53 10.4	57 53.9	45.5	116 22.5	05.4	105 14.8	20.5	240 31.2	32.0
20	68 12.8	72 54.5	44.3	131 22.9	05.1	120 16.7	20.4	255 33.8	32.1
21	83 15.3	87 55.0	. . 43.1	146 23.4	. . 04.8	135 18.5	. . 20.3	270 36.3	. . 32.1
22	98 17.8	102 55.5	41.9	161 23.8	04.4	150 20.4	20.1	285 38.9	32.2
23	113 20.2	117 56.1	40.8	176 24.2	04.1	165 22.2	20.0	300 41.5	32.2
29 00 (THURSDAY)	128 22.7	132 56.6	S 0 39.6	191 24.6	S22 03.8	180 24.1	S19 19.8	315 44.0	N 5 32.3
01	143 25.2	147 57.2	38.4	206 25.0	03.4	195 26.0	19.7	330 46.6	32.3
02	158 27.6	162 57.7	37.2	221 25.4	03.1	210 27.8	19.6	345 49.2	32.4
03	173 30.1	177 58.2	. . 36.0	236 25.9	. . 02.8	225 29.7	. . 19.4	0 51.8	. . 32.5
04	188 32.5	192 58.8	34.8	251 26.3	02.5	240 31.5	19.3	15 54.3	32.5
05	203 35.0	207 59.3	33.7	266 26.7	02.1	255 33.4	19.1	30 56.9	32.6
06	218 37.5	222 59.9	S 0 32.5	281 27.1	S22 01.8	270 35.2	S19 19.0	45 59.5	N 5 32.6
07	233 39.9	238 00.4	31.3	296 27.5	01.5	285 37.1	18.8	61 02.1	32.7
08	248 42.4	253 01.0	30.1	311 27.9	01.1	300 38.9	18.7	76 04.6	32.7
09	263 44.9	268 01.5	. . 28.9	326 28.4	. . 00.8	315 40.8	. . 18.6	91 07.2	. . 32.8
10	278 47.3	283 02.1	27.8	341 28.8	00.5	330 42.7	18.4	106 09.8	32.9
11	293 49.8	298 02.6	26.6	356 29.2	22 00.1	345 44.5	18.3	121 12.3	32.9
12	308 52.3	313 03.2	S 0 25.4	11 29.6	S21 59.8	0 46.4	S19 18.1	136 14.9	N 5 33.0
13	323 54.7	328 03.7	24.2	26 30.0	59.4	15 48.2	18.0	151 17.5	33.0
14	338 57.2	343 04.3	23.0	41 30.5	59.1	30 50.1	17.9	166 20.1	33.1
15	353 59.7	358 04.8	. . 21.9	56 30.9	. . 58.8	45 51.9	. . 17.7	181 22.6	. . 33.1
16	9 02.1	13 05.4	20.7	71 31.3	58.4	60 53.8	17.6	196 25.2	33.2
17	24 04.6	28 06.0	19.5	86 31.7	58.1	75 55.6	17.4	211 27.8	33.3
18	39 07.0	43 06.5	S 0 18.3	101 32.1	S21 57.8	90 57.5	S19 17.3	226 30.4	N 5 33.3
19	54 09.5	58 07.1	17.1	116 32.5	57.4	105 59.4	17.2	241 32.9	33.4
20	69 12.0	73 07.6	16.0	131 33.0	57.1	121 01.2	17.0	256 35.5	33.4
21	84 14.4	88 08.2	. . 14.8	146 33.4	. . 56.7	136 03.1	. . 16.9	271 38.1	. . 33.5
22	99 16.9	103 08.8	13.6	161 33.8	56.4	151 04.9	16.7	286 40.7	33.6
23	114 19.4	118 09.3	12.4	176 34.2	56.1	166 06.8	16.6	301 43.2	33.6
30 00 (FRIDAY)	129 21.8	133 09.9	S 0 11.2	191 34.6	S21 55.7	181 08.6	S19 16.4	316 45.8	N 5 33.7
01	144 24.3	148 10.5	10.1	206 35.1	55.4	196 10.5	16.3	331 48.4	33.7
02	159 26.8	163 11.0	08.9	221 35.5	55.0	211 12.4	16.2	346 51.0	33.8
03	174 29.2	178 11.6	. . 07.7	236 35.9	. . 54.7	226 14.2	. . 16.0	1 53.5	. . 33.8
04	189 31.7	193 12.2	06.5	251 36.3	54.4	241 16.1	15.9	16 56.1	33.9
05	204 34.1	208 12.8	05.4	266 36.7	54.0	256 17.9	15.7	31 58.7	34.0
06	219 36.6	223 13.3	S 0 04.2	281 37.2	S21 53.7	271 19.8	S19 15.6	47 01.3	N 5 34.0
07	234 39.1	238 13.9	03.0	296 37.6	53.3	286 21.6	15.5	62 03.9	34.1
08	249 41.5	253 14.5	01.8	311 38.0	53.0	301 23.5	15.3	77 06.4	34.1
09	264 44.0	268 15.1	S 00.7	326 38.4	. . 52.6	316 25.3	. . 15.2	92 09.0	. . 34.2
10	279 46.5	283 15.6	N 00.5	341 38.8	52.3	331 27.2	15.0	107 11.6	34.3
11	294 48.9	298 16.2	01.7	356 39.3	52.0	346 29.1	14.9	122 14.2	34.3
12	309 51.4	313 16.8	N 0 02.9	11 39.7	S21 51.6	1 30.9	S19 14.7	137 16.7	N 5 34.4
13	324 53.9	328 17.4	04.0	26 40.1	51.3	16 32.8	14.6	152 19.3	34.4
14	339 56.3	343 18.0	05.2	41 40.5	50.9	31 34.6	14.5	167 21.9	34.5
15	354 58.8	358 18.6	. . 06.4	56 40.9	. . 50.6	46 36.5	. . 14.3	182 24.5	. . 34.5
16	10 01.3	13 19.2	07.6	71 41.4	50.2	61 38.3	14.2	197 27.1	34.6
17	25 03.7	28 19.7	08.7	86 41.8	49.9	76 40.2	14.0	212 29.6	34.7
18	40 06.2	43 20.3	N 0 09.9	101 42.2	S21 49.5	91 42.1	S19 13.9	227 32.2	N 5 34.7
19	55 08.6	58 20.9	11.1	116 42.6	49.2	106 43.9	13.7	242 34.8	34.8
20	70 11.1	73 21.5	12.3	131 43.1	48.8	121 45.8	13.6	257 37.4	34.8
21	85 13.6	88 22.1	. . 13.4	146 43.5	. . 48.5	136 47.6	. . 13.5	272 39.9	. . 34.9
22	100 16.0	103 22.7	14.6	161 43.9	48.1	151 49.5	13.3	287 42.5	35.0
23	115 18.5	118 23.3	15.8	176 44.3	47.8	166 51.3	13.2	302 45.1	35.0
Mer. Pass. h m	15 24.0	v 0.6	d 1.2	v 0.4	d 0.3	v 1.9	d 0.1	v 2.6	d 0.1

STARS

Name	SHA ° ′	Dec ° ′
Acamar	315 20.6	S40 16.3
Achernar	335 29.1	S57 11.6
Acrux	173 12.9	S63 08.9
Adhara	255 14.7	S28 59.2
Aldebaran	290 52.9	N16 31.7
Alioth	166 23.0	N55 54.2
Alkaid	153 01.2	N49 15.7
Al Na'ir	27 48.1	S46 55.1
Alnilam	275 49.4	S 1 11.8
Alphard	217 59.0	S 8 42.0
Alphecca	126 13.8	N26 40.7
Alpheratz	357 47.2	N29 08.6
Altair	62 11.7	N 8 53.4
Ankaa	353 19.0	S42 15.6
Antares	112 30.5	S26 27.2
Arcturus	145 58.6	N19 07.8
Atria	107 35.7	S69 02.5
Avior	234 18.9	S59 32.4
Bellatrix	278 35.3	N 6 21.5
Betelgeuse	271 04.5	N 7 24.5
Canopus	263 57.1	S52 42.2
Capella	280 39.0	N46 00.6
Deneb	49 34.3	N45 18.7
Denebola	182 36.7	N14 31.0
Diphda	348 59.2	S17 56.3
Dubhe	193 54.8	N61 41.8
Elnath	278 16.5	N28 37.0
Eltanin	90 48.1	N51 29.0
Enif	33 50.6	N 9 55.0
Fomalhaut	15 27.8	S29 34.5
Gacrux	172 04.5	S57 09.7
Gienah	175 55.5	S17 35.6
Hadar	148 52.7	S60 24.9
Hamal	328 04.5	N23 30.5
Kaus Aust.	83 48.4	S34 22.8
Kochab	137 19.6	N74 06.6
Markab	13 41.9	N15 15.3
Menkar	314 18.4	N 4 07.6
Menkent	148 11.5	S36 24.9
Miaplacidus	221 39.8	S69 45.3
Mirfak	308 45.0	N49 53.9
Nunki	76 02.6	S26 17.2
Peacock	53 24.8	S56 42.4
Pollux	243 31.3	N28 00.2
Procyon	245 02.8	N 5 12.0
Rasalhague	96 09.7	N12 33.0
Regulus	207 46.6	N11 55.2
Rigel	281 14.9	S 8 11.5
Rigil Kent.	139 56.4	S60 52.2
Sabik	102 16.5	S15 44.3
Schedar	349 44.8	N56 35.5
Shaula	96 26.6	S37 06.6
Sirius	258 36.3	S16 43.8
Spica	158 34.6	S11 12.7
Suhail	222 54.5	S43 28.2
Vega	80 41.6	N38 47.3
Zuben'ubi	137 09.1	S16 04.9

	SHA ° ′	Mer. Pass. h m
Venus	4 33.9	15 08
Mars	63 01.9	11 14
Jupiter	52 01.4	11 57
Saturn	187 21.4	2 57

	UT	SUN GHA	SUN Dec	MOON GHA	v	MOON Dec	d	HP
	d h	° ′	° ′	° ′	′	° ′	′	′
	28 00	176 46.6	S18 13.0	158 54.8	15.0	S11 04.3	13.0	55.1
	01	191 46.5	12.4	173 28.8	14.9	10 51.3	13.1	55.1
	02	206 46.3	11.7	188 02.7	15.0	10 38.2	13.1	55.1
	03	221 46.2	. . 11.1	202 36.7	15.1	10 25.1	13.2	55.2
	04	236 46.1	10.4	217 10.8	15.0	10 11.9	13.2	55.2
	05	251 46.0	09.7	231 44.8	15.1	9 58.7	13.2	55.2
	06	266 45.9	S18 09.1	246 18.9	15.1	S 9 45.5	13.3	55.2
W	07	281 45.8	08.4	260 53.0	15.1	9 32.2	13.4	55.2
E	08	296 45.6	07.8	275 27.1	15.1	9 18.8	13.3	55.3
D	09	311 45.5	. . 07.1	290 01.2	15.2	9 05.5	13.5	55.3
N	10	326 45.4	06.4	304 35.4	15.2	8 52.0	13.4	55.3
E	11	341 45.3	05.8	319 09.6	15.2	8 38.6	13.6	55.3
S	12	356 45.2	S18 05.1	333 43.8	15.2	S 8 25.0	13.5	55.3
D	13	11 45.1	04.4	348 18.0	15.2	8 11.5	13.6	55.3
A	14	26 44.9	03.8	2 52.2	15.3	7 57.9	13.7	55.4
Y	15	41 44.8	. . 03.1	17 26.5	15.2	7 44.2	13.6	55.4
	16	56 44.7	02.4	32 00.7	15.3	7 30.6	13.8	55.4
	17	71 44.6	01.8	46 35.0	15.3	7 16.8	13.7	55.4
	18	86 44.5	S18 01.1	61 09.3	15.3	S 7 03.1	13.8	55.4
	19	101 44.4	18 00.4	75 43.6	15.3	6 49.3	13.8	55.4
	20	116 44.3	17 59.8	90 17.9	15.3	6 35.5	13.9	55.5
	21	131 44.1	. . 59.1	104 52.2	15.3	6 21.6	13.9	55.5
	22	146 44.0	58.4	119 26.5	15.3	6 07.7	13.9	55.5
	23	161 43.9	57.8	134 00.8	15.4	5 53.8	14.0	55.5
	29 00	176 43.8	S17 57.1	148 35.2	15.3	S 5 39.8	13.9	55.5
	01	191 43.7	56.4	163 09.5	15.3	5 25.9	14.1	55.6
	02	206 43.6	55.7	177 43.8	15.4	5 11.8	14.0	55.6
	03	221 43.5	. . 55.1	192 18.2	15.3	4 57.8	14.1	55.6
	04	236 43.4	54.4	206 52.5	15.4	4 43.7	14.1	55.6
	05	251 43.3	53.7	221 26.9	15.4	4 29.6	14.1	55.6
	06	266 43.2	S17 53.0	236 01.3	15.3	S 4 15.5	14.2	55.7
	07	281 43.0	52.4	250 35.6	15.4	4 01.3	14.2	55.7
T	08	296 42.9	51.7	265 10.0	15.3	3 47.1	14.2	55.7
H	09	311 42.8	. . 51.0	279 44.3	15.4	3 32.9	14.2	55.7
U	10	326 42.7	50.3	294 18.7	15.3	3 18.7	14.2	55.7
R	11	341 42.6	49.7	308 53.0	15.4	3 04.5	14.3	55.8
S	12	356 42.5	S17 49.0	323 27.4	15.3	S 2 50.2	14.3	55.8
D	13	11 42.4	48.3	338 01.7	15.3	2 35.9	14.3	55.8
A	14	26 42.3	47.6	352 36.0	15.3	2 21.6	14.3	55.8
Y	15	41 42.2	. . 46.9	7 10.3	15.3	2 07.3	14.4	55.8
	16	56 42.1	46.3	21 44.6	15.3	1 52.9	14.3	55.9
	17	71 42.0	45.6	36 18.9	15.3	1 38.6	14.4	55.9
	18	86 41.9	S17 44.9	50 53.2	15.3	S 1 24.2	14.4	55.9
	19	101 41.8	44.2	65 27.5	15.3	1 09.8	14.4	55.9
	20	116 41.7	43.5	80 01.8	15.2	0 55.4	14.4	55.9
	21	131 41.6	. . 42.9	94 36.0	15.3	0 41.0	14.5	56.0
	22	146 41.5	42.2	109 10.3	15.2	0 26.5	14.4	56.0
	23	161 41.3	41.5	123 44.5	15.2	S 0 12.1	14.5	56.0
	30 00	176 41.2	S17 40.8	138 18.7	15.1	N 0 02.4	14.4	56.0
	01	191 41.1	40.1	152 52.8	15.2	0 16.8	14.5	56.0
	02	206 41.0	39.4	167 27.0	15.1	0 31.3	14.5	56.1
	03	221 40.9	. . 38.7	182 01.1	15.1	0 45.8	14.5	56.1
	04	236 40.8	38.1	196 35.2	15.1	1 00.3	14.5	56.1
	05	251 40.7	37.4	211 09.3	15.1	1 14.8	14.5	56.1
	06	266 40.6	S17 36.7	225 43.4	15.0	N 1 29.3	14.5	56.1
	07	281 40.5	36.0	240 17.4	15.1	1 43.8	14.5	56.2
	08	296 40.4	35.3	254 51.5	14.9	1 58.3	14.5	56.2
F	09	311 40.3	. . 34.6	269 25.4	15.0	2 12.8	14.5	56.2
R	10	326 40.2	33.9	283 59.4	14.9	2 27.3	14.5	56.2
I	11	341 40.1	33.2	298 33.3	14.9	2 41.8	14.5	56.3
D	12	356 40.0	S17 32.5	313 07.2	14.9	N 2 56.3	14.5	56.3
A	13	11 39.9	31.9	327 41.1	14.8	3 10.8	14.5	56.3
Y	14	26 39.8	31.2	342 14.9	14.8	3 25.3	14.5	56.3
	15	41 39.7	. . 30.5	356 48.7	14.8	3 39.8	14.5	56.3
	16	56 39.6	29.8	11 22.5	14.7	3 54.3	14.5	56.4
	17	71 39.6	29.1	25 56.2	14.7	4 08.8	14.5	56.4
	18	86 39.5	S17 28.4	40 29.9	14.6	N 4 23.3	14.5	56.4
	19	101 39.4	27.7	55 03.5	14.6	4 37.8	14.4	56.4
	20	116 39.3	27.0	69 37.1	14.6	4 52.2	14.5	56.5
	21	131 39.2	. . 26.3	84 10.7	14.5	5 06.7	14.4	56.5
	22	146 39.1	25.6	98 44.2	14.4	5 21.1	14.5	56.5
	23	161 39.0	24.9	113 17.6	14.5	N 5 35.6	14.4	56.5
		SD 16.3	d 0.7	SD 15.1		15.2		15.3

Lat.	Twilight Naut.	Twilight Civil	Sunrise	Moonrise 28	29	30	31
°	h m	h m	h m	h m	h m	h m	h m
N 72	07 19	08 51	10 54	09 39	09 06	08 37	08 05
N 70	07 10	08 30	10 01	09 25	09 02	08 40	08 16
68	07 03	08 14	09 30	09 14	08 58	08 42	08 25
66	06 56	08 01	09 06	09 05	08 54	08 44	08 33
64	06 51	07 50	08 48	08 58	08 51	08 45	08 39
62	06 46	07 41	08 33	08 51	08 49	08 47	08 45
60	06 41	07 32	08 20	08 45	08 47	08 48	08 49
N 58	06 37	07 25	08 10	08 40	08 45	08 49	08 54
56	06 33	07 19	08 00	08 35	08 43	08 50	08 58
54	06 30	07 13	07 52	08 31	08 41	08 51	09 01
52	06 26	07 07	07 44	08 28	08 40	08 52	09 04
50	06 23	07 02	07 38	08 24	08 38	08 52	09 07
45	06 16	06 52	07 23	08 17	08 36	08 54	09 13
N 40	06 10	06 42	07 11	08 11	08 33	08 55	09 19
35	06 04	06 34	07 01	08 05	08 31	08 57	09 23
30	05 58	06 27	06 52	08 00	08 29	08 58	09 27
20	05 47	06 13	06 36	07 52	08 26	09 00	09 34
N 10	05 35	06 01	06 23	07 45	08 23	09 01	09 41
0	05 23	05 48	06 10	07 38	08 20	09 03	09 47
S 10	05 08	05 34	05 57	07 31	08 18	09 04	09 53
20	04 51	05 19	05 42	07 24	08 15	09 06	09 59
30	04 29	05 00	05 26	07 15	08 12	09 08	10 06
35	04 15	04 49	05 16	07 10	08 10	09 09	10 11
40	03 58	04 35	05 05	07 05	08 07	09 11	10 16
45	03 36	04 19	04 52	06 58	08 05	09 12	10 21
S 50	03 07	03 58	04 37	06 50	08 02	09 14	10 28
52	02 52	03 48	04 29	06 47	08 01	09 15	10 32
54	02 34	03 36	04 21	06 43	07 59	09 16	10 35
56	02 11	03 23	04 11	06 38	07 57	09 17	10 39
58	01 39	03 07	04 01	06 33	07 56	09 18	10 43
S 60	00 27	02 48	03 48	06 28	07 54	09 20	10 48

Lat.	Sunset	Twilight Civil	Twilight Naut.	Moonset 28	29	30	31
°	h m	h m	h m	h m	h m	h m	h m
N 72	13 34	15 37	17 08	18 26	20 29	22 33	24 50
N 70	14 26	15 57	17 17	18 37	20 31	22 25	24 29
68	14 58	16 13	17 25	18 46	20 32	22 19	24 12
66	15 21	16 26	17 31	18 53	20 33	22 13	23 59
64	15 39	16 37	17 37	18 59	20 33	22 09	23 49
62	15 54	16 47	17 42	19 04	20 34	22 05	23 40
60	16 07	16 55	17 46	19 09	20 35	22 02	23 32
N 58	16 17	17 02	17 50	19 13	20 35	21 59	23 25
56	16 27	17 09	17 54	19 16	20 35	21 56	23 19
54	16 35	17 14	17 57	19 19	20 36	21 54	23 14
52	16 43	17 20	18 01	19 22	20 36	21 51	23 09
50	16 49	17 25	18 04	19 25	20 37	21 49	23 05
45	17 04	17 35	18 11	19 30	20 37	21 45	22 55
N 40	17 16	17 44	18 17	19 35	20 38	21 42	22 48
35	17 26	17 53	18 23	19 39	20 38	21 39	22 41
30	17 35	18 00	18 29	19 42	20 39	21 36	22 35
20	17 50	18 13	18 40	19 48	20 39	21 31	22 25
N 10	18 04	18 26	18 51	19 54	20 40	21 27	22 16
0	18 17	18 38	19 04	19 58	20 41	21 24	22 08
S 10	18 30	18 52	19 18	20 03	20 41	21 20	22 00
20	18 44	19 07	19 35	20 08	20 42	21 16	21 51
30	19 00	19 26	19 57	20 14	20 42	21 11	21 42
35	19 09	19 37	20 11	20 17	20 43	21 09	21 36
40	19 20	19 50	20 28	20 21	20 43	21 06	21 30
45	19 33	20 07	20 49	20 25	20 44	21 02	21 22
S 50	19 49	20 27	21 17	20 30	20 44	20 58	21 13
52	19 56	20 37	21 32	20 32	20 44	20 56	21 09
54	20 04	20 48	21 50	20 35	20 45	20 54	21 05
56	20 14	21 01	22 12	20 38	20 45	20 52	21 00
58	20 24	21 17	22 42	20 41	20 45	20 49	20 55
S 60	20 36	21 36	00 00	20 44	20 45	20 47	20 48

Day	SUN Eqn. of Time 00^h	SUN Eqn. of Time 12^h	SUN Mer. Pass.	MOON Mer. Pass. Upper	MOON Mer. Pass. Lower	Age	Phase
d	m s	m s	h m	h m	h m	d	%
28	12 53	12 59	12 13	13 48	01 27	02	4
29	13 05	13 10	12 13	14 30	02 09	03	10
30	13 15	13 20	12 13	15 13	02 52	04	16

	UT	ARIES	VENUS −4.7		MARS +1.3		JUPITER −1.9		SATURN +0.8		STARS		
		GHA	GHA	Dec	GHA	Dec	GHA	Dec	GHA	Dec	Name	SHA	Dec
	d h	° ′	° ′	° ′	° ′	° ′	° ′	° ′	° ′	° ′		° ′	° ′
	31 00	130 21.0	133 23.9	N 0 16.9	191 44.7	S21 47.4	181 53.2	S19 13.0	317 47.7	N 5 35.1	Acamar	315 20.7	S40 16.3
	01	145 23.4	148 24.5	18.1	206 45.2	47.1	196 55.1	12.9	332 50.3	35.1	Achernar	335 29.1	S57 11.6
	02	160 25.9	163 25.1	19.3	221 45.6	46.7	211 56.9	12.8	347 52.8	35.2	Acrux	173 12.9	S63 08.9
	03	175 28.4	178 25.7	. . 20.4	236 46.0	. . 46.4	226 58.8	. . 12.6	2 55.4	. . 35.3	Adhara	255 14.8	S28 59.2
	04	190 30.8	193 26.3	21.6	251 46.4	46.0	242 00.6	12.5	17 58.0	35.3	Aldebaran	290 53.0	N16 31.7
	05	205 33.3	208 26.9	22.8	266 46.8	45.7	257 02.5	12.3	33 00.6	35.4			
	06	220 35.7	223 27.5	N 0 24.0	281 47.3	S21 45.3	272 04.3	S19 12.2	48 03.2	N 5 35.4	Alioth	166 23.0	N55 54.2
	07	235 38.2	238 28.1	25.1	296 47.7	45.0	287 06.2	12.0	63 05.7	35.5	Alkaid	153 01.1	N49 15.7
S	08	250 40.7	253 28.7	26.3	311 48.1	44.6	302 08.1	11.9	78 08.3	35.6	Al Na'ir	27 48.1	S46 55.1
A	09	265 43.1	268 29.3	. . 27.5	326 48.5	. . 44.3	317 09.9	. . 11.8	93 10.9	. . 35.6	Alnilam	275 49.4	S 1 11.8
T	10	280 45.6	283 30.0	28.6	341 49.0	43.9	332 11.8	11.6	108 13.5	35.7	Alphard	217 59.0	S 8 42.0
U	11	295 48.1	298 30.6	29.8	356 49.4	43.6	347 13.6	11.5	123 16.1	35.7			
R	12	310 50.5	313 31.2	N 0 31.0	11 49.8	S21 43.2	2 15.5	S19 11.3	138 18.6	N 5 35.8	Alphecca	126 13.8	N26 40.7
D	13	325 53.0	328 31.8	32.1	26 50.2	42.8	17 17.3	11.2	153 21.2	35.9	Alpheratz	357 47.2	N29 08.6
A	14	340 55.5	343 32.4	33.3	41 50.7	42.5	32 19.2	11.0	168 23.8	35.9	Altair	62 11.7	N 8 53.4
Y	15	355 57.9	358 33.0	. . 34.5	56 51.1	. . 42.1	47 21.1	. . 10.9	183 26.4	. . 36.0	Ankaa	353 19.0	S42 15.5
	16	11 00.4	13 33.6	35.6	71 51.5	41.8	62 22.9	10.8	198 29.0	36.0	Antares	112 30.4	S26 27.2
	17	26 02.9	28 34.3	36.8	86 51.9	41.4	77 24.8	10.6	213 31.5	36.1			
	18	41 05.3	43 34.9	N 0 38.0	101 52.4	S21 41.1	92 26.6	S19 10.5	228 34.1	N 5 36.2	Arcturus	145 58.6	N19 07.8
	19	56 07.8	58 35.5	39.1	116 52.8	40.7	107 28.5	10.3	243 36.7	36.2	Atria	107 35.6	S69 02.5
	20	71 10.2	73 36.1	40.3	131 53.2	40.3	122 30.3	10.2	258 39.3	36.3	Avior	234 18.9	S59 32.4
	21	86 12.7	88 36.8	. . 41.5	146 53.6	. . 40.0	137 32.2	. . 10.0	273 41.9	. . 36.3	Bellatrix	278 35.3	N 6 21.5
	22	101 15.2	103 37.4	42.6	161 54.0	39.6	152 34.1	09.9	288 44.5	36.4	Betelgeuse	271 04.6	N 7 24.5
	23	116 17.6	118 38.0	43.8	176 54.5	39.3	167 35.9	09.8	303 47.0	36.5			
	1 00	131 20.1	133 38.6	N 0 45.0	191 54.9	S21 38.9	182 37.8	S19 09.6	318 49.6	N 5 36.5	Canopus	263 57.2	S52 42.2
	01	146 22.6	148 39.3	46.1	206 55.3	38.5	197 39.6	09.5	333 52.2	36.6	Capella	280 39.0	N46 00.6
	02	161 25.0	163 39.9	47.3	221 55.7	38.2	212 41.5	09.3	348 54.8	36.6	Deneb	49 34.3	N45 18.7
	03	176 27.5	178 40.5	. . 48.5	236 56.2	. . 37.8	227 43.3	. . 09.2	3 57.4	. . 36.7	Denebola	182 36.7	N14 31.0
	04	191 30.0	193 41.2	49.6	251 56.6	37.5	242 45.2	09.0	18 59.9	36.8	Diphda	348 59.2	S17 56.3
	05	206 32.4	208 41.8	50.8	266 57.0	37.1	257 47.1	08.9	34 02.5	36.8			
	06	221 34.9	223 42.4	N 0 51.9	281 57.4	S21 36.7	272 48.9	S19 08.8	49 05.1	N 5 36.9	Dubhe	193 54.8	N61 41.8
	07	236 37.4	238 43.1	53.1	296 57.9	36.4	287 50.8	08.6	64 07.7	36.9	Elnath	278 16.5	N28 37.0
	08	251 39.8	253 43.7	54.3	311 58.3	36.0	302 52.6	08.5	79 10.3	37.0	Eltanin	90 48.0	N51 29.0
S	09	266 42.3	268 44.4	. . 55.4	326 58.7	. . 35.6	317 54.5	. . 08.3	94 12.9	. . 37.1	Enif	33 50.6	N 9 55.0
U	10	281 44.7	283 45.0	56.6	341 59.2	35.3	332 56.4	08.2	109 15.4	37.1	Fomalhaut	15 27.8	S29 34.5
N	11	296 47.2	298 45.7	57.8	356 59.6	34.9	347 58.2	08.0	124 18.0	37.2			
D	12	311 49.7	313 46.3	N 0 58.9	12 00.0	S21 34.6	3 00.1	S19 07.9	139 20.6	N 5 37.2	Gacrux	172 04.5	S57 09.8
A	13	326 52.1	328 46.9	1 00.1	27 00.4	34.2	18 01.9	07.8	154 23.2	37.3	Gienah	175 55.5	S17 35.6
Y	14	341 54.6	343 47.6	01.2	42 00.9	33.8	33 03.8	07.6	169 25.8	37.4	Hadar	148 52.7	S60 24.9
	15	356 57.1	358 48.2	. . 02.4	57 01.3	. . 33.5	48 05.6	. . 07.5	184 28.4	. . 37.4	Hamal	328 04.5	N23 30.5
	16	11 59.5	13 48.9	03.6	72 01.7	33.1	63 07.5	07.3	199 31.0	37.5	Kaus Aust.	83 48.4	S34 22.8
	17	27 02.0	28 49.5	04.7	87 02.1	32.7	78 09.4	07.2	214 33.5	37.6			
	18	42 04.5	43 50.2	N 1 05.9	102 02.6	S21 32.4	93 11.2	S19 07.0	229 36.1	N 5 37.6	Kochab	137 19.5	N74 06.6
	19	57 06.9	58 50.9	07.0	117 03.0	32.0	108 13.1	06.9	244 38.7	37.7	Markab	13 41.9	N15 15.3
	20	72 09.4	73 51.5	08.2	132 03.4	31.6	123 14.9	06.7	259 41.3	37.7	Menkar	314 18.4	N 4 07.6
	21	87 11.8	88 52.2	. . 09.3	147 03.8	. . 31.2	138 16.8	. . 06.6	274 43.9	. . 37.8	Menkent	148 11.4	S36 24.9
	22	102 14.3	103 52.8	10.5	162 04.3	30.9	153 18.7	06.5	289 46.5	37.9	Miaplacidus	221 39.8	S69 45.3
	23	117 16.8	118 53.5	11.7	177 04.7	30.5	168 20.5	06.3	304 49.0	37.9			
	2 00	132 19.2	133 54.1	N 1 12.8	192 05.1	S21 30.1	183 22.4	S19 06.2	319 51.6	N 5 38.0	Mirfak	308 45.0	N49 53.9
	01	147 21.7	148 54.8	14.0	207 05.6	29.8	198 24.2	06.0	334 54.2	38.0	Nunki	76 02.6	S26 17.2
	02	162 24.2	163 55.5	15.1	222 06.0	29.4	213 26.1	05.9	349 56.8	38.1	Peacock	53 24.8	S56 42.4
	03	177 26.6	178 56.1	. . 16.3	237 06.4	. . 29.0	228 27.9	. . 05.7	4 59.4	. . 38.2	Pollux	243 31.3	N28 00.2
	04	192 29.1	193 56.8	17.4	252 06.8	28.7	243 29.8	05.6	20 02.0	38.2	Procyon	245 02.8	N 5 12.0
	05	207 31.6	208 57.5	18.6	267 07.3	28.3	258 31.7	05.5	35 04.6	38.3			
	06	222 34.0	223 58.1	N 1 19.8	282 07.7	S21 27.9	273 33.5	S19 05.3	50 07.1	N 5 38.4	Rasalhague	96 09.7	N12 33.0
	07	237 36.5	238 58.8	20.9	297 08.1	27.5	288 35.4	05.2	65 09.7	38.4	Regulus	207 46.6	N11 55.2
	08	252 39.0	253 59.5	22.1	312 08.6	27.2	303 37.2	05.0	80 12.3	38.5	Rigel	281 15.0	S 8 11.5
M	09	267 41.4	269 00.2	. . 23.2	327 09.0	. . 26.8	318 39.1	. . 04.9	95 14.9	. . 38.5	Rigil Kent.	139 56.4	S60 52.2
O	10	282 43.9	284 00.8	24.4	342 09.4	26.4	333 41.0	04.7	110 17.5	38.6	Sabik	102 16.5	S15 44.3
N	11	297 46.3	299 01.5	25.5	357 09.8	26.0	348 42.8	04.6	125 20.1	38.7			
D	12	312 48.8	314 02.2	N 1 26.7	12 10.3	S21 25.7	3 44.7	S19 04.5	140 22.7	N 5 38.7	Schedar	349 44.8	N56 35.5
A	13	327 51.3	329 02.9	27.8	27 10.7	25.3	18 46.5	04.3	155 25.3	38.8	Shaula	96 26.6	S37 06.6
Y	14	342 53.7	344 03.6	29.0	42 11.1	24.9	33 48.4	04.2	170 27.8	38.9	Sirius	258 36.3	S16 43.8
	15	357 56.2	359 04.2	. . 30.1	57 11.6	. . 24.5	48 50.3	. . 04.0	185 30.4	. . 38.9	Spica	158 34.6	S11 12.7
	16	12 58.7	14 04.9	31.3	72 12.0	24.2	63 52.1	03.9	200 33.0	39.0	Suhail	222 54.5	S43 28.2
	17	28 01.1	29 05.6	32.4	87 12.4	23.8	78 54.0	03.7	215 35.6	39.0			
	18	43 03.6	44 06.3	N 1 33.6	102 12.8	S21 23.4	93 55.8	S19 03.6	230 38.2	N 5 39.1	Vega	80 41.5	N38 47.3
	19	58 06.1	59 07.0	34.7	117 13.3	23.0	108 57.7	03.4	245 40.8	39.2	Zuben'ubi	137 09.1	S16 04.9
	20	73 08.5	74 07.7	35.9	132 13.7	22.6	123 59.5	03.3	260 43.4	39.2		SHA	Mer. Pass.
	21	88 11.0	89 08.4	. . 37.0	147 14.1	. . 22.3	139 01.4	. . 03.2	275 46.0	. . 39.3		° ′	h m
	22	103 13.5	104 09.1	38.2	162 14.6	21.9	154 03.3	03.0	290 48.5	39.4	Venus	2 18.5	15 05
	23	118 15.9	119 09.7	39.3	177 15.0	21.5	169 05.1	02.9	305 51.1	39.4	Mars	60 34.8	11 12
		h m									Jupiter	51 17.7	11 48
	Mer. Pass.	15 12.2	*v* 0.6	*d* 1.2	*v* 0.4	*d* 0.4	*v* 1.9	*d* 0.1	*v* 2.6	*d* 0.1	Saturn	187 29.5	2 44

UT	SUN GHA	SUN Dec	MOON GHA	v	MOON Dec	d	HP
d h	° ′	° ′	° ′	′	° ′	′	′
31 00	176 38.9	S17 24.2	127 51.1	14.3	N 5 50.0	14.4	56.6
01	191 38.8	23.5	142 24.4	14.4	6 04.4	14.4	56.6
02	206 38.7	22.8	156 57.8	14.3	6 18.8	14.3	56.6
03	221 38.6	. . 22.1	171 31.1	14.2	6 33.1	14.4	56.6
04	236 38.5	21.4	186 04.3	14.2	6 47.5	14.3	56.6
05	251 38.4	20.7	200 37.5	14.1	7 01.8	14.4	56.7
06	266 38.3	S17 20.0	215 10.6	14.1	N 7 16.2	14.3	56.7
07	281 38.2	19.3	229 43.7	14.0	7 30.5	14.2	56.7
08	296 38.1	18.6	244 16.7	14.0	7 44.7	14.3	56.7
09	311 38.1	. . 17.9	258 49.7	13.9	7 59.0	14.2	56.8
10	326 38.0	17.2	273 22.6	13.8	8 13.2	14.2	56.8
11	341 37.9	16.5	287 55.4	13.8	8 27.4	14.2	56.8
12	356 37.8	S17 15.8	302 28.2	13.7	N 8 41.6	14.1	56.8
13	11 37.7	15.1	317 00.9	13.7	8 55.7	14.2	56.9
14	26 37.6	14.4	331 33.6	13.6	9 09.9	14.1	56.9
15	41 37.5	. . 13.7	346 06.2	13.5	9 24.0	14.0	56.9
16	56 37.4	13.0	0 38.7	13.5	9 38.0	14.1	56.9
17	71 37.3	12.3	15 11.2	13.4	9 52.1	14.0	57.0
18	86 37.3	S17 11.6	29 43.6	13.3	N10 06.1	13.9	57.0
19	101 37.2	10.9	44 15.9	13.3	10 20.0	14.0	57.0
20	116 37.1	10.1	58 48.2	13.2	10 34.0	13.8	57.0
21	131 37.0	. . 09.4	73 20.4	13.1	10 47.8	13.9	57.1
22	146 36.9	08.7	87 52.5	13.1	11 01.7	13.8	57.1
23	161 36.8	08.0	102 24.6	12.9	11 15.5	13.8	57.1
1 00	176 36.7	S17 07.3	116 56.5	12.9	N11 29.3	13.7	57.1
01	191 36.6	06.6	131 28.4	12.9	11 43.0	13.7	57.2
02	206 36.6	05.9	146 00.3	12.7	11 56.7	13.7	57.2
03	221 36.5	. . 05.2	160 32.0	12.7	12 10.4	13.6	57.2
04	236 36.4	04.5	175 03.7	12.6	12 24.0	13.5	57.2
05	251 36.3	03.8	189 35.3	12.5	12 37.5	13.5	57.3
06	266 36.2	S17 03.0	204 06.8	12.4	N12 51.0	13.5	57.3
07	281 36.1	02.3	218 38.2	12.3	13 04.5	13.4	57.3
08	296 36.1	01.6	233 09.5	12.3	13 17.9	13.4	57.4
09	311 36.0	. . 00.9	247 40.8	12.2	13 31.3	13.3	57.4
10	326 35.9	17 00.2	262 12.0	12.1	13 44.6	13.2	57.4
11	341 35.8	16 59.5	276 43.1	12.0	13 57.8	13.2	57.4
12	356 35.7	S16 58.7	291 14.1	11.9	N14 11.0	13.2	57.5
13	11 35.7	58.0	305 45.0	11.8	14 24.2	13.0	57.5
14	26 35.6	57.3	320 15.8	11.7	14 37.2	13.1	57.5
15	41 35.5	. . 56.6	334 46.5	11.7	14 50.3	12.9	57.5
16	56 35.4	55.9	349 17.2	11.5	15 03.2	12.9	57.6
17	71 35.3	55.2	3 47.7	11.5	15 16.1	12.8	57.6
18	86 35.3	S16 54.4	18 18.2	11.3	N15 28.9	12.8	57.6
19	101 35.2	53.7	32 48.5	11.3	15 41.7	12.7	57.6
20	116 35.1	53.0	47 18.8	11.1	15 54.4	12.6	57.7
21	131 35.0	. . 52.3	61 48.9	11.1	16 07.0	12.6	57.7
22	146 34.9	51.6	76 19.0	11.0	16 19.6	12.5	57.7
23	161 34.9	50.8	90 49.0	10.8	16 32.1	12.4	57.8
2 00	176 34.8	S16 50.1	105 18.8	10.8	N16 44.5	12.4	57.8
01	191 34.7	49.4	119 48.6	10.7	16 56.9	12.2	57.8
02	206 34.6	48.7	134 18.3	10.5	17 09.1	12.2	57.8
03	221 34.6	. . 47.9	148 47.8	10.5	17 21.3	12.1	57.9
04	236 34.5	47.2	163 17.3	10.3	17 33.4	12.1	57.9
05	251 34.4	46.5	177 46.6	10.3	17 45.5	11.9	57.9
06	266 34.3	S16 45.8	192 15.9	10.1	N17 57.4	11.9	58.0
07	281 34.3	45.0	206 45.0	10.1	18 09.3	11.8	58.0
08	296 34.2	44.3	221 14.1	9.9	18 21.1	11.7	58.0
09	311 34.1	. . 43.6	235 43.0	9.8	18 32.8	11.6	58.0
10	326 34.0	42.9	250 11.8	9.7	18 44.4	11.5	58.1
11	341 34.0	42.1	264 40.5	9.6	18 55.9	11.4	58.1
12	356 33.9	S16 41.4	279 09.1	9.5	N19 07.3	11.3	58.1
13	11 33.8	40.7	293 37.6	9.4	19 18.6	11.3	58.1
14	26 33.8	39.9	308 06.0	9.3	19 29.9	11.1	58.2
15	41 33.7	. . 39.2	322 34.3	9.1	19 41.0	11.1	58.2
16	56 33.6	38.5	337 02.4	9.1	19 52.1	10.9	58.2
17	71 33.5	37.7	351 30.5	8.9	20 03.0	10.9	58.3
18	86 33.5	S16 37.0	5 58.4	8.8	N20 13.9	10.7	58.3
19	101 33.4	36.3	20 26.2	8.7	20 24.6	10.6	58.3
20	116 33.3	35.6	34 53.9	8.6	20 35.2	10.6	58.3
21	131 33.3	. . 34.8	49 21.5	8.5	20 45.8	10.4	58.4
22	146 33.2	34.1	63 49.0	8.4	20 56.2	10.3	58.4
23	161 33.1	33.4	78 16.4	8.2	N21 06.5	10.2	58.4
	SD 16.3	*d* 0.7	SD		15.5	15.7	15.8

Lat.	Twilight Naut.	Twilight Civil	Sunrise	Moonrise 31	1	2	3
°	h m	h m	h m	h m	h m	h m	h m
N 72	07 09	08 38	10 26	08 05	07 24	05 47	☐
N 70	07 01	08 19	09 45	08 16	07 48	07 00	☐
68	06 55	08 05	09 17	08 25	08 06	07 38	☐
66	06 49	07 53	08 56	08 33	08 20	08 04	07 36
64	06 44	07 43	08 39	08 39	08 33	08 25	08 16
62	06 40	07 34	08 25	08 45	08 43	08 42	08 43
60	06 36	07 26	08 14	08 49	08 52	08 56	09 05
N 58	06 32	07 20	08 03	08 54	09 00	09 09	09 23
56	06 29	07 14	07 55	08 58	09 07	09 19	09 38
54	06 26	07 08	07 47	09 01	09 13	09 29	09 51
52	06 23	07 03	07 40	09 04	09 19	09 37	10 02
50	06 20	06 59	07 33	09 07	09 24	09 45	10 12
45	06 13	06 49	07 20	09 13	09 35	10 01	10 33
N 40	06 07	06 40	07 08	09 19	09 44	10 14	10 51
35	06 02	06 32	06 59	09 23	09 52	10 26	11 05
30	05 57	06 25	06 50	09 27	09 59	10 36	11 18
20	05 46	06 12	06 35	09 34	10 12	10 53	11 40
N 10	05 35	06 01	06 22	09 41	10 23	11 08	11 59
0	05 23	05 48	06 10	09 47	10 33	11 23	12 17
S 10	05 10	05 36	05 58	09 53	10 43	11 37	12 35
20	04 53	05 21	05 44	09 59	10 54	11 53	12 55
30	04 32	05 03	05 29	10 06	11 07	12 11	13 17
35	04 19	04 52	05 20	10 11	11 15	12 21	13 31
40	04 02	04 39	05 09	10 16	11 23	12 33	13 46
45	03 42	04 24	04 57	10 21	11 33	12 48	14 05
S 50	03 14	04 04	04 42	10 28	11 45	13 06	14 28
52	03 00	03 54	04 35	10 32	11 51	13 14	14 39
54	02 44	03 44	04 27	10 35	11 57	13 23	14 51
56	02 23	03 31	04 18	10 39	12 04	13 34	15 06
58	01 55	03 16	04 08	10 43	12 12	13 46	15 23
S 60	01 12	02 59	03 57	10 48	12 21	14 00	15 44

Lat.	Sunset	Twilight Civil	Twilight Naut.	Moonset 31	1	2	3
°	h m	h m	h m	h m	h m	h m	h m
N 72	14 03	15 51	17 20	24 50	00 50	04 09	☐
N 70	14 44	16 09	17 27	24 29	00 29	02 58	☐
68	15 12	16 24	17 34	24 12	00 12	02 22	☐
66	15 32	16 36	17 39	23 59	25 56	01 56	04 15
64	15 49	16 46	17 44	23 49	25 36	01 36	03 36
62	16 03	16 54	17 49	23 40	25 21	01 21	03 09
60	16 14	17 02	17 53	23 32	25 07	01 07	02 48
N 58	16 25	17 08	17 56	23 25	24 56	00 56	02 31
56	16 33	17 14	17 59	23 19	24 46	00 46	02 17
54	16 41	17 20	18 02	23 14	24 38	00 38	02 05
52	16 48	17 25	18 05	23 09	24 30	00 30	01 54
50	16 54	17 29	18 08	23 05	24 23	00 23	01 44
45	17 08	17 39	18 14	22 55	24 08	00 08	01 24
N 40	17 19	17 48	18 20	22 48	23 56	25 08	01 08
35	17 29	17 55	18 26	22 41	23 46	24 54	00 54
30	17 37	18 02	18 31	22 35	23 37	24 42	00 42
20	17 52	18 15	18 41	22 25	23 22	24 22	00 22
N 10	18 05	18 27	18 52	22 16	23 08	24 04	00 04
0	18 17	18 39	19 04	22 08	22 56	23 48	24 45
S 10	18 29	18 52	19 17	22 00	22 44	23 32	24 25
20	18 43	19 06	19 34	21 51	22 31	23 14	24 05
30	18 58	19 24	19 55	21 42	22 16	22 55	23 41
35	19 07	19 35	20 08	21 36	22 07	22 43	23 27
40	19 17	19 47	20 24	21 30	21 57	22 30	23 10
45	19 30	20 03	20 44	21 22	21 46	22 14	22 51
S 50	19 44	20 22	21 11	21 13	21 32	21 55	22 27
52	19 51	20 31	21 25	21 09	21 25	21 46	22 15
54	19 59	20 42	21 41	21 05	21 18	21 36	22 02
56	20 08	20 54	22 01	21 00	21 10	21 25	21 47
58	20 17	21 09	22 28	20 55	21 01	21 12	21 30
S 60	20 29	21 26	23 08	20 48	20 51	20 57	21 08

Day	SUN Eqn. of Time 00^h	SUN Eqn. of Time 12^h	SUN Mer. Pass.	MOON Mer. Pass. Upper	MOON Mer. Pass. Lower	Age	Phase
d	m s	m s	h m	h m	h m	d %	
31	13 24	13 29	12 13	15 57	03 35	05 25	
1	13 33	13 37	12 14	16 44	04 20	06 34	◐
2	13 41	13 44	12 14	17 35	05 09	07 45	

Day	UT (d h)	ARIES GHA	VENUS −4.7 GHA	VENUS Dec	MARS +1.3 GHA	MARS Dec	JUPITER −1.9 GHA	JUPITER Dec	SATURN +0.7 GHA	SATURN Dec
		° ′	° ′	° ′	° ′	° ′	° ′	° ′	° ′	° ′
TUESDAY	3 00	133 18.4	134 10.4	N 1 40.5	192 15.4	S21 21.1	184 07.0	S19 02.7	320 53.7	N 5 39.5
	01	148 20.8	149 11.1	41.6	207 15.9	20.7	199 08.8	02.6	335 56.3	39.5
	02	163 23.3	164 11.8	42.8	222 16.3	20.4	214 10.7	02.4	350 58.9	39.6
	03	178 25.8	179 12.5	. . 43.9	237 16.7	. . 20.0	229 12.6	. . 02.3	6 01.5	. . 39.7
	04	193 28.2	194 13.2	45.1	252 17.2	19.6	244 14.4	02.1	21 04.1	39.7
	05	208 30.7	209 13.9	46.2	267 17.6	19.2	259 16.3	02.0	36 06.7	39.8
	06	223 33.2	224 14.6	N 1 47.4	282 18.0	S21 18.8	274 18.1	S19 01.9	51 09.3	N 5 39.9
	07	238 35.6	239 15.3	48.5	297 18.4	18.5	289 20.0	01.7	66 11.8	39.9
	08	253 38.1	254 16.0	49.7	312 18.9	18.1	304 21.9	01.6	81 14.4	40.0
	09	268 40.6	269 16.8	. . 50.8	327 19.3	. . 17.7	319 23.7	. . 01.4	96 17.0	. . 40.1
	10	283 43.0	284 17.5	52.0	342 19.7	17.3	334 25.6	01.3	111 19.6	40.1
	11	298 45.5	299 18.2	53.1	357 20.2	16.9	349 27.4	01.1	126 22.2	40.2
	12	313 47.9	314 18.9	N 1 54.2	12 20.6	S21 16.5	4 29.3	S19 01.0	141 24.8	N 5 40.2
	13	328 50.4	329 19.6	55.4	27 21.0	16.2	19 31.2	00.8	156 27.4	40.3
	14	343 52.9	344 20.3	56.5	42 21.5	15.8	34 33.0	00.7	171 30.0	40.4
	15	358 55.3	359 21.0	. . 57.7	57 21.9	. . 15.4	49 34.9	. . 00.6	186 32.6	. . 40.4
	16	13 57.8	14 21.7	1 58.8	72 22.3	15.0	64 36.7	00.4	201 35.2	40.5
	17	29 00.3	29 22.5	2 00.0	87 22.8	14.6	79 38.6	00.3	216 37.8	40.6
	18	44 02.7	44 23.2	N 2 01.1	102 23.2	S21 14.2	94 40.5	S19 00.1	231 40.3	N 5 40.6
	19	59 05.2	59 23.9	02.2	117 23.6	13.8	109 42.3	19 00.0	246 42.9	40.7
	20	74 07.7	74 24.6	03.4	132 24.1	13.4	124 44.2	18 59.8	261 45.5	40.8
	21	89 10.1	89 25.4	. . 04.5	147 24.5	. . 13.1	139 46.0	. . 59.7	276 48.1	. . 40.8
	22	104 12.6	104 26.1	05.7	162 24.9	12.7	154 47.9	59.5	291 50.7	40.9
	23	119 15.1	119 26.8	06.8	177 25.4	12.3	169 49.8	59.4	306 53.3	40.9
WEDNESDAY	4 00	134 17.5	134 27.5	N 2 07.9	192 25.8	S21 11.9	184 51.6	S18 59.3	321 55.9	N 5 41.0
	01	149 20.0	149 28.3	09.1	207 26.2	11.5	199 53.5	59.1	336 58.5	41.1
	02	164 22.4	164 29.0	10.2	222 26.7	11.1	214 55.3	59.0	352 01.1	41.1
	03	179 24.9	179 29.7	. . 11.4	237 27.1	. . 10.7	229 57.2	. . 58.8	7 03.7	. . 41.2
	04	194 27.4	194 30.5	12.5	252 27.5	10.3	244 59.1	58.7	22 06.3	41.3
	05	209 29.8	209 31.2	13.6	267 28.0	09.9	260 00.9	58.5	37 08.9	41.3
	06	224 32.3	224 31.9	N 2 14.8	282 28.4	S21 09.5	275 02.8	S18 58.4	52 11.4	N 5 41.4
	07	239 34.8	239 32.7	15.9	297 28.8	09.1	290 04.7	58.2	67 14.0	41.5
	08	254 37.2	254 33.4	17.0	312 29.3	08.8	305 06.5	58.1	82 16.6	41.5
	09	269 39.7	269 34.2	. . 18.2	327 29.7	. . 08.4	320 08.4	. . 57.9	97 19.2	. . 41.6
	10	284 42.2	284 34.9	19.3	342 30.1	08.0	335 10.2	57.8	112 21.8	41.6
	11	299 44.6	299 35.7	20.5	357 30.6	07.6	350 12.1	57.7	127 24.4	41.7
	12	314 47.1	314 36.4	N 2 21.6	12 31.0	S21 07.2	5 14.0	S18 57.5	142 27.0	N 5 41.8
	13	329 49.6	329 37.2	22.7	27 31.4	06.8	20 15.8	57.4	157 29.6	41.8
	14	344 52.0	344 37.9	23.9	42 31.9	06.4	35 17.7	57.2	172 32.2	41.9
	15	359 54.5	359 38.7	. . 25.0	57 32.3	. . 06.0	50 19.5	. . 57.1	187 34.8	. . 42.0
	16	14 56.9	14 39.4	26.1	72 32.8	05.6	65 21.4	56.9	202 37.4	42.0
	17	29 59.4	29 40.2	27.3	87 33.2	05.2	80 23.3	56.8	217 40.0	42.1
	18	45 01.9	44 40.9	N 2 28.4	102 33.6	S21 04.8	95 25.1	S18 56.6	232 42.6	N 5 42.2
	19	60 04.3	59 41.7	29.5	117 34.1	04.4	110 27.0	56.5	247 45.2	42.2
	20	75 06.8	74 42.4	30.7	132 34.5	04.0	125 28.8	56.4	262 47.8	42.3
	21	90 09.3	89 43.2	. . 31.8	147 34.9	. . 03.6	140 30.7	. . 56.2	277 50.4	. . 42.4
	22	105 11.7	104 44.0	32.9	162 35.4	03.2	155 32.6	56.1	292 52.9	42.4
	23	120 14.2	119 44.7	34.0	177 35.8	02.8	170 34.4	55.9	307 55.5	42.5
THURSDAY	5 00	135 16.7	134 45.5	N 2 35.2	192 36.2	S21 02.4	185 36.3	S18 55.8	322 58.1	N 5 42.6
	01	150 19.1	149 46.2	36.3	207 36.7	02.0	200 38.2	55.6	338 00.7	42.6
	02	165 21.6	164 47.0	37.4	222 37.1	01.6	215 40.0	55.5	353 03.3	42.7
	03	180 24.1	179 47.8	. . 38.6	237 37.6	. . 01.2	230 41.9	. . 55.3	8 05.9	. . 42.8
	04	195 26.5	194 48.6	39.7	252 38.0	00.8	245 43.7	55.2	23 08.5	42.8
	05	210 29.0	209 49.3	40.8	267 38.4	00.4	260 45.6	55.0	38 11.1	42.9
	06	225 31.4	224 50.1	N 2 41.9	282 38.9	S21 00.0	275 47.5	S18 54.9	53 13.7	N 5 43.0
	07	240 33.9	239 50.9	43.1	297 39.3	20 59.6	290 49.3	54.8	68 16.3	43.0
	08	255 36.4	254 51.6	44.2	312 39.7	59.2	305 51.2	54.6	83 18.9	43.1
	09	270 38.8	269 52.4	. . 45.3	327 40.2	. . 58.8	320 53.0	. . 54.5	98 21.5	. . 43.1
	10	285 41.3	284 53.2	46.5	342 40.6	58.4	335 54.9	54.3	113 24.1	43.2
	11	300 43.8	299 54.0	47.6	357 41.1	58.0	350 56.8	54.2	128 26.7	43.3
	12	315 46.2	314 54.8	N 2 48.7	12 41.5	S20 57.6	5 58.6	S18 54.0	143 29.3	N 5 43.3
	13	330 48.7	329 55.6	49.8	27 41.9	57.2	21 00.5	53.9	158 31.9	43.4
	14	345 51.2	344 56.3	51.0	42 42.4	56.8	36 02.4	53.7	173 34.5	43.5
	15	0 53.6	359 57.1	. . 52.1	57 42.8	. . 56.4	51 04.2	. . 53.6	188 37.1	. . 43.5
	16	15 56.1	14 57.9	53.2	72 43.3	56.0	66 06.1	53.4	203 39.7	43.6
	17	30 58.6	29 58.7	54.3	87 43.7	55.6	81 07.9	53.3	218 42.3	43.7
	18	46 01.0	44 59.5	N 2 55.4	102 44.1	S20 55.2	96 09.8	S18 53.2	233 44.9	N 5 43.7
	19	61 03.5	60 00.3	56.6	117 44.6	54.7	111 11.7	53.0	248 47.5	43.8
	20	76 05.9	75 01.1	57.7	132 45.0	54.3	126 13.5	52.9	263 50.1	43.9
	21	91 08.4	90 01.9	. . 58.8	147 45.4	. . 53.9	141 15.4	. . 52.7	278 52.7	. . 43.9
	22	106 10.9	105 02.7	2 59.9	162 45.9	53.5	156 17.3	52.6	293 55.3	44.0
	23	121 13.3	120 03.5	N 3 01.0	177 46.3	53.1	171 19.1	52.4	308 57.9	44.1
	Mer. Pass. (h m)	15 00.4	*v* 0.7	*d* 1.1	*v* 0.4	*d* 0.4	*v* 1.9	*d* 0.1	*v* 2.6	*d* 0.1

STARS

Name	SHA	Dec
	° ′	° ′
Acamar	315 20.7	S40 16.
Achernar	335 29.1	S57 11.
Acrux	173 12.8	S63 08.
Adhara	255 14.8	S28 59.
Aldebaran	290 53.0	N16 31.
Alioth	166 23.0	N55 54.
Alkaid	153 01.1	N49 15.
Al Na'ir	27 48.1	S46 55.
Alnilam	275 49.4	S 1 11.
Alphard	217 59.0	S 8 42.
Alphecca	126 13.8	N26 40.
Alpheratz	357 47.2	N29 08.
Altair	62 11.7	N 8 53.4
Ankaa	353 19.0	S42 15.5
Antares	112 30.4	S26 27.2
Arcturus	145 58.6	N19 07.8
Atria	107 35.6	S69 02.5
Avior	234 18.9	S59 32.4
Bellatrix	278 35.3	N 6 21.5
Betelgeuse	271 04.6	N 7 24.5
Canopus	263 57.2	S52 42.2
Capella	280 39.0	N46 00.6
Deneb	49 34.3	N45 18.7
Denebola	182 36.7	N14 31.0
Diphda	348 59.3	S17 56.3
Dubhe	193 54.7	N61 41.8
Elnath	278 16.5	N28 37.0
Eltanin	90 48.0	N51 28.9
Enif	33 50.6	N 9 55.0
Fomalhaut	15 27.8	S29 34.5
Gacrux	172 04.4	S57 09.8
Gienah	175 55.5	S17 35.7
Hadar	148 52.6	S60 24.9
Hamal	328 04.5	N23 30.5
Kaus Aust.	83 48.4	S34 22.8
Kochab	137 19.4	N74 06.6
Markab	13 41.9	N15 15.3
Menkar	314 18.4	N 4 07.6
Menkent	148 11.4	S36 24.9
Miaplacidus	221 39.7	S69 45.3
Mirfak	308 45.0	N49 53.9
Nunki	76 02.6	S26 17.2
Peacock	53 24.8	S56 42.4
Pollux	243 31.3	N28 00.2
Procyon	245 02.8	N 5 12.0
Rasalhague	96 09.7	N12 33.0
Regulus	207 46.6	N11 55.2
Rigel	281 15.0	S 8 11.5
Rigil Kent.	139 56.3	S60 52.2
Sabik	102 16.5	S15 44.3
Schedar	349 44.8	N56 35.5
Shaula	96 26.6	S37 06.6
Sirius	258 36.3	S16 43.8
Spica	158 34.6	S11 12.7
Suhail	222 54.5	S43 28.2
Vega	80 41.5	N38 47.3
Zuben'ubi	137 09.0	S16 04.9

	SHA	Mer. Pass
	° ′	h m
Venus	0 10.0	15 01
Mars	58 08.3	11 10
Jupiter	50 34.1	11 39
Saturn	187 38.4	2 32

UT d h	SUN GHA ° ′	SUN Dec ° ′	MOON GHA ° ′	v ′	Dec ° ′	d ′	HP ′
3 00	176 33.1	S16 32.6	92 43.6	8.2	N21 16.7	10.1	58.5
01	191 33.0	31.9	107 10.8	8.0	21 26.8	10.0	58.5
02	206 32.9	31.1	121 37.8	7.9	21 36.8	9.9	58.5
03	221 32.9	. . 30.4	136 04.7	7.8	21 46.7	9.7	58.5
04	236 32.8	29.7	150 31.5	7.6	21 56.4	9.6	58.6
05	251 32.7	28.9	164 58.1	7.6	22 06.0	9.5	58.6
06	266 32.7	S16 28.2	179 24.7	7.4	N22 15.5	9.4	58.6
07	281 32.6	27.5	193 51.1	7.4	22 24.9	9.3	58.7
08	296 32.5	26.7	208 17.5	7.2	22 34.2	9.1	58.7
09	311 32.5	. . 26.0	222 43.7	7.1	22 43.3	9.0	58.7
10	326 32.4	25.2	237 09.8	7.0	22 52.3	8.9	58.7
11	341 32.3	24.5	251 35.8	6.8	23 01.2	8.7	58.8
12	356 32.3	S16 23.8	266 01.6	6.8	N23 09.9	8.6	58.8
13	11 32.2	23.0	280 27.4	6.6	23 18.5	8.5	58.8
14	26 32.1	22.3	294 53.0	6.5	23 27.0	8.3	58.8
15	41 32.1	. . 21.5	309 18.5	6.4	23 35.3	8.2	58.9
16	56 32.0	20.8	323 43.9	6.3	23 43.5	8.1	58.9
17	71 32.0	20.1	338 09.2	6.2	23 51.6	7.9	58.9
18	86 31.9	S16 19.3	352 34.4	6.1	N23 59.5	7.8	59.0
19	101 31.8	18.6	6 59.5	5.9	24 07.3	7.6	59.0
20	116 31.8	17.8	21 24.4	5.9	24 14.9	7.5	59.0
21	131 31.7	. . 17.1	35 49.3	5.7	24 22.4	7.3	59.0
22	146 31.7	16.3	50 14.0	5.7	24 29.7	7.2	59.1
23	161 31.6	15.6	64 38.7	5.5	24 36.9	7.1	59.1
4 00	176 31.5	S16 14.8	79 03.2	5.4	N24 44.0	6.9	59.1
01	191 31.5	14.1	93 27.6	5.3	24 50.9	6.7	59.2
02	206 31.4	13.3	107 51.9	5.2	24 57.6	6.6	59.2
03	221 31.4	. . 12.6	122 16.1	5.1	25 04.2	6.4	59.2
04	236 31.3	11.8	136 40.2	5.0	25 10.6	6.2	59.2
05	251 31.2	11.1	151 04.2	4.9	25 16.8	6.1	59.3
06	266 31.2	S16 10.3	165 28.1	4.8	N25 22.9	6.0	59.3
07	281 31.1	09.6	179 51.9	4.7	25 28.9	5.7	59.3
08	296 31.1	08.8	194 15.6	4.6	25 34.6	5.6	59.3
09	311 31.0	. . 08.1	208 39.2	4.5	25 40.2	5.5	59.4
10	326 31.0	07.3	223 02.7	4.4	25 45.7	5.2	59.4
11	341 30.9	06.6	237 26.1	4.3	25 50.9	5.1	59.4
12	356 30.8	S16 05.8	251 49.4	4.2	N25 56.0	5.0	59.4
13	11 30.8	05.1	266 12.6	4.2	26 01.0	4.7	59.5
14	26 30.7	04.3	280 35.8	4.0	26 05.7	4.6	59.5
15	41 30.7	. . 03.6	294 58.8	4.0	26 10.3	4.4	59.5
16	56 30.6	02.8	309 21.8	3.9	26 14.7	4.2	59.5
17	71 30.6	02.1	323 44.7	3.7	26 18.9	4.1	59.6
18	86 30.5	S16 01.3	338 07.4	3.8	N26 23.0	3.8	59.6
19	101 30.5	16 00.6	352 30.2	3.6	26 26.8	3.7	59.6
20	116 30.4	15 59.8	6 52.8	3.5	26 30.5	3.5	59.6
21	131 30.4	. . 59.1	21 15.3	3.5	26 34.0	3.4	59.7
22	146 30.3	58.3	35 37.8	3.4	26 37.4	3.1	59.7
23	161 30.3	57.5	50 00.2	3.4	26 40.5	3.0	59.7
5 00	176 30.2	S15 56.8	64 22.6	3.2	N26 43.5	2.8	59.7
01	191 30.2	56.0	78 44.8	3.2	26 46.3	2.5	59.8
02	206 30.1	55.3	93 07.0	3.2	26 48.8	2.4	59.8
03	221 30.1	. . 54.5	107 29.2	3.1	26 51.2	2.2	59.8
04	236 30.0	53.7	121 51.3	3.0	26 53.4	2.1	59.8
05	251 30.0	53.0	136 13.3	2.9	26 55.5	1.8	59.9
06	266 29.9	S15 52.2	150 35.2	3.0	N26 57.3	1.6	59.9
07	281 29.9	51.5	164 57.2	2.8	26 58.9	1.5	59.9
08	296 29.8	50.7	179 19.0	2.8	27 00.4	1.2	59.9
09	311 29.8	. . 49.9	193 40.8	2.8	27 01.6	1.1	59.9
10	326 29.7	49.2	208 02.6	2.7	27 02.7	0.8	60.0
11	341 29.7	48.4	222 24.3	2.7	27 03.5	0.7	60.0
12	356 29.6	S15 47.6	236 46.0	2.6	N27 04.2	0.5	60.0
13	11 29.6	46.9	251 07.6	2.6	27 04.7	0.3	60.0
14	26 29.5	46.1	265 29.2	2.6	27 05.0	0.0	60.1
15	41 29.5	. . 45.4	279 50.8	2.5	27 05.0	0.1	60.1
16	56 29.4	44.6	294 12.3	2.6	27 04.9	0.3	60.1
17	71 29.4	43.8	308 33.9	2.4	27 04.6	0.5	60.1
18	86 29.4	S15 43.1	322 55.3	2.5	N27 04.1	0.7	60.1
19	101 29.3	42.3	337 16.8	2.5	27 03.4	0.9	60.2
20	116 29.3	41.5	351 38.3	2.4	27 02.5	1.1	60.2
21	131 29.2	. . 40.8	5 59.7	2.4	27 01.4	1.4	60.2
22	146 29.2	40.0	20 21.1	2.4	27 00.0	1.5	60.2
23	161 29.1	39.2	34 42.5	2.4	N26 58.5	1.7	60.2
	SD 16.3	*d* 0.8	SD	16.0	16.2		16.4

Lat. °	Twilight Naut. h m	Twilight Civil h m	Sunrise h m	Moonrise 3 h m	Moonrise 4 h m	Moonrise 5 h m	Moonrise 6 h m
N 72	06 58	08 24	10 03	□	□	□	□
N 70	06 52	08 08	09 28	□	□	□	□
68	06 46	07 55	09 04	□	□	□	□
66	06 41	07 44	08 45	07 36	□	□	□
64	06 37	07 35	08 30	08 16	07 55	□	□
62	06 33	07 27	08 17	08 43	08 50	09 17	10 38
60	06 30	07 20	08 06	09 05	09 23	10 02	11 19
N 58	06 27	07 14	07 57	09 23	09 47	10 32	11 46
56	06 24	07 08	07 49	09 38	10 07	10 55	12 08
54	06 21	07 03	07 41	09 51	10 24	11 14	12 26
52	06 18	06 59	07 35	10 02	10 38	11 30	12 41
50	06 16	06 54	07 29	10 12	10 50	11 44	12 54
45	06 10	06 45	07 16	10 33	11 16	12 12	13 21
N 40	06 05	06 37	07 06	10 51	11 37	12 34	13 42
35	06 00	06 30	06 56	11 05	11 54	12 52	14 00
30	05 55	06 23	06 48	11 18	12 09	13 08	14 15
20	05 45	06 12	06 34	11 40	12 34	13 35	14 41
N 10	05 35	06 00	06 22	11 59	12 56	13 58	15 03
0	05 24	05 49	06 10	12 17	13 17	14 20	15 24
S 10	05 11	05 37	05 59	12 35	13 37	14 41	15 45
20	04 55	05 23	05 46	12 55	13 59	15 04	16 07
30	04 35	05 06	05 31	13 17	14 25	15 32	16 33
35	04 22	04 55	05 23	13 31	14 41	15 47	16 48
40	04 07	04 43	05 13	13 46	14 58	16 06	17 05
45	03 47	04 28	05 01	14 05	15 20	16 29	17 26
S 50	03 22	04 10	04 47	14 28	15 47	16 57	17 52
52	03 08	04 01	04 41	14 39	16 01	17 11	18 05
54	02 53	03 51	04 33	14 51	16 16	17 27	18 20
56	02 34	03 39	04 25	15 06	16 34	17 47	18 37
58	02 10	03 25	04 16	15 23	16 56	18 11	18 57
S 60	01 36	03 09	04 05	15 44	17 25	18 42	19 23

Lat. °	Sunset h m	Twilight Civil h m	Twilight Naut. h m	Moonset 3 h m	Moonset 4 h m	Moonset 5 h m	Moonset 6 h m
N 72	14 26	16 05	17 32	□	□	□	□
N 70	15 01	16 21	17 38	□	□	□	□
68	15 25	16 34	17 43	□	□	□	□
66	15 44	16 45	17 48	04 15	□	□	□
64	15 59	16 54	17 52	03 36	05 58	□	□
62	16 12	17 02	17 56	03 09	05 04	06 47	07 41
60	16 22	17 09	17 59	02 48	04 31	06 02	07 00
N 58	16 32	17 15	18 02	02 31	04 07	05 32	06 32
56	16 40	17 20	18 05	02 17	03 48	05 09	06 11
54	16 47	17 25	18 08	02 05	03 32	04 50	05 52
52	16 54	17 30	18 10	01 54	03 18	04 35	05 37
50	17 00	17 34	18 13	01 44	03 06	04 21	05 24
45	17 12	17 43	18 18	01 24	02 40	03 53	04 56
N 40	17 23	17 51	18 23	01 08	02 21	03 31	04 35
35	17 32	17 58	18 28	00 54	02 04	03 13	04 17
30	17 40	18 05	18 33	00 42	01 50	02 57	04 01
20	17 54	18 17	18 43	00 22	01 25	02 31	03 35
N 10	18 06	18 28	18 53	00 04	01 04	02 08	03 12
0	18 17	18 39	19 04	24 45	00 45	01 46	02 50
S 10	18 29	18 51	19 17	24 25	00 25	01 25	02 29
20	18 42	19 05	19 32	24 05	00 05	01 02	02 06
30	18 56	19 22	19 52	23 41	24 35	00 35	01 39
35	19 05	19 32	20 05	23 27	24 20	00 20	01 23
40	19 14	19 44	20 20	23 10	24 02	00 02	01 05
45	19 26	19 59	20 39	22 51	23 40	24 42	00 42
S 50	19 40	20 17	21 05	22 27	23 12	24 14	00 14
52	19 46	20 26	21 17	22 15	22 58	24 00	00 00
54	19 53	20 36	21 33	22 02	22 43	23 44	25 05
56	20 01	20 47	21 51	21 47	22 24	23 24	24 49
58	20 11	21 00	22 14	21 30	22 02	23 01	24 29
S 60	20 21	21 16	22 46	21 08	21 33	22 29	24 03

Day	SUN Eqn. of Time 00^h	SUN Eqn. of Time 12^h	SUN Mer. Pass.	MOON Mer. Pass. Upper	MOON Mer. Pass. Lower	MOON Age	MOON Phase
d	m s	m s	h m	h m	h m	d %	
3	13 48	13 51	12 14	18 31	06 02	08 56	
4	13 54	13 57	12 14	19 31	07 01	09 67	
5	13 59	14 01	12 14	20 35	08 03	10 78	

UT	ARIES	VENUS −4.8		MARS +1.3		JUPITER −1.9		SATURN +0.7		STARS		
	GHA	GHA	Dec	GHA	Dec	GHA	Dec	GHA	Dec	Name	SHA	Dec
d h	° ′	° ′	° ′	° ′	° ′	° ′	° ′	° ′	° ′		° ′	° ′
6 00	136 15.8	135 04.3	N 3 02.2	192 46.8	S20 52.7	186 21.0	S18 52.3	324 00.4	N 5 44.1	Acamar	315 20.7	S40 16.
01	151 18.3	150 05.1	03.3	207 47.2	52.3	201 22.8	52.1	339 03.0	44.2	Achernar	335 29.1	S57 11.
02	166 20.7	165 05.9	04.4	222 47.6	51.9	216 24.7	52.0	354 05.6	44.3	Acrux	173 12.8	S63 08.
03	181 23.2	180 06.7	05.5	237 48.1	51.5	231 26.6	51.8	9 08.2	44.3	Adhara	255 14.8	S28 59.
04	196 25.7	195 07.5	06.6	252 48.5	51.1	246 28.4	51.7	24 10.8	44.4	Aldebaran	290 53.0	N16 31.
05	211 28.1	210 08.3	07.8	267 49.0	50.7	261 30.3	51.5	39 13.4	44.5			
06	226 30.6	225 09.1	N 3 08.9	282 49.4	S20 50.2	276 32.2	S18 51.4	54 16.0	N 5 44.5	Alioth	166 22.9	N55 54.
07	241 33.1	240 09.9	10.0	297 49.9	49.8	291 34.0	51.3	69 18.6	44.6	Alkaid	153 01.1	N49 15.
08	256 35.5	255 10.8	11.1	312 50.3	49.4	306 35.9	51.1	84 21.2	44.7	Al Na'ir	27 48.1	S46 55.
F 09	271 38.0	270 11.6	12.2	327 50.7	49.0	321 37.7	51.0	99 23.8	44.7	Alnilam	275 49.4	S 1 11.
R 10	286 40.4	285 12.4	13.3	342 51.2	48.6	336 39.6	50.8	114 26.4	44.8	Alphard	217 58.9	S 8 42.
I 11	301 42.9	300 13.2	14.5	357 51.6	48.2	351 41.5	50.7	129 29.0	44.9			
D 12	316 45.4	315 14.0	N 3 15.6	12 52.1	S20 47.8	6 43.3	S18 50.5	144 31.6	N 5 44.9	Alphecca	126 13.7	N26 40.
A 13	331 47.8	330 14.8	16.7	27 52.5	47.4	21 45.2	50.4	159 34.2	45.0	Alpheratz	357 47.2	N29 08.
Y 14	346 50.3	345 15.7	17.8	42 52.9	46.9	36 47.1	50.2	174 36.8	45.1	Altair	62 11.7	N 8 53.
15	1 52.8	0 16.5	18.9	57 53.4	46.5	51 48.9	50.1	189 39.4	45.1	Ankaa	353 19.0	S42 15.
16	16 55.2	15 17.3	20.0	72 53.8	46.1	66 50.8	49.9	204 42.0	45.2	Antares	112 30.4	S26 27.
17	31 57.7	30 18.1	21.1	87 54.3	45.7	81 52.6	49.8	219 44.6	45.3			
18	47 00.2	45 19.0	N 3 22.2	102 54.7	S20 45.3	96 54.5	S18 49.6	234 47.2	N 5 45.3	Arcturus	145 58.5	N19 07.
19	62 02.6	60 19.8	23.4	117 55.2	44.9	111 56.4	49.5	249 49.8	45.4	Atria	107 35.5	S69 02.
20	77 05.1	75 20.6	24.5	132 55.6	44.4	126 58.2	49.4	264 52.4	45.5	Avior	234 18.9	S59 32.
21	92 07.5	90 21.5	25.6	147 56.0	44.0	142 00.1	49.2	279 55.0	45.5	Bellatrix	278 35.3	N 6 21.
22	107 10.0	105 22.3	26.7	162 56.5	43.6	157 02.0	49.1	294 57.6	45.6	Betelgeuse	271 04.6	N 7 24.
23	122 12.5	120 23.1	27.8	177 56.9	43.2	172 03.8	48.9	310 00.2	45.7			
7 00	137 14.9	135 24.0	N 3 28.9	192 57.4	S20 42.8	187 05.7	S18 48.8	325 02.8	N 5 45.7	Canopus	263 57.2	S52 42.
01	152 17.4	150 24.8	30.0	207 57.8	42.4	202 07.6	48.6	340 05.4	45.8	Capella	280 39.0	N46 00.
02	167 19.9	165 25.7	31.1	222 58.3	41.9	217 09.4	48.5	355 08.0	45.9	Deneb	49 34.2	N45 18.
03	182 22.3	180 26.5	32.2	237 58.7	41.5	232 11.3	48.3	10 10.6	45.9	Denebola	182 36.6	N14 31.
04	197 24.8	195 27.4	33.3	252 59.1	41.1	247 13.1	48.2	25 13.2	46.0	Diphda	348 59.3	S17 56.
05	212 27.3	210 28.2	34.4	267 59.6	40.7	262 15.0	48.0	40 15.8	46.1			
06	227 29.7	225 29.1	N 3 35.5	283 00.0	S20 40.3	277 16.9	S18 47.9	55 18.4	N 5 46.1	Dubhe	193 54.7	N61 41.
S 07	242 32.2	240 29.9	36.6	298 00.5	39.8	292 18.7	47.7	70 21.0	46.2	Elnath	278 16.5	N28 37.
A 08	257 34.7	255 30.8	37.7	313 00.9	39.4	307 20.6	47.6	85 23.6	46.3	Eltanin	90 48.0	N51 28.
T 09	272 37.1	270 31.6	38.8	328 01.4	39.0	322 22.5	47.4	100 26.2	46.3	Enif	33 50.6	N 9 55.
U 10	287 39.6	285 32.5	40.0	343 01.8	38.6	337 24.3	47.3	115 28.8	46.4	Fomalhaut	15 27.8	S29 34.
R 11	302 42.0	300 33.3	41.1	358 02.3	38.1	352 26.2	47.2	130 31.4	46.5			
D 12	317 44.5	315 34.2	N 3 42.2	13 02.7	S20 37.7	7 28.1	S18 47.0	145 34.0	N 5 46.6	Gacrux	172 04.4	S57 09.
A 13	332 47.0	330 35.0	43.3	28 03.1	37.3	22 29.9	46.9	160 36.7	46.6	Gienah	175 55.4	S17 35.
Y 14	347 49.4	345 35.9	44.4	43 03.6	36.9	37 31.8	46.7	175 39.3	46.7	Hadar	148 52.6	S60 24.
15	2 51.9	0 36.8	45.5	58 04.0	36.5	52 33.7	46.6	190 41.9	46.8	Hamal	328 04.5	N23 30.
16	17 54.4	15 37.6	46.6	73 04.5	36.0	67 35.5	46.4	205 44.5	46.8	Kaus Aust.	83 48.4	S34 22.
17	32 56.8	30 38.5	47.7	88 04.9	35.6	82 37.4	46.3	220 47.1	46.9			
18	47 59.3	45 39.4	N 3 48.8	103 05.4	S20 35.2	97 39.2	S18 46.1	235 49.7	N 5 47.0	Kochab	137 19.4	N74 06.
19	63 01.8	60 40.2	49.9	118 05.8	34.8	112 41.1	46.0	250 52.3	47.0	Markab	13 41.9	N15 15.
20	78 04.2	75 41.1	51.0	133 06.3	34.3	127 43.0	45.8	265 54.9	47.1	Menkar	314 18.4	N 4 07.
21	93 06.7	90 42.0	52.1	148 06.7	33.9	142 44.8	45.7	280 57.5	47.2	Menkent	148 11.4	S36 24.
22	108 09.2	105 42.9	53.1	163 07.2	33.5	157 46.7	45.5	296 00.1	47.2	Miaplacidus	221 39.7	S69 45.
23	123 11.6	120 43.7	54.2	178 07.6	33.0	172 48.6	45.4	311 02.7	47.3			
8 00	138 14.1	135 44.6	N 3 55.3	193 08.1	S20 32.6	187 50.4	S18 45.2	326 05.3	N 5 47.4	Mirfak	308 45.0	N49 53.
01	153 16.5	150 45.5	56.4	208 08.5	32.2	202 52.3	45.1	341 07.9	47.4	Nunki	76 02.6	S26 17.
02	168 19.0	165 46.4	57.5	223 08.9	31.8	217 54.2	45.0	356 10.5	47.5	Peacock	53 24.7	S56 42.
03	183 21.5	180 47.3	58.6	238 09.4	31.3	232 56.0	44.8	11 13.1	47.6	Pollux	243 31.3	N28 00.
04	198 23.9	195 48.1	3 59.7	253 09.8	30.9	247 57.9	44.7	26 15.7	47.6	Procyon	245 02.8	N 5 12.
05	213 26.4	210 49.0	4 00.8	268 10.3	30.5	262 59.8	44.5	41 18.3	47.7			
06	228 28.9	225 49.9	N 4 01.9	283 10.7	S20 30.0	278 01.6	S18 44.4	56 20.9	N 5 47.8	Rasalhague	96 09.6	N12 33.
07	243 31.3	240 50.8	03.0	298 11.2	29.6	293 03.5	44.2	71 23.5	47.8	Regulus	207 46.6	N11 55.
08	258 33.8	255 51.7	04.1	313 11.6	29.2	308 05.4	44.1	86 26.1	47.9	Rigel	281 15.0	S 8 11.
S 09	273 36.3	270 52.6	05.2	328 12.1	28.7	323 07.2	43.9	101 28.7	48.0	Rigil Kent.	139 56.3	S60 52.
U 10	288 38.7	285 53.5	06.3	343 12.5	28.3	338 09.1	43.8	116 31.3	48.1	Sabik	102 16.4	S15 44.
N 11	303 41.2	300 54.4	07.4	358 13.0	27.9	353 11.0	43.6	131 33.9	48.1			
D 12	318 43.7	315 55.3	N 4 08.5	13 13.4	S20 27.5	8 12.8	S18 43.5	146 36.5	N 5 48.2	Schedar	349 44.8	N56 35.
A 13	333 46.1	330 56.2	09.5	28 13.9	27.0	23 14.7	43.3	161 39.1	48.3	Shaula	96 26.6	S37 06.
Y 14	348 48.6	345 57.1	10.6	43 14.3	26.6	38 16.6	43.2	176 41.7	48.3	Sirius	258 36.3	S16 43.
15	3 51.0	0 58.0	11.7	58 14.8	26.2	53 18.4	43.0	191 44.4	48.4	Spica	158 34.5	S11 12.
16	18 53.5	15 58.9	12.8	73 15.2	25.7	68 20.3	42.9	206 47.0	48.5	Suhail	222 54.5	S43 28.
17	33 56.0	30 59.8	13.9	88 15.7	25.3	83 22.1	42.7	221 49.6	48.5			
18	48 58.4	46 00.7	N 4 15.0	103 16.1	S20 24.9	98 24.0	S18 42.6	236 52.2	N 5 48.6	Vega	80 41.5	N38 47.
19	64 00.9	61 01.6	16.1	118 16.6	24.4	113 25.9	42.5	251 54.8	48.7	Zuben'ubi	137 09.0	S16 04.
20	79 03.4	76 02.5	17.2	133 17.0	24.0	128 27.7	42.3	266 57.4	48.7		SHA	Mer. Pass
21	94 05.8	91 03.5	18.2	148 17.5	23.5	143 29.6	42.2	282 00.0	48.8		° ′	h m
22	109 08.3	106 04.4	19.3	163 17.9	23.1	158 31.5	42.0	297 02.6	48.9	Venus	358 09.0	14 58
23	124 10.8	121 05.3	20.4	178 18.4	22.7	173 33.3	41.9	312 05.2	49.0	Mars	55 42.4	11 08
	h m									Jupiter	49 50.8	11 30
Mer. Pass.	14 48.6	*v* 0.9	*d* 1.1	*v* 0.4	*d* 0.4	*v* 1.9	*d* 0.1	*v* 2.6	*d* 0.1	Saturn	187 47.9	2 19

UT	SUN GHA	SUN Dec	MOON GHA	v	MOON Dec	d	HP
d h	° ′	° ′	° ′	′	° ′	′	′
6 00	176 29.1	S15 38.4	49 03.9	2.4	N26 56.8	1.9	60.2
01	191 29.1	37.7	63 25.3	2.4	26 54.9	2.1	60.3
02	206 29.0	36.9	77 46.7	2.4	26 52.8	2.3	60.3
03	221 29.0	. . 36.1	92 08.1	2.4	26 50.5	2.5	60.3
04	236 28.9	35.4	106 29.5	2.4	26 48.0	2.7	60.3
05	251 28.9	34.6	120 50.9	2.4	26 45.3	2.9	60.3
06	266 28.9	S15 33.8	135 12.3	2.4	N26 42.4	3.1	60.3
07	281 28.8	33.1	149 33.7	2.4	26 39.3	3.3	60.4
08	296 28.8	32.3	163 55.1	2.5	26 36.0	3.5	60.4
09	311 28.7	. . 31.5	178 16.6	2.5	26 32.5	3.7	60.4
10	326 28.7	30.7	192 38.1	2.5	26 28.8	3.9	60.4
11	341 28.7	30.0	206 59.6	2.5	26 24.9	4.1	60.4
12	356 28.6	S15 29.2	221 21.1	2.5	N26 20.8	4.3	60.4
13	11 28.6	28.4	235 42.6	2.6	26 16.5	4.4	60.4
14	26 28.5	27.6	250 04.2	2.6	26 12.1	4.7	60.5
15	41 28.5	. . 26.9	264 25.8	2.7	26 07.4	4.9	60.5
16	56 28.5	26.1	278 47.5	2.7	26 02.5	5.1	60.5
17	71 28.4	25.3	293 09.2	2.7	25 57.4	5.2	60.5
18	86 28.4	S15 24.5	307 30.9	2.8	N25 52.2	5.5	60.5
19	101 28.4	23.7	321 52.7	2.8	25 46.7	5.6	60.5
20	116 28.3	23.0	336 14.5	2.8	25 41.1	5.8	60.5
21	131 28.3	. . 22.2	350 36.3	2.9	25 35.3	6.0	60.5
22	146 28.3	21.4	4 58.2	3.0	25 29.3	6.2	60.5
23	161 28.2	20.6	19 20.2	3.0	25 23.1	6.4	60.6
7 00	176 28.2	S15 19.9	33 42.2	3.1	N25 16.7	6.6	60.6
01	191 28.2	19.1	48 04.3	3.2	25 10.1	6.8	60.6
02	206 28.1	18.3	62 26.5	3.2	25 03.3	6.9	60.6
03	221 28.1	. . 17.5	76 48.7	3.2	24 56.4	7.1	60.6
04	236 28.1	16.7	91 10.9	3.4	24 49.3	7.3	60.6
05	251 28.0	15.9	105 33.3	3.3	24 42.0	7.5	60.6
06	266 28.0	S15 15.2	119 55.6	3.5	N24 34.5	7.7	60.6
07	281 28.0	14.4	134 18.1	3.6	24 26.8	7.8	60.6
08	296 27.9	13.6	148 40.7	3.6	24 19.0	8.1	60.6
09	311 27.9	. . 12.8	163 03.3	3.7	24 10.9	8.1	60.6
10	326 27.9	12.0	177 26.0	3.7	24 02.8	8.4	60.6
11	341 27.8	11.2	191 48.7	3.9	23 54.4	8.5	60.6
12	356 27.8	S15 10.5	206 11.6	3.9	N23 45.9	8.7	60.6
13	11 27.8	09.7	220 34.5	4.0	23 37.2	8.9	60.6
14	26 27.8	08.9	234 57.5	4.1	23 28.3	9.1	60.7
15	41 27.7	. . 08.1	249 20.6	4.1	23 19.2	9.2	60.7
16	56 27.7	07.3	263 43.7	4.3	23 10.0	9.3	60.7
17	71 27.7	06.5	278 07.0	4.3	23 00.7	9.6	60.7
18	86 27.6	S15 05.7	292 30.3	4.5	N22 51.1	9.7	60.7
19	101 27.6	04.9	306 53.8	4.5	22 41.4	9.8	60.7
20	116 27.6	04.2	321 17.3	4.6	22 31.6	10.0	60.7
21	131 27.6	. . 03.4	335 40.9	4.7	22 21.6	10.2	60.7
22	146 27.5	02.6	350 04.6	4.8	22 11.4	10.3	60.7
23	161 27.5	01.8	4 28.4	4.9	22 01.1	10.4	60.7
8 00	176 27.5	S15 01.0	18 52.3	5.0	N21 50.7	10.6	60.7
01	191 27.5	15 00.2	33 16.3	5.1	21 40.1	10.8	60.7
02	206 27.4	14 59.4	47 40.4	5.2	21 29.3	10.9	60.7
03	221 27.4	. . 58.6	62 04.6	5.2	21 18.4	11.0	60.6
04	236 27.4	57.8	76 28.8	5.4	21 07.4	11.2	60.6
05	251 27.4	57.0	90 53.2	5.5	20 56.2	11.4	60.6
06	266 27.3	S14 56.2	105 17.7	5.6	N20 44.8	11.4	60.6
07	281 27.3	55.5	119 42.3	5.6	20 33.4	11.6	60.6
08	296 27.3	54.7	134 06.9	5.8	20 21.8	11.8	60.6
09	311 27.3	. . 53.9	148 31.7	5.9	20 10.0	11.8	60.6
10	326 27.3	53.1	162 56.6	6.0	19 58.2	12.0	60.6
11	341 27.2	52.3	177 21.6	6.0	19 46.2	12.1	60.6
12	356 27.2	S14 51.5	191 46.6	6.2	N19 34.1	12.3	60.6
13	11 27.2	50.7	206 11.8	6.3	19 21.8	12.3	60.6
14	26 27.2	49.9	220 37.1	6.4	19 09.5	12.5	60.6
15	41 27.2	. . 49.1	235 02.5	6.5	18 57.0	12.6	60.6
16	56 27.1	48.3	249 28.0	6.6	18 44.4	12.8	60.6
17	71 27.1	47.5	263 53.6	6.6	18 31.6	12.8	60.6
18	86 27.1	S14 46.7	278 19.2	6.8	N18 18.8	13.0	60.5
19	101 27.1	45.9	292 45.0	6.9	18 05.8	13.0	60.5
20	116 27.1	45.1	307 10.9	7.0	17 52.8	13.2	60.5
21	131 27.0	. . 44.3	321 36.9	7.1	17 39.6	13.3	60.5
22	146 27.0	43.5	336 03.0	7.2	17 26.3	13.4	60.5
23	161 27.0	42.7	350 29.2	7.3	N17 12.9	13.5	60.5
	SD 16.2	*d* 0.8	SD 16.5		16.5		16.5

Lat.	Twilight Naut.	Twilight Civil	Sunrise	Moonrise 6	Moonrise 7	Moonrise 8	Moonrise 9
°	h m	h m	h m	h m	h m	h m	h m
N 72	06 47	08 11	09 42	□	□	□	15 08
N 70	06 41	07 56	09 13	□	□	12 14	15 35
68	06 37	07 45	08 51	□	□	13 19	15 55
66	06 33	07 35	08 34	□	10 46	13 55	16 11
64	06 30	07 27	08 20	□	12 01	14 20	16 24
62	06 26	07 20	08 09	10 38	12 37	14 39	16 35
60	06 24	07 13	07 59	11 19	13 03	14 55	16 44
N 58	06 21	07 08	07 50	11 46	13 24	15 09	16 52
56	06 18	07 03	07 43	12 08	13 40	15 20	16 59
54	06 16	06 58	07 36	12 26	13 55	15 30	17 05
52	06 14	06 54	07 30	12 41	14 07	15 39	17 11
50	06 12	06 50	07 24	12 54	14 18	15 47	17 16
45	06 07	06 42	07 12	13 21	14 41	16 04	17 27
N 40	06 02	06 34	07 02	13 42	14 59	16 18	17 36
35	05 58	06 28	06 54	14 00	15 14	16 30	17 43
30	05 53	06 22	06 46	14 15	15 27	16 40	17 50
20	05 44	06 10	06 33	14 41	15 50	16 57	18 02
N 10	05 35	06 00	06 22	15 03	16 09	17 12	18 12
0	05 24	05 49	06 11	15 24	16 27	17 26	18 21
S 10	05 12	05 38	06 00	15 45	16 45	17 40	18 30
20	04 57	05 25	05 48	16 07	17 04	17 55	18 40
30	04 38	05 08	05 34	16 33	17 26	18 12	18 52
35	04 26	04 59	05 26	16 48	17 39	18 22	18 58
40	04 11	04 47	05 17	17 05	17 54	18 33	19 05
45	03 53	04 33	05 06	17 26	18 11	18 46	19 14
S 50	03 29	04 16	04 52	17 52	18 33	19 02	19 24
52	03 16	04 07	04 46	18 05	18 43	19 09	19 29
54	03 02	03 58	04 39	18 20	18 54	19 18	19 34
56	02 45	03 47	04 32	18 37	19 07	19 27	19 40
58	02 24	03 34	04 23	18 57	19 23	19 37	19 46
S 60	01 55	03 20	04 13	19 23	19 41	19 49	19 53

Lat.	Sunset	Twilight Civil	Twilight Naut.	Moonset 6	Moonset 7	Moonset 8	Moonset 9
°	h m	h m	h m	h m	h m	h m	h m
N 72	14 48	16 19	17 43	□	□	□	09 34
N 70	15 17	16 33	17 49	□	□	10 28	09 05
68	15 38	16 45	17 53	□	□	09 22	08 43
66	15 55	16 55	17 57	□	09 48	08 45	08 25
64	16 09	17 03	18 00	□	08 32	08 19	08 11
62	16 20	17 10	18 03	07 41	07 55	07 58	07 59
60	16 30	17 16	18 06	07 00	07 29	07 42	07 48
N 58	16 39	17 22	18 08	06 32	07 08	07 27	07 39
56	16 46	17 26	18 11	06 11	06 50	07 15	07 31
54	16 53	17 31	18 13	05 52	06 36	07 04	07 24
52	16 59	17 35	18 15	05 37	06 23	06 54	07 17
50	17 05	17 39	18 17	05 24	06 11	06 46	07 11
45	17 17	17 47	18 22	04 56	05 48	06 27	06 59
N 40	17 26	17 55	18 27	04 35	05 28	06 12	06 48
35	17 35	18 01	18 31	04 17	05 12	05 59	06 39
30	17 42	18 07	18 36	04 01	04 58	05 48	06 31
20	17 55	18 18	18 44	03 35	04 35	05 29	06 17
N 10	18 07	18 28	18 54	03 12	04 14	05 12	06 04
0	18 18	18 39	19 04	02 50	03 54	04 56	05 53
S 10	18 29	18 50	19 16	02 29	03 35	04 39	05 41
20	18 40	19 03	19 31	02 06	03 14	04 22	05 28
30	18 54	19 19	19 49	01 39	02 49	04 02	05 14
35	19 02	19 29	20 01	01 23	02 35	03 50	05 05
40	19 11	19 40	20 16	01 05	02 18	03 36	04 55
45	19 22	19 54	20 34	00 42	01 57	03 20	04 44
S 50	19 35	20 11	20 58	00 14	01 32	03 00	04 29
52	19 41	20 20	21 10	00 00	01 19	02 50	04 23
54	19 48	20 29	21 24	25 05	01 05	02 39	04 15
56	19 55	20 40	21 41	24 49	00 49	02 27	04 07
58	20 04	20 52	22 01	24 29	00 29	02 12	03 58
S 60	20 13	21 06	22 29	24 03	00 03	01 55	03 47

Day	SUN Eqn. of Time 00^h	SUN Eqn. of Time 12^h	SUN Mer. Pass.	MOON Mer. Pass. Upper	MOON Mer. Pass. Lower	MOON Age	MOON Phase
d	m s	m s	h m	h m	h m	d %	
6	14 04	14 05	12 14	21 39	09 07	11 87	
7	14 07	14 09	12 14	22 41	10 11	12 94	
8	14 10	14 11	12 14	23 40	11 11	13 98	

UT	ARIES	VENUS −4.8		MARS +1.3		JUPITER −1.9		SATURN +0.7		STARS		
	GHA	GHA	Dec	GHA	Dec	GHA	Dec	GHA	Dec	Name	SHA	Dec
d h	° ′	° ′	° ′	° ′	° ′	° ′	° ′	° ′	° ′		° ′	° ′
9 00 MONDAY	139 13.2	136 06.2	N 4 21.5	193 18.8	S20 22.2	188 35.2	S18 41.7	327 07.8	N 5 49.0	Acamar	315 20.7	S40 16.
01	154 15.7	151 07.1	22.6	208 19.3	21.8	203 37.1	41.6	342 10.4	49.1	Achernar	335 29.2	S57 11.
02	169 18.2	166 08.0	23.7	223 19.7	21.4	218 38.9	41.4	357 13.0	49.2	Acrux	173 12.7	S63 08.
03	184 20.6	181 09.0	. . 24.7	238 20.2	. . 20.9	233 40.8	. . 41.3	12 15.6	. . 49.2	Adhara	255 14.8	S28 59.
04	199 23.1	196 09.9	25.8	253 20.6	20.5	248 42.7	41.1	27 18.2	49.3	Aldebaran	290 53.0	N16 31.
05	214 25.5	211 10.8	26.9	268 21.1	20.0	263 44.5	41.0	42 20.8	49.4			
06	229 28.0	226 11.8	N 4 28.0	283 21.5	S20 19.6	278 46.4	S18 40.8	57 23.4	N 5 49.4	Alioth	166 22.9	N55 54.
07	244 30.5	241 12.7	29.1	298 22.0	19.2	293 48.3	40.7	72 26.0	49.5	Alkaid	153 01.0	N49 15.
08	259 32.9	256 13.6	30.1	313 22.4	18.7	308 50.1	40.5	87 28.7	49.6	Al Na'ir	27 48.0	S46 55.
09	274 35.4	271 14.6	. . 31.2	328 22.9	. . 18.3	323 52.0	. . 40.4	102 31.3	. . 49.6	Alnilam	275 49.4	S 1 11.
10	289 37.9	286 15.5	32.3	343 23.3	17.8	338 53.9	40.2	117 33.9	49.7	Alphard	217 58.9	S 8 42.
11	304 40.3	301 16.4	33.4	358 23.8	17.4	353 55.7	40.1	132 36.5	49.8			
12	319 42.8	316 17.4	N 4 34.4	13 24.2	S20 17.0	8 57.6	S18 39.9	147 39.1	N 5 49.9	Alphecca	126 13.7	N26 40.
13	334 45.3	331 18.3	35.5	28 24.7	16.5	23 59.5	39.8	162 41.7	49.9	Alpheratz	357 47.2	N29 08.
14	349 47.7	346 19.3	36.6	43 25.1	16.1	39 01.3	39.7	177 44.3	50.0	Altair	62 11.7	N 8 53.
15	4 50.2	1 20.2	. . 37.7	58 25.6	. . 15.6	54 03.2	. . 39.5	192 46.9	. . 50.1	Ankaa	353 19.0	S42 15.
16	19 52.6	16 21.1	38.7	73 26.0	15.2	69 05.1	39.4	207 49.5	50.1	Antares	112 30.3	S26 27.
17	34 55.1	31 22.1	39.8	88 26.5	14.7	84 06.9	39.2	222 52.1	50.2			
18	49 57.6	46 23.0	N 4 40.9	103 26.9	S20 14.3	99 08.8	S18 39.1	237 54.7	N 5 50.3	Arcturus	145 58.5	N19 07.
19	65 00.0	61 24.0	42.0	118 27.4	13.9	114 10.7	38.9	252 57.3	50.3	Atria	107 35.4	S69 02.
20	80 02.5	76 24.9	43.0	133 27.8	13.4	129 12.5	38.8	268 00.0	50.4	Avior	234 18.9	S59 32.
21	95 05.0	91 25.9	. . 44.1	148 28.3	. . 13.0	144 14.4	. . 38.6	283 02.6	. . 50.5	Bellatrix	278 35.3	N 6 21.
22	110 07.4	106 26.9	45.2	163 28.7	12.5	159 16.3	38.5	298 05.2	50.6	Betelgeuse	271 04.6	N 7 24.
23	125 09.9	121 27.8	46.2	178 29.2	12.1	174 18.1	38.3	313 07.8	50.6			
10 00 TUESDAY	140 12.4	136 28.8	N 4 47.3	193 29.7	S20 11.6	189 20.0	S18 38.2	328 10.4	N 5 50.7	Canopus	263 57.2	S52 42.
01	155 14.8	151 29.7	48.4	208 30.1	11.2	204 21.9	38.0	343 13.0	50.8	Capella	280 39.0	N46 00.
02	170 17.3	166 30.7	49.5	223 30.6	10.7	219 23.7	37.9	358 15.6	50.8	Deneb	49 34.2	N45 18.
03	185 19.8	181 31.7	. . 50.5	238 31.0	. . 10.3	234 25.6	. . 37.7	13 18.2	. . 50.9	Denebola	182 36.6	N14 31.
04	200 22.2	196 32.6	51.6	253 31.5	09.8	249 27.5	37.6	28 20.8	51.0	Diphda	348 59.3	S17 56.
05	215 24.7	211 33.6	52.7	268 31.9	09.4	264 29.3	37.4	43 23.4	51.1			
06	230 27.1	226 34.6	N 4 53.7	283 32.4	S20 08.9	279 31.2	S18 37.3	58 26.0	N 5 51.1	Dubhe	193 54.7	N61 41.
07	245 29.6	241 35.6	54.8	298 32.8	08.5	294 33.1	37.1	73 28.7	51.2	Elnath	278 16.5	N28 37.
08	260 32.1	256 36.5	55.9	313 33.3	08.0	309 35.0	37.0	88 31.3	51.3	Eltanin	90 48.0	N51 28.
09	275 34.5	271 37.5	. . 56.9	328 33.7	. . 07.6	324 36.8	. . 36.8	103 33.9	. . 51.3	Enif	33 50.6	N 9 55.
10	290 37.0	286 38.5	58.0	343 34.2	07.1	339 38.7	36.7	118 36.5	51.4	Fomalhaut	15 27.8	S29 34.
11	305 39.5	301 39.5	4 59.0	358 34.7	06.7	354 40.6	36.5	133 39.1	51.5			
12	320 41.9	316 40.5	N 5 00.1	13 35.1	S20 06.2	9 42.4	S18 36.4	148 41.7	N 5 51.5	Gacrux	172 04.4	S57 09.
13	335 44.4	331 41.4	01.2	28 35.6	05.8	24 44.3	36.3	163 44.3	51.6	Gienah	175 55.4	S17 35.
14	350 46.9	346 42.4	02.2	43 36.0	05.3	39 46.2	36.1	178 46.9	51.7	Hadar	148 52.5	S60 24.
15	5 49.3	1 43.4	. . 03.3	58 36.5	. . 04.9	54 48.0	. . 36.0	193 49.5	. . 51.8	Hamal	328 04.6	N23 30.
16	20 51.8	16 44.4	04.3	73 36.9	04.4	69 49.9	35.8	208 52.1	51.8	Kaus Aust.	83 48.3	S34 22.
17	35 54.3	31 45.4	05.4	88 37.4	04.0	84 51.8	35.7	223 54.8	51.9			
18	50 56.7	46 46.4	N 5 06.5	103 37.8	S20 03.5	99 53.6	S18 35.5	238 57.4	N 5 52.0	Kochab	137 19.3	N74 06.
19	65 59.2	61 47.4	07.5	118 38.3	03.1	114 55.5	35.4	254 00.0	52.0	Markab	13 41.9	N15 15.
20	81 01.6	76 48.4	08.6	133 38.8	02.6	129 57.4	35.2	269 02.6	52.1	Menkar	314 18.4	N 4 07.
21	96 04.1	91 49.4	. . 09.6	148 39.2	. . 02.2	144 59.2	. . 35.1	284 05.2	. . 52.2	Menkent	148 11.4	S36 24.
22	111 06.6	106 50.4	10.7	163 39.7	01.7	160 01.1	34.9	299 07.8	52.3	Miaplacidus	221 39.8	S69 45.
23	126 09.0	121 51.4	11.7	178 40.1	01.3	175 03.0	34.8	314 10.4	52.3			
11 00 WEDNESDAY	141 11.5	136 52.4	N 5 12.8	193 40.6	S20 00.8	190 04.8	S18 34.6	329 13.0	N 5 52.4	Mirfak	308 45.1	N49 53.
01	156 14.0	151 53.4	13.9	208 41.0	20 00.4	205 06.7	34.5	344 15.6	52.5	Nunki	76 02.6	S26 17.
02	171 16.4	166 54.4	14.9	223 41.5	19 59.9	220 08.6	34.3	359 18.3	52.5	Peacock	53 24.7	S56 42.
03	186 18.9	181 55.4	. . 16.0	238 42.0	. . 59.4	235 10.4	. . 34.2	14 20.9	. . 52.6	Pollux	243 31.3	N28 00.
04	201 21.4	196 56.4	17.0	253 42.4	59.0	250 12.3	34.0	29 23.5	52.7	Procyon	245 02.8	N 5 12.
05	216 23.8	211 57.4	18.1	268 42.9	58.5	265 14.2	33.9	44 26.1	52.8			
06	231 26.3	226 58.5	N 5 19.1	283 43.3	S19 58.1	280 16.1	S18 33.7	59 28.7	N 5 52.8	Rasalhague	96 09.6	N12 33.
07	246 28.7	241 59.5	20.2	298 43.8	57.6	295 17.9	33.6	74 31.3	52.9	Regulus	207 46.6	N11 55.
08	261 31.2	257 00.5	21.2	313 44.2	57.1	310 19.8	33.4	89 33.9	53.0	Rigel	281 15.0	S 8 11.
09	276 33.7	272 01.5	. . 22.3	328 44.7	. . 56.7	325 21.7	. . 33.3	104 36.5	. . 53.0	Rigil Kent.	139 56.2	S60 52.
10	291 36.1	287 02.5	23.3	343 45.2	56.2	340 23.5	33.1	119 39.2	53.1	Sabik	102 16.4	S15 44.
11	306 38.6	302 03.6	24.4	358 45.6	55.8	355 25.4	33.0	134 41.8	53.2			
12	321 41.1	317 04.6	N 5 25.4	13 46.1	S19 55.3	10 27.3	S18 32.8	149 44.4	N 5 53.3	Schedar	349 44.8	N56 35.
13	336 43.5	332 05.6	26.5	28 46.5	54.9	25 29.1	32.7	164 47.0	53.3	Shaula	96 26.5	S37 06.
14	351 46.0	347 06.6	27.5	43 47.0	54.4	40 31.0	32.5	179 49.6	53.4	Sirius	258 36.3	S16 43.
15	6 48.5	2 07.7	. . 28.6	58 47.5	. . 53.9	55 32.9	. . 32.4	194 52.2	. . 53.5	Spica	158 34.5	S11 12.
16	21 50.9	17 08.7	29.6	73 47.9	53.5	70 34.7	32.2	209 54.8	53.5	Suhail	222 54.5	S43 28.
17	36 53.4	32 09.7	30.6	88 48.4	53.0	85 36.6	32.1	224 57.4	53.6			
18	51 55.9	47 10.8	N 5 31.7	103 48.8	S19 52.5	100 38.5	S18 31.9	240 00.1	N 5 53.7	Vega	80 41.5	N38 47.
19	66 58.3	62 11.8	32.7	118 49.3	52.1	115 40.4	31.8	255 02.7	53.8	Zuben'ubi	137 09.0	S16 04.
20	82 00.8	77 12.9	33.8	133 49.8	51.6	130 42.2	31.7	270 05.3	53.8		SHA	Mer. Pass.
21	97 03.2	92 13.9	. . 34.8	148 50.2	. . 51.2	145 44.1	. . 31.5	285 07.9	. . 53.9		° ′	h m
22	112 05.7	107 15.0	35.9	163 50.7	50.7	160 46.0	31.4	300 10.5	54.0	Venus	356 16.4	14 53
23	127 08.2	122 16.0	36.9	178 51.1	50.2	175 47.8	31.2	315 13.1	54.0	Mars	53 17.3	11 06
	h m									Jupiter	49 07.6	11 21
Mer. Pass.	14 36.8	*v* 1.0	*d* 1.1	*v* 0.5	*d* 0.5	*v* 1.9	*d* 0.1	*v* 2.6	*d* 0.1	Saturn	187 58.0	2 07

UT	SUN GHA	SUN Dec	MOON GHA	v	MOON Dec	d	HP
d h	° ′	° ′	° ′	′	° ′	′	′
9 00	176 27.0	S14 41.9	4 55.5	7.4	N16 59.4	13.5	60.5
01	191 27.0	41.1	19 21.9	7.6	16 45.9	13.7	60.5
02	206 27.0	40.3	33 48.5	7.6	16 32.2	13.8	60.4
03	221 26.9	. . 39.5	48 15.1	7.7	16 18.4	13.9	60.4
04	236 26.9	38.7	62 41.8	7.8	16 04.5	14.0	60.4
05	251 26.9	37.9	77 08.6	7.9	15 50.5	14.0	60.4
06	266 26.9	S14 37.1	91 35.5	8.0	N15 36.5	14.2	60.4
07	281 26.9	36.3	106 02.5	8.1	15 22.3	14.2	60.4
M 08	296 26.9	35.5	120 29.6	8.2	15 08.1	14.3	60.4
O 09	311 26.8	. . 34.7	134 56.8	8.3	14 53.8	14.4	60.3
N 10	326 26.8	33.9	149 24.1	8.4	14 39.4	14.5	60.3
D 11	341 26.8	33.1	163 51.5	8.5	14 24.9	14.5	60.3
A 12	356 26.8	S14 32.2	178 19.0	8.6	N14 10.4	14.7	60.3
Y 13	11 26.8	31.4	192 46.6	8.7	13 55.7	14.7	60.3
14	26 26.8	30.6	207 14.3	8.8	13 41.0	14.7	60.2
15	41 26.8	. . 29.8	221 42.1	8.9	13 26.3	14.9	60.2
16	56 26.8	29.0	236 10.0	9.0	13 11.4	14.9	60.2
17	71 26.7	28.2	250 38.0	9.0	12 56.5	15.0	60.2
18	86 26.7	S14 27.4	265 06.0	9.2	N12 41.5	15.0	60.2
19	101 26.7	26.6	279 34.2	9.2	12 26.5	15.1	60.1
20	116 26.7	25.8	294 02.4	9.4	12 11.4	15.1	60.1
21	131 26.7	. . 25.0	308 30.8	9.4	11 56.3	15.3	60.1
22	146 26.7	24.2	322 59.2	9.5	11 41.0	15.2	60.1
23	161 26.7	23.4	337 27.7	9.6	11 25.8	15.3	60.1
10 00	176 26.7	S14 22.5	351 56.3	9.7	N11 10.5	15.4	60.0
01	191 26.7	21.7	6 25.0	9.8	10 55.1	15.4	60.0
02	206 26.7	20.9	20 53.8	9.8	10 39.7	15.5	60.0
03	221 26.7	. . 20.1	35 22.6	10.0	10 24.2	15.5	60.0
04	236 26.6	19.3	49 51.6	10.0	10 08.7	15.6	59.9
05	251 26.6	18.5	64 20.6	10.1	9 53.1	15.6	59.9
06	266 26.6	S14 17.7	78 49.7	10.2	N 9 37.5	15.6	59.9
07	281 26.6	16.9	93 18.9	10.3	9 21.9	15.7	59.9
T 08	296 26.6	16.0	107 48.2	10.3	9 06.2	15.7	59.8
U 09	311 26.6	. . 15.2	122 17.5	10.4	8 50.5	15.7	59.8
E 10	326 26.6	14.4	136 46.9	10.5	8 34.8	15.8	59.8
S 11	341 26.6	13.6	151 16.4	10.6	8 19.0	15.8	59.8
D 12	356 26.6	S14 12.8	165 46.0	10.7	N 8 03.2	15.9	59.7
A 13	11 26.6	12.0	180 15.7	10.7	7 47.3	15.8	59.7
Y 14	26 26.6	11.1	194 45.4	10.8	7 31.5	15.9	59.7
15	41 26.6	. . 10.3	209 15.2	10.9	7 15.6	15.9	59.6
16	56 26.6	09.5	223 45.1	10.9	6 59.7	15.9	59.6
17	71 26.6	08.7	238 15.0	11.0	6 43.8	16.0	59.6
18	86 26.6	S14 07.9	252 45.0	11.1	N 6 27.8	15.9	59.6
19	101 26.6	07.1	267 15.1	11.1	6 11.9	16.0	59.5
20	116 26.6	06.2	281 45.2	11.2	5 55.9	16.0	59.5
21	131 26.6	. . 05.4	296 15.4	11.3	5 39.9	16.0	59.5
22	146 26.6	04.6	310 45.7	11.3	5 23.9	16.0	59.4
23	161 26.6	03.8	325 16.0	11.4	5 07.9	16.1	59.4
11 00	176 26.6	S14 02.9	339 46.4	11.5	N 4 51.8	16.0	59.4
01	191 26.6	02.1	354 16.9	11.5	4 35.8	16.0	59.3
02	206 26.6	01.3	8 47.4	11.6	4 19.8	16.1	59.3
03	221 26.6	14 00.5	23 18.0	11.6	4 03.7	16.0	59.3
04	236 26.6	13 59.7	37 48.6	11.7	3 47.7	16.1	59.2
05	251 26.6	58.8	52 19.3	11.7	3 31.6	16.0	59.2
06	266 26.6	S13 58.0	66 50.0	11.8	N 3 15.6	16.1	59.2
W 07	281 26.6	57.2	81 20.8	11.9	2 59.5	16.0	59.1
E 08	296 26.6	56.4	95 51.7	11.9	2 43.5	16.1	59.1
D 09	311 26.6	. . 55.5	110 22.6	11.9	2 27.4	16.0	59.1
N 10	326 26.6	54.7	124 53.5	12.0	2 11.4	16.0	59.0
E 11	341 26.6	53.9	139 24.5	12.1	1 55.4	16.0	59.0
S 12	356 26.6	S13 53.1	153 55.6	12.1	N 1 39.4	16.0	59.0
D 13	11 26.6	52.2	168 26.7	12.1	1 23.4	16.0	58.9
A 14	26 26.6	51.4	182 57.8	12.2	1 07.4	16.0	58.9
Y 15	41 26.6	. . 50.6	197 29.0	12.2	0 51.4	15.9	58.9
16	56 26.6	49.8	212 00.2	12.3	0 35.5	16.0	58.8
17	71 26.6	48.9	226 31.5	12.3	0 19.5	15.9	58.8
18	86 26.6	S13 48.1	241 02.8	12.3	N 0 03.6	15.9	58.8
19	101 26.6	47.3	255 34.1	12.4	S 0 12.3	15.9	58.7
20	116 26.6	46.4	270 05.5	12.5	0 28.2	15.8	58.7
21	131 26.6	. . 45.6	284 37.0	12.4	0 44.0	15.9	58.7
22	146 26.6	44.8	299 08.4	12.5	0 59.9	15.8	58.6
23	161 26.6	44.0	313 39.9	12.5	S 1 15.7	15.8	58.6
	SD 16.2	d 0.8	SD 16.4		16.3		16.1

Lat.	Twilight Naut.	Twilight Civil	Sunrise	Moonrise 9	Moonrise 10	Moonrise 11	Moonrise 12
°	h m	h m	h m	h m	h m	h m	h m
N 72	06 35	07 57	09 23	15 08	17 48	20 08	22 23
N 70	06 31	07 44	08 58	15 35	17 58	20 07	22 12
68	06 27	07 34	08 39	15 55	18 06	20 06	22 03
66	06 24	07 26	08 23	16 11	18 13	20 06	21 55
64	06 22	07 18	08 11	16 24	18 18	20 05	21 49
62	06 19	07 12	08 00	16 35	18 23	20 05	21 43
60	06 17	07 06	07 51	16 44	18 27	20 04	21 39
N 58	06 15	07 01	07 43	16 52	18 30	20 04	21 34
56	06 13	06 57	07 36	16 59	18 33	20 04	21 31
54	06 11	06 53	07 30	17 05	18 36	20 03	21 28
52	06 09	06 49	07 24	17 11	18 39	20 03	21 25
50	06 07	06 45	07 19	17 16	18 41	20 03	21 22
45	06 03	06 38	07 08	17 27	18 46	20 02	21 16
N 40	05 59	06 31	06 59	17 36	18 50	20 02	21 11
35	05 55	06 25	06 51	17 43	18 54	20 02	21 07
30	05 51	06 19	06 44	17 50	18 57	20 01	21 04
20	05 43	06 09	06 32	18 02	19 03	20 01	20 57
N 10	05 34	05 59	06 21	18 12	19 08	20 01	20 52
0	05 25	05 49	06 11	18 21	19 12	20 00	20 47
S 10	05 13	05 39	06 00	18 30	19 17	20 00	20 42
20	04 59	05 26	05 49	18 40	19 21	20 00	20 37
30	04 41	05 11	05 36	18 52	19 27	19 59	20 31
35	04 30	05 02	05 29	18 58	19 30	19 59	20 27
40	04 16	04 51	05 20	19 05	19 33	19 59	20 23
45	03 58	04 38	05 10	19 14	19 37	19 59	20 19
S 50	03 36	04 22	04 58	19 24	19 42	19 58	20 14
52	03 24	04 14	04 52	19 29	19 44	19 58	20 11
54	03 11	04 05	04 46	19 34	19 47	19 58	20 09
56	02 55	03 55	04 39	19 40	19 50	19 58	20 06
58	02 36	03 43	04 31	19 46	19 52	19 58	20 03
S 60	02 11	03 30	04 22	19 53	19 56	19 58	19 59

Lat.	Sunset	Twilight Civil	Twilight Naut.	Moonset 9	Moonset 10	Moonset 11	Moonset 12
°	h m	h m	h m	h m	h m	h m	h m
N 72	15 07	16 33	17 56	09 34	08 43	08 08	07 36
N 70	15 32	16 46	17 59	09 05	08 31	08 05	07 41
68	15 51	16 56	18 03	08 43	08 20	08 02	07 45
66	16 06	17 04	18 05	08 25	08 11	08 00	07 48
64	16 19	17 11	18 08	08 11	08 04	07 58	07 51
62	16 29	17 18	18 10	07 59	07 57	07 56	07 54
60	16 38	17 23	18 13	07 48	07 52	07 54	07 56
N 58	16 46	17 28	18 15	07 39	07 47	07 53	07 58
56	16 53	17 33	18 17	07 31	07 42	07 52	08 00
54	16 59	17 37	18 18	07 24	07 38	07 50	08 01
52	17 05	17 40	18 20	07 17	07 35	07 49	08 03
50	17 10	17 44	18 22	07 11	07 31	07 48	08 04
45	17 21	17 51	18 26	06 59	07 24	07 46	08 07
N 40	17 30	17 58	18 30	06 48	07 18	07 45	08 10
35	17 38	18 04	18 34	06 39	07 13	07 43	08 12
30	17 45	18 09	18 38	06 31	07 08	07 42	08 14
20	17 57	18 20	18 46	06 17	07 00	07 39	08 17
N 10	18 08	18 29	18 54	06 04	06 53	07 37	08 20
0	18 18	18 39	19 04	05 53	06 46	07 35	08 22
S 10	18 28	18 50	19 15	05 41	06 39	07 33	08 25
20	18 39	19 02	19 29	05 28	06 31	07 31	08 28
30	18 52	19 17	19 47	05 14	06 22	07 28	08 31
35	18 59	19 26	19 58	05 05	06 17	07 27	08 33
40	19 08	19 37	20 12	04 55	06 12	07 25	08 35
45	19 18	19 49	20 29	04 44	06 05	07 23	08 38
S 50	19 30	20 06	20 51	04 29	05 57	07 20	08 41
52	19 35	20 13	21 02	04 23	05 53	07 19	08 42
54	19 41	20 22	21 15	04 15	05 49	07 18	08 44
56	19 48	20 32	21 30	04 07	05 44	07 17	08 45
58	19 56	20 43	21 49	03 58	05 39	07 15	08 47
S 60	20 05	20 56	22 13	03 47	05 33	07 13	08 49

Day	SUN Eqn. of Time 00^h	SUN Eqn. of Time 12^h	SUN Mer. Pass.	MOON Mer. Pass. Upper	MOON Mer. Pass. Lower	MOON Age	MOON Phase
d	m s	m s	h m	h m	h m	d %	
9	14 12	14 13	12 14	24 33	12 07	14 100	○
10	14 13	14 14	12 14	00 33	12 59	15 99	
11	14 14	14 14	12 14	01 24	13 48	16 95	

UT	ARIES	VENUS −4.8		MARS +1.3		JUPITER −1.9		SATURN +0.7		STARS		
	GHA	GHA	Dec	GHA	Dec	GHA	Dec	GHA	Dec	Name	SHA	Dec
d h	° ′	° ′	° ′	° ′	° ′	° ′	° ′	° ′	° ′		° ′	° ′
12 00	142 10.6	137 17.1	N 5 37.9	193 51.6	S19 49.8	190 49.7	S18 31.1	330 15.7	N 5 54.1	Acamar	315 20.7	S40 16.3
01	157 13.1	152 18.1	39.0	208 52.1	49.3	205 51.6	30.9	345 18.4	54.2	Achernar	335 29.2	S57 11.6
02	172 15.6	167 19.2	40.0	223 52.5	48.8	220 53.4	30.8	0 21.0	54.3	Acrux	173 12.7	S63 08.9
03	187 18.0	182 20.2	. . 41.0	238 53.0	. . 48.4	235 55.3	. . 30.6	15 23.6	. . 54.3	Adhara	255 14.8	S28 59.2
04	202 20.5	197 21.3	42.1	253 53.4	47.9	250 57.2	30.5	30 26.2	54.4	Aldebaran	290 53.0	N16 31.7
05	217 23.0	212 22.3	43.1	268 53.9	47.4	265 59.1	30.3	45 28.8	54.5			
06	232 25.4	227 23.4	N 5 44.2	283 54.4	S19 47.0	281 00.9	S18 30.2	60 31.4	N 5 54.6	Alioth	166 22.9	N55 54.2
07	247 27.9	242 24.5	45.2	298 54.8	46.5	296 02.8	30.0	75 34.0	54.6	Alkaid	153 01.0	N49 15.6
T 08	262 30.3	257 25.5	46.2	313 55.3	46.0	311 04.7	29.9	90 36.7	54.7	Al Na'ir	27 48.0	S46 55.1
H 09	277 32.8	272 26.6	. . 47.3	328 55.7	. . 45.6	326 06.5	. . 29.7	105 39.3	. . 54.8	Alnilam	275 49.5	S 1 11.8
U 10	292 35.3	287 27.7	48.3	343 56.2	45.1	341 08.4	29.6	120 41.9	54.8	Alphard	217 58.9	S 8 42.0
R 11	307 37.7	302 28.7	49.3	358 56.7	44.6	356 10.3	29.4	135 44.5	54.9			
S 12	322 40.2	317 29.8	N 5 50.4	13 57.1	S19 44.2	11 12.2	S18 29.3	150 47.1	N 5 55.0	Alphecca	126 13.7	N26 40.7
D 13	337 42.7	332 30.9	51.4	28 57.6	43.7	26 14.0	29.1	165 49.7	55.1	Alpheratz	357 47.2	N29 08.6
A 14	352 45.1	347 31.9	52.4	43 58.1	43.2	41 15.9	29.0	180 52.4	55.1	Altair	62 11.7	N 8 53.4
Y 15	7 47.6	2 33.0	. . 53.4	58 58.5	. . 42.8	56 17.8	. . 28.8	195 55.0	. . 55.2	Ankaa	353 19.0	S42 15.5
16	22 50.1	17 34.1	54.5	73 59.0	42.3	71 19.6	28.7	210 57.6	55.3	Antares	112 30.3	S26 27.2
17	37 52.5	32 35.2	55.5	88 59.5	41.8	86 21.5	28.5	226 00.2	55.4			
18	52 55.0	47 36.3	N 5 56.5	103 59.9	S19 41.3	101 23.4	S18 28.4	241 02.8	N 5 55.4	Arcturus	145 58.5	N19 07.8
19	67 57.5	62 37.4	57.6	119 00.4	40.9	116 25.2	28.2	256 05.4	55.5	Atria	107 35.4	S69 02.5
20	82 59.9	77 38.4	58.6	134 00.8	40.4	131 27.1	28.1	271 08.0	55.6	Avior	234 18.9	S59 32.5
21	98 02.4	92 39.5	5 59.6	149 01.3	. . 39.9	146 29.0	. . 27.9	286 10.7	. . 55.6	Bellatrix	278 35.3	N 6 21.5
22	113 04.8	107 40.6	6 00.6	164 01.8	39.5	161 30.9	27.8	301 13.3	55.7	Betelgeuse	271 04.6	N 7 24.5
23	128 07.3	122 41.7	01.7	179 02.2	39.0	176 32.7	27.6	316 15.9	55.8			
13 00	143 09.8	137 42.8	N 6 02.7	194 02.7	S19 38.5	191 34.6	S18 27.5	331 18.5	N 5 55.9	Canopus	263 57.2	S52 42.2
01	158 12.2	152 43.9	03.7	209 03.2	38.0	206 36.5	27.3	346 21.1	55.9	Capella	280 39.0	N46 00.6
02	173 14.7	167 45.0	04.7	224 03.6	37.6	221 38.3	27.2	1 23.7	56.0	Deneb	49 34.2	N45 18.7
03	188 17.2	182 46.1	. . 05.8	239 04.1	. . 37.1	236 40.2	. . 27.0	16 26.4	. . 56.1	Denebola	182 36.6	N14 31.0
04	203 19.6	197 47.2	06.8	254 04.6	36.6	251 42.1	26.9	31 29.0	56.2	Diphda	348 59.3	S17 56.3
05	218 22.1	212 48.3	07.8	269 05.0	36.1	266 44.0	26.7	46 31.6	56.2			
06	233 24.6	227 49.4	N 6 08.8	284 05.5	S19 35.7	281 45.8	S18 26.6	61 34.2	N 5 56.3	Dubhe	193 54.7	N61 41.9
07	248 27.0	242 50.5	09.8	299 05.9	35.2	296 47.7	26.4	76 36.8	56.4	Elnath	278 16.5	N28 37.0
08	263 29.5	257 51.7	10.8	314 06.4	34.7	311 49.6	26.3	91 39.4	56.4	Eltanin	90 47.9	N51 28.9
F 09	278 31.9	272 52.8	. . 11.9	329 06.9	. . 34.2	326 51.4	. . 26.1	106 42.1	. . 56.5	Enif	33 50.6	N 9 54.9
R 10	293 34.4	287 53.9	12.9	344 07.3	33.8	341 53.3	26.0	121 44.7	56.6	Fomalhaut	15 27.8	S29 34.5
I 11	308 36.9	302 55.0	13.9	359 07.8	33.3	356 55.2	25.8	136 47.3	56.7			
D 12	323 39.3	317 56.1	N 6 14.9	14 08.3	S19 32.8	11 57.1	S18 25.7	151 49.9	N 5 56.7	Gacrux	172 04.3	S57 09.8
A 13	338 41.8	332 57.2	15.9	29 08.7	32.3	26 58.9	25.5	166 52.5	56.8	Gienah	175 55.4	S17 35.7
Y 14	353 44.3	347 58.4	16.9	44 09.2	31.8	42 00.8	25.4	181 55.2	56.9	Hadar	148 52.5	S60 24.9
15	8 46.7	2 59.5	. . 18.0	59 09.7	. . 31.4	57 02.7	. . 25.2	196 57.8	. . 57.0	Hamal	328 04.6	N23 30.5
16	23 49.2	18 00.6	19.0	74 10.1	30.9	72 04.6	25.1	212 00.4	57.0	Kaus Aust.	83 48.3	S34 22.8
17	38 51.7	33 01.7	20.0	89 10.6	30.4	87 06.4	24.9	227 03.0	57.1			
18	53 54.1	48 02.9	N 6 21.0	104 11.1	S19 29.9	102 08.3	S18 24.8	242 05.6	N 5 57.2	Kochab	137 19.2	N74 06.6
19	68 56.6	63 04.0	22.0	119 11.5	29.4	117 10.2	24.6	257 08.2	57.3	Markab	13 41.9	N15 15.3
20	83 59.1	78 05.2	23.0	134 12.0	29.0	132 12.0	24.5	272 10.9	57.3	Menkar	314 18.5	N 4 07.6
21	99 01.5	93 06.3	. . 24.0	149 12.5	. . 28.5	147 13.9	. . 24.3	287 13.5	. . 57.4	Menkent	148 11.3	S36 24.9
22	114 04.0	108 07.4	25.0	164 12.9	28.0	162 15.8	24.2	302 16.1	57.5	Miaplacidus	221 39.8	S69 45.4
23	129 06.4	123 08.6	26.0	179 13.4	27.5	177 17.7	24.0	317 18.7	57.6			
14 00	144 08.9	138 09.7	N 6 27.0	194 13.9	S19 27.0	192 19.5	S18 23.9	332 21.3	N 5 57.6	Mirfak	308 45.1	N49 53.9
01	159 11.4	153 10.9	28.0	209 14.4	26.6	207 21.4	23.7	347 24.0	57.7	Nunki	76 02.6	S26 17.2
02	174 13.8	168 12.0	29.0	224 14.8	26.1	222 23.3	23.6	2 26.6	57.8	Peacock	53 24.7	S56 42.3
03	189 16.3	183 13.2	. . 30.1	239 15.3	. . 25.6	237 25.2	. . 23.4	17 29.2	. . 57.8	Pollux	243 31.3	N28 00.2
04	204 18.8	198 14.3	31.1	254 15.8	25.1	252 27.0	23.3	32 31.8	57.9	Procyon	245 02.8	N 5 12.0
05	219 21.2	213 15.5	32.1	269 16.2	24.6	267 28.9	23.1	47 34.4	58.0			
06	234 23.7	228 16.6	N 6 33.1	284 16.7	S19 24.1	282 30.8	S18 23.0	62 37.1	N 5 58.1	Rasalhague	96 09.6	N12 33.0
07	249 26.2	243 17.8	34.1	299 17.2	23.7	297 32.6	22.8	77 39.7	58.1	Regulus	207 46.6	N11 55.2
S 08	264 28.6	258 18.9	35.1	314 17.6	23.2	312 34.5	22.7	92 42.3	58.2	Rigel	281 15.0	S 8 11.5
A 09	279 31.1	273 20.1	. . 36.1	329 18.1	. . 22.7	327 36.4	. . 22.5	107 44.9	. . 58.3	Rigil Kent.	139 56.2	S60 52.2
T 10	294 33.6	288 21.3	37.1	344 18.6	22.2	342 38.3	22.4	122 47.5	58.4	Sabik	102 16.4	S15 44.3
U 11	309 36.0	303 22.4	38.1	359 19.0	21.7	357 40.1	22.2	137 50.2	58.4			
R 12	324 38.5	318 23.6	N 6 39.1	14 19.5	S19 21.2	12 42.0	S18 22.1	152 52.8	N 5 58.5	Schedar	349 44.8	N56 35.5
D 13	339 40.9	333 24.8	40.1	29 20.0	20.7	27 43.9	21.9	167 55.4	58.6	Shaula	96 26.5	S37 06.6
A 14	354 43.4	348 25.9	41.0	44 20.4	20.2	42 45.8	21.8	182 58.0	58.7	Sirius	258 36.3	S16 43.8
Y 15	9 45.9	3 27.1	. . 42.0	59 20.9	. . 19.8	57 47.6	. . 21.6	198 00.6	. . 58.7	Spica	158 34.5	S11 12.7
16	24 48.3	18 28.3	43.0	74 21.4	19.3	72 49.5	21.5	213 03.3	58.8	Suhail	222 54.5	S43 28.3
17	39 50.8	33 29.5	44.0	89 21.9	18.8	87 51.4	21.3	228 05.9	58.9			
18	54 53.3	48 30.7	N 6 45.0	104 22.3	S19 18.3	102 53.3	S18 21.2	243 08.5	N 5 59.0	Vega	80 41.5	N38 47.2
19	69 55.7	63 31.8	46.0	119 22.8	17.8	117 55.1	21.0	258 11.1	59.0	Zuben'ubi	137 09.0	S16 04.9
20	84 58.2	78 33.0	47.0	134 23.3	17.3	132 57.0	20.9	273 13.7	59.1		SHA	Mer. Pass.
21	100 00.7	93 34.2	. . 48.0	149 23.7	. . 16.8	147 58.9	. . 20.7	288 16.4	. . 59.2		° ′	h m
22	115 03.1	108 35.4	49.0	164 24.2	16.3	163 00.8	20.6	303 19.0	59.3	Venus	354 33.0	14 48
23	130 05.6	123 36.6	50.0	179 24.7	15.8	178 02.6	20.4	318 21.6	59.3	Mars	50 52.9	11 03
	h m									Jupiter	48 24.8	11 12
Mer. Pass.	14 25.0	v 1.1	d 1.0	v 0.5	d 0.5	v 1.9	d 0.1	v 2.6	d 0.1	Saturn	188 08.7	1 54

UT	SUN GHA	SUN Dec	MOON GHA	v	MOON Dec	d	HP
d h	° ′	° ′	° ′	′	° ′	′	′
12 00	176 26.6	S13 43.1	328 11.4	12.6	S 1 31.5	15.7	58.6
01	191 26.6	42.3	342 43.0	12.6	1 47.2	15.7	58.5
02	206 26.7	41.5	357 14.6	12.6	2 02.9	15.7	58.5
03	221 26.7	. . 40.6	11 46.2	12.7	2 18.6	15.7	58.4
04	236 26.7	39.8	26 17.9	12.7	2 34.3	15.6	58.4
05	251 26.7	39.0	40 49.6	12.7	2 49.9	15.6	58.4
06	266 26.7	S13 38.1	55 21.3	12.7	S 3 05.5	15.6	58.3
07	281 26.7	37.3	69 53.0	12.8	3 21.1	15.5	58.3
08	296 26.7	36.5	84 24.8	12.8	3 36.6	15.5	58.3
09	311 26.7	. . 35.6	98 56.6	12.8	3 52.1	15.4	58.2
10	326 26.7	34.8	113 28.4	12.8	4 07.5	15.4	58.2
11	341 26.7	34.0	128 00.2	12.8	4 22.9	15.4	58.1
12	356 26.7	S13 33.1	142 32.0	12.9	S 4 38.3	15.3	58.1
13	11 26.8	32.3	157 03.9	12.9	4 53.6	15.3	58.1
14	26 26.8	31.5	171 35.8	12.9	5 08.9	15.2	58.0
15	41 26.8	. . 30.6	186 07.7	12.9	5 24.1	15.2	58.0
16	56 26.8	29.8	200 39.6	13.0	5 39.3	15.1	58.0
17	71 26.8	28.9	215 11.6	12.9	5 54.4	15.1	57.9
18	86 26.8	S13 28.1	229 43.5	13.0	S 6 09.5	15.0	57.9
19	101 26.8	27.3	244 15.5	13.0	6 24.5	15.0	57.8
20	116 26.8	26.4	258 47.5	12.9	6 39.5	14.9	57.8
21	131 26.9	. . 25.6	273 19.4	13.0	6 54.4	14.9	57.8
22	146 26.9	24.8	287 51.4	13.0	7 09.3	14.9	57.7
23	161 26.9	23.9	302 23.4	13.1	7 24.2	14.7	57.7
13 00	176 26.9	S13 23.1	316 55.5	13.0	S 7 38.9	14.7	57.7
01	191 26.9	22.2	331 27.5	13.0	7 53.6	14.7	57.6
02	206 26.9	21.4	345 59.5	13.0	8 08.3	14.6	57.6
03	221 26.9	. . 20.5	0 31.5	13.1	8 22.9	14.5	57.5
04	236 27.0	19.7	15 03.6	13.0	8 37.4	14.5	57.5
05	251 27.0	18.9	29 35.6	13.1	8 51.9	14.4	57.5
06	266 27.0	S13 18.0	44 07.7	13.0	S 9 06.3	14.4	57.4
07	281 27.0	17.2	58 39.7	13.0	9 20.7	14.3	57.4
08	296 27.0	16.3	73 11.7	13.1	9 35.0	14.2	57.4
09	311 27.0	. . 15.5	87 43.8	13.0	9 49.2	14.2	57.3
10	326 27.1	14.7	102 15.8	13.1	10 03.4	14.1	57.3
11	341 27.1	13.8	116 47.9	13.0	10 17.5	14.0	57.2
12	356 27.1	S13 13.0	131 19.9	13.0	S10 31.5	14.0	57.2
13	11 27.1	12.1	145 51.9	13.1	10 45.5	13.9	57.2
14	26 27.1	11.3	160 24.0	13.0	10 59.4	13.8	57.1
15	41 27.2	. . 10.4	174 56.0	13.0	11 13.2	13.8	57.1
16	56 27.2	09.6	189 28.0	13.0	11 27.0	13.7	57.1
17	71 27.2	08.7	204 00.0	13.0	11 40.7	13.6	57.0
18	86 27.2	S13 07.9	218 32.0	13.0	S11 54.3	13.5	57.0
19	101 27.2	07.0	233 04.0	12.9	12 07.8	13.5	57.0
20	116 27.3	06.2	247 35.9	13.0	12 21.3	13.4	56.9
21	131 27.3	. . 05.3	262 07.9	13.0	12 34.7	13.3	56.9
22	146 27.3	04.5	276 39.9	12.9	12 48.0	13.2	56.8
23	161 27.3	03.6	291 11.8	12.9	13 01.2	13.2	56.8
14 00	176 27.3	S13 02.8	305 43.7	13.0	S13 14.4	13.1	56.8
01	191 27.4	01.9	320 15.7	12.9	13 27.5	13.0	56.7
02	206 27.4	01.1	334 47.6	12.8	13 40.5	12.9	56.7
03	221 27.4	13 00.3	349 19.4	12.9	13 53.4	12.9	56.7
04	236 27.4	12 59.4	3 51.3	12.9	14 06.3	12.7	56.6
05	251 27.5	58.5	18 23.2	12.8	14 19.0	12.7	56.6
06	266 27.5	S12 57.7	32 55.0	12.8	S14 31.7	12.6	56.6
07	281 27.5	56.8	47 26.8	12.8	14 44.3	12.5	56.5
08	296 27.5	56.0	61 58.6	12.8	14 56.8	12.5	56.5
09	311 27.6	. . 55.1	76 30.4	12.7	15 09.3	12.3	56.5
10	326 27.6	54.3	91 02.1	12.8	15 21.6	12.3	56.4
11	341 27.6	53.4	105 33.9	12.7	15 33.9	12.1	56.4
12	356 27.6	S12 52.6	120 05.6	12.7	S15 46.0	12.1	56.4
13	11 27.7	51.7	134 37.3	12.6	15 58.1	12.0	56.3
14	26 27.7	50.9	149 08.9	12.7	16 10.1	11.9	56.3
15	41 27.7	. . 50.0	163 40.6	12.6	16 22.0	11.8	56.3
16	56 27.7	49.2	178 12.2	12.6	16 33.8	11.8	56.2
17	71 27.8	48.3	192 43.8	12.6	16 45.6	11.6	56.2
18	86 27.8	S12 47.5	207 15.4	12.5	S16 57.2	11.5	56.2
19	101 27.8	46.6	221 46.9	12.5	17 08.7	11.5	56.1
20	116 27.9	45.7	236 18.4	12.5	17 20.2	11.4	56.1
21	131 27.9	. . 44.9	250 49.9	12.5	17 31.6	11.2	56.1
22	146 27.9	44.0	265 21.4	12.4	17 42.8	11.2	56.0
23	161 27.9	43.2	279 52.8	12.5	S17 54.0	11.1	56.0
	SD 16.2	d 0.8	SD 15.8		15.6		15.4

Lat.	Twilight Naut.	Twilight Civil	Sunrise	Moonrise 12	Moonrise 13	Moonrise 14	Moonrise 15
°	h m	h m	h m	h m	h m	h m	h m
N 72	06 22	07 43	09 05	22 23	24 50	00 50	■
N 70	06 20	07 32	08 43	22 12	24 22	00 22	03 04
68	06 18	07 23	08 26	22 03	24 01	00 01	02 12
66	06 15	07 16	08 12	21 55	23 45	25 41	01 41
64	06 14	07 10	08 01	21 49	23 32	25 17	01 17
62	06 12	07 04	07 51	21 43	23 21	24 59	00 59
60	06 10	06 59	07 43	21 39	23 11	24 44	00 44
N 58	06 08	06 55	07 36	21 34	23 03	24 31	00 31
56	06 07	06 51	07 30	21 31	22 56	24 20	00 20
54	06 05	06 47	07 24	21 28	22 50	24 11	00 11
52	06 04	06 44	07 19	21 25	22 44	24 02	00 02
50	06 03	06 40	07 14	21 22	22 39	23 55	25 08
45	05 59	06 34	07 04	21 16	22 28	23 38	24 47
N 40	05 56	06 28	06 55	21 11	22 19	23 25	24 30
35	05 52	06 22	06 48	21 07	22 11	23 14	24 16
30	05 49	06 17	06 42	21 04	22 04	23 04	24 03
20	05 42	06 08	06 30	20 57	21 53	22 48	23 42
N 10	05 34	05 59	06 20	20 52	21 42	22 33	23 24
0	05 25	05 50	06 11	20 47	21 33	22 20	23 07
S 10	05 14	05 40	06 01	20 42	21 24	22 06	22 51
20	05 01	05 28	05 51	20 37	21 14	21 52	22 33
30	04 44	05 14	05 39	20 31	21 03	21 36	22 12
35	04 33	05 05	05 32	20 27	20 56	21 27	22 01
40	04 20	04 55	05 24	20 23	20 49	21 16	21 47
45	04 04	04 43	05 14	20 19	20 40	21 04	21 31
S 50	03 42	04 27	05 03	20 14	20 30	20 49	21 11
52	03 32	04 20	04 58	20 11	20 26	20 42	21 02
54	03 20	04 12	04 52	20 09	20 21	20 34	20 52
56	03 05	04 03	04 45	20 06	20 15	20 26	20 40
58	02 48	03 52	04 38	20 03	20 09	20 16	20 27
S 60	02 26	03 39	04 30	19 59	20 02	20 05	20 11

Lat.	Sunset	Twilight Civil	Twilight Naut.	Moonset 12	Moonset 13	Moonset 14	Moonset 15
°	h m	h m	h m	h m	h m	h m	h m
N 72	15 25	16 47	18 08	07 36	07 02	06 13	■
N 70	15 47	16 58	18 10	07 41	07 16	06 43	05 40
68	16 04	17 06	18 12	07 45	07 27	07 06	06 33
66	16 17	17 14	18 14	07 48	07 37	07 24	07 06
64	16 28	17 20	18 16	07 51	07 45	07 38	07 30
62	16 38	17 26	18 18	07 54	07 52	07 50	07 49
60	16 46	17 30	18 19	07 56	07 58	08 01	08 05
N 58	16 53	17 35	18 21	07 58	08 03	08 10	08 19
56	17 00	17 39	18 22	08 00	08 08	08 18	08 30
54	17 05	17 42	18 24	08 01	08 13	08 25	08 41
52	17 10	17 46	18 25	08 03	08 17	08 32	08 50
50	17 15	17 49	18 27	08 04	08 20	08 38	08 58
45	17 25	17 55	18 30	08 07	08 28	08 50	09 15
N 40	17 34	18 01	18 33	08 10	08 35	09 01	09 30
35	17 41	18 07	18 36	08 12	08 40	09 10	09 42
30	17 47	18 12	18 40	08 14	08 45	09 18	09 53
20	17 58	18 21	18 47	08 17	08 54	09 32	10 11
N 10	18 08	18 30	18 55	08 20	09 02	09 44	10 27
0	18 18	18 39	19 03	08 22	09 09	09 55	10 42
S 10	18 27	18 49	19 14	08 25	09 16	10 07	10 58
20	18 37	19 00	19 27	08 28	09 24	10 19	11 14
30	18 49	19 14	19 44	08 31	09 33	10 33	11 33
35	18 56	19 23	19 54	08 33	09 38	10 41	11 44
40	19 04	19 33	20 07	08 35	09 44	10 51	11 56
45	19 13	19 45	20 23	08 38	09 51	11 02	12 11
S 50	19 24	20 00	20 44	08 41	09 59	11 15	12 30
52	19 29	20 07	20 55	08 42	10 03	11 22	12 39
54	19 35	20 15	21 07	08 44	10 07	11 29	12 49
56	19 42	20 24	21 20	08 45	10 12	11 36	13 00
58	19 49	20 35	21 37	08 47	10 17	11 45	13 12
S 60	19 57	20 47	21 58	08 49	10 23	11 55	13 28

Day	SUN Eqn. of Time 00^h	SUN Eqn. of Time 12^h	SUN Mer. Pass.	MOON Mer. Pass. Upper	MOON Mer. Pass. Lower	MOON Age	MOON Phase
d	m s	m s	h m	h m	h m	d %	
12	14 13	14 13	12 14	02 11	14 35	17 89	
13	14 12	14 12	12 14	02 58	15 21	18 82	◑
14	14 11	14 09	12 14	03 44	16 07	19 73	

	UT	ARIES	VENUS −4.8		MARS +1.3		JUPITER −1.9		SATURN +0.6	
		GHA	GHA	Dec	GHA	Dec	GHA	Dec	GHA	Dec
	d h	° ′	° ′	° ′	° ′	° ′	° ′	° ′	° ′	° ′
	15 00	145 08.0	138 37.8	N 6 51.0	194 25.2	S19 15.4	193 04.5	S18 20.3	333 24.2	N 5 59.4
	01	160 10.5	153 39.0	52.0	209 25.6	14.9	208 06.4	20.1	348 26.8	59.5
	02	175 13.0	168 40.2	52.9	224 26.1	14.4	223 08.3	20.0	3 29.5	59.6
	03	190 15.4	183 41.4	. . 53.9	239 26.6	. . 13.9	238 10.1	. . 19.8	18 32.1	. . 59.6
	04	205 17.9	198 42.6	54.9	254 27.0	13.4	253 12.0	19.7	33 34.7	59.7
	05	220 20.4	213 43.8	55.9	269 27.5	12.9	268 13.9	19.5	48 37.3	59.8
	06	235 22.8	228 45.0	N 6 56.9	284 28.0	S19 12.4	283 15.8	S18 19.4	63 39.9	N 5 59.9
	07	250 25.3	243 46.2	57.9	299 28.5	11.9	298 17.6	19.2	78 42.6	5 59.9
	08	265 27.8	258 47.4	58.8	314 28.9	11.4	313 19.5	19.1	93 45.2	6 00.0
S	09	280 30.2	273 48.6	6 59.8	329 29.4	. . 10.9	328 21.4	. . 18.9	108 47.8	. . 00.1
U	10	295 32.7	288 49.8	7 00.8	344 29.9	10.4	343 23.3	18.8	123 50.4	00.2
N	11	310 35.2	303 51.1	01.8	359 30.4	09.9	358 25.1	18.6	138 53.1	00.2
D	12	325 37.6	318 52.3	N 7 02.8	14 30.8	S19 09.4	13 27.0	S18 18.5	153 55.7	N 6 00.3
A	13	340 40.1	333 53.5	03.7	29 31.3	08.9	28 28.9	18.3	168 58.3	00.4
Y	14	355 42.5	348 54.7	04.7	44 31.8	08.4	43 30.8	18.2	184 00.9	00.5
	15	10 45.0	3 56.0	. . 05.7	59 32.2	. . 08.0	58 32.6	. . 18.0	199 03.5	. . 00.5
	16	25 47.5	18 57.2	06.7	74 32.7	07.5	73 34.5	17.9	214 06.2	00.6
	17	40 49.9	33 58.4	07.6	89 33.2	07.0	88 36.4	17.7	229 08.8	00.7
	18	55 52.4	48 59.6	N 7 08.6	104 33.7	S19 06.5	103 38.3	S18 17.6	244 11.4	N 6 00.8
	19	70 54.9	64 00.9	09.6	119 34.1	06.0	118 40.1	17.4	259 14.0	00.8
	20	85 57.3	79 02.1	10.6	134 34.6	05.5	133 42.0	17.3	274 16.7	00.9
	21	100 59.8	94 03.4	. . 11.5	149 35.1	. . 05.0	148 43.9	. . 17.1	289 19.3	. . 01.0
	22	116 02.3	109 04.6	12.5	164 35.6	04.5	163 45.8	17.0	304 21.9	01.1
	23	131 04.7	124 05.8	13.5	179 36.0	04.0	178 47.6	16.8	319 24.5	01.1
	16 00	146 07.2	139 07.1	N 7 14.4	194 36.5	S19 03.5	193 49.5	S18 16.7	334 27.2	N 6 01.2
	01	161 09.7	154 08.3	15.4	209 37.0	03.0	208 51.4	16.5	349 29.8	01.3
	02	176 12.1	169 09.6	16.4	224 37.5	02.5	223 53.3	16.4	4 32.4	01.4
	03	191 14.6	184 10.8	. . 17.4	239 37.9	. . 02.0	238 55.1	. . 16.2	19 35.0	. . 01.4
	04	206 17.0	199 12.1	18.3	254 38.4	01.5	253 57.0	16.1	34 37.7	01.5
	05	221 19.5	214 13.3	19.3	269 38.9	01.0	268 58.9	15.9	49 40.3	01.6
	06	236 22.0	229 14.6	N 7 20.2	284 39.4	S19 00.5	284 00.8	S18 15.8	64 42.9	N 6 01.7
	07	251 24.4	244 15.9	21.2	299 39.9	19 00.0	299 02.6	15.6	79 45.5	01.7
	08	266 26.9	259 17.1	22.2	314 40.3	18 59.5	314 04.5	15.5	94 48.1	01.8
M	09	281 29.4	274 18.4	. . 23.1	329 40.8	. . 59.0	329 06.4	. . 15.3	109 50.8	. . 01.9
O	10	296 31.8	289 19.7	24.1	344 41.3	58.5	344 08.3	15.2	124 53.4	02.0
N	11	311 34.3	304 20.9	25.1	359 41.8	58.0	359 10.1	15.0	139 56.0	02.0
D	12	326 36.8	319 22.2	N 7 26.0	14 42.2	S18 57.4	14 12.0	S18 14.9	154 58.6	N 6 02.1
A	13	341 39.2	334 23.5	27.0	29 42.7	56.9	29 13.9	14.7	170 01.3	02.2
Y	14	356 41.7	349 24.8	27.9	44 43.2	56.4	44 15.8	14.6	185 03.9	02.3
	15	11 44.1	4 26.0	. . 28.9	59 43.7	. . 55.9	59 17.7	. . 14.4	200 06.5	. . 02.3
	16	26 46.6	19 27.3	29.8	74 44.1	55.4	74 19.5	14.3	215 09.1	02.4
	17	41 49.1	34 28.6	30.8	89 44.6	54.9	89 21.4	14.1	230 11.8	02.5
	18	56 51.5	49 29.9	N 7 31.8	104 45.1	S18 54.4	104 23.3	S18 14.0	245 14.4	N 6 02.6
	19	71 54.0	64 31.2	32.7	119 45.6	53.9	119 25.2	13.8	260 17.0	02.6
	20	86 56.5	79 32.5	33.7	134 46.1	53.4	134 27.0	13.7	275 19.6	02.7
	21	101 58.9	94 33.8	. . 34.6	149 46.5	. . 52.9	149 28.9	. . 13.5	290 22.3	. . 02.8
	22	117 01.4	109 35.1	35.6	164 47.0	52.4	164 30.8	13.4	305 24.9	02.9
	23	132 03.9	124 36.4	36.5	179 47.5	51.9	179 32.7	13.2	320 27.5	02.9
	17 00	147 06.3	139 37.7	N 7 37.5	194 48.0	S18 51.4	194 34.6	S18 13.1	335 30.1	N 6 03.0
	01	162 08.8	154 39.0	38.4	209 48.5	50.9	209 36.4	12.9	350 32.8	03.1
	02	177 11.3	169 40.3	39.4	224 48.9	50.4	224 38.3	12.8	5 35.4	03.2
	03	192 13.7	184 41.6	. . 40.3	239 49.4	. . 49.8	239 40.2	. . 12.6	20 38.0	. . 03.2
	04	207 16.2	199 42.9	41.2	254 49.9	49.3	254 42.1	12.5	35 40.6	03.3
	05	222 18.6	214 44.2	42.2	269 50.4	48.8	269 43.9	12.3	50 43.3	03.4
	06	237 21.1	229 45.5	N 7 43.1	284 50.9	S18 48.3	284 45.8	S18 12.2	65 45.9	N 6 03.5
	07	252 23.6	244 46.8	44.1	299 51.3	47.8	299 47.7	12.0	80 48.5	03.6
T	08	267 26.0	259 48.1	45.0	314 51.8	47.3	314 49.6	11.9	95 51.2	03.6
U	09	282 28.5	274 49.4	. . 46.0	329 52.3	. . 46.8	329 51.5	. . 11.7	110 53.8	. . 03.7
E	10	297 31.0	289 50.8	46.9	344 52.8	46.3	344 53.3	11.6	125 56.4	03.8
S	11	312 33.4	304 52.1	47.8	359 53.3	45.8	359 55.2	11.4	140 59.0	03.9
D	12	327 35.9	319 53.4	N 7 48.8	14 53.7	S18 45.3	14 57.1	S18 11.3	156 01.7	N 6 03.9
A	13	342 38.4	334 54.8	49.7	29 54.2	44.7	29 59.0	11.1	171 04.3	04.0
Y	14	357 40.8	349 56.1	50.7	44 54.7	44.2	45 00.8	11.0	186 06.9	04.1
	15	12 43.3	4 57.4	. . 51.6	59 55.2	. . 43.7	60 02.7	. . 10.8	201 09.5	. . 04.2
	16	27 45.8	19 58.8	52.5	74 55.7	43.2	75 04.6	10.7	216 12.2	04.2
	17	42 48.2	35 00.1	53.5	89 56.1	42.7	90 06.5	10.5	231 14.8	04.3
	18	57 50.7	50 01.4	N 7 54.4	104 56.6	S18 42.2	105 08.4	S18 10.4	246 17.4	N 6 04.4
	19	72 53.1	65 02.8	55.3	119 57.1	41.7	120 10.2	10.2	261 20.0	04.5
	20	87 55.6	80 04.1	56.3	134 57.6	41.1	135 12.1	10.1	276 22.7	04.5
	21	102 58.1	95 05.5	. . 57.2	149 58.1	. . 40.6	150 14.0	. . 09.9	291 25.3	. . 04.6
	22	118 00.5	110 06.8	58.1	164 58.6	40.1	165 15.9	09.8	306 27.9	04.7
	23	133 03.0	125 08.2	59.1	179 59.0	39.6	180 17.8	09.6	321 30.6	04.8
	Mer. Pass.	h m 14 13.2	*v* 1.3	*d* 1.0	*v* 0.5	*d* 0.5	*v* 1.9	*d* 0.2	*v* 2.6	*d* 0.1

STARS		
Name	SHA	Dec
	° ′	° ′
Acamar	315 20.7	S40 16.
Achernar	335 29.2	S57 11.
Acrux	173 12.7	S63 08.
Adhara	255 14.8	S28 59.
Aldebaran	290 53.0	N16 31.
Alioth	166 22.9	N55 54.
Alkaid	153 01.0	N49 15.
Al Na'ir	27 48.0	S46 55.
Alnilam	275 49.5	S 1 11.
Alphard	217 58.9	S 8 42.
Alphecca	126 13.7	N26 40.
Alpheratz	357 47.2	N29 08.
Altair	62 11.7	N 8 53.
Ankaa	353 19.0	S42 15.
Antares	112 30.3	S26 27.
Arcturus	145 58.5	N19 07.
Atria	107 35.3	S69 02.
Avior	234 18.9	S59 32.
Bellatrix	278 35.3	N 6 21.
Betelgeuse	271 04.6	N 7 24.
Canopus	263 57.2	S52 42.
Capella	280 39.0	N46 00.
Deneb	49 34.2	N45 18.
Denebola	182 36.6	N14 31.
Diphda	348 59.3	S17 56.
Dubhe	193 54.7	N61 41.
Elnath	278 16.5	N28 37.
Eltanin	90 47.9	N51 28.
Enif	33 50.6	N 9 54.
Fomalhaut	15 27.8	S29 34.
Gacrux	172 04.3	S57 09.
Gienah	175 55.4	S17 35.
Hadar	148 52.5	S60 24.
Hamal	328 04.6	N23 30.
Kaus Aust.	83 48.3	S34 22.
Kochab	137 19.2	N74 06.
Markab	13 41.9	N15 15.
Menkar	314 18.5	N 4 07.
Menkent	148 11.3	S36 24.
Miaplacidus	221 39.8	S69 45.
Mirfak	308 45.1	N49 53.
Nunki	76 02.5	S26 17.
Peacock	53 24.7	S56 42.
Pollux	243 31.3	N28 00.
Procyon	245 02.8	N 5 12.
Rasalhague	96 09.6	N12 33.
Regulus	207 46.6	N11 55.
Rigel	281 15.0	S 8 11.
Rigil Kent.	139 56.2	S60 52.
Sabik	102 16.4	S15 44.
Schedar	349 44.9	N56 35.
Shaula	96 26.5	S37 06.
Sirius	258 36.3	S16 43.
Spica	158 34.5	S11 12.
Suhail	222 54.5	S43 28.
Vega	80 41.4	N38 47.
Zuben'ubi	137 08.9	S16 04.

	SHA	Mer. Pass
	° ′	h m
Venus	352 59.9	14 42
Mars	48 29.3	11 01
Jupiter	47 42.3	11 03
Saturn	188 20.0	1 42

UT (d h)	SUN GHA	SUN Dec	MOON GHA	v	MOON Dec	d	HP
	° ′	° ′	° ′	′	° ′	′	′
15 00	176 28.0	S12 42.3	294 24.3	12.3	S18 05.1	10.9	56.0
01	191 28.0	41.5	308 55.6	12.4	18 16.0	10.9	55.9
02	206 28.0	40.6	323 27.0	12.3	18 26.9	10.8	55.9
03	221 28.1	. . 39.7	337 58.3	12.3	18 37.7	10.7	55.9
04	236 28.1	38.9	352 29.6	12.3	18 48.4	10.6	55.8
05	251 28.1	38.0	7 00.9	12.2	18 59.0	10.5	55.8
06	266 28.2	S12 37.2	21 32.1	12.2	S19 09.5	10.4	55.8
07	281 28.2	36.3	36 03.3	12.2	19 19.9	10.3	55.7
08	296 28.2	35.4	50 34.5	12.2	19 30.2	10.1	55.7
09	311 28.3	. . 34.6	65 05.7	12.1	19 40.3	10.1	55.7
10	326 28.3	33.7	79 36.8	12.1	19 50.4	10.0	55.7
11	341 28.3	32.9	94 07.9	12.0	20 00.4	9.9	55.6
12	356 28.4	S12 32.0	108 38.9	12.0	S20 10.3	9.8	55.6
13	11 28.4	31.1	123 09.9	12.0	20 20.1	9.6	55.6
14	26 28.4	30.3	137 40.9	12.0	20 29.7	9.6	55.5
15	41 28.5	. . 29.4	152 11.9	11.9	20 39.3	9.5	55.5
16	56 28.5	28.5	166 42.8	11.9	20 48.8	9.3	55.5
17	71 28.5	27.7	181 13.7	11.9	20 58.1	9.3	55.5
18	86 28.6	S12 26.8	195 44.6	11.8	S21 07.4	9.1	55.4
19	101 28.6	25.9	210 15.4	11.8	21 16.5	9.0	55.4
20	116 28.6	25.1	224 46.2	11.7	21 25.5	9.0	55.4
21	131 28.7	. . 24.2	239 16.9	11.8	21 34.5	8.8	55.3
22	146 28.7	23.4	253 47.7	11.6	21 43.3	8.7	55.3
23	161 28.7	22.5	268 18.3	11.7	21 52.0	8.6	55.3
16 00	176 28.8	S12 21.6	282 49.0	11.6	S22 00.6	8.5	55.3
01	191 28.8	20.8	297 19.6	11.6	22 09.1	8.4	55.2
02	206 28.9	19.9	311 50.2	11.6	22 17.5	8.2	55.2
03	221 28.9	. . 19.0	326 20.8	11.5	22 25.7	8.2	55.2
04	236 28.9	18.2	340 51.3	11.5	22 33.9	8.0	55.2
05	251 29.0	17.3	355 21.8	11.5	22 41.9	7.9	55.1
06	266 29.0	S12 16.4	9 52.3	11.4	S22 49.8	7.8	55.1
07	281 29.0	15.5	24 22.7	11.4	22 57.6	7.7	55.1
08	296 29.1	14.7	38 53.1	11.3	23 05.3	7.6	55.1
09	311 29.1	. . 13.8	53 23.4	11.4	23 12.9	7.5	55.0
10	326 29.2	12.9	67 53.8	11.3	23 20.4	7.3	55.0
11	341 29.2	12.1	82 24.1	11.2	23 27.7	7.3	55.0
12	356 29.2	S12 11.2	96 54.3	11.3	S23 35.0	7.1	55.0
13	11 29.3	10.3	111 24.6	11.2	23 42.1	7.0	55.0
14	26 29.3	09.5	125 54.8	11.1	23 49.1	6.9	54.9
15	41 29.4	. . 08.6	140 24.9	11.2	23 56.0	6.7	54.9
16	56 29.4	07.7	154 55.1	11.1	24 02.7	6.7	54.9
17	71 29.5	06.8	169 25.2	11.1	24 09.4	6.5	54.9
18	86 29.5	S12 06.0	183 55.3	11.0	S24 15.9	6.4	54.9
19	101 29.5	05.1	198 25.3	11.0	24 22.3	6.3	54.8
20	116 29.6	04.2	212 55.3	11.0	24 28.6	6.2	54.8
21	131 29.6	. . 03.3	227 25.3	11.0	24 34.8	6.0	54.8
22	146 29.7	02.5	241 55.3	10.9	24 40.8	6.0	54.8
23	161 29.7	01.6	256 25.2	10.9	24 46.8	5.8	54.8
17 00	176 29.8	S12 00.7	270 55.1	10.9	S24 52.6	5.7	54.7
01	191 29.8	11 59.9	285 25.0	10.8	24 58.3	5.5	54.7
02	206 29.8	59.0	299 54.8	10.8	25 03.8	5.5	54.7
03	221 29.9	. . 58.1	314 24.6	10.8	25 09.3	5.3	54.7
04	236 29.9	57.2	328 54.4	10.8	25 14.6	5.2	54.7
05	251 30.0	56.3	343 24.2	10.7	25 19.8	5.0	54.6
06	266 30.0	S11 55.5	357 53.9	10.8	S25 24.8	5.0	54.6
07	281 30.1	54.6	12 23.7	10.6	25 29.8	4.8	54.6
08	296 30.1	53.7	26 53.3	10.7	25 34.6	4.7	54.6
09	311 30.2	. . 52.8	41 23.0	10.7	25 39.3	4.6	54.6
10	326 30.2	52.0	55 52.7	10.6	25 43.9	4.4	54.6
11	341 30.3	51.1	70 22.3	10.6	25 48.3	4.3	54.5
12	356 30.3	S11 50.2	84 51.9	10.6	S25 52.6	4.2	54.5
13	11 30.4	49.3	99 21.5	10.5	25 56.8	4.1	54.5
14	26 30.4	48.4	113 51.0	10.5	26 00.9	4.0	54.5
15	41 30.5	. . 47.6	128 20.5	10.6	26 04.9	3.8	54.5
16	56 30.5	46.7	142 50.1	10.5	26 08.7	3.7	54.5
17	71 30.5	45.8	157 19.6	10.4	26 12.4	3.5	54.5
18	86 30.6	S11 44.9	171 49.0	10.5	S26 15.9	3.5	54.4
19	101 30.6	44.0	186 18.5	10.4	26 19.4	3.3	54.4
20	116 30.7	43.2	200 47.9	10.5	26 22.7	3.1	54.4
21	131 30.7	. . 42.3	215 17.4	10.4	26 25.8	3.1	54.4
22	146 30.8	41.4	229 46.8	10.4	26 28.9	2.9	54.4
23	161 30.8	40.5	244 16.2	10.4	S26 31.8	2.8	54.4
	SD 16.2	*d* 0.9	SD 15.1		15.0		14.9

Lat.	Twilight Naut.	Twilight Civil	Sunrise	Moonrise 15	Moonrise 16	Moonrise 17	Moonrise 18
°	h m	h m	h m	h m	h m	h m	h m
N 72	06 10	07 29	08 47	■	■	■	■
N 70	06 08	07 20	08 28	03 04	■	■	■
68	06 07	07 12	08 13	02 12	■	■	■
66	06 06	07 06	08 01	01 41	03 55	■	■
64	06 05	07 01	07 51	01 17	03 08	05 16	■
62	06 04	06 56	07 43	00 59	02 38	04 17	05 47
60	06 03	06 51	07 35	00 44	02 16	03 44	05 01
N 58	06 02	06 48	07 29	00 31	01 58	03 19	04 31
56	06 01	06 44	07 23	00 20	01 42	03 00	04 08
54	06 00	06 41	07 18	00 11	01 29	02 44	03 49
52	05 59	06 38	07 13	00 02	01 18	02 30	03 34
50	05 58	06 35	07 09	25 08	01 08	02 17	03 20
45	05 55	06 29	06 59	24 47	00 47	01 52	02 52
N 40	05 52	06 24	06 52	24 30	00 30	01 32	02 30
35	05 49	06 19	06 45	24 16	00 16	01 16	02 12
30	05 46	06 15	06 39	24 03	00 03	01 01	01 57
20	05 40	06 06	06 29	23 42	24 37	00 37	01 31
N 10	05 33	05 58	06 19	23 24	24 16	00 16	01 08
0	05 25	05 50	06 11	23 07	23 57	24 47	00 47
S 10	05 15	05 40	06 02	22 51	23 37	24 26	00 26
20	05 03	05 30	05 52	22 33	23 17	24 04	00 04
30	04 47	05 16	05 41	22 12	22 53	23 38	24 27
35	04 37	05 08	05 35	22 01	22 39	23 22	24 11
40	04 24	04 59	05 27	21 47	22 23	23 05	23 53
45	04 09	04 47	05 19	21 31	22 04	22 43	23 31
S 50	03 49	04 33	05 08	21 11	21 40	22 16	23 03
52	03 39	04 26	05 04	21 02	21 28	22 03	22 49
54	03 28	04 19	04 58	20 52	21 15	21 48	22 33
56	03 15	04 10	04 52	20 40	21 01	21 31	22 15
58	02 59	04 00	04 46	20 27	20 43	21 10	21 52
S 60	02 40	03 49	04 38	20 11	20 22	20 42	21 22

Lat.	Sunset	Twilight Civil	Twilight Naut.	Moonset 15	Moonset 16	Moonset 17	Moonset 18
°	h m	h m	h m	h m	h m	h m	h m
N 72	15 43	17 01	18 21	■	■	■	■
N 70	16 01	17 10	18 22	05 40	■	■	■
68	16 16	17 17	18 23	06 33	■	■	■
66	16 28	17 23	18 24	07 06	06 32	■	■
64	16 38	17 29	18 25	07 30	07 19	06 54	■
62	16 47	17 34	18 25	07 49	07 50	07 54	08 09
60	16 54	17 38	18 26	08 05	08 13	08 27	08 56
N 58	17 00	17 42	18 27	08 19	08 32	08 52	09 26
56	17 06	17 45	18 28	08 30	08 47	09 12	09 49
54	17 11	17 48	18 29	08 41	09 01	09 29	10 07
52	17 16	17 51	18 30	08 50	09 13	09 43	10 23
50	17 20	17 54	18 31	08 58	09 23	09 55	10 37
45	17 29	18 00	18 34	09 15	09 45	10 21	11 05
N 40	17 37	18 05	18 36	09 30	10 03	10 42	11 27
35	17 44	18 10	18 39	09 42	10 18	10 59	11 45
30	17 50	18 14	18 42	09 53	10 31	11 14	12 01
20	18 00	18 22	18 48	10 11	10 53	11 39	12 27
N 10	18 09	18 30	18 55	10 27	11 13	12 01	12 50
0	18 17	18 39	19 03	10 42	11 31	12 21	13 12
S 10	18 26	18 48	19 13	10 58	11 49	12 41	13 33
20	18 36	18 58	19 25	11 14	12 09	13 03	13 56
30	18 46	19 11	19 41	11 33	12 31	13 28	14 22
35	18 53	19 19	19 51	11 44	12 45	13 43	14 38
40	19 00	19 28	20 03	11 56	13 00	14 01	14 56
45	19 08	19 40	20 18	12 11	13 19	14 22	15 18
S 50	19 19	19 54	20 37	12 30	13 42	14 48	15 46
52	19 23	20 00	20 47	12 39	13 53	15 01	16 00
54	19 29	20 08	20 58	12 49	14 05	15 16	16 16
56	19 34	20 16	21 11	13 00	14 20	15 33	16 34
58	19 41	20 26	21 26	13 12	14 37	15 54	16 57
S 60	19 48	20 37	21 45	13 28	14 58	16 21	17 28

Day	SUN Eqn. of Time 00h	SUN Eqn. of Time 12h	SUN Mer. Pass.	MOON Mer. Pass. Upper	MOON Mer. Pass. Lower	MOON Age	MOON Phase
d	m s	m s	h m	h m	h m	d %	
15	14 08	14 07	12 14	04 31	16 55	20 64	
16	14 05	14 03	12 14	05 19	17 44	21 54	◑
17	14 01	13 59	12 14	06 09	18 34	22 44	

UT	ARIES	VENUS −4.8		MARS +1.2		JUPITER −1.9		SATURN +0.6		STARS		
d h	GHA ° ′	GHA ° ′	Dec ° ′	GHA ° ′	Dec ° ′	GHA ° ′	Dec ° ′	GHA ° ′	Dec ° ′	Name	SHA ° ′	Dec ° ′
18 00	148 05.5	140 09.5	N 8 00.0	194 59.5	S18 39.1	195 19.6	S18 09.5	336 33.2	N 6 04.9	Acamar	315 20.8	S40 16.
01	163 07.9	155 10.9	00.9	210 00.0	38.6	210 21.5	09.3	351 35.8	04.9	Achernar	335 29.2	S57 11.
02	178 10.4	170 12.2	01.8	225 00.5	38.0	225 23.4	09.2	6 38.4	05.0	Acrux	173 12.6	S63 09.
03	193 12.9	185 13.6	02.8	240 01.0	37.5	240 25.3	09.0	21 41.1	05.1	Adhara	255 14.8	S28 59.
04	208 15.3	200 15.0	03.7	255 01.5	37.0	255 27.1	08.9	36 43.7	05.2	Aldebaran	290 53.0	N16 31.
05	223 17.8	215 16.3	04.6	270 01.9	36.5	270 29.0	08.7	51 46.3	05.2			
06	238 20.3	230 17.7	N 8 05.5	285 02.4	S18 36.0	285 30.9	S18 08.6	66 49.0	N 6 05.3	Alioth	166 22.8	N55 54.
W 07	253 22.7	245 19.1	06.4	300 02.9	35.5	300 32.8	08.4	81 51.6	05.4	Alkaid	153 01.0	N49 15.
E 08	268 25.2	260 20.4	07.4	315 03.4	34.9	315 34.7	08.3	96 54.2	05.5	Al Na'ir	27 48.0	S46 55.
D 09	283 27.6	275 21.8	08.3	330 03.9	34.4	330 36.5	08.1	111 56.8	05.5	Alnilam	275 49.5	S 1 11.
N 10	298 30.1	290 23.2	09.2	345 04.4	33.9	345 38.4	08.0	126 59.5	05.6	Alphard	217 58.9	S 8 42.
E 11	313 32.6	305 24.6	10.1	0 04.8	33.4	0 40.3	07.8	142 02.1	05.7			
S 12	328 35.0	320 26.0	N 8 11.0	15 05.3	S18 32.9	15 42.2	S18 07.6	157 04.7	N 6 05.8	Alphecca	126 13.6	N26 40.
D 13	343 37.5	335 27.4	12.0	30 05.8	32.3	30 44.1	07.5	172 07.3	05.9	Alpheratz	357 47.2	N29 08.
A 14	358 40.0	350 28.7	12.9	45 06.3	31.8	45 45.9	07.3	187 10.0	05.9	Altair	62 11.6	N 8 53.
Y 15	13 42.4	5 30.1	13.8	60 06.8	31.3	60 47.8	07.2	202 12.6	06.0	Ankaa	353 19.0	S42 15.
16	28 44.9	20 31.5	14.7	75 07.3	30.8	75 49.7	07.0	217 15.2	06.1	Antares	112 30.3	S26 27.
17	43 47.4	35 32.9	15.6	90 07.8	30.2	90 51.6	06.9	232 17.9	06.2			
18	58 49.8	50 34.3	N 8 16.5	105 08.2	S18 29.7	105 53.5	S18 06.7	247 20.5	N 6 06.2	Arcturus	145 58.5	N19 07.
19	73 52.3	65 35.7	17.4	120 08.7	29.2	120 55.4	06.6	262 23.1	06.3	Atria	107 35.2	S69 02.
20	88 54.7	80 37.1	18.3	135 09.2	28.7	135 57.2	06.4	277 25.8	06.4	Avior	234 18.9	S59 32.
21	103 57.2	95 38.5	19.3	150 09.7	28.2	150 59.1	06.3	292 28.4	06.5	Bellatrix	278 35.3	N 6 21.
22	118 59.7	110 39.9	20.2	165 10.2	27.6	166 01.0	06.1	307 31.0	06.5	Betelgeuse	271 04.6	N 7 24.
23	134 02.1	125 41.3	21.1	180 10.7	27.1	181 02.9	06.0	322 33.6	06.6			
19 00	149 04.6	140 42.8	N 8 22.0	195 11.2	S18 26.6	196 04.8	S18 05.8	337 36.3	N 6 06.7	Canopus	263 57.3	S52 42.
01	164 07.1	155 44.2	22.9	210 11.6	26.1	211 06.6	05.7	352 38.9	06.8	Capella	280 39.0	N46 00.
02	179 09.5	170 45.6	23.8	225 12.1	25.5	226 08.5	05.5	7 41.5	06.9	Deneb	49 34.2	N45 18.
03	194 12.0	185 47.0	24.7	240 12.6	25.0	241 10.4	05.4	22 44.2	06.9	Denebola	182 36.6	N14 31.
04	209 14.5	200 48.4	25.6	255 13.1	24.5	256 12.3	05.2	37 46.8	07.0	Diphda	348 59.3	S17 56.
05	224 16.9	215 49.9	26.5	270 13.6	24.0	271 14.2	05.1	52 49.4	07.1			
06	239 19.4	230 51.3	N 8 27.4	285 14.1	S18 23.4	286 16.0	S18 04.9	67 52.0	N 6 07.2	Dubhe	193 54.6	N61 41.
T 07	254 21.9	245 52.7	28.3	300 14.6	22.9	301 17.9	04.8	82 54.7	07.2	Elnath	278 16.5	N28 37.
H 08	269 24.3	260 54.1	29.2	315 15.1	22.4	316 19.8	04.6	97 57.3	07.3	Eltanin	90 47.9	N51 28.
U 09	284 26.8	275 55.6	30.1	330 15.5	21.8	331 21.7	04.5	112 59.9	07.4	Enif	33 50.6	N 9 54.
R 10	299 29.2	290 57.0	31.0	345 16.0	21.3	346 23.6	04.3	128 02.6	07.5	Fomalhaut	15 27.8	S29 34.
S 11	314 31.7	305 58.5	31.9	0 16.5	20.8	1 25.4	04.2	143 05.2	07.6			
D 12	329 34.2	320 59.9	N 8 32.8	15 17.0	S18 20.3	16 27.3	S18 04.0	158 07.8	N 6 07.6	Gacrux	172 04.3	S57 09.9
A 13	344 36.6	336 01.3	33.7	30 17.5	19.7	31 29.2	03.9	173 10.5	07.7	Gienah	175 55.4	S17 35.7
Y 14	359 39.1	351 02.8	34.5	45 18.0	19.2	46 31.1	03.7	188 13.1	07.8	Hadar	148 52.4	S60 24.9
15	14 41.6	6 04.2	35.4	60 18.5	18.7	61 33.0	03.6	203 15.7	07.9	Hamal	328 04.6	N23 30.5
16	29 44.0	21 05.7	36.3	75 19.0	18.1	76 34.9	03.4	218 18.4	07.9	Kaus Aust.	83 48.3	S34 22.8
17	44 46.5	36 07.2	37.2	90 19.5	17.6	91 36.7	03.3	233 21.0	08.0			
18	59 49.0	51 08.6	N 8 38.1	105 19.9	S18 17.1	106 38.6	S18 03.1	248 23.6	N 6 08.1	Kochab	137 19.1	N74 06.6
19	74 51.4	66 10.1	39.0	120 20.4	16.6	121 40.5	03.0	263 26.2	08.2	Markab	13 41.9	N15 15.3
20	89 53.9	81 11.5	39.9	135 20.9	16.0	136 42.4	02.8	278 28.9	08.2	Menkar	314 18.5	N 4 07.5
21	104 56.4	96 13.0	40.8	150 21.4	15.5	151 44.3	02.7	293 31.5	08.3	Menkent	148 11.3	S36 24.9
22	119 58.8	111 14.5	41.6	165 21.9	15.0	166 46.2	02.5	308 34.1	08.4	Miaplacidus	221 39.8	S69 45.4
23	135 01.3	126 15.9	42.5	180 22.4	14.4	181 48.0	02.4	323 36.8	08.5			
20 00	150 03.7	141 17.4	N 8 43.4	195 22.9	S18 13.9	196 49.9	S18 02.2	338 39.4	N 6 08.6	Mirfak	308 45.1	N49 53.9
01	165 06.2	156 18.9	44.3	210 23.4	13.4	211 51.8	02.0	353 42.0	08.6	Nunki	76 02.5	S26 17.2
02	180 08.7	171 20.4	45.2	225 23.9	12.8	226 53.7	01.9	8 44.7	08.7	Peacock	53 24.6	S56 42.3
03	195 11.1	186 21.8	46.1	240 24.4	12.3	241 55.6	01.7	23 47.3	08.8	Pollux	243 31.3	N28 00.2
04	210 13.6	201 23.3	46.9	255 24.9	11.8	256 57.5	01.6	38 49.9	08.9	Procyon	245 02.8	N 5 12.0
05	225 16.1	216 24.8	47.8	270 25.3	11.2	271 59.3	01.4	53 52.6	08.9			
06	240 18.5	231 26.3	N 8 48.7	285 25.8	S18 10.7	287 01.2	S18 01.3	68 55.2	N 6 09.0	Rasalhague	96 09.6	N12 32.9
07	255 21.0	246 27.8	49.6	300 26.3	10.2	302 03.1	01.1	83 57.8	09.1	Regulus	207 46.6	N11 55.2
08	270 23.5	261 29.3	50.4	315 26.8	09.6	317 05.0	01.0	99 00.5	09.2	Rigel	281 15.0	S 8 11.5
F 09	285 25.9	276 30.8	51.3	330 27.3	09.1	332 06.9	00.8	114 03.1	09.3	Rigil Kent.	139 56.1	S60 52.2
R 10	300 28.4	291 32.3	52.2	345 27.8	08.5	347 08.8	00.7	129 05.7	09.3	Sabik	102 16.3	S15 44.3
I 11	315 30.9	306 33.8	53.0	0 28.3	08.0	2 10.6	00.5	144 08.4	09.4			
D 12	330 33.3	321 35.3	N 8 53.9	15 28.8	S18 07.5	17 12.5	S18 00.4	159 11.0	N 6 09.5	Schedar	349 44.9	N56 35.5
A 13	345 35.8	336 36.8	54.8	30 29.3	06.9	32 14.4	00.2	174 13.6	09.6	Shaula	96 26.4	S37 06.6
Y 14	0 38.2	351 38.3	55.7	45 29.8	06.4	47 16.3	18 00.1	189 16.2	09.7	Sirius	258 36.3	S16 43.9
15	15 40.7	6 39.8	56.5	60 30.3	05.9	62 18.2	17 59.9	204 18.9	09.7	Spica	158 34.5	S11 12.7
16	30 43.2	21 41.3	57.4	75 30.8	05.3	77 20.1	59.8	219 21.5	09.8	Suhail	222 54.5	S43 28.3
17	45 45.6	36 42.8	58.3	90 31.3	04.8	92 21.9	59.6	234 24.1	09.9			
18	60 48.1	51 44.3	N 8 59.1	105 31.8	S18 04.2	107 23.8	S17 59.5	249 26.8	N 6 10.0	Vega	80 41.4	N38 47.2
19	75 50.6	66 45.9	9 00.0	120 32.2	03.7	122 25.7	59.3	264 29.4	10.0	Zuben'ubi	137 08.9	S16 04.9
20	90 53.0	81 47.4	00.8	135 32.7	03.2	137 27.6	59.2	279 32.0	10.1		SHA	Mer. Pass
21	105 55.5	96 48.9	01.7	150 33.2	02.6	152 29.5	59.0	294 34.7	10.2		° ′	h m
22	120 58.0	111 50.4	02.6	165 33.7	02.1	167 31.4	58.9	309 37.3	10.3	Venus	351 38.2	14 36
23	136 00.4	126 52.0	03.4	180 34.2	01.5	182 33.2	58.7	324 39.9	10.4	Mars	46 06.6	10 59
	h m									Jupiter	47 00.1	10 54
Mer. Pass.	14 01.4	v 1.4	d 0.9	v 0.5	d 0.5	v 1.9	d 0.2	v 2.6	d 0.1	Saturn	188 31.7	1 29

UT	SUN GHA	SUN Dec	MOON GHA	v	MOON Dec	d	HP
d h	° ′	° ′	° ′	′	° ′	′	′
WEDNESDAY 18 00	176 30.9	S11 39.6	258 45.6	10.3	S26 34.6	2.7	54.4
01	191 31.0	38.8	273 14.9	10.4	26 37.3	2.5	54.4
02	206 31.0	37.9	287 44.3	10.3	26 39.8	2.4	54.3
03	221 31.1	. . 37.0	302 13.6	10.4	26 42.2	2.3	54.3
04	236 31.1	36.1	316 43.0	10.3	26 44.5	2.2	54.3
05	251 31.2	35.2	331 12.3	10.3	26 46.7	2.0	54.3
06	266 31.2	S11 34.3	345 41.6	10.3	S26 48.7	1.9	54.3
07	281 31.3	33.5	0 10.9	10.3	26 50.6	1.8	54.3
08	296 31.3	32.6	14 40.2	10.3	26 52.4	1.6	54.3
09	311 31.4	. . 31.7	29 09.5	10.3	26 54.0	1.5	54.3
10	326 31.4	30.8	43 38.8	10.3	26 55.5	1.4	54.3
11	341 31.5	29.9	58 08.1	10.3	26 56.9	1.2	54.3
12	356 31.5	S11 29.0	72 37.4	10.2	S26 58.1	1.1	54.2
13	11 31.6	28.1	87 06.6	10.3	26 59.2	1.0	54.2
14	26 31.6	27.3	101 35.9	10.3	27 00.2	0.9	54.2
15	41 31.7	. . 26.4	116 05.2	10.2	27 01.1	0.7	54.2
16	56 31.8	25.5	130 34.4	10.3	27 01.8	0.6	54.2
17	71 31.8	24.6	145 03.7	10.3	27 02.4	0.5	54.2
18	86 31.9	S11 23.7	159 33.0	10.3	S27 02.9	0.3	54.2
19	101 31.9	22.8	174 02.3	10.2	27 03.2	0.2	54.2
20	116 32.0	21.9	188 31.5	10.3	27 03.4	0.1	54.2
21	131 32.0	. . 21.0	203 00.8	10.3	27 03.5	0.1	54.2
22	146 32.1	20.1	217 30.1	10.3	27 03.4	0.1	54.2
23	161 32.2	19.3	231 59.4	10.3	27 03.3	0.4	54.2
THURSDAY 19 00	176 32.2	S11 18.4	246 28.7	10.3	S27 02.9	0.4	54.2
01	191 32.3	17.5	260 58.0	10.3	27 02.5	0.6	54.2
02	206 32.3	16.6	275 27.3	10.3	27 01.9	0.7	54.2
03	221 32.4	. . 15.7	289 56.6	10.4	27 01.2	0.8	54.2
04	236 32.5	14.8	304 26.0	10.3	27 00.4	1.0	54.1
05	251 32.5	13.9	318 55.3	10.4	26 59.4	1.0	54.1
06	266 32.6	S11 13.0	333 24.7	10.3	S26 58.4	1.3	54.1
07	281 32.6	12.1	347 54.0	10.4	26 57.1	1.3	54.1
08	296 32.7	11.2	2 23.4	10.4	26 55.8	1.5	54.1
09	311 32.8	. . 10.3	16 52.8	10.4	26 54.3	1.6	54.1
10	326 32.8	09.4	31 22.2	10.4	26 52.7	1.7	54.1
11	341 32.9	08.6	45 51.6	10.5	26 51.0	1.9	54.1
12	356 32.9	S11 07.7	60 21.1	10.4	S26 49.1	1.9	54.1
13	11 33.0	06.8	74 50.5	10.5	26 47.2	2.1	54.1
14	26 33.1	05.9	89 20.0	10.5	26 45.1	2.3	54.1
15	41 33.1	. . 05.0	103 49.5	10.5	26 42.8	2.3	54.1
16	56 33.2	04.1	118 19.0	10.6	26 40.5	2.5	54.1
17	71 33.2	03.2	132 48.6	10.5	26 38.0	2.7	54.1
18	86 33.3	S11 02.3	147 18.1	10.6	S26 35.3	2.7	54.1
19	101 33.4	01.4	161 47.7	10.6	26 32.6	2.9	54.1
20	116 33.4	11 00.5	176 17.3	10.7	26 29.7	3.0	54.1
21	131 33.5	10 59.6	190 47.0	10.6	26 26.7	3.1	54.1
22	146 33.6	58.7	205 16.6	10.7	26 23.6	3.2	54.1
23	161 33.6	57.8	219 46.3	10.7	26 20.4	3.4	54.1
FRIDAY 20 00	176 33.7	S10 56.9	234 16.0	10.7	S26 17.0	3.5	54.1
01	191 33.8	56.0	248 45.7	10.8	26 13.5	3.6	54.1
02	206 33.8	55.1	263 15.5	10.8	26 09.9	3.8	54.1
03	221 33.9	. . 54.2	277 45.3	10.8	26 06.1	3.8	54.1
04	236 34.0	53.3	292 15.1	10.9	26 02.3	4.0	54.1
05	251 34.0	52.4	306 45.0	10.9	25 58.3	4.1	54.1
06	266 34.1	S10 51.5	321 14.9	10.9	S25 54.2	4.3	54.1
07	281 34.2	50.6	335 44.8	10.9	25 49.9	4.3	54.2
08	296 34.2	49.7	350 14.7	11.0	25 45.6	4.5	54.2
09	311 34.3	. . 48.8	4 44.7	11.0	25 41.1	4.6	54.2
10	326 34.4	47.9	19 14.7	11.1	25 36.5	4.7	54.2
11	341 34.4	47.0	33 44.8	11.0	25 31.8	4.8	54.2
12	356 34.5	S10 46.1	48 14.8	11.1	S25 27.0	5.0	54.2
13	11 34.6	45.2	62 44.9	11.2	25 22.0	5.1	54.2
14	26 34.6	44.3	77 15.1	11.2	25 16.9	5.1	54.2
15	41 34.7	. . 43.4	91 45.3	11.2	25 11.8	5.4	54.2
16	56 34.8	42.5	106 15.5	11.3	25 06.4	5.4	54.2
17	71 34.8	41.6	120 45.8	11.3	25 01.0	5.5	54.2
18	86 34.9	S10 40.7	135 16.1	11.3	S24 55.5	5.7	54.2
19	101 35.0	39.8	149 46.4	11.4	24 49.8	5.8	54.2
20	116 35.1	38.9	164 16.8	11.4	24 44.0	5.9	54.2
21	131 35.1	. . 38.0	178 47.2	11.5	24 38.1	6.0	54.2
22	146 35.2	37.1	193 17.7	11.5	24 32.1	6.1	54.2
23	161 35.3	36.2	207 48.2	11.5	S24 26.0	6.2	54.2
	SD 16.2	*d* 0.9	SD 14.8		14.7		14.8

Lat.	Twilight Naut.	Twilight Civil	Sunrise	Moonrise 18	Moonrise 19	Moonrise 20	Moonrise 21
°	h m	h m	h m	h m	h m	h m	h m
N 72	05 57	07 15	08 30	▬	▬	▬	▬
N 70	05 57	07 08	08 14	▬	▬	▬	▬
68	05 57	07 01	08 01	▬	▬	▬	▬
66	05 56	06 56	07 50	▬	▬	▬	08 35
64	05 56	06 51	07 41	▬	▬	08 12	07 39
62	05 56	06 47	07 33	05 47	06 43	07 01	07 06
60	05 55	06 44	07 27	05 01	05 55	06 26	06 42
N 58	05 55	06 40	07 21	04 31	05 25	06 00	06 22
56	05 54	06 38	07 16	04 08	05 02	05 40	06 06
54	05 54	06 35	07 11	03 49	04 43	05 23	05 52
52	05 53	06 32	07 07	03 34	04 27	05 08	05 39
50	05 52	06 30	07 03	03 20	04 13	04 56	05 29
45	05 51	06 25	06 55	02 52	03 45	04 29	05 06
N 40	05 49	06 20	06 48	02 30	03 23	04 09	04 48
35	05 46	06 16	06 42	02 12	03 05	03 51	04 32
30	05 44	06 12	06 36	01 57	02 49	03 36	04 19
20	05 39	06 04	06 27	01 31	02 22	03 11	03 56
N 10	05 32	05 57	06 18	01 08	01 59	02 49	03 36
0	05 25	05 49	06 10	00 47	01 38	02 28	03 18
S 10	05 16	05 41	06 02	00 26	01 16	02 08	02 59
20	05 05	05 31	05 54	00 04	00 54	01 46	02 39
30	04 50	05 19	05 44	24 27	00 27	01 20	02 16
35	04 40	05 11	05 38	24 11	00 11	01 05	02 02
40	04 29	05 03	05 31	23 53	24 47	00 47	01 46
45	04 14	04 52	05 23	23 31	24 26	00 26	01 28
S 50	03 56	04 39	05 14	23 03	23 59	25 04	01 04
52	03 47	04 33	05 09	22 49	23 46	24 53	00 53
54	03 36	04 26	05 04	22 33	23 31	24 40	00 40
56	03 24	04 18	04 59	22 15	23 14	24 25	00 25
58	03 10	04 09	04 53	21 52	22 52	24 07	00 07
S 60	02 53	03 58	04 46	21 22	22 25	23 45	25 13

Lat.	Sunset	Twilight Civil	Twilight Naut.	Moonset 18	Moonset 19	Moonset 20	Moonset 21
°	h m	h m	h m	h m	h m	h m	h m
N 72	15 59	17 14	18 33	▬	▬	▬	▬
N 70	16 15	17 22	18 33	▬	▬	▬	▬
68	16 28	17 28	18 33	▬	▬	▬	▬
66	16 39	17 33	18 33	▬	▬	▬	10 34
64	16 48	17 38	18 33	▬	▬	09 15	11 29
62	16 55	17 42	18 33	08 09	08 59	10 25	12 02
60	17 02	17 45	18 34	08 56	09 47	11 00	12 26
N 58	17 08	17 48	18 34	09 26	10 17	11 26	12 45
56	17 13	17 51	18 34	09 49	10 40	11 46	13 01
54	17 17	17 54	18 35	10 07	10 59	12 03	13 14
52	17 22	17 56	18 35	10 23	11 15	12 17	13 26
50	17 25	17 58	18 36	10 37	11 28	12 29	13 37
45	17 34	18 04	18 38	11 05	11 56	12 55	13 58
N 40	17 41	18 08	18 40	11 27	12 18	13 15	14 16
35	17 47	18 12	18 42	11 45	12 36	13 32	14 31
30	17 52	18 16	18 44	12 01	12 52	13 47	14 43
20	18 01	18 23	18 49	12 27	13 18	14 11	15 05
N 10	18 09	18 31	18 55	12 50	13 41	14 33	15 23
0	18 17	18 38	19 03	13 12	14 02	14 52	15 41
S 10	18 25	18 47	19 12	13 33	14 23	15 12	15 58
20	18 34	18 56	19 23	13 56	14 46	15 33	16 16
30	18 44	19 08	19 37	14 22	15 12	15 57	16 37
35	18 49	19 16	19 47	14 38	15 28	16 11	16 50
40	18 56	19 24	19 58	14 56	15 45	16 28	17 04
45	19 04	19 35	20 12	15 18	16 07	16 47	17 20
S 50	19 13	19 47	20 30	15 46	16 34	17 12	17 41
52	19 17	19 54	20 39	16 00	16 47	17 23	17 50
54	19 22	20 00	20 49	16 16	17 02	17 37	18 01
56	19 27	20 08	21 01	16 34	17 20	17 52	18 13
58	19 33	20 17	21 15	16 57	17 42	18 10	18 27
S 60	19 40	20 27	21 32	17 28	18 10	18 32	18 44

Day	SUN Eqn. of Time 00^h	SUN Eqn. of Time 12^h	SUN Mer. Pass.	MOON Mer. Pass. Upper	MOON Mer. Pass. Lower	MOON Age	MOON Phase
d	m s	m s	h m	h m	h m	d %	
18	13 56	13 54	12 14	06 59	19 25	23 35	
19	13 51	13 48	12 14	07 50	20 15	24 26	
20	13 45	13 42	12 14	08 40	21 05	25 19	

UT	ARIES	VENUS −4.8		MARS +1.2		JUPITER −2.0		SATURN +0.6	
	GHA	GHA	Dec	GHA	Dec	GHA	Dec	GHA	Dec
d h	° ′	° ′	° ′	° ′	° ′	° ′	° ′	° ′	° ′
21 00 (SATURDAY)	151 02.9	141 53.5	N 9 04.3	195 34.7	S18 01.0	197 35.1	S17 58.6	339 42.6	N 6 10.4
01	166 05.3	156 55.1	05.1	210 35.2	18 00.5	212 37.0	58.4	354 45.2	10.5
02	181 07.8	171 56.6	06.0	225 35.7	17 59.9	227 38.9	58.3	9 47.8	10.6
03	196 10.3	186 58.1	. . 06.8	240 36.2	. . 59.4	242 40.8	. . 58.1	24 50.5	. . 10.7
04	211 12.7	201 59.7	07.7	255 36.7	58.8	257 42.7	58.0	39 53.1	10.7
05	226 15.2	217 01.2	08.5	270 37.2	58.3	272 44.6	57.8	54 55.7	10.8
06	241 17.7	232 02.8	N 9 09.4	285 37.7	S17 57.8	287 46.4	S17 57.7	69 58.4	N 6 10.9
07	256 20.1	247 04.3	10.2	300 38.2	57.2	302 48.3	57.5	85 01.0	11.0
08	271 22.6	262 05.9	11.1	315 38.7	56.7	317 50.2	57.3	100 03.6	11.1
09	286 25.1	277 07.4	. . 11.9	330 39.2	. . 56.1	332 52.1	. . 57.2	115 06.3	. . 11.1
10	301 27.5	292 09.0	12.8	345 39.7	55.6	347 54.0	57.0	130 08.9	11.2
11	316 30.0	307 10.6	13.6	0 40.2	55.0	2 55.9	56.9	145 11.5	11.3
12	331 32.5	322 12.1	N 9 14.5	15 40.7	S17 54.5	17 57.7	S17 56.7	160 14.2	N 6 11.4
13	346 34.9	337 13.7	15.3	30 41.2	53.9	32 59.6	56.6	175 16.8	11.5
14	1 37.4	352 15.3	16.2	45 41.7	53.4	48 01.5	56.4	190 19.4	11.5
15	16 39.8	7 16.8	. . 17.0	60 42.2	. . 52.9	63 03.4	. . 56.3	205 22.1	. . 11.6
16	31 42.3	22 18.4	17.8	75 42.7	52.3	78 05.3	56.1	220 24.7	11.7
17	46 44.8	37 20.0	18.7	90 43.2	51.8	93 07.2	56.0	235 27.4	11.8
18	61 47.2	52 21.6	N 9 19.5	105 43.7	S17 51.2	108 09.1	S17 55.8	250 30.0	N 6 11.8
19	76 49.7	67 23.2	20.4	120 44.2	50.7	123 10.9	55.7	265 32.6	11.9
20	91 52.2	82 24.8	21.2	135 44.6	50.1	138 12.8	55.5	280 35.3	12.0
21	106 54.6	97 26.4	. . 22.0	150 45.1	. . 49.6	153 14.7	. . 55.4	295 37.9	. . 12.1
22	121 57.1	112 27.9	22.9	165 45.6	49.0	168 16.6	55.2	310 40.5	12.2
23	136 59.6	127 29.5	23.7	180 46.1	48.5	183 18.5	55.1	325 43.2	12.2
22 00 (SUNDAY)	152 02.0	142 31.1	N 9 24.5	195 46.6	S17 47.9	198 20.4	S17 54.9	340 45.8	N 6 12.3
01	167 04.5	157 32.7	25.4	210 47.1	47.4	213 22.3	54.8	355 48.4	12.4
02	182 07.0	172 34.3	26.2	225 47.6	46.8	228 24.2	54.6	10 51.1	12.5
03	197 09.4	187 36.0	. . 27.0	240 48.1	. . 46.3	243 26.0	. . 54.5	25 53.7	. . 12.6
04	212 11.9	202 37.6	27.8	255 48.6	45.7	258 27.9	54.3	40 56.3	12.6
05	227 14.3	217 39.2	28.7	270 49.1	45.2	273 29.8	54.2	55 59.0	12.7
06	242 16.8	232 40.8	N 9 29.5	285 49.6	S17 44.6	288 31.7	S17 54.0	71 01.6	N 6 12.8
07	257 19.3	247 42.4	30.3	300 50.1	44.1	303 33.6	53.9	86 04.2	12.9
08	272 21.7	262 44.0	31.1	315 50.6	43.5	318 35.5	53.7	101 06.9	13.0
09	287 24.2	277 45.7	. . 32.0	330 51.1	. . 43.0	333 37.4	. . 53.5	116 09.5	. . 13.0
10	302 26.7	292 47.3	32.8	345 51.6	42.4	348 39.2	53.4	131 12.1	13.1
11	317 29.1	307 48.9	33.6	0 52.1	41.9	3 41.1	53.2	146 14.8	13.2
12	332 31.6	322 50.5	N 9 34.4	15 52.6	S17 41.3	18 43.0	S17 53.1	161 17.4	N 6 13.3
13	347 34.1	337 52.2	35.2	30 53.1	40.8	33 44.9	52.9	176 20.1	13.3
14	2 36.5	352 53.8	36.0	45 53.6	40.2	48 46.8	52.8	191 22.7	13.4
15	17 39.0	7 55.5	. . 36.9	60 54.1	. . 39.7	63 48.7	. . 52.6	206 25.3	. . 13.5
16	32 41.5	22 57.1	37.7	75 54.6	39.1	78 50.6	52.5	221 28.0	13.6
17	47 43.9	37 58.7	38.5	90 55.1	38.6	93 52.5	52.3	236 30.6	13.7
18	62 46.4	53 00.4	N 9 39.3	105 55.7	S17 38.0	108 54.4	S17 52.2	251 33.2	N 6 13.7
19	77 48.8	68 02.0	40.1	120 56.2	37.4	123 56.2	52.0	266 35.9	13.8
20	92 51.3	83 03.7	40.9	135 56.7	36.9	138 58.1	51.9	281 38.5	13.9
21	107 53.8	98 05.4	. . 41.7	150 57.2	. . 36.3	154 00.0	. . 51.7	296 41.1	. . 14.0
22	122 56.2	113 07.0	42.5	165 57.7	35.8	169 01.9	51.6	311 43.8	14.1
23	137 58.7	128 08.7	43.3	180 58.2	35.2	184 03.8	51.4	326 46.4	14.1
23 00 (MONDAY)	153 01.2	143 10.3	N 9 44.1	195 58.7	S17 34.7	199 05.7	S17 51.3	341 49.0	N 6 14.2
01	168 03.6	158 12.0	44.9	210 59.2	34.1	214 07.6	51.1	356 51.7	14.3
02	183 06.1	173 13.7	45.7	225 59.7	33.6	229 09.5	51.0	11 54.3	14.4
03	198 08.6	188 15.4	. . 46.5	241 00.2	. . 33.0	244 11.3	. . 50.8	26 57.0	. . 14.5
04	213 11.0	203 17.0	47.3	256 00.7	32.4	259 13.2	50.7	41 59.6	14.5
05	228 13.5	218 18.7	48.1	271 01.2	31.9	274 15.1	50.5	57 02.2	14.6
06	243 15.9	233 20.4	N 9 48.9	286 01.7	S17 31.3	289 17.0	S17 50.4	72 04.9	N 6 14.7
07	258 18.4	248 22.1	49.7	301 02.2	30.8	304 18.9	50.2	87 07.5	14.8
08	273 20.9	263 23.8	50.5	316 02.7	30.2	319 20.8	50.1	102 10.1	14.9
09	288 23.3	278 25.5	. . 51.3	331 03.2	. . 29.7	334 22.7	. . 49.9	117 12.8	. . 14.9
10	303 25.8	293 27.1	52.1	346 03.7	29.1	349 24.6	49.7	132 15.4	15.0
11	318 28.3	308 28.8	52.9	1 04.2	28.5	4 26.5	49.6	147 18.1	15.1
12	333 30.7	323 30.5	N 9 53.7	16 04.7	S17 28.0	19 28.3	S17 49.4	162 20.7	N 6 15.2
13	348 33.2	338 32.2	54.5	31 05.2	27.4	34 30.2	49.3	177 23.3	15.2
14	3 35.7	353 34.0	55.3	46 05.7	26.8	49 32.1	49.1	192 26.0	15.3
15	18 38.1	8 35.7	. . 56.0	61 06.2	. . 26.3	64 34.0	. . 49.0	207 28.6	. . 15.4
16	33 40.6	23 37.4	56.8	76 06.7	25.7	79 35.9	48.8	222 31.2	15.5
17	48 43.1	38 39.1	57.6	91 07.2	25.2	94 37.8	48.7	237 33.9	15.6
18	63 45.5	53 40.8	N 9 58.4	106 07.7	S17 24.6	109 39.7	S17 48.5	252 36.5	N 6 15.6
19	78 48.0	68 42.5	59.2	121 08.3	24.0	124 41.6	48.4	267 39.2	15.7
20	93 50.4	83 44.2	9 59.9	136 08.8	23.5	139 43.5	48.2	282 41.8	15.8
21	108 52.9	98 46.0	10 00.7	151 09.3	. . 22.9	154 45.4	. . 48.1	297 44.4	. . 15.9
22	123 55.4	113 47.7	01.5	166 09.8	22.4	169 47.2	47.9	312 47.1	16.0
23	138 57.8	128 49.4	02.3	181 10.3	21.8	184 49.1	47.8	327 49.7	16.0
Mer. Pass. (h m)	13 49.6	v 1.6	d 0.8	v 0.5	d 0.6	v 1.9	d 0.2	v 2.6	d 0.1

STARS

Name	SHA	Dec
	° ′	° ′
Acamar	315 20.8	S40 16.[illegible]
Achernar	335 29.3	S57 11.[illegible]
Acrux	173 12.6	S63 09.[illegible]
Adhara	255 14.8	S28 59.[illegible]
Aldebaran	290 53.0	N16 31.[illegible]
Alioth	166 22.8	N55 54.[illegible]
Alkaid	153 00.9	N49 15.[illegible]
Al Na'ir	27 48.0	S46 55.[illegible]
Alnilam	275 49.5	S 1 11.[illegible]
Alphard	217 58.9	S 8 42.[illegible]
Alphecca	126 13.6	N26 40.[illegible]
Alpheratz	357 47.2	N29 08.5
Altair	62 11.6	N 8 53.4
Ankaa	353 19.0	S42 15.5
Antares	112 30.2	S26 27.2
Arcturus	145 58.4	N19 07.8
Atria	107 35.2	S69 02.5
Avior	234 18.9	S59 32.5
Bellatrix	278 35.3	N 6 21.5
Betelgeuse	271 04.6	N 7 24.5
Canopus	263 57.3	S52 42.3
Capella	280 39.1	N46 00.6
Deneb	49 34.2	N45 18.6
Denebola	182 36.6	N14 31.0
Diphda	348 59.3	S17 56.[illegible]
Dubhe	193 54.6	N61 41.9
Elnath	278 16.5	N28 37.0
Eltanin	90 47.9	N51 28.9
Enif	33 50.6	N 9 54.9
Fomalhaut	15 27.8	S29 34.5
Gacrux	172 04.3	S57 09.9
Gienah	175 55.4	S17 35.7
Hadar	148 52.4	S60 25.0
Hamal	328 04.6	N23 30.5
Kaus Aust.	83 48.2	S34 22.8
Kochab	137 19.1	N74 06.6
Markab	13 41.9	N15 15.3
Menkar	314 18.5	N 4 07.5
Menkent	148 11.2	S36 24.9
Miaplacidus	221 39.8	S69 45.4
Mirfak	308 45.1	N49 53.9
Nunki	76 02.5	S26 17.2
Peacock	53 24.6	S56 42.3
Pollux	243 31.3	N28 00.2
Procyon	245 02.8	N 5 12.0
Rasalhague	96 09.5	N12 32.9
Regulus	207 46.5	N11 55.2
Rigel	281 15.0	S 8 11.5
Rigil Kent.	139 56.1	S60 52.3
Sabik	102 16.3	S15 44.3
Schedar	349 44.9	N56 35.5
Shaula	96 26.4	S37 06.6
Sirius	258 36.3	S16 43.9
Spica	158 34.4	S11 12.7
Suhail	222 54.5	S43 28.3
Vega	80 41.4	N38 47.2
Zuben'ubi	137 08.9	S16 04.9

	SHA	Mer. Pass
	° ′	h m
Venus	350 29.1	14 28
Mars	43 44.6	10 57
Jupiter	46 18.4	10 45
Saturn	188 43.8	1 17

UT	SUN GHA	SUN Dec	MOON GHA	v	MOON Dec	d	HP
d h	° ′	° ′	° ′	′	° ′	′	′
21 00	176 35.3	S10 35.3	222 18.7	11.6	S24 19.8	6.4	54.2
01	191 35.4	34.4	236 49.3	11.6	24 13.4	6.4	54.3
02	206 35.5	33.5	251 19.9	11.6	24 07.0	6.6	54.3
03	221 35.6	. . 32.6	265 50.5	11.7	24 00.4	6.7	54.3
04	236 35.6	31.7	280 21.2	11.8	23 53.7	6.8	54.3
05	251 35.7	30.8	294 52.0	11.8	23 46.9	6.9	54.3
06	266 35.8	S10 29.9	309 22.8	11.8	S23 40.0	7.0	54.3
07	281 35.8	29.0	323 53.6	11.9	23 33.0	7.2	54.3
08	296 35.9	28.0	338 24.5	11.9	23 25.8	7.2	54.3
09	311 36.0	. . 27.1	352 55.4	12.0	23 18.6	7.3	54.3
10	326 36.1	26.2	7 26.4	12.0	23 11.3	7.5	54.3
11	341 36.1	25.3	21 57.4	12.0	23 03.8	7.6	54.3
12	356 36.2	S10 24.4	36 28.4	12.1	S22 56.2	7.6	54.4
13	11 36.3	23.5	50 59.5	12.1	22 48.6	7.8	54.4
14	26 36.4	22.6	65 30.6	12.2	22 40.8	7.9	54.4
15	41 36.4	. . 21.7	80 01.8	12.2	22 32.9	8.0	54.4
16	56 36.5	20.8	94 33.0	12.3	22 24.9	8.0	54.4
17	71 36.6	19.9	109 04.3	12.3	22 16.9	8.2	54.4
18	86 36.7	S10 19.0	123 35.6	12.4	S22 08.7	8.3	54.4
19	101 36.7	18.1	138 07.0	12.4	22 00.4	8.4	54.4
20	116 36.8	17.1	152 38.4	12.4	21 52.0	8.5	54.4
21	131 36.9	. . 16.2	167 09.8	12.5	21 43.5	8.6	54.5
22	146 37.0	15.3	181 41.3	12.6	21 34.9	8.7	54.5
23	161 37.1	14.4	196 12.9	12.5	21 26.2	8.8	54.5
22 00	176 37.1	S10 13.5	210 44.4	12.7	S21 17.4	8.9	54.5
01	191 37.2	12.6	225 16.1	12.6	21 08.5	9.0	54.5
02	206 37.3	11.7	239 47.7	12.8	20 59.5	9.0	54.5
03	221 37.4	. . 10.8	254 19.5	12.7	20 50.5	9.2	54.5
04	236 37.4	09.9	268 51.2	12.8	20 41.3	9.3	54.5
05	251 37.5	08.9	283 23.0	12.9	20 32.0	9.4	54.6
06	266 37.6	S10 08.0	297 54.9	12.9	S20 22.6	9.4	54.6
07	281 37.7	07.1	312 26.8	12.9	20 13.2	9.6	54.6
08	296 37.8	06.2	326 58.7	13.0	20 03.6	9.6	54.6
09	311 37.8	. . 05.3	341 30.7	13.1	19 54.0	9.8	54.6
10	326 37.9	04.4	356 02.8	13.0	19 44.2	9.8	54.6
11	341 38.0	03.5	10 34.8	13.2	19 34.4	10.0	54.6
12	356 38.1	S10 02.5	25 07.0	13.1	S19 24.4	10.0	54.7
13	11 38.2	01.6	39 39.1	13.2	19 14.4	10.1	54.7
14	26 38.3	10 00.7	54 11.3	13.3	19 04.3	10.2	54.7
15	41 38.3	9 59.8	68 43.6	13.3	18 54.1	10.2	54.7
16	56 38.4	58.9	83 15.9	13.3	18 43.9	10.4	54.7
17	71 38.5	58.0	97 48.2	13.4	18 33.5	10.5	54.7
18	86 38.6	S 9 57.1	112 20.6	13.4	S18 23.0	10.5	54.7
19	101 38.7	56.1	126 53.0	13.5	18 12.5	10.6	54.8
20	116 38.8	55.2	141 25.5	13.5	18 01.9	10.7	54.8
21	131 38.8	. . 54.3	155 58.0	13.5	17 51.2	10.8	54.8
22	146 38.9	53.4	170 30.5	13.6	17 40.4	10.9	54.8
23	161 39.0	52.5	185 03.1	13.6	17 29.5	10.9	54.8
23 00	176 39.1	S 9 51.6	199 35.7	13.7	S17 18.6	11.1	54.8
01	191 39.2	50.6	214 08.4	13.7	17 07.5	11.1	54.9
02	206 39.3	49.7	228 41.1	13.7	16 56.4	11.2	54.9
03	221 39.3	. . 48.8	243 13.8	13.8	16 45.2	11.2	54.9
04	236 39.4	47.9	257 46.6	13.8	16 34.0	11.4	54.9
05	251 39.5	47.0	272 19.4	13.9	16 22.6	11.4	54.9
06	266 39.6	S 9 46.0	286 52.3	13.9	S16 11.2	11.5	54.9
07	281 39.7	45.1	301 25.2	13.9	15 59.7	11.6	55.0
08	296 39.8	44.2	315 58.1	14.0	15 48.1	11.6	55.0
09	311 39.9	. . 43.3	330 31.1	14.0	15 36.5	11.7	55.0
10	326 39.9	42.4	345 04.1	14.0	15 24.8	11.8	55.0
11	341 40.0	41.4	359 37.1	14.1	15 13.0	11.9	55.0
12	356 40.1	S 9 40.5	14 10.2	14.1	S15 01.1	11.9	55.0
13	11 40.2	39.6	28 43.3	14.1	14 49.2	12.0	55.1
14	26 40.3	38.7	43 16.4	14.2	14 37.2	12.1	55.1
15	41 40.4	. . 37.8	57 49.6	14.2	14 25.1	12.2	55.1
16	56 40.5	36.8	72 22.8	14.2	14 12.9	12.2	55.1
17	71 40.6	35.9	86 56.0	14.3	14 00.7	12.2	55.1
18	86 40.7	S 9 35.0	101 29.3	14.3	S13 48.5	12.4	55.1
19	101 40.7	34.1	116 02.6	14.4	13 36.1	12.4	55.2
20	116 40.8	33.2	130 36.0	14.3	13 23.7	12.5	55.2
21	131 40.9	. . 32.2	145 09.3	14.4	13 11.2	12.5	55.2
22	146 41.0	31.3	159 42.7	14.4	12 58.7	12.6	55.2
23	161 41.1	30.4	174 16.1	14.5	S12 46.1	12.6	55.2
	SD 16.2	*d* 0.9	SD	14.8	14.9		15.0

Lat.	Twilight Naut.	Twilight Civil	Sunrise	Moonrise 21	22	23	24
°	h m	h m	h m	h m	h m	h m	h m
N 72	05 43	07 01	08 14	▬	▬	08 58	08 05
N 70	05 44	06 55	08 00	▬	09 48	08 22	07 47
68	05 46	06 50	07 48	▬	08 32	07 56	07 34
66	05 46	06 46	07 39	08 35	07 54	07 36	07 22
64	05 47	06 42	07 31	07 39	07 28	07 19	07 12
62	05 47	06 39	07 24	07 06	07 07	07 06	07 04
60	05 48	06 36	07 18	06 42	06 50	06 54	06 57
N 58	05 48	06 33	07 13	06 22	06 35	06 44	06 51
56	05 48	06 31	07 09	06 06	06 23	06 36	06 45
54	05 47	06 28	07 04	05 52	06 12	06 28	06 40
52	05 47	06 26	07 01	05 39	06 03	06 21	06 35
50	05 47	06 24	06 57	05 29	05 54	06 14	06 31
45	05 46	06 20	06 50	05 06	05 36	06 01	06 22
N 40	05 45	06 16	06 43	04 48	05 21	05 49	06 15
35	05 43	06 12	06 38	04 32	05 08	05 40	06 08
30	05 41	06 09	06 33	04 19	04 57	05 31	06 02
20	05 37	06 03	06 25	03 56	04 38	05 16	05 52
N 10	05 31	05 56	06 17	03 36	04 21	05 03	05 43
0	05 25	05 49	06 10	03 18	04 05	04 51	05 35
S 10	05 17	05 41	06 03	02 59	03 49	04 39	05 27
20	05 06	05 32	05 55	02 39	03 32	04 25	05 18
30	04 52	05 21	05 46	02 16	03 13	04 10	05 07
35	04 43	05 14	05 41	02 02	03 01	04 01	05 01
40	04 33	05 06	05 34	01 46	02 48	03 51	04 55
45	04 19	04 57	05 27	01 28	02 33	03 39	04 47
S 50	04 02	04 45	05 19	01 04	02 13	03 25	04 37
52	03 54	04 39	05 15	00 53	02 04	03 18	04 33
54	03 44	04 32	05 10	00 40	01 54	03 10	04 28
56	03 33	04 25	05 06	00 25	01 42	03 02	04 22
58	03 20	04 17	05 00	00 07	01 29	02 52	04 16
S 60	03 05	04 07	04 54	25 13	01 13	02 41	04 09

Lat.	Sunset	Twilight Civil	Twilight Naut.	Moonset 21	22	23	24
°	h m	h m	h m	h m	h m	h m	h m
N 72	16 15	17 28	18 46	▬	▬	13 28	15 53
N 70	16 29	17 34	18 45	▬	11 01	14 02	16 08
68	16 40	17 39	18 43	▬	12 16	14 27	16 20
66	16 49	17 43	18 42	10 34	12 53	14 45	16 30
64	16 57	17 46	18 41	11 29	13 19	15 00	16 38
62	17 04	17 50	18 41	12 02	13 39	15 13	16 45
60	17 10	17 52	18 41	12 26	13 55	15 23	16 51
N 58	17 15	17 55	18 40	12 45	14 08	15 32	16 56
56	17 19	17 57	18 40	13 01	14 20	15 40	17 01
54	17 23	17 59	18 41	13 14	14 30	15 47	17 05
52	17 27	18 01	18 41	13 26	14 39	15 54	17 08
50	17 30	18 03	18 41	13 37	14 47	15 59	17 12
45	17 38	18 08	18 42	13 58	15 05	16 12	17 19
N 40	17 44	18 11	18 43	14 16	15 19	16 22	17 25
35	17 49	18 15	18 44	14 31	15 30	16 30	17 30
30	17 54	18 18	18 46	14 43	15 41	16 38	17 35
20	18 02	18 25	18 50	15 05	15 58	16 51	17 43
N 10	18 10	18 31	18 56	15 23	16 13	17 02	17 49
0	18 17	18 38	19 02	15 41	16 27	17 12	17 56
S 10	18 24	18 45	19 10	15 58	16 41	17 23	18 02
20	18 32	18 54	19 20	16 16	16 56	17 33	18 09
30	18 41	19 05	19 34	16 37	17 13	17 46	18 16
35	18 46	19 12	19 43	16 50	17 23	17 53	18 20
40	18 52	19 20	19 53	17 04	17 34	18 01	18 25
45	18 59	19 29	20 06	17 20	17 47	18 11	18 31
S 50	19 07	19 41	20 23	17 41	18 03	18 22	18 38
52	19 11	19 47	20 31	17 50	18 11	18 27	18 41
54	19 15	19 53	20 41	18 01	18 19	18 33	18 44
56	19 20	20 00	20 52	18 13	18 28	18 39	18 48
58	19 25	20 08	21 04	18 27	18 38	18 46	18 52
S 60	19 31	20 17	21 19	18 44	18 50	18 54	18 56

Day	SUN Eqn. of Time 00^h	SUN Eqn. of Time 12^h	SUN Mer. Pass.	MOON Mer. Pass. Upper	MOON Mer. Pass. Lower	Age	Phase
d	m s	m s	h m	h m	h m	d %	
21	13 39	13 35	12 14	09 29	21 53	26 12	●
22	13 32	13 28	12 13	10 16	22 39	27 6	
23	13 24	13 20	12 13	11 02	23 24	28 2	

UT		ARIES	VENUS −4.8		MARS +1.2		JUPITER −2.0		SATURN +0.6	
		GHA	GHA	Dec	GHA	Dec	GHA	Dec	GHA	Dec
d	h	° ′	° ′	° ′	° ′	° ′	° ′	° ′	° ′	° ′
24	00	154 00.3	143 51.2	N10 03.1	196 10.8	S17 21.2	199 51.0	S17 47.6	342 52.3	N 6 16.1
	01	169 02.8	158 52.9	03.8	211 11.3	20.7	214 52.9	47.5	357 55.0	16.2
	02	184 05.2	173 54.6	04.6	226 11.8	20.1	229 54.8	47.3	12 57.6	16.3
	03	199 07.7	188 56.4	. . 05.4	241 12.3	. . 19.5	244 56.7	. . 47.2	28 00.3	. . 16.4
	04	214 10.2	203 58.1	06.1	256 12.8	19.0	259 58.6	47.0	43 02.9	16.4
	05	229 12.6	218 59.9	06.9	271 13.3	18.4	275 00.5	46.9	58 05.5	16.5
	06	244 15.1	234 01.6	N10 07.7	286 13.8	S17 17.8	290 02.4	S17 46.7	73 08.2	N 6 16.6
	07	259 17.6	249 03.4	08.4	301 14.3	17.3	305 04.3	46.5	88 10.8	16.7
T	08	274 20.0	264 05.1	09.2	316 14.8	16.7	320 06.2	46.4	103 13.4	16.8
	09	289 22.5	279 06.9	. . 10.0	331 15.4	. . 16.1	335 08.0	. . 46.2	118 16.1	. . 16.8
U	10	304 24.9	294 08.7	10.7	346 15.9	15.6	350 09.9	46.1	133 18.7	16.9
E	11	319 27.4	309 10.4	11.5	1 16.4	15.0	5 11.8	45.9	148 21.4	17.0
S	12	334 29.9	324 12.2	N10 12.2	16 16.9	S17 14.4	20 13.7	S17 45.8	163 24.0	N 6 17.1
D	13	349 32.3	339 14.0	13.0	31 17.4	13.9	35 15.6	45.6	178 26.6	17.2
A	14	4 34.8	354 15.8	13.8	46 17.9	13.3	50 17.5	45.5	193 29.3	17.2
Y	15	19 37.3	9 17.5	. . 14.5	61 18.4	. . 12.7	65 19.4	. . 45.3	208 31.9	. . 17.3
	16	34 39.7	24 19.3	15.3	76 18.9	12.2	80 21.3	45.2	223 34.6	17.4
	17	49 42.2	39 21.1	16.0	91 19.4	11.6	95 23.2	45.0	238 37.2	17.5
	18	64 44.7	54 22.9	N10 16.8	106 19.9	S17 11.0	110 25.1	S17 44.9	253 39.8	N 6 17.6
	19	79 47.1	69 24.7	17.5	121 20.5	10.5	125 27.0	44.7	268 42.5	17.6
	20	94 49.6	84 26.5	18.3	136 21.0	09.9	140 28.9	44.6	283 45.1	17.7
	21	109 52.0	99 28.3	. . 19.0	151 21.5	. . 09.3	155 30.7	. . 44.4	298 47.7	. . 17.8
	22	124 54.5	114 30.1	19.8	166 22.0	08.7	170 32.6	44.3	313 50.4	17.9
	23	139 57.0	129 31.9	20.5	181 22.5	08.2	185 34.5	44.1	328 53.0	18.0
25	00	154 59.4	144 33.7	N10 21.3	196 23.0	S17 07.6	200 36.4	S17 44.0	343 55.7	N 6 18.0
	01	170 01.9	159 35.5	22.0	211 23.5	07.0	215 38.3	43.8	358 58.3	18.1
	02	185 04.4	174 37.3	22.7	226 24.0	06.5	230 40.2	43.7	14 00.9	18.2
	03	200 06.8	189 39.1	. . 23.5	241 24.5	. . 05.9	245 42.1	. . 43.5	29 03.6	. . 18.3
	04	215 09.3	204 40.9	24.2	256 25.1	05.3	260 44.0	43.4	44 06.2	18.4
	05	230 11.8	219 42.8	25.0	271 25.6	04.7	275 45.9	43.2	59 08.9	18.4
	06	245 14.2	234 44.6	N10 25.7	286 26.1	S17 04.2	290 47.8	S17 43.0	74 11.5	N 6 18.5
W	07	260 16.7	249 46.4	26.4	301 26.6	03.6	305 49.7	42.9	89 14.1	18.6
E	08	275 19.2	264 48.2	27.2	316 27.1	03.0	320 51.6	42.7	104 16.8	18.7
D	09	290 21.6	279 50.1	. . 27.9	331 27.6	. . 02.4	335 53.5	. . 42.6	119 19.4	. . 18.8
N	10	305 24.1	294 51.9	28.6	346 28.1	01.9	350 55.4	42.4	134 22.1	18.8
E	11	320 26.5	309 53.7	29.4	1 28.6	01.3	5 57.2	42.3	149 24.7	18.9
S	12	335 29.0	324 55.6	N10 30.1	16 29.2	S17 00.7	20 59.1	S17 42.1	164 27.3	N 6 19.0
D	13	350 31.5	339 57.4	30.8	31 29.7	17 00.1	36 01.0	42.0	179 30.0	19.1
	14	5 33.9	354 59.3	31.5	46 30.2	16 59.6	51 02.9	41.8	194 32.6	19.2
A	15	20 36.4	10 01.1	. . 32.3	61 30.7	. . 59.0	66 04.8	. . 41.7	209 35.3	. . 19.2
Y	16	35 38.9	25 03.0	33.0	76 31.2	58.4	81 06.7	41.5	224 37.9	19.3
	17	50 41.3	40 04.8	33.7	91 31.7	57.8	96 08.6	41.4	239 40.5	19.4
	18	65 43.8	55 06.7	N10 34.4	106 32.2	S16 57.3	111 10.5	S17 41.2	254 43.2	N 6 19.5
	19	80 46.3	70 08.6	35.1	121 32.7	56.7	126 12.4	41.1	269 45.8	19.6
	20	95 48.7	85 10.4	35.9	136 33.3	56.1	141 14.3	40.9	284 48.5	19.6
	21	110 51.2	100 12.3	. . 36.6	151 33.8	. . 55.5	156 16.2	. . 40.8	299 51.1	. . 19.7
	22	125 53.6	115 14.2	37.3	166 34.3	54.9	171 18.1	40.6	314 53.7	19.8
	23	140 56.1	130 16.1	38.0	181 34.8	54.4	186 20.0	40.5	329 56.4	19.9
26	00	155 58.6	145 17.9	N10 38.7	196 35.3	S16 53.8	201 21.9	S17 40.3	344 59.0	N 6 20.0
	01	171 01.0	160 19.8	39.4	211 35.8	53.2	216 23.8	40.1	0 01.7	20.0
	02	186 03.5	175 21.7	40.1	226 36.4	52.6	231 25.7	40.0	15 04.3	20.1
	03	201 06.0	190 23.6	. . 40.8	241 36.9	. . 52.1	246 27.6	. . 39.8	30 06.9	. . 20.2
	04	216 08.4	205 25.5	41.5	256 37.4	51.5	261 29.5	39.7	45 09.6	20.3
	05	231 10.9	220 27.4	42.2	271 37.9	50.9	276 31.3	39.5	60 12.2	20.4
	06	246 13.4	235 29.3	N10 43.0	286 38.4	S16 50.3	291 33.2	S17 39.4	75 14.9	N 6 20.4
	07	261 15.8	250 31.2	43.7	301 38.9	49.7	306 35.1	39.2	90 17.5	20.5
T	08	276 18.3	265 33.1	44.4	316 39.5	49.2	321 37.0	39.1	105 20.1	20.6
H	09	291 20.8	280 35.0	. . 45.1	331 40.0	. . 48.6	336 38.9	. . 38.9	120 22.8	. . 20.7
U	10	306 23.2	295 36.9	45.8	346 40.5	48.0	351 40.8	38.8	135 25.4	20.8
R	11	321 25.7	310 38.8	46.4	1 41.0	47.4	6 42.7	38.6	150 28.1	20.8
S	12	336 28.1	325 40.7	N10 47.1	16 41.5	S16 46.8	21 44.6	S17 38.5	165 30.7	N 6 20.9
D	13	351 30.6	340 42.7	47.8	31 42.0	46.2	36 46.5	38.3	180 33.4	21.0
A	14	6 33.1	355 44.6	48.5	46 42.6	45.7	51 48.4	38.2	195 36.0	21.1
Y	15	21 35.5	10 46.5	. . 49.2	61 43.1	. . 45.1	66 50.3	. . 38.0	210 38.6	. . 21.2
	16	36 38.0	25 48.4	49.9	76 43.6	44.5	81 52.2	37.9	225 41.3	21.2
	17	51 40.5	40 50.4	50.6	91 44.1	43.9	96 54.1	37.7	240 43.9	21.3
	18	66 42.9	55 52.3	N10 51.3	106 44.6	S16 43.3	111 56.0	S17 37.6	255 46.6	N 6 21.4
	19	81 45.4	70 54.2	52.0	121 45.1	42.7	126 57.9	37.4	270 49.2	21.5
	20	96 47.9	85 56.2	52.7	136 45.7	42.2	141 59.8	37.3	285 51.8	21.6
	21	111 50.3	100 58.1	. . 53.3	151 46.2	. . 41.6	157 01.7	. . 37.1	300 54.5	. . 21.7
	22	126 52.8	116 00.1	54.0	166 46.7	41.0	172 03.6	36.9	315 57.1	21.7
	23	141 55.2	131 02.0	54.7	181 47.2	40.4	187 05.5	36.8	330 59.8	21.8
Mer. Pass.		h m 13 37.8	*v* 1.8	*d* 0.7	*v* 0.5	*d* 0.6	*v* 1.9	*d* 0.2	*v* 2.6	*d* 0.1

STARS

Name	SHA	Dec
	° ′	° ′
Acamar	315 20.8	S40 16.
Achernar	335 29.3	S57 11
Acrux	173 12.6	S63 09.
Adhara	255 14.8	S28 59.
Aldebaran	290 53.0	N16 31.
Alioth	166 22.8	N55 54.
Alkaid	153 00.9	N49 15.
Al Na'ir	27 48.0	S46 55.
Alnilam	275 49.5	S 1 11.
Alphard	217 58.9	S 8 42.
Alphecca	126 13.6	N26 40.
Alpheratz	357 47.2	N29 08.
Altair	62 11.6	N 8 53.
Ankaa	353 19.0	S42 15.
Antares	112 30.2	S26 27.
Arcturus	145 58.4	N19 07.
Atria	107 35.1	S69 02.
Avior	234 18.9	S59 32.
Bellatrix	278 35.3	N 6 21.
Betelgeuse	271 04.6	N 7 24.
Canopus	263 57.3	S52 42.
Capella	280 39.1	N46 00.
Deneb	49 34.2	N45 18.
Denebola	182 36.6	N14 31.
Diphda	348 59.3	S17 56.
Dubhe	193 54.6	N61 41.
Elnath	278 16.6	N28 37.
Eltanin	90 47.8	N51 28.
Enif	33 50.6	N 9 54.
Fomalhaut	15 27.8	S29 34.
Gacrux	172 04.2	S57 09.
Gienah	175 55.3	S17 35.
Hadar	148 52.3	S60 25.
Hamal	328 04.6	N23 30.
Kaus Aust.	83 48.2	S34 22.
Kochab	137 19.0	N74 06.6
Markab	13 41.9	N15 15.2
Menkar	314 18.5	N 4 07.5
Menkent	148 11.2	S36 25.0
Miaplacidus	221 39.8	S69 45.4
Mirfak	308 45.2	N49 53.9
Nunki	76 02.5	S26 17.2
Peacock	53 24.6	S56 42.3
Pollux	243 31.3	N28 00.2
Procyon	245 02.8	N 5 12.0
Rasalhague	96 09.5	N12 32.9
Regulus	207 46.5	N11 55.2
Rigel	281 15.0	S 8 11.5
Rigil Kent.	139 56.0	S60 52.3
Sabik	102 16.3	S15 44.3
Schedar	349 44.9	N56 35.4
Shaula	96 26.4	S37 06.6
Sirius	258 36.3	S16 43.9
Spica	158 34.4	S11 12.7
Suhail	222 54.5	S43 28.3
Vega	80 41.4	N38 47.2
Zuben'ubi	137 08.9	S16 04.9

	SHA	Mer. Pass
	° ′	h m
Venus	349 34.2	14 20
Mars	41 23.6	10 54
Jupiter	45 37.0	10 36
Saturn	188 56.2	1 04

UT	SUN GHA	SUN Dec	MOON GHA	*v*	MOON Dec	*d*	HP
d h	° ′	° ′	° ′	′	° ′	′	′
24 00	176 41.2	S 9 29.5	188 49.6	14.4	S12 33.5	12.8	55.3
01	191 41.3	28.5	203 23.0	14.5	12 20.7	12.7	55.3
02	206 41.4	27.6	217 56.5	14.6	12 08.0	12.9	55.3
03	221 41.5	. . 26.7	232 30.1	14.5	11 55.1	12.9	55.3
04	236 41.6	25.8	247 03.6	14.6	11 42.2	12.9	55.3
05	251 41.7	24.8	261 37.2	14.6	11 29.3	13.0	55.3
06	266 41.7	S 9 23.9	276 10.8	14.6	S11 16.3	13.1	55.4
T 07	281 41.8	23.0	290 44.4	14.6	11 03.2	13.1	55.4
U 08	296 41.9	22.1	305 18.0	14.7	10 50.1	13.2	55.4
E 09	311 42.0	. . 21.1	319 51.7	14.7	10 36.9	13.2	55.4
S 10	326 42.1	20.2	334 25.4	14.7	10 23.7	13.3	55.4
D 11	341 42.2	19.3	348 59.1	14.7	10 10.4	13.3	55.5
A 12	356 42.3	S 9 18.4	3 32.8	14.8	S 9 57.1	13.4	55.5
Y 13	11 42.4	17.4	18 06.6	14.7	9 43.7	13.5	55.5
14	26 42.5	16.5	32 40.3	14.8	9 30.2	13.4	55.5
15	41 42.6	. . 15.6	47 14.1	14.8	9 16.8	13.6	55.5
16	56 42.7	14.7	61 47.9	14.8	9 03.2	13.5	55.6
17	71 42.8	13.7	76 21.7	14.8	8 49.7	13.7	55.6
18	86 42.9	S 9 12.8	90 55.5	14.9	S 8 36.0	13.6	55.6
19	101 43.0	11.9	105 29.4	14.8	8 22.4	13.7	55.6
20	116 43.1	10.9	120 03.2	14.9	8 08.7	13.8	55.6
21	131 43.2	. . 10.0	134 37.1	14.8	7 54.9	13.8	55.7
22	146 43.3	09.1	149 10.9	14.9	7 41.1	13.8	55.7
23	161 43.3	08.2	163 44.8	14.9	7 27.3	13.9	55.7
25 00	176 43.4	S 9 07.2	178 18.7	14.9	S 7 13.4	13.9	55.7
01	191 43.5	06.3	192 52.6	14.9	6 59.5	13.9	55.7
02	206 43.6	05.4	207 26.5	14.9	6 45.6	14.0	55.8
03	221 43.7	. . 04.4	222 00.4	15.0	6 31.6	14.1	55.8
04	236 43.8	03.5	236 34.4	14.9	6 17.5	14.0	55.8
05	251 43.9	02.6	251 08.3	14.9	6 03.5	14.1	55.8
06	266 44.0	S 9 01.6	265 42.2	14.9	S 5 49.4	14.2	55.8
W 07	281 44.1	9 00.7	280 16.1	15.0	5 35.2	14.1	55.9
E 08	296 44.2	8 59.8	294 50.1	14.9	5 21.1	14.2	55.9
D 09	311 44.3	. . 58.8	309 24.0	15.0	5 06.9	14.2	55.9
N 10	326 44.4	57.9	323 58.0	14.9	4 52.7	14.3	55.9
E 11	341 44.5	57.0	338 31.9	14.9	4 38.4	14.3	55.9
S 12	356 44.6	S 8 56.1	353 05.8	15.0	S 4 24.1	14.3	56.0
D 13	11 44.7	55.1	7 39.8	14.9	4 09.8	14.3	56.0
A 14	26 44.8	54.2	22 13.7	14.9	3 55.5	14.4	56.0
Y 15	41 44.9	. . 53.3	36 47.6	14.9	3 41.1	14.4	56.0
16	56 45.0	52.3	51 21.5	15.0	3 26.7	14.4	56.0
17	71 45.1	51.4	65 55.5	14.9	3 12.3	14.5	56.1
18	86 45.2	S 8 50.5	80 29.4	14.9	S 2 57.8	14.4	56.1
19	101 45.3	49.5	95 03.3	14.9	2 43.4	14.5	56.1
20	116 45.4	48.6	109 37.2	14.8	2 28.9	14.5	56.1
21	131 45.5	. . 47.7	124 11.0	14.9	2 14.4	14.5	56.1
22	146 45.6	46.7	138 44.9	14.9	1 59.9	14.6	56.2
23	161 45.7	45.8	153 18.8	14.8	1 45.3	14.5	56.2
26 00	176 45.8	S 8 44.8	167 52.6	14.9	S 1 30.8	14.6	56.2
01	191 45.9	43.9	182 26.5	14.8	1 16.2	14.6	56.2
02	206 46.0	43.0	197 00.3	14.8	1 01.6	14.6	56.3
03	221 46.2	. . 42.0	211 34.1	14.8	0 47.0	14.6	56.3
04	236 46.3	41.1	226 07.9	14.7	0 32.4	14.7	56.3
05	251 46.4	40.2	240 41.6	14.8	0 17.7	14.6	56.3
06	266 46.5	S 8 39.2	255 15.4	14.7	S 0 03.1	14.7	56.3
T 07	281 46.6	38.3	269 49.1	14.7	N 0 11.6	14.6	56.4
H 08	296 46.7	37.4	284 22.8	14.7	0 26.2	14.7	56.4
U 09	311 46.8	. . 36.4	298 56.5	14.7	0 40.9	14.7	56.4
R 10	326 46.9	35.5	313 30.2	14.6	0 55.6	14.6	56.4
S 11	341 47.0	34.5	328 03.8	14.6	1 10.2	14.7	56.4
D 12	356 47.1	S 8 33.6	342 37.4	14.6	N 1 24.9	14.7	56.5
A 13	11 47.2	32.7	357 11.0	14.6	1 39.6	14.7	56.5
Y 14	26 47.3	31.7	11 44.6	14.5	1 54.3	14.7	56.5
15	41 47.4	. . 30.8	26 18.1	14.5	2 09.0	14.7	56.5
16	56 47.5	29.9	40 51.6	14.5	2 23.7	14.7	56.5
17	71 47.6	28.9	55 25.1	14.4	2 38.4	14.7	56.6
18	86 47.7	S 8 28.0	69 58.5	14.5	N 2 53.1	14.7	56.6
19	101 47.8	27.0	84 32.0	14.3	3 07.8	14.7	56.6
20	116 47.9	26.1	99 05.3	14.4	3 22.5	14.6	56.6
21	131 48.1	. . 25.2	113 38.7	14.3	3 37.1	14.7	56.6
22	146 48.2	24.2	128 12.0	14.3	3 51.8	14.7	56.7
23	161 48.3	23.3	142 45.3	14.2	N 4 06.5	14.7	56.7
	SD 16.2	*d* 0.9	SD	15.1	15.2		15.4

Lat.	Twilight Naut.	Twilight Civil	Sunrise	Moonrise 24	Moonrise 25	Moonrise 26	Moonrise 27
°	h m	h m	h m	h m	h m	h m	h m
N 72	05 29	06 47	07 57	08 05	07 30	07 00	06 29
N 70	05 32	06 42	07 45	07 47	07 22	07 00	06 37
68	05 34	06 38	07 36	07 34	07 16	07 00	06 44
66	05 36	06 35	07 28	07 22	07 11	07 00	06 49
64	05 38	06 32	07 21	07 12	07 06	07 00	06 54
62	05 39	06 30	07 15	07 04	07 02	07 00	06 58
60	05 40	06 28	07 10	06 57	06 59	07 00	07 02
N 58	05 40	06 26	07 05	06 51	06 56	07 00	07 05
56	05 41	06 24	07 01	06 45	06 53	07 00	07 08
54	05 41	06 22	06 58	06 40	06 51	07 00	07 11
52	05 41	06 20	06 54	06 35	06 48	07 00	07 13
50	05 41	06 19	06 51	06 31	06 46	07 01	07 15
45	05 41	06 15	06 45	06 22	06 42	07 01	07 20
N 40	05 41	06 12	06 39	06 15	06 38	07 01	07 24
35	05 40	06 09	06 34	06 08	06 35	07 01	07 27
30	05 38	06 06	06 30	06 02	06 32	07 01	07 30
20	05 35	06 01	06 23	05 52	06 27	07 01	07 36
N 10	05 30	05 55	06 16	05 43	06 22	07 01	07 41
0	05 24	05 49	06 10	05 35	06 18	07 01	07 45
S 10	05 17	05 42	06 03	05 27	06 14	07 01	07 50
20	05 08	05 34	05 56	05 18	06 09	07 02	07 55
30	04 55	05 24	05 48	05 07	06 04	07 02	08 00
35	04 47	05 17	05 43	05 01	06 01	07 02	08 04
40	04 37	05 10	05 38	04 55	05 58	07 02	08 07
45	04 24	05 01	05 32	04 47	05 54	07 02	08 12
S 50	04 08	04 50	05 24	04 37	05 49	07 02	08 17
52	04 01	04 45	05 20	04 33	05 47	07 03	08 19
54	03 52	04 39	05 17	04 28	05 45	07 03	08 22
56	03 42	04 32	05 12	04 22	05 42	07 03	08 25
58	03 30	04 25	05 07	04 16	05 39	07 03	08 28
S 60	03 16	04 16	05 02	04 09	05 36	07 03	08 32

Lat.	Sunset	Twilight Civil	Twilight Naut.	Moonset 24	Moonset 25	Moonset 26	Moonset 27
°	h m	h m	h m	h m	h m	h m	h m
N 72	16 31	17 42	19 00	15 53	18 00	20 04	22 17
N 70	16 42	17 46	18 56	16 08	18 04	19 59	22 01
68	16 52	17 49	18 54	16 20	18 08	19 56	21 48
66	17 00	17 52	18 52	16 30	18 11	19 52	21 38
64	17 07	17 55	18 50	16 38	18 13	19 50	21 29
62	17 12	17 58	18 49	16 45	18 16	19 47	21 22
60	17 17	18 00	18 48	16 51	18 18	19 45	21 16
N 58	17 22	18 02	18 47	16 56	18 19	19 43	21 10
56	17 26	18 03	18 47	17 01	18 21	19 42	21 05
54	17 29	18 05	18 46	17 05	18 22	19 40	21 01
52	17 33	18 07	18 46	17 08	18 23	19 39	20 57
50	17 36	18 08	18 46	17 12	18 24	19 38	20 53
45	17 42	18 12	18 46	17 19	18 27	19 35	20 46
N 40	17 47	18 15	18 46	17 25	18 29	19 33	20 39
35	17 52	18 18	18 47	17 30	18 30	19 31	20 34
30	17 56	18 20	18 48	17 35	18 32	19 30	20 29
20	18 04	18 26	18 52	17 43	18 34	19 27	20 21
N 10	18 10	18 31	18 56	17 49	18 37	19 24	20 14
0	18 16	18 37	19 02	17 56	18 39	19 22	20 07
S 10	18 23	18 44	19 09	18 02	18 41	19 20	20 00
20	18 30	18 52	19 18	18 09	18 43	19 17	19 53
30	18 37	19 02	19 31	18 16	18 45	19 14	19 45
35	18 42	19 08	19 39	18 20	18 47	19 13	19 40
40	18 47	19 15	19 48	18 25	18 48	19 11	19 35
45	18 54	19 24	20 01	18 31	18 50	19 09	19 29
S 50	19 01	19 35	20 16	18 38	18 52	19 06	19 21
52	19 04	19 40	20 24	18 41	18 53	19 05	19 18
54	19 08	19 46	20 32	18 44	18 54	19 04	19 14
56	19 12	19 52	20 42	18 48	18 55	19 02	19 10
58	19 17	19 59	20 54	18 52	18 56	19 01	19 06
S 60	19 22	20 08	21 07	18 56	18 58	18 59	19 01

Day	SUN Eqn. of Time 00^h	SUN Eqn. of Time 12^h	SUN Mer. Pass.	MOON Mer. Pass. Upper	MOON Mer. Pass. Lower	Age	Phase
d	m s	m s	h m	h m	h m	d %	
24	13 15	13 11	12 13	11 45	24 07	29 0	●
25	13 06	13 02	12 13	12 28	00 07	00 0	
26	12 57	12 52	12 13	13 12	00 50	01 2	

	UT	ARIES	VENUS −4.8		MARS +1.2		JUPITER −2.0		SATURN +0.5	
		GHA	GHA	Dec	GHA	Dec	GHA	Dec	GHA	Dec
	d h	° ′	° ′	° ′	° ′	° ′	° ′	° ′	° ′	° ′
	27 00	156 57.7	146 04.0	N10 55.4	196 47.7	S16 39.8	202 07.4	S17 36.6	346 02.4	N 6 21.9
	01	172 00.2	161 05.9	56.0	211 48.3	39.2	217 09.3	36.5	1 05.1	22.0
	02	187 02.6	176 07.9	56.7	226 48.8	38.6	232 11.2	36.3	16 07.7	22.1
	03	202 05.1	191 09.9	. . 57.4	241 49.3	. . 38.1	247 13.1	. . 36.2	31 10.3	. . 22.1
	04	217 07.6	206 11.8	58.1	256 49.8	37.5	262 15.0	36.0	46 13.0	22.2
	05	232 10.0	221 13.8	58.7	271 50.3	36.9	277 16.9	35.9	61 15.6	22.3
	06	247 12.5	236 15.8	N10 59.4	286 50.9	S16 36.3	292 18.8	S17 35.7	76 18.3	N 6 22.4
	07	262 15.0	251 17.8	11 00.1	301 51.4	35.7	307 20.7	35.6	91 20.9	22.5
	08	277 17.4	266 19.8	00.7	316 51.9	35.1	322 22.6	35.4	106 23.5	22.5
F	09	292 19.9	281 21.7	. . 01.4	331 52.4	. . 34.5	337 24.5	. . 35.3	121 26.2	. . 22.6
R	10	307 22.4	296 23.7	02.1	346 52.9	33.9	352 26.3	35.1	136 28.8	22.7
I	11	322 24.8	311 25.7	02.7	1 53.5	33.3	7 28.2	35.0	151 31.5	22.8
D	12	337 27.3	326 27.7	N11 03.4	16 54.0	S16 32.8	22 30.1	S17 34.8	166 34.1	N 6 22.9
A	13	352 29.7	341 29.7	04.0	31 54.5	32.2	37 32.0	34.7	181 36.8	22.9
Y	14	7 32.2	356 31.7	04.7	46 55.0	31.6	52 33.9	34.5	196 39.4	23.0
	15	22 34.7	11 33.7	. . 05.4	61 55.6	. . 31.0	67 35.8	. . 34.4	211 42.0	. . 23.1
	16	37 37.1	26 35.7	06.0	76 56.1	30.4	82 37.7	34.2	226 44.7	23.2
	17	52 39.6	41 37.7	06.7	91 56.6	29.8	97 39.6	34.0	241 47.3	23.3
	18	67 42.1	56 39.7	N11 07.3	106 57.1	S16 29.2	112 41.5	S17 33.9	256 50.0	N 6 23.3
	19	82 44.5	71 41.8	08.0	121 57.6	28.6	127 43.4	33.7	271 52.6	23.4
	20	97 47.0	86 43.8	08.6	136 58.2	28.0	142 45.3	33.6	286 55.3	23.5
	21	112 49.5	101 45.8	. . 09.3	151 58.7	. . 27.4	157 47.2	. . 33.4	301 57.9	. . 23.6
	22	127 51.9	116 47.8	09.9	166 59.2	26.8	172 49.1	33.3	317 00.5	23.7
	23	142 54.4	131 49.9	10.6	181 59.7	26.3	187 51.0	33.1	332 03.2	23.7
	28 00	157 56.8	146 51.9	N11 11.2	197 00.3	S16 25.7	202 52.9	S17 33.0	347 05.8	N 6 23.8
	01	172 59.3	161 53.9	11.8	212 00.8	25.1	217 54.8	32.8	2 08.5	23.9
	02	188 01.8	176 56.0	12.5	227 01.3	24.5	232 56.7	32.7	17 11.1	24.0
	03	203 04.2	191 58.0	. . 13.1	242 01.8	. . 23.9	247 58.6	. . 32.5	32 13.7	. . 24.1
	04	218 06.7	207 00.1	13.8	257 02.4	23.3	263 00.5	32.4	47 16.4	24.2
	05	233 09.2	222 02.1	14.4	272 02.9	22.7	278 02.4	32.2	62 19.0	24.2
	06	248 11.6	237 04.2	N11 15.0	287 03.4	S16 22.1	293 04.3	S17 32.1	77 21.7	N 6 24.3
S	07	263 14.1	252 06.2	15.6	302 03.9	21.5	308 06.2	31.9	92 24.3	24.4
A	08	278 16.6	267 08.3	16.3	317 04.5	20.9	323 08.1	31.8	107 27.0	24.5
T	09	293 19.0	282 10.4	. . 16.9	332 05.0	. . 20.3	338 10.0	. . 31.6	122 29.6	. . 24.6
U	10	308 21.5	297 12.4	17.5	347 05.5	19.7	353 11.9	31.5	137 32.3	24.6
R	11	323 24.0	312 14.5	18.2	2 06.0	19.1	8 13.8	31.3	152 34.9	24.7
D	12	338 26.4	327 16.6	N11 18.8	17 06.6	S16 18.5	23 15.7	S17 31.1	167 37.5	N 6 24.8
A	13	353 28.9	342 18.7	19.4	32 07.1	17.9	38 17.6	31.0	182 40.2	24.9
Y	14	8 31.3	357 20.7	20.0	47 07.6	17.3	53 19.5	30.8	197 42.8	25.0
	15	23 33.8	12 22.8	. . 20.6	62 08.1	. . 16.7	68 21.4	. . 30.7	212 45.5	. . 25.0
	16	38 36.3	27 24.9	21.3	77 08.7	16.1	83 23.3	30.5	227 48.1	25.1
	17	53 38.7	42 27.0	21.9	92 09.2	15.5	98 25.2	30.4	242 50.8	25.2
	18	68 41.2	57 29.1	N11 22.5	107 09.7	S16 14.9	113 27.1	S17 30.2	257 53.4	N 6 25.3
	19	83 43.7	72 31.2	23.1	122 10.2	14.3	128 29.0	30.1	272 56.0	25.4
	20	98 46.1	87 33.3	23.7	137 10.8	13.7	143 30.9	29.9	287 58.7	25.4
	21	113 48.6	102 35.4	. . 24.3	152 11.3	. . 13.1	158 32.8	. . 29.8	303 01.3	. . 25.5
	22	128 51.1	117 37.5	24.9	167 11.8	12.5	173 34.7	29.6	318 04.0	25.6
	23	143 53.5	132 39.6	25.5	182 12.4	11.9	188 36.6	29.5	333 06.6	25.7
	1 00	158 56.0	147 41.7	N11 26.1	197 12.9	S16 11.3	203 38.5	S17 29.3	348 09.3	N 6 25.8
	01	173 58.5	162 43.9	26.7	212 13.4	10.7	218 40.4	29.2	3 11.9	25.8
	02	189 00.9	177 46.0	27.3	227 13.9	10.1	233 42.3	29.0	18 14.5	25.9
	03	204 03.4	192 48.1	. . 27.9	242 14.5	. . 09.5	248 44.2	. . 28.9	33 17.2	. . 26.0
	04	219 05.8	207 50.2	28.5	257 15.0	08.9	263 46.1	28.7	48 19.8	26.1
	05	234 08.3	222 52.4	29.1	272 15.5	08.3	278 48.0	28.6	63 22.5	26.2
	06	249 10.8	237 54.5	N11 29.7	287 16.1	S16 07.7	293 49.9	S17 28.4	78 25.1	N 6 26.3
	07	264 13.2	252 56.6	30.3	302 16.6	07.1	308 51.9	28.2	93 27.8	26.3
	08	279 15.7	267 58.8	30.9	317 17.1	06.5	323 53.8	28.1	108 30.4	26.4
S	09	294 18.2	283 00.9	. . 31.5	332 17.6	. . 05.9	338 55.7	. . 27.9	123 33.1	. . 26.5
U	10	309 20.6	298 03.1	32.1	347 18.2	05.3	353 57.6	27.8	138 35.7	26.6
N	11	324 23.1	313 05.2	32.7	2 18.7	04.7	8 59.5	27.6	153 38.3	26.7
D	12	339 25.6	328 07.4	N11 33.3	17 19.2	S16 04.1	24 01.4	S17 27.5	168 41.0	N 6 26.7
A	13	354 28.0	343 09.6	33.9	32 19.8	03.5	39 03.3	27.3	183 43.6	26.8
Y	14	9 30.5	358 11.7	34.4	47 20.3	02.9	54 05.2	27.2	198 46.3	26.9
	15	24 32.9	13 13.9	. . 35.0	62 20.8	. . 02.3	69 07.1	. . 27.0	213 48.9	. . 27.0
	16	39 35.4	28 16.1	35.6	77 21.4	01.7	84 09.0	26.9	228 51.6	27.1
	17	54 37.9	43 18.2	36.2	92 21.9	01.1	99 10.9	26.7	243 54.2	27.1
	18	69 40.3	58 20.4	N11 36.8	107 22.4	S16 00.5	114 12.8	S17 26.6	258 56.8	N 6 27.2
	19	84 42.8	73 22.6	37.3	122 23.0	15 59.9	129 14.7	26.4	273 59.5	27.3
	20	99 45.3	88 24.8	37.9	137 23.5	59.3	144 16.6	26.3	289 02.1	27.4
	21	114 47.7	103 27.0	. . 38.5	152 24.0	. . 58.7	159 18.5	. . 26.1	304 04.8	. . 27.5
	22	129 50.2	118 29.1	39.0	167 24.5	58.1	174 20.4	26.0	319 07.4	27.5
	23	144 52.7	133 31.3	39.6	182 25.1	57.5	189 22.3	25.8	334 10.1	27.6
	Mer. Pass.	h m 13 26.0	*v* 2.1	*d* 0.6	*v* 0.5	*d* 0.6	*v* 1.9	*d* 0.2	*v* 2.6	*d* 0.1

STARS		
Name	SHA	Dec
	° ′	° ′
Acamar	315 20.8	S40 16.
Achernar	335 29.3	S57 11.
Acrux	173 12.6	S63 09.
Adhara	255 14.8	S28 59.
Aldebaran	290 53.1	N16 31.
Alioth	166 22.8	N55 54.
Alkaid	153 00.9	N49 15.
Al Na'ir	27 48.0	S46 55.
Alnilam	275 49.5	S 1 11.
Alphard	217 58.9	S 8 42.
Alphecca	126 13.6	N26 40.
Alpheratz	357 47.2	N29 08.
Altair	62 11.6	N 8 53.
Ankaa	353 19.0	S42 15.
Antares	112 30.2	S26 27.
Arcturus	145 58.4	N19 07.
Atria	107 35.1	S69 02.
Avior	234 19.0	S59 32.
Bellatrix	278 35.4	N 6 21.
Betelgeuse	271 04.6	N 7 24.
Canopus	263 57.3	S52 42.
Capella	280 39.1	N46 00.
Deneb	49 34.2	N45 18.
Denebola	182 36.5	N14 31.
Diphda	348 59.3	S17 56.
Dubhe	193 54.6	N61 41.
Elnath	278 16.6	N28 37.
Eltanin	90 47.8	N51 28.
Enif	33 50.6	N 9 54.
Fomalhaut	15 27.8	S29 34.
Gacrux	172 04.2	S57 09.
Gienah	175 55.3	S17 35.
Hadar	148 52.3	S60 25.
Hamal	328 04.6	N23 30.
Kaus Aust.	83 48.2	S34 22.
Kochab	137 18.9	N74 06.
Markab	13 41.9	N15 15.
Menkar	314 18.5	N 4 07.
Menkent	148 11.2	S36 25.
Miaplacidus	221 39.8	S69 45.
Mirfak	308 45.2	N49 53.
Nunki	76 02.4	S26 17.
Peacock	53 24.6	S56 42.
Pollux	243 31.3	N28 00.
Procyon	245 02.8	N 5 12.
Rasalhague	96 09.5	N12 32.
Regulus	207 46.5	N11 55.
Rigel	281 15.1	S 8 11.
Rigil Kent.	139 56.0	S60 52.
Sabik	102 16.3	S15 44.
Schedar	349 44.9	N56 35.
Shaula	96 26.4	S37 06.
Sirius	258 36.4	S16 43.
Spica	158 34.4	S11 12.
Suhail	222 54.5	S43 28.
Vega	80 41.3	N38 47.
Zuben'ubi	137 08.8	S16 04.

	SHA	Mer. Pass.
	° ′	h m
Venus	348 55.1	14 11
Mars	39 03.4	10 52
Jupiter	44 56.1	10 27
Saturn	189 09.0	0 51

UT		SUN GHA	SUN Dec	MOON GHA	v	MOON Dec	d	HP
d	h	° ′	° ′	° ′	′	° ′	′	′
27	00	176 48.4	S 8 22.3	157 18.5	14.2	N 4 21.2	14.6	56.7
	01	191 48.5	21.4	171 51.7	14.2	4 35.8	14.6	56.7
	02	206 48.6	20.5	186 24.9	14.1	4 50.4	14.7	56.7
	03	221 48.7	. . 19.5	200 58.0	14.0	5 05.1	14.6	56.8
	04	236 48.8	18.6	215 31.0	14.1	5 19.7	14.6	56.8
	05	251 48.9	17.6	230 04.1	14.0	5 34.3	14.6	56.8
	06	266 49.0	S 8 16.7	244 37.1	13.9	N 5 48.9	14.6	56.8
	07	281 49.1	15.8	259 10.0	13.9	6 03.5	14.5	56.8
	08	296 49.3	14.8	273 42.9	13.9	6 18.0	14.5	56.9
F	09	311 49.4	. . 13.9	288 15.8	13.8	6 32.5	14.6	56.9
R	10	326 49.5	12.9	302 48.6	13.7	6 47.1	14.5	56.9
I	11	341 49.6	12.0	317 21.3	13.7	7 01.6	14.4	56.9
D	12	356 49.7	S 8 11.0	331 54.0	13.7	N 7 16.0	14.5	56.9
A	13	11 49.8	10.1	346 26.7	13.6	7 30.5	14.4	57.0
Y	14	26 49.9	09.2	0 59.3	13.5	7 44.9	14.4	57.0
	15	41 50.0	. . 08.2	15 31.8	13.5	7 59.3	14.4	57.0
	16	56 50.2	07.3	30 04.3	13.5	8 13.7	14.3	57.0
	17	71 50.3	06.3	44 36.8	13.4	8 28.0	14.3	57.0
	18	86 50.4	S 8 05.4	59 09.2	13.3	N 8 42.3	14.3	57.1
	19	101 50.5	04.4	73 41.5	13.3	8 56.6	14.3	57.1
	20	116 50.6	03.5	88 13.8	13.2	9 10.9	14.2	57.1
	21	131 50.7	. . 02.5	102 46.0	13.1	9 25.1	14.2	57.1
	22	146 50.8	01.6	117 18.1	13.1	9 39.3	14.1	57.1
	23	161 50.9	8 00.7	131 50.2	13.0	9 53.4	14.1	57.2
28	00	176 51.1	S 7 59.7	146 22.2	13.0	N10 07.5	14.1	57.2
	01	191 51.2	58.8	160 54.2	12.9	10 21.6	14.0	57.2
	02	206 51.3	57.8	175 26.1	12.8	10 35.6	14.0	57.2
	03	221 51.4	. . 56.9	189 57.9	12.8	10 49.6	14.0	57.2
	04	236 51.5	55.9	204 29.7	12.7	11 03.6	13.9	57.3
	05	251 51.6	55.0	219 01.4	12.6	11 17.5	13.9	57.3
	06	266 51.7	S 7 54.0	233 33.0	12.5	N11 31.4	13.8	57.3
	07	281 51.9	53.1	248 04.5	12.5	11 45.2	13.7	57.3
S	08	296 52.0	52.1	262 36.0	12.5	11 58.9	13.8	57.3
A	09	311 52.1	. . 51.2	277 07.5	12.3	12 12.7	13.6	57.4
T	10	326 52.2	50.3	291 38.8	12.3	12 26.3	13.7	57.4
U	11	341 52.3	49.3	306 10.1	12.2	12 40.0	13.5	57.4
R	12	356 52.4	S 7 48.4	320 41.3	12.1	N12 53.5	13.5	57.4
D	13	11 52.6	47.4	335 12.4	12.0	13 07.0	13.5	57.4
A	14	26 52.7	46.5	349 43.4	12.0	13 20.5	13.4	57.5
Y	15	41 52.8	. . 45.5	4 14.4	11.9	13 33.9	13.3	57.5
	16	56 52.9	44.6	18 45.3	11.8	13 47.2	13.3	57.5
	17	71 53.0	43.6	33 16.1	11.7	14 00.5	13.2	57.5
	18	86 53.2	S 7 42.7	47 46.8	11.7	N14 13.7	13.2	57.5
	19	101 53.3	41.7	62 17.5	11.5	14 26.9	13.1	57.6
	20	116 53.4	40.8	76 48.0	11.5	14 40.0	13.0	57.6
	21	131 53.5	. . 39.8	91 18.5	11.4	14 53.0	13.0	57.6
	22	146 53.6	38.9	105 48.9	11.3	15 06.0	12.9	57.6
	23	161 53.8	37.9	120 19.2	11.3	15 18.9	12.8	57.6
1	00	176 53.9	S 7 37.0	134 49.5	11.1	N15 31.7	12.8	57.7
	01	191 54.0	36.0	149 19.6	11.1	15 44.5	12.7	57.7
	02	206 54.1	35.1	163 49.7	10.9	15 57.2	12.6	57.7
	03	221 54.2	. . 34.1	178 19.6	10.9	16 09.8	12.5	57.7
	04	236 54.4	33.2	192 49.5	10.8	16 22.3	12.5	57.7
	05	251 54.5	32.2	207 19.3	10.7	16 34.8	12.3	57.8
	06	266 54.6	S 7 31.3	221 49.0	10.6	N16 47.1	12.3	57.8
	07	281 54.7	30.3	236 18.6	10.6	16 59.4	12.3	57.8
	08	296 54.8	29.4	250 48.2	10.4	17 11.7	12.1	57.8
S	09	311 55.0	. . 28.4	265 17.6	10.3	17 23.8	12.1	57.8
U	10	326 55.1	27.5	279 46.9	10.3	17 35.9	11.9	57.8
N	11	341 55.2	26.5	294 16.2	10.1	17 47.8	11.9	57.9
D	12	356 55.3	S 7 25.6	308 45.3	10.1	N17 59.7	11.8	57.9
A	13	11 55.4	24.6	323 14.4	10.0	18 11.5	11.7	57.9
Y	14	26 55.6	23.7	337 43.4	9.8	18 23.2	11.6	57.9
	15	41 55.7	. . 22.7	352 12.2	9.8	18 34.8	11.5	57.9
	16	56 55.8	21.7	6 41.0	9.7	18 46.3	11.5	58.0
	17	71 55.9	20.8	21 09.7	9.6	18 57.8	11.3	58.0
	18	86 56.1	S 7 19.8	35 38.3	9.5	N19 09.1	11.2	58.0
	19	101 56.2	18.9	50 06.8	9.4	19 20.3	11.1	58.0
	20	116 56.3	17.9	64 35.2	9.3	19 31.4	11.1	58.0
	21	131 56.4	. . 17.0	79 03.5	9.2	19 42.5	10.9	58.1
	22	146 56.6	16.0	93 31.7	9.1	19 53.4	10.8	58.1
	23	161 56.7	15.1	107 59.8	9.0	N20 04.2	10.8	58.1
		SD 16.2	*d* 0.9	SD 15.5		15.6		15.8

Lat.	Twilight Naut.	Twilight Civil	Sunrise	Moonrise 27	Moonrise 28	Moonrise 1	Moonrise 2
°	h m	h m	h m	h m	h m	h m	h m
N 72	05 14	06 32	07 41	06 29	05 53	04 50	□
N 70	05 19	06 29	07 31	06 37	06 11	05 33	□
68	05 23	06 27	07 23	06 44	06 26	06 02	05 14
66	05 25	06 24	07 16	06 49	06 38	06 24	06 02
64	05 28	06 23	07 11	06 54	06 48	06 41	06 33
62	05 30	06 21	07 06	06 58	06 57	06 56	06 57
60	05 31	06 19	07 01	07 02	07 04	07 08	07 15
N 58	05 33	06 18	06 57	07 05	07 11	07 19	07 31
56	05 34	06 16	06 54	07 08	07 17	07 28	07 45
54	05 34	06 15	06 51	07 11	07 22	07 37	07 56
52	05 35	06 14	06 48	07 13	07 27	07 44	08 07
50	05 35	06 13	06 45	07 15	07 31	07 51	08 16
45	05 36	06 10	06 40	07 20	07 41	08 06	08 36
N 40	05 36	06 08	06 35	07 24	07 49	08 18	08 52
35	05 36	06 05	06 31	07 27	07 56	08 28	09 06
30	05 35	06 03	06 27	07 30	08 02	08 37	09 18
20	05 33	05 58	06 21	07 36	08 13	08 53	09 38
N 10	05 29	05 54	06 15	07 41	08 22	09 07	09 56
0	05 24	05 48	06 09	07 45	08 31	09 20	10 13
S 10	05 18	05 42	06 04	07 50	08 40	09 33	10 30
20	05 09	05 35	05 57	07 55	08 50	09 47	10 48
30	04 57	05 26	05 50	08 00	09 01	10 04	11 09
35	04 50	05 20	05 46	08 04	09 07	10 14	11 22
40	04 41	05 14	05 41	08 07	09 15	10 25	11 36
45	04 29	05 06	05 36	08 12	09 23	10 38	11 53
S 50	04 15	04 56	05 29	08 17	09 34	10 54	12 15
52	04 07	04 51	05 26	08 19	09 39	11 01	12 25
54	03 59	04 45	05 23	08 22	09 44	11 09	12 37
56	03 50	04 39	05 19	08 25	09 50	11 19	12 50
58	03 40	04 33	05 15	08 28	09 57	11 30	13 05
S 60	03 27	04 25	05 10	08 32	10 05	11 42	13 24

Lat.	Sunset	Twilight Civil	Twilight Naut.	Moonset 27	Moonset 28	Moonset 1	Moonset 2
°	h m	h m	h m	h m	h m	h m	h m
N 72	16 46	17 55	19 14	22 17	25 00	01 00	□
N 70	16 55	17 58	19 09	22 01	24 19	00 19	□
68	17 03	18 00	19 05	21 48	23 52	26 28	02 28
66	17 10	18 02	19 01	21 38	23 31	25 40	01 40
64	17 16	18 04	18 59	21 29	23 15	25 10	01 10
62	17 21	18 06	18 57	21 22	23 01	24 48	00 48
60	17 25	18 07	18 55	21 16	22 50	24 29	00 29
N 58	17 29	18 08	18 54	21 10	22 40	24 14	00 14
56	17 32	18 10	18 53	21 05	22 32	24 02	00 02
54	17 35	18 11	18 52	21 01	22 24	23 51	25 17
52	17 38	18 12	18 51	20 57	22 18	23 41	25 04
50	17 41	18 13	18 51	20 53	22 11	23 32	24 53
45	17 46	18 16	18 50	20 46	21 58	23 14	24 29
N 40	17 51	18 18	18 49	20 39	21 48	22 59	24 10
35	17 55	18 20	18 50	20 34	21 39	22 46	23 55
30	17 58	18 22	18 50	20 29	21 31	22 35	23 41
20	18 05	18 27	18 53	20 21	21 17	22 16	23 18
N 10	18 10	18 32	18 56	20 14	21 05	22 00	22 58
0	18 16	18 37	19 01	20 07	20 54	21 45	22 40
S 10	18 21	18 43	19 07	20 00	20 43	21 30	22 21
20	18 27	18 50	19 16	19 53	20 31	21 14	22 01
30	18 34	18 59	19 27	19 45	20 18	20 55	21 39
35	18 38	19 04	19 35	19 40	20 10	20 45	21 25
40	18 43	19 11	19 44	19 35	20 01	20 32	21 10
45	18 48	19 18	19 55	19 29	19 51	20 18	20 52
S 50	18 55	19 28	20 09	19 21	19 39	20 01	20 29
52	18 58	19 33	20 16	19 18	19 33	19 53	20 19
54	19 01	19 38	20 24	19 14	19 27	19 43	20 07
56	19 05	19 44	20 33	19 10	19 20	19 33	19 53
58	19 09	19 50	20 43	19 06	19 12	19 22	19 37
S 60	19 14	19 58	20 55	19 01	19 04	19 08	19 18

Day	SUN Eqn. of Time 00^h	SUN Eqn. of Time 12^h	SUN Mer. Pass.	MOON Mer. Pass. Upper	MOON Mer. Pass. Lower	MOON Age	MOON Phase
d	m s	m s	h m	h m	h m	d %	
27	12 47	12 41	12 13	13 56	01 34	02 6	
28	12 36	12 30	12 13	14 42	02 19	03 13	
1	12 25	12 19	12 12	15 32	03 07	04 21	

UT		ARIES	VENUS −4.8		MARS +1.2		JUPITER −2.0		SATURN +0.5	
		GHA	GHA	Dec	GHA	Dec	GHA	Dec	GHA	Dec
d	h	° ′	° ′	° ′	° ′	° ′	° ′	° ′	° ′	° ′
2	00	159 55.1	148 33.5	N11 40.2	197 25.6	S15 56.8	204 24.2	S17 25.7	349 12.7	N 6 27.7
	01	174 57.6	163 35.7	40.7	212 26.1	56.2	219 26.1	25.5	4 15.4	27.8
	02	190 00.1	178 37.9	41.3	227 26.7	55.6	234 28.0	25.3	19 18.0	27.9
	03	205 02.5	193 40.2	. . 41.9	242 27.2	. . 55.0	249 29.9	. . 25.2	34 20.7	. . 28.0
	04	220 05.0	208 42.4	42.4	257 27.7	54.4	264 31.8	25.0	49 23.3	28.0
	05	235 07.4	223 44.6	43.0	272 28.3	53.8	279 33.7	24.9	64 25.9	28.1
	06	250 09.9	238 46.8	N11 43.5	287 28.8	S15 53.2	294 35.6	S17 24.7	79 28.6	N 6 28.2
	07	265 12.4	253 49.0	44.1	302 29.3	52.6	309 37.5	24.6	94 31.2	28.3
	08	280 14.8	268 51.3	44.6	317 29.9	52.0	324 39.4	24.4	109 33.9	28.4
M	09	295 17.3	283 53.5	. . 45.2	332 30.4	. . 51.4	339 41.3	. . 24.3	124 36.5	. . 28.4
O	10	310 19.8	298 55.7	45.7	347 30.9	50.8	354 43.2	24.1	139 39.2	28.5
N	11	325 22.2	313 58.0	46.3	2 31.5	50.2	9 45.2	24.0	154 41.8	28.6
D	12	340 24.7	329 00.2	N11 46.8	17 32.0	S15 49.5	24 47.1	S17 23.8	169 44.5	N 6 28.7
A	13	355 27.2	344 02.4	47.4	32 32.5	48.9	39 49.0	23.7	184 47.1	28.8
Y	14	10 29.6	359 04.7	47.9	47 33.1	48.3	54 50.9	23.5	199 49.7	28.8
	15	25 32.1	14 06.9	. . 48.4	62 33.6	. . 47.7	69 52.8	. . 23.4	214 52.4	. . 28.9
	16	40 34.6	29 09.2	49.0	77 34.1	47.1	84 54.7	23.2	229 55.0	29.0
	17	55 37.0	44 11.5	49.5	92 34.7	46.5	99 56.6	23.1	244 57.7	29.1
	18	70 39.5	59 13.7	N11 50.1	107 35.2	S15 45.9	114 58.5	S17 22.9	260 00.3	N 6 29.2
	19	85 41.9	74 16.0	50.6	122 35.8	45.3	130 00.4	22.8	275 03.0	29.2
	20	100 44.4	89 18.3	51.1	137 36.3	44.7	145 02.3	22.6	290 05.6	29.3
	21	115 46.9	104 20.5	. . 51.6	152 36.8	. . 44.0	160 04.2	. . 22.4	305 08.3	. . 29.4
	22	130 49.3	119 22.8	52.2	167 37.4	43.4	175 06.1	22.3	320 10.9	29.5
	23	145 51.8	134 25.1	52.7	182 37.9	42.8	190 08.0	22.1	335 13.5	29.6
3	00	160 54.3	149 27.4	N11 53.2	197 38.4	S15 42.2	205 09.9	S17 22.0	350 16.2	N 6 29.7
	01	175 56.7	164 29.7	53.7	212 39.0	41.6	220 11.8	21.8	5 18.8	29.7
	02	190 59.2	179 31.9	54.3	227 39.5	41.0	235 13.7	21.7	20 21.5	29.8
	03	206 01.7	194 34.2	. . 54.8	242 40.0	. . 40.4	250 15.6	. . 21.5	35 24.1	. . 29.9
	04	221 04.1	209 36.5	55.3	257 40.6	39.7	265 17.6	21.4	50 26.8	30.0
	05	236 06.6	224 38.8	55.8	272 41.1	39.1	280 19.5	21.2	65 29.4	30.1
	06	251 09.0	239 41.1	N11 56.3	287 41.7	S15 38.5	295 21.4	S17 21.1	80 32.1	N 6 30.1
	07	266 11.5	254 43.5	56.8	302 42.2	37.9	310 23.3	20.9	95 34.7	30.2
T	08	281 14.0	269 45.8	57.4	317 42.7	37.3	325 25.2	20.8	110 37.4	30.3
U	09	296 16.4	284 48.1	. . 57.9	332 43.3	. . 36.7	340 27.1	. . 20.6	125 40.0	. . 30.4
E	10	311 18.9	299 50.4	58.4	347 43.8	36.0	355 29.0	20.5	140 42.6	30.5
S	11	326 21.4	314 52.7	58.9	2 44.3	35.4	10 30.9	20.3	155 45.3	30.5
D	12	341 23.8	329 55.1	N11 59.4	17 44.9	S15 34.8	25 32.8	S17 20.2	170 47.9	N 6 30.6
A	13	356 26.3	344 57.4	11 59.9	32 45.4	34.2	40 34.7	20.0	185 50.6	30.7
Y	14	11 28.8	359 59.7	12 00.4	47 46.0	33.6	55 36.6	19.9	200 53.2	30.8
	15	26 31.2	15 02.1	. . 00.9	62 46.5	. . 33.0	70 38.5	. . 19.7	215 55.9	. . 30.9
	16	41 33.7	30 04.4	01.4	77 47.0	32.3	85 40.4	19.5	230 58.5	30.9
	17	56 36.2	45 06.7	01.9	92 47.6	31.7	100 42.4	19.4	246 01.2	31.0
	18	71 38.6	60 09.1	N12 02.4	107 48.1	S15 31.1	115 44.3	S17 19.2	261 03.8	N 6 31.1
	19	86 41.1	75 11.4	02.8	122 48.7	30.5	130 46.2	19.1	276 06.5	31.2
	20	101 43.5	90 13.8	03.3	137 49.2	29.9	145 48.1	18.9	291 09.1	31.3
	21	116 46.0	105 16.2	. . 03.8	152 49.7	. . 29.2	160 50.0	. . 18.8	306 11.7	. . 31.4
	22	131 48.5	120 18.5	04.3	167 50.3	28.6	175 51.9	18.6	321 14.4	31.4
	23	146 50.9	135 20.9	04.8	182 50.8	28.0	190 53.8	18.5	336 17.0	31.5
4	00	161 53.4	150 23.3	N12 05.3	197 51.4	S15 27.4	205 55.7	S17 18.3	351 19.7	N 6 31.6
	01	176 55.9	165 25.6	05.7	212 51.9	26.8	220 57.6	18.2	6 22.3	31.7
	02	191 58.3	180 28.0	06.2	227 52.4	26.1	235 59.5	18.0	21 25.0	31.8
	03	207 00.8	195 30.4	. . 06.7	242 53.0	. . 25.5	251 01.4	. . 17.9	36 27.6	. . 31.8
	04	222 03.3	210 32.8	07.2	257 53.5	24.9	266 03.3	17.7	51 30.3	31.9
	05	237 05.7	225 35.2	07.6	272 54.1	24.3	281 05.3	17.6	66 32.9	32.0
	06	252 08.2	240 37.6	N12 08.1	287 54.6	S15 23.7	296 07.2	S17 17.4	81 35.6	N 6 32.1
W	07	267 10.7	255 40.0	08.6	302 55.2	23.0	311 09.1	17.3	96 38.2	32.2
E	08	282 13.1	270 42.4	09.0	317 55.7	22.4	326 11.0	17.1	111 40.9	32.2
D	09	297 15.6	285 44.8	. . 09.5	332 56.2	. . 21.8	341 12.9	. . 17.0	126 43.5	. . 32.3
N	10	312 18.0	300 47.2	10.0	347 56.8	21.2	356 14.8	16.8	141 46.1	32.4
E	11	327 20.5	315 49.6	10.4	2 57.3	20.6	11 16.7	16.6	156 48.8	32.5
S	12	342 23.0	330 52.0	N12 10.9	17 57.9	S15 19.9	26 18.6	S17 16.5	171 51.4	N 6 32.6
D	13	357 25.4	345 54.4	11.3	32 58.4	19.3	41 20.5	16.3	186 54.1	32.7
A	14	12 27.9	0 56.9	11.8	47 59.0	18.7	56 22.4	16.2	201 56.7	32.7
Y	15	27 30.4	15 59.3	. . 12.2	62 59.5	. . 18.1	71 24.4	. . 16.0	216 59.4	. . 32.8
	16	42 32.8	31 01.7	12.7	78 00.0	17.4	86 26.3	15.9	232 02.0	32.9
	17	57 35.3	46 04.1	13.1	93 00.6	16.8	101 28.2	15.7	247 04.7	33.0
	18	72 37.8	61 06.6	N12 13.6	108 01.1	S15 16.2	116 30.1	S17 15.6	262 07.3	N 6 33.1
	19	87 40.2	76 09.0	14.0	123 01.7	15.6	131 32.0	15.4	277 10.0	33.1
	20	102 42.7	91 11.5	14.5	138 02.2	14.9	146 33.9	15.3	292 12.6	33.2
	21	117 45.2	106 13.9	. . 14.9	153 02.8	. . 14.3	161 35.8	. . 15.1	307 15.3	. . 33.3
	22	132 47.6	121 16.4	15.4	168 03.3	13.7	176 37.7	15.0	322 17.9	33.4
	23	147 50.1	136 18.8	15.8	183 03.8	13.1	191 39.6	14.8	337 20.5	33.5
Mer. Pass.		h m 13 14.2	v 2.3	d 0.5	v 0.5	d 0.6	v 1.9	d 0.2	v 2.6	d 0.1

STARS		
Name	SHA	Dec
	° ′	° ′
Acamar	315 20.8	S40 16.
Achernar	335 29.3	S57 11.
Acrux	173 12.6	S63 09.
Adhara	255 14.8	S28 59.
Aldebaran	290 53.1	N16 31.
Alioth	166 22.7	N55 54.
Alkaid	153 00.9	N49 15.
Al Na'ir	27 48.0	S46 55.
Alnilam	275 49.5	S 1 11.
Alphard	217 58.9	S 8 42.
Alphecca	126 13.5	N26 40.
Alpheratz	357 47.2	N29 08.
Altair	62 11.6	N 8 53.
Ankaa	353 19.1	S42 15.
Antares	112 30.2	S26 27.
Arcturus	145 58.4	N19 07.
Atria	107 35.0	S69 02.
Avior	234 19.0	S59 32.
Bellatrix	278 35.4	N 6 21.
Betelgeuse	271 04.6	N 7 24.
Canopus	263 57.4	S52 42.
Capella	280 39.1	N46 00.
Deneb	49 34.1	N45 18.
Denebola	182 36.5	N14 31.
Diphda	348 59.3	S17 56.
Dubhe	193 54.6	N61 41.
Elnath	278 16.6	N28 37.
Eltanin	90 47.8	N51 28.
Enif	33 50.6	N 9 54.9
Fomalhaut	15 27.7	S29 34.5
Gacrux	172 04.2	S57 09.9
Gienah	175 55.3	S17 35.8
Hadar	148 52.3	S60 25.0
Hamal	328 04.6	N23 30.4
Kaus Aust.	83 48.2	S34 22.8
Kochab	137 18.9	N74 06.6
Markab	13 41.9	N15 15.2
Menkar	314 18.5	N 4 07.5
Menkent	148 11.2	S36 25.0
Miaplacidus	221 39.8	S69 45.5
Mirfak	308 45.2	N49 53.9
Nunki	76 02.4	S26 17.2
Peacock	53 24.5	S56 42.2
Pollux	243 31.3	N28 00.2
Procyon	245 02.8	N 5 12.0
Rasalhague	96 09.5	N12 32.9
Regulus	207 46.5	N11 55.2
Rigel	281 15.1	S 8 11.6
Rigil Kent.	139 56.0	S60 52.3
Sabik	102 16.2	S15 44.3
Schedar	349 44.9	N56 35.4
Shaula	96 26.3	S37 06.6
Sirius	258 36.4	S16 43.9
Spica	158 34.4	S11 12.7
Suhail	222 54.5	S43 28.4
Vega	80 41.3	N38 47.2
Zuben'ubi	137 08.8	S16 04.9

	SHA	Mer. Pass
	° ′	h m
Venus	348 33.1	14 00
Mars	36 44.2	10 49
Jupiter	44 15.7	10 18
Saturn	189 21.9	0 39

UT d h	SUN GHA ° ′	SUN Dec ° ′	MOON GHA ° ′	v ′	MOON Dec ° ′	d ′	HP ′
2 00	176 56.8	S 7 14.1	122 27.8	8.9	N20 15.0	10.6	58.1
01	191 56.9	13.2	136 55.7	8.8	20 25.6	10.5	58.1
02	206 57.1	12.2	151 23.5	8.7	20 36.1	10.4	58.1
03	221 57.2	. . 11.3	165 51.2	8.6	20 46.5	10.3	58.2
04	236 57.3	10.3	180 18.8	8.5	20 56.8	10.1	58.2
05	251 57.4	09.3	194 46.3	8.4	21 06.9	10.1	58.2
06	266 57.6	S 7 08.4	209 13.7	8.3	N21 17.0	9.9	58.2
07	281 57.7	07.4	223 41.0	8.2	21 26.9	9.8	58.2
08	296 57.8	06.5	238 08.2	8.1	21 36.7	9.8	58.3
M 09	311 57.9	. . 05.5	252 35.3	8.0	21 46.5	9.5	58.3
O 10	326 58.1	04.6	267 02.3	7.9	21 56.0	9.5	58.3
N 11	341 58.2	03.6	281 29.2	7.8	22 05.5	9.3	58.3
D 12	356 58.3	S 7 02.7	295 56.0	7.7	N22 14.8	9.3	58.3
A 13	11 58.5	01.7	310 22.7	7.7	22 24.1	9.0	58.3
Y 14	26 58.6	7 00.7	324 49.4	7.5	22 33.1	9.0	58.4
15	41 58.7	6 59.8	339 15.9	7.4	22 42.1	8.8	58.4
16	56 58.8	58.8	353 42.3	7.3	22 50.9	8.7	58.4
17	71 59.0	57.9	8 08.6	7.2	22 59.6	8.6	58.4
18	86 59.1	S 6 56.9	22 34.8	7.2	N23 08.2	8.4	58.4
19	101 59.2	56.0	37 01.0	7.0	23 16.6	8.3	58.5
20	116 59.4	55.0	51 27.0	6.9	23 24.9	8.2	58.5
21	131 59.5	. . 54.0	65 52.9	6.9	23 33.1	8.0	58.5
22	146 59.6	53.1	80 18.8	6.7	23 41.1	7.9	58.5
23	161 59.7	52.1	94 44.5	6.7	23 49.0	7.8	58.5
3 00	176 59.9	S 6 51.2	109 10.2	6.5	N23 56.8	7.6	58.5
01	192 00.0	50.2	123 35.7	6.5	24 04.4	7.4	58.6
02	207 00.1	49.3	138 01.2	6.4	24 11.8	7.3	58.6
03	222 00.3	. . 48.3	152 26.6	6.3	24 19.1	7.2	58.6
04	237 00.4	47.3	166 51.9	6.2	24 26.3	7.0	58.6
05	252 00.5	46.4	181 17.1	6.1	24 33.3	6.9	58.6
06	267 00.7	S 6 45.4	195 42.2	6.0	N24 40.2	6.7	58.6
07	282 00.8	44.5	210 07.2	5.9	24 46.9	6.6	58.7
T 08	297 00.9	43.5	224 32.1	5.8	24 53.5	6.4	58.7
U 09	312 01.1	. . 42.5	238 56.9	5.8	24 59.9	6.2	58.7
E 10	327 01.2	41.6	253 21.7	5.7	25 06.1	6.1	58.7
S 11	342 01.3	40.6	267 46.4	5.6	25 12.2	6.0	58.7
D 12	357 01.5	S 6 39.7	282 11.0	5.5	N25 18.2	5.8	58.7
A 13	12 01.6	38.7	296 35.5	5.4	25 24.0	5.6	58.8
Y 14	27 01.7	37.7	310 59.9	5.4	25 29.6	5.5	58.8
15	42 01.9	. . 36.8	325 24.3	5.2	25 35.1	5.3	58.8
16	57 02.0	35.8	339 48.5	5.2	25 40.4	5.1	58.8
17	72 02.1	34.9	354 12.7	5.2	25 45.5	5.0	58.8
18	87 02.3	S 6 33.9	8 36.9	5.0	N25 50.5	4.8	58.8
19	102 02.4	32.9	23 00.9	5.0	25 55.3	4.6	58.9
20	117 02.5	32.0	37 24.9	4.9	25 59.9	4.5	58.9
21	132 02.7	. . 31.0	51 48.8	4.8	26 04.4	4.3	58.9
22	147 02.8	30.0	66 12.6	4.8	26 08.7	4.1	58.9
23	162 02.9	29.1	80 36.4	4.7	26 12.8	4.0	58.9
4 00	177 03.1	S 6 28.1	95 00.1	4.7	N26 16.8	3.8	58.9
01	192 03.2	27.2	109 23.8	4.6	26 20.6	3.6	58.9
02	207 03.3	26.2	123 47.4	4.5	26 24.2	3.5	59.0
03	222 03.5	. . 25.2	138 10.9	4.4	26 27.7	3.2	59.0
04	237 03.6	24.3	152 34.3	4.4	26 30.9	3.1	59.0
05	252 03.8	23.3	166 57.7	4.4	26 34.0	3.0	59.0
06	267 03.9	S 6 22.3	181 21.1	4.3	N26 37.0	2.7	59.0
W 07	282 04.0	21.4	195 44.4	4.3	26 39.7	2.6	59.0
E 08	297 04.2	20.4	210 07.7	4.2	26 42.3	2.4	59.1
D 09	312 04.3	. . 19.5	224 30.9	4.1	26 44.7	2.2	59.1
N 10	327 04.4	18.5	238 54.0	4.1	26 46.9	2.0	59.1
E 11	342 04.6	17.5	253 17.1	4.1	26 48.9	1.9	59.1
S 12	357 04.7	S 6 16.6	267 40.2	4.0	N26 50.8	1.6	59.1
D 13	12 04.8	15.6	282 03.2	4.0	26 52.4	1.5	59.1
A 14	27 05.0	14.6	296 26.2	4.0	26 53.9	1.3	59.1
Y 15	42 05.1	. . 13.7	310 49.2	3.9	26 55.2	1.1	59.2
16	57 05.3	12.7	325 12.1	3.9	26 56.3	1.0	59.2
17	72 05.4	11.7	339 35.0	3.9	26 57.3	0.7	59.2
18	87 05.5	S 6 10.8	353 57.9	3.8	N26 58.0	0.6	59.2
19	102 05.7	09.8	8 20.7	3.8	26 58.6	0.4	59.2
20	117 05.8	08.8	22 43.5	3.8	26 59.0	0.2	59.2
21	132 06.0	. . 07.9	37 06.3	3.8	26 59.2	0.0	59.2
22	147 06.1	06.9	51 29.1	3.7	26 59.2	0.2	59.3
23	162 06.2	06.0	65 51.8	3.8	N26 59.0	0.3	59.3
	SD 16.2	*d* 1.0	SD 15.9		16.0		16.1

Lat. °	Twilight Naut. h m	Twilight Civil h m	Sunrise h m	Moonrise 2 h m	Moonrise 3 h m	Moonrise 4 h m	Moonrise 5 h m
N 72	04 59	06 17	07 26	▭	▭	▭	▭
N 70	05 06	06 16	07 17	▭	▭	▭	▭
68	05 10	06 15	07 11	05 14	▭	▭	▭
66	05 14	06 14	07 05	06 02	▭	▭	▭
64	05 18	06 13	07 00	06 33	06 20	▭	▭
62	05 20	06 12	06 56	06 57	07 01	07 18	08 17
60	05 23	06 11	06 52	07 15	07 30	08 00	09 00
N 58	05 25	06 10	06 49	07 31	07 52	08 28	09 30
56	05 26	06 09	06 46	07 45	08 10	08 50	09 52
54	05 27	06 08	06 44	07 56	08 25	09 08	10 10
52	05 29	06 08	06 41	08 07	08 38	09 24	10 26
50	05 29	06 07	06 39	08 16	08 50	09 37	10 39
45	05 31	06 05	06 34	08 36	09 14	10 04	11 07
N 40	05 32	06 03	06 30	08 52	09 34	10 26	11 29
35	05 32	06 02	06 27	09 06	09 50	10 44	11 47
30	05 32	06 00	06 24	09 18	10 05	11 00	12 02
20	05 31	05 56	06 18	09 38	10 29	11 26	12 29
N 10	05 28	05 52	06 13	09 56	10 50	11 49	12 51
0	05 24	05 48	06 09	10 13	11 10	12 10	13 13
S 10	05 18	05 43	06 04	10 30	11 30	12 32	13 34
20	05 10	05 36	05 58	10 48	11 51	12 55	13 56
30	05 00	05 28	05 52	11 09	12 16	13 22	14 23
35	04 53	05 23	05 49	11 22	12 31	13 37	14 38
40	04 44	05 17	05 45	11 36	12 48	13 56	14 56
45	04 34	05 10	05 40	11 53	13 08	14 18	15 18
S 50	04 20	05 01	05 34	12 15	13 34	14 46	15 45
52	04 14	04 57	05 32	12 25	13 47	15 00	15 58
54	04 07	04 52	05 29	12 37	14 02	15 16	16 14
56	03 58	04 46	05 25	12 50	14 19	15 35	16 32
58	03 49	04 40	05 22	13 05	14 39	15 59	16 54
S 60	03 37	04 34	05 18	13 24	15 05	16 30	17 22

Lat. °	Sunset h m	Twilight Civil h m	Twilight Naut. h m	Moonset 2 h m	Moonset 3 h m	Moonset 4 h m	Moonset 5 h m
N 72	17 00	18 09	19 28	▭	▭	▭	▭
N 70	17 08	18 10	19 21	▭	▭	▭	▭
68	17 15	18 11	19 16	02 28	▭	▭	▭
66	17 20	18 12	19 11	01 40	▭	▭	▭
64	17 25	18 13	19 08	01 10	03 20	▭	▭
62	17 29	18 14	19 05	00 48	02 39	04 27	05 38
60	17 33	18 14	19 03	00 29	02 11	03 45	04 54
N 58	17 36	18 15	19 01	00 14	01 50	03 17	04 25
56	17 39	18 16	18 59	00 02	01 32	02 56	04 03
54	17 41	18 17	18 58	25 17	01 17	02 38	03 44
52	17 43	18 17	18 56	25 04	01 04	02 22	03 28
50	17 46	18 18	18 55	24 53	00 53	02 09	03 15
45	17 50	18 20	18 54	24 29	00 29	01 42	02 47
N 40	17 54	18 21	18 53	24 10	00 10	01 21	02 25
35	17 57	18 23	18 52	23 55	25 03	01 03	02 07
30	18 00	18 24	18 52	23 41	24 48	00 48	01 51
20	18 06	18 28	18 54	23 18	24 22	00 22	01 24
N 10	18 11	18 32	18 56	22 58	23 59	25 01	01 01
0	18 15	18 36	19 00	22 40	23 38	24 40	00 40
S 10	18 20	18 41	19 06	22 21	23 18	24 19	00 19
20	18 25	18 47	19 13	22 01	22 55	23 55	25 00
30	18 31	18 55	19 24	21 39	22 30	23 29	24 34
35	18 34	19 00	19 30	21 25	22 14	23 13	24 19
40	18 38	19 06	19 39	21 10	21 57	22 54	24 01
45	18 43	19 13	19 49	20 52	21 36	22 32	23 40
S 50	18 49	19 22	20 02	20 29	21 09	22 03	23 13
52	18 51	19 26	20 08	20 19	20 56	21 49	23 00
54	18 54	19 31	20 15	20 07	20 41	21 33	22 45
56	18 57	19 36	20 24	19 53	20 24	21 14	22 27
58	19 01	19 42	20 33	19 37	20 03	20 51	22 06
S 60	19 05	19 48	20 44	19 18	19 37	20 19	21 38

Day	SUN Eqn. of Time 00^h	SUN Eqn. of Time 12^h	SUN Mer. Pass.	MOON Mer. Pass. Upper	MOON Mer. Pass. Lower	Age	Phase
d	m s	m s	h m	h m	h m	d %	
2	12 13	12 07	12 12	16 26	03 59	05 30	
3	12 01	11 54	12 12	17 24	04 55	06 41	◐
4	11 48	11 41	12 12	18 25	05 54	07 52	

	UT	ARIES	VENUS −4.7		MARS +1.2		JUPITER −2.0		SATURN +0.5		STARS		
		GHA	GHA	Dec	GHA	Dec	GHA	Dec	GHA	Dec	Name	SHA	Dec
	d h	° ′	° ′	° ′	° ′	° ′	° ′	° ′	° ′	° ′		° ′	° ′
	5 00	162 52.5	151 21.3	N12 16.2	198 04.4	S15 12.4	206 41.6	S17 14.7	352 23.2	N 6 33.5	Acamar	315 20.8	S40 16.
	01	177 55.0	166 23.7	16.7	213 04.9	11.8	221 43.5	14.5	7 25.8	33.6	Achernar	335 29.3	S57 11.
	02	192 57.5	181 26.2	17.1	228 05.5	11.2	236 45.4	14.4	22 28.5	33.7	Acrux	173 12.5	S63 09.
	03	207 59.9	196 28.7	. . 17.5	243 06.0	. . 10.5	251 47.3	. . 14.2	37 31.1	. . 33.8	Adhara	255 14.9	S28 59.
	04	223 02.4	211 31.2	18.0	258 06.6	09.9	266 49.2	14.1	52 33.8	33.9	Aldebaran	290 53.1	N16 31.
	05	238 04.9	226 33.6	18.4	273 07.1	09.3	281 51.1	13.9	67 36.4	33.9			
	06	253 07.3	241 36.1	N12 18.8	288 07.7	S15 08.7	296 53.0	S17 13.8	82 39.1	N 6 34.0	Alioth	166 22.7	N55 54.
	07	268 09.8	256 38.6	19.2	303 08.2	08.0	311 54.9	13.6	97 41.7	34.1	Alkaid	153 00.8	N49 15.
T	08	283 12.3	271 41.1	19.6	318 08.8	07.4	326 56.9	13.4	112 44.4	34.2	Al Na'ir	27 48.0	S46 55.
H	09	298 14.7	286 43.6	. . 20.1	333 09.3	. . 06.8	341 58.8	. . 13.3	127 47.0	. . 34.3	Alnilam	275 49.5	S 1 11.
U	10	313 17.2	301 46.1	20.5	348 09.8	06.1	357 00.7	13.1	142 49.7	34.4	Alphard	217 58.9	S 8 42.
R	11	328 19.7	316 48.6	20.9	3 10.4	05.5	12 02.6	13.0	157 52.3	34.4			
S	12	343 22.1	331 51.1	N12 21.3	18 10.9	S15 04.9	27 04.5	S17 12.8	172 55.0	N 6 34.5	Alphecca	126 13.5	N26 40.
D	13	358 24.6	346 53.6	21.7	33 11.5	04.3	42 06.4	12.7	187 57.6	34.6	Alpheratz	357 47.2	N29 08.
A	14	13 27.0	1 56.1	22.1	48 12.0	03.6	57 08.3	12.5	203 00.2	34.7	Altair	62 11.6	N 8 53.
Y	15	28 29.5	16 58.6	. . 22.5	63 12.6	. . 03.0	72 10.2	. . 12.4	218 02.9	. . 34.8	Ankaa	353 19.1	S42 15.
	16	43 32.0	32 01.2	22.9	78 13.1	02.4	87 12.2	12.2	233 05.5	34.8	Antares	112 30.1	S26 27.
	17	58 34.4	47 03.7	23.3	93 13.7	01.7	102 14.1	12.1	248 08.2	34.9			
	18	73 36.9	62 06.2	N12 23.7	108 14.2	S15 01.1	117 16.0	S17 11.9	263 10.8	N 6 35.0	Arcturus	145 58.4	N19 07.
	19	88 39.4	77 08.7	24.1	123 14.8	15 00.5	132 17.9	11.8	278 13.5	35.1	Atria	107 34.9	S69 02.
	20	103 41.8	92 11.3	24.5	138 15.3	14 59.8	147 19.8	11.6	293 16.1	35.2	Avior	234 19.0	S59 32.
	21	118 44.3	107 13.8	. . 24.9	153 15.9	. . 59.2	162 21.7	. . 11.5	308 18.8	. . 35.2	Bellatrix	278 35.4	N 6 21.
	22	133 46.8	122 16.4	25.3	168 16.4	58.6	177 23.6	11.3	323 21.4	35.3	Betelgeuse	271 04.7	N 7 24.
	23	148 49.2	137 18.9	25.7	183 17.0	57.9	192 25.5	11.2	338 24.1	35.4			
	6 00	163 51.7	152 21.5	N12 26.1	198 17.5	S14 57.3	207 27.5	S17 11.0	353 26.7	N 6 35.5	Canopus	263 57.4	S52 42.
	01	178 54.1	167 24.0	26.5	213 18.1	56.7	222 29.4	10.9	8 29.4	35.6	Capella	280 39.1	N46 00.
	02	193 56.6	182 26.6	26.9	228 18.6	56.0	237 31.3	10.7	23 32.0	35.6	Deneb	49 34.1	N45 18.
	03	208 59.1	197 29.1	. . 27.2	243 19.2	. . 55.4	252 33.2	. . 10.6	38 34.7	. . 35.7	Denebola	182 36.5	N14 31.
	04	224 01.5	212 31.7	27.6	258 19.7	54.8	267 35.1	10.4	53 37.3	35.8	Diphda	348 59.3	S17 56.
	05	239 04.0	227 34.3	28.0	273 20.3	54.1	282 37.0	10.2	68 39.9	35.9			
	06	254 06.5	242 36.8	N12 28.4	288 20.8	S14 53.5	297 38.9	S17 10.1	83 42.6	N 6 36.0	Dubhe	193 54.6	N61 41.
	07	269 08.9	257 39.4	28.8	303 21.4	52.9	312 40.9	09.9	98 45.2	36.0	Elnath	278 16.6	N28 37.
	08	284 11.4	272 42.0	29.1	318 21.9	52.2	327 42.8	09.8	113 47.9	36.1	Eltanin	90 47.7	N51 28.
F	09	299 13.9	287 44.6	. . 29.5	333 22.5	. . 51.6	342 44.7	. . 09.6	128 50.5	. . 36.2	Enif	33 50.6	N 9 54.
R	10	314 16.3	302 47.2	29.9	348 23.0	51.0	357 46.6	09.5	143 53.2	36.3	Fomalhaut	15 27.7	S29 34.
I	11	329 18.8	317 49.8	30.2	3 23.6	50.3	12 48.5	09.3	158 55.8	36.4			
D	12	344 21.3	332 52.4	N12 30.6	18 24.1	S14 49.7	27 50.4	S17 09.2	173 58.5	N 6 36.5	Gacrux	172 04.2	S57 09.
A	13	359 23.7	347 55.0	31.0	33 24.7	49.1	42 52.4	09.0	189 01.1	36.5	Gienah	175 55.3	S17 35.
Y	14	14 26.2	2 57.6	31.3	48 25.2	48.4	57 54.3	08.9	204 03.8	36.6	Hadar	148 52.2	S60 25.
	15	29 28.6	18 00.2	. . 31.7	63 25.8	. . 47.8	72 56.2	. . 08.7	219 06.4	. . 36.7	Hamal	328 04.6	N23 30.
	16	44 31.1	33 02.8	32.0	78 26.3	47.1	87 58.1	08.6	234 09.1	36.8	Kaus Aust.	83 48.1	S34 22.
	17	59 33.6	48 05.4	32.4	93 26.9	46.5	103 00.0	08.4	249 11.7	36.9			
	18	74 36.0	63 08.0	N12 32.7	108 27.4	S14 45.9	118 01.9	S17 08.3	264 14.4	N 6 36.9	Kochab	137 18.8	N74 06.
	19	89 38.5	78 10.6	33.1	123 28.0	45.2	133 03.8	08.1	279 17.0	37.0	Markab	13 41.9	N15 15.
	20	104 41.0	93 13.3	33.4	138 28.5	44.6	148 05.8	08.0	294 19.7	37.1	Menkar	314 18.5	N 4 07.
	21	119 43.4	108 15.9	. . 33.8	153 29.1	. . 44.0	163 07.7	. . 07.8	309 22.3	. . 37.2	Menkent	148 11.2	S36 25.
	22	134 45.9	123 18.5	34.1	168 29.6	43.3	178 09.6	07.7	324 24.9	37.3	Miaplacidus	221 39.9	S69 45.
	23	149 48.4	138 21.2	34.5	183 30.2	42.7	193 11.5	07.5	339 27.6	37.3			
	7 00	164 50.8	153 23.8	N12 34.8	198 30.8	S14 42.0	208 13.4	S17 07.4	354 30.2	N 6 37.4	Mirfak	308 45.2	N49 53.
	01	179 53.3	168 26.5	35.1	213 31.3	41.4	223 15.3	07.2	9 32.9	37.5	Nunki	76 02.4	S26 17.
	02	194 55.8	183 29.1	35.5	228 31.9	40.8	238 17.3	07.1	24 35.5	37.6	Peacock	53 24.5	S56 42.
	03	209 58.2	198 31.8	. . 35.8	243 32.4	. . 40.1	253 19.2	. . 06.9	39 38.2	. . 37.7	Pollux	243 31.3	N28 00.
	04	225 00.7	213 34.4	36.1	258 33.0	39.5	268 21.1	06.7	54 40.8	37.7	Procyon	245 02.8	N 5 12.
	05	240 03.1	228 37.1	36.4	273 33.5	38.8	283 23.0	06.6	69 43.5	37.8			
	06	255 05.6	243 39.7	N12 36.8	288 34.1	S14 38.2	298 24.9	S17 06.4	84 46.1	N 6 37.9	Rasalhague	96 09.4	N12 32.
	07	270 08.1	258 42.4	37.1	303 34.6	37.6	313 26.8	06.3	99 48.8	38.0	Regulus	207 46.5	N11 55.
S	08	285 10.5	273 45.1	37.4	318 35.2	36.9	328 28.8	06.1	114 51.4	38.1	Rigel	281 15.1	S 8 11.
A	09	300 13.0	288 47.8	. . 37.7	333 35.7	. . 36.3	343 30.7	. . 06.0	129 54.1	. . 38.1	Rigil Kent.	139 55.9	S60 52.
T	10	315 15.5	303 50.4	38.1	348 36.3	35.6	358 32.6	05.8	144 56.7	38.2	Sabik	102 16.2	S15 44.
U	11	330 17.9	318 53.1	38.4	3 36.8	35.0	13 34.5	05.7	159 59.4	38.3			
R	12	345 20.4	333 55.8	N12 38.7	18 37.4	S14 34.4	28 36.4	S17 05.5	175 02.0	N 6 38.4	Schedar	349 44.9	N56 35.
D	13	0 22.9	348 58.5	39.0	33 38.0	33.7	43 38.4	05.4	190 04.7	38.5	Shaula	96 26.3	S37 06.
A	14	15 25.3	4 01.2	39.3	48 38.5	33.1	58 40.3	05.2	205 07.3	38.6	Sirius	258 36.4	S16 43.
Y	15	30 27.8	19 03.9	. . 39.6	63 39.1	. . 32.4	73 42.2	. . 05.1	220 09.9	. . 38.6	Spica	158 34.4	S11 12.
	16	45 30.3	34 06.6	39.9	78 39.6	31.8	88 44.1	04.9	235 12.6	38.7	Suhail	222 54.5	S43 28.
	17	60 32.7	49 09.3	40.2	93 40.2	31.1	103 46.0	04.8	250 15.2	38.8			
	18	75 35.2	64 12.0	N12 40.5	108 40.7	S14 30.5	118 47.9	S17 04.6	265 17.9	N 6 38.9	Vega	80 41.3	N38 47.2
	19	90 37.6	79 14.7	40.8	123 41.3	29.9	133 49.9	04.5	280 20.5	39.0	Zuben'ubi	137 08.8	S16 05.0
	20	105 40.1	94 17.5	41.1	138 41.9	29.2	148 51.8	04.3	295 23.2	39.0		SHA	Mer. Pass
	21	120 42.6	109 20.2	. . 41.4	153 42.4	. . 28.6	163 53.7	. . 04.2	310 25.8	. . 39.1		° ′	h m
	22	135 45.0	124 22.9	41.7	168 43.0	27.9	178 55.6	04.0	325 28.5	39.2	Venus	348 29.8	13 48
	23	150 47.5	139 25.6	42.0	183 43.5	27.3	193 57.5	03.9	340 31.1	39.3	Mars	34 25.8	10 46
		h m									Jupiter	43 35.8	10 09
	Mer. Pass.	13 02.4	*v* 2.6	*d* 0.4	*v* 0.6	*d* 0.6	*v* 1.9	*d* 0.2	*v* 2.6	*d* 0.1	Saturn	189 35.0	0 26

UT	SUN GHA	SUN Dec	MOON GHA	v	MOON Dec	d	HP
d h	° ′	° ′	° ′	′	° ′	′	′
5 00	177 06.4	S 6 05.0	80 14.6	3.7	N26 58.7	0.6	59.3
01	192 06.5	04.0	94 37.3	3.7	26 58.1	0.7	59.3
02	207 06.7	03.1	109 00.0	3.7	26 57.4	0.9	59.3
03	222 06.8	. . 02.1	123 22.7	3.7	26 56.5	1.1	59.3
04	237 06.9	01.1	137 45.4	3.7	26 55.4	1.3	59.3
05	252 07.1	6 00.2	152 08.1	3.7	26 54.1	1.5	59.3
06	267 07.2	S 5 59.2	166 30.8	3.7	N26 52.6	1.6	59.4
07	282 07.4	58.2	180 53.5	3.7	26 51.0	1.9	59.4
08	297 07.5	57.3	195 16.2	3.7	26 49.1	2.0	59.4
09	312 07.6	. . 56.3	209 38.9	3.7	26 47.1	2.2	59.4
10	327 07.8	55.3	224 01.6	3.7	26 44.9	2.4	59.4
11	342 07.9	54.4	238 24.3	3.8	26 42.5	2.6	59.4
12	357 08.1	S 5 53.4	252 47.1	3.7	N26 39.9	2.8	59.4
13	12 08.2	52.4	267 09.8	3.8	26 37.1	2.9	59.4
14	27 08.4	51.5	281 32.6	3.8	26 34.2	3.2	59.5
15	42 08.5	. . 50.5	295 55.4	3.8	26 31.0	3.3	59.5
16	57 08.6	49.5	310 18.2	3.8	26 27.7	3.5	59.5
17	72 08.8	48.6	324 41.0	3.8	26 24.2	3.7	59.5
18	87 08.9	S 5 47.6	339 03.8	3.9	N26 20.5	3.9	59.5
19	102 09.1	46.6	353 26.7	3.9	26 16.6	4.1	59.5
20	117 09.2	45.6	7 49.6	4.0	26 12.5	4.2	59.5
21	132 09.4	. . 44.7	22 12.6	3.9	26 08.3	4.4	59.5
22	147 09.5	43.7	36 35.5	4.0	26 03.9	4.6	59.5
23	162 09.7	42.7	50 58.5	4.1	25 59.3	4.8	59.5
6 00	177 09.8	S 5 41.8	65 21.6	4.1	N25 54.5	5.0	59.6
01	192 09.9	40.8	79 44.7	4.1	25 49.5	5.1	59.6
02	207 10.1	39.8	94 07.8	4.2	25 44.4	5.3	59.6
03	222 10.2	. . 38.9	108 31.0	4.2	25 39.1	5.5	59.6
04	237 10.4	37.9	122 54.2	4.3	25 33.6	5.7	59.6
05	252 10.5	36.9	137 17.5	4.3	25 27.9	5.8	59.6
06	267 10.7	S 5 36.0	151 40.8	4.3	N25 22.1	6.0	59.6
07	282 10.8	35.0	166 04.1	4.5	25 16.1	6.2	59.6
08	297 11.0	34.0	180 27.6	4.4	25 09.9	6.4	59.6
09	312 11.1	. . 33.0	194 51.0	4.6	25 03.5	6.5	59.6
10	327 11.3	32.1	209 14.6	4.6	24 57.0	6.7	59.6
11	342 11.4	31.1	223 38.2	4.6	24 50.3	6.9	59.6
12	357 11.6	S 5 30.1	238 01.8	4.7	N24 43.4	7.1	59.6
13	12 11.7	29.2	252 25.5	4.8	24 36.3	7.2	59.7
14	27 11.8	28.2	266 49.3	4.8	24 29.1	7.3	59.7
15	42 12.0	. . 27.2	281 13.1	4.9	24 21.8	7.6	59.7
16	57 12.1	26.3	295 37.0	5.0	24 14.2	7.7	59.7
17	72 12.3	25.3	310 01.0	5.0	24 06.5	7.8	59.7
18	87 12.4	S 5 24.3	324 25.0	5.1	N23 58.7	8.1	59.7
19	102 12.6	23.3	338 49.1	5.2	23 50.6	8.2	59.7
20	117 12.7	22.4	353 13.3	5.2	23 42.4	8.3	59.7
21	132 12.9	. . 21.4	7 37.5	5.4	23 34.1	8.5	59.7
22	147 13.0	20.4	22 01.9	5.4	23 25.6	8.7	59.7
23	162 13.2	19.5	36 26.3	5.4	23 16.9	8.8	59.7
7 00	177 13.3	S 5 18.5	50 50.7	5.6	N23 08.1	8.9	59.7
01	192 13.5	17.5	65 15.3	5.6	22 59.2	9.2	59.7
02	207 13.6	16.5	79 39.9	5.7	22 50.0	9.2	59.7
03	222 13.8	. . 15.6	94 04.6	5.8	22 40.8	9.5	59.7
04	237 13.9	14.6	108 29.4	5.8	22 31.3	9.5	59.7
05	252 14.1	13.6	122 54.2	6.0	22 21.8	9.7	59.7
06	267 14.2	S 5 12.6	137 19.2	6.0	N22 12.1	9.9	59.7
07	282 14.4	11.7	151 44.2	6.1	22 02.2	10.0	59.7
08	297 14.5	10.7	166 09.3	6.2	21 52.2	10.1	59.7
09	312 14.7	. . 09.7	180 34.5	6.2	21 42.1	10.3	59.7
10	327 14.8	08.8	194 59.7	6.4	21 31.8	10.4	59.7
11	342 15.0	07.8	209 25.1	6.4	21 21.4	10.6	59.7
12	357 15.1	S 5 06.8	223 50.5	6.5	N21 10.8	10.7	59.7
13	12 15.3	05.8	238 16.0	6.6	21 00.1	10.8	59.7
14	27 15.4	04.9	252 41.6	6.7	20 49.3	10.9	59.7
15	42 15.6	. . 03.9	267 07.3	6.8	20 38.4	11.1	59.7
16	57 15.7	02.9	281 33.1	6.9	20 27.3	11.2	59.7
17	72 15.9	01.9	295 59.0	6.9	20 16.1	11.4	59.7
18	87 16.0	S 5 01.0	310 24.9	7.0	N20 04.7	11.4	59.7
19	102 16.2	5 00.0	324 50.9	7.2	19 53.3	11.6	59.7
20	117 16.4	4 59.0	339 17.1	7.2	19 41.7	11.7	59.7
21	132 16.5	. . 58.0	353 43.3	7.3	19 30.0	11.9	59.7
22	147 16.7	57.1	8 09.6	7.4	19 18.1	11.9	59.7
23	162 16.8	56.1	22 36.0	7.4	N19 06.2	12.1	59.7
	SD 16.1	d 1.0	SD 16.2		16.3		16.3

Lat.	Twilight Naut.	Twilight Civil	Sunrise	Moonrise 5	Moonrise 6	Moonrise 7	Moonrise 8
°	h m	h m	h m	h m	h m	h m	h m
N 72	04 44	06 03	07 10	▭	▭	▭	11 42
N 70	04 52	06 03	07 03	▭	▭	▭	12 30
68	04 58	06 03	06 58	▭	▭	09 57	13 00
66	05 03	06 03	06 54	▭	▭	11 00	13 22
64	05 07	06 02	06 50	▭	09 13	11 35	13 40
62	05 11	06 02	06 46	08 17	10 02	12 00	13 54
60	05 14	06 02	06 44	09 00	10 33	12 19	14 06
N 58	05 16	06 02	06 41	09 30	10 56	12 35	14 16
56	05 19	06 02	06 39	09 52	11 15	12 49	14 25
54	05 20	06 01	06 37	10 10	11 30	13 01	14 33
52	05 22	06 01	06 35	10 26	11 44	13 11	14 40
50	05 23	06 01	06 33	10 39	11 56	13 20	14 47
45	05 26	06 00	06 29	11 07	12 20	13 40	15 00
N 40	05 27	05 59	06 26	11 29	12 40	13 55	15 12
35	05 28	05 58	06 23	11 47	12 56	14 09	15 21
30	05 29	05 56	06 20	12 02	13 10	14 20	15 29
20	05 28	05 54	06 16	12 29	13 34	14 40	15 44
N 10	05 26	05 51	06 12	12 51	13 55	14 57	15 56
0	05 23	05 47	06 08	13 13	14 14	15 13	16 08
S 10	05 18	05 43	06 04	13 34	14 33	15 28	16 19
20	05 11	05 37	06 00	13 56	14 54	15 45	16 32
30	05 02	05 30	05 54	14 23	15 17	16 04	16 45
35	04 56	05 26	05 51	14 38	15 31	16 16	16 54
40	04 48	05 21	05 48	14 56	15 47	16 28	17 03
45	04 39	05 14	05 44	15 18	16 06	16 43	17 13
S 50	04 26	05 06	05 39	15 45	16 29	17 02	17 26
52	04 20	05 02	05 37	15 58	16 41	17 10	17 32
54	04 14	04 58	05 34	16 14	16 53	17 20	17 39
56	04 06	04 53	05 32	16 32	17 08	17 31	17 46
58	03 57	04 48	05 29	16 54	17 25	17 43	17 54
S 60	03 47	04 42	05 25	17 22	17 46	17 57	18 03

Lat.	Sunset	Twilight Civil	Twilight Naut.	Moonset 5	Moonset 6	Moonset 7	Moonset 8
°	h m	h m	h m	h m	h m	h m	h m
N 72	17 15	18 22	19 42	▭	▭	▭	08 30
N 70	17 21	18 22	19 34	▭	▭	▭	07 40
68	17 26	18 22	19 27	▭	▭	08 14	07 08
66	17 30	18 22	19 21	▭	▭	07 10	06 45
64	17 34	18 22	19 17	▭	06 52	06 35	06 26
62	17 37	18 22	19 13	05 38	06 02	06 09	06 10
60	17 40	18 22	19 10	04 54	05 31	05 49	05 57
N 58	17 43	18 22	19 07	04 25	05 08	05 32	05 46
56	17 45	18 22	19 05	04 03	04 48	05 17	05 36
54	17 47	18 22	19 03	03 44	04 32	05 05	05 27
52	17 49	18 23	19 02	03 28	04 18	04 54	05 20
50	17 51	18 23	19 00	03 15	04 06	04 44	05 12
45	17 54	18 24	18 58	02 47	03 41	04 24	04 57
N 40	17 57	18 24	18 56	02 25	03 21	04 07	04 44
35	18 00	18 25	18 55	02 07	03 04	03 52	04 34
30	18 03	18 26	18 54	01 51	02 49	03 40	04 24
20	18 07	18 29	18 54	01 24	02 24	03 19	04 08
N 10	18 11	18 32	18 56	01 01	02 02	03 00	03 53
0	18 15	18 35	18 59	00 40	01 42	02 42	03 39
S 10	18 18	18 39	19 04	00 19	01 22	02 25	03 25
20	18 23	18 45	19 11	25 00	01 00	02 06	03 11
30	18 28	18 52	19 20	24 34	00 34	01 43	02 53
35	18 30	18 56	19 26	24 19	00 19	01 30	02 43
40	18 34	19 01	19 34	24 01	00 01	01 15	02 32
45	18 38	19 07	19 43	23 40	24 57	00 57	02 18
S 50	18 42	19 15	19 55	23 13	24 35	00 35	02 01
52	18 44	19 19	20 01	23 00	24 24	00 24	01 53
54	18 47	19 23	20 07	22 45	24 12	00 12	01 44
56	18 49	19 28	20 15	22 27	23 58	25 34	01 34
58	18 52	19 33	20 23	22 06	23 41	25 23	01 23
S 60	18 56	19 39	20 33	21 38	23 21	25 10	01 10

Day	SUN Eqn. of Time 00^h	SUN Eqn. of Time 12^h	SUN Mer. Pass.	MOON Mer. Pass. Upper	MOON Mer. Pass. Lower	MOON Age	MOON Phase
d	m s	m s	h m	h m	h m	d %	
5	11 35	11 28	12 11	19 27	06 56	08 63	
6	11 21	11 14	12 11	20 28	07 58	09 74	◐
7	11 07	11 00	12 11	21 26	08 58	10 84	

	UT	ARIES	VENUS −4.7		MARS +1.2		JUPITER −2.0		SATURN +0.5	
		GHA	GHA	Dec	GHA	Dec	GHA	Dec	GHA	Dec
	d h	° ′	° ′	° ′	° ′	° ′	° ′	° ′	° ′	° ′
	8 00	165 50.0	154 28.4	N12 42.3	198 44.1	S14 26.6	208 59.5	S17 03.7	355 33.8	N 6 39.4
	01	180 52.4	169 31.1	42.6	213 44.6	26.0	224 01.4	03.6	10 36.4	39.4
	02	195 54.9	184 33.8	42.8	228 45.2	25.3	239 03.3	03.4	25 39.1	39.5
	03	210 57.4	199 36.6	. . 43.1	243 45.8	. . 24.7	254 05.2	. . 03.2	40 41.7	. . 39.6
	04	225 59.8	214 39.3	43.4	258 46.3	24.0	269 07.1	03.1	55 44.4	39.7
	05	241 02.3	229 42.1	43.7	273 46.9	23.4	284 09.1	02.9	70 47.0	39.8
	06	256 04.7	244 44.9	N12 43.9	288 47.4	S14 22.8	299 11.0	S17 02.8	85 49.7	N 6 39.8
	07	271 07.2	259 47.6	44.2	303 48.0	22.1	314 12.9	02.6	100 52.3	39.9
	08	286 09.7	274 50.4	44.5	318 48.5	21.5	329 14.8	02.5	115 55.0	40.0
S	09	301 12.1	289 53.2	. . 44.8	333 49.1	. . 20.8	344 16.7	. . 02.3	130 57.6	. . 40.1
U	10	316 14.6	304 55.9	45.0	348 49.7	20.2	359 18.7	02.2	146 00.2	40.2
N	11	331 17.1	319 58.7	45.3	3 50.2	19.5	14 20.6	02.0	161 02.9	40.2
D	12	346 19.5	335 01.5	N12 45.5	18 50.8	S14 18.9	29 22.5	S17 01.9	176 05.5	N 6 40.3
A	13	1 22.0	350 04.3	45.8	33 51.3	18.2	44 24.4	01.7	191 08.2	40.4
Y	14	16 24.5	5 07.1	46.1	48 51.9	17.6	59 26.3	01.6	206 10.8	40.5
	15	31 26.9	20 09.8	. . 46.3	63 52.5	. . 16.9	74 28.3	. . 01.4	221 13.5	. . 40.6
	16	46 29.4	35 12.6	46.6	78 53.0	16.3	89 30.2	01.3	236 16.1	40.6
	17	61 31.9	50 15.4	46.8	93 53.6	15.6	104 32.1	01.1	251 18.8	40.7
	18	76 34.3	65 18.2	N12 47.1	108 54.1	S14 15.0	119 34.0	S17 01.0	266 21.4	N 6 40.8
	19	91 36.8	80 21.1	47.3	123 54.7	14.3	134 35.9	00.8	281 24.1	40.9
	20	106 39.2	95 23.9	47.5	138 55.3	13.7	149 37.9	00.7	296 26.7	41.0
	21	121 41.7	110 26.7	. . 47.8	153 55.8	. . 13.0	164 39.8	. . 00.5	311 29.4	. . 41.1
	22	136 44.2	125 29.5	48.0	168 56.4	12.4	179 41.7	00.4	326 32.0	41.1
	23	151 46.6	140 32.3	48.3	183 56.9	11.7	194 43.6	00.2	341 34.7	41.2
	9 00	166 49.1	155 35.1	N12 48.5	198 57.5	S14 11.1	209 45.6	S17 00.1	356 37.3	N 6 41.3
	01	181 51.6	170 38.0	48.7	213 58.1	10.4	224 47.5	16 59.9	11 40.0	41.4
	02	196 54.0	185 40.8	48.9	228 58.6	09.8	239 49.4	59.8	26 42.6	41.5
	03	211 56.5	200 43.6	. . 49.2	243 59.2	. . 09.1	254 51.3	. . 59.6	41 45.3	. . 41.5
	04	226 59.0	215 46.5	49.4	258 59.8	08.5	269 53.2	59.5	56 47.9	41.6
	05	242 01.4	230 49.3	49.6	274 00.3	07.8	284 55.2	59.3	71 50.5	41.7
	06	257 03.9	245 52.2	N12 49.8	289 00.9	S14 07.2	299 57.1	S16 59.1	86 53.2	N 6 41.8
	07	272 06.4	260 55.0	50.1	304 01.4	06.5	314 59.0	59.0	101 55.8	41.9
	08	287 08.8	275 57.9	50.3	319 02.0	05.9	330 00.9	58.8	116 58.5	41.9
M	09	302 11.3	291 00.8	. . 50.5	334 02.6	. . 05.2	345 02.9	. . 58.7	132 01.1	. . 42.0
O	10	317 13.7	306 03.6	50.7	349 03.1	04.6	0 04.8	58.5	147 03.8	42.1
N	11	332 16.2	321 06.5	50.9	4 03.7	03.9	15 06.7	58.4	162 06.4	42.2
D	12	347 18.7	336 09.4	N12 51.1	19 04.3	S14 03.2	30 08.6	S16 58.2	177 09.1	N 6 42.3
A	13	2 21.1	351 12.2	51.3	34 04.8	02.6	45 10.6	58.1	192 11.7	42.3
Y	14	17 23.6	6 15.1	51.5	49 05.4	01.9	60 12.5	57.9	207 14.4	42.4
	15	32 26.1	21 18.0	. . 51.7	64 05.9	. . 01.3	75 14.4	. . 57.8	222 17.0	. . 42.5
	16	47 28.5	36 20.9	51.9	79 06.5	00.6	90 16.3	57.6	237 19.7	42.6
	17	62 31.0	51 23.8	52.1	94 07.1	14 00.0	105 18.2	57.5	252 22.3	42.7
	18	77 33.5	66 26.7	N12 52.3	109 07.6	S13 59.3	120 20.2	S16 57.3	267 25.0	N 6 42.7
	19	92 35.9	81 29.6	52.5	124 08.2	58.7	135 22.1	57.2	282 27.6	42.8
	20	107 38.4	96 32.5	52.7	139 08.8	58.0	150 24.0	57.0	297 30.3	42.9
	21	122 40.8	111 35.4	. . 52.9	154 09.3	. . 57.3	165 25.9	. . 56.9	312 32.9	. . 43.0
	22	137 43.3	126 38.3	53.0	169 09.9	56.7	180 27.9	56.7	327 35.6	43.1
	23	152 45.8	141 41.2	53.2	184 10.5	56.0	195 29.8	56.6	342 38.2	43.1
	10 00	167 48.2	156 44.1	N12 53.4	199 11.0	S13 55.4	210 31.7	S16 56.4	357 40.9	N 6 43.2
	01	182 50.7	171 47.1	53.6	214 11.6	54.7	225 33.6	56.3	12 43.5	43.3
	02	197 53.2	186 50.0	53.8	229 12.2	54.1	240 35.6	56.1	27 46.1	43.4
	03	212 55.6	201 52.9	. . 53.9	244 12.7	. . 53.4	255 37.5	. . 56.0	42 48.8	. . 43.5
	04	227 58.1	216 55.8	54.1	259 13.3	52.7	270 39.4	55.8	57 51.4	43.5
	05	243 00.6	231 58.8	54.3	274 13.9	52.1	285 41.3	55.7	72 54.1	43.6
	06	258 03.0	247 01.7	N12 54.4	289 14.4	S13 51.4	300 43.3	S16 55.5	87 56.7	N 6 43.7
	07	273 05.5	262 04.7	54.6	304 15.0	50.8	315 45.2	55.4	102 59.4	43.8
T	08	288 08.0	277 07.6	54.7	319 15.6	50.1	330 47.1	55.2	118 02.0	43.9
U	09	303 10.4	292 10.6	. . 54.9	334 16.1	. . 49.5	345 49.0	. . 55.1	133 04.7	. . 43.9
E	10	318 12.9	307 13.5	55.1	349 16.7	48.8	0 51.0	54.9	148 07.3	44.0
S	11	333 15.3	322 16.5	55.2	4 17.3	48.1	15 52.9	54.8	163 10.0	44.1
D	12	348 17.8	337 19.5	N12 55.4	19 17.8	S13 47.5	30 54.8	S16 54.6	178 12.6	N 6 44.2
A	13	3 20.3	352 22.4	55.5	34 18.4	46.8	45 56.7	54.4	193 15.3	44.3
Y	14	18 22.7	7 25.4	55.6	49 19.0	46.2	60 58.7	54.3	208 17.9	44.3
	15	33 25.2	22 28.4	. . 55.8	64 19.5	. . 45.5	76 00.6	. . 54.1	223 20.6	. . 44.4
	16	48 27.7	37 31.4	55.9	79 20.1	44.8	91 02.5	54.0	238 23.2	44.5
	17	63 30.1	52 34.3	56.1	94 20.7	44.2	106 04.5	53.8	253 25.9	44.6
	18	78 32.6	67 37.3	N12 56.2	109 21.2	S13 43.5	121 06.4	S16 53.7	268 28.5	N 6 44.7
	19	93 35.1	82 40.3	56.3	124 21.8	42.9	136 08.3	53.5	283 31.2	44.7
	20	108 37.5	97 43.3	56.5	139 22.4	42.2	151 10.2	53.4	298 33.8	44.8
	21	123 40.0	112 46.3	. . 56.6	154 22.9	. . 41.5	166 12.2	. . 53.2	313 36.4	. . 44.9
	22	138 42.4	127 49.3	56.7	169 23.5	40.9	181 14.1	53.1	328 39.1	45.0
	23	153 44.9	142 52.3	56.8	184 24.1	40.2	196 16.0	52.9	343 41.7	45.1
	Mer. Pass.	h m 12 50.6	v 2.9	d 0.2	v 0.6	d 0.7	v 1.9	d 0.2	v 2.6	d 0.1

STARS		
Name	SHA	Dec
	° ′	° ′
Acamar	315 20.9	S40 16.
Achernar	335 29.3	S57 11.
Acrux	173 12.5	S63 09.
Adhara	255 14.9	S28 59.
Aldebaran	290 53.1	N16 31.
Alioth	166 22.7	N55 54.
Alkaid	153 00.8	N49 15.
Al Na'ir	27 48.0	S46 55.
Alnilam	275 49.5	S 1 11.
Alphard	217 58.9	S 8 42.
Alphecca	126 13.5	N26 40.
Alpheratz	357 47.2	N29 08.
Altair	62 11.5	N 8 53.
Ankaa	353 19.1	S42 15.
Antares	112 30.1	S26 27.
Arcturus	145 58.3	N19 07.
Atria	107 34.9	S69 02.
Avior	234 19.0	S59 32.
Bellatrix	278 35.4	N 6 21.
Betelgeuse	271 04.7	N 7 24.
Canopus	263 57.4	S52 42.
Capella	280 39.1	N46 00.
Deneb	49 34.1	N45 18.
Denebola	182 36.5	N14 31.
Diphda	348 59.3	S17 56.
Dubhe	193 54.6	N61 42.
Elnath	278 16.6	N28 37.
Eltanin	90 47.7	N51 28.
Enif	33 50.5	N 9 54.
Fomalhaut	15 27.7	S29 34.
Gacrux	172 04.2	S57 10.
Gienah	175 55.3	S17 35.
Hadar	148 52.2	S60 25.
Hamal	328 04.6	N23 30.
Kaus Aust.	83 48.1	S34 22.
Kochab	137 18.8	N74 06.
Markab	13 41.9	N15 15.
Menkar	314 18.5	N 4 07.
Menkent	148 11.1	S36 25.
Miaplacidus	221 39.9	S69 45.
Mirfak	308 45.2	N49 53.
Nunki	76 02.4	S26 17.
Peacock	53 24.5	S56 42.
Pollux	243 31.3	N28 00.
Procyon	245 02.8	N 5 12.
Rasalhague	96 09.4	N12 32.
Regulus	207 46.5	N11 55.
Rigel	281 15.1	S 8 11.
Rigil Kent.	139 55.9	S60 52.
Sabik	102 16.2	S15 44.
Schedar	349 44.9	N56 35.
Shaula	96 26.3	S37 06.
Sirius	258 36.4	S16 43.
Spica	158 34.4	S11 12.
Suhail	222 54.5	S43 28.
Vega	80 41.3	N38 47.
Zuben'ubi	137 08.8	S16 05.

	SHA	Mer. Pass.
	° ′	h m
Venus	348 46.0	13 35
Mars	32 08.4	10 44
Jupiter	42 56.5	10 00
Saturn	189 48.2	0 13

UT	SUN GHA	SUN Dec	MOON GHA	v	MOON Dec	d	HP
d h	° ′	° ′	° ′	′	° ′	′	′
8 00	177 17.0	S 4 55.1	37 02.4	7.6	N18 54.1	12.1	59.7
01	192 17.1	54.1	51 29.0	7.6	18 42.0	12.3	59.7
02	207 17.3	53.2	65 55.6	7.8	18 29.7	12.4	59.7
03	222 17.4	. . 52.2	80 22.4	7.8	18 17.3	12.5	59.7
04	237 17.6	51.2	94 49.2	7.9	18 04.8	12.7	59.7
05	252 17.7	50.2	109 16.1	8.0	17 52.1	12.7	59.7
06	267 17.9	S 4 49.3	123 43.1	8.0	N17 39.4	12.8	59.7
07	282 18.0	48.3	138 10.1	8.2	17 26.6	13.0	59.7
08	297 18.2	47.3	152 37.3	8.3	17 13.6	13.0	59.7
09	312 18.4	. . 46.3	167 04.6	8.3	17 00.6	13.1	59.7
10	327 18.5	45.4	181 31.9	8.4	16 47.5	13.3	59.7
11	342 18.7	44.4	195 59.3	8.5	16 34.2	13.3	59.7
12	357 18.8	S 4 43.4	210 26.8	8.6	N16 20.9	13.4	59.7
13	12 19.0	42.4	224 54.4	8.7	16 07.5	13.5	59.7
14	27 19.1	41.5	239 22.1	8.7	15 54.0	13.6	59.7
15	42 19.3	. . 40.5	253 49.8	8.9	15 40.4	13.7	59.7
16	57 19.4	39.5	268 17.7	8.9	15 26.7	13.8	59.7
17	72 19.6	38.5	282 45.6	9.0	15 12.9	13.8	59.6
18	87 19.8	S 4 37.6	297 13.6	9.1	N14 59.1	14.0	59.6
19	102 19.9	36.6	311 41.7	9.2	14 45.1	14.0	59.6
20	117 20.1	35.6	326 09.9	9.2	14 31.1	14.1	59.6
21	132 20.2	. . 34.6	340 38.1	9.3	14 17.0	14.2	59.6
22	147 20.4	33.7	355 06.4	9.4	14 02.8	14.3	59.6
23	162 20.5	32.7	9 34.8	9.5	13 48.5	14.3	59.6
9 00	177 20.7	S 4 31.7	24 03.3	9.6	N13 34.2	14.4	59.6
01	192 20.9	30.7	38 31.9	9.6	13 19.8	14.5	59.6
02	207 21.0	29.7	53 00.5	9.7	13 05.3	14.5	59.6
03	222 21.2	. . 28.8	67 29.2	9.8	12 50.8	14.6	59.6
04	237 21.3	27.8	81 58.0	9.9	12 36.2	14.7	59.5
05	252 21.5	26.8	96 26.9	9.9	12 21.5	14.7	59.5
06	267 21.6	S 4 25.8	110 55.8	10.1	N12 06.8	14.8	59.5
07	282 21.8	24.9	125 24.9	10.0	11 52.0	14.9	59.5
08	297 22.0	23.9	139 53.9	10.2	11 37.1	14.9	59.5
09	312 22.1	. . 22.9	154 23.1	10.2	11 22.2	15.0	59.5
10	327 22.3	21.9	168 52.3	10.3	11 07.2	15.0	59.5
11	342 22.4	20.9	183 21.6	10.4	10 52.2	15.1	59.5
12	357 22.6	S 4 20.0	197 51.0	10.4	N10 37.1	15.2	59.4
13	12 22.8	19.0	212 20.4	10.5	10 21.9	15.1	59.4
14	27 22.9	18.0	226 49.9	10.6	10 06.8	15.3	59.4
15	42 23.1	. . 17.0	241 19.5	10.6	9 51.5	15.2	59.4
16	57 23.2	16.1	255 49.1	10.7	9 36.3	15.4	59.4
17	72 23.4	15.1	270 18.8	10.8	9 20.9	15.3	59.4
18	87 23.6	S 4 14.1	284 48.6	10.8	N 9 05.6	15.4	59.4
19	102 23.7	13.1	299 18.4	10.9	8 50.2	15.5	59.3
20	117 23.9	12.1	313 48.3	10.9	8 34.7	15.5	59.3
21	132 24.0	. . 11.2	328 18.2	11.0	8 19.2	15.5	59.3
22	147 24.2	10.2	342 48.2	11.1	8 03.7	15.5	59.3
23	162 24.4	09.2	357 18.3	11.1	7 48.2	15.6	59.3
10 00	177 24.5	S 4 08.2	11 48.4	11.2	N 7 32.6	15.6	59.3
01	192 24.7	07.2	26 18.6	11.2	7 17.0	15.7	59.2
02	207 24.8	06.3	40 48.8	11.3	7 01.3	15.6	59.2
03	222 25.0	. . 05.3	55 19.1	11.4	6 45.7	15.7	59.2
04	237 25.2	04.3	69 49.5	11.3	6 30.0	15.7	59.2
05	252 25.3	03.3	84 19.8	11.5	6 14.3	15.8	59.2
06	267 25.5	S 4 02.3	98 50.3	11.5	N 5 58.5	15.7	59.1
07	282 25.6	01.4	113 20.8	11.5	5 42.8	15.8	59.1
08	297 25.8	4 00.4	127 51.3	11.6	5 27.0	15.8	59.1
09	312 26.0	3 59.4	142 21.9	11.7	5 11.2	15.8	59.1
10	327 26.1	58.4	156 52.6	11.6	4 55.4	15.9	59.1
11	342 26.3	57.4	171 23.2	11.8	4 39.5	15.8	59.0
12	357 26.5	S 3 56.5	185 54.0	11.7	N 4 23.7	15.9	59.0
13	12 26.6	55.5	200 24.7	11.9	4 07.8	15.8	59.0
14	27 26.8	54.5	214 55.6	11.8	3 52.0	15.9	59.0
15	42 26.9	. . 53.5	229 26.4	11.9	3 36.1	15.9	59.0
16	57 27.1	52.5	243 57.3	12.0	3 20.2	15.9	58.9
17	72 27.3	51.6	258 28.3	11.9	3 04.3	15.8	58.9
18	87 27.4	S 3 50.6	272 59.2	12.1	N 2 48.5	15.9	58.9
19	102 27.6	49.6	287 30.3	12.0	2 32.6	15.9	58.9
20	117 27.8	48.6	302 01.3	12.1	2 16.7	15.9	58.8
21	132 27.9	. . 47.6	316 32.4	12.1	2 00.8	15.9	58.8
22	147 28.1	46.7	331 03.5	12.2	1 44.9	15.9	58.8
23	162 28.3	45.7	345 34.7	12.2	N 1 29.0	15.8	58.8
	SD 16.1	*d* 1.0	SD	16.3		16.2	16.1

Lat.	Twilight Naut.	Twilight Civil	Sunrise	Moonrise 8	Moonrise 9	Moonrise 10	Moonrise 11
°	h m	h m	h m	h m	h m	h m	h m
N 72	04 27	05 47	06 54	11 42	14 46	17 11	19 27
N 70	04 37	05 49	06 50	12 30	15 03	17 15	19 21
68	04 45	05 50	06 46	13 00	15 16	17 19	19 16
66	04 52	05 51	06 42	13 22	15 26	17 21	19 12
64	04 57	05 52	06 39	13 40	15 35	17 24	19 09
62	05 01	05 53	06 37	13 54	15 43	17 26	19 06
60	05 05	05 53	06 35	14 06	15 49	17 28	19 03
N 58	05 08	05 54	06 33	14 16	15 55	17 29	19 01
56	05 11	05 54	06 31	14 25	16 00	17 31	18 59
54	05 13	05 54	06 29	14 33	16 04	17 32	18 57
52	05 15	05 54	06 28	14 40	16 08	17 33	18 56
50	05 17	05 54	06 26	14 47	16 12	17 34	18 54
45	05 20	05 54	06 24	15 00	16 20	17 36	18 51
N 40	05 23	05 54	06 21	15 12	16 26	17 38	18 49
35	05 24	05 54	06 19	15 21	16 32	17 40	18 46
30	05 25	05 53	06 17	15 29	16 37	17 41	18 44
20	05 26	05 51	06 13	15 44	16 45	17 44	18 41
N 10	05 25	05 49	06 10	15 56	16 52	17 46	18 38
0	05 22	05 46	06 07	16 08	16 59	17 48	18 36
S 10	05 18	05 43	06 04	16 19	17 06	17 50	18 33
20	05 13	05 38	06 01	16 32	17 14	17 53	18 30
30	05 04	05 32	05 56	16 45	17 22	17 55	18 27
35	04 59	05 29	05 54	16 54	17 27	17 57	18 25
40	04 52	05 24	05 51	17 03	17 32	17 58	18 23
45	04 43	05 18	05 48	17 13	17 38	18 00	18 21
S 50	04 32	05 11	05 44	17 26	17 46	18 02	18 18
52	04 27	05 08	05 42	17 32	17 49	18 03	18 17
54	04 20	05 04	05 40	17 39	17 53	18 05	18 16
56	04 14	05 00	05 38	17 46	17 57	18 06	18 14
58	04 06	04 55	05 36	17 54	18 01	18 07	18 13
S 60	03 56	04 50	05 33	18 03	18 06	18 09	18 11

Lat.	Sunset	Twilight Civil	Twilight Naut.	Moonset 8	Moonset 9	Moonset 10	Moonset 11
°	h m	h m	h m	h m	h m	h m	h m
N 72	17 29	18 36	19 57	08 30	07 16	06 37	06 05
N 70	17 33	18 34	19 47	07 40	06 57	06 29	06 05
68	17 37	18 33	19 38	07 08	06 42	06 23	06 06
66	17 40	18 31	19 32	06 45	06 29	06 17	06 06
64	17 43	18 31	19 26	06 26	06 19	06 13	06 07
62	17 46	18 30	19 22	06 10	06 10	06 09	06 07
60	17 48	18 29	19 18	05 57	06 02	06 05	06 07
N 58	17 50	18 29	19 14	05 46	05 55	06 02	06 08
56	17 51	18 28	19 12	05 36	05 49	05 59	06 08
54	17 53	18 28	19 09	05 27	05 44	05 57	06 08
52	17 54	18 28	19 07	05 20	05 39	05 54	06 08
50	17 55	18 28	19 05	05 12	05 34	05 52	06 09
45	17 58	18 27	19 02	04 57	05 24	05 48	06 09
N 40	18 01	18 28	18 59	04 44	05 16	05 44	06 09
35	18 03	18 28	18 57	04 34	05 09	05 40	06 10
30	18 04	18 28	18 56	04 24	05 03	05 37	06 10
20	18 08	18 30	18 55	04 08	04 52	05 32	06 10
N 10	18 11	18 32	18 56	03 53	04 42	05 27	06 11
0	18 14	18 34	18 59	03 39	04 33	05 23	06 11
S 10	18 17	18 38	19 02	03 25	04 23	05 18	06 11
20	18 20	18 42	19 08	03 11	04 13	05 13	06 11
30	18 24	18 48	19 16	02 53	04 02	05 08	06 12
35	18 26	18 52	19 22	02 43	03 55	05 05	06 12
40	18 29	18 56	19 28	02 32	03 47	05 01	06 12
45	18 32	19 02	19 37	02 18	03 38	04 57	06 12
S 50	18 36	19 09	19 48	02 01	03 27	04 51	06 13
52	18 38	19 12	19 53	01 53	03 22	04 49	06 13
54	18 39	19 15	19 59	01 44	03 17	04 46	06 13
56	18 42	19 19	20 06	01 34	03 10	04 43	06 13
58	18 44	19 24	20 13	01 23	03 03	04 40	06 13
S 60	18 47	19 29	20 22	01 10	02 55	04 36	06 13

Day	SUN Eqn. of Time 00^h	SUN Eqn. of Time 12^h	SUN Mer. Pass.	MOON Mer. Pass. Upper	MOON Mer. Pass. Lower	MOON Age	MOON Phase
d	m s	m s	h m	h m	h m	d	%
8	10 52	10 45	12 11	22 20	09 54	11	91
9	10 38	10 30	12 10	23 11	10 46	12	97
10	10 22	10 14	12 10	24 00	11 36	13	99

UT		ARIES	VENUS −4.6		MARS +1.2		JUPITER −2.0		SATURN +0.5		STARS		
		GHA	GHA	Dec	GHA	Dec	GHA	Dec	GHA	Dec	Name	SHA	Dec
d h		° ′	° ′	° ′	° ′	° ′	° ′	° ′	° ′	° ′		° ′	° ′
11 00		168 47.4	157 55.3	N12 57.0	199 24.7	S13 39.5	211 17.9	S16 52.8	358 44.4	N 6 45.1	Acamar	315 20.9	S40 16
01		183 49.8	172 58.3	57.1	214 25.2	38.9	226 19.9	52.6	13 47.0	45.2	Achernar	335 29.4	S57 11
02		198 52.3	188 01.4	57.2	229 25.8	38.2	241 21.8	52.5	28 49.7	45.3	Acrux	173 12.5	S63 09.
03		213 54.8	203 04.4	. . 57.3	244 26.4	. . 37.6	256 23.7	. . 52.3	43 52.3	. . 45.4	Adhara	255 14.9	S28 59
04		228 57.2	218 07.4	57.4	259 26.9	36.9	271 25.7	52.2	58 55.0	45.5	Aldebaran	290 53.1	N16 31.
05		243 59.7	233 10.4	57.5	274 27.5	36.2	286 27.6	52.0	73 57.6	45.5			
06		259 02.2	248 13.5	N12 57.6	289 28.1	S13 35.6	301 29.5	S16 51.9	89 00.3	N 6 45.6	Alioth	166 22.7	N55 54
07	W	274 04.6	263 16.5	57.7	304 28.6	34.9	316 31.4	51.7	104 02.9	45.7	Alkaid	153 00.8	N49 15
08	E	289 07.1	278 19.5	57.8	319 29.2	34.2	331 33.4	51.6	119 05.6	45.8	Al Na'ir	27 48.0	S46 55
09	D	304 09.6	293 22.6	. . 57.9	334 29.8	. . 33.6	346 35.3	. . 51.4	134 08.2	. . 45.9	Alnilam	275 49.6	S 1 11
10	N	319 12.0	308 25.6	58.0	349 30.4	32.9	1 37.2	51.3	149 10.9	45.9	Alphard	217 58.9	S 8 42
11	E	334 14.5	323 28.7	58.1	4 30.9	32.2	16 39.2	51.1	164 13.5	46.0			
12		349 16.9	338 31.7	N12 58.2	19 31.5	S13 31.6	31 41.1	S16 51.0	179 16.2	N 6 46.1	Alphecca	126 13.5	N26 40.
13	S	4 19.4	353 34.8	58.3	34 32.1	30.9	46 43.0	50.8	194 18.8	46.2	Alpheratz	357 47.2	N29 08.
14	D	19 21.9	8 37.9	58.4	49 32.6	30.2	61 44.9	50.7	209 21.5	46.3	Altair	62 11.5	N 8 53.
15	A	34 24.3	23 40.9	. . 58.5	64 33.2	. . 29.6	76 46.9	. . 50.5	224 24.1	. . 46.3	Ankaa	353 19.1	S42 15.
16	Y	49 26.8	38 44.0	58.6	79 33.8	28.9	91 48.8	50.4	239 26.7	46.4	Antares	112 30.1	S26 27.
17		64 29.3	53 47.1	58.6	94 34.4	28.2	106 50.7	50.2	254 29.4	46.5			
18		79 31.7	68 50.1	N12 58.7	109 34.9	S13 27.6	121 52.7	S16 50.1	269 32.0	N 6 46.6	Arcturus	145 58.3	N19 07.
19		94 34.2	83 53.2	58.8	124 35.5	26.9	136 54.6	49.9	284 34.7	46.7	Atria	107 34.8	S69 02.
20		109 36.7	98 56.3	58.9	139 36.1	26.2	151 56.5	49.8	299 37.3	46.7	Avior	234 19.0	S59 32.
21		124 39.1	113 59.4	. . 58.9	154 36.7	. . 25.6	166 58.5	. . 49.6	314 40.0	. . 46.8	Bellatrix	278 35.4	N 6 21.
22		139 41.6	129 02.5	59.0	169 37.2	24.9	182 00.4	49.5	329 42.6	46.9	Betelgeuse	271 04.7	N 7 24.
23		154 44.0	144 05.6	59.1	184 37.8	24.2	197 02.3	49.3	344 45.3	47.0			
12 00		169 46.5	159 08.7	N12 59.1	199 38.4	S13 23.6	212 04.2	S16 49.2	359 47.9	N 6 47.1	Canopus	263 57.4	S52 42.
01		184 49.0	174 11.8	59.2	214 38.9	22.9	227 06.2	49.0	14 50.6	47.1	Capella	280 39.2	N46 00.
02		199 51.4	189 14.9	59.3	229 39.5	22.2	242 08.1	48.9	29 53.2	47.2	Deneb	49 34.1	N45 18
03		214 53.9	204 18.0	. . 59.3	244 40.1	. . 21.6	257 10.0	. . 48.7	44 55.9	. . 47.3	Denebola	182 36.5	N14 31.
04		229 56.4	219 21.1	59.4	259 40.7	20.9	272 12.0	48.5	59 58.5	47.4	Diphda	348 59.3	S17 56.
05		244 58.8	234 24.2	59.4	274 41.2	20.2	287 13.9	48.4	75 01.2	47.5			
06		260 01.3	249 27.4	N12 59.5	289 41.8	S13 19.6	302 15.8	S16 48.2	90 03.8	N 6 47.5	Dubhe	193 54.6	N61 42.
07		275 03.8	264 30.5	59.5	304 42.4	18.9	317 17.8	48.1	105 06.5	47.6	Elnath	278 16.6	N28 37.
08	T	290 06.2	279 33.6	59.5	319 43.0	18.2	332 19.7	47.9	120 09.1	47.7	Eltanin	90 47.7	N51 28.
09	H	305 08.7	294 36.8	. . 59.6	334 43.5	. . 17.6	347 21.6	. . 47.8	135 11.8	. . 47.8	Enif	33 50.5	N 9 54.
10	U	320 11.2	309 39.9	59.6	349 44.1	16.9	2 23.6	47.6	150 14.4	47.9	Fomalhaut	15 27.7	S29 34.
11	R	335 13.6	324 43.0	59.7	4 44.7	16.2	17 25.5	47.5	165 17.0	47.9			
12	S	350 16.1	339 46.2	N12 59.7	19 45.3	S13 15.5	32 27.4	S16 47.3	180 19.7	N 6 48.0	Gacrux	172 04.2	S57 10.
13	D	5 18.5	354 49.3	59.7	34 45.8	14.9	47 29.4	47.2	195 22.3	48.1	Gienah	175 55.3	S17 35.
14	A	20 21.0	9 52.5	59.7	49 46.4	14.2	62 31.3	47.0	210 25.0	48.2	Hadar	148 52.2	S60 25.
15	Y	35 23.5	24 55.6	. . 59.8	64 47.0	. . 13.5	77 33.2	. . 46.9	225 27.6	. . 48.2	Hamal	328 04.7	N23 30.
16		50 25.9	39 58.8	59.8	79 47.6	12.9	92 35.1	46.7	240 30.3	48.3	Kaus Aust.	83 48.1	S34 22.
17		65 28.4	55 01.9	59.8	94 48.1	12.2	107 37.1	46.6	255 32.9	48.4			
18		80 30.9	70 05.1	N12 59.8	109 48.7	S13 11.5	122 39.0	S16 46.4	270 35.6	N 6 48.5	Kochab	137 18.7	N74 06.
19		95 33.3	85 08.3	59.8	124 49.3	10.8	137 40.9	46.3	285 38.2	48.6	Markab	13 41.9	N15 15.
20		110 35.8	100 11.5	59.9	139 49.9	10.2	152 42.9	46.1	300 40.9	48.6	Menkar	314 18.6	N 4 07.
21		125 38.3	115 14.6	. . 59.9	154 50.5	. . 09.5	167 44.8	. . 46.0	315 43.5	. . 48.7	Menkent	148 11.1	S36 25.
22		140 40.7	130 17.8	59.9	169 51.0	08.8	182 46.7	45.8	330 46.2	48.8	Miaplacidus	221 39.9	S69 45.
23		155 43.2	145 21.0	59.9	184 51.6	08.1	197 48.7	45.7	345 48.8	48.9			
13 00		170 45.7	160 24.2	N12 59.9	199 52.2	S13 07.5	212 50.6	S16 45.5	0 51.5	N 6 49.0	Mirfak	308 45.3	N49 53.
01		185 48.1	175 27.4	59.9	214 52.8	06.8	227 52.5	45.4	15 54.1	49.0	Nunki	76 02.3	S26 17.
02		200 50.6	190 30.6	59.9	229 53.3	06.1	242 54.5	45.2	30 56.8	49.1	Peacock	53 24.4	S56 42.
03		215 53.0	205 33.8	. . 59.9	244 53.9	. . 05.5	257 56.4	. . 45.1	45 59.4	. . 49.2	Pollux	243 31.3	N28 00.
04		230 55.5	220 37.0	59.9	259 54.5	04.8	272 58.3	44.9	61 02.0	49.3	Procyon	245 02.9	N 5 12.
05		245 58.0	235 40.2	59.9	274 55.1	04.1	288 00.3	44.8	76 04.7	49.4			
06		261 00.4	250 43.4	N12 59.8	289 55.7	S13 03.4	303 02.2	S16 44.6	91 07.3	N 6 49.4	Rasalhague	96 09.4	N12 32.
07		276 02.9	265 46.6	59.8	304 56.2	02.8	318 04.1	44.5	106 10.0	49.5	Regulus	207 46.5	N11 55.
08		291 05.4	280 49.8	59.8	319 56.8	02.1	333 06.1	44.3	121 12.6	49.6	Rigel	281 15.1	S 8 11.
09	F	306 07.8	295 53.0	. . 59.8	334 57.4	. . 01.4	348 08.0	. . 44.2	136 15.3	. . 49.7	Rigil Kent.	139 55.9	S60 52.
10	R	321 10.3	310 56.3	59.8	349 58.0	00.7	3 09.9	44.0	151 17.9	49.8	Sabik	102 16.2	S15 44.
11	I	336 12.8	325 59.5	59.7	4 58.5	13 00.1	18 11.9	43.9	166 20.6	49.8			
12	D	351 15.2	341 02.7	N12 59.7	19 59.1	S12 59.4	33 13.8	S16 43.7	181 23.2	N 6 49.9	Schedar	349 44.9	N56 35.
13	A	6 17.7	356 06.0	59.7	34 59.7	58.7	48 15.8	43.6	196 25.9	50.0	Shaula	96 26.2	S37 06.
14	Y	21 20.1	11 09.2	59.6	50 00.3	58.0	63 17.7	43.4	211 28.5	50.1	Sirius	258 36.4	S16 43.
15		36 22.6	26 12.4	. . 59.6	65 00.9	. . 57.3	78 19.6	. . 43.3	226 31.2	. . 50.1	Spica	158 34.3	S11 12.
16		51 25.1	41 15.7	59.6	80 01.4	56.7	93 21.6	43.1	241 33.8	50.2	Suhail	222 54.5	S43 28.
17		66 27.5	56 18.9	59.5	95 02.0	56.0	108 23.5	43.0	256 36.5	50.3			
18		81 30.0	71 22.2	N12 59.5	110 02.6	S12 55.3	123 25.4	S16 42.8	271 39.1	N 6 50.4	Vega	80 41.2	N38 47.
19		96 32.5	86 25.4	59.4	125 03.2	54.6	138 27.4	42.7	286 41.7	50.5	Zuben'ubi	137 08.8	S16 05.
20		111 34.9	101 28.7	59.4	140 03.8	54.0	153 29.3	42.5	301 44.4	50.5		SHA	Mer.Pass
21		126 37.4	116 31.9	. . 59.3	155 04.3	. . 53.3	168 31.2	. . 42.4	316 47.0	. . 50.6		° ′	h m
22		141 39.9	131 35.2	59.3	170 04.9	52.6	183 33.2	42.2	331 49.7	50.7	Venus	349 22.2	13 21
23		156 42.3	146 38.5	59.2	185 05.5	51.9	198 35.1	42.1	346 52.3	50.8	Mars	29 51.9	10 41
		h m									Jupiter	42 17.7	9 50
Mer. Pass.		12 38.8	*v* 3.1	*d* 0.0	*v* 0.6	*d* 0.7	*v* 1.9	*d* 0.2	*v* 2.6	*d* 0.1	Saturn	190 01.4	0 01

UT d	UT h	SUN GHA	SUN Dec	MOON GHA	v	MOON Dec	d	HP
		° ′	° ′	° ′	′	° ′	′	′
11	00	177 28.4	S 3 44.7	0 05.9	12.2	N 1 13.2	15.9	58.8
	01	192 28.6	43.7	14 37.1	12.2	0 57.3	15.9	58.7
	02	207 28.7	42.7	29 08.3	12.3	0 41.4	15.8	58.7
	03	222 28.9	. . 41.8	43 39.6	12.3	0 25.6	15.9	58.7
	04	237 29.1	40.8	58 10.9	12.4	N 0 09.7	15.8	58.7
	05	252 29.2	39.8	72 42.3	12.3	S 0 06.1	15.8	58.6
	06	267 29.4	S 3 38.8	87 13.6	12.4	S 0 21.9	15.8	58.6
	07	282 29.6	37.8	101 45.0	12.4	0 37.7	15.8	58.6
	08	297 29.7	36.8	116 16.4	12.5	0 53.5	15.7	58.6
	09	312 29.9	. . 35.9	130 47.9	12.4	1 09.2	15.8	58.5
	10	327 30.1	34.9	145 19.3	12.5	1 25.0	15.7	58.5
	11	342 30.2	33.9	159 50.8	12.5	1 40.7	15.7	58.5
	12	357 30.4	S 3 32.9	174 22.3	12.5	S 1 56.4	15.6	58.5
	13	12 30.6	31.9	188 53.8	12.5	2 12.0	15.7	58.4
	14	27 30.7	31.0	203 25.3	12.6	2 27.7	15.6	58.4
	15	42 30.9	. . 30.0	217 56.9	12.6	2 43.3	15.6	58.4
	16	57 31.1	29.0	232 28.5	12.5	2 58.9	15.6	58.3
	17	72 31.2	28.0	247 00.0	12.6	3 14.5	15.5	58.3
	18	87 31.4	S 3 27.0	261 31.6	12.6	S 3 30.0	15.5	58.3
	19	102 31.6	26.0	276 03.2	12.7	3 45.5	15.4	58.3
	20	117 31.7	25.1	290 34.9	12.6	4 00.9	15.5	58.2
	21	132 31.9	. . 24.1	305 06.5	12.6	4 16.4	15.4	58.2
	22	147 32.1	23.1	319 38.1	12.7	4 31.8	15.3	58.2
	23	162 32.2	22.1	334 09.8	12.6	4 47.1	15.3	58.1
12	00	177 32.4	S 3 21.1	348 41.4	12.7	S 5 02.4	15.3	58.1
	01	192 32.6	20.1	3 13.1	12.7	5 17.7	15.2	58.1
	02	207 32.7	19.2	17 44.8	12.6	5 32.9	15.2	58.1
	03	222 32.9	. . 18.2	32 16.4	12.7	5 48.1	15.2	58.0
	04	237 33.1	17.2	46 48.1	12.7	6 03.3	15.1	58.0
	05	252 33.2	16.2	61 19.8	12.7	6 18.4	15.0	58.0
	06	267 33.4	S 3 15.2	75 51.5	12.7	S 6 33.4	15.1	57.9
	07	282 33.6	14.2	90 23.2	12.7	6 48.5	14.9	57.9
	08	297 33.8	13.3	104 54.9	12.6	7 03.4	14.9	57.9
	09	312 33.9	. . 12.3	119 26.5	12.7	7 18.3	14.9	57.9
	10	327 34.1	11.3	133 58.2	12.7	7 33.2	14.8	57.8
	11	342 34.3	10.3	148 29.9	12.7	7 48.0	14.8	57.8
	12	357 34.4	S 3 09.3	163 01.6	12.7	S 8 02.8	14.7	57.8
	13	12 34.6	08.3	177 33.3	12.6	8 17.5	14.6	57.7
	14	27 34.8	07.4	192 04.9	12.7	8 32.1	14.6	57.7
	15	42 34.9	. . 06.4	206 36.6	12.7	8 46.7	14.5	57.7
	16	57 35.1	05.4	221 08.3	12.6	9 01.2	14.5	57.6
	17	72 35.3	04.4	235 39.9	12.7	9 15.7	14.4	57.6
	18	87 35.4	S 3 03.4	250 11.6	12.6	S 9 30.1	14.3	57.6
	19	102 35.6	02.4	264 43.2	12.6	9 44.4	14.3	57.6
	20	117 35.8	01.4	279 14.8	12.6	9 58.7	14.3	57.5
	21	132 36.0	3 00.5	293 46.4	12.6	10 13.0	14.1	57.5
	22	147 36.1	2 59.5	308 18.0	12.6	10 27.1	14.1	57.5
	23	162 36.3	58.5	322 49.6	12.6	10 41.2	14.0	57.4
13	00	177 36.5	S 2 57.5	337 21.2	12.6	S10 55.2	14.0	57.4
	01	192 36.6	56.5	351 52.8	12.5	11 09.2	13.9	57.4
	02	207 36.8	55.5	6 24.3	12.5	11 23.1	13.8	57.3
	03	222 37.0	. . 54.6	20 55.8	12.6	11 36.9	13.7	57.3
	04	237 37.1	53.6	35 27.4	12.5	11 50.6	13.7	57.3
	05	252 37.3	52.6	49 58.9	12.5	12 04.3	13.6	57.2
	06	267 37.5	S 2 51.6	64 30.4	12.4	S12 17.9	13.6	57.2
	07	282 37.7	50.6	79 01.8	12.5	12 31.5	13.4	57.2
	08	297 37.8	49.6	93 33.3	12.4	12 44.9	13.4	57.2
	09	312 38.0	. . 48.6	108 04.7	12.4	12 58.3	13.3	57.1
	10	327 38.2	47.7	122 36.1	12.4	13 11.6	13.2	57.1
	11	342 38.3	46.7	137 07.5	12.4	13 24.8	13.2	57.1
	12	357 38.5	S 2 45.7	151 38.9	12.3	S13 38.0	13.0	57.0
	13	12 38.7	44.7	166 10.2	12.4	13 51.0	13.0	57.0
	14	27 38.9	43.7	180 41.6	12.3	14 04.0	12.9	57.0
	15	42 39.0	. . 42.7	195 12.9	12.3	14 16.9	12.8	56.9
	16	57 39.2	41.7	209 44.2	12.2	14 29.7	12.8	56.9
	17	72 39.4	40.8	224 15.4	12.2	14 42.5	12.6	56.9
	18	87 39.6	S 2 39.8	238 46.6	12.3	S14 55.1	12.6	56.8
	19	102 39.7	38.8	253 17.9	12.1	15 07.7	12.4	56.8
	20	117 39.9	37.8	267 49.0	12.2	15 20.1	12.4	56.8
	21	132 40.1	. . 36.8	282 20.2	12.1	15 32.5	12.3	56.7
	22	147 40.2	35.8	296 51.3	12.1	15 44.8	12.2	56.7
	23	162 40.4	34.8	311 22.4	12.1	S15 57.0	12.2	56.7
		SD 16.1	*d* 1.0	SD		15.9	15.7	15.5

Lat.	Twilight Naut.	Twilight Civil	Sunrise	Moonrise 11	Moonrise 12	Moonrise 13	Moonrise 14
°	h m	h m	h m	h m	h m	h m	h m
N 72	04 11	05 32	06 39	19 27	21 46	24 37	00 37
N 70	04 22	05 35	06 36	19 21	21 28	23 49	■
68	04 32	05 38	06 33	19 16	21 13	23 18	25 52
66	04 40	05 40	06 31	19 12	21 02	22 56	24 59
64	04 46	05 42	06 29	19 09	20 52	22 38	24 27
62	04 51	05 43	06 27	19 06	20 44	22 23	24 04
60	04 56	05 44	06 26	19 03	20 37	22 11	23 45
N 58	05 00	05 45	06 24	19 01	20 31	22 01	23 30
56	05 03	05 46	06 23	18 59	20 26	21 52	23 17
54	05 06	05 47	06 22	18 57	20 21	21 44	23 05
52	05 08	05 47	06 21	18 56	20 17	21 37	22 55
50	05 10	05 48	06 20	18 54	20 13	21 30	22 46
45	05 15	05 49	06 18	18 51	20 05	21 17	22 28
N 40	05 18	05 49	06 16	18 49	19 58	21 06	22 12
35	05 20	05 50	06 15	18 46	19 52	20 56	22 00
30	05 22	05 50	06 13	18 44	19 46	20 48	21 48
20	05 23	05 49	06 11	18 41	19 37	20 33	21 29
N 10	05 23	05 48	06 09	18 38	19 30	20 21	21 13
0	05 22	05 46	06 06	18 36	19 22	20 10	20 58
S 10	05 19	05 43	06 04	18 33	19 15	19 58	20 43
20	05 14	05 39	06 01	18 30	19 08	19 46	20 26
30	05 06	05 34	05 58	18 27	18 59	19 32	20 08
35	05 01	05 31	05 57	18 25	18 54	19 24	19 57
40	04 55	05 27	05 54	18 23	18 49	19 15	19 45
45	04 47	05 22	05 52	18 21	18 42	19 05	19 31
S 50	04 37	05 16	05 49	18 18	18 34	18 52	19 14
52	04 33	05 14	05 48	18 17	18 31	18 47	19 05
54	04 27	05 10	05 46	18 16	18 27	18 40	18 56
56	04 21	05 07	05 44	18 14	18 23	18 33	18 46
58	04 14	05 03	05 43	18 13	18 18	18 25	18 35
S 60	04 06	04 58	05 40	18 11	18 13	18 16	18 21

Lat.	Sunset	Twilight Civil	Twilight Naut.	Moonset 11	Moonset 12	Moonset 13	Moonset 14
°	h m	h m	h m	h m	h m	h m	h m
N 72	17 43	18 50	20 13	06 05	05 32	04 53	03 42
N 70	17 46	18 46	20 00	06 05	05 42	05 14	04 33
68	17 48	18 44	19 50	06 06	05 49	05 30	05 05
66	17 50	18 41	19 42	06 06	05 55	05 43	05 29
64	17 52	18 39	19 36	06 07	06 01	05 54	05 47
62	17 54	18 38	19 30	06 07	06 05	06 04	06 03
60	17 55	18 37	19 25	06 07	06 10	06 12	06 16
N 58	17 56	18 35	19 21	06 08	06 13	06 19	06 27
56	17 57	18 34	19 18	06 08	06 16	06 26	06 37
54	17 58	18 34	19 15	06 08	06 19	06 31	06 46
52	17 59	18 33	19 12	06 08	06 22	06 37	06 54
50	18 00	18 32	19 10	06 09	06 24	06 41	07 01
45	18 02	18 31	19 05	06 09	06 30	06 52	07 16
N 40	18 04	18 31	19 02	06 09	06 34	07 00	07 28
35	18 05	18 30	19 00	06 10	06 38	07 08	07 39
30	18 06	18 30	18 58	06 10	06 42	07 14	07 49
20	18 09	18 31	18 56	06 10	06 48	07 25	08 05
N 10	18 11	18 32	18 56	06 11	06 53	07 35	08 19
0	18 13	18 34	18 58	06 11	06 58	07 45	08 32
S 10	18 15	18 36	19 01	06 11	07 03	07 54	08 46
20	18 18	18 40	19 05	06 11	07 08	08 04	09 00
30	18 21	18 45	19 13	06 12	07 14	08 16	09 17
35	18 22	18 48	19 17	06 12	07 18	08 23	09 27
40	18 24	18 51	19 23	06 12	07 22	08 30	09 38
45	18 27	18 56	19 31	06 12	07 26	08 39	09 51
S 50	18 29	19 02	19 41	06 13	07 32	08 50	10 07
52	18 31	19 05	19 45	06 13	07 35	08 55	10 15
54	18 32	19 08	19 51	06 13	07 37	09 01	10 23
56	18 34	19 11	19 57	06 13	07 41	09 07	10 32
58	18 36	19 15	20 04	06 13	07 44	09 14	10 43
S 60	18 38	19 20	20 12	06 13	07 48	09 22	10 56

Day	SUN Eqn. of Time 00^h	SUN Eqn. of Time 12^h	SUN Mer. Pass.	MOON Mer. Pass. Upper	MOON Mer. Pass. Lower	Age	%	Phase
d	m s	m s	h m	h m	h m	d	%	
11	10 07	09 59	12 10	00 00	12 23	14	100	○
12	09 51	09 43	12 10	00 47	13 10	15	97	
13	09 34	09 26	12 09	01 34	13 57	16	93	

UT		ARIES	VENUS −4.5		MARS +1.2		JUPITER −2.0		SATURN +0.5	
		GHA	GHA	Dec	GHA	Dec	GHA	Dec	GHA	Dec
d	h	° ′	° ′	° ′	° ′	° ′	° ′	° ′	° ′	° ′
14	00	171 44.8	161 41.8	N12 59.2	200 06.1	S12 51.2	213 37.0	S16 41.9	1 55.0	N 6 50.9
	01	186 47.3	176 45.0	59.1	215 06.7	50.6	228 39.0	41.8	16 57.6	50.9
	02	201 49.7	191 48.3	59.0	230 07.3	49.9	243 40.9	41.6	32 00.3	51.0
	03	216 52.2	206 51.6	. . 59.0	245 07.8	. . 49.2	258 42.9	. . 41.5	47 02.9	. . 51.1
	04	231 54.6	221 54.9	58.9	260 08.4	48.5	273 44.8	41.3	62 05.6	51.2
	05	246 57.1	236 58.2	58.8	275 09.0	47.8	288 46.7	41.2	77 08.2	51.3
	06	261 59.6	252 01.5	N12 58.8	290 09.6	S12 47.2	303 48.7	S16 41.0	92 10.9	N 6 51.3
	07	277 02.0	267 04.8	58.7	305 10.2	46.5	318 50.6	40.9	107 13.5	51.4
S	08	292 04.5	282 08.1	58.6	320 10.7	45.8	333 52.5	40.7	122 16.2	51.5
A	09	307 07.0	297 11.4	. . 58.5	335 11.3	. . 45.1	348 54.5	. . 40.6	137 18.8	. . 51.6
T	10	322 09.4	312 14.7	58.4	350 11.9	44.4	3 56.4	40.4	152 21.4	51.6
U	11	337 11.9	327 18.0	58.3	5 12.5	43.8	18 58.4	40.3	167 24.1	51.7
R	12	352 14.4	342 21.3	N12 58.3	20 13.1	S12 43.1	34 00.3	S16 40.1	182 26.7	N 6 51.8
D	13	7 16.8	357 24.6	58.2	35 13.7	42.4	49 02.2	40.0	197 29.4	51.9
A	14	22 19.3	12 27.9	58.1	50 14.2	41.7	64 04.2	39.8	212 32.0	52.0
Y	15	37 21.7	27 31.3	. . 58.0	65 14.8	. . 41.0	79 06.1	. . 39.7	227 34.7	. . 52.0
	16	52 24.2	42 34.6	57.9	80 15.4	40.4	94 08.0	39.5	242 37.3	52.1
	17	67 26.7	57 37.9	57.8	95 16.0	39.7	109 10.0	39.4	257 40.0	52.2
	18	82 29.1	72 41.2	N12 57.7	110 16.6	S12 39.0	124 11.9	S16 39.2	272 42.6	N 6 52.3
	19	97 31.6	87 44.6	57.6	125 17.2	38.3	139 13.9	39.1	287 45.3	52.4
	20	112 34.1	102 47.9	57.4	140 17.7	37.6	154 15.8	38.9	302 47.9	52.4
	21	127 36.5	117 51.3	. . 57.3	155 18.3	. . 36.9	169 17.7	. . 38.8	317 50.6	. . 52.5
	22	142 39.0	132 54.6	57.2	170 18.9	36.3	184 19.7	38.6	332 53.2	52.6
	23	157 41.5	147 58.0	57.1	185 19.5	35.6	199 21.6	38.5	347 55.8	52.7
15	00	172 43.9	163 01.3	N12 57.0	200 20.1	S12 34.9	214 23.6	S16 38.3	2 58.5	N 6 52.7
	01	187 46.4	178 04.7	56.8	215 20.7	34.2	229 25.5	38.2	18 01.1	52.8
	02	202 48.9	193 08.0	56.7	230 21.3	33.5	244 27.4	38.0	33 03.8	52.9
	03	217 51.3	208 11.4	. . 56.6	245 21.8	. . 32.8	259 29.4	. . 37.9	48 06.4	. . 53.0
	04	232 53.8	223 14.8	56.5	260 22.4	32.2	274 31.3	37.7	63 09.1	53.1
	05	247 56.2	238 18.1	56.3	275 23.0	31.5	289 33.3	37.6	78 11.7	53.1
	06	262 58.7	253 21.5	N12 56.2	290 23.6	S12 30.8	304 35.2	S16 37.4	93 14.4	N 6 53.2
	07	278 01.2	268 24.9	56.1	305 24.2	30.1	319 37.1	37.3	108 17.0	53.3
	08	293 03.6	283 28.3	55.9	320 24.8	29.4	334 39.1	37.1	123 19.7	53.4
S	09	308 06.1	298 31.7	. . 55.8	335 25.4	. . 28.7	349 41.0	. . 37.0	138 22.3	. . 53.4
U	10	323 08.6	313 35.0	55.6	350 25.9	28.0	4 43.0	36.8	153 25.0	53.5
N	11	338 11.0	328 38.4	55.5	5 26.5	27.4	19 44.9	36.7	168 27.6	53.6
D	12	353 13.5	343 41.8	N12 55.3	20 27.1	S12 26.7	34 46.8	S16 36.5	183 30.2	N 6 53.7
A	13	8 16.0	358 45.2	55.2	35 27.7	26.0	49 48.8	36.4	198 32.9	53.8
Y	14	23 18.4	13 48.6	55.0	50 28.3	25.3	64 50.7	36.2	213 35.5	53.8
	15	38 20.9	28 52.0	. . 54.8	65 28.9	. . 24.6	79 52.7	. . 36.1	228 38.2	. . 53.9
	16	53 23.4	43 55.4	54.7	80 29.5	23.9	94 54.6	35.9	243 40.8	54.0
	17	68 25.8	58 58.8	54.5	95 30.1	23.2	109 56.5	35.8	258 43.5	54.1
	18	83 28.3	74 02.3	N12 54.3	110 30.6	S12 22.5	124 58.5	S16 35.6	273 46.1	N 6 54.2
	19	98 30.7	89 05.7	54.2	125 31.2	21.9	140 00.4	35.5	288 48.8	54.2
	20	113 33.2	104 09.1	54.0	140 31.8	21.2	155 02.4	35.3	303 51.4	54.3
	21	128 35.7	119 12.5	. . 53.8	155 32.4	. . 20.5	170 04.3	. . 35.2	318 54.1	. . 54.4
	22	143 38.1	134 15.9	53.6	170 33.0	19.8	185 06.3	35.0	333 56.7	54.5
	23	158 40.6	149 19.4	53.5	185 33.6	19.1	200 08.2	34.9	348 59.4	54.5
16	00	173 43.1	164 22.8	N12 53.3	200 34.2	S12 18.4	215 10.1	S16 34.7	4 02.0	N 6 54.6
	01	188 45.5	179 26.2	53.1	215 34.8	17.7	230 12.1	34.6	19 04.6	54.7
	02	203 48.0	194 29.7	52.9	230 35.3	17.0	245 14.0	34.4	34 07.3	54.8
	03	218 50.5	209 33.1	. . 52.7	245 35.9	. . 16.3	260 16.0	. . 34.3	49 09.9	. . 54.9
	04	233 52.9	224 36.6	52.5	260 36.5	15.7	275 17.9	34.1	64 12.6	54.9
	05	248 55.4	239 40.0	52.3	275 37.1	15.0	290 19.8	34.0	79 15.2	55.0
	06	263 57.8	254 43.5	N12 52.1	290 37.7	S12 14.3	305 21.8	S16 33.8	94 17.9	N 6 55.1
	07	279 00.3	269 46.9	51.9	305 38.3	13.6	320 23.7	33.7	109 20.5	55.2
	08	294 02.8	284 50.4	51.7	320 38.9	12.9	335 25.7	33.5	124 23.2	55.2
M	09	309 05.2	299 53.8	. . 51.5	335 39.5	. . 12.2	350 27.6	. . 33.4	139 25.8	. . 55.3
O	10	324 07.7	314 57.3	51.3	350 40.1	11.5	5 29.6	33.2	154 28.5	55.4
N	11	339 10.2	330 00.8	51.1	5 40.7	10.8	20 31.5	33.1	169 31.1	55.5
D	12	354 12.6	345 04.2	N12 50.9	20 41.2	S12 10.1	35 33.5	S16 32.9	184 33.7	N 6 55.6
A	13	9 15.1	0 07.7	50.6	35 41.8	09.4	50 35.4	32.8	199 36.4	55.6
Y	14	24 17.6	15 11.2	50.4	50 42.4	08.8	65 37.3	32.6	214 39.0	55.7
	15	39 20.0	30 14.7	. . 50.2	65 43.0	. . 08.1	80 39.3	. . 32.5	229 41.7	. . 55.8
	16	54 22.5	45 18.1	50.0	80 43.6	07.4	95 41.2	32.3	244 44.3	55.9
	17	69 25.0	60 21.6	49.7	95 44.2	06.7	110 43.2	32.2	259 47.0	55.9
	18	84 27.4	75 25.1	N12 49.5	110 44.8	S12 06.0	125 45.1	S16 32.0	274 49.6	N 6 56.0
	19	99 29.9	90 28.6	49.3	125 45.4	05.3	140 47.1	31.9	289 52.3	56.1
	20	114 32.3	105 32.1	49.0	140 46.0	04.6	155 49.0	31.7	304 54.9	56.2
	21	129 34.8	120 35.6	. . 48.8	155 46.6	. . 03.9	170 51.0	. . 31.6	319 57.6	. . 56.2
	22	144 37.3	135 39.1	48.6	170 47.2	03.2	185 52.9	31.4	335 00.2	56.3
	23	159 39.7	150 42.6	48.3	185 47.8	02.5	200 54.8	31.3	350 02.8	56.4
Mer. Pass.		h m 12 27.0	v 3.4	d 0.2	v 0.6	d 0.7	v 1.9	d 0.2	v 2.6	d 0.1

STARS

Name	SHA	Dec
	° ′	° ′
Acamar	315 20.9	S40 16.2
Achernar	335 29.4	S57 11.5
Acrux	173 12.5	S63 09.1
Adhara	255 14.9	S28 59.2
Aldebaran	290 53.1	N16 31.7
Alioth	166 22.7	N55 54.1
Alkaid	153 00.8	N49 15.2
Al Na'ir	27 47.9	S46 54.9
Alnilam	275 49.6	S 1 11.8
Alphard	217 58.9	S 8 42.1
Alphecca	126 13.5	N26 40.7
Alpheratz	357 47.2	N29 08.5
Altair	62 11.5	N 8 53.4
Ankaa	353 19.1	S42 15.4
Antares	112 30.1	S26 27.2
Arcturus	145 58.3	N19 07.8
Atria	107 34.8	S69 02.5
Avior	234 19.1	S59 32.6
Bellatrix	278 35.4	N 6 21.5
Betelgeuse	271 04.7	N 7 24.5
Canopus	263 57.5	S52 42.3
Capella	280 39.2	N46 00.6
Deneb	49 34.1	N45 18.5
Denebola	182 36.5	N14 31.0
Diphda	348 59.3	S17 56.2
Dubhe	193 54.6	N61 42.0
Elnath	278 16.6	N28 37.0
Eltanin	90 47.6	N51 28.9
Enif	33 50.5	N 9 54.9
Fomalhaut	15 27.7	S29 34.4
Gacrux	172 04.1	S57 10.0
Gienah	175 55.3	S17 35.8
Hadar	148 52.2	S60 25.1
Hamal	328 04.7	N23 30.4
Kaus Aust.	83 48.1	S34 22.8
Kochab	137 18.7	N74 06.7
Markab	13 41.8	N15 15.2
Menkar	314 18.6	N 4 07.5
Menkent	148 11.1	S36 25.0
Miaplacidus	221 40.0	S69 45.5
Mirfak	308 45.3	N49 53.8
Nunki	76 02.3	S26 17.1
Peacock	53 24.4	S56 42.2
Pollux	243 31.4	N28 00.3
Procyon	245 02.9	N 5 12.0
Rasalhague	96 09.4	N12 32.9
Regulus	207 46.5	N11 55.2
Rigel	281 15.1	S 8 11.6
Rigil Kent.	139 55.8	S60 52.3
Sabik	102 16.1	S15 44.3
Schedar	349 44.9	N56 35.4
Shaula	96 26.2	S37 06.6
Sirius	258 36.4	S16 43.9
Spica	158 34.3	S11 12.8
Suhail	222 54.5	S43 28.4
Vega	80 41.2	N38 47.2
Zuben'ubi	137 08.7	S16 05.0

	SHA	Mer. Pass
	° ′	h m
Venus	350 17.4	13 05
Mars	27 36.2	10 38
Jupiter	41 39.6	9 41
Saturn	190 14.6	23 44

	UT	SUN GHA	SUN Dec	MOON GHA	v	MOON Dec	d	HP
	d h	° ′	° ′	° ′	′	° ′	′	′
	14 00	177 40.6	S 2 33.9	325 53.5	12.1	S16 09.2	12.0	56.7
	01	192 40.8	32.9	340 24.6	12.0	16 21.2	11.9	56.6
	02	207 40.9	31.9	354 55.6	12.0	16 33.1	11.9	56.6
	03	222 41.1	. . 30.9	9 26.6	12.0	16 45.0	11.7	56.6
	04	237 41.3	29.9	23 57.6	11.9	16 56.7	11.7	56.5
	05	252 41.5	28.9	38 28.5	11.9	17 08.4	11.6	56.5
	06	267 41.6	S 2 27.9	52 59.4	11.9	S17 20.0	11.4	56.5
	07	282 41.8	27.0	67 30.3	11.8	17 31.4	11.4	56.4
S	08	297 42.0	26.0	82 01.1	11.9	17 42.8	11.3	56.4
A	09	312 42.2	. . 25.0	96 32.0	11.8	17 54.1	11.1	56.4
T	10	327 42.3	24.0	111 02.8	11.7	18 05.2	11.1	56.3
U	11	342 42.5	23.0	125 33.5	11.7	18 16.3	11.0	56.3
R	12	357 42.7	S 2 22.0	140 04.2	11.7	S18 27.3	10.9	56.3
D	13	12 42.9	21.0	154 34.9	11.7	18 38.2	10.7	56.3
A	14	27 43.0	20.1	169 05.6	11.7	18 48.9	10.7	56.2
Y	15	42 43.2	. . 19.1	183 36.3	11.6	18 59.6	10.6	56.2
	16	57 43.4	18.1	198 06.9	11.5	19 10.2	10.4	56.2
	17	72 43.6	17.1	212 37.4	11.6	19 20.6	10.4	56.1
	18	87 43.7	S 2 16.1	227 08.0	11.5	S19 31.0	10.3	56.1
	19	102 43.9	15.1	241 38.5	11.4	19 41.3	10.1	56.1
	20	117 44.1	14.1	256 08.9	11.5	19 51.4	10.0	56.1
	21	132 44.3	. . 13.1	270 39.4	11.4	20 01.4	10.0	56.0
	22	147 44.4	12.2	285 09.8	11.4	20 11.4	9.8	56.0
	23	162 44.6	11.2	299 40.2	11.3	20 21.2	9.7	56.0
	15 00	177 44.8	S 2 10.2	314 10.5	11.3	S20 30.9	9.6	55.9
	01	192 45.0	09.2	328 40.8	11.3	20 40.5	9.5	55.9
	02	207 45.1	08.2	343 11.1	11.3	20 50.0	9.4	55.9
	03	222 45.3	. . 07.2	357 41.4	11.2	20 59.4	9.3	55.9
	04	237 45.5	06.2	12 11.6	11.2	21 08.7	9.2	55.8
	05	252 45.7	05.3	26 41.8	11.1	21 17.9	9.0	55.8
	06	267 45.8	S 2 04.3	41 11.9	11.1	S21 26.9	8.9	55.8
	07	282 46.0	03.3	55 42.0	11.1	21 35.8	8.9	55.7
	08	297 46.2	02.3	70 12.1	11.1	21 44.7	8.7	55.7
S	09	312 46.4	. . 01.3	84 42.2	11.0	21 53.4	8.6	55.7
U	10	327 46.5	2 00.3	99 12.2	11.0	22 02.0	8.4	55.7
N	11	342 46.7	1 59.3	113 42.2	11.0	22 10.4	8.4	55.6
D	12	357 46.9	S 1 58.3	128 12.2	10.9	S22 18.8	8.3	55.6
A	13	12 47.1	57.4	142 42.1	10.9	22 27.1	8.1	55.6
Y	14	27 47.2	56.4	157 12.0	10.9	22 35.2	8.0	55.6
	15	42 47.4	. . 55.4	171 41.9	10.8	22 43.2	7.9	55.5
	16	57 47.6	54.4	186 11.7	10.8	22 51.1	7.8	55.5
	17	72 47.8	53.4	200 41.5	10.8	22 58.9	7.6	55.5
	18	87 48.0	S 1 52.4	215 11.3	10.8	S23 06.5	7.5	55.5
	19	102 48.1	51.4	229 41.1	10.7	23 14.0	7.5	55.4
	20	117 48.3	50.4	244 10.8	10.7	23 21.5	7.3	55.4
	21	132 48.5	. . 49.5	258 40.5	10.6	23 28.8	7.1	55.4
	22	147 48.7	48.5	273 10.1	10.7	23 35.9	7.1	55.4
	23	162 48.8	47.5	287 39.8	10.6	23 43.0	6.9	55.3
	16 00	177 49.0	S 1 46.5	302 09.4	10.6	S23 49.9	6.8	55.3
	01	192 49.2	45.5	316 39.0	10.5	23 56.7	6.7	55.3
	02	207 49.4	44.5	331 08.5	10.6	24 03.4	6.5	55.3
	03	222 49.6	. . 43.5	345 38.1	10.5	24 09.9	6.5	55.2
	04	237 49.7	42.5	0 07.6	10.4	24 16.4	6.3	55.2
	05	252 49.9	41.6	14 37.0	10.5	24 22.7	6.2	55.2
	06	267 50.1	S 1 40.6	29 06.5	10.4	S24 28.9	6.0	55.2
	07	282 50.3	39.6	43 35.9	10.4	24 34.9	6.0	55.1
	08	297 50.4	38.6	58 05.3	10.4	24 40.9	5.8	55.1
M	09	312 50.6	. . 37.6	72 34.7	10.4	24 46.7	5.6	55.1
O	10	327 50.8	36.6	87 04.1	10.3	24 52.3	5.6	55.1
N	11	342 51.0	35.6	101 33.4	10.3	24 57.9	5.4	55.1
D	12	357 51.2	S 1 34.6	116 02.7	10.3	S25 03.3	5.3	55.0
A	13	12 51.3	33.6	130 32.0	10.3	25 08.6	5.2	55.0
Y	14	27 51.5	32.7	145 01.3	10.3	25 13.8	5.0	55.0
	15	42 51.7	. . 31.7	159 30.6	10.2	25 18.8	4.9	55.0
	16	57 51.9	30.7	173 59.8	10.2	25 23.7	4.8	55.0
	17	72 52.1	29.7	188 29.0	10.2	25 28.5	4.7	54.9
	18	87 52.2	S 1 28.7	202 58.2	10.2	S25 33.2	4.5	54.9
	19	102 52.4	27.7	217 27.4	10.2	25 37.7	4.4	54.9
	20	117 52.6	26.7	231 56.6	10.1	25 42.1	4.3	54.9
	21	132 52.8	. . 25.7	246 25.7	10.2	25 46.4	4.1	54.9
	22	147 53.0	24.8	260 54.9	10.1	25 50.5	4.0	54.8
	23	162 53.1	23.8	275 24.0	10.1	S25 54.5	3.9	54.8
		SD 16.1	d 1.0	SD	15.3	15.2		15.0

Lat.	Twilight Naut.	Twilight Civil	Sunrise	Moonrise 14	15	16	17
°	h m	h m	h m	h m	h m	h m	h m
N 72	03 53	05 17	06 24	00 37	■	■	■
N 70	04 07	05 21	06 22	■	■	■	■
68	04 18	05 25	06 21	25 52	01 52	■	■
66	04 27	05 29	06 19	24 59	00 59	■	■
64	04 35	05 31	06 18	24 27	00 27	02 25	■
62	04 41	05 33	06 17	24 04	00 04	01 45	03 20
60	04 46	05 35	06 17	23 45	25 17	01 17	02 40
N 58	04 51	05 37	06 16	23 30	24 56	00 56	02 13
56	04 55	05 38	06 15	23 17	24 38	00 38	01 52
54	04 58	05 40	06 15	23 05	24 23	00 23	01 34
52	05 01	05 41	06 14	22 55	24 10	00 10	01 19
50	05 04	05 41	06 14	22 46	23 59	25 06	01 06
45	05 09	05 43	06 13	22 28	23 36	24 39	00 39
N 40	05 13	05 45	06 12	22 12	23 17	24 18	00 18
35	05 16	05 45	06 11	22 00	23 02	24 01	00 01
30	05 18	05 46	06 10	21 48	22 48	23 46	24 40
20	05 21	05 46	06 08	21 29	22 25	23 20	24 13
N 10	05 22	05 46	06 07	21 13	22 06	22 58	23 51
0	05 21	05 45	06 06	20 58	21 47	22 38	23 29
S 10	05 19	05 43	06 04	20 43	21 29	22 18	23 08
20	05 15	05 40	06 02	20 26	21 10	21 56	22 45
30	05 08	05 36	06 00	20 08	20 47	21 31	22 19
35	05 04	05 34	05 59	19 57	20 34	21 16	22 04
40	04 59	05 31	05 58	19 45	20 19	20 59	21 45
45	04 52	05 26	05 56	19 31	20 02	20 39	21 24
S 50	04 43	05 21	05 54	19 14	19 40	20 13	20 56
52	04 38	05 19	05 53	19 05	19 29	20 01	20 43
54	04 34	05 16	05 52	18 56	19 18	19 47	20 27
56	04 28	05 13	05 51	18 46	19 04	19 31	20 09
58	04 22	05 10	05 49	18 35	18 49	19 11	19 47
S 60	04 14	05 06	05 48	18 21	18 30	18 47	19 18

Lat.	Sunset	Twilight Civil	Twilight Naut.	Moonset 14	15	16	17
°	h m	h m	h m	h m	h m	h m	h m
N 72	17 57	19 04	20 29	03 42	■	■	■
N 70	17 58	18 59	20 14	04 33	■	■	■
68	17 59	18 55	20 02	05 05	04 13	■	■
66	18 00	18 51	19 53	05 29	05 06	■	■
64	18 01	18 48	19 45	05 47	05 39	05 25	■
62	18 02	18 46	19 39	06 03	06 03	06 06	06 17
60	18 02	18 44	19 33	06 16	06 23	06 34	06 57
N 58	18 03	18 42	19 29	06 27	06 39	06 56	07 24
56	18 04	18 41	19 24	06 37	06 52	07 14	07 46
54	18 04	18 39	19 21	06 46	07 04	07 29	08 04
52	18 05	18 38	19 18	06 54	07 15	07 42	08 19
50	18 05	18 37	19 15	07 01	07 24	07 54	08 32
45	18 06	18 35	19 10	07 16	07 44	08 18	08 59
N 40	18 07	18 34	19 05	07 28	08 00	08 37	09 20
35	18 08	18 33	19 02	07 39	08 14	08 53	09 38
30	18 08	18 32	19 00	07 49	08 26	09 08	09 53
20	18 10	18 32	18 57	08 05	08 47	09 31	10 19
N 10	18 11	18 32	18 56	08 19	09 05	09 52	10 42
0	18 12	18 33	18 57	08 32	09 21	10 12	11 03
S 10	18 14	18 35	18 59	08 46	09 38	10 31	11 24
20	18 15	18 37	19 03	09 00	09 56	10 52	11 46
30	18 17	18 41	19 09	09 17	10 17	11 16	12 12
35	18 18	18 43	19 13	09 27	10 30	11 31	12 28
40	18 19	18 47	19 18	09 38	10 44	11 47	12 46
45	18 21	18 50	19 25	09 51	11 01	12 07	13 07
S 50	18 23	18 55	19 34	10 07	11 22	12 32	13 34
52	18 24	18 58	19 38	10 15	11 32	12 44	13 48
54	18 25	19 00	19 43	10 23	11 43	12 58	14 03
56	18 26	19 03	19 48	10 32	11 56	13 14	14 21
58	18 27	19 07	19 54	10 43	12 11	13 33	14 44
S 60	18 28	19 10	20 01	10 56	12 29	13 57	15 12

Day	SUN Eqn. of Time 00^h	SUN Eqn. of Time 12^h	SUN Mer. Pass.	MOON Mer. Pass. Upper	MOON Mer. Pass. Lower	Age	Phase
d	m s	m s	h m	h m	h m	d %	
14	09 18	09 10	12 09	02 21	14 45	17 87	
15	09 01	08 53	12 09	03 10	15 34	18 79	
16	08 44	08 36	12 09	03 59	16 25	19 71	

Day	UT d h	ARIES GHA ° ′	VENUS −4.3 GHA ° ′	VENUS Dec ° ′	MARS +1.2 GHA ° ′	MARS Dec ° ′	JUPITER −2.0 GHA ° ′	JUPITER Dec ° ′	SATURN +0.5 GHA ° ′	SATURN Dec ° ′
TUESDAY	17 00	174 42.2	165 46.1	N12 48.1	200 48.3	S12 01.8	215 56.8	S16 31.1	5 05.5	N 6 56.5
	01	189 44.7	180 49.6	47.8	215 48.9	01.1	230 58.7	31.0	20 08.1	56.6
	02	204 47.1	195 53.1	47.6	230 49.5	12 00.4	246 00.7	30.8	35 10.8	56.6
	03	219 49.6	210 56.6	. . 47.3	245 50.1	11 59.7	261 02.6	. . 30.7	50 13.4	. . 56.7
	04	234 52.1	226 00.1	47.0	260 50.7	59.0	276 04.6	30.5	65 16.1	56.8
	05	249 54.5	241 03.7	46.8	275 51.3	58.3	291 06.5	30.4	80 18.7	56.9
	06	264 57.0	256 07.2	N12 46.5	290 51.9	S11 57.7	306 08.5	S16 30.2	95 21.4	N 6 56.9
	07	279 59.5	271 10.7	46.3	305 52.5	57.0	321 10.4	30.1	110 24.0	57.0
	08	295 01.9	286 14.2	46.0	320 53.1	56.3	336 12.4	29.9	125 26.6	57.1
	09	310 04.4	301 17.8	. . 45.7	335 53.7	. . 55.6	351 14.3	. . 29.8	140 29.3	. . 57.2
	10	325 06.8	316 21.3	45.4	350 54.3	54.9	6 16.3	29.6	155 31.9	57.3
	11	340 09.3	331 24.8	45.2	5 54.9	54.2	21 18.2	29.5	170 34.6	57.3
	12	355 11.8	346 28.4	N12 44.9	20 55.5	S11 53.5	36 20.1	S16 29.3	185 37.2	N 6 57.4
	13	10 14.2	1 31.9	44.6	35 56.1	52.8	51 22.1	29.2	200 39.9	57.5
	14	25 16.7	16 35.4	44.3	50 56.7	52.1	66 24.0	29.0	215 42.5	57.6
	15	40 19.2	31 39.0	. . 44.0	65 57.3	. . 51.4	81 26.0	. . 28.9	230 45.2	. . 57.6
	16	55 21.6	46 42.5	43.7	80 57.8	50.7	96 27.9	28.7	245 47.8	57.7
	17	70 24.1	61 46.1	43.5	95 58.4	50.0	111 29.9	28.6	260 50.4	57.8
	18	85 26.6	76 49.7	N12 43.2	110 59.0	S11 49.3	126 31.8	S16 28.4	275 53.1	N 6 57.9
	19	100 29.0	91 53.2	42.9	125 59.6	48.6	141 33.8	28.3	290 55.7	57.9
	20	115 31.5	106 56.8	42.6	141 00.2	47.9	156 35.7	28.1	305 58.4	58.0
	21	130 34.0	122 00.3	. . 42.3	156 00.8	. . 47.2	171 37.7	. . 28.0	321 01.0	. . 58.1
	22	145 36.4	137 03.9	42.0	171 01.4	46.5	186 39.6	27.8	336 03.7	58.2
	23	160 38.9	152 07.5	41.6	186 02.0	45.8	201 41.6	27.7	351 06.3	58.3
WEDNESDAY	18 00	175 41.3	167 11.0	N12 41.3	201 02.6	S11 45.1	216 43.5	S16 27.5	6 09.0	N 6 58.3
	01	190 43.8	182 14.6	41.0	216 03.2	44.4	231 45.5	27.4	21 11.6	58.4
	02	205 46.3	197 18.2	40.7	231 03.8	43.7	246 47.4	27.2	36 14.2	58.5
	03	220 48.7	212 21.8	. . 40.4	246 04.4	. . 43.0	261 49.4	. . 27.1	51 16.9	. . 58.6
	04	235 51.2	227 25.4	40.1	261 05.0	42.3	276 51.3	26.9	66 19.5	58.6
	05	250 53.7	242 28.9	39.7	276 05.6	41.6	291 53.3	26.8	81 22.2	58.7
	06	265 56.1	257 32.5	N12 39.4	291 06.2	S11 40.9	306 55.2	S16 26.6	96 24.8	N 6 58.8
	07	280 58.6	272 36.1	39.1	306 06.8	40.2	321 57.2	26.5	111 27.5	58.9
	08	296 01.1	287 39.7	38.8	321 07.4	39.5	336 59.1	26.3	126 30.1	58.9
	09	311 03.5	302 43.3	. . 38.4	336 08.0	. . 38.8	352 01.1	. . 26.2	141 32.8	. . 59.0
	10	326 06.0	317 46.9	38.1	351 08.6	38.1	7 03.0	26.0	156 35.4	59.1
	11	341 08.4	332 50.5	37.7	6 09.2	37.4	22 05.0	25.9	171 38.0	59.2
	12	356 10.9	347 54.1	N12 37.4	21 09.8	S11 36.7	37 06.9	S16 25.7	186 40.7	N 6 59.2
	13	11 13.4	2 57.7	37.1	36 10.4	36.0	52 08.9	25.6	201 43.3	59.3
	14	26 15.8	18 01.3	36.7	51 11.0	35.3	67 10.8	25.4	216 46.0	59.4
	15	41 18.3	33 04.9	. . 36.4	66 11.6	. . 34.6	82 12.8	. . 25.3	231 48.6	. . 59.5
	16	56 20.8	48 08.5	36.0	81 12.2	33.9	97 14.7	25.1	246 51.3	59.6
	17	71 23.2	63 12.2	35.7	96 12.8	33.2	112 16.7	25.0	261 53.9	59.6
	18	86 25.7	78 15.8	N12 35.3	111 13.4	S11 32.5	127 18.6	S16 24.8	276 56.5	N 6 59.7
	19	101 28.2	93 19.4	34.9	126 14.0	31.8	142 20.6	24.7	291 59.2	59.8
	20	116 30.6	108 23.0	34.6	141 14.6	31.1	157 22.5	24.6	307 01.8	59.9
	21	131 33.1	123 26.7	. . 34.2	156 15.2	. . 30.4	172 24.5	. . 24.4	322 04.5	6 59.9
	22	146 35.6	138 30.3	33.8	171 15.8	29.7	187 26.4	24.3	337 07.1	7 00.0
	23	161 38.0	153 33.9	33.5	186 16.4	29.0	202 28.4	24.1	352 09.8	00.1
THURSDAY	19 00	176 40.5	168 37.5	N12 33.1	201 17.0	S11 28.3	217 30.3	S16 24.0	7 12.4	N 7 00.2
	01	191 42.9	183 41.2	32.7	216 17.6	27.6	232 32.3	23.8	22 15.1	00.2
	02	206 45.4	198 44.8	32.3	231 18.2	26.9	247 34.2	23.7	37 17.7	00.3
	03	221 47.9	213 48.5	. . 32.0	246 18.8	. . 26.2	262 36.2	. . 23.5	52 20.3	. . 00.4
	04	236 50.3	228 52.1	31.6	261 19.4	25.5	277 38.1	23.4	67 23.0	00.5
	05	251 52.8	243 55.7	31.2	276 20.0	24.8	292 40.1	23.2	82 25.6	00.5
	06	266 55.3	258 59.4	N12 30.8	291 20.6	S11 24.1	307 42.0	S16 23.1	97 28.3	N 7 00.6
	07	281 57.7	274 03.0	30.4	306 21.2	23.4	322 44.0	22.9	112 30.9	00.7
	08	297 00.2	289 06.7	30.0	321 21.8	22.7	337 45.9	22.8	127 33.6	00.8
	09	312 02.7	304 10.3	. . 29.6	336 22.4	. . 22.0	352 47.9	. . 22.6	142 36.2	. . 00.8
	10	327 05.1	319 14.0	29.2	351 23.0	21.2	7 49.9	22.5	157 38.8	00.9
	11	342 07.6	334 17.7	28.8	6 23.6	20.5	22 51.8	22.3	172 41.5	01.0
	12	357 10.1	349 21.3	N12 28.4	21 24.2	S11 19.8	37 53.8	S16 22.2	187 44.1	N 7 01.1
	13	12 12.5	4 25.0	28.0	36 24.8	19.1	52 55.7	22.0	202 46.8	01.2
	14	27 15.0	19 28.6	27.6	51 25.4	18.4	67 57.7	21.9	217 49.4	01.2
	15	42 17.4	34 32.3	. . 27.2	66 26.0	. . 17.7	82 59.6	. . 21.7	232 52.1	. . 01.3
	16	57 19.9	49 36.0	26.8	81 26.6	17.0	98 01.6	21.6	247 54.7	01.4
	17	72 22.4	64 39.7	26.4	96 27.2	16.3	113 03.5	21.4	262 57.3	01.5
	18	87 24.8	79 43.3	N12 25.9	111 27.8	S11 15.6	128 05.5	S16 21.3	278 00.0	N 7 01.5
	19	102 27.3	94 47.0	25.5	126 28.4	14.9	143 07.4	21.1	293 02.6	01.6
	20	117 29.8	109 50.7	25.1	141 29.0	14.2	158 09.4	21.0	308 05.3	01.7
	21	132 32.2	124 54.4	. . 24.7	156 29.6	. . 13.5	173 11.3	. . 20.8	323 07.9	. . 01.8
	22	147 34.7	139 58.0	24.2	171 30.2	12.8	188 13.3	20.7	338 10.6	01.8
	23	162 37.2	155 01.7	23.8	186 30.8	12.1	203 15.3	20.5	353 13.2	01.9
	Mer. Pass. (h m)	12 15.2	v 3.6	d 0.3	v 0.6	d 0.7	v 2.0	d 0.1	v 2.6	d 0.1

STARS

Name	SHA ° ′	Dec °
Acamar	315 20.9	S40 16
Achernar	335 29.4	S57 11
Acrux	173 12.5	S63 09
Adhara	255 14.9	S28 59
Aldebaran	290 53.1	N16 31
Alioth	166 22.7	N55 54
Alkaid	153 00.8	N49 15
Al Na'ir	27 47.9	S46 54
Alnilam	275 49.6	S 1 11
Alphard	217 58.9	S 8 42
Alphecca	126 13.4	N26 40
Alpheratz	357 47.2	N29 08
Altair	62 11.5	N 8 53
Ankaa	353 19.1	S42 15
Antares	112 30.0	S26 27
Arcturus	145 58.3	N19 07
Atria	107 34.7	S69 02
Avior	234 19.1	S59 32
Bellatrix	278 35.4	N 6 21
Betelgeuse	271 04.7	N 7 24
Canopus	263 57.5	S52 42
Capella	280 39.2	N46 00
Deneb	49 34.0	N45 18
Denebola	182 36.5	N14 31
Diphda	348 59.3	S17 56
Dubhe	193 54.6	N61 42
Elnath	278 16.7	N28 37
Eltanin	90 47.6	N51 28
Enif	33 50.5	N 9 54
Fomalhaut	15 27.7	S29 34
Gacrux	172 04.1	S57 10
Gienah	175 55.3	S17 35
Hadar	148 52.1	S60 25
Hamal	328 04.7	N23 30
Kaus Aust.	83 48.0	S34 22
Kochab	137 18.6	N74 06
Markab	13 41.8	N15 15
Menkar	314 18.6	N 4 07
Menkent	148 11.1	S36 25
Miaplacidus	221 40.0	S69 45
Mirfak	308 45.3	N49 53
Nunki	76 02.3	S26 17
Peacock	53 24.4	S56 42
Pollux	243 31.4	N28 00
Procyon	245 02.9	N 5 12
Rasalhague	96 09.4	N12 32
Regulus	207 46.5	N11 55
Rigel	281 15.1	S 8 11
Rigil Kent.	139 55.8	S60 52
Sabik	102 16.1	S15 44
Schedar	349 44.9	N56 35
Shaula	96 26.2	S37 06
Sirius	258 36.4	S16 43
Spica	158 34.3	S11 12
Suhail	222 54.6	S43 28
Vega	80 41.2	N38 47
Zuben'ubi	137 08.7	S16 05

	SHA ° ′	Mer. Pass. h m
Venus	351 29.7	12 48
Mars	25 21.3	10 35
Jupiter	41 02.2	9 32
Saturn	190 27.6	23 31

UT	SUN GHA	SUN Dec	MOON GHA	v	MOON Dec	d	HP
d h	° ′	° ′	° ′	′	° ′	′	′
17 00	177 53.3	S 1 22.8	289 53.1	10.1	S25 58.4	3.8	54.8
01	192 53.5	21.8	304 22.2	10.1	26 02.2	3.6	54.8
02	207 53.7	20.8	318 51.3	10.1	26 05.8	3.5	54.8
03	222 53.9	. . 19.8	333 20.4	10.1	26 09.3	3.3	54.7
04	237 54.0	18.8	347 49.5	10.0	26 12.6	3.3	54.7
05	252 54.2	17.8	2 18.5	10.1	26 15.9	3.1	54.7
06	267 54.4	S 1 16.8	16 47.6	10.0	S26 19.0	2.9	54.7
07	282 54.6	15.9	31 16.6	10.1	26 21.9	2.9	54.7
08	297 54.8	14.9	45 45.7	10.0	26 24.8	2.7	54.7
09	312 54.9	. . 13.9	60 14.7	10.0	26 27.5	2.5	54.6
10	327 55.1	12.9	74 43.7	10.1	26 30.0	2.5	54.6
11	342 55.3	11.9	89 12.8	10.0	26 32.5	2.3	54.6
12	357 55.5	S 1 10.9	103 41.8	10.0	S26 34.8	2.2	54.6
13	12 55.7	09.9	118 10.8	10.0	26 37.0	2.0	54.6
14	27 55.8	08.9	132 39.8	10.0	26 39.0	1.9	54.6
15	42 56.0	. . 07.9	147 08.8	10.0	26 40.9	1.8	54.6
16	57 56.2	07.0	161 37.8	10.1	26 42.7	1.7	54.5
17	72 56.4	06.0	176 06.9	10.0	26 44.4	1.5	54.5
18	87 56.6	S 1 05.0	190 35.9	10.0	S26 45.9	1.4	54.5
19	102 56.7	04.0	205 04.9	10.0	26 47.3	1.2	54.5
20	117 56.9	03.0	219 33.9	10.1	26 48.5	1.1	54.5
21	132 57.1	. . 02.0	234 03.0	10.0	26 49.6	1.0	54.5
22	147 57.3	01.0	248 32.0	10.0	26 50.6	0.9	54.5
23	162 57.5	1 00.0	263 01.0	10.1	26 51.5	0.7	54.5
18 00	177 57.7	S 0 59.1	277 30.1	10.0	S26 52.2	0.6	54.4
01	192 57.8	58.1	291 59.1	10.1	26 52.8	0.5	54.4
02	207 58.0	57.1	306 28.2	10.1	26 53.3	0.3	54.4
03	222 58.2	. . 56.1	320 57.3	10.1	26 53.6	0.2	54.4
04	237 58.4	55.1	335 26.4	10.1	26 53.8	0.1	54.4
05	252 58.6	54.1	349 55.5	10.1	26 53.9	0.0	54.4
06	267 58.7	S 0 53.1	4 24.6	10.1	S26 53.9	0.2	54.4
07	282 58.9	52.1	18 53.7	10.2	26 53.7	0.3	54.4
08	297 59.1	51.1	33 22.9	10.1	26 53.4	0.5	54.4
09	312 59.3	. . 50.2	47 52.0	10.2	26 52.9	0.6	54.4
10	327 59.5	49.2	62 21.2	10.2	26 52.3	0.7	54.3
11	342 59.7	48.2	76 50.4	10.2	26 51.6	0.8	54.3
12	357 59.8	S 0 47.2	91 19.6	10.2	S26 50.8	1.0	54.3
13	13 00.0	46.2	105 48.8	10.2	26 49.8	1.1	54.3
14	28 00.2	45.2	120 18.0	10.3	26 48.7	1.2	54.3
15	43 00.4	. . 44.2	134 47.3	10.3	26 47.5	1.3	54.3
16	58 00.6	43.2	149 16.6	10.3	26 46.2	1.5	54.3
17	73 00.7	42.2	163 45.9	10.3	26 44.7	1.6	54.3
18	88 00.9	S 0 41.3	178 15.2	10.4	S26 43.1	1.8	54.3
19	103 01.1	40.3	192 44.6	10.3	26 41.3	1.8	54.3
20	118 01.3	39.3	207 13.9	10.4	26 39.5	2.0	54.3
21	133 01.5	. . 38.3	221 43.3	10.5	26 37.5	2.1	54.3
22	148 01.7	37.3	236 12.8	10.4	26 35.4	2.3	54.3
23	163 01.8	36.3	250 42.2	10.5	26 33.1	2.4	54.3
19 00	178 02.0	S 0 35.3	265 11.7	10.5	S26 30.7	2.5	54.3
01	193 02.2	34.3	279 41.2	10.6	26 28.2	2.6	54.3
02	208 02.4	33.3	294 10.8	10.5	26 25.6	2.7	54.3
03	223 02.6	. . 32.4	308 40.3	10.6	26 22.9	2.9	54.3
04	238 02.8	31.4	323 09.9	10.6	26 20.0	3.0	54.2
05	253 02.9	30.4	337 39.5	10.7	26 17.0	3.1	54.2
06	268 03.1	S 0 29.4	352 09.2	10.7	S26 13.9	3.3	54.2
07	283 03.3	28.4	6 38.9	10.7	26 10.6	3.3	54.2
08	298 03.5	27.4	21 08.6	10.8	26 07.3	3.5	54.2
09	313 03.7	. . 26.4	35 38.4	10.8	26 03.8	3.6	54.2
10	328 03.9	25.4	50 08.2	10.8	26 00.2	3.8	54.2
11	343 04.0	24.4	64 38.0	10.8	25 56.4	3.8	54.2
12	358 04.2	S 0 23.5	79 07.8	10.9	S25 52.6	4.0	54.2
13	13 04.4	22.5	93 37.7	11.0	25 48.6	4.1	54.2
14	28 04.6	21.5	108 07.7	10.9	25 44.5	4.2	54.2
15	43 04.8	. . 20.5	122 37.6	11.0	25 40.3	4.4	54.2
16	58 05.0	19.5	137 07.6	11.1	25 35.9	4.4	54.2
17	73 05.2	18.5	151 37.7	11.1	25 31.5	4.6	54.2
18	88 05.3	S 0 17.5	166 07.8	11.1	S25 26.9	4.7	54.2
19	103 05.5	16.5	180 37.9	11.1	25 22.2	4.8	54.2
20	118 05.7	15.6	195 08.0	11.2	25 17.4	5.0	54.2
21	133 05.9	. . 14.6	209 38.2	11.3	25 12.4	5.0	54.2
22	148 06.1	13.6	224 08.5	11.3	25 07.4	5.2	54.2
23	163 06.3	12.6	238 38.8	11.3	S25 02.2	5.3	54.2
	SD 16.1	d 1.0	SD 14.9		14.8		14.8

Lat.	Twilight Naut.	Twilight Civil	Sunrise	Moonrise 17	Moonrise 18	Moonrise 19	Moonrise 20
°	h m	h m	h m	h m	h m	h m	h m
N 72	03 34	05 01	06 08	■	■	■	■
N 70	03 51	05 07	06 08	■	■	■	■
68	04 04	05 12	06 08	■	■	■	■
66	04 15	05 17	06 08	■	■	■	■
64	04 23	05 20	06 08	■	■	■	05 55
62	04 31	05 24	06 08	03 20	04 33	05 05	05 15
60	04 37	05 26	06 08	02 40	03 45	04 25	04 47
N 58	04 42	05 28	06 07	02 13	03 15	03 58	04 25
56	04 47	05 30	06 07	01 52	02 52	03 37	04 07
54	04 51	05 32	06 07	01 34	02 33	03 19	03 52
52	04 54	05 34	06 07	01 19	02 18	03 04	03 39
50	04 57	05 35	06 07	01 06	02 04	02 51	03 27
45	05 03	05 38	06 07	00 39	01 36	02 24	03 03
N 40	05 08	05 40	06 07	00 18	01 14	02 03	02 44
35	05 12	05 41	06 06	00 01	00 56	01 45	02 28
30	05 15	05 42	06 06	24 40	00 40	01 30	02 14
20	05 18	05 44	06 06	24 13	00 13	01 04	01 50
N 10	05 20	05 44	06 05	23 51	24 41	00 41	01 29
0	05 20	05 44	06 05	23 29	24 20	00 20	01 10
S 10	05 19	05 43	06 04	23 08	23 59	24 51	00 51
20	05 16	05 41	06 03	22 45	23 37	24 30	00 30
30	05 10	05 38	06 02	22 19	23 11	24 06	00 06
35	05 07	05 36	06 02	22 04	22 56	23 52	24 50
40	05 02	05 34	06 01	21 45	22 38	23 35	24 36
45	04 56	05 30	06 00	21 24	22 16	23 15	24 19
S 50	04 48	05 26	05 59	20 56	21 49	22 50	23 58
52	04 44	05 24	05 58	20 43	21 36	22 38	23 48
54	04 40	05 22	05 58	20 27	21 20	22 25	23 36
56	04 35	05 20	05 57	20 09	21 02	22 09	23 23
58	04 29	05 17	05 56	19 47	20 40	21 50	23 08
S 60	04 23	05 13	05 55	19 18	20 12	21 26	22 50

Lat.	Sunset	Twilight Civil	Twilight Naut.	Moonset 17	Moonset 18	Moonset 19	Moonset 20
°	h m	h m	h m	h m	h m	h m	h m
N 72	18 10	19 18	20 47	■	■	■	■
N 70	18 10	19 11	20 29	■	■	■	■
68	18 10	19 06	20 15	■	■	■	■
66	18 10	19 01	20 04	■	■	■	■
64	18 10	18 57	19 55	■	■	■	08 57
62	18 10	18 54	19 48	06 17	06 51	08 04	09 37
60	18 10	18 51	19 41	06 57	07 38	08 43	10 05
N 58	18 10	18 49	19 36	07 24	08 08	09 11	10 26
56	18 10	18 47	19 31	07 46	08 31	09 32	10 43
54	18 10	18 45	19 27	08 04	08 50	09 49	10 58
52	18 10	18 44	19 23	08 19	09 06	10 04	11 11
50	18 10	18 42	19 20	08 32	09 20	10 17	11 22
45	18 10	18 39	19 14	08 59	09 48	10 44	11 45
N 40	18 10	18 37	19 09	09 20	10 10	11 05	12 04
35	18 10	18 35	19 05	09 38	10 28	11 22	12 19
30	18 10	18 34	19 02	09 53	10 44	11 37	12 33
20	18 10	18 32	18 58	10 19	11 10	12 02	12 56
N 10	18 11	18 32	18 56	10 42	11 33	12 24	13 15
0	18 11	18 32	18 56	11 03	11 54	12 45	13 34
S 10	18 12	18 33	18 57	11 24	12 15	13 05	13 52
20	18 12	18 34	19 00	11 46	12 38	13 26	14 11
30	18 13	18 37	19 05	12 12	13 04	13 51	14 33
35	18 14	18 39	19 09	12 28	13 20	14 06	14 46
40	18 15	18 42	19 13	12 46	13 38	14 23	15 01
45	18 15	18 45	19 19	13 07	14 00	14 43	15 19
S 50	18 16	18 49	19 27	13 34	14 27	15 09	15 41
52	18 17	18 51	19 30	13 48	14 40	15 21	15 51
54	18 17	18 53	19 35	14 03	14 56	15 35	16 03
56	18 18	18 55	19 39	14 21	15 14	15 51	16 17
58	18 19	18 58	19 45	14 44	15 36	16 11	16 32
S 60	18 19	19 01	19 51	15 12	16 05	16 35	16 51

Day	SUN Eqn. of Time 00^h	SUN Eqn. of Time 12^h	SUN Mer. Pass.	MOON Mer. Pass. Upper	MOON Mer. Pass. Lower	MOON Age	MOON Phase
d	m s	m s	h m	h m	h m	d %	
17	08 27	08 18	12 08	04 50	17 16	20 62	
18	08 10	08 01	12 08	05 42	18 07	21 52	◑
19	07 52	07 43	12 08	06 32	18 57	22 43	

UT		ARIES	VENUS −4.2		MARS +1.2		JUPITER −2.0		SATURN +0.5		STARS		
		GHA	GHA	Dec	GHA	Dec	GHA	Dec	GHA	Dec	Name	SHA	Dec
d	h	° ′	° ′	° ′	° ′	° ′	° ′	° ′	° ′	° ′		° ′	° ′
20	00	177 39.6	170 05.4	N12 23.4	201 31.4	S11 11.4	218 17.2	S16 20.4	8 15.8	N 7 02.0	Acamar	315 20.9	S40 16.
	01	192 42.1	185 09.1	22.9	216 32.0	10.7	233 19.2	20.2	23 18.5	02.1	Achernar	335 29.4	S57 11.5
	02	207 44.6	200 12.8	22.5	231 32.6	09.9	248 21.1	20.1	38 21.1	02.1	Acrux	173 12.5	S63 09.
	03	222 47.0	215 16.5	22.1	246 33.2	09.2	263 23.1	20.0	53 23.8	02.2	Adhara	255 14.9	S28 59.
	04	237 49.5	230 20.2	21.6	261 33.8	08.5	278 25.0	19.8	68 26.4	02.3	Aldebaran	290 53.1	N16 31.
	05	252 51.9	245 23.9	21.2	276 34.4	07.8	293 27.0	19.7	83 29.0	02.4			
	06	267 54.4	260 27.6	N12 20.7	291 35.0	S11 07.1	308 28.9	S16 19.5	98 31.7	N 7 02.4	Alioth	166 22.6	N55 54.
	07	282 56.9	275 31.3	20.3	306 35.6	06.4	323 30.9	19.4	113 34.3	02.5	Alkaid	153 00.8	N49 15.
	08	297 59.3	290 35.0	19.8	321 36.2	05.7	338 32.9	19.2	128 37.0	02.6	Al Na'ir	27 47.9	S46 54.
F	09	313 01.8	305 38.7	19.3	336 36.8	05.0	353 34.8	19.1	143 39.6	02.7	Alnilam	275 49.6	S 1 11.8
R	10	328 04.3	320 42.4	18.9	351 37.4	04.3	8 36.8	18.9	158 42.3	02.7	Alphard	217 59.0	S 8 42.
I	11	343 06.7	335 46.1	18.4	6 38.1	03.6	23 38.7	18.8	173 44.9	02.8			
D	12	358 09.2	350 49.8	N12 18.0	21 38.7	S11 02.9	38 40.7	S16 18.6	188 47.5	N 7 02.9	Alphecca	126 13.4	N26 40.
A	13	13 11.7	5 53.5	17.5	36 39.3	02.1	53 42.6	18.5	203 50.2	03.0	Alpheratz	357 47.2	N29 08.5
Y	14	28 14.1	20 57.3	17.0	51 39.9	01.4	68 44.6	18.3	218 52.8	03.0	Altair	62 11.5	N 8 53.4
	15	43 16.6	36 01.0	16.5	66 40.5	00.7	83 46.6	18.2	233 55.5	03.1	Ankaa	353 19.1	S42 15.4
	16	58 19.0	51 04.7	16.1	81 41.1	11 00.0	98 48.5	18.0	248 58.1	03.2	Antares	112 30.0	S26 27.2
	17	73 21.5	66 08.4	15.6	96 41.7	10 59.3	113 50.5	17.9	264 00.8	03.3			
	18	88 24.0	81 12.1	N12 15.1	111 42.3	S10 58.6	128 52.4	S16 17.7	279 03.4	N 7 03.3	Arcturus	145 58.3	N19 07.8
	19	103 26.4	96 15.9	14.6	126 42.9	57.9	143 54.4	17.6	294 06.0	03.4	Atria	107 34.6	S69 02.5
	20	118 28.9	111 19.6	14.1	141 43.5	57.2	158 56.3	17.4	309 08.7	03.5	Avior	234 19.1	S59 32.6
	21	133 31.4	126 23.3	13.7	156 44.1	56.5	173 58.3	17.3	324 11.3	03.6	Bellatrix	278 35.4	N 6 21.5
	22	148 33.8	141 27.0	13.2	171 44.7	55.7	189 00.3	17.1	339 14.0	03.6	Betelgeuse	271 04.7	N 7 24.5
	23	163 36.3	156 30.8	12.7	186 45.3	55.0	204 02.2	17.0	354 16.6	03.7			
21	00	178 38.8	171 34.5	N12 12.2	201 45.9	S10 54.3	219 04.2	S16 16.8	9 19.2	N 7 03.8	Canopus	263 57.5	S52 42.3
	01	193 41.2	186 38.2	11.7	216 46.5	53.6	234 06.1	16.7	24 21.9	03.9	Capella	280 39.2	N46 00.6
	02	208 43.7	201 42.0	11.2	231 47.1	52.9	249 08.1	16.6	39 24.5	03.9	Deneb	49 34.0	N45 18.5
	03	223 46.2	216 45.7	10.7	246 47.8	52.2	264 10.1	16.4	54 27.2	04.0	Denebola	182 36.5	N14 31.0
	04	238 48.6	231 49.5	10.2	261 48.4	51.5	279 12.0	16.3	69 29.8	04.1	Diphda	348 59.3	S17 56.2
	05	253 51.1	246 53.2	09.7	276 49.0	50.8	294 14.0	16.1	84 32.4	04.2			
	06	268 53.5	261 56.9	N12 09.2	291 49.6	S10 50.0	309 15.9	S16 16.0	99 35.1	N 7 04.2	Dubhe	193 54.6	N61 42.0
S	07	283 56.0	277 00.7	08.7	306 50.2	49.3	324 17.9	15.8	114 37.7	04.3	Elnath	278 16.7	N28 37.0
	08	298 58.5	292 04.4	08.1	321 50.8	48.6	339 19.8	15.7	129 40.4	04.4	Eltanin	90 47.6	N51 28.9
A	09	314 00.9	307 08.2	07.6	336 51.4	47.9	354 21.8	15.5	144 43.0	04.5	Enif	33 50.5	N 9 54.9
T	10	329 03.4	322 11.9	07.1	351 52.0	47.2	9 23.8	15.4	159 45.7	04.5	Fomalhaut	15 27.7	S29 34.4
U	11	344 05.9	337 15.7	06.6	6 52.6	46.5	24 25.7	15.2	174 48.3	04.6			
R	12	359 08.3	352 19.4	N12 06.1	21 53.2	S10 45.8	39 27.7	S16 15.1	189 50.9	N 7 04.7	Gacrux	172 04.1	S57 10.0
D	13	14 10.8	7 23.2	05.5	36 53.8	45.1	54 29.6	14.9	204 53.6	04.8	Gienah	175 55.3	S17 35.8
A	14	29 13.3	22 27.0	05.0	51 54.4	44.3	69 31.6	14.8	219 56.2	04.8	Hadar	148 52.1	S60 25.1
Y	15	44 15.7	37 30.7	04.5	66 55.1	43.6	84 33.6	14.6	234 58.9	04.9	Hamal	328 04.7	N23 30.4
	16	59 18.2	52 34.5	03.9	81 55.7	42.9	99 35.5	14.5	250 01.5	05.0	Kaus Aust.	83 48.0	S34 22.8
	17	74 20.7	67 38.2	03.4	96 56.3	42.2	114 37.5	14.3	265 04.1	05.1			
	18	89 23.1	82 42.0	N12 02.9	111 56.9	S10 41.5	129 39.5	S16 14.2	280 06.8	N 7 05.1	Kochab	137 18.6	N74 06.7
	19	104 25.6	97 45.8	02.3	126 57.5	40.8	144 41.4	14.1	295 09.4	05.2	Markab	13 41.8	N15 15.2
	20	119 28.0	112 49.5	01.8	141 58.1	40.1	159 43.4	13.9	310 12.1	05.3	Menkar	314 18.6	N 4 07.5
	21	134 30.5	127 53.3	01.2	156 58.7	39.3	174 45.3	13.8	325 14.7	05.4	Menkent	148 11.1	S36 25.0
	22	149 33.0	142 57.1	00.7	171 59.3	38.6	189 47.3	13.6	340 17.3	05.4	Miaplacidus	221 40.0	S69 45.6
	23	164 35.4	158 00.8	12 00.1	186 59.9	37.9	204 49.3	13.5	355 20.0	05.5			
22	00	179 37.9	173 04.6	N11 59.6	202 00.5	S10 37.2	219 51.2	S16 13.3	10 22.6	N 7 05.6	Mirfak	308 45.3	N49 53.8
	01	194 40.4	188 08.4	59.0	217 01.2	36.5	234 53.2	13.2	25 25.3	05.6	Nunki	76 02.3	S26 17.1
	02	209 42.8	203 12.2	58.5	232 01.8	35.8	249 55.1	13.0	40 27.9	05.7	Peacock	53 24.4	S56 42.2
	03	224 45.3	218 15.9	57.9	247 02.4	35.0	264 57.1	12.9	55 30.5	05.8	Pollux	243 31.4	N28 00.3
	04	239 47.8	233 19.7	57.3	262 03.0	34.3	279 59.1	12.7	70 33.2	05.9	Procyon	245 02.9	N 5 12.0
	05	254 50.2	248 23.5	56.8	277 03.6	33.6	295 01.0	12.6	85 35.8	05.9			
	06	269 52.7	263 27.3	N11 56.2	292 04.2	S10 32.9	310 03.0	S16 12.4	100 38.5	N 7 06.0	Rasalhague	96 09.3	N12 32.9
	07	284 55.1	278 31.1	55.6	307 04.8	32.2	325 05.0	12.3	115 41.1	06.1	Regulus	207 46.5	N11 55.2
	08	299 57.6	293 34.8	55.1	322 05.4	31.5	340 06.9	12.1	130 43.7	06.2	Rigel	281 15.1	S 8 11.6
S	09	315 00.1	308 38.6	54.5	337 06.0	30.7	355 08.9	12.0	145 46.4	06.2	Rigil Kent.	139 55.8	S60 52.4
U	10	330 02.5	323 42.4	53.9	352 06.7	30.0	10 10.8	11.8	160 49.0	06.3	Sabik	102 16.1	S15 44.3
N	11	345 05.0	338 46.2	53.3	7 07.3	29.3	25 12.8	11.7	175 51.7	06.4			
D	12	0 07.5	353 50.0	N11 52.8	22 07.9	S10 28.6	40 14.8	S16 11.6	190 54.3	N 7 06.5	Schedar	349 44.9	N56 35.3
A	13	15 09.9	8 53.8	52.2	37 08.5	27.9	55 16.7	11.4	205 56.9	06.5	Shaula	96 26.2	S37 06.6
Y	14	30 12.4	23 57.6	51.6	52 09.1	27.1	70 18.7	11.3	220 59.6	06.6	Sirius	258 36.4	S16 43.9
	15	45 14.9	39 01.4	51.0	67 09.7	26.4	85 20.7	11.1	236 02.2	06.7	Spica	158 34.3	S11 12.8
	16	60 17.3	54 05.2	50.4	82 10.3	25.7	100 22.6	11.0	251 04.9	06.8	Suhail	222 54.6	S43 28.5
	17	75 19.8	69 09.0	49.8	97 10.9	25.0	115 24.6	10.8	266 07.5	06.8			
	18	90 22.3	84 12.7	N11 49.2	112 11.6	S10 24.3	130 26.6	S16 10.7	281 10.1	N 7 06.9	Vega	80 41.2	N38 47.2
	19	105 24.7	99 16.5	48.6	127 12.2	23.6	145 28.5	10.5	296 12.8	07.0	Zuben'ubi	137 08.7	S16 05.0
	20	120 27.2	114 20.3	48.0	142 12.8	22.8	160 30.5	10.4	311 15.4	07.1		SHA	Mer. Pass
	21	135 29.6	129 24.1	47.4	157 13.4	22.1	175 32.5	10.2	326 18.1	07.1		° ′	h m
	22	150 32.1	144 27.9	46.8	172 14.0	21.4	190 34.4	10.1	341 20.7	07.2	Venus	352 55.7	12 31
	23	165 34.6	159 31.7	46.2	187 14.6	20.7	205 36.4	09.9	356 23.3	07.3	Mars	23 07.2	10 33
		h m									Jupiter	40 25.4	9 22
Mer. Pass.		12 03.4	v 3.8	d 0.5	v 0.6	d 0.7	v 2.0	d 0.1	v 2.6	d 0.1	Saturn	190 40.5	23 19

UT	SUN		MOON				
	GHA	Dec	GHA	*v*	Dec	*d*	HP
d h	° ′	° ′	° ′	′	° ′	′	′
20 00	178 06.4	S 0 11.6	253 09.1	11.3	S24 56.9	5.4	54.3
01	193 06.6	10.6	267 39.4	11.5	24 51.5	5.5	54.3
02	208 06.8	09.6	282 09.9	11.4	24 46.0	5.6	54.3
03	223 07.0	. . 08.6	296 40.3	11.5	24 40.4	5.7	54.3
04	238 07.2	07.6	311 10.8	11.5	24 34.7	5.9	54.3
05	253 07.4	06.7	325 41.3	11.6	24 28.8	6.0	54.3
06	268 07.5	S 0 05.7	340 11.9	11.6	S24 22.8	6.0	54.3
07	283 07.7	04.7	354 42.5	11.7	24 16.8	6.2	54.3
08	298 07.9	03.7	9 13.2	11.7	24 10.6	6.3	54.3
09	313 08.1	. . 02.7	23 43.9	11.7	24 04.3	6.5	54.3
10	328 08.3	01.7	38 14.6	11.8	23 57.8	6.5	54.3
11	343 08.5	S 00.7	52 45.4	11.8	23 51.3	6.6	54.3
12	358 08.7	N 0 00.3	67 16.2	11.9	S23 44.7	6.8	54.3
13	13 08.8	01.3	81 47.1	12.0	23 37.9	6.8	54.3
14	28 09.0	02.2	96 18.1	11.9	23 31.1	7.0	54.3
15	43 09.2	. . 03.2	110 49.0	12.0	23 24.1	7.0	54.3
16	58 09.4	04.2	125 20.0	12.1	23 17.1	7.2	54.3
17	73 09.6	05.2	139 51.1	12.1	23 09.9	7.3	54.3
18	88 09.8	N 0 06.2	154 22.2	12.2	S23 02.6	7.4	54.4
19	103 10.0	07.2	168 53.4	12.2	22 55.2	7.5	54.4
20	118 10.1	08.2	183 24.6	12.2	22 47.7	7.6	54.4
21	133 10.3	. . 09.2	197 55.8	12.3	22 40.1	7.7	54.4
22	148 10.5	10.1	212 27.1	12.3	22 32.4	7.8	54.4
23	163 10.7	11.1	226 58.4	12.4	22 24.6	7.9	54.4
21 00	178 10.9	N 0 12.1	241 29.8	12.4	S22 16.7	8.0	54.4
01	193 11.1	13.1	256 01.2	12.5	22 08.7	8.1	54.4
02	208 11.3	14.1	270 32.7	12.5	22 00.6	8.2	54.4
03	223 11.4	. . 15.1	285 04.2	12.5	21 52.4	8.3	54.4
04	238 11.6	16.1	299 35.7	12.7	21 44.1	8.4	54.5
05	253 11.8	17.1	314 07.4	12.6	21 35.7	8.5	54.5
06	268 12.0	N 0 18.0	328 39.0	12.7	S21 27.2	8.6	54.5
07	283 12.2	19.0	343 10.7	12.7	21 18.6	8.7	54.5
08	298 12.4	20.0	357 42.4	12.8	21 09.9	8.8	54.5
09	313 12.6	. . 21.0	12 14.2	12.8	21 01.1	8.9	54.5
10	328 12.7	22.0	26 46.0	12.9	20 52.2	9.0	54.5
11	343 12.9	23.0	41 17.9	12.9	20 43.2	9.1	54.5
12	358 13.1	N 0 24.0	55 49.8	13.0	S20 34.1	9.1	54.6
13	13 13.3	25.0	70 21.8	13.0	20 25.0	9.3	54.6
14	28 13.5	25.9	84 53.8	13.0	20 15.7	9.4	54.6
15	43 13.7	. . 26.9	99 25.8	13.1	20 06.3	9.4	54.6
16	58 13.9	27.9	113 57.9	13.1	19 56.9	9.6	54.6
17	73 14.0	28.9	128 30.0	13.2	19 47.3	9.6	54.6
18	88 14.2	N 0 29.9	143 02.2	13.2	S19 37.7	9.7	54.6
19	103 14.4	30.9	157 34.4	13.3	19 28.0	9.8	54.7
20	118 14.6	31.9	172 06.7	13.3	19 18.2	10.0	54.7
21	133 14.8	. . 32.9	186 39.0	13.3	19 08.2	9.9	54.7
22	148 15.0	33.8	201 11.3	13.4	18 58.3	10.1	54.7
23	163 15.2	34.8	215 43.7	13.4	18 48.2	10.2	54.7
22 00	178 15.3	N 0 35.8	230 16.1	13.5	S18 38.0	10.2	54.7
01	193 15.5	36.8	244 48.6	13.5	18 27.8	10.4	54.7
02	208 15.7	37.8	259 21.1	13.5	18 17.4	10.4	54.8
03	223 15.9	. . 38.8	273 53.6	13.6	18 07.0	10.5	54.8
04	238 16.1	39.8	288 26.2	13.6	17 56.5	10.6	54.8
05	253 16.3	40.8	302 58.8	13.7	17 45.9	10.7	54.8
06	268 16.5	N 0 41.7	317 31.5	13.7	S17 35.2	10.7	54.8
07	283 16.7	42.7	332 04.2	13.7	17 24.5	10.9	54.8
08	298 16.8	43.7	346 36.9	13.7	17 13.6	10.9	54.9
09	313 17.0	. . 44.7	1 09.6	13.8	17 02.7	11.0	54.9
10	328 17.2	45.7	15 42.4	13.9	16 51.7	11.1	54.9
11	343 17.4	46.7	30 15.3	13.8	16 40.6	11.1	54.9
12	358 17.6	N 0 47.7	44 48.1	14.0	S16 29.5	11.3	54.9
13	13 17.8	48.7	59 21.1	13.9	16 18.2	11.3	55.0
14	28 18.0	49.6	73 54.0	14.0	16 06.9	11.3	55.0
15	43 18.2	. . 50.6	88 27.0	14.0	15 55.6	11.5	55.0
16	58 18.3	51.6	103 00.0	14.0	15 44.1	11.5	55.0
17	73 18.5	52.6	117 33.0	14.1	15 32.6	11.6	55.0
18	88 18.7	N 0 53.6	132 06.1	14.1	S15 21.0	11.7	55.1
19	103 18.9	54.6	146 39.2	14.1	15 09.3	11.8	55.1
20	118 19.1	55.6	161 12.3	14.2	14 57.5	11.8	55.1
21	133 19.3	. . 56.5	175 45.5	14.1	14 45.7	11.9	55.1
22	148 19.5	57.5	190 18.6	14.3	14 33.8	12.0	55.1
23	163 19.6	58.5	204 51.9	14.2	S14 21.8	12.0	55.2
	SD 16.1	*d* 1.0	SD 14.8		14.9		15.0

Lat.	Twilight Naut.	Twilight Civil	Sunrise	Moonrise 20	Moonrise 21	Moonrise 22	Moonrise 23
°	h m	h m	h m	h m	h m	h m	h m
N 72	03 14	04 44	05 53	■	■	07 50	06 34
N 70	03 34	04 53	05 54	■	■	06 52	06 11
68	03 49	04 59	05 55	■	07 10	06 18	05 53
66	04 01	05 05	05 56	■	06 15	05 53	05 39
64	04 11	05 10	05 57	05 55	05 42	05 33	05 27
62	04 20	05 14	05 58	05 15	05 17	05 17	05 16
60	04 27	05 17	05 58	04 47	04 58	05 04	05 08
N 58	04 33	05 20	05 59	04 25	04 42	04 52	05 00
56	04 38	05 22	05 59	04 07	04 28	04 42	04 53
54	04 43	05 25	06 00	03 52	04 16	04 33	04 47
52	04 47	05 27	06 00	03 39	04 05	04 25	04 41
50	04 50	05 28	06 01	03 27	03 56	04 18	04 36
45	04 58	05 32	06 01	03 03	03 36	04 02	04 25
N 40	05 03	05 35	06 02	02 44	03 19	03 50	04 16
35	05 08	05 37	06 02	02 28	03 06	03 39	04 08
30	05 11	05 39	06 03	02 14	02 54	03 29	04 01
20	05 16	05 41	06 03	01 50	02 33	03 13	03 49
N 10	05 18	05 43	06 04	01 29	02 15	02 58	03 39
0	05 19	05 43	06 04	01 10	01 58	02 44	03 29
S 10	05 19	05 43	06 04	00 51	01 41	02 31	03 19
20	05 17	05 42	06 04	00 30	01 23	02 16	03 08
30	05 12	05 40	06 04	00 06	01 02	01 59	02 56
35	05 09	05 39	06 04	24 50	00 50	01 49	02 49
40	05 05	05 37	06 04	24 36	00 36	01 38	02 41
45	05 00	05 34	06 04	24 19	00 19	01 24	02 31
S 50	04 53	05 31	06 03	23 58	25 08	01 08	02 20
52	04 50	05 30	06 03	23 48	25 00	01 00	02 14
54	04 46	05 28	06 03	23 36	24 52	00 52	02 08
56	04 42	05 26	06 03	23 23	24 42	00 42	02 02
58	04 37	05 24	06 03	23 08	24 31	00 31	01 54
S 60	04 31	05 21	06 03	22 50	24 18	00 18	01 46

Lat.	Sunset	Twilight Civil	Twilight Naut.	Moonset 20	Moonset 21	Moonset 22	Moonset 23
°	h m	h m	h m	h m	h m	h m	h m
N 72	18 24	19 33	21 05	■	■	10 21	13 10
N 70	18 22	19 24	20 44	■	■	11 17	13 31
68	18 21	19 17	20 28	■	09 22	11 50	13 47
66	18 20	19 11	20 16	■	10 17	12 14	14 00
64	18 19	19 07	20 05	08 57	10 49	12 32	14 11
62	18 18	19 02	19 57	09 37	11 13	12 47	14 20
60	18 17	18 59	19 49	10 05	11 32	13 00	14 27
N 58	18 17	18 56	19 43	10 26	11 47	13 11	14 34
56	18 16	18 53	19 38	10 43	12 01	13 20	14 40
54	18 15	18 51	19 33	10 58	12 12	13 28	14 45
52	18 15	18 49	19 29	11 11	12 22	13 36	14 50
50	18 15	18 47	19 25	11 22	12 31	13 42	14 54
45	18 14	18 43	19 18	11 45	12 50	13 56	15 03
N 40	18 13	18 40	19 12	12 04	13 05	14 08	15 11
35	18 13	18 38	19 07	12 19	13 18	14 18	15 18
30	18 12	18 36	19 04	12 33	13 30	14 27	15 23
20	18 11	18 33	18 59	12 56	13 49	14 41	15 33
N 10	18 11	18 32	18 56	13 15	14 05	14 54	15 42
0	18 10	18 31	18 55	13 34	14 21	15 06	15 50
S 10	18 10	18 31	18 55	13 52	14 36	15 18	15 58
20	18 10	18 32	18 57	14 11	14 52	15 30	16 06
30	18 10	18 34	19 01	14 33	15 11	15 45	16 16
35	18 10	18 35	19 04	14 46	15 22	15 53	16 21
40	18 10	18 37	19 08	15 01	15 34	16 02	16 27
45	18 10	18 39	19 13	15 19	15 48	16 13	16 35
S 50	18 10	18 42	19 20	15 41	16 06	16 26	16 43
52	18 10	18 44	19 23	15 51	16 14	16 32	16 47
54	18 10	18 45	19 27	16 03	16 24	16 39	16 51
56	18 10	18 47	19 31	16 17	16 34	16 46	16 56
58	18 10	18 49	19 36	16 32	16 46	16 55	17 01
S 60	18 10	18 52	19 41	16 51	16 59	17 04	17 07

Day	SUN Eqn. of Time 00^h	SUN Eqn. of Time 12^h	SUN Mer. Pass.	MOON Mer. Pass. Upper	MOON Mer. Pass. Lower	MOON Age	MOON Phase
d	m s	m s	h m	h m	h m	d	%
20	07 35	07 26	12 07	07 22	19 46	23	34
21	07 17	07 08	12 07	08 09	20 33	24	25
22	06 59	06 50	12 07	08 55	21 17	25	17

UT	ARIES	VENUS −4.2		MARS +1.2		JUPITER −2.0		SATURN +0.6		STARS		
d h	GHA	GHA	Dec	GHA	Dec	GHA	Dec	GHA	Dec	Name	SHA	Dec
	° ′	° ′	° ′	° ′	° ′	° ′	° ′	° ′	° ′		° ′	°
23 00	180 37.0	174 35.5	N11 45.6	202 15.2	S10 20.0	220 38.4	S16 09.8	11 26.0	N 7 07.3	Acamar	315 20.9	S40 16.
01	195 39.5	189 39.4	45.0	217 15.8	19.2	235 40.3	09.6	26 28.6	07.4	Achernar	335 29.4	S57 11
02	210 42.0	204 43.2	44.4	232 16.5	18.5	250 42.3	09.5	41 31.2	07.5	Acrux	173 12.5	S63 09.
03	225 44.4	219 47.0	43.8	247 17.1	17.8	265 44.2	09.4	56 33.9	07.6	Adhara	255 15.0	S28 59.
04	240 46.9	234 50.8	43.2	262 17.7	17.1	280 46.2	09.2	71 36.5	07.6	Aldebaran	290 53.1	N16 31.
05	255 49.4	249 54.6	42.5	277 18.3	16.4	295 48.2	09.1	86 39.2	07.7			
06	270 51.8	264 58.4	N11 41.9	292 18.9	S10 15.6	310 50.1	S16 08.9	101 41.8	N 7 07.8	Alioth	166 22.6	N55 54.
07	285 54.3	280 02.2	41.3	307 19.5	14.9	325 52.1	08.8	116 44.4	07.9	Alkaid	153 00.7	N49 15.
08	300 56.8	295 06.0	40.7	322 20.2	14.2	340 54.1	08.6	131 47.1	07.9	Al Na'ir	27 47.9	S46 54.
M 09	315 59.2	310 09.8	40.0	337 20.8	13.5	355 56.0	08.5	146 49.7	08.0	Alnilam	275 49.6	S 1 11
O 10	331 01.7	325 13.6	39.4	352 21.4	12.7	10 58.0	08.3	161 52.4	08.1	Alphard	217 59.0	S 8 42.
N 11	346 04.1	340 17.4	38.8	7 22.0	12.0	26 00.0	08.2	176 55.0	08.1			
D 12	1 06.6	355 21.3	N11 38.1	22 22.6	S10 11.3	41 01.9	S16 08.0	191 57.6	N 7 08.2	Alphecca	126 13.4	N26 40.
A 13	16 09.1	10 25.1	37.5	37 23.2	10.6	56 03.9	07.9	207 00.3	08.3	Alpheratz	357 47.2	N29 08.
Y 14	31 11.5	25 28.9	36.9	52 23.8	09.9	71 05.9	07.7	222 02.9	08.4	Altair	62 11.4	N 8 53.
15	46 14.0	40 32.7	36.2	67 24.5	09.1	86 07.8	07.6	237 05.6	08.4	Ankaa	353 19.1	S42 15.
16	61 16.5	55 36.5	35.6	82 25.1	08.4	101 09.8	07.5	252 08.2	08.5	Antares	112 30.0	S26 27.
17	76 18.9	70 40.3	34.9	97 25.7	07.7	116 11.8	07.3	267 10.8	08.6			
18	91 21.4	85 44.2	N11 34.3	112 26.3	S10 07.0	131 13.8	S16 07.2	282 13.5	N 7 08.7	Arcturus	145 58.3	N19 07.
19	106 23.9	100 48.0	33.6	127 26.9	06.2	146 15.7	07.0	297 16.1	08.7	Atria	107 34.6	S69 02.
20	121 26.3	115 51.8	33.0	142 27.5	05.5	161 17.7	06.9	312 18.7	08.8	Avior	234 19.1	S59 32
21	136 28.8	130 55.6	32.3	157 28.2	04.8	176 19.7	06.7	327 21.4	08.9	Bellatrix	278 35.5	N 6 21.
22	151 31.2	145 59.4	31.7	172 28.8	04.1	191 21.6	06.6	342 24.0	08.9	Betelgeuse	271 04.7	N 7 24.
23	166 33.7	161 03.3	31.0	187 29.4	03.3	206 23.6	06.4	357 26.7	09.0			
24 00	181 36.2	176 07.1	N11 30.4	202 30.0	S10 02.6	221 25.6	S16 06.3	12 29.3	N 7 09.1	Canopus	263 57.5	S52 42.
01	196 38.6	191 10.9	29.7	217 30.6	01.9	236 27.5	06.1	27 31.9	09.2	Capella	280 39.2	N46 00.
02	211 41.1	206 14.7	29.0	232 31.2	01.2	251 29.5	06.0	42 34.6	09.2	Deneb	49 34.0	N45 18.
03	226 43.6	221 18.6	28.4	247 31.9	10 00.5	266 31.5	05.9	57 37.2	09.3	Denebola	182 36.5	N14 31.
04	241 46.0	236 22.4	27.7	262 32.5	9 59.7	281 33.4	05.7	72 39.8	09.4	Diphda	348 59.3	S17 56.
05	256 48.5	251 26.2	27.0	277 33.1	59.0	296 35.4	05.6	87 42.5	09.5			
06	271 51.0	266 30.0	N11 26.3	292 33.7	S 9 58.3	311 37.4	S16 05.4	102 45.1	N 7 09.5	Dubhe	193 54.6	N61 42.
07	286 53.4	281 33.9	25.7	307 34.3	57.6	326 39.3	05.3	117 47.8	09.6	Elnath	278 16.7	N28 37.
T 08	301 55.9	296 37.7	25.0	322 35.0	56.8	341 41.3	05.1	132 50.4	09.7	Eltanin	90 47.6	N51 28.
U 09	316 58.4	311 41.5	24.3	337 35.6	56.1	356 43.3	05.0	147 53.0	09.7	Enif	33 50.5	N 9 54.
E 10	332 00.8	326 45.4	23.6	352 36.2	55.4	11 45.3	04.8	162 55.7	09.8	Fomalhaut	15 27.7	S29 34.
S 11	347 03.3	341 49.2	22.9	7 36.8	54.7	26 47.2	04.7	177 58.3	09.9			
D 12	2 05.7	356 53.0	N11 22.3	22 37.4	S 9 53.9	41 49.2	S16 04.5	193 00.9	N 7 10.0	Gacrux	172 04.1	S57 10.
A 13	17 08.2	11 56.9	21.6	37 38.0	53.2	56 51.2	04.4	208 03.6	10.0	Gienah	175 55.3	S17 35.
Y 14	32 10.7	27 00.7	20.9	52 38.7	52.5	71 53.1	04.3	223 06.2	10.1	Hadar	148 52.1	S60 25.
15	47 13.1	42 04.5	20.2	67 39.3	51.7	86 55.1	04.1	238 08.8	10.2	Hamal	328 04.7	N23 30.
16	62 15.6	57 08.4	19.5	82 39.9	51.0	101 57.1	04.0	253 11.5	10.3	Kaus Aust.	83 48.0	S34 22.
17	77 18.1	72 12.2	18.8	97 40.5	50.3	116 59.1	03.8	268 14.1	10.3			
18	92 20.5	87 16.0	N11 18.1	112 41.1	S 9 49.6	132 01.0	S16 03.7	283 16.8	N 7 10.4	Kochab	137 18.5	N74 06.
19	107 23.0	102 19.9	17.4	127 41.8	48.8	147 03.0	03.5	298 19.4	10.5	Markab	13 41.8	N15 15.
20	122 25.5	117 23.7	16.7	142 42.4	48.1	162 05.0	03.4	313 22.0	10.5	Menkar	314 18.6	N 4 07.
21	137 27.9	132 27.5	16.0	157 43.0	47.4	177 06.9	03.2	328 24.7	10.6	Menkent	148 11.1	S36 25.
22	152 30.4	147 31.4	15.3	172 43.6	46.7	192 08.9	03.1	343 27.3	10.7	Miaplacidus	221 40.1	S69 45.
23	167 32.8	162 35.2	14.6	187 44.2	45.9	207 10.9	02.9	358 29.9	10.8			
25 00	182 35.3	177 39.0	N11 13.9	202 44.9	S 9 45.2	222 12.9	S16 02.8	13 32.6	N 7 10.8	Mirfak	308 45.3	N49 53.
01	197 37.8	192 42.9	13.2	217 45.5	44.5	237 14.8	02.7	28 35.2	10.9	Nunki	76 02.2	S26 17.
02	212 40.2	207 46.7	12.4	232 46.1	43.8	252 16.8	02.5	43 37.8	11.0	Peacock	53 24.3	S56 42.
03	227 42.7	222 50.5	11.7	247 46.7	43.0	267 18.8	02.4	58 40.5	11.0	Pollux	243 31.4	N28 00.
04	242 45.2	237 54.4	11.0	262 47.4	42.3	282 20.8	02.2	73 43.1	11.1	Procyon	245 02.9	N 5 12.
05	257 47.6	252 58.2	10.3	277 48.0	41.6	297 22.7	02.1	88 45.8	11.2			
06	272 50.1	268 02.1	N11 09.6	292 48.6	S 9 40.8	312 24.7	S16 01.9	103 48.4	N 7 11.3	Rasalhague	96 09.3	N12 32.
W 07	287 52.6	283 05.9	08.9	307 49.2	40.1	327 26.7	01.8	118 51.0	11.3	Regulus	207 46.6	N11 55.
E 08	302 55.0	298 09.7	08.1	322 49.8	39.4	342 28.6	01.6	133 53.7	11.4	Rigel	281 15.2	S 8 11.
D 09	317 57.5	313 13.6	07.4	337 50.5	38.7	357 30.6	01.5	148 56.3	11.5	Rigil Kent.	139 55.7	S60 52.
N 10	333 00.0	328 17.4	06.7	352 51.1	37.9	12 32.6	01.3	163 58.9	11.5	Sabik	102 16.1	S15 44.
E 11	348 02.4	343 21.2	05.9	7 51.7	37.2	27 34.6	01.2	179 01.6	11.6			
S 12	3 04.9	358 25.1	N11 05.2	22 52.3	S 9 36.5	42 36.5	S16 01.1	194 04.2	N 7 11.7	Schedar	349 44.9	N56 35.
D 13	18 07.3	13 28.9	04.5	37 53.0	35.7	57 38.5	00.9	209 06.8	11.8	Shaula	96 26.1	S37 06.
A 14	33 09.8	28 32.8	03.7	52 53.6	35.0	72 40.5	00.8	224 09.5	11.8	Sirius	258 36.5	S16 43.
Y 15	48 12.3	43 36.6	03.0	67 54.2	34.3	87 42.5	00.6	239 12.1	11.9	Spica	158 34.3	S11 12.
16	63 14.7	58 40.4	02.3	82 54.8	33.5	102 44.4	00.5	254 14.7	12.0	Suhail	222 54.6	S43 28.
17	78 17.2	73 44.3	01.5	97 55.4	32.8	117 46.4	00.3	269 17.4	12.0			
18	93 19.7	88 48.1	N11 00.8	112 56.1	S 9 32.1	132 48.4	S16 00.2	284 20.0	N 7 12.1	Vega	80 41.1	N38 47.
19	108 22.1	103 52.0	11 00.0	127 56.7	31.4	147 50.4	16 00.0	299 22.7	12.2	Zuben'ubi	137 08.7	S16 05.
20	123 24.6	118 55.8	10 59.3	142 57.3	30.6	162 52.3	15 59.9	314 25.3	12.3		SHA	Mer. Pass.
21	138 27.1	133 59.6	58.5	157 57.9	29.9	177 54.3	59.8	329 27.9	12.3		° ′	h m
22	153 29.5	149 03.5	57.8	172 58.6	29.2	192 56.3	59.6	344 30.6	12.4	Venus	354 30.9	12 12
23	168 32.0	164 07.3	57.0	187 59.2	28.4	207 58.3	59.5	359 33.2	12.5	Mars	20 53.8	10 30
	h m									Jupiter	39 49.4	9 13
Mer. Pass.	11 51.6	*v* 3.8	*d* 0.7	*v* 0.6	*d* 0.7	*v* 2.0	*d* 0.1	*v* 2.6	*d* 0.1	Saturn	190 53.1	23 06

UT	SUN GHA	SUN Dec	MOON GHA	v	MOON Dec	d	HP
d h	° ′	° ′	° ′	′	° ′	′	′
23 00 (MONDAY)	178 19.8	N 0 59.5	219 25.1	14.3	S14 09.8	12.1	55.2
01	193 20.0	1 00.5	233 58.4	14.3	13 57.7	12.2	55.2
02	208 20.2	01.5	248 31.7	14.3	13 45.5	12.2	55.2
03	223 20.4	. . 02.5	263 05.0	14.3	13 33.3	12.3	55.2
04	238 20.6	03.4	277 38.3	14.4	13 21.0	12.4	55.3
05	253 20.8	04.4	292 11.7	14.4	13 08.6	12.5	55.3
06	268 21.0	N 1 05.4	306 45.1	14.4	S12 56.1	12.5	55.3
07	283 21.1	06.4	321 18.5	14.5	12 43.6	12.5	55.3
08	298 21.3	07.4	335 52.0	14.4	12 31.1	12.7	55.3
09	313 21.5	. . 08.4	350 25.4	14.5	12 18.4	12.6	55.4
10	328 21.7	09.4	4 58.9	14.5	12 05.8	12.8	55.4
11	343 21.9	10.3	19 32.4	14.5	11 53.0	12.8	55.4
12	358 22.1	N 1 11.3	34 05.9	14.6	S11 40.2	12.9	55.4
13	13 22.3	12.3	48 39.5	14.5	11 27.3	12.9	55.5
14	28 22.5	13.3	63 13.0	14.6	11 14.4	13.0	55.5
15	43 22.6	. . 14.3	77 46.6	14.6	11 01.4	13.0	55.5
16	58 22.8	15.3	92 20.2	14.6	10 48.4	13.1	55.5
17	73 23.0	16.3	106 53.8	14.6	10 35.3	13.2	55.6
18	88 23.2	N 1 17.2	121 27.4	14.6	S10 22.1	13.2	55.6
19	103 23.4	18.2	136 01.0	14.7	10 08.9	13.3	55.6
20	118 23.6	19.2	150 34.7	14.6	9 55.6	13.3	55.6
21	133 23.8	. . 20.2	165 08.3	14.7	9 42.3	13.4	55.6
22	148 24.0	21.2	179 42.0	14.7	9 28.9	13.4	55.7
23	163 24.1	22.2	194 15.7	14.7	9 15.5	13.5	55.7
24 00 (TUESDAY)	178 24.3	N 1 23.2	208 49.4	14.7	S 9 02.0	13.5	55.7
01	193 24.5	24.1	223 23.1	14.7	8 48.5	13.6	55.7
02	208 24.7	25.1	237 56.8	14.7	8 34.9	13.6	55.8
03	223 24.9	. . 26.1	252 30.5	14.7	8 21.3	13.7	55.8
04	238 25.1	27.1	267 04.2	14.7	8 07.6	13.7	55.8
05	253 25.3	28.1	281 37.9	14.8	7 53.9	13.7	55.8
06	268 25.5	N 1 29.1	296 11.7	14.7	S 7 40.2	13.8	55.9
07	283 25.6	30.0	310 45.4	14.7	7 26.4	13.9	55.9
08	298 25.8	31.0	325 19.1	14.8	7 12.5	13.9	55.9
09	313 26.0	. . 32.0	339 52.9	14.7	6 58.6	13.9	55.9
10	328 26.2	33.0	354 26.6	14.7	6 44.7	14.0	56.0
11	343 26.4	34.0	9 00.3	14.8	6 30.7	14.0	56.0
12	358 26.6	N 1 35.0	23 34.1	14.7	S 6 16.7	14.0	56.0
13	13 26.8	36.0	38 07.8	14.7	6 02.7	14.1	56.0
14	28 27.0	36.9	52 41.5	14.8	5 48.6	14.2	56.1
15	43 27.2	. . 37.9	67 15.3	14.7	5 34.4	14.1	56.1
16	58 27.3	38.9	81 49.0	14.7	5 20.3	14.2	56.1
17	73 27.5	39.9	96 22.7	14.7	5 06.1	14.3	56.1
18	88 27.7	N 1 40.9	110 56.4	14.7	S 4 51.8	14.2	56.2
19	103 27.9	41.9	125 30.1	14.7	4 37.6	14.3	56.2
20	118 28.1	42.8	140 03.8	14.7	4 23.3	14.4	56.2
21	133 28.3	. . 43.8	154 37.5	14.6	4 08.9	14.4	56.2
22	148 28.5	44.8	169 11.1	14.7	3 54.5	14.3	56.3
23	163 28.7	45.8	183 44.8	14.7	3 40.2	14.5	56.3
25 00 (WEDNESDAY)	178 28.8	N 1 46.8	198 18.5	14.6	S 3 25.7	14.4	56.3
01	193 29.0	47.8	212 52.1	14.6	3 11.3	14.5	56.3
02	208 29.2	48.7	227 25.7	14.6	2 56.8	14.5	56.4
03	223 29.4	. . 49.7	241 59.3	14.6	2 42.3	14.5	56.4
04	238 29.6	50.7	256 32.9	14.6	2 27.8	14.6	56.4
05	253 29.8	51.7	271 06.5	14.5	2 13.2	14.6	56.4
06	268 30.0	N 1 52.7	285 40.0	14.5	S 1 58.6	14.6	56.5
07	283 30.2	53.7	300 13.5	14.6	1 44.0	14.6	56.5
08	298 30.4	54.6	314 47.1	14.4	1 29.4	14.6	56.5
09	313 30.5	. . 55.6	329 20.5	14.5	1 14.8	14.7	56.6
10	328 30.7	56.6	343 54.0	14.5	1 00.1	14.7	56.6
11	343 30.9	57.6	358 27.5	14.4	0 45.4	14.7	56.6
12	358 31.1	N 1 58.6	13 00.9	14.4	S 0 30.7	14.7	56.6
13	13 31.3	1 59.6	27 34.3	14.3	0 16.0	14.7	56.7
14	28 31.5	2 00.5	42 07.6	14.4	S 0 01.3	14.8	56.7
15	43 31.7	. . 01.5	56 41.0	14.3	N 0 13.5	14.7	56.7
16	58 31.9	02.5	71 14.3	14.3	0 28.2	14.8	56.7
17	73 32.0	03.5	85 47.6	14.2	0 43.0	14.7	56.8
18	88 32.2	N 2 04.5	100 20.8	14.2	N 0 57.7	14.8	56.8
19	103 32.4	05.4	114 54.0	14.2	1 12.5	14.8	56.8
20	118 32.6	06.4	129 27.2	14.2	1 27.3	14.8	56.8
21	133 32.8	. . 07.4	144 00.4	14.1	1 42.1	14.8	56.9
22	148 33.0	08.4	158 33.5	14.1	1 56.9	14.8	56.9
23	163 33.2	09.4	173 06.6	14.0	N 2 11.7	14.9	56.9
	SD 16.1	d 1.0	SD 15.1		15.3		15.4

Lat.	Twilight Naut.	Twilight Civil	Sunrise	Moonrise 23	Moonrise 24	Moonrise 25	Moonrise 26
°	h m	h m	h m	h m	h m	h m	h m
N 72	02 52	04 28	05 37	06 34	05 55	05 24	04 54
N 70	03 16	04 38	05 40	06 11	05 44	05 21	04 59
68	03 34	04 46	05 43	05 53	05 35	05 19	05 03
66	03 48	04 53	05 45	05 39	05 27	05 17	05 06
64	03 59	04 59	05 47	05 27	05 21	05 15	05 09
62	04 09	05 03	05 48	05 16	05 15	05 13	05 12
60	04 17	05 08	05 49	05 08	05 10	05 12	05 14
N 58	04 24	05 11	05 51	05 00	05 06	05 11	05 16
56	04 30	05 14	05 52	04 53	05 02	05 10	05 17
54	04 35	05 17	05 52	04 47	04 58	05 09	05 19
52	04 39	05 20	05 53	04 41	04 55	05 08	05 20
50	04 43	05 22	05 54	04 36	04 52	05 07	05 22
45	04 52	05 26	05 56	04 25	04 46	05 05	05 25
N 40	04 58	05 30	05 57	04 16	04 40	05 04	05 27
35	05 03	05 33	05 58	04 08	04 36	05 02	05 29
30	05 07	05 35	05 59	04 01	04 32	05 01	05 31
20	05 13	05 39	06 01	03 49	04 25	04 59	05 34
N 10	05 16	05 41	06 02	03 39	04 18	04 58	05 37
0	05 18	05 42	06 03	03 29	04 13	04 56	05 40
S 10	05 19	05 43	06 04	03 19	04 07	04 54	05 43
20	05 17	05 43	06 05	03 08	04 00	04 53	05 46
30	05 14	05 42	06 06	02 56	03 53	04 51	05 49
35	05 12	05 41	06 06	02 49	03 49	04 49	05 51
40	05 08	05 40	06 07	02 41	03 44	04 48	05 54
45	05 04	05 38	06 08	02 31	03 39	04 47	05 56
S 50	04 58	05 36	06 08	02 20	03 32	04 45	06 00
52	04 55	05 35	06 09	02 14	03 29	04 44	06 01
54	04 52	05 34	06 09	02 08	03 25	04 43	06 03
56	04 48	05 32	06 09	02 02	03 22	04 42	06 05
58	04 44	05 30	06 10	01 54	03 17	04 41	06 07
S 60	04 39	05 28	06 10	01 46	03 13	04 40	06 09

Lat.	Sunset	Twilight Civil	Twilight Naut.	Moonset 23	Moonset 24	Moonset 25	Moonset 26
°	h m	h m	h m	h m	h m	h m	h m
N 72	18 38	19 48	21 26	13 10	15 21	17 26	19 36
N 70	18 35	19 37	21 01	13 31	15 30	17 25	19 25
68	18 32	19 29	20 42	13 47	15 36	17 24	19 16
66	18 30	19 22	20 28	14 00	15 42	17 24	19 08
64	18 28	19 16	20 16	14 11	15 47	17 23	19 02
62	18 26	19 11	20 06	14 20	15 51	17 23	18 57
60	18 25	19 07	19 58	14 27	15 54	17 22	18 53
N 58	18 23	19 03	19 51	14 34	15 57	17 22	18 49
56	18 22	19 00	19 45	14 40	16 00	17 21	18 45
54	18 21	18 57	19 39	14 45	16 03	17 21	18 42
52	18 20	18 54	19 34	14 50	16 05	17 21	18 39
50	18 19	18 52	19 30	14 54	16 07	17 20	18 36
45	18 18	18 47	19 22	15 03	16 11	17 20	18 31
N 40	18 16	18 43	19 15	15 11	16 15	17 19	18 26
35	18 15	18 40	19 10	15 18	16 18	17 19	18 22
30	18 14	18 38	19 06	15 23	16 21	17 19	18 18
20	18 12	18 34	19 00	15 33	16 25	17 18	18 12
N 10	18 11	18 32	18 56	15 42	16 29	17 17	18 07
0	18 09	18 30	18 54	15 50	16 33	17 17	18 02
S 10	18 08	18 29	18 54	15 58	16 37	17 16	17 57
20	18 07	18 29	18 55	16 06	16 41	17 16	17 51
30	18 06	18 30	18 58	16 16	16 45	17 15	17 45
35	18 06	18 31	19 00	16 21	16 48	17 15	17 42
40	18 05	18 32	19 03	16 27	16 51	17 14	17 38
45	18 04	18 33	19 07	16 35	16 54	17 13	17 33
S 50	18 03	18 35	19 13	16 43	16 58	17 13	17 28
52	18 03	18 37	19 16	16 47	17 00	17 12	17 26
54	18 03	18 38	19 19	16 51	17 02	17 12	17 23
56	18 02	18 39	19 23	16 56	17 04	17 12	17 20
58	18 02	18 41	19 27	17 01	17 07	17 11	17 17
S 60	18 01	18 43	19 32	17 07	17 09	17 11	17 13

Day	SUN Eqn. of Time 00^h	SUN Eqn. of Time 12^h	SUN Mer. Pass.	MOON Mer. Pass. Upper	MOON Mer. Pass. Lower	Age	Phase
d	m s	m s	h m	h m	h m	d %	
23	06 41	06 32	12 07	09 39	22 01	26 11	
24	06 23	06 14	12 06	10 23	22 45	27 5	
25	06 05	05 56	12 06	11 06	23 28	28 2	

UT		ARIES	VENUS −4.2		MARS +1.2		JUPITER −2.1		SATURN +0.6	
	d h	GHA	GHA	Dec	GHA	Dec	GHA	Dec	GHA	Dec
		° ′	° ′	° ′	° ′	° ′	° ′	° ′	° ′	° ′
	26 00	183 34.4	179 11.2	N10 56.3	202 59.8	S 9 27.7	223 00.2	S15 59.3	14 35.8	N 7 12.5
	01	198 36.9	194 15.0	55.5	218 00.4	27.0	238 02.2	59.2	29 38.5	12.6
	02	213 39.4	209 18.8	54.8	233 01.1	26.2	253 04.2	59.0	44 41.1	12.7
	03	228 41.8	224 22.7	. . 54.0	248 01.7	. . 25.5	268 06.2	. . 58.9	59 43.7	. . 12.7
	04	243 44.3	239 26.5	53.2	263 02.3	24.8	283 08.2	58.7	74 46.4	12.8
	05	258 46.8	254 30.3	52.5	278 02.9	24.0	298 10.1	58.6	89 49.0	12.9
	06	273 49.2	269 34.2	N10 51.7	293 03.6	S 9 23.3	313 12.1	S15 58.5	104 51.6	N 7 13.0
	07	288 51.7	284 38.0	50.9	308 04.2	22.6	328 14.1	58.3	119 54.3	13.0
T	08	303 54.2	299 41.9	50.2	323 04.8	21.8	343 16.1	58.2	134 56.9	13.1
H	09	318 56.6	314 45.7	. . 49.4	338 05.4	. . 21.1	358 18.0	. . 58.0	149 59.5	. . 13.2
U	10	333 59.1	329 49.5	48.6	353 06.1	20.4	13 20.0	57.9	165 02.2	13.2
R	11	349 01.6	344 53.4	47.9	8 06.7	19.6	28 22.0	57.7	180 04.8	13.3
S	12	4 04.0	359 57.2	N10 47.1	23 07.3	S 9 18.9	43 24.0	S15 57.6	195 07.4	N 7 13.4
D	13	19 06.5	15 01.0	46.3	38 07.9	18.2	58 26.0	57.4	210 10.1	13.5
A	14	34 08.9	30 04.9	45.5	53 08.6	17.4	73 27.9	57.3	225 12.7	13.5
Y	15	49 11.4	45 08.7	. . 44.8	68 09.2	. . 16.7	88 29.9	. . 57.2	240 15.3	. . 13.6
	16	64 13.9	60 12.6	44.0	83 09.8	16.0	103 31.9	57.0	255 18.0	13.7
	17	79 16.3	75 16.4	43.2	98 10.4	15.2	118 33.9	56.9	270 20.6	13.7
	18	94 18.8	90 20.2	N10 42.4	113 11.1	S 9 14.5	133 35.8	S15 56.7	285 23.2	N 7 13.8
	19	109 21.3	105 24.1	41.6	128 11.7	13.8	148 37.8	56.6	300 25.9	13.9
	20	124 23.7	120 27.9	40.8	143 12.3	13.0	163 39.8	56.4	315 28.5	13.9
	21	139 26.2	135 31.7	. . 40.0	158 12.9	. . 12.3	178 41.8	. . 56.3	330 31.1	. . 14.0
	22	154 28.7	150 35.6	39.3	173 13.6	11.6	193 43.8	56.2	345 33.8	14.1
	23	169 31.1	165 39.4	38.5	188 14.2	10.8	208 45.7	56.0	0 36.4	14.2
	27 00	184 33.6	180 43.2	N10 37.7	203 14.8	S 9 10.1	223 47.7	S15 55.9	15 39.0	N 7 14.2
	01	199 36.0	195 47.1	36.9	218 15.5	09.4	238 49.7	55.7	30 41.7	14.3
	02	214 38.5	210 50.9	36.1	233 16.1	08.6	253 51.7	55.6	45 44.3	14.4
	03	229 41.0	225 54.7	. . 35.3	248 16.7	. . 07.9	268 53.7	. . 55.4	60 46.9	. . 14.4
	04	244 43.4	240 58.6	34.5	263 17.3	07.2	283 55.6	55.3	75 49.6	14.5
	05	259 45.9	256 02.4	33.7	278 18.0	06.4	298 57.6	55.1	90 52.2	14.6
	06	274 48.4	271 06.2	N10 32.9	293 18.6	S 9 05.7	313 59.6	S15 55.0	105 54.8	N 7 14.6
	07	289 50.8	286 10.0	32.1	308 19.2	05.0	329 01.6	54.9	120 57.5	14.7
	08	304 53.3	301 13.9	31.3	323 19.9	04.2	344 03.6	54.7	136 00.1	14.8
F	09	319 55.8	316 17.7	. . 30.4	338 20.5	. . 03.5	359 05.5	. . 54.6	151 02.7	. . 14.8
R	10	334 58.2	331 21.5	29.6	353 21.1	02.8	14 07.5	54.4	166 05.4	14.9
I	11	350 00.7	346 25.4	28.8	8 21.7	02.0	29 09.5	54.3	181 08.0	15.0
D	12	5 03.2	1 29.2	N10 28.0	23 22.4	S 9 01.3	44 11.5	S15 54.1	196 10.6	N 7 15.1
A	13	20 05.6	16 33.0	27.2	38 23.0	9 00.5	59 13.5	54.0	211 13.3	15.1
Y	14	35 08.1	31 36.8	26.4	53 23.6	8 59.8	74 15.5	53.9	226 15.9	15.2
	15	50 10.5	46 40.7	. . 25.6	68 24.3	. . 59.1	89 17.4	. . 53.7	241 18.5	. . 15.3
	16	65 13.0	61 44.5	24.7	83 24.9	58.3	104 19.4	53.6	256 21.2	15.3
	17	80 15.5	76 48.3	23.9	98 25.5	57.6	119 21.4	53.4	271 23.8	15.4
	18	95 17.9	91 52.1	N10 23.1	113 26.1	S 8 56.9	134 23.4	S15 53.3	286 26.4	N 7 15.5
	19	110 20.4	106 56.0	22.3	128 26.8	56.1	149 25.4	53.1	301 29.1	15.5
	20	125 22.9	121 59.8	21.5	143 27.4	55.4	164 27.4	53.0	316 31.7	15.6
	21	140 25.3	137 03.6	. . 20.6	158 28.0	. . 54.6	179 29.3	. . 52.9	331 34.3	. . 15.7
	22	155 27.8	152 07.4	19.8	173 28.7	53.9	194 31.3	52.7	346 36.9	15.7
	23	170 30.3	167 11.2	19.0	188 29.3	53.2	209 33.3	52.6	1 39.6	15.8
	28 00	185 32.7	182 15.0	N10 18.2	203 29.9	S 8 52.4	224 35.3	S15 52.4	16 42.2	N 7 15.9
	01	200 35.2	197 18.9	17.3	218 30.6	51.7	239 37.3	52.3	31 44.8	16.0
	02	215 37.7	212 22.7	16.5	233 31.2	51.0	254 39.3	52.1	46 47.5	16.0
	03	230 40.1	227 26.5	. . 15.7	248 31.8	. . 50.2	269 41.2	. . 52.0	61 50.1	. . 16.1
	04	245 42.6	242 30.3	14.8	263 32.5	49.5	284 43.2	51.9	76 52.7	16.2
	05	260 45.0	257 34.1	14.0	278 33.1	48.7	299 45.2	51.7	91 55.4	16.2
	06	275 47.5	272 37.9	N10 13.1	293 33.7	S 8 48.0	314 47.2	S15 51.6	106 58.0	N 7 16.3
	07	290 50.0	287 41.8	12.3	308 34.3	47.3	329 49.2	51.4	122 00.6	16.4
S	08	305 52.4	302 45.6	11.5	323 35.0	46.5	344 51.2	51.3	137 03.3	16.4
A	09	320 54.9	317 49.4	. . 10.6	338 35.6	. . 45.8	359 53.1	. . 51.1	152 05.9	. . 16.5
T	10	335 57.4	332 53.2	09.8	353 36.2	45.1	14 55.1	51.0	167 08.5	16.6
U	11	350 59.8	347 57.0	08.9	8 36.9	44.3	29 57.1	50.9	182 11.1	16.6
R	12	6 02.3	3 00.8	N10 08.1	23 37.5	S 8 43.6	44 59.1	S15 50.7	197 13.8	N 7 16.7
D	13	21 04.8	18 04.6	07.2	38 38.1	42.8	60 01.1	50.6	212 16.4	16.8
A	14	36 07.2	33 08.4	06.4	53 38.8	42.1	75 03.1	50.4	227 19.0	16.8
Y	15	51 09.7	48 12.2	. . 05.5	68 39.4	. . 41.4	90 05.1	. . 50.3	242 21.7	. . 16.9
	16	66 12.1	63 16.0	04.7	83 40.0	40.6	105 07.0	50.1	257 24.3	17.0
	17	81 14.6	78 19.8	03.8	98 40.7	39.9	120 09.0	50.0	272 26.9	17.0
	18	96 17.1	93 23.6	N10 03.0	113 41.3	S 8 39.1	135 11.0	S15 49.9	287 29.6	N 7 17.1
	19	111 19.5	108 27.4	02.1	128 41.9	38.4	150 13.0	49.7	302 32.2	17.2
	20	126 22.0	123 31.2	01.3	143 42.6	37.7	165 15.0	49.6	317 34.8	17.3
	21	141 24.5	138 35.0	10 00.4	158 43.2	. . 36.9	180 17.0	. . 49.4	332 37.4	. . 17.3
	22	156 26.9	153 38.8	9 59.6	173 43.8	36.2	195 19.0	49.3	347 40.1	17.4
	23	171 29.4	168 42.6	N 9 58.7	188 44.5	35.4	210 21.0	49.1	2 42.7	17.5
	Mer. Pass.	h m 11 39.8	v 3.8	d 0.8	v 0.6	d 0.7	v 2.0	d 0.1	v 2.6	d 0.1

STARS		
Name	SHA	Dec
	° ′	°
Acamar	315 20.9	S40 16
Achernar	335 29.4	S57 11
Acrux	173 12.5	S63 09
Adhara	255 15.0	S28 59
Aldebaran	290 53.2	N16 31
Alioth	166 22.6	N55 54
Alkaid	153 00.7	N49 15
Al Na'ir	27 47.9	S46 54
Alnilam	275 49.6	S 1 11
Alphard	217 59.0	S 8 42
Alphecca	126 13.4	N26 40
Alpheratz	357 47.2	N29 08
Altair	62 11.4	N 8 53
Ankaa	353 19.1	S42 15
Antares	112 30.0	S26 27
Arcturus	145 58.2	N19 07
Atria	107 34.6	S69 02
Avior	234 19.2	S59 32
Bellatrix	278 35.5	N 6 21
Betelgeuse	271 04.7	N 7 24
Canopus	263 57.6	S52 42
Capella	280 39.3	N46 00
Deneb	49 34.0	N45 18
Denebola	182 36.5	N14 31
Diphda	348 59.3	S17 56
Dubhe	193 54.6	N61 42
Elnath	278 16.7	N28 37
Eltanin	90 47.5	N51 28
Enif	33 50.5	N 9 54
Fomalhaut	15 27.7	S29 34
Gacrux	172 04.1	S57 10
Gienah	175 55.3	S17 35
Hadar	148 52.1	S60 25
Hamal	328 04.7	N23 30
Kaus Aust.	83 48.0	S34 22
Kochab	137 18.5	N74 06
Markab	13 41.8	N15 15
Menkar	314 18.6	N 4 07
Menkent	148 11.0	S36 25
Miaplacidus	221 40.1	S69 45
Mirfak	308 45.3	N49 53
Nunki	76 02.2	S26 17
Peacock	53 24.3	S56 42
Pollux	243 31.4	N28 00
Procyon	245 02.9	N 5 12
Rasalhague	96 09.3	N12 32
Regulus	207 46.6	N11 55
Rigel	281 15.2	S 8 11
Rigil Kent.	139 55.7	S60 52
Sabik	102 16.0	S15 44
Schedar	349 44.9	N56 35
Shaula	96 26.1	S37 06
Sirius	258 36.5	S16 43
Spica	158 34.3	S11 12
Suhail	222 54.6	S43 28
Vega	80 41.1	N38 47
Zuben'ubi	137 08.7	S16 05

	SHA	Mer. Pas
	° ′	h m
Venus	356 09.6	11 54
Mars	18 41.2	10 27
Jupiter	39 14.1	9 04
Saturn	191 05.5	22 53

UT	SUN GHA	SUN Dec	MOON GHA	v	MOON Dec	d	HP
d h	° ′	° ′	° ′	′	° ′	′	′
26 00	178 33.4	N 2 10.4	187 39.6	14.0	N 2 26.6	14.8	56.9
01	193 33.6	11.3	202 12.6	14.0	2 41.4	14.8	57.0
02	208 33.7	12.3	216 45.6	13.9	2 56.2	14.8	57.0
03	223 33.9	. . 13.3	231 18.5	13.9	3 11.0	14.8	57.0
04	238 34.1	14.3	245 51.4	13.8	3 25.8	14.8	57.0
05	253 34.3	15.3	260 24.2	13.8	3 40.6	14.8	57.1
06	268 34.5	N 2 16.2	274 57.0	13.8	N 3 55.4	14.9	57.1
07	283 34.7	17.2	289 29.8	13.7	4 10.3	14.8	57.1
08	298 34.9	18.2	304 02.5	13.7	4 25.1	14.7	57.1
09	313 35.1	. . 19.2	318 35.2	13.6	4 39.8	14.8	57.2
10	328 35.2	20.2	333 07.8	13.5	4 54.6	14.8	57.2
11	343 35.4	21.1	347 40.3	13.6	5 09.4	14.8	57.2
12	358 35.6	N 2 22.1	2 12.9	13.4	N 5 24.2	14.7	57.2
13	13 35.8	23.1	16 45.3	13.4	5 38.9	14.8	57.3
14	28 36.0	24.1	31 17.7	13.4	5 53.7	14.7	57.3
15	43 36.2	. . 25.1	45 50.1	13.3	6 08.4	14.7	57.3
16	58 36.4	26.0	60 22.4	13.3	6 23.1	14.7	57.3
17	73 36.6	27.0	74 54.7	13.2	6 37.8	14.7	57.4
18	88 36.8	N 2 28.0	89 26.9	13.1	N 6 52.5	14.6	57.4
19	103 36.9	29.0	103 59.0	13.1	7 07.1	14.7	57.4
20	118 37.1	30.0	118 31.1	13.0	7 21.8	14.6	57.4
21	133 37.3	. . 30.9	133 03.1	13.0	7 36.4	14.6	57.5
22	148 37.5	31.9	147 35.1	12.9	7 51.0	14.5	57.5
23	163 37.7	32.9	162 07.0	12.8	8 05.5	14.6	57.5
27 00	178 37.9	N 2 33.9	176 38.8	12.8	N 8 20.1	14.5	57.5
01	193 38.1	34.9	191 10.6	12.7	8 34.6	14.4	57.6
02	208 38.3	35.8	205 42.3	12.7	8 49.0	14.5	57.6
03	223 38.4	. . 36.8	220 14.0	12.6	9 03.5	14.4	57.6
04	238 38.6	37.8	234 45.6	12.5	9 17.9	14.4	57.6
05	253 38.8	38.8	249 17.1	12.4	9 32.3	14.3	57.6
06	268 39.0	N 2 39.8	263 48.5	12.4	N 9 46.6	14.3	57.7
07	283 39.2	40.7	278 19.9	12.3	10 00.9	14.3	57.7
08	298 39.4	41.7	292 51.2	12.3	10 15.2	14.3	57.7
09	313 39.6	. . 42.7	307 22.5	12.1	10 29.5	14.1	57.7
10	328 39.8	43.7	321 53.6	12.1	10 43.6	14.2	57.8
11	343 40.0	44.7	336 24.7	12.1	10 57.8	14.1	57.8
12	358 40.1	N 2 45.6	350 55.8	11.9	N11 11.9	14.1	57.8
13	13 40.3	46.6	5 26.7	11.9	11 26.0	14.0	57.8
14	28 40.5	47.6	19 57.6	11.8	11 40.0	14.0	57.9
15	43 40.7	. . 48.6	34 28.4	11.7	11 54.0	13.9	57.9
16	58 40.9	49.5	48 59.1	11.6	12 07.9	13.8	57.9
17	73 41.1	50.5	63 29.7	11.6	12 21.7	13.9	57.9
18	88 41.3	N 2 51.5	78 00.3	11.5	N12 35.6	13.7	57.9
19	103 41.5	52.5	92 30.8	11.4	12 49.3	13.7	58.0
20	118 41.6	53.5	107 01.2	11.3	13 03.0	13.7	58.0
21	133 41.8	. . 54.4	121 31.5	11.2	13 16.7	13.6	58.0
22	148 42.0	55.4	136 01.7	11.2	13 30.3	13.5	58.0
23	163 42.2	56.4	150 31.9	11.1	13 43.8	13.5	58.0
28 00	178 42.4	N 2 57.4	165 02.0	11.0	N13 57.3	13.4	58.1
01	193 42.6	58.3	179 32.0	10.9	14 10.7	13.3	58.1
02	208 42.8	2 59.3	194 01.9	10.8	14 24.0	13.3	58.1
03	223 43.0	. . 3 00.3	208 31.7	10.7	14 37.3	13.2	58.1
04	238 43.1	01.3	223 01.4	10.6	14 50.5	13.2	58.1
05	253 43.3	02.3	237 31.0	10.6	15 03.7	13.0	58.2
06	268 43.5	N 3 03.2	252 00.6	10.5	N15 16.7	13.0	58.2
07	283 43.7	04.2	266 30.1	10.3	15 29.7	12.9	58.2
08	298 43.9	05.2	280 59.4	10.3	15 42.6	12.9	58.2
09	313 44.1	. . 06.2	295 28.7	10.2	15 55.5	12.8	58.2
10	328 44.3	07.1	309 57.9	10.1	16 08.3	12.7	58.3
11	343 44.5	08.1	324 27.0	10.0	16 21.0	12.6	58.3
12	358 44.7	N 3 09.1	338 56.0	10.0	N16 33.6	12.5	58.3
13	13 44.8	10.1	353 25.0	9.8	16 46.1	12.4	58.3
14	28 45.0	11.0	7 53.8	9.7	16 58.5	12.4	58.3
15	43 45.2	. . 12.0	22 22.5	9.7	17 10.9	12.3	58.3
16	58 45.4	13.0	36 51.2	9.5	17 23.2	12.1	58.4
17	73 45.6	14.0	51 19.7	9.5	17 35.3	12.1	58.4
18	88 45.8	N 3 14.9	65 48.2	9.3	N17 47.4	12.0	58.4
19	103 46.0	15.9	80 16.5	9.3	17 59.4	11.9	58.4
20	118 46.2	16.9	94 44.8	9.1	18 11.3	11.8	58.4
21	133 46.3	. . 17.9	109 12.9	9.1	18 23.1	11.8	58.5
22	148 46.5	18.8	123 41.0	9.0	18 34.9	11.6	58.5
23	163 46.7	19.8	138 09.0	8.9	N18 46.5	11.5	58.5
	SD 16.1	d 1.0	SD 15.6		15.8		15.9

Lat.	Twilight Naut.	Twilight Civil	Sunrise	Moonrise 26	27	28	29
°	h m	h m	h m	h m	h m	h m	h m
N 72	02 28	04 11	05 22	04 54	04 21	03 34	▭
N 70	02 57	04 23	05 26	04 59	04 35	04 03	02 56
68	03 18	04 33	05 30	05 03	04 46	04 25	03 52
66	03 34	04 41	05 33	05 06	04 56	04 43	04 26
64	03 47	04 48	05 36	05 09	05 04	04 58	04 51
62	03 57	04 53	05 38	05 12	05 10	05 10	05 10
60	04 07	04 58	05 40	05 14	05 16	05 20	05 27
N 58	04 14	05 02	05 42	05 16	05 22	05 29	05 41
56	04 21	05 06	05 44	05 17	05 26	05 37	05 52
54	04 27	05 09	05 45	05 19	05 31	05 44	06 03
52	04 32	05 12	05 46	05 20	05 34	05 51	06 12
50	04 36	05 15	05 48	05 22	05 38	05 57	06 21
45	04 46	05 21	05 50	05 25	05 46	06 10	06 39
N 40	04 53	05 25	05 52	05 27	05 52	06 20	06 53
35	04 59	05 29	05 54	05 29	05 58	06 29	07 06
30	05 03	05 31	05 55	05 31	06 03	06 37	07 17
20	05 10	05 36	05 58	05 34	06 11	06 51	07 36
N 10	05 15	05 39	06 00	05 37	06 19	07 03	07 52
0	05 17	05 41	06 02	05 40	06 26	07 15	08 08
S 10	05 19	05 43	06 04	05 43	06 33	07 27	08 24
20	05 18	05 44	06 06	05 46	06 41	07 39	08 40
30	05 16	05 44	06 08	05 49	06 50	07 54	09 00
35	05 14	05 44	06 09	05 51	06 56	08 02	09 12
40	05 12	05 43	06 10	05 54	07 02	08 12	09 25
45	05 08	05 42	06 11	05 56	07 09	08 23	09 41
S 50	05 03	05 41	06 13	06 00	07 17	08 37	10 00
52	05 01	05 40	06 14	06 01	07 21	08 44	10 09
54	04 58	05 39	06 14	06 03	07 25	08 51	10 20
56	04 55	05 38	06 15	06 05	07 30	08 59	10 32
58	04 51	05 37	06 16	06 07	07 36	09 09	10 45
S 60	04 47	05 36	06 17	06 09	07 42	09 19	11 02

Lat.	Sunset	Twilight Civil	Twilight Naut.	Moonset 26	27	28	29
°	h m	h m	h m	h m	h m	h m	h m
N 72	18 52	20 04	21 50	19 36	22 04	▭	▭
N 70	18 47	19 51	21 19	19 25	21 37	24 32	00 32
68	18 43	19 41	20 57	19 16	21 16	23 37	▭
66	18 39	19 32	20 40	19 08	21 00	23 05	25 40
64	18 37	19 25	20 27	19 02	20 47	22 41	24 46
62	18 34	19 19	20 16	18 57	20 36	22 22	24 13
60	18 32	19 14	20 06	18 53	20 27	22 07	23 50
N 58	18 30	19 10	19 58	18 49	20 19	21 54	23 31
56	18 28	19 06	19 51	18 45	20 12	21 43	23 15
54	18 27	19 02	19 45	18 42	20 06	21 33	23 01
52	18 25	18 59	19 40	18 39	20 00	21 24	22 49
50	18 24	18 57	19 36	18 36	19 55	21 16	22 39
45	18 21	18 51	19 26	18 31	19 44	21 00	22 17
N 40	18 19	18 46	19 18	18 26	19 35	20 46	22 00
35	18 17	18 43	19 13	18 22	19 27	20 35	21 45
30	18 16	18 40	19 08	18 18	19 20	20 25	21 32
20	18 13	18 35	19 01	18 12	19 09	20 08	21 11
N 10	18 11	18 32	18 56	18 07	18 59	19 53	20 52
0	18 09	18 29	18 53	18 02	18 49	19 40	20 34
S 10	18 07	18 28	18 52	17 57	18 40	19 26	20 17
20	18 05	18 27	18 52	17 51	18 30	19 11	19 58
30	18 02	18 26	18 54	17 45	18 18	18 55	19 37
35	18 01	18 27	18 56	17 42	18 12	18 45	19 25
40	18 00	18 27	18 58	17 38	18 04	18 34	19 11
45	17 59	18 28	19 02	17 33	17 55	18 21	18 54
S 50	17 57	18 29	19 06	17 28	17 45	18 06	18 33
52	17 56	18 30	19 09	17 26	17 40	17 59	18 23
54	17 55	18 30	19 11	17 23	17 35	17 51	18 12
56	17 54	18 31	19 15	17 20	17 29	17 42	18 00
58	17 53	18 32	19 18	17 17	17 23	17 32	17 45
S 60	17 52	18 33	19 22	17 13	17 16	17 20	17 28

Day	SUN Eqn. of Time 00^h	SUN Eqn. of Time 12^h	SUN Mer. Pass.	MOON Mer. Pass. Upper	MOON Mer. Pass. Lower	MOON Age	MOON Phase
d	m s	m s	h m	h m	h m	d %	
26	05 47	05 38	12 06	11 51	24 14	29 0	●
27	05 29	05 20	12 05	12 37	00 14	01 1	
28	05 11	05 02	12 05	13 27	01 02	02 4	

UT (d h)	ARIES GHA	VENUS −4.2 GHA	VENUS Dec	MARS +1.2 GHA	MARS Dec	JUPITER −2.1 GHA	JUPITER Dec	SATURN +0.6 GHA	SATURN Dec	STARS Name	SHA	Dec
	° ′	° ′	° ′	° ′	° ′	° ′	° ′	° ′	° ′		° ′	°
29 00 (SUNDAY)	186 31.9	183 46.4	N 9 57.8	203 45.1	S 8 34.7	225 22.9	S15 49.0	17 45.3	N 7 17.5	Acamar	315 21.0	S40
01	201 34.3	198 50.2	57.0	218 45.7	33.9	240 24.9	48.9	32 48.0	17.6	Achernar	335 29.4	S57
02	216 36.8	213 54.0	56.1	233 46.4	33.2	255 26.9	48.7	47 50.6	17.7	Acrux	173 12.4	S63
03	231 39.3	228 57.8	55.3	248 47.0	32.5	270 28.9	48.6	62 53.2	17.7	Adhara	255 15.0	S28
04	246 41.7	244 01.6	54.4	263 47.6	31.7	285 30.9	48.4	77 55.9	17.8	Aldebaran	290 53.2	N16
05	261 44.2	259 05.4	53.5	278 48.3	31.0	300 32.9	48.3	92 58.5	17.9			
06	276 46.6	274 09.1	N 9 52.7	293 48.9	S 8 30.2	315 34.9	S15 48.2	108 01.1	N 7 17.9	Alioth	166 22.6	N55
07	291 49.1	289 12.9	51.8	308 49.5	29.5	330 36.9	48.0	123 03.7	18.0	Alkaid	153 00.7	N49
08	306 51.6	304 16.7	50.9	323 50.2	28.8	345 38.8	47.9	138 06.4	18.1	Al Na'ir	27 47.9	S46
09	321 54.0	319 20.5	50.1	338 50.8	28.0	0 40.8	47.7	153 09.0	18.1	Alnilam	275 49.6	S 1
10	336 56.5	334 24.3	49.2	353 51.4	27.3	15 42.8	47.6	168 11.6	18.2	Alphard	217 59.0	S 8
11	351 59.0	349 28.1	48.3	8 52.1	26.5	30 44.8	47.4	183 14.3	18.3			
12	7 01.4	4 31.8	N 9 47.4	23 52.7	S 8 25.8	45 46.8	S15 47.3	198 16.9	N 7 18.3	Alphecca	126 13.4	N26
13	22 03.9	19 35.6	46.6	38 53.4	25.0	60 48.8	47.2	213 19.5	18.4	Alpheratz	357 47.2	N29
14	37 06.4	34 39.4	45.7	53 54.0	24.3	75 50.8	47.0	228 22.1	18.5	Altair	62 11.4	N 8
15	52 08.8	49 43.2	44.8	68 54.6	23.6	90 52.8	46.9	243 24.8	18.5	Ankaa	353 19.1	S42
16	67 11.3	64 46.9	43.9	83 55.3	22.8	105 54.8	46.7	258 27.4	18.6	Antares	112 30.0	S26
17	82 13.7	79 50.7	43.1	98 55.9	22.1	120 56.8	46.6	273 30.0	18.7			
18	97 16.2	94 54.5	N 9 42.2	113 56.5	S 8 21.3	135 58.7	S15 46.4	288 32.7	N 7 18.7	Arcturus	145 58.2	N19
19	112 18.7	109 58.3	41.3	128 57.2	20.6	151 00.7	46.3	303 35.3	18.8	Atria	107 34.5	S69
20	127 21.1	125 02.0	40.4	143 57.8	19.8	166 02.7	46.2	318 37.9	18.9	Avior	234 19.2	S59
21	142 23.6	140 05.8	39.5	158 58.4	19.1	181 04.7	46.0	333 40.5	18.9	Bellatrix	278 35.5	N 6
22	157 26.1	155 09.6	38.7	173 59.1	18.4	196 06.7	45.9	348 43.2	19.0	Betelgeuse	271 04.8	N 7
23	172 28.5	170 13.3	37.8	188 59.7	17.6	211 08.7	45.7	3 45.8	19.1			
30 00 (MONDAY)	187 31.0	185 17.1	N 9 36.9	204 00.4	S 8 16.9	226 10.7	S15 45.6	18 48.4	N 7 19.1	Canopus	263 57.6	S52
01	202 33.5	200 20.8	36.0	219 01.0	16.1	241 12.7	45.5	33 51.0	19.2	Capella	280 39.3	N46
02	217 35.9	215 24.6	35.1	234 01.6	15.4	256 14.7	45.3	48 53.7	19.3	Deneb	49 34.0	N45
03	232 38.4	230 28.4	34.2	249 02.3	14.6	271 16.7	45.2	63 56.3	19.3	Denebola	182 36.5	N14
04	247 40.9	245 32.1	33.3	264 02.9	13.9	286 18.7	45.0	78 58.9	19.4	Diphda	348 59.3	S17
05	262 43.3	260 35.9	32.4	279 03.5	13.1	301 20.6	44.9	94 01.6	19.5			
06	277 45.8	275 39.6	N 9 31.5	294 04.2	S 8 12.4	316 22.6	S15 44.8	109 04.2	N 7 19.5	Dubhe	193 54.6	N61
07	292 48.2	290 43.4	30.7	309 04.8	11.7	331 24.6	44.6	124 06.8	19.6	Elnath	278 16.7	N28
08	307 50.7	305 47.1	29.8	324 05.5	10.9	346 26.6	44.5	139 09.4	19.7	Eltanin	90 47.5	N51
09	322 53.2	320 50.9	28.9	339 06.1	10.2	1 28.6	44.3	154 12.1	19.7	Enif	33 50.5	N 9
10	337 55.6	335 54.6	28.0	354 06.7	09.4	16 30.6	44.2	169 14.7	19.8	Fomalhaut	15 27.7	S29
11	352 58.1	350 58.4	27.1	9 07.4	08.7	31 32.6	44.0	184 17.3	19.9			
12	8 00.6	6 02.1	N 9 26.2	24 08.0	S 8 07.9	46 34.6	S15 43.9	199 19.9	N 7 19.9	Gacrux	172 04.1	S57
13	23 03.0	21 05.8	25.3	39 08.6	07.2	61 36.6	43.8	214 22.6	20.0	Gienah	175 55.3	S17
14	38 05.5	36 09.6	24.4	54 09.3	06.4	76 38.6	43.6	229 25.2	20.1	Hadar	148 52.0	S60
15	53 08.0	51 13.3	23.5	69 09.9	05.7	91 40.6	43.5	244 27.8	20.1	Hamal	328 04.7	N23
16	68 10.4	66 17.1	22.6	84 10.6	04.9	106 42.6	43.3	259 30.4	20.2	Kaus Aust.	83 47.9	S34
17	83 12.9	81 20.8	21.7	99 11.2	04.2	121 44.6	43.2	274 33.1	20.3			
18	98 15.4	96 24.5	N 9 20.8	114 11.8	S 8 03.5	136 46.6	S15 43.1	289 35.7	N 7 20.3	Kochab	137 18.5	N74
19	113 17.8	111 28.3	19.9	129 12.5	02.7	151 48.6	42.9	304 38.3	20.4	Markab	13 41.8	N15
20	128 20.3	126 32.0	19.0	144 13.1	02.0	166 50.6	42.8	319 41.0	20.5	Menkar	314 18.6	N 4
21	143 22.7	141 35.7	18.1	159 13.8	01.2	181 52.5	42.6	334 43.6	20.5	Menkent	148 11.0	S36
22	158 25.2	156 39.4	17.2	174 14.4	8 00.5	196 54.5	42.5	349 46.2	20.6	Miaplacidus	221 40.1	S69
23	173 27.7	171 43.2	16.3	189 15.0	7 59.7	211 56.5	42.4	4 48.8	20.7			
31 00 (TUESDAY)	188 30.1	186 46.9	N 9 15.4	204 15.7	S 7 59.0	226 58.5	S15 42.2	19 51.5	N 7 20.7	Mirfak	308 45.4	N49
01	203 32.6	201 50.6	14.5	219 16.3	58.2	242 00.5	42.1	34 54.1	20.8	Nunki	76 02.2	S26
02	218 35.1	216 54.3	13.6	234 17.0	57.5	257 02.5	41.9	49 56.7	20.9	Peacock	53 24.3	S56
03	233 37.5	231 58.0	12.6	249 17.6	56.7	272 04.5	41.8	64 59.3	20.9	Pollux	243 31.4	N28
04	248 40.0	247 01.8	11.7	264 18.2	56.0	287 06.5	41.7	80 02.0	21.0	Procyon	245 02.9	N 5
05	263 42.5	262 05.5	10.8	279 18.9	55.2	302 08.5	41.5	95 04.6	21.1			
06	278 44.9	277 09.2	N 9 09.9	294 19.5	S 7 54.5	317 10.5	S15 41.4	110 07.2	N 7 21.1	Rasalhague	96 09.3	N12
07	293 47.4	292 12.9	09.0	309 20.2	53.7	332 12.5	41.2	125 09.8	21.2	Regulus	207 46.6	N11
08	308 49.9	307 16.6	08.1	324 20.8	53.0	347 14.5	41.1	140 12.5	21.3	Rigel	281 15.2	S 8
09	323 52.3	322 20.3	07.2	339 21.4	52.2	2 16.5	41.0	155 15.1	21.3	Rigil Kent.	139 55.7	S60
10	338 54.8	337 24.0	06.3	354 22.1	51.5	17 18.5	40.8	170 17.7	21.4	Sabik	102 16.0	S15
11	353 57.2	352 27.7	05.4	9 22.7	50.7	32 20.5	40.7	185 20.3	21.4			
12	8 59.7	7 31.4	N 9 04.4	24 23.4	S 7 50.0	47 22.5	S15 40.5	200 23.0	N 7 21.5	Schedar	349 44.9	N56
13	24 02.2	22 35.1	03.5	39 24.0	49.3	62 24.5	40.4	215 25.6	21.6	Shaula	96 26.1	S37
14	39 04.6	37 38.8	02.6	54 24.7	48.5	77 26.5	40.3	230 28.2	21.6	Sirius	258 36.5	S16
15	54 07.1	52 42.5	01.7	69 25.3	47.8	92 28.5	40.1	245 30.8	21.7	Spica	158 34.3	S11
16	69 09.6	67 46.2	9 00.8	84 25.9	47.0	107 30.5	40.0	260 33.5	21.8	Suhail	222 54.6	S43
17	84 12.0	82 49.9	8 59.9	99 26.6	46.3	122 32.5	39.8	275 36.1	21.8			
18	99 14.5	97 53.5	N 8 58.9	114 27.2	S 7 45.5	137 34.5	S15 39.7	290 38.7	N 7 21.9	Vega	80 41.1	N38
19	114 17.0	112 57.2	58.0	129 27.9	44.8	152 36.5	39.6	305 41.3	22.0	Zuben'ubi	137 08.7	S16
20	129 19.4	128 00.9	57.1	144 28.5	44.0	167 38.5	39.4	320 43.9	22.0			
21	144 21.9	143 04.6	56.2	159 29.2	43.3	182 40.5	39.3	335 46.6	22.1			
22	159 24.4	158 08.3	55.3	174 29.8	42.5	197 42.5	39.1	350 49.2	22.2			
23	174 26.8	173 11.9	54.4	189 30.4	41.8	212 44.5	39.0	5 51.8	22.2			
Mer. Pass.	h m 11 28.0	v 3.7	d 0.9	v 0.6	d 0.7	v 2.0	d 0.1	v 2.6	d 0.1			

	SHA ° ′	Mer. P. h
Venus	357 46.1	11
Mars	16 29.4	10
Jupiter	38 39.7	8
Saturn	191 17.4	22

	UT d h	SUN GHA ° ′	SUN Dec ° ′	MOON GHA ° ′	v ′	MOON Dec ° ′	d ′	HP ′
	29 00	178 46.9	N 3 20.8	152 36.9	8.8	N18 58.0	11.4	58.5
	01	193 47.1	21.8	167 04.7	8.6	19 09.4	11.3	58.5
	02	208 47.3	22.7	181 32.3	8.6	19 20.7	11.2	58.5
	03	223 47.5	. . 23.7	195 59.9	8.5	19 31.9	11.1	58.6
	04	238 47.7	24.7	210 27.4	8.4	19 43.0	11.0	58.6
	05	253 47.8	25.7	224 54.8	8.3	19 54.0	10.9	58.6
	06	268 48.0	N 3 26.6	239 22.1	8.2	N20 04.9	10.7	58.6
	07	283 48.2	27.6	253 49.3	8.1	20 15.6	10.7	58.6
	08	298 48.4	28.6	268 16.4	8.0	20 26.3	10.5	58.6
S	09	313 48.6	. . 29.6	282 43.4	7.9	20 36.8	10.4	58.6
U	10	328 48.8	30.5	297 10.3	7.8	20 47.2	10.3	58.7
N	11	343 49.0	31.5	311 37.1	7.8	20 57.5	10.2	58.7
D	12	358 49.2	N 3 32.5	326 03.9	7.6	N21 07.7	10.1	58.7
A	13	13 49.3	33.4	340 30.5	7.5	21 17.8	9.9	58.7
Y	14	28 49.5	34.4	354 57.0	7.4	21 27.7	9.8	58.7
	15	43 49.7	. . 35.4	9 23.4	7.4	21 37.5	9.7	58.7
	16	58 49.9	36.4	23 49.8	7.2	21 47.2	9.6	58.7
	17	73 50.1	37.3	38 16.0	7.1	21 56.8	9.4	58.8
	18	88 50.3	N 3 38.3	52 42.1	7.1	N22 06.2	9.3	58.8
	19	103 50.5	39.3	67 08.2	7.0	22 15.5	9.2	58.8
	20	118 50.7	40.3	81 34.2	6.8	22 24.7	9.0	58.8
	21	133 50.8	. . 41.2	96 00.0	6.8	22 33.7	8.9	58.8
	22	148 51.0	42.2	110 25.8	6.7	22 42.6	8.8	58.8
	23	163 51.2	43.2	124 51.5	6.6	22 51.4	8.6	58.8
	30 00	178 51.4	N 3 44.1	139 17.1	6.5	N23 00.0	8.5	58.8
	01	193 51.6	45.1	153 42.6	6.4	23 08.5	8.3	58.9
	02	208 51.8	46.1	168 08.0	6.3	23 16.8	8.2	58.9
	03	223 52.0	. . 47.1	182 33.3	6.2	23 25.0	8.0	58.9
	04	238 52.2	48.0	196 58.5	6.2	23 33.0	8.0	58.9
	05	253 52.3	49.0	211 23.7	6.0	23 41.0	7.7	58.9
	06	268 52.5	N 3 50.0	225 48.7	6.0	N23 48.7	7.6	58.9
	07	283 52.7	50.9	240 13.7	5.9	23 56.3	7.5	58.9
	08	298 52.9	51.9	254 38.6	5.8	24 03.8	7.3	58.9
M	09	313 53.1	. . 52.9	269 03.4	5.7	24 11.1	7.1	58.9
O	10	328 53.3	53.9	283 28.1	5.7	24 18.2	7.0	59.0
N	11	343 53.5	54.8	297 52.8	5.5	24 25.2	6.9	59.0
D	12	358 53.7	N 3 55.8	312 17.3	5.5	N24 32.1	6.7	59.0
A	13	13 53.8	56.8	326 41.8	5.4	24 38.8	6.5	59.0
Y	14	28 54.0	57.7	341 06.2	5.4	24 45.3	6.4	59.0
	15	43 54.2	. . 58.7	355 30.6	5.2	24 51.7	6.2	59.0
	16	58 54.4	3 59.7	9 54.8	5.2	24 57.9	6.0	59.0
	17	73 54.6	4 00.6	24 19.0	5.2	25 03.9	5.9	59.0
	18	88 54.8	N 4 01.6	38 43.2	5.0	N25 09.8	5.8	59.0
	19	103 55.0	02.6	53 07.2	5.0	25 15.6	5.5	59.0
	20	118 55.1	03.6	67 31.2	4.9	25 21.1	5.4	59.0
	21	133 55.3	. . 04.5	81 55.1	4.9	25 26.5	5.2	59.1
	22	148 55.5	05.5	96 19.0	4.8	25 31.7	5.1	59.1
	23	163 55.7	06.5	110 42.8	4.7	25 36.8	4.9	59.1
	31 00	178 55.9	N 4 07.4	125 06.5	4.6	N25 41.7	4.7	59.1
	01	193 56.1	08.4	139 30.1	4.7	25 46.4	4.5	59.1
	02	208 56.3	09.4	153 53.8	4.5	25 50.9	4.4	59.1
	03	223 56.5	. . 10.3	168 17.3	4.5	25 55.3	4.2	59.1
	04	238 56.6	11.3	182 40.8	4.5	25 59.5	4.0	59.1
	05	253 56.8	12.3	197 04.3	4.4	26 03.5	3.9	59.1
	06	268 57.0	N 4 13.2	211 27.7	4.3	N26 07.4	3.7	59.1
	07	283 57.2	14.2	225 51.0	4.3	26 11.1	3.5	59.1
T	08	298 57.4	15.2	240 14.3	4.3	26 14.6	3.3	59.1
	09	313 57.6	. . 16.1	254 37.6	4.2	26 17.9	3.1	59.1
U	10	328 57.8	17.1	269 00.8	4.2	26 21.0	3.0	59.1
E	11	343 57.9	18.1	283 24.0	4.1	26 24.0	2.8	59.2
S	12	358 58.1	N 4 19.0	297 47.1	4.1	N26 26.8	2.6	59.2
D	13	13 58.3	20.0	312 10.2	4.1	26 29.4	2.5	59.2
A	14	28 58.5	21.0	326 33.3	4.0	26 31.9	2.2	59.2
Y	15	43 58.7	. . 21.9	340 56.3	4.0	26 34.1	2.1	59.2
	16	58 58.9	22.9	355 19.3	4.0	26 36.2	1.9	59.2
	17	73 59.1	23.9	9 42.3	4.0	26 38.1	1.7	59.2
	18	88 59.2	N 4 24.8	24 05.3	3.9	N26 39.8	1.5	59.2
	19	103 59.4	25.8	38 28.2	3.9	26 41.3	1.3	59.2
	20	118 59.6	26.8	52 51.1	3.9	26 42.6	1.2	59.2
	21	133 59.8	. . 27.7	67 14.0	3.9	26 43.8	1.0	59.2
	22	149 00.0	28.7	81 36.9	3.9	26 44.8	0.8	59.2
	23	164 00.2	29.7	95 59.8	3.8	N26 45.6	0.6	59.2
		SD 16.0	*d* 1.0	SD	16.0		16.1	16.1

Lat. °	Twilight Naut. h m	Twilight Civil h m	Sunrise h m	Moonrise 29 h m	Moonrise 30 h m	Moonrise 31 h m	Moonrise 1 h m
N 72	01 59	03 53	05 06	▭	▭	▭	▭
N 70	02 36	04 07	05 12	02 56	▭	▭	▭
68	03 00	04 19	05 17	03 52	▭	▭	▭
66	03 19	04 28	05 22	04 26	03 47	▭	▭
64	03 34	04 36	05 25	04 51	04 42	04 15	▭
62	03 46	04 43	05 28	05 10	05 15	05 28	06 12
60	03 56	04 49	05 31	05 27	05 39	06 04	06 55
N 58	04 05	04 54	05 34	05 41	05 59	06 30	07 24
56	04 12	04 58	05 36	05 52	06 15	06 51	07 46
54	04 19	05 02	05 38	06 03	06 29	07 08	08 05
52	04 24	05 05	05 39	06 12	06 41	07 23	08 20
50	04 29	05 08	05 41	06 21	06 52	07 36	08 33
45	04 40	05 15	05 44	06 39	07 15	08 02	09 01
N 40	04 48	05 20	05 47	06 53	07 34	08 23	09 23
35	04 54	05 24	05 50	07 06	07 49	08 41	09 41
30	05 00	05 28	05 52	07 17	08 03	08 56	09 56
20	05 08	05 33	05 55	07 36	08 26	09 22	10 23
N 10	05 13	05 37	05 58	07 52	08 46	09 44	10 45
0	05 16	05 40	06 01	08 08	09 05	10 05	11 07
S 10	05 18	05 43	06 04	08 24	09 24	10 26	11 28
20	05 19	05 45	06 07	08 40	09 44	10 48	11 51
30	05 18	05 46	06 10	09 00	10 08	11 14	12 17
35	05 17	05 46	06 11	09 12	10 22	11 30	12 33
40	05 15	05 46	06 13	09 25	10 38	11 48	12 51
45	05 12	05 46	06 15	09 41	10 57	12 10	13 13
S 50	05 08	05 45	06 18	10 00	11 22	12 37	13 40
52	05 06	05 45	06 19	10 09	11 34	12 51	13 54
54	05 04	05 45	06 20	10 20	11 47	13 06	14 09
56	05 01	05 44	06 21	10 32	12 03	13 25	14 28
58	04 58	05 44	06 23	10 45	12 22	13 47	14 50
S 60	04 54	05 43	06 24	11 02	12 45	14 17	15 19

Lat. °	Sunset h m	Twilight Civil h m	Twilight Naut. h m	Moonset 29 h m	Moonset 30 h m	Moonset 31 h m	Moonset 1 h m
N 72	19 06	20 20	22 18	▭	▭	▭	▭
N 70	18 59	20 05	21 39	00 32	▭	▭	▭
68	18 54	19 53	21 13	▭	▭	▭	▭
66	18 49	19 43	20 53	25 40	01 40	▭	▭
64	18 45	19 35	20 38	24 46	00 46	03 16	▭
62	18 42	19 28	20 25	24 13	00 13	02 04	03 29
60	18 39	19 22	20 15	23 50	25 28	01 28	02 46
N 58	18 37	19 17	20 06	23 31	25 03	01 03	02 18
56	18 34	19 12	19 59	23 15	24 42	00 42	01 55
54	18 32	19 08	19 52	23 01	24 25	00 25	01 37
52	18 31	19 05	19 46	22 49	24 11	00 11	01 21
50	18 29	19 02	19 41	22 39	23 58	25 08	01 08
45	18 25	18 55	19 30	22 17	23 32	24 40	00 40
N 40	18 22	18 50	19 22	22 00	23 12	24 18	00 18
35	18 20	18 45	19 15	21 45	22 54	24 00	00 00
30	18 18	18 42	19 10	21 32	22 40	23 45	24 44
20	18 14	18 36	19 02	21 11	22 15	23 18	24 19
N 10	18 11	18 32	18 56	20 52	21 53	22 56	23 57
0	18 08	18 28	18 52	20 34	21 33	22 34	23 36
S 10	18 05	18 26	18 50	20 17	21 13	22 13	23 15
20	18 02	18 24	18 50	19 58	20 51	21 50	22 53
30	17 59	18 23	18 50	19 37	20 27	21 24	22 27
35	17 57	18 22	18 52	19 25	20 12	21 08	22 12
40	17 55	18 22	18 53	19 11	19 55	20 50	21 54
45	17 53	18 22	18 56	18 54	19 35	20 28	21 32
S 50	17 50	18 22	19 00	18 33	19 10	20 00	21 05
52	17 49	18 23	19 02	18 23	18 58	19 46	20 52
54	17 48	18 23	19 04	18 12	18 44	19 31	20 36
56	17 46	18 23	19 06	18 00	18 28	19 12	20 18
58	17 45	18 24	19 09	17 45	18 08	18 49	19 56
S 60	17 43	18 24	19 13	17 28	17 45	18 20	19 27

Day	SUN Eqn. of Time 00^h	SUN Eqn. of Time 12^h	SUN Mer. Pass.	MOON Mer. Pass. Upper	MOON Mer. Pass. Lower	Age	Phase
d	m s	m s	h m	h m	h m	d	%
29	04 53	04 44	12 05	14 21	01 54	03	10
30	04 35	04 26	12 04	15 19	02 49	04	17
31	04 17	04 08	12 04	16 20	03 49	05	27

	UT	ARIES	VENUS −4.1		MARS +1.2		JUPITER −2.1		SATURN +0.6		STARS		
		GHA	GHA	Dec	GHA	Dec	GHA	Dec	GHA	Dec	Name	SHA	Dec
	d h	° ′	° ′	° ′	° ′	° ′	° ′	° ′	° ′	° ′		° ′	° ′
	1 00	189 29.3	188 15.6	N 8 53.4	204 31.1	S 7 41.0	227 46.5	S15 38.9	20 54.4	N 7 22.3	Acamar	315 21.0	S40 16
	01	204 31.7	203 19.3	52.5	219 31.7	40.3	242 48.5	38.7	35 57.1	22.4	Achernar	335 29.4	S57 11
	02	219 34.2	218 22.9	51.6	234 32.4	39.5	257 50.5	38.6	50 59.7	22.4	Acrux	173 12.4	S63 09
	03	234 36.7	233 26.6	. . 50.7	249 33.0	. . 38.8	272 52.5	. . 38.4	66 02.3	. . 22.5	Adhara	255 15.0	S28 59
	04	249 39.1	248 30.3	49.7	264 33.7	38.0	287 54.5	38.3	81 04.9	22.6	Aldebaran	290 53.2	N16 31
	05	264 41.6	263 33.9	48.8	279 34.3	37.3	302 56.5	38.2	96 07.6	22.6			
	06	279 44.1	278 37.6	N 8 47.9	294 34.9	S 7 36.5	317 58.5	S15 38.0	111 10.2	N 7 22.7	Alioth	166 22.6	N55 54
W	07	294 46.5	293 41.3	47.0	309 35.6	35.8	333 00.5	37.9	126 12.8	22.7	Alkaid	153 00.7	N49 15
E	08	309 49.0	308 44.9	46.0	324 36.2	35.0	348 02.5	37.7	141 15.4	22.8	Al Na'ir	27 47.8	S46 54
D	09	324 51.5	323 48.6	. . 45.1	339 36.9	. . 34.3	3 04.5	. . 37.6	156 18.0	. . 22.9	Alnilam	275 49.6	S 1 11
N	10	339 53.9	338 52.2	44.2	354 37.5	33.5	18 06.5	37.5	171 20.7	22.9	Alphard	217 59.0	S 8 42
E	11	354 56.4	353 55.9	43.3	9 38.2	32.8	33 08.5	37.3	186 23.3	23.0			
S	12	9 58.8	8 59.5	N 8 42.3	24 38.8	S 7 32.0	48 10.5	S15 37.2	201 25.9	N 7 23.1	Alphecca	126 13.3	N26 40
D	13	25 01.3	24 03.2	41.4	39 39.5	31.3	63 12.5	37.0	216 28.5	23.1	Alpheratz	357 47.2	N29 08
A	14	40 03.8	39 06.8	40.5	54 40.1	30.5	78 14.5	36.9	231 31.2	23.2	Altair	62 11.4	N 8 53
Y	15	55 06.2	54 10.4	. . 39.6	69 40.7	. . 29.8	93 16.5	. . 36.8	246 33.8	. . 23.3	Ankaa	353 19.0	S42 15
	16	70 08.7	69 14.1	38.6	84 41.4	29.0	108 18.5	36.6	261 36.4	23.3	Antares	112 29.9	S26 27
	17	85 11.2	84 17.7	37.7	99 42.0	28.3	123 20.5	36.5	276 39.0	23.4			
	18	100 13.6	99 21.3	N 8 36.8	114 42.7	S 7 27.5	138 22.5	S15 36.3	291 41.6	N 7 23.4	Arcturus	145 58.2	N19 07
	19	115 16.1	114 25.0	35.9	129 43.3	26.7	153 24.5	36.2	306 44.3	23.5	Atria	107 34.4	S69 02
	20	130 18.6	129 28.6	34.9	144 44.0	26.0	168 26.5	36.1	321 46.9	23.6	Avior	234 19.2	S59 32
	21	145 21.0	144 32.2	. . 34.0	159 44.6	. . 25.2	183 28.5	. . 35.9	336 49.5	. . 23.6	Bellatrix	278 35.5	N 6 21
	22	160 23.5	159 35.8	33.1	174 45.3	24.5	198 30.5	35.8	351 52.1	23.7	Betelgeuse	271 04.8	N 7 24
	23	175 26.0	174 39.5	32.1	189 45.9	23.7	213 32.5	35.6	6 54.8	23.8			
	2 00	190 28.4	189 43.1	N 8 31.2	204 46.6	S 7 23.0	228 34.5	S15 35.5	21 57.4	N 7 23.8	Canopus	263 57.6	S52 42
	01	205 30.9	204 46.7	30.3	219 47.2	22.2	243 36.5	35.4	37 00.0	23.9	Capella	280 39.3	N46 00
	02	220 33.3	219 50.3	29.3	234 47.9	21.5	258 38.5	35.2	52 02.6	24.0	Deneb	49 33.9	N45 18
	03	235 35.8	234 53.9	. . 28.4	249 48.5	. . 20.7	273 40.5	. . 35.1	67 05.2	. . 24.0	Denebola	182 36.5	N14 31
	04	250 38.3	249 57.5	27.5	264 49.1	20.0	288 42.5	35.0	82 07.9	24.1	Diphda	348 59.3	S17 56
	05	265 40.7	265 01.1	26.6	279 49.8	19.2	303 44.5	34.8	97 10.5	24.1			
	06	280 43.2	280 04.7	N 8 25.6	294 50.4	S 7 18.5	318 46.5	S15 34.7	112 13.1	N 7 24.2	Dubhe	193 54.6	N61 42
T	07	295 45.7	295 08.3	24.7	309 51.1	17.7	333 48.5	34.5	127 15.7	24.3	Elnath	278 16.7	N28 37
H	08	310 48.1	310 11.9	23.8	324 51.7	17.0	348 50.5	34.4	142 18.3	24.3	Eltanin	90 47.5	N51 28
U	09	325 50.6	325 15.5	. . 22.8	339 52.4	. . 16.2	3 52.5	. . 34.3	157 21.0	. . 24.4	Enif	33 50.4	N 9 54
R	10	340 53.1	340 19.1	21.9	354 53.0	15.5	18 54.5	34.1	172 23.6	24.5	Fomalhaut	15 27.6	S29 34
S	11	355 55.5	355 22.7	21.0	9 53.7	14.7	33 56.6	34.0	187 26.2	24.5			
D	12	10 58.0	10 26.3	N 8 20.0	24 54.3	S 7 14.0	48 58.6	S15 33.8	202 28.8	N 7 24.6	Gacrux	172 04.1	S57 10
A	13	26 00.5	25 29.9	19.1	39 55.0	13.2	64 00.6	33.7	217 31.4	24.6	Gienah	175 55.3	S17 35
Y	14	41 02.9	40 33.4	18.2	54 55.6	12.4	79 02.6	33.6	232 34.1	24.7	Hadar	148 52.0	S60 25
	15	56 05.4	55 37.0	. . 17.2	69 56.3	. . 11.7	94 04.6	. . 33.4	247 36.7	. . 24.8	Hamal	328 04.7	N23 30
	16	71 07.8	70 40.6	16.3	84 56.9	10.9	109 06.6	33.3	262 39.3	24.8	Kaus Aust.	83 47.9	S34 22
	17	86 10.3	85 44.2	15.4	99 57.6	10.2	124 08.6	33.2	277 41.9	24.9			
	18	101 12.8	100 47.7	N 8 14.4	114 58.2	S 7 09.4	139 10.6	S15 33.0	292 44.5	N 7 25.0	Kochab	137 18.4	N74 06
	19	116 15.2	115 51.3	13.5	129 58.9	08.7	154 12.6	32.9	307 47.2	25.0	Markab	13 41.8	N15 15
	20	131 17.7	130 54.9	12.6	144 59.5	07.9	169 14.6	32.7	322 49.8	25.1	Menkar	314 18.6	N 4 07
	21	146 20.2	145 58.4	. . 11.6	160 00.2	. . 07.2	184 16.6	. . 32.6	337 52.4	. . 25.1	Menkent	148 11.0	S36 25
	22	161 22.6	161 02.0	10.7	175 00.8	06.4	199 18.6	32.5	352 55.0	25.2	Miaplacidus	221 40.2	S69 45
	23	176 25.1	176 05.5	09.8	190 01.5	05.7	214 20.6	32.3	7 57.6	25.3			
	3 00	191 27.6	191 09.1	N 8 08.8	205 02.1	S 7 04.9	229 22.6	S15 32.2	23 00.3	N 7 25.3	Mirfak	308 45.4	N49 53
	01	206 30.0	206 12.6	07.9	220 02.8	04.2	244 24.6	32.1	38 02.9	25.4	Nunki	76 02.2	S26 17
	02	221 32.5	221 16.2	07.0	235 03.4	03.4	259 26.7	31.9	53 05.5	25.5	Peacock	53 24.2	S56 42
	03	236 35.0	236 19.7	. . 06.0	250 04.1	. . 02.6	274 28.7	. . 31.8	68 08.1	. . 25.5	Pollux	243 31.4	N28 00
	04	251 37.4	251 23.3	05.1	265 04.7	01.9	289 30.7	31.6	83 10.7	25.6	Procyon	245 02.9	N 5 12
	05	266 39.9	266 26.8	04.2	280 05.3	01.1	304 32.7	31.5	98 13.4	25.6			
	06	281 42.3	281 30.4	N 8 03.2	295 06.0	S 7 00.4	319 34.7	S15 31.4	113 16.0	N 7 25.7	Rasalhague	96 09.2	N12 32
	07	296 44.8	296 33.9	02.3	310 06.6	6 59.6	334 36.7	31.2	128 18.6	25.8	Regulus	207 46.6	N11 55
	08	311 47.3	311 37.4	01.4	325 07.3	58.9	349 38.7	31.1	143 21.2	25.8	Rigel	281 15.2	S 8 11
F	09	326 49.7	326 40.9	8 00.4	340 07.9	. . 58.1	4 40.7	. . 31.0	158 23.8	. . 25.9	Rigil Kent.	139 55.7	S60 52
R	10	341 52.2	341 44.5	7 59.5	355 08.6	57.4	19 42.7	30.8	173 26.4	26.0	Sabik	102 16.0	S15 44
I	11	356 54.7	356 48.0	58.6	10 09.2	56.6	34 44.7	30.7	188 29.1	26.0			
D	12	11 57.1	11 51.5	N 7 57.6	25 09.9	S 6 55.8	49 46.7	S15 30.5	203 31.7	N 7 26.1	Schedar	349 44.9	N56 35
A	13	26 59.6	26 55.0	56.7	40 10.6	55.1	64 48.8	30.4	218 34.3	26.1	Shaula	96 26.0	S37 06
Y	14	42 02.1	41 58.5	55.8	55 11.2	54.3	79 50.8	30.3	233 36.9	26.2	Sirius	258 36.5	S16 43
	15	57 04.5	57 02.1	. . 54.8	70 11.9	. . 53.6	94 52.8	. . 30.1	248 39.5	. . 26.3	Spica	158 34.3	S11 12
	16	72 07.0	72 05.6	53.9	85 12.5	52.8	109 54.8	30.0	263 42.1	26.3	Suhail	222 54.6	S43 28
	17	87 09.5	87 09.1	53.0	100 13.2	52.1	124 56.8	29.9	278 44.8	26.4			
	18	102 11.9	102 12.6	N 7 52.0	115 13.8	S 6 51.3	139 58.8	S15 29.7	293 47.4	N 7 26.4	Vega	80 41.1	N38 47
	19	117 14.4	117 16.1	51.1	130 14.5	50.6	155 00.8	29.6	308 50.0	26.5	Zuben'ubi	137 08.6	S16 05
	20	132 16.8	132 19.6	50.2	145 15.1	49.8	170 02.8	29.4	323 52.6	26.6		SHA	Mer.Pass
	21	147 19.3	147 23.1	. . 49.2	160 15.8	. . 49.0	185 04.8	. . 29.3	338 55.2	. . 26.6		° ′	h m
	22	162 21.8	162 26.5	48.3	175 16.4	48.3	200 06.8	29.2	353 57.9	26.7	Venus	359 14.7	11 18
	23	177 24.2	177 30.0	47.4	190 17.1	47.5	215 08.9	29.0	9 00.5	26.8	Mars	14 18.1	10 20
		h m									Jupiter	38 06.1	8 45
	Mer. Pass.	11 16.3	*v* 3.6	*d* 0.9	*v* 0.6	*d* 0.8	*v* 2.0	*d* 0.1	*v* 2.6	*d* 0.1	Saturn	191 29.0	22 28

UT	SUN GHA	SUN Dec	MOON GHA	v	MOON Dec	d	HP
d h	° ′	° ′	° ′	′	° ′	′	′
1 00	179 00.4	N 4 30.6	110 22.6	3.9	N26 46.2	0.4	59.2
01	194 00.5	31.6	124 45.5	3.8	26 46.6	0.2	59.2
02	209 00.7	32.6	139 08.3	3.8	26 46.8	0.1	59.2
03	224 00.9	. . 33.5	153 31.1	3.9	26 46.9	0.1	59.2
04	239 01.1	34.5	167 54.0	3.8	26 46.8	0.3	59.2
05	254 01.3	35.5	182 16.8	3.8	26 46.5	0.5	59.2
06	269 01.5	N 4 36.4	196 39.6	3.9	N26 46.0	0.7	59.2
07	284 01.7	37.4	211 02.5	3.8	26 45.3	0.9	59.2
08	299 01.8	38.4	225 25.3	3.9	26 44.4	1.0	59.2
09	314 02.0	. . 39.3	239 48.2	3.8	26 43.4	1.2	59.2
10	329 02.2	40.3	254 11.0	3.9	26 42.2	1.5	59.2
11	344 02.4	41.2	268 33.9	3.9	26 40.7	1.5	59.2
12	359 02.6	N 4 42.2	282 56.8	3.9	N26 39.2	1.8	59.2
13	14 02.8	43.2	297 19.7	4.0	26 37.4	2.0	59.2
14	29 03.0	44.1	311 42.7	3.9	26 35.4	2.1	59.3
15	44 03.1	. . 45.1	326 05.6	4.0	26 33.3	2.3	59.3
16	59 03.3	46.1	340 28.6	4.0	26 31.0	2.6	59.3
17	74 03.5	47.0	354 51.6	4.0	26 28.4	2.6	59.3
18	89 03.7	N 4 48.0	9 14.6	4.1	N26 25.8	2.9	59.3
19	104 03.9	49.0	23 37.7	4.1	26 22.9	3.0	59.3
20	119 04.1	49.9	38 00.8	4.2	26 19.9	3.3	59.3
21	134 04.3	. . 50.9	52 24.0	4.1	26 16.6	3.4	59.3
22	149 04.4	51.8	66 47.1	4.2	26 13.2	3.6	59.3
23	164 04.6	52.8	81 10.3	4.3	26 09.6	3.7	59.3
2 00	179 04.8	N 4 53.8	95 33.6	4.3	N26 05.9	3.9	59.3
01	194 05.0	54.7	109 56.9	4.3	26 02.0	4.2	59.3
02	209 05.2	55.7	124 20.2	4.4	25 57.8	4.2	59.3
03	224 05.4	. . 56.6	138 43.6	4.5	25 53.6	4.5	59.3
04	239 05.5	57.6	153 07.1	4.5	25 49.1	4.6	59.3
05	254 05.7	58.6	167 30.6	4.5	25 44.5	4.8	59.3
06	269 05.9	N 4 59.5	181 54.1	4.6	N25 39.7	5.0	59.3
07	284 06.1	5 00.5	196 17.7	4.7	25 34.7	5.2	59.3
08	299 06.3	01.4	210 41.4	4.7	25 29.5	5.3	59.3
09	314 06.5	. . 02.4	225 05.1	4.8	25 24.2	5.5	59.3
10	329 06.7	03.4	239 28.9	4.8	25 18.7	5.6	59.3
11	344 06.8	04.3	253 52.7	4.9	25 13.1	5.8	59.3
12	359 07.0	N 5 05.3	268 16.6	4.9	N25 07.3	6.0	59.3
13	14 07.2	06.2	282 40.5	5.1	25 01.3	6.2	59.3
14	29 07.4	07.2	297 04.6	5.1	24 55.1	6.3	59.3
15	44 07.6	. . 08.2	311 28.7	5.1	24 48.8	6.5	59.3
16	59 07.8	09.1	325 52.8	5.2	24 42.3	6.6	59.2
17	74 07.9	10.1	340 17.0	5.3	24 35.7	6.8	59.2
18	89 08.1	N 5 11.0	354 41.3	5.4	N24 28.9	7.0	59.2
19	104 08.3	12.0	9 05.7	5.5	24 21.9	7.1	59.2
20	119 08.5	13.0	23 30.2	5.5	24 14.8	7.2	59.2
21	134 08.7	. . 13.9	37 54.7	5.6	24 07.6	7.5	59.2
22	149 08.9	14.9	52 19.3	5.6	24 00.1	7.5	59.2
23	164 09.0	15.8	66 43.9	5.8	23 52.6	7.8	59.2
3 00	179 09.2	N 5 16.8	81 08.7	5.8	N23 44.8	7.9	59.2
01	194 09.4	17.8	95 33.5	5.9	23 36.9	8.0	59.2
02	209 09.6	18.7	109 58.4	6.0	23 28.9	8.2	59.2
03	224 09.8	. . 19.7	124 23.4	6.1	23 20.7	8.3	59.2
04	239 10.0	20.6	138 48.5	6.1	23 12.4	8.5	59.2
05	254 10.1	21.6	153 13.6	6.3	23 03.9	8.6	59.2
06	269 10.3	N 5 22.5	167 38.9	6.3	N22 55.3	8.8	59.2
07	284 10.5	23.5	182 04.2	6.4	22 46.5	8.9	59.2
08	299 10.7	24.5	196 29.6	6.5	22 37.6	9.0	59.2
09	314 10.9	. . 25.4	210 55.1	6.5	22 28.6	9.2	59.2
10	329 11.1	26.4	225 20.6	6.7	22 19.4	9.4	59.2
11	344 11.2	27.3	239 46.3	6.7	22 10.0	9.4	59.2
12	359 11.4	N 5 28.3	254 12.0	6.9	N22 00.6	9.6	59.2
13	14 11.6	29.2	268 37.9	6.9	21 51.0	9.7	59.2
14	29 11.8	30.2	283 03.8	7.0	21 41.3	9.9	59.2
15	44 12.0	. . 31.1	297 29.8	7.1	21 31.4	10.0	59.2
16	59 12.2	32.1	311 55.9	7.1	21 21.4	10.1	59.2
17	74 12.3	33.1	326 22.0	7.3	21 11.3	10.3	59.2
18	89 12.5	N 5 34.0	340 48.3	7.4	N21 01.0	10.3	59.2
19	104 12.7	35.0	355 14.7	7.4	20 50.7	10.5	59.2
20	119 12.9	35.9	9 41.1	7.5	20 40.2	10.7	59.1
21	134 13.1	. . 36.9	24 07.6	7.7	20 29.5	10.7	59.1
22	149 13.3	37.8	38 34.3	7.7	20 18.8	10.9	59.1
23	164 13.4	38.8	53 01.0	7.8	N20 07.9	10.9	59.1
	SD 16.0	d 1.0	SD 16.1		16.1		16.1

Lat.	Twilight Naut.	Twilight Civil	Sunrise	Moonrise 1	Moonrise 2	Moonrise 3	Moonrise 4
°	h m	h m	h m	h m	h m	h m	h m
N 72	01 23	03 34	04 50	▭	▭	▭	▭
N 70	02 12	03 52	04 58	▭	▭	▭	09 40
68	02 42	04 05	05 05	▭	▭	▭	10 23
66	03 04	04 16	05 10	▭	▭	08 23	10 51
64	03 20	04 25	05 15	▭	06 45	09 08	11 12
62	03 34	04 33	05 19	06 12	07 44	09 37	11 29
60	03 45	04 39	05 22	06 55	08 18	09 59	11 43
N 58	03 55	04 45	05 25	07 24	08 43	10 17	11 55
56	04 03	04 50	05 28	07 46	09 02	10 32	12 06
54	04 10	04 54	05 30	08 05	09 19	10 45	12 15
52	04 17	04 58	05 33	08 20	09 33	10 56	12 23
50	04 22	05 02	05 35	08 33	09 45	11 06	12 30
45	04 34	05 09	05 39	09 01	10 11	11 27	12 46
N 40	04 43	05 15	05 42	09 23	10 31	11 44	12 58
35	04 50	05 20	05 46	09 41	10 48	11 58	13 09
30	04 56	05 24	05 48	09 56	11 02	12 10	13 18
20	05 05	05 31	05 53	10 23	11 27	12 31	13 34
N 10	05 11	05 36	05 57	10 45	11 48	12 49	13 48
0	05 15	05 40	06 00	11 07	12 08	13 06	14 01
S 10	05 18	05 43	06 04	11 28	12 28	13 23	14 14
20	05 20	05 45	06 07	11 51	12 49	13 41	14 28
30	05 20	05 47	06 11	12 17	13 13	14 02	14 44
35	05 19	05 48	06 14	12 33	13 27	14 14	14 53
40	05 18	05 49	06 16	12 51	13 44	14 27	15 03
45	05 16	05 50	06 19	13 13	14 04	14 44	15 15
S 50	05 13	05 50	06 22	13 40	14 28	15 03	15 30
52	05 11	05 50	06 24	13 54	14 40	15 13	15 36
54	05 09	05 50	06 26	14 09	14 54	15 23	15 44
56	05 07	05 50	06 27	14 28	15 09	15 35	15 52
58	05 05	05 50	06 29	14 50	15 27	15 49	16 02
S 60	05 02	05 50	06 32	15 19	15 50	16 05	16 12

Lat.	Sunset	Twilight Civil	Twilight Naut.	Moonset 1	Moonset 2	Moonset 3	Moonset 4
°	h m	h m	h m	h m	h m	h m	h m
N 72	19 20	20 37	22 58	▭	▭	▭	▭
N 70	19 12	20 19	22 02	▭	▭	▭	06 15
68	19 05	20 05	21 30	▭	▭	▭	05 31
66	18 59	19 54	21 07	▭	▭	05 32	05 01
64	18 54	19 44	20 50	▭	05 06	04 47	04 39
62	18 50	19 37	20 36	03 29	04 06	04 17	04 21
60	18 47	19 30	20 24	02 46	03 32	03 55	04 06
N 58	18 43	19 24	20 14	02 18	03 07	03 36	03 53
56	18 40	19 19	20 06	01 55	02 47	03 20	03 42
54	18 38	19 14	19 58	01 37	02 30	03 07	03 32
52	18 36	19 10	19 52	01 21	02 16	02 55	03 23
50	18 34	19 07	19 46	01 08	02 03	02 45	03 15
45	18 29	18 59	19 34	00 40	01 37	02 23	02 58
N 40	18 25	18 53	19 25	00 18	01 17	02 05	02 44
35	18 22	18 48	19 18	00 00	00 59	01 50	02 32
30	18 19	18 43	19 12	24 44	00 44	01 37	02 22
20	18 15	18 37	19 03	24 19	00 19	01 14	02 04
N 10	18 11	18 32	18 56	23 57	24 55	00 55	01 48
0	18 07	18 28	18 52	23 36	24 36	00 36	01 33
S 10	18 03	18 24	18 49	23 15	24 18	00 18	01 18
20	17 59	18 22	18 47	22 53	23 58	25 02	01 02
30	17 55	18 19	18 47	22 27	23 35	24 43	00 43
35	17 53	18 18	18 48	22 12	23 21	24 32	00 32
40	17 50	18 17	18 49	21 54	23 05	24 19	00 19
45	17 47	18 17	18 51	21 32	22 46	24 04	00 04
S 50	17 44	18 16	18 53	21 05	22 22	23 46	25 10
52	17 42	18 16	18 55	20 52	22 11	23 37	25 04
54	17 41	18 16	18 57	20 36	21 58	23 27	24 57
56	17 39	18 16	18 59	20 18	21 43	23 16	24 49
58	17 36	18 16	19 01	19 56	21 25	23 03	24 41
S 60	17 34	18 16	19 04	19 27	21 03	22 48	24 31

Day	SUN Eqn. of Time 00^h	SUN Eqn. of Time 12^h	SUN Mer. Pass.	MOON Mer. Pass. Upper	MOON Mer. Pass. Lower	MOON Age	MOON Phase
d	m s	m s	h m	h m	h m	d %	
1	03 59	03 50	12 04	17 21	04 51	06 38	
2	03 41	03 32	12 04	18 22	05 52	07 49	◐
3	03 23	03 15	12 03	19 20	06 51	08 60	

	UT	ARIES	VENUS −4.3		MARS +1.2		JUPITER −2.1		SATURN +0.6	
		GHA	GHA	Dec	GHA	Dec	GHA	Dec	GHA	Dec
	d h	° ′	° ′	° ′	° ′	° ′	° ′	° ′	° ′	° ′
	4 00	192 26.7	192 33.5	N 7 46.4	205 17.7	S 6 46.8	230 10.9	S15 28.9	24 03.1	N 7 26.8
	01	207 29.2	207 37.0	45.5	220 18.4	46.0	245 12.9	28.8	39 05.7	26.9
	02	222 31.6	222 40.5	44.6	235 19.0	45.3	260 14.9	28.6	54 08.3	26.9
	03	237 34.1	237 43.9	. . 43.6	250 19.7	. . 44.5	275 16.9	. . 28.5	69 10.9	. . 27.0
	04	252 36.6	252 47.4	42.7	265 20.3	43.7	290 18.9	28.3	84 13.5	27.1
	05	267 39.0	267 50.9	41.8	280 21.0	43.0	305 20.9	28.2	99 16.2	27.1
	06	282 41.5	282 54.4	N 7 40.8	295 21.6	S 6 42.2	320 22.9	S15 28.1	114 18.8	N 7 27.2
	07	297 43.9	297 57.8	39.9	310 22.3	41.5	335 25.0	27.9	129 21.4	27.2
S	08	312 46.4	313 01.3	39.0	325 22.9	40.7	350 27.0	27.8	144 24.0	27.3
A	09	327 48.9	328 04.7	. . 38.0	340 23.6	. . 40.0	5 29.0	. . 27.7	159 26.6	. . 27.4
T	10	342 51.3	343 08.2	37.1	355 24.2	39.2	20 31.0	27.5	174 29.2	27.4
U	11	357 53.8	358 11.6	36.2	10 24.9	38.4	35 33.0	27.4	189 31.9	27.5
R	12	12 56.3	13 15.1	N 7 35.3	25 25.5	S 6 37.7	50 35.0	S15 27.3	204 34.5	N 7 27.5
D	13	27 58.7	28 18.5	34.3	40 26.2	36.9	65 37.0	27.1	219 37.1	27.6
A	14	43 01.2	43 22.0	33.4	55 26.9	36.2	80 39.1	27.0	234 39.7	27.7
Y	15	58 03.7	58 25.4	. . 32.5	70 27.5	. . 35.4	95 41.1	. . 26.8	249 42.3	. . 27.7
	16	73 06.1	73 28.8	31.5	85 28.2	34.6	110 43.1	26.7	264 44.9	27.8
	17	88 08.6	88 32.3	30.6	100 28.8	33.9	125 45.1	26.6	279 47.6	27.8
	18	103 11.1	103 35.7	N 7 29.7	115 29.5	S 6 33.1	140 47.1	S15 26.4	294 50.2	N 7 27.9
	19	118 13.5	118 39.1	28.8	130 30.1	32.4	155 49.1	26.3	309 52.8	28.0
	20	133 16.0	133 42.5	27.8	145 30.8	31.6	170 51.1	26.2	324 55.4	28.0
	21	148 18.4	148 46.0	. . 26.9	160 31.4	. . 30.9	185 53.2	. . 26.0	339 58.0	. . 28.1
	22	163 20.9	163 49.4	26.0	175 32.1	30.1	200 55.2	25.9	355 00.6	28.1
	23	178 23.4	178 52.8	25.1	190 32.7	29.3	215 57.2	25.8	10 03.2	28.2
	5 00	193 25.8	193 56.2	N 7 24.1	205 33.4	S 6 28.6	230 59.2	S15 25.6	25 05.9	N 7 28.3
	01	208 28.3	208 59.6	23.2	220 34.0	27.8	246 01.2	25.5	40 08.5	28.3
	02	223 30.8	224 03.0	22.3	235 34.7	27.1	261 03.2	25.4	55 11.1	28.4
	03	238 33.2	239 06.4	. . 21.4	250 35.4	. . 26.3	276 05.3	. . 25.2	70 13.7	. . 28.4
	04	253 35.7	254 09.8	20.4	265 36.0	25.5	291 07.3	25.1	85 16.3	28.5
	05	268 38.2	269 13.2	19.5	280 36.7	24.8	306 09.3	24.9	100 18.9	28.6
	06	283 40.6	284 16.6	N 7 18.6	295 37.3	S 6 24.0	321 11.3	S15 24.8	115 21.5	N 7 28.6
	07	298 43.1	299 20.0	17.7	310 38.0	23.3	336 13.3	24.7	130 24.1	28.7
	08	313 45.6	314 23.3	16.7	325 38.6	22.5	351 15.3	24.5	145 26.8	28.7
S	09	328 48.0	329 26.7	. . 15.8	340 39.3	. . 21.7	6 17.4	. . 24.4	160 29.4	. . 28.8
U	10	343 50.5	344 30.1	14.9	355 39.9	21.0	21 19.4	24.3	175 32.0	28.9
N	11	358 52.9	359 33.5	14.0	10 40.6	20.2	36 21.4	24.1	190 34.6	28.9
D	12	13 55.4	14 36.8	N 7 13.1	25 41.3	S 6 19.5	51 23.4	S15 24.0	205 37.2	N 7 29.0
A	13	28 57.9	29 40.2	12.1	40 41.9	18.7	66 25.4	23.9	220 39.8	29.0
Y	14	44 00.3	44 43.6	11.2	55 42.6	17.9	81 27.4	23.7	235 42.4	29.1
	15	59 02.8	59 46.9	. . 10.3	70 43.2	. . 17.2	96 29.5	. . 23.6	250 45.1	. . 29.2
	16	74 05.3	74 50.3	09.4	85 43.9	16.4	111 31.5	23.5	265 47.7	29.2
	17	89 07.7	89 53.6	08.5	100 44.5	15.7	126 33.5	23.3	280 50.3	29.3
	18	104 10.2	104 57.0	N 7 07.5	115 45.2	S 6 14.9	141 35.5	S15 23.2	295 52.9	N 7 29.3
	19	119 12.7	120 00.3	06.6	130 45.9	14.1	156 37.5	23.0	310 55.5	29.4
	20	134 15.1	135 03.7	05.7	145 46.5	13.4	171 39.6	22.9	325 58.1	29.5
	21	149 17.6	150 07.0	. . 04.8	160 47.2	. . 12.6	186 41.6	. . 22.8	341 00.7	. . 29.5
	22	164 20.0	165 10.3	03.9	175 47.8	11.9	201 43.6	22.6	356 03.3	29.6
	23	179 22.5	180 13.7	03.0	190 48.5	11.1	216 45.6	22.5	11 06.0	29.6
	6 00	194 25.0	195 17.0	N 7 02.1	205 49.1	S 6 10.3	231 47.6	S15 22.4	26 08.6	N 7 29.7
	01	209 27.4	210 20.3	01.1	220 49.8	09.6	246 49.7	22.2	41 11.2	29.7
	02	224 29.9	225 23.6	7 00.2	235 50.5	08.8	261 51.7	22.1	56 13.8	29.8
	03	239 32.4	240 27.0	6 59.3	250 51.1	. . 08.0	276 53.7	. . 22.0	71 16.4	. . 29.9
	04	254 34.8	255 30.3	58.4	265 51.8	07.3	291 55.7	21.8	86 19.0	29.9
	05	269 37.3	270 33.6	57.5	280 52.4	06.5	306 57.7	21.7	101 21.6	30.0
	06	284 39.8	285 36.9	N 6 56.6	295 53.1	S 6 05.8	321 59.8	S15 21.6	116 24.2	N 7 30.0
	07	299 42.2	300 40.2	55.7	310 53.7	05.0	337 01.8	21.4	131 26.8	30.1
	08	314 44.7	315 43.5	54.8	325 54.4	04.2	352 03.8	21.3	146 29.5	30.2
M	09	329 47.2	330 46.8	. . 53.9	340 55.1	. . 03.5	7 05.8	. . 21.2	161 32.1	. . 30.2
O	10	344 49.6	345 50.1	53.0	355 55.7	02.7	22 07.8	21.0	176 34.7	30.3
N	11	359 52.1	0 53.4	52.1	10 56.4	02.0	37 09.9	20.9	191 37.3	30.3
D	12	14 54.5	15 56.7	N 6 51.1	25 57.0	S 6 01.2	52 11.9	S15 20.8	206 39.9	N 7 30.4
A	13	29 57.0	30 59.9	50.2	40 57.7	6 00.4	67 13.9	20.6	221 42.5	30.4
Y	14	44 59.5	46 03.2	49.3	55 58.4	5 59.7	82 15.9	20.5	236 45.1	30.5
	15	60 01.9	61 06.5	. . 48.4	70 59.0	. . 58.9	97 18.0	. . 20.4	251 47.7	. . 30.6
	16	75 04.4	76 09.8	47.5	85 59.7	58.1	112 20.0	20.2	266 50.3	30.6
	17	90 06.9	91 13.0	46.6	101 00.3	57.4	127 22.0	20.1	281 52.9	30.7
	18	105 09.3	106 16.3	N 6 45.7	116 01.0	S 5 56.6	142 24.0	S15 20.0	296 55.6	N 7 30.7
	19	120 11.8	121 19.6	44.8	131 01.6	55.9	157 26.1	19.8	311 58.2	30.8
	20	135 14.3	136 22.8	43.9	146 02.3	55.1	172 28.1	19.7	327 00.8	30.9
	21	150 16.7	151 26.1	. . 43.0	161 03.0	. . 54.3	187 30.1	. . 19.6	342 03.4	. . 30.9
	22	165 19.2	166 29.3	42.1	176 03.6	53.6	202 32.1	19.4	357 06.0	31.0
	23	180 21.6	181 32.6	41.2	191 04.3	52.8	217 34.2	19.3	12 08.6	31.0
	Mer. Pass.	h m 11 04.5	v 3.4	d 0.9	v 0.7	d 0.8	v 2.0	d 0.1	v 2.6	d 0.1

STARS		
Name	SHA	Dec
	° ′	°
Acamar	315 21.0	S40 1
Achernar	335 29.4	S57 1
Acrux	173 12.4	S63 0
Adhara	255 15.0	S28 5
Aldebaran	290 53.2	N16 3
Alioth	166 22.6	N55 5
Alkaid	153 00.7	N49 1
Al Na'ir	27 47.8	S46 5
Alnilam	275 49.7	S 1 1
Alphard	217 59.0	S 8 4
Alphecca	126 13.3	N26 4
Alpheratz	357 47.2	N29 0
Altair	62 11.4	N 8 5
Ankaa	353 19.0	S42 1
Antares	112 29.9	S26 2
Arcturus	145 58.2	N19 0
Atria	107 34.4	S69 0
Avior	234 19.3	S59 3
Bellatrix	278 35.5	N 6 2
Betelgeuse	271 04.8	N 7 2
Canopus	263 57.7	S52 4
Capella	280 39.3	N46 0
Deneb	49 33.9	N45 1
Denebola	182 36.5	N14 3
Diphda	348 59.3	S17 5
Dubhe	193 54.6	N61 4
Elnath	278 16.7	N28 3
Eltanin	90 47.4	N51 2
Enif	33 50.4	N 9 5
Fomalhaut	15 27.6	S29 3
Gacrux	172 04.1	S57 1
Gienah	175 55.2	S17 3
Hadar	148 52.0	S60 2
Hamal	328 04.7	N23 3
Kaus Aust.	83 47.9	S34 2
Kochab	137 18.4	N74 0
Markab	13 41.8	N15 1
Menkar	314 18.6	N 4 0
Menkent	148 11.0	S36 2
Miaplacidus	221 40.2	S69 4
Mirfak	308 45.4	N49 5
Nunki	76 02.1	S26 1
Peacock	53 24.2	S56 4
Pollux	243 31.4	N28 0
Procyon	245 02.9	N 5 1
Rasalhague	96 09.2	N12 3
Regulus	207 46.6	N11 5
Rigel	281 15.2	S 8 1
Rigil Kent.	139 55.6	S60 5
Sabik	102 16.0	S15 4
Schedar	349 44.9	N56 3
Shaula	96 26.0	S37 0
Sirius	258 36.5	S16 4
Spica	158 34.3	S11 1
Suhail	222 54.6	S43 2
Vega	80 41.0	N38 4
Zuben'ubi	137 08.6	S16 0

	SHA	Mer. Pa
	° ′	h
Venus	0 30.4	11 0
Mars	12 07.6	10 1
Jupiter	37 33.4	8 3
Saturn	191 40.0	22 1

UT	SUN GHA	SUN Dec	MOON GHA	v	MOON Dec	d	HP
h	° ′	° ′	° ′	′	° ′	′	′
00	179 13.6	N 5 39.7	67 27.8	7.9	N19 57.0	11.1	59.1
01	194 13.8	40.7	81 54.7	7.9	19 45.9	11.2	59.1
02	209 14.0	41.6	96 21.6	8.1	19 34.7	11.4	59.1
03	224 14.2	. . 42.6	110 48.7	8.1	19 23.3	11.4	59.1
04	239 14.3	43.5	125 15.8	8.3	19 11.9	11.6	59.1
05	254 14.5	44.5	139 43.1	8.3	19 00.3	11.6	59.1
06	269 14.7	N 5 45.5	154 10.4	8.4	N18 48.7	11.8	59.1
07	284 14.9	46.4	168 37.8	8.5	18 36.9	11.8	59.1
08	299 15.1	47.4	183 05.3	8.6	18 25.1	12.0	59.1
09	314 15.3	. . 48.3	197 32.9	8.7	18 13.1	12.1	59.1
10	329 15.4	49.3	212 00.6	8.8	18 01.0	12.2	59.1
11	344 15.6	50.2	226 28.4	8.8	17 48.8	12.2	59.0
12	359 15.8	N 5 51.2	240 56.2	9.0	N17 36.6	12.4	59.0
13	14 16.0	52.1	255 24.2	9.0	17 24.2	12.5	59.0
14	29 16.2	53.1	269 52.2	9.1	17 11.7	12.5	59.0
15	44 16.3	. . 54.0	284 20.3	9.2	16 59.2	12.7	59.0
16	59 16.5	55.0	298 48.5	9.3	16 46.5	12.7	59.0
17	74 16.7	55.9	313 16.8	9.3	16 33.8	12.9	59.0
18	89 16.9	N 5 56.9	327 45.1	9.5	N16 20.9	12.9	59.0
19	104 17.1	57.8	342 13.6	9.5	16 08.0	13.0	59.0
20	119 17.2	58.8	356 42.1	9.6	15 55.0	13.1	59.0
21	134 17.4	5 59.7	11 10.7	9.7	15 41.9	13.2	59.0
22	149 17.6	6 00.7	25 39.4	9.7	15 28.7	13.2	59.0
23	164 17.8	01.6	40 08.1	9.9	15 15.5	13.4	58.9
00	179 18.0	N 6 02.6	54 37.0	9.9	N15 02.1	13.4	58.9
01	194 18.1	03.5	69 05.9	10.0	14 48.7	13.5	58.9
02	209 18.3	04.5	83 34.9	10.1	14 35.2	13.5	58.9
03	224 18.5	. . 05.4	98 04.0	10.1	14 21.7	13.7	58.9
04	239 18.7	06.4	112 33.1	10.3	14 08.0	13.7	58.9
05	254 18.9	07.3	127 02.4	10.3	13 54.3	13.8	58.9
06	269 19.0	N 6 08.3	141 31.7	10.3	N13 40.5	13.8	58.9
07	284 19.2	09.2	156 01.0	10.5	13 26.7	14.0	58.9
08	299 19.4	10.2	170 30.5	10.5	13 12.7	14.0	58.9
09	314 19.6	. . 11.1	185 00.0	10.6	12 58.7	14.0	58.8
10	329 19.8	12.1	199 29.6	10.7	12 44.7	14.1	58.8
11	344 19.9	13.0	213 59.3	10.7	12 30.6	14.2	58.8
12	359 20.1	N 6 13.9	228 29.0	10.8	N12 16.4	14.2	58.8
13	14 20.3	14.9	242 58.8	10.9	12 02.2	14.3	58.8
14	29 20.5	15.8	257 28.7	10.9	11 47.9	14.4	58.8
15	44 20.7	. . 16.8	271 58.6	11.0	11 33.5	14.4	58.8
16	59 20.8	17.7	286 28.6	11.1	11 19.1	14.5	58.8
17	74 21.0	18.7	300 58.7	11.1	11 04.6	14.5	58.8
18	89 21.2	N 6 19.6	315 28.8	11.2	N10 50.1	14.5	58.7
19	104 21.4	20.6	329 59.0	11.3	10 35.6	14.7	58.7
20	119 21.6	21.5	344 29.3	11.3	10 20.9	14.6	58.7
21	134 21.7	. . 22.5	358 59.6	11.4	10 06.3	14.7	58.7
22	149 21.9	23.4	13 30.0	11.4	9 51.6	14.8	58.7
23	164 22.1	24.4	28 00.4	11.5	9 36.8	14.8	58.7
00	179 22.3	N 6 25.3	42 30.9	11.6	N 9 22.0	14.8	58.7
01	194 22.5	26.2	57 01.5	11.6	9 07.2	14.9	58.7
02	209 22.6	27.2	71 32.1	11.7	8 52.3	14.9	58.6
03	224 22.8	. . 28.1	86 02.8	11.7	8 37.4	15.0	58.6
04	239 23.0	29.1	100 33.5	11.8	8 22.4	15.0	58.6
05	254 23.2	30.0	115 04.3	11.8	8 07.4	15.0	58.6
06	269 23.3	N 6 31.0	129 35.1	11.9	N 7 52.4	15.1	58.6
07	284 23.5	31.9	144 06.0	11.9	7 37.3	15.1	58.6
08	299 23.7	32.9	158 36.9	12.0	7 22.2	15.1	58.6
09	314 23.9	. . 33.8	173 07.9	12.0	7 07.1	15.1	58.5
10	329 24.1	34.7	187 38.9	12.1	6 52.0	15.2	58.5
11	344 24.2	35.7	202 10.0	12.1	6 36.8	15.2	58.5
12	359 24.4	N 6 36.6	216 41.1	12.1	N 6 21.6	15.3	58.5
13	14 24.6	37.6	231 12.2	12.2	6 06.3	15.2	58.5
14	29 24.8	38.5	245 43.4	12.3	5 51.1	15.3	58.5
15	44 24.9	. . 39.5	260 14.7	12.3	5 35.8	15.3	58.5
16	59 25.1	40.4	274 46.0	12.3	5 20.5	15.3	58.4
17	74 25.3	41.3	289 17.3	12.3	5 05.2	15.3	58.4
18	89 25.5	N 6 42.3	303 48.6	12.5	N 4 49.9	15.4	58.4
19	104 25.7	43.2	318 20.1	12.4	4 34.5	15.3	58.4
20	119 25.8	44.2	332 51.5	12.5	4 19.2	15.4	58.4
21	134 26.0	. . 45.1	347 23.0	12.5	4 03.8	15.4	58.4
22	149 26.2	46.0	1 54.5	12.5	3 48.4	15.4	58.3
23	164 26.4	47.0	16 26.0	12.6	N 3 33.0	15.4	58.3
	SD 16.0	*d* 0.9	SD 16.1		16.0		15.9

Lat.	Twilight Naut.	Twilight Civil	Sunrise	Moonrise 4	Moonrise 5	Moonrise 6	Moonrise 7
°	h m	h m	h m	h m	h m	h m	h m
N 72	////	03 15	04 34	▭	12 01	14 28	16 42
N 70	01 44	03 35	04 44	09 40	12 24	14 36	16 40
68	02 22	03 51	04 52	10 23	12 41	14 43	16 39
66	02 47	04 03	04 58	10 51	12 55	14 49	16 38
64	03 06	04 13	05 04	11 12	13 06	14 54	16 37
62	03 22	04 22	05 09	11 29	13 16	14 58	16 36
60	03 34	04 29	05 13	11 43	13 24	15 01	16 35
N 58	03 45	04 36	05 17	11 55	13 31	15 04	16 35
56	03 54	04 42	05 20	12 06	13 38	15 07	16 34
54	04 02	04 47	05 23	12 15	13 43	15 10	16 34
52	04 09	04 51	05 26	12 23	13 48	15 12	16 33
50	04 15	04 55	05 28	12 30	13 53	15 14	16 33
45	04 28	05 03	05 33	12 46	14 03	15 18	16 32
N 40	04 38	05 10	05 38	12 58	14 11	15 22	16 31
35	04 46	05 16	05 41	13 09	14 18	15 25	16 31
30	04 52	05 20	05 45	13 18	14 24	15 28	16 30
20	05 02	05 28	05 50	13 34	14 35	15 33	16 29
N 10	05 09	05 34	05 55	13 48	14 44	15 37	16 28
0	05 14	05 39	05 59	14 01	14 52	15 41	16 28
S 10	05 18	05 43	06 04	14 14	15 01	15 45	16 27
20	05 20	05 46	06 08	14 28	15 10	15 49	16 26
30	05 21	05 49	06 13	14 44	15 20	15 54	16 26
35	05 21	05 51	06 16	14 53	15 26	15 57	16 25
40	05 21	05 52	06 19	15 03	15 33	16 00	16 25
45	05 19	05 53	06 23	15 15	15 41	16 03	16 24
S 50	05 17	05 55	06 27	15 30	15 50	16 08	16 23
52	05 16	05 55	06 29	15 36	15 54	16 09	16 23
54	05 15	05 56	06 31	15 44	15 59	16 12	16 23
56	05 13	05 56	06 33	15 52	16 04	16 14	16 23
58	05 11	05 57	06 36	16 02	16 10	16 17	16 22
S 60	05 09	05 57	06 39	16 12	16 17	16 19	16 22

Lat.	Sunset	Twilight Civil	Twilight Naut.	Moonset 4	Moonset 5	Moonset 6	Moonset 7
°	h m	h m	h m	h m	h m	h m	h m
N 72	19 34	20 55	////	▭	05 45	05 02	04 29
N 70	19 24	20 34	22 30	06 15	05 20	04 50	04 27
68	19 16	20 18	21 49	05 31	05 01	04 41	04 24
66	19 09	20 05	21 22	05 01	04 45	04 33	04 23
64	19 03	19 54	21 02	04 39	04 32	04 27	04 21
62	18 58	19 45	20 46	04 21	04 21	04 21	04 20
60	18 54	19 38	20 33	04 06	04 12	04 16	04 18
N 58	18 50	19 31	20 22	03 53	04 04	04 11	04 17
56	18 47	19 25	20 13	03 42	03 56	04 07	04 16
54	18 44	19 20	20 05	03 32	03 50	04 04	04 15
52	18 41	19 16	19 58	03 23	03 44	04 00	04 15
50	18 38	19 11	19 52	03 15	03 38	03 57	04 14
45	18 33	19 03	19 39	02 58	03 27	03 51	04 12
N 40	18 28	18 56	19 29	02 44	03 17	03 45	04 11
35	18 25	18 50	19 20	02 32	03 08	03 40	04 10
30	18 21	18 45	19 14	02 22	03 01	03 36	04 08
20	18 15	18 38	19 04	02 04	02 48	03 29	04 07
N 10	18 10	18 32	18 56	01 48	02 37	03 22	04 05
0	18 06	18 27	18 51	01 33	02 26	03 16	04 03
S 10	18 02	18 23	18 47	01 18	02 15	03 09	04 02
20	17 57	18 19	18 45	01 02	02 03	03 03	04 00
30	17 52	18 16	18 43	00 43	01 50	02 55	03 58
35	17 49	18 14	18 44	00 32	01 42	02 50	03 57
40	17 46	18 13	18 44	00 19	01 33	02 45	03 55
45	17 42	18 11	18 45	00 04	01 23	02 39	03 54
S 50	17 38	18 10	18 47	25 10	01 10	02 32	03 52
52	17 35	18 09	18 48	25 04	01 04	02 28	03 51
54	17 33	18 09	18 49	24 57	00 57	02 25	03 50
56	17 31	18 08	18 51	24 49	00 49	02 20	03 49
58	17 28	18 07	18 53	24 41	00 41	02 16	03 47
S 60	17 25	18 07	18 55	24 31	00 31	02 11	03 46

Day	SUN Eqn. of Time 00h	SUN Eqn. of Time 12h	SUN Mer. Pass.	MOON Mer. Pass. Upper	MOON Mer. Pass. Lower	Age	Phase
d	m s	m s	h m	h m	h m	d %	
4	03 06	02 57	12 03	20 14	07 47	09 71	
5	02 48	02 40	12 03	21 04	08 39	10 81	
6	02 31	02 23	12 02	21 52	09 28	11 89	

Day	UT d h	ARIES GHA	VENUS −4.4 GHA	VENUS Dec	MARS +1.2 GHA	MARS Dec	JUPITER −2.1 GHA	JUPITER Dec	SATURN +0.6 GHA	SATURN Dec
		° ′	° ′	° ′	° ′	° ′	° ′	° ′	° ′	° ′
TUESDAY	7 00	195 24.1	196 35.8	N 6 40.3	206 04.9	S 5 52.0	232 36.2	S15 19.2	27 11.2	N 7 31.1
	01	210 26.6	211 39.1	39.4	221 05.6	51.3	247 38.2	19.0	42 13.8	31.1
	02	225 29.0	226 42.3	38.5	236 06.3	50.5	262 40.2	18.9	57 16.4	31.2
	03	240 31.5	241 45.5	. . 37.7	251 06.9	. . 49.8	277 42.2	. . 18.7	72 19.0	. . 31.3
	04	255 34.0	256 48.8	36.8	266 07.6	49.0	292 44.3	18.6	87 21.6	31.3
	05	270 36.4	271 52.0	35.9	281 08.2	48.2	307 46.3	18.5	102 24.3	31.4
	06	285 38.9	286 55.2	N 6 35.0	296 08.9	S 5 47.5	322 48.3	S15 18.3	117 26.9	N 7 31.4
	07	300 41.4	301 58.4	34.1	311 09.6	46.7	337 50.4	18.2	132 29.5	31.5
	08	315 43.8	317 01.6	33.2	326 10.2	45.9	352 52.4	18.1	147 32.1	31.5
	09	330 46.3	332 04.8	. . 32.3	341 10.9	. . 45.2	7 54.4	. . 17.9	162 34.7	. . 31.6
	10	345 48.8	347 08.1	31.4	356 11.6	44.4	22 56.4	17.8	177 37.3	31.7
	11	0 51.2	2 11.3	30.5	11 12.2	43.6	37 58.5	17.7	192 39.9	31.7
	12	15 53.7	17 14.5	N 6 29.6	26 12.9	S 5 42.9	53 00.5	S15 17.5	207 42.5	N 7 31.8
	13	30 56.1	32 17.7	28.8	41 13.5	42.1	68 02.5	17.4	222 45.1	31.8
	14	45 58.6	47 20.8	27.9	56 14.2	41.4	83 04.5	17.3	237 47.7	31.9
	15	61 01.1	62 24.0	. . 27.0	71 14.9	. . 40.6	98 06.6	. . 17.1	252 50.3	. . 31.9
	16	76 03.5	77 27.2	26.1	86 15.5	39.8	113 08.6	17.0	267 52.9	32.0
	17	91 06.0	92 30.4	25.2	101 16.2	39.1	128 10.6	16.9	282 55.6	32.1
	18	106 08.5	107 33.6	N 6 24.3	116 16.8	S 5 38.3	143 12.6	S15 16.8	297 58.2	N 7 32.1
	19	121 10.9	122 36.8	23.5	131 17.5	37.5	158 14.7	16.6	313 00.8	32.2
	20	136 13.4	137 39.9	22.6	146 18.2	36.8	173 16.7	16.5	328 03.4	32.2
	21	151 15.9	152 43.1	. . 21.7	161 18.8	. . 36.0	188 18.7	. . 16.4	343 06.0	. . 32.3
	22	166 18.3	167 46.3	20.8	176 19.5	35.2	203 20.8	16.2	358 08.6	32.3
	23	181 20.8	182 49.4	19.9	191 20.2	34.5	218 22.8	16.1	13 11.2	32.4
WEDNESDAY	8 00	196 23.2	197 52.6	N 6 19.1	206 20.8	S 5 33.7	233 24.8	S15 16.0	28 13.8	N 7 32.4
	01	211 25.7	212 55.7	18.2	221 21.5	32.9	248 26.8	15.8	43 16.4	32.5
	02	226 28.2	227 58.9	17.3	236 22.1	32.2	263 28.9	15.7	58 19.0	32.6
	03	241 30.6	243 02.0	. . 16.4	251 22.8	. . 31.4	278 30.9	. . 15.6	73 21.6	. . 32.6
	04	256 33.1	258 05.2	15.6	266 23.5	30.6	293 32.9	15.4	88 24.2	32.7
	05	271 35.6	273 08.3	14.7	281 24.1	29.9	308 35.0	15.3	103 26.8	32.7
	06	286 38.0	288 11.4	N 6 13.8	296 24.8	S 5 29.1	323 37.0	S15 15.2	118 29.4	N 7 32.8
	07	301 40.5	303 14.6	13.0	311 25.5	28.4	338 39.0	15.0	133 32.0	32.8
	08	316 43.0	318 17.7	12.1	326 26.1	27.6	353 41.1	14.9	148 34.6	32.9
	09	331 45.4	333 20.8	. . 11.2	341 26.8	. . 26.8	8 43.1	. . 14.8	163 37.3	. . 32.9
	10	346 47.9	348 23.9	10.4	356 27.4	26.1	23 45.1	14.6	178 39.9	33.0
	11	1 50.4	3 27.0	09.5	11 28.1	25.3	38 47.1	14.5	193 42.5	33.1
	12	16 52.8	18 30.1	N 6 08.6	26 28.8	S 5 24.5	53 49.2	S15 14.4	208 45.1	N 7 33.1
	13	31 55.3	33 33.2	07.8	41 29.4	23.8	68 51.2	14.2	223 47.7	33.2
	14	46 57.7	48 36.4	06.9	56 30.1	23.0	83 53.2	14.1	238 50.3	33.2
	15	62 00.2	63 39.4	. . 06.1	71 30.8	. . 22.2	98 55.3	. . 14.0	253 52.9	. . 33.3
	16	77 02.7	78 42.5	05.2	86 31.4	21.5	113 57.3	13.8	268 55.5	33.3
	17	92 05.1	93 45.6	04.3	101 32.1	20.7	128 59.3	13.7	283 58.1	33.4
	18	107 07.6	108 48.7	N 6 03.5	116 32.8	S 5 19.9	144 01.4	S15 13.6	299 00.7	N 7 33.4
	19	122 10.1	123 51.8	02.6	131 33.4	19.2	159 03.4	13.4	314 03.3	33.5
	20	137 12.5	138 54.9	01.8	146 34.1	18.4	174 05.4	13.3	329 05.9	33.6
	21	152 15.0	153 58.0	. . 00.9	161 34.7	. . 17.6	189 07.5	. . 13.2	344 08.5	. . 33.6
	22	167 17.5	169 01.0	6 00.1	176 35.4	16.9	204 09.5	13.0	359 11.1	33.7
	23	182 19.9	184 04.1	5 59.2	191 36.1	16.1	219 11.5	12.9	14 13.7	33.7
THURSDAY	9 00	197 22.4	199 07.2	N 5 58.4	206 36.7	S 5 15.3	234 13.6	S15 12.8	29 16.3	N 7 33.8
	01	212 24.9	214 10.2	57.5	221 37.4	14.6	249 15.6	12.6	44 18.9	33.8
	02	227 27.3	229 13.3	56.7	236 38.1	13.8	264 17.6	12.5	59 21.5	33.9
	03	242 29.8	244 16.3	. . 55.8	251 38.7	. . 13.0	279 19.7	. . 12.4	74 24.1	. . 33.9
	04	257 32.2	259 19.4	55.0	266 39.4	12.3	294 21.7	12.3	89 26.7	34.0
	05	272 34.7	274 22.4	54.1	281 40.1	11.5	309 23.7	12.1	104 29.3	34.0
	06	287 37.2	289 25.5	N 5 53.3	296 40.7	S 5 10.7	324 25.8	S15 12.0	119 31.9	N 7 34.1
	07	302 39.6	304 28.5	52.4	311 41.4	10.0	339 27.8	11.9	134 34.5	34.2
	08	317 42.1	319 31.5	51.6	326 42.1	09.2	354 29.8	11.7	149 37.2	34.2
	09	332 44.6	334 34.6	. . 50.8	341 42.7	. . 08.4	9 31.9	. . 11.6	164 39.8	. . 34.3
	10	347 47.0	349 37.6	49.9	356 43.4	07.7	24 33.9	11.5	179 42.4	34.3
	11	2 49.5	4 40.6	49.1	11 44.1	06.9	39 35.9	11.3	194 45.0	34.4
	12	17 52.0	19 43.6	N 5 48.2	26 44.7	S 5 06.1	54 38.0	S15 11.2	209 47.6	N 7 34.4
	13	32 54.4	34 46.7	47.4	41 45.4	05.4	69 40.0	11.1	224 50.2	34.5
	14	47 56.9	49 49.7	46.6	56 46.1	04.6	84 42.0	10.9	239 52.8	34.5
	15	62 59.3	64 52.7	. . 45.7	71 46.7	. . 03.8	99 44.1	. . 10.8	254 55.4	. . 34.6
	16	78 01.8	79 55.7	44.9	86 47.4	03.1	114 46.1	10.7	269 58.0	34.6
	17	93 04.3	94 58.7	44.1	101 48.1	02.3	129 48.2	10.5	285 00.6	34.7
	18	108 06.7	110 01.7	N 5 43.2	116 48.7	S 5 01.5	144 50.2	S15 10.4	300 03.2	N 7 34.8
	19	123 09.2	125 04.7	42.4	131 49.4	00.8	159 52.2	10.3	315 05.8	34.8
	20	138 11.7	140 07.7	41.6	146 50.1	5 00.0	174 54.3	10.2	330 08.4	34.9
	21	153 14.1	155 10.6	. . 40.8	161 50.7	4 59.2	189 56.3	. . 10.0	345 11.0	. . 34.9
	22	168 16.6	170 13.6	39.9	176 51.4	58.5	204 58.3	09.9	0 13.6	35.0
	23	183 19.1	185 16.6	39.1	191 52.1	57.7	220 00.4	09.8	15 16.2	35.0
	Mer. Pass.	h m 10 52.7	v 3.1	d 0.9	v 0.7	d 0.8	v 2.0	d 0.1	v 2.6	d 0.1

STARS

Name	SHA	Dec
	° ′	° ′
Acamar	315 21.0	S40 16.
Achernar	335 29.4	S57 11.
Acrux	173 12.4	S63 09.
Adhara	255 15.0	S28 59.
Aldebaran	290 53.2	N16 31.7
Alioth	166 22.6	N55 54.
Alkaid	153 00.7	N49 15.8
Al Na'ir	27 47.8	S46 54.8
Alnilam	275 49.7	S 1 11.8
Alphard	217 59.0	S 8 42.
Alphecca	126 13.3	N26 40.7
Alpheratz	357 47.2	N29 08.
Altair	62 11.3	N 8 53.4
Ankaa	353 19.0	S42 15.
Antares	112 29.9	S26 27.2
Arcturus	145 58.2	N19 07.8
Atria	107 34.3	S69 02.5
Avior	234 19.3	S59 32.
Bellatrix	278 35.5	N 6 21.5
Betelgeuse	271 04.8	N 7 24.5
Canopus	263 57.7	S52 42.3
Capella	280 39.3	N46 00.
Deneb	49 33.9	N45 18.5
Denebola	182 36.5	N14 31.0
Diphda	348 59.3	S17 56.2
Dubhe	193 54.6	N61 42.1
Elnath	278 16.7	N28 37.0
Eltanin	90 47.4	N51 28.9
Enif	33 50.4	N 9 54.9
Fomalhaut	15 27.6	S29 34.3
Gacrux	172 04.1	S57 10.1
Gienah	175 55.2	S17 35.8
Hadar	148 52.0	S60 25.2
Hamal	328 04.7	N23 30.4
Kaus Aust.	83 47.8	S34 22.8
Kochab	137 18.4	N74 06.8
Markab	13 41.8	N15 15.2
Menkar	314 18.6	N 4 07.5
Menkent	148 11.0	S36 25.1
Miaplacidus	221 40.2	S69 45.6
Mirfak	308 45.4	N49 53.8
Nunki	76 02.1	S26 17.1
Peacock	53 24.1	S56 42.1
Pollux	243 31.5	N28 00.3
Procyon	245 03.0	N 5 12.0
Rasalhague	96 09.2	N12 32.9
Regulus	207 46.6	N11 55.2
Rigel	281 15.2	S 8 11.5
Rigil Kent.	139 55.6	S60 52.5
Sabik	102 16.0	S15 44.3
Schedar	349 44.9	N56 35.3
Shaula	96 26.0	S37 06.6
Sirius	258 36.5	S16 43.9
Spica	158 34.3	S11 12.8
Suhail	222 54.7	S43 28.5
Vega	80 41.0	N38 47.2
Zuben'ubi	137 08.6	S16 05.0

	SHA	Mer. Pass.
	° ′	h m
Venus	1 29.3	10 46
Mars	9 57.6	10 14
Jupiter	37 01.6	8 25
Saturn	191 50.6	22 03

UT	SUN GHA	SUN Dec	MOON GHA	*v*	MOON Dec	*d*	HP
d h	° ′	° ′	° ′	′	° ′	′	′
7 00	179 26.5	N 6 47.9	30 57.6	12.6	N 3 17.6	15.4	58.3
01	194 26.7	48.9	45 29.2	12.6	3 02.2	15.4	58.3
02	209 26.9	49.8	60 00.8	12.7	2 46.8	15.5	58.3
03	224 27.1	. . 50.7	74 32.5	12.7	2 31.3	15.4	58.3
04	239 27.2	51.7	89 04.2	12.7	2 15.9	15.4	58.2
05	254 27.4	52.6	103 35.9	12.8	2 00.5	15.4	58.2
06	269 27.6	N 6 53.6	118 07.7	12.7	N 1 45.1	15.5	58.2
07	284 27.8	54.5	132 39.4	12.8	1 29.6	15.4	58.2
08	299 28.0	55.4	147 11.2	12.8	1 14.2	15.4	58.2
09	314 28.1	. . 56.4	161 43.0	12.9	0 58.8	15.5	58.2
10	329 28.3	57.3	176 14.9	12.8	0 43.3	15.4	58.1
11	344 28.5	58.2	190 46.7	12.9	0 27.9	15.4	58.1
12	359 28.7	N 6 59.2	205 18.6	12.9	N 0 12.5	15.4	58.1
13	14 28.8	7 00.1	219 50.5	12.9	S 0 02.9	15.4	58.1
14	29 29.0	01.1	234 22.4	13.0	0 18.3	15.4	58.1
15	44 29.2	. . 02.0	248 54.4	12.9	0 33.7	15.3	58.0
16	59 29.4	02.9	263 26.3	13.0	0 49.0	15.4	58.0
17	74 29.5	03.9	277 58.3	12.9	1 04.4	15.3	58.0
18	89 29.7	N 7 04.8	292 30.2	13.0	S 1 19.7	15.4	58.0
19	104 29.9	05.7	307 02.2	13.0	1 35.1	15.3	58.0
20	119 30.1	06.7	321 34.2	13.0	1 50.4	15.3	57.9
21	134 30.2	. . 07.6	336 06.2	13.0	2 05.7	15.2	57.9
22	149 30.4	08.6	350 38.2	13.1	2 20.9	15.3	57.9
23	164 30.6	09.5	5 10.3	13.0	2 36.2	15.2	57.9
8 00	179 30.8	N 7 10.4	19 42.3	13.0	S 2 51.4	15.2	57.9
01	194 30.9	11.4	34 14.3	13.1	3 06.6	15.2	57.8
02	209 31.1	12.3	48 46.4	13.0	3 21.8	15.2	57.8
03	224 31.3	. . 13.2	63 18.4	13.1	3 37.0	15.1	57.8
04	239 31.4	14.2	77 50.5	13.0	3 52.1	15.1	57.8
05	254 31.6	15.1	92 22.5	13.1	4 07.2	15.1	57.8
06	269 31.8	N 7 16.0	106 54.6	13.0	S 4 22.3	15.0	57.7
07	284 32.0	17.0	121 26.6	13.1	4 37.3	15.0	57.7
08	299 32.1	17.9	135 58.7	13.0	4 52.3	15.0	57.7
09	314 32.3	. . 18.8	150 30.7	13.1	5 07.3	14.9	57.7
10	329 32.5	19.8	165 02.8	13.0	5 22.2	14.9	57.7
11	344 32.7	20.7	179 34.8	13.0	5 37.1	14.9	57.6
12	359 32.8	N 7 21.6	194 06.8	13.1	S 5 52.0	14.8	57.6
13	14 33.0	22.6	208 38.9	13.0	6 06.8	14.8	57.6
14	29 33.2	23.5	223 10.9	13.0	6 21.6	14.7	57.6
15	44 33.4	. . 24.4	237 42.9	13.0	6 36.3	14.7	57.6
16	59 33.5	25.4	252 14.9	13.0	6 51.0	14.7	57.5
17	74 33.7	26.3	266 46.9	13.0	7 05.7	14.6	57.5
18	89 33.9	N 7 27.2	281 18.9	13.0	S 7 20.3	14.6	57.5
19	104 34.0	28.1	295 50.9	13.0	7 34.9	14.5	57.5
20	119 34.2	29.1	310 22.9	12.9	7 49.4	14.5	57.4
21	134 34.4	. . 30.0	324 54.8	12.9	8 03.9	14.4	57.4
22	149 34.6	30.9	339 26.7	13.0	8 18.3	14.4	57.4
23	164 34.7	31.9	353 58.7	12.9	8 32.7	14.3	57.4
9 00	179 34.9	N 7 32.8	8 30.6	12.9	S 8 47.0	14.3	57.4
01	194 35.1	33.7	23 02.5	12.9	9 01.3	14.2	57.3
02	209 35.3	34.7	37 34.4	12.8	9 15.5	14.2	57.3
03	224 35.4	. . 35.6	52 06.2	12.9	9 29.7	14.1	57.3
04	239 35.6	36.5	66 38.1	12.8	9 43.8	14.0	57.3
05	254 35.8	37.4	81 09.9	12.8	9 57.8	14.0	57.2
06	269 35.9	N 7 38.4	95 41.7	12.8	S10 11.8	14.0	57.2
07	284 36.1	39.3	110 13.5	12.7	10 25.8	13.8	57.2
08	299 36.3	40.2	124 45.2	12.8	10 39.6	13.8	57.2
09	314 36.4	. . 41.2	139 17.0	12.7	10 53.4	13.8	57.1
10	329 36.6	42.1	153 48.7	12.7	11 07.2	13.7	57.1
11	344 36.8	43.0	168 20.4	12.7	11 20.9	13.6	57.1
12	359 37.0	N 7 43.9	182 52.1	12.6	S11 34.5	13.6	57.1
13	14 37.1	44.9	197 23.7	12.6	11 48.1	13.4	57.1
14	29 37.3	45.8	211 55.3	12.6	12 01.5	13.5	57.0
15	44 37.5	. . 46.7	226 26.9	12.6	12 15.0	13.3	57.0
16	59 37.6	47.7	240 58.5	12.5	12 28.3	13.3	57.0
17	74 37.8	48.6	255 30.0	12.6	12 41.6	13.2	57.0
18	89 38.0	N 7 49.5	270 01.6	12.4	S12 54.8	13.1	56.9
19	104 38.2	50.4	284 33.0	12.5	13 07.9	13.1	56.9
20	119 38.3	51.4	299 04.5	12.4	13 21.0	13.0	56.9
21	134 38.5	. . 52.3	313 35.9	12.4	13 34.0	12.9	56.9
22	149 38.7	53.2	328 07.3	12.4	13 46.9	12.9	56.8
23	164 38.8	54.1	342 38.7	12.3	S13 59.8	12.7	56.8
	SD 16.0	*d* 0.9	SD	15.8		15.7	15.6

Lat.	Twilight Naut.	Twilight Civil	Sunrise	Moonrise 7	8	9	10
°	h m	h m	h m	h m	h m	h m	h m
N 72	////	02 54	04 18	16 42	18 55	21 22	■
N 70	01 07	03 18	04 30	16 40	18 43	20 53	23 39
68	01 59	03 36	04 39	16 39	18 33	20 32	22 45
66	02 30	03 50	04 47	16 38	18 25	20 15	22 12
64	02 52	04 02	04 53	16 37	18 18	20 02	21 48
62	03 09	04 11	04 59	16 36	18 13	19 50	21 30
60	03 23	04 20	05 04	16 35	18 08	19 41	21 14
N 58	03 35	04 27	05 08	16 35	18 04	19 32	21 01
56	03 45	04 33	05 12	16 34	18 00	19 25	20 50
54	03 54	04 39	05 16	16 34	17 56	19 19	20 41
52	04 01	04 44	05 19	16 33	17 53	19 13	20 32
50	04 08	04 48	05 22	16 33	17 50	19 08	20 24
45	04 22	04 58	05 28	16 32	17 44	18 56	20 08
N 40	04 33	05 05	05 33	16 31	17 39	18 47	19 54
35	04 41	05 12	05 37	16 31	17 35	18 39	19 43
30	04 48	05 17	05 41	16 30	17 31	18 32	19 33
20	04 59	05 25	05 48	16 29	17 25	18 20	19 16
N 10	05 07	05 32	05 53	16 28	17 19	18 10	19 02
0	05 14	05 38	05 59	16 28	17 14	18 00	18 48
S 10	05 18	05 42	06 04	16 27	17 09	17 51	18 35
20	05 21	05 47	06 09	16 26	17 03	17 41	18 20
30	05 23	05 51	06 15	16 26	16 57	17 29	18 04
35	05 24	05 53	06 18	16 25	16 53	17 23	17 55
40	05 23	05 55	06 22	16 25	16 49	17 15	17 44
45	05 23	05 57	06 26	16 24	16 45	17 07	17 31
S 50	05 22	05 59	06 32	16 23	16 39	16 57	17 16
52	05 21	06 00	06 34	16 23	16 37	16 52	17 09
54	05 20	06 01	06 37	16 23	16 34	16 47	17 01
56	05 19	06 02	06 39	16 23	16 31	16 41	16 53
58	05 18	06 03	06 43	16 22	16 28	16 34	16 43
S 60	05 16	06 04	06 46	16 22	16 24	16 27	16 32

Lat.	Sunset	Twilight Civil	Twilight Naut.	Moonset 7	8	9	10
°	h m	h m	h m	h m	h m	h m	h m
N 72	19 49	21 15	////	04 29	03 59	03 25	02 37
N 70	19 37	20 50	23 12	04 27	04 04	03 40	03 08
68	19 27	20 31	22 11	04 24	04 09	03 52	03 31
66	19 19	20 16	21 39	04 23	04 12	04 01	03 49
64	19 12	20 04	21 16	04 21	04 15	04 10	04 04
62	19 06	19 54	20 58	04 20	04 18	04 17	04 16
60	19 01	19 46	20 43	04 18	04 21	04 23	04 27
N 58	18 57	19 38	20 31	04 17	04 23	04 29	04 36
56	18 53	19 32	20 21	04 16	04 25	04 34	04 44
54	18 49	19 26	20 12	04 15	04 27	04 38	04 52
52	18 46	19 21	20 04	04 15	04 28	04 42	04 58
50	18 43	19 16	19 57	04 14	04 30	04 46	05 04
45	18 37	19 07	19 43	04 12	04 33	04 54	05 17
N 40	18 31	18 59	19 32	04 11	04 36	05 01	05 28
35	18 27	18 53	19 23	04 10	04 38	05 07	05 37
30	18 23	18 47	19 16	04 08	04 40	05 12	05 45
20	18 16	18 39	19 05	04 07	04 43	05 21	05 59
N 10	18 10	18 32	18 56	04 05	04 47	05 29	06 11
0	18 05	18 26	18 50	04 03	04 50	05 36	06 23
S 10	18 00	18 21	18 46	04 02	04 53	05 43	06 35
20	17 54	18 17	18 42	04 00	04 56	05 51	06 47
30	17 48	18 12	18 40	03 58	04 59	06 00	07 02
35	17 45	18 10	18 40	03 57	05 01	06 06	07 10
40	17 41	18 08	18 39	03 55	05 04	06 12	07 19
45	17 36	18 06	18 40	03 54	05 07	06 19	07 31
S 50	17 31	18 04	18 41	03 52	05 10	06 27	07 44
52	17 29	18 03	18 42	03 51	05 11	06 31	07 51
54	17 26	18 02	18 42	03 50	05 13	06 36	07 58
56	17 23	18 00	18 43	03 49	05 15	06 40	08 06
58	17 20	17 59	18 45	03 47	05 17	06 46	08 15
S 60	17 16	17 58	18 46	03 46	05 19	06 52	08 25

Day	SUN Eqn. of Time 00^h	Eqn. of Time 12^h	Mer. Pass.	MOON Mer. Pass. Upper	Mer. Pass. Lower	Age	Phase
d	m s	m s	h m	h m	h m	d %	
7	02 14	02 06	12 02	22 39	10 15	12 95	
8	01 57	01 49	12 02	23 25	11 02	13 98	
9	01 41	01 32	12 02	24 12	11 48	14 100	

UT	ARIES	VENUS −4.5		MARS +1.2		JUPITER −2.1		SATURN +0.6	
	GHA	GHA	Dec	GHA	Dec	GHA	Dec	GHA	Dec
d h	° ′	° ′	° ′	° ′	° ′	° ′	° ′	° ′	° ′
10 00	198 21.5	200 19.6	N 5 38.3	206 52.7	S 4 56.9	235 02.4	S15 09.6	30 18.8	N 7 35.1
01	213 24.0	215 22.5	37.5	221 53.4	56.1	250 04.5	09.5	45 21.4	35.1
02	228 26.5	230 25.5	36.7	236 54.1	55.4	265 06.5	09.4	60 24.0	35.2
03	243 28.9	245 28.5	35.8	251 54.7	54.6	280 08.5	09.2	75 26.6	35.2
04	258 31.4	260 31.4	35.0	266 55.4	53.8	295 10.6	09.1	90 29.2	35.3
05	273 33.8	275 34.4	34.2	281 56.1	53.1	310 12.6	09.0	105 31.8	35.3
06	288 36.3	290 37.3	N 5 33.4	296 56.7	S 4 52.3	325 14.6	S15 08.9	120 34.4	N 7 35.4
07	303 38.8	305 40.3	32.6	311 57.4	51.5	340 16.7	08.7	135 37.0	35.4
08	318 41.2	320 43.2	31.8	326 58.1	50.8	355 18.7	08.6	150 39.6	35.5
F 09	333 43.7	335 46.1	31.0	341 58.7	50.0	10 20.8	08.5	165 42.2	35.6
R 10	348 46.2	350 49.1	30.1	356 59.4	49.2	25 22.8	08.3	180 44.8	35.6
I 11	3 48.6	5 52.0	29.3	12 00.1	48.5	40 24.8	08.2	195 47.4	35.7
D 12	18 51.1	20 54.9	N 5 28.5	27 00.7	S 4 47.7	55 26.9	S15 08.1	210 50.0	N 7 35.7
A 13	33 53.6	35 57.8	27.7	42 01.4	46.9	70 28.9	07.9	225 52.6	35.8
Y 14	48 56.0	51 00.8	26.9	57 02.1	46.2	85 31.0	07.8	240 55.2	35.8
15	63 58.5	66 03.7	26.1	72 02.7	45.4	100 33.0	07.7	255 57.8	35.9
16	79 00.9	81 06.6	25.3	87 03.4	44.6	115 35.0	07.6	271 00.4	35.9
17	94 03.4	96 09.5	24.5	102 04.1	43.8	130 37.1	07.4	286 03.0	36.0
18	109 05.9	111 12.4	N 5 23.7	117 04.7	S 4 43.1	145 39.1	S15 07.3	301 05.6	N 7 36.0
19	124 08.3	126 15.3	22.9	132 05.4	42.3	160 41.2	07.2	316 08.2	36.1
20	139 10.8	141 18.2	22.1	147 06.1	41.5	175 43.2	07.0	331 10.8	36.1
21	154 13.3	156 21.1	21.3	162 06.7	40.8	190 45.3	06.9	346 13.4	36.2
22	169 15.7	171 24.0	20.5	177 07.4	40.0	205 47.3	06.8	1 16.0	36.2
23	184 18.2	186 26.8	19.7	192 08.1	39.2	220 49.3	06.6	16 18.6	36.3
11 00	199 20.7	201 29.7	N 5 19.0	207 08.8	S 4 38.5	235 51.4	S15 06.5	31 21.2	N 7 36.3
01	214 23.1	216 32.6	18.2	222 09.4	37.7	250 53.4	06.4	46 23.8	36.4
02	229 25.6	231 35.5	17.4	237 10.1	36.9	265 55.5	06.3	61 26.4	36.4
03	244 28.1	246 38.3	16.6	252 10.8	36.2	280 57.5	06.1	76 29.0	36.5
04	259 30.5	261 41.2	15.8	267 11.4	35.4	295 59.5	06.0	91 31.6	36.5
05	274 33.0	276 44.0	15.0	282 12.1	34.6	311 01.6	05.9	106 34.2	36.6
06	289 35.4	291 46.9	N 5 14.2	297 12.8	S 4 33.8	326 03.6	S15 05.7	121 36.8	N 7 36.6
S 07	304 37.9	306 49.7	13.5	312 13.4	33.1	341 05.7	05.6	136 39.4	36.7
08	319 40.4	321 52.6	12.7	327 14.1	32.3	356 07.7	05.5	151 42.0	36.8
A 09	334 42.8	336 55.4	11.9	342 14.8	31.5	11 09.8	05.4	166 44.6	36.8
T 10	349 45.3	351 58.3	11.1	357 15.4	30.8	26 11.8	05.2	181 47.2	36.9
U 11	4 47.8	7 01.1	10.4	12 16.1	30.0	41 13.9	05.1	196 49.7	36.9
R 12	19 50.2	22 03.9	N 5 09.6	27 16.8	S 4 29.2	56 15.9	S15 05.0	211 52.3	N 7 37.0
D 13	34 52.7	37 06.8	08.8	42 17.5	28.5	71 17.9	04.8	226 54.9	37.0
A 14	49 55.2	52 09.6	08.0	57 18.1	27.7	86 20.0	04.7	241 57.5	37.1
Y 15	64 57.6	67 12.4	07.3	72 18.8	26.9	101 22.0	04.6	257 00.1	37.1
16	80 00.1	82 15.2	06.5	87 19.5	26.1	116 24.1	04.5	272 02.7	37.2
17	95 02.6	97 18.0	05.7	102 20.1	25.4	131 26.1	04.3	287 05.3	37.2
18	110 05.0	112 20.8	N 5 05.0	117 20.8	S 4 24.6	146 28.2	S15 04.2	302 07.9	N 7 37.3
19	125 07.5	127 23.6	04.2	132 21.5	23.8	161 30.2	04.1	317 10.5	37.3
20	140 09.9	142 26.4	03.5	147 22.2	23.1	176 32.3	03.9	332 13.1	37.4
21	155 12.4	157 29.2	02.7	162 22.8	22.3	191 34.3	03.8	347 15.7	37.4
22	170 14.9	172 32.0	01.9	177 23.5	21.5	206 36.4	03.7	2 18.3	37.5
23	185 17.3	187 34.8	01.2	192 24.2	20.7	221 38.4	03.6	17 20.9	37.5
12 00	200 19.8	202 37.6	N 5 00.4	207 24.8	S 4 20.0	236 40.4	S15 03.4	32 23.5	N 7 37.6
01	215 22.3	217 40.4	4 59.7	222 25.5	19.2	251 42.5	03.3	47 26.1	37.6
02	230 24.7	232 43.1	58.9	237 26.2	18.4	266 44.5	03.2	62 28.7	37.7
03	245 27.2	247 45.9	58.2	252 26.9	17.7	281 46.6	03.0	77 31.3	37.7
04	260 29.7	262 48.7	57.4	267 27.5	16.9	296 48.6	02.9	92 33.9	37.8
05	275 32.1	277 51.4	56.7	282 28.2	16.1	311 50.7	02.8	107 36.5	37.8
06	290 34.6	292 54.2	N 4 55.9	297 28.9	S 4 15.4	326 52.7	S15 02.7	122 39.1	N 7 37.9
07	305 37.0	307 57.0	55.2	312 29.5	14.6	341 54.8	02.5	137 41.7	37.9
08	320 39.5	322 59.7	54.4	327 30.2	13.8	356 56.8	02.4	152 44.3	38.0
S 09	335 42.0	338 02.5	53.7	342 30.9	13.0	11 58.9	02.3	167 46.9	38.0
U 10	350 44.4	353 05.2	52.9	357 31.6	12.3	27 00.9	02.1	182 49.4	38.1
N 11	5 46.9	8 07.9	52.2	12 32.2	11.5	42 03.0	02.0	197 52.0	38.1
D 12	20 49.4	23 10.7	N 4 51.5	27 32.9	S 4 10.7	57 05.0	S15 01.9	212 54.6	N 7 38.2
A 13	35 51.8	38 13.4	50.7	42 33.6	10.0	72 07.1	01.8	227 57.2	38.2
Y 14	50 54.3	53 16.1	50.0	57 34.2	09.2	87 09.1	01.6	242 59.8	38.3
15	65 56.8	68 18.9	49.3	72 34.9	08.4	102 11.2	01.5	258 02.4	38.3
16	80 59.2	83 21.6	48.5	87 35.6	07.6	117 13.2	01.4	273 05.0	38.4
17	96 01.7	98 24.3	47.8	102 36.3	06.9	132 15.3	01.3	288 07.6	38.4
18	111 04.2	113 27.0	N 4 47.1	117 36.9	S 4 06.1	147 17.3	S15 01.1	303 10.2	N 7 38.5
19	126 06.6	128 29.7	46.4	132 37.6	05.3	162 19.4	01.0	318 12.8	38.5
20	141 09.1	143 32.4	45.6	147 38.3	04.6	177 21.4	00.9	333 15.4	38.6
21	156 11.5	158 35.1	44.9	162 39.0	03.8	192 23.5	00.7	348 18.0	38.6
22	171 14.0	173 37.8	44.2	177 39.6	03.0	207 25.5	00.6	3 20.6	38.7
23	186 16.5	188 40.5	43.5	192 40.3	02.2	222 27.6	00.5	18 23.2	38.7
Mer. Pass.	h m 10 40.9	v 2.8	d 0.8	v 0.7	d 0.8	v 2.0	d 0.1	v 2.6	d 0.1

STARS Name	SHA	Dec
	° ′	° ′
Acamar	315 21.0	S40 16.1
Achernar	335 29.4	S57 11.4
Acrux	173 12.4	S63 09.3
Adhara	255 15.1	S28 59.3
Aldebaran	290 53.2	N16 31.7
Alioth	166 22.6	N55 54.4
Alkaid	153 00.7	N49 15.8
Al Na'ir	27 47.8	S46 54.8
Alnilam	275 49.7	S 1 11.8
Alphard	217 59.0	S 8 42.1
Alphecca	126 13.3	N26 40.7
Alpheratz	357 47.2	N29 08.4
Altair	62 11.3	N 8 53.4
Ankaa	353 19.0	S42 15.3
Antares	112 29.9	S26 27.2
Arcturus	145 58.2	N19 07.8
Atria	107 34.3	S69 02.5
Avior	234 19.3	S59 32.7
Bellatrix	278 35.5	N 6 21.5
Betelgeuse	271 04.8	N 7 24.5
Canopus	263 57.7	S52 42.3
Capella	280 39.3	N46 00.6
Deneb	49 33.8	N45 18.5
Denebola	182 36.5	N14 31.0
Diphda	348 59.3	S17 56.1
Dubhe	193 54.7	N61 42.1
Elnath	278 16.8	N28 37.0
Eltanin	90 47.4	N51 28.9
Enif	33 50.4	N 9 54.9
Fomalhaut	15 27.6	S29 34.3
Gacrux	172 04.1	S57 10.1
Gienah	175 55.3	S17 35.8
Hadar	148 52.0	S60 25.2
Hamal	328 04.7	N23 30.4
Kaus Aust.	83 47.8	S34 22.8
Kochab	137 18.4	N74 06.8
Markab	13 41.7	N15 15.2
Menkar	314 18.6	N 4 07.5
Menkent	148 11.0	S36 25.1
Miaplacidus	221 40.3	S69 45.6
Mirfak	308 45.4	N49 53.8
Nunki	76 02.1	S26 17.1
Peacock	53 24.1	S56 42.1
Pollux	243 31.5	N28 00.3
Procyon	245 03.0	N 5 12.0
Rasalhague	96 09.2	N12 32.9
Regulus	207 46.6	N11 55.2
Rigel	281 15.2	S 8 11.5
Rigil Kent.	139 55.6	S60 52.5
Sabik	102 15.9	S15 44.3
Schedar	349 44.9	N56 35.3
Shaula	96 26.0	S37 06.6
Sirius	258 36.5	S16 43.9
Spica	158 34.3	S11 12.8
Suhail	222 54.7	S43 28.5
Vega	80 41.0	N38 47.2
Zuben'ubi	137 08.6	S16 05.0

	SHA	Mer. Pass.
	° ′	h m
Venus	2 09.1	10 32
Mars	7 48.1	10 11
Jupiter	36 30.7	8 15
Saturn	192 00.5	21 51

UT d h	SUN GHA ° ′	SUN Dec ° ′	MOON GHA ° ′	v ′	MOON Dec ° ′	d ′	HP ′
(10) 00	179 39.0	N 7 55.1	357 10.0	12.3	S14 12.5	12.7	56.8
01	194 39.2	56.0	11 41.3	12.3	14 25.2	12.6	56.8
02	209 39.3	56.9	26 12.6	12.3	14 37.8	12.5	56.7
03	224 39.5	. . 57.8	40 43.9	12.2	14 50.3	12.5	56.7
04	239 39.7	58.8	55 15.1	12.1	15 02.8	12.3	56.7
05	254 39.8	7 59.7	69 46.2	12.2	15 15.1	12.3	56.7
06	269 40.0	N 8 00.6	84 17.4	12.1	S15 27.4	12.2	56.6
07	284 40.2	01.5	98 48.5	12.1	15 39.6	12.1	56.6
08	299 40.3	02.4	113 19.6	12.0	15 51.7	12.0	56.6
09	314 40.5	. . 03.4	127 50.6	12.0	16 03.7	11.9	56.6
10	329 40.7	04.3	142 21.6	12.0	16 15.6	11.9	56.5
11	344 40.9	05.2	156 52.6	11.9	16 27.5	11.7	56.5
12	359 41.0	N 8 06.1	171 23.5	11.9	S16 39.2	11.7	56.5
13	14 41.2	07.1	185 54.4	11.8	16 50.9	11.6	56.5
14	29 41.4	08.0	200 25.2	11.9	17 02.5	11.4	56.4
15	44 41.5	. . 08.9	214 56.1	11.7	17 13.9	11.4	56.4
16	59 41.7	09.8	229 26.8	11.8	17 25.3	11.3	56.4
17	74 41.9	10.7	243 57.6	11.7	17 36.6	11.2	56.4
18	89 42.0	N 8 11.7	258 28.3	11.7	S17 47.8	11.1	56.3
19	104 42.2	12.6	272 59.0	11.6	17 58.9	11.0	56.3
20	119 42.4	13.5	287 29.6	11.6	18 09.9	10.9	56.3
21	134 42.5	. . 14.4	302 00.2	11.6	18 20.8	10.8	56.3
22	149 42.7	15.3	316 30.8	11.5	18 31.6	10.7	56.2
23	164 42.9	16.3	331 01.3	11.5	18 42.3	10.6	56.2
(11) 00	179 43.0	N 8 17.2	345 31.8	11.4	S18 52.9	10.5	56.2
01	194 43.2	18.1	0 02.2	11.4	19 03.4	10.4	56.2
02	209 43.4	19.0	14 32.6	11.4	19 13.8	10.3	56.2
03	224 43.5	. . 19.9	29 03.0	11.3	19 24.1	10.2	56.1
04	239 43.7	20.8	43 33.3	11.3	19 34.3	10.1	56.1
05	254 43.9	21.8	58 03.6	11.3	19 44.4	10.0	56.1
06	269 44.0	N 8 22.7	72 33.9	11.2	S19 54.4	9.9	56.1
07	284 44.2	23.6	87 04.1	11.1	20 04.3	9.8	56.0
08	299 44.3	24.5	101 34.2	11.2	20 14.1	9.6	56.0
09	314 44.5	. . 25.4	116 04.4	11.1	20 23.7	9.6	56.0
10	329 44.7	26.4	130 34.5	11.0	20 33.3	9.4	56.0
11	344 44.8	27.3	145 04.5	11.1	20 42.7	9.4	55.9
12	359 45.0	N 8 28.2	159 34.6	11.0	S20 52.1	9.2	55.9
13	14 45.2	29.1	174 04.6	10.9	21 01.3	9.1	55.9
14	29 45.3	30.0	188 34.5	10.9	21 10.4	9.0	55.9
15	44 45.5	. . 30.9	203 04.4	10.9	21 19.4	8.9	55.8
16	59 45.7	31.8	217 34.3	10.8	21 28.3	8.8	55.8
17	74 45.8	32.8	232 04.1	10.8	21 37.1	8.6	55.8
18	89 46.0	N 8 33.7	246 33.9	10.8	S21 45.7	8.6	55.8
19	104 46.2	34.6	261 03.7	10.7	21 54.3	8.4	55.7
20	119 46.3	35.5	275 33.4	10.7	22 02.7	8.3	55.7
21	134 46.5	. . 36.4	290 03.1	10.7	22 11.0	8.2	55.7
22	149 46.6	37.3	304 32.8	10.6	22 19.2	8.1	55.7
23	164 46.8	38.2	319 02.4	10.6	22 27.3	8.0	55.7
(12) 00	179 47.0	N 8 39.2	333 32.0	10.6	S22 35.3	7.8	55.6
01	194 47.1	40.1	348 01.6	10.5	22 43.1	7.7	55.6
02	209 47.3	41.0	2 31.1	10.5	22 50.8	7.6	55.6
03	224 47.5	. . 41.9	17 00.6	10.5	22 58.4	7.5	55.6
04	239 47.6	42.8	31 30.1	10.4	23 05.9	7.4	55.5
05	254 47.8	43.7	45 59.5	10.4	23 13.3	7.2	55.5
06	269 48.0	N 8 44.6	60 28.9	10.4	S23 20.5	7.1	55.5
07	284 48.1	45.5	74 58.3	10.3	23 27.6	7.0	55.5
08	299 48.3	46.5	89 27.6	10.3	23 34.6	6.9	55.5
09	314 48.4	. . 47.4	103 56.9	10.3	23 41.5	6.7	55.4
10	329 48.6	48.3	118 26.2	10.2	23 48.2	6.7	55.4
11	344 48.8	49.2	132 55.4	10.2	23 54.9	6.5	55.4
12	359 48.9	N 8 50.1	147 24.6	10.2	S24 01.4	6.3	55.4
13	14 49.1	51.0	161 53.8	10.2	24 07.7	6.3	55.3
14	29 49.2	51.9	176 23.0	10.1	24 14.0	6.1	55.3
15	44 49.4	. . 52.8	190 52.1	10.2	24 20.1	6.0	55.3
16	59 49.6	53.7	205 21.3	10.1	24 26.1	5.8	55.3
17	74 49.7	54.6	219 50.4	10.0	24 31.9	5.8	55.3
18	89 49.9	N 8 55.6	234 19.4	10.1	S24 37.7	5.6	55.2
19	104 50.0	56.5	248 48.5	10.0	24 43.3	5.5	55.2
20	119 50.2	57.4	263 17.5	10.0	24 48.8	5.3	55.2
21	134 50.4	. . 58.3	277 46.5	10.0	24 54.1	5.2	55.2
22	149 50.5	8 59.2	292 15.5	9.9	24 59.3	5.1	55.2
23	164 50.7	N 9 00.1	306 44.4	10.0	S25 04.4	5.0	55.1
	SD 16.0	*d* 0.9	SD	15.4	15.2		15.1

Lat.	Twilight Naut.	Twilight Civil	Sunrise	Moonrise 10	Moonrise 11	Moonrise 12	Moonrise 13
°	h m	h m	h m	h m	h m	h m	h m
N 72	////	02 32	04 01	■	■	■	■
N 70	////	03 00	04 15	23 39	■	■	■
68	01 32	03 21	04 26	22 45	■	■	■
66	02 10	03 37	04 35	22 12	24 29	00 29	■
64	02 36	03 50	04 43	21 48	23 41	25 46	01 46
62	02 56	04 01	04 49	21 30	23 10	24 48	00 48
60	03 12	04 10	04 55	21 14	22 47	24 15	00 15
N 58	03 25	04 18	05 00	21 01	22 29	23 51	25 00
56	03 36	04 25	05 05	20 50	22 14	23 31	24 38
54	03 45	04 31	05 09	20 41	22 00	23 15	24 20
52	03 53	04 37	05 12	20 32	21 49	23 01	24 05
50	04 01	04 42	05 15	20 24	21 39	22 49	23 51
45	04 16	04 52	05 22	20 08	21 18	22 24	23 24
N 40	04 27	05 00	05 28	19 54	21 00	22 04	23 02
35	04 37	05 07	05 33	19 43	20 46	21 47	22 44
30	04 45	05 13	05 38	19 33	20 34	21 33	22 29
20	04 57	05 23	05 45	19 16	20 13	21 09	22 03
N 10	05 06	05 30	05 52	19 02	19 54	20 48	21 41
0	05 13	05 37	05 58	18 48	19 37	20 28	21 20
S 10	05 18	05 42	06 04	18 35	19 21	20 09	20 59
20	05 22	05 48	06 10	18 20	19 03	19 48	20 37
30	05 25	05 53	06 17	18 04	18 42	19 24	20 11
35	05 26	05 55	06 21	17 55	18 30	19 10	19 56
40	05 26	05 58	06 25	17 44	18 17	18 54	19 38
45	05 27	06 01	06 30	17 31	18 00	18 35	19 17
S 50	05 26	06 04	06 36	17 16	17 41	18 11	18 51
52	05 26	06 05	06 39	17 09	17 31	18 00	18 38
54	05 25	06 06	06 42	17 01	17 21	17 47	18 23
56	05 25	06 08	06 45	16 53	17 09	17 32	18 06
58	05 24	06 09	06 49	16 43	16 55	17 14	17 45
S 60	05 23	06 11	06 53	16 32	16 40	16 53	17 18

Lat.	Sunset	Twilight Civil	Twilight Naut.	Moonset 10	Moonset 11	Moonset 12	Moonset 13
°	h m	h m	h m	h m	h m	h m	h m
N 72	20 04	21 37	////	02 37	■	■	■
N 70	19 50	21 07	////	03 08	02 03	■	■
68	19 38	20 45	22 39	03 31	02 59	■	■
66	19 29	20 28	21 57	03 49	03 32	02 59	■
64	19 21	20 15	21 30	04 04	03 57	03 48	03 29
62	19 14	20 03	21 09	04 16	04 17	04 19	04 27
60	19 09	19 54	20 53	04 27	04 33	04 42	05 01
N 58	19 03	19 46	20 40	04 36	04 46	05 01	05 25
56	18 59	19 38	20 28	04 44	04 58	05 17	05 45
54	18 55	19 32	20 19	04 52	05 09	05 31	06 01
52	18 51	19 27	20 10	04 58	05 18	05 43	06 16
50	18 48	19 21	20 03	05 04	05 26	05 53	06 28
45	18 40	19 11	19 47	05 17	05 44	06 15	06 54
N 40	18 34	19 02	19 36	05 28	05 58	06 33	07 14
35	18 29	18 55	19 26	05 37	06 11	06 48	07 31
30	18 25	18 49	19 18	05 45	06 21	07 02	07 46
20	18 17	18 39	19 06	05 59	06 40	07 24	08 11
N 10	18 10	18 32	18 56	06 11	06 56	07 44	08 33
0	18 04	18 25	18 49	06 23	07 12	08 02	08 53
S 10	17 58	18 19	18 44	06 35	07 27	08 20	09 13
20	17 52	18 14	18 40	06 47	07 43	08 40	09 35
30	17 45	18 09	18 37	07 02	08 02	09 02	10 00
35	17 41	18 06	18 36	07 10	08 14	09 16	10 15
40	17 36	18 04	18 35	07 19	08 26	09 31	10 33
45	17 31	18 01	18 35	07 31	08 41	09 50	10 53
S 50	17 25	17 58	18 35	07 44	09 00	10 13	11 20
52	17 22	17 56	18 35	07 51	09 09	10 24	11 32
54	17 19	17 55	18 36	07 58	09 19	10 37	11 47
56	17 16	17 53	18 36	08 06	09 30	10 51	12 04
58	17 12	17 51	18 37	08 15	09 43	11 08	12 25
S 60	17 08	17 50	18 38	08 25	09 58	11 29	12 51

Day	SUN Eqn. of Time 00^h	SUN Eqn. of Time 12^h	SUN Mer. Pass.	MOON Mer. Pass. Upper	MOON Mer. Pass. Lower	Age	Phase
d	m s	m s	h m	h m	h m	d %	
10	01 24	01 16	12 01	00 12	12 36	15 99	
11	01 08	01 00	12 01	01 00	13 25	16 96	○
12	00 52	00 45	12 01	01 50	14 15	17 91	

UT d	h	ARIES GHA	VENUS −4.6 GHA	VENUS Dec	MARS +1.2 GHA	MARS Dec	JUPITER −2.1 GHA	JUPITER Dec	SATURN +0.7 GHA	SATURN Dec
		° ′	° ′	° ′	° ′	° ′	° ′	° ′	° ′	° ′
13	00	201 18.9	203 43.2	N 4 42.7	207 41.0	S 4 01.5	237 29.6	S15 00.4	33 25.8	N 7 38.8
	01	216 21.4	218 45.9	42.0	222 41.6	4 00.7	252 31.7	00.2	48 28.3	38.8
	02	231 23.9	233 48.5	41.3	237 42.3	3 59.9	267 33.7	00.1	63 30.9	38.9
	03	246 26.3	248 51.2	. . 40.6	252 43.0	. . 59.1	282 35.8	15 00.0	78 33.5	. . 38.9
	04	261 28.8	263 53.9	39.9	267 43.7	58.4	297 37.8	14 59.9	93 36.1	39.0
	05	276 31.3	278 56.6	39.2	282 44.3	57.6	312 39.9	59.7	108 38.7	39.0
	06	291 33.7	293 59.2	N 4 38.5	297 45.0	S 3 56.8	327 41.9	S14 59.6	123 41.3	N 7 39.1
	07	306 36.2	309 01.9	37.8	312 45.7	56.1	342 44.0	59.5	138 43.9	39.1
	08	321 38.7	324 04.5	37.1	327 46.4	55.3	357 46.0	59.3	153 46.5	39.2
M	09	336 41.1	339 07.2	. . 36.3	342 47.0	. . 54.5	12 48.1	. . 59.2	168 49.1	. . 39.2
O	10	351 43.6	354 09.8	35.6	357 47.7	53.7	27 50.1	59.1	183 51.7	39.3
N	11	6 46.0	9 12.5	34.9	12 48.4	53.0	42 52.2	59.0	198 54.3	39.3
D	12	21 48.5	24 15.1	N 4 34.2	27 49.1	S 3 52.2	57 54.3	S14 58.8	213 56.9	N 7 39.4
A	13	36 51.0	39 17.8	33.5	42 49.7	51.4	72 56.3	58.7	228 59.4	39.4
Y	14	51 53.4	54 20.4	32.9	57 50.4	50.7	87 58.4	58.6	244 02.0	39.5
	15	66 55.9	69 23.0	. . 32.2	72 51.1	. . 49.9	103 00.4	. . 58.5	259 04.6	. . 39.5
	16	81 58.4	84 25.6	31.5	87 51.8	49.1	118 02.5	58.3	274 07.2	39.5
	17	97 00.8	99 28.3	30.8	102 52.4	48.3	133 04.5	58.2	289 09.8	39.6
	18	112 03.3	114 30.9	N 4 30.1	117 53.1	S 3 47.6	148 06.6	S14 58.1	304 12.4	N 7 39.6
	19	127 05.8	129 33.5	29.4	132 53.8	46.8	163 08.6	58.0	319 15.0	39.7
	20	142 08.2	144 36.1	28.7	147 54.5	46.0	178 10.7	57.8	334 17.6	39.7
	21	157 10.7	159 38.7	. . 28.0	162 55.1	. . 45.2	193 12.7	. . 57.7	349 20.2	. . 39.8
	22	172 13.2	174 41.3	27.3	177 55.8	44.5	208 14.8	57.6	4 22.8	39.8
	23	187 15.6	189 43.9	26.7	192 56.5	43.7	223 16.9	57.5	19 25.3	39.9
14	00	202 18.1	204 46.5	N 4 26.0	207 57.2	S 3 42.9	238 18.9	S14 57.3	34 27.9	N 7 39.9
	01	217 20.5	219 49.1	25.3	222 57.8	42.2	253 21.0	57.2	49 30.5	40.0
	02	232 23.0	234 51.7	24.6	237 58.5	41.4	268 23.0	57.1	64 33.1	40.0
	03	247 25.5	249 54.2	. . 24.0	252 59.2	. . 40.6	283 25.1	. . 57.0	79 35.7	. . 40.1
	04	262 27.9	264 56.8	23.3	267 59.9	39.8	298 27.1	56.8	94 38.3	40.1
	05	277 30.4	279 59.4	22.6	283 00.5	39.1	313 29.2	56.7	109 40.9	40.2
	06	292 32.9	295 02.0	N 4 21.9	298 01.2	S 3 38.3	328 31.3	S14 56.6	124 43.5	N 7 40.2
	07	307 35.3	310 04.5	21.3	313 01.9	37.5	343 33.3	56.5	139 46.1	40.3
T	08	322 37.8	325 07.1	20.6	328 02.6	36.7	358 35.4	56.3	154 48.6	40.3
U	09	337 40.3	340 09.6	. . 19.9	343 03.2	. . 36.0	13 37.4	. . 56.2	169 51.2	. . 40.4
E	10	352 42.7	355 12.2	19.3	358 03.9	35.2	28 39.5	56.1	184 53.8	40.4
S	11	7 45.2	10 14.7	18.6	13 04.6	34.4	43 41.5	56.0	199 56.4	40.5
D	12	22 47.6	25 17.3	N 4 18.0	28 05.3	S 3 33.6	58 43.6	S14 55.8	214 59.0	N 7 40.5
	13	37 50.1	40 19.8	17.3	43 05.9	32.9	73 45.7	55.7	230 01.6	40.5
A	14	52 52.6	55 22.4	16.6	58 06.6	32.1	88 47.7	55.6	245 04.2	40.6
Y	15	67 55.0	70 24.9	. . 16.0	73 07.3	. . 31.3	103 49.8	. . 55.5	260 06.8	. . 40.6
	16	82 57.5	85 27.4	15.3	88 08.0	30.6	118 51.8	55.3	275 09.4	40.7
	17	98 00.0	100 30.0	14.7	103 08.6	29.8	133 53.9	55.2	290 11.9	40.7
	18	113 02.4	115 32.5	N 4 14.0	118 09.3	S 3 29.0	148 56.0	S14 55.1	305 14.5	N 7 40.8
	19	128 04.9	130 35.0	13.4	133 10.0	28.2	163 58.0	55.0	320 17.1	40.8
	20	143 07.4	145 37.5	12.7	148 10.7	27.5	179 00.1	54.8	335 19.7	40.9
	21	158 09.8	160 40.0	. . 12.1	163 11.4	. . 26.7	194 02.1	. . 54.7	350 22.3	. . 40.9
	22	173 12.3	175 42.5	11.4	178 12.0	25.9	209 04.2	54.6	5 24.9	41.0
	23	188 14.8	190 45.0	10.8	193 12.7	25.1	224 06.3	54.5	20 27.5	41.0
15	00	203 17.2	205 47.5	N 4 10.2	208 13.4	S 3 24.4	239 08.3	S14 54.3	35 30.0	N 7 41.1
	01	218 19.7	220 50.0	09.5	223 14.1	23.6	254 10.4	54.2	50 32.6	41.1
	02	233 22.1	235 52.5	08.9	238 14.7	22.8	269 12.4	54.1	65 35.2	41.2
	03	248 24.6	250 55.0	. . 08.3	253 15.4	. . 22.0	284 14.5	. . 54.0	80 37.8	. . 41.2
	04	263 27.1	265 57.5	07.6	268 16.1	21.3	299 16.6	53.8	95 40.4	41.2
	05	278 29.5	281 00.0	07.0	283 16.8	20.5	314 18.6	53.7	110 43.0	41.3
	06	293 32.0	296 02.4	N 4 06.4	298 17.4	S 3 19.7	329 20.7	S14 53.6	125 45.6	N 7 41.3
W	07	308 34.5	311 04.9	05.7	313 18.1	18.9	344 22.8	53.5	140 48.1	41.4
E	08	323 36.9	326 07.4	05.1	328 18.8	18.2	359 24.8	53.3	155 50.7	41.4
D	09	338 39.4	341 09.8	. . 04.5	343 19.5	. . 17.4	14 26.9	. . 53.2	170 53.3	. . 41.5
N	10	353 41.9	356 12.3	03.9	358 20.2	16.6	29 28.9	53.1	185 55.9	41.5
E	11	8 44.3	11 14.7	03.2	13 20.8	15.8	44 31.0	53.0	200 58.5	41.6
S	12	23 46.8	26 17.2	N 4 02.6	28 21.5	S 3 15.1	59 33.1	S14 52.8	216 01.1	N 7 41.6
D	13	38 49.3	41 19.6	02.0	43 22.2	14.3	74 35.1	52.7	231 03.7	41.7
A	14	53 51.7	56 22.1	01.4	58 22.9	13.5	89 37.2	52.6	246 06.2	41.7
Y	15	68 54.2	71 24.5	. . 00.8	73 23.5	. . 12.7	104 39.3	. . 52.5	261 08.8	. . 41.7
	16	83 56.6	86 27.0	4 00.2	88 24.2	12.0	119 41.3	52.3	276 11.4	41.8
	17	98 59.1	101 29.4	3 59.6	103 24.9	11.2	134 43.4	52.2	291 14.0	41.8
	18	114 01.6	116 31.8	N 3 59.0	118 25.6	S 3 10.4	149 45.4	S14 52.1	306 16.6	N 7 41.9
	19	129 04.0	131 34.2	58.4	133 26.3	09.7	164 47.5	52.0	321 19.2	41.9
	20	144 06.5	146 36.7	57.7	148 26.9	08.9	179 49.6	51.9	336 21.7	42.0
	21	159 09.0	161 39.1	. . 57.1	163 27.6	. . 08.1	194 51.6	. . 51.7	351 24.3	. . 42.0
	22	174 11.4	176 41.5	56.5	178 28.3	07.3	209 53.7	51.6	6 26.9	42.1
	23	189 13.9	191 43.9	55.9	193 29.0	06.6	224 55.8	51.5	21 29.5	42.1
Mer. Pass.		h m 10 29.1	v 2.5	d 0.7	v 0.7	d 0.8	v 2.1	d 0.1	v 2.6	d 0.0

STARS

Name	SHA	Dec
	° ′	° ′
Acamar	315 21.0	S40 16.
Achernar	335 29.4	S57 11.
Acrux	173 12.4	S63 09.
Adhara	255 15.1	S28 59.
Aldebaran	290 53.2	N16 31.
Alioth	166 22.6	N55 54.
Alkaid	153 00.7	N49 15.
Al Na'ir	27 47.7	S46 54.
Alnilam	275 49.7	S 1 11.
Alphard	217 59.0	S 8 42.
Alphecca	126 13.3	N26 40.
Alpheratz	357 47.1	N29 08.
Altair	62 11.3	N 8 53.
Ankaa	353 19.0	S42 15.
Antares	112 29.8	S26 27.
Arcturus	145 58.2	N19 07.
Atria	107 34.2	S69 02.6
Avior	234 19.3	S59 32.
Bellatrix	278 35.5	N 6 21.5
Betelgeuse	271 04.8	N 7 24.5
Canopus	263 57.7	S52 42.3
Capella	280 39.4	N46 00.6
Deneb	49 33.8	N45 18.5
Denebola	182 36.5	N14 31.0
Diphda	348 59.3	S17 56.1
Dubhe	193 54.7	N61 42.1
Elnath	278 16.8	N28 37.0
Eltanin	90 47.4	N51 28.9
Enif	33 50.4	N 9 54.9
Fomalhaut	15 27.6	S29 34.3
Gacrux	172 04.1	S57 10.2
Gienah	175 55.2	S17 35.9
Hadar	148 51.9	S60 25.2
Hamal	328 04.7	N23 30.4
Kaus Aust.	83 47.8	S34 22.8
Kochab	137 18.4	N74 06.8
Markab	13 41.7	N15 15.2
Menkar	314 18.6	N 4 07.6
Menkent	148 11.0	S36 25.1
Miaplacidus	221 40.3	S69 45.7
Mirfak	308 45.4	N49 53.8
Nunki	76 02.1	S26 17.1
Peacock	53 24.1	S56 42.1
Pollux	243 31.5	N28 00.3
Procyon	245 03.0	N 5 12.0
Rasalhague	96 09.2	N12 33.0
Regulus	207 46.6	N11 55.2
Rigel	281 15.2	S 8 11.5
Rigil Kent.	139 55.6	S60 52.5
Sabik	102 15.9	S15 44.3
Schedar	349 44.9	N56 35.2
Shaula	96 25.9	S37 06.6
Sirius	258 36.6	S16 43.9
Spica	158 34.2	S11 12.8
Suhail	222 54.7	S43 28.5
Vega	80 41.0	N38 47.2
Zuben'ubi	137 08.6	S16 05.0

	SHA	Mer. Pass
	° ′	h m
Venus	2 28.4	10 19
Mars	5 39.1	10 08
Jupiter	36 00.8	8 06
Saturn	192 09.9	21 38

UT	SUN GHA	SUN Dec	MOON GHA	v	MOON Dec	d	HP
d h	° ′	° ′	° ′	′	° ′	′	′
13 00	179 50.8	N 9 01.0	321 13.4	9.9	S25 09.4	4.8	55.1
01	194 51.0	01.9	335 42.3	9.9	25 14.2	4.7	55.1
02	209 51.2	02.8	350 11.2	9.9	25 18.9	4.6	55.1
03	224 51.3	. . 03.7	4 40.1	9.9	25 23.5	4.4	55.1
04	239 51.5	04.6	19 09.0	9.8	25 27.9	4.4	55.0
05	254 51.6	05.5	33 37.8	9.9	25 32.3	4.1	55.0
06	269 51.8	N 9 06.4	48 06.7	9.8	S25 36.4	4.1	55.0
07	284 52.0	07.3	62 35.5	9.8	25 40.5	3.9	55.0
08	299 52.1	08.2	77 04.3	9.8	25 44.4	3.8	55.0
09	314 52.3	. . 09.1	91 33.1	9.8	25 48.2	3.6	54.9
10	329 52.4	10.1	106 01.9	9.8	25 51.8	3.5	54.9
11	344 52.6	11.0	120 30.7	9.8	25 55.3	3.4	54.9
12	359 52.8	N 9 11.9	134 59.5	9.8	S25 58.7	3.3	54.9
13	14 52.9	12.8	149 28.3	9.8	26 02.0	3.1	54.9
14	29 53.1	13.7	163 57.1	9.7	26 05.1	3.0	54.9
15	44 53.2	. . 14.6	178 25.8	9.8	26 08.1	2.8	54.8
16	59 53.4	15.5	192 54.6	9.7	26 10.9	2.8	54.8
17	74 53.5	16.4	207 23.3	9.8	26 13.7	2.6	54.8
18	89 53.7	N 9 17.3	221 52.1	9.7	S26 16.3	2.4	54.8
19	104 53.9	18.2	236 20.8	9.8	26 18.7	2.3	54.8
20	119 54.0	19.1	250 49.6	9.7	26 21.0	2.2	54.8
21	134 54.2	. . 20.0	265 18.3	9.8	26 23.2	2.1	54.7
22	149 54.3	20.9	279 47.1	9.7	26 25.3	1.9	54.7
23	164 54.5	21.8	294 15.8	9.8	26 27.2	1.8	54.7
14 00	179 54.6	N 9 22.7	308 44.6	9.8	S26 29.0	1.6	54.7
01	194 54.8	23.6	323 13.4	9.7	26 30.6	1.6	54.7
02	209 55.0	24.5	337 42.1	9.8	26 32.2	1.3	54.7
03	224 55.1	. . 25.4	352 10.9	9.8	26 33.5	1.3	54.7
04	239 55.3	26.3	6 39.7	9.8	26 34.8	1.1	54.6
05	254 55.4	27.2	21 08.5	9.8	26 35.9	1.0	54.6
06	269 55.6	N 9 28.1	35 37.3	9.8	S26 36.9	0.9	54.6
07	284 55.7	29.0	50 06.1	9.8	26 37.8	0.7	54.6
08	299 55.9	29.9	64 34.9	9.9	26 38.5	0.6	54.6
09	314 56.0	. . 30.8	79 03.8	9.8	26 39.1	0.4	54.6
10	329 56.2	31.7	93 32.6	9.9	26 39.5	0.4	54.6
11	344 56.3	32.6	108 01.5	9.9	26 39.9	0.2	54.5
12	359 56.5	N 9 33.5	122 30.4	9.9	S26 40.1	0.0	54.5
13	14 56.7	34.4	136 59.3	9.9	26 40.1	0.0	54.5
14	29 56.8	35.3	151 28.2	9.9	26 40.1	0.2	54.5
15	44 57.0	. . 36.2	165 57.1	10.0	26 39.9	0.4	54.5
16	59 57.1	37.1	180 26.1	9.9	26 39.5	0.4	54.5
17	74 57.3	38.0	194 55.0	10.0	26 39.1	0.6	54.5
18	89 57.4	N 9 38.9	209 24.0	10.0	S26 38.5	0.7	54.5
19	104 57.6	39.8	223 53.0	10.1	26 37.8	0.9	54.5
20	119 57.7	40.6	238 22.1	10.0	26 36.9	1.0	54.4
21	134 57.9	. . 41.5	252 51.1	10.1	26 35.9	1.1	54.4
22	149 58.0	42.4	267 20.2	10.1	26 34.8	1.2	54.4
23	164 58.2	43.3	281 49.3	10.2	26 33.6	1.4	54.4
15 00	179 58.3	N 9 44.2	296 18.5	10.2	S26 32.2	1.5	54.4
01	194 58.5	45.1	310 47.7	10.2	26 30.7	1.6	54.4
02	209 58.6	46.0	325 16.9	10.2	26 29.1	1.8	54.4
03	224 58.8	. . 46.9	339 46.1	10.2	26 27.3	1.9	54.4
04	239 59.0	47.8	354 15.3	10.3	26 25.4	2.0	54.4
05	254 59.1	48.7	8 44.6	10.4	26 23.4	2.1	54.4
06	269 59.3	N 9 49.6	23 14.0	10.3	S26 21.3	2.3	54.4
07	284 59.4	50.5	37 43.3	10.4	26 19.0	2.4	54.3
08	299 59.6	51.4	52 12.7	10.4	26 16.6	2.5	54.3
09	314 59.7	. . 52.3	66 42.1	10.5	26 14.1	2.6	54.3
10	329 59.9	53.2	81 11.6	10.5	26 11.5	2.8	54.3
11	345 00.0	54.0	95 41.1	10.5	26 08.7	2.9	54.3
12	0 00.2	N 9 54.9	110 10.6	10.6	S26 05.8	3.0	54.3
13	15 00.3	55.8	124 40.2	10.6	26 02.8	3.1	54.3
14	30 00.5	56.7	139 09.8	10.6	25 59.7	3.3	54.3
15	45 00.6	. . 57.6	153 39.4	10.7	25 56.4	3.4	54.3
16	60 00.8	58.5	168 09.1	10.7	25 53.0	3.5	54.3
17	75 00.9	9 59.4	182 38.8	10.8	25 49.5	3.6	54.3
18	90 01.1	N10 00.3	197 08.6	10.8	S25 45.9	3.8	54.3
19	105 01.2	01.2	211 38.4	10.8	25 42.1	3.8	54.3
20	120 01.4	02.1	226 08.2	10.9	25 38.3	4.0	54.3
21	135 01.5	. . 02.9	240 38.1	10.9	25 34.3	4.1	54.3
22	150 01.7	03.8	255 08.0	11.0	25 30.2	4.2	54.3
23	165 01.8	04.7	269 38.0	11.0	S25 26.0	4.4	54.3
	SD 16.0	d 0.9	SD 15.0		14.9		14.8

Lat.	Twilight Naut.	Twilight Civil	Sunrise	Moonrise 13	Moonrise 14	Moonrise 15	Moonrise 16
°	h m	h m	h m	h m	h m	h m	h m
N 72	////	02 06	03 44	■	■	■	■
N 70	////	02 41	04 00	■	■	■	■
68	00 55	03 05	04 13	■	■	■	■
66	01 49	03 23	04 24	■	■	■	■
64	02 20	03 38	04 32	01 46	■	■	04 08
62	02 42	03 50	04 40	00 48	02 12	03 01	03 19
60	03 00	04 00	04 46	00 15	01 29	02 20	02 48
N 58	03 14	04 09	04 52	25 00	01 00	01 51	02 25
56	03 26	04 17	04 57	24 38	00 38	01 30	02 06
54	03 36	04 24	05 01	24 20	00 20	01 12	01 50
52	03 45	04 30	05 05	24 05	00 05	00 56	01 36
50	03 53	04 35	05 09	23 51	24 43	00 43	01 24
45	04 10	04 47	05 17	23 24	24 16	00 16	00 59
N 40	04 22	04 56	05 24	23 02	23 54	24 39	00 39
35	04 33	05 03	05 29	22 44	23 36	24 22	00 22
30	04 41	05 10	05 34	22 29	23 21	24 08	00 08
20	04 54	05 20	05 43	22 03	22 55	23 43	24 27
N 10	05 04	05 29	05 50	21 41	22 32	23 22	24 08
0	05 12	05 36	05 57	21 20	22 11	23 02	23 51
S 10	05 18	05 42	06 04	20 59	21 50	22 42	23 33
20	05 23	05 48	06 11	20 37	21 28	22 21	23 14
30	05 26	05 54	06 18	20 11	21 02	21 56	22 52
35	05 28	05 57	06 23	19 56	20 47	21 41	22 38
40	05 29	06 01	06 28	19 38	20 29	21 24	22 23
45	05 30	06 04	06 34	19 17	20 07	21 04	22 06
S 50	05 31	06 08	06 41	18 51	19 40	20 38	21 43
52	05 31	06 10	06 44	18 38	19 27	20 26	21 32
54	05 31	06 12	06 48	18 23	19 11	20 11	21 20
56	05 30	06 14	06 51	18 06	18 53	19 55	21 06
58	05 30	06 16	06 56	17 45	18 31	19 35	20 50
S 60	05 30	06 18	07 01	17 18	18 03	19 09	20 30

Lat.	Sunset	Twilight Civil	Twilight Naut.	Moonset 13	Moonset 14	Moonset 15	Moonset 16
°	h m	h m	h m	h m	h m	h m	h m
N 72	20 20	22 02	////	■	■	■	■
N 70	20 03	21 25	////	■	■	■	■
68	19 50	21 00	23 24	■	■	■	■
66	19 39	20 41	22 18	■	■	■	■
64	19 30	20 25	21 45	03 29	■	■	06 27
62	19 23	20 13	21 22	04 27	04 51	05 49	07 15
60	19 16	20 02	21 04	05 01	05 34	06 30	07 46
N 58	19 10	19 53	20 49	05 25	06 03	06 58	08 09
56	19 05	19 45	20 36	05 45	06 25	07 20	08 28
54	19 00	19 38	20 26	06 01	06 43	07 38	08 43
52	18 56	19 32	20 17	06 16	06 59	07 53	08 57
50	18 52	19 27	20 09	06 28	07 12	08 06	09 09
45	18 44	19 15	19 52	06 54	07 40	08 33	09 33
N 40	18 37	19 06	19 39	07 14	08 02	08 55	09 52
35	18 32	18 58	19 29	07 31	08 20	09 12	10 09
30	18 27	18 51	19 20	07 46	08 35	09 28	10 23
20	18 18	18 40	19 07	08 11	09 01	09 53	10 46
N 10	18 10	18 32	18 57	08 33	09 24	10 15	11 07
0	18 04	18 25	18 49	08 53	09 45	10 36	11 26
S 10	17 57	18 18	18 43	09 13	10 06	10 57	11 45
20	17 50	18 12	18 38	09 35	10 28	11 19	12 05
30	17 42	18 06	18 34	10 00	10 55	11 44	12 28
35	17 37	18 03	18 32	10 15	11 10	11 59	12 42
40	17 32	17 59	18 31	10 33	11 28	12 16	12 57
45	17 26	17 56	18 30	10 53	11 50	12 37	13 16
S 50	17 19	17 52	18 29	11 20	12 17	13 03	13 39
52	17 16	17 50	18 29	11 32	12 30	13 16	13 50
54	17 12	17 48	18 29	11 47	12 45	13 30	14 03
56	17 08	17 46	18 29	12 04	13 04	13 47	14 17
58	17 04	17 44	18 29	12 25	13 26	14 07	14 34
S 60	16 59	17 41	18 29	12 51	13 54	14 33	14 54

Day	SUN Eqn. of Time 00h	SUN Eqn. of Time 12h	SUN Mer. Pass.	MOON Mer. Pass. Upper	MOON Mer. Pass. Lower	Age	Phase
d	m s	m s	h m	h m	h m	d	%
13	00 37	00 29	12 00	02 41	15 06	18	85
14	00 22	00 14	12 00	03 32	15 58	19	78
15	00 07	00 00	12 00	04 24	16 49	20	69

	UT d h	ARIES GHA ° ′	VENUS −4.7 GHA ° ′	VENUS Dec ° ′	MARS +1.2 GHA ° ′	MARS Dec ° ′	JUPITER −2.2 GHA ° ′	JUPITER Dec ° ′	SATURN +0.7 GHA ° ′	SATURN Dec ° ′
	16 00	204 16.4	206 46.3	N 3 55.3	208 29.7	S 3 05.8	239 57.8	S14 51.4	36 32.1	N 7 42.2
	01	219 18.8	221 48.7	54.7	223 30.3	05.0	254 59.9	51.2	51 34.7	42.2
	02	234 21.3	236 51.1	54.2	238 31.0	04.2	270 02.0	51.1	66 37.2	42.2
	03	249 23.8	251 53.5	53.6	253 31.7	03.5	285 04.0	51.0	81 39.8	42.3
	04	264 26.2	266 55.9	53.0	268 32.4	02.7	300 06.1	50.9	96 42.4	42.3
	05	279 28.7	281 58.3	52.4	283 33.0	01.9	315 08.2	50.7	111 45.0	42.4
	06	294 31.1	297 00.7	N 3 51.8	298 33.7	S 3 01.1	330 10.2	S14 50.6	126 47.6	N 7 42.4
	07	309 33.6	312 03.0	51.2	313 34.4	3 00.4	345 12.3	50.5	141 50.2	42.5
T	08	324 36.1	327 05.4	50.6	328 35.1	2 59.6	0 14.4	50.4	156 52.7	42.5
H	09	339 38.5	342 07.8	50.0	343 35.8	58.8	15 16.4	50.3	171 55.3	42.6
U	10	354 41.0	357 10.1	49.5	358 36.4	58.0	30 18.5	50.1	186 57.9	42.6
R	11	9 43.5	12 12.5	48.9	13 37.1	57.3	45 20.6	50.0	202 00.5	42.6
S	12	24 45.9	27 14.9	N 3 48.3	28 37.8	S 2 56.5	60 22.6	S14 49.9	217 03.1	N 7 42.7
D	13	39 48.4	42 17.2	47.7	43 38.5	55.7	75 24.7	49.8	232 05.7	42.7
A	14	54 50.9	57 19.6	47.2	58 39.2	54.9	90 26.8	49.6	247 08.2	42.8
Y	15	69 53.3	72 21.9	46.6	73 39.8	54.2	105 28.8	49.5	262 10.8	42.8
	16	84 55.8	87 24.3	46.0	88 40.5	53.4	120 30.9	49.4	277 13.4	42.9
	17	99 58.2	102 26.6	45.5	103 41.2	52.6	135 33.0	49.3	292 16.0	42.9
	18	115 00.7	117 28.9	N 3 44.9	118 41.9	S 2 51.8	150 35.1	S14 49.2	307 18.6	N 7 43.0
	19	130 03.2	132 31.3	44.3	133 42.6	51.1	165 37.1	49.0	322 21.1	43.0
	20	145 05.6	147 33.6	43.8	148 43.2	50.3	180 39.2	48.9	337 23.7	43.0
	21	160 08.1	162 35.9	43.2	163 43.9	49.5	195 41.3	48.8	352 26.3	43.1
	22	175 10.6	177 38.2	42.6	178 44.6	48.7	210 43.3	48.7	7 28.9	43.1
	23	190 13.0	192 40.6	42.1	193 45.3	48.0	225 45.4	48.5	22 31.5	43.2
	17 00	205 15.5	207 42.9	N 3 41.5	208 46.0	S 2 47.2	240 47.5	S14 48.4	37 34.0	N 7 43.2
	01	220 18.0	222 45.2	41.0	223 46.6	46.4	255 49.5	48.3	52 36.6	43.3
	02	235 20.4	237 47.5	40.4	238 47.3	45.6	270 51.6	48.2	67 39.2	43.3
	03	250 22.9	252 49.8	39.9	253 48.0	44.9	285 53.7	48.1	82 41.8	43.3
	04	265 25.4	267 52.1	39.3	268 48.7	44.1	300 55.8	47.9	97 44.4	43.4
	05	280 27.8	282 54.4	38.8	283 49.4	43.3	315 57.8	47.8	112 46.9	43.4
	06	295 30.3	297 56.7	N 3 38.2	298 50.0	S 2 42.5	330 59.9	S14 47.7	127 49.5	N 7 43.5
	07	310 32.7	312 59.0	37.7	313 50.7	41.8	346 02.0	47.6	142 52.1	43.5
	08	325 35.2	328 01.2	37.2	328 51.4	41.0	1 04.0	47.5	157 54.7	43.6
F	09	340 37.7	343 03.5	36.6	343 52.1	40.2	16 06.1	47.3	172 57.3	43.6
R	10	355 40.1	358 05.8	36.1	358 52.8	39.4	31 08.2	47.2	187 59.8	43.6
I	11	10 42.6	13 08.1	35.5	13 53.5	38.7	46 10.3	47.1	203 02.4	43.7
D	12	25 45.1	28 10.3	N 3 35.0	28 54.1	S 2 37.9	61 12.3	S14 47.0	218 05.0	N 7 43.7
A	13	40 47.5	43 12.6	34.5	43 54.8	37.1	76 14.4	46.8	233 07.6	43.8
Y	14	55 50.0	58 14.9	34.0	58 55.5	36.3	91 16.5	46.7	248 10.1	43.8
	15	70 52.5	73 17.1	33.4	73 56.2	35.5	106 18.6	46.6	263 12.7	43.9
	16	85 54.9	88 19.4	32.9	88 56.9	34.8	121 20.6	46.5	278 15.3	43.9
	17	100 57.4	103 21.6	32.4	103 57.5	34.0	136 22.7	46.4	293 17.9	43.9
	18	115 59.9	118 23.9	N 3 31.9	118 58.2	S 2 33.2	151 24.8	S14 46.2	308 20.5	N 7 44.0
	19	131 02.3	133 26.1	31.3	133 58.9	32.4	166 26.9	46.1	323 23.0	44.0
	20	146 04.8	148 28.3	30.8	148 59.6	31.7	181 28.9	46.0	338 25.6	44.1
	21	161 07.2	163 30.6	30.3	164 00.3	30.9	196 31.0	45.9	353 28.2	44.1
	22	176 09.7	178 32.8	29.8	179 01.0	30.1	211 33.1	45.8	8 30.8	44.2
	23	191 12.2	193 35.0	29.3	194 01.6	29.3	226 35.2	45.6	23 33.3	44.2
	18 00	206 14.6	208 37.3	N 3 28.8	209 02.3	S 2 28.6	241 37.2	S14 45.5	38 35.9	N 7 44.2
	01	221 17.1	223 39.5	28.2	224 03.0	27.8	256 39.3	45.4	53 38.5	44.3
	02	236 19.6	238 41.7	27.7	239 03.7	27.0	271 41.4	45.3	68 41.1	44.3
	03	251 22.0	253 43.9	27.2	254 04.4	26.2	286 43.5	45.2	83 43.7	44.4
	04	266 24.5	268 46.1	26.7	269 05.0	25.5	301 45.5	45.0	98 46.2	44.4
	05	281 27.0	283 48.3	26.2	284 05.7	24.7	316 47.6	44.9	113 48.8	44.4
	06	296 29.4	298 50.5	N 3 25.7	299 06.4	S 2 23.9	331 49.7	S14 44.8	128 51.4	N 7 44.5
	07	311 31.9	313 52.7	25.2	314 07.1	23.1	346 51.8	44.7	143 54.0	44.5
S	08	326 34.3	328 54.9	24.7	329 07.8	22.4	1 53.8	44.6	158 56.5	44.6
A	09	341 36.8	343 57.1	24.2	344 08.5	21.6	16 55.9	44.4	173 59.1	44.6
T	10	356 39.3	358 59.3	23.7	359 09.1	20.8	31 58.0	44.3	189 01.7	44.7
U	11	11 41.7	14 01.5	23.2	14 09.8	20.0	47 00.1	44.2	204 04.3	44.7
R	12	26 44.2	29 03.7	N 3 22.8	29 10.5	S 2 19.3	62 02.1	S14 44.1	219 06.8	N 7 44.7
D	13	41 46.7	44 05.8	22.3	44 11.2	18.5	77 04.2	44.0	234 09.4	44.8
A	14	56 49.1	59 08.0	21.8	59 11.9	17.7	92 06.3	43.8	249 12.0	44.8
Y	15	71 51.6	74 10.2	21.3	74 12.6	16.9	107 08.4	43.7	264 14.6	44.9
	16	86 54.1	89 12.3	20.8	89 13.2	16.2	122 10.5	43.6	279 17.1	44.9
	17	101 56.5	104 14.5	20.3	104 13.9	15.4	137 12.5	43.5	294 19.7	44.9
	18	116 59.0	119 16.7	N 3 19.9	119 14.6	S 2 14.6	152 14.6	S14 43.4	309 22.3	N 7 45.0
	19	132 01.5	134 18.8	19.4	134 15.3	13.8	167 16.7	43.2	324 24.9	45.0
	20	147 03.9	149 21.0	18.9	149 16.0	13.0	182 18.8	43.1	339 27.4	45.1
	21	162 06.4	164 23.1	18.4	164 16.7	12.3	197 20.9	43.0	354 30.0	45.1
	22	177 08.8	179 25.2	18.0	179 17.3	11.5	212 22.9	42.9	9 32.6	45.1
	23	192 11.3	194 27.4	17.5	194 18.0	10.7	227 25.0	42.8	24 35.2	45.2
	Mer. Pass. (h m)	10 17.3	v 2.3	d 0.5	v 0.7	d 0.8	v 2.1	d 0.1	v 2.6	d 0.0

STARS

Name	SHA ° ′	Dec ° ′
Acamar	315 21.0	S40 16.1
Achernar	335 29.4	S57 11.3
Acrux	173 12.4	S63 09.3
Adhara	255 15.1	S28 59.3
Aldebaran	290 53.2	N16 31.7
Alioth	166 22.6	N55 54.5
Alkaid	153 00.7	N49 15.8
Al Na'ir	27 47.7	S46 54.8
Alnilam	275 49.7	S 1 11.8
Alphard	217 59.0	S 8 42.1
Alphecca	126 13.3	N26 40.7
Alpheratz	357 47.1	N29 08.4
Altair	62 11.3	N 8 53.4
Ankaa	353 19.0	S42 15.2
Antares	112 29.8	S26 27.3
Arcturus	145 58.2	N19 07.8
Atria	107 34.2	S69 02.6
Avior	234 19.4	S59 32.7
Bellatrix	278 35.5	N 6 21.5
Betelgeuse	271 04.8	N 7 24.5
Canopus	263 57.8	S52 42.3
Capella	280 39.4	N46 00.6
Deneb	49 33.8	N45 18.5
Denebola	182 36.5	N14 31.0
Diphda	348 59.3	S17 56.1
Dubhe	193 54.7	N61 42.1
Elnath	278 16.8	N28 37.0
Eltanin	90 47.3	N51 28.9
Enif	33 50.3	N 9 54.9
Fomalhaut	15 27.6	S29 34.3
Gacrux	172 04.1	S57 10.2
Gienah	175 55.2	S17 35.9
Hadar	148 51.9	S60 25.2
Hamal	328 04.7	N23 30.4
Kaus Aust.	83 47.8	S34 22.8
Kochab	137 18.3	N74 06.8
Markab	13 41.7	N15 15.2
Menkar	314 18.6	N 4 07.6
Menkent	148 10.9	S36 25.1
Miaplacidus	221 40.4	S69 45.7
Mirfak	308 45.4	N49 53.8
Nunki	76 02.0	S26 17.1
Peacock	53 24.0	S56 42.1
Pollux	243 31.5	N28 00.3
Procyon	245 03.0	N 5 12.0
Rasalhague	96 09.1	N12 33.0
Regulus	207 46.6	N11 55.2
Rigel	281 15.2	S 8 11.5
Rigil Kent.	139 55.6	S60 52.5
Sabik	102 15.9	S15 44.3
Schedar	349 44.9	N56 35.2
Shaula	96 25.9	S37 06.6
Sirius	258 36.6	S16 43.9
Spica	158 34.2	S11 12.8
Suhail	222 54.7	S43 28.5
Vega	80 40.9	N38 47.2
Zuben'ubi	137 08.6	S16 05.0

	SHA ° ′	Mer. Pass. h m
Venus	2 27.4	10 08
Mars	3 30.5	10 04
Jupiter	35 32.0	7 56
Saturn	192 18.5	21 26

UT	SUN GHA	SUN Dec	MOON GHA	v	MOON Dec	d	HP
d h	° ′	° ′	° ′	′	° ′	′	′
00	180 02.0	N10 05.6	284 08.0	11.1	S25 21.6	4.5	54.3
01	195 02.1	06.5	298 38.1	11.1	25 17.1	4.5	54.3
02	210 02.3	07.4	313 08.2	11.1	25 12.6	4.7	54.3
03	225 02.4	. . 08.3	327 38.3	11.2	25 07.9	4.8	54.3
04	240 02.5	09.2	342 08.5	11.3	25 03.1	5.0	54.2
05	255 02.7	10.0	356 38.8	11.3	24 58.1	5.0	54.2
06	270 02.8	N10 10.9	11 09.1	11.3	S24 53.1	5.2	54.2
07	285 03.0	11.8	25 39.4	11.4	24 47.9	5.2	54.2
08	300 03.1	12.7	40 09.8	11.4	24 42.7	5.4	54.2
09	315 03.3	. . 13.6	54 40.2	11.5	24 37.3	5.5	54.2
10	330 03.4	14.5	69 10.7	11.5	24 31.8	5.6	54.2
11	345 03.6	15.3	83 41.2	11.6	24 26.2	5.8	54.2
12	0 03.7	N10 16.2	98 11.8	11.6	S24 20.4	5.8	54.2
13	15 03.9	17.1	112 42.4	11.7	24 14.6	5.9	54.2
14	30 04.0	18.0	127 13.1	11.7	24 08.7	6.1	54.2
15	45 04.2	. . 18.9	141 43.8	11.8	24 02.6	6.1	54.2
16	60 04.3	19.8	156 14.6	11.8	23 56.5	6.3	54.3
17	75 04.5	20.6	170 45.4	11.8	23 50.2	6.4	54.3
18	90 04.6	N10 21.5	185 16.2	12.0	S23 43.8	6.5	54.3
19	105 04.7	22.4	199 47.2	11.9	23 37.3	6.6	54.3
20	120 04.9	23.3	214 18.1	12.1	23 30.7	6.7	54.3
21	135 05.0	. . 24.2	228 49.2	12.0	23 24.0	6.8	54.3
22	150 05.2	25.1	243 20.2	12.1	23 17.2	6.9	54.3
23	165 05.3	25.9	257 51.3	12.2	23 10.3	7.0	54.3
7 00	180 05.5	N10 26.8	272 22.5	12.2	S23 03.3	7.1	54.3
01	195 05.6	27.7	286 53.7	12.3	22 56.2	7.2	54.3
02	210 05.8	28.6	301 25.0	12.3	22 49.0	7.3	54.3
03	225 05.9	. . 29.5	315 56.3	12.4	22 41.7	7.5	54.3
04	240 06.0	30.3	330 27.7	12.4	22 34.2	7.5	54.3
05	255 06.2	31.2	344 59.1	12.5	22 26.7	7.6	54.3
06	270 06.3	N10 32.1	359 30.6	12.5	S22 19.1	7.7	54.3
07	285 06.5	33.0	14 02.1	12.6	22 11.4	7.9	54.3
08	300 06.6	33.9	28 33.7	12.6	22 03.5	7.9	54.3
09	315 06.8	. . 34.7	43 05.3	12.7	21 55.6	8.0	54.3
10	330 06.9	35.6	57 37.0	12.7	21 47.6	8.1	54.3
11	345 07.1	36.5	72 08.7	12.7	21 39.5	8.2	54.3
12	0 07.2	N10 37.4	86 40.4	12.9	S21 31.3	8.3	54.4
13	15 07.3	38.2	101 12.3	12.8	21 23.0	8.4	54.4
14	30 07.5	39.1	115 44.1	12.9	21 14.6	8.5	54.4
15	45 07.6	. . 40.0	130 16.0	13.0	21 06.1	8.6	54.4
16	60 07.8	40.9	144 48.0	13.0	20 57.5	8.7	54.4
17	75 07.9	41.7	159 20.0	13.1	20 48.8	8.8	54.4
18	90 08.0	N10 42.6	173 52.1	13.1	S20 40.0	8.9	54.4
19	105 08.2	43.5	188 24.2	13.1	20 31.1	8.9	54.4
20	120 08.3	44.4	202 56.3	13.2	20 22.2	9.1	54.4
21	135 08.5	. . 45.2	217 28.5	13.3	20 13.1	9.1	54.4
22	150 08.6	46.1	232 00.8	13.3	20 04.0	9.2	54.4
23	165 08.7	47.0	246 33.1	13.3	19 54.8	9.4	54.5
8 00	180 08.9	N10 47.9	261 05.4	13.4	S19 45.4	9.4	54.5
01	195 09.0	48.7	275 37.8	13.4	19 36.0	9.5	54.5
02	210 09.2	49.6	290 10.2	13.5	19 26.5	9.5	54.5
03	225 09.3	. . 50.5	304 42.7	13.5	19 17.0	9.7	54.5
04	240 09.4	51.4	319 15.2	13.6	19 07.3	9.8	54.5
05	255 09.6	52.2	333 47.8	13.6	18 57.5	9.8	54.5
06	270 09.7	N10 53.1	348 20.4	13.7	S18 47.7	9.9	54.5
07	285 09.9	54.0	2 53.1	13.7	18 37.8	10.0	54.6
08	300 10.0	54.8	17 25.8	13.7	18 27.8	10.1	54.6
09	315 10.1	. . 55.7	31 58.5	13.8	18 17.7	10.2	54.6
10	330 10.3	56.6	46 31.3	13.8	18 07.5	10.2	54.6
11	345 10.4	57.4	61 04.1	13.8	17 57.3	10.4	54.6
12	0 10.6	N10 58.3	75 36.9	13.9	S17 46.9	10.4	54.6
13	15 10.7	10 59.2	90 09.8	14.0	17 36.5	10.5	54.6
14	30 10.8	11 00.1	104 42.8	14.0	17 26.0	10.5	54.7
15	45 11.0	. . 00.9	119 15.8	14.0	17 15.5	10.7	54.7
16	60 11.1	01.8	133 48.8	14.0	17 04.8	10.7	54.7
17	75 11.2	02.7	148 21.8	14.1	16 54.1	10.8	54.7
18	90 11.4	N11 03.5	162 54.9	14.1	S16 43.3	10.9	54.7
19	105 11.5	04.4	177 28.0	14.2	16 32.4	10.9	54.7
20	120 11.7	05.3	192 01.2	14.2	16 21.5	11.1	54.8
21	135 11.8	. . 06.1	206 34.4	14.2	16 10.4	11.1	54.8
22	150 11.9	07.0	221 07.6	14.3	15 59.3	11.1	54.8
23	165 12.1	07.9	235 40.9	14.3	S15 48.2	11.3	54.8
	SD 16.0	*d* 0.9	SD 14.8		14.8		14.9

Lat.	Twilight Naut.	Twilight Civil	Sunrise	Moonrise 16	Moonrise 17	Moonrise 18	Moonrise 19
°	h m	h m	h m	h m	h m	h m	h m
N 72	////	01 36	03 27	▬	▬	▬	05 04
N 70	////	02 20	03 45	▬	▬	05 29	04 34
68	////	02 48	04 00	▬	▬	04 40	04 12
66	01 23	03 09	04 12	▬	04 36	04 09	03 54
64	02 02	03 26	04 22	04 08	03 54	03 46	03 39
62	02 28	03 39	04 30	03 19	03 25	03 27	03 27
60	02 47	03 51	04 37	02 48	03 03	03 11	03 16
N 58	03 03	04 00	04 44	02 25	02 45	02 58	03 07
56	03 17	04 09	04 49	02 06	02 30	02 47	02 59
54	03 28	04 16	04 54	01 50	02 17	02 37	02 52
52	03 38	04 23	04 59	01 36	02 06	02 28	02 45
50	03 46	04 29	05 03	01 24	01 55	02 20	02 40
45	04 04	04 41	05 12	00 59	01 34	02 03	02 27
N 40	04 17	04 51	05 19	00 39	01 17	01 49	02 16
35	04 28	04 59	05 26	00 22	01 02	01 37	02 07
30	04 37	05 06	05 31	00 08	00 49	01 26	01 59
20	04 52	05 18	05 40	24 27	00 27	01 08	01 45
N 10	05 02	05 27	05 49	24 08	00 08	00 52	01 33
0	05 11	05 35	05 56	23 51	24 37	00 37	01 22
S 10	05 18	05 42	06 04	23 33	24 22	00 22	01 10
20	05 23	05 49	06 11	23 14	24 06	00 06	00 58
30	05 28	05 56	06 20	22 52	23 48	24 44	00 44
35	05 30	06 00	06 25	22 38	23 37	24 36	00 36
40	05 32	06 04	06 31	22 23	23 25	24 26	00 26
45	05 34	06 08	06 38	22 06	23 10	24 15	00 15
S 50	05 35	06 12	06 45	21 43	22 52	24 02	00 02
52	05 35	06 15	06 49	21 32	22 43	23 56	25 09
54	05 36	06 17	06 53	21 20	22 34	23 49	25 04
56	05 36	06 19	06 57	21 06	22 23	23 41	25 00
58	05 36	06 22	07 02	20 50	22 10	23 32	24 54
S 60	05 36	06 25	07 08	20 30	21 56	23 22	24 48

Lat.	Sunset	Twilight Civil	Twilight Naut.	Moonset 16	Moonset 17	Moonset 18	Moonset 19
°	h m	h m	h m	h m	h m	h m	h m
N 72	20 36	22 33	////	▬	▬	▬	10 23
N 70	20 17	21 45	////	▬	▬	08 24	10 51
68	20 02	21 15	////	▬	▬	09 12	11 13
66	19 50	20 53	22 45	▬	07 39	09 42	11 29
64	19 39	20 36	22 03	06 27	08 21	10 04	11 42
62	19 31	20 22	21 35	07 15	08 49	10 22	11 54
60	19 23	20 11	21 15	07 46	09 10	10 37	12 03
N 58	19 17	20 01	20 58	08 09	09 28	10 49	12 11
56	19 11	19 52	20 45	08 28	09 42	11 00	12 19
54	19 06	19 44	20 33	08 43	09 55	11 09	12 25
52	19 01	19 38	20 23	08 57	10 06	11 18	12 31
50	18 57	19 32	20 14	09 09	10 16	11 25	12 36
45	18 48	19 19	19 57	09 33	10 36	11 41	12 47
N 40	18 41	19 09	19 43	09 52	10 53	11 54	12 56
35	18 34	19 00	19 32	10 09	11 07	12 05	13 04
30	18 29	18 53	19 22	10 23	11 19	12 15	13 11
20	18 19	18 41	19 08	10 46	11 39	12 31	13 23
N 10	18 11	18 32	18 57	11 07	11 57	12 46	13 33
0	18 03	18 24	18 48	11 26	12 13	12 59	13 43
S 10	17 55	18 17	18 41	11 45	12 30	13 12	13 52
20	17 47	18 10	18 35	12 05	12 47	13 26	14 02
30	17 38	18 03	18 31	12 28	13 07	13 42	14 14
35	17 33	17 59	18 28	12 42	13 19	13 51	14 20
40	17 27	17 55	18 26	12 57	13 32	14 02	14 28
45	17 21	17 51	18 25	13 16	13 48	14 14	14 37
S 50	17 13	17 46	18 23	13 39	14 07	14 29	14 47
52	17 09	17 44	18 23	13 50	14 16	14 36	14 52
54	17 05	17 41	18 22	14 03	14 26	14 43	14 57
56	17 01	17 39	18 22	14 17	14 37	14 52	15 03
58	16 56	17 36	18 22	14 34	14 50	15 01	15 09
S 60	16 50	17 33	18 21	14 54	15 06	15 12	15 16

Day	SUN Eqn. of Time 00^h	SUN Eqn. of Time 12^h	SUN Mer. Pass.	MOON Mer. Pass. Upper	MOON Mer. Pass. Lower	Age	Phase
d	m s	m s	h m	h m	h m	d %	
16	00 08	00 15	12 00	05 14	17 38	21 60	
17	00 22	00 28	12 00	06 02	18 25	22 51	
18	00 35	00 42	11 59	06 48	19 10	23 41	

UT	ARIES	VENUS −4.7		MARS +1.2		JUPITER −2.2		SATURN +0.7		STARS		
	GHA	GHA	Dec	GHA	Dec	GHA	Dec	GHA	Dec	Name	SHA	Dec
d h	° ′	° ′	° ′	° ′	° ′	° ′	° ′	° ′	° ′		° ′	° ′
19 00 SUNDAY	207 13.8	209 29.5	N 3 17.0	209 18.7	S 2 09.9	242 27.1	S14 42.7	39 37.7	N 7 45.2	Acamar	315 21.0	S40 16.[illegible]
01	222 16.2	224 31.7	16.6	224 19.4	09.2	257 29.2	42.5	54 40.3	45.3	Achernar	335 29.4	S57 11.[illegible]
02	237 18.7	239 33.8	16.1	239 20.1	08.4	272 31.3	42.4	69 42.9	45.3	Acrux	173 12.4	S63 09.[illegible]
03	252 21.2	254 35.9	. . 15.6	254 20.8	. . 07.6	287 33.3	. . 42.3	84 45.4	. . 45.3	Adhara	255 15.1	S28 59.[illegible]
04	267 23.6	269 38.0	15.2	269 21.4	06.8	302 35.4	42.2	99 48.0	45.4	Aldebaran	290 53.2	N16 31.7
05	282 26.1	284 40.1	14.7	284 22.1	06.1	317 37.5	42.1	114 50.6	45.4			
06	297 28.6	299 42.3	N 3 14.3	299 22.8	S 2 05.3	332 39.6	S14 41.9	129 53.2	N 7 45.5	Alioth	166 22.6	N55 54.5
07	312 31.0	314 44.4	13.8	314 23.5	04.5	347 41.7	41.8	144 55.7	45.5	Alkaid	153 00.7	N49 15.8
08	327 33.5	329 46.5	13.3	329 24.2	03.7	2 43.7	41.7	159 58.3	45.5	Al Na'ir	27 47.7	S46 54.8
09	342 36.0	344 48.6	. . 12.9	344 24.9	. . 03.0	17 45.8	. . 41.6	175 00.9	. . 45.6	Alnilam	275 49.7	S 1 11.8
10	357 38.4	359 50.7	12.4	359 25.5	02.2	32 47.9	41.5	190 03.5	45.6	Alphard	217 59.0	S 8 42.1
11	12 40.9	14 52.8	12.0	14 26.2	01.4	47 50.0	41.3	205 06.0	45.7			
12	27 43.3	29 54.9	N 3 11.5	29 26.9	S 2 00.6	62 52.1	S14 41.2	220 08.6	N 7 45.7	Alphecca	126 13.2	N26 40.7
13	42 45.8	44 57.0	11.1	44 27.6	1 59.9	77 54.2	41.1	235 11.2	45.7	Alpheratz	357 47.1	N29 08.4
14	57 48.3	59 59.0	10.7	59 28.3	59.1	92 56.2	41.0	250 13.7	45.8	Altair	62 11.2	N 8 53.4
15	72 50.7	75 01.1	. . 10.2	74 29.0	. . 58.3	107 58.3	. . 40.9	265 16.3	. . 45.8	Ankaa	353 19.0	S42 15.2
16	87 53.2	90 03.2	09.8	89 29.7	57.5	123 00.4	40.8	280 18.9	45.9	Antares	112 29.8	S26 27.3
17	102 55.7	105 05.3	09.3	104 30.3	56.8	138 02.5	40.6	295 21.5	45.9			
18	117 58.1	120 07.3	N 3 08.9	119 31.0	S 1 56.0	153 04.6	S14 40.5	310 24.0	N 7 45.9	Arcturus	145 58.2	N19 07.8
19	133 00.6	135 09.4	08.5	134 31.7	55.2	168 06.7	40.4	325 26.6	46.0	Atria	107 34.1	S69 02.6
20	148 03.1	150 11.5	08.0	149 32.4	54.4	183 08.7	40.3	340 29.2	46.0	Avior	234 19.4	S59 32.7
21	163 05.5	165 13.5	. . 07.6	164 33.1	. . 53.6	198 10.8	. . 40.2	355 31.7	. . 46.1	Bellatrix	278 35.5	N 6 21.5
22	178 08.0	180 15.6	07.2	179 33.8	52.9	213 12.9	40.1	10 34.3	46.1	Betelgeuse	271 04.8	N 7 24.5
23	193 10.4	195 17.6	06.8	194 34.5	52.1	228 15.0	39.9	25 36.9	46.1			
20 00 MONDAY	208 12.9	210 19.7	N 3 06.3	209 35.1	S 1 51.3	243 17.1	S14 39.8	40 39.5	N 7 46.2	Canopus	263 57.8	S52 42.3
01	223 15.4	225 21.7	05.9	224 35.8	50.5	258 19.2	39.7	55 42.0	46.2	Capella	280 39.4	N46 00.6
02	238 17.8	240 23.8	05.5	239 36.5	49.8	273 21.3	39.6	70 44.6	46.3	Deneb	49 33.8	N45 18.5
03	253 20.3	255 25.8	. . 05.1	254 37.2	. . 49.0	288 23.3	. . 39.5	85 47.2	. . 46.3	Denebola	182 36.5	N14 31.1
04	268 22.8	270 27.9	04.7	269 37.9	48.2	303 25.4	39.3	100 49.7	46.3	Diphda	348 59.3	S17 56.1
05	283 25.2	285 29.9	04.2	284 38.6	47.4	318 27.5	39.2	115 52.3	46.4			
06	298 27.7	300 31.9	N 3 03.8	299 39.2	S 1 46.7	333 29.6	S14 39.1	130 54.9	N 7 46.4	Dubhe	193 54.7	N61 42.1
07	313 30.2	315 33.9	03.4	314 39.9	45.9	348 31.7	39.0	145 57.4	46.4	Elnath	278 16.8	N28 37.0
08	328 32.6	330 36.0	03.0	329 40.6	45.1	3 33.8	38.9	161 00.0	46.5	Eltanin	90 47.3	N51 28.9
09	343 35.1	345 38.0	. . 02.6	344 41.3	. . 44.3	18 35.9	. . 38.8	176 02.6	. . 46.5	Enif	33 50.3	N 9 54.9
10	358 37.6	0 40.0	02.2	359 42.0	43.6	33 38.0	38.6	191 05.1	46.6	Fomalhaut	15 27.5	S29 34.3
11	13 40.0	15 42.0	01.8	14 42.7	42.8	48 40.0	38.5	206 07.7	46.6			
12	28 42.5	30 44.0	N 3 01.4	29 43.4	S 1 42.0	63 42.1	S14 38.4	221 10.3	N 7 46.6	Gacrux	172 04.1	S57 10.2
13	43 44.9	45 46.0	01.0	44 44.0	41.2	78 44.2	38.3	236 12.9	46.7	Gienah	175 55.2	S17 35.9
14	58 47.4	60 48.0	00.6	59 44.7	40.4	93 46.3	38.2	251 15.4	46.7	Hadar	148 51.9	S60 25.2
15	73 49.9	75 50.0	3 00.2	74 45.4	. . 39.7	108 48.4	. . 38.1	266 18.0	. . 46.8	Hamal	328 04.7	N23 30.4
16	88 52.3	90 52.0	2 59.8	89 46.1	38.9	123 50.5	37.9	281 20.6	46.8	Kaus Aust.	83 47.7	S34 22.8
17	103 54.8	105 54.0	59.4	104 46.8	38.1	138 52.6	37.8	296 23.1	46.8			
18	118 57.3	120 56.0	N 2 59.0	119 47.5	S 1 37.3	153 54.7	S14 37.7	311 25.7	N 7 46.9	Kochab	137 18.3	N74 06.8
19	133 59.7	135 58.0	58.6	134 48.2	36.6	168 56.8	37.6	326 28.3	46.9	Markab	13 41.7	N15 15.2
20	149 02.2	151 00.0	58.2	149 48.9	35.8	183 58.8	37.5	341 30.8	46.9	Menkar	314 18.6	N 4 07.6
21	164 04.7	166 01.9	. . 57.9	164 49.5	. . 35.0	199 00.9	. . 37.4	356 33.4	. . 47.0	Menkent	148 10.9	S36 25.1
22	179 07.1	181 03.9	57.5	179 50.2	34.2	214 03.0	37.2	11 36.0	47.0	Miaplacidus	221 40.4	S69 45.7
23	194 09.6	196 05.9	57.1	194 50.9	33.5	229 05.1	37.1	26 38.5	47.1			
21 00 TUESDAY	209 12.1	211 07.8	N 2 56.7	209 51.6	S 1 32.7	244 07.2	S14 37.0	41 41.1	N 7 47.1	Mirfak	308 45.4	N49 53.7
01	224 14.5	226 09.8	56.3	224 52.3	31.9	259 09.3	36.9	56 43.7	47.1	Nunki	76 02.0	S26 17.1
02	239 17.0	241 11.8	55.9	239 53.0	31.1	274 11.4	36.8	71 46.2	47.2	Peacock	53 24.0	S56 42.1
03	254 19.4	256 13.7	. . 55.6	254 53.7	. . 30.4	289 13.5	. . 36.7	86 48.8	. . 47.2	Pollux	243 31.5	N28 00.3
04	269 21.9	271 15.7	55.2	269 54.3	29.6	304 15.6	36.6	101 51.4	47.2	Procyon	245 03.0	N 5 12.0
05	284 24.4	286 17.6	54.8	284 55.0	28.8	319 17.7	36.4	116 53.9	47.3			
06	299 26.8	301 19.6	N 2 54.5	299 55.7	S 1 28.0	334 19.8	S14 36.3	131 56.5	N 7 47.3	Rasalhague	96 09.1	N12 33.0
07	314 29.3	316 21.5	54.1	314 56.4	27.3	349 21.8	36.2	146 59.1	47.3	Regulus	207 46.6	N11 55.2
08	329 31.8	331 23.4	53.7	329 57.1	26.5	4 23.9	36.1	162 01.6	47.4	Rigel	281 15.3	S 8 11.5
09	344 34.2	346 25.4	. . 53.4	344 57.8	. . 25.7	19 26.0	. . 36.0	177 04.2	. . 47.4	Rigil Kent.	139 55.5	S60 52.5
10	359 36.7	1 27.3	53.0	359 58.5	24.9	34 28.1	35.9	192 06.8	47.5	Sabik	102 15.9	S15 44.3
11	14 39.2	16 29.2	52.6	14 59.2	24.1	49 30.2	35.7	207 09.3	47.5			
12	29 41.6	31 31.2	N 2 52.3	29 59.8	S 1 23.4	64 32.3	S14 35.6	222 11.9	N 7 47.5	Schedar	349 44.9	N56 35.2
13	44 44.1	46 33.1	51.9	45 00.5	22.6	79 34.4	35.5	237 14.5	47.6	Shaula	96 25.9	S37 06.6
14	59 46.5	61 35.0	51.6	60 01.2	21.8	94 36.5	35.4	252 17.0	47.6	Sirius	258 36.6	S16 43.9
15	74 49.0	76 36.9	. . 51.2	75 01.9	. . 21.0	109 38.6	. . 35.3	267 19.6	. . 47.6	Spica	158 34.2	S11 12.8
16	89 51.5	91 38.8	50.9	90 02.6	20.3	124 40.7	35.2	282 22.1	47.7	Suhail	222 54.7	S43 28.5
17	104 53.9	106 40.7	50.5	105 03.3	19.5	139 42.8	35.1	297 24.7	47.7			
18	119 56.4	121 42.7	N 2 50.2	120 04.0	S 1 18.7	154 44.9	S14 34.9	312 27.3	N 7 47.7	Vega	80 40.9	N38 47.2
19	134 58.9	136 44.6	49.8	135 04.7	17.9	169 47.0	34.8	327 29.8	47.8	Zuben'ubi	137 08.6	S16 05.0
20	150 01.3	151 46.5	49.5	150 05.3	17.2	184 49.1	34.7	342 32.4	47.8		SHA	Mer. Pass.
21	165 03.8	166 48.3	. . 49.1	165 06.0	. . 16.4	199 51.2	. . 34.6	357 35.0	. . 47.9		° ′	h m
22	180 06.3	181 50.2	48.8	180 06.7	15.6	214 53.3	34.5	12 37.5	47.9	Venus	2 06.8	9 57
23	195 08.7	196 52.1	48.5	195 07.4	14.8	229 55.3	34.4	27 40.1	47.9	Mars	1 22.2	10 01
Mer. Pass.	h m 10 05.5	v 2.0	d 0.4	v 0.7	d 0.8	v 2.1	d 0.1	v 2.6	d 0.0	Jupiter	35 04.2	7 46
										Saturn	192 26.5	21 14

UT	SUN GHA	SUN Dec	MOON GHA	v	MOON Dec	d	HP
d h	° ′	° ′	° ′	′	° ′	′	′
9 00	180 12.2	N11 08.7	250 14.2	14.3	S15 36.9	11.3	54.8
01	195 12.3	09.6	264 47.5	14.4	15 25.6	11.4	54.8
02	210 12.5	10.5	279 20.9	14.4	15 14.2	11.5	54.9
03	225 12.6	. . 11.3	293 54.3	14.4	15 02.7	11.5	54.9
04	240 12.7	12.2	308 27.7	14.5	14 51.2	11.6	54.9
05	255 12.9	13.1	323 01.2	14.5	14 39.6	11.7	54.9
06	270 13.0	N11 13.9	337 34.7	14.5	S14 27.9	11.7	54.9
07	285 13.1	14.8	352 08.2	14.5	14 16.2	11.8	55.0
08	300 13.3	15.6	6 41.7	14.6	14 04.4	11.9	55.0
09	315 13.4	. . 16.5	21 15.3	14.6	13 52.5	11.9	55.0
10	330 13.6	17.4	35 48.9	14.6	13 40.6	12.0	55.0
11	345 13.7	18.2	50 22.5	14.6	13 28.6	12.1	55.0
12	0 13.8	N11 19.1	64 56.1	14.7	S13 16.5	12.1	55.1
13	15 14.0	20.0	79 29.8	14.7	13 04.4	12.2	55.1
14	30 14.1	20.8	94 03.5	14.7	12 52.2	12.3	55.1
15	45 14.2	. . 21.7	108 37.2	14.7	12 39.9	12.3	55.1
16	60 14.4	22.5	123 10.9	14.8	12 27.6	12.4	55.2
17	75 14.5	23.4	137 44.7	14.8	12 15.2	12.4	55.2
18	90 14.6	N11 24.3	152 18.5	14.8	S12 02.8	12.5	55.2
19	105 14.8	25.1	166 52.3	14.8	11 50.3	12.6	55.2
20	120 14.9	26.0	181 26.1	14.8	11 37.7	12.6	55.3
21	135 15.0	. . 26.8	195 59.9	14.8	11 25.1	12.7	55.3
22	150 15.1	27.7	210 33.7	14.9	11 12.4	12.7	55.3
23	165 15.3	28.6	225 07.6	14.9	10 59.7	12.8	55.3
0 00	180 15.4	N11 29.4	239 41.5	14.8	S10 46.9	12.8	55.3
01	195 15.5	30.3	254 15.3	14.9	10 34.1	12.9	55.4
02	210 15.7	31.1	268 49.2	14.9	10 21.2	13.0	55.4
03	225 15.8	. . 32.0	283 23.1	15.0	10 08.2	13.0	55.4
04	240 15.9	32.8	297 57.1	14.9	9 55.2	13.0	55.4
05	255 16.1	33.7	312 31.0	14.9	9 42.2	13.2	55.5
06	270 16.2	N11 34.6	327 04.9	15.0	S 9 29.0	13.1	55.5
07	285 16.3	35.4	341 38.9	14.9	9 15.9	13.2	55.5
08	300 16.5	36.3	356 12.8	15.0	9 02.7	13.3	55.5
09	315 16.6	. . 37.1	10 46.8	14.9	8 49.4	13.3	55.6
10	330 16.7	38.0	25 20.7	15.0	8 36.1	13.4	55.6
11	345 16.8	38.8	39 54.7	15.0	8 22.7	13.4	55.6
12	0 17.0	N11 39.7	54 28.7	14.9	S 8 09.3	13.4	55.7
13	15 17.1	40.5	69 02.6	15.0	7 55.9	13.5	55.7
14	30 17.2	41.4	83 36.6	15.0	7 42.4	13.6	55.7
15	45 17.4	. . 42.2	98 10.6	15.0	7 28.8	13.6	55.7
16	60 17.5	43.1	112 44.6	14.9	7 15.2	13.6	55.8
17	75 17.6	43.9	127 18.5	15.0	7 01.6	13.7	55.8
18	90 17.7	N11 44.8	141 52.5	15.0	S 6 47.9	13.7	55.8
19	105 17.9	45.7	156 26.5	14.9	6 34.2	13.8	55.8
20	120 18.0	46.5	171 00.4	15.0	6 20.4	13.8	55.9
21	135 18.1	. . 47.4	185 34.4	14.9	6 06.6	13.8	55.9
22	150 18.3	48.2	200 08.3	15.0	5 52.8	13.9	55.9
23	165 18.4	49.1	214 42.3	14.9	5 38.9	14.0	56.0
21 00	180 18.5	N11 49.9	229 16.2	14.9	S 5 24.9	13.9	56.0
01	195 18.6	50.8	243 50.1	14.9	5 11.0	14.0	56.0
02	210 18.8	51.6	258 24.0	14.9	4 57.0	14.0	56.0
03	225 18.9	. . 52.5	272 57.9	14.9	4 43.0	14.1	56.1
04	240 19.0	53.3	287 31.8	14.9	4 28.9	14.1	56.1
05	255 19.1	54.2	302 05.7	14.8	4 14.8	14.1	56.1
06	270 19.3	N11 55.0	316 39.5	14.9	S 4 00.7	14.2	56.2
07	285 19.4	55.8	331 13.4	14.8	3 46.5	14.2	56.2
08	300 19.5	56.7	345 47.2	14.8	3 32.3	14.2	56.2
09	315 19.6	. . 57.5	0 21.0	14.8	3 18.1	14.3	56.3
10	330 19.8	58.4	14 54.8	14.7	3 03.8	14.3	56.3
11	345 19.9	11 59.2	29 28.5	14.8	2 49.5	14.3	56.3
12	0 20.0	N12 00.1	44 02.3	14.7	S 2 35.2	14.4	56.3
13	15 20.1	00.9	58 36.0	14.7	2 20.8	14.3	56.4
14	30 20.3	01.8	73 09.7	14.6	2 06.5	14.4	56.4
15	45 20.4	. . 02.6	87 43.3	14.7	1 52.1	14.5	56.4
16	60 20.5	03.5	102 17.0	14.6	1 37.6	14.4	56.5
17	75 20.6	04.3	116 50.6	14.6	1 23.2	14.5	56.5
18	90 20.8	N12 05.1	131 24.2	14.5	S 1 08.7	14.5	56.5
19	105 20.9	06.0	145 57.7	14.5	0 54.2	14.5	56.6
20	120 21.0	06.8	160 31.2	14.5	0 39.7	14.5	56.6
21	135 21.1	. . 07.7	175 04.7	14.5	0 25.2	14.5	56.6
22	150 21.2	08.5	189 38.2	14.4	S 0 10.7	14.6	56.7
23	165 21.4	09.4	204 11.6	14.4	N 0 03.9	14.6	56.7
	SD 15.9	*d* 0.9	SD		15.0	15.2	15.4

Lat.	Twilight Naut.	Twilight Civil	Sunrise	Moonrise 19	Moonrise 20	Moonrise 21	Moonrise 22
°	h m	h m	h m	h m	h m	h m	h m
N 72	////	00 54	03 09	05 04	04 20	03 47	03 19
N 70	////	01 56	03 30	04 34	04 04	03 41	03 20
68	////	02 30	03 47	04 12	03 52	03 36	03 21
66	00 46	02 55	04 00	03 54	03 42	03 32	03 22
64	01 41	03 13	04 11	03 39	03 33	03 28	03 23
62	02 12	03 28	04 21	03 27	03 26	03 25	03 24
60	02 35	03 41	04 29	03 16	03 20	03 22	03 24
N 58	02 52	03 51	04 36	03 07	03 14	03 19	03 25
56	03 07	04 01	04 42	02 59	03 09	03 17	03 25
54	03 19	04 09	04 47	02 52	03 04	03 15	03 26
52	03 30	04 16	04 52	02 45	03 00	03 13	03 26
50	03 39	04 22	04 57	02 40	02 56	03 12	03 26
45	03 58	04 36	05 07	02 27	02 48	03 08	03 27
N 40	04 12	04 46	05 15	02 16	02 41	03 05	03 28
35	04 24	04 55	05 22	02 07	02 35	03 02	03 29
30	04 34	05 03	05 28	01 59	02 30	03 00	03 29
20	04 49	05 16	05 38	01 45	02 21	02 55	03 30
N 10	05 01	05 26	05 47	01 33	02 13	02 52	03 31
0	05 10	05 34	05 56	01 22	02 05	02 48	03 32
S 10	05 18	05 42	06 04	01 10	01 58	02 45	03 33
20	05 24	05 50	06 12	00 58	01 50	02 41	03 34
30	05 30	05 58	06 22	00 44	01 40	02 37	03 35
35	05 32	06 02	06 28	00 36	01 35	02 34	03 35
40	05 35	06 06	06 34	00 26	01 29	02 32	03 36
45	05 37	06 11	06 41	00 15	01 22	02 28	03 37
S 50	05 39	06 17	06 50	00 02	01 13	02 25	03 38
52	05 40	06 19	06 54	25 09	01 09	02 23	03 38
54	05 41	06 22	06 58	25 04	01 04	02 21	03 39
56	05 42	06 25	07 03	25 00	01 00	02 19	03 40
58	05 42	06 28	07 09	24 54	00 54	02 16	03 40
S 60	05 43	06 31	07 15	24 48	00 48	02 14	03 41

Lat.	Sunset	Twilight Civil	Twilight Naut.	Moonset 19	Moonset 20	Moonset 21	Moonset 22
°	h m	h m	h m	h m	h m	h m	h m
N 72	20 53	23 26	////	10 23	12 39	14 43	16 49
N 70	20 31	22 09	////	10 51	12 52	14 46	16 42
68	20 14	21 32	////	11 13	13 02	14 49	16 37
66	20 00	21 07	23 31	11 29	13 10	14 51	16 33
64	19 49	20 47	22 23	11 42	13 17	14 52	16 29
62	19 39	20 32	21 50	11 54	13 23	14 54	16 26
60	19 31	20 19	21 26	12 03	13 29	14 55	16 23
N 58	19 24	20 08	21 08	12 11	13 33	14 56	16 21
56	19 17	19 59	20 53	12 19	13 37	14 57	16 19
54	19 12	19 51	20 41	12 25	13 41	14 58	16 17
52	19 06	19 43	20 30	12 31	13 44	14 59	16 15
50	19 02	19 37	20 20	12 36	13 47	14 59	16 14
45	18 52	19 23	20 01	12 47	13 54	15 01	16 10
N 40	18 44	19 12	19 46	12 56	13 59	15 02	16 08
35	18 37	19 03	19 34	13 04	14 03	15 03	16 05
30	18 30	18 55	19 25	13 11	14 07	15 04	16 03
20	18 20	18 42	19 09	13 23	14 14	15 06	15 59
N 10	18 11	18 32	18 57	13 33	14 20	15 08	15 56
0	18 02	18 23	18 48	13 43	14 26	15 09	15 53
S 10	17 54	18 15	18 40	13 52	14 31	15 10	15 50
20	17 45	18 08	18 33	14 02	14 37	15 12	15 47
30	17 35	18 00	18 28	14 14	14 44	15 13	15 43
35	17 30	17 55	18 25	14 20	14 48	15 14	15 41
40	17 23	17 51	18 22	14 28	14 52	15 15	15 39
45	17 16	17 46	18 20	14 37	14 57	15 16	15 36
S 50	17 07	17 40	18 18	14 47	15 03	15 17	15 32
52	17 03	17 38	18 17	14 52	15 05	15 18	15 31
54	16 58	17 35	18 16	14 57	15 08	15 19	15 29
56	16 53	17 32	18 15	15 03	15 11	15 19	15 28
58	16 48	17 29	18 14	15 09	15 15	15 20	15 26
S 60	16 42	17 25	18 14	15 16	15 19	15 21	15 23

Day	SUN Eqn. of Time 00^h	SUN Eqn. of Time 12^h	SUN Mer. Pass.	MOON Mer. Pass. Upper	MOON Mer. Pass. Lower	Age	Phase
d	m s	m s	h m	h m	h m	d %	
19	00 49	00 55	11 59	07 32	19 54	24 32	
20	01 01	01 08	11 59	08 16	20 37	25 23	
21	01 14	01 20	11 59	08 59	21 20	26 15	

UT	ARIES	VENUS −4.7		MARS +1.2		JUPITER −2.2		SATURN +0.7		STARS		
d h	GHA ° ′	GHA ° ′	Dec ° ′	GHA ° ′	Dec ° ′	GHA ° ′	Dec ° ′	GHA ° ′	Dec ° ′	Name	SHA ° ′	Dec ° ′
22 00	210 11.2	211 54.0	N 2 48.1	210 08.1	S 1 14.1	244 57.4	S14 34.3	42 42.7	N 7 48.0	Acamar	315 21.0	S40 16.1
01	225 13.7	226 55.9	47.8	225 08.8	13.3	259 59.5	34.1	57 45.2	48.0	Achernar	335 29.4	S57 11.3
02	240 16.1	241 57.8	47.5	240 09.5	12.5	275 01.6	34.0	72 47.8	48.0	Acrux	173 12.5	S63 09.3
03	255 18.6	256 59.7	. . 47.1	255 10.2	. . 11.7	290 03.7	. . 33.9	87 50.3	. . 48.1	Adhara	255 15.1	S28 59.3
04	270 21.0	272 01.5	46.8	270 10.8	11.0	305 05.8	33.8	102 52.9	48.1	Aldebaran	290 53.2	N16 31.7
05	285 23.5	287 03.4	46.5	285 11.5	10.2	320 07.9	33.7	117 55.5	48.1			
06	300 26.0	302 05.3	N 2 46.1	300 12.2	S 1 09.4	335 10.0	S14 33.6	132 58.0	N 7 48.2	Alioth	166 22.6	N55 54.5
W 07	315 28.4	317 07.1	45.8	315 12.9	08.6	350 12.1	33.5	148 00.6	48.2	Alkaid	153 00.7	N49 15.9
E 08	330 30.9	332 09.0	45.5	330 13.6	07.8	5 14.2	33.3	163 03.2	48.2	Al Na'ir	27 47.7	S46 54.8
D 09	345 33.4	347 10.8	. . 45.2	345 14.3	. . 07.1	20 16.3	. . 33.2	178 05.7	. . 48.3	Alnilam	275 49.7	S 1 11.8
N 10	0 35.8	2 12.7	44.9	0 15.0	06.3	35 18.4	33.1	193 08.3	48.3	Alphard	217 59.1	S 8 42.1
E 11	15 38.3	17 14.5	44.5	15 15.7	05.5	50 20.5	33.0	208 10.8	48.4			
S 12	30 40.8	32 16.4	N 2 44.2	30 16.4	S 1 04.7	65 22.6	S14 32.9	223 13.4	N 7 48.4	Alphecca	126 13.2	N26 40.8
D 13	45 43.2	47 18.2	43.9	45 17.0	04.0	80 24.7	32.8	238 16.0	48.4	Alpheratz	357 47.1	N29 08.4
A 14	60 45.7	62 20.1	43.6	60 17.7	03.2	95 26.8	32.7	253 18.5	48.5	Altair	62 11.2	N 8 53.4
Y 15	75 48.1	77 21.9	. . 43.3	75 18.4	. . 02.4	110 28.9	. . 32.5	268 21.1	. . 48.5	Ankaa	353 19.0	S42 15.2
16	90 50.6	92 23.7	43.0	90 19.1	01.6	125 31.0	32.4	283 23.7	48.5	Antares	112 29.8	S26 27.3
17	105 53.1	107 25.6	42.7	105 19.8	00.9	140 33.1	32.3	298 26.2	48.6			
18	120 55.5	122 27.4	N 2 42.4	120 20.5	S 1 00.1	155 35.2	S14 32.2	313 28.8	N 7 48.6	Arcturus	145 58.2	N19 07.8
19	135 58.0	137 29.2	42.1	135 21.2	0 59.3	170 37.3	32.1	328 31.3	48.6	Atria	107 34.1	S69 02.6
20	151 00.5	152 31.0	41.8	150 21.9	58.5	185 39.4	32.0	343 33.9	48.7	Avior	234 19.4	S59 32.7
21	166 02.9	167 32.9	. . 41.5	165 22.6	. . 57.8	200 41.5	. . 31.9	358 36.5	. . 48.7	Bellatrix	278 35.6	N 6 21.5
22	181 05.4	182 34.7	41.2	180 23.2	57.0	215 43.6	31.8	13 39.0	48.7	Betelgeuse	271 04.8	N 7 24.5
23	196 07.9	197 36.5	40.9	195 23.9	56.2	230 45.7	31.6	28 41.6	48.8			
23 00	211 10.3	212 38.3	N 2 40.6	210 24.6	S 0 55.4	245 47.8	S14 31.5	43 44.1	N 7 48.8	Canopus	263 57.8	S52 42.3
01	226 12.8	227 40.1	40.3	225 25.3	54.7	260 49.9	31.4	58 46.7	48.8	Capella	280 39.4	N46 00.6
02	241 15.3	242 41.9	40.0	240 26.0	53.9	275 52.0	31.3	73 49.3	48.9	Deneb	49 33.7	N45 18.5
03	256 17.7	257 43.7	. . 39.7	255 26.7	. . 53.1	290 54.1	. . 31.2	88 51.8	. . 48.9	Denebola	182 36.5	N14 31.1
04	271 20.2	272 45.5	39.4	270 27.4	52.3	305 56.2	31.1	103 54.4	48.9	Diphda	348 59.2	S17 56.1
05	286 22.6	287 47.3	39.1	285 28.1	51.5	320 58.3	31.0	118 56.9	49.0			
06	301 25.1	302 49.1	N 2 38.9	300 28.8	S 0 50.8	336 00.4	S14 30.9	133 59.5	N 7 49.0	Dubhe	193 54.7	N61 42.1
T 07	316 27.6	317 50.8	38.6	315 29.5	50.0	351 02.5	30.7	149 02.1	49.0	Elnath	278 16.8	N28 37.0
H 08	331 30.0	332 52.6	38.3	330 30.1	49.2	6 04.6	30.6	164 04.6	49.1	Eltanin	90 47.3	N51 28.9
U 09	346 32.5	347 54.4	. . 38.0	345 30.8	. . 48.4	21 06.7	. . 30.5	179 07.2	. . 49.1	Enif	33 50.3	N 9 54.9
R 10	1 35.0	2 56.2	37.7	0 31.5	47.7	36 08.8	30.4	194 09.7	49.1	Fomalhaut	15 27.5	S29 34.3
S 11	16 37.4	17 58.0	37.5	15 32.2	46.9	51 10.9	30.3	209 12.3	49.2			
D 12	31 39.9	32 59.7	N 2 37.2	30 32.9	S 0 46.1	66 13.0	S14 30.2	224 14.8	N 7 49.2	Gacrux	172 04.1	S57 10.2
A 13	46 42.4	48 01.5	36.9	45 33.6	45.3	81 15.1	30.1	239 17.4	49.2	Gienah	175 55.3	S17 35.9
Y 14	61 44.8	63 03.2	36.7	60 34.3	44.6	96 17.2	30.0	254 20.0	49.3	Hadar	148 51.9	S60 25.2
15	76 47.3	78 05.0	. . 36.4	75 35.0	. . 43.8	111 19.4	. . 29.8	269 22.5	. . 49.3	Hamal	328 04.7	N23 30.4
16	91 49.7	93 06.8	36.1	90 35.7	43.0	126 21.5	29.7	284 25.1	49.3	Kaus Aust.	83 47.7	S34 22.8
17	106 52.2	108 08.5	35.9	105 36.3	42.2	141 23.6	29.6	299 27.6	49.4			
18	121 54.7	123 10.3	N 2 35.6	120 37.0	S 0 41.5	156 25.7	S14 29.5	314 30.2	N 7 49.4	Kochab	137 18.3	N74 06.8
19	136 57.1	138 12.0	35.3	135 37.7	40.7	171 27.8	29.4	329 32.7	49.4	Markab	13 41.7	N15 15.2
20	151 59.6	153 13.8	35.1	150 38.4	39.9	186 29.9	29.3	344 35.3	49.5	Menkar	314 18.6	N 4 07.6
21	167 02.1	168 15.5	. . 34.8	165 39.1	. . 39.1	201 32.0	. . 29.2	359 37.9	. . 49.5	Menkent	148 10.9	S36 25.1
22	182 04.5	183 17.2	34.6	180 39.8	38.4	216 34.1	29.1	14 40.4	49.5	Miaplacidus	221 40.5	S69 45.7
23	197 07.0	198 19.0	34.3	195 40.5	37.6	231 36.2	29.0	29 43.0	49.6			
24 00	212 09.5	213 20.7	N 2 34.1	210 41.2	S 0 36.8	246 38.3	S14 28.8	44 45.5	N 7 49.6	Mirfak	308 45.4	N49 53.7
01	227 11.9	228 22.4	33.8	225 41.9	36.0	261 40.4	28.7	59 48.1	49.6	Nunki	76 02.0	S26 17.1
02	242 14.4	243 24.2	33.6	240 42.6	35.3	276 42.5	28.6	74 50.6	49.7	Peacock	53 24.0	S56 42.1
03	257 16.9	258 25.9	. . 33.3	255 43.3	. . 34.5	291 44.6	. . 28.5	89 53.2	. . 49.7	Pollux	243 31.5	N28 00.3
04	272 19.3	273 27.6	33.1	270 43.9	33.7	306 46.7	28.4	104 55.8	49.7	Procyon	245 03.0	N 5 12.0
05	287 21.8	288 29.3	32.8	285 44.6	32.9	321 48.8	28.3	119 58.3	49.8			
06	302 24.2	303 31.0	N 2 32.6	300 45.3	S 0 32.2	336 50.9	S14 28.2	135 00.9	N 7 49.8	Rasalhague	96 09.1	N12 33.0
07	317 26.7	318 32.7	32.4	315 46.0	31.4	351 53.1	28.1	150 03.4	49.8	Regulus	207 46.6	N11 55.2
08	332 29.2	333 34.4	32.1	330 46.7	30.6	6 55.2	28.0	165 06.0	49.9	Rigel	281 15.3	S 8 11.5
F 09	347 31.6	348 36.1	. . 31.9	345 47.4	. . 29.8	21 57.3	. . 27.8	180 08.5	. . 49.9	Rigil Kent.	139 55.5	S60 52.5
R 10	2 34.1	3 37.8	31.6	0 48.1	29.1	36 59.4	27.7	195 11.1	49.9	Sabik	102 15.9	S15 44.3
I 11	17 36.6	18 39.5	31.4	15 48.8	28.3	52 01.5	27.6	210 13.6	50.0			
D 12	32 39.0	33 41.2	N 2 31.2	30 49.5	S 0 27.5	67 03.6	S14 27.5	225 16.2	N 7 50.0	Schedar	349 44.8	N56 35.2
A 13	47 41.5	48 42.9	31.0	45 50.2	26.7	82 05.7	27.4	240 18.8	50.0	Shaula	96 25.9	S37 06.6
Y 14	62 44.0	63 44.6	30.7	60 50.9	26.0	97 07.8	27.3	255 21.3	50.0	Sirius	258 36.6	S16 43.9
15	77 46.4	78 46.3	. . 30.5	75 51.5	. . 25.2	112 09.9	. . 27.2	270 23.9	. . 50.1	Spica	158 34.2	S11 12.8
16	92 48.9	93 48.0	30.3	90 52.2	24.4	127 12.0	27.1	285 26.4	50.1	Suhail	222 54.7	S43 28.5
17	107 51.4	108 49.7	30.1	105 52.9	23.6	142 14.1	27.0	300 29.0	50.1			
18	122 53.8	123 51.3	N 2 29.8	120 53.6	S 0 22.8	157 16.3	S14 26.9	315 31.5	N 7 50.2	Vega	80 40.9	N38 47.2
19	137 56.3	138 53.0	29.6	135 54.3	22.1	172 18.4	26.7	330 34.1	50.2	Zuben'ubi	137 08.6	S16 05.0
20	152 58.7	153 54.7	29.4	150 55.0	21.3	187 20.5	26.6	345 36.6	50.2			
21	168 01.2	168 56.4	. . 29.2	165 55.7	. . 20.5	202 22.6	. . 26.5	0 39.2	. . 50.3			
22	183 03.7	183 58.0	29.0	180 56.4	19.7	217 24.7	26.4	15 41.7	50.3			
23	198 06.1	198 59.7	28.8	195 57.1	19.0	232 26.8	26.3	30 44.3	50.3			
Mer. Pass. h m	9 53.7	*v* 1.8	*d* 0.3	*v* 0.7	*d* 0.8	*v* 2.1	*d* 0.1	*v* 2.6	*d* 0.0			

	SHA ° ′	Mer. Pass. h m
Venus	1 28.0	9 48
Mars	359 14.3	9 58
Jupiter	34 37.5	7 36
Saturn	192 33.8	21 01

UT	SUN GHA	SUN Dec	MOON GHA	v	MOON Dec	d	HP
h	° ′	° ′	° ′	′	° ′	′	′
00	180 21.5	N12 10.2	218 45.0	14.4	N 0 18.5	14.6	56.7
01	195 21.6	11.0	233 18.4	14.3	0 33.1	14.6	56.8
02	210 21.7	11.9	247 51.7	14.3	0 47.7	14.6	56.8
03	225 21.9	. . 12.7	262 25.0	14.2	1 02.3	14.7	56.8
04	240 22.0	13.6	276 58.2	14.2	1 17.0	14.6	56.9
05	255 22.1	14.4	291 31.4	14.2	1 31.6	14.7	56.9
06	270 22.2	N12 15.2	306 04.6	14.1	N 1 46.3	14.7	56.9
07	285 22.3	16.1	320 37.7	14.0	2 01.0	14.6	56.9
08	300 22.5	16.9	335 10.7	14.1	2 15.6	14.7	57.0
09	315 22.6	. . 17.8	349 43.8	13.9	2 30.3	14.7	57.0
10	330 22.7	18.6	4 16.7	14.0	2 45.0	14.7	57.0
11	345 22.8	19.4	18 49.7	13.9	2 59.7	14.7	57.1
12	0 22.9	N12 20.3	33 22.6	13.8	N 3 14.4	14.7	57.1
13	15 23.1	21.1	47 55.4	13.8	3 29.1	14.7	57.1
14	30 23.2	22.0	62 28.2	13.7	3 43.8	14.8	57.2
15	45 23.3	. . 22.8	77 00.9	13.7	3 58.6	14.7	57.2
16	60 23.4	23.6	91 33.6	13.6	4 13.3	14.7	57.2
17	75 23.5	24.5	106 06.2	13.5	4 28.0	14.7	57.3
18	90 23.7	N12 25.3	120 38.7	13.6	N 4 42.7	14.7	57.3
19	105 23.8	26.1	135 11.3	13.4	4 57.4	14.7	57.3
20	120 23.9	27.0	149 43.7	13.4	5 12.1	14.6	57.4
21	135 24.0	. . 27.8	164 16.1	13.3	5 26.7	14.7	57.4
22	150 24.1	28.6	178 48.4	13.3	5 41.4	14.7	57.4
23	165 24.2	29.5	193 20.7	13.2	5 56.1	14.7	57.5
00	180 24.4	N12 30.3	207 52.9	13.1	N 6 10.8	14.6	57.5
01	195 24.5	31.1	222 25.0	13.1	6 25.4	14.6	57.5
02	210 24.6	32.0	236 57.1	13.0	6 40.0	14.7	57.6
03	225 24.7	. . 32.8	251 29.1	13.0	6 54.7	14.6	57.6
04	240 24.8	33.6	266 01.1	12.8	7 09.3	14.6	57.6
05	255 24.9	34.5	280 32.9	12.8	7 23.9	14.5	57.7
06	270 25.1	N12 35.3	295 04.7	12.8	N 7 38.4	14.6	57.7
07	285 25.2	36.1	309 36.5	12.6	7 53.0	14.5	57.7
08	300 25.3	37.0	324 08.1	12.6	8 07.5	14.5	57.7
09	315 25.4	. . 37.8	338 39.7	12.5	8 22.0	14.5	57.8
10	330 25.5	38.6	353 11.2	12.5	8 36.5	14.5	57.8
11	345 25.6	39.5	7 42.7	12.4	8 51.0	14.4	57.8
12	0 25.8	N12 40.3	22 14.1	12.2	N 9 05.4	14.4	57.9
13	15 25.9	41.1	36 45.3	12.3	9 19.8	14.4	57.9
14	30 26.0	41.9	51 16.6	12.1	9 34.2	14.3	57.9
15	45 26.1	. . 42.8	65 47.7	12.0	9 48.5	14.3	58.0
16	60 26.2	43.6	80 18.7	12.0	10 02.8	14.3	58.0
17	75 26.3	44.4	94 49.7	11.9	10 17.1	14.2	58.0
18	90 26.4	N12 45.2	109 20.6	11.8	N10 31.3	14.2	58.1
19	105 26.6	46.1	123 51.4	11.7	10 45.5	14.2	58.1
20	120 26.7	46.9	138 22.1	11.7	10 59.7	14.1	58.1
21	135 26.8	. . 47.7	152 52.8	11.5	11 13.8	14.1	58.1
22	150 26.9	48.6	167 23.3	11.5	11 27.9	14.0	58.2
23	165 27.0	49.4	181 53.8	11.3	11 41.9	14.0	58.2
00	180 27.1	N12 50.2	196 24.1	11.3	N11 55.9	14.0	58.2
01	195 27.2	51.0	210 54.4	11.2	12 09.9	13.9	58.3
02	210 27.3	51.9	225 24.6	11.1	12 23.8	13.8	58.3
03	225 27.5	. . 52.7	239 54.7	11.0	12 37.6	13.8	58.3
04	240 27.6	53.5	254 24.7	10.9	12 51.4	13.8	58.4
05	255 27.7	54.3	268 54.6	10.9	13 05.2	13.7	58.4
06	270 27.8	N12 55.1	283 24.5	10.7	N13 18.9	13.6	58.4
07	285 27.9	56.0	297 54.2	10.6	13 32.5	13.6	58.4
08	300 28.0	56.8	312 23.8	10.5	13 46.1	13.5	58.5
09	315 28.1	. . 57.6	326 53.3	10.5	13 59.6	13.5	58.5
10	330 28.2	58.4	341 22.8	10.3	14 13.1	13.4	58.5
11	345 28.3	12 59.3	355 52.1	10.3	14 26.5	13.3	58.6
12	0 28.4	N13 00.1	10 21.4	10.1	N14 39.8	13.2	58.6
13	15 28.6	00.9	24 50.5	10.0	14 53.0	13.2	58.6
14	30 28.7	01.7	39 19.5	10.0	15 06.2	13.2	58.6
15	45 28.8	. . 02.5	53 48.5	9.8	15 19.4	13.0	58.7
16	60 28.9	03.3	68 17.3	9.8	15 32.4	13.0	58.7
17	75 29.0	04.2	82 46.1	9.6	15 45.4	12.9	58.7
18	90 29.1	N13 05.0	97 14.7	9.5	N15 58.3	12.8	58.7
19	105 29.2	05.8	111 43.2	9.5	16 11.1	12.8	58.8
20	120 29.3	06.6	126 11.7	9.3	16 23.9	12.6	58.8
21	135 29.4	. . 07.4	140 40.0	9.2	16 36.5	12.6	58.8
22	150 29.5	08.3	155 08.2	9.1	16 49.1	12.5	58.8
23	165 29.6	09.1	169 36.3	9.0	N17 01.6	12.4	58.9
	SD 15.9	d 0.8	SD 15.6		15.8		16.0

Lat.	Twilight Naut.	Twilight Civil	Sunrise	Moonrise 22	23	24	25
°	h m	h m	h m	h m	h m	h m	h m
N 72	////	////	02 50	03 19	02 49	02 10	00 53
N 70	////	01 29	03 15	03 20	02 58	02 31	01 49
68	////	02 11	03 33	03 21	03 06	02 48	02 22
66	////	02 39	03 48	03 22	03 12	03 01	02 47
64	01 17	03 00	04 01	03 23	03 18	03 12	03 07
62	01 56	03 17	04 11	03 24	03 22	03 22	03 23
60	02 22	03 31	04 20	03 24	03 27	03 30	03 36
N 58	02 41	03 43	04 28	03 25	03 30	03 38	03 48
56	02 57	03 53	04 35	03 25	03 34	03 44	03 58
54	03 10	04 01	04 41	03 26	03 37	03 50	04 07
52	03 22	04 09	04 46	03 26	03 40	03 55	04 15
50	03 32	04 16	04 51	03 26	03 42	04 00	04 22
45	03 52	04 30	05 02	03 27	03 48	04 11	04 38
N 40	04 07	04 42	05 11	03 28	03 52	04 19	04 51
35	04 20	04 52	05 18	03 29	03 56	04 27	05 02
30	04 30	05 00	05 25	03 29	04 00	04 34	05 12
20	04 47	05 13	05 36	03 30	04 06	04 45	05 28
N 10	04 59	05 24	05 46	03 31	04 12	04 55	05 43
0	05 09	05 34	05 55	03 32	04 17	05 05	05 57
S 10	05 18	05 42	06 04	03 33	04 22	05 15	06 11
20	05 25	05 51	06 13	03 34	04 28	05 25	06 26
30	05 31	06 00	06 24	03 35	04 35	05 37	06 44
35	05 35	06 04	06 30	03 35	04 38	05 45	06 54
40	05 38	06 09	06 37	03 36	04 43	05 53	07 06
45	05 40	06 15	06 45	03 37	04 48	06 02	07 20
S 50	05 43	06 21	06 55	03 38	04 54	06 14	07 37
52	05 45	06 24	06 59	03 38	04 57	06 19	07 45
54	05 46	06 27	07 04	03 39	05 00	06 25	07 54
56	05 47	06 30	07 09	03 40	05 03	06 32	08 04
58	05 48	06 34	07 15	03 40	05 07	06 39	08 16
S 60	05 49	06 38	07 22	03 41	05 12	06 48	08 29

Lat.	Sunset	Twilight Civil	Twilight Naut.	Moonset 22	23	24	25
°	h m	h m	h m	h m	h m	h m	h m
N 72	21 12	////	////	16 49	19 06	22 10	☐
N 70	20 46	22 37	////	16 42	18 47	21 16	☐
68	20 26	21 51	////	16 37	18 33	20 44	23 45
66	20 11	21 21	////	16 33	18 21	20 20	22 39
64	19 58	20 59	22 48	16 29	18 12	20 02	22 03
62	19 47	20 42	22 06	16 26	18 03	19 47	21 38
60	19 38	20 28	21 39	16 23	17 56	19 35	21 18
N 58	19 30	20 16	21 18	16 21	17 50	19 24	21 02
56	19 23	20 06	21 02	16 19	17 45	19 15	20 48
54	19 17	19 57	20 48	16 17	17 40	19 07	20 36
52	19 12	19 49	20 36	16 15	17 35	18 59	20 26
50	19 06	19 42	20 26	16 14	17 31	18 53	20 17
45	18 56	19 27	20 06	16 10	17 23	18 39	19 57
N 40	18 47	19 15	19 50	16 08	17 16	18 27	19 41
35	18 39	19 06	19 37	16 05	17 10	18 17	19 28
30	18 32	18 57	19 27	16 03	17 04	18 09	19 16
20	18 21	18 44	19 10	15 59	16 55	17 54	18 57
N 10	18 11	18 32	18 58	15 56	16 47	17 41	18 40
0	18 02	18 23	18 47	15 53	16 39	17 29	18 24
S 10	17 53	18 14	18 39	15 50	16 32	17 17	18 08
20	17 43	18 05	18 31	15 47	16 24	17 05	17 51
30	17 32	17 57	18 25	15 43	16 15	16 50	17 31
35	17 26	17 52	18 22	15 41	16 10	16 42	17 20
40	17 19	17 47	18 18	15 39	16 04	16 33	17 07
45	17 11	17 41	18 15	15 36	15 57	16 22	16 52
S 50	17 01	17 35	18 12	15 32	15 49	16 09	16 34
52	16 57	17 32	18 11	15 31	15 45	16 02	16 25
54	16 52	17 29	18 10	15 29	15 41	15 56	16 15
56	16 46	17 25	18 09	15 28	15 37	15 48	16 04
58	16 40	17 21	18 07	15 26	15 32	15 40	15 52
S 60	16 33	17 17	18 06	15 23	15 26	15 30	15 37

Day	SUN Eqn. of Time 00^h	SUN Eqn. of Time 12^h	SUN Mer. Pass.	MOON Mer. Pass. Upper	MOON Mer. Pass. Lower	Age	Phase
d	m s	m s	h m	h m	h m	d	%
22	01 26	01 32	11 58	09 42	22 05	27	9
23	01 37	01 43	11 58	10 28	22 52	28	4
24	01 48	01 54	11 58	11 17	23 43	29	1

	UT	ARIES	VENUS −4.7		MARS +1.2		JUPITER −2.2		SATURN +0.7	
		GHA	GHA	Dec	GHA	Dec	GHA	Dec	GHA	Dec
	d h	° ′	° ′	° ′	° ′	° ′	° ′	° ′	° ′	° ′
	25 00	213 08.6	214 01.3	N 2 28.6	210 57.8	S 0 18.2	247 28.9	S14 26.2	45 46.8	N 7 50.4
	01	228 11.1	229 03.0	28.4	225 58.5	17.4	262 31.0	26.1	60 49.4	50.4
	02	243 13.5	244 04.7	28.1	240 59.2	16.6	277 33.1	26.0	75 51.9	50.4
	03	258 16.0	259 06.3	. . 27.9	255 59.8	. . 15.9	292 35.3	. . 25.9	90 54.5	. . 50.5
	04	273 18.5	274 07.9	27.7	271 00.5	15.1	307 37.4	25.8	105 57.1	50.5
	05	288 20.9	289 09.6	27.5	286 01.2	14.3	322 39.5	25.6	120 59.6	50.5
	06	303 23.4	304 11.2	N 2 27.3	301 01.9	S 0 13.5	337 41.6	S14 25.5	136 02.2	N 7 50.5
	07	318 25.8	319 12.9	27.1	316 02.6	12.8	352 43.7	25.4	151 04.7	50.6
S	08	333 28.3	334 14.5	27.0	331 03.3	12.0	7 45.8	25.3	166 07.3	50.6
A	09	348 30.8	349 16.1	. . 26.8	346 04.0	. . 11.2	22 47.9	. . 25.2	181 09.8	. . 50.6
T	10	3 33.2	4 17.8	26.6	1 04.7	10.4	37 50.1	25.1	196 12.4	50.7
U	11	18 35.7	19 19.4	26.4	16 05.4	09.7	52 52.2	25.0	211 14.9	50.7
R	12	33 38.2	34 21.0	N 2 26.2	31 06.1	S 0 08.9	67 54.3	S14 24.9	226 17.5	N 7 50.7
D	13	48 40.6	49 22.6	26.0	46 06.8	08.1	82 56.4	24.8	241 20.0	50.8
A	14	63 43.1	64 24.3	25.8	61 07.5	07.3	97 58.5	24.7	256 22.6	50.8
Y	15	78 45.6	79 25.9	. . 25.6	76 08.2	. . 06.6	113 00.6	. . 24.6	271 25.1	. . 50.8
	16	93 48.0	94 27.5	25.5	91 08.8	05.8	128 02.7	24.4	286 27.7	50.8
	17	108 50.5	109 29.1	25.3	106 09.5	05.0	143 04.9	24.3	301 30.2	50.9
	18	123 53.0	124 30.7	N 2 25.1	121 10.2	S 0 04.2	158 07.0	S14 24.2	316 32.8	N 7 50.9
	19	138 55.4	139 32.3	24.9	136 10.9	03.5	173 09.1	24.1	331 35.3	50.9
	20	153 57.9	154 33.9	24.7	151 11.6	02.7	188 11.2	24.0	346 37.9	51.0
	21	169 00.3	169 35.5	. . 24.6	166 12.3	. . 01.9	203 13.3	. . 23.9	1 40.4	. . 51.0
	22	184 02.8	184 37.1	24.4	181 13.0	01.1	218 15.4	23.8	16 43.0	51.0
	23	199 05.3	199 38.7	24.2	196 13.7	S 00.4	233 17.6	23.7	31 45.5	51.0
	26 00	214 07.7	214 40.3	N 2 24.1	211 14.4	N 0 00.4	248 19.7	S14 23.6	46 48.1	N 7 51.1
	01	229 10.2	229 41.8	23.9	226 15.1	01.2	263 21.8	23.5	61 50.6	51.1
	02	244 12.7	244 43.4	23.7	241 15.8	01.9	278 23.9	23.4	76 53.2	51.1
	03	259 15.1	259 45.0	. . 23.6	256 16.5	. . 02.7	293 26.0	. . 23.3	91 55.7	. . 51.2
	04	274 17.6	274 46.6	23.4	271 17.2	03.5	308 28.1	23.2	106 58.3	51.2
	05	289 20.1	289 48.2	23.2	286 17.9	04.3	323 30.3	23.0	122 00.8	51.2
	06	304 22.5	304 49.7	N 2 23.1	301 18.5	N 0 05.0	338 32.4	S14 22.9	137 03.4	N 7 51.3
	07	319 25.0	319 51.3	22.9	316 19.2	05.8	353 34.5	22.8	152 05.9	51.3
	08	334 27.5	334 52.9	22.8	331 19.9	06.6	8 36.6	22.7	167 08.4	51.3
S	09	349 29.9	349 54.4	. . 22.6	346 20.6	. . 07.4	23 38.7	. . 22.6	182 11.0	. . 51.3
U	10	4 32.4	4 56.0	22.5	1 21.3	08.1	38 40.9	22.5	197 13.5	51.4
N	11	19 34.8	19 57.5	22.3	16 22.0	08.9	53 43.0	22.4	212 16.1	51.4
D	12	34 37.3	34 59.1	N 2 22.2	31 22.7	N 0 09.7	68 45.1	S14 22.3	227 18.6	N 7 51.4
A	13	49 39.8	50 00.7	22.0	46 23.4	10.5	83 47.2	22.2	242 21.2	51.5
Y	14	64 42.2	65 02.2	21.9	61 24.1	11.2	98 49.3	22.1	257 23.7	51.5
	15	79 44.7	80 03.7	. . 21.7	76 24.8	. . 12.0	113 51.5	. . 22.0	272 26.3	. . 51.5
	16	94 47.2	95 05.3	21.6	91 25.5	12.8	128 53.6	21.9	287 28.8	51.5
	17	109 49.6	110 06.8	21.5	106 26.2	13.6	143 55.7	21.8	302 31.4	51.6
	18	124 52.1	125 08.4	N 2 21.3	121 26.9	N 0 14.3	158 57.8	S14 21.7	317 33.9	N 7 51.6
	19	139 54.6	140 09.9	21.2	136 27.6	15.1	173 59.9	21.5	332 36.5	51.6
	20	154 57.0	155 11.4	21.1	151 28.3	15.9	189 02.1	21.4	347 39.0	51.6
	21	169 59.5	170 13.0	. . 20.9	166 28.9	. . 16.7	204 04.2	. . 21.3	2 41.6	. . 51.7
	22	185 01.9	185 14.5	20.8	181 29.6	17.4	219 06.3	21.2	17 44.1	51.7
	23	200 04.4	200 16.0	20.7	196 30.3	18.2	234 08.4	21.1	32 46.6	51.7
	27 00	215 06.9	215 17.5	N 2 20.5	211 31.0	N 0 19.0	249 10.6	S14 21.0	47 49.2	N 7 51.8
	01	230 09.3	230 19.0	20.4	226 31.7	19.8	264 12.7	20.9	62 51.7	51.8
	02	245 11.8	245 20.6	20.3	241 32.4	20.5	279 14.8	20.8	77 54.3	51.8
	03	260 14.3	260 22.1	. . 20.2	256 33.1	. . 21.3	294 16.9	. . 20.7	92 56.8	. . 51.8
	04	275 16.7	275 23.6	20.0	271 33.8	22.1	309 19.1	20.6	107 59.4	51.9
	05	290 19.2	290 25.1	19.9	286 34.5	22.8	324 21.2	20.5	123 01.9	51.9
	06	305 21.7	305 26.6	N 2 19.8	301 35.2	N 0 23.6	339 23.3	S14 20.4	138 04.5	N 7 51.9
	07	320 24.1	320 28.1	19.7	316 35.9	24.4	354 25.4	20.3	153 07.0	51.9
	08	335 26.6	335 29.6	19.6	331 36.6	25.2	9 27.6	20.2	168 09.6	52.0
M	09	350 29.1	350 31.1	. . 19.5	346 37.3	. . 25.9	24 29.7	. . 20.1	183 12.1	. . 52.0
O	10	5 31.5	5 32.6	19.4	1 38.0	26.7	39 31.8	20.0	198 14.6	52.0
N	11	20 34.0	20 34.1	19.3	16 38.7	27.5	54 33.9	19.9	213 17.2	52.1
D	12	35 36.4	35 35.6	N 2 19.1	31 39.4	N 0 28.3	69 36.1	S14 19.7	228 19.7	N 7 52.1
A	13	50 38.9	50 37.0	19.0	46 40.1	29.0	84 38.2	19.6	243 22.3	52.1
Y	14	65 41.4	65 38.5	18.9	61 40.8	29.8	99 40.3	19.5	258 24.8	52.1
	15	80 43.8	80 40.0	. . 18.8	76 41.4	. . 30.6	114 42.4	. . 19.4	273 27.4	. . 52.2
	16	95 46.3	95 41.5	18.7	91 42.1	31.4	129 44.6	19.3	288 29.9	52.2
	17	110 48.8	110 43.0	18.6	106 42.8	32.1	144 46.7	19.2	303 32.4	52.2
	18	125 51.2	125 44.4	N 2 18.5	121 43.5	N 0 32.9	159 48.8	S14 19.1	318 35.0	N 7 52.2
	19	140 53.7	140 45.9	18.4	136 44.2	33.7	174 50.9	19.0	333 37.5	52.3
	20	155 56.2	155 47.4	18.3	151 44.9	34.4	189 53.1	18.9	348 40.1	52.3
	21	170 58.6	170 48.8	. . 18.3	166 45.6	. . 35.2	204 55.2	. . 18.8	3 42.6	. . 52.3
	22	186 01.1	185 50.3	18.2	181 46.3	36.0	219 57.3	18.7	18 45.2	52.3
	23	201 03.6	200 51.7	18.1	196 47.0	36.8	234 59.4	18.6	33 47.7	52.4
		h m								
	Mer. Pass.	9 41.9	*v* 1.6	*d* 0.1	*v* 0.7	*d* 0.8	*v* 2.1	*d* 0.1	*v* 2.5	*d* 0.0

STARS

Name	SHA	Dec
	° ′	° ′
Acamar	315 21.0	S40 16.
Achernar	335 29.4	S57 11.
Acrux	173 12.5	S63 09.
Adhara	255 15.1	S28 59.
Aldebaran	290 53.2	N16 31.
Alioth	166 22.6	N55 54.5
Alkaid	153 00.7	N49 15.9
Al Na'ir	27 47.6	S46 54.8
Alnilam	275 49.7	S 1 11.8
Alphard	217 59.1	S 8 42.
Alphecca	126 13.2	N26 40.8
Alpheratz	357 47.1	N29 08.4
Altair	62 11.2	N 8 53.4
Ankaa	353 19.0	S42 15.2
Antares	112 29.8	S26 27.3
Arcturus	145 58.2	N19 07.8
Atria	107 34.1	S69 02.6
Avior	234 19.5	S59 32.7
Bellatrix	278 35.6	N 6 21.5
Betelgeuse	271 04.9	N 7 24.5
Canopus	263 57.8	S52 42.3
Capella	280 39.4	N46 00.6
Deneb	49 33.7	N45 18.5
Denebola	182 36.5	N14 31.1
Diphda	348 59.2	S17 56.1
Dubhe	193 54.8	N61 42.1
Elnath	278 16.8	N28 37.0
Eltanin	90 47.2	N51 28.9
Enif	33 50.3	N 9 54.9
Fomalhaut	15 27.5	S29 34.3
Gacrux	172 04.1	S57 10.2
Gienah	175 55.3	S17 35.9
Hadar	148 51.9	S60 25.3
Hamal	328 04.7	N23 30.4
Kaus Aust.	83 47.7	S34 22.8
Kochab	137 18.3	N74 06.9
Markab	13 41.7	N15 15.2
Menkar	314 18.6	N 4 07.6
Menkent	148 10.9	S36 25.2
Miaplacidus	221 40.5	S69 45.7
Mirfak	308 45.4	N49 53.7
Nunki	76 02.0	S26 17.1
Peacock	53 23.9	S56 42.1
Pollux	243 31.5	N28 00.3
Procyon	245 03.0	N 5 12.0
Rasalhague	96 09.1	N12 33.0
Regulus	207 46.6	N11 55.2
Rigel	281 15.3	S 8 11.5
Rigil Kent.	139 55.5	S60 52.5
Sabik	102 15.8	S15 44.3
Schedar	349 44.8	N56 35.2
Shaula	96 25.8	S37 06.6
Sirius	258 36.6	S16 43.9
Spica	158 34.2	S11 12.8
Suhail	222 54.8	S43 28.5
Vega	80 40.9	N38 47.2
Zuben'ubi	137 08.5	S16 05.0

	SHA	Mer. Pass.
	° ′	h m
Venus	0 32.5	9 40
Mars	357 06.6	9 55
Jupiter	34 11.9	7 26
Saturn	192 40.3	20 49

UT d h	SUN GHA ° ′	SUN Dec ° ′	MOON GHA ° ′	v ′	MOON Dec ° ′	d ′	HP ′
25 00	180 29.8	N13 09.9	184 04.3	8.9	N17 14.0	12.4	58.9
01	195 29.9	10.7	198 32.2	8.8	17 26.4	12.2	58.9
02	210 30.0	11.5	213 00.0	8.7	17 38.6	12.1	58.9
03	225 30.1	. . 12.3	227 27.7	8.6	17 50.7	12.1	59.0
04	240 30.2	13.1	241 55.3	8.4	18 02.8	12.0	59.0
05	255 30.3	14.0	256 22.7	8.4	18 14.8	11.8	59.0
06	270 30.4	N13 14.8	270 50.1	8.2	N18 26.6	11.8	59.0
07	285 30.5	15.6	285 17.3	8.2	18 38.4	11.6	59.1
08	300 30.6	16.4	299 44.5	8.0	18 50.0	11.6	59.1
09	315 30.7	. . 17.2	314 11.5	7.9	19 01.6	11.4	59.1
10	330 30.8	18.0	328 38.4	7.9	19 13.0	11.4	59.1
11	345 30.9	18.8	343 05.3	7.7	19 24.4	11.2	59.2
12	0 31.0	N13 19.6	357 32.0	7.6	N19 35.6	11.1	59.2
13	15 31.1	20.5	11 58.6	7.5	19 46.7	11.1	59.2
14	30 31.2	21.3	26 25.1	7.3	19 57.8	10.9	59.2
15	45 31.3	. . 22.1	40 51.4	7.3	20 08.7	10.7	59.2
16	60 31.4	22.9	55 17.7	7.2	20 19.4	10.7	59.3
17	75 31.5	23.7	69 43.9	7.0	20 30.1	10.6	59.3
18	90 31.6	N13 24.5	84 09.9	7.0	N20 40.7	10.4	59.3
19	105 31.7	25.3	98 35.9	6.8	20 51.1	10.3	59.3
20	120 31.9	26.1	113 01.7	6.8	21 01.4	10.2	59.3
21	135 32.0	. . 26.9	127 27.5	6.6	21 11.6	10.0	59.4
22	150 32.1	27.7	141 53.1	6.5	21 21.6	9.9	59.4
23	165 32.2	28.5	156 18.6	6.4	21 31.5	9.8	59.4
26 00	180 32.3	N13 29.4	170 44.0	6.3	N21 41.3	9.7	59.4
01	195 32.4	30.2	185 09.3	6.2	21 51.0	9.5	59.4
02	210 32.5	31.0	199 34.5	6.1	22 00.5	9.4	59.5
03	225 32.6	. . 31.8	213 59.6	6.0	22 09.9	9.3	59.5
04	240 32.7	32.6	228 24.6	5.9	22 19.2	9.1	59.5
05	255 32.8	33.4	242 49.5	5.8	22 28.3	8.9	59.5
06	270 32.9	N13 34.2	257 14.3	5.7	N22 37.2	8.9	59.5
07	285 33.0	35.0	271 39.0	5.6	22 46.1	8.6	59.5
08	300 33.1	35.8	286 03.6	5.5	22 54.7	8.6	59.5
09	315 33.2	. . 36.6	300 28.1	5.4	23 03.3	8.4	59.6
10	330 33.3	37.4	314 52.5	5.3	23 11.7	8.2	59.6
11	345 33.4	38.2	329 16.8	5.2	23 19.9	8.1	59.6
12	0 33.5	N13 39.0	343 41.0	5.1	N23 28.0	7.9	59.6
13	15 33.6	39.8	358 05.1	5.1	23 35.9	7.8	59.6
14	30 33.7	40.6	12 29.2	4.9	23 43.7	7.6	59.6
15	45 33.8	. . 41.4	26 53.1	4.8	23 51.3	7.5	59.6
16	60 33.9	42.2	41 16.9	4.8	23 58.8	7.3	59.7
17	75 34.0	43.0	55 40.7	4.6	24 06.1	7.2	59.7
18	90 34.1	N13 43.8	70 04.3	4.6	N24 13.3	7.0	59.7
19	105 34.2	44.6	84 27.9	4.5	24 20.3	6.8	59.7
20	120 34.3	45.4	98 51.4	4.4	24 27.1	6.6	59.7
21	135 34.4	. . 46.2	113 14.8	4.4	24 33.7	6.5	59.7
22	150 34.5	47.0	127 38.2	4.2	24 40.2	6.3	59.7
23	165 34.6	47.8	142 01.4	4.2	24 46.5	6.2	59.7
27 00	180 34.7	N13 48.6	156 24.6	4.1	N24 52.7	6.0	59.8
01	195 34.7	49.4	170 47.7	4.0	24 58.7	5.8	59.8
02	210 34.8	50.2	185 10.7	3.9	25 04.5	5.6	59.8
03	225 34.9	. . 51.0	199 33.6	3.9	25 10.1	5.5	59.8
04	240 35.0	51.8	213 56.5	3.8	25 15.6	5.3	59.8
05	255 35.1	52.6	228 19.3	3.8	25 20.9	5.1	59.8
06	270 35.2	N13 53.4	242 42.1	3.7	N25 26.0	4.9	59.8
07	285 35.3	54.2	257 04.8	3.6	25 30.9	4.8	59.8
08	300 35.4	55.0	271 27.4	3.6	25 35.7	4.6	59.8
09	315 35.5	. . 55.8	285 50.0	3.5	25 40.3	4.4	59.8
10	330 35.6	56.6	300 12.5	3.4	25 44.7	4.2	59.8
11	345 35.7	57.3	314 34.9	3.4	25 48.9	4.0	59.8
12	0 35.8	N13 58.1	328 57.3	3.4	N25 52.9	3.9	59.9
13	15 35.9	58.9	343 19.7	3.3	25 56.8	3.6	59.9
14	30 36.0	13 59.7	357 42.0	3.2	26 00.4	3.5	59.9
15	45 36.1	14 00.5	12 04.2	3.2	26 03.9	3.3	59.9
16	60 36.2	01.3	26 26.4	3.2	26 07.2	3.1	59.9
17	75 36.3	02.1	40 48.6	3.1	26 10.3	2.9	59.9
18	90 36.4	N14 02.9	55 10.7	3.1	N26 13.2	2.8	59.9
19	105 36.5	03.7	69 32.8	3.1	26 16.0	2.5	59.9
20	120 36.5	04.5	83 54.9	3.0	26 18.5	2.4	59.9
21	135 36.6	. . 05.3	98 16.9	3.0	26 20.9	2.1	59.9
22	150 36.7	06.0	112 38.9	3.0	26 23.0	2.0	59.9
23	165 36.8	06.8	127 00.9	3.0	N26 25.0	1.8	59.9
	SD 15.9	*d* 0.8	SD	16.1		16.2	16.3

Lat.	Twilight Naut.	Twilight Civil	Sunrise	Moonrise 25	26	27	28
°	h m	h m	h m	h m	h m	h m	h m
N 72	////	////	02 30	00 53	▭	▭	▭
N 70	////	00 50	02 59	01 49	▭	▭	▭
68	////	01 50	03 20	02 22	01 17	▭	▭
66	////	02 24	03 37	02 47	02 24	▭	▭
64	00 44	02 47	03 50	03 07	03 00	02 50	▭
62	01 37	03 06	04 02	03 23	03 26	03 37	04 10
60	02 08	03 21	04 12	03 36	03 47	04 08	04 50
N 58	02 30	03 34	04 20	03 48	04 04	04 31	05 18
56	02 47	03 45	04 27	03 58	04 18	04 50	05 40
54	03 02	03 54	04 34	04 07	04 31	05 06	05 57
52	03 14	04 02	04 40	04 15	04 42	05 19	06 13
50	03 25	04 10	04 45	04 22	04 52	05 32	06 26
45	03 46	04 25	04 57	04 38	05 12	05 57	06 53
N 40	04 03	04 38	05 07	04 51	05 29	06 17	07 15
35	04 16	04 48	05 15	05 02	05 44	06 34	07 33
30	04 27	04 57	05 22	05 12	05 56	06 48	07 48
20	04 44	05 11	05 34	05 28	06 17	07 13	08 14
N 10	04 58	05 23	05 45	05 43	06 36	07 34	08 37
0	05 08	05 33	05 54	05 57	06 54	07 55	08 58
S 10	05 18	05 43	06 04	06 11	07 12	08 15	09 19
20	05 26	05 52	06 14	06 26	07 31	08 37	09 42
30	05 33	06 01	06 26	06 44	07 53	09 02	10 08
35	05 37	06 06	06 33	06 54	08 06	09 17	10 24
40	05 40	06 12	06 40	07 06	08 21	09 34	10 42
45	05 44	06 18	06 49	07 20	08 39	09 55	11 04
S 50	05 47	06 25	06 59	07 37	09 01	10 22	11 31
52	05 49	06 29	07 04	07 45	09 12	10 35	11 45
54	05 51	06 32	07 09	07 54	09 24	10 50	12 01
56	05 52	06 36	07 15	08 04	09 39	11 07	12 19
58	05 54	06 40	07 22	08 16	09 55	11 28	12 41
S 60	05 56	06 45	07 29	08 29	10 15	11 55	13 10

Lat.	Sunset	Twilight Civil	Twilight Naut.	Moonset 25	26	27	28
°	h m	h m	h m	h m	h m	h m	h m
N 72	21 31	////	////	▭	▭	▭	▭
N 70	21 01	23 25	////	▭	▭	▭	▭
68	20 39	22 12	////	23 45	▭	▭	▭
66	20 21	21 36	////	22 39	▭	▭	▭
64	20 07	21 11	23 29	22 03	24 19	00 19	▭
62	19 56	20 52	22 24	21 38	23 32	25 11	01 11
60	19 46	20 37	21 52	21 18	23 02	24 31	00 31
N 58	19 37	20 24	21 29	21 02	22 39	24 03	00 03
56	19 29	20 13	21 11	20 48	22 21	23 42	24 42
54	19 23	20 03	20 56	20 36	22 05	23 24	24 25
52	19 17	19 55	20 43	20 26	21 52	23 09	24 10
50	19 11	19 47	20 32	20 17	21 40	22 56	23 58
45	18 59	19 31	20 11	19 57	21 15	22 28	23 31
N 40	18 50	19 19	19 54	19 41	20 56	22 07	23 10
35	18 41	19 08	19 40	19 28	20 40	21 49	22 52
30	18 34	18 59	19 29	19 16	20 26	21 34	22 37
20	18 22	18 45	19 12	18 57	20 02	21 08	22 11
N 10	18 11	18 33	18 58	18 40	19 41	20 46	21 49
0	18 01	18 22	18 47	18 24	19 22	20 25	21 28
S 10	17 51	18 13	18 38	18 08	19 03	20 04	21 07
20	17 41	18 04	18 30	17 51	18 43	19 41	20 45
30	17 29	17 54	18 22	17 31	18 19	19 15	20 19
35	17 23	17 49	18 18	17 20	18 06	19 00	20 03
40	17 15	17 43	18 15	17 07	17 50	18 42	19 45
45	17 06	17 37	18 11	16 52	17 31	18 21	19 24
S 50	16 56	17 29	18 07	16 34	17 07	17 54	18 56
52	16 51	17 26	18 06	16 25	16 56	17 41	18 43
54	16 45	17 23	18 04	16 15	16 43	17 26	18 27
56	16 39	17 19	18 02	16 04	16 29	17 08	18 09
58	16 33	17 14	18 01	15 52	16 12	16 47	17 46
S 60	16 25	17 10	17 59	15 37	15 51	16 20	17 17

Day	SUN Eqn. of Time 00h	12h	Mer. Pass.	MOON Mer. Pass. Upper	Lower	Age	Phase
d	m s	m s	h m	h m	h m	d	%
25	01 59	02 04	11 58	12 10	24 39	00	0
26	02 09	02 14	11 58	13 08	00 39	01	3
27	02 18	02 23	11 58	14 10	01 38	02	8

UT (d h)	ARIES GHA ° ′	VENUS −4.7 GHA ° ′	VENUS Dec ° ′	MARS +1.2 GHA ° ′	MARS Dec ° ′	JUPITER −2.2 GHA ° ′	JUPITER Dec ° ′	SATURN +0.7 GHA ° ′	SATURN Dec ° ′
28 00 TUESDAY	216 06.0	215 53.2	N 2 18.0	211 47.7	N 0 37.5	250 01.6	S14 18.5	48 50.2	N 7 52.4
01	231 08.5	230 54.7	17.9	226 48.4	38.3	265 03.7	18.4	63 52.8	52.4
02	246 10.9	245 56.1	17.8	241 49.1	39.1	280 05.8	18.3	78 55.3	52.4
03	261 13.4	260 57.5	17.7	256 49.8	39.9	295 08.0	18.2	93 57.9	52.5
04	276 15.9	275 59.0	17.6	271 50.5	40.6	310 10.1	18.1	109 00.4	52.5
05	291 18.3	291 00.4	17.6	286 51.2	41.4	325 12.2	18.0	124 02.9	52.5
06	306 20.8	306 01.9	N 2 17.5	301 51.9	N 0 42.2	340 14.4	S14 17.9	139 05.5	N 7 52.5
07	321 23.3	321 03.3	17.4	316 52.6	43.0	355 16.5	17.8	154 08.0	52.6
08	336 25.7	336 04.7	17.3	331 53.3	43.7	10 18.6	17.7	169 10.6	52.6
09	351 28.2	351 06.2	17.3	346 54.0	44.5	25 20.7	17.6	184 13.1	52.6
10	6 30.7	6 07.6	17.2	1 54.7	45.3	40 22.9	17.4	199 15.6	52.7
11	21 33.1	21 09.0	17.1	16 55.3	46.0	55 25.0	17.3	214 18.2	52.7
12	36 35.6	36 10.5	N 2 17.1	31 56.0	N 0 46.8	70 27.1	S14 17.2	229 20.7	N 7 52.7
13	51 38.1	51 11.9	17.0	46 56.7	47.6	85 29.3	17.1	244 23.3	52.7
14	66 40.5	66 13.3	16.9	61 57.4	48.4	100 31.4	17.0	259 25.8	52.8
15	81 43.0	81 14.7	16.9	76 58.1	49.1	115 33.5	16.9	274 28.3	52.8
16	96 45.4	96 16.1	16.8	91 58.8	49.9	130 35.7	16.8	289 30.9	52.8
17	111 47.9	111 17.5	16.7	106 59.5	50.7	145 37.8	16.7	304 33.4	52.8
18	126 50.4	126 18.9	N 2 16.7	122 00.2	N 0 51.4	160 39.9	S14 16.6	319 36.0	N 7 52.8
19	141 52.8	141 20.3	16.6	137 00.9	52.2	175 42.1	16.5	334 38.5	52.9
20	156 55.3	156 21.8	16.6	152 01.6	53.0	190 44.2	16.4	349 41.0	52.9
21	171 57.8	171 23.2	16.5	167 02.3	53.8	205 46.3	16.3	4 43.6	52.9
22	187 00.2	186 24.6	16.5	182 03.0	54.5	220 48.5	16.2	19 46.1	52.9
23	202 02.7	201 25.9	16.4	197 03.7	55.3	235 50.6	16.1	34 48.7	53.0
29 00 WEDNESDAY	217 05.2	216 27.3	N 2 16.4	212 04.4	N 0 56.1	250 52.7	S14 16.0	49 51.2	N 7 53.0
01	232 07.6	231 28.7	16.3	227 05.1	56.9	265 54.9	15.9	64 53.7	53.0
02	247 10.1	246 30.1	16.3	242 05.8	57.6	280 57.0	15.8	79 56.3	53.0
03	262 12.6	261 31.5	16.2	257 06.5	58.4	295 59.1	15.7	94 58.8	53.1
04	277 15.0	276 32.9	16.2	272 07.2	59.2	311 01.3	15.6	110 01.3	53.1
05	292 17.5	291 34.3	16.1	287 07.9	0 59.9	326 03.4	15.5	125 03.9	53.1
06	307 19.9	306 35.6	N 2 16.1	302 08.6	N 1 00.7	341 05.5	S14 15.4	140 06.4	N 7 53.1
07	322 22.4	321 37.0	16.1	317 09.3	01.5	356 07.7	15.3	155 09.0	53.2
08	337 24.9	336 38.4	16.0	332 10.0	02.3	11 09.8	15.2	170 11.5	53.2
09	352 27.3	351 39.8	16.0	347 10.7	03.0	26 11.9	15.1	185 14.0	53.2
10	7 29.8	6 41.1	16.0	2 11.4	03.8	41 14.1	15.0	200 16.6	53.2
11	22 32.3	21 42.5	15.9	17 12.1	04.6	56 16.2	14.9	215 19.1	53.3
12	37 34.7	36 43.9	N 2 15.9	32 12.7	N 1 05.3	71 18.4	S14 14.8	230 21.6	N 7 53.3
13	52 37.2	51 45.2	15.9	47 13.4	06.1	86 20.5	14.7	245 24.2	53.3
14	67 39.7	66 46.6	15.8	62 14.1	06.9	101 22.6	14.6	260 26.7	53.3
15	82 42.1	81 47.9	15.8	77 14.8	07.7	116 24.8	14.5	275 29.2	53.3
16	97 44.6	96 49.3	15.8	92 15.5	08.4	131 26.9	14.4	290 31.8	53.4
17	112 47.1	111 50.6	15.8	107 16.2	09.2	146 29.0	14.3	305 34.3	53.4
18	127 49.5	126 52.0	N 2 15.8	122 16.9	N 1 10.0	161 31.2	S14 14.2	320 36.9	N 7 53.4
19	142 52.0	141 53.3	15.7	137 17.6	10.7	176 33.3	14.1	335 39.4	53.4
20	157 54.4	156 54.7	15.7	152 18.3	11.5	191 35.5	14.0	350 41.9	53.5
21	172 56.9	171 56.0	15.7	167 19.0	12.3	206 37.6	13.9	5 44.5	53.5
22	187 59.4	186 57.3	15.7	182 19.7	13.1	221 39.7	13.8	20 47.0	53.5
23	203 01.8	201 58.7	15.7	197 20.4	13.8	236 41.9	13.7	35 49.5	53.5
30 00 THURSDAY	218 04.3	217 00.0	N 2 15.7	212 21.1	N 1 14.6	251 44.0	S14 13.6	50 52.1	N 7 53.6
01	233 06.8	232 01.3	15.7	227 21.8	15.4	266 46.2	13.5	65 54.6	53.6
02	248 09.2	247 02.7	15.6	242 22.5	16.1	281 48.3	13.4	80 57.1	53.6
03	263 11.7	262 04.0	15.6	257 23.2	16.9	296 50.4	13.3	95 59.7	53.6
04	278 14.2	277 05.3	15.6	272 23.9	17.7	311 52.6	13.2	111 02.2	53.6
05	293 16.6	292 06.6	15.6	287 24.6	18.4	326 54.7	13.1	126 04.7	53.7
06	308 19.1	307 07.9	N 2 15.6	302 25.3	N 1 19.2	341 56.9	S14 13.0	141 07.3	N 7 53.7
07	323 21.6	322 09.3	15.6	317 26.0	20.0	356 59.0	12.9	156 09.8	53.7
08	338 24.0	337 10.6	15.6	332 26.7	20.8	12 01.1	12.8	171 12.3	53.7
09	353 26.5	352 11.9	15.6	347 27.4	21.5	27 03.3	12.7	186 14.9	53.8
10	8 28.9	7 13.2	15.6	2 28.1	22.3	42 05.4	12.6	201 17.4	53.8
11	23 31.4	22 14.5	15.6	17 28.8	23.1	57 07.6	12.5	216 19.9	53.8
12	38 33.9	37 15.8	N 2 15.7	32 29.5	N 1 23.8	72 09.7	S14 12.4	231 22.5	N 7 53.8
13	53 36.3	52 17.1	15.7	47 30.2	24.6	87 11.9	12.3	246 25.0	53.8
14	68 38.8	67 18.4	15.7	62 30.9	25.4	102 14.0	12.2	261 27.5	53.9
15	83 41.3	82 19.7	15.7	77 31.6	26.1	117 16.1	12.1	276 30.1	53.9
16	98 43.7	97 21.0	15.7	92 32.3	26.9	132 18.3	12.0	291 32.6	53.9
17	113 46.2	112 22.3	15.7	107 33.0	27.7	147 20.4	11.9	306 35.1	53.9
18	128 48.7	127 23.6	N 2 15.7	122 33.7	N 1 28.5	162 22.6	S14 11.8	321 37.6	N 7 53.9
19	143 51.1	142 24.9	15.7	137 34.4	29.2	177 24.7	11.7	336 40.2	54.0
20	158 53.6	157 26.1	15.8	152 35.1	30.0	192 26.9	11.6	351 42.7	54.0
21	173 56.0	172 27.4	15.8	167 35.8	30.8	207 29.0	11.5	6 45.2	54.0
22	188 58.5	187 28.7	15.8	182 36.4	31.5	222 31.2	11.4	21 47.8	54.0
23	204 01.0	202 30.0	15.8	197 37.1	32.3	237 33.3	11.3	36 50.3	54.1
Mer. Pass. (h m)	9 30.1	*v* 1.4	*d* 0.0	*v* 0.7	*d* 0.8	*v* 2.1	*d* 0.1	*v* 2.5	*d* 0.0

STARS

Name	SHA ° ′	Dec ° ′
Acamar	315 21.0	S40 16.0
Achernar	335 29.4	S57 11.2
Acrux	173 12.5	S63 09.4
Adhara	255 15.1	S28 59.3
Aldebaran	290 53.2	N16 31.7
Alioth	166 22.6	N55 54.5
Alkaid	153 00.7	N49 15.9
Al Na'ir	27 47.6	S46 54.7
Alnilam	275 49.7	S 1 11.8
Alphard	217 59.1	S 8 42.1
Alphecca	126 13.2	N26 40.8
Alpheratz	357 47.1	N29 08.4
Altair	62 11.2	N 8 53.4
Ankaa	353 18.9	S42 15.2
Antares	112 29.7	S26 27.3
Arcturus	145 58.2	N19 07.8
Atria	107 34.0	S69 02.6
Avior	234 19.5	S59 32.7
Bellatrix	278 35.6	N 6 21.5
Betelgeuse	271 04.9	N 7 24.5
Canopus	263 57.8	S52 42.3
Capella	280 39.4	N46 00.6
Deneb	49 33.7	N45 18.5
Denebola	182 36.5	N14 31.1
Diphda	348 59.2	S17 56.1
Dubhe	193 54.8	N61 42.2
Elnath	278 16.8	N28 37.0
Eltanin	90 47.2	N51 28.9
Enif	33 50.2	N 9 55.0
Fomalhaut	15 27.5	S29 34.3
Gacrux	172 04.1	S57 10.2
Gienah	175 55.3	S17 35.9
Hadar	148 51.9	S60 25.3
Hamal	328 04.6	N23 30.4
Kaus Aust.	83 47.7	S34 22.8
Kochab	137 18.3	N74 06.9
Markab	13 41.6	N15 15.2
Menkar	314 18.6	N 4 07.6
Menkent	148 10.9	S36 25.2
Miaplacidus	221 40.5	S69 45.7
Mirfak	308 45.4	N49 53.7
Nunki	76 01.9	S26 17.1
Peacock	53 23.9	S56 42.1
Pollux	243 31.6	N28 00.3
Procyon	245 03.0	N 5 12.0
Rasalhague	96 09.1	N12 33.0
Regulus	207 46.6	N11 55.2
Rigel	281 15.3	S 8 11.5
Rigil Kent.	139 55.5	S60 52.5
Sabik	102 15.8	S15 44.3
Schedar	349 44.8	N56 35.2
Shaula	96 25.8	S37 06.6
Sirius	258 36.6	S16 43.9
Spica	158 34.2	S11 12.8
Suhail	222 54.8	S43 28.5
Vega	80 40.8	N38 47.2
Zuben'ubi	137 08.5	S16 05.0

	SHA ° ′	Mer. Pass. h m
Venus	359 22.2	9 33
Mars	354 59.2	9 51
Jupiter	33 47.6	7 15
Saturn	192 46.0	20 37

UT		SUN GHA	SUN Dec	MOON GHA	v	MOON Dec	d	HP
	d h	° ′	° ′	° ′	′	° ′	′	′
	28 00	180 36.9	N14 07.6	141 22.9	2.9	N26 26.8	1.6	59.9
	01	195 37.0	08.4	155 44.8	2.9	26 28.4	1.4	59.9
	02	210 37.1	09.2	170 06.7	3.0	26 29.8	1.2	59.9
	03	225 37.2	. . 10.0	184 28.7	2.9	26 31.0	1.0	59.9
	04	240 37.3	10.8	198 50.6	2.8	26 32.0	0.8	59.9
	05	255 37.4	11.5	213 12.4	2.9	26 32.8	0.7	59.9
	06	270 37.5	N14 12.3	227 34.3	2.9	N26 33.5	0.4	59.9
	07	285 37.6	13.1	241 56.2	2.9	26 33.9	0.3	59.9
T	08	300 37.6	13.9	256 18.1	2.9	26 34.2	0.0	59.9
U	09	315 37.7	. . 14.7	270 40.0	2.8	26 34.2	0.1	59.9
E	10	330 37.8	15.5	285 01.8	2.9	26 34.1	0.3	59.9
S	11	345 37.9	16.3	299 23.7	2.9	26 33.8	0.5	59.9
D	12	0 38.0	N14 17.0	313 45.6	2.9	N26 33.3	0.7	59.9
A	13	15 38.1	17.8	328 07.5	3.0	26 32.6	0.9	59.9
Y	14	30 38.2	18.6	342 29.5	2.9	26 31.7	1.1	59.9
	15	45 38.3	. . 19.4	356 51.4	3.0	26 30.6	1.3	59.9
	16	60 38.4	20.2	11 13.4	3.0	26 29.3	1.5	59.9
	17	75 38.4	20.9	25 35.4	3.0	26 27.8	1.6	59.9
	18	90 38.5	N14 21.7	39 57.4	3.0	N26 26.2	1.9	59.9
	19	105 38.6	22.5	54 19.4	3.1	26 24.3	2.0	59.9
	20	120 38.7	23.3	68 41.5	3.1	26 22.3	2.3	59.9
	21	135 38.8	. . 24.1	83 03.6	3.1	26 20.0	2.4	59.9
	22	150 38.9	24.8	97 25.7	3.2	26 17.6	2.6	59.9
	23	165 39.0	25.6	111 47.9	3.2	26 15.0	2.8	59.9
	29 00	180 39.1	N14 26.4	126 10.1	3.3	N26 12.2	3.0	59.9
	01	195 39.1	27.2	140 32.4	3.3	26 09.2	3.1	59.9
	02	210 39.2	28.0	154 54.7	3.4	26 06.1	3.4	59.8
	03	225 39.3	. . 28.7	169 17.1	3.4	26 02.7	3.5	59.8
	04	240 39.4	29.5	183 39.5	3.4	25 59.2	3.7	59.8
	05	255 39.5	30.3	198 01.9	3.6	25 55.5	3.9	59.8
	06	270 39.6	N14 31.1	212 24.5	3.5	N25 51.6	4.1	59.8
W	07	285 39.7	31.8	226 47.0	3.7	25 47.5	4.2	59.8
E	08	300 39.7	32.6	241 09.7	3.6	25 43.3	4.5	59.8
D	09	315 39.8	. . 33.4	255 32.3	3.8	25 38.8	4.6	59.8
N	10	330 39.9	34.2	269 55.1	3.8	25 34.2	4.8	59.8
E	11	345 40.0	34.9	284 17.9	3.9	25 29.4	4.9	59.8
S	12	0 40.1	N14 35.7	298 40.8	3.9	N25 24.5	5.2	59.8
D	13	15 40.2	36.5	313 03.7	4.1	25 19.3	5.3	59.8
A	14	30 40.2	37.2	327 26.8	4.1	25 14.0	5.5	59.8
Y	15	45 40.3	. . 38.0	341 49.9	4.1	25 08.5	5.6	59.8
	16	60 40.4	38.8	356 13.0	4.3	25 02.9	5.9	59.7
	17	75 40.5	39.6	10 36.3	4.3	24 57.0	6.0	59.7
	18	90 40.6	N14 40.3	24 59.6	4.4	N24 51.0	6.1	59.7
	19	105 40.7	41.1	39 23.0	4.5	24 44.9	6.4	59.7
	20	120 40.7	41.9	53 46.5	4.6	24 38.5	6.4	59.7
	21	135 40.8	. . 42.6	68 10.1	4.6	24 32.1	6.7	59.7
	22	150 40.9	43.4	82 33.7	4.7	24 25.4	6.8	59.7
	23	165 41.0	44.2	96 57.4	4.9	24 18.6	7.0	59.7

UT		SUN GHA	SUN Dec	MOON GHA	v	MOON Dec	d	HP
	30 00	180 41.1	N14 44.9	111 21.3	4.9	N24 11.6	7.2	59.7
	01	195 41.2	45.7	125 45.2	5.0	24 04.4	7.2	59.7
	02	210 41.2	46.5	140 09.2	5.0	23 57.2	7.5	59.6
	03	225 41.3	. . 47.2	154 33.2	5.2	23 49.7	7.6	59.6
	04	240 41.4	48.0	168 57.4	5.3	23 42.1	7.8	59.6
	05	255 41.5	48.8	183 21.7	5.4	23 34.3	7.9	59.6
	06	270 41.6	N14 49.5	197 46.1	5.4	N23 26.4	8.1	59.6
T	07	285 41.6	50.3	212 10.5	5.6	23 18.3	8.2	59.6
H	08	300 41.7	51.1	226 35.1	5.6	23 10.1	8.3	59.6
	09	315 41.8	. . 51.8	240 59.7	5.8	23 01.8	8.5	59.6
U	10	330 41.9	52.6	255 24.5	5.8	22 53.3	8.7	59.5
R	11	345 42.0	53.4	269 49.3	5.9	22 44.6	8.8	59.5
S	12	0 42.0	N14 54.1	284 14.2	6.1	N22 35.8	8.9	59.5
D	13	15 42.1	54.9	298 39.3	6.1	22 26.9	9.1	59.5
A	14	30 42.2	55.6	313 04.4	6.2	22 17.8	9.2	59.5
Y	15	45 42.3	. . 56.4	327 29.6	6.4	22 08.6	9.3	59.5
	16	60 42.3	57.2	341 55.0	6.4	21 59.3	9.5	59.5
	17	75 42.4	57.9	356 20.4	6.6	21 49.8	9.6	59.4
	18	90 42.5	N14 58.7	10 46.0	6.6	N21 40.2	9.8	59.4
	19	105 42.6	14 59.4	25 11.6	6.8	21 30.4	9.9	59.4
	20	120 42.7	15 00.2	39 37.4	6.8	21 20.5	10.0	59.4
	21	135 42.7	. . 01.0	54 03.2	6.9	21 10.5	10.1	59.4
	22	150 42.8	01.7	68 29.1	7.1	21 00.4	10.2	59.4
	23	165 42.9	02.5	82 55.2	7.1	N20 50.2	10.4	59.4
		SD 15.9	d 0.8	SD 16.3		16.3		16.2

Lat.	Twilight Naut.	Twilight Civil	Sunrise	Moonrise 28	29	30	1
°	h m	h m	h m	h m	h m	h m	h m
N 72	////	////	02 08	□	□	□	□
N 70	////	////	02 42	□	□	□	06 50
68	////	01 26	03 06	□	□	□	07 53
66	////	02 07	03 25	□	□	05 47	08 27
64	////	02 34	03 40	□	04 16	06 45	08 52
62	01 16	02 55	03 53	04 10	05 29	07 19	09 12
60	01 53	03 11	04 03	04 50	06 05	07 43	09 27
N 58	02 18	03 25	04 12	05 18	06 31	08 03	09 41
56	02 37	03 37	04 21	05 40	06 51	08 19	09 52
54	02 53	03 47	04 28	05 57	07 08	08 33	10 02
52	03 06	03 56	04 34	06 13	07 23	08 45	10 11
50	03 18	04 04	04 40	06 26	07 35	08 55	10 19
45	03 41	04 20	04 52	06 53	08 01	09 17	10 36
N 40	03 58	04 33	05 03	07 15	08 22	09 35	10 50
35	04 12	04 44	05 11	07 33	08 39	09 50	11 01
30	04 24	04 54	05 19	07 48	08 54	10 03	11 11
20	04 42	05 09	05 32	08 14	09 19	10 25	11 29
N 10	04 56	05 22	05 43	08 37	09 41	10 44	11 44
0	05 08	05 33	05 54	08 58	10 01	11 01	11 58
S 10	05 18	05 43	06 04	09 19	10 21	11 19	12 12
20	05 26	05 53	06 15	09 42	10 43	11 38	12 27
30	05 35	06 03	06 28	10 08	11 08	11 59	12 43
35	05 39	06 09	06 35	10 24	11 23	12 12	12 53
40	05 43	06 15	06 43	10 42	11 40	12 27	13 04
45	05 47	06 22	06 52	11 04	12 00	12 44	13 18
S 50	05 52	06 30	07 04	11 31	12 25	13 05	13 33
52	05 53	06 33	07 09	11 45	12 38	13 15	13 41
54	05 55	06 37	07 15	12 01	12 52	13 26	13 49
56	05 57	06 41	07 21	12 19	13 08	13 39	13 58
58	05 59	06 46	07 28	12 41	13 27	13 53	14 09
S 60	06 02	06 51	07 36	13 10	13 51	14 11	14 21

Lat.	Sunset	Twilight Civil	Twilight Naut.	Moonset 28	29	30	1
°	h m	h m	h m	h m	h m	h m	h m
N 72	21 53	////	////	□	□	□	□
N 70	21 17	////	////	□	□	□	04 55
68	20 52	22 37	////	□	□	□	03 52
66	20 32	21 53	////	□	□	03 57	03 16
64	20 17	21 24	////	□	03 19	02 58	02 50
62	20 04	21 03	22 46	01 11	02 05	02 24	02 30
60	19 53	20 46	22 06	00 31	01 29	01 58	02 13
N 58	19 44	20 32	21 40	00 03	01 03	01 38	01 59
56	19 36	20 20	21 20	24 42	00 42	01 22	01 47
54	19 28	20 09	21 04	24 25	00 25	01 07	01 36
52	19 22	20 00	20 50	24 10	00 10	00 55	01 26
50	19 16	19 52	20 39	23 58	24 44	00 44	01 18
45	19 03	19 35	20 15	23 31	24 21	00 21	01 00
N 40	18 53	19 22	19 57	23 10	24 02	00 02	00 45
35	18 44	19 11	19 43	22 52	23 47	24 32	00 32
30	18 36	19 01	19 31	22 37	23 33	24 21	00 21
20	18 23	18 46	19 13	22 11	23 10	24 01	00 01
N 10	18 11	18 33	18 59	21 49	22 49	23 45	24 35
0	18 01	18 22	18 47	21 28	22 30	23 29	24 23
S 10	17 50	18 12	18 37	21 07	22 11	23 13	24 11
20	17 39	18 02	18 28	20 45	21 51	22 56	23 58
30	17 27	17 51	18 20	20 19	21 27	22 36	23 43
35	17 19	17 46	18 15	20 03	21 12	22 24	23 35
40	17 11	17 39	18 11	19 45	20 56	22 10	23 25
45	17 02	17 32	18 07	19 24	20 36	21 54	23 13
S 50	16 50	17 24	18 02	18 56	20 12	21 35	22 59
52	16 45	17 21	18 00	18 43	20 00	21 25	22 52
54	16 39	17 17	17 59	18 27	19 46	21 14	22 44
56	16 33	17 12	17 56	18 09	19 30	21 02	22 36
58	16 25	17 08	17 54	17 46	19 11	20 48	22 26
S 60	16 17	17 02	17 52	17 17	18 47	20 31	22 15

Day	SUN Eqn. of Time 00^h	12^h	Mer. Pass.	MOON Mer. Pass. Upper	Lower	Age	Phase
d	m s	m s	h m	h m	h m	d %	
28	02 27	02 32	11 57	15 13	02 41	03 15	
29	02 36	02 40	11 57	16 16	03 45	04 24	
30	02 44	02 48	11 57	17 15	04 46	05 35	

	UT d h	ARIES GHA ° ′	VENUS −4.7 GHA ° ′	VENUS Dec ° ′	MARS +1.2 GHA ° ′	MARS Dec ° ′	JUPITER −2.2 GHA ° ′	JUPITER Dec ° ′	SATURN +0.8 GHA ° ′	SATURN Dec ° ′
	1 00	219 03.4	217 31.3	N 2 15.9	212 37.8	N 1 33.1	252 35.4	S14 11.2	51 52.8	N 7 54.1
	01	234 05.9	232 32.5	15.9	227 38.5	33.8	267 37.6	11.1	66 55.4	54.1
	02	249 08.4	247 33.8	15.9	242 39.2	34.6	282 39.7	11.0	81 57.9	54.1
	03	264 10.8	262 35.1	15.9	257 39.9	35.4	297 41.9	10.9	97 00.4	54.1
	04	279 13.3	277 36.3	16.0	272 40.6	36.2	312 44.0	10.8	112 03.0	54.2
	05	294 15.8	292 37.6	16.0	287 41.3	36.9	327 46.2	10.7	127 05.5	54.2
	06	309 18.2	307 38.8	N 2 16.0	302 42.0	N 1 37.7	342 48.3	S14 10.6	142 08.0	N 7 54.2
	07	324 20.7	322 40.1	16.1	317 42.7	38.5	357 50.5	10.5	157 10.5	54.2
	08	339 23.2	337 41.4	16.1	332 43.4	39.2	12 52.6	10.4	172 13.1	54.2
F	09	354 25.6	352 42.6	16.2	347 44.1	40.0	27 54.8	10.3	187 15.6	54.3
R	10	9 28.1	7 43.9	16.2	2 44.8	40.8	42 56.9	10.2	202 18.1	54.3
I	11	24 30.5	22 45.1	16.2	17 45.5	41.5	57 59.1	10.1	217 20.7	54.3
D	12	39 33.0	37 46.4	N 2 16.3	32 46.2	N 1 42.3	73 01.2	S14 10.0	232 23.2	N 7 54.3
A	13	54 35.5	52 47.6	16.3	47 46.9	43.1	88 03.4	09.9	247 25.7	54.3
Y	14	69 37.9	67 48.8	16.4	62 47.6	43.8	103 05.5	09.8	262 28.2	54.4
	15	84 40.4	82 50.1	16.4	77 48.3	44.6	118 07.7	09.7	277 30.8	54.4
	16	99 42.9	97 51.3	16.5	92 49.0	45.4	133 09.8	09.6	292 33.3	54.4
	17	114 45.3	112 52.6	16.5	107 49.7	46.1	148 12.0	09.5	307 35.8	54.4
	18	129 47.8	127 53.8	N 2 16.6	122 50.4	N 1 46.9	163 14.1	S14 09.4	322 38.4	N 7 54.4
	19	144 50.3	142 55.0	16.6	137 51.1	47.7	178 16.3	09.3	337 40.9	54.5
	20	159 52.7	157 56.2	16.7	152 51.8	48.5	193 18.4	09.2	352 43.4	54.5
	21	174 55.2	172 57.5	16.8	167 52.5	49.2	208 20.6	09.1	7 45.9	54.5
	22	189 57.7	187 58.7	16.8	182 53.2	50.0	223 22.7	09.0	22 48.5	54.5
	23	205 00.1	202 59.9	16.9	197 53.9	50.8	238 24.9	08.9	37 51.0	54.5
	2 00	220 02.6	218 01.1	N 2 16.9	212 54.6	N 1 51.5	253 27.0	S14 08.8	52 53.5	N 7 54.6
	01	235 05.0	233 02.4	17.0	227 55.3	52.3	268 29.2	08.7	67 56.0	54.6
	02	250 07.5	248 03.6	17.1	242 56.0	53.1	283 31.3	08.6	82 58.6	54.6
	03	265 10.0	263 04.8	17.1	257 56.7	53.8	298 33.5	08.5	98 01.1	54.6
	04	280 12.4	278 06.0	17.2	272 57.4	54.6	313 35.6	08.4	113 03.6	54.6
	05	295 14.9	293 07.2	17.3	287 58.1	55.4	328 37.8	08.3	128 06.1	54.6
	06	310 17.4	308 08.4	N 2 17.3	302 58.8	N 1 56.1	343 39.9	S14 08.2	143 08.7	N 7 54.7
	07	325 19.8	323 09.6	17.4	317 59.5	56.9	358 42.1	08.1	158 11.2	54.7
S	08	340 22.3	338 10.8	17.5	333 00.2	57.7	13 44.2	08.0	173 13.7	54.7
A	09	355 24.8	353 12.0	17.6	348 00.9	58.4	28 46.4	07.9	188 16.2	54.7
T	10	10 27.2	8 13.2	17.6	3 01.6	1 59.2	43 48.5	07.8	203 18.8	54.7
U	11	25 29.7	23 14.4	17.7	18 02.3	2 00.0	58 50.7	07.7	218 21.3	54.8
R	12	40 32.2	38 15.6	N 2 17.8	33 03.0	N 2 00.7	73 52.8	S14 07.6	233 23.8	N 7 54.8
D	13	55 34.6	53 16.8	17.9	48 03.7	01.5	88 55.0	07.6	248 26.3	54.8
A	14	70 37.1	68 18.0	18.0	63 04.4	02.3	103 57.2	07.5	263 28.9	54.8
Y	15	85 39.5	83 19.1	18.0	78 05.1	03.0	118 59.3	07.4	278 31.4	54.8
	16	100 42.0	98 20.3	18.1	93 05.8	03.8	134 01.5	07.3	293 33.9	54.8
	17	115 44.5	113 21.5	18.2	108 06.5	04.6	149 03.6	07.2	308 36.4	54.9
	18	130 46.9	128 22.7	N 2 18.3	123 07.2	N 2 05.3	164 05.8	S14 07.1	323 39.0	N 7 54.9
	19	145 49.4	143 23.9	18.4	138 07.9	06.1	179 07.9	07.0	338 41.5	54.9
	20	160 51.9	158 25.0	18.5	153 08.6	06.9	194 10.1	06.9	353 44.0	54.9
	21	175 54.3	173 26.2	18.6	168 09.3	07.6	209 12.2	06.8	8 46.5	54.9
	22	190 56.8	188 27.4	18.7	183 10.0	08.4	224 14.4	06.7	23 49.1	55.0
	23	205 59.3	203 28.5	18.8	198 10.7	09.2	239 16.6	06.6	38 51.6	55.0
	3 00	221 01.7	218 29.7	N 2 18.9	213 11.4	N 2 09.9	254 18.7	S14 06.5	53 54.1	N 7 55.0
	01	236 04.2	233 30.9	19.0	228 12.1	10.7	269 20.9	06.4	68 56.6	55.0
	02	251 06.6	248 32.0	19.1	243 12.8	11.5	284 23.0	06.3	83 59.1	55.0
	03	266 09.1	263 33.2	19.2	258 13.5	12.2	299 25.2	06.2	99 01.7	55.0
	04	281 11.6	278 34.3	19.3	273 14.2	13.0	314 27.3	06.1	114 04.2	55.1
	05	296 14.0	293 35.5	19.4	288 14.9	13.8	329 29.5	06.0	129 06.7	55.1
	06	311 16.5	308 36.7	N 2 19.5	303 15.6	N 2 14.5	344 31.7	S14 05.9	144 09.2	N 7 55.1
	07	326 19.0	323 37.8	19.6	318 16.3	15.3	359 33.8	05.8	159 11.8	55.1
	08	341 21.4	338 39.0	19.7	333 17.0	16.1	14 36.0	05.7	174 14.3	55.1
S	09	356 23.9	353 40.1	19.8	348 17.7	16.8	29 38.1	05.6	189 16.8	55.1
U	10	11 26.4	8 41.2	19.9	3 18.4	17.6	44 40.3	05.6	204 19.3	55.2
N	11	26 28.8	23 42.4	20.0	18 19.1	18.4	59 42.5	05.5	219 21.8	55.2
D	12	41 31.3	38 43.5	N 2 20.1	33 19.8	N 2 19.1	74 44.6	S14 05.4	234 24.4	N 7 55.2
A	13	56 33.8	53 44.7	20.2	48 20.5	19.9	89 46.8	05.3	249 26.9	55.2
Y	14	71 36.2	68 45.8	20.4	63 21.2	20.7	104 48.9	05.2	264 29.4	55.2
	15	86 38.7	83 46.9	20.5	78 21.9	21.4	119 51.1	05.1	279 31.9	55.2
	16	101 41.1	98 48.1	20.6	93 22.6	22.2	134 53.3	05.0	294 34.4	55.3
	17	116 43.6	113 49.2	20.7	108 23.3	23.0	149 55.4	04.9	309 37.0	55.3
	18	131 46.1	128 50.3	N 2 20.8	123 24.0	N 2 23.7	164 57.6	S14 04.8	324 39.5	N 7 55.3
	19	146 48.5	143 51.4	21.0	138 24.7	24.5	179 59.7	04.7	339 42.0	55.3
	20	161 51.0	158 52.6	21.1	153 25.4	25.3	195 01.9	04.6	354 44.5	55.3
	21	176 53.5	173 53.7	21.2	168 26.1	26.0	210 04.1	04.5	9 47.0	55.3
	22	191 55.9	188 54.8	21.3	183 26.8	26.8	225 06.2	04.4	24 49.6	55.4
	23	206 58.4	203 55.9	21.5	198 27.5	27.6	240 08.4	04.3	39 52.1	55.4
	Mer. Pass.	h m 9 18.3	v 1.2	d 0.1	v 0.7	d 0.8	v 2.2	d 0.1	v 2.5	d 0.0

STARS

Name	SHA ° ′	Dec ° ′
Acamar	315 21.0	S40 16.
Achernar	335 29.4	S57 11.
Acrux	173 12.5	S63 09.
Adhara	255 15.1	S28 59.
Aldebaran	290 53.2	N16 31.
Alioth	166 22.6	N55 54.
Alkaid	153 00.7	N49 15.
Al Na'ir	27 47.6	S46 54.
Alnilam	275 49.7	S 1 11.
Alphard	217 59.1	S 8 42.
Alphecca	126 13.2	N26 40.
Alpheratz	357 47.0	N29 08.
Altair	62 11.1	N 8 53.
Ankaa	353 18.9	S42 15.
Antares	112 29.7	S26 27.
Arcturus	145 58.1	N19 07.
Atria	107 34.0	S69 02.
Avior	234 19.5	S59 32.
Bellatrix	278 35.6	N 6 21.
Betelgeuse	271 04.9	N 7 24.
Canopus	263 57.9	S52 42.
Capella	280 39.4	N46 00.
Deneb	49 33.6	N45 18.
Denebola	182 36.5	N14 31.
Diphda	348 59.2	S17 56.
Dubhe	193 54.8	N61 42.
Elnath	278 16.8	N28 37.
Eltanin	90 47.2	N51 29.
Enif	33 50.2	N 9 55.
Fomalhaut	15 27.5	S29 34.
Gacrux	172 04.1	S57 10.
Gienah	175 55.3	S17 35.
Hadar	148 51.9	S60 25.
Hamal	328 04.6	N23 30.
Kaus Aust.	83 47.6	S34 22.
Kochab	137 18.3	N74 06.
Markab	13 41.6	N15 15.
Menkar	314 18.6	N 4 07.
Menkent	148 10.9	S36 25.
Miaplacidus	221 40.6	S69 45.
Mirfak	308 45.4	N49 53.
Nunki	76 01.9	S26 17.
Peacock	53 23.8	S56 42.
Pollux	243 31.6	N28 00.
Procyon	245 03.0	N 5 12.
Rasalhague	96 09.0	N12 33.
Regulus	207 46.6	N11 55.
Rigel	281 15.3	S 8 11.
Rigil Kent.	139 55.5	S60 52.
Sabik	102 15.8	S15 44.
Schedar	349 44.8	N56 35.
Shaula	96 25.8	S37 06.
Sirius	258 36.6	S16 43.
Spica	158 34.2	S11 12.
Suhail	222 54.8	S43 28.
Vega	80 40.8	N38 47.
Zuben'ubi	137 08.5	S16 05.

	SHA ° ′	Mer. Pass h m
Venus	357 58.6	9 27
Mars	352 52.0	9 48
Jupiter	33 24.4	7 05
Saturn	192 50.9	20 25

UT (d h)	SUN GHA	SUN Dec	MOON GHA	MOON v	MOON Dec	MOON d	MOON HP
	° ′	° ′	° ′	′	° ′	′	′
1 00	180 43.0	N15 03.2	97 21.3	7.3	N20 39.8	10.5	59.3
01	195 43.0	04.0	111 47.6	7.4	20 29.3	10.6	59.3
02	210 43.1	04.8	126 14.0	7.4	20 18.7	10.7	59.3
03	225 43.2	. . 05.5	140 40.4	7.6	20 08.0	10.9	59.3
04	240 43.3	06.3	155 07.0	7.6	19 57.1	10.9	59.3
05	255 43.3	07.0	169 33.6	7.8	19 46.2	11.1	59.3
06	270 43.4	N15 07.8	184 00.4	7.9	N19 35.1	11.2	59.3
07	285 43.5	08.5	198 27.3	8.0	19 23.9	11.3	59.2
08	300 43.6	09.3	212 54.3	8.0	19 12.6	11.4	59.2
09	315 43.6	. . 10.0	227 21.3	8.2	19 01.2	11.5	59.2
10	330 43.7	10.8	241 48.5	8.3	18 49.7	11.6	59.2
11	345 43.8	11.5	256 15.8	8.4	18 38.1	11.7	59.2
12	0 43.9	N15 12.3	270 43.2	8.4	N18 26.4	11.8	59.2
13	15 43.9	13.0	285 10.6	8.6	18 14.6	11.9	59.1
14	30 44.0	13.8	299 38.2	8.7	18 02.7	12.0	59.1
15	45 44.1	. . 14.5	314 05.9	8.7	17 50.7	12.1	59.1
16	60 44.1	15.3	328 33.6	8.9	17 38.6	12.2	59.1
17	75 44.2	16.0	343 01.5	9.0	17 26.4	12.3	59.1
18	90 44.3	N15 16.8	357 29.5	9.1	N17 14.1	12.4	59.1
19	105 44.4	17.5	11 57.6	9.1	17 01.7	12.5	59.0
20	120 44.4	18.3	26 25.7	9.3	16 49.2	12.5	59.0
21	135 44.5	. . 19.0	40 54.0	9.3	16 36.7	12.6	59.0
22	150 44.6	19.8	55 22.3	9.5	16 24.1	12.8	59.0
23	165 44.6	20.5	69 50.8	9.5	16 11.3	12.8	59.0
2 00	180 44.7	N15 21.3	84 19.3	9.6	N15 58.5	12.9	58.9
01	195 44.8	22.0	98 47.9	9.8	15 45.6	12.9	58.9
02	210 44.9	22.8	113 16.7	9.8	15 32.7	13.1	58.9
03	225 44.9	. . 23.5	127 45.5	9.9	15 19.6	13.1	58.9
04	240 45.0	24.3	142 14.4	10.0	15 06.5	13.2	58.9
05	255 45.1	25.0	156 43.4	10.1	14 53.3	13.2	58.9
06	270 45.1	N15 25.8	171 12.5	10.1	N14 40.1	13.4	58.8
07	285 45.2	26.5	185 41.6	10.3	14 26.7	13.4	58.8
08	300 45.3	27.2	200 10.9	10.4	14 13.3	13.5	58.8
09	315 45.3	. . 28.0	214 40.3	10.4	13 59.8	13.5	58.8
10	330 45.4	28.7	229 09.7	10.5	13 46.3	13.6	58.8
11	345 45.5	29.5	243 39.2	10.6	13 32.7	13.7	58.7
12	0 45.5	N15 30.2	258 08.8	10.7	N13 19.0	13.7	58.7
13	15 45.6	31.0	272 38.5	10.7	13 05.3	13.8	58.7
14	30 45.7	31.7	287 08.2	10.9	12 51.5	13.8	58.7
15	45 45.8	. . 32.4	301 38.1	10.9	12 37.7	13.9	58.7
16	60 45.8	33.2	316 08.0	11.0	12 23.8	14.0	58.7
17	75 45.9	33.9	330 38.0	11.1	12 09.8	14.0	58.6
18	90 46.0	N15 34.7	345 08.1	11.1	N11 55.8	14.1	58.6
19	105 46.0	35.4	359 38.2	11.2	11 41.7	14.1	58.6
20	120 46.1	36.1	14 08.4	11.3	11 27.6	14.2	58.6
21	135 46.1	. . 36.9	28 38.7	11.4	11 13.4	14.2	58.6
22	150 46.2	37.6	43 09.1	11.5	10 59.2	14.3	58.5
23	165 46.3	38.3	57 39.6	11.5	10 44.9	14.3	58.5
3 00	180 46.3	N15 39.1	72 10.1	11.6	N10 30.6	14.3	58.5
01	195 46.4	39.8	86 40.7	11.6	10 16.3	14.4	58.5
02	210 46.5	40.5	101 11.3	11.7	10 01.9	14.5	58.5
03	225 46.5	. . 41.3	115 42.0	11.8	9 47.4	14.4	58.4
04	240 46.6	42.0	130 12.8	11.9	9 33.0	14.5	58.4
05	255 46.7	42.7	144 43.7	11.9	9 18.5	14.6	58.4
06	270 46.7	N15 43.5	159 14.6	12.0	N 9 03.9	14.6	58.4
07	285 46.8	44.2	173 45.6	12.0	8 49.3	14.6	58.4
08	300 46.9	44.9	188 16.6	12.1	8 34.7	14.7	58.3
09	315 46.9	. . 45.7	202 47.7	12.1	8 20.0	14.7	58.3
10	330 47.0	46.4	217 18.8	12.3	8 05.3	14.7	58.3
11	345 47.0	47.1	231 50.1	12.2	7 50.6	14.7	58.3
12	0 47.1	N15 47.9	246 21.3	12.4	N 7 35.9	14.8	58.3
13	15 47.2	48.6	260 52.7	12.3	7 21.1	14.8	58.2
14	30 47.2	49.3	275 24.0	12.5	7 06.3	14.8	58.2
15	45 47.3	. . 50.1	289 55.5	12.4	6 51.5	14.9	58.2
16	60 47.4	50.8	304 26.9	12.6	6 36.6	14.8	58.2
17	75 47.4	51.5	318 58.5	12.6	6 21.8	14.9	58.2
18	90 47.5	N15 52.2	333 30.1	12.6	N 6 06.9	15.0	58.1
19	105 47.5	53.0	348 01.7	12.7	5 51.9	14.9	58.1
20	120 47.6	53.7	2 33.4	12.7	5 37.0	14.9	58.1
21	135 47.7	. . 54.4	17 05.1	12.8	5 22.1	15.0	58.1
22	150 47.7	55.2	31 36.9	12.8	5 07.1	15.0	58.1
23	165 47.8	55.9	46 08.7	12.8	N 4 52.1	15.0	58.0
	SD 15.9	*d* 0.7	SD 16.1		16.0		15.9

Lat.	Twilight Naut.	Twilight Civil	Sunrise	Moonrise 1	Moonrise 2	Moonrise 3	Moonrise 4
°	h m	h m	h m	h m	h m	h m	h m
N 72	////	////	01 43	□	09 27	12 00	14 12
N 70	////	////	02 24	06 50	09 56	12 11	14 14
68	////	00 54	02 52	07 53	10 17	12 21	14 15
66	////	01 48	03 13	08 27	10 34	12 28	14 16
64	////	02 20	03 30	08 52	10 48	12 35	14 17
62	00 47	02 43	03 43	09 12	10 59	12 40	14 17
60	01 36	03 01	03 55	09 27	11 09	12 45	14 18
N 58	02 05	03 16	04 05	09 41	11 17	12 49	14 18
56	02 27	03 29	04 14	09 52	11 24	12 53	14 19
54	02 44	03 40	04 21	10 02	11 31	12 56	14 19
52	02 58	03 49	04 28	10 11	11 37	12 59	14 20
50	03 11	03 58	04 35	10 19	11 42	13 02	14 20
45	03 35	04 15	04 48	10 36	11 53	13 08	14 21
N 40	03 54	04 29	04 59	10 50	12 03	13 13	14 21
35	04 08	04 41	05 08	11 01	12 10	13 17	14 22
30	04 21	04 51	05 16	11 11	12 17	13 21	14 22
20	04 40	05 07	05 30	11 29	12 29	13 27	14 23
N 10	04 55	05 20	05 42	11 44	12 40	13 33	14 24
0	05 07	05 32	05 54	11 58	12 50	13 38	14 25
S 10	05 18	05 43	06 05	12 12	12 59	13 44	14 25
20	05 27	05 54	06 16	12 27	13 10	13 49	14 26
30	05 36	06 05	06 30	12 43	13 22	13 55	14 27
35	05 41	06 11	06 37	12 53	13 28	13 59	14 28
40	05 46	06 18	06 46	13 04	13 36	14 03	14 28
45	05 50	06 25	06 56	13 18	13 45	14 08	14 29
S 50	05 55	06 34	07 08	13 33	13 56	14 14	14 30
52	05 58	06 38	07 14	13 41	14 00	14 16	14 30
54	06 00	06 42	07 20	13 49	14 06	14 19	14 31
56	06 02	06 47	07 27	13 58	14 12	14 22	14 31
58	06 05	06 52	07 35	14 09	14 18	14 26	14 32
S 60	06 08	06 58	07 44	14 21	14 26	14 29	14 32

Lat.	Sunset	Twilight Civil	Twilight Naut.	Moonset 1	Moonset 2	Moonset 3	Moonset 4
°	h m	h m	h m	h m	h m	h m	h m
N 72	22 19	////	////	□	04 12	03 23	02 49
N 70	21 34	////	////	04 55	03 40	03 08	02 44
68	21 05	23 14	////	03 52	03 17	02 57	02 40
66	20 44	22 11	////	03 16	02 59	02 47	02 37
64	20 26	21 38	////	02 50	02 44	02 39	02 34
62	20 12	21 14	23 19	02 30	02 32	02 32	02 31
60	20 01	20 55	22 23	02 13	02 21	02 25	02 29
N 58	19 50	20 40	21 52	01 59	02 11	02 20	02 27
56	19 42	20 27	21 30	01 47	02 03	02 15	02 25
54	19 34	20 16	21 12	01 36	01 56	02 11	02 23
52	19 27	20 06	20 57	01 26	01 49	02 07	02 21
50	19 20	19 57	20 45	01 18	01 43	02 03	02 20
45	19 07	19 39	20 20	01 00	01 30	01 55	02 17
N 40	18 56	19 25	20 01	00 45	01 19	01 48	02 14
35	18 46	19 13	19 46	00 32	01 10	01 43	02 12
30	18 38	19 04	19 34	00 21	01 02	01 37	02 10
20	18 24	18 47	19 14	00 01	00 47	01 29	02 07
N 10	18 12	18 34	18 59	24 35	00 35	01 21	02 03
0	18 00	18 22	18 47	24 23	00 23	01 13	02 01
S 10	17 49	18 11	18 36	24 11	00 11	01 06	01 57
20	17 37	18 00	18 26	23 58	24 57	00 57	01 54
30	17 24	17 49	18 17	23 43	24 48	00 48	01 50
35	17 16	17 43	18 13	23 35	24 43	00 43	01 48
40	17 08	17 36	18 08	23 25	24 36	00 36	01 46
45	16 57	17 28	18 03	23 13	24 29	00 29	01 43
S 50	16 45	17 19	17 58	22 59	24 20	00 20	01 39
52	16 39	17 15	17 56	22 52	24 16	00 16	01 38
54	16 33	17 11	17 53	22 44	24 12	00 12	01 36
56	16 26	17 06	17 51	22 36	24 07	00 07	01 34
58	16 18	17 01	17 48	22 26	24 01	00 01	01 32
S 60	16 09	16 55	17 45	22 15	23 55	25 29	01 29

Day	SUN Eqn. of Time 00^h	SUN Eqn. of Time 12^h	SUN Mer. Pass.	MOON Mer. Pass. Upper	MOON Mer. Pass. Lower	MOON Age	MOON Phase
d	m s	m s	h m	h m	h m	d %	
1	02 52	02 55	11 57	18 10	05 43	06 46	
2	02 59	03 02	11 57	19 02	06 36	07 57	◐
3	03 05	03 08	11 57	19 49	07 26	08 68	

UT		ARIES	VENUS −4.7		MARS +1.2		JUPITER −2.3		SATURN +0.8	
		GHA	GHA	Dec	GHA	Dec	GHA	Dec	GHA	Dec
d h		° ′	° ′	° ′	° ′	° ′	° ′	° ′	° ′	° ′
4 00		222 00.9	218 57.0	N 2 21.6	213 28.2	N 2 28.3	255 10.6	S14 04.2	54 54.6	N 7 55.4
01		237 03.3	233 58.1	21.7	228 28.9	29.1	270 12.7	04.1	69 57.1	55.4
02		252 05.8	248 59.2	21.9	243 29.6	29.8	285 14.9	04.1	84 59.6	55.4
03		267 08.2	264 00.4	. . 22.0	258 30.3	. . 30.6	300 17.1	. . 04.0	100 02.2	. . 55.4
04		282 10.7	279 01.5	22.1	273 31.0	31.4	315 19.2	03.9	115 04.7	55.4
05		297 13.2	294 02.6	22.3	288 31.7	32.1	330 21.4	03.8	130 07.2	55.5
06		312 15.6	309 03.7	N 2 22.4	303 32.4	N 2 32.9	345 23.5	S14 03.7	145 09.7	N 7 55.5
07		327 18.1	324 04.8	22.6	318 33.1	33.7	0 25.7	03.6	160 12.2	55.5
08		342 20.6	339 05.9	22.7	333 33.8	34.4	15 27.9	03.5	175 14.7	55.5
09	M	357 23.0	354 07.0	. . 22.8	348 34.5	. . 35.2	30 30.0	. . 03.4	190 17.3	. . 55.5
10	O	12 25.5	9 08.0	23.0	3 35.2	36.0	45 32.2	03.3	205 19.8	55.5
11	N	27 28.0	24 09.1	23.1	18 35.9	36.7	60 34.4	03.2	220 22.3	55.6
12	D	42 30.4	39 10.2	N 2 23.3	33 36.6	N 2 37.5	75 36.5	S14 03.1	235 24.8	N 7 55.6
13	A	57 32.9	54 11.3	23.4	48 37.3	38.3	90 38.7	03.0	250 27.3	55.6
14	Y	72 35.4	69 12.4	23.6	63 38.0	39.0	105 40.9	02.9	265 29.8	55.6
15		87 37.8	84 13.5	. . 23.7	78 38.7	. . 39.8	120 43.0	. . 02.8	280 32.4	. . 55.6
16		102 40.3	99 14.6	23.9	93 39.4	40.5	135 45.2	02.8	295 34.9	55.6
17		117 42.7	114 15.6	24.0	108 40.1	41.3	150 47.4	02.7	310 37.4	55.6
18		132 45.2	129 16.7	N 2 24.2	123 40.8	N 2 42.1	165 49.5	S14 02.6	325 39.9	N 7 55.7
19		147 47.7	144 17.8	24.4	138 41.5	42.8	180 51.7	02.5	340 42.4	55.7
20		162 50.1	159 18.9	24.5	153 42.2	43.6	195 53.9	02.4	355 44.9	55.7
21		177 52.6	174 19.9	. . 24.7	168 42.9	. . 44.4	210 56.0	. . 02.3	10 47.5	. . 55.7
22		192 55.1	189 21.0	24.8	183 43.6	45.1	225 58.2	02.2	25 50.0	55.7
23		207 57.5	204 22.1	25.0	198 44.3	45.9	241 00.4	02.1	40 52.5	55.7
5 00		223 00.0	219 23.1	N 2 25.2	213 45.0	N 2 46.7	256 02.6	S14 02.0	55 55.0	N 7 55.7
01		238 02.5	234 24.2	25.3	228 45.7	47.4	271 04.7	01.9	70 57.5	55.8
02		253 04.9	249 25.2	25.5	243 46.4	48.2	286 06.9	01.8	86 00.0	55.8
03		268 07.4	264 26.3	. . 25.7	258 47.1	. . 48.9	301 09.1	. . 01.7	101 02.5	. . 55.8
04		283 09.9	279 27.4	25.8	273 47.8	49.7	316 11.2	01.7	116 05.1	55.8
05		298 12.3	294 28.4	26.0	288 48.5	50.5	331 13.4	01.6	131 07.6	55.8
06		313 14.8	309 29.5	N 2 26.2	303 49.2	N 2 51.2	346 15.6	S14 01.5	146 10.1	N 7 55.8
07		328 17.2	324 30.5	26.3	318 49.9	52.0	1 17.7	01.4	161 12.6	55.8
08	T	343 19.7	339 31.6	26.5	333 50.6	52.8	16 19.9	01.3	176 15.1	55.9
09	U	358 22.2	354 32.6	. . 26.7	348 51.3	. . 53.5	31 22.1	. . 01.2	191 17.6	. . 55.9
10	E	13 24.6	9 33.7	26.9	3 52.0	54.3	46 24.3	01.1	206 20.1	55.9
11	S	28 27.1	24 34.7	27.1	18 52.7	55.0	61 26.4	01.0	221 22.7	55.9
12	D	43 29.6	39 35.7	N 2 27.2	33 53.4	N 2 55.8	76 28.6	S14 00.9	236 25.2	N 7 55.9
13		58 32.0	54 36.8	27.4	48 54.1	56.6	91 30.8	00.8	251 27.7	55.9
14	A	73 34.5	69 37.8	27.6	63 54.8	57.3	106 32.9	00.8	266 30.2	55.9
15	Y	88 37.0	84 38.8	. . 27.8	78 55.5	. . 58.1	121 35.1	. . 00.7	281 32.7	. . 55.9
16		103 39.4	99 39.9	28.0	93 56.2	58.9	136 37.3	00.6	296 35.2	56.0
17		118 41.9	114 40.9	28.2	108 56.9	2 59.6	151 39.5	00.5	311 37.7	56.0
18		133 44.3	129 41.9	N 2 28.3	123 57.6	N 3 00.4	166 41.6	S14 00.4	326 40.2	N 7 56.0
19		148 46.8	144 43.0	28.5	138 58.3	01.1	181 43.8	00.3	341 42.8	56.0
20		163 49.3	159 44.0	28.7	153 59.0	01.9	196 46.0	00.2	356 45.3	56.0
21		178 51.7	174 45.0	. . 28.9	168 59.7	. . 02.7	211 48.2	. . 00.1	11 47.8	. . 56.0
22		193 54.2	189 46.0	29.1	184 00.4	03.4	226 50.3	14 00.0	26 50.3	56.0
23		208 56.7	204 47.0	29.3	199 01.1	04.2	241 52.5	13 59.9	41 52.8	56.0
6 00		223 59.1	219 48.1	N 2 29.5	214 01.8	N 3 04.9	256 54.7	S13 59.8	56 55.3	N 7 56.1
01		239 01.6	234 49.1	29.7	229 02.5	05.7	271 56.9	59.8	71 57.8	56.1
02		254 04.1	249 50.1	29.9	244 03.2	06.5	286 59.0	59.7	87 00.3	56.1
03		269 06.5	264 51.1	. . 30.1	259 03.9	. . 07.2	302 01.2	. . 59.6	102 02.8	. . 56.1
04		284 09.0	279 52.1	30.3	274 04.6	08.0	317 03.4	59.5	117 05.4	56.1
05		299 11.5	294 53.1	30.5	289 05.3	08.8	332 05.6	59.4	132 07.9	56.1
06		314 13.9	309 54.1	N 2 30.7	304 06.0	N 3 09.5	347 07.7	S13 59.3	147 10.4	N 7 56.1
07	W	329 16.4	324 55.1	30.9	319 06.7	10.3	2 09.9	59.2	162 12.9	56.1
08	E	344 18.8	339 56.1	31.1	334 07.4	11.0	17 12.1	59.1	177 15.4	56.2
09	D	359 21.3	354 57.1	. . 31.3	349 08.1	. . 11.8	32 14.3	. . 59.1	192 17.9	. . 56.2
10	N	14 23.8	9 58.1	31.5	4 08.8	12.6	47 16.4	59.0	207 20.4	56.2
11	E	29 26.2	24 59.1	31.7	19 09.5	13.3	62 18.6	58.9	222 22.9	56.2
12	S	44 28.7	40 00.1	N 2 32.0	34 10.2	N 3 14.1	77 20.8	S13 58.8	237 25.4	N 7 56.2
13		59 31.2	55 01.1	32.2	49 10.9	14.8	92 23.0	58.7	252 27.9	56.2
14	D	74 33.6	70 02.1	32.4	64 11.6	15.6	107 25.2	58.6	267 30.4	56.2
15	A	89 36.1	85 03.1	. . 32.6	79 12.3	. . 16.4	122 27.3	. . 58.5	282 33.0	. . 56.2
16	Y	104 38.6	100 04.1	32.8	94 13.0	17.1	137 29.5	58.4	297 35.5	56.2
17		119 41.0	115 05.0	33.0	109 13.7	17.9	152 31.7	58.3	312 38.0	56.3
18		134 43.5	130 06.0	N 2 33.2	124 14.4	N 3 18.6	167 33.9	S13 58.3	327 40.5	N 7 56.3
19		149 45.9	145 07.0	33.5	139 15.1	19.4	182 36.1	58.2	342 43.0	56.3
20		164 48.4	160 08.0	33.7	154 15.8	20.2	197 38.2	58.1	357 45.5	56.3
21		179 50.9	175 09.0	. . 33.9	169 16.5	. . 20.9	212 40.4	. . 58.0	12 48.0	. . 56.3
22		194 53.3	190 09.9	34.1	184 17.2	21.7	227 42.6	57.9	27 50.5	56.3
23		209 55.8	205 10.9	34.4	199 17.9	22.4	242 44.8	57.8	42 53.0	56.3
Mer. Pass.		h m 9 06.5	v 1.0	d 0.2	v 0.7	d 0.8	v 2.2	d 0.1	v 2.5	d 0.0

STARS		
Name	SHA	De
	° ′	°
Acamar	315 21.0	S40 1
Achernar	335 29.4	S57 1
Acrux	173 12.5	S63 0
Adhara	255 15.2	S28 5
Aldebaran	290 53.2	N16 3
Alioth	166 22.7	N55 5
Alkaid	153 00.7	N49 1
Al Na'ir	27 47.5	S46 5
Alnilam	275 49.7	S 1 1
Alphard	217 59.1	S 8 4
Alphecca	126 13.2	N26 4
Alpheratz	357 47.0	N29 0
Altair	62 11.1	N 8 5
Ankaa	353 18.9	S42 1
Antares	112 29.7	S26 2
Arcturus	145 58.1	N19 0
Atria	107 33.9	S69 0
Avior	234 19.6	S59 3
Bellatrix	278 35.6	N 6 2
Betelgeuse	271 04.9	N 7 2
Canopus	263 57.9	S52 4
Capella	280 39.4	N46 0
Deneb	49 33.6	N45 1
Denebola	182 36.5	N14 3
Diphda	348 59.2	S17 5
Dubhe	193 54.8	N61 4
Elnath	278 16.8	N28 3
Eltanin	90 47.2	N51 2
Enif	33 50.2	N 9 5
Fomalhaut	15 27.4	S29 3
Gacrux	172 04.1	S57 1
Gienah	175 55.3	S17 3
Hadar	148 51.9	S60 2
Hamal	328 04.6	N23 3
Kaus Aust.	83 47.6	S34 2
Kochab	137 18.3	N74 0
Markab	13 41.6	N15 1
Menkar	314 18.6	N 4 0
Menkent	148 10.9	S36 2
Miaplacidus	221 40.6	S69 4
Mirfak	308 45.4	N49 5
Nunki	76 01.9	S26 1
Peacock	53 23.8	S56 4
Pollux	243 31.6	N28 0
Procyon	245 03.1	N 5 1
Rasalhague	96 09.0	N12 3
Regulus	207 46.7	N11 5
Rigel	281 15.3	S 8 1
Rigil Kent.	139 55.5	S60 5
Sabik	102 15.8	S15 4
Schedar	349 44.7	N56 3
Shaula	96 25.8	S37 0
Sirius	258 36.6	S16 4
Spica	158 34.2	S11 1
Suhail	222 54.8	S43 2
Vega	80 40.8	N38 4
Zuben'ubi	137 08.5	S16 0
	SHA	Mer. Pa
	° ′	h
Venus	356 23.1	9 2
Mars	350 45.0	9 4
Jupiter	33 02.6	6 5
Saturn	192 55.0	20 1

	UT	SUN GHA	SUN Dec	MOON GHA	v	MOON Dec	d	HP
	d h	° ′	° ′	° ′	′	° ′	′	′
	4 00	180 47.8	N15 56.6	60 40.5	12.9	N 4 37.1	15.0	58.0
	01	195 47.9	57.3	75 12.4	13.0	4 22.1	15.0	58.0
	02	210 48.0	58.1	89 44.4	12.9	4 07.1	15.0	58.0
	03	225 48.0	. . 58.8	104 16.3	13.0	3 52.1	15.0	58.0
	04	240 48.1	15 59.5	118 48.3	13.1	3 37.1	15.1	57.9
	05	255 48.1	16 00.2	133 20.4	13.1	3 22.0	15.0	57.9
	06	270 48.2	N16 00.9	147 52.5	13.1	N 3 07.0	15.1	57.9
	07	285 48.3	01.7	162 24.6	13.1	2 51.9	15.0	57.9
	08	300 48.3	02.4	176 56.7	13.2	2 36.9	15.1	57.9
M	09	315 48.4	. . 03.1	191 28.9	13.2	2 21.8	15.1	57.8
O	10	330 48.4	03.8	206 01.1	13.3	2 06.7	15.0	57.8
N	11	345 48.5	04.6	220 33.4	13.2	1 51.7	15.1	57.8
D	12	0 48.5	N16 05.3	235 05.6	13.3	N 1 36.6	15.0	57.8
A	13	15 48.6	06.0	249 37.9	13.3	1 21.6	15.1	57.8
Y	14	30 48.7	06.7	264 10.2	13.4	1 06.5	15.0	57.7
	15	45 48.7	. . 07.4	278 42.6	13.3	0 51.5	15.1	57.7
	16	60 48.8	08.1	293 14.9	13.4	0 36.4	15.0	57.7
	17	75 48.8	08.9	307 47.3	13.4	0 21.4	15.0	57.7
	18	90 48.9	N16 09.6	322 19.7	13.4	N 0 06.4	15.0	57.7
	19	105 48.9	10.3	336 52.1	13.5	S 0 08.6	15.0	57.6
	20	120 49.0	11.0	351 24.6	13.5	0 23.6	15.0	57.6
	21	135 49.0	. . 11.7	5 57.1	13.4	0 38.6	15.0	57.6
	22	150 49.1	12.4	20 29.5	13.5	0 53.6	15.0	57.6
	23	165 49.1	13.2	35 02.0	13.5	1 08.6	14.9	57.6
	5 00	180 49.2	N16 13.9	49 34.5	13.5	S 1 23.5	15.0	57.5
	01	195 49.3	14.6	64 07.0	13.6	1 38.5	14.9	57.5
	02	210 49.3	15.3	78 39.6	13.5	1 53.4	14.9	57.5
	03	225 49.4	. . 16.0	93 12.1	13.6	2 08.3	14.9	57.5
	04	240 49.4	16.7	107 44.7	13.5	2 23.2	14.8	57.5
	05	255 49.5	17.4	122 17.2	13.6	2 38.0	14.9	57.4
	06	270 49.5	N16 18.1	136 49.8	13.6	S 2 52.9	14.8	57.4
	07	285 49.6	18.9	151 22.4	13.5	3 07.7	14.8	57.4
T	08	300 49.6	19.6	165 54.9	13.6	3 22.5	14.8	57.4
U	09	315 49.7	. . 20.3	180 27.5	13.6	3 37.3	14.7	57.3
	10	330 49.7	21.0	195 00.1	13.6	3 52.0	14.7	57.3
E	11	345 49.8	21.7	209 32.7	13.6	4 06.7	14.7	57.3
S	12	0 49.8	N16 22.4	224 05.3	13.6	S 4 21.4	14.7	57.3
D	13	15 49.9	23.1	238 37.9	13.5	4 36.1	14.6	57.3
A	14	30 49.9	23.8	253 10.4	13.6	4 50.7	14.6	57.2
Y	15	45 50.0	. . 24.5	267 43.0	13.6	5 05.3	14.5	57.2
	16	60 50.0	25.2	282 15.6	13.6	5 19.8	14.6	57.2
	17	75 50.1	25.9	296 48.2	13.5	5 34.4	14.5	57.2
	18	90 50.1	N16 26.6	311 20.7	13.6	S 5 48.9	14.4	57.2
	19	105 50.2	27.3	325 53.3	13.6	6 03.3	14.4	57.1
	20	120 50.2	28.0	340 25.9	13.5	6 17.7	14.4	57.1
	21	135 50.3	. . 28.8	354 58.4	13.5	6 32.1	14.4	57.1
	22	150 50.3	29.5	9 30.9	13.6	6 46.5	14.3	57.1
	23	165 50.4	30.2	24 03.5	13.5	7 00.8	14.2	57.1
	6 00	180 50.4	N16 30.9	38 36.0	13.5	S 7 15.0	14.3	57.0
	01	195 50.5	31.6	53 08.5	13.5	7 29.3	14.1	57.0
	02	210 50.5	32.3	67 41.0	13.4	7 43.4	14.2	57.0
	03	225 50.6	. . 33.0	82 13.4	13.5	7 57.6	14.1	57.0
	04	240 50.6	33.7	96 45.9	13.5	8 11.7	14.0	57.0
	05	255 50.7	34.4	111 18.4	13.4	8 25.7	14.0	56.9
	06	270 50.7	N16 35.1	125 50.8	13.4	S 8 39.7	13.9	56.9
W	07	285 50.8	35.8	140 23.2	13.4	8 53.6	13.9	56.9
E	08	300 50.8	36.5	154 55.6	13.3	9 07.5	13.9	56.9
D	09	315 50.9	. . 37.2	169 27.9	13.4	9 21.4	13.8	56.9
	10	330 50.9	37.9	184 00.3	13.3	9 35.2	13.7	56.8
N	11	345 50.9	38.6	198 32.6	13.3	9 48.9	13.7	56.8
E	12	0 51.0	N16 39.3	213 04.9	13.3	S10 02.6	13.6	56.8
S	13	15 51.0	40.0	227 37.2	13.3	10 16.2	13.6	56.8
D	14	30 51.1	40.6	242 09.5	13.2	10 29.8	13.5	56.7
A	15	45 51.1	. . 41.3	256 41.7	13.2	10 43.3	13.5	56.7
Y	16	60 51.2	42.0	271 13.9	13.2	10 56.8	13.4	56.7
	17	75 51.2	42.7	285 46.1	13.2	11 10.2	13.3	56.7
	18	90 51.3	N16 43.4	300 18.3	13.1	S11 23.5	13.3	56.7
	19	105 51.3	44.1	314 50.4	13.1	11 36.8	13.2	56.6
	20	120 51.3	44.8	329 22.5	13.1	11 50.0	13.2	56.6
	21	135 51.4	. . 45.5	343 54.6	13.0	12 03.2	13.1	56.6
	22	150 51.4	46.2	358 26.6	13.0	12 16.3	13.0	56.6
	23	165 51.5	46.9	12 58.6	13.0	S12 29.3	13.0	56.6
		SD 15.9	d 0.7	SD 15.7		15.6		15.5

Lat.	Twilight Naut.	Twilight Civil	Sunrise	Moonrise 4	Moonrise 5	Moonrise 6	Moonrise 7
°	h m	h m	h m	h m	h m	h m	h m
N 72	////	////	01 13	14 12	16 21	18 38	21 40
N 70	////	////	02 06	14 14	16 13	18 16	20 38
68	////	////	02 38	14 15	16 06	18 00	20 03
66	////	01 28	03 01	14 16	16 01	17 47	19 38
64	////	02 06	03 20	14 17	15 56	17 36	19 19
62	////	02 31	03 35	14 17	15 52	17 27	19 04
60	01 18	02 51	03 47	14 18	15 49	17 19	18 51
N 58	01 52	03 08	03 58	14 18	15 45	17 12	18 40
56	02 17	03 21	04 07	14 19	15 43	17 06	18 30
54	02 35	03 33	04 15	14 19	15 40	17 01	18 21
52	02 51	03 43	04 23	14 20	15 38	16 56	18 14
50	03 04	03 52	04 29	14 20	15 36	16 52	18 07
45	03 30	04 11	04 44	14 21	15 32	16 42	17 53
N 40	03 49	04 26	04 55	14 21	15 28	16 35	17 41
35	04 05	04 38	05 05	14 22	15 25	16 28	17 31
30	04 18	04 48	05 14	14 22	15 22	16 22	17 22
20	04 38	05 05	05 29	14 23	15 18	16 12	17 07
N 10	04 54	05 19	05 41	14 24	15 14	16 04	16 54
0	05 07	05 32	05 53	14 25	15 10	15 55	16 42
S 10	05 18	05 43	06 05	14 25	15 06	15 47	16 30
20	05 28	05 55	06 17	14 26	15 02	15 39	16 17
30	05 38	06 07	06 32	14 27	14 58	15 29	16 03
35	05 43	06 13	06 40	14 28	14 55	15 24	15 54
40	05 48	06 20	06 49	14 28	14 53	15 18	15 45
45	05 54	06 29	07 00	14 29	14 49	15 10	15 34
S 50	05 59	06 38	07 13	14 30	14 45	15 02	15 20
52	06 02	06 42	07 19	14 30	14 44	14 58	15 14
54	06 05	06 47	07 25	14 31	14 42	14 54	15 08
56	06 07	06 52	07 33	14 31	14 40	14 49	15 00
58	06 10	06 58	07 41	14 32	14 37	14 44	14 52
S 60	06 14	07 04	07 51	14 32	14 35	14 38	14 42

Lat.	Sunset	Twilight Civil	Twilight Naut.	Moonset 4	Moonset 5	Moonset 6	Moonset 7
°	h m	h m	h m	h m	h m	h m	h m
N 72	22 52	////	////	02 49	02 20	01 49	{01 10 / 23 46}
N 70	21 53	////	////	02 44	02 23	02 00	01 33
68	21 19	////	////	02 40	02 25	02 09	01 51
66	20 55	22 32	////	02 37	02 27	02 17	02 06
64	20 36	21 52	////	02 34	02 29	02 23	02 18
62	20 21	21 25	////	02 31	02 30	02 29	02 28
60	20 08	21 04	22 41	02 29	02 31	02 34	02 37
N 58	19 57	20 48	22 05	02 27	02 32	02 38	02 45
56	19 48	20 34	21 40	02 25	02 33	02 42	02 52
54	19 39	20 22	21 20	02 23	02 34	02 46	02 58
52	19 32	20 12	21 04	02 21	02 35	02 49	03 04
50	19 25	20 02	20 51	02 20	02 36	02 52	03 09
45	19 11	19 44	20 25	02 17	02 38	02 58	03 20
N 40	18 59	19 29	20 05	02 14	02 39	03 04	03 30
35	18 49	19 16	19 49	02 12	02 40	03 08	03 37
30	18 40	19 06	19 36	02 10	02 41	03 12	03 45
20	18 25	18 48	19 16	02 07	02 43	03 19	03 57
N 10	18 12	18 34	19 00	02 03	02 45	03 26	04 07
0	18 00	18 22	18 47	02 01	02 46	03 32	04 18
S 10	17 48	18 10	18 35	01 57	02 48	03 37	04 28
20	17 36	17 59	18 25	01 54	02 49	03 44	04 38
30	17 22	17 46	18 15	01 50	02 51	03 51	04 51
35	17 13	17 40	18 10	01 48	02 52	03 55	04 58
40	17 04	17 32	18 05	01 46	02 53	04 00	05 06
45	16 53	17 24	17 59	01 43	02 55	04 06	05 16
S 50	16 40	17 15	17 53	01 39	02 56	04 12	05 28
52	16 34	17 10	17 51	01 38	02 57	04 15	05 33
54	16 27	17 06	17 48	01 36	02 58	04 19	05 39
56	16 20	17 01	17 45	01 34	02 59	04 23	05 46
58	16 11	16 55	17 42	01 32	03 00	04 27	05 54
S 60	16 02	16 49	17 39	01 29	03 01	04 32	06 02

Day	SUN Eqn. of Time 00^h	SUN Eqn. of Time 12^h	SUN Mer. Pass.	MOON Mer. Pass. Upper	MOON Mer. Pass. Lower	MOON Age	MOON Phase
d	m s	m s	h m	h m	h m	d %	
4	03 11	03 14	11 57	20 35	08 13	09 78	
5	03 17	03 19	11 57	21 21	08 58	10 86	
6	03 22	03 24	11 57	22 06	09 43	11 92	

UT	ARIES	VENUS −4.7		MARS +1.2		JUPITER −2.3		SATURN +0.8	
	GHA	GHA	Dec	GHA	Dec	GHA	Dec	GHA	Dec
d h	° ′	° ′	° ′	° ′	° ′	° ′	° ′	° ′	° ′
7 00	224 58.3	220 11.9	N 2 34.6	214 18.6	N 3 23.2	257 47.0	S13 57.7	57 55.5	N 7 56.3
01	240 00.7	235 12.9	34.8	229 19.3	24.0	272 49.1	57.6	72 58.0	56.3
02	255 03.2	250 13.8	35.1	244 20.0	24.7	287 51.3	57.6	88 00.5	56.4
03	270 05.7	265 14.8 . .	35.3	259 20.7 . .	25.5	302 53.5 . .	57.5	103 03.0 . .	56.4
04	285 08.1	280 15.7	35.5	274 21.4	26.2	317 55.7	57.4	118 05.6	56.4
05	300 10.6	295 16.7	35.7	289 22.1	27.0	332 57.9	57.3	133 08.1	56.4
06	315 13.1	310 17.7	N 2 36.0	304 22.8	N 3 27.8	348 00.1	S13 57.2	148 10.6	N 7 56.4
07	330 15.5	325 18.6	36.2	319 23.5	28.5	3 02.2	57.1	163 13.1	56.4
T 08	345 18.0	340 19.6	36.5	334 24.2	29.3	18 04.4	57.0	178 15.6	56.4
H 09	0 20.4	355 20.5 . .	36.7	349 24.9 . .	30.0	33 06.6 . .	56.9	193 18.1 . .	56.4
U 10	15 22.9	10 21.5	36.9	4 25.6	30.8	48 08.8	56.9	208 20.6	56.4
R 11	30 25.4	25 22.4	37.2	19 26.3	31.5	63 11.0	56.8	223 23.1	56.4
S 12	45 27.8	40 23.4	N 2 37.4	34 27.0	N 3 32.3	78 13.2	S13 56.7	238 25.6	N 7 56.5
D 13	60 30.3	55 24.3	37.7	49 27.7	33.1	93 15.3	56.6	253 28.1	56.5
A 14	75 32.8	70 25.3	37.9	64 28.4	33.8	108 17.5	56.5	268 30.6	56.5
Y 15	90 35.2	85 26.2 . .	38.2	79 29.1 . .	34.6	123 19.7 . .	56.4	283 33.1 . .	56.5
16	105 37.7	100 27.2	38.4	94 29.8	35.3	138 21.9	56.3	298 35.6	56.5
17	120 40.2	115 28.1	38.7	109 30.5	36.1	153 24.1	56.3	313 38.1	56.5
18	135 42.6	130 29.0	N 2 38.9	124 31.2	N 3 36.9	168 26.3	S13 56.2	328 40.6	N 7 56.5
19	150 45.1	145 30.0	39.2	139 31.9	37.6	183 28.4	56.1	343 43.1	56.5
20	165 47.6	160 30.9	39.4	154 32.6	38.4	198 30.6	56.0	358 45.6	56.5
21	180 50.0	175 31.8 . .	39.7	169 33.3 . .	39.1	213 32.8 . .	55.9	13 48.1 . .	56.5
22	195 52.5	190 32.8	39.9	184 34.0	39.9	228 35.0	55.8	28 50.6	56.6
23	210 54.9	205 33.7	40.2	199 34.7	40.6	243 37.2	55.7	43 53.1	56.6
8 00	225 57.4	220 34.6	N 2 40.4	214 35.4	N 3 41.4	258 39.4	S13 55.7	58 55.7	N 7 56.6
01	240 59.9	235 35.6	40.7	229 36.1	42.2	273 41.6	55.6	73 58.2	56.6
02	256 02.3	250 36.5	40.9	244 36.8	42.9	288 43.8	55.5	89 00.7	56.6
03	271 04.8	265 37.4 . .	41.2	259 37.5 . .	43.7	303 45.9 . .	55.4	104 03.2 . .	56.6
04	286 07.3	280 38.3	41.5	274 38.2	44.4	318 48.1	55.3	119 05.7	56.6
05	301 09.7	295 39.2	41.7	289 38.9	45.2	333 50.3	55.2	134 08.2	56.6
06	316 12.2	310 40.1	N 2 42.0	304 39.6	N 3 45.9	348 52.5	S13 55.1	149 10.7	N 7 56.6
07	331 14.7	325 41.1	42.3	319 40.3	46.7	3 54.7	55.1	164 13.2	56.6
08	346 17.1	340 42.0	42.5	334 41.0	47.5	18 56.9	55.0	179 15.7	56.6
F 09	1 19.6	355 42.9 . .	42.8	349 41.7 . .	48.2	33 59.1 . .	54.9	194 18.2 . .	56.6
R 10	16 22.0	10 43.8	43.1	4 42.4	49.0	49 01.3	54.8	209 20.7	56.7
I 11	31 24.5	25 44.7	43.3	19 43.1	49.7	64 03.5	54.7	224 23.2	56.7
D 12	46 27.0	40 45.6	N 2 43.6	34 43.8	N 3 50.5	79 05.6	S13 54.6	239 25.7	N 7 56.7
A 13	61 29.4	55 46.5	43.9	49 44.5	51.2	94 07.8	54.6	254 28.2	56.7
Y 14	76 31.9	70 47.4	44.2	64 45.2	52.0	109 10.0	54.5	269 30.7	56.7
15	91 34.4	85 48.3 . .	44.4	79 45.9 . .	52.7	124 12.2 . .	54.4	284 33.2 . .	56.7
16	106 36.8	100 49.2	44.7	94 46.6	53.5	139 14.4	54.3	299 35.7	56.7
17	121 39.3	115 50.1	45.0	109 47.3	54.3	154 16.6	54.2	314 38.2	56.7
18	136 41.8	130 51.0	N 2 45.3	124 48.0	N 3 55.0	169 18.8	S13 54.1	329 40.7	N 7 56.7
19	151 44.2	145 51.9	45.5	139 48.7	55.8	184 21.0	54.0	344 43.2	56.7
20	166 46.7	160 52.8	45.8	154 49.4	56.5	199 23.2	54.0	359 45.7	56.7
21	181 49.2	175 53.7 . .	46.1	169 50.1 . .	57.3	214 25.4 . .	53.9	14 48.2 . .	56.7
22	196 51.6	190 54.6	46.4	184 50.8	58.0	229 27.6	53.8	29 50.7	56.7
23	211 54.1	205 55.5	46.7	199 51.5	58.8	244 29.8	53.7	44 53.2	56.8
9 00	226 56.5	220 56.3	N 2 47.0	214 52.2	N 3 59.5	259 32.0	S13 53.6	59 55.7	N 7 56.8
01	241 59.0	235 57.2	47.3	229 52.9	4 00.3	274 34.1	53.5	74 58.2	56.8
02	257 01.5	250 58.1	47.5	244 53.6	01.1	289 36.3	53.5	90 00.7	56.8
03	272 03.9	265 59.0 . .	47.8	259 54.3 . .	01.8	304 38.5 . .	53.4	105 03.2 . .	56.8
04	287 06.4	280 59.9	48.1	274 55.0	02.6	319 40.7	53.3	120 05.7	56.8
05	302 08.9	296 00.7	48.4	289 55.7	03.3	334 42.9	53.2	135 08.2	56.8
06	317 11.3	311 01.6	N 2 48.7	304 56.4	N 4 04.1	349 45.1	S13 53.1	150 10.7	N 7 56.8
07	332 13.8	326 02.5	49.0	319 57.1	04.8	4 47.3	53.0	165 13.2	56.8
S 08	347 16.3	341 03.3	49.3	334 57.8	05.6	19 49.5	53.0	180 15.7	56.8
A 09	2 18.7	356 04.2 . .	49.6	349 58.5 . .	06.3	34 51.7 . .	52.9	195 18.2 . .	56.8
T 10	17 21.2	11 05.1	49.9	4 59.2	07.1	49 53.9	52.8	210 20.7	56.8
U 11	32 23.7	26 05.9	50.2	19 59.9	07.8	64 56.1	52.7	225 23.2	56.8
R 12	47 26.1	41 06.8	N 2 50.5	35 00.6	N 4 08.6	79 58.3	S13 52.6	240 25.7	N 7 56.8
D 13	62 28.6	56 07.7	50.8	50 01.3	09.4	95 00.5	52.6	255 28.2	56.8
A 14	77 31.0	71 08.5	51.1	65 02.0	10.1	110 02.7	52.5	270 30.6	56.9
Y 15	92 33.5	86 09.4 . .	51.4	80 02.7 . .	10.9	125 04.9 . .	52.4	285 33.1 . .	56.9
16	107 36.0	101 10.3	51.7	95 03.4	11.6	140 07.1	52.3	300 35.6	56.9
17	122 38.4	116 11.1	52.0	110 04.1	12.4	155 09.3	52.2	315 38.1	56.9
18	137 40.9	131 12.0	N 2 52.3	125 04.8	N 4 13.1	170 11.5	S13 52.1	330 40.6	N 7 56.9
19	152 43.4	146 12.8	52.6	140 05.5	13.9	185 13.7	52.1	345 43.1	56.9
20	167 45.8	161 13.7	52.9	155 06.2	14.6	200 15.9	52.0	0 45.6	56.9
21	182 48.3	176 14.5 . .	53.3	170 06.9 . .	15.4	215 18.1 . .	51.9	15 48.1 . .	56.9
22	197 50.8	191 15.4	53.6	185 07.6	16.1	230 20.3	51.8	30 50.6	56.9
23	212 53.2	206 16.2	53.9	200 08.3	16.9	245 22.5	51.7	45 53.1	56.9
Mer. Pass.	h m 8 54.7	*v* 0.9	*d* 0.3	*v* 0.7	*d* 0.8	*v* 2.2	*d* 0.1	*v* 2.5	*d* 0.0

STARS

Name	SHA	Dec
	° ′	°
Acamar	315 21.0	S40 16
Achernar	335 29.4	S57 11
Acrux	173 12.5	S63 09
Adhara	255 15.2	S28 59
Aldebaran	290 53.2	N16 31
Alioth	166 22.7	N55 54
Alkaid	153 00.7	N49 15
Al Na'ir	27 47.5	S46 54
Alnilam	275 49.7	S 1 11
Alphard	217 59.1	S 8 42
Alphecca	126 13.2	N26 40
Alpheratz	357 47.0	N29 08
Altair	62 11.1	N 8 53
Ankaa	353 18.9	S42 15
Antares	112 29.7	S26 27
Arcturus	145 58.1	N19 07
Atria	107 33.9	S69 02
Avior	234 19.6	S59 32
Bellatrix	278 35.6	N 6 21
Betelgeuse	271 04.9	N 7 24
Canopus	263 57.9	S52 42
Capella	280 39.4	N46 00
Deneb	49 33.6	N45 18
Denebola	182 36.5	N14 31
Diphda	348 59.2	S17 56
Dubhe	193 54.8	N61 42
Elnath	278 16.8	N28 37
Eltanin	90 47.2	N51 29
Enif	33 50.2	N 9 55
Fomalhaut	15 27.4	S29 34
Gacrux	172 04.1	S57 10
Gienah	175 55.3	S17 35
Hadar	148 51.9	S60 25
Hamal	328 04.6	N23 30
Kaus Aust.	83 47.6	S34 22
Kochab	137 18.3	N74 06
Markab	13 41.6	N15 15
Menkar	314 18.6	N 4 07
Menkent	148 10.9	S36 25
Miaplacidus	221 40.7	S69 45
Mirfak	308 45.4	N49 53
Nunki	76 01.9	S26 17
Peacock	53 23.8	S56 42
Pollux	243 31.6	N28 00
Procyon	245 03.1	N 5 12
Rasalhague	96 09.0	N12 33
Regulus	207 46.7	N11 55
Rigel	281 15.3	S 8 11
Rigil Kent.	139 55.5	S60 52
Sabik	102 15.8	S15 44
Schedar	349 44.7	N56 35
Shaula	96 25.7	S37 06
Sirius	258 36.6	S16 43
Spica	158 34.2	S11 12
Suhail	222 54.8	S43 28
Vega	80 40.8	N38 47
Zuben'ubi	137 08.5	S16 05

	SHA	Mer. Pas
	° ′	h m
Venus	354 37.2	9 17
Mars	348 38.0	9 41
Jupiter	32 42.0	6 44
Saturn	192 58.2	20 01

UT	SUN GHA	SUN Dec	MOON GHA	v	MOON Dec	d	HP
d h	° ′	° ′	° ′	′	° ′	′	′
7 00	180 51.5	N16 47.6	27 30.6	13.0	S12 42.3	12.9	56.5
01	195 51.6	48.3	42 02.6	12.9	12 55.2	12.8	56.5
02	210 51.6	49.0	56 34.5	12.9	13 08.0	12.7	56.5
03	225 51.6	. . 49.7	71 06.4	12.8	13 20.7	12.7	56.5
04	240 51.7	50.3	85 38.2	12.9	13 33.4	12.6	56.5
05	255 51.7	51.0	100 10.1	12.7	13 46.0	12.6	56.4
06	270 51.8	N16 51.7	114 41.8	12.8	S13 58.6	12.5	56.4
07	285 51.8	52.4	129 13.6	12.7	14 11.1	12.4	56.4
08	300 51.8	53.1	143 45.3	12.7	14 23.5	12.3	56.4
09	315 51.9	. . 53.8	158 17.0	12.6	14 35.8	12.2	56.4
10	330 51.9	54.5	172 48.6	12.6	14 48.0	12.2	56.3
11	345 52.0	55.1	187 20.2	12.6	15 00.2	12.1	56.3
12	0 52.0	N16 55.8	201 51.8	12.5	S15 12.3	12.0	56.3
13	15 52.0	56.5	216 23.3	12.5	15 24.3	11.9	56.3
14	30 52.1	57.2	230 54.8	12.5	15 36.2	11.9	56.3
15	45 52.1	. . 57.9	245 26.3	12.4	15 48.1	11.8	56.2
16	60 52.2	58.6	259 57.7	12.3	15 59.9	11.6	56.2
17	75 52.2	59.3	274 29.0	12.4	16 11.5	11.6	56.2
18	90 52.2	N16 59.9	289 00.4	12.3	S16 23.1	11.6	56.2
19	105 52.3	17 00.6	303 31.7	12.2	16 34.7	11.4	56.2
20	120 52.3	01.3	318 02.9	12.2	16 46.1	11.3	56.1
21	135 52.4	. . 02.0	332 34.1	12.2	16 57.4	11.3	56.1
22	150 52.4	02.7	347 05.3	12.1	17 08.7	11.2	56.1
23	165 52.4	03.3	1 36.4	12.1	17 19.9	11.1	56.1
8 00	180 52.5	N17 04.0	16 07.5	12.0	S17 31.0	11.0	56.1
01	195 52.5	04.7	30 38.5	12.0	17 42.0	10.9	56.0
02	210 52.5	05.4	45 09.5	12.0	17 52.9	10.8	56.0
03	225 52.6	. . 06.1	59 40.5	11.9	18 03.7	10.7	56.0
04	240 52.6	06.7	74 11.4	11.9	18 14.4	10.6	56.0
05	255 52.6	07.4	88 42.3	11.8	18 25.0	10.5	56.0
06	270 52.7	N17 08.1	103 13.1	11.8	S18 35.5	10.5	55.9
07	285 52.7	08.8	117 43.9	11.7	18 46.0	10.3	55.9
08	300 52.7	09.4	132 14.6	11.7	18 56.3	10.2	55.9
09	315 52.8	. . 10.1	146 45.3	11.6	19 06.5	10.2	55.9
10	330 52.8	10.8	161 15.9	11.6	19 16.7	10.0	55.9
11	345 52.9	11.5	175 46.5	11.6	19 26.7	10.0	55.8
12	0 52.9	N17 12.1	190 17.1	11.5	S19 36.7	9.8	55.8
13	15 52.9	12.8	204 47.6	11.5	19 46.5	9.7	55.8
14	30 53.0	13.5	219 18.1	11.4	19 56.2	9.7	55.8
15	45 53.0	. . 14.1	233 48.5	11.4	20 05.9	9.5	55.8
16	60 53.0	14.8	248 18.9	11.3	20 15.4	9.4	55.7
17	75 53.1	15.5	262 49.2	11.3	20 24.8	9.3	55.7
18	90 53.1	N17 16.2	277 19.5	11.3	S20 34.1	9.3	55.7
19	105 53.1	16.8	291 49.8	11.2	20 43.4	9.1	55.7
20	120 53.1	17.5	306 20.0	11.2	20 52.5	9.0	55.7
21	135 53.2	. . 18.2	320 50.2	11.1	21 01.5	8.9	55.6
22	150 53.2	18.8	335 20.3	11.1	21 10.4	8.8	55.6
23	165 53.2	19.5	349 50.4	11.0	21 19.2	8.6	55.6
9 00	180 53.3	N17 20.2	4 20.4	11.0	S21 27.8	8.6	55.6
01	195 53.3	20.8	18 50.4	11.0	21 36.4	8.4	55.6
02	210 53.3	21.5	33 20.4	10.9	21 44.8	8.4	55.5
03	225 53.4	. . 22.2	47 50.3	10.9	21 53.2	8.2	55.5
04	240 53.4	22.8	62 20.2	10.8	22 01.4	8.1	55.5
05	255 53.4	23.5	76 50.0	10.8	22 09.5	8.0	55.5
06	270 53.5	N17 24.2	91 19.8	10.8	S22 17.5	7.9	55.5
07	285 53.5	24.8	105 49.6	10.7	22 25.4	7.8	55.5
08	300 53.5	25.5	120 19.3	10.7	22 33.2	7.6	55.4
09	315 53.5	. . 26.2	134 49.0	10.6	22 40.8	7.6	55.4
10	330 53.6	26.8	149 18.6	10.6	22 48.4	7.4	55.4
11	345 53.6	27.5	163 48.2	10.6	22 55.8	7.3	55.4
12	0 53.6	N17 28.1	178 17.8	10.5	S23 03.1	7.2	55.4
13	15 53.6	28.8	192 47.3	10.5	23 10.3	7.0	55.3
14	30 53.7	29.5	207 16.8	10.4	23 17.3	7.0	55.3
15	45 53.7	. . 30.1	221 46.2	10.4	23 24.3	6.8	55.3
16	60 53.7	30.8	236 15.6	10.4	23 31.1	6.7	55.3
17	75 53.8	31.4	250 45.0	10.4	23 37.8	6.6	55.3
18	90 53.8	N17 32.1	265 14.4	10.3	S23 44.4	6.4	55.3
19	105 53.8	32.8	279 43.7	10.3	23 50.8	6.4	55.2
20	120 53.8	33.4	294 13.0	10.2	23 57.2	6.2	55.2
21	135 53.9	. . 34.1	308 42.2	10.2	24 03.4	6.1	55.2
22	150 53.9	34.7	323 11.4	10.2	24 09.5	5.9	55.2
23	165 53.9	35.4	337 40.6	10.2	S24 15.4	5.9	55.2
	SD 15.9	d 0.7	SD 15.3		15.2		15.1

Lat.	Twilight Naut.	Twilight Civil	Sunrise	Moonrise 7	Moonrise 8	Moonrise 9	Moonrise 10
°	h m	h m	h m	h m	h m	h m	h m
N 72	////	////	00 21	21 40	■	■	■
N 70	////	////	01 45	20 38	■	■	■
68	////	////	02 23	20 03	22 43	■	■
66	////	01 02	02 50	19 38	21 41	■	■
64	////	01 50	03 10	19 19	21 07	23 02	■
62	////	02 20	03 26	19 04	20 42	22 20	23 50
60	00 55	02 42	03 39	18 51	20 23	21 52	23 12
N 58	01 39	02 59	03 51	18 40	20 07	21 30	22 45
56	02 06	03 14	04 01	18 30	19 53	21 13	22 24
54	02 26	03 26	04 10	18 21	19 41	20 58	22 06
52	02 43	03 37	04 17	18 14	19 31	20 45	21 52
50	02 57	03 47	04 25	18 07	19 22	20 33	21 39
45	03 25	04 06	04 40	17 53	19 02	20 10	21 12
N 40	03 45	04 22	04 52	17 41	18 47	19 51	20 51
35	04 01	04 35	05 02	17 31	18 34	19 35	20 34
30	04 15	04 46	05 11	17 22	18 22	19 22	20 19
20	04 36	05 04	05 27	17 07	18 03	18 59	19 54
N 10	04 53	05 18	05 40	16 54	17 46	18 39	19 32
0	05 06	05 31	05 53	16 42	17 30	18 20	19 12
S 10	05 18	05 43	06 05	16 30	17 15	18 02	18 51
20	05 29	05 56	06 19	16 17	16 58	17 42	18 30
30	05 40	06 08	06 33	16 03	16 39	17 20	18 05
35	05 45	06 15	06 42	15 54	16 28	17 06	17 50
40	05 51	06 23	06 52	15 45	16 16	16 51	17 33
45	05 57	06 32	07 03	15 34	16 01	16 33	17 13
S 50	06 03	06 42	07 17	15 20	15 43	16 11	16 47
52	06 06	06 47	07 23	15 14	15 34	16 00	16 35
54	06 09	06 52	07 31	15 08	15 25	15 49	16 21
56	06 12	06 57	07 38	15 00	15 15	15 35	16 04
58	06 16	07 03	07 47	14 52	15 03	15 19	15 45
S 60	06 19	07 10	07 58	14 42	14 49	15 00	15 21

Lat.	Sunset	Twilight Civil	Twilight Naut.	Moonset 7	Moonset 8	Moonset 9	Moonset 10
°	h m	h m	h m	h m	h m	h m	h m
N 72	□	□	□	{01 23 / 10 46}	■	■	■
N 70	22 14	////	////	01 33	00 50	■	■
68	21 34	////	////	01 51	01 26	00 27	■
66	21 06	23 00	////	02 06	01 52	01 30	■
64	20 46	22 07	////	02 18	02 12	02 05	01 55
62	20 29	21 37	////	02 28	02 29	02 31	02 37
60	20 15	21 14	23 06	02 37	02 43	02 51	03 06
N 58	20 04	20 56	22 18	02 45	02 54	03 08	03 28
56	19 53	20 41	21 50	02 52	03 05	03 22	03 46
54	19 45	20 28	21 29	02 58	03 14	03 34	04 01
52	19 37	20 17	21 12	03 04	03 22	03 45	04 15
50	19 29	20 08	20 57	03 09	03 29	03 54	04 26
45	19 14	19 48	20 30	03 20	03 45	04 15	04 51
N 40	19 02	19 32	20 09	03 30	03 58	04 31	05 10
35	18 51	19 19	19 52	03 37	04 09	04 45	05 26
30	18 42	19 08	19 39	03 45	04 19	04 58	05 40
20	18 26	18 50	19 17	03 57	04 36	05 19	06 04
N 10	18 13	18 35	19 01	04 07	04 51	05 37	06 25
0	18 00	18 22	18 47	04 18	05 05	05 54	06 45
S 10	17 47	18 09	18 35	04 28	05 19	06 11	07 04
20	17 34	17 57	18 24	04 38	05 34	06 30	07 25
30	17 19	17 44	18 13	04 51	05 51	06 51	07 50
35	17 11	17 37	18 07	04 58	06 01	07 03	08 04
40	17 01	17 29	18 02	05 06	06 13	07 18	08 20
45	16 49	17 21	17 56	05 16	06 26	07 35	08 40
S 50	16 35	17 10	17 49	05 28	06 43	07 56	09 05
52	16 29	17 06	17 46	05 33	06 51	08 07	09 17
54	16 22	17 01	17 43	05 39	07 00	08 18	09 31
56	16 14	16 55	17 40	05 46	07 10	08 31	09 47
58	16 05	16 49	17 37	05 54	07 21	08 47	10 07
S 60	15 55	16 42	17 33	06 02	07 34	09 05	10 31

Day	SUN Eqn. of Time 00h	SUN Eqn. of Time 12h	SUN Mer. Pass.	MOON Mer. Pass. Upper	MOON Mer. Pass. Lower	MOON Age	MOON Phase
d	m s	m s	h m	h m	h m	d %	
7	03 26	03 28	11 57	22 53	10 30	12 97	○
8	03 30	03 31	11 56	23 42	11 17	13 99	
9	03 33	03 34	11 56	24 32	12 07	14 100	

UT d h		ARIES GHA	VENUS −4.7 GHA	VENUS Dec	MARS +1.2 GHA	MARS Dec	JUPITER −2.3 GHA	JUPITER Dec	SATURN +0.8 GHA	SATURN Dec
		° ′	° ′	° ′	° ′	° ′	° ′	° ′	° ′	° ′
10 00		227 55.7	221 17.0	N 2 54.2	215 09.0	N 4 17.6	260 24.7	S13 51.7	60 55.6	N 7 56.9
01		242 58.1	236 17.9	54.5	230 09.7	18.4	275 26.9	51.6	75 58.1	56.9
02		258 00.6	251 18.7	54.8	245 10.5	19.1	290 29.1	51.5	91 00.6	56.9
03		273 03.1	266 19.6	. . 55.2	260 11.2	. . 19.9	305 31.3	. . 51.4	106 03.1	. . 56.9
04		288 05.5	281 20.4	55.5	275 11.9	20.7	320 33.5	51.3	121 05.6	56.9
05		303 08.0	296 21.2	55.8	290 12.6	21.4	335 35.7	51.2	136 08.1	56.9
06		318 10.5	311 22.1	N 2 56.1	305 13.3	N 4 22.2	350 37.9	S13 51.2	151 10.6	N 7 56.9
07		333 12.9	326 22.9	56.4	320 14.0	22.9	5 40.1	51.1	166 13.1	57.0
08		348 15.4	341 23.7	56.8	335 14.7	23.7	20 42.3	51.0	181 15.6	57.0
09	S	3 17.9	356 24.6	. . 57.1	350 15.4	. . 24.4	35 44.5	. . 50.9	196 18.1	. . 57.0
10	U	18 20.3	11 25.4	57.4	5 16.1	25.2	50 46.7	50.8	211 20.6	57.0
11	N	33 22.8	26 26.2	57.7	20 16.8	25.9	65 48.9	50.8	226 23.0	57.0
12	D	48 25.3	41 27.0	N 2 58.1	35 17.5	N 4 26.7	80 51.1	S13 50.7	241 25.5	N 7 57.0
13	A	63 27.7	56 27.9	58.4	50 18.2	27.4	95 53.3	50.6	256 28.0	57.0
14	Y	78 30.2	71 28.7	58.7	65 18.9	28.2	110 55.5	50.5	271 30.5	57.0
15		93 32.6	86 29.5	. . 59.1	80 19.6	. . 28.9	125 57.7	. . 50.4	286 33.0	. . 57.0
16		108 35.1	101 30.3	59.4	95 20.3	29.7	140 59.9	50.4	301 35.5	57.0
17		123 37.6	116 31.1	2 59.7	110 21.0	30.4	156 02.1	50.3	316 38.0	57.0
18		138 40.0	131 32.0	N 3 00.1	125 21.7	N 4 31.2	171 04.3	S13 50.2	331 40.5	N 7 57.0
19		153 42.5	146 32.8	00.4	140 22.4	31.9	186 06.5	50.1	346 43.0	57.0
20		168 45.0	161 33.6	00.7	155 23.1	32.7	201 08.7	50.0	1 45.5	57.0
21		183 47.4	176 34.4	. . 01.1	170 23.8	. . 33.4	216 10.9	. . 50.0	16 48.0	. . 57.0
22		198 49.9	191 35.2	01.4	185 24.5	34.2	231 13.1	49.9	31 50.5	57.0
23		213 52.4	206 36.0	01.8	200 25.2	34.9	246 15.3	49.8	46 53.0	57.0
11 00		228 54.8	221 36.8	N 3 02.1	215 25.9	N 4 35.7	261 17.5	S13 49.7	61 55.4	N 7 57.0
01		243 57.3	236 37.6	02.4	230 26.6	36.4	276 19.7	49.6	76 57.9	57.0
02		258 59.8	251 38.4	02.8	245 27.3	37.2	291 22.0	49.6	92 00.4	57.0
03		274 02.2	266 39.2	. . 03.1	260 28.0	. . 37.9	306 24.2	. . 49.5	107 02.9	. . 57.0
04		289 04.7	281 40.0	03.5	275 28.7	38.7	321 26.4	49.4	122 05.4	57.0
05		304 07.1	296 40.8	03.8	290 29.4	39.4	336 28.6	49.3	137 07.9	57.0
06		319 09.6	311 41.6	N 3 04.2	305 30.1	N 4 40.2	351 30.8	S13 49.3	152 10.4	N 7 57.0
07		334 12.1	326 42.4	04.5	320 30.8	40.9	6 33.0	49.2	167 12.9	57.0
08		349 14.5	341 43.2	04.9	335 31.5	41.7	21 35.2	49.1	182 15.4	57.1
09	M	4 17.0	356 44.0	. . 05.2	350 32.2	. . 42.4	36 37.4	. . 49.0	197 17.9	. . 57.1
10	O	19 19.5	11 44.8	05.6	5 32.9	43.2	51 39.6	48.9	212 20.4	57.1
11	N	34 21.9	26 45.6	05.9	20 33.6	43.9	66 41.8	48.9	227 22.8	57.1
12	D	49 24.4	41 46.3	N 3 06.3	35 34.3	N 4 44.7	81 44.0	S13 48.8	242 25.3	N 7 57.1
13	A	64 26.9	56 47.1	06.6	50 35.0	45.4	96 46.2	48.7	257 27.8	57.1
14	Y	79 29.3	71 47.9	07.0	65 35.7	46.2	111 48.5	48.6	272 30.3	57.1
15		94 31.8	86 48.7	. . 07.4	80 36.4	. . 46.9	126 50.7	. . 48.5	287 32.8	. . 57.1
16		109 34.3	101 49.5	07.7	95 37.1	47.7	141 52.9	48.5	302 35.3	57.1
17		124 36.7	116 50.2	08.1	110 37.8	48.4	156 55.1	48.4	317 37.8	57.1
18		139 39.2	131 51.0	N 3 08.4	125 38.5	N 4 49.2	171 57.3	S13 48.3	332 40.3	N 7 57.1
19		154 41.6	146 51.8	08.8	140 39.2	49.9	186 59.5	48.2	347 42.8	57.1
20		169 44.1	161 52.6	09.2	155 39.9	50.7	202 01.7	48.2	2 45.2	57.1
21		184 46.6	176 53.3	. . 09.5	170 40.6	. . 51.4	217 03.9	. . 48.1	17 47.7	. . 57.1
22		199 49.0	191 54.1	09.9	185 41.3	52.2	232 06.1	48.0	32 50.2	57.1
23		214 51.5	206 54.9	10.3	200 42.0	52.9	247 08.3	47.9	47 52.7	57.1
12 00		229 54.0	221 55.7	N 3 10.6	215 42.7	N 4 53.7	262 10.6	S13 47.8	62 55.2	N 7 57.1
01		244 56.4	236 56.4	11.0	230 43.4	54.4	277 12.8	47.8	77 57.7	57.1
02		259 58.9	251 57.2	11.4	245 44.1	55.2	292 15.0	47.7	93 00.2	57.1
03		275 01.4	266 57.9	. . 11.7	260 44.8	. . 55.9	307 17.2	. . 47.6	108 02.7	. . 57.1
04		290 03.8	281 58.7	12.1	275 45.5	56.7	322 19.4	47.5	123 05.1	57.1
05		305 06.3	296 59.5	12.5	290 46.2	57.4	337 21.6	47.5	138 07.6	57.1
06		320 08.8	312 00.2	N 3 12.9	305 46.9	N 4 58.1	352 23.8	S13 47.4	153 10.1	N 7 57.1
07		335 11.2	327 01.0	13.2	320 47.6	58.9	7 26.1	47.3	168 12.6	57.1
08	T	350 13.7	342 01.7	13.6	335 48.3	4 59.6	22 28.3	47.2	183 15.1	57.1
09	U	5 16.1	357 02.5	. . 14.0	350 49.0	5 00.4	37 30.5	. . 47.2	198 17.6	. . 57.1
10	E	20 18.6	12 03.2	14.4	5 49.7	01.1	52 32.7	47.1	213 20.1	57.1
11	S	35 21.1	27 04.0	14.7	20 50.4	01.9	67 34.9	47.0	228 22.5	57.1
12	D	50 23.5	42 04.7	N 3 15.1	35 51.1	N 5 02.6	82 37.1	S13 46.9	243 25.0	N 7 57.1
13	A	65 26.0	57 05.5	15.5	50 51.8	03.4	97 39.3	46.9	258 27.5	57.1
14	Y	80 28.5	72 06.2	15.9	65 52.5	04.1	112 41.6	46.8	273 30.0	57.1
15		95 30.9	87 07.0	. . 16.3	80 53.2	. . 04.9	127 43.8	. . 46.7	288 32.5	. . 57.1
16		110 33.4	102 07.7	16.7	95 53.9	05.6	142 46.0	46.6	303 35.0	57.1
17		125 35.9	117 08.5	17.0	110 54.6	06.4	157 48.2	46.6	318 37.4	57.1
18		140 38.3	132 09.2	N 3 17.4	125 55.3	N 5 07.1	172 50.4	S13 46.5	333 39.9	N 7 57.1
19		155 40.8	147 09.9	17.8	140 56.0	07.9	187 52.6	46.4	348 42.4	57.1
20		170 43.2	162 10.7	18.2	155 56.7	08.6	202 54.9	46.3	3 44.9	57.1
21		185 45.7	177 11.4	. . 18.6	170 57.4	. . 09.3	217 57.1	. . 46.2	18 47.4	. . 57.1
22		200 48.2	192 12.1	19.0	185 58.1	10.1	232 59.3	46.2	33 49.9	57.1
23		215 50.6	207 12.9	19.4	200 58.8	10.8	248 01.5	46.1	48 52.4	57.1
Mer. Pass.		h m 8 42.9	v 0.8	d 0.4	v 0.7	d 0.7	v 2.2	d 0.1	v 2.5	d 0.0

STARS

Name	SHA	Dec
	° ′	°
Acamar	315 21.0	S40
Achernar	335 29.4	S57
Acrux	173 12.5	S63
Adhara	255 15.2	S28
Aldebaran	290 53.2	N16
Alioth	166 22.7	N55
Alkaid	153 00.7	N49
Al Na'ir	27 47.5	S46
Alnilam	275 49.8	S 1
Alphard	217 59.1	S 8
Alphecca	126 13.2	N26
Alpheratz	357 47.0	N29
Altair	62 11.1	N 8
Ankaa	353 18.9	S42
Antares	112 29.7	S26
Arcturus	145 58.1	N19
Atria	107 33.9	S69
Avior	234 19.6	S59
Bellatrix	278 35.6	N 6
Betelgeuse	271 04.9	N 7
Canopus	263 57.9	S52
Capella	280 39.4	N46
Deneb	49 33.6	N45
Denebola	182 36.5	N14
Diphda	348 59.2	S17
Dubhe	193 54.9	N61
Elnath	278 16.8	N28
Eltanin	90 47.1	N51
Enif	33 50.2	N 9
Fomalhaut	15 27.4	S29
Gacrux	172 04.1	S57
Gienah	175 55.3	S17
Hadar	148 51.9	S60
Hamal	328 04.6	N23
Kaus Aust.	83 47.6	S34
Kochab	137 18.3	N74
Markab	13 41.6	N15
Menkar	314 18.6	N 4
Menkent	148 10.9	S36
Miaplacidus	221 40.7	S69
Mirfak	308 45.4	N49
Nunki	76 01.8	S26
Peacock	53 23.7	S56
Pollux	243 31.6	N28
Procyon	245 03.1	N 5
Rasalhague	96 09.0	N12
Regulus	207 46.7	N11
Rigel	281 15.3	S 8
Rigil Kent.	139 55.5	S60
Sabik	102 15.7	S15
Schedar	349 44.7	N56
Shaula	96 25.7	S37
Sirius	258 36.6	S16
Spica	158 34.2	S11
Suhail	222 54.8	S43
Vega	80 40.8	N38
Zuben'ubi	137 08.5	S16

	SHA	Mer. Pass.
	° ′	h
Venus	352 42.0	9
Mars	346 31.0	9
Jupiter	32 22.7	6
Saturn	193 00.6	19

UT	SUN GHA	SUN Dec	MOON GHA	v	MOON Dec	d	HP
d h	° ′	° ′	° ′	′	° ′	′	′
10 00	180 53.9	N17 36.0	352 09.8	10.1	S24 21.3	5.7	55.1
01	195 54.0	36.7	6 38.9	10.1	24 27.0	5.6	55.1
02	210 54.0	37.3	21 08.0	10.1	24 32.6	5.4	55.1
03	225 54.0	. . 38.0	35 37.1	10.0	24 38.0	5.4	55.1
04	240 54.0	38.7	50 06.1	10.1	24 43.4	5.2	55.1
05	255 54.1	39.3	64 35.2	10.0	24 48.6	5.0	55.1
06	270 54.1	N17 40.0	79 04.2	9.9	S24 53.6	5.0	55.0
07	285 54.1	40.6	93 33.1	10.0	24 58.6	4.8	55.0
08	300 54.1	41.3	108 02.1	9.9	25 03.4	4.7	55.0
09	315 54.1	. . 41.9	122 31.0	9.9	25 08.1	4.6	55.0
10	330 54.2	42.6	136 59.9	9.9	25 12.7	4.4	55.0
11	345 54.2	43.2	151 28.8	9.9	25 17.1	4.3	55.0
12	0 54.2	N17 43.9	165 57.7	9.8	S25 21.4	4.2	54.9
13	15 54.2	44.5	180 26.5	9.9	25 25.6	4.0	54.9
14	30 54.3	45.2	194 55.4	9.8	25 29.6	3.9	54.9
15	45 54.3	. . 45.8	209 24.2	9.8	25 33.5	3.8	54.9
16	60 54.3	46.5	223 53.0	9.8	25 37.3	3.7	54.9
17	75 54.3	47.1	238 21.8	9.7	25 41.0	3.5	54.9
18	90 54.3	N17 47.7	252 50.5	9.8	S25 44.5	3.4	54.9
19	105 54.4	48.4	267 19.3	9.8	25 47.9	3.2	54.8
20	120 54.4	49.0	281 48.1	9.7	25 51.1	3.1	54.8
21	135 54.4	. . 49.7	296 16.8	9.7	25 54.2	3.0	54.8
22	150 54.4	50.3	310 45.5	9.8	25 57.2	2.9	54.8
23	165 54.4	51.0	325 14.3	9.7	26 00.1	2.7	54.8
11 00	180 54.5	N17 51.6	339 43.0	9.7	S26 02.8	2.6	54.8
01	195 54.5	52.3	354 11.7	9.7	26 05.4	2.5	54.7
02	210 54.5	52.9	8 40.4	9.7	26 07.9	2.3	54.7
03	225 54.5	. . 53.5	23 09.1	9.7	26 10.2	2.2	54.7
04	240 54.5	54.2	37 37.8	9.7	26 12.4	2.0	54.7
05	255 54.5	54.8	52 06.5	9.7	26 14.4	2.0	54.7
06	270 54.6	N17 55.5	66 35.2	9.7	S26 16.4	1.8	54.7
07	285 54.6	56.1	81 03.9	9.7	26 18.2	1.6	54.7
08	300 54.6	56.7	95 32.6	9.7	26 19.8	1.5	54.6
09	315 54.6	. . 57.4	110 01.3	9.7	26 21.3	1.4	54.6
10	330 54.6	58.0	124 30.0	9.8	26 22.7	1.3	54.6
11	345 54.6	58.6	138 58.8	9.7	26 24.0	1.1	54.6
12	0 54.7	N17 59.3	153 27.5	9.7	S26 25.1	1.0	54.6
13	15 54.7	17 59.9	167 56.2	9.7	26 26.1	0.9	54.6
14	30 54.7	18 00.6	182 24.9	9.8	26 27.0	0.7	54.6
15	45 54.7	. . 01.2	196 53.7	9.8	26 27.7	0.6	54.6
16	60 54.7	01.8	211 22.5	9.7	26 28.3	0.5	54.5
17	75 54.7	02.5	225 51.2	9.8	26 28.8	0.3	54.5
18	90 54.8	N18 03.1	240 20.0	9.8	S26 29.1	0.2	54.5
19	105 54.8	03.7	254 48.8	9.9	26 29.3	0.1	54.5
20	120 54.8	04.4	269 17.7	9.8	26 29.4	0.1	54.5
21	135 54.8	. . 05.0	283 46.5	9.8	26 29.3	0.2	54.5
22	150 54.8	05.6	298 15.3	9.9	26 29.1	0.3	54.5
23	165 54.8	06.3	312 44.2	9.9	26 28.8	0.5	54.5
12 00	180 54.8	N18 06.9	327 13.1	9.9	S26 28.3	0.5	54.5
01	195 54.8	07.5	341 42.0	10.0	26 27.8	0.8	54.4
02	210 54.9	08.1	356 11.0	9.9	26 27.0	0.8	54.4
03	225 54.9	. . 08.8	10 39.9	10.0	26 26.2	1.0	54.4
04	240 54.9	09.4	25 08.9	10.0	26 25.2	1.1	54.4
05	255 54.9	10.0	39 37.9	10.0	26 24.1	1.2	54.4
06	270 54.9	N18 10.7	54 06.9	10.1	S26 22.9	1.4	54.4
07	285 54.9	11.3	68 36.0	10.1	26 21.5	1.5	54.4
08	300 54.9	11.9	83 05.1	10.1	26 20.0	1.6	54.4
09	315 54.9	. . 12.5	97 34.2	10.2	26 18.4	1.8	54.4
10	330 54.9	13.2	112 03.4	10.2	26 16.6	1.9	54.3
11	345 55.0	13.8	126 32.6	10.2	26 14.7	2.0	54.3
12	0 55.0	N18 14.4	141 01.8	10.2	S26 12.7	2.1	54.3
13	15 55.0	15.0	155 31.0	10.3	26 10.6	2.3	54.3
14	30 55.0	15.7	170 00.3	10.3	26 08.3	2.3	54.3
15	45 55.0	. . 16.3	184 29.6	10.4	26 06.0	2.5	54.3
16	60 55.0	16.9	198 59.0	10.4	26 03.5	2.7	54.3
17	75 55.0	17.5	213 28.4	10.4	26 00.8	2.7	54.3
18	90 55.0	N18 18.1	227 57.8	10.5	S25 58.1	2.9	54.3
19	105 55.0	18.8	242 27.3	10.5	25 55.2	3.0	54.3
20	120 55.0	19.4	256 56.8	10.5	25 52.2	3.2	54.3
21	135 55.0	. . 20.0	271 26.3	10.6	25 49.0	3.2	54.3
22	150 55.1	20.6	285 55.9	10.6	25 45.8	3.4	54.3
23	165 55.1	21.2	300 25.5	10.7	S25 42.4	3.5	54.2
	SD 15.9	*d* 0.6	SD 15.0		14.9		14.8

Lat.	Twilight Naut.	Twilight Civil	Sunrise	Moonrise 10	11	12	13
°	h m	h m	h m	h m	h m	h m	h m
N 72	□	□	□	■	■	■	■
N 70	////	////	01 22	■	■	■	■
68	////	////	02 08	■	■	■	■
66	////	00 22	02 38	■	■	■	■
64	////	01 33	03 00	■	■	■	02 22
62	////	02 07	03 17	23 50	24 53	00 53	01 22
60	00 19	02 32	03 32	23 12	24 12	00 12	00 48
N 58	01 24	02 51	03 44	22 45	23 43	24 23	00 23
56	01 55	03 07	03 55	22 24	23 21	24 03	00 03
54	02 18	03 20	04 04	22 06	23 03	23 46	24 17
52	02 36	03 31	04 12	21 52	22 48	23 32	24 05
50	02 51	03 41	04 20	21 39	22 35	23 20	23 54
45	03 20	04 02	04 36	21 12	22 07	22 54	23 32
N 40	03 41	04 18	04 49	20 51	21 46	22 33	23 14
35	03 58	04 32	05 00	20 34	21 28	22 16	22 58
30	04 12	04 43	05 09	20 19	21 13	22 02	22 45
20	04 34	05 02	05 26	19 54	20 47	21 36	22 22
N 10	04 52	05 18	05 40	19 32	20 24	21 15	22 02
0	05 06	05 31	05 53	19 12	20 03	20 54	21 44
S 10	05 18	05 44	06 06	18 51	19 42	20 34	21 25
20	05 30	05 57	06 20	18 30	19 20	20 12	21 05
30	05 41	06 10	06 35	18 05	18 54	19 47	20 42
35	05 47	06 18	06 44	17 50	18 39	19 32	20 29
40	05 53	06 26	06 55	17 33	18 21	19 15	20 13
45	06 00	06 35	07 07	17 13	18 00	18 54	19 54
S 50	06 07	06 46	07 21	16 47	17 33	18 28	19 31
52	06 10	06 51	07 28	16 35	17 20	18 15	19 20
54	06 13	06 56	07 36	16 21	17 05	18 01	19 07
56	06 17	07 02	07 44	16 04	16 47	17 44	18 52
58	06 21	07 09	07 54	15 45	16 25	17 23	18 34
S 60	06 25	07 16	08 04	15 21	15 58	16 57	18 13

Lat.	Sunset	Twilight Civil	Twilight Naut.	Moonset 10	11	12	13
°	h m	h m	h m	h m	h m	h m	h m
N 72	□	□	□	■	■	■	■
N 70	22 38	////	////	■	■	■	■
68	21 49	////	////	■	■	■	■
66	21 18	////	////	■	■	■	■
64	20 55	22 25	////	01 55	■	■	03 57
62	20 37	21 49	////	02 37	02 55	03 39	04 57
60	20 23	21 23	////	03 06	03 33	04 21	05 30
N 58	20 10	21 04	22 34	03 28	04 00	04 49	05 55
56	19 59	20 48	22 01	03 46	04 22	05 11	06 15
54	19 50	20 34	21 37	04 01	04 39	05 29	06 31
52	19 41	20 23	21 19	04 15	04 54	05 44	06 45
50	19 34	20 13	21 04	04 26	05 07	05 58	06 57
45	19 18	19 52	20 34	04 51	05 34	06 25	07 22
N 40	19 05	19 35	20 12	05 10	05 55	06 46	07 43
35	18 54	19 21	19 55	05 26	06 13	07 04	07 59
30	18 44	19 10	19 41	05 40	06 28	07 19	08 14
20	18 27	18 51	19 19	06 04	06 54	07 45	08 38
N 10	18 13	18 35	19 01	06 25	07 16	08 07	08 59
0	18 00	18 22	18 47	06 45	07 37	08 28	09 19
S 10	17 47	18 09	18 34	07 04	07 57	08 49	09 38
20	17 33	17 56	18 23	07 25	08 20	09 11	09 59
30	17 17	17 42	18 11	07 50	08 45	09 37	10 23
35	17 08	17 35	18 05	08 04	09 01	09 52	10 37
40	16 58	17 27	17 59	08 20	09 18	10 09	10 53
45	16 46	17 17	17 53	08 40	09 39	10 30	11 13
S 50	16 31	17 06	17 45	09 05	10 06	10 57	11 37
52	16 24	17 01	17 42	09 17	10 19	11 10	11 48
54	16 16	16 56	17 39	09 31	10 35	11 25	12 01
56	16 08	16 50	17 35	09 47	10 52	11 42	12 17
58	15 58	16 43	17 31	10 07	11 14	12 03	12 34
S 60	15 47	16 36	17 27	10 31	11 41	12 29	12 56

Day	SUN Eqn. of Time 00^h	SUN Eqn. of Time 12^h	SUN Mer. Pass.	MOON Mer. Pass. Upper	MOON Mer. Pass. Lower	MOON Age	MOON Phase
d	m s	m s	h m	h m	h m	d	%
10	03 36	03 37	11 56	00 32	12 58	15	98
11	03 38	03 39	11 56	01 24	13 50	16	95
12	03 39	03 40	11 56	02 16	14 41	17	90

UT d h	ARIES GHA	VENUS −4.7 GHA	VENUS Dec	MARS +1.2 GHA	MARS Dec	JUPITER −2.3 GHA	JUPITER Dec	SATURN +0.8 GHA	SATURN Dec
	° ′	° ′	° ′	° ′	° ′	° ′	° ′	° ′	° ′
13 00 (WEDNESDAY)	230 53.1	222 13.6	N 3 19.8	215 59.5	N 5 11.6	263 03.7	S13 46.0	63 54.8	N 7 57.1
01	245 55.6	237 14.3	20.2	231 00.2	12.3	278 06.0	45.9	78 57.3	57.1
02	260 58.0	252 15.1	20.6	246 00.9	13.1	293 08.2	45.9	93 59.8	57.1
03	276 00.5	267 15.8	21.0	261 01.6	13.8	308 10.4	45.8	109 02.3	57.1
04	291 03.0	282 16.5	21.4	276 02.3	14.6	323 12.6	45.7	124 04.8	57.1
05	306 05.4	297 17.2	21.8	291 03.0	15.3	338 14.8	45.7	139 07.3	57.1
06	321 07.9	312 18.0	N 3 22.2	306 03.7	N 5 16.0	353 17.1	S13 45.6	154 09.7	N 7 57.1
07	336 10.4	327 18.7	22.6	321 04.4	16.8	8 19.3	45.5	169 12.2	57.1
08	351 12.8	342 19.4	23.0	336 05.1	17.5	23 21.5	45.4	184 14.7	57.1
09	6 15.3	357 20.1	23.4	351 05.8	18.3	38 23.7	45.4	199 17.2	57.1
10	21 17.7	12 20.8	23.8	6 06.5	19.0	53 25.9	45.3	214 19.7	57.1
11	36 20.2	27 21.6	24.2	21 07.2	19.8	68 28.2	45.2	229 22.1	57.1
12	51 22.7	42 22.3	N 3 24.6	36 07.9	N 5 20.5	83 30.4	S13 45.1	244 24.6	N 7 57.1
13	66 25.1	57 23.0	25.0	51 08.6	21.3	98 32.6	45.1	259 27.1	57.1
14	81 27.6	72 23.7	25.4	66 09.3	22.0	113 34.8	45.0	274 29.6	57.1
15	96 30.1	87 24.4	25.8	81 10.0	22.7	128 37.0	44.9	289 32.1	57.1
16	111 32.5	102 25.1	26.2	96 10.7	23.5	143 39.3	44.8	304 34.5	57.1
17	126 35.0	117 25.8	26.6	111 11.4	24.2	158 41.5	44.8	319 37.0	57.1
18	141 37.5	132 26.5	N 3 27.0	126 12.1	N 5 25.0	173 43.7	S13 44.7	334 39.5	N 7 57.1
19	156 39.9	147 27.2	27.4	141 12.8	25.7	188 45.9	44.6	349 42.0	57.1
20	171 42.4	162 27.9	27.9	156 13.5	26.5	203 48.2	44.5	4 44.5	57.1
21	186 44.9	177 28.6	28.3	171 14.2	27.2	218 50.4	44.5	19 46.9	57.1
22	201 47.3	192 29.3	28.7	186 14.9	27.9	233 52.6	44.4	34 49.4	57.1
23	216 49.8	207 30.0	29.1	201 15.6	28.7	248 54.8	44.3	49 51.9	57.1
14 00 (THURSDAY)	231 52.2	222 30.7	N 3 29.5	216 16.3	N 5 29.4	263 57.1	S13 44.3	64 54.4	N 7 57.1
01	246 54.7	237 31.4	29.9	231 17.0	30.2	278 59.3	44.2	79 56.9	57.1
02	261 57.2	252 32.1	30.4	246 17.7	30.9	294 01.5	44.1	94 59.3	57.1
03	276 59.6	267 32.8	30.8	261 18.4	31.7	309 03.7	44.0	110 01.8	57.1
04	292 02.1	282 33.5	31.2	276 19.1	32.4	324 06.0	44.0	125 04.3	57.1
05	307 04.6	297 34.2	31.6	291 19.8	33.1	339 08.2	43.9	140 06.8	57.1
06	322 07.0	312 34.9	N 3 32.0	306 20.5	N 5 33.9	354 10.4	S13 43.8	155 09.3	N 7 57.1
07	337 09.5	327 35.6	32.5	321 21.2	34.6	9 12.6	43.7	170 11.7	57.1
08	352 12.0	342 36.2	32.9	336 21.9	35.4	24 14.9	43.7	185 14.2	57.1
09	7 14.4	357 36.9	33.3	351 22.6	36.1	39 17.1	43.6	200 16.7	57.1
10	22 16.9	12 37.6	33.8	6 23.3	36.9	54 19.3	43.5	215 19.2	57.1
11	37 19.4	27 38.3	34.2	21 24.0	37.6	69 21.6	43.5	230 21.6	57.1
12	52 21.8	42 39.0	N 3 34.6	36 24.7	N 5 38.3	84 23.8	S13 43.4	245 24.1	N 7 57.1
13	67 24.3	57 39.6	35.0	51 25.4	39.1	99 26.0	43.3	260 26.6	57.1
14	82 26.7	72 40.3	35.5	66 26.1	39.8	114 28.2	43.2	275 29.1	57.1
15	97 29.2	87 41.0	35.9	81 26.8	40.6	129 30.5	43.2	290 31.6	57.1
16	112 31.7	102 41.7	36.3	96 27.5	41.3	144 32.7	43.1	305 34.0	57.1
17	127 34.1	117 42.3	36.8	111 28.2	42.0	159 34.9	43.0	320 36.5	57.1
18	142 36.6	132 43.0	N 3 37.2	126 28.9	N 5 42.8	174 37.2	S13 43.0	335 39.0	N 7 57.1
19	157 39.1	147 43.7	37.6	141 29.6	43.5	189 39.4	42.9	350 41.5	57.1
20	172 41.5	162 44.4	38.1	156 30.3	44.3	204 41.6	42.8	5 43.9	57.1
21	187 44.0	177 45.0	38.5	171 31.0	45.0	219 43.9	42.7	20 46.4	57.1
22	202 46.5	192 45.7	39.0	186 31.7	45.7	234 46.1	42.7	35 48.9	57.1
23	217 48.9	207 46.4	39.4	201 32.4	46.5	249 48.3	42.6	50 51.4	57.1
15 00 (FRIDAY)	232 51.4	222 47.0	N 3 39.8	216 33.1	N 5 47.2	264 50.5	S13 42.5	65 53.8	N 7 57.1
01	247 53.9	237 47.7	40.3	231 33.8	48.0	279 52.8	42.5	80 56.3	57.1
02	262 56.3	252 48.3	40.7	246 34.5	48.7	294 55.0	42.4	95 58.8	57.1
03	277 58.8	267 49.0	41.2	261 35.2	49.4	309 57.2	42.3	111 01.3	57.0
04	293 01.2	282 49.7	41.6	276 35.9	50.2	324 59.5	42.3	126 03.7	57.0
05	308 03.7	297 50.3	42.1	291 36.6	50.9	340 01.7	42.2	141 06.2	57.0
06	323 06.2	312 51.0	N 3 42.5	306 37.3	N 5 51.7	355 03.9	S13 42.1	156 08.7	N 7 57.0
07	338 08.6	327 51.6	42.9	321 38.0	52.4	10 06.2	42.0	171 11.2	57.0
08	353 11.1	342 52.3	43.4	336 38.7	53.1	25 08.4	42.0	186 13.6	57.0
09	8 13.6	357 52.9	43.8	351 39.4	53.9	40 10.6	41.9	201 16.1	57.0
10	23 16.0	12 53.6	44.3	6 40.1	54.6	55 12.9	41.8	216 18.6	57.0
11	38 18.5	27 54.2	44.7	21 40.8	55.3	70 15.1	41.8	231 21.1	57.0
12	53 21.0	42 54.9	N 3 45.2	36 41.5	N 5 56.1	85 17.3	S13 41.7	246 23.5	N 7 57.0
13	68 23.4	57 55.5	45.6	51 42.2	56.8	100 19.6	41.6	261 26.0	57.0
14	83 25.9	72 56.2	46.1	66 42.9	57.6	115 21.8	41.6	276 28.5	57.0
15	98 28.3	87 56.8	46.6	81 43.6	58.3	130 24.1	41.5	291 31.0	57.0
16	113 30.8	102 57.4	47.0	96 44.3	59.0	145 26.3	41.4	306 33.4	57.0
17	128 33.3	117 58.1	47.5	111 45.0	5 59.8	160 28.5	41.4	321 35.9	57.0
18	143 35.7	132 58.7	N 3 47.9	126 45.7	N 6 00.5	175 30.8	S13 41.3	336 38.4	N 7 57.0
19	158 38.2	147 59.4	48.4	141 46.4	01.2	190 33.0	41.2	351 40.8	57.0
20	173 40.7	163 00.0	48.8	156 47.1	02.0	205 35.2	41.1	6 43.3	57.0
21	188 43.1	178 00.6	49.3	171 47.8	02.7	220 37.5	41.1	21 45.8	57.0
22	203 45.6	193 01.3	49.8	186 48.5	03.5	235 39.7	41.0	36 48.3	57.0
23	218 48.1	208 01.9	50.2	201 49.2	04.2	250 41.9	40.9	51 50.7	57.0
Mer. Pass.	h m 8 31.1	v 0.7	d 0.4	v 0.7	d 0.7	v 2.2	d 0.1	v 2.5	d 0.0

STARS Name	SHA	Dec
	° ′	°
Acamar	315 21.0	S40 1
Achernar	335 29.3	S57 1
Acrux	173 12.5	S63 0
Adhara	255 15.2	S28 5
Aldebaran	290 53.2	N16 3
Alioth	166 22.7	N55 5
Alkaid	153 00.7	N49 1
Al Na'ir	27 47.4	S46 5
Alnilam	275 49.7	S 1 1
Alphard	217 59.1	S 8 4
Alphecca	126 13.2	N26 4
Alpheratz	357 47.0	N29 0
Altair	62 11.1	N 8 5
Ankaa	353 18.8	S42 1
Antares	112 29.7	S26 2
Arcturus	145 58.1	N19 0
Atria	107 33.8	S69 0
Avior	234 19.6	S59 3
Bellatrix	278 35.6	N 6 2
Betelgeuse	271 04.9	N 7 2
Canopus	263 57.9	S52 4
Capella	280 39.4	N46 0
Deneb	49 33.5	N45 1
Denebola	182 36.5	N14 3
Diphda	348 59.1	S17 5
Dubhe	193 54.9	N61 4
Elnath	278 16.8	N28 3
Eltanin	90 47.1	N51 2
Enif	33 50.1	N 9 5
Fomalhaut	15 27.4	S29 3
Gacrux	172 04.1	S57 1
Gienah	175 55.3	S17 3
Hadar	148 51.9	S60 2
Hamal	328 04.6	N23 3
Kaus Aust.	83 47.5	S34 2
Kochab	137 18.3	N74 07
Markab	13 41.5	N15 15
Menkar	314 18.6	N 4 07
Menkent	148 10.9	S36 25
Miaplacidus	221 40.8	S69 45
Mirfak	308 45.3	N49 53
Nunki	76 01.8	S26 17
Peacock	53 23.7	S56 42
Pollux	243 31.6	N28 00
Procyon	245 03.1	N 5 12
Rasalhague	96 09.0	N12 33
Regulus	207 46.7	N11 55
Rigel	281 15.3	S 8 11
Rigil Kent.	139 55.5	S60 52
Sabik	102 15.7	S15 44
Schedar	349 44.7	N56 35
Shaula	96 25.7	S37 06
Sirius	258 36.7	S16 43
Spica	158 34.2	S11 12
Suhail	222 54.9	S43 28
Vega	80 40.7	N38 47
Zuben'ubi	137 08.5	S16 05

	SHA	Mer. Pa[ss.]
	° ′	h m
Venus	350 38.5	9 10
Mars	344 24.1	9 34
Jupiter	32 04.8	6 23
Saturn	193 02.1	19 37

JT d h	SUN GHA ° ′	SUN Dec ° ′	MOON GHA ° ′	v ′	MOON Dec ° ′	d ′	HP ′
3 00	180 55.1	N18 21.9	314 55.2	10.7	S25 38.9	3.6	54.2
01	195 55.1	22.5	329 24.9	10.8	25 35.3	3.8	54.2
02	210 55.1	23.1	343 54.7	10.8	25 31.5	3.8	54.2
03	225 55.1	. . 23.7	358 24.5	10.8	25 27.7	4.0	54.2
04	240 55.1	24.3	12 54.3	10.9	25 23.7	4.1	54.2
05	255 55.1	24.9	27 24.2	10.9	25 19.6	4.2	54.2
06	270 55.1	N18 25.5	41 54.1	11.0	S25 15.4	4.3	54.2
07	285 55.1	26.2	56 24.1	11.1	25 11.1	4.5	54.2
08	300 55.1	26.8	70 54.2	11.0	25 06.6	4.6	54.2
09	315 55.1	. . 27.4	85 24.2	11.2	25 02.0	4.6	54.2
10	330 55.1	28.0	99 54.4	11.1	24 57.4	4.8	54.2
11	345 55.1	28.6	114 24.5	11.3	24 52.6	5.0	54.2
12	0 55.1	N18 29.2	128 54.8	11.3	S24 47.6	5.0	54.2
13	15 55.1	29.8	143 25.1	11.3	24 42.6	5.1	54.2
14	30 55.1	30.4	157 55.4	11.4	24 37.5	5.3	54.2
15	45 55.1	. . 31.1	172 25.8	11.4	24 32.2	5.3	54.2
16	60 55.1	31.7	186 56.2	11.5	24 26.9	5.5	54.2
17	75 55.1	32.3	201 26.7	11.5	24 21.4	5.6	54.2
18	90 55.1	N18 32.9	215 57.2	11.6	S24 15.8	5.7	54.2
19	105 55.1	33.5	230 27.8	11.6	24 10.1	5.8	54.2
20	120 55.1	34.1	244 58.4	11.7	24 04.3	5.9	54.2
21	135 55.1	. . 34.7	259 29.1	11.8	23 58.4	6.1	54.2
22	150 55.1	35.3	273 59.9	11.8	23 52.3	6.1	54.2
23	165 55.1	35.9	288 30.7	11.8	23 46.2	6.2	54.2
4 00	180 55.1	N18 36.5	303 01.5	12.0	S23 40.0	6.4	54.2
01	195 55.1	37.1	317 32.5	11.9	23 33.6	6.4	54.2
02	210 55.1	37.7	332 03.4	12.0	23 27.2	6.6	54.2
03	225 55.1	. . 38.3	346 34.4	12.1	23 20.6	6.7	54.2
04	240 55.1	38.9	1 05.5	12.1	23 13.9	6.7	54.2
05	255 55.1	39.5	15 36.6	12.2	23 07.2	6.9	54.2
06	270 55.1	N18 40.1	30 07.8	12.3	S23 00.3	7.0	54.2
07	285 55.1	40.7	44 39.1	12.3	22 53.3	7.0	54.2
08	300 55.1	41.3	59 10.4	12.3	22 46.3	7.2	54.2
09	315 55.1	. . 41.9	73 41.7	12.4	22 39.1	7.3	54.2
10	330 55.1	42.5	88 13.1	12.5	22 31.8	7.3	54.2
11	345 55.1	43.1	102 44.6	12.5	22 24.5	7.5	54.2
12	0 55.1	N18 43.7	117 16.1	12.5	S22 17.0	7.6	54.2
13	15 55.1	44.3	131 47.6	12.7	22 09.4	7.6	54.2
14	30 55.1	44.9	146 19.3	12.7	22 01.8	7.8	54.2
15	45 55.1	. . 45.5	160 51.0	12.7	21 54.0	7.9	54.2
16	60 55.1	46.1	175 22.7	12.8	21 46.1	7.9	54.2
17	75 55.1	46.7	189 54.5	12.8	21 38.2	8.0	54.2
18	90 55.1	N18 47.3	204 26.3	12.9	S21 30.2	8.2	54.2
19	105 55.1	47.9	218 58.2	13.0	21 22.0	8.2	54.2
20	120 55.1	48.5	233 30.2	13.0	21 13.8	8.3	54.2
21	135 55.1	. . 49.1	248 02.2	13.0	21 05.5	8.5	54.2
22	150 55.1	49.7	262 34.2	13.2	20 57.0	8.5	54.2
23	165 55.1	50.3	277 06.4	13.1	20 48.5	8.6	54.2
5 00	180 55.1	N18 50.9	291 38.5	13.2	S20 39.9	8.7	54.2
01	195 55.1	51.5	306 10.7	13.3	20 31.2	8.7	54.2
02	210 55.1	52.0	320 43.0	13.3	20 22.5	8.9	54.2
03	225 55.1	. . 52.6	335 15.3	13.4	20 13.6	8.9	54.2
04	240 55.1	53.2	349 47.7	13.5	20 04.7	9.1	54.2
05	255 55.1	53.8	4 20.2	13.4	19 55.6	9.1	54.2
06	270 55.1	N18 54.4	18 52.6	13.6	S19 46.5	9.2	54.2
07	285 55.0	55.0	33 25.2	13.6	19 37.3	9.3	54.3
08	300 55.0	55.6	47 57.8	13.6	19 28.0	9.4	54.3
09	315 55.0	. . 56.2	62 30.4	13.7	19 18.6	9.4	54.3
10	330 55.0	56.7	77 03.1	13.7	19 09.2	9.6	54.3
11	345 55.0	57.3	91 35.8	13.8	18 59.6	9.6	54.3
12	0 55.0	N18 57.9	106 08.6	13.8	S18 50.0	9.7	54.3
13	15 55.0	58.5	120 41.4	13.9	18 40.3	9.8	54.3
14	30 55.0	59.1	135 14.3	13.9	18 30.5	9.8	54.3
15	45 55.0	18 59.7	149 47.2	14.0	18 20.7	10.0	54.3
16	60 55.0	19 00.3	164 20.2	14.0	18 10.7	10.0	54.3
17	75 55.0	00.8	178 53.2	14.1	18 00.7	10.1	54.3
18	90 54.9	N19 01.4	193 26.3	14.1	S17 50.6	10.2	54.4
19	105 54.9	02.0	207 59.4	14.2	17 40.4	10.2	54.4
20	120 54.9	02.6	222 32.6	14.2	17 30.2	10.4	54.4
21	135 54.9	. . 03.2	237 05.8	14.3	17 19.8	10.4	54.4
22	150 54.9	03.7	251 39.1	14.2	17 09.4	10.4	54.4
23	165 54.9	04.3	266 12.3	14.4	S16 59.0	10.6	54.4
	SD 15.8	*d* 0.6	SD 14.8		14.8		14.8

Lat. °	Twilight Naut. h m	Twilight Civil h m	Sunrise h m	Moonrise 13 h m	Moonrise 14 h m	Moonrise 15 h m	Moonrise 16 h m
N 72	□	□	□	■	■	■	03 39
N 70	////	////	00 52	■	■	■	02 58
68	////	////	01 52	■	■	03 04	02 29
66	////	////	02 26	■	03 01	02 25	02 08
64	////	01 14	02 50	02 22	02 05	01 57	01 51
62	////	01 55	03 09	01 22	01 32	01 35	01 36
60	////	02 22	03 25	00 48	01 07	01 18	01 24
N 58	01 07	02 43	03 38	00 23	00 48	01 03	01 14
56	01 44	02 59	03 49	00 03	00 31	00 51	01 05
54	02 09	03 14	03 59	24 17	00 17	00 40	00 57
52	02 28	03 26	04 08	24 05	00 05	00 30	00 49
50	02 44	03 36	04 15	23 54	24 21	00 21	00 43
45	03 15	03 58	04 32	23 32	24 03	00 03	00 28
N 40	03 37	04 15	04 46	23 14	23 48	24 17	00 17
35	03 55	04 29	04 57	22 58	23 35	24 06	00 06
30	04 10	04 41	05 07	22 45	23 23	23 58	24 29
20	04 33	05 01	05 24	22 22	23 04	23 42	24 18
N 10	04 51	05 17	05 39	22 02	22 47	23 29	24 08
0	05 06	05 31	05 53	21 44	22 31	23 16	23 59
S 10	05 19	05 44	06 06	21 25	22 15	23 03	23 50
20	05 31	05 58	06 21	21 05	21 58	22 50	23 41
30	05 43	06 12	06 37	20 42	21 38	22 34	23 29
35	05 49	06 20	06 47	20 29	21 27	22 25	23 23
40	05 56	06 28	06 57	20 13	21 13	22 14	23 16
45	06 03	06 38	07 10	19 54	20 58	22 02	23 07
S 50	06 10	06 50	07 25	19 31	20 38	21 47	22 57
52	06 14	06 55	07 33	19 20	20 29	21 40	22 52
54	06 17	07 01	07 41	19 07	20 18	21 32	22 46
56	06 21	07 07	07 49	18 52	20 06	21 23	22 40
58	06 25	07 14	08 00	18 34	19 53	21 13	22 34
S 60	06 30	07 22	08 11	18 13	19 37	21 02	22 26

Lat. °	Sunset h m	Twilight Civil h m	Twilight Naut. h m	Moonset 13 h m	Moonset 14 h m	Moonset 15 h m	Moonset 16 h m
N 72	□	□	□	■	■	■	07 35
N 70	23 13	////	////	■	■	■	08 15
68	22 05	////	////	■	■	06 34	08 41
66	21 30	////	////	■	05 00	07 13	09 02
64	21 05	22 45	////	03 57	05 55	07 40	09 18
62	20 46	22 01	////	04 57	06 28	08 00	09 31
60	20 30	21 33	////	05 30	06 52	08 17	09 42
N 58	20 16	21 12	22 52	05 55	07 11	08 31	09 52
56	20 05	20 55	22 12	06 15	07 27	08 43	10 00
54	19 55	20 41	21 46	06 31	07 40	08 54	10 08
52	19 46	20 28	21 26	06 45	07 52	09 03	10 14
50	19 38	20 17	21 10	06 57	08 03	09 11	10 20
45	19 21	19 55	20 39	07 22	08 24	09 29	10 33
N 40	19 08	19 38	20 16	07 43	08 42	09 43	10 44
35	18 56	19 24	19 58	07 59	08 57	09 55	10 53
30	18 46	19 12	19 43	08 14	09 09	10 05	11 01
20	18 29	18 52	19 20	08 38	09 31	10 23	11 14
N 10	18 14	18 36	19 02	08 59	09 50	10 39	11 26
0	18 00	18 22	18 47	09 19	10 07	10 53	11 37
S 10	17 46	18 08	18 34	09 38	10 24	11 07	11 48
20	17 32	17 55	18 22	09 59	10 43	11 23	11 59
30	17 15	17 41	18 09	10 23	11 04	11 40	12 12
35	17 06	17 33	18 03	10 37	11 16	11 50	12 20
40	16 55	17 24	17 57	10 53	11 30	12 01	12 29
45	16 42	17 14	17 50	11 13	11 47	12 15	12 39
S 50	16 27	17 02	17 42	11 37	12 07	12 31	12 50
52	16 19	16 57	17 38	11 48	12 17	12 39	12 56
54	16 11	16 51	17 35	12 01	12 28	12 47	13 02
56	16 03	16 45	17 31	12 17	12 40	12 57	13 09
58	15 52	16 38	17 26	12 34	12 54	13 07	13 16
S 60	15 41	16 30	17 22	12 56	13 11	13 20	13 25

Day	SUN Eqn. of Time 00ʰ	SUN Eqn. of Time 12ʰ	SUN Mer. Pass.	MOON Mer. Pass. Upper	MOON Mer. Pass. Lower	MOON Age	MOON Phase
d	m s	m s	h m	h m	h m	d %	
13	03 40	03 40	11 56	03 07	15 31	18 83	
14	03 41	03 41	11 56	03 55	16 19	19 76	
15	03 40	03 40	11 56	04 42	17 05	20 67	

UT	ARIES	VENUS −4.6		MARS +1.2		JUPITER −2.3		SATURN +0.8	
	GHA	GHA	Dec	GHA	Dec	GHA	Dec	GHA	Dec
d h	° ′	° ′	° ′	° ′	° ′	° ′	° ′	° ′	° ′
16 00	233 50.5	223 02.5	N 3 50.7	216 49.9	N 6 04.9	265 44.2	S13 40.9	66 53.2	N 7 57.0
01	248 53.0	238 03.2	51.2	231 50.6	05.7	280 46.4	40.8	81 55.7	57.0
02	263 55.5	253 03.8	51.6	246 51.3	06.4	295 48.7	40.7	96 58.1	57.0
03	278 57.9	268 04.4	. . 52.1	261 52.0	. . 07.1	310 50.9	. . 40.7	112 00.6	. . 56.9
04	294 00.4	283 05.0	52.6	276 52.7	07.9	325 53.1	40.6	127 03.1	56.9
05	309 02.8	298 05.7	53.0	291 53.4	08.6	340 55.4	40.5	142 05.6	56.9
06	324 05.3	313 06.3	N 3 53.5	306 54.1	N 6 09.3	355 57.6	S13 40.5	157 08.0	N 7 56.9
07	339 07.8	328 06.9	54.0	321 54.8	10.1	10 59.9	40.4	172 10.5	56.9
S 08	354 10.2	343 07.5	54.4	336 55.5	10.8	26 02.1	40.3	187 13.0	56.9
A 09	9 12.7	358 08.1	. . 54.9	351 56.2	. . 11.6	41 04.3	. . 40.3	202 15.4	. . 56.9
T 10	24 15.2	13 08.8	55.4	6 56.9	12.3	56 06.6	40.2	217 17.9	56.9
U 11	39 17.6	28 09.4	55.8	21 57.6	13.0	71 08.8	40.1	232 20.4	56.9
R 12	54 20.1	43 10.0	N 3 56.3	36 58.3	N 6 13.8	86 11.1	S13 40.1	247 22.8	N 7 56.9
D 13	69 22.6	58 10.6	56.8	51 59.0	14.5	101 13.3	40.0	262 25.3	56.9
A 14	84 25.0	73 11.2	57.3	66 59.7	15.2	116 15.5	39.9	277 27.8	56.9
Y 15	99 27.5	88 11.8	. . 57.7	82 00.4	. . 16.0	131 17.8	. . 39.9	292 30.3	. . 56.9
16	114 30.0	103 12.4	58.2	97 01.1	16.7	146 20.0	39.8	307 32.7	56.9
17	129 32.4	118 13.1	58.7	112 01.8	17.4	161 22.3	39.7	322 35.2	56.9
18	144 34.9	133 13.7	N 3 59.2	127 02.5	N 6 18.2	176 24.5	S13 39.7	337 37.7	N 7 56.9
19	159 37.3	148 14.3	3 59.7	142 03.2	18.9	191 26.8	39.6	352 40.1	56.9
20	174 39.8	163 14.9	4 00.1	157 03.9	19.6	206 29.0	39.5	7 42.6	56.9
21	189 42.3	178 15.5	. . 00.6	172 04.6	. . 20.4	221 31.3	. . 39.5	22 45.1	. . 56.8
22	204 44.7	193 16.1	01.1	187 05.3	21.1	236 33.5	39.4	37 47.5	56.8
23	219 47.2	208 16.7	01.6	202 06.0	21.8	251 35.7	39.3	52 50.0	56.8
17 00	234 49.7	223 17.3	N 4 02.1	217 06.7	N 6 22.6	266 38.0	S13 39.3	67 52.5	N 7 56.8
01	249 52.1	238 17.9	02.6	232 07.4	23.3	281 40.2	39.2	82 54.9	56.8
02	264 54.6	253 18.5	03.0	247 08.1	24.0	296 42.5	39.1	97 57.4	56.8
03	279 57.1	268 19.1	. . 03.5	262 08.8	. . 24.8	311 44.7	. . 39.1	112 59.9	. . 56.8
04	294 59.5	283 19.7	04.0	277 09.5	25.5	326 47.0	39.0	128 02.3	56.8
05	310 02.0	298 20.3	04.5	292 10.2	26.2	341 49.2	38.9	143 04.8	56.8
06	325 04.4	313 20.9	N 4 05.0	307 10.9	N 6 27.0	356 51.5	S13 38.9	158 07.3	N 7 56.8
07	340 06.9	328 21.4	05.5	322 11.6	27.7	11 53.7	38.8	173 09.7	56.8
08	355 09.4	343 22.0	06.0	337 12.3	28.4	26 56.0	38.7	188 12.2	56.8
S 09	10 11.8	358 22.6	. . 06.5	352 13.0	. . 29.2	41 58.2	. . 38.7	203 14.7	. . 56.8
U 10	25 14.3	13 23.2	07.0	7 13.7	29.9	57 00.4	38.6	218 17.1	56.8
N 11	40 16.8	28 23.8	07.5	22 14.4	30.6	72 02.7	38.5	233 19.6	56.8
D 12	55 19.2	43 24.4	N 4 07.9	37 15.1	N 6 31.4	87 04.9	S13 38.5	248 22.1	N 7 56.7
A 13	70 21.7	58 25.0	08.4	52 15.8	32.1	102 07.2	38.4	263 24.5	56.7
Y 14	85 24.2	73 25.6	08.9	67 16.5	32.8	117 09.4	38.3	278 27.0	56.7
15	100 26.6	88 26.1	. . 09.4	82 17.2	. . 33.6	132 11.7	. . 38.3	293 29.5	. . 56.7
16	115 29.1	103 26.7	09.9	97 17.9	34.3	147 13.9	38.2	308 31.9	56.7
17	130 31.6	118 27.3	10.4	112 18.6	35.0	162 16.2	38.2	323 34.4	56.7
18	145 34.0	133 27.9	N 4 10.9	127 19.3	N 6 35.7	177 18.4	S13 38.1	338 36.8	N 7 56.7
19	160 36.5	148 28.4	11.4	142 20.0	36.5	192 20.7	38.0	353 39.3	56.7
20	175 38.9	163 29.0	11.9	157 20.7	37.2	207 22.9	38.0	8 41.8	56.7
21	190 41.4	178 29.6	. . 12.4	172 21.4	. . 37.9	222 25.2	. . 37.9	23 44.2	. . 56.7
22	205 43.9	193 30.2	12.9	187 22.1	38.7	237 27.4	37.8	38 46.7	56.7
23	220 46.3	208 30.7	13.4	202 22.8	39.4	252 29.7	37.8	53 49.2	56.7
18 00	235 48.8	223 31.3	N 4 14.0	217 23.5	N 6 40.1	267 31.9	S13 37.7	68 51.6	N 7 56.7
01	250 51.3	238 31.9	14.5	232 24.2	40.9	282 34.2	37.6	83 54.1	56.6
02	265 53.7	253 32.4	15.0	247 24.9	41.6	297 36.5	37.6	98 56.6	56.6
03	280 56.2	268 33.0	. . 15.5	262 25.6	. . 42.3	312 38.7	. . 37.5	113 59.0	. . 56.6
04	295 58.7	283 33.6	16.0	277 26.3	43.1	327 41.0	37.5	129 01.5	56.6
05	311 01.1	298 34.1	16.5	292 27.0	43.8	342 43.2	37.4	144 03.9	56.6
06	326 03.6	313 34.7	N 4 17.0	307 27.7	N 6 44.5	357 45.5	S13 37.3	159 06.4	N 7 56.6
07	341 06.1	328 35.3	17.5	322 28.4	45.2	12 47.7	37.3	174 08.9	56.6
08	356 08.5	343 35.8	18.0	337 29.1	46.0	27 50.0	37.2	189 11.3	56.6
M 09	11 11.0	358 36.4	. . 18.5	352 29.8	. . 46.7	42 52.2	. . 37.1	204 13.8	. . 56.6
O 10	26 13.4	13 36.9	19.0	7 30.5	47.4	57 54.5	37.1	219 16.3	56.6
N 11	41 15.9	28 37.5	19.6	22 31.2	48.2	72 56.7	37.0	234 18.7	56.6
D 12	56 18.4	43 38.1	N 4 20.1	37 31.9	N 6 48.9	87 59.0	S13 37.0	249 21.2	N 7 56.5
A 13	71 20.8	58 38.6	20.6	52 32.6	49.6	103 01.2	36.9	264 23.6	56.5
Y 14	86 23.3	73 39.2	21.1	67 33.3	50.3	118 03.5	36.8	279 26.1	56.5
15	101 25.8	88 39.7	. . 21.6	82 34.0	. . 51.1	133 05.8	. . 36.8	294 28.6	. . 56.5
16	116 28.2	103 40.3	22.1	97 34.7	51.8	148 08.0	36.7	309 31.0	56.5
17	131 30.7	118 40.8	22.7	112 35.4	52.5	163 10.3	36.6	324 33.5	56.5
18	146 33.2	133 41.4	N 4 23.2	127 36.1	N 6 53.3	178 12.5	S13 36.6	339 35.9	N 7 56.5
19	161 35.6	148 41.9	23.7	142 36.8	54.0	193 14.8	36.5	354 38.4	56.5
20	176 38.1	163 42.5	24.2	157 37.5	54.7	208 17.0	36.5	9 40.9	56.5
21	191 40.5	178 43.0	. . 24.7	172 38.2	. . 55.4	223 19.3	. . 36.4	24 43.3	. . 56.5
22	206 43.0	193 43.6	25.3	187 38.9	56.2	238 21.6	36.3	39 45.8	56.5
23	221 45.5	208 44.1	25.8	202 39.6	56.9	253 23.8	36.3	54 48.2	56.4
Mer. Pass.	h m 8 19.3	v 0.6	d 0.5	v 0.7	d 0.7	v 2.2	d 0.1	v 2.5	d 0.0

STARS

Name	SHA	Dec
	° ′	° ′
Acamar	315 21.0	S40 15.
Achernar	335 29.3	S57 11.
Acrux	173 12.5	S63 09
Adhara	255 15.2	S28 59
Aldebaran	290 53.2	N16 31.
Alioth	166 22.7	N55 54
Alkaid	153 00.7	N49 16.
Al Na'ir	27 47.4	S46 54
Alnilam	275 49.8	S 1 11.
Alphard	217 59.1	S 8 42.
Alphecca	126 13.2	N26 40.
Alpheratz	357 46.9	N29 08.
Altair	62 11.0	N 8 53.
Ankaa	353 18.8	S42 15.
Antares	112 29.7	S26 27.
Arcturus	145 58.1	N19 07.
Atria	107 33.8	S69 02.
Avior	234 19.7	S59 32.
Bellatrix	278 35.6	N 6 21.
Betelgeuse	271 04.9	N 7 24.
Canopus	263 58.0	S52 42.
Capella	280 39.4	N46 00.
Deneb	49 33.5	N45 18.
Denebola	182 36.6	N14 31.
Diphda	348 59.1	S17 56.
Dubhe	193 54.9	N61 42.
Elnath	278 16.8	N28 37.
Eltanin	90 47.1	N51 29.
Enif	33 50.1	N 9 55.
Fomalhaut	15 27.3	S29 34.
Gacrux	172 04.2	S57 10.
Gienah	175 55.3	S17 35.
Hadar	148 51.9	S60 25.
Hamal	328 04.6	N23 30.
Kaus Aust.	83 47.5	S34 22.
Kochab	137 18.3	N74 07.
Markab	13 41.5	N15 15.
Menkar	314 18.6	N 4 07.
Menkent	148 10.9	S36 25.
Miaplacidus	221 40.8	S69 45.
Mirfak	308 45.3	N49 53.
Nunki	76 01.8	S26 17.
Peacock	53 23.6	S56 42.
Pollux	243 31.6	N28 00.
Procyon	245 03.1	N 5 12.
Rasalhague	96 09.0	N12 33.
Regulus	207 46.7	N11 55.
Rigel	281 15.3	S 8 11.
Rigil Kent.	139 55.5	S60 52.
Sabik	102 15.7	S15 44.
Schedar	349 44.6	N56 35.
Shaula	96 25.7	S37 06.
Sirius	258 36.7	S16 43.
Spica	158 34.2	S11 12.
Suhail	222 54.9	S43 28.
Vega	80 40.7	N38 47.
Zuben'ubi	137 08.5	S16 05.

	SHA	Mer. Pas
	° ′	h m
Venus	348 27.6	9 06
Mars	342 17.1	9 31
Jupiter	31 48.3	6 13
Saturn	193 02.8	19 25

UT	SUN GHA	SUN Dec	MOON GHA	*v*	MOON Dec	*d*	HP
d h	° ′	° ′	° ′	′	° ′	′	′
6 00	180 54.9	N19 04.9	280 45.7	14.4	S16 48.4	10.6	54.4
01	195 54.9	05.5	295 19.1	14.4	16 37.8	10.7	54.4
02	210 54.9	06.1	309 52.5	14.4	16 27.1	10.8	54.5
03	225 54.8	. . 06.6	324 25.9	14.5	16 16.3	10.8	54.5
04	240 54.8	07.2	338 59.4	14.6	16 05.5	10.9	54.5
05	255 54.8	07.8	353 33.0	14.6	15 54.6	11.0	54.5
06	270 54.8	N19 08.4	8 06.6	14.6	S15 43.6	11.0	54.5
07	285 54.8	08.9	22 40.2	14.6	15 32.6	11.1	54.5
08	300 54.8	09.5	37 13.8	14.7	15 21.5	11.2	54.5
09	315 54.8	. . 10.1	51 47.5	14.8	15 10.3	11.3	54.5
10	330 54.8	10.6	66 21.3	14.7	14 59.0	11.3	54.6
11	345 54.7	11.2	80 55.0	14.8	14 47.7	11.3	54.6
12	0 54.7	N19 11.8	95 28.8	14.8	S14 36.4	11.5	54.6
13	15 54.7	12.4	110 02.6	14.9	14 24.9	11.5	54.6
14	30 54.7	12.9	124 36.5	14.9	14 13.4	11.6	54.6
15	45 54.7	. . 13.5	139 10.4	14.9	14 01.8	11.6	54.6
16	60 54.7	14.1	153 44.3	15.0	13 50.2	11.7	54.7
17	75 54.6	14.6	168 18.3	15.0	13 38.5	11.7	54.7
18	90 54.6	N19 15.2	182 52.3	15.0	S13 26.8	11.9	54.7
19	105 54.6	15.8	197 26.3	15.0	13 14.9	11.8	54.7
20	120 54.6	16.3	212 00.3	15.1	13 03.1	12.0	54.7
21	135 54.6	. . 16.9	226 34.4	15.1	12 51.1	12.0	54.7
22	150 54.6	17.5	241 08.5	15.1	12 39.1	12.0	54.8
23	165 54.5	18.0	255 42.6	15.2	12 27.1	12.1	54.8
7 00	180 54.5	N19 18.6	270 16.8	15.2	S12 15.0	12.2	54.8
01	195 54.5	19.2	284 51.0	15.1	12 02.8	12.2	54.8
02	210 54.5	19.7	299 25.1	15.3	11 50.6	12.3	54.8
03	225 54.5	. . 20.3	313 59.4	15.2	11 38.3	12.3	54.9
04	240 54.5	20.9	328 33.6	15.3	11 26.0	12.4	54.9
05	255 54.4	21.4	343 07.9	15.2	11 13.6	12.4	54.9
06	270 54.4	N19 22.0	357 42.1	15.3	S11 01.2	12.5	54.9
07	285 54.4	22.5	12 16.4	15.3	10 48.7	12.6	54.9
08	300 54.4	23.1	26 50.7	15.4	10 36.1	12.6	55.0
09	315 54.4	. . 23.7	41 25.1	15.3	10 23.5	12.6	55.0
10	330 54.3	24.2	55 59.4	15.4	10 10.9	12.7	55.0
11	345 54.3	24.8	70 33.8	15.4	9 58.2	12.8	55.0
12	0 54.3	N19 25.3	85 08.2	15.3	S 9 45.4	12.8	55.1
13	15 54.3	25.9	99 42.5	15.4	9 32.6	12.8	55.1
14	30 54.3	26.5	114 16.9	15.4	9 19.8	12.9	55.1
15	45 54.2	. . 27.0	128 51.3	15.5	9 06.9	12.9	55.1
16	60 54.2	27.6	143 25.8	15.4	8 54.0	13.0	55.1
17	75 54.2	28.1	158 00.2	15.4	8 41.0	13.0	55.2
18	90 54.2	N19 28.7	172 34.6	15.5	S 8 28.0	13.1	55.2
19	105 54.1	29.2	187 09.1	15.4	8 14.9	13.1	55.2
20	120 54.1	29.8	201 43.5	15.5	8 01.8	13.2	55.2
21	135 54.1	. . 30.3	216 18.0	15.4	7 48.6	13.2	55.3
22	150 54.1	30.9	230 52.4	15.5	7 35.4	13.2	55.3
23	165 54.1	31.4	245 26.9	15.4	7 22.2	13.3	55.3
8 00	180 54.0	N19 32.0	260 01.3	15.5	S 7 08.9	13.4	55.3
01	195 54.0	32.5	274 35.8	15.5	6 55.5	13.3	55.4
02	210 54.0	33.1	289 10.3	15.4	6 42.2	13.4	55.4
03	225 54.0	. . 33.6	303 44.7	15.5	6 28.8	13.5	55.4
04	240 53.9	34.2	318 19.2	15.4	6 15.3	13.5	55.5
05	255 53.9	34.7	332 53.6	15.5	6 01.8	13.5	55.5
06	270 53.9	N19 35.3	347 28.1	15.4	S 5 48.3	13.6	55.5
07	285 53.9	35.8	2 02.5	15.4	5 34.7	13.6	55.5
08	300 53.8	36.4	16 36.9	15.5	5 21.1	13.6	55.6
09	315 53.8	. . 36.9	31 11.4	15.4	5 07.5	13.7	55.6
10	330 53.8	37.5	45 45.8	15.4	4 53.8	13.7	55.6
11	345 53.8	38.0	60 20.2	15.4	4 40.1	13.7	55.6
12	0 53.7	N19 38.6	74 54.6	15.3	S 4 26.4	13.8	55.7
13	15 53.7	39.1	89 28.9	15.4	4 12.6	13.8	55.7
14	30 53.7	39.6	104 03.3	15.4	3 58.8	13.8	55.7
15	45 53.7	. . 40.2	118 37.7	15.3	3 45.0	13.8	55.8
16	60 53.6	40.7	133 12.0	15.3	3 31.2	13.9	55.8
17	75 53.6	41.3	147 46.3	15.3	3 17.3	13.9	55.8
18	90 53.6	N19 41.8	162 20.6	15.3	S 3 03.4	14.0	55.9
19	105 53.5	42.3	176 54.9	15.2	2 49.4	14.0	55.9
20	120 53.5	42.9	191 29.1	15.3	2 35.4	14.0	55.9
21	135 53.5	. . 43.4	206 03.4	15.2	2 21.4	14.0	55.9
22	150 53.5	44.0	220 37.6	15.2	2 07.4	14.0	56.0
23	165 53.4	44.5	235 11.8	15.2	S 1 53.4	14.1	56.0
	SD 15.8	*d* 0.6	SD 14.9		15.0		15.2

Lat.	Twilight Naut.	Twilight Civil	Sunrise	Moonrise 16	Moonrise 17	Moonrise 18	Moonrise 19
°	h m	h m	h m	h m	h m	h m	h m
N 72	▭	▭	▭	03 39	02 44	02 09	01 41
N 70	▭	▭	▭	02 58	02 24	02 00	01 39
68	////	////	01 35	02 29	02 09	01 52	01 38
66	////	////	02 14	02 08	01 56	01 46	01 36
64	////	00 51	02 40	01 51	01 45	01 40	01 35
62	////	01 42	03 01	01 36	01 36	01 35	01 34
60	////	02 12	03 18	01 24	01 28	01 31	01 33
N 58	00 46	02 35	03 31	01 14	01 21	01 27	01 33
56	01 32	02 53	03 43	01 05	01 15	01 24	01 32
54	02 00	03 08	03 54	00 57	01 10	01 21	01 31
52	02 21	03 20	04 03	00 49	01 05	01 18	01 31
50	02 38	03 32	04 11	00 43	01 00	01 16	01 30
45	03 10	03 54	04 29	00 28	00 50	01 10	01 29
N 40	03 34	04 12	04 43	00 17	00 42	01 06	01 28
35	03 52	04 27	04 55	00 06	00 35	01 02	01 28
30	04 08	04 39	05 05	24 29	00 29	00 58	01 27
20	04 31	04 59	05 23	24 18	00 18	00 52	01 26
N 10	04 50	05 16	05 39	24 08	00 08	00 47	01 25
0	05 05	05 31	05 53	23 59	24 42	00 42	01 24
S 10	05 19	05 45	06 07	23 50	24 36	00 36	01 23
20	05 32	05 59	06 22	23 41	24 31	00 31	01 22
30	05 45	06 14	06 39	23 29	24 25	00 25	01 21
35	05 51	06 22	06 49	23 23	24 21	00 21	01 20
40	05 58	06 31	07 00	23 16	24 17	00 17	01 19
45	06 06	06 41	07 13	23 07	24 12	00 12	01 18
S 50	06 14	06 54	07 29	22 57	24 06	00 06	01 17
52	06 17	06 59	07 37	22 52	24 04	00 04	01 17
54	06 21	07 05	07 45	22 46	24 01	00 01	01 16
56	06 26	07 12	07 55	22 40	23 58	25 16	01 16
58	06 30	07 19	08 05	22 34	23 54	25 15	01 15
S 60	06 35	07 28	08 18	22 26	23 50	25 15	01 15

Lat.	Sunset	Twilight Civil	Twilight Naut.	Moonset 16	Moonset 17	Moonset 18	Moonset 19
°	h m	h m	h m	h m	h m	h m	h m
N 72	▭	▭	▭	07 35	10 00	12 04	14 05
N 70	▭	▭	▭	08 15	10 18	12 11	14 02
68	22 23	////	////	08 41	10 31	12 16	14 01
66	21 42	////	////	09 02	10 43	12 20	13 59
64	21 15	23 11	////	09 18	10 52	12 24	13 58
62	20 54	22 15	////	09 31	11 00	12 27	13 57
60	20 37	21 43	////	09 42	11 06	12 30	13 56
N 58	20 23	21 20	23 15	09 52	11 12	12 33	13 55
56	20 11	21 02	22 24	10 00	11 17	12 35	13 54
54	20 00	20 47	21 55	10 08	11 22	12 37	13 53
52	19 51	20 34	21 33	10 14	11 26	12 39	13 53
50	19 42	20 22	21 16	10 20	11 30	12 40	13 52
45	19 25	19 59	20 43	10 33	11 38	12 44	13 51
N 40	19 10	19 41	20 20	10 44	11 45	12 47	13 50
35	18 58	19 27	20 01	10 53	11 51	12 49	13 49
30	18 48	19 14	19 46	11 01	11 56	12 52	13 48
20	18 30	18 54	19 22	11 14	12 05	12 55	13 47
N 10	18 14	18 37	19 03	11 26	12 12	12 59	13 46
0	18 00	18 22	18 47	11 37	12 20	13 02	13 44
S 10	17 46	18 08	18 34	11 48	12 27	13 05	13 43
20	17 31	17 54	18 21	11 59	12 34	13 08	13 42
30	17 13	17 39	18 08	12 12	12 43	13 12	13 40
35	17 04	17 31	18 01	12 20	12 47	13 14	13 40
40	16 52	17 21	17 54	12 29	12 53	13 16	13 39
45	16 39	17 11	17 47	12 39	12 59	13 19	13 38
S 50	16 23	16 59	17 38	12 50	13 07	13 22	13 36
52	16 15	16 53	17 35	12 56	13 10	13 23	13 36
54	16 07	16 47	17 31	13 02	13 14	13 25	13 35
56	15 57	16 40	17 27	13 09	13 18	13 26	13 34
58	15 47	16 33	17 22	13 16	13 23	13 28	13 34
S 60	15 34	16 24	17 17	13 25	13 28	13 31	13 33

Day	SUN Eqn. of Time 00^h	SUN Eqn. of Time 12^h	SUN Mer. Pass.	MOON Mer. Pass. Upper	MOON Mer. Pass. Lower	MOON Age	MOON Phase
d	m s	m s	h m	h m	h m	d %	
16	03 40	03 39	11 56	05 27	17 48	21 58	
17	03 38	03 37	11 56	06 09	18 31	22 48	◐
18	03 36	03 35	11 56	06 52	19 13	23 38	

	UT	ARIES	VENUS −4.6		MARS +1.2		JUPITER −2.4		SATURN +0.9	
	d h	GHA ° ′	GHA ° ′	Dec ° ′	GHA ° ′	Dec ° ′	GHA ° ′	Dec ° ′	GHA ° ′	Dec ° ′
	19 00	236 47.9	223 44.6	N 4 26.3	217 40.3	N 6 57.6	268 26.1	S13 36.2	69 50.7	N 7 56.4
	01	251 50.4	238 45.2	26.8	232 41.0	58.3	283 28.3	36.2	84 53.2	56.4
	02	266 52.9	253 45.7	27.4	247 41.7	59.1	298 30.6	36.1	99 55.6	56.4
	03	281 55.3	268 46.3	. . 27.9	262 42.4	6 59.8	313 32.8	. . 36.0	114 58.1	. . 56.4
	04	296 57.8	283 46.8	28.4	277 43.1	7 00.5	328 35.1	36.0	130 00.5	56.4
	05	312 00.3	298 47.3	28.9	292 43.8	01.2	343 37.4	35.9	145 03.0	56.4
	06	327 02.7	313 47.9	N 4 29.5	307 44.5	N 7 02.0	358 39.6	S13 35.8	160 05.5	N 7 56.4
	07	342 05.2	328 48.4	30.0	322 45.2	02.7	13 41.9	35.8	175 07.9	56.4
T	08	357 07.7	343 48.9	30.5	337 45.9	03.4	28 44.1	35.7	190 10.4	56.4
U	09	12 10.1	358 49.5	. . 31.1	352 46.6	. . 04.2	43 46.4	. . 35.7	205 12.8	. . 56.3
E	10	27 12.6	13 50.0	31.6	7 47.3	04.9	58 48.7	35.6	220 15.3	56.3
	11	42 15.0	28 50.5	32.1	22 48.0	05.6	73 50.9	35.5	235 17.7	56.3
S	12	57 17.5	43 51.0	N 4 32.7	37 48.7	N 7 06.3	88 53.2	S13 35.5	250 20.2	N 7 56.3
D	13	72 20.0	58 51.6	33.2	52 49.4	07.1	103 55.5	35.4	265 22.7	56.3
A	14	87 22.4	73 52.1	33.7	67 50.1	07.8	118 57.7	35.4	280 25.1	56.3
Y	15	102 24.9	88 52.6	. . 34.3	82 50.8	. . 08.5	134 00.0	. . 35.3	295 27.6	. . 56.3
	16	117 27.4	103 53.1	34.8	97 51.5	09.2	149 02.2	35.2	310 30.0	56.3
	17	132 29.8	118 53.7	35.3	112 52.2	10.0	164 04.5	35.2	325 32.5	56.3
	18	147 32.3	133 54.2	N 4 35.9	127 52.9	N 7 10.7	179 06.8	S13 35.1	340 34.9	N 7 56.2
	19	162 34.8	148 54.7	36.4	142 53.6	11.4	194 09.0	35.1	355 37.4	56.2
	20	177 37.2	163 55.2	37.0	157 54.3	12.1	209 11.3	35.0	10 39.9	56.2
	21	192 39.7	178 55.7	. . 37.5	172 54.9	. . 12.8	224 13.6	. . 34.9	25 42.3	. . 56.2
	22	207 42.1	193 56.3	38.0	187 55.6	13.6	239 15.8	34.9	40 44.8	56.2
	23	222 44.6	208 56.8	38.6	202 56.3	14.3	254 18.1	34.8	55 47.2	56.2
	20 00	237 47.1	223 57.3	N 4 39.1	217 57.0	N 7 15.0	269 20.4	S13 34.8	70 49.7	N 7 56.2
	01	252 49.5	238 57.8	39.7	232 57.7	15.7	284 22.6	34.7	85 52.1	56.2
	02	267 52.0	253 58.3	40.2	247 58.4	16.5	299 24.9	34.7	100 54.6	56.2
	03	282 54.5	268 58.8	. . 40.8	262 59.1	. . 17.2	314 27.2	. . 34.6	115 57.0	. . 56.1
	04	297 56.9	283 59.3	41.3	277 59.8	17.9	329 29.4	34.5	130 59.5	56.1
	05	312 59.4	298 59.8	41.8	293 00.5	18.6	344 31.7	34.5	146 01.9	56.1
	06	328 01.9	314 00.4	N 4 42.4	308 01.2	N 7 19.4	359 34.0	S13 34.4	161 04.4	N 7 56.1
W	07	343 04.3	329 00.9	42.9	323 01.9	20.1	14 36.2	34.4	176 06.9	56.1
E	08	358 06.8	344 01.4	43.5	338 02.6	20.8	29 38.5	34.3	191 09.3	56.1
D	09	13 09.3	359 01.9	. . 44.0	353 03.3	. . 21.5	44 40.8	. . 34.2	206 11.8	. . 56.1
N	10	28 11.7	14 02.4	44.6	8 04.0	22.2	59 43.0	34.2	221 14.2	56.1
E	11	43 14.2	29 02.9	45.1	23 04.7	23.0	74 45.3	34.1	236 16.7	56.0
S	12	58 16.6	44 03.4	N 4 45.7	38 05.4	N 7 23.7	89 47.6	S13 34.1	251 19.1	N 7 56.0
D	13	73 19.1	59 03.9	46.2	53 06.1	24.4	104 49.8	34.0	266 21.6	56.0
A	14	88 21.6	74 04.4	46.8	68 06.8	25.1	119 52.1	34.0	281 24.0	56.0
Y	15	103 24.0	89 04.9	. . 47.3	83 07.5	. . 25.8	134 54.4	. . 33.9	296 26.5	. . 56.0
	16	118 26.5	104 05.4	47.9	98 08.2	26.6	149 56.6	33.8	311 28.9	56.0
	17	133 29.0	119 05.9	48.5	113 08.9	27.3	164 58.9	33.8	326 31.4	56.0
	18	148 31.4	134 06.4	N 4 49.0	128 09.6	N 7 28.0	180 01.2	S13 33.7	341 33.8	N 7 56.0
	19	163 33.9	149 06.8	49.6	143 10.3	28.7	195 03.4	33.7	356 36.3	55.9
	20	178 36.4	164 07.3	50.1	158 11.0	29.5	210 05.7	33.6	11 38.7	55.9
	21	193 38.8	179 07.8	. . 50.7	173 11.7	. . 30.2	225 08.0	. . 33.6	26 41.2	. . 55.9
	22	208 41.3	194 08.3	51.2	188 12.4	30.9	240 10.3	33.5	41 43.6	55.9
	23	223 43.8	209 08.8	51.8	203 13.1	31.6	255 12.5	33.4	56 46.1	55.9
	21 00	238 46.2	224 09.3	N 4 52.4	218 13.8	N 7 32.3	270 14.8	S13 33.4	71 48.5	N 7 55.9
	01	253 48.7	239 09.8	52.9	233 14.5	33.1	285 17.1	33.3	86 51.0	55.9
	02	268 51.1	254 10.3	53.5	248 15.2	33.8	300 19.4	33.3	101 53.5	55.9
	03	283 53.6	269 10.8	. . 54.0	263 15.9	. . 34.5	315 21.6	. . 33.2	116 55.9	. . 55.8
	04	298 56.1	284 11.2	54.6	278 16.6	35.2	330 23.9	33.2	131 58.4	55.8
	05	313 58.5	299 11.7	55.2	293 17.3	35.9	345 26.2	33.1	147 00.8	55.8
	06	329 01.0	314 12.2	N 4 55.7	308 18.0	N 7 36.6	0 28.4	S13 33.1	162 03.3	N 7 55.8
	07	344 03.5	329 12.7	56.3	323 18.7	37.4	15 30.7	33.0	177 05.7	55.8
T	08	359 05.9	344 13.2	56.9	338 19.4	38.1	30 33.0	32.9	192 08.2	55.8
H	09	14 08.4	359 13.6	. . 57.4	353 20.1	. . 38.8	45 35.3	. . 32.9	207 10.6	. . 55.8
U	10	29 10.9	14 14.1	58.0	8 20.8	39.5	60 37.5	32.8	222 13.1	55.7
R	11	44 13.3	29 14.6	58.6	23 21.5	40.2	75 39.8	32.8	237 15.5	55.7
S	12	59 15.8	44 15.1	N 4 59.1	38 22.1	N 7 41.0	90 42.1	S13 32.7	252 18.0	N 7 55.7
D	13	74 18.2	59 15.5	4 59.7	53 22.8	41.7	105 44.4	32.7	267 20.4	55.7
A	14	89 20.7	74 16.0	5 00.3	68 23.5	42.4	120 46.6	32.6	282 22.8	55.7
Y	15	104 23.2	89 16.5	. . 00.9	83 24.2	. . 43.1	135 48.9	. . 32.6	297 25.3	. . 55.7
	16	119 25.6	104 17.0	01.4	98 24.9	43.8	150 51.2	32.5	312 27.7	55.7
	17	134 28.1	119 17.4	02.0	113 25.6	44.5	165 53.5	32.4	327 30.2	55.6
	18	149 30.6	134 17.9	N 5 02.6	128 26.3	N 7 45.3	180 55.8	S13 32.4	342 32.6	N 7 55.6
	19	164 33.0	149 18.4	03.1	143 27.0	46.0	195 58.0	32.3	357 35.1	55.6
	20	179 35.5	164 18.8	03.7	158 27.7	46.7	211 00.3	32.3	12 37.5	55.6
	21	194 38.0	179 19.3	. . 04.3	173 28.4	. . 47.4	226 02.6	. . 32.2	27 40.0	. . 55.6
	22	209 40.4	194 19.8	04.9	188 29.1	48.1	241 04.9	32.2	42 42.4	55.6
	23	224 42.9	209 20.2	05.4	203 29.8	48.8	256 07.1	32.1	57 44.9	55.6
	Mer. Pass.	h m 8 07.5	*v* 0.5	*d* 0.6	*v* 0.7	*d* 0.7	*v* 2.3	*d* 0.1	*v* 2.5	*d* 0.0

STARS

Name	SHA ° ′	Dec ° ′
Acamar	315 21.0	S40 15.
Achernar	335 29.3	S57 11.
Acrux	173 12.6	S63 09.
Adhara	255 15.2	S28 59.
Aldebaran	290 53.2	N16 31.
Alioth	166 22.7	N55 54.
Alkaid	153 00.7	N49 16.
Al Na'ir	27 47.4	S46 54.
Alnilam	275 49.8	S 1 11.
Alphard	217 59.1	S 8 42.
Alphecca	126 13.2	N26 40.
Alpheratz	357 46.9	N29 08.
Altair	62 11.0	N 8 53.
Ankaa	353 18.8	S42 15.
Antares	112 29.6	S26 27.
Arcturus	145 58.1	N19 07.
Atria	107 33.8	S69 02.
Avior	234 19.7	S59 32.
Bellatrix	278 35.6	N 6 21.
Betelgeuse	271 04.9	N 7 24.
Canopus	263 58.0	S52 42.
Capella	280 39.4	N46 00.
Deneb	49 33.5	N45 18.
Denebola	182 36.6	N14 31.
Diphda	348 59.1	S17 56.
Dubhe	193 54.9	N61 42.
Elnath	278 16.8	N28 37.
Eltanin	90 47.1	N51 29.
Enif	33 50.1	N 9 55.
Fomalhaut	15 27.3	S29 34.
Gacrux	172 04.2	S57 10.
Gienah	175 55.3	S17 35.
Hadar	148 51.9	S60 25.
Hamal	328 04.6	N23 30.
Kaus Aust.	83 47.5	S34 22.
Kochab	137 18.3	N74 07.
Markab	13 41.5	N15 15.
Menkar	314 18.6	N 4 07.
Menkent	148 10.9	S36 25.
Miaplacidus	221 40.8	S69 45.
Mirfak	308 45.3	N49 53.
Nunki	76 01.8	S26 17.
Peacock	53 23.6	S56 42.
Pollux	243 31.6	N28 00.
Procyon	245 03.1	N 5 12.
Rasalhague	96 09.0	N12 33.
Regulus	207 46.7	N11 55.
Rigel	281 15.3	S 8 11.
Rigil Kent.	139 55.5	S60 52.
Sabik	102 15.7	S15 44.
Schedar	349 44.6	N56 35.
Shaula	96 25.7	S37 06.
Sirius	258 36.7	S16 43.
Spica	158 34.2	S11 12.
Suhail	222 54.9	S43 28.
Vega	80 40.7	N38 47.
Zuben'ubi	137 08.5	S16 05.

	SHA ° ′	Mer. Pas h m
Venus	346 10.2	9 04
Mars	340 10.0	9 28
Jupiter	31 33.3	6 02
Saturn	193 02.6	19 14

UT	SUN GHA	SUN Dec	MOON GHA	v	MOON Dec	d	HP
d h	° ′	° ′	° ′	′	° ′	′	′
19 00	180 53.4	N19 45.0	249 46.0	15.1	S 1 39.3	14.1	56.0
01	195 53.4	45.6	264 20.1	15.1	1 25.2	14.1	56.1
02	210 53.3	46.1	278 54.2	15.1	1 11.1	14.1	56.1
03	225 53.3	. . 46.6	293 28.3	15.0	0 57.0	14.2	56.1
04	240 53.3	47.2	308 02.3	15.0	0 42.8	14.2	56.2
05	255 53.2	47.7	322 36.3	15.0	0 28.6	14.2	56.2
06	270 53.2	N19 48.2	337 10.3	15.0	S 0 14.4	14.2	56.2
07	285 53.2	48.8	351 44.3	14.9	S 0 00.2	14.2	56.3
08	300 53.2	49.3	6 18.2	14.9	N 0 14.0	14.3	56.3
09	315 53.1	. . 49.8	20 52.1	14.8	0 28.3	14.2	56.3
10	330 53.1	50.4	35 25.9	14.8	0 42.5	14.3	56.4
11	345 53.1	50.9	49 59.7	14.8	0 56.8	14.3	56.4
12	0 53.0	N19 51.4	64 33.5	14.7	N 1 11.1	14.3	56.4
13	15 53.0	52.0	79 07.2	14.7	1 25.4	14.4	56.5
14	30 53.0	52.5	93 40.9	14.6	1 39.8	14.3	56.5
15	45 52.9	. . 53.0	108 14.5	14.6	1 54.1	14.3	56.5
16	60 52.9	53.6	122 48.1	14.6	2 08.4	14.4	56.6
17	75 52.9	54.1	137 21.7	14.5	2 22.8	14.3	56.6
18	90 52.8	N19 54.6	151 55.2	14.4	N 2 37.1	14.4	56.6
19	105 52.8	55.1	166 28.6	14.4	2 51.5	14.4	56.7
20	120 52.8	55.7	181 02.0	14.4	3 05.9	14.3	56.7
21	135 52.7	. . 56.2	195 35.4	14.3	3 20.2	14.4	56.7
22	150 52.7	56.7	210 08.7	14.2	3 34.6	14.4	56.8
23	165 52.7	57.2	224 41.9	14.2	3 49.0	14.4	56.8
20 00	180 52.6	N19 57.8	239 15.1	14.1	N 4 03.4	14.4	56.8
01	195 52.6	58.3	253 48.2	14.1	4 17.8	14.4	56.9
02	210 52.5	58.8	268 21.3	14.0	4 32.2	14.4	56.9
03	225 52.5	. . 59.3	282 54.3	14.0	4 46.6	14.3	57.0
04	240 52.5	19 59.8	297 27.3	13.9	5 00.9	14.4	57.0
05	255 52.4	20 00.4	312 00.2	13.8	5 15.3	14.4	57.0
06	270 52.4	N20 00.9	326 33.0	13.8	N 5 29.7	14.4	57.1
07	285 52.4	01.4	341 05.8	13.7	5 44.1	14.3	57.1
08	300 52.3	01.9	355 38.5	13.6	5 58.4	14.4	57.1
09	315 52.3	. . 02.4	10 11.1	13.6	6 12.8	14.3	57.2
10	330 52.2	03.0	24 43.7	13.5	6 27.1	14.4	57.2
11	345 52.2	03.5	39 16.2	13.4	6 41.5	14.3	57.2
12	0 52.2	N20 04.0	53 48.6	13.4	N 6 55.8	14.3	57.3
13	15 52.1	04.5	68 21.0	13.3	7 10.1	14.3	57.3
14	30 52.1	05.0	82 53.3	13.2	7 24.4	14.3	57.4
15	45 52.1	. . 05.5	97 25.5	13.1	7 38.7	14.3	57.4
16	60 52.0	06.0	111 57.6	13.1	7 53.0	14.2	57.4
17	75 52.0	06.6	126 29.7	13.0	8 07.2	14.3	57.5
18	90 51.9	N20 07.1	141 01.7	12.9	N 8 21.5	14.2	57.5
19	105 51.9	07.6	155 33.6	12.8	8 35.7	14.2	57.5
20	120 51.9	08.1	170 05.4	12.8	8 49.9	14.2	57.6
21	135 51.8	. . 08.6	184 37.2	12.6	9 04.1	14.1	57.6
22	150 51.8	09.1	199 08.8	12.6	9 18.2	14.1	57.7
23	165 51.7	09.6	213 40.4	12.5	9 32.3	14.1	57.7
21 00	180 51.7	N20 10.1	228 11.9	12.4	N 9 46.4	14.1	57.7
01	195 51.7	10.6	242 43.3	12.3	10 00.5	14.0	57.8
02	210 51.6	11.1	257 14.6	12.2	10 14.5	14.0	57.8
03	225 51.6	. . 11.7	271 45.8	12.2	10 28.5	14.0	57.8
04	240 51.5	12.2	286 17.0	12.0	10 42.5	14.0	57.9
05	255 51.5	12.7	300 48.0	12.0	10 56.5	13.9	57.9
06	270 51.4	N20 13.2	315 19.0	11.8	N11 10.4	13.8	58.0
07	285 51.4	13.7	329 49.8	11.8	11 24.2	13.9	58.0
08	300 51.4	14.2	344 20.6	11.7	11 38.1	13.8	58.0
09	315 51.3	. . 14.7	358 51.3	11.6	11 51.9	13.7	58.1
10	330 51.3	15.2	13 21.9	11.4	12 05.6	13.7	58.1
11	345 51.2	15.7	27 52.3	11.4	12 19.3	13.7	58.1
12	0 51.2	N20 16.2	42 22.7	11.3	N12 33.0	13.6	58.2
13	15 51.1	16.7	56 53.0	11.2	12 46.6	13.6	58.2
14	30 51.1	17.2	71 23.2	11.0	13 00.2	13.5	58.2
15	45 51.0	. . 17.7	85 53.2	11.0	13 13.7	13.5	58.3
16	60 51.0	18.2	100 23.2	10.9	13 27.2	13.4	58.3
17	75 51.0	18.7	114 53.1	10.7	13 40.6	13.4	58.4
18	90 50.9	N20 19.2	129 22.8	10.7	N13 54.0	13.3	58.4
19	105 50.9	19.7	143 52.5	10.6	14 07.3	13.3	58.4
20	120 50.8	20.2	158 22.1	10.4	14 20.6	13.2	58.5
21	135 50.8	. . 20.7	172 51.5	10.3	14 33.8	13.1	58.5
22	150 50.7	21.2	187 20.8	10.2	14 46.9	13.1	58.5
23	165 50.7	21.7	201 50.0	10.2	N15 00.0	13.0	58.6
	SD 15.8	d 0.5	SD 15.4		15.6		15.9

Lat.	Twilight Naut.	Twilight Civil	Sunrise	Moonrise 19	Moonrise 20	Moonrise 21	Moonrise 22
°	h m	h m	h m	h m	h m	h m	h m
N 72	▭	▭	▭	01 41	01 13	{00 40 / 23 51}	▭
N 70	▭	▭	▭	01 39	01 18	00 55	{00 23 / 23 06}
68	////	////	01 16	01 38	01 23	01 07	00 47
66	////	////	02 02	01 36	01 27	01 17	01 05
64	////	00 09	02 31	01 35	01 30	01 25	01 20
62	////	01 28	02 53	01 34	01 33	01 33	01 33
60	////	02 03	03 11	01 33	01 36	01 39	01 44
N 58	00 08	02 27	03 26	01 33	01 38	01 45	01 54
56	01 20	02 46	03 38	01 32	01 40	01 50	02 02
54	01 51	03 02	03 49	01 31	01 42	01 54	02 09
52	02 14	03 15	03 59	01 31	01 44	01 58	02 16
50	02 32	03 27	04 08	01 30	01 45	02 02	02 22
45	03 06	03 51	04 26	01 29	01 49	02 10	02 35
N 40	03 31	04 09	04 40	01 28	01 52	02 17	02 46
35	03 50	04 25	04 53	01 28	01 54	02 23	02 56
30	04 06	04 37	05 04	01 27	01 57	02 28	03 04
20	04 30	04 58	05 22	01 26	02 00	02 37	03 18
N 10	04 49	05 16	05 38	01 25	02 04	02 46	03 31
0	05 05	05 31	05 53	01 24	02 07	02 53	03 43
S 10	05 20	05 45	06 08	01 23	02 11	03 01	03 55
20	05 33	06 00	06 23	01 22	02 14	03 09	04 08
30	05 46	06 15	06 41	01 21	02 18	03 19	04 23
35	05 53	06 24	06 51	01 20	02 21	03 24	04 32
40	06 00	06 33	07 03	01 19	02 24	03 31	04 42
45	06 08	06 44	07 17	01 18	02 27	03 38	04 54
S 50	06 17	06 57	07 33	01 17	02 31	03 47	05 08
52	06 21	07 03	07 41	01 17	02 32	03 51	05 15
54	06 25	07 09	07 50	01 16	02 34	03 56	05 22
56	06 30	07 16	08 00	01 16	02 36	04 01	05 31
58	06 35	07 24	08 11	01 15	02 39	04 07	05 40
S 60	06 40	07 33	08 24	01 15	02 42	04 13	05 52

Lat.	Sunset	Twilight Civil	Twilight Naut.	Moonset 19	Moonset 20	Moonset 21	Moonset 22
°	h m	h m	h m	h m	h m	h m	h m
N 72	▭	▭	▭	14 05	16 12	18 43	▭
N 70	▭	▭	▭	14 02	16 00	18 13	21 21
68	22 44	////	////	14 01	15 50	17 51	20 16
66	21 55	////	////	13 59	15 42	17 34	19 40
64	21 24	////	////	13 58	15 35	17 20	19 15
62	21 02	22 29	////	13 57	15 29	17 09	18 56
60	20 44	21 53	////	13 56	15 24	16 59	18 40
N 58	20 29	21 28	////	13 55	15 20	16 50	18 26
56	20 16	21 08	22 37	13 54	15 16	16 43	18 15
54	20 05	20 52	22 04	13 53	15 13	16 36	18 05
52	19 55	20 39	21 41	13 53	15 10	16 30	17 56
50	19 46	20 27	21 22	13 52	15 07	16 25	17 48
45	19 28	20 03	20 48	13 51	15 00	16 14	17 31
N 40	19 13	19 44	20 23	13 50	14 55	16 04	17 17
35	19 01	19 29	20 04	13 49	14 51	15 56	17 05
30	18 50	19 16	19 48	13 48	14 47	15 49	16 55
20	18 31	18 55	19 23	13 47	14 40	15 37	16 38
N 10	18 15	18 37	19 04	13 46	14 34	15 26	16 22
0	18 00	18 22	18 48	13 44	14 29	15 16	16 08
S 10	17 45	18 08	18 33	13 43	14 23	15 07	15 54
20	17 30	17 53	18 20	13 42	14 17	14 56	15 39
30	17 12	17 38	18 07	13 40	14 11	14 44	15 22
35	17 02	17 29	18 00	13 40	14 07	14 37	15 12
40	16 50	17 19	17 52	13 39	14 03	14 30	15 01
45	16 36	17 08	17 44	13 38	13 58	14 21	14 48
S 50	16 19	16 56	17 35	13 36	13 52	14 10	14 32
52	16 11	16 50	17 31	13 36	13 49	14 05	14 25
54	16 02	16 43	17 27	13 35	13 46	13 59	14 16
56	15 53	16 36	17 23	13 34	13 43	13 53	14 07
58	15 41	16 28	17 18	13 34	13 39	13 46	13 57
S 60	15 28	16 19	17 12	13 33	13 35	13 39	13 45

Day	SUN Eqn. of Time 00h	SUN Eqn. of Time 12h	SUN Mer. Pass.	MOON Mer. Pass. Upper	MOON Mer. Pass. Lower	MOON Age	MOON Phase
d	m s	m s	h m	h m	h m	d %	
19	03 34	03 32	11 56	07 34	19 56	24 29	
20	03 31	03 29	11 57	08 18	20 41	25 20	
21	03 27	03 25	11 57	09 05	21 30	26 12	

UT	ARIES	VENUS −4.6		MARS +1.2		JUPITER −2.4		SATURN +0.9	
d h	GHA ° ′	GHA ° ′	Dec ° ′	GHA ° ′	Dec ° ′	GHA ° ′	Dec ° ′	GHA ° ′	Dec ° ′
22 00	239 45.4	224 20.7	N 5 06.0	218 30.5	N 7 49.6	271 09.4	S13 32.1	72 47.3	N 7 55.5
01	254 47.8	239 21.1	06.6	233 31.2	50.3	286 11.7	32.0	87 49.8	55.5
02	269 50.3	254 21.6	07.2	248 31.9	51.0	301 14.0	32.0	102 52.2	55.5
03	284 52.7	269 22.1	07.8	263 32.6	51.7	316 16.3	31.9	117 54.7	55.5
04	299 55.2	284 22.5	08.3	278 33.3	52.4	331 18.5	31.8	132 57.1	55.5
05	314 57.7	299 23.0	08.9	293 34.0	53.1	346 20.8	31.8	147 59.6	55.5
06	330 00.1	314 23.4	N 5 09.5	308 34.7	N 7 53.9	1 23.1	S13 31.7	163 02.0	N 7 55.4
07	345 02.6	329 23.9	10.1	323 35.4	54.6	16 25.4	31.7	178 04.5	55.4
08	0 05.1	344 24.3	10.7	338 36.1	55.3	31 27.7	31.6	193 06.9	55.4
F 09	15 07.5	359 24.8	11.2	353 36.8	56.0	46 29.9	31.6	208 09.3	55.4
R 10	30 10.0	14 25.2	11.8	8 37.5	56.7	61 32.2	31.5	223 11.8	55.4
I 11	45 12.5	29 25.7	12.4	23 38.2	57.4	76 34.5	31.5	238 14.2	55.4
D 12	60 14.9	44 26.1	N 5 13.0	38 38.9	N 7 58.1	91 36.8	S13 31.4	253 16.7	N 7 55.4
A 13	75 17.4	59 26.6	13.6	53 39.6	58.9	106 39.1	31.4	268 19.1	55.3
Y 14	90 19.9	74 27.0	14.2	68 40.3	7 59.6	121 41.4	31.3	283 21.6	55.3
15	105 22.3	89 27.5	14.8	83 41.0	8 00.3	136 43.6	31.3	298 24.0	55.3
16	120 24.8	104 27.9	15.3	98 41.7	01.0	151 45.9	31.2	313 26.5	55.3
17	135 27.2	119 28.4	15.9	113 42.3	01.7	166 48.2	31.2	328 28.9	55.3
18	150 29.7	134 28.8	N 5 16.5	128 43.0	N 8 02.4	181 50.5	S13 31.1	343 31.3	N 7 55.3
19	165 32.2	149 29.3	17.1	143 43.7	03.1	196 52.8	31.1	358 33.8	55.2
20	180 34.6	164 29.7	17.7	158 44.4	03.8	211 55.1	31.0	13 36.2	55.2
21	195 37.1	179 30.1	18.3	173 45.1	04.6	226 57.4	31.0	28 38.7	55.2
22	210 39.6	194 30.6	18.9	188 45.8	05.3	241 59.6	30.9	43 41.1	55.2
23	225 42.0	209 31.0	19.5	203 46.5	06.0	257 01.9	30.8	58 43.6	55.2
23 00	240 44.5	224 31.4	N 5 20.1	218 47.2	N 8 06.7	272 04.2	S13 30.8	73 46.0	N 7 55.2
01	255 47.0	239 31.9	20.7	233 47.9	07.4	287 06.5	30.7	88 48.5	55.1
02	270 49.4	254 32.3	21.3	248 48.6	08.1	302 08.8	30.7	103 50.9	55.1
03	285 51.9	269 32.8	21.9	263 49.3	08.8	317 11.1	30.6	118 53.3	55.1
04	300 54.3	284 33.2	22.4	278 50.0	09.5	332 13.4	30.6	133 55.8	55.1
05	315 56.8	299 33.6	23.0	293 50.7	10.3	347 15.6	30.5	148 58.2	55.1
06	330 59.3	314 34.1	N 5 23.6	308 51.4	N 8 11.0	2 17.9	S13 30.5	164 00.7	N 7 55.1
S 07	346 01.7	329 34.5	24.2	323 52.1	11.7	17 20.2	30.4	179 03.1	55.0
A 08	1 04.2	344 34.9	24.8	338 52.8	12.4	32 22.5	30.4	194 05.6	55.0
T 09	16 06.7	359 35.3	25.4	353 53.5	13.1	47 24.8	30.3	209 08.0	55.0
U 10	31 09.1	14 35.8	26.0	8 54.2	13.8	62 27.1	30.3	224 10.4	55.0
R 11	46 11.6	29 36.2	26.6	23 54.9	14.5	77 29.4	30.2	239 12.9	55.0
D 12	61 14.1	44 36.6	N 5 27.2	38 55.6	N 8 15.2	92 31.7	S13 30.2	254 15.3	N 7 55.0
A 13	76 16.5	59 37.0	27.8	53 56.3	15.9	107 34.0	30.1	269 17.8	54.9
Y 14	91 19.0	74 37.5	28.4	68 57.0	16.6	122 36.2	30.1	284 20.2	54.9
15	106 21.5	89 37.9	29.0	83 57.7	17.4	137 38.5	30.0	299 22.6	54.9
16	121 23.9	104 38.3	29.6	98 58.4	18.1	152 40.8	30.0	314 25.1	54.9
17	136 26.4	119 38.7	30.2	113 59.0	18.8	167 43.1	29.9	329 27.5	54.9
18	151 28.8	134 39.1	N 5 30.9	128 59.7	N 8 19.5	182 45.4	S13 29.9	344 30.0	N 7 54.8
19	166 31.3	149 39.6	31.5	144 00.4	20.2	197 47.7	29.8	359 32.4	54.8
20	181 33.8	164 40.0	32.1	159 01.1	20.9	212 50.0	29.8	14 34.8	54.8
21	196 36.2	179 40.4	32.7	174 01.8	21.6	227 52.3	29.7	29 37.3	54.8
22	211 38.7	194 40.8	33.3	189 02.5	22.3	242 54.6	29.7	44 39.7	54.8
23	226 41.2	209 41.2	33.9	204 03.2	23.0	257 56.9	29.6	59 42.2	54.8
24 00	241 43.6	224 41.6	N 5 34.5	219 03.9	N 8 23.7	272 59.2	S13 29.6	74 44.6	N 7 54.7
01	256 46.1	239 42.0	35.1	234 04.6	24.4	288 01.5	29.5	89 47.0	54.7
02	271 48.6	254 42.5	35.7	249 05.3	25.2	303 03.8	29.5	104 49.5	54.7
03	286 51.0	269 42.9	36.3	264 06.0	25.9	318 06.0	29.4	119 51.9	54.7
04	301 53.5	284 43.3	36.9	279 06.7	26.6	333 08.3	29.4	134 54.4	54.7
05	316 56.0	299 43.7	37.5	294 07.4	27.3	348 10.6	29.3	149 56.8	54.6
06	331 58.4	314 44.1	N 5 38.2	309 08.1	N 8 28.0	3 12.9	S13 29.3	164 59.2	N 7 54.6
07	347 00.9	329 44.5	38.8	324 08.8	28.7	18 15.2	29.2	180 01.7	54.6
08	2 03.3	344 44.9	39.4	339 09.5	29.4	33 17.5	29.2	195 04.1	54.6
S 09	17 05.8	359 45.3	40.0	354 10.2	30.1	48 19.8	29.2	210 06.5	54.6
U 10	32 08.3	14 45.7	40.6	9 10.9	30.8	63 22.1	29.1	225 09.0	54.5
N 11	47 10.7	29 46.1	41.2	24 11.6	31.5	78 24.4	29.1	240 11.4	54.5
D 12	62 13.2	44 46.5	N 5 41.8	39 12.3	N 8 32.2	93 26.7	S13 29.0	255 13.9	N 7 54.5
A 13	77 15.7	59 46.9	42.5	54 13.0	32.9	108 29.0	29.0	270 16.3	54.5
Y 14	92 18.1	74 47.3	43.1	69 13.6	33.6	123 31.3	28.9	285 18.7	54.5
15	107 20.6	89 47.7	43.7	84 14.3	34.3	138 33.6	28.9	300 21.2	54.5
16	122 23.1	104 48.1	44.3	99 15.0	35.0	153 35.9	28.8	315 23.6	54.4
17	137 25.5	119 48.5	44.9	114 15.7	35.8	168 38.2	28.8	330 26.0	54.4
18	152 28.0	134 48.9	N 5 45.5	129 16.4	N 8 36.5	183 40.5	S13 28.7	345 28.5	N 7 54.4
19	167 30.5	149 49.3	46.2	144 17.1	37.2	198 42.8	28.7	0 30.9	54.4
20	182 32.9	164 49.7	46.8	159 17.8	37.9	213 45.1	28.6	15 33.3	54.4
21	197 35.4	179 50.1	47.4	174 18.5	38.6	228 47.4	28.6	30 35.8	54.3
22	212 37.8	194 50.5	48.0	189 19.2	39.3	243 49.7	28.5	45 38.2	54.3
23	227 40.3	209 50.9	48.6	204 19.9	40.0	258 52.0	28.5	60 40.7	54.3
Mer. Pass.	h m 7 55.7	v 0.4	d 0.6	v 0.7	d 0.7	v 2.3	d 0.1	v 2.4	d 0.0

STARS

Name	SHA ° ′	Dec ° ′
Acamar	315 21.0	S40 15.
Achernar	335 29.3	S57 11.
Acrux	173 12.6	S63 09.
Adhara	255 15.2	S28 59.
Aldebaran	290 53.2	N16 31.
Alioth	166 22.7	N55 54.
Alkaid	153 00.7	N49 16.
Al Na'ir	27 47.4	S46 54.
Alnilam	275 49.8	S 1 11.
Alphard	217 59.2	S 8 42.
Alphecca	126 13.2	N26 40.
Alpheratz	357 46.9	N29 08.
Altair	62 11.0	N 8 53.
Ankaa	353 18.8	S42 15.
Antares	112 29.6	S26 27.
Arcturus	145 58.1	N19 07.
Atria	107 33.7	S69 02.
Avior	234 19.7	S59 32.
Bellatrix	278 35.6	N 6 21.
Betelgeuse	271 04.9	N 7 24.
Canopus	263 58.0	S52 42.
Capella	280 39.4	N46 00.
Deneb	49 33.4	N45 18.
Denebola	182 36.6	N14 31.
Diphda	348 59.1	S17 56.
Dubhe	193 55.0	N61 42.
Elnath	278 16.8	N28 37.
Eltanin	90 47.1	N51 29.
Enif	33 50.1	N 9 55.
Fomalhaut	15 27.3	S29 34.
Gacrux	172 04.2	S57 10.
Gienah	175 55.3	S17 35.
Hadar	148 51.9	S60 25.
Hamal	328 04.5	N23 30.
Kaus Aust.	83 47.5	S34 22.
Kochab	137 18.4	N74 07.
Markab	13 41.5	N15 15.
Menkar	314 18.5	N 4 07.
Menkent	148 10.9	S36 25.
Miaplacidus	221 40.9	S69 45.
Mirfak	308 45.3	N49 53.
Nunki	76 01.8	S26 17.
Peacock	53 23.6	S56 42.
Pollux	243 31.6	N28 00.
Procyon	245 03.1	N 5 12.
Rasalhague	96 08.9	N12 33.
Regulus	207 46.7	N11 55.
Rigel	281 15.3	S 8 11.
Rigil Kent.	139 55.5	S60 52.
Sabik	102 15.7	S15 44.
Schedar	349 44.6	N56 35.
Shaula	96 25.6	S37 06.
Sirius	258 36.7	S16 43.
Spica	158 34.2	S11 12.
Suhail	222 54.9	S43 28.
Vega	80 40.7	N38 47.
Zuben'ubi	137 08.5	S16 05.

	SHA ° ′	Mer. Pas h m
Venus	343 47.0	9 02
Mars	338 02.7	9 24
Jupiter	31 19.7	5 51
Saturn	193 01.5	19 02

	UT	SUN GHA	SUN Dec	MOON GHA	v	MOON Dec	d	HP
	d h	° ′	° ′	° ′	′	° ′	′	′
	22 00	180 50.6	N20 22.2	216 19.2	9.9	N15 13.0	13.0	58.6
	01	195 50.6	22.7	230 48.1	9.9	15 26.0	12.8	58.6
	02	210 50.5	23.1	245 17.0	9.8	15 38.8	12.8	58.7
	03	225 50.5	. . 23.6	259 45.8	9.7	15 51.6	12.8	58.7
	04	240 50.4	24.1	274 14.5	9.5	16 04.4	12.6	58.8
	05	255 50.4	24.6	288 43.0	9.4	16 17.0	12.6	58.8
	06	270 50.4	N20 25.1	303 11.4	9.3	N16 29.6	12.5	58.8
	07	285 50.3	25.6	317 39.7	9.2	16 42.1	12.5	58.9
	08	300 50.3	26.1	332 07.9	9.1	16 54.6	12.3	58.9
F	09	315 50.2	. . 26.6	346 36.0	8.9	17 06.9	12.3	58.9
R	10	330 50.2	27.1	1 03.9	8.9	17 19.2	12.2	59.0
I	11	345 50.1	27.6	15 31.8	8.7	17 31.4	12.1	59.0
D	12	0 50.1	N20 28.0	29 59.5	8.6	N17 43.5	12.0	59.0
A	13	15 50.0	28.5	44 27.1	8.5	17 55.5	11.9	59.1
Y	14	30 50.0	29.0	58 54.6	8.3	18 07.4	11.9	59.1
	15	45 49.9	. . 29.5	73 21.9	8.2	18 19.3	11.7	59.1
	16	60 49.9	30.0	87 49.1	8.1	18 31.0	11.7	59.2
	17	75 49.8	30.5	102 16.2	8.0	18 42.7	11.5	59.2
	18	90 49.8	N20 31.0	116 43.2	7.9	N18 54.2	11.5	59.2
	19	105 49.7	31.4	131 10.1	7.7	19 05.7	11.3	59.3
	20	120 49.7	31.9	145 36.8	7.6	19 17.0	11.2	59.3
	21	135 49.6	. . 32.4	160 03.4	7.5	19 28.2	11.2	59.3
	22	150 49.5	32.9	174 29.9	7.4	19 39.4	11.0	59.4
	23	165 49.5	33.4	188 56.3	7.3	19 50.4	10.9	59.4
	23 00	180 49.4	N20 33.8	203 22.6	7.1	N20 01.3	10.8	59.4
	01	195 49.4	34.3	217 48.7	7.0	20 12.1	10.7	59.5
	02	210 49.3	34.8	232 14.7	6.8	20 22.8	10.6	59.5
	03	225 49.3	. . 35.3	246 40.5	6.8	20 33.4	10.5	59.5
	04	240 49.2	35.8	261 06.3	6.6	20 43.9	10.3	59.6
	05	255 49.2	36.2	275 31.9	6.5	20 54.2	10.2	59.6
	06	270 49.1	N20 36.7	289 57.4	6.4	N21 04.4	10.1	59.6
	07	285 49.1	37.2	304 22.8	6.3	21 14.5	10.0	59.6
S	08	300 49.0	37.7	318 48.1	6.1	21 24.5	9.8	59.7
A	09	315 49.0	. . 38.1	333 13.2	6.0	21 34.3	9.8	59.7
T	10	330 48.9	38.6	347 38.2	5.9	21 44.1	9.5	59.7
U	11	345 48.9	39.1	2 03.1	5.8	21 53.6	9.5	59.8
R	12	0 48.8	N20 39.5	16 27.9	5.7	N22 03.1	9.3	59.8
D	13	15 48.7	40.0	30 52.6	5.5	22 12.4	9.2	59.8
A	14	30 48.7	40.5	45 17.1	5.4	22 21.6	9.0	59.8
Y	15	45 48.6	. . 41.0	59 41.5	5.3	22 30.6	8.9	59.9
	16	60 48.6	41.4	74 05.8	5.2	22 39.5	8.7	59.9
	17	75 48.5	41.9	88 30.0	5.1	22 48.2	8.7	59.9
	18	90 48.5	N20 42.4	102 54.1	4.9	N22 56.9	8.4	59.9
	19	105 48.4	42.8	117 18.0	4.8	23 05.3	8.3	60.0
	20	120 48.3	43.3	131 41.8	4.8	23 13.6	8.2	60.0
	21	135 48.3	. . 43.8	146 05.6	4.6	23 21.8	8.0	60.0
	22	150 48.2	44.2	160 29.2	4.5	23 29.8	7.8	60.0
	23	165 48.2	44.7	174 52.7	4.4	23 37.6	7.7	60.1
	24 00	180 48.1	N20 45.2	189 16.1	4.3	N23 45.3	7.6	60.1
	01	195 48.1	45.6	203 39.4	4.2	23 52.9	7.3	60.1
	02	210 48.0	46.1	218 02.6	4.0	24 00.2	7.2	60.1
	03	225 47.9	. . 46.6	232 25.6	4.0	24 07.4	7.1	60.2
	04	240 47.9	47.0	246 48.6	3.9	24 14.5	6.9	60.2
	05	255 47.8	47.5	261 11.5	3.8	24 21.4	6.7	60.2
	06	270 47.8	N20 47.9	275 34.3	3.6	N24 28.1	6.5	60.2
	07	285 47.7	48.4	289 56.9	3.6	24 34.6	6.4	60.2
	08	300 47.6	48.9	304 19.5	3.5	24 41.0	6.2	60.3
S	09	315 47.6	. . 49.3	318 42.0	3.4	24 47.2	6.1	60.3
U	10	330 47.5	49.8	333 04.4	3.3	24 53.3	5.8	60.3
N	11	345 47.5	50.2	347 26.7	3.2	24 59.1	5.7	60.3
D	12	0 47.4	N20 50.7	1 48.9	3.1	N25 04.8	5.5	60.3
A	13	15 47.3	51.2	16 11.0	3.1	25 10.3	5.3	60.4
Y	14	30 47.3	51.6	30 33.1	2.9	25 15.6	5.1	60.4
	15	45 47.2	. . 52.1	44 55.0	2.9	25 20.7	5.0	60.4
	16	60 47.2	52.5	59 16.9	2.8	25 25.7	4.8	60.4
	17	75 47.1	53.0	73 38.7	2.8	25 30.5	4.6	60.4
	18	90 47.0	N20 53.4	88 00.5	2.6	N25 35.1	4.4	60.4
	19	105 47.0	53.9	102 22.1	2.6	25 39.5	4.2	60.5
	20	120 46.9	54.3	116 43.7	2.5	25 43.7	4.0	60.5
	21	135 46.9	. . 54.8	131 05.2	2.5	25 47.7	3.8	60.5
	22	150 46.8	55.2	145 26.7	2.4	25 51.5	3.7	60.5
	23	165 46.7	55.7	159 48.1	2.3	N25 55.2	3.4	60.5
		SD 15.8	d 0.5	SD 16.1		16.3		16.4

Lat.	Twilight Naut.	Twilight Civil	Sunrise	Moonrise 22	Moonrise 23	Moonrise 24	Moonrise 25
°	h m	h m	h m	h m	h m	h m	h m
N 72	□	□	□	□	□	□	□
N 70	□	□	□	00 23 / 23 06	□	□	□
68	////	////	00 53	00 47	00 12	□	□
66	////	////	01 49	01 05	00 49	00 07	□
64	////	////	02 22	01 20	01 15	01 08	00 51
62	////	01 14	02 46	01 33	01 36	01 43	02 05
60	////	01 53	03 05	01 44	01 52	02 08	02 41
N 58	////	02 20	03 20	01 54	02 07	02 28	03 07
56	01 07	02 40	03 34	02 02	02 19	02 45	03 27
54	01 43	02 57	03 45	02 09	02 30	03 00	03 45
52	02 08	03 11	03 55	02 16	02 39	03 12	03 59
50	02 27	03 23	04 04	02 22	02 48	03 23	04 12
45	03 02	03 48	04 23	02 35	03 06	03 46	04 39
N 40	03 28	04 07	04 38	02 46	03 21	04 05	05 00
35	03 48	04 23	04 51	02 56	03 34	04 21	05 17
30	04 04	04 36	05 02	03 04	03 45	04 35	05 33
20	04 29	04 57	05 21	03 18	04 05	04 58	05 58
N 10	04 49	05 15	05 38	03 31	04 22	05 18	06 21
0	05 05	05 31	05 53	03 43	04 38	05 38	06 41
S 10	05 20	05 46	06 08	03 55	04 54	05 57	07 02
20	05 34	06 01	06 24	04 08	05 11	06 17	07 25
30	05 48	06 17	06 43	04 23	05 31	06 41	07 51
35	05 55	06 26	06 53	04 32	05 43	06 56	08 06
40	06 03	06 36	07 05	04 42	05 56	07 12	08 24
45	06 11	06 47	07 20	04 54	06 13	07 32	08 46
S 50	06 20	07 00	07 37	05 08	06 33	07 57	09 14
52	06 24	07 07	07 45	05 15	06 42	08 09	09 27
54	06 29	07 13	07 55	05 22	06 53	08 23	09 42
56	06 34	07 21	08 05	05 31	07 05	08 39	10 01
58	06 39	07 29	08 17	05 40	07 19	08 58	10 23
S 60	06 45	07 38	08 30	05 52	07 36	09 22	10 52

Lat.	Sunset	Twilight Civil	Twilight Naut.	Moonset 22	Moonset 23	Moonset 24	Moonset 25
°	h m	h m	h m	h m	h m	h m	h m
N 72	□	□	□	□	□	□	□
N 70	□	□	□	21 21	□	□	□
68	23 10	////	////	20 16	□	□	□
66	22 08	////	////	19 40	22 24	□	□
64	21 34	////	////	19 15	21 24	23 53	□
62	21 09	22 44	////	18 56	20 50	22 40	23 57
60	20 50	22 03	////	18 40	20 25	22 04	23 18
N 58	20 34	21 36	////	18 26	20 05	21 38	22 51
56	20 21	21 15	22 51	18 15	19 49	21 18	22 30
54	20 09	20 58	22 13	18 05	19 35	21 01	22 12
52	19 59	20 44	21 48	17 56	19 23	20 46	21 57
50	19 50	20 31	21 28	17 48	19 12	20 34	21 44
45	19 31	20 07	20 52	17 31	18 50	20 07	21 17
N 40	19 16	19 47	20 26	17 17	18 32	19 47	20 55
35	19 03	19 31	20 06	17 05	18 17	19 30	20 38
30	18 51	19 18	19 50	16 55	18 04	19 15	20 22
20	18 32	18 56	19 25	16 38	17 42	18 50	19 56
N 10	18 16	18 38	19 05	16 22	17 23	18 28	19 34
0	18 00	18 22	18 48	16 08	17 06	18 08	19 13
S 10	17 45	18 08	18 33	15 54	16 48	17 47	18 52
20	17 29	17 52	18 19	15 39	16 29	17 26	18 29
30	17 11	17 36	18 06	15 22	16 07	17 01	18 03
35	17 00	17 27	17 58	15 12	15 55	16 46	17 47
40	16 48	17 17	17 51	15 01	15 40	16 29	17 29
45	16 33	17 06	17 42	14 48	15 23	16 08	17 07
S 50	16 16	16 53	17 33	14 32	15 02	15 43	16 40
52	16 08	16 46	17 29	14 25	14 52	15 31	16 26
54	15 59	16 40	17 24	14 16	14 40	15 16	16 11
56	15 48	16 32	17 19	14 07	14 27	15 00	15 53
58	15 36	16 24	17 14	13 57	14 13	14 41	15 30
S 60	15 23	16 15	17 08	13 45	13 55	14 16	15 01

Day	SUN Eqn. of Time 00^h	SUN Eqn. of Time 12^h	SUN Mer. Pass.	MOON Mer. Pass. Upper	MOON Mer. Pass. Lower	MOON Age	MOON Phase
d	m s	m s	h m	h m	h m	d %	
22	03 23	03 20	11 57	09 56	22 23	27 6	●
23	03 18	03 15	11 57	10 51	23 21	28 2	
24	03 13	03 10	11 57	11 52	24 24	29 0	

	UT	ARIES	VENUS −4.5		MARS +1.2		JUPITER −2.4		SATURN +0.9	
		GHA	GHA	Dec	GHA	Dec	GHA	Dec	GHA	Dec
	d h	° ′	° ′	° ′	° ′	° ′	° ′	° ′	° ′	° ′
	25 00	242 42.8	224 51.3	N 5 49.3	219 20.6	N 8 40.7	273 54.3	S13 28.4	75 43.1	N 7 54.3
	01	257 45.2	239 51.7	49.9	234 21.3	41.4	288 56.6	28.4	90 45.5	54.3
	02	272 47.7	254 52.0	50.5	249 22.0	42.1	303 58.9	28.3	105 48.0	54.2
	03	287 50.2	269 52.4	. . 51.1	264 22.7	. . 42.8	319 01.2	. . 28.3	120 50.4	. . 54.2
	04	302 52.6	284 52.8	51.8	279 23.4	43.5	334 03.5	28.3	135 52.8	54.2
	05	317 55.1	299 53.2	52.4	294 24.1	44.2	349 05.8	28.2	150 55.3	54.2
	06	332 57.6	314 53.6	N 5 53.0	309 24.8	N 8 44.9	4 08.1	S13 28.2	165 57.7	N 7 54.2
	07	348 00.0	329 54.0	53.6	324 25.5	45.6	19 10.4	28.1	181 00.1	54.1
	08	3 02.5	344 54.3	54.3	339 26.1	46.3	34 12.7	28.1	196 02.6	54.1
M	09	18 05.0	359 54.7	. . 54.9	354 26.8	. . 47.0	49 15.0	. . 28.0	211 05.0	. . 54.1
O	10	33 07.4	14 55.1	55.5	9 27.5	47.7	64 17.3	28.0	226 07.4	54.1
N	11	48 09.9	29 55.5	56.1	24 28.2	48.4	79 19.6	27.9	241 09.9	54.0
D	12	63 12.3	44 55.9	N 5 56.8	39 28.9	N 8 49.1	94 21.9	S13 27.9	256 12.3	N 7 54.0
A	13	78 14.8	59 56.2	57.4	54 29.6	49.8	109 24.2	27.8	271 14.7	54.0
Y	14	93 17.3	74 56.6	58.0	69 30.3	50.5	124 26.5	27.8	286 17.2	54.0
	15	108 19.7	89 57.0	. . 58.7	84 31.0	. . 51.2	139 28.8	. . 27.8	301 19.6	. . 54.0
	16	123 22.2	104 57.4	59.3	99 31.7	51.9	154 31.1	27.7	316 22.0	53.9
	17	138 24.7	119 57.7	5 59.9	114 32.4	52.6	169 33.4	27.7	331 24.5	53.9
	18	153 27.1	134 58.1	N 6 00.6	129 33.1	N 8 53.3	184 35.7	S13 27.6	346 26.9	N 7 53.9
	19	168 29.6	149 58.5	01.2	144 33.8	54.0	199 38.1	27.6	1 29.3	53.9
	20	183 32.1	164 58.9	01.8	159 34.5	54.7	214 40.4	27.5	16 31.8	53.9
	21	198 34.5	179 59.2	. . 02.5	174 35.2	. . 55.4	229 42.7	. . 27.5	31 34.2	. . 53.8
	22	213 37.0	194 59.6	03.1	189 35.9	56.1	244 45.0	27.4	46 36.6	53.8
	23	228 39.5	210 00.0	03.7	204 36.6	56.8	259 47.3	27.4	61 39.0	53.8
	26 00	243 41.9	225 00.3	N 6 04.4	219 37.3	N 8 57.5	274 49.6	S13 27.4	76 41.5	N 7 53.8
	01	258 44.4	240 00.7	05.0	234 37.9	58.2	289 51.9	27.3	91 43.9	53.7
	02	273 46.8	255 01.1	05.6	249 38.6	58.9	304 54.2	27.3	106 46.3	53.7
	03	288 49.3	270 01.4	. . 06.3	264 39.3	8 59.6	319 56.5	. . 27.2	121 48.8	. . 53.7
	04	303 51.8	285 01.8	06.9	279 40.0	9 00.3	334 58.8	27.2	136 51.2	53.7
	05	318 54.2	300 02.2	07.6	294 40.7	01.0	350 01.1	27.1	151 53.6	53.7
	06	333 56.7	315 02.5	N 6 08.2	309 41.4	N 9 01.7	5 03.4	S13 27.1	166 56.1	N 7 53.6
	07	348 59.2	330 02.9	08.8	324 42.1	02.4	20 05.7	27.0	181 58.5	53.6
T	08	4 01.6	345 03.2	09.5	339 42.8	03.1	35 08.1	27.0	197 00.9	53.6
U	09	19 04.1	0 03.6	. . 10.1	354 43.5	. . 03.8	50 10.4	. . 27.0	212 03.4	. . 53.6
E	10	34 06.6	15 04.0	10.8	9 44.2	04.5	65 12.7	26.9	227 05.8	53.5
S	11	49 09.0	30 04.3	11.4	24 44.9	05.2	80 15.0	26.9	242 08.2	53.5
D	12	64 11.5	45 04.7	N 6 12.0	39 45.6	N 9 05.9	95 17.3	S13 26.8	257 10.6	N 7 53.5
A	13	79 14.0	60 05.0	12.7	54 46.3	06.6	110 19.6	26.8	272 13.1	53.5
Y	14	94 16.4	75 05.4	13.3	69 47.0	07.3	125 21.9	26.7	287 15.5	53.5
	15	109 18.9	90 05.7	. . 14.0	84 47.7	. . 08.0	140 24.2	. . 26.7	302 17.9	. . 53.4
	16	124 21.3	105 06.1	14.6	99 48.3	08.7	155 26.5	26.7	317 20.4	53.4
	17	139 23.8	120 06.4	15.3	114 49.0	09.4	170 28.9	26.6	332 22.8	53.4
	18	154 26.3	135 06.8	N 6 15.9	129 49.7	N 9 10.1	185 31.2	S13 26.6	347 25.2	N 7 53.4
	19	169 28.7	150 07.1	16.5	144 50.4	10.8	200 33.5	26.5	2 27.6	53.3
	20	184 31.2	165 07.5	17.2	159 51.1	11.5	215 35.8	26.5	17 30.1	53.3
	21	199 33.7	180 07.8	. . 17.8	174 51.8	. . 12.2	230 38.1	. . 26.5	32 32.5	. . 53.3
	22	214 36.1	195 08.2	18.5	189 52.5	12.9	245 40.4	26.4	47 34.9	53.3
	23	229 38.6	210 08.5	19.1	204 53.2	13.6	260 42.7	26.4	62 37.3	53.2
	27 00	244 41.1	225 08.9	N 6 19.8	219 53.9	N 9 14.3	275 45.1	S13 26.3	77 39.8	N 7 53.2
	01	259 43.5	240 09.2	20.4	234 54.6	15.0	290 47.4	26.3	92 42.2	53.2
	02	274 46.0	255 09.6	21.1	249 55.3	15.7	305 49.7	26.2	107 44.6	53.2
	03	289 48.4	270 09.9	. . 21.7	264 56.0	. . 16.4	320 52.0	. . 26.2	122 47.1	. . 53.2
	04	304 50.9	285 10.3	22.4	279 56.7	17.1	335 54.3	26.2	137 49.5	53.1
	05	319 53.4	300 10.6	23.0	294 57.4	17.8	350 56.6	26.1	152 51.9	53.1
	06	334 55.8	315 10.9	N 6 23.7	309 58.1	N 9 18.4	5 59.0	S13 26.1	167 54.3	N 7 53.1
W	07	349 58.3	330 11.3	24.3	324 58.7	19.1	21 01.3	26.0	182 56.8	53.1
E	08	5 00.8	345 11.6	25.0	339 59.4	19.8	36 03.6	26.0	197 59.2	53.0
D	09	20 03.2	0 12.0	. . 25.6	355 00.1	. . 20.5	51 05.9	. . 26.0	213 01.6	. . 53.0
N	10	35 05.7	15 12.3	26.3	10 00.8	21.2	66 08.2	25.9	228 04.0	53.0
E	11	50 08.2	30 12.6	26.9	25 01.5	21.9	81 10.5	25.9	243 06.5	53.0
S	12	65 10.6	45 13.0	N 6 27.6	40 02.2	N 9 22.6	96 12.9	S13 25.8	258 08.9	N 7 52.9
D	13	80 13.1	60 13.3	28.3	55 02.9	23.3	111 15.2	25.8	273 11.3	52.9
A	14	95 15.6	75 13.6	28.9	70 03.6	24.0	126 17.5	25.8	288 13.7	52.9
Y	15	110 18.0	90 14.0	. . 29.6	85 04.3	. . 24.7	141 19.8	. . 25.7	303 16.2	. . 52.9
	16	125 20.5	105 14.3	30.2	100 05.0	25.4	156 22.1	25.7	318 18.6	52.8
	17	140 22.9	120 14.6	30.9	115 05.7	26.1	171 24.5	25.6	333 21.0	52.8
	18	155 25.4	135 15.0	N 6 31.5	130 06.4	N 9 26.8	186 26.8	S13 25.6	348 23.4	N 7 52.8
	19	170 27.9	150 15.3	32.2	145 07.1	27.5	201 29.1	25.6	3 25.9	52.8
	20	185 30.3	165 15.6	32.8	160 07.7	28.2	216 31.4	25.5	18 28.3	52.7
	21	200 32.8	180 15.9	. . 33.5	175 08.4	. . 28.8	231 33.7	. . 25.5	33 30.7	. . 52.7
	22	215 35.3	195 16.3	34.2	190 09.1	29.5	246 36.1	25.4	48 33.1	52.7
	23	230 37.7	210 16.6	34.8	205 09.8	30.2	261 38.4	25.4	63 35.6	52.7
	Mer. Pass.	h m 7 43.9	*v* 0.4	*d* 0.6	*v* 0.7	*d* 0.7	*v* 2.3	*d* 0.0	*v* 2.4	*d* 0.0

STARS

Name	SHA	Dec
	° ′	° ′
Acamar	315 20.9	S40 15.
Achernar	335 29.3	S57 11.
Acrux	173 12.6	S63 09.
Adhara	255 15.2	S28 59.
Aldebaran	290 53.2	N16 31.
Alioth	166 22.8	N55 54.
Alkaid	153 00.7	N49 16.
Al Na'ir	27 47.3	S46 54.
Alnilam	275 49.8	S 1 11.
Alphard	217 59.2	S 8 42.
Alphecca	126 13.2	N26 40.
Alpheratz	357 46.9	N29 08.
Altair	62 11.0	N 8 53.
Ankaa	353 18.8	S42 15.
Antares	112 29.6	S26 27.
Arcturus	145 58.1	N19 07.
Atria	107 33.7	S69 02.
Avior	234 19.7	S59 32.
Bellatrix	278 35.6	N 6 21.
Betelgeuse	271 04.9	N 7 24.
Canopus	263 58.0	S52 42.
Capella	280 39.4	N46 00.
Deneb	49 33.4	N45 18.
Denebola	182 36.6	N14 31.
Diphda	348 59.1	S17 56.
Dubhe	193 55.0	N61 42.
Elnath	278 16.8	N28 37.
Eltanin	90 47.0	N51 29.
Enif	33 50.0	N 9 55.
Fomalhaut	15 27.3	S29 34.
Gacrux	172 04.2	S57 10.
Gienah	175 55.3	S17 35.
Hadar	148 51.9	S60 25.
Hamal	328 04.5	N23 30.
Kaus Aust.	83 47.4	S34 22.
Kochab	137 18.4	N74 07.
Markab	13 41.4	N15 15.
Menkar	314 18.5	N 4 07.
Menkent	148 10.9	S36 25.
Miaplacidus	221 40.9	S69 45.
Mirfak	308 45.3	N49 53.
Nunki	76 01.7	S26 17.
Peacock	53 23.5	S56 42.
Pollux	243 31.6	N28 00.
Procyon	245 03.1	N 5 12.
Rasalhague	96 08.9	N12 33.
Regulus	207 46.7	N11 55.
Rigel	281 15.3	S 8 11.
Rigil Kent.	139 55.5	S60 52.
Sabik	102 15.7	S15 44.
Schedar	349 44.5	N56 35.
Shaula	96 25.6	S37 06.
Sirius	258 36.7	S16 43.
Spica	158 34.2	S11 12.
Suhail	222 54.9	S43 28.
Vega	80 40.7	N38 47.
Zuben'ubi	137 08.5	S16 05.

	SHA	Mer. Pass.
	° ′	h m
Venus	341 18.4	9 00
Mars	335 55.3	9 21
Jupiter	31 07.7	5 40
Saturn	192 59.6	18 50

UT	SUN GHA	SUN Dec	MOON GHA	v	MOON Dec	d	HP
d h	° ′	° ′	° ′	′	° ′	′	′
25 00	180 46.7	N20 56.1	174 09.4	2.3	N25 58.6	3.3	60.5
01	195 46.6	56.6	188 30.7	2.2	26 01.9	3.1	60.5
02	210 46.5	57.0	202 51.9	2.2	26 05.0	2.8	60.6
03	225 46.5	. . 57.5	217 13.1	2.1	26 07.8	2.7	60.6
04	240 46.4	57.9	231 34.2	2.1	26 10.5	2.5	60.6
05	255 46.3	58.4	245 55.3	2.0	26 13.0	2.3	60.6
06	270 46.3	N20 58.8	260 16.3	2.0	N26 15.3	2.0	60.6
07	285 46.2	59.3	274 37.3	2.0	26 17.3	1.9	60.6
08	300 46.2	20 59.7	288 58.3	1.9	26 19.2	1.7	60.6
09	315 46.1	21 00.2	303 19.2	1.9	26 20.9	1.5	60.6
10	330 46.0	00.6	317 40.1	1.9	26 22.4	1.3	60.6
11	345 46.0	01.0	332 01.0	1.9	26 23.7	1.1	60.6
12	0 45.9	N21 01.5	346 21.9	1.8	N26 24.8	0.9	60.7
13	15 45.8	01.9	0 42.7	1.8	26 25.7	0.6	60.7
14	30 45.8	02.4	15 03.5	1.8	26 26.3	0.5	60.7
15	45 45.7	. . 02.8	29 24.3	1.8	26 26.8	0.3	60.7
16	60 45.6	03.2	43 45.1	1.8	26 27.1	0.1	60.7
17	75 45.6	03.7	58 05.9	1.8	26 27.2	0.1	60.7
18	90 45.5	N21 04.1	72 26.7	1.7	N26 27.1	0.4	60.7
19	105 45.4	04.6	86 47.4	1.8	26 26.7	0.5	60.7
20	120 45.4	05.0	101 08.2	1.8	26 26.2	0.7	60.7
21	135 45.3	. . 05.4	115 29.0	1.8	26 25.5	1.0	60.7
22	150 45.2	05.9	129 49.8	1.7	26 24.5	1.1	60.7
23	165 45.2	06.3	144 10.5	1.9	26 23.4	1.3	60.7
26 00	180 45.1	N21 06.7	158 31.4	1.8	N26 22.1	1.6	60.7
01	195 45.0	07.2	172 52.2	1.8	26 20.5	1.7	60.7
02	210 45.0	07.6	187 13.0	1.9	26 18.8	1.9	60.7
03	225 44.9	. . 08.0	201 33.9	1.9	26 16.9	2.2	60.7
04	240 44.8	08.5	215 54.8	1.9	26 14.7	2.3	60.7
05	255 44.7	08.9	230 15.7	2.0	26 12.4	2.5	60.7
06	270 44.7	N21 09.3	244 36.7	2.0	N26 09.9	2.8	60.7
07	285 44.6	09.8	258 57.7	2.0	26 07.1	2.9	60.7
08	300 44.5	10.2	273 18.7	2.0	26 04.2	3.1	60.7
09	315 44.5	. . 10.6	287 39.7	2.2	26 01.1	3.4	60.7
10	330 44.4	11.1	302 00.9	2.1	25 57.7	3.5	60.7
11	345 44.3	11.5	316 22.0	2.2	25 54.2	3.7	60.7
12	0 44.3	N21 11.9	330 43.2	2.3	N25 50.5	3.9	60.7
13	15 44.2	12.3	345 04.5	2.3	25 46.6	4.1	60.7
14	30 44.1	12.8	359 25.8	2.4	25 42.5	4.3	60.7
15	45 44.0	. . 13.2	13 47.2	2.4	25 38.2	4.5	60.7
16	60 44.0	13.6	28 08.6	2.5	25 33.7	4.7	60.7
17	75 43.9	14.0	42 30.1	2.5	25 29.0	4.8	60.7
18	90 43.8	N21 14.5	56 51.6	2.7	N25 24.2	5.1	60.7
19	105 43.8	14.9	71 13.3	2.6	25 19.1	5.2	60.7
20	120 43.7	15.3	85 34.9	2.8	25 13.9	5.4	60.7
21	135 43.6	. . 15.7	99 56.7	2.9	25 08.5	5.6	60.6
22	150 43.5	16.1	114 18.6	2.9	25 02.9	5.8	60.6
23	165 43.5	16.6	128 40.5	3.0	24 57.1	6.0	60.6
27 00	180 43.4	N21 17.0	143 02.5	3.0	N24 51.1	6.1	60.6
01	195 43.3	17.4	157 24.5	3.2	24 45.0	6.3	60.6
02	210 43.3	17.8	171 46.7	3.2	24 38.7	6.5	60.6
03	225 43.2	. . 18.2	186 08.9	3.4	24 32.2	6.7	60.6
04	240 43.1	18.7	200 31.3	3.4	24 25.5	6.9	60.6
05	255 43.0	19.1	214 53.7	3.5	24 18.6	7.0	60.6
06	270 43.0	N21 19.5	229 16.2	3.6	N24 11.6	7.2	60.6
07	285 42.9	19.9	243 38.8	3.7	24 04.4	7.3	60.5
08	300 42.8	20.3	258 01.5	3.8	23 57.1	7.6	60.5
09	315 42.7	. . 20.7	272 24.3	3.9	23 49.5	7.7	60.5
10	330 42.7	21.1	286 47.2	3.9	23 41.8	7.8	60.5
11	345 42.6	21.6	301 10.1	4.1	23 34.0	8.0	60.5
12	0 42.5	N21 22.0	315 33.2	4.2	N23 26.0	8.2	60.5
13	15 42.4	22.4	329 56.4	4.3	23 17.8	8.3	60.5
14	30 42.4	22.8	344 19.7	4.4	23 09.5	8.5	60.5
15	45 42.3	. . 23.2	358 43.1	4.5	23 01.0	8.7	60.4
16	60 42.2	23.6	13 06.6	4.6	22 52.3	8.8	60.4
17	75 42.1	24.0	27 30.2	4.7	22 43.5	8.9	60.4
18	90 42.1	N21 24.4	41 53.9	4.8	N22 34.6	9.1	60.4
19	105 42.0	24.8	56 17.7	4.9	22 25.5	9.3	60.4
20	120 41.9	25.2	70 41.6	5.1	22 16.2	9.4	60.4
21	135 41.8	. . 25.6	85 05.7	5.1	22 06.8	9.5	60.3
22	150 41.7	26.1	99 29.8	5.3	21 57.3	9.7	60.3
23	165 41.7	26.5	113 54.1	5.3	N21 47.6	9.8	60.3
	SD 15.8	d 0.4	SD 16.5		16.5		16.5

Lat.	Twilight Naut.	Twilight Civil	Sunrise	Moonrise 25	Moonrise 26	Moonrise 27	Moonrise 28
°	h m	h m	h m	h m	h m	h m	h m
N 72	▭	▭	▭	▭	▭	▭	▭
N 70	▭	▭	▭	▭	▭	▭	▭
68	////	////	00 14	▭	▭	▭	05 05
66	////	////	01 37	▭	▭	▭	05 53
64	////	////	02 13	00 51	▭	04 09	06 24
62	////	00 58	02 39	02 05	03 05	04 50	06 47
60	////	01 44	02 59	02 41	03 44	05 18	07 05
N 58	////	02 13	03 15	03 07	04 11	05 40	07 20
56	00 52	02 34	03 29	03 27	04 32	05 58	07 33
54	01 34	02 52	03 41	03 45	04 50	06 12	07 44
52	02 01	03 07	03 52	03 59	05 05	06 25	07 54
50	02 22	03 19	04 01	04 12	05 18	06 37	08 03
45	02 59	03 45	04 20	04 39	05 44	07 00	08 21
N 40	03 25	04 05	04 36	05 00	06 06	07 19	08 36
35	03 45	04 21	04 50	05 17	06 23	07 35	08 49
30	04 02	04 34	05 01	05 33	06 38	07 49	09 00
20	04 28	04 57	05 21	05 58	07 04	08 12	09 19
N 10	04 48	05 15	05 38	06 21	07 26	08 32	09 35
0	05 05	05 31	05 53	06 41	07 47	08 51	09 50
S 10	05 21	05 47	06 09	07 02	08 08	09 09	10 06
20	05 35	06 02	06 26	07 25	08 30	09 29	10 22
30	05 49	06 19	06 44	07 51	08 56	09 52	10 40
35	05 57	06 28	06 55	08 06	09 11	10 06	10 51
40	06 05	06 38	07 08	08 24	09 28	10 21	11 03
45	06 13	06 50	07 23	08 46	09 49	10 39	11 18
S 50	06 23	07 04	07 41	09 14	10 16	11 02	11 35
52	06 28	07 10	07 49	09 27	10 29	11 13	11 43
54	06 32	07 17	07 59	09 42	10 43	11 25	11 53
56	06 37	07 25	08 09	10 01	11 01	11 39	12 03
58	06 43	07 33	08 22	10 23	11 21	11 55	12 14
S 60	06 49	07 43	08 36	10 52	11 48	12 15	12 28

Lat.	Sunset	Twilight Civil	Twilight Naut.	Moonset 25	Moonset 26	Moonset 27	Moonset 28
°	h m	h m	h m	h m	h m	h m	h m
N 72	▭	▭	▭	▭	▭	▭	▭
N 70	▭	▭	▭	▭	▭	▭	▭
68	▭	▭	▭	▭	▭	▭	02 22
66	22 21	////	////	▭	▭	▭	01 33
64	21 43	////	////	▭	▭	01 09	01 01
62	21 17	23 02	////	23 57	24 27	00 27	00 38
60	20 56	22 13	////	23 18	23 59	24 19	00 19
N 58	20 40	21 43	////	22 51	23 37	24 03	00 03
56	20 26	21 21	23 07	22 30	23 18	23 49	24 09
54	20 14	21 03	22 22	22 12	23 03	23 37	24 00
52	20 03	20 49	21 55	21 57	22 50	23 27	23 53
50	19 54	20 36	21 34	21 44	22 38	23 17	23 46
45	19 34	20 10	20 56	21 17	22 13	22 58	23 32
N 40	19 18	19 50	20 30	20 55	21 54	22 41	23 19
35	19 05	19 34	20 09	20 38	21 37	22 28	23 09
30	18 53	19 20	19 52	20 22	21 23	22 16	23 00
20	18 33	18 58	19 26	19 56	20 59	21 55	22 44
N 10	18 16	18 39	19 06	19 34	20 38	21 37	22 30
0	18 01	18 23	18 49	19 13	20 18	21 20	22 17
S 10	17 45	18 07	18 33	18 52	19 58	21 03	22 04
20	17 28	17 52	18 19	18 29	19 36	20 44	21 50
30	17 09	17 35	18 05	18 03	19 11	20 23	21 33
35	16 58	17 26	17 57	17 47	18 57	20 10	21 23
40	16 46	17 16	17 49	17 29	18 39	19 56	21 12
45	16 31	17 04	17 40	17 07	18 19	19 38	20 59
S 50	16 13	16 50	17 30	16 40	17 53	19 17	20 43
52	16 04	16 44	17 26	16 26	17 40	19 06	20 36
54	15 55	16 37	17 21	16 11	17 26	18 55	20 27
56	15 44	16 29	17 16	15 53	17 09	18 41	20 18
58	15 32	16 20	17 11	15 30	16 48	18 25	20 07
S 60	15 17	16 10	17 05	15 01	16 22	18 06	19 54

Day	SUN Eqn. of Time 00^h	SUN Eqn. of Time 12^h	SUN Mer. Pass.	MOON Mer. Pass. Upper	MOON Mer. Pass. Lower	Age	Phase
d	m s	m s	h m	h m	h m	d %	
25	03 07	03 04	11 57	12 57	00 24	01 2	
26	03 01	02 57	11 57	14 02	01 30	02 6	●
27	02 54	02 50	11 57	15 05	02 34	03 13	

	UT	ARIES	VENUS −4.5		MARS +1.2		JUPITER −2.4		SATURN +0.9		STARS		
		GHA	GHA	Dec	GHA	Dec	GHA	Dec	GHA	Dec	Name	SHA	Dec
	d h	° ′	° ′	° ′	° ′	° ′	° ′	° ′	° ′	° ′		° ′	° ′
	28 00	245 40.2	225 16.9	N 6 35.5	220 10.5	N 9 30.9	276 40.7	S13 25.4	78 38.0	N 7 52.6	Acamar	315 20.9	S40 15.
	01	260 42.7	240 17.2	36.1	235 11.2	31.6	291 43.0	25.3	93 40.4	52.6	Achernar	335 29.2	S57 11.
	02	275 45.1	255 17.6	36.8	250 11.9	32.3	306 45.3	25.3	108 42.8	52.6	Acrux	173 12.6	S63 09.
	03	290 47.6	270 17.9	. . 37.5	265 12.6	. . 33.0	321 47.7	. . 25.2	123 45.2	. . 52.6	Adhara	255 15.2	S28 59.
	04	305 50.1	285 18.2	38.1	280 13.3	33.7	336 50.0	25.2	138 47.7	52.5	Aldebaran	290 53.2	N16 31.
	05	320 52.5	300 18.5	38.8	295 14.0	34.4	351 52.3	25.2	153 50.1	52.5			
	06	335 55.0	315 18.8	N 6 39.5	310 14.7	N 9 35.1	6 54.6	S13 25.1	168 52.5	N 7 52.5	Alioth	166 22.8	N55 54.
	07	350 57.4	330 19.2	40.1	325 15.4	35.7	21 57.0	25.1	183 54.9	52.5	Alkaid	153 00.7	N49 16.
T	08	5 59.9	345 19.5	40.8	340 16.1	36.4	36 59.3	25.1	198 57.4	52.4	Al Na'ir	27 47.3	S46 54.
H	09	21 02.4	0 19.8	. . 41.4	355 16.7	. . 37.1	52 01.6	. . 25.0	213 59.8	. . 52.4	Alnilam	275 49.7	S 1 11.
U	10	36 04.8	15 20.1	42.1	10 17.4	37.8	67 03.9	25.0	229 02.2	52.4	Alphard	217 59.2	S 8 42.
R	11	51 07.3	30 20.4	42.8	25 18.1	38.5	82 06.3	24.9	244 04.6	52.4			
S	12	66 09.8	45 20.7	N 6 43.4	40 18.8	N 9 39.2	97 08.6	S13 24.9	259 07.0	N 7 52.3	Alphecca	126 13.1	N26 40.
D	13	81 12.2	60 21.1	44.1	55 19.5	39.9	112 10.9	24.9	274 09.5	52.3	Alpheratz	357 46.8	N29 08.
A	14	96 14.7	75 21.4	44.8	70 20.2	40.6	127 13.2	24.8	289 11.9	52.3	Altair	62 11.0	N 8 53.
Y	15	111 17.2	90 21.7	. . 45.4	85 20.9	. . 41.3	142 15.6	. . 24.8	304 14.3	. . 52.3	Ankaa	353 18.7	S42 15.
	16	126 19.6	105 22.0	46.1	100 21.6	42.0	157 17.9	24.8	319 16.7	52.2	Antares	112 29.6	S26 27.
	17	141 22.1	120 22.3	46.8	115 22.3	42.6	172 20.2	24.7	334 19.1	52.2			
	18	156 24.6	135 22.6	N 6 47.4	130 23.0	N 9 43.3	187 22.5	S13 24.7	349 21.6	N 7 52.2	Arcturus	145 58.1	N19 07.
	19	171 27.0	150 22.9	48.1	145 23.7	44.0	202 24.9	24.6	4 24.0	52.2	Atria	107 33.7	S69 02.
	20	186 29.5	165 23.2	48.8	160 24.4	44.7	217 27.2	24.6	19 26.4	52.1	Avior	234 19.8	S59 32.
	21	201 31.9	180 23.5	. . 49.5	175 25.0	. . 45.4	232 29.5	. . 24.6	34 28.8	. . 52.1	Bellatrix	278 35.6	N 6 21.
	22	216 34.4	195 23.8	50.1	190 25.7	46.1	247 31.9	24.5	49 31.2	52.1	Betelgeuse	271 04.9	N 7 24.
	23	231 36.9	210 24.1	50.8	205 26.4	46.8	262 34.2	24.5	64 33.7	52.0			
	29 00	246 39.3	225 24.4	N 6 51.5	220 27.1	N 9 47.4	277 36.5	S13 24.5	79 36.1	N 7 52.0	Canopus	263 58.0	S52 42.
	01	261 41.8	240 24.7	52.1	235 27.8	48.1	292 38.8	24.4	94 38.5	52.0	Capella	280 39.4	N46 00.
	02	276 44.3	255 25.0	52.8	250 28.5	48.8	307 41.2	24.4	109 40.9	52.0	Deneb	49 33.4	N45 18.
	03	291 46.7	270 25.3	. . 53.5	265 29.2	. . 49.5	322 43.5	. . 24.4	124 43.3	. . 51.9	Denebola	182 36.6	N14 31.
	04	306 49.2	285 25.6	54.2	280 29.9	50.2	337 45.8	24.3	139 45.8	51.9	Diphda	348 59.0	S17 56.
	05	321 51.7	300 25.9	54.8	295 30.6	50.9	352 48.2	24.3	154 48.2	51.9			
	06	336 54.1	315 26.2	N 6 55.5	310 31.3	N 9 51.6	7 50.5	S13 24.2	169 50.6	N 7 51.9	Dubhe	193 55.0	N61 42.
	07	351 56.6	330 26.5	56.2	325 32.0	52.3	22 52.8	24.2	184 53.0	51.8	Elnath	278 16.8	N28 37.
	08	6 59.1	345 26.8	56.9	340 32.7	52.9	37 55.2	24.2	199 55.4	51.8	Eltanin	90 47.0	N51 29.
F	09	22 01.5	0 27.1	. . 57.5	355 33.3	. . 53.6	52 57.5	. . 24.1	214 57.8	. . 51.8	Enif	33 50.0	N 9 55.
R	10	37 04.0	15 27.4	58.2	10 34.0	54.3	67 59.8	24.1	230 00.3	51.7	Fomalhaut	15 27.2	S29 34.
I	11	52 06.4	30 27.7	58.9	25 34.7	55.0	83 02.2	24.1	245 02.7	51.7			
D	12	67 08.9	45 28.0	N 6 59.6	40 35.4	N 9 55.7	98 04.5	S13 24.0	260 05.1	N 7 51.7	Gacrux	172 04.2	S57 10.
A	13	82 11.4	60 28.3	7 00.2	55 36.1	56.4	113 06.8	24.0	275 07.5	51.7	Gienah	175 55.3	S17 35.
Y	14	97 13.8	75 28.6	00.9	70 36.8	57.0	128 09.2	24.0	290 09.9	51.6	Hadar	148 51.9	S60 25.
	15	112 16.3	90 28.9	. . 01.6	85 37.5	. . 57.7	143 11.5	. . 23.9	305 12.3	. . 51.6	Hamal	328 04.5	N23 30.
	16	127 18.8	105 29.2	02.3	100 38.2	58.4	158 13.8	23.9	320 14.8	51.6	Kaus Aust.	83 47.4	S34 22.
	17	142 21.2	120 29.5	02.9	115 38.9	59.1	173 16.2	23.9	335 17.2	51.6			
	18	157 23.7	135 29.8	N 7 03.6	130 39.6	N 9 59.8	188 18.5	S13 23.8	350 19.6	N 7 51.5	Kochab	137 18.4	N74 07.
	19	172 26.2	150 30.1	04.3	145 40.3	10 00.5	203 20.8	23.8	5 22.0	51.5	Markab	13 41.4	N15 15.
	20	187 28.6	165 30.3	05.0	160 40.9	01.1	218 23.2	23.8	20 24.4	51.5	Menkar	314 18.5	N 4 07.
	21	202 31.1	180 30.6	. . 05.7	175 41.6	. . 01.8	233 25.5	. . 23.7	35 26.8	. . 51.4	Menkent	148 10.9	S36 25.
	22	217 33.5	195 30.9	06.3	190 42.3	02.5	248 27.8	23.7	50 29.3	51.4	Miaplacidus	221 41.0	S69 45.
	23	232 36.0	210 31.2	07.0	205 43.0	03.2	263 30.2	23.7	65 31.7	51.4			
	30 00	247 38.5	225 31.5	N 7 07.7	220 43.7	N10 03.9	278 32.5	S13 23.6	80 34.1	N 7 51.4	Mirfak	308 45.3	N49 53.
	01	262 40.9	240 31.8	08.4	235 44.4	04.6	293 34.8	23.6	95 36.5	51.3	Nunki	76 01.7	S26 17.
	02	277 43.4	255 32.0	09.1	250 45.1	05.2	308 37.2	23.6	110 38.9	51.3	Peacock	53 23.5	S56 42.
	03	292 45.9	270 32.3	. . 09.8	265 45.8	. . 05.9	323 39.5	. . 23.5	125 41.3	. . 51.3	Pollux	243 31.6	N28 00.
	04	307 48.3	285 32.6	10.4	280 46.5	06.6	338 41.9	23.5	140 43.7	51.2	Procyon	245 03.1	N 5 12.
	05	322 50.8	300 32.9	11.1	295 47.2	07.3	353 44.2	23.5	155 46.2	51.2			
	06	337 53.3	315 33.2	N 7 11.8	310 47.9	N10 08.0	8 46.5	S13 23.4	170 48.6	N 7 51.2	Rasalhague	96 08.9	N12 33.
	07	352 55.7	330 33.4	12.5	325 48.5	08.6	23 48.9	23.4	185 51.0	51.2	Regulus	207 46.7	N11 55.
S	08	7 58.2	345 33.7	13.2	340 49.2	09.3	38 51.2	23.4	200 53.4	51.1	Rigel	281 15.3	S 8 11.
A	09	23 00.7	0 34.0	. . 13.9	355 49.9	. . 10.0	53 53.6	. . 23.3	215 55.8	. . 51.1	Rigil Kent.	139 55.5	S60 52.
T	10	38 03.1	15 34.3	14.5	10 50.6	10.7	68 55.9	23.3	230 58.2	51.1	Sabik	102 15.7	S15 44.
U	11	53 05.6	30 34.5	15.2	25 51.3	11.4	83 58.2	23.3	246 00.6	51.0			
R	12	68 08.0	45 34.8	N 7 15.9	40 52.0	N10 12.0	99 00.6	S13 23.2	261 03.1	N 7 51.0	Schedar	349 44.5	N56 35.
D	13	83 10.5	60 35.1	16.6	55 52.7	12.7	114 02.9	23.2	276 05.5	51.0	Shaula	96 25.6	S37 06.
A	14	98 13.0	75 35.4	17.3	70 53.4	13.4	129 05.3	23.2	291 07.9	51.0	Sirius	258 36.7	S16 43.
Y	15	113 15.4	90 35.6	. . 18.0	85 54.1	. . 14.1	144 07.6	. . 23.1	306 10.3	. . 50.9	Spica	158 34.2	S11 12.
	16	128 17.9	105 35.9	18.7	100 54.8	14.8	159 09.9	23.1	321 12.7	50.9	Suhail	222 54.9	S43 28.
	17	143 20.4	120 36.2	19.4	115 55.4	15.4	174 12.3	23.1	336 15.1	50.9			
	18	158 22.8	135 36.4	N 7 20.0	130 56.1	N10 16.1	189 14.6	S13 23.0	351 17.5	N 7 50.8	Vega	80 40.6	N38 47.
	19	173 25.3	150 36.7	20.7	145 56.8	16.8	204 17.0	23.0	6 19.9	50.8	Zuben'ubi	137 08.5	S16 05.
	20	188 27.8	165 37.0	21.4	160 57.5	17.5	219 19.3	23.0	21 22.4	50.8		SHA	Mer. Pass
	21	203 30.2	180 37.2	. . 22.1	175 58.2	. . 18.1	234 21.6	. . 22.9	36 24.8	. . 50.7		° ′	h m
	22	218 32.7	195 37.5	22.8	190 58.9	18.8	249 24.0	22.9	51 27.2	50.7	Venus	338 45.1	8 58
	23	233 35.2	210 37.8	23.5	205 59.6	19.5	264 26.3	22.9	66 29.6	50.7	Mars	333 47.8	9 18
	Mer. Pass.	h m 7 32.1	*v* 0.3	*d* 0.7	*v* 0.7	*d* 0.7	*v* 2.3	*d* 0.0	*v* 2.4	*d* 0.0	Jupiter	30 57.2	5 29
											Saturn	192 56.7	18 39

UT (d h)	SUN GHA (° ′)	SUN Dec (° ′)	MOON GHA (° ′)	v (′)	MOON Dec (° ′)	d (′)	HP (′)
28 00	180 41.6	N21 26.9	128 18.4	5.5	N21 37.8	10.0	60.3
01	195 41.5	27.3	142 42.9	5.6	21 27.8	10.1	60.3
02	210 41.4	27.7	157 07.5	5.7	21 17.7	10.2	60.3
03	225 41.4	. . 28.1	171 32.2	5.8	21 07.5	10.4	60.2
04	240 41.3	28.5	185 57.0	6.0	20 57.1	10.4	60.2
05	255 41.2	28.9	200 22.0	6.0	20 46.7	10.7	60.2
06	270 41.1	N21 29.3	214 47.0	6.2	N20 36.0	10.7	60.2
07	285 41.0	29.7	229 12.2	6.3	20 25.3	10.9	60.2
08	300 41.0	30.1	243 37.5	6.4	20 14.4	11.0	60.1
09	315 40.9	. . 30.5	258 02.9	6.5	20 03.4	11.1	60.1
10	330 40.8	30.9	272 28.4	6.6	19 52.3	11.2	60.1
11	345 40.7	31.3	286 54.0	6.7	19 41.1	11.3	60.1
12	0 40.6	N21 31.7	301 19.7	6.9	N19 29.8	11.5	60.1
13	15 40.6	32.1	315 45.6	6.9	19 18.3	11.6	60.0
14	30 40.5	32.5	330 11.5	7.1	19 06.7	11.7	60.0
15	45 40.4	. . 32.8	344 37.6	7.2	18 55.0	11.7	60.0
16	60 40.3	33.2	359 03.8	7.3	18 43.3	11.9	60.0
17	75 40.2	33.6	13 30.1	7.5	18 31.4	12.0	59.9
18	90 40.2	N21 34.0	27 56.6	7.5	N18 19.4	12.1	59.9
19	105 40.1	34.4	42 23.1	7.6	18 07.3	12.2	59.9
20	120 40.0	34.8	56 49.7	7.8	17 55.1	12.3	59.9
21	135 39.9	. . 35.2	71 16.5	7.9	17 42.8	12.4	59.8
22	150 39.8	35.6	85 43.4	8.0	17 30.4	12.5	59.8
23	165 39.8	36.0	100 10.4	8.1	17 17.9	12.6	59.8
29 00	180 39.7	N21 36.4	114 37.5	8.2	N17 05.3	12.7	59.8
01	195 39.6	36.8	129 04.7	8.3	16 52.6	12.8	59.8
02	210 39.5	37.1	143 32.0	8.4	16 39.8	12.8	59.7
03	225 39.4	. . 37.5	157 59.4	8.6	16 27.0	13.0	59.7
04	240 39.3	37.9	172 27.0	8.6	16 14.0	13.0	59.7
05	255 39.3	38.3	186 54.6	8.8	16 01.0	13.1	59.7
06	270 39.2	N21 38.7	201 22.4	8.8	N15 47.9	13.2	59.6
07	285 39.1	39.1	215 50.2	9.0	15 34.7	13.2	59.6
08	300 39.0	39.4	230 18.2	9.1	15 21.5	13.4	59.6
09	315 38.9	. . 39.8	244 46.3	9.2	15 08.1	13.4	59.5
10	330 38.8	40.2	259 14.5	9.2	14 54.7	13.5	59.5
11	345 38.8	40.6	273 42.7	9.4	14 41.2	13.5	59.5
12	0 38.7	N21 41.0	288 11.1	9.5	N14 27.7	13.6	59.5
13	15 38.6	41.4	302 39.6	9.6	14 14.1	13.7	59.4
14	30 38.5	41.7	317 08.2	9.7	14 00.4	13.8	59.4
15	45 38.4	. . 42.1	331 36.9	9.7	13 46.6	13.8	59.4
16	60 38.3	42.5	346 05.6	9.9	13 32.8	13.9	59.4
17	75 38.2	42.9	0 34.5	10.0	13 18.9	13.9	59.3
18	90 38.2	N21 43.2	15 03.5	10.1	N13 05.0	14.0	59.3
19	105 38.1	43.6	29 32.6	10.1	12 51.0	14.1	59.3
20	120 38.0	44.0	44 01.7	10.3	12 36.9	14.1	59.3
21	135 37.9	. . 44.4	58 31.0	10.4	12 22.8	14.2	59.2
22	150 37.8	44.7	73 00.4	10.4	12 08.6	14.2	59.2
23	165 37.7	45.1	87 29.8	10.5	11 54.4	14.3	59.2
30 00	180 37.6	N21 45.5	101 59.3	10.6	N11 40.1	14.3	59.1
01	195 37.6	45.9	116 28.9	10.8	11 25.8	14.3	59.1
02	210 37.5	46.2	130 58.7	10.8	11 11.5	14.4	59.1
03	225 37.4	. . 46.6	145 28.5	10.8	10 57.1	14.5	59.1
04	240 37.3	47.0	159 58.3	11.0	10 42.6	14.5	59.0
05	255 37.2	47.3	174 28.3	11.0	10 28.1	14.5	59.0
06	270 37.1	N21 47.7	188 58.3	11.2	N10 13.6	14.6	59.0
07	285 37.0	48.1	203 28.5	11.2	9 59.0	14.6	58.9
08	300 37.0	48.5	217 58.7	11.3	9 44.4	14.6	58.9
09	315 36.9	. . 48.8	232 29.0	11.3	9 29.8	14.7	58.9
10	330 36.8	49.2	246 59.3	11.5	9 15.1	14.7	58.9
11	345 36.7	49.5	261 29.8	11.5	9 00.4	14.8	58.8
12	0 36.6	N21 49.9	276 00.3	11.6	N 8 45.6	14.7	58.8
13	15 36.5	50.3	290 30.9	11.7	8 30.9	14.8	58.8
14	30 36.4	50.6	305 01.6	11.7	8 16.1	14.9	58.7
15	45 36.3	. . 51.0	319 32.3	11.8	8 01.2	14.8	58.7
16	60 36.2	51.4	334 03.1	11.9	7 46.4	14.9	58.7
17	75 36.2	51.7	348 34.0	11.9	7 31.5	14.9	58.7
18	90 36.1	N21 52.1	3 04.9	12.0	N 7 16.6	14.9	58.6
19	105 36.0	52.4	17 35.9	12.1	7 01.7	15.0	58.6
20	120 35.9	52.8	32 07.0	12.2	6 46.7	14.9	58.6
21	135 35.8	. . 53.2	46 38.2	12.2	6 31.8	15.0	58.5
22	150 35.7	53.5	61 09.4	12.2	6 16.8	15.0	58.5
23	165 35.6	53.9	75 40.6	12.4	N 6 01.8	15.0	58.5
	SD 15.8	*d* 0.4	SD 16.4		16.2		16.0

Lat.	Twilight Naut.	Twilight Civil	Sunrise	Moonrise 28	29	30	31
°	h m	h m	h m	h m	h m	h m	h m
N 72	▭	▭	▭	▭	06 42	09 31	11 48
N 70	▭	▭	▭	▭	07 22	09 47	11 52
68	▭	▭	▭	05 05	07 49	09 59	11 56
66	////	////	01 25	05 53	08 09	10 08	11 58
64	////	////	02 05	06 24	08 25	10 17	12 00
62	////	00 38	02 33	06 47	08 39	10 24	12 02
60	////	01 35	02 54	07 05	08 50	10 30	12 04
N 58	////	02 06	03 11	07 20	09 00	10 35	12 05
56	00 34	02 29	03 25	07 33	09 08	10 40	12 07
54	01 26	02 47	03 38	07 44	09 16	10 44	12 08
52	01 55	03 03	03 49	07 54	09 22	10 47	12 09
50	02 17	03 16	03 58	08 03	09 28	10 51	12 10
45	02 56	03 42	04 18	08 21	09 41	10 58	12 12
N 40	03 23	04 03	04 35	08 36	09 52	11 04	12 14
35	03 44	04 19	04 48	08 49	10 01	11 10	12 15
30	04 01	04 33	05 00	09 00	10 09	11 14	12 17
20	04 27	04 56	05 20	09 19	10 22	11 22	12 19
N 10	04 48	05 15	05 38	09 35	10 34	11 29	12 21
0	05 06	05 32	05 54	09 50	10 45	11 36	12 23
S 10	05 21	05 47	06 10	10 06	10 56	11 42	12 25
20	05 36	06 03	06 27	10 22	11 08	11 49	12 27
30	05 51	06 20	06 46	10 40	11 21	11 57	12 30
35	05 58	06 30	06 57	10 51	11 29	12 02	12 31
40	06 07	06 40	07 10	11 03	11 38	12 07	12 32
45	06 16	06 52	07 25	11 18	11 48	12 12	12 34
S 50	06 26	07 07	07 44	11 35	12 00	12 19	12 36
52	06 31	07 13	07 53	11 43	12 06	12 23	12 37
54	06 36	07 21	08 03	11 53	12 12	12 26	12 38
56	06 41	07 29	08 14	12 03	12 19	12 30	12 40
58	06 47	07 38	08 27	12 14	12 26	12 34	12 41
S 60	06 53	07 48	08 42	12 28	12 35	12 39	12 42

Lat.	Sunset	Twilight Civil	Twilight Naut.	Moonset 28	29	30	31
°	h m	h m	h m	h m	h m	h m	h m
N 72	▭	▭	▭	▭	02 45	01 44	01 08
N 70	▭	▭	▭	▭	02 04	01 26	01 01
68	▭	▭	▭	02 22	01 35	01 12	00 55
66	22 34	////	////	01 33	01 13	01 00	00 50
64	21 52	////	////	01 01	00 56	00 51	00 46
62	21 24	23 25	////	00 38	00 41	00 42	00 42
60	21 02	22 23	////	00 19	00 29	00 35	00 39
N 58	20 45	21 51	////	00 03	00 18	00 28	00 36
56	20 30	21 27	23 28	24 09	00 09	00 22	00 33
54	20 18	21 09	22 31	24 00	00 00	00 17	00 31
52	20 07	20 53	22 01	23 53	24 12	00 12	00 28
50	19 57	20 40	21 39	23 46	24 08	00 08	00 26
45	19 37	20 13	21 00	23 32	23 59	24 22	00 22
N 40	19 21	19 53	20 33	23 19	23 51	24 18	00 18
35	19 07	19 36	20 12	23 09	23 44	24 15	00 15
30	18 55	19 22	19 54	23 00	23 38	24 12	00 12
20	18 35	18 59	19 28	22 44	23 28	24 07	00 07
N 10	18 17	18 40	19 07	22 30	23 18	24 03	00 03
0	18 01	18 23	18 49	22 17	23 10	23 59	24 45
S 10	17 45	18 08	18 34	22 04	23 01	23 54	24 45
20	17 28	17 52	18 19	21 50	22 51	23 50	24 46
30	17 09	17 34	18 04	21 33	22 40	23 44	24 46
35	16 57	17 25	17 56	21 23	22 34	23 41	24 46
40	16 44	17 14	17 48	21 12	22 27	23 38	24 46
45	16 29	17 02	17 39	20 59	22 18	23 34	24 47
S 50	16 10	16 48	17 28	20 43	22 08	23 29	24 47
52	16 02	16 41	17 24	20 36	22 03	23 27	24 47
54	15 52	16 34	17 19	20 27	21 58	23 24	24 47
56	15 41	16 26	17 14	20 18	21 52	23 21	24 47
58	15 28	16 17	17 08	20 07	21 45	23 18	24 47
S 60	15 13	16 07	17 01	19 54	21 38	23 15	24 48

Day	SUN Eqn. of Time 00^h	SUN Eqn. of Time 12^h	SUN Mer. Pass.	MOON Mer. Pass. Upper	MOON Mer. Pass. Lower	MOON Age	MOON Phase
d	m s	m s	h m	h m	h m	d %	
28	02 47	02 43	11 57	16 04	03 35	04 22	
29	02 39	02 35	11 57	16 58	04 31	05 32	
30	02 31	02 27	11 58	17 47	05 23	06 43	

	UT	ARIES	VENUS −4.5		MARS +1.2		JUPITER −2.5		SATURN +0.9	
		GHA	GHA	Dec	GHA	Dec	GHA	Dec	GHA	Dec
	d h	° ′	° ′	° ′	° ′	° ′	° ′	° ′	° ′	° ′
	31 00	248 37.6	225 38.0	N 7 24.2	221 00.3	N10 20.2	279 28.7	S13 22.8	81 32.0	N 7 50.7
	01	263 40.1	240 38.3	24.9	236 01.0	20.9	294 31.0	22.8	96 34.4	50.6
	02	278 42.5	255 38.6	25.6	251 01.6	21.5	309 33.4	22.8	111 36.8	50.6
	03	293 45.0	270 38.8	. . 26.3	266 02.3	. . 22.2	324 35.7	. . 22.8	126 39.2	. . 50.6
	04	308 47.5	285 39.1	27.0	281 03.0	22.9	339 38.1	22.7	141 41.6	50.5
	05	323 49.9	300 39.3	27.6	296 03.7	23.6	354 40.4	22.7	156 44.1	50.5
	06	338 52.4	315 39.6	N 7 28.3	311 04.4	N10 24.2	9 42.7	S13 22.7	171 46.5	N 7 50.5
	07	353 54.9	330 39.9	29.0	326 05.1	24.9	24 45.1	22.6	186 48.9	50.4
	08	8 57.3	345 40.1	29.7	341 05.8	25.6	39 47.4	22.6	201 51.3	50.4
S	09	23 59.8	0 40.4	. . 30.4	356 06.5	. . 26.3	54 49.8	. . 22.6	216 53.7	. . 50.4
U	10	39 02.3	15 40.6	31.1	11 07.2	26.9	69 52.1	22.5	231 56.1	50.4
N	11	54 04.7	30 40.9	31.8	26 07.9	27.6	84 54.5	22.5	246 58.5	50.3
D	12	69 07.2	45 41.1	N 7 32.5	41 08.5	N10 28.3	99 56.8	S13 22.5	262 00.9	N 7 50.3
A	13	84 09.6	60 41.4	33.2	56 09.2	29.0	114 59.2	22.5	277 03.3	50.3
Y	14	99 12.1	75 41.6	33.9	71 09.9	29.6	130 01.5	22.4	292 05.7	50.2
	15	114 14.6	90 41.9	. . 34.6	86 10.6	. . 30.3	145 03.9	. . 22.4	307 08.2	. . 50.2
	16	129 17.0	105 42.1	35.3	101 11.3	31.0	160 06.2	22.4	322 10.6	50.2
	17	144 19.5	120 42.4	36.0	116 12.0	31.7	175 08.6	22.3	337 13.0	50.1
	18	159 22.0	135 42.6	N 7 36.7	131 12.7	N10 32.3	190 10.9	S13 22.3	352 15.4	N 7 50.1
	19	174 24.4	150 42.9	37.4	146 13.4	33.0	205 13.3	22.3	7 17.8	50.1
	20	189 26.9	165 43.1	38.1	161 14.1	33.7	220 15.6	22.3	22 20.2	50.0
	21	204 29.4	180 43.4	. . 38.8	176 14.7	. . 34.4	235 18.0	. . 22.2	37 22.6	. . 50.0
	22	219 31.8	195 43.6	39.5	191 15.4	35.0	250 20.3	22.2	52 25.0	50.0
	23	234 34.3	210 43.9	40.2	206 16.1	35.7	265 22.7	22.2	67 27.4	49.9
	1 00	249 36.8	225 44.1	N 7 40.9	221 16.8	N10 36.4	280 25.0	S13 22.1	82 29.8	N 7 49.9
	01	264 39.2	240 44.4	41.6	236 17.5	37.0	295 27.4	22.1	97 32.2	49.9
	02	279 41.7	255 44.6	42.3	251 18.2	37.7	310 29.7	22.1	112 34.6	49.9
	03	294 44.1	270 44.8	. . 43.0	266 18.9	. . 38.4	325 32.1	. . 22.1	127 37.0	. . 49.8
	04	309 46.6	285 45.1	43.7	281 19.6	39.1	340 34.4	22.0	142 39.5	49.8
	05	324 49.1	300 45.3	44.4	296 20.3	39.7	355 36.8	22.0	157 41.9	49.8
	06	339 51.5	315 45.6	N 7 45.1	311 20.9	N10 40.4	10 39.1	S13 22.0	172 44.3	N 7 49.7
	07	354 54.0	330 45.8	45.8	326 21.6	41.1	25 41.5	21.9	187 46.7	49.7
	08	9 56.5	345 46.0	46.5	341 22.3	41.7	40 43.8	21.9	202 49.1	49.7
M	09	24 58.9	0 46.3	. . 47.2	356 23.0	. . 42.4	55 46.2	. . 21.9	217 51.5	. . 49.6
O	10	40 01.4	15 46.5	47.9	11 23.7	43.1	70 48.6	21.9	232 53.9	49.6
N	11	55 03.9	30 46.8	48.6	26 24.4	43.8	85 50.9	21.8	247 56.3	49.6
D	12	70 06.3	45 47.0	N 7 49.3	41 25.1	N10 44.4	100 53.3	S13 21.8	262 58.7	N 7 49.5
A	13	85 08.8	60 47.2	50.0	56 25.8	45.1	115 55.6	21.8	278 01.1	49.5
Y	14	100 11.3	75 47.5	50.7	71 26.4	45.8	130 58.0	21.7	293 03.5	49.5
	15	115 13.7	90 47.7	. . 51.4	86 27.1	. . 46.4	146 00.3	. . 21.7	308 05.9	. . 49.4
	16	130 16.2	105 47.9	52.1	101 27.8	47.1	161 02.7	21.7	323 08.3	49.4
	17	145 18.6	120 48.1	52.8	116 28.5	47.8	176 05.0	21.7	338 10.7	49.4
	18	160 21.1	135 48.4	N 7 53.5	131 29.2	N10 48.4	191 07.4	S13 21.6	353 13.1	N 7 49.3
	19	175 23.6	150 48.6	54.3	146 29.9	49.1	206 09.8	21.6	8 15.5	49.3
	20	190 26.0	165 48.8	55.0	161 30.6	49.8	221 12.1	21.6	23 17.9	49.3
	21	205 28.5	180 49.1	. . 55.7	176 31.3	. . 50.5	236 14.5	. . 21.6	38 20.3	. . 49.2
	22	220 31.0	195 49.3	56.4	191 32.0	51.1	251 16.8	21.5	53 22.7	49.2
	23	235 33.4	210 49.5	57.1	206 32.6	51.8	266 19.2	21.5	68 25.1	49.2
	2 00	250 35.9	225 49.7	N 7 57.8	221 33.3	N10 52.5	281 21.5	S13 21.5	83 27.6	N 7 49.1
	01	265 38.4	240 50.0	58.5	236 34.0	53.1	296 23.9	21.5	98 30.0	49.1
	02	280 40.8	255 50.2	59.2	251 34.7	53.8	311 26.3	21.4	113 32.4	49.1
	03	295 43.3	270 50.4	7 59.9	266 35.4	. . 54.5	326 28.6	. . 21.4	128 34.8	. . 49.0
	04	310 45.7	285 50.6	8 00.6	281 36.1	55.1	341 31.0	21.4	143 37.2	49.0
	05	325 48.2	300 50.9	01.3	296 36.8	55.8	356 33.3	21.4	158 39.6	49.0
	06	340 50.7	315 51.1	N 8 02.1	311 37.5	N10 56.5	11 35.7	S13 21.3	173 42.0	N 7 48.9
	07	355 53.1	330 51.3	02.8	326 38.1	57.1	26 38.1	21.3	188 44.4	48.9
T	08	10 55.6	345 51.5	03.5	341 38.8	57.8	41 40.4	21.3	203 46.8	48.9
U	09	25 58.1	0 51.7	. . 04.2	356 39.5	. . 58.5	56 42.8	. . 21.3	218 49.2	. . 48.8
E	10	41 00.5	15 51.9	04.9	11 40.2	59.1	71 45.2	21.2	233 51.6	48.8
S	11	56 03.0	30 52.2	05.6	26 40.9	10 59.8	86 47.5	21.2	248 54.0	48.8
D	12	71 05.5	45 52.4	N 8 06.3	41 41.6	N11 00.4	101 49.9	S13 21.2	263 56.4	N 7 48.7
A	13	86 07.9	60 52.6	07.0	56 42.3	01.1	116 52.2	21.2	278 58.8	48.7
Y	14	101 10.4	75 52.8	07.7	71 42.9	01.8	131 54.6	21.1	294 01.2	48.7
	15	116 12.9	90 53.0	. . 08.5	86 43.6	. . 02.4	146 57.0	. . 21.1	309 03.6	. . 48.6
	16	131 15.3	105 53.2	09.2	101 44.3	03.1	161 59.3	21.1	324 06.0	48.6
	17	146 17.8	120 53.4	09.9	116 45.0	03.8	177 01.7	21.1	339 08.4	48.6
	18	161 20.2	135 53.7	N 8 10.6	131 45.7	N11 04.4	192 04.1	S13 21.0	354 10.8	N 7 48.5
	19	176 22.7	150 53.9	11.3	146 46.4	05.1	207 06.4	21.0	9 13.2	48.5
	20	191 25.2	165 54.1	12.0	161 47.1	05.8	222 08.8	21.0	24 15.6	48.5
	21	206 27.6	180 54.3	. . 12.7	176 47.8	. . 06.4	237 11.2	. . 21.0	39 18.0	. . 48.4
	22	221 30.1	195 54.5	13.5	191 48.4	07.1	252 13.5	20.9	54 20.4	48.4
	23	236 32.6	210 54.7	14.2	206 49.1	07.8	267 15.9	20.9	69 22.8	48.4
		h m								
	Mer. Pass.	7 20.3	v 0.2	d 0.7	v 0.7	d 0.7	v 2.4	d 0.0	v 2.4	d 0.0

STARS		
Name	SHA	Dec
	° ′	° ′
Acamar	315 20.9	S40 15.
Achernar	335 29.2	S57 11.
Acrux	173 12.6	S63 09.9
Adhara	255 15.2	S28 59.2
Aldebaran	290 53.2	N16 31.
Alioth	166 22.8	N55 54.6
Alkaid	153 00.7	N49 16.
Al Na'ir	27 47.3	S46 54.6
Alnilam	275 49.7	S 1 11.8
Alphard	217 59.2	S 8 42.1
Alphecca	126 13.1	N26 40.9
Alpheratz	357 46.8	N29 08.4
Altair	62 10.9	N 8 53.5
Ankaa	353 18.7	S42 15.6
Antares	112 29.6	S26 27.1
Arcturus	145 58.1	N19 07.9
Atria	107 33.7	S69 02.1
Avior	234 19.8	S59 32.6
Bellatrix	278 35.6	N 6 21.5
Betelgeuse	271 04.9	N 7 24.5
Canopus	263 58.0	S52 42.1
Capella	280 39.4	N46 00.5
Deneb	49 33.4	N45 18.6
Denebola	182 36.6	N14 31.1
Diphda	348 59.0	S17 56.0
Dubhe	193 55.0	N61 42.2
Elnath	278 16.8	N28 37.0
Eltanin	90 47.0	N51 29.1
Enif	33 50.0	N 9 55.0
Fomalhaut	15 27.2	S29 34.1
Gacrux	172 04.2	S57 10.3
Gienah	175 55.3	S17 35.9
Hadar	148 51.9	S60 25.4
Hamal	328 04.5	N23 30.4
Kaus Aust.	83 47.4	S34 22.8
Kochab	137 18.4	N74 07.0
Markab	13 41.4	N15 15.3
Menkar	314 18.5	N 4 07.6
Menkent	148 10.9	S36 25.2
Miaplacidus	221 41.0	S69 45.7
Mirfak	308 45.3	N49 53.6
Nunki	76 01.7	S26 17.1
Peacock	53 23.5	S56 42.1
Pollux	243 31.6	N28 00.3
Procyon	245 03.1	N 5 12.0
Rasalhague	96 08.9	N12 33.1
Regulus	207 46.7	N11 55.2
Rigel	281 15.3	S 8 11.4
Rigil Kent.	139 55.5	S60 52.7
Sabik	102 15.6	S15 44.3
Schedar	349 44.5	N56 35.2
Shaula	96 25.6	S37 06.7
Sirius	258 36.7	S16 43.8
Spica	158 34.2	S11 12.8
Suhail	222 55.0	S43 28.5
Vega	80 40.6	N38 47.4
Zuben'ubi	137 08.5	S16 05.0

	SHA	Mer. Pass.
	° ′	h m
Venus	336 07.4	8 57
Mars	331 40.1	9 14
Jupiter	30 48.3	5 18
Saturn	192 53.1	18 27

JT	SUN GHA	SUN Dec	MOON GHA	v	MOON Dec	d	HP
d h	° ′	° ′	° ′	′	° ′	′	′
1 00	180 35.5	N21 54.2	90 12.0	12.3	N 5 46.8	15.0	58.5
01	195 35.4	54.6	104 43.3	12.5	5 31.8	15.0	58.4
02	210 35.3	55.0	119 14.8	12.5	5 16.8	15.1	58.4
03	225 35.2	. . 55.3	133 46.3	12.5	5 01.7	15.0	58.4
04	240 35.2	55.7	148 17.8	12.6	4 46.7	15.1	58.3
05	255 35.1	56.0	162 49.4	12.7	4 31.6	15.1	58.3
06	270 35.0	N21 56.4	177 21.1	12.7	N 4 16.5	15.0	58.3
07	285 34.9	56.7	191 52.8	12.8	4 01.5	15.1	58.2
08	300 34.8	57.1	206 24.6	12.8	3 46.4	15.1	58.2
09	315 34.7	. . 57.4	220 56.4	12.8	3 31.3	15.1	58.2
10	330 34.6	57.8	235 28.2	12.9	3 16.2	15.1	58.2
11	345 34.5	58.1	250 00.1	13.0	3 01.1	15.0	58.1
12	0 34.4	N21 58.5	264 32.1	13.0	N 2 46.1	15.1	58.1
13	15 34.3	58.8	279 04.1	13.0	2 31.0	15.1	58.1
14	30 34.2	59.2	293 36.1	13.0	2 15.9	15.1	58.0
15	45 34.1	. . 59.5	308 08.1	13.2	2 00.8	15.0	58.0
16	60 34.1	21 59.9	322 40.3	13.1	1 45.8	15.1	58.0
17	75 34.0	22 00.2	337 12.4	13.2	1 30.7	15.0	58.0
18	90 33.9	N22 00.6	351 44.6	13.2	N 1 15.7	15.1	57.9
19	105 33.8	00.9	6 16.8	13.3	1 00.6	15.0	57.9
20	120 33.7	01.2	20 49.1	13.2	0 45.6	15.0	57.9
21	135 33.6	. . 01.6	35 21.3	13.4	0 30.6	15.1	57.8
22	150 33.5	01.9	49 53.7	13.3	0 15.5	15.0	57.8
23	165 33.4	02.3	64 26.0	13.4	N 0 00.5	14.9	57.8
1 00	180 33.3	N22 02.6	78 58.4	13.4	S 0 14.4	15.0	57.8
01	195 33.2	02.9	93 30.8	13.4	0 29.4	15.0	57.7
02	210 33.1	03.3	108 03.2	13.5	0 44.4	14.9	57.7
03	225 33.0	. . 03.6	122 35.7	13.4	0 59.3	14.9	57.7
04	240 32.9	04.0	137 08.1	13.5	1 14.2	14.9	57.6
05	255 32.8	04.3	151 40.6	13.6	1 29.1	14.9	57.6
06	270 32.7	N22 04.6	166 13.2	13.5	S 1 44.0	14.8	57.6
07	285 32.6	05.0	180 45.7	13.6	1 58.8	14.8	57.6
08	300 32.5	05.3	195 18.3	13.5	2 13.6	14.9	57.5
09	315 32.4	. . 05.6	209 50.8	13.6	2 28.5	14.7	57.5
10	330 32.3	06.0	224 23.4	13.7	2 43.2	14.8	57.5
11	345 32.2	06.3	238 56.1	13.6	2 58.0	14.7	57.4
12	0 32.2	N22 06.6	253 28.7	13.6	S 3 12.7	14.7	57.4
13	15 32.1	07.0	268 01.3	13.7	3 27.4	14.7	57.4
14	30 32.0	07.3	282 34.0	13.6	3 42.1	14.6	57.4
15	45 31.9	. . 07.6	297 06.6	13.7	3 56.7	14.6	57.3
16	60 31.8	08.0	311 39.3	13.7	4 11.3	14.6	57.3
17	75 31.7	08.3	326 12.0	13.7	4 25.9	14.5	57.3
18	90 31.6	N22 08.6	340 44.7	13.7	S 4 40.4	14.5	57.3
19	105 31.5	09.0	355 17.4	13.7	4 54.9	14.5	57.2
20	120 31.4	09.3	9 50.1	13.7	5 09.4	14.5	57.2
21	135 31.3	. . 09.6	24 22.8	13.7	5 23.9	14.4	57.2
22	150 31.2	09.9	38 55.5	13.7	5 38.3	14.3	57.1
23	165 31.1	10.3	53 28.2	13.7	5 52.6	14.3	57.1
2 00	180 31.0	N22 10.6	68 00.9	13.7	S 6 06.9	14.3	57.1
01	195 30.9	10.9	82 33.6	13.7	6 21.2	14.3	57.1
02	210 30.8	11.2	97 06.3	13.7	6 35.5	14.2	57.0
03	225 30.7	. . 11.6	111 39.0	13.7	6 49.7	14.1	57.0
04	240 30.6	11.9	126 11.7	13.7	7 03.8	14.1	57.0
05	255 30.5	12.2	140 44.4	13.7	7 17.9	14.1	57.0
06	270 30.4	N22 12.5	155 17.1	13.7	S 7 32.0	14.0	56.9
07	285 30.3	12.8	169 49.8	13.6	7 46.0	14.0	56.9
08	300 30.2	13.2	184 22.4	13.7	8 00.0	13.9	56.9
09	315 30.1	. . 13.5	198 55.1	13.7	8 13.9	13.9	56.9
10	330 30.0	13.8	213 27.8	13.6	8 27.8	13.8	56.8
11	345 29.9	14.1	228 00.4	13.7	8 41.6	13.8	56.8
12	0 29.8	N22 14.4	242 33.1	13.6	S 8 55.4	13.7	56.8
13	15 29.7	14.8	257 05.7	13.6	9 09.1	13.7	56.8
14	30 29.6	15.1	271 38.3	13.6	9 22.8	13.6	56.7
15	45 29.5	. . 15.4	286 10.9	13.6	9 36.4	13.6	56.7
16	60 29.4	15.7	300 43.5	13.5	9 50.0	13.5	56.7
17	75 29.3	16.0	315 16.0	13.6	10 03.5	13.5	56.7
18	90 29.2	N22 16.3	329 48.6	13.5	S10 17.0	13.4	56.6
19	105 29.1	16.6	344 21.1	13.5	10 30.4	13.3	56.6
20	120 29.0	16.9	358 53.6	13.5	10 43.7	13.3	56.6
21	135 28.9	. . 17.3	13 26.1	13.4	10 57.0	13.2	56.6
22	150 28.8	17.6	27 58.5	13.5	11 10.2	13.2	56.5
23	165 28.7	17.9	42 31.0	13.4	S11 23.4	13.1	56.5
	SD 15.8	*d* 0.3	SD 15.8		15.6		15.5

Lat.	Twilight Naut.	Twilight Civil	Sunrise	Moonrise 31	Moonrise 1	Moonrise 2	Moonrise 3
°	h m	h m	h m	h m	h m	h m	h m
N 72	▭	▭	▭	11 48	13 57	16 09	18 44
N 70	▭	▭	▭	11 52	13 52	15 52	18 03
68	▭	▭	▭	11 56	13 47	15 39	17 36
66	////	////	01 12	11 58	13 43	15 28	17 16
64	////	////	01 58	12 00	13 40	15 19	16 59
62	////	////	02 27	12 02	13 37	15 11	16 46
60	////	01 26	02 49	12 04	13 35	15 05	16 35
N 58	////	02 00	03 07	12 05	13 33	14 59	16 25
56	////	02 24	03 22	12 07	13 31	14 54	16 17
54	01 18	02 44	03 35	12 08	13 29	14 49	16 09
52	01 50	02 59	03 46	12 09	13 28	14 45	16 02
50	02 13	03 13	03 56	12 10	13 27	14 42	15 56
45	02 53	03 40	04 17	12 12	13 24	14 34	15 43
N 40	03 21	04 01	04 33	12 14	13 21	14 27	15 33
35	03 42	04 18	04 47	12 15	13 19	14 22	15 24
30	04 00	04 32	04 59	12 17	13 17	14 17	15 16
20	04 27	04 56	05 20	12 19	13 14	14 08	15 02
N 10	04 48	05 15	05 38	12 21	13 11	14 01	14 50
0	05 06	05 32	05 54	12 23	13 09	13 54	14 39
S 10	05 22	05 48	06 11	12 25	13 06	13 47	14 29
20	05 37	06 04	06 28	12 27	13 04	13 40	14 17
30	05 52	06 22	06 48	12 30	13 01	13 32	14 04
35	06 00	06 31	06 59	12 31	12 59	13 27	13 56
40	06 09	06 42	07 12	12 32	12 57	13 22	13 48
45	06 18	06 55	07 28	12 34	12 55	13 15	13 38
S 50	06 29	07 10	07 47	12 36	12 52	13 08	13 26
52	06 33	07 16	07 56	12 37	12 51	13 05	13 20
54	06 39	07 24	08 06	12 38	12 50	13 01	13 14
56	06 44	07 32	08 18	12 40	12 48	12 57	13 08
58	06 50	07 41	08 31	12 41	12 47	12 53	13 00
S 60	06 57	07 52	08 47	12 42	12 45	12 48	12 52

Lat.	Sunset	Twilight Civil	Twilight Naut.	Moonset 31	Moonset 1	Moonset 2	Moonset 3
°	h m	h m	h m	h m	h m	h m	h m
N 72	▭	▭	▭	01 08	00 39	{00 09 / 23 34}	22 36
N 70	▭	▭	▭	01 01	00 39	{00 18 / 23 53}	23 18
68	▭	▭	▭	00 55	00 40	00 25	{00 08 / 23 46}
66	22 48	////	////	00 50	00 40	00 31	00 20
64	22 00	////	////	00 46	00 41	00 36	00 31
62	21 30	////	////	00 42	00 41	00 40	00 40
60	21 08	22 33	////	00 39	00 41	00 44	00 47
N 58	20 50	21 57	////	00 36	00 42	00 48	00 54
56	20 35	21 33	////	00 33	00 42	00 51	01 00
54	20 22	21 13	22 40	00 31	00 42	00 54	01 06
52	20 10	20 57	22 07	00 28	00 42	00 56	01 11
50	20 00	20 43	21 44	00 26	00 43	00 58	01 15
45	19 40	20 16	21 04	00 22	00 43	01 04	01 25
N 40	19 23	19 55	20 36	00 18	00 43	01 08	01 33
35	19 09	19 38	20 14	00 15	00 44	01 11	01 40
30	18 57	19 24	19 56	00 12	00 44	01 15	01 46
20	18 36	19 00	19 29	00 07	00 44	01 20	01 57
N 10	18 18	18 41	19 08	00 03	00 45	01 25	02 07
0	18 01	18 24	18 50	24 45	00 45	01 30	02 15
S 10	17 45	18 08	18 34	24 45	00 45	01 35	02 24
20	17 28	17 51	18 19	24 46	00 46	01 40	02 34
30	17 08	17 34	18 03	24 46	00 46	01 46	02 45
35	16 56	17 24	17 55	24 46	00 46	01 49	02 51
40	16 43	17 13	17 47	24 46	00 46	01 53	02 59
45	16 27	17 01	17 37	24 47	00 47	01 57	03 07
S 50	16 08	16 46	17 27	24 47	00 47	02 03	03 17
52	15 59	16 39	17 22	24 47	00 47	02 05	03 22
54	15 49	16 32	17 17	24 47	00 47	02 08	03 27
56	15 38	16 23	17 11	24 47	00 47	02 11	03 33
58	15 24	16 14	17 05	24 47	00 47	02 14	03 40
S 60	15 09	16 03	16 59	24 48	00 48	02 18	03 47

Day	SUN Eqn. of Time 00^h	SUN Eqn. of Time 12^h	SUN Mer. Pass.	MOON Mer. Pass. Upper	MOON Mer. Pass. Lower	MOON Age	MOON Phase
d	m s	m s	h m	h m	h m	d %	
31	02 22	02 18	11 58	18 34	06 11	07 54	
1	02 13	02 09	11 58	19 19	06 57	08 65	
2	02 04	01 59	11 58	20 05	07 42	09 74	

	UT	ARIES	VENUS −4.5		MARS +1.2		JUPITER −2.5		SATURN +0.9	
		GHA	GHA	Dec	GHA	Dec	GHA	Dec	GHA	Dec
	d h	° ′	° ′	° ′	° ′	° ′	° ′	° ′	° ′	° ′
	3 00	251 35.0	225 54.9	N 8 14.9	221 49.8	N11 08.4	282 18.3	S13 20.9	84 25.2	N 7 48.3
	01	266 37.5	240 55.1	15.6	236 50.5	09.1	297 20.6	20.9	99 27.6	48.3
	02	281 40.0	255 55.3	16.3	251 51.2	09.7	312 23.0	20.9	114 30.0	48.2
	03	296 42.4	270 55.5	. . 17.0	266 51.9	. . 10.4	327 25.4	. . 20.8	129 32.4	. . 48.2
	04	311 44.9	285 55.7	17.8	281 52.6	11.1	342 27.7	20.8	144 34.8	48.2
	05	326 47.3	300 55.9	18.5	296 53.2	11.7	357 30.1	20.8	159 37.2	48.1
	06	341 49.8	315 56.1	N 8 19.2	311 53.9	N11 12.4	12 32.5	S13 20.8	174 39.6	N 7 48.1
W	07	356 52.3	330 56.3	19.9	326 54.6	13.0	27 34.8	20.7	189 42.0	48.1
E	08	11 54.7	345 56.5	20.6	341 55.3	13.7	42 37.2	20.7	204 44.4	48.0
D	09	26 57.2	0 56.7	. . 21.4	356 56.0	. . 14.4	57 39.6	. . 20.7	219 46.8	. . 48.0
N	10	41 59.7	15 56.9	22.1	11 56.7	15.0	72 41.9	20.7	234 49.2	48.0
E	11	57 02.1	30 57.1	22.8	26 57.4	15.7	87 44.3	20.7	249 51.6	47.9
S	12	72 04.6	45 57.3	N 8 23.5	41 58.1	N11 16.3	102 46.7	S13 20.6	264 54.0	N 7 47.9
D	13	87 07.1	60 57.5	24.2	56 58.7	17.0	117 49.0	20.6	279 56.4	47.9
A	14	102 09.5	75 57.7	24.9	71 59.4	17.7	132 51.4	20.6	294 58.8	47.8
Y	15	117 12.0	90 57.9	. . 25.7	87 00.1	. . 18.3	147 53.8	. . 20.6	310 01.2	. . 47.8
	16	132 14.5	105 58.1	26.4	102 00.8	19.0	162 56.2	20.5	325 03.6	47.8
	17	147 16.9	120 58.3	27.1	117 01.5	19.6	177 58.5	20.5	340 06.0	47.7
	18	162 19.4	135 58.5	N 8 27.8	132 02.2	N11 20.3	193 00.9	S13 20.5	355 08.4	N 7 47.7
	19	177 21.8	150 58.7	28.6	147 02.9	21.0	208 03.3	20.5	10 10.7	47.6
	20	192 24.3	165 58.9	29.3	162 03.5	21.6	223 05.6	20.5	25 13.1	47.6
	21	207 26.8	180 59.1	. . 30.0	177 04.2	. . 22.3	238 08.0	. . 20.4	40 15.5	. . 47.6
	22	222 29.2	195 59.2	30.7	192 04.9	22.9	253 10.4	20.4	55 17.9	47.5
	23	237 31.7	210 59.4	31.4	207 05.6	23.6	268 12.8	20.4	70 20.3	47.5
	4 00	252 34.2	225 59.6	N 8 32.2	222 06.3	N11 24.3	283 15.1	S13 20.4	85 22.7	N 7 47.5
	01	267 36.6	240 59.8	32.9	237 07.0	24.9	298 17.5	20.4	100 25.1	47.4
	02	282 39.1	256 00.0	33.6	252 07.7	25.6	313 19.9	20.3	115 27.5	47.4
	03	297 41.6	271 00.2	. . 34.3	267 08.3	. . 26.2	328 22.3	. . 20.3	130 29.9	. . 47.4
	04	312 44.0	286 00.4	35.1	282 09.0	26.9	343 24.6	20.3	145 32.3	47.3
	05	327 46.5	301 00.6	35.8	297 09.7	27.5	358 27.0	20.3	160 34.7	47.3
	06	342 49.0	316 00.7	N 8 36.5	312 10.4	N11 28.2	13 29.4	S13 20.3	175 37.1	N 7 47.2
	07	357 51.4	331 00.9	37.2	327 11.1	28.8	28 31.8	20.2	190 39.5	47.2
T	08	12 53.9	346 01.1	38.0	342 11.8	29.5	43 34.1	20.2	205 41.9	47.2
H	09	27 56.3	1 01.3	. . 38.7	357 12.4	. . 30.2	58 36.5	. . 20.2	220 44.3	. . 47.1
U	10	42 58.8	16 01.5	39.4	12 13.1	30.8	73 38.9	20.2	235 46.7	47.1
R	11	58 01.3	31 01.6	40.1	27 13.8	31.5	88 41.3	20.2	250 49.1	47.1
S	12	73 03.7	46 01.8	N 8 40.9	42 14.5	N11 32.1	103 43.6	S13 20.1	265 51.5	N 7 47.0
D	13	88 06.2	61 02.0	41.6	57 15.2	32.8	118 46.0	20.1	280 53.9	47.0
A	14	103 08.7	76 02.2	42.3	72 15.9	33.4	133 48.4	20.1	295 56.3	46.9
Y	15	118 11.1	91 02.3	. . 43.0	87 16.6	. . 34.1	148 50.8	. . 20.1	310 58.6	. . 46.9
	16	133 13.6	106 02.5	43.8	102 17.2	34.7	163 53.2	20.1	326 01.0	46.9
	17	148 16.1	121 02.7	44.5	117 17.9	35.4	178 55.5	20.0	341 03.4	46.8
	18	163 18.5	136 02.9	N 8 45.2	132 18.6	N11 36.0	193 57.9	S13 20.0	356 05.8	N 7 46.8
	19	178 21.0	151 03.0	46.0	147 19.3	36.7	209 00.3	20.0	11 08.2	46.8
	20	193 23.5	166 03.2	46.7	162 20.0	37.4	224 02.7	20.0	26 10.6	46.7
	21	208 25.9	181 03.4	. . 47.4	177 20.7	. . 38.0	239 05.0	. . 20.0	41 13.0	. . 46.7
	22	223 28.4	196 03.6	48.1	192 21.4	38.7	254 07.4	20.0	56 15.4	46.6
	23	238 30.8	211 03.7	48.9	207 22.0	39.3	269 09.8	19.9	71 17.8	46.6
	5 00	253 33.3	226 03.9	N 8 49.6	222 22.7	N11 40.0	284 12.2	S13 19.9	86 20.2	N 7 46.6
	01	268 35.8	241 04.1	50.3	237 23.4	40.6	299 14.6	19.9	101 22.6	46.5
	02	283 38.2	256 04.2	51.1	252 24.1	41.3	314 17.0	19.9	116 25.0	46.5
	03	298 40.7	271 04.4	. . 51.8	267 24.8	. . 41.9	329 19.3	. . 19.9	131 27.4	. . 46.5
	04	313 43.2	286 04.6	52.5	282 25.5	42.6	344 21.7	19.8	146 29.7	46.4
	05	328 45.6	301 04.7	53.2	297 26.1	43.2	359 24.1	19.8	161 32.1	46.4
	06	343 48.1	316 04.9	N 8 54.0	312 26.8	N11 43.9	14 26.5	S13 19.8	176 34.5	N 7 46.3
	07	358 50.6	331 05.1	54.7	327 27.5	44.5	29 28.9	19.8	191 36.9	46.3
	08	13 53.0	346 05.2	55.4	342 28.2	45.2	44 31.3	19.8	206 39.3	46.3
F	09	28 55.5	1 05.4	. . 56.2	357 28.9	. . 45.8	59 33.6	. . 19.8	221 41.7	. . 46.2
R	10	43 57.9	16 05.6	56.9	12 29.6	46.5	74 36.0	19.7	236 44.1	46.2
I	11	59 00.4	31 05.7	57.6	27 30.2	47.1	89 38.4	19.7	251 46.5	46.1
D	12	74 02.9	46 05.9	N 8 58.4	42 30.9	N11 47.8	104 40.8	S13 19.7	266 48.9	N 7 46.1
A	13	89 05.3	61 06.0	59.1	57 31.6	48.4	119 43.2	19.7	281 51.3	46.1
Y	14	104 07.8	76 06.2	8 59.8	72 32.3	49.1	134 45.6	19.7	296 53.6	46.0
	15	119 10.3	91 06.4	9 00.6	87 33.0	. . 49.7	149 47.9	. . 19.7	311 56.0	. . 46.0
	16	134 12.7	106 06.5	01.3	102 33.7	50.4	164 50.3	19.6	326 58.4	45.9
	17	149 15.2	121 06.7	02.0	117 34.3	51.0	179 52.7	19.6	342 00.8	45.9
	18	164 17.7	136 06.8	N 9 02.8	132 35.0	N11 51.7	194 55.1	S13 19.6	357 03.2	N 7 45.9
	19	179 20.1	151 07.0	03.5	147 35.7	52.3	209 57.5	19.6	12 05.6	45.8
	20	194 22.6	166 07.1	04.2	162 36.4	53.0	224 59.9	19.6	27 08.0	45.8
	21	209 25.1	181 07.3	. . 05.0	177 37.1	. . 53.6	240 02.3	. . 19.6	42 10.4	. . 45.8
	22	224 27.5	196 07.4	05.7	192 37.8	54.3	255 04.7	19.5	57 12.8	45.7
	23	239 30.0	211 07.6	06.4	207 38.4	54.9	270 07.0	19.5	72 15.1	45.7
	Mer. Pass.	h m 7 08.5	v 0.2	d 0.7	v 0.7	d 0.7	v 2.4	d 0.0	v 2.4	d 0.0

STARS		
Name	SHA	Dec
	° ′	° ′
Acamar	315 20.9	S40 15.
Achernar	335 29.2	S57 11.
Acrux	173 12.7	S63 09.
Adhara	255 15.2	S28 59.
Aldebaran	290 53.2	N16 31.
Alioth	166 22.8	N55 54.
Alkaid	153 00.8	N49 16.
Al Na'ir	27 47.2	S46 54.
Alnilam	275 49.7	S 1 11.
Alphard	217 59.2	S 8 42.
Alphecca	126 13.2	N26 40.
Alpheratz	357 46.8	N29 08.
Altair	62 10.9	N 8 53.
Ankaa	353 18.7	S42 15.
Antares	112 29.6	S26 27.
Arcturus	145 58.2	N19 07.
Atria	107 33.6	S69 02.
Avior	234 19.8	S59 32.
Bellatrix	278 35.6	N 6 21.
Betelgeuse	271 04.9	N 7 24.
Canopus	263 58.0	S52 42.
Capella	280 39.4	N46 00.
Deneb	49 33.3	N45 18.
Denebola	182 36.6	N14 31.
Diphda	348 59.0	S17 55.
Dubhe	193 55.1	N61 42.
Elnath	278 16.8	N28 37.
Eltanin	90 47.0	N51 29.
Enif	33 50.0	N 9 55.
Fomalhaut	15 27.2	S29 34.
Gacrux	172 04.2	S57 10.
Gienah	175 55.3	S17 35.
Hadar	148 51.9	S60 25.
Hamal	328 04.5	N23 30.
Kaus Aust.	83 47.4	S34 22.
Kochab	137 18.5	N74 07.
Markab	13 41.4	N15 15.
Menkar	314 18.5	N 4 07.
Menkent	148 10.9	S36 25.
Miaplacidus	221 41.0	S69 45.
Mirfak	308 45.2	N49 53.
Nunki	76 01.7	S26 17.
Peacock	53 23.4	S56 42.
Pollux	243 31.6	N28 00.
Procyon	245 03.1	N 5 12.
Rasalhague	96 08.9	N12 33.
Regulus	207 46.8	N11 55.
Rigel	281 15.3	S 8 11.
Rigil Kent.	139 55.5	S60 52.
Sabik	102 15.6	S15 44.
Schedar	349 44.4	N56 35.
Shaula	96 25.6	S37 06.
Sirius	258 36.7	S16 43.
Spica	158 34.2	S11 12.
Suhail	222 55.0	S43 28.
Vega	80 40.6	N38 47.
Zuben'ubi	137 08.5	S16 05.

	SHA	Mer. Pass.
	° ′	h m
Venus	333 25.5	8 56
Mars	329 32.1	9 11
Jupiter	30 41.0	5 06
Saturn	192 48.6	18 16

Day	UT	SUN GHA	SUN Dec	MOON GHA	v	MOON Dec	d	HP
	d h	° ′	° ′	° ′	′	° ′	′	′
WEDNESDAY	3 00	180 28.6	N22 18.2	57 03.4	13.4	S11 36.5	13.1	56.5
	01	195 28.5	18.5	71 35.8	13.4	11 49.6	12.9	56.5
	02	210 28.4	18.8	86 08.2	13.3	12 02.5	13.0	56.4
	03	225 28.3	. . 19.1	100 40.5	13.3	12 15.5	12.8	56.4
	04	240 28.2	19.4	115 12.8	13.3	12 28.3	12.8	56.4
	05	255 28.1	19.7	129 45.1	13.3	12 41.1	12.7	56.4
	06	270 28.0	N22 20.0	144 17.4	13.2	S12 53.8	12.7	56.3
	07	285 27.9	20.3	158 49.6	13.3	13 06.5	12.5	56.3
	08	300 27.7	20.6	173 21.9	13.1	13 19.0	12.5	56.3
	09	315 27.6	. . 20.9	187 54.0	13.2	13 31.5	12.5	56.3
	10	330 27.5	21.2	202 26.2	13.1	13 44.0	12.4	56.2
	11	345 27.4	21.5	216 58.3	13.1	13 56.4	12.3	56.2
	12	0 27.3	N22 21.8	231 30.4	13.0	S14 08.7	12.2	56.2
	13	15 27.2	22.1	246 02.4	13.1	14 20.9	12.1	56.2
	14	30 27.1	22.4	260 34.5	13.0	14 33.0	12.1	56.2
	15	45 27.0	. . 22.7	275 06.5	12.9	14 45.1	12.0	56.1
	16	60 26.9	23.0	289 38.4	12.9	14 57.1	11.9	56.1
	17	75 26.8	23.3	304 10.3	12.9	15 09.0	11.9	56.1
	18	90 26.7	N22 23.6	318 42.2	12.9	S15 20.9	11.8	56.1
	19	105 26.6	23.9	333 14.1	12.8	15 32.7	11.6	56.0
	20	120 26.5	24.2	347 45.9	12.8	15 44.3	11.7	56.0
	21	135 26.4	. . 24.5	2 17.7	12.7	15 56.0	11.5	56.0
	22	150 26.3	24.8	16 49.4	12.7	16 07.5	11.4	56.0
	23	165 26.2	25.1	31 21.1	12.7	16 18.9	11.4	56.0
THURSDAY	4 00	180 26.1	N22 25.4	45 52.8	12.6	S16 30.3	11.3	55.9
	01	195 26.0	25.7	60 24.4	12.6	16 41.6	11.2	55.9
	02	210 25.9	26.0	74 56.0	12.5	16 52.8	11.1	55.9
	03	225 25.8	. . 26.3	89 27.5	12.5	17 03.9	11.0	55.9
	04	240 25.7	26.5	103 59.0	12.5	17 14.9	11.0	55.8
	05	255 25.5	26.8	118 30.5	12.4	17 25.9	10.8	55.8
	06	270 25.4	N22 27.1	133 01.9	12.4	S17 36.7	10.8	55.8
	07	285 25.3	27.4	147 33.3	12.3	17 47.5	10.7	55.8
	08	300 25.2	27.7	162 04.6	12.3	17 58.2	10.5	55.8
	09	315 25.1	. . 28.0	176 35.9	12.3	18 08.7	10.5	55.7
	10	330 25.0	28.3	191 07.2	12.2	18 19.2	10.4	55.7
	11	345 24.9	28.6	205 38.4	12.1	18 29.6	10.4	55.7
	12	0 24.8	N22 28.8	220 09.5	12.2	S18 40.0	10.2	55.7
	13	15 24.7	29.1	234 40.7	12.0	18 50.2	10.1	55.7
	14	30 24.6	29.4	249 11.7	12.1	19 00.3	10.0	55.6
	15	45 24.5	. . 29.7	263 42.8	12.0	19 10.3	10.0	55.6
	16	60 24.4	30.0	278 13.8	11.9	19 20.3	9.8	55.6
	17	75 24.3	30.3	292 44.7	11.9	19 30.1	9.7	55.6
	18	90 24.1	N22 30.5	307 15.6	11.9	S19 39.8	9.7	55.6
	19	105 24.0	30.8	321 46.5	11.8	19 49.5	9.5	55.5
	20	120 23.9	31.1	336 17.3	11.8	19 59.0	9.4	55.5
	21	135 23.8	. . 31.4	350 48.1	11.7	20 08.4	9.4	55.5
	22	150 23.7	31.6	5 18.8	11.7	20 17.8	9.2	55.5
	23	165 23.6	31.9	19 49.5	11.7	20 27.0	9.2	55.5
FRIDAY	5 00	180 23.5	N22 32.2	34 20.2	11.6	S20 36.2	9.0	55.4
	01	195 23.4	32.5	48 50.8	11.5	20 45.2	8.9	55.4
	02	210 23.3	32.7	63 21.3	11.5	20 54.1	8.8	55.4
	03	225 23.2	. . 33.0	77 51.8	11.5	21 02.9	8.8	55.4
	04	240 23.1	33.3	92 22.3	11.4	21 11.7	8.6	55.4
	05	255 22.9	33.6	106 52.7	11.4	21 20.3	8.5	55.4
	06	270 22.8	N22 33.8	121 23.1	11.4	S21 28.8	8.4	55.3
	07	285 22.7	34.1	135 53.5	11.2	21 37.2	8.3	55.3
	08	300 22.6	34.4	150 23.7	11.3	21 45.5	8.2	55.3
	09	315 22.5	. . 34.7	164 54.0	11.2	21 53.7	8.0	55.3
	10	330 22.4	34.9	179 24.2	11.2	22 01.7	8.0	55.3
	11	345 22.3	35.2	193 54.4	11.1	22 09.7	7.9	55.2
	12	0 22.2	N22 35.5	208 24.5	11.1	S22 17.6	7.7	55.2
	13	15 22.1	35.7	222 54.6	11.0	22 25.3	7.6	55.2
	14	30 22.0	36.0	237 24.6	11.0	22 32.9	7.5	55.2
	15	45 21.8	. . 36.3	251 54.6	11.0	22 40.4	7.4	55.2
	16	60 21.7	36.5	266 24.6	10.9	22 47.8	7.3	55.2
	17	75 21.6	36.8	280 54.5	10.9	22 55.1	7.2	55.1
	18	90 21.5	N22 37.1	295 24.4	10.8	S23 02.3	7.1	55.1
	19	105 21.4	37.3	309 54.2	10.8	23 09.4	6.9	55.1
	20	120 21.3	37.6	324 24.0	10.8	23 16.3	6.8	55.1
	21	135 21.2	. . 37.8	338 53.8	10.7	23 23.1	6.7	55.1
	22	150 21.1	38.1	353 23.5	10.7	23 29.8	6.6	55.1
	23	165 20.9	38.4	7 53.2	10.6	S23 36.4	6.5	55.0
		SD 15.8	*d* 0.3	SD 15.3		15.2		15.0

Lat.	Twilight Naut.	Twilight Civil	Sunrise	Moonrise 3	Moonrise 4	Moonrise 5	Moonrise 6
°	h m	h m	h m	h m	h m	h m	h m
N 72	□	□	□	18 44	■	■	■
N 70	□	□	□	18 03	■	■	■
68	□	□	□	17 36	19 54	■	■
66	////	////	00 58	17 16	19 12	21 39	■
64	////	////	01 51	16 59	18 44	20 35	22 40
62	////	////	02 22	16 46	18 23	20 00	21 33
60	////	01 17	02 45	16 35	18 05	19 35	20 58
N 58	////	01 55	03 04	16 25	17 51	19 15	20 33
56	////	02 20	03 19	16 17	17 39	18 59	20 12
54	01 10	02 40	03 32	16 09	17 28	18 45	19 56
52	01 45	02 57	03 44	16 02	17 19	18 33	19 42
50	02 09	03 11	03 54	15 56	17 10	18 22	19 29
45	02 50	03 38	04 15	15 43	16 52	18 00	19 03
N 40	03 19	04 00	04 32	15 33	16 38	17 42	18 43
35	03 41	04 17	04 46	15 24	16 26	17 27	18 26
30	03 59	04 32	04 59	15 16	16 15	17 14	18 12
20	04 27	04 55	05 20	15 02	15 57	16 52	17 47
N 10	04 48	05 15	05 38	14 50	15 41	16 33	17 26
0	05 06	05 32	05 55	14 39	15 26	16 15	17 06
S 10	05 22	05 49	06 11	14 29	15 12	15 58	16 46
20	05 38	06 05	06 29	14 17	14 56	15 39	16 25
30	05 53	06 23	06 49	14 04	14 39	15 18	16 01
35	06 02	06 33	07 01	13 56	14 29	15 05	15 46
40	06 10	06 44	07 14	13 48	14 17	14 51	15 30
45	06 20	06 57	07 30	13 38	14 03	14 34	15 10
S 50	06 31	07 12	07 50	13 26	13 47	14 13	14 46
52	06 36	07 19	07 59	13 20	13 39	14 03	14 34
54	06 41	07 27	08 10	13 14	13 31	13 52	14 21
56	06 47	07 35	08 21	13 08	13 21	13 39	14 05
58	06 53	07 45	08 35	13 00	13 10	13 24	13 47
S 60	07 00	07 56	08 51	12 52	12 58	13 07	13 24

Lat.	Sunset	Twilight Civil	Twilight Naut.	Moonset 3	Moonset 4	Moonset 5	Moonset 6
°	h m	h m	h m	h m	h m	h m	h m
N 72	□	□	□	22 36	■	■	■
N 70	□	□	□	23 18	■	■	■
68	□	□	□	{00 08 / 23 46}	23 08	■	■
66	23 03	////	////	00 20	{00 08 / 23 50}	23 06	■
64	22 08	////	////	00 31	00 25	00 19	{00 11 / 23 52}
62	21 36	////	////	00 40	00 40	00 41	00 46
60	21 13	22 42	////	00 47	00 52	00 59	01 12
N 58	20 54	22 04	////	00 54	01 03	01 14	01 32
56	20 38	21 38	////	01 00	01 12	01 27	01 49
54	20 25	21 17	22 49	01 06	01 20	01 39	02 03
52	20 14	21 01	22 13	01 11	01 28	01 48	02 16
50	20 03	20 47	21 49	01 15	01 34	01 57	02 27
45	19 42	20 19	21 07	01 25	01 49	02 16	02 50
N 40	19 25	19 57	20 38	01 33	02 01	02 32	03 08
35	19 11	19 40	20 16	01 40	02 11	02 45	03 24
30	18 58	19 25	19 58	01 46	02 20	02 57	03 37
20	18 37	19 01	19 30	01 57	02 35	03 16	04 01
N 10	18 19	18 42	19 09	02 07	02 49	03 34	04 21
0	18 02	18 24	18 50	02 15	03 02	03 50	04 40
S 10	17 45	18 08	18 34	02 24	03 15	04 06	04 58
20	17 28	17 51	18 19	02 34	03 28	04 23	05 19
30	17 07	17 33	18 03	02 45	03 44	04 43	05 42
35	16 56	17 23	17 55	02 51	03 53	04 55	05 56
40	16 42	17 12	17 46	02 59	04 04	05 09	06 12
45	16 26	16 59	17 36	03 07	04 16	05 25	06 31
S 50	16 06	16 44	17 25	03 17	04 32	05 45	06 55
52	15 57	16 37	17 20	03 22	04 39	05 54	07 06
54	15 47	16 30	17 15	03 27	04 47	06 05	07 19
56	15 35	16 21	17 09	03 33	04 56	06 17	07 35
58	15 21	16 12	17 03	03 40	05 06	06 31	07 53
S 60	15 05	16 01	16 56	03 47	05 18	06 48	08 15

Day	SUN Eqn. of Time 00^h	SUN Eqn. of Time 12^h	SUN Mer. Pass.	MOON Mer. Pass. Upper	MOON Mer. Pass. Lower	MOON Age	MOON Phase
d	m s	m s	h m	h m	h m	d %	
3	01 54	01 50	11 58	20 51	08 27	10 83	
4	01 45	01 39	11 58	21 38	09 14	11 90	
5	01 34	01 29	11 59	22 27	10 02	12 95	

Day	d h	ARIES GHA	VENUS −4.4 GHA	VENUS Dec	MARS +1.1 GHA	MARS Dec	JUPITER −2.5 GHA	JUPITER Dec	SATURN +1.0 GHA	SATURN Dec
		° ′	° ′	° ′	° ′	° ′	° ′	° ′	° ′	° ′
SATURDAY	6 00	254 32.4	226 07.7	N 9 07.2	222 39.1	N11 55.6	285 09.4	S13 19.5	87 17.5	N 7 45.6
	01	269 34.9	241 07.9	07.9	237 39.8	56.2	300 11.8	19.5	102 19.9	45.6
	02	284 37.4	256 08.0	08.6	252 40.5	56.8	315 14.2	19.5	117 22.3	45.6
	03	299 39.8	271 08.2	. . 09.4	267 41.2	. . 57.5	330 16.6	. . 19.5	132 24.7	. . 45.5
	04	314 42.3	286 08.3	10.1	282 41.9	58.1	345 19.0	19.5	147 27.1	45.5
	05	329 44.8	301 08.5	10.8	297 42.5	58.8	0 21.4	19.4	162 29.5	45.4
	06	344 47.2	316 08.6	N 9 11.6	312 43.2	N11 59.4	15 23.8	S13 19.4	177 31.9	N 7 45.4
	07	359 49.7	331 08.8	12.3	327 43.9	12 00.1	30 26.2	19.4	192 34.2	45.4
	08	14 52.2	346 08.9	13.1	342 44.6	00.7	45 28.6	19.4	207 36.6	45.3
	09	29 54.6	1 09.1	. . 13.8	357 45.3	. . 01.4	60 30.9	. . 19.4	222 39.0	. . 45.3
	10	44 57.1	16 09.2	14.5	12 46.0	02.0	75 33.3	19.4	237 41.4	45.2
	11	59 59.6	31 09.4	15.3	27 46.6	02.7	90 35.7	19.4	252 43.8	45.2
	12	75 02.0	46 09.5	N 9 16.0	42 47.3	N12 03.3	105 38.1	S13 19.3	267 46.2	N 7 45.2
	13	90 04.5	61 09.6	16.7	57 48.0	03.9	120 40.5	19.3	282 48.6	45.1
	14	105 06.9	76 09.8	17.5	72 48.7	04.6	135 42.9	19.3	297 51.0	45.1
	15	120 09.4	91 09.9	. . 18.2	87 49.4	. . 05.2	150 45.3	. . 19.3	312 53.3	. . 45.0
	16	135 11.9	106 10.1	19.0	102 50.1	05.9	165 47.7	19.3	327 55.7	45.0
	17	150 14.3	121 10.2	19.7	117 50.7	06.5	180 50.1	19.3	342 58.1	45.0
	18	165 16.8	136 10.3	N 9 20.4	132 51.4	N12 07.2	195 52.5	S13 19.3	358 00.5	N 7 44.9
	19	180 19.3	151 10.5	21.2	147 52.1	07.8	210 54.9	19.3	13 02.9	44.9
	20	195 21.7	166 10.6	21.9	162 52.8	08.5	225 57.3	19.2	28 05.3	44.8
	21	210 24.2	181 10.7	. . 22.6	177 53.5	. . 09.1	240 59.7	. . 19.2	43 07.7	. . 44.8
	22	225 26.7	196 10.9	23.4	192 54.1	09.7	256 02.1	19.2	58 10.0	44.7
	23	240 29.1	211 11.0	24.1	207 54.8	10.4	271 04.5	19.2	73 12.4	44.7
SUNDAY	7 00	255 31.6	226 11.1	N 9 24.9	222 55.5	N12 11.0	286 06.9	S13 19.2	88 14.8	N 7 44.7
	01	270 34.1	241 11.3	25.6	237 56.2	11.7	301 09.2	19.2	103 17.2	44.6
	02	285 36.5	256 11.4	26.3	252 56.9	12.3	316 11.6	19.2	118 19.6	44.6
	03	300 39.0	271 11.5	. . 27.1	267 57.6	. . 12.9	331 14.0	. . 19.2	133 22.0	. . 44.5
	04	315 41.4	286 11.7	27.8	282 58.2	13.6	346 16.4	19.1	148 24.3	44.5
	05	330 43.9	301 11.8	28.6	297 58.9	14.2	1 18.8	19.1	163 26.7	44.5
	06	345 46.4	316 11.9	N 9 29.3	312 59.6	N12 14.9	16 21.2	S13 19.1	178 29.1	N 7 44.4
	07	0 48.8	331 12.1	30.0	328 00.3	15.5	31 23.6	19.1	193 31.5	44.4
	08	15 51.3	346 12.2	30.8	343 01.0	16.1	46 26.0	19.1	208 33.9	44.3
	09	30 53.8	1 12.3	. . 31.5	358 01.6	. . 16.8	61 28.4	. . 19.1	223 36.3	. . 44.3
	10	45 56.2	16 12.4	32.3	13 02.3	17.4	76 30.8	19.1	238 38.6	44.2
	11	60 58.7	31 12.6	33.0	28 03.0	18.1	91 33.2	19.1	253 41.0	44.2
	12	76 01.2	46 12.7	N 9 33.8	43 03.7	N12 18.7	106 35.6	S13 19.0	268 43.4	N 7 44.2
	13	91 03.6	61 12.8	34.5	58 04.4	19.3	121 38.0	19.0	283 45.8	44.1
	14	106 06.1	76 12.9	35.2	73 05.0	20.0	136 40.4	19.0	298 48.2	44.1
	15	121 08.5	91 13.1	. . 36.0	88 05.7	. . 20.6	151 42.8	. . 19.0	313 50.5	. . 44.0
	16	136 11.0	106 13.2	36.7	103 06.4	21.3	166 45.2	19.0	328 52.9	44.0
	17	151 13.5	121 13.3	37.5	118 07.1	21.9	181 47.6	19.0	343 55.3	44.0
	18	166 15.9	136 13.4	N 9 38.2	133 07.8	N12 22.5	196 50.0	S13 19.0	358 57.7	N 7 43.9
	19	181 18.4	151 13.5	39.0	148 08.5	23.2	211 52.4	19.0	14 00.1	43.9
	20	196 20.9	166 13.7	39.7	163 09.1	23.8	226 54.8	19.0	29 02.5	43.8
	21	211 23.3	181 13.8	. . 40.4	178 09.8	. . 24.4	241 57.2	. . 18.9	44 04.8	. . 43.8
	22	226 25.8	196 13.9	41.2	193 10.5	25.1	256 59.6	18.9	59 07.2	43.7
	23	241 28.3	211 14.0	41.9	208 11.2	25.7	272 02.0	18.9	74 09.6	43.7
MONDAY	8 00	256 30.7	226 14.1	N 9 42.7	223 11.9	N12 26.4	287 04.5	S13 18.9	89 12.0	N 7 43.7
	01	271 33.2	241 14.2	43.4	238 12.5	27.0	302 06.9	18.9	104 14.4	43.6
	02	286 35.7	256 14.4	44.2	253 13.2	27.6	317 09.3	18.9	119 16.7	43.6
	03	301 38.1	271 14.5	. . 44.9	268 13.9	. . 28.3	332 11.7	. . 18.9	134 19.1	. . 43.5
	04	316 40.6	286 14.6	45.7	283 14.6	28.9	347 14.1	18.9	149 21.5	43.5
	05	331 43.0	301 14.7	46.4	298 15.3	29.5	2 16.5	18.9	164 23.9	43.4
	06	346 45.5	316 14.8	N 9 47.1	313 15.9	N12 30.2	17 18.9	S13 18.9	179 26.3	N 7 43.4
	07	1 48.0	331 14.9	47.9	328 16.6	30.8	32 21.3	18.9	194 28.6	43.4
	08	16 50.4	346 15.0	48.6	343 17.3	31.4	47 23.7	18.8	209 31.0	43.3
	09	31 52.9	1 15.1	. . 49.4	358 18.0	. . 32.1	62 26.1	. . 18.8	224 33.4	. . 43.3
	10	46 55.4	16 15.3	50.1	13 18.7	32.7	77 28.5	18.8	239 35.8	43.2
	11	61 57.8	31 15.4	50.9	28 19.3	33.3	92 30.9	18.8	254 38.2	43.2
	12	77 00.3	46 15.5	N 9 51.6	43 20.0	N12 34.0	107 33.3	S13 18.8	269 40.5	N 7 43.1
	13	92 02.8	61 15.6	52.4	58 20.7	34.6	122 35.7	18.8	284 42.9	43.1
	14	107 05.2	76 15.7	53.1	73 21.4	35.2	137 38.1	18.8	299 45.3	43.1
	15	122 07.7	91 15.8	. . 53.9	88 22.1	. . 35.9	152 40.5	. . 18.8	314 47.7	. . 43.0
	16	137 10.2	106 15.9	54.6	103 22.7	36.5	167 43.0	18.8	329 50.0	43.0
	17	152 12.6	121 16.0	55.3	118 23.4	37.1	182 45.4	18.8	344 52.4	42.9
	18	167 15.1	136 16.1	N 9 56.1	133 24.1	N12 37.8	197 47.8	S13 18.8	359 54.8	N 7 42.9
	19	182 17.5	151 16.2	56.8	148 24.8	38.4	212 50.2	18.8	14 57.2	42.8
	20	197 20.0	166 16.3	57.6	163 25.5	39.0	227 52.6	18.7	29 59.6	42.8
	21	212 22.5	181 16.4	. . 58.3	178 26.1	. . 39.7	242 55.0	. . 18.7	45 01.9	. . 42.7
	22	227 24.9	196 16.5	59.1	193 26.8	40.3	257 57.4	18.7	60 04.3	42.7
	23	242 27.4	211 16.6	59.8	208 27.5	40.9	272 59.8	18.7	75 06.7	42.7
	Mer. Pass.	h m 6 56.8	*v* 0.1	*d* 0.7	*v* 0.7	*d* 0.6	*v* 2.4	*d* 0.0	*v* 2.4	*d* 0.0

STARS

Name	SHA	Dec
	° ′	° ′
Acamar	315 20.9	S40 15.
Achernar	335 29.2	S57 11.
Acrux	173 12.7	S63 09.
Adhara	255 15.2	S28 59.
Aldebaran	290 53.2	N16 31.
Alioth	166 22.8	N55 54.
Alkaid	153 00.8	N49 16.
Al Na'ir	27 47.2	S46 54.
Alnilam	275 49.7	S 1 11.
Alphard	217 59.2	S 8 42.
Alphecca	126 13.2	N26 40.
Alpheratz	357 46.8	N29 08.
Altair	62 10.9	N 8 53.
Ankaa	353 18.6	S42 15.
Antares	112 29.6	S26 27.
Arcturus	145 58.2	N19 07.
Atria	107 33.6	S69 02.
Avior	234 19.8	S59 32.
Bellatrix	278 35.6	N 6 21.
Betelgeuse	271 04.9	N 7 24.
Canopus	263 58.0	S52 42.1
Capella	280 39.4	N46 00.
Deneb	49 33.3	N45 18.
Denebola	182 36.6	N14 31.1
Diphda	348 59.0	S17 55.
Dubhe	193 55.1	N61 42.
Elnath	278 16.8	N28 36.
Eltanin	90 47.0	N51 29.1
Enif	33 49.9	N 9 55.1
Fomalhaut	15 27.2	S29 34.1
Gacrux	172 04.3	S57 10.
Gienah	175 55.3	S17 35.
Hadar	148 51.9	S60 25.
Hamal	328 04.4	N23 30.
Kaus Aust.	83 47.4	S34 22.
Kochab	137 18.5	N74 07.1
Markab	13 41.3	N15 15.
Menkar	314 18.5	N 4 07.
Menkent	148 10.9	S36 25.
Miaplacidus	221 41.1	S69 45.
Mirfak	308 45.2	N49 53.
Nunki	76 01.7	S26 17.
Peacock	53 23.4	S56 42.
Pollux	243 31.6	N28 00.
Procyon	245 03.1	N 5 12.
Rasalhague	96 08.9	N12 33.1
Regulus	207 46.8	N11 55.
Rigel	281 15.3	S 8 11.
Rigil Kent.	139 55.5	S60 52.
Sabik	102 15.6	S15 44.
Schedar	349 44.4	N56 35.
Shaula	96 25.6	S37 06.
Sirius	258 36.7	S16 43.
Spica	158 34.2	S11 12.
Suhail	222 55.0	S43 28.5
Vega	80 40.6	N38 47.4
Zuben'ubi	137 08.5	S16 05.

	SHA	Mer. Pass.
	° ′	h m
Venus	330 39.6	8 55
Mars	327 23.9	9 08
Jupiter	30 35.3	4 55
Saturn	192 43.2	18 04

UT	SUN GHA	SUN Dec	MOON GHA	v	MOON Dec	d	HP
d h	° ′	° ′	° ′	′	° ′	′	′
6 00	180 20.8	N22 38.6	22 22.8	10.7	S23 42.9	6.3	55.0
01	195 20.7	38.9	36 52.5	10.5	23 49.2	6.3	55.0
02	210 20.6	39.1	51 22.0	10.6	23 55.5	6.1	55.0
03	225 20.5	. . 39.4	65 51.6	10.5	24 01.6	6.0	55.0
04	240 20.4	39.6	80 21.1	10.5	24 07.6	5.8	55.0
05	255 20.3	39.9	94 50.6	10.4	24 13.4	5.8	54.9
06	270 20.2	N22 40.2	109 20.0	10.4	S24 19.2	5.6	54.9
S 07	285 20.0	40.4	123 49.4	10.4	24 24.8	5.5	54.9
A 08	300 19.9	40.7	138 18.8	10.3	24 30.3	5.4	54.9
T 09	315 19.8	. . 40.9	152 48.1	10.3	24 35.7	5.2	54.9
U 10	330 19.7	41.2	167 17.4	10.3	24 40.9	5.2	54.9
R 11	345 19.6	41.4	181 46.7	10.3	24 46.1	5.0	54.9
D 12	0 19.5	N22 41.7	196 16.0	10.2	S24 51.1	4.9	54.8
A 13	15 19.4	41.9	210 45.2	10.2	24 56.0	4.7	54.8
Y 14	30 19.3	42.2	225 14.4	10.2	25 00.7	4.6	54.8
15	45 19.1	. . 42.4	239 43.6	10.1	25 05.3	4.5	54.8
16	60 19.0	42.7	254 12.7	10.2	25 09.8	4.4	54.8
17	75 18.9	42.9	268 41.9	10.1	25 14.2	4.3	54.8
18	90 18.8	N22 43.2	283 11.0	10.1	S25 18.5	4.1	54.8
19	105 18.7	43.4	297 40.1	10.0	25 22.6	4.0	54.7
20	120 18.6	43.7	312 09.1	10.1	25 26.6	3.8	54.7
21	135 18.4	. . 43.9	326 38.2	10.0	25 30.4	3.8	54.7
22	150 18.3	44.2	341 07.2	10.0	25 34.2	3.6	54.7
23	165 18.2	44.4	355 36.2	10.0	25 37.8	3.5	54.7
7 00	180 18.1	N22 44.6	10 05.2	9.9	S25 41.3	3.3	54.7
01	195 18.0	44.9	24 34.1	10.0	25 44.6	3.2	54.7
02	210 17.9	45.1	39 03.1	9.9	25 47.8	3.1	54.6
03	225 17.8	. . 45.4	53 32.0	9.9	25 50.9	3.0	54.6
04	240 17.6	45.6	68 00.9	10.0	25 53.9	2.8	54.6
05	255 17.5	45.8	82 29.9	9.9	25 56.7	2.7	54.6
06	270 17.4	N22 46.1	96 58.8	9.8	S25 59.4	2.6	54.6
07	285 17.3	46.3	111 27.6	9.9	26 02.0	2.4	54.6
S 08	300 17.2	46.6	125 56.5	9.9	26 04.4	2.3	54.6
U 09	315 17.1	. . 46.8	140 25.4	9.9	26 06.7	2.2	54.6
N 10	330 16.9	47.0	154 54.3	9.8	26 08.9	2.0	54.5
D 11	345 16.8	47.3	169 23.1	9.9	26 10.9	2.0	54.5
A 12	0 16.7	N22 47.5	183 52.0	9.8	S26 12.9	1.7	54.5
Y 13	15 16.6	47.7	198 20.8	9.9	26 14.6	1.7	54.5
14	30 16.5	48.0	212 49.7	9.8	26 16.3	1.5	54.5
15	45 16.4	. . 48.2	227 18.5	9.9	26 17.8	1.4	54.5
16	60 16.2	48.4	241 47.4	9.8	26 19.2	1.2	54.5
17	75 16.1	48.7	256 16.2	9.9	26 20.4	1.2	54.5
18	90 16.0	N22 48.9	270 45.1	9.8	S26 21.6	1.0	54.5
19	105 15.9	49.1	285 13.9	9.9	26 22.6	0.8	54.4
20	120 15.8	49.4	299 42.8	9.9	26 23.4	0.8	54.4
21	135 15.7	. . 49.6	314 11.7	9.8	26 24.2	0.6	54.4
22	150 15.5	49.8	328 40.5	9.9	26 24.8	0.4	54.4
23	165 15.4	50.0	343 09.4	9.9	26 25.2	0.4	54.4
8 00	180 15.3	N22 50.3	357 38.3	9.9	S26 25.6	0.2	54.4
01	195 15.2	50.5	12 07.2	9.9	26 25.8	0.0	54.4
02	210 15.1	50.7	26 36.1	9.9	26 25.8	0.0	54.4
03	225 14.9	. . 50.9	41 05.0	10.0	26 25.8	0.2	54.4
04	240 14.8	51.2	55 34.0	9.9	26 25.6	0.3	54.3
05	255 14.7	51.4	70 02.9	10.0	26 25.3	0.5	54.3
06	270 14.6	N22 51.6	84 31.9	10.0	S26 24.8	0.6	54.3
07	285 14.5	51.8	99 00.9	10.0	26 24.2	0.7	54.3
08	300 14.3	52.0	113 29.9	10.0	26 23.5	0.8	54.3
M 09	315 14.2	. . 52.3	127 58.9	10.0	26 22.7	1.0	54.3
O 10	330 14.1	52.5	142 27.9	10.1	26 21.7	1.1	54.3
N 11	345 14.0	52.7	156 57.0	10.1	26 20.6	1.2	54.3
D 12	0 13.9	N22 52.9	171 26.1	10.1	S26 19.4	1.3	54.3
A 13	15 13.8	53.1	185 55.2	10.2	26 18.1	1.5	54.3
Y 14	30 13.6	53.4	200 24.4	10.1	26 16.6	1.6	54.3
15	45 13.5	. . 53.6	214 53.5	10.2	26 15.0	1.8	54.2
16	60 13.4	53.8	229 22.7	10.2	26 13.2	1.8	54.2
17	75 13.3	54.0	243 51.9	10.3	26 11.4	2.0	54.2
18	90 13.1	N22 54.2	258 21.2	10.2	S26 09.4	2.1	54.2
19	105 13.0	54.4	272 50.4	10.4	26 07.3	2.3	54.2
20	120 12.9	54.6	287 19.8	10.3	26 05.0	2.4	54.2
21	135 12.8	. . 54.9	301 49.1	10.4	26 02.6	2.4	54.2
22	150 12.7	55.1	316 18.5	10.4	26 00.2	2.7	54.2
23	165 12.5	55.3	330 47.9	10.4	S25 57.5	2.7	54.2
	SD 15.8	*d* 0.2	SD 14.9		14.9		14.8

Lat.	Twilight Naut.	Twilight Civil	Sunrise	Moonrise 6	Moonrise 7	Moonrise 8	Moonrise 9
°	h m	h m	h m	h m	h m	h m	h m
N 72	□	□	□	■	■	■	■
N 70	□	□	□	■	■	■	■
68	□	□	□	■	■	■	■
66	////	////	00 44	■	■	■	■
64	////	////	01 44	22 40	■	■	00 48
62	////	////	02 18	21 33	22 46	23 25	23 39
60	////	01 09	02 42	20 58	22 04	22 48	23 12
N 58	////	01 50	03 01	20 33	21 36	22 22	22 51
56	////	02 17	03 17	20 12	21 14	22 01	22 34
54	01 03	02 37	03 30	19 56	20 56	21 44	22 19
52	01 41	02 54	03 42	19 42	20 41	21 29	22 06
50	02 06	03 09	03 52	19 29	20 28	21 16	21 54
45	02 48	03 37	04 14	19 03	20 01	20 50	21 31
N 40	03 18	03 59	04 31	18 43	19 39	20 29	21 12
35	03 40	04 16	04 46	18 26	19 22	20 12	20 56
30	03 58	04 31	04 58	18 12	19 06	19 57	20 42
20	04 26	04 55	05 20	17 47	18 40	19 31	20 18
N 10	04 48	05 15	05 38	17 26	18 18	19 09	19 58
0	05 07	05 33	05 55	17 06	17 57	18 49	19 39
S 10	05 23	05 49	06 12	16 46	17 37	18 28	19 20
20	05 39	06 06	06 30	16 25	17 14	18 06	18 59
30	05 55	06 24	06 51	16 01	16 49	17 40	18 35
35	06 03	06 35	07 03	15 46	16 33	17 25	18 21
40	06 12	06 46	07 16	15 30	16 16	17 08	18 05
45	06 22	06 59	07 33	15 10	15 55	16 47	17 45
S 50	06 33	07 14	07 53	14 46	15 28	16 20	17 21
52	06 38	07 22	08 02	14 34	15 15	16 07	17 09
54	06 44	07 29	08 13	14 21	15 00	15 52	16 56
56	06 49	07 38	08 25	14 05	14 43	15 35	16 40
58	06 56	07 48	08 39	13 47	14 22	15 14	16 21
S 60	07 03	07 59	08 55	13 24	13 55	14 47	15 58

Lat.	Sunset	Twilight Civil	Twilight Naut.	Moonset 6	Moonset 7	Moonset 8	Moonset 9
°	h m	h m	h m	h m	h m	h m	h m
N 72	□	□	□	■	■	■	■
N 70	□	□	□	■	■	■	■
68	□	□	□	■	■	■	■
66	23 19	////	////	■	■	■	■
64	22 15	////	////	{00 11 / 23 52}	■	■	01 18
62	21 41	////	////	00 46	00 59	01 33	02 41
60	21 17	22 51	////	01 12	01 34	02 15	03 17
N 58	20 58	22 09	////	01 32	02 00	02 43	03 43
56	20 42	21 42	////	01 49	02 20	03 05	04 04
54	20 28	21 21	22 57	02 03	02 37	03 23	04 21
52	20 16	21 04	22 18	02 16	02 51	03 38	04 35
50	20 06	20 50	21 53	02 27	03 04	03 51	04 48
45	19 44	20 21	21 10	02 50	03 30	04 19	05 14
N 40	19 27	19 59	20 40	03 08	03 51	04 40	05 35
35	19 12	19 42	20 18	03 24	04 08	04 58	05 52
30	18 59	19 27	20 00	03 37	04 23	05 13	06 07
20	18 38	19 03	19 32	04 01	04 48	05 39	06 32
N 10	18 20	18 43	19 10	04 21	05 10	06 01	06 53
0	18 03	18 25	18 51	04 40	05 31	06 22	07 13
S 10	17 46	18 08	18 35	04 58	05 51	06 43	07 33
20	17 28	17 51	18 19	05 19	06 13	07 05	07 54
30	17 07	17 33	18 03	05 42	06 38	07 31	08 19
35	16 55	17 23	17 55	05 56	06 53	07 46	08 33
40	16 41	17 12	17 46	06 12	07 11	08 04	08 50
45	16 25	16 59	17 36	06 31	07 32	08 25	09 10
S 50	16 05	16 43	17 24	06 55	07 58	08 52	09 35
52	15 56	16 36	17 19	07 06	08 11	09 05	09 47
54	15 45	16 28	17 14	07 19	08 26	09 20	10 01
56	15 33	16 19	17 08	07 35	08 43	09 37	10 17
58	15 19	16 10	17 02	07 53	09 04	09 59	10 36
S 60	15 02	15 59	16 55	08 15	09 30	10 26	10 59

Day	SUN Eqn. of Time 00^h	SUN Eqn. of Time 12^h	SUN Mer. Pass.	MOON Mer. Pass. Upper	MOON Mer. Pass. Lower	MOON Age	MOON Phase
d	m s	m s	h m	h m	h m	d %	
6	01 24	01 18	11 59	23 18	10 53	13 98	
7	01 13	01 07	11 59	24 10	11 44	14 100	○
8	01 01	00 56	11 59	00 10	12 35	15 99	

	UT	ARIES	VENUS −4.4		MARS +1.1		JUPITER −2.5		SATURN +1.0	
		GHA	GHA	Dec	GHA	Dec	GHA	Dec	GHA	Dec
	d h	° ′	° ′	° ′	° ′	° ′	° ′	° ′	° ′	° ′
	9 00	257 29.9	226 16.7	N10 00.6	223 28.2	N12 41.6	288 02.2	S13 18.7	90 09.1	N 7 42.6
	01	272 32.3	241 16.8	01.3	238 28.9	42.2	303 04.6	18.7	105 11.4	42.6
	02	287 34.8	256 16.9	02.1	253 29.5	42.8	318 07.1	18.7	120 13.8	42.5
	03	302 37.3	271 17.0	. . 02.8	268 30.2	. . 43.4	333 09.5	. . 18.7	135 16.2	. . 42.5
	04	317 39.7	286 17.1	03.6	283 30.9	44.1	348 11.9	18.7	150 18.6	42.4
	05	332 42.2	301 17.2	04.3	298 31.6	44.7	3 14.3	18.7	165 20.9	42.4
	06	347 44.7	316 17.3	N10 05.1	313 32.2	N12 45.3	18 16.7	S13 18.7	180 23.3	N 7 42.3
	07	2 47.1	331 17.4	05.8	328 32.9	46.0	33 19.1	18.7	195 25.7	42.3
T	08	17 49.6	346 17.5	06.6	343 33.6	46.6	48 21.5	18.7	210 28.1	42.3
U	09	32 52.0	1 17.5	. . 07.3	358 34.3	. . 47.2	63 24.0	. . 18.7	225 30.5	. . 42.2
E	10	47 54.5	16 17.6	08.1	13 35.0	47.8	78 26.4	18.7	240 32.8	42.2
S	11	62 57.0	31 17.7	08.8	28 35.6	48.5	93 28.8	18.6	255 35.2	42.1
D	12	77 59.4	46 17.8	N10 09.6	43 36.3	N12 49.1	108 31.2	S13 18.6	270 37.6	N 7 42.1
A	13	93 01.9	61 17.9	10.3	58 37.0	49.7	123 33.6	18.6	285 40.0	42.0
Y	14	108 04.4	76 18.0	11.1	73 37.7	50.4	138 36.0	18.6	300 42.3	42.0
	15	123 06.8	91 18.1	. . 11.8	88 38.4	. . 51.0	153 38.4	. . 18.6	315 44.7	. . 41.9
	16	138 09.3	106 18.2	12.6	103 39.0	51.6	168 40.9	18.6	330 47.1	41.9
	17	153 11.8	121 18.3	13.3	118 39.7	52.2	183 43.3	18.6	345 49.4	41.9
	18	168 14.2	136 18.3	N10 14.1	133 40.4	N12 52.9	198 45.7	S13 18.6	0 51.8	N 7 41.8
	19	183 16.7	151 18.4	14.8	148 41.1	53.5	213 48.1	18.6	15 54.2	41.8
	20	198 19.2	166 18.5	15.6	163 41.7	54.1	228 50.5	18.6	30 56.6	41.7
	21	213 21.6	181 18.6	. . 16.3	178 42.4	. . 54.7	243 52.9	. . 18.6	45 58.9	. . 41.7
	22	228 24.1	196 18.7	17.1	193 43.1	55.4	258 55.4	18.6	61 01.3	41.6
	23	243 26.5	211 18.8	17.8	208 43.8	56.0	273 57.8	18.6	76 03.7	41.6
	10 00	258 29.0	226 18.8	N10 18.6	223 44.5	N12 56.6	289 00.2	S13 18.6	91 06.1	N 7 41.5
	01	273 31.5	241 18.9	19.3	238 45.1	57.2	304 02.6	18.6	106 08.4	41.5
	02	288 33.9	256 19.0	20.1	253 45.8	57.9	319 05.0	18.6	121 10.8	41.4
	03	303 36.4	271 19.1	. . 20.8	268 46.5	. . 58.5	334 07.5	. . 18.6	136 13.2	. . 41.4
	04	318 38.9	286 19.2	21.6	283 47.2	59.1	349 09.9	18.6	151 15.6	41.3
	05	333 41.3	301 19.2	22.3	298 47.8	12 59.7	4 12.3	18.6	166 17.9	41.3
	06	348 43.8	316 19.3	N10 23.1	313 48.5	N13 00.4	19 14.7	S13 18.6	181 20.3	N 7 41.3
W	07	3 46.3	331 19.4	23.8	328 49.2	01.0	34 17.1	18.6	196 22.7	41.2
E	08	18 48.7	346 19.5	24.6	343 49.9	01.6	49 19.6	18.6	211 25.0	41.2
D	09	33 51.2	1 19.5	. . 25.3	358 50.6	. . 02.2	64 22.0	. . 18.5	226 27.4	. . 41.1
N	10	48 53.7	16 19.6	26.1	13 51.2	02.9	79 24.4	18.5	241 29.8	41.1
E	11	63 56.1	31 19.7	26.8	28 51.9	03.5	94 26.8	18.5	256 32.2	41.0
S	12	78 58.6	46 19.8	N10 27.6	43 52.6	N13 04.1	109 29.3	S13 18.5	271 34.5	N 7 41.0
D	13	94 01.0	61 19.8	28.3	58 53.3	04.7	124 31.7	18.5	286 36.9	40.9
A	14	109 03.5	76 19.9	29.1	73 53.9	05.3	139 34.1	18.5	301 39.3	40.9
Y	15	124 06.0	91 20.0	. . 29.8	88 54.6	. . 06.0	154 36.5	. . 18.5	316 41.6	. . 40.8
	16	139 08.4	106 20.0	30.6	103 55.3	06.6	169 39.0	18.5	331 44.0	40.8
	17	154 10.9	121 20.1	31.3	118 56.0	07.2	184 41.4	18.5	346 46.4	40.7
	18	169 13.4	136 20.2	N10 32.1	133 56.7	N13 07.8	199 43.8	S13 18.5	1 48.8	N 7 40.7
	19	184 15.8	151 20.2	32.8	148 57.3	08.5	214 46.2	18.5	16 51.1	40.7
	20	199 18.3	166 20.3	33.6	163 58.0	09.1	229 48.7	18.5	31 53.5	40.6
	21	214 20.8	181 20.4	. . 34.3	178 58.7	. . 09.7	244 51.1	. . 18.5	46 55.9	. . 40.6
	22	229 23.2	196 20.4	35.1	193 59.4	10.3	259 53.5	18.5	61 58.2	40.5
	23	244 25.7	211 20.5	35.8	209 00.0	10.9	274 55.9	18.5	77 00.6	40.5
	11 00	259 28.1	226 20.6	N10 36.6	224 00.7	N13 11.5	289 58.4	S13 18.5	92 03.0	N 7 40.4
	01	274 30.6	241 20.6	37.3	239 01.4	12.2	305 00.8	18.5	107 05.3	40.4
	02	289 33.1	256 20.7	38.1	254 02.1	12.8	320 03.2	18.5	122 07.7	40.3
	03	304 35.5	271 20.8	. . 38.9	269 02.7	. . 13.4	335 05.6	. . 18.5	137 10.1	. . 40.3
	04	319 38.0	286 20.8	39.6	284 03.4	14.0	350 08.1	18.5	152 12.4	40.2
	05	334 40.5	301 20.9	40.4	299 04.1	14.6	5 10.5	18.5	167 14.8	40.2
	06	349 42.9	316 20.9	N10 41.1	314 04.8	N13 15.3	20 12.9	S13 18.5	182 17.2	N 7 40.1
	07	4 45.4	331 21.0	41.9	329 05.4	15.9	35 15.4	18.5	197 19.6	40.1
T	08	19 47.9	346 21.1	42.6	344 06.1	16.5	50 17.8	18.5	212 21.9	40.0
H	09	34 50.3	1 21.1	. . 43.4	359 06.8	. . 17.1	65 20.2	. . 18.5	227 24.3	. . 40.0
U	10	49 52.8	16 21.2	44.1	14 07.5	17.7	80 22.6	18.5	242 26.7	39.9
R	11	64 55.3	31 21.2	44.9	29 08.2	18.3	95 25.1	18.5	257 29.0	39.9
S	12	79 57.7	46 21.3	N10 45.6	44 08.8	N13 19.0	110 27.5	S13 18.5	272 31.4	N 7 39.8
D	13	95 00.2	61 21.3	46.4	59 09.5	19.6	125 29.9	18.5	287 33.8	39.8
A	14	110 02.6	76 21.4	47.1	74 10.2	20.2	140 32.4	18.5	302 36.1	39.7
Y	15	125 05.1	91 21.4	. . 47.9	89 10.9	. . 20.8	155 34.8	. . 18.5	317 38.5	. . 39.7
	16	140 07.6	106 21.5	48.6	104 11.5	21.4	170 37.2	18.5	332 40.9	39.7
	17	155 10.0	121 21.6	49.4	119 12.2	22.0	185 39.7	18.5	347 43.2	39.6
	18	170 12.5	136 21.6	N10 50.2	134 12.9	N13 22.7	200 42.1	S13 18.5	2 45.6	N 7 39.6
	19	185 15.0	151 21.7	50.9	149 13.6	23.3	215 44.5	18.5	17 48.0	39.5
	20	200 17.4	166 21.7	51.7	164 14.2	23.9	230 47.0	18.5	32 50.3	39.5
	21	215 19.9	181 21.8	. . 52.4	179 14.9	. . 24.5	245 49.4	. . 18.5	47 52.7	. . 39.4
	22	230 22.4	196 21.8	53.2	194 15.6	25.1	260 51.8	18.5	62 55.1	39.4
	23	245 24.8	211 21.9	53.9	209 16.3	25.7	275 54.3	18.5	77 57.4	39.3
	Mer. Pass.	h m 6 45.0	v 0.1	d 0.8	v 0.7	d 0.6	v 2.4	d 0.0	v 2.4	d 0.0

STARS Name	SHA	Dec
	° ′	° ′
Acamar	315 20.9	S40 15.
Achernar	335 29.1	S57 11.
Acrux	173 12.7	S63 09.
Adhara	255 15.2	S28 59.
Aldebaran	290 53.2	N16 31.
Alioth	166 22.8	N55 54.
Alkaid	153 00.8	N49 16.
Al Na'ir	27 47.2	S46 54
Alnilam	275 49.7	S 1 11.
Alphard	217 59.2	S 8 42.
Alphecca	126 13.1	N26 40.
Alpheratz	357 46.7	N29 08.
Altair	62 10.9	N 8 53.
Ankaa	353 18.6	S42 15.
Antares	112 29.6	S26 27.
Arcturus	145 58.2	N19 08.
Atria	107 33.6	S69 02.
Avior	234 19.8	S59 32.
Bellatrix	278 35.5	N 6 21.
Betelgeuse	271 04.9	N 7 24.
Canopus	263 58.0	S52 42.
Capella	280 39.4	N46 00.
Deneb	49 33.3	N45 18.
Denebola	182 36.6	N14 31.
Diphda	348 58.9	S17 55.
Dubhe	193 55.1	N61 42.
Elnath	278 16.8	N28 36.
Eltanin	90 47.0	N51 29.
Enif	33 49.9	N 9 55.
Fomalhaut	15 27.1	S29 34.
Gacrux	172 04.3	S57 10.
Gienah	175 55.3	S17 35.
Hadar	148 51.9	S60 25.
Hamal	328 04.4	N23 30.
Kaus Aust.	83 47.4	S34 22.
Kochab	137 18.5	N74 07.
Markab	13 41.3	N15 15.
Menkar	314 18.5	N 4 07.
Menkent	148 10.9	S36 25.
Miaplacidus	221 41.1	S69 45.
Mirfak	308 45.2	N49 53.
Nunki	76 01.6	S26 17.
Peacock	53 23.3	S56 42.
Pollux	243 31.6	N28 00.
Procyon	245 03.1	N 5 12.
Rasalhague	96 08.9	N12 33.
Regulus	207 46.8	N11 55.
Rigel	281 15.3	S 8 11.
Rigil Kent.	139 55.5	S60 52.
Sabik	102 15.6	S15 44.
Schedar	349 44.4	N56 35.
Shaula	96 25.5	S37 06.
Sirius	258 36.7	S16 43.
Spica	158 34.2	S11 12.
Suhail	222 55.0	S43 28.
Vega	80 40.6	N38 47.
Zuben'ubi	137 08.5	S16 05.

	SHA	Mer. Pas
	° ′	h m
Venus	327 49.8	8 55
Mars	325 15.5	9 05
Jupiter	30 31.2	4 43
Saturn	192 37.1	17 53

UT	SUN GHA	SUN Dec	MOON GHA	v	MOON Dec	d	HP
d h	° ′	° ′	° ′	′	° ′	′	′
9 00	180 12.4	N22 55.5	345 17.3	10.5	S25 54.8	2.9	54.2
01	195 12.3	55.7	359 46.8	10.5	25 51.9	3.0	54.2
02	210 12.2	55.9	14 16.3	10.6	25 48.9	3.1	54.2
03	225 12.1	. . 56.1	28 45.9	10.6	25 45.8	3.2	54.2
04	240 11.9	56.3	43 15.5	10.6	25 42.6	3.4	54.2
05	255 11.8	56.5	57 45.1	10.7	25 39.2	3.5	54.1
06	270 11.7	N22 56.7	72 14.8	10.7	S25 35.7	3.6	54.1
T 07	285 11.6	56.9	86 44.5	10.7	25 32.1	3.7	54.1
U 08	300 11.5	57.1	101 14.2	10.8	25 28.4	3.8	54.1
E 09	315 11.3	. . 57.3	115 44.0	10.9	25 24.6	4.0	54.1
S 10	330 11.2	57.5	130 13.9	10.9	25 20.6	4.1	54.1
D 11	345 11.1	57.7	144 43.8	10.9	25 16.5	4.2	54.1
A 12	0 11.0	N22 57.9	159 13.7	11.0	S25 12.3	4.3	54.1
Y 13	15 10.8	58.1	173 43.7	11.0	25 08.0	4.4	54.1
14	30 10.7	58.3	188 13.7	11.0	25 03.6	4.6	54.1
15	45 10.6	. . 58.5	202 43.7	11.2	24 59.0	4.6	54.1
16	60 10.5	58.7	217 13.9	11.1	24 54.4	4.8	54.1
17	75 10.4	58.9	231 44.0	11.2	24 49.6	4.9	54.1
18	90 10.2	N22 59.1	246 14.2	11.3	S24 44.7	5.0	54.1
19	105 10.1	59.3	260 44.5	11.3	24 39.7	5.2	54.1
20	120 10.0	59.5	275 14.8	11.4	24 34.5	5.2	54.1
21	135 09.9	. . 59.7	289 45.2	11.4	24 29.3	5.4	54.1
22	150 09.7	22 59.9	304 15.6	11.5	24 23.9	5.4	54.1
23	165 09.6	23 00.1	318 46.1	11.5	24 18.5	5.6	54.1
10 00	180 09.5	N23 00.3	333 16.6	11.6	S24 12.9	5.7	54.1
01	195 09.4	00.5	347 47.2	11.6	24 07.2	5.8	54.1
02	210 09.2	00.7	2 17.8	11.7	24 01.4	5.9	54.1
03	225 09.1	. . 00.9	16 48.5	11.7	23 55.5	6.0	54.1
04	240 09.0	01.1	31 19.2	11.8	23 49.5	6.1	54.1
05	255 08.9	01.3	45 50.0	11.8	23 43.4	6.2	54.0
06	270 08.7	N23 01.4	60 20.8	11.9	S23 37.2	6.3	54.0
W 07	285 08.6	01.6	74 51.7	12.0	23 30.9	6.5	54.0
E 08	300 08.5	01.8	89 22.7	12.0	23 24.4	6.5	54.0
D 09	315 08.4	. . 02.0	103 53.7	12.0	23 17.9	6.6	54.0
N 10	330 08.2	02.2	118 24.7	12.2	23 11.3	6.8	54.0
E 11	345 08.1	02.4	132 55.9	12.1	23 04.5	6.8	54.0
S 12	0 08.0	N23 02.6	147 27.0	12.3	S22 57.7	7.0	54.0
D 13	15 07.9	02.7	161 58.3	12.2	22 50.7	7.0	54.0
A 14	30 07.8	02.9	176 29.5	12.4	22 43.7	7.2	54.0
Y 15	45 07.6	. . 03.1	191 00.9	12.4	22 36.5	7.2	54.0
16	60 07.5	03.3	205 32.3	12.5	22 29.3	7.4	54.0
17	75 07.4	03.5	220 03.8	12.5	22 21.9	7.4	54.0
18	90 07.2	N23 03.7	234 35.3	12.5	S22 14.5	7.6	54.0
19	105 07.1	03.8	249 06.8	12.7	22 06.9	7.6	54.0
20	120 07.0	04.0	263 38.5	12.7	21 59.3	7.8	54.0
21	135 06.9	. . 04.2	278 10.2	12.7	21 51.5	7.8	54.0
22	150 06.7	04.4	292 41.9	12.8	21 43.7	7.9	54.0
23	165 06.6	04.5	307 13.7	12.9	21 35.8	8.0	54.0
11 00	180 06.5	N23 04.7	321 45.6	12.9	S21 27.8	8.1	54.0
01	195 06.4	04.9	336 17.5	12.9	21 19.7	8.2	54.0
02	210 06.2	05.1	350 49.4	13.1	21 11.5	8.3	54.0
03	225 06.1	. . 05.2	5 21.5	13.1	21 03.2	8.4	54.0
04	240 06.0	05.4	19 53.6	13.1	20 54.8	8.5	54.1
05	255 05.9	05.6	34 25.7	13.2	20 46.3	8.6	54.1
06	270 05.7	N23 05.8	48 57.9	13.3	S20 37.7	8.6	54.1
T 07	285 05.6	05.9	63 30.2	13.3	20 29.1	8.8	54.1
H 08	300 05.5	06.1	78 02.5	13.3	20 20.3	8.8	54.1
U 09	315 05.4	. . 06.3	92 34.8	13.5	20 11.5	8.9	54.1
R 10	330 05.2	06.4	107 07.3	13.4	20 02.6	9.0	54.1
S 11	345 05.1	06.6	121 39.7	13.6	19 53.6	9.1	54.1
D 12	0 05.0	N23 06.8	136 12.3	13.6	S19 44.5	9.1	54.1
A 13	15 04.9	06.9	150 44.9	13.6	19 35.4	9.3	54.1
Y 14	30 04.7	07.1	165 17.5	13.7	19 26.1	9.3	54.1
15	45 04.6	. . 07.3	179 50.2	13.8	19 16.8	9.4	54.1
16	60 04.5	07.4	194 23.0	13.8	19 07.4	9.5	54.1
17	75 04.3	07.6	208 55.8	13.8	18 57.9	9.6	54.1
18	90 04.2	N23 07.8	223 28.6	13.9	S18 48.3	9.6	54.1
19	105 04.1	07.9	238 01.5	14.0	18 38.7	9.7	54.1
20	120 04.0	08.1	252 34.5	14.0	18 29.0	9.9	54.1
21	135 03.8	. . 08.3	267 07.5	14.1	18 19.1	9.8	54.1
22	150 03.7	08.4	281 40.6	14.1	18 09.3	10.0	54.1
23	165 03.6	08.6	296 13.7	14.2	S17 59.3	10.0	54.1
	SD 15.8	d 0.2	SD 14.7		14.7		14.7

Lat.	Twilight Naut.	Twilight Civil	Sunrise	Moonrise 9	10	11	12
°	h m	h m	h m	h m	h m	h m	h m
N 72	□	□	□	■	■	■	02 25
N 70	□	□	□	■	■	■	01 24
68	□	□	□	■	■	01 37	00 48
66	////	////	00 27	■	■	00 42	00 23
64	////	////	01 39	00 48	00 18	00 09	{00 03 / 23 57}
62	////	////	02 14	23 39	23 45	23 47	23 47
60	////	01 02	02 39	23 12	23 25	23 33	23 38
N 58	////	01 46	02 59	22 51	23 09	23 21	23 30
56	////	02 14	03 15	22 34	22 56	23 11	23 23
54	00 57	02 35	03 29	22 19	22 44	23 02	23 16
52	01 37	02 52	03 41	22 06	22 33	22 54	23 10
50	02 03	03 07	03 51	21 54	22 24	22 47	23 05
45	02 47	03 36	04 13	21 31	22 04	22 31	22 54
N 40	03 17	03 58	04 31	21 12	21 48	22 18	22 44
35	03 39	04 16	04 46	20 56	21 34	22 07	22 36
30	03 58	04 31	04 58	20 42	21 22	21 57	22 29
20	04 26	04 55	05 20	20 18	21 01	21 41	22 17
N 10	04 49	05 16	05 39	19 58	20 43	21 26	22 06
0	05 07	05 33	05 56	19 39	20 27	21 12	21 56
S 10	05 24	05 50	06 13	19 20	20 10	20 58	21 45
20	05 40	06 07	06 31	18 59	19 52	20 44	21 34
30	05 56	06 26	06 52	18 35	19 31	20 27	21 22
35	06 04	06 36	07 04	18 21	19 19	20 17	21 14
40	06 14	06 48	07 18	18 05	19 05	20 05	21 06
45	06 24	07 01	07 34	17 45	18 48	19 52	20 56
S 50	06 35	07 16	07 55	17 21	18 27	19 35	20 44
52	06 40	07 24	08 04	17 09	18 17	19 28	20 39
54	06 46	07 32	08 15	16 56	18 06	19 19	20 33
56	06 52	07 41	08 27	16 40	17 53	19 09	20 26
58	06 58	07 51	08 42	16 21	17 38	18 58	20 18
S 60	07 06	08 02	08 59	15 58	17 20	18 45	20 09

Lat.	Sunset	Twilight Civil	Twilight Naut.	Moonset 9	10	11	12
°	h m	h m	h m	h m	h m	h m	h m
N 72	□	□	□	■	■	■	04 40
N 70	□	□	□	■	■	■	05 40
68	□	□	□	■	■	03 52	06 14
66	23 39	////	////	■	■	04 46	06 38
64	22 21	////	////	01 18	03 31	05 18	06 57
62	21 46	////	////	02 41	04 09	05 42	07 12
60	21 21	22 59	////	03 17	04 36	06 00	07 25
N 58	21 01	22 14	////	03 43	04 57	06 16	07 36
56	20 45	21 46	////	04 04	05 14	06 29	07 46
54	20 31	21 24	23 04	04 21	05 28	06 41	07 54
52	20 19	21 07	22 23	04 35	05 41	06 51	08 02
50	20 08	20 52	21 56	04 48	05 52	07 00	08 08
45	19 46	20 23	21 12	05 14	06 15	07 18	08 23
N 40	19 28	20 01	20 42	05 35	06 33	07 34	08 35
35	19 14	19 43	20 20	05 52	06 49	07 47	08 45
30	19 01	19 28	20 01	06 07	07 02	07 58	08 53
20	18 39	19 04	19 33	06 32	07 25	08 17	09 08
N 10	18 20	18 43	19 10	06 53	07 44	08 34	09 21
0	18 03	18 26	18 52	07 13	08 02	08 49	09 33
S 10	17 46	18 09	18 35	07 33	08 20	09 04	09 45
20	17 28	17 52	18 19	07 54	08 39	09 21	09 58
30	17 07	17 33	18 03	08 19	09 02	09 39	10 13
35	16 55	17 23	17 54	08 33	09 14	09 50	10 21
40	16 41	17 11	17 45	08 50	09 29	10 02	10 31
45	16 24	16 58	17 35	09 10	09 47	10 17	10 42
S 50	16 04	16 42	17 24	09 35	10 09	10 34	10 55
52	15 54	16 35	17 19	09 47	10 19	10 43	11 01
54	15 44	16 27	17 13	10 01	10 31	10 52	11 08
56	15 31	16 18	17 07	10 17	10 44	11 02	11 16
58	15 17	16 08	17 00	10 36	10 59	11 14	11 24
S 60	15 00	15 57	16 53	10 59	11 18	11 28	11 34

Day	SUN Eqn. of Time 00^h	SUN Eqn. of Time 12^h	SUN Mer. Pass.	MOON Mer. Pass. Upper	MOON Mer. Pass. Lower	Age	Phase
d	m s	m s	h m	h m	h m	d	%
9	00 50	00 44	11 59	01 01	13 26	16	97
10	00 38	00 32	11 59	01 51	14 14	17	93
11	00 26	00 20	12 00	02 38	15 01	18	88

UT d h		ARIES GHA ° ′	VENUS −4.4 GHA ° ′	VENUS Dec ° ′	MARS +1.1 GHA ° ′	MARS Dec ° ′	JUPITER −2.5 GHA ° ′	JUPITER Dec ° ′	SATURN +1.0 GHA ° ′	SATURN Dec ° ′
12 00		260 27.3	226 21.9	N10 54.7	224 16.9	N13 26.3	290 56.7	S13 18.5	92 59.8	N 7 39.3
01		275 29.8	241 21.9	55.4	239 17.6	27.0	305 59.1	18.5	108 02.2	39.2
02		290 32.2	256 22.0	56.2	254 18.3	27.6	321 01.6	18.5	123 04.5	39.2
03		305 34.7	271 22.0	. . 56.9	269 19.0	. . 28.2	336 04.0	. . 18.5	138 06.9	. . 39.1
04		320 37.1	286 22.1	57.7	284 19.6	28.8	351 06.4	18.5	153 09.3	39.1
05		335 39.6	301 22.1	58.5	299 20.3	29.4	6 08.9	18.5	168 11.6	39.0
06		350 42.1	316 22.2	N10 59.2	314 21.0	N13 30.0	21 11.3	S13 18.5	183 14.0	N 7 39.0
07		5 44.5	331 22.2	11 00.0	329 21.7	30.6	36 13.7	18.5	198 16.3	38.9
08		20 47.0	346 22.2	00.7	344 22.3	31.2	51 16.2	18.5	213 18.7	38.9
09	F	35 49.5	1 22.3	. . 01.5	359 23.0	. . 31.9	66 18.6	. . 18.5	228 21.1	. . 38.8
10	R	50 51.9	16 22.3	02.2	14 23.7	32.5	81 21.1	18.5	243 23.4	38.8
11	I	65 54.4	31 22.4	03.0	29 24.4	33.1	96 23.5	18.5	258 25.8	38.7
12	D	80 56.9	46 22.4	N11 03.7	44 25.0	N13 33.7	111 25.9	S13 18.5	273 28.2	N 7 38.7
13	A	95 59.3	61 22.4	04.5	59 25.7	34.3	126 28.4	18.5	288 30.5	38.6
14	Y	111 01.8	76 22.5	05.2	74 26.4	34.9	141 30.8	18.5	303 32.9	38.6
15		126 04.3	91 22.5	. . 06.0	89 27.1	. . 35.5	156 33.2	. . 18.5	318 35.3	. . 38.5
16		141 06.7	106 22.6	06.8	104 27.7	36.1	171 35.7	18.5	333 37.6	38.5
17		156 09.2	121 22.6	07.5	119 28.4	36.7	186 38.1	18.5	348 40.0	38.4
18		171 11.6	136 22.6	N11 08.3	134 29.1	N13 37.3	201 40.6	S13 18.5	3 42.4	N 7 38.4
19		186 14.1	151 22.7	09.0	149 29.8	38.0	216 43.0	18.6	18 44.7	38.3
20		201 16.6	166 22.7	09.8	164 30.4	38.6	231 45.4	18.6	33 47.1	38.3
21		216 19.0	181 22.7	. . 10.5	179 31.1	. . 39.2	246 47.9	. . 18.6	48 49.4	. . 38.2
22		231 21.5	196 22.8	11.3	194 31.8	39.8	261 50.3	18.6	63 51.8	38.2
23		246 24.0	211 22.8	12.0	209 32.4	40.4	276 52.8	18.6	78 54.2	38.1
13 00		261 26.4	226 22.8	N11 12.8	224 33.1	N13 41.0	291 55.2	S13 18.6	93 56.5	N 7 38.1
01		276 28.9	241 22.8	13.6	239 33.8	41.6	306 57.7	18.6	108 58.9	38.0
02		291 31.4	256 22.9	14.3	254 34.5	42.2	322 00.1	18.6	124 01.2	38.0
03		306 33.8	271 22.9	. . 15.1	269 35.1	. . 42.8	337 02.5	. . 18.6	139 03.6	. . 37.9
04		321 36.3	286 22.9	15.8	284 35.8	43.4	352 05.0	18.6	154 06.0	37.9
05		336 38.7	301 23.0	16.6	299 36.5	44.0	7 07.4	18.6	169 08.3	37.8
06		351 41.2	316 23.0	N11 17.3	314 37.2	N13 44.6	22 09.9	S13 18.6	184 10.7	N 7 37.8
07		6 43.7	331 23.0	18.1	329 37.8	45.2	37 12.3	18.6	199 13.1	37.7
08	S	21 46.1	346 23.0	18.8	344 38.5	45.8	52 14.8	18.6	214 15.4	37.7
09	A	36 48.6	1 23.1	. . 19.6	359 39.2	. . 46.4	67 17.2	. . 18.6	229 17.8	. . 37.6
10	T	51 51.1	16 23.1	20.3	14 39.9	47.1	82 19.6	18.6	244 20.1	37.6
11	U	66 53.5	31 23.1	21.1	29 40.5	47.7	97 22.1	18.6	259 22.5	37.5
12	R	81 56.0	46 23.1	N11 21.9	44 41.2	N13 48.3	112 24.5	S13 18.6	274 24.9	N 7 37.5
13	D	96 58.5	61 23.1	22.6	59 41.9	48.9	127 27.0	18.6	289 27.2	37.4
14	A	112 00.9	76 23.2	23.4	74 42.5	49.5	142 29.4	18.6	304 29.6	37.4
15	Y	127 03.4	91 23.2	. . 24.1	89 43.2	. . 50.1	157 31.9	. . 18.6	319 31.9	. . 37.3
16		142 05.9	106 23.2	24.9	104 43.9	50.7	172 34.3	18.6	334 34.3	37.3
17		157 08.3	121 23.2	25.6	119 44.6	51.3	187 36.8	18.7	349 36.7	37.2
18		172 10.8	136 23.2	N11 26.4	134 45.2	N13 51.9	202 39.2	S13 18.7	4 39.0	N 7 37.2
19		187 13.2	151 23.3	27.1	149 45.9	52.5	217 41.7	18.7	19 41.4	37.1
20		202 15.7	166 23.3	27.9	164 46.6	53.1	232 44.1	18.7	34 43.7	37.1
21		217 18.2	181 23.3	. . 28.7	179 47.3	. . 53.7	247 46.6	. . 18.7	49 46.1	. . 37.0
22		232 20.6	196 23.3	29.4	194 47.9	54.3	262 49.0	18.7	64 48.5	37.0
23		247 23.1	211 23.3	30.2	209 48.6	54.9	277 51.5	18.7	79 50.8	36.9
14 00		262 25.6	226 23.3	N11 30.9	224 49.3	N13 55.5	292 53.9	S13 18.7	94 53.2	N 7 36.9
01		277 28.0	241 23.3	31.7	239 49.9	56.1	307 56.4	18.7	109 55.5	36.8
02		292 30.5	256 23.4	32.4	254 50.6	56.7	322 58.8	18.7	124 57.9	36.7
03		307 33.0	271 23.4	. . 33.2	269 51.3	. . 57.3	338 01.3	. . 18.7	140 00.2	. . 36.7
04		322 35.4	286 23.4	33.9	284 52.0	57.9	353 03.7	18.7	155 02.6	36.6
05		337 37.9	301 23.4	34.7	299 52.6	58.5	8 06.2	18.7	170 05.0	36.6
06		352 40.4	316 23.4	N11 35.5	314 53.3	N13 59.1	23 08.6	S13 18.7	185 07.3	N 7 36.5
07		7 42.8	331 23.4	36.2	329 54.0	13 59.7	38 11.1	18.7	200 09.7	36.5
08		22 45.3	346 23.4	37.0	344 54.7	14 00.3	53 13.5	18.8	215 12.0	36.4
09	S	37 47.7	1 23.4	. . 37.7	359 55.3	. . 00.9	68 16.0	. . 18.8	230 14.4	. . 36.4
10	U	52 50.2	16 23.4	38.5	14 56.0	01.5	83 18.4	18.8	245 16.7	36.3
11	N	67 52.7	31 23.4	39.2	29 56.7	02.1	98 20.9	18.8	260 19.1	36.3
12	D	82 55.1	46 23.4	N11 40.0	44 57.3	N14 02.7	113 23.3	S13 18.8	275 21.5	N 7 36.2
13	A	97 57.6	61 23.4	40.7	59 58.0	03.3	128 25.8	18.8	290 23.8	36.2
14	Y	113 00.1	76 23.5	41.5	74 58.7	03.9	143 28.2	18.8	305 26.2	36.1
15		128 02.5	91 23.5	. . 42.3	89 59.4	. . 04.5	158 30.7	. . 18.8	320 28.5	. . 36.1
16		143 05.0	106 23.5	43.0	105 00.0	05.1	173 33.2	18.8	335 30.9	36.0
17		158 07.5	121 23.5	43.8	120 00.7	05.7	188 35.6	18.8	350 33.2	36.0
18		173 09.9	136 23.5	N11 44.5	135 01.4	N14 06.3	203 38.1	S13 18.8	5 35.6	N 7 35.9
19		188 12.4	151 23.5	45.3	150 02.0	06.9	218 40.5	18.8	20 38.0	35.9
20		203 14.8	166 23.5	46.0	165 02.7	07.5	233 43.0	18.9	35 40.3	35.8
21		218 17.3	181 23.5	. . 46.8	180 03.4	. . 08.1	248 45.4	. . 18.9	50 42.7	. . 35.8
22		233 19.8	196 23.5	47.5	195 04.1	08.7	263 47.9	18.9	65 45.0	35.7
23		248 22.2	211 23.5	48.3	210 04.7	09.3	278 50.3	18.9	80 47.4	35.6
Mer. Pass.		h m 6 33.2	v 0.0	d 0.8	v 0.7	d 0.6	v 2.4	d 0.0	v 2.4	d 0.1

STARS

Name	SHA ° ′	Dec ° ′
Acamar	315 20.8	S40 15.8
Achernar	335 29.1	S57 11.0
Acrux	173 12.7	S63 09.5
Adhara	255 15.2	S28 59.2
Aldebaran	290 53.1	N16 31.7
Alioth	166 22.9	N55 54.7
Alkaid	153 00.8	N49 16.1
Al Na'ir	27 47.1	S46 54.6
Alnilam	275 49.7	S 1 11.7
Alphard	217 59.2	S 8 42.1
Alphecca	126 13.1	N26 40.9
Alpheratz	357 46.7	N29 08.5
Altair	62 10.9	N 8 53.6
Ankaa	353 18.6	S42 15.0
Antares	112 29.6	S26 27.3
Arcturus	145 58.2	N19 08.0
Atria	107 33.6	S69 02.8
Avior	234 19.9	S59 32.6
Bellatrix	278 35.5	N 6 21.5
Betelgeuse	271 04.9	N 7 24.6
Canopus	263 58.0	S52 42.1
Capella	280 39.4	N46 00.5
Deneb	49 33.3	N45 18.7
Denebola	182 36.6	N14 31.1
Diphda	348 58.9	S17 55.9
Dubhe	193 55.1	N61 42.2
Elnath	278 16.8	N28 36.9
Eltanin	90 47.0	N51 29.2
Enif	33 49.9	N 9 55.1
Fomalhaut	15 27.1	S29 34.1
Gacrux	172 04.3	S57 10.3
Gienah	175 55.3	S17 35.9
Hadar	148 51.9	S60 25.4
Hamal	328 04.4	N23 30.4
Kaus Aust.	83 47.3	S34 22.8
Kochab	137 18.6	N74 07.1
Markab	13 41.3	N15 15.3
Menkar	314 18.4	N 4 07.7
Menkent	148 10.9	S36 25.3
Miaplacidus	221 41.2	S69 45.7
Mirfak	308 45.2	N49 53.6
Nunki	76 01.6	S26 17.1
Peacock	53 23.3	S56 42.1
Pollux	243 31.6	N28 00.3
Procyon	245 03.1	N 5 12.0
Rasalhague	96 08.9	N12 33.1
Regulus	207 46.8	N11 55.2
Rigel	281 15.2	S 8 11.4
Rigil Kent.	139 55.5	S60 52.7
Sabik	102 15.6	S15 44.3
Schedar	349 44.3	N56 35.2
Shaula	96 25.5	S37 06.7
Sirius	258 36.7	S16 43.8
Spica	158 34.2	S11 12.8
Suhail	222 55.0	S43 28.5
Vega	80 40.6	N38 47.4
Zuben'ubi	137 08.5	S16 05.0

	SHA ° ′	Mer. Pass h m
Venus	324 56.4	8 54
Mars	323 06.7	9 01
Jupiter	30 28.8	4 32
Saturn	192 30.1	17 41

UT d h	SUN GHA ° ′	SUN Dec ° ′	MOON GHA ° ′	v ′	Dec ° ′	d ′	HP ′
12 00	180 03.4	N23 08.7	310 46.9	14.2	S17 49.3	10.1	54.1
01	195 03.3	08.9	325 20.1	14.3	17 39.2	10.2	54.1
02	210 03.2	09.1	339 53.4	14.3	17 29.0	10.3	54.2
03	225 03.1	. . 09.2	354 26.7	14.4	17 18.7	10.3	54.2
04	240 02.9	09.4	9 00.1	14.4	17 08.4	10.4	54.2
05	255 02.8	09.5	23 33.5	14.5	16 58.0	10.5	54.2
06	270 02.7	N23 09.7	38 07.0	14.5	S16 47.5	10.5	54.2
07	285 02.5	09.8	52 40.5	14.5	16 37.0	10.6	54.2
08	300 02.4	10.0	67 14.0	14.6	16 26.4	10.7	54.2
09	315 02.3	. . 10.1	81 47.6	14.7	16 15.7	10.7	54.2
10	330 02.2	10.3	96 21.3	14.7	16 05.0	10.8	54.2
11	345 02.0	10.4	110 55.0	14.7	15 54.2	10.9	54.2
12	0 01.9	N23 10.6	125 28.7	14.8	S15 43.3	11.0	54.2
13	15 01.8	10.7	140 02.5	14.8	15 32.3	11.0	54.3
14	30 01.6	10.9	154 36.3	14.9	15 21.3	11.0	54.3
15	45 01.5	. . 11.0	169 10.2	14.9	15 10.3	11.2	54.3
16	60 01.4	11.2	183 44.1	14.9	14 59.1	11.2	54.3
17	75 01.3	11.3	198 18.0	15.0	14 47.9	11.2	54.3
18	90 01.1	N23 11.5	212 52.0	15.1	S14 36.7	11.4	54.3
19	105 01.0	11.6	227 26.1	15.0	14 25.3	11.3	54.3
20	120 00.9	11.8	242 00.1	15.1	14 14.0	11.5	54.3
21	135 00.7	. . 11.9	256 34.2	15.2	14 02.5	11.5	54.3
22	150 00.6	12.1	271 08.4	15.2	13 51.0	11.5	54.4
23	165 00.5	12.2	285 42.6	15.2	13 39.5	11.7	54.4
13 00	180 00.4	N23 12.3	300 16.8	15.2	S13 27.8	11.6	54.4
01	195 00.2	12.5	314 51.0	15.3	13 16.2	11.8	54.4
02	210 00.1	12.6	329 25.3	15.3	13 04.4	11.8	54.4
03	225 00.0	. . 12.8	343 59.6	15.4	12 52.6	11.8	54.4
04	239 59.8	12.9	358 34.0	15.4	12 40.8	11.9	54.4
05	254 59.7	13.0	13 08.4	15.4	12 28.9	11.9	54.4
06	269 59.6	N23 13.2	27 42.8	15.4	S12 17.0	12.0	54.5
07	284 59.4	13.3	42 17.2	15.5	12 05.0	12.1	54.5
08	299 59.3	13.4	56 51.7	15.5	11 52.9	12.1	54.5
09	314 59.2	. . 13.6	71 26.2	15.5	11 40.8	12.2	54.5
10	329 59.0	13.7	86 00.7	15.6	11 28.6	12.2	54.5
11	344 58.9	13.9	100 35.3	15.6	11 16.4	12.2	54.5
12	359 58.8	N23 14.0	115 09.9	15.6	S11 04.2	12.3	54.5
13	14 58.7	14.1	129 44.5	15.6	10 51.9	12.4	54.6
14	29 58.5	14.3	144 19.1	15.6	10 39.5	12.4	54.6
15	44 58.4	. . 14.4	158 53.7	15.7	10 27.1	12.5	54.6
16	59 58.3	14.5	173 28.4	15.7	10 14.6	12.5	54.6
17	74 58.1	14.6	188 03.1	15.7	10 02.1	12.5	54.6
18	89 58.0	N23 14.8	202 37.8	15.8	S 9 49.6	12.6	54.6
19	104 57.9	14.9	217 12.6	15.7	9 37.0	12.6	54.7
20	119 57.7	15.0	231 47.3	15.8	9 24.4	12.7	54.7
21	134 57.6	. . 15.2	246 22.1	15.8	9 11.7	12.7	54.7
22	149 57.5	15.3	260 56.9	15.8	8 59.0	12.8	54.7
23	164 57.3	15.4	275 31.7	15.8	8 46.2	12.8	54.7
14 00	179 57.2	N23 15.5	290 06.5	15.8	S 8 33.4	12.8	54.8
01	194 57.1	15.7	304 41.3	15.9	8 20.6	12.9	54.8
02	209 56.9	15.8	319 16.2	15.8	8 07.7	12.9	54.8
03	224 56.8	. . 15.9	333 51.0	15.9	7 54.8	13.0	54.8
04	239 56.7	16.0	348 25.9	15.9	7 41.8	13.0	54.8
05	254 56.6	16.1	3 00.8	15.9	7 28.8	13.0	54.8
06	269 56.4	N23 16.3	17 35.7	15.9	S 7 15.8	13.1	54.9
07	284 56.3	16.4	32 10.6	15.9	7 02.7	13.1	54.9
08	299 56.2	16.5	46 45.5	15.9	6 49.6	13.1	54.9
09	314 56.0	. . 16.6	61 20.4	15.9	6 36.5	13.2	54.9
10	329 55.9	16.7	75 55.3	15.9	6 23.3	13.2	55.0
11	344 55.8	16.9	90 30.2	15.9	6 10.1	13.3	55.0
12	359 55.6	N23 17.0	105 05.1	15.9	S 5 56.8	13.3	55.0
13	14 55.5	17.1	119 40.0	15.9	5 43.5	13.3	55.0
14	29 55.4	17.2	134 14.9	16.0	5 30.2	13.3	55.0
15	44 55.2	. . 17.3	148 49.9	15.9	5 16.9	13.4	55.1
16	59 55.1	17.4	163 24.8	15.9	5 03.5	13.4	55.1
17	74 55.0	17.5	177 59.7	15.9	4 50.1	13.4	55.1
18	89 54.8	N23 17.7	192 34.6	15.9	S 4 36.7	13.5	55.1
19	104 54.7	17.8	207 09.5	15.9	4 23.2	13.5	55.2
20	119 54.6	17.9	221 44.4	15.8	4 09.7	13.5	55.2
21	134 54.4	. . 18.0	236 19.3	15.8	3 56.2	13.6	55.2
22	149 54.3	18.1	250 54.1	15.9	3 42.6	13.5	55.2
23	164 54.2	18.2	265 29.0	15.9	S 3 29.1	13.6	55.3
	SD 15.8	*d* 0.1	SD	14.8	14.9		15.0

Lat. °	Twilight Naut. h m	Twilight Civil h m	Sunrise h m	Moonrise 12 h m	13 h m	14 h m	15 h m
N 72	□	□	□	02 25	01 08	00 31	{00 01 / 23 34}
N 70	□	□	□	01 24	00 44	{00 18 / 23 57}	23 37
68	□	□	□	00 48	00 25	{00 08 / 23 53}	23 39
66	□	□	□	00 23	00 10	{00 00 / 23 50}	23 41
64	////	////	01 35	{00 03 / 23 57}	23 52	23 48	23 43
62	////	////	02 11	23 47	23 46	23 45	23 44
60	////	00 56	02 37	23 38	23 41	23 43	23 45
N 58	////	01 43	02 57	23 30	23 36	23 41	23 47
56	////	02 12	03 14	23 23	23 32	23 40	23 48
54	00 51	02 34	03 28	23 16	23 28	23 38	23 49
52	01 34	02 51	03 40	23 10	23 24	23 37	23 49
50	02 01	03 06	03 50	23 05	23 21	23 36	23 50
45	02 46	03 35	04 13	22 54	23 14	23 33	23 52
N 40	03 16	03 58	04 31	22 44	23 08	23 31	23 53
35	03 39	04 16	04 45	22 36	23 03	23 29	23 54
30	03 58	04 31	04 58	22 29	22 59	23 27	23 56
20	04 26	04 56	05 20	22 17	22 51	23 24	23 57
N 10	04 49	05 16	05 39	22 06	22 44	23 21	23 59
0	05 08	05 34	05 56	21 56	22 38	23 19	24 01
S 10	05 25	05 51	06 14	21 45	22 31	23 17	24 03
20	05 41	06 08	06 32	21 34	22 24	23 14	24 04
30	05 57	06 27	06 53	21 22	22 16	23 11	24 06
35	06 06	06 37	07 05	21 14	22 12	23 09	24 08
40	06 15	06 49	07 19	21 06	22 06	23 07	24 09
45	06 25	07 02	07 36	20 56	22 00	23 05	24 11
S 50	06 37	07 18	07 57	20 44	21 53	23 02	24 13
52	06 42	07 26	08 06	20 39	21 50	23 01	24 13
54	06 47	07 34	08 17	20 33	21 46	23 00	24 14
56	06 54	07 43	08 30	20 26	21 42	22 58	24 16
58	07 00	07 53	08 44	20 18	21 37	22 56	24 17
S 60	07 08	08 04	09 02	20 09	21 32	22 54	24 18

Lat. °	Sunset h m	Twilight Civil h m	Twilight Naut. h m	Moonset 12 h m	13 h m	14 h m	15 h m
N 72	□	□	□	04 40	07 27	09 32	11 30
N 70	□	□	□	05 40	07 49	09 42	11 31
68	□	□	□	06 14	08 06	09 50	11 32
66	□	□	□	06 38	08 20	09 57	11 33
64	22 26	////	////	06 57	08 31	10 02	11 33
62	21 50	////	////	07 12	08 41	10 07	11 34
60	21 24	23 06	////	07 25	08 49	10 11	11 34
N 58	21 04	22 18	////	07 36	08 56	10 15	11 35
56	20 47	21 49	////	07 46	09 02	10 18	11 35
54	20 33	21 27	23 10	07 54	09 08	10 21	11 35
52	20 21	21 09	22 26	08 02	09 13	10 24	11 35
50	20 10	20 54	21 59	08 08	09 17	10 26	11 36
45	19 48	20 25	21 14	08 23	09 27	10 31	11 36
N 40	19 30	20 03	20 44	08 35	09 35	10 36	11 37
35	19 15	19 44	20 21	08 45	09 42	10 39	11 37
30	19 02	19 29	20 03	08 53	09 48	10 43	11 37
20	18 40	19 05	19 34	09 08	09 59	10 48	11 38
N 10	18 21	18 44	19 11	09 21	10 08	10 53	11 38
0	18 04	18 26	18 52	09 33	10 16	10 58	11 39
S 10	17 46	18 09	18 36	09 45	10 24	11 02	11 39
20	17 28	17 52	18 20	09 58	10 33	11 07	11 40
30	17 07	17 33	18 03	10 13	10 43	11 12	11 40
35	16 55	17 23	17 55	10 21	10 49	11 15	11 40
40	16 41	17 11	17 45	10 31	10 56	11 19	11 41
45	16 24	16 58	17 35	10 42	11 03	11 23	11 41
S 50	16 03	16 42	17 23	10 55	11 12	11 27	11 42
52	15 54	16 34	17 18	11 01	11 16	11 29	11 42
54	15 43	16 26	17 13	11 08	11 21	11 32	11 42
56	15 30	16 17	17 06	11 16	11 26	11 34	11 42
58	15 16	16 07	17 00	11 24	11 31	11 37	11 42
S 60	14 58	15 56	16 52	11 34	11 38	11 40	11 43

Day	SUN Eqn. of Time 00^h	SUN Eqn. of Time 12^h	SUN Mer. Pass.	MOON Mer. Pass. Upper	MOON Mer. Pass. Lower	Age	Phase
d	m s	m s	h m	h m	h m	d	%
12	00 14	00 08	12 00	03 23	15 45	19	81
13	00 02	00 05	12 00	04 06	16 27	20	73
14	00 11	00 17	12 00	04 48	17 08	21	64

UT		ARIES	VENUS −4.3		MARS +1.1		JUPITER −2.6		SATURN +1.0	
		GHA	GHA	Dec	GHA	Dec	GHA	Dec	GHA	Dec
d	h	° ′	° ′	° ′	° ′	° ′	° ′	° ′	° ′	° ′
15	00	263 24.7	226 23.5	N11 49.1	225 05.4	N14 09.9	293 52.8	S13 18.9	95 49.7	N 7 35.6
	01	278 27.2	241 23.4	49.8	240 06.1	10.5	308 55.3	18.9	110 52.1	35.5
	02	293 29.6	256 23.4	50.6	255 06.7	11.0	323 57.7	18.9	125 54.4	35.5
	03	308 32.1	271 23.4	. . 51.3	270 07.4	. . 11.6	339 00.2	. . 18.9	140 56.8	. . 35.4
	04	323 34.6	286 23.4	52.1	285 08.1	12.2	354 02.6	18.9	155 59.1	35.4
	05	338 37.0	301 23.4	52.8	300 08.8	12.8	9 05.1	18.9	171 01.5	35.3
	06	353 39.5	316 23.4	N11 53.6	315 09.4	N14 13.4	24 07.6	S13 19.0	186 03.9	N 7 35.3
	07	8 42.0	331 23.4	54.3	330 10.1	14.0	39 10.0	19.0	201 06.2	35.2
	08	23 44.4	346 23.4	55.1	345 10.8	14.6	54 12.5	19.0	216 08.6	35.2
M	09	38 46.9	1 23.4	. . 55.9	0 11.4	. . 15.2	69 14.9	. . 19.0	231 10.9	. . 35.1
O	10	53 49.3	16 23.4	56.6	15 12.1	15.8	84 17.4	19.0	246 13.3	35.1
N	11	68 51.8	31 23.4	57.4	30 12.8	16.4	99 19.9	19.0	261 15.6	35.0
D	12	83 54.3	46 23.4	N11 58.1	45 13.4	N14 17.0	114 22.3	S13 19.0	276 18.0	N 7 34.9
A	13	98 56.7	61 23.4	58.9	60 14.1	17.6	129 24.8	19.0	291 20.3	34.9
Y	14	113 59.2	76 23.3	11 59.6	75 14.8	18.2	144 27.2	19.0	306 22.7	34.8
	15	129 01.7	91 23.3	12 00.4	90 15.5	. . 18.8	159 29.7	. . 19.0	321 25.0	. . 34.8
	16	144 04.1	106 23.3	01.1	105 16.1	19.4	174 32.2	19.1	336 27.4	34.7
	17	159 06.6	121 23.3	01.9	120 16.8	19.9	189 34.6	19.1	351 29.7	34.7
	18	174 09.1	136 23.3	N12 02.6	135 17.5	N14 20.5	204 37.1	S13 19.1	6 32.1	N 7 34.6
	19	189 11.5	151 23.3	03.4	150 18.1	21.1	219 39.6	19.1	21 34.4	34.6
	20	204 14.0	166 23.2	04.2	165 18.8	21.7	234 42.0	19.1	36 36.8	34.5
	21	219 16.5	181 23.2	. . 04.9	180 19.5	. . 22.3	249 44.5	. . 19.1	51 39.1	. . 34.5
	22	234 18.9	196 23.2	05.7	195 20.1	22.9	264 47.0	19.1	66 41.5	34.4
	23	249 21.4	211 23.2	06.4	210 20.8	23.5	279 49.4	19.1	81 43.9	34.4
16	00	264 23.8	226 23.2	N12 07.2	225 21.5	N14 24.1	294 51.9	S13 19.2	96 46.2	N 7 34.3
	01	279 26.3	241 23.2	07.9	240 22.2	24.7	309 54.3	19.2	111 48.6	34.2
	02	294 28.8	256 23.1	08.7	255 22.8	25.3	324 56.8	19.2	126 50.9	34.2
	03	309 31.2	271 23.1	. . 09.4	270 23.5	. . 25.8	339 59.3	. . 19.2	141 53.3	. . 34.1
	04	324 33.7	286 23.1	10.2	285 24.2	26.4	355 01.7	19.2	156 55.6	34.1
	05	339 36.2	301 23.1	10.9	300 24.8	27.0	10 04.2	19.2	171 58.0	34.0
	06	354 38.6	316 23.0	N12 11.7	315 25.5	N14 27.6	25 06.7	S13 19.2	187 00.3	N 7 34.0
	07	9 41.1	331 23.0	12.4	330 26.2	28.2	40 09.1	19.2	202 02.7	33.9
T	08	24 43.6	346 23.0	13.2	345 26.8	28.8	55 11.6	19.3	217 05.0	33.9
U	09	39 46.0	1 23.0	. . 14.0	0 27.5	. . 29.4	70 14.1	. . 19.3	232 07.4	. . 33.8
E	10	54 48.5	16 22.9	14.7	15 28.2	30.0	85 16.6	19.3	247 09.7	33.7
S	11	69 50.9	31 22.9	15.5	30 28.8	30.5	100 19.0	19.3	262 12.1	33.7
D	12	84 53.4	46 22.9	N12 16.2	45 29.5	N14 31.1	115 21.5	S13 19.3	277 14.4	N 7 33.6
A	13	99 55.9	61 22.9	17.0	60 30.2	31.7	130 24.0	19.3	292 16.8	33.6
Y	14	114 58.3	76 22.8	17.7	75 30.9	32.3	145 26.4	19.3	307 19.1	33.5
	15	130 00.8	91 22.8	. . 18.5	90 31.5	. . 32.9	160 28.9	. . 19.4	322 21.5	. . 33.5
	16	145 03.3	106 22.8	19.2	105 32.2	33.5	175 31.4	19.4	337 23.8	33.4
	17	160 05.7	121 22.7	20.0	120 32.9	34.1	190 33.8	19.4	352 26.2	33.4
	18	175 08.2	136 22.7	N12 20.7	135 33.5	N14 34.6	205 36.3	S13 19.4	7 28.5	N 7 33.3
	19	190 10.7	151 22.7	21.5	150 34.2	35.2	220 38.8	19.4	22 30.8	33.2
	20	205 13.1	166 22.6	22.2	165 34.9	35.8	235 41.3	19.4	37 33.2	33.2
	21	220 15.6	181 22.6	. . 23.0	180 35.5	. . 36.4	250 43.7	. . 19.4	52 35.5	. . 33.1
	22	235 18.1	196 22.6	23.7	195 36.2	37.0	265 46.2	19.5	67 37.9	33.1
	23	250 20.5	211 22.5	24.5	210 36.9	37.6	280 48.7	19.5	82 40.2	33.0
17	00	265 23.0	226 22.5	N12 25.3	225 37.5	N14 38.1	295 51.1	S13 19.5	97 42.6	N 7 33.0
	01	280 25.4	241 22.5	26.0	240 38.2	38.7	310 53.6	19.5	112 44.9	32.9
	02	295 27.9	256 22.4	26.8	255 38.9	39.3	325 56.1	19.5	127 47.3	32.9
	03	310 30.4	271 22.4	. . 27.5	270 39.5	. . 39.9	340 58.6	. . 19.5	142 49.6	. . 32.8
	04	325 32.8	286 22.4	28.3	285 40.2	40.5	356 01.0	19.5	157 52.0	32.7
	05	340 35.3	301 22.3	29.0	300 40.9	41.0	11 03.5	19.6	172 54.3	32.7
	06	355 37.8	316 22.3	N12 29.8	315 41.5	N14 41.6	26 06.0	S13 19.6	187 56.7	N 7 32.6
W	07	10 40.2	331 22.2	30.5	330 42.2	42.2	41 08.5	19.6	202 59.0	32.6
E	08	25 42.7	346 22.2	31.3	345 42.9	42.8	56 10.9	19.6	218 01.4	32.5
D	09	40 45.2	1 22.2	. . 32.0	0 43.5	. . 43.4	71 13.4	. . 19.6	233 03.7	. . 32.5
N	10	55 47.6	16 22.1	32.8	15 44.2	44.0	86 15.9	19.6	248 06.1	32.4
E	11	70 50.1	31 22.1	33.5	30 44.9	44.5	101 18.4	19.7	263 08.4	32.3
S	12	85 52.6	46 22.0	N12 34.3	45 45.6	N14 45.1	116 20.8	S13 19.7	278 10.8	N 7 32.3
D	13	100 55.0	61 22.0	35.0	60 46.2	45.7	131 23.3	19.7	293 13.1	32.2
A	14	115 57.5	76 21.9	35.8	75 46.9	46.3	146 25.8	19.7	308 15.4	32.2
Y	15	130 59.9	91 21.9	. . 36.5	90 47.6	. . 46.8	161 28.3	. . 19.7	323 17.8	. . 32.1
	16	146 02.4	106 21.8	37.3	105 48.2	47.4	176 30.8	19.7	338 20.1	32.1
	17	161 04.9	121 21.8	38.0	120 48.9	48.0	191 33.2	19.8	353 22.5	32.0
	18	176 07.3	136 21.7	N12 38.8	135 49.6	N14 48.6	206 35.7	S13 19.8	8 24.8	N 7 31.9
	19	191 09.8	151 21.7	39.5	150 50.2	49.2	221 38.2	19.8	23 27.2	31.9
	20	206 12.3	166 21.7	40.3	165 50.9	49.7	236 40.7	19.8	38 29.5	31.8
	21	221 14.7	181 21.6	. . 41.0	180 51.6	. . 50.3	251 43.2	. . 19.8	53 31.9	. . 31.8
	22	236 17.2	196 21.6	41.8	195 52.2	50.9	266 45.6	19.8	68 34.2	31.7
	23	251 19.7	211 21.5	42.5	210 52.9	51.5	281 48.1	19.9	83 36.5	31.7
Mer. Pass.		h m 6 21.4	v 0.0	d 0.8	v 0.7	d 0.6	v 2.5	d 0.0	v 2.3	d 0.1

STARS		
Name	SHA	Dec
	° ′	° ′
Acamar	315 20.8	S40 15.8
Achernar	335 29.1	S57 11.0
Acrux	173 12.8	S63 09.5
Adhara	255 15.2	S28 59.1
Aldebaran	290 53.1	N16 31.7
Alioth	166 22.9	N55 54.7
Alkaid	153 00.8	N49 16.1
Al Na'ir	27 47.1	S46 54.6
Alnilam	275 49.7	S 1 11.7
Alphard	217 59.2	S 8 42.1
Alphecca	126 13.2	N26 41.0
Alpheratz	357 46.7	N29 08.5
Altair	62 10.8	N 8 53.6
Ankaa	353 18.6	S42 15.0
Antares	112 29.6	S26 27.3
Arcturus	145 58.2	N19 08.0
Atria	107 33.6	S69 02.8
Avior	234 19.9	S59 32.6
Bellatrix	278 35.5	N 6 21.5
Betelgeuse	271 04.9	N 7 24.6
Canopus	263 58.0	S52 42.1
Capella	280 39.3	N46 00.5
Deneb	49 33.2	N45 18.7
Denebola	182 36.6	N14 31.1
Diphda	348 58.9	S17 55.9
Dubhe	193 55.2	N61 42.2
Elnath	278 16.8	N28 36.9
Eltanin	90 47.0	N51 29.2
Enif	33 49.9	N 9 55.1
Fomalhaut	15 27.1	S29 34.1
Gacrux	172 04.3	S57 10.3
Gienah	175 55.3	S17 35.9
Hadar	148 51.9	S60 25.5
Hamal	328 04.4	N23 30.4
Kaus Aust.	83 47.3	S34 22.8
Kochab	137 18.6	N74 07.1
Markab	13 41.3	N15 15.4
Menkar	314 18.4	N 4 07.7
Menkent	148 10.9	S36 25.3
Miaplacidus	221 41.2	S69 45.6
Mirfak	308 45.1	N49 53.6
Nunki	76 01.6	S26 17.1
Peacock	53 23.3	S56 42.1
Pollux	243 31.6	N28 00.3
Procyon	245 03.1	N 5 12.0
Rasalhague	96 08.9	N12 33.1
Regulus	207 46.8	N11 55.2
Rigel	281 15.2	S 8 11.4
Rigil Kent.	139 55.5	S60 52.7
Sabik	102 15.6	S15 44.3
Schedar	349 44.3	N56 35.2
Shaula	96 25.5	S37 06.7
Sirius	258 36.7	S16 43.8
Spica	158 34.3	S11 12.8
Suhail	222 55.0	S43 28.5
Vega	80 40.6	N38 47.5
Zuben'ubi	137 08.5	S16 05.0

	SHA	Mer. Pass
	° ′	h m
Venus	321 59.3	8 54
Mars	320 57.6	8 58
Jupiter	30 28.0	4 20
Saturn	192 22.4	17 30

UT	SUN GHA	SUN Dec	MOON GHA	v	Dec	d	HP
d h	° ′	° ′	° ′	′	° ′	′	′
15 00	179 54.0	N23 18.3	280 03.9	15.8	S 3 15.5	13.7	55.3
01	194 53.9	18.4	294 38.7	15.9	3 01.8	13.6	55.3
02	209 53.8	18.5	309 13.6	15.8	2 48.2	13.7	55.3
03	224 53.6 . .	18.6	323 48.4	15.8	2 34.5	13.7	55.4
04	239 53.5	18.7	338 23.2	15.8	2 20.8	13.7	55.4
05	254 53.4	18.8	352 58.0	15.7	2 07.1	13.7	55.4
06	269 53.2	N23 18.9	7 32.7	15.8	S 1 53.4	13.8	55.4
07	284 53.1	19.1	22 07.5	15.7	1 39.6	13.7	55.5
08	299 53.0	19.2	36 42.2	15.7	1 25.9	13.8	55.5
09	314 52.8 . .	19.3	51 16.9	15.7	1 12.1	13.8	55.5
10	329 52.7	19.4	65 51.6	15.7	0 58.3	13.9	55.5
11	344 52.6	19.5	80 26.3	15.6	0 44.4	13.8	55.6
12	359 52.4	N23 19.6	95 00.9	15.7	S 0 30.6	13.9	55.6
13	14 52.3	19.7	109 35.6	15.5	0 16.7	13.9	55.6
14	29 52.2	19.7	124 10.1	15.6	S 0 02.8	13.8	55.7
15	44 52.0 . .	19.8	138 44.7	15.5	N 0 11.0	13.9	55.7
16	59 51.9	19.9	153 19.2	15.5	0 24.9	14.0	55.7
17	74 51.8	20.0	167 53.7	15.5	0 38.9	13.9	55.7
18	89 51.6	N23 20.1	182 28.2	15.4	N 0 52.8	13.9	55.8
19	104 51.5	20.2	197 02.6	15.4	1 06.7	14.0	55.8
20	119 51.4	20.3	211 37.0	15.4	1 20.7	13.9	55.8
21	134 51.2 . .	20.4	226 11.4	15.3	1 34.6	14.0	55.9
22	149 51.1	20.5	240 45.7	15.3	1 48.6	14.0	55.9
23	164 51.0	20.6	255 20.0	15.3	2 02.6	14.0	55.9
16 00	179 50.8	N23 20.7	269 54.3	15.2	N 2 16.6	14.0	56.0
01	194 50.7	20.8	284 28.5	15.1	2 30.6	14.0	56.0
02	209 50.5	20.9	299 02.6	15.2	2 44.6	14.0	56.0
03	224 50.4 . .	21.0	313 36.8	15.1	2 58.6	14.0	56.1
04	239 50.3	21.0	328 10.9	15.0	3 12.6	14.0	56.1
05	254 50.1	21.1	342 44.9	15.0	3 26.6	14.0	56.1
06	269 50.0	N23 21.2	357 18.9	14.9	N 3 40.6	14.0	56.1
07	284 49.9	21.3	11 52.8	14.9	3 54.6	14.0	56.2
08	299 49.7	21.4	26 26.7	14.8	4 08.6	14.0	56.2
09	314 49.6 . .	21.5	41 00.5	14.8	4 22.6	14.0	56.2
10	329 49.5	21.6	55 34.3	14.8	4 36.6	14.0	56.3
11	344 49.3	21.6	70 08.1	14.6	4 50.6	14.0	56.3
12	359 49.2	N23 21.7	84 41.7	14.7	N 5 04.6	14.0	56.3
13	14 49.1	21.8	99 15.4	14.5	5 18.6	14.0	56.4
14	29 48.9	21.9	113 48.9	14.5	5 32.6	14.0	56.4
15	44 48.8 . .	22.0	128 22.4	14.5	5 46.6	14.0	56.4
16	59 48.7	22.0	142 55.9	14.4	6 00.6	14.0	56.5
17	74 48.5	22.1	157 29.3	14.3	6 14.6	13.9	56.5
18	89 48.4	N23 22.2	172 02.6	14.3	N 6 28.5	14.0	56.6
19	104 48.3	22.3	186 35.9	14.2	6 42.5	13.9	56.6
20	119 48.1	22.3	201 09.1	14.1	6 56.4	14.0	56.6
21	134 48.0 . .	22.4	215 42.2	14.1	7 10.4	13.9	56.7
22	149 47.8	22.5	230 15.3	14.0	7 24.3	13.9	56.7
23	164 47.7	22.6	244 48.3	13.9	7 38.2	13.9	56.7
17 00	179 47.6	N23 22.6	259 21.2	13.8	N 7 52.1	13.8	56.8
01	194 47.4	22.7	273 54.0	13.8	8 05.9	13.9	56.8
02	209 47.3	22.8	288 26.8	13.7	8 19.8	13.8	56.8
03	224 47.2 . .	22.9	302 59.5	13.6	8 33.6	13.8	56.9
04	239 47.0	22.9	317 32.1	13.6	8 47.4	13.8	56.9
05	254 46.9	23.0	332 04.7	13.5	9 01.2	13.8	56.9
06	269 46.8	N23 23.1	346 37.2	13.4	N 9 15.0	13.8	57.0
07	284 46.6	23.1	1 09.6	13.3	9 28.8	13.7	57.0
08	299 46.5	23.2	15 41.9	13.2	9 42.5	13.7	57.1
09	314 46.4 . .	23.3	30 14.1	13.2	9 56.2	13.6	57.1
10	329 46.2	23.3	44 46.3	13.0	10 09.8	13.7	57.1
11	344 46.1	23.4	59 18.3	13.0	10 23.5	13.6	57.2
12	359 45.9	N23 23.5	73 50.3	12.9	N10 37.1	13.6	57.2
13	14 45.8	23.5	88 22.2	12.8	10 50.7	13.5	57.2
14	29 45.7	23.6	102 54.0	12.7	11 04.2	13.6	57.3
15	44 45.5 . .	23.7	117 25.7	12.6	11 17.8	13.4	57.3
16	59 45.4	23.7	131 57.3	12.6	11 31.2	13.5	57.4
17	74 45.3	23.8	146 28.9	12.4	11 44.7	13.4	57.4
18	89 45.1	N23 23.8	161 00.3	12.4	N11 58.1	13.4	57.4
19	104 45.0	23.9	175 31.7	12.2	12 11.5	13.3	57.5
20	119 44.9	24.0	190 02.9	12.2	12 24.8	13.3	57.5
21	134 44.7 . .	24.0	204 34.1	12.0	12 38.1	13.2	57.6
22	149 44.6	24.1	219 05.1	12.0	12 51.3	13.2	57.6
23	164 44.4	24.1	233 36.1	11.8	N13 04.5	13.2	57.6
	SD 15.8	*d* 0.1	SD 15.1		15.4		15.6

Lat.	Twilight Naut.	Twilight Civil	Sunrise	Moonrise 15	16	17	18
°	h m	h m	h m	h m	h m	h m	h m
N 72	▭	▭	▭	{00 01 / 23 34}	23 05	22 27	21 03
N 70	▭	▭	▭	23 37	23 16	22 49	22 06
68	▭	▭	▭	23 39	23 24	23 07	22 42
66	▭	▭	▭	23 41	23 32	23 21	23 08
64	////	////	01 32	23 43	23 38	23 33	23 28
62	////	////	02 10	23 44	23 43	23 43	23 45
60	////	00 52	02 36	23 45	23 48	23 52	23 58
N 58	////	01 41	02 56	23 47	23 52	24 00	00 00
56	////	02 11	03 13	23 48	23 56	24 07	00 07
54	00 47	02 33	03 27	23 49	24 00	00 00	00 13
52	01 33	02 51	03 39	23 49	24 03	00 03	00 18
50	02 00	03 06	03 50	23 50	24 06	00 06	00 23
45	02 46	03 35	04 13	23 52	24 12	00 12	00 34
N 40	03 16	03 58	04 31	23 53	24 17	00 17	00 43
35	03 39	04 16	04 46	23 54	24 21	00 21	00 51
30	03 58	04 31	04 59	23 56	24 25	00 25	00 58
20	04 27	04 56	05 21	23 57	24 32	00 32	01 10
N 10	04 49	05 16	05 39	23 59	24 39	00 39	01 21
0	05 08	05 35	05 57	24 01	00 01	00 44	01 31
S 10	05 25	05 52	06 14	24 03	00 03	00 50	01 41
20	05 41	06 09	06 33	24 04	00 04	00 57	01 52
30	05 58	06 28	06 54	24 06	00 06	01 04	02 05
35	06 07	06 38	07 06	24 08	00 08	01 08	02 12
40	06 16	06 50	07 21	24 09	00 09	01 13	02 20
45	06 26	07 04	07 37	24 11	00 11	01 19	02 30
S 50	06 38	07 20	07 58	24 13	00 13	01 25	02 42
52	06 43	07 27	08 08	24 13	00 13	01 29	02 48
54	06 49	07 35	08 19	24 14	00 14	01 32	02 54
56	06 55	07 44	08 32	24 16	00 16	01 36	03 01
58	07 02	07 54	08 46	24 17	00 17	01 40	03 09
S 60	07 09	08 06	09 04	24 18	00 18	01 45	03 18

Lat.	Sunset	Twilight Civil	Twilight Naut.	Moonset 15	16	17	18
°	h m	h m	h m	h m	h m	h m	h m
N 72	▭	▭	▭	11 30	13 30	15 44	18 51
N 70	▭	▭	▭	11 31	13 22	15 24	17 50
68	▭	▭	▭	11 32	13 16	15 08	17 15
66	▭	▭	▭	11 33	13 11	14 55	16 51
64	22 30	////	////	11 33	13 06	14 45	16 32
62	21 52	////	////	11 34	13 02	14 36	16 16
60	21 26	23 11	////	11 34	12 59	14 28	16 04
N 58	21 06	22 21	////	11 35	12 56	14 22	15 52
56	20 49	21 51	////	11 35	12 53	14 16	15 43
54	20 35	21 29	23 15	11 35	12 51	14 11	15 34
52	20 22	21 11	22 29	11 35	12 49	14 06	15 27
50	20 11	20 56	22 01	11 36	12 47	14 02	15 20
45	19 49	20 26	21 16	11 36	12 43	13 52	15 06
N 40	19 31	20 04	20 45	11 37	12 39	13 45	14 54
35	19 16	19 46	20 22	11 37	12 36	13 38	14 44
30	19 03	19 30	20 04	11 37	12 34	13 33	14 35
20	18 41	19 05	19 35	11 38	12 29	13 23	14 20
N 10	18 22	18 45	19 12	11 38	12 25	13 14	14 07
0	18 04	18 27	18 53	11 39	12 21	13 06	13 55
S 10	17 47	18 10	18 36	11 39	12 18	12 58	13 43
20	17 28	17 52	18 20	11 40	12 14	12 50	13 30
30	17 07	17 34	18 04	11 40	12 09	12 40	13 15
35	16 55	17 23	17 55	11 40	12 06	12 35	13 06
40	16 41	17 11	17 45	11 41	12 04	12 28	12 57
45	16 24	16 58	17 35	11 41	12 00	12 21	12 45
S 50	16 03	16 42	17 23	11 42	11 56	12 12	12 32
52	15 53	16 34	17 18	11 42	11 54	12 08	12 25
54	15 42	16 26	17 12	11 42	11 52	12 04	12 18
56	15 30	16 17	17 06	11 42	11 50	11 59	12 11
58	15 15	16 07	17 00	11 42	11 48	11 54	12 02
S 60	14 58	15 55	16 52	11 43	11 45	11 48	11 52

Day	SUN Eqn. of Time 00^h	SUN Eqn. of Time 12^h	SUN Mer. Pass.	MOON Mer. Pass. Upper	MOON Mer. Pass. Lower	MOON Age	MOON Phase
d	m s	m s	h m	h m	h m	d	%
15	00 24	00 30	12 01	05 29	17 50	22	54
16	00 36	00 43	12 01	06 11	18 33	23	44
17	00 49	00 56	12 01	06 55	19 18	24	34

UT	ARIES	VENUS −4.3		MARS +1.1		JUPITER −2.6		SATURN +1.0		STARS		
	GHA	GHA	Dec	GHA	Dec	GHA	Dec	GHA	Dec	Name	SHA	Dec
d h	° ′	° ′	° ′	° ′	° ′	° ′	° ′	° ′	° ′		° ′	°
18 00	266 22.1	226 21.4	N12 43.3	225 53.6	N14 52.0	296 50.6	S13 19.9	98 38.9	N 7 31.6	Acamar	315 20.8	S40 15
01	281 24.6	241 21.4	44.0	240 54.2	52.6	311 53.1	19.9	113 41.2	31.5	Achernar	335 29.0	S57 11
02	296 27.0	256 21.3	44.8	255 54.9	53.2	326 55.6	19.9	128 43.6	31.5	Acrux	173 12.8	S63 09
03	311 29.5	271 21.3	45.5	270 55.6	53.8	341 58.0	19.9	143 45.9	31.4	Adhara	255 15.2	S28 59
04	326 32.0	286 21.2	46.3	285 56.2	54.4	357 00.5	19.9	158 48.3	31.4	Aldebaran	290 53.1	N16 31
05	341 34.4	301 21.2	47.0	300 56.9	54.9	12 03.0	20.0	173 50.6	31.3			
06	356 36.9	316 21.1	N12 47.8	315 57.6	N14 55.5	27 05.5	S13 20.0	188 53.0	N 7 31.3	Alioth	166 22.9	N55 54
07	11 39.4	331 21.1	48.5	330 58.2	56.1	42 08.0	20.0	203 55.3	31.2	Alkaid	153 00.8	N49 16
T 08	26 41.8	346 21.0	49.3	345 58.9	56.7	57 10.5	20.0	218 57.6	31.1	Al Na'ir	27 47.1	S46 54
H 09	41 44.3	1 21.0	50.0	0 59.6	57.2	72 12.9	20.0	234 00.0	31.1	Alnilam	275 49.7	S 1 11
U 10	56 46.8	16 20.9	50.8	16 00.2	57.8	87 15.4	20.1	249 02.3	31.0	Alphard	217 59.2	S 8 42
R 11	71 49.2	31 20.8	51.5	31 00.9	58.4	102 17.9	20.1	264 04.7	31.0			
S 12	86 51.7	46 20.8	N12 52.3	46 01.6	N14 58.9	117 20.4	S13 20.1	279 07.0	N 7 30.9	Alphecca	126 13.2	N26 41
D 13	101 54.2	61 20.7	53.0	61 02.2	14 59.5	132 22.9	20.1	294 09.4	30.9	Alpheratz	357 46.7	N29 08
A 14	116 56.6	76 20.7	53.8	76 02.9	15 00.1	147 25.4	20.1	309 11.7	30.8	Altair	62 10.8	N 8 53
Y 15	131 59.1	91 20.6	54.5	91 03.6	00.7	162 27.8	20.2	324 14.0	30.7	Ankaa	353 18.5	S42 15
16	147 01.5	106 20.5	55.3	106 04.2	01.2	177 30.3	20.2	339 16.4	30.7	Antares	112 29.6	S26 27
17	162 04.0	121 20.5	56.0	121 04.9	01.8	192 32.8	20.2	354 18.7	30.6			
18	177 06.5	136 20.4	N12 56.8	136 05.6	N15 02.4	207 35.3	S13 20.2	9 21.1	N 7 30.6	Arcturus	145 58.2	N19 08
19	192 08.9	151 20.3	57.5	151 06.2	03.0	222 37.8	20.2	24 23.4	30.5	Atria	107 33.6	S69 02
20	207 11.4	166 20.3	58.3	166 06.9	03.5	237 40.3	20.3	39 25.7	30.4	Avior	234 19.9	S59 32
21	222 13.9	181 20.2	59.0	181 07.6	04.1	252 42.8	20.3	54 28.1	30.4	Bellatrix	278 35.5	N 6 21
22	237 16.3	196 20.1	12 59.8	196 08.2	04.7	267 45.3	20.3	69 30.4	30.3	Betelgeuse	271 04.8	N 7 24
23	252 18.8	211 20.1	13 00.5	211 08.9	05.2	282 47.7	20.3	84 32.8	30.3			
19 00	267 21.3	226 20.0	N13 01.3	226 09.5	N15 05.8	297 50.2	S13 20.3	99 35.1	N 7 30.2	Canopus	263 58.0	S52 42
01	282 23.7	241 19.9	02.0	241 10.2	06.4	312 52.7	20.4	114 37.5	30.1	Capella	280 39.3	N46 00
02	297 26.2	256 19.9	02.8	256 10.9	06.9	327 55.2	20.4	129 39.8	30.1	Deneb	49 33.2	N45 18
03	312 28.7	271 19.8	03.5	271 11.5	07.5	342 57.7	20.4	144 42.1	30.0	Denebola	182 36.6	N14 31
04	327 31.1	286 19.7	04.3	286 12.2	08.1	358 00.2	20.4	159 44.5	30.0	Diphda	348 58.9	S17 55
05	342 33.6	301 19.7	05.0	301 12.9	08.7	13 02.7	20.4	174 46.8	29.9			
06	357 36.0	316 19.6	N13 05.7	316 13.5	N15 09.2	28 05.2	S13 20.5	189 49.2	N 7 29.8	Dubhe	193 55.2	N61 42
07	12 38.5	331 19.5	06.5	331 14.2	09.8	43 07.7	20.5	204 51.5	29.8	Elnath	278 16.8	N28 36
08	27 41.0	346 19.4	07.2	346 14.9	10.4	58 10.2	20.5	219 53.8	29.7	Eltanin	90 47.0	N51 29
F 09	42 43.4	1 19.4	08.0	1 15.5	10.9	73 12.6	20.5	234 56.2	29.7	Enif	33 49.9	N 9 55
R 10	57 45.9	16 19.3	08.7	16 16.2	11.5	88 15.1	20.5	249 58.5	29.6	Fomalhaut	15 27.1	S29 34
I 11	72 48.4	31 19.2	09.5	31 16.9	12.1	103 17.6	20.6	265 00.8	29.6			
D 12	87 50.8	46 19.1	N13 10.2	46 17.5	N15 12.6	118 20.1	S13 20.6	280 03.2	N 7 29.5	Gacrux	172 04.3	S57 10
A 13	102 53.3	61 19.1	11.0	61 18.2	13.2	133 22.6	20.6	295 05.5	29.4	Gienah	175 55.4	S17 35
Y 14	117 55.8	76 19.0	11.7	76 18.9	13.8	148 25.1	20.6	310 07.9	29.4	Hadar	148 52.0	S60 25
15	132 58.2	91 18.9	12.5	91 19.5	14.3	163 27.6	20.7	325 10.2	29.3	Hamal	328 04.4	N23 30
16	148 00.7	106 18.8	13.2	106 20.2	14.9	178 30.1	20.7	340 12.5	29.3	Kaus Aust.	83 47.3	S34 22
17	163 03.1	121 18.7	13.9	121 20.9	15.5	193 32.6	20.7	355 14.9	29.2			
18	178 05.6	136 18.7	N13 14.7	136 21.5	N15 16.0	208 35.1	S13 20.7	10 17.2	N 7 29.1	Kochab	137 18.7	N74 07
19	193 08.1	151 18.6	15.4	151 22.2	16.6	223 37.6	20.7	25 19.6	29.1	Markab	13 41.2	N15 15
20	208 10.5	166 18.5	16.2	166 22.8	17.2	238 40.1	20.8	40 21.9	29.0	Menkar	314 18.4	N 4 07
21	223 13.0	181 18.4	16.9	181 23.5	17.7	253 42.6	20.8	55 24.2	29.0	Menkent	148 10.9	S36 25
22	238 15.5	196 18.3	17.7	196 24.2	18.3	268 45.1	20.8	70 26.6	28.9	Miaplacidus	221 41.2	S69 45
23	253 17.9	211 18.3	18.4	211 24.8	18.8	283 47.6	20.8	85 28.9	28.8			
20 00	268 20.4	226 18.2	N13 19.2	226 25.5	N15 19.4	298 50.1	S13 20.9	100 31.2	N 7 28.8	Mirfak	308 45.1	N49 53
01	283 22.9	241 18.1	19.9	241 26.2	20.0	313 52.6	20.9	115 33.6	28.7	Nunki	76 01.6	S26 17
02	298 25.3	256 18.0	20.6	256 26.8	20.5	328 55.0	20.9	130 35.9	28.7	Peacock	53 23.3	S56 42
03	313 27.8	271 17.9	21.4	271 27.5	21.1	343 57.5	20.9	145 38.3	28.6	Pollux	243 31.7	N28 00
04	328 30.3	286 17.8	22.1	286 28.2	21.7	359 00.0	21.0	160 40.6	28.5	Procyon	245 03.1	N 5 12
05	343 32.7	301 17.7	22.9	301 28.8	22.2	14 02.5	21.0	175 42.9	28.5			
06	358 35.2	316 17.7	N13 23.6	316 29.5	N15 22.8	29 05.0	S13 21.0	190 45.3	N 7 28.4	Rasalhague	96 08.9	N12 33
07	13 37.6	331 17.6	24.4	331 30.2	23.3	44 07.5	21.0	205 47.6	28.3	Regulus	207 46.8	N11 55
S 08	28 40.1	346 17.5	25.1	346 30.8	23.9	59 10.0	21.1	220 49.9	28.3	Rigel	281 15.2	S 8 11
A 09	43 42.6	1 17.4	25.8	1 31.5	24.5	74 12.5	21.1	235 52.3	28.2	Rigil Kent.	139 55.5	S60 52
T 10	58 45.0	16 17.3	26.6	16 32.1	25.0	89 15.0	21.1	250 54.6	28.2	Sabik	102 15.6	S15 44
U 11	73 47.5	31 17.2	27.3	31 32.8	25.6	104 17.5	21.1	265 56.9	28.1			
R 12	88 50.0	46 17.1	N13 28.1	46 33.5	N15 26.1	119 20.0	S13 21.1	280 59.3	N 7 28.0	Schedar	349 44.2	N56 35
D 13	103 52.4	61 17.0	28.8	61 34.1	26.7	134 22.5	21.2	296 01.6	28.0	Shaula	96 25.5	S37 06
A 14	118 54.9	76 16.9	29.6	76 34.8	27.3	149 25.0	21.2	311 04.0	27.9	Sirius	258 36.7	S16 43
Y 15	133 57.4	91 16.8	30.3	91 35.5	27.8	164 27.5	21.2	326 06.3	27.9	Spica	158 34.3	S11 12
16	148 59.8	106 16.7	31.0	106 36.1	28.4	179 30.0	21.2	341 08.6	27.8	Suhail	222 55.0	S43 28
17	164 02.3	121 16.7	31.8	121 36.8	28.9	194 32.5	21.3	356 11.0	27.7			
18	179 04.8	136 16.6	N13 32.5	136 37.5	N15 29.5	209 35.0	S13 21.3	11 13.3	N 7 27.7	Vega	80 40.6	N38 47
19	194 07.2	151 16.5	33.3	151 38.1	30.1	224 37.5	21.3	26 15.6	27.6	Zuben'ubi	137 08.5	S16 05
20	209 09.7	166 16.4	34.0	166 38.8	30.6	239 40.1	21.3	41 18.0	27.6		SHA	Mer. Pass
21	224 12.1	181 16.3	34.7	181 39.4	31.2	254 42.6	21.4	56 20.3	27.5		° ′	h m
22	239 14.6	196 16.2	35.5	196 40.1	31.7	269 45.1	21.4	71 22.6	27.4	Venus	318 58.7	8 55
23	254 17.1	211 16.1	36.2	211 40.8	32.3	284 47.6	21.4	86 25.0	27.4	Mars	318 48.3	8 55
	h m									Jupiter	30 29.0	4 08
Mer. Pass.	6 09.6	v −0.1	d 0.7	v 0.7	d 0.6	v 2.5	d 0.0	v 2.3	d 0.1	Saturn	192 13.9	17 19

UT d h	SUN GHA ° ′	SUN Dec ° ′	MOON GHA ° ′	v ′	MOON Dec ° ′	d ′	HP ′
18 00	179 44.3	N23 24.2	248 06.9	11.8	N13 17.7	13.1	57.7
01	194 44.2	24.2	262 37.7	11.6	13 30.8	13.1	57.7
02	209 44.0	24.3	277 08.3	11.6	13 43.9	13.0	57.7
03	224 43.9	. . 24.4	291 38.9	11.4	13 56.9	13.0	57.8
04	239 43.8	24.4	306 09.3	11.3	14 09.9	12.9	57.8
05	254 43.6	24.5	320 39.6	11.2	14 22.8	12.8	57.9
06	269 43.5	N23 24.5	335 09.8	11.2	N14 35.6	12.8	57.9
07	284 43.4	24.6	349 40.0	11.0	14 48.4	12.8	57.9
08	299 43.2	24.6	4 10.0	10.9	15 01.2	12.6	58.0
09	314 43.1	. . 24.7	18 39.9	10.7	15 13.8	12.7	58.0
10	329 42.9	24.7	33 09.6	10.7	15 26.5	12.5	58.1
11	344 42.8	24.8	47 39.3	10.6	15 39.0	12.5	58.1
12	359 42.7	N23 24.8	62 08.9	10.4	N15 51.5	12.5	58.1
13	14 42.5	24.9	76 38.3	10.3	16 04.0	12.3	58.2
14	29 42.4	24.9	91 07.6	10.2	16 16.3	12.3	58.2
15	44 42.3	. . 24.9	105 36.8	10.1	16 28.6	12.2	58.3
16	59 42.1	25.0	120 05.9	10.0	16 40.8	12.2	58.3
17	74 42.0	25.0	134 34.9	9.8	16 53.0	12.1	58.3
18	89 41.9	N23 25.1	149 03.7	9.8	N17 05.1	12.0	58.4
19	104 41.7	25.1	163 32.5	9.6	17 17.1	11.9	58.4
20	119 41.6	25.2	178 01.1	9.5	17 29.0	11.9	58.5
21	134 41.4	. . 25.2	192 29.6	9.3	17 40.9	11.7	58.5
22	149 41.3	25.2	206 57.9	9.3	17 52.6	11.7	58.5
23	164 41.2	25.3	221 26.2	9.1	18 04.3	11.6	58.6
19 00	179 41.0	N23 25.3	235 54.3	9.0	N18 15.9	11.5	58.6
01	194 40.9	25.4	250 22.3	8.8	18 27.4	11.5	58.7
02	209 40.8	25.4	264 50.1	8.8	18 38.9	11.3	58.7
03	224 40.6	. . 25.4	279 17.9	8.6	18 50.2	11.2	58.7
04	239 40.5	25.5	293 45.5	8.5	19 01.4	11.2	58.8
05	254 40.3	25.5	308 13.0	8.3	19 12.6	11.0	58.8
06	269 40.2	N23 25.5	322 40.3	8.2	N19 23.6	11.0	58.9
07	284 40.1	25.6	337 07.5	8.1	19 34.6	10.8	58.9
08	299 39.9	25.6	351 34.6	8.0	19 45.4	10.8	58.9
09	314 39.8	. . 25.6	6 01.6	7.8	19 56.2	10.7	59.0
10	329 39.7	25.7	20 28.4	7.8	20 06.9	10.5	59.0
11	344 39.5	25.7	34 55.2	7.5	20 17.4	10.4	59.0
12	359 39.4	N23 25.7	49 21.7	7.5	N20 27.8	10.4	59.1
13	14 39.3	25.8	63 48.2	7.3	20 38.2	10.2	59.1
14	29 39.1	25.8	78 14.5	7.2	20 48.4	10.1	59.2
15	44 39.0	. . 25.8	92 40.7	7.0	20 58.5	10.0	59.2
16	59 38.8	25.9	107 06.7	7.0	21 08.5	9.9	59.2
17	74 38.7	25.9	121 32.7	6.8	21 18.4	9.7	59.3
18	89 38.6	N23 25.9	135 58.5	6.6	N21 28.1	9.7	59.3
19	104 38.4	25.9	150 24.1	6.6	21 37.8	9.5	59.4
20	119 38.3	26.0	164 49.7	6.4	21 47.3	9.4	59.4
21	134 38.2	. . 26.0	179 15.1	6.3	21 56.7	9.2	59.4
22	149 38.0	26.0	193 40.4	6.1	22 05.9	9.2	59.5
23	164 37.9	26.0	208 05.5	6.0	22 15.1	9.0	59.5
20 00	179 37.7	N23 26.0	222 30.5	5.9	N22 24.1	8.8	59.5
01	194 37.6	26.1	236 55.4	5.8	22 32.9	8.8	59.6
02	209 37.5	26.1	251 20.2	5.6	22 41.7	8.5	59.6
03	224 37.3	. . 26.1	265 44.8	5.5	22 50.2	8.5	59.6
04	239 37.2	26.1	280 09.3	5.4	22 58.7	8.3	59.7
05	254 37.1	26.1	294 33.7	5.2	23 07.0	8.2	59.7
06	269 36.9	N23 26.2	308 57.9	5.2	N23 15.2	8.0	59.7
07	284 36.8	26.2	323 22.1	5.0	23 23.2	7.9	59.8
08	299 36.6	26.2	337 46.1	4.8	23 31.1	7.7	59.8
09	314 36.5	. . 26.2	352 09.9	4.8	23 38.8	7.6	59.9
10	329 36.4	26.2	6 33.7	4.6	23 46.4	7.5	59.9
11	344 36.2	26.2	20 57.3	4.5	23 53.9	7.2	59.9
12	359 36.1	N23 26.2	35 20.8	4.4	N24 01.1	7.1	60.0
13	14 36.0	26.3	49 44.2	4.3	24 08.2	7.0	60.0
14	29 35.8	26.3	64 07.5	4.2	24 15.2	6.8	60.0
15	44 35.7	. . 26.3	78 30.7	4.0	24 22.0	6.7	60.1
16	59 35.5	26.3	92 53.7	4.0	24 28.7	6.4	60.1
17	74 35.4	26.3	107 16.7	3.8	24 35.1	6.3	60.1
18	89 35.3	N23 26.3	121 39.5	3.7	N24 41.4	6.2	60.1
19	104 35.1	26.3	136 02.2	3.6	24 47.6	6.0	60.2
20	119 35.0	26.3	150 24.8	3.5	24 53.6	5.8	60.2
21	134 34.9	. . 26.3	164 47.3	3.4	24 59.4	5.6	60.2
22	149 34.7	26.3	179 09.7	3.3	25 05.0	5.4	60.3
23	164 34.6	26.3	193 32.0	3.2	N25 10.4	5.3	60.3
	SD 15.8	*d* 0.0	SD	15.8	16.1		16.3

Lat. °	Twilight Naut. h m	Twilight Civil h m	Sunrise h m	Moonrise 18 h m	Moonrise 19 h m	Moonrise 20 h m	Moonrise 21 h m
N 72	□	□	□	21 03	□	□	□
N 70	□	□	□	22 06	□	□	□
68	□	□	□	22 42	21 31	□	□
66	□	□	□	23 08	22 45	□	□
64	////	////	01 31	23 28	23 22	23 14	□
62	////	////	02 09	23 45	23 49	24 01	00 01
60	////	00 49	02 35	23 58	24 10	00 10	00 32
N 58	////	01 40	02 56	00 00	00 10	00 27	00 56
56	////	02 10	03 13	00 07	00 21	00 42	01 15
54	00 45	02 33	03 27	00 13	00 30	00 54	01 31
52	01 32	02 51	03 39	00 18	00 38	01 05	01 45
50	02 00	03 06	03 50	00 23	00 46	01 15	01 57
45	02 46	03 35	04 13	00 34	01 01	01 36	02 22
N 40	03 16	03 58	04 31	00 43	01 15	01 53	02 42
35	03 39	04 16	04 46	00 51	01 26	02 08	02 59
30	03 58	04 32	04 59	00 58	01 36	02 20	03 13
20	04 27	04 57	05 21	01 10	01 53	02 42	03 38
N 10	04 50	05 17	05 40	01 21	02 08	03 01	04 00
0	05 09	05 35	05 58	01 31	02 22	03 18	04 20
S 10	05 26	05 52	06 15	01 41	02 36	03 36	04 40
20	05 42	06 10	06 34	01 52	02 52	03 55	05 02
30	05 59	06 29	06 55	02 05	03 09	04 18	05 28
35	06 07	06 39	07 07	02 12	03 20	04 31	05 43
40	06 17	06 51	07 22	02 20	03 32	04 46	06 00
45	06 27	07 04	07 38	02 30	03 46	05 04	06 21
S 50	06 39	07 21	07 59	02 42	04 03	05 27	06 48
52	06 44	07 28	08 09	02 48	04 12	05 38	07 01
54	06 50	07 36	08 20	02 54	04 21	05 50	07 16
56	06 56	07 45	08 33	03 01	04 31	06 04	07 33
58	07 03	07 56	08 48	03 09	04 43	06 21	07 54
S 60	07 11	08 07	09 05	03 18	04 57	06 41	08 21

Lat. °	Sunset h m	Twilight Civil h m	Twilight Naut. h m	Moonset 18 h m	Moonset 19 h m	Moonset 20 h m	Moonset 21 h m
N 72	□	□	□	18 51	□	□	□
N 70	□	□	□	17 50	□	□	□
68	□	□	□	17 15	20 21	□	□
66	□	□	□	16 51	19 07	□	□
64	22 32	////	////	16 32	18 31	20 45	□
62	21 54	////	////	16 16	18 05	19 58	21 36
60	21 27	23 14	////	16 04	17 45	19 28	20 56
N 58	21 07	22 23	////	15 52	17 29	19 05	20 29
56	20 50	21 53	////	15 43	17 15	18 46	20 08
54	20 36	21 30	23 18	15 34	17 02	18 30	19 50
52	20 23	21 12	22 31	15 27	16 52	18 17	19 35
50	20 13	20 57	22 03	15 20	16 42	18 05	19 22
45	19 50	20 27	21 17	15 06	16 23	17 41	18 55
N 40	19 32	20 05	20 46	14 54	16 07	17 21	18 33
35	19 17	19 47	20 23	14 44	15 53	17 05	18 15
30	19 04	19 31	20 05	14 35	15 42	16 51	18 00
20	18 42	19 06	19 35	14 20	15 22	16 27	17 34
N 10	18 23	18 46	19 13	14 07	15 04	16 06	17 12
0	18 05	18 28	18 54	13 55	14 48	15 47	16 51
S 10	17 48	18 10	18 37	13 43	14 32	15 28	16 30
20	17 29	17 53	18 21	13 30	14 15	15 08	16 08
30	17 08	17 34	18 04	13 15	13 55	14 44	15 42
35	16 55	17 24	17 55	13 06	13 44	14 30	15 26
40	16 41	17 12	17 46	12 57	13 31	14 14	15 09
45	16 24	16 58	17 35	12 45	13 16	13 55	14 47
S 50	16 04	16 42	17 24	12 32	12 57	13 32	14 20
52	15 54	16 35	17 18	12 25	12 48	13 20	14 07
54	15 43	16 26	17 13	12 18	12 38	13 07	13 52
56	15 30	16 17	17 07	12 11	12 27	12 53	13 34
58	15 15	16 07	17 00	12 02	12 15	12 35	13 13
S 60	14 58	15 55	16 52	11 52	12 00	12 15	12 46

Day d	SUN Eqn. of Time 00^h m s	SUN Eqn. of Time 12^h m s	SUN Mer. Pass. h m	MOON Mer. Pass. Upper h m	MOON Mer. Pass. Lower h m	MOON Age d	MOON Phase %
18	01 02	01 09	12 01	07 43	20 08	25	24
19	01 16	01 22	12 01	08 35	21 03	26	15
20	01 29	01 35	12 02	09 33	22 03	27	8

UT (d h)	ARIES GHA	VENUS −4.3 GHA	VENUS Dec	MARS +1.1 GHA	MARS Dec	JUPITER −2.6 GHA	JUPITER Dec	SATURN +1.0 GHA	SATURN Dec
	° ′	° ′	° ′	° ′	° ′	° ′	° ′	° ′	° ′
21 00 (SUNDAY)	269 19.5	226 16.0	N13 37.0	226 41.4	N15 32.8	299 50.1	S13 21.5	101 27.3	N 7 27.3
01	284 22.0	241 15.9	37.7	241 42.1	33.4	314 52.6	21.5	116 29.6	27.2
02	299 24.5	256 15.8	38.4	256 42.8	34.0	329 55.1	21.5	131 32.0	27.2
03	314 26.9	271 15.7	39.2	271 43.4	34.5	344 57.6	21.5	146 34.3	27.1
04	329 29.4	286 15.6	39.9	286 44.1	35.1	0 00.1	21.6	161 36.6	27.1
05	344 31.9	301 15.5	40.7	301 44.7	35.6	15 02.6	21.6	176 39.0	27.0
06	359 34.3	316 15.4	N13 41.4	316 45.4	N15 36.2	30 05.1	S13 21.6	191 41.3	N 7 26.9
07	14 36.8	331 15.3	42.1	331 46.1	36.7	45 07.6	21.6	206 43.6	26.9
08	29 39.3	346 15.1	42.9	346 46.7	37.3	60 10.1	21.7	221 46.0	26.8
09	44 41.7	1 15.0	43.6	1 47.4	37.8	75 12.6	21.7	236 48.3	26.7
10	59 44.2	16 14.9	44.4	16 48.1	38.4	90 15.1	21.7	251 50.6	26.7
11	74 46.6	31 14.8	45.1	31 48.7	39.0	105 17.6	21.7	266 53.0	26.6
12	89 49.1	46 14.7	N13 45.8	46 49.4	N15 39.5	120 20.1	S13 21.8	281 55.3	N 7 26.6
13	104 51.6	61 14.6	46.6	61 50.0	40.1	135 22.6	21.8	296 57.6	26.5
14	119 54.0	76 14.5	47.3	76 50.7	40.6	150 25.2	21.8	312 00.0	26.4
15	134 56.5	91 14.4	48.0	91 51.4	41.2	165 27.7	21.9	327 02.3	26.4
16	149 59.0	106 14.3	48.8	106 52.0	41.7	180 30.2	21.9	342 04.6	26.3
17	165 01.4	121 14.2	49.5	121 52.7	42.3	195 32.7	21.9	357 06.9	26.2
18	180 03.9	136 14.1	N13 50.3	136 53.4	N15 42.8	210 35.2	S13 21.9	12 09.3	N 7 26.2
19	195 06.4	151 14.0	51.0	151 54.0	43.4	225 37.7	22.0	27 11.6	26.1
20	210 08.8	166 13.8	51.7	166 54.7	43.9	240 40.2	22.0	42 13.9	26.1
21	225 11.3	181 13.7	52.5	181 55.3	44.5	255 42.7	22.0	57 16.3	26.0
22	240 13.8	196 13.6	53.2	196 56.0	45.0	270 45.2	22.1	72 18.6	25.9
23	255 16.2	211 13.5	53.9	211 56.7	45.6	285 47.8	22.1	87 20.9	25.9
22 00 (MONDAY)	270 18.7	226 13.4	N13 54.7	226 57.3	N15 46.1	300 50.3	S13 22.1	102 23.3	N 7 25.8
01	285 21.1	241 13.3	55.4	241 58.0	46.7	315 52.8	22.1	117 25.6	25.7
02	300 23.6	256 13.2	56.1	256 58.7	47.2	330 55.3	22.2	132 27.9	25.7
03	315 26.1	271 13.0	56.9	271 59.3	47.8	345 57.8	22.2	147 30.3	25.6
04	330 28.5	286 12.9	57.6	287 00.0	48.3	1 00.3	22.2	162 32.6	25.5
05	345 31.0	301 12.8	58.3	302 00.6	48.9	16 02.8	22.3	177 34.9	25.5
06	0 33.5	316 12.7	N13 59.1	317 01.3	N15 49.4	31 05.3	S13 22.3	192 37.2	N 7 25.4
07	15 35.9	331 12.6	13 59.8	332 02.0	50.0	46 07.9	22.3	207 39.6	25.4
08	30 38.4	346 12.4	14 00.5	347 02.6	50.5	61 10.4	22.3	222 41.9	25.3
09	45 40.9	1 12.3	01.3	2 03.3	51.1	76 12.9	22.4	237 44.2	25.2
10	60 43.3	16 12.2	02.0	17 03.9	51.6	91 15.4	22.4	252 46.6	25.2
11	75 45.8	31 12.1	02.7	32 04.6	52.2	106 17.9	22.4	267 48.9	25.1
12	90 48.3	46 12.0	N14 03.5	47 05.3	N15 52.7	121 20.4	S13 22.5	282 51.2	N 7 25.0
13	105 50.7	61 11.8	04.2	62 05.9	53.3	136 22.9	22.5	297 53.5	25.0
14	120 53.2	76 11.7	04.9	77 06.6	53.8	151 25.5	22.5	312 55.9	24.9
15	135 55.6	91 11.6	05.7	92 07.2	54.4	166 28.0	22.6	327 58.2	24.8
16	150 58.1	106 11.5	06.4	107 07.9	54.9	181 30.5	22.6	343 00.5	24.8
17	166 00.6	121 11.3	07.1	122 08.6	55.4	196 33.0	22.6	358 02.9	24.7
18	181 03.0	136 11.2	N14 07.9	137 09.2	N15 56.0	211 35.5	S13 22.6	13 05.2	N 7 24.7
19	196 05.5	151 11.1	08.6	152 09.9	56.5	226 38.1	22.7	28 07.5	24.6
20	211 08.0	166 10.9	09.3	167 10.6	57.1	241 40.6	22.7	43 09.8	24.5
21	226 10.4	181 10.8	10.1	182 11.2	57.6	256 43.1	22.7	58 12.2	24.5
22	241 12.9	196 10.7	10.8	197 11.9	58.2	271 45.6	22.8	73 14.5	24.4
23	256 15.4	211 10.6	11.5	212 12.5	58.7	286 48.1	22.8	88 16.8	24.3
23 00 (TUESDAY)	271 17.8	226 10.4	N14 12.2	227 13.2	N15 59.3	301 50.6	S13 22.8	103 19.2	N 7 24.3
01	286 20.3	241 10.3	13.0	242 13.9	15 59.8	316 53.2	22.9	118 21.5	24.2
02	301 22.8	256 10.2	13.7	257 14.5	16 00.3	331 55.7	22.9	133 23.8	24.1
03	316 25.2	271 10.0	14.4	272 15.2	00.9	346 58.2	22.9	148 26.1	24.1
04	331 27.7	286 09.9	15.2	287 15.8	01.4	2 00.7	23.0	163 28.5	24.0
05	346 30.1	301 09.8	15.9	302 16.5	02.0	17 03.2	23.0	178 30.8	23.9
06	1 32.6	316 09.6	N14 16.6	317 17.2	N16 02.5	32 05.8	S13 23.0	193 33.1	N 7 23.9
07	16 35.1	331 09.5	17.4	332 17.8	03.1	47 08.3	23.1	208 35.4	23.8
08	31 37.5	346 09.4	18.1	347 18.5	03.6	62 10.8	23.1	223 37.8	23.7
09	46 40.0	1 09.2	18.8	2 19.1	04.1	77 13.3	23.1	238 40.1	23.7
10	61 42.5	16 09.1	19.5	17 19.8	04.7	92 15.9	23.2	253 42.4	23.6
11	76 44.9	31 08.9	20.3	32 20.5	05.2	107 18.4	23.2	268 44.7	23.6
12	91 47.4	46 08.8	N14 21.0	47 21.1	N16 05.8	122 20.9	S13 23.2	283 47.1	N 7 23.5
13	106 49.9	61 08.7	21.7	62 21.8	06.3	137 23.4	23.3	298 49.4	23.4
14	121 52.3	76 08.5	22.4	77 22.4	06.8	152 26.0	23.3	313 51.7	23.4
15	136 54.8	91 08.4	23.2	92 23.1	07.4	167 28.5	23.3	328 54.0	23.3
16	151 57.3	106 08.2	23.9	107 23.8	07.9	182 31.0	23.4	343 56.4	23.2
17	166 59.7	121 08.1	24.6	122 24.4	08.4	197 33.5	23.4	358 58.7	23.2
18	182 02.2	136 08.0	N14 25.3	137 25.1	N16 09.0	212 36.1	S13 23.4	14 01.0	N 7 23.1
19	197 04.6	151 07.8	26.1	152 25.7	09.5	227 38.6	23.5	29 03.3	23.0
20	212 07.1	166 07.7	26.8	167 26.4	10.1	242 41.1	23.5	44 05.7	23.0
21	227 09.6	181 07.5	27.5	182 27.1	10.6	257 43.6	23.5	59 08.0	22.9
22	242 12.0	196 07.4	28.2	197 27.7	11.1	272 46.2	23.6	74 10.3	22.8
23	257 14.5	211 07.2	29.0	212 28.4	11.7	287 48.7	23.6	89 12.6	22.8
Mer. Pass. (h m)	5 57.8	v −0.1	d 0.7	v 0.7	d 0.5	v 2.5	d 0.0	v 2.3	d 0.1

STARS Name	SHA (° ′)	Dec (° ′)
Acamar	315 20.8	S40 15
Achernar	335 29.0	S57 11
Acrux	173 12.8	S63 09
Adhara	255 15.2	S28 59
Aldebaran	290 53.1	N16 31
Alioth	166 22.9	N55 54
Alkaid	153 00.8	N49 16
Al Na'ir	27 47.0	S46 54
Alnilam	275 49.7	S 1 11
Alphard	217 59.2	S 8 42
Alphecca	126 13.2	N26 41
Alpheratz	357 46.6	N29 08
Altair	62 10.8	N 8 53
Ankaa	353 18.5	S42 14
Antares	112 29.6	S26 27
Arcturus	145 58.2	N19 08
Atria	107 33.6	S69 02
Avior	234 19.9	S59 32
Bellatrix	278 35.5	N 6 21
Betelgeuse	271 04.8	N 7 24
Canopus	263 58.0	S52 42
Capella	280 39.3	N46 00
Deneb	49 33.2	N45 18
Denebola	182 36.6	N14 31
Diphda	348 58.9	S17 55
Dubhe	193 55.2	N61 42
Elnath	278 16.7	N28 36
Eltanin	90 47.0	N51 29
Enif	33 49.8	N 9 55
Fomalhaut	15 27.0	S29 34
Gacrux	172 04.3	S57 10
Gienah	175 55.4	S17 35
Hadar	148 52.0	S60 25
Hamal	328 04.3	N23 30
Kaus Aust.	83 47.3	S34 22
Kochab	137 18.7	N74 07
Markab	13 41.2	N15 15
Menkar	314 18.4	N 4 07
Menkent	148 10.9	S36 25
Miaplacidus	221 41.2	S69 45
Mirfak	308 45.1	N49 53
Nunki	76 01.6	S26 17
Peacock	53 23.2	S56 42
Pollux	243 31.6	N28 00
Procyon	245 03.1	N 5 12
Rasalhague	96 08.8	N12 33
Regulus	207 46.8	N11 55
Rigel	281 15.2	S 8 11
Rigil Kent.	139 55.5	S60 52.
Sabik	102 15.6	S15 44.
Schedar	349 44.2	N56 35.
Shaula	96 25.5	S37 06.
Sirius	258 36.7	S16 43.
Spica	158 34.3	S11 12
Suhail	222 55.0	S43 28
Vega	80 40.5	N38 47.
Zuben'ubi	137 08.5	S16 05

	SHA (° ′)	Mer. Pas (h m)
Venus	315 54.7	8 55
Mars	316 38.6	8 52
Jupiter	30 31.6	3 56
Saturn	192 04.6	17 08

UT	SUN GHA	SUN Dec	MOON GHA	v	MOON Dec	d	HP
d h	° ′	° ′	° ′	′	° ′	′	′
21 00	179 34.5	N23 26.3	207 54.2	3.1	N25 15.7	5.1	60.3
01	194 34.3	26.4	222 16.3	3.0	25 20.8	5.0	60.4
02	209 34.2	26.4	236 38.3	2.9	25 25.8	4.7	60.4
03	224 34.0	. . 26.4	251 00.2	2.8	25 30.5	4.6	60.4
04	239 33.9	26.4	265 22.0	2.7	25 35.1	4.3	60.4
05	254 33.8	26.4	279 43.7	2.7	25 39.4	4.2	60.5
06	269 33.6	N23 26.4	294 05.4	2.5	N25 43.6	4.0	60.5
07	284 33.5	26.4	308 26.9	2.5	25 47.6	3.9	60.5
08	299 33.4	26.4	322 48.4	2.4	25 51.5	3.6	60.6
09	314 33.2	. . 26.4	337 09.8	2.3	25 55.1	3.4	60.6
10	329 33.1	26.4	351 31.1	2.2	25 58.5	3.3	60.6
11	344 32.9	26.3	5 52.3	2.2	26 01.8	3.0	60.6
12	359 32.8	N23 26.3	20 13.5	2.0	N26 04.8	2.9	60.7
13	14 32.7	26.3	34 34.5	2.1	26 07.7	2.7	60.7
14	29 32.5	26.3	48 55.6	1.9	26 10.4	2.4	60.7
15	44 32.4	. . 26.3	63 16.5	1.9	26 12.8	2.3	60.7
16	59 32.3	26.3	77 37.4	1.9	26 15.1	2.1	60.7
17	74 32.1	26.3	91 58.3	1.7	26 17.2	1.9	60.8
18	89 32.0	N23 26.3	106 19.0	1.8	N26 19.1	1.6	60.8
19	104 31.8	26.3	120 39.8	1.6	26 20.7	1.5	60.8
20	119 31.7	26.3	135 00.4	1.7	26 22.2	1.3	60.8
21	134 31.6	. . 26.3	149 21.1	1.6	26 23.5	1.0	60.9
22	149 31.4	26.3	163 41.7	1.5	26 24.5	0.9	60.9
23	164 31.3	26.3	178 02.2	1.5	26 25.4	0.7	60.9
22 00	179 31.2	N23 26.2	192 22.7	1.5	N26 26.1	0.4	60.9
01	194 31.0	26.2	206 43.2	1.4	26 26.5	0.3	60.9
02	209 30.9	26.2	221 03.6	1.5	26 26.8	0.0	61.0
03	224 30.8	. . 26.2	235 24.1	1.4	26 26.8	0.1	61.0
04	239 30.6	26.2	249 44.5	1.3	26 26.7	0.4	61.0
05	254 30.5	26.2	264 04.8	1.4	26 26.3	0.5	61.0
06	269 30.3	N23 26.1	278 25.2	1.3	N26 25.8	0.8	61.0
07	284 30.2	26.1	292 45.5	1.4	26 25.0	1.0	61.0
08	299 30.1	26.1	307 05.9	1.3	26 24.0	1.2	61.0
09	314 29.9	. . 26.1	321 26.2	1.3	26 22.8	1.4	61.1
10	329 29.8	26.1	335 46.5	1.3	26 21.4	1.6	61.1
11	344 29.7	26.1	350 06.8	1.3	26 19.8	1.8	61.1
12	359 29.5	N23 26.0	4 27.1	1.4	N26 18.0	2.0	61.1
13	14 29.4	26.0	18 47.5	1.3	26 16.0	2.2	61.1
14	29 29.2	26.0	33 07.8	1.3	26 13.8	2.4	61.1
15	44 29.1	. . 26.0	47 28.1	1.4	26 11.4	2.7	61.1
16	59 29.0	25.9	61 48.5	1.4	26 08.7	2.8	61.1
17	74 28.8	25.9	76 08.9	1.4	26 05.9	3.0	61.2
18	89 28.7	N23 25.9	90 29.3	1.4	N26 02.9	3.3	61.2
19	104 28.6	25.9	104 49.7	1.4	25 59.6	3.4	61.2
20	119 28.4	25.8	119 10.1	1.5	25 56.2	3.7	61.2
21	134 28.3	. . 25.8	133 30.6	1.5	25 52.5	3.8	61.2
22	149 28.2	25.8	147 51.1	1.6	25 48.7	4.1	61.2
23	164 28.0	25.7	162 11.7	1.6	25 44.6	4.2	61.2
23 00	179 27.9	N23 25.7	176 32.3	1.6	N25 40.4	4.5	61.2
01	194 27.7	25.7	190 52.9	1.7	25 35.9	4.6	61.2
02	209 27.6	25.7	205 13.6	1.7	25 31.3	4.9	61.2
03	224 27.5	. . 25.6	219 34.3	1.8	25 26.4	5.0	61.2
04	239 27.3	25.6	233 55.1	1.8	25 21.4	5.3	61.2
05	254 27.2	25.6	248 15.9	1.9	25 16.1	5.4	61.2
06	269 27.1	N23 25.5	262 36.8	2.0	N25 10.7	5.7	61.2
07	284 26.9	25.5	276 57.8	2.0	25 05.0	5.8	61.2
08	299 26.8	25.4	291 18.8	2.1	24 59.2	6.0	61.2
09	314 26.7	. . 25.4	305 39.9	2.1	24 53.2	6.2	61.2
10	329 26.5	25.4	320 01.0	2.3	24 47.0	6.4	61.2
11	344 26.4	25.3	334 22.3	2.3	24 40.6	6.6	61.2
12	359 26.3	N23 25.3	348 43.6	2.3	N24 34.0	6.8	61.2
13	14 26.1	25.3	3 04.9	2.5	24 27.2	6.9	61.2
14	29 26.0	25.2	17 26.4	2.5	24 20.3	7.1	61.2
15	44 25.8	. . 25.2	31 47.9	2.6	24 13.2	7.4	61.2
16	59 25.7	25.1	46 09.5	2.7	24 05.8	7.5	61.2
17	74 25.6	25.1	60 31.2	2.8	23 58.3	7.6	61.2
18	89 25.4	N23 25.1	74 53.0	2.8	N23 50.7	7.9	61.2
19	104 25.3	25.0	89 14.8	3.0	23 42.8	8.0	61.2
20	119 25.2	25.0	103 36.8	3.0	23 34.8	8.2	61.2
21	134 25.0	. . 24.9	117 58.8	3.2	23 26.6	8.4	61.2
22	149 24.9	24.9	132 21.0	3.2	23 18.2	8.5	61.2
23	164 24.8	24.8	146 43.2	3.3	N23 09.7	8.8	61.2
	SD 15.8	d 0.0	SD 16.5		16.6		16.7

Lat.	Twilight Naut.	Twilight Civil	Sunrise	Moonrise 21	Moonrise 22	Moonrise 23	Moonrise 24
°	h m	h m	h m	h m	h m	h m	h m
N 72	▭	▭	▭	▭	▭	▭	▭
N 70	▭	▭	▭	▭	▭	▭	▭
68	▭	▭	▭	▭	▭	▭	▭
66	▭	▭	▭	▭	▭	▭	02 52
64	////	////	01 31	▭	▭	01 09	03 36
62	////	////	02 09	00 01	00 40	02 08	04 05
60	////	00 49	02 36	00 32	01 19	02 41	04 27
N 58	////	01 41	02 57	00 56	01 47	03 06	04 45
56	////	02 11	03 13	01 15	02 08	03 26	05 01
54	00 45	02 33	03 28	01 31	02 26	03 42	05 13
52	01 32	02 51	03 40	01 45	02 41	03 56	05 25
50	02 00	03 06	03 51	01 57	02 54	04 09	05 35
45	02 46	03 36	04 13	02 22	03 21	04 34	05 56
N 40	03 17	03 59	04 32	02 42	03 43	04 54	06 13
35	03 40	04 17	04 47	02 59	04 01	05 11	06 27
30	03 59	04 32	05 00	03 13	04 16	05 26	06 39
20	04 28	04 57	05 22	03 38	04 42	05 51	07 00
N 10	04 51	05 18	05 41	04 00	05 05	06 12	07 18
0	05 10	05 36	05 58	04 20	05 26	06 32	07 35
S 10	05 27	05 53	06 16	04 40	05 47	06 52	07 52
20	05 43	06 10	06 34	05 02	06 09	07 13	08 10
30	05 59	06 29	06 56	05 28	06 36	07 37	08 31
35	06 08	06 40	07 08	05 43	06 51	07 52	08 43
40	06 18	06 52	07 22	06 00	07 09	08 08	08 56
45	06 28	07 05	07 39	06 21	07 31	08 28	09 13
S 50	06 40	07 21	08 00	06 48	07 58	08 53	09 33
52	06 45	07 29	08 10	07 01	08 12	09 05	09 42
54	06 51	07 37	08 21	07 16	08 27	09 18	09 52
56	06 57	07 46	08 34	07 33	08 45	09 34	10 04
58	07 04	07 56	08 48	07 54	09 07	09 52	10 18
S 60	07 11	08 08	09 06	08 21	09 36	10 15	10 34

Lat.	Sunset	Twilight Civil	Twilight Naut.	Moonset 21	Moonset 22	Moonset 23	Moonset 24
°	h m	h m	h m	h m	h m	h m	h m
N 72	▭	▭	▭	▭	▭	▭	▭
N 70	▭	▭	▭	▭	▭	▭	▭
68	▭	▭	▭	▭	▭	▭	23 59
66	▭	▭	▭	▭	▭	▭	{00 00, 23 31}
64	22 33	////	////	▭	23 26	23 15	23 08
62	21 54	////	////	21 36	22 27	22 45	22 51
60	21 28	23 14	////	20 56	21 53	22 22	22 36
N 58	21 07	22 23	////	20 29	21 28	22 03	22 23
56	20 51	21 53	////	20 08	21 08	21 47	22 12
54	20 36	21 31	23 19	19 50	20 51	21 34	22 02
52	20 24	21 13	22 31	19 35	20 37	21 22	21 53
50	20 13	20 58	22 03	19 22	20 24	21 11	21 46
45	19 51	20 28	21 18	18 55	19 58	20 49	21 29
N 40	19 33	20 05	20 47	18 33	19 38	20 31	21 15
35	19 17	19 47	20 24	18 15	19 20	20 16	21 03
30	19 04	19 32	20 05	18 00	19 05	20 03	20 53
20	18 42	19 07	19 36	17 34	18 40	19 41	20 35
N 10	18 23	18 46	19 13	17 12	18 18	19 21	20 19
0	18 06	18 28	18 54	16 51	17 57	19 02	20 04
S 10	17 48	18 11	18 37	16 30	17 36	18 44	19 49
20	17 30	17 54	18 21	16 08	17 14	18 24	19 33
30	17 08	17 35	18 05	15 42	16 48	18 00	19 14
35	16 56	17 24	17 56	15 26	16 33	17 47	19 03
40	16 42	17 12	17 47	15 09	16 15	17 31	18 50
45	16 25	16 59	17 36	14 47	15 54	17 12	18 35
S 50	16 04	16 43	17 24	14 20	15 26	16 48	18 17
52	15 54	16 35	17 19	14 07	15 13	16 36	18 08
54	15 43	16 27	17 13	13 52	14 58	16 23	17 58
56	15 30	16 18	17 07	13 34	14 40	16 08	17 47
58	15 16	16 08	17 00	13 13	14 18	15 50	17 34
S 60	14 58	15 56	16 53	12 46	13 50	15 28	17 19

Day	SUN Eqn. of Time 00ʰ	SUN Eqn. of Time 12ʰ	SUN Mer. Pass.	MOON Mer. Pass. Upper	MOON Mer. Pass. Lower	MOON Age	MOON Phase
d	m s	m s	h m	h m	h m	d	%
21	01 42	01 49	12 02	10 35	23 08	28	3
22	01 55	02 02	12 02	11 41	24 14	29	0
23	02 08	02 15	12 02	12 47	00 14	01	1

UT	ARIES	VENUS −4.3		MARS +1.1		JUPITER −2.6		SATURN +1.0		STARS		
	GHA	GHA	Dec	GHA	Dec	GHA	Dec	GHA	Dec	Name	SHA	Dec
d h	° ′	° ′	° ′	° ′	° ′	° ′	° ′	° ′	° ′		° ′	° ′
24 00	272 17.0	226 07.1	N14 29.7	227 29.0	N16 12.2	302 51.2	S13 23.6	104 15.0	N 7 22.7	Acamar	315 20.8	S40 15.
01	287 19.4	241 06.9	30.4	242 29.7	12.7	317 53.7	23.7	119 17.3	22.6	Achernar	335 29.0	S57 10.
02	302 21.9	256 06.8	31.1	257 30.4	13.3	332 56.3	23.7	134 19.6	22.6	Acrux	173 12.8	S63 09.
03	317 24.4	271 06.7	. . 31.9	272 31.0	. . 13.8	347 58.8	. . 23.7	149 21.9	. . 22.5	Adhara	255 15.2	S28 59.
04	332 26.8	286 06.5	32.6	287 31.7	14.4	3 01.3	23.8	164 24.3	22.4	Aldebaran	290 53.1	N16 31.
05	347 29.3	301 06.4	33.3	302 32.3	14.9	18 03.9	23.8	179 26.6	22.4			
06	2 31.8	316 06.2	N14 34.0	317 33.0	N16 15.4	33 06.4	S13 23.8	194 28.9	N 7 22.3	Alioth	166 22.9	N55 54.
W 07	17 34.2	331 06.1	34.8	332 33.7	16.0	48 08.9	23.9	209 31.2	22.2	Alkaid	153 00.8	N49 16.
E 08	32 36.7	346 05.9	35.5	347 34.3	16.5	63 11.4	23.9	224 33.5	22.2	Al Na'ir	27 47.0	S46 54.
D 09	47 39.1	1 05.8	. . 36.2	2 35.0	. . 17.0	78 14.0	. . 23.9	239 35.9	. . 22.1	Alnilam	275 49.7	S 1 11.
N 10	62 41.6	16 05.6	36.9	17 35.6	17.6	93 16.5	24.0	254 38.2	22.0	Alphard	217 59.2	S 8 42.
E 11	77 44.1	31 05.4	37.6	32 36.3	18.1	108 19.0	24.0	269 40.5	22.0			
S 12	92 46.5	46 05.3	N14 38.4	47 36.9	N16 18.6	123 21.6	S13 24.0	284 42.8	N 7 21.9	Alphecca	126 13.2	N26 41.
D 13	107 49.0	61 05.1	39.1	62 37.6	19.2	138 24.1	24.1	299 45.2	21.8	Alpheratz	357 46.6	N29 08.
A 14	122 51.5	76 05.0	39.8	77 38.3	19.7	153 26.6	24.1	314 47.5	21.8	Altair	62 10.8	N 8 53.
Y 15	137 53.9	91 04.8	. . 40.5	92 38.9	. . 20.2	168 29.2	. . 24.1	329 49.8	. . 21.7	Ankaa	353 18.5	S42 14.
16	152 56.4	106 04.7	41.2	107 39.6	20.8	183 31.7	24.2	344 52.1	21.6	Antares	112 29.5	S26 27.
17	167 58.9	121 04.5	42.0	122 40.2	21.3	198 34.2	24.2	359 54.4	21.6			
18	183 01.3	136 04.4	N14 42.7	137 40.9	N16 21.8	213 36.8	S13 24.3	14 56.8	N 7 21.5	Arcturus	145 58.2	N19 08.
19	198 03.8	151 04.2	43.4	152 41.6	22.4	228 39.3	24.3	29 59.1	21.4	Atria	107 33.6	S69 02.
20	213 06.2	166 04.0	44.1	167 42.2	22.9	243 41.8	24.3	45 01.4	21.4	Avior	234 19.9	S59 32.
21	228 08.7	181 03.9	. . 44.8	182 42.9	. . 23.4	258 44.4	. . 24.4	60 03.7	. . 21.3	Bellatrix	278 35.5	N 6 21.
22	243 11.2	196 03.7	45.6	197 43.5	23.9	273 46.9	24.4	75 06.0	21.2	Betelgeuse	271 04.8	N 7 24.
23	258 13.6	211 03.6	46.3	212 44.2	24.5	288 49.4	24.4	90 08.4	21.2			
25 00	273 16.1	226 03.4	N14 47.0	227 44.9	N16 25.0	303 52.0	S13 24.5	105 10.7	N 7 21.1	Canopus	263 58.0	S52 42.
01	288 18.6	241 03.2	47.7	242 45.5	25.5	318 54.5	24.5	120 13.0	21.0	Capella	280 39.3	N46 00.
02	303 21.0	256 03.1	48.4	257 46.2	26.1	333 57.0	24.6	135 15.3	21.0	Deneb	49 33.2	N45 18.
03	318 23.5	271 02.9	. . 49.1	272 46.8	. . 26.6	348 59.6	. . 24.6	150 17.6	. . 20.9	Denebola	182 36.7	N14 31.
04	333 26.0	286 02.7	49.9	287 47.5	27.1	4 02.1	24.6	165 20.0	20.8	Diphda	348 58.8	S17 55.
05	348 28.4	301 02.6	50.6	302 48.1	27.6	19 04.7	24.7	180 22.3	20.8			
06	3 30.9	316 02.4	N14 51.3	317 48.8	N16 28.2	34 07.2	S13 24.7	195 24.6	N 7 20.7	Dubhe	193 55.2	N61 42.
07	18 33.4	331 02.2	52.0	332 49.5	28.7	49 09.7	24.7	210 26.9	20.6	Elnath	278 16.7	N28 36.
T 08	33 35.8	346 02.1	52.7	347 50.1	29.2	64 12.3	24.8	225 29.2	20.6	Eltanin	90 47.0	N51 29.
H 09	48 38.3	1 01.9	. . 53.4	2 50.8	. . 29.8	79 14.8	. . 24.8	240 31.6	. . 20.5	Enif	33 49.8	N 9 55.
U 10	63 40.7	16 01.7	54.2	17 51.4	30.3	94 17.3	24.9	255 33.9	20.4	Fomalhaut	15 27.0	S29 34.
R 11	78 43.2	31 01.6	54.9	32 52.1	30.8	109 19.9	24.9	270 36.2	20.4			
S 12	93 45.7	46 01.4	N14 55.6	47 52.7	N16 31.3	124 22.4	S13 24.9	285 38.5	N 7 20.3	Gacrux	172 04.4	S57 10.
D 13	108 48.1	61 01.2	56.3	62 53.4	31.9	139 25.0	25.0	300 40.8	20.2	Gienah	175 55.4	S17 35.
A 14	123 50.6	76 01.1	57.0	77 54.1	32.4	154 27.5	25.0	315 43.2	20.2	Hadar	148 52.0	S60 25.
Y 15	138 53.1	91 00.9	. . 57.7	92 54.7	. . 32.9	169 30.0	. . 25.0	330 45.5	. . 20.1	Hamal	328 04.3	N23 30.
16	153 55.5	106 00.7	58.4	107 55.4	33.4	184 32.6	25.1	345 47.8	20.0	Kaus Aust.	83 47.3	S34 22.
17	168 58.0	121 00.6	59.2	122 56.0	34.0	199 35.1	25.1	0 50.1	19.9			
18	184 00.5	136 00.4	N14 59.9	137 56.7	N16 34.5	214 37.7	S13 25.2	15 52.4	N 7 19.9	Kochab	137 18.8	N74 07.
19	199 02.9	151 00.2	15 00.6	152 57.4	35.0	229 40.2	25.2	30 54.7	19.8	Markab	13 41.2	N15 15.
20	214 05.4	166 00.0	01.3	167 58.0	35.5	244 42.7	25.2	45 57.1	19.7	Menkar	314 18.4	N 4 07.
21	229 07.9	180 59.9	. . 02.0	182 58.7	. . 36.1	259 45.3	. . 25.3	60 59.4	. . 19.7	Menkent	148 10.9	S36 25.
22	244 10.3	195 59.7	02.7	197 59.3	36.6	274 47.8	25.3	76 01.7	19.6	Miaplacidus	221 41.3	S69 45.
23	259 12.8	210 59.5	03.4	213 00.0	37.1	289 50.4	25.4	91 04.0	19.5			
26 00	274 15.2	225 59.3	N15 04.1	228 00.6	N16 37.6	304 52.9	S13 25.4	106 06.3	N 7 19.5	Mirfak	308 45.1	N49 53.
01	289 17.7	240 59.2	04.8	243 01.3	38.1	319 55.5	25.4	121 08.6	19.4	Nunki	76 01.6	S26 17.
02	304 20.2	255 59.0	05.6	258 02.0	38.7	334 58.0	25.5	136 11.0	19.3	Peacock	53 23.2	S56 42.
03	319 22.6	270 58.8	. . 06.3	273 02.6	. . 39.2	350 00.5	. . 25.5	151 13.3	. . 19.3	Pollux	243 31.6	N28 00.
04	334 25.1	285 58.6	07.0	288 03.3	39.7	5 03.1	25.6	166 15.6	19.2	Procyon	245 03.1	N 5 12.
05	349 27.6	300 58.4	07.7	303 03.9	40.2	20 05.6	25.6	181 17.9	19.1			
06	4 30.0	315 58.3	N15 08.4	318 04.6	N16 40.8	35 08.2	S13 25.6	196 20.2	N 7 19.1	Rasalhague	96 08.8	N12 33.
07	19 32.5	330 58.1	09.1	333 05.2	41.3	50 10.7	25.7	211 22.5	19.0	Regulus	207 46.8	N11 55.
08	34 35.0	345 57.9	09.8	348 05.9	41.8	65 13.3	25.7	226 24.9	18.9	Rigel	281 15.2	S 8 11.
F 09	49 37.4	0 57.7	. . 10.5	3 06.6	. . 42.3	80 15.8	. . 25.8	241 27.2	. . 18.8	Rigil Kent.	139 55.5	S60 52.
R 10	64 39.9	15 57.5	11.2	18 07.2	42.8	95 18.4	25.8	256 29.5	18.8	Sabik	102 15.6	S15 44.
I 11	79 42.4	30 57.3	11.9	33 07.9	43.4	110 20.9	25.8	271 31.8	18.7			
D 12	94 44.8	45 57.2	N15 12.6	48 08.5	N16 43.9	125 23.4	S13 25.9	286 34.1	N 7 18.6	Schedar	349 44.2	N56 35.
A 13	109 47.3	60 57.0	13.3	63 09.2	44.4	140 26.0	25.9	301 36.4	18.6	Shaula	96 25.5	S37 06.
Y 14	124 49.7	75 56.8	14.0	78 09.8	44.9	155 28.5	26.0	316 38.8	18.5	Sirius	258 36.7	S16 43.
15	139 52.2	90 56.6	. . 14.8	93 10.5	. . 45.4	170 31.1	. . 26.0	331 41.1	. . 18.4	Spica	158 34.3	S11 12.
16	154 54.7	105 56.4	15.5	108 11.2	45.9	185 33.6	26.0	346 43.4	18.4	Suhail	222 55.0	S43 28.
17	169 57.1	120 56.2	16.2	123 11.8	46.5	200 36.2	26.1	1 45.7	18.3			
18	184 59.6	135 56.0	N15 16.9	138 12.5	N16 47.0	215 38.7	S13 26.1	16 48.0	N 7 18.2	Vega	80 40.5	N38 47.
19	200 02.1	150 55.9	17.6	153 13.1	47.5	230 41.3	26.2	31 50.3	18.2	Zuben'ubi	137 08.5	S16 05.
20	215 04.5	165 55.7	18.3	168 13.8	48.0	245 43.8	26.2	46 52.6	18.1		SHA	Mer. Pas
21	230 07.0	180 55.5	. . 19.0	183 14.4	. . 48.5	260 46.4	. . 26.3	61 55.0	. . 18.0		° ′	h m
22	245 09.5	195 55.3	19.7	198 15.1	49.0	275 48.9	26.3	76 57.3	17.9	Venus	312 47.3	8 56
23	260 11.9	210 55.1	20.4	213 15.7	49.6	290 51.5	26.3	91 59.6	17.9	Mars	314 28.7	8 49
Mer. Pass.	h m 5 46.0	*v* −0.2	*d* 0.7	*v* 0.7	*d* 0.5	*v* 2.5	*d* 0.0	*v* 2.3	*d* 0.1	Jupiter	30 35.9	3 44
										Saturn	191 54.6	16 57

UT	SUN GHA	SUN Dec	MOON GHA	v	MOON Dec	d	HP
d h	° ′	° ′	° ′	′	° ′	′	′
24 00	179 24.6	N23 24.8	161 05.5	3.5	N23 00.9	8.8	61.2
01	194 24.5	24.7	175 28.0	3.5	22 52.1	9.1	61.2
02	209 24.4	24.7	189 50.5	3.6	22 43.0	9.2	61.2
03	224 24.2	. . 24.6	204 13.1	3.7	22 33.8	9.3	61.2
04	239 24.1	24.6	218 35.8	3.9	22 24.5	9.6	61.2
05	254 24.0	24.5	232 58.7	3.9	22 14.9	9.6	61.2
06	269 23.8	N23 24.5	247 21.6	4.1	N22 05.3	9.9	61.1
07	284 23.7	24.4	261 44.7	4.1	21 55.4	10.0	61.1
08	299 23.5	24.4	276 07.8	4.3	21 45.4	10.1	61.1
09	314 23.4	. . 24.3	290 31.1	4.4	21 35.3	10.3	61.1
10	329 23.3	24.3	304 54.5	4.4	21 25.0	10.4	61.1
11	344 23.1	24.2	319 17.9	4.6	21 14.6	10.6	61.1
12	359 23.0	N23 24.2	333 41.5	4.8	N21 04.0	10.7	61.1
13	14 22.9	24.1	348 05.3	4.8	20 53.3	10.8	61.1
14	29 22.7	24.0	2 29.1	4.9	20 42.5	11.0	61.0
15	44 22.6	. . 24.0	16 53.0	5.0	20 31.5	11.2	61.0
16	59 22.5	23.9	31 17.0	5.2	20 20.3	11.2	61.0
17	74 22.3	23.9	45 41.2	5.3	20 09.1	11.4	61.0
18	89 22.2	N23 23.8	60 05.5	5.4	N19 57.7	11.5	61.0
19	104 22.1	23.7	74 29.9	5.5	19 46.2	11.7	61.0
20	119 21.9	23.7	88 54.4	5.6	19 34.5	11.8	60.9
21	134 21.8	. . 23.6	103 19.0	5.7	19 22.7	11.8	60.9
22	149 21.7	23.6	117 43.7	5.9	19 10.9	12.1	60.9
23	164 21.5	23.5	132 08.6	6.0	18 58.8	12.1	60.9
25 00	179 21.4	N23 23.4	146 33.6	6.1	N18 46.7	12.2	60.9
01	194 21.3	23.4	160 58.7	6.2	18 34.5	12.4	60.9
02	209 21.1	23.3	175 23.9	6.3	18 22.1	12.5	60.8
03	224 21.0	. . 23.2	189 49.2	6.4	18 09.6	12.6	60.8
04	239 20.9	23.2	204 14.6	6.6	17 57.0	12.7	60.8
05	254 20.7	23.1	218 40.2	6.6	17 44.3	12.8	60.8
06	269 20.6	N23 23.0	233 05.8	6.8	N17 31.5	12.9	60.7
07	284 20.5	23.0	247 31.6	6.9	17 18.6	13.0	60.7
08	299 20.3	22.9	261 57.5	7.1	17 05.6	13.0	60.7
09	314 20.2	. . 22.8	276 23.6	7.1	16 52.6	13.2	60.7
10	329 20.1	22.7	290 49.7	7.2	16 39.4	13.3	60.7
11	344 19.9	22.7	305 15.9	7.4	16 26.1	13.4	60.6
12	359 19.8	N23 22.6	319 42.3	7.5	N16 12.7	13.5	60.6
13	14 19.7	22.5	334 08.8	7.5	15 59.2	13.5	60.6
14	29 19.5	22.5	348 35.3	7.7	15 45.7	13.7	60.6
15	44 19.4	. . 22.4	3 02.0	7.8	15 32.0	13.7	60.5
16	59 19.3	22.3	17 28.8	8.0	15 18.3	13.8	60.5
17	74 19.1	22.2	31 55.8	8.0	15 04.5	13.9	60.5
18	89 19.0	N23 22.1	46 22.8	8.1	N14 50.6	14.0	60.5
19	104 18.9	22.1	60 49.9	8.3	14 36.6	14.0	60.4
20	119 18.7	22.0	75 17.2	8.3	14 22.6	14.1	60.4
21	134 18.6	. . 21.9	89 44.5	8.5	14 08.5	14.2	60.4
22	149 18.5	21.8	104 12.0	8.6	13 54.3	14.2	60.3
23	164 18.3	21.8	118 39.6	8.7	13 40.1	14.4	60.3
26 00	179 18.2	N23 21.7	133 07.3	8.7	N13 25.7	14.3	60.3
01	194 18.1	21.6	147 35.0	8.9	13 11.4	14.5	60.3
02	209 17.9	21.5	162 02.9	9.0	12 56.9	14.5	60.2
03	224 17.8	. . 21.4	176 30.9	9.1	12 42.4	14.6	60.2
04	239 17.7	21.3	190 59.0	9.2	12 27.8	14.6	60.2
05	254 17.5	21.3	205 27.2	9.3	12 13.2	14.7	60.1
06	269 17.4	N23 21.2	219 55.5	9.4	N11 58.5	14.7	60.1
07	284 17.3	21.1	234 23.9	9.4	11 43.8	14.8	60.1
08	299 17.1	21.0	248 52.3	9.6	11 29.0	14.8	60.0
09	314 17.0	. . 20.9	263 20.9	9.7	11 14.2	14.9	60.0
10	329 16.9	20.8	277 49.6	9.8	10 59.3	14.9	60.0
11	344 16.7	20.7	292 18.4	9.8	10 44.4	15.0	60.0
12	359 16.6	N23 20.6	306 47.2	10.0	N10 29.4	15.0	59.9
13	14 16.5	20.5	321 16.2	10.1	10 14.4	15.0	59.9
14	29 16.3	20.5	335 45.3	10.1	9 59.4	15.1	59.9
15	44 16.2	. . 20.4	350 14.4	10.2	9 44.3	15.1	59.8
16	59 16.1	20.3	4 43.6	10.3	9 29.2	15.2	59.8
17	74 16.0	20.2	19 12.9	10.4	9 14.0	15.2	59.8
18	89 15.8	N23 20.1	33 42.3	10.5	N 8 58.8	15.2	59.7
19	104 15.7	20.0	48 11.8	10.6	8 43.6	15.3	59.7
20	119 15.6	19.9	62 41.4	10.6	8 28.3	15.2	59.7
21	134 15.4	. . 19.8	77 11.0	10.8	8 13.1	15.3	59.6
22	149 15.3	19.7	91 40.8	10.8	7 57.8	15.4	59.6
23	164 15.2	19.6	106 10.6	10.9	N 7 42.4	15.3	59.6
	SD 15.8	*d* 0.1	SD	16.6		16.5	16.3

Lat.	Twilight Naut.	Twilight Civil	Sunrise	Moonrise 24	Moonrise 25	Moonrise 26	Moonrise 27
°	h m	h m	h m	h m	h m	h m	h m
N 72	▭	▭	▭	▭	▭	06 45	09 14
N 70	▭	▭	▭	▭	04 18	07 06	09 22
68	▭	▭	▭	▭	04 59	07 23	09 28
66	▭	▭	▭	02 52	05 26	07 36	09 32
64	////	////	01 33	03 36	05 47	07 47	09 37
62	////	////	02 11	04 05	06 04	07 56	09 40
60	////	00 52	02 37	04 27	06 18	08 04	09 43
N 58	////	01 42	02 58	04 45	06 30	08 11	09 46
56	////	02 12	03 14	05 01	06 40	08 17	09 48
54	00 48	02 34	03 29	05 13	06 49	08 22	09 51
52	01 34	02 52	03 41	05 25	06 57	08 27	09 53
50	02 02	03 07	03 52	05 35	07 04	08 31	09 54
45	02 47	03 37	04 14	05 56	07 19	08 41	09 58
N 40	03 18	04 00	04 32	06 13	07 32	08 49	10 02
35	03 41	04 18	04 47	06 27	07 42	08 55	10 04
30	04 00	04 33	05 00	06 39	07 52	09 01	10 07
20	04 29	04 58	05 22	07 00	08 08	09 11	10 11
N 10	04 51	05 18	05 41	07 18	08 21	09 20	10 15
0	05 10	05 36	05 59	07 35	08 34	09 28	10 18
S 10	05 27	05 54	06 16	07 52	08 47	09 36	10 22
20	05 43	06 11	06 35	08 10	09 01	09 45	10 26
30	06 00	06 30	06 56	08 31	09 16	09 55	10 30
35	06 09	06 40	07 08	08 43	09 25	10 01	10 32
40	06 18	06 52	07 23	08 56	09 35	10 07	10 35
45	06 28	07 06	07 39	09 13	09 47	10 15	10 38
S 50	06 40	07 22	08 00	09 33	10 01	10 24	10 42
52	06 45	07 29	08 10	09 42	10 08	10 28	10 44
54	06 51	07 37	08 21	09 52	10 15	10 32	10 46
56	06 57	07 46	08 34	10 04	10 24	10 37	10 48
58	07 04	07 57	08 48	10 18	10 33	10 43	10 50
S 60	07 11	08 08	09 06	10 34	10 43	10 49	10 52

Lat.	Sunset	Twilight Civil	Twilight Naut.	Moonset 24	Moonset 25	Moonset 26	Moonset 27
°	h m	h m	h m	h m	h m	h m	h m
N 72	▭	▭	▭	▭	▭	{00 13 / 23 31}	22 59
N 70	▭	▭	▭	▭	{00 42 / 23 49}	23 20	22 57
68	▭	▭	▭	23 59	23 31	23 12	22 56
66	▭	▭	▭	{00 00 / 23 31}	23 16	23 04	22 55
64	22 32	////	////	23 08	23 03	22 58	22 53
62	21 54	////	////	22 51	22 53	22 53	22 52
60	21 28	23 12	////	22 36	22 43	22 48	22 52
N 58	21 07	22 23	////	22 23	22 35	22 44	22 51
56	20 51	21 53	////	22 12	22 28	22 40	22 50
54	20 36	21 31	23 17	22 02	22 22	22 37	22 50
52	20 24	21 13	22 31	21 53	22 16	22 34	22 49
50	20 13	20 58	22 03	21 46	22 11	22 31	22 48
45	19 51	20 28	21 18	21 29	22 00	22 25	22 47
N 40	19 33	20 06	20 47	21 15	21 50	22 20	22 46
35	19 18	19 48	20 24	21 03	21 42	22 15	22 45
30	19 05	19 32	20 06	20 53	21 35	22 11	22 45
20	18 43	19 07	19 37	20 35	21 22	22 04	22 43
N 10	18 24	18 47	19 14	20 19	21 11	21 58	22 42
0	18 06	18 29	18 55	20 04	21 00	21 52	22 41
S 10	17 49	18 12	18 38	19 49	20 50	21 46	22 40
20	17 30	17 54	18 22	19 33	20 38	21 40	22 39
30	17 09	17 36	18 05	19 14	20 25	21 33	22 37
35	16 57	17 25	17 57	19 03	20 17	21 29	22 36
40	16 43	17 13	17 47	18 50	20 09	21 24	22 35
45	16 26	17 00	17 37	18 35	19 58	21 18	22 34
S 50	16 05	16 44	17 25	18 17	19 46	21 11	22 33
52	15 55	16 36	17 20	18 08	19 40	21 08	22 32
54	15 44	16 28	17 14	17 58	19 33	21 05	22 31
56	15 32	16 19	17 08	17 47	19 26	21 01	22 30
58	15 17	16 09	17 01	17 34	19 18	20 56	22 30
S 60	15 00	15 57	16 54	17 19	19 08	20 51	22 29

Day	SUN Eqn. of Time 00^h	SUN Eqn. of Time 12^h	SUN Mer. Pass.	MOON Mer. Pass. Upper	MOON Mer. Pass. Lower	Age	Phase
d	m s	m s	h m	h m	h m	d	%
24	02 21	02 28	12 02	13 50	01 19	02	4
25	02 34	02 41	12 03	14 47	02 19	03	11
26	02 47	02 53	12 03	15 40	03 14	04	19

	UT	ARIES	VENUS −4.2		MARS +1.1		JUPITER −2.7		SATURN +1.0		STARS		
	d h	GHA ° ′	GHA ° ′	Dec ° ′	GHA ° ′	Dec ° ′	GHA ° ′	Dec ° ′	GHA ° ′	Dec ° ′	Name	SHA ° ′	Dec °
	27 00	275 14.4	225 54.9	N15 21.1	228 16.4	N16 50.1	305 54.0	S13 26.4	107 01.9	N 7 17.8	Acamar	315 20.7	S40 15
	01	290 16.8	240 54.7	21.8	243 17.1	50.6	320 56.6	26.4	122 04.2	17.7	Achernar	335 28.9	S57 10
	02	305 19.3	255 54.5	22.5	258 17.7	51.1	335 59.1	26.5	137 06.5	17.7	Acrux	173 12.8	S63 09
	03	320 21.8	270 54.3	. . 23.2	273 18.4	. . 51.6	351 01.7	. . 26.5	152 08.8	. . 17.6	Adhara	255 15.2	S28 59
	04	335 24.2	285 54.1	23.9	288 19.0	52.1	6 04.2	26.5	167 11.2	17.5	Aldebaran	290 53.1	N16 31
	05	350 26.7	300 53.9	24.6	303 19.7	52.7	21 06.8	26.6	182 13.5	17.5			
	06	5 29.2	315 53.7	N15 25.3	318 20.3	N16 53.2	36 09.3	S13 26.6	197 15.8	N 7 17.4	Alioth	166 23.0	N55 54
	07	20 31.6	330 53.5	26.0	333 21.0	53.7	51 11.9	26.7	212 18.1	17.3	Alkaid	153 00.9	N49 16
S	08	35 34.1	345 53.3	26.7	348 21.7	54.2	66 14.4	26.7	227 20.4	17.2	Al Na'ir	27 47.0	S46 54
A	09	50 36.6	0 53.1	. . 27.4	3 22.3	. . 54.7	81 17.0	. . 26.8	242 22.7	. . 17.2	Alnilam	275 49.7	S 1 11
T	10	65 39.0	15 52.9	28.1	18 23.0	55.2	96 19.5	26.8	257 25.0	17.1	Alphard	217 59.2	S 8 42
U	11	80 41.5	30 52.8	28.8	33 23.6	55.7	111 22.1	26.9	272 27.3	17.0			
R	12	95 44.0	45 52.6	N15 29.5	48 24.3	N16 56.2	126 24.7	S13 26.9	287 29.7	N 7 17.0	Alphecca	126 13.2	N26 41
D	13	110 46.4	60 52.4	30.2	63 24.9	56.8	141 27.2	26.9	302 32.0	16.9	Alpheratz	357 46.6	N29 08
A	14	125 48.9	75 52.1	30.9	78 25.6	57.3	156 29.8	27.0	317 34.3	16.8	Altair	62 10.8	N 8 53
Y	15	140 51.3	90 51.9	. . 31.6	93 26.2	. . 57.8	171 32.3	. . 27.0	332 36.6	. . 16.8	Ankaa	353 18.4	S42 14
	16	155 53.8	105 51.7	32.3	108 26.9	58.3	186 34.9	27.1	347 38.9	16.7	Antares	112 29.5	S26 27
	17	170 56.3	120 51.5	33.0	123 27.6	58.8	201 37.4	27.1	2 41.2	16.6			
	18	185 58.7	135 51.3	N15 33.7	138 28.2	N16 59.3	216 40.0	S13 27.2	17 43.5	N 7 16.5	Arcturus	145 58.2	N19 08
	19	201 01.2	150 51.1	34.4	153 28.9	16 59.8	231 42.5	27.2	32 45.8	16.5	Atria	107 33.6	S69 02
	20	216 03.7	165 50.9	35.1	168 29.5	17 00.3	246 45.1	27.2	47 48.1	16.4	Avior	234 19.9	S59 32
	21	231 06.1	180 50.7	. . 35.8	183 30.2	. . 00.8	261 47.7	. . 27.3	62 50.5	. . 16.3	Bellatrix	278 35.5	N 6 21
	22	246 08.6	195 50.5	36.5	198 30.8	01.3	276 50.2	27.3	77 52.8	16.3	Betelgeuse	271 04.8	N 7 24
	23	261 11.1	210 50.3	37.2	213 31.5	01.9	291 52.8	27.4	92 55.1	16.2			
	28 00	276 13.5	225 50.1	N15 37.9	228 32.1	N17 02.4	306 55.3	S13 27.4	107 57.4	N 7 16.1	Canopus	263 58.0	S52 42
	01	291 16.0	240 49.9	38.6	243 32.8	02.9	321 57.9	27.5	122 59.7	16.0	Capella	280 39.3	N46 00
	02	306 18.5	255 49.7	39.3	258 33.5	03.4	337 00.4	27.5	138 02.0	16.0	Deneb	49 33.2	N45 18
	03	321 20.9	270 49.5	. . 39.9	273 34.1	. . 03.9	352 03.0	. . 27.6	153 04.3	. . 15.9	Denebola	182 36.7	N14 31
	04	336 23.4	285 49.3	40.6	288 34.8	04.4	7 05.6	27.6	168 06.6	15.8	Diphda	348 58.8	S17 55
	05	351 25.8	300 49.1	41.3	303 35.4	04.9	22 08.1	27.7	183 08.9	15.8			
	06	6 28.3	315 48.9	N15 42.0	318 36.1	N17 05.4	37 10.7	S13 27.7	198 11.3	N 7 15.7	Dubhe	193 55.2	N61 42
	07	21 30.8	330 48.6	42.7	333 36.7	05.9	52 13.2	27.7	213 13.6	15.6	Elnath	278 16.7	N28 36
	08	36 33.2	345 48.4	43.4	348 37.4	06.4	67 15.8	27.8	228 15.9	15.5	Eltanin	90 47.0	N51 29
S	09	51 35.7	0 48.2	. . 44.1	3 38.0	. . 06.9	82 18.4	. . 27.8	243 18.2	. . 15.5	Enif	33 49.8	N 9 55
U	10	66 38.2	15 48.0	44.8	18 38.7	07.4	97 20.9	27.9	258 20.5	15.4	Fomalhaut	15 27.0	S29 34
N	11	81 40.6	30 47.8	45.5	33 39.4	07.9	112 23.5	27.9	273 22.8	15.3			
D	12	96 43.1	45 47.6	N15 46.2	48 40.0	N17 08.4	127 26.1	S13 28.0	288 25.1	N 7 15.3	Gacrux	172 04.4	S57 10
A	13	111 45.6	60 47.4	46.9	63 40.7	08.9	142 28.6	28.0	303 27.4	15.2	Gienah	175 55.4	S17 35
Y	14	126 48.0	75 47.1	47.6	78 41.3	09.4	157 31.2	28.1	318 29.7	15.1	Hadar	148 52.0	S60 25
	15	141 50.5	90 46.9	. . 48.2	93 42.0	. . 09.9	172 33.7	. . 28.1	333 32.0	. . 15.0	Hamal	328 04.3	N23 30
	16	156 52.9	105 46.7	48.9	108 42.6	10.5	187 36.3	28.2	348 34.3	15.0	Kaus Aust.	83 47.3	S34 22
	17	171 55.4	120 46.5	49.6	123 43.3	11.0	202 38.9	28.2	3 36.7	14.9			
	18	186 57.9	135 46.3	N15 50.3	138 43.9	N17 11.5	217 41.4	S13 28.3	18 39.0	N 7 14.8	Kochab	137 18.8	N74 07
	19	202 00.3	150 46.1	51.0	153 44.6	12.0	232 44.0	28.3	33 41.3	14.7	Markab	13 41.2	N15 15
	20	217 02.8	165 45.8	51.7	168 45.2	12.5	247 46.6	28.3	48 43.6	14.7	Menkar	314 18.3	N 4 07
	21	232 05.3	180 45.6	. . 52.4	183 45.9	. . 13.0	262 49.1	. . 28.4	63 45.9	. . 14.6	Menkent	148 10.9	S36 25
	22	247 07.7	195 45.4	53.1	198 46.6	13.5	277 51.7	28.4	78 48.2	14.5	Miaplacidus	221 41.3	S69 45
	23	262 10.2	210 45.2	53.8	213 47.2	14.0	292 54.2	28.5	93 50.5	14.5			
	29 00	277 12.7	225 44.9	N15 54.4	228 47.9	N17 14.5	307 56.8	S13 28.5	108 52.8	N 7 14.4	Mirfak	308 45.0	N49 53
	01	292 15.1	240 44.7	55.1	243 48.5	15.0	322 59.4	28.6	123 55.1	14.3	Nunki	76 01.5	S26 17
	02	307 17.6	255 44.5	55.8	258 49.2	15.5	338 01.9	28.6	138 57.4	14.2	Peacock	53 23.2	S56 42
	03	322 20.1	270 44.3	. . 56.5	273 49.8	. . 16.0	353 04.5	. . 28.7	153 59.7	. . 14.2	Pollux	243 31.6	N28 00
	04	337 22.5	285 44.1	57.2	288 50.5	16.5	8 07.1	28.7	169 02.0	14.1	Procyon	245 03.1	N 5 12
	05	352 25.0	300 43.8	57.9	303 51.1	17.0	23 09.6	28.8	184 04.3	14.0			
	06	7 27.4	315 43.6	N15 58.5	318 51.8	N17 17.5	38 12.2	S13 28.8	199 06.6	N 7 14.0	Rasalhague	96 08.8	N12 33
	07	22 29.9	330 43.4	59.2	333 52.4	18.0	53 14.8	28.9	214 09.0	13.9	Regulus	207 46.8	N11 55
	08	37 32.4	345 43.1	15 59.9	348 53.1	18.5	68 17.3	28.9	229 11.3	13.8	Rigel	281 15.2	S 8 11
M	09	52 34.8	0 42.9	16 00.6	3 53.7	. . 19.0	83 19.9	. . 29.0	244 13.6	. . 13.7	Rigil Kent.	139 55.6	S60 52
O	10	67 37.3	15 42.7	01.3	18 54.4	19.5	98 22.5	29.0	259 15.9	13.7	Sabik	102 15.6	S15 44
N	11	82 39.8	30 42.5	02.0	33 55.1	20.0	113 25.1	29.1	274 18.2	13.6			
D	12	97 42.2	45 42.2	N16 02.6	48 55.7	N17 20.5	128 27.6	S13 29.1	289 20.5	N 7 13.5	Schedar	349 44.1	N56 35
A	13	112 44.7	60 42.0	03.3	63 56.4	21.0	143 30.2	29.2	304 22.8	13.4	Shaula	96 25.5	S37 06
Y	14	127 47.2	75 41.8	04.0	78 57.0	21.5	158 32.8	29.2	319 25.1	13.4	Sirius	258 36.7	S16 43
	15	142 49.6	90 41.5	. . 04.7	93 57.7	. . 21.9	173 35.3	. . 29.3	334 27.4	. . 13.3	Spica	158 34.3	S11 12
	16	157 52.1	105 41.3	05.4	108 58.3	22.4	188 37.9	29.3	349 29.7	13.2	Suhail	222 55.1	S43 28
	17	172 54.5	120 41.1	06.0	123 59.0	22.9	203 40.5	29.4	4 32.0	13.1			
	18	187 57.0	135 40.8	N16 06.7	138 59.6	N17 23.4	218 43.0	S13 29.4	19 34.3	N 7 13.1	Vega	80 40.5	N38 47
	19	202 59.5	150 40.6	07.4	154 00.3	23.9	233 45.6	29.5	34 36.6	13.0	Zuben'ubi	137 08.5	S16 05
	20	218 01.9	165 40.4	08.1	169 00.9	24.4	248 48.2	29.5	49 38.9	12.9			
	21	233 04.4	180 40.1	. . 08.8	184 01.6	. . 24.9	263 50.8	. . 29.6	64 41.2	. . 12.9			
	22	248 06.9	195 39.9	09.4	199 02.2	25.4	278 53.3	29.6	79 43.5	12.8			
	23	263 09.3	210 39.7	10.1	214 02.9	25.9	293 55.9	29.7	94 45.8	12.7			
	Mer. Pass.	h m 5 34.2	v −0.2	d 0.7	v 0.7	d 0.5	v 2.6	d 0.0	v 2.3	d 0.1			

	SHA ° ′	Mer. Pass. h m
Venus	309 36.6	8 57
Mars	312 18.6	8 45
Jupiter	30 41.8	3 32
Saturn	191 43.9	16 46

UT	SUN GHA	SUN Dec	MOON GHA	v	MOON Dec	d	HP
d h	° ′	° ′	° ′	′	° ′	′	′
27 00	179 15.0	N23 19.5	120 40.5	11.0	N 7 27.1	15.4	59.5
01	194 14.9	19.4	135 10.5	11.0	7 11.7	15.4	59.5
02	209 14.8	19.3	149 40.5	11.1	6 56.3	15.4	59.5
03	224 14.6	. . 19.2	164 10.6	11.2	6 40.9	15.4	59.4
04	239 14.5	19.1	178 40.8	11.3	6 25.5	15.5	59.4
05	254 14.4	19.0	193 11.1	11.3	6 10.0	15.4	59.4
06	269 14.3	N23 18.9	207 41.4	11.4	N 5 54.6	15.5	59.3
07	284 14.1	18.8	222 11.8	11.5	5 39.1	15.5	59.3
08	299 14.0	18.7	236 42.3	11.6	5 23.6	15.5	59.2
09	314 13.9	. . 18.6	251 12.9	11.6	5 08.1	15.4	59.2
10	329 13.7	18.5	265 43.5	11.6	4 52.7	15.5	59.2
11	344 13.6	18.4	280 14.1	11.8	4 37.2	15.5	59.1
12	359 13.5	N23 18.3	294 44.9	11.8	N 4 21.7	15.6	59.1
13	14 13.3	18.2	309 15.7	11.8	4 06.1	15.5	59.1
14	29 13.2	18.0	323 46.5	11.9	3 50.6	15.5	59.0
15	44 13.1	. . 17.9	338 17.4	12.0	3 35.1	15.5	59.0
16	59 13.0	17.8	352 48.4	12.1	3 19.6	15.5	59.0
17	74 12.8	17.7	7 19.5	12.0	3 04.1	15.5	58.9
18	89 12.7	N23 17.6	21 50.5	12.2	N 2 48.6	15.5	58.9
19	104 12.6	17.5	36 21.7	12.2	2 33.1	15.4	58.9
20	119 12.4	17.4	50 52.9	12.2	2 17.7	15.5	58.8
21	134 12.3	. . 17.3	65 24.1	12.3	2 02.2	15.5	58.8
22	149 12.2	17.1	79 55.4	12.4	1 46.7	15.4	58.7
23	164 12.1	17.0	94 26.8	12.4	1 31.3	15.5	58.7
28 00	179 11.9	N23 16.9	108 58.2	12.4	N 1 15.8	15.4	58.7
01	194 11.8	16.8	123 29.6	12.5	1 00.4	15.4	58.6
02	209 11.7	16.7	138 01.1	12.5	0 45.0	15.4	58.6
03	224 11.5	. . 16.6	152 32.6	12.6	0 29.6	15.4	58.6
04	239 11.4	16.4	167 04.2	12.6	N 0 14.2	15.4	58.5
05	254 11.3	16.3	181 35.8	12.6	S 0 01.2	15.3	58.5
06	269 11.2	N23 16.2	196 07.4	12.7	S 0 16.5	15.3	58.5
07	284 11.0	16.1	210 39.1	12.7	0 31.8	15.3	58.4
08	299 10.9	16.0	225 10.8	12.8	0 47.1	15.3	58.4
09	314 10.8	. . 15.8	239 42.6	12.8	1 02.4	15.2	58.3
10	329 10.6	15.7	254 14.4	12.8	1 17.6	15.3	58.3
11	344 10.5	15.6	268 46.2	12.9	1 32.9	15.2	58.3
12	359 10.4	N23 15.5	283 18.1	12.9	S 1 48.1	15.1	58.2
13	14 10.3	15.3	297 50.0	12.9	2 03.2	15.2	58.2
14	29 10.1	15.2	312 21.9	12.9	2 18.4	15.1	58.2
15	44 10.0	. . 15.1	326 53.8	13.0	2 33.5	15.1	58.1
16	59 09.9	15.0	341 25.8	13.0	2 48.6	15.0	58.1
17	74 09.8	14.8	355 57.8	13.0	3 03.6	15.0	58.1
18	89 09.6	N23 14.7	10 29.8	13.1	S 3 18.6	15.0	58.0
19	104 09.5	14.6	25 01.9	13.0	3 33.6	15.0	58.0
20	119 09.4	14.5	39 33.9	13.1	3 48.6	14.9	57.9
21	134 09.2	. . 14.3	54 06.0	13.1	4 03.5	14.9	57.9
22	149 09.1	14.2	68 38.1	13.1	4 18.4	14.8	57.9
23	164 09.0	14.1	83 10.2	13.2	4 33.2	14.8	57.8
29 00	179 08.9	N23 13.9	97 42.4	13.1	S 4 48.0	14.7	57.8
01	194 08.7	13.8	112 14.5	13.2	5 02.7	14.7	57.8
02	209 08.6	13.7	126 46.7	13.2	5 17.4	14.7	57.7
03	224 08.5	. . 13.5	141 18.9	13.2	5 32.1	14.6	57.7
04	239 08.4	13.4	155 51.1	13.2	5 46.7	14.6	57.7
05	254 08.2	13.3	170 23.3	13.2	6 01.3	14.5	57.6
06	269 08.1	N23 13.1	184 55.5	13.3	S 6 15.8	14.5	57.6
07	284 08.0	13.0	199 27.8	13.2	6 30.3	14.5	57.6
08	299 07.9	12.8	214 00.0	13.3	6 44.8	14.4	57.5
09	314 07.7	. . 12.7	228 32.3	13.2	6 59.2	14.3	57.5
10	329 07.6	12.6	243 04.5	13.3	7 13.5	14.3	57.5
11	344 07.5	12.4	257 36.8	13.2	7 27.8	14.2	57.4
12	359 07.4	N23 12.3	272 09.0	13.3	S 7 42.0	14.2	57.4
13	14 07.2	12.1	286 41.3	13.2	7 56.2	14.2	57.4
14	29 07.1	12.0	301 13.5	13.3	8 10.4	14.0	57.3
15	44 07.0	. . 11.9	315 45.8	13.3	8 24.4	14.1	57.3
16	59 06.9	11.7	330 18.1	13.2	8 38.5	13.9	57.3
17	74 06.7	11.6	344 50.3	13.3	8 52.4	13.9	57.2
18	89 06.6	N23 11.4	359 22.6	13.2	S 9 06.3	13.9	57.2
19	104 06.5	11.3	13 54.8	13.3	9 20.2	13.8	57.1
20	119 06.4	11.1	28 27.1	13.2	9 34.0	13.7	57.1
21	134 06.2	. . 11.0	42 59.3	13.3	9 47.7	13.7	57.1
22	149 06.1	10.8	57 31.6	13.2	10 01.4	13.6	57.0
23	164 06.0	10.7	72 03.8	13.2	S10 15.0	13.6	57.0
	SD 15.8	d 0.1	SD 16.1		15.9		15.6

Lat.	Twilight Naut.	Twilight Civil	Sunrise	Moonrise 27	28	29	30
°	h m	h m	h m	h m	h m	h m	h m
N 72	▭	▭	▭	09 14	11 28	13 40	16 06
N 70	▭	▭	▭	09 22	11 26	13 27	15 35
68	▭	▭	▭	09 28	11 23	13 17	15 13
66	▭	▭	▭	09 32	11 22	13 08	14 56
64	////	////	01 37	09 37	11 20	13 01	14 42
62	////	////	02 13	09 40	11 19	12 55	14 30
60	////	00 57	02 39	09 43	11 18	12 50	14 20
N 58	////	01 45	03 00	09 46	11 17	12 45	14 12
56	////	02 14	03 16	09 48	11 16	12 41	14 04
54	00 52	02 36	03 30	09 51	11 15	12 37	13 58
52	01 36	02 54	03 42	09 53	11 15	12 34	13 52
50	02 04	03 09	03 53	09 54	11 14	12 31	13 46
45	02 49	03 38	04 16	09 58	11 13	12 24	13 35
N 40	03 19	04 01	04 34	10 02	11 11	12 19	13 25
35	03 42	04 19	04 48	10 04	11 10	12 14	13 17
30	04 01	04 34	05 01	10 07	11 10	12 10	13 10
20	04 30	04 59	05 23	10 11	11 08	12 03	12 58
N 10	04 52	05 19	05 42	10 15	11 07	11 57	12 47
0	05 11	05 37	06 00	10 18	11 06	11 52	12 38
S 10	05 28	05 54	06 17	10 22	11 05	11 46	12 28
20	05 44	06 11	06 35	10 26	11 03	11 40	12 17
30	06 00	06 30	06 56	10 30	11 02	11 34	12 06
35	06 09	06 41	07 09	10 32	11 01	11 30	11 59
40	06 18	06 52	07 23	10 35	11 00	11 25	11 51
45	06 29	07 06	07 39	10 38	11 00	11 20	11 43
S 50	06 40	07 22	08 00	10 42	10 58	11 15	11 32
52	06 45	07 29	08 10	10 44	10 58	11 12	11 27
54	06 51	07 37	08 21	10 46	10 57	11 09	11 22
56	06 57	07 46	08 33	10 48	10 57	11 06	11 16
58	07 04	07 56	08 48	10 50	10 56	11 02	11 09
S 60	07 11	08 08	09 05	10 52	10 55	10 58	11 02

Lat.	Sunset	Twilight Civil	Twilight Naut.	Moonset 27	28	29	30
°	h m	h m	h m	h m	h m	h m	h m
N 72	▭	▭	▭	22 59	22 29	21 56	21 08
N 70	▭	▭	▭	22 57	22 36	22 12	21 41
68	▭	▭	▭	22 56	22 41	22 24	22 04
66	▭	▭	▭	22 55	22 45	22 35	22 23
64	22 29	////	////	22 53	22 49	22 44	22 38
62	21 52	////	////	22 52	22 52	22 51	22 51
60	21 27	23 08	////	22 52	22 55	22 58	23 02
N 58	21 07	22 21	////	22 51	22 57	23 03	23 11
56	20 50	21 52	////	22 50	22 59	23 09	23 20
54	20 36	21 30	23 13	22 50	23 01	23 13	23 27
52	20 24	21 12	22 29	22 49	23 03	23 18	23 34
50	20 13	20 57	22 02	22 48	23 05	23 21	23 40
45	19 51	20 28	21 18	22 47	23 08	23 30	23 53
N 40	19 33	20 06	20 47	22 46	23 11	23 37	24 04
35	19 18	19 48	20 24	22 45	23 14	23 43	24 13
30	19 05	19 33	20 06	22 45	23 17	23 48	24 21
20	18 43	19 08	19 37	22 43	23 21	23 58	24 36
N 10	18 24	18 47	19 15	22 42	23 24	24 06	00 06
0	18 07	18 29	18 56	22 41	23 28	24 14	00 14
S 10	17 50	18 12	18 39	22 40	23 31	24 21	00 21
20	17 31	17 55	18 23	22 39	23 35	24 30	00 30
30	17 10	17 37	18 06	22 37	23 39	24 39	00 39
35	16 58	17 26	17 58	22 36	23 41	24 45	00 45
40	16 44	17 14	17 48	22 35	23 44	24 51	00 51
45	16 27	17 01	17 38	22 34	23 47	24 58	00 58
S 50	16 07	16 45	17 27	22 33	23 51	25 07	01 07
52	15 57	16 38	17 21	22 32	23 53	25 11	01 11
54	15 46	16 30	17 16	22 31	23 54	25 16	01 16
56	15 33	16 21	17 10	22 30	23 57	25 21	01 21
58	15 19	16 10	17 03	22 30	23 59	25 26	01 26
S 60	15 02	15 59	16 56	22 29	24 02	00 02	01 33

Day	SUN Eqn. of Time 00h	SUN Eqn. of Time 12h	SUN Mer. Pass.	MOON Mer. Pass. Upper	MOON Mer. Pass. Lower	MOON Age	MOON Phase
d	m s	m s	h m	h m	h m	d %	
27	03 00	03 06	12 03	16 30	04 05	05 29	
28	03 12	03 18	12 03	17 17	04 53	06 40	◑
29	03 24	03 30	12 04	18 03	05 40	07 50	

UT	ARIES GHA	VENUS −4.2 GHA	VENUS Dec	MARS +1.1 GHA	MARS Dec	JUPITER −2.7 GHA	JUPITER Dec	SATURN +1.0 GHA	SATURN Dec
d h	° ′	° ′	° ′	° ′	° ′	° ′	° ′	° ′	° ′
30 00	278 11.8	225 39.4	N16 10.8	229 03.6	N17 26.4	308 58.5	S13 29.7	109 48.1	N 7 12.6
01	293 14.3	240 39.2	11.5	244 04.2	26.9	324 01.1	29.8	124 50.4	12.6
02	308 16.7	255 39.0	12.1	259 04.9	27.4	339 03.6	29.8	139 52.8	12.5
03	323 19.2	270 38.7	. . 12.8	274 05.5	. . 27.9	354 06.2	. . 29.9	154 55.1	. . 12.4
04	338 21.7	285 38.5	13.5	289 06.2	28.4	9 08.8	29.9	169 57.4	12.3
05	353 24.1	300 38.2	14.2	304 06.8	28.9	24 11.3	30.0	184 59.7	12.3
06	8 26.6	315 38.0	N16 14.8	319 07.5	N17 29.4	39 13.9	S13 30.0	200 02.0	N 7 12.2
07	23 29.0	330 37.7	15.5	334 08.1	29.9	54 16.5	30.1	215 04.3	12.1
T 08	38 31.5	345 37.5	16.2	349 08.8	30.3	69 19.1	30.1	230 06.6	12.0
U 09	53 34.0	0 37.3	. . 16.9	4 09.4	. . 30.8	84 21.6	. . 30.2	245 08.9	. . 12.0
E 10	68 36.4	15 37.0	17.5	19 10.1	31.3	99 24.2	30.2	260 11.2	11.9
S 11	83 38.9	30 36.8	18.2	34 10.7	31.8	114 26.8	30.3	275 13.5	11.8
D 12	98 41.4	45 36.5	N16 18.9	49 11.4	N17 32.3	129 29.4	S13 30.3	290 15.8	N 7 11.7
13	113 43.8	60 36.3	19.6	64 12.0	32.8	144 32.0	30.4	305 18.1	11.7
A 14	128 46.3	75 36.0	20.2	79 12.7	33.3	159 34.5	30.4	320 20.4	11.6
Y 15	143 48.8	90 35.8	. . 20.9	94 13.4	. . 33.8	174 37.1	. . 30.5	335 22.7	. . 11.5
16	158 51.2	105 35.5	21.6	109 14.0	34.3	189 39.7	30.5	350 25.0	11.4
17	173 53.7	120 35.3	22.2	124 14.7	34.8	204 42.3	30.6	5 27.3	11.4
18	188 56.2	135 35.0	N16 22.9	139 15.3	N17 35.2	219 44.8	S13 30.6	20 29.6	N 7 11.3
19	203 58.6	150 34.8	23.6	154 16.0	35.7	234 47.4	30.7	35 31.9	11.2
20	219 01.1	165 34.6	24.2	169 16.6	36.2	249 50.0	30.7	50 34.2	11.1
21	234 03.5	180 34.3	. . 24.9	184 17.3	. . 36.7	264 52.6	. . 30.8	65 36.5	. . 11.1
22	249 06.0	195 34.0	25.6	199 17.9	37.2	279 55.2	30.8	80 38.8	11.0
23	264 08.5	210 33.8	26.3	214 18.6	37.7	294 57.7	30.9	95 41.1	10.9
1 00	279 10.9	225 33.5	N16 26.9	229 19.2	N17 38.2	310 00.3	S13 30.9	110 43.4	N 7 10.8
01	294 13.4	240 33.3	27.6	244 19.9	38.6	325 02.9	31.0	125 45.7	10.8
02	309 15.9	255 33.0	28.3	259 20.5	39.1	340 05.5	31.0	140 48.0	10.7
03	324 18.3	270 32.8	. . 28.9	274 21.2	. . 39.6	355 08.1	. . 31.1	155 50.3	. . 10.6
04	339 20.8	285 32.5	29.6	289 21.8	40.1	10 10.6	31.1	170 52.6	10.5
05	354 23.3	300 32.3	30.3	304 22.5	40.6	25 13.2	31.2	185 54.9	10.5
06	9 25.7	315 32.0	N16 30.9	319 23.1	N17 41.1	40 15.8	S13 31.3	200 57.2	N 7 10.4
W 07	24 28.2	330 31.8	31.6	334 23.8	41.6	55 18.4	31.3	215 59.5	10.3
E 08	39 30.6	345 31.5	32.2	349 24.4	42.0	70 21.0	31.4	231 01.8	10.2
D 09	54 33.1	0 31.2	. . 32.9	4 25.1	. . 42.5	85 23.6	. . 31.4	246 04.1	. . 10.2
N 10	69 35.6	15 31.0	33.6	19 25.7	43.0	100 26.1	31.5	261 06.4	10.1
E 11	84 38.0	30 30.7	34.2	34 26.4	43.5	115 28.7	31.5	276 08.7	10.0
S 12	99 40.5	45 30.5	N16 34.9	49 27.1	N17 44.0	130 31.3	S13 31.6	291 11.0	N 7 09.9
D 13	114 43.0	60 30.2	35.6	64 27.7	44.5	145 33.9	31.6	306 13.3	09.9
A 14	129 45.4	75 29.9	36.2	79 28.4	44.9	160 36.5	31.7	321 15.6	09.8
Y 15	144 47.9	90 29.7	. . 36.9	94 29.0	. . 45.4	175 39.1	. . 31.7	336 17.9	. . 09.7
16	159 50.4	105 29.4	37.5	109 29.7	45.9	190 41.6	31.8	351 20.2	09.6
17	174 52.8	120 29.2	38.2	124 30.3	46.4	205 44.2	31.8	6 22.5	09.6
18	189 55.3	135 28.9	N16 38.9	139 31.0	N17 46.9	220 46.8	S13 31.9	21 24.8	N 7 09.5
19	204 57.8	150 28.6	39.5	154 31.6	47.3	235 49.4	32.0	36 27.1	09.4
20	220 00.2	165 28.4	40.2	169 32.3	47.8	250 52.0	32.0	51 29.4	09.3
21	235 02.7	180 28.1	. . 40.8	184 32.9	. . 48.3	265 54.6	. . 32.1	66 31.7	. . 09.3
22	250 05.1	195 27.8	41.5	199 33.6	48.8	280 57.2	32.1	81 34.0	09.2
23	265 07.6	210 27.6	42.2	214 34.2	49.3	295 59.7	32.2	96 36.3	09.1
2 00	280 10.1	225 27.3	N16 42.8	229 34.9	N17 49.7	311 02.3	S13 32.2	111 38.6	N 7 09.0
01	295 12.5	240 27.0	43.5	244 35.5	50.2	326 04.9	32.3	126 40.9	09.0
02	310 15.0	255 26.8	44.1	259 36.2	50.7	341 07.5	32.3	141 43.2	08.9
03	325 17.5	270 26.5	. . 44.8	274 36.8	. . 51.2	356 10.1	. . 32.4	156 45.5	. . 08.8
04	340 19.9	285 26.2	45.4	289 37.5	51.7	11 12.7	32.4	171 47.8	08.7
05	355 22.4	300 26.0	46.1	304 38.1	52.1	26 15.3	32.5	186 50.1	08.7
06	10 24.9	315 25.7	N16 46.7	319 38.8	N17 52.6	41 17.9	S13 32.6	201 52.4	N 7 08.6
07	25 27.3	330 25.4	47.4	334 39.4	53.1	56 20.5	32.6	216 54.7	08.5
T 08	40 29.8	345 25.1	48.0	349 40.1	53.6	71 23.0	32.7	231 57.0	08.4
H 09	55 32.3	0 24.9	. . 48.7	4 40.7	. . 54.0	86 25.6	. . 32.7	246 59.3	. . 08.3
U 10	70 34.7	15 24.6	49.4	19 41.4	54.5	101 28.2	32.8	262 01.6	08.3
R 11	85 37.2	30 24.3	50.0	34 42.0	55.0	116 30.8	32.8	277 03.9	08.2
S 12	100 39.6	45 24.0	N16 50.7	49 42.7	N17 55.5	131 33.4	S13 32.9	292 06.2	N 7 08.1
D 13	115 42.1	60 23.8	51.3	64 43.3	55.9	146 36.0	32.9	307 08.5	08.0
A 14	130 44.6	75 23.5	52.0	79 44.0	56.4	161 38.6	33.0	322 10.8	08.0
Y 15	145 47.0	90 23.2	. . 52.6	94 44.6	. . 56.9	176 41.2	. . 33.1	337 13.1	. . 07.9
16	160 49.5	105 22.9	53.3	109 45.3	57.4	191 43.8	33.1	352 15.4	07.8
17	175 52.0	120 22.7	53.9	124 45.9	57.8	206 46.4	33.2	7 17.7	07.7
18	190 54.4	135 22.4	N16 54.6	139 46.6	N17 58.3	221 49.0	S13 33.2	22 19.9	N 7 07.7
19	205 56.9	150 22.1	55.2	154 47.2	58.8	236 51.6	33.3	37 22.2	07.6
20	220 59.4	165 21.8	55.9	169 47.9	59.3	251 54.1	33.3	52 24.5	07.5
21	236 01.8	180 21.5	. . 56.5	184 48.5	17 59.7	266 56.7	. . 33.4	67 26.8	. . 07.4
22	251 04.3	195 21.3	57.1	199 49.2	18 00.2	281 59.3	33.5	82 29.1	07.3
23	266 06.8	210 21.0	57.8	214 49.8	N18 00.7	297 01.9	33.5	97 31.4	07.3
Mer. Pass.	h m 5 22.4	v −0.3	d 0.7	v 0.7	d 0.5	v 2.6	d 0.1	v 2.3	d 0.1

STARS

Name	SHA	Dec
	° ′	°
Acamar	315 20.7	S40 15
Achernar	335 28.9	S57 10
Acrux	173 12.9	S63 09
Adhara	255 15.2	S28 59
Aldebaran	290 53.1	N16 31
Alioth	166 23.0	N55 54
Alkaid	153 00.9	N49 16
Al Na'ir	27 46.9	S46 54
Alnilam	275 49.7	S 1 11
Alphard	217 59.2	S 8 42
Alphecca	126 13.2	N26 41
Alpheratz	357 46.6	N29 08
Altair	62 10.8	N 8 53
Ankaa	353 18.4	S42 14
Antares	112 29.6	S26 27
Arcturus	145 58.2	N19 08
Atria	107 33.6	S69 02
Avior	234 19.9	S59 32
Bellatrix	278 35.5	N 6 21
Betelgeuse	271 04.8	N 7 24
Canopus	263 58.0	S52 42
Capella	280 39.3	N46 00
Deneb	49 33.2	N45 18
Denebola	182 36.7	N14 31
Diphda	348 58.8	S17 55
Dubhe	193 55.3	N61 42
Elnath	278 16.7	N28 36
Eltanin	90 47.0	N51 29
Enif	33 49.8	N 9 55
Fomalhaut	15 27.0	S29 34
Gacrux	172 04.4	S57 10
Gienah	175 55.4	S17 35
Hadar	148 52.0	S60 25
Hamal	328 04.3	N23 30
Kaus Aust.	83 47.3	S34 22
Kochab	137 18.8	N74 07
Markab	13 41.1	N15 15
Menkar	314 18.3	N 4 07
Menkent	148 11.0	S36 25
Miaplacidus	221 41.3	S69 45
Mirfak	308 45.0	N49 53
Nunki	76 01.5	S26 17
Peacock	53 23.1	S56 42
Pollux	243 31.6	N28 00
Procyon	245 03.1	N 5 12
Rasalhague	96 08.8	N12 33
Regulus	207 46.8	N11 55
Rigel	281 15.2	S 8 11
Rigil Kent.	139 55.6	S60 52
Sabik	102 15.6	S15 44
Schedar	349 44.1	N56 35
Shaula	96 25.5	S37 06
Sirius	258 36.7	S16 43
Spica	158 34.3	S11 12
Suhail	222 55.1	S43 28
Vega	80 40.5	N38 47
Zuben'ubi	137 08.5	S16 05

	SHA	Mer. Pass
	° ′	h m
Venus	306 22.6	8 58
Mars	310 08.3	8 42
Jupiter	30 49.4	3 19
Saturn	191 32.5	16 35

UT	SUN GHA	SUN Dec	MOON GHA	v	MOON Dec	d	HP
d h	° ′	° ′	° ′	′	° ′	′	′
30 00	179 05.9	N23 10.5	86 36.0	13.2	S10 28.6	13.5	57.0
01	194 05.7	10.4	101 08.2	13.2	10 42.1	13.4	57.0
02	209 05.6	10.2	115 40.4	13.2	10 55.5	13.4	56.9
03	224 05.5	. . 10.1	130 12.6	13.2	11 08.9	13.3	56.9
04	239 05.4	09.9	144 44.8	13.1	11 22.2	13.2	56.9
05	254 05.3	09.8	159 16.9	13.2	11 35.4	13.2	56.8
06	269 05.1	N23 09.6	173 49.1	13.1	S11 48.6	13.0	56.8
07	284 05.0	09.5	188 21.2	13.1	12 01.6	13.1	56.8
08	299 04.9	09.3	202 53.3	13.1	12 14.7	12.9	56.7
09	314 04.8	. . 09.2	217 25.4	13.1	12 27.6	12.9	56.7
10	329 04.6	09.0	231 57.5	13.1	12 40.5	12.8	56.7
11	344 04.5	08.8	246 29.6	13.0	12 53.3	12.8	56.6
12	359 04.4	N23 08.7	261 01.6	13.0	S13 06.1	12.6	56.6
13	14 04.3	08.5	275 33.6	13.0	13 18.7	12.6	56.6
14	29 04.1	08.4	290 05.6	13.0	13 31.3	12.6	56.5
15	44 04.0	. . 08.2	304 37.6	13.0	13 43.9	12.4	56.5
16	59 03.9	08.0	319 09.6	12.9	13 56.3	12.4	56.5
17	74 03.8	07.9	333 41.5	12.9	14 08.7	12.3	56.4
18	89 03.7	N23 07.7	348 13.4	12.9	S14 21.0	12.2	56.4
19	104 03.5	07.6	2 45.3	12.9	14 33.2	12.1	56.4
20	119 03.4	07.4	17 17.2	12.8	14 45.3	12.1	56.4
21	134 03.3	. . 07.2	31 49.0	12.8	14 57.4	12.0	56.3
22	149 03.2	07.1	46 20.8	12.8	15 09.4	11.8	56.3
23	164 03.1	06.9	60 52.6	12.7	15 21.2	11.9	56.3
1 00	179 02.9	N23 06.7	75 24.3	12.8	S15 33.1	11.7	56.2
01	194 02.8	06.6	89 56.1	12.6	15 44.8	11.6	56.2
02	209 02.7	06.4	104 27.7	12.7	15 56.4	11.6	56.2
03	224 02.6	. . 06.2	118 59.4	12.6	16 08.0	11.5	56.2
04	239 02.5	06.1	133 31.0	12.7	16 19.5	11.4	56.1
05	254 02.3	05.9	148 02.7	12.5	16 30.9	11.3	56.1
06	269 02.2	N23 05.7	162 34.2	12.6	S16 42.2	11.2	56.1
07	284 02.1	05.5	177 05.8	12.5	16 53.4	11.1	56.0
08	299 02.0	05.4	191 37.3	12.4	17 04.5	11.1	56.0
09	314 01.8	. . 05.2	206 08.7	12.5	17 15.6	10.9	56.0
10	329 01.7	05.0	220 40.2	12.4	17 26.5	10.9	56.0
11	344 01.6	04.9	235 11.6	12.4	17 37.4	10.8	55.9
12	359 01.5	N23 04.7	249 43.0	12.3	S17 48.2	10.7	55.9
13	14 01.4	04.5	264 14.3	12.3	17 58.9	10.6	55.9
14	29 01.3	04.3	278 45.6	12.3	18 09.5	10.4	55.9
15	44 01.1	. . 04.1	293 16.9	12.2	18 19.9	10.5	55.8
16	59 01.0	04.0	307 48.1	12.2	18 30.4	10.3	55.8
17	74 00.9	03.8	322 19.3	12.2	18 40.7	10.2	55.8
18	89 00.8	N23 03.6	336 50.5	12.1	S18 50.9	10.1	55.8
19	104 00.7	03.4	351 21.6	12.1	19 01.0	10.0	55.7
20	119 00.5	03.3	5 52.7	12.0	19 11.0	9.9	55.7
21	134 00.4	. . 03.1	20 23.7	12.0	19 20.9	9.9	55.7
22	149 00.3	02.9	34 54.7	12.0	19 30.8	9.7	55.7
23	164 00.2	02.7	49 25.7	12.0	19 40.5	9.6	55.6
2 00	179 00.1	N23 02.5	63 56.7	11.8	S19 50.1	9.5	55.6
01	193 59.9	02.3	78 27.5	11.9	19 59.6	9.5	55.6
02	208 59.8	02.2	92 58.4	11.8	20 09.1	9.3	55.6
03	223 59.7	. . 02.0	107 29.2	11.8	20 18.4	9.2	55.5
04	238 59.6	01.8	122 00.0	11.8	20 27.6	9.1	55.5
05	253 59.5	01.6	136 30.8	11.7	20 36.7	9.1	55.5
06	268 59.4	N23 01.4	151 01.5	11.6	S20 45.8	8.9	55.5
07	283 59.2	01.2	165 32.1	11.7	20 54.7	8.8	55.4
08	298 59.1	01.0	180 02.8	11.5	21 03.5	8.7	55.4
09	313 59.0	. . 00.8	194 33.3	11.6	21 12.2	8.6	55.4
10	328 58.9	00.7	209 03.9	11.5	21 20.8	8.5	55.4
11	343 58.8	00.5	223 34.4	11.5	21 29.3	8.3	55.4
12	358 58.7	N23 00.3	238 04.9	11.4	S21 37.6	8.3	55.3
13	13 58.5	23 00.1	252 35.3	11.4	21 45.9	8.2	55.3
14	28 58.4	22 59.9	267 05.7	11.4	21 54.1	8.0	55.3
15	43 58.3	. . 59.7	281 36.1	11.3	22 02.1	8.0	55.3
16	58 58.2	59.5	296 06.4	11.3	22 10.1	7.8	55.2
17	73 58.1	59.3	310 36.7	11.2	22 17.9	7.7	55.2
18	88 58.0	N22 59.1	325 06.9	11.2	S22 25.6	7.6	55.2
19	103 57.8	58.9	339 37.1	11.2	22 33.2	7.5	55.2
20	118 57.7	58.7	354 07.3	11.1	22 40.7	7.4	55.2
21	133 57.6	. . 58.5	8 37.4	11.1	22 48.1	7.3	55.1
22	148 57.5	58.3	23 07.5	11.1	22 55.4	7.2	55.1
23	163 57.4	58.1	37 37.6	11.0	S23 02.6	7.0	55.1
	SD 15.8	d 0.2	SD		15.4	15.2	15.1

Lat.	Twilight Naut.	Twilight Civil	Sunrise	Moonrise 30	Moonrise 1	Moonrise 2	Moonrise 3
°	h m	h m	h m	h m	h m	h m	h m
N 72	□	□	□	16 06	■	■	■
N 70	□	□	□	15 35	18 18	■	■
68	□	□	□	15 13	17 22	■	■
66	////	////	00 20	14 56	16 49	19 00	■
64	////	////	01 41	14 42	16 25	18 14	20 11
62	////	////	02 17	14 30	16 07	17 44	19 18
60	////	01 03	02 42	14 20	15 51	17 21	18 46
N 58	////	01 49	03 02	14 12	15 38	17 03	18 22
56	////	02 17	03 18	14 04	15 27	16 48	18 03
54	00 58	02 39	03 32	13 58	15 17	16 35	17 47
52	01 40	02 56	03 44	13 52	15 09	16 23	17 34
50	02 06	03 11	03 55	13 46	15 01	16 13	17 22
45	02 51	03 40	04 17	13 35	14 44	15 52	16 57
N 40	03 21	04 02	04 35	13 25	14 31	15 35	16 37
35	03 44	04 20	04 50	13 17	14 19	15 21	16 21
30	04 02	04 35	05 02	13 10	14 10	15 09	16 06
20	04 31	05 00	05 24	12 58	13 53	14 48	15 42
N 10	04 53	05 20	05 43	12 47	13 38	14 29	15 22
0	05 12	05 38	06 00	12 38	13 24	14 12	15 02
S 10	05 28	05 55	06 17	12 28	13 11	13 56	14 43
20	05 44	06 12	06 36	12 17	12 56	13 38	14 23
30	06 00	06 30	06 57	12 06	12 40	13 17	13 59
35	06 09	06 41	07 09	11 59	12 30	13 05	13 45
40	06 18	06 52	07 23	11 51	12 20	12 52	13 29
45	06 28	07 05	07 39	11 43	12 07	12 36	13 10
S 50	06 40	07 21	08 00	11 32	11 52	12 16	12 47
52	06 45	07 29	08 09	11 27	11 45	12 07	12 35
54	06 51	07 37	08 20	11 22	11 37	11 56	12 22
56	06 57	07 45	08 32	11 16	11 28	11 44	12 08
58	07 03	07 55	08 47	11 09	11 18	11 31	11 50
S 60	07 10	08 07	09 04	11 02	11 07	11 15	11 29

Lat.	Sunset	Twilight Civil	Twilight Naut.	Moonset 30	Moonset 1	Moonset 2	Moonset 3
°	h m	h m	h m	h m	h m	h m	h m
N 72	□	□	□	21 08	■	■	■
N 70	□	□	□	21 41	20 37	■	■
68	□	□	□	22 04	21 33	■	■
66	23 39	////	////	22 23	22 07	21 38	■
64	22 25	////	////	22 38	22 32	22 25	22 11
62	21 50	////	////	22 51	22 52	22 55	23 05
60	21 25	23 02	////	23 02	23 08	23 19	23 37
N 58	21 05	22 18	////	23 11	23 22	23 37	24 01
56	20 49	21 50	////	23 20	23 34	23 53	24 21
54	20 35	21 29	23 08	23 27	23 44	24 06	00 06
52	20 23	21 11	22 27	23 34	23 53	24 18	00 18
50	20 12	20 57	22 01	23 40	24 02	00 02	00 29
45	19 50	20 28	21 17	23 53	24 19	00 19	00 51
N 40	19 33	20 05	20 47	24 04	00 04	00 34	01 08
35	19 18	19 48	20 24	24 13	00 13	00 46	01 23
30	19 05	19 33	20 06	24 21	00 21	00 57	01 36
20	18 44	19 08	19 37	24 36	00 36	01 16	01 59
N 10	18 25	18 48	19 15	00 06	00 48	01 32	02 18
0	18 08	18 30	18 56	00 14	01 00	01 47	02 36
S 10	17 50	18 13	18 39	00 21	01 12	02 03	02 54
20	17 32	17 56	18 24	00 30	01 24	02 19	03 14
30	17 11	17 38	18 07	00 39	01 39	02 38	03 37
35	16 59	17 27	17 59	00 45	01 47	02 49	03 50
40	16 45	17 16	17 50	00 51	01 57	03 02	04 05
45	16 29	17 02	17 40	00 58	02 08	03 17	04 24
S 50	16 08	16 47	17 28	01 07	02 22	03 36	04 46
52	15 59	16 39	17 23	01 11	02 29	03 44	04 57
54	15 48	16 31	17 17	01 16	02 36	03 54	05 10
56	15 36	16 23	17 11	01 21	02 44	04 06	05 24
58	15 21	16 13	17 05	01 26	02 53	04 19	05 41
S 60	15 04	16 01	16 58	01 33	03 03	04 34	06 02

Day	SUN Eqn. of Time 00^h	SUN Eqn. of Time 12^h	SUN Mer. Pass.	MOON Mer. Pass. Upper	MOON Mer. Pass. Lower	MOON Age	MOON Phase
d	m s	m s	h m	h m	h m	d %	
30	03 36	03 42	12 04	18 49	06 26	08 61	
1	03 48	03 54	12 04	19 36	07 12	09 71	
2	04 00	04 05	12 04	20 24	08 00	10 79	

UT		ARIES	VENUS −4.2		MARS +1.1		JUPITER −2.7		SATURN +1.1	
d h		GHA	GHA	Dec	GHA	Dec	GHA	Dec	GHA	Dec
		° ′	° ′	° ′	° ′	° ′	° ′	° ′	° ′	° ′
3 00	FRIDAY	281 09.2	225 20.7	N16 58.4	229 50.5	N18 01.2	312 04.5	S13 33.6	112 33.7	N 7 07.2
01		296 11.7	240 20.4	59.1	244 51.1	01.6	327 07.1	33.6	127 36.0	07.1
02		311 14.1	255 20.1	16 59.7	259 51.8	02.1	342 09.7	33.7	142 38.3	07.0
03		326 16.6	270 19.8	17 00.4	274 52.5	02.6	357 12.3	33.7	157 40.6	07.0
04		341 19.1	285 19.6	01.0	289 53.1	03.0	12 14.9	33.8	172 42.9	06.9
05		356 21.5	300 19.3	01.7	304 53.8	03.5	27 17.5	33.9	187 45.2	06.8
06		11 24.0	315 19.0	N17 02.3	319 54.4	N18 04.0	42 20.1	S13 33.9	202 47.5	N 7 06.7
07		26 26.5	330 18.7	03.0	334 55.1	04.4	57 22.7	34.0	217 49.8	06.6
08		41 28.9	345 18.4	03.6	349 55.7	04.9	72 25.3	34.0	232 52.1	06.6
09		56 31.4	0 18.1	04.2	4 56.4	05.4	87 27.9	34.1	247 54.4	06.5
10		71 33.9	15 17.8	04.9	19 57.0	05.9	102 30.5	34.2	262 56.7	06.4
11		86 36.3	30 17.5	05.5	34 57.7	06.3	117 33.1	34.2	277 59.0	06.3
12		101 38.8	45 17.3	N17 06.2	49 58.3	N18 06.8	132 35.7	S13 34.3	293 01.3	N 7 06.3
13		116 41.2	60 17.0	06.8	64 59.0	07.3	147 38.3	34.3	308 03.5	06.2
14		131 43.7	75 16.7	07.4	79 59.6	07.7	162 40.9	34.4	323 05.8	06.1
15		146 46.2	90 16.4	08.1	95 00.3	08.2	177 43.5	34.4	338 08.1	06.0
16		161 48.6	105 16.1	08.7	110 00.9	08.7	192 46.1	34.5	353 10.4	05.9
17		176 51.1	120 15.8	09.4	125 01.6	09.1	207 48.7	34.6	8 12.7	05.9
18		191 53.6	135 15.5	N17 10.0	140 02.2	N18 09.6	222 51.3	S13 34.6	23 15.0	N 7 05.8
19		206 56.0	150 15.2	10.6	155 02.9	10.1	237 53.9	34.7	38 17.3	05.7
20		221 58.5	165 14.9	11.3	170 03.5	10.5	252 56.5	34.7	53 19.6	05.6
21		237 01.0	180 14.6	11.9	185 04.2	11.0	267 59.1	34.8	68 21.9	05.5
22		252 03.4	195 14.3	12.5	200 04.8	11.4	283 01.7	34.9	83 24.2	05.5
23		267 05.9	210 14.0	13.2	215 05.5	11.9	298 04.3	34.9	98 26.5	05.4
4 00	SATURDAY	282 08.4	225 13.7	N17 13.8	230 06.1	N18 12.4	313 06.9	S13 35.0	113 28.8	N 7 05.3
01		297 10.8	240 13.4	14.4	245 06.8	12.8	328 09.5	35.0	128 31.1	05.2
02		312 13.3	255 13.1	15.1	260 07.4	13.3	343 12.1	35.1	143 33.4	05.2
03		327 15.7	270 12.8	15.7	275 08.1	13.8	358 14.7	35.2	158 35.6	05.1
04		342 18.2	285 12.5	16.3	290 08.7	14.2	13 17.3	35.2	173 37.9	05.0
05		357 20.7	300 12.2	17.0	305 09.4	14.7	28 19.9	35.3	188 40.2	04.9
06		12 23.1	315 11.9	N17 17.6	320 10.0	N18 15.2	43 22.5	S13 35.3	203 42.5	N 7 04.8
07		27 25.6	330 11.6	18.2	335 10.6	15.6	58 25.1	35.4	218 44.8	04.8
08		42 28.1	345 11.3	18.9	350 11.3	16.1	73 27.7	35.5	233 47.1	04.7
09		57 30.5	0 11.0	19.5	5 11.9	16.5	88 30.3	35.5	248 49.4	04.6
10		72 33.0	15 10.7	20.1	20 12.6	17.0	103 32.9	35.6	263 51.7	04.5
11		87 35.5	30 10.4	20.8	35 13.2	17.5	118 35.5	35.7	278 54.0	04.4
12		102 37.9	45 10.1	N17 21.4	50 13.9	N18 17.9	133 38.1	S13 35.7	293 56.3	N 7 04.4
13		117 40.4	60 09.8	22.0	65 14.5	18.4	148 40.7	35.8	308 58.6	04.3
14		132 42.9	75 09.5	22.6	80 15.2	18.8	163 43.4	35.8	324 00.8	04.2
15		147 45.3	90 09.2	23.3	95 15.8	19.3	178 46.0	35.9	339 03.1	04.1
16		162 47.8	105 08.9	23.9	110 16.5	19.8	193 48.6	36.0	354 05.4	04.0
17		177 50.2	120 08.6	24.5	125 17.1	20.2	208 51.2	36.0	9 07.7	04.0
18		192 52.7	135 08.3	N17 25.2	140 17.8	N18 20.7	223 53.8	S13 36.1	24 10.0	N 7 03.9
19		207 55.2	150 08.0	25.8	155 18.4	21.1	238 56.4	36.1	39 12.3	03.8
20		222 57.6	165 07.6	26.4	170 19.1	21.6	253 59.0	36.2	54 14.6	03.7
21		238 00.1	180 07.3	27.0	185 19.7	22.1	269 01.6	36.3	69 16.9	03.7
22		253 02.6	195 07.0	27.7	200 20.4	22.5	284 04.2	36.3	84 19.2	03.6
23		268 05.0	210 06.7	28.3	215 21.0	23.0	299 06.8	36.4	99 21.5	03.5
5 00	SUNDAY	283 07.5	225 06.4	N17 28.9	230 21.7	N18 23.4	314 09.4	S13 36.5	114 23.7	N 7 03.4
01		298 10.0	240 06.1	29.5	245 22.3	23.9	329 12.0	36.5	129 26.0	03.3
02		313 12.4	255 05.8	30.1	260 23.0	24.3	344 14.6	36.6	144 28.3	03.3
03		328 14.9	270 05.5	30.8	275 23.6	24.8	359 17.3	36.6	159 30.6	03.2
04		343 17.4	285 05.1	31.4	290 24.3	25.2	14 19.9	36.7	174 32.9	03.1
05		358 19.8	300 04.8	32.0	305 24.9	25.7	29 22.5	36.8	189 35.2	03.0
06		13 22.3	315 04.5	N17 32.6	320 25.6	N18 26.2	44 25.1	S13 36.8	204 37.5	N 7 02.9
07		28 24.7	330 04.2	33.2	335 26.2	26.6	59 27.7	36.9	219 39.8	02.9
08		43 27.2	345 03.9	33.9	350 26.9	27.1	74 30.3	37.0	234 42.1	02.8
09		58 29.7	0 03.6	34.5	5 27.5	27.5	89 32.9	37.0	249 44.3	02.7
10		73 32.1	15 03.2	35.1	20 28.2	28.0	104 35.5	37.1	264 46.6	02.6
11		88 34.6	30 02.9	35.7	35 28.8	28.4	119 38.1	37.1	279 48.9	02.5
12		103 37.1	45 02.6	N17 36.3	50 29.5	N18 28.9	134 40.8	S13 37.2	294 51.2	N 7 02.5
13		118 39.5	60 02.3	37.0	65 30.1	29.3	149 43.4	37.3	309 53.5	02.4
14		133 42.0	75 02.0	37.6	80 30.8	29.8	164 46.0	37.3	324 55.8	02.3
15		148 44.5	90 01.6	38.2	95 31.4	30.2	179 48.6	37.4	339 58.1	02.2
16		163 46.9	105 01.3	38.8	110 32.1	30.7	194 51.2	37.5	355 00.4	02.1
17		178 49.4	120 01.0	39.4	125 32.7	31.1	209 53.8	37.5	10 02.6	02.0
18		193 51.9	135 00.7	N17 40.0	140 33.4	N18 31.6	224 56.4	S13 37.6	25 04.9	N 7 02.0
19		208 54.3	150 00.4	40.6	155 34.0	32.0	239 59.1	37.7	40 07.2	01.9
20		223 56.8	165 00.0	41.3	170 34.7	32.5	255 01.7	37.7	55 09.5	01.8
21		238 59.2	179 59.7	41.9	185 35.3	32.9	270 04.3	37.8	70 11.8	01.7
22		254 01.7	194 59.4	42.5	200 36.0	33.4	285 06.9	37.8	85 14.1	01.6
23		269 04.2	209 59.1	43.1	215 36.6	33.8	300 09.5	37.9	100 16.4	01.6
Mer. Pass.		h m 5 10.6	v −0.3	d 0.6	v 0.6	d 0.5	v 2.6	d 0.1	v 2.3	d 0.1

STARS		
Name	SHA	Dec
	° ′	°
Acamar	315 20.7	S40 15
Achernar	335 28.9	S57 10
Acrux	173 12.9	S63 09
Adhara	255 15.2	S28 59
Aldebaran	290 53.0	N16 31
Alioth	166 23.0	N55 54
Alkaid	153 00.9	N49 16
Al Na'ir	27 46.9	S46 54
Alnilam	275 49.6	S 1 11
Alphard	217 59.2	S 8 42
Alphecca	126 13.2	N26 41
Alpheratz	357 46.5	N29 08
Altair	62 10.8	N 8 53
Ankaa	353 18.4	S42 14
Antares	112 29.6	S26 27
Arcturus	145 58.2	N19 08
Atria	107 33.6	S69 02
Avior	234 20.0	S59 32
Bellatrix	278 35.5	N 6 21
Betelgeuse	271 04.8	N 7 24
Canopus	263 58.0	S52 42
Capella	280 39.2	N46 00
Deneb	49 33.1	N45 18
Denebola	182 36.7	N14 31
Diphda	348 58.8	S17 55
Dubhe	193 55.3	N61 42
Elnath	278 16.7	N28 36
Eltanin	90 47.0	N51 29
Enif	33 49.8	N 9 55
Fomalhaut	15 26.9	S29 34
Gacrux	172 04.4	S57 10
Gienah	175 55.4	S17 35
Hadar	148 52.0	S60 25
Hamal	328 04.2	N23 30
Kaus Aust.	83 47.3	S34 22
Kochab	137 18.9	N74 07
Markab	13 41.1	N15 15
Menkar	314 18.3	N 4 07
Menkent	148 11.0	S36 25
Miaplacidus	221 41.4	S69 45
Mirfak	308 45.0	N49 53
Nunki	76 01.5	S26 17
Peacock	53 23.1	S56 42
Pollux	243 31.6	N28 00
Procyon	245 03.1	N 5 12
Rasalhague	96 08.8	N12 33
Regulus	207 46.8	N11 55
Rigel	281 15.2	S 8 11
Rigil Kent.	139 55.6	S60 52
Sabik	102 15.6	S15 44
Schedar	349 44.1	N56 35
Shaula	96 25.5	S37 06
Sirius	258 36.6	S16 43
Spica	158 34.3	S11 12
Suhail	222 55.1	S43 28
Vega	80 40.5	N38 47
Zuben'ubi	137 08.5	S16 05

	SHA	Mer. Pas
	° ′	h m
Venus	303 05.4	8 59
Mars	307 57.7	8 39
Jupiter	30 58.5	3 07
Saturn	191 20.4	16 24

UT	SUN GHA	SUN Dec	MOON GHA	v	MOON Dec	d	HP
d h	° ′	° ′	° ′	′	° ′	′	′
3 00	178 57.3	N22 57.9	52 07.6	11.0	S23 09.6	6.9	55.1
01	193 57.2	57.7	66 37.6	10.9	23 16.5	6.8	55.1
02	208 57.0	57.5	81 07.5	10.9	23 23.3	6.7	55.0
03	223 56.9	. . 57.3	95 37.4	10.9	23 30.0	6.6	55.0
04	238 56.8	57.1	110 07.3	10.8	23 36.6	6.5	55.0
05	253 56.7	56.9	124 37.1	10.8	23 43.1	6.3	55.0
06	268 56.6	N22 56.7	139 06.9	10.8	S23 49.4	6.2	55.0
07	283 56.5	56.5	153 36.7	10.8	23 55.6	6.1	54.9
08	298 56.4	56.3	168 06.5	10.7	24 01.7	6.0	54.9
09	313 56.2	. . 56.1	182 36.2	10.7	24 07.7	5.9	54.9
10	328 56.1	55.9	197 05.9	10.6	24 13.6	5.7	54.9
11	343 56.0	55.7	211 35.5	10.6	24 19.3	5.7	54.9
12	358 55.9	N22 55.5	226 05.1	10.6	S24 25.0	5.5	54.9
13	13 55.8	55.3	240 34.7	10.5	24 30.5	5.3	54.8
14	28 55.7	55.0	255 04.3	10.5	24 35.8	5.3	54.8
15	43 55.6	. . 54.8	269 33.8	10.5	24 41.1	5.1	54.8
16	58 55.5	54.6	284 03.3	10.5	24 46.2	5.0	54.8
17	73 55.3	54.4	298 32.8	10.4	24 51.2	4.9	54.8
18	88 55.2	N22 54.2	313 02.2	10.4	S24 56.1	4.8	54.8
19	103 55.1	54.0	327 31.6	10.4	25 00.9	4.6	54.7
20	118 55.0	53.8	342 01.0	10.4	25 05.5	4.6	54.7
21	133 54.9	. . 53.6	356 30.4	10.3	25 10.1	4.3	54.7
22	148 54.8	53.3	10 59.7	10.3	25 14.4	4.3	54.7
23	163 54.7	53.1	25 29.0	10.3	25 18.7	4.2	54.7
4 00	178 54.6	N22 52.9	39 58.3	10.3	S25 22.9	4.0	54.7
01	193 54.4	52.7	54 27.6	10.3	25 26.9	3.9	54.6
02	208 54.3	52.5	68 56.9	10.2	25 30.8	3.7	54.6
03	223 54.2	. . 52.3	83 26.1	10.2	25 34.5	3.7	54.6
04	238 54.1	52.0	97 55.3	10.2	25 38.2	3.5	54.6
05	253 54.0	51.8	112 24.5	10.2	25 41.7	3.3	54.6
06	268 53.9	N22 51.6	126 53.7	10.2	S25 45.0	3.3	54.6
07	283 53.8	51.4	141 22.9	10.1	25 48.3	3.1	54.6
08	298 53.7	51.2	155 52.0	10.1	25 51.4	3.0	54.5
09	313 53.6	. . 50.9	170 21.1	10.2	25 54.4	2.9	54.5
10	328 53.4	50.7	184 50.3	10.1	25 57.3	2.7	54.5
11	343 53.3	50.5	199 19.4	10.1	26 00.0	2.6	54.5
12	358 53.2	N22 50.3	213 48.5	10.0	S26 02.6	2.5	54.5
13	13 53.1	50.0	228 17.5	10.1	26 05.1	2.4	54.5
14	28 53.0	49.8	242 46.6	10.1	26 07.5	2.2	54.5
15	43 52.9	. . 49.6	257 15.7	10.0	26 09.7	2.1	54.5
16	58 52.8	49.4	271 44.7	10.1	26 11.8	1.9	54.4
17	73 52.7	49.1	286 13.8	10.0	26 13.7	1.9	54.4
18	88 52.6	N22 48.9	300 42.8	10.1	S26 15.6	1.7	54.4
19	103 52.5	48.7	315 11.9	10.0	26 17.3	1.5	54.4
20	118 52.4	48.4	329 40.9	10.1	26 18.8	1.5	54.4
21	133 52.2	. . 48.2	344 10.0	10.0	26 20.3	1.3	54.4
22	148 52.1	48.0	358 39.0	10.0	26 21.6	1.2	54.4
23	163 52.0	47.7	13 08.0	10.1	26 22.8	1.0	54.4
5 00	178 51.9	N22 47.5	27 37.1	10.0	S26 23.8	1.0	54.4
01	193 51.8	47.3	42 06.1	10.0	26 24.8	0.8	54.3
02	208 51.7	47.0	56 35.1	10.1	26 25.6	0.6	54.3
03	223 51.6	. . 46.8	71 04.2	10.0	26 26.2	0.6	54.3
04	238 51.5	46.6	85 33.2	10.1	26 26.8	0.4	54.3
05	253 51.4	46.3	100 02.3	10.0	26 27.2	0.2	54.3
06	268 51.3	N22 46.1	114 31.3	10.1	S26 27.4	0.2	54.3
07	283 51.2	45.9	129 00.4	10.1	26 27.6	0.0	54.3
08	298 51.1	45.6	143 29.5	10.1	26 27.6	0.1	54.3
09	313 51.0	. . 45.4	157 58.6	10.1	26 27.5	0.2	54.3
10	328 50.9	45.1	172 27.7	10.1	26 27.3	0.4	54.3
11	343 50.7	44.9	186 56.8	10.1	26 26.9	0.5	54.2
12	358 50.6	N22 44.7	201 25.9	10.1	S26 26.4	0.6	54.2
13	13 50.5	44.4	215 55.0	10.2	26 25.8	0.8	54.2
14	28 50.4	44.2	230 24.2	10.2	26 25.0	0.9	54.2
15	43 50.3	. . 43.9	244 53.4	10.2	26 24.1	1.0	54.2
16	58 50.2	43.7	259 22.6	10.2	26 23.1	1.1	54.2
17	73 50.1	43.4	273 51.8	10.2	26 22.0	1.3	54.2
18	88 50.0	N22 43.2	288 21.0	10.2	S26 20.7	1.4	54.2
19	103 49.9	42.9	302 50.2	10.3	26 19.3	1.5	54.2
20	118 49.8	42.7	317 19.5	10.3	26 17.8	1.7	54.2
21	133 49.7	. . 42.5	331 48.8	10.3	26 16.1	1.8	54.2
22	148 49.6	42.2	346 18.1	10.4	26 14.3	1.9	54.2
23	163 49.5	42.0	0 47.5	10.3	S26 12.4	2.0	54.1
	SD 15.8	d 0.2	SD 14.9		14.8		14.8

Lat.	Twilight Naut.	Twilight Civil	Sunrise	Moonrise 3	Moonrise 4	Moonrise 5	Moonrise 6
°	h m	h m	h m	h m	h m	h m	h m
N 72	□	□	□	■	■	■	■
N 70	□	□	□	■	■	■	■
68	□	□	□	■	■	■	■
66	////	////	00 42	■	■	■	■
64	////	////	01 47	20 11	■	■	22 35
62	////	////	02 21	19 18	20 39	21 28	21 48
60	////	01 11	02 46	18 46	19 59	20 49	21 18
N 58	////	01 53	03 05	18 22	19 31	20 22	20 55
56	////	02 21	03 21	18 03	19 09	20 00	20 37
54	01 05	02 42	03 35	17 47	18 51	19 43	20 21
52	01 44	02 59	03 47	17 34	18 36	19 27	20 07
50	02 10	03 13	03 57	17 22	18 23	19 14	19 55
45	02 53	03 42	04 19	16 57	17 56	18 48	19 31
N 40	03 23	04 04	04 36	16 37	17 35	18 26	19 11
35	03 45	04 22	04 51	16 21	17 17	18 09	18 54
30	04 03	04 36	05 04	16 06	17 02	17 53	18 40
20	04 32	05 01	05 25	15 42	16 36	17 27	18 16
N 10	04 54	05 21	05 44	15 22	16 14	17 05	17 55
0	05 12	05 38	06 01	15 02	15 53	16 44	17 35
S 10	05 29	05 55	06 18	14 43	15 33	16 24	17 15
20	05 45	06 12	06 36	14 23	15 11	16 01	16 54
30	06 00	06 30	06 56	13 59	14 45	15 36	16 29
35	06 09	06 40	07 08	13 45	14 30	15 20	16 15
40	06 18	06 52	07 22	13 29	14 13	15 03	15 58
45	06 28	07 05	07 39	13 10	13 52	14 41	15 38
S 50	06 39	07 20	07 59	12 47	13 26	14 14	15 13
52	06 44	07 28	08 08	12 35	13 13	14 01	15 00
54	06 50	07 36	08 19	12 22	12 58	13 46	14 46
56	06 56	07 44	08 31	12 08	12 41	13 29	14 30
58	07 02	07 54	08 45	11 50	12 21	13 07	14 10
S 60	07 09	08 05	09 02	11 29	11 55	12 39	13 45

Lat.	Sunset	Twilight Civil	Twilight Naut.	Moonset 3	Moonset 4	Moonset 5	Moonset 6
°	h m	h m	h m	h m	h m	h m	h m
N 72	□	□	□	■	■	■	■
N 70	□	□	□	■	■	■	■
68	□	□	□	■	■	■	■
66	23 22	////	////	■	■	■	■
64	22 20	////	////	22 11	■	■	■
62	21 47	////	////	23 05	23 30	24 28	00 28
60	21 22	22 55	////	23 37	24 11	00 11	01 07
N 58	21 03	22 14	////	24 01	00 01	00 39	01 34
56	20 47	21 47	////	24 21	00 21	01 01	01 55
54	20 34	21 26	23 01	00 06	00 37	01 19	02 13
52	20 22	21 09	22 23	00 18	00 51	01 34	02 28
50	20 11	20 55	21 58	00 29	01 03	01 47	02 41
45	19 50	20 27	21 15	00 51	01 29	02 14	03 08
N 40	19 32	20 05	20 46	01 08	01 49	02 36	03 29
35	19 18	19 47	20 23	01 23	02 06	02 53	03 46
30	19 05	19 32	20 05	01 36	02 20	03 09	04 01
20	18 44	19 08	19 37	01 59	02 45	03 35	04 27
N 10	18 25	18 48	19 15	02 18	03 07	03 57	04 49
0	18 08	18 31	18 57	02 36	03 27	04 18	05 09
S 10	17 51	18 14	18 40	02 54	03 47	04 39	05 29
20	17 33	17 57	18 24	03 14	04 08	05 01	05 51
30	17 13	17 39	18 09	03 37	04 33	05 27	06 16
35	17 01	17 29	18 00	03 50	04 48	05 42	06 31
40	16 47	17 17	17 51	04 05	05 05	06 00	06 48
45	16 31	17 04	17 41	04 24	05 26	06 21	07 09
S 50	16 11	16 49	17 30	04 46	05 52	06 48	07 35
52	16 01	16 41	17 25	04 57	06 04	07 01	07 47
54	15 50	16 34	17 19	05 10	06 19	07 17	08 01
56	15 38	16 25	17 14	05 24	06 36	07 34	08 18
58	15 24	16 15	17 07	05 41	06 56	07 56	08 38
S 60	15 08	16 04	17 00	06 02	07 21	08 23	09 03

Day	SUN Eqn. of Time 00^h	SUN Eqn. of Time 12^h	SUN Mer. Pass.	MOON Mer. Pass. Upper	MOON Mer. Pass. Lower	Age	Phase
d	m s	m s	h m	h m	h m	d %	
3	04 11	04 16	12 04	21 14	08 49	11 87	
4	04 22	04 27	12 04	22 06	09 40	12 93	
5	04 32	04 37	12 05	22 57	10 31	13 97	

	UT d h	ARIES GHA	VENUS −4.2 GHA	VENUS Dec	MARS +1.1 GHA	MARS Dec	JUPITER −2.7 GHA	JUPITER Dec	SATURN +1.1 GHA	SATURN Dec	STARS Name	SHA	Dec
		° ′	° ′	° ′	° ′	° ′	° ′	° ′	° ′	° ′		° ′	°
	6 00	284 06.6	224 58.7	N17 43.7	230 37.2	N18 34.3	315 12.1	S13 38.0	115 18.6	N 7 01.5	Acamar	315 20.7	S40 15
	01	299 09.1	239 58.4	44.3	245 37.9	34.7	330 14.8	38.0	130 20.9	01.4	Achernar	335 28.8	S57 10
	02	314 11.6	254 58.1	44.9	260 38.5	35.2	345 17.4	38.1	145 23.2	01.3	Acrux	173 12.9	S63 09
	03	329 14.0	269 57.7	. . 45.5	275 39.2	. . 35.6	0 20.0	. . 38.2	160 25.5	. . 01.2	Adhara	255 15.2	S28 59
	04	344 16.5	284 57.4	46.1	290 39.8	36.1	15 22.6	38.2	175 27.8	01.2	Aldebaran	290 53.0	N16 31
	05	359 19.0	299 57.1	46.7	305 40.5	36.5	30 25.2	38.3	190 30.1	01.1			
	06	14 21.4	314 56.7	N17 47.4	320 41.1	N18 37.0	45 27.8	S13 38.4	205 32.4	N 7 01.0	Alioth	166 23.0	N55 54
	07	29 23.9	329 56.4	48.0	335 41.8	37.4	60 30.5	38.4	220 34.6	00.9	Alkaid	153 00.9	N49 16
	08	44 26.4	344 56.1	48.6	350 42.4	37.9	75 33.1	38.5	235 36.9	00.8	Al Na'ir	27 46.9	S46 54
M	09	59 28.8	359 55.8	. . 49.2	5 43.1	. . 38.3	90 35.7	. . 38.6	250 39.2	. . 00.8	Alnilam	275 49.6	S 1 11
O	10	74 31.3	14 55.4	49.8	20 43.7	38.8	105 38.3	38.6	265 41.5	00.7	Alphard	217 59.2	S 8 42
N	11	89 33.7	29 55.1	50.4	35 44.4	39.2	120 40.9	38.7	280 43.8	00.6			
D	12	104 36.2	44 54.8	N17 51.0	50 45.0	N18 39.7	135 43.6	S13 38.8	295 46.1	N 7 00.5	Alphecca	126 13.2	N26 41
A	13	119 38.7	59 54.4	51.6	65 45.7	40.1	150 46.2	38.8	310 48.4	00.4	Alpheratz	357 46.5	N29 08
Y	14	134 41.1	74 54.1	52.2	80 46.3	40.5	165 48.8	38.9	325 50.6	00.3	Altair	62 10.7	N 8 53
	15	149 43.6	89 53.7	. . 52.8	95 47.0	. . 41.0	180 51.4	. . 39.0	340 52.9	. . 00.3	Ankaa	353 18.4	S42 14
	16	164 46.1	104 53.4	53.4	110 47.6	41.4	195 54.0	39.0	355 55.2	00.2	Antares	112 29.5	S26 27
	17	179 48.5	119 53.1	54.0	125 48.3	41.9	210 56.7	39.1	10 57.5	00.1			
	18	194 51.0	134 52.7	N17 54.6	140 48.9	N18 42.3	225 59.3	S13 39.2	25 59.8	N 7 00.0	Arcturus	145 58.2	N19 08
	19	209 53.5	149 52.4	55.2	155 49.6	42.8	241 01.9	39.2	41 02.1	6 59.9	Atria	107 33.6	S69 02
	20	224 55.9	164 52.1	55.8	170 50.2	43.2	256 04.5	39.3	56 04.3	59.9	Avior	234 20.0	S59 32
	21	239 58.4	179 51.7	. . 56.4	185 50.9	. . 43.6	271 07.1	. . 39.4	71 06.6	. . 59.8	Bellatrix	278 35.4	N 6 21
	22	255 00.8	194 51.4	57.0	200 51.5	44.1	286 09.8	39.4	86 08.9	59.7	Betelgeuse	271 04.8	N 7 24
	23	270 03.3	209 51.0	57.6	215 52.1	44.5	301 12.4	39.5	101 11.2	59.6			
	7 00	285 05.8	224 50.7	N17 58.2	230 52.8	N18 45.0	316 15.0	S13 39.6	116 13.5	N 6 59.5	Canopus	263 58.0	S52 41
	01	300 08.2	239 50.3	58.8	245 53.4	45.4	331 17.6	39.6	131 15.8	59.4	Capella	280 39.2	N46 00
	02	315 10.7	254 50.0	17 59.4	260 54.1	45.9	346 20.3	39.7	146 18.1	59.4	Deneb	49 33.1	N45 18
	03	330 13.2	269 49.7	18 00.0	275 54.7	. . 46.3	1 22.9	. . 39.8	161 20.3	. . 59.3	Denebola	182 36.7	N14 31
	04	345 15.6	284 49.3	00.6	290 55.4	46.7	16 25.5	39.8	176 22.6	59.2	Diphda	348 58.7	S17 55
	05	0 18.1	299 49.0	01.2	305 56.0	47.2	31 28.1	39.9	191 24.9	59.1			
	06	15 20.6	314 48.6	N18 01.8	320 56.7	N18 47.6	46 30.8	S13 40.0	206 27.2	N 6 59.0	Dubhe	193 55.3	N61 42
	07	30 23.0	329 48.3	02.4	335 57.3	48.1	61 33.4	40.0	221 29.5	58.9	Elnath	278 16.7	N28 36
T	08	45 25.5	344 47.9	03.0	350 58.0	48.5	76 36.0	40.1	236 31.7	58.9	Eltanin	90 47.0	N51 29
U	09	60 28.0	359 47.6	. . 03.5	5 58.6	. . 48.9	91 38.6	. . 40.2	251 34.0	. . 58.8	Enif	33 49.7	N 9 55
E	10	75 30.4	14 47.2	04.1	20 59.3	49.4	106 41.3	40.2	266 36.3	58.7	Fomalhaut	15 26.9	S29 34
S	11	90 32.9	29 46.9	04.7	35 59.9	49.8	121 43.9	40.3	281 38.6	58.6			
D	12	105 35.3	44 46.5	N18 05.3	51 00.6	N18 50.2	136 46.5	S13 40.4	296 40.9	N 6 58.5	Gacrux	172 04.4	S57 10
	13	120 37.8	59 46.2	05.9	66 01.2	50.7	151 49.1	40.4	311 43.2	58.5	Gienah	175 55.4	S17 35
A	14	135 40.3	74 45.8	06.5	81 01.9	51.1	166 51.8	40.5	326 45.4	58.4	Hadar	148 52.0	S60 25
Y	15	150 42.7	89 45.5	. . 07.1	96 02.5	. . 51.6	181 54.4	. . 40.6	341 47.7	. . 58.3	Hamal	328 04.2	N23 30
	16	165 45.2	104 45.1	07.7	111 03.1	52.0	196 57.0	40.6	356 50.0	58.2	Kaus Aust.	83 47.2	S34 22
	17	180 47.7	119 44.8	08.3	126 03.8	52.4	211 59.6	40.7	11 52.3	58.1			
	18	195 50.1	134 44.4	N18 08.9	141 04.4	N18 52.9	227 02.3	S13 40.8	26 54.6	N 6 58.0	Kochab	137 19.0	N74 07
	19	210 52.6	149 44.1	09.4	156 05.1	53.3	242 04.9	40.8	41 56.8	58.0	Markab	13 41.1	N15 15
	20	225 55.1	164 43.7	10.0	171 05.7	53.7	257 07.5	40.9	56 59.1	57.9	Menkar	314 18.3	N 4 07
	21	240 57.5	179 43.4	. . 10.6	186 06.4	. . 54.2	272 10.2	. . 41.0	72 01.4	. . 57.8	Menkent	148 11.0	S36 25
	22	256 00.0	194 43.0	11.2	201 07.0	54.6	287 12.8	41.1	87 03.7	57.7	Miaplacidus	221 41.4	S69 45
	23	271 02.5	209 42.7	11.8	216 07.7	55.0	302 15.4	41.1	102 06.0	57.6			
	8 00	286 04.9	224 42.3	N18 12.4	231 08.3	N18 55.5	317 18.1	S13 41.2	117 08.3	N 6 57.5	Mirfak	308 44.9	N49 53
	01	301 07.4	239 41.9	13.0	246 09.0	55.9	332 20.7	41.3	132 10.5	57.5	Nunki	76 01.5	S26 17
	02	316 09.8	254 41.6	13.5	261 09.6	56.3	347 23.3	41.3	147 12.8	57.4	Peacock	53 23.1	S56 42
	03	331 12.3	269 41.2	. . 14.1	276 10.3	. . 56.8	2 25.9	. . 41.4	162 15.1	. . 57.3	Pollux	243 31.6	N28 00
	04	346 14.8	284 40.9	14.7	291 10.9	57.2	17 28.6	41.5	177 17.4	57.2	Procyon	245 03.1	N 5 12
	05	1 17.2	299 40.5	15.3	306 11.6	57.6	32 31.2	41.5	192 19.7	57.1			
	06	16 19.7	314 40.2	N18 15.9	321 12.2	N18 58.1	47 33.8	S13 41.6	207 21.9	N 6 57.0	Rasalhague	96 08.8	N12 33
W	07	31 22.2	329 39.8	16.4	336 12.8	58.5	62 36.5	41.7	222 24.2	57.0	Regulus	207 46.8	N11 55
E	08	46 24.6	344 39.4	17.0	351 13.5	58.9	77 39.1	41.8	237 26.5	56.9	Rigel	281 15.1	S 8 11
D	09	61 27.1	359 39.1	. . 17.6	6 14.1	. . 59.4	92 41.7	. . 41.8	252 28.8	. . 56.8	Rigil Kent.	139 55.6	S60 52
N	10	76 29.6	14 38.7	18.2	21 14.8	18 59.8	107 44.4	41.9	267 31.1	56.7	Sabik	102 15.6	S15 44
E	11	91 32.0	29 38.3	18.8	36 15.4	19 00.2	122 47.0	42.0	282 33.3	56.6			
S	12	106 34.5	44 38.0	N18 19.3	51 16.1	N19 00.6	137 49.6	S13 42.0	297 35.6	N 6 56.5	Schedar	349 44.0	N56 35
D	13	121 37.0	59 37.6	19.9	66 16.7	01.1	152 52.3	42.1	312 37.9	56.5	Shaula	96 25.5	S37 06
A	14	136 39.4	74 37.2	20.5	81 17.4	01.5	167 54.9	42.2	327 40.2	56.4	Sirius	258 36.6	S16 43
Y	15	151 41.9	89 36.9	. . 21.1	96 18.0	. . 01.9	182 57.5	. . 42.2	342 42.4	. . 56.3	Spica	158 34.3	S11 12
	16	166 44.3	104 36.5	21.6	111 18.7	02.4	198 00.2	42.3	357 44.7	56.2	Suhail	222 55.1	S43 28
	17	181 46.8	119 36.2	22.2	126 19.3	02.8	213 02.8	42.4	12 47.0	56.1			
	18	196 49.3	134 35.8	N18 22.8	141 20.0	N19 03.2	228 05.4	S13 42.5	27 49.3	N 6 56.0	Vega	80 40.5	N38 47
	19	211 51.7	149 35.4	23.4	156 20.6	03.6	243 08.1	42.5	42 51.6	55.9	Zuben'ubi	137 08.5	S16 05
	20	226 54.2	164 35.0	23.9	171 21.2	04.1	258 10.7	42.6	57 53.8	55.9			
	21	241 56.7	179 34.7	. . 24.5	186 21.9	. . 04.5	273 13.3	. . 42.7	72 56.1	. . 55.8			
	22	256 59.1	194 34.3	25.1	201 22.5	04.9	288 16.0	42.7	87 58.4	55.7			
	23	272 01.6	209 33.9	25.7	216 23.2	05.4	303 18.6	42.8	103 00.7	55.6			
	Mer. Pass.	h m 4 58.8	*v* −0.3	*d* 0.6	*v* 0.6	*d* 0.4	*v* 2.6	*d* 0.1	*v* 2.3	*d* 0.1			

	SHA	Mer. Pass.
	° ′	h m
Venus	299 44.9	9 01
Mars	305 47.0	8 36
Jupiter	31 09.2	2 54
Saturn	191 07.7	16 13

Day	UT	SUN GHA	SUN Dec	MOON GHA	v	MOON Dec	d	HP
	d h	° ′	° ′	° ′	′	° ′	′	′
MONDAY	6 00	178 49.4	N22 41.7	15 16.8	10.4	S26 10.4	2.2	54.1
	01	193 49.3	41.5	29 46.2	10.4	26 08.2	2.2	54.1
	02	208 49.2	41.2	44 15.6	10.5	26 06.0	2.4	54.1
	03	223 49.1	. . 41.0	58 45.1	10.5	26 03.6	2.6	54.1
	04	238 49.0	40.7	73 14.6	10.5	26 01.0	2.6	54.1
	05	253 48.9	40.5	87 44.1	10.5	25 58.4	2.8	54.1
	06	268 48.8	N22 40.2	102 13.6	10.6	S25 55.6	2.9	54.1
	07	283 48.6	39.9	116 43.2	10.6	25 52.7	3.0	54.1
	08	298 48.5	39.7	131 12.8	10.7	25 49.7	3.2	54.1
	09	313 48.4	. . 39.4	145 42.5	10.7	25 46.5	3.3	54.1
	10	328 48.3	39.2	160 12.2	10.7	25 43.2	3.4	54.1
	11	343 48.2	38.9	174 41.9	10.7	25 39.8	3.5	54.1
	12	358 48.1	N22 38.7	189 11.6	10.8	S25 36.3	3.6	54.1
	13	13 48.0	38.4	203 41.4	10.9	25 32.7	3.8	54.1
	14	28 47.9	38.1	218 11.3	10.8	25 28.9	3.8	54.1
	15	43 47.8	. . 37.9	232 41.1	10.9	25 25.1	4.0	54.0
	16	58 47.7	37.6	247 11.0	11.0	25 21.1	4.1	54.0
	17	73 47.6	37.4	261 41.0	11.0	25 17.0	4.3	54.0
	18	88 47.5	N22 37.1	276 11.0	11.0	S25 12.7	4.3	54.0
	19	103 47.4	36.8	290 41.0	11.1	25 08.4	4.5	54.0
	20	118 47.3	36.6	305 11.1	11.1	25 03.9	4.6	54.0
	21	133 47.2	. . 36.3	319 41.2	11.2	24 59.3	4.7	54.0
	22	148 47.1	36.0	334 11.4	11.2	24 54.6	4.8	54.0
	23	163 47.0	35.8	348 41.6	11.3	24 49.8	4.9	54.0
TUESDAY	7 00	178 46.9	N22 35.5	3 11.9	11.3	S24 44.9	5.0	54.0
	01	193 46.8	35.3	17 42.2	11.3	24 39.9	5.2	54.0
	02	208 46.7	35.0	32 12.5	11.4	24 34.7	5.3	54.0
	03	223 46.6	. . 34.7	46 42.9	11.5	24 29.4	5.3	54.0
	04	238 46.5	34.4	61 13.4	11.5	24 24.1	5.5	54.0
	05	253 46.4	34.2	75 43.9	11.5	24 18.6	5.6	54.0
	06	268 46.3	N22 33.9	90 14.4	11.6	S24 13.0	5.7	54.0
	07	283 46.2	33.6	104 45.0	11.7	24 07.3	5.9	54.0
	08	298 46.1	33.4	119 15.7	11.7	24 01.4	5.9	54.0
	09	313 46.0	. . 33.1	133 46.4	11.7	23 55.5	6.0	54.0
	10	328 45.9	32.8	148 17.1	11.8	23 49.5	6.2	54.0
	11	343 45.8	32.5	162 47.9	11.9	23 43.3	6.2	54.0
	12	358 45.7	N22 32.3	177 18.8	11.9	S23 37.1	6.4	54.0
	13	13 45.6	32.0	191 49.7	11.9	23 30.7	6.5	54.0
	14	28 45.5	31.7	206 20.6	12.0	23 24.2	6.5	54.0
	15	43 45.4	. . 31.4	220 51.6	12.1	23 17.7	6.7	54.0
	16	58 45.3	31.2	235 22.7	12.1	23 11.0	6.8	54.0
	17	73 45.2	30.9	249 53.8	12.2	23 04.2	6.9	54.0
	18	88 45.1	N22 30.6	264 25.0	12.2	S22 57.3	6.9	54.0
	19	103 45.0	30.3	278 56.2	12.3	22 50.4	7.1	54.0
	20	118 44.9	30.1	293 27.5	12.4	22 43.3	7.2	54.0
	21	133 44.8	. . 29.8	307 58.9	12.4	22 36.1	7.3	54.0
	22	148 44.7	29.5	322 30.3	12.4	22 28.8	7.4	54.0
	23	163 44.7	29.2	337 01.7	12.5	22 21.4	7.5	54.0
WEDNESDAY	8 00	178 44.6	N22 28.9	351 33.2	12.6	S22 13.9	7.5	54.0
	01	193 44.5	28.7	6 04.8	12.6	22 06.4	7.7	54.0
	02	208 44.4	28.4	20 36.4	12.7	21 58.7	7.8	54.0
	03	223 44.3	. . 28.1	35 08.1	12.7	21 50.9	7.9	54.0
	04	238 44.2	27.8	49 39.8	12.8	21 43.0	7.9	54.0
	05	253 44.1	27.5	64 11.6	12.8	21 35.1	8.1	54.0
	06	268 44.0	N22 27.2	78 43.4	12.9	S21 27.0	8.1	54.0
	07	283 43.9	26.9	93 15.3	13.0	21 18.9	8.3	54.0
	08	298 43.8	26.7	107 47.3	13.0	21 10.6	8.3	54.0
	09	313 43.7	. . 26.4	122 19.3	13.0	21 02.3	8.4	54.0
	10	328 43.6	26.1	136 51.3	13.1	20 53.9	8.5	54.0
	11	343 43.5	25.8	151 23.4	13.2	20 45.4	8.6	54.0
	12	358 43.4	N22 25.5	165 55.6	13.2	S20 36.8	8.7	54.0
	13	13 43.3	25.2	180 27.8	13.3	20 28.1	8.8	54.0
	14	28 43.2	24.9	195 00.1	13.4	20 19.3	8.9	54.0
	15	43 43.1	. . 24.6	209 32.5	13.4	20 10.4	8.9	54.0
	16	58 43.0	24.3	224 04.9	13.4	20 01.5	9.1	54.0
	17	73 42.9	24.0	238 37.3	13.5	19 52.4	9.1	54.0
	18	88 42.8	N22 23.7	253 09.8	13.6	S19 43.3	9.2	54.0
	19	103 42.8	23.4	267 42.4	13.6	19 34.1	9.3	54.0
	20	118 42.7	23.2	282 15.0	13.7	19 24.8	9.3	54.0
	21	133 42.6	. . 22.9	296 47.7	13.7	19 15.5	9.5	54.0
	22	148 42.5	22.6	311 20.4	13.8	19 06.0	9.5	54.0
	23	163 42.4	22.3	325 53.2	13.8	S18 56.5	9.6	54.0
		SD 15.8	*d* 0.3	SD 14.7		14.7		14.7

Lat.	Twilight Naut.	Twilight Civil	Sunrise	Moonrise 6	Moonrise 7	Moonrise 8	Moonrise 9
°	h m	h m	h m	h m	h m	h m	h m
N 72	□	□	□	■	■	■	23 36
N 70	□	□	□	■	■	23 58	23 06
68	□	□	□	■	■	23 11	22 43
66	////	////	00 58	■	23 04	22 40	22 26
64	////	////	01 54	22 35	22 24	22 17	22 11
62	////	////	02 26	21 48	21 56	21 58	21 59
60	////	01 20	02 50	21 18	21 34	21 43	21 48
N 58	////	01 59	03 09	20 55	21 16	21 30	21 39
56	////	02 25	03 24	20 37	21 01	21 19	21 31
54	01 13	02 45	03 38	20 21	20 49	21 09	21 24
52	01 49	03 02	03 49	20 07	20 37	21 00	21 17
50	02 14	03 16	04 00	19 55	20 27	20 52	21 11
45	02 56	03 44	04 21	19 31	20 06	20 35	20 59
N 40	03 25	04 06	04 38	19 11	19 49	20 21	20 48
35	03 47	04 23	04 53	18 54	19 34	20 09	20 39
30	04 05	04 38	05 05	18 40	19 22	19 58	20 31
20	04 33	05 02	05 26	18 16	19 00	19 40	20 17
N 10	04 55	05 22	05 44	17 55	18 41	19 25	20 05
0	05 13	05 39	06 01	17 35	18 23	19 10	19 54
S 10	05 29	05 55	06 18	17 15	18 06	18 55	19 43
20	05 45	06 12	06 36	16 54	17 47	18 39	19 30
30	06 00	06 30	06 56	16 29	17 25	18 21	19 16
35	06 09	06 40	07 08	16 15	17 12	18 10	19 08
40	06 17	06 51	07 22	15 58	16 57	17 58	18 59
45	06 27	07 04	07 38	15 38	16 39	17 43	18 48
S 50	06 38	07 19	07 57	15 13	16 17	17 25	18 34
52	06 43	07 26	08 07	15 00	16 07	17 17	18 28
54	06 48	07 34	08 17	14 46	15 55	17 07	18 21
56	06 54	07 43	08 29	14 30	15 41	16 56	18 13
58	07 00	07 52	08 43	14 10	15 25	16 44	18 04
S 60	07 07	08 03	08 59	13 45	15 05	16 30	17 54

Lat.	Sunset	Twilight Civil	Twilight Naut.	Moonset 6	Moonset 7	Moonset 8	Moonset 9
°	h m	h m	h m	h m	h m	h m	h m
N 72	□	□	□	■	■	■	■
N 70	□	□	□	■	■	■	03 00
68	□	□	□	■	■	■	03 46
66	23 07	////	////	■	■	02 17	04 16
64	22 14	////	////	■	01 06	02 57	04 38
62	21 42	////	////	00 28	01 52	03 24	04 56
60	21 19	22 47	////	01 07	02 21	03 45	05 10
N 58	21 00	22 10	////	01 34	02 44	04 02	05 23
56	20 45	21 44	////	01 55	03 02	04 17	05 33
54	20 32	21 24	22 54	02 13	03 18	04 29	05 43
52	20 20	21 07	22 19	02 28	03 31	04 40	05 51
50	20 10	20 53	21 55	02 41	03 43	04 50	05 58
45	19 49	20 25	21 13	03 08	04 07	05 10	06 14
N 40	19 31	20 04	20 45	03 29	04 26	05 26	06 27
35	19 17	19 46	20 23	03 46	04 42	05 40	06 38
30	19 05	19 32	20 05	04 01	04 56	05 52	06 48
20	18 44	19 08	19 37	04 27	05 20	06 12	07 04
N 10	18 25	18 48	19 15	04 49	05 40	06 30	07 18
0	18 09	18 31	18 57	05 09	05 59	06 46	07 31
S 10	17 52	18 15	18 41	05 29	06 17	07 02	07 45
20	17 34	17 58	18 25	05 51	06 37	07 20	07 58
30	17 14	17 40	18 10	06 16	07 00	07 40	08 14
35	17 02	17 30	18 01	06 31	07 14	07 51	08 24
40	16 49	17 19	17 53	06 48	07 29	08 04	08 34
45	16 33	17 06	17 43	07 09	07 48	08 20	08 46
S 50	16 13	16 51	17 32	07 35	08 11	08 39	09 01
52	16 04	16 44	17 27	07 47	08 22	08 48	09 08
54	15 53	16 36	17 22	08 01	08 34	08 58	09 15
56	15 41	16 28	17 16	08 18	08 48	09 09	09 24
58	15 28	16 18	17 10	08 38	09 05	09 22	09 33
S 60	15 11	16 07	17 03	09 03	09 25	09 37	09 44

Day	SUN Eqn. of Time 00^h	SUN Eqn. of Time 12^h	SUN Mer. Pass.	MOON Mer. Pass. Upper	MOON Mer. Pass. Lower	Age	Phase
d	m s	m s	h m	h m	h m	d %	
6	04 42	04 47	12 05	23 47	11 22	14 99	
7	04 52	04 57	12 05	24 35	12 11	15 100	○
8	05 02	05 06	12 05	00 35	12 58	16 99	

Day	UT d h	ARIES GHA	VENUS −4.1 GHA	VENUS Dec	MARS +1.1 GHA	MARS Dec	JUPITER −2.7 GHA	JUPITER Dec	SATURN +1.1 GHA	SATURN Dec
		° ′	° ′	° ′	° ′	° ′	° ′	° ′	° ′	° ′
THURSDAY	9 00	287 04.1	224 33.6	N18 26.2	231 23.8	N19 05.8	318 21.3	S13 42.9	118 03.0	N 6 55.5
	01	302 06.5	239 33.2	26.8	246 24.5	06.2	333 23.9	43.0	133 05.2	55.4
	02	317 09.0	254 32.8	27.4	261 25.1	06.6	348 26.5	43.0	148 07.5	55.4
	03	332 11.5	269 32.4	27.9	276 25.8	07.1	3 29.2	43.1	163 09.8	55.3
	04	347 13.9	284 32.1	28.5	291 26.4	07.5	18 31.8	43.2	178 12.1	55.2
	05	2 16.4	299 31.7	29.1	306 27.1	07.9	33 34.4	43.2	193 14.3	55.1
	06	17 18.8	314 31.3	N18 29.6	321 27.7	N19 08.3	48 37.1	S13 43.3	208 16.6	N 6 55.0
	07	32 21.3	329 31.0	30.2	336 28.3	08.8	63 39.7	43.4	223 18.9	54.9
	08	47 23.8	344 30.6	30.8	351 29.0	09.2	78 42.4	43.5	238 21.2	54.8
	09	62 26.2	359 30.2	31.3	6 29.6	09.6	93 45.0	43.5	253 23.4	54.8
	10	77 28.7	14 29.8	31.9	21 30.3	10.0	108 47.6	43.6	268 25.7	54.7
	11	92 31.2	29 29.4	32.5	36 30.9	10.4	123 50.3	43.7	283 28.0	54.6
	12	107 33.6	44 29.1	N18 33.0	51 31.6	N19 10.9	138 52.9	S13 43.8	298 30.3	N 6 54.5
	13	122 36.1	59 28.7	33.6	66 32.2	11.3	153 55.6	43.8	313 32.6	54.4
	14	137 38.6	74 28.3	34.1	81 32.9	11.7	168 58.2	43.9	328 34.8	54.3
	15	152 41.0	89 27.9	34.7	96 33.5	12.1	184 00.8	44.0	343 37.1	54.3
	16	167 43.5	104 27.5	35.3	111 34.2	12.6	199 03.5	44.0	358 39.4	54.2
	17	182 45.9	119 27.2	35.8	126 34.8	13.0	214 06.1	44.1	13 41.7	54.1
	18	197 48.4	134 26.8	N18 36.4	141 35.4	N19 13.4	229 08.8	S13 44.2	28 43.9	N 6 54.0
	19	212 50.9	149 26.4	36.9	156 36.1	13.8	244 11.4	44.3	43 46.2	53.9
	20	227 53.3	164 26.0	37.5	171 36.7	14.2	259 14.1	44.3	58 48.5	53.8
	21	242 55.8	179 25.6	38.1	186 37.4	14.7	274 16.7	44.4	73 50.8	53.7
	22	257 58.3	194 25.3	38.6	201 38.0	15.1	289 19.3	44.5	88 53.0	53.7
	23	273 00.7	209 24.9	39.2	216 38.7	15.5	304 22.0	44.6	103 55.3	53.6
FRIDAY	10 00	288 03.2	224 24.5	N18 39.7	231 39.3	N19 15.9	319 24.6	S13 44.6	118 57.6	N 6 53.5
	01	303 05.7	239 24.1	40.3	246 40.0	16.3	334 27.3	44.7	133 59.9	53.4
	02	318 08.1	254 23.7	40.8	261 40.6	16.7	349 29.9	44.8	149 02.1	53.3
	03	333 10.6	269 23.3	41.4	276 41.3	17.2	4 32.6	44.9	164 04.4	53.2
	04	348 13.1	284 22.9	41.9	291 41.9	17.6	19 35.2	44.9	179 06.7	53.1
	05	3 15.5	299 22.5	42.5	306 42.5	18.0	34 37.9	45.0	194 09.0	53.1
	06	18 18.0	314 22.2	N18 43.1	321 43.2	N19 18.4	49 40.5	S13 45.1	209 11.2	N 6 53.0
	07	33 20.4	329 21.8	43.6	336 43.8	18.8	64 43.1	45.2	224 13.5	52.9
	08	48 22.9	344 21.4	44.2	351 44.5	19.2	79 45.8	45.2	239 15.8	52.8
	09	63 25.4	359 21.0	44.7	6 45.1	19.7	94 48.4	45.3	254 18.1	52.7
	10	78 27.8	14 20.6	45.3	21 45.8	20.1	109 51.1	45.4	269 20.3	52.6
	11	93 30.3	29 20.2	45.8	36 46.4	20.5	124 53.7	45.5	284 22.6	52.5
	12	108 32.8	44 19.8	N18 46.4	51 47.1	N19 20.9	139 56.4	S13 45.5	299 24.9	N 6 52.5
	13	123 35.2	59 19.4	46.9	66 47.7	21.3	154 59.0	45.6	314 27.2	52.4
	14	138 37.7	74 19.0	47.4	81 48.4	21.7	170 01.7	45.7	329 29.4	52.3
	15	153 40.2	89 18.6	48.0	96 49.0	22.1	185 04.3	45.8	344 31.7	52.2
	16	168 42.6	104 18.2	48.5	111 49.6	22.6	200 07.0	45.8	359 34.0	52.1
	17	183 45.1	119 17.8	49.1	126 50.3	23.0	215 09.6	45.9	14 36.2	52.0
	18	198 47.6	134 17.4	N18 49.6	141 50.9	N19 23.4	230 12.3	S13 46.0	29 38.5	N 6 51.9
	19	213 50.0	149 17.0	50.2	156 51.6	23.8	245 14.9	46.1	44 40.8	51.9
	20	228 52.5	164 16.7	50.7	171 52.2	24.2	260 17.6	46.1	59 43.1	51.8
	21	243 54.9	179 16.3	51.3	186 52.9	24.6	275 20.2	46.2	74 45.3	51.7
	22	258 57.4	194 15.9	51.8	201 53.5	25.0	290 22.9	46.3	89 47.6	51.6
	23	273 59.9	209 15.5	52.3	216 54.2	25.4	305 25.5	46.4	104 49.9	51.5
SATURDAY	11 00	289 02.3	224 15.1	N18 52.9	231 54.8	N19 25.8	320 28.2	S13 46.4	119 52.2	N 6 51.4
	01	304 04.8	239 14.7	53.4	246 55.4	26.3	335 30.8	46.5	134 54.4	51.3
	02	319 07.3	254 14.3	54.0	261 56.1	26.7	350 33.5	46.6	149 56.7	51.2
	03	334 09.7	269 13.9	54.5	276 56.7	27.1	5 36.1	46.7	164 59.0	51.2
	04	349 12.2	284 13.5	55.0	291 57.4	27.5	20 38.8	46.7	180 01.2	51.1
	05	4 14.7	299 13.1	55.6	306 58.0	27.9	35 41.4	46.8	195 03.5	51.0
	06	19 17.1	314 12.6	N18 56.1	321 58.7	N19 28.3	50 44.1	S13 46.9	210 05.8	N 6 50.9
	07	34 19.6	329 12.2	56.7	336 59.3	28.7	65 46.7	47.0	225 08.1	50.8
	08	49 22.0	344 11.8	57.2	352 00.0	29.1	80 49.4	47.0	240 10.3	50.7
	09	64 24.5	359 11.4	57.7	7 00.6	29.5	95 52.0	47.1	255 12.6	50.6
	10	79 27.0	14 11.0	58.3	22 01.2	29.9	110 54.7	47.2	270 14.9	50.6
	11	94 29.4	29 10.6	58.8	37 01.9	30.3	125 57.3	47.3	285 17.1	50.5
	12	109 31.9	44 10.2	N18 59.3	52 02.5	N19 30.7	141 00.0	S13 47.4	300 19.4	N 6 50.4
	13	124 34.4	59 09.8	18 59.9	67 03.2	31.1	156 02.6	47.4	315 21.7	50.3
	14	139 36.8	74 09.4	19 00.4	82 03.8	31.6	171 05.3	47.5	330 24.0	50.2
	15	154 39.3	89 09.0	00.9	97 04.5	32.0	186 07.9	47.6	345 26.2	50.1
	16	169 41.8	104 08.6	01.5	112 05.1	32.4	201 10.6	47.7	0 28.5	50.0
	17	184 44.2	119 08.2	02.0	127 05.8	32.8	216 13.3	47.7	15 30.8	49.9
	18	199 46.7	134 07.8	N19 02.5	142 06.4	N19 33.2	231 15.9	S13 47.8	30 33.0	N 6 49.9
	19	214 49.2	149 07.4	03.0	157 07.0	33.6	246 18.6	47.9	45 35.3	49.8
	20	229 51.6	164 06.9	03.6	172 07.7	34.0	261 21.2	48.0	60 37.6	49.7
	21	244 54.1	179 06.5	04.1	187 08.3	34.4	276 23.9	48.0	75 39.8	49.6
	22	259 56.5	194 06.1	04.6	202 09.0	34.8	291 26.5	48.1	90 42.1	49.5
	23	274 59.0	209 05.7	05.2	217 09.6	35.2	306 29.2	48.2	105 44.4	49.4
	Mer. Pass. (h m)	4 47.0	v −0.4	d 0.5	v 0.6	d 0.4	v 2.6	d 0.1	v 2.3	d 0.1

STARS

Name	SHA	Dec
	° ′	° ′
Acamar	315 20.6	S40 15.
Achernar	335 28.8	S57 10.
Acrux	173 12.9	S63 09.
Adhara	255 15.2	S28 59.
Aldebaran	290 53.0	N16 31.
Alioth	166 23.0	N55 54.
Alkaid	153 00.9	N49 16.
Al Na'ir	27 46.9	S46 54.
Alnilam	275 49.6	S 1 11.
Alphard	217 59.2	S 8 42.
Alphecca	126 13.2	N26 41.
Alpheratz	357 46.5	N29 08.
Altair	62 10.7	N 8 53.
Ankaa	353 18.3	S42 14.
Antares	112 29.5	S26 27.
Arcturus	145 58.2	N19 08.
Atria	107 33.6	S69 02.
Avior	234 20.0	S59 32.
Bellatrix	278 35.4	N 6 21.
Betelgeuse	271 04.8	N 7 24.
Canopus	263 58.0	S52 41.
Capella	280 39.2	N46 00.
Deneb	49 33.1	N45 18.
Denebola	182 36.7	N14 31.
Diphda	348 58.7	S17 55.
Dubhe	193 55.3	N61 42.
Elnath	278 16.6	N28 36.
Eltanin	90 47.0	N51 29.
Enif	33 49.7	N 9 55.
Fomalhaut	15 26.9	S29 34.
Gacrux	172 04.5	S57 10.
Gienah	175 55.4	S17 35.
Hadar	148 52.1	S60 25.
Hamal	328 04.2	N23 30.
Kaus Aust.	83 47.2	S34 22.
Kochab	137 19.0	N74 07.
Markab	13 41.1	N15 15.
Menkar	314 18.2	N 4 07.
Menkent	148 11.0	S36 25.
Miaplacidus	221 41.4	S69 45.
Mirfak	308 44.9	N49 53.
Nunki	76 01.5	S26 17.
Peacock	53 23.1	S56 42.
Pollux	243 31.6	N28 00.
Procyon	245 03.1	N 5 12.
Rasalhague	96 08.8	N12 33.
Regulus	207 46.8	N11 55.
Rigel	281 15.1	S 8 11.
Rigil Kent.	139 55.6	S60 52.
Sabik	102 15.6	S15 44.
Schedar	349 44.0	N56 35.
Shaula	96 25.5	S37 06.
Sirius	258 36.6	S16 43.
Spica	158 34.3	S11 12.
Suhail	222 55.1	S43 28.
Vega	80 40.5	N38 47.
Zuben'ubi	137 08.5	S16 05.

	SHA	Mer. Pass.
	° ′	h m
Venus	296 21.3	9 03
Mars	303 36.1	8 33
Jupiter	31 21.4	2 42
Saturn	190 54.4	16 02

UT	SUN GHA	SUN Dec	MOON GHA	v	MOON Dec	d	HP
d h	° ′	° ′	° ′	′	° ′	′	′
9 00	178 42.3	N22 22.0	340 26.0	13.9	S18 46.9	9.7	54.0
01	193 42.2	21.7	354 58.9	14.0	18 37.2	9.8	54.0
02	208 42.1	21.4	9 31.9	14.0	18 27.4	9.8	54.0
03	223 42.0	. . 21.1	24 04.9	14.0	18 17.6	9.9	54.0
04	238 41.9	20.8	38 37.9	14.1	18 07.7	10.0	54.1
05	253 41.8	20.5	53 11.0	14.1	17 57.7	10.1	54.1
06	268 41.7	N22 20.2	67 44.1	14.2	S17 47.6	10.1	54.1
07	283 41.7	19.9	82 17.3	14.3	17 37.5	10.3	54.1
08	298 41.6	19.6	96 50.6	14.3	17 27.2	10.3	54.1
09	313 41.5	. . 19.3	111 23.9	14.3	17 16.9	10.3	54.1
10	328 41.4	18.9	125 57.2	14.4	17 06.6	10.5	54.1
11	343 41.3	18.6	140 30.6	14.5	16 56.1	10.5	54.1
12	358 41.2	N22 18.3	155 04.1	14.5	S16 45.6	10.5	54.1
13	13 41.1	18.0	169 37.6	14.5	16 35.1	10.7	54.1
14	28 41.0	17.7	184 11.1	14.6	16 24.4	10.7	54.1
15	43 40.9	. . 17.4	198 44.7	14.7	16 13.7	10.8	54.1
16	58 40.8	17.1	213 18.4	14.7	16 02.9	10.8	54.1
17	73 40.7	16.8	227 52.1	14.7	15 52.1	10.9	54.1
18	88 40.7	N22 16.5	242 25.8	14.8	S15 41.2	11.0	54.1
19	103 40.6	16.2	256 59.6	14.8	15 30.2	11.0	54.1
20	118 40.5	15.9	271 33.4	14.9	15 19.2	11.1	54.2
21	133 40.4	. . 15.5	286 07.3	14.9	15 08.1	11.2	54.2
22	148 40.3	15.2	300 41.2	14.9	14 56.9	11.2	54.2
23	163 40.2	14.9	315 15.1	15.0	14 45.7	11.3	54.2
10 00	178 40.1	N22 14.6	329 49.1	15.1	S14 34.4	11.3	54.2
01	193 40.0	14.3	344 23.2	15.1	14 23.1	11.4	54.2
02	208 40.0	14.0	358 57.3	15.1	14 11.7	11.5	54.2
03	223 39.9	. . 13.7	13 31.4	15.2	14 00.2	11.5	54.2
04	238 39.8	13.3	28 05.6	15.2	13 48.7	11.6	54.2
05	253 39.7	13.0	42 39.8	15.2	13 37.1	11.6	54.2
06	268 39.6	N22 12.7	57 14.0	15.3	S13 25.5	11.7	54.2
07	283 39.5	12.4	71 48.3	15.3	13 13.8	11.8	54.2
08	298 39.4	12.1	86 22.6	15.3	13 02.0	11.8	54.3
09	313 39.3	. . 11.8	100 56.9	15.4	12 50.2	11.8	54.3
10	328 39.3	11.4	115 31.3	15.5	12 38.4	11.9	54.3
11	343 39.2	11.1	130 05.8	15.4	12 26.5	12.0	54.3
12	358 39.1	N22 10.8	144 40.2	15.5	S12 14.5	12.0	54.3
13	13 39.0	10.5	159 14.7	15.5	12 02.5	12.0	54.3
14	28 38.9	10.1	173 49.2	15.6	11 50.5	12.1	54.3
15	43 38.8	. . 09.8	188 23.8	15.6	11 38.4	12.2	54.3
16	58 38.7	09.5	202 58.4	15.6	11 26.2	12.2	54.3
17	73 38.7	09.2	217 33.0	15.6	11 14.0	12.3	54.4
18	88 38.6	N22 08.8	232 07.6	15.7	S11 01.7	12.2	54.4
19	103 38.5	08.5	246 42.3	15.7	10 49.5	12.4	54.4
20	118 38.4	08.2	261 17.0	15.8	10 37.1	12.4	54.4
21	133 38.3	. . 07.9	275 51.8	15.7	10 24.7	12.4	54.4
22	148 38.2	07.5	290 26.5	15.8	10 12.3	12.5	54.4
23	163 38.2	07.2	305 01.3	15.8	9 59.8	12.5	54.4
11 00	178 38.1	N22 06.9	319 36.1	15.8	S 9 47.3	12.6	54.4
01	193 38.0	06.5	334 10.9	15.9	9 34.7	12.6	54.5
02	208 37.9	06.2	348 45.8	15.9	9 22.1	12.6	54.5
03	223 37.8	. . 05.9	3 20.7	15.9	9 09.5	12.7	54.5
04	238 37.7	05.5	17 55.6	15.9	8 56.8	12.7	54.5
05	253 37.7	05.2	32 30.5	15.9	8 44.1	12.8	54.5
06	268 37.6	N22 04.9	47 05.4	16.0	S 8 31.3	12.8	54.5
07	283 37.5	04.5	61 40.4	16.0	8 18.5	12.8	54.5
08	298 37.4	04.2	76 15.4	16.0	8 05.7	12.9	54.5
09	313 37.3	. . 03.9	90 50.4	16.0	7 52.8	12.9	54.6
10	328 37.2	03.5	105 25.4	16.0	7 39.9	13.0	54.6
11	343 37.2	03.2	120 00.4	16.0	7 26.9	12.9	54.6
12	358 37.1	N22 02.9	134 35.4	16.1	S 7 14.0	13.0	54.6
13	13 37.0	02.5	149 10.5	16.1	7 01.0	13.1	54.6
14	28 36.9	02.2	163 45.6	16.0	6 47.9	13.1	54.6
15	43 36.8	. . 01.8	178 20.6	16.1	6 34.8	13.1	54.7
16	58 36.8	01.5	192 55.7	16.1	6 21.7	13.1	54.7
17	73 36.7	01.2	207 30.8	16.1	6 08.6	13.2	54.7
18	88 36.6	N22 00.8	222 05.9	16.1	S 5 55.4	13.2	54.7
19	103 36.5	00.5	236 41.0	16.1	5 42.2	13.2	54.7
20	118 36.4	22 00.1	251 16.1	16.2	5 29.0	13.3	54.7
21	133 36.4	21 59.8	265 51.3	16.1	5 15.7	13.2	54.8
22	148 36.3	59.4	280 26.4	16.1	5 02.5	13.4	54.8
23	163 36.2	59.1	295 01.5	16.2	S 4 49.1	13.3	54.8
	SD 15.8	*d* 0.3	SD	14.7	14.8		14.9

Lat.	Twilight Naut.	Twilight Civil	Sunrise	Moonrise 9	10	11	12
°	h m	h m	h m	h m	h m	h m	h m
N 72	▭	▭	▭	23 36	22 53	22 22	21 55
N 70	▭	▭	▭	23 06	22 37	22 15	21 55
68	▭	▭	▭	22 43	22 25	22 10	21 56
66	////	////	01 13	22 26	22 15	22 05	21 56
64	////	////	02 02	22 11	22 06	22 01	21 56
62	////	////	02 32	21 59	21 58	21 57	21 56
60	////	01 29	02 55	21 48	21 52	21 54	21 56
N 58	////	02 05	03 13	21 39	21 46	21 51	21 56
56	////	02 30	03 28	21 31	21 41	21 49	21 57
54	01 22	02 49	03 41	21 24	21 36	21 47	21 57
52	01 55	03 06	03 52	21 17	21 32	21 45	21 57
50	02 18	03 19	04 02	21 11	21 28	21 43	21 57
45	02 59	03 47	04 23	20 59	21 20	21 39	21 57
N 40	03 27	04 08	04 40	20 48	21 13	21 35	21 57
35	03 49	04 25	04 54	20 39	21 07	21 32	21 58
30	04 07	04 39	05 07	20 31	21 01	21 30	21 58
20	04 34	05 03	05 27	20 17	20 52	21 25	21 58
N 10	04 56	05 22	05 45	20 05	20 44	21 21	21 58
0	05 13	05 39	06 02	19 54	20 36	21 17	21 59
S 10	05 29	05 56	06 18	19 43	20 29	21 14	21 59
20	05 45	06 12	06 36	19 30	20 20	21 10	21 59
30	06 00	06 30	06 56	19 16	20 11	21 05	21 59
35	06 08	06 39	07 07	19 08	20 05	21 02	22 00
40	06 17	06 50	07 21	18 59	19 59	20 59	22 00
45	06 26	07 03	07 36	18 48	19 52	20 56	22 00
S 50	06 37	07 18	07 55	18 34	19 43	20 52	22 01
52	06 42	07 25	08 05	18 28	19 39	20 50	22 01
54	06 47	07 32	08 15	18 21	19 34	20 48	22 01
56	06 52	07 40	08 26	18 13	19 29	20 45	22 01
58	06 58	07 50	08 40	18 04	19 24	20 43	22 01
S 60	07 05	08 00	08 55	17 54	19 18	20 40	22 02

Lat.	Sunset	Twilight Civil	Twilight Naut.	Moonset 9	10	11	12
°	h m	h m	h m	h m	h m	h m	h m
N 72	▭	▭	▭	■	04 54	07 06	09 04
N 70	▭	▭	▭	03 00	05 23	07 19	09 07
68	▭	▭	▭	03 46	05 44	07 29	09 10
66	22 54	////	////	04 16	06 00	07 38	09 13
64	22 07	////	////	04 38	06 14	07 45	09 15
62	21 37	////	////	04 56	06 25	07 52	09 17
60	21 15	22 39	////	05 10	06 34	07 57	09 19
N 58	20 57	22 04	////	05 23	06 43	08 02	09 20
56	20 42	21 40	////	05 33	06 50	08 06	09 22
54	20 29	21 20	22 47	05 43	06 56	08 10	09 23
52	20 18	21 04	22 14	05 51	07 02	08 13	09 24
50	20 08	20 51	21 51	05 58	07 07	08 16	09 25
45	19 47	20 23	21 11	06 14	07 19	08 23	09 27
N 40	19 30	20 02	20 43	06 27	07 28	08 28	09 28
35	19 16	19 45	20 21	06 38	07 36	08 33	09 30
30	19 04	19 31	20 04	06 48	07 43	08 37	09 31
20	18 43	19 08	19 36	07 04	07 55	08 44	09 33
N 10	18 26	18 48	19 15	07 18	08 05	08 50	09 35
0	18 09	18 31	18 57	07 31	08 15	08 56	09 37
S 10	17 53	18 15	18 41	07 45	08 24	09 02	09 39
20	17 35	17 59	18 26	07 58	08 34	09 08	09 41
30	17 15	17 41	18 11	08 14	08 46	09 15	09 43
35	17 04	17 32	18 03	08 24	08 52	09 19	09 44
40	16 50	17 21	17 54	08 34	09 00	09 23	09 45
45	16 35	17 08	17 45	08 46	09 08	09 28	09 47
S 50	16 16	16 53	17 34	09 01	09 19	09 34	09 49
52	16 07	16 46	17 29	09 08	09 24	09 37	09 49
54	15 56	16 39	17 24	09 15	09 29	09 40	09 50
56	15 45	16 31	17 19	09 24	09 35	09 44	09 51
58	15 32	16 21	17 13	09 33	09 41	09 47	09 52
S 60	15 16	16 11	17 06	09 44	09 48	09 51	09 54

Day	SUN Eqn. of Time 00^h	SUN Eqn. of Time 12^h	SUN Mer. Pass.	MOON Mer. Pass. Upper	MOON Mer. Pass. Lower	Age	Phase
d	m s	m s	h m	h m	h m	d %	
9	05 11	05 15	12 05	01 21	13 43	17 96	
10	05 19	05 23	12 05	02 04	14 25	18 92	◐
11	05 28	05 32	12 06	02 46	15 07	19 85	

	UT	ARIES	VENUS −4.1		MARS +1.1		JUPITER −2.7		SATURN +1.1	
		GHA	GHA	Dec	GHA	Dec	GHA	Dec	GHA	Dec
	d h	° ′	° ′	° ′	° ′	° ′	° ′	° ′	° ′	° ′
SUNDAY	12 00	290 01.5	224 05.3	N19 05.7	232 10.3	N19 35.6	321 31.8	S13 48.3	120 46.7	N 6 49.3
	01	305 03.9	239 04.9	06.2	247 10.9	36.0	336 34.5	48.4	135 48.9	49.2
	02	320 06.4	254 04.5	06.7	262 11.6	36.4	351 37.2	48.4	150 51.2	49.2
	03	335 08.9	269 04.0	. . 07.2	277 12.2	. . 36.8	6 39.8	. . 48.5	165 53.5	. . 49.1
	04	350 11.3	284 03.6	07.8	292 12.8	37.2	21 42.5	48.6	180 55.7	49.0
	05	5 13.8	299 03.2	08.3	307 13.5	37.6	36 45.1	48.7	195 58.0	48.9
	06	20 16.3	314 02.8	N19 08.8	322 14.1	N19 38.0	51 47.8	S13 48.8	211 00.3	N 6 48.8
	07	35 18.7	329 02.4	09.3	337 14.8	38.4	66 50.5	48.8	226 02.5	48.7
	08	50 21.2	344 02.0	09.9	352 15.4	38.8	81 53.1	48.9	241 04.8	48.6
	09	65 23.7	359 01.5	. . 10.4	7 16.1	. . 39.2	96 55.8	. . 49.0	256 07.1	. . 48.5
	10	80 26.1	14 01.1	10.9	22 16.7	39.6	111 58.4	49.1	271 09.3	48.5
	11	95 28.6	29 00.7	11.4	37 17.3	40.0	127 01.1	49.1	286 11.6	48.4
	12	110 31.0	44 00.3	N19 11.9	52 18.0	N19 40.4	142 03.8	S13 49.2	301 13.9	N 6 48.3
	13	125 33.5	58 59.9	12.4	67 18.6	40.8	157 06.4	49.3	316 16.2	48.2
	14	140 36.0	73 59.4	13.0	82 19.3	41.2	172 09.1	49.4	331 18.4	48.1
	15	155 38.4	88 59.0	. . 13.5	97 19.9	. . 41.6	187 11.7	. . 49.5	346 20.7	. . 48.0
	16	170 40.9	103 58.6	14.0	112 20.6	42.0	202 14.4	49.5	1 23.0	47.9
	17	185 43.4	118 58.2	14.5	127 21.2	42.4	217 17.1	49.6	16 25.2	47.8
	18	200 45.8	133 57.7	N19 15.0	142 21.9	N19 42.8	232 19.7	S13 49.7	31 27.5	N 6 47.7
	19	215 48.3	148 57.3	15.5	157 22.5	43.2	247 22.4	49.8	46 29.8	47.7
	20	230 50.8	163 56.9	16.0	172 23.1	43.6	262 25.0	49.9	61 32.0	47.6
	21	245 53.2	178 56.5	. . 16.6	187 23.8	. . 44.0	277 27.7	. . 49.9	76 34.3	. . 47.5
	22	260 55.7	193 56.0	17.1	202 24.4	44.4	292 30.4	50.0	91 36.6	47.4
	23	275 58.1	208 55.6	17.6	217 25.1	44.8	307 33.0	50.1	106 38.8	47.3
MONDAY	13 00	291 00.6	223 55.2	N19 18.1	232 25.7	N19 45.2	322 35.7	S13 50.2	121 41.1	N 6 47.2
	01	306 03.1	238 54.8	18.6	247 26.4	45.5	337 38.4	50.3	136 43.4	47.1
	02	321 05.5	253 54.3	19.1	262 27.0	45.9	352 41.0	50.3	151 45.6	47.0
	03	336 08.0	268 53.9	. . 19.6	277 27.6	. . 46.3	7 43.7	. . 50.4	166 47.9	. . 46.9
	04	351 10.5	283 53.5	20.1	292 28.3	46.7	22 46.3	50.5	181 50.2	46.9
	05	6 12.9	298 53.0	20.6	307 28.9	47.1	37 49.0	50.6	196 52.4	46.8
	06	21 15.4	313 52.6	N19 21.1	322 29.6	N19 47.5	52 51.7	S13 50.7	211 54.7	N 6 46.7
	07	36 17.9	328 52.2	21.6	337 30.2	47.9	67 54.3	50.7	226 57.0	46.6
	08	51 20.3	343 51.7	22.1	352 30.9	48.3	82 57.0	50.8	241 59.2	46.5
	09	66 22.8	358 51.3	. . 22.6	7 31.5	. . 48.7	97 59.7	. . 50.9	257 01.5	. . 46.4
	10	81 25.3	13 50.9	23.1	22 32.2	49.1	113 02.3	51.0	272 03.8	46.3
	11	96 27.7	28 50.5	23.6	37 32.8	49.5	128 05.0	51.1	287 06.0	46.2
	12	111 30.2	43 50.0	N19 24.1	52 33.4	N19 49.9	143 07.7	S13 51.2	302 08.3	N 6 46.1
	13	126 32.6	58 49.6	24.6	67 34.1	50.3	158 10.3	51.2	317 10.6	46.1
	14	141 35.1	73 49.1	25.1	82 34.7	50.6	173 13.0	51.3	332 12.8	46.0
	15	156 37.6	88 48.7	. . 25.6	97 35.4	. . 51.0	188 15.7	. . 51.4	347 15.1	. . 45.9
	16	171 40.0	103 48.3	26.1	112 36.0	51.4	203 18.3	51.5	2 17.4	45.8
	17	186 42.5	118 47.8	26.6	127 36.7	51.8	218 21.0	51.6	17 19.6	45.7
	18	201 45.0	133 47.4	N19 27.1	142 37.3	N19 52.2	233 23.7	S13 51.6	32 21.9	N 6 45.6
	19	216 47.4	148 47.0	27.6	157 37.9	52.6	248 26.3	51.7	47 24.1	45.5
	20	231 49.9	163 46.5	28.1	172 38.6	53.0	263 29.0	51.8	62 26.4	45.4
	21	246 52.4	178 46.1	. . 28.6	187 39.2	. . 53.4	278 31.7	. . 51.9	77 28.7	. . 45.3
	22	261 54.8	193 45.6	29.1	202 39.9	53.7	293 34.4	52.0	92 30.9	45.3
	23	276 57.3	208 45.2	29.6	217 40.5	54.1	308 37.0	52.1	107 33.2	45.2
TUESDAY	14 00	291 59.7	223 44.8	N19 30.1	232 41.2	N19 54.5	323 39.7	S13 52.1	122 35.5	N 6 45.1
	01	307 02.2	238 44.3	30.6	247 41.8	54.9	338 42.4	52.2	137 37.7	45.0
	02	322 04.7	253 43.9	31.1	262 42.4	55.3	353 45.0	52.3	152 40.0	44.9
	03	337 07.1	268 43.4	. . 31.6	277 43.1	. . 55.7	8 47.7	. . 52.4	167 42.3	. . 44.8
	04	352 09.6	283 43.0	32.1	292 43.7	56.1	23 50.4	52.5	182 44.5	44.7
	05	7 12.1	298 42.5	32.6	307 44.4	56.4	38 53.0	52.5	197 46.8	44.6
	06	22 14.5	313 42.1	N19 33.1	322 45.0	N19 56.8	53 55.7	S13 52.6	212 49.1	N 6 44.5
	07	37 17.0	328 41.7	33.5	337 45.7	57.2	68 58.4	52.7	227 51.3	44.4
	08	52 19.5	343 41.2	34.0	352 46.3	57.6	84 01.1	52.8	242 53.6	44.4
	09	67 21.9	358 40.8	. . 34.5	7 47.0	. . 58.0	99 03.7	. . 52.9	257 55.8	. . 44.3
	10	82 24.4	13 40.3	35.0	22 47.6	58.4	114 06.4	53.0	272 58.1	44.2
	11	97 26.9	28 39.9	35.5	37 48.2	58.8	129 09.1	53.0	288 00.4	44.1
	12	112 29.3	43 39.4	N19 36.0	52 48.9	N19 59.1	144 11.7	S13 53.1	303 02.6	N 6 44.0
	13	127 31.8	58 39.0	36.5	67 49.5	59.5	159 14.4	53.2	318 04.9	43.9
	14	142 34.2	73 38.5	36.9	82 50.2	19 59.9	174 17.1	53.3	333 07.2	43.8
	15	157 36.7	88 38.1	. . 37.4	97 50.8	20 00.3	189 19.8	. . 53.4	348 09.4	. . 43.7
	16	172 39.2	103 37.6	37.9	112 51.5	00.7	204 22.4	53.5	3 11.7	43.6
	17	187 41.6	118 37.2	38.4	127 52.1	01.0	219 25.1	53.5	18 13.9	43.5
	18	202 44.1	133 36.7	N19 38.9	142 52.7	N20 01.4	234 27.8	S13 53.6	33 16.2	N 6 43.4
	19	217 46.6	148 36.3	39.3	157 53.4	01.8	249 30.5	53.7	48 18.5	43.4
	20	232 49.0	163 35.8	39.8	172 54.0	02.2	264 33.1	53.8	63 20.7	43.3
	21	247 51.5	178 35.4	. . 40.3	187 54.7	. . 02.6	279 35.8	. . 53.9	78 23.0	. . 43.2
	22	262 54.0	193 34.9	40.8	202 55.3	02.9	294 38.5	54.0	93 25.3	43.1
	23	277 56.4	208 34.5	41.3	217 56.0	03.3	309 41.2	54.0	108 27.5	43.0
	Mer. Pass.	h m 4 35.2	*v* −0.4	*d* 0.5	*v* 0.6	*d* 0.4	*v* 2.7	*d* 0.1	*v* 2.3	*d* 0.1

STARS		
Name	SHA	Dec
	° ′	° ′
Acamar	315 20.6	S40 15.
Achernar	335 28.8	S57 10.
Acrux	173 13.0	S63 09.
Adhara	255 15.2	S28 59.
Aldebaran	290 53.0	N16 31.
Alioth	166 23.1	N55 54.
Alkaid	153 00.9	N49 16.
Al Na'ir	27 46.8	S46 54.
Alnilam	275 49.6	S 1 11.
Alphard	217 59.2	S 8 42.
Alphecca	126 13.2	N26 41.
Alpheratz	357 46.5	N29 08.
Altair	62 10.7	N 8 53.
Ankaa	353 18.3	S42 14.
Antares	112 29.6	S26 27.
Arcturus	145 58.2	N19 08.
Atria	107 33.6	S69 02.
Avior	234 20.0	S59 32.
Bellatrix	278 35.4	N 6 21.
Betelgeuse	271 04.7	N 7 24.
Canopus	263 58.0	S52 41.
Capella	280 39.2	N46 00.
Deneb	49 33.1	N45 18.
Denebola	182 36.7	N14 31.
Diphda	348 58.7	S17 55.
Dubhe	193 55.3	N61 42.
Elnath	278 16.6	N28 36.
Eltanin	90 47.0	N51 29.
Enif	33 49.7	N 9 55.
Fomalhaut	15 26.9	S29 34.
Gacrux	172 04.5	S57 10.3
Gienah	175 55.4	S17 35.8
Hadar	148 52.1	S60 25.5
Hamal	328 04.2	N23 30.5
Kaus Aust.	83 47.2	S34 22.8
Kochab	137 19.0	N74 07.2
Markab	13 41.1	N15 15.5
Menkar	314 18.2	N 4 07.7
Menkent	148 11.0	S36 25.3
Miaplacidus	221 41.4	S69 45.6
Mirfak	308 44.9	N49 53.6
Nunki	76 01.5	S26 17.1
Peacock	53 23.1	S56 42.1
Pollux	243 31.6	N28 00.2
Procyon	245 03.1	N 5 12.1
Rasalhague	96 08.8	N12 33.2
Regulus	207 46.8	N11 55.3
Rigel	281 15.1	S 8 11.3
Rigil Kent.	139 55.7	S60 52.8
Sabik	102 15.6	S15 44.3
Schedar	349 43.9	N56 35.2
Shaula	96 25.5	S37 06.7
Sirius	258 36.6	S16 43.7
Spica	158 34.3	S11 12.8
Suhail	222 55.1	S43 28.4
Vega	80 40.5	N38 47.6
Zuben'ubi	137 08.5	S16 05.0
	SHA	Mer. Pass.
	° ′	h m
Venus	292 54.6	9 05
Mars	301 25.1	8 30
Jupiter	31 35.1	2 29
Saturn	190 40.5	15 51

JT	SUN GHA	SUN Dec	MOON GHA	v	MOON Dec	d	HP
d h	° ′	° ′	° ′	′	° ′	′	′
2 00	178 36.1	N21 58.8	309 36.7	16.1	S 4 35.8	13.4	54.8
01	193 36.0	58.4	324 11.8	16.1	4 22.4	13.3	54.8
02	208 36.0	58.1	338 46.9	16.2	4 09.1	13.4	54.8
03	223 35.9	. . 57.7	353 22.1	16.1	3 55.7	13.5	54.9
04	238 35.8	57.4	7 57.2	16.1	3 42.2	13.4	54.9
05	253 35.7	57.0	22 32.3	16.2	3 28.8	13.5	54.9
06	268 35.7	N21 56.7	37 07.5	16.1	S 3 15.3	13.5	54.9
07	283 35.6	56.3	51 42.6	16.1	3 01.8	13.5	54.9
08	298 35.5	56.0	66 17.7	16.1	2 48.3	13.5	55.0
09	313 35.4	. . 55.6	80 52.8	16.1	2 34.8	13.6	55.0
10	328 35.3	55.3	95 27.9	16.1	2 21.2	13.6	55.0
11	343 35.3	54.9	110 03.0	16.1	2 07.6	13.6	55.0
12	358 35.2	N21 54.5	124 38.1	16.1	S 1 54.0	13.6	55.0
13	13 35.1	54.2	139 13.2	16.0	1 40.4	13.6	55.1
14	28 35.0	53.8	153 48.2	16.1	1 26.8	13.6	55.1
15	43 35.0	. . 53.5	168 23.3	16.0	1 13.2	13.7	55.1
16	58 34.9	53.1	182 58.3	16.0	0 59.5	13.6	55.1
17	73 34.8	52.8	197 33.3	16.0	0 45.9	13.7	55.1
18	88 34.7	N21 52.4	212 08.3	16.0	S 0 32.2	13.7	55.2
19	103 34.7	52.1	226 43.3	16.0	0 18.5	13.7	55.2
20	118 34.6	51.7	241 18.3	15.9	S 0 04.8	13.7	55.2
21	133 34.5	. . 51.3	255 53.2	15.9	N 0 08.9	13.7	55.2
22	148 34.4	51.0	270 28.1	15.9	0 22.6	13.7	55.3
23	163 34.4	50.6	285 03.0	15.9	0 36.3	13.8	55.3
3 00	178 34.3	N21 50.3	299 37.9	15.9	N 0 50.1	13.7	55.3
01	193 34.2	49.9	314 12.8	15.8	1 03.8	13.8	55.3
02	208 34.1	49.5	328 47.6	15.8	1 17.6	13.7	55.3
03	223 34.1	. . 49.2	343 22.4	15.8	1 31.3	13.8	55.4
04	238 34.0	48.8	357 57.2	15.7	1 45.1	13.7	55.4
05	253 33.9	48.4	12 31.9	15.7	1 58.8	13.8	55.4
06	268 33.8	N21 48.1	27 06.6	15.7	N 2 12.6	13.8	55.4
07	283 33.8	47.7	41 41.3	15.7	2 26.4	13.7	55.5
08	298 33.7	47.3	56 16.0	15.6	2 40.1	13.8	55.5
09	313 33.6	. . 47.0	70 50.6	15.6	2 53.9	13.8	55.5
10	328 33.6	46.6	85 25.2	15.5	3 07.7	13.8	55.5
11	343 33.5	46.2	99 59.7	15.5	3 21.5	13.7	55.6
12	358 33.4	N21 45.9	114 34.2	15.5	N 3 35.2	13.8	55.6
13	13 33.3	45.5	129 08.7	15.4	3 49.0	13.8	55.6
14	28 33.3	45.1	143 43.1	15.4	4 02.8	13.7	55.6
15	43 33.2	. . 44.8	158 17.5	15.4	4 16.5	13.8	55.7
16	58 33.1	44.4	172 51.9	15.3	4 30.3	13.7	55.7
17	73 33.1	44.0	187 26.2	15.2	4 44.0	13.8	55.7
18	88 33.0	N21 43.6	202 00.4	15.2	N 4 57.8	13.7	55.7
19	103 32.9	43.3	216 34.6	15.2	5 11.5	13.7	55.8
20	118 32.8	42.9	231 08.8	15.1	5 25.2	13.8	55.8
21	133 32.8	. . 42.5	245 42.9	15.1	5 39.0	13.7	55.8
22	148 32.7	42.1	260 17.0	15.0	5 52.7	13.7	55.9
23	163 32.6	41.8	274 51.0	15.0	6 06.4	13.7	55.9
4 00	178 32.6	N21 41.4	289 25.0	14.9	N 6 20.1	13.6	55.9
01	193 32.5	41.0	303 58.9	14.9	6 33.7	13.7	55.9
02	208 32.4	40.6	318 32.8	14.8	6 47.4	13.6	56.0
03	223 32.4	. . 40.2	333 06.6	14.7	7 01.0	13.7	56.0
04	238 32.3	39.9	347 40.3	14.7	7 14.7	13.6	56.0
05	253 32.2	39.5	2 14.0	14.7	7 28.3	13.6	56.1
06	268 32.2	N21 39.1	16 47.7	14.5	N 7 41.9	13.6	56.1
07	283 32.1	38.7	31 21.2	14.6	7 55.5	13.5	56.1
08	298 32.0	38.3	45 54.8	14.4	8 09.0	13.6	56.1
09	313 31.9	. . 38.0	60 28.2	14.4	8 22.6	13.5	56.2
10	328 31.9	37.6	75 01.6	14.3	8 36.1	13.5	56.2
11	343 31.8	37.2	89 34.9	14.3	8 49.6	13.4	56.2
12	358 31.7	N21 36.8	104 08.2	14.2	N 9 03.0	13.5	56.3
13	13 31.7	36.4	118 41.4	14.1	9 16.5	13.4	56.3
14	28 31.6	36.0	133 14.5	14.0	9 29.9	13.4	56.3
15	43 31.5	. . 35.6	147 47.5	14.0	9 43.3	13.4	56.4
16	58 31.5	35.3	162 20.5	13.9	9 56.7	13.3	56.4
17	73 31.4	34.9	176 53.4	13.8	10 10.0	13.3	56.4
18	88 31.3	N21 34.5	191 26.2	13.8	N10 23.3	13.3	56.5
19	103 31.3	34.1	205 59.0	13.7	10 36.6	13.3	56.5
20	118 31.2	33.7	220 31.7	13.6	10 49.9	13.2	56.5
21	133 31.2	. . 33.3	235 04.3	13.5	11 03.1	13.2	56.6
22	148 31.1	32.9	249 36.8	13.5	11 16.3	13.1	56.6
23	163 31.0	32.5	264 09.3	13.3	N11 29.4	13.1	56.6
	SD 15.8	d 0.4	SD 15.0		15.1		15.3

Lat.	Twilight Naut.	Twilight Civil	Sunrise	Moonrise 12	13	14	15
°	h m	h m	h m	h m	h m	h m	h m
N 72	▭	▭	▭	21 55	21 27	20 54	20 01
N 70	▭	▭	▭	21 55	21 35	21 11	20 38
68	▭	▭	▭	21 56	21 41	21 25	21 05
66	////	////	01 27	21 56	21 47	21 37	21 25
64	////	////	02 10	21 56	21 51	21 46	21 41
62	////	00 32	02 38	21 56	21 55	21 55	21 55
60	////	01 39	03 00	21 56	21 59	22 02	22 07
N 58	////	02 11	03 17	21 56	22 02	22 08	22 17
56	00 30	02 35	03 32	21 57	22 05	22 14	22 26
54	01 31	02 54	03 45	21 57	22 07	22 19	22 34
52	02 01	03 10	03 56	21 57	22 09	22 24	22 41
50	02 23	03 23	04 05	21 57	22 12	22 28	22 47
45	03 02	03 50	04 26	21 57	22 16	22 37	23 01
N 40	03 30	04 10	04 42	21 57	22 20	22 44	23 13
35	03 51	04 27	04 56	21 58	22 23	22 51	23 23
30	04 08	04 41	05 08	21 58	22 26	22 57	23 31
20	04 35	05 04	05 28	21 58	22 31	23 07	23 46
N 10	04 56	05 23	05 46	21 58	22 36	23 16	24 00
0	05 14	05 40	06 02	21 59	22 40	23 25	24 12
S 10	05 30	05 56	06 18	21 59	22 45	23 33	24 25
20	05 44	06 12	06 35	21 59	22 50	23 42	24 38
30	05 59	06 29	06 55	21 59	22 55	23 53	24 54
35	06 07	06 39	07 06	22 00	22 58	23 59	25 03
40	06 16	06 49	07 19	22 00	23 02	24 06	00 06
45	06 25	07 01	07 35	22 00	23 06	24 14	00 14
S 50	06 35	07 16	07 53	22 01	23 11	24 24	00 24
52	06 40	07 23	08 02	22 01	23 13	24 29	00 29
54	06 45	07 30	08 12	22 01	23 16	24 34	00 34
56	06 50	07 38	08 23	22 01	23 19	24 40	00 40
58	06 56	07 47	08 36	22 01	23 22	24 46	00 46
S 60	07 02	07 57	08 51	22 02	23 26	24 54	00 54

Lat.	Sunset	Twilight Civil	Twilight Naut.	Moonset 12	13	14	15
°	h m	h m	h m	h m	h m	h m	h m
N 72	▭	▭	▭	09 04	11 00	13 05	15 34
N 70	▭	▭	▭	09 07	10 56	12 50	14 59
68	▭	▭	▭	09 10	10 52	12 38	14 34
66	22 41	////	////	09 13	10 49	12 28	14 15
64	21 59	////	////	09 15	10 46	12 20	14 00
62	21 32	23 30	////	09 17	10 43	12 13	13 48
60	21 10	22 30	////	09 19	10 41	12 07	13 37
N 58	20 53	21 58	////	09 20	10 40	12 02	13 28
56	20 39	21 35	23 33	09 22	10 38	11 57	13 20
54	20 26	21 17	22 39	09 23	10 37	11 53	13 13
52	20 15	21 01	22 09	09 24	10 35	11 49	13 06
50	20 06	20 48	21 47	09 25	10 34	11 46	13 01
45	19 45	20 21	21 08	09 27	10 32	11 38	12 48
N 40	19 29	20 01	20 41	09 28	10 29	11 32	12 38
35	19 15	19 44	20 20	09 30	10 28	11 27	12 29
30	19 03	19 30	20 03	09 31	10 26	11 23	12 22
20	18 43	19 07	19 36	09 33	10 23	11 15	12 09
N 10	18 26	18 48	19 15	09 35	10 21	11 08	11 57
0	18 09	18 32	18 58	09 37	10 18	11 01	11 47
S 10	17 53	18 16	18 42	09 39	10 16	10 55	11 36
20	17 36	18 00	18 27	09 41	10 14	10 48	11 25
30	17 17	17 43	18 12	09 43	10 11	10 40	11 12
35	17 06	17 33	18 04	09 44	10 09	10 36	11 05
40	16 53	17 23	17 56	09 45	10 07	10 31	10 57
45	16 37	17 10	17 47	09 47	10 05	10 25	10 47
S 50	16 19	16 56	17 37	09 49	10 03	10 18	10 35
52	16 10	16 49	17 32	09 49	10 02	10 15	10 30
54	16 00	16 42	17 27	09 50	10 00	10 11	10 24
56	15 49	16 34	17 22	09 51	09 59	10 07	10 17
58	15 36	16 25	17 16	09 52	09 57	10 03	10 10
S 60	15 21	16 15	17 10	09 54	09 56	09 58	10 02

Day	SUN Eqn. of Time 00^h	SUN Eqn. of Time 12^h	SUN Mer. Pass.	MOON Mer. Pass. Upper	MOON Mer. Pass. Lower	MOON Age	MOON Phase
d	m s	m s	h m	h m	h m	d %	
12	05 35	05 39	12 06	03 27	15 48	20 78	◐
13	05 43	05 46	12 06	04 08	16 29	21 69	
14	05 50	05 53	12 06	04 51	17 13	22 60	

UT d	h	ARIES GHA	VENUS −4.1 GHA	VENUS Dec	MARS +1.1 GHA	MARS Dec	JUPITER −2.8 GHA	JUPITER Dec	SATURN +1.1 GHA	SATURN Dec
		° ′	° ′	° ′	° ′	° ′	° ′	° ′	° ′	° ′
15	00	292 58.9	223 34.0	N19 41.7	232 56.6	N20 03.7	324 43.8	S13 54.1	123 29.8	N 6 42.9
	01	308 01.4	238 33.5	42.2	247 57.2	04.1	339 46.5	54.2	138 32.0	42.8
	02	323 03.8	253 33.1	42.7	262 57.9	04.5	354 49.2	54.3	153 34.3	42.7
	03	338 06.3	268 32.6	. . 43.2	277 58.5	. . 04.8	9 51.9	. . 54.4	168 36.6	. . 42.6
	04	353 08.7	283 32.2	43.6	292 59.2	05.2	24 54.5	54.5	183 38.8	42.5
	05	8 11.2	298 31.7	44.1	307 59.8	05.6	39 57.2	54.6	198 41.1	42.4
WEDNESDAY	06	23 13.7	313 31.3	N19 44.6	323 00.5	N20 06.0	54 59.9	S13 54.6	213 43.3	N 6 42.4
	07	38 16.1	328 30.8	45.0	338 01.1	06.3	70 02.6	54.7	228 45.6	42.3
	08	53 18.6	343 30.3	45.5	353 01.7	06.7	85 05.3	54.8	243 47.9	42.2
	09	68 21.1	358 29.9	. . 46.0	8 02.4	. . 07.1	100 07.9	. . 54.9	258 50.1	. . 42.1
	10	83 23.5	13 29.4	46.5	23 03.0	07.5	115 10.6	55.0	273 52.4	42.0
	11	98 26.0	28 29.0	46.9	38 03.7	07.8	130 13.3	55.1	288 54.7	41.9
	12	113 28.5	43 28.5	N19 47.4	53 04.3	N20 08.2	145 16.0	S13 55.1	303 56.9	N 6 41.8
	13	128 30.9	58 28.0	47.9	68 05.0	08.6	160 18.7	55.2	318 59.2	41.7
	14	143 33.4	73 27.6	48.3	83 05.6	09.0	175 21.3	55.3	334 01.4	41.6
	15	158 35.8	88 27.1	. . 48.8	98 06.2	. . 09.3	190 24.0	. . 55.4	349 03.7	. . 41.5
	16	173 38.3	103 26.7	49.3	113 06.9	09.7	205 26.7	55.5	4 06.0	41.4
	17	188 40.8	118 26.2	49.7	128 07.5	10.1	220 29.4	55.6	19 08.2	41.4
	18	203 43.2	133 25.7	N19 50.2	143 08.2	N20 10.5	235 32.1	S13 55.7	34 10.5	N 6 41.3
	19	218 45.7	148 25.3	50.6	158 08.8	10.8	250 34.7	55.7	49 12.7	41.2
	20	233 48.2	163 24.8	51.1	173 09.5	11.2	265 37.4	55.8	64 15.0	41.1
	21	248 50.6	178 24.3	. . 51.6	188 10.1	. . 11.6	280 40.1	. . 55.9	79 17.3	. . 41.0
	22	263 53.1	193 23.9	52.0	203 10.7	11.9	295 42.8	56.0	94 19.5	40.9
	23	278 55.6	208 23.4	52.5	218 11.4	12.3	310 45.5	56.1	109 21.8	40.8
16	00	293 58.0	223 22.9	N19 52.9	233 12.0	N20 12.7	325 48.1	S13 56.2	124 24.0	N 6 40.7
	01	309 00.5	238 22.5	53.4	248 12.7	13.0	340 50.8	56.3	139 26.3	40.6
	02	324 03.0	253 22.0	53.9	263 13.3	13.4	355 53.5	56.4	154 28.5	40.5
	03	339 05.4	268 21.5	. . 54.3	278 14.0	. . 13.8	10 56.2	. . 56.4	169 30.8	. . 40.4
	04	354 07.9	283 21.1	54.8	293 14.6	14.2	25 58.9	56.5	184 33.1	40.3
	05	9 10.3	298 20.6	55.2	308 15.2	14.5	41 01.6	56.6	199 35.3	40.3
THURSDAY	06	24 12.8	313 20.1	N19 55.7	323 15.9	N20 14.9	56 04.2	S13 56.7	214 37.6	N 6 40.2
	07	39 15.3	328 19.6	56.1	338 16.5	15.3	71 06.9	56.8	229 39.8	40.1
	08	54 17.7	343 19.2	56.6	353 17.2	15.6	86 09.6	56.9	244 42.1	40.0
	09	69 20.2	358 18.7	. . 57.0	8 17.8	. . 16.0	101 12.3	. . 57.0	259 44.4	. . 39.9
	10	84 22.7	13 18.2	57.5	23 18.5	16.4	116 15.0	57.0	274 46.6	39.8
	11	99 25.1	28 17.7	57.9	38 19.1	16.7	131 17.7	57.1	289 48.9	39.7
	12	114 27.6	43 17.3	N19 58.4	53 19.7	N20 17.1	146 20.3	S13 57.2	304 51.1	N 6 39.6
	13	129 30.1	58 16.8	58.8	68 20.4	17.5	161 23.0	57.3	319 53.4	39.5
	14	144 32.5	73 16.3	59.3	83 21.0	17.8	176 25.7	57.4	334 55.6	39.4
	15	159 35.0	88 15.8	19 59.7	98 21.7	. . 18.2	191 28.4	. . 57.5	349 57.9	. . 39.3
	16	174 37.5	103 15.4	20 00.2	113 22.3	18.6	206 31.1	57.6	5 00.2	39.2
	17	189 39.9	118 14.9	00.6	128 23.0	18.9	221 33.8	57.7	20 02.4	39.1
	18	204 42.4	133 14.4	N20 01.1	143 23.6	N20 19.3	236 36.5	S13 57.7	35 04.7	N 6 39.0
	19	219 44.8	148 13.9	01.5	158 24.2	19.7	251 39.1	57.8	50 06.9	39.0
	20	234 47.3	163 13.5	02.0	173 24.9	20.0	266 41.8	57.9	65 09.2	38.9
	21	249 49.8	178 13.0	. . 02.4	188 25.5	. . 20.4	281 44.5	. . 58.0	80 11.4	. . 38.8
	22	264 52.2	193 12.5	02.8	203 26.2	20.7	296 47.2	58.1	95 13.7	38.7
	23	279 54.7	208 12.0	03.3	218 26.8	21.1	311 49.9	58.2	110 16.0	38.6
17	00	294 57.2	223 11.5	N20 03.7	233 27.5	N20 21.5	326 52.6	S13 58.3	125 18.2	N 6 38.5
	01	309 59.6	238 11.1	04.2	248 28.1	21.8	341 55.3	58.4	140 20.5	38.4
	02	325 02.1	253 10.6	04.6	263 28.7	22.2	356 58.0	58.4	155 22.7	38.3
	03	340 04.6	268 10.1	. . 05.0	278 29.4	. . 22.6	12 00.6	. . 58.5	170 25.0	. . 38.2
	04	355 07.0	283 09.6	05.5	293 30.0	22.9	27 03.3	58.6	185 27.2	38.1
	05	10 09.5	298 09.1	05.9	308 30.7	23.3	42 06.0	58.7	200 29.5	38.0
FRIDAY	06	25 11.9	313 08.6	N20 06.4	323 31.3	N20 23.6	57 08.7	S13 58.8	215 31.8	N 6 37.9
	07	40 14.4	328 08.2	06.8	338 32.0	24.0	72 11.4	58.9	230 34.0	37.8
	08	55 16.9	343 07.7	07.2	353 32.6	24.4	87 14.1	59.0	245 36.3	37.7
	09	70 19.3	358 07.2	. . 07.7	8 33.2	. . 24.7	102 16.8	. . 59.1	260 38.5	. . 37.7
	10	85 21.8	13 06.7	08.1	23 33.9	25.1	117 19.5	59.2	275 40.8	37.6
	11	100 24.3	28 06.2	08.5	38 34.5	25.4	132 22.2	59.2	290 43.0	37.5
	12	115 26.7	43 05.7	N20 09.0	53 35.2	N20 25.8	147 24.9	S13 59.3	305 45.3	N 6 37.4
	13	130 29.2	58 05.2	09.4	68 35.8	26.1	162 27.6	59.4	320 47.5	37.3
	14	145 31.7	73 04.8	09.8	83 36.4	26.5	177 30.2	59.5	335 49.8	37.2
	15	160 34.1	88 04.3	. . 10.3	98 37.1	. . 26.9	192 32.9	. . 59.6	350 52.1	. . 37.1
	16	175 36.6	103 03.8	10.7	113 37.7	27.2	207 35.6	59.7	5 54.3	37.0
	17	190 39.1	118 03.3	11.1	128 38.4	27.6	222 38.3	59.8	20 56.6	36.9
	18	205 41.5	133 02.8	N20 11.5	143 39.0	N20 27.9	237 41.0	S13 59.9	35 58.8	N 6 36.8
	19	220 44.0	148 02.3	12.0	158 39.7	28.3	252 43.7	14 00.0	51 01.1	36.7
	20	235 46.4	163 01.8	12.4	173 40.3	28.6	267 46.4	00.0	66 03.3	36.6
	21	250 48.9	178 01.3	. . 12.8	188 40.9	. . 29.0	282 49.1	. . 00.1	81 05.6	. . 36.5
	22	265 51.4	193 00.8	13.2	203 41.6	29.4	297 51.8	00.2	96 07.8	36.4
	23	280 53.8	208 00.3	13.7	218 42.2	29.7	312 54.5	00.3	111 10.1	36.3
Mer. Pass.		h m 4 23.4	v −0.5	d 0.4	v 0.6	d 0.4	v 2.7	d 0.1	v 2.3	d 0.1

STARS

Name	SHA	Dec
	° ′	°
Acamar	315 20.6	S40 15.
Achernar	335 28.7	S57 10.
Acrux	173 13.0	S63 09.
Adhara	255 15.2	S28 59.
Aldebaran	290 53.0	N16 31.
Alioth	166 23.1	N55 54.
Alkaid	153 01.0	N49 16.
Al Na'ir	27 46.8	S46 54
Alnilam	275 49.6	S 1 11.
Alphard	217 59.2	S 8 42.
Alphecca	126 13.2	N26 41.
Alpheratz	357 46.4	N29 08.
Altair	62 10.7	N 8 53.
Ankaa	353 18.3	S42 14.
Antares	112 29.6	S26 27.
Arcturus	145 58.2	N19 08.
Atria	107 33.6	S69 02.
Avior	234 20.0	S59 32.
Bellatrix	278 35.4	N 6 21.
Betelgeuse	271 04.7	N 7 24.
Canopus	263 58.0	S52 41.
Capella	280 39.1	N46 00.
Deneb	49 33.1	N45 18.
Denebola	182 36.7	N14 31.
Diphda	348 58.7	S17 55.
Dubhe	193 55.4	N61 42.
Elnath	278 16.6	N28 36.
Eltanin	90 47.0	N51 29.
Enif	33 49.7	N 9 55.
Fomalhaut	15 26.8	S29 34.
Gacrux	172 04.5	S57 10.
Gienah	175 55.4	S17 35.
Hadar	148 52.1	S60 25.
Hamal	328 04.1	N23 30.
Kaus Aust.	83 47.2	S34 22.
Kochab	137 19.1	N74 07.
Markab	13 41.0	N15 15.
Menkar	314 18.2	N 4 07.
Menkent	148 11.0	S36 25.
Miaplacidus	221 41.4	S69 45.
Mirfak	308 44.8	N49 53.
Nunki	76 01.5	S26 17.
Peacock	53 23.0	S56 42.
Pollux	243 31.6	N28 00.
Procyon	245 03.1	N 5 12.
Rasalhague	96 08.8	N12 33.
Regulus	207 46.8	N11 55.
Rigel	281 15.1	S 8 11.
Rigil Kent.	139 55.7	S60 52.
Sabik	102 15.6	S15 44.
Schedar	349 43.9	N56 35.
Shaula	96 25.5	S37 06.
Sirius	258 36.6	S16 43.
Spica	158 34.3	S11 12.
Suhail	222 55.1	S43 28.
Vega	80 40.5	N38 47.
Zuben'ubi	137 08.5	S16 05.

	SHA	Mer. Pass
	° ′	h m
Venus	289 24.9	9 07
Mars	299 14.0	8 27
Jupiter	31 50.1	2 16
Saturn	190 26.0	15 40

UT	SUN GHA	SUN Dec	MOON GHA	v	MOON Dec	d	HP
d h	° ′	° ′	° ′	′	° ′	′	′
15 00	178 31.0	N21 32.1	278 41.6	13.3	N11 42.5	13.1	56.7
01	193 30.9	31.7	293 13.9	13.2	11 55.6	13.0	56.7
02	208 30.8	31.4	307 46.1	13.1	12 08.6	13.0	56.7
03	223 30.8	. . 31.0	322 18.2	13.0	12 21.6	13.0	56.8
04	238 30.7	30.6	336 50.2	13.0	12 34.6	12.9	56.8
05	253 30.6	30.2	351 22.2	12.8	12 47.5	12.8	56.8
06	268 30.6	N21 29.8	5 54.0	12.8	N13 00.3	12.9	56.9
07	283 30.5	29.4	20 25.8	12.7	13 13.2	12.7	56.9
08	298 30.4	29.0	34 57.5	12.5	13 25.9	12.8	56.9
09	313 30.4	. . 28.6	49 29.0	12.5	13 38.7	12.7	57.0
10	328 30.3	28.2	64 00.5	12.4	13 51.4	12.6	57.0
11	343 30.3	27.8	78 31.9	12.3	14 04.0	12.6	57.0
12	358 30.2	N21 27.4	93 03.2	12.2	N14 16.6	12.5	57.1
13	13 30.1	27.0	107 34.4	12.1	14 29.1	12.5	57.1
14	28 30.1	26.6	122 05.5	12.0	14 41.6	12.4	57.1
15	43 30.0	. . 26.2	136 36.5	11.9	14 54.0	12.4	57.2
16	58 30.0	25.8	151 07.4	11.8	15 06.4	12.3	57.2
17	73 29.9	25.4	165 38.2	11.8	15 18.7	12.2	57.2
18	88 29.8	N21 25.0	180 09.0	11.6	N15 30.9	12.2	57.3
19	103 29.8	24.6	194 39.6	11.5	15 43.1	12.1	57.3
20	118 29.7	24.2	209 10.1	11.3	15 55.2	12.1	57.3
21	133 29.7	. . 23.7	223 40.4	11.3	16 07.3	12.0	57.4
22	148 29.6	23.3	238 10.7	11.2	16 19.3	11.9	57.4
23	163 29.5	22.9	252 40.9	11.1	16 31.2	11.9	57.5
16 00	178 29.5	N21 22.5	267 11.0	10.9	N16 43.1	11.8	57.5
01	193 29.4	22.1	281 40.9	10.9	16 54.9	11.7	57.5
02	208 29.4	21.7	296 10.8	10.7	17 06.6	11.6	57.6
03	223 29.3	. . 21.3	310 40.5	10.7	17 18.2	11.6	57.6
04	238 29.2	20.9	325 10.2	10.5	17 29.8	11.5	57.6
05	253 29.2	20.5	339 39.7	10.4	17 41.3	11.4	57.7
06	268 29.1	N21 20.1	354 09.1	10.3	N17 52.7	11.4	57.7
07	283 29.1	19.7	8 38.4	10.1	18 04.1	11.3	57.7
08	298 29.0	19.2	23 07.5	10.1	18 15.4	11.1	57.8
09	313 28.9	. . 18.8	37 36.6	9.9	18 26.5	11.1	57.8
10	328 28.9	18.4	52 05.5	9.9	18 37.6	11.1	57.9
11	343 28.8	18.0	66 34.4	9.7	18 48.7	10.9	57.9
12	358 28.8	N21 17.6	81 03.1	9.6	N18 59.6	10.8	57.9
13	13 28.7	17.2	95 31.7	9.4	19 10.4	10.8	58.0
14	28 28.7	16.8	110 00.1	9.4	19 21.2	10.7	58.0
15	43 28.6	. . 16.3	124 28.5	9.2	19 31.9	10.5	58.0
16	58 28.6	15.9	138 56.7	9.1	19 42.4	10.5	58.1
17	73 28.5	15.5	153 24.8	9.0	19 52.9	10.4	58.1
18	88 28.4	N21 15.1	167 52.8	8.9	N20 03.3	10.3	58.2
19	103 28.4	14.7	182 20.7	8.7	20 13.6	10.2	58.2
20	118 28.3	14.2	196 48.4	8.6	20 23.8	10.0	58.2
21	133 28.3	. . 13.8	211 16.0	8.5	20 33.8	10.0	58.3
22	148 28.2	13.4	225 43.5	8.4	20 43.8	9.9	58.3
23	163 28.2	13.0	240 10.9	8.2	20 53.7	9.7	58.4
17 00	178 28.1	N21 12.6	254 38.1	8.1	N21 03.4	9.7	58.4
01	193 28.1	12.1	269 05.2	8.0	21 13.1	9.6	58.4
02	208 28.0	11.7	283 32.2	7.9	21 22.7	9.4	58.5
03	223 28.0	. . 11.3	297 59.1	7.7	21 32.1	9.3	58.5
04	238 27.9	10.9	312 25.8	7.7	21 41.4	9.2	58.5
05	253 27.8	10.4	326 52.5	7.4	21 50.6	9.1	58.6
06	268 27.8	N21 10.0	341 18.9	7.4	N21 59.7	9.0	58.6
07	283 27.7	09.6	355 45.3	7.3	22 08.7	8.8	58.7
08	298 27.7	09.1	10 11.6	7.1	22 17.5	8.8	58.7
09	313 27.6	. . 08.7	24 37.7	7.0	22 26.3	8.6	58.7
10	328 27.6	08.3	39 03.7	6.8	22 34.9	8.4	58.8
11	343 27.5	07.9	53 29.5	6.8	22 43.3	8.4	58.8
12	358 27.5	N21 07.4	67 55.3	6.6	N22 51.7	8.2	58.8
13	13 27.4	07.0	82 20.9	6.5	22 59.9	8.1	58.9
14	28 27.4	06.6	96 46.4	6.4	23 08.0	8.0	58.9
15	43 27.3	. . 06.1	111 11.8	6.2	23 16.0	7.8	59.0
16	58 27.3	05.7	125 37.0	6.2	23 23.8	7.7	59.0
17	73 27.2	05.3	140 02.2	6.0	23 31.5	7.5	59.0
18	88 27.2	N21 04.8	154 27.2	5.8	N23 39.0	7.4	59.1
19	103 27.1	04.4	168 52.0	5.8	23 46.4	7.3	59.1
20	118 27.1	04.0	183 16.8	5.7	23 53.7	7.1	59.1
21	133 27.0	. . 03.5	197 41.5	5.5	24 00.8	7.0	59.2
22	148 27.0	03.1	212 06.0	5.4	24 07.8	6.8	59.2
23	163 26.9	02.7	226 30.4	5.3	N24 14.6	6.7	59.3
	SD 15.8	d 0.4	SD	15.5	15.8		16.0

Lat.	Twilight Naut.	Twilight Civil	Sunrise	Moonrise 15	Moonrise 16	Moonrise 17	Moonrise 18
°	h m	h m	h m	h m	h m	h m	h m
N 72	▭	▭	▭	20 01	▭	▭	▭
N 70	▭	▭	▭	20 38	▭	▭	▭
68	▭	▭	▭	21 05	20 28	▭	▭
66	////	////	01 40	21 25	21 08	20 21	▭
64	////	////	02 19	21 41	21 36	21 29	21 09
62	////	00 57	02 45	21 55	21 57	22 05	22 26
60	////	01 48	03 06	22 07	22 15	22 31	23 03
N 58	////	02 18	03 22	22 17	22 30	22 51	23 29
56	00 53	02 41	03 36	22 26	22 42	23 08	23 50
54	01 39	02 59	03 49	22 34	22 54	23 23	24 07
52	02 07	03 14	03 59	22 41	23 03	23 35	24 22
50	02 28	03 27	04 09	22 47	23 12	23 47	24 35
45	03 06	03 53	04 29	23 01	23 31	24 10	00 10
N 40	03 33	04 13	04 45	23 13	23 47	24 29	00 29
35	03 54	04 29	04 58	23 23	24 00	00 00	00 45
30	04 10	04 43	05 10	23 31	24 11	00 11	00 59
20	04 37	05 05	05 30	23 46	24 31	00 31	01 22
N 10	04 57	05 24	05 47	24 00	00 00	00 48	01 43
0	05 14	05 40	06 02	24 12	00 12	01 04	02 02
S 10	05 30	05 56	06 18	24 25	00 25	01 21	02 21
20	05 44	06 11	06 35	24 38	00 38	01 38	02 42
30	05 59	06 28	06 54	24 54	00 54	01 59	03 06
35	06 06	06 37	07 05	25 03	01 03	02 11	03 20
40	06 14	06 48	07 18	00 06	01 14	02 25	03 37
45	06 23	07 00	07 33	00 14	01 26	02 41	03 57
S 50	06 33	07 14	07 51	00 24	01 41	03 01	04 22
52	06 38	07 20	07 59	00 29	01 48	03 11	04 34
54	06 42	07 27	08 09	00 34	01 56	03 22	04 48
56	06 47	07 35	08 20	00 40	02 05	03 35	05 04
58	06 53	07 44	08 32	00 46	02 15	03 49	05 23
S 60	06 59	07 53	08 47	00 54	02 27	04 07	05 47

Lat.	Sunset	Twilight Civil	Twilight Naut.	Moonset 15	Moonset 16	Moonset 17	Moonset 18
°	h m	h m	h m	h m	h m	h m	h m
N 72	▭	▭	▭	15 34	▭	▭	▭
N 70	▭	▭	▭	14 59	▭	▭	▭
68	▭	▭	▭	14 34	16 57	▭	▭
66	22 28	////	////	14 15	16 18	19 01	▭
64	21 51	////	////	14 00	15 51	17 54	20 22
62	21 25	23 09	////	13 48	15 30	17 19	19 05
60	21 05	22 21	////	13 37	15 13	16 53	18 28
N 58	20 49	21 52	////	13 28	14 59	16 33	18 02
56	20 35	21 30	23 14	13 20	14 47	16 17	17 42
54	20 23	21 12	22 30	13 13	14 37	16 02	17 25
52	20 12	20 57	22 03	13 06	14 27	15 50	17 10
50	20 03	20 45	21 42	13 01	14 19	15 39	16 57
45	19 43	20 19	21 05	12 48	14 01	15 17	16 31
N 40	19 27	19 59	20 39	12 38	13 47	14 59	16 11
35	19 14	19 43	20 18	12 29	13 35	14 44	15 53
30	19 02	19 29	20 01	12 22	13 25	14 31	15 38
20	18 42	19 07	19 35	12 09	13 07	14 08	15 13
N 10	18 25	18 48	19 15	11 57	12 51	13 49	14 51
0	18 10	18 32	18 58	11 47	12 36	13 31	14 31
S 10	17 54	18 16	18 42	11 36	12 22	13 13	14 11
20	17 37	18 01	18 28	11 25	12 07	12 54	13 49
30	17 18	17 44	18 14	11 12	11 49	12 32	13 24
35	17 07	17 35	18 06	11 05	11 39	12 19	13 09
40	16 55	17 25	17 58	10 57	11 27	12 05	12 52
45	16 40	17 13	17 49	10 47	11 13	11 47	12 32
S 50	16 22	16 59	17 39	10 35	10 57	11 26	12 06
52	16 13	16 52	17 35	10 30	10 49	11 15	11 53
54	16 04	16 45	17 30	10 24	10 40	11 04	11 39
56	15 53	16 38	17 25	10 17	10 31	10 51	11 23
58	15 40	16 29	17 20	10 10	10 20	10 36	11 03
S 60	15 26	16 19	17 13	10 02	10 07	10 18	10 39

Day	SUN Eqn. of Time 00^h	SUN Eqn. of Time 12^h	SUN Mer. Pass.	MOON Mer. Pass. Upper	MOON Mer. Pass. Lower	MOON Age	MOON Phase
d	m s	m s	h m	h m	h m	d %	
15	05 56	05 59	12 06	05 36	17 59	23 49	
16	06 02	06 05	12 06	06 24	18 50	24 39	
17	06 07	06 10	12 06	07 18	19 46	25 28	

	UT	ARIES	VENUS −4.1		MARS +1.1		JUPITER −2.8		SATURN +1.1	
		GHA	GHA	Dec	GHA	Dec	GHA	Dec	GHA	Dec
	d h	° ′	° ′	° ′	° ′	° ′	° ′	° ′	° ′	° ′
	18 00	295 56.3	222 59.8	N20 14.1	233 42.9	N20 30.1	327 57.2	S14 00.4	126 12.3	N 6 36.3
	01	310 58.8	237 59.4	14.5	248 43.5	30.4	342 59.9	00.5	141 14.6	36.2
	02	326 01.2	252 58.9	14.9	263 44.2	30.8	358 02.6	00.6	156 16.9	36.1
	03	341 03.7	267 58.4	. . 15.3	278 44.8	. . 31.1	13 05.3	. . 00.7	171 19.1	. . 36.0
	04	356 06.2	282 57.9	15.8	293 45.4	31.5	28 07.9	00.8	186 21.4	35.9
	05	11 08.6	297 57.4	16.2	308 46.1	31.8	43 10.6	00.9	201 23.6	35.8
	06	26 11.1	312 56.9	N20 16.6	323 46.7	N20 32.2	58 13.3	S14 00.9	216 25.9	N 6 35.7
	07	41 13.6	327 56.4	17.0	338 47.4	32.5	73 16.0	01.0	231 28.1	35.6
S	08	56 16.0	342 55.9	17.4	353 48.0	32.9	88 18.7	01.1	246 30.4	35.5
A	09	71 18.5	357 55.4	. . 17.9	8 48.7	. . 33.2	103 21.4	. . 01.2	261 32.6	. . 35.4
T	10	86 20.9	12 54.9	18.3	23 49.3	33.6	118 24.1	01.3	276 34.9	35.3
U	11	101 23.4	27 54.4	18.7	38 49.9	33.9	133 26.8	01.4	291 37.1	35.2
R	12	116 25.9	42 53.9	N20 19.1	53 50.6	N20 34.3	148 29.5	S14 01.5	306 39.4	N 6 35.1
D	13	131 28.3	57 53.4	19.5	68 51.2	34.6	163 32.2	01.6	321 41.6	35.0
A	14	146 30.8	72 52.9	19.9	83 51.9	35.0	178 34.9	01.7	336 43.9	34.9
Y	15	161 33.3	87 52.4	. . 20.3	98 52.5	. . 35.3	193 37.6	. . 01.8	351 46.1	. . 34.8
	16	176 35.7	102 51.9	20.7	113 53.2	35.7	208 40.3	01.9	6 48.4	34.7
	17	191 38.2	117 51.4	21.2	128 53.8	36.0	223 43.0	01.9	21 50.7	34.6
	18	206 40.7	132 50.9	N20 21.6	143 54.4	N20 36.4	238 45.7	S14 02.0	36 52.9	N 6 34.6
	19	221 43.1	147 50.4	22.0	158 55.1	36.7	253 48.4	02.1	51 55.2	34.5
	20	236 45.6	162 49.9	22.4	173 55.7	37.1	268 51.1	02.2	66 57.4	34.4
	21	251 48.1	177 49.4	. . 22.8	188 56.4	. . 37.4	283 53.8	. . 02.3	81 59.7	. . 34.3
	22	266 50.5	192 48.9	23.2	203 57.0	37.8	298 56.5	02.4	97 01.9	34.2
	23	281 53.0	207 48.4	23.6	218 57.7	38.1	313 59.2	02.5	112 04.2	34.1
	19 00	296 55.4	222 47.9	N20 24.0	233 58.3	N20 38.5	329 01.9	S14 02.6	127 06.4	N 6 34.0
	01	311 57.9	237 47.3	24.4	248 58.9	38.8	344 04.6	02.7	142 08.7	33.9
	02	327 00.4	252 46.8	24.8	263 59.6	39.2	359 07.3	02.8	157 10.9	33.8
	03	342 02.8	267 46.3	. . 25.2	279 00.2	. . 39.5	14 10.0	. . 02.9	172 13.2	. . 33.7
	04	357 05.3	282 45.8	25.6	294 00.9	39.8	29 12.7	02.9	187 15.4	33.6
	05	12 07.8	297 45.3	26.0	309 01.5	40.2	44 15.4	03.0	202 17.7	33.5
	06	27 10.2	312 44.8	N20 26.4	324 02.2	N20 40.5	59 18.1	S14 03.1	217 19.9	N 6 33.4
	07	42 12.7	327 44.3	26.8	339 02.8	40.9	74 20.8	03.2	232 22.2	33.3
	08	57 15.2	342 43.8	27.2	354 03.4	41.2	89 23.5	03.3	247 24.4	33.2
S	09	72 17.6	357 43.3	. . 27.6	9 04.1	. . 41.6	104 26.2	. . 03.4	262 26.7	. . 33.1
U	10	87 20.1	12 42.8	28.0	24 04.7	41.9	119 28.9	03.5	277 28.9	33.0
N	11	102 22.6	27 42.3	28.4	39 05.4	42.2	134 31.6	03.6	292 31.2	32.9
D	12	117 25.0	42 41.7	N20 28.8	54 06.0	N20 42.6	149 34.3	S14 03.7	307 33.4	N 6 32.8
A	13	132 27.5	57 41.2	29.2	69 06.7	42.9	164 37.0	03.8	322 35.7	32.7
Y	14	147 29.9	72 40.7	29.6	84 07.3	43.3	179 39.7	03.9	337 37.9	32.7
	15	162 32.4	87 40.2	. . 30.0	99 07.9	. . 43.6	194 42.4	. . 04.0	352 40.2	. . 32.6
	16	177 34.9	102 39.7	30.4	114 08.6	44.0	209 45.1	04.1	7 42.4	32.5
	17	192 37.3	117 39.2	30.7	129 09.2	44.3	224 47.8	04.1	22 44.7	32.4
	18	207 39.8	132 38.7	N20 31.1	144 09.9	N20 44.6	239 50.5	S14 04.2	37 46.9	N 6 32.3
	19	222 42.3	147 38.2	31.5	159 10.5	45.0	254 53.2	04.3	52 49.2	32.2
	20	237 44.7	162 37.6	31.9	174 11.2	45.3	269 55.9	04.4	67 51.4	32.1
	21	252 47.2	177 37.1	. . 32.3	189 11.8	. . 45.7	284 58.6	. . 04.5	82 53.7	. . 32.0
	22	267 49.7	192 36.6	32.7	204 12.4	46.0	300 01.3	04.6	97 55.9	31.9
	23	282 52.1	207 36.1	33.1	219 13.1	46.3	315 04.1	04.7	112 58.2	31.8
	20 00	297 54.6	222 35.6	N20 33.5	234 13.7	N20 46.7	330 06.8	S14 04.8	128 00.4	N 6 31.7
	01	312 57.1	237 35.1	33.8	249 14.4	47.0	345 09.5	04.9	143 02.7	31.6
	02	327 59.5	252 34.5	34.2	264 15.0	47.3	0 12.2	05.0	158 04.9	31.5
	03	343 02.0	267 34.0	. . 34.6	279 15.7	. . 47.7	15 14.9	. . 05.1	173 07.2	. . 31.4
	04	358 04.4	282 33.5	35.0	294 16.3	48.0	30 17.6	05.2	188 09.4	31.3
	05	13 06.9	297 33.0	35.4	309 16.9	48.3	45 20.3	05.3	203 11.7	31.2
	06	28 09.4	312 32.5	N20 35.7	324 17.6	N20 48.7	60 23.0	S14 05.4	218 13.9	N 6 31.1
	07	43 11.8	327 31.9	36.1	339 18.2	49.0	75 25.7	05.5	233 16.2	31.0
	08	58 14.3	342 31.4	36.5	354 18.9	49.4	90 28.4	05.5	248 18.4	30.9
M	09	73 16.8	357 30.9	. . 36.9	9 19.5	. . 49.7	105 31.1	. . 05.6	263 20.7	. . 30.8
O	10	88 19.2	12 30.4	37.3	24 20.2	50.0	120 33.8	05.7	278 22.9	30.7
N	11	103 21.7	27 29.8	37.6	39 20.8	50.4	135 36.5	05.8	293 25.2	30.6
D	12	118 24.2	42 29.3	N20 38.0	54 21.5	N20 50.7	150 39.2	S14 05.9	308 27.4	N 6 30.5
A	13	133 26.6	57 28.8	38.4	69 22.1	51.0	165 41.9	06.0	323 29.7	30.4
Y	14	148 29.1	72 28.3	38.8	84 22.7	51.4	180 44.7	06.1	338 31.9	30.3
	15	163 31.6	87 27.7	. . 39.1	99 23.4	. . 51.7	195 47.4	. . 06.2	353 34.2	. . 30.3
	16	178 34.0	102 27.2	39.5	114 24.0	52.0	210 50.1	06.3	8 36.4	30.2
	17	193 36.5	117 26.7	39.9	129 24.7	52.4	225 52.8	06.4	23 38.6	30.1
	18	208 38.9	132 26.2	N20 40.2	144 25.3	N20 52.7	240 55.5	S14 06.5	38 40.9	N 6 30.0
	19	223 41.4	147 25.6	40.6	159 26.0	53.0	255 58.2	06.6	53 43.1	29.9
	20	238 43.9	162 25.1	41.0	174 26.6	53.4	271 00.9	06.7	68 45.4	29.8
	21	253 46.3	177 24.6	. . 41.3	189 27.2	. . 53.7	286 03.6	. . 06.8	83 47.6	. . 29.7
	22	268 48.8	192 24.1	41.7	204 27.9	54.0	301 06.3	06.9	98 49.9	29.6
	23	283 51.3	207 23.5	42.1	219 28.5	54.3	316 09.0	07.0	113 52.1	29.5
	Mer. Pass.	h m 4 11.6	*v* −0.5	*d* 0.4	*v* 0.6	*d* 0.3	*v* 2.7	*d* 0.1	*v* 2.3	*d* 0.1

STARS

Name	SHA	Dec
	° ′	° ′
Acamar	315 20.6	S40 15.
Achernar	335 28.7	S57 10.
Acrux	173 13.0	S63 09.
Adhara	255 15.2	S28 59.
Aldebaran	290 52.9	N16 31.
Alioth	166 23.1	N55 54.
Alkaid	153 01.0	N49 16.
Al Na'ir	27 46.8	S46 54.
Alnilam	275 49.6	S 1 11.
Alphard	217 59.2	S 8 42.
Alphecca	126 13.2	N26 41.
Alpheratz	357 46.4	N29 08.
Altair	62 10.7	N 8 53.
Ankaa	353 18.2	S42 14.
Antares	112 29.6	S26 27.
Arcturus	145 58.3	N19 08.
Atria	107 33.6	S69 02.
Avior	234 20.0	S59 32.
Bellatrix	278 35.4	N 6 21.
Betelgeuse	271 04.7	N 7 24.6
Canopus	263 58.0	S52 41.9
Capella	280 39.1	N46 00.4
Deneb	49 33.1	N45 18.9
Denebola	182 36.7	N14 31.2
Diphda	348 58.6	S17 55.8
Dubhe	193 55.4	N61 42.1
Elnath	278 16.6	N28 36.9
Eltanin	90 47.0	N51 29.4
Enif	33 49.7	N 9 55.2
Fomalhaut	15 26.8	S29 34.0
Gacrux	172 04.5	S57 10.3
Gienah	175 55.4	S17 35.8
Hadar	148 52.1	S60 25.5
Hamal	328 04.1	N23 30.5
Kaus Aust.	83 47.2	S34 22.8
Kochab	137 19.2	N74 07.2
Markab	13 41.0	N15 15.5
Menkar	314 18.2	N 4 07.8
Menkent	148 11.0	S36 25.3
Miaplacidus	221 41.5	S69 45.5
Mirfak	308 44.8	N49 53.6
Nunki	76 01.5	S26 17.1
Peacock	53 23.0	S56 42.1
Pollux	243 31.6	N28 00.2
Procyon	245 03.1	N 5 12.1
Rasalhague	96 08.8	N12 33.2
Regulus	207 46.8	N11 55.3
Rigel	281 15.1	S 8 11.3
Rigil Kent.	139 55.7	S60 52.8
Sabik	102 15.6	S15 44.2
Schedar	349 43.9	N56 35.3
Shaula	96 25.5	S37 06.7
Sirius	258 36.6	S16 43.7
Spica	158 34.3	S11 12.8
Suhail	222 55.1	S43 28.4
Vega	80 40.5	N38 47.6
Zuben'ubi	137 08.5	S16 05.0

	SHA	Mer. Pass
	° ′	h m
Venus	285 52.4	9 09
Mars	297 02.9	8 24
Jupiter	32 06.4	2 04
Saturn	190 11.0	15 29

JT	SUN GHA	SUN Dec	MOON GHA	v	MOON Dec	d	HP
d h	° ′	° ′	° ′	′	° ′	′	′
8 00	178 26.9	N21 02.2	240 54.7	5.1	N24 21.3	6.5	59.3
01	193 26.8	01.8	255 18.8	5.1	24 27.8	6.3	59.3
02	208 26.8	01.3	269 42.9	4.9	24 34.1	6.3	59.4
03	223 26.7	. . 00.9	284 06.8	4.9	24 40.4	6.0	59.4
04	238 26.7	00.5	298 30.7	4.7	24 46.4	5.9	59.4
05	253 26.6	21 00.0	312 54.4	4.6	24 52.3	5.8	59.5
06	268 26.6	N20 59.6	327 18.0	4.5	N24 58.1	5.5	59.5
07	283 26.5	59.1	341 41.5	4.4	25 03.6	5.5	59.6
08	298 26.5	58.7	356 04.9	4.3	25 09.1	5.2	59.6
09	313 26.4	. . 58.2	10 28.2	4.1	25 14.3	5.1	59.6
10	328 26.4	57.8	24 51.3	4.1	25 19.4	4.9	59.7
11	343 26.4	57.4	39 14.4	4.0	25 24.3	4.8	59.7
12	358 26.3	N20 56.9	53 37.4	3.9	N25 29.1	4.5	59.7
13	13 26.3	56.5	68 00.3	3.7	25 33.6	4.4	59.8
14	28 26.2	56.0	82 23.0	3.7	25 38.0	4.3	59.8
15	43 26.2	. . 55.6	96 45.7	3.6	25 42.3	4.0	59.8
16	58 26.1	55.1	111 08.3	3.5	25 46.3	3.9	59.9
17	73 26.1	54.7	125 30.8	3.4	25 50.2	3.7	59.9
18	88 26.0	N20 54.2	139 53.2	3.3	N25 53.9	3.5	59.9
19	103 26.0	53.8	154 15.5	3.2	25 57.4	3.4	60.0
20	118 25.9	53.3	168 37.7	3.2	26 00.8	3.1	60.0
21	133 25.9	. . 52.9	182 59.9	3.0	26 03.9	3.0	60.0
22	148 25.9	52.4	197 21.9	3.0	26 06.9	2.8	60.1
23	163 25.8	52.0	211 43.9	2.9	26 09.7	2.6	60.1
9 00	178 25.8	N20 51.5	226 05.8	2.8	N26 12.3	2.4	60.1
01	193 25.7	51.1	240 27.6	2.8	26 14.7	2.3	60.2
02	208 25.7	50.6	254 49.4	2.7	26 17.0	2.0	60.2
03	223 25.6	. . 50.2	269 11.1	2.6	26 19.0	1.8	60.2
04	238 25.6	49.7	283 32.7	2.5	26 20.8	1.7	60.3
05	253 25.6	49.2	297 54.2	2.5	26 22.5	1.5	60.3
06	268 25.5	N20 48.8	312 15.7	2.4	N26 24.0	1.2	60.3
07	283 25.5	48.3	326 37.1	2.4	26 25.2	1.1	60.3
08	298 25.4	47.9	340 58.5	2.3	26 26.3	0.9	60.4
09	313 25.4	. . 47.4	355 19.8	2.3	26 27.2	0.7	60.4
10	328 25.4	47.0	9 41.1	2.2	26 27.9	0.5	60.4
11	343 25.3	46.5	24 02.3	2.2	26 28.4	0.3	60.5
12	358 25.3	N20 46.0	38 23.5	2.1	N26 28.7	0.1	60.5
13	13 25.2	45.6	52 44.6	2.1	26 28.8	0.1	60.5
14	28 25.2	45.1	67 05.7	2.0	26 28.7	0.3	60.5
15	43 25.2	. . 44.7	81 26.7	2.0	26 28.4	0.5	60.6
16	58 25.1	44.2	95 47.7	2.0	26 27.9	0.7	60.6
17	73 25.1	43.7	110 08.7	2.0	26 27.2	1.0	60.6
18	88 25.0	N20 43.3	124 29.7	1.9	N26 26.2	1.1	60.7
19	103 25.0	42.8	138 50.6	1.9	26 25.1	1.3	60.7
20	118 25.0	42.3	153 11.5	1.9	26 23.8	1.5	60.7
21	133 24.9	. . 41.9	167 32.4	1.8	26 22.3	1.7	60.7
22	148 24.9	41.4	181 53.2	1.9	26 20.6	1.9	60.8
23	163 24.8	40.9	196 14.1	1.8	26 18.7	2.1	60.8
20 00	178 24.8	N20 40.5	210 34.9	1.9	N26 16.6	2.3	60.8
01	193 24.8	40.0	224 55.8	1.8	26 14.3	2.5	60.8
02	208 24.7	39.5	239 16.6	1.8	26 11.8	2.8	60.8
03	223 24.7	. . 39.1	253 37.4	1.8	26 09.0	2.9	60.9
04	238 24.7	38.6	267 58.2	1.9	26 06.1	3.1	60.9
05	253 24.6	38.1	282 19.1	1.8	26 03.0	3.3	60.9
06	268 24.6	N20 37.7	296 39.9	1.9	N25 59.7	3.5	60.9
07	283 24.5	37.2	311 00.8	1.8	25 56.2	3.8	61.0
08	298 24.5	36.7	325 21.6	1.9	25 52.4	3.9	61.0
09	313 24.5	. . 36.2	339 42.5	1.9	25 48.5	4.1	61.0
10	328 24.4	35.8	354 03.4	1.9	25 44.4	4.3	61.0
11	343 24.4	35.3	8 24.3	1.9	25 40.1	4.6	61.0
12	358 24.4	N20 34.8	22 45.2	2.0	N25 35.5	4.7	61.0
13	13 24.3	34.3	37 06.2	2.0	25 30.8	4.9	61.1
14	28 24.3	33.9	51 27.2	2.0	25 25.9	5.1	61.1
15	43 24.3	. . 33.4	65 48.2	2.1	25 20.8	5.3	61.1
16	58 24.2	32.9	80 09.3	2.1	25 15.5	5.5	61.1
17	73 24.2	32.4	94 30.4	2.2	25 10.0	5.7	61.1
18	88 24.2	N20 32.0	108 51.6	2.1	N25 04.3	5.9	61.1
19	103 24.1	31.5	123 12.7	2.3	24 58.4	6.1	61.2
20	118 24.1	31.0	137 34.0	2.3	24 52.3	6.3	61.2
21	133 24.1	. . 30.5	151 55.3	2.3	24 46.0	6.5	61.2
22	148 24.0	30.0	166 16.6	2.4	24 39.5	6.6	61.2
23	163 24.0	29.6	180 38.0	2.4	N24 32.9	6.9	61.2
	SD 15.8	*d* 0.5	SD	16.3	16.5		16.6

Lat.	Twilight Naut.	Twilight Civil	Sunrise	Moonrise 18	Moonrise 19	Moonrise 20	Moonrise 21
°	h m	h m	h m	h m	h m	h m	h m
N 72	□	□	□	□	□	□	□
N 70	□	□	□	□	□	□	□
68	////	////	00 47	□	□	□	□
66	////	////	01 53	□	□	□	□
64	////	////	02 28	21 09	□	□	00 33
62	////	01 16	02 52	22 26	23 27	25 14	01 14
60	////	01 58	03 12	23 03	24 06	00 06	01 42
N 58	////	02 26	03 28	23 29	24 33	00 33	02 04
56	01 09	02 47	03 41	23 50	24 54	00 54	02 21
54	01 48	03 04	03 53	24 07	00 07	01 12	02 36
52	02 14	03 18	04 03	24 22	00 22	01 27	02 49
50	02 34	03 31	04 12	24 35	00 35	01 40	03 01
45	03 10	03 56	04 31	00 10	01 01	02 07	03 24
N 40	03 36	04 15	04 47	00 29	01 22	02 28	03 43
35	03 56	04 31	05 00	00 45	01 40	02 45	03 59
30	04 12	04 45	05 11	00 59	01 55	03 01	04 12
20	04 38	05 07	05 31	01 22	02 21	03 26	04 36
N 10	04 58	05 25	05 47	01 43	02 43	03 49	04 56
0	05 15	05 41	06 03	02 02	03 04	04 09	05 14
S 10	05 30	05 56	06 18	02 21	03 25	04 30	05 33
20	05 44	06 11	06 34	02 42	03 48	04 52	05 53
30	05 58	06 27	06 53	03 06	04 14	05 18	06 16
35	06 05	06 36	07 04	03 20	04 29	05 33	06 29
40	06 13	06 46	07 16	03 37	04 47	05 51	06 44
45	06 21	06 58	07 30	03 57	05 09	06 12	07 03
S 50	06 31	07 11	07 48	04 22	05 36	06 38	07 25
52	06 35	07 17	07 56	04 34	05 50	06 51	07 36
54	06 40	07 24	08 05	04 48	06 05	07 06	07 48
56	06 44	07 32	08 16	05 04	06 24	07 23	08 02
58	06 50	07 40	08 28	05 23	06 46	07 44	08 18
S 60	06 56	07 49	08 41	05 47	07 15	08 10	08 38

Lat.	Sunset	Twilight Civil	Twilight Naut.	Moonset 18	Moonset 19	Moonset 20	Moonset 21
°	h m	h m	h m	h m	h m	h m	h m
N 72	□	□	□	□	□	□	□
N 70	□	□	□	□	□	□	□
68	23 16	////	////	□	□	□	22 40
66	22 16	////	////	□	□	□	21 54
64	21 43	////	////	20 22	□	21 31	21 23
62	21 18	22 52	////	19 05	20 19	20 50	21 00
60	20 59	22 12	////	18 28	19 41	20 21	20 41
N 58	20 44	21 45	////	18 02	19 13	19 59	20 26
56	20 30	21 24	22 59	17 42	18 52	19 41	20 12
54	20 19	21 07	22 22	17 25	18 34	19 26	20 00
52	20 09	20 53	21 57	17 10	18 19	19 12	19 50
50	20 00	20 41	21 37	16 57	18 06	19 01	19 41
45	19 41	20 16	21 02	16 31	17 39	18 36	19 21
N 40	19 25	19 57	20 36	16 11	17 18	18 17	19 05
35	19 12	19 41	20 16	15 53	17 00	18 00	18 51
30	19 01	19 28	20 00	15 38	16 45	17 46	18 40
20	18 42	19 06	19 34	15 13	16 19	17 21	18 19
N 10	18 25	18 48	19 14	14 51	15 56	17 00	18 01
0	18 10	18 32	18 58	14 31	15 35	16 40	17 44
S 10	17 55	18 17	18 43	14 11	15 14	16 20	17 27
20	17 38	18 02	18 29	13 49	14 51	15 59	17 09
30	17 20	17 46	18 15	13 24	14 25	15 34	16 47
35	17 09	17 37	18 08	13 09	14 09	15 19	16 35
40	16 57	17 27	18 00	12 52	13 51	15 02	16 20
45	16 43	17 15	17 52	12 32	13 30	14 41	16 03
S 50	16 25	17 02	17 42	12 06	13 02	14 15	15 42
52	16 17	16 56	17 38	11 53	12 49	14 03	15 31
54	16 08	16 49	17 34	11 39	12 33	13 48	15 20
56	15 57	16 42	17 29	11 23	12 15	13 31	15 06
58	15 45	16 33	17 23	11 03	11 52	13 11	14 51
S 60	15 32	16 24	17 18	10 39	11 23	12 45	14 32

Day	SUN Eqn. of Time 00^h	SUN Eqn. of Time 12^h	SUN Mer. Pass.	MOON Mer. Pass. Upper	MOON Mer. Pass. Lower	MOON Age	MOON Phase
d	m s	m s	h m	h m	h m	d	%
18	06 12	06 15	12 06	08 16	20 47	26	18
19	06 17	06 19	12 06	09 20	21 52	27	10
20	06 21	06 22	12 06	10 25	22 57	28	4

	UT	ARIES	VENUS −4.1		MARS +1.1		JUPITER −2.8		SATURN +1.1		STARS		
		GHA	GHA	Dec	GHA	Dec	GHA	Dec	GHA	Dec	Name	SHA	Dec
	d h	° ′	° ′	° ′	° ′	° ′	° ′	° ′	° ′	° ′		° ′	° ′
	21 00	298 53.7	222 23.0	N20 42.4	234 29.2	N20 54.7	331 11.8	S14 07.1	128 54.4	N 6 29.4	Acamar	315 20.5	S40 15.4
	01	313 56.2	237 22.5	42.8	249 29.8	55.0	346 14.5	07.2	143 56.6	29.3	Achernar	335 28.7	S57 10.[illegible]
	02	328 58.7	252 21.9	43.2	264 30.5	55.3	1 17.2	07.2	158 58.9	29.2	Acrux	173 13.0	S63 09.5
	03	344 01.1	267 21.4	. . 43.5	279 31.1	. . 55.7	16 19.9	. . 07.3	174 01.1	. . 29.1	Adhara	255 15.2	S28 59.[illegible]
	04	359 03.6	282 20.9	43.9	294 31.7	56.0	31 22.6	07.4	189 03.4	29.0	Aldebaran	290 52.9	N16 31.2
	05	14 06.1	297 20.4	44.3	309 32.4	56.3	46 25.3	07.5	204 05.6	28.9			
	06	29 08.5	312 19.8	N20 44.6	324 33.0	N20 56.6	61 28.0	S14 07.6	219 07.9	N 6 28.8	Alioth	166 23.1	N55 54.[illegible]
	07	44 11.0	327 19.3	45.0	339 33.7	57.0	76 30.7	07.7	234 10.1	28.7	Alkaid	153 01.0	N49 16.7
T	08	59 13.4	342 18.8	45.3	354 34.3	57.3	91 33.4	07.8	249 12.4	28.6	Al Na'ir	27 46.8	S46 54.[illegible]
	09	74 15.9	357 18.2	. . 45.7	9 35.0	. . 57.6	106 36.2	. . 07.9	264 14.6	. . 28.5	Alnilam	275 49.5	S 1 11.[illegible]
U	10	89 18.4	12 17.7	46.1	24 35.6	57.9	121 38.9	08.0	279 16.8	28.4	Alphard	217 59.2	S 8 42.[illegible]
E	11	104 20.8	27 17.2	46.4	39 36.3	58.3	136 41.6	08.1	294 19.1	28.3			
S	12	119 23.3	42 16.6	N20 46.8	54 36.9	N20 58.6	151 44.3	S14 08.2	309 21.3	N 6 28.2	Alphecca	126 13.2	N26 41.[illegible]
D	13	134 25.8	57 16.1	47.1	69 37.5	58.9	166 47.0	08.3	324 23.6	28.1	Alpheratz	357 46.4	N29 08.[illegible]
A	14	149 28.2	72 15.5	47.5	84 38.2	59.3	181 49.7	08.4	339 25.8	28.0	Altair	62 10.7	N 8 53.7
Y	15	164 30.7	87 15.0	. . 47.8	99 38.8	. . 59.6	196 52.4	. . 08.5	354 28.1	. . 27.9	Ankaa	353 18.2	S42 14.9
	16	179 33.2	102 14.5	48.2	114 39.5	20 59.9	211 55.2	08.6	9 30.3	27.8	Antares	112 29.6	S26 27.[illegible]
	17	194 35.6	117 13.9	48.5	129 40.1	21 00.2	226 57.9	08.7	24 32.6	27.7			
	18	209 38.1	132 13.4	N20 48.9	144 40.8	N21 00.5	242 00.6	S14 08.8	39 34.8	N 6 27.6	Arcturus	145 58.3	N19 08.0
	19	224 40.5	147 12.9	49.2	159 41.4	00.9	257 03.3	08.9	54 37.1	27.5	Atria	107 33.6	S69 02.9
	20	239 43.0	162 12.3	49.6	174 42.0	01.2	272 06.0	09.0	69 39.3	27.4	Avior	234 20.0	S59 32.4
	21	254 45.5	177 11.8	. . 49.9	189 42.7	. . 01.5	287 08.7	. . 09.1	84 41.5	. . 27.3	Bellatrix	278 35.3	N 6 21.6
	22	269 47.9	192 11.2	50.3	204 43.3	01.8	302 11.4	09.2	99 43.8	27.2	Betelgeuse	271 04.7	N 7 24.6
	23	284 50.4	207 10.7	50.6	219 44.0	02.2	317 14.2	09.3	114 46.0	27.1			
	22 00	299 52.9	222 10.2	N20 51.0	234 44.6	N21 02.5	332 16.9	S14 09.4	129 48.3	N 6 27.0	Canopus	263 57.9	S52 41.9
	01	314 55.3	237 09.6	51.3	249 45.3	02.8	347 19.6	09.5	144 50.5	26.9	Capella	280 39.1	N46 00.4
	02	329 57.8	252 09.1	51.7	264 45.9	03.1	2 22.3	09.6	159 52.8	26.9	Deneb	49 33.1	N45 18.9
	03	345 00.3	267 08.5	. . 52.0	279 46.6	. . 03.4	17 25.0	. . 09.6	174 55.0	. . 26.8	Denebola	182 36.7	N14 31.2
	04	0 02.7	282 08.0	52.3	294 47.2	03.8	32 27.7	09.7	189 57.3	26.7	Diphda	348 58.6	S17 55.8
	05	15 05.2	297 07.5	52.7	309 47.8	04.1	47 30.5	09.8	204 59.5	26.6			
	06	30 07.7	312 06.9	N20 53.0	324 48.5	N21 04.4	62 33.2	S14 09.9	220 01.7	N 6 26.5	Dubhe	193 55.4	N61 42.1
W	07	45 10.1	327 06.4	53.4	339 49.1	04.7	77 35.9	10.0	235 04.0	26.4	Elnath	278 16.6	N28 36.9
E	08	60 12.6	342 05.8	53.7	354 49.8	05.0	92 38.6	10.1	250 06.2	26.3	Eltanin	90 47.0	N51 29.4
D	09	75 15.0	357 05.3	. . 54.0	9 50.4	. . 05.4	107 41.3	. . 10.2	265 08.5	. . 26.2	Enif	33 49.7	N 9 55.2
N	10	90 17.5	12 04.7	54.4	24 51.1	05.7	122 44.0	10.3	280 10.7	26.1	Fomalhaut	15 26.8	S29 34.0
E	11	105 20.0	27 04.2	54.7	39 51.7	06.0	137 46.8	10.4	295 13.0	26.0			
S	12	120 22.4	42 03.6	N20 55.0	54 52.3	N21 06.3	152 49.5	S14 10.5	310 15.2	N 6 25.9	Gacrux	172 04.5	S57 10.3
D	13	135 24.9	57 03.1	55.4	69 53.0	06.6	167 52.2	10.6	325 17.4	25.8	Gienah	175 55.4	S17 35.8
A	14	150 27.4	72 02.6	55.7	84 53.6	06.9	182 54.9	10.7	340 19.7	25.7	Hadar	148 52.1	S60 25.5
Y	15	165 29.8	87 02.0	. . 56.0	99 54.3	. . 07.3	197 57.6	. . 10.8	355 21.9	. . 25.6	Hamal	328 04.1	N23 30.5
	16	180 32.3	102 01.5	56.4	114 54.9	07.6	213 00.4	10.9	10 24.2	25.5	Kaus Aust.	83 47.2	S34 22.8
	17	195 34.8	117 00.9	56.7	129 55.6	07.9	228 03.1	11.0	25 26.4	25.4			
	18	210 37.2	132 00.4	N20 57.0	144 56.2	N21 08.2	243 05.8	S14 11.1	40 28.7	N 6 25.3	Kochab	137 19.2	N74 07.2
	19	225 39.7	146 59.8	57.4	159 56.9	08.5	258 08.5	11.2	55 30.9	25.2	Markab	13 41.0	N15 15.5
	20	240 42.2	161 59.3	57.7	174 57.5	08.8	273 11.2	11.3	70 33.1	25.1	Menkar	314 18.2	N 4 07.8
	21	255 44.6	176 58.7	. . 58.0	189 58.1	. . 09.2	288 14.0	. . 11.4	85 35.4	. . 25.0	Menkent	148 11.0	S36 25.3
	22	270 47.1	191 58.2	58.4	204 58.8	09.5	303 16.7	11.5	100 37.6	24.9	Miaplacidus	221 41.5	S69 45.5
	23	285 49.5	206 57.6	58.7	219 59.4	09.8	318 19.4	11.6	115 39.9	24.8			
	23 00	300 52.0	221 57.1	N20 59.0	235 00.1	N21 10.1	333 22.1	S14 11.7	130 42.1	N 6 24.7	Mirfak	308 44.8	N49 53.6
	01	315 54.5	236 56.5	59.3	250 00.7	10.4	348 24.8	11.8	145 44.4	24.6	Nunki	76 01.5	S26 17.1
	02	330 56.9	251 56.0	20 59.7	265 01.4	10.7	3 27.6	11.9	160 46.6	24.5	Peacock	53 23.0	S56 42.1
	03	345 59.4	266 55.4	21 00.0	280 02.0	. . 11.0	18 30.3	. . 12.0	175 48.8	. . 24.4	Pollux	243 31.6	N28 00.2
	04	1 01.9	281 54.9	00.3	295 02.7	11.3	33 33.0	12.1	190 51.1	24.3	Procyon	245 03.0	N 5 12.1
	05	16 04.3	296 54.3	00.6	310 03.3	11.7	48 35.7	12.2	205 53.3	24.2			
	06	31 06.8	311 53.7	N21 00.9	325 03.9	N21 12.0	63 38.4	S14 12.3	220 55.6	N 6 24.1	Rasalhague	96 08.8	N12 33.2
	07	46 09.3	326 53.2	01.3	340 04.6	12.3	78 41.2	12.4	235 57.8	24.0	Regulus	207 46.8	N11 55.3
T	08	61 11.7	341 52.6	01.6	355 05.2	12.6	93 43.9	12.5	251 00.0	23.9	Rigel	281 15.1	S 8 11.3
H	09	76 14.2	356 52.1	. . 01.9	10 05.9	. . 12.9	108 46.6	. . 12.6	266 02.3	. . 23.8	Rigil Kent.	139 55.7	S60 52.8
U	10	91 16.7	11 51.5	02.2	25 06.5	13.2	123 49.3	12.7	281 04.5	23.7	Sabik	102 15.6	S15 44.2
R	11	106 19.1	26 51.0	02.5	40 07.2	13.5	138 52.1	12.8	296 06.8	23.6			
S	12	121 21.6	41 50.4	N21 02.8	55 07.8	N21 13.8	153 54.8	S14 12.9	311 09.0	N 6 23.5	Schedar	349 43.8	N56 35.3
D	13	136 24.0	56 49.9	03.2	70 08.5	14.1	168 57.5	13.0	326 11.3	23.4	Shaula	96 25.5	S37 06.7
A	14	151 26.5	71 49.3	03.5	85 09.1	14.4	184 00.2	13.1	341 13.5	23.3	Sirius	258 36.6	S16 43.6
Y	15	166 29.0	86 48.7	. . 03.8	100 09.8	. . 14.8	199 03.0	. . 13.2	356 15.7	. . 23.2	Spica	158 34.3	S11 12.8
	16	181 31.4	101 48.2	04.1	115 10.4	15.1	214 05.7	13.3	11 18.0	23.1	Suhail	222 55.1	S43 28.3
	17	196 33.9	116 47.6	04.4	130 11.0	15.4	229 08.4	13.4	26 20.2	23.0			
	18	211 36.4	131 47.1	N21 04.7	145 11.7	N21 15.7	244 11.1	S14 13.5	41 22.5	N 6 22.9	Vega	80 40.5	N38 47.6
	19	226 38.8	146 46.5	05.0	160 12.3	16.0	259 13.9	13.6	56 24.7	22.8	Zuben'ubi	137 08.5	S16 05.0
	20	241 41.3	161 46.0	05.3	175 13.0	16.3	274 16.6	13.7	71 26.9	22.7			
	21	256 43.8	176 45.4	. . 05.6	190 13.6	. . 16.6	289 19.3	. . 13.8	86 29.2	. . 22.6			
	22	271 46.2	191 44.8	05.9	205 14.3	16.9	304 22.0	13.9	101 31.4	22.5			
	23	286 48.7	206 44.3	06.2	220 14.9	17.2	319 24.8	14.0	116 33.7	22.4			
	Mer. Pass.	h m 3 59.8	v −0.5	d 0.3	v 0.6	d 0.3	v 2.7	d 0.1	v 2.2	d 0.1			

	SHA	Mer. Pass.
	° ′	h m
Venus	282 17.3	9 12
Mars	294 51.7	8 21
Jupiter	32 24.0	1 51
Saturn	189 55.4	15 18

UT	SUN GHA	SUN Dec	MOON GHA	v	MOON Dec	d	HP
d h	° ′	° ′	° ′	′	° ′	′	′
21 00	178 24.0	N20 29.1	194 59.4	2.5	N24 26.0	7.0	61.2
01	193 23.9	28.6	209 20.9	2.6	24 19.0	7.2	61.2
02	208 23.9	28.1	223 42.5	2.6	24 11.8	7.4	61.2
03	223 23.9	. . 27.6	238 04.1	2.7	24 04.4	7.6	61.3
04	238 23.8	27.1	252 25.8	2.8	23 56.8	7.8	61.3
05	253 23.8	26.7	266 47.6	2.8	23 49.0	7.9	61.3
06	268 23.8	N20 26.2	281 09.4	2.9	N23 41.1	8.2	61.3
07	283 23.8	25.7	295 31.3	3.0	23 32.9	8.3	61.3
08	298 23.7	25.2	309 53.3	3.0	23 24.6	8.4	61.3
09	313 23.7	. . 24.7	324 15.3	3.2	23 16.2	8.7	61.3
10	328 23.7	24.2	338 37.5	3.2	23 07.5	8.8	61.3
11	343 23.6	23.7	352 59.7	3.3	22 58.7	9.0	61.3
12	358 23.6	N20 23.2	7 22.0	3.3	N22 49.7	9.2	61.3
13	13 23.6	22.8	21 44.3	3.5	22 40.5	9.3	61.3
14	28 23.5	22.3	36 06.8	3.5	22 31.2	9.5	61.3
15	43 23.5	. . 21.8	50 29.3	3.6	22 21.7	9.7	61.3
16	58 23.5	21.3	64 51.9	3.7	22 12.0	9.8	61.3
17	73 23.5	20.8	79 14.6	3.8	22 02.2	10.0	61.3
18	88 23.4	N20 20.3	93 37.4	3.9	N21 52.2	10.1	61.3
19	103 23.4	19.8	108 00.3	4.0	21 42.1	10.3	61.3
20	118 23.4	19.3	122 23.3	4.1	21 31.8	10.5	61.3
21	133 23.4	. . 18.8	136 46.4	4.2	21 21.3	10.6	61.3
22	148 23.3	18.3	151 09.6	4.2	21 10.7	10.8	61.3
23	163 23.3	17.8	165 32.8	4.4	N20 59.9	10.9	61.3
22 00	178 23.3	N20 17.3	A total eclipse of the Sun occurs on this date. See page 5.				
01	193 23.2	16.8					
02	208 23.2	16.3					
03	223 23.2	. . 15.8					
04	238 23.2	15.3					
05	253 23.1	14.8					
06	268 23.1	N20 14.3	266 18.5	5.0	N19 40.6	11.9	61.3
07	283 23.1	13.8	280 42.5	5.2	19 28.7	12.0	61.3
08	298 23.1	13.3	295 06.7	5.3	19 16.7	12.2	61.3
09	313 23.1	. . 12.8	309 31.0	5.4	19 04.5	12.3	61.3
10	328 23.0	12.3	323 55.4	5.4	18 52.2	12.4	61.3
11	343 23.0	11.8	338 19.8	5.6	18 39.8	12.5	61.3
12	358 23.0	N20 11.3	352 44.4	5.7	N18 27.3	12.7	61.3
13	13 23.0	10.8	7 09.1	5.8	18 14.6	12.7	61.3
14	28 22.9	10.3	21 33.9	5.9	18 01.9	12.9	61.3
15	43 22.9	. . 09.8	35 58.8	6.0	17 49.0	13.0	61.2
16	58 22.9	09.3	50 23.8	6.1	17 36.0	13.1	61.2
17	73 22.9	08.8	64 48.9	6.2	17 22.9	13.2	61.2
18	88 22.8	N20 08.3	79 14.1	6.3	N17 09.7	13.4	61.2
19	103 22.8	07.8	93 39.4	6.4	16 56.3	13.4	61.2
20	118 22.8	07.3	108 04.8	6.5	16 42.9	13.5	61.2
21	133 22.8	. . 06.8	122 30.3	6.6	16 29.4	13.7	61.2
22	148 22.8	06.3	136 55.9	6.8	16 15.7	13.7	61.2
23	163 22.7	05.8	151 21.7	6.8	16 02.0	13.9	61.1
23 00	178 22.7	N20 05.3	165 47.5	6.9	N15 48.1	13.9	61.1
01	193 22.7	04.7	180 13.4	7.1	15 34.2	14.0	61.1
02	208 22.7	04.2	194 39.5	7.1	15 20.2	14.1	61.1
03	223 22.7	. . 03.7	209 05.6	7.3	15 06.1	14.2	61.1
04	238 22.6	03.2	223 31.9	7.3	14 51.9	14.3	61.1
05	253 22.6	02.7	237 58.2	7.5	14 37.6	14.4	61.0
06	268 22.6	N20 02.2	252 24.7	7.5	N14 23.2	14.4	61.0
07	283 22.6	01.7	266 51.2	7.7	14 08.8	14.6	61.0
08	298 22.6	01.2	281 17.9	7.7	13 54.2	14.6	61.0
09	313 22.6	. . 00.6	295 44.6	7.9	13 39.6	14.7	61.0
10	328 22.5	20 00.1	310 11.5	7.9	13 24.9	14.7	60.9
11	343 22.5	19 59.6	324 38.4	8.0	13 10.2	14.8	60.9
12	358 22.5	N19 59.1	339 05.4	8.2	N12 55.4	14.9	60.9
13	13 22.5	58.6	353 32.6	8.2	12 40.5	15.0	60.9
14	28 22.5	58.1	7 59.8	8.3	12 25.5	15.0	60.9
15	43 22.5	. . 57.5	22 27.1	8.5	12 10.5	15.1	60.8
16	58 22.4	57.0	36 54.6	8.5	11 55.4	15.1	60.8
17	73 22.4	56.5	51 22.1	8.6	11 40.3	15.2	60.8
18	88 22.4	N19 56.0	65 49.7	8.7	N11 25.1	15.3	60.8
19	103 22.4	55.5	80 17.4	8.8	11 09.8	15.3	60.7
20	118 22.4	54.9	94 45.2	8.9	10 54.5	15.3	60.7
21	133 22.4	. . 54.4	109 13.1	9.0	10 39.2	15.5	60.7
22	148 22.3	53.9	123 41.1	9.0	10 23.7	15.4	60.7
23	163 22.3	53.4	138 09.1	9.2	N10 08.3	15.5	60.6
	SD 15.8	*d* 0.5	SD 16.7		16.7		16.6

Lat.	Twilight Naut.	Twilight Civil	Sunrise	Moonrise 21	22	23	24
°	h m	h m	h m	h m	h m	h m	h m
N 72	▭	▭	▭	▭	▭	03 27	06 20
N 70	▭	▭	▭	▭	▭	04 02	06 32
68	////	////	01 15	▭	01 38	04 27	06 43
66	////	////	02 06	▭	02 23	04 46	06 51
64	////	////	02 37	00 33	02 53	05 01	06 58
62	////	01 31	03 00	01 14	03 15	05 13	07 04
60	////	02 08	03 18	01 42	03 33	05 24	07 10
N 58	////	02 33	03 33	02 04	03 48	05 33	07 14
56	01 23	02 53	03 46	02 21	04 00	05 41	07 18
54	01 57	03 09	03 57	02 36	04 11	05 48	07 22
52	02 21	03 23	04 07	02 49	04 21	05 55	07 25
50	02 40	03 35	04 16	03 01	04 30	06 00	07 28
45	03 14	03 59	04 34	03 24	04 48	06 13	07 34
N 40	03 39	04 18	04 49	03 43	05 03	06 23	07 40
35	03 59	04 34	05 02	03 59	05 15	06 31	07 44
30	04 15	04 47	05 13	04 12	05 26	06 39	07 48
20	04 40	05 08	05 32	04 36	05 45	06 52	07 55
N 10	04 59	05 25	05 48	04 56	06 01	07 03	08 02
0	05 15	05 41	06 03	05 14	06 16	07 14	08 07
S 10	05 30	05 55	06 18	05 33	06 31	07 24	08 13
20	05 43	06 10	06 33	05 53	06 47	07 35	08 19
30	05 57	06 26	06 51	06 16	07 05	07 48	08 26
35	06 04	06 35	07 02	06 29	07 16	07 55	08 30
40	06 11	06 44	07 14	06 44	07 28	08 04	08 34
45	06 19	06 55	07 28	07 03	07 42	08 13	08 39
S 50	06 28	07 08	07 45	07 25	07 59	08 25	08 45
52	06 32	07 14	07 52	07 36	08 07	08 30	08 48
54	06 36	07 21	08 01	07 48	08 16	08 36	08 51
56	06 41	07 28	08 11	08 02	08 26	08 43	08 55
58	06 46	07 36	08 23	08 18	08 38	08 50	08 58
S 60	06 51	07 45	08 36	08 38	08 51	08 58	09 03

Lat.	Sunset	Twilight Civil	Twilight Naut.	Moonset 21	22	23	24
°	h m	h m	h m	h m	h m	h m	h m
N 72	▭	▭	▭	▭	22 58	22 00	21 24
N 70	▭	▭	▭	▭	22 21	21 44	21 19
68	22 52	////	////	22 40	21 55	21 32	21 14
66	22 04	////	////	21 54	21 34	21 21	21 10
64	21 34	////	////	21 23	21 17	21 12	21 07
62	21 11	22 37	////	21 00	21 03	21 04	21 04
60	20 53	22 02	////	20 41	20 52	20 58	21 02
N 58	20 38	21 37	////	20 26	20 41	20 52	21 00
56	20 26	21 18	22 46	20 12	20 32	20 47	20 58
54	20 15	21 02	22 13	20 00	20 24	20 42	20 56
52	20 05	20 49	21 50	19 50	20 17	20 37	20 54
50	19 56	20 37	21 32	19 41	20 11	20 33	20 53
45	19 38	20 13	20 58	19 21	19 56	20 25	20 49
N 40	19 23	19 54	20 33	19 05	19 45	20 18	20 46
35	19 10	19 39	20 14	18 51	19 35	20 11	20 44
30	19 00	19 26	19 58	18 40	19 26	20 06	20 42
20	18 41	19 05	19 33	18 19	19 10	19 56	20 38
N 10	18 25	18 47	19 14	18 01	18 57	19 48	20 35
0	18 10	18 32	18 58	17 44	18 44	19 40	20 31
S 10	17 55	18 18	18 43	17 27	18 31	19 31	20 28
20	17 40	18 03	18 30	17 09	18 17	19 23	20 24
30	17 22	17 47	18 17	16 47	18 01	19 12	20 20
35	17 11	17 39	18 10	16 35	17 52	19 07	20 18
40	17 00	17 29	18 02	16 20	17 41	19 00	20 15
45	16 46	17 18	17 54	16 03	17 28	18 52	20 12
S 50	16 29	17 05	17 45	15 42	17 13	18 42	20 08
52	16 21	16 59	17 41	15 31	17 05	18 38	20 07
54	16 12	16 53	17 37	15 20	16 57	18 33	20 05
56	16 02	16 46	17 32	15 06	16 48	18 27	20 02
58	15 51	16 38	17 28	14 51	16 37	18 21	20 00
S 60	15 38	16 29	17 22	14 32	16 25	18 14	19 57

Day	SUN Eqn. of Time 00^h	SUN Eqn. of Time 12^h	SUN Mer. Pass.	MOON Mer. Pass. Upper	MOON Mer. Pass. Lower	Age	Phase
d	m s	m s	h m	h m	h m	d	%
21	06 24	06 26	12 06	11 29	24 00	29	1
22	06 27	06 28	12 06	12 30	00 00	00	0
23	06 29	06 30	12 06	13 27	00 59	01	3

	UT	ARIES	VENUS −4.0		MARS +1.1		JUPITER −2.8		SATURN +1.1	
		GHA	GHA	Dec	GHA	Dec	GHA	Dec	GHA	Dec
	d h	° ′	° ′	° ′	° ′	° ′	° ′	° ′	° ′	° ′
	24 00	301 51.1	221 43.7	N21 06.5	235 15.6	N21 17.5	334 27.5	S14 14.1	131 35.9	N 6 22.3
	01	316 53.6	236 43.1	06.9	250 16.2	17.8	349 30.2	14.2	146 38.1	22.2
	02	331 56.1	251 42.6	07.2	265 16.9	18.1	4 32.9	14.3	161 40.4	22.1
	03	346 58.5	266 42.0	. . 07.5	280 17.5	. . 18.4	19 35.7	. . 14.4	176 42.6	. . 22.0
	04	2 01.0	281 41.5	07.8	295 18.1	18.7	34 38.4	14.5	191 44.9	21.9
	05	17 03.5	296 40.9	08.1	310 18.8	19.0	49 41.1	14.6	206 47.1	21.8
	06	32 05.9	311 40.3	N21 08.4	325 19.4	N21 19.3	64 43.8	S14 14.7	221 49.3	N 6 21.7
	07	47 08.4	326 39.8	08.7	340 20.1	19.6	79 46.6	14.8	236 51.6	21.6
	08	62 10.9	341 39.2	09.0	355 20.7	19.9	94 49.3	14.9	251 53.8	21.5
F	09	77 13.3	356 38.6	. . 09.3	10 21.4	. . 20.2	109 52.0	. . 15.0	266 56.1	. . 21.4
R	10	92 15.8	11 38.1	09.5	25 22.0	20.5	124 54.8	15.1	281 58.3	21.3
I	11	107 18.3	26 37.5	09.8	40 22.7	20.8	139 57.5	15.2	297 00.5	21.2
D	12	122 20.7	41 36.9	N21 10.1	55 23.3	N21 21.1	155 00.2	S14 15.3	312 02.8	N 6 21.1
A	13	137 23.2	56 36.4	10.4	70 24.0	21.4	170 02.9	15.4	327 05.0	21.0
Y	14	152 25.6	71 35.8	10.7	85 24.6	21.7	185 05.7	15.5	342 07.3	20.9
	15	167 28.1	86 35.2	. . 11.0	100 25.2	. . 22.0	200 08.4	. . 15.6	357 09.5	. . 20.8
	16	182 30.6	101 34.7	11.3	115 25.9	22.3	215 11.1	15.7	12 11.7	20.7
	17	197 33.0	116 34.1	11.6	130 26.5	22.6	230 13.9	15.8	27 14.0	20.6
	18	212 35.5	131 33.5	N21 11.9	145 27.2	N21 22.9	245 16.6	S14 15.9	42 16.2	N 6 20.5
	19	227 38.0	146 33.0	12.2	160 27.8	23.2	260 19.3	16.0	57 18.4	20.4
	20	242 40.4	161 32.4	12.5	175 28.5	23.5	275 22.0	16.1	72 20.7	20.3
	21	257 42.9	176 31.8	. . 12.7	190 29.1	. . 23.8	290 24.8	. . 16.2	87 22.9	. . 20.2
	22	272 45.4	191 31.2	13.0	205 29.8	24.1	305 27.5	16.3	102 25.2	20.1
	23	287 47.8	206 30.7	13.3	220 30.4	24.4	320 30.2	16.4	117 27.4	20.0
	25 00	302 50.3	221 30.1	N21 13.6	235 31.1	N21 24.7	335 33.0	S14 16.5	132 29.6	N 6 19.9
	01	317 52.8	236 29.5	13.9	250 31.7	25.0	350 35.7	16.6	147 31.9	19.8
	02	332 55.2	251 29.0	14.2	265 32.3	25.3	5 38.4	16.7	162 34.1	19.7
	03	347 57.7	266 28.4	. . 14.4	280 33.0	. . 25.6	20 41.2	. . 16.8	177 36.3	. . 19.6
	04	3 00.1	281 27.8	14.7	295 33.6	25.9	35 43.9	16.9	192 38.6	19.5
	05	18 02.6	296 27.2	15.0	310 34.3	26.2	50 46.6	17.0	207 40.8	19.4
	06	33 05.1	311 26.7	N21 15.3	325 34.9	N21 26.5	65 49.4	S14 17.1	222 43.1	N 6 19.3
S	07	48 07.5	326 26.1	15.6	340 35.6	26.8	80 52.1	17.2	237 45.3	19.2
	08	63 10.0	341 25.5	15.8	355 36.2	27.1	95 54.8	17.3	252 47.5	19.1
A	09	78 12.5	356 24.9	. . 16.1	10 36.9	. . 27.4	110 57.6	. . 17.4	267 49.8	. . 19.0
T	10	93 14.9	11 24.4	16.4	25 37.5	27.7	126 00.3	17.5	282 52.0	18.9
U	11	108 17.4	26 23.8	16.7	40 38.2	28.0	141 03.0	17.6	297 54.2	18.8
R	12	123 19.9	41 23.2	N21 16.9	55 38.8	N21 28.3	156 05.7	S14 17.7	312 56.5	N 6 18.7
D	13	138 22.3	56 22.6	17.2	70 39.5	28.5	171 08.5	17.8	327 58.7	18.6
A	14	153 24.8	71 22.1	17.5	85 40.1	28.8	186 11.2	17.9	343 01.0	18.5
Y	15	168 27.2	86 21.5	. . 17.7	100 40.8	. . 29.1	201 13.9	. . 18.0	358 03.2	. . 18.4
	16	183 29.7	101 20.9	18.0	115 41.4	29.4	216 16.7	18.1	13 05.4	18.3
	17	198 32.2	116 20.3	18.3	130 42.0	29.7	231 19.4	18.2	28 07.7	18.2
	18	213 34.6	131 19.7	N21 18.5	145 42.7	N21 30.0	246 22.2	S14 18.3	43 09.9	N 6 18.1
	19	228 37.1	146 19.2	18.8	160 43.3	30.3	261 24.9	18.4	58 12.1	18.0
	20	243 39.6	161 18.6	19.1	175 44.0	30.6	276 27.6	18.5	73 14.4	17.9
	21	258 42.0	176 18.0	. . 19.3	190 44.6	. . 30.9	291 30.4	. . 18.6	88 16.6	. . 17.8
	22	273 44.5	191 17.4	19.6	205 45.3	31.2	306 33.1	18.7	103 18.8	17.7
	23	288 47.0	206 16.8	19.9	220 45.9	31.4	321 35.8	18.8	118 21.1	17.6
	26 00	303 49.4	221 16.3	N21 20.1	235 46.6	N21 31.7	336 38.6	S14 18.9	133 23.3	N 6 17.5
	01	318 51.9	236 15.7	20.4	250 47.2	32.0	351 41.3	19.0	148 25.5	17.4
	02	333 54.4	251 15.1	20.7	265 47.9	32.3	6 44.0	19.1	163 27.8	17.3
	03	348 56.8	266 14.5	. . 20.9	280 48.5	. . 32.6	21 46.8	. . 19.2	178 30.0	. . 17.2
	04	3 59.3	281 13.9	21.2	295 49.2	32.9	36 49.5	19.3	193 32.3	17.1
	05	19 01.7	296 13.4	21.4	310 49.8	33.2	51 52.2	19.4	208 34.5	17.0
	06	34 04.2	311 12.8	N21 21.7	325 50.5	N21 33.5	66 55.0	S14 19.5	223 36.7	N 6 16.9
	07	49 06.7	326 12.2	21.9	340 51.1	33.7	81 57.7	19.6	238 39.0	16.8
	08	64 09.1	341 11.6	22.2	355 51.8	34.0	97 00.4	19.7	253 41.2	16.7
S	09	79 11.6	356 11.0	. . 22.5	10 52.4	. . 34.3	112 03.2	. . 19.8	268 43.4	. . 16.6
U	10	94 14.1	11 10.4	22.7	25 53.0	34.6	127 05.9	19.9	283 45.7	16.5
N	11	109 16.5	26 09.8	23.0	40 53.7	34.9	142 08.7	20.0	298 47.9	16.4
D	12	124 19.0	41 09.3	N21 23.2	55 54.3	N21 35.2	157 11.4	S14 20.1	313 50.1	N 6 16.3
A	13	139 21.5	56 08.7	23.5	70 55.0	35.5	172 14.1	20.2	328 52.4	16.2
Y	14	154 23.9	71 08.1	23.7	85 55.6	35.7	187 16.9	20.3	343 54.6	16.1
	15	169 26.4	86 07.5	. . 24.0	100 56.3	. . 36.0	202 19.6	. . 20.4	358 56.8	. . 16.0
	16	184 28.8	101 06.9	24.2	115 56.9	36.3	217 22.3	20.5	13 59.1	15.9
	17	199 31.3	116 06.3	24.5	130 57.6	36.6	232 25.1	20.6	29 01.3	15.8
	18	214 33.8	131 05.7	N21 24.7	145 58.2	N21 36.9	247 27.8	S14 20.7	44 03.5	N 6 15.7
	19	229 36.2	146 05.1	24.9	160 58.9	37.1	262 30.6	20.8	59 05.8	15.6
	20	244 38.7	161 04.6	25.2	175 59.5	37.4	277 33.3	20.9	74 08.0	15.5
	21	259 41.2	176 04.0	. . 25.4	191 00.2	. . 37.7	292 36.0	. . 21.1	89 10.2	. . 15.4
	22	274 43.6	191 03.4	25.7	206 00.8	38.0	307 38.8	21.2	104 12.5	15.3
	23	289 46.1	206 02.8	25.9	221 01.5	38.3	322 41.5	21.3	119 14.7	15.2
	Mer. Pass.	h m 3 48.0	v −0.6	d 0.3	v 0.6	d 0.3	v 2.7	d 0.1	v 2.2	d 0.1

STARS

Name	SHA	Dec
	° ′	° ′
Acamar	315 20.5	S40 15.6
Achernar	335 28.6	S57 10.9
Acrux	173 13.1	S63 09.5
Adhara	255 15.2	S28 59.0
Aldebaran	290 52.9	N16 31.8
Alioth	166 23.1	N55 54.7
Alkaid	153 01.0	N49 16.1
Al Na'ir	27 46.7	S46 54.6
Alnilam	275 49.5	S 1 11.6
Alphard	217 59.2	S 8 42.0
Alphecca	126 13.2	N26 41.1
Alpheratz	357 46.4	N29 08.6
Altair	62 10.7	N 8 53.7
Ankaa	353 18.2	S42 14.9
Antares	112 29.6	S26 27.3
Arcturus	145 58.3	N19 08.0
Atria	107 33.7	S69 02.9
Avior	234 20.0	S59 32.4
Bellatrix	278 35.3	N 6 21.6
Betelgeuse	271 04.7	N 7 24.6
Canopus	263 57.9	S52 41.9
Capella	280 39.1	N46 00.4
Deneb	49 33.1	N45 18.9
Denebola	182 36.7	N14 31.2
Diphda	348 58.6	S17 55.8
Dubhe	193 55.4	N61 42.1
Elnath	278 16.5	N28 36.9
Eltanin	90 47.0	N51 29.4
Enif	33 49.6	N 9 55.2
Fomalhaut	15 26.8	S29 34.0
Gacrux	172 04.6	S57 10.3
Gienah	175 55.4	S17 35.8
Hadar	148 52.2	S60 25.5
Hamal	328 04.0	N23 30.5
Kaus Aust.	83 47.2	S34 22.8
Kochab	137 19.3	N74 07.2
Markab	13 41.0	N15 15.5
Menkar	314 18.1	N 4 07.8
Menkent	148 11.0	S36 25.3
Miaplacidus	221 41.5	S69 45.5
Mirfak	308 44.7	N49 53.6
Nunki	76 01.5	S26 17.1
Peacock	53 23.0	S56 42.1
Pollux	243 31.6	N28 00.2
Procyon	245 03.0	N 5 12.1
Rasalhague	96 08.8	N12 33.2
Regulus	207 46.8	N11 55.3
Rigel	281 15.0	S 8 11.3
Rigil Kent.	139 55.7	S60 52.8
Sabik	102 15.6	S15 44.2
Schedar	349 43.8	N56 35.3
Shaula	96 25.5	S37 06.7
Sirius	258 36.6	S16 43.6
Spica	158 34.3	S11 12.8
Suhail	222 55.1	S43 28.3
Vega	80 40.5	N38 47.7
Zuben'ubi	137 08.5	S16 05.0

	SHA	Mer. Pass.
	° ′	h m
Venus	278 39.8	9 14
Mars	292 40.8	8 18
Jupiter	32 42.7	1 38
Saturn	189 39.3	15 08

UT	SUN GHA	SUN Dec	MOON GHA	v	MOON Dec	d	HP
d h	° ′	° ′	° ′	′	° ′	′	′
24 00	178 22.3	N19 52.8	152 37.3	9.2	N 9 52.8	15.6	60.6
01	193 22.3	52.3	167 05.5	9.3	9 37.2	15.5	60.6
02	208 22.3	51.8	181 33.8	9.4	9 21.7	15.7	60.5
03	223 22.3	. . 51.3	196 02.2	9.5	9 06.0	15.6	60.5
04	238 22.3	50.7	210 30.7	9.6	8 50.4	15.7	60.5
05	253 22.3	50.2	224 59.3	9.6	8 34.7	15.8	60.5
06	268 22.2	N19 49.7	239 27.9	9.8	N 8 18.9	15.7	60.4
07	283 22.2	49.2	253 56.7	9.8	8 03.2	15.8	60.4
08	298 22.2	48.6	268 25.5	9.9	7 47.4	15.8	60.4
09	313 22.2	. . 48.1	282 54.4	9.9	7 31.6	15.9	60.3
10	328 22.2	47.6	297 23.3	10.1	7 15.7	15.8	60.3
11	343 22.2	47.0	311 52.4	10.1	6 59.9	15.9	60.3
12	358 22.2	N19 46.5	326 21.5	10.2	N 6 44.0	15.9	60.3
13	13 22.2	46.0	340 50.7	10.3	6 28.1	16.0	60.2
14	28 22.2	45.4	355 20.0	10.3	6 12.1	15.9	60.2
15	43 22.1	. . 44.9	9 49.3	10.4	5 56.2	16.0	60.2
16	58 22.1	44.4	24 18.7	10.5	5 40.2	15.9	60.1
17	73 22.1	43.8	38 48.2	10.5	5 24.3	16.0	60.1
18	88 22.1	N19 43.3	53 17.7	10.6	N 5 08.3	16.0	60.1
19	103 22.1	42.8	67 47.3	10.7	4 52.3	16.0	60.0
20	118 22.1	42.2	82 17.0	10.7	4 36.3	16.0	60.0
21	133 22.1	. . 41.7	96 46.7	10.8	4 20.3	16.0	60.0
22	148 22.1	41.2	111 16.5	10.9	4 04.3	16.0	59.9
23	163 22.1	40.6	125 46.4	10.9	3 48.3	16.0	59.9
25 00	178 22.1	N19 40.1	140 16.3	11.0	N 3 32.3	16.1	59.9
01	193 22.1	39.6	154 46.3	11.0	3 16.2	16.0	59.8
02	208 22.1	39.0	169 16.3	11.1	3 00.2	16.0	59.8
03	223 22.0	. . 38.5	183 46.4	11.1	2 44.2	16.0	59.7
04	238 22.0	37.9	198 16.5	11.2	2 28.2	16.0	59.7
05	253 22.0	37.4	212 46.7	11.3	2 12.2	15.9	59.7
06	268 22.0	N19 36.9	227 17.0	11.3	N 1 56.3	16.0	59.6
07	283 22.0	36.3	241 47.3	11.3	1 40.3	16.0	59.6
08	298 22.0	35.8	256 17.6	11.4	1 24.3	15.9	59.6
09	313 22.0	. . 35.2	270 48.0	11.5	1 08.4	15.9	59.5
10	328 22.0	34.7	285 18.5	11.5	0 52.5	16.0	59.5
11	343 22.0	34.1	299 49.0	11.5	0 36.5	15.9	59.5
12	358 22.0	N19 33.6	314 19.5	11.6	N 0 20.6	15.8	59.4
13	13 22.0	33.1	328 50.1	11.6	N 0 04.8	15.9	59.4
14	28 22.0	32.5	343 20.7	11.7	S 0 11.1	15.8	59.3
15	43 22.0	. . 32.0	357 51.4	11.7	0 26.9	15.8	59.3
16	58 22.0	31.4	12 22.1	11.8	0 42.7	15.8	59.3
17	73 22.0	30.9	26 52.9	11.8	0 58.5	15.8	59.2
18	88 22.0	N19 30.3	41 23.7	11.8	S 1 14.3	15.7	59.2
19	103 22.0	29.8	55 54.5	11.9	1 30.0	15.7	59.2
20	118 22.0	29.2	70 25.4	11.9	1 45.7	15.7	59.1
21	133 22.0	. . 28.7	84 56.3	11.9	2 01.4	15.6	59.1
22	148 22.0	28.1	99 27.2	12.0	2 17.0	15.6	59.0
23	163 22.0	27.6	113 58.2	11.9	2 32.6	15.6	59.0
26 00	178 22.0	N19 27.0	128 29.1	12.1	S 2 48.2	15.6	59.0
01	193 22.0	26.5	143 00.2	12.0	3 03.8	15.5	58.9
02	208 22.0	25.9	157 31.2	12.1	3 19.3	15.4	58.9
03	223 22.0	. . 25.4	172 02.3	12.1	3 34.7	15.4	58.8
04	238 22.0	24.8	186 33.4	12.1	3 50.1	15.4	58.8
05	253 22.0	24.3	201 04.5	12.2	4 05.5	15.4	58.8
06	268 22.0	N19 23.7	215 35.7	12.2	S 4 20.9	15.3	58.7
07	283 22.0	23.2	230 06.9	12.2	4 36.2	15.2	58.7
08	298 22.0	22.6	244 38.1	12.2	4 51.4	15.2	58.6
09	313 22.0	. . 22.0	259 09.3	12.2	5 06.6	15.2	58.6
10	328 22.0	21.5	273 40.5	12.3	5 21.8	15.1	58.6
11	343 22.0	20.9	288 11.8	12.3	5 36.9	15.1	58.5
12	358 22.0	N19 20.4	302 43.1	12.3	S 5 52.0	15.0	58.5
13	13 22.0	19.8	317 14.4	12.3	6 07.0	14.9	58.4
14	28 22.0	19.3	331 45.7	12.3	6 21.9	14.9	58.4
15	43 22.0	. . 18.7	346 17.0	12.3	6 36.8	14.9	58.4
16	58 22.0	18.1	0 48.3	12.4	6 51.7	14.8	58.3
17	73 22.0	17.6	15 19.7	12.3	7 06.5	14.7	58.3
18	88 22.0	N19 17.0	29 51.0	12.4	S 7 21.2	14.7	58.2
19	103 22.0	16.5	44 22.4	12.4	7 35.9	14.7	58.2
20	118 22.0	15.9	58 53.8	12.3	7 50.6	14.5	58.2
21	133 22.0	. . 15.3	73 25.1	12.4	8 05.1	14.6	58.1
22	148 22.0	14.8	87 56.5	12.4	8 19.7	14.4	58.1
23	163 22.0	14.2	102 27.9	12.4	S 8 34.1	14.4	58.1
	SD 15.8	*d* 0.5	SD 16.4		16.2		15.9

Lat.	Twilight Naut.	Twilight Civil	Sunrise	Moonrise 24	Moonrise 25	Moonrise 26	Moonrise 27
°	h m	h m	h m	h m	h m	h m	h m
N 72	▭	▭	▭	06 20	08 43	11 00	13 22
N 70	▭	▭	▭	06 32	08 45	10 51	12 59
68	////	////	01 36	06 43	08 46	10 44	12 42
66	////	////	02 18	06 51	08 47	10 38	12 28
64	////	00 45	02 46	06 58	08 48	10 33	12 17
62	////	01 46	03 08	07 04	08 49	10 29	12 07
60	////	02 18	03 25	07 10	08 49	10 25	11 59
N 58	00 41	02 41	03 39	07 14	08 50	10 22	11 52
56	01 36	03 00	03 51	07 18	08 51	10 19	11 46
54	02 06	03 15	04 02	07 22	08 51	10 17	11 40
52	02 28	03 28	04 11	07 25	08 52	10 15	11 35
50	02 46	03 40	04 20	07 28	08 52	10 13	11 31
45	03 19	04 03	04 38	07 34	08 53	10 08	11 21
N 40	03 43	04 21	04 52	07 40	08 53	10 04	11 13
35	04 01	04 36	05 04	07 44	08 54	10 01	11 06
30	04 17	04 48	05 15	07 48	08 55	09 58	11 00
20	04 41	05 09	05 33	07 55	08 56	09 53	10 50
N 10	05 00	05 26	05 49	08 02	08 56	09 49	10 41
0	05 16	05 41	06 03	08 07	08 57	09 45	10 33
S 10	05 29	05 55	06 17	08 13	08 58	09 42	10 24
20	05 42	06 09	06 33	08 19	08 59	09 37	10 16
30	05 55	06 24	06 50	08 26	09 00	09 33	10 06
35	06 02	06 33	07 00	08 30	09 01	09 30	10 00
40	06 09	06 42	07 11	08 34	09 01	09 27	09 54
45	06 17	06 53	07 25	08 39	09 02	09 24	09 46
S 50	06 25	07 05	07 41	08 45	09 03	09 20	09 37
52	06 29	07 11	07 49	08 48	09 04	09 18	09 33
54	06 33	07 17	07 57	08 51	09 04	09 16	09 29
56	06 37	07 23	08 07	08 55	09 05	09 14	09 24
58	06 42	07 31	08 17	08 58	09 05	09 12	09 18
S 60	06 47	07 40	08 30	09 03	09 06	09 09	09 12

Lat.	Sunset	Twilight Civil	Twilight Naut.	Moonset 24	Moonset 25	Moonset 26	Moonset 27
°	h m	h m	h m	h m	h m	h m	h m
N 72	▭	▭	▭	21 24	20 54	20 22	19 41
N 70	▭	▭	▭	21 19	20 56	20 33	20 05
68	22 32	////	////	21 14	20 59	20 43	20 24
66	21 52	////	////	21 10	21 01	20 51	20 39
64	21 24	23 18	////	21 07	21 02	20 57	20 52
62	21 04	22 24	////	21 04	21 04	21 03	21 03
60	20 47	21 53	////	21 02	21 05	21 08	21 12
N 58	20 33	21 30	23 22	21 00	21 06	21 13	21 20
56	20 21	21 12	22 33	20 58	21 07	21 17	21 27
54	20 10	20 56	22 05	20 56	21 08	21 20	21 34
52	20 01	20 44	21 43	20 54	21 09	21 24	21 40
50	19 52	20 32	21 26	20 53	21 10	21 27	21 45
45	19 35	20 09	20 53	20 49	21 11	21 33	21 56
N 40	19 21	19 51	20 30	20 46	21 13	21 39	22 06
35	19 08	19 37	20 11	20 44	21 14	21 44	22 14
30	18 58	19 24	19 56	20 42	21 15	21 48	22 21
20	18 40	19 04	19 32	20 38	21 17	21 55	22 34
N 10	18 24	18 47	19 13	20 35	21 19	22 02	22 45
0	18 10	18 32	18 58	20 31	21 20	22 08	22 55
S 10	17 56	18 18	18 44	20 28	21 22	22 14	23 06
20	17 41	18 04	18 31	20 24	21 23	22 21	23 17
30	17 23	17 49	18 18	20 20	21 25	22 28	23 30
35	17 14	17 41	18 11	20 18	21 26	22 32	23 37
40	17 02	17 31	18 04	20 15	21 28	22 37	23 46
45	16 49	17 21	17 57	20 12	21 29	22 43	23 56
S 50	16 33	17 09	17 48	20 08	21 31	22 50	24 08
52	16 25	17 03	17 45	20 07	21 31	22 53	24 13
54	16 17	16 57	17 41	20 05	21 32	22 57	24 20
56	16 07	16 50	17 36	20 02	21 33	23 01	24 27
58	15 56	16 43	17 32	20 00	21 34	23 05	24 34
S 60	15 44	16 34	17 27	19 57	21 35	23 10	24 43

Day	SUN Eqn. of Time 00^h	SUN Eqn. of Time 12^h	SUN Mer. Pass.	MOON Mer. Pass. Upper	MOON Mer. Pass. Lower	MOON Age	MOON Phase
d	m s	m s	h m	h m	h m	d	%
24	06 31	06 31	12 07	14 19	01 54	02	8
25	06 32	06 32	12 07	15 09	02 44	03	16
26	06 32	06 32	12 07	15 57	03 33	04	25

UT d h	ARIES GHA	VENUS −4.0 GHA	VENUS Dec	MARS +1.1 GHA	MARS Dec	JUPITER −2.8 GHA	JUPITER Dec	SATURN +1.1 GHA	SATURN Dec
	° ′	° ′	° ′	° ′	° ′	° ′	° ′	° ′	° ′
27 00 (MONDAY)	304 48.6	221 02.2	N21 26.2	236 02.1	N21 38.6	337 44.3	S14 21.4	134 16.9	N 6 15.1
01	319 51.0	236 01.6	26.4	251 02.8	38.8	352 47.0	21.5	149 19.2	15.0
02	334 53.5	251 01.0	26.6	266 03.4	39.1	7 49.7	21.6	164 21.4	14.8
03	349 56.0	266 00.4	26.9	281 04.1	39.4	22 52.5	21.7	179 23.6	14.7
04	4 58.4	280 59.8	27.1	296 04.7	39.7	37 55.2	21.8	194 25.9	14.6
05	20 00.9	295 59.2	27.3	311 05.4	39.9	52 58.0	21.9	209 28.1	14.5
06	35 03.3	310 58.6	N21 27.6	326 06.0	N21 40.2	68 00.7	S14 22.0	224 30.3	N 6 14.4
07	50 05.8	325 58.1	27.8	341 06.7	40.5	83 03.4	22.1	239 32.6	14.3
08	65 08.3	340 57.5	28.0	356 07.3	40.8	98 06.2	22.2	254 34.8	14.2
09	80 10.7	355 56.9	28.3	11 07.9	41.1	113 08.9	22.3	269 37.0	14.1
10	95 13.2	10 56.3	28.5	26 08.6	41.3	128 11.7	22.4	284 39.3	14.0
11	110 15.7	25 55.7	28.7	41 09.2	41.6	143 14.4	22.5	299 41.5	13.9
12	125 18.1	40 55.1	N21 29.0	56 09.9	N21 41.9	158 17.1	S14 22.6	314 43.7	N 6 13.8
13	140 20.6	55 54.5	29.2	71 10.5	42.2	173 19.9	22.7	329 46.0	13.7
14	155 23.1	70 53.9	29.4	86 11.2	42.4	188 22.6	22.8	344 48.2	13.6
15	170 25.5	85 53.3	29.7	101 11.8	42.7	203 25.4	22.9	359 50.4	13.5
16	185 28.0	100 52.7	29.9	116 12.5	43.0	218 28.1	23.0	14 52.7	13.4
17	200 30.5	115 52.1	30.1	131 13.1	43.3	233 30.9	23.1	29 54.9	13.3
18	215 32.9	130 51.5	N21 30.3	146 13.8	N21 43.5	248 33.6	S14 23.2	44 57.1	N 6 13.2
19	230 35.4	145 50.9	30.5	161 14.4	43.8	263 36.3	23.3	59 59.4	13.1
20	245 37.8	160 50.3	30.8	176 15.1	44.1	278 39.1	23.4	75 01.6	13.0
21	260 40.3	175 49.7	31.0	191 15.7	44.4	293 41.8	23.5	90 03.8	12.9
22	275 42.8	190 49.1	31.2	206 16.4	44.6	308 44.6	23.6	105 06.1	12.8
23	290 45.2	205 48.5	31.4	221 17.0	44.9	323 47.3	23.7	120 08.3	12.7
28 00 (TUESDAY)	305 47.7	220 47.9	N21 31.7	236 17.7	N21 45.2	338 50.1	S14 23.9	135 10.5	N 6 12.6
01	320 50.2	235 47.3	31.9	251 18.3	45.4	353 52.8	24.0	150 12.7	12.5
02	335 52.6	250 46.7	32.1	266 19.0	45.7	8 55.5	24.1	165 15.0	12.4
03	350 55.1	265 46.1	32.3	281 19.6	46.0	23 58.3	24.2	180 17.2	12.3
04	5 57.6	280 45.5	32.5	296 20.3	46.3	39 01.0	24.3	195 19.4	12.2
05	21 00.0	295 44.9	32.7	311 20.9	46.5	54 03.8	24.4	210 21.7	12.1
06	36 02.5	310 44.3	N21 32.9	326 21.6	N21 46.8	69 06.5	S14 24.5	225 23.9	N 6 12.0
07	51 04.9	325 43.7	33.2	341 22.2	47.1	84 09.3	24.6	240 26.1	11.9
08	66 07.4	340 43.1	33.4	356 22.9	47.3	99 12.0	24.7	255 28.4	11.8
09	81 09.9	355 42.5	33.6	11 23.5	47.6	114 14.8	24.8	270 30.6	11.7
10	96 12.3	10 41.9	33.8	26 24.2	47.9	129 17.5	24.9	285 32.8	11.6
11	111 14.8	25 41.3	34.0	41 24.8	48.1	144 20.3	25.0	300 35.1	11.5
12	126 17.3	40 40.7	N21 34.2	56 25.5	N21 48.4	159 23.0	S14 25.1	315 37.3	N 6 11.4
13	141 19.7	55 40.1	34.4	71 26.1	48.7	174 25.7	25.2	330 39.5	11.3
14	156 22.2	70 39.5	34.6	86 26.8	48.9	189 28.5	25.3	345 41.7	11.1
15	171 24.7	85 38.9	34.8	101 27.4	49.2	204 31.2	25.4	0 44.0	11.0
16	186 27.1	100 38.3	35.0	116 28.1	49.5	219 34.0	25.5	15 46.2	10.9
17	201 29.6	115 37.7	35.2	131 28.7	49.7	234 36.7	25.6	30 48.4	10.8
18	216 32.1	130 37.1	N21 35.4	146 29.4	N21 50.0	249 39.5	S14 25.7	45 50.7	N 6 10.7
19	231 34.5	145 36.4	35.6	161 30.0	50.3	264 42.2	25.8	60 52.9	10.6
20	246 37.0	160 35.8	35.8	176 30.7	50.5	279 45.0	25.9	75 55.1	10.5
21	261 39.4	175 35.2	36.0	191 31.3	50.8	294 47.7	26.1	90 57.4	10.4
22	276 41.9	190 34.6	36.2	206 32.0	51.1	309 50.5	26.2	105 59.6	10.3
23	291 44.4	205 34.0	36.4	221 32.6	51.3	324 53.2	26.3	121 01.8	10.2
29 00 (WEDNESDAY)	306 46.8	220 33.4	N21 36.6	236 33.3	N21 51.6	339 56.0	S14 26.4	136 04.0	N 6 10.1
01	321 49.3	235 32.8	36.8	251 33.9	51.9	354 58.7	26.5	151 06.3	10.0
02	336 51.8	250 32.2	37.0	266 34.6	52.1	10 01.5	26.6	166 08.5	09.9
03	351 54.2	265 31.6	37.2	281 35.2	52.4	25 04.2	26.7	181 10.7	09.8
04	6 56.7	280 31.0	37.4	296 35.9	52.6	40 07.0	26.8	196 13.0	09.7
05	21 59.2	295 30.4	37.6	311 36.5	52.9	55 09.7	26.9	211 15.2	09.6
06	37 01.6	310 29.8	N21 37.8	326 37.2	N21 53.2	70 12.4	S14 27.0	226 17.4	N 6 09.5
07	52 04.1	325 29.1	38.0	341 37.8	53.4	85 15.2	27.1	241 19.6	09.4
08	67 06.6	340 28.5	38.2	356 38.5	53.7	100 17.9	27.2	256 21.9	09.3
09	82 09.0	355 27.9	38.3	11 39.1	53.9	115 20.7	27.3	271 24.1	09.2
10	97 11.5	10 27.3	38.5	26 39.8	54.2	130 23.4	27.4	286 26.3	09.1
11	112 13.9	25 26.7	38.7	41 40.4	54.5	145 26.2	27.5	301 28.6	09.0
12	127 16.4	40 26.1	N21 38.9	56 41.1	N21 54.7	160 28.9	S14 27.6	316 30.8	N 6 08.9
13	142 18.9	55 25.5	39.1	71 41.7	55.0	175 31.7	27.7	331 33.0	08.8
14	157 21.3	70 24.9	39.3	86 42.4	55.2	190 34.4	27.8	346 35.2	08.7
15	172 23.8	85 24.2	39.4	101 43.0	55.5	205 37.2	27.9	1 37.5	08.6
16	187 26.3	100 23.6	39.6	116 43.7	55.8	220 39.9	28.1	16 39.7	08.5
17	202 28.7	115 23.0	39.8	131 44.3	56.0	235 42.7	28.2	31 41.9	08.3
18	217 31.2	130 22.4	N21 40.0	146 45.0	N21 56.3	250 45.4	S14 28.3	46 44.2	N 6 08.2
19	232 33.7	145 21.8	40.2	161 45.6	56.5	265 48.2	28.4	61 46.4	08.1
20	247 36.1	160 21.2	40.3	176 46.3	56.8	280 50.9	28.5	76 48.6	08.0
21	262 38.6	175 20.6	40.5	191 46.9	57.0	295 53.7	28.6	91 50.8	07.9
22	277 41.0	190 19.9	40.7	206 47.6	57.3	310 56.4	28.7	106 53.1	07.8
23	292 43.5	205 19.3	40.9	221 48.2	57.6	325 59.2	28.8	121 55.3	07.7
Mer. Pass.	h m 3 36.2	v −0.6	d 0.2	v 0.6	d 0.3	v 2.7	d 0.1	v 2.2	d 0.1

STARS

Name	SHA	Dec
	° ′	°
Acamar	315 20.5	S40 15
Achernar	335 28.6	S57 10
Acrux	173 13.1	S63 09
Adhara	255 15.1	S28 59
Aldebaran	290 52.9	N16 31
Alioth	166 23.2	N55 54
Alkaid	153 01.0	N49 16
Al Na'ir	27 46.7	S46 54
Alnilam	275 49.5	S 1 11
Alphard	217 59.2	S 8 42
Alphecca	126 13.3	N26 41
Alpheratz	357 46.3	N29 08
Altair	62 10.7	N 8 53
Ankaa	353 18.2	S42 14
Antares	112 29.6	S26 27
Arcturus	145 58.3	N19 08
Atria	107 33.7	S69 03
Avior	234 20.0	S59 32
Bellatrix	278 35.3	N 6 21
Betelgeuse	271 04.7	N 7 24
Canopus	263 57.9	S52 41
Capella	280 39.0	N46 00
Deneb	49 33.0	N45 18
Denebola	182 36.7	N14 31
Diphda	348 58.6	S17 55
Dubhe	193 55.4	N61 42
Elnath	278 16.5	N28 36
Eltanin	90 47.0	N51 29
Enif	33 49.6	N 9 55
Fomalhaut	15 26.8	S29 34
Gacrux	172 04.6	S57 10
Gienah	175 55.5	S17 35
Hadar	148 52.2	S60 25
Hamal	328 04.0	N23 30
Kaus Aust.	83 47.2	S34 22
Kochab	137 19.3	N74 07
Markab	13 41.0	N15 15
Menkar	314 18.1	N 4 07
Menkent	148 11.0	S36 25
Miaplacidus	221 41.5	S69 45
Mirfak	308 44.7	N49 53
Nunki	76 01.5	S26 17
Peacock	53 23.0	S56 42
Pollux	243 31.6	N28 00
Procyon	245 03.0	N 5 12
Rasalhague	96 08.8	N12 33
Regulus	207 46.8	N11 55
Rigel	281 15.0	S 8 11
Rigil Kent.	139 55.8	S60 52
Sabik	102 15.6	S15 44
Schedar	349 43.8	N56 35
Shaula	96 25.5	S37 06
Sirius	258 36.6	S16 43
Spica	158 34.4	S11 12
Suhail	222 55.1	S43 28
Vega	80 40.5	N38 47
Zuben'ubi	137 08.6	S16 05

	SHA	Mer. Pas
	° ′	h m
Venus	275 00.2	9 17
Mars	290 30.0	8 14
Jupiter	33 02.4	1 24
Saturn	189 22.8	14 57

UT	SUN GHA	SUN Dec	MOON GHA	v	MOON Dec	d	HP
d h	° ′	° ′	° ′	′	° ′	′	′
7 00	178 22.0	N19 13.6	116 59.3	12.4	S 8 48.5	14.3	58.0
01	193 22.0	13.1	131 30.7	12.4	9 02.8	14.3	58.0
02	208 22.0	12.5	146 02.1	12.4	9 17.1	14.2	57.9
03	223 22.0	. . 11.9	160 33.5	12.4	9 31.3	14.1	57.9
04	238 22.0	11.4	175 04.9	12.4	9 45.4	14.1	57.9
05	253 22.0	10.8	189 36.3	12.4	9 59.5	14.0	57.8
06	268 22.1	N19 10.2	204 07.7	12.5	S10 13.5	13.9	57.8
07	283 22.1	09.7	218 39.2	12.4	10 27.4	13.9	57.7
08	298 22.1	09.1	233 10.6	12.3	10 41.3	13.8	57.7
09	313 22.1	. . 08.5	247 41.9	12.4	10 55.1	13.7	57.7
10	328 22.1	08.0	262 13.3	12.4	11 08.8	13.7	57.6
11	343 22.1	07.4	276 44.7	12.4	11 22.5	13.5	57.6
12	358 22.1	N19 06.8	291 16.1	12.4	S11 36.0	13.5	57.5
13	13 22.1	06.3	305 47.5	12.3	11 49.5	13.5	57.5
14	28 22.1	05.7	320 18.8	12.4	12 03.0	13.3	57.5
15	43 22.1	. . 05.1	334 50.2	12.3	12 16.3	13.3	57.4
16	58 22.1	04.5	349 21.5	12.4	12 29.6	13.2	57.4
17	73 22.1	04.0	3 52.9	12.3	12 42.8	13.1	57.4
18	88 22.2	N19 03.4	18 24.2	12.3	S12 55.9	13.0	57.3
19	103 22.2	02.8	32 55.5	12.3	13 08.9	13.0	57.3
20	118 22.2	02.2	47 26.8	12.3	13 21.9	12.9	57.2
21	133 22.2	. . 01.7	61 58.1	12.3	13 34.8	12.8	57.2
22	148 22.2	01.1	76 29.4	12.2	13 47.6	12.7	57.2
23	163 22.2	19 00.5	91 00.6	12.2	14 00.3	12.6	57.1
8 00	178 22.2	N18 59.9	105 31.8	12.3	S14 12.9	12.6	57.1
01	193 22.2	59.4	120 03.1	12.2	14 25.5	12.4	57.1
02	208 22.2	58.8	134 34.3	12.2	14 37.9	12.4	57.0
03	223 22.3	. . 58.2	149 05.5	12.1	14 50.3	12.3	57.0
04	238 22.3	57.6	163 36.6	12.2	15 02.6	12.2	56.9
05	253 22.3	57.0	178 07.8	12.1	15 14.8	12.1	56.9
06	268 22.3	N18 56.5	192 38.9	12.1	S15 26.9	12.0	56.9
07	283 22.3	55.9	207 10.0	12.1	15 38.9	12.0	56.8
08	298 22.3	55.3	221 41.1	12.1	15 50.9	11.8	56.8
09	313 22.3	. . 54.7	236 12.2	12.0	16 02.7	11.8	56.8
10	328 22.4	54.1	250 43.2	12.0	16 14.5	11.6	56.7
11	343 22.4	53.5	265 14.2	12.0	16 26.1	11.6	56.7
12	358 22.4	N18 53.0	279 45.2	12.0	S16 37.7	11.5	56.7
13	13 22.4	52.4	294 16.2	11.9	16 49.2	11.3	56.6
14	28 22.4	51.8	308 47.1	12.0	17 00.5	11.3	56.6
15	43 22.4	. . 51.2	323 18.1	11.9	17 11.8	11.2	56.6
16	58 22.4	50.6	337 49.0	11.8	17 23.0	11.1	56.5
17	73 22.5	50.0	352 19.8	11.9	17 34.1	11.0	56.5
18	88 22.5	N18 49.4	6 50.7	11.8	S17 45.1	10.9	56.5
19	103 22.5	48.9	21 21.5	11.8	17 56.0	10.8	56.4
20	118 22.5	48.3	35 52.3	11.8	18 06.8	10.7	56.4
21	133 22.5	. . 47.7	50 23.1	11.7	18 17.5	10.6	56.4
22	148 22.5	47.1	64 53.8	11.7	18 28.1	10.5	56.3
23	163 22.6	46.5	79 24.5	11.7	18 38.6	10.4	56.3
9 00	178 22.6	N18 45.9	93 55.2	11.6	S18 49.0	10.3	56.3
01	193 22.6	45.3	108 25.8	11.7	18 59.3	10.2	56.2
02	208 22.6	44.7	122 56.5	11.6	19 09.5	10.1	56.2
03	223 22.6	. . 44.1	137 27.1	11.5	19 19.6	10.0	56.2
04	238 22.7	43.5	151 57.6	11.6	19 29.6	9.9	56.1
05	253 22.7	42.9	166 28.2	11.5	19 39.5	9.8	56.1
06	268 22.7	N18 42.4	180 58.7	11.4	S19 49.3	9.6	56.1
07	283 22.7	41.8	195 29.1	11.5	19 58.9	9.6	56.0
08	298 22.7	41.2	209 59.6	11.4	20 08.5	9.5	56.0
09	313 22.8	. . 40.6	224 30.0	11.4	20 18.0	9.3	56.0
10	328 22.8	40.0	239 00.4	11.3	20 27.3	9.3	55.9
11	343 22.8	39.4	253 30.7	11.4	20 36.6	9.1	55.9
12	358 22.8	N18 38.8	268 01.1	11.3	S20 45.7	9.0	55.9
13	13 22.8	38.2	282 31.4	11.2	20 54.7	8.9	55.9
14	28 22.9	37.6	297 01.6	11.3	21 03.6	8.9	55.8
15	43 22.9	. . 37.0	311 31.9	11.2	21 12.5	8.7	55.8
16	58 22.9	36.4	326 02.1	11.1	21 21.2	8.5	55.8
17	73 22.9	35.8	340 32.2	11.2	21 29.7	8.5	55.7
18	88 23.0	N18 35.2	355 02.4	11.1	S21 38.2	8.4	55.7
19	103 23.0	34.6	9 32.5	11.1	21 46.6	8.2	55.7
20	118 23.0	34.0	24 02.6	11.0	21 54.8	8.2	55.6
21	133 23.0	. . 33.4	38 32.6	11.0	22 03.0	8.0	55.6
22	148 23.1	32.8	53 02.6	11.0	22 11.0	7.9	55.6
23	163 23.1	32.2	67 32.6	11.0	S22 18.9	7.8	55.6
	SD 15.8	d 0.6	SD 15.7		15.4		15.2

Lat.	Twilight Naut.	Twilight Civil	Sunrise	Moonrise 27	28	29	30
°	h m	h m	h m	h m	h m	h m	h m
N 72	□	□	□	13 22	16 42	■	■
N 70	////	////	00 38	12 59	15 25	■	■
68	////	////	01 54	12 42	14 48	17 39	■
66	////	////	02 30	12 28	14 22	16 26	■
64	////	01 13	02 56	12 17	14 02	15 50	17 45
62	////	01 59	03 15	12 07	13 46	15 24	17 01
60	////	02 28	03 32	11 59	13 32	15 04	16 33
N 58	01 07	02 49	03 45	11 52	13 21	14 48	16 10
56	01 48	03 06	03 57	11 46	13 11	14 34	15 52
54	02 15	03 21	04 07	11 40	13 02	14 22	15 37
52	02 35	03 33	04 16	11 35	12 55	14 11	15 24
50	02 52	03 44	04 24	11 31	12 48	14 02	15 13
45	03 23	04 07	04 41	11 21	12 33	13 42	14 49
N 40	03 46	04 24	04 55	11 13	12 21	13 27	14 30
35	04 04	04 38	05 06	11 06	12 10	13 13	14 14
30	04 19	04 50	05 17	11 00	12 01	13 02	14 00
20	04 42	05 10	05 34	10 50	11 46	12 42	13 37
N 10	05 01	05 27	05 49	10 41	11 33	12 25	13 17
0	05 16	05 41	06 03	10 33	11 20	12 09	12 58
S 10	05 29	05 55	06 17	10 24	11 08	11 53	12 40
20	05 41	06 08	06 31	10 16	10 55	11 36	12 20
30	05 54	06 23	06 48	10 06	10 40	11 17	11 57
35	06 00	06 31	06 58	10 00	10 31	11 06	11 44
40	06 07	06 40	07 09	09 54	10 22	10 53	11 29
45	06 14	06 50	07 21	09 46	10 10	10 38	11 11
S 50	06 22	07 01	07 37	09 37	09 57	10 20	10 48
52	06 25	07 07	07 44	09 33	09 50	10 11	10 37
54	06 29	07 13	07 52	09 29	09 43	10 01	10 25
56	06 33	07 19	08 01	09 24	09 36	09 51	10 11
58	06 37	07 26	08 12	09 18	09 27	09 38	09 55
S 60	06 42	07 34	08 24	09 12	09 17	09 24	09 36

Lat.	Sunset	Twilight Civil	Twilight Naut.	Moonset 27	28	29	30
°	h m	h m	h m	h m	h m	h m	h m
N 72	□	□	□	19 41	18 01	■	■
N 70	23 19	////	////	20 05	19 19	■	■
68	22 14	////	////	20 24	19 58	18 49	■
66	21 40	////	////	20 39	20 25	20 03	■
64	21 15	22 53	////	20 52	20 46	20 39	20 29
62	20 56	22 11	////	21 03	21 03	21 06	21 13
60	20 40	21 43	////	21 12	21 18	21 27	21 42
N 58	20 27	21 22	23 00	21 20	21 30	21 44	22 04
56	20 15	21 05	22 22	21 27	21 40	21 58	22 23
54	20 05	20 51	21 56	21 34	21 50	22 10	22 38
52	19 56	20 38	21 36	21 40	21 58	22 21	22 52
50	19 48	20 28	21 20	21 45	22 06	22 31	23 03
45	19 31	20 06	20 49	21 56	22 22	22 52	23 28
N 40	19 18	19 48	20 26	22 06	22 35	23 09	23 47
35	19 06	19 34	20 08	22 14	22 47	23 23	24 04
30	18 56	19 22	19 54	22 21	22 57	23 35	24 18
20	18 39	19 02	19 30	22 34	23 14	23 57	24 42
N 10	18 24	18 46	19 12	22 45	23 29	24 15	00 15
0	18 10	18 32	18 57	22 55	23 43	24 32	00 32
S 10	17 56	18 18	18 44	23 06	23 58	24 50	00 50
20	17 42	18 05	18 32	23 17	24 13	00 13	01 08
30	17 25	17 51	18 20	23 30	24 30	00 30	01 30
35	17 16	17 43	18 13	23 37	24 41	00 41	01 43
40	17 05	17 34	18 07	23 46	24 52	00 52	01 57
45	16 52	17 24	18 00	23 56	25 06	01 06	02 15
S 50	16 37	17 12	17 52	24 08	00 08	01 23	02 36
52	16 29	17 07	17 48	24 13	00 13	01 32	02 47
54	16 21	17 01	17 45	24 20	00 20	01 41	02 58
56	16 12	16 55	17 41	24 27	00 27	01 51	03 12
58	16 02	16 48	17 36	24 34	00 34	02 02	03 28
S 60	15 50	16 40	17 32	24 43	00 43	02 16	03 46

Day	SUN Eqn. of Time 00^h	SUN Eqn. of Time 12^h	SUN Mer. Pass.	MOON Mer. Pass. Upper	MOON Mer. Pass. Lower	Age	Phase
d	m s	m s	h m	h m	h m	d %	
27	06 32	06 32	12 07	16 44	04 20	05 35	
28	06 31	06 30	12 07	17 32	05 08	06 46	
29	06 30	06 29	12 06	18 21	05 56	07 56	

Day	UT d h	ARIES GHA ° ′	VENUS −4.0 GHA ° ′	VENUS Dec ° ′	MARS +1.1 GHA ° ′	MARS Dec ° ′	JUPITER −2.8 GHA ° ′	JUPITER Dec ° ′	SATURN +1.1 GHA ° ′	SATURN Dec ° ′
	30 00	307 46.0	220 18.7	N21 41.0	236 48.9	N21 57.8	341 01.9	S14 28.9	136 57.5	N 6 07.6
	01	322 48.4	235 18.1	41.2	251 49.5	58.1	356 04.7	29.0	151 59.7	07.5
	02	337 50.9	250 17.5	41.4	266 50.2	58.3	11 07.5	29.1	167 02.0	07.4
	03	352 53.4	265 16.9	. . 41.6	281 50.9	. . 58.6	26 10.2	. . 29.2	182 04.2	. . 07.3
	04	7 55.8	280 16.2	41.7	296 51.5	58.8	41 13.0	29.3	197 06.4	07.2
	05	22 58.3	295 15.6	41.9	311 52.2	59.1	56 15.7	29.4	212 08.7	07.1
	06	38 00.8	310 15.0	N21 42.1	326 52.8	N21 59.3	71 18.5	S14 29.5	227 10.9	N 6 07.0
	07	53 03.2	325 14.4	42.2	341 53.5	59.6	86 21.2	29.6	242 13.1	06.9
T	08	68 05.7	340 13.8	42.4	356 54.1	21 59.8	101 24.0	29.8	257 15.3	06.8
H	09	83 08.2	355 13.2	. . 42.6	11 54.8	22 00.1	116 26.7	. . 29.9	272 17.6	. . 06.7
U	10	98 10.6	10 12.5	42.7	26 55.4	00.3	131 29.5	30.0	287 19.8	06.6
R	11	113 13.1	25 11.9	42.9	41 56.1	00.6	146 32.2	30.1	302 22.0	06.5
S	12	128 15.5	40 11.3	N21 43.1	56 56.7	N22 00.8	161 35.0	S14 30.2	317 24.2	N 6 06.4
D	13	143 18.0	55 10.7	43.2	71 57.4	01.1	176 37.7	30.3	332 26.5	06.3
A	14	158 20.5	70 10.1	43.4	86 58.0	01.3	191 40.5	30.4	347 28.7	06.1
Y	15	173 22.9	85 09.4	. . 43.5	101 58.7	. . 01.6	206 43.2	. . 30.5	2 30.9	. . 06.0
	16	188 25.4	100 08.8	43.7	116 59.3	01.8	221 46.0	30.6	17 33.1	05.9
	17	203 27.9	115 08.2	43.9	132 00.0	02.1	236 48.7	30.7	32 35.4	05.8
	18	218 30.3	130 07.6	N21 44.0	147 00.6	N22 02.3	251 51.5	S14 30.8	47 37.6	N 6 05.7
	19	233 32.8	145 06.9	44.2	162 01.3	02.6	266 54.3	30.9	62 39.8	05.6
	20	248 35.3	160 06.3	44.3	177 01.9	02.8	281 57.0	31.0	77 42.1	05.5
	21	263 37.7	175 05.7	. . 44.5	192 02.6	. . 03.1	296 59.8	. . 31.1	92 44.3	. . 05.4
	22	278 40.2	190 05.1	44.6	207 03.2	03.3	312 02.5	31.2	107 46.5	05.3
	23	293 42.7	205 04.5	44.8	222 03.9	03.6	327 05.3	31.4	122 48.7	05.2
	31 00	308 45.1	220 03.8	N21 44.9	237 04.6	N22 03.8	342 08.0	S14 31.5	137 51.0	N 6 05.1
	01	323 47.6	235 03.2	45.1	252 05.2	04.1	357 10.8	31.6	152 53.2	05.0
	02	338 50.0	250 02.6	45.2	267 05.9	04.3	12 13.5	31.7	167 55.4	04.9
	03	353 52.5	265 02.0	. . 45.4	282 06.5	. . 04.6	27 16.3	. . 31.8	182 57.6	. . 04.8
	04	8 55.0	280 01.3	45.5	297 07.2	04.8	42 19.0	31.9	197 59.9	04.7
	05	23 57.4	295 00.7	45.7	312 07.8	05.1	57 21.8	32.0	213 02.1	04.6
	06	38 59.9	310 00.1	N21 45.8	327 08.5	N22 05.3	72 24.6	S14 32.1	228 04.3	N 6 04.5
	07	54 02.4	324 59.4	46.0	342 09.1	05.5	87 27.3	32.2	243 06.5	04.4
	08	69 04.8	339 58.8	46.1	357 09.8	05.8	102 30.1	32.3	258 08.8	04.3
F	09	84 07.3	354 58.2	. . 46.2	12 10.4	. . 06.0	117 32.8	. . 32.4	273 11.0	. . 04.1
R	10	99 09.8	9 57.6	46.4	27 11.1	06.3	132 35.6	32.5	288 13.2	04.0
I	11	114 12.2	24 56.9	46.5	42 11.7	06.5	147 38.3	32.6	303 15.4	03.9
D	12	129 14.7	39 56.3	N21 46.7	57 12.4	N22 06.8	162 41.1	S14 32.7	318 17.7	N 6 03.8
A	13	144 17.2	54 55.7	46.8	72 13.0	07.0	177 43.9	32.9	333 19.9	03.7
Y	14	159 19.6	69 55.1	46.9	87 13.7	07.2	192 46.6	33.0	348 22.1	03.6
	15	174 22.1	84 54.4	. . 47.1	102 14.4	. . 07.5	207 49.4	. . 33.1	3 24.3	. . 03.5
	16	189 24.5	99 53.8	47.2	117 15.0	07.7	222 52.1	33.2	18 26.5	03.4
	17	204 27.0	114 53.2	47.3	132 15.7	08.0	237 54.9	33.3	33 28.8	03.3
	18	219 29.5	129 52.5	N21 47.5	147 16.3	N22 08.2	252 57.6	S14 33.4	48 31.0	N 6 03.2
	19	234 31.9	144 51.9	47.6	162 17.0	08.5	268 00.4	33.5	63 33.2	03.1
	20	249 34.4	159 51.3	47.7	177 17.6	08.7	283 03.2	33.6	78 35.4	03.0
	21	264 36.9	174 50.7	. . 47.9	192 18.3	. . 08.9	298 05.9	. . 33.7	93 37.7	. . 02.9
	22	279 39.3	189 50.0	48.0	207 18.9	09.2	313 08.7	33.8	108 39.9	02.8
	23	294 41.8	204 49.4	48.1	222 19.6	09.4	328 11.4	33.9	123 42.1	02.7
	1 00	309 44.3	219 48.8	N21 48.3	237 20.2	N22 09.6	343 14.2	S14 34.0	138 44.3	N 6 02.6
	01	324 46.7	234 48.1	48.4	252 20.9	09.9	358 16.9	34.1	153 46.6	02.5
	02	339 49.2	249 47.5	48.5	267 21.5	10.1	13 19.7	34.3	168 48.8	02.3
	03	354 51.7	264 46.9	. . 48.6	282 22.2	. . 10.4	28 22.5	. . 34.4	183 51.0	. . 02.2
	04	9 54.1	279 46.2	48.8	297 22.9	10.6	43 25.2	34.5	198 53.2	02.1
	05	24 56.6	294 45.6	48.9	312 23.5	10.8	58 28.0	34.6	213 55.5	02.0
	06	39 59.0	309 45.0	N21 49.0	327 24.2	N22 11.1	73 30.7	S14 34.7	228 57.7	N 6 01.9
	07	55 01.5	324 44.3	49.1	342 24.8	11.3	88 33.5	34.8	243 59.9	01.8
S	08	70 04.0	339 43.7	49.3	357 25.5	11.5	103 36.3	34.9	259 02.1	01.7
A	09	85 06.4	354 43.1	. . 49.4	12 26.1	. . 11.8	118 39.0	. . 35.0	274 04.3	. . 01.6
T	10	100 08.9	9 42.4	49.5	27 26.8	12.0	133 41.8	35.1	289 06.6	01.5
U	11	115 11.4	24 41.8	49.6	42 27.4	12.3	148 44.5	35.2	304 08.8	01.4
R	12	130 13.8	39 41.2	N21 49.7	57 28.1	N22 12.5	163 47.3	S14 35.3	319 11.0	N 6 01.3
D	13	145 16.3	54 40.5	49.8	72 28.8	12.7	178 50.1	35.4	334 13.2	01.2
A	14	160 18.8	69 39.9	50.0	87 29.4	13.0	193 52.8	35.6	349 15.5	01.1
Y	15	175 21.2	84 39.3	. . 50.1	102 30.1	. . 13.2	208 55.6	. . 35.7	4 17.7	. . 01.0
	16	190 23.7	99 38.6	50.2	117 30.7	13.4	223 58.3	35.8	19 19.9	00.9
	17	205 26.1	114 38.0	50.3	132 31.4	13.7	239 01.1	35.9	34 22.1	00.8
	18	220 28.6	129 37.3	N21 50.4	147 32.0	N22 13.9	254 03.9	S14 36.0	49 24.3	N 6 00.6
	19	235 31.1	144 36.7	50.5	162 32.7	14.1	269 06.6	36.1	64 26.6	00.5
	20	250 33.5	159 36.1	50.6	177 33.3	14.3	284 09.4	36.2	79 28.8	00.4
	21	265 36.0	174 35.4	. . 50.7	192 34.0	. . 14.6	299 12.1	. . 36.3	94 31.0	. . 00.3
	22	280 38.5	189 34.8	50.8	207 34.6	14.8	314 14.9	36.4	109 33.2	00.2
	23	295 40.9	204 34.2	50.9	222 35.3	15.0	329 17.7	36.5	124 35.5	00.1
	Mer. Pass.	h m 3 24.4	v −0.6	d 0.1	v 0.7	d 0.2	v 2.8	d 0.1	v 2.2	d 0.1

STARS

Name	SHA ° ′	Dec ° ′
Acamar	315 20.5	S40 15.
Achernar	335 28.6	S57 10.
Acrux	173 13.1	S63 09.
Adhara	255 15.1	S28 59.
Aldebaran	290 52.8	N16 31.
Alioth	166 23.2	N55 54.
Alkaid	153 01.1	N49 16.
Al Na'ir	27 46.7	S46 54.
Alnilam	275 49.5	S 1 11.
Alphard	217 59.2	S 8 42.
Alphecca	126 13.3	N26 41.
Alpheratz	357 46.3	N29 08.
Altair	62 10.7	N 8 53.
Ankaa	353 18.1	S42 14.
Antares	112 29.6	S26 27.
Arcturus	145 58.3	N19 08.
Atria	107 33.7	S69 03.
Avior	234 20.0	S59 32.
Bellatrix	278 35.3	N 6 21.
Betelgeuse	271 04.6	N 7 24.
Canopus	263 57.9	S52 41.
Capella	280 39.0	N46 00.
Deneb	49 33.0	N45 18.
Denebola	182 36.7	N14 31.
Diphda	348 58.5	S17 55.
Dubhe	193 55.4	N61 42.
Elnath	278 16.5	N28 36.
Eltanin	90 47.0	N51 29.
Enif	33 49.6	N 9 55.
Fomalhaut	15 26.7	S29 34.
Gacrux	172 04.6	S57 10.
Gienah	175 55.5	S17 35.
Hadar	148 52.2	S60 25.
Hamal	328 04.0	N23 30.
Kaus Aust.	83 47.2	S34 22.
Kochab	137 19.4	N74 07.
Markab	13 40.9	N15 15.
Menkar	314 18.1	N 4 07.
Menkent	148 11.1	S36 25.
Miaplacidus	221 41.5	S69 45.
Mirfak	308 44.7	N49 53.
Nunki	76 01.5	S26 17.
Peacock	53 23.0	S56 42.
Pollux	243 31.5	N28 00.
Procyon	245 03.0	N 5 12.
Rasalhague	96 08.8	N12 33.
Regulus	207 46.8	N11 55.
Rigel	281 15.0	S 8 11.
Rigil Kent.	139 55.8	S60 52.
Sabik	102 15.6	S15 44.
Schedar	349 43.7	N56 35.
Shaula	96 25.5	S37 06.
Sirius	258 36.5	S16 43.
Spica	158 34.4	S11 12.
Suhail	222 55.1	S43 28.
Vega	80 40.5	N38 47.
Zuben'ubi	137 08.6	S16 05.

	SHA ° ′	Mer. Pass h m
Venus	271 18.7	9 20
Mars	288 19.4	8 11
Jupiter	33 22.9	1 11
Saturn	189 05.8	14 46

UT	SUN GHA	SUN Dec	MOON GHA	*v*	MOON Dec	*d*	HP
d h	° ′	° ′	° ′	′	° ′	′	′
0 00	178 23.1	N18 31.6	82 02.6	10.9	S22 26.7	7.6	55.5
01	193 23.1	31.0	96 32.5	10.9	22 34.3	7.6	55.5
02	208 23.2	30.4	111 02.4	10.9	22 41.9	7.4	55.5
03	223 23.2	. . 29.8	125 32.3	10.8	22 49.3	7.4	55.5
04	238 23.2	29.2	140 02.1	10.8	22 56.7	7.2	55.4
05	253 23.2	28.6	154 31.9	10.8	23 03.9	7.1	55.4
06	268 23.3	N18 28.0	169 01.7	10.8	S23 11.0	6.9	55.4
07	283 23.3	27.3	183 31.5	10.7	23 17.9	6.9	55.4
08	298 23.3	26.7	198 01.2	10.7	23 24.8	6.7	55.3
09	313 23.3	. . 26.1	212 30.9	10.7	23 31.5	6.6	55.3
10	328 23.4	25.5	227 00.6	10.6	23 38.1	6.5	55.3
11	343 23.4	24.9	241 30.2	10.6	23 44.6	6.4	55.3
12	358 23.4	N18 24.3	255 59.8	10.6	S23 51.0	6.2	55.2
13	13 23.4	23.7	270 29.4	10.6	23 57.2	6.2	55.2
14	28 23.5	23.1	284 59.0	10.5	24 03.4	6.0	55.2
15	43 23.5	. . 22.5	299 28.5	10.5	24 09.4	5.8	55.2
16	58 23.5	21.9	313 58.0	10.5	24 15.2	5.8	55.1
17	73 23.6	21.2	328 27.5	10.5	24 21.0	5.6	55.1
18	88 23.6	N18 20.6	342 57.0	10.4	S24 26.6	5.5	55.1
19	103 23.6	20.0	357 26.4	10.5	24 32.1	5.4	55.1
20	118 23.7	19.4	11 55.9	10.3	24 37.5	5.3	55.0
21	133 23.7	. . 18.8	26 25.2	10.4	24 42.8	5.1	55.0
22	148 23.7	18.2	40 54.6	10.4	24 47.9	5.0	55.0
23	163 23.7	17.6	55 24.0	10.3	24 52.9	4.9	55.0
1 00	178 23.8	N18 17.0	69 53.3	10.3	S24 57.8	4.8	55.0
01	193 23.8	16.3	84 22.6	10.3	25 02.6	4.6	54.9
02	208 23.8	15.7	98 51.9	10.3	25 07.2	4.5	54.9
03	223 23.9	. . 15.1	113 21.2	10.3	25 11.7	4.4	54.9
04	238 23.9	14.5	127 50.5	10.2	25 16.1	4.3	54.9
05	253 23.9	13.9	142 19.7	10.2	25 20.4	4.1	54.9
06	268 24.0	N18 13.2	156 48.9	10.2	S25 24.5	4.0	54.8
07	283 24.0	12.6	171 18.1	10.2	25 28.5	3.9	54.8
08	298 24.0	12.0	185 47.3	10.2	25 32.4	3.7	54.8
09	313 24.1	. . 11.4	200 16.5	10.2	25 36.1	3.7	54.8
10	328 24.1	10.8	214 45.7	10.1	25 39.8	3.5	54.8
11	343 24.1	10.1	229 14.8	10.2	25 43.3	3.3	54.7
12	358 24.2	N18 09.5	243 44.0	10.1	S25 46.6	3.3	54.7
13	13 24.2	08.9	258 13.1	10.1	25 49.9	3.1	54.7
14	28 24.2	08.3	272 42.2	10.1	25 53.0	2.9	54.7
15	43 24.3	. . 07.7	287 11.3	10.1	25 55.9	2.9	54.7
16	58 24.3	07.0	301 40.4	10.1	25 58.8	2.7	54.7
17	73 24.3	06.4	316 09.5	10.1	26 01.5	2.6	54.6
18	88 24.4	N18 05.8	330 38.6	10.1	S26 04.1	2.5	54.6
19	103 24.4	05.2	345 07.7	10.1	26 06.6	2.3	54.6
20	118 24.5	04.5	359 36.8	10.0	26 08.9	2.2	54.6
21	133 24.5	. . 03.9	14 05.8	10.1	26 11.1	2.1	54.6
22	148 24.5	03.3	28 34.9	10.1	26 13.2	1.9	54.6
23	163 24.6	02.7	43 04.0	10.0	26 15.1	1.8	54.5
1 00	178 24.6	N18 02.0	57 33.0	10.1	S26 16.9	1.7	54.5
01	193 24.6	01.4	72 02.1	10.0	26 18.6	1.6	54.5
02	208 24.7	00.8	86 31.1	10.1	26 20.2	1.4	54.5
03	223 24.7	18 00.1	101 00.2	10.1	26 21.6	1.3	54.5
04	238 24.8	17 59.5	115 29.3	10.0	26 22.9	1.1	54.5
05	253 24.8	58.9	129 58.3	10.1	26 24.0	1.1	54.5
06	268 24.8	N17 58.2	144 27.4	10.1	S26 25.1	0.9	54.4
07	283 24.9	57.6	158 56.5	10.1	26 26.0	0.8	54.4
08	298 24.9	57.0	173 25.6	10.0	26 26.8	0.6	54.4
09	313 25.0	. . 56.4	187 54.6	10.1	26 27.4	0.5	54.4
10	328 25.0	55.7	202 23.7	10.1	26 27.9	0.4	54.4
11	343 25.0	55.1	216 52.8	10.2	26 28.3	0.3	54.4
12	358 25.1	N17 54.5	231 22.0	10.1	S26 28.6	0.1	54.4
13	13 25.1	53.8	245 51.1	10.1	26 28.7	0.0	54.3
14	28 25.2	53.2	260 20.2	10.2	26 28.7	0.1	54.3
15	43 25.2	. . 52.5	274 49.4	10.1	26 28.6	0.3	54.3
16	58 25.2	51.9	289 18.5	10.2	26 28.3	0.4	54.3
17	73 25.3	51.3	303 47.7	10.2	26 27.9	0.5	54.3
18	88 25.3	N17 50.6	318 16.9	10.2	S26 27.4	0.6	54.3
19	103 25.4	50.0	332 46.1	10.2	26 26.8	0.8	54.3
20	118 25.4	49.4	347 15.3	10.3	26 26.0	0.9	54.3
21	133 25.5	. . 48.7	1 44.6	10.2	26 25.1	1.0	54.3
22	148 25.5	48.1	16 13.8	10.3	26 24.1	1.2	54.2
23	163 25.5	47.4	30 43.1	10.3	S26 22.9	1.2	54.2
	SD 15.8	*d* 0.6	SD 15.0		14.9		14.8

Lat.	Twilight Naut.	Twilight Civil	Sunrise	Moonrise 30	Moonrise 31	Moonrise 1	Moonrise 2
°	h m	h m	h m	h m	h m	h m	h m
N 72	▭	▭	▭	▬	▬	▬	▬
N 70	////	////	01 18	▬	▬	▬	▬
68	////	////	02 10	▬	▬	▬	▬
66	////	////	02 42	▬	▬	▬	▬
64	////	01 34	03 05	17 45	▬	▬	20 55
62	////	02 11	03 23	17 01	18 29	19 30	19 57
60	////	02 37	03 38	16 33	17 50	18 48	19 23
N 58	01 25	02 57	03 51	16 10	17 23	18 20	18 59
56	01 59	03 13	04 02	15 52	17 02	17 58	18 39
54	02 23	03 27	04 12	15 37	16 45	17 40	18 22
52	02 42	03 39	04 20	15 24	16 30	17 25	18 08
50	02 58	03 49	04 28	15 13	16 17	17 12	17 56
45	03 27	04 10	04 44	14 49	15 50	16 44	17 30
N 40	03 50	04 27	04 57	14 30	15 29	16 23	17 10
35	04 07	04 41	05 09	14 14	15 12	16 05	16 53
30	04 21	04 52	05 18	14 00	14 57	15 50	16 38
20	04 44	05 12	05 35	13 37	14 31	15 24	16 13
N 10	05 01	05 27	05 50	13 17	14 10	15 01	15 51
0	05 16	05 41	06 03	12 58	13 49	14 40	15 31
S 10	05 29	05 54	06 16	12 40	13 29	14 20	15 11
20	05 40	06 07	06 30	12 20	13 07	13 57	14 49
30	05 52	06 21	06 46	11 57	12 42	13 31	14 24
35	05 58	06 28	06 55	11 44	12 27	13 16	14 09
40	06 04	06 37	07 06	11 29	12 10	12 58	13 52
45	06 11	06 46	07 18	11 11	11 50	12 37	13 31
S 50	06 18	06 57	07 33	10 48	11 24	12 10	13 05
52	06 21	07 02	07 40	10 37	11 12	11 57	12 52
54	06 25	07 08	07 47	10 25	10 58	11 42	12 38
56	06 29	07 14	07 56	10 11	10 41	11 24	12 21
58	06 32	07 21	08 06	09 55	10 22	11 03	12 00
S 60	06 37	07 28	08 17	09 36	09 57	10 35	11 34

Lat.	Sunset	Twilight Civil	Twilight Naut.	Moonset 30	Moonset 31	Moonset 1	Moonset 2
°	h m	h m	h m	h m	h m	h m	h m
N 72	▭	▭	▭	▬	▬	▬	▬
N 70	22 47	////	////	▬	▬	▬	▬
68	21 58	////	////	▬	▬	▬	▬
66	21 28	////	////	▬	▬	▬	▬
64	21 05	22 33	////	20 29	▬	▬	22 37
62	20 47	21 58	////	21 13	21 31	22 17	23 35
60	20 33	21 33	////	21 42	22 10	22 59	24 08
N 58	20 20	21 14	22 43	22 04	22 37	23 27	24 32
56	20 09	20 58	22 10	22 23	22 59	23 49	24 52
54	20 00	20 44	21 47	22 38	23 16	24 06	00 06
52	19 51	20 33	21 29	22 52	23 31	24 22	00 22
50	19 44	20 23	21 14	23 03	23 44	24 35	00 35
45	19 28	20 02	20 44	23 28	24 11	00 11	01 02
N 40	19 15	19 45	20 22	23 47	24 32	00 32	01 23
35	19 04	19 31	20 05	24 04	00 04	00 50	01 41
30	18 54	19 20	19 51	24 18	00 18	01 05	01 56
20	18 37	19 01	19 29	24 42	00 42	01 31	02 22
N 10	18 23	18 45	19 11	00 15	01 03	01 53	02 44
0	18 10	18 32	18 57	00 32	01 23	02 14	03 05
S 10	17 57	18 19	18 44	00 50	01 42	02 35	03 26
20	17 43	18 06	18 33	01 08	02 03	02 57	03 48
30	17 27	17 52	18 21	01 30	02 28	03 23	04 13
35	17 18	17 45	18 15	01 43	02 42	03 38	04 29
40	17 07	17 36	18 09	01 57	02 59	03 56	04 46
45	16 55	17 27	18 02	02 15	03 19	04 17	05 07
S 50	16 41	17 16	17 55	02 36	03 44	04 44	05 33
52	16 34	17 11	17 52	02 47	03 56	04 57	05 46
54	16 26	17 05	17 49	02 58	04 10	05 12	06 01
56	16 17	16 59	17 45	03 12	04 26	05 30	06 18
58	16 08	16 53	17 41	03 28	04 46	05 51	06 39
S 60	15 57	16 45	17 37	03 46	05 10	06 19	07 05

Day	SUN Eqn. of Time 00^h	SUN Eqn. of Time 12^h	SUN Mer. Pass.	MOON Mer. Pass. Upper	MOON Mer. Pass. Lower	MOON Age	MOON Phase
d	m s	m s	h m	h m	h m	d %	
30	06 28	06 26	12 06	19 11	06 45	08 66	
31	06 25	06 23	12 06	20 02	07 36	09 75	◐
1	06 22	06 20	12 06	20 53	08 27	10 83	

UT		ARIES	VENUS −4.0		MARS +1.1		JUPITER −2.8		SATURN +1.1	
		GHA	GHA	Dec	GHA	Dec	GHA	Dec	GHA	Dec
d	h	° ′	° ′	° ′	° ′	° ′	° ′	° ′	° ′	° ′
2	00	310 43.4	219 33.5	N21 51.0	237 36.0	N22 15.3	344 20.4	S14 36.6	139 37.7	N 6 00.0
	01	325 45.9	234 32.9	51.1	252 36.6	15.5	359 23.2	36.7	154 39.9	5 59.9
	02	340 48.3	249 32.2	51.2	267 37.3	15.7	14 26.0	36.9	169 42.1	59.8
	03	355 50.8	264 31.6	. . 51.4	282 37.9	. . 16.0	29 28.7	. . 37.0	184 44.3	. . 59.7
	04	10 53.3	279 31.0	51.5	297 38.6	16.2	44 31.5	37.1	199 46.6	59.6
	05	25 55.7	294 30.3	51.6	312 39.2	16.4	59 34.2	37.2	214 48.8	59.5
	06	40 58.2	309 29.7	N21 51.6	327 39.9	N22 16.6	74 37.0	S14 37.3	229 51.0	N 5 59.4
	07	56 00.6	324 29.0	51.7	342 40.6	16.9	89 39.8	37.4	244 53.2	59.3
	08	71 03.1	339 28.4	51.8	357 41.2	17.1	104 42.5	37.5	259 55.4	59.2
S	09	86 05.6	354 27.8	. . 51.9	12 41.9	. . 17.3	119 45.3	. . 37.6	274 57.7	. . 59.0
U	10	101 08.0	9 27.1	52.0	27 42.5	17.6	134 48.1	37.7	289 59.9	58.9
N	11	116 10.5	24 26.5	52.1	42 43.2	17.8	149 50.8	37.8	305 02.1	58.8
D	12	131 13.0	39 25.8	N21 52.2	57 43.8	N22 18.0	164 53.6	S14 37.9	320 04.3	N 5 58.7
A	13	146 15.4	54 25.2	52.3	72 44.5	18.2	179 56.3	38.1	335 06.5	58.6
Y	14	161 17.9	69 24.6	52.4	87 45.2	18.5	194 59.1	38.2	350 08.8	58.5
	15	176 20.4	84 23.9	. . 52.5	102 45.8	. . 18.7	210 01.9	. . 38.3	5 11.0	. . 58.4
	16	191 22.8	99 23.3	52.6	117 46.5	18.9	225 04.6	38.4	20 13.2	58.3
	17	206 25.3	114 22.6	52.7	132 47.1	19.1	240 07.4	38.5	35 15.4	58.2
	18	221 27.8	129 22.0	N21 52.8	147 47.8	N22 19.4	255 10.2	S14 38.6	50 17.7	N 5 58.1
	19	236 30.2	144 21.3	52.8	162 48.4	19.6	270 12.9	38.7	65 19.9	58.0
	20	251 32.7	159 20.7	52.9	177 49.1	19.8	285 15.7	38.8	80 22.1	57.9
	21	266 35.1	174 20.0	. . 53.0	192 49.8	. . 20.0	300 18.4	. . 38.9	95 24.3	. . 57.8
	22	281 37.6	189 19.4	53.1	207 50.4	20.3	315 21.2	39.0	110 26.5	57.7
	23	296 40.1	204 18.8	53.2	222 51.1	20.5	330 24.0	39.1	125 28.8	57.5
3	00	311 42.5	219 18.1	N21 53.3	237 51.7	N22 20.7	345 26.7	S14 39.3	140 31.0	N 5 57.4
	01	326 45.0	234 17.5	53.3	252 52.4	20.9	0 29.5	39.4	155 33.2	57.3
	02	341 47.5	249 16.8	53.4	267 53.0	21.1	15 32.3	39.5	170 35.4	57.2
	03	356 49.9	264 16.2	. . 53.5	282 53.7	. . 21.4	30 35.0	. . 39.6	185 37.6	. . 57.1
	04	11 52.4	279 15.5	53.6	297 54.4	21.6	45 37.8	39.7	200 39.8	57.0
	05	26 54.9	294 14.9	53.6	312 55.0	21.8	60 40.6	39.8	215 42.1	56.9
	06	41 57.3	309 14.2	N21 53.7	327 55.7	N22 22.0	75 43.3	S14 39.9	230 44.3	N 5 56.8
	07	56 59.8	324 13.6	53.8	342 56.3	22.2	90 46.1	40.0	245 46.5	56.7
	08	72 02.3	339 12.9	53.9	357 57.0	22.5	105 48.9	40.1	260 48.7	56.6
M	09	87 04.7	354 12.3	. . 53.9	12 57.6	. . 22.7	120 51.6	. . 40.2	275 50.9	. . 56.5
O	10	102 07.2	9 11.7	54.0	27 58.3	22.9	135 54.4	40.3	290 53.2	56.4
N	11	117 09.6	24 11.0	54.1	42 59.0	23.1	150 57.2	40.5	305 55.4	56.3
D	12	132 12.1	39 10.4	N21 54.2	57 59.6	N22 23.3	165 59.9	S14 40.6	320 57.6	N 5 56.1
A	13	147 14.6	54 09.7	54.2	73 00.3	23.6	181 02.7	40.7	335 59.8	56.0
Y	14	162 17.0	69 09.1	54.3	88 00.9	23.8	196 05.5	40.8	351 02.0	55.9
	15	177 19.5	84 08.4	. . 54.4	103 01.6	. . 24.0	211 08.2	. . 40.9	6 04.3	. . 55.8
	16	192 22.0	99 07.8	54.4	118 02.3	24.2	226 11.0	41.0	21 06.5	55.7
	17	207 24.4	114 07.1	54.5	133 02.9	24.4	241 13.8	41.1	36 08.7	55.6
	18	222 26.9	129 06.5	N21 54.5	148 03.6	N22 24.6	256 16.5	S14 41.2	51 10.9	N 5 55.5
	19	237 29.4	144 05.8	54.6	163 04.2	24.9	271 19.3	41.3	66 13.1	55.4
	20	252 31.8	159 05.2	54.7	178 04.9	25.1	286 22.1	41.4	81 15.3	55.3
	21	267 34.3	174 04.5	. . 54.7	193 05.5	. . 25.3	301 24.8	. . 41.5	96 17.6	. . 55.2
	22	282 36.8	189 03.9	54.8	208 06.2	25.5	316 27.6	41.7	111 19.8	55.1
	23	297 39.2	204 03.2	54.9	223 06.9	25.7	331 30.4	41.8	126 22.0	55.0
4	00	312 41.7	219 02.6	N21 54.9	238 07.5	N22 25.9	346 33.1	S14 41.9	141 24.2	N 5 54.8
	01	327 44.1	234 01.9	55.0	253 08.2	26.1	1 35.9	42.0	156 26.4	54.7
	02	342 46.6	249 01.3	55.0	268 08.8	26.4	16 38.7	42.1	171 28.7	54.6
	03	357 49.1	264 00.6	. . 55.1	283 09.5	. . 26.6	31 41.4	. . 42.2	186 30.9	. . 54.5
	04	12 51.5	279 00.0	55.1	298 10.2	26.8	46 44.2	42.3	201 33.1	54.4
	05	27 54.0	293 59.3	55.2	313 10.8	27.0	61 47.0	42.4	216 35.3	54.3
	06	42 56.5	308 58.7	N21 55.2	328 11.5	N22 27.2	76 49.7	S14 42.5	231 37.5	N 5 54.2
	07	57 58.9	323 58.0	55.3	343 12.1	27.4	91 52.5	42.6	246 39.7	54.1
T	08	73 01.4	338 57.4	55.3	358 12.8	27.6	106 55.3	42.8	261 42.0	54.0
U	09	88 03.9	353 56.7	. . 55.4	13 13.5	. . 27.8	121 58.0	. . 42.9	276 44.2	. . 53.9
E	10	103 06.3	8 56.0	55.4	28 14.1	28.0	137 00.8	43.0	291 46.4	53.8
S	11	118 08.8	23 55.4	55.5	43 14.8	28.3	152 03.6	43.1	306 48.6	53.7
D	12	133 11.2	38 54.7	N21 55.5	58 15.4	N22 28.5	167 06.3	S14 43.2	321 50.8	N 5 53.5
A	13	148 13.7	53 54.1	55.6	73 16.1	28.7	182 09.1	43.3	336 53.0	53.4
Y	14	163 16.2	68 53.4	55.6	88 16.8	28.9	197 11.9	43.4	351 55.3	53.3
	15	178 18.6	83 52.8	. . 55.6	103 17.4	. . 29.1	212 14.6	. . 43.5	6 57.5	. . 53.2
	16	193 21.1	98 52.1	55.7	118 18.1	29.3	227 17.4	43.6	21 59.7	53.1
	17	208 23.6	113 51.5	55.7	133 18.7	29.5	242 20.2	43.7	37 01.9	53.0
	18	223 26.0	128 50.8	N21 55.8	148 19.4	N22 29.7	257 23.0	S14 43.9	52 04.1	N 5 52.9
	19	238 28.5	143 50.2	55.8	163 20.1	29.9	272 25.7	44.0	67 06.3	52.8
	20	253 31.0	158 49.5	55.8	178 20.7	30.1	287 28.5	44.1	82 08.6	52.7
	21	268 33.4	173 48.8	. . 55.9	193 21.4	. . 30.3	302 31.3	. . 44.2	97 10.8	. . 52.6
	22	283 35.9	188 48.2	55.9	208 22.0	30.5	317 34.0	44.3	112 13.0	52.5
	23	298 38.4	203 47.5	56.0	223 22.7	30.7	332 36.8	44.4	127 15.2	52.4
Mer. Pass.		h m 3 12.6	v −0.6	d 0.1	v 0.7	d 0.2	v 2.8	d 0.1	v 2.2	d 0.1

STARS

Name	SHA	Dec
	° ′	° ′
Acamar	315 20.4	S40 15.
Achernar	335 28.5	S57 10.
Acrux	173 13.1	S63 09.
Adhara	255 15.1	S28 59.
Aldebaran	290 52.8	N16 31.
Alioth	166 23.2	N55 54.
Alkaid	153 01.1	N49 16.
Al Na'ir	27 46.7	S46 54.
Alnilam	275 49.5	S 1 11.
Alphard	217 59.2	S 8 42.
Alphecca	126 13.3	N26 41.
Alpheratz	357 46.3	N29 08.
Altair	62 10.7	N 8 53.
Ankaa	353 18.1	S42 14.
Antares	112 29.6	S26 27.
Arcturus	145 58.3	N19 08.
Atria	107 33.7	S69 03.
Avior	234 20.0	S59 32.
Bellatrix	278 35.3	N 6 21.
Betelgeuse	271 04.6	N 7 24.
Canopus	263 57.9	S52 41.
Capella	280 39.0	N46 00.
Deneb	49 33.0	N45 19.
Denebola	182 36.7	N14 31.
Diphda	348 58.5	S17 55.
Dubhe	193 55.4	N61 42.
Elnath	278 16.5	N28 36.
Eltanin	90 47.1	N51 29.
Enif	33 49.6	N 9 55.
Fomalhaut	15 26.7	S29 34.
Gacrux	172 04.6	S57 10.
Gienah	175 55.5	S17 35.
Hadar	148 52.2	S60 25.
Hamal	328 04.0	N23 30.
Kaus Aust.	83 47.2	S34 22.
Kochab	137 19.5	N74 07.
Markab	13 40.9	N15 15.
Menkar	314 18.1	N 4 07.
Menkent	148 11.1	S36 25.
Miaplacidus	221 41.5	S69 45.
Mirfak	308 44.6	N49 53.
Nunki	76 01.5	S26 17.
Peacock	53 23.0	S56 42.
Pollux	243 31.5	N28 00.
Procyon	245 03.0	N 5 12.
Rasalhague	96 08.8	N12 33.
Regulus	207 46.8	N11 55.
Rigel	281 15.0	S 8 11.
Rigil Kent.	139 55.8	S60 52.
Sabik	102 15.6	S15 44.
Schedar	349 43.7	N56 35.
Shaula	96 25.5	S37 06.
Sirius	258 36.5	S16 43.
Spica	158 34.4	S11 12.
Suhail	222 55.1	S43 28.
Vega	80 40.5	N38 47.
Zuben'ubi	137 08.6	S16 05.

	SHA	Mer. Pass
	° ′	h m
Venus	267 35.6	9 23
Mars	286 09.2	8 08
Jupiter	33 44.2	0 58
Saturn	188 48.4	14 36

JT	SUN		MOON				
	GHA	Dec	GHA	*v*	Dec	*d*	HP
d h	° ′	° ′	° ′	′	° ′	′	′
2 00	178 25.6	N17 46.8	45 12.4	10.3	S26 21.7	1.4	54.2
01	193 25.6	46.2	59 41.7	10.4	26 20.3	1.6	54.2
02	208 25.7	45.5	74 11.1	10.4	26 18.7	1.6	54.2
03	223 25.7	. . 44.9	88 40.5	10.4	26 17.1	1.8	54.2
04	238 25.8	44.2	103 09.9	10.4	26 15.3	1.9	54.2
05	253 25.8	43.6	117 39.3	10.4	26 13.4	2.1	54.2
06	268 25.9	N17 43.0	132 08.7	10.5	S26 11.3	2.1	54.2
07	283 25.9	42.3	146 38.2	10.5	26 09.2	2.3	54.2
08	298 26.0	41.7	161 07.7	10.5	26 06.9	2.4	54.2
09	313 26.0	. . 41.0	175 37.2	10.6	26 04.5	2.5	54.2
10	328 26.0	40.4	190 06.8	10.6	26 02.0	2.7	54.1
11	343 26.1	39.7	204 36.4	10.6	25 59.3	2.8	54.1
12	358 26.1	N17 39.1	219 06.0	10.6	S25 56.5	2.9	54.1
13	13 26.2	38.4	233 35.6	10.7	25 53.6	3.0	54.1
14	28 26.2	37.8	248 05.3	10.8	25 50.6	3.2	54.1
15	43 26.3	. . 37.1	262 35.1	10.7	25 47.4	3.2	54.1
16	58 26.3	36.5	277 04.8	10.8	25 44.2	3.4	54.1
17	73 26.4	35.9	291 34.6	10.8	25 40.8	3.5	54.1
18	88 26.4	N17 35.2	306 04.4	10.9	S25 37.3	3.7	54.1
19	103 26.5	34.6	320 34.3	10.9	25 33.6	3.7	54.1
20	118 26.5	33.9	335 04.2	10.9	25 29.9	3.9	54.1
21	133 26.6	. . 33.3	349 34.1	11.0	25 26.0	4.0	54.1
22	148 26.6	32.6	4 04.1	11.0	25 22.0	4.1	54.1
23	163 26.7	32.0	18 34.1	11.1	25 17.9	4.2	54.1
3 00	178 26.7	N17 31.3	33 04.2	11.1	S25 13.7	4.3	54.1
01	193 26.8	30.6	47 34.3	11.1	25 09.4	4.5	54.1
02	208 26.8	30.0	62 04.4	11.2	25 04.9	4.6	54.1
03	223 26.9	. . 29.3	76 34.6	11.2	25 00.3	4.6	54.0
04	238 26.9	28.7	91 04.8	11.3	24 55.7	4.8	54.0
05	253 27.0	28.0	105 35.1	11.3	24 50.9	5.0	54.0
06	268 27.0	N17 27.4	120 05.4	11.4	S24 45.9	5.0	54.0
07	283 27.1	26.7	134 35.8	11.4	24 40.9	5.1	54.0
08	298 27.1	26.1	149 06.2	11.4	24 35.8	5.3	54.0
09	313 27.2	. . 25.4	163 36.6	11.5	24 30.5	5.4	54.0
10	328 27.3	24.8	178 07.1	11.6	24 25.1	5.5	54.0
11	343 27.3	24.1	192 37.7	11.6	24 19.6	5.5	54.0
12	358 27.4	N17 23.4	207 08.3	11.6	S24 14.1	5.7	54.0
13	13 27.4	22.8	221 38.9	11.7	24 08.4	5.9	54.0
14	28 27.5	22.1	236 09.6	11.7	24 02.5	5.9	54.0
15	43 27.5	. . 21.5	250 40.3	11.8	23 56.6	6.0	54.0
16	58 27.6	20.8	265 11.1	11.8	23 50.6	6.2	54.0
17	73 27.6	20.1	279 41.9	11.9	23 44.4	6.2	54.0
18	88 27.7	N17 19.5	294 12.8	11.9	S23 38.2	6.3	54.0
19	103 27.7	18.8	308 43.7	12.0	23 31.9	6.5	54.0
20	118 27.8	18.2	323 14.7	12.0	23 25.4	6.6	54.0
21	133 27.9	. . 17.5	337 45.7	12.1	23 18.8	6.6	54.0
22	148 27.9	16.8	352 16.8	12.2	23 12.2	6.8	54.0
23	163 28.0	16.2	6 48.0	12.1	23 05.4	6.9	54.0
4 00	178 28.0	N17 15.5	21 19.1	12.3	S22 58.5	7.0	54.0
01	193 28.1	14.8	35 50.4	12.3	22 51.5	7.0	54.0
02	208 28.1	14.2	50 21.7	12.3	22 44.5	7.2	54.0
03	223 28.2	. . 13.5	64 53.0	12.4	22 37.3	7.3	54.0
04	238 28.3	12.8	79 24.4	12.4	22 30.0	7.4	54.0
05	253 28.3	12.2	93 55.8	12.5	22 22.6	7.4	54.0
06	268 28.4	N17 11.5	108 27.3	12.6	S22 15.2	7.6	54.0
07	283 28.4	10.9	122 58.9	12.6	22 07.6	7.7	54.0
08	298 28.5	10.2	137 30.5	12.7	21 59.9	7.8	54.0
09	313 28.5	. . 09.5	152 02.2	12.7	21 52.1	7.8	54.0
10	328 28.6	08.8	166 33.9	12.7	21 44.3	8.0	54.0
11	343 28.7	08.2	181 05.6	12.8	21 36.3	8.0	54.0
12	358 28.7	N17 07.5	195 37.4	12.9	S21 28.3	8.2	54.0
13	13 28.8	06.8	210 09.3	12.9	21 20.1	8.2	54.0
14	28 28.8	06.2	224 41.2	13.0	21 11.9	8.4	54.0
15	43 28.9	. . 05.5	239 13.2	13.0	21 03.5	8.4	54.0
16	58 29.0	04.8	253 45.2	13.1	20 55.1	8.5	54.0
17	73 29.0	04.2	268 17.3	13.1	20 46.6	8.6	54.0
18	88 29.1	N17 03.5	282 49.4	13.2	S20 38.0	8.7	54.0
19	103 29.2	02.8	297 21.6	13.3	20 29.3	8.8	54.0
20	118 29.2	02.1	311 53.9	13.3	20 20.5	8.9	54.0
21	133 29.3	. . 01.5	326 26.2	13.3	20 11.6	8.9	54.0
22	148 29.3	00.8	340 58.5	13.4	20 02.7	9.1	54.0
23	163 29.4	00.1	355 30.9	13.5	S19 53.6	9.1	54.0
	SD 15.8	*d* 0.7	SD 14.7		14.7		14.7

Lat.	Twilight		Sunrise	Moonrise			
	Naut.	Civil		2	3	4	5
°	h m	h m	h m	h m	h m	h m	h m
N 72	□	□	□	■	■	■	22 11
N 70	////	////	01 44	■	■	23 05	21 31
68	////	////	02 26	■	■	21 38	21 04
66	////	00 54	02 53	■	21 33	20 59	20 42
64	////	01 51	03 15	20 55	20 39	20 32	20 25
62	////	02 23	03 32	19 57	20 07	20 10	20 11
60	00 49	02 47	03 46	19 23	19 43	19 53	19 59
N 58	01 41	03 05	03 57	18 59	19 23	19 39	19 49
56	02 10	03 20	04 08	18 39	19 07	19 26	19 40
54	02 32	03 33	04 17	18 22	18 53	19 15	19 32
52	02 49	03 44	04 25	18 08	18 41	19 06	19 25
50	03 04	03 54	04 32	17 56	18 30	18 57	19 18
45	03 32	04 14	04 48	17 30	18 08	18 38	19 04
N 40	03 53	04 30	05 00	17 10	17 50	18 23	18 52
35	04 10	04 43	05 11	16 53	17 34	18 11	18 42
30	04 23	04 54	05 20	16 38	17 21	17 59	18 34
20	04 45	05 13	05 36	16 13	16 58	17 40	18 18
N 10	05 02	05 28	05 50	15 51	16 39	17 23	18 05
0	05 16	05 41	06 03	15 31	16 20	17 07	17 52
S 10	05 28	05 53	06 15	15 11	16 02	16 52	17 40
20	05 39	06 06	06 29	14 49	15 42	16 35	17 26
30	05 50	06 19	06 44	14 24	15 19	16 15	17 11
35	05 56	06 26	06 53	14 09	15 06	16 04	17 02
40	06 01	06 34	07 03	13 52	14 50	15 50	16 51
45	06 08	06 43	07 14	13 31	14 31	15 35	16 39
S 50	06 14	06 53	07 28	13 05	14 08	15 15	16 24
52	06 17	06 58	07 35	12 52	13 57	15 06	16 17
54	06 20	07 03	07 42	12 38	13 44	14 56	16 10
56	06 24	07 09	07 50	12 21	13 29	14 44	16 01
58	06 27	07 15	07 59	12 00	13 12	14 30	15 51
S 60	06 31	07 22	08 10	11 34	12 50	14 14	15 39

Lat.	Sunset	Twilight		Moonset			
		Civil	Naut.	2	3	4	5
°	h m	h m	h m	h m	h m	h m	h m
N 72	□	□	□	■	■	■	■
N 70	22 22	////	////	■	■	23 47	26 54
68	21 43	////	////	■	■	■	01 14
66	21 16	23 07	////	■	23 41	25 52	01 52
64	20 55	22 16	////	22 37	24 34	00 34	02 18
62	20 39	21 46	////	23 35	25 06	01 06	02 38
60	20 25	21 23	23 14	24 08	00 08	01 30	02 55
N 58	20 13	21 05	22 27	24 32	00 32	01 48	03 09
56	20 03	20 51	21 59	24 52	00 52	02 04	03 21
54	19 54	20 38	21 38	00 06	01 08	02 18	03 31
52	19 46	20 27	21 21	00 22	01 22	02 29	03 40
50	19 39	20 17	21 07	00 35	01 34	02 40	03 48
45	19 24	19 57	20 39	01 02	01 59	03 01	04 06
N 40	19 12	19 42	20 18	01 23	02 19	03 19	04 20
35	19 01	19 29	20 02	01 41	02 36	03 33	04 32
30	18 52	19 18	19 48	01 56	02 50	03 46	04 42
20	18 36	18 59	19 27	02 22	03 15	04 08	05 00
N 10	18 22	18 44	19 10	02 44	03 36	04 26	05 15
0	18 10	18 31	18 56	03 05	03 55	04 43	05 29
S 10	17 57	18 19	18 44	03 26	04 15	05 01	05 44
20	17 44	18 07	18 33	03 48	04 35	05 19	05 59
30	17 29	17 54	18 23	04 13	04 59	05 40	06 16
35	17 20	17 47	18 17	04 29	05 13	05 52	06 26
40	17 10	17 39	18 11	04 46	05 29	06 06	06 37
45	16 59	17 30	18 05	05 07	05 49	06 23	06 51
S 50	16 45	17 20	17 59	05 33	06 13	06 43	07 07
52	16 38	17 15	17 56	05 46	06 24	06 53	07 14
54	16 31	17 10	17 53	06 01	06 37	07 03	07 23
56	16 23	17 04	17 49	06 18	06 52	07 16	07 32
58	16 14	16 58	17 46	06 39	07 10	07 30	07 43
S 60	16 03	16 51	17 42	07 05	07 32	07 46	07 55

Day	SUN			MOON			
	Eqn. of Time		Mer. Pass.	Mer. Pass.		Age	Phase
	00^h	12^h		Upper	Lower		
d	m s	m s	h m	h m	h m	d	%
2	06 18	06 16	12 06	21 43	09 18	11	89
3	06 13	06 11	12 06	22 32	10 08	12	94
4	06 08	06 05	12 06	23 19	10 55	13	98

UT	ARIES	VENUS −4.0		MARS +1.1		JUPITER −2.8		SATURN +1.1		STARS		
	GHA	GHA	Dec	GHA	Dec	GHA	Dec	GHA	Dec	Name	SHA	Dec
d h	° ′	° ′	° ′	° ′	° ′	° ′	° ′	° ′	° ′		° ′	°
5 00	313 40.8	218 46.9	N21 56.0	238 23.4	N22 31.0	347 39.6	S14 44.5	142 17.4	N 5 52.2	Acamar	315 20.4	S40 1
01	328 43.3	233 46.2	56.0	253 24.0	31.2	2 42.3	44.6	157 19.6	52.1	Achernar	335 28.5	S57 1
02	343 45.7	248 45.6	56.0	268 24.7	31.4	17 45.1	44.7	172 21.9	52.0	Acrux	173 13.2	S63 0
03	358 48.2	263 44.9	. . 56.1	283 25.4	. . 31.6	32 47.9	. . 44.8	187 24.1	. . 51.9	Adhara	255 15.1	S28 5
04	13 50.7	278 44.2	56.1	298 26.0	31.8	47 50.6	45.0	202 26.3	51.8	Aldebaran	290 52.8	N16 31
05	28 53.1	293 43.6	56.1	313 26.7	32.0	62 53.4	45.1	217 28.5	51.7			
06	43 55.6	308 42.9	N21 56.2	328 27.3	N22 32.2	77 56.2	S14 45.2	232 30.7	N 5 51.6	Alioth	166 23.2	N55 5
W 07	58 58.1	323 42.3	56.2	343 28.0	32.4	92 59.0	45.3	247 32.9	51.5	Alkaid	153 01.1	N49 1
E 08	74 00.5	338 41.6	56.2	358 28.7	32.6	108 01.7	45.4	262 35.2	51.4	Al Na'ir	27 46.7	S46 5
D 09	89 03.0	353 41.0	. . 56.2	13 29.3	. . 32.8	123 04.5	. . 45.5	277 37.4	. . 51.3	Alnilam	275 49.5	S 1 1
N 10	104 05.5	8 40.3	56.3	28 30.0	33.0	138 07.3	45.6	292 39.6	51.2	Alphard	217 59.2	S 8 4
E 11	119 07.9	23 39.6	56.3	43 30.6	33.2	153 10.0	45.7	307 41.8	51.0			
S 12	134 10.4	38 39.0	N21 56.3	58 31.3	N22 33.4	168 12.8	S14 45.8	322 44.0	N 5 50.9	Alphecca	126 13.3	N26 4
D 13	149 12.9	53 38.3	56.3	73 32.0	33.6	183 15.6	46.0	337 46.2	50.8	Alpheratz	357 46.3	N29 0
A 14	164 15.3	68 37.7	56.3	88 32.6	33.8	198 18.4	46.1	352 48.4	50.7	Altair	62 10.7	N 8 5
Y 15	179 17.8	83 37.0	. . 56.4	103 33.3	. . 34.0	213 21.1	. . 46.2	7 50.7	. . 50.6	Ankaa	353 18.1	S42 1
16	194 20.2	98 36.3	56.4	118 34.0	34.2	228 23.9	46.3	22 52.9	50.5	Antares	112 29.6	S26 2
17	209 22.7	113 35.7	56.4	133 34.6	34.4	243 26.7	46.4	37 55.1	50.4			
18	224 25.2	128 35.0	N21 56.4	148 35.3	N22 34.6	258 29.4	S14 46.5	52 57.3	N 5 50.3	Arcturus	145 58.3	N19 0
19	239 27.6	143 34.4	56.4	163 35.9	34.8	273 32.2	46.6	67 59.5	50.2	Atria	107 33.8	S69 0
20	254 30.1	158 33.7	56.4	178 36.6	35.0	288 35.0	46.7	83 01.7	50.1	Avior	234 20.0	S59 3
21	269 32.6	173 33.0	. . 56.5	193 37.3	. . 35.2	303 37.8	. . 46.8	98 04.0	. . 50.0	Bellatrix	278 35.3	N 6 21
22	284 35.0	188 32.4	56.5	208 37.9	35.4	318 40.5	46.9	113 06.2	49.8	Betelgeuse	271 04.6	N 7 24
23	299 37.5	203 31.7	56.5	223 38.6	35.6	333 43.3	47.1	128 08.4	49.7			
6 00	314 40.0	218 31.1	N21 56.5	238 39.3	N22 35.8	348 46.1	S14 47.2	143 10.6	N 5 49.6	Canopus	263 57.9	S52 4
01	329 42.4	233 30.4	56.5	253 39.9	36.0	3 48.8	47.3	158 12.8	49.5	Capella	280 38.9	N46 0
02	344 44.9	248 29.7	56.5	268 40.6	36.2	18 51.6	47.4	173 15.0	49.4	Deneb	49 33.0	N45 1
03	359 47.3	263 29.1	. . 56.5	283 41.2	. . 36.4	33 54.4	. . 47.5	188 17.2	. . 49.3	Denebola	182 36.7	N14 31
04	14 49.8	278 28.4	56.5	298 41.9	36.6	48 57.2	47.6	203 19.5	49.2	Diphda	348 58.5	S17 5
05	29 52.3	293 27.7	56.5	313 42.6	36.8	63 59.9	47.7	218 21.7	49.1			
06	44 54.7	308 27.1	N21 56.5	328 43.2	N22 37.0	79 02.7	S14 47.8	233 23.9	N 5 49.0	Dubhe	193 55.4	N61 4
07	59 57.2	323 26.4	56.5	343 43.9	37.2	94 05.5	47.9	248 26.1	48.9	Elnath	278 16.4	N28 3
T 08	74 59.7	338 25.8	56.5	358 44.6	37.4	109 08.2	48.1	263 28.3	48.7	Eltanin	90 47.1	N51 2
H 09	90 02.1	353 25.1	. . 56.5	13 45.2	. . 37.5	124 11.0	. . 48.2	278 30.5	. . 48.6	Enif	33 49.6	N 9 55
U 10	105 04.6	8 24.4	56.5	28 45.9	37.7	139 13.8	48.3	293 32.7	48.5	Fomalhaut	15 26.7	S29 34
R 11	120 07.1	23 23.8	56.5	43 46.6	37.9	154 16.6	48.4	308 34.9	48.4			
S 12	135 09.5	38 23.1	N21 56.5	58 47.2	N22 38.1	169 19.3	S14 48.5	323 37.2	N 5 48.3	Gacrux	172 04.6	S57 1
D 13	150 12.0	53 22.4	56.5	73 47.9	38.3	184 22.1	48.6	338 39.4	48.2	Gienah	175 55.5	S17 35
A 14	165 14.5	68 21.8	56.5	88 48.5	38.5	199 24.9	48.7	353 41.6	48.1	Hadar	148 52.3	S60 25
Y 15	180 16.9	83 21.1	. . 56.5	103 49.2	. . 38.7	214 27.7	. . 48.8	8 43.8	. . 48.0	Hamal	328 03.9	N23 3
16	195 19.4	98 20.4	56.5	118 49.9	38.9	229 30.4	48.9	23 46.0	47.9	Kaus Aust.	83 47.2	S34 22
17	210 21.8	113 19.8	56.5	133 50.5	39.1	244 33.2	49.1	38 48.2	47.8			
18	225 24.3	128 19.1	N21 56.5	148 51.2	N22 39.3	259 36.0	S14 49.2	53 50.4	N 5 47.6	Kochab	137 19.5	N74 07
19	240 26.8	143 18.5	56.5	163 51.9	39.5	274 38.8	49.3	68 52.7	47.5	Markab	13 40.9	N15 15
20	255 29.2	158 17.8	56.5	178 52.5	39.7	289 41.5	49.4	83 54.9	47.4	Menkar	314 18.0	N 4 07
21	270 31.7	173 17.1	. . 56.4	193 53.2	. . 39.8	304 44.3	. . 49.5	98 57.1	. . 47.3	Menkent	148 11.1	S36 2
22	285 34.2	188 16.5	56.4	208 53.9	40.0	319 47.1	49.6	113 59.3	47.2	Miaplacidus	221 41.5	S69 45
23	300 36.6	203 15.8	56.4	223 54.5	40.2	334 49.9	49.7	129 01.5	47.1			
7 00	315 39.1	218 15.1	N21 56.4	238 55.2	N22 40.4	349 52.6	S14 49.8	144 03.7	N 5 47.0	Mirfak	308 44.6	N49 53
01	330 41.6	233 14.5	56.4	253 55.9	40.6	4 55.4	49.9	159 05.9	46.9	Nunki	76 01.5	S26 17
02	345 44.0	248 13.8	56.4	268 56.5	40.8	19 58.2	50.1	174 08.1	46.8	Peacock	53 22.9	S56 42
03	0 46.5	263 13.1	. . 56.4	283 57.2	. . 41.0	35 00.9	. . 50.2	189 10.4	. . 46.7	Pollux	243 31.5	N28 00
04	15 49.0	278 12.5	56.3	298 57.9	41.2	50 03.7	50.3	204 12.6	46.5	Procyon	245 03.0	N 5 12
05	30 51.4	293 11.8	56.3	313 58.5	41.4	65 06.5	50.4	219 14.8	46.4			
06	45 53.9	308 11.1	N21 56.3	328 59.2	N22 41.5	80 09.3	S14 50.5	234 17.0	N 5 46.3	Rasalhague	96 08.9	N12 33
07	60 56.3	323 10.5	56.3	343 59.8	41.7	95 12.0	50.6	249 19.2	46.2	Regulus	207 46.8	N11 55
08	75 58.8	338 09.8	56.2	359 00.5	41.9	110 14.8	50.7	264 21.4	46.1	Rigel	281 15.0	S 8 11
F 09	91 01.3	353 09.1	. . 56.2	14 01.2	. . 42.1	125 17.6	. . 50.8	279 23.6	. . 46.0	Rigil Kent.	139 55.8	S60 52
R 10	106 03.7	8 08.4	56.2	29 01.8	42.3	140 20.4	50.9	294 25.8	45.9	Sabik	102 15.6	S15 44
I 11	121 06.2	23 07.8	56.2	44 02.5	42.5	155 23.1	51.1	309 28.1	45.8			
D 12	136 08.7	38 07.1	N21 56.1	59 03.2	N22 42.7	170 25.9	S14 51.2	324 30.3	N 5 45.7	Schedar	349 43.7	N56 35
A 13	151 11.1	53 06.4	56.1	74 03.8	42.8	185 28.7	51.3	339 32.5	45.6	Shaula	96 25.5	S37 06
Y 14	166 13.6	68 05.8	56.1	89 04.5	43.0	200 31.5	51.4	354 34.7	45.4	Sirius	258 36.5	S16 43
15	181 16.1	83 05.1	. . 56.1	104 05.2	. . 43.2	215 34.2	. . 51.5	9 36.9	. . 45.3	Spica	158 34.4	S11 12
16	196 18.5	98 04.4	56.0	119 05.8	43.4	230 37.0	51.6	24 39.1	45.2	Suhail	222 55.1	S43 28
17	211 21.0	113 03.8	56.0	134 06.5	43.6	245 39.8	51.7	39 41.3	45.1			
18	226 23.4	128 03.1	N21 56.0	149 07.2	N22 43.8	260 42.6	S14 51.8	54 43.5	N 5 45.0	Vega	80 40.5	N38 47
19	241 25.9	143 02.4	55.9	164 07.8	43.9	275 45.4	51.9	69 45.7	44.9	Zuben'ubi	137 08.6	S16 05
20	256 28.4	158 01.8	55.9	179 08.5	44.1	290 48.1	52.1	84 48.0	44.8		SHA	Mer. Pass.
21	271 30.8	173 01.1	. . 55.9	194 09.2	. . 44.3	305 50.9	. . 52.2	99 50.2	. . 44.7		° ′	h m
22	286 33.3	188 00.4	55.8	209 09.8	44.5	320 53.7	52.3	114 52.4	44.6	Venus	263 51.1	9 26
23	301 35.8	202 59.8	55.8	224 10.5	44.7	335 56.5	52.4	129 54.6	44.4	Mars	283 59.3	8 05
	h m									Jupiter	34 06.1	0 45
Mer. Pass.	3 00.8	v −0.7	d 0.0	v 0.7	d 0.2	v 2.8	d 0.1	v 2.2	d 0.1	Saturn	188 30.6	14 25

UT	SUN GHA	SUN Dec	MOON GHA	v	MOON Dec	d	HP
d h	° ′	° ′	° ′	′	° ′	′	′
5 00	178 29.5	N16 59.4	10 03.4	13.5	S19 44.5	9.2	54.0
01	193 29.5	58.8	24 35.9	13.5	19 35.3	9.3	54.1
02	208 29.6	58.1	39 08.4	13.6	19 26.0	9.4	54.1
03	223 29.7	. . 57.4	53 41.0	13.7	19 16.6	9.5	54.1
04	238 29.7	56.7	68 13.7	13.7	19 07.1	9.5	54.1
05	253 29.8	56.1	82 46.4	13.7	18 57.6	9.7	54.1
06	268 29.9	N16 55.4	97 19.1	13.9	S18 47.9	9.7	54.1
07	283 29.9	54.7	111 52.0	13.8	18 38.2	9.7	54.1
08	298 30.0	54.0	126 24.8	13.9	18 28.5	9.9	54.1
09	313 30.1	. . 53.3	140 57.7	14.0	18 18.6	10.0	54.1
10	328 30.1	52.7	155 30.7	14.0	18 08.6	10.0	54.1
11	343 30.2	52.0	170 03.7	14.1	17 58.6	10.1	54.1
12	358 30.2	N16 51.3	184 36.8	14.1	S17 48.5	10.1	54.1
13	13 30.3	50.6	199 09.9	14.1	17 38.4	10.3	54.1
14	28 30.4	49.9	213 43.0	14.2	17 28.1	10.3	54.1
15	43 30.4	. . 49.3	228 16.2	14.3	17 17.8	10.4	54.1
16	58 30.5	48.6	242 49.5	14.3	17 07.4	10.5	54.1
17	73 30.6	47.9	257 22.8	14.3	16 56.9	10.5	54.1
18	88 30.7	N16 47.2	271 56.1	14.4	S16 46.4	10.6	54.1
19	103 30.7	46.5	286 29.5	14.5	16 35.8	10.7	54.1
20	118 30.8	45.8	301 03.0	14.5	16 25.1	10.7	54.2
21	133 30.9	. . 45.2	315 36.5	14.5	16 14.4	10.8	54.2
22	148 30.9	44.5	330 10.0	14.6	16 03.6	10.9	54.2
23	163 31.0	43.8	344 43.6	14.6	15 52.7	11.0	54.2
6 00	178 31.1	N16 43.1	359 17.2	14.7	S15 41.7	11.0	54.2
01	193 31.1	42.4	13 50.9	14.7	15 30.7	11.1	54.2
02	208 31.2	41.7	28 24.6	14.7	15 19.6	11.1	54.2
03	223 31.3	. . 41.0	42 58.3	14.8	15 08.5	11.2	54.2
04	238 31.3	40.3	57 32.1	14.8	14 57.3	11.3	54.2
05	253 31.4	39.7	72 05.9	14.9	14 46.0	11.3	54.2
06	268 31.5	N16 39.0	86 39.8	14.9	S14 34.7	11.4	54.2
07	283 31.6	38.3	101 13.7	15.0	14 23.3	11.5	54.2
08	298 31.6	37.6	115 47.7	15.0	14 11.8	11.5	54.2
09	313 31.7	. . 36.9	130 21.7	15.0	14 00.3	11.5	54.3
10	328 31.8	36.2	144 55.7	15.1	13 48.8	11.7	54.3
11	343 31.8	35.5	159 29.8	15.1	13 37.1	11.6	54.3
12	358 31.9	N16 34.8	174 03.9	15.2	S13 25.5	11.8	54.3
13	13 32.0	34.1	188 38.1	15.1	13 13.7	11.8	54.3
14	28 32.1	33.4	203 12.2	15.3	13 01.9	11.8	54.3
15	43 32.1	. . 32.7	217 46.5	15.2	12 50.1	11.9	54.3
16	58 32.2	32.1	232 20.7	15.3	12 38.2	12.0	54.3
17	73 32.3	31.4	246 55.0	15.3	12 26.2	12.0	54.3
18	88 32.4	N16 30.7	261 29.3	15.4	S12 14.2	12.1	54.3
19	103 32.4	30.0	276 03.7	15.4	12 02.1	12.1	54.4
20	118 32.5	29.3	290 38.1	15.4	11 50.0	12.2	54.4
21	133 32.6	. . 28.6	305 12.5	15.5	11 37.8	12.2	54.4
22	148 32.7	27.9	319 47.0	15.4	11 25.6	12.2	54.4
23	163 32.7	27.2	334 21.4	15.5	11 13.4	12.3	54.4
7 00	178 32.8	N16 26.5	348 55.9	15.6	S11 01.1	12.4	54.4
01	193 32.9	25.8	3 30.5	15.6	10 48.7	12.4	54.4
02	208 33.0	25.1	18 05.1	15.5	10 36.3	12.5	54.4
03	223 33.0	. . 24.4	32 39.6	15.7	10 23.8	12.4	54.4
04	238 33.1	23.7	47 14.3	15.6	10 11.4	12.6	54.4
05	253 33.2	23.0	61 48.9	15.7	9 58.8	12.6	54.5
06	268 33.3	N16 22.3	76 23.6	15.7	S 9 46.2	12.6	54.5
07	283 33.3	21.6	90 58.3	15.7	9 33.6	12.7	54.5
08	298 33.4	20.9	105 33.0	15.8	9 20.9	12.7	54.5
09	313 33.5	. . 20.2	120 07.8	15.7	9 08.2	12.7	54.5
10	328 33.6	19.5	134 42.5	15.8	8 55.5	12.8	54.5
11	343 33.7	18.8	149 17.3	15.8	8 42.7	12.8	54.5
12	358 33.7	N16 18.1	163 52.1	15.8	S 8 29.9	12.9	54.5
13	13 33.8	17.4	178 26.9	15.9	8 17.0	12.9	54.6
14	28 33.9	16.7	193 01.8	15.9	8 04.1	12.9	54.6
15	43 34.0	. . 16.0	207 36.7	15.8	7 51.2	13.0	54.6
16	58 34.1	15.3	222 11.5	15.9	7 38.2	13.0	54.6
17	73 34.1	14.6	236 46.4	15.9	7 25.2	13.0	54.6
18	88 34.2	N16 13.9	251 21.3	16.0	S 7 12.2	13.1	54.6
19	103 34.3	13.1	265 56.3	15.9	6 59.1	13.1	54.6
20	118 34.4	12.4	280 31.2	16.0	6 46.0	13.1	54.6
21	133 34.5	. . 11.7	295 06.2	15.9	6 32.9	13.2	54.7
22	148 34.5	11.0	309 41.1	16.0	6 19.7	13.2	54.7
23	163 34.6	10.3	324 16.1	16.0	S 6 06.5	13.2	54.7
	SD 15.8	d 0.7	SD 14.7		14.8		14.9

Lat.	Twilight Naut.	Twilight Civil	Sunrise	Moonrise 5	6	7	8
°	h m	h m	h m	h m	h m	h m	h m
N 72	////	////	00 59	22 11	21 18	20 44	20 16
N 70	////	////	02 05	21 31	20 58	20 35	20 14
68	////	////	02 40	21 04	20 43	20 27	20 13
66	////	01 24	03 05	20 42	20 31	20 21	20 11
64	////	02 07	03 24	20 25	20 20	20 15	20 10
62	////	02 35	03 40	20 11	20 11	20 10	20 09
60	01 16	02 56	03 53	19 59	20 03	20 06	20 08
N 58	01 55	03 13	04 04	19 49	19 57	20 02	20 08
56	02 20	03 27	04 13	19 40	19 51	19 59	20 07
54	02 40	03 39	04 22	19 32	19 45	19 56	20 06
52	02 56	03 50	04 30	19 25	19 40	19 53	20 06
50	03 10	03 59	04 36	19 18	19 36	19 51	20 05
45	03 37	04 18	04 51	19 04	19 26	19 46	20 04
N 40	03 57	04 33	05 03	18 52	19 18	19 41	20 03
35	04 13	04 46	05 13	18 42	19 11	19 37	20 02
30	04 26	04 56	05 22	18 34	19 05	19 34	20 02
20	04 46	05 14	05 37	18 18	18 54	19 28	20 00
N 10	05 03	05 28	05 50	18 05	18 44	19 22	19 59
0	05 16	05 41	06 02	17 52	18 35	19 17	19 58
S 10	05 27	05 53	06 14	17 40	18 26	19 12	19 57
20	05 38	06 04	06 27	17 26	18 17	19 07	19 56
30	05 48	06 17	06 42	17 11	18 06	19 00	19 55
35	05 53	06 23	06 50	17 02	18 00	18 57	19 54
40	05 58	06 31	06 59	16 51	17 52	18 53	19 53
45	06 04	06 39	07 10	16 39	17 44	18 48	19 53
S 50	06 10	06 49	07 24	16 24	17 34	18 42	19 51
52	06 13	06 53	07 30	16 17	17 29	18 40	19 51
54	06 16	06 58	07 37	16 10	17 23	18 37	19 50
56	06 19	07 03	07 44	16 01	17 18	18 34	19 50
58	06 22	07 09	07 53	15 51	17 11	18 30	19 49
S 60	06 25	07 16	08 02	15 39	17 04	18 26	19 48

Lat.	Sunset	Twilight Civil	Twilight Naut.	Moonset 5	6	7	8
°	h m	h m	h m	h m	h m	h m	h m
N 72	22 59	////	////	■	02 16	04 38	06 39
N 70	22 01	////	////	26 54	02 54	04 55	06 46
68	21 28	////	////	01 14	03 20	05 09	06 51
66	21 04	22 40	////	01 52	03 40	05 20	06 56
64	20 45	22 01	////	02 18	03 56	05 29	07 00
62	20 30	21 34	////	02 38	04 09	05 37	07 03
60	20 17	21 13	22 50	02 55	04 20	05 43	07 06
N 58	20 06	20 57	22 13	03 09	04 29	05 49	07 08
56	19 57	20 43	21 49	03 21	04 38	05 54	07 10
54	19 49	20 31	21 29	03 31	04 45	05 59	07 12
52	19 41	20 21	21 14	03 40	04 52	06 03	07 14
50	19 34	20 12	21 00	03 48	04 58	06 07	07 16
45	19 20	19 53	20 34	04 06	05 10	06 15	07 19
N 40	19 08	19 38	20 14	04 20	05 21	06 22	07 22
35	18 58	19 25	19 58	04 32	05 30	06 27	07 25
30	18 49	19 15	19 46	04 42	05 38	06 32	07 27
20	18 34	18 58	19 25	05 00	05 51	06 41	07 31
N 10	18 21	18 43	19 09	05 15	06 03	06 49	07 34
0	18 09	18 31	18 56	05 29	06 13	06 56	07 37
S 10	17 57	18 19	18 45	05 44	06 24	07 03	07 40
20	17 45	18 08	18 34	05 59	06 36	07 10	07 43
30	17 30	17 55	18 24	06 16	06 48	07 18	07 47
35	17 22	17 49	18 19	06 26	06 56	07 23	07 49
40	17 13	17 41	18 14	06 37	07 04	07 29	07 51
45	17 02	17 33	18 08	06 51	07 14	07 35	07 54
S 50	16 49	17 23	18 02	07 07	07 26	07 42	07 57
52	16 43	17 19	18 00	07 14	07 31	07 46	07 58
54	16 36	17 14	17 57	07 23	07 37	07 49	08 00
56	16 28	17 09	17 54	07 32	07 44	07 54	08 02
58	16 20	17 03	17 51	07 43	07 52	07 58	08 04
S 60	16 10	16 57	17 48	07 55	08 00	08 03	08 06

Day	SUN Eqn. of Time 00^h	SUN Eqn. of Time 12^h	SUN Mer. Pass.	MOON Mer. Pass. Upper	MOON Mer. Pass. Lower	Age	Phase
d	m s	m s	h m	h m	h m	d %	
5	06 02	05 59	12 06	24 03	11 41	14 100	○
6	05 56	05 52	12 06	00 03	12 24	15 100	
7	05 49	05 45	12 06	00 46	13 06	16 98	

UT	ARIES	VENUS −4.0		MARS +1.1		JUPITER −2.9		SATURN +1.1		STARS		
d h	GHA ° ′	GHA ° ′	Dec ° ′	GHA ° ′	Dec ° ′	GHA ° ′	Dec ° ′	GHA ° ′	Dec ° ′	Name	SHA ° ′	Dec ° ′
8 00 (Saturday)	316 38.2	217 59.1	N21 55.7	239 11.2	N22 44.9	350 59.2	S14 52.5	144 56.8	N 5 44.3	Acamar	315 20.4	S40 15
01	331 40.7	232 58.4	55.7	254 11.8	45.0	6 02.0	52.6	159 59.0	44.2	Achernar	335 28.4	S57 10.
02	346 43.2	247 57.7	55.7	269 12.5	45.2	21 04.8	52.7	175 01.2	44.1	Acrux	173 13.2	S63 09.
03	1 45.6	262 57.1	55.6	284 13.2	45.4	36 07.6	52.8	190 03.4	44.0	Adhara	255 15.1	S28 58.
04	16 48.1	277 56.4	55.6	299 13.8	45.6	51 10.3	52.9	205 05.6	43.9	Aldebaran	290 52.8	N16 31.
05	31 50.6	292 55.7	55.5	314 14.5	45.8	66 13.1	53.1	220 07.9	43.8			
06	46 53.0	307 55.1	N21 55.5	329 15.2	N22 45.9	81 15.9	S14 53.2	235 10.1	N 5 43.7	Alioth	166 23.2	N55 54.
07	61 55.5	322 54.4	55.4	344 15.8	46.1	96 18.7	53.3	250 12.3	43.6	Alkaid	153 01.1	N49 16.
08	76 57.9	337 53.7	55.4	359 16.5	46.3	111 21.4	53.4	265 14.5	43.4	Al Na'ir	27 46.7	S46 54.
09	92 00.4	352 53.0	55.3	14 17.2	46.5	126 24.2	53.5	280 16.7	43.3	Alnilam	275 49.4	S 1 11.
10	107 02.9	7 52.4	55.3	29 17.8	46.6	141 27.0	53.6	295 18.9	43.2	Alphard	217 59.2	S 8 42.
11	122 05.3	22 51.7	55.2	44 18.5	46.8	156 29.8	53.7	310 21.1	43.1			
12	137 07.8	37 51.0	N21 55.2	59 19.2	N22 47.0	171 32.6	S14 53.8	325 23.3	N 5 43.0	Alphecca	126 13.3	N26 41.
13	152 10.3	52 50.4	55.1	74 19.9	47.2	186 35.3	53.9	340 25.5	42.9	Alpheratz	357 46.3	N29 08.
14	167 12.7	67 49.7	55.1	89 20.5	47.4	201 38.1	54.1	355 27.8	42.8	Altair	62 10.7	N 8 53.
15	182 15.2	82 49.0	55.0	104 21.2	47.5	216 40.9	54.2	10 30.0	42.7	Ankaa	353 18.1	S42 14.
16	197 17.7	97 48.3	55.0	119 21.9	47.7	231 43.7	54.3	25 32.2	42.6	Antares	112 29.6	S26 27.
17	212 20.1	112 47.7	54.9	134 22.5	47.9	246 46.4	54.4	40 34.4	42.4			
18	227 22.6	127 47.0	N21 54.9	149 23.2	N22 48.1	261 49.2	S14 54.5	55 36.6	N 5 42.3	Arcturus	145 58.3	N19 08.
19	242 25.0	142 46.3	54.8	164 23.9	48.2	276 52.0	54.6	70 38.8	42.2	Atria	107 33.8	S69 03.
20	257 27.5	157 45.6	54.7	179 24.5	48.4	291 54.8	54.7	85 41.0	42.1	Avior	234 20.0	S59 32.
21	272 30.0	172 45.0	54.7	194 25.2	48.6	306 57.5	54.8	100 43.2	42.0	Bellatrix	278 35.2	N 6 21.
22	287 32.4	187 44.3	54.6	209 25.9	48.7	322 00.3	54.9	115 45.4	41.9	Betelgeuse	271 04.6	N 7 24.
23	302 34.9	202 43.6	54.6	224 26.5	48.9	337 03.1	55.1	130 47.6	41.8			
9 00 (Sunday)	317 37.4	217 42.9	N21 54.5	239 27.2	N22 49.1	352 05.9	S14 55.2	145 49.9	N 5 41.7	Canopus	263 57.8	S52 41.
01	332 39.8	232 42.3	54.4	254 27.9	49.3	7 08.7	55.3	160 52.1	41.6	Capella	280 38.9	N46 00.
02	347 42.3	247 41.6	54.4	269 28.6	49.4	22 11.4	55.4	175 54.3	41.4	Deneb	49 33.0	N45 19.
03	2 44.8	262 40.9	54.3	284 29.2	49.6	37 14.2	55.5	190 56.5	41.3	Denebola	182 36.7	N14 31.
04	17 47.2	277 40.2	54.2	299 29.9	49.8	52 17.0	55.6	205 58.7	41.2	Diphda	348 58.5	S17 55.
05	32 49.7	292 39.6	54.2	314 30.6	50.0	67 19.8	55.7	221 00.9	41.1			
06	47 52.2	307 38.9	N21 54.1	329 31.2	N22 50.1	82 22.5	S14 55.8	236 03.1	N 5 41.0	Dubhe	193 55.5	N61 42.
07	62 54.6	322 38.2	54.0	344 31.9	50.3	97 25.3	55.9	251 05.3	40.9	Elnath	278 16.4	N28 36.
08	77 57.1	337 37.5	53.9	359 32.6	50.5	112 28.1	56.1	266 07.5	40.8	Eltanin	90 47.1	N51 29.
09	92 59.5	352 36.9	53.9	14 33.2	50.6	127 30.9	56.2	281 09.7	40.7	Enif	33 49.6	N 9 55.
10	108 02.0	7 36.2	53.8	29 33.9	50.8	142 33.7	56.3	296 11.9	40.6	Fomalhaut	15 26.7	S29 34
11	123 04.5	22 35.5	53.7	44 34.6	51.0	157 36.4	56.4	311 14.1	40.4			
12	138 06.9	37 34.8	N21 53.6	59 35.3	N22 51.1	172 39.2	S14 56.5	326 16.4	N 5 40.3	Gacrux	172 04.7	S57 10.
13	153 09.4	52 34.2	53.6	74 35.9	51.3	187 42.0	56.6	341 18.6	40.2	Gienah	175 55.5	S17 35.
14	168 11.9	67 33.5	53.5	89 36.6	51.5	202 44.8	56.7	356 20.8	40.1	Hadar	148 52.3	S60 25.
15	183 14.3	82 32.8	53.4	104 37.3	51.6	217 47.6	56.8	11 23.0	40.0	Hamal	328 03.9	N23 30.
16	198 16.8	97 32.1	53.3	119 37.9	51.8	232 50.3	57.0	26 25.2	39.9	Kaus Aust.	83 47.2	S34 22.
17	213 19.3	112 31.5	53.2	134 38.6	52.0	247 53.1	57.1	41 27.4	39.8			
18	228 21.7	127 30.8	N21 53.2	149 39.3	N22 52.1	262 55.9	S14 57.2	56 29.6	N 5 39.7	Kochab	137 19.6	N74 07.
19	243 24.2	142 30.1	53.1	164 39.9	52.3	277 58.7	57.3	71 31.8	39.5	Markab	13 40.9	N15 15
20	258 26.7	157 29.4	53.0	179 40.6	52.5	293 01.4	57.4	86 34.0	39.4	Menkar	314 18.0	N 4 07.
21	273 29.1	172 28.8	52.9	194 41.3	52.6	308 04.2	57.5	101 36.2	39.3	Menkent	148 11.1	S36 25.
22	288 31.6	187 28.1	52.8	209 42.0	52.8	323 07.0	57.6	116 38.4	39.2	Miaplacidus	221 41.5	S69 45.
23	303 34.0	202 27.4	52.7	224 42.6	53.0	338 09.8	57.7	131 40.6	39.1			
10 00 (Monday)	318 36.5	217 26.7	N21 52.6	239 43.3	N22 53.1	353 12.6	S14 57.8	146 42.9	N 5 39.0	Mirfak	308 44.6	N49 53
01	333 39.0	232 26.0	52.6	254 44.0	53.3	8 15.3	58.0	161 45.1	38.9	Nunki	76 01.5	S26 17.
02	348 41.4	247 25.4	52.5	269 44.6	53.5	23 18.1	58.1	176 47.3	38.8	Peacock	53 22.9	S56 42.
03	3 43.9	262 24.7	52.4	284 45.3	53.6	38 20.9	58.2	191 49.5	38.7	Pollux	243 31.5	N28 00.
04	18 46.4	277 24.0	52.3	299 46.0	53.8	53 23.7	58.3	206 51.7	38.5	Procyon	245 03.0	N 5 12.
05	33 48.8	292 23.3	52.2	314 46.7	54.0	68 26.5	58.4	221 53.9	38.4			
06	48 51.3	307 22.7	N21 52.1	329 47.3	N22 54.1	83 29.2	S14 58.5	236 56.1	N 5 38.3	Rasalhague	96 08.9	N12 33.
07	63 53.8	322 22.0	52.0	344 48.0	54.3	98 32.0	58.6	251 58.3	38.2	Regulus	207 46.8	N11 55.
08	78 56.2	337 21.3	51.9	359 48.7	54.4	113 34.8	58.7	267 00.5	38.1	Rigel	281 15.0	S 8 11.
09	93 58.7	352 20.6	51.8	14 49.4	54.6	128 37.6	58.8	282 02.7	38.0	Rigil Kent.	139 55.9	S60 52.
10	109 01.1	7 19.9	51.7	29 50.0	54.8	143 40.4	59.0	297 04.9	37.9	Sabik	102 15.6	S15 44.
11	124 03.6	22 19.3	51.6	44 50.7	54.9	158 43.1	59.1	312 07.1	37.8			
12	139 06.1	37 18.6	N21 51.5	59 51.4	N22 55.1	173 45.9	S14 59.2	327 09.3	N 5 37.6	Schedar	349 43.6	N56 35.
13	154 08.5	52 17.9	51.4	74 52.0	55.2	188 48.7	59.3	342 11.6	37.5	Shaula	96 25.5	S37 06.
14	169 11.0	67 17.2	51.3	89 52.7	55.4	203 51.5	59.4	357 13.8	37.4	Sirius	258 36.5	S16 43.
15	184 13.5	82 16.6	51.2	104 53.4	55.6	218 54.3	59.5	12 16.0	37.3	Spica	158 34.4	S11 12.
16	199 15.9	97 15.9	51.1	119 54.1	55.7	233 57.0	59.6	27 18.2	37.2	Suhail	222 55.1	S43 28.
17	214 18.4	112 15.2	51.0	134 54.7	55.9	248 59.8	59.7	42 20.4	37.1			
18	229 20.9	127 14.5	N21 50.9	149 55.4	N22 56.0	264 02.6	S14 59.8	57 22.6	N 5 37.0	Vega	80 40.6	N38 47.
19	244 23.3	142 13.8	50.8	164 56.1	56.2	279 05.4	15 00.0	72 24.8	36.9	Zuben'ubi	137 08.6	S16 05.
20	259 25.8	157 13.2	50.7	179 56.8	56.4	294 08.2	00.1	87 27.0	36.7			
21	274 28.3	172 12.5	50.5	194 57.4	56.5	309 10.9	00.2	102 29.2	36.6			
22	289 30.7	187 11.8	50.4	209 58.1	56.7	324 13.7	00.3	117 31.4	36.5			
23	304 33.2	202 11.1	50.3	224 58.8	56.8	339 16.5	00.4	132 33.6	36.4			
Mer. Pass.	h m 2 49.0	v −0.7	d 0.1	v 0.7	d 0.2	v 2.8	d 0.1	v 2.2	d 0.1			

	SHA ° ′	Mer. Pass h m
Venus	260 05.6	9 30
Mars	281 49.8	8 02
Jupiter	34 28.5	0 32
Saturn	188 12.5	14 15

UT	SUN GHA	SUN Dec	MOON GHA	v	Dec	d	HP
d h	° ′	° ′	° ′	′	° ′	′	′
8 00	178 34.7	N16 09.6	338 51.1	16.0	S 5 53.3	13.2	54.7
01	193 34.8	08.9	353 26.1	16.0	5 40.1	13.3	54.7
02	208 34.9	08.2	8 01.1	16.0	5 26.8	13.3	54.7
03	223 35.0	. . 07.5	22 36.1	16.0	5 13.5	13.4	54.7
04	238 35.0	06.8	37 11.1	16.0	5 00.1	13.3	54.8
05	253 35.1	06.1	51 46.1	16.1	4 46.8	13.4	54.8
06	268 35.2	N16 05.3	66 21.2	16.0	S 4 33.4	13.4	54.8
07	283 35.3	04.6	80 56.2	16.0	4 20.0	13.4	54.8
08	298 35.4	03.9	95 31.2	16.1	4 06.6	13.5	54.8
09	313 35.5	. . 03.2	110 06.3	16.0	3 53.1	13.4	54.8
10	328 35.5	02.5	124 41.3	16.1	3 39.7	13.5	54.8
11	343 35.6	01.8	139 16.4	16.0	3 26.2	13.5	54.9
12	358 35.7	N16 01.1	153 51.4	16.0	S 3 12.7	13.6	54.9
13	13 35.8	16 00.4	168 26.4	16.1	2 59.1	13.5	54.9
14	28 35.9	15 59.6	183 01.5	16.0	2 45.6	13.6	54.9
15	43 36.0	. . 58.9	197 36.5	16.0	2 32.0	13.6	54.9
16	58 36.0	58.2	212 11.5	16.0	2 18.4	13.6	54.9
17	73 36.1	57.5	226 46.5	16.1	2 04.8	13.6	55.0
18	88 36.2	N15 56.8	241 21.6	16.0	S 1 51.2	13.6	55.0
19	103 36.3	56.1	255 56.6	16.0	1 37.6	13.6	55.0
20	118 36.4	55.3	270 31.6	15.9	1 24.0	13.7	55.0
21	133 36.5	. . 54.6	285 06.5	16.0	1 10.3	13.7	55.0
22	148 36.6	53.9	299 41.5	16.0	0 56.6	13.6	55.0
23	163 36.7	53.2	314 16.5	15.9	0 43.0	13.7	55.1
9 00	178 36.7	N15 52.5	328 51.4	16.0	S 0 29.3	13.7	55.1
01	193 36.8	51.7	343 26.4	15.9	0 15.6	13.7	55.1
02	208 36.9	51.0	358 01.3	15.9	S 0 01.9	13.7	55.1
03	223 37.0	. . 50.3	12 36.2	15.9	N 0 11.8	13.8	55.1
04	238 37.1	49.6	27 11.1	15.9	0 25.6	13.7	55.1
05	253 37.2	48.9	41 46.0	15.8	0 39.3	13.7	55.2
06	268 37.3	N15 48.1	56 20.8	15.9	N 0 53.0	13.7	55.2
07	283 37.4	47.4	70 55.7	15.8	1 06.7	13.8	55.2
08	298 37.5	46.7	85 30.5	15.8	1 20.5	13.7	55.2
09	313 37.5	. . 46.0	100 05.3	15.8	1 34.2	13.8	55.2
10	328 37.6	45.3	114 40.1	15.7	1 48.0	13.7	55.3
11	343 37.7	44.5	129 14.8	15.7	2 01.7	13.7	55.3
12	358 37.8	N15 43.8	143 49.5	15.7	N 2 15.4	13.8	55.3
13	13 37.9	43.1	158 24.2	15.7	2 29.2	13.7	55.3
14	28 38.0	42.4	172 58.9	15.7	2 42.9	13.8	55.3
15	43 38.1	. . 41.6	187 33.6	15.6	2 56.7	13.7	55.4
16	58 38.2	40.9	202 08.2	15.6	3 10.4	13.7	55.4
17	73 38.3	40.2	216 42.8	15.5	3 24.1	13.8	55.4
18	88 38.4	N15 39.4	231 17.3	15.5	N 3 37.9	13.7	55.4
19	103 38.5	38.7	245 51.8	15.5	3 51.6	13.7	55.4
20	118 38.5	38.0	260 26.3	15.5	4 05.3	13.7	55.5
21	133 38.6	. . 37.3	275 00.8	15.4	4 19.0	13.7	55.5
22	148 38.7	36.5	289 35.2	15.4	4 32.7	13.7	55.5
23	163 38.8	35.8	304 09.6	15.3	4 46.4	13.7	55.5
10 00	178 38.9	N15 35.1	318 43.9	15.3	N 5 00.1	13.7	55.5
01	193 39.0	34.3	333 18.2	15.3	5 13.8	13.6	55.6
02	208 39.1	33.6	347 52.5	15.2	5 27.4	13.7	55.6
03	223 39.2	. . 32.9	2 26.7	15.2	5 41.1	13.6	55.6
04	238 39.3	32.2	17 00.9	15.2	5 54.7	13.6	55.6
05	253 39.4	31.4	31 35.1	15.1	6 08.3	13.6	55.6
06	268 39.5	N15 30.7	46 09.2	15.0	N 6 21.9	13.6	55.7
07	283 39.6	30.0	60 43.2	15.0	6 35.5	13.6	55.7
08	298 39.7	29.2	75 17.2	15.0	6 49.1	13.5	55.7
09	313 39.8	. . 28.5	89 51.2	14.9	7 02.6	13.6	55.7
10	328 39.9	27.8	104 25.1	14.9	7 16.2	13.5	55.7
11	343 40.0	27.0	118 59.0	14.8	7 29.7	13.5	55.8
12	358 40.1	N15 26.3	133 32.8	14.8	N 7 43.2	13.4	55.8
13	13 40.2	25.6	148 06.6	14.7	7 56.6	13.5	55.8
14	28 40.3	24.8	162 40.3	14.6	8 10.1	13.4	55.8
15	43 40.4	. . 24.1	177 13.9	14.6	8 23.5	13.4	55.9
16	58 40.5	23.3	191 47.5	14.6	8 36.9	13.3	55.9
17	73 40.6	22.6	206 21.1	14.5	8 50.2	13.4	55.9
18	88 40.6	N15 21.9	220 54.6	14.4	N 9 03.6	13.3	55.9
19	103 40.7	21.1	235 28.0	14.4	9 16.9	13.3	56.0
20	118 40.8	20.4	250 01.4	14.3	9 30.2	13.2	56.0
21	133 40.9	. . 19.7	264 34.7	14.2	9 43.4	13.3	56.0
22	148 41.0	18.9	279 07.9	14.2	9 56.7	13.2	56.0
23	163 41.1	18.2	293 41.1	14.1	N10 09.9	13.1	56.1
	SD 15.8	*d* 0.7	SD	15.0	15.1		15.2

Lat.	Twilight Naut.	Twilight Civil	Sunrise	Moonrise 8	9	10	11
°	h m	h m	h m	h m	h m	h m	h m
N 72	////	////	01 36	20 16	19 49	19 19	18 36
N 70	////	////	02 24	20 14	19 54	19 32	19 04
68	////	00 31	02 54	20 13	19 59	19 44	19 25
66	////	01 46	03 16	20 11	20 02	19 53	19 42
64	////	02 21	03 33	20 10	20 06	20 01	19 56
62	00 28	02 46	03 48	20 09	20 08	20 08	20 08
60	01 35	03 05	04 00	20 08	20 11	20 13	20 18
N 58	02 07	03 21	04 10	20 08	20 13	20 19	20 26
56	02 30	03 34	04 19	20 07	20 15	20 23	20 34
54	02 48	03 45	04 27	20 06	20 17	20 28	20 41
52	03 03	03 55	04 34	20 06	20 18	20 32	20 47
50	03 16	04 04	04 41	20 05	20 20	20 35	20 53
45	03 41	04 22	04 55	20 04	20 23	20 43	21 05
N 40	04 00	04 36	05 06	20 03	20 25	20 49	21 15
35	04 15	04 48	05 15	20 02	20 28	20 55	21 24
30	04 28	04 58	05 24	20 02	20 30	21 00	21 32
20	04 48	05 15	05 38	20 00	20 34	21 08	21 45
N 10	05 03	05 29	05 51	19 59	20 37	21 16	21 57
0	05 16	05 41	06 02	19 58	20 40	21 23	22 08
S 10	05 26	05 52	06 13	19 57	20 43	21 30	22 20
20	05 36	06 03	06 25	19 56	20 46	21 38	22 32
30	05 46	06 14	06 39	19 55	20 50	21 47	22 46
35	05 50	06 21	06 47	19 54	20 52	21 52	22 54
40	05 55	06 27	06 56	19 53	20 55	21 58	23 03
45	06 00	06 35	07 06	19 53	20 58	22 05	23 14
S 50	06 06	06 44	07 19	19 51	21 01	22 13	23 28
52	06 08	06 48	07 24	19 51	21 03	22 17	23 34
54	06 11	06 53	07 31	19 50	21 05	22 21	23 41
56	06 13	06 58	07 38	19 50	21 07	22 26	23 49
58	06 16	07 03	07 46	19 49	21 09	22 31	23 57
S 60	06 19	07 09	07 55	19 48	21 11	22 37	24 07

Lat.	Sunset	Twilight Civil	Twilight Naut.	Moonset 8	9	10	11
°	h m	h m	h m	h m	h m	h m	h m
N 72	22 27	////	////	06 39	08 36	10 36	12 52
N 70	21 43	////	////	06 46	08 34	10 25	12 26
68	21 14	23 20	////	06 51	08 32	10 16	12 07
66	20 52	22 19	////	06 56	08 31	10 08	11 52
64	20 35	21 46	////	07 00	08 30	10 02	11 39
62	20 21	21 22	23 25	07 03	08 29	09 57	11 28
60	20 09	21 04	22 30	07 06	08 28	09 52	11 19
N 58	19 59	20 48	22 00	07 08	08 27	09 48	11 12
56	19 50	20 35	21 38	07 10	08 27	09 44	11 05
54	19 42	20 24	21 20	07 12	08 26	09 41	10 59
52	19 35	20 15	21 06	07 14	08 25	09 38	10 53
50	19 29	20 06	20 53	07 16	08 25	09 35	10 48
45	19 16	19 48	20 29	07 19	08 24	09 30	10 38
N 40	19 04	19 34	20 10	07 22	08 23	09 25	10 29
35	18 55	19 22	19 55	07 25	08 22	09 21	10 21
30	18 47	19 12	19 42	07 27	08 21	09 17	10 15
20	18 33	18 56	19 23	07 31	08 20	09 11	10 03
N 10	18 20	18 42	19 08	07 34	08 19	09 05	09 54
0	18 09	18 30	18 55	07 37	08 18	09 00	09 44
S 10	17 58	18 19	18 45	07 40	08 17	08 55	09 35
20	17 46	18 09	18 35	07 43	08 16	08 50	09 25
30	17 32	17 57	18 26	07 47	08 15	08 43	09 14
35	17 24	17 51	18 21	07 49	08 14	08 40	09 08
40	17 16	17 44	18 16	07 51	08 13	08 36	09 00
45	17 05	17 36	18 11	07 54	08 12	08 31	08 52
S 50	16 53	17 27	18 06	07 57	08 11	08 26	08 42
52	16 47	17 23	18 04	07 58	08 11	08 23	08 37
54	16 41	17 19	18 01	08 00	08 10	08 20	08 32
56	16 34	17 14	17 59	08 02	08 09	08 17	08 27
58	16 26	17 09	17 56	08 04	08 09	08 14	08 20
S 60	16 17	17 03	17 53	08 06	08 08	08 10	08 13

Day	SUN Eqn. of Time 00ʰ	Eqn. of Time 12ʰ	Mer. Pass.	MOON Mer. Pass. Upper	Mer. Pass. Lower	Age	Phase
d	m s	m s	h m	h m	h m	d %	
8	05 41	05 37	12 06	01 27	13 48	17 94	
9	05 33	05 29	12 05	02 08	14 29	18 89	
10	05 25	05 20	12 05	02 50	15 11	19 82	

UT d	h	ARIES GHA	VENUS −4.0 GHA	Dec	MARS +1.0 GHA	Dec	JUPITER −2.9 GHA	Dec	SATURN +1.1 GHA	Dec
		° ′	° ′	° ′	° ′	° ′	° ′	° ′	° ′	° ′
11	00	319 35.6	217 10.4	N21 50.2	239 59.5	N22 57.0	354 19.3	S15 00.5	147 35.8	N 5 36.3
	01	334 38.1	232 09.8	50.1	255 00.1	57.1	9 22.1	00.6	162 38.0	36.2
	02	349 40.6	247 09.1	50.0	270 00.8	57.3	24 24.8	00.7	177 40.2	36.1
	03	4 43.0	262 08.4 . .	49.9	285 01.5 . .	57.5	39 27.6 . .	00.9	192 42.4 . .	36.0
	04	19 45.5	277 07.7	49.7	300 02.2	57.6	54 30.4	01.0	207 44.7	35.8
	05	34 48.0	292 07.0	49.6	315 02.8	57.8	69 33.2	01.1	222 46.9	35.7
	06	49 50.4	307 06.4	N21 49.5	330 03.5	N22 57.9	84 36.0	S15 01.2	237 49.1	N 5 35.6
	07	64 52.9	322 05.7	49.4	345 04.2	58.1	99 38.7	01.3	252 51.3	35.5
T	08	79 55.4	337 05.0	49.3	0 04.9	58.2	114 41.5	01.4	267 53.5	35.4
U	09	94 57.8	352 04.3 . .	49.1	15 05.5 . .	58.4	129 44.3 . .	01.5	282 55.7 . .	35.3
E	10	110 00.3	7 03.6	49.0	30 06.2	58.5	144 47.1	01.6	297 57.9	35.2
	11	125 02.7	22 03.0	48.9	45 06.9	58.7	159 49.9	01.7	313 00.1	35.1
S	12	140 05.2	37 02.3	N21 48.8	60 07.6	N22 58.8	174 52.7	S15 01.9	328 02.3	N 5 34.9
D	13	155 07.7	52 01.6	48.6	75 08.2	59.0	189 55.4	02.0	343 04.5	34.8
A	14	170 10.1	67 00.9	48.5	90 08.9	59.1	204 58.2	02.1	358 06.7	34.7
Y	15	185 12.6	82 00.2 . .	48.4	105 09.6 . .	59.3	220 01.0 . .	02.2	13 08.9 . .	34.6
	16	200 15.1	96 59.5	48.3	120 10.3	59.4	235 03.8	02.3	28 11.1	34.5
	17	215 17.5	111 58.9	48.1	135 10.9	59.6	250 06.6	02.4	43 13.3	34.4
	18	230 20.0	126 58.2	N21 48.0	150 11.6	N22 59.7	265 09.3	S15 02.5	58 15.5	N 5 34.3
	19	245 22.5	141 57.5	47.9	165 12.3	22 59.9	280 12.1	02.6	73 17.7	34.1
	20	260 24.9	156 56.8	47.7	180 13.0	23 00.0	295 14.9	02.7	88 19.9	34.0
	21	275 27.4	171 56.1 . .	47.6	195 13.6 . .	00.2	310 17.7 . .	02.9	103 22.1 . .	33.9
	22	290 29.9	186 55.5	47.4	210 14.3	00.3	325 20.5	03.0	118 24.3	33.8
	23	305 32.3	201 54.8	47.3	225 15.0	00.5	340 23.2	03.1	133 26.6	33.7
12	00	320 34.8	216 54.1	N21 47.2	240 15.7	N23 00.6	355 26.0	S15 03.2	148 28.8	N 5 33.6
	01	335 37.2	231 53.4	47.0	255 16.3	00.8	10 28.8	03.3	163 31.0	33.5
	02	350 39.7	246 52.7	46.9	270 17.0	00.9	25 31.6	03.4	178 33.2	33.4
	03	5 42.2	261 52.0 . .	46.8	285 17.7 . .	01.1	40 34.4 . .	03.5	193 35.4 . .	33.2
	04	20 44.6	276 51.4	46.6	300 18.4	01.2	55 37.2	03.6	208 37.6	33.1
	05	35 47.1	291 50.7	46.5	315 19.0	01.4	70 39.9	03.8	223 39.8	33.0
	06	50 49.6	306 50.0	N21 46.3	330 19.7	N23 01.5	85 42.7	S15 03.9	238 42.0	N 5 32.9
W	07	65 52.0	321 49.3	46.2	345 20.4	01.7	100 45.5	04.0	253 44.2	32.8
E	08	80 54.5	336 48.6	46.0	0 21.1	01.8	115 48.3	04.1	268 46.4	32.7
D	09	95 57.0	351 47.9 . .	45.9	15 21.8 . .	02.0	130 51.1 . .	04.2	283 48.6 . .	32.6
N	10	110 59.4	6 47.3	45.7	30 22.4	02.1	145 53.8	04.3	298 50.8	32.4
	11	126 01.9	21 46.6	45.6	45 23.1	02.2	160 56.6	04.4	313 53.0	32.3
E	12	141 04.4	36 45.9	N21 45.4	60 23.8	N23 02.4	175 59.4	S15 04.5	328 55.2	N 5 32.2
S	13	156 06.8	51 45.2	45.3	75 24.5	02.5	191 02.2	04.6	343 57.4	32.1
D	14	171 09.3	66 44.5	45.1	90 25.1	02.7	206 05.0	04.8	358 59.6	32.0
A	15	186 11.7	81 43.8 . .	45.0	105 25.8 . .	02.8	221 07.8 . .	04.9	14 01.8 . .	31.9
Y	16	201 14.2	96 43.2	44.8	120 26.5	03.0	236 10.5	05.0	29 04.0	31.8
	17	216 16.7	111 42.5	44.7	135 27.2	03.1	251 13.3	05.1	44 06.2	31.7
	18	231 19.1	126 41.8	N21 44.5	150 27.9	N23 03.2	266 16.1	S15 05.2	59 08.4	N 5 31.5
	19	246 21.6	141 41.1	44.4	165 28.5	03.4	281 18.9	05.3	74 10.6	31.4
	20	261 24.1	156 40.4	44.2	180 29.2	03.5	296 21.7	05.4	89 12.8	31.3
	21	276 26.5	171 39.7 . .	44.0	195 29.9 . .	03.7	311 24.5 . .	05.5	104 15.0 . .	31.2
	22	291 29.0	186 39.1	43.9	210 30.6	03.8	326 27.2	05.6	119 17.2	31.1
	23	306 31.5	201 38.4	43.7	225 31.3	04.0	341 30.0	05.8	134 19.5	31.0
13	00	321 33.9	216 37.7	N21 43.5	240 31.9	N23 04.1	356 32.8	S15 05.9	149 21.7	N 5 30.9
	01	336 36.4	231 37.0	43.4	255 32.6	04.2	11 35.6	06.0	164 23.9	30.7
	02	351 38.8	246 36.3	43.2	270 33.3	04.4	26 38.4	06.1	179 26.1	30.6
	03	6 41.3	261 35.6 . .	43.1	285 34.0 . .	04.5	41 41.1 . .	06.2	194 28.3 . .	30.5
	04	21 43.8	276 35.0	42.9	300 34.7	04.7	56 43.9	06.3	209 30.5	30.4
	05	36 46.2	291 34.3	42.7	315 35.3	04.8	71 46.7	06.4	224 32.7	30.3
	06	51 48.7	306 33.6	N21 42.5	330 36.0	N23 04.9	86 49.5	S15 06.5	239 34.9	N 5 30.2
	07	66 51.2	321 32.9	42.4	345 36.7	05.1	101 52.3	06.6	254 37.1	30.1
T	08	81 53.6	336 32.2	42.2	0 37.4	05.2	116 55.1	06.8	269 39.3	29.9
H	09	96 56.1	351 31.5 . .	42.0	15 38.1 . .	05.3	131 57.8 . .	06.9	284 41.5 . .	29.8
U	10	111 58.6	6 30.9	41.9	30 38.7	05.5	147 00.6	07.0	299 43.7	29.7
R	11	127 01.0	21 30.2	41.7	45 39.4	05.6	162 03.4	07.1	314 45.9	29.6
S	12	142 03.5	36 29.5	N21 41.5	60 40.1	N23 05.7	177 06.2	S15 07.2	329 48.1	N 5 29.5
D	13	157 06.0	51 28.8	41.3	75 40.8	05.9	192 09.0	07.3	344 50.3	29.4
A	14	172 08.4	66 28.1	41.2	90 41.5	06.0	207 11.8	07.4	359 52.5	29.3
Y	15	187 10.9	81 27.4 . .	41.0	105 42.1 . .	06.2	222 14.5 . .	07.5	14 54.7 . .	29.1
	16	202 13.3	96 26.7	40.8	120 42.8	06.3	237 17.3	07.6	29 56.9	29.0
	17	217 15.8	111 26.1	40.6	135 43.5	06.4	252 20.1	07.8	44 59.1	28.9
	18	232 18.3	126 25.4	N21 40.4	150 44.2	N23 06.6	267 22.9	S15 07.9	60 01.3	N 5 28.8
	19	247 20.7	141 24.7	40.3	165 44.9	06.7	282 25.7	08.0	75 03.5	28.7
	20	262 23.2	156 24.0	40.1	180 45.6	06.8	297 28.5	08.1	90 05.7	28.6
	21	277 25.7	171 23.3 . .	39.9	195 46.2 . .	07.0	312 31.2 . .	08.2	105 07.9 . .	28.5
	22	292 28.1	186 22.6	39.7	210 46.9	07.1	327 34.0	08.3	120 10.1	28.4
	23	307 30.6	201 22.0	39.5	225 47.6	07.2	342 36.8	08.4	135 12.3	28.2
Mer. Pass.	h m	2 37.3	v −0.7	d 0.2	v 0.7	d 0.1	v 2.8	d 0.1	v 2.2	d 0.1

STARS

Name	SHA	Dec
	° ′	° ′
Acamar	315 20.4	S40 15.6
Achernar	335 28.4	S57 10.9
Acrux	173 13.2	S63 09.4
Adhara	255 15.1	S28 58.9
Aldebaran	290 52.8	N16 31.8
Alioth	166 23.3	N55 54.6
Alkaid	153 01.1	N49 16.1
Al Na'ir	27 46.7	S46 54.7
Alnilam	275 49.4	S 1 11.6
Alphard	217 59.2	S 8 42.0
Alphecca	126 13.3	N26 41.1
Alpheratz	357 46.2	N29 08.7
Altair	62 10.7	N 8 53.8
Ankaa	353 18.0	S42 14.9
Antares	112 29.6	S26 27.3
Arcturus	145 58.3	N19 08.0
Atria	107 33.8	S69 03.0
Avior	234 19.9	S59 32.3
Bellatrix	278 35.2	N 6 21.6
Betelgeuse	271 04.6	N 7 24.
Canopus	263 57.8	S52 41.8
Capella	280 38.9	N46 00.4
Deneb	49 33.0	N45 19.
Denebola	182 36.8	N14 31.
Diphda	348 58.5	S17 55.
Dubhe	193 55.5	N61 42.
Elnath	278 16.4	N28 36.
Eltanin	90 47.1	N51 29.5
Enif	33 49.6	N 9 55.
Fomalhaut	15 26.7	S29 34.
Gacrux	172 04.7	S57 10.
Gienah	175 55.5	S17 35.
Hadar	148 52.3	S60 25.
Hamal	328 03.9	N23 30.
Kaus Aust.	83 47.2	S34 22.
Kochab	137 19.6	N74 07.
Markab	13 40.9	N15 15.
Menkar	314 18.0	N 4 07.
Menkent	148 11.1	S36 25.
Miaplacidus	221 41.5	S69 45.
Mirfak	308 44.5	N49 53.
Nunki	76 01.5	S26 17.
Peacock	53 23.0	S56 42.
Pollux	243 31.5	N28 00.
Procyon	245 03.0	N 5 12.
Rasalhague	96 08.9	N12 33.
Regulus	207 46.8	N11 55.
Rigel	281 14.9	S 8 11.
Rigil Kent.	139 55.9	S60 52.
Sabik	102 15.6	S15 44.
Schedar	349 43.6	N56 35.
Shaula	96 25.5	S37 06.
Sirius	258 36.5	S16 43.
Spica	158 34.4	S11 12.
Suhail	222 55.1	S43 28.
Vega	80 40.6	N38 47.
Zuben'ubi	137 08.6	S16 05.

	SHA	Mer. Pas
	° ′	h m
Venus	256 19.3	9 33
Mars	279 40.9	7 59
Jupiter	34 51.2	0 18
Saturn	187 54.0	14 04

	UT	SUN GHA	SUN Dec	MOON GHA	v	MOON Dec	d	HP
	d h	° ′	° ′	° ′	′	° ′	′	′
	11 00	178 41.2	N15 17.4	308 14.2	14.1	N10 23.0	13.2	56.1
	01	193 41.3	16.7	322 47.3	14.0	10 36.2	13.0	56.1
	02	208 41.4	16.0	337 20.3	13.9	10 49.2	13.1	56.1
	03	223 41.5	. . 15.2	351 53.2	13.9	11 02.3	13.0	56.1
	04	238 41.6	14.5	6 26.1	13.8	11 15.3	13.0	56.2
	05	253 41.7	13.7	20 58.9	13.7	11 28.3	12.9	56.2
	06	268 41.8	N15 13.0	35 31.6	13.6	N11 41.2	12.9	56.2
	07	283 41.9	12.2	50 04.2	13.6	11 54.1	12.9	56.3
T	08	298 42.0	11.5	64 36.8	13.5	12 07.0	12.8	56.3
U	09	313 42.1	. . 10.8	79 09.3	13.4	12 19.8	12.8	56.3
E	10	328 42.2	10.0	93 41.7	13.4	12 32.6	12.7	56.3
S	11	343 42.4	09.3	108 14.1	13.3	12 45.3	12.7	56.4
D	12	358 42.5	N15 08.5	122 46.4	13.2	N12 58.0	12.6	56.4
	13	13 42.6	07.8	137 18.6	13.1	13 10.6	12.6	56.4
A	14	28 42.7	07.0	151 50.7	13.0	13 23.2	12.5	56.4
Y	15	43 42.8	. . 06.3	166 22.7	13.0	13 35.7	12.5	56.5
	16	58 42.9	05.5	180 54.7	12.9	13 48.2	12.4	56.5
	17	73 43.0	04.8	195 26.6	12.8	14 00.6	12.4	56.5
	18	88 43.1	N15 04.0	209 58.4	12.7	N14 13.0	12.3	56.5
	19	103 43.2	03.3	224 30.1	12.6	14 25.3	12.3	56.6
	20	118 43.3	02.5	239 01.7	12.6	14 37.6	12.2	56.6
	21	133 43.4	. . 01.8	253 33.3	12.4	14 49.8	12.2	56.6
	22	148 43.5	01.0	268 04.7	12.4	15 02.0	12.1	56.6
	23	163 43.6	15 00.3	282 36.1	12.3	15 14.1	12.0	56.7
	12 00	178 43.7	N14 59.5	297 07.4	12.2	N15 26.1	12.0	56.7
	01	193 43.8	58.8	311 38.6	12.1	15 38.1	11.9	56.7
	02	208 43.9	58.0	326 09.7	12.0	15 50.0	11.8	56.8
	03	223 44.0	. . 57.3	340 40.7	11.9	16 01.8	11.8	56.8
	04	238 44.1	56.5	355 11.6	11.8	16 13.6	11.7	56.8
	05	253 44.2	55.8	9 42.4	11.8	16 25.3	11.7	56.8
	06	268 44.3	N14 55.0	24 13.2	11.6	N16 37.0	11.5	56.9
W	07	283 44.4	54.3	38 43.8	11.6	16 48.5	11.5	56.9
E	08	298 44.6	53.5	53 14.4	11.4	17 00.0	11.5	56.9
D	09	313 44.7	. . 52.8	67 44.8	11.4	17 11.5	11.3	57.0
N	10	328 44.8	52.0	82 15.2	11.2	17 22.8	11.3	57.0
E	11	343 44.9	51.3	96 45.4	11.2	17 34.1	11.2	57.0
S	12	358 45.0	N14 50.5	111 15.6	11.0	N17 45.3	11.2	57.0
D	13	13 45.1	49.8	125 45.6	11.0	17 56.5	11.0	57.1
A	14	28 45.2	49.0	140 15.6	10.9	18 07.5	11.0	57.1
Y	15	43 45.3	. . 48.2	154 45.5	10.7	18 18.5	10.9	57.1
	16	58 45.4	47.5	169 15.2	10.7	18 29.4	10.8	57.2
	17	73 45.5	46.7	183 44.9	10.5	18 40.2	10.7	57.2
	18	88 45.6	N14 46.0	198 14.4	10.5	N18 50.9	10.6	57.2
	19	103 45.7	45.2	212 43.9	10.3	19 01.5	10.6	57.3
	20	118 45.9	44.4	227 13.2	10.2	19 12.1	10.4	57.3
	21	133 46.0	. . 43.7	241 42.4	10.2	19 22.5	10.4	57.3
	22	148 46.1	42.9	256 11.6	10.0	19 32.9	10.3	57.3
	23	163 46.2	42.2	270 40.6	9.9	19 43.2	10.2	57.4
	13 00	178 46.3	N14 41.4	285 09.5	9.8	N19 53.4	10.1	57.4
	01	193 46.4	40.7	299 38.3	9.7	20 03.5	10.0	57.4
	02	208 46.5	39.9	314 07.0	9.6	20 13.5	9.9	57.5
	03	223 46.6	. . 39.1	328 35.6	9.5	20 23.4	9.8	57.5
	04	238 46.7	38.4	343 04.1	9.4	20 33.2	9.7	57.5
	05	253 46.9	37.6	357 32.5	9.3	20 42.9	9.6	57.6
	06	268 47.0	N14 36.8	12 00.8	9.2	N20 52.5	9.5	57.6
T	07	283 47.1	36.1	26 29.0	9.0	21 02.0	9.3	57.6
	08	298 47.2	35.3	40 57.0	9.0	21 11.3	9.3	57.7
H	09	313 47.3	. . 34.6	55 25.0	8.8	21 20.6	9.2	57.7
U	10	328 47.4	33.8	69 52.8	8.7	21 29.8	9.1	57.7
R	11	343 47.5	33.0	84 20.5	8.6	21 38.9	8.9	57.7
S	12	358 47.6	N14 32.3	98 48.1	8.6	N21 47.8	8.9	57.8
D	13	13 47.8	31.5	113 15.7	8.4	21 56.7	8.7	57.8
A	14	28 47.9	30.7	127 43.1	8.2	22 05.4	8.6	57.8
Y	15	43 48.0	. . 30.0	142 10.3	8.2	22 14.0	8.5	57.9
	16	58 48.1	29.2	156 37.5	8.1	22 22.5	8.4	57.9
	17	73 48.2	28.4	171 04.6	7.9	22 30.9	8.2	57.9
	18	88 48.3	N14 27.7	185 31.5	7.9	N22 39.1	8.2	58.0
	19	103 48.4	26.9	199 58.4	7.7	22 47.3	8.0	58.0
	20	118 48.6	26.1	214 25.1	7.6	22 55.3	7.9	58.0
	21	133 48.7	. . 25.4	228 51.7	7.6	23 03.2	7.7	58.1
	22	148 48.8	24.6	243 18.3	7.4	23 10.9	7.6	58.1
	23	163 48.9	23.8	257 44.7	7.3	N23 18.5	7.5	58.1
		SD 15.8	d 0.8	SD 15.4		15.5		15.7

Lat.	Twilight Naut.	Twilight Civil	Sunrise	Moonrise 11	Moonrise 12	Moonrise 13	Moonrise 14
°	h m	h m	h m	h m	h m	h m	h m
N 72	////	////	02 03	18 36	□	□	□
N 70	////	////	02 41	19 04	18 12	□	□
68	////	01 17	03 07	19 25	18 58	□	□
66	////	02 05	03 27	19 42	19 28	19 02	□
64	////	02 35	03 43	19 56	19 51	19 45	19 35
62	01 09	02 57	03 56	20 08	20 09	20 14	20 27
60	01 52	03 14	04 07	20 18	20 24	20 36	20 59
N 58	02 19	03 28	04 17	20 26	20 37	20 54	21 23
56	02 40	03 41	04 25	20 34	20 48	21 10	21 43
54	02 56	03 51	04 33	20 41	20 58	21 23	21 59
52	03 10	04 00	04 39	20 47	21 07	21 34	22 13
50	03 22	04 09	04 45	20 53	21 15	21 45	22 25
45	03 46	04 26	04 58	21 05	21 32	22 06	22 51
N 40	04 04	04 39	05 09	21 15	21 46	22 24	23 11
35	04 18	04 51	05 18	21 24	21 58	22 39	23 28
30	04 30	05 00	05 26	21 32	22 09	22 52	23 43
20	04 49	05 16	05 39	21 45	22 27	23 14	24 08
N 10	05 03	05 29	05 51	21 57	22 43	23 33	24 30
0	05 15	05 40	06 02	22 08	22 58	23 52	24 50
S 10	05 26	05 51	06 12	22 20	23 13	24 10	00 10
20	05 35	06 01	06 24	22 32	23 29	24 29	00 29
30	05 43	06 12	06 36	22 46	23 48	24 52	00 52
35	05 48	06 18	06 44	22 54	23 59	25 06	01 06
40	05 52	06 24	06 52	23 03	24 11	00 11	01 21
45	05 56	06 31	07 02	23 14	24 26	00 26	01 40
S 50	06 01	06 39	07 13	23 28	24 45	00 45	02 03
52	06 03	06 43	07 19	23 34	24 54	00 54	02 14
54	06 05	06 47	07 25	23 41	25 03	01 03	02 27
56	06 07	06 51	07 31	23 49	25 15	01 15	02 42
58	06 10	06 56	07 39	23 57	25 27	01 27	02 59
S 60	06 12	07 02	07 47	24 07	00 07	01 42	03 20

Lat.	Sunset	Twilight Civil	Twilight Naut.	Moonset 11	Moonset 12	Moonset 13	Moonset 14
°	h m	h m	h m	h m	h m	h m	h m
N 72	22 01	////	////	12 52	□	□	□
N 70	21 25	////	////	12 26	14 59	□	□
68	21 00	22 44	////	12 07	14 14	□	□
66	20 40	22 00	////	11 52	13 45	16 00	□
64	20 25	21 32	////	11 39	13 23	15 18	17 27
62	20 12	21 11	22 53	11 28	13 06	14 50	16 35
60	20 01	20 54	22 14	11 19	12 52	14 28	16 03
N 58	19 52	20 40	21 48	11 12	12 39	14 10	15 40
56	19 44	20 28	21 28	11 05	12 29	13 56	15 21
54	19 36	20 17	21 12	10 59	12 20	13 43	15 05
52	19 30	20 08	20 58	10 53	12 11	13 32	14 51
50	19 24	20 00	20 46	10 48	12 04	13 22	14 39
45	19 11	19 43	20 23	10 38	11 48	13 01	14 14
N 40	19 01	19 30	20 05	10 29	11 35	12 44	13 54
35	18 52	19 19	19 51	10 21	11 25	12 30	13 38
30	18 44	19 09	19 39	10 15	11 15	12 18	13 23
20	18 31	18 54	19 21	10 03	10 59	11 58	12 59
N 10	18 19	18 41	19 06	09 54	10 45	11 40	12 38
0	18 08	18 30	18 55	09 44	10 32	11 23	12 19
S 10	17 58	18 19	18 45	09 35	10 18	11 06	11 59
20	17 47	18 09	18 36	09 25	10 04	10 48	11 39
30	17 34	17 59	18 27	09 14	09 48	10 28	11 15
35	17 27	17 53	18 23	09 08	09 39	10 16	11 01
40	17 18	17 47	18 19	09 00	09 29	10 03	10 44
45	17 09	17 40	18 14	08 52	09 16	09 47	10 25
S 50	16 57	17 31	18 10	08 42	09 02	09 27	10 01
52	16 52	17 28	18 08	08 37	08 55	09 18	09 49
54	16 46	17 24	18 06	08 32	08 47	09 07	09 36
56	16 40	17 19	18 04	08 27	08 39	08 55	09 21
58	16 32	17 15	18 01	08 20	08 29	08 42	09 03
S 60	16 24	17 09	17 59	08 13	08 18	08 26	08 42

Day	SUN Eqn. of Time 00h	SUN Eqn. of Time 12h	SUN Mer. Pass.	MOON Mer. Pass. Upper	MOON Mer. Pass. Lower	Age	Phase
d	m s	m s	h m	h m	h m	d %	
11	05 15	05 10	12 05	03 33	15 56	20 74	
12	05 05	05 00	12 05	04 20	16 44	21 64	◐
13	04 55	04 50	12 05	05 10	17 37	22 53	

	UT	ARIES	VENUS −4.0		MARS +1.0		JUPITER −2.9		SATURN +1.1	
		GHA	GHA	Dec	GHA	Dec	GHA	Dec	GHA	Dec
	d h	° ′	° ′	° ′	° ′	° ′	° ′	° ′	° ′	° ′
	14 00	322 33.1	216 21.3	N21 39.3	240 48.3	N23 07.4	357 39.6	S15 08.5	150 14.5	N 5 28.1
	01	337 35.5	231 20.6	39.1	255 49.0	07.5	12 42.4	08.6	165 16.7	28.0
	02	352 38.0	246 19.9	38.9	270 49.6	07.6	27 45.2	08.8	180 18.9	27.9
	03	7 40.5	261 19.2	. . 38.8	285 50.3	. . 07.8	42 47.9	. . 08.9	195 21.1	. . 27.8
	04	22 42.9	276 18.5	38.6	300 51.0	07.9	57 50.7	09.0	210 23.3	27.7
	05	37 45.4	291 17.8	38.4	315 51.7	08.0	72 53.5	09.1	225 25.5	27.5
	06	52 47.8	306 17.2	N21 38.2	330 52.4	N23 08.1	87 56.3	S15 09.2	240 27.7	N 5 27.4
	07	67 50.3	321 16.5	38.0	345 53.1	08.3	102 59.1	09.3	255 29.9	27.3
	08	82 52.8	336 15.8	37.8	0 53.7	08.4	118 01.8	09.4	270 32.1	27.2
F	09	97 55.2	351 15.1	. . 37.6	15 54.4	. . 08.5	133 04.6	. . 09.5	285 34.3	. . 27.1
R	10	112 57.7	6 14.4	37.4	30 55.1	08.7	148 07.4	09.6	300 36.5	27.0
I	11	128 00.2	21 13.7	37.2	45 55.8	08.8	163 10.2	09.8	315 38.7	26.9
D	12	143 02.6	36 13.0	N21 37.0	60 56.5	N23 08.9	178 13.0	S15 09.9	330 40.9	N 5 26.7
A	13	158 05.1	51 12.4	36.8	75 57.2	09.0	193 15.8	10.0	345 43.1	26.6
Y	14	173 07.6	66 11.7	36.6	90 57.8	09.2	208 18.5	10.1	0 45.3	26.5
	15	188 10.0	81 11.0	. . 36.4	105 58.5	. . 09.3	223 21.3	. . 10.2	15 47.5	. . 26.4
	16	203 12.5	96 10.3	36.2	120 59.2	09.4	238 24.1	10.3	30 49.7	26.3
	17	218 15.0	111 09.6	36.0	135 59.9	09.6	253 26.9	10.4	45 51.9	26.2
	18	233 17.4	126 08.9	N21 35.8	151 00.6	N23 09.7	268 29.7	S15 10.5	60 54.1	N 5 26.1
	19	248 19.9	141 08.2	35.6	166 01.3	09.8	283 32.5	10.6	75 56.3	25.9
	20	263 22.3	156 07.6	35.3	181 02.0	09.9	298 35.2	10.7	90 58.5	25.8
	21	278 24.8	171 06.9	. . 35.1	196 02.6	. . 10.1	313 38.0	. . 10.9	106 00.7	. . 25.7
	22	293 27.3	186 06.2	34.9	211 03.3	10.2	328 40.8	11.0	121 02.9	25.6
	23	308 29.7	201 05.5	34.7	226 04.0	10.3	343 43.6	11.1	136 05.1	25.5
	15 00	323 32.2	216 04.8	N21 34.5	241 04.7	N23 10.4	358 46.4	S15 11.2	151 07.3	N 5 25.4
	01	338 34.7	231 04.1	34.3	256 05.4	10.6	13 49.2	11.3	166 09.5	25.3
	02	353 37.1	246 03.4	34.1	271 06.1	10.7	28 51.9	11.4	181 11.7	25.1
	03	8 39.6	261 02.8	. . 33.9	286 06.8	. . 10.8	43 54.7	. . 11.5	196 13.9	. . 25.0
	04	23 42.1	276 02.1	33.6	301 07.4	10.9	58 57.5	11.6	211 16.1	24.9
	05	38 44.5	291 01.4	33.4	316 08.1	11.0	74 00.3	11.7	226 18.3	24.8
	06	53 47.0	306 00.7	N21 33.2	331 08.8	N23 11.2	89 03.1	S15 11.9	241 20.5	N 5 24.7
	07	68 49.4	321 00.0	33.0	346 09.5	11.3	104 05.9	12.0	256 22.7	24.6
S	08	83 51.9	335 59.3	32.8	1 10.2	11.4	119 08.6	12.1	271 24.9	24.5
A	09	98 54.4	350 58.6	. . 32.5	16 10.9	. . 11.5	134 11.4	. . 12.2	286 27.1	. . 24.3
T	10	113 56.8	5 58.0	32.3	31 11.6	11.7	149 14.2	12.3	301 29.3	24.2
U	11	128 59.3	20 57.3	32.1	46 12.2	11.8	164 17.0	12.4	316 31.5	24.1
R	12	144 01.8	35 56.6	N21 31.9	61 12.9	N23 11.9	179 19.8	S15 12.5	331 33.7	N 5 24.0
D	13	159 04.2	50 55.9	31.6	76 13.6	12.0	194 22.6	12.6	346 35.9	23.9
A	14	174 06.7	65 55.2	31.4	91 14.3	12.1	209 25.3	12.7	1 38.1	23.8
Y	15	189 09.2	80 54.5	. . 31.2	106 15.0	. . 12.3	224 28.1	. . 12.9	16 40.3	. . 23.6
	16	204 11.6	95 53.8	31.0	121 15.7	12.4	239 30.9	13.0	31 42.5	23.5
	17	219 14.1	110 53.2	30.7	136 16.4	12.5	254 33.7	13.1	46 44.7	23.4
	18	234 16.6	125 52.5	N21 30.5	151 17.1	N23 12.6	269 36.5	S15 13.2	61 46.9	N 5 23.3
	19	249 19.0	140 51.8	30.3	166 17.7	12.7	284 39.3	13.3	76 49.1	23.2
	20	264 21.5	155 51.1	30.0	181 18.4	12.8	299 42.0	13.4	91 51.3	23.1
	21	279 23.9	170 50.4	. . 29.8	196 19.1	. . 13.0	314 44.8	. . 13.5	106 53.5	. . 23.0
	22	294 26.4	185 49.7	29.6	211 19.8	13.1	329 47.6	13.6	121 55.7	22.8
	23	309 28.9	200 49.0	29.3	226 20.5	13.2	344 50.4	13.7	136 57.9	22.7
	16 00	324 31.3	215 48.4	N21 29.1	241 21.2	N23 13.3	359 53.2	S15 13.8	152 00.1	N 5 22.6
	01	339 33.8	230 47.7	28.8	256 21.9	13.4	14 56.0	14.0	167 02.3	22.5
	02	354 36.3	245 47.0	28.6	271 22.6	13.5	29 58.7	14.1	182 04.5	22.4
	03	9 38.7	260 46.3	. . 28.4	286 23.2	. . 13.7	45 01.5	. . 14.2	197 06.7	. . 22.3
	04	24 41.2	275 45.6	28.1	301 23.9	13.8	60 04.3	14.3	212 08.9	22.1
	05	39 43.7	290 44.9	27.9	316 24.6	13.9	75 07.1	14.4	227 11.1	22.0
	06	54 46.1	305 44.2	N21 27.6	331 25.3	N23 14.0	90 09.9	S15 14.5	242 13.3	N 5 21.9
	07	69 48.6	320 43.5	27.4	346 26.0	14.1	105 12.7	14.6	257 15.5	21.8
	08	84 51.1	335 42.9	27.1	1 26.7	14.2	120 15.4	14.7	272 17.7	21.7
S	09	99 53.5	350 42.2	. . 26.9	16 27.4	. . 14.3	135 18.2	. . 14.8	287 19.9	. . 21.6
U	10	114 56.0	5 41.5	26.6	31 28.1	14.5	150 21.0	14.9	302 22.1	21.5
N	11	129 58.4	20 40.8	26.4	46 28.8	14.6	165 23.8	15.1	317 24.3	21.3
D	12	145 00.9	35 40.1	N21 26.1	61 29.5	N23 14.7	180 26.6	S15 15.2	332 26.5	N 5 21.2
A	13	160 03.4	50 39.4	25.9	76 30.1	14.8	195 29.4	15.3	347 28.7	21.1
Y	14	175 05.8	65 38.7	25.6	91 30.8	14.9	210 32.1	15.4	2 30.9	21.0
	15	190 08.3	80 38.1	. . 25.4	106 31.5	. . 15.0	225 34.9	. . 15.5	17 33.1	. . 20.9
	16	205 10.8	95 37.4	25.1	121 32.2	15.1	240 37.7	15.6	32 35.3	20.8
	17	220 13.2	110 36.7	24.9	136 32.9	15.2	255 40.5	15.7	47 37.5	20.6
	18	235 15.7	125 36.0	N21 24.6	151 33.6	N23 15.3	270 43.3	S15 15.8	62 39.7	N 5 20.5
	19	250 18.2	140 35.3	24.4	166 34.3	15.5	285 46.1	15.9	77 41.9	20.4
	20	265 20.6	155 34.6	24.1	181 35.0	15.6	300 48.8	16.0	92 44.1	20.3
	21	280 23.1	170 33.9	. . 23.8	196 35.7	. . 15.7	315 51.6	. . 16.2	107 46.3	. . 20.2
	22	295 25.6	185 33.3	23.6	211 36.4	15.8	330 54.4	16.3	122 48.5	20.1
	23	310 28.0	200 32.6	23.3	226 37.1	15.9	345 57.2	16.4	137 50.7	19.9
	Mer. Pass.	h m 2 25.5	*v* −0.7	*d* 0.2	*v* 0.7	*d* 0.1	*v* 2.8	*d* 0.1	*v* 2.2	*d* 0.1

STARS

Name	SHA	Dec
	° ′	° ′
Acamar	315 20.3	S40 15.6
Achernar	335 28.4	S57 10.9
Acrux	173 13.2	S63 09.4
Adhara	255 15.1	S28 58.9
Aldebaran	290 52.7	N16 31.8
Alioth	166 23.3	N55 54.6
Alkaid	153 01.2	N49 16.1
Al Na'ir	27 46.6	S46 54.7
Alnilam	275 49.4	S 1 11.6
Alphard	217 59.2	S 8 42.0
Alphecca	126 13.3	N26 41.1
Alpheratz	357 46.2	N29 08.7
Altair	62 10.7	N 8 53.8
Ankaa	353 18.0	S42 14.9
Antares	112 29.6	S26 27.1
Arcturus	145 58.4	N19 08.
Atria	107 33.9	S69 03.
Avior	234 19.9	S59 32.
Bellatrix	278 35.2	N 6 21.
Betelgeuse	271 04.5	N 7 24.
Canopus	263 57.8	S52 41.
Capella	280 38.9	N46 00.
Deneb	49 33.0	N45 19.
Denebola	182 36.8	N14 31.
Diphda	348 58.4	S17 55.
Dubhe	193 55.5	N61 42.
Elnath	278 16.4	N28 36.
Eltanin	90 47.1	N51 29.
Enif	33 49.6	N 9 55.
Fomalhaut	15 26.7	S29 34.
Gacrux	172 04.7	S57 10.
Gienah	175 55.5	S17 35.
Hadar	148 52.3	S60 25.
Hamal	328 03.9	N23 30.
Kaus Aust.	83 47.3	S34 22.
Kochab	137 19.7	N74 07.
Markab	13 40.9	N15 15.
Menkar	314 18.0	N 4 07.
Menkent	148 11.1	S36 25.
Miaplacidus	221 41.5	S69 45.
Mirfak	308 44.5	N49 53.
Nunki	76 01.5	S26 17.
Peacock	53 22.9	S56 42.
Pollux	243 31.5	N28 00.
Procyon	245 03.0	N 5 12.
Rasalhague	96 08.9	N12 33.
Regulus	207 46.8	N11 55.
Rigel	281 14.9	S 8 11.
Rigil Kent.	139 55.9	S60 52.
Sabik	102 15.6	S15 44.
Schedar	349 43.6	N56 35.
Shaula	96 25.5	S37 06.
Sirius	258 36.5	S16 43.
Spica	158 34.4	S11 12.
Suhail	222 55.1	S43 28.
Vega	80 40.6	N38 47
Zuben'ubi	137 08.6	S16 05

	SHA	Mer. Pas
	° ′	h m
Venus	252 32.6	9 36
Mars	277 32.5	7 55
Jupiter	35 14.2	0 05
Saturn	187 35.1	13 53

UT	SUN GHA	SUN Dec	MOON GHA	v	MOON Dec	d	HP
d h	° ′	° ′	° ′	′	° ′	′	′
14 00	178 49.0	N14 23.0	272 11.0	7.2	N23 26.0	7.4	58.2
01	193 49.1	22.3	286 37.2	7.0	23 33.4	7.3	58.2
02	208 49.3	21.5	301 03.2	7.0	23 40.7	7.1	58.2
03	223 49.4	. . 20.7	315 29.2	6.9	23 47.8	6.9	58.3
04	238 49.5	20.0	329 55.1	6.8	23 54.7	6.9	58.3
05	253 49.6	19.2	344 20.9	6.6	24 01.6	6.7	58.3
06	268 49.7	N14 18.4	358 46.5	6.6	N24 08.3	6.5	58.4
07	283 49.8	17.6	13 12.1	6.4	24 14.8	6.4	58.4
08	298 50.0	16.9	27 37.5	6.4	24 21.2	6.3	58.4
09	313 50.1	. . 16.1	42 02.9	6.2	24 27.5	6.1	58.5
10	328 50.2	15.3	56 28.1	6.1	24 33.6	6.0	58.5
11	343 50.3	14.5	70 53.2	6.1	24 39.6	5.8	58.5
12	358 50.4	N14 13.8	85 18.3	5.9	N24 45.4	5.7	58.6
13	13 50.6	13.0	99 43.2	5.9	24 51.1	5.5	58.6
14	28 50.7	12.2	114 08.1	5.7	24 56.6	5.4	58.6
15	43 50.8	. . 11.4	128 32.8	5.6	25 02.0	5.2	58.6
16	58 50.9	10.7	142 57.4	5.6	25 07.2	5.1	58.7
17	73 51.0	09.9	157 22.0	5.4	25 12.3	4.9	58.7
18	88 51.2	N14 09.1	171 46.4	5.4	N25 17.2	4.8	58.7
19	103 51.3	08.3	186 10.8	5.3	25 22.0	4.6	58.8
20	118 51.4	07.6	200 35.1	5.1	25 26.6	4.4	58.8
21	133 51.5	. . 06.8	214 59.2	5.1	25 31.0	4.3	58.8
22	148 51.6	06.0	229 23.3	5.0	25 35.3	4.1	58.9
23	163 51.8	05.2	243 47.3	4.9	25 39.4	4.0	58.9
15 00	178 51.9	N14 04.4	258 11.2	4.9	N25 43.4	3.8	58.9
01	193 52.0	03.7	272 35.1	4.7	25 47.2	3.6	59.0
02	208 52.1	02.9	286 58.8	4.7	25 50.8	3.5	59.0
03	223 52.2	. . 02.1	301 22.5	4.5	25 54.3	3.2	59.0
04	238 52.4	01.3	315 46.0	4.5	25 57.5	3.2	59.1
05	253 52.5	14 00.5	330 09.5	4.5	26 00.7	2.9	59.1
06	268 52.6	N13 59.8	344 33.0	4.3	N26 03.6	2.8	59.1
07	283 52.7	59.0	358 56.3	4.3	26 06.4	2.6	59.2
08	298 52.9	58.2	13 19.6	4.2	26 09.0	2.4	59.2
09	313 53.0	. . 57.4	27 42.8	4.1	26 11.4	2.3	59.2
10	328 53.1	56.6	42 05.9	4.1	26 13.7	2.0	59.3
11	343 53.2	55.8	56 29.0	4.0	26 15.7	1.9	59.3
12	358 53.4	N13 55.1	70 52.0	3.9	N26 17.6	1.8	59.3
13	13 53.5	54.3	85 14.9	3.9	26 19.4	1.5	59.3
14	28 53.6	53.5	99 37.8	3.8	26 20.9	1.4	59.4
15	43 53.7	. . 52.7	114 00.6	3.8	26 22.3	1.1	59.4
16	58 53.9	51.9	128 23.4	3.7	26 23.4	1.0	59.4
17	73 54.0	51.1	142 46.1	3.6	26 24.4	0.8	59.5
18	88 54.1	N13 50.3	157 08.7	3.6	N26 25.2	0.7	59.5
19	103 54.2	49.6	171 31.3	3.6	26 25.9	0.4	59.5
20	118 54.4	48.8	185 53.9	3.5	26 26.3	0.3	59.6
21	133 54.5	. . 48.0	200 16.4	3.4	26 26.6	0.1	59.6
22	148 54.6	47.2	214 38.8	3.5	26 26.7	0.1	59.6
23	163 54.7	46.4	229 01.3	3.3	26 26.6	0.3	59.6
16 00	178 54.9	N13 45.6	243 23.6	3.4	N26 26.3	0.5	59.7
01	193 55.0	44.8	257 46.0	3.3	26 25.8	0.7	59.7
02	208 55.1	44.0	272 08.3	3.3	26 25.1	0.8	59.7
03	223 55.3	. . 43.3	286 30.6	3.2	26 24.3	1.1	59.8
04	238 55.4	42.5	300 52.8	3.2	26 23.2	1.2	59.8
05	253 55.5	41.7	315 15.0	3.2	26 22.0	1.4	59.8
06	268 55.6	N13 40.9	329 37.2	3.2	N26 20.6	1.6	59.8
07	283 55.8	40.1	343 59.4	3.1	26 19.0	1.8	59.9
08	298 55.9	39.3	358 21.5	3.2	26 17.2	2.0	59.9
09	313 56.0	. . 38.5	12 43.7	3.1	26 15.2	2.2	59.9
10	328 56.2	37.7	27 05.8	3.1	26 13.0	2.4	60.0
11	343 56.3	36.9	41 27.9	3.1	26 10.6	2.5	60.0
12	358 56.4	N13 36.1	55 50.0	3.1	N26 08.1	2.8	60.0
13	13 56.5	35.3	70 12.1	3.0	26 05.3	2.9	60.0
14	28 56.7	34.5	84 34.1	3.1	26 02.4	3.1	60.1
15	43 56.8	. . 33.7	98 56.2	3.1	25 59.3	3.4	60.1
16	58 56.9	32.9	113 18.3	3.1	25 55.9	3.5	60.1
17	73 57.1	32.2	127 40.4	3.1	25 52.4	3.7	60.1
18	88 57.2	N13 31.4	142 02.5	3.1	N25 48.7	3.9	60.2
19	103 57.3	30.6	156 24.6	3.1	25 44.8	4.0	60.2
20	118 57.5	29.8	170 46.7	3.1	25 40.8	4.3	60.2
21	133 57.6	. . 29.0	185 08.8	3.1	25 36.5	4.4	60.2
22	148 57.7	28.2	199 30.9	3.1	25 32.1	4.7	60.3
23	163 57.8	27.4	213 53.0	3.2	N25 27.4	4.8	60.3
	SD 15.8	d 0.8	SD 16.0		16.2		16.4

Lat.	Twilight Naut.	Twilight Civil	Sunrise	Moonrise 14	Moonrise 15	Moonrise 16	Moonrise 17
°	h m	h m	h m	h m	h m	h m	h m
N 72	////	////	02 25	▭	▭	▭	▭
N 70	////	////	02 57	▭	▭	▭	▭
68	////	01 44	03 20	▭	▭	▭	▭
66	////	02 21	03 38	▭	▭	▭	23 11
64	////	02 47	03 52	19 35	▭	21 34	23 56
62	01 32	03 07	04 04	20 27	21 06	22 32	24 26
60	02 06	03 23	04 14	20 59	21 46	23 05	24 48
N 58	02 30	03 36	04 23	21 23	22 13	23 30	25 07
56	02 49	03 47	04 31	21 43	22 35	23 49	25 22
54	03 04	03 57	04 38	21 59	22 52	24 06	00 06
52	03 17	04 06	04 44	22 13	23 08	24 20	00 20
50	03 28	04 14	04 50	22 25	23 21	24 32	00 32
45	03 50	04 30	05 02	22 51	23 48	24 58	00 58
N 40	04 07	04 43	05 12	23 11	24 09	00 09	01 18
35	04 21	04 53	05 20	23 28	24 27	00 27	01 35
30	04 32	05 02	05 27	23 43	24 42	00 42	01 49
20	04 50	05 17	05 40	24 08	00 08	01 08	02 14
N 10	05 04	05 29	05 51	24 30	00 30	01 31	02 35
0	05 15	05 40	06 01	24 50	00 50	01 52	02 55
S 10	05 25	05 49	06 11	00 10	01 10	02 13	03 15
20	05 33	05 59	06 22	00 29	01 32	02 35	03 36
30	05 41	06 09	06 34	00 52	01 58	03 02	04 01
35	05 45	06 14	06 40	01 06	02 13	03 17	04 15
40	05 48	06 20	06 48	01 21	02 30	03 35	04 31
45	05 52	06 27	06 57	01 40	02 51	03 56	04 51
S 50	05 56	06 34	07 08	02 03	03 18	04 24	05 16
52	05 58	06 38	07 13	02 14	03 31	04 37	05 28
54	06 00	06 41	07 18	02 27	03 46	04 52	05 41
56	06 01	06 45	07 25	02 42	04 04	05 10	05 57
58	06 03	06 50	07 31	02 59	04 25	05 32	06 16
S 60	06 05	06 54	07 39	03 20	04 52	06 01	06 38

Lat.	Sunset	Twilight Civil	Twilight Naut.	Moonset 14	Moonset 15	Moonset 16	Moonset 17
°	h m	h m	h m	h m	h m	h m	h m
N 72	21 38	////	////	▭	▭	▭	▭
N 70	21 08	23 29	////	▭	▭	▭	▭
68	20 46	22 18	////	▭	▭	▭	▭
66	20 29	21 43	////	▭	▭	▭	20 24
64	20 15	21 18	23 33	17 27	▭	19 50	19 38
62	20 03	20 59	22 30	16 35	18 04	18 51	19 08
60	19 53	20 44	21 59	16 03	17 25	18 17	18 45
N 58	19 44	20 31	21 36	15 40	16 57	17 52	18 26
56	19 37	20 20	21 18	15 21	16 36	17 32	18 10
54	19 30	20 10	21 03	15 05	16 18	17 16	17 57
52	19 24	20 02	20 50	14 51	16 03	17 01	17 45
50	19 18	19 54	20 39	14 39	15 50	16 49	17 34
45	19 06	19 38	20 17	14 14	15 23	16 23	17 12
N 40	18 57	19 26	20 01	13 54	15 01	16 02	16 54
35	18 48	19 15	19 47	13 38	14 44	15 45	16 39
30	18 41	19 06	19 36	13 23	14 28	15 30	16 26
20	18 29	18 52	19 18	12 59	14 02	15 05	16 03
N 10	18 18	18 40	19 05	12 38	13 40	14 43	15 43
0	18 08	18 29	18 54	12 19	13 19	14 22	15 25
S 10	17 58	18 20	18 44	11 59	12 58	14 01	15 06
20	17 47	18 10	18 36	11 39	12 36	13 39	14 46
30	17 36	18 00	18 29	11 15	12 10	13 13	14 23
35	17 29	17 55	18 25	11 01	11 54	12 58	14 09
40	17 21	17 49	18 21	10 44	11 37	12 40	13 53
45	17 12	17 43	18 17	10 25	11 15	12 19	13 34
S 50	17 02	17 35	18 13	10 01	10 48	11 52	13 10
52	16 57	17 32	18 12	09 49	10 35	11 39	12 59
54	16 51	17 28	18 10	09 36	10 20	11 23	12 46
56	16 45	17 25	18 08	09 21	10 02	11 05	12 30
58	16 38	17 20	18 07	09 03	09 41	10 44	12 12
S 60	16 31	17 16	18 05	08 42	09 13	10 16	11 50

Day	SUN Eqn. of Time 00^h	SUN Eqn. of Time 12^h	SUN Mer. Pass.	MOON Mer. Pass. Upper	MOON Mer. Pass. Lower	MOON Age	MOON Phase
d	m s	m s	h m	h m	h m	d %	
14	04 44	04 38	12 05	06 05	18 34	23 42	
15	04 33	04 27	12 04	07 04	19 35	24 31	
16	04 21	04 15	12 04	08 07	20 38	25 21	

UT	ARIES	VENUS −3.9		MARS +1.0		JUPITER −2.9		SATURN +1.1	
	GHA	GHA	Dec	GHA	Dec	GHA	Dec	GHA	Dec
d h	° ′	° ′	° ′	° ′	° ′	° ′	° ′	° ′	° ′
17 00	325 30.5	215 31.9	N21 23.1	241 37.7	N23 16.0	1 00.0	S15 16.5	152 52.9	N 5 19.8
01	340 32.9	230 31.2	22.8	256 38.4	16.1	16 02.8	16.6	167 55.1	19.7
02	355 35.4	245 30.5	22.5	271 39.1	16.2	31 05.5	16.7	182 57.3	19.6
03	10 37.9	260 29.8	. . 22.3	286 39.8	. . 16.3	46 08.3	. . 16.8	197 59.5	. . 19.5
04	25 40.3	275 29.1	22.0	301 40.5	16.4	61 11.1	16.9	213 01.7	19.4
05	40 42.8	290 28.5	21.7	316 41.2	16.5	76 13.9	17.0	228 03.9	19.3
06	55 45.3	305 27.8	N21 21.5	331 41.9	N23 16.6	91 16.7	S15 17.1	243 06.1	N 5 19.1
07	70 47.7	320 27.1	21.2	346 42.6	16.7	106 19.5	17.3	258 08.3	19.0
08	85 50.2	335 26.4	20.9	1 43.3	16.9	121 22.2	17.4	273 10.5	18.9
M 09	100 52.7	350 25.7	. . 20.7	16 44.0	. . 17.0	136 25.0	. . 17.5	288 12.7	. . 18.8
O 10	115 55.1	5 25.0	20.4	31 44.7	17.1	151 27.8	17.6	303 14.9	18.7
N 11	130 57.6	20 24.3	20.1	46 45.4	17.2	166 30.6	17.7	318 17.1	18.6
D 12	146 00.1	35 23.7	N21 19.8	61 46.1	N23 17.3	181 33.4	S15 17.8	333 19.3	N 5 18.4
A 13	161 02.5	50 23.0	19.6	76 46.8	17.4	196 36.2	17.9	348 21.5	18.3
Y 14	176 05.0	65 22.3	19.3	91 47.5	17.5	211 38.9	18.0	3 23.7	18.2
15	191 07.4	80 21.6	. . 19.0	106 48.1	. . 17.6	226 41.7	. . 18.1	18 25.9	. . 18.1
16	206 09.9	95 20.9	18.7	121 48.8	17.7	241 44.5	18.2	33 28.1	18.0
17	221 12.4	110 20.2	18.4	136 49.5	17.8	256 47.3	18.3	48 30.3	17.9
18	236 14.8	125 19.6	N21 18.2	151 50.2	N23 17.9	271 50.1	S15 18.5	63 32.4	N 5 17.7
19	251 17.3	140 18.9	17.9	166 50.9	18.0	286 52.9	18.6	78 34.6	17.6
20	266 19.8	155 18.2	17.6	181 51.6	18.1	301 55.6	18.7	93 36.8	17.5
21	281 22.2	170 17.5	. . 17.3	196 52.3	. . 18.2	316 58.4	. . 18.8	108 39.0	. . 17.4
22	296 24.7	185 16.8	17.0	211 53.0	18.3	332 01.2	18.9	123 41.2	17.3
23	311 27.2	200 16.1	16.7	226 53.7	18.4	347 04.0	19.0	138 43.4	17.2
18 00	326 29.6	215 15.4	N21 16.5	241 54.4	N23 18.5	2 06.8	S15 19.1	153 45.6	N 5 17.0
01	341 32.1	230 14.8	16.2	256 55.1	18.6	17 09.6	19.2	168 47.8	16.9
02	356 34.6	245 14.1	15.9	271 55.8	18.7	32 12.3	19.3	183 50.0	16.8
03	11 37.0	260 13.4	. . 15.6	286 56.5	. . 18.8	47 15.1	. . 19.4	198 52.2	. . 16.7
04	26 39.5	275 12.7	15.3	301 57.2	18.9	62 17.9	19.6	213 54.4	16.6
05	41 41.9	290 12.0	15.0	316 57.9	19.0	77 20.7	19.7	228 56.6	16.5
06	56 44.4	305 11.3	N21 14.7	331 58.6	N23 19.1	92 23.5	S15 19.8	243 58.8	N 5 16.3
07	71 46.9	320 10.7	14.4	346 59.3	19.2	107 26.3	19.9	259 01.0	16.2
T 08	86 49.3	335 10.0	14.1	2 00.0	19.3	122 29.0	20.0	274 03.2	16.1
U 09	101 51.8	350 09.3	. . 13.8	17 00.7	. . 19.4	137 31.8	. . 20.1	289 05.4	. . 16.0
E 10	116 54.3	5 08.6	13.5	32 01.4	19.5	152 34.6	20.2	304 07.6	15.9
S 11	131 56.7	20 07.9	13.2	47 02.1	19.6	167 37.4	20.3	319 09.8	15.8
D 12	146 59.2	35 07.2	N21 12.9	62 02.8	N23 19.7	182 40.2	S15 20.4	334 12.0	N 5 15.6
A 13	162 01.7	50 06.5	12.6	77 03.5	19.8	197 42.9	20.5	349 14.2	15.5
Y 14	177 04.1	65 05.9	12.3	92 04.2	19.9	212 45.7	20.6	4 16.4	15.4
15	192 06.6	80 05.2	. . 12.0	107 04.8	. . 20.0	227 48.5	. . 20.8	19 18.6	. . 15.3
16	207 09.0	95 04.5	11.7	122 05.5	20.1	242 51.3	20.9	34 20.8	15.2
17	222 11.5	110 03.8	11.4	137 06.2	20.2	257 54.1	21.0	49 23.0	15.1
18	237 14.0	125 03.1	N21 11.1	152 06.9	N23 20.2	272 56.9	S15 21.1	64 25.2	N 5 14.9
19	252 16.4	140 02.4	10.8	167 07.6	20.3	287 59.6	21.2	79 27.4	14.8
20	267 18.9	155 01.8	10.5	182 08.3	20.4	303 02.4	21.3	94 29.5	14.7
21	282 21.4	170 01.1	. . 10.2	197 09.0	. . 20.5	318 05.2	. . 21.4	109 31.7	. . 14.6
22	297 23.8	185 00.4	09.9	212 09.7	20.6	333 08.0	21.5	124 33.9	14.5
23	312 26.3	199 59.7	09.6	227 10.4	20.7	348 10.8	21.6	139 36.1	14.4
19 00	327 28.8	214 59.0	N21 09.2	242 11.1	N23 20.8	3 13.6	S15 21.7	154 38.3	N 5 14.2
01	342 31.2	229 58.3	08.9	257 11.8	20.9	18 16.3	21.8	169 40.5	14.1
02	357 33.7	244 57.7	08.6	272 12.5	21.0	33 19.1	21.9	184 42.7	14.0
03	12 36.2	259 57.0	. . 08.3	287 13.2	. . 21.1	48 21.9	. . 22.1	199 44.9	. . 13.9
04	27 38.6	274 56.3	08.0	302 13.9	21.2	63 24.7	22.2	214 47.1	13.8
05	42 41.1	289 55.6	07.7	317 14.6	21.3	78 27.5	22.3	229 49.3	13.7
06	57 43.5	304 54.9	N21 07.4	332 15.3	N23 21.4	93 30.2	S15 22.4	244 51.5	N 5 13.5
W 07	72 46.0	319 54.2	07.0	347 16.0	21.4	108 33.0	22.5	259 53.7	13.4
E 08	87 48.5	334 53.6	06.7	2 16.7	21.5	123 35.8	22.6	274 55.9	13.3
D 09	102 50.9	349 52.9	. . 06.4	17 17.4	. . 21.6	138 38.6	. . 22.7	289 58.1	. . 13.2
N 10	117 53.4	4 52.2	06.1	32 18.1	21.7	153 41.4	22.8	305 00.3	13.1
E 11	132 55.9	19 51.5	05.7	47 18.8	21.8	168 44.2	22.9	320 02.5	13.0
S 12	147 58.3	34 50.8	N21 05.4	62 19.5	N23 21.9	183 46.9	S15 23.0	335 04.7	N 5 12.8
D 13	163 00.8	49 50.1	05.1	77 20.2	22.0	198 49.7	23.1	350 06.9	12.7
A 14	178 03.3	64 49.5	04.8	92 20.9	22.1	213 52.5	23.2	5 09.1	12.6
Y 15	193 05.7	79 48.8	. . 04.4	107 21.6	. . 22.2	228 55.3	. . 23.4	20 11.2	. . 12.5
16	208 08.2	94 48.1	04.1	122 22.3	22.2	243 58.1	23.5	35 13.4	12.4
17	223 10.7	109 47.4	03.8	137 23.0	22.3	259 00.8	23.6	50 15.6	12.3
18	238 13.1	124 46.7	N21 03.5	152 23.7	N23 22.4	274 03.6	S15 23.7	65 17.8	N 5 12.1
19	253 15.6	139 46.1	03.1	167 24.4	22.5	289 06.4	23.8	80 20.0	12.0
20	268 18.0	154 45.4	02.8	182 25.2	22.6	304 09.2	23.9	95 22.2	11.9
21	283 20.5	169 44.7	. . 02.5	197 25.9	. . 22.7	319 12.0	. . 24.0	110 24.4	. . 11.8
22	298 23.0	184 44.0	02.1	212 26.6	22.8	334 14.8	24.1	125 26.6	11.7
23	313 25.4	199 43.3	01.8	227 27.3	22.8	349 17.5	24.2	140 28.8	11.6
Mer. Pass.	h m 2 13.7	*v* −0.7	*d* 0.3	*v* 0.7	*d* 0.1	*v* 2.8	*d* 0.1	*v* 2.2	*d* 0.1

STARS

Name	SHA	Dec
	° ′	° ′
Acamar	315 20.3	S40 15.
Achernar	335 28.4	S57 10.
Acrux	173 13.2	S63 09.
Adhara	255 15.0	S28 58.
Aldebaran	290 52.7	N16 31.
Alioth	166 23.3	N55 54.
Alkaid	153 01.2	N49 16.
Al Na'ir	27 46.6	S46 54.
Alnilam	275 49.4	S 1 11.
Alphard	217 59.2	S 8 42.
Alphecca	126 13.3	N26 41.
Alpheratz	357 46.2	N29 08.
Altair	62 10.7	N 8 53.
Ankaa	353 18.0	S42 14.
Antares	112 29.6	S26 27.
Arcturus	145 58.4	N19 08.
Atria	107 33.9	S69 03.
Avior	234 19.9	S59 32.
Bellatrix	278 35.2	N 6 21.
Betelgeuse	271 04.5	N 7 24.
Canopus	263 57.8	S52 41.
Capella	280 38.8	N46 00.
Deneb	49 33.0	N45 19.
Denebola	182 36.8	N14 31.
Diphda	348 58.4	S17 55.
Dubhe	193 55.5	N61 42.
Elnath	278 16.3	N28 36.
Eltanin	90 47.1	N51 29.
Enif	33 49.6	N 9 55.
Fomalhaut	15 26.6	S29 34.
Gacrux	172 04.7	S57 10.
Gienah	175 55.5	S17 35.
Hadar	148 52.4	S60 25.
Hamal	328 03.8	N23 30.
Kaus Aust.	83 47.3	S34 22.
Kochab	137 19.7	N74 07.
Markab	13 40.9	N15 15.
Menkar	314 17.9	N 4 07.
Menkent	148 11.1	S36 25.
Miaplacidus	221 41.5	S69 45.
Mirfak	308 44.5	N49 53.
Nunki	76 01.5	S26 17.
Peacock	53 22.9	S56 42.
Pollux	243 31.4	N28 00.
Procyon	245 02.9	N 5 12.
Rasalhague	96 08.9	N12 33.
Regulus	207 46.8	N11 55.
Rigel	281 14.9	S 8 11.
Rigil Kent.	139 55.9	S60 52.
Sabik	102 15.6	S15 44.
Schedar	349 43.5	N56 35.
Shaula	96 25.5	S37 06.
Sirius	258 36.4	S16 43.
Spica	158 34.4	S11 12.
Suhail	222 55.0	S43 28.
Vega	80 40.6	N38 47.
Zuben'ubi	137 08.6	S16 05.

	SHA	Mer. Pas
	° ′	h m
Venus	248 45.8	9 39
Mars	275 24.8	7 52
Jupiter	35 37.2	23 47
Saturn	187 16.0	13 43

UT	SUN GHA	SUN Dec	MOON GHA	v	MOON Dec	d	HP
d h	° ′	° ′	° ′	′	° ′	′	′
17 00	178 58.0	N13 26.6	228 15.2	3.2	N25 22.6	5.0	60.3
01	193 58.1	25.8	242 37.4	3.2	25 17.6	5.2	60.3
02	208 58.2	25.0	256 59.6	3.2	25 12.4	5.4	60.4
03	223 58.4	. . 24.2	271 21.8	3.3	25 07.0	5.6	60.4
04	238 58.5	23.4	285 44.1	3.3	25 01.4	5.7	60.4
05	253 58.6	22.6	300 06.4	3.3	24 55.7	6.0	60.4
06	268 58.8	N13 21.8	314 28.7	3.3	N24 49.7	6.1	60.4
07	283 58.9	21.0	328 51.0	3.4	24 43.6	6.3	60.5
08	298 59.0	20.2	343 13.4	3.5	24 37.3	6.5	60.5
09	313 59.2	. . 19.4	357 35.9	3.4	24 30.8	6.6	60.5
10	328 59.3	18.6	11 58.3	3.6	24 24.2	6.9	60.5
11	343 59.5	17.8	26 20.9	3.5	24 17.3	7.0	60.5
12	358 59.6	N13 17.0	40 43.4	3.6	N24 10.3	7.2	60.6
13	13 59.7	16.2	55 06.0	3.7	24 03.1	7.4	60.6
14	28 59.9	15.4	69 28.7	3.7	23 55.7	7.5	60.6
15	44 00.0	. . 14.6	83 51.4	3.7	23 48.2	7.8	60.6
16	59 00.1	13.8	98 14.1	3.8	23 40.4	7.9	60.6
17	74 00.3	13.0	112 36.9	3.9	23 32.5	8.0	60.7
18	89 00.4	N13 12.2	126 59.8	3.9	N23 24.5	8.3	60.7
19	104 00.5	11.3	141 22.7	4.0	23 16.2	8.4	60.7
20	119 00.7	10.5	155 45.7	4.0	23 07.8	8.6	60.7
21	134 00.8	. . 09.7	170 08.7	4.1	22 59.2	8.7	60.7
22	149 00.9	08.9	184 31.8	4.2	22 50.5	8.9	60.7
23	164 01.1	08.1	198 55.0	4.2	22 41.6	9.1	60.7
18 00	179 01.2	N13 07.3	213 18.2	4.3	N22 32.5	9.3	60.8
01	194 01.4	06.5	227 41.5	4.4	22 23.2	9.4	60.8
02	209 01.5	05.7	242 04.9	4.4	22 13.8	9.5	60.8
03	224 01.6	. . 04.9	256 28.3	4.5	22 04.3	9.8	60.8
04	239 01.8	04.1	270 51.8	4.6	21 54.5	9.8	60.8
05	254 01.9	03.3	285 15.4	4.6	21 44.7	10.1	60.8
06	269 02.0	N13 02.5	299 39.0	4.8	N21 34.6	10.2	60.8
07	284 02.2	01.7	314 02.8	4.8	21 24.4	10.3	60.8
08	299 02.3	00.9	328 26.6	4.8	21 14.1	10.5	60.9
09	314 02.5	13 00.0	342 50.4	5.0	21 03.6	10.6	60.9
10	329 02.6	12 59.2	357 14.4	5.0	20 53.0	10.8	60.9
11	344 02.7	58.4	11 38.4	5.1	20 42.2	11.0	60.9
12	359 02.9	N12 57.6	26 02.5	5.2	N20 31.2	11.0	60.9
13	14 03.0	56.8	40 26.7	5.2	20 20.2	11.2	60.9
14	29 03.2	56.0	54 50.9	5.4	20 09.0	11.4	60.9
15	44 03.3	. . 55.2	69 15.3	5.4	19 57.6	11.5	60.9
16	59 03.4	54.4	83 39.7	5.5	19 46.1	11.6	60.9
17	74 03.6	53.6	98 04.2	5.6	19 34.5	11.8	60.9
18	89 03.7	N12 52.7	112 28.8	5.6	N19 22.7	11.9	60.9
19	104 03.9	51.9	126 53.4	5.8	19 10.8	12.0	60.9
20	119 04.0	51.1	141 18.2	5.8	18 58.8	12.2	61.0
21	134 04.2	. . 50.3	155 43.0	6.0	18 46.6	12.3	61.0
22	149 04.3	49.5	170 08.0	6.0	18 34.3	12.4	61.0
23	164 04.4	48.7	184 33.0	6.0	18 21.9	12.5	61.0
19 00	179 04.6	N12 47.9	198 58.0	6.2	N18 09.4	12.7	61.0
01	194 04.7	47.0	213 23.2	6.3	17 56.7	12.8	61.0
02	209 04.9	46.2	227 48.5	6.3	17 43.9	12.9	61.0
03	224 05.0	. . 45.4	242 13.8	6.4	17 31.0	13.0	61.0
04	239 05.2	44.6	256 39.2	6.6	17 18.0	13.1	61.0
05	254 05.3	43.8	271 04.8	6.6	17 04.9	13.2	61.0
06	269 05.4	N12 43.0	285 30.4	6.6	N16 51.7	13.4	61.0
07	284 05.6	42.1	299 56.0	6.8	16 38.3	13.4	61.0
08	299 05.7	41.3	314 21.8	6.9	16 24.9	13.6	61.0
09	314 05.9	. . 40.5	328 47.7	6.9	16 11.3	13.7	61.0
10	329 06.0	39.7	343 13.6	7.0	15 57.6	13.7	61.0
11	344 06.2	38.9	357 39.6	7.2	15 43.9	13.9	61.0
12	359 06.3	N12 38.0	12 05.8	7.2	N15 30.0	14.0	61.0
13	14 06.4	37.2	26 32.0	7.2	15 16.0	14.0	61.0
14	29 06.6	36.4	40 58.2	7.4	15 02.0	14.2	60.9
15	44 06.7	. . 35.6	55 24.6	7.4	14 47.8	14.2	60.9
16	59 06.9	34.8	69 51.0	7.6	14 33.6	14.3	60.9
17	74 07.0	33.9	84 17.6	7.6	14 19.3	14.5	60.9
18	89 07.2	N12 33.1	98 44.2	7.7	N14 04.8	14.5	60.9
19	104 07.3	32.3	113 10.9	7.8	13 50.3	14.5	60.9
20	119 07.5	31.5	127 37.7	7.8	13 35.8	14.7	60.9
21	134 07.6	. . 30.7	142 04.5	8.0	13 21.1	14.7	60.9
22	149 07.8	29.8	156 31.5	8.0	13 06.4	14.9	60.9
23	164 07.9	29.0	170 58.5	8.1	N12 51.5	14.9	60.9
	SD 15.8	d 0.8	SD 16.5		16.6		16.6

Lat.	Twilight Naut.	Twilight Civil	Sunrise	Moonrise 17	18	19	20
°	h m	h m	h m	h m	h m	h m	h m
N 72	////	////	02 45	▭	▭	▭	03 08
N 70	////	01 14	03 12	▭	▭	00 37	03 29
68	////	02 06	03 32	▭	▭	01 18	03 46
66	////	02 37	03 48	23 11	25 47	01 47	03 59
64	01 05	02 59	04 01	23 56	26 08	02 08	04 10
62	01 51	03 17	04 12	24 26	00 26	02 25	04 19
60	02 20	03 32	04 22	24 48	00 48	02 39	04 27
N 58	02 41	03 44	04 30	25 07	01 07	02 51	04 34
56	02 58	03 54	04 37	25 22	01 22	03 01	04 40
54	03 12	04 03	04 43	00 06	01 35	03 10	04 45
52	03 24	04 11	04 49	00 20	01 46	03 18	04 50
50	03 34	04 19	04 54	00 32	01 56	03 25	04 55
45	03 55	04 34	05 05	00 58	02 17	03 41	05 04
N 40	04 11	04 46	05 14	01 18	02 34	03 53	05 12
35	04 24	04 56	05 22	01 35	02 49	04 04	05 19
30	04 34	05 04	05 29	01 49	03 01	04 13	05 25
20	04 51	05 18	05 41	02 14	03 22	04 29	05 35
N 10	05 04	05 29	05 51	02 35	03 40	04 43	05 43
0	05 15	05 39	06 00	02 55	03 57	04 56	05 52
S 10	05 23	05 48	06 10	03 15	04 14	05 09	06 00
20	05 31	05 57	06 20	03 36	04 32	05 23	06 09
30	05 38	06 06	06 31	04 01	04 53	05 39	06 19
35	05 41	06 11	06 37	04 15	05 05	05 48	06 24
40	05 45	06 16	06 44	04 31	05 19	05 58	06 31
45	05 48	06 22	06 52	04 51	05 35	06 10	06 38
S 50	05 51	06 29	07 02	05 16	05 55	06 24	06 47
52	05 52	06 32	07 07	05 28	06 04	06 31	06 51
54	05 54	06 35	07 12	05 41	06 15	06 38	06 56
56	05 55	06 39	07 18	05 57	06 27	06 47	07 01
58	05 57	06 43	07 24	06 16	06 41	06 56	07 06
S 60	05 58	06 47	07 31	06 38	06 57	07 07	07 12

Lat.	Sunset	Twilight Civil	Twilight Naut.	Moonset 17	18	19	20
°	h m	h m	h m	h m	h m	h m	h m
N 72	21 18	////	////	▭	▭	20 37	19 54
N 70	20 52	22 43	////	▭	21 07	20 13	19 43
68	20 32	21 56	////	▭	20 24	19 54	19 35
66	20 17	21 27	////	20 24	19 54	19 39	19 28
64	20 04	21 05	22 53	19 38	19 32	19 27	19 22
62	19 54	20 48	22 11	19 08	19 14	19 16	19 17
60	19 44	20 34	21 44	18 45	18 59	19 07	19 12
N 58	19 36	20 22	21 24	18 26	18 46	18 59	19 08
56	19 29	20 12	21 08	18 10	18 35	18 52	19 04
54	19 23	20 03	20 54	17 57	18 25	18 45	19 01
52	19 18	19 55	20 42	17 45	18 16	18 39	18 58
50	19 12	19 48	20 32	17 34	18 08	18 34	18 55
45	19 02	19 33	20 12	17 12	17 51	18 23	18 49
N 40	18 53	19 21	19 56	16 54	17 37	18 13	18 44
35	18 45	19 11	19 43	16 39	17 25	18 05	18 39
30	18 38	19 03	19 33	16 26	17 15	17 57	18 36
20	18 26	18 49	19 16	16 03	16 57	17 45	18 29
N 10	18 16	18 38	19 03	15 43	16 41	17 34	18 23
0	18 07	18 28	18 53	15 25	16 26	17 23	18 17
S 10	17 58	18 19	18 44	15 06	16 11	17 12	18 11
20	17 48	18 11	18 37	14 46	15 54	17 01	18 05
30	17 37	18 02	18 30	14 23	15 35	16 48	17 57
35	17 31	17 57	18 27	14 09	15 24	16 40	17 53
40	17 24	17 52	18 24	13 53	15 12	16 31	17 48
45	17 16	17 46	18 21	13 34	14 56	16 20	17 43
S 50	17 06	17 39	18 17	13 10	14 38	16 08	17 36
52	17 01	17 37	18 16	12 59	14 29	16 02	17 33
54	16 56	17 33	18 15	12 46	14 19	15 55	17 29
56	16 51	17 30	18 13	12 30	14 08	15 48	17 26
58	16 45	17 26	18 12	12 12	13 55	15 39	17 21
S 60	16 38	17 22	18 11	11 50	13 39	15 30	17 16

Day	SUN Eqn. of Time 00h	SUN Eqn. of Time 12h	SUN Mer. Pass.	MOON Mer. Pass. Upper	MOON Mer. Pass. Lower	Age	Phase
d	m s	m s	h m	h m	h m	d	%
17	04 08	04 02	12 04	09 10	21 41	26	12
18	03 55	03 49	12 04	10 12	22 41	27	5
19	03 42	03 35	12 04	11 10	23 37	28	1

	UT	ARIES	VENUS −3.9		MARS +1.0		JUPITER −2.9		SATURN +1.1	
		GHA	GHA	Dec	GHA	Dec	GHA	Dec	GHA	Dec
	d h	° ′	° ′	° ′	° ′	° ′	° ′	° ′	° ′	° ′
	20 00	328 27.9	214 42.6	N21 01.5	242 28.0	N23 22.9	4 20.3	S15 24.3	155 31.0	N 5 11.4
	01	343 30.4	229 42.0	01.1	257 28.7	23.0	19 23.1	24.4	170 33.2	11.3
	02	358 32.8	244 41.3	00.8	272 29.4	23.1	34 25.9	24.5	185 35.4	11.2
	03	13 35.3	259 40.6	. . 00.4	287 30.1	. . 23.2	49 28.7	. . 24.7	200 37.6	. . 11.1
	04	28 37.8	274 39.9	21 00.1	302 30.8	23.3	64 31.4	24.8	215 39.8	11.0
	05	43 40.2	289 39.2	20 59.8	317 31.5	23.4	79 34.2	24.9	230 42.0	10.8
	06	58 42.7	304 38.6	N20 59.4	332 32.2	N23 23.4	94 37.0	S15 25.0	245 44.2	N 5 10.7
	07	73 45.1	319 37.9	59.1	347 32.9	23.5	109 39.8	25.1	260 46.4	10.6
T	08	88 47.6	334 37.2	58.7	2 33.6	23.6	124 42.6	25.2	275 48.5	10.5
H	09	103 50.1	349 36.5	. . 58.4	17 34.3	. . 23.7	139 45.3	. . 25.3	290 50.7	. . 10.4
U	10	118 52.5	4 35.8	58.0	32 35.0	23.8	154 48.1	25.4	305 52.9	10.3
R	11	133 55.0	19 35.1	57.7	47 35.7	23.8	169 50.9	25.5	320 55.1	10.1
S	12	148 57.5	34 34.5	N20 57.3	62 36.4	N23 23.9	184 53.7	S15 25.6	335 57.3	N 5 10.0
D	13	163 59.9	49 33.8	57.0	77 37.1	24.0	199 56.5	25.7	350 59.5	09.9
A	14	179 02.4	64 33.1	56.6	92 37.8	24.1	214 59.3	25.8	6 01.7	09.8
Y	15	194 04.9	79 32.4	. . 56.3	107 38.5	. . 24.2	230 02.0	. . 25.9	21 03.9	. . 09.7
	16	209 07.3	94 31.7	55.9	122 39.2	24.2	245 04.8	26.1	36 06.1	09.6
	17	224 09.8	109 31.1	55.6	137 39.9	24.3	260 07.6	26.2	51 08.3	09.4
	18	239 12.3	124 30.4	N20 55.2	152 40.6	N23 24.4	275 10.4	S15 26.3	66 10.5	N 5 09.3
	19	254 14.7	139 29.7	54.9	167 41.4	24.5	290 13.2	26.4	81 12.7	09.2
	20	269 17.2	154 29.0	54.5	182 42.1	24.6	305 15.9	26.5	96 14.9	09.1
	21	284 19.6	169 28.3	. . 54.1	197 42.8	. . 24.6	320 18.7	. . 26.6	111 17.1	. . 09.0
	22	299 22.1	184 27.7	53.8	212 43.5	24.7	335 21.5	26.7	126 19.3	08.9
	23	314 24.6	199 27.0	53.4	227 44.2	24.8	350 24.3	26.8	141 21.4	08.7
	21 00	329 27.0	214 26.3	N20 53.1	242 44.9	N23 24.9	5 27.1	S15 26.9	156 23.6	N 5 08.6
	01	344 29.5	229 25.6	52.7	257 45.6	24.9	20 29.8	27.0	171 25.8	08.5
	02	359 32.0	244 25.0	52.3	272 46.3	25.0	35 32.6	27.1	186 28.0	08.4
	03	14 34.4	259 24.3	. . 52.0	287 47.0	. . 25.1	50 35.4	. . 27.2	201 30.2	. . 08.3
	04	29 36.9	274 23.6	51.6	302 47.7	25.2	65 38.2	27.3	216 32.4	08.1
	05	44 39.4	289 22.9	51.2	317 48.4	25.3	80 41.0	27.4	231 34.6	08.0
	06	59 41.8	304 22.2	N20 50.9	332 49.1	N23 25.3	95 43.7	S15 27.6	246 36.8	N 5 07.9
	07	74 44.3	319 21.6	50.5	347 49.8	25.4	110 46.5	27.7	261 39.0	07.8
	08	89 46.8	334 20.9	50.1	2 50.5	25.5	125 49.3	27.8	276 41.2	07.7
F	09	104 49.2	349 20.2	. . 49.8	17 51.3	. . 25.6	140 52.1	. . 27.9	291 43.4	. . 07.6
R	10	119 51.7	4 19.5	49.4	32 52.0	25.6	155 54.9	28.0	306 45.6	07.4
I	11	134 54.1	19 18.8	49.0	47 52.7	25.7	170 57.6	28.1	321 47.8	07.3
D	12	149 56.6	34 18.2	N20 48.7	62 53.4	N23 25.8	186 00.4	S15 28.2	336 49.9	N 5 07.2
A	13	164 59.1	49 17.5	48.3	77 54.1	25.8	201 03.2	28.3	351 52.1	07.1
Y	14	180 01.5	64 16.8	47.9	92 54.8	25.9	216 06.0	28.4	6 54.3	07.0
	15	195 04.0	79 16.1	. . 47.5	107 55.5	. . 26.0	231 08.8	. . 28.5	21 56.5	. . 06.8
	16	210 06.5	94 15.5	47.2	122 56.2	26.1	246 11.5	28.6	36 58.7	06.7
	17	225 08.9	109 14.8	46.8	137 56.9	26.1	261 14.3	28.7	52 00.9	06.6
	18	240 11.4	124 14.1	N20 46.4	152 57.6	N23 26.2	276 17.1	S15 28.8	67 03.1	N 5 06.5
	19	255 13.9	139 13.4	46.0	167 58.3	26.3	291 19.9	28.9	82 05.3	06.4
	20	270 16.3	154 12.7	45.6	182 59.1	26.3	306 22.6	29.0	97 07.5	06.3
	21	285 18.8	169 12.1	. . 45.2	197 59.8	. . 26.4	321 25.4	. . 29.2	112 09.7	. . 06.1
	22	300 21.2	184 11.4	44.9	213 00.5	26.5	336 28.2	29.3	127 11.9	06.0
	23	315 23.7	199 10.7	44.5	228 01.2	26.6	351 31.0	29.4	142 14.1	05.9
	22 00	330 26.2	214 10.0	N20 44.1	243 01.9	N23 26.6	6 33.8	S15 29.5	157 16.3	N 5 05.8
	01	345 28.6	229 09.4	43.7	258 02.6	26.7	21 36.5	29.6	172 18.4	05.7
	02	0 31.1	244 08.7	43.3	273 03.3	26.8	36 39.3	29.7	187 20.6	05.6
	03	15 33.6	259 08.0	. . 42.9	288 04.0	. . 26.8	51 42.1	. . 29.8	202 22.8	. . 05.4
	04	30 36.0	274 07.3	42.5	303 04.8	26.9	66 44.9	29.9	217 25.0	05.3
	05	45 38.5	289 06.7	42.2	318 05.5	27.0	81 47.7	30.0	232 27.2	05.2
	06	60 41.0	304 06.0	N20 41.8	333 06.2	N23 27.0	96 50.4	S15 30.1	247 29.4	N 5 05.1
	07	75 43.4	319 05.3	41.4	348 06.9	27.1	111 53.2	30.2	262 31.6	05.0
S	08	90 45.9	334 04.6	41.0	3 07.6	27.2	126 56.0	30.3	277 33.8	04.8
A	09	105 48.4	349 04.0	. . 40.6	18 08.3	. . 27.2	141 58.8	. . 30.4	292 36.0	. . 04.7
T	10	120 50.8	4 03.3	40.2	33 09.0	27.3	157 01.5	30.5	307 38.2	04.6
U	11	135 53.3	19 02.6	39.8	48 09.7	27.4	172 04.3	30.6	322 40.4	04.5
R	12	150 55.7	34 01.9	N20 39.4	63 10.5	N23 27.4	187 07.1	S15 30.7	337 42.5	N 5 04.4
D	13	165 58.2	49 01.3	39.0	78 11.2	27.5	202 09.9	30.8	352 44.7	04.3
A	14	181 00.7	64 00.6	38.6	93 11.9	27.6	217 12.7	31.0	7 46.9	04.1
Y	15	196 03.1	78 59.9	. . 38.2	108 12.6	. . 27.6	232 15.4	. . 31.1	22 49.1	. . 04.0
	16	211 05.6	93 59.2	37.8	123 13.3	27.7	247 18.2	31.2	37 51.3	03.9
	17	226 08.1	108 58.6	37.4	138 14.0	27.8	262 21.0	31.3	52 53.5	03.8
	18	241 10.5	123 57.9	N20 37.0	153 14.7	N23 27.8	277 23.8	S15 31.4	67 55.7	N 5 03.7
	19	256 13.0	138 57.2	36.6	168 15.4	27.9	292 26.5	31.5	82 57.9	03.5
	20	271 15.5	153 56.5	36.2	183 16.2	27.9	307 29.3	31.6	98 00.1	03.4
	21	286 17.9	168 55.9	. . 35.8	198 16.9	. . 28.0	322 32.1	. . 31.7	113 02.3	. . 03.3
	22	301 20.4	183 55.2	35.4	213 17.6	28.1	337 34.9	31.8	128 04.5	03.2
	23	316 22.8	198 54.5	34.9	228 18.3	28.1	352 37.7	31.9	143 06.6	03.1
	Mer. Pass.	h m 2 01.9	v −0.7	d 0.4	v 0.7	d 0.1	v 2.8	d 0.1	v 2.2	d 0.1

STARS Name	SHA	Dec
	° ′	° ′
Acamar	315 20.3	S40 15.
Achernar	335 28.3	S57 10.
Acrux	173 13.3	S63 09.
Adhara	255 15.0	S28 58.
Aldebaran	290 52.7	N16 31.
Alioth	166 23.3	N55 54.
Alkaid	153 01.2	N49 16.
Al Na'ir	27 46.6	S46 54.
Alnilam	275 49.4	S 1 11.
Alphard	217 59.2	S 8 42.
Alphecca	126 13.4	N26 41.
Alpheratz	357 46.2	N29 08.
Altair	62 10.7	N 8 53.
Ankaa	353 18.0	S42 14.
Antares	112 29.7	S26 27.
Arcturus	145 58.4	N19 08.
Atria	107 33.9	S69 03.
Avior	234 19.9	S59 32.
Bellatrix	278 35.1	N 6 21.
Betelgeuse	271 04.5	N 7 24.
Canopus	263 57.7	S52 41.
Capella	280 38.8	N46 00.
Deneb	49 33.0	N45 19.
Denebola	182 36.8	N14 31.
Diphda	348 58.4	S17 55.
Dubhe	193 55.5	N61 42.
Elnath	278 16.3	N28 36.
Eltanin	90 47.2	N51 29.
Enif	33 49.6	N 9 55.
Fomalhaut	15 26.6	S29 34.
Gacrux	172 04.7	S57 10.
Gienah	175 55.5	S17 35.
Hadar	148 52.4	S60 25.
Hamal	328 03.8	N23 30.
Kaus Aust.	83 47.3	S34 22.
Kochab	137 19.8	N74 07.
Markab	13 40.8	N15 15.
Menkar	314 17.9	N 4 07.
Menkent	148 11.1	S36 25.
Miaplacidus	221 41.5	S69 45.
Mirfak	308 44.4	N49 53.
Nunki	76 01.5	S26 17.
Peacock	53 22.9	S56 42.
Pollux	243 31.4	N28 00.
Procyon	245 02.9	N 5 12.
Rasalhague	96 08.9	N12 33.
Regulus	207 46.8	N11 55.
Rigel	281 14.9	S 8 11.
Rigil Kent.	139 56.0	S60 52.
Sabik	102 15.6	S15 44.
Schedar	349 43.5	N56 35.
Shaula	96 25.5	S37 06.
Sirius	258 36.4	S16 43.
Spica	158 34.4	S11 12.
Suhail	222 55.0	S43 28.
Vega	80 40.6	N38 47.
Zuben'ubi	137 08.6	S16 05.

	SHA	Mer. Pas
	° ′	h m
Venus	244 59.3	9 43
Mars	273 17.8	7 49
Jupiter	36 00.0	23 34
Saturn	186 56.6	13 32

UT	SUN GHA	SUN Dec	MOON GHA	v	MOON Dec	d	HP
d h	° ′	° ′	° ′	′	° ′	′	′
20 00	179 08.1	N12 28.2	185 25.6	8.2	N12 36.6	14.9	60.9
01	194 08.2	27.4	199 52.8	8.2	12 21.7	15.1	60.9
02	209 08.4	26.5	214 20.0	8.4	12 06.6	15.1	60.9
03	224 08.5	. . 25.7	228 47.4	8.4	11 51.5	15.1	60.8
04	239 08.7	24.9	243 14.8	8.5	11 36.4	15.3	60.8
05	254 08.8	24.1	257 42.3	8.6	11 21.1	15.3	60.8
06	269 08.9	N12 23.2	272 09.9	8.6	N11 05.8	15.3	60.8
07	284 09.1	22.4	286 37.5	8.7	10 50.5	15.5	60.8
08	299 09.2	21.6	301 05.2	8.8	10 35.0	15.4	60.8
09	314 09.4	. . 20.8	315 33.0	8.9	10 19.6	15.6	60.8
10	329 09.5	19.9	330 00.9	8.9	10 04.0	15.5	60.8
11	344 09.7	19.1	344 28.8	9.1	9 48.5	15.7	60.7
12	359 09.8	N12 18.3	358 56.9	9.0	N 9 32.8	15.6	60.7
13	14 10.0	17.4	13 24.9	9.2	9 17.2	15.8	60.7
14	29 10.1	16.6	27 53.1	9.2	9 01.4	15.7	60.7
15	44 10.3	. . 15.8	42 21.3	9.3	8 45.7	15.8	60.7
16	59 10.4	15.0	56 49.6	9.4	8 29.9	15.9	60.7
17	74 10.6	14.1	71 18.0	9.4	8 14.0	15.9	60.6
18	89 10.7	N12 13.3	85 46.4	9.5	N 7 58.1	15.9	60.6
19	104 10.9	12.5	100 14.9	9.5	7 42.2	16.0	60.6
20	119 11.1	11.6	114 43.4	9.6	7 26.2	15.9	60.6
21	134 11.2	. . 10.8	129 12.0	9.7	7 10.3	16.1	60.6
22	149 11.4	10.0	143 40.7	9.8	6 54.2	16.0	60.5
23	164 11.5	09.1	158 09.5	9.8	6 38.2	16.1	60.5
21 00	179 11.7	N12 08.3	172 38.3	9.8	N 6 22.1	16.1	60.5
01	194 11.8	07.5	187 07.1	10.0	6 06.0	16.1	60.5
02	209 12.0	06.7	201 36.1	9.9	5 49.9	16.1	60.5
03	224 12.1	. . 05.8	216 05.0	10.1	5 33.8	16.2	60.4
04	239 12.3	05.0	230 34.1	10.1	5 17.6	16.2	60.4
05	254 12.4	04.2	245 03.2	10.1	5 01.4	16.2	60.4
06	269 12.6	N12 03.3	259 32.3	10.2	N 4 45.2	16.2	60.4
07	284 12.7	02.5	274 01.5	10.2	4 29.0	16.2	60.3
08	299 12.9	01.7	288 30.7	10.3	4 12.8	16.2	60.3
09	314 13.0	. . 00.8	303 00.0	10.4	3 56.6	16.2	60.3
10	329 13.2	12 00.0	317 29.4	10.4	3 40.4	16.3	60.3
11	344 13.3	11 59.1	331 58.8	10.4	3 24.1	16.2	60.2
12	359 13.5	N11 58.3	346 28.2	10.5	N 3 07.9	16.3	60.2
13	14 13.7	57.5	0 57.7	10.6	2 51.6	16.2	60.2
14	29 13.8	56.6	15 27.3	10.6	2 35.4	16.3	60.2
15	44 14.0	. . 55.8	29 56.9	10.6	2 19.1	16.2	60.1
16	59 14.1	55.0	44 26.5	10.7	2 02.9	16.2	60.1
17	74 14.3	54.1	58 56.2	10.7	1 46.7	16.3	60.1
18	89 14.4	N11 53.3	73 25.9	10.7	N 1 30.4	16.2	60.0
19	104 14.6	52.5	87 55.6	10.8	1 14.2	16.2	60.0
20	119 14.7	51.6	102 25.4	10.9	0 58.0	16.2	60.0
21	134 14.9	. . 50.8	116 55.3	10.8	0 41.8	16.2	60.0
22	149 15.1	49.9	131 25.1	11.0	0 25.6	16.2	59.9
23	164 15.2	49.1	145 55.1	10.9	N 0 09.4	16.1	59.9
22 00	179 15.4	N11 48.3	160 25.0	11.0	S 0 06.7	16.1	59.9
01	194 15.5	47.4	174 55.0	11.0	0 22.8	16.2	59.8
02	209 15.7	46.6	189 25.0	11.0	0 39.0	16.1	59.8
03	224 15.8	. . 45.7	203 55.0	11.1	0 55.1	16.0	59.8
04	239 16.0	44.9	218 25.1	11.1	1 11.1	16.1	59.7
05	254 16.2	44.1	232 55.2	11.1	1 27.2	16.0	59.7
06	269 16.3	N11 43.2	247 25.3	11.2	S 1 43.2	16.0	59.7
07	284 16.5	42.4	261 55.5	11.2	1 59.2	16.0	59.6
08	299 16.6	41.5	276 25.7	11.2	2 15.2	15.9	59.6
09	314 16.8	. . 40.7	290 55.9	11.2	2 31.1	15.9	59.6
10	329 17.0	39.8	305 26.1	11.3	2 47.0	15.9	59.5
11	344 17.1	39.0	319 56.4	11.3	3 02.9	15.8	59.5
12	359 17.3	N11 38.2	334 26.7	11.3	S 3 18.7	15.8	59.5
13	14 17.4	37.3	348 57.0	11.3	3 34.5	15.8	59.4
14	29 17.6	36.5	3 27.3	11.3	3 50.3	15.7	59.4
15	44 17.8	. . 35.6	17 57.6	11.4	4 06.0	15.7	59.4
16	59 17.9	34.8	32 28.0	11.4	4 21.7	15.6	59.3
17	74 18.1	33.9	46 58.4	11.4	4 37.3	15.6	59.3
18	89 18.2	N11 33.1	61 28.8	11.4	S 4 52.9	15.5	59.3
19	104 18.4	32.2	75 59.2	11.4	5 08.4	15.5	59.2
20	119 18.6	31.4	90 29.6	11.5	5 23.9	15.5	59.2
21	134 18.7	. . 30.6	105 00.1	11.4	5 39.4	15.4	59.2
22	149 18.9	29.7	119 30.5	11.5	5 54.8	15.4	59.1
23	164 19.0	28.9	134 01.0	11.5	S 6 10.2	15.3	59.1
	SD 15.8	*d* 0.8	SD	16.5		16.4	16.2

Lat.	Twilight Naut.	Twilight Civil	Sunrise	Moonrise 20	Moonrise 21	Moonrise 22	Moonrise 23
°	h m	h m	h m	h m	h m	h m	h m
N 72	////	////	03 03	03 08	05 43	08 05	10 26
N 70	////	01 45	03 26	03 29	05 50	08 01	10 11
68	////	02 24	03 44	03 46	05 56	07 58	09 59
66	////	02 51	03 59	03 59	06 01	07 56	09 49
64	01 32	03 11	04 10	04 10	06 04	07 54	09 41
62	02 08	03 27	04 20	04 19	06 08	07 52	09 34
60	02 32	03 40	04 29	04 27	06 11	07 51	09 28
N 58	02 51	03 51	04 36	04 34	06 13	07 49	09 23
56	03 06	04 01	04 43	04 40	06 16	07 48	09 18
54	03 19	04 09	04 49	04 45	06 18	07 47	09 14
52	03 30	04 17	04 54	04 50	06 20	07 46	09 10
50	03 40	04 24	04 59	04 55	06 21	07 45	09 07
45	03 59	04 38	05 09	05 04	06 25	07 43	09 00
N 40	04 14	04 49	05 17	05 12	06 28	07 42	08 53
35	04 27	04 58	05 25	05 19	06 31	07 40	08 48
30	04 37	05 06	05 31	05 25	06 33	07 39	08 44
20	04 52	05 19	05 42	05 35	06 37	07 37	08 36
N 10	05 04	05 30	05 51	05 43	06 41	07 36	08 29
0	05 14	05 39	06 00	05 52	06 44	07 34	08 23
S 10	05 22	05 47	06 08	06 00	06 47	07 33	08 17
20	05 29	05 55	06 17	06 09	06 51	07 31	08 10
30	05 35	06 03	06 28	06 19	06 55	07 29	08 03
35	05 38	06 08	06 33	06 24	06 57	07 28	07 58
40	05 41	06 12	06 40	06 31	07 00	07 27	07 54
45	05 43	06 17	06 48	06 38	07 03	07 26	07 48
S 50	05 46	06 23	06 57	06 47	07 06	07 24	07 42
52	05 47	06 26	07 01	06 51	07 08	07 23	07 39
54	05 48	06 29	07 05	06 56	07 10	07 23	07 35
56	05 49	06 32	07 10	07 01	07 12	07 22	07 32
58	05 50	06 35	07 16	07 06	07 14	07 21	07 28
S 60	05 51	06 39	07 23	07 12	07 16	07 20	07 23

Lat.	Sunset	Twilight Civil	Twilight Naut.	Moonset 20	Moonset 21	Moonset 22	Moonset 23
°	h m	h m	h m	h m	h m	h m	h m
N 72	20 59	23 29	////	19 54	19 21	18 50	18 14
N 70	20 36	22 13	////	19 43	19 20	18 57	18 32
68	20 19	21 37	////	19 35	19 19	19 03	18 46
66	20 05	21 12	23 34	19 28	19 18	19 08	18 57
64	19 54	20 52	22 27	19 22	19 17	19 12	19 07
62	19 44	20 37	21 54	19 17	19 16	19 16	19 16
60	19 36	20 24	21 31	19 12	19 16	19 19	19 23
N 58	19 28	20 13	21 13	19 08	19 15	19 22	19 29
56	19 22	20 04	20 58	19 04	19 15	19 24	19 35
54	19 16	19 55	20 45	19 01	19 14	19 27	19 40
52	19 11	19 48	20 34	18 58	19 14	19 29	19 45
50	19 07	19 41	20 25	18 55	19 13	19 31	19 49
45	18 57	19 28	20 06	18 49	19 13	19 35	19 58
N 40	18 48	19 17	19 51	18 44	19 12	19 39	20 06
35	18 41	19 07	19 39	18 39	19 11	19 42	20 13
30	18 35	19 00	19 29	18 36	19 11	19 44	20 18
20	18 24	18 47	19 13	18 29	19 10	19 49	20 29
N 10	18 15	18 37	19 02	18 23	19 09	19 53	20 38
0	18 06	18 28	18 52	18 17	19 08	19 57	20 46
S 10	17 58	18 19	18 44	18 11	19 07	20 01	20 55
20	17 49	18 12	18 37	18 05	19 06	20 05	21 04
30	17 39	18 03	18 31	17 57	19 05	20 10	21 14
35	17 33	17 59	18 29	17 53	19 04	20 13	21 20
40	17 27	17 55	18 26	17 48	19 03	20 16	21 27
45	17 19	17 49	18 24	17 43	19 03	20 20	21 35
S 50	17 10	17 44	18 21	17 36	19 01	20 24	21 45
52	17 06	17 41	18 20	17 33	19 01	20 26	21 50
54	17 02	17 38	18 20	17 29	19 00	20 28	21 55
56	16 57	17 35	18 19	17 26	19 00	20 31	22 00
58	16 51	17 32	18 18	17 21	18 59	20 34	22 06
S 60	16 45	17 28	18 17	17 16	18 58	20 37	22 13

Day	SUN Eqn. of Time 00^h	SUN Eqn. of Time 12^h	SUN Mer. Pass.	MOON Mer. Pass. Upper	MOON Mer. Pass. Lower	Age	Phase
d	m s	m s	h m	h m	h m	d	%
20	03 28	03 21	12 03	12 04	24 31	00	0
21	03 14	03 06	12 03	12 56	00 31	01	2
22	02 59	02 51	12 03	13 46	01 21	02	6

	UT	ARIES	VENUS −3.9		MARS +1.0		JUPITER −2.8		SATURN +1.1		STARS		
		GHA	GHA	Dec	GHA	Dec	GHA	Dec	GHA	Dec	Name	SHA	Dec
	d h	° ′	° ′	° ′	° ′	° ′	° ′	° ′	° ′	° ′		° ′	° ′
	23 00	331 25.3	213 53.8	N20 34.5	243 19.0	N23 28.2	7 40.4	S15 32.0	158 08.8	N 5 02.9	Acamar	315 20.3	S40 15.6
	01	346 27.8	228 53.2	34.1	258 19.7	28.3	22 43.2	32.1	173 11.0	02.8	Achernar	335 28.3	S57 10.9
	02	1 30.2	243 52.5	33.7	273 20.5	28.3	37 46.0	32.2	188 13.2	02.7	Acrux	173 13.3	S63 09.4
	03	16 32.7	258 51.8	. . 33.3	288 21.2	. . 28.4	52 48.8	. . 32.3	203 15.4	. . 02.6	Adhara	255 15.0	S28 58.9
	04	31 35.2	273 51.1	32.9	303 21.9	28.4	67 51.5	32.4	218 17.6	02.5	Aldebaran	290 52.7	N16 31.8
	05	46 37.6	288 50.5	32.5	318 22.6	28.5	82 54.3	32.5	233 19.8	02.4			
	06	61 40.1	303 49.8	N20 32.1	333 23.3	N23 28.6	97 57.1	S15 32.6	248 22.0	N 5 02.2	Alioth	166 23.3	N55 54.6
	07	76 42.6	318 49.1	31.6	348 24.0	28.6	112 59.9	32.7	263 24.2	02.1	Alkaid	153 01.2	N49 16.1
	08	91 45.0	333 48.5	31.2	3 24.8	28.7	128 02.6	32.9	278 26.4	02.0	Al Na'ir	27 46.6	S46 54.7
S	09	106 47.5	348 47.8	. . 30.8	18 25.5	. . 28.7	143 05.4	. . 33.0	293 28.6	. . 01.9	Alnilam	275 49.3	S 1 11.6
U	10	121 50.0	3 47.1	30.4	33 26.2	28.8	158 08.2	33.1	308 30.7	01.8	Alphard	217 59.2	S 8 42.0
N	11	136 52.4	18 46.4	30.0	48 26.9	28.9	173 11.0	33.2	323 32.9	01.6			
D	12	151 54.9	33 45.8	N20 29.5	63 27.6	N23 28.9	188 13.8	S15 33.3	338 35.1	N 5 01.5	Alphecca	126 13.4	N26 41.1
A	13	166 57.3	48 45.1	29.1	78 28.3	29.0	203 16.5	33.4	353 37.3	01.4	Alpheratz	357 46.2	N29 08.7
Y	14	181 59.8	63 44.4	28.7	93 29.1	29.0	218 19.3	33.5	8 39.5	01.3	Altair	62 10.7	N 8 53.8
	15	197 02.3	78 43.8	. . 28.3	108 29.8	. . 29.1	233 22.1	. . 33.6	23 41.7	. . 01.2	Ankaa	353 18.0	S42 14.9
	16	212 04.7	93 43.1	27.8	123 30.5	29.1	248 24.9	33.7	38 43.9	01.1	Antares	112 29.7	S26 27.3
	17	227 07.2	108 42.4	27.4	138 31.2	29.2	263 27.6	33.8	53 46.1	00.9			
	18	242 09.7	123 41.7	N20 27.0	153 31.9	N23 29.3	278 30.4	S15 33.9	68 48.3	N 5 00.8	Arcturus	145 58.4	N19 08.0
	19	257 12.1	138 41.1	26.6	168 32.7	29.3	293 33.2	34.0	83 50.5	00.7	Atria	107 34.0	S69 03.0
	20	272 14.6	153 40.4	26.1	183 33.4	29.4	308 36.0	34.1	98 52.6	00.6	Avior	234 19.9	S59 32.3
	21	287 17.1	168 39.7	. . 25.7	198 34.1	. . 29.4	323 38.7	. . 34.2	113 54.8	. . 00.5	Bellatrix	278 35.1	N 6 21.6
	22	302 19.5	183 39.1	25.3	213 34.8	29.5	338 41.5	34.3	128 57.0	00.3	Betelgeuse	271 04.5	N 7 24.7
	23	317 22.0	198 38.4	24.8	228 35.5	29.5	353 44.3	34.4	143 59.2	00.2			
	24 00	332 24.5	213 37.7	N20 24.4	243 36.2	N23 29.6	8 47.1	S15 34.5	159 01.4	N 5 00.1	Canopus	263 57.7	S52 41.7
	01	347 26.9	228 37.1	24.0	258 37.0	29.6	23 49.8	34.6	174 03.6	5 00.0	Capella	280 38.8	N46 00.4
	02	2 29.4	243 36.4	23.5	273 37.7	29.7	38 52.6	34.7	189 05.8	4 59.9	Deneb	49 33.1	N45 19.1
	03	17 31.8	258 35.7	. . 23.1	288 38.4	. . 29.7	53 55.4	. . 34.8	204 08.0	. . 59.7	Denebola	182 36.8	N14 31.1
	04	32 34.3	273 35.0	22.7	303 39.1	29.8	68 58.2	34.9	219 10.2	59.6	Diphda	348 58.4	S17 55.7
	05	47 36.8	288 34.4	22.2	318 39.9	29.8	84 00.9	35.0	234 12.3	59.5			
	06	62 39.2	303 33.7	N20 21.8	333 40.6	N23 29.9	99 03.7	S15 35.1	249 14.5	N 4 59.4	Dubhe	193 55.5	N61 42.0
	07	77 41.7	318 33.0	21.3	348 41.3	30.0	114 06.5	35.3	264 16.7	59.3	Elnath	278 16.3	N28 36.9
	08	92 44.2	333 32.4	20.9	3 42.0	30.0	129 09.3	35.4	279 18.9	59.2	Eltanin	90 47.2	N51 29.5
M	09	107 46.6	348 31.7	. . 20.5	18 42.7	. . 30.1	144 12.0	. . 35.5	294 21.1	. . 59.0	Enif	33 49.6	N 9 55.3
O	10	122 49.1	3 31.0	20.0	33 43.5	30.1	159 14.8	35.6	309 23.3	58.9	Fomalhaut	15 26.6	S29 34.0
N	11	137 51.6	18 30.4	19.6	48 44.2	30.2	174 17.6	35.7	324 25.5	58.8			
D	12	152 54.0	33 29.7	N20 19.1	63 44.9	N23 30.2	189 20.4	S15 35.8	339 27.7	N 4 58.7	Gacrux	172 04.7	S57 10.2
A	13	167 56.5	48 29.0	18.7	78 45.6	30.3	204 23.1	35.9	354 29.9	58.6	Gienah	175 55.5	S17 35.8
Y	14	182 58.9	63 28.4	18.2	93 46.3	30.3	219 25.9	36.0	9 32.1	58.4	Hadar	148 52.4	S60 25.5
	15	198 01.4	78 27.7	. . 17.8	108 47.1	. . 30.4	234 28.7	. . 36.1	24 34.2	. . 58.3	Hamal	328 03.8	N23 30.6
	16	213 03.9	93 27.0	17.3	123 47.8	30.4	249 31.5	36.2	39 36.4	58.2	Kaus Aust.	83 47.3	S34 22.9
	17	228 06.3	108 26.4	16.9	138 48.5	30.5	264 34.2	36.3	54 38.6	58.1			
	18	243 08.8	123 25.7	N20 16.4	153 49.2	N23 30.5	279 37.0	S15 36.4	69 40.8	N 4 58.0	Kochab	137 19.9	N74 07.2
	19	258 11.3	138 25.0	16.0	168 50.0	30.6	294 39.8	36.5	84 43.0	57.8	Markab	13 40.8	N15 15.6
	20	273 13.7	153 24.4	15.5	183 50.7	30.6	309 42.5	36.6	99 45.2	57.7	Menkar	314 17.9	N 4 07.9
	21	288 16.2	168 23.7	. . 15.1	198 51.4	. . 30.6	324 45.3	. . 36.7	114 47.4	. . 57.6	Menkent	148 11.2	S36 25.2
	22	303 18.7	183 23.0	14.6	213 52.1	30.7	339 48.1	36.8	129 49.6	57.5	Miaplacidus	221 41.5	S69 45.3
	23	318 21.1	198 22.4	14.1	228 52.9	30.7	354 50.9	36.9	144 51.7	57.4			
	25 00	333 23.6	213 21.7	N20 13.7	243 53.6	N23 30.8	9 53.6	S15 37.0	159 53.9	N 4 57.2	Mirfak	308 44.4	N49 53.7
	01	348 26.1	228 21.0	13.2	258 54.3	30.8	24 56.4	37.1	174 56.1	57.1	Nunki	76 01.5	S26 17.1
	02	3 28.5	243 20.4	12.8	273 55.0	30.9	39 59.2	37.2	189 58.3	57.0	Peacock	53 23.0	S56 42.2
	03	18 31.0	258 19.7	. . 12.3	288 55.8	. . 30.9	55 02.0	. . 37.3	205 00.5	. . 56.9	Pollux	243 31.4	N28 00.2
	04	33 33.4	273 19.0	11.9	303 56.5	31.0	70 04.7	37.4	220 02.7	56.8	Procyon	245 02.9	N 5 12.1
	05	48 35.9	288 18.4	11.4	318 57.2	31.0	85 07.5	37.5	235 04.9	56.7			
	06	63 38.4	303 17.7	N20 10.9	333 57.9	N23 31.1	100 10.3	S15 37.6	250 07.1	N 4 56.5	Rasalhague	96 08.9	N12 33.3
	07	78 40.8	318 17.0	10.5	348 58.7	31.1	115 13.1	37.7	265 09.3	56.4	Regulus	207 46.8	N11 55.3
T	08	93 43.3	333 16.4	10.0	3 59.4	31.2	130 15.8	37.8	280 11.4	56.3	Rigel	281 14.8	S 8 11.2
U	09	108 45.8	348 15.7	. . 09.5	19 00.1	. . 31.2	145 18.6	. . 37.9	295 13.6	. . 56.2	Rigil Kent.	139 56.0	S60 52.8
E	10	123 48.2	3 15.0	09.1	34 00.8	31.2	160 21.4	38.0	310 15.8	56.1	Sabik	102 15.7	S15 44.2
S	11	138 50.7	18 14.4	08.6	49 01.6	31.3	175 24.1	38.1	325 18.0	55.9			
D	12	153 53.2	33 13.7	N20 08.1	64 02.3	N23 31.3	190 26.9	S15 38.2	340 20.2	N 4 55.8	Schedar	349 43.5	N56 35.4
A	13	168 55.6	48 13.1	07.7	79 03.0	31.4	205 29.7	38.3	355 22.4	55.7	Shaula	96 25.6	S37 06.8
Y	14	183 58.1	63 12.4	07.2	94 03.7	31.4	220 32.5	38.4	10 24.6	55.6	Sirius	258 36.4	S16 43.6
	15	199 00.5	78 11.7	. . 06.7	109 04.5	. . 31.5	235 35.2	. . 38.6	25 26.8	. . 55.5	Spica	158 34.4	S11 12.7
	16	214 03.0	93 11.1	06.2	124 05.2	31.5	250 38.0	38.7	40 29.0	55.3	Suhail	222 55.0	S43 28.2
	17	229 05.5	108 10.4	05.8	139 05.9	31.5	265 40.8	38.8	55 31.1	55.2			
	18	244 07.9	123 09.7	N20 05.3	154 06.7	N23 31.6	280 43.5	S15 38.9	70 33.3	N 4 55.1	Vega	80 40.6	N38 47.8
	19	259 10.4	138 09.1	04.8	169 07.4	31.6	295 46.3	39.0	85 35.5	55.0	Zuben'ubi	137 08.6	S16 05.[illegible]
	20	274 12.9	153 08.4	04.3	184 08.1	31.7	310 49.1	39.1	100 37.7	54.9		SHA	Mer. Pass
	21	289 15.3	168 07.8	. . 03.8	199 08.8	. . 31.7	325 51.9	. . 39.2	115 39.9	. . 54.7		° ′	h m
	22	304 17.8	183 07.1	03.4	214 09.6	31.7	340 54.6	39.3	130 42.1	54.6	Venus	241 13.3	9 46
	23	319 20.3	198 06.4	02.9	229 10.3	31.8	355 57.4	39.4	145 44.3	54.5	Mars	271 11.8	7 45
		h m									Jupiter	36 22.6	23 21
	Mer. Pass.	1 50.1	v −0.7	d 0.4	v 0.7	d 0.1	v 2.8	d 0.1	v 2.2	d 0.1	Saturn	186 37.0	13 22

UT	SUN GHA	SUN Dec	MOON GHA	v	MOON Dec	d	HP
d h	° ′	° ′	° ′	′	° ′	′	′
23 00	179 19.2	N11 28.0	148 31.5	11.4	S 6 25.5	15.2	59.1
01	194 19.4	27.2	163 01.9	11.5	6 40.7	15.2	59.0
02	209 19.5	26.3	177 32.4	11.5	6 55.9	15.1	59.0
03	224 19.7	. . 25.5	192 02.9	11.5	7 11.0	15.1	58.9
04	239 19.9	24.6	206 33.4	11.6	7 26.1	15.0	58.9
05	254 20.0	23.8	221 04.0	11.5	7 41.1	15.0	58.9
06	269 20.2	N11 22.9	235 34.5	11.5	S 7 56.1	14.9	58.8
07	284 20.3	22.1	250 05.0	11.5	8 11.0	14.9	58.8
08	299 20.5	21.2	264 35.5	11.6	8 25.9	14.7	58.8
09	314 20.7	. . 20.4	279 06.1	11.5	8 40.6	14.7	58.7
10	329 20.8	19.5	293 36.6	11.5	8 55.3	14.7	58.7
11	344 21.0	18.7	308 07.1	11.5	9 10.0	14.6	58.6
12	359 21.2	N11 17.8	322 37.6	11.6	S 9 24.6	14.5	58.6
13	14 21.3	17.0	337 08.2	11.5	9 39.1	14.4	58.6
14	29 21.5	16.1	351 38.7	11.5	9 53.5	14.4	58.5
15	44 21.7	. . 15.3	6 09.2	11.5	10 07.9	14.3	58.5
16	59 21.8	14.4	20 39.7	11.6	10 22.2	14.2	58.5
17	74 22.0	13.6	35 10.3	11.5	10 36.4	14.2	58.4
18	89 22.2	N11 12.7	49 40.8	11.5	S10 50.6	14.1	58.4
19	104 22.3	11.9	64 11.3	11.5	11 04.7	14.0	58.3
20	119 22.5	11.0	78 41.8	11.5	11 18.7	13.9	58.3
21	134 22.7	. . 10.2	93 12.3	11.4	11 32.6	13.9	58.3
22	149 22.8	09.3	107 42.7	11.5	11 46.5	13.7	58.2
23	164 23.0	08.4	122 13.2	11.5	12 00.2	13.7	58.2
24 00	179 23.1	N11 07.6	136 43.7	11.4	S12 13.9	13.6	58.1
01	194 23.3	06.7	151 14.1	11.5	12 27.5	13.6	58.1
02	209 23.5	05.9	165 44.6	11.4	12 41.1	13.4	58.1
03	224 23.6	. . 05.0	180 15.0	11.4	12 54.5	13.4	58.0
04	239 23.8	04.2	194 45.4	11.4	13 07.9	13.3	58.0
05	254 24.0	03.3	209 15.8	11.4	13 21.2	13.1	58.0
06	269 24.2	N11 02.5	223 46.2	11.4	S13 34.3	13.2	57.9
07	284 24.3	01.6	238 16.6	11.4	13 47.5	13.0	57.9
08	299 24.5	11 00.7	252 47.0	11.3	14 00.5	12.9	57.8
09	314 24.7	10 59.9	267 17.3	11.4	14 13.4	12.8	57.8
10	329 24.8	59.0	281 47.7	11.3	14 26.2	12.8	57.8
11	344 25.0	58.2	296 18.0	11.3	14 39.0	12.6	57.7
12	359 25.2	N10 57.3	310 48.3	11.3	S14 51.6	12.6	57.7
13	14 25.3	56.5	325 18.6	11.2	15 04.2	12.5	57.6
14	29 25.5	55.6	339 48.8	11.3	15 16.7	12.3	57.6
15	44 25.7	. . 54.7	354 19.1	11.2	15 29.0	12.3	57.6
16	59 25.8	53.9	8 49.3	11.2	15 41.3	12.2	57.5
17	74 26.0	53.0	23 19.5	11.2	15 53.5	12.1	57.5
18	89 26.2	N10 52.2	37 49.7	11.2	S16 05.6	12.0	57.5
19	104 26.3	51.3	52 19.9	11.1	16 17.6	11.9	57.4
20	119 26.5	50.4	66 50.0	11.1	16 29.5	11.8	57.4
21	134 26.7	. . 49.6	81 20.1	11.1	16 41.3	11.7	57.3
22	149 26.9	48.7	95 50.2	11.1	16 53.0	11.6	57.3
23	164 27.0	47.9	110 20.3	11.1	17 04.6	11.4	57.3
25 00	179 27.2	N10 47.0	124 50.4	11.0	S17 16.0	11.4	57.2
01	194 27.4	46.1	139 20.4	11.0	17 27.4	11.3	57.2
02	209 27.5	45.3	153 50.4	11.0	17 38.7	11.2	57.1
03	224 27.7	. . 44.4	168 20.4	11.0	17 49.9	11.1	57.1
04	239 27.9	43.5	182 50.4	10.9	18 01.0	10.9	57.1
05	254 28.1	42.7	197 20.3	11.0	18 11.9	10.9	57.0
06	269 28.2	N10 41.8	211 50.3	10.9	S18 22.8	10.8	57.0
07	284 28.4	40.9	226 20.2	10.8	18 33.6	10.6	57.0
08	299 28.6	40.1	240 50.0	10.9	18 44.2	10.5	56.9
09	314 28.8	. . 39.2	255 19.9	10.8	18 54.7	10.5	56.9
10	329 28.9	38.4	269 49.7	10.8	19 05.2	10.3	56.9
11	344 29.1	37.5	284 19.5	10.8	19 15.5	10.2	56.8
12	359 29.3	N10 36.6	298 49.3	10.7	S19 25.7	10.1	56.8
13	14 29.4	35.8	313 19.0	10.8	19 35.8	10.0	56.7
14	29 29.6	34.9	327 48.8	10.7	19 45.8	9.9	56.7
15	44 29.8	. . 34.0	342 18.5	10.7	19 55.7	9.7	56.7
16	59 30.0	33.2	356 48.2	10.6	20 05.4	9.7	56.6
17	74 30.1	32.3	11 17.8	10.6	20 15.1	9.5	56.6
18	89 30.3	N10 31.4	25 47.4	10.6	S20 24.6	9.4	56.6
19	104 30.5	30.6	40 17.0	10.6	20 34.0	9.3	56.5
20	119 30.7	29.7	54 46.6	10.6	20 43.3	9.2	56.5
21	134 30.8	. . 28.8	69 16.2	10.5	20 52.5	9.1	56.5
22	149 31.0	28.0	83 45.7	10.5	21 01.6	8.9	56.4
23	164 31.2	27.1	98 15.2	10.5	S21 10.5	8.9	56.4
	SD 15.8	*d* 0.9	SD 16.0		15.7		15.5

Lat.	Twilight Naut.	Twilight Civil	Sunrise	Moonrise 23	24	25	26
°	h m	h m	h m	h m	h m	h m	h m
N 72	////	01 16	03 20	10 26	13 07	▬	▬
N 70	////	02 09	03 40	10 11	12 29	▬	▬
68	////	02 41	03 56	09 59	12 03	14 26	▬
66	01 06	03 04	04 09	09 49	11 44	13 45	16 15
64	01 53	03 22	04 19	09 41	11 28	13 17	15 11
62	02 22	03 36	04 28	09 34	11 15	12 56	14 36
60	02 43	03 48	04 36	09 28	11 04	12 39	14 11
N 58	03 00	03 59	04 43	09 23	10 55	12 25	13 51
56	03 14	04 08	04 49	09 18	10 46	12 13	13 35
54	03 26	04 15	04 54	09 14	10 39	12 02	13 21
52	03 36	04 22	04 59	09 10	10 32	11 53	13 09
50	03 45	04 28	05 03	09 07	10 27	11 44	12 58
45	04 04	04 41	05 12	09 00	10 14	11 27	12 36
N 40	04 18	04 52	05 20	08 53	10 04	11 12	12 18
35	04 29	05 01	05 27	08 48	09 55	11 00	12 03
30	04 39	05 08	05 33	08 44	09 47	10 49	11 50
20	04 53	05 20	05 42	08 36	09 34	10 31	11 28
N 10	05 05	05 30	05 51	08 29	09 22	10 16	11 09
0	05 13	05 38	05 59	08 23	09 12	10 01	10 52
S 10	05 21	05 45	06 07	08 17	09 01	09 47	10 34
20	05 27	05 53	06 15	08 10	08 50	09 31	10 15
30	05 32	06 00	06 24	08 03	08 37	09 14	09 54
35	05 34	06 04	06 30	07 58	08 30	09 04	09 42
40	05 37	06 08	06 36	07 54	08 22	08 52	09 27
45	05 38	06 13	06 42	07 48	08 12	08 39	09 10
S 50	05 40	06 18	06 51	07 42	08 01	08 23	08 49
52	05 41	06 20	06 54	07 39	07 55	08 15	08 39
54	05 41	06 22	06 59	07 35	07 49	08 06	08 28
56	05 42	06 25	07 03	07 32	07 43	07 57	08 16
58	05 42	06 28	07 08	07 28	07 36	07 46	08 01
S 60	05 43	06 31	07 14	07 23	07 28	07 34	07 44

Lat.	Sunset	Twilight Civil	Twilight Naut.	Moonset 23	24	25	26
°	h m	h m	h m	h m	h m	h m	h m
N 72	20 40	22 37	////	18 14	17 17	▬	▬
N 70	20 21	21 49	////	18 32	17 56	▬	▬
68	20 06	21 19	////	18 46	18 24	17 46	▬
66	19 53	20 57	22 49	18 57	18 45	18 27	17 43
64	19 43	20 40	22 06	19 07	19 02	18 56	18 48
62	19 35	20 26	21 39	19 16	19 16	19 18	19 23
60	19 27	20 14	21 18	19 23	19 28	19 36	19 49
N 58	19 20	20 04	21 02	19 29	19 38	19 51	20 09
56	19 15	19 55	20 48	19 35	19 47	20 03	20 26
54	19 09	19 48	20 37	19 40	19 55	20 15	20 40
52	19 05	19 41	20 27	19 45	20 03	20 25	20 52
50	19 01	19 35	20 18	19 49	20 09	20 33	21 03
45	18 51	19 22	20 00	19 58	20 23	20 52	21 27
N 40	18 44	19 12	19 46	20 06	20 35	21 08	21 45
35	18 37	19 04	19 35	20 13	20 45	21 21	22 01
30	18 32	18 56	19 25	20 18	20 54	21 32	22 14
20	18 22	18 44	19 11	20 29	21 09	21 52	22 37
N 10	18 13	18 35	19 00	20 38	21 23	22 09	22 58
0	18 06	18 27	18 51	20 46	21 35	22 25	23 16
S 10	17 58	18 19	18 44	20 55	21 48	22 41	23 35
20	17 50	18 12	18 38	21 04	22 02	22 59	23 55
30	17 41	18 05	18 33	21 14	22 17	23 19	24 19
35	17 36	18 01	18 31	21 20	22 26	23 31	24 33
40	17 30	17 57	18 29	21 27	22 37	23 44	24 48
45	17 23	17 53	18 27	21 35	22 49	24 00	00 00
S 50	17 15	17 48	18 25	21 45	23 04	24 20	00 20
52	17 11	17 46	18 25	21 50	23 11	24 29	00 29
54	17 07	17 43	18 24	21 55	23 19	24 40	00 40
56	17 02	17 41	18 24	22 00	23 27	24 52	00 52
58	16 57	17 38	18 23	22 06	23 37	25 06	01 06
S 60	16 52	17 35	18 23	22 13	23 49	25 23	01 23

Day	SUN Eqn. of Time 00^h	SUN Eqn. of Time 12^h	SUN Mer. Pass.	MOON Mer. Pass. Upper	MOON Mer. Pass. Lower	Age	Phase
d	m s	m s	h m	h m	h m	d %	
23	02 44	02 36	12 03	14 35	02 10	03 13	
24	02 28	02 20	12 02	15 24	02 59	04 21	
25	02 12	02 03	12 02	16 13	03 48	05 30	

UT		ARIES	VENUS −3.9		MARS +1.0		JUPITER −2.8		SATURN +1.1		STARS		
		GHA	GHA	Dec	GHA	Dec	GHA	Dec	GHA	Dec	Name	SHA	Dec
d	h	° ′	° ′	° ′	° ′	° ′	° ′	° ′	° ′	° ′		° ′	° ′
26	00	334 22.7	213 05.8	N20 02.4	244 11.0	N23 31.8	11 00.2	S15 39.5	160 46.5	N 4 54.4	Acamar	315 20.2	S40 15.
	01	349 25.2	228 05.1	01.9	259 11.8	31.9	26 02.9	39.6	175 48.6	54.3	Achernar	335 28.3	S57 10.
	02	4 27.7	243 04.4	01.4	274 12.5	31.9	41 05.7	39.7	190 50.8	54.1	Acrux	173 13.3	S63 09.
	03	19 30.1	258 03.8	. . 01.0	289 13.2	. . 31.9	56 08.5	. . 39.8	205 53.0	. . 54.0	Adhara	255 15.0	S28 58.
	04	34 32.6	273 03.1	00.5	304 13.9	32.0	71 11.2	39.9	220 55.2	53.9	Aldebaran	290 52.6	N16 31.
	05	49 35.0	288 02.5	20 00.0	319 14.7	32.0	86 14.0	40.0	235 57.4	53.8			
W	06	64 37.5	303 01.8	N19 59.5	334 15.4	N23 32.0	101 16.8	S15 40.1	250 59.6	N 4 53.7	Alioth	166 23.3	N55 54.
E	07	79 40.0	318 01.1	59.0	349 16.1	32.1	116 19.6	40.2	266 01.8	53.5	Alkaid	153 01.2	N49 16.
D	08	94 42.4	333 00.5	58.5	4 16.9	32.1	131 22.3	40.3	281 04.0	53.4	Al Na'ir	27 46.6	S46 54.
N	09	109 44.9	347 59.8	. . 58.0	19 17.6	. . 32.2	146 25.1	. . 40.4	296 06.1	. . 53.3	Alnilam	275 49.3	S 1 11.
E	10	124 47.4	2 59.2	57.5	34 18.3	32.2	161 27.9	40.5	311 08.3	53.2	Alphard	217 59.2	S 8 42.
S	11	139 49.8	17 58.5	57.0	49 19.1	32.2	176 30.6	40.6	326 10.5	53.1			
D	12	154 52.3	32 57.8	N19 56.6	64 19.8	N23 32.3	191 33.4	S15 40.7	341 12.7	N 4 52.9	Alphecca	126 13.4	N26 41.
A	13	169 54.8	47 57.2	56.1	79 20.5	32.3	206 36.2	40.8	356 14.9	52.8	Alpheratz	357 46.2	N29 08.
Y	14	184 57.2	62 56.5	55.6	94 21.2	32.3	221 38.9	40.9	11 17.1	52.7	Altair	62 10.7	N 8 53.
	15	199 59.7	77 55.9	. . 55.1	109 22.0	. . 32.4	236 41.7	. . 41.0	26 19.3	. . 52.6	Ankaa	353 17.9	S42 14.
	16	215 02.2	92 55.2	54.6	124 22.7	32.4	251 44.5	41.1	41 21.5	52.5	Antares	112 29.7	S26 27.
	17	230 04.6	107 54.6	54.1	139 23.4	32.4	266 47.3	41.2	56 23.6	52.3			
	18	245 07.1	122 53.9	N19 53.6	154 24.2	N23 32.5	281 50.0	S15 41.3	71 25.8	N 4 52.2	Arcturus	145 58.4	N19 08.
	19	260 09.5	137 53.2	53.1	169 24.9	32.5	296 52.8	41.4	86 28.0	52.1	Atria	107 34.0	S69 03.
	20	275 12.0	152 52.6	52.6	184 25.6	32.5	311 55.6	41.5	101 30.2	52.0	Avior	234 19.9	S59 32.
	21	290 14.5	167 51.9	. . 52.1	199 26.4	. . 32.6	326 58.3	. . 41.6	116 32.4	. . 51.9	Bellatrix	278 35.1	N 6 21.
	22	305 16.9	182 51.3	51.6	214 27.1	32.6	342 01.1	41.7	131 34.6	51.8	Betelgeuse	271 04.5	N 7 24.
	23	320 19.4	197 50.6	51.1	229 27.8	32.6	357 03.9	41.8	146 36.8	51.6			
27	00	335 21.9	212 50.0	N19 50.6	244 28.6	N23 32.7	12 06.6	S15 41.9	161 38.9	N 4 51.5	Canopus	263 57.7	S52 41.
	01	350 24.3	227 49.3	50.1	259 29.3	32.7	27 09.4	42.0	176 41.1	51.4	Capella	280 38.7	N46 00.
	02	5 26.8	242 48.6	49.5	274 30.0	32.7	42 12.2	42.1	191 43.3	51.3	Deneb	49 33.1	N45 19.
	03	20 29.3	257 48.0	. . 49.0	289 30.8	. . 32.8	57 14.9	. . 42.2	206 45.5	. . 51.2	Denebola	182 36.8	N14 31.
	04	35 31.7	272 47.3	48.5	304 31.5	32.8	72 17.7	42.3	221 47.7	51.0	Diphda	348 58.4	S17 55.
	05	50 34.2	287 46.7	48.0	319 32.2	32.8	87 20.5	42.4	236 49.9	50.9			
	06	65 36.7	302 46.0	N19 47.5	334 33.0	N23 32.9	102 23.2	S15 42.5	251 52.1	N 4 50.8	Dubhe	193 55.5	N61 42.
T	07	80 39.1	317 45.4	47.0	349 33.7	32.9	117 26.0	42.6	266 54.3	50.7	Elnath	278 16.3	N28 36.
H	08	95 41.6	332 44.7	46.5	4 34.5	32.9	132 28.8	42.7	281 56.4	50.6	Eltanin	90 47.2	N51 29.
U	09	110 44.0	347 44.0	. . 46.0	19 35.2	. . 32.9	147 31.5	. . 42.8	296 58.6	. . 50.4	Enif	33 49.6	N 9 55.
R	10	125 46.5	2 43.4	45.5	34 35.9	33.0	162 34.3	42.9	312 00.8	50.3	Fomalhaut	15 26.6	S29 34.
S	11	140 49.0	17 42.7	44.9	49 36.7	33.0	177 37.1	43.0	327 03.0	50.2			
D	12	155 51.4	32 42.1	N19 44.4	64 37.4	N23 33.0	192 39.8	S15 43.1	342 05.2	N 4 50.1	Gacrux	172 04.7	S57 10.
A	13	170 53.9	47 41.4	43.9	79 38.1	33.1	207 42.6	43.2	357 07.4	50.0	Gienah	175 55.5	S17 35.
Y	14	185 56.4	62 40.8	43.4	94 38.9	33.1	222 45.4	43.3	12 09.6	49.8	Hadar	148 52.4	S60 25.
	15	200 58.8	77 40.1	. . 42.9	109 39.6	. . 33.1	237 48.1	. . 43.4	27 11.7	. . 49.7	Hamal	328 03.8	N23 30
	16	216 01.3	92 39.5	42.3	124 40.3	33.1	252 50.9	43.5	42 13.9	49.6	Kaus Aust.	83 47.3	S34 22.
	17	231 03.8	107 38.8	41.8	139 41.1	33.2	267 53.7	43.6	57 16.1	49.5			
	18	246 06.2	122 38.2	N19 41.3	154 41.8	N23 33.2	282 56.4	S15 43.7	72 18.3	N 4 49.4	Kochab	137 19.9	N74 07.
	19	261 08.7	137 37.5	40.8	169 42.6	33.2	297 59.2	43.8	87 20.5	49.2	Markab	13 40.8	N15 15.
	20	276 11.1	152 36.9	40.3	184 43.3	33.2	313 02.0	43.9	102 22.7	49.1	Menkar	314 17.9	N 4 07.
	21	291 13.6	167 36.2	. . 39.7	199 44.0	. . 33.3	328 04.7	. . 44.0	117 24.9	. . 49.0	Menkent	148 11.2	S36 25.
	22	306 16.1	182 35.6	39.2	214 44.8	33.3	343 07.5	44.1	132 27.0	48.9	Miaplacidus	221 41.5	S69 45.
	23	321 18.5	197 34.9	38.7	229 45.5	33.3	358 10.3	44.2	147 29.2	48.8			
28	00	336 21.0	212 34.2	N19 38.1	244 46.2	N23 33.3	13 13.0	S15 44.3	162 31.4	N 4 48.6	Mirfak	308 44.4	N49 53.
	01	351 23.5	227 33.6	37.6	259 47.0	33.4	28 15.8	44.4	177 33.6	48.5	Nunki	76 01.5	S26 17.
	02	6 25.9	242 32.9	37.1	274 47.7	33.4	43 18.6	44.5	192 35.8	48.4	Peacock	53 23.0	S56 42.
	03	21 28.4	257 32.3	. . 36.6	289 48.5	. . 33.4	58 21.3	. . 44.6	207 38.0	. . 48.3	Pollux	243 31.4	N28 00.
	04	36 30.9	272 31.6	36.0	304 49.2	33.4	73 24.1	44.7	222 40.2	48.2	Procyon	245 02.9	N 5 12.
	05	51 33.3	287 31.0	35.5	319 49.9	33.5	88 26.9	44.8	237 42.3	48.0			
	06	66 35.8	302 30.3	N19 35.0	334 50.7	N23 33.5	103 29.6	S15 44.9	252 44.5	N 4 47.9	Rasalhague	96 08.9	N12 33.
	07	81 38.3	317 29.7	34.4	349 51.4	33.5	118 32.4	45.0	267 46.7	47.8	Regulus	207 46.8	N11 55.
	08	96 40.7	332 29.0	33.9	4 52.2	33.5	133 35.2	45.1	282 48.9	47.7	Rigel	281 14.8	S 8 11.
F	09	111 43.2	347 28.4	. . 33.3	19 52.9	. . 33.6	148 37.9	. . 45.2	297 51.1	. . 47.6	Rigil Kent.	139 56.0	S60 52.
R	10	126 45.6	2 27.7	32.8	34 53.6	33.6	163 40.7	45.3	312 53.3	47.4	Sabik	102 15.7	S15 44.
I	11	141 48.1	17 27.1	32.3	49 54.4	33.6	178 43.4	45.4	327 55.5	47.3			
D	12	156 50.6	32 26.4	N19 31.7	64 55.1	N23 33.6	193 46.2	S15 45.5	342 57.6	N 4 47.2	Schedar	349 43.5	N56 35.
A	13	171 53.0	47 25.8	31.2	79 55.9	33.6	208 49.0	45.6	357 59.8	47.1	Shaula	96 25.6	S37 06.
Y	14	186 55.5	62 25.1	30.6	94 56.6	33.7	223 51.7	45.7	13 02.0	46.9	Sirius	258 36.4	S16 43.
	15	201 58.0	77 24.5	. . 30.1	109 57.3	. . 33.7	238 54.5	. . 45.8	28 04.2	. . 46.8	Spica	158 34.4	S11 12.
	16	217 00.4	92 23.8	29.6	124 58.1	33.7	253 57.3	45.9	43 06.4	46.7	Suhail	222 55.0	S43 28.
	17	232 02.9	107 23.2	29.0	139 58.8	33.7	269 00.0	46.0	58 08.6	46.6			
	18	247 05.4	122 22.6	N19 28.5	154 59.6	N23 33.7	284 02.8	S15 46.1	73 10.8	N 4 46.5	Vega	80 40.6	N38 47.
	19	262 07.8	137 21.9	27.9	170 00.3	33.8	299 05.6	46.2	88 12.9	46.3	Zuben'ubi	137 08.7	S16 05.
	20	277 10.3	152 21.3	27.4	185 01.1	33.8	314 08.3	46.3	103 15.1	46.2			
	21	292 12.8	167 20.6	. . 26.8	200 01.8	. . 33.8	329 11.1	. . 46.4	118 17.3	. . 46.1			
	22	307 15.2	182 20.0	26.3	215 02.5	33.8	344 13.8	46.5	133 19.5	46.0			
	23	322 17.7	197 19.3	25.7	230 03.3	33.8	359 16.6	46.6	148 21.7	45.9			
Mer. Pass.		1 38.3 (h m)	v −0.7	d 0.5	v 0.7	d 0.0	v 2.8	d 0.1	v 2.2	d 0.1			

	SHA	Mer. Pas
	° ′	h m
Venus	237 28.1	9 49
Mars	269 06.7	7 42
Jupiter	36 44.8	23 07
Saturn	186 17.1	13 11

UT	SUN GHA	SUN Dec	MOON GHA	v	MOON Dec	d	HP
d h	° ′	° ′	° ′	′	° ′	′	′
26 00	179 31.4	N10 26.2	112 44.7	10.4	S21 19.4	8.7	56.4
01	194 31.5	25.4	127 14.1	10.4	21 28.1	8.6	56.3
02	209 31.7	24.5	141 43.5	10.5	21 36.7	8.5	56.3
03	224 31.9	. . 23.6	156 13.0	10.3	21 45.2	8.3	56.3
04	239 32.1	22.7	170 42.3	10.4	21 53.5	8.3	56.2
05	254 32.2	21.9	185 11.7	10.3	22 01.8	8.1	56.2
06	269 32.4	N10 21.0	199 41.0	10.3	S22 09.9	8.0	56.2
07	284 32.6	20.1	214 10.3	10.3	22 17.9	7.9	56.1
08	299 32.8	19.3	228 39.6	10.3	22 25.8	7.7	56.1
09	314 32.9	. . 18.4	243 08.9	10.2	22 33.5	7.6	56.1
10	329 33.1	17.5	257 38.1	10.2	22 41.1	7.5	56.0
11	344 33.3	16.6	272 07.3	10.2	22 48.6	7.4	56.0
12	359 33.5	N10 15.8	286 36.5	10.2	S22 56.0	7.3	56.0
13	14 33.7	14.9	301 05.7	10.2	23 03.3	7.1	55.9
14	29 33.8	14.0	315 34.9	10.1	23 10.4	7.0	55.9
15	44 34.0	. . 13.2	330 04.0	10.1	23 17.4	6.9	55.9
16	59 34.2	12.3	344 33.1	10.1	23 24.3	6.8	55.8
17	74 34.4	11.4	359 02.2	10.1	23 31.1	6.6	55.8
18	89 34.5	N10 10.5	13 31.3	10.0	S23 37.7	6.5	55.8
19	104 34.7	09.7	28 00.3	10.1	23 44.2	6.4	55.8
20	119 34.9	08.8	42 29.4	10.0	23 50.6	6.2	55.7
21	134 35.1	. . 07.9	56 58.4	10.0	23 56.8	6.2	55.7
22	149 35.3	07.0	71 27.4	10.0	24 03.0	6.0	55.7
23	164 35.4	06.2	85 56.4	9.9	24 09.0	5.8	55.6
27 00	179 35.6	N10 05.3	100 25.3	10.0	S24 14.8	5.8	55.6
01	194 35.8	04.4	114 54.3	9.9	24 20.6	5.6	55.6
02	209 36.0	03.5	129 23.2	9.9	24 26.2	5.5	55.6
03	224 36.2	. . 02.7	143 52.1	9.9	24 31.7	5.3	55.5
04	239 36.3	01.8	158 21.0	9.9	24 37.0	5.2	55.5
05	254 36.5	00.9	172 49.9	9.9	24 42.2	5.1	55.5
06	269 36.7	N10 00.0	187 18.8	9.9	S24 47.3	5.0	55.4
07	284 36.9	9 59.2	201 47.7	9.8	24 52.3	4.8	55.4
08	299 37.1	58.3	216 16.5	9.9	24 57.1	4.7	55.4
09	314 37.2	. . 57.4	230 45.4	9.8	25 01.8	4.6	55.4
10	329 37.4	56.5	245 14.2	9.8	25 06.4	4.4	55.3
11	344 37.6	55.6	259 43.0	9.8	25 10.8	4.4	55.3
12	359 37.8	N 9 54.8	274 11.8	9.8	S25 15.2	4.1	55.3
13	14 38.0	53.9	288 40.6	9.8	25 19.3	4.1	55.3
14	29 38.2	53.0	303 09.4	9.8	25 23.4	3.9	55.2
15	44 38.3	. . 52.1	317 38.2	9.8	25 27.3	3.8	55.2
16	59 38.5	51.2	332 07.0	9.8	25 31.1	3.6	55.2
17	74 38.7	50.4	346 35.8	9.8	25 34.7	3.6	55.2
18	89 38.9	N 9 49.5	1 04.6	9.7	S25 38.3	3.4	55.1
19	104 39.1	48.6	15 33.3	9.8	25 41.7	3.2	55.1
20	119 39.3	47.7	30 02.1	9.8	25 44.9	3.1	55.1
21	134 39.4	. . 46.8	44 30.9	9.7	25 48.0	3.0	55.1
22	149 39.6	46.0	58 59.6	9.8	25 51.0	2.9	55.0
23	164 39.8	45.1	73 28.4	9.8	25 53.9	2.7	55.0
28 00	179 40.0	N 9 44.2	87 57.2	9.7	S25 56.6	2.6	55.0
01	194 40.2	43.3	102 25.9	9.8	25 59.2	2.5	55.0
02	209 40.4	42.4	116 54.7	9.8	26 01.7	2.3	55.0
03	224 40.5	. . 41.6	131 23.5	9.8	26 04.0	2.2	54.9
04	239 40.7	40.7	145 52.3	9.7	26 06.2	2.1	54.9
05	254 40.9	39.8	160 21.0	9.8	26 08.3	2.0	54.9
06	269 41.1	N 9 38.9	174 49.8	9.8	S26 10.3	1.8	54.9
07	284 41.3	38.0	189 18.6	9.8	26 12.1	1.6	54.9
08	299 41.5	37.1	203 47.4	9.8	26 13.7	1.6	54.8
09	314 41.6	. . 36.3	218 16.2	9.8	26 15.3	1.4	54.8
10	329 41.8	35.4	232 45.0	9.9	26 16.7	1.3	54.8
11	344 42.0	34.5	247 13.9	9.8	26 18.0	1.1	54.8
12	359 42.2	N 9 33.6	261 42.7	9.8	S26 19.1	1.1	54.8
13	14 42.4	32.7	276 11.5	9.9	26 20.2	0.8	54.7
14	29 42.6	31.8	290 40.4	9.9	26 21.0	0.8	54.7
15	44 42.8	. . 30.9	305 09.3	9.9	26 21.8	0.6	54.7
16	59 42.9	30.1	319 38.2	9.9	26 22.4	0.5	54.7
17	74 43.1	29.2	334 07.1	9.9	26 22.9	0.4	54.7
18	89 43.3	N 9 28.3	348 36.0	9.9	S26 23.3	0.2	54.6
19	104 43.5	27.4	3 04.9	10.0	26 23.5	0.1	54.6
20	119 43.7	26.5	17 33.9	10.0	26 23.6	0.0	54.6
21	134 43.9	. . 25.6	32 02.9	9.9	26 23.6	0.2	54.6
22	149 44.1	24.7	46 31.8	10.1	26 23.4	0.3	54.6
23	164 44.3	23.8	61 00.9	10.0	S26 23.1	0.4	54.6
	SD 15.9	*d* 0.9	SD 15.3		15.1		14.9

Lat.	Twilight Naut.	Twilight Civil	Sunrise	Moonrise 26	27	28	29
°	h m	h m	h m	h m	h m	h m	h m
N 72	////	01 50	03 36	■	■	■	■
N 70	////	02 30	03 54	■	■	■	■
68	////	02 57	04 08	■	■	■	■
66	01 35	03 17	04 19	16 15	■	■	■
64	02 11	03 33	04 28	15 11	17 17	■	19 19
62	02 35	03 46	04 36	14 36	16 10	17 22	18 01
60	02 54	03 57	04 43	14 11	15 35	16 41	17 24
N 58	03 10	04 06	04 49	13 51	15 10	16 13	16 58
56	03 22	04 14	04 55	13 35	14 50	15 52	16 38
54	03 33	04 21	04 59	13 21	14 33	15 34	16 21
52	03 43	04 28	05 04	13 09	14 19	15 18	16 06
50	03 51	04 33	05 08	12 58	14 06	15 05	15 53
45	04 08	04 45	05 16	12 36	13 41	14 38	15 27
N 40	04 21	04 55	05 23	12 18	13 20	14 17	15 06
35	04 32	05 03	05 29	12 03	13 03	13 59	14 49
30	04 41	05 10	05 34	11 50	12 49	13 44	14 34
20	04 54	05 21	05 43	11 28	12 24	13 18	14 08
N 10	05 05	05 30	05 51	11 09	12 03	12 56	13 46
0	05 13	05 37	05 58	10 52	11 43	12 35	13 26
S 10	05 19	05 44	06 05	10 34	11 23	12 14	13 05
20	05 25	05 50	06 13	10 15	11 02	11 52	12 43
30	05 29	05 57	06 21	09 54	10 38	11 26	12 18
35	05 31	06 00	06 26	09 42	10 24	11 11	12 03
40	05 32	06 04	06 31	09 27	10 07	10 53	11 45
45	05 34	06 08	06 37	09 10	09 48	10 32	11 25
S 50	05 34	06 12	06 45	08 49	09 23	10 06	10 58
52	05 35	06 14	06 48	08 39	09 11	09 53	10 45
54	05 35	06 16	06 52	08 28	08 58	09 38	10 30
56	05 35	06 18	06 56	08 16	08 42	09 21	10 13
58	05 35	06 20	07 00	08 01	08 24	09 00	09 52
S 60	05 35	06 23	07 05	07 44	08 02	08 33	09 25

Lat.	Sunset	Twilight Civil	Twilight Naut.	Moonset 26	27	28	29
°	h m	h m	h m	h m	h m	h m	h m
N 72	20 23	22 05	////	■	■	■	■
N 70	20 06	21 28	////	■	■	■	■
68	19 53	21 02	23 26	■	■	■	■
66	19 42	20 43	22 21	17 43	■	■	■
64	19 33	20 28	21 48	18 48	18 30	■	20 02
62	19 25	20 15	21 24	19 23	19 37	20 12	21 20
60	19 18	20 04	21 06	19 49	20 12	20 53	21 56
N 58	19 12	19 55	20 51	20 09	20 37	21 21	22 21
56	19 07	19 47	20 38	20 26	20 58	21 43	22 42
54	19 02	19 40	20 28	20 40	21 14	22 01	22 59
52	18 58	19 34	20 19	20 52	21 29	22 16	23 13
50	18 54	19 28	20 10	21 03	21 42	22 29	23 26
45	18 46	19 17	19 54	21 27	22 08	22 56	23 52
N 40	18 39	19 07	19 41	21 45	22 28	23 18	24 12
35	18 33	18 59	19 30	22 01	22 46	23 35	24 29
30	18 28	18 53	19 22	22 14	23 00	23 51	24 44
20	18 19	18 42	19 08	22 37	23 26	24 17	00 17
N 10	18 12	18 33	18 58	22 58	23 48	24 39	00 39
0	18 05	18 26	18 50	23 16	24 08	00 08	01 00
S 10	17 58	18 19	18 44	23 35	24 28	00 28	01 20
20	17 51	18 13	18 39	23 55	24 50	00 50	01 43
30	17 42	18 07	18 34	24 19	00 19	01 16	02 08
35	17 38	18 03	18 33	24 33	00 33	01 31	02 24
40	17 32	18 00	18 31	24 48	00 48	01 48	02 41
45	17 26	17 56	18 30	00 00	01 08	02 09	03 02
S 50	17 19	17 52	18 29	00 20	01 31	02 35	03 29
52	17 16	17 50	18 29	00 29	01 43	02 48	03 42
54	17 12	17 48	18 29	00 40	01 56	03 03	03 57
56	17 08	17 46	18 29	00 52	02 11	03 20	04 14
58	17 04	17 44	18 29	01 06	02 29	03 41	04 36
S 60	16 59	17 41	18 30	01 23	02 51	04 07	05 02

Day	SUN Eqn. of Time 00^h	SUN Eqn. of Time 12^h	SUN Mer. Pass.	MOON Mer. Pass. Upper	MOON Mer. Pass. Lower	Age	Phase
d	m s	m s	h m	h m	h m	d	%
26	01 55	01 46	12 02	17 04	04 38	06	40
27	01 38	01 29	12 01	17 56	05 30	07	50
28	01 20	01 12	12 01	18 47	06 21	08	60

	UT	ARIES	VENUS −3.9		MARS +1.0		JUPITER −2.8		SATURN +1.1	
	d h	GHA ° ′	GHA ° ′	Dec ° ′	GHA ° ′	Dec ° ′	GHA ° ′	Dec ° ′	GHA ° ′	Dec ° ′
	29 00	337 20.1	212 18.7	N19 25.2	245 04.0	N23 33.8	14 19.4	S15 46.7	163 23.9	N 4 45.7
	01	352 22.6	227 18.0	24.6	260 04.8	33.9	29 22.1	46.8	178 26.1	45.6
	02	7 25.1	242 17.4	24.1	275 05.5	33.9	44 24.9	46.9	193 28.2	45.5
	03	22 27.5	257 16.7	. . 23.5	290 06.3	. . 33.9	59 27.7	. . 47.0	208 30.4	. . 45.4
	04	37 30.0	272 16.1	23.0	305 07.0	33.9	74 30.4	47.1	223 32.6	45.3
	05	52 32.5	287 15.4	22.4	320 07.8	33.9	89 33.2	47.2	238 34.8	45.1
	06	67 34.9	302 14.8	N19 21.8	335 08.5	N23 33.9	104 35.9	S15 47.2	253 37.0	N 4 45.0
	07	82 37.4	317 14.1	21.3	350 09.2	34.0	119 38.7	47.3	268 39.2	44.9
S	08	97 39.9	332 13.5	20.7	5 10.0	34.0	134 41.5	47.4	283 41.3	44.8
A	09	112 42.3	347 12.9	. . 20.2	20 10.7	. . 34.0	149 44.2	. . 47.5	298 43.5	. . 44.7
T	10	127 44.8	2 12.2	19.6	35 11.5	34.0	164 47.0	47.6	313 45.7	44.5
U	11	142 47.3	17 11.6	19.0	50 12.2	34.0	179 49.7	47.7	328 47.9	44.4
R	12	157 49.7	32 10.9	N19 18.5	65 13.0	N23 34.0	194 52.5	S15 47.8	343 50.1	N 4 44.3
D	13	172 52.2	47 10.3	17.9	80 13.7	34.0	209 55.3	47.9	358 52.3	44.2
A	14	187 54.6	62 09.6	17.4	95 14.5	34.1	224 58.0	48.0	13 54.5	44.1
Y	15	202 57.1	77 09.0	. . 16.8	110 15.2	. . 34.1	240 00.8	. . 48.1	28 56.6	. . 43.9
	16	217 59.6	92 08.4	16.2	125 16.0	34.1	255 03.5	48.2	43 58.8	43.8
	17	233 02.0	107 07.7	15.7	140 16.7	34.1	270 06.3	48.3	59 01.0	43.7
	18	248 04.5	122 07.1	N19 15.1	155 17.4	N23 34.1	285 09.1	S15 48.4	74 03.2	N 4 43.6
	19	263 07.0	137 06.4	14.5	170 18.2	34.1	300 11.8	48.5	89 05.4	43.5
	20	278 09.4	152 05.8	13.9	185 18.9	34.1	315 14.6	48.6	104 07.6	43.3
	21	293 11.9	167 05.1	. . 13.4	200 19.7	. . 34.1	330 17.3	. . 48.7	119 09.7	. . 43.2
	22	308 14.4	182 04.5	12.8	215 20.4	34.2	345 20.1	48.8	134 11.9	43.1
	23	323 16.8	197 03.9	12.2	230 21.2	34.2	0 22.9	48.9	149 14.1	43.0
	30 00	338 19.3	212 03.2	N19 11.7	245 21.9	N23 34.2	15 25.6	S15 49.0	164 16.3	N 4 42.9
	01	353 21.7	227 02.6	11.1	260 22.7	34.2	30 28.4	49.1	179 18.5	42.7
	02	8 24.2	242 01.9	10.5	275 23.4	34.2	45 31.1	49.2	194 20.7	42.6
	03	23 26.7	257 01.3	. . 09.9	290 24.2	. . 34.2	60 33.9	. . 49.3	209 22.9	. . 42.5
	04	38 29.1	272 00.7	09.4	305 24.9	34.2	75 36.7	49.4	224 25.0	42.4
	05	53 31.6	287 00.0	08.8	320 25.7	34.2	90 39.4	49.5	239 27.2	42.3
	06	68 34.1	301 59.4	N19 08.2	335 26.4	N23 34.2	105 42.2	S15 49.6	254 29.4	N 4 42.1
	07	83 36.5	316 58.7	07.6	350 27.2	34.2	120 44.9	49.7	269 31.6	42.0
	08	98 39.0	331 58.1	07.0	5 27.9	34.2	135 47.7	49.8	284 33.8	41.9
S	09	113 41.5	346 57.5	. . 06.4	20 28.7	. . 34.3	150 50.4	. . 49.9	299 36.0	. . 41.8
U	10	128 43.9	1 56.8	05.9	35 29.4	34.3	165 53.2	50.0	314 38.1	41.6
N	11	143 46.4	16 56.2	05.3	50 30.2	34.3	180 56.0	50.1	329 40.3	41.5
D	12	158 48.9	31 55.5	N19 04.7	65 30.9	N23 34.3	195 58.7	S15 50.1	344 42.5	N 4 41.4
A	13	173 51.3	46 54.9	04.1	80 31.7	34.3	211 01.5	50.2	359 44.7	41.3
Y	14	188 53.8	61 54.3	03.5	95 32.4	34.3	226 04.2	50.3	14 46.9	41.2
	15	203 56.2	76 53.6	. . 02.9	110 33.2	. . 34.3	241 07.0	. . 50.4	29 49.1	. . 41.0
	16	218 58.7	91 53.0	02.3	125 33.9	34.3	256 09.7	50.5	44 51.2	40.9
	17	234 01.2	106 52.4	01.7	140 34.7	34.3	271 12.5	50.6	59 53.4	40.8
	18	249 03.6	121 51.7	N19 01.2	155 35.4	N23 34.3	286 15.3	S15 50.7	74 55.6	N 4 40.7
	19	264 06.1	136 51.1	00.6	170 36.2	34.3	301 18.0	50.8	89 57.8	40.6
	20	279 08.6	151 50.4	19 00.0	185 37.0	34.3	316 20.8	50.9	105 00.0	40.4
	21	294 11.0	166 49.8	18 59.4	200 37.7	. . 34.3	331 23.5	. . 51.0	120 02.2	. . 40.3
	22	309 13.5	181 49.2	58.8	215 38.5	34.3	346 26.3	51.1	135 04.3	40.2
	23	324 16.0	196 48.5	58.2	230 39.2	34.3	1 29.0	51.2	150 06.5	40.1
	31 00	339 18.4	211 47.9	N18 57.6	245 40.0	N23 34.3	16 31.8	S15 51.3	165 08.7	N 4 40.0
	01	354 20.9	226 47.3	57.0	260 40.7	34.3	31 34.5	51.4	180 10.9	39.8
	02	9 23.4	241 46.6	56.4	275 41.5	34.3	46 37.3	51.5	195 13.1	39.7
	03	24 25.8	256 46.0	. . 55.8	290 42.2	. . 34.3	61 40.1	. . 51.6	210 15.3	. . 39.6
	04	39 28.3	271 45.4	55.2	305 43.0	34.3	76 42.8	51.7	225 17.5	39.5
	05	54 30.7	286 44.7	54.6	320 43.7	34.3	91 45.6	51.8	240 19.6	39.4
	06	69 33.2	301 44.1	N18 54.0	335 44.5	N23 34.3	106 48.3	S15 51.9	255 21.8	N 4 39.2
	07	84 35.7	316 43.5	53.4	350 45.2	34.3	121 51.1	52.0	270 24.0	39.1
	08	99 38.1	331 42.8	52.8	5 46.0	34.3	136 53.8	52.0	285 26.2	39.0
M	09	114 40.6	346 42.2	. . 52.2	20 46.8	. . 34.3	151 56.6	. . 52.1	300 28.4	. . 38.9
O	10	129 43.1	1 41.6	51.6	35 47.5	34.3	166 59.3	52.2	315 30.6	38.7
N	11	144 45.5	16 40.9	51.0	50 48.3	34.3	182 02.1	52.3	330 32.7	38.6
D	12	159 48.0	31 40.3	N18 50.4	65 49.0	N23 34.3	197 04.8	S15 52.4	345 34.9	N 4 38.5
A	13	174 50.5	46 39.7	49.7	80 49.8	34.3	212 07.6	52.5	0 37.1	38.4
Y	14	189 52.9	61 39.0	49.1	95 50.5	34.3	227 10.4	52.6	15 39.3	38.3
	15	204 55.4	76 38.4	. . 48.5	110 51.3	. . 34.3	242 13.1	. . 52.7	30 41.5	. . 38.1
	16	219 57.9	91 37.8	47.9	125 52.1	34.3	257 15.9	52.8	45 43.6	38.0
	17	235 00.3	106 37.1	47.3	140 52.8	34.3	272 18.6	52.9	60 45.8	37.9
	18	250 02.8	121 36.5	N18 46.7	155 53.6	N23 34.3	287 21.4	S15 53.0	75 48.0	N 4 37.8
	19	265 05.2	136 35.9	46.1	170 54.3	34.3	302 24.1	53.1	90 50.2	37.7
	20	280 07.7	151 35.2	45.5	185 55.1	34.3	317 26.9	53.2	105 52.4	37.5
	21	295 10.2	166 34.6	. . 44.8	200 55.8	. . 34.3	332 29.6	. . 53.3	120 54.6	. . 37.4
	22	310 12.6	181 34.0	44.2	215 56.6	34.3	347 32.4	53.4	135 56.7	37.3
	23	325 15.1	196 33.4	43.6	230 57.4	34.3	2 35.1	53.5	150 58.9	37.2
	Mer. Pass.	h m 1 26.5	v −0.6	d 0.6	v 0.8	d 0.0	v 2.8	d 0.1	v 2.2	d 0.1

STARS		
Name	SHA ° ′	Dec ° ′
Acamar	315 20.2	S40 15.
Achernar	335 28.2	S57 10.
Acrux	173 13.3	S63 09.
Adhara	255 15.0	S28 58.
Aldebaran	290 52.6	N16 31.
Alioth	166 23.3	N55 54.
Alkaid	153 01.2	N49 16.
Al Na'ir	27 46.6	S46 54.
Alnilam	275 49.3	S 1 11.
Alphard	217 59.2	S 8 42.
Alphecca	126 13.4	N26 41.
Alpheratz	357 46.1	N29 08.
Altair	62 10.7	N 8 53.
Ankaa	353 17.9	S42 14.
Antares	112 29.7	S26 27.
Arcturus	145 58.4	N19 08.
Atria	107 34.0	S69 03.
Avior	234 19.8	S59 32.
Bellatrix	278 35.1	N 6 21.
Betelgeuse	271 04.4	N 7 24.
Canopus	263 57.7	S52 41.
Capella	280 38.7	N46 00.
Deneb	49 33.1	N45 19.
Denebola	182 36.8	N14 31.
Diphda	348 58.4	S17 55.
Dubhe	193 55.5	N61 41.
Elnath	278 16.2	N28 36.
Eltanin	90 47.2	N51 29.
Enif	33 49.6	N 9 55.
Fomalhaut	15 26.6	S29 34.
Gacrux	172 04.7	S57 10.
Gienah	175 55.5	S17 35.
Hadar	148 52.4	S60 25.
Hamal	328 03.8	N23 30.
Kaus Aust.	83 47.3	S34 22.
Kochab	137 20.0	N74 07.
Markab	13 40.8	N15 15.
Menkar	314 17.9	N 4 07.
Menkent	148 11.2	S36 25.
Miaplacidus	221 41.4	S69 45.
Mirfak	308 44.3	N49 53.
Nunki	76 01.5	S26 17.
Peacock	53 23.0	S56 42.
Pollux	243 31.4	N28 00.
Procyon	245 02.9	N 5 12.
Rasalhague	96 08.9	N12 33.
Regulus	207 46.8	N11 55.
Rigel	281 14.8	S 8 11.
Rigil Kent.	139 56.0	S60 52.
Sabik	102 15.7	S15 44.
Schedar	349 43.5	N56 35.
Shaula	96 25.6	S37 06.
Sirius	258 36.4	S16 43.
Spica	158 34.4	S11 12.
Suhail	222 55.0	S43 28.
Vega	80 40.6	N38 47.
Zuben'ubi	137 08.7	S16 05.

	SHA ° ′	Mer. Pass h m
Venus	233 43.9	9 52
Mars	267 02.7	7 38
Jupiter	37 06.3	22 54
Saturn	185 57.0	13 01

UT		SUN GHA	SUN Dec	MOON GHA	v	MOON Dec	d	HP
d	h	° ′	° ′	° ′	′	° ′	′	′
29	00	179 44.4	N 9 23.0	75 29.9	10.0	S26 22.7	0.5	54.5
	01	194 44.6	22.1	89 58.9	10.1	26 22.2	0.7	54.5
	02	209 44.8	21.2	104 28.0	10.1	26 21.5	0.8	54.5
	03	224 45.0	. . 20.3	118 57.1	10.1	26 20.7	0.9	54.5
	04	239 45.2	19.4	133 26.2	10.2	26 19.8	1.1	54.5
	05	254 45.4	18.5	147 55.4	10.2	26 18.7	1.2	54.5
	06	269 45.6	N 9 17.6	162 24.6	10.2	S26 17.5	1.3	54.5
	07	284 45.8	16.7	176 53.8	10.2	26 16.2	1.4	54.4
S	08	299 45.9	15.8	191 23.0	10.3	26 14.8	1.6	54.4
A	09	314 46.1	. . 15.0	205 52.3	10.3	26 13.2	1.7	54.4
T	10	329 46.3	14.1	220 21.6	10.3	26 11.5	1.8	54.4
U	11	344 46.5	13.2	234 50.9	10.3	26 09.7	1.9	54.4
R	12	359 46.7	N 9 12.3	249 20.2	10.4	S26 07.8	2.1	54.4
D	13	14 46.9	11.4	263 49.6	10.4	26 05.7	2.2	54.4
A	14	29 47.1	10.5	278 19.0	10.5	26 03.5	2.3	54.4
Y	15	44 47.3	. . 09.6	292 48.5	10.5	26 01.2	2.4	54.3
	16	59 47.5	08.7	307 18.0	10.5	25 58.8	2.6	54.3
	17	74 47.7	07.8	321 47.5	10.5	25 56.2	2.7	54.3
	18	89 47.8	N 9 06.9	336 17.0	10.6	S25 53.5	2.8	54.3
	19	104 48.0	06.0	350 46.6	10.6	25 50.7	2.9	54.3
	20	119 48.2	05.1	5 16.2	10.7	25 47.8	3.1	54.3
	21	134 48.4	. . 04.2	19 45.9	10.7	25 44.7	3.2	54.3
	22	149 48.6	03.4	34 15.6	10.7	25 41.5	3.3	54.3
	23	164 48.8	02.5	48 45.3	10.8	25 38.2	3.4	54.3
30	00	179 49.0	N 9 01.6	63 15.1	10.8	S25 34.8	3.5	54.3
	01	194 49.2	9 00.7	77 44.9	10.9	25 31.3	3.7	54.2
	02	209 49.4	8 59.8	92 14.8	10.9	25 27.6	3.7	54.2
	03	224 49.6	. . 58.9	106 44.7	10.9	25 23.9	3.9	54.2
	04	239 49.8	58.0	121 14.6	11.0	25 20.0	4.1	54.2
	05	254 49.9	57.1	135 44.6	11.0	25 15.9	4.1	54.2
	06	269 50.1	N 8 56.2	150 14.6	11.1	S25 11.8	4.2	54.2
	07	284 50.3	55.3	164 44.7	11.1	25 07.6	4.4	54.2
	08	299 50.5	54.4	179 14.8	11.2	25 03.2	4.5	54.2
S	09	314 50.7	. . 53.5	193 45.0	11.2	24 58.7	4.6	54.2
U	10	329 50.9	52.6	208 15.2	11.2	24 54.1	4.7	54.2
N	11	344 51.1	51.7	222 45.4	11.3	24 49.4	4.8	54.2
D	12	359 51.3	N 8 50.8	237 15.7	11.4	S24 44.6	4.9	54.2
A	13	14 51.5	49.9	251 46.1	11.4	24 39.7	5.1	54.2
Y	14	29 51.7	49.0	266 16.5	11.4	24 34.6	5.1	54.2
	15	44 51.9	. . 48.1	280 46.9	11.5	24 29.5	5.3	54.2
	16	59 52.1	47.2	295 17.4	11.5	24 24.2	5.4	54.1
	17	74 52.3	46.3	309 47.9	11.6	24 18.8	5.5	54.1
	18	89 52.5	N 8 45.4	324 18.5	11.6	S24 13.3	5.6	54.1
	19	104 52.7	44.5	338 49.1	11.7	24 07.7	5.7	54.1
	20	119 52.8	43.6	353 19.8	11.7	24 02.0	5.8	54.1
	21	134 53.0	. . 42.7	7 50.5	11.8	23 56.2	5.9	54.1
	22	149 53.2	41.8	22 21.3	11.8	23 50.3	6.1	54.1
	23	164 53.4	40.9	36 52.1	11.9	23 44.2	6.1	54.1
31	00	179 53.6	N 8 40.0	51 23.0	11.9	S23 38.1	6.3	54.1
	01	194 53.8	39.1	65 53.9	11.9	23 31.8	6.3	54.1
	02	209 54.0	38.2	80 24.8	12.1	23 25.5	6.5	54.1
	03	224 54.2	. . 37.3	94 55.9	12.0	23 19.0	6.6	54.1
	04	239 54.4	36.4	109 26.9	12.2	23 12.4	6.6	54.1
	05	254 54.6	35.5	123 58.1	12.1	23 05.8	6.8	54.1
	06	269 54.8	N 8 34.6	138 29.2	12.3	S22 59.0	6.9	54.1
	07	284 55.0	33.7	153 00.5	12.2	22 52.1	6.9	54.1
	08	299 55.2	32.8	167 31.7	12.4	22 45.2	7.1	54.1
M	09	314 55.4	. . 31.9	182 03.1	12.4	22 38.1	7.2	54.1
O	10	329 55.6	31.0	196 34.5	12.4	22 30.9	7.3	54.1
N	11	344 55.8	30.1	211 05.9	12.5	22 23.6	7.3	54.1
D	12	359 56.0	N 8 29.2	225 37.4	12.5	S22 16.3	7.5	54.1
A	13	14 56.2	28.3	240 08.9	12.6	22 08.8	7.6	54.1
Y	14	29 56.4	27.4	254 40.5	12.6	22 01.2	7.7	54.1
	15	44 56.6	. . 26.5	269 12.1	12.7	21 53.5	7.7	54.1
	16	59 56.8	25.6	283 43.8	12.8	21 45.8	7.9	54.1
	17	74 57.0	24.7	298 15.6	12.8	21 37.9	7.9	54.1
	18	89 57.2	N 8 23.8	312 47.4	12.8	S21 30.0	8.1	54.1
	19	104 57.4	22.9	327 19.2	12.9	21 21.9	8.1	54.1
	20	119 57.6	22.0	341 51.1	13.0	21 13.8	8.3	54.1
	21	134 57.7	. . 21.1	356 23.1	13.0	21 05.5	8.3	54.1
	22	149 57.9	20.2	10 55.1	13.0	20 57.2	8.4	54.1
	23	164 58.1	19.3	25 27.1	13.1	S20 48.8	8.5	54.1
		SD 15.9	*d* 0.9	SD 14.8		14.8		14.7

Lat.	Twilight Naut.	Twilight Civil	Sunrise	Moonrise 29	Moonrise 30	Moonrise 31	Moonrise 1
°	h m	h m	h m	h m	h m	h m	h m
N 72	////	02 17	03 52	▬	▬	▬	21 00
N 70	////	02 48	04 07	▬	▬	▬	20 00
68	01 10	03 11	04 19	▬	▬	20 12	19 25
66	01 57	03 29	04 29	▬	▬	19 18	18 59
64	02 26	03 43	04 37	19 19	18 54	18 46	18 40
62	02 48	03 55	04 44	18 01	18 16	18 21	18 23
60	03 05	04 05	04 50	17 24	17 49	18 02	18 10
N 58	03 19	04 13	04 56	16 58	17 28	17 46	17 58
56	03 30	04 21	05 00	16 38	17 10	17 32	17 48
54	03 40	04 27	05 05	16 21	16 55	17 20	17 39
52	03 49	04 33	05 09	16 06	16 43	17 10	17 31
50	03 57	04 38	05 12	15 53	16 31	17 00	17 24
45	04 12	04 49	05 20	15 27	16 08	16 41	17 08
N 40	04 25	04 58	05 26	15 06	15 49	16 24	16 55
35	04 34	05 05	05 31	14 49	15 33	16 11	16 44
30	04 43	05 11	05 36	14 34	15 19	15 59	16 34
20	04 55	05 22	05 44	14 08	14 55	15 38	16 18
N 10	05 05	05 30	05 51	13 46	14 35	15 20	16 03
0	05 12	05 36	05 57	13 26	14 16	15 04	15 49
S 10	05 18	05 42	06 04	13 05	13 57	14 47	15 36
20	05 22	05 48	06 10	12 43	13 36	14 29	15 21
30	05 26	05 53	06 18	12 18	13 12	14 08	15 04
35	05 27	05 56	06 22	12 03	12 58	13 56	14 54
40	05 28	05 59	06 27	11 45	12 42	13 42	14 43
45	05 28	06 02	06 32	11 25	12 23	13 25	14 29
S 50	05 29	06 06	06 38	10 58	11 58	13 04	14 13
52	05 28	06 07	06 41	10 45	11 47	12 54	14 05
54	05 28	06 09	06 45	10 30	11 33	12 43	13 56
56	05 28	06 11	06 48	10 13	11 18	12 30	13 47
58	05 27	06 13	06 52	09 52	10 59	12 15	13 35
S 60	05 27	06 15	06 57	09 25	10 36	11 58	13 23

Lat.	Sunset	Twilight Civil	Twilight Naut.	Moonset 29	Moonset 30	Moonset 31	Moonset 1
°	h m	h m	h m	h m	h m	h m	h m
N 72	20 06	21 38	////	▬	▬	▬	23 19
N 70	19 51	21 08	////	▬	▬	▬	▬
68	19 40	20 46	22 40	▬	▬	22 31	24 52
66	19 30	20 29	21 58	▬	▬	23 24	25 16
64	19 22	20 16	21 31	20 02	22 09	23 56	25 35
62	19 15	20 04	21 10	21 20	22 47	24 19	00 19
60	19 09	19 55	20 54	21 56	23 14	24 38	00 38
N 58	19 04	19 46	20 40	22 21	23 34	24 53	00 53
56	18 59	19 39	20 29	22 42	23 51	25 07	01 07
54	18 55	19 33	20 19	22 59	24 06	00 06	01 18
52	18 52	19 27	20 11	23 13	24 18	00 18	01 28
50	18 48	19 22	20 03	23 26	24 29	00 29	01 37
45	18 41	19 11	19 48	23 52	24 52	00 52	01 56
N 40	18 35	19 02	19 36	24 12	00 12	01 11	02 11
35	18 29	18 55	19 26	24 29	00 29	01 26	02 24
30	18 25	18 49	19 18	24 44	00 44	01 39	02 35
20	18 17	18 39	19 05	00 17	01 09	02 02	02 54
N 10	18 10	18 31	18 56	00 39	01 30	02 21	03 11
0	18 04	18 25	18 49	01 00	01 50	02 39	03 26
S 10	17 58	18 19	18 43	01 20	02 10	02 57	03 41
20	17 51	18 13	18 39	01 43	02 31	03 16	03 58
30	17 44	18 08	18 36	02 08	02 56	03 38	04 16
35	17 40	18 05	18 35	02 24	03 10	03 51	04 27
40	17 35	18 03	18 34	02 41	03 27	04 06	04 39
45	17 30	17 59	18 34	03 02	03 47	04 24	04 54
S 50	17 23	17 56	18 34	03 29	04 12	04 45	05 11
52	17 21	17 55	18 34	03 42	04 24	04 56	05 20
54	17 17	17 53	18 34	03 57	04 38	05 07	05 29
56	17 14	17 52	18 35	04 14	04 53	05 20	05 39
58	17 10	17 50	18 35	04 36	05 12	05 36	05 51
S 60	17 05	17 48	18 36	05 02	05 36	05 54	06 04

Day	SUN Eqn. of Time 00^h	SUN Eqn. of Time 12^h	SUN Mer. Pass.	MOON Mer. Pass. Upper	MOON Mer. Pass. Lower	MOON Age	MOON Phase
d	m s	m s	h m	h m	h m	d %	
29	01 03	00 54	12 01	19 38	07 13	09 69	
30	00 44	00 35	12 01	20 28	08 03	10 77	
31	00 26	00 16	12 00	21 15	08 52	11 85	

	UT	ARIES	VENUS −3.9		MARS +1.0		JUPITER −2.8		SATURN +1.1	
		GHA	GHA	Dec	GHA	Dec	GHA	Dec	GHA	Dec
	d h	° ′	° ′	° ′	° ′	° ′	° ′	° ′	° ′	° ′
	1 00	340 17.6	211 32.7	N18 43.0	245 58.1	N23 34.3	17 37.9	S15 53.6	166 01.1	N 4 37.1
	01	355 20.0	226 32.1	42.4	260 58.9	34.3	32 40.6	53.6	181 03.3	36.9
	02	10 22.5	241 31.5	41.7	275 59.6	34.3	47 43.4	53.7	196 05.5	36.8
	03	25 25.0	256 30.8	. . 41.1	291 00.4	. . 34.3	62 46.1	. . 53.8	211 07.7	. . 36.7
	04	40 27.4	271 30.2	40.5	306 01.2	34.3	77 48.9	53.9	226 09.8	36.6
	05	55 29.9	286 29.6	39.9	321 01.9	34.3	92 51.6	54.0	241 12.0	36.4
	06	70 32.3	301 29.0	N18 39.3	336 02.7	N23 34.3	107 54.4	S15 54.1	256 14.2	N 4 36.3
	07	85 34.8	316 28.3	38.6	351 03.4	34.3	122 57.1	54.2	271 16.4	36.2
T	08	100 37.3	331 27.7	38.0	6 04.2	34.3	137 59.9	54.3	286 18.6	36.1
U	09	115 39.7	346 27.1	. . 37.4	21 05.0	. . 34.3	153 02.6	. . 54.4	301 20.8	. . 36.0
E	10	130 42.2	1 26.4	36.7	36 05.7	34.3	168 05.4	54.5	316 22.9	35.8
S	11	145 44.7	16 25.8	36.1	51 06.5	34.3	183 08.1	54.6	331 25.1	35.7
D	12	160 47.1	31 25.2	N18 35.5	66 07.2	N23 34.3	198 10.9	S15 54.7	346 27.3	N 4 35.6
A	13	175 49.6	46 24.6	34.9	81 08.0	34.2	213 13.6	54.8	1 29.5	35.5
Y	14	190 52.1	61 23.9	34.2	96 08.8	34.2	228 16.4	54.9	16 31.7	35.4
	15	205 54.5	76 23.3	. . 33.6	111 09.5	. . 34.2	243 19.1	. . 54.9	31 33.9	. . 35.2
	16	220 57.0	91 22.7	33.0	126 10.3	34.2	258 21.9	55.0	46 36.0	35.1
	17	235 59.5	106 22.1	32.3	141 11.1	34.2	273 24.6	55.1	61 38.2	35.0
	18	251 01.9	121 21.4	N18 31.7	156 11.8	N23 34.2	288 27.4	S15 55.2	76 40.4	N 4 34.9
	19	266 04.4	136 20.8	31.0	171 12.6	34.2	303 30.1	55.3	91 42.6	34.8
	20	281 06.8	151 20.2	30.4	186 13.3	34.2	318 32.9	55.4	106 44.8	34.6
	21	296 09.3	166 19.6	. . 29.8	201 14.1	. . 34.2	333 35.6	. . 55.5	121 46.9	. . 34.5
	22	311 11.8	181 18.9	29.1	216 14.9	34.2	348 38.4	55.6	136 49.1	34.4
	23	326 14.2	196 18.3	28.5	231 15.6	34.2	3 41.1	55.7	151 51.3	34.3
	2 00	341 16.7	211 17.7	N18 27.9	246 16.4	N23 34.1	18 43.9	S15 55.8	166 53.5	N 4 34.1
	01	356 19.2	226 17.1	27.2	261 17.2	34.1	33 46.6	55.9	181 55.7	34.0
	02	11 21.6	241 16.5	26.6	276 17.9	34.1	48 49.4	56.0	196 57.9	33.9
	03	26 24.1	256 15.8	. . 25.9	291 18.7	. . 34.1	63 52.1	. . 56.0	212 00.0	. . 33.8
	04	41 26.6	271 15.2	25.3	306 19.5	34.1	78 54.9	56.1	227 02.2	33.7
	05	56 29.0	286 14.6	24.6	321 20.2	34.1	93 57.6	56.2	242 04.4	33.5
	06	71 31.5	301 14.0	N18 24.0	336 21.0	N23 34.1	109 00.4	S15 56.3	257 06.6	N 4 33.4
W	07	86 34.0	316 13.3	23.3	351 21.8	34.1	124 03.1	56.4	272 08.8	33.3
E	08	101 36.4	331 12.7	22.7	6 22.5	34.1	139 05.9	56.5	287 10.9	33.2
D	09	116 38.9	346 12.1	. . 22.0	21 23.3	. . 34.0	154 08.6	. . 56.6	302 13.1	. . 33.1
N	10	131 41.3	1 11.5	21.4	36 24.1	34.0	169 11.3	56.7	317 15.3	32.9
E	11	146 43.8	16 10.9	20.7	51 24.8	34.0	184 14.1	56.8	332 17.5	32.8
S	12	161 46.3	31 10.2	N18 20.1	66 25.6	N23 34.0	199 16.8	S15 56.9	347 19.7	N 4 32.7
D	13	176 48.7	46 09.6	19.4	81 26.4	34.0	214 19.6	57.0	2 21.9	32.6
A	14	191 51.2	61 09.0	18.8	96 27.1	34.0	229 22.3	57.1	17 24.0	32.4
Y	15	206 53.7	76 08.4	. . 18.1	111 27.9	. . 34.0	244 25.1	. . 57.1	32 26.2	. . 32.3
	16	221 56.1	91 07.8	17.5	126 28.7	33.9	259 27.8	57.2	47 28.4	32.2
	17	236 58.6	106 07.1	16.8	141 29.4	33.9	274 30.6	57.3	62 30.6	32.1
	18	252 01.1	121 06.5	N18 16.2	156 30.2	N23 33.9	289 33.3	S15 57.4	77 32.8	N 4 32.0
	19	267 03.5	136 05.9	15.5	171 31.0	33.9	304 36.1	57.5	92 34.9	31.8
	20	282 06.0	151 05.3	14.8	186 31.7	33.9	319 38.8	57.6	107 37.1	31.7
	21	297 08.4	166 04.7	. . 14.2	201 32.5	. . 33.9	334 41.5	. . 57.7	122 39.3	. . 31.6
	22	312 10.9	181 04.1	13.5	216 33.3	33.8	349 44.3	57.8	137 41.5	31.5
	23	327 13.4	196 03.4	12.9	231 34.0	33.8	4 47.0	57.9	152 43.7	31.4
	3 00	342 15.8	211 02.8	N18 12.2	246 34.8	N23 33.8	19 49.8	S15 58.0	167 45.9	N 4 31.2
	01	357 18.3	226 02.2	11.5	261 35.6	33.8	34 52.5	58.0	182 48.0	31.1
	02	12 20.8	241 01.6	10.9	276 36.3	33.8	49 55.3	58.1	197 50.2	31.0
	03	27 23.2	256 01.0	. . 10.2	291 37.1	. . 33.8	64 58.0	. . 58.2	212 52.4	. . 30.9
	04	42 25.7	271 00.4	09.5	306 37.9	33.7	80 00.8	58.3	227 54.6	30.7
	05	57 28.2	285 59.7	08.9	321 38.7	33.7	95 03.5	58.4	242 56.8	30.6
	06	72 30.6	300 59.1	N18 08.2	336 39.4	N23 33.7	110 06.2	S15 58.5	257 58.9	N 4 30.5
	07	87 33.1	315 58.5	07.5	351 40.2	33.7	125 09.0	58.6	273 01.1	30.4
T	08	102 35.6	330 57.9	06.9	6 41.0	33.7	140 11.7	58.7	288 03.3	30.3
H	09	117 38.0	345 57.3	. . 06.2	21 41.7	. . 33.6	155 14.5	. . 58.8	303 05.5	. . 30.1
U	10	132 40.5	0 56.7	05.5	36 42.5	33.6	170 17.2	58.9	318 07.7	30.0
R	11	147 42.9	15 56.1	04.8	51 43.3	33.6	185 20.0	58.9	333 09.9	29.9
S	12	162 45.4	30 55.4	N18 04.2	66 44.1	N23 33.6	200 22.7	S15 59.0	348 12.0	N 4 29.8
D	13	177 47.9	45 54.8	03.5	81 44.8	33.6	215 25.4	59.1	3 14.2	29.7
A	14	192 50.3	60 54.2	02.8	96 45.6	33.5	230 28.2	59.2	18 16.4	29.5
Y	15	207 52.8	75 53.6	. . 02.1	111 46.4	. . 33.5	245 30.9	. . 59.3	33 18.6	. . 29.4
	16	222 55.3	90 53.0	01.5	126 47.2	33.5	260 33.7	59.4	48 20.8	29.3
	17	237 57.7	105 52.4	00.8	141 47.9	33.5	275 36.4	59.5	63 22.9	29.2
	18	253 00.2	120 51.8	N18 00.1	156 48.7	N23 33.4	290 39.1	S15 59.6	78 25.1	N 4 29.0
	19	268 02.7	135 51.2	17 59.4	171 49.5	33.4	305 41.9	59.7	93 27.3	28.9
	20	283 05.1	150 50.6	58.7	186 50.3	33.4	320 44.6	59.7	108 29.5	28.8
	21	298 07.6	165 49.9	. . 58.1	201 51.0	. . 33.4	335 47.4	. . 59.8	123 31.7	. . 28.7
	22	313 10.0	180 49.3	57.4	216 51.8	33.3	350 50.1	15 59.9	138 33.8	28.6
	23	328 12.5	195 48.7	56.7	231 52.6	33.3	5 52.8	S16 00.0	153 36.0	28.4
	Mer. Pass.	h m 1 14.7	*v* −0.6	*d* 0.7	*v* 0.8	*d* 0.0	*v* 2.7	*d* 0.1	*v* 2.2	*d* 0.1

STARS		
Name	SHA	Dec
	° ′	° ′
Acamar	315 20.2	S40 15.6
Achernar	335 28.2	S57 10.9
Acrux	173 13.3	S63 09.3
Adhara	255 14.9	S28 58.9
Aldebaran	290 52.6	N16 31.8
Alioth	166 23.4	N55 54.5
Alkaid	153 01.2	N49 16.0
Al Na'ir	27 46.6	S46 54.7
Alnilam	275 49.3	S 1 11.6
Alphard	217 59.1	S 8 41.9
Alphecca	126 13.4	N26 41.1
Alpheratz	357 46.1	N29 08.8
Altair	62 10.7	N 8 53.8
Ankaa	353 17.9	S42 14.9
Antares	112 29.7	S26 27.3
Arcturus	145 58.4	N19 08.0
Atria	107 34.1	S69 03.0
Avior	234 19.8	S59 32.2
Bellatrix	278 35.1	N 6 21.7
Betelgeuse	271 04.4	N 7 24.7
Canopus	263 57.6	S52 41.7
Capella	280 38.7	N46 00.4
Deneb	49 33.1	N45 19.1
Denebola	182 36.8	N14 31.1
Diphda	348 58.3	S17 55.7
Dubhe	193 55.5	N61 41.9
Elnath	278 16.2	N28 36.9
Eltanin	90 47.3	N51 29.5
Enif	33 49.6	N 9 55.4
Fomalhaut	15 26.6	S29 34.0
Gacrux	172 04.7	S57 10.2
Gienah	175 55.5	S17 35.8
Hadar	148 52.5	S60 25.4
Hamal	328 03.7	N23 30.6
Kaus Aust.	83 47.3	S34 22.9
Kochab	137 20.0	N74 07.1
Markab	13 40.8	N15 15.6
Menkar	314 17.8	N 4 07.9
Menkent	148 11.2	S36 25.2
Miaplacidus	221 41.4	S69 45.3
Mirfak	308 44.3	N49 53.7
Nunki	76 01.5	S26 17.1
Peacock	53 23.0	S56 42.3
Pollux	243 31.3	N28 00.2
Procyon	245 02.9	N 5 12.1
Rasalhague	96 08.9	N12 33.3
Regulus	207 46.8	N11 55.2
Rigel	281 14.8	S 8 11.2
Rigil Kent.	139 56.1	S60 52.8
Sabik	102 15.7	S15 44.2
Schedar	349 43.4	N56 35.5
Shaula	96 25.6	S37 06.8
Sirius	258 36.3	S16 43.5
Spica	158 34.4	S11 12.7
Suhail	222 55.0	S43 28.2
Vega	80 40.7	N38 47.8
Zuben'ubi	137 08.7	S16 05.0

	SHA	Mer. Pass
	° ′	h m
Venus	230 01.0	9 55
Mars	264 59.7	7 35
Jupiter	37 27.2	22 41
Saturn	185 36.8	12 51

UT d h	SUN GHA ° ′	SUN Dec ° ′	MOON GHA ° ′	v ′	MOON Dec ° ′	d ′	HP ′
1 00	179 58.3	N 8 18.4	39 59.2	13.2	S20 40.3	8.6	54.1
01	194 58.5	17.5	54 31.4	13.2	20 31.7	8.7	54.1
02	209 58.7	16.6	69 03.6	13.2	20 23.0	8.8	54.1
03	224 58.9	. . 15.6	83 35.8	13.4	20 14.2	8.8	54.1
04	239 59.1	14.7	98 08.2	13.3	20 05.4	9.0	54.1
05	254 59.3	13.8	112 40.5	13.4	19 56.4	9.0	54.1
06	269 59.5	N 8 12.9	127 12.9	13.5	S19 47.4	9.1	54.1
T 07	284 59.7	12.0	141 45.4	13.5	19 38.3	9.3	54.1
U 08	299 59.9	11.1	156 17.9	13.5	19 29.0	9.2	54.2
E 09	315 00.1	. . 10.2	170 50.4	13.6	19 19.8	9.4	54.2
S 10	330 00.3	09.3	185 23.0	13.7	19 10.4	9.5	54.2
D 11	345 00.5	08.4	199 55.7	13.7	19 00.9	9.5	54.2
A 12	0 00.7	N 8 07.5	214 28.4	13.8	S18 51.4	9.6	54.2
Y 13	15 00.9	06.6	229 01.2	13.8	18 41.8	9.7	54.2
14	30 01.1	05.7	243 34.0	13.8	18 32.1	9.8	54.2
15	45 01.3	. . 04.8	258 06.8	13.9	18 22.3	9.9	54.2
16	60 01.5	03.8	272 39.7	13.9	18 12.4	9.9	54.2
17	75 01.7	02.9	287 12.6	14.0	18 02.5	10.0	54.2
18	90 01.9	N 8 02.0	301 45.6	14.1	S17 52.5	10.1	54.2
19	105 02.1	01.1	316 18.7	14.0	17 42.4	10.2	54.2
20	120 02.3	8 00.2	330 51.7	14.2	17 32.2	10.2	54.2
21	135 02.5	7 59.3	345 24.9	14.1	17 22.0	10.3	54.2
22	150 02.7	58.4	359 58.0	14.3	17 11.7	10.4	54.2
23	165 02.9	57.5	14 31.3	14.2	17 01.3	10.5	54.2
2 00	180 03.1	N 7 56.6	29 04.5	14.3	S16 50.8	10.5	54.3
01	195 03.3	55.7	43 37.8	14.4	16 40.3	10.6	54.3
02	210 03.5	54.7	58 11.2	14.3	16 29.7	10.7	54.3
03	225 03.7	. . 53.8	72 44.5	14.5	16 19.0	10.8	54.3
04	240 03.9	52.9	87 18.0	14.4	16 08.2	10.8	54.3
05	255 04.1	52.0	101 51.4	14.6	15 57.4	10.9	54.3
06	270 04.3	N 7 51.1	116 25.0	14.5	S15 46.5	10.9	54.3
W 07	285 04.5	50.2	130 58.5	14.6	15 35.6	11.0	54.3
E 08	300 04.8	49.3	145 32.1	14.6	15 24.6	11.1	54.3
D 09	315 05.0	. . 48.4	160 05.7	14.7	15 13.5	11.2	54.3
N 10	330 05.2	47.4	174 39.4	14.7	15 02.3	11.2	54.3
E 11	345 05.4	46.5	189 13.1	14.8	14 51.1	11.3	54.3
S 12	0 05.6	N 7 45.6	203 46.9	14.8	S14 39.8	11.3	54.4
D 13	15 05.8	44.7	218 20.7	14.8	14 28.5	11.4	54.4
A 14	30 06.0	43.8	232 54.5	14.8	14 17.1	11.5	54.4
Y 15	45 06.2	. . 42.9	247 28.3	14.9	14 05.6	11.5	54.4
16	60 06.4	42.0	262 02.2	15.0	13 54.1	11.6	54.4
17	75 06.6	41.0	276 36.2	14.9	13 42.5	11.6	54.4
18	90 06.8	N 7 40.1	291 10.1	15.0	S13 30.9	11.7	54.4
19	105 07.0	39.2	305 44.1	15.1	13 19.2	11.8	54.4
20	120 07.2	38.3	320 18.2	15.0	13 07.4	11.8	54.4
21	135 07.4	. . 37.4	334 52.2	15.1	12 55.6	11.9	54.4
22	150 07.6	36.5	349 26.3	15.1	12 43.7	11.9	54.5
23	165 07.8	35.6	4 00.4	15.2	12 31.8	12.0	54.5
3 00	180 08.0	N 7 34.6	18 34.6	15.2	S12 19.8	12.1	54.5
01	195 08.2	33.7	33 08.8	15.2	12 07.7	12.0	54.5
02	210 08.4	32.8	47 43.0	15.2	11 55.7	12.2	54.5
03	225 08.6	. . 31.9	62 17.2	15.3	11 43.5	12.2	54.5
04	240 08.8	31.0	76 51.5	15.3	11 31.3	12.2	54.5
05	255 09.0	30.1	91 25.8	15.3	11 19.1	12.3	54.5
06	270 09.2	N 7 29.1	106 00.1	15.4	S11 06.8	12.4	54.6
T 07	285 09.4	28.2	120 34.5	15.4	10 54.4	12.4	54.6
H 08	300 09.6	27.3	135 08.9	15.4	10 42.0	12.4	54.6
U 09	315 09.8	. . 26.4	149 43.3	15.4	10 29.6	12.5	54.6
R 10	330 10.1	25.5	164 17.7	15.4	10 17.1	12.6	54.6
S 11	345 10.3	24.5	178 52.1	15.5	10 04.5	12.5	54.6
D 12	0 10.5	N 7 23.6	193 26.6	15.5	S 9 52.0	12.7	54.6
A 13	15 10.7	22.7	208 01.1	15.5	9 39.3	12.6	54.6
Y 14	30 10.9	21.8	222 35.6	15.5	9 26.7	12.8	54.7
15	45 11.1	. . 20.9	237 10.1	15.6	9 13.9	12.7	54.7
16	60 11.3	20.0	251 44.7	15.6	9 01.2	12.8	54.7
17	75 11.5	19.0	266 19.3	15.5	8 48.4	12.8	54.7
18	90 11.7	N 7 18.1	280 53.8	15.6	S 8 35.6	12.9	54.7
19	105 11.9	17.2	295 28.4	15.7	8 22.7	12.9	54.7
20	120 12.1	16.3	310 03.1	15.6	8 09.8	13.0	54.7
21	135 12.3	. . 15.4	324 37.7	15.6	7 56.8	13.0	54.7
22	150 12.5	14.4	339 12.3	15.7	7 43.8	13.0	54.8
23	165 12.7	13.5	353 47.0	15.7	S 7 30.8	13.1	54.8
	SD 15.9	*d* 0.9	SD 14.8		14.8		14.9

Lat. °	Twilight Naut. h m	Twilight Civil h m	Sunrise h m	Moonrise 1 h m	Moonrise 2 h m	Moonrise 3 h m	Moonrise 4 h m
N 72	////	02 39	04 07	21 00	19 45	19 07	18 38
N 70	00 18	03 05	04 20	20 00	19 20	18 55	18 34
68	01 41	03 25	04 30	19 25	19 02	18 45	18 30
66	02 16	03 41	04 39	18 59	18 47	18 37	18 27
64	02 41	03 53	04 46	18 40	18 34	18 29	18 25
62	02 59	04 04	04 52	18 23	18 24	18 23	18 23
60	03 15	04 13	04 57	18 10	18 15	18 18	18 21
N 58	03 27	04 20	05 02	17 58	18 07	18 13	18 19
56	03 38	04 27	05 06	17 48	18 00	18 09	18 17
54	03 47	04 33	05 10	17 39	17 53	18 05	18 16
52	03 55	04 38	05 13	17 31	17 48	18 02	18 15
50	04 02	04 43	05 17	17 24	17 42	17 59	18 13
45	04 17	04 53	05 23	17 08	17 31	17 52	18 11
N 40	04 28	05 01	05 29	16 55	17 22	17 46	18 09
35	04 37	05 08	05 33	16 44	17 14	17 41	18 07
30	04 45	05 13	05 38	16 34	17 07	17 36	18 05
20	04 56	05 22	05 45	16 18	16 54	17 29	18 02
N 10	05 05	05 30	05 51	16 03	16 43	17 22	18 00
0	05 11	05 35	05 56	15 49	16 33	17 16	17 57
S 10	05 16	05 41	06 02	15 36	16 23	17 09	17 55
20	05 20	05 45	06 08	15 21	16 12	17 02	17 52
30	05 22	05 50	06 14	15 04	15 59	16 54	17 49
35	05 23	05 52	06 18	14 54	15 52	16 50	17 48
40	05 23	05 55	06 22	14 43	15 44	16 45	17 46
45	05 23	05 57	06 27	14 29	15 34	16 39	17 44
S 50	05 22	06 00	06 32	14 13	15 22	16 31	17 41
52	05 22	06 01	06 35	14 05	15 17	16 28	17 40
54	05 21	06 02	06 38	13 56	15 10	16 24	17 38
56	05 20	06 03	06 41	13 47	15 04	16 20	17 37
58	05 19	06 05	06 44	13 35	14 56	16 16	17 35
S 60	05 18	06 06	06 48	13 23	14 47	16 11	17 33

Lat. °	Sunset h m	Twilight Civil h m	Twilight Naut. h m	Moonset 1 h m	Moonset 2 h m	Moonset 3 h m	Moonset 4 h m
N 72	19 49	21 15	////	23 19	26 06	02 06	04 12
N 70	19 37	20 50	23 13	■	00 18	02 28	04 22
68	19 27	20 31	22 11	24 52	00 52	02 45	04 30
66	19 18	20 16	21 38	25 16	01 16	02 59	04 36
64	19 12	20 04	21 15	25 35	01 35	03 10	04 42
62	19 06	19 53	20 57	00 19	01 50	03 19	04 46
60	19 00	19 45	20 42	00 38	02 03	03 27	04 50
N 58	18 56	19 37	20 30	00 53	02 14	03 34	04 54
56	18 52	19 31	20 20	01 07	02 24	03 40	04 57
54	18 48	19 25	20 11	01 18	02 32	03 46	05 00
52	18 45	19 20	20 03	01 28	02 39	03 51	05 02
50	18 42	19 15	19 56	01 37	02 46	03 56	05 05
45	18 35	19 05	19 42	01 56	03 00	04 05	05 10
N 40	18 30	18 58	19 30	02 11	03 12	04 13	05 14
35	18 25	18 51	19 22	02 24	03 22	04 20	05 18
30	18 21	18 46	19 14	02 35	03 31	04 26	05 21
20	18 14	18 37	19 03	02 54	03 46	04 36	05 26
N 10	18 08	18 30	18 54	03 11	03 59	04 45	05 31
0	18 03	18 24	18 48	03 26	04 11	04 54	05 36
S 10	17 58	18 19	18 43	03 41	04 23	05 02	05 40
20	17 52	18 14	18 40	03 58	04 35	05 11	05 45
30	17 46	18 10	18 37	04 16	04 50	05 21	05 50
35	17 42	18 07	18 37	04 27	04 58	05 27	05 53
40	17 38	18 05	18 37	04 39	05 08	05 33	05 56
45	17 33	18 03	18 37	04 54	05 19	05 40	06 00
S 50	17 28	18 00	18 38	05 11	05 32	05 49	06 05
52	17 25	17 59	18 38	05 20	05 38	05 54	06 07
54	17 23	17 58	18 39	05 29	05 45	05 58	06 09
56	17 20	17 57	18 40	05 39	05 53	06 03	06 12
58	17 16	17 56	18 41	05 51	06 01	06 09	06 15
S 60	17 12	17 54	18 43	06 04	06 11	06 15	06 18

Day	SUN Eqn. of Time 00^h	SUN Eqn. of Time 12^h	SUN Mer. Pass.	MOON Mer. Pass. Upper	MOON Mer. Pass. Lower	Age	Phase
d	m s	m s	h m	h m	h m	d %	
1	00 07	00 03	12 00	22 00	09 38	12 91	
2	00 12	00 22	12 00	22 43	10 22	13 96	○
3	00 32	00 41	11 59	23 26	11 05	14 99	

UT	ARIES	VENUS −3.9		MARS +0.9		JUPITER −2.8		SATURN +1.1		STARS		
	GHA	GHA	Dec	GHA	Dec	GHA	Dec	GHA	Dec	Name	SHA	Dec
d h	° ′	° ′	° ′	° ′	° ′	° ′	° ′	° ′	° ′		° ′	° ′
4 00	343 15.0	210 48.1	N17 56.0	246 53.4	N23 33.3	20 55.6	S16 00.1	168 38.2	N 4 28.3	Acamar	315 20.2	S40 15.
01	358 17.4	225 47.5	55.3	261 54.1	33.3	35 58.3	00.2	183 40.4	28.2	Achernar	335 28.2	S57 10.
02	13 19.9	240 46.9	54.6	276 54.9	33.3	51 01.1	00.3	198 42.6	28.1	Acrux	173 13.3	S63 09.
03	28 22.4	255 46.3	. . 53.9	291 55.7	. . 33.2	66 03.8	. . 00.4	213 44.8	. . 28.0	Adhara	255 14.9	S28 58.
04	43 24.8	270 45.7	53.3	306 56.5	33.2	81 06.5	00.5	228 46.9	27.8	Aldebaran	290 52.6	N16 31.
05	58 27.3	285 45.1	52.6	321 57.2	33.2	96 09.3	00.5	243 49.1	27.7			
06	73 29.8	300 44.5	N17 51.9	336 58.0	N23 33.1	111 12.0	S16 00.6	258 51.3	N 4 27.6	Alioth	166 23.4	N55 54.
07	88 32.2	315 43.8	51.2	351 58.8	33.1	126 14.8	00.7	273 53.5	27.5	Alkaid	153 01.3	N49 16.
08	103 34.7	330 43.2	50.5	6 59.6	33.1	141 17.5	00.8	288 55.7	27.3	Al Na'ir	27 46.6	S46 54.
F 09	118 37.2	345 42.6	. . 49.8	22 00.3	. . 33.1	156 20.2	. . 00.9	303 57.8	. . 27.2	Alnilam	275 49.3	S 1 11.
R 10	133 39.6	0 42.0	49.1	37 01.1	33.0	171 23.0	01.0	319 00.0	27.1	Alphard	217 59.1	S 8 41.
I 11	148 42.1	15 41.4	48.4	52 01.9	33.0	186 25.7	01.1	334 02.2	27.0			
D 12	163 44.5	30 40.8	N17 47.7	67 02.7	N23 33.0	201 28.4	S16 01.2	349 04.4	N 4 26.9	Alphecca	126 13.4	N26 41.
A 13	178 47.0	45 40.2	47.0	82 03.5	33.0	216 31.2	01.2	4 06.6	26.7	Alpheratz	357 46.1	N29 08.
Y 14	193 49.5	60 39.6	46.3	97 04.2	32.9	231 33.9	01.3	19 08.7	26.6	Altair	62 10.7	N 8 53.
15	208 51.9	75 39.0	. . 45.6	112 05.0	. . 32.9	246 36.7	. . 01.4	34 10.9	. . 26.5	Ankaa	353 17.9	S42 14.
16	223 54.4	90 38.4	44.9	127 05.8	32.9	261 39.4	01.5	49 13.1	26.4	Antares	112 29.7	S26 27.
17	238 56.9	105 37.8	44.2	142 06.6	32.8	276 42.1	01.6	64 15.3	26.2			
18	253 59.3	120 37.2	N17 43.5	157 07.3	N23 32.8	291 44.9	S16 01.7	79 17.5	N 4 26.1	Arcturus	145 58.4	N19 08.
19	269 01.8	135 36.6	42.8	172 08.1	32.8	306 47.6	01.8	94 19.6	26.0	Atria	107 34.1	S69 03.
20	284 04.3	150 36.0	42.1	187 08.9	32.8	321 50.3	01.9	109 21.8	25.9	Avior	234 19.8	S59 32.
21	299 06.7	165 35.4	. . 41.4	202 09.7	. . 32.7	336 53.1	. . 01.9	124 24.0	. . 25.8	Bellatrix	278 35.0	N 6 21.
22	314 09.2	180 34.8	40.7	217 10.5	32.7	351 55.8	02.0	139 26.2	25.6	Betelgeuse	271 04.4	N 7 24.
23	329 11.6	195 34.2	40.0	232 11.2	32.7	6 58.5	02.1	154 28.4	25.5			
5 00	344 14.1	210 33.6	N17 39.3	247 12.0	N23 32.6	22 01.3	S16 02.2	169 30.5	N 4 25.4	Canopus	263 57.6	S52 41.
01	359 16.6	225 33.0	38.6	262 12.8	32.6	37 04.0	02.3	184 32.7	25.3	Capella	280 38.6	N46 00.
02	14 19.0	240 32.3	37.9	277 13.6	32.6	52 06.8	02.4	199 34.9	25.2	Deneb	49 33.1	N45 19.
03	29 21.5	255 31.7	. . 37.2	292 14.4	. . 32.5	67 09.5	. . 02.5	214 37.1	. . 25.0	Denebola	182 36.8	N14 31.
04	44 24.0	270 31.1	36.5	307 15.2	32.5	82 12.2	02.5	229 39.3	24.9	Diphda	348 58.3	S17 55.
05	59 26.4	285 30.5	35.8	322 15.9	32.5	97 15.0	02.6	244 41.4	24.8			
06	74 28.9	300 29.9	N17 35.1	337 16.7	N23 32.4	112 17.7	S16 02.7	259 43.6	N 4 24.7	Dubhe	193 55.5	N61 41.
S 07	89 31.4	315 29.3	34.3	352 17.5	32.4	127 20.4	02.8	274 45.8	24.5	Elnath	278 16.2	N28 36.
A 08	104 33.8	330 28.7	33.6	7 18.3	32.4	142 23.2	02.9	289 48.0	24.4	Eltanin	90 47.3	N51 29.
T 09	119 36.3	345 28.1	. . 32.9	22 19.1	. . 32.3	157 25.9	. . 03.0	304 50.2	. . 24.3	Enif	33 49.6	N 9 55.
U 10	134 38.8	0 27.5	32.2	37 19.9	32.3	172 28.6	03.1	319 52.3	24.2	Fomalhaut	15 26.6	S29 34.
R 11	149 41.2	15 26.9	31.5	52 20.6	32.3	187 31.4	03.1	334 54.5	24.1			
D 12	164 43.7	30 26.3	N17 30.8	67 21.4	N23 32.2	202 34.1	S16 03.2	349 56.7	N 4 23.9	Gacrux	172 04.8	S57 10.
A 13	179 46.1	45 25.7	30.1	82 22.2	32.2	217 36.8	03.3	4 58.9	23.8	Gienah	175 55.5	S17 35.
Y 14	194 48.6	60 25.1	29.3	97 23.0	32.2	232 39.6	03.4	20 01.1	23.7	Hadar	148 52.5	S60 25.
15	209 51.1	75 24.5	. . 28.6	112 23.8	. . 32.1	247 42.3	. . 03.5	35 03.2	. . 23.6	Hamal	328 03.7	N23 30.
16	224 53.5	90 23.9	27.9	127 24.6	32.1	262 45.0	03.6	50 05.4	23.4	Kaus Aust.	83 47.3	S34 22.
17	239 56.0	105 23.3	27.2	142 25.3	32.1	277 47.8	03.7	65 07.6	23.3			
18	254 58.5	120 22.7	N17 26.5	157 26.1	N23 32.0	292 50.5	S16 03.7	80 09.8	N 4 23.2	Kochab	137 20.1	N74 07.
19	270 00.9	135 22.1	25.7	172 26.9	32.0	307 53.2	03.8	95 12.0	23.1	Markab	13 40.8	N15 15.
20	285 03.4	150 21.5	25.0	187 27.7	32.0	322 55.9	03.9	110 14.1	23.0	Menkar	314 17.8	N 4 07.
21	300 05.9	165 21.0	. . 24.3	202 28.5	. . 31.9	337 58.7	. . 04.0	125 16.3	. . 22.8	Menkent	148 11.2	S36 25.
22	315 08.3	180 20.4	23.6	217 29.3	31.9	353 01.4	04.1	140 18.5	22.7	Miaplacidus	221 41.4	S69 45.
23	330 10.8	195 19.8	22.8	232 30.1	31.8	8 04.1	04.2	155 20.7	22.6			
6 00	345 13.3	210 19.2	N17 22.1	247 30.8	N23 31.8	23 06.9	S16 04.3	170 22.9	N 4 22.5	Mirfak	308 44.3	N49 53.
01	0 15.7	225 18.6	21.4	262 31.6	31.8	38 09.6	04.3	185 25.0	22.4	Nunki	76 01.5	S26 17.
02	15 18.2	240 18.0	20.7	277 32.4	31.7	53 12.3	04.4	200 27.2	22.2	Peacock	53 23.0	S56 42.
03	30 20.6	255 17.4	. . 19.9	292 33.2	. . 31.7	68 15.1	. . 04.5	215 29.4	. . 22.1	Pollux	243 31.3	N28 00.
04	45 23.1	270 16.8	19.2	307 34.0	31.7	83 17.8	04.6	230 31.6	22.0	Procyon	245 02.8	N 5 12.
05	60 25.6	285 16.2	18.5	322 34.8	31.6	98 20.5	04.7	245 33.8	21.9			
06	75 28.0	300 15.6	N17 17.7	337 35.6	N23 31.6	113 23.2	S16 04.8	260 35.9	N 4 21.7	Rasalhague	96 09.0	N12 33.
07	90 30.5	315 15.0	17.0	352 36.4	31.5	128 26.0	04.8	275 38.1	21.6	Regulus	207 46.8	N11 55.
08	105 33.0	330 14.4	16.3	7 37.1	31.5	143 28.7	04.9	290 40.3	21.5	Rigel	281 14.8	S 8 11.
S 09	120 35.4	345 13.8	. . 15.5	22 37.9	. . 31.5	158 31.4	. . 05.0	305 42.5	. . 21.4	Rigil Kent.	139 56.1	S60 52.
U 10	135 37.9	0 13.2	14.8	37 38.7	31.4	173 34.2	05.1	320 44.7	21.3	Sabik	102 15.7	S15 44.
N 11	150 40.4	15 12.6	14.1	52 39.5	31.4	188 36.9	05.2	335 46.8	21.1			
D 12	165 42.8	30 12.0	N17 13.3	67 40.3	N23 31.3	203 39.6	S16 05.3	350 49.0	N 4 21.0	Schedar	349 43.4	N56 35.
A 13	180 45.3	45 11.4	12.6	82 41.1	31.3	218 42.4	05.4	5 51.2	20.9	Shaula	96 25.6	S37 06.
Y 14	195 47.7	60 10.8	11.8	97 41.9	31.3	233 45.1	05.4	20 53.4	20.8	Sirius	258 36.3	S16 43.
15	210 50.2	75 10.3	. . 11.1	112 42.7	. . 31.2	248 47.8	. . 05.5	35 55.6	. . 20.6	Spica	158 34.5	S11 12.
16	225 52.7	90 09.7	10.4	127 43.5	31.2	263 50.5	05.6	50 57.7	20.5	Suhail	222 55.0	S43 28.
17	240 55.1	105 09.1	09.6	142 44.3	31.1	278 53.3	05.7	65 59.9	20.4			
18	255 57.6	120 08.5	N17 08.9	157 45.0	N23 31.1	293 56.0	S16 05.8	81 02.1	N 4 20.3	Vega	80 40.7	N38 47.
19	271 00.1	135 07.9	08.1	172 45.8	31.0	308 58.7	05.9	96 04.3	20.2	Zuben'ubi	137 08.7	S16 05.
20	286 02.5	150 07.3	07.4	187 46.6	31.0	324 01.4	05.9	111 06.5	20.0			
21	301 05.0	165 06.7	. . 06.7	202 47.4	. . 31.0	339 04.2	. . 06.0	126 08.6	. . 19.9			
22	316 07.5	180 06.1	05.9	217 48.2	30.9	354 06.9	06.1	141 10.8	19.8			
23	331 09.9	195 05.5	05.2	232 49.0	30.9	9 09.6	06.2	156 13.0	19.7			
Mer. Pass.	h m 1 02.9	v −0.6	d 0.7	v 0.8	d 0.0	v 2.7	d 0.1	v 2.2	d 0.1			

	SHA	Mer. Pass.
	° ′	h m
Venus	226 19.4	9 58
Mars	262 57.9	7 31
Jupiter	37 47.2	22 28
Saturn	185 16.4	12 40

	UT	SUN GHA	SUN Dec	MOON GHA	v	MOON Dec	d	HP
	d h	° ′	° ′	° ′	′	° ′	′	′
	4 00	180 12.9	N 7 12.6	8 21.7	15.6	S 7 17.7	13.1	54.8
	01	195 13.1	11.7	22 56.3	15.7	7 04.6	13.1	54.8
	02	210 13.4	10.7	37 31.0	15.7	6 51.5	13.2	54.8
	03	225 13.6	. . 09.8	52 05.7	15.8	6 38.3	13.2	54.8
	04	240 13.8	08.9	66 40.5	15.7	6 25.1	13.2	54.9
	05	255 14.0	08.0	81 15.2	15.7	6 11.9	13.3	54.9
	06	270 14.2	N 7 07.1	95 49.9	15.7	S 5 58.6	13.3	54.9
	07	285 14.4	06.1	110 24.6	15.8	5 45.3	13.3	54.9
	08	300 14.6	05.2	124 59.4	15.7	5 32.0	13.4	54.9
F	09	315 14.8	. . 04.3	139 34.1	15.8	5 18.6	13.4	54.9
R	10	330 15.0	03.4	154 08.9	15.7	5 05.2	13.4	54.9
I	11	345 15.2	02.4	168 43.6	15.8	4 51.8	13.4	55.0
D	12	0 15.4	N 7 01.5	183 18.4	15.8	S 4 38.4	13.5	55.0
A	13	15 15.6	7 00.6	197 53.2	15.7	4 24.9	13.4	55.0
Y	14	30 15.8	6 59.7	212 27.9	15.8	4 11.5	13.6	55.0
	15	45 16.1	. . 58.7	227 02.7	15.7	3 57.9	13.5	55.0
	16	60 16.3	57.8	241 37.4	15.8	3 44.4	13.5	55.0
	17	75 16.5	56.9	256 12.2	15.8	3 30.9	13.6	55.1
	18	90 16.7	N 6 56.0	270 47.0	15.7	S 3 17.3	13.6	55.1
	19	105 16.9	55.1	285 21.7	15.7	3 03.7	13.6	55.1
	20	120 17.1	54.1	299 56.4	15.8	2 50.1	13.7	55.1
	21	135 17.3	. . 53.2	314 31.2	15.7	2 36.4	13.6	55.1
	22	150 17.5	52.3	329 05.9	15.7	2 22.8	13.7	55.1
	23	165 17.7	51.3	343 40.6	15.8	2 09.1	13.7	55.1
	5 00	180 17.9	N 6 50.4	358 15.4	15.7	S 1 55.4	13.7	55.2
	01	195 18.1	49.5	12 50.1	15.7	1 41.7	13.7	55.2
	02	210 18.4	48.6	27 24.8	15.7	1 28.0	13.7	55.2
	03	225 18.6	. . 47.6	41 59.5	15.6	1 14.3	13.8	55.2
	04	240 18.8	46.7	56 34.1	15.7	1 00.5	13.7	55.2
	05	255 19.0	45.8	71 08.8	15.6	0 46.8	13.8	55.3
	06	270 19.2	N 6 44.9	85 43.4	15.7	S 0 33.0	13.8	55.3
	07	285 19.4	43.9	100 18.1	15.6	0 19.2	13.7	55.3
S	08	300 19.6	43.0	114 52.7	15.6	S 0 05.5	13.8	55.3
A	09	315 19.8	. . 42.1	129 27.3	15.6	N 0 08.3	13.8	55.3
T	10	330 20.0	41.2	144 01.9	15.5	0 22.1	13.8	55.3
U	11	345 20.2	40.2	158 36.4	15.6	0 35.9	13.8	55.4
R	12	0 20.5	N 6 39.3	173 11.0	15.5	N 0 49.7	13.9	55.4
D	13	15 20.7	38.4	187 45.5	15.5	1 03.6	13.8	55.4
A	14	30 20.9	37.4	202 20.0	15.5	1 17.4	13.8	55.4
Y	15	45 21.1	. . 36.5	216 54.5	15.5	1 31.2	13.8	55.4
	16	60 21.3	35.6	231 29.0	15.4	1 45.0	13.9	55.4
	17	75 21.5	34.7	246 03.4	15.4	1 58.9	13.8	55.5
	18	90 21.7	N 6 33.7	260 37.8	15.4	N 2 12.7	13.8	55.5
	19	105 21.9	32.8	275 12.2	15.3	2 26.5	13.9	55.5
	20	120 22.1	31.9	289 46.5	15.4	2 40.4	13.8	55.5
	21	135 22.4	. . 30.9	304 20.9	15.3	2 54.2	13.8	55.5
	22	150 22.6	30.0	318 55.2	15.2	3 08.0	13.8	55.6
	23	165 22.8	29.1	333 29.4	15.3	3 21.8	13.8	55.6
	6 00	180 23.0	N 6 28.1	348 03.7	15.2	N 3 35.6	13.8	55.6
	01	195 23.2	27.2	2 37.9	15.1	3 49.4	13.8	55.6
	02	210 23.4	26.3	17 12.0	15.2	4 03.2	13.8	55.6
	03	225 23.6	. . 25.4	31 46.2	15.1	4 17.0	13.8	55.6
	04	240 23.8	24.4	46 20.3	15.1	4 30.8	13.8	55.7
	05	255 24.0	23.5	60 54.4	15.0	4 44.6	13.7	55.7
	06	270 24.3	N 6 22.6	75 28.4	15.0	N 4 58.3	13.8	55.7
	07	285 24.5	21.6	90 02.4	14.9	5 12.1	13.7	55.7
	08	300 24.7	20.7	104 36.3	15.0	5 25.8	13.7	55.7
S	09	315 24.9	. . 19.8	119 10.3	14.8	5 39.5	13.7	55.8
U	10	330 25.1	18.8	133 44.1	14.9	5 53.2	13.7	55.8
N	11	345 25.3	17.9	148 18.0	14.7	6 06.9	13.7	55.8
D	12	0 25.5	N 6 17.0	162 51.7	14.8	N 6 20.6	13.6	55.8
A	13	15 25.7	16.0	177 25.5	14.7	6 34.2	13.7	55.8
Y	14	30 26.0	15.1	191 59.2	14.7	6 47.9	13.6	55.9
	15	45 26.2	. . 14.2	206 32.9	14.6	7 01.5	13.6	55.9
	16	60 26.4	13.2	221 06.5	14.5	7 15.1	13.6	55.9
	17	75 26.6	12.3	235 40.0	14.5	7 28.7	13.5	55.9
	18	90 26.8	N 6 11.4	250 13.5	14.5	N 7 42.2	13.5	55.9
	19	105 27.0	10.4	264 47.0	14.4	7 55.7	13.5	56.0
	20	120 27.2	09.5	279 20.4	14.4	8 09.2	13.5	56.0
	21	135 27.5	. . 08.6	293 53.8	14.3	8 22.7	13.5	56.0
	22	150 27.7	07.6	308 27.1	14.2	8 36.2	13.4	56.0
	23	165 27.9	06.7	323 00.3	14.2	N 8 49.6	13.4	56.0
		SD 15.9	*d* 0.9	SD	15.0	15.1		15.2

Lat.	Twilight Naut.	Twilight Civil	Sunrise	Moonrise 4	Moonrise 5	Moonrise 6	Moonrise 7
°	h m	h m	h m	h m	h m	h m	h m
N 72	////	02 59	04 21	18 38	18 11	17 43	17 06
N 70	01 18	03 21	04 32	18 34	18 14	17 53	17 28
68	02 04	03 38	04 41	18 30	18 17	18 02	17 45
66	02 33	03 52	04 48	18 27	18 19	18 09	18 00
64	02 54	04 03	04 55	18 25	18 20	18 16	18 11
62	03 11	04 12	05 00	18 23	18 22	18 21	18 21
60	03 24	04 20	05 04	18 21	18 23	18 26	18 30
N 58	03 35	04 27	05 09	18 19	18 24	18 30	18 37
56	03 45	04 33	05 12	18 17	18 25	18 34	18 44
54	03 53	04 39	05 15	18 16	18 26	18 37	18 50
52	04 01	04 43	05 18	18 15	18 27	18 40	18 56
50	04 07	04 48	05 21	18 13	18 28	18 43	19 01
45	04 21	04 57	05 27	18 11	18 30	18 49	19 11
N 40	04 31	05 04	05 32	18 09	18 31	18 55	19 20
35	04 40	05 10	05 36	18 07	18 32	18 59	19 28
30	04 46	05 15	05 39	18 05	18 34	19 03	19 35
20	04 57	05 23	05 45	18 02	18 36	19 10	19 47
N 10	05 05	05 29	05 51	18 00	18 37	19 16	19 57
0	05 10	05 35	05 55	17 57	18 39	19 22	20 07
S 10	05 15	05 39	06 00	17 55	18 41	19 28	20 17
20	05 17	05 43	06 05	17 52	18 43	19 34	20 28
30	05 19	05 47	06 11	17 49	18 45	19 41	20 40
35	05 19	05 48	06 14	17 48	18 46	19 46	20 47
40	05 19	05 50	06 17	17 46	18 47	19 50	20 55
45	05 18	05 52	06 21	17 44	18 49	19 56	21 05
S 50	05 16	05 54	06 26	17 41	18 51	20 03	21 17
52	05 15	05 54	06 28	17 40	18 52	20 06	21 22
54	05 14	05 55	06 30	17 38	18 53	20 09	21 28
56	05 13	05 56	06 33	17 37	18 54	20 13	21 35
58	05 11	05 57	06 36	17 35	18 55	20 17	21 43
S 60	05 09	05 57	06 39	17 33	18 57	20 22	21 51

Lat.	Sunset	Twilight Civil	Twilight Naut.	Moonset 4	Moonset 5	Moonset 6	Moonset 7
°	h m	h m	h m	h m	h m	h m	h m
N 72	19 33	20 54	////	04 12	06 10	08 09	10 19
N 70	19 22	20 32	22 29	04 22	06 11	08 01	09 59
68	19 14	20 16	21 47	04 30	06 11	07 54	09 43
66	19 07	20 03	21 20	04 36	06 12	07 49	09 31
64	19 01	19 52	21 00	04 42	06 12	07 45	09 20
62	18 56	19 43	20 44	04 46	06 13	07 41	09 12
60	18 51	19 35	20 31	04 50	06 13	07 37	09 04
N 58	18 47	19 28	20 20	04 54	06 14	07 34	08 58
56	18 44	19 22	20 10	04 57	06 14	07 32	08 52
54	18 41	19 17	20 02	05 00	06 14	07 29	08 47
52	18 38	19 13	19 55	05 02	06 14	07 27	08 42
50	18 35	19 08	19 49	05 05	06 15	07 25	08 38
45	18 30	19 00	19 35	05 10	06 15	07 21	08 29
N 40	18 25	18 53	19 25	05 14	06 15	07 18	08 21
35	18 21	18 47	19 17	05 18	06 16	07 14	08 15
30	18 18	18 42	19 10	05 21	06 16	07 12	08 09
20	18 12	18 34	19 00	05 26	06 16	07 07	08 00
N 10	18 07	18 28	18 52	05 31	06 17	07 03	07 51
0	18 02	18 23	18 47	05 36	06 17	06 59	07 43
S 10	17 57	18 18	18 43	05 40	06 18	06 56	07 35
20	17 53	18 15	18 40	05 45	06 18	06 52	07 27
30	17 47	18 11	18 39	05 50	06 18	06 47	07 17
35	17 44	18 10	18 39	05 53	06 18	06 44	07 12
40	17 41	18 08	18 39	05 56	06 19	06 41	07 06
45	17 37	18 06	18 40	06 00	06 19	06 38	06 59
S 50	17 32	18 05	18 42	06 05	06 19	06 34	06 50
52	17 30	18 04	18 43	06 07	06 19	06 32	06 46
54	17 28	18 03	18 44	06 09	06 20	06 30	06 42
56	17 25	18 03	18 46	06 12	06 20	06 28	06 37
58	17 22	18 02	18 47	06 15	06 20	06 25	06 32
S 60	17 19	18 01	18 49	06 18	06 20	06 23	06 26

Day	SUN Eqn. of Time 00h	SUN Eqn. of Time 12h	SUN Mer. Pass.	MOON Mer. Pass. Upper	MOON Mer. Pass. Lower	MOON Age	MOON Phase
d	m s	m s	h m	h m	h m	d %	
4	00 51	01 01	11 59	24 07	11 46	15 100	○
5	01 11	01 21	11 59	00 07	12 28	16 99	
6	01 32	01 42	11 58	00 49	13 11	17 97	

UT		ARIES	VENUS −3.9		MARS +0.9		JUPITER −2.8		SATURN +1.1	
		GHA	GHA	Dec	GHA	Dec	GHA	Dec	GHA	Dec
d	h	° ′	° ′	° ′	° ′	° ′	° ′	° ′	° ′	° ′
7	00	346 12.4	210 04.9	N17 04.4	247 49.8	N23 30.8	24 12.3	S16 06.3	171 15.2	N 4 19.5
	01	1 14.9	225 04.3	03.7	262 50.6	30.8	39 15.1	06.4	186 17.4	19.4
	02	16 17.3	240 03.8	02.9	277 51.4	30.7	54 17.8	06.4	201 19.5	19.3
	03	31 19.8	255 03.2	. . 02.2	292 52.2	. . 30.7	69 20.5	. . 06.5	216 21.7	. . 19.2
	04	46 22.2	270 02.6	01.4	307 53.0	30.6	84 23.2	06.6	231 23.9	19.1
	05	61 24.7	285 02.0	17 00.7	322 53.8	30.6	99 26.0	06.7	246 26.1	18.9
	06	76 27.2	300 01.4	N16 59.9	337 54.6	N23 30.6	114 28.7	S16 06.8	261 28.3	N 4 18.8
	07	91 29.6	315 00.8	59.2	352 55.4	30.5	129 31.4	06.8	276 30.4	18.7
	08	106 32.1	330 00.2	58.4	7 56.2	30.5	144 34.1	06.9	291 32.6	18.6
MONDAY	09	121 34.6	344 59.6	. . 57.7	22 56.9	. . 30.4	159 36.9	. . 07.0	306 34.8	. . 18.4
	10	136 37.0	359 59.1	56.9	37 57.7	30.4	174 39.6	07.1	321 37.0	18.3
	11	151 39.5	14 58.5	56.1	52 58.5	30.3	189 42.3	07.2	336 39.2	18.2
	12	166 42.0	29 57.9	N16 55.4	67 59.3	N23 30.3	204 45.0	S16 07.3	351 41.3	N 4 18.1
	13	181 44.4	44 57.3	54.6	83 00.1	30.2	219 47.8	07.3	6 43.5	18.0
	14	196 46.9	59 56.7	53.9	98 00.9	30.2	234 50.5	07.4	21 45.7	17.8
	15	211 49.3	74 56.1	. . 53.1	113 01.7	. . 30.1	249 53.2	. . 07.5	36 47.9	. . 17.7
	16	226 51.8	89 55.5	52.3	128 02.5	30.1	264 55.9	07.6	51 50.0	17.6
	17	241 54.3	104 55.0	51.6	143 03.3	30.0	279 58.7	07.7	66 52.2	17.5
	18	256 56.7	119 54.4	N16 50.8	158 04.1	N23 30.0	295 01.4	S16 07.7	81 54.4	N 4 17.4
	19	271 59.2	134 53.8	50.1	173 04.9	29.9	310 04.1	07.8	96 56.6	17.2
	20	287 01.7	149 53.2	49.3	188 05.7	29.9	325 06.8	07.9	111 58.8	17.1
	21	302 04.1	164 52.6	. . 48.5	203 06.5	. . 29.8	340 09.5	. . 08.0	127 00.9	. . 17.0
	22	317 06.6	179 52.0	47.8	218 07.3	29.8	355 12.3	08.1	142 03.1	16.9
	23	332 09.1	194 51.5	47.0	233 08.1	29.7	10 15.0	08.2	157 05.3	16.7
8	00	347 11.5	209 50.9	N16 46.2	248 08.9	N23 29.7	25 17.7	S16 08.2	172 07.5	N 4 16.6
	01	2 14.0	224 50.3	45.5	263 09.7	29.6	40 20.4	08.3	187 09.7	16.5
	02	17 16.5	239 49.7	44.7	278 10.5	29.6	55 23.1	08.4	202 11.8	16.4
	03	32 18.9	254 49.1	. . 43.9	293 11.3	. . 29.5	70 25.9	. . 08.5	217 14.0	. . 16.3
	04	47 21.4	269 48.6	43.2	308 12.1	29.5	85 28.6	08.6	232 16.2	16.1
	05	62 23.8	284 48.0	42.4	323 12.9	29.4	100 31.3	08.6	247 18.4	16.0
	06	77 26.3	299 47.4	N16 41.6	338 13.7	N23 29.4	115 34.0	S16 08.7	262 20.6	N 4 15.9
	07	92 28.8	314 46.8	40.8	353 14.5	29.3	130 36.7	08.8	277 22.7	15.8
TUESDAY	08	107 31.2	329 46.2	40.1	8 15.3	29.3	145 39.5	08.9	292 24.9	15.6
	09	122 33.7	344 45.7	. . 39.3	23 16.1	. . 29.2	160 42.2	. . 09.0	307 27.1	. . 15.5
	10	137 36.2	359 45.1	38.5	38 16.9	29.2	175 44.9	09.0	322 29.3	15.4
	11	152 38.6	14 44.5	37.7	53 17.7	29.1	190 47.6	09.1	337 31.4	15.3
	12	167 41.1	29 43.9	N16 37.0	68 18.5	N23 29.1	205 50.3	S16 09.2	352 33.6	N 4 15.2
	13	182 43.6	44 43.3	36.2	83 19.3	29.0	220 53.1	09.3	7 35.8	15.0
	14	197 46.0	59 42.8	35.4	98 20.1	28.9	235 55.8	09.4	22 38.0	14.9
	15	212 48.5	74 42.2	. . 34.6	113 20.9	. . 28.9	250 58.5	. . 09.4	37 40.2	. . 14.8
	16	227 50.9	89 41.6	33.8	128 21.7	28.8	266 01.2	09.5	52 42.3	14.7
	17	242 53.4	104 41.0	33.1	143 22.5	28.8	281 03.9	09.6	67 44.5	14.5
	18	257 55.9	119 40.5	N16 32.3	158 23.3	N23 28.7	296 06.6	S16 09.7	82 46.7	N 4 14.4
	19	272 58.3	134 39.9	31.5	173 24.1	28.7	311 09.4	09.8	97 48.9	14.3
	20	288 00.8	149 39.3	30.7	188 24.9	28.6	326 12.1	09.8	112 51.1	14.2
	21	303 03.3	164 38.7	. . 29.9	203 25.7	. . 28.6	341 14.8	. . 09.9	127 53.2	. . 14.1
	22	318 05.7	179 38.2	29.1	218 26.5	28.5	356 17.5	10.0	142 55.4	13.9
	23	333 08.2	194 37.6	28.4	233 27.3	28.4	11 20.2	10.1	157 57.6	13.8
9	00	348 10.7	209 37.0	N16 27.6	248 28.1	N23 28.4	26 22.9	S16 10.2	172 59.8	N 4 13.7
	01	3 13.1	224 36.4	26.8	263 29.0	28.3	41 25.7	10.2	188 02.0	13.6
	02	18 15.6	239 35.9	26.0	278 29.8	28.3	56 28.4	10.3	203 04.1	13.4
	03	33 18.1	254 35.3	. . 25.2	293 30.6	. . 28.2	71 31.1	. . 10.4	218 06.3	. . 13.3
	04	48 20.5	269 34.7	24.4	308 31.4	28.2	86 33.8	10.5	233 08.5	13.2
	05	63 23.0	284 34.1	23.6	323 32.2	28.1	101 36.5	10.6	248 10.7	13.1
	06	78 25.4	299 33.6	N16 22.8	338 33.0	N23 28.0	116 39.2	S16 10.6	263 12.8	N 4 13.0
WEDNESDAY	07	93 27.9	314 33.0	22.0	353 33.8	28.0	131 41.9	10.7	278 15.0	12.8
	08	108 30.4	329 32.4	21.2	8 34.6	27.9	146 44.7	10.8	293 17.2	12.7
	09	123 32.8	344 31.8	. . 20.4	23 35.4	. . 27.9	161 47.4	. . 10.9	308 19.4	. . 12.6
	10	138 35.3	359 31.3	19.6	38 36.2	27.8	176 50.1	10.9	323 21.6	12.5
	11	153 37.8	14 30.7	18.9	53 37.0	27.7	191 52.8	11.0	338 23.7	12.3
	12	168 40.2	29 30.1	N16 18.1	68 37.8	N23 27.7	206 55.5	S16 11.1	353 25.9	N 4 12.2
	13	183 42.7	44 29.6	17.3	83 38.6	27.6	221 58.2	11.2	8 28.1	12.1
	14	198 45.2	59 29.0	16.5	98 39.4	27.6	237 00.9	11.3	23 30.3	12.0
	15	213 47.6	74 28.4	. . 15.7	113 40.3	. . 27.5	252 03.6	. . 11.3	38 32.4	. . 11.9
	16	228 50.1	89 27.8	14.9	128 41.1	27.4	267 06.4	11.4	53 34.6	11.7
	17	243 52.6	104 27.3	14.1	143 41.9	27.4	282 09.1	11.5	68 36.8	11.6
	18	258 55.0	119 26.7	N16 13.3	158 42.7	N23 27.3	297 11.8	S16 11.6	83 39.0	N 4 11.5
	19	273 57.5	134 26.1	12.5	173 43.5	27.3	312 14.5	11.6	98 41.2	11.4
	20	288 59.9	149 25.6	11.7	188 44.3	27.2	327 17.2	11.7	113 43.3	11.2
	21	304 02.4	164 25.0	. . 10.8	203 45.1	. . 27.1	342 19.9	. . 11.8	128 45.5	. . 11.1
	22	319 04.9	179 24.4	10.0	218 45.9	27.1	357 22.6	11.9	143 47.7	11.0
	23	334 07.3	194 23.9	09.2	233 46.7	27.0	12 25.3	12.0	158 49.9	10.9
Mer. Pass.		h m 0 51.1	v −0.6	d 0.8	v 0.8	d 0.1	v 2.7	d 0.1	v 2.2	d 0.1

STARS		
Name	SHA	Dec
	° ′	° ′
Acamar	315 20.1	S40 15.6
Achernar	335 28.2	S57 10.9
Acrux	173 13.3	S63 09.3
Adhara	255 14.9	S28 58.8
Aldebaran	290 52.5	N16 31.8
Alioth	166 23.4	N55 54.5
Alkaid	153 01.3	N49 16.0
Al Na'ir	27 46.6	S46 54.7
Alnilam	275 49.2	S 1 11.6
Alphard	217 59.1	S 8 41.9
Alphecca	126 13.4	N26 41.1
Alpheratz	357 46.1	N29 08.8
Altair	62 10.7	N 8 53.8
Ankaa	353 17.9	S42 14.9
Antares	112 29.7	S26 27.3
Arcturus	145 58.4	N19 08.0
Atria	107 34.2	S69 03.0
Avior	234 19.8	S59 32.2
Bellatrix	278 35.0	N 6 21.7
Betelgeuse	271 04.4	N 7 24.7
Canopus	263 57.6	S52 41.7
Capella	280 38.6	N46 00.4
Deneb	49 33.1	N45 19.1
Denebola	182 36.8	N14 31.1
Diphda	348 58.3	S17 55.7
Dubhe	193 55.5	N61 41.9
Elnath	278 16.2	N28 36.9
Eltanin	90 47.3	N51 29.5
Enif	33 49.6	N 9 55.4
Fomalhaut	15 26.6	S29 34.1
Gacrux	172 04.8	S57 10.2
Gienah	175 55.5	S17 35.8
Hadar	148 52.5	S60 25.4
Hamal	328 03.7	N23 30.6
Kaus Aust.	83 47.3	S34 22.9
Kochab	137 20.1	N74 07.1
Markab	13 40.8	N15 15.7
Menkar	314 17.8	N 4 07.9
Menkent	148 11.2	S36 25.2
Miaplacidus	221 41.4	S69 45.3
Mirfak	308 44.2	N49 53.7
Nunki	76 01.6	S26 17.1
Peacock	53 23.0	S56 42.3
Pollux	243 31.3	N28 00.2
Procyon	245 02.8	N 5 12.1
Rasalhague	96 09.0	N12 33.3
Regulus	207 46.8	N11 55.2
Rigel	281 14.7	S 8 11.2
Rigil Kent.	139 56.1	S60 52.7
Sabik	102 15.7	S15 44.2
Schedar	349 43.4	N56 35.5
Shaula	96 25.6	S37 06.8
Sirius	258 36.3	S16 43.5
Spica	158 34.5	S11 12.7
Suhail	222 55.0	S43 28.1
Vega	80 40.7	N38 47.8
Zuben'ubi	137 08.7	S16 05.0

	SHA	Mer. Pass.
	° ′	h m
Venus	222 39.4	10 01
Mars	260 57.4	7 27
Jupiter	38 06.2	22 15
Saturn	184 56.0	12 30

UT	SUN GHA	SUN Dec	MOON GHA	v	MOON Dec	d	HP
d h	° ′	° ′	° ′	′	° ′	′	′
7 00	180 28.1	N 6 05.8	337 33.5	14.2	N 9 03.0	13.3	56.1
01	195 28.3	04.8	352 06.7	14.1	9 16.3	13.4	56.1
02	210 28.5	03.9	6 39.8	14.0	9 29.7	13.3	56.1
03	225 28.7 . .	03.0	21 12.8	14.0	9 43.0	13.2	56.1
04	240 28.9	02.0	35 45.8	13.9	9 56.2	13.3	56.1
05	255 29.2	01.1	50 18.7	13.8	10 09.5	13.1	56.2
06	270 29.4	N 6 00.1	64 51.5	13.8	N10 22.6	13.2	56.2
07	285 29.6	5 59.2	79 24.3	13.8	10 35.8	13.1	56.2
08	300 29.8	58.3	93 57.1	13.6	10 48.9	13.1	56.2
M 09	315 30.0 . .	57.3	108 29.7	13.6	11 02.0	13.0	56.2
O 10	330 30.2	56.4	123 02.3	13.6	11 15.0	13.0	56.3
N 11	345 30.4	55.5	137 34.9	13.5	11 28.0	13.0	56.3
D 12	0 30.7	N 5 54.5	152 07.4	13.4	N11 41.0	12.9	56.3
A 13	15 30.9	53.6	166 39.8	13.3	11 53.9	12.8	56.3
Y 14	30 31.1	52.7	181 12.1	13.3	12 06.7	12.9	56.3
15	45 31.3 . .	51.7	195 44.4	13.2	12 19.6	12.7	56.4
16	60 31.5	50.8	210 16.6	13.1	12 32.3	12.8	56.4
17	75 31.7	49.8	224 48.7	13.1	12 45.1	12.6	56.4
18	90 32.0	N 5 48.9	239 20.8	13.0	N12 57.7	12.7	56.4
19	105 32.2	48.0	253 52.8	12.9	13 10.4	12.5	56.4
20	120 32.4	47.0	268 24.7	12.8	13 22.9	12.5	56.5
21	135 32.6 . .	46.1	282 56.5	12.8	13 35.4	12.5	56.5
22	150 32.8	45.1	297 28.3	12.7	13 47.9	12.4	56.5
23	165 33.0	44.2	312 00.0	12.6	14 00.3	12.4	56.5
8 00	180 33.2	N 5 43.3	326 31.6	12.6	N14 12.7	12.3	56.5
01	195 33.5	42.3	341 03.2	12.4	14 25.0	12.2	56.6
02	210 33.7	41.4	355 34.6	12.4	14 37.2	12.2	56.6
03	225 33.9 . .	40.5	10 06.0	12.3	14 49.4	12.1	56.6
04	240 34.1	39.5	24 37.3	12.2	15 01.5	12.0	56.6
05	255 34.3	38.6	39 08.5	12.2	15 13.5	12.0	56.7
06	270 34.5	N 5 37.6	53 39.7	12.0	N15 25.5	11.9	56.7
07	285 34.8	36.7	68 10.7	12.0	15 37.4	11.9	56.7
T 08	300 35.0	35.8	82 41.7	11.9	15 49.3	11.8	56.7
U 09	315 35.2 . .	34.8	97 12.6	11.8	16 01.1	11.7	56.7
E 10	330 35.4	33.9	111 43.4	11.8	16 12.8	11.6	56.8
S 11	345 35.6	32.9	126 14.2	11.6	16 24.4	11.6	56.8
D 12	0 35.8	N 5 32.0	140 44.8	11.6	N16 36.0	11.5	56.8
A 13	15 36.1	31.1	155 15.4	11.5	16 47.5	11.4	56.8
Y 14	30 36.3	30.1	169 45.9	11.3	16 58.9	11.3	56.8
15	45 36.5 . .	29.2	184 16.2	11.3	17 10.2	11.3	56.9
16	60 36.7	28.2	198 46.5	11.3	17 21.5	11.2	56.9
17	75 36.9	27.3	213 16.8	11.1	17 32.7	11.1	56.9
18	90 37.1	N 5 26.3	227 46.9	11.0	N17 43.8	11.1	56.9
19	105 37.4	25.4	242 16.9	10.9	17 54.9	10.9	57.0
20	120 37.6	24.5	256 46.8	10.9	18 05.8	10.9	57.0
21	135 37.8 . .	23.5	271 16.7	10.8	18 16.7	10.8	57.0
22	150 38.0	22.6	285 46.5	10.6	18 27.5	10.6	57.0
23	165 38.2	21.6	300 16.1	10.6	18 38.1	10.7	57.0
9 00	180 38.4	N 5 20.7	314 45.7	10.5	N18 48.8	10.5	57.1
01	195 38.7	19.7	329 15.2	10.4	18 59.3	10.4	57.1
02	210 38.9	18.8	343 44.6	10.3	19 09.7	10.3	57.1
03	225 39.1 . .	17.9	358 13.9	10.2	19 20.0	10.3	57.1
04	240 39.3	16.9	12 43.1	10.1	19 30.3	10.1	57.2
05	255 39.5	16.0	27 12.2	10.0	19 40.4	10.1	57.2
06	270 39.7	N 5 15.0	41 41.2	10.0	N19 50.5	9.9	57.2
W 07	285 40.0	14.1	56 10.2	9.8	20 00.4	9.9	57.2
E 08	300 40.2	13.1	70 39.0	9.7	20 10.3	9.7	57.2
D 09	315 40.4 . .	12.2	85 07.7	9.7	20 20.0	9.7	57.3
N 10	330 40.6	11.2	99 36.4	9.5	20 29.7	9.6	57.3
E 11	345 40.8	10.3	114 04.9	9.5	20 39.3	9.4	57.3
S 12	0 41.0	N 5 09.4	128 33.4	9.3	N20 48.7	9.3	57.3
D 13	15 41.3	08.4	143 01.7	9.3	20 58.0	9.3	57.4
A 14	30 41.5	07.5	157 30.0	9.2	21 07.3	9.1	57.4
Y 15	45 41.7 . .	06.5	171 58.2	9.0	21 16.4	9.0	57.4
16	60 41.9	05.6	186 26.2	9.0	21 25.4	8.9	57.4
17	75 42.1	04.6	200 54.2	8.9	21 34.3	8.8	57.4
18	90 42.4	N 5 03.7	215 22.1	8.8	N21 43.1	8.7	57.5
19	105 42.6	02.7	229 49.9	8.7	21 51.8	8.6	57.5
20	120 42.8	01.8	244 17.6	8.6	22 00.4	8.4	57.5
21	135 43.0	5 00.8	258 45.2	8.4	22 08.8	8.4	57.5
22	150 43.2	4 59.9	273 12.6	8.5	22 17.2	8.2	57.6
23	165 43.4	N 4 59.0	287 40.1	8.3	N22 25.4	8.1	57.6
	SD 15.9	d 0.9	SD	15.3		15.5	15.6

Lat.	Twilight Naut.	Twilight Civil	Sunrise	Moonrise 7	Moonrise 8	Moonrise 9	Moonrise 10
°	h m	h m	h m	h m	h m	h m	h m
N 72	00 39	03 17	04 35	17 06	15 55	▭	▭
N 70	01 49	03 36	04 44	17 28	16 49	▭	▭
68	02 24	03 51	04 52	17 45	17 23	16 33	▭
66	02 48	04 03	04 58	18 00	17 47	17 28	▭
64	03 06	04 13	05 03	18 11	18 07	18 02	17 56
62	03 21	04 21	05 08	18 21	18 22	18 26	18 37
60	03 33	04 28	05 12	18 30	18 36	18 46	19 05
N 58	03 43	04 34	05 15	18 37	18 47	19 02	19 27
56	03 52	04 40	05 18	18 44	18 57	19 16	19 45
54	04 00	04 44	05 21	18 50	19 06	19 28	20 00
52	04 07	04 49	05 23	18 56	19 14	19 39	20 13
50	04 13	04 52	05 25	19 01	19 21	19 48	20 25
45	04 25	05 00	05 30	19 11	19 37	20 09	20 49
N 40	04 34	05 07	05 34	19 20	19 50	20 25	21 08
35	04 42	05 12	05 38	19 28	20 01	20 39	21 24
30	04 48	05 17	05 41	19 35	20 10	20 51	21 39
20	04 58	05 24	05 46	19 47	20 27	21 12	22 03
N 10	05 05	05 29	05 50	19 57	20 41	21 30	22 24
0	05 09	05 34	05 54	20 07	20 55	21 47	22 43
S 10	05 13	05 37	05 58	20 17	21 09	22 04	23 03
20	05 15	05 40	06 02	20 28	21 24	22 23	23 24
30	05 15	05 43	06 07	20 40	21 41	22 44	23 49
35	05 15	05 44	06 09	20 47	21 51	22 57	24 03
40	05 14	05 45	06 12	20 55	22 03	23 12	24 20
45	05 12	05 46	06 16	21 05	22 16	23 29	24 40
S 50	05 10	05 47	06 19	21 17	22 33	23 51	25 06
52	05 09	05 48	06 21	21 22	22 41	24 01	00 01
54	05 07	05 48	06 23	21 28	22 50	24 13	00 13
56	05 05	05 48	06 25	21 35	23 00	24 26	00 26
58	05 03	05 48	06 28	21 43	23 11	24 42	00 42
S 60	05 01	05 49	06 30	21 51	23 25	25 01	01 01

Lat.	Sunset	Twilight Civil	Twilight Naut.	Moonset 7	Moonset 8	Moonset 9	Moonset 10
°	h m	h m	h m	h m	h m	h m	h m
N 72	19 17	20 34	22 56	10 19	13 09	▭	▭
N 70	19 08	20 16	21 59	09 59	12 16	▭	▭
68	19 01	20 01	21 27	09 43	11 44	14 19	▭
66	18 55	19 50	21 03	09 31	11 21	13 25	▭
64	18 50	19 40	20 46	09 20	11 02	12 52	14 52
62	18 46	19 32	20 32	09 12	10 47	12 29	14 12
60	18 42	19 25	20 20	09 04	10 35	12 10	13 45
N 58	18 39	19 19	20 10	08 58	10 24	11 54	13 23
56	18 36	19 14	20 01	08 52	10 15	11 41	13 06
54	18 33	19 10	19 54	08 47	10 07	11 29	12 51
52	18 31	19 05	19 47	08 42	10 00	11 19	12 38
50	18 29	19 02	19 41	08 38	09 53	11 10	12 27
45	18 24	18 54	19 29	08 29	09 39	10 51	12 03
N 40	18 20	18 48	19 20	08 21	09 27	10 36	11 44
35	18 17	18 42	19 13	08 15	09 18	10 23	11 29
30	18 14	18 38	19 06	08 09	09 09	10 11	11 15
20	18 09	18 31	18 57	08 00	08 54	09 52	10 52
N 10	18 05	18 26	18 50	07 51	08 42	09 35	10 32
0	18 01	18 22	18 46	07 43	08 30	09 20	10 14
S 10	17 57	18 18	18 43	07 35	08 18	09 04	09 55
20	17 53	18 15	18 41	07 27	08 05	08 48	09 35
30	17 49	18 13	18 40	07 17	07 51	08 29	09 12
35	17 46	18 12	18 41	07 12	07 42	08 18	08 59
40	17 44	18 11	18 42	07 06	07 33	08 05	08 44
45	17 40	18 10	18 44	06 59	07 22	07 50	08 26
S 50	17 37	18 09	18 46	06 50	07 09	07 32	08 03
52	17 35	18 09	18 48	06 46	07 03	07 24	07 52
54	17 33	18 08	18 49	06 42	06 56	07 14	07 40
56	17 31	18 08	18 51	06 37	06 48	07 04	07 26
58	17 29	18 08	18 54	06 32	06 40	06 51	07 10
S 60	17 26	18 08	18 56	06 26	06 30	06 37	06 50

Day	SUN Eqn. of Time 00^h	SUN Eqn. of Time 12^h	SUN Mer. Pass.	MOON Mer. Pass. Upper	MOON Mer. Pass. Lower	MOON Age	MOON Phase
d	m s	m s	h m	h m	h m	d %	
7	01 52	02 02	11 58	01 33	13 55	18 92	
8	02 13	02 23	11 58	02 18	14 42	19 85	
9	02 33	02 44	11 57	03 07	15 33	20 77	

UT	ARIES	VENUS −3.9		MARS +0.9		JUPITER −2.8		SATURN +1.1	
	GHA	GHA	Dec	GHA	Dec	GHA	Dec	GHA	Dec
d h	° ′	° ′	° ′	° ′	° ′	° ′	° ′	° ′	° ′
10 00 THURSDAY	349 09.8	209 23.3	N16 08.4	248 47.5	N23 26.9	27 28.0	S16 12.0	173 52.1	N 4 10.8
01	4 12.3	224 22.7	07.6	263 48.4	26.9	42 30.8	12.1	188 54.2	10.6
02	19 14.7	239 22.2	06.8	278 49.2	26.8	57 33.5	12.2	203 56.4	10.5
03	34 17.2	254 21.6	06.0	293 50.0	26.8	72 36.2	12.3	218 58.6	10.4
04	49 19.7	269 21.0	05.2	308 50.8	26.7	87 38.9	12.3	234 00.8	10.3
05	64 22.1	284 20.5	04.4	323 51.6	26.6	102 41.6	12.4	249 02.9	10.1
06	79 24.6	299 19.9	N16 03.6	338 52.4	N23 26.6	117 44.3	S16 12.5	264 05.1	N 4 10.0
07	94 27.0	314 19.3	02.8	353 53.2	26.5	132 47.0	12.6	279 07.3	09.9
08	109 29.5	329 18.8	01.9	8 54.0	26.4	147 49.7	12.6	294 09.5	09.8
09	124 32.0	344 18.2	01.1	23 54.9	26.4	162 52.4	12.7	309 11.7	09.7
10	139 34.4	359 17.6	16 00.3	38 55.7	26.3	177 55.1	12.8	324 13.8	09.5
11	154 36.9	14 17.1	15 59.5	53 56.5	26.2	192 57.8	12.9	339 16.0	09.4
12	169 39.4	29 16.5	N15 58.7	68 57.3	N23 26.2	208 00.6	S16 12.9	354 18.2	N 4 09.3
13	184 41.8	44 15.9	57.9	83 58.1	26.1	223 03.3	13.0	9 20.4	09.2
14	199 44.3	59 15.4	57.1	98 58.9	26.0	238 06.0	13.1	24 22.5	09.0
15	214 46.8	74 14.8	56.2	113 59.7	26.0	253 08.7	13.2	39 24.7	08.9
16	229 49.2	89 14.3	55.4	129 00.6	25.9	268 11.4	13.3	54 26.9	08.8
17	244 51.7	104 13.7	54.6	144 01.4	25.8	283 14.1	13.3	69 29.1	08.7
18	259 54.2	119 13.1	N15 53.8	159 02.2	N23 25.8	298 16.8	S16 13.4	84 31.3	N 4 08.6
19	274 56.6	134 12.6	52.9	174 03.0	25.7	313 19.5	13.5	99 33.4	08.4
20	289 59.1	149 12.0	52.1	189 03.8	25.6	328 22.2	13.6	114 35.6	08.3
21	305 01.5	164 11.4	51.3	204 04.6	25.6	343 24.9	13.6	129 37.8	08.2
22	320 04.0	179 10.9	50.5	219 05.5	25.5	358 27.6	13.7	144 40.0	08.1
23	335 06.5	194 10.3	49.7	234 06.3	25.4	13 30.3	13.8	159 42.1	07.9
11 00 FRIDAY	350 08.9	209 09.8	N15 48.8	249 07.1	N23 25.4	28 33.0	S16 13.9	174 44.3	N 4 07.8
01	5 11.4	224 09.2	48.0	264 07.9	25.3	43 35.7	13.9	189 46.5	07.7
02	20 13.9	239 08.6	47.2	279 08.7	25.2	58 38.4	14.0	204 48.7	07.6
03	35 16.3	254 08.1	46.3	294 09.5	25.1	73 41.1	14.1	219 50.9	07.4
04	50 18.8	269 07.5	45.5	309 10.4	25.1	88 43.8	14.2	234 53.0	07.3
05	65 21.3	284 07.0	44.7	324 11.2	25.0	103 46.5	14.2	249 55.2	07.2
06	80 23.7	299 06.4	N15 43.9	339 12.0	N23 24.9	118 49.3	S16 14.3	264 57.4	N 4 07.1
07	95 26.2	314 05.8	43.0	354 12.8	24.9	133 52.0	14.4	279 59.6	07.0
08	110 28.7	329 05.3	42.2	9 13.6	24.8	148 54.7	14.4	295 01.7	06.8
09	125 31.1	344 04.7	41.4	24 14.5	24.7	163 57.4	14.5	310 03.9	06.7
10	140 33.6	359 04.2	40.5	39 15.3	24.6	179 00.1	14.6	325 06.1	06.6
11	155 36.0	14 03.6	39.7	54 16.1	24.6	194 02.8	14.7	340 08.3	06.5
12	170 38.5	29 03.1	N15 38.9	69 16.9	N23 24.5	209 05.5	S16 14.7	355 10.5	N 4 06.3
13	185 41.0	44 02.5	38.0	84 17.7	24.4	224 08.2	14.8	10 12.6	06.2
14	200 43.4	59 01.9	37.2	99 18.6	24.4	239 10.9	14.9	25 14.8	06.1
15	215 45.9	74 01.4	36.3	114 19.4	24.3	254 13.6	15.0	40 17.0	06.0
16	230 48.4	89 00.8	35.5	129 20.2	24.2	269 16.3	15.0	55 19.2	05.9
17	245 50.8	104 00.3	34.7	144 21.0	24.1	284 19.0	15.1	70 21.3	05.7
18	260 53.3	118 59.7	N15 33.8	159 21.9	N23 24.1	299 21.7	S16 15.2	85 23.5	N 4 05.6
19	275 55.8	133 59.2	33.0	174 22.7	24.0	314 24.4	15.3	100 25.7	05.5
20	290 58.2	148 58.6	32.1	189 23.5	23.9	329 27.1	15.3	115 27.9	05.4
21	306 00.7	163 58.1	31.3	204 24.3	23.8	344 29.8	15.4	130 30.1	05.2
22	321 03.2	178 57.5	30.5	219 25.2	23.8	359 32.5	15.5	145 32.2	05.1
23	336 05.6	193 57.0	29.6	234 26.0	23.7	14 35.2	15.6	160 34.4	05.0
12 00 SATURDAY	351 08.1	208 56.4	N15 28.8	249 26.8	N23 23.6	29 37.9	S16 15.6	175 36.6	N 4 04.9
01	6 10.5	223 55.9	27.9	264 27.6	23.5	44 40.6	15.7	190 38.8	04.8
02	21 13.0	238 55.3	27.1	279 28.4	23.5	59 43.3	15.8	205 40.9	04.6
03	36 15.5	253 54.7	26.2	294 29.3	23.4	74 46.0	15.8	220 43.1	04.5
04	51 17.9	268 54.2	25.4	309 30.1	23.3	89 48.7	15.9	235 45.3	04.4
05	66 20.4	283 53.6	24.5	324 30.9	23.2	104 51.4	16.0	250 47.5	04.3
06	81 22.9	298 53.1	N15 23.7	339 31.8	N23 23.2	119 54.1	S16 16.1	265 49.7	N 4 04.1
07	96 25.3	313 52.5	22.8	354 32.6	23.1	134 56.8	16.1	280 51.8	04.0
08	111 27.8	328 52.0	22.0	9 33.4	23.0	149 59.5	16.2	295 54.0	03.9
09	126 30.3	343 51.4	21.1	24 34.2	22.9	165 02.2	16.3	310 56.2	03.8
10	141 32.7	358 50.9	20.3	39 35.1	22.8	180 04.9	16.4	325 58.4	03.7
11	156 35.2	13 50.3	19.4	54 35.9	22.8	195 07.5	16.4	341 00.5	03.5
12	171 37.7	28 49.8	N15 18.6	69 36.7	N23 22.7	210 10.2	S16 16.5	356 02.7	N 4 03.4
13	186 40.1	43 49.2	17.7	84 37.5	22.6	225 12.9	16.6	11 04.9	03.3
14	201 42.6	58 48.7	16.9	99 38.4	22.5	240 15.6	16.6	26 07.1	03.2
15	216 45.0	73 48.1	16.0	114 39.2	22.5	255 18.3	16.7	41 09.3	03.0
16	231 47.5	88 47.6	15.2	129 40.0	22.4	270 21.0	16.8	56 11.4	02.9
17	246 50.0	103 47.1	14.3	144 40.9	22.3	285 23.7	16.9	71 13.6	02.8
18	261 52.4	118 46.5	N15 13.4	159 41.7	N23 22.2	300 26.4	S16 16.9	86 15.8	N 4 02.7
19	276 54.9	133 46.0	12.6	174 42.5	22.1	315 29.1	17.0	101 18.0	02.6
20	291 57.4	148 45.4	11.7	189 43.3	22.1	330 31.8	17.1	116 20.1	02.4
21	306 59.8	163 44.9	10.9	204 44.2	22.0	345 34.5	17.1	131 22.3	02.3
22	322 02.3	178 44.3	10.0	219 45.0	21.9	0 37.2	17.2	146 24.5	02.2
23	337 04.8	193 43.8	09.1	234 45.8	21.8	15 39.9	17.3	161 26.7	02.1
Mer. Pass.	h m 0 39.3	v −0.6	d 0.8	v 0.8	d 0.1	v 2.7	d 0.1	v 2.2	d 0.1

STARS

Name	SHA	Dec
	° ′	° ′
Acamar	315 20.1	S40 15.
Achernar	335 28.1	S57 10.
Acrux	173 13.3	S63 09.
Adhara	255 14.9	S28 58.
Aldebaran	290 52.5	N16 31.
Alioth	166 23.4	N55 54.
Alkaid	153 01.3	N49 16.
Al Na'ir	27 46.6	S46 54.
Alnilam	275 49.2	S 1 11.
Alphard	217 59.1	S 8 41.
Alphecca	126 13.5	N26 41.
Alpheratz	357 46.1	N29 08.
Altair	62 10.7	N 8 53.
Ankaa	353 17.9	S42 14.
Antares	112 29.8	S26 27.
Arcturus	145 58.4	N19 08.
Atria	107 34.2	S69 03.
Avior	234 19.8	S59 32.
Bellatrix	278 35.0	N 6 21.
Betelgeuse	271 04.4	N 7 24.
Canopus	263 57.6	S52 41.
Capella	280 38.6	N46 00.
Deneb	49 33.1	N45 19.
Denebola	182 36.8	N14 31.
Diphda	348 58.3	S17 55.
Dubhe	193 55.5	N61 41.
Elnath	278 16.2	N28 36.
Eltanin	90 47.3	N51 29.
Enif	33 49.6	N 9 55.
Fomalhaut	15 26.6	S29 34.
Gacrux	172 04.8	S57 10.
Gienah	175 55.5	S17 35.
Hadar	148 52.5	S60 25.
Hamal	328 03.7	N23 30.
Kaus Aust.	83 47.4	S34 22.
Kochab	137 20.2	N74 07.
Markab	13 40.8	N15 15.
Menkar	314 17.8	N 4 07.
Menkent	148 11.2	S36 25.
Miaplacidus	221 41.3	S69 45.
Mirfak	308 44.2	N49 53.
Nunki	76 01.6	S26 17.
Peacock	53 23.0	S56 42.
Pollux	243 31.3	N28 00.
Procyon	245 02.8	N 5 12.
Rasalhague	96 09.0	N12 33.
Regulus	207 46.7	N11 55.
Rigel	281 14.7	S 8 11.
Rigil Kent.	139 56.1	S60 52.
Sabik	102 15.7	S15 44.
Schedar	349 43.4	N56 35.
Shaula	96 25.7	S37 06.
Sirius	258 36.3	S16 43.
Spica	158 34.5	S11 12.
Suhail	222 55.0	S43 28.
Vega	80 40.7	N38 47.
Zuben'ubi	137 08.7	S16 05.

	SHA	Mer. Pass.
	° ′	h m
Venus	219 00.8	10 04
Mars	258 58.2	7 23
Jupiter	38 24.1	22 02
Saturn	184 35.4	12 19

UT	SUN GHA	SUN Dec	MOON GHA	v	Dec	d	HP
d h	° ′	° ′	° ′	′	° ′	′	′
0 00	180 43.7	N 4 58.0	302 07.4	8.2	N22 33.5	7.9	57.6
01	195 43.9	57.1	316 34.6	8.1	22 41.4	7.9	57.6
02	210 44.1	56.1	331 01.7	8.0	22 49.3	7.7	57.7
03	225 44.3	. . 55.2	345 28.7	8.0	22 57.0	7.6	57.7
04	240 44.5	54.2	359 55.7	7.8	23 04.6	7.5	57.7
05	255 44.8	53.3	14 22.5	7.7	23 12.1	7.3	57.7
06	270 45.0	N 4 52.3	28 49.2	7.7	N23 19.4	7.2	57.7
07	285 45.2	51.4	43 15.9	7.6	23 26.6	7.1	57.8
08	300 45.4	50.4	57 42.5	7.4	23 33.7	6.9	57.8
09	315 45.6	. . 49.5	72 08.9	7.4	23 40.6	6.8	57.8
10	330 45.9	48.5	86 35.3	7.3	23 47.4	6.7	57.8
11	345 46.1	47.6	101 01.6	7.2	23 54.1	6.6	57.9
12	0 46.3	N 4 46.6	115 27.8	7.1	N24 00.7	6.4	57.9
13	15 46.5	45.7	129 53.9	7.0	24 07.1	6.2	57.9
14	30 46.7	44.7	144 19.9	7.0	24 13.3	6.2	57.9
15	45 47.0	. . 43.8	158 45.9	6.8	24 19.5	5.9	58.0
16	60 47.2	42.8	173 11.7	6.8	24 25.4	5.9	58.0
17	75 47.4	41.9	187 37.5	6.7	24 31.3	5.7	58.0
18	90 47.6	N 4 40.9	202 03.2	6.6	N24 37.0	5.5	58.0
19	105 47.8	40.0	216 28.8	6.5	24 42.5	5.4	58.1
20	120 48.0	39.0	230 54.3	6.5	24 47.9	5.3	58.1
21	135 48.3	. . 38.1	245 19.8	6.3	24 53.2	5.1	58.1
22	150 48.5	37.1	259 45.1	6.3	24 58.3	4.9	58.1
23	165 48.7	36.2	274 10.4	6.2	25 03.2	4.9	58.1
1 00	180 48.9	N 4 35.2	288 35.6	6.1	N25 08.1	4.6	58.2
01	195 49.1	34.3	303 00.7	6.1	25 12.7	4.5	58.2
02	210 49.4	33.3	317 25.8	5.9	25 17.2	4.4	58.2
03	225 49.6	. . 32.4	331 50.7	5.9	25 21.6	4.1	58.2
04	240 49.8	31.4	346 15.6	5.9	25 25.7	4.1	58.3
05	255 50.0	30.5	0 40.5	5.7	25 29.8	3.9	58.3
06	270 50.2	N 4 29.5	15 05.2	5.7	N25 33.7	3.7	58.3
07	285 50.5	28.6	29 29.9	5.6	25 37.4	3.5	58.3
08	300 50.7	27.6	43 54.5	5.6	25 40.9	3.4	58.4
09	315 50.9	. . 26.7	58 19.1	5.5	25 44.3	3.3	58.4
10	330 51.1	25.7	72 43.6	5.4	25 47.6	3.1	58.4
11	345 51.3	24.8	87 08.0	5.4	25 50.7	2.9	58.4
12	0 51.6	N 4 23.8	101 32.4	5.3	N25 53.6	2.7	58.4
13	15 51.8	22.9	115 56.7	5.2	25 56.3	2.6	58.5
14	30 52.0	21.9	130 20.9	5.2	25 58.9	2.4	58.5
15	45 52.2	. . 21.0	144 45.1	5.2	26 01.3	2.3	58.5
16	60 52.4	20.0	159 09.3	5.0	26 03.6	2.1	58.5
17	75 52.7	19.1	173 33.3	5.1	26 05.7	1.9	58.6
18	90 52.9	N 4 18.1	187 57.4	4.9	N26 07.6	1.7	58.6
19	105 53.1	17.2	202 21.3	5.0	26 09.3	1.6	58.6
20	120 53.3	16.2	216 45.3	4.9	26 10.9	1.4	58.6
21	135 53.6	. . 15.3	231 09.2	4.8	26 12.3	1.3	58.7
22	150 53.8	14.3	245 33.0	4.8	26 13.6	1.0	58.7
23	165 54.0	13.4	259 56.8	4.8	26 14.6	0.9	58.7
2 00	180 54.2	N 4 12.4	274 20.6	4.7	N26 15.5	0.7	58.7
01	195 54.4	11.4	288 44.3	4.7	26 16.2	0.6	58.7
02	210 54.7	10.5	303 08.0	4.6	26 16.8	0.4	58.8
03	225 54.9	. . 09.5	317 31.6	4.6	26 17.2	0.2	58.8
04	240 55.1	08.6	331 55.2	4.6	26 17.4	0.0	58.8
05	255 55.3	07.6	346 18.8	4.5	26 17.4	0.1	58.8
06	270 55.5	N 4 06.7	0 42.3	4.6	N26 17.3	0.4	58.9
07	285 55.8	05.7	15 05.9	4.5	26 16.9	0.5	58.9
08	300 56.0	04.8	29 29.4	4.4	26 16.4	0.6	58.9
09	315 56.2	. . 03.8	43 52.8	4.5	26 15.8	0.9	58.9
10	330 56.4	02.9	58 16.3	4.4	26 14.9	1.0	59.0
11	345 56.6	01.9	72 39.7	4.5	26 13.9	1.2	59.0
12	0 56.9	N 4 00.9	87 03.2	4.4	N26 12.7	1.4	59.0
13	15 57.1	4 00.0	101 26.6	4.3	26 11.3	1.6	59.0
14	30 57.3	3 59.0	115 49.9	4.4	26 09.7	1.7	59.0
15	45 57.5	. . 58.1	130 13.3	4.4	26 08.0	1.9	59.1
16	60 57.7	57.1	144 36.7	4.4	26 06.1	2.1	59.1
17	75 58.0	56.2	159 00.1	4.3	26 04.0	2.3	59.1
18	90 58.2	N 3 55.2	173 23.4	4.4	N26 01.7	2.4	59.1
19	105 58.4	54.3	187 46.8	4.3	25 59.3	2.6	59.1
20	120 58.6	53.3	202 10.1	4.4	25 56.7	2.8	59.2
21	135 58.9	. . 52.3	216 33.5	4.3	25 53.9	3.0	59.2
22	150 59.1	51.4	230 56.8	4.4	25 50.9	3.1	59.2
23	165 59.3	50.4	245 20.2	4.4	N25 47.8	3.4	59.2
	SD 15.9	d 1.0	SD		15.8	15.9	16.1

Lat.	Twilight Naut.	Twilight Civil	Sunrise	Moonrise 10	11	12	13
°	h m	h m	h m	h m	h m	h m	h m
N 72	01 29	03 34	04 49	▭	▭	▭	▭
N 70	02 13	03 50	04 57	▭	▭	▭	▭
68	02 41	04 03	05 03	▭	▭	▭	▭
66	03 02	04 13	05 08	▭	▭	▭	19 54
64	03 18	04 22	05 12	17 56	17 39	18 56	21 16
62	03 31	04 29	05 15	18 37	19 04	20 11	21 52
60	03 42	04 36	05 19	19 05	19 41	20 47	22 19
N 58	03 51	04 41	05 21	19 27	20 08	21 12	22 39
56	03 59	04 46	05 24	19 45	20 28	21 33	22 56
54	04 06	04 50	05 26	20 00	20 46	21 50	23 10
52	04 12	04 54	05 28	20 13	21 01	22 05	23 23
50	04 18	04 57	05 30	20 25	21 13	22 17	23 34
45	04 29	05 04	05 34	20 49	21 40	22 43	23 57
N 40	04 38	05 10	05 37	21 08	22 01	23 04	24 15
35	04 44	05 14	05 40	21 24	22 19	23 21	24 31
30	04 50	05 18	05 42	21 39	22 34	23 36	24 44
20	04 59	05 24	05 47	22 03	23 00	24 01	00 01
N 10	05 05	05 29	05 50	22 24	23 22	24 23	00 23
0	05 09	05 33	05 53	22 43	23 43	24 44	00 44
S 10	05 11	05 35	05 56	23 03	24 03	00 03	01 04
20	05 12	05 38	06 00	23 24	24 26	00 26	01 26
30	05 12	05 39	06 03	23 49	24 52	00 52	01 51
35	05 11	05 40	06 05	24 03	00 03	01 07	02 06
40	05 09	05 40	06 07	24 20	00 20	01 25	02 23
45	05 07	05 41	06 10	24 40	00 40	01 46	02 43
S 50	05 03	05 41	06 13	25 06	01 06	02 13	03 09
52	05 02	05 41	06 14	00 01	01 18	02 27	03 21
54	05 00	05 41	06 16	00 13	01 33	02 42	03 36
56	04 57	05 40	06 17	00 26	01 49	03 00	03 52
58	04 55	05 40	06 19	00 42	02 09	03 22	04 12
S 60	04 52	05 40	06 21	01 01	02 34	03 50	04 37

Lat.	Sunset	Twilight Civil	Twilight Naut.	Moonset 10	11	12	13
°	h m	h m	h m	h m	h m	h m	h m
N 72	19 01	20 15	22 14	▭	▭	▭	▭
N 70	18 54	20 00	21 34	▭	▭	▭	▭
68	18 48	19 47	21 08	▭	▭	▭	▭
66	18 44	19 37	20 48	▭	▭	▭	19 13
64	18 40	19 29	20 32	14 52	17 12	18 03	17 51
62	18 36	19 22	20 19	14 12	15 47	16 48	17 13
60	18 33	19 16	20 09	13 45	15 10	16 12	16 47
N 58	18 30	19 11	20 00	13 23	14 44	15 46	16 26
56	18 28	19 06	19 52	13 06	14 23	15 25	16 08
54	18 26	19 02	19 45	12 51	14 06	15 08	15 53
52	18 24	18 58	19 39	12 38	13 51	14 53	15 41
50	18 22	18 55	19 34	12 27	13 38	14 40	15 29
45	18 18	18 48	19 23	12 03	13 12	14 14	15 05
N 40	18 15	18 43	19 15	11 44	12 51	13 53	14 46
35	18 13	18 38	19 08	11 29	12 34	13 35	14 30
30	18 10	18 34	19 02	11 15	12 19	13 20	14 16
20	18 06	18 28	18 54	10 52	11 54	12 55	13 53
N 10	18 03	18 24	18 48	10 32	11 32	12 32	13 32
0	18 00	18 21	18 45	10 14	11 11	12 12	13 13
S 10	17 57	18 18	18 42	09 55	10 51	11 51	12 53
20	17 54	18 16	18 41	09 35	10 29	11 28	12 32
30	17 50	18 14	18 42	09 12	10 04	11 03	12 08
35	17 48	18 14	18 43	08 59	09 49	10 47	11 53
40	17 46	18 13	18 45	08 44	09 31	10 29	11 37
45	17 44	18 13	18 47	08 26	09 11	10 08	11 17
S 50	17 41	18 13	18 51	08 03	08 45	09 41	10 52
52	17 40	18 13	18 53	07 52	08 32	09 28	10 39
54	17 38	18 14	18 55	07 40	08 18	09 12	10 25
56	17 37	18 14	18 57	07 26	08 01	08 54	10 09
58	17 35	18 14	19 00	07 10	07 41	08 33	09 50
S 60	17 33	18 15	19 03	06 50	07 15	08 05	09 25

Day	SUN Eqn. of Time 00^h	12^h	Mer. Pass.	MOON Mer. Pass. Upper	Lower	Age	Phase
d	m s	m s	h m	h m	h m	d	%
10	02 54	03 05	11 57	04 00	16 28	21	67
11	03 15	03 26	11 57	04 57	17 27	22	57 (phase symbol)
12	03 36	03 47	11 56	05 57	18 28	23	46

UT d h	ARIES GHA ° ′	VENUS −3.9 GHA ° ′	VENUS Dec ° ′	MARS +0.9 GHA ° ′	MARS Dec ° ′	JUPITER −2.8 GHA ° ′	JUPITER Dec ° ′	SATURN +1.1 GHA ° ′	SATURN Dec ° ′	STARS Name	SHA ° ′	Dec °
13 00	352 07.2	208 43.2	N15 08.3	249 46.7	N23 21.7	30 42.6	S16 17.4	176 28.8	N 4 01.9	Acamar	315 20.1	S40 15
01	7 09.7	223 42.7	07.4	264 47.5	21.7	45 45.3	17.4	191 31.0	01.8	Achernar	335 28.1	S57 10
02	22 12.1	238 42.1	06.5	279 48.3	21.6	60 48.0	17.5	206 33.2	01.7	Acrux	173 13.3	S63 09
03	37 14.6	253 41.6	05.7	294 49.2	21.5	75 50.7	17.6	221 35.4	01.6	Adhara	255 14.9	S28 58
04	52 17.1	268 41.0	04.8	309 50.0	21.4	90 53.3	17.6	236 37.6	01.5	Aldebaran	290 52.5	N16 31
05	67 19.5	283 40.5	04.0	324 50.8	21.3	105 56.0	17.7	251 39.7	01.3			
06	82 22.0	298 40.0	N15 03.1	339 51.7	N23 21.2	120 58.7	S16 17.8	266 41.9	N 4 01.2	Alioth	166 23.4	N55 54
07	97 24.5	313 39.4	02.2	354 52.5	21.2	136 01.4	17.8	281 44.1	01.1	Alkaid	153 01.3	N49 16
08	112 26.9	328 38.9	01.3	9 53.3	21.1	151 04.1	17.9	296 46.3	01.0	Al Na'ir	27 46.6	S46 54
S 09	127 29.4	343 38.3	15 00.5	24 54.2	21.0	166 06.8	18.0	311 48.4	00.8	Alnilam	275 49.2	S 1 11
U 10	142 31.9	358 37.8	14 59.6	39 55.0	20.9	181 09.5	18.1	326 50.6	00.7	Alphard	217 59.1	S 8 41
N 11	157 34.3	13 37.2	58.7	54 55.8	20.8	196 12.2	18.1	341 52.8	00.6			
D 12	172 36.8	28 36.7	N14 57.9	69 56.7	N23 20.7	211 14.9	S16 18.2	356 55.0	N 4 00.5	Alphecca	126 13.5	N26 41
A 13	187 39.3	43 36.2	57.0	84 57.5	20.7	226 17.6	18.3	11 57.2	00.3	Alpheratz	357 46.1	N29 08
Y 14	202 41.7	58 35.6	56.1	99 58.3	20.6	241 20.3	18.3	26 59.3	00.2	Altair	62 10.8	N 8 53
15	217 44.2	73 35.1	55.2	114 59.2	20.5	256 22.9	18.4	42 01.5	00.1	Ankaa	353 17.9	S42 14
16	232 46.6	88 34.5	54.4	130 00.0	20.4	271 25.6	18.5	57 03.7	4 00.0	Antares	112 29.8	S26 27
17	247 49.1	103 34.0	53.5	145 00.8	20.3	286 28.3	18.5	72 05.9	3 59.9			
18	262 51.6	118 33.5	N14 52.6	160 01.7	N23 20.2	301 31.0	S16 18.6	87 08.0	N 3 59.7	Arcturus	145 58.4	N19 08
19	277 54.0	133 32.9	51.7	175 02.5	20.1	316 33.7	18.7	102 10.2	59.6	Atria	107 34.2	S69 03
20	292 56.5	148 32.4	50.9	190 03.3	20.1	331 36.4	18.7	117 12.4	59.5	Avior	234 19.7	S59 32
21	307 59.0	163 31.8	50.0	205 04.2	20.0	346 39.1	18.8	132 14.6	59.4	Bellatrix	278 35.0	N 6 21
22	323 01.4	178 31.3	49.1	220 05.0	19.9	1 41.8	18.9	147 16.7	59.2	Betelgeuse	271 04.3	N 7 24
23	338 03.9	193 30.8	48.2	235 05.9	19.8	16 44.5	19.0	162 18.9	59.1			
14 00	353 06.4	208 30.2	N14 47.3	250 06.7	N23 19.7	31 47.1	S16 19.0	177 21.1	N 3 59.0	Canopus	263 57.5	S52 41
01	8 08.8	223 29.7	46.5	265 07.5	19.6	46 49.8	19.1	192 23.3	58.9	Capella	280 38.5	N46 00
02	23 11.3	238 29.2	45.6	280 08.4	19.5	61 52.5	19.2	207 25.5	58.8	Deneb	49 33.1	N45 19
03	38 13.8	253 28.6	44.7	295 09.2	19.4	76 55.2	19.2	222 27.6	58.6	Denebola	182 36.8	N14 31
04	53 16.2	268 28.1	43.8	310 10.1	19.4	91 57.9	19.3	237 29.8	58.5	Diphda	348 58.3	S17 55
05	68 18.7	283 27.5	42.9	325 10.9	19.3	107 00.6	19.4	252 32.0	58.4			
06	83 21.1	298 27.0	N14 42.0	340 11.7	N23 19.2	122 03.3	S16 19.4	267 34.2	N 3 58.3	Dubhe	193 55.4	N61 41
07	98 23.6	313 26.5	41.2	355 12.6	19.1	137 05.9	19.5	282 36.3	58.1	Elnath	278 16.1	N28 37
08	113 26.1	328 25.9	40.3	10 13.4	19.0	152 08.6	19.6	297 38.5	58.0	Eltanin	90 47.4	N51 29
M 09	128 28.5	343 25.4	39.4	25 14.3	18.9	167 11.3	19.6	312 40.7	57.9	Enif	33 49.6	N 9 55
O 10	143 31.0	358 24.9	38.5	40 15.1	18.8	182 14.0	19.7	327 42.9	57.8	Fomalhaut	15 26.6	S29 34
N 11	158 33.5	13 24.3	37.6	55 15.9	18.7	197 16.7	19.8	342 45.0	57.7			
D 12	173 35.9	28 23.8	N14 36.7	70 16.8	N23 18.6	212 19.4	S16 19.8	357 47.2	N 3 57.5	Gacrux	172 04.8	S57 10
A 13	188 38.4	43 23.3	35.8	85 17.6	18.6	227 22.1	19.9	12 49.4	57.4	Gienah	175 55.5	S17 35
Y 14	203 40.9	58 22.7	34.9	100 18.5	18.5	242 24.7	20.0	27 51.6	57.3	Hadar	148 52.5	S60 25
15	218 43.3	73 22.2	34.0	115 19.3	18.4	257 27.4	20.0	42 53.8	57.2	Hamal	328 03.7	N23 30
16	233 45.8	88 21.7	33.2	130 20.1	18.3	272 30.1	20.1	57 55.9	57.0	Kaus Aust.	83 47.4	S34 22
17	248 48.3	103 21.1	32.3	145 21.0	18.2	287 32.8	20.2	72 58.1	56.9			
18	263 50.7	118 20.6	N14 31.4	160 21.8	N23 18.1	302 35.5	S16 20.2	88 00.3	N 3 56.8	Kochab	137 20.2	N74 07
19	278 53.2	133 20.1	30.5	175 22.7	18.0	317 38.1	20.3	103 02.5	56.7	Markab	13 40.8	N15 15
20	293 55.6	148 19.5	29.6	190 23.5	17.9	332 40.8	20.4	118 04.6	56.6	Menkar	314 17.8	N 4 07
21	308 58.1	163 19.0	28.7	205 24.4	17.8	347 43.5	20.4	133 06.8	56.4	Menkent	148 11.2	S36 25
22	324 00.6	178 18.5	27.8	220 25.2	17.7	2 46.2	20.5	148 09.0	56.3	Miaplacidus	221 41.3	S69 45
23	339 03.0	193 17.9	26.9	235 26.1	17.6	17 48.9	20.6	163 11.2	56.2			
15 00	354 05.5	208 17.4	N14 26.0	250 26.9	N23 17.6	32 51.6	S16 20.6	178 13.3	N 3 56.1	Mirfak	308 44.2	N49 53
01	9 08.0	223 16.9	25.1	265 27.7	17.5	47 54.2	20.7	193 15.5	55.9	Nunki	76 01.6	S26 17
02	24 10.4	238 16.3	24.2	280 28.6	17.4	62 56.9	20.8	208 17.7	55.8	Peacock	53 23.0	S56 42
03	39 12.9	253 15.8	23.3	295 29.4	17.3	77 59.6	20.8	223 19.9	55.7	Pollux	243 31.3	N28 00
04	54 15.4	268 15.3	22.4	310 30.3	17.2	93 02.3	20.9	238 22.1	55.6	Procyon	245 02.8	N 5 12
05	69 17.8	283 14.8	21.5	325 31.1	17.1	108 05.0	21.0	253 24.2	55.5			
06	84 20.3	298 14.2	N14 20.6	340 32.0	N23 17.0	123 07.6	S16 21.0	268 26.4	N 3 55.3	Rasalhague	96 09.0	N12 33
07	99 22.7	313 13.7	19.7	355 32.8	16.9	138 10.3	21.1	283 28.6	55.2	Regulus	207 46.7	N11 55
T 08	114 25.2	328 13.2	18.8	10 33.7	16.8	153 13.0	21.2	298 30.8	55.1	Rigel	281 14.7	S 8 11
U 09	129 27.7	343 12.6	17.9	25 34.5	16.7	168 15.7	21.2	313 32.9	55.0	Rigil Kent.	139 56.2	S60 52
E 10	144 30.1	358 12.1	17.0	40 35.4	16.6	183 18.4	21.3	328 35.1	54.8	Sabik	102 15.7	S15 44
S 11	159 32.6	13 11.6	16.1	55 36.2	16.5	198 21.0	21.4	343 37.3	54.7			
D 12	174 35.1	28 11.1	N14 15.2	70 37.1	N23 16.4	213 23.7	S16 21.4	358 39.5	N 3 54.6	Schedar	349 43.4	N56 35
A 13	189 37.5	43 10.5	14.2	85 37.9	16.3	228 26.4	21.5	13 41.6	54.5	Shaula	96 25.7	S37 06
Y 14	204 40.0	58 10.0	13.3	100 38.8	16.2	243 29.1	21.6	28 43.8	54.4	Sirius	258 36.3	S16 43
15	219 42.5	73 09.5	12.4	115 39.6	16.1	258 31.7	21.6	43 46.0	54.2	Spica	158 34.5	S11 12
16	234 44.9	88 09.0	11.5	130 40.5	16.0	273 34.4	21.7	58 48.2	54.1	Suhail	222 54.9	S43 28
17	249 47.4	103 08.4	10.6	145 41.3	15.9	288 37.1	21.8	73 50.4	54.0			
18	264 49.9	118 07.9	N14 09.7	160 42.2	N23 15.8	303 39.8	S16 21.8	88 52.5	N 3 53.9	Vega	80 40.7	N38 47
19	279 52.3	133 07.4	08.8	175 43.0	15.7	318 42.4	21.9	103 54.7	53.7	Zuben'ubi	137 08.7	S16 05
20	294 54.8	148 06.9	07.9	190 43.9	15.6	333 45.1	22.0	118 56.9	53.6			
21	309 57.2	163 06.3	07.0	205 44.7	15.5	348 47.8	22.0	133 59.1	53.5			
22	324 59.7	178 05.8	06.0	220 45.6	15.5	3 50.5	22.1	149 01.2	53.4			
23	340 02.2	193 05.3	05.1	235 46.4	15.4	18 53.1	22.1	164 03.4	53.3			
Mer. Pass.	h m 0 27.5	v −0.5	d 0.9	v 0.8	d 0.1	v 2.7	d 0.1	v 2.2	d 0.1			

	SHA ° ′	Mer. Pass. h m
Venus	215 23.9	10 06
Mars	257 00.3	7 19
Jupiter	38 40.8	21 49
Saturn	184 14.7	12 09

UT	SUN GHA	SUN Dec	MOON GHA	v	MOON Dec	d	HP
d h	° ′	° ′	° ′	′	° ′	′	′
13 00	180 59.5	N 3 49.5	259 43.6	4.3	N25 44.4	3.5	59.3
01	195 59.7	48.5	274 06.9	4.4	25 40.9	3.6	59.3
02	211 00.0	47.6	288 30.3	4.4	25 37.3	3.9	59.3
03	226 00.2	· · 46.6	302 53.7	4.4	25 33.4	4.0	59.3
04	241 00.4	45.7	317 17.1	4.5	25 29.4	4.2	59.3
05	256 00.6	44.7	331 40.6	4.4	25 25.2	4.4	59.4
06	271 00.8	N 3 43.7	346 04.0	4.5	N25 20.8	4.6	59.4
07	286 01.1	42.8	0 27.5	4.5	25 16.2	4.7	59.4
S 08	301 01.3	41.8	14 51.0	4.5	25 11.5	4.9	59.4
U 09	316 01.5	· · 40.9	29 14.5	4.5	25 06.6	5.1	59.4
N 10	331 01.7	39.9	43 38.0	4.6	25 01.5	5.2	59.5
D 11	346 02.0	38.9	58 01.6	4.6	24 56.3	5.4	59.5
A 12	1 02.2	N 3 38.0	72 25.2	4.6	N24 50.9	5.6	59.5
Y 13	16 02.4	37.0	86 48.8	4.7	24 45.3	5.8	59.5
14	31 02.6	36.1	101 12.5	4.7	24 39.5	5.9	59.5
15	46 02.8	· · 35.1	115 36.2	4.7	24 33.6	6.1	59.6
16	61 03.1	34.2	129 59.9	4.8	24 27.5	6.3	59.6
17	76 03.3	33.2	144 23.7	4.8	24 21.2	6.4	59.6
18	91 03.5	N 3 32.2	158 47.5	4.8	N24 14.8	6.6	59.6
19	106 03.7	31.3	173 11.3	4.9	24 08.2	6.8	59.6
20	121 03.9	30.3	187 35.2	5.0	24 01.4	6.9	59.6
21	136 04.2	· · 29.4	201 59.2	5.0	23 54.5	7.1	59.7
22	151 04.4	28.4	216 23.2	5.0	23 47.4	7.3	59.7
23	166 04.6	27.4	230 47.2	5.0	23 40.1	7.4	59.7
14 00	181 04.8	N 3 26.5	245 11.2	5.2	N23 32.7	7.6	59.7
01	196 05.1	25.5	259 35.4	5.1	23 25.1	7.7	59.7
02	211 05.3	24.6	273 59.5	5.3	23 17.4	7.9	59.7
03	226 05.5	· · 23.6	288 23.8	5.2	23 09.5	8.1	59.8
04	241 05.7	22.6	302 48.0	5.4	23 01.4	8.2	59.8
05	256 05.9	21.7	317 12.4	5.4	22 53.2	8.4	59.8
06	271 06.2	N 3 20.7	331 36.8	5.4	N22 44.8	8.5	59.8
07	286 06.4	19.8	346 01.2	5.5	22 36.3	8.7	59.8
08	301 06.6	18.8	0 25.7	5.6	22 27.6	8.8	59.8
M 09	316 06.8	· · 17.8	14 50.3	5.6	22 18.8	9.0	59.9
O 10	331 07.1	16.9	29 14.9	5.7	22 09.8	9.1	59.9
N 11	346 07.3	15.9	43 39.6	5.7	22 00.7	9.3	59.9
D 12	1 07.5	N 3 15.0	58 04.3	5.8	N21 51.4	9.4	59.9
A 13	16 07.7	14.0	72 29.1	5.9	21 42.0	9.6	59.9
Y 14	31 07.9	13.0	86 54.0	5.9	21 32.4	9.8	59.9
15	46 08.2	· · 12.1	101 18.9	6.0	21 22.6	9.8	59.9
16	61 08.4	11.1	115 43.9	6.0	21 12.8	10.0	60.0
17	76 08.6	10.2	130 08.9	6.2	21 02.8	10.2	60.0
18	91 08.8	N 3 09.2	144 34.1	6.1	N20 52.6	10.3	60.0
19	106 09.1	08.2	158 59.2	6.3	20 42.3	10.4	60.0
20	121 09.3	07.3	173 24.5	6.3	20 31.9	10.6	60.0
21	136 09.5	· · 06.3	187 49.8	6.4	20 21.3	10.7	60.0
22	151 09.7	05.4	202 15.2	6.5	20 10.6	10.8	60.0
23	166 09.9	04.4	216 40.7	6.5	19 59.8	11.0	60.0
15 00	181 10.2	N 3 03.4	231 06.2	6.6	N19 48.8	11.1	60.0
01	196 10.4	02.5	245 31.8	6.7	19 37.7	11.2	60.1
02	211 10.6	01.5	259 57.5	6.7	19 26.5	11.4	60.1
03	226 10.8	3 00.5	274 23.2	6.8	19 15.1	11.4	60.1
04	241 11.1	2 59.6	288 49.0	6.9	19 03.7	11.6	60.1
05	256 11.3	58.6	303 14.9	6.9	18 52.1	11.8	60.1
06	271 11.5	N 2 57.7	317 40.8	7.1	N18 40.3	11.8	60.1
T 07	286 11.7	56.7	332 06.9	7.0	18 28.5	12.0	60.1
08	301 11.9	55.7	346 32.9	7.2	18 16.5	12.1	60.1
U 09	316 12.2	· · 54.8	0 59.1	7.2	18 04.4	12.2	60.1
E 10	331 12.4	53.8	15 25.3	7.3	17 52.2	12.3	60.1
S 11	346 12.6	52.8	29 51.6	7.4	17 39.9	12.5	60.2
D 12	1 12.8	N 2 51.9	44 18.0	7.5	N17 27.4	12.5	60.2
A 13	16 13.1	50.9	58 44.5	7.5	17 14.9	12.7	60.2
Y 14	31 13.3	50.0	73 11.0	7.6	17 02.2	12.7	60.2
15	46 13.5	· · 49.0	87 37.6	7.6	16 49.5	12.9	60.2
16	61 13.7	48.0	102 04.2	7.8	16 36.6	13.0	60.2
17	76 13.9	47.1	116 31.0	7.8	16 23.6	13.1	60.2
18	91 14.2	N 2 46.1	130 57.8	7.8	N16 10.5	13.1	60.2
19	106 14.4	45.1	145 24.6	8.0	15 57.4	13.3	60.2
20	121 14.6	44.2	159 51.6	8.0	15 44.1	13.4	60.2
21	136 14.8	· · 43.2	174 18.6	8.1	15 30.7	13.5	60.2
22	151 15.1	42.2	188 45.7	8.1	15 17.2	13.6	60.2
23	166 15.3	41.3	203 12.8	8.2	N15 03.6	13.6	60.2
	SD 15.9	*d* 1.0	SD		16.2	16.3	16.4

Lat.	Twilight Naut.	Twilight Civil	Sunrise	Moonrise 13	14	15	16
°	h m	h m	h m	h m	h m	h m	h m
N 72	02 00	03 50	05 03	▭	▭	23 54	26 42
N 70	02 33	04 04	05 09	▭	▭	▭	00 28
68	02 57	04 15	05 13	▭	22 15	24 53	00 53
66	03 15	04 24	05 17	19 54	22 57	25 11	01 11
64	03 29	04 31	05 20	21 16	23 25	25 26	01 26
62	03 41	04 38	05 23	21 52	23 46	25 39	01 39
60	03 51	04 43	05 26	22 19	24 03	00 03	01 49
N 58	03 59	04 48	05 28	22 39	24 17	00 17	01 58
56	04 06	04 52	05 30	22 56	24 30	00 30	02 06
54	04 12	04 56	05 31	23 10	24 40	00 40	02 13
52	04 18	04 59	05 33	23 23	24 50	00 50	02 19
50	04 23	05 02	05 34	23 34	24 58	00 58	02 25
45	04 33	05 08	05 37	23 57	25 16	01 16	02 37
N 40	04 41	05 13	05 40	24 15	00 15	01 31	02 47
35	04 47	05 17	05 42	24 31	00 31	01 43	02 56
30	04 52	05 20	05 44	24 44	00 44	01 54	03 03
20	04 59	05 25	05 47	00 01	01 06	02 12	03 16
N 10	05 04	05 29	05 50	00 23	01 26	02 28	03 27
0	05 08	05 32	05 52	00 44	01 44	02 42	03 38
S 10	05 09	05 34	05 55	01 04	02 02	02 57	03 48
20	05 09	05 35	05 57	01 26	02 22	03 13	03 59
30	05 08	05 36	05 59	01 51	02 44	03 31	04 12
35	05 06	05 36	06 01	02 06	02 57	03 41	04 19
40	05 04	05 36	06 02	02 23	03 12	03 53	04 27
45	05 01	05 35	06 04	02 43	03 30	04 07	04 37
S 50	04 57	05 34	06 06	03 09	03 52	04 24	04 48
52	04 55	05 34	06 07	03 21	04 02	04 31	04 54
54	04 52	05 33	06 08	03 36	04 14	04 40	05 00
56	04 49	05 33	06 09	03 52	04 27	04 50	05 06
58	04 46	05 32	06 11	04 12	04 43	05 01	05 13
S 60	04 42	05 31	06 12	04 37	05 01	05 14	05 21

Lat.	Sunset	Twilight Civil	Twilight Naut.	Moonset 13	14	15	16
°	h m	h m	h m	h m	h m	h m	h m
N 72	18 45	19 57	21 44	▭	▭	19 19	18 24
N 70	18 40	19 44	21 12	▭	▭	18 43	18 08
68	18 36	19 33	20 50	▭	18 56	18 17	17 55
66	18 32	19 25	20 33	19 13	18 14	17 57	17 45
64	18 29	19 18	20 19	17 51	17 45	17 40	17 36
62	18 26	19 11	20 08	17 13	17 23	17 27	17 28
60	18 24	19 06	19 58	16 47	17 05	17 15	17 21
N 58	18 22	19 02	19 50	16 26	16 50	17 05	17 15
56	18 20	18 58	19 43	16 08	16 37	16 56	17 10
54	18 18	18 54	19 37	15 53	16 25	16 48	17 05
52	18 17	18 51	19 32	15 41	16 15	16 41	17 01
50	18 16	18 48	19 27	15 29	16 06	16 34	16 57
45	18 13	18 42	19 17	15 05	15 47	16 21	16 48
N 40	18 10	18 38	19 10	14 46	15 31	16 09	16 41
35	18 08	18 34	19 03	14 30	15 18	15 59	16 35
30	18 07	18 31	18 59	14 16	15 06	15 50	16 29
20	18 04	18 26	18 51	13 53	14 46	15 35	16 20
N 10	18 01	18 22	18 47	13 32	14 29	15 22	16 11
0	17 59	18 19	18 43	13 13	14 12	15 09	16 03
S 10	17 57	18 18	18 42	12 53	13 56	14 56	15 55
20	17 54	18 16	18 42	12 32	13 38	14 43	15 46
30	17 52	18 16	18 44	12 08	13 17	14 27	15 36
35	17 51	18 16	18 45	11 53	13 05	14 18	15 30
40	17 49	18 16	18 48	11 37	12 51	14 07	15 23
45	17 48	18 17	18 51	11 17	12 34	13 55	15 16
S 50	17 46	18 18	18 55	10 52	12 13	13 39	15 06
52	17 45	18 18	18 58	10 39	12 03	13 32	15 02
54	17 44	18 19	19 00	10 25	11 52	13 24	14 57
56	17 43	18 20	19 03	10 09	11 39	13 15	14 51
58	17 41	18 21	19 06	09 50	11 24	13 05	14 45
S 60	17 40	18 22	19 10	09 25	11 06	12 53	14 38

Day	SUN Eqn. of Time 00^h	SUN Eqn. of Time 12^h	SUN Mer. Pass.	MOON Mer. Pass. Upper	MOON Mer. Pass. Lower	Age	Phase
d	m s	m s	h m	h m	h m	d %	
13	03 58	04 08	11 56	06 58	19 28	24 34	
14	04 19	04 30	11 56	07 58	20 27	25 24	
15	04 40	04 51	11 55	08 56	21 24	26 15	

UT	ARIES	VENUS −3.9		MARS +0.9		JUPITER −2.7		SATURN +1.1		STARS		
	GHA	GHA	Dec	GHA	Dec	GHA	Dec	GHA	Dec	Name	SHA	Dec
d h	° ′	° ′	° ′	° ′	° ′	° ′	° ′	° ′	° ′		° ′	°
16 00	355 04.6	208 04.8	N14 04.2	250 47.3	N23 15.3	33 55.8	S16 22.2	179 05.6	N 3 53.1	Acamar	315 20.1	S40 15
01	10 07.1	223 04.2	03.3	265 48.1	15.2	48 58.5	22.3	194 07.8	53.0	Achernar	335 28.1	S57 10.
02	25 09.6	238 03.7	02.4	280 49.0	15.1	64 01.2	22.3	209 09.9	52.9	Acrux	173 13.4	S63 09.
03	40 12.0	253 03.2	. . 01.5	295 49.8	. . 15.0	79 03.8	. . 22.4	224 12.1	. . 52.8	Adhara	255 14.8	S28 58
04	55 14.5	268 02.7	14 00.6	310 50.7	14.9	94 06.5	22.5	239 14.3	52.6	Aldebaran	290 52.5	N16 31
05	70 17.0	283 02.2	13 59.6	325 51.5	14.8	109 09.2	22.5	254 16.5	52.5			
06	85 19.4	298 01.6	N13 58.7	340 52.4	N23 14.7	124 11.9	S16 22.6	269 18.7	N 3 52.4	Alioth	166 23.4	N55 54.
07	100 21.9	313 01.1	57.8	355 53.2	14.6	139 14.5	22.7	284 20.8	52.3	Alkaid	153 01.3	N49 16.
08	115 24.4	328 00.6	56.9	10 54.1	14.5	154 17.2	22.7	299 23.0	52.2	Al Na'ir	27 46.6	S46 54
09	130 26.8	343 00.1	. . 55.9	25 55.0	. . 14.4	169 19.9	. . 22.8	314 25.2	. . 52.0	Alnilam	275 49.2	S 1 11.
10	145 29.3	357 59.6	55.0	40 55.8	14.3	184 22.6	22.8	329 27.4	51.9	Alphard	217 59.1	S 8 41.
11	160 31.7	12 59.0	54.1	55 56.7	14.2	199 25.2	22.9	344 29.5	51.8			
12	175 34.2	27 58.5	N13 53.2	70 57.5	N23 14.1	214 27.9	S16 23.0	359 31.7	N 3 51.7	Alphecca	126 13.5	N26 41
13	190 36.7	42 58.0	52.3	85 58.4	14.0	229 30.6	23.0	14 33.9	51.5	Alpheratz	357 46.1	N29 08
14	205 39.1	57 57.5	51.3	100 59.2	13.9	244 33.2	23.1	29 36.1	51.4	Altair	62 10.8	N 8 53
15	220 41.6	72 57.0	. . 50.4	116 00.1	. . 13.8	259 35.9	. . 23.2	44 38.2	. . 51.3	Ankaa	353 17.8	S42 14
16	235 44.1	87 56.4	49.5	131 00.9	13.6	274 38.6	23.2	59 40.4	51.2	Antares	112 29.8	S26 27
17	250 46.5	102 55.9	48.6	146 01.8	13.5	289 41.3	23.3	74 42.6	51.1			
18	265 49.0	117 55.4	N13 47.6	161 02.7	N23 13.4	304 43.9	S16 23.3	89 44.8	N 3 50.9	Arcturus	145 58.5	N19 08.
19	280 51.5	132 54.9	46.7	176 03.5	13.3	319 46.6	23.4	104 46.9	50.8	Atria	107 34.3	S69 03
20	295 53.9	147 54.4	45.8	191 04.4	13.2	334 49.3	23.5	119 49.1	50.7	Avior	234 19.7	S59 32.
21	310 56.4	162 53.9	. . 44.8	206 05.2	. . 13.1	349 51.9	. . 23.5	134 51.3	. . 50.6	Bellatrix	278 34.9	N 6 21.
22	325 58.8	177 53.3	43.9	221 06.1	13.0	4 54.6	23.6	149 53.5	50.4	Betelgeuse	271 04.3	N 7 24.
23	341 01.3	192 52.8	43.0	236 07.0	12.9	19 57.3	23.7	164 55.7	50.3			
17 00	356 03.8	207 52.3	N13 42.0	251 07.8	N23 12.8	34 59.9	S16 23.7	179 57.8	N 3 50.2	Canopus	263 57.5	S52 41.
01	11 06.2	222 51.8	41.1	266 08.7	12.7	50 02.6	23.8	195 00.0	50.1	Capella	280 38.5	N46 00.
02	26 08.7	237 51.3	40.2	281 09.5	12.6	65 05.3	23.8	210 02.2	50.0	Deneb	49 33.1	N45 19.
03	41 11.2	252 50.8	. . 39.2	296 10.4	. . 12.5	80 07.9	. . 23.9	225 04.4	. . 49.8	Denebola	182 36.7	N14 31.
04	56 13.6	267 50.2	38.3	311 11.3	12.4	95 10.6	24.0	240 06.5	49.7	Diphda	348 58.3	S17 55.
05	71 16.1	282 49.7	37.4	326 12.1	12.3	110 13.3	24.0	255 08.7	49.6			
06	86 18.6	297 49.2	N13 36.4	341 13.0	N23 12.2	125 15.9	S16 24.1	270 10.9	N 3 49.5	Dubhe	193 55.4	N61 41.
07	101 21.0	312 48.7	35.5	356 13.8	12.1	140 18.6	24.2	285 13.1	49.3	Elnath	278 16.1	N28 37.
08	116 23.5	327 48.2	34.6	11 14.7	12.0	155 21.3	24.2	300 15.2	49.2	Eltanin	90 47.4	N51 29.
09	131 26.0	342 47.7	. . 33.6	26 15.6	. . 11.9	170 23.9	. . 24.3	315 17.4	. . 49.1	Enif	33 49.6	N 9 55.
10	146 28.4	357 47.2	32.7	41 16.4	11.8	185 26.6	24.3	330 19.6	49.0	Fomalhaut	15 26.6	S29 34.
11	161 30.9	12 46.7	31.7	56 17.3	11.7	200 29.3	24.4	345 21.8	48.9			
12	176 33.3	27 46.1	N13 30.8	71 18.2	N23 11.6	215 31.9	S16 24.5	0 24.0	N 3 48.7	Gacrux	172 04.8	S57 10.
13	191 35.8	42 45.6	29.9	86 19.0	11.5	230 34.6	24.5	15 26.1	48.6	Gienah	175 55.5	S17 35.
14	206 38.3	57 45.1	28.9	101 19.9	11.3	245 37.3	24.6	30 28.3	48.5	Hadar	148 52.6	S60 25.
15	221 40.7	72 44.6	. . 28.0	116 20.8	. . 11.2	260 39.9	. . 24.6	45 30.5	. . 48.4	Hamal	328 03.7	N23 30.
16	236 43.2	87 44.1	27.0	131 21.6	11.1	275 42.6	24.7	60 32.7	48.2	Kaus Aust.	83 47.4	S34 22.
17	251 45.7	102 43.6	26.1	146 22.5	11.0	290 45.3	24.8	75 34.8	48.1			
18	266 48.1	117 43.1	N13 25.2	161 23.3	N23 10.9	305 47.9	S16 24.8	90 37.0	N 3 48.0	Kochab	137 20.3	N74 07.
19	281 50.6	132 42.6	24.2	176 24.2	10.8	320 50.6	24.9	105 39.2	47.9	Markab	13 40.8	N15 15.
20	296 53.1	147 42.1	23.3	191 25.1	10.7	335 53.2	24.9	120 41.4	47.8	Menkar	314 17.7	N 4 07.
21	311 55.5	162 41.6	. . 22.3	206 25.9	. . 10.6	350 55.9	. . 25.0	135 43.5	. . 47.6	Menkent	148 11.2	S36 25.
22	326 58.0	177 41.0	21.4	221 26.8	10.5	5 58.6	25.1	150 45.7	47.5	Miaplacidus	221 41.3	S69 45.
23	342 00.4	192 40.5	20.4	236 27.7	10.4	21 01.2	25.1	165 47.9	47.4			
18 00	357 02.9	207 40.0	N13 19.5	251 28.5	N23 10.3	36 03.9	S16 25.2	180 50.1	N 3 47.3	Mirfak	308 44.1	N49 53.
01	12 05.4	222 39.5	18.5	266 29.4	10.2	51 06.6	25.2	195 52.3	47.1	Nunki	76 01.6	S26 17.
02	27 07.8	237 39.0	17.6	281 30.3	10.0	66 09.2	25.3	210 54.4	47.0	Peacock	53 23.0	S56 42.
03	42 10.3	252 38.5	. . 16.6	296 31.1	. . 09.9	81 11.9	. . 25.4	225 56.6	. . 46.9	Pollux	243 31.2	N28 00.
04	57 12.8	267 38.0	15.7	311 32.0	09.8	96 14.5	25.4	240 58.8	46.8	Procyon	245 02.8	N 5 12.
05	72 15.2	282 37.5	14.7	326 32.9	09.7	111 17.2	25.5	256 01.0	46.7			
06	87 17.7	297 37.0	N13 13.8	341 33.8	N23 09.6	126 19.9	S16 25.5	271 03.1	N 3 46.5	Rasalhague	96 09.0	N12 33.
07	102 20.2	312 36.5	12.8	356 34.6	09.5	141 22.5	25.6	286 05.3	46.4	Regulus	207 46.7	N11 55.
08	117 22.6	327 36.0	11.9	11 35.5	09.4	156 25.2	25.7	301 07.5	46.3	Rigel	281 14.7	S 8 11.
09	132 25.1	342 35.5	. . 10.9	26 36.4	. . 09.3	171 27.8	. . 25.7	316 09.7	. . 46.2	Rigil Kent.	139 56.2	S60 52.
10	147 27.6	357 35.0	10.0	41 37.2	09.2	186 30.5	25.8	331 11.8	46.0	Sabik	102 15.7	S15 44.
11	162 30.0	12 34.5	09.0	56 38.1	09.0	201 33.1	25.8	346 14.0	45.9			
12	177 32.5	27 33.9	N13 08.0	71 39.0	N23 08.9	216 35.8	S16 25.9	1 16.2	N 3 45.8	Schedar	349 43.4	N56 35.
13	192 34.9	42 33.4	07.1	86 39.8	08.8	231 38.5	25.9	16 18.4	45.7	Shaula	96 25.7	S37 06.
14	207 37.4	57 32.9	06.1	101 40.7	08.7	246 41.1	26.0	31 20.5	45.6	Sirius	258 36.2	S16 43.
15	222 39.9	72 32.4	. . 05.2	116 41.6	. . 08.6	261 43.8	. . 26.1	46 22.7	. . 45.4	Spica	158 34.5	S11 12.
16	237 42.3	87 31.9	04.2	131 42.5	08.5	276 46.4	26.1	61 24.9	45.3	Suhail	222 54.9	S43 28.
17	252 44.8	102 31.4	03.3	146 43.3	08.4	291 49.1	26.2	76 27.1	45.2			
18	267 47.3	117 30.9	N13 02.3	161 44.2	N23 08.3	306 51.7	S16 26.2	91 29.3	N 3 45.1	Vega	80 40.7	N38 47.
19	282 49.7	132 30.4	01.3	176 45.1	08.1	321 54.4	26.3	106 31.4	44.9	Zuben'ubi	137 08.7	S16 05.
20	297 52.2	147 29.9	13 00.4	191 46.0	08.0	336 57.1	26.3	121 33.6	44.8			
21	312 54.7	162 29.4	12 59.4	206 46.8	. . 07.9	351 59.7	. . 26.4	136 35.8	. . 44.7			
22	327 57.1	177 28.9	58.5	221 47.7	07.8	7 02.4	26.5	151 38.0	44.6			
23	342 59.6	192 28.4	57.5	236 48.6	07.7	22 05.0	26.5	166 40.1	44.5			
Mer. Pass.	h m 0 15.7	v −0.5	d 0.9	v 0.9	d 0.1	v 2.7	d 0.1	v 2.2	d 0.1			

	SHA	Mer. Pas
	° ′	h m
Venus	211 48.5	10 09
Mars	255 04.0	7 15
Jupiter	38 56.2	21 36
Saturn	183 54.1	11 58

UT (d h)	SUN GHA (° ′)	SUN Dec (° ′)	MOON GHA (° ′)	*v* (′)	MOON Dec (° ′)	*d* (′)	HP (′)
16 00	181 15.5	N 2 40.3	217 40.0	8.3	N14 50.0	13.8	60.2
01	196 15.7	39.3	232 07.3	8.4	14 36.2	13.8	60.2
02	211 15.9	38.4	246 34.7	8.4	14 22.4	14.0	60.2
03	226 16.2	. . 37.4	261 02.1	8.5	14 08.4	14.0	60.2
04	241 16.4	36.5	275 29.6	8.5	13 54.4	14.1	60.2
05	256 16.6	35.5	289 57.1	8.6	13 40.3	14.2	60.2
06	271 16.8	N 2 34.5	304 24.7	8.7	N13 26.1	14.2	60.2
07	286 17.1	33.6	318 52.4	8.8	13 11.9	14.4	60.2
08	301 17.3	32.6	333 20.2	8.8	12 57.5	14.4	60.2
09	316 17.5	. . 31.6	347 48.0	8.9	12 43.1	14.5	60.2
10	331 17.7	30.7	2 15.9	8.9	12 28.6	14.6	60.2
11	346 17.9	29.7	16 43.8	9.0	12 14.0	14.6	60.2
12	1 18.2	N 2 28.7	31 11.8	9.1	N11 59.4	14.7	60.2
13	16 18.4	27.8	45 39.9	9.1	11 44.7	14.8	60.2
14	31 18.6	26.8	60 08.0	9.2	11 29.9	14.9	60.2
15	46 18.8	. . 25.8	74 36.2	9.3	11 15.0	14.9	60.2
16	61 19.1	24.9	89 04.5	9.3	11 00.1	15.0	60.2
17	76 19.3	23.9	103 32.8	9.3	10 45.1	15.0	60.2
18	91 19.5	N 2 22.9	118 01.1	9.5	N10 30.1	15.1	60.2
19	106 19.7	22.0	132 29.6	9.5	10 15.0	15.1	60.2
20	121 19.9	21.0	146 58.1	9.5	9 59.9	15.2	60.2
21	136 20.2	. . 20.0	161 26.6	9.6	9 44.7	15.3	60.2
22	151 20.4	19.1	175 55.2	9.6	9 29.4	15.3	60.2
23	166 20.6	18.1	190 23.8	9.7	9 14.1	15.4	60.2
17 00	181 20.8	N 2 17.1	204 52.5	9.8	N 8 58.7	15.4	60.2
01	196 21.1	16.2	219 21.3	9.8	8 43.3	15.5	60.2
02	211 21.3	15.2	233 50.1	9.8	8 27.8	15.5	60.2
03	226 21.5	. . 14.2	248 18.9	10.0	8 12.3	15.5	60.2
04	241 21.7	13.3	262 47.9	9.9	7 56.8	15.6	60.2
05	256 21.9	12.3	277 16.8	10.0	7 41.2	15.6	60.1
06	271 22.2	N 2 11.3	291 45.8	10.1	N 7 25.6	15.7	60.1
07	286 22.4	10.4	306 14.9	10.1	7 09.9	15.7	60.1
08	301 22.6	09.4	320 44.0	10.1	6 54.2	15.7	60.1
09	316 22.8	. . 08.4	335 13.1	10.2	6 38.5	15.8	60.1
10	331 23.1	07.5	349 42.3	10.2	6 22.7	15.8	60.1
11	346 23.3	06.5	4 11.5	10.3	6 06.9	15.8	60.1
12	1 23.5	N 2 05.5	18 40.8	10.3	N 5 51.1	15.9	60.1
13	16 23.7	04.6	33 10.1	10.4	5 35.2	15.8	60.1
14	31 23.9	03.6	47 39.5	10.4	5 19.4	15.9	60.1
15	46 24.2	. . 02.6	62 08.9	10.4	5 03.5	16.0	60.0
16	61 24.4	01.7	76 38.3	10.5	4 47.5	15.9	60.0
17	76 24.6	2 00.7	91 07.8	10.5	4 31.6	16.0	60.0
18	91 24.8	N 1 59.7	105 37.3	10.6	N 4 15.6	15.9	60.0
19	106 25.1	58.8	120 06.9	10.6	3 59.7	16.0	60.0
20	121 25.3	57.8	134 36.5	10.6	3 43.7	16.0	60.0
21	136 25.5	. . 56.8	149 06.1	10.6	3 27.7	16.1	60.0
22	151 25.7	55.9	163 35.7	10.7	3 11.6	16.0	59.9
23	166 25.9	54.9	178 05.4	10.8	2 55.6	16.0	59.9
18 00	181 26.2	N 1 53.9	192 35.2	10.7	N 2 39.6	16.1	59.9
01	196 26.4	53.0	207 04.9	10.8	2 23.5	16.0	59.9
02	211 26.6	52.0	221 34.7	10.8	2 07.5	16.0	59.9
03	226 26.8	. . 51.0	236 04.5	10.8	1 51.5	16.1	59.9
04	241 27.1	50.1	250 34.3	10.9	1 35.4	16.0	59.9
05	256 27.3	49.1	265 04.2	10.9	1 19.4	16.1	59.8
06	271 27.5	N 1 48.1	279 34.1	10.9	N 1 03.3	16.0	59.8
07	286 27.7	47.2	294 04.0	10.9	0 47.3	16.1	59.8
08	301 27.9	46.2	308 33.9	11.0	0 31.2	16.0	59.8
09	316 28.2	. . 45.2	323 03.9	10.9	N 0 15.2	16.0	59.8
10	331 28.4	44.2	337 33.8	11.0	S 0 00.8	16.0	59.7
11	346 28.6	43.3	352 03.8	11.1	0 16.8	16.0	59.7
12	1 28.8	N 1 42.3	6 33.9	11.0	S 0 32.8	16.0	59.7
13	16 29.1	41.3	21 03.9	11.0	0 48.8	15.9	59.7
14	31 29.3	40.4	35 33.9	11.1	1 04.7	16.0	59.7
15	46 29.5	. . 39.4	50 04.0	11.1	1 20.7	15.9	59.6
16	61 29.7	38.4	64 34.1	11.1	1 36.6	15.9	59.6
17	76 29.9	37.5	79 04.2	11.1	1 52.5	15.9	59.6
18	91 30.2	N 1 36.5	93 34.3	11.1	S 2 08.4	15.9	59.6
19	106 30.4	35.5	108 04.4	11.2	2 24.3	15.8	59.6
20	121 30.6	34.6	122 34.6	11.1	2 40.1	15.8	59.5
21	136 30.8	. . 33.6	137 04.7	11.2	2 55.9	15.8	59.5
22	151 31.1	32.6	151 34.9	11.2	3 11.7	15.8	59.5
23	166 31.3	31.6	166 05.1	11.1	S 3 27.5	15.7	59.5
	SD 15.9	*d* 1.0	SD 16.4		16.4		16.3

Lat. (°)	Twilight Naut. (h m)	Twilight Civil (h m)	Sunrise (h m)	Moonrise 16 (h m)	Moonrise 17 (h m)	Moonrise 18 (h m)	Moonrise 19 (h m)
N 72	02 24	04 06	05 16	26 42	02 42	05 06	07 25
N 70	02 52	04 17	05 20	00 28	02 55	05 08	07 17
68	03 12	04 26	05 24	00 53	03 05	05 09	07 10
66	03 27	04 34	05 27	01 11	03 14	05 10	07 04
64	03 40	04 40	05 29	01 26	03 21	05 11	06 59
62	03 50	04 46	05 31	01 39	03 27	05 12	06 55
60	03 59	04 50	05 33	01 49	03 32	05 13	06 51
N 58	04 06	04 54	05 34	01 58	03 37	05 13	06 48
56	04 13	04 58	05 36	02 06	03 41	05 14	06 45
54	04 18	05 01	05 37	02 13	03 45	05 14	06 42
52	04 23	05 04	05 38	02 19	03 48	05 15	06 40
50	04 28	05 06	05 39	02 25	03 51	05 15	06 38
45	04 37	05 12	05 41	02 37	03 58	05 16	06 33
N 40	04 44	05 16	05 43	02 47	04 03	05 17	06 30
35	04 49	05 19	05 44	02 56	04 08	05 18	06 26
30	04 54	05 22	05 46	03 03	04 12	05 18	06 24
20	05 00	05 26	05 48	03 16	04 19	05 19	06 19
N 10	05 04	05 29	05 50	03 27	04 25	05 20	06 14
0	05 07	05 31	05 51	03 38	04 31	05 21	06 11
S 10	05 07	05 32	05 53	03 48	04 36	05 22	06 07
20	05 07	05 32	05 54	03 59	04 42	05 23	06 03
30	05 04	05 32	05 56	04 12	04 49	05 24	05 58
35	05 02	05 31	05 57	04 19	04 53	05 25	05 55
40	04 59	05 31	05 58	04 27	04 58	05 25	05 52
45	04 55	05 29	05 59	04 37	05 03	05 26	05 49
S 50	04 50	05 28	06 00	04 48	05 09	05 27	05 45
52	04 47	05 27	06 00	04 54	05 12	05 28	05 43
54	04 45	05 26	06 01	05 00	05 15	05 28	05 41
56	04 41	05 25	06 02	05 06	05 18	05 29	05 39
58	04 37	05 23	06 02	05 13	05 22	05 30	05 37
S 60	04 33	05 22	06 03	05 21	05 26	05 30	05 34

Lat. (°)	Sunset (h m)	Twilight Civil (h m)	Twilight Naut. (h m)	Moonset 16 (h m)	Moonset 17 (h m)	Moonset 18 (h m)	Moonset 19 (h m)
N 72	18 30	19 40	21 18	18 24	17 49	17 18	16 46
N 70	18 26	19 29	20 53	18 08	17 43	17 21	16 58
68	18 23	19 20	20 33	17 55	17 39	17 23	17 07
66	18 20	19 13	20 18	17 45	17 35	17 25	17 16
64	18 18	19 06	20 06	17 36	17 31	17 27	17 22
62	18 16	19 01	19 56	17 28	17 28	17 28	17 28
60	18 15	18 57	19 48	17 21	17 26	17 30	17 33
N 58	18 13	18 53	19 41	17 15	17 24	17 31	17 38
56	18 12	18 49	19 34	17 10	17 21	17 32	17 42
54	18 11	18 46	19 29	17 05	17 20	17 33	17 46
52	18 10	18 44	19 24	17 01	17 18	17 33	17 49
50	18 09	18 41	19 20	16 57	17 16	17 34	17 52
45	18 07	18 36	19 11	16 48	17 13	17 36	17 59
N 40	18 05	18 32	19 04	16 41	17 10	17 37	18 05
35	18 04	18 29	18 59	16 35	17 08	17 39	18 09
30	18 03	18 27	18 55	16 29	17 05	17 40	18 14
20	18 01	18 23	18 48	16 20	17 01	17 41	18 21
N 10	17 59	18 20	18 45	16 11	16 58	17 43	18 28
0	17 58	18 18	18 42	16 03	16 55	17 45	18 34
S 10	17 56	18 17	18 42	15 55	16 51	17 46	18 40
20	17 55	18 17	18 43	15 46	16 48	17 48	18 47
30	17 54	18 17	18 45	15 36	16 44	17 49	18 54
35	17 53	18 18	18 47	15 30	16 41	17 51	18 59
40	17 52	18 19	18 50	15 23	16 38	17 52	19 04
45	17 51	18 20	18 55	15 16	16 35	17 53	19 10
S 50	17 50	18 22	19 00	15 06	16 31	17 55	19 17
52	17 50	18 23	19 03	15 02	16 29	17 55	19 20
54	17 49	18 24	19 06	14 57	16 27	17 56	19 24
56	17 48	18 26	19 09	14 51	16 25	17 57	19 28
58	17 48	18 27	19 13	14 45	16 23	17 58	19 32
S 60	17 47	18 29	19 18	14 38	16 20	17 59	19 37

Day (d)	SUN Eqn. of Time 00^h (m s)	SUN Eqn. of Time 12^h (m s)	SUN Mer. Pass. (h m)	MOON Mer. Pass. Upper (h m)	MOON Mer. Pass. Lower (h m)	MOON Age (d)	MOON Phase (%)
16	05 02	05 12	11 55	09 51	22 17	27	7
17	05 23	05 34	11 54	10 43	23 08	28	2
18	05 44	05 55	11 54	11 33	23 58	29	0

UT	ARIES	VENUS −3.9		MARS +0.9		JUPITER −2.7		SATURN +1.1	
	GHA	GHA	Dec	GHA	Dec	GHA	Dec	GHA	Dec
d h	° ′	° ′	° ′	° ′	° ′	° ′	° ′	° ′	° ′
19 00	358 02.1	207 27.9	N12 56.5	251 49.5	N23 07.6	37 07.7	S16 26.6	181 42.3	N 3 44.3
01	13 04.5	222 27.4	55.6	266 50.3	07.5	52 10.3	26.6	196 44.5	44.2
02	28 07.0	237 26.9	54.6	281 51.2	07.3	67 13.0	26.7	211 46.7	44.1
03	43 09.4	252 26.4	. . 53.6	296 52.1	. . 07.2	82 15.6	. . 26.8	226 48.8	. . 44.0
04	58 11.9	267 25.9	52.7	311 53.0	07.1	97 18.3	26.8	241 51.0	43.8
05	73 14.4	282 25.4	51.7	326 53.8	07.0	112 21.0	26.9	256 53.2	43.7
06	88 16.8	297 24.9	N12 50.7	341 54.7	N23 06.9	127 23.6	S16 26.9	271 55.4	N 3 43.6
07	103 19.3	312 24.4	49.8	356 55.6	06.8	142 26.3	27.0	286 57.6	43.5
S 08	118 21.8	327 23.9	48.8	11 56.5	06.7	157 28.9	27.0	301 59.7	43.4
A 09	133 24.2	342 23.4	. . 47.8	26 57.3	. . 06.5	172 31.6	. . 27.1	317 01.9	. . 43.2
T 10	148 26.7	357 22.9	46.9	41 58.2	06.4	187 34.2	27.1	332 04.1	43.1
U 11	163 29.2	12 22.4	45.9	56 59.1	06.3	202 36.9	27.2	347 06.3	43.0
R 12	178 31.6	27 21.9	N12 44.9	72 00.0	N23 06.2	217 39.5	S16 27.3	2 08.4	N 3 42.9
D 13	193 34.1	42 21.4	43.9	87 00.9	06.1	232 42.2	27.3	17 10.6	42.7
A 14	208 36.5	57 20.9	43.0	102 01.7	06.0	247 44.8	27.4	32 12.8	42.6
Y 15	223 39.0	72 20.4	. . 42.0	117 02.6	. . 05.8	262 47.5	. . 27.4	47 15.0	. . 42.5
16	238 41.5	87 19.9	41.0	132 03.5	05.7	277 50.1	27.5	62 17.1	42.4
17	253 43.9	102 19.4	40.0	147 04.4	05.6	292 52.8	27.5	77 19.3	42.3
18	268 46.4	117 19.0	N12 39.1	162 05.3	N23 05.5	307 55.4	S16 27.6	92 21.5	N 3 42.1
19	283 48.9	132 18.5	38.1	177 06.1	05.4	322 58.1	27.6	107 23.7	42.0
20	298 51.3	147 18.0	37.1	192 07.0	05.2	338 00.7	27.7	122 25.8	41.9
21	313 53.8	162 17.5	. . 36.1	207 07.9	. . 05.1	353 03.4	. . 27.8	137 28.0	. . 41.8
22	328 56.3	177 17.0	35.2	222 08.8	05.0	8 06.0	27.8	152 30.2	41.6
23	343 58.7	192 16.5	34.2	237 09.7	04.9	23 08.7	27.9	167 32.4	41.5
20 00	359 01.2	207 16.0	N12 33.2	252 10.6	N23 04.8	38 11.3	S16 27.9	182 34.6	N 3 41.4
01	14 03.7	222 15.5	32.2	267 11.4	04.6	53 14.0	28.0	197 36.7	41.3
02	29 06.1	237 15.0	31.2	282 12.3	04.5	68 16.6	28.0	212 38.9	41.2
03	44 08.6	252 14.5	. . 30.3	297 13.2	. . 04.4	83 19.3	. . 28.1	227 41.1	. . 41.0
04	59 11.0	267 14.0	29.3	312 14.1	04.3	98 21.9	28.1	242 43.3	40.9
05	74 13.5	282 13.5	28.3	327 15.0	04.2	113 24.5	28.2	257 45.4	40.8
06	89 16.0	297 13.0	N12 27.3	342 15.9	N23 04.0	128 27.2	S16 28.2	272 47.6	N 3 40.7
07	104 18.4	312 12.5	26.3	357 16.7	03.9	143 29.8	28.3	287 49.8	40.5
08	119 20.9	327 12.0	25.4	12 17.6	03.8	158 32.5	28.4	302 52.0	40.4
S 09	134 23.4	342 11.6	. . 24.4	27 18.5	. . 03.7	173 35.1	. . 28.4	317 54.1	. . 40.3
U 10	149 25.8	357 11.1	23.4	42 19.4	03.6	188 37.8	28.5	332 56.3	40.2
N 11	164 28.3	12 10.6	22.4	57 20.3	03.4	203 40.4	28.5	347 58.5	40.1
D 12	179 30.8	27 10.1	N12 21.4	72 21.2	N23 03.3	218 43.1	S16 28.6	3 00.7	N 3 39.9
A 13	194 33.2	42 09.6	20.4	87 22.1	03.2	233 45.7	28.6	18 02.9	39.8
Y 14	209 35.7	57 09.1	19.4	102 23.0	03.1	248 48.3	28.7	33 05.0	39.7
15	224 38.1	72 08.6	. . 18.4	117 23.8	. . 02.9	263 51.0	. . 28.7	48 07.2	. . 39.6
16	239 40.6	87 08.1	17.5	132 24.7	02.8	278 53.6	28.8	63 09.4	39.5
17	254 43.1	102 07.6	16.5	147 25.6	02.7	293 56.3	28.8	78 11.6	39.3
18	269 45.5	117 07.2	N12 15.5	162 26.5	N23 02.6	308 58.9	S16 28.9	93 13.7	N 3 39.2
19	284 48.0	132 06.7	14.5	177 27.4	02.5	324 01.6	28.9	108 15.9	39.1
20	299 50.5	147 06.2	13.5	192 28.3	02.3	339 04.2	29.0	123 18.1	39.0
21	314 52.9	162 05.7	. . 12.5	207 29.2	. . 02.2	354 06.8	. . 29.1	138 20.3	. . 38.8
22	329 55.4	177 05.2	11.5	222 30.1	02.1	9 09.5	29.1	153 22.4	38.7
23	344 57.9	192 04.7	10.5	237 31.0	02.0	24 12.1	29.2	168 24.6	38.6
21 00	0 00.3	207 04.2	N12 09.5	252 31.8	N23 01.8	39 14.8	S16 29.2	183 26.8	N 3 38.5
01	15 02.8	222 03.7	08.5	267 32.7	01.7	54 17.4	29.3	198 29.0	38.4
02	30 05.3	237 03.3	07.5	282 33.6	01.6	69 20.1	29.3	213 31.2	38.2
03	45 07.7	252 02.8	. . 06.5	297 34.5	. . 01.5	84 22.7	. . 29.4	228 33.3	. . 38.1
04	60 10.2	267 02.3	05.5	312 35.4	01.3	99 25.3	29.4	243 35.5	38.0
05	75 12.6	282 01.8	04.5	327 36.3	01.2	114 28.0	29.5	258 37.7	37.9
06	90 15.1	297 01.3	N12 03.5	342 37.2	N23 01.1	129 30.6	S16 29.5	273 39.9	N 3 37.7
07	105 17.6	312 00.8	02.5	357 38.1	01.0	144 33.2	29.6	288 42.0	37.6
08	120 20.0	327 00.3	01.5	12 39.0	00.8	159 35.9	29.6	303 44.2	37.5
M 09	135 22.5	341 59.9	12 00.6	27 39.9	. . 00.7	174 38.5	. . 29.7	318 46.4	. . 37.4
O 10	150 25.0	356 59.4	11 59.6	42 40.8	00.6	189 41.2	29.7	333 48.6	37.3
N 11	165 27.4	11 58.9	58.5	57 41.7	00.4	204 43.8	29.8	348 50.7	37.1
D 12	180 29.9	26 58.4	N11 57.5	72 42.6	N23 00.3	219 46.4	S16 29.8	3 52.9	N 3 37.0
A 13	195 32.4	41 57.9	56.5	87 43.5	00.2	234 49.1	29.9	18 55.1	36.9
Y 14	210 34.8	56 57.4	55.5	102 44.4	23 00.1	249 51.7	29.9	33 57.3	36.8
15	225 37.3	71 57.0	. . 54.5	117 45.2	22 59.9	264 54.3	. . 30.0	48 59.4	. . 36.6
16	240 39.7	86 56.5	53.5	132 46.1	59.8	279 57.0	30.0	64 01.6	36.5
17	255 42.2	101 56.0	52.5	147 47.0	59.7	294 59.6	30.1	79 03.8	36.4
18	270 44.7	116 55.5	N11 51.5	162 47.9	N22 59.6	310 02.3	S16 30.1	94 06.0	N 3 36.3
19	285 47.1	131 55.0	50.5	177 48.8	59.4	325 04.9	30.2	109 08.2	36.2
20	300 49.6	146 54.6	49.5	192 49.7	59.3	340 07.5	30.2	124 10.3	36.0
21	315 52.1	161 54.1	. . 48.5	207 50.6	. . 59.2	355 10.2	. . 30.3	139 12.5	. . 35.9
22	330 54.5	176 53.6	47.5	222 51.5	59.0	10 12.8	30.3	154 14.7	35.8
23	345 57.0	191 53.1	46.5	237 52.4	58.9	25 15.4	30.4	169 16.9	35.7
Mer. Pass.	h m 0 03.9	*v* −0.5	*d* 1.0	*v* 0.9	*d* 0.1	*v* 2.6	*d* 0.1	*v* 2.2	*d* 0.1

STARS		
Name	SHA	Dec
	° ′	°
Acamar	315 20.1	S40 15
Achernar	335 28.1	S57 10
Acrux	173 13.4	S63 09
Adhara	255 14.8	S28 58
Aldebaran	290 52.5	N16 31
Alioth	166 23.4	N55 54
Alkaid	153 01.3	N49 16
Al Na'ir	27 46.6	S46 54
Alnilam	275 49.1	S 1 11
Alphard	217 59.1	S 8 41
Alphecca	126 13.5	N26 41
Alpheratz	357 46.1	N29 08
Altair	62 10.8	N 8 53
Ankaa	353 17.8	S42 15
Antares	112 29.8	S26 27.
Arcturus	145 58.5	N19 08.
Atria	107 34.3	S69 03
Avior	234 19.7	S59 32
Bellatrix	278 34.9	N 6 21.
Betelgeuse	271 04.3	N 7 24
Canopus	263 57.5	S52 41
Capella	280 38.5	N46 00.
Deneb	49 33.2	N45 19
Denebola	182 36.7	N14 31
Diphda	348 58.3	S17 55.
Dubhe	193 55.4	N61 41
Elnath	278 16.1	N28 37
Eltanin	90 47.4	N51 29
Enif	33 49.6	N 9 55
Fomalhaut	15 26.6	S29 34.
Gacrux	172 04.8	S57 10.
Gienah	175 55.5	S17 35.
Hadar	148 52.6	S60 25
Hamal	328 03.6	N23 30.
Kaus Aust.	83 47.4	S34 22.
Kochab	137 20.3	N74 07.
Markab	13 40.8	N15 15.
Menkar	314 17.7	N 4 07.
Menkent	148 11.2	S36 25.
Miaplacidus	221 41.3	S69 45.
Mirfak	308 44.1	N49 53.
Nunki	76 01.6	S26 17.
Peacock	53 23.1	S56 42.
Pollux	243 31.2	N28 00.
Procyon	245 02.7	N 5 12.
Rasalhague	96 09.0	N12 33.
Regulus	207 46.7	N11 55.
Rigel	281 14.6	S 8 11.
Rigil Kent.	139 56.2	S60 52.
Sabik	102 15.8	S15 44.
Schedar	349 43.3	N56 35.
Shaula	96 25.7	S37 06.
Sirius	258 36.2	S16 43.
Spica	158 34.5	S11 12.
Suhail	222 54.9	S43 28.
Vega	80 40.8	N38 47.
Zuben'ubi	137 08.7	S16 05.

	SHA	Mer. Pas
	° ′	h m
Venus	208 14.8	10 11
Mars	253 09.4	7 11
Jupiter	39 10.1	21 23
Saturn	183 33.4	11 48

UT	SUN GHA	SUN Dec	MOON GHA	v	MOON Dec	d	HP
d h	° ′	° ′	° ′	′	° ′	′	′
19 00	181 31.5	N 1 30.7	180 35.2	11.2	S 3 43.2	15.7	59.4
01	196 31.7	29.7	195 05.4	11.2	3 58.9	15.6	59.4
02	211 31.9	28.7	209 35.6	11.2	4 14.5	15.7	59.4
03	226 32.2	. . 27.8	224 05.8	11.2	4 30.2	15.6	59.4
04	241 32.4	26.8	238 36.0	11.2	4 45.8	15.5	59.3
05	256 32.6	25.8	253 06.2	11.2	5 01.3	15.5	59.3
06	271 32.8	N 1 24.9	267 36.4	11.2	S 5 16.8	15.5	59.3
07	286 33.1	23.9	282 06.6	11.2	5 32.3	15.4	59.3
08	301 33.3	22.9	296 36.8	11.2	5 47.7	15.4	59.2
09	316 33.5	. . 21.9	311 07.0	11.3	6 03.1	15.3	59.2
10	331 33.7	21.0	325 37.3	11.2	6 18.4	15.3	59.2
11	346 33.9	20.0	340 07.5	11.2	6 33.7	15.2	59.2
12	1 34.2	N 1 19.0	354 37.7	11.2	S 6 48.9	15.2	59.1
13	16 34.4	18.1	9 07.9	11.2	7 04.1	15.1	59.1
14	31 34.6	17.1	23 38.1	11.1	7 19.2	15.1	59.1
15	46 34.8	. . 16.1	38 08.2	11.2	7 34.3	15.0	59.0
16	61 35.0	15.2	52 38.4	11.2	7 49.3	14.9	59.0
17	76 35.3	14.2	67 08.6	11.2	8 04.2	15.0	59.0
18	91 35.5	N 1 13.2	81 38.8	11.1	S 8 19.2	14.8	59.0
19	106 35.7	12.2	96 08.9	11.2	8 34.0	14.8	58.9
20	121 35.9	11.3	110 39.1	11.1	8 48.8	14.7	58.9
21	136 36.2	. . 10.3	125 09.2	11.2	9 03.5	14.7	58.9
22	151 36.4	09.3	139 39.4	11.1	9 18.2	14.6	58.8
23	166 36.6	08.4	154 09.5	11.1	9 32.8	14.5	58.8
20 00	181 36.8	N 1 07.4	168 39.6	11.1	S 9 47.3	14.5	58.8
01	196 37.0	06.4	183 09.7	11.1	10 01.8	14.4	58.7
02	211 37.3	05.4	197 39.8	11.1	10 16.2	14.3	58.7
03	226 37.5	. . 04.5	212 09.9	11.0	10 30.5	14.2	58.7
04	241 37.7	03.5	226 39.9	11.1	10 44.7	14.2	58.7
05	256 37.9	02.5	241 10.0	11.0	10 58.9	14.1	58.6
06	271 38.1	N 1 01.6	255 40.0	11.0	S11 13.0	14.1	58.6
07	286 38.4	1 00.6	270 10.0	11.0	11 27.1	13.9	58.6
08	301 38.6	0 59.6	284 40.0	11.0	11 41.0	13.9	58.5
09	316 38.8	. . 58.6	299 10.0	10.9	11 54.9	13.8	58.5
10	331 39.0	57.7	313 39.9	11.0	12 08.7	13.8	58.5
11	346 39.3	56.7	328 09.9	10.9	12 22.5	13.6	58.4
12	1 39.5	N 0 55.7	342 39.8	10.9	S12 36.1	13.6	58.4
13	16 39.7	54.8	357 09.7	10.8	12 49.7	13.4	58.4
14	31 39.9	53.8	11 39.5	10.9	13 03.1	13.4	58.3
15	46 40.1	. . 52.8	26 09.4	10.8	13 16.5	13.3	58.3
16	61 40.4	51.8	40 39.2	10.9	13 29.8	13.3	58.3
17	76 40.6	50.9	55 09.1	10.7	13 43.1	13.1	58.2
18	91 40.8	N 0 49.9	69 38.8	10.8	S13 56.2	13.1	58.2
19	106 41.0	48.9	84 08.6	10.8	14 09.3	12.9	58.2
20	121 41.2	48.0	98 38.4	10.7	14 22.2	12.9	58.1
21	136 41.5	. . 47.0	113 08.1	10.7	14 35.1	12.8	58.1
22	151 41.7	46.0	127 37.8	10.7	14 47.9	12.6	58.1
23	166 41.9	45.0	142 07.5	10.6	15 00.5	12.6	58.0
21 00	181 42.1	N 0 44.1	156 37.1	10.6	S15 13.1	12.5	58.0
01	196 42.3	43.1	171 06.7	10.6	15 25.6	12.4	58.0
02	211 42.6	42.1	185 36.3	10.6	15 38.0	12.3	57.9
03	226 42.8	. . 41.2	200 05.9	10.6	15 50.3	12.2	57.9
04	241 43.0	40.2	214 35.5	10.5	16 02.5	12.1	57.9
05	256 43.2	39.2	229 05.0	10.5	16 14.6	12.0	57.8
06	271 43.4	N 0 38.2	243 34.5	10.5	S16 26.6	11.9	57.8
07	286 43.7	37.3	258 04.0	10.4	16 38.5	11.8	57.8
08	301 43.9	36.3	272 33.4	10.4	16 50.3	11.7	57.7
09	316 44.1	. . 35.3	287 02.8	10.4	17 02.0	11.6	57.7
10	331 44.3	34.4	301 32.2	10.4	17 13.6	11.5	57.7
11	346 44.5	33.4	316 01.6	10.3	17 25.1	11.4	57.6
12	1 44.8	N 0 32.4	330 30.9	10.3	S17 36.5	11.2	57.6
13	16 45.0	31.4	345 00.2	10.3	17 47.7	11.2	57.6
14	31 45.2	30.5	359 29.5	10.2	17 58.9	11.1	57.5
15	46 45.4	. . 29.5	13 58.7	10.3	18 10.0	10.9	57.5
16	61 45.6	28.5	28 28.0	10.2	18 20.9	10.9	57.5
17	76 45.9	27.5	42 57.2	10.1	18 31.8	10.7	57.4
18	91 46.1	N 0 26.6	57 26.3	10.2	S18 42.5	10.6	57.4
19	106 46.3	25.6	71 55.5	10.1	18 53.1	10.5	57.3
20	121 46.5	24.6	86 24.6	10.1	19 03.6	10.4	57.3
21	136 46.7	. . 23.7	100 53.7	10.0	19 14.0	10.3	57.3
22	151 47.0	22.7	115 22.7	10.1	19 24.3	10.1	57.2
23	166 47.2	21.7	129 51.8	10.0	S19 34.4	10.1	57.2
	SD 16.0	d 1.0	SD		16.1	15.9	15.7

Lat.	Twilight Naut.	Twilight Civil	Sunrise	Moonrise 19	Moonrise 20	Moonrise 21	Moonrise 22
°	h m	h m	h m	h m	h m	h m	h m
N 72	02 46	04 20	05 30	07 25	09 53	▬	▬
N 70	03 08	04 30	05 32	07 17	09 30	12 05	▬
68	03 26	04 38	05 34	07 10	09 12	11 24	▬
66	03 39	04 44	05 36	07 04	08 58	10 57	13 07
64	03 50	04 49	05 37	06 59	08 46	10 36	12 29
62	03 59	04 54	05 39	06 55	08 37	10 19	12 02
60	04 07	04 58	05 40	06 51	08 28	10 05	11 41
N 58	04 14	05 01	05 41	06 48	08 21	09 54	11 24
56	04 19	05 04	05 41	06 45	08 15	09 43	11 10
54	04 24	05 07	05 42	06 42	08 09	09 35	10 57
52	04 29	05 09	05 43	06 40	08 04	09 27	10 47
50	04 33	05 11	05 43	06 38	07 59	09 19	10 37
45	04 41	05 15	05 45	06 33	07 49	09 04	10 17
N 40	04 47	05 19	05 46	06 30	07 41	08 52	10 01
35	04 52	05 21	05 47	06 26	07 34	08 41	09 47
30	04 55	05 23	05 47	06 24	07 28	08 32	09 35
20	05 01	05 26	05 48	06 19	07 18	08 16	09 15
N 10	05 04	05 28	05 49	06 14	07 08	08 03	08 57
0	05 05	05 29	05 50	06 11	07 00	07 50	08 41
S 10	05 05	05 30	05 51	06 07	06 51	07 37	08 25
20	05 04	05 29	05 51	06 03	06 43	07 24	08 08
30	05 00	05 28	05 52	05 58	06 32	07 09	07 48
35	04 58	05 27	05 52	05 55	06 27	07 00	07 37
40	04 54	05 26	05 53	05 52	06 20	06 50	07 24
45	04 49	05 24	05 53	05 49	06 13	06 39	07 09
S 50	04 43	05 21	05 53	05 45	06 04	06 25	06 50
52	04 40	05 20	05 53	05 43	05 59	06 18	06 41
54	04 37	05 18	05 53	05 41	05 55	06 11	06 31
56	04 33	05 17	05 54	05 39	05 50	06 03	06 20
58	04 28	05 15	05 54	05 37	05 44	05 54	06 08
S 60	04 23	05 13	05 54	05 34	05 38	05 44	05 53

Lat.	Sunset	Twilight Civil	Twilight Naut.	Moonset 19	Moonset 20	Moonset 21	Moonset 22
°	h m	h m	h m	h m	h m	h m	h m
N 72	18 14	19 23	20 56	16 46	16 04	▬	▬
N 70	18 12	19 14	20 34	16 58	16 29	15 41	▬
68	18 10	19 07	20 18	17 07	16 49	16 22	▬
66	18 09	19 00	20 05	17 16	17 05	16 51	16 28
64	18 08	18 55	19 54	17 22	17 18	17 13	17 07
62	18 07	18 51	19 45	17 28	17 29	17 30	17 35
60	18 06	18 47	19 38	17 33	17 38	17 45	17 56
N 58	18 05	18 44	19 31	17 38	17 47	17 58	18 14
56	18 04	18 41	19 26	17 42	17 54	18 09	18 29
54	18 04	18 39	19 21	17 46	18 01	18 18	18 41
52	18 03	18 37	19 17	17 49	18 06	18 27	18 53
50	18 02	18 35	19 13	17 52	18 12	18 35	19 03
45	18 01	18 31	19 05	17 59	18 24	18 51	19 24
N 40	18 00	18 27	18 59	18 05	18 33	19 05	19 41
35	18 00	18 25	18 54	18 09	18 42	19 17	19 56
30	17 59	18 23	18 51	18 14	18 49	19 27	20 08
20	17 58	18 20	18 46	18 21	19 02	19 44	20 30
N 10	17 57	18 18	18 43	18 28	19 13	20 00	20 48
0	17 57	18 17	18 41	18 34	19 24	20 14	21 06
S 10	17 56	18 17	18 41	18 40	19 34	20 29	21 24
20	17 56	18 18	18 43	18 47	19 46	20 44	21 43
30	17 55	18 19	18 47	18 54	19 59	21 02	22 05
35	17 55	18 20	18 50	18 59	20 06	21 13	22 17
40	17 55	18 22	18 53	19 04	20 15	21 25	22 32
45	17 55	18 24	18 58	19 10	20 25	21 39	22 50
S 50	17 55	18 27	19 05	19 17	20 38	21 57	23 12
52	17 54	18 28	19 08	19 20	20 43	22 05	23 23
54	17 54	18 30	19 11	19 24	20 50	22 14	23 35
56	17 54	18 31	19 15	19 28	20 57	22 25	23 48
58	17 54	18 33	19 20	19 32	21 05	22 37	24 04
S 60	17 54	18 36	19 25	19 37	21 14	22 51	24 24

Day	SUN Eqn. of Time 00^h	SUN Eqn. of Time 12^h	SUN Mer. Pass.	MOON Mer. Pass. Upper	MOON Mer. Pass. Lower	Age	Phase
d	m s	m s	h m	h m	h m	d %	
19	06 06	06 16	11 54	12 22	24 47	01 1	●
20	06 27	06 37	11 53	13 12	00 47	02 4	
21	06 48	06 59	11 53	14 02	01 37	03 9	

UT	ARIES	VENUS −3.9		MARS +0.8		JUPITER −2.7		SATURN +1.1	
	GHA	GHA	Dec	GHA	Dec	GHA	Dec	GHA	Dec
d h	° ′	° ′	° ′	° ′	° ′	° ′	° ′	° ′	° ′
22 00 (TUESDAY)	0 59.5	206 52.6	N11 45.5	252 53.3	N22 58.8	40 18.1	S16 30.4	184 19.0	N 3 35.6
01	16 01.9	221 52.2	44.5	267 54.2	58.6	55 20.7	30.5	199 21.2	35.4
02	31 04.4	236 51.7	43.5	282 55.1	58.5	70 23.3	30.5	214 23.4	35.3
03	46 06.9	251 51.2	42.5	297 56.0	58.4	85 26.0	30.6	229 25.6	35.2
04	61 09.3	266 50.7	41.4	312 56.9	58.3	100 28.6	30.6	244 27.7	35.1
05	76 11.8	281 50.2	40.4	327 57.8	58.1	115 31.2	30.7	259 29.9	34.9
06	91 14.2	296 49.8	N11 39.4	342 58.7	N22 58.0	130 33.9	S16 30.7	274 32.1	N 3 34.8
07	106 16.7	311 49.3	38.4	357 59.6	57.9	145 36.5	30.8	289 34.3	34.7
08	121 19.2	326 48.8	37.4	13 00.5	57.7	160 39.1	30.8	304 36.5	34.6
09	136 21.6	341 48.3	36.4	28 01.4	57.6	175 41.7	30.9	319 38.6	34.5
10	151 24.1	356 47.9	35.4	43 02.3	57.5	190 44.4	30.9	334 40.8	34.3
11	166 26.6	11 47.4	34.4	58 03.2	57.3	205 47.0	31.0	349 43.0	34.2
12	181 29.0	26 46.9	N11 33.3	73 04.1	N22 57.2	220 49.6	S16 31.0	4 45.2	N 3 34.1
13	196 31.5	41 46.4	32.3	88 05.0	57.1	235 52.3	31.1	19 47.3	34.0
14	211 34.0	56 46.0	31.3	103 05.9	56.9	250 54.9	31.1	34 49.5	33.9
15	226 36.4	71 45.5	30.3	118 06.9	56.8	265 57.5	31.2	49 51.7	33.7
16	241 38.9	86 45.0	29.3	133 07.8	56.7	281 00.2	31.2	64 53.9	33.6
17	256 41.4	101 44.5	28.3	148 08.7	56.5	296 02.8	31.3	79 56.1	33.5
18	271 43.8	116 44.1	N11 27.2	163 09.6	N22 56.4	311 05.4	S16 31.3	94 58.2	N 3 33.4
19	286 46.3	131 43.6	26.2	178 10.5	56.3	326 08.0	31.4	110 00.4	33.2
20	301 48.7	146 43.1	25.2	193 11.4	56.1	341 10.7	31.4	125 02.6	33.1
21	316 51.2	161 42.6	24.2	208 12.3	56.0	356 13.3	31.5	140 04.8	33.0
22	331 53.7	176 42.2	23.2	223 13.2	55.9	11 15.9	31.5	155 06.9	32.9
23	346 56.1	191 41.7	22.1	238 14.1	55.7	26 18.5	31.6	170 09.1	32.8
23 00 (WEDNESDAY)	1 58.6	206 41.2	N11 21.1	253 15.0	N22 55.6	41 21.2	S16 31.6	185 11.3	N 3 32.6
01	17 01.1	221 40.7	20.1	268 15.9	55.5	56 23.8	31.6	200 13.5	32.5
02	32 03.5	236 40.3	19.1	283 16.8	55.3	71 26.4	31.7	215 15.6	32.4
03	47 06.0	251 39.8	18.0	298 17.7	55.2	86 29.1	31.7	230 17.8	32.3
04	62 08.5	266 39.3	17.0	313 18.6	55.1	101 31.7	31.8	245 20.0	32.2
05	77 10.9	281 38.9	16.0	328 19.5	54.9	116 34.3	31.8	260 22.2	32.0
06	92 13.4	296 38.4	N11 15.0	343 20.5	N22 54.8	131 36.9	S16 31.9	275 24.4	N 3 31.9
07	107 15.8	311 37.9	13.9	358 21.4	54.7	146 39.5	31.9	290 26.5	31.8
08	122 18.3	326 37.4	12.9	13 22.3	54.5	161 42.2	32.0	305 28.7	31.7
09	137 20.8	341 37.0	11.9	28 23.2	54.4	176 44.8	32.0	320 30.9	31.5
10	152 23.2	356 36.5	10.9	43 24.1	54.3	191 47.4	32.1	335 33.1	31.4
11	167 25.7	11 36.0	09.8	58 25.0	54.1	206 50.0	32.1	350 35.2	31.3
12	182 28.2	26 35.6	N11 08.8	73 25.9	N22 54.0	221 52.7	S16 32.2	5 37.4	N 3 31.2
13	197 30.6	41 35.1	07.8	88 26.8	53.8	236 55.3	32.2	20 39.6	31.1
14	212 33.1	56 34.6	06.7	103 27.7	53.7	251 57.9	32.3	35 41.8	30.9
15	227 35.6	71 34.2	05.7	118 28.7	53.6	267 00.5	32.3	50 43.9	30.8
16	242 38.0	86 33.7	04.7	133 29.6	53.4	282 03.2	32.3	65 46.1	30.7
17	257 40.5	101 33.2	03.6	148 30.5	53.3	297 05.8	32.4	80 48.3	30.6
18	272 43.0	116 32.8	N11 02.6	163 31.4	N22 53.2	312 08.4	S16 32.4	95 50.5	N 3 30.5
19	287 45.4	131 32.3	01.6	178 32.3	53.0	327 11.0	32.5	110 52.7	30.3
20	302 47.9	146 31.8	11 00.6	193 33.2	52.9	342 13.6	32.5	125 54.8	30.2
21	317 50.3	161 31.4	10 59.5	208 34.1	52.7	357 16.3	32.6	140 57.0	30.1
22	332 52.8	176 30.9	58.5	223 35.0	52.6	12 18.9	32.6	155 59.2	30.0
23	347 55.3	191 30.4	57.4	238 36.0	52.5	27 21.5	32.7	171 01.4	29.8
24 00 (THURSDAY)	2 57.7	206 30.0	N10 56.4	253 36.9	N22 52.3	42 24.1	S16 32.7	186 03.5	N 3 29.7
01	18 00.2	221 29.5	55.4	268 37.8	52.2	57 26.7	32.8	201 05.7	29.6
02	33 02.7	236 29.0	54.3	283 38.7	52.1	72 29.3	32.8	216 07.9	29.5
03	48 05.1	251 28.6	53.3	298 39.6	51.9	87 32.0	32.8	231 10.1	29.4
04	63 07.6	266 28.1	52.3	313 40.5	51.8	102 34.6	32.9	246 12.3	29.2
05	78 10.1	281 27.6	51.2	328 41.5	51.6	117 37.2	32.9	261 14.4	29.1
06	93 12.5	296 27.2	N10 50.2	343 42.4	N22 51.5	132 39.8	S16 33.0	276 16.6	N 3 29.0
07	108 15.0	311 26.7	49.1	358 43.3	51.4	147 42.4	33.0	291 18.8	28.9
08	123 17.5	326 26.2	48.1	13 44.2	51.2	162 45.1	33.1	306 21.0	28.8
09	138 19.9	341 25.8	47.1	28 45.1	51.1	177 47.7	33.1	321 23.1	28.6
10	153 22.4	356 25.3	46.0	43 46.1	50.9	192 50.3	33.2	336 25.3	28.5
11	168 24.8	11 24.8	45.0	58 47.0	50.8	207 52.9	33.2	351 27.5	28.4
12	183 27.3	26 24.4	N10 43.9	73 47.9	N22 50.6	222 55.5	S16 33.2	6 29.7	N 3 28.3
13	198 29.8	41 23.9	42.9	88 48.8	50.5	237 58.1	33.3	21 31.8	28.1
14	213 32.2	56 23.5	41.9	103 49.7	50.4	253 00.7	33.3	36 34.0	28.0
15	228 34.7	71 23.0	40.8	118 50.7	50.2	268 03.4	33.4	51 36.2	27.9
16	243 37.2	86 22.5	39.8	133 51.6	50.1	283 06.0	33.4	66 38.4	27.8
17	258 39.6	101 22.1	38.7	148 52.5	49.9	298 08.6	33.5	81 40.6	27.7
18	273 42.1	116 21.6	N10 37.7	163 53.4	N22 49.8	313 11.2	S16 33.5	96 42.7	N 3 27.5
19	288 44.6	131 21.2	36.6	178 54.3	49.7	328 13.8	33.5	111 44.9	27.4
20	303 47.0	146 20.7	35.6	193 55.3	49.5	343 16.4	33.6	126 47.1	27.3
21	318 49.5	161 20.2	34.5	208 56.2	49.4	358 19.0	33.6	141 49.3	27.2
22	333 52.0	176 19.8	33.5	223 57.1	49.2	13 21.6	33.7	156 51.4	27.1
23	348 54.4	191 19.3	32.4	238 58.0	49.1	28 24.3	33.7	171 53.6	26.9
Mer. Pass.	h m 23 48.2	*v* −0.5	*d* 1.0	*v* 0.9	*d* 0.1	*v* 2.6	*d* 0.0	*v* 2.2	*d* 0.1

STARS

Name	SHA	Dec
	° ′	° ′
Acamar	315 20.0	S40 15.
Achernar	335 28.1	S57 11.
Acrux	173 13.4	S63 09.
Adhara	255 14.8	S28 58.
Aldebaran	290 52.4	N16 31.
Alioth	166 23.4	N55 54.
Alkaid	153 01.3	N49 16.
Al Na'ir	27 46.6	S46 54.
Alnilam	275 49.1	S 1 11.
Alphard	217 59.1	S 8 41.
Alphecca	126 13.5	N26 41.
Alpheratz	357 46.1	N29 08.
Altair	62 10.8	N 8 53.
Ankaa	353 17.8	S42 15.
Antares	112 29.8	S26 27.
Arcturus	145 58.5	N19 08.
Atria	107 34.3	S69 03.
Avior	234 19.6	S59 32.
Bellatrix	278 34.9	N 6 21.
Betelgeuse	271 04.3	N 7 24.
Canopus	263 57.4	S52 41.
Capella	280 38.4	N46 00.
Deneb	49 33.2	N45 19.
Denebola	182 36.7	N14 31.
Diphda	348 58.3	S17 55.
Dubhe	193 55.4	N61 41.
Elnath	278 16.0	N28 37.
Eltanin	90 47.4	N51 29.
Enif	33 49.6	N 9 55.
Fomalhaut	15 26.6	S29 34.
Gacrux	172 04.8	S57 10.
Gienah	175 55.5	S17 35.
Hadar	148 52.6	S60 25.
Hamal	328 03.6	N23 30.
Kaus Aust.	83 47.4	S34 22.
Kochab	137 20.4	N74 07.
Markab	13 40.8	N15 15.
Menkar	314 17.7	N 4 07.
Menkent	148 11.2	S36 25.
Miaplacidus	221 41.2	S69 45.
Mirfak	308 44.1	N49 53.
Nunki	76 01.6	S26 17.
Peacock	53 23.1	S56 42.
Pollux	243 31.2	N28 00.
Procyon	245 02.7	N 5 12.
Rasalhague	96 09.0	N12 33.
Regulus	207 46.7	N11 55.
Rigel	281 14.6	S 8 11.
Rigil Kent.	139 56.2	S60 52.
Sabik	102 15.8	S15 44.
Schedar	349 43.3	N56 35.
Shaula	96 25.7	S37 06.
Sirius	258 36.2	S16 43.
Spica	158 34.5	S11 12.
Suhail	222 54.9	S43 28.
Vega	80 40.8	N38 47.
Zuben'ubi	137 08.7	S16 05.

	SHA	Mer. Pass
	° ′	h m
Venus	204 42.6	10 14
Mars	251 16.4	7 07
Jupiter	39 22.6	21 11
Saturn	183 12.7	11 38

JT d h	SUN GHA ° ′	SUN Dec ° ′	MOON GHA ° ′	v ′	Dec ° ′	d ′	HP ′
2 00	181 47.4	N 0 20.7	144 20.8	10.0	S19 44.5	9.9	57.2
01	196 47.6	19.8	158 49.8	9.9	19 54.4	9.8	57.1
02	211 47.8	18.8	173 18.7	9.9	20 04.2	9.7	57.1
03	226 48.1	. . 17.8	187 47.6	9.9	20 13.9	9.6	57.1
04	241 48.3	16.8	202 16.5	9.9	20 23.5	9.4	57.0
05	256 48.5	15.9	216 45.4	9.8	20 32.9	9.3	57.0
06	271 48.7	N 0 14.9	231 14.2	9.9	S20 42.2	9.3	57.0
07	286 48.9	13.9	245 43.1	9.8	20 51.5	9.0	56.9
08	301 49.2	12.9	260 11.9	9.7	21 00.5	9.0	56.9
09	316 49.4	. . 12.0	274 40.6	9.8	21 09.5	8.9	56.9
10	331 49.6	11.0	289 09.4	9.7	21 18.4	8.7	56.8
11	346 49.8	10.0	303 38.1	9.7	21 27.1	8.6	56.8
12	1 50.0	N 0 09.1	318 06.8	9.7	S21 35.7	8.4	56.8
13	16 50.3	08.1	332 35.5	9.6	21 44.1	8.4	56.7
14	31 50.5	07.1	347 04.1	9.6	21 52.5	8.2	56.7
15	46 50.7	. . 06.1	1 32.7	9.6	22 00.7	8.1	56.7
16	61 50.9	05.2	16 01.3	9.6	22 08.8	8.0	56.6
17	76 51.1	04.2	30 29.9	9.6	22 16.8	7.8	56.6
18	91 51.4	N 0 03.2	44 58.5	9.5	S22 24.6	7.7	56.6
19	106 51.6	02.2	59 27.0	9.5	22 32.3	7.6	56.5
20	121 51.8	01.3	73 55.5	9.5	22 39.9	7.5	56.5
21	136 52.0	N 00.3	88 24.0	9.5	22 47.4	7.3	56.5
22	151 52.2	S 00.7	102 52.5	9.4	22 54.7	7.2	56.4
23	166 52.5	01.6	117 20.9	9.5	23 01.9	7.1	56.4
3 00	181 52.7	S 0 02.6	131 49.4	9.4	S23 09.0	6.9	56.4
01	196 52.9	03.6	146 17.8	9.4	23 15.9	6.8	56.3
02	211 53.1	04.6	160 46.2	9.4	23 22.7	6.7	56.3
03	226 53.3	. . 05.5	175 14.6	9.4	23 29.4	6.5	56.3
04	241 53.5	06.5	189 43.0	9.3	23 35.9	6.4	56.2
05	256 53.8	07.5	204 11.3	9.4	23 42.3	6.3	56.2
06	271 54.0	S 0 08.5	218 39.7	9.3	S23 48.6	6.2	56.2
07	286 54.2	09.4	233 08.0	9.3	23 54.8	6.0	56.2
08	301 54.4	10.4	247 36.3	9.3	24 00.8	5.9	56.1
09	316 54.6	. . 11.4	262 04.6	9.3	24 06.7	5.7	56.1
10	331 54.9	12.4	276 32.9	9.3	24 12.4	5.6	56.1
11	346 55.1	13.3	291 01.2	9.2	24 18.0	5.5	56.0
12	1 55.3	S 0 14.3	305 29.4	9.3	S24 23.5	5.3	56.0
13	16 55.5	15.3	319 57.7	9.2	24 28.8	5.2	56.0
14	31 55.7	16.3	334 25.9	9.3	24 34.0	5.1	55.9
15	46 55.9	. . 17.2	348 54.2	9.2	24 39.1	4.9	55.9
16	61 56.2	18.2	3 22.4	9.2	24 44.0	4.9	55.9
17	76 56.4	19.2	17 50.6	9.3	24 48.9	4.6	55.9
18	91 56.6	S 0 20.1	32 18.9	9.2	S24 53.5	4.6	55.8
19	106 56.8	21.1	46 47.1	9.2	24 58.1	4.3	55.8
20	121 57.0	22.1	61 15.3	9.2	25 02.4	4.3	55.8
21	136 57.3	. . 23.1	75 43.5	9.2	25 06.7	4.1	55.7
22	151 57.5	24.0	90 11.7	9.2	25 10.8	4.0	55.7
23	166 57.7	25.0	104 39.9	9.2	25 14.8	3.9	55.7
4 00	181 57.9	S 0 26.0	119 08.1	9.2	S25 18.7	3.7	55.7
01	196 58.1	27.0	133 36.3	9.2	25 22.4	3.6	55.6
02	211 58.3	27.9	148 04.5	9.3	25 26.0	3.4	55.6
03	226 58.6	. . 28.9	162 32.8	9.2	25 29.4	3.3	55.6
04	241 58.8	29.9	177 01.0	9.2	25 32.7	3.2	55.5
05	256 59.0	30.9	191 29.2	9.2	25 35.9	3.0	55.5
06	271 59.2	S 0 31.8	205 57.4	9.2	S25 38.9	2.9	55.5
07	286 59.4	32.8	220 25.6	9.3	25 41.8	2.8	55.5
08	301 59.6	33.8	234 53.9	9.2	25 44.6	2.6	55.4
09	316 59.9	. . 34.8	249 22.1	9.3	25 47.2	2.5	55.4
10	332 00.1	35.7	263 50.4	9.2	25 49.7	2.4	55.4
11	347 00.3	36.7	278 18.6	9.3	25 52.1	2.2	55.4
12	2 00.5	S 0 37.7	292 46.9	9.3	S25 54.3	2.1	55.3
13	17 00.7	38.6	307 15.2	9.3	25 56.4	1.9	55.3
14	32 00.9	39.6	321 43.5	9.3	25 58.3	1.8	55.3
15	47 01.2	. . 40.6	336 11.8	9.4	26 00.1	1.7	55.3
16	62 01.4	41.6	350 40.2	9.3	26 01.8	1.5	55.2
17	77 01.6	42.5	5 08.5	9.4	26 03.3	1.5	55.2
18	92 01.8	S 0 43.5	19 36.9	9.3	S26 04.8	1.2	55.2
19	107 02.0	44.5	34 05.2	9.4	26 06.0	1.2	55.2
20	122 02.2	45.5	48 33.6	9.5	26 07.2	1.0	55.1
21	137 02.5	. . 46.4	63 02.1	9.4	26 08.2	0.8	55.1
22	152 02.7	47.4	77 30.5	9.5	26 09.0	0.8	55.1
23	167 02.9	48.4	91 59.0	9.4	S26 09.8	0.6	55.1
	SD 16.0	d 1.0	SD	15.5	15.3		15.1

Lat. °	Twilight Naut. h m	Twilight Civil h m	Sunrise h m	Moonrise 22 h m	Moonrise 23 h m	Moonrise 24 h m	Moonrise 25 h m
N 72	03 05	04 35	05 43	▬	▬	▬	▬
N 70	03 24	04 43	05 44	▬	▬	▬	▬
68	03 39	04 49	05 45	▬	▬	▬	▬
66	03 51	04 54	05 45	13 07	▬	▬	▬
64	04 00	04 58	05 46	12 29	14 26	▬	▬
62	04 08	05 02	05 46	12 02	13 40	15 04	15 57
60	04 15	05 05	05 47	11 41	13 11	14 26	15 19
N 58	04 21	05 08	05 47	11 24	12 48	13 59	14 53
56	04 26	05 10	05 47	11 10	12 30	13 38	14 32
54	04 30	05 12	05 47	10 57	12 14	13 21	14 14
52	04 34	05 14	05 48	10 47	12 01	13 06	13 59
50	04 38	05 16	05 48	10 37	11 49	12 53	13 47
45	04 45	05 19	05 48	10 17	11 25	12 27	13 20
N 40	04 50	05 21	05 49	10 01	11 06	12 06	12 59
35	04 54	05 23	05 49	09 47	10 50	11 49	12 42
30	04 57	05 25	05 49	09 35	10 36	11 34	12 26
20	05 01	05 27	05 49	09 15	10 12	11 08	12 01
N 10	05 04	05 28	05 49	08 57	09 52	10 46	11 39
0	05 04	05 28	05 49	08 41	09 33	10 26	11 18
S 10	05 04	05 28	05 49	08 25	09 15	10 06	10 58
20	05 01	05 27	05 49	08 08	08 55	09 44	10 36
30	04 57	05 24	05 48	07 48	08 32	09 19	10 10
35	04 53	05 23	05 48	07 37	08 18	09 04	09 55
40	04 49	05 21	05 48	07 24	08 03	08 47	09 38
45	04 44	05 18	05 47	07 09	07 44	08 27	09 17
S 50	04 36	05 14	05 47	06 50	07 21	08 01	08 50
52	04 33	05 13	05 46	06 41	07 10	07 49	08 37
54	04 29	05 11	05 46	06 31	06 58	07 34	08 22
56	04 24	05 09	05 46	06 20	06 44	07 18	08 05
58	04 19	05 06	05 45	06 08	06 27	06 58	07 44
S 60	04 13	05 03	05 45	05 53	06 08	06 34	07 18

Lat. °	Sunset h m	Twilight Civil h m	Twilight Naut. h m	Moonset 22 h m	Moonset 23 h m	Moonset 24 h m	Moonset 25 h m
N 72	17 59	19 07	20 35	▬	▬	▬	▬
N 70	17 58	18 59	20 17	▬	▬	▬	▬
68	17 58	18 53	20 03	▬	▬	▬	▬
66	17 57	18 49	19 51	16 28	▬	▬	▬
64	17 57	18 44	19 42	17 07	16 59	▬	▬
62	17 57	18 41	19 34	17 35	17 45	18 11	19 07
60	17 57	18 38	19 28	17 56	18 15	18 49	19 44
N 58	17 56	18 35	19 22	18 14	18 38	19 16	20 11
56	17 56	18 33	19 17	18 29	18 57	19 37	20 32
54	17 56	18 31	19 13	18 41	19 13	19 55	20 49
52	17 56	18 30	19 09	18 53	19 26	20 10	21 04
50	17 56	18 28	19 06	19 03	19 38	20 23	21 16
45	17 56	18 25	18 59	19 24	20 03	20 49	21 43
N 40	17 55	18 22	18 54	19 41	20 23	21 10	22 04
35	17 55	18 21	18 50	19 56	20 39	21 28	22 21
30	17 55	18 19	18 47	20 08	20 54	21 43	22 36
20	17 55	18 17	18 43	20 30	21 18	22 09	23 01
N 10	17 55	18 16	18 41	20 48	21 39	22 31	23 23
0	17 56	18 16	18 40	21 06	21 59	22 51	23 43
S 10	17 56	18 17	18 41	21 24	22 18	23 12	24 03
20	17 56	18 18	18 44	21 43	22 40	23 34	24 25
30	17 57	18 21	18 49	22 05	23 04	24 00	00 00
35	17 57	18 23	18 52	22 17	23 19	24 15	00 15
40	17 58	18 25	18 56	22 32	23 35	24 32	00 32
45	17 58	18 28	19 02	22 50	23 55	24 53	00 53
S 50	17 59	18 31	19 09	23 12	24 21	00 21	01 20
52	17 59	18 33	19 13	23 23	24 33	00 33	01 32
54	18 00	18 35	19 17	23 35	24 47	00 47	01 47
56	18 00	18 37	19 22	23 48	25 03	01 03	02 05
58	18 01	18 40	19 27	24 04	00 04	01 23	02 26
S 60	18 01	18 43	19 33	24 24	00 24	01 47	02 52

Day	SUN Eqn. of Time 00h	SUN Eqn. of Time 12h	SUN Mer. Pass.	MOON Mer. Pass. Upper	MOON Mer. Pass. Lower	MOON Age	MOON Phase
d	m s	m s	h m	h m	h m	d %	
22	07 09	07 20	11 53	14 54	02 28	04 17	
23	07 30	07 41	11 52	15 46	03 20	05 25	
24	07 51	08 02	11 52	16 39	04 12	06 34	

UT	ARIES	VENUS −3.9		MARS +0.8		JUPITER −2.7		SATURN +1.1		STARS		
	GHA	GHA	Dec	GHA	Dec	GHA	Dec	GHA	Dec	Name	SHA	Dec
d h	° ′	° ′	° ′	° ′	° ′	° ′	° ′	° ′	° ′		° ′	°
25 00	3 56.9	206 18.9	N10 31.4	253 59.0	N22 48.9	43 26.9	S16 33.8	186 55.8	N 3 26.8	Acamar	315 20.0	S40 15
01	18 59.3	221 18.4	30.3	268 59.9	48.8	58 29.5	33.8	201 58.0	26.7	Achernar	335 28.1	S57 11
02	34 01.8	236 17.9	29.3	284 00.8	48.6	73 32.1	33.8	217 00.2	26.6	Acrux	173 13.3	S63 09
03	49 04.3	251 17.5	. . 28.2	299 01.7	. . 48.5	88 34.7	. . 33.9	232 02.3	. . 26.5	Adhara	255 14.8	S28 58
04	64 06.7	266 17.0	27.2	314 02.7	48.4	103 37.3	33.9	247 04.5	26.3	Aldebaran	290 52.4	N16 31
05	79 09.2	281 16.6	26.1	329 03.6	48.2	118 39.9	34.0	262 06.7	26.2			
06	94 11.7	296 16.1	N10 25.1	344 04.5	N22 48.1	133 42.5	S16 34.0	277 08.9	N 3 26.1	Alioth	166 23.4	N55 54
07	109 14.1	311 15.6	24.0	359 05.4	47.9	148 45.1	34.1	292 11.0	26.0	Alkaid	153 01.3	N49 15
08	124 16.6	326 15.2	23.0	14 06.4	47.8	163 47.8	34.1	307 13.2	25.8	Al Na'ir	27 46.6	S46 54
F 09	139 19.1	341 14.7	. . 21.9	29 07.3	. . 47.6	178 50.4	. . 34.1	322 15.4	. . 25.7	Alnilam	275 49.1	S 1 11
R 10	154 21.5	356 14.3	20.9	44 08.2	47.5	193 53.0	34.2	337 17.6	25.6	Alphard	217 59.0	S 8 41
I 11	169 24.0	11 13.8	19.8	59 09.1	47.3	208 55.6	34.2	352 19.8	25.5			
D 12	184 26.4	26 13.4	N10 18.8	74 10.1	N22 47.2	223 58.2	S16 34.3	7 21.9	N 3 25.4	Alphecca	126 13.5	N26 41
A 13	199 28.9	41 12.9	17.7	89 11.0	47.1	239 00.8	34.3	22 24.1	25.2	Alpheratz	357 46.1	N29 08
Y 14	214 31.4	56 12.4	16.7	104 11.9	46.9	254 03.4	34.3	37 26.3	25.1	Altair	62 10.8	N 8 53
15	229 33.8	71 12.0	. . 15.6	119 12.9	. . 46.8	269 06.0	. . 34.4	52 28.5	. . 25.0	Ankaa	353 17.8	S42 15
16	244 36.3	86 11.5	14.5	134 13.8	46.6	284 08.6	34.4	67 30.6	24.9	Antares	112 29.8	S26 27
17	259 38.8	101 11.1	13.5	149 14.7	46.5	299 11.2	34.5	82 32.8	24.8			
18	274 41.2	116 10.6	N10 12.4	164 15.6	N22 46.3	314 13.8	S16 34.5	97 35.0	N 3 24.6	Arcturus	145 58.5	N19 08
19	289 43.7	131 10.2	11.4	179 16.6	46.2	329 16.4	34.5	112 37.2	24.5	Atria	107 34.4	S69 03
20	304 46.2	146 09.7	10.3	194 17.5	46.0	344 19.0	34.6	127 39.4	24.4	Avior	234 19.6	S59 32
21	319 48.6	161 09.3	. . 09.2	209 18.4	. . 45.9	359 21.6	. . 34.6	142 41.5	. . 24.3	Bellatrix	278 34.9	N 6 21
22	334 51.1	176 08.8	08.2	224 19.4	45.7	14 24.2	34.7	157 43.7	24.2	Betelgeuse	271 04.2	N 7 24
23	349 53.6	191 08.4	07.1	239 20.3	45.6	29 26.8	34.7	172 45.9	24.0			
26 00	4 56.0	206 07.9	N10 06.1	254 21.2	N22 45.4	44 29.5	S16 34.7	187 48.1	N 3 23.9	Canopus	263 57.4	S52 41
01	19 58.5	221 07.5	05.0	269 22.2	45.3	59 32.1	34.8	202 50.2	23.8	Capella	280 38.4	N46 00
02	35 00.9	236 07.0	03.9	284 23.1	45.1	74 34.7	34.8	217 52.4	23.7	Deneb	49 33.2	N45 19
03	50 03.4	251 06.5	. . 02.9	299 24.0	. . 45.0	89 37.3	. . 34.9	232 54.6	. . 23.5	Denebola	182 36.7	N14 31
04	65 05.9	266 06.1	01.8	314 25.0	44.8	104 39.9	34.9	247 56.8	23.4	Diphda	348 58.3	S17 55
05	80 08.3	281 05.6	10 00.7	329 25.9	44.7	119 42.5	34.9	262 59.0	23.3			
06	95 10.8	296 05.2	N 9 59.7	344 26.8	N22 44.5	134 45.1	S16 35.0	278 01.1	N 3 23.2	Dubhe	193 55.4	N61 41
S 07	110 13.3	311 04.7	58.6	359 27.8	44.4	149 47.7	35.0	293 03.3	23.1	Elnath	278 16.0	N28 37
A 08	125 15.7	326 04.3	57.6	14 28.7	44.2	164 50.3	35.1	308 05.5	22.9	Eltanin	90 47.5	N51 29
T 09	140 18.2	341 03.8	. . 56.5	29 29.6	. . 44.1	179 52.9	. . 35.1	323 07.7	. . 22.8	Enif	33 49.6	N 9 55
U 10	155 20.7	356 03.4	55.4	44 30.6	43.9	194 55.5	35.1	338 09.8	22.7	Fomalhaut	15 26.6	S29 34
R 11	170 23.1	11 02.9	54.4	59 31.5	43.8	209 58.1	35.2	353 12.0	22.6			
D 12	185 25.6	26 02.5	N 9 53.3	74 32.4	N22 43.6	225 00.7	S16 35.2	8 14.2	N 3 22.5	Gacrux	172 04.8	S57 10
A 13	200 28.1	41 02.0	52.2	89 33.4	43.5	240 03.3	35.3	23 16.4	22.3	Gienah	175 55.5	S17 35
Y 14	215 30.5	56 01.6	51.2	104 34.3	43.3	255 05.9	35.3	38 18.6	22.2	Hadar	148 52.6	S60 25
15	230 33.0	71 01.1	. . 50.1	119 35.3	. . 43.2	270 08.5	. . 35.3	53 20.7	. . 22.1	Hamal	328 03.6	N23 30
16	245 35.4	86 00.7	49.0	134 36.2	43.0	285 11.1	35.4	68 22.9	22.0	Kaus Aust.	83 47.4	S34 22
17	260 37.9	101 00.2	47.9	149 37.1	42.9	300 13.7	35.4	83 25.1	21.9			
18	275 40.4	115 59.8	N 9 46.9	164 38.1	N22 42.7	315 16.3	S16 35.4	98 27.3	N 3 21.7	Kochab	137 20.4	N74 07
19	290 42.8	130 59.3	45.8	179 39.0	42.6	330 18.9	35.5	113 29.5	21.6	Markab	13 40.8	N15 15
20	305 45.3	145 58.9	44.7	194 40.0	42.4	345 21.5	35.5	128 31.6	21.5	Menkar	314 17.7	N 4 07
21	320 47.8	160 58.4	. . 43.7	209 40.9	. . 42.3	0 24.1	. . 35.6	143 33.8	. . 21.4	Menkent	148 11.2	S36 25
22	335 50.2	175 58.0	42.6	224 41.8	42.1	15 26.7	35.6	158 36.0	21.3	Miaplacidus	221 41.2	S69 45
23	350 52.7	190 57.6	41.5	239 42.8	42.0	30 29.3	35.6	173 38.2	21.1			
27 00	5 55.2	205 57.1	N 9 40.4	254 43.7	N22 41.8	45 31.8	S16 35.7	188 40.3	N 3 21.0	Mirfak	308 44.1	N49 53
01	20 57.6	220 56.7	39.4	269 44.7	41.7	60 34.4	35.7	203 42.5	20.9	Nunki	76 01.6	S26 17
02	36 00.1	235 56.2	38.3	284 45.6	41.5	75 37.0	35.8	218 44.7	20.8	Peacock	53 23.1	S56 42
03	51 02.6	250 55.8	. . 37.2	299 46.5	. . 41.4	90 39.6	. . 35.8	233 46.9	. . 20.7	Pollux	243 31.2	N28 00
04	66 05.0	265 55.3	36.1	314 47.5	41.2	105 42.2	35.8	248 49.1	20.5	Procyon	245 02.7	N 5 12
05	81 07.5	280 54.9	35.1	329 48.4	41.1	120 44.8	35.9	263 51.2	20.4			
06	96 09.9	295 54.4	N 9 34.0	344 49.4	N22 40.9	135 47.4	S16 35.9	278 53.4	N 3 20.3	Rasalhague	96 09.1	N12 33
07	111 12.4	310 54.0	32.9	359 50.3	40.8	150 50.0	35.9	293 55.6	20.2	Regulus	207 46.7	N11 55
08	126 14.9	325 53.5	31.8	14 51.3	40.6	165 52.6	36.0	308 57.8	20.1	Rigel	281 14.6	S 8 11
S 09	141 17.3	340 53.1	. . 30.8	29 52.2	. . 40.5	180 55.2	. . 36.0	323 59.9	. . 19.9	Rigil Kent.	139 56.2	S60 52
U 10	156 19.8	355 52.6	29.7	44 53.1	40.3	195 57.8	36.0	339 02.1	19.8	Sabik	102 15.8	S15 44
N 11	171 22.3	10 52.2	28.6	59 54.1	40.1	211 00.4	36.1	354 04.3	19.7			
D 12	186 24.7	25 51.8	N 9 27.5	74 55.0	N22 40.0	226 03.0	S16 36.1	9 06.5	N 3 19.6	Schedar	349 43.3	N56 35
A 13	201 27.2	40 51.3	26.4	89 56.0	39.8	241 05.6	36.2	24 08.7	19.4	Shaula	96 25.7	S37 06
Y 14	216 29.7	55 50.9	25.4	104 56.9	39.7	256 08.2	36.2	39 10.8	19.3	Sirius	258 36.2	S16 43
15	231 32.1	70 50.4	. . 24.3	119 57.9	. . 39.5	271 10.8	. . 36.2	54 13.0	. . 19.2	Spica	158 34.5	S11 12
16	246 34.6	85 50.0	23.2	134 58.8	39.4	286 13.3	36.3	69 15.2	19.1	Suhail	222 54.9	S43 28
17	261 37.0	100 49.5	22.1	149 59.8	39.2	301 15.9	36.3	84 17.4	19.0			
18	276 39.5	115 49.1	N 9 21.0	165 00.7	N22 39.1	316 18.5	S16 36.3	99 19.6	N 3 18.8	Vega	80 40.8	N38 47
19	291 42.0	130 48.7	20.0	180 01.7	38.9	331 21.1	36.4	114 21.7	18.7	Zuben'ubi	137 08.7	S16 05
20	306 44.4	145 48.2	18.9	195 02.6	38.8	346 23.7	36.4	129 23.9	18.6		SHA	Mer.Pas
21	321 46.9	160 47.8	. . 17.8	210 03.6	. . 38.6	1 26.3	. . 36.4	144 26.1	. . 18.5		° ′	h m
22	336 49.4	175 47.3	16.7	225 04.5	38.4	16 28.9	36.5	159 28.3	18.4	Venus	201 11.9	10 16
23	351 51.8	190 46.9	15.6	240 05.5	38.3	31 31.5	36.5	174 30.4	18.2	Mars	249 25.2	7 02
Mer. Pass.	h m 23 36.4	v −0.5	d 1.1	v 0.9	d 0.2	v 2.6	d 0.0	v 2.2	d 0.1	Jupiter	39 33.4	20 58
										Saturn	182 52.1	11 27

UT	SUN GHA	SUN Dec	MOON GHA	v	MOON Dec	d	HP
d h	° ′	° ′	° ′	′	° ′	′	′
25 00	182 03.1	S 0 49.4	106 27.4	9.5	S26 10.4	0.4	55.1
01	197 03.3	50.3	120 55.9	9.6	26 10.8	0.4	55.0
02	212 03.5	51.3	135 24.5	9.5	26 11.2	0.2	55.0
03	227 03.8	. . 52.3	149 53.0	9.6	26 11.4	0.0	55.0
04	242 04.0	53.3	164 21.6	9.6	26 11.4	0.0	55.0
05	257 04.2	54.2	178 50.2	9.7	26 11.4	0.2	54.9
06	272 04.4	S 0 55.2	193 18.9	9.6	S26 11.2	0.3	54.9
07	287 04.6	56.2	207 47.5	9.7	26 10.9	0.5	54.9
08	302 04.8	57.2	222 16.2	9.7	26 10.4	0.6	54.9
09	317 05.1	. . 58.1	236 44.9	9.8	26 09.8	0.7	54.9
10	332 05.3	0 59.1	251 13.7	9.8	26 09.1	0.9	54.8
11	347 05.5	1 00.1	265 42.5	9.8	26 08.2	0.9	54.8
12	2 05.7	S 1 01.0	280 11.3	9.8	S26 07.3	1.1	54.8
13	17 05.9	02.0	294 40.1	9.9	26 06.2	1.3	54.8
14	32 06.1	03.0	309 09.0	9.9	26 04.9	1.3	54.8
15	47 06.3	. . 04.0	323 37.9	10.0	26 03.6	1.5	54.8
16	62 06.6	04.9	338 06.9	10.0	26 02.1	1.7	54.7
17	77 06.8	05.9	352 35.9	10.0	26 00.4	1.7	54.7
18	92 07.0	S 1 06.9	7 04.9	10.1	S25 58.7	1.9	54.7
19	107 07.2	07.9	21 34.0	10.1	25 56.8	2.0	54.7
20	122 07.4	08.8	36 03.1	10.2	25 54.8	2.1	54.7
21	137 07.6	. . 09.8	50 32.3	10.2	25 52.7	2.3	54.7
22	152 07.8	10.8	65 01.5	10.2	25 50.4	2.3	54.6
23	167 08.1	11.8	79 30.7	10.3	25 48.1	2.6	54.6
26 00	182 08.3	S 1 12.7	94 00.0	10.3	S25 45.5	2.6	54.6
01	197 08.5	13.7	108 29.3	10.3	25 42.9	2.7	54.6
02	212 08.7	14.7	122 58.6	10.4	25 40.2	2.9	54.6
03	227 08.9	. . 15.7	137 28.0	10.5	25 37.3	3.0	54.6
04	242 09.1	16.6	151 57.5	10.5	25 34.3	3.1	54.5
05	257 09.3	17.6	166 27.0	10.5	25 31.2	3.2	54.5
06	272 09.6	S 1 18.6	180 56.5	10.6	S25 28.0	3.4	54.5
07	287 09.8	19.5	195 26.1	10.6	25 24.6	3.5	54.5
08	302 10.0	20.5	209 55.7	10.7	25 21.1	3.6	54.5
09	317 10.2	. . 21.5	224 25.4	10.7	25 17.5	3.7	54.5
10	332 10.4	22.5	238 55.1	10.8	25 13.8	3.8	54.5
11	347 10.6	23.4	253 24.9	10.8	25 10.0	4.0	54.5
12	2 10.8	S 1 24.4	267 54.7	10.9	S25 06.0	4.1	54.4
13	17 11.1	25.4	282 24.6	10.9	25 01.9	4.1	54.4
14	32 11.3	26.4	296 54.5	11.0	24 57.8	4.3	54.4
15	47 11.5	. . 27.3	311 24.5	11.0	24 53.5	4.5	54.4
16	62 11.7	28.3	325 54.5	11.1	24 49.0	4.5	54.4
17	77 11.9	29.3	340 24.6	11.1	24 44.5	4.6	54.4
18	92 12.1	S 1 30.3	354 54.7	11.2	S24 39.9	4.8	54.4
19	107 12.3	31.2	9 24.9	11.2	24 35.1	4.9	54.4
20	122 12.6	32.2	23 55.1	11.3	24 30.2	5.0	54.4
21	137 12.8	. . 33.2	38 25.4	11.3	24 25.2	5.1	54.4
22	152 13.0	34.1	52 55.7	11.4	24 20.1	5.2	54.3
23	167 13.2	35.1	67 26.1	11.5	24 14.9	5.3	54.3
27 00	182 13.4	S 1 36.1	81 56.6	11.5	S24 09.6	5.4	54.3
01	197 13.6	37.1	96 27.1	11.5	24 04.2	5.6	54.3
02	212 13.8	38.0	110 57.6	11.6	23 58.6	5.6	54.3
03	227 14.0	. . 39.0	125 28.2	11.7	23 53.0	5.7	54.3
04	242 14.3	40.0	139 58.9	11.7	23 47.3	5.9	54.3
05	257 14.5	41.0	154 29.6	11.7	23 41.4	6.0	54.3
06	272 14.7	S 1 41.9	169 00.3	11.9	S23 35.4	6.0	54.3
07	287 14.9	42.9	183 31.2	11.8	23 29.4	6.2	54.3
08	302 15.1	43.9	198 02.0	12.0	23 23.2	6.3	54.3
09	317 15.3	. . 44.9	212 33.0	11.9	23 16.9	6.4	54.3
10	332 15.5	45.8	227 03.9	12.1	23 10.5	6.5	54.3
11	347 15.7	46.8	241 35.0	12.1	23 04.0	6.6	54.3
12	2 15.9	S 1 47.8	256 06.1	12.1	S22 57.4	6.7	54.3
13	17 16.2	48.7	270 37.2	12.2	22 50.7	6.8	54.2
14	32 16.4	49.7	285 08.4	12.3	22 43.9	6.9	54.2
15	47 16.6	. . 50.7	299 39.7	12.3	22 37.0	7.0	54.2
16	62 16.8	51.7	314 11.0	12.4	22 30.0	7.0	54.2
17	77 17.0	52.6	328 42.4	12.4	22 23.0	7.2	54.2
18	92 17.2	S 1 53.6	343 13.8	12.5	S22 15.8	7.3	54.2
19	107 17.4	54.6	357 45.3	12.5	22 08.5	7.4	54.2
20	122 17.6	55.6	12 16.8	12.6	22 01.1	7.5	54.2
21	137 17.8	. . 56.5	26 48.4	12.7	21 53.6	7.6	54.2
22	152 18.1	57.5	41 20.1	12.7	21 46.0	7.6	54.2
23	167 18.3	58.5	55 51.8	12.7	S21 38.4	7.8	54.2
	SD 16.0	*d* 1.0	SD 14.9		14.8		14.8

Lat.	Twilight Naut.	Twilight Civil	Sunrise	Moonrise 25	Moonrise 26	Moonrise 27	Moonrise 28
°	h m	h m	h m	h m	h m	h m	h m
N 72	03 23	04 49	05 57	▬▬	▬▬	▬▬	▬▬
N 70	03 39	04 55	05 56	▬▬	▬▬	▬▬	18 34
68	03 51	05 00	05 55	▬▬	▬▬	▬▬	17 46
66	04 02	05 04	05 55	▬▬	▬▬	17 38	17 15
64	04 10	05 07	05 54	▬▬	17 05	16 57	16 52
62	04 17	05 10	05 54	15 57	16 20	16 29	16 33
60	04 23	05 12	05 54	15 19	15 50	16 08	16 18
N 58	04 28	05 14	05 53	14 53	15 28	15 50	16 05
56	04 32	05 16	05 53	14 32	15 09	15 35	15 53
54	04 36	05 18	05 53	14 14	14 54	15 23	15 43
52	04 39	05 19	05 53	13 59	14 40	15 11	15 35
50	04 42	05 20	05 52	13 47	14 29	15 01	15 27
45	04 48	05 23	05 52	13 20	14 04	14 40	15 09
N 40	04 53	05 24	05 51	12 59	13 44	14 23	14 55
35	04 56	05 26	05 51	12 42	13 28	14 08	14 43
30	04 59	05 27	05 51	12 26	13 14	13 56	14 33
20	05 02	05 28	05 50	12 01	12 50	13 34	14 15
N 10	05 04	05 28	05 49	11 39	12 29	13 15	13 59
0	05 03	05 27	05 48	11 18	12 09	12 58	13 44
S 10	05 02	05 26	05 47	10 58	11 49	12 40	13 29
20	04 58	05 24	05 46	10 36	11 28	12 21	13 14
30	04 53	05 21	05 45	10 10	11 04	11 59	12 55
35	04 49	05 18	05 44	09 55	10 50	11 47	12 44
40	04 44	05 16	05 43	09 38	10 33	11 32	12 32
45	04 38	05 12	05 41	09 17	10 13	11 14	12 17
S 50	04 29	05 08	05 40	08 50	09 48	10 52	12 00
52	04 25	05 06	05 39	08 37	09 36	10 41	11 51
54	04 21	05 03	05 39	08 22	09 22	10 30	11 42
56	04 16	05 00	05 38	08 05	09 06	10 16	11 31
58	04 10	04 57	05 37	07 44	08 46	10 00	11 18
S 60	04 03	04 54	05 36	07 18	08 22	09 40	11 04

Lat.	Sunset	Twilight Civil	Twilight Naut.	Moonset 25	Moonset 26	Moonset 27	Moonset 28
°	h m	h m	h m	h m	h m	h m	h m
N 72	17 44	18 51	20 16	▬▬	▬▬	▬▬	▬▬
N 70	17 44	18 45	20 00	▬▬	▬▬	▬▬	21 32
68	17 45	18 41	19 48	▬▬	▬▬	▬▬	22 19
66	17 46	18 37	19 38	▬▬	▬▬	20 52	22 49
64	17 46	18 34	19 30	▬▬	19 44	21 31	23 12
62	17 47	18 31	19 24	19 07	20 28	21 59	23 29
60	17 47	18 29	19 18	19 44	20 57	22 20	23 44
N 58	17 48	18 27	19 13	20 11	21 20	22 37	23 57
56	17 48	18 25	19 09	20 32	21 38	22 51	24 07
54	17 49	18 24	19 05	20 49	21 53	23 04	24 17
52	17 49	18 22	19 02	21 04	22 06	23 14	24 25
50	17 49	18 21	18 59	21 16	22 18	23 24	24 32
45	17 50	18 19	18 53	21 43	22 42	23 44	24 48
N 40	17 51	18 17	18 49	22 04	23 01	24 01	00 01
35	17 51	18 16	18 46	22 21	23 17	24 14	00 14
30	17 52	18 15	18 43	22 36	23 31	24 26	00 26
20	17 53	18 15	18 40	23 01	23 54	24 47	00 47
N 10	17 54	18 15	18 39	23 23	24 14	00 14	01 04
0	17 55	18 15	18 39	23 43	24 33	00 33	01 21
S 10	17 56	18 17	18 41	24 03	00 03	00 52	01 37
20	17 57	18 19	18 45	24 25	00 25	01 12	01 54
30	17 59	18 22	18 50	00 00	00 50	01 34	02 14
35	17 59	18 25	18 54	00 15	01 04	01 48	02 25
40	18 01	18 28	19 00	00 32	01 21	02 03	02 39
45	18 02	18 31	19 06	00 53	01 42	02 22	02 54
S 50	18 04	18 36	19 14	01 20	02 07	02 44	03 13
52	18 04	18 38	19 19	01 32	02 20	02 55	03 22
54	18 05	18 41	19 23	01 47	02 34	03 08	03 32
56	18 06	18 43	19 28	02 05	02 50	03 22	03 43
58	18 07	18 47	19 34	02 26	03 10	03 38	03 56
S 60	18 08	18 50	19 41	02 52	03 34	03 58	04 11

Day	SUN Eqn. of Time 00^h	SUN Eqn. of Time 12^h	SUN Mer. Pass.	MOON Mer. Pass. Upper	MOON Mer. Pass. Lower	MOON Age	MOON Phase
d	m s	m s	h m	h m	h m	d %	
25	08 12	08 22	11 52	17 31	05 05	07 43	
26	08 33	08 43	11 51	18 21	05 56	08 53	◐
27	08 53	09 03	11 51	19 09	06 45	09 62	

	UT	ARIES	VENUS −3.9		MARS +0.8		JUPITER −2.7		SATURN +1.1		STARS		
		GHA	GHA	Dec	GHA	Dec	GHA	Dec	GHA	Dec	Name	SHA	Dec
	d h	° ′	° ′	° ′	° ′	° ′	° ′	° ′	° ′	° ′		° ′	° ′
	28 00	6 54.3	205 46.4	N 9 14.5	255 06.4	N22 38.1	46 34.1	S16 36.5	189 32.6	N 3 18.1	Acamar	315 20.0	S40 15.
	01	21 56.8	220 46.0	13.4	270 07.4	38.0	61 36.6	36.6	204 34.8	18.0	Achernar	335 28.0	S57 11.
	02	36 59.2	235 45.6	12.4	285 08.3	37.8	76 39.2	36.6	219 37.0	17.9	Acrux	173 13.3	S63 09.
	03	52 01.7	250 45.1	11.3	300 09.3	37.7	91 41.8	36.6	234 39.2	17.8	Adhara	255 14.7	S28 58.
	04	67 04.2	265 44.7	10.2	315 10.2	37.5	106 44.4	36.7	249 41.3	17.6	Aldebaran	290 52.4	N16 31.
	05	82 06.6	280 44.2	09.1	330 11.2	37.3	121 47.0	36.7	264 43.5	17.5			
	06	97 09.1	295 43.8	N 9 08.0	345 12.1	N22 37.2	136 49.6	S16 36.8	279 45.7	N 3 17.4	Alioth	166 23.4	N55 54.
	07	112 11.5	310 43.4	06.9	0 13.1	37.0	151 52.2	36.8	294 47.9	17.3	Alkaid	153 01.3	N49 15.
	08	127 14.0	325 42.9	05.8	15 14.0	36.9	166 54.8	36.8	309 50.1	17.2	Al Na'ir	27 46.6	S46 54.
M	09	142 16.5	340 42.5	04.7	30 15.0	36.7	181 57.3	36.9	324 52.2	17.0	Alnilam	275 49.1	S 1 11.
O	10	157 18.9	355 42.0	03.7	45 15.9	36.6	196 59.9	36.9	339 54.4	16.9	Alphard	217 59.0	S 8 41.
N	11	172 21.4	10 41.6	02.6	60 16.9	36.4	212 02.5	36.9	354 56.6	16.8			
D	12	187 23.9	25 41.2	N 9 01.5	75 17.8	N22 36.2	227 05.1	S16 37.0	9 58.8	N 3 16.7	Alphecca	126 13.5	N26 41.
A	13	202 26.3	40 40.7	9 00.4	90 18.8	36.1	242 07.7	37.0	25 01.0	16.6	Alpheratz	357 46.1	N29 08.
Y	14	217 28.8	55 40.3	8 59.3	105 19.7	35.9	257 10.3	37.0	40 03.1	16.4	Altair	62 10.8	N 8 53.
	15	232 31.3	70 39.9	58.2	120 20.7	35.8	272 12.8	37.1	55 05.3	16.3	Ankaa	353 17.8	S42 15.
	16	247 33.7	85 39.4	57.1	135 21.6	35.6	287 15.4	37.1	70 07.5	16.2	Antares	112 29.8	S26 27.
	17	262 36.2	100 39.0	56.0	150 22.6	35.4	302 18.0	37.1	85 09.7	16.1			
	18	277 38.7	115 38.5	N 8 54.9	165 23.6	N22 35.3	317 20.6	S16 37.2	100 11.8	N 3 16.0	Arcturus	145 58.5	N19 08.
	19	292 41.1	130 38.1	53.8	180 24.5	35.1	332 23.2	37.2	115 14.0	15.8	Atria	107 34.4	S69 03.
	20	307 43.6	145 37.7	52.7	195 25.5	35.0	347 25.8	37.2	130 16.2	15.7	Avior	234 19.6	S59 32.
	21	322 46.0	160 37.2	51.6	210 26.4	34.8	2 28.3	37.3	145 18.4	15.6	Bellatrix	278 34.9	N 6 21.
	22	337 48.5	175 36.8	50.5	225 27.4	34.6	17 30.9	37.3	160 20.6	15.5	Betelgeuse	271 04.2	N 7 24.
	23	352 51.0	190 36.4	49.4	240 28.3	34.5	32 33.5	37.3	175 22.7	15.4			
	29 00	7 53.4	205 35.9	N 8 48.4	255 29.3	N22 34.3	47 36.1	S16 37.4	190 24.9	N 3 15.2	Canopus	263 57.4	S52 41.
	01	22 55.9	220 35.5	47.3	270 30.3	34.2	62 38.7	37.4	205 27.1	15.1	Capella	280 38.4	N46 00.
	02	37 58.4	235 35.1	46.2	285 31.2	34.0	77 41.2	37.4	220 29.3	15.0	Deneb	49 33.2	N45 19.
	03	53 00.8	250 34.6	45.1	300 32.2	33.8	92 43.8	37.4	235 31.5	14.9	Denebola	182 36.7	N14 31.
	04	68 03.3	265 34.2	44.0	315 33.1	33.7	107 46.4	37.5	250 33.6	14.8	Diphda	348 58.3	S17 55.
	05	83 05.8	280 33.7	42.9	330 34.1	33.5	122 49.0	37.5	265 35.8	14.6			
	06	98 08.2	295 33.3	N 8 41.8	345 35.1	N22 33.4	137 51.6	S16 37.5	280 38.0	N 3 14.5	Dubhe	193 55.4	N61 41.
	07	113 10.7	310 32.9	40.7	0 36.0	33.2	152 54.1	37.6	295 40.2	14.4	Elnath	278 16.0	N28 37.
T	08	128 13.1	325 32.4	39.6	15 37.0	33.0	167 56.7	37.6	310 42.4	14.3	Eltanin	90 47.5	N51 29.
U	09	143 15.6	340 32.0	38.5	30 37.9	32.9	182 59.3	37.6	325 44.5	14.2	Enif	33 49.6	N 9 55.
E	10	158 18.1	355 31.6	37.4	45 38.9	32.7	198 01.9	37.7	340 46.7	14.0	Fomalhaut	15 26.6	S29 34.
S	11	173 20.5	10 31.1	36.3	60 39.9	32.6	213 04.5	37.7	355 48.9	13.9			
D	12	188 23.0	25 30.7	N 8 35.2	75 40.8	N22 32.4	228 07.0	S16 37.7	10 51.1	N 3 13.8	Gacrux	172 04.8	S57 10.
A	13	203 25.5	40 30.3	34.1	90 41.8	32.2	243 09.6	37.8	25 53.3	13.7	Gienah	175 55.5	S17 35.
Y	14	218 27.9	55 29.8	33.0	105 42.8	32.1	258 12.2	37.8	40 55.4	13.6	Hadar	148 52.6	S60 25.
	15	233 30.4	70 29.4	31.9	120 43.7	31.9	273 14.8	37.8	55 57.6	13.4	Hamal	328 03.6	N23 30.
	16	248 32.9	85 29.0	30.7	135 44.7	31.7	288 17.3	37.9	70 59.8	13.3	Kaus Aust.	83 47.4	S34 22.
	17	263 35.3	100 28.5	29.6	150 45.7	31.6	303 19.9	37.9	86 02.0	13.2			
	18	278 37.8	115 28.1	N 8 28.5	165 46.6	N22 31.4	318 22.5	S16 37.9	101 04.1	N 3 13.1	Kochab	137 20.5	N74 07.
	19	293 40.3	130 27.7	27.4	180 47.6	31.2	333 25.1	37.9	116 06.3	13.0	Markab	13 40.8	N15 15.
	20	308 42.7	145 27.3	26.3	195 48.5	31.1	348 27.6	38.0	131 08.5	12.8	Menkar	314 17.7	N 4 07.
	21	323 45.2	160 26.8	25.2	210 49.5	30.9	3 30.2	38.0	146 10.7	12.7	Menkent	148 11.3	S36 25.
	22	338 47.6	175 26.4	24.1	225 50.5	30.8	18 32.8	38.0	161 12.9	12.6	Miaplacidus	221 41.1	S69 45.
	23	353 50.1	190 26.0	23.0	240 51.4	30.6	33 35.4	38.1	176 15.0	12.5			
	30 00	8 52.6	205 25.5	N 8 21.9	255 52.4	N22 30.4	48 37.9	S16 38.1	191 17.2	N 3 12.4	Mirfak	308 44.0	N49 53.
	01	23 55.0	220 25.1	20.8	270 53.4	30.3	63 40.5	38.1	206 19.4	12.2	Nunki	76 01.6	S26 17.
	02	38 57.5	235 24.7	19.7	285 54.4	30.1	78 43.1	38.2	221 21.6	12.1	Peacock	53 23.1	S56 42.
	03	54 00.0	250 24.2	18.6	300 55.3	29.9	93 45.6	38.2	236 23.8	12.0	Pollux	243 31.2	N28 00.
	04	69 02.4	265 23.8	17.5	315 56.3	29.8	108 48.2	38.2	251 25.9	11.9	Procyon	245 02.7	N 5 12.
	05	84 04.9	280 23.4	16.4	330 57.3	29.6	123 50.8	38.2	266 28.1	11.8			
	06	99 07.4	295 23.0	N 8 15.3	345 58.2	N22 29.4	138 53.4	S16 38.3	281 30.3	N 3 11.6	Rasalhague	96 09.1	N12 33.
W	07	114 09.8	310 22.5	14.1	0 59.2	29.3	153 55.9	38.3	296 32.5	11.5	Regulus	207 46.7	N11 55.
E	08	129 12.3	325 22.1	13.0	16 00.2	29.1	168 58.5	38.3	311 34.7	11.4	Rigel	281 14.6	S 8 11.
D	09	144 14.8	340 21.7	11.9	31 01.1	28.9	184 01.1	38.4	326 36.8	11.3	Rigil Kent.	139 56.2	S60 52.
N	10	159 17.2	355 21.2	10.8	46 02.1	28.8	199 03.6	38.4	341 39.0	11.2	Sabik	102 15.8	S15 44.
E	11	174 19.7	10 20.8	09.7	61 03.1	28.6	214 06.2	38.4	356 41.2	11.0			
S	12	189 22.1	25 20.4	N 8 08.6	76 04.1	N22 28.4	229 08.8	S16 38.4	11 43.4	N 3 10.9	Schedar	349 43.3	N56 35.
D	13	204 24.6	40 20.0	07.5	91 05.0	28.3	244 11.3	38.5	26 45.6	10.8	Shaula	96 25.7	S37 06.
A	14	219 27.1	55 19.5	06.4	106 06.0	28.1	259 13.9	38.5	41 47.7	10.7	Sirius	258 36.1	S16 43.
Y	15	234 29.5	70 19.1	05.3	121 07.0	27.9	274 16.5	38.5	56 49.9	10.6	Spica	158 34.5	S11 12.
	16	249 32.0	85 18.7	04.1	136 07.9	27.8	289 19.0	38.6	71 52.1	10.5	Suhail	222 54.8	S43 28.
	17	264 34.5	100 18.2	03.0	151 08.9	27.6	304 21.6	38.6	86 54.3	10.3			
	18	279 36.9	115 17.8	N 8 01.9	166 09.9	N22 27.4	319 24.2	S16 38.6	101 56.5	N 3 10.2	Vega	80 40.8	N38 47.
	19	294 39.4	130 17.4	8 00.8	181 10.9	27.3	334 26.7	38.6	116 58.6	10.1	Zuben'ubi	137 08.7	S16 05.
	20	309 41.9	145 17.0	7 59.7	196 11.8	27.1	349 29.3	38.7	132 00.8	10.0		SHA	Mer. Pass.
	21	324 44.3	160 16.5	58.6	211 12.8	26.9	4 31.9	38.7	147 03.0	09.9		° ′	h m
	22	339 46.8	175 16.1	57.4	226 13.8	26.8	19 34.4	38.7	162 05.2	09.7	Venus	197 42.5	10 18
	23	354 49.2	190 15.7	56.3	241 14.8	26.6	34 37.0	38.8	177 07.4	09.6	Mars	247 35.9	6 58
		h m									Jupiter	39 42.7	20 46
	Mer. Pass.	23 24.6	v −0.4	d 1.1	v 1.0	d 0.2	v 2.6	d 0.0	v 2.2	d 0.1	Saturn	182 31.5	11 17

UT	SUN GHA	SUN Dec	MOON GHA	v	MOON Dec	d	HP
d h	° ′	° ′	° ′	′	° ′	′	′
28 00	182 18.5	S 1 59.4	70 23.5	12.8	S21 30.6	7.9	54.2
01	197 18.7	2 00.4	84 55.3	12.9	21 22.7	7.9	54.2
02	212 18.9	01.4	99 27.2	12.9	21 14.8	8.1	54.2
03	227 19.1 . .	02.4	113 59.1	13.0	21 06.7	8.1	54.2
04	242 19.3	03.3	128 31.1	13.0	20 58.6	8.2	54.2
05	257 19.5	04.3	143 03.1	13.1	20 50.4	8.3	54.2
06	272 19.7	S 2 05.3	157 35.2	13.1	S20 42.1	8.4	54.2
07	287 20.0	06.3	172 07.3	13.2	20 33.7	8.5	54.2
08	302 20.2	07.2	186 39.5	13.2	20 25.2	8.6	54.2
09	317 20.4 . .	08.2	201 11.7	13.3	20 16.6	8.7	54.2
10	332 20.6	09.2	215 44.0	13.3	20 07.9	8.8	54.2
11	347 20.8	10.1	230 16.3	13.4	19 59.1	8.8	54.2
12	2 21.0	S 2 11.1	244 48.7	13.5	S19 50.3	8.9	54.2
13	17 21.2	12.1	259 21.2	13.5	19 41.4	9.0	54.2
14	32 21.4	13.1	273 53.7	13.5	19 32.4	9.1	54.2
15	47 21.6 . .	14.0	288 26.2	13.6	19 23.3	9.2	54.2
16	62 21.8	15.0	302 58.8	13.6	19 14.1	9.3	54.2
17	77 22.0	16.0	317 31.4	13.7	19 04.8	9.3	54.2
18	92 22.3	S 2 16.9	332 04.1	13.8	S18 55.5	9.4	54.2
19	107 22.5	17.9	346 36.9	13.8	18 46.1	9.5	54.3
20	122 22.7	18.9	1 09.7	13.8	18 36.6	9.6	54.3
21	137 22.9 . .	19.9	15 42.5	13.9	18 27.0	9.7	54.3
22	152 23.1	20.8	30 15.4	13.9	18 17.3	9.7	54.3
23	167 23.3	21.8	44 48.3	14.0	18 07.6	9.8	54.3
29 00	182 23.5	S 2 22.8	59 21.3	14.0	S17 57.8	9.9	54.3
01	197 23.7	23.8	73 54.3	14.1	17 47.9	10.0	54.3
02	212 23.9	24.7	88 27.4	14.1	17 37.9	10.0	54.3
03	227 24.1 . .	25.7	103 00.5	14.2	17 27.9	10.1	54.3
04	242 24.3	26.7	117 33.7	14.2	17 17.8	10.2	54.3
05	257 24.5	27.6	132 06.9	14.3	17 07.6	10.3	54.3
06	272 24.8	S 2 28.6	146 40.2	14.3	S16 57.3	10.3	54.3
07	287 25.0	29.6	161 13.5	14.3	16 47.0	10.4	54.3
08	302 25.2	30.6	175 46.8	14.4	16 36.6	10.5	54.3
09	317 25.4 . .	31.5	190 20.2	14.4	16 26.1	10.6	54.3
10	332 25.6	32.5	204 53.6	14.5	16 15.5	10.6	54.3
11	347 25.8	33.5	219 27.1	14.5	16 04.9	10.7	54.4
12	2 26.0	S 2 34.4	234 00.6	14.5	S15 54.2	10.7	54.4
13	17 26.2	35.4	248 34.1	14.6	15 43.5	10.9	54.4
14	32 26.4	36.4	263 07.7	14.6	15 32.6	10.9	54.4
15	47 26.6 . .	37.4	277 41.3	14.7	15 21.7	10.9	54.4
16	62 26.8	38.3	292 15.0	14.7	15 10.8	11.0	54.4
17	77 27.0	39.3	306 48.7	14.7	14 59.8	11.1	54.4
18	92 27.2	S 2 40.3	321 22.4	14.8	S14 48.7	11.2	54.4
19	107 27.4	41.2	335 56.2	14.8	14 37.5	11.2	54.4
20	122 27.6	42.2	350 30.0	14.8	14 26.3	11.3	54.4
21	137 27.9 . .	43.2	5 03.8	14.9	14 15.0	11.3	54.4
22	152 28.1	44.1	19 37.7	14.9	14 03.7	11.4	54.5
23	167 28.3	45.1	34 11.6	14.9	13 52.3	11.5	54.5
30 00	182 28.5	S 2 46.1	48 45.5	15.0	S13 40.8	11.5	54.5
01	197 28.7	47.1	63 19.5	15.0	13 29.3	11.6	54.5
02	212 28.9	48.0	77 53.5	15.0	13 17.7	11.6	54.5
03	227 29.1 . .	49.0	92 27.5	15.1	13 06.1	11.7	54.5
04	242 29.3	50.0	107 01.6	15.1	12 54.4	11.8	54.5
05	257 29.5	50.9	121 35.7	15.1	12 42.6	11.8	54.5
06	272 29.7	S 2 51.9	136 09.8	15.2	S12 30.8	11.9	54.6
07	287 29.9	52.9	150 44.0	15.1	12 18.9	11.9	54.6
08	302 30.1	53.9	165 18.1	15.2	12 07.0	11.9	54.6
09	317 30.3 . .	54.8	179 52.3	15.3	11 55.1	12.1	54.6
10	332 30.5	55.8	194 26.6	15.2	11 43.0	12.1	54.6
11	347 30.7	56.8	209 00.8	15.3	11 30.9	12.1	54.6
12	2 30.9	S 2 57.7	223 35.1	15.3	S11 18.8	12.2	54.6
13	17 31.1	58.7	238 09.4	15.3	11 06.6	12.2	54.6
14	32 31.3	2 59.7	252 43.7	15.3	10 54.4	12.3	54.7
15	47 31.5 . .	3 00.6	267 18.0	15.4	10 42.1	12.3	54.7
16	62 31.7	01.6	281 52.4	15.4	10 29.8	12.4	54.7
17	77 32.0	02.6	296 26.8	15.4	10 17.4	12.4	54.7
18	92 32.2	S 3 03.6	311 01.2	15.4	S10 05.0	12.5	54.7
19	107 32.4	04.5	325 35.6	15.4	9 52.5	12.5	54.7
20	122 32.6	05.5	340 10.0	15.5	9 40.0	12.6	54.7
21	137 32.8 . .	06.5	354 44.5	15.5	9 27.4	12.6	54.8
22	152 33.0	07.4	9 19.0	15.4	9 14.8	12.7	54.8
23	167 33.2	08.4	23 53.4	15.5	S 9 02.1	12.7	54.8
	SD 16.0	d 1.0	SD 14.8		14.8		14.9

Lat.	Twilight Naut.	Twilight Civil	Sunrise	Moonrise 28	Moonrise 29	Moonrise 30	Moonrise 1
°	h m	h m	h m	h m	h m	h m	h m
N 72	03 39	05 03	06 10	■	18 14	17 31	17 00
N 70	03 53	05 07	06 08	18 34	17 43	17 15	16 53
68	04 03	05 11	06 06	17 46	17 20	17 02	16 48
66	04 12	05 13	06 04	17 15	17 02	16 52	16 43
64	04 19	05 16	06 03	16 52	16 47	16 43	16 38
62	04 25	05 18	06 02	16 33	16 35	16 35	16 35
60	04 30	05 19	06 01	16 18	16 24	16 28	16 31
N 58	04 35	05 21	06 00	16 05	16 15	16 22	16 29
56	04 38	05 22	05 59	15 53	16 07	16 17	16 26
54	04 42	05 23	05 58	15 43	15 59	16 12	16 24
52	04 45	05 24	05 58	15 35	15 53	16 08	16 22
50	04 47	05 25	05 57	15 27	15 47	16 04	16 20
45	04 52	05 26	05 56	15 09	15 34	15 56	16 16
N 40	04 56	05 27	05 54	14 55	15 24	15 49	16 12
35	04 59	05 28	05 53	14 43	15 14	15 42	16 09
30	05 01	05 28	05 52	14 33	15 06	15 37	16 06
20	05 03	05 28	05 50	14 15	14 52	15 28	16 02
N 10	05 03	05 28	05 49	13 59	14 40	15 19	15 57
0	05 02	05 26	05 47	13 44	14 29	15 12	15 53
S 10	05 00	05 24	05 45	13 29	14 17	15 04	15 50
20	04 56	05 21	05 43	13 14	14 05	14 55	15 45
30	04 49	05 17	05 41	12 55	13 51	14 46	15 41
35	04 44	05 14	05 39	12 44	13 42	14 40	15 38
40	04 39	05 11	05 38	12 32	13 33	14 34	15 35
45	04 32	05 06	05 36	12 17	13 22	14 26	15 31
S 50	04 22	05 01	05 33	12 00	13 08	14 17	15 27
52	04 18	04 58	05 32	11 51	13 02	14 13	15 25
54	04 13	04 56	05 31	11 42	12 55	14 09	15 23
56	04 07	04 52	05 30	11 31	12 47	14 04	15 20
58	04 01	04 49	05 28	11 18	12 38	13 58	15 17
S 60	03 53	04 44	05 27	11 04	12 28	13 51	15 14

Lat.	Sunset	Twilight Civil	Twilight Naut.	Moonset 28	Moonset 29	Moonset 30	Moonset 1
°	h m	h m	h m	h m	h m	h m	h m
N 72	17 28	18 35	19 58	■	23 26	25 38	01 38
N 70	17 31	18 31	19 45	21 32	23 55	25 52	01 52
68	17 33	18 28	19 34	22 19	24 17	00 17	02 03
66	17 34	18 25	19 26	22 49	24 33	00 33	02 12
64	17 36	18 23	19 19	23 12	24 47	00 47	02 19
62	17 37	18 21	19 13	23 29	24 58	00 58	02 26
60	17 38	18 20	19 08	23 44	25 08	01 08	02 31
N 58	17 39	18 18	19 04	23 57	25 16	01 16	02 36
56	17 40	18 17	19 01	24 07	00 07	01 24	02 40
54	17 41	18 16	18 57	24 17	00 17	01 30	02 44
52	17 42	18 15	18 55	24 25	00 25	01 36	02 48
50	17 43	18 15	18 52	24 32	00 32	01 42	02 51
45	17 44	18 13	18 47	24 48	00 48	01 53	02 57
N 40	17 46	18 13	18 44	00 01	01 01	02 02	03 03
35	17 47	18 12	18 41	00 14	01 12	02 10	03 08
30	17 48	18 12	18 39	00 26	01 22	02 17	03 12
20	17 50	18 12	18 37	00 47	01 38	02 29	03 19
N 10	17 52	18 13	18 37	01 04	01 53	02 40	03 26
0	17 54	18 14	18 38	01 21	02 06	02 49	03 32
S 10	17 55	18 16	18 41	01 37	02 19	02 59	03 38
20	17 58	18 20	18 45	01 54	02 33	03 09	03 44
30	18 00	18 24	18 52	02 14	02 49	03 21	03 51
35	18 02	18 27	18 57	02 25	02 58	03 28	03 55
40	18 04	18 31	19 03	02 39	03 09	03 35	03 59
45	18 06	18 35	19 10	02 54	03 21	03 44	04 05
S 50	18 08	18 41	19 19	03 13	03 36	03 55	04 11
52	18 09	18 43	19 24	03 22	03 43	03 59	04 14
54	18 11	18 46	19 29	03 32	03 50	04 05	04 17
56	18 12	18 50	19 35	03 43	03 59	04 11	04 20
58	18 14	18 53	19 42	03 56	04 09	04 17	04 24
S 60	18 15	18 58	19 50	04 11	04 19	04 25	04 28

Day	SUN Eqn. of Time 00^h	SUN Eqn. of Time 12^h	SUN Mer. Pass.	MOON Mer. Pass. Upper	MOON Mer. Pass. Lower	MOON Age	MOON Phase
d	m s	m s	h m	h m	h m	d	%
28	09 13	09 24	11 51	19 55	07 33	10	71
29	09 34	09 44	11 50	20 39	08 17	11	79
30	09 53	10 03	11 50	21 22	09 01	12	86

UT d	UT h	Day	ARIES GHA	VENUS −3.9 GHA	VENUS Dec	MARS +0.8 GHA	MARS Dec	JUPITER −2.7 GHA	JUPITER Dec	SATURN +1.1 GHA	SATURN Dec
d	h		° ′	° ′	° ′	° ′	° ′	° ′	° ′	° ′	° ′
1	00		9 51.7	205 15.3	N 7 55.2	256 15.7	N22 26.4	49 39.6	S16 38.8	192 09.5	N 3 09.5
	01		24 54.2	220 14.8	54.1	271 16.7	26.3	64 42.1	38.8	207 11.7	09.4
	02		39 56.6	235 14.4	53.0	286 17.7	26.1	79 44.7	38.8	222 13.9	09.3
	03		54 59.1	250 14.0	51.9	301 18.7	25.9	94 47.3	38.9	237 16.1	09.1
	04		70 01.6	265 13.6	50.7	316 19.7	25.8	109 49.8	38.9	252 18.3	09.0
	05		85 04.0	280 13.1	49.6	331 20.6	25.6	124 52.4	38.9	267 20.4	08.9
	06		100 06.5	295 12.7	N 7 48.5	346 21.6	N22 25.4	139 55.0	S16 38.9	282 22.6	N 3 08.8
	07		115 09.0	310 12.3	47.4	1 22.6	25.3	154 57.5	39.0	297 24.8	08.7
	08	T	130 11.4	325 11.9	46.3	16 23.6	25.1	170 00.1	39.0	312 27.0	08.5
	09	H	145 13.9	340 11.4	45.1	31 24.5	24.9	185 02.6	39.0	327 29.2	08.4
	10	U	160 16.4	355 11.0	44.0	46 25.5	24.7	200 05.2	39.1	342 31.3	08.3
	11	R	175 18.8	10 10.6	42.9	61 26.5	24.6	215 07.8	39.1	357 33.5	08.2
	12	S	190 21.3	25 10.2	N 7 41.8	76 27.5	N22 24.4	230 10.3	S16 39.1	12 35.7	N 3 08.1
	13	D	205 23.7	40 09.7	40.6	91 28.5	24.2	245 12.9	39.1	27 37.9	07.9
	14	A	220 26.2	55 09.3	39.5	106 29.5	24.1	260 15.4	39.2	42 40.1	07.8
	15	Y	235 28.7	70 08.9	38.4	121 30.4	23.9	275 18.0	39.2	57 42.2	07.7
	16		250 31.1	85 08.5	37.3	136 31.4	23.7	290 20.6	39.2	72 44.4	07.6
	17		265 33.6	100 08.1	36.2	151 32.4	23.6	305 23.1	39.2	87 46.6	07.5
	18		280 36.1	115 07.6	N 7 35.0	166 33.4	N22 23.4	320 25.7	S16 39.3	102 48.8	N 3 07.4
	19		295 38.5	130 07.2	33.9	181 34.4	23.2	335 28.2	39.3	117 51.0	07.2
	20		310 41.0	145 06.8	32.8	196 35.3	23.0	350 30.8	39.3	132 53.1	07.1
	21		325 43.5	160 06.4	31.7	211 36.3	22.9	5 33.4	39.3	147 55.3	07.0
	22		340 45.9	175 06.0	30.5	226 37.3	22.7	20 35.9	39.4	162 57.5	06.9
	23		355 48.4	190 05.5	29.4	241 38.3	22.5	35 38.5	39.4	177 59.7	06.8
2	00		10 50.8	205 05.1	N 7 28.3	256 39.3	N22 22.4	50 41.0	S16 39.4	193 01.9	N 3 06.6
	01		25 53.3	220 04.7	27.1	271 40.3	22.2	65 43.6	39.4	208 04.0	06.5
	02		40 55.8	235 04.3	26.0	286 41.3	22.0	80 46.1	39.5	223 06.2	06.4
	03		55 58.2	250 03.8	24.9	301 42.2	21.8	95 48.7	39.5	238 08.4	06.3
	04		71 00.7	265 03.4	23.8	316 43.2	21.7	110 51.3	39.5	253 10.6	06.2
	05		86 03.2	280 03.0	22.6	331 44.2	21.5	125 53.8	39.5	268 12.8	06.0
	06		101 05.6	295 02.6	N 7 21.5	346 45.2	N22 21.3	140 56.4	S16 39.6	283 15.0	N 3 05.9
	07		116 08.1	310 02.2	20.4	1 46.2	21.2	155 58.9	39.6	298 17.1	05.8
	08		131 10.6	325 01.7	19.2	16 47.2	21.0	171 01.5	39.6	313 19.3	05.7
	09	F	146 13.0	340 01.3	18.1	31 48.2	20.8	186 04.0	39.6	328 21.5	05.6
	10	R	161 15.5	355 00.9	17.0	46 49.2	20.6	201 06.6	39.6	343 23.7	05.4
	11	I	176 18.0	10 00.5	15.9	61 50.2	20.5	216 09.1	39.7	358 25.9	05.3
	12	D	191 20.4	25 00.1	N 7 14.7	76 51.1	N22 20.3	231 11.7	S16 39.7	13 28.0	N 3 05.2
	13	A	206 22.9	39 59.7	13.6	91 52.1	20.1	246 14.2	39.7	28 30.2	05.1
	14	Y	221 25.3	54 59.2	12.5	106 53.1	19.9	261 16.8	39.7	43 32.4	05.0
	15		236 27.8	69 58.8	11.3	121 54.1	19.8	276 19.3	39.8	58 34.6	04.9
	16		251 30.3	84 58.4	10.2	136 55.1	19.6	291 21.9	39.8	73 36.8	04.7
	17		266 32.7	99 58.0	09.1	151 56.1	19.4	306 24.5	39.8	88 38.9	04.6
	18		281 35.2	114 57.6	N 7 07.9	166 57.1	N22 19.2	321 27.0	S16 39.8	103 41.1	N 3 04.5
	19		296 37.7	129 57.2	06.8	181 58.1	19.1	336 29.6	39.9	118 43.3	04.4
	20		311 40.1	144 56.7	05.7	196 59.1	18.9	351 32.1	39.9	133 45.5	04.3
	21		326 42.6	159 56.3	04.5	212 00.1	18.7	6 34.7	39.9	148 47.7	04.1
	22		341 45.1	174 55.9	03.4	227 01.1	18.5	21 37.2	39.9	163 49.8	04.0
	23		356 47.5	189 55.5	02.2	242 02.1	18.4	36 39.8	39.9	178 52.0	03.9
3	00		11 50.0	204 55.1	N 7 01.1	257 03.1	N22 18.2	51 42.3	S16 40.0	193 54.2	N 3 03.8
	01		26 52.4	219 54.7	7 00.0	272 04.0	18.0	66 44.9	40.0	208 56.4	03.7
	02		41 54.9	234 54.2	6 58.8	287 05.0	17.8	81 47.4	40.0	223 58.6	03.6
	03		56 57.4	249 53.8	57.7	302 06.0	17.7	96 49.9	40.0	239 00.8	03.4
	04		71 59.8	264 53.4	56.6	317 07.0	17.5	111 52.5	40.1	254 02.9	03.3
	05		87 02.3	279 53.0	55.4	332 08.0	17.3	126 55.0	40.1	269 05.1	03.2
	06		102 04.8	294 52.6	N 6 54.3	347 09.0	N22 17.1	141 57.6	S16 40.1	284 07.3	N 3 03.1
	07		117 07.2	309 52.2	53.1	2 10.0	17.0	157 00.1	40.1	299 09.5	03.0
	08	S	132 09.7	324 51.7	52.0	17 11.0	16.8	172 02.7	40.1	314 11.7	02.8
	09	A	147 12.2	339 51.3	50.9	32 12.0	16.6	187 05.2	40.2	329 13.8	02.7
	10	T	162 14.6	354 50.9	49.7	47 13.0	16.4	202 07.8	40.2	344 16.0	02.6
	11	U	177 17.1	9 50.5	48.6	62 14.0	16.3	217 10.3	40.2	359 18.2	02.5
	12	R	192 19.6	24 50.1	N 6 47.4	77 15.0	N22 16.1	232 12.9	S16 40.2	14 20.4	N 3 02.4
	13	D	207 22.0	39 49.7	46.3	92 16.0	15.9	247 15.4	40.2	29 22.6	02.2
	14	A	222 24.5	54 49.3	45.2	107 17.0	15.7	262 18.0	40.3	44 24.8	02.1
	15	Y	237 26.9	69 48.8	44.0	122 18.0	15.5	277 20.5	40.3	59 26.9	02.0
	16		252 29.4	84 48.4	42.9	137 19.0	15.4	292 23.0	40.3	74 29.1	01.9
	17		267 31.9	99 48.0	41.7	152 20.0	15.2	307 25.6	40.3	89 31.3	01.8
	18		282 34.3	114 47.6	N 6 40.6	167 21.0	N22 15.0	322 28.1	S16 40.4	104 33.5	N 3 01.7
	19		297 36.8	129 47.2	39.4	182 22.0	14.8	337 30.7	40.4	119 35.7	01.5
	20		312 39.3	144 46.8	38.3	197 23.0	14.7	352 33.2	40.4	134 37.8	01.4
	21		327 41.7	159 46.4	37.2	212 24.0	14.5	7 35.8	40.4	149 40.0	01.3
	22		342 44.2	174 46.0	36.0	227 25.0	14.3	22 38.3	40.4	164 42.2	01.2
	23		357 46.7	189 45.5	34.9	242 26.0	14.1	37 40.8	40.4	179 44.4	01.1
Mer. Pass.			h m 23 12.8	*v* −0.4	*d* 1.1	*v* 1.0	*d* 0.2	*v* 2.6	*d* 0.0	*v* 2.2	*d* 0.1

STARS

Name	SHA	Dec
	° ′	°
Acamar	315 20.0	S40 15
Achernar	335 28.0	S57 11
Acrux	173 13.3	S63 09
Adhara	255 14.7	S28 58
Aldebaran	290 52.4	N16 31
Alioth	166 23.4	N55 54
Alkaid	153 01.3	N49 15
Al Na'ir	27 46.6	S46 54
Alnilam	275 49.1	S 1 11
Alphard	217 59.0	S 8 41
Alphecca	126 13.5	N26 41
Alpheratz	357 46.1	N29 08
Altair	62 10.8	N 8 53
Ankaa	353 17.8	S42 15
Antares	112 29.8	S26 27
Arcturus	145 58.5	N19 08
Atria	107 34.4	S69 03
Avior	234 19.5	S59 32
Bellatrix	278 34.8	N 6 21
Betelgeuse	271 04.2	N 7 24
Canopus	263 57.3	S52 41
Capella	280 38.4	N46 00
Deneb	49 33.2	N45 19
Denebola	182 36.7	N14 31
Diphda	348 58.2	S17 55
Dubhe	193 55.4	N61 41
Elnath	278 16.0	N28 37
Eltanin	90 47.5	N51 29
Enif	33 49.6	N 9 55
Fomalhaut	15 26.6	S29 34
Gacrux	172 04.8	S57 10
Gienah	175 55.5	S17 35
Hadar	148 52.6	S60 25
Hamal	328 03.6	N23 30
Kaus Aust.	83 47.5	S34 22
Kochab	137 20.5	N74 07
Markab	13 40.8	N15 15
Menkar	314 17.7	N 4 07
Menkent	148 11.3	S36 25
Miaplacidus	221 41.1	S69 45
Mirfak	308 44.0	N49 53
Nunki	76 01.7	S26 17
Peacock	53 23.1	S56 42
Pollux	243 31.1	N28 00
Procyon	245 02.7	N 5 12
Rasalhague	96 09.1	N12 33
Regulus	207 46.6	N11 55
Rigel	281 14.6	S 8 11
Rigil Kent.	139 56.3	S60 52
Sabik	102 15.8	S15 44
Schedar	349 43.3	N56 35
Shaula	96 25.8	S37 06
Sirius	258 36.1	S16 43
Spica	158 34.5	S11 12
Suhail	222 54.8	S43 28
Vega	80 40.8	N38 47
Zuben'ubi	137 08.8	S16 05

	SHA	Mer. Pas
	° ′	h m
Venus	194 14.3	10 20
Mars	245 48.4	6 53
Jupiter	39 50.2	20 34
Saturn	182 11.0	11 06

UT d	UT h	SUN GHA ° ′	SUN Dec ° ′	MOON GHA ° ′	MOON v ′	MOON Dec ° ′	MOON d ′	MOON HP ′
1 THURSDAY	00	182 33.4	S 3 09.4	38 27.9	15.6	S 8 49.4	12.7	54.8
	01	197 33.6	10.3	53 02.5	15.5	8 36.7	12.8	54.8
	02	212 33.8	11.3	67 37.0	15.5	8 23.9	12.8	54.8
	03	227 34.0	. . 12.3	82 11.5	15.6	8 11.1	12.9	54.9
	04	242 34.2	13.2	96 46.1	15.5	7 58.2	12.9	54.9
	05	257 34.4	14.2	111 20.6	15.6	7 45.3	13.0	54.9
	06	272 34.6	S 3 15.2	125 55.2	15.6	S 7 32.3	13.0	54.9
	07	287 34.8	16.2	140 29.8	15.5	7 19.3	13.0	54.9
	08	302 35.0	17.1	155 04.3	15.6	7 06.3	13.0	54.9
	09	317 35.2	. . 18.1	169 38.9	15.6	6 53.3	13.1	55.0
	10	332 35.4	19.1	184 13.5	15.6	6 40.2	13.2	55.0
	11	347 35.6	20.0	198 48.1	15.6	6 27.0	13.1	55.0
	12	2 35.8	S 3 21.0	213 22.7	15.6	S 6 13.9	13.2	55.0
	13	17 36.0	22.0	227 57.3	15.6	6 00.7	13.3	55.0
	14	32 36.2	22.9	242 31.9	15.6	5 47.4	13.2	55.1
	15	47 36.4	. . 23.9	257 06.5	15.6	5 34.2	13.3	55.1
	16	62 36.6	24.9	271 41.1	15.6	5 20.9	13.3	55.1
	17	77 36.8	25.8	286 15.7	15.6	5 07.6	13.4	55.1
	18	92 37.0	S 3 26.8	300 50.3	15.6	S 4 54.2	13.4	55.1
	19	107 37.2	27.8	315 24.9	15.6	4 40.8	13.4	55.1
	20	122 37.4	28.7	329 59.5	15.6	4 27.4	13.4	55.2
	21	137 37.6	. . 29.7	344 34.1	15.6	4 14.0	13.5	55.2
	22	152 37.8	30.7	359 08.7	15.6	4 00.5	13.5	55.2
	23	167 38.0	31.6	13 43.3	15.6	3 47.0	13.5	55.2
2 FRIDAY	00	182 38.2	S 3 32.6	28 17.9	15.6	S 3 33.5	13.5	55.2
	01	197 38.4	33.6	42 52.5	15.5	3 20.0	13.6	55.3
	02	212 38.6	34.5	57 27.0	15.6	3 06.4	13.6	55.3
	03	227 38.8	. . 35.5	72 01.6	15.5	2 52.8	13.6	55.3
	04	242 39.0	36.5	86 36.1	15.5	2 39.2	13.6	55.3
	05	257 39.2	37.5	101 10.6	15.5	2 25.6	13.7	55.3
	06	272 39.4	S 3 38.4	115 45.1	15.5	S 2 11.9	13.7	55.4
	07	287 39.6	39.4	130 19.6	15.5	1 58.2	13.6	55.4
	08	302 39.8	40.4	144 54.1	15.5	1 44.6	13.7	55.4
	09	317 40.0	. . 41.3	159 28.6	15.4	1 30.9	13.8	55.4
	10	332 40.2	42.3	174 03.0	15.5	1 17.1	13.7	55.4
	11	347 40.4	43.3	188 37.5	15.4	1 03.4	13.7	55.5
	12	2 40.6	S 3 44.2	203 11.9	15.4	S 0 49.7	13.8	55.5
	13	17 40.8	45.2	217 46.3	15.4	0 35.9	13.8	55.5
	14	32 41.0	46.2	232 20.7	15.3	0 22.1	13.8	55.5
	15	47 41.2	. . 47.1	246 55.0	15.4	S 0 08.3	13.8	55.5
	16	62 41.4	48.1	261 29.4	15.3	N 0 05.5	13.8	55.6
	17	77 41.6	49.1	276 03.7	15.3	0 19.3	13.8	55.6
	18	92 41.8	S 3 50.0	290 38.0	15.2	N 0 33.1	13.8	55.6
	19	107 42.0	51.0	305 12.2	15.3	0 46.9	13.9	55.6
	20	122 42.2	52.0	319 46.5	15.2	1 00.8	13.8	55.6
	21	137 42.4	. . 52.9	334 20.7	15.1	1 14.6	13.9	55.7
	22	152 42.6	53.9	348 54.8	15.2	1 28.5	13.8	55.7
	23	167 42.8	54.9	3 29.0	15.1	1 42.3	13.9	55.7
3 SATURDAY	00	182 43.0	S 3 55.8	18 03.1	15.1	N 1 56.2	13.9	55.7
	01	197 43.2	56.8	32 37.2	15.1	2 10.1	13.8	55.8
	02	212 43.4	57.8	47 11.3	15.0	2 23.9	13.9	55.8
	03	227 43.6	. . 58.7	61 45.3	15.0	2 37.8	13.8	55.8
	04	242 43.8	3 59.7	76 19.3	14.9	2 51.6	13.9	55.8
	05	257 44.0	4 00.6	90 53.2	14.9	3 05.5	13.9	55.8
	06	272 44.1	S 4 01.6	105 27.1	14.9	N 3 19.4	13.8	55.9
	07	287 44.3	02.6	120 01.0	14.9	3 33.2	13.9	55.9
	08	302 44.5	03.5	134 34.9	14.8	3 47.1	13.8	55.9
	09	317 44.7	. . 04.5	149 08.7	14.7	4 00.9	13.9	55.9
	10	332 44.9	05.5	163 42.4	14.8	4 14.8	13.8	56.0
	11	347 45.1	06.4	178 16.2	14.7	4 28.6	13.9	56.0
	12	2 45.3	S 4 07.4	192 49.9	14.6	N 4 42.5	13.8	56.0
	13	17 45.5	08.4	207 23.5	14.6	4 56.3	13.8	56.0
	14	32 45.7	09.3	221 57.1	14.6	5 10.1	13.8	56.0
	15	47 45.9	. . 10.3	236 30.7	14.5	5 23.9	13.8	56.1
	16	62 46.1	11.3	251 04.2	14.4	5 37.7	13.7	56.1
	17	77 46.3	12.2	265 37.6	14.4	5 51.4	13.8	56.1
	18	92 46.5	S 4 13.2	280 11.0	14.4	N 6 05.2	13.8	56.1
	19	107 46.7	14.2	294 44.4	14.3	6 19.0	13.7	56.2
	20	122 46.9	15.1	309 17.7	14.3	6 32.7	13.7	56.2
	21	137 47.1	. . 16.1	323 51.0	14.2	6 46.4	13.7	56.2
	22	152 47.3	17.0	338 24.2	14.2	7 00.1	13.7	56.2
	23	167 47.5	18.0	352 57.4	14.1	N 7 13.8	13.6	56.2
		SD 16.0	*d* 1.0	SD 15.0		15.1		15.3

Lat. °	Twilight Naut. h m	Twilight Civil h m	Sunrise h m	Moonrise 1 h m	Moonrise 2 h m	Moonrise 3 h m	Moonrise 4 h m
N 72	03 55	05 16	06 24	17 00	16 34	16 07	15 34
N 70	04 06	05 19	06 20	16 53	16 34	16 14	15 51
68	04 15	05 21	06 17	16 48	16 34	16 20	16 05
66	04 22	05 23	06 14	16 43	16 34	16 26	16 17
64	04 29	05 24	06 12	16 38	16 34	16 30	16 26
62	04 34	05 26	06 10	16 35	16 34	16 34	16 34
60	04 38	05 27	06 08	16 31	16 34	16 37	16 41
N 58	04 42	05 27	06 06	16 29	16 34	16 40	16 48
56	04 45	05 28	06 05	16 26	16 34	16 43	16 53
54	04 47	05 29	06 04	16 24	16 35	16 46	16 58
52	04 50	05 29	06 03	16 22	16 35	16 48	17 03
50	04 52	05 29	06 02	16 20	16 35	16 50	17 07
45	04 56	05 30	05 59	16 16	16 35	16 55	17 16
N 40	04 59	05 30	05 57	16 12	16 35	16 58	17 24
35	05 01	05 30	05 56	16 09	16 35	17 02	17 30
30	05 02	05 30	05 54	16 06	16 35	17 05	17 36
20	05 04	05 29	05 51	16 02	16 35	17 10	17 46
N 10	05 03	05 28	05 49	15 57	16 35	17 14	17 55
0	05 01	05 25	05 46	15 53	16 35	17 19	18 04
S 10	04 58	05 22	05 43	15 50	16 36	17 23	18 12
20	04 53	05 18	05 41	15 45	16 36	17 28	18 21
30	04 45	05 13	05 37	15 41	16 36	17 33	18 32
35	04 40	05 10	05 35	15 38	16 36	17 36	18 38
40	04 34	05 06	05 33	15 35	16 36	17 40	18 45
45	04 26	05 01	05 30	15 31	16 37	17 44	18 53
S 50	04 15	04 54	05 27	15 27	16 37	17 49	19 03
52	04 10	04 51	05 25	15 25	16 37	17 51	19 07
54	04 05	04 48	05 24	15 23	16 37	17 53	19 13
56	03 58	04 44	05 22	15 20	16 37	17 56	19 18
58	03 51	04 40	05 20	15 17	16 37	17 59	19 25
S 60	03 43	04 35	05 17	15 14	16 38	18 03	19 32

Lat. °	Sunset h m	Twilight Civil h m	Twilight Naut. h m	Moonset 1 h m	Moonset 2 h m	Moonset 3 h m	Moonset 4 h m
N 72	17 13	18 20	19 40	01 38	03 38	05 36	07 42
N 70	17 17	18 17	19 30	01 52	03 42	05 32	07 27
68	17 20	18 15	19 21	02 03	03 45	05 28	07 15
66	17 23	18 14	19 14	02 12	03 48	05 25	07 05
64	17 25	18 13	19 08	02 19	03 50	05 22	06 57
62	17 27	18 11	19 03	02 26	03 52	05 20	06 51
60	17 29	18 11	18 59	02 31	03 54	05 18	06 45
N 58	17 31	18 10	18 56	02 36	03 56	05 16	06 40
56	17 32	18 09	18 53	02 40	03 57	05 15	06 35
54	17 34	18 09	18 50	02 44	03 58	05 13	06 31
52	17 35	18 09	18 48	02 48	03 59	05 12	06 27
50	17 36	18 08	18 46	02 51	04 00	05 11	06 24
45	17 39	18 08	18 42	02 57	04 02	05 09	06 17
N 40	17 41	18 08	18 39	03 03	04 04	05 07	06 11
35	17 43	18 08	18 37	03 08	04 06	05 05	06 05
30	17 44	18 08	18 36	03 12	04 07	05 03	06 01
20	17 47	18 09	18 35	03 19	04 10	05 00	05 53
N 10	17 50	18 11	18 35	03 26	04 12	04 58	05 46
0	17 53	18 13	18 37	03 32	04 14	04 56	05 40
S 10	17 55	18 16	18 41	03 38	04 15	04 54	05 33
20	17 58	18 20	18 46	03 44	04 17	04 51	05 27
30	18 02	18 26	18 54	03 51	04 20	04 49	05 19
35	18 04	18 30	18 59	03 55	04 21	04 47	05 14
40	18 06	18 34	19 06	03 59	04 22	04 45	05 09
45	18 09	18 39	19 14	04 05	04 24	04 43	05 04
S 50	18 13	18 45	19 25	04 11	04 26	04 41	04 57
52	18 14	18 49	19 30	04 14	04 27	04 40	04 54
54	18 16	18 52	19 36	04 17	04 28	04 38	04 50
56	18 18	18 56	19 42	04 20	04 29	04 37	04 46
58	18 20	19 00	19 50	04 24	04 30	04 36	04 42
S 60	18 23	19 05	19 58	04 28	04 31	04 34	04 37

Day	SUN Eqn. of Time 00^h	SUN Eqn. of Time 12^h	SUN Mer. Pass.	MOON Mer. Pass. Upper	MOON Mer. Pass. Lower	MOON Age	MOON Phase
d	m s	m s	h m	h m	h m	d %	
1	10 13	10 23	11 50	22 04	09 43	13 92	○
2	10 32	10 42	11 49	22 46	10 24	14 97	
3	10 51	11 01	11 49	23 29	11 07	15 99	

UT	ARIES	VENUS −3.9		MARS +0.7		JUPITER −2.6		SATURN +1.1		STARS		
	GHA	GHA	Dec	GHA	Dec	GHA	Dec	GHA	Dec	Name	SHA	Dec
d h	° ′	° ′	° ′	° ′	° ′	° ′	° ′	° ′	° ′		° ′	°
4 00	12 49.1	204 45.1	N 6 33.7	257 27.0	N22 13.9	52 43.4	S16 40.5	194 46.6	N 3 00.9	Acamar	315 20.0	S40 15
01	27 51.6	219 44.7	32.6	272 28.0	13.8	67 45.9	40.5	209 48.8	00.8	Achernar	335 28.0	S57 11
02	42 54.0	234 44.3	31.4	287 29.0	13.6	82 48.5	40.5	224 50.9	00.7	Acrux	173 13.3	S63 09
03	57 56.5	249 43.9	. . 30.3	302 30.1	. . 13.4	97 51.0	. . 40.5	239 53.1	. . 00.6	Adhara	255 14.7	S28 58
04	72 59.0	264 43.5	29.1	317 31.1	13.2	112 53.5	40.5	254 55.3	00.5	Aldebaran	290 52.3	N16 31
05	88 01.4	279 43.1	28.0	332 32.1	13.0	127 56.1	40.6	269 57.5	00.4			
06	103 03.9	294 42.7	N 6 26.8	347 33.1	N22 12.9	142 58.6	S16 40.6	284 59.7	N 3 00.2	Alioth	166 23.4	N55 54
07	118 06.4	309 42.2	25.7	2 34.1	12.7	158 01.2	40.6	300 01.8	00.1	Alkaid	153 01.4	N49 15
08	133 08.8	324 41.8	24.6	17 35.1	12.5	173 03.7	40.6	315 04.0	3 00.0	Al Na'ir	27 46.7	S46 54
S 09	148 11.3	339 41.4	. . 23.4	32 36.1	. . 12.3	188 06.2	. . 40.6	330 06.2	2 59.9	Alnilam	275 49.0	S 1 11
U 10	163 13.8	354 41.0	22.3	47 37.1	12.1	203 08.8	40.7	345 08.4	59.8	Alphard	217 59.0	S 8 41
N 11	178 16.2	9 40.6	21.1	62 38.1	12.0	218 11.3	40.7	0 10.6	59.6			
D 12	193 18.7	24 40.2	N 6 20.0	77 39.1	N22 11.8	233 13.8	S16 40.7	15 12.8	N 2 59.5	Alphecca	126 13.5	N26 41
A 13	208 21.2	39 39.8	18.8	92 40.1	11.6	248 16.4	40.7	30 14.9	59.4	Alpheratz	357 46.1	N29 08
Y 14	223 23.6	54 39.4	17.7	107 41.1	11.4	263 18.9	40.7	45 17.1	59.3	Altair	62 10.8	N 8 53
15	238 26.1	69 39.0	. . 16.5	122 42.1	. . 11.2	278 21.5	. . 40.8	60 19.3	. . 59.2	Ankaa	353 17.8	S42 15
16	253 28.5	84 38.6	15.4	137 43.2	11.1	293 24.0	40.8	75 21.5	59.1	Antares	112 29.9	S26 27
17	268 31.0	99 38.1	14.2	152 44.2	10.9	308 26.5	40.8	90 23.7	58.9			
18	283 33.5	114 37.7	N 6 13.1	167 45.2	N22 10.7	323 29.1	S16 40.8	105 25.8	N 2 58.8	Arcturus	145 58.5	N19 08
19	298 35.9	129 37.3	11.9	182 46.2	10.5	338 31.6	40.8	120 28.0	58.7	Atria	107 34.5	S69 03
20	313 38.4	144 36.9	10.7	197 47.2	10.3	353 34.1	40.8	135 30.2	58.6	Avior	234 19.5	S59 32
21	328 40.9	159 36.5	. . 09.6	212 48.2	. . 10.2	8 36.7	. . 40.9	150 32.4	. . 58.5	Bellatrix	278 34.8	N 6 21
22	343 43.3	174 36.1	08.4	227 49.2	10.0	23 39.2	40.9	165 34.6	58.3	Betelgeuse	271 04.2	N 7 24
23	358 45.8	189 35.7	07.3	242 50.2	09.8	38 41.7	40.9	180 36.8	58.2			
5 00	13 48.3	204 35.3	N 6 06.1	257 51.3	N22 09.6	53 44.3	S16 40.9	195 38.9	N 2 58.1	Canopus	263 57.3	S52 41
01	28 50.7	219 34.9	05.0	272 52.3	09.4	68 46.8	40.9	210 41.1	58.0	Capella	280 38.3	N46 00
02	43 53.2	234 34.5	03.8	287 53.3	09.2	83 49.3	40.9	225 43.3	57.9	Deneb	49 33.2	N45 19
03	58 55.7	249 34.1	. . 02.7	302 54.3	. . 09.1	98 51.9	. . 41.0	240 45.5	. . 57.8	Denebola	182 36.7	N14 31
04	73 58.1	264 33.6	01.5	317 55.3	08.9	113 54.4	41.0	255 47.7	57.6	Diphda	348 58.2	S17 55
05	89 00.6	279 33.2	6 00.4	332 56.3	08.7	128 56.9	41.0	270 49.9	57.5			
06	104 03.0	294 32.8	N 5 59.2	347 57.3	N22 08.5	143 59.5	S16 41.0	285 52.0	N 2 57.4	Dubhe	193 55.3	N61 41
07	119 05.5	309 32.4	58.1	2 58.4	08.3	159 02.0	41.0	300 54.2	57.3	Elnath	278 16.0	N28 37
08	134 08.0	324 32.0	56.9	17 59.4	08.1	174 04.5	41.0	315 56.4	57.2	Eltanin	90 47.5	N51 29
M 09	149 10.4	339 31.6	. . 55.7	33 00.4	. . 08.0	189 07.0	. . 41.1	330 58.6	. . 57.1	Enif	33 49.6	N 9 55
O 10	164 12.9	354 31.2	54.6	48 01.4	07.8	204 09.6	41.1	346 00.8	56.9	Fomalhaut	15 26.6	S29 34
N 11	179 15.4	9 30.8	53.4	63 02.4	07.6	219 12.1	41.1	1 03.0	56.8			
D 12	194 17.8	24 30.4	N 5 52.3	78 03.4	N22 07.4	234 14.6	S16 41.1	16 05.1	N 2 56.7	Gacrux	172 04.8	S57 10
A 13	209 20.3	39 30.0	51.1	93 04.5	07.2	249 17.2	41.1	31 07.3	56.6	Gienah	175 55.5	S17 35
Y 14	224 22.8	54 29.6	50.0	108 05.5	07.0	264 19.7	41.1	46 09.5	56.5	Hadar	148 52.6	S60 25
15	239 25.2	69 29.2	. . 48.8	123 06.5	. . 06.9	279 22.2	. . 41.2	61 11.7	. . 56.3	Hamal	328 03.6	N23 30
16	254 27.7	84 28.8	47.6	138 07.5	06.7	294 24.7	41.2	76 13.9	56.2	Kaus Aust.	83 47.5	S34 22
17	269 30.1	99 28.4	46.5	153 08.5	06.5	309 27.3	41.2	91 16.0	56.1			
18	284 32.6	114 28.0	N 5 45.3	168 09.6	N22 06.3	324 29.8	S16 41.2	106 18.2	N 2 56.0	Kochab	137 20.5	N74 07
19	299 35.1	129 27.5	44.2	183 10.6	06.1	339 32.3	41.2	121 20.4	55.9	Markab	13 40.8	N15 15
20	314 37.5	144 27.1	43.0	198 11.6	05.9	354 34.8	41.2	136 22.6	55.8	Menkar	314 17.6	N 4 07
21	329 40.0	159 26.7	. . 41.8	213 12.6	. . 05.8	9 37.4	. . 41.2	151 24.8	. . 55.6	Menkent	148 11.3	S36 25
22	344 42.5	174 26.3	40.7	228 13.6	05.6	24 39.9	41.3	166 27.0	55.5	Miaplacidus	221 41.1	S69 45
23	359 44.9	189 25.9	39.5	243 14.7	05.4	39 42.4	41.3	181 29.1	55.4			
6 00	14 47.4	204 25.5	N 5 38.4	258 15.7	N22 05.2	54 44.9	S16 41.3	196 31.3	N 2 55.3	Mirfak	308 44.0	N49 53
01	29 49.9	219 25.1	37.2	273 16.7	05.0	69 47.5	41.3	211 33.5	55.2	Nunki	76 01.7	S26 17
02	44 52.3	234 24.7	36.0	288 17.7	04.8	84 50.0	41.3	226 35.7	55.1	Peacock	53 23.2	S56 42
03	59 54.8	249 24.3	. . 34.9	303 18.8	. . 04.6	99 52.5	. . 41.3	241 37.9	. . 54.9	Pollux	243 31.1	N28 00
04	74 57.3	264 23.9	33.7	318 19.8	04.5	114 55.0	41.3	256 40.1	54.8	Procyon	245 02.6	N 5 12
05	89 59.7	279 23.5	32.5	333 20.8	04.3	129 57.6	41.4	271 42.2	54.7			
06	105 02.2	294 23.1	N 5 31.4	348 21.8	N22 04.1	145 00.1	S16 41.4	286 44.4	N 2 54.6	Rasalhague	96 09.1	N12 33
07	120 04.6	309 22.7	30.2	3 22.9	03.9	160 02.6	41.4	301 46.6	54.5	Regulus	207 46.6	N11 55
08	135 07.1	324 22.3	29.0	18 23.9	03.7	175 05.1	41.4	316 48.8	54.4	Rigel	281 14.5	S 8 11
T 09	150 09.6	339 21.9	. . 27.9	33 24.9	. . 03.5	190 07.7	. . 41.4	331 51.0	. . 54.2	Rigil Kent.	139 56.3	S60 52
U 10	165 12.0	354 21.5	26.7	48 25.9	03.3	205 10.2	41.4	346 53.2	54.1	Sabik	102 15.8	S15 44
E 11	180 14.5	9 21.1	25.6	63 27.0	03.2	220 12.7	41.4	1 55.3	54.0			
S 12	195 17.0	24 20.7	N 5 24.4	78 28.0	N22 03.0	235 15.2	S16 41.4	16 57.5	N 2 53.9	Schedar	349 43.3	N56 35
D 13	210 19.4	39 20.3	23.2	93 29.0	02.8	250 17.7	41.5	31 59.7	53.8	Shaula	96 25.8	S37 06
A 14	225 21.9	54 19.9	22.1	108 30.1	02.6	265 20.3	41.5	47 01.9	53.6	Sirius	258 36.1	S16 43
Y 15	240 24.4	69 19.5	. . 20.9	123 31.1	. . 02.4	280 22.8	. . 41.5	62 04.1	. . 53.5	Spica	158 34.5	S11 12
16	255 26.8	84 19.1	19.7	138 32.1	02.2	295 25.3	41.5	77 06.3	53.4	Suhail	222 54.8	S43 28
17	270 29.3	99 18.7	18.6	153 33.1	02.0	310 27.8	41.5	92 08.5	53.3			
18	285 31.7	114 18.3	N 5 17.4	168 34.2	N22 01.8	325 30.3	S16 41.5	107 10.6	N 2 53.2	Vega	80 40.9	N38 47
19	300 34.2	129 17.8	16.2	183 35.2	01.7	340 32.9	41.5	122 12.8	53.1	Zuben'ubi	137 08.8	S16 04
20	315 36.7	144 17.4	15.1	198 36.2	01.5	355 35.4	41.6	137 15.0	52.9		SHA	Mer. Pas
21	330 39.1	159 17.0	. . 13.9	213 37.3	. . 01.3	10 37.9	. . 41.6	152 17.2	. . 52.8		° ′	h m
22	345 41.6	174 16.6	12.7	228 38.3	01.1	25 40.4	41.6	167 19.4	52.7	Venus	190 47.0	10 22
23	0 44.1	189 16.2	11.6	243 39.3	00.9	40 42.9	41.6	182 21.6	52.6	Mars	244 03.0	6 48
	h m									Jupiter	39 56.0	20 22
Mer. Pass.	23 01.0	v −0.4	d 1.2	v 1.0	d 0.2	v 2.5	d 0.0	v 2.2	d 0.1	Saturn	181 50.7	10 56

UT	SUN GHA	SUN Dec	MOON GHA	*v*	MOON Dec	*d*	HP
d h	° ′	° ′	° ′	′	° ′	′	′
4 00	182 47.7	S 4 19.0	7 30.5	14.0	N 7 27.4	13.7	56.3
01	197 47.8	19.9	22 03.5	14.0	7 41.1	13.6	56.3
02	212 48.0	20.9	36 36.5	14.0	7 54.7	13.5	56.3
03	227 48.2	. . 21.9	51 09.5	13.8	8 08.2	13.6	56.3
04	242 48.4	22.8	65 42.3	13.9	8 21.8	13.5	56.4
05	257 48.6	23.8	80 15.2	13.7	8 35.3	13.5	56.4
06	272 48.8	S 4 24.8	94 47.9	13.7	N 8 48.8	13.5	56.4
07	287 49.0	25.7	109 20.6	13.7	9 02.3	13.5	56.4
08	302 49.2	26.7	123 53.3	13.6	9 15.8	13.4	56.4
S 09	317 49.4	. . 27.6	138 25.9	13.5	9 29.2	13.4	56.5
U 10	332 49.6	28.6	152 58.4	13.5	9 42.6	13.3	56.5
N 11	347 49.8	29.6	167 30.9	13.4	9 55.9	13.3	56.5
D 12	2 50.0	S 4 30.5	182 03.3	13.3	N10 09.2	13.3	56.5
A 13	17 50.2	31.5	196 35.6	13.3	10 22.5	13.3	56.6
Y 14	32 50.3	32.5	211 07.9	13.2	10 35.8	13.2	56.6
15	47 50.5	. . 33.4	225 40.1	13.1	10 49.0	13.1	56.6
16	62 50.7	34.4	240 12.2	13.0	11 02.1	13.2	56.6
17	77 50.9	35.4	254 44.2	13.0	11 15.3	13.0	56.6
18	92 51.1	S 4 36.3	269 16.2	13.0	N11 28.3	13.1	56.7
19	107 51.3	37.3	283 48.2	12.8	11 41.4	13.0	56.7
20	122 51.5	38.2	298 20.0	12.8	11 54.4	12.9	56.7
21	137 51.7	. . 39.2	312 51.8	12.7	12 07.3	12.9	56.7
22	152 51.9	40.2	327 23.5	12.6	12 20.2	12.9	56.8
23	167 52.1	41.1	341 55.1	12.6	12 33.1	12.8	56.8
5 00	182 52.2	S 4 42.1	356 26.7	12.5	N12 45.9	12.7	56.8
01	197 52.4	43.0	10 58.2	12.4	12 58.6	12.7	56.8
02	212 52.6	44.0	25 29.6	12.3	13 11.3	12.7	56.8
03	227 52.8	. . 45.0	40 00.9	12.2	13 24.0	12.6	56.9
04	242 53.0	45.9	54 32.1	12.2	13 36.6	12.5	56.9
05	257 53.2	46.9	69 03.3	12.1	13 49.1	12.5	56.9
06	272 53.4	S 4 47.9	83 34.4	12.0	N14 01.6	12.4	56.9
07	287 53.6	48.8	98 05.4	11.9	14 14.0	12.4	56.9
08	302 53.8	49.8	112 36.3	11.9	14 26.4	12.3	57.0
M 09	317 53.9	. . 50.7	127 07.2	11.8	14 38.7	12.2	57.0
O 10	332 54.1	51.7	141 38.0	11.7	14 50.9	12.2	57.0
N 11	347 54.3	52.7	156 08.7	11.6	15 03.1	12.1	57.0
D 12	2 54.5	S 4 53.6	170 39.3	11.5	N15 15.2	12.0	57.1
A 13	17 54.7	54.6	185 09.8	11.4	15 27.2	12.0	57.1
Y 14	32 54.9	55.5	199 40.2	11.4	15 39.2	11.9	57.1
15	47 55.1	. . 56.5	214 10.6	11.2	15 51.1	11.8	57.1
16	62 55.3	57.5	228 40.8	11.2	16 02.9	11.8	57.1
17	77 55.4	58.4	243 11.0	11.1	16 14.7	11.7	57.2
18	92 55.6	S 4 59.4	257 41.1	11.0	N16 26.4	11.6	57.2
19	107 55.8	5 00.3	272 11.1	10.9	16 38.0	11.5	57.2
20	122 56.0	01.3	286 41.0	10.9	16 49.5	11.5	57.2
21	137 56.2	. . 02.3	301 10.9	10.7	17 01.0	11.3	57.2
22	152 56.4	03.2	315 40.6	10.6	17 12.3	11.3	57.3
23	167 56.6	04.2	330 10.2	10.6	17 23.6	11.2	57.3
6 00	182 56.7	S 5 05.1	344 39.8	10.5	N17 34.8	11.2	57.3
01	197 56.9	06.1	359 09.3	10.4	17 46.0	11.0	57.3
02	212 57.1	07.1	13 38.7	10.3	17 57.0	10.9	57.3
03	227 57.3	. . 08.0	28 08.0	10.2	18 07.9	10.9	57.4
04	242 57.5	09.0	42 37.2	10.1	18 18.8	10.8	57.4
05	257 57.7	09.9	57 06.3	10.0	18 29.6	10.7	57.4
06	272 57.9	S 5 10.9	71 35.3	9.9	N18 40.3	10.6	57.4
07	287 58.0	11.8	86 04.2	9.9	18 50.9	10.5	57.4
T 08	302 58.2	12.8	100 33.1	9.7	19 01.4	10.4	57.5
U 09	317 58.4	. . 13.8	115 01.8	9.7	19 11.8	10.3	57.5
E 10	332 58.6	14.7	129 30.5	9.5	19 22.1	10.2	57.5
S 11	347 58.8	15.7	143 59.0	9.5	19 32.3	10.1	57.5
D 12	2 59.0	S 5 16.6	158 27.5	9.4	N19 42.4	10.0	57.5
A 13	17 59.1	17.6	172 55.9	9.2	19 52.4	9.9	57.6
Y 14	32 59.3	18.6	187 24.1	9.2	20 02.3	9.8	57.6
15	47 59.5	. . 19.5	201 52.3	9.1	20 12.1	9.7	57.6
16	62 59.7	20.5	216 20.4	9.0	20 21.8	9.6	57.6
17	77 59.9	21.4	230 48.4	8.9	20 31.4	9.5	57.6
18	93 00.1	S 5 22.4	245 16.3	8.9	N20 40.9	9.3	57.7
19	108 00.2	23.3	259 44.2	8.7	20 50.2	9.3	57.7
20	123 00.4	24.3	274 11.9	8.6	20 59.5	9.1	57.7
21	138 00.6	. . 25.3	288 39.5	8.6	21 08.6	9.1	57.7
22	153 00.8	26.2	303 07.1	8.4	21 17.7	8.9	57.7
23	168 01.0	27.2	317 34.5	8.4	N21 26.6	8.8	57.8
	SD 16.0	*d* 1.0	SD 15.4		15.5		15.7

Lat.	Twilight Naut.	Twilight Civil	Sunrise	Moonrise 4	Moonrise 5	Moonrise 6	Moonrise 7
°	h m	h m	h m	h m	h m	h m	h m
N 72	04 10	05 30	06 37	15 34	14 44	▭	▭
N 70	04 19	05 31	06 32	15 51	15 20	13 55	▭
68	04 26	05 32	06 27	16 05	15 46	15 13	▭
66	04 33	05 32	06 24	16 17	16 06	15 51	15 19
64	04 38	05 33	06 20	16 26	16 22	16 18	16 15
62	04 42	05 33	06 18	16 34	16 36	16 39	16 48
60	04 45	05 34	06 15	16 41	16 47	16 57	17 13
N 58	04 48	05 34	06 13	16 48	16 57	17 11	17 33
56	04 51	05 34	06 11	16 53	17 06	17 24	17 49
54	04 53	05 34	06 09	16 58	17 14	17 34	18 03
52	04 55	05 34	06 08	17 03	17 21	17 44	18 16
50	04 56	05 34	06 06	17 07	17 27	17 53	18 27
45	05 00	05 34	06 03	17 16	17 41	18 11	18 49
N 40	05 02	05 33	06 00	17 24	17 52	18 27	19 08
35	05 03	05 33	05 58	17 30	18 02	18 39	19 23
30	05 04	05 32	05 56	17 36	18 11	18 51	19 37
20	05 04	05 30	05 52	17 46	18 26	19 10	20 00
N 10	05 03	05 27	05 48	17 55	18 39	19 27	20 20
0	05 00	05 24	05 45	18 04	18 52	19 43	20 39
S 10	04 56	05 21	05 42	18 12	19 04	19 59	20 58
20	04 50	05 16	05 38	18 21	19 17	20 17	21 18
30	04 41	05 09	05 33	18 32	19 33	20 37	21 42
35	04 35	05 05	05 31	18 38	19 42	20 48	21 55
40	04 28	05 01	05 28	18 45	19 52	21 02	22 12
45	04 20	04 55	05 24	18 53	20 05	21 18	22 31
S 50	04 08	04 48	05 20	19 03	20 20	21 38	22 55
52	04 03	04 44	05 18	19 07	20 27	21 48	23 07
54	03 56	04 40	05 16	19 13	20 35	21 58	23 20
56	03 49	04 36	05 14	19 18	20 43	22 11	23 36
58	03 41	04 31	05 11	19 25	20 53	22 25	23 54
S 60	03 32	04 25	05 08	19 32	21 05	22 42	24 17

Lat.	Sunset	Twilight Civil	Twilight Naut.	Moonset 4	Moonset 5	Moonset 6	Moonset 7
°	h m	h m	h m	h m	h m	h m	h m
N 72	16 57	18 05	19 24	07 42	10 10	▭	▭
N 70	17 03	18 04	19 15	07 27	09 36	12 47	▭
68	17 08	18 03	19 08	07 15	09 12	11 30	▭
66	17 12	18 02	19 02	07 05	08 53	10 53	13 19
64	17 15	18 02	18 57	06 57	08 38	10 27	12 23
62	17 18	18 02	18 53	06 51	08 26	10 06	11 50
60	17 20	18 02	18 50	06 45	08 15	09 50	11 26
N 58	17 23	18 02	18 47	06 40	08 06	09 36	11 07
56	17 25	18 02	18 45	06 35	07 58	09 24	10 51
54	17 26	18 02	18 43	06 31	07 51	09 14	10 37
52	17 28	18 02	18 41	06 27	07 45	09 05	10 25
50	17 30	18 02	18 39	06 24	07 39	08 57	10 15
45	17 33	18 02	18 36	06 17	07 27	08 40	09 53
N 40	17 36	18 03	18 34	06 11	07 17	08 25	09 35
35	17 38	18 04	18 33	06 05	07 08	08 14	09 20
30	17 41	18 05	18 32	06 01	07 01	08 03	09 08
20	17 45	18 07	18 32	05 53	06 48	07 45	08 46
N 10	17 48	18 09	18 34	05 46	06 37	07 30	08 27
0	17 52	18 12	18 36	05 40	06 26	07 16	08 09
S 10	17 55	18 16	18 41	05 33	06 16	07 01	07 52
20	17 59	18 21	18 47	05 27	06 04	06 46	07 33
30	18 04	18 28	18 56	05 19	05 52	06 29	07 11
35	18 06	18 32	19 02	05 14	05 44	06 19	06 59
40	18 09	18 37	19 09	05 09	05 36	06 07	06 44
45	18 13	18 43	19 18	05 04	05 27	05 54	06 27
S 50	18 17	18 50	19 30	04 57	05 15	05 37	06 06
52	18 19	18 54	19 36	04 54	05 10	05 30	05 56
54	18 22	18 58	19 42	04 50	05 04	05 21	05 45
56	18 24	19 02	19 49	04 46	04 57	05 12	05 32
58	18 27	19 07	19 58	04 42	04 50	05 01	05 18
S 60	18 30	19 13	20 07	04 37	04 42	04 49	05 00

Day	SUN Eqn. of Time 00^h	SUN Eqn. of Time 12^h	SUN Mer. Pass.	MOON Mer. Pass. Upper	MOON Mer. Pass. Lower	MOON Age	MOON Phase
d	m s	m s	h m	h m	h m	d %	
4	11 10	11 19	11 49	24 15	11 52	16 100	
5	11 29	11 38	11 48	00 15	12 39	17 98	○
6	11 47	11 55	11 48	01 03	13 29	18 94	

UT		ARIES	VENUS −3.9		MARS +0.7		JUPITER −2.6		SATURN +1.1	
		GHA	GHA	Dec	GHA	Dec	GHA	Dec	GHA	Dec
d h		° ′	° ′	° ′	° ′	° ′	° ′	° ′	° ′	° ′
7 00		15 46.5	204 15.8	N 5 10.4	258 40.4	N22 00.7	55 45.4	S16 41.6	197 23.7	N 2 52.5
01		30 49.0	219 15.4	09.2	273 41.4	00.5	70 48.0	41.6	212 25.9	52.4
02		45 51.5	234 15.0	08.0	288 42.4	00.3	85 50.5	41.6	227 28.1	52.2
03		60 53.9	249 14.6	. . 06.9	303 43.5	. . 00.2	100 53.0	. . 41.6	242 30.3	. . 52.1
04		75 56.4	264 14.2	05.7	318 44.5	22 00.0	115 55.5	41.6	257 32.5	52.0
05		90 58.9	279 13.8	04.5	333 45.5	21 59.8	130 58.0	41.7	272 34.7	51.9
06		106 01.3	294 13.4	N 5 03.4	348 46.6	N21 59.6	146 00.5	S16 41.7	287 36.8	N 2 51.8
07	W	121 03.8	309 13.0	02.2	3 47.6	59.4	161 03.0	41.7	302 39.0	51.7
08	E	136 06.2	324 12.6	5 01.0	18 48.6	59.2	176 05.6	41.7	317 41.2	51.5
09	D	151 08.7	339 12.2	4 59.8	33 49.7	. . 59.0	191 08.1	. . 41.7	332 43.4	. . 51.4
10	N	166 11.2	354 11.8	58.7	48 50.7	58.8	206 10.6	41.7	347 45.6	51.3
11	E	181 13.6	9 11.4	57.5	63 51.7	58.6	221 13.1	41.7	2 47.8	51.2
12	S	196 16.1	24 11.0	N 4 56.3	78 52.8	N21 58.4	236 15.6	S16 41.7	17 49.9	N 2 51.1
13	D	211 18.6	39 10.6	55.2	93 53.8	58.3	251 18.1	41.7	32 52.1	51.0
14	A	226 21.0	54 10.2	54.0	108 54.9	58.1	266 20.6	41.8	47 54.3	50.8
15	Y	241 23.5	69 09.8	. . 52.8	123 55.9	. . 57.9	281 23.2	. . 41.8	62 56.5	. . 50.7
16		256 26.0	84 09.4	51.6	138 56.9	57.7	296 25.7	41.8	77 58.7	50.6
17		271 28.4	99 09.0	50.5	153 58.0	57.5	311 28.2	41.8	93 00.9	50.5
18		286 30.9	114 08.6	N 4 49.3	168 59.0	N21 57.3	326 30.7	S16 41.8	108 03.1	N 2 50.4
19		301 33.4	129 08.2	48.1	184 00.1	57.1	341 33.2	41.8	123 05.2	50.3
20		316 35.8	144 07.8	46.9	199 01.1	56.9	356 35.7	41.8	138 07.4	50.1
21		331 38.3	159 07.4	. . 45.8	214 02.1	. . 56.7	11 38.2	. . 41.8	153 09.6	. . 50.0
22		346 40.7	174 07.0	44.6	229 03.2	56.5	26 40.7	41.8	168 11.8	49.9
23		1 43.2	189 06.6	43.4	244 04.2	56.4	41 43.2	41.8	183 14.0	49.8
8 00		16 45.7	204 06.2	N 4 42.2	259 05.3	N21 56.2	56 45.7	S16 41.9	198 16.2	N 2 49.7
01		31 48.1	219 05.8	41.1	274 06.3	56.0	71 48.2	41.9	213 18.3	49.6
02		46 50.6	234 05.4	39.9	289 07.4	55.8	86 50.8	41.9	228 20.5	49.4
03		61 53.1	249 05.0	. . 38.7	304 08.4	. . 55.6	101 53.3	. . 41.9	243 22.7	. . 49.3
04		76 55.5	264 04.6	37.5	319 09.4	55.4	116 55.8	41.9	258 24.9	49.2
05		91 58.0	279 04.2	36.4	334 10.5	55.2	131 58.3	41.9	273 27.1	49.1
06		107 00.5	294 03.8	N 4 35.2	349 11.5	N21 55.0	147 00.8	S16 41.9	288 29.3	N 2 49.0
07		122 02.9	309 03.4	34.0	4 12.6	54.8	162 03.3	41.9	303 31.5	48.9
08	T	137 05.4	324 03.0	32.8	19 13.6	54.6	177 05.8	41.9	318 33.6	48.7
09	H	152 07.9	339 02.6	. . 31.6	34 14.7	. . 54.4	192 08.3	. . 41.9	333 35.8	. . 48.6
10	U	167 10.3	354 02.2	30.5	49 15.7	54.2	207 10.8	41.9	348 38.0	48.5
11	R	182 12.8	9 01.8	29.3	64 16.8	54.1	222 13.3	41.9	3 40.2	48.4
12	S	197 15.2	24 01.4	N 4 28.1	79 17.8	N21 53.9	237 15.8	S16 42.0	18 42.4	N 2 48.3
13	D	212 17.7	39 01.0	26.9	94 18.9	53.7	252 18.3	42.0	33 44.6	48.2
14	A	227 20.2	54 00.6	25.7	109 19.9	53.5	267 20.8	42.0	48 46.8	48.0
15	Y	242 22.6	69 00.2	. . 24.6	124 21.0	. . 53.3	282 23.3	. . 42.0	63 48.9	. . 47.9
16		257 25.1	83 59.9	23.4	139 22.0	53.1	297 25.8	42.0	78 51.1	47.8
17		272 27.6	98 59.5	22.2	154 23.0	52.9	312 28.3	42.0	93 53.3	47.7
18		287 30.0	113 59.1	N 4 21.0	169 24.1	N21 52.7	327 30.8	S16 42.0	108 55.5	N 2 47.6
19		302 32.5	128 58.7	19.8	184 25.1	52.5	342 33.3	42.0	123 57.7	47.5
20		317 35.0	143 58.3	18.7	199 26.2	52.3	357 35.8	42.0	138 59.9	47.3
21		332 37.4	158 57.9	. . 17.5	214 27.3	. . 52.1	12 38.3	. . 42.0	154 02.0	. . 47.2
22		347 39.9	173 57.5	16.3	229 28.3	51.9	27 40.8	42.0	169 04.2	47.1
23		2 42.3	188 57.1	15.1	244 29.4	51.7	42 43.3	42.0	184 06.4	47.0
9 00		17 44.8	203 56.7	N 4 13.9	259 30.4	N21 51.5	57 45.8	S16 42.0	199 08.6	N 2 46.9
01		32 47.3	218 56.3	12.8	274 31.5	51.4	72 48.3	42.0	214 10.8	46.8
02		47 49.7	233 55.9	11.6	289 32.5	51.2	87 50.8	42.1	229 13.0	46.6
03		62 52.2	248 55.5	. . 10.4	304 33.6	. . 51.0	102 53.3	. . 42.1	244 15.2	. . 46.5
04		77 54.7	263 55.1	09.2	319 34.6	50.8	117 55.8	42.1	259 17.3	46.4
05		92 57.1	278 54.7	08.0	334 35.7	50.6	132 58.3	42.1	274 19.5	46.3
06		107 59.6	293 54.3	N 4 06.8	349 36.7	N21 50.4	148 00.8	S16 42.1	289 21.7	N 2 46.2
07		123 02.1	308 53.9	05.7	4 37.8	50.2	163 03.3	42.1	304 23.9	46.1
08		138 04.5	323 53.5	04.5	19 38.8	50.0	178 05.8	42.1	319 26.1	46.0
09	F	153 07.0	338 53.1	. . 03.3	34 39.9	. . 49.8	193 08.3	. . 42.1	334 28.3	. . 45.8
10	R	168 09.5	353 52.7	02.1	49 41.0	49.6	208 10.8	42.1	349 30.5	45.7
11	I	183 11.9	8 52.3	4 00.9	64 42.0	49.4	223 13.3	42.1	4 32.6	45.6
12	D	198 14.4	23 51.9	N 3 59.7	79 43.1	N21 49.2	238 15.8	S16 42.1	19 34.8	N 2 45.5
13	A	213 16.8	38 51.5	58.5	94 44.1	49.0	253 18.3	42.1	34 37.0	45.4
14	Y	228 19.3	53 51.1	57.4	109 45.2	48.8	268 20.8	42.1	49 39.2	45.3
15		243 21.8	68 50.7	. . 56.2	124 46.2	. . 48.6	283 23.3	. . 42.1	64 41.4	. . 45.1
16		258 24.2	83 50.3	55.0	139 47.3	48.4	298 25.8	42.1	79 43.6	45.0
17		273 26.7	98 49.9	53.8	154 48.4	48.2	313 28.3	42.1	94 45.8	44.9
18		288 29.2	113 49.5	N 3 52.6	169 49.4	N21 48.0	328 30.8	S16 42.1	109 48.0	N 2 44.8
19		303 31.6	128 49.2	51.4	184 50.5	47.8	343 33.3	42.1	124 50.1	44.7
20		318 34.1	143 48.8	50.2	199 51.5	47.6	358 35.8	42.2	139 52.3	44.6
21		333 36.6	158 48.4	. . 49.1	214 52.6	. . 47.5	13 38.3	. . 42.2	154 54.5	. . 44.4
22		348 39.0	173 48.0	47.9	229 53.7	47.3	28 40.8	42.2	169 56.7	44.3
23		3 41.5	188 47.6	46.7	244 54.7	47.1	43 43.3	42.2	184 58.9	44.2
Mer. Pass.		h m 22 49.2	v −0.4	d 1.2	v 1.0	d 0.2	v 2.5	d 0.0	v 2.2	d 0.1

STARS

Name	SHA	Dec
	° ′	° ′
Acamar	315 19.9	S40 15.
Achernar	335 28.0	S57 11.
Acrux	173 13.3	S63 09.
Adhara	255 14.7	S28 58.
Aldebaran	290 52.3	N16 31.
Alioth	166 23.4	N55 54.
Alkaid	153 01.4	N49 15.
Al Na'ir	27 46.7	S46 54.
Alnilam	275 49.0	S 1 11.
Alphard	217 59.0	S 8 41.
Alphecca	126 13.6	N26 41.
Alpheratz	357 46.1	N29 08.
Altair	62 10.9	N 8 53.
Ankaa	353 17.8	S42 15.
Antares	112 29.9	S26 27.
Arcturus	145 58.5	N19 07.
Atria	107 34.5	S69 03.
Avior	234 19.5	S59 32.
Bellatrix	278 34.8	N 6 21.
Betelgeuse	271 04.2	N 7 24.
Canopus	263 57.3	S52 41.
Capella	280 38.3	N46 00.
Deneb	49 33.3	N45 19.
Denebola	182 36.7	N14 31.
Diphda	348 58.2	S17 55.
Dubhe	193 55.3	N61 41.
Elnath	278 15.9	N28 37.
Eltanin	90 47.6	N51 29.
Enif	33 49.6	N 9 55.
Fomalhaut	15 26.6	S29 34.
Gacrux	172 04.8	S57 10.
Gienah	175 55.5	S17 35.
Hadar	148 52.6	S60 25.
Hamal	328 03.6	N23 30.
Kaus Aust.	83 47.5	S34 22.
Kochab	137 20.6	N74 07.
Markab	13 40.8	N15 15.
Menkar	314 17.6	N 4 07.
Menkent	148 11.3	S36 25.
Miaplacidus	221 41.0	S69 45.
Mirfak	308 44.0	N49 53.
Nunki	76 01.7	S26 17.
Peacock	53 23.2	S56 42.
Pollux	243 31.1	N28 00.
Procyon	245 02.6	N 5 12.
Rasalhague	96 09.1	N12 33.
Regulus	207 46.6	N11 55.
Rigel	281 14.5	S 8 11.
Rigil Kent.	139 56.3	S60 52.
Sabik	102 15.8	S15 44.
Schedar	349 43.3	N56 35.
Shaula	96 25.8	S37 06.
Sirius	258 36.1	S16 43.
Spica	158 34.5	S11 12.
Suhail	222 54.8	S43 28.
Vega	80 40.9	N38 47.
Zuben'ubi	137 08.8	S16 04.

	SHA	Mer. Pass
	° ′	h m
Venus	187 20.6	10 24
Mars	242 19.6	6 43
Jupiter	40 00.1	20 10
Saturn	181 30.5	10 45

UT	SUN GHA	SUN Dec	MOON GHA	v	MOON Dec	d	HP
d h	° ′	° ′	° ′	′	° ′	′	′
7 00	183 01.2	S 5 28.1	332 01.9	8.3	N21 35.4	8.6	57.8
01	198 01.3	29.1	346 29.2	8.1	21 44.0	8.6	57.8
02	213 01.5	30.0	0 56.3	8.1	21 52.6	8.4	57.8
03	228 01.7	. . 31.0	15 23.4	8.0	22 01.0	8.4	57.8
04	243 01.9	31.9	29 50.4	7.9	22 09.4	8.2	57.8
05	258 02.1	32.9	44 17.3	7.9	22 17.6	8.0	57.9
06	273 02.2	S 5 33.9	58 44.2	7.7	N22 25.6	8.0	57.9
07	288 02.4	34.8	73 10.9	7.6	22 33.6	7.8	57.9
08	303 02.6	35.8	87 37.5	7.6	22 41.4	7.7	57.9
09	318 02.8	. . 36.7	102 04.1	7.5	22 49.1	7.5	57.9
10	333 03.0	37.7	116 30.6	7.4	22 56.6	7.4	58.0
11	348 03.1	38.6	130 57.0	7.3	23 04.0	7.3	58.0
12	3 03.3	S 5 39.6	145 23.3	7.2	N23 11.3	7.2	58.0
13	18 03.5	40.5	159 49.5	7.1	23 18.5	7.0	58.0
14	33 03.7	41.5	174 15.6	7.1	23 25.5	6.9	58.0
15	48 03.8	. . 42.5	188 41.7	7.0	23 32.4	6.7	58.0
16	63 04.0	43.4	203 07.7	6.8	23 39.1	6.6	58.1
17	78 04.2	44.4	217 33.5	6.9	23 45.7	6.5	58.1
18	93 04.4	S 5 45.3	231 59.4	6.7	N23 52.2	6.3	58.1
19	108 04.6	46.3	246 25.1	6.6	23 58.5	6.2	58.1
20	123 04.7	47.2	260 50.7	6.6	24 04.7	6.0	58.1
21	138 04.9	. . 48.2	275 16.3	6.5	24 10.7	5.9	58.1
22	153 05.1	49.1	289 41.8	6.4	24 16.6	5.7	58.2
23	168 05.3	50.1	304 07.2	6.4	24 22.3	5.6	58.2
8 00	183 05.4	S 5 51.0	318 32.6	6.3	N24 27.9	5.5	58.2
01	198 05.6	52.0	332 57.9	6.2	24 33.4	5.3	58.2
02	213 05.8	53.0	347 23.1	6.1	24 38.7	5.1	58.2
03	228 06.0	. . 53.9	1 48.2	6.1	24 43.8	5.0	58.2
04	243 06.2	54.9	16 13.3	6.0	24 48.8	4.8	58.3
05	258 06.3	55.8	30 38.3	5.9	24 53.6	4.7	58.3
06	273 06.5	S 5 56.8	45 03.2	5.9	N24 58.3	4.5	58.3
07	288 06.7	57.7	59 28.1	5.8	25 02.8	4.4	58.3
08	303 06.9	58.7	73 52.9	5.8	25 07.2	4.2	58.3
09	318 07.0	5 59.6	88 17.7	5.7	25 11.4	4.1	58.3
10	333 07.2	6 00.6	102 42.4	5.6	25 15.5	3.9	58.4
11	348 07.4	01.5	117 07.0	5.6	25 19.4	3.7	58.4
12	3 07.6	S 6 02.5	131 31.6	5.5	N25 23.1	3.6	58.4
13	18 07.7	03.4	145 56.1	5.4	25 26.7	3.4	58.4
14	33 07.9	04.4	160 20.5	5.5	25 30.1	3.2	58.4
15	48 08.1	. . 05.3	174 45.0	5.3	25 33.3	3.1	58.4
16	63 08.3	06.3	189 09.3	5.3	25 36.4	3.0	58.4
17	78 08.4	07.2	203 33.6	5.3	25 39.4	2.7	58.5
18	93 08.6	S 6 08.2	217 57.9	5.2	N25 42.1	2.6	58.5
19	108 08.8	09.1	232 22.1	5.2	25 44.7	2.5	58.5
20	123 08.9	10.1	246 46.3	5.1	25 47.2	2.2	58.5
21	138 09.1	. . 11.0	261 10.4	5.1	25 49.4	2.1	58.5
22	153 09.3	12.0	275 34.5	5.1	25 51.5	1.9	58.5
23	168 09.5	12.9	289 58.6	5.0	25 53.4	1.8	58.5
9 00	183 09.6	S 6 13.9	304 22.6	5.0	N25 55.2	1.6	58.6
01	198 09.8	14.8	318 46.6	5.0	25 56.8	1.4	58.6
02	213 10.0	15.8	333 10.6	4.9	25 58.2	1.3	58.6
03	228 10.2	. . 16.7	347 34.5	4.9	25 59.5	1.1	58.6
04	243 10.3	17.7	1 58.4	4.9	26 00.6	0.9	58.6
05	258 10.5	18.6	16 22.3	4.8	26 01.5	0.7	58.6
06	273 10.7	S 6 19.6	30 46.1	4.8	N26 02.2	0.6	58.6
07	288 10.8	20.5	45 09.9	4.8	26 02.8	0.4	58.7
08	303 11.0	21.5	59 33.7	4.8	26 03.2	0.2	58.7
09	318 11.2	. . 22.4	73 57.5	4.8	26 03.4	0.1	58.7
10	333 11.4	23.4	88 21.3	4.8	26 03.5	0.2	58.7
11	348 11.5	24.3	102 45.1	4.7	26 03.3	0.2	58.7
12	3 11.7	S 6 25.3	117 08.8	4.7	N26 03.1	0.5	58.7
13	18 11.9	26.2	131 32.5	4.8	26 02.6	0.6	58.7
14	33 12.0	27.2	145 56.3	4.7	26 02.0	0.8	58.7
15	48 12.2	. . 28.1	160 20.0	4.7	26 01.2	1.0	58.8
16	63 12.4	29.1	174 43.7	4.7	26 00.2	1.2	58.8
17	78 12.5	30.0	189 07.4	4.7	25 59.0	1.3	58.8
18	93 12.7	S 6 31.0	203 31.1	4.7	N25 57.7	1.5	58.8
19	108 12.9	31.9	217 54.8	4.7	25 56.2	1.7	58.8
20	123 13.0	32.9	232 18.5	4.7	25 54.5	1.8	58.8
21	138 13.2	. . 33.8	246 42.2	4.8	25 52.7	2.0	58.8
22	153 13.4	34.8	261 06.0	4.7	25 50.7	2.2	58.8
23	168 13.5	35.7	275 29.7	4.7	N25 48.5	2.4	58.9
	SD 16.0	*d* 1.0	SD 15.8		15.9		16.0

Lat.	Twilight Naut.	Twilight Civil	Sunrise	Moonrise 7	Moonrise 8	Moonrise 9	Moonrise 10
°	h m	h m	h m	h m	h m	h m	h m
N 72	04 24	05 43	06 51	▭	▭	▭	▭
N 70	04 31	05 43	06 44	▭	▭	▭	▭
68	04 37	05 42	06 38	▭	▭	▭	▭
66	04 42	05 42	06 33	15 19	▭	▭	▭
64	04 46	05 42	06 29	16 15	16 13	16 50	18 53
62	04 50	05 41	06 26	16 48	17 11	18 04	19 35
60	04 52	05 41	06 22	17 13	17 44	18 40	20 04
N 58	04 55	05 40	06 20	17 33	18 09	19 06	20 26
56	04 57	05 40	06 17	17 49	18 29	19 27	20 43
54	04 58	05 39	06 15	18 03	18 45	19 44	20 58
52	05 00	05 39	06 13	18 16	18 59	19 58	21 12
50	05 01	05 39	06 11	18 27	19 12	20 11	21 23
45	05 03	05 37	06 07	18 49	19 38	20 37	21 47
N 40	05 05	05 36	06 03	19 08	19 58	20 58	22 06
35	05 06	05 35	06 00	19 23	20 15	21 15	22 22
30	05 06	05 34	05 57	19 37	20 30	21 30	22 35
20	05 05	05 31	05 53	20 00	20 55	21 55	22 59
N 10	05 03	05 27	05 48	20 20	21 17	22 17	23 19
0	04 59	05 23	05 44	20 39	21 37	22 38	23 38
S 10	04 54	05 19	05 40	20 58	21 58	22 58	23 56
20	04 47	05 13	05 35	21 18	22 20	23 20	24 16
30	04 37	05 06	05 30	21 42	22 45	23 45	24 40
35	04 31	05 01	05 27	21 55	23 00	24 00	00 00
40	04 23	04 56	05 23	22 12	23 18	24 18	00 18
45	04 13	04 49	05 19	22 31	23 39	24 38	00 38
S 50	04 01	04 41	05 14	22 55	24 06	00 06	01 04
52	03 55	04 37	05 12	23 07	24 18	00 18	01 17
54	03 48	04 33	05 09	23 20	24 33	00 33	01 32
56	03 40	04 28	05 06	23 36	24 51	00 51	01 48
58	03 31	04 22	05 03	23 54	25 12	01 12	02 09
S 60	03 21	04 16	04 59	24 17	00 17	01 39	02 34

Lat.	Sunset	Twilight Civil	Twilight Naut.	Moonset 7	Moonset 8	Moonset 9	Moonset 10
°	h m	h m	h m	h m	h m	h m	h m
N 72	16 42	17 50	19 08	▭	▭	▭	▭
N 70	16 49	17 50	19 01	▭	▭	▭	▭
68	16 55	17 51	18 55	▭	▭	▭	▭
66	17 00	17 51	18 51	13 19	▭	▭	▭
64	17 04	17 52	18 47	12 23	14 26	15 55	15 58
62	17 08	17 52	18 44	11 50	13 29	14 41	15 16
60	17 11	17 53	18 41	11 26	12 56	14 05	14 47
N 58	17 14	17 53	18 39	11 07	12 31	13 39	14 25
56	17 17	17 54	18 37	10 51	12 12	13 19	14 07
54	17 19	17 55	18 35	10 37	11 55	13 01	13 51
52	17 21	17 55	18 34	10 25	11 41	12 47	13 38
50	17 23	17 56	18 33	10 15	11 29	12 34	13 26
45	17 28	17 57	18 31	09 53	11 04	12 08	13 02
N 40	17 31	17 58	18 30	09 35	10 44	11 47	12 42
35	17 34	18 00	18 29	09 20	10 27	11 29	12 26
30	17 37	18 01	18 29	09 08	10 12	11 14	12 12
20	17 42	18 04	18 30	08 46	09 48	10 49	11 47
N 10	17 46	18 08	18 32	08 27	09 26	10 27	11 26
0	17 51	18 12	18 36	08 09	09 06	10 06	11 07
S 10	17 55	18 16	18 41	07 52	08 47	09 46	10 47
20	18 00	18 22	18 48	07 33	08 25	09 23	10 25
30	18 06	18 30	18 58	07 11	08 01	08 58	10 01
35	18 09	18 34	19 05	06 59	07 47	08 43	09 46
40	18 13	18 40	19 13	06 44	07 30	08 25	09 29
45	18 17	18 47	19 23	06 27	07 10	08 04	09 09
S 50	18 22	18 55	19 35	06 06	06 45	07 37	08 43
52	18 25	18 59	19 42	05 56	06 33	07 24	08 31
54	18 27	19 04	19 49	05 45	06 19	07 09	08 16
56	18 30	19 09	19 57	05 32	06 04	06 52	08 00
58	18 34	19 15	20 06	05 18	05 45	06 30	07 40
S 60	18 37	19 21	20 16	05 00	05 22	06 03	07 14

Day	SUN Eqn. of Time 00^h	SUN Eqn. of Time 12^h	SUN Mer. Pass.	MOON Mer. Pass. Upper	MOON Mer. Pass. Lower	MOON Age	MOON Phase
d	m s	m s	h m	h m	h m	d %	
7	12 04	12 13	11 48	01 56	14 24	19 88	
8	12 21	12 30	11 48	02 53	15 22	20 80	
9	12 38	12 46	11 47	03 52	16 22	21 71	

	UT	ARIES	VENUS −3.9		MARS +0.7		JUPITER −2.6		SATURN +1.1		STARS		
		GHA	GHA	Dec	GHA	Dec	GHA	Dec	GHA	Dec	Name	SHA	Dec
	d h	° ′	° ′	° ′	° ′	° ′	° ′	° ′	° ′	° ′		° ′	°
	10 00	18 44.0	203 47.2	N 3 45.5	259 55.8	N21 46.9	58 45.7	S16 42.2	200 01.1	N 2 44.1	Acamar	315 19.9	S40 15
	01	33 46.4	218 46.8	44.3	274 56.9	46.7	73 48.2	42.2	215 03.3	44.0	Achernar	335 28.0	S57 11
	02	48 48.9	233 46.4	43.1	289 57.9	46.5	88 50.7	42.2	230 05.4	43.9	Acrux	173 13.3	S63 09
	03	63 51.3	248 46.0	. . 41.9	304 59.0	. . 46.3	103 53.2	. . 42.2	245 07.6	. . 43.8	Adhara	255 14.6	S28 58
	04	78 53.8	263 45.6	40.7	320 00.0	46.1	118 55.7	42.2	260 09.8	43.6	Aldebaran	290 52.3	N16 31
	05	93 56.3	278 45.2	39.5	335 01.1	45.9	133 58.2	42.2	275 12.0	43.5			
	06	108 58.7	293 44.8	N 3 38.4	350 02.2	N21 45.7	149 00.7	S16 42.2	290 14.2	N 2 43.4	Alioth	166 23.4	N55 54
	07	124 01.2	308 44.4	37.2	5 03.2	45.5	164 03.2	42.2	305 16.4	43.3	Alkaid	153 01.4	N49 15
S	08	139 03.7	323 44.0	36.0	20 04.3	45.3	179 05.7	42.2	320 18.6	43.2	Al Na'ir	27 46.7	S46 54
A	09	154 06.1	338 43.6	. . 34.8	35 05.4	. . 45.1	194 08.2	. . 42.2	335 20.7	. . 43.1	Alnilam	275 49.0	S 1 11
T	10	169 08.6	353 43.2	33.6	50 06.4	44.9	209 10.6	42.2	350 22.9	42.9	Alphard	217 58.9	S 8 41
U	11	184 11.1	8 42.8	32.4	65 07.5	44.7	224 13.1	42.2	5 25.1	42.8			
R	12	199 13.5	23 42.4	N 3 31.2	80 08.6	N21 44.5	239 15.6	S16 42.2	20 27.3	N 2 42.7	Alphecca	126 13.6	N26 41
D	13	214 16.0	38 42.1	30.0	95 09.6	44.3	254 18.1	42.2	35 29.5	42.6	Alpheratz	357 46.0	N29 08
A	14	229 18.5	53 41.7	28.8	110 10.7	44.1	269 20.6	42.2	50 31.7	42.5	Altair	62 10.9	N 8 53
Y	15	244 20.9	68 41.3	. . 27.6	125 11.8	. . 43.9	284 23.1	. . 42.2	65 33.9	. . 42.4	Ankaa	353 17.8	S42 15
	16	259 23.4	83 40.9	26.4	140 12.8	43.7	299 25.6	42.2	80 36.1	42.3	Antares	112 29.9	S26 27
	17	274 25.8	98 40.5	25.2	155 13.9	43.5	314 28.1	42.2	95 38.2	42.1			
	18	289 28.3	113 40.1	N 3 24.1	170 15.0	N21 43.3	329 30.5	S16 42.2	110 40.4	N 2 42.0	Arcturus	145 58.5	N19 07
	19	304 30.8	128 39.7	22.9	185 16.1	43.1	344 33.0	42.2	125 42.6	41.9	Atria	107 34.5	S69 02
	20	319 33.2	143 39.3	21.7	200 17.1	42.9	359 35.5	42.2	140 44.8	41.8	Avior	234 19.4	S59 32
	21	334 35.7	158 38.9	. . 20.5	215 18.2	. . 42.7	14 38.0	. . 42.2	155 47.0	. . 41.7	Bellatrix	278 34.8	N 6 21
	22	349 38.2	173 38.5	19.3	230 19.3	42.5	29 40.5	42.2	170 49.2	41.6	Betelgeuse	271 04.1	N 7 24
	23	4 40.6	188 38.1	18.1	245 20.3	42.3	44 43.0	42.2	185 51.4	41.4			
	11 00	19 43.1	203 37.7	N 3 16.9	260 21.4	N21 42.1	59 45.5	S16 42.2	200 53.6	N 2 41.3	Canopus	263 57.2	S52 41
	01	34 45.6	218 37.3	15.7	275 22.5	41.9	74 47.9	42.2	215 55.7	41.2	Capella	280 38.3	N46 00
	02	49 48.0	233 36.9	14.5	290 23.6	41.7	89 50.4	42.2	230 57.9	41.1	Deneb	49 33.3	N45 19
	03	64 50.5	248 36.5	. . 13.3	305 24.6	. . 41.5	104 52.9	. . 42.2	246 00.1	. . 41.0	Denebola	182 36.7	N14 31
	04	79 53.0	263 36.2	12.1	320 25.7	41.3	119 55.4	42.2	261 02.3	40.9	Diphda	348 58.2	S17 55
	05	94 55.4	278 35.8	10.9	335 26.8	41.1	134 57.9	42.2	276 04.5	40.8			
	06	109 57.9	293 35.4	N 3 09.7	350 27.9	N21 40.9	150 00.3	S16 42.2	291 06.7	N 2 40.6	Dubhe	193 55.3	N61 41
	07	125 00.3	308 35.0	08.5	5 28.9	40.7	165 02.8	42.2	306 08.9	40.5	Elnath	278 15.9	N28 37
	08	140 02.8	323 34.6	07.3	20 30.0	40.5	180 05.3	42.2	321 11.1	40.4	Eltanin	90 47.6	N51 29
S	09	155 05.3	338 34.2	. . 06.1	35 31.1	. . 40.3	195 07.8	. . 42.2	336 13.2	. . 40.3	Enif	33 49.6	N 9 55
U	10	170 07.7	353 33.8	04.9	50 32.2	40.1	210 10.3	42.2	351 15.4	40.2	Fomalhaut	15 26.6	S29 34
N	11	185 10.2	8 33.4	03.8	65 33.2	39.9	225 12.7	42.2	6 17.6	40.1			
D	12	200 12.7	23 33.0	N 3 02.6	80 34.3	N21 39.7	240 15.2	S16 42.2	21 19.8	N 2 40.0	Gacrux	172 04.7	S57 10
A	13	215 15.1	38 32.6	01.4	95 35.4	39.5	255 17.7	42.2	36 22.0	39.8	Gienah	175 55.5	S17 35
Y	14	230 17.6	53 32.2	3 00.2	110 36.5	39.3	270 20.2	42.2	51 24.2	39.7	Hadar	148 52.6	S60 25
	15	245 20.1	68 31.8	. . 2 59.0	125 37.6	. . 39.1	285 22.7	. . 42.2	66 26.4	. . 39.6	Hamal	328 03.5	N23 30
	16	260 22.5	83 31.4	57.8	140 38.6	38.9	300 25.1	42.2	81 28.6	39.5	Kaus Aust.	83 47.5	S34 22
	17	275 25.0	98 31.0	56.6	155 39.7	38.7	315 27.6	42.2	96 30.7	39.4			
	18	290 27.4	113 30.7	N 2 55.4	170 40.8	N21 38.5	330 30.1	S16 42.2	111 32.9	N 2 39.3	Kochab	137 20.6	N74 07
	19	305 29.9	128 30.3	54.2	185 41.9	38.3	345 32.6	42.2	126 35.1	39.1	Markab	13 40.8	N15 15
	20	320 32.4	143 29.9	53.0	200 43.0	38.1	0 35.1	42.2	141 37.3	39.0	Menkar	314 17.6	N 4 07
	21	335 34.8	158 29.5	. . 51.8	215 44.0	. . 37.9	15 37.5	. . 42.2	156 39.5	. . 38.9	Menkent	148 11.3	S36 25
	22	350 37.3	173 29.1	50.6	230 45.1	37.7	30 40.0	42.2	171 41.7	38.8	Miaplacidus	221 41.0	S69 45
	23	5 39.8	188 28.7	49.4	245 46.2	37.5	45 42.5	42.2	186 43.9	38.7			
	12 00	20 42.2	203 28.3	N 2 48.2	260 47.3	N21 37.3	60 45.0	S16 42.2	201 46.1	N 2 38.6	Mirfak	308 43.9	N49 53
	01	35 44.7	218 27.9	47.0	275 48.4	37.1	75 47.4	42.2	216 48.3	38.5	Nunki	76 01.7	S26 17
	02	50 47.2	233 27.5	45.8	290 49.5	36.9	90 49.9	42.2	231 50.4	38.3	Peacock	53 23.2	S56 42
	03	65 49.6	248 27.1	. . 44.6	305 50.6	. . 36.7	105 52.4	. . 42.2	246 52.6	. . 38.2	Pollux	243 31.1	N28 00
	04	80 52.1	263 26.7	43.4	320 51.6	36.5	120 54.9	42.2	261 54.8	38.1	Procyon	245 02.6	N 5 12
	05	95 54.6	278 26.3	42.2	335 52.7	36.3	135 57.3	42.2	276 57.0	38.0			
	06	110 57.0	293 26.0	N 2 41.0	350 53.8	N21 36.1	150 59.8	S16 42.2	291 59.2	N 2 37.9	Rasalhague	96 09.1	N12 33
	07	125 59.5	308 25.6	39.8	5 54.9	35.9	166 02.3	42.2	307 01.4	37.8	Regulus	207 46.6	N11 55
	08	141 01.9	323 25.2	38.6	20 56.0	35.7	181 04.8	42.2	322 03.6	37.7	Rigel	281 14.5	S 8 11
M	09	156 04.4	338 24.8	. . 37.4	35 57.1	. . 35.5	196 07.2	. . 42.2	337 05.8	. . 37.5	Rigil Kent.	139 56.3	S60 52
O	10	171 06.9	353 24.4	36.2	50 58.2	35.3	211 09.7	42.2	352 07.9	37.4	Sabik	102 15.8	S15 44
N	11	186 09.3	8 24.0	35.0	65 59.2	35.1	226 12.2	42.2	7 10.1	37.3			
D	12	201 11.8	23 23.6	N 2 33.8	81 00.3	N21 34.9	241 14.6	S16 42.2	22 12.3	N 2 37.2	Schedar	349 43.3	N56 35
A	13	216 14.3	38 23.2	32.6	96 01.4	34.7	256 17.1	42.2	37 14.5	37.1	Shaula	96 25.8	S37 06
Y	14	231 16.7	53 22.8	31.4	111 02.5	34.5	271 19.6	42.2	52 16.7	37.0	Sirius	258 36.1	S16 43
	15	246 19.2	68 22.4	. . 30.2	126 03.6	. . 34.3	286 22.1	. . 42.2	67 18.9	. . 36.9	Spica	158 34.5	S11 12
	16	261 21.7	83 22.0	29.0	141 04.7	34.1	301 24.5	42.2	82 21.1	36.7	Suhail	222 54.7	S43 28
	17	276 24.1	98 21.7	27.8	156 05.8	33.9	316 27.0	42.2	97 23.3	36.6			
	18	291 26.6	113 21.3	N 2 26.6	171 06.9	N21 33.7	331 29.5	S16 42.2	112 25.5	N 2 36.5	Vega	80 40.9	N38 47
	19	306 29.1	128 20.9	25.4	186 08.0	33.5	346 31.9	42.2	127 27.6	36.4	Zuben'ubi	137 08.8	S16 04
	20	321 31.5	143 20.5	24.2	201 09.1	33.3	1 34.4	42.2	142 29.8	36.3		SHA	Mer. Pas
	21	336 34.0	158 20.1	. . 23.0	216 10.2	. . 33.1	16 36.9	. . 42.2	157 32.0	. . 36.2		° ′	h m
	22	351 36.4	173 19.7	21.8	231 11.2	32.9	31 39.3	42.2	172 34.2	36.1	Venus	183 54.6	10 26
	23	6 38.9	188 19.3	20.6	246 12.3	32.7	46 41.8	42.2	187 36.4	35.9	Mars	240 38.3	6 38
		h m									Jupiter	40 02.4	19 58
	Mer. Pass.	22 37.4	v −0.4	d 1.2	v 1.1	d 0.2	v 2.5	d 0.0	v 2.2	d 0.1	Saturn	181 10.5	10 35

	UT	SUN GHA	SUN Dec	MOON GHA	v	MOON Dec	d	HP
	d h	° ′	° ′	° ′	′	° ′	′	′
	10 00	183 13.7	S 6 36.7	289 53.4	4.8	N25 46.1	2.5	58.9
	01	198 13.9	37.6	304 17.2	4.8	25 43.6	2.7	58.9
	02	213 14.0	38.6	318 41.0	4.8	25 40.9	2.9	58.9
	03	228 14.2	. . 39.5	333 04.8	4.8	25 38.0	3.0	58.9
	04	243 14.4	40.4	347 28.6	4.8	25 35.0	3.2	58.9
	05	258 14.5	41.4	1 52.4	4.9	25 31.8	3.4	58.9
	06	273 14.7	S 6 42.3	16 16.3	4.8	N25 28.4	3.6	58.9
	07	288 14.9	43.3	30 40.1	4.9	25 24.8	3.7	58.9
S	08	303 15.0	44.2	45 04.0	5.0	25 21.1	3.9	59.0
A	09	318 15.2	. . 45.2	59 28.0	4.9	25 17.2	4.0	59.0
T	10	333 15.4	46.1	73 51.9	5.0	25 13.2	4.3	59.0
U	11	348 15.5	47.1	88 15.9	5.0	25 08.9	4.3	59.0
R	12	3 15.7	S 6 48.0	102 39.9	5.1	N25 04.6	4.6	59.0
D	13	18 15.9	49.0	117 04.0	5.1	25 00.0	4.7	59.0
A	14	33 16.0	49.9	131 28.1	5.1	24 55.3	4.9	59.0
Y	15	48 16.2	. . 50.8	145 52.2	5.2	24 50.4	5.1	59.0
	16	63 16.4	51.8	160 16.4	5.2	24 45.3	5.2	59.0
	17	78 16.5	52.7	174 40.6	5.3	24 40.1	5.4	59.0
	18	93 16.7	S 6 53.7	189 04.9	5.3	N24 34.7	5.5	59.1
	19	108 16.9	54.6	203 29.2	5.3	24 29.2	5.7	59.1
	20	123 17.0	55.6	217 53.5	5.4	24 23.5	5.9	59.1
	21	138 17.2	. . 56.5	232 17.9	5.4	24 17.6	6.0	59.1
	22	153 17.3	57.5	246 42.3	5.5	24 11.6	6.2	59.1
	23	168 17.5	58.4	261 06.8	5.5	24 05.4	6.3	59.1
	11 00	183 17.7	S 6 59.3	275 31.3	5.6	N23 59.1	6.5	59.1
	01	198 17.8	7 00.3	289 55.9	5.7	23 52.6	6.7	59.1
	02	213 18.0	01.2	304 20.6	5.7	23 45.9	6.8	59.1
	03	228 18.2	. . 02.2	318 45.3	5.7	23 39.1	7.0	59.1
	04	243 18.3	03.1	333 10.0	5.8	23 32.1	7.1	59.2
	05	258 18.5	04.1	347 34.8	5.9	23 25.0	7.3	59.2
	06	273 18.6	S 7 05.0	1 59.7	5.9	N23 17.7	7.4	59.2
	07	288 18.8	05.9	16 24.6	6.0	23 10.3	7.6	59.2
	08	303 19.0	06.9	30 49.6	6.0	23 02.7	7.7	59.2
S	09	318 19.1	. . 07.8	45 14.6	6.2	22 55.0	7.9	59.2
U	10	333 19.3	08.8	59 39.8	6.1	22 47.1	8.0	59.2
N	11	348 19.4	09.7	74 04.9	6.3	22 39.1	8.1	59.2
D	12	3 19.6	S 7 10.6	88 30.2	6.3	N22 31.0	8.3	59.2
A	13	18 19.8	11.6	102 55.5	6.3	22 22.7	8.5	59.2
Y	14	33 19.9	12.5	117 20.8	6.4	22 14.2	8.6	59.2
	15	48 20.1	. . 13.5	131 46.2	6.5	22 05.6	8.7	59.2
	16	63 20.2	14.4	146 11.7	6.6	21 56.9	8.9	59.2
	17	78 20.4	15.4	160 37.3	6.6	21 48.0	9.0	59.3
	18	93 20.6	S 7 16.3	175 02.9	6.7	N21 39.0	9.1	59.3
	19	108 20.7	17.2	189 28.6	6.8	21 29.9	9.3	59.3
	20	123 20.9	18.2	203 54.4	6.8	21 20.6	9.5	59.3
	21	138 21.0	. . 19.1	218 20.2	6.9	21 11.1	9.5	59.3
	22	153 21.2	20.1	232 46.1	7.0	21 01.6	9.7	59.3
	23	168 21.3	21.0	247 12.1	7.0	20 51.9	9.8	59.3
	12 00	183 21.5	S 7 21.9	261 38.1	7.1	N20 42.1	10.0	59.3
	01	198 21.7	22.9	276 04.2	7.2	20 32.1	10.0	59.3
	02	213 21.8	23.8	290 30.4	7.3	20 22.1	10.2	59.3
	03	228 22.0	. . 24.7	304 56.7	7.3	20 11.9	10.4	59.3
	04	243 22.1	25.7	319 23.0	7.4	20 01.5	10.4	59.3
	05	258 22.3	26.6	333 49.4	7.5	19 51.1	10.6	59.3
	06	273 22.4	S 7 27.6	348 15.9	7.5	N19 40.5	10.7	59.3
	07	288 22.6	28.5	2 42.4	7.6	19 29.8	10.8	59.3
	08	303 22.7	29.4	17 09.0	7.7	19 19.0	10.9	59.3
M	09	318 22.9	. . 30.4	31 35.7	7.8	19 08.1	11.1	59.3
O	10	333 23.1	31.3	46 02.5	7.8	18 57.0	11.1	59.4
N	11	348 23.2	32.3	60 29.3	7.9	18 45.9	11.3	59.4
D	12	3 23.4	S 7 33.2	74 56.2	8.0	N18 34.6	11.4	59.4
A	13	18 23.5	34.1	89 23.2	8.1	18 23.2	11.5	59.4
Y	14	33 23.7	35.1	103 50.3	8.1	18 11.7	11.6	59.4
	15	48 23.8	. . 36.0	118 17.4	8.2	18 00.1	11.7	59.4
	16	63 24.0	36.9	132 44.6	8.3	17 48.4	11.9	59.4
	17	78 24.1	37.9	147 11.9	8.3	17 36.5	11.9	59.4
	18	93 24.3	S 7 38.8	161 39.2	8.4	N17 24.6	12.0	59.4
	19	108 24.4	39.7	176 06.6	8.5	17 12.6	12.2	59.4
	20	123 24.6	40.7	190 34.1	8.5	17 00.4	12.2	59.4
	21	138 24.7	. . 41.6	205 01.6	8.7	16 48.2	12.4	59.4
	22	153 24.9	42.6	219 29.3	8.6	16 35.8	12.4	59.4
	23	168 25.0	43.5	233 56.9	8.8	N16 23.4	12.5	59.4
		SD 16.0	d 0.9	SD		16.1	16.1	16.2

Lat.	Twilight Naut.	Twilight Civil	Sunrise	Moonrise 10	Moonrise 11	Moonrise 12	Moonrise 13
°	h m	h m	h m	h m	h m	h m	h m
N 72	04 38	05 56	07 05	▭	▭	20 50	23 56
N 70	04 44	05 54	06 56	▭	▭	21 44	24 14
68	04 48	05 53	06 49	▭	19 25	22 16	24 29
66	04 52	05 51	06 43	▭	20 24	22 39	24 40
64	04 55	05 50	06 38	18 53	20 58	22 57	24 50
62	04 58	05 49	06 34	19 35	21 23	23 12	24 58
60	05 00	05 48	06 30	20 04	21 43	23 25	25 05
N 58	05 01	05 47	06 26	20 26	21 59	23 35	25 12
56	05 03	05 46	06 23	20 43	22 12	23 45	25 17
54	05 04	05 45	06 20	20 58	22 24	23 53	25 22
52	05 05	05 44	06 18	21 12	22 34	24 00	00 00
50	05 06	05 43	06 16	21 23	22 43	24 07	00 07
45	05 07	05 41	06 11	21 47	23 03	24 21	00 21
N 40	05 08	05 39	06 06	22 06	23 18	24 32	00 32
35	05 08	05 37	06 03	22 22	23 32	24 42	00 42
30	05 08	05 35	05 59	22 35	23 43	24 51	00 51
20	05 06	05 31	05 54	22 59	24 03	00 03	01 05
N 10	05 03	05 27	05 48	23 19	24 20	00 20	01 18
0	04 58	05 23	05 43	23 38	24 35	00 35	01 30
S 10	04 53	05 17	05 38	23 56	24 51	00 51	01 42
20	04 45	05 11	05 33	24 16	00 16	01 08	01 55
30	04 34	05 02	05 26	24 40	00 40	01 27	02 09
35	04 27	04 57	05 23	00 00	00 53	01 38	02 17
40	04 18	04 51	05 18	00 18	01 09	01 51	02 27
45	04 07	04 43	05 13	00 38	01 27	02 06	02 38
S 50	03 54	04 34	05 07	01 04	01 50	02 25	02 51
52	03 47	04 30	05 05	01 17	02 01	02 33	02 57
54	03 40	04 25	05 02	01 32	02 14	02 43	03 04
56	03 31	04 19	04 58	01 48	02 28	02 54	03 12
58	03 21	04 13	04 55	02 09	02 44	03 06	03 20
S 60	03 10	04 06	04 50	02 34	03 04	03 20	03 30

Lat.	Sunset	Twilight Civil	Twilight Naut.	Moonset 10	Moonset 11	Moonset 12	Moonset 13
°	h m	h m	h m	h m	h m	h m	h m
N 72	16 26	17 35	18 53	▭	▭	18 05	16 50
N 70	16 35	17 37	18 47	▭	▭	17 10	16 29
68	16 43	17 39	18 43	▭	17 31	16 36	16 13
66	16 49	17 40	18 39	▭	16 30	16 11	15 59
64	16 54	17 42	18 37	15 58	15 55	15 52	15 48
62	16 59	17 43	18 34	15 16	15 30	15 36	15 38
60	17 03	17 44	18 32	14 47	15 10	15 22	15 30
N 58	17 06	17 45	18 31	14 25	14 53	15 11	15 23
56	17 09	17 46	18 29	14 07	14 39	15 01	15 16
54	17 12	17 47	18 28	13 51	14 27	14 52	15 10
52	17 15	17 48	18 27	13 38	14 16	14 44	15 05
50	17 17	17 49	18 27	13 26	14 06	14 36	15 00
45	17 22	17 52	18 25	13 02	13 46	14 21	14 50
N 40	17 26	17 54	18 25	12 42	13 29	14 08	14 41
35	17 30	17 56	18 25	12 26	13 15	13 57	14 33
30	17 34	17 58	18 25	12 12	13 03	13 47	14 27
20	17 40	18 02	18 27	11 47	12 41	13 31	14 15
N 10	17 45	18 06	18 31	11 26	12 23	13 16	14 05
0	17 50	18 11	18 35	11 07	12 06	13 02	13 55
S 10	17 55	18 16	18 41	10 47	11 48	12 48	13 45
20	18 01	18 23	18 49	10 25	11 29	12 33	13 35
30	18 07	18 32	19 00	10 01	11 07	12 15	13 23
35	18 11	18 37	19 07	09 46	10 55	12 05	13 16
40	18 16	18 43	19 16	09 29	10 40	11 54	13 08
45	18 21	18 51	19 27	09 09	10 22	11 40	12 58
S 50	18 27	19 00	19 41	08 43	10 00	11 23	12 47
52	18 30	19 05	19 48	08 31	09 49	11 15	12 41
54	18 33	19 10	19 55	08 16	09 37	11 06	12 35
56	18 36	19 15	20 04	08 00	09 24	10 55	12 29
58	18 40	19 22	20 14	07 40	09 08	10 44	12 21
S 60	18 45	19 29	20 26	07 14	08 48	10 30	12 12

Day	SUN Eqn. of Time 00^h	SUN Eqn. of Time 12^h	SUN Mer. Pass.	MOON Mer. Pass. Upper	MOON Mer. Pass. Lower	MOON Age	MOON Phase
d	m s	m s	h m	h m	h m	d	%
10	12 55	13 02	11 47	04 52	17 22	22	60
11	13 10	13 18	11 47	05 52	18 21	23	49
12	13 26	13 33	11 46	06 49	19 16	24	37

	UT	ARIES	VENUS −3.9		MARS +0.7		JUPITER −2.6		SATURN +1.1		STARS		
		GHA	GHA	Dec	GHA	Dec	GHA	Dec	GHA	Dec	Name	SHA	Dec
	d h	° ′	° ′	° ′	° ′	° ′	° ′	° ′	° ′	° ′		° ′	°
	13 00	21 41.4	203 18.9	N 2 19.4	261 13.4	N21 32.5	61 44.3	S16 42.2	202 38.6	N 2 35.8	Acamar	315 19.9	S40 15.
	01	36 43.8	218 18.5	18.2	276 14.5	32.3	76 46.7	42.2	217 40.8	35.7	Achernar	335 28.0	S57 11
	02	51 46.3	233 18.1	17.0	291 15.6	32.1	91 49.2	42.2	232 43.0	35.6	Acrux	173 13.3	S63 09
	03	66 48.8	248 17.7	. . 15.7	306 16.7	. . 31.9	106 51.7	. . 42.2	247 45.2	. . 35.5	Adhara	255 14.6	S28 58.
	04	81 51.2	263 17.4	14.5	321 17.8	31.7	121 54.1	42.2	262 47.4	35.4	Aldebaran	290 52.3	N16 31
	05	96 53.7	278 17.0	13.3	336 18.9	31.5	136 56.6	42.2	277 49.5	35.3			
	06	111 56.2	293 16.6	N 2 12.1	351 20.0	N21 31.3	151 59.1	S16 42.2	292 51.7	N 2 35.1	Alioth	166 23.4	N55 54.
	07	126 58.6	308 16.2	10.9	6 21.1	31.1	167 01.5	42.2	307 53.9	35.0	Alkaid	153 01.4	N49 15
T	08	142 01.1	323 15.8	09.7	21 22.2	30.9	182 04.0	42.1	322 56.1	34.9	Al Na'ir	27 46.7	S46 54.
U	09	157 03.6	338 15.4	. . 08.5	36 23.3	. . 30.6	197 06.5	. . 42.1	337 58.3	. . 34.8	Alnilam	275 49.0	S 1 11
E	10	172 06.0	353 15.0	07.3	51 24.4	30.4	212 08.9	42.1	353 00.5	34.7	Alphard	217 58.9	S 8 41
S	11	187 08.5	8 14.6	06.1	66 25.5	30.2	227 11.4	42.1	8 02.7	34.6			
D	12	202 10.9	23 14.2	N 2 04.9	81 26.6	N21 30.0	242 13.8	S16 42.1	23 04.9	N 2 34.5	Alphecca	126 13.6	N26 41.
	13	217 13.4	38 13.8	03.7	96 27.7	29.8	257 16.3	42.1	38 07.1	34.4	Alpheratz	357 46.0	N29 08
A	14	232 15.9	53 13.4	02.5	111 28.8	29.6	272 18.8	42.1	53 09.2	34.2	Altair	62 10.9	N 8 53.
Y	15	247 18.3	68 13.1	. . 01.3	126 29.9	. . 29.4	287 21.2	. . 42.1	68 11.4	. . 34.1	Ankaa	353 17.8	S42 15.
	16	262 20.8	83 12.7	2 00.1	141 31.0	29.2	302 23.7	42.1	83 13.6	34.0	Antares	112 29.9	S26 27
	17	277 23.3	98 12.3	1 58.9	156 32.1	29.0	317 26.2	42.1	98 15.8	33.9			
	18	292 25.7	113 11.9	N 1 57.7	171 33.2	N21 28.8	332 28.6	S16 42.1	113 18.0	N 2 33.8	Arcturus	145 58.5	N19 07.
	19	307 28.2	128 11.5	56.5	186 34.3	28.6	347 31.1	42.1	128 20.2	33.7	Atria	107 34.6	S69 02.
	20	322 30.7	143 11.1	55.3	201 35.4	28.4	2 33.5	42.1	143 22.4	33.6	Avior	234 19.4	S59 32.
	21	337 33.1	158 10.7	. . 54.1	216 36.5	. . 28.2	17 36.0	. . 42.1	158 24.6	. . 33.4	Bellatrix	278 34.7	N 6 21.
	22	352 35.6	173 10.3	52.8	231 37.6	28.0	32 38.5	42.1	173 26.8	33.3	Betelgeuse	271 04.1	N 7 24.
	23	7 38.0	188 09.9	51.6	246 38.7	27.8	47 40.9	42.1	188 29.0	33.2			
	14 00	22 40.5	203 09.5	N 1 50.4	261 39.8	N21 27.6	62 43.4	S16 42.1	203 31.1	N 2 33.1	Canopus	263 57.2	S52 41.
	01	37 43.0	218 09.2	49.2	276 40.9	27.4	77 45.8	42.1	218 33.3	33.0	Capella	280 38.2	N46 00.
	02	52 45.4	233 08.8	48.0	291 42.0	27.2	92 48.3	42.0	233 35.5	32.9	Deneb	49 33.3	N45 19.
	03	67 47.9	248 08.4	. . 46.8	306 43.1	. . 27.0	107 50.8	. . 42.0	248 37.7	. . 32.8	Denebola	182 36.7	N14 31.
	04	82 50.4	263 08.0	45.6	321 44.2	26.8	122 53.2	42.0	263 39.9	32.7	Diphda	348 58.2	S17 55.
	05	97 52.8	278 07.6	44.4	336 45.4	26.6	137 55.7	42.0	278 42.1	32.5			
	06	112 55.3	293 07.2	N 1 43.2	351 46.5	N21 26.4	152 58.1	S16 42.0	293 44.3	N 2 32.4	Dubhe	193 55.3	N61 41.
W	07	127 57.8	308 06.8	42.0	6 47.6	26.1	168 00.6	42.0	308 46.5	32.3	Elnath	278 15.9	N28 37.
E	08	143 00.2	323 06.4	40.8	21 48.7	25.9	183 03.0	42.0	323 48.7	32.2	Eltanin	90 47.6	N51 29.
D	09	158 02.7	338 06.0	. . 39.6	36 49.8	. . 25.7	198 05.5	. . 42.0	338 50.9	. . 32.1	Enif	33 49.6	N 9 55.
N	10	173 05.2	353 05.6	38.4	51 50.9	25.5	213 07.9	42.0	353 53.1	32.0	Fomalhaut	15 26.6	S29 34.
E	11	188 07.6	8 05.3	37.1	66 52.0	25.3	228 10.4	42.0	8 55.2	31.9			
S	12	203 10.1	23 04.9	N 1 35.9	81 53.1	N21 25.1	243 12.9	S16 42.0	23 57.4	N 2 31.7	Gacrux	172 04.7	S57 10.
D	13	218 12.5	38 04.5	34.7	96 54.2	24.9	258 15.3	42.0	38 59.6	31.6	Gienah	175 55.5	S17 35.
A	14	233 15.0	53 04.1	33.5	111 55.3	24.7	273 17.8	42.0	54 01.8	31.5	Hadar	148 52.6	S60 25.
Y	15	248 17.5	68 03.7	. . 32.3	126 56.5	. . 24.5	288 20.2	. . 42.0	69 04.0	. . 31.4	Hamal	328 03.5	N23 30.
	16	263 19.9	83 03.3	31.1	141 57.6	24.3	303 22.7	41.9	84 06.2	31.3	Kaus Aust.	83 47.5	S34 22.
	17	278 22.4	98 02.9	29.9	156 58.7	24.1	318 25.1	41.9	99 08.4	31.2			
	18	293 24.9	113 02.5	N 1 28.7	171 59.8	N21 23.9	333 27.6	S16 41.9	114 10.6	N 2 31.1	Kochab	137 20.6	N74 06.
	19	308 27.3	128 02.1	27.5	187 00.9	23.7	348 30.0	41.9	129 12.8	31.0	Markab	13 40.8	N15 15.
	20	323 29.8	143 01.7	26.3	202 02.0	23.5	3 32.5	41.9	144 15.0	30.8	Menkar	314 17.6	N 4 07.
	21	338 32.3	158 01.4	. . 25.1	217 03.1	. . 23.3	18 34.9	. . 41.9	159 17.2	. . 30.7	Menkent	148 11.3	S36 25.
	22	353 34.7	173 01.0	23.8	232 04.2	23.1	33 37.4	41.9	174 19.3	30.6	Miaplacidus	221 40.9	S69 45.
	23	8 37.2	188 00.6	22.6	247 05.4	22.9	48 39.8	41.9	189 21.5	30.5			
	15 00	23 39.6	203 00.2	N 1 21.4	262 06.5	N21 22.6	63 42.3	S16 41.9	204 23.7	N 2 30.4	Mirfak	308 43.9	N49 53.
	01	38 42.1	217 59.8	20.2	277 07.6	22.4	78 44.7	41.9	219 25.9	30.3	Nunki	76 01.7	S26 17.
	02	53 44.6	232 59.4	19.0	292 08.7	22.2	93 47.2	41.9	234 28.1	30.2	Peacock	53 23.2	S56 42.
	03	68 47.0	247 59.0	. . 17.8	307 09.8	. . 22.0	108 49.6	. . 41.9	249 30.3	. . 30.1	Pollux	243 31.0	N28 00.
	04	83 49.5	262 58.6	16.6	322 10.9	21.8	123 52.1	41.8	264 32.5	29.9	Procyon	245 02.6	N 5 12.
	05	98 52.0	277 58.2	15.4	337 12.1	21.6	138 54.5	41.8	279 34.7	29.8			
	06	113 54.4	292 57.8	N 1 14.2	352 13.2	N21 21.4	153 57.0	S16 41.8	294 36.9	N 2 29.7	Rasalhague	96 09.1	N12 33.
	07	128 56.9	307 57.5	12.9	7 14.3	21.2	168 59.4	41.8	309 39.1	29.6	Regulus	207 46.6	N11 55.
T	08	143 59.4	322 57.1	11.7	22 15.4	21.0	184 01.9	41.8	324 41.3	29.5	Rigel	281 14.5	S 8 11.
H	09	159 01.8	337 56.7	. . 10.5	37 16.5	. . 20.8	199 04.3	. . 41.8	339 43.5	. . 29.4	Rigil Kent.	139 56.3	S60 52.
U	10	174 04.3	352 56.3	09.3	52 17.7	20.6	214 06.8	41.8	354 45.6	29.3	Sabik	102 15.9	S15 44.
R	11	189 06.8	7 55.9	08.1	67 18.8	20.4	229 09.2	41.8	9 47.8	29.1			
S	12	204 09.2	22 55.5	N 1 06.9	82 19.9	N21 20.2	244 11.7	S16 41.8	24 50.0	N 2 29.0	Schedar	349 43.3	N56 35.
D	13	219 11.7	37 55.1	05.7	97 21.0	20.0	259 14.1	41.8	39 52.2	28.9	Shaula	96 25.8	S37 06.
A	14	234 14.1	52 54.7	04.5	112 22.2	19.7	274 16.6	41.7	54 54.4	28.8	Sirius	258 36.0	S16 43.
Y	15	249 16.6	67 54.3	. . 03.2	127 23.3	. . 19.5	289 19.0	. . 41.7	69 56.6	. . 28.7	Spica	158 34.5	S11 12.
	16	264 19.1	82 53.9	02.0	142 24.4	19.3	304 21.5	41.7	84 58.8	28.6	Suhail	222 54.7	S43 28.
	17	279 21.5	97 53.6	1 00.8	157 25.5	19.1	319 23.9	41.7	100 01.0	28.5			
	18	294 24.0	112 53.2	N 0 59.6	172 26.6	N21 18.9	334 26.3	S16 41.7	115 03.2	N 2 28.4	Vega	80 40.9	N38 47.
	19	309 26.5	127 52.8	58.4	187 27.8	18.7	349 28.8	41.7	130 05.4	28.2	Zuben'ubi	137 08.8	S16 04.
	20	324 28.9	142 52.4	57.2	202 28.9	18.5	4 31.2	41.7	145 07.6	28.1		SHA	Mer. Pas
	21	339 31.4	157 52.0	. . 56.0	217 30.0	. . 18.3	19 33.7	. . 41.7	160 09.8	. . 28.0		° ′	h m
	22	354 33.9	172 51.6	54.8	232 31.2	18.1	34 36.1	41.7	175 12.0	27.9	Venus	180 29.0	10 28
	23	9 36.3	187 51.2	53.5	247 32.3	17.9	49 38.6	41.6	190 14.1	27.8	Mars	238 59.3	6 33
		h m									Jupiter	40 02.9	19 46
	Mer. Pass.	22 25.6	v −0.4	d 1.2	v 1.1	d 0.2	v 2.5	d 0.0	v 2.2	d 0.1	Saturn	180 50.6	10 24

UT	SUN GHA	SUN Dec	MOON GHA	*v*	MOON Dec	*d*	HP
d h	° ′	° ′	° ′	′	° ′	′	′
13 00	183 25.2	S 7 44.4	248 24.7	8.8	N16 10.9	12.7	59.4
01	198 25.3	45.4	262 52.5	8.9	15 58.2	12.7	59.4
02	213 25.5	46.3	277 20.4	9.0	15 45.5	12.8	59.4
03	228 25.6	. . 47.2	291 48.4	9.0	15 32.7	12.9	59.4
04	243 25.8	48.2	306 16.4	9.1	15 19.8	13.0	59.4
05	258 25.9	49.1	320 44.5	9.2	15 06.8	13.1	59.4
06	273 26.1	S 7 50.0	335 12.7	9.2	N14 53.7	13.2	59.4
07	288 26.2	51.0	349 40.9	9.3	14 40.5	13.2	59.4
08	303 26.4	51.9	4 09.2	9.4	14 27.3	13.3	59.4
09	318 26.5	. . 52.8	18 37.6	9.4	14 14.0	13.5	59.4
10	333 26.7	53.8	33 06.0	9.5	14 00.5	13.5	59.4
11	348 26.8	54.7	47 34.5	9.6	13 47.0	13.5	59.4
12	3 27.0	S 7 55.6	62 03.1	9.6	N13 33.5	13.7	59.4
13	18 27.1	56.6	76 31.7	9.7	13 19.8	13.7	59.4
14	33 27.3	57.5	91 00.4	9.7	13 06.1	13.8	59.4
15	48 27.4	. . 58.4	105 29.1	9.8	12 52.3	13.9	59.4
16	63 27.6	7 59.4	119 57.9	9.9	12 38.4	13.9	59.4
17	78 27.7	8 00.3	134 26.8	9.9	12 24.5	14.0	59.4
18	93 27.9	S 8 01.2	148 55.7	10.0	N12 10.5	14.1	59.4
19	108 28.0	02.2	163 24.7	10.0	11 56.4	14.2	59.4
20	123 28.2	03.1	177 53.7	10.1	11 42.2	14.2	59.4
21	138 28.3	. . 04.0	192 22.8	10.2	11 28.0	14.3	59.4
22	153 28.5	04.9	206 52.0	10.2	11 13.7	14.3	59.4
23	168 28.6	05.9	221 21.2	10.2	10 59.4	14.4	59.4
14 00	183 28.8	S 8 06.8	235 50.4	10.3	N10 45.0	14.4	59.4
01	198 28.9	07.7	250 19.7	10.4	10 30.6	14.6	59.4
02	213 29.0	08.7	264 49.1	10.4	10 16.0	14.5	59.4
03	228 29.2	. . 09.6	279 18.5	10.5	10 01.5	14.6	59.4
04	243 29.3	10.5	293 48.0	10.5	9 46.9	14.7	59.4
05	258 29.5	11.5	308 17.5	10.5	9 32.2	14.7	59.4
06	273 29.6	S 8 12.4	322 47.0	10.7	N 9 17.5	14.8	59.4
07	288 29.8	13.3	337 16.7	10.6	9 02.7	14.8	59.4
08	303 29.9	14.2	351 46.3	10.7	8 47.9	14.9	59.4
09	318 30.1	. . 15.2	6 16.0	10.8	8 33.0	14.9	59.4
10	333 30.2	16.1	20 45.8	10.7	8 18.1	14.9	59.4
11	348 30.3	17.0	35 15.5	10.9	8 03.2	15.0	59.4
12	3 30.5	S 8 18.0	49 45.4	10.9	N 7 48.2	15.1	59.4
13	18 30.6	18.9	64 15.3	10.9	7 33.1	15.0	59.4
14	33 30.8	19.8	78 45.2	10.9	7 18.1	15.2	59.3
15	48 30.9	. . 20.7	93 15.1	11.0	7 02.9	15.1	59.3
16	63 31.1	21.7	107 45.1	11.1	6 47.8	15.2	59.3
17	78 31.2	22.6	122 15.2	11.0	6 32.6	15.2	59.3
18	93 31.3	S 8 23.5	136 45.2	11.1	N 6 17.4	15.2	59.3
19	108 31.5	24.4	151 15.3	11.2	6 02.2	15.3	59.3
20	123 31.6	25.4	165 45.5	11.2	5 46.9	15.3	59.3
21	138 31.8	. . 26.3	180 15.7	11.2	5 31.6	15.3	59.3
22	153 31.9	27.2	194 45.9	11.2	5 16.3	15.4	59.3
23	168 32.0	28.2	209 16.1	11.3	5 00.9	15.3	59.3
15 00	183 32.2	S 8 29.1	223 46.4	11.3	N 4 45.6	15.4	59.3
01	198 32.3	30.0	238 16.7	11.3	4 30.2	15.5	59.3
02	213 32.5	30.9	252 47.0	11.4	4 14.7	15.4	59.3
03	228 32.6	. . 31.9	267 17.4	11.3	3 59.3	15.4	59.3
04	243 32.7	32.8	281 47.7	11.5	3 43.9	15.5	59.2
05	258 32.9	33.7	296 18.2	11.4	3 28.4	15.5	59.2
06	273 33.0	S 8 34.6	310 48.6	11.4	N 3 12.9	15.5	59.2
07	288 33.2	35.6	325 19.0	11.5	2 57.4	15.5	59.2
08	303 33.3	36.5	339 49.5	11.5	2 41.9	15.5	59.2
09	318 33.4	. . 37.4	354 20.0	11.6	2 26.4	15.5	59.2
10	333 33.6	38.3	8 50.6	11.5	2 10.9	15.5	59.2
11	348 33.7	39.2	23 21.1	11.6	1 55.4	15.6	59.2
12	3 33.8	S 8 40.2	37 51.7	11.5	N 1 39.8	15.5	59.2
13	18 34.0	41.1	52 22.2	11.6	1 24.3	15.6	59.2
14	33 34.1	42.0	66 52.8	11.6	1 08.7	15.5	59.1
15	48 34.2	. . 42.9	81 23.4	11.7	0 53.2	15.5	59.1
16	63 34.4	43.9	95 54.1	11.6	0 37.7	15.6	59.1
17	78 34.5	44.8	110 24.7	11.7	0 22.1	15.5	59.1
18	93 34.7	S 8 45.7	124 55.4	11.6	N 0 06.6	15.5	59.1
19	108 34.8	46.6	139 26.0	11.7	S 0 08.9	15.6	59.1
20	123 34.9	47.6	153 56.7	11.7	0 24.5	15.5	59.1
21	138 35.1	. . 48.5	168 27.4	11.7	0 40.0	15.5	59.1
22	153 35.2	49.4	182 58.1	11.7	0 55.5	15.5	59.1
23	168 35.3	50.3	197 28.8	11.7	S 1 11.0	15.5	59.0
	SD 16.1	*d* 0.9	SD 16.2		16.2		16.1

Lat.	Twilight Naut.	Twilight Civil	Sunrise	Moonrise 13	Moonrise 14	Moonrise 15	Moonrise 16
°	h m	h m	h m	h m	h m	h m	h m
N 72	04 51	06 09	07 20	23 56	26 20	02 20	04 35
N 70	04 56	06 06	07 09	24 14	00 14	02 26	04 31
68	04 59	06 03	07 00	24 29	00 29	02 31	04 28
66	05 01	06 01	06 53	24 40	00 40	02 34	04 25
64	05 04	05 59	06 47	24 50	00 50	02 38	04 23
62	05 05	05 57	06 42	24 58	00 58	02 41	04 21
60	05 07	05 55	06 37	25 05	01 05	02 43	04 19
N 58	05 08	05 53	06 33	25 12	01 12	02 46	04 18
56	05 09	05 52	06 29	25 17	01 17	02 47	04 17
54	05 09	05 50	06 26	25 22	01 22	02 49	04 15
52	05 10	05 49	06 23	00 00	01 26	02 51	04 14
50	05 10	05 48	06 20	00 07	01 30	02 52	04 13
45	05 11	05 45	06 14	00 21	01 39	02 55	04 11
N 40	05 11	05 42	06 09	00 32	01 46	02 58	04 09
35	05 10	05 40	06 05	00 42	01 52	03 00	04 08
30	05 09	05 37	06 01	00 51	01 57	03 02	04 07
20	05 07	05 32	05 54	01 05	02 07	03 06	04 04
N 10	05 03	05 27	05 48	01 18	02 15	03 09	04 02
0	04 58	05 22	05 43	01 30	02 22	03 12	04 00
S 10	04 51	05 16	05 37	01 42	02 30	03 15	03 59
20	04 42	05 08	05 30	01 55	02 38	03 18	03 57
30	04 30	04 59	05 23	02 09	02 47	03 21	03 55
35	04 22	04 53	05 19	02 17	02 52	03 23	03 53
40	04 13	04 46	05 14	02 27	02 58	03 26	03 52
45	04 01	04 38	05 08	02 38	03 04	03 28	03 51
S 50	03 47	04 28	05 01	02 51	03 13	03 31	03 49
52	03 39	04 23	04 58	02 57	03 16	03 33	03 48
54	03 31	04 17	04 54	03 04	03 20	03 34	03 47
56	03 22	04 11	04 51	03 12	03 25	03 36	03 46
58	03 11	04 04	04 46	03 20	03 30	03 38	03 45
S 60	02 58	03 56	04 41	03 30	03 35	03 40	03 44

Lat.	Sunset	Twilight Civil	Twilight Naut.	Moonset 13	Moonset 14	Moonset 15	Moonset 16
°	h m	h m	h m	h m	h m	h m	h m
N 72	16 10	17 20	18 38	16 50	16 13	15 43	15 13
N 70	16 21	17 24	18 34	16 29	16 04	15 42	15 21
68	16 30	17 27	18 31	16 13	15 56	15 41	15 27
66	16 37	17 30	18 29	15 59	15 50	15 41	15 32
64	16 44	17 32	18 27	15 48	15 44	15 40	15 36
62	16 49	17 34	18 25	15 38	15 39	15 40	15 40
60	16 54	17 36	18 24	15 30	15 35	15 40	15 44
N 58	16 58	17 38	18 23	15 23	15 32	15 39	15 47
56	17 02	17 39	18 22	15 16	15 28	15 39	15 49
54	17 05	17 41	18 21	15 10	15 25	15 39	15 52
52	17 08	17 42	18 21	15 05	15 23	15 38	15 54
50	17 11	17 43	18 21	15 00	15 20	15 38	15 56
45	17 17	17 46	18 20	14 50	15 15	15 38	16 00
N 40	17 22	17 49	18 21	14 41	15 10	15 37	16 04
35	17 26	17 52	18 21	14 33	15 06	15 37	16 07
30	17 30	17 54	18 22	14 27	15 03	15 37	16 10
20	17 37	17 59	18 25	14 15	14 57	15 36	16 15
N 10	17 43	18 05	18 29	14 05	14 51	15 36	16 19
0	17 49	18 10	18 34	13 55	14 46	15 35	16 23
S 10	17 55	18 17	18 41	13 45	14 41	15 34	16 28
20	18 02	18 24	18 50	13 35	14 35	15 34	16 32
30	18 09	18 34	19 03	13 23	14 29	15 33	16 37
35	18 14	18 40	19 10	13 16	14 25	15 33	16 40
40	18 19	18 47	19 20	13 08	14 20	15 32	16 43
45	18 25	18 55	19 32	12 58	14 15	15 32	16 47
S 50	18 32	19 06	19 47	12 47	14 09	15 31	16 52
52	18 35	19 11	19 54	12 41	14 07	15 31	16 54
54	18 39	19 16	20 03	12 35	14 03	15 30	16 56
56	18 43	19 22	20 12	12 29	14 00	15 30	16 59
58	18 47	19 29	20 23	12 21	13 56	15 29	17 01
S 60	18 52	19 37	20 36	12 12	13 52	15 29	17 05

Day	SUN Eqn. of Time 00^h	SUN Eqn. of Time 12^h	SUN Mer. Pass.	MOON Mer. Pass. Upper	MOON Mer. Pass. Lower	MOON Age	MOON Phase
d	m s	m s	h m	h m	h m	d	%
13	13 40	13 48	11 46	07 43	20 09	25	27
14	13 55	14 02	11 46	08 34	20 59	26	17
15	14 08	14 15	11 46	09 23	21 48	27	9

UT		ARIES	VENUS −3.9		MARS +0.6		JUPITER −2.6		SATURN +1.1		STARS		
d	h	GHA	GHA	Dec	GHA	Dec	GHA	Dec	GHA	Dec	Name	SHA	Dec
		° ′	° ′	° ′	° ′	° ′	° ′	° ′	° ′	° ′		° ′	° ′
16	00	24 38.8	202 50.8	N 0 52.3	262 33.4	N21 17.7	64 41.0	S16 41.6	205 16.3	N 2 27.7	Acamar	315 19.9	S40 15.7
	01	39 41.2	217 50.4	51.1	277 34.5	17.5	79 43.4	41.6	220 18.5	27.6	Achernar	335 28.0	S57 11.1
	02	54 43.7	232 50.0	49.9	292 35.7	17.3	94 45.9	41.6	235 20.7	27.5	Acrux	173 13.3	S63 09.2
	03	69 46.2	247 49.7	. . 48.7	307 36.8	. . 17.0	109 48.3	. . 41.6	250 22.9	. . 27.4	Adhara	255 14.6	S28 58.8
	04	84 48.6	262 49.3	47.5	322 37.9	16.8	124 50.8	41.6	265 25.1	27.2	Aldebaran	290 52.3	N16 31.9
	05	99 51.1	277 48.9	46.3	337 39.0	16.6	139 53.2	41.6	280 27.3	27.1			
	06	114 53.6	292 48.5	N 0 45.1	352 40.2	N21 16.4	154 55.6	S16 41.6	295 29.5	N 2 27.0	Alioth	166 23.4	N55 54.3
	07	129 56.0	307 48.1	43.8	7 41.3	16.2	169 58.1	41.6	310 31.7	26.9	Alkaid	153 01.4	N49 15.8
	08	144 58.5	322 47.7	42.6	22 42.4	16.0	185 00.5	41.5	325 33.9	26.8	Al Na'ir	27 46.7	S46 54.9
F	09	160 01.0	337 47.3	. . 41.4	37 43.6	. . 15.8	200 03.0	. . 41.5	340 36.1	. . 26.7	Alnilam	275 48.9	S 1 11.6
R	10	175 03.4	352 46.9	40.2	52 44.7	15.6	215 05.4	41.5	355 38.3	26.6	Alphard	217 58.9	S 8 41.9
I	11	190 05.9	7 46.5	39.0	67 45.8	15.4	230 07.8	41.5	10 40.5	26.5			
D	12	205 08.4	22 46.1	N 0 37.8	82 47.0	N21 15.2	245 10.3	S16 41.5	25 42.6	N 2 26.3	Alphecca	126 13.6	N26 41.0
A	13	220 10.8	37 45.8	36.6	97 48.1	15.0	260 12.7	41.5	40 44.8	26.2	Alpheratz	357 46.1	N29 08.9
Y	14	235 13.3	52 45.4	35.3	112 49.2	14.8	275 15.2	41.5	55 47.0	26.1	Altair	62 10.9	N 8 53.8
	15	250 15.7	67 45.0	. . 34.1	127 50.4	. . 14.5	290 17.6	. . 41.4	70 49.2	. . 26.0	Ankaa	353 17.8	S42 15.1
	16	265 18.2	82 44.6	32.9	142 51.5	14.3	305 20.0	41.4	85 51.4	25.9	Antares	112 29.9	S26 27.3
	17	280 20.7	97 44.2	31.7	157 52.6	14.1	320 22.5	41.4	100 53.6	25.8			
	18	295 23.1	112 43.8	N 0 30.5	172 53.8	N21 13.9	335 24.9	S16 41.4	115 55.8	N 2 25.7	Arcturus	145 58.5	N19 07.9
	19	310 25.6	127 43.4	29.3	187 54.9	13.7	350 27.3	41.4	130 58.0	25.6	Atria	107 34.6	S69 02.9
	20	325 28.1	142 43.0	28.0	202 56.1	13.5	5 29.8	41.4	146 00.2	25.4	Avior	234 19.4	S59 32.1
	21	340 30.5	157 42.6	. . 26.8	217 57.2	. . 13.3	20 32.2	. . 41.4	161 02.4	. . 25.3	Bellatrix	278 34.7	N 6 21.7
	22	355 33.0	172 42.2	25.6	232 58.3	13.1	35 34.6	41.4	176 04.6	25.2	Betelgeuse	271 04.1	N 7 24.7
	23	10 35.5	187 41.9	24.4	247 59.5	12.9	50 37.1	41.3	191 06.8	25.1			
17	00	25 37.9	202 41.5	N 0 23.2	263 00.6	N21 12.7	65 39.5	S16 41.3	206 09.0	N 2 25.0	Canopus	263 57.2	S52 41.7
	01	40 40.4	217 41.1	22.0	278 01.7	12.4	80 41.9	41.3	221 11.2	24.9	Capella	280 38.2	N46 00.4
	02	55 42.9	232 40.7	20.8	293 02.9	12.2	95 44.4	41.3	236 13.4	24.8	Deneb	49 33.3	N45 19.3
	03	70 45.3	247 40.3	. . 19.5	308 04.0	. . 12.0	110 46.8	. . 41.3	251 15.6	. . 24.7	Denebola	182 36.7	N14 31.0
	04	85 47.8	262 39.9	18.3	323 05.2	11.8	125 49.2	41.3	266 17.7	24.6	Diphda	348 58.2	S17 55.8
	05	100 50.2	277 39.5	17.1	338 06.3	11.6	140 51.7	41.3	281 19.9	24.4			
	06	115 52.7	292 39.1	N 0 15.9	353 07.4	N21 11.4	155 54.1	S16 41.2	296 22.1	N 2 24.3	Dubhe	193 55.2	N61 41.7
	07	130 55.2	307 38.7	14.7	8 08.6	11.2	170 56.5	41.2	311 24.3	24.2	Elnath	278 15.9	N28 37.0
S	08	145 57.6	322 38.3	13.5	23 09.7	11.0	185 59.0	41.2	326 26.5	24.1	Eltanin	90 47.6	N51 29.5
A	09	161 00.1	337 37.9	. . 12.2	38 10.9	. . 10.8	201 01.4	. . 41.2	341 28.7	. . 24.0	Enif	33 49.6	N 9 55.4
T	10	176 02.6	352 37.6	11.0	53 12.0	10.6	216 03.8	41.2	356 30.9	23.9	Fomalhaut	15 26.6	S29 34.1
U	11	191 05.0	7 37.2	09.8	68 13.2	10.3	231 06.3	41.2	11 33.1	23.8			
R	12	206 07.5	22 36.8	N 0 08.6	83 14.3	N21 10.1	246 08.7	S16 41.1	26 35.3	N 2 23.7	Gacrux	172 04.7	S57 10.0
D	13	221 10.0	37 36.4	07.4	98 15.4	09.9	261 11.1	41.1	41 37.5	23.6	Gienah	175 55.4	S17 35.7
A	14	236 12.4	52 36.0	06.2	113 16.6	09.7	276 13.6	41.1	56 39.7	23.4	Hadar	148 52.6	S60 25.3
Y	15	251 14.9	67 35.6	. . 04.9	128 17.7	. . 09.5	291 16.0	. . 41.1	71 41.9	. . 23.3	Hamal	328 03.5	N23 30.7
	16	266 17.3	82 35.2	03.7	143 18.9	09.3	306 18.4	41.1	86 44.1	23.2	Kaus Aust.	83 47.5	S34 22.9
	17	281 19.8	97 34.8	02.5	158 20.0	09.1	321 20.8	41.1	101 46.3	23.1			
	18	296 22.3	112 34.4	N 0 01.3	173 21.2	N21 08.9	336 23.3	S16 41.1	116 48.5	N 2 23.0	Kochab	137 20.7	N74 06.9
	19	311 24.7	127 34.0	N 00.1	188 22.3	08.7	351 25.7	41.0	131 50.7	22.9	Markab	13 40.8	N15 15.7
	20	326 27.2	142 33.6	S 01.1	203 23.5	08.5	6 28.1	41.0	146 52.9	22.8	Menkar	314 17.6	N 4 07.9
	21	341 29.7	157 33.3	. . 02.4	218 24.6	. . 08.2	21 30.5	. . 41.0	161 55.1	. . 22.7	Menkent	148 11.3	S36 25.1
	22	356 32.1	172 32.9	03.6	233 25.8	08.0	36 33.0	41.0	176 57.2	22.6	Miaplacidus	221 40.9	S69 45.2
	23	11 34.6	187 32.5	04.8	248 26.9	07.8	51 35.4	41.0	191 59.4	22.4			
18	00	26 37.1	202 32.1	S 0 06.0	263 28.1	N21 07.6	66 37.8	S16 41.0	207 01.6	N 2 22.3	Mirfak	308 43.9	N49 53.8
	01	41 39.5	217 31.7	07.2	278 29.2	07.4	81 40.2	40.9	222 03.8	22.2	Nunki	76 01.7	S26 17.1
	02	56 42.0	232 31.3	08.5	293 30.4	07.2	96 42.7	40.9	237 06.0	22.1	Peacock	53 23.3	S56 42.4
	03	71 44.5	247 30.9	. . 09.7	308 31.5	. . 07.0	111 45.1	. . 40.9	252 08.2	. . 22.0	Pollux	243 31.0	N28 00.1
	04	86 46.9	262 30.5	10.9	323 32.7	06.8	126 47.5	40.9	267 10.4	21.9	Procyon	245 02.5	N 5 12.1
	05	101 49.4	277 30.1	12.1	338 33.8	06.6	141 49.9	40.9	282 12.6	21.8			
	06	116 51.8	292 29.7	S 0 13.3	353 35.0	N21 06.3	156 52.4	S16 40.9	297 14.8	N 2 21.7	Rasalhague	96 09.1	N12 33.3
	07	131 54.3	307 29.3	14.5	8 36.1	06.1	171 54.8	40.8	312 17.0	21.6	Regulus	207 46.6	N11 55.2
	08	146 56.8	322 29.0	15.8	23 37.3	05.9	186 57.2	40.8	327 19.2	21.4	Rigel	281 14.5	S 8 11.2
S	09	161 59.2	337 28.6	. . 17.0	38 38.4	. . 05.7	201 59.6	. . 40.8	342 21.4	. . 21.3	Rigil Kent.	139 56.3	S60 52.6
U	10	177 01.7	352 28.2	18.2	53 39.6	05.5	217 02.1	40.8	357 23.6	21.2	Sabik	102 15.9	S15 44.2
N	11	192 04.2	7 27.8	19.4	68 40.8	05.3	232 04.5	40.8	12 25.8	21.1			
D	12	207 06.6	22 27.4	S 0 20.6	83 41.9	N21 05.1	247 06.9	S16 40.7	27 28.0	N 2 21.0	Schedar	349 43.3	N56 35.7
A	13	222 09.1	37 27.0	21.9	98 43.1	04.9	262 09.3	40.7	42 30.2	20.9	Shaula	96 25.8	S37 06.8
Y	14	237 11.6	52 26.6	23.1	113 44.2	04.7	277 11.7	40.7	57 32.4	20.8	Sirius	258 36.0	S16 43.5
	15	252 14.0	67 26.2	. . 24.3	128 45.4	. . 04.4	292 14.2	. . 40.7	72 34.6	. . 20.7	Spica	158 34.5	S11 12.7
	16	267 16.5	82 25.8	25.5	143 46.5	04.2	307 16.6	40.7	87 36.8	20.6	Suhail	222 54.7	S43 28.1
	17	282 18.9	97 25.4	26.7	158 47.7	04.0	322 19.0	40.7	102 39.0	20.5			
	18	297 21.4	112 25.0	S 0 27.9	173 48.9	N21 03.8	337 21.4	S16 40.6	117 41.2	N 2 20.3	Vega	80 40.9	N38 47.8
	19	312 23.9	127 24.6	29.2	188 50.0	03.6	352 23.8	40.6	132 43.3	20.2	Zuben'ubi	137 08.8	S16 04.9
	20	327 26.3	142 24.3	30.4	203 51.2	03.4	7 26.3	40.6	147 45.5	20.1		SHA	Mer. Pass.
	21	342 28.8	157 23.9	. . 31.6	218 52.3	. . 03.2	22 28.7	. . 40.6	162 47.7	. . 20.0		° ′	h m
	22	357 31.3	172 23.5	32.8	233 53.5	03.0	37 31.1	40.6	177 49.9	19.9	Venus	177 03.5	10 30
	23	12 33.7	187 23.1	34.0	248 54.7	02.8	52 33.5	40.5	192 52.1	19.8	Mars	237 22.7	6 27
Mer. Pass.		h m 22 13.8	v −0.4	d 1.2	v 1.1	d 0.2	v 2.4	d 0.0	v 2.2	d 0.1	Jupiter	40 01.6	19 34
											Saturn	180 31.0	10 14

UT	SUN GHA	SUN Dec	MOON GHA	v	MOON Dec	d	HP
d h	° ′	° ′	° ′	′	° ′	′	′
16 00	183 35.5	S 8 51.2	211 59.5	11.7	S 1 26.5	15.4	59.0
01	198 35.6	52.2	226 30.2	11.7	1 41.9	15.5	59.0
02	213 35.7	53.1	241 00.9	11.7	1 57.4	15.4	59.0
03	228 35.9	. . 54.0	255 31.6	11.7	2 12.8	15.4	59.0
04	243 36.0	54.9	270 02.3	11.7	2 28.2	15.4	59.0
05	258 36.1	55.8	284 33.0	11.7	2 43.6	15.4	59.0
06	273 36.3	S 8 56.8	299 03.7	11.8	S 2 59.0	15.4	58.9
07	288 36.4	57.7	313 34.5	11.7	3 14.4	15.3	58.9
08	303 36.5	58.6	328 05.2	11.7	3 29.7	15.3	58.9
09	318 36.7	8 59.5	342 35.9	11.7	3 45.0	15.3	58.9
10	333 36.8	9 00.4	357 06.6	11.7	4 00.3	15.2	58.9
11	348 36.9	01.3	11 37.3	11.7	4 15.5	15.2	58.9
12	3 37.0	S 9 02.3	26 08.0	11.7	S 4 30.7	15.2	58.8
13	18 37.2	03.2	40 38.7	11.7	4 45.9	15.2	58.8
14	33 37.3	04.1	55 09.4	11.7	5 01.1	15.1	58.8
15	48 37.4	. . 05.0	69 40.1	11.6	5 16.2	15.1	58.8
16	63 37.6	05.9	84 10.7	11.7	5 31.3	15.0	58.8
17	78 37.7	06.9	98 41.4	11.7	5 46.3	15.1	58.8
18	93 37.8	S 9 07.8	113 12.1	11.6	S 6 01.4	14.9	58.7
19	108 38.0	08.7	127 42.7	11.6	6 16.3	15.0	58.7
20	123 38.1	09.6	142 13.3	11.6	6 31.3	14.8	58.7
21	138 38.2	. . 10.5	156 43.9	11.6	6 46.1	14.9	58.7
22	153 38.3	11.4	171 14.5	11.6	7 01.0	14.8	58.7
23	168 38.5	12.3	185 45.1	11.6	7 15.8	14.7	58.6
17 00	183 38.6	S 9 13.3	200 15.7	11.5	S 7 30.5	14.7	58.6
01	198 38.7	14.2	214 46.2	11.6	7 45.2	14.7	58.6
02	213 38.8	15.1	229 16.8	11.5	7 59.9	14.6	58.6
03	228 39.0	. . 16.0	243 47.3	11.5	8 14.5	14.6	58.6
04	243 39.1	16.9	258 17.8	11.4	8 29.1	14.5	58.6
05	258 39.2	17.8	272 48.2	11.5	8 43.6	14.4	58.5
06	273 39.4	S 9 18.8	287 18.7	11.4	S 8 58.0	14.4	58.5
07	288 39.5	19.7	301 49.1	11.5	9 12.4	14.3	58.5
08	303 39.6	20.6	316 19.6	11.3	9 26.7	14.3	58.5
09	318 39.7	. . 21.5	330 49.9	11.4	9 41.0	14.2	58.4
10	333 39.9	22.4	345 20.3	11.4	9 55.2	14.2	58.4
11	348 40.0	23.3	359 50.7	11.3	10 09.4	14.1	58.4
12	3 40.1	S 9 24.2	14 21.0	11.3	S10 23.5	14.0	58.4
13	18 40.2	25.1	28 51.3	11.2	10 37.5	14.0	58.4
14	33 40.4	26.1	43 21.5	11.3	10 51.5	13.9	58.3
15	48 40.5	. . 27.0	57 51.8	11.2	11 05.4	13.8	58.3
16	63 40.6	27.9	72 22.0	11.2	11 19.2	13.7	58.3
17	78 40.7	28.8	86 52.2	11.1	11 32.9	13.7	58.3
18	93 40.8	S 9 29.7	101 22.3	11.2	S11 46.6	13.6	58.3
19	108 41.0	30.6	115 52.5	11.1	12 00.2	13.6	58.2
20	123 41.1	31.5	130 22.6	11.0	12 13.8	13.5	58.2
21	138 41.2	. . 32.4	144 52.6	11.1	12 27.3	13.4	58.2
22	153 41.3	33.3	159 22.7	11.0	12 40.7	13.3	58.2
23	168 41.5	34.3	173 52.7	10.9	12 54.0	13.2	58.1
18 00	183 41.6	S 9 35.2	188 22.6	11.0	S13 07.2	13.2	58.1
01	198 41.7	36.1	202 52.6	10.9	13 20.4	13.0	58.1
02	213 41.8	37.0	217 22.5	10.9	13 33.4	13.0	58.1
03	228 41.9	. . 37.9	231 52.4	10.8	13 46.4	13.0	58.0
04	243 42.1	38.8	246 22.2	10.8	13 59.4	12.8	58.0
05	258 42.2	39.7	260 52.0	10.8	14 12.2	12.7	58.0
06	273 42.3	S 9 40.6	275 21.8	10.8	S14 24.9	12.7	58.0
07	288 42.4	41.5	289 51.6	10.7	14 37.6	12.6	57.9
08	303 42.5	42.4	304 21.3	10.7	14 50.2	12.5	57.9
09	318 42.7	. . 43.3	318 51.0	10.6	15 02.7	12.4	57.9
10	333 42.8	44.2	333 20.6	10.6	15 15.1	12.3	57.9
11	348 42.9	45.2	347 50.2	10.6	15 27.4	12.2	57.8
12	3 43.0	S 9 46.1	2 19.8	10.5	S15 39.6	12.1	57.8
13	18 43.1	47.0	16 49.3	10.5	15 51.7	12.0	57.8
14	33 43.3	47.9	31 18.8	10.5	16 03.7	11.9	57.8
15	48 43.4	. . 48.8	45 48.3	10.4	16 15.6	11.9	57.7
16	63 43.5	49.7	60 17.7	10.4	16 27.5	11.7	57.7
17	78 43.6	50.6	74 47.1	10.3	16 39.2	11.6	57.7
18	93 43.7	S 9 51.5	89 16.4	10.3	S16 50.8	11.6	57.7
19	108 43.8	52.4	103 45.7	10.3	17 02.4	11.4	57.6
20	123 44.0	53.3	118 15.0	10.3	17 13.8	11.3	57.6
21	138 44.1	. . 54.2	132 44.3	10.2	17 25.1	11.3	57.6
22	153 44.2	55.1	147 13.5	10.1	17 36.4	11.1	57.5
23	168 44.3	56.0	161 42.6	10.2	S17 47.5	11.0	57.5
	SD 16.1	*d* 0.9	SD 16.0		15.9		15.8

Lat.	Twilight Naut.	Twilight Civil	Sunrise	Moonrise 16	Moonrise 17	Moonrise 18	Moonrise 19
°	h m	h m	h m	h m	h m	h m	h m
N 72	05 04	06 23	07 34	04 35	06 54	09 35	■
N 70	05 07	06 18	07 22	04 31	06 39	08 57	■
68	05 09	06 14	07 12	04 28	06 26	08 31	10 55
66	05 11	06 10	07 03	04 25	06 17	08 11	10 14
64	05 12	06 07	06 56	04 23	06 08	07 55	09 46
62	05 13	06 04	06 50	04 21	06 01	07 42	09 25
60	05 14	06 02	06 44	04 19	05 55	07 31	09 07
N 58	05 14	06 00	06 40	04 18	05 50	07 22	08 53
56	05 15	05 58	06 36	04 17	05 45	07 13	08 41
54	05 15	05 56	06 32	04 15	05 41	07 06	08 30
52	05 15	05 54	06 28	04 14	05 37	07 00	08 21
50	05 15	05 52	06 25	04 13	05 34	06 54	08 12
45	05 15	05 49	06 18	04 11	05 26	06 41	07 55
N 40	05 14	05 45	06 13	04 09	05 20	06 31	07 40
35	05 13	05 42	06 08	04 08	05 15	06 22	07 28
30	05 11	05 39	06 03	04 07	05 10	06 14	07 17
20	05 07	05 33	05 55	04 04	05 02	06 01	06 59
N 10	05 03	05 27	05 49	04 02	04 55	05 49	06 44
0	04 57	05 21	05 42	04 00	04 49	05 38	06 29
S 10	04 49	05 14	05 35	03 59	04 43	05 28	06 15
20	04 39	05 06	05 28	03 57	04 36	05 16	05 59
30	04 26	04 55	05 20	03 55	04 28	05 04	05 42
35	04 18	04 49	05 15	03 53	04 24	04 56	05 32
40	04 08	04 41	05 09	03 52	04 19	04 48	05 20
45	03 56	04 32	05 03	03 51	04 14	04 38	05 07
S 50	03 39	04 21	04 55	03 49	04 07	04 27	04 50
52	03 32	04 16	04 51	03 48	04 04	04 21	04 42
54	03 23	04 10	04 47	03 47	04 00	04 15	04 34
56	03 13	04 03	04 43	03 46	03 57	04 09	04 24
58	03 01	03 55	04 38	03 45	03 53	04 02	04 14
S 60	02 46	03 47	04 32	03 44	03 48	03 53	04 01

Lat.	Sunset	Twilight Civil	Twilight Naut.	Moonset 16	Moonset 17	Moonset 18	Moonset 19
°	h m	h m	h m	h m	h m	h m	h m
N 72	15 54	17 06	18 24	15 13	14 38	13 42	■
N 70	16 07	17 11	18 21	15 21	14 56	14 22	■
68	16 18	17 15	18 19	15 27	15 11	14 50	14 13
66	16 26	17 19	18 18	15 32	15 22	15 11	14 55
64	16 33	17 22	18 17	15 36	15 32	15 28	15 24
62	16 40	17 25	18 16	15 40	15 41	15 42	15 46
60	16 45	17 28	18 16	15 44	15 48	15 54	16 04
N 58	16 50	17 30	18 15	15 47	15 55	16 05	16 19
56	16 54	17 32	18 15	15 49	16 00	16 14	16 32
54	16 58	17 34	18 15	15 52	16 06	16 22	16 43
52	17 02	17 36	18 15	15 54	16 10	16 29	16 53
50	17 05	17 37	18 15	15 56	16 15	16 36	17 02
45	17 12	17 41	18 15	16 00	16 24	16 50	17 21
N 40	17 18	17 45	18 16	16 04	16 32	17 02	17 36
35	17 23	17 48	18 18	16 07	16 38	17 12	17 49
30	17 27	17 51	18 19	16 10	16 44	17 21	18 01
20	17 35	17 57	18 23	16 15	16 55	17 36	18 21
N 10	17 42	18 03	18 28	16 19	17 04	17 50	18 38
0	17 49	18 10	18 34	16 23	17 12	18 02	18 54
S 10	17 55	18 17	18 42	16 28	17 21	18 15	19 10
20	18 03	18 25	18 52	16 32	17 30	18 29	19 28
30	18 11	18 36	19 05	16 37	17 41	18 44	19 48
35	18 16	18 42	19 13	16 40	17 47	18 54	19 59
40	18 22	18 50	19 24	16 43	17 54	19 04	20 13
45	18 29	18 59	19 36	16 47	18 02	19 16	20 29
S 50	18 37	19 11	19 53	16 52	18 12	19 31	20 49
52	18 40	19 16	20 01	16 54	18 16	19 38	20 58
54	18 45	19 22	20 10	16 56	18 21	19 46	21 09
56	18 49	19 29	20 20	16 59	18 27	19 55	21 21
58	18 54	19 37	20 32	17 01	18 33	20 05	21 35
S 60	19 00	19 46	20 47	17 05	18 40	20 17	21 52

Day	SUN Eqn. of Time 00^h	SUN Eqn. of Time 12^h	SUN Mer. Pass.	MOON Mer. Pass. Upper	MOON Mer. Pass. Lower	Age	Phase
d	m s	m s	h m	h m	h m	d	%
16	14 22	14 28	11 46	10 12	22 36	28	4
17	14 34	14 40	11 45	11 01	23 25	29	1
18	14 46	14 52	11 45	11 50	24 16	00	0

2009 OCTOBER 19, 20, 21 (MON., TUES., WED.)

UT d h	ARIES GHA	VENUS −3.9 GHA	VENUS Dec	MARS +0.6 GHA	MARS Dec	JUPITER −2.5 GHA	JUPITER Dec	SATURN +1.1 GHA	SATURN Dec
	° ′	° ′	° ′	° ′	° ′	° ′	° ′	° ′	° ′
19 00 (MONDAY)	27 36.2	202 22.7	S 0 35.3	263 55.8	N21 02.5	67 35.9	S16 40.5	207 54.3	N 2 19.7
01	42 38.7	217 22.3	36.5	278 57.0	02.3	82 38.4	40.5	222 56.5	19.6
02	57 41.1	232 21.9	37.7	293 58.2	02.1	97 40.8	40.5	237 58.7	19.5
03	72 43.6	247 21.5	. . 38.9	308 59.3	. . 01.9	112 43.2	. . 40.5	253 00.9	. . 19.3
04	87 46.1	262 21.1	40.1	324 00.5	01.7	127 45.6	40.4	268 03.1	19.2
05	102 48.5	277 20.7	41.4	339 01.6	01.5	142 48.0	40.4	283 05.3	19.1
06	117 51.0	292 20.3	S 0 42.6	354 02.8	N21 01.3	157 50.4	S16 40.4	298 07.5	N 2 19.0
07	132 53.4	307 19.9	43.8	9 04.0	01.1	172 52.8	40.4	313 09.7	18.9
08	147 55.9	322 19.5	45.0	24 05.1	00.8	187 55.3	40.4	328 11.9	18.8
09	162 58.4	337 19.1	. . 46.2	39 06.3	. . 00.6	202 57.7	. . 40.3	343 14.1	. . 18.7
10	178 00.8	352 18.8	47.5	54 07.5	00.4	218 00.1	40.3	358 16.3	18.6
11	193 03.3	7 18.4	48.7	69 08.6	00.2	233 02.5	40.3	13 18.5	18.5
12	208 05.8	22 18.0	S 0 49.9	84 09.8	N21 00.0	248 04.9	S16 40.3	28 20.7	N 2 18.4
13	223 08.2	37 17.6	51.1	99 11.0	20 59.8	263 07.3	40.3	43 22.9	18.2
14	238 10.7	52 17.2	52.3	114 12.1	59.6	278 09.7	40.2	58 25.1	18.1
15	253 13.2	67 16.8	. . 53.5	129 13.3	. . 59.4	293 12.2	. . 40.2	73 27.3	. . 18.0
16	268 15.6	82 16.4	54.8	144 14.5	59.1	308 14.6	40.2	88 29.5	17.9
17	283 18.1	97 16.0	56.0	159 15.7	58.9	323 17.0	40.2	103 31.7	17.8
18	298 20.6	112 15.6	S 0 57.2	174 16.8	N20 58.7	338 19.4	S16 40.2	118 33.9	N 2 17.7
19	313 23.0	127 15.2	58.4	189 18.0	58.5	353 21.8	40.1	133 36.1	17.6
20	328 25.5	142 14.8	0 59.6	204 19.2	58.3	8 24.2	40.1	148 38.3	17.5
21	343 27.9	157 14.4	1 00.9	219 20.3	. . 58.1	23 26.6	. . 40.1	163 40.5	. . 17.4
22	358 30.4	172 14.0	02.1	234 21.5	57.9	38 29.0	40.1	178 42.7	17.3
23	13 32.9	187 13.6	03.3	249 22.7	57.7	53 31.4	40.0	193 44.9	17.2
20 00 (TUESDAY)	28 35.3	202 13.2	S 1 04.5	264 23.9	N20 57.4	68 33.8	S16 40.0	208 47.1	N 2 17.0
01	43 37.8	217 12.9	05.7	279 25.0	57.2	83 36.3	40.0	223 49.3	16.9
02	58 40.3	232 12.5	07.0	294 26.2	57.0	98 38.7	40.0	238 51.4	16.8
03	73 42.7	247 12.1	. . 08.2	309 27.4	. . 56.8	113 41.1	. . 40.0	253 53.6	. . 16.7
04	88 45.2	262 11.7	09.4	324 28.6	56.6	128 43.5	39.9	268 55.8	16.6
05	103 47.7	277 11.3	10.6	339 29.7	56.4	143 45.9	39.9	283 58.0	16.5
06	118 50.1	292 10.9	S 1 11.8	354 30.9	N20 56.2	158 48.3	S16 39.9	299 00.2	N 2 16.4
07	133 52.6	307 10.5	13.1	9 32.1	56.0	173 50.7	39.9	314 02.4	16.3
08	148 55.0	322 10.1	14.3	24 33.3	55.7	188 53.1	39.8	329 04.6	16.2
09	163 57.5	337 09.7	. . 15.5	39 34.5	. . 55.5	203 55.5	. . 39.8	344 06.8	. . 16.1
10	179 00.0	352 09.3	16.7	54 35.6	55.3	218 57.9	39.8	359 09.0	15.9
11	194 02.4	7 08.9	17.9	69 36.8	55.1	234 00.3	39.8	14 11.2	15.8
12	209 04.9	22 08.5	S 1 19.2	84 38.0	N20 54.9	249 02.7	S16 39.8	29 13.4	N 2 15.7
13	224 07.4	37 08.1	20.4	99 39.2	54.7	264 05.1	39.7	44 15.6	15.6
14	239 09.8	52 07.7	21.6	114 40.4	54.5	279 07.5	39.7	59 17.8	15.5
15	254 12.3	67 07.3	. . 22.8	129 41.5	. . 54.3	294 09.9	. . 39.7	74 20.0	. . 15.4
16	269 14.8	82 06.9	24.0	144 42.7	54.0	309 12.3	39.7	89 22.2	15.3
17	284 17.2	97 06.5	25.3	159 43.9	53.8	324 14.8	39.6	104 24.4	15.2
18	299 19.7	112 06.1	S 1 26.5	174 45.1	N20 53.6	339 17.2	S16 39.6	119 26.6	N 2 15.1
19	314 22.2	127 05.7	27.7	189 46.3	53.4	354 19.6	39.6	134 28.8	15.0
20	329 24.6	142 05.4	28.9	204 47.5	53.2	9 22.0	39.6	149 31.0	14.9
21	344 27.1	157 05.0	. . 30.1	219 48.6	. . 53.0	24 24.4	. . 39.5	164 33.2	. . 14.7
22	359 29.5	172 04.6	31.4	234 49.8	52.8	39 26.8	39.5	179 35.4	14.6
23	14 32.0	187 04.2	32.6	249 51.0	52.5	54 29.2	39.5	194 37.6	14.5
21 00 (WEDNESDAY)	29 34.5	202 03.8	S 1 33.8	264 52.2	N20 52.3	69 31.6	S16 39.5	209 39.8	N 2 14.4
01	44 36.9	217 03.4	35.0	279 53.4	52.1	84 34.0	39.4	224 42.0	14.3
02	59 39.4	232 03.0	36.2	294 54.6	51.9	99 36.4	39.4	239 44.2	14.2
03	74 41.9	247 02.6	. . 37.5	309 55.8	. . 51.7	114 38.8	. . 39.4	254 46.4	. . 14.1
04	89 44.3	262 02.2	38.7	324 56.9	51.5	129 41.2	39.4	269 48.6	14.0
05	104 46.8	277 01.8	39.9	339 58.1	51.3	144 43.6	39.3	284 50.8	13.9
06	119 49.3	292 01.4	S 1 41.1	354 59.3	N20 51.0	159 46.0	S16 39.3	299 53.0	N 2 13.8
07	134 51.7	307 01.0	42.3	10 00.5	50.8	174 48.4	39.3	314 55.2	13.7
08	149 54.2	322 00.6	43.6	25 01.7	50.6	189 50.8	39.3	329 57.4	13.6
09	164 56.7	337 00.2	. . 44.8	40 02.9	. . 50.4	204 53.2	. . 39.2	344 59.6	. . 13.4
10	179 59.1	351 59.8	46.0	55 04.1	50.2	219 55.6	39.2	0 01.8	13.3
11	195 01.6	6 59.4	47.2	70 05.3	50.0	234 57.9	39.2	15 04.0	13.2
12	210 04.0	21 59.0	S 1 48.4	85 06.5	N20 49.8	250 00.3	S16 39.2	30 06.2	N 2 13.1
13	225 06.5	36 58.6	49.7	100 07.7	49.6	265 02.7	39.1	45 08.4	13.0
14	240 09.0	51 58.2	50.9	115 08.8	49.3	280 05.1	39.1	60 10.6	12.9
15	255 11.4	66 57.8	. . 52.1	130 10.0	. . 49.1	295 07.5	. . 39.1	75 12.8	. . 12.8
16	270 13.9	81 57.4	53.3	145 11.2	48.9	310 09.9	39.1	90 15.0	12.7
17	285 16.4	96 57.0	54.5	160 12.4	48.7	325 12.3	39.0	105 17.2	12.6
18	300 18.8	111 56.6	S 1 55.8	175 13.6	N20 48.5	340 14.7	S16 39.0	120 19.4	N 2 12.5
19	315 21.3	126 56.2	57.0	190 14.8	48.3	355 17.1	39.0	135 21.6	12.4
20	330 23.8	141 55.8	58.2	205 16.0	48.1	10 19.5	39.0	150 23.8	12.2
21	345 26.2	156 55.4	1 59.4	220 17.2	. . 47.8	25 21.9	. . 38.9	165 26.0	. . 12.1
22	0 28.7	171 55.0	2 00.6	235 18.4	47.6	40 24.3	38.9	180 28.2	12.0
23	15 31.2	186 54.6	S 2 01.9	250 19.6	47.4	55 26.7	38.9	195 30.4	11.9
Mer. Pass. (h m)	22 02.0	*v* −0.4	*d* 1.2	*v* 1.2	*d* 0.2	*v* 2.4	*d* 0.0	*v* 2.2	*d* 0.1

STARS

Name	SHA	Dec
	° ′	° ′
Acamar	315 19.9	S40 15.7
Achernar	335 28.0	S57 11.1
Acrux	173 13.3	S63 09.1
Adhara	255 14.6	S28 58.9
Aldebaran	290 52.2	N16 31.9
Alioth	166 23.4	N55 54.3
Alkaid	153 01.4	N49 15.8
Al Na'ir	27 46.7	S46 54.9
Alnilam	275 48.9	S 1 11.6
Alphard	217 58.9	S 8 42.0
Alphecca	126 13.6	N26 41.0
Alpheratz	357 46.1	N29 08.9
Altair	62 10.9	N 8 53.8
Ankaa	353 17.8	S42 15.1
Antares	112 29.9	S26 27.3
Arcturus	145 58.5	N19 07.9
Atria	107 34.6	S69 02.9
Avior	234 19.3	S59 32.1
Bellatrix	278 34.7	N 6 21.6
Betelgeuse	271 04.1	N 7 24.6
Canopus	263 57.2	S52 41.7
Capella	280 38.2	N46 00.4
Deneb	49 33.3	N45 19.3
Denebola	182 36.7	N14 31.0
Diphda	348 58.2	S17 55.8
Dubhe	193 55.2	N61 41.7
Elnath	278 15.8	N28 37.0
Eltanin	90 47.7	N51 29.5
Enif	33 49.7	N 9 55.4
Fomalhaut	15 26.6	S29 34.1
Gacrux	172 04.7	S57 10.0
Gienah	175 55.4	S17 35.7
Hadar	148 52.6	S60 25.3
Hamal	328 03.5	N23 30.7
Kaus Aust.	83 47.5	S34 22.9
Kochab	137 20.7	N74 06.9
Markab	13 40.8	N15 15.7
Menkar	314 17.6	N 4 07.9
Menkent	148 11.3	S36 25.1
Miaplacidus	221 40.8	S69 45.2
Mirfak	308 43.9	N49 53.8
Nunki	76 01.7	S26 17.1
Peacock	53 23.3	S56 42.4
Pollux	243 31.0	N28 00.1
Procyon	245 02.5	N 5 12.1
Rasalhague	96 09.2	N12 33.3
Regulus	207 46.5	N11 55.2
Rigel	281 14.4	S 8 11.2
Rigil Kent.	139 56.3	S60 52.6
Sabik	102 15.9	S15 44.2
Schedar	349 43.3	N56 35.7
Shaula	96 25.8	S37 06.7
Sirius	258 36.0	S16 43.6
Spica	158 34.5	S11 12.7
Suhail	222 54.7	S43 28.1
Vega	80 41.0	N38 47.8
Zuben'ubi	137 08.8	S16 04.9

	SHA	Mer. Pass.
	° ′	h m
Venus	173 37.9	10 31
Mars	235 48.5	6 22
Jupiter	39 58.5	19 23
Saturn	180 11.7	10 03

UT (d h)	SUN GHA	SUN Dec	MOON GHA	v	MOON Dec	d	HP
	° ′	° ′	° ′	′	° ′	′	′
19 00 (MONDAY)	183 44.4	S 9 56.9	176 11.8	10.1	S17 58.5	10.9	57.5
01	198 44.5	57.8	190 40.9	10.0	18 09.4	10.8	57.5
02	213 44.6	58.7	205 09.9	10.0	18 20.2	10.7	57.4
03	228 44.8	9 59.6	219 38.9	10.0	18 30.9	10.6	57.4
04	243 44.9	10 00.5	234 07.9	10.0	18 41.5	10.5	57.4
05	258 45.0	01.4	248 36.9	9.9	18 52.0	10.4	57.4
06	273 45.1	S10 02.3	263 05.8	9.8	S19 02.4	10.2	57.3
07	288 45.2	03.2	277 34.6	9.9	19 12.6	10.2	57.3
08	303 45.3	04.1	292 03.5	9.8	19 22.8	10.0	57.3
09	318 45.4	. . 05.0	306 32.3	9.7	19 32.8	9.9	57.2
10	333 45.5	05.9	321 01.0	9.8	19 42.7	9.8	57.2
11	348 45.7	06.8	335 29.8	9.7	19 52.5	9.7	57.2
12	3 45.8	S10 07.7	349 58.5	9.6	S20 02.2	9.6	57.2
13	18 45.9	08.6	4 27.1	9.7	20 11.8	9.4	57.1
14	33 46.0	09.5	18 55.8	9.6	20 21.2	9.3	57.1
15	48 46.1	. . 10.4	33 24.4	9.5	20 30.5	9.3	57.1
16	63 46.2	11.3	47 52.9	9.5	20 39.8	9.0	57.1
17	78 46.3	12.2	62 21.4	9.5	20 48.8	9.0	57.0
18	93 46.4	S10 13.1	76 49.9	9.5	S20 57.8	8.9	57.0
19	108 46.5	14.0	91 18.4	9.4	21 06.7	8.7	57.0
20	123 46.7	14.9	105 46.8	9.4	21 15.4	8.6	56.9
21	138 46.8	. . 15.8	120 15.2	9.4	21 24.0	8.5	56.9
22	153 46.9	16.7	134 43.6	9.3	21 32.5	8.3	56.9
23	168 47.0	17.6	149 11.9	9.3	21 40.8	8.2	56.9
20 00 (TUESDAY)	183 47.1	S10 18.5	163 40.2	9.3	S21 49.0	8.1	56.8
01	198 47.2	19.4	178 08.5	9.3	21 57.1	8.0	56.8
02	213 47.3	20.3	192 36.8	9.2	22 05.1	7.9	56.8
03	228 47.4	. . 21.2	207 05.0	9.2	22 13.0	7.7	56.7
04	243 47.5	22.1	221 33.2	9.1	22 20.7	7.6	56.7
05	258 47.6	23.0	236 01.3	9.2	22 28.3	7.4	56.7
06	273 47.7	S10 23.9	250 29.5	9.1	S22 35.7	7.4	56.7
07	288 47.8	24.8	264 57.6	9.1	22 43.1	7.2	56.6
08	303 47.9	25.7	279 25.7	9.1	22 50.3	7.0	56.6
09	318 48.1	. . 26.6	293 53.8	9.0	22 57.3	7.0	56.6
10	333 48.2	27.5	308 21.8	9.0	23 04.3	6.8	56.5
11	348 48.3	28.4	322 49.8	9.0	23 11.1	6.6	56.5
12	3 48.4	S10 29.3	337 17.8	9.0	S23 17.7	6.6	56.5
13	18 48.5	30.2	351 45.8	9.0	23 24.3	6.4	56.5
14	33 48.6	31.1	6 13.8	8.9	23 30.7	6.3	56.4
15	48 48.7	. . 32.0	20 41.7	9.0	23 37.0	6.1	56.4
16	63 48.8	32.9	35 09.7	8.9	23 43.1	6.0	56.4
17	78 48.9	33.7	49 37.6	8.9	23 49.1	5.9	56.3
18	93 49.0	S10 34.6	64 05.5	8.8	S23 55.0	5.7	56.3
19	108 49.1	35.5	78 33.3	8.9	24 00.7	5.6	56.3
20	123 49.2	36.4	93 01.2	8.8	24 06.3	5.5	56.3
21	138 49.3	. . 37.3	107 29.0	8.9	24 11.8	5.3	56.2
22	153 49.4	38.2	121 56.9	8.8	24 17.1	5.2	56.2
23	168 49.5	39.1	136 24.7	8.8	24 22.3	5.0	56.2
21 00 (WEDNESDAY)	183 49.6	S10 40.0	150 52.5	8.8	S24 27.3	4.9	56.2
01	198 49.7	40.9	165 20.3	8.8	24 32.2	4.8	56.1
02	213 49.8	41.8	179 48.1	8.8	24 37.0	4.7	56.1
03	228 49.9	. . 42.7	194 15.9	8.8	24 41.7	4.5	56.1
04	243 50.0	43.6	208 43.7	8.8	24 46.2	4.3	56.0
05	258 50.1	44.4	223 11.5	8.8	24 50.5	4.2	56.0
06	273 50.2	S10 45.3	237 39.3	8.7	S24 54.7	4.1	56.0
07	288 50.3	46.2	252 07.0	8.8	24 58.8	4.0	56.0
08	303 50.4	47.1	266 34.8	8.8	25 02.8	3.8	55.9
09	318 50.5	. . 48.0	281 02.6	8.7	25 06.6	3.7	55.9
10	333 50.6	48.9	295 30.3	8.8	25 10.3	3.5	55.9
11	348 50.7	49.8	309 58.1	8.8	25 13.8	3.4	55.9
12	3 50.8	S10 50.7	324 25.9	8.8	S25 17.2	3.2	55.8
13	18 50.9	51.5	338 53.7	8.8	25 20.4	3.1	55.8
14	33 51.0	52.4	353 21.5	8.7	25 23.5	3.0	55.8
15	48 51.1	. . 53.3	7 49.2	8.8	25 26.5	2.9	55.8
16	63 51.2	54.2	22 17.0	8.8	25 29.4	2.7	55.7
17	78 51.3	55.1	36 44.8	8.9	25 32.1	2.5	55.7
18	93 51.4	S10 56.0	51 12.7	8.8	S25 34.6	2.4	55.7
19	108 51.5	56.9	65 40.5	8.8	25 37.0	2.3	55.7
20	123 51.6	57.8	80 08.3	8.9	25 39.3	2.2	55.6
21	138 51.7	. . 58.6	94 36.2	8.8	25 41.5	2.0	55.6
22	153 51.8	10 59.5	109 04.0	8.9	25 43.5	1.8	55.6
23	168 51.9	S11 00.4	123 31.9	8.9	S25 45.3	1.8	55.6
	SD 16.1	*d* 0.9	SD 15.6		15.4		15.2

Lat.	Twilight Naut.	Twilight Civil	Sunrise	Moonrise 19	Moonrise 20	Moonrise 21	Moonrise 22
°	h m	h m	h m	h m	h m	h m	h m
N 72	05 17	06 36	07 50	■	■	■	■
N 70	05 19	06 30	07 35	■	■	■	■
68	05 19	06 24	07 23	10 55	■	■	■
66	05 20	06 20	07 13	10 14	12 42	■	■
64	05 20	06 16	07 05	09 46	11 40	13 37	■
62	05 21	06 12	06 58	09 25	11 06	12 37	13 45
60	05 21	06 09	06 52	09 07	10 41	12 04	13 07
N 58	05 21	06 06	06 47	08 53	10 21	11 39	12 41
56	05 20	06 04	06 42	08 41	10 05	11 19	12 20
54	05 20	06 01	06 37	08 30	09 51	11 03	12 03
52	05 20	05 59	06 34	08 21	09 39	10 49	11 48
50	05 19	05 57	06 30	08 12	09 28	10 37	11 35
45	05 18	05 52	06 22	07 55	09 06	10 11	11 09
N 40	05 17	05 48	06 16	07 40	08 48	09 51	10 48
35	05 15	05 45	06 10	07 28	08 33	09 34	10 31
30	05 13	05 41	06 05	07 17	08 20	09 20	10 16
20	05 08	05 34	05 57	06 59	07 58	08 55	09 50
N 10	05 03	05 27	05 49	06 44	07 39	08 34	09 28
0	04 56	05 20	05 41	06 29	07 21	08 15	09 08
S 10	04 48	05 13	05 34	06 15	07 04	07 55	08 48
20	04 37	05 03	05 26	05 59	06 45	07 34	08 26
30	04 23	04 52	05 17	05 42	06 24	07 10	08 00
35	04 14	04 45	05 11	05 32	06 11	06 56	07 46
40	04 03	04 37	05 05	05 20	05 57	06 40	07 28
45	03 50	04 27	04 58	05 07	05 40	06 20	07 08
S 50	03 32	04 15	04 49	04 50	05 19	05 56	06 42
52	03 24	04 09	04 45	04 42	05 09	05 44	06 29
54	03 14	04 02	04 40	04 34	04 58	05 31	06 15
56	03 03	03 55	04 35	04 24	04 46	05 15	05 58
58	02 50	03 47	04 30	04 14	04 31	04 57	05 37
S 60	02 34	03 37	04 24	04 01	04 14	04 35	05 12

Lat.	Sunset	Twilight Civil	Twilight Naut.	Moonset 19	Moonset 20	Moonset 21	Moonset 22
°	h m	h m	h m	h m	h m	h m	h m
N 72	15 38	16 52	18 10	■	■	■	■
N 70	15 53	16 58	18 09	■	■	■	■
68	16 05	17 04	18 08	14 13	■	■	■
66	16 15	17 09	18 08	14 55	14 17	■	■
64	16 23	17 13	18 08	15 24	15 20	15 14	■
62	16 30	17 16	18 08	15 46	15 54	16 14	16 57
60	16 37	17 19	18 08	16 04	16 20	16 48	17 34
N 58	16 42	17 22	18 08	16 19	16 40	17 12	18 01
56	16 47	17 25	18 08	16 32	16 57	17 32	18 22
54	16 51	17 27	18 08	16 43	17 11	17 49	18 39
52	16 55	17 30	18 09	16 53	17 23	18 03	18 53
50	16 59	17 32	18 09	17 02	17 34	18 16	19 06
45	17 07	17 36	18 11	17 21	17 57	18 41	19 33
N 40	17 13	17 41	18 12	17 36	18 16	19 02	19 53
35	17 19	17 45	18 14	17 49	18 32	19 19	20 11
30	17 24	17 48	18 16	18 01	18 45	19 34	20 26
20	17 33	17 55	18 21	18 21	19 08	19 59	20 51
N 10	17 41	18 02	18 27	18 38	19 28	20 20	21 13
0	17 48	18 09	18 34	18 54	19 47	20 40	21 33
S 10	17 56	18 17	18 42	19 10	20 06	21 01	21 54
20	18 04	18 26	18 53	19 28	20 26	21 22	22 15
30	18 13	18 38	19 07	19 48	20 49	21 47	22 40
35	18 19	18 45	19 16	19 59	21 03	22 02	22 55
40	18 25	18 54	19 27	20 13	21 19	22 19	23 12
45	18 33	19 04	19 41	20 29	21 38	22 40	23 33
S 50	18 42	19 16	19 59	20 49	22 01	23 06	23 59
52	18 46	19 22	20 07	20 58	22 13	23 18	24 11
54	18 50	19 29	20 17	21 09	22 26	23 33	24 25
56	18 55	19 36	20 29	21 21	22 41	23 49	24 42
58	19 01	19 45	20 42	21 35	22 59	24 10	00 10
S 60	19 08	19 55	20 58	21 52	23 21	24 35	00 35

Day	SUN Eqn. of Time 00^h	SUN Eqn. of Time 12^h	SUN Mer. Pass.	MOON Mer. Pass. Upper	MOON Mer. Pass. Lower	MOON Age	MOON Phase
d	m s	m s	h m	h m	h m	d %	
19	14 57	15 03	11 45	12 42	00 16	01 2	●
20	15 08	15 13	11 45	13 34	01 08	02 6	
21	15 18	15 23	11 45	14 28	02 01	03 12	

	UT	ARIES	VENUS −3.9		MARS +0.6		JUPITER −2.5		SATURN +1.1	
		GHA	GHA	Dec	GHA	Dec	GHA	Dec	GHA	Dec
	d h	° ′	° ′	° ′	° ′	° ′	° ′	° ′	° ′	° ′
	22 00	30 33.6	201 54.2	S 2 03.1	265 20.8	N20 47.2	70 29.1	S16 38.9	210 32.6	N 2 11.8
	01	45 36.1	216 53.8	04.3	280 22.0	47.0	85 31.5	38.8	225 34.8	11.7
	02	60 38.5	231 53.4	05.5	295 23.2	46.8	100 33.9	38.8	240 37.0	11.6
	03	75 41.0	246 53.1	06.7	310 24.4	46.6	115 36.3	38.8	255 39.2	11.5
	04	90 43.5	261 52.7	08.0	325 25.6	46.3	130 38.6	38.7	270 41.4	11.4
	05	105 45.9	276 52.3	09.2	340 26.8	46.1	145 41.0	38.7	285 43.6	11.3
	06	120 48.4	291 51.9	S 2 10.4	355 28.0	N20 45.9	160 43.4	S16 38.7	300 45.8	N 2 11.2
	07	135 50.9	306 51.5	11.6	10 29.2	45.7	175 45.8	38.7	315 48.0	11.1
T	08	150 53.3	321 51.1	12.8	25 30.4	45.5	190 48.2	38.6	330 50.2	11.0
H	09	165 55.8	336 50.7	14.1	40 31.6	45.3	205 50.6	38.6	345 52.4	10.8
U	10	180 58.3	351 50.3	15.3	55 32.8	45.1	220 53.0	38.6	0 54.6	10.7
R	11	196 00.7	6 49.9	16.5	70 34.0	44.8	235 55.4	38.5	15 56.8	10.6
S	12	211 03.2	21 49.5	S 2 17.7	85 35.2	N20 44.6	250 57.8	S16 38.5	30 59.0	N 2 10.5
D	13	226 05.6	36 49.1	18.9	100 36.4	44.4	266 00.2	38.5	46 01.2	10.4
A	14	241 08.1	51 48.7	20.2	115 37.6	44.2	281 02.5	38.5	61 03.4	10.3
Y	15	256 10.6	66 48.3	21.4	130 38.8	44.0	296 04.9	38.4	76 05.6	10.2
	16	271 13.0	81 47.9	22.6	145 40.0	43.8	311 07.3	38.4	91 07.8	10.1
	17	286 15.5	96 47.5	23.8	160 41.3	43.6	326 09.7	38.4	106 10.0	10.0
	18	301 18.0	111 47.1	S 2 25.0	175 42.5	N20 43.3	341 12.1	S16 38.3	121 12.2	N 2 09.9
	19	316 20.4	126 46.7	26.3	190 43.7	43.1	356 14.5	38.3	136 14.4	09.8
	20	331 22.9	141 46.3	27.5	205 44.9	42.9	11 16.9	38.3	151 16.6	09.7
	21	346 25.4	156 45.9	28.7	220 46.1	42.7	26 19.2	38.3	166 18.8	09.6
	22	1 27.8	171 45.5	29.9	235 47.3	42.5	41 21.6	38.2	181 21.0	09.4
	23	16 30.3	186 45.1	31.1	250 48.5	42.3	56 24.0	38.2	196 23.2	09.3
	23 00	31 32.8	201 44.7	S 2 32.4	265 49.7	N20 42.1	71 26.4	S16 38.2	211 25.4	N 2 09.2
	01	46 35.2	216 44.3	33.6	280 50.9	41.8	86 28.8	38.1	226 27.6	09.1
	02	61 37.7	231 43.9	34.8	295 52.1	41.6	101 31.2	38.1	241 29.8	09.0
	03	76 40.1	246 43.5	36.0	310 53.4	41.4	116 33.5	38.1	256 32.0	08.9
	04	91 42.6	261 43.1	37.2	325 54.6	41.2	131 35.9	38.1	271 34.2	08.8
	05	106 45.1	276 42.7	38.5	340 55.8	41.0	146 38.3	38.0	286 36.4	08.7
	06	121 47.5	291 42.3	S 2 39.7	355 57.0	N20 40.8	161 40.7	S16 38.0	301 38.6	N 2 08.6
	07	136 50.0	306 41.9	40.9	10 58.2	40.6	176 43.1	38.0	316 40.8	08.5
	08	151 52.5	321 41.4	42.1	25 59.4	40.3	191 45.5	37.9	331 43.0	08.4
F	09	166 54.9	336 41.0	43.3	41 00.6	40.1	206 47.8	37.9	346 45.2	08.3
R	10	181 57.4	351 40.6	44.5	56 01.9	39.9	221 50.2	37.9	1 47.5	08.2
I	11	196 59.9	6 40.2	45.8	71 03.1	39.7	236 52.6	37.8	16 49.7	08.1
D	12	212 02.3	21 39.8	S 2 47.0	86 04.3	N20 39.5	251 55.0	S16 37.8	31 51.9	N 2 07.9
A	13	227 04.8	36 39.4	48.2	101 05.5	39.3	266 57.4	37.8	46 54.1	07.8
Y	14	242 07.3	51 39.0	49.4	116 06.7	39.1	281 59.7	37.7	61 56.3	07.7
	15	257 09.7	66 38.6	50.6	131 07.9	38.8	297 02.1	37.7	76 58.5	07.6
	16	272 12.2	81 38.2	51.9	146 09.2	38.6	312 04.5	37.7	92 00.7	07.5
	17	287 14.6	96 37.8	53.1	161 10.4	38.4	327 06.9	37.7	107 02.9	07.4
	18	302 17.1	111 37.4	S 2 54.3	176 11.6	N20 38.2	342 09.3	S16 37.6	122 05.1	N 2 07.3
	19	317 19.6	126 37.0	55.5	191 12.8	38.0	357 11.6	37.6	137 07.3	07.2
	20	332 22.0	141 36.6	56.7	206 14.0	37.8	12 14.0	37.6	152 09.5	07.1
	21	347 24.5	156 36.2	58.0	221 15.3	37.6	27 16.4	37.5	167 11.7	07.0
	22	2 27.0	171 35.8	2 59.2	236 16.5	37.3	42 18.8	37.5	182 13.9	06.9
	23	17 29.4	186 35.4	3 00.4	251 17.7	37.1	57 21.1	37.5	197 16.1	06.8
	24 00	32 31.9	201 35.0	S 3 01.6	266 18.9	N20 36.9	72 23.5	S16 37.4	212 18.3	N 2 06.7
	01	47 34.4	216 34.6	02.8	281 20.2	36.7	87 25.9	37.4	227 20.5	06.6
	02	62 36.8	231 34.2	04.0	296 21.4	36.5	102 28.3	37.4	242 22.7	06.5
	03	77 39.3	246 33.8	05.3	311 22.6	36.3	117 30.6	37.3	257 24.9	06.3
	04	92 41.8	261 33.4	06.5	326 23.8	36.0	132 33.0	37.3	272 27.1	06.2
	05	107 44.2	276 33.0	07.7	341 25.1	35.8	147 35.4	37.3	287 29.3	06.1
	06	122 46.7	291 32.6	S 3 08.9	356 26.3	N20 35.6	162 37.8	S16 37.2	302 31.5	N 2 06.0
	07	137 49.1	306 32.2	10.1	11 27.5	35.4	177 40.1	37.2	317 33.7	05.9
S	08	152 51.6	321 31.8	11.3	26 28.7	35.2	192 42.5	37.2	332 35.9	05.8
A	09	167 54.1	336 31.4	12.6	41 30.0	35.0	207 44.9	37.1	347 38.1	05.7
T	10	182 56.5	351 31.0	13.8	56 31.2	34.8	222 47.3	37.1	2 40.3	05.6
U	11	197 59.0	6 30.6	15.0	71 32.4	34.5	237 49.6	37.1	17 42.5	05.5
R	12	213 01.5	21 30.1	S 3 16.2	86 33.6	N20 34.3	252 52.0	S16 37.0	32 44.7	N 2 05.4
D	13	228 03.9	36 29.7	17.4	101 34.9	34.1	267 54.4	37.0	47 46.9	05.3
A	14	243 06.4	51 29.3	18.6	116 36.1	33.9	282 56.7	37.0	62 49.1	05.2
Y	15	258 08.9	66 28.9	19.9	131 37.3	33.7	297 59.1	36.9	77 51.3	05.1
	16	273 11.3	81 28.5	21.1	146 38.6	33.5	313 01.5	36.9	92 53.6	05.0
	17	288 13.8	96 28.1	22.3	161 39.8	33.3	328 03.9	36.9	107 55.8	04.9
	18	303 16.2	111 27.7	S 3 23.5	176 41.0	N20 33.0	343 06.2	S16 36.8	122 58.0	N 2 04.8
	19	318 18.7	126 27.3	24.7	191 42.3	32.8	358 08.6	36.8	138 00.2	04.6
	20	333 21.2	141 26.9	25.9	206 43.5	32.6	13 11.0	36.8	153 02.4	04.5
	21	348 23.6	156 26.5	27.2	221 44.7	32.4	28 13.3	36.7	168 04.6	04.4
	22	3 26.1	171 26.1	28.4	236 46.0	32.2	43 15.7	36.7	183 06.8	04.3
	23	18 28.6	186 25.7	29.6	251 47.2	32.0	58 18.1	36.7	198 09.0	04.2
	Mer. Pass.	h m 21 50.2	v −0.4	d 1.2	v 1.2	d 0.2	v 2.4	d 0.0	v 2.2	d 0.1

STARS		
Name	SHA	Dec
	° ′	° ′
Acamar	315 19.9	S40 15.7
Achernar	335 28.0	S57 11.1
Acrux	173 13.2	S63 09.1
Adhara	255 14.6	S28 58.9
Aldebaran	290 52.2	N16 31.9
Alioth	166 23.4	N55 54.3
Alkaid	153 01.4	N49 15.8
Al Na'ir	27 46.7	S46 54.9
Alnilam	275 48.9	S 1 11.6
Alphard	217 58.9	S 8 42.0
Alphecca	126 13.6	N26 41.0
Alpheratz	357 46.1	N29 09.0
Altair	62 10.9	N 8 53.8
Ankaa	353 17.8	S42 15.1
Antares	112 29.9	S26 27.3
Arcturus	145 58.5	N19 07.9
Atria	107 34.6	S69 02.9
Avior	234 19.3	S59 32.1
Bellatrix	278 34.7	N 6 21.6
Betelgeuse	271 04.0	N 7 24.6
Canopus	263 57.1	S52 41.7
Capella	280 38.1	N46 00.4
Deneb	49 33.3	N45 19.3
Denebola	182 36.6	N14 31.0
Diphda	348 58.2	S17 55.8
Dubhe	193 55.2	N61 41.6
Elnath	278 15.8	N28 37.0
Eltanin	90 47.7	N51 29.5
Enif	33 49.7	N 9 55.4
Fomalhaut	15 26.6	S29 34.2
Gacrux	172 04.7	S57 10.0
Gienah	175 55.4	S17 35.7
Hadar	148 52.6	S60 25.2
Hamal	328 03.5	N23 30.7
Kaus Aust.	83 47.6	S34 22.9
Kochab	137 20.7	N74 06.9
Markab	13 40.8	N15 15.7
Menkar	314 17.6	N 4 07.9
Menkent	148 11.2	S36 25.1
Miaplacidus	221 40.8	S69 45.2
Mirfak	308 43.9	N49 53.8
Nunki	76 01.8	S26 17.1
Peacock	53 23.3	S56 42.4
Pollux	243 31.0	N28 00.1
Procyon	245 02.5	N 5 12.1
Rasalhague	96 09.2	N12 33.3
Regulus	207 46.5	N11 55.1
Rigel	281 14.4	S 8 11.2
Rigil Kent.	139 56.3	S60 52.6
Sabik	102 15.9	S15 44.2
Schedar	349 43.3	N56 35.7
Shaula	96 25.8	S37 06.7
Sirius	258 36.0	S16 43.6
Spica	158 34.4	S11 12.7
Suhail	222 54.6	S43 28.1
Vega	80 41.0	N38 47.8
Zuben'ubi	137 08.8	S16 04.9

	SHA	Mer. Pass
	° ′	h m
Venus	170 11.9	10 33
Mars	234 17.0	6 16
Jupiter	39 53.6	19 11
Saturn	179 52.7	9 53

UT (d h)	SUN GHA ° ′	SUN Dec ° ′	MOON GHA ° ′	v ′	MOON Dec ° ′	d ′	HP ′
22 00 (Thursday)	183 52.0	S11 01.3	137 59.8	8.9	S25 47.1	1.5	55.5
01	198 52.1	02.2	152 27.7	8.9	25 48.6	1.5	55.5
02	213 52.1	03.1	166 55.6	9.0	25 50.1	1.3	55.5
03	228 52.2	. . 03.9	181 23.6	8.9	25 51.4	1.2	55.5
04	243 52.3	04.8	195 51.5	9.0	25 52.6	1.0	55.4
05	258 52.4	05.7	210 19.5	9.0	25 53.6	0.9	55.4
06	273 52.5	S11 06.6	224 47.5	9.1	S25 54.5	0.8	55.4
07	288 52.6	07.5	239 15.6	9.0	25 55.3	0.6	55.4
08	303 52.7	08.4	253 43.6	9.1	25 55.9	0.5	55.3
09	318 52.8	. . 09.2	268 11.7	9.1	25 56.4	0.4	55.3
10	333 52.9	10.1	282 39.8	9.2	25 56.8	0.2	55.3
11	348 53.0	11.0	297 08.0	9.2	25 57.0	0.1	55.3
12	3 53.1	S11 11.9	311 36.2	9.2	S25 57.1	0.1	55.2
13	18 53.2	12.8	326 04.4	9.2	25 57.0	0.2	55.2
14	33 53.3	13.6	340 32.6	9.3	25 56.8	0.3	55.2
15	48 53.3	. . 14.5	355 00.9	9.3	25 56.5	0.4	55.2
16	63 53.4	15.4	9 29.2	9.3	25 56.1	0.6	55.2
17	78 53.5	16.3	23 57.5	9.4	25 55.5	0.7	55.1
18	93 53.6	S11 17.2	38 25.9	9.4	S25 54.8	0.9	55.1
19	108 53.7	18.0	52 54.3	9.4	25 53.9	1.0	55.1
20	123 53.8	18.9	67 22.7	9.5	25 52.9	1.1	55.1
21	138 53.9	. . 19.8	81 51.2	9.5	25 51.8	1.2	55.1
22	153 54.0	20.7	96 19.7	9.6	25 50.6	1.4	55.0
23	168 54.1	21.5	110 48.3	9.6	25 49.2	1.5	55.0
23 00 (Friday)	183 54.2	S11 22.4	125 16.9	9.6	S25 47.7	1.6	55.0
01	198 54.2	23.3	139 45.5	9.7	25 46.1	1.8	55.0
02	213 54.3	24.2	154 14.2	9.8	25 44.3	1.9	55.0
03	228 54.4	. . 25.1	168 43.0	9.7	25 42.4	2.0	54.9
04	243 54.5	25.9	183 11.7	9.9	25 40.4	2.1	54.9
05	258 54.6	26.8	197 40.6	9.8	25 38.3	2.3	54.9
06	273 54.7	S11 27.7	212 09.4	10.0	S25 36.0	2.4	54.9
07	288 54.8	28.6	226 38.4	9.9	25 33.6	2.5	54.9
08	303 54.8	29.4	241 07.3	10.0	25 31.1	2.7	54.8
09	318 54.9	. . 30.3	255 36.3	10.1	25 28.4	2.8	54.8
10	333 55.0	31.2	270 05.4	10.1	25 25.6	2.9	54.8
11	348 55.1	32.1	284 34.5	10.2	25 22.7	3.0	54.8
12	3 55.2	S11 32.9	299 03.7	10.2	S25 19.7	3.1	54.8
13	18 55.3	33.8	313 32.9	10.2	25 16.6	3.3	54.8
14	33 55.4	34.7	328 02.1	10.4	25 13.3	3.4	54.7
15	48 55.4	. . 35.5	342 31.5	10.3	25 09.9	3.5	54.7
16	63 55.5	36.4	357 00.8	10.5	25 06.4	3.6	54.7
17	78 55.6	37.3	11 30.3	10.4	25 02.8	3.8	54.7
18	93 55.7	S11 38.2	25 59.7	10.6	S24 59.0	3.8	54.7
19	108 55.8	39.0	40 29.3	10.6	24 55.2	4.0	54.7
20	123 55.8	39.9	54 58.9	10.6	24 51.2	4.1	54.6
21	138 55.9	. . 40.8	69 28.5	10.7	24 47.1	4.2	54.6
22	153 56.0	41.6	83 58.2	10.8	24 42.9	4.4	54.6
23	168 56.1	42.5	98 28.0	10.8	24 38.5	4.4	54.6
24 00 (Saturday)	183 56.2	S11 43.4	112 57.8	10.9	S24 34.1	4.6	54.6
01	198 56.3	44.3	127 27.7	10.9	24 29.5	4.7	54.6
02	213 56.3	45.1	141 57.6	11.0	24 24.8	4.7	54.6
03	228 56.4	. . 46.0	156 27.6	11.0	24 20.1	4.9	54.5
04	243 56.5	46.9	170 57.6	11.1	24 15.2	5.1	54.5
05	258 56.6	47.7	185 27.7	11.2	24 10.1	5.1	54.5
06	273 56.7	S11 48.6	199 57.9	11.2	S24 05.0	5.2	54.5
07	288 56.7	49.5	214 28.1	11.3	23 59.8	5.4	54.5
08	303 56.8	50.3	228 58.4	11.4	23 54.4	5.4	54.5
09	318 56.9	. . 51.2	243 28.8	11.4	23 49.0	5.6	54.5
10	333 57.0	52.1	257 59.2	11.4	23 43.4	5.7	54.5
11	348 57.0	52.9	272 29.6	11.6	23 37.7	5.7	54.5
12	3 57.1	S11 53.8	287 00.2	11.6	S23 32.0	5.9	54.4
13	18 57.2	54.7	301 30.8	11.6	23 26.1	6.0	54.4
14	33 57.3	55.5	316 01.4	11.7	23 20.1	6.1	54.4
15	48 57.3	. . 56.4	330 32.1	11.8	23 14.0	6.2	54.4
16	63 57.4	57.3	345 02.9	11.8	23 07.8	6.3	54.4
17	78 57.5	58.1	359 33.7	11.9	23 01.5	6.4	54.4
18	93 57.6	S11 59.0	14 04.6	12.0	S22 55.1	6.5	54.4
19	108 57.7	11 59.9	28 35.6	12.0	22 48.6	6.6	54.4
20	123 57.7	12 00.7	43 06.6	12.1	22 42.0	6.7	54.4
21	138 57.8	. . 01.6	57 37.7	12.1	22 35.3	6.8	54.4
22	153 57.9	02.4	72 08.8	12.2	22 28.5	6.9	54.4
23	168 57.9	03.3	86 40.0	12.3	S22 21.6	7.0	54.3
	SD 16.1	*d* 0.9	SD	15.1	14.9		14.8

Lat. °	Twilight Naut. h m	Twilight Civil h m	Sunrise h m	Moonrise 22 h m	Moonrise 23 h m	Moonrise 24 h m	Moonrise 25 h m
N 72	05 30	06 49	08 05	■■	■■	■■	■■
N 70	05 30	06 41	07 48	■■	■■	■■	17 48
68	05 29	06 35	07 35	■■	■■	■■	16 09
66	05 29	06 29	07 24	■■	■■	15 58	15 30
64	05 29	06 24	07 14	■■	15 11	15 06	15 02
62	05 28	06 20	07 06	13 45	14 20	14 35	14 41
60	05 28	06 16	06 59	13 07	13 48	14 11	14 24
N 58	05 27	06 13	06 53	12 41	13 24	13 51	14 09
56	05 26	06 10	06 48	12 20	13 05	13 35	13 57
54	05 26	06 07	06 43	12 03	12 49	13 22	13 46
52	05 25	06 04	06 39	11 48	12 35	13 10	13 36
50	05 24	06 02	06 35	11 35	12 23	12 59	13 27
45	05 22	05 56	06 26	11 09	11 57	12 37	13 09
N 40	05 20	05 51	06 19	10 48	11 37	12 19	12 54
35	05 17	05 47	06 13	10 31	11 21	12 04	12 41
30	05 15	05 43	06 07	10 16	11 06	11 50	12 30
20	05 09	05 35	05 58	09 50	10 41	11 28	12 10
N 10	05 03	05 28	05 49	09 28	10 20	11 08	11 53
0	04 55	05 20	05 41	09 08	10 00	10 50	11 38
S 10	04 46	05 11	05 33	08 48	09 40	10 32	11 22
20	04 35	05 01	05 24	08 26	09 19	10 12	11 05
30	04 20	04 49	05 14	08 00	08 54	09 49	10 45
35	04 10	04 41	05 08	07 46	08 39	09 36	10 34
40	03 58	04 32	05 01	07 28	08 23	09 21	10 20
45	03 44	04 22	04 53	07 08	08 02	09 02	10 05
S 50	03 25	04 08	04 43	06 42	07 37	08 39	09 45
52	03 16	04 02	04 39	06 29	07 24	08 28	09 36
54	03 06	03 55	04 34	06 15	07 10	08 15	09 26
56	02 53	03 47	04 28	05 58	06 54	08 01	09 14
58	02 39	03 38	04 22	05 37	06 34	07 44	09 01
S 60	02 22	03 27	04 15	05 12	06 09	07 23	08 45

Lat. °	Sunset h m	Twilight Civil h m	Twilight Naut. h m	Moonset 22 h m	Moonset 23 h m	Moonset 24 h m	Moonset 25 h m
N 72	15 21	16 38	17 56	■■	■■	■■	■■
N 70	15 39	16 46	17 57	■■	■■	■■	18 04
68	15 53	16 53	17 57	■■	■■	■■	19 42
66	16 04	16 58	17 58	■■	■■	18 15	20 20
64	16 13	17 03	17 59	■■	17 18	19 06	20 47
62	16 21	17 08	17 59	16 57	18 10	19 37	21 08
60	16 28	17 11	18 00	17 34	18 41	20 01	21 24
N 58	16 34	17 15	18 01	18 01	19 05	20 19	21 38
56	16 40	17 18	18 01	18 22	19 24	20 35	21 50
54	16 45	17 21	18 02	18 39	19 40	20 48	22 00
52	16 49	17 24	18 03	18 53	19 53	21 00	22 10
50	16 53	17 26	18 04	19 06	20 05	21 10	22 18
45	17 02	17 32	18 06	19 33	20 30	21 32	22 35
N 40	17 09	17 37	18 08	19 53	20 50	21 49	22 50
35	17 15	17 41	18 11	20 11	21 06	22 04	23 02
30	17 21	17 45	18 13	20 26	21 21	22 16	23 12
20	17 31	17 53	18 19	20 51	21 45	22 38	23 30
N 10	17 39	18 01	18 26	21 13	22 05	22 56	23 45
0	17 48	18 09	18 33	21 33	22 25	23 13	24 00
S 10	17 56	18 18	18 43	21 54	22 44	23 30	24 14
20	18 05	18 28	18 54	22 15	23 04	23 49	24 29
30	18 16	18 40	19 10	22 40	23 28	24 10	00 10
35	18 22	18 48	19 19	22 55	23 42	24 22	00 22
40	18 29	18 57	19 31	23 12	23 58	24 36	00 36
45	18 37	19 08	19 46	23 33	24 17	00 17	00 52
S 50	18 47	19 22	20 05	23 59	24 40	00 40	01 13
52	18 51	19 28	20 14	24 11	00 11	00 52	01 22
54	18 56	19 35	20 25	24 25	00 25	01 05	01 33
56	19 02	19 43	20 38	24 42	00 42	01 19	01 45
58	19 08	19 53	20 52	00 10	01 02	01 37	01 59
S 60	19 15	20 04	21 10	00 35	01 27	01 58	02 16

Day d	SUN Eqn. of Time 00^h m s	SUN Eqn. of Time 12^h m s	SUN Mer. Pass. h m	MOON Mer. Pass. Upper h m	MOON Mer. Pass. Lower h m	MOON Age d	MOON Phase %
22	15 28	15 32	11 44	15 21	02 54	04	19
23	15 36	15 41	11 44	16 12	03 47	05	27
24	15 45	15 48	11 44	17 02	04 37	06	36

	UT	ARIES	VENUS −3.9		MARS +0.5		JUPITER −2.5		SATURN +1.1	
		GHA	GHA	Dec	GHA	Dec	GHA	Dec	GHA	Dec
	d h	° ′	° ′	° ′	° ′	° ′	° ′	° ′	° ′	° ′
	25 00	33 31.0	201 25.3	S 3 30.8	266 48.4	N20 31.8	73 20.4	S16 36.6	213 11.2	N 2 04.1
	01	48 33.5	216 24.9	32.0	281 49.7	31.5	88 22.8	36.6	228 13.4	04.0
	02	63 36.0	231 24.5	33.2	296 50.9	31.3	103 25.2	36.6	243 15.6	03.9
	03	78 38.4	246 24.0	. . 34.5	311 52.2	. . 31.1	118 27.5	. . 36.5	258 17.8	. . 03.8
	04	93 40.9	261 23.6	35.7	326 53.4	30.9	133 29.9	36.5	273 20.0	03.7
	05	108 43.4	276 23.2	36.9	341 54.6	30.7	148 32.3	36.5	288 22.2	03.6
	06	123 45.8	291 22.8	S 3 38.1	356 55.9	N20 30.5	163 34.6	S16 36.4	303 24.4	N 2 03.5
	07	138 48.3	306 22.4	39.3	11 57.1	30.2	178 37.0	36.4	318 26.6	03.4
	08	153 50.7	321 22.0	40.5	26 58.3	30.0	193 39.4	36.3	333 28.8	03.3
S	09	168 53.2	336 21.6	. . 41.8	41 59.6	. . 29.8	208 41.7	. . 36.3	348 31.0	. . 03.2
U	10	183 55.7	351 21.2	43.0	57 00.8	29.6	223 44.1	36.3	3 33.2	03.1
N	11	198 58.1	6 20.8	44.2	72 02.1	29.4	238 46.5	36.2	18 35.4	03.0
D	12	214 00.6	21 20.4	S 3 45.4	87 03.3	N20 29.2	253 48.8	S16 36.2	33 37.7	N 2 02.8
A	13	229 03.1	36 20.0	46.6	102 04.6	29.0	268 51.2	36.2	48 39.9	02.7
Y	14	244 05.5	51 19.6	47.8	117 05.8	28.7	283 53.6	36.1	63 42.1	02.6
	15	259 08.0	66 19.1	. . 49.0	132 07.0	. . 28.5	298 55.9	. . 36.1	78 44.3	. . 02.5
	16	274 10.5	81 18.7	50.3	147 08.3	28.3	313 58.3	36.1	93 46.5	02.4
	17	289 12.9	96 18.3	51.5	162 09.5	28.1	329 00.6	36.0	108 48.7	02.3
	18	304 15.4	111 17.9	S 3 52.7	177 10.8	N20 27.9	344 03.0	S16 36.0	123 50.9	N 2 02.2
	19	319 17.9	126 17.5	53.9	192 12.0	27.7	359 05.4	35.9	138 53.1	02.1
	20	334 20.3	141 17.1	55.1	207 13.3	27.5	14 07.7	35.9	153 55.3	02.0
	21	349 22.8	156 16.7	. . 56.3	222 14.5	. . 27.2	29 10.1	. . 35.9	168 57.5	. . 01.9
	22	4 25.2	171 16.3	57.5	237 15.8	27.0	44 12.5	35.8	183 59.7	01.8
	23	19 27.7	186 15.9	3 58.8	252 17.0	26.8	59 14.8	35.8	199 01.9	01.7
	26 00	34 30.2	201 15.4	S 4 00.0	267 18.3	N20 26.6	74 17.2	S16 35.8	214 04.1	N 2 01.6
	01	49 32.6	216 15.0	01.2	282 19.5	26.4	89 19.5	35.7	229 06.3	01.5
	02	64 35.1	231 14.6	02.4	297 20.8	26.2	104 21.9	35.7	244 08.5	01.4
	03	79 37.6	246 14.2	. . 03.6	312 22.0	. . 26.0	119 24.3	. . 35.7	259 10.7	. . 01.3
	04	94 40.0	261 13.8	04.8	327 23.3	25.7	134 26.6	35.6	274 13.0	01.2
	05	109 42.5	276 13.4	06.0	342 24.5	25.5	149 29.0	35.6	289 15.2	01.1
	06	124 45.0	291 13.0	S 4 07.3	357 25.8	N20 25.3	164 31.3	S16 35.5	304 17.4	N 2 01.0
	07	139 47.4	306 12.6	08.5	12 27.0	25.1	179 33.7	35.5	319 19.6	00.9
	08	154 49.9	321 12.2	09.7	27 28.3	24.9	194 36.0	35.5	334 21.8	00.7
M	09	169 52.4	336 11.7	. . 10.9	42 29.5	. . 24.7	209 38.4	. . 35.4	349 24.0	. . 00.6
O	10	184 54.8	351 11.3	12.1	57 30.8	24.5	224 40.8	35.4	4 26.2	00.5
N	11	199 57.3	6 10.9	13.3	72 32.0	24.2	239 43.1	35.3	19 28.4	00.4
D	12	214 59.7	21 10.5	S 4 14.5	87 33.3	N20 24.0	254 45.5	S16 35.3	34 30.6	N 2 00.3
A	13	230 02.2	36 10.1	15.7	102 34.5	23.8	269 47.8	35.3	49 32.8	00.2
Y	14	245 04.7	51 09.7	17.0	117 35.8	23.6	284 50.2	35.2	64 35.0	00.1
	15	260 07.1	66 09.3	. . 18.2	132 37.1	. . 23.4	299 52.5	. . 35.2	79 37.2	2 00.0
	16	275 09.6	81 08.8	19.4	147 38.3	23.2	314 54.9	35.2	94 39.4	1 59.9
	17	290 12.1	96 08.4	20.6	162 39.6	23.0	329 57.2	35.1	109 41.6	59.8
	18	305 14.5	111 08.0	S 4 21.8	177 40.8	N20 22.7	344 59.6	S16 35.1	124 43.9	N 1 59.7
	19	320 17.0	126 07.6	23.0	192 42.1	22.5	0 02.0	35.0	139 46.1	59.6
	20	335 19.5	141 07.2	24.2	207 43.3	22.3	15 04.3	35.0	154 48.3	59.5
	21	350 21.9	156 06.8	. . 25.4	222 44.6	. . 22.1	30 06.7	. . 35.0	169 50.5	. . 59.4
	22	5 24.4	171 06.4	26.7	237 45.9	21.9	45 09.0	34.9	184 52.7	59.3
	23	20 26.8	186 05.9	27.9	252 47.1	21.7	60 11.4	34.9	199 54.9	59.2
	27 00	35 29.3	201 05.5	S 4 29.1	267 48.4	N20 21.4	75 13.7	S16 34.8	214 57.1	N 1 59.1
	01	50 31.8	216 05.1	30.3	282 49.7	21.2	90 16.1	34.8	229 59.3	59.0
	02	65 34.2	231 04.7	31.5	297 50.9	21.0	105 18.4	34.8	245 01.5	58.9
	03	80 36.7	246 04.3	. . 32.7	312 52.2	. . 20.8	120 20.8	. . 34.7	260 03.7	. . 58.8
	04	95 39.2	261 03.9	33.9	327 53.4	20.6	135 23.1	34.7	275 05.9	58.7
	05	110 41.6	276 03.4	35.1	342 54.7	20.4	150 25.5	34.6	290 08.1	58.6
	06	125 44.1	291 03.0	S 4 36.3	357 56.0	N20 20.2	165 27.8	S16 34.6	305 10.3	N 1 58.5
	07	140 46.6	306 02.6	37.5	12 57.2	19.9	180 30.2	34.6	320 12.6	58.4
T	08	155 49.0	321 02.2	38.8	27 58.5	19.7	195 32.5	34.5	335 14.8	58.2
U	09	170 51.5	336 01.8	. . 40.0	42 59.8	. . 19.5	210 34.9	. . 34.5	350 17.0	. . 58.1
E	10	185 54.0	351 01.4	41.2	58 01.0	19.3	225 37.2	34.4	5 19.2	58.0
S	11	200 56.4	6 00.9	42.4	73 02.3	19.1	240 39.6	34.4	20 21.4	57.9
D	12	215 58.9	21 00.5	S 4 43.6	88 03.6	N20 18.9	255 41.9	S16 34.4	35 23.6	N 1 57.8
A	13	231 01.3	36 00.1	44.8	103 04.8	18.7	270 44.3	34.3	50 25.8	57.7
Y	14	246 03.8	50 59.7	46.0	118 06.1	18.4	285 46.6	34.3	65 28.0	57.6
	15	261 06.3	65 59.3	. . 47.2	133 07.4	. . 18.2	300 49.0	. . 34.2	80 30.2	. . 57.5
	16	276 08.7	80 58.9	48.4	148 08.6	18.0	315 51.3	34.2	95 32.4	57.4
	17	291 11.2	95 58.4	49.6	163 09.9	17.8	330 53.6	34.2	110 34.6	57.3
	18	306 13.7	110 58.0	S 4 50.9	178 11.2	N20 17.6	345 56.0	S16 34.1	125 36.9	N 1 57.2
	19	321 16.1	125 57.6	52.1	193 12.5	17.4	0 58.3	34.1	140 39.1	57.1
	20	336 18.6	140 57.2	53.3	208 13.7	17.2	16 00.7	34.0	155 41.3	57.0
	21	351 21.1	155 56.8	. . 54.5	223 15.0	. . 17.0	31 03.0	. . 34.0	170 43.5	. . 56.9
	22	6 23.5	170 56.3	55.7	238 16.3	16.7	46 05.4	33.9	185 45.7	56.8
	23	21 26.0	185 55.9	56.9	253 17.6	16.5	61 07.7	33.9	200 47.9	56.7
	Mer. Pass.	h m 21 38.4	*v* −0.4	*d* 1.2	*v* 1.3	*d* 0.2	*v* 2.4	*d* 0.0	*v* 2.2	*d* 0.1

STARS		
Name	SHA	Dec
	° ′	° ′
Acamar	315 19.9	S40 15.7
Achernar	335 28.0	S57 11.1
Acrux	173 13.2	S63 09.1
Adhara	255 14.5	S28 58.9
Aldebaran	290 52.2	N16 31.9
Alioth	166 23.4	N55 54.3
Alkaid	153 01.3	N49 15.8
Al Na'ir	27 46.7	S46 54.9
Alnilam	275 48.9	S 1 11.6
Alphard	217 58.8	S 8 42.0
Alphecca	126 13.6	N26 41.0
Alpheratz	357 46.1	N29 09.0
Altair	62 10.9	N 8 53.8
Ankaa	353 17.8	S42 15.1
Antares	112 29.9	S26 27.3
Arcturus	145 58.5	N19 07.9
Atria	107 34.7	S69 02.9
Avior	234 19.3	S59 32.1
Bellatrix	278 34.7	N 6 21.6
Betelgeuse	271 04.0	N 7 24.6
Canopus	263 57.1	S52 41.7
Capella	280 38.1	N46 00.4
Deneb	49 33.4	N45 19.3
Denebola	182 36.6	N14 31.0
Diphda	348 58.2	S17 55.8
Dubhe	193 55.2	N61 41.6
Elnath	278 15.8	N28 37.0
Eltanin	90 47.7	N51 29.5
Enif	33 49.7	N 9 55.4
Fomalhaut	15 26.6	S29 34.2
Gacrux	172 04.7	S57 10.0
Gienah	175 55.4	S17 35.7
Hadar	148 52.6	S60 25.2
Hamal	328 03.5	N23 30.7
Kaus Aust.	83 47.6	S34 22.9
Kochab	137 20.7	N74 06.9
Markab	13 40.8	N15 15.7
Menkar	314 17.6	N 4 07.9
Menkent	148 11.2	S36 25.1
Miaplacidus	221 40.7	S69 45.2
Mirfak	308 43.8	N49 53.8
Nunki	76 01.8	S26 17.1
Peacock	53 23.3	S56 42.4
Pollux	243 30.9	N28 00.1
Procyon	245 02.5	N 5 12.1
Rasalhague	96 09.2	N12 33.3
Regulus	207 46.5	N11 55.1
Rigel	281 14.4	S 8 11.2
Rigil Kent.	139 56.3	S60 52.6
Sabik	102 15.9	S15 44.2
Schedar	349 43.3	N56 35.7
Shaula	96 25.9	S37 06.7
Sirius	258 35.9	S16 43.6
Spica	158 34.4	S11 12.7
Suhail	222 54.6	S43 28.1
Vega	80 41.0	N38 47.8
Zuben'ubi	137 08.8	S16 04.9

	SHA	Mer. Pass.
	° ′	h m
Venus	166 45.3	10 35
Mars	232 48.1	6 10
Jupiter	39 47.0	19 00
Saturn	179 34.0	9 42

UT	SUN GHA	SUN Dec	MOON GHA	*v*	MOON Dec	*d*	HP
d h	° ′	° ′	° ′	′	° ′	′	′
25 00	183 58.0	S12 04.2	101 11.3	12.3	S22 14.6	7.1	54.3
01	198 58.1	05.0	115 42.6	12.4	22 07.5	7.2	54.3
02	213 58.2	05.9	130 14.0	12.5	22 00.3	7.3	54.3
03	228 58.2	. . 06.8	144 45.5	12.5	21 53.0	7.4	54.3
04	243 58.3	07.6	159 17.0	12.6	21 45.6	7.5	54.3
05	258 58.4	08.5	173 48.6	12.6	21 38.1	7.6	54.3
06	273 58.5	S12 09.3	188 20.2	12.7	S21 30.5	7.6	54.3
07	288 58.5	10.2	202 51.9	12.8	21 22.9	7.8	54.3
S 08	303 58.6	11.1	217 23.7	12.8	21 15.1	7.8	54.3
U 09	318 58.7	. . 11.9	231 55.5	12.8	21 07.3	8.0	54.3
N 10	333 58.7	12.8	246 27.3	13.0	20 59.3	8.0	54.3
D 11	348 58.8	13.6	260 59.3	13.0	20 51.3	8.1	54.3
A 12	3 58.9	S12 14.5	275 31.3	13.0	S20 43.2	8.2	54.3
Y 13	18 59.0	15.3	290 03.3	13.1	20 35.0	8.3	54.3
14	33 59.0	16.2	304 35.4	13.2	20 26.7	8.4	54.3
15	48 59.1	. . 17.1	319 07.6	13.2	20 18.3	8.4	54.3
16	63 59.2	17.9	333 39.8	13.3	20 09.9	8.6	54.3
17	78 59.2	18.8	348 12.1	13.4	20 01.3	8.6	54.3
18	93 59.3	S12 19.6	2 44.5	13.4	S19 52.7	8.7	54.3
19	108 59.4	20.5	17 16.9	13.4	19 44.0	8.8	54.3
20	123 59.4	21.3	31 49.3	13.5	19 35.2	8.9	54.3
21	138 59.5	. . 22.2	46 21.8	13.6	19 26.3	9.0	54.3
22	153 59.6	23.1	60 54.4	13.6	19 17.3	9.0	54.3
23	168 59.6	23.9	75 27.0	13.7	19 08.3	9.2	54.3
26 00	183 59.7	S12 24.8	89 59.7	13.7	S18 59.1	9.2	54.3
01	198 59.8	25.6	104 32.4	13.8	18 49.9	9.2	54.3
02	213 59.8	26.5	119 05.2	13.9	18 40.7	9.4	54.3
03	228 59.9	. . 27.3	133 38.1	13.9	18 31.3	9.4	54.3
04	244 00.0	28.2	148 11.0	13.9	18 21.9	9.5	54.3
05	259 00.0	29.0	162 43.9	14.0	18 12.4	9.6	54.3
06	274 00.1	S12 29.9	177 16.9	14.1	S18 02.8	9.7	54.3
07	289 00.2	30.7	191 50.0	14.1	17 53.1	9.7	54.3
08	304 00.2	31.6	206 23.1	14.1	17 43.4	9.9	54.3
M 09	319 00.3	. . 32.4	220 56.2	14.2	17 33.5	9.8	54.3
O 10	334 00.3	33.3	235 29.4	14.3	17 23.7	10.0	54.3
N 11	349 00.4	34.1	250 02.7	14.3	17 13.7	10.0	54.3
D 12	4 00.5	S12 35.0	264 36.0	14.3	S17 03.7	10.1	54.3
A 13	19 00.5	35.8	279 09.3	14.4	16 53.6	10.2	54.3
Y 14	34 00.6	36.7	293 42.7	14.4	16 43.4	10.2	54.3
15	49 00.7	. . 37.5	308 16.1	14.5	16 33.2	10.3	54.3
16	64 00.7	38.4	322 49.6	14.5	16 22.9	10.4	54.3
17	79 00.8	39.2	337 23.1	14.6	16 12.5	10.5	54.3
18	94 00.8	S12 40.1	351 56.7	14.6	S16 02.0	10.5	54.3
19	109 00.9	40.9	6 30.3	14.7	15 51.5	10.5	54.3
20	124 01.0	41.8	21 04.0	14.7	15 41.0	10.7	54.3
21	139 01.0	. . 42.6	35 37.7	14.7	15 30.3	10.7	54.3
22	154 01.1	43.5	50 11.4	14.8	15 19.6	10.8	54.3
23	169 01.1	44.3	64 45.2	14.9	15 08.8	10.8	54.3
27 00	184 01.2	S12 45.2	79 19.1	14.8	S14 58.0	10.9	54.3
01	199 01.3	46.0	93 52.9	14.9	14 47.1	11.0	54.4
02	214 01.3	46.9	108 26.8	15.0	14 36.1	11.0	54.4
03	229 01.4	. . 47.7	123 00.8	15.0	14 25.1	11.1	54.4
04	244 01.4	48.5	137 34.8	15.0	14 14.0	11.1	54.4
05	259 01.5	49.4	152 08.8	15.0	14 02.9	11.2	54.4
06	274 01.5	S12 50.2	166 42.8	15.1	S13 51.7	11.3	54.4
07	289 01.6	51.1	181 16.9	15.1	13 40.4	11.3	54.4
T 08	304 01.7	51.9	195 51.0	15.2	13 29.1	11.3	54.4
U 09	319 01.7	. . 52.8	210 25.2	15.2	13 17.8	11.5	54.4
E 10	334 01.8	53.6	224 59.4	15.2	13 06.3	11.5	54.4
S 11	349 01.8	54.5	239 33.6	15.2	12 54.8	11.5	54.4
D 12	4 01.9	S12 55.3	254 07.8	15.3	S12 43.3	11.6	54.5
A 13	19 01.9	56.1	268 42.1	15.3	12 31.7	11.7	54.5
Y 14	34 02.0	57.0	283 16.4	15.4	12 20.0	11.7	54.5
15	49 02.0	. . 57.8	297 50.8	15.3	12 08.3	11.7	54.5
16	64 02.1	58.7	312 25.1	15.4	11 56.6	11.8	54.5
17	79 02.1	12 59.5	326 59.5	15.4	11 44.8	11.9	54.5
18	94 02.2	S13 00.3	341 33.9	15.5	S11 32.9	11.9	54.5
19	109 02.2	01.2	356 08.4	15.4	11 21.0	12.0	54.5
20	124 02.3	02.0	10 42.8	15.5	11 09.0	12.0	54.6
21	139 02.4	. . 02.9	25 17.3	15.5	10 57.0	12.0	54.6
22	154 02.4	03.7	39 51.8	15.6	10 45.0	12.2	54.6
23	169 02.5	04.5	54 26.4	15.5	S10 32.8	12.1	54.6
	SD 16.1	*d* 0.9	SD 14.8		14.8		14.8

Lat.	Twilight Naut.	Twilight Civil	Sunrise	Moonrise 25	26	27	28
°	h m	h m	h m	h m	h m	h m	h m
N 72	05 42	07 03	08 22	■	16 48	15 55	15 22
N 70	05 41	06 53	08 02	17 48	16 06	15 34	15 12
68	05 39	06 45	07 47	16 09	15 37	15 18	15 03
66	05 38	06 38	07 34	15 30	15 16	15 05	14 56
64	05 37	06 33	07 24	15 02	14 58	14 54	14 50
62	05 36	06 28	07 15	14 41	14 44	14 45	14 45
60	05 34	06 23	07 07	14 24	14 32	14 37	14 41
N 58	05 33	06 19	07 00	14 09	14 21	14 30	14 37
56	05 32	06 16	06 54	13 57	14 12	14 23	14 33
54	05 31	06 12	06 49	13 46	14 04	14 18	14 30
52	05 30	06 09	06 44	13 36	13 56	14 13	14 27
50	05 29	06 06	06 40	13 27	13 50	14 08	14 24
45	05 26	06 00	06 30	13 09	13 35	13 58	14 18
N 40	05 23	05 55	06 22	12 54	13 23	13 50	14 13
35	05 20	05 50	06 16	12 41	13 13	13 42	14 09
30	05 17	05 45	06 09	12 30	13 04	13 36	14 05
20	05 10	05 36	05 59	12 10	12 49	13 25	13 59
N 10	05 03	05 28	05 49	11 53	12 35	13 15	13 53
0	04 55	05 19	05 41	11 38	12 23	13 06	13 48
S 10	04 45	05 10	05 32	11 22	12 10	12 56	13 42
20	04 32	04 59	05 22	11 05	11 56	12 47	13 36
30	04 16	04 46	05 11	10 45	11 41	12 35	13 30
35	04 06	04 38	05 04	10 34	11 31	12 29	13 26
40	03 54	04 28	04 57	10 20	11 21	12 21	13 21
45	03 38	04 16	04 48	10 05	11 08	12 12	13 16
S 50	03 18	04 02	04 37	09 45	10 53	12 02	13 10
52	03 08	03 55	04 32	09 36	10 46	11 57	13 07
54	02 57	03 48	04 27	09 26	10 38	11 51	13 04
56	02 44	03 39	04 21	09 14	10 30	11 45	13 01
58	02 28	03 29	04 14	09 01	10 20	11 38	12 57
S 60	02 08	03 17	04 06	08 45	10 08	11 31	12 53

Lat.	Sunset	Twilight Civil	Twilight Naut.	Moonset 25	26	27	28
°	h m	h m	h m	h m	h m	h m	h m
N 72	15 04	16 24	17 43	■	20 38	23 01	25 02
N 70	15 24	16 33	17 45	18 04	21 19	23 19	25 09
68	15 40	16 41	17 47	19 42	21 46	23 34	25 15
66	15 53	16 48	17 48	20 20	22 06	23 45	25 21
64	16 03	16 54	17 50	20 47	22 23	23 55	25 25
62	16 12	16 59	17 51	21 08	22 36	24 03	00 03
60	16 20	17 04	17 52	21 24	22 48	24 10	00 10
N 58	16 27	17 08	17 54	21 38	22 57	24 16	00 16
56	16 33	17 11	17 55	21 50	23 06	24 21	00 21
54	16 38	17 15	17 56	22 00	23 13	24 26	00 26
52	16 43	17 18	17 57	22 10	23 20	24 31	00 31
50	16 47	17 21	17 59	22 18	23 26	24 35	00 35
45	16 57	17 27	18 02	22 35	23 39	24 43	00 43
N 40	17 05	17 33	18 04	22 50	23 50	24 50	00 50
35	17 12	17 38	18 08	23 02	23 59	24 56	00 56
30	17 18	17 43	18 11	23 12	24 07	00 07	01 02
20	17 29	17 51	18 17	23 30	24 21	00 21	01 11
N 10	17 38	18 00	18 25	23 45	24 33	00 33	01 19
0	17 47	18 09	18 33	24 00	00 00	00 44	01 26
S 10	17 57	18 18	18 43	24 14	00 14	00 55	01 33
20	18 06	18 29	18 56	24 29	00 29	01 06	01 41
30	18 18	18 43	19 12	00 10	00 46	01 19	01 50
35	18 24	18 51	19 23	00 22	00 57	01 27	01 55
40	18 32	19 01	19 35	00 36	01 08	01 36	02 01
45	18 41	19 12	19 51	00 52	01 21	01 46	02 07
S 50	18 52	19 27	20 11	01 13	01 38	01 58	02 15
52	18 57	19 34	20 22	01 22	01 45	02 03	02 19
54	19 02	19 42	20 33	01 33	01 54	02 10	02 23
56	19 08	19 51	20 47	01 45	02 03	02 16	02 27
58	19 15	20 01	21 03	01 59	02 14	02 24	02 32
S 60	19 23	20 13	21 23	02 16	02 26	02 33	02 37

Day	SUN Eqn. of Time 00^h	SUN Eqn. of Time 12^h	SUN Mer. Pass.	MOON Mer. Pass. Upper	MOON Mer. Pass. Lower	Age	Phase
d	m s	m s	h m	h m	h m	d %	
25	15 52	15 55	11 44	17 49	05 26	07 45	◐
26	15 59	16 02	11 44	18 33	06 11	08 55	
27	16 05	16 07	11 44	19 16	06 55	09 64	

Day	UT d h	ARIES GHA ° ′	VENUS −3.9 GHA ° ′	VENUS Dec ° ′	MARS +0.5 GHA ° ′	MARS Dec ° ′	JUPITER −2.5 GHA ° ′	JUPITER Dec ° ′	SATURN +1.1 GHA ° ′	SATURN Dec ° ′	STARS Name	SHA ° ′	Dec ° ′
	28 00	36 28.5	200 55.5	S 4 58.1	268 18.8	N20 16.3	76 10.1	S16 33.9	215 50.1	N 1 56.6	Acamar	315 19.9	S40 15.7
	01	51 30.9	215 55.1	4 59.3	283 20.1	16.1	91 12.4	33.8	230 52.3	56.5	Achernar	335 28.0	S57 11.1
	02	66 33.4	230 54.7	5 00.5	298 21.4	15.9	106 14.7	33.8	245 54.5	56.4	Acrux	173 13.2	S63 09.1
	03	81 35.8	245 54.2	. . 01.7	313 22.7	. . 15.7	121 17.1	. . 33.7	260 56.7	. . 56.3	Adhara	255 14.5	S28 58.9
	04	96 38.3	260 53.8	02.9	328 23.9	15.5	136 19.4	33.7	275 59.0	56.2	Aldebaran	290 52.2	N16 31.9
	05	111 40.8	275 53.4	04.1	343 25.2	15.2	151 21.8	33.6	291 01.2	56.1			
	06	126 43.2	290 53.0	S 5 05.3	358 26.5	N20 15.0	166 24.1	S16 33.6	306 03.4	N 1 56.0	Alioth	166 23.4	N55 54.2
W	07	141 45.7	305 52.6	06.6	13 27.8	14.8	181 26.5	33.6	321 05.6	55.9	Alkaid	153 01.3	N49 15.8
E	08	156 48.2	320 52.1	07.8	28 29.0	14.6	196 28.8	33.5	336 07.8	55.8	Al Na'ir	27 46.8	S46 54.9
D	09	171 50.6	335 51.7	. . 09.0	43 30.3	. . 14.4	211 31.1	. . 33.5	351 10.0	. . 55.7	Alnilam	275 48.9	S 1 11.6
N	10	186 53.1	350 51.3	10.2	58 31.6	14.2	226 33.5	33.4	6 12.2	55.6	Alphard	217 58.8	S 8 42.0
E	11	201 55.6	5 50.9	11.4	73 32.9	14.0	241 35.8	33.4	21 14.4	55.5			
S	12	216 58.0	20 50.4	S 5 12.6	88 34.2	N20 13.7	256 38.2	S16 33.3	36 16.6	N 1 55.4	Alphecca	126 13.6	N26 40.9
D	13	232 00.5	35 50.0	13.8	103 35.5	13.5	271 40.5	33.3	51 18.8	55.3	Alpheratz	357 46.1	N29 09.0
A	14	247 02.9	50 49.6	15.0	118 36.7	13.3	286 42.8	33.3	66 21.1	55.1	Altair	62 10.9	N 8 53.8
Y	15	262 05.4	65 49.2	. . 16.2	133 38.0	. . 13.1	301 45.2	. . 33.2	81 23.3	. . 55.0	Ankaa	353 17.8	S42 15.1
	16	277 07.9	80 48.8	17.4	148 39.3	12.9	316 47.5	33.2	96 25.5	54.9	Antares	112 29.9	S26 27.3
	17	292 10.3	95 48.3	18.6	163 40.6	12.7	331 49.9	33.1	111 27.7	54.8			
	18	307 12.8	110 47.9	S 5 19.8	178 41.9	N20 12.5	346 52.2	S16 33.1	126 29.9	N 1 54.7	Arcturus	145 58.5	N19 07.9
	19	322 15.3	125 47.5	21.0	193 43.2	12.2	1 54.5	33.0	141 32.1	54.6	Atria	107 34.7	S69 02.9
	20	337 17.7	140 47.1	22.2	208 44.4	12.0	16 56.9	33.0	156 34.3	54.5	Avior	234 19.2	S59 32.1
	21	352 20.2	155 46.6	. . 23.4	223 45.7	. . 11.8	31 59.2	. . 32.9	171 36.5	. . 54.4	Bellatrix	278 34.7	N 6 21.6
	22	7 22.7	170 46.2	24.6	238 47.0	11.6	47 01.5	32.9	186 38.7	54.3	Betelgeuse	271 04.0	N 7 24.6
	23	22 25.1	185 45.8	25.8	253 48.3	11.4	62 03.9	32.9	201 41.0	54.2			
	29 00	37 27.6	200 45.4	S 5 27.0	268 49.6	N20 11.2	77 06.2	S16 32.8	216 43.2	N 1 54.1	Canopus	263 57.1	S52 41.7
	01	52 30.1	215 44.9	28.2	283 50.9	11.0	92 08.6	32.8	231 45.4	54.0	Capella	280 38.1	N46 00.4
	02	67 32.5	230 44.5	29.4	298 52.2	10.8	107 10.9	32.7	246 47.6	53.9	Deneb	49 33.4	N45 19.3
	03	82 35.0	245 44.1	. . 30.7	313 53.5	. . 10.5	122 13.2	. . 32.7	261 49.8	. . 53.8	Denebola	182 36.6	N14 31.0
	04	97 37.4	260 43.7	31.9	328 54.7	10.3	137 15.6	32.6	276 52.0	53.7	Diphda	348 58.2	S17 55.8
	05	112 39.9	275 43.2	33.1	343 56.0	10.1	152 17.9	32.6	291 54.2	53.6			
	06	127 42.4	290 42.8	S 5 34.3	358 57.3	N20 09.9	167 20.2	S16 32.5	306 56.4	N 1 53.5	Dubhe	193 55.1	N61 41.6
T	07	142 44.8	305 42.4	35.5	13 58.6	09.7	182 22.6	32.5	321 58.7	53.4	Elnath	278 15.8	N28 37.0
H	08	157 47.3	320 41.9	36.7	28 59.9	09.5	197 24.9	32.5	337 00.9	53.3	Eltanin	90 47.7	N51 29.5
U	09	172 49.8	335 41.5	. . 37.9	44 01.2	. . 09.3	212 27.2	. . 32.4	352 03.1	. . 53.2	Enif	33 49.7	N 9 55.4
R	10	187 52.2	350 41.1	39.1	59 02.5	09.0	227 29.6	32.4	7 05.3	53.1	Fomalhaut	15 26.6	S29 34.2
S	11	202 54.7	5 40.7	40.3	74 03.8	08.8	242 31.9	32.3	22 07.5	53.0			
D	12	217 57.2	20 40.2	S 5 41.5	89 05.1	N20 08.6	257 34.2	S16 32.3	37 09.7	N 1 52.9	Gacrux	172 04.6	S57 10.0
A	13	232 59.6	35 39.8	42.7	104 06.4	08.4	272 36.6	32.2	52 11.9	52.8	Gienah	175 55.4	S17 35.7
Y	14	248 02.1	50 39.4	43.9	119 07.7	08.2	287 38.9	32.2	67 14.1	52.7	Hadar	148 52.6	S60 25.2
	15	263 04.5	65 39.0	. . 45.1	134 09.0	. . 08.0	302 41.2	. . 32.1	82 16.4	. . 52.6	Hamal	328 03.5	N23 30.7
	16	278 07.0	80 38.5	46.3	149 10.3	07.8	317 43.5	32.1	97 18.6	52.5	Kaus Aust.	83 47.6	S34 22.9
	17	293 09.5	95 38.1	47.5	164 11.6	07.6	332 45.9	32.0	112 20.8	52.4			
	18	308 11.9	110 37.7	S 5 48.7	179 12.9	N20 07.3	347 48.2	S16 32.0	127 23.0	N 1 52.3	Kochab	137 20.7	N74 06.8
	19	323 14.4	125 37.2	49.9	194 14.2	07.1	2 50.5	31.9	142 25.2	52.2	Markab	13 40.8	N15 15.7
	20	338 16.9	140 36.8	51.1	209 15.5	06.9	17 52.9	31.9	157 27.4	52.1	Menkar	314 17.5	N 4 07.9
	21	353 19.3	155 36.4	. . 52.3	224 16.8	. . 06.7	32 55.2	. . 31.9	172 29.6	. . 52.0	Menkent	148 11.2	S36 25.1
	22	8 21.8	170 35.9	53.5	239 18.1	06.5	47 57.5	31.8	187 31.8	51.9	Miaplacidus	221 40.7	S69 45.1
	23	23 24.3	185 35.5	54.7	254 19.4	06.3	62 59.9	31.8	202 34.1	51.8			
	30 00	38 26.7	200 35.1	S 5 55.9	269 20.7	N20 06.1	78 02.2	S16 31.7	217 36.3	N 1 51.7	Mirfak	308 43.8	N49 53.9
	01	53 29.2	215 34.7	57.1	284 22.0	05.9	93 04.5	31.7	232 38.5	51.6	Nunki	76 01.8	S26 17.1
	02	68 31.7	230 34.2	58.3	299 23.3	05.6	108 06.8	31.6	247 40.7	51.5	Peacock	53 23.3	S56 42.4
	03	83 34.1	245 33.8	5 59.5	314 24.6	. . 05.4	123 09.2	. . 31.6	262 42.9	. . 51.4	Pollux	243 30.9	N28 00.1
	04	98 36.6	260 33.4	6 00.7	329 25.9	05.2	138 11.5	31.5	277 45.1	51.3	Procyon	245 02.5	N 5 12.1
	05	113 39.0	275 32.9	01.9	344 27.2	05.0	153 13.8	31.5	292 47.3	51.2			
	06	128 41.5	290 32.5	S 6 03.1	359 28.5	N20 04.8	168 16.1	S16 31.4	307 49.5	N 1 51.1	Rasalhague	96 09.2	N12 33.3
	07	143 44.0	305 32.1	04.3	14 29.8	04.6	183 18.5	31.4	322 51.8	51.0	Regulus	207 46.5	N11 55.1
	08	158 46.4	320 31.6	05.5	29 31.1	04.4	198 20.8	31.3	337 54.0	50.9	Rigel	281 14.4	S 8 11.2
F	09	173 48.9	335 31.2	. . 06.7	44 32.4	. . 04.2	213 23.1	. . 31.3	352 56.2	. . 50.8	Rigil Kent.	139 56.3	S60 52.5
R	10	188 51.4	350 30.8	07.9	59 33.7	03.9	228 25.4	31.2	7 58.4	50.7	Sabik	102 15.9	S15 44.2
I	11	203 53.8	5 30.3	09.1	74 35.0	03.7	243 27.8	31.2	23 00.6	50.6			
D	12	218 56.3	20 29.9	S 6 10.3	89 36.3	N20 03.5	258 30.1	S16 31.1	38 02.8	N 1 50.5	Schedar	349 43.3	N56 35.8
A	13	233 58.8	35 29.5	11.5	104 37.6	03.3	273 32.4	31.1	53 05.0	50.4	Shaula	96 25.9	S37 06.7
Y	14	249 01.2	50 29.0	12.7	119 38.9	03.1	288 34.7	31.0	68 07.3	50.3	Sirius	258 35.9	S16 43.6
	15	264 03.7	65 28.6	. . 13.9	134 40.3	. . 02.9	303 37.1	. . 31.0	83 09.5	. . 50.2	Spica	158 34.4	S11 12.7
	16	279 06.1	80 28.2	15.0	149 41.6	02.7	318 39.4	30.9	98 11.7	50.1	Suhail	222 54.6	S43 28.1
	17	294 08.6	95 27.7	16.2	164 42.9	02.5	333 41.7	30.9	113 13.9	50.0			
	18	309 11.1	110 27.3	S 6 17.4	179 44.2	N20 02.2	348 44.0	S16 30.8	128 16.1	N 1 49.9	Vega	80 41.0	N38 47.8
	19	324 13.5	125 26.9	18.6	194 45.5	02.0	3 46.4	30.8	143 18.3	49.8	Zuben'ubi	137 08.8	S16 04.9
	20	339 16.0	140 26.4	19.8	209 46.8	01.8	18 48.7	30.7	158 20.5	49.7		SHA	Mer. Pass
	21	354 18.5	155 26.0	. . 21.0	224 48.1	. . 01.6	33 51.0	. . 30.7	173 22.8	. . 49.6		° ′	h m
	22	9 20.9	170 25.6	22.2	239 49.4	01.4	48 53.3	30.7	188 25.0	49.5	Venus	163 17.8	10 37
	23	24 23.4	185 25.1	23.4	254 50.8	01.2	63 55.6	30.6	203 27.2	49.4	Mars	231 22.0	6 04
	Mer. Pass. h m	21 26.6	*v* −0.4	*d* 1.2	*v* 1.3	*d* 0.2	*v* 2.3	*d* 0.0	*v* 2.2	*d* 0.1	Jupiter	39 38.6	18 49
											Saturn	179 15.6	9 32

UT	SUN GHA	SUN Dec	MOON GHA	v	MOON Dec	d	HP
d h	° ′	° ′	° ′	′	° ′	′	′
28 00	184 02.5	S13 05.4	69 00.9	15.6	S10 20.7	12.2	54.6
01	199 02.6	06.2	83 35.5	15.6	10 08.5	12.3	54.6
02	214 02.6	07.0	98 10.1	15.6	9 56.2	12.3	54.6
03	229 02.7	. . 07.9	112 44.7	15.6	9 43.9	12.3	54.7
04	244 02.7	08.7	127 19.3	15.7	9 31.6	12.4	54.7
05	259 02.8	09.6	141 54.0	15.6	9 19.2	12.4	54.7
06	274 02.8	S13 10.4	156 28.6	15.7	S 9 06.8	12.5	54.7
W 07	289 02.9	11.2	171 03.3	15.7	8 54.3	12.5	54.7
E 08	304 02.9	12.1	185 38.0	15.7	8 41.8	12.6	54.7
D 09	319 03.0	. . 12.9	200 12.7	15.7	8 29.2	12.6	54.7
N 10	334 03.0	13.7	214 47.4	15.7	8 16.6	12.6	54.8
E 11	349 03.1	14.6	229 22.1	15.8	8 04.0	12.7	54.8
S 12	4 03.1	S13 15.4	243 56.9	15.7	S 7 51.3	12.7	54.8
D 13	19 03.1	16.2	258 31.6	15.7	7 38.6	12.7	54.8
A 14	34 03.2	17.1	273 06.3	15.8	7 25.9	12.8	54.8
Y 15	49 03.2	. . 17.9	287 41.1	15.8	7 13.1	12.9	54.9
16	64 03.3	18.7	302 15.9	15.7	7 00.2	12.8	54.9
17	79 03.3	19.6	316 50.6	15.8	6 47.4	12.9	54.9
18	94 03.4	S13 20.4	331 25.4	15.8	S 6 34.5	13.0	54.9
19	109 03.4	21.2	346 00.2	15.7	6 21.5	12.9	54.9
20	124 03.5	22.1	0 34.9	15.8	6 08.6	13.0	54.9
21	139 03.5	. . 22.9	15 09.7	15.8	5 55.6	13.1	55.0
22	154 03.6	23.7	29 44.5	15.7	5 42.5	13.1	55.0
23	169 03.6	24.5	44 19.2	15.8	5 29.4	13.1	55.0
29 00	184 03.6	S13 25.4	58 54.0	15.8	S 5 16.3	13.1	55.0
01	199 03.7	26.2	73 28.8	15.7	5 03.2	13.1	55.0
02	214 03.7	27.0	88 03.5	15.8	4 50.1	13.2	55.1
03	229 03.8	. . 27.9	102 38.3	15.8	4 36.9	13.3	55.1
04	244 03.8	28.7	117 13.1	15.7	4 23.6	13.2	55.1
05	259 03.9	29.5	131 47.8	15.7	4 10.4	13.3	55.1
06	274 03.9	S13 30.3	146 22.5	15.8	S 3 57.1	13.3	55.2
07	289 03.9	31.2	160 57.3	15.7	3 43.8	13.3	55.2
T 08	304 04.0	32.0	175 32.0	15.7	3 30.5	13.4	55.2
H 09	319 04.0	. . 32.8	190 06.7	15.7	3 17.1	13.4	55.2
U 10	334 04.1	33.6	204 41.4	15.6	3 03.7	13.4	55.2
R 11	349 04.1	34.5	219 16.0	15.7	2 50.3	13.4	55.3
S 12	4 04.1	S13 35.3	233 50.7	15.7	S 2 36.9	13.4	55.3
D 13	19 04.2	36.1	248 25.4	15.6	2 23.5	13.5	55.3
A 14	34 04.2	36.9	263 00.0	15.6	2 10.0	13.5	55.3
Y 15	49 04.3	. . 37.8	277 34.6	15.6	1 56.5	13.5	55.4
16	64 04.3	38.6	292 09.2	15.6	1 43.0	13.5	55.4
17	79 04.3	39.4	306 43.8	15.5	1 29.5	13.6	55.4
18	94 04.4	S13 40.2	321 18.3	15.6	S 1 15.9	13.5	55.4
19	109 04.4	41.1	335 52.9	15.5	1 02.4	13.6	55.4
20	124 04.4	41.9	350 27.4	15.5	0 48.8	13.6	55.5
21	139 04.5	. . 42.7	5 01.9	15.4	0 35.2	13.6	55.5
22	154 04.5	43.5	19 36.3	15.5	0 21.6	13.7	55.5
23	169 04.5	44.3	34 10.8	15.4	S 0 07.9	13.6	55.5
30 00	184 04.6	S13 45.2	48 45.2	15.3	N 0 05.7	13.6	55.6
01	199 04.6	46.0	63 19.5	15.4	0 19.3	13.7	55.6
02	214 04.7	46.8	77 53.9	15.3	0 33.0	13.7	55.6
03	229 04.7	. . 47.6	92 28.2	15.3	0 46.7	13.7	55.6
04	244 04.7	48.4	107 02.5	15.3	1 00.4	13.7	55.7
05	259 04.8	49.3	121 36.8	15.2	1 14.1	13.7	55.7
06	274 04.8	S13 50.1	136 11.0	15.2	N 1 27.8	13.7	55.7
07	289 04.8	50.9	150 45.2	15.1	1 41.5	13.7	55.7
08	304 04.9	51.7	165 19.3	15.2	1 55.2	13.7	55.8
F 09	319 04.9	. . 52.5	179 53.5	15.0	2 08.9	13.8	55.8
R 10	334 04.9	53.3	194 27.5	15.1	2 22.7	13.7	55.8
I 11	349 05.0	54.2	209 01.6	15.0	2 36.4	13.7	55.9
D 12	4 05.0	S13 55.0	223 35.6	15.0	N 2 50.1	13.8	55.9
A 13	19 05.0	55.8	238 09.6	14.9	3 03.9	13.7	55.9
Y 14	34 05.0	56.6	252 43.5	14.8	3 17.6	13.7	55.9
15	49 05.1	. . 57.4	267 17.3	14.9	3 31.3	13.8	56.0
16	64 05.1	58.2	281 51.2	14.8	3 45.1	13.7	56.0
17	79 05.1	59.0	296 25.0	14.7	3 58.8	13.8	56.0
18	94 05.2	S13 59.9	310 58.7	14.7	N 4 12.6	13.7	56.0
19	109 05.2	14 00.7	325 32.4	14.6	4 26.3	13.7	56.1
20	124 05.2	01.5	340 06.0	14.6	4 40.0	13.7	56.1
21	139 05.3	. . 02.3	354 39.6	14.6	4 53.7	13.8	56.1
22	154 05.3	03.1	9 13.2	14.5	5 07.5	13.7	56.2
23	169 05.3	03.9	23 46.7	14.4	N 5 21.2	13.7	56.2
	SD 16.1	d 0.8	SD 14.9		15.1		15.2

Lat.	Twilight Naut.	Twilight Civil	Sunrise	Moonrise 28	Moonrise 29	Moonrise 30	Moonrise 31
°	h m	h m	h m	h m	h m	h m	h m
N 72	05 55	07 16	08 39	15 22	14 55	14 29	14 00
N 70	05 52	07 05	08 16	15 12	14 52	14 34	14 13
68	05 49	06 56	07 59	15 03	14 50	14 37	14 23
66	05 47	06 48	07 45	14 56	14 48	14 40	14 32
64	05 45	06 41	07 33	14 50	14 47	14 43	14 39
62	05 43	06 35	07 23	14 45	14 45	14 45	14 46
60	05 41	06 30	07 15	14 41	14 44	14 47	14 51
N 58	05 39	06 26	07 07	14 37	14 43	14 49	14 56
56	05 38	06 21	07 01	14 33	14 42	14 51	15 00
54	05 36	06 18	06 55	14 30	14 41	14 52	15 04
52	05 35	06 14	06 50	14 27	14 40	14 53	15 08
50	05 33	06 11	06 45	14 24	14 39	14 55	15 11
45	05 29	06 04	06 34	14 18	14 38	14 57	15 18
N 40	05 26	05 58	06 26	14 13	14 36	15 00	15 24
35	05 22	05 52	06 18	14 09	14 35	15 02	15 29
30	05 19	05 47	06 12	14 05	14 34	15 03	15 34
20	05 12	05 38	06 00	13 59	14 32	15 06	15 42
N 10	05 03	05 28	05 50	13 53	14 31	15 09	15 49
0	04 54	05 19	05 40	13 48	14 29	15 12	15 56
S 10	04 44	05 09	05 31	13 42	14 28	15 14	16 03
20	04 30	04 57	05 20	13 36	14 26	15 17	16 10
30	04 13	04 43	05 08	13 30	14 24	15 20	16 18
35	04 02	04 34	05 01	13 26	14 23	15 22	16 23
40	03 49	04 24	04 53	13 21	14 22	15 24	16 29
45	03 32	04 11	04 43	13 16	14 21	15 27	16 35
S 50	03 11	03 56	04 32	13 10	14 19	15 30	16 43
52	03 00	03 49	04 26	13 07	14 18	15 31	16 47
54	02 48	03 40	04 20	13 04	14 18	15 33	16 51
56	02 34	03 31	04 14	13 01	14 17	15 34	16 55
58	02 16	03 20	04 06	12 57	14 16	15 36	17 00
S 60	01 54	03 08	03 58	12 53	14 15	15 38	17 06

Lat.	Sunset	Twilight Civil	Twilight Naut.	Moonset 28	Moonset 29	Moonset 30	Moonset 31
°	h m	h m	h m	h m	h m	h m	h m
N 72	14 47	16 10	17 31	25 02	01 02	02 58	04 59
N 70	15 10	16 21	17 34	25 09	01 09	02 57	04 49
68	15 27	16 31	17 37	25 15	01 15	02 56	04 41
66	15 42	16 38	17 39	25 21	01 21	02 56	04 34
64	15 53	16 45	17 41	25 25	01 25	02 55	04 28
62	16 03	16 51	17 43	00 03	01 29	02 55	04 24
60	16 12	16 56	17 45	00 10	01 32	02 54	04 19
N 58	16 19	17 01	17 47	00 16	01 35	02 54	04 16
56	16 26	17 05	17 49	00 21	01 37	02 54	04 12
54	16 32	17 09	17 50	00 26	01 39	02 53	04 10
52	16 37	17 12	17 52	00 31	01 41	02 53	04 07
50	16 42	17 16	17 54	00 35	01 43	02 53	04 04
45	16 52	17 23	17 57	00 43	01 47	02 52	03 59
N 40	17 01	17 29	18 01	00 50	01 51	02 52	03 55
35	17 09	17 35	18 05	00 56	01 53	02 51	03 51
30	17 15	17 40	18 08	01 02	01 56	02 51	03 48
20	17 27	17 50	18 16	01 11	02 00	02 50	03 42
N 10	17 37	17 59	18 24	01 19	02 04	02 50	03 37
0	17 47	18 08	18 33	01 26	02 08	02 49	03 32
S 10	17 57	18 19	18 44	01 33	02 11	02 49	03 28
20	18 08	18 31	18 57	01 41	02 15	02 48	03 23
30	18 20	18 45	19 15	01 50	02 19	02 48	03 17
35	18 27	18 54	19 26	01 55	02 21	02 47	03 14
40	18 35	19 04	19 39	02 01	02 24	02 47	03 10
45	18 45	19 17	19 56	02 07	02 27	02 46	03 06
S 50	18 57	19 33	20 18	02 15	02 31	02 46	03 01
52	19 02	19 40	20 29	02 19	02 32	02 45	02 59
54	19 08	19 49	20 41	02 23	02 34	02 45	02 56
56	19 15	19 58	20 56	02 27	02 36	02 45	02 54
58	19 23	20 09	21 14	02 32	02 38	02 44	02 51
S 60	19 31	20 22	21 37	02 37	02 41	02 44	02 47

Day	SUN Eqn. of Time 00^h	SUN Eqn. of Time 12^h	SUN Mer. Pass.	MOON Mer. Pass. Upper	MOON Mer. Pass. Lower	MOON Age	MOON Phase
d	m s	m s	h m	h m	h m	d	%
28	16 10	16 12	11 44	19 58	07 37	10	73
29	16 14	16 16	11 44	20 39	08 18	11	81
30	16 18	16 20	11 44	21 22	09 00	12	88

	UT	ARIES	VENUS −3.9		MARS +0.4		JUPITER −2.4		SATURN +1.1	
	d h	GHA ° ′	GHA ° ′	Dec ° ′	GHA ° ′	Dec ° ′	GHA ° ′	Dec ° ′	GHA ° ′	Dec ° ′
	31 00	39 25.9	200 24.7	S 6 24.6	269 52.1	N20 01.0	78 58.0	S16 30.6	218 29.4	N 1 49.3
	01	54 28.3	215 24.2	25.8	284 53.4	00.8	94 00.3	30.5	233 31.6	49.2
	02	69 30.8	230 23.8	27.0	299 54.7	00.5	109 02.6	30.5	248 33.8	49.1
	03	84 33.3	245 23.4	. . 28.2	314 56.0	. . 00.3	124 04.9	. . 30.4	263 36.1	. . 49.0
	04	99 35.7	260 22.9	29.4	329 57.3	20 00.1	139 07.2	30.4	278 38.3	48.9
	05	114 38.2	275 22.5	30.6	344 58.7	19 59.9	154 09.6	30.3	293 40.5	48.7
	06	129 40.6	290 22.1	S 6 31.8	0 00.0	N19 59.7	169 11.9	S16 30.3	308 42.7	N 1 48.6
	07	144 43.1	305 21.6	33.0	15 01.3	59.5	184 14.2	30.2	323 44.9	48.5
S	08	159 45.6	320 21.2	34.2	30 02.6	59.3	199 16.5	30.2	338 47.1	48.4
A	09	174 48.0	335 20.7	. . 35.4	45 03.9	. . 59.1	214 18.8	. . 30.1	353 49.3	. . 48.3
T	10	189 50.5	350 20.3	36.5	60 05.3	58.9	229 21.1	30.1	8 51.6	48.2
U	11	204 53.0	5 19.9	37.7	75 06.6	58.6	244 23.5	30.0	23 53.8	48.1
R	12	219 55.4	20 19.4	S 6 38.9	90 07.9	N19 58.4	259 25.8	S16 29.9	38 56.0	N 1 48.0
D	13	234 57.9	35 19.0	40.1	105 09.2	58.2	274 28.1	29.9	53 58.2	47.9
A	14	250 00.4	50 18.5	41.3	120 10.5	58.0	289 30.4	29.8	69 00.4	47.8
Y	15	265 02.8	65 18.1	. . 42.5	135 11.9	. . 57.8	304 32.7	. . 29.8	84 02.6	. . 47.7
	16	280 05.3	80 17.7	43.7	150 13.2	57.6	319 35.0	29.7	99 04.9	47.6
	17	295 07.8	95 17.2	44.9	165 14.5	57.4	334 37.4	29.7	114 07.1	47.5
	18	310 10.2	110 16.8	S 6 46.1	180 15.8	N19 57.2	349 39.7	S16 29.6	129 09.3	N 1 47.4
	19	325 12.7	125 16.3	47.3	195 17.2	57.0	4 42.0	29.6	144 11.5	47.3
	20	340 15.1	140 15.9	48.5	210 18.5	56.7	19 44.3	29.5	159 13.7	47.2
	21	355 17.6	155 15.5	. . 49.6	225 19.8	. . 56.5	34 46.6	. . 29.5	174 15.9	. . 47.1
	22	10 20.1	170 15.0	50.8	240 21.1	56.3	49 48.9	29.4	189 18.2	47.0
	23	25 22.5	185 14.6	52.0	255 22.5	56.1	64 51.2	29.4	204 20.4	46.9
	1 00	40 25.0	200 14.1	S 6 53.2	270 23.8	N19 55.9	79 53.5	S16 29.3	219 22.6	N 1 46.8
	01	55 27.5	215 13.7	54.4	285 25.1	55.7	94 55.9	29.3	234 24.8	46.7
	02	70 29.9	230 13.2	55.6	300 26.5	55.5	109 58.2	29.2	249 27.0	46.7
	03	85 32.4	245 12.8	. . 56.8	315 27.8	. . 55.3	125 00.5	. . 29.2	264 29.2	. . 46.6
	04	100 34.9	260 12.4	58.0	330 29.1	55.1	140 02.8	29.1	279 31.5	46.5
	05	115 37.3	275 11.9	6 59.2	345 30.5	54.8	155 05.1	29.1	294 33.7	46.4
	06	130 39.8	290 11.5	S 7 00.3	0 31.8	N19 54.6	170 07.4	S16 29.0	309 35.9	N 1 46.3
	07	145 42.2	305 11.0	01.5	15 33.1	54.4	185 09.7	29.0	324 38.1	46.2
	08	160 44.7	320 10.6	02.7	30 34.5	54.2	200 12.0	28.9	339 40.3	46.1
S	09	175 47.2	335 10.1	. . 03.9	45 35.8	. . 54.0	215 14.3	. . 28.9	354 42.5	. . 46.0
U	10	190 49.6	350 09.7	05.1	60 37.1	53.8	230 16.7	28.8	9 44.8	45.9
N	11	205 52.1	5 09.2	06.3	75 38.5	53.6	245 19.0	28.8	24 47.0	45.8
D	12	220 54.6	20 08.8	S 7 07.5	90 39.8	N19 53.4	260 21.3	S16 28.7	39 49.2	N 1 45.7
A	13	235 57.0	35 08.3	08.7	105 41.1	53.2	275 23.6	28.6	54 51.4	45.6
Y	14	250 59.5	50 07.9	09.8	120 42.5	53.0	290 25.9	28.6	69 53.6	45.5
	15	266 02.0	65 07.5	. . 11.0	135 43.8	. . 52.7	305 28.2	. . 28.5	84 55.9	. . 45.4
	16	281 04.4	80 07.0	12.2	150 45.1	52.5	320 30.5	28.5	99 58.1	45.3
	17	296 06.9	95 06.6	13.4	165 46.5	52.3	335 32.8	28.4	115 00.3	45.2
	18	311 09.4	110 06.1	S 7 14.6	180 47.8	N19 52.1	350 35.1	S16 28.4	130 02.5	N 1 45.1
	19	326 11.8	125 05.7	15.8	195 49.2	51.9	5 37.4	28.3	145 04.7	45.0
	20	341 14.3	140 05.2	17.0	210 50.5	51.7	20 39.7	28.3	160 06.9	44.9
	21	356 16.7	155 04.8	. . 18.1	225 51.8	. . 51.5	35 42.0	. . 28.2	175 09.2	. . 44.8
	22	11 19.2	170 04.3	19.3	240 53.2	51.3	50 44.3	28.2	190 11.4	44.7
	23	26 21.7	185 03.9	20.5	255 54.5	51.1	65 46.7	28.1	205 13.6	44.6
	2 00	41 24.1	200 03.4	S 7 21.7	270 55.9	N19 50.9	80 49.0	S16 28.1	220 15.8	N 1 44.5
	01	56 26.6	215 03.0	22.9	285 57.2	50.6	95 51.3	28.0	235 18.0	44.4
	02	71 29.1	230 02.5	24.1	300 58.6	50.4	110 53.6	27.9	250 20.3	44.3
	03	86 31.5	245 02.1	. . 25.2	315 59.9	. . 50.2	125 55.9	. . 27.9	265 22.5	. . 44.2
	04	101 34.0	260 01.6	26.4	331 01.3	50.0	140 58.2	27.8	280 24.7	44.1
	05	116 36.5	275 01.2	27.6	346 02.6	49.8	156 00.5	27.8	295 26.9	44.0
	06	131 38.9	290 00.7	S 7 28.8	1 03.9	N19 49.6	171 02.8	S16 27.7	310 29.1	N 1 43.9
	07	146 41.4	305 00.3	30.0	16 05.3	49.4	186 05.1	27.7	325 31.4	43.8
	08	161 43.8	319 59.8	31.1	31 06.6	49.2	201 07.4	27.6	340 33.6	43.7
M	09	176 46.3	334 59.4	. . 32.3	46 08.0	. . 49.0	216 09.7	. . 27.6	355 35.8	. . 43.6
O	10	191 48.8	349 58.9	33.5	61 09.3	48.8	231 12.0	27.5	10 38.0	43.5
N	11	206 51.2	4 58.5	34.7	76 10.7	48.6	246 14.3	27.4	25 40.2	43.4
D	12	221 53.7	19 58.0	S 7 35.9	91 12.0	N19 48.3	261 16.6	S16 27.4	40 42.4	N 1 43.3
A	13	236 56.2	34 57.6	37.0	106 13.4	48.1	276 18.9	27.3	55 44.7	43.2
Y	14	251 58.6	49 57.1	38.2	121 14.7	47.9	291 21.2	27.3	70 46.9	43.1
	15	267 01.1	64 56.6	. . 39.4	136 16.1	. . 47.7	306 23.5	. . 27.2	85 49.1	. . 43.0
	16	282 03.6	79 56.2	40.6	151 17.4	47.5	321 25.8	27.2	100 51.3	42.9
	17	297 06.0	94 55.7	41.8	166 18.8	47.3	336 28.1	27.1	115 53.5	42.8
	18	312 08.5	109 55.3	S 7 42.9	181 20.1	N19 47.1	351 30.4	S16 27.1	130 55.8	N 1 42.7
	19	327 11.0	124 54.8	44.1	196 21.5	46.9	6 32.7	27.0	145 58.0	42.6
	20	342 13.4	139 54.4	45.3	211 22.9	46.7	21 35.0	26.9	161 00.2	42.5
	21	357 15.9	154 53.9	. . 46.5	226 24.2	. . 46.5	36 37.3	. . 26.9	176 02.4	. . 42.4
	22	12 18.3	169 53.5	47.7	241 25.6	46.3	51 39.6	26.8	191 04.7	42.3
	23	27 20.8	184 53.0	48.8	256 26.9	46.1	66 41.9	26.8	206 06.9	42.2
	Mer. Pass.	h m 21 14.8	v −0.4	d 1.2	v 1.3	d 0.2	v 2.3	d 0.1	v 2.2	d 0.1

STARS

Name	SHA ° ′	Dec ° ′
Acamar	315 19.9	S40 15.7
Achernar	335 28.0	S57 11.1
Acrux	173 13.2	S63 09.1
Adhara	255 14.5	S28 58.9
Aldebaran	290 52.2	N16 31.9
Alioth	166 23.3	N55 54.2
Alkaid	153 01.3	N49 15.8
Al Na'ir	27 46.8	S46 54.9
Alnilam	275 48.9	S 1 11.6
Alphard	217 58.8	S 8 42.0
Alphecca	126 13.6	N26 40.9
Alpheratz	357 46.1	N29 09.0
Altair	62 10.9	N 8 53.8
Ankaa	353 17.8	S42 15.1
Antares	112 29.9	S26 27.3
Arcturus	145 58.5	N19 07.9
Atria	107 34.7	S69 02.9
Avior	234 19.2	S59 32.1
Bellatrix	278 34.6	N 6 21.6
Betelgeuse	271 04.0	N 7 24.6
Canopus	263 57.0	S52 41.7
Capella	280 38.1	N46 00.4
Deneb	49 33.4	N45 19.3
Denebola	182 36.6	N14 31.0
Diphda	348 58.2	S17 55.8
Dubhe	193 55.1	N61 41.6
Elnath	278 15.7	N28 37.0
Eltanin	90 47.7	N51 29.5
Enif	33 49.7	N 9 55.4
Fomalhaut	15 26.7	S29 34.2
Gacrux	172 04.6	S57 10.0
Gienah	175 55.4	S17 35.7
Hadar	148 52.6	S60 25.2
Hamal	328 03.5	N23 30.7
Kaus Aust.	83 47.6	S34 22.9
Kochab	137 20.7	N74 06.8
Markab	13 40.9	N15 15.7
Menkar	314 17.5	N 4 07.9
Menkent	148 11.2	S36 25.1
Miaplacidus	221 40.6	S69 45.1
Mirfak	308 43.8	N49 53.[illegible]
Nunki	76 01.8	S26 17.1
Peacock	53 23.4	S56 42.4
Pollux	243 30.9	N28 00.1
Procyon	245 02.4	N 5 12.0
Rasalhague	96 09.2	N12 33.3
Regulus	207 46.5	N11 55.1
Rigel	281 14.4	S 8 11.2
Rigil Kent.	139 56.3	S60 52.5
Sabik	102 15.9	S15 44.2
Schedar	349 43.3	N56 35.8
Shaula	96 25.9	S37 06.7
Sirius	258 35.9	S16 43.6
Spica	158 34.4	S11 12.7
Suhail	222 54.6	S43 28.2
Vega	80 41.0	N38 47.8
Zuben'ubi	137 08.8	S16 04.9
	SHA ° ′	**Mer. Pass** h m
Venus	159 49.1	10 39
Mars	229 58.8	5 58
Jupiter	39 28.6	18 38
Saturn	178 57.6	9 21

	UT	SUN GHA	SUN Dec	MOON GHA	v	MOON Dec	d	HP
	d h	° ′	° ′	° ′	′	° ′	′	′
	31 00	184 05.3	S14 04.7	38 20.1	14.4	N 5 34.9	13.7	56.2
	01	199 05.4	05.5	52 53.5	14.3	5 48.6	13.6	56.2
	02	214 05.4	06.3	67 26.8	14.3	6 02.2	13.7	56.3
	03	229 05.4	. . 07.2	82 00.1	14.2	6 15.9	13.7	56.3
	04	244 05.4	08.0	96 33.3	14.2	6 29.6	13.6	56.3
	05	259 05.5	08.8	111 06.5	14.1	6 43.2	13.6	56.3
	06	274 05.5	S14 09.6	125 39.6	14.0	N 6 56.8	13.6	56.4
S	07	289 05.5	10.4	140 12.6	14.0	7 10.4	13.6	56.4
A	08	304 05.5	11.2	154 45.6	13.9	7 24.0	13.6	56.4
T	09	319 05.6	. . 12.0	169 18.5	13.9	7 37.6	13.6	56.5
U	10	334 05.6	12.8	183 51.4	13.8	7 51.2	13.5	56.5
R	11	349 05.6	13.6	198 24.2	13.7	8 04.7	13.5	56.5
D	12	4 05.6	S14 14.4	212 56.9	13.7	N 8 18.2	13.5	56.5
A	13	19 05.7	15.2	227 29.6	13.6	8 31.7	13.5	56.6
Y	14	34 05.7	16.0	242 02.2	13.5	8 45.2	13.4	56.6
	15	49 05.7	. . 16.8	256 34.7	13.4	8 58.6	13.4	56.6
	16	64 05.7	17.6	271 07.1	13.4	9 12.0	13.4	56.7
	17	79 05.8	18.4	285 39.5	13.3	9 25.4	13.4	56.7
	18	94 05.8	S14 19.3	300 11.8	13.3	N 9 38.8	13.3	56.7
	19	109 05.8	20.1	314 44.1	13.2	9 52.1	13.3	56.7
	20	124 05.8	20.9	329 16.3	13.0	10 05.4	13.3	56.8
	21	139 05.8	. . 21.7	343 48.3	13.1	10 18.7	13.2	56.8
	22	154 05.9	22.5	358 20.4	12.9	10 31.9	13.2	56.8
	23	169 05.9	23.3	12 52.3	12.9	10 45.1	13.2	56.9
	1 00	184 05.9	S14 24.1	27 24.2	12.8	N10 58.3	13.1	56.9
	01	199 05.9	24.9	41 56.0	12.7	11 11.4	13.1	56.9
	02	214 05.9	25.7	56 27.7	12.6	11 24.5	13.0	56.9
	03	229 06.0	. . 26.5	70 59.3	12.5	11 37.5	13.0	57.0
	04	244 06.0	27.3	85 30.8	12.5	11 50.5	13.0	57.0
	05	259 06.0	28.1	100 02.3	12.4	12 03.5	12.9	57.0
	06	274 06.0	S14 28.9	114 33.7	12.3	N12 16.4	12.9	57.1
	07	289 06.0	29.7	129 05.0	12.2	12 29.3	12.8	57.1
	08	304 06.0	30.5	143 36.2	12.1	12 42.1	12.8	57.1
S	09	319 06.1	. . 31.3	158 07.3	12.1	12 54.9	12.7	57.1
U	10	334 06.1	32.1	172 38.4	11.9	13 07.6	12.7	57.2
N	11	349 06.1	32.9	187 09.3	11.9	13 20.3	12.6	57.2
D	12	4 06.1	S14 33.6	201 40.2	11.8	N13 32.9	12.6	57.2
A	13	19 06.1	34.4	216 11.0	11.6	13 45.5	12.5	57.3
Y	14	34 06.1	35.2	230 41.6	11.6	13 58.0	12.5	57.3
	15	49 06.1	. . 36.0	245 12.2	11.5	14 10.5	12.4	57.3
	16	64 06.2	36.8	259 42.7	11.5	14 22.9	12.3	57.3
	17	79 06.2	37.6	274 13.2	11.3	14 35.2	12.3	57.4
	18	94 06.2	S14 38.4	288 43.5	11.2	N14 47.5	12.2	57.4
	19	109 06.2	39.2	303 13.7	11.1	14 59.7	12.2	57.4
	20	124 06.2	40.0	317 43.8	11.1	15 11.9	12.1	57.5
	21	139 06.2	. . 40.8	332 13.9	10.9	15 24.0	12.0	57.5
	22	154 06.2	41.6	346 43.8	10.9	15 36.0	12.0	57.5
	23	169 06.2	42.4	1 13.7	10.7	15 48.0	11.8	57.5
	2 00	184 06.3	S14 43.2	15 43.4	10.7	N15 59.8	11.9	57.6
	01	199 06.3	44.0	30 13.1	10.6	16 11.7	11.7	57.6
	02	214 06.3	44.8	44 42.7	10.4	16 23.4	11.7	57.6
	03	229 06.3	. . 45.5	59 12.1	10.4	16 35.1	11.6	57.6
	04	244 06.3	46.3	73 41.5	10.2	16 46.7	11.5	57.7
	05	259 06.3	47.1	88 10.7	10.2	16 58.2	11.4	57.7
	06	274 06.3	S14 47.9	102 39.9	10.1	N17 09.6	11.3	57.7
	07	289 06.3	48.7	117 09.0	9.9	17 20.9	11.3	57.8
	08	304 06.3	49.5	131 37.9	9.9	17 32.2	11.2	57.8
M	09	319 06.3	. . 50.3	146 06.8	9.8	17 43.4	11.1	57.8
O	10	334 06.3	51.1	160 35.6	9.6	17 54.5	11.0	57.8
N	11	349 06.4	51.9	175 04.2	9.6	18 05.5	10.9	57.9
D	12	4 06.4	S14 52.6	189 32.8	9.4	N18 16.4	10.9	57.9
A	13	19 06.4	53.4	204 01.2	9.4	18 27.3	10.7	57.9
Y	14	34 06.4	54.2	218 29.6	9.3	18 38.0	10.7	57.9
	15	49 06.4	. . 55.0	232 57.9	9.1	18 48.7	10.5	58.0
	16	64 06.4	55.8	247 26.0	9.1	18 59.2	10.5	58.0
	17	79 06.4	56.6	261 54.1	8.9	19 09.7	10.3	58.0
	18	94 06.4	S14 57.4	276 22.0	8.9	N19 20.0	10.3	58.0
	19	109 06.4	58.1	290 49.9	8.7	19 30.3	10.1	58.1
	20	124 06.4	58.9	305 17.6	8.6	19 40.4	10.1	58.1
	21	139 06.4	14 59.7	319 45.2	8.6	19 50.5	9.9	58.1
	22	154 06.4	15 00.5	334 12.8	8.4	20 00.4	9.9	58.1
	23	169 06.4	S15 01.3	348 40.2	8.3	N20 10.3	9.7	58.2
		SD 16.1	d 0.8	SD	15.4		15.6	15.8

Lat.	Twilight Naut.	Twilight Civil	Sunrise	Moonrise 31	Moonrise 1	Moonrise 2	Moonrise 3
°	h m	h m	h m	h m	h m	h m	h m
N 72	06 07	07 30	08 57	14 00	13 21	▭	▭
N 70	06 02	07 17	08 31	14 13	13 47	13 02	▭
68	05 59	07 06	08 11	14 23	14 07	13 43	▭
66	05 56	06 57	07 55	14 32	14 23	14 11	13 52
64	05 53	06 50	07 43	14 39	14 36	14 33	14 31
62	05 50	06 43	07 32	14 46	14 47	14 50	14 58
60	05 48	06 37	07 22	14 51	14 57	15 05	15 19
N 58	05 45	06 32	07 14	14 56	15 05	15 17	15 37
56	05 43	06 27	07 07	15 00	15 12	15 28	15 52
54	05 41	06 23	07 01	15 04	15 19	15 38	16 04
52	05 39	06 19	06 55	15 08	15 25	15 47	16 16
50	05 38	06 16	06 50	15 11	15 30	15 54	16 26
45	05 33	06 08	06 38	15 18	15 42	16 11	16 47
N 40	05 29	06 01	06 29	15 24	15 52	16 25	17 04
35	05 25	05 55	06 21	15 29	16 00	16 36	17 18
30	05 21	05 49	06 14	15 34	16 08	16 46	17 31
20	05 13	05 39	06 01	15 42	16 21	17 04	17 53
N 10	05 04	05 29	05 51	15 49	16 32	17 20	18 12
0	04 54	05 19	05 40	15 56	16 43	17 34	18 30
S 10	04 42	05 08	05 30	16 03	16 54	17 49	18 47
20	04 28	04 55	05 18	16 10	17 06	18 05	19 07
30	04 10	04 40	05 05	16 18	17 19	18 23	19 29
35	03 59	04 31	04 58	16 23	17 27	18 34	19 42
40	03 45	04 20	04 49	16 29	17 36	18 46	19 57
45	03 27	04 07	04 39	16 35	17 46	19 01	20 15
S 50	03 04	03 50	04 26	16 43	17 59	19 19	20 38
52	02 53	03 42	04 21	16 47	18 05	19 27	20 49
54	02 40	03 33	04 14	16 51	18 12	19 37	21 01
56	02 24	03 23	04 07	16 55	18 19	19 47	21 16
58	02 05	03 12	03 59	17 00	18 28	20 00	21 32
S 60	01 40	02 58	03 50	17 06	18 37	20 14	21 53

Lat.	Sunset	Twilight Civil	Twilight Naut.	Moonset 31	Moonset 1	Moonset 2	Moonset 3
°	h m	h m	h m	h m	h m	h m	h m
N 72	14 28	15 56	17 19	04 59	07 15	▭	▭
N 70	14 55	16 09	17 23	04 49	06 51	09 20	▭
68	15 15	16 20	17 27	04 41	06 33	08 41	▭
66	15 31	16 29	17 30	04 34	06 19	08 13	10 26
64	15 44	16 37	17 33	04 28	06 07	07 53	09 47
62	15 54	16 43	17 36	04 24	05 57	07 36	09 21
60	16 04	16 49	17 38	04 19	05 49	07 23	09 00
N 58	16 12	16 54	17 41	04 16	05 41	07 11	08 43
56	16 19	16 59	17 43	04 12	05 35	07 01	08 29
54	16 26	17 03	17 45	04 10	05 29	06 52	08 17
52	16 31	17 07	17 47	04 07	05 24	06 44	08 06
50	16 37	17 11	17 49	04 04	05 19	06 37	07 56
45	16 48	17 19	17 53	03 59	05 09	06 22	07 36
N 40	16 58	17 26	17 58	03 55	05 00	06 09	07 20
35	17 06	17 32	18 02	03 51	04 53	05 59	07 06
30	17 13	17 38	18 06	03 48	04 47	05 49	06 54
20	17 25	17 48	18 14	03 42	04 36	05 34	06 34
N 10	17 37	17 58	18 23	03 37	04 27	05 20	06 16
0	17 47	18 08	18 33	03 32	04 18	05 07	06 00
S 10	17 58	18 19	18 45	03 28	04 09	04 54	05 44
20	18 09	18 32	18 59	03 23	04 00	04 41	05 26
30	18 22	18 48	19 18	03 17	03 49	04 25	05 06
35	18 30	18 57	19 29	03 14	03 43	04 16	04 55
40	18 39	19 08	19 44	03 10	03 36	04 06	04 42
45	18 49	19 22	20 01	03 06	03 28	03 54	04 26
S 50	19 02	19 38	20 25	03 01	03 19	03 40	04 07
52	19 08	19 46	20 36	02 59	03 14	03 33	03 58
54	19 14	19 55	20 50	02 56	03 09	03 26	03 48
56	19 22	20 06	21 06	02 54	03 04	03 17	03 36
58	19 30	20 18	21 26	02 51	02 58	03 08	03 23
S 60	19 39	20 32	21 52	02 47	02 51	02 58	03 08

Day	SUN Eqn. of Time 00h	SUN Eqn. of Time 12h	SUN Mer. Pass.	MOON Mer. Pass. Upper	MOON Mer. Pass. Lower	Age	Phase
d	m s	m s	h m	h m	h m	d %	
31	16 21	16 23	11 44	22 07	09 44	13 94	
1	16 24	16 24	11 44	22 55	10 30	14 98	○
2	16 25	16 25	11 44	23 47	11 20	15 100	

UT	ARIES	VENUS −3.9		MARS +0.4		JUPITER −2.4		SATURN +1.1	
	GHA	GHA	Dec	GHA	Dec	GHA	Dec	GHA	Dec
d h	° ′	° ′	° ′	° ′	° ′	° ′	° ′	° ′	° ′
3 00 (TUESDAY)	42 23.3	199 52.5	S 7 50.0	271 28.3	N19 45.8	81 44.2	S16 26.7	221 09.1	N 1 42.1
01	57 25.7	214 52.1	51.2	286 29.6	45.6	96 46.5	26.7	236 11.3	42.0
02	72 28.2	229 51.6	52.4	301 31.0	45.4	111 48.8	26.6	251 13.5	41.9
03	87 30.7	244 51.2	. . 53.5	316 32.4	. . 45.2	126 51.1	. . 26.5	266 15.8	. . 41.8
04	102 33.1	259 50.7	54.7	331 33.7	45.0	141 53.4	26.5	281 18.0	41.7
05	117 35.6	274 50.3	55.9	346 35.1	44.8	156 55.7	26.4	296 20.2	41.6
06	132 38.1	289 49.8	S 7 57.1	1 36.4	N19 44.6	171 58.0	S16 26.4	311 22.4	N 1 41.5
07	147 40.5	304 49.3	58.2	16 37.8	44.4	187 00.2	26.3	326 24.6	41.4
08	162 43.0	319 48.9	7 59.4	31 39.2	44.2	202 02.5	26.3	341 26.9	41.3
09	177 45.5	334 48.4	8 00.6	46 40.5	. . 44.0	217 04.8	. . 26.2	356 29.1	. . 41.2
10	192 47.9	349 48.0	01.8	61 41.9	43.8	232 07.1	26.1	11 31.3	41.1
11	207 50.4	4 47.5	02.9	76 43.2	43.6	247 09.4	26.1	26 33.5	41.0
12	222 52.8	19 47.0	S 8 04.1	91 44.6	N19 43.4	262 11.7	S16 26.0	41 35.7	N 1 40.9
13	237 55.3	34 46.6	05.3	106 46.0	43.2	277 14.0	26.0	56 38.0	40.8
14	252 57.8	49 46.1	06.5	121 47.3	42.9	292 16.3	25.9	71 40.2	40.7
15	268 00.2	64 45.7	. . 07.6	136 48.7	. . 42.7	307 18.6	. . 25.9	86 42.4	. . 40.6
16	283 02.7	79 45.2	08.8	151 50.1	42.5	322 20.9	25.8	101 44.6	40.6
17	298 05.2	94 44.7	10.0	166 51.4	42.3	337 23.2	25.7	116 46.9	40.5
18	313 07.6	109 44.3	S 8 11.1	181 52.8	N19 42.1	352 25.5	S16 25.7	131 49.1	N 1 40.4
19	328 10.1	124 43.8	12.3	196 54.2	41.9	7 27.8	25.6	146 51.3	40.3
20	343 12.6	139 43.4	13.5	211 55.5	41.7	22 30.1	25.6	161 53.5	40.2
21	358 15.0	154 42.9	. . 14.7	226 56.9	. . 41.5	37 32.3	. . 25.5	176 55.7	. . 40.1
22	13 17.5	169 42.4	15.8	241 58.3	41.3	52 34.6	25.4	191 58.0	40.0
23	28 20.0	184 42.0	17.0	256 59.7	41.1	67 36.9	25.4	207 00.2	39.9
4 00 (WEDNESDAY)	43 22.4	199 41.5	S 8 18.2	272 01.0	N19 40.9	82 39.2	S16 25.3	222 02.4	N 1 39.8
01	58 24.9	214 41.0	19.3	287 02.4	40.7	97 41.5	25.3	237 04.6	39.7
02	73 27.3	229 40.6	20.5	302 03.8	40.5	112 43.8	25.2	252 06.9	39.6
03	88 29.8	244 40.1	. . 21.7	317 05.2	. . 40.3	127 46.1	. . 25.1	267 09.1	. . 39.5
04	103 32.3	259 39.6	22.8	332 06.5	40.1	142 48.4	25.1	282 11.3	39.4
05	118 34.7	274 39.2	24.0	347 07.9	39.8	157 50.7	25.0	297 13.5	39.3
06	133 37.2	289 38.7	S 8 25.2	2 09.3	N19 39.6	172 52.9	S16 25.0	312 15.7	N 1 39.2
07	148 39.7	304 38.2	26.3	17 10.6	39.4	187 55.2	24.9	327 18.0	39.1
08	163 42.1	319 37.8	27.5	32 12.0	39.2	202 57.5	24.8	342 20.2	39.0
09	178 44.6	334 37.3	. . 28.7	47 13.4	. . 39.0	217 59.8	. . 24.8	357 22.4	. . 38.9
10	193 47.1	349 36.8	29.8	62 14.8	38.8	233 02.1	24.7	12 24.6	38.8
11	208 49.5	4 36.4	31.0	77 16.2	38.6	248 04.4	24.7	27 26.9	38.7
12	223 52.0	19 35.9	S 8 32.2	92 17.5	N19 38.4	263 06.7	S16 24.6	42 29.1	N 1 38.6
13	238 54.4	34 35.4	33.3	107 18.9	38.2	278 09.0	24.5	57 31.3	38.5
14	253 56.9	49 35.0	34.5	122 20.3	38.0	293 11.2	24.5	72 33.5	38.4
15	268 59.4	64 34.5	. . 35.7	137 21.7	. . 37.8	308 13.5	. . 24.4	87 35.8	. . 38.3
16	284 01.8	79 34.0	36.8	152 23.1	37.6	323 15.8	24.4	102 38.0	38.2
17	299 04.3	94 33.6	38.0	167 24.4	37.4	338 18.1	24.3	117 40.2	38.1
18	314 06.8	109 33.1	S 8 39.2	182 25.8	N19 37.2	353 20.4	S16 24.2	132 42.4	N 1 38.0
19	329 09.2	124 32.6	40.3	197 27.2	37.0	8 22.7	24.2	147 44.7	37.9
20	344 11.7	139 32.2	41.5	212 28.6	36.8	23 24.9	24.1	162 46.9	37.8
21	359 14.2	154 31.7	. . 42.7	227 30.0	. . 36.6	38 27.2	. . 24.1	177 49.1	. . 37.8
22	14 16.6	169 31.2	43.8	242 31.4	36.4	53 29.5	24.0	192 51.3	37.7
23	29 19.1	184 30.7	45.0	257 32.7	36.1	68 31.8	23.9	207 53.6	37.6
5 00 (THURSDAY)	44 21.6	199 30.3	S 8 46.2	272 34.1	N19 35.9	83 34.1	S16 23.9	222 55.8	N 1 37.5
01	59 24.0	214 29.8	47.3	287 35.5	35.7	98 36.4	23.8	237 58.0	37.4
02	74 26.5	229 29.3	48.5	302 36.9	35.5	113 38.6	23.7	253 00.2	37.3
03	89 28.9	244 28.9	. . 49.6	317 38.3	. . 35.3	128 40.9	. . 23.7	268 02.5	. . 37.2
04	104 31.4	259 28.4	50.8	332 39.7	35.1	143 43.2	23.6	283 04.7	37.1
05	119 33.9	274 27.9	52.0	347 41.1	34.9	158 45.5	23.6	298 06.9	37.0
06	134 36.3	289 27.4	S 8 53.1	2 42.5	N19 34.7	173 47.8	S16 23.5	313 09.1	N 1 36.9
07	149 38.8	304 27.0	54.3	17 43.9	34.5	188 50.0	23.4	328 11.4	36.8
08	164 41.3	319 26.5	55.4	32 45.2	34.3	203 52.3	23.4	343 13.6	36.7
09	179 43.7	334 26.0	. . 56.6	47 46.6	. . 34.1	218 54.6	. . 23.3	358 15.8	. . 36.6
10	194 46.2	349 25.5	57.8	62 48.0	33.9	233 56.9	23.2	13 18.0	36.5
11	209 48.7	4 25.1	8 58.9	77 49.4	33.7	248 59.2	23.2	28 20.3	36.4
12	224 51.1	19 24.6	S 9 00.1	92 50.8	N19 33.5	264 01.4	S16 23.1	43 22.5	N 1 36.3
13	239 53.6	34 24.1	01.2	107 52.2	33.3	279 03.7	23.1	58 24.7	36.2
14	254 56.1	49 23.6	02.4	122 53.6	33.1	294 06.0	23.0	73 26.9	36.1
15	269 58.5	64 23.2	. . 03.5	137 55.0	. . 32.9	309 08.3	. . 22.9	88 29.2	. . 36.0
16	285 01.0	79 22.7	04.7	152 56.4	32.7	324 10.5	22.9	103 31.4	35.9
17	300 03.4	94 22.2	05.9	167 57.8	32.5	339 12.8	22.8	118 33.6	35.8
18	315 05.9	109 21.7	S 9 07.0	182 59.2	N19 32.3	354 15.1	S16 22.7	133 35.8	N 1 35.7
19	330 08.4	124 21.2	08.2	198 00.6	32.1	9 17.4	22.7	148 38.1	35.6
20	345 10.8	139 20.8	09.3	213 02.0	31.9	24 19.7	22.6	163 40.3	35.6
21	0 13.3	154 20.3	. . 10.5	228 03.4	. . 31.7	39 21.9	. . 22.5	178 42.5	. . 35.5
22	15 15.8	169 19.8	11.6	243 04.8	31.5	54 24.2	22.5	193 44.7	35.4
23	30 18.2	184 19.3	12.8	258 06.2	31.3	69 26.5	22.4	208 47.0	35.3
Mer. Pass.	h m 21 03.0	v −0.5	d 1.2	v 1.4	d 0.2	v 2.3	d 0.1	v 2.2	d 0.1

STARS

Name	SHA	Dec
	° ′	° ′
Acamar	315 19.9	S40 15.8
Achernar	335 28.0	S57 11.2
Acrux	173 13.1	S63 09.1
Adhara	255 14.5	S28 58.9
Aldebaran	290 52.2	N16 31.9
Alioth	166 23.3	N55 54.2
Alkaid	153 01.3	N49 15.7
Al Na'ir	27 46.8	S46 54.9
Alnilam	275 48.8	S 1 11.[illegible]
Alphard	217 58.8	S 8 42.0
Alphecca	126 13.6	N26 40.9
Alpheratz	357 46.1	N29 09.0
Altair	62 11.0	N 8 53.8
Ankaa	353 17.9	S42 15.3
Antares	112 29.9	S26 27.2
Arcturus	145 58.5	N19 07.9
Atria	107 34.7	S69 02.9
Avior	234 19.1	S59 32.2
Bellatrix	278 34.6	N 6 21.6
Betelgeuse	271 04.0	N 7 24.6
Canopus	263 57.0	S52 41.8
Capella	280 38.0	N46 00.4
Deneb	49 33.4	N45 19.3
Denebola	182 36.6	N14 31.0
Diphda	348 58.2	S17 55.8
Dubhe	193 55.1	N61 41.6
Elnath	278 15.7	N28 37.0
Eltanin	90 47.8	N51 29.5
Enif	33 49.7	N 9 55.4
Fomalhaut	15 26.7	S29 34.2
Gacrux	172 04.6	S57 10.0
Gienah	175 55.4	S17 35.7
Hadar	148 52.6	S60 25.2
Hamal	328 03.5	N23 30.7
Kaus Aust.	83 47.6	S34 22.9
Kochab	137 20.8	N74 06.8
Markab	13 40.9	N15 15.7
Menkar	314 17.5	N 4 07.9
Menkent	148 11.2	S36 25.1
Miaplacidus	221 40.6	S69 45.2
Mirfak	308 43.8	N49 53.9
Nunki	76 01.8	S26 17.1
Peacock	53 23.4	S56 42.4
Pollux	243 30.9	N28 00.1
Procyon	245 02.4	N 5 12.0
Rasalhague	96 09.2	N12 33.3
Regulus	207 46.4	N11 55.1
Rigel	281 14.4	S 8 11.2
Rigil Kent.	139 56.3	S60 52.5
Sabik	102 15.9	S15 44.2
Schedar	349 43.3	N56 35.8
Shaula	96 25.9	S37 06.7
Sirius	258 35.9	S16 43.6
Spica	158 34.4	S11 12.7
Suhail	222 54.5	S43 28.1
Vega	80 41.0	N38 47.8
Zuben'ubi	137 08.8	S16 04.9

	SHA	Mer. Pass
	° ′	h m
Venus	156 19.1	10 42
Mars	228 38.6	5 51
Jupiter	39 16.8	18 27
Saturn	178 40.0	9 10

UT	SUN GHA	SUN Dec	MOON GHA	v	MOON Dec	d	HP
d h	° ′	° ′	° ′	′	° ′	′	′
3 00	184 06.4	S15 02.0	3 07.5	8.3	N20 20.0	9.6	58.2
01	199 06.4	02.8	17 34.8	8.1	20 29.6	9.5	58.2
02	214 06.4	03.6	32 01.9	8.0	20 39.1	9.4	58.2
03	229 06.4	. . 04.4	46 28.9	7.9	20 48.5	9.3	58.3
04	244 06.4	05.2	60 55.8	7.9	20 57.8	9.2	58.3
05	259 06.4	06.0	75 22.7	7.7	21 07.0	9.0	58.3
06	274 06.4	S15 06.7	89 49.4	7.6	N21 16.0	8.9	58.3
07	289 06.4	07.5	104 16.0	7.5	21 24.9	8.8	58.3
08	304 06.4	08.3	118 42.5	7.4	21 33.7	8.7	58.4
09	319 06.4	. . 09.1	133 08.9	7.4	21 42.4	8.6	58.4
10	334 06.4	09.8	147 35.3	7.2	21 51.0	8.4	58.4
11	349 06.4	10.6	162 01.5	7.1	21 59.4	8.3	58.4
12	4 06.4	S15 11.4	176 27.6	7.0	N22 07.7	8.2	58.5
13	19 06.4	12.2	190 53.6	7.0	22 15.9	8.1	58.5
14	34 06.4	12.9	205 19.6	6.8	22 24.0	7.9	58.5
15	49 06.4	. . 13.7	219 45.4	6.7	22 31.9	7.8	58.5
16	64 06.4	14.5	234 11.1	6.7	22 39.7	7.6	58.5
17	79 06.4	15.3	248 36.8	6.5	22 47.3	7.5	58.6
18	94 06.4	S15 16.0	263 02.3	6.5	N22 54.8	7.4	58.6
19	109 06.4	16.8	277 27.8	6.4	23 02.2	7.2	58.6
20	124 06.4	17.6	291 53.2	6.3	23 09.4	7.1	58.6
21	139 06.4	. . 18.4	306 18.5	6.2	23 16.5	7.0	58.6
22	154 06.4	19.1	320 43.7	6.1	23 23.5	6.8	58.7
23	169 06.4	19.9	335 08.8	6.0	23 30.3	6.6	58.7
4 00	184 06.4	S15 20.7	349 33.8	5.9	N23 36.9	6.5	58.7
01	199 06.4	21.4	3 58.7	5.8	23 43.4	6.4	58.7
02	214 06.4	22.2	18 23.5	5.8	23 49.8	6.2	58.7
03	229 06.4	. . 23.0	32 48.3	5.7	23 56.0	6.1	58.8
04	244 06.3	23.8	47 13.0	5.6	24 02.1	5.9	58.8
05	259 06.3	24.5	61 37.6	5.5	24 08.0	5.8	58.8
06	274 06.3	S15 25.3	76 02.1	5.4	N24 13.8	5.6	58.8
07	289 06.3	26.1	90 26.5	5.4	24 19.4	5.4	58.8
08	304 06.3	26.8	104 50.9	5.3	24 24.8	5.3	58.8
09	319 06.3	. . 27.6	119 15.2	5.2	24 30.1	5.2	58.9
10	334 06.3	28.4	133 39.4	5.2	24 35.3	4.9	58.9
11	349 06.3	29.1	148 03.6	5.1	24 40.2	4.9	58.9
12	4 06.3	S15 29.9	162 27.7	5.0	N24 45.1	4.6	58.9
13	19 06.3	30.7	176 51.7	4.9	24 49.7	4.5	58.9
14	34 06.3	31.4	191 15.6	4.9	24 54.2	4.3	58.9
15	49 06.2	. . 32.2	205 39.5	4.8	24 58.5	4.2	59.0
16	64 06.2	33.0	220 03.3	4.8	25 02.7	4.0	59.0
17	79 06.2	33.7	234 27.1	4.7	25 06.7	3.9	59.0
18	94 06.2	S15 34.5	248 50.8	4.6	N25 10.6	3.6	59.0
19	109 06.2	35.2	263 14.4	4.6	25 14.2	3.5	59.0
20	124 06.2	36.0	277 38.0	4.5	25 17.7	3.3	59.0
21	139 06.2	. . 36.8	292 01.5	4.5	25 21.0	3.2	59.0
22	154 06.2	37.5	306 25.0	4.4	25 24.2	3.0	59.1
23	169 06.1	38.3	320 48.4	4.4	25 27.2	2.8	59.1
5 00	184 06.1	S15 39.1	335 11.8	4.4	N25 30.0	2.7	59.1
01	199 06.1	39.8	349 35.2	4.3	25 32.7	2.4	59.1
02	214 06.1	40.6	3 58.5	4.2	25 35.1	2.3	59.1
03	229 06.1	. . 41.3	18 21.7	4.2	25 37.4	2.2	59.1
04	244 06.1	42.1	32 44.9	4.2	25 39.6	1.9	59.1
05	259 06.0	42.9	47 08.1	4.2	25 41.5	1.8	59.2
06	274 06.0	S15 43.6	61 31.3	4.1	N25 43.3	1.6	59.2
07	289 06.0	44.4	75 54.4	4.1	25 44.9	1.4	59.2
08	304 06.0	45.1	90 17.5	4.0	25 46.3	1.3	59.2
09	319 06.0	. . 45.9	104 40.5	4.1	25 47.6	1.0	59.2
10	334 06.0	46.6	119 03.6	4.0	25 48.6	0.9	59.2
11	349 05.9	47.4	133 26.6	4.0	25 49.5	0.7	59.2
12	4 05.9	S15 48.2	147 49.6	4.0	N25 50.2	0.6	59.2
13	19 05.9	48.9	162 12.6	3.9	25 50.8	0.3	59.2
14	34 05.9	49.7	176 35.5	4.0	25 51.1	0.2	59.2
15	49 05.9	. . 50.4	190 58.5	3.9	25 51.3	0.0	59.3
16	64 05.8	51.2	205 21.4	4.0	25 51.3	0.2	59.3
17	79 05.8	51.9	219 44.4	3.9	25 51.1	0.3	59.3
18	94 05.8	S15 52.7	234 07.3	3.9	N25 50.8	0.5	59.3
19	109 05.8	53.4	248 30.2	3.9	25 50.3	0.8	59.3
20	124 05.8	54.2	262 53.1	4.0	25 49.5	0.9	59.3
21	139 05.7	. . 54.9	277 16.1	3.9	25 48.6	1.0	59.3
22	154 05.7	55.7	291 39.0	3.9	25 47.6	1.3	59.3
23	169 05.7	56.4	306 01.9	3.9	N25 46.3	1.4	59.3
	SD 16.2	d 0.8	SD	15.9	16.1		16.1

Lat.	Twilight Naut.	Twilight Civil	Sunrise	Moonrise 3	Moonrise 4	Moonrise 5	Moonrise 6
°	h m	h m	h m	h m	h m	h m	h m
N 72	06 18	07 43	09 17	▭	▭	▭	▭
N 70	06 13	07 28	08 46	▭	▭	▭	▭
68	06 08	07 16	08 24	▭	▭	▭	▭
66	06 04	07 06	08 06	13 52	▭	▭	▭
64	06 00	06 58	07 52	14 31	14 31	14 53	16 35
62	05 57	06 51	07 40	14 58	15 16	15 59	17 21
60	05 54	06 44	07 30	15 19	15 46	16 34	17 51
N 58	05 51	06 38	07 21	15 37	16 08	17 00	18 14
56	05 49	06 33	07 14	15 52	16 27	17 20	18 33
54	05 46	06 29	07 07	16 04	16 42	17 37	18 48
52	05 44	06 24	07 00	16 16	16 56	17 51	19 02
50	05 42	06 20	06 55	16 26	17 08	18 04	19 14
45	05 37	06 12	06 43	16 47	17 33	18 30	19 38
N 40	05 32	06 04	06 33	17 04	17 52	18 51	19 58
35	05 27	05 58	06 24	17 18	18 09	19 08	20 14
30	05 23	05 51	06 16	17 31	18 23	19 23	20 28
20	05 14	05 40	06 03	17 53	18 48	19 48	20 52
N 10	05 04	05 29	05 51	18 12	19 09	20 10	21 13
0	04 54	05 19	05 40	18 30	19 29	20 30	21 32
S 10	04 41	05 07	05 29	18 47	19 49	20 51	21 51
20	04 27	04 54	05 17	19 07	20 10	21 13	22 12
30	04 07	04 38	05 03	19 29	20 35	21 38	22 35
35	03 55	04 28	04 55	19 42	20 50	21 53	22 49
40	03 40	04 16	04 45	19 57	21 07	22 10	23 05
45	03 22	04 02	04 35	20 15	21 27	22 31	23 24
S 50	02 57	03 44	04 21	20 38	21 53	22 58	23 48
52	02 45	03 36	04 15	20 49	22 06	23 10	24 00
54	02 31	03 26	04 08	21 01	22 20	23 25	24 13
56	02 14	03 16	04 00	21 16	22 37	23 42	24 28
58	01 52	03 03	03 52	21 32	22 57	24 03	00 03
S 60	01 23	02 48	03 42	21 53	23 22	24 28	00 28

Lat.	Sunset	Twilight Civil	Twilight Naut.	Moonset 3	Moonset 4	Moonset 5	Moonset 6
°	h m	h m	h m	h m	h m	h m	h m
N 72	14 09	15 42	17 07	▭	▭	▭	▭
N 70	14 40	15 57	17 13	▭	▭	▭	▭
68	15 02	16 09	17 18	▭	▭	▭	▭
66	15 20	16 20	17 22	10 26	▭	▭	▭
64	15 34	16 28	17 25	09 47	11 49	13 35	14 03
62	15 46	16 36	17 29	09 21	11 04	12 29	13 16
60	15 56	16 42	17 32	09 00	10 35	11 54	12 46
N 58	16 05	16 48	17 35	08 43	10 13	11 29	12 22
56	16 13	16 53	17 37	08 29	09 54	11 08	12 04
54	16 20	16 58	17 40	08 17	09 39	10 51	11 48
52	16 26	17 02	17 42	08 06	09 26	10 37	11 34
50	16 32	17 06	17 44	07 56	09 14	10 24	11 22
45	16 44	17 15	17 50	07 36	08 50	09 58	10 57
N 40	16 54	17 22	17 55	07 20	08 31	09 38	10 37
35	17 03	17 29	17 59	07 06	08 15	09 20	10 20
30	17 11	17 35	18 04	06 54	08 01	09 06	10 06
20	17 24	17 47	18 13	06 34	07 37	08 40	09 41
N 10	17 36	17 58	18 23	06 16	07 17	08 18	09 20
0	17 47	18 09	18 33	06 00	06 57	07 58	09 00
S 10	17 58	18 20	18 46	05 44	06 38	07 38	08 40
20	18 11	18 34	19 01	05 26	06 18	07 16	08 18
30	18 25	18 50	19 20	05 06	05 55	06 51	07 53
35	18 33	19 00	19 33	04 55	05 41	06 36	07 38
40	18 42	19 12	19 48	04 42	05 25	06 18	07 21
45	18 53	19 26	20 07	04 26	05 06	05 58	07 01
S 50	19 07	19 44	20 31	04 07	04 43	05 31	06 35
52	19 13	19 53	20 44	03 58	04 31	05 19	06 22
54	19 20	20 02	20 59	03 48	04 19	05 04	06 07
56	19 28	20 13	21 16	03 36	04 04	04 47	05 51
58	19 37	20 26	21 39	03 23	03 47	04 27	05 30
S 60	19 47	20 42	22 09	03 08	03 26	04 01	05 04

Day	SUN Eqn. of Time 00^h	SUN Eqn. of Time 12^h	SUN Mer. Pass.	MOON Mer. Pass. Upper	MOON Mer. Pass. Lower	Age	Phase
d	m s	m s	h m	h m	h m	d %	
3	16 26	16 26	11 44	24 43	12 15	16 99	○
4	16 26	16 25	11 44	00 43	13 13	17 96	
5	16 25	16 24	11 44	01 43	14 14	18 91	

	UT	ARIES	VENUS −3.9		MARS +0.4		JUPITER −2.4		SATURN +1.1		STARS		
		GHA	GHA	Dec	GHA	Dec	GHA	Dec	GHA	Dec	Name	SHA	Dec
	d h	° ′	° ′	° ′	° ′	° ′	° ′	° ′	° ′	° ′		° ′	°
	6 00	45 20.7	199 18.8	S 9 14.0	273 07.6	N19 31.1	84 28.8	S16 22.4	223 49.2	N 1 35.2	Acamar	315 19.8	S40 15
	01	60 23.2	214 18.4	15.1	288 09.0	30.9	99 31.0	22.3	238 51.4	35.1	Achernar	335 28.0	S57 11
	02	75 25.6	229 17.9	16.3	303 10.4	30.7	114 33.3	22.2	253 53.7	35.0	Acrux	173 13.1	S63 09
	03	90 28.1	244 17.4	17.4	318 11.8	30.5	129 35.6	22.2	268 55.9	34.9	Adhara	255 14.4	S28 58
	04	105 30.6	259 16.9	18.6	333 13.2	30.3	144 37.8	22.1	283 58.1	34.8	Aldebaran	290 52.1	N16 31
	05	120 33.0	274 16.4	19.7	348 14.6	30.0	159 40.1	22.0	299 00.3	34.7			
	06	135 35.5	289 16.0	S 9 20.9	3 16.0	N19 29.8	174 42.4	S16 22.0	314 02.6	N 1 34.6	Alioth	166 23.3	N55 54
	07	150 37.9	304 15.5	22.0	18 17.4	29.6	189 44.7	21.9	329 04.8	34.5	Alkaid	153 01.3	N49 15
	08	165 40.4	319 15.0	23.2	33 18.8	29.4	204 46.9	21.8	344 07.0	34.4	Al Na'ir	27 46.8	S46 54
F	09	180 42.9	334 14.5	24.3	48 20.2	29.2	219 49.2	21.8	359 09.2	34.3	Alnilam	275 48.8	S 1 11
R	10	195 45.3	349 14.0	25.5	63 21.6	29.0	234 51.5	21.7	14 11.5	34.2	Alphard	217 58.7	S 8 42
I	11	210 47.8	4 13.5	26.6	78 23.0	28.8	249 53.8	21.6	29 13.7	34.1			
D	12	225 50.3	19 13.1	S 9 27.8	93 24.4	N19 28.6	264 56.0	S16 21.6	44 15.9	N 1 34.0	Alphecca	126 13.6	N26 40
A	13	240 52.7	34 12.6	28.9	108 25.9	28.4	279 58.3	21.5	59 18.2	33.9	Alpheratz	357 46.1	N29 09
Y	14	255 55.2	49 12.1	30.1	123 27.3	28.2	295 00.6	21.5	74 20.4	33.8	Altair	62 11.0	N 8 53
	15	270 57.7	64 11.6	31.2	138 28.7	28.0	310 02.8	21.4	89 22.6	33.8	Ankaa	353 17.9	S42 15
	16	286 00.1	79 11.1	32.4	153 30.1	27.8	325 05.1	21.3	104 24.8	33.7	Antares	112 29.9	S26 27
	17	301 02.6	94 10.6	33.5	168 31.5	27.6	340 07.4	21.3	119 27.1	33.6			
	18	316 05.1	109 10.1	S 9 34.7	183 32.9	N19 27.4	355 09.6	S16 21.2	134 29.3	N 1 33.5	Arcturus	145 58.5	N19 07
	19	331 07.5	124 09.7	35.8	198 34.3	27.2	10 11.9	21.1	149 31.5	33.4	Atria	107 34.7	S69 02
	20	346 10.0	139 09.2	37.0	213 35.7	27.0	25 14.2	21.1	164 33.7	33.3	Avior	234 19.1	S59 32
	21	1 12.4	154 08.7	38.1	228 37.2	26.8	40 16.5	21.0	179 36.0	33.2	Bellatrix	278 34.6	N 6 21
	22	16 14.9	169 08.2	39.3	243 38.6	26.6	55 18.7	20.9	194 38.2	33.1	Betelgeuse	271 03.9	N 7 24
	23	31 17.4	184 07.7	40.4	258 40.0	26.4	70 21.0	20.9	209 40.4	33.0			
	7 00	46 19.8	199 07.2	S 9 41.5	273 41.4	N19 26.2	85 23.3	S16 20.8	224 42.7	N 1 32.9	Canopus	263 57.0	S52 41
	01	61 22.3	214 06.7	42.7	288 42.8	26.0	100 25.5	20.7	239 44.9	32.8	Capella	280 38.0	N46 00
	02	76 24.8	229 06.2	43.8	303 44.2	25.8	115 27.8	20.7	254 47.1	32.7	Deneb	49 33.4	N45 19
	03	91 27.2	244 05.8	45.0	318 45.7	25.6	130 30.1	20.6	269 49.3	32.6	Denebola	182 36.6	N14 31
	04	106 29.7	259 05.3	46.1	333 47.1	25.4	145 32.3	20.5	284 51.6	32.5	Diphda	348 58.2	S17 55
	05	121 32.2	274 04.8	47.3	348 48.5	25.2	160 34.6	20.5	299 53.8	32.4			
	06	136 34.6	289 04.3	S 9 48.4	3 49.9	N19 25.0	175 36.9	S16 20.4	314 56.0	N 1 32.3	Dubhe	193 55.0	N61 41
S	07	151 37.1	304 03.8	49.6	18 51.3	24.8	190 39.1	20.3	329 58.3	32.3	Elnath	278 15.7	N28 37
A	08	166 39.6	319 03.3	50.7	33 52.8	24.6	205 41.4	20.3	345 00.5	32.2	Eltanin	90 47.8	N51 29
T	09	181 42.0	334 02.8	51.8	48 54.2	24.4	220 43.6	20.2	0 02.7	32.1	Enif	33 49.7	N 9 55
U	10	196 44.5	349 02.3	53.0	63 55.6	24.2	235 45.9	20.1	15 05.0	32.0	Fomalhaut	15 26.7	S29 34
R	11	211 46.9	4 01.8	54.1	78 57.0	24.0	250 48.2	20.1	30 07.2	31.9			
D	12	226 49.4	19 01.3	S 9 55.3	93 58.5	N19 23.8	265 50.4	S16 20.0	45 09.4	N 1 31.8	Gacrux	172 04.6	S57 10
A	13	241 51.9	34 00.8	56.4	108 59.9	23.6	280 52.7	19.9	60 11.6	31.7	Gienah	175 55.3	S17 35
Y	14	256 54.3	49 00.3	57.5	124 01.3	23.4	295 55.0	19.9	75 13.9	31.6	Hadar	148 52.5	S60 25
	15	271 56.8	63 59.9	58.7	139 02.7	23.2	310 57.2	19.8	90 16.1	31.5	Hamal	328 03.5	N23 30
	16	286 59.3	78 59.4	9 59.8	154 04.2	23.0	325 59.5	19.7	105 18.3	31.4	Kaus Aust.	83 47.6	S34 22
	17	302 01.7	93 58.9	10 01.0	169 05.6	22.8	341 01.8	19.6	120 20.6	31.3			
	18	317 04.2	108 58.4	S10 02.1	184 07.0	N19 22.6	356 04.0	S16 19.6	135 22.8	N 1 31.2	Kochab	137 20.8	N74 06
	19	332 06.7	123 57.9	03.2	199 08.4	22.4	11 06.3	19.5	150 25.0	31.1	Markab	13 40.9	N15 15
	20	347 09.1	138 57.4	04.4	214 09.9	22.2	26 08.5	19.4	165 27.3	31.0	Menkar	314 17.5	N 4 07
	21	2 11.6	153 56.9	05.5	229 11.3	22.0	41 10.8	19.4	180 29.5	30.9	Menkent	148 11.2	S36 25
	22	17 14.0	168 56.4	06.7	244 12.7	21.8	56 13.1	19.3	195 31.7	30.9	Miaplacidus	221 40.5	S69 45
	23	32 16.5	183 55.9	07.8	259 14.2	21.7	71 15.3	19.2	210 33.9	30.8			
	8 00	47 19.0	198 55.4	S10 08.9	274 15.6	N19 21.5	86 17.6	S16 19.2	225 36.2	N 1 30.7	Mirfak	308 43.8	N49 53
	01	62 21.4	213 54.9	10.1	289 17.0	21.3	101 19.8	19.1	240 38.4	30.6	Nunki	76 01.8	S26 17
	02	77 23.9	228 54.4	11.2	304 18.5	21.1	116 22.1	19.0	255 40.6	30.5	Peacock	53 23.4	S56 42
	03	92 26.4	243 53.9	12.3	319 19.9	20.9	131 24.4	19.0	270 42.9	30.4	Pollux	243 30.8	N28 00
	04	107 28.8	258 53.4	13.5	334 21.3	20.7	146 26.6	18.9	285 45.1	30.3	Procyon	245 02.4	N 5 12
	05	122 31.3	273 52.9	14.6	349 22.8	20.5	161 28.9	18.8	300 47.3	30.2			
	06	137 33.8	288 52.4	S10 15.7	4 24.2	N19 20.3	176 31.1	S16 18.8	315 49.6	N 1 30.1	Rasalhague	96 09.2	N12 33
	07	152 36.2	303 51.9	16.9	19 25.6	20.1	191 33.4	18.7	330 51.8	30.0	Regulus	207 46.4	N11 55
	08	167 38.7	318 51.4	18.0	34 27.1	19.9	206 35.6	18.6	345 54.0	29.9	Rigel	281 14.3	S 8 11
S	09	182 41.2	333 50.9	19.1	49 28.5	19.7	221 37.9	18.5	0 56.3	29.8	Rigil Kent.	139 56.2	S60 52
U	10	197 43.6	348 50.4	20.3	64 29.9	19.5	236 40.2	18.5	15 58.5	29.7	Sabik	102 15.9	S15 44
N	11	212 46.1	3 49.9	21.4	79 31.4	19.3	251 42.4	18.4	31 00.7	29.7			
D	12	227 48.5	18 49.4	S10 22.5	94 32.8	N19 19.1	266 44.7	S16 18.3	46 03.0	N 1 29.6	Schedar	349 43.3	N56 35
A	13	242 51.0	33 48.9	23.7	109 34.3	18.9	281 46.9	18.3	61 05.2	29.5	Shaula	96 25.9	S37 06
Y	14	257 53.5	48 48.4	24.8	124 35.7	18.7	296 49.2	18.2	76 07.4	29.4	Sirius	258 35.9	S16 43
	15	272 55.9	63 47.9	25.9	139 37.1	18.5	311 51.4	18.1	91 09.7	29.3	Spica	158 34.4	S11 12
	16	287 58.4	78 47.4	27.1	154 38.6	18.3	326 53.7	18.1	106 11.9	29.2	Suhail	222 54.5	S43 28
	17	303 00.9	93 46.9	28.2	169 40.0	18.1	341 55.9	18.0	121 14.1	29.1			
	18	318 03.3	108 46.4	S10 29.3	184 41.5	N19 17.9	356 58.2	S16 17.9	136 16.4	N 1 29.0	Vega	80 41.0	N38 47
	19	333 05.8	123 45.9	30.5	199 42.9	17.7	12 00.5	17.8	151 18.6	28.9	Zuben'ubi	137 08.7	S16 04
	20	348 08.3	138 45.4	31.6	214 44.4	17.5	27 02.7	17.8	166 20.8	28.8		SHA	Mer. Pas
	21	3 10.7	153 44.9	32.7	229 45.8	17.3	42 05.0	17.7	181 23.0	28.7		° ′	h m
	22	18 13.2	168 44.3	33.8	244 47.3	17.1	57 07.2	17.6	196 25.3	28.6	Venus	152 47.4	10 44
	23	33 15.7	183 43.8	35.0	259 48.7	16.9	72 09.5	17.6	211 27.5	28.5	Mars	227 21.6	5 45
		h m									Jupiter	39 03.4	18 16
	Mer. Pass.	20 51.3	*v* −0.5	*d* 1.1	*v* 1.4	*d* 0.2	*v* 2.3	*d* 0.1	*v* 2.2	*d* 0.1	Saturn	178 22.8	9 00

UT	SUN GHA	SUN Dec	MOON GHA	v	MOON Dec	d	HP
d h	° ′	° ′	° ′	′	° ′	′	′
6 00	184 05.7	S15 57.2	320 24.8	4.0	N25 44.9	1.6	59.3
01	199 05.6	57.9	334 47.8	4.0	25 43.3	1.8	59.3
02	214 05.6	58.7	349 10.8	3.9	25 41.5	2.0	59.3
03	229 05.6	15 59.4	3 33.7	4.0	25 39.5	2.1	59.4
04	244 05.6	16 00.2	17 56.7	4.0	25 37.4	2.3	59.4
05	259 05.5	00.9	32 19.7	4.1	25 35.1	2.5	59.4
06	274 05.5	S16 01.7	46 42.8	4.0	N25 32.6	2.7	59.4
07	289 05.5	02.4	61 05.8	4.1	25 29.9	2.8	59.4
08	304 05.5	03.2	75 28.9	4.1	25 27.1	3.1	59.4
09	319 05.4	. . 03.9	89 52.0	4.2	25 24.0	3.1	59.4
10	334 05.4	04.7	104 15.2	4.1	25 20.9	3.4	59.4
11	349 05.4	05.4	118 38.3	4.3	25 17.5	3.6	59.4
12	4 05.4	S16 06.1	133 01.6	4.2	N25 13.9	3.7	59.4
13	19 05.3	06.9	147 24.8	4.3	25 10.2	3.9	59.4
14	34 05.3	07.6	161 48.1	4.3	25 06.3	4.0	59.4
15	49 05.3	. . 08.4	176 11.4	4.4	25 02.3	4.3	59.4
16	64 05.2	09.1	190 34.8	4.4	24 58.0	4.4	59.4
17	79 05.2	09.9	204 58.2	4.4	24 53.6	4.6	59.4
18	94 05.2	S16 10.6	219 21.6	4.6	N24 49.0	4.7	59.4
19	109 05.2	11.3	233 45.2	4.5	24 44.3	4.9	59.4
20	124 05.1	12.1	248 08.7	4.6	24 39.4	5.1	59.4
21	139 05.1	. . 12.8	262 32.3	4.7	24 34.3	5.2	59.4
22	154 05.1	13.6	276 56.0	4.7	24 29.1	5.4	59.4
23	169 05.0	14.3	291 19.7	4.8	24 23.7	5.6	59.4
7 00	184 05.0	S16 15.0	305 43.5	4.8	N24 18.1	5.8	59.4
01	199 05.0	15.8	320 07.3	4.9	24 12.3	5.9	59.4
02	214 04.9	16.5	334 31.2	4.9	24 06.4	6.0	59.4
03	229 04.9	. . 17.3	348 55.1	5.0	24 00.4	6.2	59.4
04	244 04.9	18.0	3 19.1	5.1	23 54.2	6.4	59.4
05	259 04.8	18.7	17 43.2	5.2	23 47.8	6.6	59.4
06	274 04.8	S16 19.5	32 07.4	5.2	N23 41.2	6.6	59.4
07	289 04.8	20.2	46 31.6	5.3	23 34.6	6.9	59.4
08	304 04.7	20.9	60 55.9	5.3	23 27.7	7.0	59.4
09	319 04.7	. . 21.7	75 20.2	5.4	23 20.7	7.2	59.4
10	334 04.7	22.4	89 44.6	5.5	23 13.5	7.3	59.4
11	349 04.6	23.1	104 09.1	5.6	23 06.2	7.4	59.4
12	4 04.6	S16 23.9	118 33.7	5.6	N22 58.8	7.6	59.4
13	19 04.5	24.6	132 58.3	5.7	22 51.2	7.8	59.4
14	34 04.5	25.3	147 23.0	5.8	22 43.4	7.9	59.4
15	49 04.5	. . 26.1	161 47.8	5.9	22 35.5	8.0	59.4
16	64 04.4	26.8	176 12.7	5.9	22 27.5	8.2	59.4
17	79 04.4	27.5	190 37.6	6.0	22 19.3	8.4	59.4
18	94 04.4	S16 28.3	205 02.6	6.1	N22 10.9	8.5	59.4
19	109 04.3	29.0	219 27.7	6.2	22 02.4	8.6	59.4
20	124 04.3	29.7	233 52.9	6.3	21 53.8	8.7	59.4
21	139 04.2	. . 30.4	248 18.2	6.3	21 45.1	8.9	59.4
22	154 04.2	31.2	262 43.5	6.5	21 36.2	9.1	59.4
23	169 04.2	31.9	277 09.0	6.5	21 27.1	9.1	59.4
8 00	184 04.1	S16 32.6	291 34.5	6.6	N21 18.0	9.3	59.4
01	199 04.1	33.4	306 00.1	6.6	21 08.7	9.4	59.4
02	214 04.0	34.1	320 25.7	6.8	20 59.3	9.6	59.4
03	229 04.0	. . 34.8	334 51.5	6.8	20 49.7	9.7	59.4
04	244 03.9	35.5	349 17.3	7.0	20 40.0	9.8	59.4
05	259 03.9	36.3	3 43.3	7.0	20 30.2	9.9	59.4
06	274 03.9	S16 37.0	18 09.3	7.1	N20 20.3	10.1	59.4
07	289 03.8	37.7	32 35.4	7.2	20 10.2	10.2	59.4
08	304 03.8	38.4	47 01.6	7.2	20 00.0	10.3	59.4
09	319 03.7	. . 39.2	61 27.8	7.4	19 49.7	10.4	59.4
10	334 03.7	39.9	75 54.2	7.4	19 39.3	10.6	59.4
11	349 03.6	40.6	90 20.6	7.6	19 28.7	10.6	59.4
12	4 03.6	S16 41.3	104 47.2	7.6	N19 18.1	10.8	59.4
13	19 03.5	42.0	119 13.8	7.7	19 07.3	10.9	59.4
14	34 03.5	42.8	133 40.5	7.8	18 56.4	11.0	59.4
15	49 03.4	. . 43.5	148 07.3	7.8	18 45.4	11.1	59.3
16	64 03.4	44.2	162 34.1	8.0	18 34.3	11.2	59.3
17	79 03.4	44.9	177 01.1	8.0	18 23.1	11.3	59.3
18	94 03.3	S16 45.6	191 28.1	8.2	N18 11.8	11.5	59.3
19	109 03.3	46.4	205 55.3	8.2	18 00.3	11.5	59.3
20	124 03.2	47.1	220 22.5	8.3	17 48.8	11.6	59.3
21	139 03.2	. . 47.8	234 49.8	8.4	17 37.2	11.8	59.3
22	154 03.1	48.5	249 17.2	8.4	17 25.4	11.8	59.3
23	169 03.1	49.2	263 44.6	8.6	N17 13.6	11.9	59.3
	SD 16.2	*d* 0.7	SD 16.2		16.2		16.2

Lat.	Twilight Naut.	Twilight Civil	Sunrise	Moonrise 6	7	8	9
°	h m	h m	h m	h m	h m	h m	h m
N 72	06 30	07 57	09 39	▭	▭	▭	21 24
N 70	06 23	07 40	09 02	▭	▭	19 08	21 47
68	06 17	07 27	08 37	▭	▭	19 50	22 05
66	06 12	07 16	08 17	▭	17 58	20 17	22 19
64	06 08	07 06	08 02	16 35	18 39	20 38	22 31
62	06 04	06 58	07 49	17 21	19 06	20 55	22 41
60	06 01	06 51	07 38	17 51	19 28	21 09	22 49
N 58	05 57	06 45	07 28	18 14	19 45	21 21	22 56
56	05 54	06 39	07 20	18 33	20 00	21 31	23 03
54	05 52	06 34	07 12	18 48	20 12	21 41	23 09
52	05 49	06 29	07 06	19 02	20 23	21 49	23 14
50	05 46	06 25	07 00	19 14	20 33	21 56	23 18
45	05 41	06 16	06 47	19 38	20 53	22 11	23 28
N 40	05 35	06 07	06 36	19 58	21 10	22 24	23 37
35	05 30	06 00	06 27	20 14	21 24	22 34	23 44
30	05 25	05 54	06 19	20 28	21 36	22 44	23 50
20	05 15	05 42	06 04	20 52	21 57	23 00	24 01
N 10	05 05	05 30	05 52	21 13	22 14	23 14	24 10
0	04 54	05 19	05 40	21 32	22 31	23 27	24 19
S 10	04 41	05 06	05 28	21 51	22 48	23 40	24 27
20	04 25	04 52	05 15	22 12	23 05	23 53	24 37
30	04 05	04 35	05 01	22 35	23 25	24 09	00 09
35	03 52	04 25	04 52	22 49	23 37	24 18	00 18
40	03 36	04 12	04 42	23 05	23 51	24 28	00 28
45	03 17	03 58	04 30	23 24	24 07	00 07	00 40
S 50	02 51	03 39	04 16	23 48	24 26	00 26	00 55
52	02 38	03 30	04 10	24 00	00 00	00 35	01 02
54	02 22	03 20	04 02	24 13	00 13	00 46	01 09
56	02 03	03 08	03 54	24 28	00 28	00 57	01 17
58	01 39	02 54	03 45	00 03	00 45	01 11	01 27
S 60	01 04	02 38	03 34	00 28	01 06	01 26	01 37

Lat.	Sunset	Twilight Civil	Twilight Naut.	Moonset 6	7	8	9
°	h m	h m	h m	h m	h m	h m	h m
N 72	13 47	15 29	16 56	▭	▭	▭	15 13
N 70	14 24	15 46	17 03	▭	▭	15 36	14 48
68	14 49	15 59	17 09	▭	▭	14 53	14 28
66	15 09	16 11	17 14	▭	14 45	14 24	14 12
64	15 25	16 20	17 18	14 03	14 04	14 02	13 59
62	15 38	16 28	17 22	13 16	13 36	13 44	13 48
60	15 49	16 35	17 26	12 46	13 14	13 29	13 38
N 58	15 58	16 42	17 29	12 22	12 56	13 17	13 30
56	16 07	16 47	17 32	12 04	12 41	13 06	13 23
54	16 14	16 53	17 35	11 48	12 28	12 56	13 16
52	16 21	16 57	17 38	11 34	12 16	12 47	13 10
50	16 27	17 02	17 40	11 22	12 06	12 39	13 05
45	16 40	17 11	17 46	10 57	11 45	12 22	12 53
N 40	16 51	17 19	17 52	10 37	11 27	12 09	12 43
35	17 00	17 27	17 57	10 20	11 12	11 57	12 35
30	17 08	17 33	18 02	10 06	11 00	11 46	12 27
20	17 23	17 46	18 12	09 41	10 38	11 28	12 14
N 10	17 35	17 57	18 22	09 20	10 18	11 13	12 03
0	17 47	18 09	18 34	09 00	10 00	10 58	11 52
S 10	17 59	18 21	18 47	08 40	09 42	10 43	11 41
20	18 12	18 35	19 03	08 18	09 23	10 27	11 29
30	18 27	18 53	19 23	07 53	09 00	10 08	11 16
35	18 36	19 03	19 36	07 38	08 47	09 58	11 08
40	18 46	19 16	19 52	07 21	08 31	09 45	10 59
45	18 58	19 31	20 12	07 01	08 13	09 30	10 48
S 50	19 12	19 50	20 38	06 35	07 50	09 12	10 35
52	19 19	19 59	20 52	06 22	07 39	09 03	10 29
54	19 26	20 09	21 08	06 07	07 26	08 53	10 22
56	19 35	20 21	21 27	05 51	07 12	08 42	10 15
58	19 44	20 35	21 52	05 30	06 55	08 30	10 06
S 60	19 55	20 52	22 29	05 04	06 34	08 15	09 57

Day	SUN Eqn. of Time 00^h	SUN Eqn. of Time 12^h	SUN Mer. Pass.	MOON Mer. Pass. Upper	MOON Mer. Pass. Lower	Age	Phase
d	m s	m s	h m	h m	h m	d %	
6	16 23	16 21	11 44	02 45	15 16	19 83	
7	16 20	16 18	11 44	03 46	16 16	20 74	◑
8	16 17	16 14	11 44	04 45	17 12	21 63	

	UT	ARIES	VENUS −3.9		MARS +0.3		JUPITER −2.4		SATURN +1.1	
		GHA	GHA	Dec	GHA	Dec	GHA	Dec	GHA	Dec
	d h	° ′	° ′	° ′	° ′	° ′	° ′	° ′	° ′	° ′
	9 00	48 18.1	198 43.3	S10 36.1	274 50.2	N19 16.7	87 11.7	S16 17.5	226 29.7	N 1 28.5
	01	63 20.6	213 42.8	37.2	289 51.6	16.5	102 14.0	17.4	241 32.0	28.4
	02	78 23.0	228 42.3	38.3	304 53.0	16.4	117 16.2	17.3	256 34.2	28.3
	03	93 25.5	243 41.8	. . 39.5	319 54.5	. . 16.2	132 18.5	. . 17.3	271 36.4	. . 28.2
	04	108 28.0	258 41.3	40.6	334 55.9	16.0	147 20.7	17.2	286 38.7	28.1
	05	123 30.4	273 40.8	41.7	349 57.4	15.8	162 23.0	17.1	301 40.9	28.0
	06	138 32.9	288 40.3	S10 42.8	4 58.9	N19 15.6	177 25.2	S16 17.1	316 43.1	N 1 27.9
	07	153 35.4	303 39.8	44.0	20 00.3	15.4	192 27.5	17.0	331 45.4	27.8
	08	168 37.8	318 39.3	45.1	35 01.8	15.2	207 29.7	16.9	346 47.6	27.7
M	09	183 40.3	333 38.8	. . 46.2	50 03.2	. . 15.0	222 32.0	. . 16.8	1 49.8	. . 27.6
O	10	198 42.8	348 38.2	47.3	65 04.7	14.8	237 34.2	16.8	16 52.1	27.5
N	11	213 45.2	3 37.7	48.5	80 06.1	14.6	252 36.5	16.7	31 54.3	27.5
D	12	228 47.7	18 37.2	S10 49.6	95 07.6	N19 14.4	267 38.7	S16 16.6	46 56.6	N 1 27.4
A	13	243 50.2	33 36.7	50.7	110 09.0	14.2	282 41.0	16.6	61 58.8	27.3
Y	14	258 52.6	48 36.2	51.8	125 10.5	14.0	297 43.2	16.5	77 01.0	27.2
	15	273 55.1	63 35.7	. . 52.9	140 11.9	. . 13.8	312 45.5	. . 16.4	92 03.3	. . 27.1
	16	288 57.5	78 35.2	54.1	155 13.4	13.6	327 47.7	16.3	107 05.5	27.0
	17	304 00.0	93 34.7	55.2	170 14.9	13.4	342 50.0	16.3	122 07.7	26.9
	18	319 02.5	108 34.1	S10 56.3	185 16.3	N19 13.3	357 52.2	S16 16.2	137 10.0	N 1 26.8
	19	334 04.9	123 33.6	57.4	200 17.8	13.1	12 54.5	16.1	152 12.2	26.7
	20	349 07.4	138 33.1	58.5	215 19.2	12.9	27 56.7	16.0	167 14.4	26.6
	21	4 09.9	153 32.6	10 59.7	230 20.7	. . 12.7	42 59.0	. . 16.0	182 16.7	. . 26.5
	22	19 12.3	168 32.1	11 00.8	245 22.2	12.5	58 01.2	15.9	197 18.9	26.5
	23	34 14.8	183 31.6	01.9	260 23.6	12.3	73 03.5	15.8	212 21.1	26.4
	10 00	49 17.3	198 31.1	S11 03.0	275 25.1	N19 12.1	88 05.7	S16 15.8	227 23.4	N 1 26.3
	01	64 19.7	213 30.5	04.1	290 26.6	11.9	103 07.9	15.7	242 25.6	26.2
	02	79 22.2	228 30.0	05.2	305 28.0	11.7	118 10.2	15.6	257 27.8	26.1
	03	94 24.6	243 29.5	. . 06.4	320 29.5	. . 11.5	133 12.4	. . 15.5	272 30.1	. . 26.0
	04	109 27.1	258 29.0	07.5	335 31.0	11.3	148 14.7	15.5	287 32.3	25.9
	05	124 29.6	273 28.5	08.6	350 32.4	11.1	163 16.9	15.4	302 34.5	25.8
	06	139 32.0	288 28.0	S11 09.7	5 33.9	N19 10.9	178 19.2	S16 15.3	317 36.8	N 1 25.7
	07	154 34.5	303 27.4	10.8	20 35.4	10.8	193 21.4	15.2	332 39.0	25.6
T	08	169 37.0	318 26.9	11.9	35 36.8	10.6	208 23.7	15.2	347 41.3	25.5
U	09	184 39.4	333 26.4	. . 13.0	50 38.3	. . 10.4	223 25.9	. . 15.1	2 43.5	. . 25.5
E	10	199 41.9	348 25.9	14.1	65 39.8	10.2	238 28.1	15.0	17 45.7	25.4
S	11	214 44.4	3 25.4	15.3	80 41.2	10.0	253 30.4	14.9	32 48.0	25.3
D	12	229 46.8	18 24.8	S11 16.4	95 42.7	N19 09.8	268 32.6	S16 14.9	47 50.2	N 1 25.2
A	13	244 49.3	33 24.3	17.5	110 44.2	09.6	283 34.9	14.8	62 52.4	25.1
Y	14	259 51.8	48 23.8	18.6	125 45.7	09.4	298 37.1	14.7	77 54.7	25.0
	15	274 54.2	63 23.3	. . 19.7	140 47.1	. . 09.2	313 39.3	. . 14.6	92 56.9	. . 24.9
	16	289 56.7	78 22.7	20.8	155 48.6	09.0	328 41.6	14.6	107 59.1	24.8
	17	304 59.1	93 22.2	21.9	170 50.1	08.8	343 43.8	14.5	123 01.4	24.7
	18	320 01.6	108 21.7	S11 23.0	185 51.6	N19 08.7	358 46.1	S16 14.4	138 03.6	N 1 24.6
	19	335 04.1	123 21.2	24.1	200 53.0	08.5	13 48.3	14.3	153 05.9	24.6
	20	350 06.5	138 20.6	25.2	215 54.5	08.3	28 50.5	14.3	168 08.1	24.5
	21	5 09.0	153 20.1	. . 26.3	230 56.0	. . 08.1	43 52.8	. . 14.2	183 10.3	. . 24.4
	22	20 11.5	168 19.6	27.5	245 57.5	07.9	58 55.0	14.1	198 12.6	24.3
	23	35 13.9	183 19.1	28.6	260 58.9	07.7	73 57.3	14.0	213 14.8	24.2
	11 00	50 16.4	198 18.5	S11 29.7	276 00.4	N19 07.5	88 59.5	S16 14.0	228 17.0	N 1 24.1
	01	65 18.9	213 18.0	30.8	291 01.9	07.3	104 01.7	13.9	243 19.3	24.0
	02	80 21.3	228 17.5	31.9	306 03.4	07.1	119 04.0	13.8	258 21.5	23.9
	03	95 23.8	243 17.0	. . 33.0	321 04.9	. . 07.0	134 06.2	. . 13.7	273 23.8	. . 23.8
	04	110 26.3	258 16.4	34.1	336 06.4	06.8	149 08.4	13.7	288 26.0	23.8
	05	125 28.7	273 15.9	35.2	351 07.8	06.6	164 10.7	13.6	303 28.2	23.7
	06	140 31.2	288 15.4	S11 36.3	6 09.3	N19 06.4	179 12.9	S16 13.5	318 30.5	N 1 23.6
W	07	155 33.6	303 14.8	37.4	21 10.8	06.2	194 15.2	13.4	333 32.7	23.5
E	08	170 36.1	318 14.3	38.5	36 12.3	06.0	209 17.4	13.4	348 34.9	23.4
D	09	185 38.6	333 13.8	. . 39.6	51 13.8	. . 05.8	224 19.6	. . 13.3	3 37.2	. . 23.3
N	10	200 41.0	348 13.3	40.7	66 15.3	05.6	239 21.9	13.2	18 39.4	23.2
E	11	215 43.5	3 12.7	41.8	81 16.8	05.5	254 24.1	13.1	33 41.7	23.1
S	12	230 46.0	18 12.2	S11 42.9	96 18.2	N19 05.3	269 26.3	S16 13.0	48 43.9	N 1 23.0
D	13	245 48.4	33 11.7	44.0	111 19.7	05.1	284 28.6	13.0	63 46.1	23.0
A	14	260 50.9	48 11.1	45.1	126 21.2	04.9	299 30.8	12.9	78 48.4	22.9
Y	15	275 53.4	63 10.6	. . 46.2	141 22.7	. . 04.7	314 33.0	. . 12.8	93 50.6	. . 22.8
	16	290 55.8	78 10.1	47.3	156 24.2	04.5	329 35.3	12.7	108 52.9	22.7
	17	305 58.3	93 09.5	48.4	171 25.7	04.3	344 37.5	12.7	123 55.1	22.6
	18	321 00.7	108 09.0	S11 49.5	186 27.2	N19 04.1	359 39.7	S16 12.6	138 57.3	N 1 22.5
	19	336 03.2	123 08.5	50.6	201 28.7	04.0	14 42.0	12.5	153 59.6	22.4
	20	351 05.7	138 07.9	51.7	216 30.2	03.8	29 44.2	12.4	169 01.8	22.3
	21	6 08.1	153 07.4	. . 52.8	231 31.7	. . 03.6	44 46.4	. . 12.3	184 04.1	. . 22.2
	22	21 10.6	168 06.9	53.9	246 33.2	03.4	59 48.7	12.3	199 06.3	22.2
	23	36 13.1	183 06.3	55.0	261 34.7	03.2	74 50.9	12.2	214 08.5	22.1
	Mer. Pass.	h m 20 39.5	*v* −0.5	*d* 1.1	*v* 1.5	*d* 0.2	*v* 2.2	*d* 0.1	*v* 2.2	*d* 0.1

STARS		
Name	SHA	Dec
	° ′	°
Acamar	315 19.8	S40 15
Achernar	335 28.0	S57 11
Acrux	173 13.1	S63 09
Adhara	255 14.4	S28 58
Aldebaran	290 52.1	N16 31
Alioth	166 23.3	N55 54
Alkaid	153 01.3	N49 15
Al Na'ir	27 46.8	S46 54
Alnilam	275 48.8	S 1 11
Alphard	217 58.7	S 8 42
Alphecca	126 13.6	N26 40
Alpheratz	357 46.1	N29 09
Altair	62 11.0	N 8 53
Ankaa	353 17.9	S42 15
Antares	112 29.9	S26 27
Arcturus	145 58.4	N19 07
Atria	107 34.7	S69 02
Avior	234 19.1	S59 32
Bellatrix	278 34.6	N 6 21
Betelgeuse	271 03.9	N 7 24
Canopus	263 57.0	S52 41
Capella	280 38.0	N46 00
Deneb	49 33.5	N45 19
Denebola	182 36.5	N14 30
Diphda	348 58.2	S17 55
Dubhe	193 55.0	N61 41
Elnath	278 15.7	N28 37
Eltanin	90 47.8	N51 29
Enif	33 49.7	N 9 55
Fomalhaut	15 26.7	S29 34
Gacrux	172 04.5	S57 10
Gienah	175 55.3	S17 35
Hadar	148 52.5	S60 25
Hamal	328 03.5	N23 30
Kaus Aust.	83 47.6	S34 22
Kochab	137 20.8	N74 06
Markab	13 40.9	N15 15
Menkar	314 17.5	N 4 07
Menkent	148 11.2	S36 25
Miaplacidus	221 40.5	S69 45
Mirfak	308 43.8	N49 53
Nunki	76 01.8	S26 17
Peacock	53 23.4	S56 42
Pollux	243 30.8	N28 00
Procyon	245 02.4	N 5 12
Rasalhague	96 09.2	N12 33
Regulus	207 46.4	N11 55
Rigel	281 14.3	S 8 11
Rigil Kent.	139 56.2	S60 52
Sabik	102 15.9	S15 44
Schedar	349 43.3	N56 35
Shaula	96 25.9	S37 06
Sirius	258 35.8	S16 43
Spica	158 34.4	S11 12
Suhail	222 54.5	S43 28
Vega	80 41.1	N38 47
Zuben'ubi	137 08.7	S16 04

	SHA	Mer. Pas
	° ′	h m
Venus	149 13.8	10 46
Mars	226 07.8	5 38
Jupiter	38 48.4	18 05
Saturn	178 06.1	8 49

UT		SUN GHA	SUN Dec	MOON GHA	v	MOON Dec	d	HP
d	h	° ′	° ′	° ′	′	° ′	′	′
9	00	184 03.0	S16 49.9	278 12.2	8.6	N17 01.7	12.1	59.3
	01	199 03.0	50.7	292 39.8	8.7	16 49.6	12.1	59.3
	02	214 02.9	51.4	307 07.5	8.8	16 37.5	12.2	59.3
	03	229 02.9	. . 52.1	321 35.3	8.9	16 25.3	12.3	59.3
	04	244 02.8	52.8	336 03.2	9.0	16 13.0	12.4	59.3
	05	259 02.8	53.5	350 31.2	9.0	16 00.6	12.5	59.3
	06	274 02.7	S16 54.2	4 59.2	9.2	N15 48.1	12.6	59.2
	07	289 02.6	54.9	19 27.4	9.2	15 35.5	12.6	59.2
	08	304 02.6	55.6	33 55.6	9.2	15 22.9	12.8	59.2
MONDAY	09	319 02.5	. . 56.4	48 23.8	9.4	15 10.1	12.8	59.2
	10	334 02.5	57.1	62 52.2	9.4	14 57.3	12.9	59.2
	11	349 02.4	57.8	77 20.6	9.6	14 44.4	13.0	59.2
	12	4 02.4	S16 58.5	91 49.2	9.6	N14 31.4	13.0	59.2
	13	19 02.3	59.2	106 17.8	9.6	14 18.4	13.2	59.2
	14	34 02.3	16 59.9	120 46.4	9.8	14 05.2	13.2	59.2
	15	49 02.2	17 00.6	135 15.2	9.8	13 52.0	13.3	59.2
	16	64 02.2	01.3	149 44.0	9.9	13 38.7	13.3	59.2
	17	79 02.1	02.0	164 12.9	9.9	13 25.4	13.5	59.2
	18	94 02.0	S17 02.7	178 41.8	10.1	N13 11.9	13.4	59.1
	19	109 02.0	03.4	193 10.9	10.1	12 58.5	13.6	59.1
	20	124 01.9	04.1	207 40.0	10.2	12 44.9	13.6	59.1
	21	139 01.9	. . 04.9	222 09.2	10.2	12 31.3	13.7	59.1
	22	154 01.8	05.6	236 38.4	10.3	12 17.6	13.8	59.1
	23	169 01.8	06.3	251 07.7	10.4	12 03.8	13.8	59.1
10	00	184 01.7	S17 07.0	265 37.1	10.5	N11 50.0	13.9	59.1
	01	199 01.6	07.7	280 06.6	10.5	11 36.1	13.9	59.1
	02	214 01.6	08.4	294 36.1	10.5	11 22.2	14.0	59.1
	03	229 01.5	. . 09.1	309 05.6	10.7	11 08.2	14.0	59.1
	04	244 01.5	09.8	323 35.3	10.7	10 54.2	14.1	59.0
	05	259 01.4	10.5	338 05.0	10.8	10 40.1	14.1	59.0
	06	274 01.3	S17 11.2	352 34.8	10.8	N10 26.0	14.2	59.0
	07	289 01.3	11.9	7 04.6	10.9	10 11.8	14.3	59.0
	08	304 01.2	12.6	21 34.5	10.9	9 57.5	14.3	59.0
TUESDAY	09	319 01.1	. . 13.3	36 04.4	11.0	9 43.2	14.3	59.0
	10	334 01.1	14.0	50 34.4	11.1	9 28.9	14.4	59.0
	11	349 01.0	14.7	65 04.5	11.1	9 14.5	14.4	59.0
	12	4 01.0	S17 15.4	79 34.6	11.1	N 9 00.1	14.5	59.0
	13	19 00.9	16.1	94 04.7	11.3	8 45.6	14.5	59.0
	14	34 00.8	16.8	108 35.0	11.2	8 31.1	14.6	58.9
	15	49 00.8	. . 17.5	123 05.2	11.4	8 16.5	14.5	58.9
	16	64 00.7	18.2	137 35.6	11.3	8 02.0	14.7	58.9
	17	79 00.6	18.9	152 05.9	11.5	7 47.3	14.6	58.9
	18	94 00.6	S17 19.5	166 36.4	11.4	N 7 32.7	14.7	58.9
	19	109 00.5	20.2	181 06.8	11.5	7 18.0	14.7	58.9
	20	124 00.4	20.9	195 37.3	11.6	7 03.3	14.8	58.9
	21	139 00.4	. . 21.6	210 07.9	11.6	6 48.5	14.8	58.9
	22	154 00.3	22.3	224 38.5	11.7	6 33.7	14.8	58.8
	23	169 00.2	23.0	239 09.2	11.7	6 18.9	14.8	58.8
11	00	184 00.2	S17 23.7	253 39.9	11.7	N 6 04.1	14.9	58.8
	01	199 00.1	24.4	268 10.6	11.8	5 49.2	14.9	58.8
	02	214 00.0	25.1	282 41.4	11.8	5 34.3	14.9	58.8
	03	229 00.0	. . 25.8	297 12.2	11.8	5 19.4	14.9	58.8
	04	243 59.9	26.5	311 43.0	11.9	5 04.5	15.0	58.8
	05	258 59.8	27.2	326 13.9	11.9	4 49.5	14.9	58.8
	06	273 59.7	S17 27.8	340 44.8	12.0	N 4 34.6	15.0	58.7
	07	288 59.7	28.5	355 15.8	12.0	4 19.6	15.0	58.7
	08	303 59.6	29.2	9 46.8	12.0	4 04.6	15.0	58.7
	09	318 59.5	. . 29.9	24 17.8	12.0	3 49.6	15.1	58.7
	10	333 59.5	30.6	38 48.8	12.1	3 34.5	15.0	58.7
WEDNESDAY	11	348 59.4	31.3	53 19.9	12.1	3 19.5	15.0	58.7
	12	3 59.3	S17 32.0	67 51.0	12.1	N 3 04.5	15.1	58.7
	13	18 59.2	32.6	82 22.1	12.2	2 49.4	15.1	58.7
	14	33 59.2	33.3	96 53.3	12.2	2 34.3	15.0	58.6
	15	48 59.1	. . 34.0	111 24.5	12.2	2 19.3	15.1	58.6
	16	63 59.0	34.7	125 55.7	12.2	2 04.2	15.1	58.6
	17	78 58.9	35.4	140 26.9	12.3	1 49.1	15.1	58.6
	18	93 58.9	S17 36.1	154 58.2	12.2	N 1 34.0	15.1	58.6
	19	108 58.8	36.7	169 29.4	12.3	1 18.9	15.1	58.6
	20	123 58.7	37.4	184 00.7	12.3	1 03.8	15.1	58.6
	21	138 58.6	. . 38.1	198 32.0	12.4	0 48.7	15.0	58.5
	22	153 58.6	38.8	213 03.4	12.3	0 33.7	15.1	58.5
	23	168 58.5	39.5	227 34.7	12.4	N 0 18.6	15.1	58.5
		SD 16.2	d 0.7	SD 16.1		16.1		16.0

Lat.	Twilight Naut.	Twilight Civil	Sunrise	Moonrise 9	Moonrise 10	Moonrise 11	Moonrise 12
°	h m	h m	h m	h m	h m	h m	h m
N 72	06 41	08 11	10 03	21 24	23 50	26 03	02 03
N 70	06 33	07 52	09 19	21 47	23 59	26 02	02 02
68	06 27	07 37	08 50	22 05	24 06	00 06	02 01
66	06 21	07 25	08 29	22 19	24 12	00 12	02 01
64	06 16	07 15	08 12	22 31	24 18	00 18	02 00
62	06 11	07 06	07 57	22 41	24 22	00 22	02 00
60	06 07	06 58	07 45	22 49	24 26	00 26	02 00
N 58	06 03	06 51	07 35	22 56	24 29	00 29	01 59
56	06 00	06 45	07 26	23 03	24 32	00 32	01 59
54	05 57	06 39	07 18	23 09	24 35	00 35	01 59
52	05 54	06 34	07 11	23 14	24 37	00 37	01 59
50	05 51	06 30	07 05	23 18	24 39	00 39	01 58
45	05 44	06 19	06 51	23 28	24 44	00 44	01 58
N 40	05 38	06 11	06 39	23 37	24 48	00 48	01 58
35	05 33	06 03	06 30	23 44	24 51	00 51	01 57
30	05 27	05 56	06 21	23 50	24 54	00 54	01 57
20	05 17	05 43	06 06	24 01	00 01	00 59	01 57
N 10	05 06	05 31	05 53	24 10	00 10	01 04	01 56
0	04 54	05 19	05 40	24 19	00 19	01 08	01 56
S 10	04 40	05 06	05 28	24 27	00 27	01 12	01 56
20	04 23	04 51	05 14	24 37	00 37	01 17	01 55
30	04 02	04 33	04 59	00 09	00 47	01 22	01 55
35	03 49	04 22	04 50	00 18	00 53	01 25	01 55
40	03 32	04 09	04 39	00 28	01 00	01 28	01 55
45	03 12	03 53	04 27	00 40	01 08	01 32	01 54
S 50	02 44	03 34	04 12	00 55	01 17	01 37	01 54
52	02 30	03 24	04 05	01 02	01 22	01 39	01 54
54	02 13	03 13	03 57	01 09	01 27	01 41	01 54
56	01 53	03 01	03 48	01 17	01 32	01 43	01 54
58	01 25	02 46	03 38	01 27	01 38	01 46	01 54
S 60	00 41	02 28	03 26	01 37	01 44	01 49	01 53

Lat.	Sunset	Twilight Civil	Twilight Naut.	Moonset 9	Moonset 10	Moonset 11	Moonset 12
°	h m	h m	h m	h m	h m	h m	h m
N 72	13 23	15 15	16 45	15 13	14 33	14 02	13 34
N 70	14 07	15 34	16 53	14 48	14 21	13 59	13 39
68	14 37	15 50	17 00	14 28	14 11	13 56	13 43
66	14 58	16 02	17 06	14 12	14 03	13 54	13 46
64	15 15	16 12	17 11	13 59	13 56	13 52	13 49
62	15 30	16 21	17 16	13 48	13 50	13 51	13 51
60	15 42	16 29	17 20	13 38	13 45	13 49	13 53
N 58	15 52	16 36	17 24	13 30	13 40	13 48	13 55
56	16 01	16 42	17 27	13 23	13 36	13 47	13 57
54	16 09	16 48	17 30	13 16	13 32	13 46	13 58
52	16 16	16 53	17 34	13 10	13 29	13 45	14 00
50	16 23	16 58	17 36	13 05	13 25	13 44	14 01
45	16 36	17 08	17 43	12 53	13 19	13 42	14 04
N 40	16 48	17 17	17 49	12 43	13 13	13 40	14 06
35	16 58	17 25	17 55	12 35	13 08	13 39	14 08
30	17 07	17 32	18 00	12 27	13 03	13 37	14 10
20	17 22	17 45	18 11	12 14	12 56	13 35	14 13
N 10	17 35	17 57	18 22	12 03	12 49	13 33	14 16
0	17 47	18 09	18 34	11 52	12 43	13 31	14 18
S 10	18 00	18 22	18 48	11 41	12 36	13 29	14 21
20	18 14	18 37	19 05	11 29	12 29	13 27	14 24
30	18 30	18 55	19 26	11 16	12 21	13 24	14 27
35	18 39	19 06	19 40	11 08	12 16	13 23	14 28
40	18 49	19 20	19 56	10 59	12 11	13 21	14 30
45	19 02	19 35	20 17	10 48	12 05	13 19	14 33
S 50	19 17	19 55	20 45	10 35	11 57	13 17	14 36
52	19 24	20 05	21 00	10 29	11 54	13 16	14 37
54	19 32	20 16	21 17	10 22	11 50	13 15	14 38
56	19 41	20 29	21 38	10 15	11 45	13 13	14 40
58	19 52	20 44	22 07	10 06	11 41	13 12	14 42
S 60	20 03	21 02	22 56	09 57	11 35	13 10	14 44

Day	SUN Eqn. of Time 00h	SUN Eqn. of Time 12h	SUN Mer. Pass.	MOON Mer. Pass. Upper	MOON Mer. Pass. Lower	MOON Age	MOON Phase
d	m s	m s	h m	h m	h m	d %	
9	16 12	16 10	11 44	05 39	18 05	22 52	
10	16 07	16 04	11 44	06 31	18 55	23 41	
11	16 01	15 57	11 44	07 20	19 43	24 30	

UT d	h	ARIES GHA ° ′	VENUS −3.9 GHA ° ′	VENUS Dec ° ′	MARS +0.3 GHA ° ′	MARS Dec ° ′	JUPITER −2.4 GHA ° ′	JUPITER Dec ° ′	SATURN +1.1 GHA ° ′	SATURN Dec ° ′
12	00	51 15.5	198 05.8	S11 56.1	276 36.2	N19 03.0	89 53.1	S16 12.1	229 10.8	N 1 22.0
	01	66 18.0	213 05.3	57.2	291 37.7	02.8	104 55.4	12.0	244 13.0	21.9
	02	81 20.5	228 04.7	58.2	306 39.2	02.7	119 57.6	12.0	259 15.3	21.8
	03	96 22.9	243 04.2	11 59.3	321 40.7	02.5	134 59.8	11.9	274 17.5	21.7
	04	111 25.4	258 03.6	12 00.4	336 42.2	02.3	150 02.1	11.8	289 19.7	21.6
	05	126 27.9	273 03.1	01.5	351 43.7	02.1	165 04.3	11.7	304 22.0	21.5
	06	141 30.3	288 02.6	S12 02.6	6 45.2	N19 01.9	180 06.5	S16 11.6	319 24.2	N 1 21.4
	07	156 32.8	303 02.0	03.7	21 46.7	01.7	195 08.7	11.6	334 26.5	21.4
T	08	171 35.2	318 01.5	04.8	36 48.2	01.5	210 11.0	11.5	349 28.7	21.3
H	09	186 37.7	333 00.9	05.9	51 49.7	01.4	225 13.2	11.4	4 30.9	21.2
U	10	201 40.2	348 00.4	07.0	66 51.2	01.2	240 15.4	11.3	19 33.2	21.1
R	11	216 42.6	2 59.9	08.1	81 52.7	01.0	255 17.7	11.3	34 35.4	21.0
S	12	231 45.1	17 59.3	S12 09.2	96 54.2	N19 00.8	270 19.9	S16 11.2	49 37.7	N 1 20.9
D	13	246 47.6	32 58.8	10.2	111 55.7	00.6	285 22.1	11.1	64 39.9	20.8
A	14	261 50.0	47 58.2	11.3	126 57.2	00.4	300 24.3	11.0	79 42.1	20.7
Y	15	276 52.5	62 57.7	12.4	141 58.7	00.3	315 26.6	10.9	94 44.4	20.7
	16	291 55.0	77 57.2	13.5	157 00.2	19 00.1	330 28.8	10.9	109 46.6	20.6
	17	306 57.4	92 56.6	14.6	172 01.7	18 59.9	345 31.0	10.8	124 48.9	20.5
	18	321 59.9	107 56.1	S12 15.7	187 03.2	N18 59.7	0 33.2	S16 10.7	139 51.1	N 1 20.4
	19	337 02.3	122 55.5	16.8	202 04.7	59.5	15 35.5	10.6	154 53.3	20.3
	20	352 04.8	137 55.0	17.8	217 06.3	59.3	30 37.7	10.5	169 55.6	20.2
	21	7 07.3	152 54.4	18.9	232 07.8	59.2	45 39.9	10.5	184 57.8	20.1
	22	22 09.7	167 53.9	20.0	247 09.3	59.0	60 42.1	10.4	200 00.1	20.0
	23	37 12.2	182 53.3	21.1	262 10.8	58.8	75 44.4	10.3	215 02.3	20.0
13	00	52 14.7	197 52.8	S12 22.2	277 12.3	N18 58.6	90 46.6	S16 10.2	230 04.6	N 1 19.9
	01	67 17.1	212 52.2	23.3	292 13.8	58.4	105 48.8	10.1	245 06.8	19.8
	02	82 19.6	227 51.7	24.3	307 15.3	58.3	120 51.0	10.1	260 09.0	19.7
	03	97 22.1	242 51.1	25.4	322 16.9	58.1	135 53.3	10.0	275 11.3	19.6
	04	112 24.5	257 50.6	26.5	337 18.4	57.9	150 55.5	09.9	290 13.5	19.5
	05	127 27.0	272 50.1	27.6	352 19.9	57.7	165 57.7	09.8	305 15.8	19.4
	06	142 29.5	287 49.5	S12 28.7	7 21.4	N18 57.5	180 59.9	S16 09.7	320 18.0	N 1 19.3
	07	157 31.9	302 49.0	29.7	22 22.9	57.3	196 02.1	09.6	335 20.3	19.3
	08	172 34.4	317 48.4	30.8	37 24.4	57.2	211 04.4	09.6	350 22.5	19.2
F	09	187 36.8	332 47.9	31.9	52 26.0	57.0	226 06.6	09.5	5 24.7	19.1
R	10	202 39.3	347 47.3	33.0	67 27.5	56.8	241 08.8	09.4	20 27.0	19.0
I	11	217 41.8	2 46.7	34.0	82 29.0	56.6	256 11.0	09.3	35 29.2	18.9
D	12	232 44.2	17 46.2	S12 35.1	97 30.5	N18 56.4	271 13.3	S16 09.2	50 31.5	N 1 18.8
A	13	247 46.7	32 45.6	36.2	112 32.1	56.3	286 15.5	09.2	65 33.7	18.7
Y	14	262 49.2	47 45.1	37.3	127 33.6	56.1	301 17.7	09.1	80 36.0	18.7
	15	277 51.6	62 44.5	38.3	142 35.1	55.9	316 19.9	09.0	95 38.2	18.6
	16	292 54.1	77 44.0	39.4	157 36.6	55.7	331 22.1	08.9	110 40.4	18.5
	17	307 56.6	92 43.4	40.5	172 38.2	55.5	346 24.4	08.8	125 42.7	18.4
	18	322 59.0	107 42.9	S12 41.6	187 39.7	N18 55.4	1 26.6	S16 08.7	140 44.9	N 1 18.3
	19	338 01.5	122 42.3	42.6	202 41.2	55.2	16 28.8	08.7	155 47.2	18.2
	20	353 04.0	137 41.8	43.7	217 42.7	55.0	31 31.0	08.6	170 49.4	18.1
	21	8 06.4	152 41.2	44.8	232 44.3	54.8	46 33.2	08.5	185 51.7	18.1
	22	23 08.9	167 40.7	45.9	247 45.8	54.6	61 35.4	08.4	200 53.9	18.0
	23	38 11.3	182 40.1	46.9	262 47.3	54.5	76 37.7	08.3	215 56.2	17.9
14	00	53 13.8	197 39.5	S12 48.0	277 48.9	N18 54.3	91 39.9	S16 08.3	230 58.4	N 1 17.8
	01	68 16.3	212 39.0	49.1	292 50.4	54.1	106 42.1	08.2	246 00.6	17.7
	02	83 18.7	227 38.4	50.1	307 51.9	53.9	121 44.3	08.1	261 02.9	17.6
	03	98 21.2	242 37.9	51.2	322 53.5	53.8	136 46.5	08.0	276 05.1	17.5
	04	113 23.7	257 37.3	52.3	337 55.0	53.6	151 48.7	07.9	291 07.4	17.4
	05	128 26.1	272 36.7	53.3	352 56.5	53.4	166 51.0	07.8	306 09.6	17.4
	06	143 28.6	287 36.2	S12 54.4	7 58.1	N18 53.2	181 53.2	S16 07.8	321 11.9	N 1 17.3
	07	158 31.1	302 35.6	55.5	22 59.6	53.0	196 55.4	07.7	336 14.1	17.2
S	08	173 33.5	317 35.1	56.5	38 01.1	52.9	211 57.6	07.6	351 16.4	17.1
A	09	188 36.0	332 34.5	57.6	53 02.7	52.7	226 59.8	07.5	6 18.6	17.0
T	10	203 38.4	347 33.9	58.7	68 04.2	52.5	242 02.0	07.4	21 20.9	16.9
U	11	218 40.9	2 33.4	12 59.7	83 05.8	52.3	257 04.2	07.3	36 23.1	16.8
R	12	233 43.4	17 32.8	S13 00.8	98 07.3	N18 52.2	272 06.5	S16 07.3	51 25.3	N 1 16.8
D	13	248 45.8	32 32.3	01.9	113 08.8	52.0	287 08.7	07.2	66 27.6	16.7
A	14	263 48.3	47 31.7	02.9	128 10.4	51.8	302 10.9	07.1	81 29.8	16.6
Y	15	278 50.8	62 31.1	04.0	143 11.9	51.6	317 13.1	07.0	96 32.1	16.5
	16	293 53.2	77 30.6	05.0	158 13.5	51.5	332 15.3	06.9	111 34.3	16.4
	17	308 55.7	92 30.0	06.1	173 15.0	51.3	347 17.5	06.8	126 36.6	16.3
	18	323 58.2	107 29.4	S13 07.2	188 16.6	N18 51.1	2 19.7	S16 06.7	141 38.8	N 1 16.3
	19	339 00.6	122 28.9	08.2	203 18.1	50.9	17 22.0	06.7	156 41.1	16.2
	20	354 03.1	137 28.3	09.3	218 19.6	50.8	32 24.2	06.6	171 43.3	16.1
	21	9 05.6	152 27.7	10.3	233 21.2	50.6	47 26.4	06.5	186 45.6	16.0
	22	24 08.0	167 27.2	11.4	248 22.7	50.4	62 28.6	06.4	201 47.8	15.9
	23	39 10.5	182 26.6	12.4	263 24.3	50.2	77 30.8	06.3	216 50.1	15.8
Mer. Pass.		h m 20 27.7	v −0.6	d 1.1	v 1.5	d 0.2	v 2.2	d 0.1	v 2.2	d 0.1

STARS

Name	SHA ° ′	Dec ° ′
Acamar	315 19.8	S40 15.
Achernar	335 28.0	S57 11.
Acrux	173 13.0	S63 09.
Adhara	255 14.4	S28 58.
Aldebaran	290 52.1	N16 31.
Alioth	166 23.3	N55 54.
Alkaid	153 01.3	N49 15.
Al Na'ir	27 46.9	S46 54.
Alnilam	275 48.8	S 1 11.
Alphard	217 58.7	S 8 42.
Alphecca	126 13.6	N26 40.
Alpheratz	357 46.1	N29 09.
Altair	62 11.0	N 8 53.
Ankaa	353 17.9	S42 15.
Antares	112 29.9	S26 27.
Arcturus	145 58.4	N19 07.
Atria	107 34.7	S69 02.
Avior	234 19.0	S59 32.
Bellatrix	278 34.6	N 6 21.
Betelgeuse	271 03.9	N 7 24.
Canopus	263 56.9	S52 41.
Capella	280 38.0	N46 00.
Deneb	49 33.5	N45 19.
Denebola	182 36.5	N14 30.
Diphda	348 58.2	S17 55.
Dubhe	193 55.0	N61 41.
Elnath	278 15.7	N28 37.
Eltanin	90 47.8	N51 29.
Enif	33 49.7	N 9 55.
Fomalhaut	15 26.7	S29 34.
Gacrux	172 04.5	S57 10.
Gienah	175 55.3	S17 35.
Hadar	148 52.5	S60 25.
Hamal	328 03.5	N23 30.
Kaus Aust.	83 47.6	S34 22.
Kochab	137 20.8	N74 06.
Markab	13 40.9	N15 15.
Menkar	314 17.5	N 4 07.
Menkent	148 11.2	S36 25.
Miaplacidus	221 40.4	S69 45
Mirfak	308 43.7	N49 53.
Nunki	76 01.8	S26 17.
Peacock	53 23.5	S56 42.
Pollux	243 30.8	N28 00.
Procyon	245 02.3	N 5 12.
Rasalhague	96 09.2	N12 33.
Regulus	207 46.4	N11 55.
Rigel	281 14.3	S 8 11.
Rigil Kent.	139 56.2	S60 52.
Sabik	102 15.9	S15 44.
Schedar	349 43.3	N56 35.
Shaula	96 25.9	S37 06.
Sirius	258 35.8	S16 43.
Spica	158 34.4	S11 12.
Suhail	222 54.4	S43 28.
Vega	80 41.1	N38 47.
Zuben'ubi	137 08.7	S16 04.

	SHA ° ′	Mer. Pass h m
Venus	145 38.1	10 49
Mars	224 57.6	5 31
Jupiter	38 31.9	17 54
Saturn	177 49.9	8 38

UT	SUN GHA	SUN Dec	MOON GHA	v	MOON Dec	d	HP
d h	° ′	° ′	° ′	′	° ′	′	′
12 00	183 58.4	S17 40.1	242 06.1	12.3	N 0 03.5	15.1	58.5
01	198 58.3	40.8	256 37.4	12.4	S 0 11.6	15.0	58.5
02	213 58.3	41.5	271 08.8	12.4	0 26.6	15.1	58.5
03	228 58.2	. . 42.2	285 40.2	12.4	0 41.7	15.0	58.5
04	243 58.1	42.8	300 11.6	12.4	0 56.7	15.0	58.4
05	258 58.0	43.5	314 43.0	12.4	1 11.7	15.1	58.4
06	273 57.9	S17 44.2	329 14.4	12.5	S 1 26.8	15.0	58.4
07	288 57.9	44.9	343 45.9	12.4	1 41.8	15.0	58.4
08	303 57.8	45.6	358 17.3	12.4	1 56.8	14.9	58.4
09	318 57.7	. . 46.2	12 48.7	12.5	2 11.7	15.0	58.4
10	333 57.6	46.9	27 20.2	12.4	2 26.7	14.9	58.4
11	348 57.5	47.6	41 51.6	12.5	2 41.6	14.9	58.3
12	3 57.5	S17 48.2	56 23.1	12.4	S 2 56.5	14.9	58.3
13	18 57.4	48.9	70 54.5	12.4	3 11.4	14.9	58.3
14	33 57.3	49.6	85 25.9	12.5	3 26.3	14.9	58.3
15	48 57.2	. . 50.3	99 57.4	12.4	3 41.2	14.8	58.3
16	63 57.1	50.9	114 28.8	12.4	3 56.0	14.8	58.3
17	78 57.0	51.6	129 00.2	12.5	4 10.8	14.7	58.2
18	93 57.0	S17 52.3	143 31.7	12.4	S 4 25.5	14.8	58.2
19	108 56.9	52.9	158 03.1	12.4	4 40.3	14.7	58.2
20	123 56.8	53.6	172 34.5	12.4	4 55.0	14.7	58.2
21	138 56.7	. . 54.3	187 05.9	12.4	5 09.7	14.6	58.2
22	153 56.6	54.9	201 37.3	12.4	5 24.3	14.6	58.2
23	168 56.5	55.6	216 08.7	12.4	5 38.9	14.6	58.1
13 00	183 56.4	S17 56.3	230 40.1	12.3	S 5 53.5	14.6	58.1
01	198 56.4	56.9	245 11.4	12.4	6 08.1	14.5	58.1
02	213 56.3	57.6	259 42.8	12.3	6 22.6	14.4	58.1
03	228 56.2	. . 58.3	274 14.1	12.3	6 37.0	14.5	58.1
04	243 56.1	58.9	288 45.4	12.4	6 51.5	14.4	58.1
05	258 56.0	17 59.6	303 16.8	12.2	7 05.9	14.3	58.0
06	273 55.9	S18 00.3	317 48.0	12.3	S 7 20.2	14.3	58.0
07	288 55.8	00.9	332 19.3	12.3	7 34.5	14.3	58.0
08	303 55.7	01.6	346 50.6	12.2	7 48.8	14.2	58.0
09	318 55.6	. . 02.2	1 21.8	12.2	8 03.0	14.1	58.0
10	333 55.6	02.9	15 53.0	12.2	8 17.1	14.1	58.0
11	348 55.5	03.6	30 24.2	12.2	8 31.2	14.1	57.9
12	3 55.4	S18 04.2	44 55.4	12.1	S 8 45.3	14.0	57.9
13	18 55.3	04.9	59 26.5	12.1	8 59.3	14.0	57.9
14	33 55.2	05.5	73 57.6	12.1	9 13.3	13.9	57.9
15	48 55.1	. . 06.2	88 28.7	12.1	9 27.2	13.9	57.9
16	63 55.0	06.9	102 59.8	12.0	9 41.1	13.8	57.9
17	78 54.9	07.5	117 30.8	12.1	9 54.9	13.7	57.8
18	93 54.8	S18 08.2	132 01.9	11.9	S10 08.6	13.7	57.8
19	108 54.7	08.8	146 32.8	12.0	10 22.3	13.6	57.8
20	123 54.6	09.5	161 03.8	11.9	10 35.9	13.6	57.8
21	138 54.5	. . 10.1	175 34.7	11.9	10 49.5	13.5	57.8
22	153 54.4	10.8	190 05.6	11.9	11 03.0	13.4	57.7
23	168 54.4	11.4	204 36.5	11.9	11 16.4	13.4	57.7
14 00	183 54.3	S18 12.1	219 07.4	11.8	S11 29.8	13.3	57.7
01	198 54.2	12.7	233 38.2	11.7	11 43.1	13.3	57.7
02	213 54.1	13.4	248 08.9	11.8	11 56.4	13.2	57.7
03	228 54.0	. . 14.0	262 39.7	11.7	12 09.6	13.1	57.7
04	243 53.9	14.7	277 10.4	11.7	12 22.7	13.0	57.6
05	258 53.8	15.3	291 41.1	11.6	12 35.7	13.0	57.6
06	273 53.7	S18 16.0	306 11.7	11.6	S12 48.7	12.9	57.6
07	288 53.6	16.6	320 42.3	11.6	13 01.6	12.8	57.6
08	303 53.5	17.3	335 12.9	11.5	13 14.4	12.8	57.6
09	318 53.4	. . 17.9	349 43.4	11.5	13 27.2	12.7	57.5
10	333 53.3	18.6	4 13.9	11.4	13 39.9	12.6	57.5
11	348 53.2	19.2	18 44.3	11.4	13 52.5	12.5	57.5
12	3 53.1	S18 19.9	33 14.7	11.4	S14 05.0	12.4	57.5
13	18 53.0	20.5	47 45.1	11.3	14 17.4	12.4	57.5
14	33 52.9	21.2	62 15.4	11.3	14 29.8	12.3	57.4
15	48 52.8	. . 21.8	76 45.7	11.3	14 42.1	12.2	57.4
16	63 52.7	22.5	91 16.0	11.2	14 54.3	12.1	57.4
17	78 52.6	23.1	105 46.2	11.2	15 06.4	12.1	57.4
18	93 52.5	S18 23.8	120 16.4	11.1	S15 18.5	11.9	57.4
19	108 52.4	24.4	134 46.5	11.1	15 30.4	11.9	57.3
20	123 52.3	25.0	149 16.6	11.0	15 42.3	11.8	57.3
21	138 52.2	. . 25.7	163 46.6	11.0	15 54.1	11.7	57.3
22	153 52.1	26.3	178 16.6	11.0	16 05.8	11.6	57.3
23	168 52.0	27.0	192 46.6	10.9	S16 17.4	11.5	57.3
	SD 16.2	d 0.7	SD 15.9		15.8		15.7

Lat.	Twilight Naut.	Twilight Civil	Sunrise	Moonrise 12	Moonrise 13	Moonrise 14	Moonrise 15
°	h m	h m	h m	h m	h m	h m	h m
N 72	06 53	08 25	10 34	02 03	04 15	06 39	■
N 70	06 43	08 04	09 37	02 02	04 04	06 13	08 48
68	06 35	07 47	09 04	02 01	03 55	05 53	08 03
66	06 29	07 34	08 40	02 01	03 48	05 38	07 34
64	06 23	07 23	08 21	02 00	03 42	05 25	07 12
62	06 18	07 13	08 06	02 00	03 37	05 15	06 54
60	06 13	07 05	07 53	02 00	03 33	05 06	06 40
N 58	06 09	06 57	07 42	01 59	03 29	04 58	06 28
56	06 05	06 51	07 33	01 59	03 25	04 51	06 17
54	06 02	06 45	07 24	01 59	03 22	04 45	06 08
52	05 58	06 39	07 16	01 59	03 19	04 40	06 00
50	05 55	06 34	07 10	01 58	03 17	04 35	05 52
45	05 48	06 23	06 55	01 58	03 11	04 24	05 37
N 40	05 41	06 14	06 43	01 58	03 06	04 15	05 24
35	05 35	06 06	06 33	01 57	03 02	04 08	05 13
30	05 29	05 58	06 23	01 57	02 59	04 01	05 03
20	05 18	05 45	06 08	01 57	02 53	03 50	04 47
N 10	05 06	05 32	05 54	01 56	02 48	03 40	04 33
0	04 54	05 19	05 41	01 56	02 43	03 31	04 20
S 10	04 39	05 05	05 28	01 56	02 38	03 22	04 07
20	04 22	04 50	05 13	01 55	02 33	03 12	03 53
30	04 00	04 31	04 57	01 55	02 28	03 01	03 38
35	03 46	04 19	04 47	01 55	02 24	02 55	03 29
40	03 29	04 06	04 36	01 55	02 21	02 48	03 18
45	03 07	03 50	04 23	01 54	02 17	02 40	03 06
S 50	02 38	03 29	04 07	01 54	02 12	02 30	02 52
52	02 23	03 19	04 00	01 54	02 09	02 26	02 45
54	02 05	03 07	03 51	01 54	02 07	02 21	02 38
56	01 42	02 54	03 42	01 54	02 04	02 15	02 29
58	01 10	02 38	03 31	01 54	02 01	02 09	02 20
S 60	00 00	02 19	03 19	01 53	01 58	02 03	02 10

Lat.	Sunset	Twilight Civil	Twilight Naut.	Moonset 12	Moonset 13	Moonset 14	Moonset 15
°	h m	h m	h m	h m	h m	h m	h m
N 72	12 53	15 02	16 35	13 34	13 04	12 21	■
N 70	13 50	15 23	16 44	13 39	13 17	12 49	11 59
68	14 24	15 40	16 52	13 43	13 28	13 11	12 45
66	14 48	15 54	16 59	13 46	13 37	13 28	13 15
64	15 07	16 05	17 05	13 49	13 45	13 42	13 38
62	15 22	16 15	17 10	13 51	13 52	13 53	13 56
60	15 35	16 23	17 15	13 53	13 58	14 03	14 12
N 58	15 46	16 31	17 19	13 55	14 03	14 12	14 25
56	15 55	16 37	17 23	13 57	14 08	14 20	14 36
54	16 04	16 43	17 26	13 58	14 12	14 27	14 46
52	16 12	16 49	17 30	14 00	14 15	14 33	14 55
50	16 18	16 54	17 33	14 01	14 19	14 39	15 03
45	16 33	17 05	17 40	14 04	14 26	14 51	15 20
N 40	16 45	17 14	17 47	14 06	14 33	15 01	15 34
35	16 56	17 23	17 53	14 08	14 38	15 10	15 45
30	17 05	17 30	17 59	14 10	14 43	15 18	15 56
20	17 21	17 44	18 10	14 13	14 51	15 31	16 14
N 10	17 35	17 57	18 22	14 16	14 59	15 43	16 30
0	17 48	18 10	18 35	14 18	15 06	15 54	16 44
S 10	18 01	18 23	18 49	14 21	15 13	16 05	16 59
20	18 15	18 39	19 07	14 24	15 20	16 17	17 15
30	18 32	18 58	19 29	14 27	15 29	16 31	17 33
35	18 42	19 10	19 43	14 28	15 34	16 39	17 44
40	18 53	19 23	20 01	14 30	15 39	16 48	17 56
45	19 06	19 40	20 23	14 33	15 46	16 59	18 11
S 50	19 22	20 01	20 52	14 36	15 54	17 12	18 29
52	19 30	20 11	21 08	14 37	15 57	17 18	18 37
54	19 38	20 23	21 26	14 38	16 01	17 25	18 47
56	19 48	20 37	21 50	14 40	16 06	17 32	18 57
58	19 59	20 53	22 24	14 42	16 11	17 41	19 10
S 60	20 11	21 13	////	14 44	16 17	17 50	19 24

Day	SUN Eqn. of Time 00h	SUN Eqn. of Time 12h	SUN Mer. Pass.	MOON Mer. Pass. Upper	MOON Mer. Pass. Lower	MOON Age	MOON Phase
d	m s	m s	h m	h m	h m	d %	
12	15 54	15 50	11 44	08 07	20 31	25 20	
13	15 46	15 42	11 44	08 54	21 18	26 12	
14	15 37	15 33	11 44	09 42	22 07	27 6	

UT	ARIES	VENUS −3.9		MARS +0.2		JUPITER −2.3		SATURN +1.0	
	GHA	GHA	Dec	GHA	Dec	GHA	Dec	GHA	Dec
d h	° ′	° ′	° ′	° ′	° ′	° ′	° ′	° ′	° ′
15 00	54 12.9	197 26.0	S13 13.5	278 25.8	N18 50.1	92 33.0	S16 06.2	231 52.3	N 1 15.7
01	69 15.4	212 25.5	14.6	293 27.4	49.9	107 35.2	06.2	246 54.6	15.7
02	84 17.9	227 24.9	15.6	308 28.9	49.7	122 37.4	06.1	261 56.8	15.6
03	99 20.3	242 24.3	. . 16.7	323 30.5	. . 49.5	137 39.6	. . 06.0	276 59.0	. . 15.5
04	114 22.8	257 23.7	17.7	338 32.0	49.4	152 41.8	05.9	292 01.3	15.4
05	129 25.3	272 23.2	18.8	353 33.6	49.2	167 44.1	05.8	307 03.5	15.3
06	144 27.7	287 22.6	S13 19.8	8 35.1	N18 49.0	182 46.3	S16 05.7	322 05.8	N 1 15.2
07	159 30.2	302 22.0	20.9	23 36.7	48.8	197 48.5	05.6	337 08.0	15.2
08	174 32.7	317 21.5	21.9	38 38.3	48.7	212 50.7	05.6	352 10.3	15.1
S 09	189 35.1	332 20.9	. . 23.0	53 39.8	. . 48.5	227 52.9	. . 05.5	7 12.5	. . 15.0
U 10	204 37.6	347 20.3	24.0	68 41.4	48.3	242 55.1	05.4	22 14.8	14.9
N 11	219 40.1	2 19.7	25.1	83 42.9	48.2	257 57.3	05.3	37 17.0	14.8
D 12	234 42.5	17 19.2	S13 26.1	98 44.5	N18 48.0	272 59.5	S16 05.2	52 19.3	N 1 14.7
A 13	249 45.0	32 18.6	27.2	113 46.0	47.8	288 01.7	05.1	67 21.5	14.6
Y 14	264 47.4	47 18.0	28.2	128 47.6	47.6	303 03.9	05.0	82 23.8	14.6
15	279 49.9	62 17.4	. . 29.3	143 49.2	. . 47.5	318 06.1	. . 05.0	97 26.0	. . 14.5
16	294 52.4	77 16.9	30.3	158 50.7	47.3	333 08.3	04.9	112 28.3	14.4
17	309 54.8	92 16.3	31.4	173 52.3	47.1	348 10.5	04.8	127 30.5	14.3
18	324 57.3	107 15.7	S13 32.4	188 53.8	N18 47.0	3 12.7	S16 04.7	142 32.8	N 1 14.2
19	339 59.8	122 15.1	33.5	203 55.4	46.8	18 14.9	04.6	157 35.0	14.1
20	355 02.2	137 14.6	34.5	218 57.0	46.6	33 17.1	04.5	172 37.3	14.1
21	10 04.7	152 14.0	. . 35.6	233 58.5	. . 46.4	48 19.4	. . 04.4	187 39.5	. . 14.0
22	25 07.2	167 13.4	36.6	249 00.1	46.3	63 21.6	04.3	202 41.8	13.9
23	40 09.6	182 12.8	37.7	264 01.7	46.1	78 23.8	04.3	217 44.0	13.8
16 00	55 12.1	197 12.2	S13 38.7	279 03.2	N18 45.9	93 26.0	S16 04.2	232 46.3	N 1 13.7
01	70 14.5	212 11.7	39.7	294 04.8	45.8	108 28.2	04.1	247 48.5	13.6
02	85 17.0	227 11.1	40.8	309 06.4	45.6	123 30.4	04.0	262 50.8	13.6
03	100 19.5	242 10.5	. . 41.8	324 07.9	. . 45.4	138 32.6	. . 03.9	277 53.0	. . 13.5
04	115 21.9	257 09.9	42.9	339 09.5	45.2	153 34.8	03.8	292 55.3	13.4
05	130 24.4	272 09.3	43.9	354 11.1	45.1	168 37.0	03.7	307 57.5	13.3
06	145 26.9	287 08.8	S13 44.9	9 12.7	N18 44.9	183 39.2	S16 03.6	322 59.8	N 1 13.2
07	160 29.3	302 08.2	46.0	24 14.2	44.7	198 41.4	03.6	338 02.0	13.1
08	175 31.8	317 07.6	47.0	39 15.8	44.6	213 43.6	03.5	353 04.3	13.1
M 09	190 34.3	332 07.0	. . 48.1	54 17.4	. . 44.4	228 45.8	. . 03.4	8 06.5	. . 13.0
O 10	205 36.7	347 06.4	49.1	69 19.0	44.2	243 48.0	03.3	23 08.8	12.9
N 11	220 39.2	2 05.8	50.1	84 20.5	44.1	258 50.2	03.2	38 11.0	12.8
D 12	235 41.7	17 05.3	S13 51.2	99 22.1	N18 43.9	273 52.4	S16 03.1	53 13.3	N 1 12.7
A 13	250 44.1	32 04.7	52.2	114 23.7	43.7	288 54.6	03.0	68 15.5	12.6
Y 14	265 46.6	47 04.1	53.2	129 25.3	43.6	303 56.8	02.9	83 17.8	12.6
15	280 49.0	62 03.5	. . 54.3	144 26.8	. . 43.4	318 59.0	. . 02.8	98 20.0	. . 12.5
16	295 51.5	77 02.9	55.3	159 28.4	43.2	334 01.2	02.8	113 22.3	12.4
17	310 54.0	92 02.3	56.3	174 30.0	43.1	349 03.4	02.7	128 24.5	12.3
18	325 56.4	107 01.7	S13 57.4	189 31.6	N18 42.9	4 05.6	S16 02.6	143 26.8	N 1 12.2
19	340 58.9	122 01.2	58.4	204 33.2	42.7	19 07.8	02.5	158 29.0	12.1
20	356 01.4	137 00.6	13 59.4	219 34.7	42.6	34 10.0	02.4	173 31.3	12.1
21	11 03.8	152 00.0	14 00.5	234 36.3	. . 42.4	49 12.2	. . 02.3	188 33.5	. . 12.0
22	26 06.3	166 59.4	01.5	249 37.9	42.2	64 14.4	02.2	203 35.8	11.9
23	41 08.8	181 58.8	02.5	264 39.5	42.1	79 16.6	02.1	218 38.0	11.8
17 00	56 11.2	196 58.2	S14 03.5	279 41.1	N18 41.9	94 18.8	S16 02.0	233 40.3	N 1 11.7
01	71 13.7	211 57.6	04.6	294 42.7	41.7	109 21.0	02.0	248 42.5	11.6
02	86 16.2	226 57.0	05.6	309 44.3	41.6	124 23.2	01.9	263 44.8	11.6
03	101 18.6	241 56.4	. . 06.6	324 45.8	. . 41.4	139 25.3	. . 01.8	278 47.0	. . 11.5
04	116 21.1	256 55.8	07.7	339 47.4	41.2	154 27.5	01.7	293 49.3	11.4
05	131 23.5	271 55.2	08.7	354 49.0	41.1	169 29.7	01.6	308 51.6	11.3
06	146 26.0	286 54.6	S14 09.7	9 50.6	N18 40.9	184 31.9	S16 01.5	323 53.8	N 1 11.2
07	161 28.5	301 54.1	10.7	24 52.2	40.7	199 34.1	01.4	338 56.1	11.2
T 08	176 30.9	316 53.5	11.8	39 53.8	40.6	214 36.3	01.3	353 58.3	11.1
U 09	191 33.4	331 52.9	. . 12.8	54 55.4	. . 40.4	229 38.5	. . 01.2	9 00.6	. . 11.0
E 10	206 35.9	346 52.3	13.8	69 57.0	40.3	244 40.7	01.1	24 02.8	10.9
S 11	221 38.3	1 51.7	14.8	84 58.6	40.1	259 42.9	01.1	39 05.1	10.8
D 12	236 40.8	16 51.1	S14 15.8	100 00.2	N18 39.9	274 45.1	S16 01.0	54 07.3	N 1 10.7
A 13	251 43.3	31 50.5	16.9	115 01.8	39.8	289 47.3	00.9	69 09.6	10.7
Y 14	266 45.7	46 49.9	17.9	130 03.4	39.6	304 49.5	00.8	84 11.8	10.6
15	281 48.2	61 49.3	. . 18.9	145 05.0	. . 39.4	319 51.7	. . 00.7	99 14.1	. . 10.5
16	296 50.7	76 48.7	19.9	160 06.6	39.3	334 53.9	00.6	114 16.3	10.4
17	311 53.1	91 48.1	20.9	175 08.2	39.1	349 56.1	00.5	129 18.6	10.3
18	326 55.6	106 47.5	S14 22.0	190 09.8	N18 38.9	4 58.2	S16 00.4	144 20.8	N 1 10.3
19	341 58.0	121 46.9	23.0	205 11.4	38.8	20 00.4	00.3	159 23.1	10.2
20	357 00.5	136 46.3	24.0	220 13.0	38.6	35 02.6	00.2	174 25.4	10.1
21	12 03.0	151 45.7	. . 25.0	235 14.6	. . 38.5	50 04.8	. . 00.1	189 27.6	. . 10.0
22	27 05.4	166 45.1	26.0	250 16.2	38.3	65 07.0	00.1	204 29.9	09.9
23	42 07.9	181 44.5	27.0	265 17.8	38.1	80 09.2	00.0	219 32.1	09.8
Mer. Pass.	h m 20 15.9	v −0.6	d 1.0	v 1.6	d 0.2	v 2.2	d 0.1	v 2.3	d 0.1

STARS

Name	SHA	Dec
	° ′	° ′
Acamar	315 19.8	S40 15
Achernar	335 28.0	S57 11
Acrux	173 13.0	S63 09
Adhara	255 14.4	S28 58
Aldebaran	290 52.1	N16 31
Alioth	166 23.2	N55 54
Alkaid	153 01.3	N49 15
Al Na'ir	27 46.9	S46 54
Alnilam	275 48.8	S 1 11
Alphard	217 58.7	S 8 42
Alphecca	126 13.6	N26 40
Alpheratz	357 46.1	N29 09
Altair	62 11.0	N 8 53
Ankaa	353 17.9	S42 15
Antares	112 29.9	S26 27
Arcturus	145 58.4	N19 07
Atria	107 34.7	S69 02
Avior	234 19.0	S59 32
Bellatrix	278 34.5	N 6 21
Betelgeuse	271 03.9	N 7 24
Canopus	263 56.9	S52 41
Capella	280 37.9	N46 00
Deneb	49 33.5	N45 19
Denebola	182 36.5	N14 30
Diphda	348 58.3	S17 55
Dubhe	193 54.9	N61 41
Elnath	278 15.6	N28 37
Eltanin	90 47.8	N51 29
Enif	33 49.7	N 9 55
Fomalhaut	15 26.7	S29 34
Gacrux	172 04.5	S57 09
Gienah	175 55.3	S17 35
Hadar	148 52.5	S60 25
Hamal	328 03.5	N23 30
Kaus Aust.	83 47.6	S34 22
Kochab	137 20.8	N74 06
Markab	13 40.9	N15 15
Menkar	314 17.5	N 4 07
Menkent	148 11.2	S36 25
Miaplacidus	221 40.4	S69 45
Mirfak	308 43.7	N49 53
Nunki	76 01.8	S26 17
Peacock	53 23.5	S56 42
Pollux	243 30.7	N28 00
Procyon	245 02.3	N 5 12
Rasalhague	96 09.2	N12 33
Regulus	207 46.3	N11 55
Rigel	281 14.3	S 8 11
Rigil Kent.	139 56.2	S60 52
Sabik	102 15.9	S15 44
Schedar	349 43.3	N56 35
Shaula	96 25.9	S37 06
Sirius	258 35.8	S16 43
Spica	158 34.4	S11 12
Suhail	222 54.4	S43 28
Vega	80 41.1	N38 47
Zuben'ubi	137 08.7	S16 04

	SHA	Mer. Pas
	° ′	h m
Venus	142 00.2	10 52
Mars	223 51.2	5 23
Jupiter	38 13.9	17 44
Saturn	177 34.2	8 28

UT	SUN GHA	SUN Dec	MOON GHA	v	MOON Dec	d	HP
d h	° ′	° ′	° ′	′	° ′	′	′
15 00	183 51.9	S18 27.6	207 16.5	10.9	S16 28.9	11.4	57.2
01	198 51.8	28.2	221 46.4	10.8	16 40.3	11.3	57.2
02	213 51.6	28.9	236 16.2	10.8	16 51.6	11.3	57.2
03	228 51.5	. . 29.5	250 46.0	10.8	17 02.9	11.1	57.2
04	243 51.4	30.1	265 15.8	10.7	17 14.0	11.0	57.2
05	258 51.3	30.8	279 45.5	10.6	17 25.0	11.0	57.1
06	273 51.2	S18 31.4	294 15.1	10.6	S17 36.0	10.8	57.1
07	288 51.1	32.1	308 44.7	10.6	17 46.8	10.8	57.1
08	303 51.0	32.7	323 14.3	10.5	17 57.6	10.6	57.1
09	318 50.9	. . 33.3	337 43.8	10.5	18 08.2	10.6	57.1
10	333 50.8	34.0	352 13.3	10.5	18 18.8	10.4	57.0
11	348 50.7	34.6	6 42.8	10.4	18 29.2	10.3	57.0
12	3 50.6	S18 35.2	21 12.2	10.3	S18 39.5	10.3	57.0
13	18 50.5	35.9	35 41.5	10.3	18 49.8	10.1	57.0
14	33 50.4	36.5	50 10.8	10.3	18 59.9	10.0	57.0
15	48 50.2	. . 37.1	64 40.1	10.2	19 09.9	10.0	56.9
16	63 50.1	37.8	79 09.3	10.2	19 19.9	9.8	56.9
17	78 50.0	38.4	93 38.5	10.1	19 29.7	9.7	56.9
18	93 49.9	S18 39.0	108 07.6	10.1	S19 39.4	9.6	56.9
19	108 49.8	39.6	122 36.7	10.0	19 49.0	9.4	56.8
20	123 49.7	40.3	137 05.7	10.0	19 58.4	9.4	56.8
21	138 49.6	. . 40.9	151 34.7	10.0	20 07.8	9.3	56.8
22	153 49.5	41.5	166 03.7	9.9	20 17.1	9.1	56.8
23	168 49.4	42.1	180 32.6	9.9	20 26.2	9.0	56.8
16 00	183 49.2	S18 42.8	195 01.5	9.8	S20 35.2	9.0	56.7
01	198 49.1	43.4	209 30.3	9.8	20 44.2	8.7	56.7
02	213 49.0	44.0	223 59.1	9.8	20 52.9	8.7	56.7
03	228 48.9	. . 44.6	238 27.9	9.7	21 01.6	8.6	56.7
04	243 48.8	45.3	252 56.6	9.7	21 10.2	8.4	56.7
05	258 48.7	45.9	267 25.3	9.6	21 18.6	8.4	56.6
06	273 48.6	S18 46.5	281 53.9	9.6	S21 27.0	8.2	56.6
07	288 48.4	47.1	296 22.5	9.6	21 35.2	8.1	56.6
08	303 48.3	47.8	310 51.1	9.5	21 43.3	7.9	56.6
09	318 48.2	. . 48.4	325 19.6	9.5	21 51.2	7.9	56.5
10	333 48.1	49.0	339 48.1	9.4	21 59.1	7.7	56.5
11	348 48.0	49.6	354 16.5	9.4	22 06.8	7.6	56.5
12	3 47.9	S18 50.2	8 44.9	9.4	S22 14.4	7.4	56.5
13	18 47.7	50.9	23 13.3	9.3	22 21.8	7.4	56.5
14	33 47.6	51.5	37 41.6	9.4	22 29.2	7.2	56.4
15	48 47.5	. . 52.1	52 10.0	9.2	22 36.4	7.1	56.4
16	63 47.4	52.7	66 38.2	9.3	22 43.5	7.0	56.4
17	78 47.3	53.3	81 06.5	9.2	22 50.5	6.8	56.4
18	93 47.1	S18 53.9	95 34.7	9.2	S22 57.3	6.7	56.3
19	108 47.0	54.6	110 02.9	9.1	23 04.0	6.6	56.3
20	123 46.9	55.2	124 31.0	9.2	23 10.6	6.4	56.3
21	138 46.8	. . 55.8	138 59.2	9.1	23 17.0	6.4	56.3
22	153 46.7	56.4	153 27.3	9.0	23 23.4	6.1	56.3
23	168 46.5	57.0	167 55.3	9.1	23 29.5	6.1	56.2
17 00	183 46.4	S18 57.6	182 23.4	9.0	S23 35.6	5.9	56.2
01	198 46.3	58.2	196 51.4	9.0	23 41.5	5.8	56.2
02	213 46.2	58.8	211 19.4	8.9	23 47.3	5.7	56.2
03	228 46.1	18 59.5	225 47.3	9.0	23 53.0	5.5	56.1
04	243 45.9	19 00.1	240 15.3	8.9	23 58.5	5.4	56.1
05	258 45.8	00.7	254 43.2	8.9	24 03.9	5.3	56.1
06	273 45.7	S19 01.3	269 11.1	8.9	S24 09.2	5.1	56.1
07	288 45.6	01.9	283 39.0	8.9	24 14.3	5.0	56.1
08	303 45.4	02.5	298 06.9	8.8	24 19.3	4.9	56.0
09	318 45.3	. . 03.1	312 34.7	8.9	24 24.2	4.7	56.0
10	333 45.2	03.7	327 02.6	8.8	24 28.9	4.6	56.0
11	348 45.1	04.3	341 30.4	8.8	24 33.5	4.4	56.0
12	3 44.9	S19 04.9	355 58.2	8.8	S24 37.9	4.3	55.9
13	18 44.8	05.5	10 26.0	8.8	24 42.2	4.2	55.9
14	33 44.7	06.1	24 53.8	8.8	24 46.4	4.0	55.9
15	48 44.5	. . 06.7	39 21.6	8.7	24 50.4	3.9	55.9
16	63 44.4	07.3	53 49.3	8.8	24 54.3	3.8	55.9
17	78 44.3	07.9	68 17.1	8.7	24 58.1	3.6	55.8
18	93 44.2	S19 08.5	82 44.8	8.8	S25 01.7	3.5	55.8
19	108 44.0	09.1	97 12.6	8.7	25 05.2	3.4	55.8
20	123 43.9	09.7	111 40.3	8.7	25 08.6	3.2	55.8
21	138 43.8	. . 10.3	126 08.0	8.8	25 11.8	3.1	55.8
22	153 43.6	10.9	140 35.8	8.7	25 14.9	2.9	55.7
23	168 43.5	11.5	155 03.5	8.8	S25 17.8	2.8	55.7
	SD 16.2	d 0.6	SD	15.5	15.4		15.2

Lat.	Twilight Naut.	Twilight Civil	Sunrise	Moonrise 15	16	17	18
°	h m	h m	h m	h m	h m	h m	h m
N 72	07 03	08 39	■	■	■	■	■
N 70	06 53	08 16	09 57	08 48	■	■	■
68	06 44	07 58	09 18	08 03	■	■	■
66	06 37	07 43	08 51	07 34	09 43	■	■
64	06 30	07 31	08 31	07 12	09 02	10 55	12 42
62	06 24	07 20	08 14	06 54	08 35	10 10	11 29
60	06 19	07 11	08 01	06 40	08 13	09 41	10 53
N 58	06 15	07 03	07 49	06 28	07 56	09 18	10 27
56	06 10	06 56	07 39	06 17	07 42	09 00	10 07
54	06 06	06 50	07 30	06 08	07 29	08 45	09 50
52	06 03	06 44	07 22	06 00	07 18	08 31	09 35
50	05 59	06 39	07 14	05 52	07 08	08 20	09 23
45	05 51	06 27	06 59	05 37	06 48	07 56	08 57
N 40	05 44	06 17	06 46	05 24	06 32	07 36	08 36
35	05 38	06 08	06 35	05 13	06 18	07 20	08 19
30	05 32	06 01	06 26	05 03	06 06	07 07	08 04
20	05 19	05 46	06 09	04 47	05 45	06 43	07 39
N 10	05 07	05 33	05 55	04 33	05 28	06 23	07 17
0	04 54	05 19	05 41	04 20	05 11	06 04	06 57
S 10	04 39	05 05	05 27	04 07	04 55	05 45	06 37
20	04 21	04 49	05 13	03 53	04 38	05 25	06 16
30	03 58	04 29	04 55	03 38	04 18	05 02	05 51
35	03 43	04 17	04 45	03 29	04 06	04 49	05 36
40	03 25	04 03	04 34	03 18	03 53	04 33	05 19
45	03 03	03 46	04 20	03 06	03 37	04 15	04 59
S 50	02 32	03 24	04 03	02 52	03 18	03 52	04 34
52	02 16	03 13	03 55	02 45	03 09	03 41	04 22
54	01 56	03 01	03 46	02 38	02 59	03 28	04 08
56	01 31	02 47	03 37	02 29	02 48	03 14	03 51
58	00 52	02 30	03 25	02 20	02 35	02 58	03 32
S 60	////	02 09	03 12	02 10	02 20	02 38	03 08

Lat.	Sunset	Twilight Civil	Twilight Naut.	Moonset 15	16	17	18
°	h m	h m	h m	h m	h m	h m	h m
N 72	■	14 49	16 25	■	■	■	■
N 70	13 32	15 13	16 36	11 59	■	■	■
68	14 11	15 31	16 45	12 45	■	■	■
66	14 37	15 46	16 52	13 15	12 54	■	■
64	14 58	15 58	16 59	13 38	13 35	13 32	13 37
62	15 14	16 09	17 04	13 56	14 03	14 18	14 51
60	15 28	16 18	17 10	14 12	14 25	14 47	15 26
N 58	15 40	16 26	17 14	14 25	14 43	15 10	15 52
56	15 50	16 33	17 19	14 36	14 58	15 29	16 13
54	15 59	16 39	17 23	14 46	15 11	15 44	16 30
52	16 07	16 45	17 26	14 55	15 22	15 58	16 44
50	16 15	16 50	17 30	15 03	15 32	16 10	16 57
45	16 30	17 02	17 38	15 20	15 53	16 34	17 23
N 40	16 43	17 12	17 45	15 34	16 11	16 54	17 44
35	16 54	17 21	17 51	15 45	16 25	17 10	18 01
30	17 03	17 29	17 58	15 56	16 38	17 25	18 16
20	17 20	17 43	18 10	16 14	17 00	17 49	18 41
N 10	17 35	17 57	18 22	16 30	17 19	18 10	19 03
0	17 48	18 10	18 36	16 44	17 36	18 30	19 23
S 10	18 02	18 25	18 51	16 59	17 54	18 49	19 43
20	18 17	18 41	19 09	17 15	18 13	19 10	20 05
30	18 35	19 01	19 32	17 33	18 35	19 35	20 30
35	18 45	19 13	19 47	17 44	18 48	19 49	20 45
40	18 56	19 27	20 05	17 56	19 03	20 06	21 02
45	19 10	19 45	20 28	18 11	19 21	20 26	21 23
S 50	19 27	20 07	20 59	18 29	19 43	20 51	21 48
52	19 35	20 18	21 16	18 37	19 54	21 03	22 01
54	19 44	20 30	21 36	18 47	20 06	21 17	22 15
56	19 54	20 45	22 02	18 57	20 19	21 33	22 32
58	20 06	21 02	22 43	19 10	20 36	21 52	22 52
S 60	20 19	21 23	////	19 24	20 55	22 16	23 17

Day	SUN Eqn. of Time 00^h	SUN Eqn. of Time 12^h	SUN Mer. Pass.	MOON Mer. Pass. Upper	MOON Mer. Pass. Lower	Age	Phase
d	m s	m s	h m	h m	h m	d	%
15	15 28	15 23	11 45	10 32	22 58	28	2
16	15 17	15 12	11 45	11 24	23 50	29	0 ●
17	15 06	15 00	11 45	12 17	24 43	01	1

UT	ARIES	VENUS −3.9		MARS +0.2		JUPITER −2.3		SATURN +1.0		STARS		
	GHA	GHA	Dec	GHA	Dec	GHA	Dec	GHA	Dec	Name	SHA	Dec
d h	° ′	° ′	° ′	° ′	° ′	° ′	° ′	° ′	° ′		° ′	°
18 00	57 10.4	196 43.9	S14 28.1	280 19.4	N18 38.0	95 11.4	S15 59.9	234 34.4	N 1 09.8	Acamar	315 19.8	S40 15
01	72 12.8	211 43.3	29.1	295 21.0	37.8	110 13.6	59.8	249 36.6	09.7	Achernar	335 28.0	S57 11
02	87 15.3	226 42.7	30.1	310 22.6	37.7	125 15.8	59.7	264 38.9	09.6	Acrux	173 12.9	S63 09
03	102 17.8	241 42.1	. . 31.1	325 24.2	. . 37.5	140 18.0	. . 59.6	279 41.1	. . 09.5	Adhara	255 14.3	S28 59
04	117 20.2	256 41.5	32.1	340 25.8	37.3	155 20.1	59.5	294 43.4	09.4	Aldebaran	290 52.1	N16 31
05	132 22.7	271 40.9	33.1	355 27.4	37.2	170 22.3	59.4	309 45.7	09.4			
06	147 25.1	286 40.3	S14 34.1	10 29.0	N18 37.0	185 24.5	S15 59.3	324 47.9	N 1 09.3	Alioth	166 23.2	N55 54
W 07	162 27.6	301 39.7	35.1	25 30.6	36.9	200 26.7	59.2	339 50.2	09.2	Alkaid	153 01.3	N49 15
E 08	177 30.1	316 39.0	36.1	40 32.2	36.7	215 28.9	59.1	354 52.4	09.1	Al Na'ir	27 46.9	S46 54
D 09	192 32.5	331 38.4	. . 37.2	55 33.8	. . 36.5	230 31.1	. . 59.0	9 54.7	. . 09.0	Alnilam	275 48.7	S 1 11
N 10	207 35.0	346 37.8	38.2	70 35.5	36.4	245 33.3	58.9	24 56.9	09.0	Alphard	217 58.7	S 8 42
E 11	222 37.5	1 37.2	39.2	85 37.1	36.2	260 35.5	58.9	39 59.2	08.9			
S 12	237 39.9	16 36.6	S14 40.2	100 38.7	N18 36.1	275 37.6	S15 58.8	55 01.4	N 1 08.8	Alphecca	126 13.6	N26 40
D 13	252 42.4	31 36.0	41.2	115 40.3	35.9	290 39.8	58.7	70 03.7	08.7	Alpheratz	357 46.1	N29 09
A 14	267 44.9	46 35.4	42.2	130 41.9	35.7	305 42.0	58.6	85 06.0	08.6	Altair	62 11.0	N 8 53
Y 15	282 47.3	61 34.8	. . 43.2	145 43.5	. . 35.6	320 44.2	. . 58.5	100 08.2	. . 08.6	Ankaa	353 17.9	S42 15
16	297 49.8	76 34.2	44.2	160 45.1	35.4	335 46.4	58.4	115 10.5	08.5	Antares	112 29.9	S26 27
17	312 52.3	91 33.6	45.2	175 46.8	35.3	350 48.6	58.3	130 12.7	08.4			
18	327 54.7	106 33.0	S14 46.2	190 48.4	N18 35.1	5 50.8	S15 58.2	145 15.0	N 1 08.3	Arcturus	145 58.4	N19 07
19	342 57.2	121 32.3	47.2	205 50.0	35.0	20 52.9	58.1	160 17.2	08.2	Atria	107 34.7	S69 02
20	357 59.6	136 31.7	48.2	220 51.6	34.8	35 55.1	58.0	175 19.5	08.2	Avior	234 19.0	S59 32
21	13 02.1	151 31.1	. . 49.2	235 53.2	. . 34.6	50 57.3	. . 57.9	190 21.8	. . 08.1	Bellatrix	278 34.5	N 6 21
22	28 04.6	166 30.5	50.2	250 54.9	34.5	65 59.5	57.8	205 24.0	08.0	Betelgeuse	271 03.9	N 7 24
23	43 07.0	181 29.9	51.2	265 56.5	34.3	81 01.7	57.7	220 26.3	07.9			
19 00	58 09.5	196 29.3	S14 52.2	280 58.1	N18 34.2	96 03.9	S15 57.6	235 28.5	N 1 07.8	Canopus	263 56.9	S52 41
01	73 12.0	211 28.7	53.2	295 59.7	34.0	111 06.0	57.5	250 30.8	07.8	Capella	280 37.9	N46 00
02	88 14.4	226 28.1	54.2	311 01.4	33.9	126 08.2	57.4	265 33.0	07.7	Deneb	49 33.5	N45 19
03	103 16.9	241 27.4	. . 55.2	326 03.0	. . 33.7	141 10.4	. . 57.4	280 35.3	. . 07.6	Denebola	182 36.5	N14 30
04	118 19.4	256 26.8	56.2	341 04.6	33.5	156 12.6	57.3	295 37.6	07.5	Diphda	348 58.3	S17 55
05	133 21.8	271 26.2	57.2	356 06.2	33.4	171 14.8	57.2	310 39.8	07.4			
06	148 24.3	286 25.6	S14 58.2	11 07.9	N18 33.2	186 17.0	S15 57.1	325 42.1	N 1 07.4	Dubhe	193 54.9	N61 41
T 07	163 26.8	301 25.0	14 59.2	26 09.5	33.1	201 19.1	57.0	340 44.3	07.3	Elnath	278 15.6	N28 37
08	178 29.2	316 24.4	15 00.2	41 11.1	32.9	216 21.3	56.9	355 46.6	07.2	Eltanin	90 47.8	N51 29
H 09	193 31.7	331 23.7	. . 01.2	56 12.8	. . 32.8	231 23.5	. . 56.8	10 48.8	. . 07.1	Enif	33 49.8	N 9 55
U 10	208 34.1	346 23.1	02.1	71 14.4	32.6	246 25.7	56.7	25 51.1	07.0	Fomalhaut	15 26.7	S29 34
R 11	223 36.6	1 22.5	03.1	86 16.0	32.5	261 27.9	56.6	40 53.4	07.0			
S 12	238 39.1	16 21.9	S15 04.1	101 17.6	N18 32.3	276 30.0	S15 56.5	55 55.6	N 1 06.9	Gacrux	172 04.4	S57 09
D 13	253 41.5	31 21.3	05.1	116 19.3	32.2	291 32.2	56.4	70 57.9	06.8	Gienah	175 55.3	S17 35
A 14	268 44.0	46 20.6	06.1	131 20.9	32.0	306 34.4	56.3	86 00.1	06.7	Hadar	148 52.4	S60 25
Y 15	283 46.5	61 20.0	. . 07.1	146 22.6	. . 31.9	321 36.6	. . 56.2	101 02.4	. . 06.6	Hamal	328 03.5	N23 30
16	298 48.9	76 19.4	08.1	161 24.2	31.7	336 38.8	56.1	116 04.7	06.6	Kaus Aust.	83 47.6	S34 22
17	313 51.4	91 18.8	09.1	176 25.8	31.5	351 40.9	56.0	131 06.9	06.5			
18	328 53.9	106 18.2	S15 10.1	191 27.5	N18 31.4	6 43.1	S15 55.9	146 09.2	N 1 06.4	Kochab	137 20.7	N74 06
19	343 56.3	121 17.5	11.0	206 29.1	31.2	21 45.3	55.8	161 11.4	06.3	Markab	13 40.9	N15 15
20	358 58.8	136 16.9	12.0	221 30.7	31.1	36 47.5	55.7	176 13.7	06.2	Menkar	314 17.5	N 4 07
21	14 01.3	151 16.3	. . 13.0	236 32.4	. . 30.9	51 49.7	. . 55.6	191 16.0	. . 06.2	Menkent	148 11.1	S36 25
22	29 03.7	166 15.7	14.0	251 34.0	30.8	66 51.8	55.5	206 18.2	06.1	Miaplacidus	221 40.3	S69 45
23	44 06.2	181 15.0	15.0	266 35.7	30.6	81 54.0	55.4	221 20.5	06.0			
20 00	59 08.6	196 14.4	S15 16.0	281 37.3	N18 30.5	96 56.2	S15 55.4	236 22.7	N 1 05.9	Mirfak	308 43.7	N49 53
01	74 11.1	211 13.8	16.9	296 38.9	30.3	111 58.4	55.3	251 25.0	05.9	Nunki	76 01.8	S26 17
02	89 13.6	226 13.2	17.9	311 40.6	30.2	127 00.5	55.2	266 27.3	05.8	Peacock	53 23.5	S56 42
03	104 16.0	241 12.5	. . 18.9	326 42.2	. . 30.0	142 02.7	. . 55.1	281 29.5	. . 05.7	Pollux	243 30.7	N28 00
04	119 18.5	256 11.9	19.9	341 43.9	29.9	157 04.9	55.0	296 31.8	05.6	Procyon	245 02.3	N 5 12
05	134 21.0	271 11.3	20.9	356 45.5	29.7	172 07.1	54.9	311 34.0	05.5			
06	149 23.4	286 10.6	S15 21.8	11 47.2	N18 29.6	187 09.2	S15 54.8	326 36.3	N 1 05.5	Rasalhague	96 09.2	N12 33
07	164 25.9	301 10.0	22.8	26 48.8	29.4	202 11.4	54.7	341 38.6	05.4	Regulus	207 46.3	N11 55
08	179 28.4	316 09.4	23.8	41 50.5	29.3	217 13.6	54.6	356 40.8	05.3	Rigel	281 14.3	S 8 11
F 09	194 30.8	331 08.8	. . 24.8	56 52.1	. . 29.1	232 15.8	. . 54.5	11 43.1	. . 05.2	Rigil Kent.	139 56.2	S60 52
R 10	209 33.3	346 08.1	25.8	71 53.8	29.0	247 17.9	54.4	26 45.3	05.1	Sabik	102 15.9	S15 44
I 11	224 35.8	1 07.5	26.7	86 55.4	28.8	262 20.1	54.3	41 47.6	05.1			
D 12	239 38.2	16 06.9	S15 27.7	101 57.1	N18 28.7	277 22.3	S15 54.2	56 49.9	N 1 05.0	Schedar	349 43.3	N56 35
A 13	254 40.7	31 06.2	28.7	116 58.7	28.5	292 24.5	54.1	71 52.1	04.9	Shaula	96 25.9	S37 06
Y 14	269 43.1	46 05.6	29.7	132 00.4	28.4	307 26.6	54.0	86 54.4	04.8	Sirius	258 35.8	S16 43
15	284 45.6	61 05.0	. . 30.6	147 02.0	. . 28.2	322 28.8	. . 53.9	101 56.7	. . 04.8	Spica	158 34.3	S11 12
16	299 48.1	76 04.3	31.6	162 03.7	28.1	337 31.0	53.8	116 58.9	04.7	Suhail	222 54.4	S43 28
17	314 50.5	91 03.7	32.6	177 05.4	27.9	352 33.1	53.7	132 01.2	04.6			
18	329 53.0	106 03.1	S15 33.5	192 07.0	N18 27.8	7 35.3	S15 53.6	147 03.4	N 1 04.5	Vega	80 41.1	N38 47
19	344 55.5	121 02.4	34.5	207 08.7	27.7	22 37.5	53.5	162 05.7	04.4	Zuben'ubi	137 08.7	S16 05
20	359 57.9	136 01.8	35.5	222 10.3	27.5	37 39.7	53.4	177 08.0	04.4		SHA	Mer.Pas
21	15 00.4	151 01.1	. . 36.4	237 12.0	. . 27.4	52 41.8	. . 53.3	192 10.2	. . 04.3		° ′	h m
22	30 02.9	166 00.5	37.4	252 13.6	27.2	67 44.0	53.2	207 12.5	04.2	Venus	138 19.8	10 54
23	45 05.3	180 59.9	38.4	267 15.3	27.1	82 46.2	53.1	222 14.8	04.1	Mars	222 48.6	5 16
	h m									Jupiter	37 54.4	17 33
Mer. Pass.	20 04.1	*v* −0.6	*d* 1.0	*v* 1.6	*d* 0.2	*v* 2.2	*d* 0.1	*v* 2.3	*d* 0.1	Saturn	177 19.0	8 17

UT	SUN GHA	SUN Dec	MOON GHA	v	MOON Dec	d	HP
d h	° ′	° ′	° ′	′	° ′	′	′
18 00	183 43.4	S19 12.1	169 31.3	8.7	S25 20.6	2.7	55.7
01	198 43.3	12.7	183 59.0	8.7	25 23.3	2.5	55.7
02	213 43.1	13.3	198 26.7	8.8	25 25.8	2.4	55.6
03	228 43.0	. . 13.9	212 54.5	8.7	25 28.2	2.2	55.6
04	243 42.9	14.5	227 22.2	8.8	25 30.4	2.1	55.6
05	258 42.7	15.1	241 50.0	8.8	25 32.5	2.0	55.6
06	273 42.6	S19 15.7	256 17.8	8.7	S25 34.5	1.8	55.6
W 07	288 42.5	16.3	270 45.5	8.8	25 36.3	1.7	55.5
E 08	303 42.3	16.9	285 13.3	8.8	25 38.0	1.6	55.5
D 09	318 42.2	. . 17.5	299 41.1	8.8	25 39.6	1.4	55.5
N 10	333 42.1	18.1	314 08.9	8.8	25 41.0	1.3	55.5
E 11	348 41.9	18.7	328 36.7	8.9	25 42.3	1.1	55.5
S 12	3 41.8	S19 19.3	343 04.6	8.8	S25 43.4	1.0	55.4
D 13	18 41.7	19.8	357 32.4	8.9	25 44.4	0.9	55.4
A 14	33 41.5	20.4	12 00.3	8.9	25 45.3	0.7	55.4
Y 15	48 41.4	. . 21.0	26 28.2	8.9	25 46.0	0.6	55.4
16	63 41.2	21.6	40 56.1	9.0	25 46.6	0.5	55.4
17	78 41.1	22.2	55 24.1	8.9	25 47.1	0.3	55.3
18	93 41.0	S19 22.8	69 52.0	9.0	S25 47.4	0.2	55.3
19	108 40.8	23.4	84 20.0	9.1	25 47.6	0.0	55.3
20	123 40.7	24.0	98 48.1	9.0	25 47.6	0.0	55.3
21	138 40.6	. . 24.5	113 16.1	9.1	25 47.6	0.3	55.3
22	153 40.4	25.1	127 44.2	9.1	25 47.3	0.3	55.2
23	168 40.3	25.7	142 12.3	9.1	25 47.0	0.5	55.2
19 00	183 40.1	S19 26.3	156 40.4	9.2	S25 46.5	0.6	55.2
01	198 40.0	26.9	171 08.6	9.2	25 45.9	0.8	55.2
02	213 39.9	27.5	185 36.8	9.2	25 45.1	0.9	55.2
03	228 39.7	. . 28.0	200 05.0	9.2	25 44.2	1.0	55.1
04	243 39.6	28.6	214 33.2	9.3	25 43.2	1.2	55.1
05	258 39.4	29.2	229 01.5	9.4	25 42.0	1.3	55.1
06	273 39.3	S19 29.8	243 29.9	9.4	S25 40.7	1.4	55.1
T 07	288 39.2	30.4	257 58.3	9.4	25 39.3	1.5	55.1
H 08	303 39.0	30.9	272 26.7	9.4	25 37.8	1.7	55.0
U 09	318 38.9	. . 31.5	286 55.1	9.5	25 36.1	1.8	55.0
R 10	333 38.7	32.1	301 23.6	9.6	25 34.3	2.0	55.0
S 11	348 38.6	32.7	315 52.2	9.6	25 32.3	2.0	55.0
D 12	3 38.5	S19 33.2	330 20.8	9.6	S25 30.3	2.2	55.0
A 13	18 38.3	33.8	344 49.4	9.7	25 28.1	2.4	55.0
Y 14	33 38.2	34.4	359 18.1	9.7	25 25.7	2.4	54.9
15	48 38.0	. . 35.0	13 46.8	9.8	25 23.3	2.6	54.9
16	63 37.9	35.5	28 15.6	9.8	25 20.7	2.7	54.9
17	78 37.7	36.1	42 44.4	9.9	25 18.0	2.8	54.9
18	93 37.6	S19 36.7	57 13.3	9.9	S25 15.2	3.0	54.9
19	108 37.4	37.3	71 42.2	9.9	25 12.2	3.1	54.8
20	123 37.3	37.8	86 11.1	10.1	25 09.1	3.2	54.8
21	138 37.1	. . 38.4	100 40.2	10.0	25 05.9	3.3	54.8
22	153 37.0	39.0	115 09.2	10.2	25 02.6	3.5	54.8
23	168 36.9	39.5	129 38.4	10.1	24 59.1	3.5	54.8
20 00	183 36.7	S19 40.1	144 07.5	10.3	S24 55.6	3.7	54.8
01	198 36.6	40.7	158 36.8	10.3	24 51.9	3.8	54.8
02	213 36.4	41.2	173 06.1	10.3	24 48.1	4.0	54.7
03	228 36.3	. . 41.8	187 35.4	10.4	24 44.1	4.0	54.7
04	243 36.1	42.4	202 04.8	10.5	24 40.1	4.2	54.7
05	258 36.0	42.9	216 34.3	10.5	24 35.9	4.3	54.7
06	273 35.8	S19 43.5	231 03.8	10.6	S24 31.6	4.4	54.7
07	288 35.7	44.1	245 33.4	10.6	24 27.2	4.5	54.7
08	303 35.5	44.6	260 03.0	10.7	24 22.7	4.6	54.6
F 09	318 35.4	. . 45.2	274 32.7	10.8	24 18.1	4.8	54.6
R 10	333 35.2	45.8	289 02.5	10.8	24 13.3	4.8	54.6
I 11	348 35.1	46.3	303 32.3	10.9	24 08.5	5.0	54.6
D 12	3 34.9	S19 46.9	318 02.2	11.0	S24 03.5	5.1	54.6
A 13	18 34.8	47.4	332 32.2	11.0	23 58.4	5.2	54.6
Y 14	33 34.6	48.0	347 02.2	11.0	23 53.2	5.3	54.6
15	48 34.4	. . 48.6	1 32.2	11.2	23 47.9	5.4	54.5
16	63 34.3	49.1	16 02.4	11.2	23 42.5	5.6	54.5
17	78 34.1	49.7	30 32.6	11.2	23 36.9	5.6	54.5
18	93 34.0	S19 50.2	45 02.8	11.4	S23 31.3	5.7	54.5
19	108 33.8	50.8	59 33.2	11.3	23 25.6	5.9	54.5
20	123 33.7	51.3	74 03.5	11.5	23 19.7	5.9	54.5
21	138 33.5	. . 51.9	88 34.0	11.5	23 13.8	6.1	54.5
22	153 33.4	52.5	103 04.5	11.6	23 07.7	6.1	54.5
23	168 33.2	53.0	117 35.1	11.7	S23 01.6	6.3	54.4
	SD 16.2	d 0.6	SD 15.1		15.0		14.9

Lat.	Twilight Naut.	Twilight Civil	Sunrise	Moonrise 18	Moonrise 19	Moonrise 20	Moonrise 21
°	h m	h m	h m	h m	h m	h m	h m
N 72	07 14	08 54	▬	▬	▬	▬	▬
N 70	07 02	08 27	10 19	▬	▬	▬	▬
68	06 52	08 08	09 33	▬	▬	▬	14 39
66	06 44	07 52	09 03	▬	▬	14 35	13 45
64	06 37	07 38	08 40	12 42	13 16	13 15	13 12
62	06 31	07 27	08 23	11 29	12 16	12 38	12 48
60	06 25	07 18	08 08	10 53	11 43	12 12	12 29
N 58	06 20	07 09	07 56	10 27	11 18	11 51	12 13
56	06 15	07 02	07 45	10 07	10 58	11 34	11 59
54	06 11	06 55	07 35	09 50	10 41	11 20	11 47
52	06 07	06 49	07 27	09 35	10 27	11 07	11 36
50	06 03	06 43	07 19	09 23	10 15	10 56	11 27
45	05 55	06 31	07 03	08 57	09 49	10 32	11 07
N 40	05 47	06 20	06 50	08 36	09 29	10 14	10 51
35	05 40	06 11	06 38	08 19	09 12	09 58	10 37
30	05 34	06 03	06 28	08 04	08 57	09 44	10 25
20	05 21	05 48	06 11	07 39	08 32	09 21	10 05
N 10	05 08	05 34	05 56	07 17	08 10	09 01	09 47
0	04 54	05 20	05 42	06 57	07 50	08 42	09 30
S 10	04 39	05 05	05 27	06 37	07 30	08 23	09 14
20	04 20	04 48	05 12	06 16	07 09	08 02	08 56
30	03 56	04 28	04 54	05 51	06 44	07 39	08 35
35	03 41	04 15	04 44	05 36	06 29	07 25	08 23
40	03 22	04 01	04 31	05 19	06 12	07 09	08 09
45	02 59	03 43	04 17	04 59	05 51	06 50	07 52
S 50	02 26	03 20	04 00	04 34	05 26	06 26	07 31
52	02 09	03 08	03 51	04 22	05 13	06 14	07 21
54	01 47	02 55	03 42	04 08	04 59	06 01	07 10
56	01 19	02 40	03 31	03 51	04 42	05 46	06 58
58	00 30	02 22	03 19	03 32	04 22	05 28	06 43
S 60	////	01 59	03 05	03 08	03 57	05 06	06 25

Lat.	Sunset	Twilight Civil	Twilight Naut.	Moonset 18	Moonset 19	Moonset 20	Moonset 21
°	h m	h m	h m	h m	h m	h m	h m
N 72	▬	14 36	16 16	▬	▬	▬	▬
N 70	13 11	15 02	16 28	▬	▬	▬	▬
68	13 57	15 22	16 38	▬	▬	▬	16 57
66	14 27	15 38	16 46	▬	▬	15 20	17 50
64	14 50	15 52	16 53	13 37	14 53	16 39	18 22
62	15 07	16 03	16 59	14 51	15 53	17 16	18 46
60	15 22	16 12	17 05	15 26	16 26	17 42	19 04
N 58	15 35	16 21	17 10	15 52	16 51	18 02	19 20
56	15 45	16 29	17 15	16 13	17 10	18 19	19 33
54	15 55	16 35	17 19	16 30	17 27	18 33	19 44
52	16 03	16 42	17 23	16 44	17 41	18 46	19 55
50	16 11	16 47	17 27	16 57	17 53	18 57	20 04
45	16 27	17 00	17 35	17 23	18 19	19 19	20 22
N 40	16 41	17 10	17 43	17 44	18 39	19 37	20 38
35	16 52	17 19	17 50	18 01	18 55	19 53	20 51
30	17 02	17 28	17 57	18 16	19 10	20 06	21 02
20	17 19	17 43	18 10	18 41	19 34	20 28	21 21
N 10	17 35	17 57	18 23	19 03	19 56	20 48	21 38
0	17 49	18 11	18 37	19 23	20 15	21 06	21 53
S 10	18 04	18 26	18 52	19 43	20 35	21 23	22 08
20	18 19	18 43	19 11	20 05	20 56	21 43	22 25
30	18 37	19 04	19 35	20 30	21 20	22 04	22 43
35	18 48	19 16	19 51	20 45	21 34	22 17	22 54
40	19 00	19 31	20 09	21 02	21 51	22 32	23 06
45	19 14	19 49	20 33	21 23	22 10	22 49	23 21
S 50	19 32	20 12	21 06	21 48	22 35	23 11	23 39
52	19 41	20 24	21 24	22 01	22 47	23 21	23 47
54	19 50	20 37	21 46	22 15	23 00	23 33	23 56
56	20 01	20 52	22 16	22 32	23 16	23 46	24 07
58	20 13	21 11	23 11	22 52	23 34	24 01	00 01
S 60	20 27	21 34	////	23 17	23 56	24 19	00 19

Day	SUN Eqn. of Time 00^h	SUN Eqn. of Time 12^h	SUN Mer. Pass.	MOON Mer. Pass. Upper	MOON Mer. Pass. Lower	MOON Age	MOON Phase
d	m s	m s	h m	h m	h m	d %	
18	14 54	14 47	11 45	13 10	00 43	02 3	
19	14 41	14 34	11 45	14 03	01 37	03 7	
20	14 27	14 20	11 46	14 54	02 29	04 13	

UT	ARIES	VENUS −3.9		MARS +0.1		JUPITER −2.3		SATURN +1.0		STARS		
	GHA	GHA	Dec	GHA	Dec	GHA	Dec	GHA	Dec	Name	SHA	Dec
d h	° ′	° ′	° ′	° ′	° ′	° ′	° ′	° ′	° ′		° ′	° ′
21 00	60 07.8	195 59.2	S15 39.3	282 17.0	N18 26.9	97 48.3	S15 53.0	237 17.0	N 1 04.1	Acamar	315 19.8	S40 15.
01	75 10.3	210 58.6	40.3	297 18.6	26.8	112 50.5	52.9	252 19.3	04.0	Achernar	335 28.1	S57 11.
02	90 12.7	225 58.0	41.3	312 20.3	26.6	127 52.7	52.8	267 21.5	03.9	Acrux	173 12.9	S63 09.
03	105 15.2	240 57.3	42.2	327 22.0	26.5	142 54.9	52.7	282 23.8	03.8	Adhara	255 14.3	S28 59.
04	120 17.6	255 56.7	43.2	342 23.6	26.3	157 57.0	52.6	297 26.1	03.8	Aldebaran	290 52.1	N16 31.
05	135 20.1	270 56.0	44.2	357 25.3	26.2	172 59.2	52.5	312 28.3	03.7			
06	150 22.6	285 55.4	S15 45.1	12 27.0	N18 26.0	188 01.4	S15 52.4	327 30.6	N 1 03.6	Alioth	166 23.2	N55 54.
07	165 25.0	300 54.8	46.1	27 28.6	25.9	203 03.5	52.3	342 32.9	03.5	Alkaid	153 01.2	N49 15.
S 08	180 27.5	315 54.1	47.0	42 30.3	25.8	218 05.7	52.2	357 35.1	03.4	Al Na'ir	27 46.9	S46 54.
A 09	195 30.0	330 53.5	48.0	57 32.0	25.6	233 07.9	52.1	12 37.4	03.4	Alnilam	275 48.7	S 1 11.
T 10	210 32.4	345 52.8	49.0	72 33.6	25.5	248 10.0	52.0	27 39.6	03.3	Alphard	217 58.6	S 8 42.
U 11	225 34.9	0 52.2	49.9	87 35.3	25.3	263 12.2	51.9	42 41.9	03.2			
R 12	240 37.4	15 51.5	S15 50.9	102 37.0	N18 25.2	278 14.4	S15 51.8	57 44.2	N 1 03.1	Alphecca	126 13.6	N26 40.
D 13	255 39.8	30 50.9	51.8	117 38.7	25.0	293 16.5	51.7	72 46.4	03.1	Alpheratz	357 46.1	N29 09.
A 14	270 42.3	45 50.3	52.8	132 40.3	24.9	308 18.7	51.6	87 48.7	03.0	Altair	62 11.0	N 8 53.
Y 15	285 44.7	60 49.6	53.8	147 42.0	24.8	323 20.9	51.5	102 51.0	02.9	Ankaa	353 17.9	S42 15.
16	300 47.2	75 49.0	54.7	162 43.7	24.6	338 23.0	51.4	117 53.2	02.8	Antares	112 29.9	S26 27.
17	315 49.7	90 48.3	55.7	177 45.4	24.5	353 25.2	51.3	132 55.5	02.8			
18	330 52.1	105 47.7	S15 56.6	192 47.0	N18 24.3	8 27.4	S15 51.2	147 57.8	N 1 02.7	Arcturus	145 58.4	N19 07.
19	345 54.6	120 47.0	57.6	207 48.7	24.2	23 29.5	51.1	163 00.0	02.6	Atria	107 34.7	S69 02.
20	0 57.1	135 46.4	58.5	222 50.4	24.0	38 31.7	51.0	178 02.3	02.5	Avior	234 18.9	S59 32.
21	15 59.5	150 45.7	15 59.5	237 52.1	23.9	53 33.9	50.9	193 04.6	02.5	Bellatrix	278 34.5	N 6 21.
22	31 02.0	165 45.1	16 00.4	252 53.7	23.8	68 36.0	50.8	208 06.8	02.4	Betelgeuse	271 03.8	N 7 24.
23	46 04.5	180 44.4	01.4	267 55.4	23.6	83 38.2	50.7	223 09.1	02.3			
22 00	61 06.9	195 43.8	S16 02.3	282 57.1	N18 23.5	98 40.3	S15 50.6	238 11.4	N 1 02.2	Canopus	263 56.9	S52 41.
01	76 09.4	210 43.1	03.3	297 58.8	23.3	113 42.5	50.5	253 13.6	02.1	Capella	280 37.9	N46 00.
02	91 11.9	225 42.5	04.2	313 00.5	23.2	128 44.7	50.4	268 15.9	02.1	Deneb	49 33.5	N45 19.
03	106 14.3	240 41.8	05.2	328 02.2	23.1	143 46.8	50.3	283 18.2	02.0	Denebola	182 36.5	N14 30.
04	121 16.8	255 41.2	06.1	343 03.8	22.9	158 49.0	50.2	298 20.4	01.9	Diphda	348 58.3	S17 55.
05	136 19.2	270 40.5	07.1	358 05.5	22.8	173 51.2	50.1	313 22.7	01.8			
06	151 21.7	285 39.9	S16 08.0	13 07.2	N18 22.6	188 53.3	S15 50.0	328 25.0	N 1 01.8	Dubhe	193 54.8	N61 41.
07	166 24.2	300 39.2	08.9	28 08.9	22.5	203 55.5	49.9	343 27.2	01.7	Elnath	278 15.6	N28 37.
08	181 26.6	315 38.6	09.9	43 10.6	22.4	218 57.7	49.8	358 29.5	01.6	Eltanin	90 47.8	N51 29.
S 09	196 29.1	330 37.9	10.8	58 12.3	22.2	233 59.8	49.7	13 31.8	01.5	Enif	33 49.8	N 9 55.
U 10	211 31.6	345 37.3	11.8	73 14.0	22.1	249 02.0	49.6	28 34.0	01.5	Fomalhaut	15 26.7	S29 34.
N 11	226 34.0	0 36.6	12.7	88 15.7	22.0	264 04.1	49.5	43 36.3	01.4			
D 12	241 36.5	15 35.9	S16 13.7	103 17.4	N18 21.8	279 06.3	S15 49.4	58 38.6	N 1 01.3	Gacrux	172 04.4	S57 09.
A 13	256 39.0	30 35.3	14.6	118 19.1	21.7	294 08.5	49.3	73 40.8	01.2	Gienah	175 55.2	S17 35.
Y 14	271 41.4	45 34.6	15.5	133 20.7	21.5	309 10.6	49.2	88 43.1	01.2	Hadar	148 52.4	S60 25.
15	286 43.9	60 34.0	16.5	148 22.4	21.4	324 12.8	49.1	103 45.4	01.1	Hamal	328 03.5	N23 30.
16	301 46.4	75 33.3	17.4	163 24.1	21.3	339 14.9	49.0	118 47.6	01.0	Kaus Aust.	83 47.6	S34 22.
17	316 48.8	90 32.7	18.3	178 25.8	21.1	354 17.1	48.9	133 49.9	00.9			
18	331 51.3	105 32.0	S16 19.3	193 27.5	N18 21.0	9 19.3	S15 48.8	148 52.2	N 1 00.9	Kochab	137 20.7	N74 06.
19	346 53.7	120 31.3	20.2	208 29.2	20.9	24 21.4	48.7	163 54.4	00.8	Markab	13 40.9	N15 15.
20	1 56.2	135 30.7	21.2	223 30.9	20.7	39 23.6	48.6	178 56.7	00.7	Menkar	314 17.5	N 4 07.
21	16 58.7	150 30.0	22.1	238 32.6	20.6	54 25.7	48.5	193 59.0	00.6	Menkent	148 11.1	S36 25.
22	32 01.1	165 29.4	23.0	253 34.3	20.5	69 27.9	48.4	209 01.2	00.6	Miaplacidus	221 40.2	S69 45.
23	47 03.6	180 28.7	23.9	268 36.0	20.3	84 30.0	48.3	224 03.5	00.5			
23 00	62 06.1	195 28.0	S16 24.9	283 37.7	N18 20.2	99 32.2	S15 48.2	239 05.8	N 1 00.4	Mirfak	308 43.7	N49 53.
01	77 08.5	210 27.4	25.8	298 39.4	20.0	114 34.4	48.1	254 08.0	00.3	Nunki	76 01.8	S26 17.
02	92 11.0	225 26.7	26.7	313 41.1	19.9	129 36.5	48.0	269 10.3	00.3	Peacock	53 23.5	S56 42.
03	107 13.5	240 26.0	27.7	328 42.8	19.8	144 38.7	47.9	284 12.6	00.2	Pollux	243 30.7	N28 00.
04	122 15.9	255 25.4	28.6	343 44.6	19.6	159 40.8	47.8	299 14.8	00.1	Procyon	245 02.3	N 5 12.
05	137 18.4	270 24.7	29.5	358 46.3	19.5	174 43.0	47.7	314 17.1	00.0			
06	152 20.9	285 24.1	S16 30.5	13 48.0	N18 19.4	189 45.1	S15 47.6	329 19.4	N 1 00.0	Rasalhague	96 09.2	N12 33.
07	167 23.3	300 23.4	31.4	28 49.7	19.2	204 47.3	47.5	344 21.7	0 59.9	Regulus	207 46.3	N11 55.
08	182 25.8	315 22.7	32.3	43 51.4	19.1	219 49.5	47.4	359 23.9	59.8	Rigel	281 14.3	S 8 11.
M 09	197 28.2	330 22.1	33.2	58 53.1	19.0	234 51.6	47.3	14 26.2	59.7	Rigil Kent.	139 56.1	S60 52.
O 10	212 30.7	345 21.4	34.2	73 54.8	18.8	249 53.8	47.2	29 28.5	59.7	Sabik	102 15.9	S15 44.
N 11	227 33.2	0 20.7	35.1	88 56.5	18.7	264 55.9	47.0	44 30.7	59.6			
D 12	242 35.6	15 20.1	S16 36.0	103 58.2	N18 18.6	279 58.1	S15 46.9	59 33.0	N 0 59.5	Schedar	349 43.3	N56 35.
A 13	257 38.1	30 19.4	36.9	118 59.9	18.5	295 00.2	46.8	74 35.3	59.5	Shaula	96 25.9	S37 06.
Y 14	272 40.6	45 18.7	37.8	134 01.7	18.3	310 02.4	46.7	89 37.5	59.4	Sirius	258 35.8	S16 43.
15	287 43.0	60 18.0	38.8	149 03.4	18.2	325 04.5	46.6	104 39.8	59.3	Spica	158 34.3	S11 12.
16	302 45.5	75 17.4	39.7	164 05.1	18.1	340 06.7	46.5	119 42.1	59.2	Suhail	222 54.3	S43 28.
17	317 48.0	90 16.7	40.6	179 06.8	17.9	355 08.8	46.4	134 44.4	59.2			
18	332 50.4	105 16.0	S16 41.5	194 08.5	N18 17.8	10 11.0	S15 46.3	149 46.6	N 0 59.1	Vega	80 41.1	N38 47.
19	347 52.9	120 15.4	42.4	209 10.2	17.7	25 13.2	46.2	164 48.9	59.0	Zuben'ubi	137 08.7	S16 05.
20	2 55.3	135 14.7	43.4	224 12.0	17.5	40 15.3	46.1	179 51.2	58.9			
21	17 57.8	150 14.0	44.3	239 13.7	17.4	55 17.5	46.0	194 53.4	58.9			
22	33 00.3	165 13.3	45.2	254 15.4	17.3	70 19.6	45.9	209 55.7	58.8			
23	48 02.7	180 12.7	46.1	269 17.1	17.1	85 21.8	45.8	224 58.0	58.7			
Mer. Pass.	h m 19 52.3	v −0.7	d 0.9	v 1.7	d 0.1	v 2.2	d 0.1	v 2.3	d 0.1			

	SHA	Mer. Pas
	° ′	h m
Venus	134 36.9	10 58
Mars	221 50.2	5 08
Jupiter	37 33.4	17 23
Saturn	177 04.4	8 06

	UT	SUN GHA	SUN Dec	MOON GHA	MOON v	MOON Dec	MOON d	MOON HP
	d h	° ′	° ′	° ′	′	° ′	′	′
	21 00	183 33.1	S19 53.6	132 05.8	11.7	S22 55.3	6.4	54.4
	01	198 32.9	54.1	146 36.5	11.8	22 48.9	6.5	54.4
	02	213 32.7	54.7	161 07.3	11.8	22 42.4	6.5	54.4
	03	228 32.6	. . 55.2	175 38.1	11.9	22 35.9	6.7	54.4
	04	243 32.4	55.8	190 09.0	12.0	22 29.2	6.8	54.4
	05	258 32.3	56.3	204 40.0	12.0	22 22.4	6.8	54.4
	06	273 32.1	S19 56.9	219 11.0	12.2	S22 15.6	7.0	54.4
	07	288 32.0	57.4	233 42.2	12.1	22 08.6	7.1	54.4
S	08	303 31.8	58.0	248 13.3	12.3	22 01.5	7.1	54.4
A	09	318 31.6	. . 58.5	262 44.6	12.3	21 54.4	7.3	54.3
T	10	333 31.5	59.1	277 15.9	12.4	21 47.1	7.3	54.3
U	11	348 31.3	19 59.6	291 47.3	12.4	21 39.8	7.5	54.3
R	12	3 31.2	S20 00.2	306 18.7	12.5	S21 32.3	7.5	54.3
D	13	18 31.0	00.7	320 50.2	12.6	21 24.8	7.6	54.3
A	14	33 30.8	01.2	335 21.8	12.6	21 17.2	7.7	54.3
Y	15	48 30.7	. . 01.8	349 53.4	12.7	21 09.5	7.8	54.3
	16	63 30.5	02.3	4 25.1	12.8	21 01.7	7.9	54.3
	17	78 30.4	02.9	18 56.9	12.8	20 53.8	8.0	54.3
	18	93 30.2	S20 03.4	33 28.7	12.9	S20 45.8	8.1	54.3
	19	108 30.0	04.0	48 00.6	13.0	20 37.7	8.2	54.3
	20	123 29.9	04.5	62 32.6	13.0	20 29.5	8.2	54.3
	21	138 29.7	. . 05.0	77 04.6	13.1	20 21.3	8.3	54.3
	22	153 29.5	05.6	91 36.7	13.2	20 13.0	8.4	54.2
	23	168 29.4	06.1	106 08.9	13.2	20 04.6	8.5	54.2
	22 00	183 29.2	S20 06.7	120 41.1	13.3	S19 56.1	8.6	54.2
	01	198 29.1	07.2	135 13.4	13.3	19 47.5	8.7	54.2
	02	213 28.9	07.7	149 45.7	13.4	19 38.8	8.7	54.2
	03	228 28.7	. . 08.3	164 18.1	13.5	19 30.1	8.9	54.2
	04	243 28.6	08.8	178 50.6	13.5	19 21.2	8.9	54.2
	05	258 28.4	09.3	193 23.1	13.6	19 12.3	9.0	54.2
	06	273 28.2	S20 09.9	207 55.7	13.7	S19 03.3	9.0	54.2
	07	288 28.1	10.4	222 28.4	13.7	18 54.3	9.2	54.2
	08	303 27.9	10.9	237 01.1	13.8	18 45.1	9.2	54.2
S	09	318 27.7	. . 11.5	251 33.9	13.8	18 35.9	9.3	54.2
U	10	333 27.6	12.0	266 06.7	13.9	18 26.6	9.3	54.2
N	11	348 27.4	12.5	280 39.6	13.9	18 17.3	9.5	54.2
D	12	3 27.2	S20 13.1	295 12.5	14.0	S18 07.8	9.5	54.2
A	13	18 27.1	13.6	309 45.5	14.1	17 58.3	9.6	54.2
Y	14	33 26.9	14.1	324 18.6	14.1	17 48.7	9.6	54.2
	15	48 26.7	. . 14.6	338 51.7	14.2	17 39.1	9.8	54.2
	16	63 26.5	15.2	353 24.9	14.3	17 29.3	9.8	54.2
	17	78 26.4	15.7	7 58.2	14.2	17 19.5	9.8	54.2
	18	93 26.2	S20 16.2	22 31.4	14.4	S17 09.7	10.0	54.2
	19	108 26.0	16.8	37 04.8	14.4	16 59.7	10.0	54.2
	20	123 25.9	17.3	51 38.2	14.4	16 49.7	10.1	54.2
	21	138 25.7	. . 17.8	66 11.6	14.5	16 39.6	10.1	54.2
	22	153 25.5	18.3	80 45.1	14.6	16 29.5	10.2	54.2
	23	168 25.4	18.9	95 18.7	14.6	16 19.3	10.3	54.2
	23 00	183 25.2	S20 19.4	109 52.3	14.7	S16 09.0	10.3	54.2
	01	198 25.0	19.9	124 26.0	14.7	15 58.7	10.4	54.2
	02	213 24.8	20.4	138 59.7	14.7	15 48.3	10.5	54.2
	03	228 24.7	. . 20.9	153 33.4	14.8	15 37.8	10.5	54.2
	04	243 24.5	21.5	168 07.2	14.9	15 27.3	10.6	54.2
	05	258 24.3	22.0	182 41.1	14.9	15 16.7	10.6	54.2
	06	273 24.1	S20 22.5	197 15.0	14.9	S15 06.1	10.7	54.2
	07	288 24.0	23.0	211 48.9	15.0	14 55.4	10.8	54.2
	08	303 23.8	23.5	226 22.9	15.1	14 44.6	10.8	54.2
M	09	318 23.6	. . 24.1	240 57.0	15.1	14 33.8	10.9	54.2
O	10	333 23.4	24.6	255 31.1	15.1	14 22.9	11.0	54.2
N	11	348 23.3	25.1	270 05.2	15.1	14 11.9	11.0	54.2
D	12	3 23.1	S20 25.6	284 39.3	15.3	S14 00.9	11.0	54.2
A	13	18 22.9	26.1	299 13.6	15.2	13 49.9	11.1	54.2
Y	14	33 22.7	26.6	313 47.8	15.3	13 38.8	11.2	54.2
	15	48 22.6	. . 27.1	328 22.1	15.3	13 27.6	11.2	54.2
	16	63 22.4	27.7	342 56.4	15.4	13 16.4	11.3	54.2
	17	78 22.2	28.2	357 30.8	15.4	13 05.1	11.3	54.2
	18	93 22.0	S20 28.7	12 05.2	15.5	S12 53.8	11.4	54.2
	19	108 21.9	29.2	26 39.7	15.4	12 42.4	11.4	54.3
	20	123 21.7	29.7	41 14.1	15.5	12 31.0	11.5	54.3
	21	138 21.5	. . 30.2	55 48.6	15.6	12 19.5	11.6	54.3
	22	153 21.3	30.7	70 23.2	15.6	12 07.9	11.5	54.3
	23	168 21.1	31.2	84 57.8	15.6	S11 56.4	11.7	54.3
		SD 16.2	*d* 0.5	SD 14.8		14.8		14.8

Lat.	Twilight Naut.	Twilight Civil	Sunrise	Moonrise 21	Moonrise 22	Moonrise 23	Moonrise 24
°	h m	h m	h m	h m	h m	h m	h m
N 72	07 24	09 08	■■■■	■■■■	15 38	14 19	13 43
N 70	07 11	08 39	10 46	■■■■	14 32	13 54	13 30
68	07 00	08 17	09 48	14 39	13 55	13 34	13 19
66	06 51	08 00	09 14	13 45	13 29	13 18	13 10
64	06 44	07 46	08 50	13 12	13 09	13 05	13 02
62	06 37	07 34	08 31	12 48	12 52	12 54	12 55
60	06 31	07 24	08 15	12 29	12 39	12 45	12 49
N 58	06 25	07 15	08 02	12 13	12 27	12 37	12 44
56	06 20	07 07	07 51	11 59	12 16	12 29	12 40
54	06 16	07 00	07 41	11 47	12 07	12 23	12 35
52	06 11	06 53	07 32	11 36	11 59	12 17	12 32
50	06 07	06 47	07 24	11 27	11 52	12 11	12 28
45	05 58	06 35	07 07	11 07	11 36	12 00	12 21
N 40	05 50	06 24	06 53	10 51	11 23	11 50	12 15
35	05 43	06 14	06 41	10 37	11 12	11 42	12 09
30	05 36	06 05	06 31	10 25	11 02	11 34	12 04
20	05 23	05 50	06 13	10 05	10 45	11 22	11 56
N 10	05 09	05 35	05 57	09 47	10 30	11 11	11 49
0	04 55	05 21	05 43	09 30	10 16	11 00	11 42
S 10	04 39	05 05	05 28	09 14	10 03	10 49	11 35
20	04 19	04 48	05 12	08 56	09 48	10 38	11 28
30	03 55	04 26	04 53	08 35	09 31	10 25	11 19
35	03 39	04 14	04 42	08 23	09 20	10 18	11 14
40	03 20	03 58	04 29	08 09	09 09	10 09	11 09
45	02 55	03 40	04 15	07 52	08 55	09 59	11 02
S 50	02 21	03 16	03 56	07 31	08 39	09 47	10 54
52	02 02	03 04	03 47	07 21	08 31	09 41	10 51
54	01 39	02 50	03 38	07 10	08 22	09 35	10 47
56	01 06	02 34	03 27	06 58	08 12	09 28	10 42
58	////	02 15	03 14	06 43	08 01	09 20	10 37
S 60	////	01 50	02 59	06 25	07 48	09 11	10 32

Lat.	Sunset	Twilight Civil	Twilight Naut.	Moonset 21	Moonset 22	Moonset 23	Moonset 24
°	h m	h m	h m	h m	h m	h m	h m
N 72	■■■■	14 23	16 07	■■■■	17 34	20 23	22 26
N 70	12 45	14 52	16 20	■■■■	18 39	20 47	22 38
68	13 44	15 14	16 31	16 57	19 15	21 05	22 47
66	14 17	15 31	16 40	17 50	19 40	21 19	22 54
64	14 42	15 46	16 48	18 22	19 59	21 31	23 00
62	15 01	15 58	16 55	18 46	20 15	21 41	23 06
60	15 16	16 08	17 01	19 04	20 28	21 50	23 10
N 58	15 30	16 17	17 06	19 20	20 39	21 57	23 14
56	15 41	16 25	17 11	19 33	20 49	22 04	23 18
54	15 51	16 32	17 16	19 44	20 57	22 09	23 21
52	16 00	16 38	17 20	19 55	21 05	22 15	23 24
50	16 08	16 44	17 24	20 04	21 12	22 19	23 27
45	16 25	16 57	17 33	20 22	21 26	22 29	23 33
N 40	16 39	17 08	17 42	20 38	21 38	22 38	23 37
35	16 51	17 18	17 49	20 51	21 48	22 45	23 41
30	17 01	17 27	17 56	21 02	21 57	22 51	23 45
20	17 19	17 43	18 10	21 21	22 12	23 02	23 51
N 10	17 35	17 57	18 23	21 38	22 25	23 12	23 57
0	17 50	18 12	18 37	21 53	22 38	23 20	24 02
S 10	18 05	18 27	18 54	22 08	22 50	23 29	24 07
20	18 21	18 45	19 13	22 25	23 03	23 38	24 12
30	18 40	19 06	19 38	22 43	23 18	23 49	24 18
35	18 51	19 19	19 54	22 54	23 26	23 55	24 21
40	19 03	19 35	20 14	23 06	23 36	24 02	00 02
45	19 18	19 54	20 39	23 21	23 47	24 09	00 09
S 50	19 37	20 18	21 13	23 39	24 01	00 01	00 19
52	19 46	20 30	21 32	23 47	24 07	00 07	00 23
54	19 56	20 44	21 56	23 56	24 14	00 14	00 28
56	20 07	21 00	22 30	24 07	00 07	00 22	00 33
58	20 20	21 20	////	00 01	00 18	00 30	00 39
S 60	20 35	21 45	////	00 19	00 32	00 40	00 46

Day	SUN Eqn. of Time 00^h	SUN Eqn. of Time 12^h	SUN Mer. Pass.	MOON Mer. Pass. Upper	MOON Mer. Pass. Lower	MOON Age	MOON Phase
d	m s	m s	h m	h m	h m	d %	
21	14 13	14 05	11 46	15 42	03 18	05 20	
22	13 57	13 49	11 46	16 27	04 05	06 28	
23	13 41	13 33	11 46	17 10	04 49	07 37	

Day	UT (d h)	ARIES GHA	VENUS −3.9 GHA	VENUS Dec	MARS +0.0 GHA	MARS Dec	JUPITER −2.3 GHA	JUPITER Dec	SATURN +1.0 GHA	SATURN Dec	STARS Name	SHA	Dec
		° ′	° ′	° ′	° ′	° ′	° ′	° ′	° ′	° ′		° ′	° ′
	24 00	63 05.2	195 12.0	S16 47.0	284 18.8	N18 17.0	100 23.9	S15 45.7	240 00.2	N 0 58.6	Acamar	315 19.8	S40 15.8
	01	78 07.7	210 11.3	47.9	299 20.6	16.9	115 26.1	45.6	255 02.5	58.6	Achernar	335 28.1	S57 11.2
	02	93 10.1	225 10.6	48.8	314 22.3	16.8	130 28.2	45.5	270 04.8	58.5	Acrux	173 12.9	S63 09.1
	03	108 12.6	240 10.0	. . 49.7	329 24.0	. . 16.6	145 30.4	. . 45.4	285 07.1	. . 58.4	Adhara	255 14.3	S28 59.0
	04	123 15.1	255 09.3	50.7	344 25.7	16.5	160 32.5	45.3	300 09.3	58.4	Aldebaran	290 52.1	N16 31.8
	05	138 17.5	270 08.6	51.6	359 27.5	16.4	175 34.7	45.2	315 11.6	58.3			
	06	153 20.0	285 07.9	S16 52.5	14 29.2	N18 16.2	190 36.8	S15 45.1	330 13.9	N 0 58.2	Alioth	166 23.2	N55 54.1
	07	168 22.5	300 07.3	53.4	29 30.9	16.1	205 39.0	45.0	345 16.2	58.1	Alkaid	153 01.2	N49 15.6
T	08	183 24.9	315 06.6	54.3	44 32.7	16.0	220 41.1	44.8	0 18.4	58.1	Al Na'ir	27 46.9	S46 54.9
	09	198 27.4	330 05.9	. . 55.2	59 34.4	. . 15.9	235 43.3	. . 44.7	15 20.7	. . 58.0	Alnilam	275 48.7	S 1 11.6
U	10	213 29.8	345 05.2	56.1	74 36.1	15.7	250 45.4	44.6	30 23.0	57.9	Alphard	217 58.6	S 8 42.0
E	11	228 32.3	0 04.5	57.0	89 37.9	15.6	265 47.6	44.5	45 25.2	57.8			
S	12	243 34.8	15 03.9	S16 57.9	104 39.6	N18 15.5	280 49.7	S15 44.4	60 27.5	N 0 57.8	Alphecca	126 13.6	N26 40.8
D	13	258 37.2	30 03.2	58.8	119 41.3	15.4	295 51.9	44.3	75 29.8	57.7	Alpheratz	357 46.1	N29 09.0
A	14	273 39.7	45 02.5	16 59.7	134 43.1	15.2	310 54.0	44.2	90 32.1	57.6	Altair	62 11.0	N 8 53.8
Y	15	288 42.2	60 01.8	17 00.6	149 44.8	. . 15.1	325 56.2	. . 44.1	105 34.3	. . 57.6	Ankaa	353 17.9	S42 15.2
	16	303 44.6	75 01.1	01.5	164 46.5	15.0	340 58.3	44.0	120 36.6	57.5	Antares	112 29.9	S26 27.2
	17	318 47.1	90 00.5	02.4	179 48.3	14.9	356 00.4	43.9	135 38.9	57.4			
	18	333 49.6	104 59.8	S17 03.3	194 50.0	N18 14.7	11 02.6	S15 43.8	150 41.2	N 0 57.3	Arcturus	145 58.4	N19 07.8
	19	348 52.0	119 59.1	04.2	209 51.7	14.6	26 04.7	43.7	165 43.4	57.3	Atria	107 34.7	S69 02.8
	20	3 54.5	134 58.4	05.1	224 53.5	14.5	41 06.9	43.6	180 45.7	57.2	Avior	234 18.9	S59 32.2
	21	18 57.0	149 57.7	. . 06.0	239 55.2	. . 14.4	56 09.0	. . 43.5	195 48.0	. . 57.1	Bellatrix	278 34.5	N 6 21.6
	22	33 59.4	164 57.0	06.9	254 57.0	14.2	71 11.2	43.4	210 50.2	57.0	Betelgeuse	271 03.8	N 7 24.6
	23	49 01.9	179 56.3	07.8	269 58.7	14.1	86 13.3	43.3	225 52.5	57.0			
	25 00	64 04.3	194 55.7	S17 08.7	285 00.5	N18 14.0	101 15.5	S15 43.1	240 54.8	N 0 56.9	Canopus	263 56.8	S52 41.9
	01	79 06.8	209 55.0	09.6	300 02.2	13.9	116 17.6	43.0	255 57.1	56.8	Capella	280 37.9	N46 00.5
	02	94 09.3	224 54.3	10.5	315 03.9	13.8	131 19.8	42.9	270 59.3	56.8	Deneb	49 33.6	N45 19.2
	03	109 11.7	239 53.6	. . 11.4	330 05.7	. . 13.6	146 21.9	. . 42.8	286 01.6	. . 56.7	Denebola	182 36.4	N14 30.9
	04	124 14.2	254 52.9	12.3	345 07.4	13.5	161 24.0	42.7	301 03.9	56.6	Diphda	348 58.3	S17 55.9
	05	139 16.7	269 52.2	13.2	0 09.2	13.4	176 26.2	42.6	316 06.2	56.5			
	06	154 19.1	284 51.5	S17 14.1	15 10.9	N18 13.3	191 28.3	S15 42.5	331 08.4	N 0 56.5	Dubhe	193 54.8	N61 41.5
W	07	169 21.6	299 50.8	14.9	30 12.7	13.1	206 30.5	42.4	346 10.7	56.4	Elnath	278 15.6	N28 37.0
E	08	184 24.1	314 50.1	15.8	45 14.4	13.0	221 32.6	42.3	1 13.0	56.3	Eltanin	90 47.9	N51 29.4
D	09	199 26.5	329 49.5	. . 16.7	60 16.2	. . 12.9	236 34.8	. . 42.2	16 15.3	. . 56.3	Enif	33 49.8	N 9 55.4
N	10	214 29.0	344 48.8	17.6	75 17.9	12.8	251 36.9	42.1	31 17.5	56.2	Fomalhaut	15 26.7	S29 34.2
E	11	229 31.4	359 48.1	18.5	90 19.7	12.7	266 39.1	42.0	46 19.8	56.1			
S	12	244 33.9	14 47.4	S17 19.4	105 21.5	N18 12.5	281 41.2	S15 41.9	61 22.1	N 0 56.0	Gacrux	172 04.4	S57 09.9
	13	259 36.4	29 46.7	20.3	120 23.2	12.4	296 43.3	41.7	76 24.4	56.0	Gienah	175 55.2	S17 35.8
D	14	274 38.8	44 46.0	21.1	135 25.0	12.3	311 45.5	41.6	91 26.6	55.9	Hadar	148 52.4	S60 25.1
A	15	289 41.3	59 45.3	. . 22.0	150 26.7	. . 12.2	326 47.6	. . 41.5	106 28.9	. . 55.8	Hamal	328 03.5	N23 30.8
Y	16	304 43.8	74 44.6	22.9	165 28.5	12.1	341 49.8	41.4	121 31.2	55.8	Kaus Aust.	83 47.6	S34 22.8
	17	319 46.2	89 43.9	23.8	180 30.2	11.9	356 51.9	41.3	136 33.5	55.7			
	18	334 48.7	104 43.2	S17 24.7	195 32.0	N18 11.8	11 54.0	S15 41.2	151 35.8	N 0 55.6	Kochab	137 20.7	N74 06.7
	19	349 51.2	119 42.5	25.6	210 33.8	11.7	26 56.2	41.1	166 38.0	55.5	Markab	13 40.9	N15 15.7
	20	4 53.6	134 41.8	26.4	225 35.5	11.6	41 58.3	41.0	181 40.3	55.5	Menkar	314 17.5	N 4 07.9
	21	19 56.1	149 41.1	. . 27.3	240 37.3	. . 11.5	57 00.5	. . 40.9	196 42.6	. . 55.4	Menkent	148 11.1	S36 25.0
	22	34 58.6	164 40.4	28.2	255 39.0	11.4	72 02.6	40.8	211 44.9	55.3	Miaplacidus	221 40.2	S69 45.2
	23	50 01.0	179 39.7	29.1	270 40.8	11.2	87 04.7	40.7	226 47.1	55.3			
	26 00	65 03.5	194 39.0	S17 29.9	285 42.6	N18 11.1	102 06.9	S15 40.6	241 49.4	N 0 55.2	Mirfak	308 43.7	N49 54.0
	01	80 05.9	209 38.3	30.8	300 44.3	11.0	117 09.0	40.4	256 51.7	55.1	Nunki	76 01.8	S26 17.1
	02	95 08.4	224 37.6	31.7	315 46.1	10.9	132 11.2	40.3	271 54.0	55.1	Peacock	53 23.5	S56 42.4
	03	110 10.9	239 36.9	. . 32.6	330 47.9	. . 10.8	147 13.3	. . 40.2	286 56.2	. . 55.0	Pollux	243 30.7	N28 00.0
	04	125 13.3	254 36.2	33.4	345 49.6	10.7	162 15.4	40.1	301 58.5	54.9	Procyon	245 02.2	N 5 12.0
	05	140 15.8	269 35.5	34.3	0 51.4	10.5	177 17.6	40.0	317 00.8	54.8			
	06	155 18.3	284 34.8	S17 35.2	15 53.2	N18 10.4	192 19.7	S15 39.9	332 03.1	N 0 54.8	Rasalhague	96 09.2	N12 33.2
	07	170 20.7	299 34.1	36.0	30 55.0	10.3	207 21.8	39.8	347 05.4	54.7	Regulus	207 46.3	N11 55.0
T	08	185 23.2	314 33.4	36.9	45 56.7	10.2	222 24.0	39.7	2 07.6	54.6	Rigel	281 14.2	S 8 11.3
H	09	200 25.7	329 32.7	. . 37.8	60 58.5	. . 10.1	237 26.1	. . 39.6	17 09.9	. . 54.6	Rigil Kent.	139 56.1	S60 52.4
U	10	215 28.1	344 32.0	38.6	76 00.3	10.0	252 28.3	39.5	32 12.2	54.5	Sabik	102 15.9	S15 44.2
R	11	230 30.6	359 31.3	39.5	91 02.1	09.9	267 30.4	39.3	47 14.5	54.4			
S	12	245 33.0	14 30.6	S17 40.4	106 03.8	N18 09.7	282 32.5	S15 39.2	62 16.8	N 0 54.4	Schedar	349 43.4	N56 35.9
D	13	260 35.5	29 29.9	41.2	121 05.6	09.6	297 34.7	39.1	77 19.0	54.3	Shaula	96 25.9	S37 06.7
A	14	275 38.0	44 29.2	42.1	136 07.4	09.5	312 36.8	39.0	92 21.3	54.2	Sirius	258 35.7	S16 43.7
Y	15	290 40.4	59 28.5	. . 43.0	151 09.2	. . 09.4	327 38.9	. . 38.9	107 23.6	. . 54.1	Spica	158 34.3	S11 12.8
	16	305 42.9	74 27.8	43.8	166 10.9	09.3	342 41.1	38.8	122 25.9	54.1	Suhail	222 54.3	S43 28.1
	17	320 45.4	89 27.1	44.7	181 12.7	09.2	357 43.2	38.7	137 28.1	54.0			
	18	335 47.8	104 26.3	S17 45.6	196 14.5	N18 09.1	12 45.3	S15 38.6	152 30.4	N 0 53.9	Vega	80 41.1	N38 47.7
	19	350 50.3	119 25.6	46.4	211 16.3	09.0	27 47.5	38.5	167 32.7	53.9	Zuben'ubi	137 08.7	S16 05.0
	20	5 52.8	134 24.9	47.3	226 18.1	08.9	42 49.6	38.4	182 35.0	53.8			
	21	20 55.2	149 24.2	. . 48.1	241 19.9	. . 08.7	57 51.8	. . 38.2	197 37.3	. . 53.7			
	22	35 57.7	164 23.5	49.0	256 21.6	08.6	72 53.9	38.1	212 39.5	53.7			
	23	51 00.2	179 22.8	49.8	271 23.4	08.5	87 56.0	38.0	227 41.8	53.6			
	Mer. Pass.	h m 19 40.5	v −0.7	d 0.9	v 1.8	d 0.1	v 2.1	d 0.1	v 2.3	d 0.1			

	SHA	Mer. Pass.
	° ′	h m
Venus	130 51.3	11 01
Mars	220 56.1	4 59
Jupiter	37 11.1	17 13
Saturn	176 50.5	7 55

JT	SUN GHA	SUN Dec	MOON GHA	v	MOON Dec	d	HP
d h	° ′	° ′	° ′	′	° ′	′	′
4 00	183 21.0	S20 31.7	99 32.4	15.6	S11 44.7	11.6	54.3
01	198 20.8	32.2	114 07.0	15.7	11 33.1	11.8	54.3
02	213 20.6	32.7	128 41.7	15.7	11 21.3	11.7	54.3
03	228 20.4	. . 33.2	143 16.4	15.8	11 09.6	11.9	54.3
04	243 20.2	33.7	157 51.2	15.7	10 57.7	11.8	54.3
05	258 20.1	34.2	172 25.9	15.8	10 45.9	11.9	54.3
06	273 19.9	S20 34.8	187 00.7	15.8	S10 34.0	12.0	54.3
07	288 19.7	35.3	201 35.5	15.9	10 22.0	12.0	54.4
08	303 19.5	35.8	216 10.4	15.8	10 10.0	12.0	54.4
09	318 19.3	. . 36.3	230 45.2	15.9	9 58.0	12.1	54.4
10	333 19.1	36.8	245 20.1	16.0	9 45.9	12.1	54.4
11	348 19.0	37.3	259 55.1	15.9	9 33.8	12.2	54.4
12	3 18.8	S20 37.8	274 30.0	15.9	S 9 21.6	12.2	54.4
13	18 18.6	38.3	289 04.9	16.0	9 09.4	12.2	54.4
14	33 18.4	38.8	303 39.9	16.0	8 57.2	12.3	54.4
15	48 18.2	. . 39.2	318 14.9	16.0	8 44.9	12.4	54.4
16	63 18.0	39.7	332 49.9	16.0	8 32.5	12.3	54.5
17	78 17.8	40.2	347 24.9	16.1	8 20.2	12.4	54.5
18	93 17.7	S20 40.7	2 00.0	16.0	S 8 07.8	12.5	54.5
19	108 17.5	41.2	16 35.0	16.1	7 55.3	12.4	54.5
20	123 17.3	41.7	31 10.1	16.1	7 42.9	12.5	54.5
21	138 17.1	. . 42.2	45 45.2	16.1	7 30.4	12.6	54.5
22	153 16.9	42.7	60 20.3	16.1	7 17.8	12.6	54.5
23	168 16.7	43.2	74 55.4	16.1	7 05.2	12.6	54.6
5 00	183 16.5	S20 43.7	89 30.5	16.1	S 6 52.6	12.6	54.6
01	198 16.4	44.2	104 05.6	16.2	6 40.0	12.7	54.6
02	213 16.2	44.7	118 40.8	16.1	6 27.3	12.7	54.6
03	228 16.0	. . 45.2	133 15.9	16.2	6 14.6	12.7	54.6
04	243 15.8	45.6	147 51.1	16.1	6 01.9	12.8	54.6
05	258 15.6	46.1	162 26.2	16.2	5 49.1	12.8	54.7
06	273 15.4	S20 46.6	177 01.4	16.1	S 5 36.3	12.8	54.7
07	288 15.2	47.1	191 36.5	16.2	5 23.5	12.9	54.7
08	303 15.0	47.6	206 11.7	16.1	5 10.6	12.9	54.7
09	318 14.8	. . 48.1	220 46.8	16.2	4 57.7	12.9	54.7
10	333 14.6	48.6	235 22.0	16.1	4 44.8	12.9	54.7
11	348 14.5	49.0	249 57.1	16.2	4 31.9	13.0	54.8
12	3 14.3	S20 49.5	264 32.3	16.1	S 4 18.9	13.0	54.8
13	18 14.1	50.0	279 07.4	16.2	4 05.9	13.0	54.8
14	33 13.9	50.5	293 42.6	16.1	3 52.9	13.0	54.8
15	48 13.7	. . 51.0	308 17.7	16.1	3 39.9	13.1	54.8
16	63 13.5	51.5	322 52.8	16.1	3 26.8	13.1	54.9
17	78 13.3	51.9	337 27.9	16.2	3 13.7	13.1	54.9
18	93 13.1	S20 52.4	352 03.1	16.1	S 3 00.6	13.1	54.9
19	108 12.9	52.9	6 38.2	16.0	2 47.5	13.2	54.9
20	123 12.7	53.4	21 13.2	16.1	2 34.3	13.2	54.9
21	138 12.5	. . 53.8	35 48.3	16.1	2 21.1	13.2	55.0
22	153 12.3	54.3	50 23.4	16.0	2 07.9	13.2	55.0
23	168 12.1	54.8	64 58.4	16.0	1 54.7	13.2	55.0
6 00	183 11.9	S20 55.3	79 33.4	16.1	S 1 41.5	13.2	55.0
01	198 11.7	55.7	94 08.5	16.0	1 28.3	13.3	55.1
02	213 11.6	56.2	108 43.5	15.9	1 15.0	13.3	55.1
03	228 11.4	. . 56.7	123 18.4	16.0	1 01.7	13.3	55.1
04	243 11.2	57.2	137 53.4	15.9	0 48.4	13.3	55.1
05	258 11.0	57.6	152 28.3	15.9	0 35.1	13.3	55.2
06	273 10.8	S20 58.1	167 03.2	15.9	S 0 21.8	13.4	55.2
07	288 10.6	58.6	181 38.1	15.9	S 0 08.4	13.3	55.2
08	303 10.4	59.0	196 13.0	15.8	N 0 04.9	13.4	55.2
09	318 10.2	20 59.5	210 47.8	15.8	0 18.3	13.4	55.2
10	333 10.0	21 00.0	225 22.6	15.8	0 31.7	13.4	55.3
11	348 09.8	00.5	239 57.4	15.7	0 45.1	13.3	55.3
12	3 09.6	S21 00.9	254 32.1	15.7	N 0 58.4	13.5	55.3
13	18 09.4	01.4	269 06.8	15.7	1 11.9	13.4	55.3
14	33 09.2	01.8	283 41.5	15.7	1 25.3	13.4	55.4
15	48 09.0	. . 02.3	298 16.2	15.6	1 38.7	13.4	55.4
16	63 08.8	02.8	312 50.8	15.6	1 52.1	13.5	55.4
17	78 08.6	03.2	327 25.4	15.5	2 05.6	13.4	55.5
18	93 08.4	S21 03.7	341 59.9	15.5	N 2 19.0	13.4	55.5
19	108 08.2	04.2	356 34.4	15.5	2 32.4	13.5	55.5
20	123 08.0	04.6	11 08.9	15.4	2 45.9	13.4	55.5
21	138 07.8	. . 05.1	25 43.3	15.4	2 59.3	13.5	55.6
22	153 07.6	05.5	40 17.7	15.4	3 12.8	13.5	55.6
23	168 07.4	06.0	54 52.1	15.3	N 3 26.3	13.4	55.6
	SD 16.2	d 0.5	SD	14.8	14.9		15.1

Lat.	Twilight Naut.	Twilight Civil	Sunrise	Moonrise 24	25	26	27
°	h m	h m	h m	h m	h m	h m	h m
N 72	07 34	09 22	▬	13 43	13 15	12 50	12 24
N 70	07 19	08 50	11 39	13 30	13 10	12 52	12 33
68	07 08	08 27	10 03	13 19	13 05	12 53	12 40
66	06 58	08 08	09 25	13 10	13 02	12 54	12 46
64	06 50	07 53	08 59	13 02	12 58	12 55	12 51
62	06 43	07 41	08 39	12 55	12 55	12 55	12 56
60	06 36	07 30	08 22	12 49	12 53	12 56	13 00
N 58	06 30	07 20	08 08	12 44	12 51	12 57	13 03
56	06 25	07 12	07 56	12 40	12 49	12 57	13 06
54	06 20	07 05	07 46	12 35	12 47	12 58	13 09
52	06 16	06 58	07 37	12 32	12 45	12 58	13 12
50	06 11	06 52	07 28	12 28	12 44	12 58	13 14
45	06 02	06 38	07 11	12 21	12 40	12 59	13 19
N 40	05 53	06 27	06 56	12 15	12 37	13 00	13 24
35	05 46	06 17	06 44	12 09	12 35	13 01	13 27
30	05 38	06 08	06 33	12 04	12 33	13 01	13 31
20	05 24	05 51	06 15	11 56	12 29	13 02	13 37
N 10	05 10	05 36	05 59	11 49	12 26	13 03	13 42
0	04 56	05 21	05 43	11 42	12 23	13 04	13 47
S 10	04 39	05 05	05 28	11 35	12 20	13 05	13 52
20	04 19	04 47	05 11	11 28	12 17	13 06	13 57
30	03 53	04 25	04 52	11 19	12 13	13 07	14 03
35	03 37	04 12	04 41	11 14	12 11	13 08	14 07
40	03 17	03 56	04 28	11 09	12 08	13 08	14 11
45	02 51	03 37	04 12	11 02	12 05	13 09	14 15
S 50	02 15	03 12	03 53	10 54	12 02	13 10	14 21
52	01 56	03 00	03 44	10 51	12 00	13 11	14 24
54	01 31	02 45	03 34	10 47	11 59	13 11	14 26
56	00 53	02 29	03 22	10 42	11 57	13 12	14 30
58	////	02 08	03 09	10 37	11 55	13 13	14 33
S 60	////	01 41	02 53	10 32	11 52	13 13	14 37

Lat.	Sunset	Twilight Civil	Twilight Naut.	Moonset 24	25	26	27
°	h m	h m	h m	h m	h m	h m	h m
N 72	▬	14 11	15 59	22 26	24 21	00 21	02 17
N 70	11 54	14 43	16 14	22 38	24 24	00 24	02 11
68	13 30	15 07	16 25	22 47	24 26	00 26	02 06
66	14 08	15 25	16 35	22 54	24 27	00 27	02 02
64	14 34	15 40	16 44	23 00	24 29	00 29	01 58
62	14 55	15 53	16 51	23 06	24 30	00 30	01 56
60	15 11	16 04	16 57	23 10	24 31	00 31	01 53
N 58	15 25	16 13	17 03	23 14	24 32	00 32	01 51
56	15 37	16 21	17 09	23 18	24 33	00 33	01 49
54	15 48	16 29	17 13	23 21	24 33	00 33	01 47
52	15 57	16 36	17 18	23 24	24 34	00 34	01 45
50	16 05	16 42	17 22	23 27	24 35	00 35	01 44
45	16 23	16 56	17 32	23 33	24 36	00 36	01 41
N 40	16 37	17 07	17 40	23 37	24 37	00 37	01 38
35	16 50	17 17	17 48	23 41	24 38	00 38	01 36
30	17 00	17 26	17 56	23 45	24 39	00 39	01 34
20	17 19	17 43	18 10	23 51	24 40	00 40	01 30
N 10	17 35	17 58	18 24	23 57	24 41	00 41	01 27
0	17 51	18 13	18 39	24 02	00 02	00 43	01 24
S 10	18 06	18 29	18 55	24 07	00 07	00 44	01 21
20	18 23	18 47	19 15	24 12	00 12	00 45	01 18
30	18 42	19 09	19 41	24 18	00 18	00 46	01 15
35	18 54	19 22	19 58	24 21	00 21	00 47	01 13
40	19 07	19 38	20 18	00 02	00 25	00 48	01 10
45	19 22	19 58	20 44	00 09	00 30	00 49	01 08
S 50	19 42	20 23	21 20	00 19	00 35	00 50	01 05
52	19 51	20 36	21 40	00 23	00 37	00 50	01 03
54	20 01	20 50	22 06	00 28	00 40	00 51	01 02
56	20 13	21 07	22 46	00 33	00 43	00 51	01 00
58	20 26	21 28	////	00 39	00 46	00 52	00 58
S 60	20 42	21 56	////	00 46	00 50	00 53	00 56

Day	SUN Eqn. of Time 00^h	SUN Eqn. of Time 12^h	SUN Mer. Pass.	MOON Mer. Pass. Upper	MOON Mer. Pass. Lower	MOON Age		MOON Phase
d	m s	m s	h m	h m	h m	d	%	
24	13 24	13 15	11 47	17 52	05 31	08	46	
25	13 07	12 57	11 47	18 33	06 12	09	56	◑
26	12 48	12 39	11 47	19 14	06 53	10	65	

UT		ARIES	VENUS −3.9		MARS +0.0		JUPITER −2.3		SATURN +1.0	
		GHA	GHA	Dec	GHA	Dec	GHA	Dec	GHA	Dec
d	h	° ′	° ′	° ′	° ′	° ′	° ′	° ′	° ′	° ′
27	00	66 02.6	194 22.1	S17 50.7	286 25.2	N18 08.4	102 58.2	S15 37.9	242 44.1	N 0 53.5
	01	81 05.1	209 21.4	51.6	301 27.0	08.3	118 00.3	37.8	257 46.4	53.4
	02	96 07.5	224 20.7	52.4	316 28.8	08.2	133 02.4	37.7	272 48.7	53.4
	03	111 10.0	239 20.0	. . 53.3	331 30.6	. . 08.1	148 04.5	. . 37.6	287 50.9	. . 53.3
	04	126 12.5	254 19.2	54.1	346 32.4	08.0	163 06.7	37.5	302 53.2	53.2
	05	141 14.9	269 18.5	55.0	1 34.2	07.9	178 08.8	37.3	317 55.5	53.2
	06	156 17.4	284 17.8	S17 55.8	16 36.0	N18 07.8	193 10.9	S15 37.2	332 57.8	N 0 53.1
	07	171 19.9	299 17.1	56.7	31 37.7	07.6	208 13.1	37.1	348 00.1	53.0
	08	186 22.3	314 16.4	57.5	46 39.5	07.5	223 15.2	37.0	3 02.4	53.0
F	09	201 24.8	329 15.7	. . 58.4	61 41.3	. . 07.4	238 17.3	. . 36.9	18 04.6	. . 52.9
R	10	216 27.3	344 14.9	17 59.2	76 43.1	07.3	253 19.5	36.8	33 06.9	52.8
I	11	231 29.7	359 14.2	18 00.1	91 44.9	07.2	268 21.6	36.7	48 09.2	52.8
D	12	246 32.2	14 13.5	S18 00.9	106 46.7	N18 07.1	283 23.7	S15 36.6	63 11.5	N 0 52.7
A	13	261 34.7	29 12.8	01.8	121 48.5	07.0	298 25.9	36.5	78 13.8	52.6
Y	14	276 37.1	44 12.1	02.6	136 50.3	06.9	313 28.0	36.3	93 16.0	52.6
	15	291 39.6	59 11.4	. . 03.4	151 52.1	. . 06.8	328 30.1	. . 36.2	108 18.3	. . 52.5
	16	306 42.0	74 10.6	04.3	166 53.9	06.7	343 32.3	36.1	123 20.6	52.4
	17	321 44.5	89 09.9	05.1	181 55.7	06.6	358 34.4	36.0	138 22.9	52.4
	18	336 47.0	104 09.2	S18 06.0	196 57.5	N18 06.5	13 36.5	S15 35.9	153 25.2	N 0 52.3
	19	351 49.4	119 08.5	06.8	211 59.3	06.4	28 38.6	35.8	168 27.5	52.2
	20	6 51.9	134 07.7	07.6	227 01.1	06.3	43 40.8	35.7	183 29.7	52.1
	21	21 54.4	149 07.0	. . 08.5	242 02.9	. . 06.2	58 42.9	. . 35.6	198 32.0	. . 52.1
	22	36 56.8	164 06.3	09.3	257 04.8	06.1	73 45.0	35.4	213 34.3	52.0
	23	51 59.3	179 05.6	10.2	272 06.6	05.9	88 47.2	35.3	228 36.6	51.9
28	00	67 01.8	194 04.9	S18 11.0	287 08.4	N18 05.8	103 49.3	S15 35.2	243 38.9	N 0 51.9
	01	82 04.2	209 04.1	11.8	302 10.2	05.7	118 51.4	35.1	258 41.2	51.8
	02	97 06.7	224 03.4	12.7	317 12.0	05.6	133 53.5	35.0	273 43.4	51.7
	03	112 09.1	239 02.7	. . 13.5	332 13.8	. . 05.5	148 55.7	. . 34.9	288 45.7	. . 51.7
	04	127 11.6	254 02.0	14.3	347 15.6	05.4	163 57.8	34.8	303 48.0	51.6
	05	142 14.1	269 01.2	15.2	2 17.4	05.3	178 59.9	34.6	318 50.3	51.5
	06	157 16.5	284 00.5	S18 16.0	17 19.2	N18 05.2	194 02.0	S15 34.5	333 52.6	N 0 51.5
S	07	172 19.0	298 59.8	16.8	32 21.1	05.1	209 04.2	34.4	348 54.9	51.4
A	08	187 21.5	313 59.0	17.6	47 22.9	05.0	224 06.3	34.3	3 57.1	51.3
T	09	202 23.9	328 58.3	. . 18.5	62 24.7	. . 04.9	239 08.4	. . 34.2	18 59.4	. . 51.3
U	10	217 26.4	343 57.6	19.3	77 26.5	04.8	254 10.5	34.1	34 01.7	51.2
R	11	232 28.9	358 56.9	20.1	92 28.3	04.7	269 12.7	34.0	49 04.0	51.1
D	12	247 31.3	13 56.1	S18 20.9	107 30.2	N18 04.6	284 14.8	S15 33.8	64 06.3	N 0 51.1
A	13	262 33.8	28 55.4	21.8	122 32.0	04.5	299 16.9	33.7	79 08.6	51.0
Y	14	277 36.3	43 54.7	22.6	137 33.8	04.4	314 19.0	33.6	94 10.8	50.9
	15	292 38.7	58 53.9	. . 23.4	152 35.6	. . 04.3	329 21.2	. . 33.5	109 13.1	. . 50.9
	16	307 41.2	73 53.2	24.2	167 37.4	04.2	344 23.3	33.4	124 15.4	50.8
	17	322 43.6	88 52.5	25.1	182 39.3	04.1	359 25.4	33.3	139 17.7	50.7
	18	337 46.1	103 51.7	S18 25.9	197 41.1	N18 04.0	14 27.5	S15 33.2	154 20.0	N 0 50.7
	19	352 48.6	118 51.0	26.7	212 42.9	03.9	29 29.7	33.0	169 22.3	50.6
	20	7 51.0	133 50.3	27.5	227 44.7	03.8	44 31.8	32.9	184 24.6	50.5
	21	22 53.5	148 49.5	. . 28.3	242 46.6	. . 03.7	59 33.9	. . 32.8	199 26.8	. . 50.5
	22	37 56.0	163 48.8	29.2	257 48.4	03.6	74 36.0	32.7	214 29.1	50.4
	23	52 58.4	178 48.1	30.0	272 50.2	03.5	89 38.1	32.6	229 31.4	50.3
29	00	68 00.9	193 47.3	S18 30.8	287 52.1	N18 03.4	104 40.3	S15 32.5	244 33.7	N 0 50.3
	01	83 03.4	208 46.6	31.6	302 53.9	03.4	119 42.4	32.4	259 36.0	50.2
	02	98 05.8	223 45.8	32.4	317 55.7	03.3	134 44.5	32.2	274 38.3	50.1
	03	113 08.3	238 45.1	. . 33.2	332 57.6	. . 03.2	149 46.6	. . 32.1	289 40.6	. . 50.1
	04	128 10.8	253 44.4	34.0	347 59.4	03.1	164 48.8	32.0	304 42.8	50.0
	05	143 13.2	268 43.6	34.8	3 01.2	03.0	179 50.9	31.9	319 45.1	49.9
	06	158 15.7	283 42.9	S18 35.7	18 03.1	N18 02.9	194 53.0	S15 31.8	334 47.4	N 0 49.9
	07	173 18.1	298 42.2	36.5	33 04.9	02.8	209 55.1	31.7	349 49.7	49.8
	08	188 20.6	313 41.4	37.3	48 06.7	02.7	224 57.2	31.5	4 52.0	49.7
S	09	203 23.1	328 40.7	. . 38.1	63 08.6	. . 02.6	239 59.4	. . 31.4	19 54.3	. . 49.7
U	10	218 25.5	343 39.9	38.9	78 10.4	02.5	255 01.5	31.3	34 56.6	49.6
N	11	233 28.0	358 39.2	39.7	93 12.3	02.4	270 03.6	31.2	49 58.9	49.5
D	12	248 30.5	13 38.4	S18 40.5	108 14.1	N18 02.3	285 05.7	S15 31.1	65 01.1	N 0 49.5
A	13	263 32.9	28 37.7	41.3	123 16.0	02.2	300 07.8	31.0	80 03.4	49.4
Y	14	278 35.4	43 37.0	42.1	138 17.8	02.1	315 10.0	30.8	95 05.7	49.3
	15	293 37.9	58 36.2	. . 42.9	153 19.6	. . 02.0	330 12.1	. . 30.7	110 08.0	. . 49.3
	16	308 40.3	73 35.5	43.7	168 21.5	01.9	345 14.2	30.6	125 10.3	49.2
	17	323 42.8	88 34.7	44.5	183 23.3	01.8	0 16.3	30.5	140 12.6	49.2
	18	338 45.2	103 34.0	S18 45.3	198 25.2	N18 01.8	15 18.4	S15 30.4	155 14.9	N 0 49.1
	19	353 47.7	118 33.2	46.1	213 27.0	01.7	30 20.5	30.3	170 17.2	49.0
	20	8 50.2	133 32.5	46.9	228 28.9	01.6	45 22.7	30.1	185 19.5	49.0
	21	23 52.6	148 31.7	. . 47.7	243 30.7	. . 01.5	60 24.8	. . 30.0	200 21.7	. . 48.9
	22	38 55.1	163 31.0	48.5	258 32.6	01.4	75 26.9	29.9	215 24.0	48.8
	23	53 57.6	178 30.2	49.3	273 34.4	01.3	90 29.0	29.8	230 26.3	48.8
Mer. Pass.		h m 19 28.7	v −0.7	d 0.8	v 1.8	d 0.1	v 2.1	d 0.1	v 2.3	d 0.1

STARS		
Name	SHA	Dec
	° ′	° ′
Acamar	315 19.8	S40 15.
Achernar	335 28.1	S57 11.
Acrux	173 12.8	S63 09.
Adhara	255 14.3	S28 59.
Aldebaran	290 52.1	N16 31.
Alioth	166 23.1	N55 54.
Alkaid	153 01.2	N49 15.
Al Na'ir	27 46.9	S46 54.
Alnilam	275 48.7	S 1 11.
Alphard	217 58.6	S 8 42.
Alphecca	126 13.6	N26 40.
Alpheratz	357 46.1	N29 09.
Altair	62 11.0	N 8 53.
Ankaa	353 17.9	S42 15.
Antares	112 29.9	S26 27.
Arcturus	145 58.4	N19 07.
Atria	107 34.7	S69 02.
Avior	234 18.9	S59 32.
Bellatrix	278 34.5	N 6 21.
Betelgeuse	271 03.8	N 7 24.
Canopus	263 56.8	S52 41.
Capella	280 37.9	N46 00.
Deneb	49 33.6	N45 19.
Denebola	182 36.4	N14 30.
Diphda	348 58.3	S17 55.
Dubhe	193 54.8	N61 41.
Elnath	278 15.6	N28 37.
Eltanin	90 47.9	N51 29.
Enif	33 49.8	N 9 55.
Fomalhaut	15 26.8	S29 34.
Gacrux	172 04.3	S57 09.
Gienah	175 55.2	S17 35.
Hadar	148 52.4	S60 25.
Hamal	328 03.5	N23 30.
Kaus Aust.	83 47.6	S34 22.
Kochab	137 20.7	N74 06.
Markab	13 40.9	N15 15.
Menkar	314 17.5	N 4 07.
Menkent	148 11.1	S36 25.
Miaplacidus	221 40.1	S69 45.
Mirfak	308 43.7	N49 54.
Nunki	76 01.9	S26 17.
Peacock	53 23.6	S56 42.
Pollux	243 30.6	N28 00.
Procyon	245 02.2	N 5 12.
Rasalhague	96 09.2	N12 33.
Regulus	207 46.2	N11 55.
Rigel	281 14.2	S 8 11.
Rigil Kent.	139 56.1	S60 52.
Sabik	102 15.9	S15 44.
Schedar	349 43.4	N56 35.
Shaula	96 25.9	S37 06.
Sirius	258 35.7	S16 43.
Spica	158 34.3	S11 12.
Suhail	222 54.3	S43 28.
Vega	80 41.1	N38 47.
Zuben'ubi	137 08.7	S16 05.

	SHA	Mer. Pas
	° ′	h m
Venus	127 03.1	11 04
Mars	220 06.6	4 51
Jupiter	36 47.5	17 02
Saturn	176 37.1	7 44

	UT	SUN GHA	SUN Dec	MOON GHA	MOON v	MOON Dec	MOON d	MOON HP
	d h	° ′	° ′	° ′	′	° ′	′	′
	27 00	183 07.2	S21 06.5	69 26.4	15.2	N 3 39.7	13.5	55.6
	01	198 07.0	06.9	84 00.6	15.2	3 53.2	13.4	55.7
	02	213 06.8	07.4	98 34.8	15.2	4 06.6	13.5	55.7
	03	228 06.6	. . 07.8	113 09.0	15.1	4 20.1	13.4	55.7
	04	243 06.4	08.3	127 43.1	15.1	4 33.5	13.5	55.8
	05	258 06.1	08.7	142 17.2	15.0	4 47.0	13.4	55.8
	06	273 05.9	S21 09.2	156 51.2	15.0	N 5 00.4	13.5	55.8
	07	288 05.7	09.7	171 25.2	14.9	5 13.9	13.4	55.9
	08	303 05.5	10.1	185 59.1	14.8	5 27.3	13.4	55.9
F	09	318 05.3	. . 10.6	200 32.9	14.8	5 40.7	13.4	55.9
R	10	333 05.1	11.0	215 06.7	14.8	5 54.1	13.4	55.9
I	11	348 04.9	11.5	229 40.5	14.7	6 07.5	13.4	56.0
D	12	3 04.7	S21 11.9	244 14.2	14.6	N 6 20.9	13.4	56.0
A	13	18 04.5	12.4	258 47.8	14.5	6 34.3	13.4	56.0
Y	14	33 04.3	12.8	273 21.3	14.5	6 47.7	13.3	56.1
	15	48 04.1	. . 13.3	287 54.8	14.5	7 01.0	13.4	56.1
	16	63 03.9	13.7	302 28.3	14.4	7 14.4	13.3	56.1
	17	78 03.7	14.1	317 01.7	14.3	7 27.7	13.3	56.2
	18	93 03.5	S21 14.6	331 35.0	14.2	N 7 41.0	13.3	56.2
	19	108 03.3	15.0	346 08.2	14.2	7 54.3	13.3	56.2
	20	123 03.0	15.5	0 41.4	14.1	8 07.6	13.3	56.3
	21	138 02.8	. . 15.9	15 14.5	14.1	8 20.9	13.2	56.3
	22	153 02.6	16.4	29 47.6	13.9	8 34.1	13.2	56.3
	23	168 02.4	16.8	44 20.5	13.9	8 47.3	13.2	56.4
	28 00	183 02.2	S21 17.3	58 53.4	13.9	N 9 00.5	13.2	56.4
	01	198 02.0	17.7	73 26.3	13.7	9 13.7	13.2	56.4
	02	213 01.8	18.1	87 59.0	13.7	9 26.9	13.1	56.5
	03	228 01.6	. . 18.6	102 31.7	13.6	9 40.0	13.1	56.5
	04	243 01.4	19.0	117 04.3	13.5	9 53.1	13.1	56.5
	05	258 01.2	19.5	131 36.8	13.5	10 06.2	13.0	56.6
	06	273 00.9	S21 19.9	146 09.3	13.3	N10 19.2	13.0	56.6
	07	288 00.7	20.3	160 41.6	13.3	10 32.2	13.0	56.6
S	08	303 00.5	20.8	175 13.9	13.2	10 45.2	13.0	56.7
A	09	318 00.3	. . 21.2	189 46.1	13.1	10 58.2	12.9	56.7
T	10	333 00.1	21.6	204 18.2	13.1	11 11.1	12.9	56.7
U	11	347 59.9	22.1	218 50.3	12.9	11 24.0	12.8	56.8
R	12	2 59.7	S21 22.5	233 22.2	12.9	N11 36.8	12.8	56.8
D	13	17 59.5	22.9	247 54.1	12.7	11 49.6	12.8	56.8
A	14	32 59.2	23.4	262 25.8	12.7	12 02.4	12.7	56.9
Y	15	47 59.0	. . 23.8	276 57.5	12.6	12 15.1	12.7	56.9
	16	62 58.8	24.2	291 29.1	12.5	12 27.8	12.7	56.9
	17	77 58.6	24.7	306 00.6	12.4	12 40.5	12.6	57.0
	18	92 58.4	S21 25.1	320 32.0	12.3	N12 53.1	12.5	57.0
	19	107 58.2	25.5	335 03.3	12.3	13 05.6	12.6	57.0
	20	122 57.9	25.9	349 34.6	12.1	13 18.2	12.4	57.1
	21	137 57.7	. . 26.4	4 05.7	12.0	13 30.6	12.4	57.1
	22	152 57.5	26.8	18 36.7	12.0	13 43.0	12.4	57.1
	23	167 57.3	27.2	33 07.7	11.8	13 55.4	12.3	57.2
	29 00	182 57.1	S21 27.6	47 38.5	11.8	N14 07.7	12.3	57.2
	01	197 56.9	28.1	62 09.3	11.6	14 20.0	12.2	57.2
	02	212 56.6	28.5	76 39.9	11.5	14 32.2	12.1	57.3
	03	227 56.4	. . 28.9	91 10.4	11.5	14 44.3	12.1	57.3
	04	242 56.2	29.3	105 40.9	11.3	14 56.4	12.1	57.4
	05	257 56.0	29.8	120 11.2	11.3	15 08.5	11.9	57.4
	06	272 55.8	S21 30.2	134 41.5	11.1	N15 20.4	12.0	57.4
	07	287 55.6	30.6	149 11.6	11.0	15 32.4	11.8	57.5
	08	302 55.3	31.0	163 41.6	10.9	15 44.2	11.8	57.5
S	09	317 55.1	. . 31.4	178 11.5	10.8	15 56.0	11.7	57.5
U	10	332 54.9	31.9	192 41.3	10.7	16 07.7	11.7	57.6
N	11	347 54.7	32.3	207 11.0	10.6	16 19.4	11.5	57.6
D	12	2 54.5	S21 32.7	221 40.6	10.5	N16 30.9	11.6	57.6
A	13	17 54.2	33.1	236 10.1	10.4	16 42.5	11.4	57.7
Y	14	32 54.0	33.5	250 39.5	10.3	16 53.9	11.3	57.7
	15	47 53.8	. . 33.9	265 08.8	10.1	17 05.2	11.3	57.7
	16	62 53.6	34.3	279 37.9	10.1	17 16.5	11.2	57.8
	17	77 53.3	34.8	294 07.0	9.9	17 27.7	11.2	57.8
	18	92 53.1	S21 35.2	308 35.9	9.9	N17 38.9	11.0	57.8
	19	107 52.9	35.6	323 04.8	9.7	17 49.9	11.0	57.9
	20	122 52.7	36.0	337 33.5	9.6	18 00.9	10.9	57.9
	21	137 52.5	. . 36.4	352 02.1	9.4	18 11.8	10.8	58.0
	22	152 52.2	36.8	6 30.5	9.4	18 22.6	10.7	58.0
	23	167 52.0	37.2	20 58.9	9.3	N18 33.3	10.6	58.0
		SD 16.2	*d* 0.4	SD 15.3		15.5		15.7

Lat.	Twilight Naut.	Twilight Civil	Sunrise	Moonrise 27	Moonrise 28	Moonrise 29	Moonrise 30
°	h m	h m	h m	h m	h m	h m	h m
N 72	07 43	09 37	■	12 24	11 52	10 58	□
N 70	07 28	09 01	■	12 33	12 11	11 38	□
68	07 15	08 36	10 20	12 40	12 25	12 06	11 32
66	07 05	08 16	09 37	12 46	12 38	12 28	12 14
64	06 56	08 00	09 08	12 51	12 48	12 45	12 43
62	06 48	07 47	08 46	12 56	12 57	12 59	13 05
60	06 41	07 36	08 29	13 00	13 04	13 11	13 23
N 58	06 35	07 26	08 14	13 03	13 11	13 22	13 38
56	06 29	07 17	08 02	13 06	13 17	13 31	13 51
54	06 24	07 09	07 51	13 09	13 23	13 39	14 02
52	06 20	07 02	07 41	13 12	13 27	13 47	14 12
50	06 15	06 56	07 33	13 14	13 32	13 53	14 21
45	06 05	06 42	07 14	13 19	13 41	14 08	14 40
N 40	05 56	06 30	07 00	13 24	13 50	14 19	14 56
35	05 48	06 19	06 47	13 27	13 56	14 30	15 09
30	05 41	06 10	06 36	13 31	14 03	14 39	15 20
20	05 26	05 53	06 17	13 37	14 13	14 54	15 40
N 10	05 12	05 38	06 00	13 42	14 23	15 08	15 58
0	04 56	05 22	05 44	13 47	14 32	15 21	16 14
S 10	04 39	05 06	05 28	13 52	14 41	15 34	16 31
20	04 19	04 47	05 11	13 57	14 50	15 48	16 48
30	03 52	04 25	04 52	14 03	15 02	16 04	17 09
35	03 36	04 11	04 40	14 07	15 08	16 13	17 21
40	03 15	03 55	04 26	14 11	15 16	16 24	17 35
45	02 48	03 35	04 10	14 15	15 24	16 37	17 52
S 50	02 11	03 09	03 51	14 21	15 35	16 52	18 12
52	01 50	02 56	03 41	14 24	15 40	16 59	18 22
54	01 22	02 41	03 31	14 26	15 45	17 08	18 33
56	00 37	02 23	03 19	14 30	15 51	17 17	18 46
58	////	02 01	03 05	14 33	15 58	17 27	19 00
S 60	////	01 31	02 48	14 37	16 06	17 39	19 18

Lat.	Sunset	Twilight Civil	Twilight Naut.	Moonset 27	Moonset 28	Moonset 29	Moonset 30
°	h m	h m	h m	h m	h m	h m	h m
N 72	■	13 59	15 52	02 17	04 21	06 55	□
N 70	■	14 34	16 08	02 11	04 05	06 16	□
68	13 16	15 00	16 20	02 06	03 52	05 50	08 13
66	13 59	15 19	16 31	02 02	03 41	05 29	07 31
64	14 28	15 35	16 40	01 58	03 33	05 14	07 04
62	14 49	15 49	16 47	01 56	03 25	05 00	06 42
60	15 07	16 00	16 54	01 53	03 19	04 49	06 25
N 58	15 21	16 10	17 01	01 51	03 13	04 40	06 11
56	15 34	16 19	17 06	01 49	03 08	04 31	05 58
54	15 45	16 26	17 11	01 47	03 03	04 24	05 48
52	15 54	16 34	17 16	01 45	02 59	04 17	05 38
50	16 03	16 40	17 21	01 44	02 56	04 11	05 30
45	16 21	16 54	17 31	01 41	02 48	03 58	05 12
N 40	16 36	17 06	17 39	01 38	02 41	03 48	04 57
35	16 49	17 16	17 48	01 36	02 36	03 39	04 45
30	17 00	17 26	17 55	01 34	02 31	03 31	04 35
20	17 19	17 43	18 10	01 30	02 22	03 17	04 16
N 10	17 36	17 58	18 24	01 27	02 15	03 06	04 00
0	17 52	18 14	18 40	01 24	02 08	02 55	03 46
S 10	18 08	18 30	18 57	01 21	02 01	02 44	03 31
20	18 25	18 49	19 18	01 18	01 53	02 32	03 15
30	18 45	19 12	19 44	01 15	01 45	02 19	02 57
35	18 56	19 25	20 01	01 13	01 40	02 11	02 47
40	19 10	19 42	20 22	01 10	01 35	02 02	02 35
45	19 26	20 02	20 49	01 08	01 29	01 52	02 21
S 50	19 46	20 28	21 27	01 05	01 21	01 40	02 04
52	19 56	20 41	21 48	01 03	01 18	01 35	01 56
54	20 06	20 56	22 16	01 02	01 14	01 28	01 47
56	20 18	21 14	23 05	01 00	01 10	01 22	01 38
58	20 33	21 37	////	00 58	01 05	01 14	01 26
S 60	20 50	22 07	////	00 56	01 00	01 05	01 13

Day	SUN Eqn. of Time 00^h	SUN Eqn. of Time 12^h	SUN Mer. Pass.	MOON Mer. Pass. Upper	MOON Mer. Pass. Lower	MOON Age	MOON Phase
d	m s	m s	h m	h m	h m	d	%
27	12 29	12 19	11 48	19 57	07 35	11	74
28	12 09	11 59	11 48	20 43	08 20	12	83
29	11 49	11 38	11 48	21 33	09 07	13	90

UT d	h	ARIES GHA	VENUS −3.9 GHA	VENUS Dec	MARS −0.1 GHA	MARS Dec	JUPITER −2.3 GHA	JUPITER Dec	SATURN +1.0 GHA	SATURN Dec
		° ′	° ′	° ′	° ′	° ′	° ′	° ′	° ′	° ′
30 MONDAY	00	69 00.0	193 29.5	S18 50.1	288 36.3	N18 01.2	105 31.1	S15 29.7	245 28.6	N 0 48.7
	01	84 02.5	208 28.7	50.9	303 38.2	01.1	120 33.2	29.6	260 30.9	48.6
	02	99 05.0	223 28.0	51.7	318 40.0	01.0	135 35.4	29.4	275 33.2	48.6
	03	114 07.4	238 27.2	. . 52.4	333 41.9	. . 00.9	150 37.5	. . 29.3	290 35.5	. . 48.5
	04	129 09.9	253 26.5	53.2	348 43.7	00.9	165 39.6	29.2	305 37.8	48.4
	05	144 12.4	268 25.7	54.0	3 45.6	00.8	180 41.7	29.1	320 40.1	48.4
	06	159 14.8	283 25.0	S18 54.8	18 47.4	N18 00.7	195 43.8	S15 29.0	335 42.3	N 0 48.3
	07	174 17.3	298 24.2	55.6	33 49.3	00.6	210 45.9	28.8	350 44.6	48.2
	08	189 19.7	313 23.5	56.4	48 51.2	00.5	225 48.0	28.7	5 46.9	48.2
	09	204 22.2	328 22.7	. . 57.2	63 53.0	. . 00.4	240 50.2	. . 28.6	20 49.2	. . 48.1
	10	219 24.7	343 22.0	58.0	78 54.9	00.3	255 52.3	28.5	35 51.5	48.1
	11	234 27.1	358 21.2	58.7	93 56.8	00.3	270 54.4	28.4	50 53.8	48.0
	12	249 29.6	13 20.5	S18 59.5	108 58.6	N18 00.2	285 56.5	S15 28.3	65 56.1	N 0 47.9
	13	264 32.1	28 19.7	19 00.3	124 00.5	00.1	300 58.6	28.1	80 58.4	47.9
	14	279 34.5	43 19.0	01.1	139 02.4	18 00.0	316 00.7	28.0	96 00.7	47.8
	15	294 37.0	58 18.2	. . 01.9	154 04.2	17 59.9	331 02.8	. . 27.9	111 03.0	. . 47.7
	16	309 39.5	73 17.4	02.6	169 06.1	59.8	346 04.9	27.8	126 05.3	47.7
	17	324 41.9	88 16.7	03.4	184 08.0	59.7	1 07.1	27.7	141 07.5	47.6
	18	339 44.4	103 15.9	S19 04.2	199 09.8	N17 59.7	16 09.2	S15 27.5	156 09.8	N 0 47.5
	19	354 46.9	118 15.2	05.0	214 11.7	59.6	31 11.3	27.4	171 12.1	47.5
	20	9 49.3	133 14.4	05.7	229 13.6	59.5	46 13.4	27.3	186 14.4	47.4
	21	24 51.8	148 13.6	. . 06.5	244 15.5	. . 59.4	61 15.5	. . 27.2	201 16.7	. . 47.4
	22	39 54.2	163 12.9	07.3	259 17.3	59.3	76 17.6	27.1	216 19.0	47.3
	23	54 56.7	178 12.1	08.1	274 19.2	59.2	91 19.7	27.0	231 21.3	47.2
1 TUESDAY	00	69 59.2	193 11.4	S19 08.8	289 21.1	N17 59.2	106 21.8	S15 26.8	246 23.6	N 0 47.2
	01	85 01.6	208 10.6	09.6	304 23.0	59.1	121 24.0	26.7	261 25.9	47.1
	02	100 04.1	223 09.8	10.4	319 24.8	59.0	136 26.1	26.6	276 28.2	47.0
	03	115 06.6	238 09.1	. . 11.2	334 26.7	. . 58.9	151 28.2	. . 26.5	291 30.5	. . 47.0
	04	130 09.0	253 08.3	11.9	349 28.6	58.8	166 30.3	26.4	306 32.8	46.9
	05	145 11.5	268 07.5	12.7	4 30.5	58.7	181 32.4	26.2	321 35.1	46.8
	06	160 14.0	283 06.8	S19 13.5	19 32.4	N17 58.7	196 34.5	S15 26.1	336 37.4	N 0 46.8
	07	175 16.4	298 06.0	14.2	34 34.3	58.6	211 36.6	26.0	351 39.6	46.7
	08	190 18.9	313 05.3	15.0	49 36.1	58.5	226 38.7	25.9	6 41.9	46.7
	09	205 21.4	328 04.5	. . 15.7	64 38.0	. . 58.4	241 40.8	. . 25.8	21 44.2	. . 46.6
	10	220 23.8	343 03.7	16.5	79 39.9	58.4	256 42.9	25.6	36 46.5	46.5
	11	235 26.3	358 03.0	17.3	94 41.8	58.3	271 45.0	25.5	51 48.8	46.5
	12	250 28.7	13 02.2	S19 18.0	109 43.7	N17 58.2	286 47.1	S15 25.4	66 51.1	N 0 46.4
	13	265 31.2	28 01.4	18.8	124 45.6	58.1	301 49.3	25.3	81 53.4	46.3
	14	280 33.7	43 00.6	19.5	139 47.5	58.0	316 51.4	25.1	96 55.7	46.3
	15	295 36.1	57 59.9	. . 20.3	154 49.4	. . 58.0	331 53.5	. . 25.0	111 58.0	. . 46.2
	16	310 38.6	72 59.1	21.1	169 51.3	57.9	346 55.6	24.9	127 00.3	46.2
	17	325 41.1	87 58.3	21.8	184 53.1	57.8	1 57.7	24.8	142 02.6	46.1
	18	340 43.5	102 57.6	S19 22.6	199 55.0	N17 57.7	16 59.8	S15 24.7	157 04.9	N 0 46.0
	19	355 46.0	117 56.8	23.3	214 56.9	57.7	32 01.9	24.5	172 07.2	46.0
	20	10 48.5	132 56.0	24.1	229 58.8	57.6	47 04.0	24.4	187 09.5	45.9
	21	25 50.9	147 55.3	. . 24.8	245 00.7	. . 57.5	62 06.1	. . 24.3	202 11.8	. . 45.8
	22	40 53.4	162 54.5	25.6	260 02.6	57.4	77 08.2	24.2	217 14.1	45.8
	23	55 55.8	177 53.7	26.3	275 04.5	57.3	92 10.3	24.1	232 16.4	45.7
2 WEDNESDAY	00	70 58.3	192 52.9	S19 27.1	290 06.4	N17 57.3	107 12.4	S15 23.9	247 18.7	N 0 45.7
	01	86 00.8	207 52.2	27.8	305 08.3	57.2	122 14.5	23.8	262 20.9	45.6
	02	101 03.2	222 51.4	28.6	320 10.2	57.1	137 16.6	23.7	277 23.2	45.5
	03	116 05.7	237 50.6	. . 29.3	335 12.1	. . 57.1	152 18.7	. . 23.6	292 25.5	. . 45.5
	04	131 08.2	252 49.8	30.1	350 14.0	57.0	167 20.8	23.5	307 27.8	45.4
	05	146 10.6	267 49.1	30.8	5 15.9	56.9	182 22.9	23.3	322 30.1	45.3
	06	161 13.1	282 48.3	S19 31.6	20 17.9	N17 56.8	197 25.1	S15 23.2	337 32.4	N 0 45.3
	07	176 15.6	297 47.5	32.3	35 19.8	56.8	212 27.2	23.1	352 34.7	45.2
	08	191 18.0	312 46.7	33.0	50 21.7	56.7	227 29.3	23.0	7 37.0	45.2
	09	206 20.5	327 45.9	. . 33.8	65 23.6	. . 56.6	242 31.4	. . 22.8	22 39.3	. . 45.1
	10	221 23.0	342 45.2	34.5	80 25.5	56.5	257 33.5	22.7	37 41.6	45.0
	11	236 25.4	357 44.4	35.3	95 27.4	56.5	272 35.6	22.6	52 43.9	45.0
	12	251 27.9	12 43.6	S19 36.0	110 29.3	N17 56.4	287 37.7	S15 22.5	67 46.2	N 0 44.9
	13	266 30.3	27 42.8	36.7	125 31.2	56.3	302 39.8	22.4	82 48.5	44.9
	14	281 32.8	42 42.0	37.5	140 33.1	56.3	317 41.9	22.2	97 50.8	44.8
	15	296 35.3	57 41.3	. . 38.2	155 35.1	. . 56.2	332 44.0	. . 22.1	112 53.1	. . 44.7
	16	311 37.7	72 40.5	38.9	170 37.0	56.1	347 46.1	22.0	127 55.4	44.7
	17	326 40.2	87 39.7	39.7	185 38.9	56.1	2 48.2	21.9	142 57.7	44.6
	18	341 42.7	102 38.9	S19 40.4	200 40.8	N17 56.0	17 50.3	S15 21.7	158 00.0	N 0 44.6
	19	356 45.1	117 38.1	41.1	215 42.7	55.9	32 52.4	21.6	173 02.3	44.5
	20	11 47.6	132 37.4	41.9	230 44.7	55.8	47 54.5	21.5	188 04.6	44.4
	21	26 50.1	147 36.6	. . 42.6	245 46.6	. . 55.8	62 56.6	. . 21.4	203 06.9	. . 44.4
	22	41 52.5	162 35.8	43.3	260 48.5	55.7	77 58.7	21.2	218 09.2	44.3
	23	56 55.0	177 35.0	44.1	275 50.4	55.6	93 00.8	21.1	233 11.5	44.2
Mer. Pass.		h m 19 16.9	*v* −0.8	*d* 0.8	*v* 1.9	*d* 0.1	*v* 2.1	*d* 0.1	*v* 2.3	*d* 0.1

STARS Name	SHA	Dec
	° ′	° ′
Acamar	315 19.8	S40 15.9
Achernar	335 28.1	S57 11.3
Acrux	173 12.8	S63 09.0
Adhara	255 14.3	S28 59.0
Aldebaran	290 52.0	N16 31.8
Alioth	166 23.1	N55 54.0
Alkaid	153 01.2	N49 15.6
Al Na'ir	27 47.0	S46 54.9
Alnilam	275 48.7	S 1 11.7
Alphard	217 58.6	S 8 42.1
Alphecca	126 13.5	N26 40.8
Alpheratz	357 46.1	N29 09.0
Altair	62 11.0	N 8 53.8
Ankaa	353 18.0	S42 15.2
Antares	112 29.9	S26 27.2
Arcturus	145 58.4	N19 07.7
Atria	107 34.7	S69 02.7
Avior	234 18.8	S59 32.2
Bellatrix	278 34.5	N 6 21.6
Betelgeuse	271 03.8	N 7 24.6
Canopus	263 56.8	S52 41.9
Capella	280 37.9	N46 00.5
Deneb	49 33.6	N45 19.2
Denebola	182 36.4	N14 30.9
Diphda	348 58.3	S17 55.9
Dubhe	193 54.7	N61 41.5
Elnath	278 15.6	N28 37.0
Eltanin	90 47.9	N51 29.3
Enif	33 49.8	N 9 55.4
Fomalhaut	15 26.8	S29 34.2
Gacrux	172 04.3	S57 09.9
Gienah	175 55.2	S17 35.8
Hadar	148 52.3	S60 25.1
Hamal	328 03.5	N23 30.8
Kaus Aust.	83 47.6	S34 22.8
Kochab	137 20.7	N74 06.6
Markab	13 40.9	N15 15.7
Menkar	314 17.5	N 4 07.8
Menkent	148 11.1	S36 25.0
Miaplacidus	221 40.1	S69 45.2
Mirfak	308 43.7	N49 54.0
Nunki	76 01.9	S26 17.1
Peacock	53 23.6	S56 42.3
Pollux	243 30.6	N28 00.0
Procyon	245 02.2	N 5 12.0
Rasalhague	96 09.2	N12 33.2
Regulus	207 46.2	N11 55.0
Rigel	281 14.2	S 8 11.3
Rigil Kent.	139 56.1	S60 52.4
Sabik	102 15.9	S15 44.2
Schedar	349 43.4	N56 35.9
Shaula	96 25.9	S37 06.7
Sirius	258 35.7	S16 43.7
Spica	158 34.3	S11 12.8
Suhail	222 54.3	S43 28.2
Vega	80 41.1	N38 47.7
Zuben'ubi	137 08.7	S16 05.0

	SHA	Mer. Pass.
	° ′	h m
Venus	123 12.2	11 08
Mars	219 21.9	4 42
Jupiter	36 22.7	16 52
Saturn	176 24.4	7 33

	UT	SUN GHA	SUN Dec	MOON GHA	v	MOON Dec	d	HP
	d h	° ′	° ′	° ′	′	° ′	′	′
	30 00	182 51.8	S21 37.6	35 27.2	9.1	N18 43.9	10.5	58.1
	01	197 51.6	38.0	49 55.3	9.0	18 54.4	10.5	58.1
	02	212 51.3	38.4	64 23.3	8.9	19 04.9	10.3	58.1
	03	227 51.1	. . 38.8	78 51.2	8.8	19 15.2	10.3	58.2
	04	242 50.9	39.3	93 19.0	8.7	19 25.5	10.1	58.2
	05	257 50.7	39.7	107 46.7	8.6	19 35.6	10.1	58.2
	06	272 50.4	S21 40.1	122 14.3	8.4	N19 45.7	9.9	58.3
	07	287 50.2	40.5	136 41.7	8.3	19 55.6	9.9	58.3
	08	302 50.0	40.9	151 09.0	8.3	20 05.5	9.7	58.3
M	09	317 49.8	. . 41.3	165 36.3	8.0	20 15.2	9.6	58.4
O	10	332 49.5	41.7	180 03.3	8.0	20 24.8	9.6	58.4
N	11	347 49.3	42.1	194 30.3	7.9	20 34.4	9.4	58.4
D	12	2 49.1	S21 42.5	208 57.2	7.7	N20 43.8	9.3	58.5
A	13	17 48.8	42.9	223 23.9	7.7	20 53.1	9.2	58.5
Y	14	32 48.6	43.3	237 50.6	7.5	21 02.3	9.1	58.5
	15	47 48.4	. . 43.7	252 17.1	7.4	21 11.4	8.9	58.6
	16	62 48.2	44.1	266 43.5	7.2	21 20.3	8.9	58.6
	17	77 47.9	44.5	281 09.7	7.2	21 29.2	8.7	58.6
	18	92 47.7	S21 44.8	295 35.9	7.1	N21 37.9	8.6	58.7
	19	107 47.5	45.2	310 02.0	6.9	21 46.5	8.5	58.7
	20	122 47.2	45.6	324 27.9	6.8	21 55.0	8.3	58.7
	21	137 47.0	. . 46.0	338 53.7	6.7	22 03.3	8.2	58.8
	22	152 46.8	46.4	353 19.4	6.6	22 11.5	8.1	58.8
	23	167 46.6	46.8	7 45.0	6.5	22 19.6	8.0	58.8
	1 00	182 46.3	S21 47.2	22 10.5	6.4	N22 27.6	7.8	58.9
	01	197 46.1	47.6	36 35.9	6.2	22 35.4	7.7	58.9
	02	212 45.9	48.0	51 01.1	6.2	22 43.1	7.6	58.9
	03	227 45.6	. . 48.4	65 26.3	6.0	22 50.7	7.4	58.9
	04	242 45.4	48.8	79 51.3	6.0	22 58.1	7.3	59.0
	05	257 45.2	49.1	94 16.3	5.8	23 05.4	7.1	59.0
	06	272 44.9	S21 49.5	108 41.1	5.7	N23 12.5	7.0	59.0
	07	287 44.7	49.9	123 05.8	5.6	23 19.5	6.9	59.1
T	08	302 44.5	50.3	137 30.4	5.6	23 26.4	6.7	59.1
U	09	317 44.2	. . 50.7	151 55.0	5.4	23 33.1	6.6	59.1
E	10	332 44.0	51.1	166 19.4	5.3	23 39.7	6.4	59.2
S	11	347 43.8	51.4	180 43.7	5.2	23 46.1	6.3	59.2
D	12	2 43.5	S21 51.8	195 07.9	5.1	N23 52.4	6.1	59.2
A	13	17 43.3	52.2	209 32.0	5.0	23 58.5	5.9	59.2
Y	14	32 43.1	52.6	223 56.0	4.9	24 04.4	5.8	59.3
	15	47 42.8	. . 53.0	238 19.9	4.9	24 10.2	5.7	59.3
	16	62 42.6	53.3	252 43.8	4.7	24 15.9	5.5	59.3
	17	77 42.4	53.7	267 07.5	4.6	24 21.4	5.3	59.4
	18	92 42.1	S21 54.1	281 31.1	4.6	N24 26.7	5.2	59.4
	19	107 41.9	54.5	295 54.7	4.4	24 31.9	5.0	59.4
	20	122 41.7	54.9	310 18.1	4.4	24 36.9	4.8	59.4
	21	137 41.4	. . 55.2	324 41.5	4.3	24 41.7	4.7	59.5
	22	152 41.2	55.6	339 04.8	4.2	24 46.4	4.5	59.5
	23	167 40.9	56.0	353 28.0	4.1	24 50.9	4.3	59.5
	2 00	182 40.7	S21 56.4	7 51.1	4.1	N24 55.2	4.2	59.5
	01	197 40.5	56.7	22 14.2	4.0	24 59.4	4.0	59.6
	02	212 40.2	57.1	36 37.2	3.9	25 03.4	3.8	59.6
	03	227 40.0	. . 57.5	51 00.1	3.8	25 07.2	3.7	59.6
	04	242 39.8	57.8	65 22.9	3.8	25 10.9	3.5	59.6
	05	257 39.5	58.2	79 45.7	3.6	25 14.4	3.3	59.7
	06	272 39.3	S21 58.6	94 08.3	3.7	N25 17.7	3.1	59.7
W	07	287 39.0	58.9	108 31.0	3.5	25 20.8	3.0	59.7
E	08	302 38.8	59.3	122 53.5	3.5	25 23.8	2.8	59.7
D	09	317 38.6	21 59.7	137 16.0	3.5	25 26.6	2.6	59.7
N	10	332 38.3	22 00.0	151 38.5	3.3	25 29.2	2.4	59.8
E	11	347 38.1	00.4	166 00.8	3.4	25 31.6	2.2	59.8
S	12	2 37.8	S22 00.8	180 23.2	3.3	N25 33.8	2.1	59.8
D	13	17 37.6	01.1	194 45.5	3.2	25 35.9	1.9	59.8
A	14	32 37.4	01.5	209 07.7	3.2	25 37.8	1.6	59.8
Y	15	47 37.1	. . 01.9	223 29.9	3.1	25 39.4	1.6	59.9
	16	62 36.9	02.2	237 52.0	3.1	25 41.0	1.3	59.9
	17	77 36.6	02.6	252 14.1	3.1	25 42.3	1.1	59.9
	18	92 36.4	S22 02.9	266 36.2	3.0	N25 43.4	1.0	59.9
	19	107 36.1	03.3	280 58.2	3.0	25 44.4	0.7	59.9
	20	122 35.9	03.7	295 20.2	3.0	25 45.1	0.6	60.0
	21	137 35.7	. . 04.0	309 42.2	2.9	25 45.7	0.4	60.0
	22	152 35.4	04.4	324 04.1	2.9	25 46.1	0.2	60.0
	23	167 35.2	04.7	338 26.0	2.9	N25 46.3	0.0	60.0
		SD 16.2	*d* 0.4	SD 15.9		16.1		16.3

Lat.	Twilight Naut.	Twilight Civil	Sunrise	Moonrise 30	Moonrise 1	Moonrise 2	Moonrise 3
°	h m	h m	h m	h m	h m	h m	h m
N 72	07 51	09 51	■	□	□	□	□
N 70	07 35	09 11	■	□	□	□	□
68	07 22	08 44	10 37	11 32	□	□	□
66	07 11	08 23	09 47	12 14	11 38	□	□
64	07 01	08 07	09 16	12 43	12 42	12 52	14 02
62	06 53	07 53	08 53	13 05	13 18	13 49	14 57
60	06 46	07 41	08 35	13 23	13 43	14 22	15 30
N 58	06 39	07 31	08 20	13 38	14 04	14 47	15 54
56	06 34	07 21	08 07	13 51	14 20	15 06	16 14
54	06 28	07 13	07 56	14 02	14 35	15 23	16 30
52	06 23	07 06	07 46	14 12	14 47	15 37	16 44
50	06 19	06 59	07 37	14 21	14 59	15 50	16 56
45	06 08	06 45	07 18	14 40	15 22	16 15	17 21
N 40	05 59	06 33	07 03	14 56	15 40	16 36	17 42
35	05 51	06 22	06 50	15 09	15 56	16 53	17 58
30	05 43	06 12	06 38	15 20	16 10	17 08	18 13
20	05 28	05 55	06 19	15 40	16 33	17 33	18 37
N 10	05 13	05 39	06 02	15 58	16 54	17 55	18 59
0	04 57	05 23	05 45	16 14	17 13	18 15	19 19
S 10	04 40	05 06	05 29	16 31	17 32	18 35	19 38
20	04 19	04 47	05 12	16 48	17 52	18 57	20 00
30	03 52	04 24	04 51	17 09	18 16	19 23	20 24
35	03 35	04 10	04 39	17 21	18 30	19 38	20 39
40	03 13	03 53	04 25	17 35	18 47	19 55	20 55
45	02 46	03 33	04 09	17 52	19 06	20 16	21 15
S 50	02 07	03 06	03 49	18 12	19 31	20 42	21 40
52	01 44	02 53	03 39	18 22	19 43	20 55	21 52
54	01 14	02 37	03 28	18 33	19 56	21 10	22 06
56	00 16	02 19	03 15	18 46	20 12	21 27	22 22
58	////	01 55	03 01	19 00	20 31	21 48	22 41
S 60	////	01 23	02 43	19 18	20 54	22 14	23 04

Lat.	Sunset	Twilight Civil	Twilight Naut.	Moonset 30	Moonset 1	Moonset 2	Moonset 3
°	h m	h m	h m	h m	h m	h m	h m
N 72	■	13 47	15 46	□	□	□	□
N 70	■	14 26	16 02	□	□	□	□
68	13 01	14 54	16 16	08 13	□	□	□
66	13 50	15 14	16 27	07 31	10 06	□	□
64	14 21	15 31	16 36	07 04	09 03	11 02	12 05
62	14 44	15 45	16 44	06 42	08 28	10 05	11 10
60	15 03	15 57	16 52	06 25	08 03	09 32	10 37
N 58	15 18	16 07	16 58	06 11	07 43	09 08	10 13
56	15 31	16 16	17 04	05 58	07 27	08 48	09 53
54	15 42	16 24	17 10	05 48	07 13	08 32	09 37
52	15 52	16 32	17 15	05 38	07 00	08 18	09 23
50	16 01	16 38	17 19	05 30	06 50	08 05	09 10
45	16 20	16 53	17 30	05 12	06 27	07 40	08 45
N 40	16 35	17 05	17 39	04 57	06 09	07 20	08 24
35	16 48	17 16	17 47	04 45	05 54	07 03	08 07
30	17 00	17 26	17 55	04 35	05 41	06 48	07 52
20	17 19	17 43	18 10	04 16	05 19	06 23	07 27
N 10	17 37	17 59	18 25	04 00	04 59	06 02	07 06
0	17 53	18 15	18 41	03 46	04 42	05 42	06 45
S 10	18 09	18 32	18 59	03 31	04 24	05 22	06 25
20	18 27	18 51	19 20	03 15	04 05	05 01	06 03
30	18 47	19 14	19 47	02 57	03 43	04 36	05 38
35	18 59	19 28	20 04	02 47	03 30	04 22	05 23
40	19 13	19 45	20 25	02 35	03 15	04 05	05 05
45	19 30	20 06	20 53	02 21	02 58	03 45	04 44
S 50	19 50	20 33	21 33	02 04	02 36	03 19	04 18
52	20 00	20 46	21 55	01 56	02 26	03 07	04 05
54	20 11	21 02	22 26	01 47	02 14	02 53	03 51
56	20 24	21 21	23 34	01 38	02 01	02 37	03 33
58	20 39	21 45	////	01 26	01 46	02 18	03 13
S 60	20 56	22 19	////	01 13	01 28	01 55	02 46

Day	SUN Eqn. of Time 00^h	SUN Eqn. of Time 12^h	SUN Mer. Pass.	MOON Mer. Pass. Upper	MOON Mer. Pass. Lower	Age	Phase
d	m s	m s	h m	h m	h m	d %	
30	11 28	11 17	11 49	22 28	10 00	14 96	
1	11 06	10 55	11 49	23 27	10 57	15 99	○
2	10 43	10 32	11 49	24 30	11 58	16 100	

UT	ARIES	VENUS −3.9		MARS −0.1		JUPITER −2.2		SATURN +1.0	
	GHA	GHA	Dec	GHA	Dec	GHA	Dec	GHA	Dec
d h	° ′	° ′	° ′	° ′	° ′	° ′	° ′	° ′	° ′
3 00	71 57.5	192 34.2	S19 44.8	290 52.3	N17 55.6	108 02.9	S15 21.0	248 13.8	N 0 44.2
01	86 59.9	207 33.4	45.5	305 54.3	55.5	123 05.0	20.9	263 16.1	44.1
02	102 02.4	222 32.6	46.2	320 56.2	55.4	138 07.1	20.8	278 18.4	44.1
03	117 04.8	237 31.9	47.0	335 58.1	55.4	153 09.2	20.6	293 20.7	44.0
04	132 07.3	252 31.1	47.7	351 00.1	55.3	168 11.3	20.5	308 23.0	43.9
05	147 09.8	267 30.3	48.4	6 02.0	55.2	183 13.4	20.4	323 25.3	43.9
06	162 12.2	282 29.5	S19 49.1	21 03.9	N17 55.2	198 15.5	S15 20.3	338 27.6	N 0 43.8
07	177 14.7	297 28.7	49.8	36 05.8	55.1	213 17.6	20.1	353 29.9	43.8
T 08	192 17.2	312 27.9	50.6	51 07.8	55.1	228 19.7	20.0	8 32.2	43.7
H 09	207 19.6	327 27.1	51.3	66 09.7	55.0	243 21.8	19.9	23 34.5	43.6
U 10	222 22.1	342 26.3	52.0	81 11.7	54.9	258 23.9	19.8	38 36.8	43.6
R 11	237 24.6	357 25.5	52.7	96 13.6	54.9	273 26.0	19.6	53 39.1	43.5
S 12	252 27.0	12 24.7	S19 53.4	111 15.5	N17 54.8	288 28.1	S15 19.5	68 41.4	N 0 43.5
D 13	267 29.5	27 24.0	54.1	126 17.5	54.7	303 30.2	19.4	83 43.7	43.4
A 14	282 32.0	42 23.2	54.8	141 19.4	54.7	318 32.3	19.3	98 46.0	43.3
Y 15	297 34.4	57 22.4	55.6	156 21.3	54.6	333 34.3	19.1	113 48.3	43.3
16	312 36.9	72 21.6	56.3	171 23.3	54.5	348 36.4	19.0	128 50.6	43.2
17	327 39.3	87 20.8	57.0	186 25.2	54.5	3 38.5	18.9	143 52.9	43.2
18	342 41.8	102 20.0	S19 57.7	201 27.2	N17 54.4	18 40.6	S15 18.8	158 55.2	N 0 43.1
19	357 44.3	117 19.2	58.4	216 29.1	54.4	33 42.7	18.6	173 57.5	43.0
20	12 46.7	132 18.4	59.1	231 31.1	54.3	48 44.8	18.5	188 59.8	43.0
21	27 49.2	147 17.6	19 59.8	246 33.0	54.2	63 46.9	18.4	204 02.1	42.9
22	42 51.7	162 16.8	20 00.5	261 35.0	54.2	78 49.0	18.3	219 04.4	42.9
23	57 54.1	177 16.0	01.2	276 36.9	54.1	93 51.1	18.1	234 06.7	42.8
4 00	72 56.6	192 15.2	S20 01.9	291 38.8	N17 54.1	108 53.2	S15 18.0	249 09.0	N 0 42.8
01	87 59.1	207 14.4	02.6	306 40.8	54.0	123 55.3	17.9	264 11.3	42.7
02	103 01.5	222 13.6	03.3	321 42.7	53.9	138 57.4	17.8	279 13.6	42.6
03	118 04.0	237 12.8	04.0	336 44.7	53.9	153 59.5	17.6	294 15.9	42.6
04	133 06.5	252 12.0	04.7	351 46.7	53.8	169 01.6	17.5	309 18.2	42.5
05	148 08.9	267 11.2	05.4	6 48.6	53.8	184 03.7	17.4	324 20.5	42.5
06	163 11.4	282 10.4	S20 06.1	21 50.6	N17 53.7	199 05.8	S15 17.3	339 22.8	N 0 42.4
07	178 13.8	297 09.6	06.8	36 52.5	53.7	214 07.9	17.1	354 25.1	42.3
08	193 16.3	312 08.8	07.5	51 54.5	53.6	229 10.0	17.0	9 27.4	42.3
F 09	208 18.8	327 08.0	08.2	66 56.4	53.5	244 12.0	16.9	24 29.7	42.2
R 10	223 21.2	342 07.2	08.9	81 58.4	53.5	259 14.1	16.8	39 32.0	42.2
I 11	238 23.7	357 06.4	09.6	97 00.4	53.4	274 16.2	16.6	54 34.3	42.1
D 12	253 26.2	12 05.6	S20 10.3	112 02.3	N17 53.4	289 18.3	S15 16.5	69 36.6	N 0 42.1
A 13	268 28.6	27 04.8	11.0	127 04.3	53.3	304 20.4	16.4	84 38.9	42.0
Y 14	283 31.1	42 04.0	11.7	142 06.2	53.3	319 22.5	16.2	99 41.2	41.9
15	298 33.6	57 03.2	12.4	157 08.2	53.2	334 24.6	16.1	114 43.5	41.9
16	313 36.0	72 02.4	13.0	172 10.2	53.2	349 26.7	16.0	129 45.9	41.8
17	328 38.5	87 01.6	13.7	187 12.1	53.1	4 28.8	15.9	144 48.2	41.8
18	343 41.0	102 00.8	S20 14.4	202 14.1	N17 53.1	19 30.9	S15 15.7	159 50.5	N 0 41.7
19	358 43.4	116 59.9	15.1	217 16.1	53.0	34 33.0	15.6	174 52.8	41.6
20	13 45.9	131 59.1	15.8	232 18.1	53.0	49 35.0	15.5	189 55.1	41.6
21	28 48.3	146 58.3	16.5	247 20.0	52.9	64 37.1	15.4	204 57.4	41.5
22	43 50.8	161 57.5	17.1	262 22.0	52.9	79 39.2	15.2	219 59.7	41.5
23	58 53.3	176 56.7	17.8	277 24.0	52.8	94 41.3	15.1	235 02.0	41.4
5 00	73 55.7	191 55.9	S20 18.5	292 25.9	N17 52.8	109 43.4	S15 15.0	250 04.3	N 0 41.4
01	88 58.2	206 55.1	19.2	307 27.9	52.7	124 45.5	14.9	265 06.6	41.3
02	104 00.7	221 54.3	19.9	322 29.9	52.7	139 47.6	14.7	280 08.9	41.2
03	119 03.1	236 53.5	20.5	337 31.9	52.6	154 49.7	14.6	295 11.2	41.2
04	134 05.6	251 52.7	21.2	352 33.9	52.6	169 51.8	14.5	310 13.5	41.1
05	149 08.1	266 51.8	21.9	7 35.8	52.5	184 53.8	14.3	325 15.8	41.1
06	164 10.5	281 51.0	S20 22.6	22 37.8	N17 52.5	199 55.9	S15 14.2	340 18.1	N 0 41.0
07	179 13.0	296 50.2	23.2	37 39.8	52.4	214 58.0	14.1	355 20.4	41.0
S 08	194 15.5	311 49.4	23.9	52 41.8	52.4	230 00.1	14.0	10 22.7	40.9
A 09	209 17.9	326 48.6	24.6	67 43.8	52.3	245 02.2	13.8	25 25.0	40.8
T 10	224 20.4	341 47.8	25.2	82 45.7	52.3	260 04.3	13.7	40 27.4	40.8
U 11	239 22.8	356 47.0	25.9	97 47.7	52.2	275 06.4	13.6	55 29.7	40.7
R 12	254 25.3	11 46.1	S20 26.6	112 49.7	N17 52.2	290 08.5	S15 13.4	70 32.0	N 0 40.7
D 13	269 27.8	26 45.3	27.2	127 51.7	52.1	305 10.5	13.3	85 34.3	40.6
A 14	284 30.2	41 44.5	27.9	142 53.7	52.1	320 12.6	13.2	100 36.6	40.6
Y 15	299 32.7	56 43.7	28.6	157 55.7	52.0	335 14.7	13.1	115 38.9	40.5
16	314 35.2	71 42.9	29.2	172 57.7	52.0	350 16.8	12.9	130 41.2	40.4
17	329 37.6	86 42.1	29.9	187 59.7	51.9	5 18.9	12.8	145 43.5	40.4
18	344 40.1	101 41.2	S20 30.6	203 01.7	N17 51.9	20 21.0	S15 12.7	160 45.8	N 0 40.3
19	359 42.6	116 40.4	31.2	218 03.7	51.9	35 23.1	12.5	175 48.1	40.3
20	14 45.0	131 39.6	31.9	233 05.7	51.8	50 25.1	12.4	190 50.4	40.2
21	29 47.5	146 38.8	32.5	248 07.6	51.8	65 27.2	12.3	205 52.7	40.2
22	44 50.0	161 38.0	33.2	263 09.6	51.7	80 29.3	12.2	220 55.0	40.1
23	59 52.4	176 37.1	33.9	278 11.6	51.7	95 31.4	12.0	235 57.3	40.1
Mer. Pass.	h m 19 05.1	v −0.8	d 0.7	v 2.0	d 0.1	v 2.1	d 0.1	v 2.3	d 0.1

STARS Name	SHA	Dec
	° ′	° ′
Acamar	315 19.8	S40 15.9
Achernar	335 28.1	S57 11.3
Acrux	173 12.7	S63 09.1
Adhara	255 14.3	S28 59.0
Aldebaran	290 52.0	N16 31.8
Alioth	166 23.1	N55 54.0
Alkaid	153 01.2	N49 15.6
Al Na'ir	27 47.0	S46 54.9
Alnilam	275 48.7	S 1 11.7
Alphard	217 58.5	S 8 42.1
Alphecca	126 13.5	N26 40.8
Alpheratz	357 46.1	N29 09.0
Altair	62 11.0	N 8 53.8
Ankaa	353 18.0	S42 15.2
Antares	112 29.9	S26 27.2
Arcturus	145 58.3	N19 07.7
Atria	107 34.7	S69 02.7
Avior	234 18.8	S59 32.3
Bellatrix	278 34.5	N 6 21.6
Betelgeuse	271 03.8	N 7 24.6
Canopus	263 56.8	S52 41.9
Capella	280 37.8	N46 00.5
Deneb	49 33.6	N45 19.2
Denebola	182 36.4	N14 30.9
Diphda	348 58.3	S17 55.9
Dubhe	193 54.7	N61 41.5
Elnath	278 15.5	N28 37.0
Eltanin	90 47.9	N51 29.3
Enif	33 49.8	N 9 55.4
Fomalhaut	15 26.8	S29 34.2
Gacrux	172 04.3	S57 09.9
Gienah	175 55.1	S17 35.8
Hadar	148 52.3	S60 25.1
Hamal	328 03.5	N23 30.8
Kaus Aust.	83 47.6	S34 22.8
Kochab	137 20.7	N74 06.6
Markab	13 40.9	N15 15.7
Menkar	314 17.5	N 4 07.8
Menkent	148 11.0	S36 25.0
Miaplacidus	221 40.0	S69 45.2
Mirfak	308 43.7	N49 54.0
Nunki	76 01.8	S26 17.1
Peacock	53 23.6	S56 42.3
Pollux	243 30.6	N28 00.0
Procyon	245 02.2	N 5 12.0
Rasalhague	96 09.2	N12 33.2
Regulus	207 46.2	N11 55.0
Rigel	281 14.2	S 8 11.3
Rigil Kent.	139 56.0	S60 52.4
Sabik	102 15.9	S15 44.2
Schedar	349 43.4	N56 35.9
Shaula	96 25.9	S37 06.7
Sirius	258 35.7	S16 43.7
Spica	158 34.2	S11 12.8
Suhail	222 54.2	S43 28.2
Vega	80 41.1	N38 47.7
Zuben'ubi	137 08.6	S16 05.0

	SHA	Mer. Pass.
	° ′	h m
Venus	119 18.6	11 12
Mars	218 42.3	4 33
Jupiter	35 56.6	16 42
Saturn	176 12.4	7 22

UT	SUN GHA	SUN Dec	MOON GHA	v	MOON Dec	d	HP
d h	° ′	° ′	° ′	′	° ′	′	′
3 00	182 34.9	S22 05.1	352 47.9	2.9	N25 46.3	0.1	60.0
01	197 34.7	05.4	7 09.8	2.8	25 46.2	0.4	60.0
02	212 34.4	05.8	21 31.6	2.8	25 45.8	0.5	60.1
03	227 34.2	. . 06.2	35 53.4	2.9	25 45.3	0.8	60.1
04	242 33.9	06.5	50 15.3	2.8	25 44.5	0.9	60.1
05	257 33.7	06.9	64 37.1	2.8	25 43.6	1.1	60.1
06	272 33.5	S22 07.2	78 58.9	2.8	N25 42.5	1.3	60.1
07	287 33.2	07.6	93 20.7	2.8	25 41.2	1.5	60.1
08	302 33.0	07.9	107 42.5	2.8	25 39.7	1.7	60.1
09	317 32.7	. . 08.3	122 04.3	2.8	25 38.0	1.8	60.1
10	332 32.5	08.6	136 26.1	2.8	25 36.2	2.1	60.2
11	347 32.2	09.0	150 47.9	2.9	25 34.1	2.2	60.2
12	2 32.0	S22 09.3	165 09.8	2.8	N25 31.9	2.5	60.2
13	17 31.7	09.6	179 31.6	2.9	25 29.4	2.6	60.2
14	32 31.5	10.0	193 53.5	2.9	25 26.8	2.8	60.2
15	47 31.2	. . 10.3	208 15.4	2.9	25 24.0	3.0	60.2
16	62 31.0	10.7	222 37.3	2.9	25 21.0	3.1	60.2
17	77 30.7	11.0	236 59.2	2.9	25 17.9	3.4	60.2
18	92 30.5	S22 11.4	251 21.1	3.0	N25 14.5	3.6	60.2
19	107 30.2	11.7	265 43.1	3.0	25 10.9	3.7	60.3
20	122 30.0	12.0	280 05.1	3.1	25 07.2	3.9	60.3
21	137 29.7	. . 12.4	294 27.2	3.1	25 03.3	4.1	60.3
22	152 29.5	12.7	308 49.3	3.1	24 59.2	4.3	60.3
23	167 29.2	13.1	323 11.4	3.1	24 54.9	4.5	60.3
4 00	182 29.0	S22 13.4	337 33.5	3.2	N24 50.4	4.6	60.3
01	197 28.7	13.7	351 55.7	3.3	24 45.8	4.8	60.3
02	212 28.5	14.1	6 18.0	3.3	24 41.0	5.0	60.3
03	227 28.2	. . 14.4	20 40.3	3.4	24 36.0	5.2	60.3
04	242 28.0	14.7	35 02.7	3.4	24 30.8	5.4	60.3
05	257 27.7	15.1	49 25.1	3.4	24 25.4	5.5	60.3
06	272 27.5	S22 15.4	63 47.5	3.5	N24 19.9	5.7	60.3
07	287 27.2	15.7	78 10.0	3.6	24 14.2	5.9	60.3
08	302 27.0	16.1	92 32.6	3.6	24 08.3	6.1	60.3
09	317 26.7	. . 16.4	106 55.2	3.7	24 02.2	6.2	60.3
10	332 26.5	16.7	121 17.9	3.8	23 56.0	6.4	60.3
11	347 26.2	17.1	135 40.7	3.8	23 49.6	6.6	60.3
12	2 26.0	S22 17.4	150 03.5	3.9	N23 43.0	6.7	60.3
13	17 25.7	17.7	164 26.4	4.0	23 36.3	6.9	60.3
14	32 25.5	18.1	178 49.4	4.0	23 29.4	7.1	60.3
15	47 25.2	. . 18.4	193 12.4	4.2	23 22.3	7.2	60.3
16	62 25.0	18.7	207 35.6	4.2	23 15.1	7.4	60.3
17	77 24.7	19.0	221 58.8	4.2	23 07.7	7.5	60.3
18	92 24.5	S22 19.4	236 22.0	4.4	N23 00.2	7.8	60.3
19	107 24.2	19.7	250 45.4	4.4	22 52.4	7.8	60.3
20	122 23.9	20.0	265 08.8	4.5	22 44.6	8.1	60.3
21	137 23.7	. . 20.3	279 32.3	4.6	22 36.5	8.2	60.3
22	152 23.4	20.6	293 55.9	4.7	22 28.3	8.3	60.3
23	167 23.2	21.0	308 19.6	4.7	22 20.0	8.5	60.3
5 00	182 22.9	S22 21.3	322 43.3	4.9	N22 11.5	8.6	60.3
01	197 22.7	21.6	337 07.2	4.9	22 02.9	8.8	60.3
02	212 22.4	21.9	351 31.1	5.0	21 54.1	9.0	60.3
03	227 22.2	. . 22.2	5 55.1	5.1	21 45.1	9.1	60.3
04	242 21.9	22.6	20 19.2	5.3	21 36.0	9.2	60.3
05	257 21.6	22.9	34 43.5	5.2	21 26.8	9.4	60.3
06	272 21.4	S22 23.2	49 07.7	5.4	N21 17.4	9.5	60.3
07	287 21.1	23.5	63 32.1	5.5	21 07.9	9.6	60.3
08	302 20.9	23.8	77 56.6	5.6	20 58.3	9.8	60.3
09	317 20.6	. . 24.1	92 21.2	5.6	20 48.5	9.9	60.3
10	332 20.3	24.4	106 45.8	5.8	20 38.6	10.1	60.3
11	347 20.1	24.8	121 10.6	5.9	20 28.5	10.2	60.2
12	2 19.8	S22 25.1	135 35.5	5.9	N20 18.3	10.3	60.2
13	17 19.6	25.4	150 00.4	6.1	20 08.0	10.5	60.2
14	32 19.3	25.7	164 25.5	6.1	19 57.5	10.5	60.2
15	47 19.1	. . 26.0	178 50.6	6.3	19 47.0	10.7	60.2
16	62 18.8	26.3	193 15.9	6.3	19 36.3	10.9	60.2
17	77 18.5	26.6	207 41.2	6.4	19 25.4	10.9	60.2
18	92 18.3	S22 26.9	222 06.6	6.6	N19 14.5	11.1	60.2
19	107 18.0	27.2	236 32.2	6.6	19 03.4	11.1	60.2
20	122 17.8	27.5	250 57.8	6.7	18 52.3	11.3	60.2
21	137 17.5	. . 27.8	265 23.5	6.9	18 41.0	11.5	60.1
22	152 17.2	28.1	279 49.4	6.9	18 29.5	11.5	60.1
23	167 17.0	28.4	294 15.3	7.0	N18 18.0	11.6	60.1
	SD 16.3	d 0.3	SD 16.4		16.4		16.4

Lat.	Twilight Naut.	Twilight Civil	Sunrise	Moonrise 3	Moonrise 4	Moonrise 5	Moonrise 6
°	h m	h m	h m	h m	h m	h m	h m
N 72	07 59	10 04	■	☐	☐	☐	18 46
N 70	07 42	09 21	■	☐	☐	16 04	19 16
68	07 28	08 52	10 56	☐	☐	17 11	19 39
66	07 16	08 30	09 58	☐	15 11	17 47	19 56
64	07 06	08 13	09 24	14 02	16 06	18 12	20 10
62	06 58	07 58	09 00	14 57	16 39	18 32	20 21
60	06 50	07 46	08 41	15 30	17 04	18 48	20 31
N 58	06 44	07 35	08 25	15 54	17 23	19 01	20 40
56	06 38	07 26	08 12	16 14	17 39	19 13	20 47
54	06 32	07 17	08 00	16 30	17 53	19 23	20 54
52	06 27	07 10	07 50	16 44	18 04	19 32	21 00
50	06 22	07 03	07 41	16 56	18 15	19 40	21 05
45	06 11	06 48	07 21	17 21	18 37	19 57	21 17
N 40	06 02	06 35	07 06	17 42	18 55	20 11	21 26
35	05 53	06 24	06 52	17 58	19 10	20 23	21 35
30	05 45	06 14	06 41	18 13	19 22	20 33	21 42
20	05 29	05 57	06 21	18 37	19 44	20 50	21 54
N 10	05 14	05 40	06 03	18 59	20 03	21 06	22 05
0	04 58	05 24	05 47	19 19	20 21	21 20	22 15
S 10	04 40	05 07	05 30	19 38	20 39	21 34	22 25
20	04 19	04 48	05 12	20 00	20 57	21 49	22 35
30	03 51	04 24	04 51	20 24	21 19	22 06	22 47
35	03 34	04 10	04 39	20 39	21 32	22 16	22 54
40	03 12	03 53	04 25	20 55	21 46	22 28	23 02
45	02 44	03 31	04 08	21 15	22 03	22 41	23 11
S 50	02 03	03 04	03 47	21 40	22 24	22 57	23 22
52	01 40	02 50	03 37	21 52	22 34	23 04	23 27
54	01 07	02 34	03 26	22 06	22 46	23 13	23 33
56	////	02 14	03 13	22 22	22 58	23 22	23 39
58	////	01 50	02 57	22 41	23 13	23 33	23 45
S 60	////	01 14	02 39	23 04	23 31	23 45	23 53

Lat.	Sunset	Twilight Civil	Twilight Naut.	Moonset 3	Moonset 4	Moonset 5	Moonset 6
°	h m	h m	h m	h m	h m	h m	h m
N 72	■	13 35	15 41	☐	☐	☐	13 41
N 70	■	14 19	15 58	☐	☐	14 24	13 08
68	12 44	14 48	16 12	☐	☐	13 16	12 44
66	13 43	15 10	16 24	☐	13 09	12 39	12 26
64	14 16	15 28	16 34	12 05	12 13	12 13	12 10
62	14 40	15 42	16 42	11 10	11 40	11 52	11 57
60	14 59	15 54	16 50	10 37	11 15	11 35	11 46
N 58	15 15	16 05	16 56	10 13	10 55	11 21	11 37
56	15 28	16 14	17 03	09 53	10 39	11 08	11 28
54	15 40	16 23	17 08	09 37	10 25	10 58	11 21
52	15 50	16 30	17 13	09 23	10 12	10 48	11 14
50	16 00	16 37	17 18	09 10	10 01	10 39	11 08
45	16 19	16 52	17 29	08 45	09 38	10 21	10 55
N 40	16 35	17 05	17 39	08 24	09 20	10 06	10 44
35	16 48	17 16	17 47	08 07	09 04	09 53	10 34
30	17 00	17 26	17 56	07 52	08 51	09 42	10 26
20	17 20	17 44	18 11	07 27	08 28	09 22	10 11
N 10	17 37	18 00	18 26	07 06	08 08	09 05	09 58
0	17 54	18 16	18 42	06 45	07 49	08 49	09 46
S 10	18 11	18 34	19 00	06 25	07 30	08 33	09 34
20	18 29	18 53	19 22	06 03	07 09	08 16	09 21
30	18 50	19 17	19 50	05 38	06 46	07 56	09 06
35	19 02	19 31	20 07	05 23	06 32	07 44	08 57
40	19 16	19 48	20 29	05 05	06 15	07 31	08 47
45	19 33	20 10	20 58	04 44	05 56	07 15	08 35
S 50	19 54	20 37	21 39	04 18	05 32	06 55	08 21
52	20 04	20 51	22 03	04 05	05 20	06 45	08 14
54	20 16	21 08	22 36	03 51	05 06	06 34	08 06
56	20 29	21 27	////	03 33	04 51	06 22	07 58
58	20 44	21 53	////	03 13	04 32	06 08	07 48
S 60	21 02	22 29	////	02 46	04 09	05 51	07 37

Day	SUN Eqn. of Time 00^h	SUN Eqn. of Time 12^h	SUN Mer. Pass.	MOON Mer. Pass. Upper	MOON Mer. Pass. Lower	MOON Age	MOON Phase
d	m s	m s	h m	h m	h m	d %	
3	10 20	10 08	11 50	00 30	13 02	17 98	
4	09 56	09 44	11 50	01 34	14 05	18 93	
5	09 32	09 20	11 51	02 35	15 05	19 86	

	UT	ARIES	VENUS −3.9		MARS −0.2		JUPITER −2.2		SATURN +1.0	
		GHA	GHA	Dec	GHA	Dec	GHA	Dec	GHA	Dec
	d h	° ′	° ′	° ′	° ′	° ′	° ′	° ′	° ′	° ′
	6 00	74 54.9	191 36.3	S20 34.5	293 13.6	N17 51.6	110 33.5	S15 11.9	250 59.7	N 0 40.0
	01	89 57.3	206 35.5	35.2	308 15.6	51.6	125 35.6	11.8	266 02.0	39.9
	02	104 59.8	221 34.7	35.8	323 17.6	51.6	140 37.6	11.6	281 04.3	39.9
	03	120 02.3	236 33.8	. . 36.5	338 19.7	. . 51.5	155 39.7	. . 11.5	296 06.6	. . 39.8
	04	135 04.7	251 33.0	37.1	353 21.7	51.5	170 41.8	11.4	311 08.9	39.8
	05	150 07.2	266 32.2	37.8	8 23.7	51.4	185 43.9	11.3	326 11.2	39.7
	06	165 09.7	281 31.4	S20 38.4	23 25.7	N17 51.4	200 46.0	S15 11.1	341 13.5	N 0 39.7
	07	180 12.1	296 30.6	39.1	38 27.7	51.3	215 48.1	11.0	356 15.8	39.6
	08	195 14.6	311 29.7	39.7	53 29.7	51.3	230 50.1	10.9	11 18.1	39.6
S	09	210 17.1	326 28.9	. . 40.4	68 31.7	. . 51.3	245 52.2	. . 10.7	26 20.4	. . 39.5
U	10	225 19.5	341 28.1	41.0	83 33.7	51.2	260 54.3	10.6	41 22.8	39.4
N	11	240 22.0	356 27.2	41.6	98 35.7	51.2	275 56.4	10.5	56 25.1	39.4
D	12	255 24.4	11 26.4	S20 42.3	113 37.7	N17 51.2	290 58.5	S15 10.3	71 27.4	N 0 39.3
A	13	270 26.9	26 25.6	42.9	128 39.7	51.1	306 00.5	10.2	86 29.7	39.3
Y	14	285 29.4	41 24.8	43.6	143 41.8	51.1	321 02.6	10.1	101 32.0	39.2
	15	300 31.8	56 23.9	. . 44.2	158 43.8	. . 51.0	336 04.7	. . 10.0	116 34.3	. . 39.2
	16	315 34.3	71 23.1	44.9	173 45.8	51.0	351 06.8	09.8	131 36.6	39.1
	17	330 36.8	86 22.3	45.5	188 47.8	51.0	6 08.9	09.7	146 38.9	39.1
	18	345 39.2	101 21.4	S20 46.1	203 49.8	N17 50.9	21 10.9	S15 09.6	161 41.2	N 0 39.0
	19	0 41.7	116 20.6	46.8	218 51.8	50.9	36 13.0	09.4	176 43.5	38.9
	20	15 44.2	131 19.8	47.4	233 53.9	50.9	51 15.1	09.3	191 45.9	38.9
	21	30 46.6	146 19.0	. . 48.0	248 55.9	. . 50.8	66 17.2	. . 09.2	206 48.2	. . 38.8
	22	45 49.1	161 18.1	48.7	263 57.9	50.8	81 19.3	09.0	221 50.5	38.8
	23	60 51.6	176 17.3	49.3	278 59.9	50.8	96 21.3	08.9	236 52.8	38.7
	7 00	75 54.0	191 16.5	S20 49.9	294 02.0	N17 50.7	111 23.4	S15 08.8	251 55.1	N 0 38.7
	01	90 56.5	206 15.6	50.6	309 04.0	50.7	126 25.5	08.6	266 57.4	38.6
	02	105 58.9	221 14.8	51.2	324 06.0	50.7	141 27.6	08.5	281 59.7	38.6
	03	121 01.4	236 13.9	. . 51.8	339 08.0	. . 50.6	156 29.7	. . 08.4	297 02.0	. . 38.5
	04	136 03.9	251 13.1	52.4	354 10.1	50.6	171 31.7	08.2	312 04.4	38.5
	05	151 06.3	266 12.3	53.1	9 12.1	50.6	186 33.8	08.1	327 06.7	38.4
	06	166 08.8	281 11.4	S20 53.7	24 14.1	N17 50.5	201 35.9	S15 08.0	342 09.0	N 0 38.3
	07	181 11.3	296 10.6	54.3	39 16.2	50.5	216 38.0	07.9	357 11.3	38.3
	08	196 13.7	311 09.8	54.9	54 18.2	50.5	231 40.0	07.7	12 13.6	38.2
M	09	211 16.2	326 08.9	. . 55.6	69 20.2	. . 50.4	246 42.1	. . 07.6	27 15.9	. . 38.2
O	10	226 18.7	341 08.1	56.2	84 22.3	50.4	261 44.2	07.5	42 18.2	38.1
N	11	241 21.1	356 07.3	56.8	99 24.3	50.4	276 46.3	07.3	57 20.5	38.1
D	12	256 23.6	11 06.4	S20 57.4	114 26.4	N17 50.3	291 48.3	S15 07.2	72 22.9	N 0 38.0
A	13	271 26.1	26 05.6	58.0	129 28.4	50.3	306 50.4	07.1	87 25.2	38.0
Y	14	286 28.5	41 04.7	58.6	144 30.4	50.3	321 52.5	06.9	102 27.5	37.9
	15	301 31.0	56 03.9	. . 59.3	159 32.5	. . 50.3	336 54.6	. . 06.8	117 29.8	. . 37.9
	16	316 33.4	71 03.1	20 59.9	174 34.5	50.2	351 56.7	06.7	132 32.1	37.8
	17	331 35.9	86 02.2	21 00.5	189 36.6	50.2	6 58.7	06.5	147 34.4	37.8
	18	346 38.4	101 01.4	S21 01.1	204 38.6	N17 50.2	22 00.8	S15 06.4	162 36.7	N 0 37.7
	19	1 40.8	116 00.5	01.7	219 40.7	50.2	37 02.9	06.3	177 39.1	37.7
	20	16 43.3	130 59.7	02.3	234 42.7	50.1	52 05.0	06.1	192 41.4	37.6
	21	31 45.8	145 58.8	. . 02.9	249 44.8	. . 50.1	67 07.0	. . 06.0	207 43.7	. . 37.5
	22	46 48.2	160 58.0	03.5	264 46.8	50.1	82 09.1	05.9	222 46.0	37.5
	23	61 50.7	175 57.2	04.1	279 48.9	50.1	97 11.2	05.7	237 48.3	37.4
	8 00	76 53.2	190 56.3	S21 04.7	294 50.9	N17 50.0	112 13.2	S15 05.6	252 50.6	N 0 37.4
	01	91 55.6	205 55.5	05.3	309 53.0	50.0	127 15.3	05.5	267 53.0	37.3
	02	106 58.1	220 54.6	05.9	324 55.0	50.0	142 17.4	05.3	282 55.3	37.3
	03	122 00.6	235 53.8	. . 06.5	339 57.1	. . 50.0	157 19.5	. . 05.2	297 57.6	. . 37.2
	04	137 03.0	250 52.9	07.2	354 59.1	49.9	172 21.5	05.1	312 59.9	37.2
	05	152 05.5	265 52.1	07.8	10 01.2	49.9	187 23.6	04.9	328 02.2	37.1
	06	167 07.9	280 51.2	S21 08.3	25 03.3	N17 49.9	202 25.7	S15 04.8	343 04.5	N 0 37.1
	07	182 10.4	295 50.4	08.9	40 05.3	49.9	217 27.8	04.7	358 06.8	37.0
T	08	197 12.9	310 49.5	09.5	55 07.4	49.8	232 29.8	04.5	13 09.2	37.0
U	09	212 15.3	325 48.7	. . 10.1	70 09.4	. . 49.8	247 31.9	. . 04.4	28 11.5	. . 36.9
E	10	227 17.8	340 47.8	10.7	85 11.5	49.8	262 34.0	04.3	43 13.8	36.9
S	11	242 20.3	355 47.0	11.3	100 13.6	49.8	277 36.0	04.1	58 16.1	36.8
D	12	257 22.7	10 46.1	S21 11.9	115 15.6	N17 49.8	292 38.1	S15 04.0	73 18.4	N 0 36.8
A	13	272 25.2	25 45.3	12.5	130 17.7	49.7	307 40.2	03.9	88 20.8	36.7
Y	14	287 27.7	40 44.4	13.1	145 19.8	49.7	322 42.3	03.7	103 23.1	36.7
	15	302 30.1	55 43.6	. . 13.7	160 21.8	. . 49.7	337 44.3	. . 03.6	118 25.4	. . 36.6
	16	317 32.6	70 42.7	14.3	175 23.9	49.7	352 46.4	03.5	133 27.7	36.6
	17	332 35.0	85 41.9	14.9	190 26.0	49.7	7 48.5	03.3	148 30.0	36.5
	18	347 37.5	100 41.0	S21 15.5	205 28.1	N17 49.6	22 50.5	S15 03.2	163 32.3	N 0 36.5
	19	2 40.0	115 40.2	16.0	220 30.1	49.6	37 52.6	03.1	178 34.7	36.4
	20	17 42.4	130 39.3	16.6	235 32.2	49.6	52 54.7	02.9	193 37.0	36.3
	21	32 44.9	145 38.5	. . 17.2	250 34.3	. . 49.6	67 56.7	. . 02.8	208 39.3	. . 36.3
	22	47 47.4	160 37.6	17.8	265 36.4	49.6	82 58.8	02.7	223 41.6	36.2
	23	62 49.8	175 36.8	18.4	280 38.4	49.6	98 00.9	02.5	238 43.9	36.2
	Mer. Pass.	h m 18 53.3	*v* −0.8	*d* 0.6	*v* 2.0	*d* 0.0	*v* 2.1	*d* 0.1	*v* 2.3	*d* 0.1

STARS

Name	SHA	Dec
	° ′	° ′
Acamar	315 19.8	S40 15.
Achernar	335 28.1	S57 11.
Acrux	173 12.7	S63 09.
Adhara	255 14.2	S28 59.
Aldebaran	290 52.0	N16 31.
Alioth	166 23.0	N55 54.
Alkaid	153 01.1	N49 15.
Al Na'ir	27 47.0	S46 54.
Alnilam	275 48.7	S 1 11.
Alphard	217 58.5	S 8 42.
Alphecca	126 13.5	N26 40.
Alpheratz	357 46.1	N29 09.
Altair	62 11.0	N 8 53.
Ankaa	353 18.0	S42 15.
Antares	112 29.8	S26 27.
Arcturus	145 58.3	N19 07.
Atria	107 34.7	S69 02.
Avior	234 18.8	S59 32.
Bellatrix	278 34.4	N 6 21.
Betelgeuse	271 03.8	N 7 24.
Canopus	263 56.8	S52 41.
Capella	280 37.8	N46 00.
Deneb	49 33.6	N45 19.
Denebola	182 36.3	N14 30.
Diphda	348 58.3	S17 55.
Dubhe	193 54.6	N61 41.
Elnath	278 15.5	N28 37.
Eltanin	90 47.9	N51 29.
Enif	33 49.8	N 9 55.
Fomalhaut	15 26.8	S29 34.
Gacrux	172 04.2	S57 09.
Gienah	175 55.1	S17 35.
Hadar	148 52.3	S60 25.
Hamal	328 03.5	N23 30.
Kaus Aust.	83 47.6	S34 22.
Kochab	137 20.6	N74 06.
Markab	13 40.9	N15 15.
Menkar	314 17.5	N 4 07.
Menkent	148 11.0	S36 25.
Miaplacidus	221 40.0	S69 45.
Mirfak	308 43.7	N49 54.
Nunki	76 01.8	S26 17.
Peacock	53 23.6	S56 42.
Pollux	243 30.6	N28 00.
Procyon	245 02.2	N 5 12.
Rasalhague	96 09.2	N12 33.
Regulus	207 46.2	N11 55.
Rigel	281 14.2	S 8 11.
Rigil Kent.	139 56.0	S60 52.
Sabik	102 15.9	S15 44.
Schedar	349 43.4	N56 35.
Shaula	96 25.9	S37 06.
Sirius	258 35.7	S16 43.
Spica	158 34.2	S11 12.
Suhail	222 54.2	S43 28.
Vega	80 41.1	N38 47.
Zuben'ubi	137 08.6	S16 05.

	SHA	Mer. Pass.
	° ′	h m
Venus	115 22.4	11 16
Mars	218 07.9	4 23
Jupiter	35 29.4	16 32
Saturn	176 01.1	7 11

	UT	SUN		MOON				
	d h	GHA ° ′	Dec ° ′	GHA ° ′	v ′	Dec ° ′	d ′	HP ′
	6 00	182 16.7	S22 28.7	308 41.3	7.2	N18 06.4	11.8	60.1
	01	197 16.4	29.0	323 07.5	7.2	17 54.6	11.8	60.1
	02	212 16.2	29.3	337 33.7	7.3	17 42.8	12.0	60.1
	03	227 15.9	. . 29.6	352 00.0	7.4	17 30.8	12.0	60.1
	04	242 15.7	29.9	6 26.4	7.6	17 18.8	12.2	60.1
	05	257 15.4	30.2	20 53.0	7.6	17 06.6	12.2	60.0
	06	272 15.1	S22 30.5	35 19.6	7.7	N16 54.4	12.4	60.0
	07	287 14.9	30.8	49 46.3	7.8	16 42.0	12.4	60.0
	08	302 14.6	31.1	64 13.1	7.9	16 29.6	12.6	60.0
SUNDAY	09	317 14.3	. . 31.4	78 40.0	8.0	16 17.0	12.6	60.0
	10	332 14.1	31.7	93 07.0	8.1	16 04.4	12.7	60.0
	11	347 13.8	32.0	107 34.1	8.2	15 51.7	12.8	60.0
	12	2 13.5	S22 32.3	122 01.3	8.3	N15 38.9	12.9	59.9
	13	17 13.3	32.6	136 28.6	8.4	15 26.0	13.0	59.9
	14	32 13.0	32.9	150 56.0	8.4	15 13.0	13.1	59.9
	15	47 12.7	. . 33.2	165 23.4	8.6	14 59.9	13.1	59.9
	16	62 12.5	33.5	179 51.0	8.7	14 46.8	13.3	59.9
	17	77 12.2	33.7	194 18.7	8.7	14 33.5	13.3	59.9
	18	92 12.0	S22 34.0	208 46.4	8.9	N14 20.2	13.3	59.8
	19	107 11.7	34.3	223 14.3	8.9	14 06.9	13.5	59.8
	20	122 11.4	34.6	237 42.2	9.0	13 53.4	13.5	59.8
	21	137 11.2	. . 34.9	252 10.2	9.1	13 39.9	13.6	59.8
	22	152 10.9	35.2	266 38.3	9.2	13 26.3	13.7	59.8
	23	167 10.6	35.5	281 06.5	9.3	13 12.6	13.7	59.8
	7 00	182 10.4	S22 35.7	295 34.8	9.4	N12 58.9	13.8	59.7
	01	197 10.1	36.0	310 03.2	9.4	12 45.1	13.9	59.7
	02	212 09.8	36.3	324 31.6	9.6	12 31.2	13.9	59.7
	03	227 09.5	. . 36.6	339 00.2	9.6	12 17.3	14.0	59.7
	04	242 09.3	36.9	353 28.8	9.7	12 03.3	14.0	59.7
	05	257 09.0	37.2	7 57.5	9.8	11 49.3	14.1	59.6
	06	272 08.7	S22 37.4	22 26.3	9.9	N11 35.2	14.1	59.6
	07	287 08.5	37.7	36 55.2	9.9	11 21.1	14.3	59.6
	08	302 08.2	38.0	51 24.1	10.0	11 06.8	14.2	59.6
MONDAY	09	317 07.9	. . 38.3	65 53.1	10.1	10 52.6	14.3	59.6
	10	332 07.7	38.5	80 22.2	10.2	10 38.3	14.4	59.6
	11	347 07.4	38.8	94 51.4	10.3	10 23.9	14.4	59.5
	12	2 07.1	S22 39.1	109 20.7	10.3	N10 09.5	14.4	59.5
	13	17 06.9	39.4	123 50.0	10.4	9 55.1	14.5	59.5
	14	32 06.6	39.6	138 19.4	10.5	9 40.6	14.5	59.5
	15	47 06.3	. . 39.9	152 48.9	10.5	9 26.1	14.6	59.4
	16	62 06.0	40.2	167 18.4	10.7	9 11.5	14.6	59.4
	17	77 05.8	40.4	181 48.1	10.6	8 56.9	14.6	59.4
	18	92 05.5	S22 40.7	196 17.7	10.8	N 8 42.3	14.7	59.4
	19	107 05.2	41.0	210 47.5	10.8	8 27.6	14.7	59.4
	20	122 05.0	41.3	225 17.3	10.9	8 12.9	14.8	59.3
	21	137 04.7	. . 41.5	239 47.2	10.9	7 58.1	14.8	59.3
	22	152 04.4	41.8	254 17.1	11.1	7 43.3	14.8	59.3
	23	167 04.1	42.1	268 47.2	11.0	7 28.5	14.8	59.3
	8 00	182 03.9	S22 42.3	283 17.2	11.2	N 7 13.7	14.9	59.3
	01	197 03.6	42.6	297 47.4	11.2	6 58.8	14.9	59.2
	02	212 03.3	42.8	312 17.6	11.2	6 43.9	14.9	59.2
	03	227 03.0	. . 43.1	326 47.8	11.3	6 29.0	14.9	59.2
	04	242 02.8	43.4	341 18.1	11.4	6 14.1	15.0	59.2
	05	257 02.5	43.6	355 48.5	11.4	5 59.1	14.9	59.1
	06	272 02.2	S22 43.9	10 18.9	11.5	N 5 44.2	15.0	59.1
	07	287 02.0	44.2	24 49.4	11.5	5 29.2	15.0	59.1
TUESDAY	08	302 01.7	44.4	39 19.9	11.6	5 14.2	15.1	59.1
	09	317 01.4	. . 44.7	53 50.5	11.6	4 59.1	15.0	59.1
	10	332 01.1	44.9	68 21.1	11.7	4 44.1	15.0	59.0
	11	347 00.9	45.2	82 51.8	11.7	4 29.1	15.1	59.0
	12	2 00.6	S22 45.4	97 22.5	11.8	N 4 14.0	15.1	59.0
	13	17 00.3	45.7	111 53.3	11.8	3 58.9	15.0	59.0
	14	32 00.0	45.9	126 24.1	11.9	3 43.9	15.1	58.9
	15	46 59.7	. . 46.2	140 55.0	11.9	3 28.8	15.1	58.9
	16	61 59.5	46.5	155 25.9	11.9	3 13.7	15.1	58.9
	17	76 59.2	46.7	169 56.8	12.0	2 58.6	15.1	58.9
	18	91 58.9	S22 47.0	184 27.8	12.0	N 2 43.5	15.1	58.8
	19	106 58.6	47.2	198 58.8	12.1	2 28.4	15.1	58.8
	20	121 58.4	47.5	213 29.9	12.1	2 13.3	15.1	58.8
	21	136 58.1	. . 47.7	228 01.0	12.1	1 58.2	15.1	58.8
	22	151 57.8	47.9	242 32.1	12.2	1 43.1	15.1	58.7
	23	166 57.5	48.2	257 03.3	12.2	N 1 28.0	15.1	58.7
		SD 16.3	*d* 0.3	SD		16.3	16.2	16.1

Lat.	Twilight Naut.	Twilight Civil	Sunrise	Moonrise 6	Moonrise 7	Moonrise 8	Moonrise 9
°	h m	h m	h m	h m	h m	h m	h m
N 72	08 06	10 18	▬	18 46	21 23	23 39	25 50
N 70	07 48	09 30	▬	19 16	21 35	23 41	25 42
68	07 33	08 59	11 20	19 39	21 45	23 42	25 35
66	07 21	08 36	10 07	19 56	21 53	23 43	25 30
64	07 11	08 18	09 31	20 10	22 00	23 44	25 25
62	07 02	08 03	09 06	20 21	22 05	23 45	25 22
60	06 54	07 50	08 46	20 31	22 10	23 46	25 18
N 58	06 47	07 39	08 30	20 40	22 15	23 46	25 15
56	06 41	07 30	08 16	20 47	22 19	23 47	25 13
54	06 35	07 21	08 04	20 54	22 22	23 47	25 10
52	06 30	07 13	07 54	21 00	22 25	23 48	25 08
50	06 25	07 06	07 44	21 05	22 28	23 48	25 06
45	06 14	06 51	07 24	21 17	22 34	23 49	25 02
N 40	06 04	06 38	07 08	21 26	22 39	23 50	24 59
35	05 55	06 27	06 55	21 35	22 44	23 51	24 56
30	05 47	06 17	06 43	21 42	22 48	23 51	24 53
20	05 31	05 59	06 22	21 54	22 54	23 52	24 49
N 10	05 16	05 42	06 05	22 05	23 00	23 53	24 45
0	04 59	05 25	05 48	22 15	23 06	23 54	24 41
S 10	04 41	05 08	05 31	22 25	23 11	23 55	24 38
20	04 19	04 48	05 13	22 35	23 17	23 56	24 34
30	03 51	04 24	04 51	22 47	23 24	23 57	24 30
35	03 33	04 10	04 39	22 54	23 27	23 58	24 27
40	03 11	03 52	04 24	23 02	23 32	23 59	24 25
45	02 42	03 31	04 07	23 11	23 36	23 59	24 21
S 50	02 00	03 02	03 46	23 22	23 42	24 00	00 00
52	01 35	02 48	03 35	23 27	23 45	24 01	00 01
54	00 59	02 31	03 24	23 33	23 48	24 01	00 01
56	////	02 11	03 11	23 39	23 51	24 02	00 02
58	////	01 45	02 55	23 45	23 55	24 03	00 03
S 60	////	01 06	02 36	23 53	23 59	24 03	00 03

Lat.	Sunset	Twilight Civil	Twilight Naut.	Moonset 6	Moonset 7	Moonset 8	Moonset 9
°	h m	h m	h m	h m	h m	h m	h m
N 72	▬	13 25	15 36	13 41	12 53	12 21	11 53
N 70	▬	14 13	15 55	13 08	12 38	12 15	11 55
68	12 22	14 44	16 09	12 44	12 26	12 11	11 57
66	13 35	15 07	16 21	12 26	12 16	12 07	11 59
64	14 11	15 25	16 31	12 10	12 07	12 04	12 01
62	14 37	15 40	16 40	11 57	12 00	12 01	12 02
60	14 56	15 52	16 48	11 46	11 54	11 59	12 03
N 58	15 13	16 03	16 55	11 37	11 48	11 57	12 04
56	15 27	16 13	17 02	11 28	11 43	11 55	12 05
54	15 39	16 22	17 07	11 21	11 38	11 53	12 06
52	15 49	16 29	17 13	11 14	11 34	11 51	12 07
50	15 59	16 37	17 18	11 08	11 31	11 50	12 07
45	16 18	16 52	17 29	10 55	11 22	11 46	12 09
N 40	16 35	17 05	17 39	10 44	11 15	11 44	12 10
35	16 48	17 16	17 48	10 34	11 10	11 41	12 11
30	17 00	17 26	17 56	10 26	11 04	11 39	12 12
20	17 21	17 44	18 12	10 11	10 55	11 35	12 14
N 10	17 38	18 01	18 27	09 58	10 47	11 32	12 15
0	17 55	18 18	18 44	09 46	10 39	11 29	12 17
S 10	18 12	18 35	19 02	09 34	10 31	11 26	12 18
20	18 31	18 55	19 24	09 21	10 23	11 22	12 19
30	18 52	19 19	19 52	09 06	10 13	11 18	12 21
35	19 04	19 34	20 10	08 57	10 08	11 16	12 22
40	19 19	19 51	20 32	08 47	10 01	11 13	12 23
45	19 36	20 13	21 02	08 35	09 54	11 10	12 24
S 50	19 58	20 41	21 44	08 21	09 45	11 07	12 26
52	20 08	20 56	22 09	08 14	09 41	11 05	12 26
54	20 20	21 12	22 46	08 06	09 36	11 03	12 27
56	20 33	21 33	////	07 58	09 31	11 01	12 28
58	20 49	22 00	////	07 48	09 25	10 58	12 29
S 60	21 08	22 40	////	07 37	09 19	10 56	12 30

Day	SUN Eqn. of Time 00^h	SUN Eqn. of Time 12^h	SUN Mer. Pass.	MOON Mer. Pass. Upper	MOON Mer. Pass. Lower	MOON Age	MOON Phase
d	m s	m s	h m	h m	h m	d	%
6	09 07	08 55	11 51	03 33	16 01	20	77
7	08 42	08 29	11 52	04 27	16 53	21	67
8	08 16	08 03	11 52	05 17	17 42	22	56

	UT	ARIES	VENUS −3.9		MARS −0.3		JUPITER −2.2		SATURN +1.0	
		GHA	GHA	Dec	GHA	Dec	GHA	Dec	GHA	Dec
	d h	° ′	° ′	° ′	° ′	° ′	° ′	° ′	° ′	° ′
	9 00	77 52.3	190 35.9	S21 18.9	295 40.5	N17 49.5	113 03.0	S15 02.4	253 46.2	N 0 36.1
	01	92 54.8	205 35.1	19.5	310 42.6	49.5	128 05.0	02.2	268 48.6	36.1
	02	107 57.2	220 34.2	20.1	325 44.7	49.5	143 07.1	02.1	283 50.9	36.0
	03	122 59.7	235 33.3	. . 20.7	340 46.8	. . 49.5	158 09.2	. . 02.0	298 53.2	. . 36.0
	04	138 02.2	250 32.5	21.3	355 48.8	49.5	173 11.2	01.8	313 55.5	35.9
	05	153 04.6	265 31.6	21.8	10 50.9	49.5	188 13.3	01.7	328 57.8	35.9
	06	168 07.1	280 30.8	S21 22.4	25 53.0	N17 49.5	203 15.4	S15 01.6	344 00.2	N 0 35.8
W	07	183 09.5	295 29.9	23.0	40 55.1	49.4	218 17.4	01.4	359 02.5	35.8
E	08	198 12.0	310 29.0	23.5	55 57.2	49.4	233 19.5	01.3	14 04.8	35.7
D	09	213 14.5	325 28.2	. . 24.1	70 59.3	. . 49.4	248 21.6	. . 01.2	29 07.1	. . 35.7
N	10	228 16.9	340 27.3	24.7	86 01.4	49.4	263 23.6	01.0	44 09.4	35.6
E	11	243 19.4	355 26.5	25.2	101 03.5	49.4	278 25.7	00.9	59 11.8	35.6
S	12	258 21.9	10 25.6	S21 25.8	116 05.6	N17 49.4	293 27.8	S15 00.8	74 14.1	N 0 35.5
D	13	273 24.3	25 24.7	26.4	131 07.7	49.4	308 29.8	00.6	89 16.4	35.5
A	14	288 26.8	40 23.9	26.9	146 09.7	49.4	323 31.9	00.5	104 18.7	35.4
Y	15	303 29.3	55 23.0	. . 27.5	161 11.8	. . 49.4	338 34.0	. . 00.3	119 21.0	. . 35.4
	16	318 31.7	70 22.2	28.1	176 13.9	49.3	353 36.0	00.2	134 23.4	35.3
	17	333 34.2	85 21.3	28.6	191 16.0	49.3	8 38.1	15 00.1	149 25.7	35.3
	18	348 36.6	100 20.4	S21 29.2	206 18.1	N17 49.3	23 40.2	S14 59.9	164 28.0	N 0 35.2
	19	3 39.1	115 19.6	29.8	221 20.2	49.3	38 42.2	59.8	179 30.3	35.2
	20	18 41.6	130 18.7	30.3	236 22.3	49.3	53 44.3	59.7	194 32.7	35.1
	21	33 44.0	145 17.8	. . 30.9	251 24.4	. . 49.3	68 46.3	. . 59.5	209 35.0	. . 35.1
	22	48 46.5	160 17.0	31.4	266 26.6	49.3	83 48.4	59.4	224 37.3	35.0
	23	63 49.0	175 16.1	32.0	281 28.7	49.3	98 50.5	59.3	239 39.6	35.0
	10 00	78 51.4	190 15.2	S21 32.5	296 30.8	N17 49.3	113 52.5	S14 59.1	254 41.9	N 0 34.9
	01	93 53.9	205 14.4	33.1	311 32.9	49.3	128 54.6	59.0	269 44.3	34.9
	02	108 56.4	220 13.5	33.6	326 35.0	49.3	143 56.7	58.8	284 46.6	34.8
	03	123 58.8	235 12.6	. . 34.2	341 37.1	. . 49.3	158 58.7	. . 58.7	299 48.9	. . 34.8
	04	139 01.3	250 11.8	34.7	356 39.2	49.3	174 00.8	58.6	314 51.2	34.7
	05	154 03.8	265 10.9	35.3	11 41.3	49.3	189 02.9	58.4	329 53.6	34.7
	06	169 06.2	280 10.0	S21 35.8	26 43.4	N17 49.2	204 04.9	S14 58.3	344 55.9	N 0 34.6
T	07	184 08.7	295 09.2	36.4	41 45.5	49.2	219 07.0	58.2	359 58.2	34.6
H	08	199 11.1	310 08.3	36.9	56 47.7	49.2	234 09.0	58.0	15 00.5	34.5
U	09	214 13.6	325 07.4	. . 37.5	71 49.8	. . 49.2	249 11.1	. . 57.9	30 02.9	. . 34.5
R	10	229 16.1	340 06.6	38.0	86 51.9	49.2	264 13.2	57.7	45 05.2	34.4
S	11	244 18.5	355 05.7	38.5	101 54.0	49.2	279 15.2	57.6	60 07.5	34.4
D	12	259 21.0	10 04.8	S21 39.1	116 56.1	N17 49.2	294 17.3	S14 57.5	75 09.8	N 0 34.3
A	13	274 23.5	25 03.9	39.6	131 58.3	49.2	309 19.4	57.3	90 12.1	34.3
Y	14	289 25.9	40 03.1	40.2	147 00.4	49.2	324 21.4	57.2	105 14.5	34.2
	15	304 28.4	55 02.2	. . 40.7	162 02.5	. . 49.2	339 23.5	. . 57.1	120 16.8	. . 34.2
	16	319 30.9	70 01.3	41.2	177 04.6	49.2	354 25.5	56.9	135 19.1	34.1
	17	334 33.3	85 00.5	41.8	192 06.8	49.2	9 27.6	56.8	150 21.4	34.1
	18	349 35.8	99 59.6	S21 42.3	207 08.9	N17 49.2	24 29.7	S14 56.6	165 23.8	N 0 34.1
	19	4 38.3	114 58.7	42.8	222 11.0	49.2	39 31.7	56.5	180 26.1	34.0
	20	19 40.7	129 57.8	43.4	237 13.2	49.2	54 33.8	56.4	195 28.4	34.0
	21	34 43.2	144 57.0	. . 43.9	252 15.3	. . 49.2	69 35.8	. . 56.2	210 30.7	. . 33.9
	22	49 45.6	159 56.1	44.4	267 17.4	49.2	84 37.9	56.1	225 33.1	33.9
	23	64 48.1	174 55.2	45.0	282 19.6	49.2	99 40.0	55.9	240 35.4	33.8
	11 00	79 50.6	189 54.3	S21 45.5	297 21.7	N17 49.2	114 42.0	S14 55.8	255 37.7	N 0 33.8
	01	94 53.0	204 53.5	46.0	312 23.8	49.2	129 44.1	55.7	270 40.0	33.7
	02	109 55.5	219 52.6	46.5	327 26.0	49.2	144 46.1	55.5	285 42.4	33.7
	03	124 58.0	234 51.7	. . 47.1	342 28.1	. . 49.3	159 48.2	. . 55.4	300 44.7	. . 33.6
	04	140 00.4	249 50.8	47.6	357 30.2	49.3	174 50.2	55.3	315 47.0	33.6
	05	155 02.9	264 49.9	48.1	12 32.4	49.3	189 52.3	55.1	330 49.4	33.5
	06	170 05.4	279 49.1	S21 48.6	27 34.5	N17 49.3	204 54.4	S14 55.0	345 51.7	N 0 33.5
	07	185 07.8	294 48.2	49.1	42 36.7	49.3	219 56.4	54.8	0 54.0	33.4
	08	200 10.3	309 47.3	49.7	57 38.8	49.3	234 58.5	54.7	15 56.3	33.4
F	09	215 12.7	324 46.4	. . 50.2	72 41.0	. . 49.3	250 00.5	. . 54.6	30 58.7	. . 33.3
R	10	230 15.2	339 45.5	50.7	87 43.1	49.3	265 02.6	54.4	46 01.0	33.3
I	11	245 17.7	354 44.7	51.2	102 45.3	49.3	280 04.6	54.3	61 03.3	33.2
D	12	260 20.1	9 43.8	S21 51.7	117 47.4	N17 49.3	295 06.7	S14 54.1	76 05.6	N 0 33.2
A	13	275 22.6	24 42.9	52.2	132 49.6	49.3	310 08.8	54.0	91 08.0	33.1
Y	14	290 25.1	39 42.0	52.7	147 51.7	49.3	325 10.8	53.9	106 10.3	33.1
	15	305 27.5	54 41.1	. . 53.3	162 53.9	. . 49.3	340 12.9	. . 53.7	121 12.6	. . 33.1
	16	320 30.0	69 40.2	53.8	177 56.0	49.3	355 14.9	53.6	136 15.0	33.0
	17	335 32.5	84 39.4	54.3	192 58.2	49.4	10 17.0	53.4	151 17.3	33.0
	18	350 34.9	99 38.5	S21 54.8	208 00.3	N17 49.4	25 19.0	S14 53.3	166 19.6	N 0 32.9
	19	5 37.4	114 37.6	55.3	223 02.5	49.4	40 21.1	53.2	181 21.9	32.9
	20	20 39.9	129 36.7	55.8	238 04.6	49.4	55 23.1	53.0	196 24.3	32.8
	21	35 42.3	144 35.8	. . 56.3	253 06.8	. . 49.4	70 25.2	. . 52.9	211 26.6	. . 32.8
	22	50 44.8	159 34.9	56.8	268 09.0	49.4	85 27.3	52.7	226 28.9	32.7
	23	65 47.2	174 34.1	57.3	283 11.1	49.4	100 29.3	52.6	241 31.3	32.7
	Mer. Pass.	h m 18 41.5	*v* −0.9	*d* 0.5	*v* 2.1	*d* 0.0	*v* 2.1	*d* 0.1	*v* 2.3	*d* 0.0

STARS		
Name	SHA	Dec
	° ′	° ′
Acamar	315 19.9	S40 15.9
Achernar	335 28.2	S57 11.3
Acrux	173 12.7	S63 09.1
Adhara	255 14.2	S28 59.0
Aldebaran	290 52.0	N16 31.8
Alioth	166 23.0	N55 54.0
Alkaid	153 01.1	N49 15.5
Al Na'ir	27 47.0	S46 54.9
Alnilam	275 48.6	S 1 11.7
Alphard	217 58.5	S 8 42.1
Alphecca	126 13.5	N26 40.8
Alpheratz	357 46.2	N29 09.0
Altair	62 11.0	N 8 53.8
Ankaa	353 18.0	S42 15.2
Antares	112 29.8	S26 27.2
Arcturus	145 58.3	N19 07.7
Atria	107 34.6	S69 02.7
Avior	234 18.7	S59 32.3
Bellatrix	278 34.4	N 6 21.6
Betelgeuse	271 03.8	N 7 24.6
Canopus	263 56.8	S52 41.9
Capella	280 37.8	N46 00.5
Deneb	49 33.6	N45 19.2
Denebola	182 36.3	N14 30.8
Diphda	348 58.3	S17 55.9
Dubhe	193 54.6	N61 41.5
Elnath	278 15.5	N28 37.0
Eltanin	90 47.9	N51 29.3
Enif	33 49.8	N 9 55.4
Fomalhaut	15 26.8	S29 34.2
Gacrux	172 04.2	S57 09.9
Gienah	175 55.1	S17 35.8
Hadar	148 52.2	S60 25.1
Hamal	328 03.5	N23 30.8
Kaus Aust.	83 47.6	S34 22.8
Kochab	137 20.6	N74 06.6
Markab	13 41.0	N15 15.7
Menkar	314 17.5	N 4 07.8
Menkent	148 11.0	S36 25.0
Miaplacidus	221 40.0	S69 45.2
Mirfak	308 43.7	N49 54.0
Nunki	76 01.8	S26 17.1
Peacock	53 23.6	S56 42.3
Pollux	243 30.5	N28 00.0
Procyon	245 02.1	N 5 11.9
Rasalhague	96 09.2	N12 33.2
Regulus	207 46.1	N11 55.0
Rigel	281 14.2	S 8 11.4
Rigil Kent.	139 56.0	S60 52.4
Sabik	102 15.9	S15 44.2
Schedar	349 43.4	N56 35.9
Shaula	96 25.9	S37 06.7
Sirius	258 35.7	S16 43.7
Spica	158 34.2	S11 12.8
Suhail	222 54.2	S43 28.2
Vega	80 41.1	N38 47.7
Zuben'ubi	137 08.6	S16 05.0

	SHA	Mer. Pass.
	° ′	h m
Venus	111 23.8	11 20
Mars	217 39.3	4 13
Jupiter	35 01.1	16 22
Saturn	175 50.5	7 00

UT	SUN		MOON				
	GHA	Dec	GHA	v	Dec	d	HP
d h	° ′	° ′	° ′	′	° ′	′	′
9 00 WEDNESDAY	181 57.3	S22 48.4	271 34.5	12.2	N 1 12.9	15.0	58.7
01	196 57.0	48.7	286 05.7	12.2	0 57.9	15.1	58.7
02	211 56.7	48.9	300 36.9	12.3	0 42.8	15.1	58.7
03	226 56.4	. . 49.2	315 08.2	12.3	0 27.7	15.0	58.6
04	241 56.1	49.4	329 39.5	12.4	N 0 12.7	15.1	58.6
05	256 55.9	49.7	344 10.9	12.4	S 0 02.4	15.0	58.6
06	271 55.6	S22 49.9	358 42.3	12.3	S 0 17.4	15.0	58.6
07	286 55.3	50.1	13 13.6	12.5	0 32.4	15.0	58.5
08	301 55.0	50.4	27 45.1	12.4	0 47.4	15.0	58.5
09	316 54.7	. . 50.6	42 16.5	12.4	1 02.4	15.0	58.5
10	331 54.5	50.9	56 47.9	12.5	1 17.4	14.9	58.5
11	346 54.2	51.1	71 19.4	12.5	1 32.3	14.9	58.4
12	1 53.9	S22 51.3	85 50.9	12.5	S 1 47.2	14.9	58.4
13	16 53.6	51.6	100 22.4	12.5	2 02.1	14.9	58.4
14	31 53.3	51.8	114 53.9	12.6	2 17.0	14.9	58.4
15	46 53.1	. . 52.0	129 25.5	12.5	2 31.9	14.8	58.3
16	61 52.8	52.3	143 57.0	12.6	2 46.7	14.8	58.3
17	76 52.5	52.5	158 28.6	12.6	3 01.5	14.8	58.3
18	91 52.2	S22 52.7	173 00.2	12.6	S 3 16.3	14.8	58.3
19	106 51.9	53.0	187 31.8	12.5	3 31.1	14.7	58.2
20	121 51.7	53.2	202 03.3	12.6	3 45.8	14.7	58.2
21	136 51.4	. . 53.4	216 34.9	12.7	4 00.5	14.7	58.2
22	151 51.1	53.7	231 06.6	12.6	4 15.2	14.6	58.2
23	166 50.8	53.9	245 38.2	12.6	4 29.8	14.6	58.1
10 00 THURSDAY	181 50.5	S22 54.1	260 09.8	12.6	S 4 44.4	14.6	58.1
01	196 50.2	54.3	274 41.4	12.6	4 59.0	14.5	58.1
02	211 50.0	54.6	289 13.0	12.6	5 13.5	14.5	58.1
03	226 49.7	. . 54.8	303 44.6	12.7	5 28.0	14.5	58.0
04	241 49.4	55.0	318 16.3	12.6	5 42.5	14.4	58.0
05	256 49.1	55.2	332 47.9	12.6	5 56.9	14.4	58.0
06	271 48.8	S22 55.5	347 19.5	12.6	S 6 11.3	14.3	58.0
07	286 48.5	55.7	1 51.1	12.6	6 25.6	14.3	57.9
08	301 48.3	55.9	16 22.7	12.6	6 39.9	14.2	57.9
09	316 48.0	. . 56.1	30 54.3	12.6	6 54.1	14.3	57.9
10	331 47.7	56.4	45 25.9	12.6	7 08.4	14.1	57.9
11	346 47.4	56.6	59 57.5	12.6	7 22.5	14.1	57.9
12	1 47.1	S22 56.8	74 29.1	12.5	S 7 36.6	14.1	57.8
13	16 46.8	57.0	89 00.6	12.6	7 50.7	14.0	57.8
14	31 46.6	57.2	103 32.2	12.6	8 04.7	14.0	57.8
15	46 46.3	. . 57.4	118 03.8	12.5	8 18.7	13.9	57.8
16	61 46.0	57.7	132 35.3	12.5	8 32.6	13.9	57.7
17	76 45.7	57.9	147 06.8	12.5	8 46.5	13.8	57.7
18	91 45.4	S22 58.1	161 38.3	12.5	S 9 00.3	13.8	57.7
19	106 45.1	58.3	176 09.8	12.5	9 14.1	13.7	57.7
20	121 44.8	58.5	190 41.3	12.4	9 27.8	13.6	57.6
21	136 44.6	. . 58.7	205 12.7	12.5	9 41.4	13.6	57.6
22	151 44.3	58.9	219 44.2	12.4	9 55.0	13.5	57.6
23	166 44.0	59.1	234 15.6	12.4	10 08.5	13.5	57.6
11 00 FRIDAY	181 43.7	S22 59.3	248 47.0	12.4	S10 22.0	13.4	57.5
01	196 43.4	59.6	263 18.4	12.3	10 35.4	13.4	57.5
02	211 43.1	22 59.8	277 49.7	12.3	10 48.8	13.3	57.5
03	226 42.8	23 00.0	292 21.0	12.3	11 02.1	13.2	57.5
04	241 42.5	00.2	306 52.3	12.3	11 15.3	13.2	57.5
05	256 42.3	00.4	321 23.6	12.3	11 28.5	13.1	57.4
06	271 42.0	S23 00.6	335 54.9	12.2	S11 41.6	13.0	57.4
07	286 41.7	00.8	350 26.1	12.2	11 54.6	13.0	57.4
08	301 41.4	01.0	4 57.3	12.2	12 07.6	12.9	57.4
09	316 41.1	. . 01.2	19 28.5	12.1	12 20.5	12.8	57.3
10	331 40.8	01.4	33 59.6	12.2	12 33.3	12.8	57.3
11	346 40.5	01.6	48 30.8	12.1	12 46.1	12.7	57.3
12	1 40.2	S23 01.8	63 01.9	12.0	S12 58.8	12.6	57.3
13	16 39.9	02.0	77 32.9	12.0	13 11.4	12.5	57.2
14	31 39.7	02.2	92 03.9	12.0	13 23.9	12.5	57.2
15	46 39.4	. . 02.4	106 34.9	12.0	13 36.4	12.4	57.2
16	61 39.1	02.6	121 05.9	11.9	13 48.8	12.3	57.2
17	76 38.8	02.8	135 36.8	11.9	14 01.1	12.3	57.2
18	91 38.5	S23 03.0	150 07.7	11.9	S14 13.4	12.1	57.1
19	106 38.2	03.2	164 38.6	11.8	14 25.5	12.1	57.1
20	121 37.9	03.3	179 09.4	11.8	14 37.6	12.0	57.1
21	136 37.6	. . 03.5	193 40.2	11.8	14 49.6	12.0	57.1
22	151 37.3	03.7	208 11.0	11.7	15 01.6	11.8	57.0
23	166 37.0	03.9	222 41.7	11.7	S15 13.4	11.8	57.0
	SD 16.3	*d* 0.2	SD 15.9		15.8		15.6

Lat.	Twilight Naut.	Twilight Civil	Sunrise	Moonrise 9	Moonrise 10	Moonrise 11	Moonrise 12
°	h m	h m	h m	h m	h m	h m	h m
N 72	08 12	10 30	■	25 50	01 50	04 06	07 01
N 70	07 53	09 38	■	25 42	01 42	03 46	06 05
68	07 38	09 05	■	25 35	01 35	03 30	05 32
66	07 26	08 41	10 16	25 30	01 30	03 17	05 09
64	07 15	08 23	09 38	25 25	01 25	03 07	04 50
62	07 06	08 07	09 11	25 22	01 22	02 58	04 35
60	06 58	07 54	08 51	25 18	01 18	02 50	04 22
N 58	06 51	07 43	08 34	25 15	01 15	02 43	04 12
56	06 44	07 33	08 20	25 13	01 13	02 38	04 02
54	06 38	07 24	08 08	25 10	01 10	02 32	03 54
52	06 33	07 16	07 57	25 08	01 08	02 28	03 46
50	06 28	07 09	07 47	25 06	01 06	02 23	03 40
45	06 17	06 54	07 27	25 02	01 02	02 14	03 26
N 40	06 07	06 40	07 11	24 59	00 59	02 07	03 14
35	05 57	06 29	06 57	24 56	00 56	02 00	03 04
30	05 49	06 19	06 45	24 53	00 53	01 54	02 56
20	05 33	06 00	06 24	24 49	00 49	01 45	02 41
N 10	05 17	05 43	06 06	24 45	00 45	01 36	02 28
0	05 01	05 27	05 49	24 41	00 41	01 28	02 16
S 10	04 42	05 09	05 32	24 38	00 38	01 20	02 04
20	04 20	04 49	05 13	24 34	00 34	01 12	01 52
30	03 51	04 25	04 52	24 30	00 30	01 03	01 38
35	03 33	04 10	04 39	24 27	00 27	00 57	01 29
40	03 11	03 52	04 25	24 25	00 25	00 51	01 20
45	02 41	03 30	04 07	24 21	00 21	00 44	01 09
S 50	01 58	03 01	03 45	00 00	00 18	00 36	00 56
52	01 32	02 47	03 35	00 01	00 16	00 32	00 50
54	00 53	02 30	03 23	00 01	00 14	00 28	00 43
56	////	02 08	03 09	00 02	00 12	00 23	00 36
58	////	01 41	02 53	00 03	00 10	00 18	00 28
S 60	////	00 59	02 34	00 03	00 07	00 12	00 18

Lat.	Sunset	Twilight Civil	Twilight Naut.	Moonset 9	Moonset 10	Moonset 11	Moonset 12
°	h m	h m	h m	h m	h m	h m	h m
N 72	■	13 15	15 33	11 53	11 24	10 47	09 34
N 70	■	14 08	15 52	11 55	11 35	11 10	10 31
68	■	14 40	16 07	11 57	11 43	11 27	11 06
66	13 30	15 04	16 19	11 59	11 51	11 42	11 31
64	14 08	15 23	16 30	12 01	11 57	11 54	11 50
62	14 34	15 38	16 39	12 02	12 03	12 04	12 06
60	14 55	15 51	16 47	12 03	12 07	12 13	12 20
N 58	15 11	16 02	16 55	12 04	12 12	12 20	12 31
56	15 25	16 12	17 01	12 05	12 15	12 27	12 42
54	15 38	16 21	17 07	12 06	12 19	12 33	12 51
52	15 49	16 29	17 12	12 07	12 22	12 39	12 59
50	15 58	16 36	17 18	12 07	12 25	12 44	13 06
45	16 18	16 52	17 29	12 09	12 31	12 55	13 21
N 40	16 35	17 05	17 39	12 10	12 36	13 04	13 34
35	16 49	17 17	17 48	12 11	12 41	13 12	13 45
30	17 01	17 27	17 57	12 12	12 45	13 19	13 55
20	17 21	17 45	18 13	12 14	12 52	13 30	14 11
N 10	17 39	18 02	18 29	12 15	12 58	13 41	14 26
0	17 57	18 19	18 45	12 17	13 04	13 51	14 40
S 10	18 14	18 37	19 04	12 18	13 09	14 01	14 53
20	18 32	18 57	19 26	12 19	13 15	14 12	15 08
30	18 54	19 21	19 54	12 21	13 22	14 24	15 25
35	19 07	19 36	20 13	12 22	13 27	14 31	15 35
40	19 21	19 54	20 35	12 23	13 31	14 39	15 46
45	19 39	20 16	21 05	12 24	13 37	14 48	16 00
S 50	20 01	20 45	21 49	12 26	13 43	15 00	16 16
52	20 12	21 00	22 15	12 26	13 46	15 05	16 24
54	20 23	21 17	22 55	12 27	13 49	15 11	16 32
56	20 37	21 38	////	12 28	13 53	15 18	16 42
58	20 53	22 06	////	12 29	13 57	15 25	16 53
S 60	21 13	22 49	////	12 30	14 02	15 34	17 06

Day	SUN Eqn. of Time 00^h	SUN Eqn. of Time 12^h	SUN Mer. Pass.	MOON Mer. Pass. Upper	MOON Mer. Pass. Lower	MOON Age	MOON Phase
d	m s	m s	h m	h m	h m	d %	
9	07 50	07 36	11 52	06 05	18 29	23 45	
10	07 23	07 09	11 53	06 52	19 16	24 34	
11	06 55	06 42	11 53	07 40	20 03	25 24	

UT	ARIES	VENUS −3.9		MARS −0.3		JUPITER −2.2		SATURN +1.0	
	GHA	GHA	Dec	GHA	Dec	GHA	Dec	GHA	Dec
d h	° ′	° ′	° ′	° ′	° ′	° ′	° ′	° ′	° ′
12 00	80 49.7	189 33.2	S21 57.8	298 13.3	N17 49.4	115 31.4	S14 52.5	256 33.6	N 0 32.6
01	95 52.2	204 32.3	58.3	313 15.5	49.4	130 33.4	52.3	271 35.9	32.6
02	110 54.6	219 31.4	58.8	328 17.6	49.5	145 35.5	52.2	286 38.2	32.5
03	125 57.1	234 30.5	59.3	343 19.8	49.5	160 37.5	52.0	301 40.6	32.5
04	140 59.6	249 29.6	21 59.8	358 22.0	49.5	175 39.6	51.9	316 42.9	32.4
05	156 02.0	264 28.7	22 00.3	13 24.1	49.5	190 41.6	51.7	331 45.2	32.4
06	171 04.5	279 27.8	S22 00.8	28 26.3	N17 49.5	205 43.7	S14 51.6	346 47.6	N 0 32.4
S 07	186 07.0	294 27.0	01.3	43 28.5	49.5	220 45.7	51.5	1 49.9	32.3
A 08	201 09.4	309 26.1	01.8	58 30.6	49.5	235 47.8	51.3	16 52.2	32.3
T 09	216 11.9	324 25.2	02.2	73 32.8	49.6	250 49.8	51.2	31 54.6	32.2
U 10	231 14.4	339 24.3	02.7	88 35.0	49.6	265 51.9	51.0	46 56.9	32.2
R 11	246 16.8	354 23.4	03.2	103 37.2	49.6	280 53.9	50.9	61 59.2	32.1
D 12	261 19.3	9 22.5	S22 03.7	118 39.4	N17 49.6	295 56.0	S14 50.8	77 01.6	N 0 32.1
A 13	276 21.7	24 21.6	04.2	133 41.5	49.6	310 58.1	50.6	92 03.9	32.0
Y 14	291 24.2	39 20.7	04.7	148 43.7	49.7	326 00.1	50.5	107 06.2	32.0
15	306 26.7	54 19.8	05.2	163 45.9	49.7	341 02.2	50.3	122 08.5	31.9
16	321 29.1	69 18.9	05.6	178 48.1	49.7	356 04.2	50.2	137 10.9	31.9
17	336 31.6	84 18.0	06.1	193 50.3	49.7	11 06.3	50.0	152 13.2	31.9
18	351 34.1	99 17.1	S22 06.6	208 52.5	N17 49.7	26 08.3	S14 49.9	167 15.5	N 0 31.8
19	6 36.5	114 16.3	07.1	223 54.6	49.8	41 10.4	49.8	182 17.9	31.8
20	21 39.0	129 15.4	07.6	238 56.8	49.8	56 12.4	49.6	197 20.2	31.7
21	36 41.5	144 14.5	08.0	253 59.0	49.8	71 14.5	49.5	212 22.5	31.7
22	51 43.9	159 13.6	08.5	269 01.2	49.8	86 16.5	49.3	227 24.9	31.6
23	66 46.4	174 12.7	09.0	284 03.4	49.8	101 18.6	49.2	242 27.2	31.6
13 00	81 48.9	189 11.8	S22 09.5	299 05.6	N17 49.9	116 20.6	S14 49.1	257 29.5	N 0 31.5
01	96 51.3	204 10.9	09.9	314 07.8	49.9	131 22.7	48.9	272 31.9	31.5
02	111 53.8	219 10.0	10.4	329 10.0	49.9	146 24.7	48.8	287 34.2	31.4
03	126 56.2	234 09.1	10.9	344 12.2	49.9	161 26.8	48.6	302 36.5	31.4
04	141 58.7	249 08.2	11.3	359 14.4	50.0	176 28.8	48.5	317 38.9	31.4
05	157 01.2	264 07.3	11.8	14 16.6	50.0	191 30.8	48.3	332 41.2	31.3
06	172 03.6	279 06.4	S22 12.3	29 18.8	N17 50.0	206 32.9	S14 48.2	347 43.5	N 0 31.3
07	187 06.1	294 05.5	12.7	44 21.0	50.0	221 34.9	48.1	2 45.9	31.2
08	202 08.6	309 04.6	13.2	59 23.2	50.1	236 37.0	47.9	17 48.2	31.2
S 09	217 11.0	324 03.7	13.7	74 25.4	50.1	251 39.0	47.8	32 50.5	31.1
U 10	232 13.5	339 02.8	14.1	89 27.6	50.1	266 41.1	47.6	47 52.9	31.1
N 11	247 16.0	354 01.9	14.6	104 29.8	50.1	281 43.1	47.5	62 55.2	31.1
D 12	262 18.4	9 01.0	S22 15.0	119 32.0	N17 50.2	296 45.2	S14 47.3	77 57.5	N 0 31.0
A 13	277 20.9	24 00.1	15.5	134 34.2	50.2	311 47.2	47.2	92 59.9	31.0
Y 14	292 23.3	38 59.2	16.0	149 36.4	50.2	326 49.3	47.0	108 02.2	30.9
15	307 25.8	53 58.3	16.4	164 38.6	50.2	341 51.3	46.9	123 04.5	30.9
16	322 28.3	68 57.4	16.9	179 40.9	50.3	356 53.4	46.8	138 06.9	30.8
17	337 30.7	83 56.5	17.3	194 43.1	50.3	11 55.4	46.6	153 09.2	30.8
18	352 33.2	98 55.6	S22 17.8	209 45.3	N17 50.3	26 57.5	S14 46.5	168 11.6	N 0 30.7
19	7 35.7	113 54.7	18.2	224 47.5	50.4	41 59.5	46.3	183 13.9	30.7
20	22 38.1	128 53.8	18.7	239 49.7	50.4	57 01.6	46.2	198 16.2	30.7
21	37 40.6	143 52.9	19.1	254 51.9	50.4	72 03.6	46.0	213 18.6	30.6
22	52 43.1	158 52.0	19.6	269 54.2	50.5	87 05.6	45.9	228 20.9	30.6
23	67 45.5	173 51.1	20.0	284 56.4	50.5	102 07.7	45.8	243 23.2	30.5
14 00	82 48.0	188 50.2	S22 20.5	299 58.6	N17 50.5	117 09.7	S14 45.6	258 25.6	N 0 30.5
01	97 50.5	203 49.3	20.9	315 00.8	50.6	132 11.8	45.5	273 27.9	30.4
02	112 52.9	218 48.4	21.3	330 03.0	50.6	147 13.8	45.3	288 30.2	30.4
03	127 55.4	233 47.4	21.8	345 05.3	50.6	162 15.9	45.2	303 32.6	30.4
04	142 57.8	248 46.5	22.2	0 07.5	50.7	177 17.9	45.0	318 34.9	30.3
05	158 00.3	263 45.6	22.7	15 09.7	50.7	192 20.0	44.9	333 37.3	30.3
06	173 02.8	278 44.7	S22 23.1	30 12.0	N17 50.7	207 22.0	S14 44.7	348 39.6	N 0 30.2
07	188 05.2	293 43.8	23.5	45 14.2	50.8	222 24.0	44.6	3 41.9	30.2
08	203 07.7	308 42.9	24.0	60 16.4	50.8	237 26.1	44.4	18 44.3	30.1
M 09	218 10.2	323 42.0	24.4	75 18.7	50.8	252 28.1	44.3	33 46.6	30.1
O 10	233 12.6	338 41.1	24.8	90 20.9	50.9	267 30.2	44.2	48 48.9	30.1
N 11	248 15.1	353 40.2	25.3	105 23.1	50.9	282 32.2	44.0	63 51.3	30.0
D 12	263 17.6	8 39.3	S22 25.7	120 25.4	N17 50.9	297 34.3	S14 43.9	78 53.6	N 0 30.0
A 13	278 20.0	23 38.4	26.1	135 27.6	51.0	312 36.3	43.7	93 56.0	29.9
Y 14	293 22.5	38 37.5	26.6	150 29.8	51.0	327 38.3	43.6	108 58.3	29.9
15	308 25.0	53 36.5	27.0	165 32.1	51.1	342 40.4	43.4	124 00.6	29.8
16	323 27.4	68 35.6	27.4	180 34.3	51.1	357 42.4	43.3	139 03.0	29.8
17	338 29.9	83 34.7	27.9	195 36.6	51.1	12 44.5	43.1	154 05.3	29.8
18	353 32.3	98 33.8	S22 28.3	210 38.8	N17 51.2	27 46.5	S14 43.0	169 07.7	N 0 29.7
19	8 34.8	113 32.9	28.7	225 41.1	51.2	42 48.6	42.8	184 10.0	29.7
20	23 37.3	128 32.0	29.1	240 43.3	51.3	57 50.6	42.7	199 12.3	29.6
21	38 39.7	143 31.1	29.5	255 45.6	51.3	72 52.6	42.6	214 14.7	29.6
22	53 42.2	158 30.2	30.0	270 47.8	51.3	87 54.7	42.4	229 17.0	29.6
23	68 44.7	173 29.3	30.4	285 50.1	51.4	102 56.7	42.3	244 19.4	29.5
Mer. Pass.	h m 18 29.7	v −0.9	d 0.5	v 2.2	d 0.0	v 2.0	d 0.1	v 2.3	d 0.0

STARS

Name	SHA	Dec
	° ′	° ′
Acamar	315 19.9	S40 15.[illegible]
Achernar	335 28.2	S57 11.3
Acrux	173 12.6	S63 09.1
Adhara	255 14.2	S28 59.1
Aldebaran	290 52.0	N16 31.8
Alioth	166 23.0	N55 54.0
Alkaid	153 01.1	N49 15.5
Al Na'ir	27 47.0	S46 54.9
Alnilam	275 48.6	S 1 11.7
Alphard	217 58.5	S 8 42.1
Alphecca	126 13.5	N26 40.7
Alpheratz	357 46.2	N29 09.0
Altair	62 11.0	N 8 53.8
Ankaa	353 18.0	S42 15.2
Antares	112 29.8	S26 27.2
Arcturus	145 58.3	N19 07.7
Atria	107 34.6	S69 02.7
Avior	234 18.7	S59 32.3
Bellatrix	278 34.4	N 6 21.6
Betelgeuse	271 03.7	N 7 24.6
Canopus	263 56.8	S52 42.0
Capella	280 37.8	N46 00.5
Deneb	49 33.6	N45 19.2
Denebola	182 36.3	N14 30.8
Diphda	348 58.3	S17 55.9
Dubhe	193 54.5	N61 41.5
Elnath	278 15.5	N28 37.0
Eltanin	90 47.9	N51 29.3
Enif	33 49.8	N 9 55.4
Fomalhaut	15 26.8	S29 34.2
Gacrux	172 04.1	S57 09.9
Gienah	175 55.1	S17 35.8
Hadar	148 52.2	S60 25.1
Hamal	328 03.5	N23 30.8
Kaus Aust.	83 47.6	S34 22.8
Kochab	137 20.6	N74 06.6
Markab	13 41.0	N15 15.7
Menkar	314 17.5	N 4 07.8
Menkent	148 11.0	S36 25.0
Miaplacidus	221 39.9	S69 45.3
Mirfak	308 43.7	N49 54.0
Nunki	76 01.8	S26 17.1
Peacock	53 23.6	S56 42.3
Pollux	243 30.5	N28 00.0
Procyon	245 02.1	N 5 11.9
Rasalhague	96 09.2	N12 33.2
Regulus	207 46.1	N11 55.0
Rigel	281 14.2	S 8 11.4
Rigil Kent.	139 55.9	S60 52.4
Sabik	102 15.9	S15 44.2
Schedar	349 43.5	N56 35.9
Shaula	96 25.9	S37 06.7
Sirius	258 35.7	S16 43.7
Spica	158 34.2	S11 12.8
Suhail	222 54.2	S43 28.2
Vega	80 41.1	N38 47.6
Zuben'ubi	137 08.6	S16 05.0

	SHA	Mer. Pass.
	° ′	h m
Venus	107 22.9	11 24
Mars	217 16.7	4 03
Jupiter	34 31.8	16 12
Saturn	175 40.7	6 49

UT	SUN GHA	SUN Dec	MOON GHA	v	MOON Dec	d	HP
d h	° ′	° ′	° ′	′	° ′	′	′
12 00	181 36.8	S23 04.1	237 12.4	11.6	S15 25.2	11.6	57.0
01	196 36.5	04.3	251 43.0	11.6	15 36.8	11.6	57.0
02	211 36.2	04.5	266 13.6	11.6	15 48.4	11.5	56.9
03	226 35.9	. . 04.7	280 44.2	11.5	15 59.9	11.5	56.9
04	241 35.6	04.9	295 14.7	11.5	16 11.4	11.3	56.9
05	256 35.3	05.0	309 45.2	11.5	16 22.7	11.2	56.9
06	271 35.0	S23 05.2	324 15.7	11.4	S16 33.9	11.2	56.9
S 07	286 34.7	05.4	338 46.1	11.4	16 45.1	11.1	56.8
A 08	301 34.4	05.6	353 16.5	11.3	16 56.2	10.9	56.8
T 09	316 34.1	. . 05.8	7 46.8	11.3	17 07.1	10.9	56.8
U 10	331 33.8	06.0	22 17.1	11.2	17 18.0	10.8	56.8
R 11	346 33.5	06.1	36 47.3	11.2	17 28.8	10.7	56.8
D 12	1 33.2	S23 06.3	51 17.5	11.2	S17 39.5	10.6	56.7
A 13	16 33.0	06.5	65 47.7	11.1	17 50.1	10.5	56.7
Y 14	31 32.7	06.7	80 17.8	11.1	18 00.6	10.4	56.7
15	46 32.4	. . 06.9	94 47.9	11.0	18 11.0	10.3	56.7
16	61 32.1	07.0	109 17.9	11.0	18 21.3	10.2	56.6
17	76 31.8	07.2	123 47.9	11.0	18 31.5	10.1	56.6
18	91 31.5	S23 07.4	138 17.9	10.9	S18 41.6	10.0	56.6
19	106 31.2	07.6	152 47.8	10.9	18 51.6	9.9	56.6
20	121 30.9	07.7	167 17.7	10.8	19 01.5	9.9	56.6
21	136 30.6	. . 07.9	181 47.5	10.8	19 11.4	9.7	56.5
22	151 30.3	08.1	196 17.3	10.7	19 21.1	9.6	56.5
23	166 30.0	08.3	210 47.0	10.7	19 30.7	9.4	56.5
13 00	181 29.7	S23 08.4	225 16.7	10.7	S19 40.1	9.4	56.5
01	196 29.4	08.6	239 46.4	10.6	19 49.5	9.3	56.5
02	211 29.1	08.8	254 16.0	10.5	19 58.8	9.2	56.4
03	226 28.8	. . 08.9	268 45.5	10.6	20 08.0	9.1	56.4
04	241 28.5	09.1	283 15.1	10.4	20 17.1	8.9	56.4
05	256 28.2	09.3	297 44.5	10.5	20 26.0	8.9	56.4
06	271 27.9	S23 09.4	312 14.0	10.4	S20 34.9	8.7	56.3
07	286 27.7	09.6	326 43.4	10.3	20 43.6	8.6	56.3
08	301 27.4	09.8	341 12.7	10.4	20 52.2	8.6	56.3
S 09	316 27.1	. . 09.9	355 42.1	10.2	21 00.8	8.4	56.3
U 10	331 26.8	10.1	10 11.3	10.3	21 09.2	8.2	56.3
N 11	346 26.5	10.2	24 40.6	10.2	21 17.4	8.2	56.2
D 12	1 26.2	S23 10.4	39 09.8	10.1	S21 25.6	8.1	56.2
A 13	16 25.9	10.6	53 38.9	10.1	21 33.7	7.9	56.2
Y 14	31 25.6	10.7	68 08.0	10.1	21 41.6	7.9	56.2
15	46 25.3	. . 10.9	82 37.1	10.0	21 49.5	7.7	56.2
16	61 25.0	11.0	97 06.1	10.0	21 57.2	7.6	56.1
17	76 24.7	11.2	111 35.1	10.0	22 04.8	7.4	56.1
18	91 24.4	S23 11.4	126 04.1	9.9	S22 12.2	7.4	56.1
19	106 24.1	11.5	140 33.0	9.9	22 19.6	7.2	56.1
20	121 23.8	11.7	155 01.9	9.8	22 26.8	7.1	56.1
21	136 23.5	. . 11.8	169 30.7	9.8	22 33.9	7.0	56.0
22	151 23.2	12.0	183 59.5	9.8	22 40.9	6.9	56.0
23	166 22.9	12.1	198 28.3	9.7	22 47.8	6.7	56.0
14 00	181 22.6	S23 12.3	212 57.0	9.7	S22 54.5	6.7	56.0
01	196 22.3	12.4	227 25.7	9.7	23 01.2	6.5	56.0
02	211 22.0	12.6	241 54.4	9.6	23 07.7	6.3	55.9
03	226 21.7	. . 12.7	256 23.0	9.6	23 14.0	6.3	55.9
04	241 21.4	12.9	270 51.6	9.6	23 20.3	6.1	55.9
05	256 21.1	13.0	285 20.2	9.6	23 26.4	6.0	55.9
06	271 20.8	S23 13.2	299 48.8	9.5	S23 32.4	5.9	55.9
07	286 20.5	13.3	314 17.3	9.4	23 38.3	5.7	55.8
08	301 20.2	13.5	328 45.7	9.5	23 44.0	5.7	55.8
M 09	316 19.9	. . 13.6	343 14.2	9.4	23 49.7	5.5	55.8
O 10	331 19.6	13.7	357 42.6	9.4	23 55.2	5.3	55.8
N 11	346 19.3	13.9	12 11.0	9.4	24 00.5	5.2	55.8
D 12	1 19.0	S23 14.0	26 39.4	9.3	S24 05.7	5.2	55.8
A 13	16 18.7	14.2	41 07.7	9.4	24 10.9	4.9	55.7
Y 14	31 18.4	14.3	55 36.1	9.3	24 15.8	4.9	55.7
15	46 18.1	. . 14.4	70 04.4	9.2	24 20.7	4.7	55.7
16	61 17.8	14.6	84 32.6	9.3	24 25.4	4.6	55.7
17	76 17.5	14.7	99 00.9	9.2	24 30.0	4.4	55.7
18	91 17.2	S23 14.9	113 29.1	9.3	S24 34.4	4.3	55.6
19	106 16.9	15.0	127 57.4	9.2	24 38.7	4.2	55.6
20	121 16.6	15.1	142 25.6	9.1	24 42.9	4.1	55.6
21	136 16.3	. . 15.3	156 53.7	9.2	24 47.0	3.9	55.6
22	151 16.0	15.4	171 21.9	9.2	24 50.9	3.8	55.6
23	166 15.7	15.5	185 50.1	9.1	S24 54.7	3.6	55.5
	SD 16.3	*d* 0.2	SD 15.5		15.3		15.2

Lat.	Twilight Naut.	Twilight Civil	Sunrise	Moonrise 12	Moonrise 13	Moonrise 14	Moonrise 15
°	h m	h m	h m	h m	h m	h m	h m
N 72	08 17	10 41	▬	07 01	▬	▬	▬
N 70	07 58	09 44	▬	06 05	▬	▬	▬
68	07 42	09 10	▬	05 32	08 02	▬	▬
66	07 30	08 46	10 23	05 09	07 09	▬	▬
64	07 19	08 27	09 43	04 50	06 37	08 27	10 16
62	07 09	08 11	09 16	04 35	06 13	07 49	09 14
60	07 01	07 58	08 55	04 22	05 54	07 23	08 40
N 58	06 54	07 46	08 38	04 12	05 39	07 02	08 15
56	06 47	07 36	08 23	04 02	05 26	06 45	07 56
54	06 41	07 27	08 11	03 54	05 14	06 31	07 39
52	06 36	07 19	08 00	03 46	05 04	06 18	07 25
50	06 30	07 12	07 50	03 40	04 55	06 07	07 12
45	06 19	06 56	07 30	03 26	04 36	05 44	06 47
N 40	06 09	06 43	07 13	03 14	04 21	05 26	06 27
35	05 59	06 31	06 59	03 04	04 08	05 10	06 10
30	05 51	06 21	06 47	02 56	03 57	04 57	05 55
20	05 35	06 02	06 26	02 41	03 38	04 35	05 31
N 10	05 19	05 45	06 08	02 28	03 21	04 15	05 09
0	05 02	05 28	05 50	02 16	03 06	03 57	04 50
S 10	04 43	05 10	05 33	02 04	02 50	03 39	04 30
20	04 21	04 50	05 14	01 52	02 34	03 20	04 09
30	03 52	04 25	04 53	01 38	02 15	02 58	03 44
35	03 34	04 10	04 40	01 29	02 05	02 45	03 30
40	03 11	03 52	04 25	01 20	01 52	02 30	03 13
45	02 41	03 30	04 07	01 09	01 38	02 12	02 54
S 50	01 56	03 01	03 45	00 56	01 20	01 51	02 29
52	01 30	02 46	03 34	00 50	01 12	01 40	02 17
54	00 47	02 28	03 22	00 43	01 03	01 29	02 04
56	////	02 07	03 08	00 36	00 53	01 16	01 48
58	////	01 38	02 52	00 28	00 41	01 00	01 30
S 60	////	00 52	02 32	00 18	00 27	00 42	01 07

Lat.	Sunset	Twilight Civil	Twilight Naut.	Moonset 12	Moonset 13	Moonset 14	Moonset 15
°	h m	h m	h m	h m	h m	h m	h m
N 72	▬	13 07	15 31	09 34	▬	▬	▬
N 70	▬	14 04	15 50	10 31	▬	▬	▬
68	▬	14 38	16 06	11 06	10 20	▬	▬
66	13 25	15 02	16 19	11 31	11 14	▬	▬
64	14 05	15 21	16 29	11 50	11 47	11 45	11 46
62	14 32	15 37	16 39	12 06	12 12	12 23	12 48
60	14 53	15 51	16 47	12 20	12 31	12 50	13 22
N 58	15 10	16 02	16 54	12 31	12 47	13 11	13 47
56	15 25	16 12	17 01	12 42	13 01	13 28	14 07
54	15 37	16 21	17 07	12 51	13 13	13 43	14 23
52	15 48	16 29	17 13	12 59	13 23	13 56	14 38
50	15 58	16 36	17 18	13 06	13 33	14 07	14 50
45	16 19	16 52	17 29	13 21	13 53	14 31	15 16
N 40	16 35	17 06	17 40	13 34	14 09	14 49	15 36
35	16 49	17 17	17 49	13 45	14 23	15 05	15 53
30	17 01	17 28	17 58	13 55	14 35	15 19	16 08
20	17 22	17 46	18 14	14 11	14 55	15 43	16 33
N 10	17 41	18 04	18 30	14 26	15 13	16 03	16 55
0	17 58	18 20	18 47	14 40	15 30	16 22	17 15
S 10	18 15	18 38	19 05	14 53	15 47	16 41	17 35
20	18 34	18 59	19 28	15 08	16 05	17 02	17 57
30	18 56	19 23	19 57	15 25	16 26	17 25	18 22
35	19 09	19 39	20 15	15 35	16 38	17 39	18 37
40	19 24	19 57	20 38	15 46	16 52	17 55	18 54
45	19 42	20 19	21 08	16 00	17 09	18 15	19 14
S 50	20 04	20 48	21 53	16 16	17 30	18 39	19 40
52	20 15	21 03	22 20	16 24	17 40	18 51	19 52
54	20 27	21 21	23 03	16 32	17 51	19 04	20 07
56	20 41	21 42	////	16 42	18 04	19 19	20 23
58	20 57	22 11	////	16 53	18 19	19 38	20 43
S 60	21 17	22 58	////	17 06	18 36	20 00	21 09

Day	SUN Eqn. of Time 00^h	SUN Eqn. of Time 12^h	SUN Mer. Pass.	MOON Mer. Pass. Upper	MOON Mer. Pass. Lower	Age	Phase
d	m s	m s	h m	h m	h m	d %	
12	06 28	06 14	11 54	08 28	20 53	26 16	
13	05 59	05 45	11 54	09 18	21 43	27 9	
14	05 31	05 17	11 55	10 09	22 36	28 4	

Day	UT d h	ARIES GHA ° ′	VENUS −3.9 GHA ° ′	VENUS Dec ° ′	MARS −0.4 GHA ° ′	MARS Dec ° ′	JUPITER −2.2 GHA ° ′	JUPITER Dec ° ′	SATURN +1.0 GHA ° ′	SATURN Dec ° ′
TUESDAY	15 00	83 47.1	188 28.3	S22 30.8	300 52.3	N17 51.4	117 58.8	S14 42.1	259 21.7	N 0 29.5
	01	98 49.6	203 27.4	31.2	315 54.6	51.5	133 00.8	42.0	274 24.0	29.4
	02	113 52.1	218 26.5	31.6	330 56.8	51.5	148 02.8	41.8	289 26.4	29.4
	03	128 54.5	233 25.6	32.0	345 59.1	51.6	163 04.9	41.7	304 28.7	29.3
	04	143 57.0	248 24.7	32.5	1 01.3	51.6	178 06.9	41.5	319 31.1	29.3
	05	158 59.5	263 23.8	32.9	16 03.6	51.6	193 09.0	41.4	334 33.4	29.3
	06	174 01.9	278 22.9	S22 33.3	31 05.9	N17 51.7	208 11.0	S14 41.2	349 35.7	N 0 29.2
	07	189 04.4	293 21.9	33.7	46 08.1	51.7	223 13.0	41.1	4 38.1	29.2
	08	204 06.8	308 21.0	34.1	61 10.4	51.8	238 15.1	40.9	19 40.4	29.1
	09	219 09.3	323 20.1	34.5	76 12.7	51.8	253 17.1	40.8	34 42.8	29.1
	10	234 11.8	338 19.2	34.9	91 14.9	51.9	268 19.1	40.7	49 45.1	29.1
	11	249 14.2	353 18.3	35.3	106 17.2	51.9	283 21.2	40.5	64 47.5	29.0
	12	264 16.7	8 17.4	S22 35.7	121 19.5	N17 52.0	298 23.2	S14 40.4	79 49.8	N 0 29.0
	13	279 19.2	23 16.4	36.1	136 21.7	52.0	313 25.3	40.2	94 52.1	28.9
	14	294 21.6	38 15.5	36.5	151 24.0	52.1	328 27.3	40.1	109 54.5	28.9
	15	309 24.1	53 14.6	36.9	166 26.3	52.1	343 29.3	39.9	124 56.8	28.9
	16	324 26.6	68 13.7	37.3	181 28.5	52.2	358 31.4	39.8	139 59.2	28.8
	17	339 29.0	83 12.8	37.7	196 30.8	52.2	13 33.4	39.6	155 01.5	28.8
	18	354 31.5	98 11.8	S22 38.1	211 33.1	N17 52.3	28 35.4	S14 39.5	170 03.8	N 0 28.7
	19	9 34.0	113 10.9	38.5	226 35.4	52.3	43 37.5	39.3	185 06.2	28.7
	20	24 36.4	128 10.0	38.9	241 37.6	52.4	58 39.5	39.2	200 08.5	28.7
	21	39 38.9	143 09.1	39.3	256 39.9	52.4	73 41.6	39.0	215 10.9	28.6
	22	54 41.3	158 08.2	39.7	271 42.2	52.5	88 43.6	38.9	230 13.2	28.6
	23	69 43.8	173 07.2	40.1	286 44.5	52.5	103 45.6	38.7	245 15.6	28.5
WEDNESDAY	16 00	84 46.3	188 06.3	S22 40.5	301 46.8	N17 52.6	118 47.7	S14 38.6	260 17.9	N 0 28.5
	01	99 48.7	203 05.4	40.8	316 49.1	52.6	133 49.7	38.4	275 20.3	28.5
	02	114 51.2	218 04.5	41.2	331 51.3	52.7	148 51.7	38.3	290 22.6	28.4
	03	129 53.7	233 03.5	41.6	346 53.6	52.7	163 53.8	38.1	305 24.9	28.4
	04	144 56.1	248 02.6	42.0	1 55.9	52.8	178 55.8	38.0	320 27.3	28.3
	05	159 58.6	263 01.7	42.4	16 58.2	52.8	193 57.8	37.8	335 29.6	28.3
	06	175 01.1	278 00.8	S22 42.8	32 00.5	N17 52.9	208 59.9	S14 37.7	350 32.0	N 0 28.3
	07	190 03.5	292 59.9	43.1	47 02.8	53.0	224 01.9	37.5	5 34.3	28.2
	08	205 06.0	307 58.9	43.5	62 05.1	53.0	239 03.9	37.4	20 36.7	28.2
	09	220 08.4	322 58.0	43.9	77 07.4	53.1	254 06.0	37.2	35 39.0	28.1
	10	235 10.9	337 57.1	44.3	92 09.7	53.1	269 08.0	37.1	50 41.4	28.1
	11	250 13.4	352 56.2	44.7	107 12.0	53.2	284 10.0	37.0	65 43.7	28.1
	12	265 15.8	7 55.2	S22 45.0	122 14.3	N17 53.2	299 12.1	S14 36.8	80 46.0	N 0 28.0
	13	280 18.3	22 54.3	45.4	137 16.6	53.3	314 14.1	36.7	95 48.4	28.0
	14	295 20.8	37 53.4	45.8	152 18.9	53.4	329 16.1	36.5	110 50.7	27.9
	15	310 23.2	52 52.5	46.2	167 21.2	53.4	344 18.2	36.4	125 53.1	27.9
	16	325 25.7	67 51.5	46.5	182 23.5	53.5	359 20.2	36.2	140 55.4	27.9
	17	340 28.2	82 50.6	46.9	197 25.8	53.5	14 22.2	36.1	155 57.8	27.8
	18	355 30.6	97 49.7	S22 47.3	212 28.1	N17 53.6	29 24.3	S14 35.9	171 00.1	N 0 27.8
	19	10 33.1	112 48.7	47.6	227 30.4	53.7	44 26.3	35.8	186 02.5	27.8
	20	25 35.6	127 47.8	48.0	242 32.7	53.7	59 28.3	35.6	201 04.8	27.7
	21	40 38.0	142 46.9	48.4	257 35.0	53.8	74 30.4	35.5	216 07.2	27.7
	22	55 40.5	157 46.0	48.7	272 37.3	53.8	89 32.4	35.3	231 09.5	27.6
	23	70 42.9	172 45.0	49.1	287 39.6	53.9	104 34.4	35.2	246 11.9	27.6
THURSDAY	17 00	85 45.4	187 44.1	S22 49.4	302 42.0	N17 54.0	119 36.5	S14 35.0	261 14.2	N 0 27.6
	01	100 47.9	202 43.2	49.8	317 44.3	54.0	134 38.5	34.9	276 16.6	27.5
	02	115 50.3	217 42.2	50.2	332 46.6	54.1	149 40.5	34.7	291 18.9	27.5
	03	130 52.8	232 41.3	50.5	347 48.9	54.2	164 42.6	34.6	306 21.3	27.5
	04	145 55.3	247 40.4	50.9	2 51.2	54.2	179 44.6	34.4	321 23.6	27.4
	05	160 57.7	262 39.5	51.2	17 53.5	54.3	194 46.6	34.3	336 26.0	27.4
	06	176 00.2	277 38.5	S22 51.6	32 55.9	N17 54.4	209 48.7	S14 34.1	351 28.3	N 0 27.3
	07	191 02.7	292 37.6	51.9	47 58.2	54.4	224 50.7	34.0	6 30.6	27.3
	08	206 05.1	307 36.7	52.3	63 00.5	54.5	239 52.7	33.8	21 33.0	27.3
	09	221 07.6	322 35.7	52.6	78 02.8	54.6	254 54.7	33.7	36 35.3	27.2
	10	236 10.1	337 34.8	53.0	93 05.2	54.6	269 56.8	33.5	51 37.7	27.2
	11	251 12.5	352 33.9	53.3	108 07.5	54.7	284 58.8	33.4	66 40.0	27.1
	12	266 15.0	7 32.9	S22 53.7	123 09.8	N17 54.8	300 00.8	S14 33.2	81 42.4	N 0 27.1
	13	281 17.4	22 32.0	54.0	138 12.2	54.8	315 02.9	33.1	96 44.7	27.1
	14	296 19.9	37 31.1	54.4	153 14.5	54.9	330 04.9	32.9	111 47.1	27.0
	15	311 22.4	52 30.1	54.7	168 16.8	55.0	345 06.9	32.8	126 49.4	27.0
	16	326 24.8	67 29.2	55.0	183 19.2	55.0	0 09.0	32.6	141 51.8	27.0
	17	341 27.3	82 28.3	55.4	198 21.5	55.1	15 11.0	32.5	156 54.1	26.9
	18	356 29.8	97 27.3	S22 55.7	213 23.8	N17 55.2	30 13.0	S14 32.3	171 56.5	N 0 26.9
	19	11 32.2	112 26.4	56.1	228 26.2	55.3	45 15.0	32.1	186 58.8	26.9
	20	26 34.7	127 25.5	56.4	243 28.5	55.3	60 17.1	32.0	202 01.2	26.8
	21	41 37.2	142 24.5	56.7	258 30.9	55.4	75 19.1	31.8	217 03.5	26.8
	22	56 39.6	157 23.6	57.1	273 33.2	55.5	90 21.1	31.7	232 05.9	26.7
	23	71 42.1	172 22.7	57.4	288 35.5	55.5	105 23.1	31.5	247 08.3	26.7
	Mer. Pass. h m	18 17.9	v −0.9	d 0.4	v 2.3	d 0.1	v 2.0	d 0.1	v 2.3	d 0.0

STARS

Name	SHA ° ′	Dec ° ′
Acamar	315 19.9	S40 15.
Achernar	335 28.2	S57 11.
Acrux	173 12.6	S63 09.
Adhara	255 14.2	S28 59.
Aldebaran	290 52.0	N16 31.
Alioth	166 23.0	N55 54.
Alkaid	153 01.1	N49 15.
Al Na'ir	27 47.0	S46 54.
Alnilam	275 48.6	S 1 11.
Alphard	217 58.4	S 8 42.
Alphecca	126 13.5	N26 40.
Alpheratz	357 46.2	N29 09.
Altair	62 11.0	N 8 53.
Ankaa	353 18.0	S42 15.
Antares	112 29.8	S26 27.
Arcturus	145 58.3	N19 07.
Atria	107 34.6	S69 02.
Avior	234 18.7	S59 32.
Bellatrix	278 34.4	N 6 21.
Betelgeuse	271 03.7	N 7 24.
Canopus	263 56.7	S52 42.
Capella	280 37.8	N46 00.
Deneb	49 33.6	N45 19.
Denebola	182 36.3	N14 30.
Diphda	348 58.3	S17 55.
Dubhe	193 54.5	N61 41.
Elnath	278 15.5	N28 37.
Eltanin	90 47.9	N51 29.
Enif	33 49.8	N 9 55.
Fomalhaut	15 26.8	S29 34.
Gacrux	172 04.1	S57 09.
Gienah	175 55.0	S17 35.
Hadar	148 52.1	S60 25.
Hamal	328 03.5	N23 30.
Kaus Aust.	83 47.6	S34 22.
Kochab	137 20.5	N74 06.
Markab	13 41.0	N15 15.
Menkar	314 17.5	N 4 07.
Menkent	148 10.9	S36 25.
Miaplacidus	221 39.9	S69 45.
Mirfak	308 43.7	N49 54.
Nunki	76 01.8	S26 17.
Peacock	53 23.6	S56 42.
Pollux	243 30.5	N28 00.
Procyon	245 02.1	N 5 11.
Rasalhague	96 09.2	N12 33.
Regulus	207 46.1	N11 55.
Rigel	281 14.2	S 8 11.
Rigil Kent.	139 55.9	S60 52.
Sabik	102 15.8	S15 44.
Schedar	349 43.5	N56 35.
Shaula	96 25.8	S37 06.
Sirius	258 35.6	S16 43.
Spica	158 34.1	S11 12.
Suhail	222 54.1	S43 28.
Vega	80 41.1	N38 47.
Zuben'ubi	137 08.6	S16 05.

	SHA ° ′	Mer. Pass. h m
Venus	103 20.0	11 28
Mars	217 00.5	3 52
Jupiter	34 01.4	16 03
Saturn	175 31.6	6 38

UT	SUN GHA	SUN Dec	MOON GHA	v	MOON Dec	d	HP
d h	° ′	° ′	° ′	′	° ′	′	′
15 00	181 15.4	S23 15.7	200 18.2	9.1	S24 58.3	3.6	55.5
01	196 15.1	15.8	214 46.3	9.1	25 01.9	3.3	55.5
02	211 14.8	15.9	229 14.4	9.1	25 05.2	3.3	55.5
03	226 14.5	. . 16.1	243 42.5	9.1	25 08.5	3.1	55.5
04	241 14.2	16.2	258 10.6	9.1	25 11.6	3.0	55.5
05	256 13.9	16.3	272 38.7	9.1	25 14.6	2.8	55.4
06	271 13.6	S23 16.4	287 06.8	9.1	S25 17.4	2.8	55.4
07	286 13.3	16.6	301 34.9	9.0	25 20.2	2.5	55.4
08	301 13.0	16.7	316 02.9	9.1	25 22.7	2.5	55.4
09	316 12.7	. . 16.8	330 31.0	9.1	25 25.2	2.3	55.4
10	331 12.4	16.9	344 59.1	9.0	25 27.5	2.2	55.4
11	346 12.1	17.1	359 27.1	9.1	25 29.7	2.0	55.3
12	1 11.8	S23 17.2	13 55.2	9.1	S25 31.7	1.9	55.3
13	16 11.5	17.3	28 23.3	9.0	25 33.6	1.8	55.3
14	31 11.2	17.4	42 51.3	9.1	25 35.4	1.6	55.3
15	46 10.9	. . 17.5	57 19.4	9.1	25 37.0	1.5	55.3
16	61 10.6	17.7	71 47.5	9.1	25 38.5	1.4	55.2
17	76 10.3	17.8	86 15.6	9.1	25 39.9	1.2	55.2
18	91 10.0	S23 17.9	100 43.7	9.1	S25 41.1	1.1	55.2
19	106 09.7	18.0	115 11.8	9.1	25 42.2	1.0	55.2
20	121 09.4	18.1	129 39.9	9.1	25 43.2	0.8	55.2
21	136 09.1	. . 18.2	144 08.0	9.1	25 44.0	0.7	55.2
22	151 08.8	18.4	158 36.1	9.2	25 44.7	0.5	55.1
23	166 08.5	18.5	173 04.3	9.1	25 45.2	0.4	55.1
16 00	181 08.1	S23 18.6	187 32.4	9.2	S25 45.6	0.3	55.1
01	196 07.8	18.7	202 00.6	9.2	25 45.9	0.2	55.1
02	211 07.5	18.8	216 28.8	9.2	25 46.1	0.0	55.1
03	226 07.2	. . 18.9	230 57.0	9.3	25 46.1	0.1	55.1
04	241 06.9	19.0	245 25.3	9.2	25 46.0	0.3	55.1
05	256 06.6	19.1	259 53.5	9.3	25 45.7	0.4	55.0
06	271 06.3	S23 19.2	274 21.8	9.3	S25 45.3	0.5	55.0
07	286 06.0	19.4	288 50.1	9.3	25 44.8	0.6	55.0
08	301 05.7	19.5	303 18.4	9.4	25 44.2	0.8	55.0
09	316 05.4	. . 19.6	317 46.8	9.4	25 43.4	0.9	55.0
10	331 05.1	19.7	332 15.2	9.4	25 42.5	1.1	55.0
11	346 04.8	19.8	346 43.6	9.4	25 41.4	1.1	54.9
12	1 04.5	S23 19.9	1 12.0	9.5	S25 40.3	1.4	54.9
13	16 04.2	20.0	15 40.5	9.5	25 38.9	1.4	54.9
14	31 03.9	20.1	30 09.0	9.5	25 37.5	1.6	54.9
15	46 03.6	. . 20.2	44 37.5	9.6	25 35.9	1.7	54.9
16	61 03.3	20.3	59 06.1	9.6	25 34.2	1.8	54.9
17	76 03.0	20.4	73 34.7	9.6	25 32.4	1.9	54.8
18	91 02.7	S23 20.5	88 03.3	9.7	S25 30.5	2.1	54.8
19	106 02.4	20.6	102 32.0	9.7	25 28.4	2.2	54.8
20	121 02.1	20.7	117 00.7	9.8	25 26.2	2.4	54.8
21	136 01.7	. . 20.8	131 29.5	9.8	25 23.8	2.4	54.8
22	151 01.4	20.9	145 58.3	9.8	25 21.4	2.6	54.8
23	166 01.1	21.0	160 27.1	9.9	25 18.8	2.8	54.8
17 00	181 00.8	S23 21.0	174 56.0	10.0	S25 16.0	2.8	54.7
01	196 00.5	21.1	189 25.0	9.9	25 13.2	3.0	54.7
02	211 00.2	21.2	203 53.9	10.0	25 10.2	3.1	54.7
03	225 59.9	. . 21.3	218 22.9	10.1	25 07.1	3.2	54.7
04	240 59.6	21.4	232 52.0	10.1	25 03.9	3.3	54.7
05	255 59.3	21.5	247 21.1	10.2	25 00.6	3.5	54.7
06	270 59.0	S23 21.6	261 50.3	10.2	S24 57.1	3.6	54.7
07	285 58.7	21.7	276 19.5	10.3	24 53.5	3.7	54.6
08	300 58.4	21.8	290 48.8	10.3	24 49.8	3.8	54.6
09	315 58.1	. . 21.8	305 18.1	10.3	24 46.0	3.9	54.6
10	330 57.8	21.9	319 47.4	10.4	24 42.1	4.1	54.6
11	345 57.5	22.0	334 16.8	10.5	24 38.0	4.2	54.6
12	0 57.1	S23 22.1	348 46.3	10.5	S24 33.8	4.3	54.6
13	15 56.8	22.2	3 15.8	10.6	24 29.5	4.4	54.6
14	30 56.5	22.3	17 45.4	10.6	24 25.1	4.5	54.6
15	45 56.2	. . 22.3	32 15.0	10.7	24 20.6	4.7	54.5
16	60 55.9	22.4	46 44.7	10.8	24 15.9	4.7	54.5
17	75 55.6	22.5	61 14.5	10.8	24 11.2	4.9	54.5
18	90 55.3	S23 22.6	75 44.3	10.8	S24 06.3	5.0	54.5
19	105 55.0	22.7	90 14.1	10.9	24 01.3	5.1	54.5
20	120 54.7	22.7	104 44.0	11.0	23 56.2	5.2	54.5
21	135 54.4	. . 22.8	119 14.0	11.1	23 51.0	5.3	54.5
22	150 54.1	22.9	133 44.1	11.1	23 45.7	5.4	54.5
23	165 53.8	23.0	148 14.2	11.1	S23 40.3	5.6	54.4
	SD 16.3	d 0.1	SD 15.1		15.0		14.9

Lat.	Twilight Naut.	Twilight Civil	Sunrise	Moonrise 15	Moonrise 16	Moonrise 17	Moonrise 18
°	h m	h m	h m	h m	h m	h m	h m
N 72	08 21	10 50	▬	▬	▬	▬	▬
N 70	08 02	09 49	▬	▬	▬	▬	▬
68	07 46	09 15	▬	▬	▬	▬	▬
66	07 33	08 50	10 29	▬	▬	▬	12 04
64	07 22	08 30	09 48	10 16	11 22	11 26	11 24
62	07 12	08 14	09 20	09 14	10 13	10 43	10 56
60	07 04	08 01	08 58	08 40	09 38	10 14	10 34
N 58	06 57	07 49	08 41	08 15	09 12	09 51	10 17
56	06 50	07 39	08 26	07 56	08 52	09 33	10 02
54	06 44	07 30	08 14	07 39	08 35	09 18	09 49
52	06 38	07 22	08 03	07 25	08 21	09 05	09 38
50	06 33	07 14	07 53	07 12	08 08	08 53	09 27
45	06 21	06 58	07 32	06 47	07 42	08 28	09 06
N 40	06 11	06 45	07 15	06 27	07 22	08 09	08 49
35	06 01	06 33	07 01	06 10	07 04	07 53	08 35
30	05 53	06 23	06 49	05 55	06 50	07 39	08 22
20	05 36	06 04	06 28	05 31	06 24	07 15	08 00
N 10	05 20	05 46	06 09	05 09	06 03	06 54	07 42
0	05 03	05 29	05 52	04 50	05 42	06 34	07 24
S 10	04 44	05 11	05 34	04 30	05 22	06 15	07 06
20	04 22	04 51	05 16	04 09	05 00	05 54	06 48
30	03 53	04 26	04 54	03 44	04 35	05 30	06 26
35	03 34	04 11	04 41	03 30	04 21	05 15	06 13
40	03 11	03 53	04 26	03 13	04 04	04 59	05 58
45	02 41	03 30	04 08	02 54	03 43	04 39	05 40
S 50	01 56	03 01	03 45	02 29	03 17	04 14	05 18
52	01 28	02 46	03 34	02 17	03 05	04 02	05 08
54	00 43	02 28	03 22	02 04	02 50	03 49	04 56
56	////	02 06	03 08	01 48	02 34	03 33	04 42
58	////	01 36	02 51	01 30	02 14	03 14	04 26
S 60	////	00 48	02 31	01 07	01 48	02 50	04 07

Lat.	Sunset	Twilight Civil	Twilight Naut.	Moonset 15	Moonset 16	Moonset 17	Moonset 18
°	h m	h m	h m	h m	h m	h m	h m
N 72	▬	13 01	15 30	▬	▬	▬	▬
N 70	▬	14 02	15 50	▬	▬	▬	▬
68	▬	14 37	16 05	▬	▬	▬	▬
66	13 22	15 02	16 18	▬	▬	▬	15 17
64	14 04	15 21	16 29	11 46	12 30	14 13	15 57
62	14 32	15 37	16 39	12 48	13 39	14 56	16 24
60	14 53	15 51	16 47	13 22	14 14	15 25	16 46
N 58	15 10	16 02	16 55	13 47	14 39	15 47	17 03
56	15 25	16 12	17 01	14 07	14 59	16 04	17 17
54	15 38	16 21	17 08	14 23	15 16	16 20	17 30
52	15 49	16 30	17 13	14 38	15 30	16 33	17 41
50	15 59	16 37	17 19	14 50	15 43	16 44	17 50
45	16 19	16 53	17 30	15 16	16 09	17 08	18 11
N 40	16 36	17 06	17 41	15 36	16 29	17 27	18 27
35	16 50	17 18	17 50	15 53	16 46	17 43	18 41
30	17 02	17 29	17 59	16 08	17 01	17 57	18 53
20	17 24	17 48	18 15	16 33	17 26	18 20	19 13
N 10	17 42	18 05	18 31	16 55	17 48	18 40	19 31
0	17 59	18 22	18 48	17 15	18 08	18 59	19 47
S 10	18 17	18 40	19 07	17 35	18 28	19 17	20 03
20	18 36	19 00	19 30	17 57	18 49	19 37	20 21
30	18 58	19 25	19 59	18 22	19 14	20 00	20 41
35	19 11	19 41	20 17	18 37	19 28	20 13	20 52
40	19 26	19 59	20 40	18 54	19 45	20 29	21 06
45	19 44	20 21	21 11	19 14	20 05	20 47	21 21
S 50	20 06	20 51	21 56	19 40	20 30	21 10	21 40
52	20 17	21 06	22 24	19 52	20 42	21 21	21 49
54	20 29	21 24	23 09	20 07	20 56	21 33	21 59
56	20 44	21 46	////	20 23	21 12	21 47	22 11
58	21 00	22 16	////	20 43	21 31	22 03	22 24
S 60	21 21	23 05	////	21 09	21 55	22 23	22 39

Day	SUN Eqn. of Time 00^h	SUN Eqn. of Time 12^h	SUN Mer. Pass.	MOON Mer. Pass. Upper	MOON Mer. Pass. Lower	MOON Age	MOON Phase
d	m s	m s	h m	h m	h m	d	%
15	05 02	04 48	11 55	11 02	23 29	29	1
16	04 33	04 19	11 56	11 55	24 21	30	0 ●
17	04 04	03 49	11 56	12 46	00 21	01	1

UT	ARIES	VENUS −3.9		MARS −0.5		JUPITER −2.2		SATURN +1.0	
	GHA	GHA	Dec	GHA	Dec	GHA	Dec	GHA	Dec
d h	° ′	° ′	° ′	° ′	° ′	° ′	° ′	° ′	° ′
18 00	86 44.6	187 21.7	S22 57.7	303 37.9	N17 55.6	120 25.2	S14 31.4	262 10.6	N 0 26.7
01	101 47.0	202 20.8	58.1	318 40.2	55.7	135 27.2	31.2	277 13.0	26.6
02	116 49.5	217 19.8	58.4	333 42.6	55.8	150 29.2	31.1	292 15.3	26.6
03	131 51.9	232 18.9	. . 58.7	348 44.9	. . 55.8	165 31.3	. . 30.9	307 17.7	. . 26.6
04	146 54.4	247 18.0	59.0	3 47.3	55.9	180 33.3	30.8	322 20.0	26.5
05	161 56.9	262 17.0	59.4	18 49.6	56.0	195 35.3	30.6	337 22.4	26.5
06	176 59.3	277 16.1	S22 59.7	33 52.0	N17 56.1	210 37.3	S14 30.5	352 24.7	N 0 26.5
07	192 01.8	292 15.2	23 00.0	48 54.3	56.2	225 39.4	30.3	7 27.1	26.4
08	207 04.3	307 14.2	00.3	63 56.7	56.2	240 41.4	30.2	22 29.4	26.4
F 09	222 06.7	322 13.3	. . 00.7	78 59.1	. . 56.3	255 43.4	. . 30.0	37 31.8	. . 26.3
R 10	237 09.2	337 12.3	01.0	94 01.4	56.4	270 45.4	29.9	52 34.1	26.3
I 11	252 11.7	352 11.4	01.3	109 03.8	56.5	285 47.5	29.7	67 36.5	26.3
D 12	267 14.1	7 10.5	S23 01.6	124 06.1	N17 56.5	300 49.5	S14 29.6	82 38.8	N 0 26.2
A 13	282 16.6	22 09.5	01.9	139 08.5	56.6	315 51.5	29.4	97 41.2	26.2
Y 14	297 19.1	37 08.6	02.2	154 10.9	56.7	330 53.5	29.3	112 43.5	26.2
15	312 21.5	52 07.6	. . 02.6	169 13.2	. . 56.8	345 55.6	. . 29.1	127 45.9	. . 26.1
16	327 24.0	67 06.7	02.9	184 15.6	56.9	0 57.6	29.0	142 48.2	26.1
17	342 26.4	82 05.8	03.2	199 18.0	56.9	15 59.6	28.8	157 50.6	26.1
18	357 28.9	97 04.8	S23 03.5	214 20.3	N17 57.0	31 01.6	S14 28.6	172 53.0	N 0 26.0
19	12 31.4	112 03.9	03.8	229 22.7	57.1	46 03.7	28.5	187 55.3	26.0
20	27 33.8	127 02.9	04.1	244 25.1	57.2	61 05.7	28.3	202 57.7	26.0
21	42 36.3	142 02.0	. . 04.4	259 27.5	. . 57.3	76 07.7	. . 28.2	218 00.0	. . 25.9
22	57 38.8	157 01.1	04.7	274 29.8	57.4	91 09.7	28.0	233 02.4	25.9
23	72 41.2	172 00.1	05.0	289 32.2	57.4	106 11.8	27.9	248 04.7	25.9
19 00	87 43.7	186 59.2	S23 05.3	304 34.6	N17 57.5	121 13.8	S14 27.7	263 07.1	N 0 25.8
01	102 46.2	201 58.2	05.6	319 37.0	57.6	136 15.8	27.6	278 09.4	25.8
02	117 48.6	216 57.3	05.9	334 39.3	57.7	151 17.8	27.4	293 11.8	25.8
03	132 51.1	231 56.3	. . 06.2	349 41.7	. . 57.8	166 19.8	. . 27.3	308 14.2	. . 25.7
04	147 53.6	246 55.4	06.5	4 44.1	57.9	181 21.9	27.1	323 16.5	25.7
05	162 56.0	261 54.5	06.8	19 46.5	57.9	196 23.9	27.0	338 18.9	25.6
06	177 58.5	276 53.5	S23 07.1	34 48.9	N17 58.0	211 25.9	S14 26.8	353 21.2	N 0 25.6
S 07	193 00.9	291 52.6	07.4	49 51.3	58.1	226 27.9	26.7	8 23.6	25.6
08	208 03.4	306 51.6	07.7	64 53.7	58.2	241 30.0	26.5	23 25.9	25.5
A 09	223 05.9	321 50.7	. . 08.0	79 56.0	. . 58.3	256 32.0	. . 26.3	38 28.3	. . 25.5
T 10	238 08.3	336 49.7	08.3	94 58.4	58.4	271 34.0	26.2	53 30.6	25.5
U 11	253 10.8	351 48.8	08.6	110 00.8	58.5	286 36.0	26.0	68 33.0	25.4
R 12	268 13.3	6 47.8	S23 08.9	125 03.2	N17 58.6	301 38.0	S14 25.9	83 35.4	N 0 25.4
D 13	283 15.7	21 46.9	09.1	140 05.6	58.7	316 40.1	25.7	98 37.7	25.4
A 14	298 18.2	36 46.0	09.4	155 08.0	58.7	331 42.1	25.6	113 40.1	25.3
Y 15	313 20.7	51 45.0	. . 09.7	170 10.4	. . 58.8	346 44.1	. . 25.4	128 42.4	. . 25.3
16	328 23.1	66 44.1	10.0	185 12.8	58.9	1 46.1	25.3	143 44.8	25.3
17	343 25.6	81 43.1	10.3	200 15.2	59.0	16 48.1	25.1	158 47.2	25.2
18	358 28.0	96 42.2	S23 10.6	215 17.6	N17 59.1	31 50.2	S14 25.0	173 49.5	N 0 25.2
19	13 30.5	111 41.2	10.8	230 20.0	59.2	46 52.2	24.8	188 51.9	25.2
20	28 33.0	126 40.3	11.1	245 22.4	59.3	61 54.2	24.6	203 54.2	25.1
21	43 35.4	141 39.3	. . 11.4	260 24.8	. . 59.4	76 56.2	. . 24.5	218 56.6	. . 25.1
22	58 37.9	156 38.4	11.7	275 27.2	59.5	91 58.2	24.3	233 58.9	25.1
23	73 40.4	171 37.4	11.9	290 29.6	59.6	107 00.3	24.2	249 01.3	25.0
20 00	88 42.8	186 36.5	S23 12.2	305 32.0	N17 59.7	122 02.3	S14 24.0	264 03.7	N 0 25.0
01	103 45.3	201 35.5	12.5	320 34.5	59.8	137 04.3	23.9	279 06.0	25.0
02	118 47.8	216 34.6	12.7	335 36.9	17 59.9	152 06.3	23.7	294 08.4	24.9
03	133 50.2	231 33.6	. . 13.0	350 39.3	18 00.0	167 08.3	. . 23.6	309 10.7	. . 24.9
04	148 52.7	246 32.7	13.3	5 41.7	00.1	182 10.3	23.4	324 13.1	24.9
05	163 55.2	261 31.7	13.6	20 44.1	00.2	197 12.4	23.3	339 15.5	24.8
06	178 57.6	276 30.8	S23 13.8	35 46.5	N18 00.3	212 14.4	S14 23.1	354 17.8	N 0 24.8
07	194 00.1	291 29.8	14.1	50 48.9	00.3	227 16.4	22.9	9 20.2	24.8
08	209 02.5	306 28.9	14.3	65 51.4	00.4	242 18.4	22.8	24 22.5	24.7
S 09	224 05.0	321 27.9	. . 14.6	80 53.8	. . 00.5	257 20.4	. . 22.6	39 24.9	. . 24.7
U 10	239 07.5	336 27.0	14.9	95 56.2	00.6	272 22.5	22.5	54 27.3	24.7
N 11	254 09.9	351 26.0	15.1	110 58.6	00.7	287 24.5	22.3	69 29.6	24.7
D 12	269 12.4	6 25.1	S23 15.4	126 01.1	N18 00.8	302 26.5	S14 22.2	84 32.0	N 0 24.6
A 13	284 14.9	21 24.1	15.6	141 03.5	00.9	317 28.5	22.0	99 34.3	24.6
Y 14	299 17.3	36 23.2	15.9	156 05.9	01.0	332 30.5	21.9	114 36.7	24.6
15	314 19.8	51 22.2	. . 16.2	171 08.4	. . 01.1	347 32.5	. . 21.7	129 39.1	. . 24.5
16	329 22.3	66 21.3	16.4	186 10.8	01.2	2 34.6	21.5	144 41.4	24.5
17	344 24.7	81 20.3	16.7	201 13.2	01.4	17 36.6	21.4	159 43.8	24.5
18	359 27.2	96 19.4	S23 16.9	216 15.6	N18 01.5	32 38.6	S14 21.2	174 46.2	N 0 24.4
19	14 29.7	111 18.4	17.2	231 18.1	01.6	47 40.6	21.1	189 48.5	24.4
20	29 32.1	126 17.5	17.4	246 20.5	01.7	62 42.6	20.9	204 50.9	24.4
21	44 34.6	141 16.5	. . 17.7	261 23.0	. . 01.8	77 44.6	. . 20.8	219 53.2	. . 24.3
22	59 37.0	156 15.6	17.9	276 25.4	01.9	92 46.6	20.6	234 55.6	24.3
23	74 39.5	171 14.6	18.1	291 27.8	02.0	107 48.7	20.4	249 58.0	24.3
Mer. Pass.	h m 18 06.1	*v* −0.9	*d* 0.3	*v* 2.4	*d* 0.1	*v* 2.0	*d* 0.2	*v* 2.4	*d* 0.0

STARS		
Name	SHA	Dec
	° ′	° ′
Acamar	315 19.9	S40 15.
Achernar	335 28.2	S57 11.
Acrux	173 12.5	S63 09.
Adhara	255 14.2	S28 59.
Aldebaran	290 52.0	N16 31.
Alioth	166 22.9	N55 54.
Alkaid	153 01.0	N49 15.
Al Na'ir	27 47.0	S46 54.
Alnilam	275 48.6	S 1 11.
Alphard	217 58.4	S 8 42.
Alphecca	126 13.5	N26 40.
Alpheratz	357 46.2	N29 09.
Altair	62 11.0	N 8 53.
Ankaa	353 18.0	S42 15.
Antares	112 29.8	S26 27.
Arcturus	145 58.2	N19 07.
Atria	107 34.6	S69 02.
Avior	234 18.6	S59 32.
Bellatrix	278 34.4	N 6 21.
Betelgeuse	271 03.7	N 7 24.
Canopus	263 56.7	S52 42.
Capella	280 37.8	N46 00.
Deneb	49 33.7	N45 19.
Denebola	182 36.2	N14 30.
Diphda	348 58.3	S17 55.
Dubhe	193 54.4	N61 41.
Elnath	278 15.5	N28 37.
Eltanin	90 47.9	N51 29.
Enif	33 49.8	N 9 55.
Fomalhaut	15 26.8	S29 34.
Gacrux	172 04.1	S57 10.
Gienah	175 55.0	S17 35.
Hadar	148 52.1	S60 25.
Hamal	328 03.5	N23 30.
Kaus Aust.	83 47.6	S34 22.
Kochab	137 20.5	N74 06.
Markab	13 41.0	N15 15.
Menkar	314 17.5	N 4 07.
Menkent	148 10.9	S36 25.
Miaplacidus	221 39.8	S69 45.
Mirfak	308 43.7	N49 54.
Nunki	76 01.8	S26 17.
Peacock	53 23.6	S56 42.
Pollux	243 30.5	N28 00
Procyon	245 02.1	N 5 11.
Rasalhague	96 09.2	N12 33.
Regulus	207 46.1	N11 55.
Rigel	281 14.2	S 8 11
Rigil Kent.	139 55.8	S60 52
Sabik	102 15.8	S15 44
Schedar	349 43.5	N56 35
Shaula	96 25.8	S37 06
Sirius	258 35.6	S16 43
Spica	158 34.1	S11 12
Suhail	222 54.1	S43 28
Vega	80 41.1	N38 47
Zuben'ubi	137 08.5	S16 05
	SHA	Mer. Pas
	° ′	h m
Venus	99 15.5	11 33
Mars	216 50.9	3 41
Jupiter	33 30.1	15 53
Saturn	175 23.4	6 27

	UT	SUN GHA	SUN Dec	MOON GHA	*v*	MOON Dec	*d*	HP
	d h	° ′	° ′	° ′	′	° ′	′	′
FRIDAY	18 00	180 53.5	S23 23.0	162 44.3	11.2	S23 34.7	5.6	54.4
	01	195 53.2	23.1	177 14.5	11.3	23 29.1	5.8	54.4
	02	210 52.8	23.2	191 44.8	11.4	23 23.3	5.9	54.4
	03	225 52.5	. . 23.3	206 15.2	11.4	23 17.4	5.9	54.4
	04	240 52.2	23.3	220 45.6	11.4	23 11.5	6.1	54.4
	05	255 51.9	23.4	235 16.0	11.6	23 05.4	6.2	54.4
	06	270 51.6	S23 23.5	249 46.6	11.6	S22 59.2	6.3	54.4
	07	285 51.3	23.5	264 17.2	11.6	22 52.9	6.4	54.4
	08	300 51.0	23.6	278 47.8	11.7	22 46.5	6.5	54.3
	09	315 50.7	. . 23.7	293 18.5	11.8	22 40.0	6.6	54.3
	10	330 50.4	23.7	307 49.3	11.9	22 33.4	6.7	54.3
	11	345 50.1	23.8	322 20.2	11.9	22 26.7	6.8	54.3
	12	0 49.8	S23 23.9	336 51.1	12.0	S22 19.9	6.8	54.3
	13	15 49.5	23.9	351 22.1	12.0	22 13.1	7.0	54.3
	14	30 49.1	24.0	5 53.1	12.1	22 06.1	7.1	54.3
	15	45 48.8	. . 24.0	20 24.2	12.2	21 59.0	7.2	54.3
	16	60 48.5	24.1	34 55.4	12.3	21 51.8	7.3	54.3
	17	75 48.2	24.2	49 26.7	12.3	21 44.5	7.4	54.3
	18	90 47.9	S23 24.2	63 58.0	12.3	S21 37.1	7.4	54.3
	19	105 47.6	24.3	78 29.3	12.5	21 29.7	7.6	54.2
	20	120 47.3	24.3	93 00.8	12.5	21 22.1	7.6	54.2
	21	135 47.0	. . 24.4	107 32.3	12.5	21 14.5	7.8	54.2
	22	150 46.7	24.4	122 03.8	12.7	21 06.7	7.8	54.2
	23	165 46.4	24.5	136 35.5	12.7	20 58.9	7.9	54.2
SATURDAY	19 00	180 46.1	S23 24.6	151 07.2	12.7	S20 51.0	8.1	54.2
	01	195 45.7	24.6	165 38.9	12.9	20 42.9	8.1	54.2
	02	210 45.4	24.7	180 10.8	12.9	20 34.8	8.1	54.2
	03	225 45.1	. . 24.7	194 42.7	12.9	20 26.7	8.3	54.2
	04	240 44.8	24.8	209 14.6	13.1	20 18.4	8.4	54.2
	05	255 44.5	24.8	223 46.7	13.1	20 10.0	8.4	54.2
	06	270 44.2	S23 24.9	238 18.8	13.1	S20 01.6	8.6	54.2
	07	285 43.9	24.9	252 50.9	13.2	19 53.0	8.6	54.2
	08	300 43.6	25.0	267 23.1	13.3	19 44.4	8.7	54.1
	09	315 43.3	. . 25.0	281 55.4	13.4	19 35.7	8.7	54.1
	10	330 43.0	25.1	296 27.8	13.4	19 27.0	8.9	54.1
	11	345 42.6	25.1	311 00.2	13.5	19 18.1	8.9	54.1
	12	0 42.3	S23 25.1	325 32.7	13.5	S19 09.2	9.0	54.1
	13	15 42.0	25.2	340 05.2	13.6	19 00.2	9.1	54.1
	14	30 41.7	25.2	354 37.8	13.7	18 51.1	9.2	54.1
	15	45 41.4	. . 25.3	9 10.5	13.7	18 41.9	9.2	54.1
	16	60 41.1	25.3	23 43.2	13.8	18 32.7	9.3	54.1
	17	75 40.8	25.4	38 16.0	13.8	18 23.4	9.4	54.1
	18	90 40.5	S23 25.4	52 48.8	14.0	S18 14.0	9.5	54.1
	19	105 40.2	25.4	67 21.8	13.9	18 04.5	9.5	54.1
	20	120 39.9	25.5	81 54.7	14.1	17 55.0	9.6	54.1
	21	135 39.5	. . 25.5	96 27.8	14.1	17 45.4	9.7	54.1
	22	150 39.2	25.5	111 00.9	14.1	17 35.7	9.7	54.1
	23	165 38.9	25.6	125 34.0	14.2	17 26.0	9.9	54.1
SUNDAY	20 00	180 38.6	S23 25.6	140 07.2	14.3	S17 16.1	9.8	54.1
	01	195 38.3	25.6	154 40.5	14.3	17 06.3	10.0	54.1
	02	210 38.0	25.7	169 13.8	14.4	16 56.3	10.0	54.1
	03	225 37.7	. . 25.7	183 47.2	14.4	16 46.3	10.1	54.1
	04	240 37.4	25.7	198 20.6	14.5	16 36.2	10.2	54.1
	05	255 37.1	25.8	212 54.1	14.6	16 26.0	10.2	54.1
	06	270 36.8	S23 25.8	227 27.7	14.6	S16 15.8	10.3	54.1
	07	285 36.4	25.8	242 01.3	14.7	16 05.5	10.3	54.1
	08	300 36.1	25.9	256 35.0	14.7	15 55.2	10.4	54.0
	09	315 35.8	. . 25.9	271 08.7	14.7	15 44.8	10.5	54.0
	10	330 35.5	25.9	285 42.4	14.9	15 34.3	10.5	54.0
	11	345 35.2	25.9	300 16.3	14.8	15 23.8	10.6	54.0
	12	0 34.9	S23 26.0	314 50.1	15.0	S15 13.2	10.6	54.0
	13	15 34.6	26.0	329 24.1	14.9	15 02.6	10.7	54.0
	14	30 34.3	26.0	343 58.0	15.1	14 51.9	10.8	54.0
	15	45 34.0	. . 26.0	358 32.1	15.0	14 41.1	10.8	54.0
	16	60 33.6	26.1	13 06.1	15.2	14 30.3	10.9	54.0
	17	75 33.3	26.1	27 40.3	15.1	14 19.4	10.9	54.0
	18	90 33.0	S23 26.1	42 14.4	15.3	S14 08.5	11.0	54.0
	19	105 32.7	26.1	56 48.7	15.2	13 57.5	11.0	54.0
	20	120 32.4	26.1	71 22.9	15.3	13 46.5	11.1	54.0
	21	135 32.1	. . 26.1	85 57.2	15.4	13 35.4	11.2	54.0
	22	150 31.8	26.2	100 31.6	15.4	13 24.2	11.2	54.0
	23	165 31.5	26.2	115 06.0	15.5	S13 13.0	11.2	54.1
		SD 16.3	*d* 0.0	SD 14.8		14.7		14.7

Lat.	Twilight Naut.	Twilight Civil	Sunrise	Moonrise 18	19	20	21
°	h m	h m	h m	h m	h m	h m	h m
N 72	08 24	10 56	■	■	■	12 48	12 05
N 70	08 04	09 53	■	■	13 07	12 15	11 48
68	07 48	09 18	■	■	12 17	11 51	11 35
66	07 35	08 52	10 33	12 04	11 45	11 33	11 23
64	07 24	08 33	09 51	11 24	11 21	11 17	11 14
62	07 15	08 17	09 22	10 56	11 02	11 05	11 06
60	07 06	08 03	09 01	10 34	10 46	10 54	10 59
N 58	06 59	07 51	08 43	10 17	10 33	10 44	10 52
56	06 52	07 41	08 29	10 02	10 21	10 36	10 47
54	06 46	07 32	08 16	09 49	10 11	10 28	10 42
52	06 40	07 24	08 05	09 38	10 02	10 22	10 37
50	06 35	07 16	07 55	09 27	09 54	10 16	10 33
45	06 23	07 00	07 34	09 06	09 37	10 02	10 24
N 40	06 13	06 47	07 17	08 49	09 23	09 52	10 17
35	06 03	06 35	07 03	08 35	09 11	09 42	10 10
30	05 54	06 24	06 51	08 22	09 00	09 34	10 05
20	05 38	06 05	06 29	08 00	08 42	09 20	09 55
N 10	05 22	05 48	06 11	07 42	08 26	09 07	09 46
0	05 05	05 31	05 53	07 24	08 11	08 56	09 38
S 10	04 46	05 13	05 36	07 06	07 56	08 44	09 30
20	04 23	04 52	05 17	06 48	07 40	08 31	09 21
30	03 54	04 27	04 55	06 26	07 22	08 17	09 11
35	03 35	04 12	04 42	06 13	07 11	08 08	09 05
40	03 12	03 54	04 27	05 58	06 58	07 59	08 58
45	02 41	03 31	04 09	05 40	06 44	07 47	08 51
S 50	01 56	03 01	03 46	05 18	06 26	07 34	08 41
52	01 28	02 46	03 35	05 08	06 17	07 27	08 37
54	00 41	02 28	03 23	04 56	06 07	07 20	08 32
56	////	02 06	03 09	04 42	05 57	07 12	08 27
58	////	01 36	02 52	04 26	05 44	07 03	08 21
S 60	////	00 45	02 31	04 07	05 29	06 52	08 14

Lat.	Sunset	Twilight Civil	Twilight Naut.	Moonset 18	19	20	21
°	h m	h m	h m	h m	h m	h m	h m
N 72	■	12 58	15 30	■	■	17 44	19 55
N 70	■	14 01	15 50	■	15 53	18 16	20 10
68	■	14 37	16 06	■	16 42	18 38	20 22
66	13 21	15 02	16 19	15 17	17 13	18 56	20 31
64	14 04	15 22	16 30	15 57	17 36	19 10	20 39
62	14 32	15 38	16 40	16 24	17 54	19 22	20 46
60	14 53	15 51	16 48	16 46	18 09	19 32	20 52
N 58	15 11	16 03	16 56	17 03	18 22	19 40	20 57
56	15 26	16 13	17 02	17 17	18 33	19 48	21 02
54	15 38	16 22	17 09	17 30	18 42	19 55	21 06
52	15 50	16 31	17 14	17 41	18 51	20 01	21 10
50	15 59	16 38	17 20	17 50	18 58	20 06	21 13
45	16 20	16 54	17 31	18 11	19 14	20 18	21 21
N 40	16 37	17 08	17 42	18 27	19 28	20 28	21 27
35	16 51	17 19	17 51	18 41	19 39	20 36	21 32
30	17 04	17 30	18 00	18 53	19 48	20 43	21 37
20	17 25	17 49	18 16	19 13	20 05	20 55	21 44
N 10	17 43	18 06	18 33	19 31	20 20	21 06	21 51
0	18 01	18 23	18 50	19 47	20 33	21 16	21 58
S 10	18 19	18 42	19 09	20 03	20 46	21 26	22 04
20	18 38	19 02	19 31	20 21	21 01	21 37	22 11
30	19 00	19 27	20 00	20 41	21 17	21 49	22 18
35	19 13	19 42	20 19	20 52	21 26	21 56	22 23
40	19 28	20 01	20 42	21 06	21 37	22 04	22 28
45	19 46	20 23	21 13	21 21	21 49	22 13	22 33
S 50	20 08	20 53	21 59	21 40	22 04	22 24	22 40
52	20 19	21 08	22 27	21 49	22 11	22 29	22 43
54	20 32	21 26	23 14	21 59	22 19	22 34	22 47
56	20 46	21 49	////	22 11	22 28	22 40	22 50
58	21 03	22 18	////	22 24	22 37	22 47	22 55
S 60	21 23	23 10	////	22 39	22 49	22 55	22 59

Day	SUN Eqn. of Time 00^h	SUN Eqn. of Time 12^h	SUN Mer. Pass.	MOON Mer. Pass. Upper	MOON Mer. Pass. Lower	Age	Phase
d	m s	m s	h m	h m	h m	d	%
18	03 34	03 20	11 57	13 36	01 11	02	4
19	03 05	02 50	11 57	14 22	01 59	03	8
20	02 35	02 20	11 58	15 06	02 44	04	14

UT d h		ARIES GHA	VENUS −3.9 GHA	VENUS Dec	MARS −0.5 GHA	MARS Dec	JUPITER −2.1 GHA	JUPITER Dec	SATURN +0.9 GHA	SATURN Dec
		° ′	° ′	° ′	° ′	° ′	° ′	° ′	° ′	° ′
21 00		89 42.0	186 13.7	S23 18.4	306 30.3	N18 02.1	122 50.7	S14 20.3	265 00.3	N 0 24.2
01		104 44.4	201 12.7	18.6	321 32.7	02.2	137 52.7	20.1	280 02.7	24.2
02		119 46.9	216 11.7	18.9	336 35.2	02.3	152 54.7	20.0	295 05.1	24.2
03		134 49.4	231 10.8	. . 19.1	351 37.6	. . 02.4	167 56.7	. . 19.8	310 07.4	. . 24.1
04		149 51.8	246 09.8	19.3	6 40.1	02.5	182 58.7	19.7	325 09.8	24.1
05		164 54.3	261 08.9	19.6	21 42.5	02.6	198 00.8	19.5	340 12.1	24.1
06		179 56.8	276 07.9	S23 19.8	36 45.0	N18 02.7	213 02.8	S14 19.3	355 14.5	N 0 24.1
07		194 59.2	291 07.0	20.1	51 47.4	02.8	228 04.8	19.2	10 16.9	24.0
08		210 01.7	306 06.0	20.3	66 49.9	02.9	243 06.8	19.0	25 19.2	24.0
09	M	225 04.2	321 05.1	. . 20.5	81 52.3	. . 03.0	258 08.8	. . 18.9	40 21.6	. . 24.0
10	O	240 06.6	336 04.1	20.8	96 54.8	03.2	273 10.8	18.7	55 24.0	23.9
11	N	255 09.1	351 03.1	21.0	111 57.2	03.3	288 12.8	18.6	70 26.3	23.9
12	D	270 11.5	6 02.2	S23 21.2	126 59.7	N18 03.4	303 14.8	S14 18.4	85 28.7	N 0 23.9
13	A	285 14.0	21 01.2	21.4	142 02.2	03.5	318 16.9	18.2	100 31.1	23.8
14	Y	300 16.5	36 00.3	21.7	157 04.6	03.6	333 18.9	18.1	115 33.4	23.8
15		315 18.9	50 59.3	. . 21.9	172 07.1	. . 03.7	348 20.9	. . 17.9	130 35.8	. . 23.8
16		330 21.4	65 58.4	22.1	187 09.5	03.8	3 22.9	17.8	145 38.2	23.8
17		345 23.9	80 57.4	22.3	202 12.0	03.9	18 24.9	17.6	160 40.5	23.7
18		0 26.3	95 56.5	S23 22.6	217 14.5	N18 04.1	33 26.9	S14 17.5	175 42.9	N 0 23.7
19		15 28.8	110 55.5	22.8	232 16.9	04.2	48 28.9	17.3	190 45.3	23.7
20		30 31.3	125 54.5	23.0	247 19.4	04.3	63 30.9	17.1	205 47.6	23.6
21		45 33.7	140 53.6	. . 23.2	262 21.9	. . 04.4	78 33.0	. . 17.0	220 50.0	. . 23.6
22		60 36.2	155 52.6	23.4	277 24.4	04.5	93 35.0	16.8	235 52.4	23.6
23		75 38.6	170 51.7	23.6	292 26.8	04.6	108 37.0	16.7	250 54.7	23.5
22 00		90 41.1	185 50.7	S23 23.9	307 29.3	N18 04.7	123 39.0	S14 16.5	265 57.1	N 0 23.5
01		105 43.6	200 49.7	24.1	322 31.8	04.9	138 41.0	16.3	280 59.5	23.5
02		120 46.0	215 48.8	24.3	337 34.3	05.0	153 43.0	16.2	296 01.8	23.5
03		135 48.5	230 47.8	. . 24.5	352 36.7	. . 05.1	168 45.0	. . 16.0	311 04.2	. . 23.4
04		150 51.0	245 46.9	24.7	7 39.2	05.2	183 47.0	15.9	326 06.6	23.4
05		165 53.4	260 45.9	24.9	22 41.7	05.3	198 49.0	15.7	341 08.9	23.4
06		180 55.9	275 45.0	S23 25.1	37 44.2	N18 05.4	213 51.0	S14 15.6	356 11.3	N 0 23.3
07		195 58.4	290 44.0	25.3	52 46.7	05.6	228 53.1	15.4	11 13.7	23.3
08	T	211 00.8	305 43.0	25.5	67 49.1	05.7	243 55.1	15.2	26 16.0	23.3
09	U	226 03.3	320 42.1	. . 25.7	82 51.6	. . 05.8	258 57.1	. . 15.1	41 18.4	. . 23.3
10	E	241 05.8	335 41.1	25.9	97 54.1	05.9	273 59.1	14.9	56 20.8	23.2
11	S	256 08.2	350 40.2	26.1	112 56.6	06.0	289 01.1	14.8	71 23.1	23.2
12	D	271 10.7	5 39.2	S23 26.3	127 59.1	N18 06.2	304 03.1	S14 14.6	86 25.5	N 0 23.2
13	A	286 13.1	20 38.2	26.5	143 01.6	06.3	319 05.1	14.4	101 27.9	23.1
14	Y	301 15.6	35 37.3	26.7	158 04.1	06.4	334 07.1	14.3	116 30.2	23.1
15		316 18.1	50 36.3	. . 26.9	173 06.6	. . 06.5	349 09.1	. . 14.1	131 32.6	. . 23.1
16		331 20.5	65 35.4	27.1	188 09.1	06.7	4 11.1	14.0	146 35.0	23.1
17		346 23.0	80 34.4	27.3	203 11.6	06.8	19 13.1	13.8	161 37.3	23.0
18		1 25.5	95 33.4	S23 27.5	218 14.1	N18 06.9	34 15.2	S14 13.6	176 39.7	N 0 23.0
19		16 27.9	110 32.5	27.7	233 16.6	07.0	49 17.2	13.5	191 42.1	23.0
20		31 30.4	125 31.5	27.9	248 19.1	07.2	64 19.2	13.3	206 44.5	22.9
21		46 32.9	140 30.5	. . 28.0	263 21.6	. . 07.3	79 21.2	. . 13.2	221 46.8	. . 22.9
22		61 35.3	155 29.6	28.2	278 24.1	07.4	94 23.2	13.0	236 49.2	22.9
23		76 37.8	170 28.6	28.4	293 26.6	07.5	109 25.2	12.8	251 51.6	22.9
23 00		91 40.3	185 27.7	S23 28.6	308 29.1	N18 07.7	124 27.2	S14 12.7	266 53.9	N 0 22.8
01		106 42.7	200 26.7	28.8	323 31.6	07.8	139 29.2	12.5	281 56.3	22.8
02		121 45.2	215 25.7	29.0	338 34.1	07.9	154 31.2	12.4	296 58.7	22.8
03		136 47.6	230 24.8	. . 29.1	353 36.6	. . 08.0	169 33.2	. . 12.2	312 01.1	. . 22.7
04		151 50.1	245 23.8	29.3	8 39.2	08.2	184 35.2	12.0	327 03.4	22.7
05		166 52.6	260 22.8	29.5	23 41.7	08.3	199 37.2	11.9	342 05.8	22.7
06		181 55.0	275 21.9	S23 29.7	38 44.2	N18 08.4	214 39.2	S14 11.7	357 08.2	N 0 22.7
07	W	196 57.5	290 20.9	29.8	53 46.7	08.6	229 41.2	11.6	12 10.5	22.6
08	E	212 00.0	305 20.0	30.0	68 49.2	08.7	244 43.3	11.4	27 12.9	22.6
09	D	227 02.4	320 19.0	. . 30.2	83 51.7	. . 08.8	259 45.3	. . 11.2	42 15.3	. . 22.6
10	N	242 04.9	335 18.0	30.4	98 54.3	08.9	274 47.3	11.1	57 17.7	22.6
11	E	257 07.4	350 17.1	30.5	113 56.8	09.1	289 49.3	10.9	72 20.0	22.5
12	S	272 09.8	5 16.1	S23 30.7	128 59.3	N18 09.2	304 51.3	S14 10.8	87 22.4	N 0 22.5
13	D	287 12.3	20 15.1	30.9	144 01.8	09.3	319 53.3	10.6	102 24.8	22.5
14	A	302 14.7	35 14.2	31.0	159 04.4	09.5	334 55.3	10.4	117 27.1	22.4
15	Y	317 17.2	50 13.2	. . 31.2	174 06.9	. . 09.6	349 57.3	. . 10.3	132 29.5	. . 22.4
16		332 19.7	65 12.2	31.4	189 09.4	09.7	4 59.3	10.1	147 31.9	22.4
17		347 22.1	80 11.3	31.5	204 12.0	09.9	20 01.3	10.0	162 34.3	22.4
18		2 24.6	95 10.3	S23 31.7	219 14.5	N18 10.0	35 03.3	S14 09.8	177 36.6	N 0 22.3
19		17 27.1	110 09.3	31.8	234 17.0	10.1	50 05.3	09.6	192 39.0	22.3
20		32 29.5	125 08.4	32.0	249 19.6	10.3	65 07.3	09.5	207 41.4	22.3
21		47 32.0	140 07.4	. . 32.2	264 22.1	. . 10.4	80 09.3	. . 09.3	222 43.8	. . 22.3
22		62 34.5	155 06.5	32.3	279 24.6	10.6	95 11.3	09.1	237 46.1	22.2
23		77 36.9	170 05.5	32.5	294 27.2	10.7	110 13.3	09.0	252 48.5	22.2
Mer. Pass.		h m 17 54.3	v −1.0	d 0.2	v 2.5	d 0.1	v 2.0	d 0.2	v 2.4	d 0.0

STARS

Name	SHA	Dec
	° ′	° ′
Acamar	315 19.9	S40 16.0
Achernar	335 28.2	S57 11.3
Acrux	173 12.5	S63 09.1
Adhara	255 14.2	S28 59.1
Aldebaran	290 52.0	N16 31.8
Alioth	166 22.9	N55 54.0
Alkaid	153 01.0	N49 15.5
Al Na'ir	27 47.0	S46 54.9
Alnilam	275 48.6	S 1 11.7
Alphard	217 58.4	S 8 42.1
Alphecca	126 13.4	N26 40.7
Alpheratz	357 46.2	N29 09.0
Altair	62 11.0	N 8 53.7
Ankaa	353 18.0	S42 15.2
Antares	112 29.8	S26 27.2
Arcturus	145 58.2	N19 07.6
Atria	107 34.5	S69 02.7
Avior	234 18.6	S59 32.4
Bellatrix	278 34.4	N 6 21.6
Betelgeuse	271 03.7	N 7 24.5
Canopus	263 56.7	S52 42.0
Capella	280 37.8	N46 00.5
Deneb	49 33.7	N45 19.2
Denebola	182 36.2	N14 30.8
Diphda	348 58.3	S17 55.9
Dubhe	193 54.4	N61 41.4
Elnath	278 15.5	N28 37.0
Eltanin	90 47.9	N51 29.2
Enif	33 49.8	N 9 55.4
Fomalhaut	15 26.8	S29 34.2
Gacrux	172 04.0	S57 10.0
Gienah	175 55.0	S17 35.8
Hadar	148 52.1	S60 25.1
Hamal	328 03.5	N23 30.8
Kaus Aust.	83 47.6	S34 22.8
Kochab	137 20.4	N74 06.5
Markab	13 41.0	N15 15.7
Menkar	314 17.5	N 4 07.8
Menkent	148 10.9	S36 25.1
Miaplacidus	221 39.8	S69 45.3
Mirfak	308 43.7	N49 54.0
Nunki	76 01.8	S26 17.1
Peacock	53 23.6	S56 42.3
Pollux	243 30.5	N28 00.0
Procyon	245 02.1	N 5 11.9
Rasalhague	96 09.2	N12 33.1
Regulus	207 46.0	N11 55.0
Rigel	281 14.2	S 8 11.4
Rigil Kent.	139 55.8	S60 52.4
Sabik	102 15.8	S15 44.2
Schedar	349 43.5	N56 35.9
Shaula	96 25.8	S37 06.6
Sirius	258 35.6	S16 43.8
Spica	158 34.1	S11 12.8
Suhail	222 54.1	S43 28.3
Vega	80 41.1	N38 47.6
Zuben'ubi	137 08.5	S16 05.0

	SHA	Mer. Pass
	° ′	h m
Venus	95 09.6	11 37
Mars	216 48.2	3 29
Jupiter	32 57.9	15 43
Saturn	175 16.0	6 15

UT	SUN GHA	SUN Dec	MOON GHA	v	MOON Dec	d	HP
d h	° ′	° ′	° ′	′	° ′	′	′
1 00	180 31.2	S23 26.2	129 40.5	15.4	S13 01.8	11.3	54.1
01	195 30.9	26.2	144 14.9	15.6	12 50.5	11.3	54.1
02	210 30.5	26.2	158 49.5	15.6	12 39.2	11.4	54.1
03	225 30.2	. . 26.2	173 24.1	15.6	12 27.8	11.5	54.1
04	240 29.9	26.2	187 58.7	15.6	12 16.3	11.5	54.1
05	255 29.6	26.3	202 33.3	15.7	12 04.8	11.5	54.1
06	270 29.3	S23 26.3	217 08.0	15.7	S11 53.3	11.6	54.1
07	285 29.0	26.3	231 42.7	15.8	11 41.7	11.6	54.1
08	300 28.7	26.3	246 17.5	15.8	11 30.1	11.7	54.1
09	315 28.4	. . 26.3	260 52.3	15.9	11 18.4	11.7	54.1
10	330 28.1	26.3	275 27.2	15.8	11 06.7	11.7	54.1
11	345 27.7	26.3	290 02.0	15.9	10 55.0	11.8	54.1
12	0 27.4	S23 26.3	304 36.9	16.0	S10 43.2	11.9	54.1
13	15 27.1	26.3	319 11.9	15.9	10 31.3	11.9	54.1
14	30 26.8	26.3	333 46.8	16.1	10 19.4	11.9	54.1
15	45 26.5	. . 26.3	348 21.9	16.0	10 07.5	11.9	54.1
16	60 26.2	26.3	2 56.9	16.1	9 55.6	12.0	54.1
17	75 25.9	26.3	17 32.0	16.0	9 43.6	12.1	54.1
18	90 25.6	S23 26.3	32 07.0	16.2	S 9 31.5	12.0	54.1
19	105 25.2	26.3	46 42.2	16.1	9 19.5	12.2	54.1
20	120 24.9	26.3	61 17.3	16.2	9 07.3	12.1	54.1
21	135 24.6	. . 26.3	75 52.5	16.2	8 55.2	12.2	54.2
22	150 24.3	26.3	90 27.7	16.2	8 43.0	12.2	54.2
23	165 24.0	26.3	105 02.9	16.3	8 30.8	12.3	54.2
2 00	180 23.7	S23 26.3	119 38.2	16.2	S 8 18.5	12.3	54.2
01	195 23.4	26.3	134 13.4	16.3	8 06.2	12.3	54.2
02	210 23.1	26.3	148 48.7	16.3	7 53.9	12.3	54.2
03	225 22.8	. . 26.3	163 24.0	16.4	7 41.6	12.4	54.2
04	240 22.4	26.3	177 59.4	16.3	7 29.2	12.4	54.2
05	255 22.1	26.3	192 34.7	16.4	7 16.8	12.5	54.2
06	270 21.8	S23 26.3	207 10.1	16.4	S 7 04.3	12.4	54.2
07	285 21.5	26.3	221 45.5	16.4	6 51.9	12.5	54.2
08	300 21.2	26.2	236 20.9	16.4	6 39.4	12.6	54.3
09	315 20.9	. . 26.2	250 56.3	16.4	6 26.8	12.5	54.3
10	330 20.6	26.2	265 31.7	16.4	6 14.3	12.6	54.3
11	345 20.3	26.2	280 07.1	16.5	6 01.7	12.6	54.3
12	0 20.0	S23 26.2	294 42.6	16.4	S 5 49.1	12.7	54.3
13	15 19.6	26.2	309 18.0	16.5	5 36.4	12.7	54.3
14	30 19.3	26.2	323 53.5	16.5	5 23.7	12.6	54.3
15	45 19.0	. . 26.1	338 29.0	16.5	5 11.1	12.8	54.3
16	60 18.7	26.1	353 04.5	16.5	4 58.3	12.7	54.3
17	75 18.4	26.1	7 40.0	16.5	4 45.6	12.8	54.4
18	90 18.1	S23 26.1	22 15.5	16.5	S 4 32.8	12.7	54.4
19	105 17.8	26.1	36 51.0	16.5	4 20.1	12.8	54.4
20	120 17.5	26.0	51 26.5	16.5	4 07.3	12.9	54.4
21	135 17.2	. . 26.0	66 02.0	16.5	3 54.4	12.8	54.4
22	150 16.8	26.0	80 37.5	16.5	3 41.6	12.9	54.4
23	165 16.5	26.0	95 13.0	16.5	3 28.7	12.9	54.4
3 00	180 16.2	S23 25.9	109 48.5	16.5	S 3 15.8	12.9	54.5
01	195 15.9	25.9	124 24.0	16.5	3 02.9	12.9	54.5
02	210 15.6	25.9	138 59.5	16.5	2 50.0	12.9	54.5
03	225 15.3	. . 25.9	153 35.0	16.5	2 37.1	13.0	54.5
04	240 15.0	25.8	168 10.5	16.5	2 24.1	13.0	54.5
05	255 14.7	25.8	182 46.0	16.5	2 11.1	12.9	54.5
06	270 14.4	S23 25.8	197 21.5	16.5	S 1 58.2	13.0	54.6
07	285 14.0	25.8	211 57.0	16.4	1 45.2	13.1	54.6
08	300 13.7	25.7	226 32.4	16.5	1 32.1	13.0	54.6
09	315 13.4	. . 25.7	241 07.9	16.4	1 19.1	13.0	54.6
10	330 13.1	25.7	255 43.3	16.5	1 06.1	13.1	54.6
11	345 12.8	25.6	270 18.8	16.4	0 53.0	13.1	54.6
12	0 12.5	S23 25.6	284 54.2	16.4	S 0 39.9	13.0	54.7
13	15 12.2	25.6	299 29.6	16.4	0 26.9	13.1	54.7
14	30 11.9	25.5	314 05.0	16.4	0 13.8	13.1	54.7
15	45 11.6	. . 25.5	328 40.4	16.3	S 0 00.7	13.1	54.7
16	60 11.2	25.4	343 15.7	16.3	N 0 12.4	13.1	54.7
17	75 10.9	25.4	357 51.0	16.3	0 25.5	13.1	54.8
18	90 10.6	S23 25.4	12 26.3	16.3	N 0 38.6	13.2	54.8
19	105 10.3	25.3	27 01.6	16.3	0 51.8	13.1	54.8
20	120 10.0	25.3	41 36.9	16.2	1 04.9	13.1	54.8
21	135 09.7	. . 25.2	56 12.1	16.3	1 18.0	13.2	54.8
22	150 09.4	25.2	70 47.4	16.2	1 31.2	13.1	54.9
23	165 09.1	25.2	85 22.6	16.1	N 1 44.3	13.2	54.9
	SD 16.3	*d* 0.0	SD 14.7		14.8		14.9

Lat.	Twilight Naut.	Twilight Civil	Sunrise	Moonrise 21	22	23	24
°	h m	h m	h m	h m	h m	h m	h m
N 72	08 26	10 58	▬	12 05	11 36	11 11	10 46
N 70	08 06	09 55	▬	11 48	11 28	11 09	10 51
68	07 50	09 20	▬	11 35	11 21	11 08	10 56
66	07 37	08 54	10 35	11 23	11 15	11 08	11 00
64	07 26	08 34	09 53	11 14	11 10	11 07	11 03
62	07 16	08 18	09 24	11 06	11 06	11 06	11 06
60	07 08	08 05	09 03	10 59	11 02	11 06	11 09
N 58	07 00	07 53	08 45	10 52	10 59	11 05	11 11
56	06 54	07 43	08 30	10 47	10 56	11 05	11 13
54	06 47	07 34	08 18	10 42	10 54	11 04	11 15
52	06 42	07 25	08 06	10 37	10 51	11 04	11 17
50	06 36	07 18	07 56	10 33	10 49	11 04	11 18
45	06 25	07 02	07 36	10 24	10 44	11 03	11 22
N 40	06 14	06 48	07 19	10 17	10 40	11 02	11 25
35	06 05	06 36	07 05	10 10	10 37	11 02	11 27
30	05 56	06 26	06 52	10 05	10 33	11 01	11 29
20	05 39	06 07	06 31	09 55	10 28	11 00	11 33
N 10	05 23	05 50	06 12	09 46	10 23	11 00	11 37
0	05 06	05 32	05 55	09 38	10 19	10 59	11 40
S 10	04 47	05 14	05 37	09 30	10 14	10 58	11 43
20	04 24	04 54	05 18	09 21	10 09	10 58	11 47
30	03 55	04 29	04 56	09 11	10 04	10 57	11 51
35	03 37	04 13	04 43	09 05	10 01	10 57	11 53
40	03 13	03 55	04 28	08 58	09 57	10 56	11 56
45	02 43	03 33	04 10	08 51	09 53	10 56	11 59
S 50	01 57	03 03	03 47	08 41	09 48	10 55	12 03
52	01 29	02 48	03 37	08 37	09 46	10 55	12 05
54	00 42	02 30	03 24	08 32	09 43	10 54	12 07
56	////	02 07	03 10	08 27	09 40	10 54	12 09
58	////	01 37	02 53	08 21	09 37	10 53	12 11
S 60	////	00 46	02 32	08 14	09 34	10 53	12 14

Lat.	Sunset	Twilight Civil	Twilight Naut.	Moonset 21	22	23	24
°	h m	h m	h m	h m	h m	h m	h m
N 72	▬	12 59	15 31	19 55	21 50	23 42	25 38
N 70	▬	14 02	15 51	20 10	21 55	23 39	25 27
68	▬	14 38	16 07	20 22	22 00	23 37	25 18
66	13 22	15 03	16 20	20 31	22 03	23 35	25 10
64	14 05	15 23	16 31	20 39	22 07	23 34	25 04
62	14 33	15 39	16 41	20 46	22 09	23 33	24 58
60	14 55	15 53	16 50	20 52	22 12	23 31	24 53
N 58	15 12	16 04	16 57	20 57	22 14	23 30	24 49
56	15 27	16 15	17 04	21 02	22 16	23 30	24 45
54	15 40	16 24	17 10	21 06	22 17	23 29	24 42
52	15 51	16 32	17 16	21 10	22 19	23 28	24 39
50	16 01	16 39	17 21	21 13	22 20	23 27	24 36
45	16 22	16 56	17 33	21 21	22 23	23 26	24 30
N 40	16 39	17 09	17 43	21 27	22 26	23 25	24 25
35	16 53	17 21	17 53	21 32	22 28	23 24	24 21
30	17 05	17 32	18 01	21 37	22 30	23 23	24 18
20	17 26	17 50	18 18	21 44	22 33	23 21	24 11
N 10	17 45	18 08	18 34	21 51	22 36	23 20	24 05
0	18 02	18 25	18 51	21 58	22 38	23 19	24 00
S 10	18 20	18 43	19 10	22 04	22 41	23 17	23 55
20	18 39	19 04	19 33	22 11	22 43	23 16	23 49
30	19 01	19 29	20 02	22 18	22 46	23 14	23 43
35	19 14	19 44	20 21	22 23	22 48	23 13	23 39
40	19 29	20 02	20 44	22 28	22 50	23 12	23 35
45	19 47	20 25	21 15	22 33	22 52	23 11	23 31
S 50	20 10	20 55	22 00	22 40	22 55	23 10	23 25
52	20 21	21 10	22 28	22 43	22 56	23 09	23 22
54	20 33	21 28	23 15	22 47	22 58	23 08	23 19
56	20 47	21 50	////	22 50	22 59	23 08	23 16
58	21 04	22 20	////	22 55	23 01	23 07	23 13
S 60	21 25	23 11	////	22 59	23 03	23 06	23 09

Day	SUN Eqn. of Time 00^h	Eqn. of Time 12^h	Mer. Pass.	MOON Mer. Pass. Upper	Mer. Pass. Lower	Age	Phase
d	m s	m s	h m	h m	h m	d	%
21	02 05	01 50	11 58	15 48	03 27	05	21
22	01 35	01 20	11 59	16 28	04 08	06	29
23	01 06	00 51	11 59	17 09	04 49	07	38

	UT	ARIES	VENUS −3.9		MARS −0.6		JUPITER −2.1		SATURN +0.9	
	d h	GHA	GHA	Dec	GHA	Dec	GHA	Dec	GHA	Dec
		° ′	° ′	° ′	° ′	° ′	° ′	° ′	° ′	° ′
	24 00	92 39.4	185 04.5	S23 32.6	309 29.7	N18 10.8	125 15.3	S14 08.8	267 50.9	N 0 22.2
	01	107 41.9	200 03.6	32.8	324 32.3	11.0	140 17.3	08.7	282 53.3	22.2
	02	122 44.3	215 02.6	32.9	339 34.8	11.1	155 19.3	08.5	297 55.6	22.1
	03	137 46.8	230 01.6	33.1	354 37.4	11.2	170 21.3	08.3	312 58.0	22.1
	04	152 49.2	245 00.7	33.2	9 39.9	11.4	185 23.3	08.2	328 00.4	22.1
	05	167 51.7	259 59.7	33.4	24 42.4	11.5	200 25.3	08.0	343 02.8	22.1
	06	182 54.2	274 58.7	S23 33.5	39 45.0	N18 11.7	215 27.3	S14 07.8	358 05.1	N 0 22.0
	07	197 56.6	289 57.8	33.7	54 47.5	11.8	230 29.3	07.7	13 07.5	22.0
T	08	212 59.1	304 56.8	33.8	69 50.1	11.9	245 31.3	07.5	28 09.9	22.0
H	09	228 01.6	319 55.8	33.9	84 52.7	12.1	260 33.3	07.4	43 12.3	22.0
U	10	243 04.0	334 54.9	34.1	99 55.2	12.2	275 35.3	07.2	58 14.6	21.9
R	11	258 06.5	349 53.9	34.2	114 57.8	12.4	290 37.3	07.0	73 17.0	21.9
S	12	273 09.0	4 52.9	S23 34.4	130 00.3	N18 12.5	305 39.3	S14 06.9	88 19.4	N 0 21.9
D	13	288 11.4	19 52.0	34.5	145 02.9	12.6	320 41.4	06.7	103 21.8	21.9
A	14	303 13.9	34 51.0	34.6	160 05.4	12.8	335 43.4	06.5	118 24.1	21.8
Y	15	318 16.3	49 50.0	34.8	175 08.0	12.9	350 45.4	06.4	133 26.5	21.8
	16	333 18.8	64 49.1	34.9	190 10.6	13.1	5 47.4	06.2	148 28.9	21.8
	17	348 21.3	79 48.1	35.0	205 13.1	13.2	20 49.4	06.1	163 31.3	21.8
	18	3 23.7	94 47.1	S23 35.2	220 15.7	N18 13.4	35 51.4	S14 05.9	178 33.7	N 0 21.7
	19	18 26.2	109 46.1	35.3	235 18.3	13.5	50 53.4	05.7	193 36.0	21.7
	20	33 28.7	124 45.2	35.4	250 20.8	13.7	65 55.4	05.6	208 38.4	21.7
	21	48 31.1	139 44.2	35.6	265 23.4	13.8	80 57.4	05.4	223 40.8	21.7
	22	63 33.6	154 43.2	35.7	280 26.0	14.0	95 59.4	05.2	238 43.2	21.6
	23	78 36.1	169 42.3	35.8	295 28.6	14.1	111 01.4	05.1	253 45.5	21.6
	25 00	93 38.5	184 41.3	S23 35.9	310 31.1	N18 14.2	126 03.4	S14 04.9	268 47.9	N 0 21.6
	01	108 41.0	199 40.3	36.0	325 33.7	14.4	141 05.4	04.8	283 50.3	21.6
	02	123 43.5	214 39.4	36.2	340 36.3	14.5	156 07.4	04.6	298 52.7	21.5
	03	138 45.9	229 38.4	36.3	355 38.9	14.7	171 09.4	04.4	313 55.1	21.5
	04	153 48.4	244 37.4	36.4	10 41.4	14.8	186 11.4	04.3	328 57.4	21.5
	05	168 50.8	259 36.5	36.5	25 44.0	15.0	201 13.3	04.1	343 59.8	21.5
	06	183 53.3	274 35.5	S23 36.6	40 46.6	N18 15.1	216 15.3	S14 03.9	359 02.2	N 0 21.4
	07	198 55.8	289 34.5	36.8	55 49.2	15.3	231 17.3	03.8	14 04.6	21.4
	08	213 58.2	304 33.6	36.9	70 51.8	15.4	246 19.3	03.6	29 07.0	21.4
F	09	229 00.7	319 32.6	37.0	85 54.4	15.6	261 21.3	03.4	44 09.3	21.4
R	10	244 03.2	334 31.6	37.1	100 57.0	15.8	276 23.3	03.3	59 11.7	21.3
I	11	259 05.6	349 30.6	37.2	115 59.6	15.9	291 25.3	03.1	74 14.1	21.3
D	12	274 08.1	4 29.7	S23 37.3	131 02.1	N18 16.1	306 27.3	S14 03.0	89 16.5	N 0 21.3
A	13	289 10.6	19 28.7	37.4	146 04.7	16.2	321 29.3	02.8	104 18.9	21.3
Y	14	304 13.0	34 27.7	37.5	161 07.3	16.4	336 31.3	02.6	119 21.2	21.3
	15	319 15.5	49 26.8	37.6	176 09.9	16.5	351 33.3	02.5	134 23.6	21.2
	16	334 18.0	64 25.8	37.7	191 12.5	16.7	6 35.3	02.3	149 26.0	21.2
	17	349 20.4	79 24.8	37.8	206 15.1	16.8	21 37.3	02.1	164 28.4	21.2
	18	4 22.9	94 23.9	S23 37.9	221 17.7	N18 17.0	36 39.3	S14 02.0	179 30.8	N 0 21.2
	19	19 25.3	109 22.9	38.0	236 20.3	17.1	51 41.3	01.8	194 33.2	21.1
	20	34 27.8	124 21.9	38.1	251 22.9	17.3	66 43.3	01.6	209 35.5	21.1
	21	49 30.3	139 20.9	38.2	266 25.5	17.5	81 45.3	01.5	224 37.9	21.1
	22	64 32.7	154 20.0	38.3	281 28.1	17.6	96 47.3	01.3	239 40.3	21.1
	23	79 35.2	169 19.0	38.4	296 30.8	17.8	111 49.3	01.1	254 42.7	21.0
	26 00	94 37.7	184 18.0	S23 38.5	311 33.4	N18 17.9	126 51.3	S14 01.0	269 45.1	N 0 21.0
	01	109 40.1	199 17.1	38.6	326 36.0	18.1	141 53.3	00.8	284 47.4	21.0
	02	124 42.6	214 16.1	38.7	341 38.6	18.2	156 55.3	00.6	299 49.8	21.0
	03	139 45.1	229 15.1	38.8	356 41.2	18.4	171 57.3	00.5	314 52.2	21.0
	04	154 47.5	244 14.2	38.9	11 43.8	18.6	186 59.3	00.3	329 54.6	20.9
	05	169 50.0	259 13.2	38.9	26 46.4	18.7	202 01.3	00.2	344 57.0	20.9
	06	184 52.4	274 12.2	S23 39.0	41 49.0	N18 18.9	217 03.3	S14 00.0	359 59.4	N 0 20.9
	07	199 54.9	289 11.2	39.1	56 51.7	19.0	232 05.3	13 59.8	15 01.7	20.9
S	08	214 57.4	304 10.3	39.2	71 54.3	19.2	247 07.3	59.7	30 04.1	20.8
A	09	229 59.8	319 09.3	39.3	86 56.9	19.4	262 09.3	59.5	45 06.5	20.8
T	10	245 02.3	334 08.3	39.4	101 59.5	19.5	277 11.3	59.3	60 08.9	20.8
U	11	260 04.8	349 07.4	39.4	117 02.2	19.7	292 13.2	59.2	75 11.3	20.8
R	12	275 07.2	4 06.4	S23 39.5	132 04.8	N18 19.9	307 15.2	S13 59.0	90 13.7	N 0 20.8
D	13	290 09.7	19 05.4	39.6	147 07.4	20.0	322 17.2	58.8	105 16.1	20.7
A	14	305 12.2	34 04.4	39.7	162 10.0	20.2	337 19.2	58.7	120 18.4	20.7
Y	15	320 14.6	49 03.5	39.7	177 12.7	20.3	352 21.2	58.5	135 20.8	20.7
	16	335 17.1	64 02.5	39.8	192 15.3	20.5	7 23.2	58.3	150 23.2	20.7
	17	350 19.6	79 01.5	39.9	207 17.9	20.7	22 25.2	58.2	165 25.6	20.7
	18	5 22.0	94 00.6	S23 39.9	222 20.6	N18 20.8	37 27.2	S13 58.0	180 28.0	N 0 20.6
	19	20 24.5	108 59.6	40.0	237 23.2	21.0	52 29.2	57.8	195 30.4	20.6
	20	35 26.9	123 58.6	40.1	252 25.8	21.2	67 31.2	57.7	210 32.8	20.6
	21	50 29.4	138 57.6	40.1	267 28.5	21.3	82 33.2	57.5	225 35.1	20.6
	22	65 31.9	153 56.7	40.2	282 31.1	21.5	97 35.2	57.3	240 37.5	20.5
	23	80 34.3	168 55.7	40.3	297 33.8	21.7	112 37.2	57.2	255 39.9	20.5
	Mer. Pass.	h m 17 42.5	v −1.0	d 0.1	v 2.6	d 0.2	v 2.0	d 0.2	v 2.4	d 0.0

STARS		
Name	SHA	Dec
	° ′	°
Acamar	315 19.9	S40 16
Achernar	335 28.3	S57 11
Acrux	173 12.4	S63 09
Adhara	255 14.2	S28 59
Aldebaran	290 52.0	N16 31
Alioth	166 22.8	N55 53
Alkaid	153 01.0	N49 15
Al Na'ir	27 47.1	S46 54
Alnilam	275 48.6	S 1 11
Alphard	217 58.4	S 8 42
Alphecca	126 13.4	N26 40
Alpheratz	357 46.2	N29 09
Altair	62 11.0	N 8 53
Ankaa	353 18.1	S42 15
Antares	112 29.8	S26 27
Arcturus	145 58.2	N19 07
Atria	107 34.5	S69 02
Avior	234 18.6	S59 32
Bellatrix	278 34.4	N 6 21
Betelgeuse	271 03.7	N 7 24
Canopus	263 56.7	S52 42.
Capella	280 37.8	N46 00
Deneb	49 33.7	N45 19
Denebola	182 36.2	N14 30
Diphda	348 58.4	S17 55
Dubhe	193 54.4	N61 41.
Elnath	278 15.5	N28 37.
Eltanin	90 47.9	N51 29
Enif	33 49.8	N 9 55.
Fomalhaut	15 26.8	S29 34.
Gacrux	172 04.0	S57 10.
Gienah	175 55.0	S17 35
Hadar	148 52.0	S60 25
Hamal	328 03.5	N23 30
Kaus Aust.	83 47.6	S34 22.
Kochab	137 20.4	N74 06.
Markab	13 41.0	N15 15.
Menkar	314 17.5	N 4 07.
Menkent	148 10.9	S36 25.
Miaplacidus	221 39.7	S69 45.
Mirfak	308 43.7	N49 54.
Nunki	76 01.8	S26 17.
Peacock	53 23.6	S56 42.
Pollux	243 30.5	N28 00.
Procyon	245 02.1	N 5 11.
Rasalhague	96 09.2	N12 33.
Regulus	207 46.0	N11 54.
Rigel	281 14.2	S 8 11.
Rigil Kent.	139 55.8	S60 52.
Sabik	102 15.8	S15 44.
Schedar	349 43.5	N56 35
Shaula	96 25.8	S37 06.
Sirius	258 35.6	S16 43
Spica	158 34.1	S11 12
Suhail	222 54.1	S43 28
Vega	80 41.1	N38 47
Zuben'ubi	137 08.5	S16 05

	SHA	Mer. Pas
	° ′	h m
Venus	91 02.8	11 42
Mars	216 52.6	3 17
Jupiter	32 24.8	15 34
Saturn	175 09.4	6 04

UT (d h)	SUN GHA (° ′)	SUN Dec (° ′)	MOON GHA (° ′)	v (′)	MOON Dec (° ′)	d (′)	HP (′)
24 00	180 08.8	S23 25.1	99 57.7	16.2	N 1 57.5	13.1	54.9
01	195 08.4	25.1	114 32.9	16.1	2 10.6	13.2	54.9
02	210 08.1	25.0	129 08.0	16.0	2 23.8	13.2	54.9
03	225 07.8	. . 25.0	143 43.0	16.1	2 37.0	13.1	55.0
04	240 07.5	24.9	158 18.1	16.0	2 50.1	13.2	55.0
05	255 07.2	24.9	172 53.1	16.0	3 03.3	13.1	55.0
06	270 06.9	S23 24.8	187 28.1	15.9	N 3 16.4	13.2	55.0
07	285 06.6	24.8	202 03.0	15.9	3 29.6	13.1	55.1
08	300 06.3	24.7	216 37.9	15.9	3 42.7	13.2	55.1
09	315 06.0	. . 24.7	231 12.8	15.8	3 55.9	13.1	55.1
10	330 05.7	24.6	245 47.6	15.8	4 09.0	13.2	55.1
11	345 05.3	24.6	260 22.4	15.8	4 22.2	13.1	55.2
12	0 05.0	S23 24.5	274 57.2	15.7	N 4 35.3	13.2	55.2
13	15 04.7	24.5	289 31.9	15.6	4 48.5	13.1	55.2
14	30 04.4	24.4	304 06.5	15.7	5 01.6	13.1	55.2
15	45 04.1	. . 24.4	318 41.2	15.5	5 14.7	13.1	55.3
16	60 03.8	24.3	333 15.7	15.6	5 27.8	13.1	55.3
17	75 03.5	24.2	347 50.3	15.4	5 40.9	13.1	55.3
18	90 03.2	S23 24.2	2 24.7	15.5	N 5 54.0	13.1	55.3
19	105 02.9	24.1	16 59.2	15.3	6 07.1	13.1	55.4
20	120 02.6	24.1	31 33.5	15.4	6 20.2	13.0	55.4
21	135 02.2	. . 24.0	46 07.9	15.2	6 33.2	13.1	55.4
22	150 01.9	23.9	60 42.1	15.3	6 46.3	13.0	55.5
23	165 01.6	23.9	75 16.4	15.1	6 59.3	13.0	55.5
25 00	180 01.3	S23 23.8	89 50.5	15.1	N 7 12.3	13.0	55.5
01	195 01.0	23.8	104 24.6	15.1	7 25.3	13.0	55.5
02	210 00.7	23.7	118 58.7	15.0	7 38.3	13.0	55.6
03	225 00.4	. . 23.6	133 32.7	14.9	7 51.3	12.9	55.6
04	240 00.1	23.6	148 06.6	14.9	8 04.2	12.9	55.6
05	254 59.8	23.5	162 40.5	14.8	8 17.1	12.9	55.7
06	269 59.5	S23 23.4	177 14.3	14.7	N 8 30.0	12.9	55.7
07	284 59.1	23.4	191 48.0	14.7	8 42.9	12.9	55.7
08	299 58.8	23.3	206 21.7	14.6	8 55.8	12.8	55.8
09	314 58.5	. . 23.2	220 55.3	14.5	9 08.6	12.9	55.8
10	329 58.2	23.1	235 28.8	14.5	9 21.5	12.8	55.8
11	344 57.9	23.1	250 02.3	14.4	9 34.3	12.7	55.8
12	359 57.6	S23 23.0	264 35.7	14.3	N 9 47.0	12.8	55.9
13	14 57.3	22.9	279 09.0	14.3	9 59.8	12.7	55.9
14	29 57.0	22.8	293 42.3	14.2	10 12.5	12.7	55.9
15	44 56.7	. . 22.8	308 15.5	14.1	10 25.2	12.6	56.0
16	59 56.4	22.7	322 48.6	14.0	10 37.8	12.7	56.0
17	74 56.1	22.6	337 21.6	14.0	10 50.5	12.6	56.0
18	89 55.7	S23 22.5	351 54.6	13.9	N11 03.1	12.5	56.1
19	104 55.4	22.5	6 27.5	13.8	11 15.6	12.6	56.1
20	119 55.1	22.4	21 00.3	13.7	11 28.2	12.5	56.1
21	134 54.8	. . 22.3	35 33.0	13.6	11 40.7	12.4	56.2
22	149 54.5	22.2	50 05.6	13.6	11 53.1	12.5	56.2
23	164 54.2	22.1	64 38.2	13.5	12 05.6	12.3	56.2
26 00	179 53.9	S23 22.0	79 10.7	13.3	N12 17.9	12.4	56.3
01	194 53.6	22.0	93 43.0	13.3	12 30.3	12.3	56.3
02	209 53.3	21.9	108 15.3	13.3	12 42.6	12.3	56.3
03	224 53.0	. . 21.8	122 47.6	13.1	12 54.9	12.2	56.4
04	239 52.7	21.7	137 19.7	13.0	13 07.1	12.2	56.4
05	254 52.3	21.6	151 51.7	13.0	13 19.3	12.1	56.4
06	269 52.0	S23 21.5	166 23.7	12.8	N13 31.4	12.1	56.5
07	284 51.7	21.4	180 55.5	12.8	13 43.5	12.0	56.5
08	299 51.4	21.4	195 27.3	12.6	13 55.5	12.0	56.6
09	314 51.1	. . 21.3	209 58.9	12.6	14 07.5	12.0	56.6
10	329 50.8	21.2	224 30.5	12.5	14 19.5	11.9	56.6
11	344 50.5	21.1	239 02.0	12.3	14 31.4	11.8	56.7
12	359 50.2	S23 21.0	253 33.3	12.3	N14 43.2	11.8	56.7
13	14 49.9	20.9	268 04.6	12.2	14 55.0	11.7	56.7
14	29 49.6	20.8	282 35.8	12.1	15 06.7	11.7	56.8
15	44 49.3	. . 20.7	297 06.9	12.0	15 18.4	11.6	56.8
16	59 49.0	20.6	311 37.9	11.8	15 30.0	11.6	56.9
17	74 48.6	20.5	326 08.7	11.8	15 41.6	11.5	56.9
18	89 48.3	S23 20.4	340 39.5	11.7	N15 53.1	11.5	56.9
19	104 48.0	20.3	355 10.2	11.5	16 04.6	11.3	57.0
20	119 47.7	20.2	9 40.7	11.5	16 15.9	11.3	57.0
21	134 47.4	. . 20.1	24 11.2	11.3	16 27.2	11.3	57.0
22	149 47.1	20.0	38 41.5	11.3	16 38.5	11.2	57.1
23	164 46.8	19.9	53 11.8	11.1	N16 49.7	11.1	57.1
	SD 16.3	*d* 0.1	SD 15.0		15.2		15.4

Lat.	Twilight Naut. (h m)	Twilight Civil (h m)	Sunrise (h m)	Moonrise 24 (h m)	Moonrise 25 (h m)	Moonrise 26 (h m)	Moonrise 27 (h m)
N 72	08 27	10 57	■	10 46	10 17	09 38	□
N 70	08 07	09 55	■	10 51	10 31	10 06	09 19
68	07 51	09 20	■	10 56	10 43	10 27	10 03
66	07 38	08 55	10 35	11 00	10 52	10 43	10 32
64	07 27	08 35	09 53	11 03	11 00	10 57	10 54
62	07 17	08 19	09 25	11 06	11 07	11 08	11 12
60	07 09	08 06	09 03	11 09	11 13	11 18	11 27
N 58	07 01	07 54	08 46	11 11	11 18	11 27	11 40
56	06 55	07 44	08 31	11 13	11 23	11 35	11 51
54	06 49	07 35	08 19	11 15	11 27	11 42	12 01
52	06 43	07 27	08 08	11 17	11 31	11 48	12 09
50	06 38	07 19	07 58	11 18	11 35	11 54	12 17
45	06 26	07 03	07 37	11 22	11 42	12 06	12 34
N 40	06 15	06 50	07 20	11 25	11 49	12 16	12 48
35	06 06	06 38	07 06	11 27	11 54	12 25	13 00
30	05 57	06 27	06 54	11 29	11 59	12 32	13 10
20	05 41	06 08	06 32	11 33	12 08	12 46	13 28
N 10	05 25	05 51	06 14	11 37	12 16	12 57	13 44
0	05 08	05 34	05 56	11 40	12 23	13 08	13 58
S 10	04 49	05 16	05 39	11 43	12 30	13 20	14 13
20	04 26	04 55	05 20	11 47	12 38	13 32	14 29
30	03 57	04 30	04 58	11 51	12 47	13 45	14 48
35	03 38	04 15	04 45	11 53	12 52	13 54	14 59
40	03 15	03 57	04 30	11 56	12 58	14 03	15 11
45	02 45	03 34	04 12	11 59	13 05	14 14	15 26
S 50	01 59	03 05	03 49	12 03	13 13	14 27	15 44
52	01 31	02 49	03 38	12 05	13 17	14 33	15 53
54	00 45	02 32	03 26	12 07	13 21	14 40	16 02
56	////	02 09	03 12	12 09	13 26	14 48	16 13
58	////	01 39	02 55	12 11	13 32	14 56	16 26
S 60	////	00 49	02 35	12 14	13 38	15 06	16 40

Lat.	Sunset (h m)	Twilight Civil (h m)	Twilight Naut. (h m)	Moonset 24 (h m)	Moonset 25 (h m)	Moonset 26 (h m)	Moonset 27 (h m)
N 72	■	13 03	15 33	25 38	01 38	03 51	□
N 70	■	14 05	15 53	25 27	01 27	03 25	05 53
68	■	14 40	16 09	25 18	01 18	03 06	05 11
66	13 25	15 05	16 22	25 10	01 10	02 51	04 42
64	14 07	15 25	16 33	25 04	01 04	02 38	04 21
62	14 35	15 41	16 43	24 58	00 58	02 28	04 04
60	14 57	15 55	16 51	24 53	00 53	02 19	03 50
N 58	15 14	16 06	16 59	24 49	00 49	02 11	03 38
56	15 29	16 17	17 06	24 45	00 45	02 04	03 28
54	15 42	16 26	17 12	24 42	00 42	01 58	03 19
52	15 53	16 34	17 18	24 39	00 39	01 53	03 10
50	16 03	16 41	17 23	24 36	00 36	01 48	03 03
45	16 23	16 57	17 35	24 30	00 30	01 37	02 48
N 40	16 40	17 11	17 45	24 25	00 25	01 29	02 35
35	16 54	17 23	17 54	24 21	00 21	01 21	02 24
30	17 07	17 33	18 03	24 18	00 18	01 15	02 15
20	17 28	17 52	18 19	24 11	00 11	01 03	01 59
N 10	17 46	18 09	18 36	24 05	00 05	00 53	01 45
0	18 04	18 26	18 53	24 00	00 00	00 44	01 32
S 10	18 21	18 45	19 12	23 55	24 35	00 35	01 19
20	18 40	19 05	19 34	23 49	24 25	00 25	01 05
30	19 02	19 30	20 03	23 43	24 14	00 14	00 49
35	19 15	19 45	20 22	23 39	24 08	00 08	00 40
40	19 30	20 03	20 45	23 35	24 01	00 01	00 30
45	19 48	20 26	21 16	23 31	23 52	24 17	00 17
S 50	20 11	20 56	22 01	23 25	23 42	24 03	00 03
52	20 22	21 11	22 29	23 22	23 37	23 56	24 21
54	20 34	21 29	23 15	23 19	23 32	23 48	24 10
56	20 48	21 51	////	23 16	23 27	23 40	23 59
58	21 05	22 20	////	23 13	23 20	23 31	23 46
S 60	21 25	23 10	////	23 09	23 13	23 20	23 30

Day	SUN Eqn. of Time 00^h (m s)	SUN Eqn. of Time 12^h (m s)	SUN Mer. Pass. (h m)	MOON Mer. Pass. Upper (h m)	MOON Mer. Pass. Lower (h m)	MOON Age (d)	MOON Phase (%)
24	00 36	00 21	12 00	17 50	05 29	08	48
25	00 06	00 09	12 00	18 33	06 11	09	58
26	00 24	00 39	12 01	19 20	06 56	10	68

	UT	ARIES	VENUS −4.0		MARS −0.7		JUPITER −2.1		SATURN +0.9		STARS		
		GHA	GHA	Dec	GHA	Dec	GHA	Dec	GHA	Dec	Name	SHA	Dec
	d h	° ′	° ′	° ′	° ′	° ′	° ′	° ′	° ′	° ′		° ′	°
	27 00	95 36.8	183 54.7	S23 40.3	312 36.4	N18 21.9	127 39.2	S13 57.0	270 42.3	N 0 20.5	Acamar	315 19.9	S40 16
	01	110 39.3	198 53.8	40.4	327 39.0	22.0	142 41.2	56.8	285 44.7	20.5	Achernar	335 28.3	S57 11
	02	125 41.7	213 52.8	40.5	342 41.7	22.2	157 43.1	56.7	300 47.1	20.5	Acrux	173 12.4	S63 09
	03	140 44.2	228 51.8	. . 40.5	357 44.3	. . 22.4	172 45.1	. . 56.5	315 49.5	. . 20.4	Adhara	255 14.1	S28 59
	04	155 46.7	243 50.8	40.6	12 47.0	22.5	187 47.1	56.3	330 51.8	20.4	Aldebaran	290 52.0	N16 31
	05	170 49.1	258 49.9	40.6	27 49.6	22.7	202 49.1	56.2	345 54.2	20.4			
	06	185 51.6	273 48.9	S23 40.7	42 52.3	N18 22.9	217 51.1	S13 56.0	0 56.6	N 0 20.4	Alioth	166 22.8	N55 53
	07	200 54.1	288 47.9	40.7	57 54.9	23.0	232 53.1	55.8	15 59.0	20.4	Alkaid	153 00.9	N49 15
	08	215 56.5	303 46.9	40.8	72 57.6	23.2	247 55.1	55.7	31 01.4	20.3	Al Na'ir	27 47.1	S46 54.
S	09	230 59.0	318 46.0	. . 40.8	88 00.2	. . 23.4	262 57.1	. . 55.5	46 03.8	. . 20.3	Alnilam	275 48.6	S 1 11
U	10	246 01.4	333 45.0	40.9	103 02.9	23.6	277 59.1	55.3	61 06.2	20.3	Alphard	217 58.4	S 8 42
N	11	261 03.9	348 44.0	40.9	118 05.6	23.7	293 01.1	55.2	76 08.6	20.3			
D	12	276 06.4	3 43.1	S23 41.0	133 08.2	N18 23.9	308 03.1	S13 55.0	91 10.9	N 0 20.3	Alphecca	126 13.4	N26 40.
A	13	291 08.8	18 42.1	41.0	148 10.9	24.1	323 05.0	54.8	106 13.3	20.2	Alpheratz	357 46.2	N29 09.
Y	14	306 11.3	33 41.1	41.1	163 13.6	24.3	338 07.0	54.7	121 15.7	20.2	Altair	62 11.0	N 8 53
	15	321 13.8	48 40.1	. . 41.1	178 16.2	. . 24.4	353 09.0	. . 54.5	136 18.1	. . 20.2	Ankaa	353 18.1	S42 15.
	16	336 16.2	63 39.2	41.2	193 18.9	24.6	8 11.0	54.3	151 20.5	20.2	Antares	112 29.7	S26 27.
	17	351 18.7	78 38.2	41.2	208 21.5	24.8	23 13.0	54.2	166 22.9	20.2			
	18	6 21.2	93 37.2	S23 41.2	223 24.2	N18 25.0	38 15.0	S13 54.0	181 25.3	N 0 20.1	Arcturus	145 58.2	N19 07
	19	21 23.6	108 36.2	41.3	238 26.9	25.1	53 17.0	53.8	196 27.7	20.1	Atria	107 34.5	S69 02
	20	36 26.1	123 35.3	41.3	253 29.6	25.3	68 19.0	53.7	211 30.1	20.1	Avior	234 18.6	S59 32.
	21	51 28.6	138 34.3	. . 41.3	268 32.2	. . 25.5	83 21.0	. . 53.5	226 32.5	. . 20.1	Bellatrix	278 34.4	N 6 21
	22	66 31.0	153 33.3	41.4	283 34.9	25.7	98 23.0	53.3	241 34.8	20.1	Betelgeuse	271 03.7	N 7 24.
	23	81 33.5	168 32.4	41.4	298 37.6	25.8	113 24.9	53.1	256 37.2	20.1			
	28 00	96 35.9	183 31.4	S23 41.4	313 40.3	N18 26.0	128 26.9	S13 53.0	271 39.6	N 0 20.0	Canopus	263 56.7	S52 42.
	01	111 38.4	198 30.4	41.5	328 42.9	26.2	143 28.9	52.8	286 42.0	20.0	Capella	280 37.8	N46 00
	02	126 40.9	213 29.4	41.5	343 45.6	26.4	158 30.9	52.6	301 44.4	20.0	Deneb	49 33.7	N45 19
	03	141 43.3	228 28.5	. . 41.5	358 48.3	. . 26.6	173 32.9	. . 52.5	316 46.8	. . 20.0	Denebola	182 36.2	N14 30.
	04	156 45.8	243 27.5	41.6	13 51.0	26.7	188 34.9	52.3	331 49.2	20.0	Diphda	348 58.4	S17 55
	05	171 48.3	258 26.5	41.6	28 53.7	26.9	203 36.9	52.1	346 51.6	19.9			
	06	186 50.7	273 25.5	S23 41.6	43 56.4	N18 27.1	218 38.9	S13 52.0	1 54.0	N 0 19.9	Dubhe	193 54.3	N61 41
	07	201 53.2	288 24.6	41.6	58 59.0	27.3	233 40.9	51.8	16 56.4	19.9	Elnath	278 15.5	N28 37
	08	216 55.7	303 23.6	41.6	74 01.7	27.5	248 42.8	51.6	31 58.8	19.9	Eltanin	90 47.9	N51 29
M	09	231 58.1	318 22.6	. . 41.7	89 04.4	. . 27.7	263 44.8	. . 51.5	47 01.1	. . 19.9	Enif	33 49.9	N 9 55.
O	10	247 00.6	333 21.7	41.7	104 07.1	27.8	278 46.8	51.3	62 03.5	19.8	Fomalhaut	15 26.9	S29 34.
N	11	262 03.0	348 20.7	41.7	119 09.8	28.0	293 48.8	51.1	77 05.9	19.8			
D	12	277 05.5	3 19.7	S23 41.7	134 12.5	N18 28.2	308 50.8	S13 51.0	92 08.3	N 0 19.8	Gacrux	172 03.9	S57 10.
A	13	292 08.0	18 18.7	41.7	149 15.2	28.4	323 52.8	50.8	107 10.7	19.8	Gienah	175 54.9	S17 35.
Y	14	307 10.4	33 17.8	41.7	164 17.9	28.6	338 54.8	50.6	122 13.1	19.8	Hadar	148 52.0	S60 25.
	15	322 12.9	48 16.8	. . 41.8	179 20.6	. . 28.8	353 56.8	. . 50.5	137 15.5	. . 19.8	Hamal	328 03.5	N23 30.
	16	337 15.4	63 15.8	41.8	194 23.3	28.9	8 58.7	50.3	152 17.9	19.7	Kaus Aust.	83 47.6	S34 22.
	17	352 17.8	78 14.8	41.8	209 26.0	29.1	24 00.7	50.1	167 20.3	19.7			
	18	7 20.3	93 13.9	S23 41.8	224 28.7	N18 29.3	39 02.7	S13 49.9	182 22.7	N 0 19.7	Kochab	137 20.3	N74 06.
	19	22 22.8	108 12.9	41.8	239 31.4	29.5	54 04.7	49.8	197 25.1	19.7	Markab	13 41.0	N15 15.
	20	37 25.2	123 11.9	41.8	254 34.1	29.7	69 06.7	49.6	212 27.5	19.7	Menkar	314 17.5	N 4 07.
	21	52 27.7	138 11.0	. . 41.8	269 36.8	. . 29.9	84 08.7	. . 49.4	227 29.9	. . 19.7	Menkent	148 10.8	S36 25.
	22	67 30.2	153 10.0	41.8	284 39.5	30.1	99 10.7	49.3	242 32.3	19.6	Miaplacidus	221 39.7	S69 45.
	23	82 32.6	168 09.0	41.8	299 42.2	30.3	114 12.6	49.1	257 34.7	19.6			
	29 00	97 35.1	183 08.0	S23 41.8	314 44.9	N18 30.4	129 14.6	S13 48.9	272 37.0	N 0 19.6	Mirfak	308 43.7	N49 54.
	01	112 37.5	198 07.1	41.8	329 47.7	30.6	144 16.6	48.8	287 39.4	19.6	Nunki	76 01.8	S26 17.
	02	127 40.0	213 06.1	41.8	344 50.4	30.8	159 18.6	48.6	302 41.8	19.6	Peacock	53 23.7	S56 42.
	03	142 42.5	228 05.1	. . 41.8	359 53.1	. . 31.0	174 20.6	. . 48.4	317 44.2	. . 19.5	Pollux	243 30.4	N28 00.
	04	157 44.9	243 04.1	41.8	14 55.8	31.2	189 22.6	48.2	332 46.6	19.5	Procyon	245 02.1	N 5 11.
	05	172 47.4	258 03.2	41.8	29 58.5	31.4	204 24.6	48.1	347 49.0	19.5			
	06	187 49.9	273 02.2	S23 41.8	45 01.2	N18 31.6	219 26.5	S13 47.9	2 51.4	N 0 19.5	Rasalhague	96 09.1	N12 33.
	07	202 52.3	288 01.2	41.8	60 04.0	31.8	234 28.5	47.7	17 53.8	19.5	Regulus	207 46.0	N11 54.
	08	217 54.8	303 00.2	41.8	75 06.7	32.0	249 30.5	47.6	32 56.2	19.5	Rigel	281 14.2	S 8 11
T	09	232 57.3	317 59.3	. . 41.8	90 09.4	. . 32.2	264 32.5	. . 47.4	47 58.6	. . 19.4	Rigil Kent.	139 55.7	S60 52.
U	10	247 59.7	332 58.3	41.8	105 12.1	32.4	279 34.5	47.2	63 01.0	19.4	Sabik	102 15.8	S15 44.
E	11	263 02.2	347 57.3	41.7	120 14.9	32.5	294 36.5	47.1	78 03.4	19.4			
S	12	278 04.7	2 56.4	S23 41.7	135 17.6	N18 32.7	309 38.4	S13 46.9	93 05.8	N 0 19.4	Schedar	349 43.6	N56 35.
D	13	293 07.1	17 55.4	41.7	150 20.3	32.9	324 40.4	46.7	108 08.2	19.4	Shaula	96 25.8	S37 06.
A	14	308 09.6	32 54.4	41.7	165 23.1	33.1	339 42.4	46.5	123 10.6	19.4	Sirius	258 35.6	S16 43.
Y	15	323 12.0	47 53.4	. . 41.7	180 25.8	. . 33.3	354 44.4	. . 46.4	138 13.0	. . 19.3	Spica	158 34.0	S11 12.
	16	338 14.5	62 52.5	41.7	195 28.5	33.5	9 46.4	46.2	153 15.4	19.3	Suhail	222 54.0	S43 28.
	17	353 17.0	77 51.5	41.6	210 31.3	33.7	24 48.4	46.0	168 17.8	19.3			
	18	8 19.4	92 50.5	S23 41.6	225 34.0	N18 33.9	39 50.3	S13 45.9	183 20.2	N 0 19.3	Vega	80 41.1	N38 47.
	19	23 21.9	107 49.5	41.6	240 36.7	34.1	54 52.3	45.7	198 22.6	19.3	Zuben'ubi	137 08.5	S16 05.
	20	38 24.4	122 48.6	41.6	255 39.5	34.3	69 54.3	45.5	213 25.0	19.3			
	21	53 26.8	137 47.6	. . 41.5	270 42.2	. . 34.5	84 56.3	. . 45.3	228 27.4	. . 19.3			
	22	68 29.3	152 46.6	41.5	285 45.0	34.7	99 58.3	45.2	243 29.8	19.2			
	23	83 31.8	167 45.7	41.5	300 47.7	34.9	115 00.3	45.0	258 32.2	19.2			
	Mer. Pass.	h m 17 30.7	*v* −1.0	*d* 0.0	*v* 2.7	*d* 0.2	*v* 2.0	*d* 0.2	*v* 2.4	*d* 0.0			

	SHA	Mer. Pas
	° ′	h m
Venus	86 55.4	11 47
Mars	217 04.3	3 05
Jupiter	31 51.0	15 24
Saturn	175 03.7	5 52

	UT	SUN GHA	SUN Dec	MOON GHA	v	MOON Dec	d	HP
	d h	° ′	° ′	° ′	′	° ′	′	′
	27 00	179 46.5	S23 19.8	67 41.9	11.0	N17 00.8	11.0	57.2
	01	194 46.2	19.7	82 11.9	10.9	17 11.8	11.0	57.2
	02	209 45.9	19.6	96 41.8	10.8	17 22.8	10.9	57.2
	03	224 45.6	. . 19.5	111 11.6	10.7	17 33.7	10.8	57.3
	04	239 45.3	19.4	125 41.3	10.6	17 44.5	10.8	57.3
	05	254 45.0	19.3	140 10.9	10.4	17 55.3	10.7	57.3
	06	269 44.7	S23 19.2	154 40.3	10.4	N18 06.0	10.6	57.4
	07	284 44.3	19.1	169 09.7	10.2	18 16.6	10.5	57.4
	08	299 44.0	19.0	183 38.9	10.1	18 27.1	10.4	57.5
S	09	314 43.7	. . 18.9	198 08.0	10.0	18 37.5	10.4	57.5
U	10	329 43.4	18.7	212 37.0	9.9	18 47.9	10.2	57.5
N	11	344 43.1	18.6	227 05.9	9.8	18 58.1	10.2	57.6
D	12	359 42.8	S23 18.5	241 34.7	9.6	N19 08.3	10.1	57.6
A	13	14 42.5	18.4	256 03.3	9.5	19 18.4	10.0	57.7
Y	14	29 42.2	18.3	270 31.8	9.4	19 28.4	9.9	57.7
	15	44 41.9	. . 18.2	285 00.2	9.3	19 38.3	9.8	57.7
	16	59 41.6	18.1	299 28.5	9.2	19 48.1	9.7	57.8
	17	74 41.3	17.9	313 56.7	9.0	19 57.8	9.7	57.8
	18	89 41.0	S23 17.8	328 24.7	9.0	N20 07.5	9.5	57.9
	19	104 40.7	17.7	342 52.7	8.8	20 17.0	9.4	57.9
	20	119 40.4	17.6	357 20.5	8.6	20 26.4	9.3	57.9
	21	134 40.1	. . 17.5	11 48.1	8.6	20 35.7	9.3	58.0
	22	149 39.8	17.3	26 15.7	8.4	20 45.0	9.1	58.0
	23	164 39.5	17.2	40 43.1	8.4	20 54.1	9.0	58.1
	28 00	179 39.1	S23 17.1	55 10.5	8.1	N21 03.1	8.9	58.1
	01	194 38.8	17.0	69 37.6	8.1	21 12.0	8.8	58.1
	02	209 38.5	16.9	84 04.7	8.0	21 20.8	8.7	58.2
	03	224 38.2	. . 16.7	98 31.7	7.8	21 29.5	8.6	58.2
	04	239 37.9	16.6	112 58.5	7.7	21 38.1	8.4	58.3
	05	254 37.6	16.5	127 25.2	7.6	21 46.5	8.4	58.3
	06	269 37.3	S23 16.4	141 51.8	7.4	N21 54.9	8.2	58.3
	07	284 37.0	16.2	156 18.2	7.4	22 03.1	8.1	58.4
	08	299 36.7	16.1	170 44.6	7.2	22 11.2	8.0	58.4
M	09	314 36.4	. . 16.0	185 10.8	7.1	22 19.2	7.8	58.5
O	10	329 36.1	15.8	199 36.9	7.0	22 27.0	7.8	58.5
N	11	344 35.8	15.7	214 02.9	6.8	22 34.8	7.6	58.5
D	12	359 35.5	S23 15.6	228 28.7	6.8	N22 42.4	7.5	58.6
A	13	14 35.2	15.4	242 54.5	6.6	22 49.9	7.3	58.6
Y	14	29 34.9	15.3	257 20.1	6.5	22 57.2	7.2	58.7
	15	44 34.6	. . 15.2	271 45.6	6.4	23 04.4	7.1	58.7
	16	59 34.3	15.0	286 11.0	6.2	23 11.5	6.9	58.7
	17	74 34.0	14.9	300 36.2	6.2	23 18.4	6.9	58.8
	18	89 33.7	S23 14.8	315 01.4	6.0	N23 25.3	6.6	58.8
	19	104 33.4	14.6	329 26.4	5.9	23 31.9	6.6	58.9
	20	119 33.1	14.5	343 51.3	5.8	23 38.5	6.3	58.9
	21	134 32.8	. . 14.4	358 16.1	5.7	23 44.8	6.3	58.9
	22	149 32.5	14.2	12 40.8	5.6	23 51.1	6.1	59.0
	23	164 32.1	14.1	27 05.4	5.4	23 57.2	5.9	59.0
	29 00	179 31.8	S23 13.9	41 29.8	5.4	N24 03.1	5.8	59.1
	01	194 31.5	13.8	55 54.2	5.2	24 08.9	5.7	59.1
	02	209 31.2	13.6	70 18.4	5.1	24 14.6	5.5	59.1
	03	224 30.9	. . 13.5	84 42.5	5.1	24 20.1	5.3	59.2
	04	239 30.6	13.4	99 06.6	4.9	24 25.4	5.2	59.2
	05	254 30.3	13.2	113 30.5	4.8	24 30.6	5.1	59.2
	06	269 30.0	S23 13.1	127 54.3	4.7	N24 35.7	4.8	59.3
	07	284 29.7	12.9	142 18.0	4.6	24 40.5	4.8	59.3
T	08	299 29.4	12.8	156 41.6	4.5	24 45.3	4.5	59.4
	09	314 29.1	. . 12.6	171 05.1	4.5	24 49.8	4.4	59.4
U	10	329 28.8	12.5	185 28.6	4.3	24 54.2	4.2	59.4
E	11	344 28.5	12.3	199 51.9	4.2	24 58.4	4.1	59.5
S	12	359 28.2	S23 12.2	214 15.1	4.1	N25 02.5	3.9	59.5
D	13	14 27.9	12.0	228 38.2	4.1	25 06.4	3.7	59.5
A	14	29 27.6	11.9	243 01.3	3.9	25 10.1	3.6	59.6
Y	15	44 27.3	. . 11.7	257 24.2	3.9	25 13.7	3.4	59.6
	16	59 27.0	11.6	271 47.1	3.8	25 17.1	3.2	59.6
	17	74 26.7	11.4	286 09.9	3.7	25 20.3	3.0	59.7
	18	89 26.4	S23 11.2	300 32.6	3.6	N25 23.3	2.9	59.7
	19	104 26.1	11.1	314 55.2	3.5	25 26.2	2.7	59.7
	20	119 25.8	10.9	329 17.7	3.5	25 28.9	2.5	59.8
	21	134 25.5	. . 10.8	343 40.2	3.3	25 31.4	2.3	59.8
	22	149 25.2	10.6	358 02.5	3.4	25 33.7	2.1	59.9
	23	164 24.9	10.5	12 24.9	3.2	N25 35.8	2.0	59.9
		SD 16.3	d 0.1	SD 15.7		16.0		16.2

Lat.	Twilight Naut.	Twilight Civil	Sunrise	Moonrise 27	Moonrise 28	Moonrise 29	Moonrise 30
°	h m	h m	h m	h m	h m	h m	h m
N 72	08 26	10 52	■	▭	▭	▭	▭
N 70	08 07	09 54	■	09 19	▭	▭	▭
68	07 51	09 20	■	10 03	▭	▭	▭
66	07 38	08 55	10 33	10 32	10 13	▭	▭
64	07 27	08 35	09 52	10 54	10 53	10 55	11 24
62	07 18	08 20	09 25	11 12	11 20	11 40	12 27
60	07 09	08 06	09 04	11 27	11 42	12 09	13 02
N 58	07 02	07 55	08 46	11 40	11 59	12 32	13 27
56	06 55	07 44	08 32	11 51	12 14	12 51	13 47
54	06 49	07 35	08 19	12 01	12 27	13 07	14 04
52	06 44	07 27	08 08	12 09	12 39	13 20	14 18
50	06 38	07 20	07 58	12 17	12 49	13 32	14 31
45	06 27	07 04	07 38	12 34	13 10	13 57	14 57
N 40	06 17	06 51	07 21	12 48	13 27	14 17	15 17
35	06 07	06 39	07 07	13 00	13 42	14 33	15 35
30	05 59	06 28	06 55	13 10	13 55	14 48	15 49
20	05 42	06 10	06 34	13 28	14 16	15 12	16 15
N 10	05 26	05 52	06 15	13 44	14 35	15 33	16 36
0	05 09	05 35	05 58	13 58	14 53	15 53	16 57
S 10	04 50	05 17	05 40	14 13	15 11	16 13	17 17
20	04 28	04 57	05 22	14 29	15 31	16 35	17 39
30	03 59	04 32	05 00	14 48	15 53	17 00	18 04
35	03 40	04 17	04 47	14 59	16 06	17 14	18 19
40	03 17	03 59	04 32	15 11	16 21	17 31	18 37
45	02 47	03 37	04 14	15 26	16 40	17 52	18 57
S 50	02 02	03 07	03 51	15 44	17 02	18 18	19 23
52	01 35	02 52	03 41	15 53	17 13	18 30	19 36
54	00 50	02 34	03 28	16 02	17 26	18 45	19 51
56	////	02 12	03 14	16 13	17 40	19 01	20 07
58	////	01 43	02 58	16 26	17 57	19 22	20 28
S 60	////	00 55	02 38	16 40	18 17	19 47	20 53

Lat.	Sunset	Twilight Civil	Twilight Naut.	Moonset 27	Moonset 28	Moonset 29	Moonset 30
°	h m	h m	h m	h m	h m	h m	h m
N 72	■	13 11	15 37	▭	▭	▭	▭
N 70	■	14 10	15 57	05 53	▭	▭	▭
68	■	14 44	16 12	05 11	▭	▭	▭
66	13 30	15 09	16 25	04 42	06 53	▭	▭
64	14 11	15 28	16 36	04 21	06 13	08 14	09 56
62	14 39	15 44	16 46	04 04	05 46	07 29	08 53
60	15 00	15 57	16 54	03 50	05 25	07 00	08 19
N 58	15 17	16 09	17 01	03 38	05 08	06 37	07 53
56	15 32	16 19	17 08	03 28	04 54	06 19	07 33
54	15 44	16 28	17 14	03 19	04 42	06 03	07 16
52	15 55	16 36	17 20	03 10	04 31	05 50	07 02
50	16 05	16 43	17 25	03 03	04 21	05 38	06 49
45	16 26	16 59	17 37	02 48	04 01	05 14	06 23
N 40	16 42	17 13	17 47	02 35	03 44	04 55	06 03
35	16 56	17 24	17 56	02 24	03 31	04 39	05 46
30	17 09	17 35	18 05	02 15	03 19	04 25	05 31
20	17 30	17 54	18 21	01 59	02 58	04 01	05 05
N 10	17 48	18 11	18 37	01 45	02 40	03 40	04 44
0	18 05	18 28	18 54	01 32	02 24	03 21	04 23
S 10	18 23	18 46	19 13	01 19	02 08	03 02	04 03
20	18 42	19 06	19 35	01 05	01 50	02 42	03 41
30	19 04	19 31	20 04	00 49	01 30	02 18	03 16
35	19 16	19 46	20 23	00 40	01 18	02 05	03 01
40	19 31	20 04	20 46	00 30	01 05	01 49	02 43
45	19 49	20 27	21 16	00 17	00 49	01 30	02 23
S 50	20 12	20 56	22 01	00 03	00 30	01 06	01 57
52	20 22	21 11	22 28	24 21	00 21	00 55	01 44
54	20 34	21 29	23 11	24 10	00 10	00 42	01 29
56	20 49	21 50	////	23 59	24 27	00 27	01 12
58	21 05	22 19	////	23 46	24 10	00 10	00 52
S 60	21 25	23 07	////	23 30	23 49	24 26	00 26

Day	SUN Eqn. of Time 00^h	SUN Eqn. of Time 12^h	SUN Mer. Pass.	MOON Mer. Pass. Upper	MOON Mer. Pass. Lower	MOON Age	MOON Phase
d	m s	m s	h m	h m	h m	d %	
27	00 53	01 08	12 01	20 11	07 45	11 77	
28	01 23	01 37	12 02	21 07	08 38	12 86	
29	01 52	02 07	12 02	22 08	09 37	13 93	

	UT	ARIES	VENUS −4.0		MARS −0.8		JUPITER −2.1		SATURN +0.9	
		GHA	GHA	Dec	GHA	Dec	GHA	Dec	GHA	Dec
	d h	° ′	° ′	° ′	° ′	° ′	° ′	° ′	° ′	° ′
	30 00	98 34.2	182 44.7	S23 41.4	315 50.4	N18 35.1	130 02.2	S13 44.8	273 34.6	N 0 19.2
	01	113 36.7	197 43.7	41.4	330 53.2	35.3	145 04.2	44.7	288 37.0	19.2
	02	128 39.2	212 42.7	41.4	345 55.9	35.5	160 06.2	44.5	303 39.4	19.2
	03	143 41.6	227 41.8	. . 41.4	0 58.7	. . 35.7	175 08.2	. . 44.3	318 41.8	. . 19.2
	04	158 44.1	242 40.8	41.3	16 01.4	35.9	190 10.2	44.1	333 44.2	19.1
	05	173 46.5	257 39.8	41.3	31 04.2	36.1	205 12.1	44.0	348 46.6	19.1
	06	188 49.0	272 38.9	S23 41.2	46 07.0	N18 36.3	220 14.1	S13 43.8	3 49.0	N 0 19.1
W	07	203 51.5	287 37.9	41.2	61 09.7	36.5	235 16.1	43.6	18 51.4	19.1
E	08	218 53.9	302 36.9	41.2	76 12.5	36.7	250 18.1	43.5	33 53.8	19.1
D	09	233 56.4	317 35.9	. . 41.1	91 15.2	. . 36.9	265 20.1	. . 43.3	48 56.2	. . 19.1
N	10	248 58.9	332 35.0	41.1	106 18.0	37.1	280 22.1	43.1	63 58.6	19.1
E	11	264 01.3	347 34.0	41.0	121 20.7	37.3	295 24.0	42.9	79 01.0	19.0
	12	279 03.8	2 33.0	S23 41.0	136 23.5	N18 37.5	310 26.0	S13 42.8	94 03.4	N 0 19.0
S	13	294 06.3	17 32.0	40.9	151 26.3	37.7	325 28.0	42.6	109 05.8	19.0
D	14	309 08.7	32 31.1	40.9	166 29.0	37.9	340 30.0	42.4	124 08.2	19.0
A	15	324 11.2	47 30.1	. . 40.8	181 31.8	. . 38.1	355 32.0	. . 42.3	139 10.6	. . 19.0
Y	16	339 13.7	62 29.1	40.8	196 34.6	38.3	10 33.9	42.1	154 13.0	19.0
	17	354 16.1	77 28.2	40.7	211 37.4	38.6	25 35.9	41.9	169 15.4	19.0
	18	9 18.6	92 27.2	S23 40.7	226 40.1	N18 38.8	40 37.9	S13 41.7	184 17.8	N 0 18.9
	19	24 21.0	107 26.2	40.6	241 42.9	39.0	55 39.9	41.6	199 20.2	18.9
	20	39 23.5	122 25.2	40.6	256 45.7	39.2	70 41.9	41.4	214 22.6	18.9
	21	54 26.0	137 24.3	. . 40.5	271 48.4	. . 39.4	85 43.8	. . 41.2	229 25.0	. . 18.9
	22	69 28.4	152 23.3	40.5	286 51.2	39.6	100 45.8	41.1	244 27.4	18.9
	23	84 30.9	167 22.3	40.4	301 54.0	39.8	115 47.8	40.9	259 29.8	18.9
	31 00	99 33.4	182 21.4	S23 40.4	316 56.8	N18 40.0	130 49.8	S13 40.7	274 32.2	N 0 18.9
	01	114 35.8	197 20.4	40.3	331 59.6	40.2	145 51.8	40.5	289 34.6	18.8
	02	129 38.3	212 19.4	40.2	347 02.3	40.4	160 53.7	40.4	304 37.0	18.8
	03	144 40.8	227 18.4	. . 40.2	2 05.1	. . 40.6	175 55.7	. . 40.2	319 39.4	. . 18.8
	04	159 43.2	242 17.5	40.1	17 07.9	40.8	190 57.7	40.0	334 41.8	18.8
	05	174 45.7	257 16.5	40.0	32 10.7	41.1	205 59.7	39.8	349 44.2	18.8
	06	189 48.2	272 15.5	S23 40.0	47 13.5	N18 41.3	221 01.6	S13 39.7	4 46.6	N 0 18.8
	07	204 50.6	287 14.6	39.9	62 16.3	41.5	236 03.6	39.5	19 49.0	18.8
T	08	219 53.1	302 13.6	39.8	77 19.1	41.7	251 05.6	39.3	34 51.4	18.8
H	09	234 55.5	317 12.6	. . 39.8	92 21.9	. . 41.9	266 07.6	. . 39.1	49 53.8	. . 18.7
U	10	249 58.0	332 11.6	39.7	107 24.7	42.1	281 09.6	39.0	64 56.2	18.7
R	11	265 00.5	347 10.7	39.6	122 27.5	42.3	296 11.5	38.8	79 58.6	18.7
S	12	280 02.9	2 09.7	S23 39.5	137 30.3	N18 42.5	311 13.5	S13 38.6	95 01.0	N 0 18.7
D	13	295 05.4	17 08.7	39.5	152 33.1	42.8	326 15.5	38.5	110 03.4	18.7
A	14	310 07.9	32 07.8	39.4	167 35.9	43.0	341 17.5	38.3	125 05.8	18.7
Y	15	325 10.3	47 06.8	. . 39.3	182 38.7	. . 43.2	356 19.4	. . 38.1	140 08.2	. . 18.7
	16	340 12.8	62 05.8	39.2	197 41.5	43.4	11 21.4	37.9	155 10.6	18.6
	17	355 15.3	77 04.9	39.1	212 44.3	43.6	26 23.4	37.8	170 13.1	18.6
	18	10 17.7	92 03.9	S23 39.0	227 47.1	N18 43.8	41 25.4	S13 37.6	185 15.5	N 0 18.6
	19	25 20.2	107 02.9	39.0	242 49.9	44.0	56 27.3	37.4	200 17.9	18.6
	20	40 22.7	122 01.9	38.9	257 52.7	44.3	71 29.3	37.2	215 20.3	18.6
	21	55 25.1	137 01.0	. . 38.8	272 55.5	. . 44.5	86 31.3	. . 37.1	230 22.7	. . 18.6
	22	70 27.6	152 00.0	38.7	287 58.3	44.7	101 33.3	36.9	245 25.1	18.6
	23	85 30.0	166 59.0	38.6	303 01.1	44.9	116 35.3	36.7	260 27.5	18.6
	1 00	100 32.5	181 58.1	S23 38.5	318 03.9	N18 45.1	131 37.2	S13 36.5	275 29.9	N 0 18.6
	01	115 35.0	196 57.1	38.4	333 06.8	45.4	146 39.2	36.4	290 32.3	18.5
	02	130 37.4	211 56.1	38.3	348 09.6	45.6	161 41.2	36.2	305 34.7	18.5
	03	145 39.9	226 55.2	. . 38.2	3 12.4	. . 45.8	176 43.2	. . 36.0	320 37.1	. . 18.5
	04	160 42.4	241 54.2	38.1	18 15.2	46.0	191 45.1	35.8	335 39.5	18.5
	05	175 44.8	256 53.2	38.0	33 18.0	46.2	206 47.1	35.7	350 41.9	18.5
	06	190 47.3	271 52.2	S23 37.9	48 20.9	N18 46.5	221 49.1	S13 35.5	5 44.3	N 0 18.5
	07	205 49.8	286 51.3	37.8	63 23.7	46.7	236 51.1	35.3	20 46.8	18.5
	08	220 52.2	301 50.3	37.7	78 26.5	46.9	251 53.0	35.1	35 49.2	18.5
F	09	235 54.7	316 49.3	. . 37.6	93 29.4	. . 47.1	266 55.0	. . 35.0	50 51.6	. . 18.4
R	10	250 57.2	331 48.4	37.5	108 32.2	47.3	281 57.0	34.8	65 54.0	18.4
I	11	265 59.6	346 47.4	37.4	123 35.0	47.6	296 59.0	34.6	80 56.4	18.4
D	12	281 02.1	1 46.4	S23 37.3	138 37.8	N18 47.8	312 00.9	S13 34.4	95 58.8	N 0 18.4
A	13	296 04.5	16 45.5	37.2	153 40.7	48.0	327 02.9	34.3	111 01.2	18.4
Y	14	311 07.0	31 44.5	37.1	168 43.5	48.2	342 04.9	34.1	126 03.6	18.4
	15	326 09.5	46 43.5	. . 37.0	183 46.4	. . 48.5	357 06.9	. . 33.9	141 06.0	. . 18.4
	16	341 11.9	61 42.6	36.9	198 49.2	48.7	12 08.8	33.7	156 08.4	18.4
	17	356 14.4	76 41.6	36.8	213 52.0	48.9	27 10.8	33.6	171 10.8	18.4
	18	11 16.9	91 40.6	S23 36.7	228 54.9	N18 49.1	42 12.8	S13 33.4	186 13.2	N 0 18.4
	19	26 19.3	106 39.7	36.6	243 57.7	49.4	57 14.7	33.2	201 15.7	18.3
	20	41 21.8	121 38.7	36.4	259 00.6	49.6	72 16.7	33.0	216 18.1	18.3
	21	56 24.3	136 37.7	. . 36.3	274 03.4	. . 49.8	87 18.7	. . 32.9	231 20.5	. . 18.3
	22	71 26.7	151 36.8	36.2	289 06.2	50.0	102 20.7	32.7	246 22.9	18.3
	23	86 29.2	166 35.8	36.1	304 09.1	50.3	117 22.6	32.5	261 25.3	18.3
	Mer. Pass.	h m 17 18.9	*v* −1.0	*d* 0.1	*v* 2.8	*d* 0.2	*v* 2.0	*d* 0.2	*v* 2.4	*d* 0.0

STARS Name	SHA	Dec
	° ′	° ′
Acamar	315 19.9	S40 16.0
Achernar	335 28.3	S57 11.3
Acrux	173 12.3	S63 09.1
Adhara	255 14.1	S28 59.2
Aldebaran	290 52.0	N16 31.8
Alioth	166 22.8	N55 53.9
Alkaid	153 00.9	N49 15.4
Al Na'ir	27 47.1	S46 54.9
Alnilam	275 48.6	S 1 11.7
Alphard	217 58.3	S 8 42.2
Alphecca	126 13.4	N26 40.7
Alpheratz	357 46.2	N29 09.0
Altair	62 11.0	N 8 53.7
Ankaa	353 18.1	S42 15.2
Antares	112 29.7	S26 27.2
Arcturus	145 58.1	N19 07.6
Atria	107 34.4	S69 02.6
Avior	234 18.6	S59 32.4
Bellatrix	278 34.4	N 6 21.5
Betelgeuse	271 03.7	N 7 24.5
Canopus	263 56.7	S52 42.1
Capella	280 37.7	N46 00.6
Deneb	49 33.7	N45 19.1
Denebola	182 36.1	N14 30.8
Diphda	348 58.4	S17 55.9
Dubhe	193 54.3	N61 41.4
Elnath	278 15.4	N28 37.0
Eltanin	90 47.9	N51 29.2
Enif	33 49.9	N 9 55.4
Fomalhaut	15 26.9	S29 34.2
Gacrux	172 03.9	S57 10.0
Gienah	175 54.9	S17 35.9
Hadar	148 51.9	S60 25.1
Hamal	328 03.5	N23 30.8
Kaus Aust.	83 47.6	S34 22.8
Kochab	137 20.3	N74 06.5
Markab	13 41.0	N15 15.7
Menkar	314 17.5	N 4 07.8
Menkent	148 10.8	S36 25.1
Miaplacidus	221 39.7	S69 45.4
Mirfak	308 43.7	N49 54.0
Nunki	76 01.8	S26 17.1
Peacock	53 23.6	S56 42.3
Pollux	243 30.4	N28 00.0
Procyon	245 02.0	N 5 11.9
Rasalhague	96 09.1	N12 33.1
Regulus	207 46.0	N11 54.9
Rigel	281 14.2	S 8 11.4
Rigil Kent.	139 55.7	S60 52.4
Sabik	102 15.8	S15 44.2
Schedar	349 43.6	N56 35.9
Shaula	96 25.8	S37 06.6
Sirius	258 35.6	S16 43.8
Spica	158 34.0	S11 12.9
Suhail	222 54.0	S43 28.3
Vega	80 41.1	N38 47.6
Zuben'ubi	137 08.4	S16 05.0

	SHA	Mer. Pass.
	° ′	h m
Venus	82 48.0	11 51
Mars	217 23.4	2 52
Jupiter	31 16.4	15 15
Saturn	174 58.8	5 41

UT d h	SUN GHA	SUN Dec	MOON GHA	v	MOON Dec	d	HP
	° ′	° ′	° ′	′	° ′	′	′
[30] 00	179 24.6	S23 10.3	26 47.1	3.2	N25 37.8	1.8	59.9
01	194 24.3	10.1	41 09.3	3.1	25 39.6	1.6	59.9
02	209 24.0	10.0	55 31.4	3.0	25 41.2	1.4	60.0
03	224 23.7	. . 09.8	69 53.4	3.0	25 42.6	1.3	60.0
04	239 23.4	09.6	84 15.4	3.0	25 43.9	1.0	60.0
05	254 23.1	09.5	98 37.4	2.8	25 44.9	0.9	60.1
06	269 22.8	S23 09.3	112 59.2	2.8	N25 45.8	0.7	60.1
07	284 22.5	09.1	127 21.0	2.8	25 46.5	0.5	60.1
08	299 22.2	09.0	141 42.8	2.7	25 47.0	0.3	60.2
09	314 21.9	. . 08.8	156 04.5	2.7	25 47.3	0.1	60.2
10	329 21.6	08.6	170 26.2	2.7	25 47.4	0.1	60.2
11	344 21.3	08.5	184 47.9	2.6	25 47.3	0.3	60.3
12	359 21.0	S23 08.3	199 09.5	2.5	N25 47.0	0.4	60.3
13	14 20.7	08.1	213 31.0	2.6	25 46.6	0.7	60.3
14	29 20.4	08.0	227 52.6	2.5	25 45.9	0.8	60.3
15	44 20.1	. . 07.8	242 14.1	2.4	25 45.1	1.0	60.4
16	59 19.8	07.6	256 35.5	2.5	25 44.1	1.3	60.4
17	74 19.5	07.4	270 57.0	2.4	25 42.8	1.4	60.4
18	89 19.2	S23 07.3	285 18.4	2.4	N25 41.4	1.6	60.4
19	104 18.9	07.1	299 39.8	2.4	25 39.8	1.8	60.5
20	119 18.6	06.9	314 01.2	2.4	25 38.0	2.0	60.5
21	134 18.3	. . 06.7	328 22.6	2.3	25 36.0	2.2	60.5
22	149 18.0	06.6	342 43.9	2.4	25 33.8	2.4	60.5
23	164 17.7	06.4	357 05.3	2.3	25 31.4	2.6	60.6
[31] 00	179 17.4	S23 06.2	11 26.6	2.4	N25 28.8	2.7	60.6
01	194 17.1	06.0	25 48.0	2.3	25 26.1	3.0	60.6
02	209 16.8	05.8	40 09.3	2.4	25 23.1	3.2	60.6
03	224 16.5	. . 05.7	54 30.7	2.3	25 19.9	3.3	60.7
04	239 16.2	05.5	68 52.0	2.4	25 16.6	3.5	60.7
05	254 15.9	05.3	83 13.4	2.4	25 13.1	3.8	60.7
06	269 15.6	S23 05.1	97 34.8	2.4	N25 09.3	3.9	60.7
07	284 15.3	04.9	111 56.2	2.4	25 05.4	4.1	60.7
08	299 15.0	04.7	126 17.6	2.4	25 01.3	4.3	60.8
09	314 14.7	. . 04.5	140 39.0	2.5	24 57.0	4.5	60.8
10	329 14.4	04.4	155 00.5	2.4	24 52.5	4.7	60.8
11	344 14.1	04.2	169 21.9	2.5	24 47.8	4.9	60.8
12	359 13.9	S23 04.0	183 43.4	2.6	N24 42.9	5.1	60.8
13	14 13.6	03.8	198 05.0	2.5	24 37.8	5.2	60.9
14	29 13.3	03.6	212 26.5	2.6	24 32.6	5.4	60.9
15	44 13.0	. . 03.4	226 48.1	2.7	24 27.2	5.7	60.9
16	59 12.7	03.2	241 09.8	2.7	24 21.5	5.8	60.9
17	74 12.4	03.0	255 31.5	2.7	24 15.7	6.0	60.9
18	89 12.1	S23 02.8	269 53.2	2.8	N24 09.7	6.1	60.9
19	104 11.8	02.6	284 15.0	2.8	24 03.6	6.4	61.0
20	119 11.5	02.4	298 36.8	2.8	23 57.2	6.5	61.0
21	134 11.2	. . 02.2	312 58.6	3.0	23 50.7	6.8	61.0
22	149 10.9	02.0	327 20.6	2.9	23 43.9	6.9	61.0
23	164 10.6	01.8	341 42.5	3.1	23 37.0	7.0	61.0
[1] 00	179 10.3	S23 01.6	356 04.6	3.0	N23 30.0	7.3	61.0
01	194 10.0	01.4	10 26.6	3.2	23 22.7	7.4	61.0
02	209 09.7	01.2	24 48.8	3.2	23 15.3	7.6	61.0
03	224 09.4	. . 01.0	39 11.0	3.3	23 07.7	7.8	61.0
04	239 09.1	00.8	53 33.3	3.3	22 59.9	7.9	61.1
05	254 08.8	00.6	67 55.6	3.4	22 52.0	8.2	61.1
06	269 08.5	S23 00.4	82 18.0	3.5	N22 43.8	8.2	61.1
07	284 08.2	00.2	96 40.5	3.6	22 35.6	8.5	61.1
08	299 08.0	23 00.0	111 03.1	3.6	22 27.1	8.6	61.1
09	314 07.7	22 59.8	125 25.7	3.7	22 18.5	8.8	61.1
10	329 07.4	59.6	139 48.4	3.8	22 09.7	8.9	61.1
11	344 07.1	59.4	154 11.2	3.8	22 00.8	9.1	61.1
12	359 06.8	S22 59.2	168 34.0	4.0	N21 51.7	9.3	61.1
13	14 06.5	59.0	182 57.0	4.0	21 42.4	9.4	61.1
14	29 06.2	58.8	197 20.0	4.1	21 33.0	9.6	61.1
15	44 05.9	. . 58.6	211 43.1	4.2	21 23.4	9.7	61.1
16	59 05.6	58.3	226 06.3	4.2	21 13.7	9.9	61.1
17	74 05.3	58.1	240 29.5	4.4	21 03.8	10.0	61.1
18	89 05.0	S22 57.9	254 52.9	4.4	N20 53.8	10.2	61.1
19	104 04.7	57.7	269 16.3	4.5	20 43.6	10.3	61.1
20	119 04.4	57.5	283 39.8	4.7	20 33.3	10.5	61.1
21	134 04.1	. . 57.3	298 03.5	4.7	20 22.8	10.6	61.1
22	149 03.9	57.1	312 27.2	4.8	20 12.2	10.7	61.1
23	164 03.6	56.8	326 51.0	4.9	N20 01.5	10.9	61.1
	SD 16.3	*d* 0.2	SD	16.4		16.6	16.7

Lat.	Twilight Naut.	Twilight Civil	Sunrise	Moonrise 30	Moonrise 31	Moonrise 1	Moonrise 2
°	h m	h m	h m	h m	h m	h m	h m
N 72	08 25	10 45	■	□	□	□	15 25
N 70	08 06	09 51	■	□	□	□	16 17
68	07 50	09 18	■	□	□	13 46	16 49
66	07 38	08 54	10 29	□	□	14 49	17 12
64	07 27	08 35	09 50	11 24	13 13	15 24	17 30
62	07 18	08 19	09 24	12 27	13 56	15 49	17 45
60	07 09	08 06	09 03	13 02	14 25	16 09	17 58
N 58	07 02	07 54	08 46	13 27	14 47	16 25	18 08
56	06 56	07 44	08 31	13 47	15 05	16 39	18 18
54	06 50	07 36	08 19	14 04	15 21	16 51	18 26
52	06 44	07 28	08 08	14 18	15 34	17 01	18 33
50	06 39	07 20	07 59	14 31	15 45	17 11	18 40
45	06 27	07 05	07 38	14 57	16 09	17 30	18 54
N 40	06 17	06 51	07 22	15 17	16 28	17 46	19 05
35	06 08	06 40	07 08	15 35	16 45	17 59	19 15
30	06 00	06 29	06 56	15 49	16 58	18 11	19 23
20	05 43	06 11	06 35	16 15	17 22	18 31	19 38
N 10	05 27	05 54	06 17	16 36	17 42	18 48	19 51
0	05 11	05 37	05 59	16 57	18 01	19 04	20 03
S 10	04 52	05 19	05 42	17 17	18 20	19 20	20 14
20	04 30	04 59	05 23	17 39	18 40	19 37	20 27
30	04 01	04 34	05 02	18 04	19 04	19 56	20 41
35	03 43	04 19	04 49	18 19	19 17	20 07	20 50
40	03 20	04 01	04 34	18 37	19 33	20 20	20 59
45	02 50	03 39	04 16	18 57	19 52	20 36	21 10
S 50	02 06	03 10	03 54	19 23	20 15	20 54	21 23
52	01 39	02 55	03 44	19 36	20 26	21 03	21 29
54	00 58	02 38	03 31	19 51	20 39	21 13	21 36
56	////	02 16	03 18	20 07	20 53	21 23	21 44
58	////	01 48	03 01	20 28	21 10	21 36	21 52
S 60	////	01 03	02 41	20 53	21 31	21 50	22 01

Lat.	Sunset	Twilight Civil	Twilight Naut.	Moonset 30	Moonset 31	Moonset 1	Moonset 2
°	h m	h m	h m	h m	h m	h m	h m
N 72	■	13 21	15 42	□	□	□	12 34
N 70	■	14 16	16 01	□	□	□	11 40
68	■	14 49	16 16	□	□	12 05	11 07
66	13 37	15 13	16 29	□	□	11 00	10 42
64	14 16	15 32	16 39	09 56	10 22	10 24	10 22
62	14 43	15 47	16 49	08 53	09 39	09 58	10 07
60	15 04	16 01	16 57	08 19	09 10	09 38	09 53
N 58	15 21	16 12	17 04	07 53	08 47	09 21	09 41
56	15 35	16 22	17 11	07 33	08 29	09 07	09 31
54	15 47	16 31	17 17	07 16	08 14	08 54	09 22
52	15 58	16 39	17 22	07 02	08 00	08 43	09 14
50	16 08	16 46	17 27	06 49	07 48	08 33	09 07
45	16 28	17 02	17 39	06 23	07 24	08 13	08 51
N 40	16 45	17 15	17 49	06 03	07 04	07 56	08 39
35	16 58	17 27	17 58	05 46	06 47	07 41	08 27
30	17 11	17 37	18 07	05 31	06 33	07 29	08 18
20	17 31	17 55	18 23	05 05	06 09	07 08	08 01
N 10	17 50	18 12	18 39	04 44	05 47	06 49	07 46
0	18 07	18 29	18 55	04 23	05 28	06 31	07 32
S 10	18 24	18 47	19 14	04 03	05 08	06 14	07 18
20	18 43	19 07	19 36	03 41	04 46	05 55	07 03
30	19 04	19 32	20 05	03 16	04 21	05 33	06 46
35	19 17	19 47	20 23	03 01	04 07	05 20	06 35
40	19 32	20 05	20 46	02 43	03 50	05 05	06 24
45	19 50	20 27	21 16	02 23	03 29	04 47	06 10
S 50	20 12	20 56	22 00	01 57	03 03	04 24	05 52
52	20 22	21 10	22 26	01 44	02 51	04 14	05 44
54	20 34	21 28	23 06	01 29	02 36	04 01	05 35
56	20 48	21 49	////	01 12	02 20	03 47	05 25
58	21 04	22 17	////	00 52	02 00	03 31	05 13
S 60	21 24	23 01	////	00 26	01 34	03 11	05 00

Day	SUN Eqn. of Time 00^h	SUN Eqn. of Time 12^h	SUN Mer. Pass.	MOON Mer. Pass. Upper	MOON Mer. Pass. Lower	Age	Phase
d	m s	m s	h m	h m	h m	d %	
30	02 21	02 35	12 03	23 12	10 40	14 98	
31	02 50	03 04	12 03	24 16	11 44	15 100	○
1	03 18	03 32	12 04	00 16	12 48	16 99	

EXPLANATION

PRINCIPLE AND ARRANGEMENT

1. *Object.* The object of this Almanac is to provide, in a convenient form, the data required for the practice of astronomical navigation at sea.

2. *Principle.* The main contents of the Almanac consist of data from which the *Greenwich Hour Angle* (GHA) and the *Declination* (Dec) of all the bodies used for navigation can be obtained for any instant of *Universal Time* (UT), or *Greenwich Mean Time* (GMT). The *Local Hour Angle* (LHA) can then be obtained by means of the formula:

$$\text{LHA} = \text{GHA} \begin{matrix} -\text{ west} \\ +\text{ east} \end{matrix} \text{ longitude}$$

The remaining data consist of: times of rising and setting of the Sun and Moon, and times of twilight; miscellaneous calendarial and planning data and auxiliary tables, including a list of Standard Times; corrections to be applied to observed altitude.

For the Sun, Moon, and planets the GHA and Dec are tabulated directly for each hour of UT throughout the year. For the stars the *Sidereal Hour Angle* (SHA) is given, and the GHA is obtained from:

$$\text{GHA Star} = \text{GHA Aries} + \text{SHA Star}$$

The SHA and Dec of the stars change slowly and may be regarded as constant over periods of several days. GHA Aries, or the Greenwich Hour Angle of the first point of Aries (the Vernal Equinox), is tabulated for each hour. Permanent tables give the appropriate increments and corrections to the tabulated hourly values of GHA and Dec for the minutes and seconds of UT.

The six-volume series of *Sight Reduction Tables for Marine Navigation* (published in U.S.A. as Pub. No. 229 and in U.K. as N.P. 401) has been designed for the solution of the navigational triangle and is intended for use with *The Nautical Almanac*.

Two alternative procedures for sight reduction are described on pages 277–318. The first requires the use of programmable calculators or computers, while the second uses a set of concise tables that is given on pages 286–317.

The tabular accuracy is 0'.1 throughout. The time argument on the daily pages of this Almanac is $12^h +$ the Greenwich Hour Angle of the mean sun and is here denoted by UT, although it is also known as GMT. This scale may differ from the broadcast time signals (UTC) by an amount which, if ignored, will introduce an error of up to 0'.2 in longitude determined from astronomical observations. (The difference arises because the time argument depends on the variable rate of rotation of the Earth while the broadcast time signals are now based on an atomic time-scale.) Step adjustments of exactly one second are made to the time signals as required (normally at 24^h on December 31 and June 30) so that the difference between the time signals and UT, as used in this Almanac, may not exceed 0ˢ.9. Those who require to reduce observations to a precision of better than 1^s must therefore obtain the correction (DUT1) to the time signals from coding in the signal, or from other sources; the required time is given by UT1=UTC+DUT1 to a precision of 0ˢ.1. Alternatively, the longitude, when determined from astronomical observations, may be corrected by the corresponding amount shown in the following table:

Correction to time signals	Correction to longitude
−0ˢ.9 to −0ˢ.7	0'.2 to east
−0ˢ.6 to −0ˢ.3	0'.1 to east
−0ˢ.2 to +0ˢ.2	no correction
+0ˢ.3 to +0ˢ.6	0'.1 to west
+0ˢ.7 to +0ˢ.9	0'.2 to west

3. *Lay-out.* The ephemeral data for three days are presented on an opening of two pages: the left-hand page contains the data for the planets and stars; the right-hand page contains the data for the Sun and Moon, together with times of twilight, sunrise, sunset, moonrise and moonset.

The remaining contents are arranged as follows: for ease of reference the altitude-correction tables are given on pages A2, A3, A4, xxxiv and xxxv; calendar, Moon's phases, eclipses, and planet notes (i.e. data of general interest) precede the main tabulations. The Explanation is followed by information on standard times, star charts and list of star positions, sight reduction procedures and concise sight reduction tables, tables of increments and corrections and other auxiliary tables that are frequently used.

MAIN DATA

4. *Daily pages.* The daily pages give the GHA of Aries, the GHA and Dec of the Sun, Moon, and the four navigational planets, for each hour of UT. For the Moon, values of v and d are also tabulated for each hour to facilitate the correction of GHA and Dec to intermediate times; v and d for the Sun and planets change so slowly that they are given, at the foot of the appropriate columns, once only on the page; v is zero for Aries and negligible for the Sun, and is omitted. The SHA and Dec of the 57 selected stars, arranged in alphabetical order of proper name, are also given.

5. *Stars.* The SHA and Dec of 173 stars, including the 57 selected stars, are tabulated for each month on pages 268–273; no interpolation is required and the data can be used in precisely the same way as those for the selected stars on the daily pages. The stars are arranged in order of SHA.

The list of 173 includes all stars down to magnitude 3·0, together with a few fainter ones to fill the larger gaps. The 57 selected stars have been chosen from amongst these on account of brightness and distribution in the sky; they will suffice for the majority of observations.

The 57 selected stars are known by their proper names, but they are also numbered in descending order of SHA. In the list of 173 stars, the constellation names are always given on the left-hand page; on the facing page proper names are given where well-known names exist. Numbers for the selected stars are given in both columns.

An index to the selected stars, containing lists in both alphabetical and numerical order, is given on page xxxiii and is also reprinted on the bookmark.

6. *Increments and corrections.* The tables printed on tinted paper (pages ii–xxxi) at the back of the Almanac provide the increments and corrections for minutes and seconds to be applied to the hourly values of GHA and Dec. They consist of sixty tables, one for each minute, separated into two parts: increments to GHA for Sun and planets, Aries, and Moon for every minute and second; and, for each minute, corrections to be applied to GHA and Dec corresponding to the values of v and d given on the daily pages.

The increments are based on the following adopted hourly rates of increase of the GHA: Sun and planets, 15° precisely; Aries, 15° 02′·46; Moon, 14° 19′·0. The values of v on the daily pages are the excesses of the actual hourly motions over the adopted values; they are generally positive, except for Venus. The tabulated hourly values of the Sun's GHA have been adjusted to reduce to a minimum the error caused by treating v as negligible. The values of d on the daily pages are the hourly differences of the Dec. For the Moon, the true values of v and d are given for each hour; otherwise mean values are given for the three days on the page.

7. *Method of entry.* The UT of an observation is expressed as a day and hour, followed by a number of minutes and seconds. The tabular values of GHA and Dec, and, where necessary, the corresponding values of v and d, are taken directly from the daily pages for the day and hour of UT; this hour is always *before* the time of observation. SHA and Dec of the selected stars are also taken from the daily pages.

The table of Increments and Corrections for the minute of UT is then selected. For the GHA, the increment for minutes and seconds is taken from the appropriate column opposite the seconds of UT; the v-correction is taken from the second part of the same table opposite the value of v as given on the daily pages. Both increment and v-correction are to be added to the GHA, except for Venus when v is prefixed by a minus sign and the v-correction is to be subtracted. For the Dec there is no increment, but a d-correction is applied in the same way as the v-correction; d is given without sign on the daily pages and the sign of the correction is to be supplied by inspection of the Dec column. In many cases the correction may be applied mentally.

8. *Examples.* (a) Sun and Moon. Required the GHA and Dec of the Sun and Moon on 2009 July 31 at 15^h 47^m 13^s UT.

		SUN			MOON			
		GHA	Dec	d	GHA	v	Dec	d
		° ′	° ′	′	° ′	′	° ′	′
Daily page, July 31^d 15^h		43 24·3	N 18 07·7	0·6	287 11·3	10·1	S 25 55·9	2·9
Increments for	47^m 13^s	11 48·3			11 16·0			
v or d corrections for	47^m		−0·5		+8·0		+2·3	
Sum for July 31^d 15^h 47^m 13^s		55 12·6	N 18 07·2		298 35·3		S 25 58·2	

(b) Planets. Required the LHA and Dec of (i) Venus on 2009 July 31 at 10^h 08^m 17^s UT in longitude W 75° 30′; (ii) Saturn on 2009 July 31 at 13^h 12^m 11^s UT in longitude E 79° 26′.

		VENUS					SATURN			
		GHA	v	Dec	d		GHA	v	Dec	d
		° ′	′	° ′	′		° ′	′	° ′	′
Daily page, July 31^d	(10^h)	9 57·6	−0·6	N21 46·4	0·1	(13^h)	333 19·9	2·2	N6 03·7	0·1
Increments (planets)	(08^m 17^s)	2 04·3				(12^m 11^s)	3 02·8			
v or d corrections	(08^m)	−0·1		0·0		(12^m)	+0·5		0·0	
Sum = GHA and Dec.		12 01·8		N21 46·4			336 23·2		N6 03·7	
Longitude	(west)	− 75 30·0				(east)	+ 79 26·0			
Multiples of 360°		+360					−360			
LHA planet		296 31·8					55 49·2			

(c) Stars. Required the GHA and Dec of (i) *Aldebaran* on 2009 July 31 at 8^h 01^m 32^s UT; (ii) *Vega* on 2009 July 31 at 22^h 36^m 29^s UT.

		Aldebaran			*Vega*	
		GHA	Dec		GHA	Dec
		° ′	° ′		° ′	° ′
Daily page (SHA and Dec)		290 52·8	N 16 31.8		80 40·5	N 38 47.7
Daily page (GHA Aries)	(8^h)	69 04·8		(22^h)	279 39·3	
Increments (Aries)	(01^m 32^s)	0 23·1		(36^m 29^s)	9 08·7	
Sum = GHA star		360 20·7			369 28·5	
Multiples of 360°		−360			−360	
GHA star		0 20·7			9 28·5	

9. *Polaris (Pole Star) tables.* The tables on pages 274–276 provide means by which the latitude can be deduced from an observed altitude of *Polaris*, and they also give its azimuth; their use is explained and illustrated on those pages. They are based on the following formula:

$$\text{Latitude} - H_O = -p\cos h + \tfrac{1}{2}p\ \sin p\ \sin^2 h\ \tan(\text{latitude})$$

where
H_O = Apparent altitude (corrected for refraction)

p = polar distance of *Polaris* = 90° − Dec

h = local hour angle of *Polaris* = LHA Aries + SHA

a_0, which is a function of LHA Aries only, is the value of both terms of the above formula calculated for mean values of the SHA (319° 11′) and Dec (N 89° 18′.5) of *Polaris*, for a mean latitude of 50°, and adjusted by the addition of a constant (58′.8).

a_1, which is a function of LHA Aries and latitude, is the excess of the value of the second term over its mean value for latitude 50°, increased by a constant (0'.6) to make it always positive. a_2, which is a function of LHA Aries and date, is the correction to the first term for the variation of *Polaris* from its adopted mean position; it is increased by a constant (0'.6) to make it positive. The sum of the added constants is 1°, so that:

$$\text{Latitude} = \text{Apparent altitude (corrected for refraction)} - 1° + a_0 + a_1 + a_2$$

RISING AND SETTING PHENOMENA

10. *General.* On the right-hand daily pages are given the times of sunrise and sunset, of the beginning and end of civil and nautical twilights, and of moonrise and moonset for a range of latitudes from N 72° to S 60°. These times, which are given to the nearest minute, are strictly the UT of the phenomena on the Greenwich meridian; they are given for every day for moonrise and moonset, but only for the middle day of the three on each page for the solar phenomena.

They are approximately the Local Mean Times (LMT) of the corresponding phenomena on other meridians; they can be formally interpolated if desired. The UT of a phenomenon is obtained from the LMT by:

$$\text{UT} = \text{LMT} \begin{matrix} + \text{ west} \\ - \text{ east} \end{matrix} \text{ longitude}$$

in which the longitude must first be converted to time by the table on page i or otherwise.

Interpolation for latitude can be done mentally or with the aid of Table I on page xxxii.

The following symbols are used to indicate the conditions under which, in high latitudes, some of the phenomena do not occur:

□ Sun or Moon remains continuously above the horizon;

■ Sun or Moon remains continuously below the horizon;

//// twilight lasts all night.

Basis of the tabulations. At sunrise and sunset 16′ is allowed for semi-diameter and 34′ for horizontal refraction, so that at the times given the Sun's upper limb is on the visible horizon; all times refer to phenomena as seen from sea level with a clear horizon.

At the times given for the beginning and end of twilight, the Sun's zenith distance is 96° for civil, and 102° for nautical twilight. The degree of illumination at the times given for civil twilight (in good conditions and in the absence of other illumination) is such that the brightest stars are visible and the horizon is clearly defined. At the times given for nautical twilight the horizon is in general not visible, and it is too dark for observation with a marine sextant.

Times corresponding to other depressions of the Sun may be obtained by interpolation or, for depressions of more than 12°, less reliably, by extrapolation; times so obtained will be subject to considerable uncertainty near extreme conditions.

At moonrise and moonset allowance is made for semi-diameter, parallax, and refraction (34′), so that at the times given the Moon's upper limb is on the visible horizon as seen from sea level.

11. *Sunrise, sunset, twilight.* The tabulated times may be regarded, without serious error, as the LMT of the phenomena on any of the three days on the page and in any longitude. Precise times may normally be obtained by interpolating the tabular values for latitude and to the correct day and longitude, the latter being expressed as a fraction of a day by dividing it by 360°, positive for west and negative for east longitudes. In the extreme conditions near □, ■ or //// interpolation may not be possible in one direction, but accurate times are of little value in these circumstances.

Examples. Required the UT of (a) the beginning of morning twilights and sunrise on 2009 January 7 for latitude S 48° 55′, longitude E 75° 18′; (b) sunset and the end of evening twilights on 2009 January 9 for latitude N 67° 10′, longitude W 168° 05′.

	(a)	Twilight Nautical	Twilight Civil	Sunrise	(b)	Sunset	Twilight Civil	Twilight Nautical
From p. 15		d h m	d h m	d h m		d h m	d h m	d h m
LMT for Lat	S 45°	7 03 00	7 03 48	7 04 25	N 66°	9 14 02	9 15 28	9 16 42
Corr. to (p. xxxii, Table I)	S 48° 55′	−30	−20	−16	N 67° 10′	−29	−12	−6
Long (p. i)	E 75° 18′	−5 01	−5 01	−5 01	W 168° 05′	+11 12	+11 12	+11 12
UT		6 21 29	6 22 27	6 23 08		10 00 45	10 02 28	10 03 48

The LMT are strictly for January 8 (middle date on page) and 0° longitude; for more precise times it is necessary to interpolate, but rounding errors may accumulate to about 2^m.

(a) to January $7^d - 75°/360° =$ Jan. $6^d_{\cdot}8$, i.e. $\frac{1}{3}(1{\cdot}2) = 0{\cdot}4$ backwards towards the data for the same latitude interpolated similarly from page 13; the corrections are -2^m to nautical twilight, -2^m to civil twilight and -2^m to sunrise.

(b) to January $9^d + 168°/360° =$ Jan. $9^d_{\cdot}5$, i.e. $\frac{1}{3}(1{\cdot}5) = 0{\cdot}5$ forwards towards the data for the same latitude interpolated similarly from page 17; the corrections are $+8^m$ to sunset, $+4^m$ to civil twilight, and $+3^m$ to nautical twilight.

12. *Moonrise, moonset.* Precise times of moonrise and moonset are rarely needed; a glance at the tables will generally give sufficient indication of whether the Moon is available for observation and of the hours of rising and setting. If needed, precise times may be obtained as follows. Interpolate for latitude, using Table I on page xxxii, on the day wanted and also on the preceding day in east longitudes or the following day in west longitudes; take the difference between these times and interpolate for longitude by applying to the time for the day wanted the correction from Table II on page xxxii, so that the resulting time is between the two times used. In extreme conditions near □ or ■ interpolation for latitude or longitude may be possible only in one direction; accurate times are of little value in these circumstances.

To facilitate this interpolation the times of moonrise and moonset are given for four days on each page; where no phenomenon occurs during a particular day (as happens once a month) the time of the phenomenon on the following day, increased by 24^h, is given; extra care must be taken when interpolating between two values, when one of those values exceeds 24^h. In practice it suffices to use the daily difference between the times for the nearest tabular latitude, and generally, to enter Table II with the nearest tabular arguments as in the examples below.

Examples. Required the UT of moonrise and moonset in latitude S 47° 10′, longitudes E 124° 00′ and W 78° 31′ on 2009 January 9.

	Longitude E 124° 00′ Moonrise	Longitude E 124° 00′ Moonset	Longitude W 78° 31′ Moonrise	Longitude W 78° 31′ Moonset
	d h m	d h m	d h m	d h m
LMT for Lat. S 45°	9 18 46	9 01 55	9 18 46	9 01 55
Lat correction (p. xxxii, Table I)	+13	−13	+13	−13
Long correction (p. xxxii, Table II)	−23	−20	+11	+16
Correct LMT	9 18 36	9 01 22	9 19 10	9 01 58
Longitude (p. i)	−8 16	−8 16	+5 14	+5 14
UT	9 10 20	8 17 06	10 00 24	9 07 12

ALTITUDE CORRECTION TABLES

13. *General.* In general two corrections are given for application to altitudes observed with a marine sextant; additional corrections are required for Venus and Mars and also for very low altitudes.

Tables of the correction for dip of the horizon, due to height of eye above sea level, are given on pages A2 and xxxiv. Strictly this correction should be applied first and subtracted from the sextant altitude to give apparent altitude, which is the correct argument for the other tables.

Separate tables are given of the second correction for the Sun, for stars and planets (on pages A2 and A3), and for the Moon (on pages xxxiv and xxxv). For the Sun, values are given for both lower and upper limbs, for two periods of the year. The star tables are used for the planets, but additional corrections for parallax (page A2) are required for Venus and Mars. The Moon tables are in two parts: the main correction is a function of apparent altitude only and is tabulated for the lower limb (30′ must be subtracted to obtain the correction for the upper limb); the other, which is given for both lower and upper limbs, depends also on the horizontal parallax, which has to be taken from the daily pages.

An additional correction, given on page A4, is required for the change in the refraction, due to variations of pressure and temperature from the adopted standard conditions; it may generally be ignored for altitudes greater than 10°, except possibly in extreme conditions. The correction tables for the Sun, stars, and planets are in two parts; only those for altitudes greater than 10° are reprinted on the bookmark.

14. *Critical tables.* Some of the altitude correction tables are arranged as critical tables. In these an interval of apparent altitude (or height of eye) corresponds to a single value of the correction; no interpolation is required. At a "critical" entry the upper of the two possible values of the correction is to be taken. For example, in the table of dip, a correction of −4′·1 corresponds to all values of the height of eye from 5·3 to 5·5 metres (17·5 to 18·3 feet) inclusive.

15. *Examples.* The following examples illustrate the use of the altitude correction tables; the sextant altitudes given are assumed to be taken on 2009 January 8 with a marine sextant at height 5·4 metres (18 feet), temperature −3°C and pressure 982 mb, the Moon sights being taken at about 10^{h} UT.

	SUN lower limb	SUN upper limb	MOON lower limb	MOON upper limb	VENUS	*Polaris*
	° ′	° ′	° ′	° ′	° ′	° ′
Sextant altitude	21 19·7	3 20·2	33 27·6	26 06·7	4 32·6	49 36·5
Dip, height 5·4 metres (18 feet)	−4·1	−4·1	−4·1	−4·1	−4·1	−4·1
Main correction	+13·8	−29·6	+57·4	+60·5	−10·8	−0·8
−30′ for upper limb (Moon)	—	—	—	−30·0	—	—
L, U correction for Moon	—	—	+8·3	+5·5	—	—
Additional correction for Venus	—	—	—	—	+0·2	—
Additional refraction correction	−0·1	−0·6	−0·1	−0·1	−0·5	0·0
Corrected sextant altitude	21 29·3	2 45·9	34 29·1	26 38·5	4 17·4	49 31·6

The main corrections have been taken out with apparent altitude (sextant altitude corrected for index error and dip) as argument, interpolating where possible. These refinements are rarely necessary.

16. *Composition of the Corrections.* The table for the dip of the sea horizon is based on the formula:

$$\text{Correction for dip} = -1'.76\sqrt{(\text{height of eye in metres})} = -0'.97\sqrt{(\text{height of eye in feet})}$$

The correction table for the Sun includes the effects of semi-diameter, parallax and mean refraction.

The correction tables for the stars and planets allow for the effect of mean refraction.

The phase correction for Venus has been incorporated in the tabulations for GHA and Dec, and no correction for phase is required. The additional corrections for Venus and Mars allow for parallax. Alternatively, the correction for parallax may be calculated from $p \cos H$, where p is the parallax and H is the altitude. In 2009 the values for p are:

	Jan. 1		Jan. 28		Feb. 21		Mar. 9		Apr. 14		Apr. 30		May 22		July 10		Dec. 31
Venus		0′·2		0′·3		0′·4		0′·5		0′·4		0′·3		0′·2		0′·1	

	Jan. 1		Nov. 27		Dec. 31
Mars		0′·1		0′·2	

The correction table for the Moon includes the effect of semi-diameter, parallax, augmentation and mean refraction.

Mean refraction is calculated for a temperature of 10°C (50°F), a pressure of 1010 mb (29·83 inches), humidity of 80% and wavelength 0·50169 μm.

17. *Bubble sextant observations.* When observing with a bubble sextant no correction is necessary for dip, semi-diameter, or augmentation. The altitude corrections for the stars and planets on page A2 and on the bookmark should be used for the Sun as well as for the stars and planets; for the Moon it is easiest to take the mean of the corrections for lower and upper limbs and subtract 15′ from the altitude; the correction for dip must not be applied.

AUXILIARY AND PLANNING DATA

18. *Sun and Moon.* On the daily pages are given: hourly values of the horizontal parallax of the Moon; the semi-diameters and the times of meridian passage of both Sun and Moon over the Greenwich meridian; the equation of time; the age of the Moon, the percent (%) illuminated and a symbol indicating the phase. The times of the phases of the Moon are given in UT on page 4. For the Moon, the semi-diameters for each of the three days are given at the foot of the column; for the Sun a single value is sufficient. Table II on page xxxii may be used for interpolating the time of the Moon's meridian passage for longitude. The equation of time is given daily at 00^h and 12^h UT. The sign is *positive* for unshaded values and *negative* for shaded values. To obtain apparent time add the equation of time to mean time when the sign is *positive*. Subtract the equation of time from mean time when the sign is *negative*. At 12^h UT, when the sign is *positive*, meridian passage of the Sun occurs *before* 12^h UT, otherwise it occurs *after* 12^h UT.

19. *Planets.* The magnitudes of the planets are given immediately following their names in the headings on the daily pages; also given, for the middle day of the three on the page, are their SHA at 00^h UT and their times of meridian passage.

The planet notes and diagram on pages 8 and 9 provide descriptive information as to the suitability of the planets for observation during the year, and of their positions and movements.

20. *Stars.* The time of meridian passage of the first point of Aries over the Greenwich meridian is given on the daily pages, for the middle day of the three on the page, to $0^{m}\!\!.1$. The interval between successive meridian passages is $23^h\ 56^{m}\!\!.1$ (24^h less $3^{m}\!\!.9$) so that times for intermediate days and other meridians can readily be derived. If a precise time is required it may be obtained by finding the UT at which LHA Aries is zero.

The meridian passage of a star occurs when its LHA is zero, that is when LHA Aries + SHA = 360°. An approximate time can be obtained from the planet diagram on page 9.

The star charts on pages 266 and 267 are intended to assist identification. They show the relative positions of the stars in the sky as seen from the Earth and include all 173 stars used in the Almanac, together with a few others to complete the main constellation configurations. The local meridian at any time may be located on the chart by means of its SHA which is 360° – LHA Aries, or west longitude – GHA Aries.

21. *Star globe.* To set a star globe on which is printed a scale of LHA Aries, first set the globe for latitude and then rotate about the polar axis until the scale under the edge of the meridian circle reads LHA Aries.

To mark the positions of the Sun, Moon, and planets on the star globe, take the difference GHA Aries – GHA body and use this along the LHA Aries scale, in conjunction with the declination, to plot the position. GHA Aries – GHA body is most conveniently found by taking the difference when the GHA of the body is small (less than 15°), which happens once a day.

22. *Calendar.* On page 4 are given lists of ecclesiastical festivals, and of the principal anniversaries and holidays in the United Kingdom and the United States of America. The calendar on page 5 includes the day of the year as well as the day of the week.

Brief particulars are given, at the foot of page 5, of the solar and lunar eclipses occurring during the year; the times given are in UT. The principal features of the more important solar eclipses are shown on the maps on pages 6 and 7.

23. *Standard times.* The lists on pages 262–265 give the standard times used in most countries. In general no attempt is made to give details of the beginning and end of summer time, since they are liable to frequent changes at short notice. For the latest information consult Admiralty List of Radio Signals Volume 2 (NP 282) corrected by Section VI of the weekly edition of Admiralty Notices to Mariners.

The Date or Calendar Line is an arbitrary line, on either side of which the date differs by one day; when crossing this line on a westerly course, the date must be advanced one day; when crossing it on an easterly course, the date must be put back one day. The line is a modification of the line of the 180th meridian, and is drawn so as to include, as far as possible, islands of any one group, etc., on the same side of the line. It may be traced by starting at the South Pole and joining up to the following positions:

	°	°	°	°	°	°	°
Lat	S 51·0	S 45·0	S 15·0	S 5·0	N 48·0	N 53·0	N 65·5
Long	180·0	W 172·5	W 172·5	180·0	180·0	E 170·0	W 169·0

thence through the middle of the Diomede Islands to Lat N 68°.0, Long W 169°.0, passing east of Ostrov Vrangelya (Wrangel Island) to Lat N 75°.0, Long 180°.0, and thence to the North Pole.

ACCURACY

24. *Main data.* The quantities tabulated in this Almanac are generally correct to the nearest 0′.1; the exception is the Sun's GHA which is deliberately adjusted by up to 0′.15 to reduce the error due to ignoring the v-correction. The GHA and Dec at intermediate times cannot be obtained to this precision, since at least two quantities must be added; moreover, the v- and d-corrections are based on mean values of v and d and are taken from tables for the whole minute only. The largest error that can occur in the GHA or Dec of any body other than the Sun or Moon is less than 0′.2; it may reach 0′.25 for the GHA of the Sun and 0′.3 for that of the Moon.

In practice it may be expected that only one third of the values of GHA and Dec taken out will have errors larger than 0′.05 and less than one tenth will have errors larger than 0′.1.

25. *Altitude corrections.* The errors in the altitude corrections are nominally of the same order as those in GHA and Dec, as they result from the addition of several quantities each correctly rounded off to 0′.1. But the actual values of the dip and of the refraction at low altitudes may, in extreme atmospheric conditions, differ considerably from the mean values used in the tables.

USE OF THIS ALMANAC IN 2010

This Almanac may be used for the Sun and stars in 2010 in the following manner.

For the Sun, take out the GHA and Dec for the same date but for a time $5^h 48^m 00^s$ *earlier* than the UT of observation; add 87° 00′ to the GHA so obtained. The error, mainly due to planetary perturbations of the Earth, is unlikely to exceed 0′.4.

For the stars, calculate the GHA and Dec for the same date and the same time, but *subtract* 15′.1 from the GHA so found. The error, due to incomplete correction for precession and nutation, is unlikely to exceed 0′.4. If preferred, the same result can be obtained by using a time $5^h 48^m 00^s$ earlier than the UT of observation (as for the Sun) and adding 86° 59′.2 to the GHA (or adding 87° as for the Sun and subtracting 0′.8, for precession, from the SHA of the star).

The Almanac cannot be so used for the Moon or planets.

LIST I — PLACES FAST ON UTC (mainly those EAST OF GREENWICH)

The times given below should be } *added* to UTC to give Standard Time / *subtracted* from Standard Time to give UTC.

Place	h	m
Admiralty Islands	10	
Afghanistan	04	30
Albania*	01	
Algeria	01	
Amirante Islands	04	
Andaman Islands	05	30
Angola	01	
Armenia*	04	
Australia		
Australian Capital Territory*	10	
New South Wales*[1]	10	
Northern Territory	09	30
Queensland	10	
South Australia*	09	30
Tasmania*	10	
Victoria*	10	
Western Australia*	08	
Whitsunday Islands	10	
Austria*†	01	
Azerbaijan*	04	
Bahrain	03	
Balearic Islands*†	01	
Bangladesh	06	
Belarus*	02	
Belgium*†	01	
Benin	01	
Bosnia and Herzegovina*	01	
Botswana, Republic of	02	
Brunei	08	
Bulgaria*†	02	
Burma (Myanmar)	06	30
Burundi	02	
Cambodia	07	
Cameroon Republic	01	
Caroline Islands[2]	10	
Central African Republic	01	
Chad	01	
Chagos Archipelago & Diego Garcia	06	
Chatham Islands*	12	45
China, People's Republic of	08	
Christmas Island, Indian Ocean	07	
Cocos (Keeling) Islands	06	30
Comoro Islands (Comoros)	03	
Congo, Democratic Republic		
Kinshasa, Mbandaka	01	
Haut-Zaire, Kasai, Kivu, Shaba	02	
Congo Republic	01	
Corsica*†	01	
Crete*†	02	
Croatia*	01	
Cyprus†: Ercan*, Larnaca*	02	
Czech Republic*†	01	
Denmark*†	01	
Djibouti	03	
Egypt, Arab Republic of*	02	
Equatorial Guinea, Republic of	01	
Eritrea	03	
Estonia*†	02	
Ethiopia	03	
Fiji	12	
Finland*†	02	
France*†	01	
Gabon	01	
Georgia	04	
Germany*†	01	
Gibraltar*	01	
Greece*†	02	
Guam	10	
Hong Kong	08	
Hungary*†	01	
India	05	30
Indonesia, Republic of		
Bangka, Billiton, Java, West and Central Kalimantan, Madura, Sumatra	07	
Bali, Flores, South and East Kalimantan, Lombok, Sulawesi, Sumba, Sumbawa, West Timor	08	
Aru, Irian Jaya, Kai, Moluccas, Tanimbar	09	
Iran*	03	30
Iraq*	03	
Israel*	02	
Italy*†	01	
Jan Mayen Island*	01	
Japan	09	
Jordan*	02	
Kazakhstan		
Western: Aktau, Uralsk, Atyrau	05	
Eastern: Kzyl-Orda, Astana	06	
Kenya	03	
Kerguelen Islands	05	
Kiribati Republic		
Gilbert Islands	12	
Phoenix Islands[3]	13	
Line Islands[3]	14	
Korea, North	09	
Republic of (South)	09	
Kuril Islands	11	
Kuwait	03	
Kyrgyzstan	06	
Laccadive Islands	05	30
Laos	07	
Latvia*†	02	

* Daylight-saving time may be kept in these places. † For Summer time dates see List II footnotes.

1 Except Broken Hill Area which keeps 09^{h} 30^{m}.

2 Except Pohnpei, Pingelap and Kosrae which keep 11^{h} and Palau which keeps 09^{h}.

3 The Line and Phoenix Is. not part of the Kiribati Republic keep 10^{h} and 11^{h}, respectively, slow on UTC.

LIST I — (*continued*)

	h	m
Lebanon*	02	
Lesotho	02	
Libya	02	
Liechtenstein*	01	
Lithuania*†	02	
Lord Howe Island*	10	30
Luxembourg*†	01	
Macau	08	
Macedonia*, former Yugoslav Republic	01	
Macias Nguema (Fernando Póo)	01	
Madagascar, Democratic Republic of	03	
Malawi	02	
Malaysia, Malaya, Sabah, Sarawak	08	
Maldives, Republic of The	05	
Malta*†	01	
Mariana Islands	10	
Marshall Islands[1]	12	
Mauritius	04	
Moldova*	02	
Monaco*	01	
Mongolia	08	
Montenegro*	01	
Mozambique	02	
Namibia*	01	
Nauru	12	
Nepal	05	45
Netherlands, The*†	01	
New Caledonia	11	
New Zealand*	12	
Nicobar Islands	05	30
Niger	01	
Nigeria, Republic of	01	
Norfolk Island	11	30
Norway*	01	
Novaya Zemlya	03	
Okinawa	09	
Oman	04	
Pagalu (Annobon Islands)	01	
Pakistan	05	
Palau Islands	09	
Papua New Guinea	10	
Pescadores Islands	08	
Philippine Republic	08	
Poland*†	01	
Qatar	03	
Reunion	04	
Romania*†	02	
Russia[2]*		
Zone 1 Kaliningrad	02	
Zone 2 Moscow, St Petersburg, Arkhangelsk, Astrakhan	03	
Zone 3 Samara, Izhevsk	04	
Zone 4 Perm, Amderna, Novyy Port	05	
Zone 5 Omsk, Novosibirsk	06	

	h	m
Russia (*continued*)		
Zone 6 Norilsk, Kyzyl, Dikson	07	
Zone 7 Bratsk, Irkutsk, Ulan-Ude	08	
Zone 8 Yakutsk, Chita, Tiksi	09	
Zone 9 Vladivostok, Khabarovsk, Okhotsk	10	
Zone 10 Magadan	11	
Zone 11 Petropavlovsk, Pevek	12	
Rwanda	02	
Ryukyu Islands	09	
Sakhalin Island*	10	
Santa Cruz Islands	11	
Sardinia*†	01	
Saudi Arabia	03	
Schouten Islands	09	
Serbia*	01	
Seychelles	04	
Sicily*†	01	
Singapore	08	
Slovakia*†	01	
Slovenia*†	01	
Socotra	03	
Solomon Islands	11	
Somalia Republic	03	
South Africa, Republic of	02	
Spain*†	01	
Spanish Possessions in North Africa*	01	
Spitsbergen (Svalbard)*	01	
Sri Lanka	05	30
Sudan, Republic of	03	
Swaziland	02	
Sweden*†	01	
Switzerland*	01	
Syria (Syrian Arab Republic)*	02	
Taiwan	08	
Tajikistan	05	
Tanzania	03	
Thailand	07	
Timor-Leste	09	
Tonga	13	
Tunisia*	01	
Turkey*	02	
Turkmenistan	05	
Tuvalu	12	
Uganda	03	
Ukraine*	02	
United Arab Emirates	04	
Uzbekistan	05	
Vanuatu, Republic of	11	
Vietnam, Socialist Republic of	07	
Yemen	03	
Zambia, Republic of	02	
Zimbabwe	02	

* Daylight-saving time may be kept in these places. † For Summer time dates see List II footnotes.

1 Except the Ebon Atoll which keeps time 24^h slow on that of the rest of the islands.

2 The boundaries between the zones are irregular; listed are chief towns in each zone.

LIST II — PLACES NORMALLY KEEPING UTC

Ascension Island	Ghana	Irish Republic*†	Morocco	Sierra Leone
Burkina-Faso	Great Britain†	Ivory Coast	Portugal*†	Togo Republic
Canary Islands*†	Guinea-Bissau	Liberia	Principe	Tristan da Cunha
Channel Islands†	Guinea Republic	Madeira*	St. Helena	
Faeroes*, The	Iceland	Mali	São Tomé	
Gambia, The	Ireland, Northern†	Mauritania	Senegal	

* Daylight-saving time may be kept in these places.

† Summer time (daylight-saving time), one hour in advance of UTC, will be kept from 2009 March 29^d 01^h to October 25^d 01^h UTC (Ninth Summer Time Directive of the European Union). Ratification by member countries has not been verified.

LIST III — PLACES SLOW ON UTC (WEST OF GREENWICH)

The times given below should be *subtracted* from UTC to give Standard Time / *added* to Standard Time to give UTC.

	h m
Argentina	03
Austral (Tubuai) Islands[1]	10
Azores*	01
Bahamas*	05
Barbados	04
Belize	06
Bermuda*	04
Bolivia	04
Brazil	
Fernando de Noronha I., Trindade I., Oceanic Is.	02
S and E coastal states*, Goiás*, Brasilia*, Minas Gerais*, Bahia, Tocantins, Para (eastern), N and NE coastal states	03
Mato Grosso do Sul*, Mato Grosso*, Para (western), Roraima, Rondonia, most of Amazonas	04
Amazonas (south western), Acre	05
British Antarctic Territory[2,3]	03
Canada[3]	
Alberta*	07
British Columbia*	08
Labrador*	04
Manitoba*	06
New Brunswick*	04
Newfoundland*	03 30
Nunavut*	
east of long. W. 85°	05
long. W. 85° to W. 102°	06
west of long. W. 102°	07
Northwest Territories*	07
Nova Scotia*	04
Ontario, east of long. W. 90°*	05
Ontario, west of long. W. 90°*	06
Canada (*continued*)	
Prince Edward Island*	04
Quebec, east of long. W. 63°	04
Quebec, west of long. W. 63°*	05
Saskatchewan	06
Yukon*	08
Cape Verde Islands	01
Cayman Islands	05
Chile*	04
Colombia	05
Cook Islands	10
Costa Rica	06
Cuba*	05
Curaçao Island	04
Dominican Republic	04
Easter Island (I. de Pascua)*	06
Ecuador	05
El Salvador	06
Falkland Islands*	04
Fanning Island	10
Fernando de Noronha Island	02
French Guiana	03
Galápagos Islands	06
Greenland	
General*	03
Scoresby Sound*	01
Thule, Danmarkshavn, Mesters Vig	00
Grenada	04
Guadeloupe	04
Guatemala	06
Guyana, Republic of	04
Haiti	05
Honduras	06

* Daylight-saving time may be kept in these places.

1 This is the legal standard time, but local mean time is generally used.

2 Stations may use UTC.

3 Some areas may keep another time zone.

LIST III — (*continued*)

	h m
Jamaica	05
Johnston Island	10
Juan Fernandez Islands*	04
Leeward Islands	04
Marquesas Islands	09 30
Martinique	04
Mexico*[1]	06
Midway Islands	11
Nicaragua	06
Niue	11
Panama, Republic of	05
Paraguay*	04
Peru	05
Pitcairn Island	08
Puerto Rico	04
St. Pierre and Miquelon*	03
Samoa	11
Society Islands	10
South Georgia	02
Suriname	03
Trindade Island, South Atlantic	02
Trinidad and Tobago	04
Tuamotu Archipelago	10
Tubuai (Austral) Islands	10
Turks and Caicos Islands*	05
United States of America[2]	
Alabama	06
Alaska	09
Aleutian Islands, east of W. 169° 30′	09
Aleutian Islands, west of W. 169° 30′	10
Arizona[3,4]	07
Arkansas	06
California	08
Colorado	07
Connecticut	05
Delaware	05
District of Columbia	05
Florida[4]	05
Georgia	05
Hawaii[3]	10
Idaho, southern part	07
northern part	08
Illinois	06
United States of America[2] (*continued*)	
Indiana[4]	05
Iowa	06
Kansas[4]	06
Kentucky, eastern part	05
western part	06
Louisiana	06
Maine	05
Maryland	05
Massachusetts	05
Michigan[4]	05
Minnesota	06
Mississippi	06
Missouri	06
Montana	07
Nebraska, eastern part	06
western part	07
Nevada	08
New Hampshire	05
New Jersey	05
New Mexico	07
New York	05
North Carolina	05
North Dakota, eastern part	06
western part	07
Ohio	05
Oklahoma	06
Oregon[4]	08
Pennsylvania	05
Rhode Island	05
South Carolina	05
South Dakota, eastern part	06
western part	07
Tennessee, eastern part	05
western part	06
Texas[4]	06
Utah	07
Vermont	05
Virginia	05
Washington D.C.	05
Washington	08
West Virginia	05
Wisconsin	06
Wyoming	07
Uruguay*	03
Venezuela	04
Virgin Islands	04
Windward Islands	04

* Daylight-saving time may be kept in these places.

1 Except the states of Sonora, Sinaloa*, Nayarit*, Chihuahua* and the Southern District of Lower California* which keep 07^h, and the Northern District of Lower California* which keeps 08^h.

2 Daylight-saving (Summer) time, one hour fast on the time given, is kept during 2009 from the March 8 (second Sunday) to November 1 (first Sunday), changing at $02^h\ 00^m$ local clock time.

3 Exempt from keeping daylight-saving time.

4 A small portion of the state is in another time zone.

NORTHERN STARS

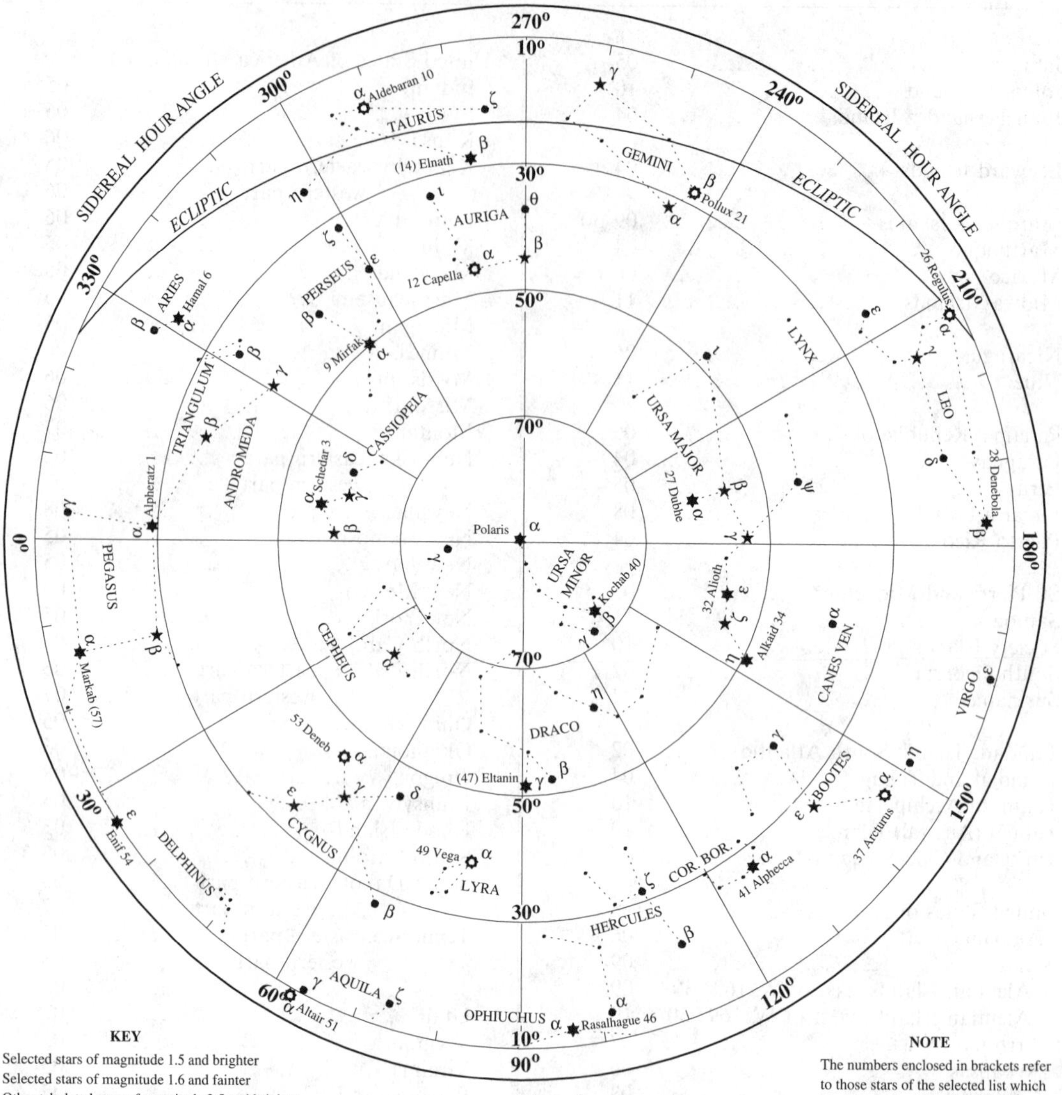

KEY

- Selected stars of magnitude 1.5 and brighter
- Selected stars of magnitude 1.6 and fainter
- Other tabulated stars of magnitude 2.5 and brighter
- Other tabulated stars of magnitude 2.6 and fainter
- Untabulated stars

NOTE

The numbers enclosed in brackets refer to those stars of the selected list which are not used in Sight Reduction Tables H.O. 249, A.P. 3270, N.P. 303.

EQUATORIAL STARS (SHA 0° to 180°)

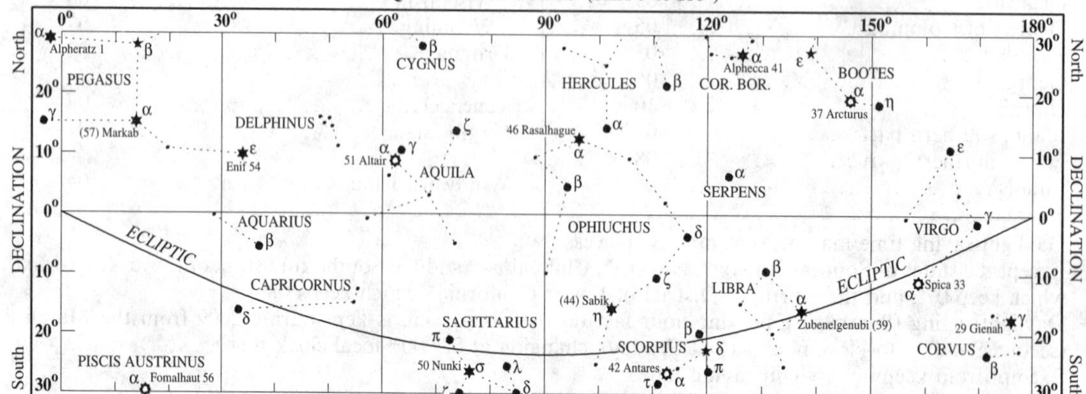

SOUTHERN STARS

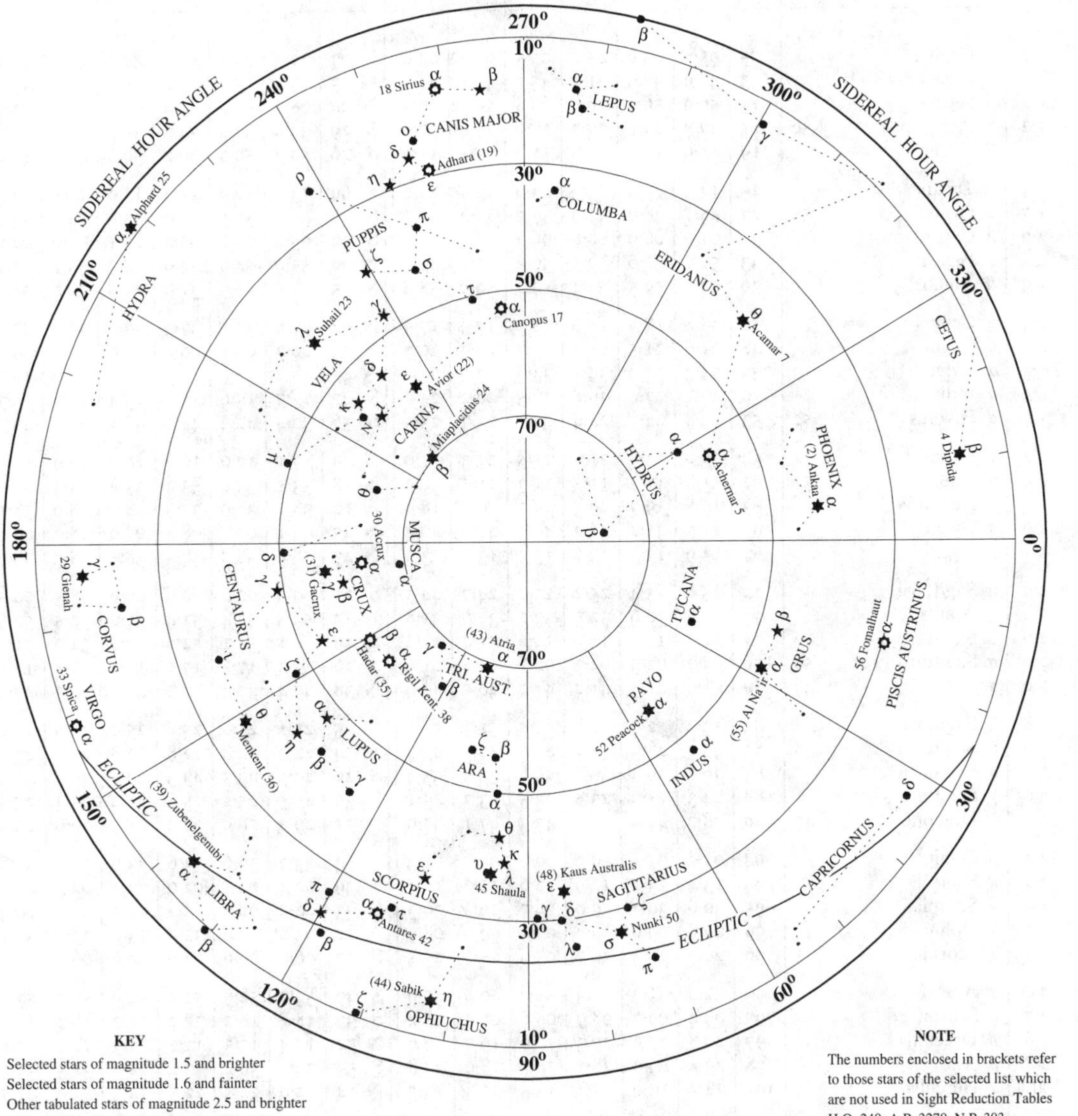

KEY

- ✪ Selected stars of magnitude 1.5 and brighter
- ✸ Selected stars of magnitude 1.6 and fainter
- ★ Other tabulated stars of magnitude 2.5 and brighter
- ● Other tabulated stars of magnitude 2.6 and fainter
- • Untabulated stars

NOTE

The numbers enclosed in brackets refer to those stars of the selected list which are not used in Sight Reduction Tables H.O. 249, A.P. 3270, N.P. 303.

EQUATORIAL STARS (SHA 180° to 360°)

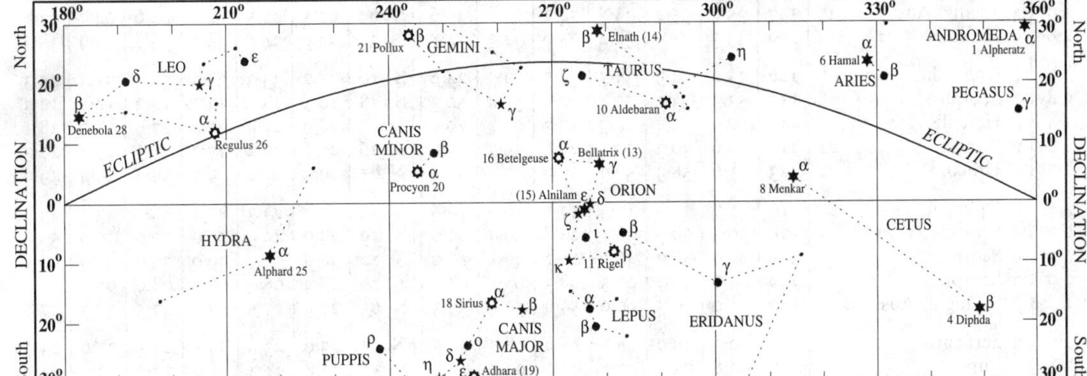

Mag.	Name and Number		SHA							Declination						
				JAN.	FEB.	MAR.	APR.	MAY	JUNE		JAN.	FEB.	MAR.	APR.	MAY	JUNE
			°	′	′	′	′	′	′	°	′	′	′	′	′	′
3·2	γ Cephei		**5**	05·1	05·6	05·8	05·5	04·8	04·0	**N 77**	41·3	41·2	41·1	40·9	40·9	40·9
2·5	α Pegasi	57	**13**	41·9	41·9	41·9	41·7	41·5	41·3	**N 15**	15·3	15·3	15·2	15·2	15·3	15·3
2·4	β Pegasi		**13**	56·9	56·9	56·9	56·8	56·5	56·3	**N 28**	08·1	08·0	07·9	07·9	07·9	08·0
1·2	α Piscis Aust.	56	**15**	27·7	27·8	27·7	27·6	27·4	27·1	**S 29**	34·6	34·5	34·4	34·3	34·2	34·1
2·1	β Gruis		**19**	11·9	12·0	11·9	11·7	11·5	11·1	**S 46**	50·4	50·3	50·2	50·0	49·9	49·8
2·9	α Tucanæ		**25**	13·4	13·4	13·3	13·0	12·7	12·2	**S 60**	13·0	12·9	12·7	12·6	12·5	12·4
1·7	α Gruis	55	**27**	48·1	48·0	47·9	47·7	47·4	47·1	**S 46**	55·2	55·1	54·9	54·8	54·7	54·6
2·9	δ Capricorni		**33**	07·0	06·9	06·8	06·7	06·4	06·2	**S 16**	05·2	05·2	05·2	05·1	05·0	04·9
2·4	ε Pegasi	54	**33**	50·6	50·6	50·5	50·4	50·1	49·9	**N 9**	55·0	54·9	54·9	54·9	55·0	55·1
2·9	β Aquarii		**36**	59·6	59·5	59·4	59·2	59·0	58·8	**S 5**	31·9	31·9	31·9	31·9	31·8	31·7
2·4	α Cephei		**40**	18·8	18·8	18·6	18·3	17·9	17·5	**N 62**	37·6	37·4	37·3	37·2	37·2	37·3
2·5	ε Cygni		**48**	21·6	21·5	21·4	21·2	20·9	20·7	**N 34**	00·3	00·1	00·1	00·0	00·1	00·2
1·3	α Cygni	53	**49**	34·3	34·2	34·1	33·8	33·5	33·3	**N 45**	18·8	18·7	18·5	18·5	18·6	18·7
3·1	α Indi		**50**	27·1	27·0	26·8	26·5	26·2	25·9	**S 47**	15·7	15·6	15·5	15·4	15·3	15·3
1·9	α Pavonis	52	**53**	24·8	24·7	24·4	24·1	23·7	23·3	**S 56**	42·4	42·3	42·2	42·1	42·1	42·1
2·2	γ Cygni		**54**	22·0	22·0	21·8	21·5	21·3	21·0	**N 40**	17·1	17·0	16·9	16·9	16·9	17·1
0·8	α Aquilæ	51	**62**	11·8	11·7	11·5	11·3	11·1	10·9	**N 8**	53·5	53·4	53·4	53·4	53·5	53·6
2·7	γ Aquilæ		**63**	19·8	19·7	19·5	19·3	19·1	18·9	**N 10**	38·1	38·0	37·9	38·0	38·0	38·1
2·9	δ Cygni		**63**	41·5	41·4	41·2	40·9	40·7	40·4	**N 45**	09·1	09·0	08·9	08·9	08·9	09·1
3·1	β Cygni		**67**	13·9	13·8	13·6	13·4	13·2	13·0	**N 27**	58·6	58·5	58·5	58·5	58·5	58·7
2·9	π Sagittarii		**72**	25·6	25·4	25·2	25·0	24·7	24·5	**S 21**	00·6	00·6	00·6	00·6	00·5	00·5
3·0	ζ Aquilæ		**73**	32·8	32·6	32·5	32·2	32·0	31·8	**N 13**	52·5	52·4	52·4	52·4	52·5	52·6
2·6	ζ Sagittarii		**74**	12·3	12·1	11·9	11·6	11·4	11·2	**S 29**	52·1	52·1	52·0	52·0	52·0	51·9
2·0	σ Sagittarii	50	**76**	02·7	02·5	02·3	02·1	01·8	01·6	**S 26**	17·2	17·2	17·1	17·1	17·1	17·1
0·0	α Lyræ	49	**80**	41·6	41·5	41·2	41·0	40·7	40·6	**N 38**	47·4	47·2	47·2	47·2	47·3	47·5
2·8	λ Sagittarii		**82**	52·2	52·0	51·8	51·5	51·3	51·1	**S 25**	25·1	25·0	25·0	25·0	25·0	25·0
1·9	ε Sagittarii	48	**83**	48·5	48·3	48·1	47·8	47·5	47·3	**S 34**	22·9	22·8	22·8	22·8	22·8	22·8
2·7	δ Sagittarii		**84**	36·5	36·3	36·1	35·8	35·5	35·4	**S 29**	49·5	49·5	49·5	49·4	49·4	49·4
3·0	γ Sagittarii		**88**	24·3	24·0	23·8	23·5	23·3	23·1	**S 30**	25·5	25·5	25·5	25·4	25·4	25·5
2·2	γ Draconis	47	**90**	48·2	47·9	47·7	47·3	47·1	47·0	**N 51**	29·0	28·9	28·9	28·9	29·0	29·2
2·8	β Ophiuchi		**94**	01·3	01·1	00·9	00·7	00·5	00·4	**N 4**	33·7	33·6	33·6	33·6	33·7	33·7
2·4	κ Scorpii		**94**	13·4	13·1	12·9	12·6	12·3	12·1	**S 39**	02·1	02·1	02·0	02·1	02·1	02·1
1·9	θ Scorpii		**95**	30·6	30·3	30·0	29·7	29·4	29·3	**S 43**	00·2	00·2	00·2	00·2	00·2	00·3
2·1	α Ophiuchi	46	**96**	09·8	09·6	09·4	09·2	09·0	08·9	**N 12**	33·0	33·0	32·9	33·0	33·0	33·1
1·6	λ Scorpii	45	**96**	26·7	26·5	26·2	25·9	25·7	25·5	**S 37**	06·6	06·6	06·6	06·6	06·7	06·7
3·0	α Aræ		**96**	52·0	51·7	51·4	51·0	50·8	50·6	**S 49**	53·0	52·9	52·9	52·9	53·0	53·1
2·7	υ Scorpii		**97**	09·4	09·2	08·9	08·6	08·3	08·2	**S 37**	18·2	18·2	18·2	18·2	18·2	18·3
2·8	β Draconis		**97**	20·8	20·6	20·3	20·0	19·7	19·6	**N 52**	17·4	17·3	17·2	17·3	17·4	17·6
2·8	β Aræ		**98**	29·4	29·0	28·7	28·3	28·0	27·8	**S 55**	32·2	32·2	32·2	32·2	32·3	32·4
Var.‡	α Herculis		**101**	14·2	14·0	13·8	13·5	13·4	13·3	**N 14**	22·6	22·5	22·5	22·5	22·6	22·7
2·4	η Ophiuchi	44	**102**	16·6	16·4	16·1	15·9	15·7	15·6	**S 15**	44·2	44·3	44·3	44·3	44·3	44·3
3·1	ζ Aræ		**105**	09·6	09·2	08·9	08·5	08·2	08·0	**S 56**	00·2	00·1	00·2	00·2	00·3	00·4
2·3	ε Scorpii		**107**	18·8	18·5	18·3	18·0	17·8	17·7	**S 34**	18·6	18·6	18·6	18·6	18·7	18·7
1·9	α Triang. Aust.	43	**107**	35·9	35·4	34·8	34·2	33·8	33·6	**S 69**	02·5	02·5	02·5	02·6	02·7	02·8
2·8	ζ Herculis		**109**	35·7	35·5	35·2	35·0	34·9	34·8	**N 31**	34·9	34·8	34·8	34·8	34·9	35·1
2·6	ζ Ophiuchi		**110**	35·2	34·9	34·7	34·5	34·3	34·2	**S 10**	35·2	35·3	35·3	35·3	35·3	35·3
2·8	τ Scorpii		**110**	53·3	53·1	52·8	52·6	52·4	52·3	**S 28**	14·1	14·1	14·2	14·2	14·2	14·3
2·8	β Herculis		**112**	20·9	20·7	20·5	20·3	20·1	20·1	**N 21**	28·0	27·9	27·8	27·9	28·0	28·1
1·0	α Scorpii	42	**112**	30·6	30·3	30·1	29·8	29·7	29·6	**S 26**	27·1	27·2	27·2	27·3	27·3	27·3
2·7	η Draconis		**113**	58·6	58·3	57·9	57·6	57·4	57·4	**N 61**	29·3	29·2	29·2	29·2	29·4	29·6
2·7	δ Ophiuchi		**116**	17·7	17·5	17·3	17·1	16·9	16·8	**S 3**	43·2	43·3	43·3	43·3	43·3	43·2
2·6	β Scorpii		**118**	30·5	30·3	30·1	29·8	29·7	29·6	**S 19**	49·9	49·9	50·0	50·0	50·0	50·0
2·3	δ Scorpii		**119**	47·0	46·7	46·5	46·2	46·1	46·0	**S 22**	38·9	38·9	39·0	39·0	39·0	39·1
2·9	π Scorpii		**120**	09·0	08·7	08·5	08·3	08·1	08·0	**S 26**	08·4	08·5	08·5	08·6	08·6	08·6
2·8	β Trianguli Aust.		**120**	61·0	60·5	60·0	59·6	59·4	59·3	**S 63**	27·4	27·4	27·4	27·5	27·7	27·8
2·6	α Serpentis		**123**	49·3	49·0	48·8	48·6	48·5	48·5	**N 6**	23·6	23·6	23·5	23·5	23·6	23·7
2·8	γ Lupi		**126**	03·8	03·5	03·2	03·0	02·8	02·8	**S 41**	11·8	11·8	11·9	12·0	12·0	12·1
2·2	α Coronæ Bor.	41	**126**	13·9	13·7	13·5	13·3	13·2	13·2	**N 26**	40·8	40·7	40·7	40·7	40·8	40·9

‡ 2·9 — 3·6

Mag.	Name and Number		SHA							Declination						
				JULY	AUG.	SEPT.	OCT.	NOV.	DEC.		JULY	AUG.	SEPT.	OCT.	NOV.	DEC.
			°	′	′	′	′	′	′	°	′	′	′	′	′	′
3·2	γ Cephei		**5**	03·2	02·6	02·4	02·5	02·9	03·5	N **77**	40·9	41·1	41·3	41·5	41·6	41·7
2·5	*Markab*	57	**13**	41·0	40·9	40·8	40·8	40·9	41·0	N **15**	15·5	15·6	15·7	15·7	15·7	15·7
2·4	*Scheat*		**13**	56·0	55·9	55·8	55·8	55·9	56·0	N **28**	08·1	08·2	08·4	08·4	08·5	08·5
1·2	*Fomalhaut*	56	**15**	26·8	26·7	26·6	26·6	26·7	26·8	S **29**	34·0	34·0	34·1	34·1	34·2	34·2
2·1	β Gruis		**19**	10·8	10·6	10·6	10·6	10·8	10·9	S **46**	49·8	49·8	49·9	50·0	50·1	50·1
2·9	α Tucanæ		**25**	11·9	11·6	11·6	11·7	12·0	12·2	S **60**	12·4	12·5	12·6	12·7	12·8	12·8
1·7	*Al Na'ir*	55	**27**	46·8	46·6	46·6	46·7	46·9	47·0	S **46**	54·6	54·7	54·8	54·8	54·9	54·9
2·9	δ Capricorni		**33**	06·0	05·8	05·8	05·9	06·0	06·1	S **16**	04·9	04·8	04·8	04·9	04·9	04·9
2·4	*Enif*	54	**33**	49·7	49·6	49·6	49·6	49·7	49·8	N **9**	55·2	55·3	55·4	55·4	55·4	55·4
2·9	β Aquarii		**36**	58·6	58·5	58·4	58·5	58·6	58·7	S **5**	31·6	31·6	31·5	31·5	31·6	31·6
2·4	*Alderamin*		**40**	17·3	17·2	17·3	17·5	17·9	18·1	N **62**	37·5	37·7	37·9	38·0	38·0	38·0
2·5	ε Cygni		**48**	20·5	20·4	20·5	20·6	20·8	20·9	N **34**	00·4	00·5	00·7	00·7	00·7	00·7
1·3	*Deneb*	53	**49**	33·1	33·0	33·1	33·3	33·5	33·6	N **45**	18·9	19·0	19·2	19·2	19·3	19·2
3·1	α Indi		**50**	25·7	25·6	25·6	25·8	25·9	26·1	S **47**	15·3	15·4	15·5	15·5	15·6	15·5
1·9	*Peacock*	52	**53**	23·0	23·0	23·0	23·2	23·5	23·6	S **56**	42·1	42·2	42·3	42·4	42·4	42·3
2·2	γ Cygni		**54**	20·9	20·9	20·9	21·1	21·3	21·4	N **40**	17·2	17·4	17·5	17·6	17·6	17·5
0·8	*Altair*	51	**62**	10·7	10·7	10·8	10·9	11·0	11·0	N **8**	53·7	53·8	53·8	53·8	53·8	53·8
2·7	γ Aquilæ		**63**	18·8	18·7	18·8	18·9	19·0	19·1	N **10**	38·2	38·3	38·4	38·4	38·4	38·3
2·9	δ Cygni		**63**	40·3	40·3	40·5	40·7	40·9	41·0	N **45**	09·3	09·4	09·5	09·6	09·6	09·5
3·1	*Albireo*		**67**	12·9	12·9	13·0	13·1	13·3	13·3	N **27**	58·8	59·0	59·0	59·1	59·0	59·0
2·9	π Sagittarii		**72**	24·4	24·4	24·5	24·6	24·7	24·7	S **21**	00·5	00·5	00·5	00·5	00·5	00·5
3·0	ζ Aquilæ		**73**	31·7	31·7	31·8	32·0	32·1	32·1	N **13**	52·7	52·8	52·9	52·9	52·8	52·8
2·6	ζ Sagittarii		**74**	11·1	11·0	11·1	11·3	11·4	11·4	S **29**	52·0	52·0	52·0	52·0	52·0	52·0
2·0	*Nunki*	50	**76**	01·5	01·5	01·6	01·7	01·8	01·8	S **26**	17·1	17·1	17·1	17·1	17·1	17·1
0·0	*Vega*	49	**80**	40·5	40·6	40·7	40·9	41·1	41·1	N **38**	47·6	47·7	47·8	47·8	47·8	47·6
2·8	λ Sagittarii		**82**	51·0	51·0	51·1	51·2	51·3	51·3	S **25**	25·0	25·0	25·0	25·0	25·0	25·0
1·9	*Kaus Australis*	48	**83**	47·2	47·3	47·4	47·5	47·6	47·6	S **34**	22·8	22·9	22·9	22·9	22·9	22·8
2·7	δ Sagittarii		**84**	35·3	35·3	35·4	35·5	35·6	35·6	S **29**	49·5	49·5	49·5	49·5	49·5	49·4
3·0	γ Sagittarii		**88**	23·0	23·1	23·2	23·3	23·4	23·4	S **30**	25·5	25·5	25·5	25·5	25·5	25·4
2·2	*Eltanin*	47	**90**	47·0	47·1	47·4	47·6	47·8	47·9	N **51**	29·3	29·5	29·5	29·5	29·4	29·3
2·8	β Ophiuchi		**94**	00·3	00·4	00·5	00·6	00·7	00·7	N **4**	33·8	33·9	33·9	33·9	33·9	33·8
2·4	κ Scorpii		**94**	12·1	12·1	12·3	12·4	12·5	12·5	S **39**	02·2	02·2	02·2	02·2	02·2	02·1
1·9	θ Scorpii		**95**	29·2	29·3	29·4	29·6	29·7	29·6	S **43**	00·3	00·4	00·4	00·4	00·3	00·2
2·1	*Rasalhague*	46	**96**	08·8	08·9	09·0	09·1	09·2	09·2	N **12**	33·2	33·3	33·3	33·3	33·3	33·2
1·6	*Shaula*	45	**96**	25·5	25·5	25·7	25·8	25·9	25·9	S **37**	06·7	06·8	06·8	06·8	06·7	06·7
3·0	α Aræ		**96**	50·5	50·6	50·8	50·9	51·1	51·0	S **49**	53·1	53·2	53·2	53·2	53·1	53·0
2·7	υ Scorpii		**97**	08·1	08·2	08·3	08·5	08·6	08·5	S **37**	18·3	18·3	18·3	18·3	18·3	18·2
2·8	β Draconis		**97**	19·7	19·8	20·1	20·3	20·5	20·6	N **52**	17·7	17·9	17·9	17·9	17·8	17·6
2·8	β Aræ		**98**	27·7	27·8	28·0	28·2	28·4	28·3	S **55**	32·5	32·5	32·6	32·5	32·4	32·3
Var.‡	α Herculis		**101**	13·3	13·3	13·4	13·6	13·7	13·6	N **14**	22·8	22·9	22·9	22·9	22·8	22·7
2·4	*Sabik*	44	**102**	15·6	15·6	15·7	15·9	15·9	15·9	S **15**	44·3	44·2	44·2	44·2	44·2	44·2
3·1	ζ Aræ		**105**	08·0	08·1	08·3	08·5	08·6	08·5	S **56**	00·5	00·5	00·6	00·5	00·4	00·3
2·3	ε Scorpii		**107**	17·7	17·7	17·9	18·0	18·1	18·0	S **34**	18·8	18·8	18·8	18·7	18·7	18·7
1·9	*Atria*	43	**107**	33·6	33·9	34·2	34·6	34·7	34·6	S **69**	02·9	03·0	03·0	02·9	02·8	02·7
2·8	ζ Herculis		**109**	34·8	34·9	35·1	35·2	35·3	35·3	N **31**	35·2	35·3	35·3	35·2	35·1	35·0
2·6	ζ Ophiuchi		**110**	34·2	34·3	34·4	34·5	34·6	34·5	S **10**	35·2	35·2	35·2	35·2	35·2	35·2
2·8	τ Scorpii		**110**	52·3	52·4	52·5	52·6	52·6	52·6	S **28**	14·3	14·3	14·3	14·2	14·2	14·2
2·8	β Herculis		**112**	20·1	20·2	20·3	20·4	20·5	20·4	N **21**	28·2	28·3	28·3	28·2	28·1	28·0
1·0	*Antares*	42	**112**	29·6	29·6	29·8	29·9	29·9	29·8	S **26**	27·3	27·3	27·3	27·3	27·2	27·2
2·7	η Draconis		**113**	57·6	57·8	58·2	58·5	58·7	58·7	N **61**	29·7	29·8	29·8	29·7	29·5	29·3
2·7	δ Ophiuchi		**116**	16·8	16·9	17·0	17·1	17·2	17·1	S **3**	43·2	43·1	43·1	43·1	43·2	43·2
2·6	β Scorpii		**118**	29·6	29·7	29·8	29·9	29·9	29·8	S **19**	50·0	50·0	50·0	50·0	49·9	50·0
2·3	*Dschubba*		**119**	46·0	46·1	46·2	46·3	46·3	46·2	S **22**	39·1	39·0	39·0	39·0	39·0	39·0
2·9	π Scorpii		**120**	08·1	08·1	08·3	08·4	08·4	08·3	S **26**	08·6	08·6	08·6	08·6	08·5	08·5
2·8	β Trianguli Aust.		**120**	59·4	59·6	59·9	60·1	60·1	59·9	S **63**	27·9	27·9	27·9	27·8	27·7	27·6
2·6	α Serpentis		**123**	48·5	48·6	48·7	48·8	48·8	48·7	N **6**	23·7	23·8	23·8	23·8	23·7	23·6
2·8	γ Lupi		**126**	02·8	02·9	03·1	03·2	03·2	03·0	S **41**	12·1	12·2	12·1	12·1	12·0	11·9
2·2	*Alphecca*	41	**126**	13·2	13·3	13·5	13·6	13·6	13·5	N **26**	41·0	41·1	41·1	41·0	40·9	40·7

‡ 2·9 — 3·6

Mag.	Name and Number		SHA						Declination						
				JAN.	FEB.	MAR.	APR.	MAY	JUNE		JAN.	FEB.	MAR.	APR.	MAY	JUNE
			°	′	′	′	′	′	′	°	′	′	′	′	′	′
3·1	γ Ursæ Minoris		**129**	49·5	49·0	48·4	48·1	48·0	48·2	**N 71**	47·7	47·6	47·6	47·8	47·9	48·1
2·9	γ Trianguli Aust.		**130**	03·7	03·1	02·6	02·2	01·9	01·9	**S 68**	42·6	42·6	42·7	42·8	43·0	43·1
2·6	β Libræ		**130**	37·5	37·3	37·1	36·9	36·8	36·7	**S 9**	25·1	25·2	25·2	25·2	25·2	25·2
2·7	β Lupi		**135**	13·0	12·7	12·5	12·2	12·1	12·1	**S 43**	10·1	10·2	10·3	10·4	10·5	10·6
2·8	α Libræ	39	**137**	09·2	09·0	08·7	08·6	08·5	08·5	**S 16**	04·8	04·9	05·0	05·0	05·0	05·0
2·1	β Ursæ Minoris	40	**137**	19·9	19·2	18·7	18·3	18·3	18·6	**N 74**	06·7	06·6	06·7	06·8	07·0	07·1
2·4	ε Bootis		**138**	39·2	38·9	38·7	38·6	38·5	38·5	**N 27**	01·9	01·8	01·8	01·9	02·0	02·1
2·3	α Lupi		**139**	21·9	21·6	21·3	21·1	21·0	21·0	**S 47**	25·5	25·6	25·7	25·8	25·9	26·0
−0·3	α Centauri	38	**139**	56·6	56·2	55·8	55·6	55·5	55·5	**S 60**	52·2	52·2	52·3	52·5	52·6	52·7
2·3	η Centauri		**140**	58·7	58·3	58·1	57·9	57·8	57·8	**S 42**	11·8	11·9	11·9	12·1	12·2	12·2
3·0	γ Bootis		**141**	53·3	53·0	52·8	52·6	52·6	52·6	**N 38**	15·8	15·7	15·7	15·8	15·9	16·1
0·0	α Bootis	37	**145**	58·7	58·5	58·3	58·2	58·1	58·2	**N 19**	07·8	07·8	07·8	07·8	07·9	08·0
2·1	θ Centauri	36	**148**	11·6	11·3	11·1	11·0	10·9	10·9	**S 36**	24·8	24·9	25·0	25·1	25·2	25·3
0·6	β Centauri	35	**148**	52·9	52·5	52·2	51·9	51·9	51·9	**S 60**	24·8	24·9	25·0	25·2	25·3	25·5
2·6	ζ Centauri		**150**	58·3	57·9	57·7	57·5	57·5	57·5	**S 47**	19·9	20·0	20·1	20·2	20·3	20·4
2·7	η Bootis		**151**	13·1	12·8	12·7	12·6	12·5	12·6	**N 18**	20·9	20·8	20·8	20·8	20·9	21·0
1·9	η Ursæ Majoris	34	**153**	01·3	01·0	00·8	00·7	00·7	00·8	**N 49**	15·7	15·6	15·7	15·8	16·0	16·1
2·3	ε Centauri		**154**	52·9	52·5	52·3	52·1	52·1	52·1	**S 53**	30·6	30·7	30·8	31·0	31·1	31·2
1·0	α Virginis	33	**158**	34·7	34·5	34·3	34·2	34·2	34·3	**S 11**	12·6	12·7	12·8	12·8	12·8	12·8
2·3	ζ Ursæ Majoris		**158**	55·4	55·0	54·8	54·7	54·7	54·9	**N 54**	52·3	52·3	52·3	52·5	52·6	52·7
2·8	ι Centauri		**159**	43·2	42·9	42·7	42·6	42·6	42·7	**S 36**	45·5	45·7	45·8	45·9	46·0	46·0
2·8	ε Virginis		**164**	20·3	20·1	20·0	19·9	19·9	19·9	**N 10**	54·4	54·3	54·3	54·3	54·4	54·4
2·9	α Canum Venat.		**165**	52·9	52·7	52·5	52·4	52·5	52·6	**N 38**	15·8	15·8	15·8	15·9	16·0	16·1
1·8	ε Ursæ Majoris	32	**166**	23·2	22·9	22·7	22·6	22·7	22·9	**N 55**	54·2	54·2	54·3	54·5	54·6	54·7
1·3	β Crucis		**167**	56·0	55·6	55·4	55·3	55·3	55·5	**S 59**	44·1	44·3	44·4	44·6	44·7	44·8
2·9	γ Virginis		**169**	27·9	27·7	27·6	27·5	27·6	27·6	**S 1**	30·1	30·2	30·2	30·2	30·2	30·2
2·2	γ Centauri		**169**	29·5	29·2	29·0	28·9	29·0	29·1	**S 49**	00·5	00·6	00·7	00·9	01·0	01·1
2·7	α Muscæ		**170**	33·8	33·3	33·0	33·0	33·1	33·3	**S 69**	10·9	11·1	11·2	11·4	11·6	11·7
2·7	β Corvi		**171**	16·8	16·5	16·4	16·4	16·4	16·4	**S 23**	26·8	26·9	27·0	27·1	27·2	27·2
1·6	γ Crucis	31	**172**	04·7	04·3	04·1	04·1	04·1	04·3	**S 57**	09·7	09·8	10·0	10·2	10·3	10·3
1·3	α Crucis	30	**173**	13·1	12·7	12·5	12·4	12·5	12·7	**S 63**	08·8	08·9	09·1	09·3	09·4	09·5
2·6	γ Corvi	29	**175**	55·6	55·4	55·3	55·2	55·3	55·3	**S 17**	35·6	35·7	35·8	35·9	35·9	35·9
2·6	δ Centauri		**177**	47·3	47·0	46·8	46·8	46·9	47·0	**S 50**	46·2	46·4	46·6	46·7	46·8	46·9
2·4	γ Ursæ Majoris		**181**	24·9	24·6	24·4	24·5	24·6	24·8	**N 53**	38·3	38·3	38·4	38·5	38·6	38·7
2·1	β Leonis	28	**182**	36·8	36·6	36·5	36·5	36·5	36·6	**N 14**	31·1	31·0	31·0	31·0	31·1	31·1
2·6	δ Leonis		**191**	20·7	20·5	20·5	20·5	20·5	20·6	**N 20**	28·2	28·2	28·2	28·2	28·3	28·3
3·0	ψ Ursæ Majoris		**192**	26·9	26·6	26·6	26·6	26·7	26·9	**N 44**	26·6	26·7	26·8	26·9	26·9	27·0
1·8	α Ursæ Majoris	27	**193**	55·0	54·7	54·6	54·7	54·9	55·1	**N 61**	41·8	41·9	42·0	42·1	42·2	42·2
2·4	β Ursæ Majoris		**194**	23·5	23·3	23·2	23·3	23·4	23·6	**N 56**	19·7	19·8	19·9	20·0	20·1	20·1
2·7	μ Velorum		**198**	12·2	12·0	12·0	12·0	12·2	12·3	**S 49**	28·0	28·2	28·3	28·5	28·5	28·5
2·8	θ Carinæ		**199**	10·3	10·1	10·0	10·2	10·4	10·7	**S 64**	26·4	26·6	26·7	26·9	27·0	27·0
2·3	γ Leonis		**204**	52·4	52·3	52·2	52·3	52·4	52·5	**N 19**	47·6	47·5	47·5	47·6	47·6	47·6
1·4	α Leonis	26	**207**	46·7	46·6	46·5	46·6	46·7	46·8	**N 11**	55·2	55·2	55·2	55·2	55·2	55·2
3·0	ε Leonis		**213**	24·0	23·9	23·8	23·9	24·0	24·1	**N 23**	43·8	43·8	43·8	43·8	43·9	43·9
3·1	N Velorum		**217**	07·0	06·9	07·0	07·2	07·4	07·6	**S 57**	04·4	04·6	04·7	04·9	04·9	04·9
2·0	α Hydræ	25	**217**	59·0	58·9	58·9	59·0	59·1	59·2	**S 8**	41·9	42·0	42·1	42·1	42·1	42·1
2·5	κ Velorum		**219**	23·5	23·5	23·5	23·7	24·0	24·2	**S 55**	02·9	03·1	03·2	03·3	03·4	03·3
2·2	ι Carinæ		**220**	39·4	39·4	39·5	39·7	40·0	40·2	**S 59**	18·7	18·9	19·1	19·2	19·2	19·2
1·7	β Carinæ	24	**221**	39·8	39·8	40·0	40·3	40·8	41·2	**S 69**	45·2	45·4	45·5	45·7	45·7	45·7
2·2	λ Velorum	23	**222**	54·5	54·5	54·5	54·7	54·9	55·0	**S 43**	28·1	28·3	28·4	28·5	28·5	28·5
3·1	ι Ursæ Majoris		**225**	01·8	01·7	01·8	01·9	02·1	02·2	**N 48**	00·2	00·3	00·4	00·4	00·5	00·4
2·0	δ Velorum		**228**	45·1	45·1	45·2	45·4	45·7	45·9	**S 54**	44·5	44·7	44·8	44·9	44·9	44·8
1·9	ε Carinæ	22	**234**	18·9	18·9	19·1	19·4	19·6	19·9	**S 59**	32·3	32·5	32·6	32·7	32·7	32·6
1·8	γ Velorum		**237**	32·3	32·3	32·4	32·6	32·8	32·9	**S 47**	21·8	22·0	22·1	22·1	22·1	22·0
2·8	ρ Puppis		**238**	00·5	00·5	00·6	00·7	00·8	00·9	**S 24**	19·8	20·0	20·1	20·1	20·1	20·0
2·3	ζ Puppis		**239**	00·9	00·9	01·0	01·2	01·4	01·5	**S 40**	01·7	01·9	02·0	02·0	02·0	01·9
1·1	β Geminorum	21	**243**	31·3	31·3	31·4	31·5	31·6	31·6	**N 28**	00·2	00·2	00·3	00·3	00·3	00·3
0·4	α Canis Minoris	20	**245**	02·8	02·8	02·9	03·0	03·1	03·1	**N 5**	12·1	12·0	12·0	12·0	12·0	12·0

Mag.	Name and Number		SHA							Declination						
				JULY	AUG.	SEPT.	OCT.	NOV.	DEC.		JULY	AUG.	SEPT.	OCT.	NOV.	DEC.
			°	′	′	′	′	′	′	°	′	′	′	′	′	′
3·1	γ Ursæ Minoris		**129**	48·5	49·1	49·6	50·0	50·2	50·0	**N 71**	48·2	48·2	48·1	48·0	47·8	47·6
2·9	γ Trianguli Aust.		**130**	02·1	02·4	02·7	03·0	03·0	02·7	**S 68**	43·2	43·2	43·2	43·1	42·9	42·8
2·6	β Libræ		**130**	36·8	36·9	37·0	37·0	37·0	36·9	**S 9**	25·2	25·1	25·1	25·1	25·1	25·2
2·7	β Lupi		**135**	12·2	12·3	12·4	12·5	12·5	12·3	**S 43**	10·6	10·6	10·5	10·5	10·4	10·4
2·8	*Zubenelgenubi*	39	**137**	08·5	08·6	08·7	08·8	08·7	08·6	**S 16**	05·0	05·0	05·0	04·9	04·9	05·0
2·1	*Kochab*	40	**137**	19·1	19·7	20·2	20·6	20·8	20·5	**N 74**	07·2	07·2	07·1	06·9	06·7	06·6
2·4	ε Bootis		**138**	38·6	38·7	38·9	38·9	38·9	38·8	**N 27**	02·2	02·2	02·2	02·1	01·9	01·8
2·3	α Lupi		**139**	21·1	21·2	21·4	21·5	21·4	21·2	**S 47**	26·0	26·0	26·0	25·9	25·8	25·7
−0·3	*Rigil Kent.*	38	**139**	55·7	55·9	56·2	56·3	56·2	55·9	**S 60**	52·8	52·8	52·7	52·6	52·5	52·4
2·3	η Centauri		**140**	57·9	58·0	58·1	58·2	58·1	57·9	**S 42**	12·2	12·2	12·2	12·1	12·0	12·0
3·0	γ Bootis		**141**	52·7	52·9	53·0	53·1	53·1	52·9	**N 38**	16·1	16·1	16·1	16·0	15·8	15·7
0·0	*Arcturus*	37	**145**	58·2	58·4	58·4	58·5	58·4	58·3	**N 19**	08·0	08·0	08·0	07·9	07·8	07·7
2·1	*Menkent*	36	**148**	11·0	11·1	11·2	11·3	11·2	11·0	**S 36**	25·3	25·2	25·2	25·1	25·1	25·0
0·6	*Hadar*	35	**148**	52·1	52·3	52·5	52·6	52·5	52·2	**S 60**	25·5	25·5	25·4	25·3	25·2	25·1
2·6	ζ Centauri		**150**	57·6	57·8	57·9	57·9	57·8	57·6	**S 47**	20·4	20·4	20·3	20·2	20·1	20·1
2·7	η Bootis		**151**	12·6	12·7	12·8	12·8	12·8	12·6	**N 18**	21·0	21·1	21·0	20·9	20·8	20·7
1·9	*Alkaid*	34	**153**	01·0	01·1	01·3	01·4	01·3	01·1	**N 49**	16·1	16·1	16·0	15·9	15·7	15·5
2·3	ε Centauri		**154**	52·3	52·5	52·6	52·6	52·5	52·2	**S 53**	31·2	31·2	31·1	31·0	30·9	30·9
1·0	*Spica*	33	**158**	34·3	34·4	34·5	34·5	34·4	34·2	**S 11**	12·8	12·8	12·7	12·7	12·7	12·8
2·3	*Mizar*		**158**	55·1	55·3	55·5	55·5	55·4	55·2	**N 54**	52·7	52·7	52·6	52·4	52·2	52·1
2·8	ι Centauri		**159**	42·8	42·9	43·0	43·0	42·8	42·6	**S 36**	46·0	46·0	45·9	45·8	45·8	45·8
2·8	ε Virginis		**164**	20·0	20·1	20·1	20·1	20·0	19·8	**N 10**	54·5	54·5	54·5	54·4	54·3	54·2
2·9	*Cor Caroli*		**165**	52·7	52·8	52·9	52·9	52·8	52·5	**N 38**	16·2	16·1	16·0	15·9	15·7	15·6
1·8	*Alioth*	32	**166**	23·1	23·3	23·4	23·4	23·3	23·0	**N 55**	54·7	54·6	54·5	54·3	54·1	54·0
1·3	*Mimosa*		**167**	55·7	55·9	56·0	56·0	55·8	55·4	**S 59**	44·8	44·8	44·6	44·5	44·4	44·4
2·9	γ Virginis		**169**	27·7	27·8	27·8	27·8	27·6	27·4	**S 1**	30·2	30·1	30·1	30·2	30·2	30·3
2·2	*Muhlifain*		**169**	29·2	29·4	29·4	29·4	29·2	28·9	**S 49**	01·1	01·0	00·9	00·8	00·7	00·7
2·7	α Muscæ		**170**	33·7	34·0	34·2	34·1	33·8	33·3	**S 69**	11·7	11·6	11·5	11·3	11·2	11·2
2·7	β Corvi		**171**	16·5	16·6	16·6	16·6	16·4	16·2	**S 23**	27·2	27·1	27·0	27·0	27·0	27·1
1·6	*Gacrux*	31	**172**	04·5	04·7	04·8	04·7	04·5	04·1	**S 57**	10·3	10·3	10·2	10·0	09·9	09·9
1·3	*Acrux*	30	**173**	13·0	13·2	13·3	13·3	13·0	12·6	**S 63**	09·5	09·4	09·3	09·2	09·1	09·1
2·6	*Gienah*	29	**175**	55·4	55·5	55·5	55·5	55·3	55·1	**S 17**	35·8	35·8	35·7	35·7	35·7	35·8
2·6	δ Centauri		**177**	47·2	47·3	47·4	47·3	47·1	46·7	**S 50**	46·8	46·8	46·6	46·5	46·5	46·5
2·4	*Phecda*		**181**	24·9	25·1	25·1	25·0	24·8	24·5	**N 53**	38·7	38·6	38·5	38·3	38·1	38·0
2·1	*Denebola*	28	**182**	36·7	36·8	36·8	36·7	36·5	36·3	**N 14**	31·2	31·1	31·1	31·0	30·9	30·8
2·6	δ Leonis		**191**	20·7	20·7	20·7	20·6	20·4	20·2	**N 20**	28·3	28·3	28·3	28·2	28·1	27·9
3·0	ψ Ursæ Majoris		**192**	27·0	27·0	27·0	26·9	26·7	26·3	**N 44**	26·9	26·9	26·7	26·6	26·4	26·3
1·8	*Dubhe*	27	**193**	55·4	55·5	55·4	55·3	54·9	54·5	**N 61**	42·1	42·0	41·9	41·7	41·5	41·4
2·4	*Merak*		**194**	23·8	23·9	23·8	23·7	23·4	23·0	**N 56**	20·0	19·9	19·8	19·6	19·5	19·4
2·7	μ Velorum		**198**	12·5	12·6	12·5	12·4	12·1	11·7	**S 49**	28·5	28·4	28·2	28·1	28·1	28·2
2·8	θ Carinæ		**199**	11·0	11·1	11·1	10·9	10·5	10·0	**S 64**	26·9	26·8	26·7	26·6	26·5	26·6
2·3	*Algeiba*		**204**	52·5	52·5	52·4	52·3	52·1	51·8	**N 19**	47·6	47·6	47·6	47·5	47·4	47·3
1·4	*Regulus*	26	**207**	46·8	46·8	46·7	46·6	46·4	46·1	**N 11**	55·3	55·3	55·2	55·2	55·1	55·0
3·0	ε Leonis		**213**	24·1	24·1	24·0	23·8	23·6	23·3	**N 23**	43·9	43·8	43·8	43·7	43·6	43·5
3·1	*N* Velorum		**217**	07·8	07·8	07·7	07·4	07·1	06·7	**S 57**	04·8	04·6	04·5	04·4	04·4	04·5
2·0	*Alphard*	25	**217**	59·2	59·2	59·1	58·9	58·7	58·4	**S 8**	42·0	42·0	41·9	41·9	42·0	42·1
2·5	κ Velorum		**219**	24·3	24·3	24·2	23·9	23·6	23·2	**S 55**	03·2	03·1	03·0	02·9	02·9	03·0
2·2	ι Carinæ		**220**	40·4	40·4	40·2	40·0	39·6	39·2	**S 59**	19·1	18·9	18·8	18·7	18·7	18·8
1·7	*Miaplacidus*	24	**221**	41·4	41·5	41·3	40·9	40·4	39·9	**S 69**	45·5	45·4	45·2	45·2	45·2	45·3
2·2	*Suhail*	23	**222**	55·1	55·1	54·9	54·7	54·4	54·1	**S 43**	28·4	28·2	28·1	28·1	28·1	28·2
3·1	ι Ursæ Majoris		**225**	02·2	02·2	02·0	01·7	01·4	01·0	**N 47**	60·3	60·2	60·1	60·0	59·9	59·9
2·0	δ Velorum		**228**	46·0	45·9	45·8	45·5	45·1	44·8	**S 54**	44·7	44·6	44·4	44·4	44·4	44·6
1·9	*Avior*	22	**234**	20·0	19·9	19·7	19·4	19·0	18·7	**S 59**	32·5	32·3	32·2	32·1	32·2	32·3
1·8	γ Velorum		**237**	33·0	32·9	32·7	32·5	32·2	31·9	**S 47**	21·9	21·8	21·7	21·6	21·7	21·8
2·8	ρ Puppis		**238**	00·9	00·8	00·7	00·5	00·2	00·0	**S 24**	19·9	19·8	19·7	19·7	19·8	19·9
2·3	ζ Puppis		**239**	01·5	01·4	01·2	01·0	00·7	00·5	**S 40**	01·8	01·7	01·6	01·6	01·6	01·8
1·1	*Pollux*	21	**243**	31·6	31·5	31·3	31·0	30·8	30·5	**N 28**	00·2	00·2	00·1	00·1	00·0	00·0
0·4	*Procyon*	20	**245**	03·1	03·0	02·8	02·6	02·3	02·1	**N 5**	12·1	12·1	12·1	12·1	12·0	11·9

Mag.	Name and Number		SHA							Declination						
				JAN.	FEB.	MAR.	APR.	MAY	JUNE		JAN.	FEB.	MAR.	APR.	MAY	JUNE
			°	′	′	′	′	′	′	°	′	′	′	′	′	′
1·6	α Geminorum		**246**	11·7	11·7	11·8	11·9	12·0	12·1	**N 31**	52·1	52·1	52·1	52·1	52·1	52·1
3·3	σ Puppis		**247**	36·7	36·7	36·9	37·1	37·2	37·3	**S 43**	19·2	19·4	19·5	19·5	19·4	19·3
2·9	β Canis Minoris		**248**	04·8	04·8	04·9	05·0	05·1	05·1	**N 8**	16·2	16·2	16·2	16·2	16·2	16·2
2·4	η Canis Majoris		**248**	52·7	52·7	52·8	53·0	53·1	53·2	**S 29**	19·3	19·4	19·5	19·5	19·5	19·4
2·7	π Puppis		**250**	37·5	37·6	37·7	37·9	38·0	38·1	**S 37**	06·8	07·0	07·1	07·1	07·1	07·0
1·8	δ Canis Majoris		**252**	48·1	48·1	48·2	48·4	48·5	48·5	**S 26**	24·5	24·6	24·7	24·7	24·6	24·5
3·0	o Canis Majoris		**254**	08·4	08·5	08·6	08·7	08·8	08·9	**S 23**	50·8	50·9	51·0	51·0	51·0	50·9
1·5	ε Canis Majoris	19	**255**	14·7	14·8	14·9	15·1	15·2	15·2	**S 28**	59·1	59·2	59·3	59·3	59·3	59·1
2·9	τ Puppis		**257**	27·0	27·1	27·3	27·5	27·7	27·9	**S 50**	37·6	37·7	37·8	37·8	37·7	37·6
−1·5	α Canis Majoris	18	**258**	36·3	36·3	36·4	36·6	36·7	36·7	**S 16**	43·7	43·8	43·9	43·9	43·9	43·8
1·9	γ Geminorum		**260**	25·9	25·9	26·0	26·2	26·2	26·2	**N 16**	23·5	23·5	23·5	23·5	23·5	23·5
−0·7	α Carinæ	17	**263**	57·1	57·2	57·5	57·7	57·9	58·0	**S 52**	42·1	42·2	42·3	42·3	42·2	42·1
2·0	β Canis Majoris		**264**	13·0	13·1	13·2	13·3	13·4	13·4	**S 17**	57·7	57·7	57·8	57·8	57·7	57·6
2·6	θ Aurigæ		**269**	54·2	54·3	54·4	54·6	54·7	54·6	**N 37**	12·9	12·9	12·9	12·9	12·9	12·8
1·9	β Aurigæ		**269**	56·4	56·4	56·6	56·8	56·9	56·8	**N 44**	57·0	57·1	57·1	57·1	57·0	56·9
Var.‡	α Orionis	16	**271**	04·5	04·6	04·7	04·8	04·9	04·9	**N 7**	24·5	24·5	24·5	24·5	24·5	24·6
2·1	κ Orionis		**272**	56·7	56·8	56·9	57·0	57·1	57·1	**S 9**	40·0	40·1	40·1	40·1	40·0	39·9
1·9	ζ Orionis		**274**	41·3	41·3	41·4	41·6	41·6	41·6	**S 1**	56·3	56·3	56·3	56·3	56·3	56·2
2·6	α Columbæ		**274**	59·8	59·9	60·1	60·3	60·4	60·4	**S 34**	04·2	04·3	04·4	04·3	04·2	04·1
3·0	ζ Tauri		**275**	26·6	26·7	26·8	26·9	27·0	27·0	**N 21**	08·9	08·9	08·9	08·9	08·9	08·9
1·7	ε Orionis	15	**275**	49·4	49·5	49·6	49·7	49·7	49·7	**S 1**	11·8	11·8	11·8	11·8	11·8	11·7
2·8	ι Orionis		**276**	01·4	01·4	01·5	01·7	01·7	01·7	**S 5**	54·2	54·3	54·3	54·3	54·3	54·2
2·6	α Leporis		**276**	42·6	42·6	42·8	42·9	43·0	43·0	**S 17**	49·0	49·1	49·1	49·1	49·0	48·9
2·2	δ Orionis		**276**	52·5	52·5	52·6	52·8	52·8	52·8	**S 0**	17·5	17·6	17·6	17·6	17·6	17·5
2·8	β Leporis		**277**	50·0	50·1	50·2	50·4	50·4	50·4	**S 20**	45·2	45·3	45·3	45·3	45·2	45·1
1·7	β Tauri	14	**278**	16·4	16·5	16·6	16·8	16·8	16·8	**N 28**	37·0	37·0	37·0	37·0	37·0	36·9
1·6	γ Orionis	13	**278**	35·2	35·3	35·4	35·5	35·6	35·5	**N 6**	21·5	21·5	21·5	21·5	21·5	21·5
0·1	α Aurigæ	12	**280**	38·9	39·0	39·2	39·4	39·4	39·3	**N 46**	00·6	00·6	00·6	00·6	00·5	00·5
0·1	β Orionis	11	**281**	14·9	15·0	15·1	15·2	15·3	15·2	**S 8**	11·5	11·5	11·6	11·5	11·5	11·4
2·8	β Eridani		**282**	55·1	55·2	55·3	55·4	55·4	55·4	**S 5**	04·5	04·5	04·6	04·5	04·5	04·4
2·7	ι Aurigæ		**285**	35·7	35·8	35·9	36·0	36·1	36·0	**N 33**	11·0	11·0	11·0	10·9	10·9	10·9
0·9	α Tauri	10	**290**	52·9	53·0	53·1	53·2	53·2	53·1	**N 16**	31·7	31·7	31·7	31·7	31·7	31·7
2·9	ε Persei		**300**	22·5	22·7	22·8	22·9	22·9	22·8	**N 40**	02·4	02·4	02·4	02·3	02·3	02·2
3·0	γ Eridani		**300**	22·8	22·9	23·0	23·1	23·1	23·0	**S 13**	29·0	29·0	29·0	29·0	28·9	28·8
2·9	ζ Persei		**301**	19·0	19·1	19·2	19·3	19·3	19·2	**N 31**	54·8	54·8	54·8	54·7	54·7	54·7
2·9	η Tauri		**302**	59·2	59·3	59·4	59·5	59·5	59·4	**N 24**	08·1	08·1	08·1	08·1	08·1	08·1
1·8	α Persei	9	**308**	44·9	45·1	45·3	45·4	45·3	45·2	**N 49**	53·9	53·9	53·8	53·8	53·7	53·6
Var.§	β Persei		**312**	48·2	48·3	48·5	48·5	48·5	48·3	**N 40**	59·7	59·7	59·6	59·6	59·5	59·5
2·5	α Ceti	8	**314**	18·4	18·5	18·6	18·6	18·6	18·4	**N 4**	07·6	07·6	07·5	07·6	07·6	07·7
3·2	θ Eridani	7	**315**	20·6	20·7	20·9	21·0	21·0	20·8	**S 40**	16·2	16·3	16·2	16·1	15·9	15·8
2·0	α Ursæ Minoris		**319**	15·5	27·6	37·9	43·6	41·5	33·1	**N 89**	18·6	18·6	18·6	18·4	18·3	18·2
3·0	β Trianguli		**327**	28·5	28·6	28·7	28·7	28·6	28·4	**N 35**	02·1	02·0	02·0	01·9	01·8	01·9
2·0	α Arietis	6	**328**	04·5	04·6	04·7	04·7	04·6	04·4	**N 23**	30·5	30·5	30·4	30·4	30·4	30·4
2·3	γ Andromedæ		**328**	52·9	53·0	53·1	53·2	53·1	52·8	**N 42**	22·7	22·6	22·6	22·5	22·4	22·4
2·9	α Hydri		**330**	13·8	14·1	14·3	14·5	14·4	14·1	**S 61**	31·8	31·7	31·6	31·4	31·3	31·1
2·6	β Arietis		**331**	12·6	12·7	12·8	12·8	12·7	12·5	**N 20**	51·3	51·3	51·2	51·2	51·2	51·2
0·5	α Eridani	5	**335**	28·9	29·2	29·4	29·4	29·3	29·1	**S 57**	11·7	11·6	11·5	11·3	11·2	11·0
2·7	δ Cassiopeiæ		**338**	23·7	23·9	24·1	24·1	23·9	23·5	**N 60**	17·3	17·3	17·2	17·0	16·9	16·9
2·1	β Andromedæ		**342**	26·3	26·4	26·4	26·4	26·3	26·0	**N 35**	40·4	40·3	40·2	40·2	40·1	40·1
Var.‖	γ Cassiopeiæ		**345**	41·1	41·3	41·5	41·4	41·2	40·8	**N 60**	46·3	46·2	46·1	46·0	45·9	45·9
2·0	β Ceti	4	**348**	59·2	59·3	59·3	59·3	59·1	58·9	**S 17**	56·3	56·3	56·2	56·1	56·0	55·9
2·2	α Cassiopeiæ	3	**349**	44·7	44·8	44·9	44·9	44·7	44·3	**N 56**	35·6	35·5	35·4	35·2	35·2	35·2
2·4	α Phœnicis	2	**353**	18·9	19·0	19·1	19·0	18·8	18·6	**S 42**	15·6	15·5	15·4	15·3	15·1	15·0
2·8	β Hydri		**353**	26·6	27·2	27·4	27·4	27·0	26·3	**S 77**	12·4	12·3	12·1	11·9	11·8	11·6
2·8	γ Pegasi		**356**	34·4	34·5	34·5	34·4	34·2	34·0	**N 15**	14·2	14·1	14·1	14·1	14·1	14·2
2·3	β Cassiopeiæ		**357**	35·2	35·4	35·5	35·3	35·1	34·7	**N 59**	12·3	12·2	12·1	12·0	11·9	11·9
2·1	α Andromedæ	1	**357**	47·1	47·2	47·2	47·1	47·0	46·7	**N 29**	08·6	08·5	08·5	08·4	08·4	08·5

‡ 0·1 — 1·2 § 2·1 — 3·4 ‖ Irregular variable; 2007 mag. 2·2

Mag.	Name and Number		SHA							Declination						
				JULY	AUG.	SEPT.	OCT.	NOV.	DEC.		JULY	AUG.	SEPT.	OCT.	NOV.	DEC.
			°	′	′	′	′	′	′	°	′	′	′	′	′	′
1·6	*Castor*		**246**	12·0	11·9	11·6	11·4	11·1	10·9	N **31**	52·1	52·0	52·0	51·9	51·9	51·9
3·3	σ Puppis		**247**	37·3	37·2	37·0	36·8	36·5	36·3	S **43**	19·2	19·1	19·0	19·0	19·0	19·2
2·9	β Canis Minoris		**248**	05·1	04·9	04·7	04·5	04·3	04·1	N **8**	16·2	16·3	16·3	16·2	16·2	16·1
2·4	η Canis Majoris		**248**	53·1	53·0	52·8	52·6	52·3	52·1	S **29**	19·3	19·1	19·1	19·1	19·1	19·3
2·7	π Puppis		**250**	38·1	38·0	37·7	37·5	37·2	37·0	S **37**	06·8	06·7	06·6	06·6	06·7	06·8
1·8	*Wezen*		**252**	48·5	48·4	48·2	47·9	47·7	47·5	S **26**	24·4	24·3	24·3	24·3	24·3	24·5
3·0	o Canis Majoris		**254**	08·8	08·7	08·5	08·3	08·0	07·9	S **23**	50·8	50·7	50·6	50·6	50·7	50·8
1·5	*Adhara*	19	**255**	15·2	15·1	14·9	14·6	14·4	14·2	S **28**	59·0	58·9	58·8	58·8	58·9	59·1
2·9	τ Puppis		**257**	27·8	27·7	27·4	27·1	26·8	26·6	S **50**	37·5	37·3	37·2	37·2	37·3	37·5
−1·5	*Sirius*	18	**258**	36·6	36·5	36·3	36·0	35·8	35·6	S **16**	43·7	43·6	43·5	43·5	43·6	43·8
1·9	*Alhena*		**260**	26·1	26·0	25·7	25·5	25·3	25·1	N **16**	23·5	23·5	23·5	23·5	23·5	23·4
−0·7	*Canopus*	17	**263**	58·0	57·8	57·5	57·2	56·9	56·8	S **52**	41·9	41·8	41·7	41·7	41·8	42·0
2·0	*Mirzam*		**264**	13·3	13·2	13·0	12·7	12·5	12·4	S **17**	57·5	57·4	57·4	57·4	57·5	57·6
2·6	θ Aurigæ		**269**	54·5	54·3	54·0	53·7	53·4	53·3	N **37**	12·8	12·8	12·7	12·7	12·8	12·8
1·9	*Menkalinan*		**269**	56·7	56·4	56·1	55·8	55·5	55·3	N **44**	56·9	56·8	56·8	56·8	56·8	56·9
Var.‡	*Betelgeuse*	16	**271**	04·7	04·6	04·3	04·1	03·9	03·7	N **7**	24·6	24·6	24·7	24·7	24·6	24·6
2·1	κ Orionis		**272**	57·0	56·8	56·5	56·3	56·1	56·0	S **9**	39·9	39·8	39·7	39·8	39·8	39·9
1·9	*Alnitak*		**274**	41·5	41·3	41·0	40·8	40·6	40·5	S **1**	56·2	56·1	56·1	56·1	56·1	56·2
2·6	*Phact*		**274**	60·3	60·1	59·8	59·6	59·4	59·3	S **34**	04·0	03·8	03·8	03·8	03·9	04·1
3·0	ζ Tauri		**275**	26·8	26·6	26·4	26·1	25·9	25·8	N **21**	08·9	08·9	09·0	09·0	09·0	08·9
1·7	*Alnilam*	15	**275**	49·6	49·4	49·2	49·0	48·8	48·6	S **1**	11·7	11·6	11·6	11·6	11·6	11·7
2·8	ι Orionis		**276**	01·6	01·4	01·2	00·9	00·8	00·6	S **5**	54·1	54·0	54·0	54·0	54·1	54·2
2·6	α Leporis		**276**	42·8	42·6	42·4	42·2	42·0	41·9	S **17**	48·8	48·7	48·6	48·7	48·8	48·9
2·2	δ Orionis		**276**	52·6	52·5	52·2	52·0	51·8	51·7	S **0**	17·4	17·4	17·3	17·3	17·4	17·5
2·8	β Leporis		**277**	50·3	50·1	49·9	49·7	49·5	49·3	S **20**	45·0	44·9	44·8	44·8	44·9	45·1
1·7	*Elnath*	14	**278**	16·6	16·4	16·1	15·9	15·6	15·5	N **28**	36·9	36·9	37·0	37·0	37·0	37·0
1·6	*Bellatrix*	13	**278**	35·4	35·2	35·0	34·7	34·6	34·4	N **6**	21·6	21·6	21·7	21·7	21·6	21·6
0·1	*Capella*	12	**280**	39·2	38·9	38·5	38·2	38·0	37·8	N **46**	00·4	00·4	00·4	00·4	00·5	00·5
0·1	*Rigel*	11	**281**	15·1	14·9	14·7	14·5	14·3	14·2	S **8**	11·3	11·2	11·2	11·2	11·3	11·4
2·8	β Eridani		**282**	55·3	55·1	54·8	54·6	54·4	54·3	S **5**	04·3	04·2	04·2	04·2	04·3	04·4
2·7	ι Aurigæ		**285**	35·8	35·5	35·3	35·0	34·8	34·7	N **33**	10·9	10·9	10·9	10·9	10·9	11·0
0·9	*Aldebaran*	10	**290**	53·0	52·7	52·5	52·3	52·1	52·0	N **16**	31·8	31·8	31·8	31·9	31·9	31·8
2·9	ε Persei		**300**	22·5	22·2	21·9	21·7	21·5	21·4	N **40**	02·2	02·2	02·3	02·4	02·4	02·5
3·0	γ Eridani		**300**	22·9	22·6	22·4	22·2	22·1	22·0	S **13**	28·7	28·6	28·6	28·6	28·7	28·8
2·9	ζ Persei		**301**	19·0	18·7	18·4	18·2	18·0	18·0	N **31**	54·7	54·7	54·8	54·8	54·9	54·9
2·9	*Alcyone*		**302**	59·2	58·9	58·6	58·4	58·3	58·2	N **24**	08·1	08·1	08·2	08·2	08·3	08·3
1·8	*Mirfak*	9	**308**	44·9	44·5	44·2	43·9	43·7	43·7	N **49**	53·6	53·6	53·7	53·8	53·9	54·0
Var.§	*Algol*		**312**	48·0	47·7	47·4	47·2	47·1	47·1	N **40**	59·5	59·5	59·6	59·7	59·8	59·9
2·5	*Menkar*	8	**314**	18·2	18·0	17·8	17·6	17·5	17·5	N **4**	07·8	07·8	07·9	07·9	07·9	07·8
3·2	*Acamar*	7	**315**	20·6	20·3	20·1	19·9	19·8	19·9	S **40**	15·6	15·6	15·6	15·7	15·8	15·9
2·0	*Polaris*		**318**	80·0	65·2	52·1	42·9	38·8	42·6	N **89**	18·1	18·2	18·3	18·4	18·6	18·8
3·0	β Trianguli		**327**	28·1	27·9	27·6	27·5	27·4	27·4	N **35**	01·9	02·0	02·1	02·2	02·3	02·3
2·0	*Hamal*	6	**328**	04·1	03·9	03·7	03·5	03·5	03·5	N **23**	30·5	30·6	30·6	30·7	30·8	30·8
2·3	*Almak*		**328**	52·5	52·2	52·0	51·8	51·7	51·8	N **42**	22·4	22·5	22·6	22·8	22·9	22·9
2·9	α Hydri		**330**	13·8	13·4	13·1	12·9	12·9	13·1	S **61**	31·0	31·0	31·0	31·1	31·3	31·4
2·6	*Sheratan*		**331**	12·3	12·0	11·8	11·7	11·7	11·7	N **20**	51·3	51·4	51·5	51·6	51·6	51·6
0·5	*Achernar*	5	**335**	28·7	28·4	28·1	28·0	28·0	28·2	S **57**	10·9	10·9	10·9	11·1	11·2	11·3
2·7	*Ruchbah*		**338**	23·1	22·7	22·4	22·2	22·2	22·3	N **60**	16·9	17·0	17·2	17·3	17·5	17·6
2·1	*Mirach*		**342**	25·7	25·5	25·3	25·2	25·2	25·2	N **35**	40·2	40·3	40·5	40·6	40·7	40·7
Var.‖	γ Cassiopeiæ		**345**	40·4	40·0	39·7	39·6	39·7	39·8	N **60**	45·9	46·1	46·2	46·4	46·5	46·6
2·0	*Diphda*	4	**348**	58·7	58·4	58·3	58·2	58·3	58·3	S **17**	55·8	55·7	55·7	55·8	55·8	55·9
2·2	*Schedar*	3	**349**	43·9	43·6	43·4	43·3	43·3	43·5	N **56**	35·2	35·4	35·5	35·7	35·8	35·9
2·4	*Ankaa*	2	**353**	18·3	18·0	17·9	17·8	17·9	18·0	S **42**	14·9	14·9	14·9	15·0	15·2	15·2
2·8	β Hydri		**353**	25·5	24·8	24·4	24·3	24·7	25·3	S **77**	11·6	11·7	11·8	11·9	12·1	12·1
2·8	*Algenib*		**356**	33·8	33·5	33·4	33·4	33·4	33·5	N **15**	14·3	14·4	14·5	14·5	14·6	14·5
2·3	*Caph*		**357**	34·3	34·0	33·8	33·7	33·8	34·0	N **59**	12·0	12·1	12·3	12·5	12·6	12·7
2·1	*Alpheratz*	1	**357**	46·4	46·2	46·1	46·1	46·1	46·2	N **29**	08·6	08·7	08·8	08·9	09·0	09·0

‡ 0·1 — 1·2 § 2·1 — 3·4 ‖ Irregular variable; 2007 mag. 2·2

POLARIS (POLE STAR) TABLES, 2009

FOR DETERMINING LATITUDE FROM SEXTANT ALTITUDE AND FOR AZIMUTH

LHA ARIES	0° – 9°	10° – 19°	20° – 29°	30° – 39°	40° – 49°	50° – 59°	60° – 69°	70° – 79°	80° – 89°	90° – 99°	100° – 109°	110° – 119°
	a_0	a_0	a_0	a_0	a_0	a_0	a_0	a_0	a_0	a_0	a_0	a_0
°	° ′	° ′	° ′	° ′	° ′	° ′	° ′	° ′	° ′	° ′	° ′	° ′
0	0 27·5	0 23·2	0 20·0	0 18·0	0 17·3	0 17·8	0 19·6	0 22·6	0 26·8	0 31·8	0 37·8	0 44·3
1	27·0	22·9	19·8	17·9	17·3	18·0	19·9	23·0	27·2	32·4	38·4	45·0
2	26·6	22·5	19·5	17·8	17·3	18·1	20·1	23·4	27·7	33·0	39·0	45·7
3	26·1	22·2	19·3	17·7	17·3	18·2	20·4	23·8	28·2	33·5	39·7	46·4
4	25·7	21·8	19·1	17·6	17·4	18·4	20·7	24·2	28·7	34·1	40·3	47·1
5	0 25·3	0 21·5	0 18·9	0 17·5	0 17·4	0 18·6	0 21·0	0 24·6	0 29·2	0 34·7	0 41·0	0 47·8
6	24·8	21·2	18·7	17·4	17·5	18·8	21·3	25·0	29·7	35·3	41·6	48·5
7	24·4	20·9	18·5	17·4	17·5	19·0	21·6	25·4	30·2	35·9	42·3	49·2
8	24·0	20·6	18·3	17·4	17·6	19·2	21·9	25·8	30·8	36·5	43·0	49·9
9	23·6	20·3	18·2	17·3	17·7	19·4	22·3	26·3	31·3	37·1	43·6	50·6
10	0 23·2	0 20·0	0 18·0	0 17·3	0 17·8	0 19·6	0 22·6	0 26·8	0 31·8	0 37·8	0 44·3	0 51·3
Lat.	a_1	a_1	a_1	a_1	a_1	a_1	a_1	a_1	a_1	a_1	a_1	a_1
°	′	′	′	′	′	′	′	′	′	′	′	′
0	0·5	0·5	0·6	0·6	0·6	0·6	0·5	0·5	0·5	0·4	0·4	0·3
10	·5	·6	·6	·6	·6	·6	·6	·5	·5	·4	·4	·4
20	·5	·6	·6	·6	·6	·6	·6	·5	·5	·5	·4	·4
30	·5	·6	·6	·6	·6	·6	·6	·6	·5	·5	·5	·5
40	0·6	0·6	0·6	0·6	0·6	0·6	0·6	0·6	0·6	0·5	0·5	0·5
45	·6	·6	·6	·6	·6	·6	·6	·6	·6	·6	·6	·6
50	·6	·6	·6	·6	·6	·6	·6	·6	·6	·6	·6	·6
55	·6	·6	·6	·6	·6	·6	·6	·6	·6	·6	·6	·7
60	·6	·6	·6	·6	·6	·6	·6	·6	·7	·7	·7	·7
62	0·7	0·6	0·6	0·6	0·6	0·6	0·6	0·7	0·7	0·7	0·7	0·8
64	·7	·6	·6	·6	·6	·6	·6	·7	·7	·7	·8	·8
66	·7	·7	·6	·6	·6	·6	·6	·7	·7	·8	·8	·8
68	0·7	0·7	0·6	0·6	0·6	0·6	0·7	0·7	0·8	0·8	0·9	0·9
Month	a_2	a_2	a_2	a_2	a_2	a_2	a_2	a_2	a_2	a_2	a_2	a_2
	′	′	′	′	′	′	′	′	′	′	′	′
Jan.	0·7	0·7	0·7	0·7	0·7	0·7	0·7	0·7	0·7	0·7	0·7	0·7
Feb.	·6	·6	·7	·7	·8	·8	·8	·8	·8	·8	·8	·8
Mar.	·5	·5	·6	·6	·7	·8	·8	·8	·9	·9	·9	0·9
Apr.	0·3	0·4	0·4	0·5	0·6	0·6	0·7	0·8	0·8	0·9	0·9	1·0
May	·2	·3	·3	·4	·4	·5	·6	·6	·7	·8	·8	0·9
June	·2	·2	·2	·3	·3	·4	·4	·5	·6	·6	·7	·8
July	0·2	0·2	0·2	0·2	0·2	0·3	0·3	0·4	0·4	0·5	0·5	0·6
Aug.	·4	·3	·3	·3	·3	·3	·3	·3	·3	·3	·4	·4
Sept.	·5	·5	·4	·4	·3	·3	·3	·3	·3	·3	·3	·3
Oct.	0·7	0·7	0·6	0·5	0·5	0·4	0·4	0·3	0·3	0·3	0·3	0·3
Nov.	0·9	0·8	·8	·7	·7	·6	·5	·5	·4	·3	·3	·3
Dec.	1·0	1·0	0·9	0·9	0·8	0·8	0·7	0·6	0·5	0·5	0·4	0·3
Lat.	AZIMUTH											
°	°	°	°	°	°	°	°	°	°	°	°	°
0	0·4	0·3	0·2	0·1	359·9	359·8	359·7	359·6	359·5	359·4	359·4	359·3
20	0·4	0·3	0·2	0·1	359·9	359·8	359·7	359·6	359·5	359·4	359·3	359·3
40	0·5	0·4	0·2	0·1	359·9	359·8	359·6	359·5	359·4	359·3	359·2	359·1
50	0·6	0·5	0·3	0·1	359·9	359·7	359·6	359·4	359·2	359·1	359·0	359·0
55	0·7	0·5	0·3	0·1	359·9	359·7	359·5	359·3	359·1	359·0	358·9	358·8
60	0·8	0·6	0·4	0·1	359·9	359·7	359·4	359·2	359·0	358·9	358·7	358·7
65	1·0	0·7	0·5	0·2	359·9	359·6	359·3	359·1	358·8	358·7	358·5	358·4

Latitude = Apparent altitude (corrected for refraction) $-1^\circ + a_0 + a_1 + a_2$

The table is entered with LHA Aries to determine the column to be used; each column refers to a range of 10°. a_0 is taken, with mental interpolation, from the upper table with the units of LHA Aries in degrees as argument; a_1, a_2 are taken, without interpolation, from the second and third tables with arguments latitude and month respectively. a_0, a_1, a_2, are always positive. The final table gives the azimuth of *Polaris.*

POLARIS (POLE STAR) TABLES, 2009

FOR DETERMINING LATITUDE FROM SEXTANT ALTITUDE AND FOR AZIMUTH

LHA ARIES	120° – 129°	130° – 139°	140° – 149°	150° – 159°	160° – 169°	170° – 179°	180° – 189°	190° – 199°	200° – 209°	210° – 219°	220° – 229°	230° – 239°
	a_0	a_0	a_0	a_0	a_0	a_0	a_0	a_0	a_0	a_0	a_0	a_0
°	° ′	° ′	° ′	° ′	° ′	° ′	° ′	° ′	° ′	° ′	° ′	° ′
0	0 51·3	0 58·5	1 05·7	1 12·7	1 19·3	1 25·2	1 30·3	1 34·5	1 37·6	1 39·6	1 40·3	1 39·8
1	52·0	0 59·2	06·4	13·4	19·9	25·8	30·8	34·9	37·9	39·7	40·3	39·7
2	52·7	1 00·0	07·1	14·1	20·5	26·3	31·3	35·2	38·1	39·8	40·3	39·5
3	53·4	00·7	07·8	14·7	21·1	26·8	31·7	35·6	38·3	39·9	40·3	39·4
4	54·2	01·4	08·5	15·4	21·7	27·4	32·1	35·9	38·6	40·0	40·2	39·2
5	0 54·9	1 02·1	1 09·2	1 16·0	1 22·3	1 27·9	1 32·6	1 36·2	1 38·8	1 40·1	1 40·2	1 39·1
6	55·6	02·8	09·9	16·7	22·9	28·4	33·0	36·5	38·9	40·2	40·1	38·9
7	56·3	03·6	10·6	17·4	23·5	28·9	33·4	36·8	39·1	40·2	40·1	38·7
8	57·1	04·3	11·3	18·0	24·1	29·4	33·8	37·1	39·3	40·3	40·0	38·5
9	57·8	05·0	12·0	18·6	24·6	29·9	34·1	37·4	39·4	40·3	39·9	38·3
10	0 58·5	1 05·7	1 12·7	1 19·3	1 25·2	1 30·3	1 34·5	1 37·6	1 39·6	1 40·3	1 39·8	1 38·0
Lat.	a_1	a_1	a_1	a_1	a_1	a_1	a_1	a_1	a_1	a_1	a_1	a_1
°	′	′	′	′	′	′	′	′	′	′	′	′
0	0·3	0·3	0·3	0·4	0·4	0·4	0·5	0·5	0·6	0·6	0·6	0·6
10	·3	·3	·4	·4	·4	·5	·5	·6	·6	·6	·6	·6
20	·4	·4	·4	·4	·5	·5	·5	·6	·6	·6	·6	·6
30	·4	·4	·5	·5	·5	·5	·5	·6	·6	·6	·6	·6
40	0·5	0·5	0·5	0·5	0·5	0·6	0·6	0·6	0·6	0·6	0·6	0·6
45	·6	·6	·6	·6	·6	·6	·6	·6	·6	·6	·6	·6
50	·6	·6	·6	·6	·6	·6	·6	·6	·6	·6	·6	·6
55	·7	·7	·7	·6	·6	·6	·6	·6	·6	·6	·6	·6
60	·7	·7	·7	·7	·7	·7	·6	·6	·6	·6	·6	·6
62	0·8	0·8	0·8	0·7	0·7	0·7	0·7	0·6	0·6	0·6	0·6	0·6
64	·8	·8	·8	·8	·7	·7	·7	·6	·6	·6	·6	·6
66	·9	·9	·8	·8	·8	·7	·7	·7	·6	·6	·6	·6
68	0·9	0·9	0·9	0·9	0·8	0·8	0·7	0·7	0·6	0·6	0·6	0·6
Month	a_2	a_2	a_2	a_2	a_2	a_2	a_2	a_2	a_2	a_2	a_2	a_2
	′	′	′	′	′	′	′	′	′	′	′	′
Jan.	0·7	0·6	0·6	0·6	0·6	0·5	0·5	0·5	0·5	0·5	0·5	0·5
Feb.	·8	·8	·8	·7	·7	·6	·6	·6	·5	·5	·4	·4
Mar.	0·9	0·9	0·9	0·9	0·8	·8	·7	·7	·6	·6	·5	·4
Apr.	1·0	1·0	1·0	1·0	1·0	0·9	0·9	0·8	0·8	0·7	0·6	0·6
May	0·9	1·0	1·0	1·0	1·0	1·0	1·0	0·9	0·9	·8	·8	·7
June	·8	0·9	0·9	1·0	1·0	1·0	1·0	1·0	1·0	0·9	0·9	·8
July	0·7	0·7	0·8	0·9	0·9	0·9	1·0	1·0	1·0	1·0	1·0	0·9
Aug.	·5	·6	·6	·7	·7	·8	0·8	0·9	0·9	0·9	0·9	·9
Sept.	·3	·4	·4	·5	·5	·6	·7	·7	·8	·8	·9	·9
Oct.	0·3	0·3	0·3	0·3	0·4	0·4	0·5	0·5	0·6	0·7	0·7	0·8
Nov.	·2	·2	·2	·2	·2	·3	·3	·4	·4	·5	·5	·6
Dec.	0·3	0·2	0·2	0·2	0·2	0·2	0·2	0·2	0·3	0·3	0·4	0·4
Lat.	AZIMUTH											
°	°	°	°	°	°	°	°	°	°	°	°	°
0	359·3	359·3	359·3	359·4	359·4	359·5	359·6	359·7	359·8	359·9	0·1	0·2
20	359·3	359·3	359·3	359·3	359·4	359·5	359·6	359·7	359·8	359·9	0·1	0·2
40	359·1	359·1	359·1	359·2	359·3	359·4	359·5	359·6	359·8	359·9	0·1	0·2
50	358·9	358·9	359·0	359·0	359·1	359·2	359·4	359·5	359·7	359·9	0·1	0·3
55	358·8	358·8	358·8	358·9	359·0	359·1	359·3	359·5	359·7	359·9	0·1	0·3
60	358·6	358·6	358·7	358·7	358·9	359·0	359·2	359·4	359·6	359·9	0·1	0·3
65	358·4	358·4	358·4	358·5	358·7	358·8	359·1	359·3	359·6	359·8	0·1	0·4

ILLUSTRATION

On 2009 April 21 at 23^h 18^m 56^s UT in longitude W 37° 14′ the apparent altitude (corrected for refraction), H_O, of *Polaris* was 49° 31′.6

From the daily pages:	° ′
GHA Aries (23^h)	195 08·7
Increment (18^m 56^s)	4 44·8
Longitude (west)	−37 14
LHA Aries	162 40

	° ′
H_O	49 31·6
a_0 (argument 162° 40′)	1 20·9
a_1 (Lat 50° approx.)	0·6
a_2 (April)	1·0
Sum − 1° = Lat =	49 54·1

POLARIS (POLE STAR) TABLES, 2009

FOR DETERMINING LATITUDE FROM SEXTANT ALTITUDE AND FOR AZIMUTH

LHA ARIES	240° – 249°	250° – 259°	260° – 269°	270° – 279°	280° – 289°	290° – 299°	300° – 309°	310° – 319°	320° – 329°	330° – 339°	340° – 349°	350° – 359°
	a_0	a_0	a_0	a_0	a_0	a_0	a_0	a_0	a_0	a_0	a_0	a_0
°	° ′	° ′	° ′	° ′	° ′	° ′	° ′	° ′	° ′	° ′	° ′	° ′
0	1 38·0	1 35·1	1 31·1	1 26·1	1 20·3	1 13·8	1 06·9	0 59·7	0 52·5	0 45·4	0 38·8	0 32·8
1	37·8	34·7	30·6	25·5	19·7	13·1	06·2	59·0	51·8	44·7	38·2	32·2
2	37·5	34·4	30·2	25·0	19·0	12·5	05·5	58·2	51·0	44·1	37·5	31·6
3	37·3	34·0	29·7	24·4	18·4	11·8	04·7	57·5	50·3	43·4	36·9	31·1
4	37·0	33·6	29·2	23·9	17·8	11·1	04·0	56·8	49·6	42·7	36·3	30·6
5	1 36·7	1 33·2	1 28·7	1 23·3	1 17·1	1 10·4	1 03·3	0 56·1	0 48·9	0 42·0	0 35·7	0 30·0
6	36·4	32·8	28·2	22·7	16·5	09·7	02·6	55·3	48·2	41·4	35·1	29·5
7	36·1	32·4	27·7	22·1	15·8	09·0	01·9	54·6	47·5	40·7	34·5	29·0
8	35·8	32·0	27·2	21·5	15·1	08·3	01·1	53·9	46·8	40·1	33·9	28·5
9	35·4	31·5	26·6	20·9	14·5	07·6	1 00·4	53·2	46·1	39·4	33·3	28·0
10	1 35·1	1 31·1	1 26·1	1 20·3	1 13·8	1 06·9	0 59·7	0 52·5	0 45·4	0 38·8	0 32·8	0 27·5
Lat.	a_1	a_1	a_1	a_1	a_1	a_1	a_1	a_1	a_1	a_1	a_1	a_1
°	′	′	′	′	′	′	′	′	′	′	′	′
0	0·5	0·5	0·5	0·4	0·4	0·3	0·3	0·3	0·3	0·4	0·4	0·4
10	·6	·5	·5	·4	·4	·4	·3	·3	·4	·4	·4	·5
20	·6	·5	·5	·5	·4	·4	·4	·4	·4	·4	·5	·5
30	·6	·6	·5	·5	·5	·5	·4	·4	·5	·5	·5	·5
40	0·6	0·6	0·6	0·5	0·5	0·5	0·5	0·5	0·5	0·5	0·5	0·6
45	·6	·6	·6	·6	·6	·6	·6	·6	·6	·6	·6	·6
50	·6	·6	·6	·6	·6	·6	·6	·6	·6	·6	·6	·6
55	·6	·6	·6	·6	·6	·7	·7	·7	·7	·6	·6	·6
60	·6	·6	·7	·7	·7	·7	·7	·7	·7	·7	·7	·7
62	0·6	0·7	0·7	0·7	0·7	0·8	0·8	0·8	0·8	0·7	0·7	0·7
64	·6	·7	·7	·7	·8	·8	·8	·8	·8	·8	·7	·7
66	·6	·7	·7	·8	·8	·8	·9	·9	·8	·8	·8	·7
68	0·7	0·7	0·8	0·8	0·9	0·9	0·9	0·9	0·9	0·9	0·8	0·8
Month	a_2	a_2	a_2	a_2	a_2	a_2	a_2	a_2	a_2	a_2	a_2	a_2
	′	′	′	′	′	′	′	′	′	′	′	′
Jan.	0·5	0·5	0·5	0·5	0·5	0·5	0·5	0·6	0·6	0·6	0·6	0·7
Feb.	·4	·4	·4	·4	·4	·4	·4	·4	·4	·5	·5	·6
Mar.	·4	·4	·3	·3	·3	·3	·3	·3	·3	·3	·4	·4
Apr.	0·5	0·4	0·4	0·3	0·3	0·2	0·2	0·2	0·2	0·2	0·2	0·3
May	·6	·6	·5	·4	·4	·3	·3	·2	·2	·2	·2	·2
June	·8	·7	·6	·6	·5	·4	·4	·3	·3	·2	·2	·2
July	0·9	0·8	0·8	0·7	0·7	0·6	0·5	0·5	0·4	0·3	0·3	0·3
Aug.	·9	·9	·9	·9	·8	·8	·7	·6	·6	·5	·5	·4
Sept.	·9	·9	·9	·9	·9	·9	·9	·8	·8	·7	·7	·6
Oct.	0·8	0·9	0·9	0·9	0·9	0·9	0·9	0·9	0·9	0·9	0·8	0·8
Nov.	·7	·7	·8	·9	·9	·9	1·0	1·0	1·0	1·0	1·0	0·9
Dec.	0·5	0·6	0·7	0·7	0·8	0·9	0·9	1·0	1·0	1·0	1·0	1·0
Lat.	AZIMUTH											
°	°	°	°	°	°	°	°	°	°	°	°	°
0	0·3	0·4	0·5	0·6	0·6	0·7	0·7	0·7	0·7	0·6	0·6	0·5
20	0·3	0·4	0·5	0·6	0·7	0·7	0·7	0·7	0·7	0·7	0·6	0·5
40	0·4	0·5	0·6	0·7	0·8	0·9	0·9	0·9	0·9	0·8	0·8	0·7
50	0·4	0·6	0·7	0·9	1·0	1·0	1·1	1·1	1·0	1·0	0·9	0·8
55	0·5	0·7	0·8	1·0	1·1	1·2	1·2	1·2	1·2	1·1	1·0	0·9
60	0·6	0·8	0·9	1·1	1·2	1·3	1·4	1·4	1·3	1·3	1·2	1·0
65	0·7	0·9	1·1	1·3	1·5	1·6	1·6	1·6	1·6	1·5	1·4	1·2

$$\text{Latitude} = \text{Apparent altitude (corrected for refraction)} - 1° + a_0 + a_1 + a_2$$

The table is entered with LHA Aries to determine the column to be used; each column refers to a range of 10°. a_0 is taken, with mental interpolation, from the upper table with the units of LHA Aries in degrees as argument; a_1, a_2 are taken, without interpolation, from the second and third tables with arguments latitude and month respectively. a_0, a_1, a_2, are always positive. The final table gives the azimuth of *Polaris*.

SIGHT REDUCTION PROCEDURES

METHODS AND FORMULAE FOR DIRECT COMPUTATION

1. *Introduction.* In this section formulae and methods are provided for *calculating* position at sea from observed altitudes taken with a marine sextant using a computer or programmable calculator.

The method uses analogous concepts and similar terminology as that used in *manual* methods of astro-navigation, where position is found by plotting position lines from their intercept and azimuth on a marine chart.

The algorithms are presented in standard algebra suitable for translating into the programming language of the user's computer. The basic ephemeris data may be taken directly from the main tabular pages of a current version of *The Nautical Almanac.* Formulae are given for calculating altitude and azimuth from the *GHA* and *Dec* of a body, and the estimated position of the observer. Formulae are also given for reducing sextant observations to observed altitudes by applying the corrections for dip, refraction, parallax and semi-diameter.

The intercept and azimuth obtained from each observation determine a position line, and the observer should lie on or close to each position line. The method of least squares is used to calculate the fix by finding the position where the sum of the squares of the distances from the position lines is a minimum. The use of least squares has other advantages. For example it is possible to improve the estimated position at the time of fix by repeating the calculation. It is also possible to include more observations in the solution and to reject doubtful ones.

2. *Notation.*

GHA = Greenwich hour angle. The range of GHA is from 0° to 360° starting at 0° on the Greenwich meridian increasing to the west, back to 360° on the Greenwich meridian.

SHA = sidereal hour angle. The range is 0° to 360°.

Dec = declination. The sign convention for declination is north is positive, south is negative. The range is from −90° at the south celestial pole to +90° at the north celestial pole.

$Long$ = longitude. The sign convention is east is positive, west is negative. The range is −180° to +180°.

Lat = latitude. The sign convention is north is positive, south is negative. The range is from −90° to +90°.

LHA = $GHA + Long$ = local hour angle. The LHA increases to the west from 0° on the local meridian to 360°.

H_C = calculated altitude. Above the horizon is positive, below the horizon is negative. The range is from −90° in the nadir to +90° in the zenith.

H_S = sextant altitude.

H = apparent altitude = sextant altitude corrected for instrumental error and dip.

H_O = observed altitude = apparent altitude corrected for refraction and, in appropriate cases, corrected for parallax and semi-diameter.

Z = Z_n= true azimuth. Z is measured from true north through east, south, west and back to north. The range is from 0° to 360°.

I = sextant index error.

D = dip of horizon.

R = atmospheric refraction.

HP = horizontal parallax of the Sun, Moon, Venus or Mars.
PA = parallax in altitude of the Sun, Moon, Venus or Mars.
S = semi-diameter of the Sun or Moon.
p = intercept = $H_O - H_C$. Towards is positive, away is negative.
T = course or track, measured as for azimuth from the north.
V = speed in knots.

3. *Entering Basic Data.* When quantities such as GHA are entered, which in *The Nautical Almanac* are given in degrees and minutes, convert them to degrees and decimals of a degree by dividing the minutes by 60 and adding to the degrees; for example, if $GHA = 123° 45'.6$, enter the two numbers 123 and 45·6 into the memory and set $GHA = 123 + 45{\cdot}6/60 = 123°7600$. Although four decimal places of a degree are shown in the examples, it is assumed that full precision is maintained in the calculations.

When using a computer or programmable calculator, write a subroutine to convert degrees and minutes to degrees and decimals. Scientific calculators usually have a special key for this purpose. For quantities like Dec which require a minus sign for southern declination, change the sign from plus to minus after the value has been converted to degrees and decimals, *e.g.* $Dec = \mathrm{S}\,0°\ 12'.3 = \mathrm{S}\,0°2050 = -0°2050$. Other quantities which require conversion are semi-diameter, horizontal parallax, longitude and latitude.

4. *Interpolation of GHA and Dec* The GHA and Dec of the Sun, Moon and planets are interpolated to the time of observation by direct calculation as follows: If the universal time is $a^h\ b^m\ c^s$, form the interpolation factor $x = b/60 + c/3600$. Enter the tabular value GHA_0 for the preceding hour (a) and the tabular value GHA_1 for the following hour ($a + 1$) then the interpolated value GHA is given by

$$GHA = GHA_0 + x(GHA_1 - GHA_0)$$

If the GHA passes through 360° between tabular values add 360° to GHA_1 before interpolation. If the interpolated value exceeds 360°, subtract 360° from GHA.

Similarly for declination, enter the tabular value Dec_0 for the preceding hour (a) and the tabular value Dec_1 for the following hour ($a+1$), then the interpolated value Dec is given by

$$Dec = Dec_0 + x(Dec_1 - Dec_0)$$

5. *Example.* (a) Find the GHA and Dec of the Sun on 2009 January 8 at $16^h\ 57^m\ 44^s$ UT.

The interpolation factor $x = 57/60 + 44/3600 = 0^h9622$

page 15 16^h $GHA_0 = 58°\ 17'.0 = 58°2833$

17^h $GHA_1 = 73°\ 16'.7 = 73°2783$

16^h9622 $GHA = 58{\cdot}2833 + 0{\cdot}9622(73{\cdot}2783 - 58{\cdot}2833) = 72°7119$

16^h $Dec_0 = \mathrm{S}\,22°\ 09'.6 = -22°1600$

17^h $Dec_1 = \mathrm{S}\,22°\ 09'.3 = -22°1550$

16^h9622 $Dec = -22{\cdot}1600 + 0{\cdot}9622(-22{\cdot}1550 + 22{\cdot}1600) = -22°1552$

GHA Aries is interpolated in the same way as GHA of a body. For a star the SHA and Dec are taken from the tabular page and do not require interpolation, then

$$GHA = GHA\ \text{Aries} + SHA$$

where GHA Aries is interpolated to the time of observation.

(b) Find the *GHA* and *Dec* of *Vega* on 2009 January 8 at $16^h\ 57^m\ 44^s$ UT.

The interpolation factor $x = 0^h\!.9622$ as in the previous example

page 14 $\quad 16^h$ GHA $Aries_0 = 348°\ 20'\!.2 = 348°\!.3367$

17^h GHA $Aries_1 = 3°\ 22'\!.6 = 363°\!.3767$ (360° added)

$16^h\!.9622$ GHA $Aries = 348{\cdot}3367 + 0{\cdot}9622(363{\cdot}3767 - 348{\cdot}3367) = 362°\!.8085$

$SHA = 80°\ 41'\!.6 = 80°\!.6933$

$GHA = GHA$ $Aries + SHA = 83°\!.5018$ (multiple of 360° removed)

$Dec = \mathrm{N}\,38°\ 47'\!.4 = +38°\!.7900$

6. *The calculated altitude and azimuth.* The calculated altitude H_C and true azimuth Z are determined from the GHA and Dec interpolated to the time of observation and from the $Long$ and Lat estimated at the time of observation as follows:

Step 1. Calculate the local hour angle

$$LHA = GHA + Long$$

Add or subtract multiples of 360° to set LHA in the range 0° to 360°.

Step 2. Calculate S, C and the altitude H_C from

$$S = \sin Dec$$
$$C = \cos Dec \cos LHA$$
$$H_C = \sin^{-1}(S \sin Lat + C \cos Lat)$$

where $\sin^{-1}$ is the inverse function of sine.

Step 3. Calculate X and A from

$$X = (S \cos Lat - C \sin Lat)/\cos H_C$$
$$\text{If } X > +1 \text{ set } X = +1$$
$$\text{If } X < -1 \text{ set } X = -1$$
$$A = \cos^{-1} X$$

where $\cos^{-1}$ is the inverse function of cosine.

Step 4. Determine the azimuth Z

$$\text{If } LHA > 180° \text{ then } Z = A$$
$$\text{Otherwise } Z = 360° - A$$

7. *Example.* Find the calculated altitude H_C and azimuth Z when

$$GHA = 53° \quad Dec = \mathrm{S}\,15° \quad Lat = \mathrm{N}\,32° \quad Long = \mathrm{W}\,16°$$

For the calculation

$$GHA = 53°\!.0000 \quad Dec = -15°\!.0000 \quad Lat = +32°\!.0000 \quad Long = -16°\!.0000$$

Step 1. $\quad LHA = 53{\cdot}0000 - 16{\cdot}0000 = 37{\cdot}0000$

Step 2. $\quad S = -0{\cdot}2588$

$C = +0{\cdot}9659 \times 0{\cdot}7986 = 0{\cdot}7714$

$\sin H_C = -0{\cdot}2588 \times 0{\cdot}5299 + 0{\cdot}7714 \times 0{\cdot}8480 = 0{\cdot}5171$

$H_C = 31°\!.1346$

Step 3. $X = (-0{\cdot}2588 \times 0{\cdot}8480 - 0{\cdot}7714 \times 0{\cdot}5299)/0{\cdot}8560 = -0{\cdot}7340$

$A = 137^\circ\!.2239$

Step 4. Since $LHA \leq 180^\circ$ then $Z = 360^\circ - A = 222^\circ\!.7761$

8. *Reduction from sextant altitude to observed altitude.* The sextant altitude H_S is corrected for both dip and index error to produce the apparent altitude. The observed altitude H_O is calculated by applying a correction for refraction. For the Sun, Moon, Venus and Mars a correction for parallax is also applied to H, and for the Sun and Moon a further correction for semi-diameter is required. The corrections are calculated as follows:

Step 1. Calculate dip

$$D = 0^\circ\!.0293\sqrt{h}$$

where h is the height of eye above the horizon in metres.

Step 2. Calculate apparent altitude

$$H = H_S + I - D$$

where I is the sextant index error.

Step 3. Calculate refraction (R) at a standard temperature of 10° Celsius (C) and pressure of 1010 millibars (mb)

$$R_0 = 0^\circ\!.0167/\tan\left(H + 7{\cdot}32/(H + 4{\cdot}32)\right)$$

If the temperature T° C and pressure P mb are known calculate the refraction from

$$R = fR_0 \qquad \text{where} \qquad f = 0{\cdot}28P/(T + 273)$$

$$\text{otherwise set} \qquad R = R_0$$

Step 4. Calculate the parallax in altitude (PA) from the horizontal parallax (HP) and the apparent altitude (H) for the Sun, Moon, Venus and Mars as follows:

$$PA = HP\cos H$$

For the Sun $HP = 0^\circ\!.0024$. This correction is very small and could be ignored.

For the Moon HP is taken for the nearest hour from the main tabular page and converted to degrees.

For Venus and Mars the HP is taken from the critical table at the bottom of page 259 and converted to degrees.

For the navigational stars and the remaining planets, Jupiter and Saturn set $PA = 0$.

If an error of $0'\!.2$ is significant the expression for the parallax in altitude for the Moon should include a small correction OB for the oblateness of the Earth as follows:

$$PA = HP\cos H + OB$$

$$\text{where} \quad OB = -0^\circ\!.0032\sin^2 Lat\cos H + 0^\circ\!.0032\sin(2Lat)\cos Z\sin H$$

At mid-latitudes and for altitudes of the Moon below 60° a simple approximation to OB is

$$OB = -0^\circ\!.0017\cos H$$

Step 5. Calculate the semi-diameter for the Sun and Moon as follows:

Sun: S is taken from the main tabular page and converted to degrees.

Moon: $S = 0°2724HP$ where HP is taken for the nearest hour from the main tabular page and converted to degrees.

Step 6. Calculate the observed altitude

$$H_O = H - R + PA \pm S$$

where the plus sign is used if the lower limb of the Sun or Moon was observed and the minus sign if the upper limb was observed.

9. *Example.* The following example illustrates how to use a calculator to reduce the sextant altitude (H_S) to observed altitude (H_O); the sextant altitudes given are assumed to be taken on 2009 January 8 with a marine sextant, zero index error, at height 5·4 m, temperature −3° C and pressure 982 mb, the Moon sights are assumed to be taken at 10^h UT.

Body limb	Sun lower	Sun upper	Moon lower	Moon upper	Venus –	*Polaris* –
Sextant altitude:						
H_S	21·3283	3·3367	33·4600	26·1117	4·5433	49·6083
Step 1. Dip:						
$D = 0{\cdot}0293\sqrt{h}$	0·0681	0·0681	0·0681	0·0681	0·0681	0·0681
Step 2. Apparent altitude:						
$H = H_S + I - D$	21·2602	3·2686	33·3919	26·0436	4·4752	49·5402
Step 3. Refraction:						
R_0	0·0423	0·2256	0·0251	0·0338	0·1798	0·0142
f	1·0184	1·0184	1·0184	1·0184	1·0184	1·0184
$R = fR_0$	0·0431	0·2298	0·0256	0·0344	0·1831	0·0144
Step 4. Parallax:			(60′7)	(60′7)	(0′2)	
HP	0·0024	0·0024	1·0117	1·0117	0·0033	–
Parallax in altitude:						
$PA = HP \cos H$	0·0022	0·0024	0·8447	0·9089	0·0033	–
Step 5. Semi-diameter:						
Sun : $S = 16{\cdot}3/60$	0·2717	0·2717	–	–	–	–
Moon : $S = 0{\cdot}2724HP$	–	–	0·2756	0·2756	–	–
Step 6. Observed altitude:						
$H_O = H - R + PA \pm S$	21·4910	2·7696	34·4865	26·6425	4·2955	49·5258

Note that for the Moon the correction for the oblateness of the Earth of about $-0°0017 \cos H$, which equals $-0°0014$ for the lower limb and $-0°0015$ for the upper limb, has been ignored in the above calculation.

10. *Position from intercept and azimuth using a chart.* An estimate is made of the position at the adopted time of fix. The position at the time of observation is then calculated by dead reckoning from the time of fix. For example if the course (track) T and the speed V (in knots) of the observer are constant then *Long* and *Lat* at the time of observation are calculated from

$$Long = L_F + t\,(V/60)\sin T / \cos B_F$$
$$Lat = B_F + t\,(V/60)\cos T$$

where L_F and B_F are the estimated longitude and latitude at the time of fix and t is the time interval in hours from the time of fix to the time of observation, t is positive if the time of observation is after the time of fix and negative if it was before.

The position line of an observation is plotted on a chart using the intercept

$$p = H_O - H_C$$

and azimuth Z with origin at the calculated position (*Long*, *Lat*) at the time of observation, where H_C and Z are calculated using the method in section 6, page 279. Starting from this calculated position a line is drawn on the chart along the direction of the azimuth to the body. Convert p to nautical miles by multiplying by 60. The position line is drawn at right angles to the azimuth line, distance p from (*Long*, *Lat*) towards the body if p is positive and distance p away from the body if p is negative. Provided there are no gross errors the navigator should be somewhere on or near the position line at the time of observation. Two or more position lines are required to determine a fix.

11. *Position from intercept and azimuth by calculation.* The position of the fix may be calculated from two or more sextant observations as follows.

If p_1, Z_1, are the intercept and azimuth of the first observation, p_2, Z_2, of the second observation and so on, form the summations

$$A = \cos^2 Z_1 + \cos^2 Z_2 + \cdots$$
$$B = \cos Z_1 \sin Z_1 + \cos Z_2 \sin Z_2 + \cdots$$
$$C = \sin^2 Z_1 + \sin^2 Z_2 + \cdots$$
$$D = p_1 \cos Z_1 + p_2 \cos Z_2 + \cdots$$
$$E = p_1 \sin Z_1 + p_2 \sin Z_2 + \cdots$$

where the number of terms in each summation is equal to the number of observations.

With $G = A\,C - B^2$, an improved estimate of the position at the time of fix (L_I, B_I) is given by

$$L_I = L_F + (A\,E - B\,D)/(G \cos B_F), \qquad B_I = B_F + (C\,D - B\,E)/G$$

Calculate the distance d between the initial estimated position (L_F, B_F) at the time of fix and the improved estimated position (L_I, B_I) in nautical miles from

$$d = 60\sqrt{((L_I - L_F)^2 \cos^2 B_F + (B_I - B_F)^2)}$$

If d exceeds about 20 nautical miles set $L_F = L_I$, $B_F = B_I$ and repeat the calculation until d, the distance between the position at the previous estimate and the improved estimate, is less than about 20 nautical miles.

12. *Example of direct computation.* Using the method described above, calculate the position of a ship on 2009 July 19 at $21^h\,00^m\,00^s$ UT from the marine sextant observations of the three stars *Regulus* (No. 26) at $20^h\,39^m\,23^s$ UT, *Antares* (No. 42) at $20^h\,45^m\,47^s$ UT and *Kochab* (No. 40) at $21^h\,02^m\,34^s$ UT, where the observed altitudes of the three stars corrected for the effects of refraction, dip and instrumental error, are $14^\circ\!.3006$, $30^\circ\!.5683$ and $46^\circ\!.7534$ respectively. The ship was travelling at a constant speed of 20 knots on a course of 325° during the period of observation, and the position of the ship at the time of fix $21^h\,00^m\,00^s$ UT is only known to the nearest whole degree W 15°, N 32°.

Intermediate values for the first iteration are shown in the table. *GHA* Aries was interpolated from the nearest tabular values on page 142. For the first iteration set $L_F = -15^\circ\!.0000$, $B_F = +32^\circ\!.0000$ at the time of fix at $21^h\ 00^m\ 00^s$ UT.

First Iteration

Body	*Regulus*	*Antares*	*Kochab*
No.	26	42	40
time of observation	$20^h\ 39^m\ 23^s$	$20^h\ 45^m\ 47^s$	$21^h\ 02^m\ 34^s$
H_O	14·3006	30·5683	46·7534
interpolation factor	0·6564	0·7631	0·0428
GHA Aries	247·6182	249·2226	253·4301
SHA (page 142)	207·7800	112·4933	137·3200
GHA	95·3982	1·7160	30·7501
Dec (page 142)	+11·9217	−26·4550	+74·1200
t	−0·3436	−0·2369	+0·0428
Long	−14·9225	−14·9466	−15·0096
Lat	+31·9062	+31·9353	+32·0117
Z	275·3165	166·2767	353·7335
H_C	14·2775	30·2625	47·1535
p	+0·0231	+0·3058	−0·4001

$$A = 1{\cdot}9404 \quad B = -0{\cdot}4312 \quad C = 1{\cdot}0596 \quad D = -0{\cdot}6926 \quad E = 0{\cdot}0932 \quad G = 1{\cdot}8701$$
$$(A\,E - B\,D)/(G\cos B_F) = -0{\cdot}0743, \qquad (C\,D - B\,E)/G = -0{\cdot}3710$$

An improved estimate of the position at the time of fix is

$$L_I = L_F - 0{\cdot}0743 = -15{\cdot}0743 \quad \text{and} \quad B_I = B_F - 0{\cdot}3710 = +31{\cdot}6290$$

Since the distance between the previous estimated position and the improved estimate $d = 22{\cdot}6$ nautical miles set $L_F = -15{\cdot}0743$, and $B_F = +31{\cdot}6290$ and repeat the calculation. The table shows the intermediate values of the calculation for the second iteration. In each iteration the quantities H_O, *GHA*, *Dec* and t do not change.

Second Iteration

Body	*Regulus*	*Antares*	*Kochab*
No.	26	42	40
Long	−14·9971	−15·0211	−15·0839
Lat	+31·5352	+31·5643	+31·6407
Z	275·3729	166·1494	353·8046
H_C	14·3061	30·6077	46·7916
p	−0·0055	−0·0394	−0·0382

$$A = 1{\cdot}9398 \quad B = -0{\cdot}4329 \quad C = 1{\cdot}0602 \quad D = -0{\cdot}0002 \quad E = 0{\cdot}0002 \quad G = 1{\cdot}8691$$
$$(A\,E - B\,D)/(G\cos B_F) = +0{\cdot}0002, \qquad (C\,D - B\,E)/G = -0{\cdot}0001$$

An improved estimate of the position at the time of fix is

$$L_I = L_F + 0{\cdot}0002 = -15{\cdot}0741 \quad \text{and} \quad B_I = B_F - 0{\cdot}0001 = +31{\cdot}6289$$

The distance between the previous estimated position and the improved estimated position $d = 0{\cdot}01$ nautical miles is so small that a third iteration would produce a negligible improvement to the estimate of the position.

USE OF CONCISE SIGHT REDUCTION TABLES

1. *Introduction.* The concise sight reduction tables given on pages 286 to 317 are intended for use when neither more extensive tables nor electronic computing aids are available. These "NAO sight reduction tables" provide for the reduction of the local hour angle and declination of a celestial object to azimuth and altitude, referred to an assumed position on the Earth, for use in the intercept method of celestial navigation which is now standard practice.

2. *Form of tables.* Entries in the reduction table are at a fixed interval of one degree for all latitudes and hour angles. A compact arrangement results from division of the navigational triangle into two right spherical triangles, so that the table has to be entered twice. Assumed latitude and local hour angle are the arguments for the first entry. The reduction table responds with the intermediate arguments A, B, and Z_1, where A is used as one of the arguments for the second entry to the table, B has to be incremented by the declination to produce the quantity F, and Z_1 is a component of the azimuth angle. The reduction table is then reentered with A and F and yields H, P, and Z_2 where H is the altitude, P is the complement of the parallactic angle, and Z_2 is the second component of the azimuth angle. It is usually necessary to adjust the tabular altitude for the fractional parts of the intermediate entering arguments to derive computed altitude, and an auxiliary table is provided for the purpose. Rules governing signs of the quantities which must be added or subtracted are given in the instructions and summarized on each tabular page. Azimuth angle is the sum of two components and is converted to true azimuth by familiar rules, repeated at the bottom of the tabular pages.

Tabular altitude and intermediate quantities are given to the nearest minute of arc, although errors of $2'$ in computed altitude may accrue during adjustment for the minutes parts of entering arguments. Components of azimuth angle are stated to $0^\circ\!.1$; for derived true azimuth, only whole degrees are warranted. Since objects near the zenith are difficult to observe with a marine sextant, they should be avoided; altitudes greater than about 80° are not suited to reduction by this method.

In many circumstances the accuracy provided by these tables is sufficient. However, to maintain the full accuracy ($0'\!.1$) of the ephemeral data in the almanac throughout their reduction to altitude and azimuth, more extensive tables or a calculator should be used.

3. *Use of Tables.*

Step 1. Determine the Greenwich hour angle (GHA) and Declination (Dec) of the body from the almanac. Select an assumed latitude (Lat) of integral degrees nearest to the estimated latitude. Choose an assumed longitude nearest to the estimated longitude such that the local hour angle

$$LHA = GHA \begin{array}{c} -\text{ west} \\ +\text{ east} \end{array} \text{longitude}$$

has integral degrees.

Step 2. Enter the reduction table with Lat and LHA as arguments. Record the quantities A, B and Z_1. Apply the rules for the sign of B and Z_1: B is minus if $90^\circ < LHA < 270^\circ$: Z_1 has the same sign as B. Set A° = nearest whole degree of A and A' = minutes part of A. This step may be repeated for all reductions before leaving the latitude opening of the table.

Step 3. Record the declination Dec. Apply the rules for the sign of Dec: Dec is minus if the name of Dec (*i.e.* N or S) is contrary to latitude. Add B and Dec algebraically to produce F. If F is negative, the object is below the horizon (in sight reduction, this can occur when the objects are close to the horizon). Regard F as positive until step 7. Set F° = nearest whole degree of F and F' = minutes part of F.

Step 4. Enter the reduction table a second time with $A°$ and $F°$ as arguments and record H, P, and Z_2. Set $P° =$ nearest whole degree of P and $Z_2° =$ nearest whole degree of Z_2.

Step 5. Enter the auxiliary table with F' and $P°$ as arguments to obtain $corr_1$ to H for F'. Apply the rule for the sign of $corr_1$: $corr_1$ is minus if $F < 90°$ and $F' > 29'$ or if $F > 90°$ and $F' < 30'$, otherwise $corr_1$ is plus.

Step 6. Enter the auxiliary table with A' and $Z_2°$ as arguments to obtain $corr_2$ to H for A'. Apply the rule for the sign of $corr_2$: $corr_2$ is minus if $A' < 30'$, otherwise $corr_2$ is plus.

Step 7. Calculate the computed altitude H_C as the sum of H, $corr_1$ and $corr_2$. Apply the rule for the sign of H_C: H_C is minus if F is negative.

Step 8. Apply the rule for the sign of Z_2: Z_2 is minus if $F > 90°$. If F is negative, replace Z_2 by $180° - Z_2$. Set the azimuth angle Z equal to the algebraic sum of Z_1 and Z_2 and ignore the resulting sign. Obtain the true azimuth Z_n from the rules

$$\begin{array}{lll} \text{For N latitude, if} & LHA > 180° & Z_n = Z \\ \text{if} & LHA < 180° & Z_n = 360° - Z \\ & & \\ \text{For S latitude, if} & LHA > 180° & Z_n = 180° - Z \\ \text{if} & LHA < 180° & Z_n = 180° + Z \end{array}$$

Observed altitude H_O is compared with H_C to obtain the altitude difference, which, with Z_n, is used to plot the position line.

4. *Example.* (a) Required the altitude and azimuth of *Schedar* on 2009 February 5 at UT $06^h\ 28^m$ from the estimated position 5° east, 53° north.

1. Assumed latitude	$Lat =$ 53° N		
From the almanac	$GHA =$ 222° 17′		
Assumed longitude	4° 43′ E		
Local hour angle	$LHA =$ 227		
2. Reduction table, 1st entry $(Lat, LHA) = (53, 227)$	$A =$ 26 07	$A° = 26$, $A' = 7$	
	$B = -27$ 12	$Z_1 = -49{\cdot}4$,	$90° < LHA < 270°$
3. From the almanac	$Dec = +56$ 36		*Lat* and *Dec* same
Sum $= B + Dec$	$F = +29$ 24	$F° = 29$, $F' = 24$	
4. Reduction table, 2nd entry $(A°, F°) = (26, 29)$	$H =$ 25 50	$P° = 61$	
		$Z_2 = 76{\cdot}3, Z_2° = 76$	
5. Auxiliary table, 1st entry $(F', P°) = (24, 61)$	$corr_1 =$ +21		$F < 90°$, $F' < 29'$
Sum	26 11		
6. Auxiliary table, 2nd entry $(A', Z_2°) = (7, 76)$	$corr_2 =$ −2		$A' < 30'$
7. Sum = computed altitude	$H_C = +26°$ 09′		$F > 0°$
8. Azimuth, first component	$Z_1 = -49{\cdot}4$		same sign as B
second component	$Z_2 = +76{\cdot}3$		$F < 90°$, $F > 0°$
Sum = azimuth angle	$Z =$ 26·9		
True azimuth	$Z_n =$ 027°		N *Lat*, $LHA > 180°$

continued on page 318

SIGHT REDUCTION TABLE

B: (–) for 90° < LHA < 270°
Dec: (–) for Lat. contrary name

Z_1: same sign as B
Z_2: (–) for F > 90°

Lat. / A		0°			1°			2°			3°			4°			5°			Lat. / A	
LHA/F		A/H	B/P	Z_1/Z_2	A/H	B/P	Z_1/Z_2	A/H	B/P	Z_1/Z_2	A/H	B/P	Z_1/Z_2	A/H	B/P	Z_1/Z_2	A/H	B/P	Z_1/Z_2	LHA	
°	°	° ′	° ′	°	° ′	° ′	°	° ′	° ′	°	° ′	° ′	°	° ′	° ′	°	° ′	° ′	°	°	°
0	180	0 00	90 00	90·0	0 00	89 00	90·0	0 00	88 00	90·0	0 00	87 00	90·0	0 00	86 00	90·0	0 00	85 00	90·0	180	360
1	179	1 00	90 00	90·0	1 00	89 00	90·0	1 00	88 00	90·0	1 00	87 00	89·9	1 00	86 00	89·9	1 00	85 00	89·9	181	359
2	178	2 00	90 00	90·0	2 00	89 00	90·0	2 00	88 00	89·9	2 00	87 00	89·9	2 00	86 00	89·9	2 00	85 00	89·8	182	358
3	177	3 00	90 00	90·0	3 00	89 00	89·9	3 00	88 00	89·9	3 00	87 00	89·8	3 00	86 00	89·8	2 59	85 00	89·7	183	357
4	176	4 00	90 00	90·0	4 00	89 00	89·9	4 00	88 00	89·9	4 00	87 00	89·8	3 59	85 59	89·7	3 59	84 59	89·7	184	356
5	175	5 00	90 00	90·0	5 00	89 00	89·9	5 00	88 00	89·8	5 00	86 59	89·7	4 59	85 59	89·7	4 59	84 59	89·6	185	355
6	174	6 00	90 00	90·0	6 00	89 00	89·9	6 00	87 59	89·8	6 00	86 59	89·7	5 59	85 59	89·6	5 59	84 58	89·5	186	354
7	173	7 00	90 00	90·0	7 00	89 00	89·9	7 00	87 59	89·8	6 59	86 59	89·6	6 59	85 58	89·5	6 58	84 58	89·4	187	353
8	172	8 00	90 00	90·0	8 00	88 59	89·9	8 00	87 59	89·7	7 59	86 58	89·6	7 59	85 58	89·4	7 58	84 57	89·3	188	352
9	171	9 00	90 00	90·0	9 00	88 59	89·8	9 00	87 59	89·7	8 59	86 58	89·5	8 59	85 57	89·4	8 58	84 56	89·2	189	351
10	170	10 00	90 00	90·0	10 00	88 59	89·8	10 00	87 58	89·6	9 59	86 57	89·5	9 59	85 56	89·3	9 58	84 55	89·1	190	350
11	169	11 00	90 00	90·0	11 00	88 59	89·8	11 00	87 58	89·6	10 59	86 57	89·4	10 58	85 56	89·2	10 57	84 54	89·0	191	349
12	168	12 00	90 00	90·0	12 00	88 59	89·8	12 00	87 57	89·6	11 59	86 56	89·4	11 58	85 55	89·2	11 57	84 53	88·9	192	348
13	167	13 00	90 00	90·0	13 00	88 58	89·8	13 00	87 57	89·5	12 59	86 55	89·3	12 58	85 54	89·1	12 57	84 52	88·8	193	347
14	166	14 00	90 00	90·0	14 00	88 58	89·8	13 59	87 56	89·5	13 59	86 55	89·3	13 58	85 53	89·0	13 57	84 51	88·8	194	346
15	165	15 00	90 00	90·0	15 00	88 58	89·7	14 59	87 56	89·5	14 59	86 54	89·2	14 58	85 52	88·9	14 56	84 49	88·7	195	345
16	164	16 00	90 00	90·0	16 00	88 58	89·7	15 59	87 55	89·4	15 59	86 53	89·1	15 58	85 50	88·9	15 56	84 48	88·6	196	344
17	163	17 00	90 00	90·0	17 00	88 57	89·7	16 59	87 55	89·4	16 59	86 52	89·1	16 57	85 49	88·8	16 56	84 46	88·5	197	343
18	162	18 00	90 00	90·0	18 00	88 57	89·7	17 59	87 54	89·4	17 58	86 51	89·0	17 57	85 48	88·7	17 56	84 45	88·4	198	342
19	161	19 00	90 00	90·0	19 00	88 57	89·7	18 59	87 53	89·3	18 58	86 50	89·0	18 57	85 46	88·6	18 55	84 43	88·3	199	341
20	160	20 00	90 00	90·0	20 00	88 56	89·6	19 59	87 52	89·3	19 58	86 48	88·9	19 57	85 45	88·5	19 55	84 41	88·2	200	340
21	159	21 00	90 00	90·0	21 00	88 56	89·6	20 59	87 51	89·2	20 58	86 47	88·8	20 57	85 43	88·5	20 55	84 39	88·1	201	339
22	158	22 00	90 00	90·0	22 00	88 55	89·6	21 59	87 51	89·2	21 58	86 46	88·8	21 57	85 41	88·4	21 55	84 37	88·0	202	338
23	157	23 00	90 00	90·0	23 00	88 55	89·6	22 59	87 50	89·2	22 58	86 44	88·7	22 56	85 39	88·3	22 54	84 34	87·9	203	337
24	156	24 00	90 00	90·0	24 00	88 54	89·6	23 59	87 49	89·1	23 58	86 43	88·7	23 56	85 37	88·2	23 54	84 32	87·8	204	336
25	155	25 00	90 00	90·0	25 00	88 54	89·5	24 59	87 48	89·1	24 58	86 41	88·6	24 56	85 35	88·1	24 54	84 29	87·7	205	335
26	154	26 00	90 00	90·0	26 00	88 53	89·5	25 59	87 47	89·0	25 58	86 40	88·5	25 56	85 33	88·1	25 54	84 26	87·6	206	334
27	153	27 00	90 00	90·0	27 00	88 53	89·5	26 59	87 45	89·0	26 58	86 38	88·5	26 56	85 31	88·0	26 53	84 24	87·5	207	333
28	152	28 00	90 00	90·0	28 00	88 52	89·5	27 59	87 44	88·9	27 57	86 36	88·4	27 56	85 28	87·9	27 53	84 20	87·3	208	332
29	151	29 00	90 00	90·0	29 00	88 51	89·4	28 59	87 43	88·9	28 57	86 34	88·3	28 55	85 26	87·8	28 53	84 17	87·2	209	331
30	150	30 00	90 00	90·0	30 00	88 51	89·4	29 59	87 41	88·8	29 57	86 32	88·3	29 55	85 23	87·7	29 52	84 14	87·1	210	330
31	149	31 00	90 00	90·0	31 00	88 50	89·4	30 59	87 40	88·8	30 57	86 30	88·2	30 55	85 20	87·6	30 52	84 10	87·0	211	329
32	148	32 00	90 00	90·0	32 00	88 49	89·4	31 59	87 39	88·8	31 57	86 28	88·1	31 55	85 17	87·5	31 52	84 07	86·9	212	328
33	147	33 00	90 00	90·0	33 00	88 48	89·4	32 59	87 37	88·7	32 57	86 25	88·1	32 55	85 14	87·4	32 52	84 03	86·8	213	327
34	146	34 00	90 00	90·0	34 00	88 48	89·3	33 59	87 35	88·7	33 57	86 23	88·0	33 54	85 11	87·3	33 51	83 59	86·6	214	326
35	145	35 00	90 00	90·0	35 00	88 47	89·3	34 59	87 34	88·6	34 57	86 20	87·9	34 54	85 07	87·2	34 51	83 54	86·5	215	325
36	144	36 00	90 00	90·0	36 00	88 46	89·3	35 58	87 32	88·5	35 57	86 18	87·8	35 54	85 04	87·1	35 51	83 50	86·4	216	324
37	143	37 00	90 00	90·0	37 00	88 45	89·2	36 58	87 30	88·5	36 56	86 15	87·7	36 54	85 00	87·0	36 50	83 45	86·2	217	323
38	142	38 00	90 00	90·0	38 00	88 44	89·2	37 58	87 28	88·4	37 56	86 12	87·7	37 53	84 56	86·9	37 50	83 40	86·1	218	322
39	141	39 00	90 00	90·0	39 00	88 43	89·2	38 58	87 26	88·4	38 56	86 09	87·6	38 53	84 52	86·8	38 49	83 35	86·0	219	321
40	140	40 00	90 00	90·0	40 00	88 42	89·2	39 58	87 23	88·3	39 56	86 05	87·5	39 53	84 47	86·7	39 49	83 29	85·8	220	320
41	139	41 00	90 00	90·0	41 00	88 41	89·1	40 58	87 21	88·3	40 56	86 02	87·4	40 53	84 42	86·5	40 49	83 23	85·7	221	319
42	138	42 00	90 00	90·0	42 00	88 39	89·1	41 58	87 19	88·2	41 56	85 58	87·3	41 52	84 37	86·4	41 48	83 17	85·5	222	318
43	137	43 00	90 00	90·0	43 00	88 38	89·1	42 58	87 16	88·1	42 56	85 54	87·2	42 52	84 32	86·3	42 48	83 11	85·4	223	317
44	136	44 00	90 00	90·0	43 59	88 37	89·0	43 58	87 13	88·1	43 55	85 50	87·1	43 52	84 27	86·1	43 47	83 04	85·2	224	316
45	135	45 00	90 00	90·0	44 59	88 35	89·0	44 58	87 10	88·0	44 55	85 46	87·0	44 52	84 21	86·0	44 47	82 57	85·0	225	315

Lat. / A		0°			1°			2°			3°			4°			5°			Lat. / A	
LHA/F		A/H	B/P	Z_1/Z_2	A/H	B/P	Z_1/Z_2	A/H	B/P	Z_1/Z_2	A/H	B/P	Z_1/Z_2	A/H	B/P	Z_1/Z_2	A/H	B/P	Z_1/Z_2	LHA	
°	°	° ′	° ′	°	° ′	° ′	°	° ′	° ′	°	° ′	° ′	°	° ′	° ′	°	° ′	° ′	°	°	°
45	135	45 00	90 00	90·0	44 59	88 35	89·0	44 58	87 10	88·0	44 55	85 46	87·0	44 52	84 21	86·0	44 47	82 57	85·0	225	315
46	134	46 00	90 00	90·0	45 59	88 34	89·0	45 58	87 07	87·9	45 55	85 41	86·9	45 51	84 15	85·9	45 46	82 49	84·8	226	314
47	133	47 00	90 00	90·0	46 59	88 32	88·9	46 58	87 04	87·9	46 55	85 36	86·8	46 51	84 09	85·7	46 46	82 41	84·7	227	313
48	132	48 00	90 00	90·0	47 59	88 30	88·9	47 58	87 01	87·8	47 55	85 31	86·7	47 51	84 02	85·6	47 46	82 33	84·5	228	312
49	131	49 00	90 00	90·0	48 59	88 29	88·8	48 58	86 57	87·7	48 55	85 26	86·6	48 50	83 55	85·4	48 45	82 24	84·3	229	311
50	130	50 00	90 00	90·0	49 59	88 27	88·8	49 58	86 53	87·6	49 54	85 20	86·4	49 50	83 47	85·2	49 44	82 15	84·1	230	310
51	129	51 00	90 00	90·0	50 59	88 25	88·8	50 57	86 49	87·5	50 54	85 14	86·3	50 50	83 40	85·1	50 44	82 05	83·9	231	309
52	128	52 00	90 00	90·0	51 59	88 23	88·7	51 57	86 45	87·4	51 54	85 08	86·2	51 49	83 31	84·9	51 43	81 55	83·6	232	308
53	127	53 00	90 00	90·0	52 59	88 20	88·7	52 57	86 41	87·3	52 54	85 01	86·0	52 49	83 22	84·7	52 43	81 44	83·4	233	307
54	126	54 00	90 00	90·0	53 59	88 18	88·6	53 57	86 36	87·2	53 54	84 54	85·9	53 49	83 13	84·5	53 42	81 32	83·2	234	306
55	125	55 00	90 00	90·0	54 59	88 15	88·6	54 57	86 31	87·1	54 53	84 47	85·7	54 48	83 03	84·3	54 41	81 20	82·9	235	305
56	124	56 00	90 00	90·0	55 59	88 13	88·5	55 57	86 26	87·0	55 53	84 39	85·6	55 48	82 52	84·1	55 41	81 06	82·6	236	304
57	123	57 00	90 00	90·0	56 59	88 10	88·5	56 57	86 20	86·9	56 53	84 30	85·4	56 47	82 41	83·9	56 40	80 52	82·4	237	303
58	122	58 00	90 00	90·0	57 59	88 07	88·4	57 57	86 14	86·8	57 52	84 21	85·2	57 47	82 29	83·6	57 39	80 38	82·1	238	302
59	121	59 00	90 00	90·0	58 59	88 04	88·3	58 57	86 07	86·7	58 52	84 11	85·0	58 46	82 16	83·4	58 38	80 22	81·7	239	301
60	120	60 00	90 00	90·0	59 59	88 00	88·3	59 56	86 00	86·5	59 52	84 01	84·8	59 46	82 02	83·1	59 37	80 05	81·4	240	300
61	119	61 00	90 00	90·0	60 59	87 56	88·2	60 56	85 53	86·4	60 52	83 50	84·6	60 45	81 48	82·8	60 37	79 46	81·1	241	299
62	118	62 00	90 00	90·0	61 59	87 52	88·1	61 56	85 45	86·2	61 51	83 38	84·4	61 44	81 32	82·5	61 36	79 27	80·7	242	298
63	117	63 00	90 00	90·0	62 59	87 48	88·0	62 56	85 36	86·1	62 51	83 25	84·1	62 44	81 15	82·2	62 35	79 06	80·3	243	297
64	116	64 00	90 00	90·0	63 59	87 43	88·0	63 56	85 27	85·9	63 50	83 11	83·9	63 43	80 56	81·9	63 33	78 43	79·9	244	296
65	115	65 00	90 00	90·0	64 59	87 38	87·9	64 56	85 17	85·7	64 50	82 56	83·6	64 42	80 36	81·5	64 32	78 18	79·4	245	295
66	114	66 00	90 00	90·0	65 59	87 33	87·8	65 55	85 06	85·5	65 49	82 39	83·3	65 41	80 15	81·1	65 31	77 52	78·9	246	294
67	113	67 00	90 00	90·0	66 59	87 27	87·6	66 55	84 54	85·3	66 49	82 22	83·0	66 40	79 51	80·7	66 29	77 23	78·4	247	293
68	112	68 00	90 00	90·0	67 59	87 20	87·5	67 55	84 40	85·1	67 48	82 02	82·6	67 39	79 26	80·2	67 28	76 51	77·8	248	292
69	111	69 00	90 00	90·0	68 59	87 13	87·4	68 55	84 26	84·8	68 48	81 41	82·2	68 38	78 58	79·7	68 26	76 17	77·2	249	291
70	110	70 00	90 00	90·0	69 59	87 05	87·3	69 54	84 10	84·5	69 47	81 17	81·8	69 37	78 27	79·2	69 25	75 39	76·5	250	290
71	109	71 00	90 00	90·0	70 58	86 56	87·1	70 54	83 53	84·2	70 46	80 51	81·4	70 36	77 53	78·5	70 23	74 58	75·8	251	289
72	108	72 00	90 00	90·0	71 58	86 46	86·9	71 54	83 33	83·9	71 46	80 22	80·8	71 35	77 15	77·9	71 20	74 12	75·0	252	288
73	107	73 00	90 00	90·0	72 58	86 35	86·7	72 53	83 11	83·5	72 45	79 50	80·3	72 33	76 33	77·1	72 18	73 20	74·1	253	287
74	106	74 00	90 00	90·0	73 58	86 23	86·5	73 53	82 47	83·1	73 44	79 14	79·7	73 31	75 46	76·3	73 15	72 23	73·1	254	286
75	105	75 00	90 00	90·0	74 58	86 09	86·3	74 52	82 19	82·6	74 43	78 33	78·9	74 29	74 53	75·4	74 12	71 19	72·0	255	285
76	104	76 00	90 00	90·0	75 58	85 52	86·0	75 52	81 47	82·0	75 41	77 47	78·1	75 27	73 53	74·4	75 09	70 07	70·7	256	284
77	103	77 00	90 00	90·0	76 58	85 34	85·7	76 51	81 11	81·4	76 40	76 53	77·2	76 25	72 44	73·2	76 05	68 45	69·3	257	283
78	102	78 00	90 00	90·0	77 58	85 12	85·3	77 50	80 28	80·7	77 38	75 51	76·2	77 22	71 25	71·8	77 01	67 11	67·7	258	282
79	101	79 00	90 00	90·0	78 57	84 46	84·9	78 49	79 38	79·8	78 36	74 39	74·9	78 18	69 52	70·3	77 56	65 22	65·8	259	281
80	100	80 00	90 00	90·0	79 57	84 16	84·3	79 48	78 38	78·8	79 34	73 12	73·5	79 14	68 04	68·4	78 50	63 16	63·7	260	280
81	99	81 00	90 00	90·0	80 57	83 38	83·7	80 47	77 25	77·6	80 31	71 29	71·7	80 09	65 55	66·2	79 43	60 47	61·2	261	279
82	98	82 00	90 00	90·0	81 56	82 51	82·9	81 45	75 55	76·1	81 28	69 22	69·6	81 04	63 19	63·6	80 34	57 51	58·2	262	278
83	97	83 00	90 00	90·0	82 56	81 51	81·9	82 43	74 01	74·1	82 23	66 44	66·9	81 57	60 09	60·4	81 24	54 20	54·6	263	277
84	96	84 00	90 00	90·0	83 55	80 31	80·6	83 41	71 32	71·6	83 18	63 22	63·5	82 48	56 13	56·4	82 12	50 04	50·3	264	276
85	95	85 00	90 00	90·0	84 54	78 40	78·7	84 37	68 10	68·3	84 10	58 59	59·1	83 36	51 16	51·4	82 56	44 53	45·1	265	275
86	94	86 00	90 00	90·0	85 53	75 57	76·0	85 32	63 24	63·5	85 00	53 05	53·2	84 21	44 56	45·1	83 36	38 34	38·7	266	274
87	93	87 00	90 00	90·0	86 50	71 33	71·6	86 24	56 17	56·3	85 45	44 58	45·0	85 00	36 49	36·9	84 10	30 53	31·0	267	273
88	92	88 00	90 00	90·0	87 46	63 26	63·4	87 10	44 59	45·0	86 24	33 40	33·7	85 32	26 31	26·6	84 37	21 45	21·8	268	272
89	91	89 00	90 00	90·0	88 35	45 00	45·0	87 46	26 33	26·6	86 50	18 25	18·4	85 53	14 01	14·0	84 54	11 17	11·3	269	271
90	90	90 00	0 00	0·0	89 00	0 00	0·0	88 00	0 00	0·0	87 00	0 00	0·0	86 00	0 00	0·0	85 00	0 00	0·0	270	270

N. Lat: for LHA > 180° ... $Z_n = Z$
for LHA < 180° ... $Z_n = 360° - Z$

S. Lat.: for LHA > 180° ... $Z_n = 180° - Z$
for LHA < 180° ... $Z_n = 180° + Z$

SIGHT REDUCTION TABLE

B: (–) for 90° < LHA < 270°
Dec: (–) for Lat. contrary name

Z_1: same sign as B
Z_2: (–) for F > 90°

Lat. / A		6°			7°			8°			9°			10°			11°			Lat. / A	
LHA/F		A/H	B/P	Z_1/Z_2	A/H	B/P	Z_1/Z_2	A/H	B/P	Z_1/Z_2	A/H	B/P	Z_1/Z_2	A/H	B/P	Z_1/Z_2	A/H	B/P	Z_1/Z_2	LHA	
°	°	° ′	° ′	°	° ′	° ′	°	° ′	° ′	°	° ′	° ′	°	° ′	° ′	°	° ′	° ′	°	°	°
0	180	0 00	84 00	90·0	0 00	83 00	90·0	0 00	82 00	90·0	0 00	81 00	90·0	0 00	80 00	90·0	0 00	79 00	90·0	180	360
1	179	1 00	84 00	89·9	1 00	83 00	89·9	0 59	82 00	89·9	0 59	81 00	89·8	0 59	80 00	89·8	0 59	79 00	89·8	181	359
2	178	1 59	84 00	89·8	1 59	83 00	89·8	1 59	82 00	89·7	1 59	81 00	89·7	1 58	80 00	89·7	1 58	79 00	89·6	182	358
3	177	2 59	84 00	89·7	2 59	82 59	89·6	2 58	81 59	89·6	2 58	80 59	89·5	2 57	79 59	89·5	2 57	78 59	89·4	183	357
4	176	3 59	83 59	89·6	3 58	82 59	89·5	3 58	81 59	89·4	3 57	80 59	89·4	3 56	79 59	89·3	3 56	78 58	89·2	184	356
5	175	4 58	83 59	89·5	4 58	82 58	89·4	4 57	81 58	89·3	4 56	80 58	89·2	4 55	79 58	89·1	4 54	78 58	89·0	185	355
6	174	5 58	83 58	89·4	5 57	82 58	89·3	5 56	81 57	89·2	5 56	80 57	89·1	5 55	79 57	89·0	5 53	78 56	88·9	186	354
7	173	6 58	83 57	89·3	6 57	82 57	89·1	6 56	81 56	89·0	6 55	80 56	88·9	6 54	79 56	88·8	6 52	78 55	88·7	187	353
8	172	7 57	83 56	89·2	7 56	82 56	89·0	7 55	81 55	88·9	7 54	80 55	88·7	7 53	79 54	88·6	7 51	78 54	88·5	188	352
9	171	8 57	83 56	89·1	8 56	82 55	88·9	8 55	81 54	88·7	8 53	80 53	88·6	8 52	79 53	88·4	8 50	78 52	88·3	189	351
10	170	9 57	83 54	88·9	9 55	82 54	88·8	9 54	81 53	88·6	9 53	80 52	88·4	9 51	79 51	88·2	9 49	78 50	88·1	190	350
11	169	10 56	83 53	88·8	10 55	82 52	88·6	10 53	81 51	88·5	10 52	80 50	88·3	10 50	79 49	88·1	10 48	78 48	87·9	191	349
12	168	11 56	83 52	88·7	11 55	82 51	88·5	11 53	81 49	88·3	11 51	80 48	88·1	11 49	79 47	87·9	11 47	78 46	87·7	192	348
13	167	12 56	83 51	88·6	12 54	82 49	88·4	12 52	81 48	88·2	12 50	80 46	87·9	12 48	79 45	87·7	12 45	78 43	87·5	193	347
14	166	13 55	83 49	88·5	13 54	82 47	88·3	13 52	81 46	88·0	13 49	80 44	87·8	13 47	79 42	87·5	13 44	78 40	87·3	194	346
15	165	14 55	83 47	88·4	14 53	82 45	88·1	14 51	81 43	87·9	14 49	80 41	87·6	14 46	79 39	87·3	14 43	78 37	87·1	195	345
16	164	15 55	83 46	88·3	15 53	82 43	88·0	15 50	81 41	87·7	15 48	80 39	87·4	15 45	79 36	87·1	15 42	78 34	86·9	196	344
17	163	16 54	83 44	88·2	16 52	82 41	87·9	16 50	81 38	87·6	16 47	80 36	87·3	16 44	79 33	87·0	16 41	78 31	86·7	197	343
18	162	17 54	83 42	88·1	17 52	82 39	87·7	17 49	81 36	87·4	17 46	80 33	87·1	17 43	79 30	86·8	17 39	78 27	86·5	198	342
19	161	18 54	83 39	87·9	18 51	82 36	87·6	18 48	81 33	87·3	18 45	80 29	86·9	18 42	79 26	86·6	18 38	78 23	86·2	199	341
20	160	19 53	83 37	87·8	19 51	82 33	87·5	19 48	81 30	87·1	19 45	80 26	86·7	19 41	79 22	86·4	19 37	78 19	86·0	200	340
21	159	20 53	83 35	87·7	20 50	82 30	87·3	20 47	81 26	86·9	20 44	80 22	86·6	20 40	79 18	86·2	20 36	78 14	85·8	201	339
22	158	21 52	83 32	87·6	21 50	82 27	87·2	21 46	81 23	86·8	21 43	80 18	86·4	21 39	79 14	86·0	21 35	78 10	85·6	202	338
23	157	22 52	83 29	87·5	22 49	82 24	87·0	22 46	81 19	86·6	22 42	80 14	86·2	22 38	79 09	85·8	22 33	78 05	85·4	203	337
24	156	23 52	83 26	87·3	23 49	82 21	86·9	23 45	81 15	86·5	23 41	80 10	86·0	23 37	79 05	85·6	23 32	77 59	85·1	204	336
25	155	24 51	83 23	87·2	24 48	82 17	86·7	24 44	81 11	86·3	24 40	80 05	85·8	24 36	78 59	85·4	24 31	77 54	84·9	205	335
26	154	25 51	83 20	87·1	25 48	82 13	86·6	25 44	81 07	86·1	25 39	80 00	85·6	25 35	78 54	85·2	25 29	77 48	84·7	206	334
27	153	26 50	83 16	87·0	26 47	82 09	86·4	26 43	81 02	85·9	26 38	79 55	85·4	26 33	78 48	84·9	26 28	77 42	84·4	207	333
28	152	27 50	83 13	86·8	27 46	82 05	86·3	27 42	80 57	85·8	27 38	79 50	85·2	27 32	78 42	84·7	27 27	77 35	84·2	208	332
29	151	28 50	83 09	86·7	28 46	82 01	86·1	28 41	80 52	85·6	28 37	79 44	85·0	28 31	78 36	84·5	28 25	77 28	84·0	209	331
30	150	29 49	83 05	86·5	29 45	81 56	86·0	29 41	80 47	85·4	29 36	79 38	84·8	29 30	78 29	84·3	29 24	77 21	83·7	210	330
31	149	30 49	83 01	86·4	30 45	81 51	85·8	30 40	80 41	85·2	30 35	79 32	84·6	30 29	78 23	84·0	30 22	77 13	83·5	211	329
32	148	31 48	82 56	86·3	31 44	81 46	85·6	31 39	80 35	85·0	31 34	79 25	84·4	31 27	78 15	83·8	31 21	77 05	83·2	212	328
33	147	32 48	82 51	86·1	32 43	81 40	85·5	32 38	80 29	84·8	32 33	79 18	84·2	32 26	78 08	83·6	32 19	76 57	82·9	213	327
34	146	33 47	82 46	86·0	33 43	81 35	85·3	33 37	80 23	84·6	33 32	79 11	84·0	33 25	78 00	83·3	33 18	76 48	82·7	214	326
35	145	34 47	82 41	85·8	34 42	81 29	85·1	34 37	80 16	84·4	34 30	79 03	83·7	34 24	77 51	83·1	34 16	76 39	82·4	215	325
36	144	35 46	82 36	85·7	35 41	81 22	84·9	35 36	80 09	84·2	35 29	78 55	83·5	35 22	77 42	82·8	35 14	76 29	82·1	216	324
37	143	36 46	82 30	85·5	36 41	81 16	84·8	36 35	80 01	84·0	36 28	78 47	83·3	36 21	77 33	82·5	36 13	76 19	81·8	217	323
38	142	37 45	82 24	85·3	37 40	81 09	84·6	37 34	79 53	83·8	37 27	78 38	83·0	37 19	77 23	82·3	37 11	76 09	81·5	218	322
39	141	38 45	82 18	85·2	38 39	81 01	84·4	38 33	79 45	83·6	38 26	78 29	82·8	38 18	77 13	82·0	38 09	75 57	81·2	219	321
40	140	39 44	82 11	85·0	39 39	80 54	84·2	39 32	79 36	83·3	39 25	78 19	82·5	39 16	77 02	81·7	39 07	75 46	80·9	220	320
41	139	40 44	82 04	84·8	40 38	80 46	84·0	40 31	79 27	83·1	40 23	78 09	82·3	40 15	76 51	81·4	40 05	75 33	80·6	221	319
42	138	41 43	81 57	84·6	41 37	80 37	83·7	41 30	79 17	82·9	41 22	77 58	82·0	41 13	76 39	81·1	41 04	75 21	80·3	222	318
43	137	42 42	81 49	84·4	42 36	80 28	83·5	42 29	79 07	82·6	42 21	77 47	81·7	42 12	76 27	80·8	42 02	75 07	79·9	223	317
44	136	43 42	81 41	84·2	43 35	80 19	83·3	43 28	78 57	82·3	43 19	77 35	81·4	43 10	76 14	80·5	43 00	74 53	79·6	224	316
45	135	44 41	81 33	84·0	44 34	80 09	83·1	44 27	78 46	82·1	44 18	77 22	81·1	44 08	76 00	80·1	43 57	74 38	79·2	225	315

Lat. / A		6°			7°			8°			9°			10°			11°			Lat. / A	
LHA/F		A/H	B/P	Z_1/Z_2	A/H	B/P	Z_1/Z_2	A/H	B/P	Z_1/Z_2	A/H	B/P	Z_1/Z_2	A/H	B/P	Z_1/Z_2	A/H	B/P	Z_1/Z_2	LHA	
°	°	° ′	° ′	°	° ′	° ′	°	° ′	° ′	°	° ′	° ′	°	° ′	° ′	°	° ′	° ′	°	°	°
45	135	44 41	81 33	84·0	44 34	80 09	83·1	44 27	78 46	82·1	44 18	77 22	81·1	44 08	76 00	80·1	43 57	74 38	79·2	225	315
46	134	45 41	81 24	83·8	45 34	79 59	82·8	45 26	78 34	81·8	45 16	77 09	80·8	45 06	75 45	79·8	44 55	74 22	78·8	226	314
47	133	46 40	81 14	83·6	46 33	79 48	82·6	46 24	78 21	81·5	46 15	76 56	80·5	46 04	75 30	79·5	45 53	74 05	78·4	227	313
48	132	47 39	81 04	83·4	47 32	79 36	82·3	47 23	78 08	81·2	47 13	76 41	80·1	47 03	75 14	79·1	46 51	73 48	78·0	228	312
49	131	48 38	80 54	83·1	48 31	79 24	82·0	48 22	77 55	80·9	48 12	76 26	79·8	48 01	74 57	78·7	47 48	73 30	77·6	229	311
50	130	49 38	80 43	82·9	49 30	79 11	81·7	49 20	77 40	80·6	49 10	76 09	79·4	48 58	74 40	78·3	48 46	73 10	77·2	230	310
51	129	50 37	80 31	82·6	50 29	78 58	81·4	50 19	77 25	80·2	50 08	75 52	79·1	49 56	74 21	77·9	49 43	72 50	76·7	231	309
52	128	51 36	80 19	82·4	51 27	78 43	81·1	51 18	77 08	79·9	51 06	75 34	78·7	50 54	74 01	77·5	50 40	72 29	76·3	232	308
53	127	52 35	80 06	82·1	52 26	78 28	80·8	52 16	76 51	79·5	52 04	75 15	78·3	51 52	73 40	77·0	51 37	72 06	75·8	233	307
54	126	53 34	79 52	81·8	53 25	78 12	80·5	53 14	76 33	79·2	53 02	74 55	77·8	52 49	73 18	76·6	52 35	71 42	75·3	234	306
55	125	54 33	79 37	81·5	54 24	77 55	80·1	54 13	76 14	78·8	54 00	74 34	77·4	53 47	72 55	76·1	53 31	71 17	74·8	235	305
56	124	55 32	79 21	81·2	55 22	77 37	79·8	55 11	75 54	78·3	54 58	74 11	76·9	54 44	72 30	75·6	54 28	70 50	74·2	236	304
57	123	56 31	79 05	80·9	56 21	77 18	79·4	56 09	75 32	77·9	55 56	73 47	76·5	55 41	72 04	75·0	55 25	70 22	73·6	237	303
58	122	57 30	78 47	80·5	57 19	76 57	79·0	57 07	75 09	77·4	56 53	73 22	75·9	56 38	71 36	74·5	56 21	69 51	73·0	238	302
59	121	58 29	78 28	80·1	58 18	76 35	78·5	58 05	74 44	77·0	57 51	72 54	75·4	57 35	71 06	73·9	57 17	69 19	72·4	239	301
60	120	59 28	78 08	79·7	59 16	76 12	78·1	59 03	74 18	76·4	58 48	72 25	74·8	58 32	70 34	73·3	58 13	68 45	71·7	240	300
61	119	60 26	77 46	79·3	60 14	75 47	77·6	60 01	73 50	75·9	59 45	71 54	74·2	59 28	70 01	72·6	59 09	68 09	71·0	241	299
62	118	61 25	77 23	78·9	61 12	75 21	77·1	60 58	73 20	75·3	60 42	71 21	73·6	60 24	69 25	71·9	60 05	67 31	70·3	242	298
63	117	62 23	76 58	78·4	62 10	74 52	76·5	61 56	72 48	74·7	61 39	70 46	72·9	61 20	68 46	71·2	61 00	66 49	69·5	243	297
64	116	63 22	76 31	77·9	63 08	74 21	76·0	62 53	72 13	74·1	62 35	70 08	72·2	62 16	68 05	70·4	61 55	66 05	68·6	244	296
65	115	64 20	76 02	77·4	64 06	73 48	75·4	63 50	71 36	73·4	63 32	69 27	71·5	63 12	67 21	69·6	62 50	65 18	67·7	245	295
66	114	65 18	75 31	76·8	65 03	73 12	74·7	64 47	70 56	72·6	64 28	68 43	70·6	64 07	66 34	68·7	63 44	64 27	66·8	246	294
67	113	66 16	74 57	76·2	66 01	72 33	74·0	65 43	70 13	71·8	65 23	67 56	69·8	65 02	65 43	67·8	64 38	63 33	65·8	247	293
68	112	67 14	74 20	75·5	66 58	71 51	73·2	66 40	69 26	71·0	66 19	67 05	68·8	65 56	64 48	66·7	65 32	62 35	64·7	248	292
69	111	68 12	73 39	74·8	67 55	71 05	72·4	67 36	68 35	70·1	67 14	66 09	67·8	66 50	63 48	65·7	66 25	61 31	63·6	249	291
70	110	69 09	72 55	74·0	68 51	70 15	71·5	68 31	67 40	69·1	68 09	65 09	66·7	67 44	62 44	64·5	67 17	60 23	62·3	250	290
71	109	70 07	72 06	73·1	69 48	69 20	70·5	69 27	66 39	68·0	69 03	64 03	65·6	68 37	61 34	63·2	68 09	59 10	61·0	251	289
72	108	71 03	71 13	72·2	70 44	68 20	69·4	70 21	65 33	66·8	69 57	62 52	64·3	69 29	60 17	61·9	69 00	57 50	59·6	252	288
73	107	72 00	70 14	71·1	71 39	67 13	68·3	71 16	64 20	65·5	70 50	61 33	62·9	70 21	58 54	60·4	69 50	56 23	58·0	253	287
74	106	72 56	69 08	70·0	72 34	65 59	67·0	72 09	62 59	64·1	71 42	60 07	61·4	71 12	57 24	58·8	70 40	54 49	56·4	254	286
75	105	73 52	67 54	68·7	73 29	64 37	65·5	73 03	61 30	62·6	72 34	58 32	59·7	72 02	55 44	57·1	71 28	53 06	54·5	255	285
76	104	74 48	66 31	67·3	74 23	63 05	64·0	73 55	59 51	60·8	73 24	56 47	57·9	72 51	53 55	55·1	72 16	51 13	52·6	256	284
77	103	75 42	64 57	65·6	75 16	61 22	62·2	74 46	58 00	58·9	74 14	54 51	55·9	73 39	51 55	53·1	73 02	49 10	50·4	257	283
78	102	76 36	63 11	63·8	76 08	59 26	60·2	75 37	55 57	56·8	75 02	52 42	53·6	74 26	49 42	50·8	73 47	46 56	48·1	258	282
79	101	77 29	61 09	61·7	76 59	57 14	57·9	76 26	53 38	54·4	75 49	50 18	51·2	75 11	47 16	48·2	74 30	44 28	45·5	259	281
80	100	78 21	58 49	59·3	77 49	54 44	55·3	77 13	51 01	51·7	76 35	47 38	48·4	75 54	44 34	45·4	75 11	41 47	42·7	260	280
81	99	79 12	56 06	56·6	78 37	51 52	52·4	77 59	48 04	48·7	77 18	44 39	45·4	76 35	41 35	42·4	75 49	38 50	39·7	261	279
82	98	80 01	52 56	53·4	79 23	48 35	49·1	78 42	44 43	45·3	77 59	41 18	41·9	77 13	38 17	39·0	76 26	35 36	36·4	262	278
83	97	80 47	49 13	49·6	80 07	44 47	45·2	79 23	40 56	41·4	78 37	37 35	38·1	77 49	34 39	35·3	76 59	32 05	32·8	263	277
84	96	81 31	44 51	45·2	80 47	40 24	40·8	80 01	36 38	37·1	79 12	33 25	33·9	78 21	30 40	31·2	77 29	28 16	28·8	264	276
85	95	82 12	39 40	39·9	81 24	35 22	35·7	80 34	31 48	32·2	79 43	28 49	29·2	78 50	26 18	26·7	77 56	24 09	24·6	265	275
86	94	82 48	33 34	33·8	81 57	29 36	29·8	81 04	26 24	26·7	80 09	23 46	24·1	79 14	21 35	21·9	78 18	19 44	20·1	266	274
87	93	83 18	26 28	26·6	82 23	23 05	23·3	81 28	20 25	20·6	80 31	18 17	18·5	79 34	16 32	16·8	78 36	15 04	15·4	267	273
88	92	83 41	18 22	18·5	82 43	15 52	16·0	81 45	13 57	14·1	80 47	12 26	12·6	79 48	11 12	11·4	78 49	10 11	10·4	268	272
89	91	83 55	9 26	9·5	82 56	8 05	8·2	81 56	7 05	7·1	80 57	6 17	6·4	79 57	5 39	5·7	78 57	5 08	5·2	269	271
90	90	84 00	0 00	0·0	83 00	0 00	0·0	82 00	0 00	0·0	81 00	0 00	0·0	80 00	0 00	0·0	79 00	0 00	0·0	270	270

N. Lat: for LHA > 180° … $Z_n = Z$
for LHA < 180° … $Z_n = 360° - Z$

S. Lat.: for LHA > 180° … $Z_n = 180° - Z$
for LHA < 180° … $Z_n = 180° + Z$

SIGHT REDUCTION TABLE

B: (−) for 90° < LHA < 270°
Dec:(−) for Lat. contrary name

Z₁ : same sign as B
Z₂: (−) for F > 90°

Lat. / A		12°			13°			14°			15°			16°			17°			Lat. / A	
LHA/F		A/H	B/P	Z_1/Z_2	A/H	B/P	Z_1/Z_2	A/H	B/P	Z_1/Z_2	A/H	B/P	Z_1/Z_2	A/H	B/P	Z_1/Z_2	A/H	B/P	Z_1/Z_2	LHA	
°	°	° ′	° ′	°	° ′	° ′	°	° ′	° ′	°	° ′	° ′	°	° ′	° ′	°	° ′	° ′	°	°	°
0	180	0 00	78 00	90·0	0 00	77 00	90·0	0 00	76 00	90·0	0 00	75 00	90·0	0 00	74 00	90·0	0 00	73 00	90·0	180	360
1	179	0 59	78 00	89·8	0 58	77 00	89·8	0 58	76 00	89·8	0 58	75 00	89·7	0 58	74 00	89·7	0 57	73 00	89·7	181	359
2	178	1 57	78 00	89·6	1 57	77 00	89·5	1 56	76 00	89·5	1 56	74 59	89·5	1 55	73 59	89·4	1 55	72 59	89·4	182	358
3	177	2 56	77 59	89·4	2 55	76 59	89·3	2 55	75 59	89·3	2 54	74 59	89·2	2 53	73 59	89·2	2 52	72 59	89·1	183	357
4	176	3 55	77 58	89·2	3 54	76 58	89·1	3 53	75 58	89·0	3 52	74 58	89·0	3 51	73 58	88·9	3 49	72 58	88·8	184	356
5	175	4 53	77 57	89·0	4 52	76 57	88·9	4 51	75 57	88·8	4 50	74 57	88·7	4 48	73 57	88·6	4 47	72 56	88·5	185	355
6	174	5 52	77 56	88·7	5 51	76 56	88·6	5 49	75 56	88·5	5 48	74 55	88·4	5 46	73 55	88·3	5 44	72 55	88·2	186	354
7	173	6 51	77 55	88·5	6 49	76 54	88·4	6 47	75 54	88·3	6 46	74 54	88·2	6 44	73 53	88·1	6 42	72 53	87·9	187	353
8	172	7 49	77 53	88·3	7 48	76 53	88·2	7 46	75 52	88·1	7 44	74 52	87·9	7 41	73 51	87·8	7 39	72 51	87·6	188	352
9	171	8 48	77 51	88·1	8 46	76 51	88·0	8 44	75 50	87·8	8 41	74 49	87·7	8 39	73 49	87·5	8 36	72 48	87·3	189	351
10	170	9 47	77 49	87·9	9 44	76 48	87·7	9 42	75 48	87·6	9 39	74 47	87·4	9 37	73 46	87·2	9 34	72 45	87·0	190	350
11	169	10 45	77 47	87·7	10 43	76 46	87·5	10 40	75 45	87·3	10 37	74 44	87·1	10 34	73 43	86·9	10 31	72 42	86·7	191	349
12	168	11 44	77 44	87·5	11 41	76 43	87·3	11 38	75 42	87·1	11 35	74 41	86·9	11 32	73 40	86·6	11 28	72 39	86·4	192	348
13	167	12 43	77 42	87·3	12 40	76 40	87·0	12 36	75 39	86·8	12 33	74 37	86·6	12 29	73 36	86·4	12 25	72 35	86·1	193	347
14	166	13 41	77 39	87·0	13 38	76 37	86·8	13 35	75 35	86·5	13 31	74 34	86·3	13 27	73 32	86·1	13 23	72 31	85·8	194	346
15	165	14 40	77 35	86·8	14 36	76 33	86·6	14 33	75 32	86·3	14 29	74 30	86·0	14 24	73 28	85·8	14 20	72 26	85·5	195	345
16	164	15 38	77 32	86·6	15 35	76 30	86·3	15 31	75 28	86·0	15 26	74 25	85·8	15 22	73 23	85·5	15 17	72 21	85·2	196	344
17	163	16 37	77 28	86·4	16 33	76 26	86·1	16 29	75 23	85·8	16 24	74 21	85·5	16 19	73 19	85·2	16 14	72 16	84·9	197	343
18	162	17 36	77 24	86·1	17 31	76 21	85·8	17 27	75 19	85·5	17 22	74 16	85·2	17 17	73 13	84·9	17 11	72 11	84·6	198	342
19	161	18 34	77 20	85·9	18 30	76 17	85·6	18 25	75 14	85·2	18 20	74 11	84·9	18 14	73 08	84·6	18 08	72 05	84·3	199	341
20	160	19 33	77 15	85·7	19 28	76 12	85·3	19 23	75 08	85·0	19 17	74 05	84·6	19 12	73 02	84·3	19 05	71 59	83·9	200	340
21	159	20 31	77 10	85·4	20 26	76 07	85·1	20 21	75 03	84·7	20 15	73 59	84·3	20 09	72 56	84·0	20 03	71 52	83·6	201	339
22	158	21 30	77 05	85·2	21 24	76 01	84·8	21 19	74 57	84·4	21 13	73 53	84·0	21 06	72 49	83·6	21 00	71 45	83·3	202	338
23	157	22 28	77 00	85·0	22 23	75 55	84·5	22 17	74 51	84·1	22 10	73 46	83·7	22 04	72 42	83·3	21 56	71 38	82·9	203	337
24	156	23 27	76 54	84·7	23 21	75 49	84·3	23 15	74 44	83·9	23 08	73 39	83·4	23 01	72 34	83·0	22 53	71 30	82·6	204	336
25	155	24 25	76 48	84·5	24 19	75 43	84·0	24 13	74 37	83·6	24 06	73 32	83·1	23 58	72 27	82·7	23 50	71 22	82·2	205	335
26	154	25 23	76 42	84·2	25 17	75 36	83·7	25 10	74 30	83·3	25 03	73 24	82·8	24 55	72 18	82·3	24 47	71 13	81·9	206	334
27	153	26 22	76 35	84·0	26 15	75 28	83·5	26 08	74 22	83·0	26 01	73 16	82·5	25 52	72 10	82·0	25 44	71 04	81·5	207	333
28	152	27 20	76 28	83·7	27 13	75 21	83·2	27 06	74 14	82·7	26 58	73 07	82·2	26 50	72 00	81·7	26 41	70 54	81·2	208	332
29	151	28 18	76 20	83·4	28 11	75 13	82·9	28 04	74 05	82·4	27 55	72 58	81·8	27 47	71 51	81·3	27 37	70 44	80·8	209	331
30	150	29 17	76 13	83·2	29 09	75 04	82·6	29 01	73 56	82·0	28 53	72 48	81·5	28 44	71 41	81·0	28 34	70 33	80·4	210	330
31	149	30 15	76 04	82·9	30 07	74 56	82·3	29 59	73 47	81·7	29 50	72 38	81·2	29 41	71 30	80·6	29 30	70 22	80·0	211	329
32	148	31 13	75 56	82·6	31 05	74 46	82·0	30 57	73 37	81·4	30 47	72 28	80·8	30 37	71 19	80·2	30 27	70 11	79·6	212	328
33	147	32 11	75 47	82·3	32 03	74 37	81·7	31 54	73 27	81·1	31 44	72 17	80·5	31 34	71 07	79·9	31 23	69 58	79·2	213	327
34	146	33 10	75 37	82·0	33 01	74 26	81·4	32 52	73 16	80·7	32 42	72 05	80·1	32 31	70 55	79·5	32 20	69 45	78·8	214	326
35	145	34 08	75 27	81·7	33 59	74 16	81·0	33 49	73 04	80·4	33 39	71 53	79·7	33 28	70 42	79·1	33 16	69 32	78·4	215	325
36	144	35 06	75 17	81·4	34 56	74 04	80·7	34 46	72 52	80·0	34 36	71 40	79·4	34 24	70 29	78·7	34 12	69 18	78·0	216	324
37	143	36 04	75 06	81·1	35 54	73 53	80·4	35 44	72 40	79·7	35 33	71 27	79·0	35 21	70 15	78·3	35 08	69 03	77·6	217	323
38	142	37 02	74 54	80·8	36 52	73 40	80·0	36 41	72 27	79·3	36 29	71 13	78·6	36 17	70 00	77·8	36 04	68 48	77·1	218	322
39	141	38 00	74 42	80·4	37 49	73 27	79·7	37 38	72 13	78·9	37 26	70 59	78·2	37 13	69 45	77·4	37 00	68 32	76·7	219	321
40	140	38 57	74 30	80·1	38 47	73 14	79·3	38 35	71 58	78·5	38 23	70 43	77·7	38 10	69 29	77·0	37 56	68 15	76·2	220	320
41	139	39 55	74 16	79·8	39 44	72 59	78·9	39 32	71 43	78·1	39 19	70 27	77·3	39 06	69 12	76·5	38 51	67 57	75·7	221	319
42	138	40 53	74 02	79·4	40 41	72 45	78·5	40 29	71 27	77·7	40 16	70 10	76·9	40 02	68 54	76·1	39 47	67 38	75·3	222	318
43	137	41 51	73 48	79·0	41 39	72 29	78·2	41 26	71 11	77·3	41 12	69 53	76·4	40 58	68 35	75·6	40 42	67 19	74·7	223	317
44	136	42 48	73 32	78·6	42 36	72 12	77·7	42 23	70 53	76·9	42 09	69 34	76·0	41 54	68 16	75·1	41 38	66 58	74·2	224	316
45	135	43 46	73 16	78·3	43 33	71 55	77·3	43 19	70 35	76·4	43 05	69 15	75·5	42 49	67 56	74·6	42 33	66 37	73·7	225	315

Lat. / A		12°			13°			14°			15°			16°			17°			Lat. / A	
LHA/F		A/H	B/P	Z_1/Z_2	A/H	B/P	Z_1/Z_2	A/H	B/P	Z_1/Z_2	A/H	B/P	Z_1/Z_2	A/H	B/P	Z_1/Z_2	A/H	B/P	Z_1/Z_2	LHA	
°	°	° ′	° ′	°	° ′	° ′	°	° ′	° ′	°	° ′	° ′	°	° ′	° ′	°	° ′	° ′	°	°	°
45	135	43 46	73 16	78·3	43 33	71 55	77·3	43 19	70 35	76·4	43 05	69 15	75·5	42 49	67 56	74·6	42 33	66 37	73·7	225	315
46	134	44 43	72 59	77·8	44 30	71 37	76·9	44 16	70 15	75·9	44 01	68 54	75·0	43 45	67 34	74·1	43 28	66 15	73·2	226	314
47	133	45 40	72 41	77·4	45 27	71 18	76·4	45 12	69 55	75·5	44 57	68 33	74·5	44 40	67 12	73·5	44 23	65 51	72·6	227	313
48	132	46 38	72 23	77·0	46 24	70 58	76·0	46 09	69 34	75·0	45 53	68 11	74·0	45 35	66 48	73·0	45 17	65 27	72·0	228	312
49	131	47 35	72 03	76·5	47 20	70 37	75·5	47 05	69 11	74·4	46 48	67 47	73·4	46 30	66 23	72·4	46 12	65 01	71·4	229	311
50	130	48 32	71 42	76·1	48 17	70 15	75·0	48 01	68 48	73·9	47 44	67 22	72·9	47 25	65 58	71·8	47 06	64 34	70·8	230	310
51	129	49 29	71 20	75·6	49 13	69 51	74·5	48 57	68 23	73·4	48 39	66 56	72·3	48 20	65 30	71·2	48 00	64 05	70·1	231	309
52	128	50 25	70 57	75·1	50 09	69 27	73·9	49 52	67 57	72·8	49 34	66 29	71·7	49 15	65 02	70·6	48 54	63 35	69·5	232	308
53	127	51 22	70 33	74·6	51 06	69 01	73·4	50 48	67 30	72·2	50 29	66 00	71·0	50 09	64 31	69·9	49 48	63 04	68·8	233	307
54	126	52 19	70 07	74·0	52 02	68 33	72·8	51 43	67 01	71·6	51 24	65 30	70·4	51 03	64 00	69·2	50 41	62 31	68·1	234	306
55	125	53 15	69 40	73·5	52 57	68 04	72·2	52 38	66 30	70·9	52 18	64 58	69·7	51 57	63 26	68·5	51 34	61 56	67·3	235	305
56	124	54 11	69 11	72·9	53 53	67 34	71·6	53 33	65 58	70·3	53 12	64 24	69·0	52 50	62 51	67·8	52 27	61 20	66·6	236	304
57	123	55 07	68 41	72·2	54 48	67 02	70·9	54 28	65 24	69·6	54 06	63 48	68·3	53 43	62 14	67·0	53 19	60 42	65·8	237	303
58	122	56 03	68 09	71·6	55 43	66 28	70·2	55 22	64 48	68·8	55 00	63 11	67·5	54 36	61 35	66·2	54 12	60 01	64·9	238	302
59	121	56 59	67 34	70·9	56 38	65 51	69·5	56 16	64 10	68·1	55 53	62 31	66·7	55 29	60 54	65·4	55 03	59 18	64·1	239	301
60	120	57 54	66 58	70·2	57 33	65 13	68·7	57 10	63 30	67·3	56 46	61 49	65·9	56 21	60 10	64·5	55 55	58 33	63·1	240	300
61	119	58 49	66 20	69·4	58 27	64 32	67·9	58 04	62 47	66·4	57 39	61 04	65·0	57 13	59 24	63·6	56 46	57 46	62·2	241	299
62	118	59 44	65 38	68·6	59 21	63 49	67·1	58 57	62 02	65·5	58 31	60 17	64·0	58 05	58 35	62·6	57 36	56 56	61·2	242	298
63	117	60 38	64 55	67·8	60 15	63 03	66·2	59 50	61 13	64·6	59 23	59 27	63·1	58 55	57 43	61·6	58 26	56 03	60·2	243	297
64	116	61 32	64 08	66·9	61 08	62 14	65·2	60 42	60 22	63·6	60 15	58 34	62·0	59 46	56 49	60·5	59 16	55 06	59·1	244	296
65	115	62 26	63 18	66·0	62 01	61 21	64·2	61 34	59 28	62·6	61 06	57 37	61·0	60 36	55 51	59·4	60 05	54 07	57·9	245	295
66	114	63 20	62 25	65·0	62 53	60 25	63·2	62 26	58 30	61·5	61 56	56 37	59·8	61 25	54 49	58·2	60 53	53 04	56·7	246	294
67	113	64 13	61 27	63·9	63 45	59 25	62·1	63 16	57 27	60·3	62 46	55 34	58·6	62 14	53 44	57·0	61 41	51 57	55·4	247	293
68	112	65 05	60 26	62·8	64 37	58 21	60·9	64 07	56 21	59·1	63 35	54 25	57·4	63 02	52 34	55·7	62 27	50 47	54·1	248	292
69	111	65 57	59 20	61·6	65 27	57 13	59·6	64 56	55 10	57·8	64 23	53 13	56·0	63 49	51 20	54·3	63 14	49 32	52·7	249	291
70	110	66 48	58 08	60·3	66 18	55 59	58·3	65 45	53 55	56·4	65 11	51 55	54·6	64 36	50 01	52·9	63 59	48 12	51·2	250	290
71	109	67 39	56 52	58·9	67 07	54 40	56·8	66 33	52 33	54·9	65 58	50 33	53·1	65 21	48 38	51·3	64 43	46 48	49·7	251	289
72	108	68 29	55 29	57·4	67 55	53 14	55·3	67 20	51 06	53·3	66 44	49 04	51·5	66 06	47 08	49·7	65 26	45 18	48·0	252	288
73	107	69 18	53 59	55·8	68 43	51 42	53·7	68 07	49 33	51·6	67 29	47 30	49·8	66 49	45 33	48·0	66 08	43 43	46·3	253	287
74	106	70 06	52 22	54·1	69 30	50 03	51·9	68 52	47 52	49·8	68 12	45 49	47·9	67 31	43 52	46·1	66 49	42 02	44·4	254	286
75	105	70 53	50 36	52·2	70 15	48 16	50·0	69 36	46 04	47·9	68 55	44 00	46·0	68 12	42 04	44·2	67 29	40 15	42·5	255	285
76	104	71 38	48 42	50·2	70 59	46 20	47·9	70 18	44 08	45·9	69 36	42 05	43·9	68 52	40 09	42·1	68 07	38 21	40·5	256	284
77	103	72 23	46 37	48·0	71 42	44 15	45·7	70 59	42 03	43·7	70 15	40 01	41·7	69 30	38 07	39·9	68 43	36 21	38·3	257	283
78	102	73 06	44 22	45·6	72 23	42 00	43·4	71 38	39 49	41·3	70 53	37 49	39·4	70 06	35 57	37·6	69 18	34 13	36·0	258	282
79	101	73 47	41 55	43·1	73 02	39 34	40·8	72 16	37 26	38·8	71 28	35 27	36·9	70 40	33 38	35·2	69 50	31 58	33·6	259	281
80	100	74 26	39 15	40·3	73 39	36 57	38·1	72 51	34 51	36·1	72 02	32 57	34·3	71 12	31 12	32·6	70 21	29 36	31·1	260	280
81	99	75 02	36 21	37·3	74 14	34 07	35·1	73 24	32 06	33·2	72 34	30 17	31·5	71 42	28 37	29·9	70 50	27 06	28·4	261	279
82	98	75 37	33 13	34·1	74 46	31 05	32·0	73 55	29 10	30·2	73 03	27 27	28·5	72 09	25 53	27·0	71 16	24 29	25·7	262	278
83	97	76 08	29 50	30·6	75 16	27 50	28·6	74 23	26 03	26·9	73 29	24 27	25·4	72 34	23 02	24·0	71 39	21 44	22·8	263	277
84	96	76 36	26 11	26·8	75 42	24 22	25·0	74 48	22 45	23·5	73 52	21 19	22·1	72 56	20 02	20·9	72 00	18 53	19·8	264	276
85	95	77 01	22 18	22·8	76 05	20 41	21·3	75 09	19 16	19·9	74 12	18 01	18·7	73 15	16 54	17·6	72 18	15 55	16·7	265	275
86	94	77 22	18 10	18·6	76 25	16 49	17·3	75 27	15 38	16·1	74 29	14 36	15·1	73 31	13 40	14·2	72 33	12 51	13·5	266	274
87	93	77 38	13 50	14·1	76 40	12 46	13·1	75 41	11 51	12·2	74 43	11 03	11·4	73 44	10 21	10·8	72 45	9 43	10·2	267	273
88	92	77 50	9 19	9·5	76 51	8 36	8·8	75 52	7 58	8·2	74 52	7 25	7·7	73 53	6 56	7·2	72 53	6 31	6·8	268	272
89	91	77 58	4 42	4·8	76 58	4 19	4·4	75 58	4 00	4·1	74 58	3 44	3·9	73 58	3 29	3·6	72 58	3 16	3·4	269	271
90	90	78 00	0 00	0·0	77 00	0 00	0·0	76 00	0 00	0·0	75 00	0 00	0·0	74 00	0 00	0·0	73 00	0 00	0·0	270	270

N. Lat: for LHA > 180° … $Z_n = Z$
for LHA < 180° … $Z_n = 360° - Z$

S. Lat.: for LHA > 180° … $Z_n = 180° - Z$
for LHA < 180° … $Z_n = 180° + Z$

SIGHT REDUCTION TABLE

B: (–) for 90° < LHA < 270°
Dec: (–) for Lat. contrary name

Z_1: same sign as B
Z_2: (–) for F > 90°

Lat. / A		18°			19°			20°			21°			22°			23°			Lat. / A	
LHA/F		A/H	B/P	Z_1/Z_2	A/H	B/P	Z_1/Z_2	A/H	B/P	Z_1/Z_2	A/H	B/P	Z_1/Z_2	A/H	B/P	Z_1/Z_2	A/H	B/P	Z_1/Z_2	LHA	
°	°	° ′	° ′	°	° ′	° ′	°	° ′	° ′	°	° ′	° ′	°	° ′	° ′	°	° ′	° ′	°	°	°
0	180	0 00	72 00	90·0	0 00	71 00	90·0	0 00	70 00	90·0	0 00	69 00	90·0	0 00	68 00	90·0	0 00	67 00	90·0	180	360
1	179	0 57	72 00	89·7	0 57	71 00	89·7	0 56	70 00	89·7	0 56	69 00	89·6	0 56	68 00	89·6	0 55	67 00	89·6	181	359
2	178	1 54	71 59	89·4	1 53	70 59	89·3	1 53	69 59	89·3	1 52	68 59	89·3	1 51	67 59	89·3	1 50	66 59	89·2	182	358
3	177	2 51	71 59	89·1	2 50	70 59	89·0	2 49	69 58	89·0	2 48	68 58	88·9	2 47	67 58	88·9	2 46	66 58	88·8	183	357
4	176	3 48	71 58	88·8	3 47	70 57	88·7	3 46	69 57	88·6	3 44	68 57	88·6	3 42	67 57	88·5	3 41	66 57	88·4	184	356
5	175	4 45	71 56	88·5	4 44	70 56	88·4	4 42	69 56	88·3	4 40	68 56	88·2	4 38	67 55	88·1	4 36	66 55	88·0	185	355
6	174	5 42	71 54	88·1	5 40	70 54	88·0	5 38	69 54	87·9	5 36	68 54	87·8	5 34	67 53	87·7	5 31	66 53	87·6	186	354
7	173	6 39	71 52	87·8	6 37	70 52	87·7	6 35	69 52	87·6	6 32	68 51	87·5	6 29	67 51	87·4	6 26	66 51	87·3	187	353
8	172	7 36	71 50	87·5	7 34	70 50	87·4	7 31	69 49	87·2	7 28	68 49	87·1	7 25	67 48	87·0	7 22	66 48	86·9	188	352
9	171	8 33	71 47	87·2	8 30	70 47	87·0	8 27	69 46	86·9	8 24	68 46	86·8	8 20	67 45	86·6	8 17	66 45	86·5	189	351
10	170	9 30	71 44	86·9	9 27	70 44	86·7	9 23	69 43	86·5	9 20	68 42	86·4	9 16	67 42	86·2	9 12	66 41	86·1	190	350
11	169	10 27	71 41	86·6	10 24	70 40	86·4	10 20	69 39	86·2	10 16	68 39	86·0	10 11	67 38	85·8	10 07	66 37	85·7	191	349
12	168	11 24	71 37	86·2	11 20	70 36	86·0	11 16	69 35	85·8	11 12	68 34	85·6	11 07	67 33	85·4	11 02	66 32	85·3	192	348
13	167	12 21	71 33	85·9	12 17	70 32	85·7	12 12	69 31	85·5	12 07	68 30	85·3	12 02	67 29	85·1	11 57	66 28	84·8	193	347
14	166	13 18	71 29	85·6	13 13	70 28	85·4	13 08	69 26	85·1	13 03	68 25	84·9	12 58	67 24	84·7	12 52	66 22	84·4	194	346
15	165	14 15	71 24	85·3	14 10	70 23	85·0	14 05	69 21	84·8	13 59	68 20	84·5	13 53	67 18	84·3	13 47	66 17	84·0	195	345
16	164	15 12	71 19	84·9	15 06	70 18	84·7	15 01	69 16	84·4	14 55	68 14	84·1	14 48	67 12	83·9	14 42	66 10	83·6	196	344
17	163	16 09	71 14	84·6	16 03	70 12	84·3	15 57	69 10	84·0	15 50	68 08	83·7	15 44	67 06	83·5	15 37	66 04	83·2	197	343
18	162	17 05	71 08	84·3	16 59	70 06	84·0	16 53	69 03	83·7	16 46	68 01	83·4	16 39	66 59	83·1	16 32	65 57	82·8	198	342
19	161	18 02	71 02	83·9	17 56	69 59	83·6	17 49	68 57	83·3	17 42	67 54	83·0	17 34	66 52	82·7	17 26	65 49	82·3	199	341
20	160	18 59	70 56	83·6	18 52	69 53	83·2	18 45	68 50	82·9	18 37	67 47	82·6	18 29	66 44	82·2	18 21	65 41	81·9	200	340
21	159	19 56	70 49	83·2	19 48	69 45	82·9	19 41	68 42	82·5	19 33	67 39	82·2	19 24	66 36	81·8	19 16	65 33	81·5	201	339
22	158	20 52	70 41	82·9	20 45	69 38	82·5	20 37	68 34	82·1	20 28	67 31	81·8	20 19	66 27	81·4	20 10	65 24	81·0	202	338
23	157	21 49	70 33	82·5	21 41	69 29	82·1	21 32	68 26	81·7	21 24	67 22	81·4	21 14	66 18	81·0	21 05	65 15	80·6	203	337
24	156	22 45	70 25	82·2	22 37	69 21	81·8	22 28	68 17	81·3	22 19	67 12	80·9	22 09	66 09	80·5	21 59	65 05	80·1	204	336
25	155	23 42	70 17	81·8	23 33	69 12	81·4	23 24	68 07	80·9	23 14	67 03	80·5	23 04	65 58	80·1	22 54	64 54	79·7	205	335
26	154	24 38	70 07	81·4	24 29	69 02	81·0	24 20	67 57	80·5	24 09	66 52	80·1	23 59	65 48	79·6	23 48	64 43	79·2	206	334
27	153	25 35	69 58	81·1	25 25	68 52	80·6	25 15	67 47	80·1	25 05	66 42	79·7	24 54	65 36	79·2	24 42	64 32	78·7	207	333
28	152	26 31	69 48	80·7	26 21	68 42	80·2	26 11	67 36	79·7	26 00	66 30	79·2	25 48	65 25	78·7	25 36	64 19	78·3	208	332
29	151	27 27	69 37	80·3	27 17	68 31	79·8	27 06	67 24	79·3	26 55	66 18	78·8	26 43	65 12	78·3	26 30	64 07	77·8	209	331
30	150	28 24	69 26	79·9	28 13	68 19	79·4	28 01	67 12	78·8	27 50	66 06	78·3	27 37	64 59	77·8	27 24	63 53	77·3	210	330
31	149	29 20	69 14	79·5	29 09	68 07	78·9	28 57	67 00	78·4	28 44	65 53	77·8	28 31	64 46	77·3	28 18	63 39	76·8	211	329
32	148	30 16	69 02	79·1	30 04	67 54	78·5	29 52	66 46	77·9	29 39	65 39	77·4	29 26	64 32	76·8	29 12	63 25	76·3	212	328
33	147	31 12	68 49	78·7	31 00	67 41	78·1	30 47	66 32	77·5	30 34	65 24	76·9	30 20	64 17	76·3	30 05	63 09	75·8	213	327
34	146	32 08	68 36	78·2	31 55	67 27	77·6	31 42	66 18	77·0	31 28	65 09	76·4	31 14	64 01	75·8	30 59	62 53	75·2	214	326
35	145	33 04	68 22	77·8	32 51	67 12	77·2	32 37	66 03	76·5	32 23	64 54	75·9	32 08	63 45	75·3	31 52	62 36	74·7	215	325
36	144	33 59	68 07	77·3	33 46	66 57	76·7	33 32	65 47	76·0	33 17	64 37	75·4	33 01	63 28	74·8	32 45	62 19	74·2	216	324
37	143	34 55	67 52	76·9	34 41	66 41	76·2	34 26	65 30	75·5	34 11	64 20	74·9	33 55	63 10	74·2	33 38	62 01	73·6	217	323
38	142	35 50	67 36	76·4	35 36	66 24	75·7	35 21	65 13	75·0	35 05	64 02	74·4	34 48	62 51	73·7	34 31	61 41	73·0	218	322
39	141	36 46	67 19	76·0	36 31	66 06	75·2	36 15	64 54	74·5	35 59	63 43	73·8	35 42	62 32	73·1	35 24	61 21	72·4	219	321
40	140	37 41	67 01	75·5	37 26	65 48	74·7	37 10	64 35	74·0	36 53	63 23	73·3	36 35	62 12	72·6	36 17	61 01	71·8	220	320
41	139	38 36	66 42	75·0	38 20	65 29	74·2	38 04	64 15	73·4	37 46	63 02	72·7	37 28	61 50	72·0	37 09	60 39	71·2	221	319
42	138	39 31	66 23	74·5	39 15	65 08	73·7	38 58	63 54	72·9	38 40	62 41	72·1	38 21	61 28	71·4	38 01	60 16	70·6	222	318
43	137	40 26	66 03	73·9	40 09	64 47	73·1	39 51	63 33	72·3	39 33	62 18	71·5	39 13	61 05	70·7	38 53	59 52	70·0	223	317
44	136	41 21	65 42	73·4	41 03	64 25	72·5	40 45	63 10	71·7	40 26	61 55	70·9	40 06	60 41	70·1	39 45	59 27	69·3	224	316
45	135	42 16	65 19	72·8	41 57	64 02	72·0	41 38	62 46	71·1	41 19	61 30	70·3	40 58	60 15	69·5	40 37	59 01	68·7	225	315

Lat. / A		18°			19°			20°			21°			22°			23°			Lat. / A	
LHA/F		A/H	B/P	Z_1/Z_2	A/H	B/P	Z_1/Z_2	A/H	B/P	Z_1/Z_2	A/H	B/P	Z_1/Z_2	A/H	B/P	Z_1/Z_2	A/H	B/P	Z_1/Z_2	LHA	
°	°	° ′	° ′	°	° ′	° ′	°	° ′	° ′	°	° ′	° ′	°	° ′	° ′	°	° ′	° ′	°	°	°
45	135	42 16	65 19	72·8	41 57	64 02	72·0	41 38	62 46	71·1	41 19	61 30	70·3	40 58	60 15	69·5	40 37	59 01	68·7	225	315
46	134	43 10	64 56	72·3	42 51	63 38	71·4	42 32	62 21	70·5	42 11	61 05	69·6	41 50	59 49	68·8	41 28	58 34	68·0	226	314
47	133	44 04	64 32	71·7	43 45	63 13	70·8	43 25	61 55	69·9	43 04	60 38	69·0	42 42	59 21	68·1	42 19	58 06	67·3	227	313
48	132	44 58	64 06	71·1	44 38	62 46	70·1	44 18	61 27	69·2	43 56	60 09	68·3	43 33	58 53	67·4	43 10	57 37	66·5	228	312
49	131	45 52	63 39	70·4	45 32	62 18	69·5	45 10	60 59	68·5	44 48	59 40	67·6	44 24	58 22	66·7	44 00	57 06	65·8	229	311
50	130	46 46	63 11	69·8	46 25	61 49	68·8	46 03	60 29	67·8	45 39	59 09	66·9	45 15	57 51	65·9	44 50	56 34	65·0	230	310
51	129	47 39	62 42	69·1	47 17	61 19	68·1	46 55	59 57	67·1	46 31	58 37	66·1	46 06	57 18	65·2	45 40	56 00	64·2	231	309
52	128	48 33	62 11	68·4	48 10	60 47	67·4	47 46	59 25	66·4	47 22	58 03	65·4	46 56	56 44	64·4	46 30	55 25	63·4	232	308
53	127	49 25	61 38	67·7	49 02	60 13	66·6	48 38	58 50	65·6	48 13	57 28	64·6	47 46	56 07	63·6	47 19	54 48	62·6	233	307
54	126	50 18	61 04	67·0	49 54	59 38	65·9	49 29	58 14	64·8	49 03	56 51	63·7	48 36	55 30	62·7	48 08	54 10	61·7	234	306
55	125	51 10	60 28	66·2	50 46	59 01	65·1	50 20	57 36	64·0	49 53	56 12	62·9	49 25	54 50	61·9	48 56	53 30	60·8	235	305
56	124	52 03	59 50	65·4	51 37	58 23	64·2	51 10	56 56	63·1	50 43	55 32	62·0	50 14	54 09	61·0	49 44	52 48	59·9	236	304
57	123	52 54	59 11	64·6	52 28	57 42	63·4	52 00	56 15	62·2	51 32	54 49	61·1	51 02	53 26	60·0	50 32	52 04	59·0	237	303
58	122	53 46	58 29	63·7	53 18	56 59	62·5	52 50	55 31	61·3	52 21	54 05	60·2	51 50	52 41	59·1	51 19	51 18	58·0	238	302
59	121	54 37	57 45	62·8	54 08	56 14	61·5	53 39	54 45	60·4	53 09	53 18	59·2	52 38	51 53	58·1	52 06	50 30	57·0	239	301
60	120	55 27	56 59	61·8	54 58	55 27	60·6	54 28	53 57	59·4	53 57	52 29	58·2	53 25	51 04	57·0	52 52	49 40	55·9	240	300
61	119	56 17	56 10	60·9	55 47	54 37	59·6	55 16	53 06	58·3	54 44	51 38	57·1	54 11	50 12	55·9	53 37	48 48	54·8	241	299
62	118	57 07	55 19	59·8	56 36	53 45	58·5	56 04	52 13	57·2	55 31	50 44	56·0	54 57	49 17	54·8	54 22	47 53	53·7	242	298
63	117	57 56	54 25	58·8	57 24	52 49	57·4	56 51	51 17	56·1	56 17	49 47	54·9	55 42	48 20	53·7	55 06	46 55	52·5	243	297
64	116	58 44	53 27	57·6	58 12	51 51	56·3	57 38	50 18	55·0	57 03	48 48	53·7	56 27	47 20	52·5	55 50	45 55	51·3	244	296
65	115	59 32	52 27	56·5	58 58	50 50	55·1	58 24	49 16	53·7	57 47	47 45	52·5	57 10	46 17	51·2	56 32	44 52	50·0	245	295
66	114	60 19	51 23	55·2	59 45	49 45	53·8	59 09	48 11	52·5	58 32	46 39	51·2	57 53	45 11	49·9	57 14	43 47	48·7	246	294
67	113	61 06	50 15	53·9	60 30	48 37	52·5	59 53	47 02	51·1	59 15	45 30	49·8	58 36	44 02	48·6	57 55	42 38	47·4	247	293
68	112	61 52	49 04	52·6	61 15	47 25	51·1	60 36	45 50	49·8	59 57	44 18	48·4	59 17	42 50	47·2	58 36	41 26	46·0	248	292
69	111	62 37	47 48	51·2	61 58	46 09	49·7	61 19	44 33	48·3	60 39	43 02	47·0	59 57	41 34	45·7	59 15	40 10	44·5	249	291
70	110	63 21	46 28	49·7	62 41	44 48	48·2	62 01	43 13	46·8	61 19	41 42	45·4	60 36	40 15	44·2	59 53	38 52	43·0	250	290
71	109	64 04	45 03	48·1	63 23	43 24	46·6	62 41	41 49	45·2	61 58	40 18	43·9	61 15	38 52	42·6	60 30	37 29	41·4	251	289
72	108	64 45	43 34	46·4	64 04	41 54	44·9	63 21	40 20	43·5	62 37	38 50	42·2	61 52	37 25	40·9	61 06	36 03	39·7	252	288
73	107	65 26	41 59	44·7	64 43	40 20	43·2	63 59	38 46	41·8	63 14	37 18	40·5	62 27	35 53	39·2	61 41	34 34	38·0	253	287
74	106	66 06	40 19	42·9	65 21	38 41	41·4	64 36	37 08	40·0	63 49	35 41	38·7	63 02	34 18	37·4	62 14	33 00	36·3	254	286
75	105	66 44	38 32	40·9	65 58	36 56	39·5	65 11	35 25	38·1	64 23	33 59	36·8	63 35	32 39	35·6	62 46	31 22	34·4	255	285
76	104	67 20	36 40	38·9	66 33	35 05	37·4	65 45	33 37	36·1	64 56	32 13	34·8	64 07	30 55	33·6	63 16	29 41	32·5	256	284
77	103	67 55	34 42	36·8	67 07	33 09	35·3	66 18	31 43	34·0	65 27	30 22	32·8	64 37	29 06	31·6	63 45	27 55	30·6	257	283
78	102	68 29	32 37	34·5	67 39	31 07	33·1	66 48	29 44	31·9	65 57	28 26	30·7	65 05	27 14	29·6	64 13	26 06	28·5	258	282
79	101	69 00	30 25	32·2	68 09	29 00	30·8	67 17	27 40	29·6	66 25	26 26	28·5	65 32	25 17	27·4	64 38	24 12	26·4	259	281
80	100	69 29	28 07	29·7	68 37	26 46	28·4	67 44	25 30	27·3	66 50	24 20	26·2	65 56	23 15	25·2	65 02	22 15	24·3	260	280
81	99	69 57	25 43	27·1	69 03	24 26	25·9	68 09	23 15	24·8	67 14	22 10	23·8	66 19	21 10	22·9	65 23	20 14	22·1	261	279
82	98	70 21	23 11	24·5	69 27	22 00	23·3	68 31	20 56	22·3	67 36	19 56	21·4	66 40	19 00	20·6	65 43	18 09	19·8	262	278
83	97	70 44	20 34	21·7	69 48	19 29	20·7	68 51	18 31	19·7	67 55	17 37	18·9	66 58	16 47	18·1	66 01	16 01	17·4	263	277
84	96	71 03	17 50	18·8	70 07	16 53	17·9	69 09	16 01	17·1	68 12	15 14	16·3	67 14	14 30	15·7	66 16	13 50	15·1	264	276
85	95	71 20	15 01	15·8	70 23	14 12	15·0	69 25	13 28	14·3	68 26	12 48	13·7	67 28	12 10	13·1	66 29	11 36	12·6	265	275
86	94	71 35	12 07	12·8	70 36	11 27	12·1	69 37	10 51	11·6	68 38	10 18	11·0	67 39	9 48	10·6	66 40	9 20	10·1	266	274
87	93	71 46	9 09	9·6	70 46	8 39	9·1	69 47	8 11	8·7	68 48	7 46	8·3	67 48	7 23	8·0	66 49	7 02	7·6	267	273
88	92	71 54	6 08	6·4	70 54	5 47	6·1	69 54	5 29	5·8	68 55	5 12	5·6	67 55	4 56	5·3	66 55	4 42	5·1	268	272
89	91	71 58	3 04	3·2	70 58	2 54	3·1	69 59	2 45	2·9	68 59	2 36	2·8	67 59	2 28	2·7	66 59	2 21	2·6	269	271
90	90	72 00	0 00	0·0	71 00	0 00	0·0	70 00	0 00	0·0	69 00	0 00	0·0	68 00	0 00	0·0	67 00	0 00	0·0	270	270

N. Lat: for LHA > 180° … $Z_n = Z$
for LHA < 180° … $Z_n = 360° - Z$

S. Lat.: for LHA > 180° … $Z_n = 180° - Z$
for LHA < 180° … $Z_n = 180° + Z$

SIGHT REDUCTION TABLE

B: (–) for 90° < LHA < 270°
Dec:(–) for Lat. contrary name

Z_1: same sign as B
Z_2: (–) for F > 90°

Lat. / A		24°			25°			26°			27°			28°			29°			Lat. / A	
LHA/F		A/H	B/P	Z_1/Z_2	A/H	B/P	Z_1/Z_2	A/H	B/P	Z_1/Z_2	A/H	B/P	Z_1/Z_2	A/H	B/P	Z_1/Z_2	A/H	B/P	Z_1/Z_2	LHA	
°	°	° ′	° ′	°	° ′	° ′	°	° ′	° ′	°	° ′	° ′	°	° ′	° ′	°	° ′	° ′	°	°	°
0	180	0 00	66 00	90·0	0 00	65 00	90·0	0 00	64 00	90·0	0 00	63 00	90·0	0 00	62 00	90·0	0 00	61 00	90·0	180	360
1	179	0 55	66 00	89·6	0 54	65 00	89·6	0 54	64 00	89·6	0 53	63 00	89·5	0 53	62 00	89·5	0 52	61 00	89·5	181	359
2	178	1 50	65 59	89·2	1 49	64 59	89·2	1 48	63 59	89·1	1 47	62 59	89·1	1 46	61 59	89·1	1 45	60 59	89·0	182	358
3	177	2 44	65 58	88·8	2 43	64 58	88·7	2 42	63 58	88·7	2 40	62 58	88·6	2 39	61 58	88·6	2 37	60 58	88·5	183	357
4	176	3 39	65 57	88·4	3 37	64 57	88·3	3 36	63 57	88·2	3 34	62 57	88·2	3 32	61 57	88·1	3 30	60 56	88·1	184	356
5	175	4 34	65 55	88·0	4 32	64 55	87·9	4 30	63 55	87·8	4 27	62 55	87·7	4 25	61 55	87·6	4 22	60 54	87·6	185	355
6	174	5 29	65 53	87·6	5 26	64 53	87·5	5 23	63 53	87·4	5 21	62 52	87·3	5 18	61 52	87·2	5 15	60 52	87·1	186	354
7	173	6 24	65 50	87·1	6 20	64 50	87·0	6 17	63 50	86·9	6 14	62 50	86·8	6 11	61 49	86·7	6 07	60 49	86·6	187	353
8	172	7 18	65 47	86·7	7 15	64 47	86·6	7 11	63 47	86·5	7 07	62 46	86·3	7 04	61 46	86·2	6 59	60 46	86·1	188	352
9	171	8 13	65 44	86·3	8 09	64 44	86·2	8 05	63 43	86·0	8 01	62 43	85·9	7 56	61 42	85·7	7 52	60 42	85·6	189	351
10	170	9 08	65 40	85·9	9 03	64 40	85·7	8 59	63 39	85·6	8 54	62 39	85·4	8 49	61 38	85·3	8 44	60 38	85·1	190	350
11	169	10 02	65 36	85·5	9 57	64 35	85·3	9 52	63 35	85·1	9 47	62 34	85·0	9 42	61 33	84·8	9 36	60 33	84·6	191	349
12	168	10 57	65 32	85·1	10 52	64 31	84·9	10 46	63 30	84·7	10 41	62 29	84·5	10 35	61 28	84·3	10 29	60 28	84·1	192	348
13	167	11 52	65 27	84·6	11 46	64 26	84·4	11 40	63 25	84·2	11 34	62 24	84·0	11 27	61 23	83·8	11 21	60 22	83·6	193	347
14	166	12 46	65 21	84·2	12 40	64 20	84·0	12 34	63 19	83·8	12 27	62 18	83·5	12 20	61 17	83·3	12 13	60 16	83·1	194	346
15	165	13 41	65 15	83·8	13 34	64 14	83·5	13 27	63 13	83·3	13 20	62 11	83·1	13 13	61 10	82·8	13 05	60 09	82·6	195	345
16	164	14 35	65 09	83·3	14 28	64 07	83·1	14 21	63 06	82·8	14 13	62 04	82·6	14 05	61 03	82·3	13 57	60 02	82·1	196	344
17	163	15 29	65 02	82·9	15 22	64 00	82·6	15 14	62 59	82·4	15 06	61 57	82·1	14 58	60 56	81·8	14 49	59 54	81·6	197	343
18	162	16 24	64 55	82·5	16 16	63 53	82·2	16 08	62 51	81·9	15 59	61 49	81·6	15 50	60 47	81·3	15 41	59 46	81·0	198	342
19	161	17 18	64 47	82·0	17 10	63 45	81·7	17 01	62 43	81·4	16 52	61 41	81·1	16 42	60 39	80·8	16 33	59 37	80·5	199	341
20	160	18 12	64 39	81·6	18 03	63 36	81·3	17 54	62 34	80·9	17 45	61 32	80·6	17 35	60 30	80·3	17 24	59 28	80·0	200	340
21	159	19 07	64 30	81·1	18 57	63 28	80·8	18 47	62 25	80·4	18 37	61 23	80·1	18 27	60 20	79·8	18 16	59 18	79·5	201	339
22	158	20 01	64 21	80·7	19 51	63 18	80·3	19 41	62 15	80·0	19 30	61 13	79·6	19 19	60 10	79·3	19 08	59 08	78·9	202	338
23	157	20 55	64 11	80·2	20 44	63 08	79·8	20 34	62 05	79·5	20 22	61 02	79·1	20 11	59 59	78·7	19 59	58 57	78·4	203	337
24	156	21 49	64 01	79·7	21 38	62 58	79·3	21 27	61 54	79·0	21 15	60 51	78·6	21 03	59 48	78·2	20 50	58 45	77·8	204	336
25	155	22 43	63 50	79·3	22 31	62 46	78·9	22 19	61 43	78·4	22 07	60 39	78·0	21 55	59 36	77·7	21 42	58 33	77·3	205	335
26	154	23 36	63 39	78·8	23 25	62 35	78·4	23 12	61 31	77·9	22 59	60 27	77·5	22 46	59 24	77·1	22 33	58 20	76·7	206	334
27	153	24 30	63 27	78·3	24 18	62 22	77·8	24 05	61 18	77·4	23 52	60 14	77·0	23 38	59 10	76·5	23 24	58 07	76·1	207	333
28	152	25 24	63 14	77·8	25 11	62 10	77·3	24 57	61 05	76·9	24 44	60 01	76·4	24 29	58 57	76·0	24 15	57 53	75·5	208	332
29	151	26 17	63 01	77·3	26 04	61 56	76·8	25 50	60 51	76·3	25 36	59 47	75·9	25 21	58 42	75·4	25 05	57 38	75·0	209	331
30	150	27 11	62 48	76·8	26 57	61 42	76·3	26 42	60 37	75·8	26 27	59 32	75·3	26 12	58 27	74·8	25 56	57 23	74·4	210	330
31	149	28 04	62 33	76·3	27 50	61 27	75·8	27 35	60 22	75·2	27 19	59 16	74·7	27 03	58 11	74·2	26 46	57 07	73·8	211	329
32	148	28 57	62 18	75·7	28 42	61 12	75·2	28 27	60 06	74·7	28 10	59 00	74·2	27 54	57 55	73·7	27 37	56 50	73·1	212	328
33	147	29 50	62 02	75·2	29 35	60 56	74·7	29 19	59 49	74·1	29 02	58 43	73·6	28 45	57 38	73·0	28 27	56 32	72·5	213	327
34	146	30 43	61 46	74·7	30 27	60 39	74·1	30 10	59 32	73·5	29 53	58 26	73·0	29 35	57 20	72·4	29 17	56 14	71·9	214	326
35	145	31 36	61 28	74·1	31 19	60 21	73·5	31 02	59 14	72·9	30 44	58 07	72·4	30 26	57 01	71·8	30 07	55 55	71·2	215	325
36	144	32 29	61 10	73·5	32 11	60 02	72·9	31 53	58 55	72·3	31 35	57 48	71·7	31 16	56 41	71·2	30 56	55 35	70·6	216	324
37	143	33 21	60 52	73·0	33 03	59 43	72·3	32 45	58 35	71·7	32 26	57 28	71·1	32 06	56 21	70·5	31 46	55 14	69·9	217	323
38	142	34 13	60 32	72·4	33 55	59 23	71·7	33 36	58 15	71·1	33 16	57 07	70·5	32 56	55 59	69·9	32 35	54 53	69·3	218	322
39	141	35 06	60 11	71·8	34 47	59 02	71·1	34 27	57 53	70·5	34 06	56 45	69·8	33 45	55 37	69·2	33 24	54 30	68·6	219	321
40	140	35 58	59 50	71·2	35 38	58 40	70·5	35 17	57 31	69·8	34 56	56 22	69·1	34 35	55 14	68·5	34 12	54 07	67·9	220	320
41	139	36 49	59 28	70·5	36 29	58 17	69·8	36 08	57 08	69·1	35 46	55 59	68·5	35 24	54 50	67·8	35 01	53 42	67·1	221	319
42	138	37 41	59 04	69·9	37 20	57 54	69·2	36 58	56 43	68·5	36 36	55 34	67·8	36 13	54 25	67·1	35 49	53 17	66·4	222	318
43	137	38 32	58 40	69·2	38 11	57 29	68·5	37 48	56 18	67·8	37 25	55 08	67·1	37 02	53 59	66·4	36 37	52 50	65·7	223	317
44	136	39 23	58 15	68·6	39 01	57 03	67·8	38 38	55 52	67·1	38 14	54 41	66·3	37 50	53 32	65·6	37 25	52 23	64·9	224	316
45	135	40 14	57 48	67·9	39 51	56 36	67·1	39 28	55 24	66·3	39 03	54 13	65·6	38 38	53 04	64·9	38 12	51 54	64·1	225	315

Lat. / A		24°			25°			26°			27°			28°			29°			Lat. / A	
LHA/F		A/H	B/P	Z_1/Z_2	A/H	B/P	Z_1/Z_2	A/H	B/P	Z_1/Z_2	A/H	B/P	Z_1/Z_2	A/H	B/P	Z_1/Z_2	A/H	B/P	Z_1/Z_2	LHA	
°	°	° ′	° ′	°	° ′	° ′	°	° ′	° ′	°	° ′	° ′	°	° ′	° ′	°	° ′	° ′	°	°	°
45	135	40 14	57 48	67·9	39 51	56 36	67·1	39 28	55 24	66·3	39 03	54 13	65·6	38 38	53 04	64·9	38 12	51 54	64·1	225	315
46	134	41 05	57 21	67·2	40 41	56 08	66·4	40 17	54 56	65·6	39 52	53 44	64·8	39 26	52 34	64·1	38 59	51 25	63·3	226	314
47	133	41 55	56 52	66·4	41 31	55 38	65·6	41 06	54 26	64·8	40 40	53 14	64·0	40 13	52 04	63·3	39 46	50 54	62·5	227	313
48	132	42 45	56 22	65·7	42 20	55 08	64·9	41 54	53 55	64·0	41 28	52 43	63·2	41 00	51 32	62·5	40 32	50 22	61·7	228	312
49	131	43 35	55 50	64·9	43 09	54 36	64·1	42 43	53 22	63·2	42 15	52 10	62·4	41 47	50 59	61·6	41 18	49 48	60·9	229	311
50	130	44 25	55 17	64·1	43 58	54 02	63·3	43 31	52 49	62·4	43 03	51 36	61·6	42 34	50 24	60·8	42 04	49 14	60·0	230	310
51	129	45 14	54 43	63·3	44 47	53 28	62·4	44 18	52 13	61·6	43 49	51 00	60·7	43 20	49 48	59·9	42 49	48 38	59·1	231	309
52	128	46 03	54 08	62·5	45 35	52 52	61·6	45 06	51 37	60·7	44 36	50 23	59·8	44 05	49 11	59·0	43 34	48 00	58·2	232	308
53	127	46 51	53 30	61·6	46 22	52 14	60·7	45 52	50 59	59·8	45 22	49 45	58·9	44 51	48 32	58·1	44 18	47 21	57·2	233	307
54	126	47 39	52 51	60·8	47 09	51 34	59·8	46 39	50 19	58·9	46 07	49 05	58·0	45 35	47 52	57·1	45 02	46 41	56·3	234	306
55	125	48 27	52 11	59·8	47 56	50 53	58·9	47 25	49 37	58·0	46 53	48 23	57·0	46 19	47 10	56·2	45 46	45 59	55·3	235	305
56	124	49 14	51 28	58·9	48 43	50 11	57·9	48 10	48 54	57·0	47 37	47 40	56·1	47 03	46 27	55·2	46 29	45 15	54·3	236	304
57	123	50 01	50 44	57·9	49 28	49 26	56·9	48 55	48 09	56·0	48 21	46 54	55·0	47 46	45 41	54·1	47 11	44 30	53·3	237	303
58	122	50 47	49 58	56·9	50 14	48 39	55·9	49 40	47 22	54·9	49 05	46 07	54·0	48 29	44 54	53·1	47 53	43 43	52·2	238	302
59	121	51 33	49 09	55·9	50 58	47 51	54·9	50 23	46 34	53·9	49 48	45 18	52·9	49 11	44 05	52·0	48 34	42 54	51·1	239	301
60	120	52 18	48 19	54·8	51 43	47 00	53·8	51 07	45 43	52·8	50 30	44 28	51·8	49 53	43 14	50·9	49 14	42 03	50·0	240	300
61	119	53 02	47 26	53·7	52 26	46 07	52·7	51 49	44 50	51·7	51 12	43 35	50·7	50 33	42 22	49·7	49 54	41 10	48·8	241	299
62	118	53 46	46 31	52·6	53 09	45 12	51·5	52 31	43 54	50·5	51 53	42 39	49·5	51 13	41 27	48·6	50 33	40 16	47·6	242	298
63	117	54 29	45 33	51·4	53 51	44 14	50·3	53 13	42 57	49·3	52 33	41 42	48·3	51 53	40 30	47·3	51 12	39 19	46·4	243	297
64	116	55 12	44 33	50·2	54 33	43 14	49·1	53 53	41 57	48·1	53 13	40 42	47·1	52 31	39 30	46·1	51 49	38 20	45·2	244	296
65	115	55 53	43 30	48·9	55 13	42 11	47·8	54 33	40 55	46·8	53 51	39 40	45·8	53 09	38 29	44·8	52 26	37 19	43·9	245	295
66	114	56 34	42 25	47·6	55 53	41 06	46·5	55 12	39 50	45·4	54 29	38 36	44·4	53 46	37 25	43·5	53 02	36 16	42·6	246	294
67	113	57 14	41 16	46·2	56 32	39 58	45·1	55 50	38 42	44·1	55 06	37 29	43·1	54 22	36 19	42·1	53 37	35 11	41·2	247	293
68	112	57 53	40 05	44·8	57 10	38 47	43·7	56 27	37 32	42·7	55 42	36 19	41·7	54 57	35 10	40·7	54 11	34 03	39·8	248	292
69	111	58 32	38 50	43·3	57 47	37 33	42·2	57 03	36 18	41·2	56 17	35 07	40·2	55 31	33 59	39·3	54 44	32 53	38·4	249	291
70	110	59 09	37 32	41·8	58 24	36 16	40·7	57 38	35 02	39·7	56 51	33 52	38·7	56 04	32 45	37·8	55 16	31 41	36·9	250	290
71	109	59 45	36 11	40·2	58 58	34 55	39·2	58 12	33 43	38·1	57 24	32 35	37·2	56 36	31 29	36·3	55 47	30 26	35·4	251	289
72	108	60 19	34 46	38·6	59 32	33 32	37·6	58 44	32 21	36·5	57 56	31 14	35·6	57 07	30 10	34·7	56 17	29 08	33·8	252	288
73	107	60 53	33 18	36·9	60 05	32 05	35·9	59 16	30 56	34·9	58 26	29 51	34·0	57 36	28 48	33·1	56 46	27 49	32·2	253	287
74	106	61 25	31 46	35·2	60 36	30 35	34·2	59 46	29 28	33·2	58 55	28 25	32·3	58 05	27 24	31·4	57 13	26 26	30·6	254	286
75	105	61 56	30 10	33·4	61 06	29 02	32·4	60 15	27 57	31·4	59 23	26 56	30·5	58 31	25 57	29·7	57 39	25 02	28·9	255	285
76	104	62 26	28 31	31·5	61 34	27 25	30·5	60 42	26 23	29·6	59 50	25 24	28·8	58 57	24 28	28·0	58 04	23 35	27·2	256	284
77	103	62 53	26 48	29·6	62 01	25 45	28·6	61 08	24 46	27·8	60 15	23 49	27·0	59 21	22 56	26·2	58 27	22 05	25·5	257	283
78	102	63 20	25 02	27·6	62 26	24 02	26·7	61 32	23 05	25·9	60 38	22 12	25·1	59 44	21 21	24·4	58 49	20 34	23·7	258	282
79	101	63 44	23 12	25·5	62 50	22 15	24·7	61 55	21 22	23·9	61 00	20 32	23·2	60 05	19 44	22·5	59 09	19 00	21·8	259	281
80	100	64 07	21 18	23·4	63 12	20 25	22·6	62 16	19 36	21·9	61 20	18 49	21·2	60 24	18 05	20·6	59 28	17 24	20·0	260	280
81	99	64 28	19 22	21·3	63 32	18 33	20·5	62 35	17 47	19·9	61 39	17 04	19·2	60 42	16 24	18·6	59 45	15 46	18·1	261	279
82	98	64 47	17 22	19·1	63 50	16 37	18·4	62 53	15 56	17·8	61 56	15 17	17·2	60 58	14 40	16·7	60 01	14 06	16·2	262	278
83	97	65 03	15 18	16·8	64 06	14 39	16·2	63 08	14 02	15·6	62 10	13 27	15·1	61 12	12 55	14·7	60 14	12 24	14·2	263	277
84	96	65 18	13 13	14·5	64 20	12 38	14·0	63 22	12 06	13·5	62 23	11 36	13·0	61 25	11 07	12·6	60 26	10 41	12·2	264	276
85	95	65 31	11 05	12·1	64 32	10 35	11·7	63 33	10 08	11·3	62 35	9 42	10·9	61 36	9 19	10·6	60 37	8 56	10·2	265	275
86	94	65 41	8 54	9·8	64 42	8 30	9·4	63 43	8 08	9·1	62 44	7 48	8·8	61 44	7 28	8·5	60 45	7 10	8·2	266	274
87	93	65 49	6 42	7·3	64 50	6 24	7·1	63 50	6 07	6·8	62 51	5 52	6·6	61 51	5 37	6·4	60 52	5 24	6·2	267	273
88	92	65 55	4 29	4·9	64 56	4 17	4·7	63 56	4 06	4·6	62 56	3 55	4·4	61 56	3 45	4·3	60 56	3 36	4·1	268	272
89	91	65 59	2 15	2·5	64 59	2 09	2·4	63 59	2 03	2·3	62 59	1 58	2·2	61 59	1 53	2·1	60 59	1 48	2·1	269	271
90	90	66 00	0 00	0·0	65 00	0 00	0·0	64 00	0 00	0·0	63 00	0 00	0·0	62 00	0 00	0·0	61 00	0 00	0·0	270	270

N. Lat: for LHA > 180° ... Z_n = Z
for LHA < 180° ... Z_n = 360° − Z

S. Lat.: for LHA > 180° ... Z_n = 180° − Z
for LHA < 180° ... Z_n = 180° + Z

SIGHT REDUCTION TABLE

B: (–) for 90° < LHA < 270°
Dec:(–) for Lat. contrary name

Z_1: same sign as B
Z_2: (–) for F > 90°

Lat. / A		30°			31°			32°			33°			34°			35°			Lat. / A	
LHA/F		A/H	B/P	Z_1/Z_2	A/H	B/P	Z_1/Z_2	A/H	B/P	Z_1/Z_2	A/H	B/P	Z_1/Z_2	A/H	B/P	Z_1/Z_2	A/H	B/P	Z_1/Z_2	LHA	
°	°	° ′	° ′	°	° ′	° ′	°	° ′	° ′	°	° ′	° ′	°	° ′	° ′	°	° ′	° ′	°	°	°
0	180	0 00	60 00	90·0	0 00	59 00	90·0	0 00	58 00	90·0	0 00	57 00	90·0	0 00	56 00	90·0	0 00	55 00	90·0	180	360
1	179	0 52	60 00	89·5	0 51	59 00	89·5	0 51	58 00	89·5	0 50	57 00	89·5	0 50	56 00	89·4	0 49	55 00	89·4	181	359
2	178	1 44	59 59	89·0	1 43	58 59	89·0	1 42	57 59	88·9	1 41	56 59	88·9	1 39	55 59	88·9	1 38	54 59	88·9	182	358
3	177	2 36	59 58	88·5	2 34	58 58	88·5	2 33	57 58	88·4	2 31	56 58	88·4	2 29	55 58	88·3	2 27	54 58	88·3	183	357
4	176	3 28	59 56	88·0	3 26	58 56	87·9	3 23	57 56	87·9	3 21	56 56	87·8	3 19	55 56	87·8	3 17	54 56	87·7	184	356
5	175	4 20	59 54	87·5	4 17	58 54	87·4	4 14	57 54	87·3	4 12	56 54	87·3	4 09	55 54	87·2	4 06	54 54	87·1	185	355
6	174	5 12	59 52	87·0	5 08	58 52	86·9	5 05	57 52	86·8	5 02	56 51	86·7	4 58	55 51	86·6	4 55	54 51	86·6	186	354
7	173	6 04	59 49	86·5	6 00	58 49	86·4	5 56	57 48	86·3	5 52	56 48	86·2	5 48	55 48	86·1	5 44	54 48	86·0	187	353
8	172	6 55	59 45	86·0	6 51	58 45	85·9	6 47	57 45	85·7	6 42	56 45	85·6	6 38	55 44	85·5	6 33	54 44	85·4	188	352
9	171	7 47	59 42	85·5	7 42	58 41	85·3	7 37	57 41	85·2	7 32	56 40	85·1	7 27	55 40	84·9	7 22	54 40	84·8	189	351
10	170	8 39	59 37	85·0	8 34	58 37	84·8	8 28	57 36	84·7	8 22	56 36	84·5	8 17	55 36	84·4	8 11	54 35	84·2	190	350
11	169	9 31	59 32	84·4	9 25	58 32	84·3	9 19	57 31	84·1	9 13	56 31	84·0	9 06	55 30	83·8	9 00	54 30	83·6	191	349
12	168	10 22	59 27	83·9	10 16	58 26	83·8	10 09	57 26	83·6	10 03	56 25	83·4	9 56	55 25	83·2	9 48	54 24	83·0	192	348
13	167	11 14	59 21	83·4	11 07	58 20	83·2	11 00	57 20	83·0	10 52	56 19	82·8	10 45	55 18	82·6	10 37	54 18	82·5	193	347
14	166	12 06	59 15	82·9	11 58	58 14	82·7	11 50	57 13	82·5	11 42	56 12	82·3	11 34	55 12	82·1	11 26	54 11	81·9	194	346
15	165	12 57	59 08	82·4	12 49	58 07	82·1	12 41	57 06	81·9	12 32	56 05	81·7	12 23	55 04	81·5	12 14	54 04	81·3	195	345
16	164	13 49	59 01	81·8	13 40	57 59	81·6	13 31	56 58	81·4	13 22	55 57	81·1	13 13	54 57	80·9	13 03	53 56	80·7	196	344
17	163	14 40	58 53	81·3	14 31	57 51	81·1	14 21	56 50	80·8	14 12	55 49	80·5	14 02	54 48	80·3	13 51	53 47	80·1	197	343
18	162	15 31	58 44	80·8	15 22	57 43	80·5	15 12	56 42	80·2	15 01	55 40	80·0	14 51	54 39	79·7	14 40	53 38	79·4	198	342
19	161	16 23	58 35	80·2	16 12	57 34	79·9	16 02	56 32	79·7	15 51	55 31	79·4	15 40	54 30	79·1	15 28	53 29	78·8	199	341
20	160	17 14	58 26	79·7	17 03	57 24	79·4	16 52	56 23	79·1	16 40	55 21	78·8	16 28	54 20	78·5	16 16	53 19	78·2	200	340
21	159	18 05	58 16	79·1	17 53	57 14	78·8	17 42	56 12	78·5	17 29	55 11	78·2	17 17	54 09	77·9	17 04	53 08	77·6	201	339
22	158	18 56	58 05	78·6	18 44	57 03	78·2	18 31	56 01	77·9	18 19	55 00	77·6	18 06	53 58	77·3	17 52	52 56	77·0	202	338
23	157	19 47	57 54	78·0	19 34	56 52	77·7	19 21	55 50	77·3	19 08	54 48	77·0	18 54	53 46	76·6	18 40	52 44	76·3	203	337
24	156	20 37	57 42	77·4	20 24	56 40	77·1	20 11	55 38	76·7	19 57	54 36	76·4	19 42	53 34	76·0	19 28	52 32	75·7	204	336
25	155	21 28	57 30	76·9	21 14	56 27	76·5	21 00	55 25	76·1	20 46	54 23	75·7	20 31	53 21	75·4	20 15	52 19	75·0	205	335
26	154	22 19	57 17	76·3	22 04	56 14	75·9	21 49	55 12	75·5	21 34	54 09	75·1	21 19	53 07	74·7	21 03	52 05	74·4	206	334
27	153	23 09	57 03	75·7	22 54	56 00	75·3	22 39	54 57	74·9	22 23	53 55	74·5	22 07	52 52	74·1	21 50	51 50	73·7	207	333
28	152	23 59	56 49	75·1	23 44	55 46	74·7	23 28	54 43	74·3	23 11	53 40	73·8	22 54	52 37	73·4	22 37	51 35	73·0	208	332
29	151	24 50	56 34	74·5	24 33	55 31	74·1	24 17	54 27	73·6	23 59	53 24	73·2	23 42	52 22	72·8	23 24	51 19	72·4	209	331
30	150	25 40	56 19	73·9	25 23	55 15	73·4	25 05	54 11	73·0	24 48	53 08	72·5	24 29	52 05	72·1	24 11	51 03	71·7	210	330
31	149	26 29	56 02	73·3	26 12	54 58	72·8	25 54	53 54	72·3	25 35	52 51	71·9	25 17	51 48	71·4	24 57	50 45	71·0	211	329
32	148	27 19	55 45	72·6	27 01	54 41	72·2	26 42	53 37	71·7	26 23	52 33	71·2	26 04	51 30	70·7	25 44	50 27	70·3	212	328
33	147	28 09	55 27	72·0	27 50	54 23	71·5	27 31	53 19	71·0	27 11	52 15	70·5	26 50	51 12	70·0	26 30	50 08	69·6	213	327
34	146	28 58	55 09	71·4	28 38	54 04	70·8	28 19	53 00	70·3	27 58	51 56	69·8	27 37	50 52	69·3	27 16	49 49	68·8	214	326
35	145	29 47	54 49	70·7	29 27	53 44	70·2	29 06	52 40	69·6	28 45	51 36	69·1	28 24	50 32	68·6	28 01	49 29	68·1	215	325
36	144	30 36	54 29	70·0	30 15	53 24	69·5	29 54	52 19	68·9	29 32	51 15	68·4	29 10	50 11	67·9	28 47	49 07	67·4	216	324
37	143	31 25	54 08	69·4	31 03	53 03	68·8	30 41	51 58	68·2	30 19	50 53	67·7	29 56	49 49	67·2	29 32	48 45	66·6	217	323
38	142	32 13	53 46	68·7	31 51	52 40	68·1	31 28	51 35	67·5	31 05	50 30	66·9	30 41	49 26	66·4	30 17	48 23	65·9	218	322
39	141	33 02	53 23	68·0	32 39	52 17	67·4	32 15	51 12	66·8	31 51	50 07	66·2	31 27	49 03	65·6	31 02	47 59	65·1	219	321
40	140	33 50	53 00	67·2	33 26	51 53	66·6	33 02	50 48	66·0	32 37	49 43	65·4	32 12	48 38	64·9	31 46	47 34	64·3	220	320
41	139	34 37	52 35	66·5	34 13	51 29	65·9	33 48	50 23	65·3	33 23	49 17	64·7	32 57	48 13	64·1	32 30	47 09	63·5	221	319
42	138	35 25	52 09	65·8	35 00	51 03	65·1	34 34	49 56	64·5	34 08	48 51	63·9	33 42	47 46	63·3	33 14	46 42	62·7	222	318
43	137	36 12	51 43	65·0	35 46	50 36	64·3	35 20	49 29	63·7	34 53	48 24	63·1	34 26	47 19	62·5	33 58	46 15	61·9	223	317
44	136	36 59	51 15	64·2	36 33	50 08	63·6	36 06	49 01	62·9	35 38	47 55	62·3	35 10	46 51	61·6	34 41	45 46	61·0	224	316
45	135	37 46	50 46	63·4	37 19	49 39	62·7	36 51	48 32	62·1	36 22	47 26	61·4	35 53	46 21	60·8	35 24	45 17	60·2	225	315

Lat. / A		30°			31°			32°			33°			34°			35°			Lat. / A	
LHA/F		A/H	B/P	Z_1/Z_2	A/H	B/P	Z_1/Z_2	A/H	B/P	Z_1/Z_2	A/H	B/P	Z_1/Z_2	A/H	B/P	Z_1/Z_2	A/H	B/P	Z_1/Z_2	LHA	
°	°	° ′	° ′	°	° ′	° ′	°	° ′	° ′	°	° ′	° ′	°	° ′	° ′	°	° ′	° ′	°	°	°
45	135	37 46	50 46	63·4	37 19	49 39	62·7	36 51	48 32	62·1	36 22	47 26	61·4	35 53	46 21	60·8	35 24	45 17	60·2	225	315
46	134	38 32	50 16	62·6	38 04	49 08	61·9	37 36	48 02	61·2	37 06	46 56	60·6	36 37	45 51	59·9	36 06	44 46	59·3	226	314
47	133	39 18	49 45	61·8	38 49	48 37	61·1	38 20	47 30	60·4	37 50	46 24	59·7	37 19	45 19	59·1	36 48	44 15	58·4	227	313
48	132	40 04	49 13	61·0	39 34	48 05	60·2	39 04	46 58	59·5	38 33	45 51	58·8	38 02	44 46	58·2	37 30	43 42	57·5	228	312
49	131	40 49	48 39	60·1	40 19	47 31	59·4	39 48	46 24	58·6	39 16	45 18	57·9	38 44	44 12	57·2	38 11	43 08	56·6	229	311
50	130	41 34	48 04	59·2	41 03	46 56	58·5	40 31	45 49	57·7	39 59	44 42	57·0	39 26	43 37	56·3	38 52	42 33	55·6	230	310
51	129	42 18	47 28	58·3	41 46	46 20	57·5	41 14	45 12	56·8	40 41	44 06	56·1	40 07	43 01	55·4	39 32	41 57	54·7	231	309
52	128	43 02	46 50	57·4	42 29	45 42	56·6	41 56	44 34	55·9	41 22	43 28	55·1	40 47	42 23	54·4	40 12	41 19	53·7	232	308
53	127	43 46	46 11	56·4	43 12	45 03	55·6	42 38	43 55	54·9	42 03	42 49	54·1	41 28	41 44	53·4	40 52	40 41	52·7	233	307
54	126	44 29	45 31	55·5	43 54	44 22	54·7	43 19	43 15	53·9	42 44	42 09	53·1	42 07	41 04	52·4	41 30	40 01	51·7	234	306
55	125	45 11	44 49	54·5	44 36	43 40	53·7	44 00	42 33	52·9	43 24	41 27	52·1	42 46	40 23	51·4	42 09	39 19	50·7	235	305
56	124	45 53	44 05	53·5	45 17	42 57	52·6	44 40	41 50	51·8	44 03	40 44	51·1	43 25	39 40	50·3	42 46	38 37	49·6	236	304
57	123	46 35	43 20	52·4	45 58	42 11	51·6	45 20	41 05	50·8	44 42	39 59	50·0	44 03	38 55	49·3	43 24	37 53	48·5	237	303
58	122	47 16	42 33	51·3	46 38	41 25	50·5	45 59	40 18	49·7	45 20	39 13	48·9	44 40	38 09	48·2	44 00	37 07	47·5	238	302
59	121	47 56	41 44	50·2	47 17	40 36	49·4	46 38	39 30	48·6	45 58	38 25	47·8	45 17	37 22	47·1	44 36	36 20	46·3	239	301
60	120	48 35	40 54	49·1	47 56	39 46	48·3	47 16	38 40	47·5	46 35	37 36	46·7	45 53	36 33	45·9	45 11	35 32	45·2	240	300
61	119	49 14	40 01	47·9	48 34	38 54	47·1	47 53	37 48	46·3	47 11	36 45	45·5	46 29	35 42	44·7	45 46	34 42	44·0	241	299
62	118	49 53	39 07	46·8	49 11	38 00	45·9	48 29	36 55	45·1	47 46	35 52	44·3	47 03	34 50	43·6	46 19	33 50	42·8	242	298
63	117	50 30	38 11	45·5	49 48	37 04	44·7	49 05	36 00	43·9	48 21	34 57	43·1	47 37	33 57	42·3	46 53	32 57	41·6	243	297
64	116	51 07	37 13	44·3	50 23	36 07	43·4	49 40	35 03	42·6	48 55	34 01	41·8	48 10	33 01	41·1	47 25	32 03	40·4	244	296
65	115	51 43	36 12	43·0	50 58	35 07	42·2	50 14	34 04	41·3	49 28	33 03	40·6	48 43	32 04	39·8	47 56	31 07	39·1	245	295
66	114	52 18	35 10	41·7	51 33	34 06	40·8	50 47	33 04	40·0	50 01	32 04	39·3	49 14	31 05	38·5	48 27	30 09	37·8	246	294
67	113	52 52	34 05	40·3	52 06	33 02	39·5	51 19	32 01	38·7	50 32	31 02	37·9	49 44	30 05	37·2	48 56	29 10	36·5	247	293
68	112	53 25	32 59	38·9	52 38	31 56	38·1	51 50	30 57	37·3	51 02	29 59	36·6	50 14	29 03	35·8	49 25	28 09	35·2	248	292
69	111	53 57	31 50	37·5	53 09	30 49	36·7	52 21	29 50	35·9	51 32	28 53	35·2	50 43	27 59	34·5	49 53	27 06	33·8	249	291
70	110	54 28	30 39	36·1	53 39	29 39	35·2	52 50	28 42	34·5	52 00	27 46	33·8	51 10	26 53	33·1	50 20	26 02	32·4	250	290
71	109	54 58	29 25	34·6	54 08	28 27	33·8	53 18	27 31	33·0	52 28	26 38	32·3	51 37	25 46	31·6	50 46	24 56	31·0	251	289
72	108	55 27	28 09	33·0	54 37	27 13	32·2	53 46	26 19	31·5	52 54	25 27	30·8	52 03	24 37	30·2	51 10	23 49	29·5	252	288
73	107	55 55	26 51	31·4	55 03	25 57	30·7	54 12	25 04	30·0	53 19	24 14	29·3	52 27	23 26	28·7	51 34	22 40	28·1	253	287
74	106	56 21	25 31	29·8	55 29	24 39	29·1	54 36	23 48	28·4	53 43	23 00	27·8	52 50	22 14	27·1	51 57	21 29	26·6	254	286
75	105	56 46	24 09	28·2	55 53	23 18	27·5	55 00	22 30	26·8	54 06	21 44	26·2	53 12	21 00	25·6	52 18	20 17	25·0	255	285
76	104	57 10	22 44	26·5	56 16	21 56	25·8	55 22	21 10	25·2	54 28	20 26	24·6	53 33	19 44	24·0	52 38	19 04	23·5	256	284
77	103	57 33	21 17	24·8	56 38	20 31	24·1	55 43	19 48	23·5	54 48	19 06	23·0	53 53	18 27	22·4	52 57	17 49	21·9	257	283
78	102	57 54	19 48	23·0	56 59	19 05	22·4	56 03	18 24	21·9	55 07	17 45	21·3	54 11	17 08	20·8	53 15	16 32	20·3	258	282
79	101	58 13	18 17	21·2	57 17	17 37	20·7	56 21	16 59	20·1	55 25	16 22	19·6	54 28	15 48	19·2	53 31	15 15	18·7	259	281
80	100	58 32	16 44	19·4	57 35	16 07	18·9	56 38	15 32	18·4	55 41	14 58	17·9	54 44	14 26	17·5	53 47	13 56	17·1	260	280
81	99	58 48	15 10	17·6	57 51	14 36	17·1	56 53	14 03	16·6	55 56	13 33	16·2	54 58	13 03	15·8	54 00	12 36	15·4	261	279
82	98	59 03	13 33	15·7	58 05	13 02	15·3	57 07	12 33	14·9	56 09	12 06	14·5	55 11	11 40	14·1	54 13	11 14	13·8	262	278
83	97	59 16	11 55	13·8	58 18	11 28	13·4	57 19	11 02	13·0	56 21	10 38	12·7	55 22	10 14	12·4	54 24	9 52	12·1	263	277
84	96	59 28	10 16	11·9	58 29	9 52	11·5	57 30	9 30	11·2	56 31	9 09	10·9	55 32	8 49	10·6	54 33	8 29	10·4	264	276
85	95	59 37	8 35	9·9	58 38	8 15	9·6	57 39	7 56	9·4	56 40	7 39	9·1	55 41	7 22	8·9	54 41	7 06	8·7	265	275
86	94	59 46	6 53	8·0	58 46	6 37	7·7	57 47	6 22	7·5	56 47	6 08	7·3	55 48	5 54	7·1	54 48	5 41	7·0	266	274
87	93	59 52	5 11	6·0	58 52	4 59	5·8	57 52	4 47	5·6	56 53	4 36	5·5	55 53	4 26	5·4	54 53	4 16	5·2	267	273
88	92	59 56	3 28	4·0	58 57	3 19	3·9	57 57	3 12	3·8	56 57	3 05	3·7	55 57	2 58	3·6	54 57	2 51	3·5	268	272
89	91	59 59	1 44	2·0	58 59	1 40	1·9	57 59	1 36	1·9	56 59	1 32	1·8	55 59	1 29	1·8	54 59	1 26	1·7	269	271
90	90	60 00	0 00	0·0	59 00	0 00	0·0	58 00	0 00	0·0	57 00	0 00	0·0	56 00	0 00	0·0	55 00	0 00	0·0	270	270

N. Lat: for LHA > 180° ... $Z_n = Z$
for LHA < 180° ... $Z_n = 360° - Z$

S. Lat.: for LHA > 180° ... $Z_n = 180° - Z$
for LHA < 180° ... $Z_n = 180° + Z$

B: (–) for 90° < LHA < 270°
Dec: (–) for Lat. contrary name

SIGHT REDUCTION TABLE

Z_1: same sign as B
Z_2: (–) for F > 90°

Lat. / A		36°			37°			38°			39°			40°			41°			Lat. / A	
LHA/F		A/H	B/P	Z_1/Z_2	A/H	B/P	Z_1/Z_2	A/H	B/P	Z_1/Z_2	A/H	B/P	Z_1/Z_2	A/H	B/P	Z_1/Z_2	A/H	B/P	Z_1/Z_2	LHA	
°	°	° ′	° ′	°	° ′	° ′	°	° ′	° ′	°	° ′	° ′	°	° ′	° ′	°	° ′	° ′	°	°	°
0	180	0 00	54 00	90·0	0 00	53 00	90·0	0 00	52 00	90·0	0 00	51 00	90·0	0 00	50 00	90·0	0 00	49 00	90·0	180	360
1	179	0 49	54 00	89·4	0 48	53 00	89·4	0 47	52 00	89·4	0 47	51 00	89·4	0 46	50 00	89·4	0 45	49 00	89·3	181	359
2	178	1 37	53 59	88·8	1 36	52 59	88·8	1 35	51 59	88·8	1 33	50 59	88·7	1 32	49 59	88·7	1 31	48 59	88·7	182	358
3	177	2 26	53 58	88·2	2 24	52 58	88·2	2 22	51 58	88·2	2 20	50 58	88·1	2 18	49 58	88·1	2 16	48 58	88·0	183	357
4	176	3 14	53 56	87·6	3 12	52 56	87·6	3 09	51 56	87·5	3 06	50 56	87·5	3 04	49 56	87·4	3 01	48 56	87·4	184	356
5	175	4 03	53 54	87·1	3 59	52 54	87·0	3 56	51 54	86·9	3 53	50 54	86·8	3 50	49 54	86·8	3 46	48 54	86·7	185	355
6	174	4 51	53 51	86·5	4 47	52 51	86·4	4 43	51 51	86·3	4 40	50 51	86·2	4 36	49 51	86·1	4 31	48 51	86·1	186	354
7	173	5 39	53 48	85·9	5 35	52 48	85·8	5 31	51 48	85·7	5 26	50 47	85·6	5 21	49 47	85·5	5 17	48 47	85·4	187	353
8	172	6 28	53 44	85·3	6 23	52 44	85·2	6 18	51 44	85·1	6 13	50 44	84·9	6 07	49 43	84·8	6 02	48 43	84·7	188	352
9	171	7 16	53 40	84·7	7 11	52 39	84·6	7 05	51 39	84·4	6 59	50 39	84·3	6 53	49 39	84·2	6 47	48 39	84·1	189	351
10	170	8 05	53 35	84·1	7 58	52 35	83·9	7 52	51 34	83·8	7 45	50 34	83·7	7 39	49 34	83·5	7 32	48 34	83·4	190	350
11	169	8 53	53 30	83·5	8 46	52 29	83·3	8 39	51 29	83·2	8 32	50 29	83·0	8 24	49 29	82·9	8 17	48 28	82·7	191	349
12	168	9 41	53 24	82·9	9 33	52 23	82·7	9 26	51 23	82·5	9 18	50 23	82·4	9 10	49 23	82·2	9 02	48 22	82·1	192	348
13	167	10 29	53 17	82·3	10 21	52 17	82·1	10 13	51 17	81·9	10 04	50 16	81·7	9 55	49 16	81·6	9 46	48 16	81·4	193	347
14	166	11 17	53 10	81·7	11 08	52 10	81·5	10 59	51 10	81·3	10 50	50 09	81·1	10 41	49 09	80·9	10 31	48 09	80·7	194	346
15	165	12 05	53 03	81·0	11 56	52 02	80·8	11 46	51 02	80·6	11 36	50 02	80·4	11 26	49 01	80·2	11 16	48 01	80·0	195	345
16	164	12 53	52 55	80·4	12 43	51 54	80·2	12 33	50 54	80·0	12 22	49 53	79·8	12 11	48 53	79·6	12 00	47 53	79·3	196	344
17	163	13 41	52 46	79·8	13 30	51 46	79·6	13 19	50 45	79·3	13 08	49 45	79·1	12 57	48 44	78·9	12 45	47 44	78·7	197	343
18	162	14 29	52 37	79·2	14 17	51 37	78·9	14 06	50 36	78·7	13 54	49 35	78·4	13 42	48 35	78·2	13 29	47 34	78·0	198	342
19	161	15 16	52 28	78·6	15 04	51 27	78·3	14 52	50 26	78·0	14 39	49 25	77·8	14 27	48 25	77·5	14 13	47 24	77·3	199	341
20	160	16 04	52 17	77·9	15 51	51 16	77·6	15 38	50 16	77·4	15 25	49 15	77·1	15 11	48 14	76·8	14 58	47 14	76·6	200	340
21	159	16 51	52 07	77·3	16 38	51 05	77·0	16 24	50 05	76·7	16 10	49 04	76·4	15 56	48 03	76·1	15 42	47 03	75·9	201	339
22	158	17 39	51 55	76·6	17 24	50 54	76·3	17 10	49 53	76·0	16 56	48 52	75·7	16 41	47 51	75·4	16 25	46 51	75·2	202	338
23	157	18 26	51 43	76·0	18 11	50 42	75·7	17 56	49 41	75·4	17 41	48 40	75·0	17 25	47 39	74·7	17 09	46 38	74·4	203	337
24	156	19 13	51 30	75·3	18 57	50 29	75·0	18 42	49 28	74·7	18 26	48 27	74·3	18 09	47 26	74·0	17 53	46 25	73·7	204	336
25	155	20 00	51 17	74·7	19 44	50 15	74·3	19 27	49 14	74·0	19 10	48 13	73·6	18 53	47 12	73·3	18 36	46 12	73·0	205	335
26	154	20 46	51 03	74·0	20 30	50 01	73·6	20 13	49 00	73·3	19 55	47 59	72·9	19 37	46 58	72·6	19 19	45 57	72·3	206	334
27	153	21 33	50 48	73·3	21 15	49 47	73·0	20 58	48 45	72·6	20 40	47 44	72·2	20 21	46 43	71·9	20 02	45 42	71·5	207	333
28	152	22 19	50 33	72·6	22 01	49 31	72·3	21 43	48 30	71·9	21 24	47 28	71·5	21 05	46 28	71·1	20 45	45 27	70·8	208	332
29	151	23 06	50 17	72·0	22 47	49 15	71·6	22 28	48 14	71·2	22 08	47 12	70·8	21 48	46 11	70·4	21 28	45 11	70·0	209	331
30	150	23 52	50 00	71·3	23 32	48 58	70·8	23 12	47 57	70·4	22 52	46 55	70·0	22 31	45 54	69·6	22 10	44 54	69·3	210	330
31	149	24 37	49 43	70·5	24 17	48 41	70·1	23 57	47 39	69·7	23 36	46 38	69·3	23 14	45 37	68·9	22 52	44 36	68·5	211	329
32	148	25 23	49 25	69·8	25 02	48 23	69·4	24 41	47 21	69·0	24 19	46 19	68·5	23 57	45 18	68·1	23 34	44 17	67·7	212	328
33	147	26 09	49 06	69·1	25 47	48 04	68·7	25 25	47 02	68·2	25 02	46 00	67·8	24 40	44 59	67·3	24 16	43 58	66·9	213	327
34	146	26 54	48 46	68·4	26 32	47 44	67·9	26 09	46 42	67·4	25 45	45 40	67·0	25 22	44 39	66·6	24 58	43 39	66·1	214	326
35	145	27 39	48 26	67·6	27 16	47 23	67·1	26 52	46 21	66·7	26 28	45 20	66·2	26 04	44 19	65·8	25 39	43 18	65·3	215	325
36	144	28 24	48 04	66·9	28 00	47 02	66·4	27 36	46 00	65·9	27 11	44 58	65·4	26 46	43 57	65·0	26 20	42 57	64·5	216	324
37	143	29 08	47 42	66·1	28 44	46 40	65·6	28 19	45 38	65·1	27 53	44 36	64·6	27 27	43 35	64·2	27 01	42 34	63·7	217	323
38	142	29 52	47 19	65·3	29 27	46 17	64·8	29 01	45 15	64·3	28 35	44 13	63·8	28 08	43 12	63·3	27 41	42 12	62·9	218	322
39	141	30 36	46 56	64·5	30 10	45 53	64·0	29 44	44 51	63·5	29 17	43 49	63·0	28 49	42 48	62·5	28 21	41 48	62·0	219	321
40	140	31 20	46 31	63·7	30 53	45 28	63·2	30 26	44 26	62·7	29 58	43 25	62·2	29 30	42 24	61·7	29 01	41 23	61·2	220	320
41	139	32 03	46 05	62·9	31 36	45 03	62·4	31 08	44 01	61·8	30 39	42 59	61·3	30 10	41 58	60·8	29 41	40 58	60·3	221	319
42	138	32 46	45 39	62·1	32 18	44 36	61·5	31 49	43 34	61·0	31 20	42 33	60·5	30 50	41 32	59·9	30 20	40 32	59·4	222	318
43	137	33 29	45 11	61·3	33 00	44 09	60·7	32 30	43 07	60·1	32 00	42 05	59·6	31 30	41 05	59·1	30 59	40 04	58·5	223	317
44	136	34 12	44 43	60·4	33 42	43 40	59·8	33 11	42 38	59·3	32 40	41 37	58·7	32 09	40 36	58·2	31 37	39 36	57·6	224	316
45	135	34 54	44 13	59·6	34 23	43 11	59·0	33 52	42 09	58·4	33 20	41 08	57·8	32 48	40 07	57·3	32 15	39 08	56·7	225	315

Lat. / A		36°			37°			38°			39°			40°			41°			Lat. / A	
LHA/F		A/H	B/P	Z_1/Z_2	A/H	B/P	Z_1/Z_2	A/H	B/P	Z_1/Z_2	A/H	B/P	Z_1/Z_2	A/H	B/P	Z_1/Z_2	A/H	B/P	Z_1/Z_2	LHA	
°	°	° ′	° ′	°	° ′	° ′	°	° ′	° ′	°	° ′	° ′	°	° ′	° ′	°	° ′	° ′	°	°	°
45	135	34 54	44 13	59·6	34 23	43 11	59·0	33 52	42 09	58·4	33 20	41 08	57·8	32 48	40 07	57·3	32 15	39 08	56·7	225	315
46	134	35 35	43 43	58·7	35 04	42 40	58·1	34 32	41 38	57·5	33 59	40 37	56·9	33 26	39 37	56·4	32 53	38 38	55·8	226	314
47	133	36 17	43 11	57·8	35 44	42 09	57·2	35 12	41 07	56·6	34 38	40 06	56·0	34 04	39 06	55·4	33 30	38 07	54·9	227	313
48	132	36 57	42 39	56·9	36 24	41 36	56·2	35 51	40 35	55·6	35 17	39 34	55·0	34 42	38 34	54·5	34 07	37 35	53·9	228	312
49	131	37 38	42 05	55·9	37 04	41 03	55·3	36 30	40 01	54·7	35 55	39 01	54·1	35 19	38 01	53·5	34 43	37 03	53·0	229	311
50	130	38 18	41 30	55·0	37 43	40 28	54·4	37 08	39 27	53·7	36 32	38 27	53·1	35 56	37 27	52·5	35 19	36 29	52·0	230	310
51	129	38 57	40 54	54·0	38 22	39 52	53·4	37 46	38 51	52·8	37 09	37 51	52·1	36 32	36 52	51·6	35 55	35 54	51·0	231	309
52	128	39 36	40 17	53·0	39 00	39 15	52·4	38 23	38 14	51·8	37 46	37 15	51·1	37 08	36 16	50·6	36 30	35 18	50·0	232	308
53	127	40 15	39 38	52·0	39 38	38 37	51·4	39 00	37 36	50·8	38 22	36 37	50·1	37 43	35 39	49·5	37 04	34 42	49·0	233	307
54	126	40 53	38 58	51·0	40 15	37 57	50·4	39 36	36 57	49·7	38 57	35 58	49·1	38 18	35 01	48·5	37 38	34 04	47·9	234	306
55	125	41 30	38 17	50·0	40 52	37 17	49·3	40 12	36 17	48·7	39 32	35 19	48·1	38 52	34 21	47·4	38 11	33 25	46·9	235	305
56	124	42 07	37 35	48·9	41 28	36 35	48·3	40 47	35 36	47·6	40 07	34 38	47·0	39 26	33 41	46·4	38 44	32 45	45·8	236	304
57	123	42 44	36 51	47·9	42 03	35 51	47·2	41 22	34 53	46·5	40 41	33 55	45·9	39 59	32 59	45·3	39 16	32 04	44·7	237	303
58	122	43 19	36 06	46·8	42 38	35 07	46·1	41 56	34 09	45·4	41 14	33 12	44·8	40 31	32 16	44·2	39 48	31 22	43·6	238	302
59	121	43 54	35 20	45·6	43 12	34 21	45·0	42 29	33 24	44·3	41 46	32 27	43·7	41 03	31 32	43·1	40 19	30 39	42·5	239	301
60	120	44 29	34 32	44·5	43 46	33 34	43·8	43 02	32 37	43·2	42 18	31 42	42·5	41 34	30 47	41·9	40 49	29 54	41·3	240	300
61	119	45 02	33 43	43·3	44 18	32 45	42·6	43 34	31 49	42·0	42 49	30 55	41·4	42 04	30 01	40·8	41 18	29 09	40·2	241	299
62	118	45 35	32 52	42·1	44 51	31 55	41·5	44 05	31 00	40·8	43 20	30 06	40·2	42 34	29 14	39·6	41 47	28 22	39·0	242	298
63	117	46 07	32 00	40·9	45 22	31 04	40·3	44 36	30 10	39·6	43 49	29 17	39·0	43 03	28 25	38·4	42 15	27 35	37·8	243	297
64	116	46 39	31 06	39·7	45 52	30 11	39·0	45 06	29 18	38·4	44 18	28 26	37·8	43 31	27 35	37·2	42 43	26 46	36·6	244	296
65	115	47 09	30 11	38·4	46 22	29 17	37·8	45 35	28 25	37·1	44 47	27 34	36·5	43 58	26 44	36·0	43 09	25 56	35·4	245	295
66	114	47 39	29 14	37·1	46 51	28 21	36·5	46 03	27 30	35·9	45 14	26 40	35·3	44 25	25 52	34·7	43 35	25 04	34·2	246	294
67	113	48 08	28 16	35·8	47 19	27 24	35·2	46 30	26 34	34·6	45 40	25 45	34·0	44 50	24 58	33·4	44 00	24 12	32·9	247	293
68	112	48 36	27 17	34·5	47 46	26 26	33·9	46 56	25 37	33·3	46 06	24 50	32·7	45 15	24 03	32·2	44 24	23 19	31·6	248	292
69	111	49 03	26 15	33·1	48 13	25 26	32·5	47 22	24 38	31·9	46 31	23 52	31·4	45 39	23 08	30·8	44 48	22 24	30·3	249	291
70	110	49 29	25 13	31·8	48 38	24 25	31·2	47 46	23 39	30·6	46 55	22 54	30·0	46 03	22 11	29·5	45 10	21 29	29·0	250	290
71	109	49 54	24 08	30·4	49 02	23 22	29·8	48 10	22 37	29·2	47 17	21 54	28·7	46 25	21 12	28·2	45 32	20 32	27·7	251	289
72	108	50 18	23 02	28·9	49 25	22 18	28·4	48 33	21 35	27·8	47 39	20 53	27·3	46 46	20 13	26·8	45 52	19 34	26·3	252	288
73	107	50 41	21 55	27·5	49 48	21 12	26·9	48 54	20 31	26·4	48 00	19 51	25·9	47 06	19 13	25·4	46 12	18 35	25·0	253	287
74	106	51 03	20 47	26·0	50 09	20 06	25·5	49 15	19 26	25·0	48 20	18 48	24·5	47 25	18 11	24·0	46 30	17 36	23·6	254	286
75	105	51 24	19 36	24·5	50 29	18 57	24·0	49 34	18 20	23·5	48 39	17 43	23·1	47 44	17 09	22·6	46 48	16 35	22·2	255	285
76	104	51 43	18 25	23·0	50 48	17 48	22·5	49 52	17 12	22·0	48 57	16 38	21·6	48 01	16 05	21·2	47 05	15 33	20·8	256	284
77	103	52 02	17 12	21·4	51 06	16 37	21·0	50 09	16 04	20·6	49 13	15 31	20·1	48 17	15 00	19·8	47 20	14 31	19·4	257	283
78	102	52 19	15 58	19·9	51 22	15 25	19·5	50 25	14 54	19·0	49 29	14 24	18·7	48 32	13 55	18·3	47 35	13 27	18·0	258	282
79	101	52 35	14 43	18·3	51 37	14 13	17·9	50 40	13 43	17·5	49 43	13 16	17·2	48 46	12 49	16·8	47 48	12 23	16·5	259	281
80	100	52 49	13 27	16·7	51 52	12 59	16·3	50 54	12 32	16·0	49 56	12 06	15·7	48 58	11 42	15·3	48 01	11 18	15·0	260	280
81	99	53 02	12 09	15·1	52 04	11 44	14·7	51 06	11 19	14·4	50 08	10 56	14·1	49 10	10 34	13·8	48 12	10 12	13·6	261	279
82	98	53 14	10 51	13·4	52 16	10 28	13·1	51 18	10 06	12·9	50 19	9 45	12·6	49 20	9 25	12·3	48 22	9 06	12·1	262	278
83	97	53 25	9 31	11·8	52 26	9 11	11·5	51 27	8 52	11·3	50 29	8 34	11·0	49 30	8 16	10·8	48 31	7 59	10·6	263	277
84	96	53 34	8 11	10·1	52 35	7 54	9·9	51 36	7 37	9·7	50 37	7 21	9·5	49 38	7 06	9·3	48 38	6 51	9·1	264	276
85	95	53 42	6 50	8·5	52 43	6 36	8·3	51 43	6 22	8·1	50 44	6 09	7·9	49 44	5 56	7·8	48 45	5 44	7·6	265	275
86	94	53 49	5 29	6·8	52 49	5 17	6·6	51 49	5 06	6·5	50 50	4 55	6·3	49 50	4 45	6·2	48 50	4 35	6·1	266	274
87	93	53 54	4 07	5·1	52 54	3 58	5·0	51 54	3 50	4·9	50 54	3 42	4·8	49 54	3 34	4·7	48 55	3 27	4·6	267	273
88	92	53 57	2 45	3·4	52 57	2 39	3·3	51 57	2 33	3·2	50 57	2 28	3·2	49 58	2 23	3·1	48 58	2 18	3·0	268	272
89	91	53 59	1 23	1·7	52 59	1 20	1·7	51 59	1 17	1·6	50 59	1 14	1·6	49 59	1 11	1·6	48 59	1 09	1·5	269	271
90	90	54 00	0 00	0·0	53 00	0 00	0·0	52 00	0 00	0·0	51 00	0 00	0·0	50 00	0 00	0·0	49 00	0 00	0·0	270	270

N. Lat: for LHA > 180° … $Z_n = Z$
for LHA < 180° … $Z_n = 360° - Z$

S. Lat.: for LHA > 180° … $Z_n = 180° - Z$
for LHA < 180° … $Z_n = 180° + Z$

SIGHT REDUCTION TABLE

B: (–) for 90° < LHA < 270°
Dec:(–) for Lat. contrary name

Z_1: same sign as B
Z_2: (–) for F > 90°

Lat. / A		42°			43°			44°			45°			46°			47°			Lat. / A	
LHA/F		A/H	B/P	Z_1/Z_2	A/H	B/P	Z_1/Z_2	A/H	B/P	Z_1/Z_2	A/H	B/P	Z_1/Z_2	A/H	B/P	Z_1/Z_2	A/H	B/P	Z_1/Z_2	LHA	
°	°	° ′	° ′	°	° ′	° ′	°	° ′	° ′	°	° ′	° ′	°	° ′	° ′	°	° ′	° ′	°	°	°
0	180	0 00	48 00	90·0	0 00	47 00	90·0	0 00	46 00	90·0	0 00	45 00	90·0	0 00	44 00	90·0	0 00	43 00	90·0	180	360
1	179	0 45	48 00	89·3	0 44	47 00	89·3	0 43	46 00	89·3	0 42	45 00	89·3	0 42	44 00	89·3	0 41	43 00	89·3	181	359
2	178	1 29	47 59	88·7	1 28	46 59	88·6	1 26	45 59	88·6	1 25	44 59	88·6	1 23	43 59	88·6	1 22	42 59	88·5	182	358
3	177	2 14	47 58	88·0	2 12	46 58	88·0	2 09	45 58	87·9	2 07	44 58	87·9	2 05	43 58	87·8	2 03	42 58	87·8	183	357
4	176	2 58	47 56	87·3	2 55	46 56	87·3	2 53	45 56	87·2	2 50	44 56	87·2	2 47	43 56	87·1	2 44	42 56	87·1	184	356
5	175	3 43	47 53	86·6	3 39	46 53	86·6	3 36	45 53	86·5	3 32	44 53	86·5	3 28	43 53	86·4	3 24	42 53	86·3	185	355
6	174	4 27	47 51	86·0	4 23	46 51	85·9	4 19	45 51	85·8	4 14	44 51	85·7	4 10	43 51	85·7	4 05	42 51	85·6	186	354
7	173	5 12	47 47	85·3	5 07	46 47	85·2	5 02	45 47	85·1	4 57	44 47	85·0	4 51	43 47	85·0	4 46	42 47	84·9	187	353
8	172	5 56	47 43	84·6	5 51	46 43	84·5	5 45	45 43	84·4	5 39	44 43	84·3	5 33	43 43	84·2	5 27	42 43	84·1	188	352
9	171	6 41	47 39	84·0	6 34	46 39	83·8	6 28	45 39	83·7	6 21	44 39	83·6	6 14	43 39	83·5	6 07	42 39	83·4	189	351
10	170	7 25	47 34	83·3	7 18	46 34	83·1	7 11	45 34	83·0	7 03	44 34	82·9	6 56	43 34	82·8	6 48	42 34	82·7	190	350
11	169	8 09	47 28	82·6	8 01	46 28	82·4	7 53	45 28	82·3	7 45	44 28	82·2	7 37	43 28	82·0	7 29	42 28	81·9	191	349
12	168	8 53	47 22	81·9	8 45	46 22	81·8	8 36	45 22	81·6	8 27	44 22	81·5	8 18	43 22	81·3	8 09	42 22	81·2	192	348
13	167	9 37	47 16	81·2	9 28	46 15	81·1	9 19	45 15	80·9	9 09	44 15	80·7	8 59	43 15	80·6	8 49	42 16	80·4	193	347
14	166	10 21	47 08	80·5	10 11	46 08	80·3	10 01	45 08	80·2	9 51	44 08	80·0	9 40	43 08	79·8	9 30	42 08	79·7	194	346
15	165	11 05	47 01	79·8	10 55	46 00	79·6	10 44	45 00	79·5	10 33	44 00	79·3	10 21	43 00	79·1	10 10	42 01	78·9	195	345
16	164	11 49	46 52	79·1	11 38	45 52	78·9	11 26	44 52	78·7	11 14	43 52	78·5	11 02	42 52	78·3	10 50	41 52	78·2	196	344
17	163	12 33	46 43	78·4	12 21	45 43	78·2	12 08	44 43	78·0	11 56	43 43	77·8	11 43	42 43	77·6	11 30	41 44	77·4	197	343
18	162	13 17	46 34	77·7	13 04	45 34	77·5	12 51	44 34	77·3	12 37	43 34	77·1	12 24	42 34	76·8	12 10	41 34	76·6	198	342
19	161	14 00	46 24	77·0	13 46	45 24	76·8	13 33	44 24	76·5	13 19	43 24	76·3	13 04	42 24	76·1	12 50	41 24	75·9	199	341
20	160	14 43	46 13	76·3	14 29	45 13	76·1	14 15	44 13	75·8	14 00	43 13	75·6	13 45	42 13	75·3	13 29	41 14	75·1	200	340
21	159	15 27	46 02	75·6	15 12	45 02	75·3	14 56	44 02	75·1	14 41	43 02	74·8	14 25	42 02	74·6	14 09	41 03	74·3	201	339
22	158	16 10	45 50	74·9	15 54	44 50	74·6	15 38	43 50	74·3	15 22	42 50	74·1	15 05	41 50	73·8	14 48	40 51	73·5	202	338
23	157	16 53	45 38	74·1	16 36	44 38	73·9	16 19	43 38	73·6	16 02	42 38	73·3	15 45	41 38	73·0	15 27	40 39	72·8	203	337
24	156	17 36	45 25	73·4	17 18	44 25	73·1	17 01	43 25	72·8	16 43	42 25	72·5	16 25	41 25	72·2	16 06	40 26	72·0	204	336
25	155	18 18	45 11	72·7	18 00	44 11	72·4	17 42	43 11	72·1	17 23	42 11	71·8	17 04	41 12	71·5	16 45	40 12	71·2	205	335
26	154	19 01	44 57	71·9	18 42	43 57	71·6	18 23	42 57	71·3	18 03	41 57	71·0	17 44	40 57	70·7	17 24	39 58	70·4	206	334
27	153	19 43	44 42	71·2	19 24	43 42	70·8	19 04	42 42	70·5	18 43	41 42	70·2	18 23	40 43	69·9	18 02	39 43	69·6	207	333
28	152	20 25	44 26	70·4	20 05	43 26	70·1	19 44	42 26	69·7	19 23	41 27	69·4	19 02	40 27	69·1	18 40	39 28	68·8	208	332
29	151	21 07	44 10	69·6	20 46	43 10	69·3	20 25	42 10	68·9	20 03	41 10	68·6	19 41	40 11	68·3	19 18	39 12	67·9	209	331
30	150	21 49	43 53	68·9	21 27	42 53	68·5	21 05	41 53	68·1	20 42	40 54	67·8	20 19	39 54	67·4	19 56	38 55	67·1	210	330
31	149	22 30	43 35	68·1	22 08	42 35	67·7	21 45	41 36	67·3	21 21	40 36	67·0	20 58	39 37	66·6	20 34	38 38	66·3	211	329
32	148	23 11	43 17	67·3	22 48	42 17	66·9	22 24	41 17	66·5	22 00	40 18	66·2	21 36	39 19	65·8	21 11	38 20	65·4	212	328
33	147	23 53	42 58	66·5	23 28	41 58	66·1	23 04	40 58	65·7	22 39	39 59	65·3	22 14	39 00	65·0	21 48	38 02	64·6	213	327
34	146	24 33	42 38	65·7	24 08	41 38	65·3	23 43	40 39	64·9	23 17	39 40	64·5	22 51	38 41	64·1	22 25	37 42	63·7	214	326
35	145	25 14	42 18	64·9	24 48	41 18	64·5	24 22	40 18	64·1	23 56	39 19	63·7	23 29	38 21	63·3	23 02	37 23	62·9	215	325
36	144	25 54	41 56	64·1	25 28	40 57	63·6	25 01	39 57	63·2	24 34	38 58	62·8	24 06	38 00	62·4	23 38	37 02	62·0	216	324
37	143	26 34	41 34	63·2	26 07	40 35	62·8	25 39	39 35	62·4	25 11	38 37	61·9	24 43	37 38	61·5	24 14	36 41	61·1	217	323
38	142	27 14	41 11	62·4	26 46	40 12	61·9	26 17	39 13	61·5	25 48	38 14	61·1	25 19	37 16	60·7	24 50	36 19	60·3	218	322
39	141	27 53	40 48	61·5	27 24	39 48	61·1	26 55	38 50	60·6	26 25	37 51	60·2	25 55	36 53	59·8	25 25	35 56	59·4	219	321
40	140	28 32	40 23	60·7	28 02	39 24	60·2	27 32	38 25	59·8	27 02	37 27	59·3	26 31	36 30	58·9	26 00	35 32	58·5	220	320
41	139	29 11	39 58	59·8	28 40	38 59	59·3	28 10	38 01	58·9	27 38	37 03	58·4	27 07	36 05	58·0	26 35	35 08	57·6	221	319
42	138	29 49	39 32	58·9	29 18	38 33	58·4	28 46	37 35	58·0	28 14	36 37	57·5	27 42	35 40	57·1	27 09	34 43	56·6	222	318
43	137	30 27	39 05	58·0	29 55	38 06	57·5	29 23	37 08	57·1	28 50	36 11	56·6	28 17	35 14	56·1	27 43	34 18	55·7	223	317
44	136	31 05	38 37	57·1	30 32	37 39	56·6	29 59	36 41	56·1	29 25	35 44	55·7	28 51	34 47	55·2	28 17	33 51	54·8	224	316
45	135	31 42	38 09	56·2	31 08	37 10	55·7	30 34	36 13	55·2	30 00	35 16	54·7	29 25	34 20	54·3	28 50	33 24	53·8	225	315

Lat. / A		42°			43°			44°			45°			46°			47°			Lat. / A	
LHA/F		A/H	B/P	Z_1/Z_2	A/H	B/P	Z_1/Z_2	A/H	B/P	Z_1/Z_2	A/H	B/P	Z_1/Z_2	A/H	B/P	Z_1/Z_2	A/H	B/P	Z_1/Z_2	LHA	
°	°	° ′	° ′	°	° ′	° ′	°	° ′	° ′	°	° ′	° ′	°	° ′	° ′	°	° ′	° ′	°	°	°
45	135	31 42	38 09	56·2	31 08	37 10	55·7	30 34	36 13	55·2	30 00	35 16	54·7	29 25	34 20	54·3	28 50	33 24	53·8	225	315
46	134	32 19	37 39	55·3	31 45	36 41	54·8	31 10	35 44	54·3	30 34	34 47	53·8	29 59	33 51	53·3	29 23	32 56	52·9	226	314
47	133	32 55	37 08	54·3	32 20	36 11	53·8	31 45	35 14	53·3	31 08	34 18	52·8	30 32	33 22	52·4	29 55	32 27	51·9	227	313
48	132	33 31	36 37	53·4	32 55	35 40	52·9	32 19	34 43	52·3	31 42	33 47	51·9	31 05	32 52	51·4	30 27	31 58	50·9	228	312
49	131	34 07	36 05	52·4	33 30	35 08	51·9	32 53	34 11	51·4	32 15	33 16	50·9	31 37	32 21	50·4	30 59	31 27	49·9	229	311
50	130	34 42	35 31	51·4	34 04	34 35	50·9	33 26	33 39	50·4	32 48	32 44	49·9	32 09	31 50	49·4	31 30	30 56	48·9	230	310
51	129	35 17	34 57	50·4	34 38	34 01	49·9	33 59	33 05	49·4	33 20	32 11	48·9	32 40	31 17	48·4	32 00	30 24	47·9	231	309
52	128	35 51	34 22	49·4	35 12	33 26	48·9	34 32	32 31	48·4	33 52	31 37	47·9	33 11	30 44	47·4	32 30	29 52	46·9	232	308
53	127	36 24	33 45	48·4	35 44	32 50	47·9	35 04	31 56	47·3	34 23	31 02	46·8	33 42	30 10	46·3	33 00	29 18	45·9	233	307
54	126	36 57	33 08	47·4	36 17	32 13	46·8	35 35	31 20	46·3	34 54	30 27	45·8	34 12	29 35	45·3	33 29	28 44	44·8	234	306
55	125	37 30	32 30	46·3	36 48	31 36	45·8	36 06	30 43	45·2	35 24	29 50	44·7	34 41	28 59	44·2	33 58	28 08	43·8	235	305
56	124	38 02	31 51	45·2	37 19	30 57	44·7	36 37	30 04	44·2	35 53	29 13	43·6	35 10	28 22	43·2	34 26	27 32	42·7	236	304
57	123	38 33	31 10	44·1	37 50	30 17	43·6	37 06	29 25	43·1	36 22	28 34	42·6	35 38	27 45	42·1	34 53	26 56	41·6	237	303
58	122	39 04	30 29	43·0	38 20	29 36	42·5	37 36	28 45	42·0	36 51	27 55	41·5	36 06	27 06	41·0	35 20	26 18	40·5	238	302
59	121	39 34	29 46	41·9	38 49	28 55	41·4	38 04	28 04	40·9	37 19	27 15	40·4	36 33	26 27	39·9	35 46	25 39	39·4	239	301
60	120	40 04	29 03	40·8	39 18	28 12	40·2	38 32	27 22	39·7	37 46	26 34	39·2	36 59	25 46	38·8	36 12	25 00	38·3	240	300
61	119	40 32	28 18	39·6	39 46	27 28	39·1	38 59	26 39	38·6	38 12	25 52	38·1	37 25	25 05	37·6	36 37	24 20	37·2	241	299
62	118	41 00	27 32	38·5	40 13	26 43	37·9	39 26	25 56	37·4	38 38	25 09	36·9	37 50	24 23	36·5	37 02	23 39	36·0	242	298
63	117	41 28	26 45	37·3	40 40	25 58	36·8	39 52	25 11	36·3	39 03	24 25	35·8	38 14	23 40	35·3	37 25	22 57	34·9	243	297
64	116	41 54	25 58	36·1	41 06	25 11	35·6	40 17	24 25	35·1	39 28	23 40	34·6	38 38	22 57	34·1	37 48	22 14	33·7	244	296
65	115	42 20	25 09	34·9	41 31	24 23	34·4	40 41	23 38	33·9	39 51	22 55	33·4	39 01	22 12	33·0	38 11	21 31	32·5	245	295
66	114	42 45	24 19	33·6	41 55	23 34	33·1	41 05	22 50	32·7	40 14	22 08	32·2	39 23	21 27	31·8	38 32	20 46	31·3	246	294
67	113	43 10	23 28	32·4	42 19	22 44	31·9	41 28	22 02	31·4	40 37	21 21	31·0	39 45	20 40	30·5	38 53	20 01	30·1	247	293
68	112	43 33	22 35	31·1	42 42	21 53	30·6	41 50	21 12	30·2	40 58	20 32	29·7	40 06	19 53	29·3	39 13	19 15	28·9	248	292
69	111	43 56	21 42	29·8	43 04	21 01	29·4	42 11	20 22	28·9	41 19	19 43	28·5	40 26	19 05	28·1	39 33	18 29	27·7	249	291
70	110	44 18	20 48	28·5	43 25	20 08	28·1	42 32	19 30	27·7	41 38	18 53	27·2	40 45	18 17	26·8	39 51	17 41	26·5	250	290
71	109	44 38	19 53	27·2	43 45	19 15	26·8	42 51	18 38	26·4	41 57	18 02	26·0	41 03	17 27	25·6	40 09	16 53	25·2	251	289
72	108	44 58	18 57	25·9	44 04	18 20	25·5	43 10	17 45	25·1	42 16	17 10	24·7	41 21	16 37	24·3	40 26	16 05	24·0	252	288
73	107	45 17	17 59	24·6	44 23	17 24	24·1	43 28	16 51	23·8	42 33	16 18	23·4	41 38	15 46	23·0	40 42	15 15	22·7	253	287
74	106	45 35	17 01	23·2	44 40	16 28	22·8	43 45	15 56	22·4	42 49	15 25	22·1	41 54	14 54	21·7	40 58	14 25	21·4	254	286
75	105	45 53	16 02	21·8	44 57	15 31	21·4	44 01	15 00	21·1	43 05	14 31	20·8	42 09	14 02	20·4	41 12	13 34	20·1	255	285
76	104	46 09	15 02	20·4	45 12	14 33	20·1	44 16	14 04	19·7	43 19	13 36	19·4	42 23	13 09	19·1	41 26	12 43	18·8	256	284
77	103	46 24	14 02	19·0	45 27	13 34	18·7	44 30	13 07	18·4	43 33	12 41	18·1	42 36	12 15	17·8	41 39	11 51	17·5	257	283
78	102	46 38	13 00	17·6	45 40	12 34	17·3	44 43	12 09	17·0	43 46	11 45	16·7	42 48	11 21	16·5	41 51	10 58	16·2	258	282
79	101	46 51	11 58	16·2	45 53	11 34	15·9	44 55	11 11	15·6	43 57	10 48	15·4	43 00	10 26	15·1	42 02	10 05	14·9	259	281
80	100	47 03	10 55	14·8	46 04	10 33	14·5	45 06	10 12	14·2	44 08	9 51	14·0	43 10	9 31	13·8	42 12	9 12	13·6	260	280
81	99	47 13	9 51	13·3	46 15	9 31	13·1	45 16	9 12	12·8	44 18	8 53	12·6	43 19	8 35	12·4	42 21	8 18	12·2	261	279
82	98	47 23	8 47	11·9	46 24	8 29	11·6	45 26	8 12	11·4	44 27	7 55	11·2	43 28	7 39	11·1	42 29	7 24	10·9	262	278
83	97	47 32	7 42	10·4	46 33	7 27	10·2	45 34	7 12	10·0	44 34	6 57	9·9	43 35	6 43	9·7	42 36	6 29	9·5	263	277
84	96	47 39	6 37	8·9	46 40	6 24	8·8	45 41	6 11	8·6	44 41	5 58	8·5	43 42	5 46	8·3	42 42	5 34	8·2	264	276
85	95	47 46	5 32	7·4	46 46	5 20	7·3	45 46	5 09	7·2	44 47	4 59	7·1	43 47	4 49	6·9	42 48	4 39	6·8	265	275
86	94	47 51	4 26	6·0	46 51	4 17	5·9	45 51	4 08	5·7	44 52	3 59	5·6	43 52	3 51	5·6	42 52	3 43	5·5	266	274
87	93	47 55	3 20	4·5	46 55	3 13	4·4	45 55	3 06	4·3	44 55	3 00	4·2	43 55	2 54	4·2	42 56	2 48	4·1	267	273
88	92	47 58	2 13	3·0	46 58	2 09	2·9	45 58	2 04	2·9	44 58	2 00	2·8	43 58	1 56	2·8	42 58	1 52	2·7	268	272
89	91	47 59	1 07	1·5	46 59	1 04	1·5	45 59	1 02	1·4	44 59	1 00	1·4	43 59	0 58	1·4	43 00	0 56	1·4	269	271
90	90	48 00	0 00	0·0	47 00	0 00	0·0	46 00	0 00	0·0	45 00	0 00	0·0	44 00	0 00	0·0	43 00	0 00	0·0	270	270

N. Lat: for LHA > 180° … $Z_n = Z$
for LHA < 180° … $Z_n = 360° - Z$

S. Lat.: for LHA > 180° … $Z_n = 180° - Z$
for LHA < 180° … $Z_n = 180° + Z$

SIGHT REDUCTION TABLE

B: (−) for 90° < LHA < 270°
Dec:(−) for Lat. contrary name

Z_1: same sign as B
Z_2: (−) for F > 90°

Lat. / A		48°			49°			50°			51°			52°			53°			Lat. / A	
LHA/F		A/H	B/P	Z_1/Z_2	A/H	B/P	Z_1/Z_2	A/H	B/P	Z_1/Z_2	A/H	B/P	Z_1/Z_2	A/H	B/P	Z_1/Z_2	A/H	B/P	Z_1/Z_2	LHA	
°	°	° ′	° ′	°	° ′	° ′	°	° ′	° ′	°	° ′	° ′	°	° ′	° ′	°	° ′	° ′	°	°	°
0	180	0 00	42 00	90·0	0 00	41 00	90·0	0 00	40 00	90·0	0 00	39 00	90·0	0 00	38 00	90·0	0 00	37 00	90·0	180	360
1	179	0 40	42 00	89·3	0 39	41 00	89·2	0 39	40 00	89·2	0 38	39 00	89·2	0 37	38 00	89·2	0 36	37 00	89·2	181	359
2	178	1 20	41 59	88·5	1 19	40 59	88·5	1 17	39 59	88·5	1 16	38 59	88·4	1 14	37 59	88·4	1 12	36 59	88·4	182	358
3	177	2 00	41 58	87·8	1 58	40 58	87·7	1 56	39 58	87·7	1 53	38 58	87·7	1 51	37 58	87·6	1 48	36 58	87·6	183	357
4	176	2 41	41 56	87·0	2 37	40 56	87·0	2 34	39 56	86·9	2 31	38 56	86·9	2 28	37 56	86·8	2 24	36 56	86·8	184	356
5	175	3 21	41 53	86·3	3 17	40 54	86·2	3 13	39 54	86·2	3 09	38 54	86·1	3 05	37 54	86·1	3 00	36 54	86·0	185	355
6	174	4 01	41 51	85·5	3 56	40 51	85·5	3 51	39 51	85·4	3 46	38 51	85·3	3 41	37 51	85·3	3 36	36 51	85·2	186	354
7	173	4 41	41 47	84·8	4 35	40 47	84·7	4 30	39 47	84·6	4 24	38 47	84·5	4 18	37 48	84·5	4 12	36 48	84·4	187	353
8	172	5 21	41 43	84·0	5 14	40 43	83·9	5 08	39 43	83·9	5 01	38 44	83·8	4 55	37 44	83·7	4 48	36 44	83·6	188	352
9	171	6 01	41 39	83·3	5 53	40 39	83·2	5 46	39 39	83·1	5 39	38 39	83·0	5 32	37 39	82·9	5 24	36 40	82·8	189	351
10	170	6 40	41 34	82·5	6 32	40 34	82·4	6 25	39 34	82·3	6 16	38 34	82·2	6 08	37 35	82·1	6 00	36 35	82·0	190	350
11	169	7 20	41 28	81·8	7 11	40 28	81·7	7 03	39 29	81·5	6 54	38 29	81·4	6 45	37 29	81·3	6 36	36 29	81·2	191	349
12	168	8 00	41 22	81·0	7 50	40 22	80·9	7 41	39 23	80·8	7 31	38 23	80·6	7 21	37 23	80·5	7 11	36 24	80·4	192	348
13	167	8 39	41 16	80·3	8 29	40 16	80·1	8 19	39 16	80·0	8 08	38 16	79·8	7 58	37 17	79·7	7 47	36 17	79·6	193	347
14	166	9 19	41 09	79·5	9 08	40 09	79·3	8 57	39 09	79·2	8 45	38 09	79·0	8 34	37 10	78·9	8 22	36 10	78·7	194	346
15	165	9 58	41 01	78·7	9 47	40 01	78·6	9 35	39 02	78·4	9 22	38 02	78·2	9 10	37 02	78·1	8 58	36 03	77·9	195	345
16	164	10 38	40 53	78·0	10 25	39 53	77·8	10 12	38 53	77·6	9 59	37 54	77·4	9 46	36 54	77·3	9 33	35 55	77·1	196	344
17	163	11 17	40 44	77·2	11 04	39 44	77·0	10 50	38 45	76·8	10 36	37 45	76·6	10 22	36 46	76·5	10 08	35 47	76·3	197	343
18	162	11 56	40 34	76·4	11 42	39 35	76·2	11 27	38 35	76·0	11 13	37 36	75·8	10 58	36 37	75·6	10 43	35 38	75·5	198	342
19	161	12 35	40 25	75·6	12 20	39 25	75·4	12 05	38 26	75·2	11 49	37 26	75·0	11 34	36 27	74·8	11 18	35 28	74·6	199	341
20	160	13 14	40 14	74·9	12 58	39 15	74·6	12 42	38 15	74·4	12 26	37 16	74·2	12 09	36 17	74·0	11 53	35 18	73·8	200	340
21	159	13 52	40 03	74·1	13 36	39 04	73·8	13 19	38 04	73·6	13 02	37 05	73·4	12 45	36 06	73·2	12 27	35 08	73·0	201	339
22	158	14 31	39 51	73·3	14 14	38 52	73·0	13 56	37 53	72·8	13 38	36 54	72·6	13 20	35 55	72·3	13 02	34 56	72·1	202	338
23	157	15 09	39 39	72·5	14 51	38 40	72·2	14 33	37 41	72·0	14 14	36 42	71·7	13 55	35 43	71·5	13 36	34 45	71·3	203	337
24	156	15 48	39 26	71·7	15 29	38 27	71·4	15 09	37 28	71·2	14 50	36 30	70·9	14 30	35 31	70·7	14 10	34 33	70·4	204	336
25	155	16 26	39 13	70·9	16 06	38 14	70·6	15 46	37 15	70·3	15 25	36 17	70·1	15 05	35 18	69·8	14 44	34 20	69·6	205	335
26	154	17 03	38 59	70·1	16 43	38 00	69·8	16 22	37 01	69·5	16 01	36 03	69·2	15 39	35 05	69·0	15 18	34 07	68·7	206	334
27	153	17 41	38 44	69·3	17 20	37 46	69·0	16 58	36 47	68·7	16 36	35 49	68·4	16 14	34 51	68·1	15 51	33 53	67·9	207	333
28	152	18 19	38 29	68·4	17 56	37 30	68·1	17 34	36 32	67·8	17 11	35 34	67·5	16 48	34 36	67·3	16 25	33 38	67·0	208	332
29	151	18 56	38 13	67·6	18 33	37 15	67·3	18 09	36 16	67·0	17 46	35 18	66·7	17 22	34 21	66·4	16 58	33 23	66·1	209	331
30	150	19 33	37 57	66·8	19 09	36 58	66·5	18 45	36 00	66·1	18 20	35 03	65·8	17 56	34 05	65·5	17 31	33 08	65·2	210	330
31	149	20 10	37 40	65·9	19 45	36 41	65·6	19 20	35 44	65·3	18 55	34 46	65·0	18 29	33 49	64·7	18 03	32 52	64·4	211	329
32	148	20 46	37 22	65·1	20 21	36 24	64·8	19 55	35 26	64·4	19 29	34 29	64·1	19 02	33 32	63·8	18 36	32 35	63·5	212	328
33	147	21 22	37 03	64·2	20 56	36 06	63·9	20 30	35 08	63·6	20 03	34 11	63·2	19 35	33 14	62·9	19 08	32 18	62·6	213	327
34	146	21 58	36 44	63·4	21 31	35 47	63·0	21 04	34 49	62·7	20 36	33 53	62·3	20 08	32 56	62·0	19 40	32 00	61·7	214	326
35	145	22 34	36 25	62·5	22 06	35 27	62·1	21 38	34 30	61·8	21 10	33 33	61·4	20 41	32 37	61·1	20 12	31 41	60·8	215	325
36	144	23 10	36 04	61·6	22 41	35 07	61·3	22 12	34 10	60·9	21 43	33 14	60·5	21 13	32 18	60·2	20 43	31 22	59·9	216	324
37	143	23 45	35 43	60·8	23 15	34 46	60·4	22 45	33 50	60·0	22 15	32 53	59·6	21 45	31 58	59·3	21 14	31 02	59·0	217	323
38	142	24 20	35 21	59·9	23 49	34 25	59·5	23 19	33 28	59·1	22 48	32 33	58·7	22 16	31 37	58·4	21 45	30 42	58·0	218	322
39	141	24 54	34 59	59·0	24 23	34 02	58·6	23 52	33 07	58·2	23 20	32 11	57·8	22 48	31 16	57·5	22 15	30 21	57·1	219	321
40	140	25 28	34 36	58·1	24 57	33 40	57·7	24 24	32 44	57·3	23 52	31 49	56·9	23 19	30 54	56·5	22 45	30 00	56·2	220	320
41	139	26 02	34 12	57·1	25 30	33 16	56·7	24 57	32 21	56·3	24 23	31 26	56·0	23 49	30 32	55·6	23 15	29 38	55·2	221	319
42	138	26 36	33 47	56·2	26 02	32 52	55·8	25 28	31 57	55·4	24 54	31 02	55·0	24 20	30 08	54·6	23 45	29 15	54·3	222	318
43	137	27 09	33 22	55·3	26 35	32 27	54·9	26 00	31 32	54·5	25 25	30 38	54·1	24 50	29 45	53·7	24 14	28 52	53·3	223	317
44	136	27 42	32 56	54·3	27 07	32 01	53·9	26 31	31 07	53·5	25 55	30 13	53·1	25 19	29 20	52·7	24 43	28 28	52·4	224	316
45	135	28 14	32 29	53·4	27 38	31 35	53·0	27 02	30 41	52·5	26 25	29 48	52·1	25 48	28 55	51·8	25 11	28 03	51·4	225	315

Lat. / A		48°			49°			50°			51°			52°			53°			Lat. / A	
LHA/F		A/H	B/P	Z_1/Z_2	A/H	B/P	Z_1/Z_2	A/H	B/P	Z_1/Z_2	A/H	B/P	Z_1/Z_2	A/H	B/P	Z_1/Z_2	A/H	B/P	Z_1/Z_2	LHA	
°	°	° ′	° ′	°	° ′	° ′	°	° ′	° ′	°	° ′	° ′	°	° ′	° ′	°	° ′	° ′	°	°	°
45	135	28 14	32 29	53·4	27 38	31 35	53·0	27 02	30 41	52·5	26 25	29 48	52·1	25 48	28 55	51·8	25 11	28 03	51·4	225	315
46	134	28 46	32 01	52·4	28 10	31 08	52·0	27 32	30 14	51·6	26 55	29 22	51·2	26 17	28 29	50·8	25 39	27 38	50·4	226	314
47	133	29 18	31 33	51·4	28 40	30 40	51·0	28 02	29 47	50·6	27 24	28 55	50·2	26 46	28 03	49·8	26 07	27 12	49·4	227	313
48	132	29 49	31 04	50·5	29 11	30 11	50·0	28 32	29 19	49·6	27 53	28 27	49·2	27 14	27 36	48·8	26 34	26 46	48·4	228	312
49	131	30 20	30 34	49·5	29 41	29 42	49·0	29 01	28 50	48·6	28 21	27 59	48·2	27 41	27 08	47·8	27 01	26 18	47·4	229	311
50	130	30 50	30 04	48·5	30 10	29 12	48·0	29 30	28 20	47·6	28 49	27 30	47·2	28 08	26 40	46·8	27 27	25 51	46·4	230	310
51	129	31 20	29 32	47·5	30 39	28 41	47·0	29 58	27 50	46·6	29 17	27 00	46·2	28 35	26 11	45·8	27 53	25 22	45·4	231	309
52	128	31 49	29 00	46·4	31 08	28 09	46·0	30 26	27 19	45·6	29 44	26 30	45·2	29 01	25 41	44·8	28 19	24 53	44·4	232	308
53	127	32 18	28 27	45·4	31 36	27 37	45·0	30 53	26 48	44·5	30 10	25 59	44·1	29 27	25 11	43·7	28 44	24 24	43·3	233	307
54	126	32 46	27 53	44·4	32 03	27 04	43·9	31 20	26 15	43·5	30 36	25 27	43·1	29 52	24 40	42·7	29 08	23 53	42·3	234	306
55	125	33 14	27 19	43·3	32 30	26 30	42·9	31 46	25 42	42·4	31 02	24 55	42·0	30 17	24 08	41·6	29 32	23 23	41·2	235	305
56	124	33 42	26 44	42·2	32 57	25 55	41·8	32 12	25 08	41·4	31 27	24 22	41·0	30 41	23 36	40·6	29 56	22 51	40·2	236	304
57	123	34 08	26 07	41·1	33 23	25 20	40·7	32 37	24 34	40·3	31 51	23 48	39·9	31 05	23 03	39·5	30 19	22 19	39·1	237	303
58	122	34 34	25 30	40·1	33 48	24 44	39·6	33 02	23 58	39·2	32 15	23 14	38·8	31 28	22 29	38·4	30 41	21 46	38·0	238	302
59	121	35 00	24 53	39·0	34 13	24 07	38·5	33 26	23 22	38·1	32 39	22 38	37·7	31 51	21 55	37·3	31 03	21 13	37·0	239	301
60	120	35 25	24 14	37·8	34 37	23 30	37·4	33 50	22 46	37·0	33 02	22 03	36·6	32 13	21 20	36·2	31 25	20 39	35·9	240	300
61	119	35 49	23 35	36·7	35 01	22 51	36·3	34 12	22 08	35·9	33 24	21 26	35·5	32 35	20 45	35·1	31 46	20 04	34·8	241	299
62	118	36 13	22 55	35·6	35 24	22 12	35·2	34 35	21 30	34·8	33 45	20 49	34·4	32 56	20 09	34·0	32 06	19 29	33·7	242	298
63	117	36 36	22 14	34·4	35 46	21 32	34·0	34 56	20 51	33·6	34 06	20 11	33·3	33 16	19 32	32·9	32 26	18 53	32·5	243	297
64	116	36 58	21 32	33·3	36 08	20 52	32·9	35 17	20 12	32·5	34 27	19 33	32·1	33 36	18 54	31·8	32 45	18 17	31·4	244	296
65	115	37 20	20 50	32·1	36 29	20 10	31·7	35 38	19 32	31·3	34 47	18 54	31·0	33 55	18 16	30·6	33 03	17 40	30·3	245	295
66	114	37 41	20 07	30·9	36 49	19 28	30·5	35 58	18 51	30·2	35 06	18 14	29·8	34 13	17 38	29·5	33 21	17 02	29·1	246	294
67	113	38 01	19 23	29·7	37 09	18 46	29·4	36 17	18 09	29·0	35 24	17 33	28·6	34 31	16 59	28·3	33 38	16 24	28·0	247	293
68	112	38 21	18 38	28·5	37 28	18 02	28·2	36 35	17 27	27·8	35 42	16 53	27·5	34 48	16 19	27·1	33 55	15 46	26·8	248	292
69	111	38 40	17 53	27·3	37 46	17 18	27·0	36 53	16 44	26·6	35 59	16 11	26·3	35 05	15 38	26·0	34 11	15 07	25·7	249	291
70	110	38 58	17 07	26·1	38 04	16 33	25·7	37 10	16 01	25·4	36 15	15 29	25·1	35 21	14 58	24·8	34 26	14 27	24·5	250	290
71	109	39 15	16 20	24·9	38 20	15 48	24·5	37 26	15 17	24·2	36 31	14 46	23·9	35 36	14 16	23·6	34 41	13 47	23·3	251	289
72	108	39 31	15 33	23·6	38 36	15 02	23·3	37 41	14 32	23·0	36 46	14 03	22·7	35 50	13 34	22·4	34 55	13 07	22·1	252	288
73	107	39 47	14 45	22·4	38 51	14 16	22·1	37 56	13 47	21·8	37 00	13 19	21·5	36 04	12 52	21·2	35 08	12 25	20·9	253	287
74	106	40 02	13 56	21·1	39 06	13 28	20·8	38 10	13 01	20·5	37 13	12 35	20·3	36 17	12 09	20·0	35 21	11 44	19·8	254	286
75	105	40 16	13 07	19·8	39 19	12 41	19·5	38 23	12 15	19·3	37 26	11 50	19·0	36 29	11 26	18·8	35 33	11 02	18·5	255	285
76	104	40 29	12 17	18·5	39 32	11 53	18·3	38 35	11 28	18·0	37 38	11 05	17·8	36 41	10 42	17·6	35 44	10 20	17·3	256	284
77	103	40 41	11 27	17·3	39 44	11 04	17·0	38 47	10 41	16·8	37 49	10 19	16·5	36 52	9 58	16·3	35 54	9 37	16·1	257	283
78	102	40 53	10 36	16·0	39 55	10 15	15·7	38 57	9 54	15·5	38 00	9 33	15·3	37 02	9 14	15·1	36 04	8 54	14·9	258	282
79	101	41 04	9 45	14·7	40 05	9 25	14·4	39 07	9 06	14·2	38 09	8 47	14·0	37 11	8 29	13·9	36 13	8 11	13·7	259	281
80	100	41 13	8 53	13·3	40 15	8 35	13·2	39 16	8 17	13·0	38 18	8 00	12·8	37 19	7 44	12·6	36 21	7 27	12·5	260	280
81	99	41 22	8 01	12·0	40 23	7 45	11·9	39 25	7 29	11·7	38 26	7 13	11·5	37 27	6 58	11·4	36 28	6 43	11·2	261	279
82	98	41 30	7 09	10·7	40 31	6 54	10·5	39 32	6 40	10·4	38 33	6 26	10·3	37 34	6 12	10·1	36 35	5 59	10·0	262	278
83	97	41 37	6 16	9·4	40 38	6 03	9·2	39 39	5 50	9·1	38 39	5 38	9·0	37 40	5 26	8·9	36 41	5 15	8·7	263	277
84	96	41 43	5 23	8·1	40 44	5 12	7·9	39 44	5 01	7·8	38 45	4 50	7·7	37 45	4 40	7·6	36 46	4 30	7·5	264	276
85	95	41 48	4 29	6·7	40 49	4 20	6·6	39 49	4 11	6·5	38 49	4 02	6·4	37 50	3 54	6·3	36 50	3 45	6·3	265	275
86	94	41 52	3 36	5·4	40 53	3 28	5·3	39 53	3 21	5·2	38 53	3 14	5·1	37 53	3 07	5·1	36 54	3 01	5·0	266	274
87	93	41 56	2 42	4·0	40 56	2 36	4·0	39 56	2 31	3·9	38 56	2 26	3·9	37 56	2 20	3·8	36 56	2 16	3·8	267	273
88	92	41 58	1 48	2·7	40 58	1 44	2·6	39 58	1 41	2·6	38 58	1 37	2·6	37 58	1 34	2·5	36 58	1 30	2·5	268	272
89	91	42 00	0 54	1·3	41 00	0 52	1·3	40 00	0 50	1·3	39 00	0 49	1·3	38 00	0 47	1·3	37 00	0 45	1·3	269	271
90	90	42 00	0 00	0·0	41 00	0 00	0·0	40 00	0 00	0·0	39 00	0 00	0·0	38 00	0 00	0·0	37 00	0 00	0·0	270	270

N. Lat: for LHA > 180° … $Z_n = Z$
for LHA < 180° … $Z_n = 360° - Z$

S. Lat.: for LHA > 180° … $Z_n = 180° - Z$
for LHA < 180° … $Z_n = 180° + Z$

SIGHT REDUCTION TABLE

B: (–) for 90° < LHA < 270°
Dec: (–) for Lat. contrary name

Z_1: same sign as B
Z_2: (–) for F > 90°

Lat. / A		54°			55°			56°			57°			58°			59°			Lat. / A	
LHA/F		A/H	B/P	Z_1/Z_2	A/H	B/P	Z_1/Z_2	A/H	B/P	Z_1/Z_2	A/H	B/P	Z_1/Z_2	A/H	B/P	Z_1/Z_2	A/H	B/P	Z_1/Z_2	LHA	
°	°	° ′	° ′	°	° ′	° ′	°	° ′	° ′	°	° ′	° ′	°	° ′	° ′	°	° ′	° ′	°	°	°
0	180	0 00	36 00	90·0	0 00	35 00	90·0	0 00	34 00	90·0	0 00	33 00	90·0	0 00	32 00	90·0	0 00	31 00	90·0	180	360
1	179	0 35	36 00	89·2	0 34	35 00	89·2	0 34	34 00	89·2	0 33	33 00	89·2	0 32	32 00	89·2	0 31	31 00	89·1	181	359
2	178	1 11	35 59	88·4	1 09	34 59	88·4	1 07	33 59	88·3	1 05	32 59	88·3	1 04	31 59	88·3	1 02	30 59	88·3	182	358
3	177	1 46	35 58	87·6	1 43	34 58	87·5	1 41	33 58	87·5	1 38	32 58	87·5	1 35	31 58	87·5	1 33	30 58	87·4	183	357
4	176	2 21	35 56	86·8	2 18	34 56	86·7	2 14	33 56	86·7	2 11	32 56	86·6	2 07	31 56	86·6	2 04	30 56	86·6	184	356
5	175	2 56	35 54	86·0	2 52	34 54	85·9	2 48	33 54	85·9	2 43	32 54	85·8	2 39	31 54	85·8	2 34	30 54	85·7	185	355
6	174	3 31	35 51	85·1	3 26	34 51	85·1	3 21	33 51	85·0	3 16	32 51	85·0	3 11	31 52	84·9	3 05	30 52	84·9	186	354
7	173	4 06	35 48	84·3	4 00	34 48	84·3	3 54	33 48	84·2	3 48	32 48	84·1	3 42	31 48	84·1	3 36	30 49	84·0	187	353
8	172	4 42	35 44	83·5	4 35	34 44	83·4	4 28	33 44	83·4	4 21	32 45	83·3	4 14	31 45	83·2	4 07	30 45	83·1	188	352
9	171	5 17	35 40	82·7	5 09	34 40	82·6	5 01	33 40	82·5	4 53	32 41	82·4	4 45	31 41	82·3	4 37	30 41	82·3	189	351
10	170	5 51	35 35	81·9	5 43	34 35	81·8	5 34	33 36	81·7	5 26	32 36	81·6	5 17	31 36	81·5	5 08	30 37	81·4	190	350
11	169	6 26	35 30	81·1	6 17	34 30	81·0	6 08	33 31	80·8	5 58	32 31	80·7	5 48	31 31	80·6	5 38	30 32	80·5	191	349
12	168	7 01	35 24	80·2	6 51	34 24	80·1	6 41	33 25	80·0	6 30	32 25	79·9	6 20	31 26	79·8	6 09	30 27	79·7	192	348
13	167	7 36	35 18	79·4	7 25	34 18	79·3	7 14	33 19	79·2	7 02	32 19	79·0	6 51	31 20	78·9	6 39	30 21	78·8	193	347
14	166	8 11	35 11	78·6	7 59	34 12	78·5	7 46	33 12	78·3	7 34	32 13	78·2	7 22	31 14	78·1	7 09	30 15	77·9	194	346
15	165	8 45	35 04	77·8	8 32	34 04	77·6	8 19	33 05	77·5	8 06	32 06	77·3	7 53	31 07	77·2	7 40	30 08	77·1	195	345
16	164	9 19	34 56	76·9	9 06	33 57	76·8	8 52	32 58	76·6	8 38	31 58	76·5	8 24	31 00	76·3	8 10	30 01	76·2	196	344
17	163	9 54	34 47	76·1	9 39	33 48	75·9	9 25	32 49	75·8	9 10	31 50	75·6	8 55	30 52	75·5	8 40	29 53	75·3	197	343
18	162	10 28	34 39	75·3	10 13	33 40	75·1	9 57	32 41	74·9	9 41	31 42	74·8	9 25	30 43	74·6	9 09	29 45	74·4	198	342
19	161	11 02	34 29	74·4	10 46	33 30	74·2	10 29	32 32	74·1	10 13	31 33	73·9	9 56	30 35	73·7	9 39	29 36	73·6	199	341
20	160	11 36	34 19	73·6	11 19	33 21	73·4	11 02	32 22	73·2	10 44	31 24	73·0	10 27	30 25	72·8	10 09	29 27	72·7	200	340
21	159	12 10	34 09	72·7	11 52	33 10	72·5	11 34	32 12	72·3	11 15	31 14	72·2	10 57	30 15	72·0	10 38	29 17	71·8	201	339
22	158	12 43	33 58	71·9	12 24	33 00	71·7	12 06	32 01	71·5	11 46	31 03	71·3	11 27	30 05	71·1	11 07	29 07	70·9	202	338
23	157	13 17	33 46	71·0	12 57	32 48	70·8	12 37	31 50	70·6	12 17	30 52	70·4	11 57	29 54	70·2	11 37	28 57	70·0	203	337
24	156	13 50	33 34	70·2	13 29	32 36	70·0	13 09	31 38	69·7	12 48	30 41	69·5	12 27	29 43	69·3	12 06	28 46	69·1	204	336
25	155	14 23	33 22	69·3	14 02	32 24	69·1	13 40	31 26	68·9	13 18	30 29	68·6	12 56	29 31	68·4	12 34	28 34	68·2	205	335
26	154	14 56	33 09	68·5	14 34	32 11	68·2	14 11	31 14	68·0	13 49	30 16	67·8	13 26	29 19	67·5	13 03	28 22	67·3	206	334
27	153	15 29	32 55	67·6	15 06	31 58	67·3	14 42	31 00	67·1	14 19	30 03	66·9	13 55	29 06	66·6	13 31	28 10	66·4	207	333
28	152	16 01	32 41	66·7	15 37	31 44	66·5	15 13	30 47	66·2	14 49	29 50	66·0	14 24	28 53	65·7	14 00	27 57	65·5	208	332
29	151	16 33	32 26	65·8	16 09	31 29	65·6	15 44	30 32	65·3	15 19	29 36	65·1	14 53	28 39	64·8	14 28	27 43	64·6	209	331
30	150	17 05	32 11	65·0	16 40	31 14	64·7	16 14	30 17	64·4	15 48	29 21	64·2	15 22	28 25	63·9	14 55	27 29	63·7	210	330
31	149	17 37	31 55	64·1	17 11	30 58	63·8	16 44	30 02	63·5	16 17	29 06	63·3	15 50	28 10	63·0	15 23	27 15	62·7	211	329
32	148	18 09	31 38	63·2	17 42	30 42	62·9	17 14	29 46	62·6	16 47	28 51	62·3	16 19	27 55	62·1	15 50	27 00	61·8	212	328
33	147	18 40	31 21	62·3	18 12	30 25	62·0	17 44	29 30	61·7	17 15	28 34	61·4	16 47	27 39	61·2	16 17	26 45	60·9	213	327
34	146	19 11	31 04	61·4	18 42	30 08	61·1	18 13	29 13	60·8	17 44	28 18	60·5	17 14	27 23	60·2	16 44	26 29	60·0	214	326
35	145	19 42	30 46	60·5	19 12	29 50	60·2	18 42	28 55	59·9	18 12	28 01	59·6	17 42	27 06	59·3	17 11	26 12	59·0	215	325
36	144	20 13	30 27	59·6	19 42	29 32	59·2	19 11	28 37	58·9	18 40	27 43	58·6	18 09	26 49	58·4	17 37	25 55	58·1	216	324
37	143	20 43	30 07	58·6	20 12	29 13	58·3	19 40	28 19	58·0	19 08	27 25	57·7	18 36	26 31	57·4	18 03	25 38	57·1	217	323
38	142	21 13	29 48	57·7	20 41	28 53	57·4	20 08	27 59	57·1	19 35	27 06	56·8	19 02	26 13	56·5	18 29	25 20	56·2	218	322
39	141	21 43	29 27	56·8	21 10	28 33	56·4	20 36	27 40	56·1	20 03	26 47	55·8	19 29	25 54	55·5	18 55	25 02	55·2	219	321
40	140	22 12	29 06	55·8	21 38	28 13	55·5	21 04	27 20	55·2	20 30	26 27	54·9	19 55	25 35	54·6	19 20	24 43	54·3	220	320
41	139	22 41	28 44	54·9	22 06	27 51	54·5	21 31	26 59	54·2	20 56	26 07	53·9	20 21	25 15	53·6	19 45	24 24	53·3	221	319
42	138	23 10	28 22	53·9	22 34	27 29	53·6	21 58	26 37	53·3	21 22	25 46	52·9	20 46	24 55	52·6	20 10	24 04	52·3	222	318
43	137	23 38	27 59	53·0	23 02	27 07	52·6	22 25	26 15	52·3	21 48	25 24	52·0	21 11	24 34	51·7	20 34	23 43	51·4	223	317
44	136	24 06	27 36	52·0	23 29	26 44	51·7	22 51	25 53	51·3	22 14	25 02	51·0	21 36	24 12	50·7	20 58	23 23	50·4	224	316
45	135	24 34	27 11	51·0	23 56	26 20	50·7	23 17	25 30	50·3	22 39	24 40	50·0	22 00	23 50	49·7	21 21	23 01	49·4	225	315

Lat. / A		54°			55°			56°			57°			58°			59°			Lat. / A	
LHA/F		A/H	B/P	Z_1/Z_2	A/H	B/P	Z_1/Z_2	A/H	B/P	Z_1/Z_2	A/H	B/P	Z_1/Z_2	A/H	B/P	Z_1/Z_2	A/H	B/P	Z_1/Z_2	LHA	
°	°	° ′	° ′	°	° ′	° ′	°	° ′	° ′	°	° ′	° ′	°	° ′	° ′	°	° ′	° ′	°	°	°
45	135	24 34	27 11	51·0	23 56	26 20	50·7	23 17	25 30	50·3	22 39	24 40	50·0	22 00	23 50	49·7	21 21	23 01	49·4	225	315
46	134	25 01	26 47	50·0	24 22	25 56	49·7	23 43	25 06	49·4	23 04	24 17	49·0	22 24	23 28	48·7	21 45	22 39	48·4	226	314
47	133	25 28	26 22	49·1	24 48	25 32	48·7	24 08	24 42	48·4	23 28	23 53	48·0	22 48	23 05	47·7	22 08	22 17	47·4	227	313
48	132	25 54	25 56	48·1	25 14	25 06	47·7	24 33	24 17	47·4	23 53	23 29	47·0	23 11	22 41	46·7	22 30	21 54	46·4	228	312
49	131	26 20	25 29	47·1	25 39	24 40	46·7	24 58	23 52	46·4	24 16	23 05	46·0	23 34	22 17	45·7	22 52	21 31	45·4	229	311
50	130	26 46	25 02	46·0	26 04	24 14	45·7	25 22	23 26	45·3	24 40	22 39	45·0	23 57	21 53	44·7	23 14	21 07	44·4	230	310
51	129	27 11	24 34	45·0	26 28	23 47	44·7	25 45	23 00	44·3	25 02	22 14	44·0	24 19	21 28	43·7	23 36	20 43	43·4	231	309
52	128	27 36	24 06	44·0	26 52	23 19	43·6	26 09	22 33	43·3	25 25	21 48	43·0	24 41	21 03	42·7	23 57	20 18	42·3	232	308
53	127	28 00	23 37	43·0	27 16	22 51	42·6	26 32	22 06	42·3	25 47	21 21	41·9	25 02	20 37	41·6	24 17	19 53	41·3	233	307
54	126	28 24	23 07	41·9	27 39	22 22	41·6	26 54	21 38	41·2	26 09	20 54	40·9	25 23	20 10	40·6	24 37	19 27	40·3	234	306
55	125	28 47	22 37	40·9	28 01	21 53	40·5	27 16	21 09	40·2	26 30	20 26	39·9	25 44	19 43	39·5	24 57	19 01	39·2	235	305
56	124	29 10	22 07	39·8	28 24	21 23	39·5	27 37	20 40	39·1	26 50	19 57	38·8	26 04	19 16	38·5	25 17	18 34	38·2	236	304
57	123	29 32	21 35	38·8	28 45	20 52	38·4	27 58	20 10	38·1	27 11	19 29	37·8	26 23	18 48	37·4	25 35	18 07	37·1	237	303
58	122	29 54	21 03	37·7	29 06	20 21	37·3	28 19	19 40	37·0	27 31	18 59	36·7	26 42	18 19	36·4	25 54	17 40	36·1	238	302
59	121	30 15	20 31	36·6	29 27	19 50	36·3	28 38	19 09	35·9	27 50	18 30	35·6	27 01	17 50	35·3	26 12	17 12	35·0	239	301
60	120	30 36	19 58	35·5	29 47	19 18	35·2	28 58	18 38	34·9	28 09	17 59	34·5	27 19	17 21	34·2	26 29	16 43	34·0	240	300
61	119	30 56	19 24	34·4	30 07	18 45	34·1	29 17	18 06	33·8	28 27	17 29	33·5	27 37	16 51	33·2	26 46	16 14	32·9	241	299
62	118	31 16	18 50	33·3	30 26	18 12	33·0	29 35	17 34	32·7	28 45	16 57	32·4	27 54	16 21	32·1	27 03	15 45	31·8	242	298
63	117	31 35	18 15	32·2	30 44	17 38	31·9	29 53	17 02	31·6	29 02	16 26	31·3	28 10	15 50	31·0	27 19	15 15	30·7	243	297
64	116	31 53	17 40	31·1	31 02	17 04	30·8	30 10	16 28	30·5	29 19	15 53	30·2	28 27	15 19	29·9	27 35	14 45	29·6	244	296
65	115	32 11	17 04	30·0	31 19	16 29	29·7	30 27	15 55	29·4	29 35	15 21	29·1	28 42	14 48	28·8	27 50	14 15	28·5	245	295
66	114	32 29	16 28	28·8	31 36	15 54	28·5	30 43	15 20	28·2	29 50	14 48	28·0	28 57	14 16	27·7	28 04	13 44	27·4	246	294
67	113	32 45	15 51	27·7	31 52	15 18	27·4	30 59	14 46	27·1	30 05	14 14	26·8	29 12	13 43	26·6	28 18	13 13	26·3	247	293
68	112	33 01	15 14	26·5	32 08	14 42	26·3	31 14	14 11	26·0	30 20	13 40	25·7	29 26	13 10	25·5	28 31	12 41	25·2	248	292
69	111	33 17	14 36	25·4	32 23	14 05	25·1	31 28	13 35	24·8	30 34	13 06	24·6	29 39	12 37	24·4	28 44	12 09	24·1	249	291
70	110	33 32	13 57	24·2	32 37	13 28	24·0	31 42	12 59	23·7	30 47	12 31	23·5	29 52	12 04	23·2	28 57	11 37	23·0	250	290
71	109	33 46	13 18	23·1	32 51	12 51	22·8	31 55	12 23	22·6	31 00	11 56	22·3	30 04	11 30	22·1	29 09	11 04	21·9	251	289
72	108	33 59	12 39	21·9	33 04	12 13	21·6	32 08	11 46	21·4	31 12	11 21	21·2	30 16	10 56	21·0	29 20	10 31	20·8	252	288
73	107	34 12	12 00	20·7	33 16	11 34	20·5	32 20	11 09	20·2	31 23	10 45	20·0	30 27	10 21	19·8	29 30	9 58	19·6	253	287
74	106	34 24	11 19	19·5	33 28	10 55	19·3	32 31	10 32	19·1	31 34	10 09	18·9	30 37	9 46	18·7	29 41	9 24	18·5	254	286
75	105	34 36	10 39	18·3	33 39	10 16	18·1	32 42	9 54	17·9	31 44	9 32	17·7	30 47	9 11	17·5	29 50	8 50	17·4	255	285
76	104	34 46	9 58	17·1	33 49	9 37	16·9	32 52	9 16	16·7	31 54	8 56	16·6	30 57	8 36	16·4	29 59	8 16	16·2	256	284
77	103	34 56	9 17	15·9	33 59	8 57	15·7	33 01	8 38	15·6	32 03	8 19	15·4	31 05	8 00	15·2	30 07	7 42	15·1	257	283
78	102	35 06	8 35	14·7	34 08	8 17	14·5	33 10	7 59	14·4	32 11	7 41	14·2	31 13	7 24	14·1	30 15	7 07	13·9	258	282
79	101	35 14	7 54	13·5	34 16	7 37	13·3	33 18	7 20	13·2	32 19	7 04	13·0	31 21	6 48	12·9	30 22	6 32	12·8	259	281
80	100	35 22	7 11	12·3	34 24	6 56	12·1	33 25	6 41	12·0	32 26	6 26	11·9	31 27	6 12	11·7	30 29	5 57	11·6	260	280
81	99	35 29	6 29	11·1	34 30	6 15	10·9	33 32	6 01	10·8	32 33	5 48	10·7	31 34	5 35	10·6	30 35	5 22	10·5	261	279
82	98	35 36	5 46	9·9	34 37	5 34	9·7	33 37	5 22	9·6	32 38	5 10	9·5	31 39	4 58	9·4	30 40	4 47	9·3	262	278
83	97	35 41	5 04	8·6	34 42	4 53	8·5	33 43	4 42	8·4	32 43	4 32	8·3	31 44	4 21	8·2	30 45	4 11	8·2	263	277
84	96	35 46	4 21	7·4	34 47	4 11	7·3	33 47	4 02	7·2	32 48	3 53	7·1	31 48	3 44	7·1	30 49	3 36	7·0	264	276
85	95	35 51	3 37	6·2	34 51	3 30	6·1	33 51	3 22	6·0	32 52	3 14	6·0	31 52	3 07	5·9	30 52	3 00	5·8	265	275
86	94	35 54	2 54	4·9	34 54	2 48	4·9	33 54	2 42	4·8	32 55	2 36	4·8	31 55	2 30	4·7	30 55	2 24	4·7	266	274
87	93	35 57	2 11	3·7	34 57	2 06	3·7	33 57	2 01	3·6	32 57	1 57	3·6	31 57	1 52	3·5	30 57	1 48	3·5	267	273
88	92	35 58	1 27	2·5	34 59	1 24	2·4	33 59	1 21	2·4	32 59	1 18	2·4	31 59	1 15	2·4	30 59	1 12	2·3	268	272
89	91	36 00	0 44	1·2	35 00	0 42	1·2	34 00	0 40	1·2	33 00	0 39	1·2	32 00	0 37	1·2	31 00	0 36	1·2	269	271
90	90	36 00	0 00	0·0	35 00	0 00	0·0	34 00	0 00	0·0	33 00	0 00	0·0	32 00	0 00	0·0	31 00	0 00	0·0	270	270

N. Lat: for LHA > 180° … $Z_n = Z$
for LHA < 180° … $Z_n = 360° - Z$

S. Lat.: for LHA > 180° … $Z_n = 180° - Z$
for LHA < 180° … $Z_n = 180° + Z$

SIGHT REDUCTION TABLE

B: (–) for 90° < LHA < 270°
Dec: (–) for Lat. contrary name

Z_1: same sign as B
Z_2: (–) for F > 90°

Lat. / A		60°			61°			62°			63°			64°			65°			Lat. / A	
LHA/F		A/H	B/P	Z_1/Z_2	A/H	B/P	Z_1/Z_2	A/H	B/P	Z_1/Z_2	A/H	B/P	Z_1/Z_2	A/H	B/P	Z_1/Z_2	A/H	B/P	Z_1/Z_2	LHA	
°	°	° ′	° ′	°	° ′	° ′	°	° ′	° ′	°	° ′	° ′	°	° ′	° ′	°	° ′	° ′	°	°	°
0	180	0 00	30 00	90·0	0 00	29 00	90·0	0 00	28 00	90·0	0 00	27 00	90·0	0 00	26 00	90·0	0 00	25 00	90·0	180	360
1	179	0 30	30 00	89·1	0 29	29 00	89·1	0 28	28 00	89·1	0 27	27 00	89·1	0 26	26 00	89·1	0 25	25 00	89·1	181	359
2	178	1 00	29 59	88·3	0 58	28 59	88·3	0 56	27 59	88·2	0 54	26 59	88·2	0 53	25 59	88·2	0 51	24 59	88·2	182	358
3	177	1 30	29 58	87·4	1 27	28 58	87·4	1 24	27 58	87·4	1 22	26 58	87·3	1 19	25 58	87·3	1 16	24 58	87·3	183	357
4	176	2 00	29 56	86·5	1 56	28 56	86·5	1 53	27 57	86·5	1 49	26 57	86·4	1 45	25 57	86·4	1 41	24 57	86·4	184	356
5	175	2 30	29 54	85·7	2 25	28 54	85·6	2 21	27 55	85·6	2 16	26 55	85·5	2 11	25 55	85·5	2 07	24 55	85·5	185	355
6	174	3 00	29 52	84·8	2 54	28 52	84·7	2 49	27 52	84·7	2 43	26 52	84·6	2 38	25 53	84·6	2 32	24 53	84·6	186	354
7	173	3 30	29 49	83·9	3 23	28 49	83·9	3 17	27 49	83·8	3 10	26 50	83·8	3 04	25 50	83·7	2 57	24 50	83·7	187	353
8	172	3 59	29 45	83·1	3 52	28 46	83·0	3 45	27 46	82·9	3 37	26 46	82·9	3 30	25 47	82·8	3 22	24 47	82·7	188	352
9	171	4 29	29 42	82·2	4 21	28 42	82·1	4 13	27 42	82·0	4 04	26 43	82·0	3 56	25 43	81·9	3 47	24 44	81·8	189	351
10	170	4 59	29 37	81·3	4 50	28 38	81·2	4 41	27 38	81·2	4 31	26 39	81·1	4 22	25 39	81·0	4 13	24 40	80·9	190	350
11	169	5 28	29 33	80·4	5 18	28 33	80·4	5 08	27 34	80·3	4 58	26 34	80·2	4 48	25 35	80·1	4 38	24 36	80·0	191	349
12	168	5 58	29 27	79·6	5 47	28 28	79·5	5 36	27 29	79·4	5 25	26 29	79·3	5 14	25 30	79·2	5 02	24 31	79·1	192	348
13	167	6 27	29 22	78·7	6 16	28 22	78·6	6 04	27 23	78·5	5 52	26 24	78·4	5 40	25 25	78·3	5 27	24 26	78·2	193	347
14	166	6 57	29 15	77·8	6 44	28 16	77·7	6 31	27 17	77·6	6 18	26 18	77·5	6 05	25 20	77·4	5 52	24 21	77·3	194	346
15	165	7 26	29 09	76·9	7 13	28 10	76·8	6 59	27 11	76·7	6 45	26 12	76·6	6 31	25 14	76·5	6 17	24 15	76·4	195	345
16	164	7 55	29 02	76·1	7 41	28 03	75·9	7 26	27 04	75·8	7 11	26 06	75·7	6 56	25 07	75·5	6 41	24 09	75·4	196	344
17	163	8 24	28 54	75·2	8 09	27 56	75·0	7 53	26 57	74·9	7 38	25 59	74·8	7 22	25 00	74·6	7 06	24 02	74·5	197	343
18	162	8 53	28 46	74·3	8 37	27 48	74·1	8 20	26 50	74·0	8 04	25 51	73·9	7 47	24 53	73·7	7 30	23 55	73·6	198	342
19	161	9 22	28 38	73·4	9 05	27 40	73·2	8 48	26 41	73·1	8 30	25 43	72·9	8 12	24 45	72·8	7 55	23 48	72·7	199	341
20	160	9 51	28 29	72·5	9 33	27 31	72·3	9 14	26 33	72·2	8 56	25 35	72·0	8 37	24 37	71·9	8 19	23 40	71·7	200	340
21	159	10 19	28 19	71·6	10 00	27 22	71·4	9 41	26 24	71·3	9 22	25 26	71·1	9 02	24 29	71·0	8 43	23 32	70·8	201	339
22	158	10 48	28 10	70·7	10 28	27 12	70·5	10 08	26 15	70·4	9 48	25 17	70·2	9 27	24 20	70·0	9 07	23 23	69·9	202	338
23	157	11 16	27 59	69·8	10 55	27 02	69·6	10 34	26 05	69·5	10 13	25 08	69·3	9 52	24 11	69·1	9 30	23 14	69·0	203	337
24	156	11 44	27 49	68·9	11 22	26 51	68·7	11 00	25 54	68·5	10 38	24 58	68·4	10 16	24 01	68·2	9 54	23 04	68·0	204	336
25	155	12 12	27 37	68·0	11 49	26 40	67·8	11 27	25 44	67·6	11 04	24 47	67·4	10 41	23 51	67·3	10 17	22 55	67·1	205	335
26	154	12 40	27 26	67·1	12 16	26 29	66·9	11 53	25 33	66·7	11 29	24 36	66·5	11 05	23 40	66·3	10 41	22 44	66·2	206	334
27	153	13 07	27 13	66·2	12 43	26 17	66·0	12 18	25 21	65·8	11 54	24 25	65·6	11 29	23 29	65·4	11 04	22 34	65·2	207	333
28	152	13 35	27 01	65·3	13 09	26 05	65·1	12 44	25 09	64·9	12 18	24 13	64·7	11 53	23 18	64·5	11 27	22 23	64·3	208	332
29	151	14 02	26 48	64·4	13 36	25 52	64·1	13 09	24 56	63·9	12 43	24 01	63·7	12 16	23 06	63·5	11 49	22 11	63·3	209	331
30	150	14 29	26 34	63·4	14 02	25 39	63·2	13 35	24 43	63·0	13 07	23 49	62·8	12 40	22 54	62·6	12 12	21 59	62·4	210	330
31	149	14 55	26 20	62·5	14 28	25 25	62·3	14 00	24 30	62·1	13 31	23 36	61·8	13 03	22 41	61·6	12 34	21 47	61·4	211	329
32	148	15 22	26 05	61·6	14 53	25 11	61·3	14 24	24 16	61·1	13 55	23 22	60·9	13 26	22 28	60·7	12 56	21 35	60·5	212	328
33	147	15 48	25 50	60·6	15 19	24 56	60·4	14 49	24 02	60·2	14 19	23 08	59·9	13 49	22 15	59·7	13 18	21 22	59·5	213	327
34	146	16 14	25 35	59·7	15 44	24 41	59·5	15 13	23 47	59·2	14 42	22 54	59·0	14 11	22 01	58·8	13 40	21 08	58·6	214	326
35	145	16 40	25 19	58·8	16 09	24 25	58·5	15 37	23 32	58·3	15 06	22 39	58·0	14 34	21 47	57·8	14 02	20 54	57·6	215	325
36	144	17 05	25 02	57·8	16 33	24 09	57·6	16 01	23 17	57·3	15 29	22 24	57·1	14 56	21 32	56·9	14 23	20 40	56·6	216	324
37	143	17 31	24 45	56·9	16 58	23 53	56·6	16 25	23 00	56·4	15 51	22 09	56·1	15 18	21 17	55·9	14 44	20 26	55·7	217	323
38	142	17 56	24 28	55·9	17 22	23 36	55·7	16 48	22 44	55·4	16 14	21 53	55·2	15 39	21 01	54·9	15 05	20 11	54·7	218	322
39	141	18 20	24 10	55·0	17 46	23 18	54·7	17 11	22 27	54·4	16 36	21 36	54·2	16 01	20 46	54·0	15 25	19 55	53·7	219	321
40	140	18 45	23 52	54·0	18 09	23 00	53·7	17 34	22 10	53·5	16 58	21 19	53·2	16 22	20 29	53·0	15 46	19 39	52·7	220	320
41	139	19 09	23 33	53·0	18 33	22 42	52·8	17 56	21 52	52·5	17 20	21 02	52·2	16 43	20 13	52·0	16 06	19 23	51·8	221	319
42	138	19 33	23 13	52·1	18 56	22 23	51·8	18 19	21 34	51·5	17 41	20 44	51·3	17 03	19 55	51·0	16 26	19 07	50·8	222	318
43	137	19 56	22 54	51·1	19 18	22 04	50·8	18 40	21 15	50·5	18 02	20 26	50·3	17 24	19 38	50·0	16 45	18 50	49·8	223	317
44	136	20 19	22 33	50·1	19 41	21 44	49·8	19 02	20 56	49·5	18 23	20 08	49·3	17 44	19 20	49·0	17 04	18 33	48·8	224	316
45	135	20 42	22 12	49·1	20 03	21 24	48·8	19 23	20 36	48·6	18 43	19 49	48·3	18 03	19 02	48·1	17 23	18 15	47·8	225	315

Lat. / A		60°			61°			62°			63°			64°			65°			Lat. / A	
LHA/F		A/H	B/P	Z_1/Z_2	A/H	B/P	Z_1/Z_2	A/H	B/P	Z_1/Z_2	A/H	B/P	Z_1/Z_2	A/H	B/P	Z_1/Z_2	A/H	B/P	Z_1/Z_2	LHA	
°	°	° ′	° ′	°	° ′	° ′	°	° ′	° ′	°	° ′	° ′	°	° ′	° ′	°	° ′	° ′	°	°	°
45	135	20 42	22 12	49·1	20 03	21 24	48·8	19 23	20 36	48·6	18 43	19 49	48·3	18 03	19 02	48·1	17 23	18 15	47·8	225	315
46	134	21 05	21 51	48·1	20 25	21 04	47·8	19 44	20 16	47·6	19 04	19 29	47·3	18 23	18 43	47·1	17 42	17 57	46·8	226	314
47	133	21 27	21 30	47·1	20 46	20 43	46·8	20 05	19 56	46·6	19 24	19 10	46·3	18 42	18 24	46·1	18 00	17 39	45·8	227	313
48	132	21 49	21 07	46·1	21 07	20 21	45·8	20 25	19 35	45·6	19 43	18 50	45·3	19 01	18 04	45·1	18 18	17 20	44·8	228	312
49	131	22 10	20 45	45·1	21 28	19 59	44·8	20 45	19 14	44·6	20 02	18 29	44·3	19 19	17 45	44·0	18 36	17 01	43·8	229	311
50	130	22 31	20 22	44·1	21 48	19 37	43·8	21 05	18 52	43·5	20 21	18 08	43·3	19 37	17 24	43·0	18 53	16 41	42·8	230	310
51	129	22 52	19 58	43·1	22 08	19 14	42·8	21 24	18 30	42·5	20 40	17 47	42·3	19 55	17 04	42·0	19 10	16 21	41·8	231	309
52	128	23 12	19 34	42·1	22 28	18 51	41·8	21 43	18 08	41·5	20 58	17 25	41·2	20 13	16 43	41·0	19 27	16 01	40·8	232	308
53	127	23 32	19 10	41·0	22 47	18 27	40·7	22 01	17 45	40·5	21 15	17 03	40·2	20 30	16 21	40·0	19 44	15 41	39·7	233	307
54	126	23 52	18 45	40·0	23 06	18 03	39·7	22 19	17 21	39·4	21 33	16 40	39·2	20 46	16 00	39·0	20 00	15 20	38·7	234	306
55	125	24 11	18 19	39·0	23 24	17 38	38·7	22 37	16 58	38·4	21 50	16 17	38·2	21 03	15 38	37·9	20 15	14 58	37·7	235	305
56	124	24 29	17 54	37·9	23 42	17 13	37·6	22 54	16 34	37·4	22 07	15 54	37·1	21 19	15 15	36·9	20 31	14 37	36·7	236	304
57	123	24 48	17 27	36·9	23 59	16 48	36·6	23 11	16 09	36·3	22 23	15 31	36·1	21 34	14 53	35·8	20 46	14 15	35·6	237	303
58	122	25 05	17 01	35·8	24 17	16 22	35·5	23 28	15 44	35·3	22 39	15 07	35·0	21 49	14 29	34·8	21 00	13 53	34·6	238	302
59	121	25 23	16 34	34·8	24 33	15 56	34·5	23 44	15 19	34·2	22 54	14 42	34·0	22 04	14 06	33·8	21 14	13 30	33·5	239	301
60	120	25 40	16 06	33·7	24 50	15 29	33·4	23 59	14 53	33·2	23 09	14 18	32·9	22 19	13 42	32·7	21 28	13 07	32·5	240	300
61	119	25 56	15 38	32·6	25 05	15 03	32·4	24 15	14 27	32·1	23 24	13 53	31·9	22 33	13 18	31·7	21 42	12 44	31·5	241	299
62	118	26 12	15 10	31·5	25 21	14 35	31·3	24 29	14 01	31·1	23 38	13 27	30·8	22 46	12 54	30·6	21 55	12 21	30·4	242	298
63	117	26 27	14 41	30·5	25 36	14 08	30·2	24 44	13 34	30·0	23 52	13 01	29·8	22 59	12 29	29·5	22 07	11 57	29·3	243	297
64	116	26 42	14 12	29·4	25 50	13 39	29·1	24 57	13 07	28·9	24 05	12 35	28·7	23 12	12 04	28·5	22 19	11 33	28·3	244	296
65	115	26 57	13 43	28·3	26 04	13 11	28·1	25 11	12 40	27·8	24 18	12 09	27·6	23 25	11 39	27·4	22 31	11 09	27·2	245	295
66	114	27 11	13 13	27·2	26 17	12 42	27·0	25 24	12 12	26·8	24 30	11 43	26·6	23 36	11 13	26·4	22 43	10 44	26·2	246	294
67	113	27 24	12 43	26·1	26 30	12 13	25·9	25 36	11 44	25·7	24 42	11 16	25·5	23 48	10 47	25·3	22 54	10 20	25·1	247	293
68	112	27 37	12 12	25·0	26 43	11 44	24·8	25 48	11 16	24·6	24 54	10 48	24·4	23 59	10 21	24·2	23 04	9 55	24·0	248	292
69	111	27 50	11 41	23·9	26 55	11 14	23·7	26 00	10 47	23·5	25 05	10 21	23·3	24 09	9 55	23·1	23 14	9 29	23·0	249	291
70	110	28 01	11 10	22·8	27 06	10 44	22·6	26 11	10 18	22·4	25 15	9 53	22·2	24 20	9 28	22·0	23 24	9 04	21·9	250	290
71	109	28 13	10 39	21·7	27 17	10 14	21·5	26 21	9 49	21·3	25 25	9 25	21·1	24 29	9 01	21·0	23 33	8 38	20·8	251	289
72	108	28 24	10 07	20·6	27 27	9 43	20·4	26 31	9 20	20·2	25 35	8 57	20·0	24 38	8 34	19·9	23 42	8 12	19·7	252	288
73	107	28 34	9 35	19·4	27 37	9 12	19·3	26 41	8 50	19·1	25 44	8 28	18·9	24 47	8 07	18·8	23 50	7 46	18·6	253	287
74	106	28 44	9 03	18·3	27 47	8 41	18·2	26 50	8 20	18·0	25 52	8 00	17·8	24 55	7 39	17·7	23 58	7 19	17·6	254	286
75	105	28 53	8 30	17·2	27 55	8 10	17·0	26 58	7 50	16·9	26 01	7 31	16·7	25 03	7 12	16·6	24 06	6 53	16·5	255	285
76	104	29 01	7 57	16·1	28 04	7 38	15·9	27 06	7 20	15·8	26 08	7 02	15·6	25 10	6 44	15·5	24 13	6 26	15·4	256	284
77	103	29 09	7 24	14·9	28 11	7 06	14·8	27 13	6 49	14·7	26 15	6 32	14·5	25 17	6 16	14·4	24 19	5 59	14·3	257	283
78	102	29 17	6 51	13·8	28 18	6 34	13·7	27 20	6 19	13·5	26 22	6 03	13·4	25 23	5 47	13·3	24 25	5 32	13·2	258	282
79	101	29 24	6 17	12·7	28 25	6 02	12·5	27 27	5 48	12·4	26 28	5 33	12·3	25 29	5 19	12·2	24 31	5 05	12·1	259	281
80	100	29 30	5 44	11·5	28 31	5 30	11·4	27 32	5 17	11·3	26 33	5 03	11·2	25 35	4 50	11·1	24 36	4 38	11·0	260	280
81	99	29 36	5 10	10·4	28 37	4 57	10·3	27 38	4 45	10·2	26 38	4 33	10·1	25 39	4 22	10·0	24 40	4 10	9·9	261	279
82	98	29 41	4 36	9·2	28 41	4 25	9·1	27 42	4 14	9·0	26 43	4 03	9·0	25 44	3 53	8·9	24 44	3 43	8·8	262	278
83	97	29 45	4 01	8·1	28 46	3 52	8·0	27 46	3 42	7·9	26 47	3 33	7·8	25 48	3 24	7·8	24 48	3 15	7·7	263	277
84	96	29 49	3 27	6·9	28 50	3 19	6·9	27 50	3 11	6·8	26 50	3 03	6·7	25 51	2 55	6·7	24 51	2 47	6·6	264	276
85	95	29 52	2 53	5·8	28 53	2 46	5·7	27 53	2 39	5·7	26 53	2 33	5·6	25 54	2 26	5·6	24 54	2 20	5·5	265	275
86	94	29 55	2 18	4·6	28 55	2 13	4·6	27 56	2 07	4·5	26 56	2 02	4·5	25 56	1 57	4·4	24 56	1 52	4·4	266	274
87	93	29 57	1 44	3·5	28 57	1 40	3·4	27 57	1 36	3·4	26 58	1 32	3·4	25 58	1 28	3·3	24 58	1 24	3·3	267	273
88	92	29 59	1 09	2·3	28 59	1 06	2·3	27 59	1 04	2·3	26 59	1 01	2·2	25 59	0 59	2·2	24 59	0 56	2·2	268	272
89	91	30 00	0 35	1·2	29 00	0 33	1·1	28 00	0 32	1·1	27 00	0 31	1·1	26 00	0 29	1·1	25 00	0 28	1·1	269	271
90	90	30 00	0 00	0·0	29 00	0 00	0·0	28 00	0 00	0·0	27 00	0 00	0·0	26 00	0 00	0·0	25 00	0 00	0·0	270	270

N. Lat: for LHA > 180° … $Z_n = Z$
for LHA < 180° … $Z_n = 360° - Z$

S. Lat.: for LHA > 180° … $Z_n = 180° - Z$
for LHA < 180° … $Z_n = 180° + Z$

SIGHT REDUCTION TABLE

B: (–) for 90° < LHA < 270°
Dec: (–) for Lat. contrary name

Z_1: same sign as B
Z_2: (–) for F > 90°

Lat. / A		66°			67°			68°			69°			70°			71°			Lat. / A	
LHA/F		A/H	B/P	Z_1/Z_2	A/H	B/P	Z_1/Z_2	A/H	B/P	Z_1/Z_2	A/H	B/P	Z_1/Z_2	A/H	B/P	Z_1/Z_2	A/H	B/P	Z_1/Z_2	LHA	
°	°	° ′	° ′	°	° ′	° ′	°	° ′	° ′	°	° ′	° ′	°	° ′	° ′	°	° ′	° ′	°	°	°
0	180	0 00	24 00	90·0	0 00	23 00	90·0	0 00	22 00	90·0	0 00	21 00	90·0	0 00	20 00	90·0	0 00	19 00	90·0	180	360
1	179	0 24	24 00	89·1	0 23	23 00	89·1	0 22	22 00	89·1	0 22	21 00	89·1	0 21	20 00	89·1	0 20	19 00	89·1	181	359
2	178	0 49	23 59	88·2	0 47	22 59	88·2	0 45	21 59	88·1	0 43	20 59	88·1	0 41	19 59	88·1	0 39	18 59	88·1	182	358
3	177	1 13	23 58	87·3	1 10	22 58	87·2	1 07	21 58	87·2	1 04	20 58	87·2	1 02	19 58	87·2	0 59	18 59	87·2	183	357
4	176	1 38	23 57	86·3	1 34	22 57	86·3	1 30	21 57	86·3	1 26	20 57	86·3	1 22	19 57	86·2	1 18	18 57	86·2	184	356
5	175	2 02	23 55	85·4	1 57	22 55	85·4	1 52	21 55	85·4	1 47	20 56	85·3	1 42	19 56	85·3	1 38	18 56	85·3	185	355
6	174	2 26	23 53	84·5	2 20	22 53	84·5	2 15	21 53	84·4	2 09	20 54	84·4	2 03	19 54	84·4	1 57	18 54	84·3	186	354
7	173	2 50	23 50	83·6	2 44	22 51	83·6	2 37	21 51	83·5	2 30	20 51	83·5	2 23	19 52	83·4	2 16	18 52	83·4	187	353
8	172	3 15	23 48	82·7	3 07	22 48	82·6	2 59	21 48	82·6	2 52	20 49	82·5	2 44	19 49	82·5	2 36	18 50	82·4	188	352
9	171	3 39	23 44	81·8	3 30	22 45	81·7	3 22	21 45	81·6	3 13	20 46	81·6	3 04	19 46	81·5	2 55	18 47	81·5	189	351
10	170	4 03	23 41	80·8	3 53	22 41	80·8	3 44	21 42	80·7	3 34	20 42	80·7	3 24	19 43	80·6	3 14	18 44	80·5	190	350
11	169	4 27	23 36	79·9	4 17	22 37	79·9	4 06	21 38	79·8	3 55	20 39	79·7	3 45	19 40	79·6	3 34	18 41	79·6	191	349
12	168	4 51	23 32	79·0	4 40	22 33	78·9	4 28	21 34	78·9	4 16	20 35	78·8	4 05	19 36	78·7	3 53	18 37	78·6	192	348
13	167	5 15	23 27	78·1	5 03	22 28	78·0	4 50	21 29	77·9	4 37	20 30	77·8	4 25	19 32	77·8	4 12	18 33	77·7	193	347
14	166	5 39	23 22	77·2	5 25	22 23	77·1	5 12	21 24	77·0	4 58	20 26	76·9	4 45	19 27	76·8	4 31	18 28	76·7	194	346
15	165	6 03	23 16	76·2	5 48	22 18	76·1	5 34	21 19	76·0	5 19	20 21	76·0	5 05	19 22	75·9	4 50	18 24	75·8	195	345
16	164	6 26	23 10	75·3	6 11	22 12	75·2	5 56	21 13	75·1	5 40	20 15	75·0	5 25	19 17	74·9	5 09	18 19	74·8	196	344
17	163	6 50	23 04	74·4	6 34	22 06	74·3	6 17	21 08	74·2	6 01	20 09	74·1	5 44	19 11	74·0	5 28	18 14	73·9	197	343
18	162	7 13	22 57	73·5	6 56	21 59	73·3	6 39	21 01	73·2	6 21	20 03	73·1	6 04	19 06	73·0	5 46	18 08	72·9	198	342
19	161	7 37	22 50	72·5	7 19	21 52	72·4	7 00	20 54	72·3	6 42	19 57	72·2	6 24	18 59	72·1	6 05	18 02	72·0	199	341
20	160	8 00	22 42	71·6	7 41	21 45	71·5	7 22	20 47	71·4	7 02	19 50	71·2	6 43	18 53	71·1	6 24	17 56	71·0	200	340
21	159	8 23	22 34	70·7	8 03	21 37	70·5	7 43	20 40	70·4	7 23	19 43	70·3	7 02	18 46	70·2	6 42	17 49	70·1	201	339
22	158	8 46	22 26	69·7	8 25	21 29	69·6	8 04	20 32	69·5	7 43	19 35	69·3	7 22	18 39	69·2	7 00	17 42	69·1	202	338
23	157	9 09	22 17	68·8	8 47	21 21	68·7	8 25	20 24	68·5	8 03	19 28	68·4	7 41	18 31	68·3	7 19	17 35	68·1	203	337
24	156	9 31	22 08	67·9	9 09	21 12	67·7	8 46	20 16	67·6	8 23	19 19	67·4	8 00	18 24	67·3	7 37	17 28	67·2	204	336
25	155	9 54	21 58	66·9	9 30	21 03	66·8	9 07	20 07	66·6	8 43	19 11	66·5	8 19	18 15	66·3	7 55	17 20	66·2	205	335
26	154	10 16	21 49	66·0	9 52	20 53	65·8	9 27	19 57	65·7	9 02	19 02	65·5	8 37	18 07	65·4	8 12	17 12	65·2	206	334
27	153	10 38	21 38	65·0	10 13	20 43	64·9	9 48	19 48	64·7	9 22	18 53	64·6	8 56	17 58	64·4	8 30	17 03	64·3	207	333
28	152	11 00	21 28	64·1	10 34	20 33	63·9	10 08	19 38	63·8	9 41	18 43	63·6	9 14	17 49	63·5	8 48	16 55	63·3	208	332
29	151	11 22	21 17	63·1	10 55	20 22	63·0	10 28	19 28	62·8	10 00	18 34	62·6	9 33	17 39	62·5	9 05	16 46	62·3	209	331
30	150	11 44	21 05	62·2	11 16	20 11	62·0	10 48	19 17	61·8	10 19	18 23	61·7	9 51	17 30	61·5	9 22	16 36	61·4	210	330
31	149	12 06	20 53	61·2	11 37	20 00	61·1	11 07	19 06	60·9	10 38	18 13	60·7	10 09	17 20	60·5	9 39	16 27	60·4	211	329
32	148	12 27	20 41	60·3	11 57	19 48	60·1	11 27	18 55	59·9	10 57	18 02	59·7	10 27	17 09	59·6	9 56	16 17	59·4	212	328
33	147	12 48	20 29	59·3	12 17	19 36	59·1	11 46	18 43	58·9	11 15	17 51	58·8	10 44	16 58	58·6	10 13	16 06	58·4	213	327
34	146	13 09	20 16	58·4	12 37	19 23	58·2	12 06	18 31	58·0	11 34	17 39	57·8	11 02	16 47	57·6	10 29	15 56	57·5	214	326
35	145	13 29	20 02	57·4	12 57	19 10	57·2	12 24	18 19	57·0	11 52	17 27	56·8	11 19	16 36	56·7	10 46	15 45	56·5	215	325
36	144	13 50	19 49	56·4	13 17	18 57	56·2	12 43	18 06	56·0	12 10	17 15	55·9	11 36	16 24	55·7	11 02	15 34	55·5	216	324
37	143	14 10	19 34	55·5	13 36	18 44	55·3	13 02	17 53	55·1	12 27	17 03	54·9	11 53	16 12	54·7	11 18	15 23	54·5	217	323
38	142	14 30	19 20	54·5	13 55	18 30	54·3	13 20	17 40	54·1	12 45	16 50	53·9	12 09	16 00	53·7	11 34	15 11	53·5	218	322
39	141	14 50	19 05	53·5	14 14	18 15	53·3	13 38	17 26	53·1	13 02	16 37	52·9	12 26	15 48	52·7	11 49	14 59	52·6	219	321
40	140	15 09	18 50	52·5	14 33	18 01	52·3	13 56	17 12	52·1	13 19	16 23	51·9	12 42	15 35	51·7	12 05	14 47	51·6	220	320
41	139	15 29	18 34	51·5	14 51	17 46	51·3	14 14	16 57	51·1	13 36	16 09	50·9	12 58	15 22	50·8	12 20	14 34	50·6	221	319
42	138	15 48	18 18	50·6	15 09	17 30	50·3	14 31	16 43	50·1	13 52	15 55	49·9	13 14	15 08	49·8	12 35	14 21	49·6	222	318
43	137	16 06	18 02	49·6	15 27	17 15	49·4	14 48	16 28	49·2	14 09	15 41	49·0	13 29	14 54	48·8	12 50	14 08	48·6	223	317
44	136	16 25	17 46	48·6	15 45	16 59	48·4	15 05	16 12	48·2	14 25	15 26	48·0	13 45	14 40	47·8	13 04	13 55	47·6	224	316
45	135	16 43	17 29	47·6	16 02	16 42	47·4	15 22	15 57	47·2	14 41	15 11	47·0	14 00	14 26	46·8	13 19	13 41	46·6	225	315

Lat. / A		66°			67°			68°			69°			70°			71°			Lat. / A	
LHA/F		A/H	B/P	Z_1/Z_2	A/H	B/P	Z_1/Z_2	A/H	B/P	Z_1/Z_2	A/H	B/P	Z_1/Z_2	A/H	B/P	Z_1/Z_2	A/H	B/P	Z_1/Z_2	LHA	
°	°	° ′	° ′	°	° ′	° ′	°	° ′	° ′	°	° ′	° ′	°	° ′	° ′	°	° ′	° ′	°	°	°
45	135	16 43	17 29	47·6	16 02	16 42	47·4	15 22	15 57	47·2	14 41	15 11	47·0	14 00	14 26	46·8	13 19	13 41	46·6	225	315
46	134	17 01	17 11	46·6	16 19	16 26	46·4	15 38	15 41	46·2	14 56	14 56	46·0	14 15	14 11	45·8	13 33	13 27	45·6	226	314
47	133	17 18	16 53	45·6	16 36	16 09	45·4	15 54	15 24	45·2	15 12	14 40	45·0	14 29	13 56	44·8	13 46	13 13	44·6	227	313
48	132	17 36	16 35	44·6	16 53	15 51	44·4	16 10	15 08	44·2	15 27	14 24	44·0	14 43	13 41	43·8	14 00	12 58	43·6	228	312
49	131	17 53	16 17	43·6	17 09	15 34	43·4	16 25	14 51	43·2	15 42	14 08	43·0	14 58	13 26	42·8	14 13	12 44	42·6	229	311
50	130	18 09	15 58	42·6	17 25	15 16	42·4	16 41	14 33	42·1	15 56	13 52	41·9	15 11	13 10	41·8	14 27	12 29	41·6	230	310
51	129	18 26	15 39	41·6	17 41	14 57	41·3	16 56	14 16	41·1	16 10	13 35	40·9	15 25	12 54	40·8	14 39	12 14	40·6	231	309
52	128	18 42	15 20	40·5	17 56	14 39	40·3	17 10	13 58	40·1	16 24	13 18	39·9	15 38	12 38	39·7	14 52	11 58	39·6	232	308
53	127	18 57	15 00	39·5	18 11	14 20	39·3	17 24	13 40	39·1	16 38	13 00	38·9	15 51	12 21	38·7	15 04	11 42	38·6	233	307
54	126	19 13	14 40	38·5	18 26	14 01	38·3	17 39	13 22	38·1	16 51	12 43	37·9	16 04	12 05	37·7	15 16	11 26	37·5	234	306
55	125	19 28	14 20	37·5	18 40	13 41	37·3	17 52	13 03	37·1	17 04	12 25	36·9	16 16	11 48	36·7	15 28	11 10	36·5	235	305
56	124	19 42	13 59	36·4	18 54	13 21	36·2	18 06	12 44	36·0	17 17	12 07	35·8	16 28	11 30	35·7	15 40	10 54	35·5	236	304
57	123	19 57	13 38	35·4	19 08	13 01	35·2	18 19	12 25	35·0	17 29	11 49	34·8	16 40	11 13	34·6	15 51	10 37	34·5	237	303
58	122	20 11	13 17	34·4	19 21	12 41	34·2	18 31	12 05	34·0	17 42	11 30	33·8	16 52	10 55	33·6	16 02	10 20	33·5	238	302
59	121	20 24	12 55	33·3	19 34	12 20	33·1	18 44	11 45	32·9	17 53	11 11	32·8	17 03	10 37	32·6	16 12	10 03	32·4	239	301
60	120	20 37	12 33	32·3	19 47	11 59	32·1	18 56	11 25	31·9	18 05	10 52	31·7	17 14	10 19	31·6	16 23	9 46	31·4	240	300
61	119	20 50	12 11	31·2	19 59	11 38	31·1	19 08	11 05	30·9	18 16	10 33	30·7	17 24	10 00	30·5	16 33	9 29	30·4	241	299
62	118	21 03	11 48	30·2	20 11	11 16	30·0	19 19	10 44	29·8	18 27	10 13	29·7	17 35	9 42	29·5	16 42	9 11	29·4	242	298
63	117	21 15	11 26	29·2	20 22	10 54	29·0	19 30	10 24	28·8	18 37	9 53	28·6	17 45	9 23	28·5	16 52	8 53	28·3	243	297
64	116	21 27	11 03	28·1	20 34	10 32	27·9	19 41	10 03	27·7	18 47	9 33	27·6	17 54	9 04	27·4	17 01	8 35	27·3	244	296
65	115	21 38	10 39	27·0	20 44	10 10	26·9	19 51	9 41	26·7	18 57	9 13	26·5	18 03	8 45	26·4	17 10	8 17	26·3	245	295
66	114	21 49	10 16	26·0	20 55	9 48	25·8	20 01	9 20	25·7	19 07	8 52	25·5	18 12	8 25	25·4	17 18	7 58	25·2	246	294
67	113	21 59	9 52	24·9	21 05	9 25	24·8	20 10	8 58	24·6	19 16	8 32	24·5	18 21	8 06	24·3	17 26	7 40	24·2	247	293
68	112	22 09	9 28	23·9	21 14	9 02	23·7	20 19	8 36	23·5	19 24	8 11	23·4	18 29	7 46	23·3	17 34	7 21	23·1	248	292
69	111	22 19	9 04	22·8	21 24	8 39	22·6	20 28	8 14	22·5	19 33	7 50	22·4	18 37	7 26	22·2	17 42	7 02	22·1	249	291
70	110	22 28	8 39	21·7	21 32	8 16	21·6	20 37	7 52	21·4	19 41	7 29	21·3	18 45	7 06	21·2	17 49	6 43	21·1	250	290
71	109	22 37	8 15	20·7	21 41	7 52	20·5	20 45	7 30	20·4	19 48	7 07	20·2	18 52	6 45	20·1	17 56	6 24	20·0	251	289
72	108	22 45	7 50	19·6	21 49	7 28	19·4	20 52	7 07	19·3	19 56	6 46	19·2	18 59	6 25	19·1	18 02	6 04	19·0	252	288
73	107	22 53	7 25	18·5	21 56	7 04	18·4	21 00	6 44	18·2	20 03	6 24	18·1	19 05	6 04	18·0	18 08	5 45	17·9	253	287
74	106	23 01	7 00	17·4	22 04	6 40	17·3	21 06	6 21	17·2	20 09	6 02	17·1	19 12	5 44	17·0	18 14	5 25	16·9	254	286
75	105	23 08	6 34	16·3	22 10	6 16	16·2	21 13	5 58	16·1	20 15	5 40	16·0	19 17	5 23	15·9	18 20	5 06	15·8	255	285
76	104	23 15	6 09	15·3	22 17	5 52	15·2	21 19	5 35	15·1	20 21	5 18	15·0	19 23	5 02	14·9	18 25	4 46	14·8	256	284
77	103	23 21	5 43	14·2	22 23	5 27	14·1	21 24	5 12	14·0	20 26	4 56	13·9	19 28	4 41	13·8	18 30	4 26	13·7	257	283
78	102	23 27	5 17	13·1	22 28	5 03	13·0	21 30	4 48	12·9	20 31	4 34	12·8	19 33	4 20	12·7	18 34	4 06	12·7	258	282
79	101	23 32	4 51	12·0	22 33	4 38	11·9	21 35	4 24	11·8	20 36	4 11	11·8	19 37	3 58	11·7	18 38	3 46	11·6	259	281
80	100	23 37	4 25	10·9	22 38	4 13	10·8	21 39	4 01	10·8	20 40	3 49	10·7	19 41	3 37	10·6	18 42	3 25	10·6	260	280
81	99	23 41	3 59	9·8	22 42	3 48	9·8	21 43	3 37	9·7	20 44	3 26	9·6	19 45	3 16	9·6	18 45	3 05	9·5	261	279
82	98	23 45	3 33	8·7	22 46	3 23	8·7	21 46	3 13	8·6	20 47	3 03	8·6	19 48	2 54	8·5	18 48	2 45	8·5	262	278
83	97	23 49	3 06	7·7	22 49	2 58	7·6	21 50	2 49	7·5	20 50	2 41	7·5	19 51	2 32	7·4	18 51	2 24	7·4	263	277
84	96	23 52	2 40	6·6	22 52	2 32	6·5	21 52	2 25	6·5	20 53	2 18	6·4	19 53	2 11	6·4	18 54	2 04	6·3	264	276
85	95	23 54	2 13	5·5	22 54	2 07	5·4	21 55	2 01	5·4	20 55	1 55	5·4	19 55	1 49	5·3	18 55	1 43	5·3	265	275
86	94	23 56	1 47	4·4	22 56	1 42	4·3	21 57	1 37	4·3	20 57	1 32	4·3	19 57	1 27	4·3	18 57	1 23	4·2	266	274
87	93	23 58	1 20	3·3	22 58	1 16	3·3	21 58	1 13	3·2	20 58	1 09	3·2	19 58	1 05	3·2	18 58	1 02	3·2	267	273
88	92	23 59	0 53	2·2	22 59	0 51	2·2	21 59	0 48	2·2	20 59	0 46	2·1	19 59	0 44	2·1	18 59	0 41	2·1	268	272
89	91	24 00	0 27	1·1	23 00	0 25	1·1	22 00	0 24	1·1	21 00	0 23	1·1	20 00	0 22	1·1	19 00	0 21	1·1	269	271
90	90	24 00	0 00	0·0	23 00	0 00	0·0	22 00	0 00	0·0	21 00	0 00	0·0	20 00	0 00	0·0	19 00	0 00	0·0	270	270

N. Lat: for LHA > 180° … $Z_n = Z$
for LHA < 180° … $Z_n = 360° - Z$

S. Lat.: for LHA > 180° … $Z_n = 180° - Z$
for LHA < 180° … $Z_n = 180° + Z$

SIGHT REDUCTION TABLE

B: (–) for 90° < LHA < 270°
Dec:(–) for Lat. contrary name

Z_1: same sign as B
Z_2: (–) for F > 90°

Lat. / A		72°			73°			74°			75°			76°			77°			Lat. / A	
LHA/F		A/H	B/P	Z_1/Z_2	A/H	B/P	Z_1/Z_2	A/H	B/P	Z_1/Z_2	A/H	B/P	Z_1/Z_2	A/H	B/P	Z_1/Z_2	A/H	B/P	Z_1/Z_2	LHA	
°	°	° ′	° ′	°	° ′	° ′	°	° ′	° ′	°	° ′	° ′	°	° ′	° ′	°	° ′	° ′	°	°	°
0	180	0 00	18 00	90·0	0 00	17 00	90·0	0 00	16 00	90·0	0 00	15 00	90·0	0 00	14 00	90·0	0 00	13 00	90·0	180	360
1	179	0 19	18 00	89·0	0 18	17 00	89·0	0 17	16 00	89·0	0 16	15 00	89·0	0 15	14 00	89·0	0 13	13 00	89·0	181	359
2	178	0 37	17 59	88·1	0 35	16 59	88·1	0 33	15 59	88·1	0 31	14 59	88·1	0 29	14 00	88·1	0 27	13 00	88·1	182	358
3	177	0 56	17 59	87·1	0 53	16 59	87·1	0 50	15 59	87·1	0 47	14 59	87·1	0 44	13 59	87·1	0 40	12 59	87·1	183	357
4	176	1 14	17 58	86·2	1 10	16 58	86·2	1 06	15 58	86·2	1 02	14 58	86·1	0 58	13 58	86·1	0 54	12 58	86·1	184	356
5	175	1 33	17 56	85·2	1 28	16 56	85·2	1 23	15 57	85·2	1 18	14 57	85·2	1 12	13 57	85·1	1 07	12 57	85·1	185	355
6	174	1 51	17 54	84·3	1 45	16 55	84·3	1 39	15 55	84·2	1 33	14 55	84·2	1 27	13 56	84·2	1 21	12 56	84·2	186	354
7	173	2 09	17 52	83·3	2 03	16 53	83·3	1 56	15 53	83·3	1 48	14 54	83·2	1 41	13 54	83·2	1 34	12 54	83·2	187	353
8	172	2 28	17 50	82·4	2 20	16 51	82·3	2 12	15 51	82·3	2 04	14 52	82·3	1 56	13 52	82·2	1 48	12 53	82·2	188	352
9	171	2 46	17 48	81·4	2 37	16 48	81·4	2 28	15 49	81·3	2 19	14 49	81·3	2 10	13 50	81·3	2 01	12 51	81·2	189	351
10	170	3 05	17 45	80·5	2 55	16 45	80·4	2 45	15 46	80·4	2 35	14 47	80·3	2 24	13 48	80·3	2 14	12 49	80·3	190	350
11	169	3 23	17 41	79·5	3 12	16 42	79·5	3 01	15 43	79·4	2 50	14 44	79·4	2 39	13 45	79·3	2 28	12 46	79·3	191	349
12	168	3 41	17 38	78·6	3 29	16 39	78·5	3 17	15 40	78·5	3 05	14 41	78·4	2 53	13 42	78·3	2 41	12 44	78·3	192	348
13	167	3 59	17 34	77·6	3 46	16 35	77·5	3 33	15 37	77·5	3 20	14 38	77·4	3 07	13 39	77·4	2 54	12 41	77·3	193	347
14	166	4 17	17 30	76·7	4 03	16 31	76·6	3 49	15 33	76·5	3 35	14 34	76·5	3 21	13 36	76·4	3 07	12 38	76·3	194	346
15	165	4 35	17 25	75·7	4 20	16 27	75·6	4 05	15 29	75·6	3 50	14 31	75·5	3 35	13 32	75·4	3 20	12 34	75·4	195	345
16	164	4 53	17 21	74·7	4 37	16 23	74·7	4 21	15 25	74·6	4 05	14 27	74·5	3 49	13 29	74·5	3 33	12 31	74·4	196	344
17	163	5 11	17 16	73·8	4 54	16 18	73·7	4 37	15 20	73·6	4 20	14 22	73·5	4 03	13 25	73·5	3 46	12 27	73·4	197	343
18	162	5 29	17 10	72·8	5 11	16 13	72·7	4 53	15 15	72·7	4 35	14 18	72·6	4 17	13 20	72·5	3 59	12 23	72·4	198	342
19	161	5 46	17 05	71·9	5 28	16 07	71·8	5 09	15 10	71·7	4 50	14 13	71·6	4 31	13 16	71·5	4 12	12 19	71·5	199	341
20	160	6 04	16 59	70·9	5 44	16 02	70·8	5 25	15 05	70·7	5 05	14 08	70·6	4 45	13 11	70·5	4 25	12 14	70·5	200	340
21	159	6 21	16 52	69·9	6 01	15 56	69·8	5 40	14 59	69·7	5 19	14 03	69·7	4 58	13 06	69·6	4 37	12 10	69·5	201	339
22	158	6 39	16 46	69·0	6 17	15 50	68·9	5 56	14 53	68·8	5 34	13 57	68·7	5 12	13 01	68·6	4 50	12 05	68·5	202	338
23	157	6 56	16 39	68·0	6 34	15 43	67·9	6 11	14 47	67·8	5 48	13 51	67·7	5 25	12 56	67·6	5 03	12 00	67·5	203	337
24	156	7 13	16 32	67·1	6 50	15 36	66·9	6 26	14 41	66·8	6 03	13 45	66·7	5 39	12 50	66·6	5 15	11 55	66·5	204	336
25	155	7 30	16 25	66·1	7 06	15 29	66·0	6 41	14 34	65·9	6 17	13 39	65·8	5 52	12 44	65·7	5 27	11 49	65·6	205	335
26	154	7 47	16 17	65·1	7 22	15 22	65·0	6 56	14 27	64·9	6 31	13 32	64·8	6 05	12 38	64·7	5 40	11 43	64·6	206	334
27	153	8 04	16 09	64·1	7 38	15 14	64·0	7 11	14 20	63·9	6 45	13 26	63·8	6 18	12 32	63·7	5 52	11 37	63·6	207	333
28	152	8 20	16 00	63·2	7 53	15 06	63·0	7 26	14 12	62·9	6 59	13 19	62·8	6 31	12 25	62·7	6 04	11 31	62·6	208	332
29	151	8 37	15 52	62·2	8 09	14 58	62·1	7 41	14 05	61·9	7 13	13 11	61·8	6 44	12 18	61·7	6 16	11 25	61·6	209	331
30	150	8 53	15 43	61·2	8 24	14 50	61·1	7 55	13 57	61·0	7 26	13 04	60·9	6 57	12 11	60·7	6 27	11 18	60·6	210	330
31	149	9 09	15 34	60·3	8 40	14 41	60·1	8 10	13 49	60·0	7 40	12 56	59·9	7 09	12 04	59·8	6 39	11 12	59·7	211	329
32	148	9 25	15 24	59·3	8 55	14 32	59·1	8 24	13 40	59·0	7 53	12 48	58·9	7 22	11 56	58·8	6 51	11 05	58·7	212	328
33	147	9 41	15 15	58·3	9 10	14 23	58·2	8 38	13 31	58·0	8 06	12 40	57·9	7 34	11 49	57·8	7 02	10 57	57·7	213	327
34	146	9 57	15 05	57·3	9 25	14 13	57·2	8 52	13 22	57·0	8 19	12 31	56·9	7 46	11 41	56·8	7 14	10 50	56·7	214	326
35	145	10 13	14 54	56·3	9 39	14 04	56·2	9 06	13 13	56·1	8 32	12 23	55·9	7 59	11 33	55·8	7 25	10 43	55·7	215	325
36	144	10 28	14 44	55·4	9 54	13 54	55·2	9 19	13 04	55·1	8 45	12 14	54·9	8 11	11 24	54·8	7 36	10 35	54·7	216	324
37	143	10 43	14 33	54·4	10 08	13 43	54·2	9 33	12 54	54·1	8 58	12 05	53·9	8 22	11 16	53·8	7 47	10 27	53·7	217	323
38	142	10 58	14 22	53·4	10 22	13 33	53·2	9 46	12 44	53·1	9 10	11 55	53·0	8 34	11 07	52·8	7 58	10 19	52·7	218	322
39	141	11 13	14 10	52·4	10 36	13 22	52·2	9 59	12 34	52·1	9 22	11 46	52·0	8 45	10 58	51·8	8 08	10 10	51·7	219	321
40	140	11 27	13 59	51·4	10 50	13 11	51·3	10 12	12 23	51·1	9 35	11 36	51·0	8 57	10 49	50·8	8 19	10 02	50·7	220	320
41	139	11 42	13 47	50·4	11 04	13 00	50·3	10 25	12 13	50·1	9 47	11 26	50·0	9 08	10 39	49·9	8 29	9 53	49·7	221	319
42	138	11 56	13 34	49·4	11 17	12 48	49·3	10 38	12 02	49·1	9 58	11 16	49·0	9 19	10 30	48·9	8 39	9 44	48·7	222	318
43	137	12 10	13 22	48·4	11 30	12 36	48·3	10 50	11 51	48·1	10 10	11 05	48·0	9 30	10 20	47·9	8 49	9 35	47·7	223	317
44	136	12 24	13 09	47·4	11 43	12 24	47·3	11 02	11 39	47·1	10 21	10 55	47·0	9 40	10 10	46·9	8 59	9 26	46·7	224	316
45	135	12 37	12 56	46·4	11 56	12 12	46·3	11 14	11 28	46·1	10 33	10 44	46·0	9 51	10 00	45·9	9 09	9 16	45·7	225	315

Lat. / A		72°			73°			74°			75°			76°			77°			Lat. / A	
LHA/F		A/H	B/P	Z_1/Z_2	A/H	B/P	Z_1/Z_2	A/H	B/P	Z_1/Z_2	A/H	B/P	Z_1/Z_2	A/H	B/P	Z_1/Z_2	A/H	B/P	Z_1/Z_2	LHA	
°	°	° ′	° ′	°	° ′	° ′	°	° ′	° ′	°	° ′	° ′	°	° ′	° ′	°	° ′	° ′	°	°	°
45	135	12 37	12 56	46·4	11 56	12 12	46·3	11 14	11 28	46·1	10 33	10 44	46·0	9 51	10 00	45·9	9 09	9 16	45·7	225	315
46	134	12 51	12 43	45·4	12 08	11 59	45·3	11 26	11 16	45·1	10 44	10 33	45·0	10 01	9 50	44·9	9 19	9 07	44·7	226	314
47	133	13 04	12 30	44·4	12 21	11 47	44·3	11 38	11 04	44·1	10 55	10 21	44·0	10 11	9 39	43·9	9 28	8 57	43·7	227	313
48	132	13 17	12 16	43·4	12 33	11 34	43·3	11 49	10 52	43·1	11 05	10 10	43·0	10 21	9 28	42·9	9 37	8 47	42·7	228	312
49	131	13 29	12 02	42·4	12 45	11 21	42·3	12 00	10 39	42·1	11 16	9 58	42·0	10 31	9 17	41·9	9 46	8 37	41·7	229	311
50	130	13 42	11 48	41·4	12 57	11 07	41·3	12 11	10 27	41·1	11 26	9 46	41·0	10 41	9 06	40·9	9 55	8 26	40·7	230	310
51	129	13 54	11 33	40·4	13 08	10 53	40·3	12 22	10 14	40·1	11 36	9 34	40·0	10 50	8 55	39·8	10 04	8 16	39·7	231	309
52	128	14 06	11 19	39·4	13 19	10 40	39·2	12 33	10 01	39·1	11 46	9 22	39·0	10 59	8 44	38·8	10 13	8 05	38·7	232	308
53	127	14 17	11 04	38·4	13 30	10 26	38·2	12 43	9 47	38·1	11 56	9 10	38·0	11 08	8 32	37·8	10 21	7 55	37·7	233	307
54	126	14 29	10 49	37·4	13 41	10 11	37·2	12 53	9 34	37·1	12 05	8 57	36·9	11 17	8 20	36·8	10 29	7 44	36·7	234	306
55	125	14 40	10 33	36·4	13 51	9 57	36·2	13 03	9 20	36·1	12 14	8 44	35·9	11 26	8 08	35·8	10 37	7 33	35·7	235	305
56	124	14 51	10 18	35·3	14 02	9 42	35·2	13 13	9 07	35·1	12 23	8 31	34·9	11 34	7 56	34·8	10 45	7 21	34·7	236	304
57	123	15 01	10 02	34·3	14 12	9 27	34·2	13 22	8 53	34·0	12 32	8 18	33·9	11 42	7 44	33·8	10 52	7 10	33·7	237	303
58	122	15 12	9 46	33·3	14 21	9 12	33·2	13 31	8 38	33·0	12 41	8 05	32·9	11 50	7 32	32·8	11 00	6 58	32·7	238	302
59	121	15 22	9 30	32·3	14 31	8 57	32·1	13 40	8 24	32·0	12 49	7 51	31·9	11 58	7 19	31·8	11 07	6 47	31·7	239	301
60	120	15 31	9 14	31·3	14 40	8 41	31·1	13 49	8 10	31·0	12 57	7 38	30·9	12 06	7 06	30·8	11 14	6 35	30·6	240	300
61	119	15 41	8 57	30·2	14 49	8 26	30·1	13 57	7 55	30·0	13 05	7 24	29·8	12 13	6 54	29·7	11 21	6 23	29·6	241	299
62	118	15 50	8 40	29·2	14 58	8 10	29·1	14 05	7 40	28·9	13 13	7 10	28·8	12 20	6 41	28·7	11 27	6 11	28·6	242	298
63	117	15 59	8 23	28·2	15 06	7 54	28·0	14 13	7 25	27·9	13 20	6 56	27·8	12 27	6 27	27·7	11 34	5 59	27·6	243	297
64	116	16 08	8 06	27·2	15 14	7 38	27·0	14 21	7 10	26·9	13 27	6 42	26·8	12 34	6 14	26·7	11 40	5 47	26·6	244	296
65	115	16 16	7 49	26·1	15 22	7 22	26·0	14 28	6 55	25·9	13 34	6 28	25·8	12 40	6 01	25·7	11 46	5 34	25·6	245	295
66	114	16 24	7 32	25·1	15 29	7 05	25·0	14 35	6 39	24·9	13 41	6 13	24·7	12 46	5 47	24·6	11 52	5 22	24·6	246	294
67	113	16 32	7 14	24·1	15 37	6 49	23·9	14 42	6 24	23·8	13 47	5 59	23·7	12 52	5 34	23·6	11 57	5 09	23·5	247	293
68	112	16 39	6 56	23·0	15 44	6 32	22·9	14 48	6 08	22·8	13 53	5 44	22·7	12 58	5 20	22·6	12 02	4 57	22·5	248	292
69	111	16 46	6 38	22·0	15 50	6 15	21·9	14 55	5 52	21·8	13 59	5 29	21·7	13 03	5 06	21·6	12 07	4 44	21·5	249	291
70	110	16 53	6 20	20·9	15 57	5 58	20·8	15 01	5 36	20·7	14 05	5 14	20·6	13 08	4 52	20·6	12 12	4 31	20·5	250	290
71	109	16 59	6 02	19·9	16 03	5 41	19·8	15 06	5 20	19·7	14 10	4 59	19·6	13 13	4 38	19·5	12 17	4 18	19·5	251	289
72	108	17 05	5 44	18·9	16 09	5 24	18·8	15 12	5 04	18·7	14 15	4 44	18·6	13 18	4 24	18·5	12 21	4 05	18·4	252	288
73	107	17 11	5 26	17·8	16 14	5 06	17·7	15 17	4 48	17·6	14 20	4 29	17·6	13 23	4 10	17·5	12 25	3 52	17·4	253	287
74	106	17 17	5 07	16·8	16 19	4 49	16·7	15 22	4 31	16·6	14 24	4 13	16·5	13 27	3 56	16·5	12 29	3 38	16·4	254	286
75	105	17 22	4 48	15·7	16 24	4 31	15·7	15 26	4 15	15·6	14 29	3 58	15·5	13 31	3 42	15·4	12 33	3 25	15·4	255	285
76	104	17 27	4 30	14·7	16 29	4 14	14·6	15 31	3 58	14·5	14 33	3 43	14·5	13 35	3 27	14·4	12 36	3 12	14·4	256	284
77	103	17 31	4 11	13·6	16 33	3 56	13·6	15 35	3 41	13·5	14 36	3 27	13·4	13 38	3 13	13·4	12 40	2 58	13·3	257	283
78	102	17 36	3 52	12·6	16 37	3 38	12·5	15 38	3 25	12·5	14 40	3 11	12·4	13 41	2 58	12·4	12 43	2 45	12·3	258	282
79	101	17 39	3 33	11·6	16 41	3 20	11·5	15 42	3 08	11·4	14 43	2 56	11·4	13 44	2 43	11·3	12 45	2 31	11·3	259	281
80	100	17 43	3 14	10·5	16 44	3 02	10·4	15 45	2 51	10·4	14 46	2 40	10·3	13 47	2 29	10·3	12 48	2 18	10·3	260	280
81	99	17 46	2 55	9·5	16 47	2 44	9·4	15 48	2 34	9·4	14 49	2 24	9·3	13 49	2 14	9·3	12 50	2 04	9·2	261	279
82	98	17 49	2 35	8·4	16 50	2 26	8·4	15 50	2 17	8·3	14 51	2 08	8·3	13 52	1 59	8·2	12 52	1 50	8·2	262	278
83	97	17 52	2 16	7·4	16 52	2 08	7·3	15 53	2 00	7·3	14 53	1 52	7·2	13 54	1 44	7·2	12 54	1 37	7·2	263	277
84	96	17 54	1 57	6·3	16 54	1 50	6·3	15 55	1 43	6·2	14 55	1 36	6·2	13 55	1 30	6·2	12 56	1 23	6·2	264	276
85	95	17 56	1 37	5·3	16 56	1 32	5·2	15 56	1 26	5·2	14 56	1 20	5·2	13 57	1 15	5·2	12 57	1 09	5·1	265	275
86	94	17 57	1 18	4·2	16 57	1 13	4·2	15 58	1 09	4·2	14 58	1 04	4·1	13 58	1 00	4·1	12 58	0 55	4·1	266	274
87	93	17 58	0 58	3·2	16 59	0 55	3·1	15 59	0 52	3·1	14 59	0 48	3·1	13 59	0 45	3·1	12 59	0 42	3·1	267	273
88	92	17 59	0 39	2·1	16 59	0 37	2·1	15 59	0 34	2·1	14 59	0 32	2·1	13 59	0 30	2·1	13 00	0 28	2·1	268	272
89	91	18 00	0 19	1·1	17 00	0 18	1·0	16 00	0 17	1·0	15 00	0 16	1·0	14 00	0 15	1·0	13 00	0 14	1·0	269	271
90	90	18 00	0 00	0·0	17 00	0 00	0·0	16 00	0 00	0·0	15 00	0 00	0·0	14 00	0 00	0·0	13 00	0 00	0·0	270	270

N. Lat: for LHA > 180° … $Z_n = Z$
for LHA < 180° … $Z_n = 360° - Z$

S. Lat.: for LHA > 180° … $Z_n = 180° - Z$
for LHA < 180° … $Z_n = 180° + Z$

SIGHT REDUCTION TABLE

B: (–) for 90° < LHA < 270°
Dec:(–) for Lat. contrary name

Z_1: same sign as B
Z_2: (–) for F > 90°

Lat. / A		78°			79°			80°			81°			82°			83°			Lat. / A	
LHA/F		A/H	B/P	Z_1/Z_2	A/H	B/P	Z_1/Z_2	A/H	B/P	Z_1/Z_2	A/H	B/P	Z_1/Z_2	A/H	B/P	Z_1/Z_2	A/H	B/P	Z_1/Z_2	LHA	
°	°	° ′	° ′	°	° ′	° ′	°	° ′	° ′	°	° ′	° ′	°	° ′	° ′	°	° ′	° ′	°	°	°
0	180	0 00	12 00	90·0	0 00	11 00	90·0	0 00	10 00	90·0	0 00	9 00	90·0	0 00	8 00	90·0	0 00	7 00	90·0	180	360
1	179	0 12	12 00	89·0	0 11	11 00	89·0	0 10	10 00	89·0	0 09	9 00	89·0	0 08	8 00	89·0	0 07	7 00	89·0	181	359
2	178	0 25	12 00	88·0	0 23	11 00	88·0	0 21	10 00	88·0	0 19	9 00	88·0	0 17	8 00	88·0	0 15	7 00	88·0	182	358
3	177	0 37	11 59	87·1	0 34	10 59	87·1	0 31	9 59	87·0	0 28	8 59	87·0	0 25	7 59	87·0	0 22	6 59	87·0	183	357
4	176	0 50	11 58	86·1	0 46	10 58	86·1	0 42	9 59	86·1	0 38	8 59	86·0	0 33	7 59	86·0	0 29	6 59	86·0	184	356
5	175	1 02	11 57	85·1	0 57	10 58	85·1	0 52	9 58	85·1	0 47	8 58	85·1	0 42	7 58	85·0	0 37	6 58	85·0	185	355
6	174	1 15	11 56	84·1	1 09	10 56	84·1	1 02	9 57	84·1	0 56	8 57	84·1	0 50	7 57	84·1	0 44	6 58	84·0	186	354
7	173	1 27	11 55	83·2	1 20	10 55	83·1	1 13	9 56	83·1	1 06	8 56	83·1	0 58	7 56	83·1	0 51	6 57	83·1	187	353
8	172	1 39	11 53	82·2	1 31	10 54	82·1	1 23	9 54	82·1	1 15	8 55	82·1	1 07	7 55	82·1	0 58	6 56	82·1	188	352
9	171	1 52	11 51	81·2	1 43	10 52	81·2	1 33	9 53	81·1	1 24	8 53	81·1	1 15	7 54	81·1	1 06	6 55	81·1	189	351
10	170	2 04	11 49	80·2	1 54	10 50	80·2	1 44	9 51	80·1	1 33	8 52	80·1	1 23	7 53	80·1	1 13	6 54	80·1	190	350
11	169	2 16	11 47	79·2	2 05	10 48	79·2	1 54	9 49	79·2	1 43	8 50	79·1	1 31	7 51	79·1	1 20	6 52	79·1	191	349
12	168	2 29	11 45	78·3	2 16	10 46	78·2	2 04	9 47	78·2	1 52	8 48	78·1	1 39	7 50	78·1	1 27	6 51	78·1	192	348
13	167	2 41	11 42	77·3	2 28	10 43	77·2	2 14	9 45	77·2	2 01	8 46	77·2	1 48	7 48	77·1	1 34	6 49	77·1	193	347
14	166	2 53	11 39	76·3	2 39	10 41	76·2	2 24	9 43	76·2	2 10	8 44	76·2	1 56	7 46	76·1	1 41	6 48	76·1	194	346
15	165	3 05	11 36	75·3	2 50	10 38	75·3	2 35	9 40	75·2	2 19	8 42	75·2	2 04	7 44	75·1	1 48	6 46	75·1	195	345
16	164	3 17	11 33	74·3	3 01	10 35	74·3	2 45	9 37	74·2	2 28	8 39	74·2	2 12	7 42	74·1	1 56	6 44	74·1	196	344
17	163	3 29	11 29	73·4	3 12	10 32	73·3	2 55	9 34	73·2	2 37	8 37	73·2	2 20	7 39	73·2	2 03	6 42	73·1	197	343
18	162	3 41	11 26	72·4	3 23	10 28	72·3	3 05	9 31	72·3	2 46	8 34	72·2	2 28	7 37	72·2	2 09	6 40	72·1	198	342
19	161	3 53	11 22	71·4	3 34	10 25	71·3	3 14	9 28	71·3	2 55	8 31	71·2	2 36	7 34	71·2	2 16	6 37	71·1	199	341
20	160	4 05	11 18	70·4	3 45	10 21	70·3	3 24	9 24	70·3	3 04	8 28	70·2	2 44	7 31	70·2	2 23	6 35	70·1	200	340
21	159	4 16	11 13	69·4	3 55	10 17	69·4	3 34	9 21	69·3	3 13	8 25	69·2	2 52	7 28	69·2	2 30	6 32	69·1	201	339
22	158	4 28	11 09	68·4	4 06	10 13	68·4	3 44	9 17	68·3	3 22	8 21	68·2	2 59	7 25	68·2	2 37	6 30	68·1	202	338
23	157	4 40	11 04	67·5	4 17	10 09	67·4	3 53	9 13	67·3	3 30	8 18	67·3	3 07	7 22	67·2	2 44	6 27	67·2	203	337
24	156	4 51	10 59	66·5	4 27	10 04	66·4	4 03	9 09	66·3	3 39	8 14	66·3	3 15	7 19	66·2	2 50	6 24	66·2	204	336
25	155	5 02	10 54	65·5	4 38	9 59	65·4	4 13	9 05	65·3	3 47	8 10	65·3	3 22	7 16	65·2	2 57	6 21	65·2	205	335
26	154	5 14	10 49	64·5	4 48	9 55	64·4	4 22	9 00	64·3	3 56	8 06	64·3	3 30	7 12	64·2	3 04	6 18	64·2	206	334
27	153	5 25	10 43	63·5	4 58	9 50	63·4	4 31	8 56	63·4	4 04	8 02	63·3	3 37	7 08	63·2	3 10	6 15	63·2	207	333
28	152	5 36	10 38	62·5	5 08	9 44	62·4	4 41	8 51	62·4	4 13	7 58	62·3	3 45	7 04	62·2	3 17	6 11	62·2	208	332
29	151	5 47	10 32	61·5	5 18	9 39	61·4	4 50	8 46	61·4	4 21	7 53	61·3	3 52	7 00	61·2	3 23	6 08	61·2	209	331
30	150	5 58	10 26	60·5	5 28	9 33	60·5	4 59	8 41	60·4	4 29	7 49	60·3	3 59	6 56	60·2	3 30	6 04	60·2	210	330
31	149	6 09	10 20	59·6	5 38	9 28	59·5	5 08	8 36	59·4	4 37	7 44	59·3	4 07	6 52	59·2	3 36	6 00	59·2	211	329
32	148	6 20	10 13	58·6	5 48	9 22	58·5	5 17	8 30	58·4	4 45	7 39	58·3	4 14	6 48	58·3	3 42	5 57	58·2	212	328
33	147	6 30	10 06	57·6	5 58	9 16	57·5	5 26	8 25	57·4	4 53	7 34	57·3	4 21	6 43	57·3	3 48	5 53	57·2	213	327
34	146	6 41	10 00	56·6	6 08	9 09	56·5	5 34	8 19	56·4	5 01	7 29	56·3	4 28	6 39	56·3	3 54	5 49	56·2	214	326
35	145	6 51	9 53	55·6	6 17	9 03	55·5	5 43	8 13	55·4	5 09	7 24	55·3	4 35	6 34	55·3	4 00	5 45	55·2	215	325
36	144	7 01	9 45	54·6	6 26	8 56	54·5	5 51	8 07	54·4	5 17	7 18	54·3	4 42	6 29	54·3	4 06	5 40	54·2	216	324
37	143	7 11	9 38	53·6	6 36	8 49	53·5	6 00	8 01	53·4	5 24	7 13	53·3	4 48	6 24	53·3	4 12	5 36	53·2	217	323
38	142	7 21	9 31	52·6	6 45	8 43	52·5	6 08	7 55	52·4	5 32	7 07	52·3	4 55	6 19	52·3	4 18	5 32	52·2	218	322
39	141	7 31	9 23	51·6	6 54	8 35	51·5	6 16	7 48	51·4	5 39	7 01	51·3	5 01	6 14	51·3	4 24	5 27	51·2	219	321
40	140	7 41	9 15	50·6	7 03	8 28	50·5	6 25	7 42	50·4	5 46	6 55	50·3	5 08	6 09	50·3	4 30	5 22	50·2	220	320
41	139	7 50	9 07	49·6	7 11	8 21	49·5	6 32	7 35	49·4	5 53	6 49	49·4	5 14	6 03	49·3	4 35	5 18	49·2	221	319
42	138	8 00	8 59	48·6	7 20	8 13	48·5	6 40	7 28	48·4	6 01	6 43	48·4	5 21	5 58	48·3	4 41	5 13	48·2	222	318
43	137	8 09	8 50	47·6	7 29	8 05	47·5	6 48	7 21	47·4	6 07	6 36	47·4	5 27	5 52	47·3	4 46	5 08	47·2	223	317
44	136	8 18	8 42	46·6	7 37	7 58	46·5	6 56	7 14	46·4	6 14	6 30	46·4	5 33	5 46	46·3	4 51	5 03	46·2	224	316
45	135	8 27	8 33	45·6	7 45	7 50	45·5	7 03	7 06	45·4	6 21	6 23	45·4	5 39	5 41	45·3	4 57	4 58	45·2	225	315

Lat. / A		78°			79°			80°			81°			82°			83°			Lat. / A	
LHA/F		A/H	B/P	Z_1/Z_2	A/H	B/P	Z_1/Z_2	A/H	B/P	Z_1/Z_2	A/H	B/P	Z_1/Z_2	A/H	B/P	Z_1/Z_2	A/H	B/P	Z_1/Z_2	LHA	
°	°	° ′	° ′	°	° ′	° ′	°	° ′	° ′	°	° ′	° ′	°	° ′	° ′	°	° ′	° ′	°	°	°
45	135	8 27	8 33	45·6	7 45	7 50	45·5	7 03	7 06	45·4	6 21	6 23	45·4	5 39	5 41	45·3	4 57	4 58	45·2	225	315
46	134	8 36	8 24	44·6	7 53	7 41	44·5	7 11	6 59	44·4	6 28	6 17	44·4	5 45	5 35	44·3	5 02	4 53	44·2	226	314
47	133	8 45	8 15	43·6	8 01	7 33	43·5	7 18	6 51	43·4	6 34	6 10	43·4	5 51	5 28	43·3	5 07	4 47	43·2	227	313
48	132	8 53	8 06	42·6	8 09	7 25	42·5	7 25	6 44	42·4	6 41	6 03	42·4	5 56	5 22	42·3	5 12	4 42	42·2	228	312
49	131	9 02	7 56	41·6	8 17	7 16	41·5	7 32	6 36	41·4	6 47	5 56	41·4	6 02	5 16	41·3	5 17	4 36	41·2	229	311
50	130	9 10	7 47	40·6	8 24	7 07	40·5	7 39	6 28	40·4	6 53	5 49	40·3	6 07	5 10	40·3	5 21	4 31	40·2	230	310
51	129	9 18	7 37	39·6	8 32	6 58	39·5	7 45	6 20	39·4	6 59	5 42	39·3	6 13	5 03	39·3	5 26	4 25	39·2	231	309
52	128	9 26	7 27	38·6	8 39	6 49	38·5	7 52	6 12	38·4	7 05	5 34	38·3	6 18	4 57	38·3	5 31	4 19	38·2	232	308
53	127	9 33	7 17	37·6	8 46	6 40	37·5	7 58	6 03	37·4	7 11	5 27	37·3	6 23	4 50	37·3	5 35	4 14	37·2	233	307
54	126	9 41	7 07	36·6	8 53	6 31	36·5	8 05	5 55	36·4	7 16	5 19	36·3	6 28	4 43	36·3	5 39	4 08	36·2	234	306
55	125	9 48	6 57	35·6	9 00	6 22	35·5	8 11	5 47	35·4	7 22	5 11	35·3	6 33	4 37	35·3	5 44	4 02	35·2	235	305
56	124	9 56	6 47	34·6	9 06	6 12	34·5	8 17	5 38	34·4	7 27	5 04	34·3	6 38	4 30	34·3	5 48	3 56	34·2	236	304
57	123	10 03	6 36	33·6	9 13	6 03	33·5	8 22	5 29	33·4	7 32	4 56	33·3	6 42	4 23	33·3	5 52	3 50	33·2	237	303
58	122	10 09	6 26	32·6	9 19	5 53	32·5	8 28	5 20	32·4	7 37	4 48	32·3	6 47	4 16	32·3	5 56	3 43	32·2	238	302
59	121	10 16	6 15	31·6	9 25	5 43	31·5	8 34	5 11	31·4	7 42	4 40	31·3	6 51	4 08	31·2	6 00	3 37	31·2	239	301
60	120	10 22	6 04	30·6	9 31	5 33	30·5	8 39	5 02	30·4	7 47	4 32	30·3	6 55	4 01	30·2	6 04	3 31	30·2	240	300
61	119	10 29	5 53	29·5	9 36	5 23	29·5	8 44	4 53	29·4	7 52	4 23	29·3	6 59	3 54	29·2	6 07	3 24	29·2	241	299
62	118	10 35	5 42	28·5	9 42	5 13	28·4	8 49	4 44	28·4	7 56	4 15	28·3	7 04	3 46	28·2	6 11	3 18	28·2	242	298
63	117	10 41	5 31	27·5	9 47	5 03	27·4	8 54	4 35	27·4	8 01	4 07	27·3	7 07	3 39	27·2	6 14	3 11	27·2	243	297
64	116	10 46	5 19	26·5	9 52	4 52	26·4	8 59	4 25	26·3	8 05	3 58	26·3	7 11	3 32	26·2	6 17	3 05	26·2	244	296
65	115	10 52	5 08	25·5	9 57	4 42	25·4	9 03	4 16	25·3	8 09	3 50	25·3	7 15	3 24	25·2	6 20	2 58	25·2	245	295
66	114	10 57	4 56	24·5	10 02	4 31	24·4	9 08	4 06	24·3	8 13	3 41	24·3	7 18	3 16	24·2	6 24	2 52	24·2	246	294
67	113	11 02	4 45	23·5	10 07	4 21	23·4	9 12	3 56	23·3	8 17	3 32	23·3	7 22	3 09	23·2	6 26	2 45	23·2	247	293
68	112	11 07	4 33	22·4	10 11	4 10	22·4	9 16	3 47	22·3	8 20	3 24	22·2	7 25	3 01	22·2	6 29	2 38	22·1	248	292
69	111	11 12	4 21	21·4	10 16	3 59	21·4	9 20	3 37	21·3	8 24	3 15	21·2	7 28	2 53	21·2	6 32	2 31	21·1	249	291
70	110	11 16	4 09	20·4	10 20	3 48	20·3	9 23	3 27	20·3	8 27	3 06	20·2	7 31	2 45	20·2	6 35	2 24	20·1	250	290
71	109	11 20	3 58	19·4	10 24	3 37	19·3	9 27	3 17	19·3	8 30	2 57	19·2	7 34	2 37	19·2	6 37	2 17	19·1	251	289
72	108	11 24	3 45	18·4	10 27	3 26	18·3	9 30	3 07	18·3	8 33	2 48	18·2	7 36	2 29	18·2	6 39	2 10	18·1	252	288
73	107	11 28	3 33	17·4	10 31	3 15	17·3	9 34	2 57	17·2	8 36	2 39	17·2	7 39	2 21	17·2	6 42	2 03	17·1	253	287
74	106	11 32	3 21	16·3	10 34	3 04	16·3	9 37	2 47	16·2	8 39	2 30	16·2	7 41	2 13	16·1	6 44	1 56	16·1	254	286
75	105	11 35	3 09	15·3	10 37	2 53	15·3	9 39	2 37	15·2	8 41	2 21	15·2	7 44	2 05	15·1	6 46	1 49	15·1	255	285
76	104	11 38	2 57	14·3	10 40	2 42	14·3	9 42	2 27	14·2	8 44	2 12	14·2	7 46	1 57	14·1	6 47	1 42	14·1	256	284
77	103	11 41	2 44	13·3	10 43	2 30	13·2	9 44	2 16	13·2	8 46	2 02	13·2	7 48	1 49	13·1	6 49	1 35	13·1	257	283
78	102	11 44	2 32	12·3	10 45	2 19	12·2	9 47	2 06	12·2	8 48	1 53	12·1	7 49	1 40	12·1	6 51	1 28	12·1	258	282
79	101	11 47	2 19	11·2	10 48	2 07	11·2	9 49	1 56	11·2	8 50	1 44	11·1	7 51	1 32	11·1	6 52	1 21	11·1	259	281
80	100	11 49	2 07	10·2	10 50	1 56	10·2	9 51	1 45	10·2	8 52	1 35	10·1	7 53	1 24	10·1	6 54	1 13	10·1	260	280
81	99	11 51	1 54	9·2	10 52	1 45	9·2	9 53	1 35	9·1	8 53	1 25	9·1	7 54	1 16	9·1	6 55	1 06	9·1	261	279
82	98	11 53	1 42	8·2	10 53	1 33	8·1	9 54	1 24	8·1	8 55	1 16	8·1	7 55	1 07	8·1	6 56	0 59	8·1	262	278
83	97	11 55	1 29	7·2	10 55	1 21	7·1	9 55	1 14	7·1	8 56	1 06	7·1	7 56	0 59	7·1	6 57	0 51	7·1	263	277
84	96	11 56	1 16	6·1	10 56	1 10	6·1	9 57	1 03	6·1	8 57	0 57	6·1	7 57	0 50	6·1	6 58	0 44	6·0	264	276
85	95	11 57	1 04	5·1	10 57	0 58	5·1	9 58	0 53	5·1	8 58	0 47	5·1	7 58	0 42	5·0	6 58	0 37	5·0	265	275
86	94	11 58	0 51	4·1	10 58	0 47	4·1	9 59	0 42	4·1	8 59	0 38	4·0	7 59	0 34	4·0	6 59	0 29	4·0	266	274
87	93	11 59	0 38	3·1	10 59	0 35	3·1	9 59	0 32	3·0	8 59	0 28	3·0	7 59	0 25	3·0	6 59	0 22	3·0	267	273
88	92	12 00	0 26	2·0	11 00	0 23	2·0	10 00	0 21	2·0	9 00	0 19	2·0	8 00	0 17	2·0	7 00	0 15	2·0	268	272
89	91	12 00	0 13	1·0	11 00	0 12	1·0	10 00	0 11	1·0	9 00	0 10	1·0	8 00	0 08	1·0	7 00	0 07	1·0	269	271
90	90	12 00	0 00	0·0	11 00	0 00	0·0	10 00	0 00	0·0	9 00	0 00	0·0	8 00	0 00	0·0	7 00	0 00	0·0	270	270

N. Lat: for LHA > 180° … $Z_n = Z$
for LHA < 180° … $Z_n = 360° - Z$

S. Lat.: for LHA > 180° … $Z_n = 180° - Z$
for LHA < 180° … $Z_n = 180° + Z$

SIGHT REDUCTION TABLE

B: (−) for 90° < LHA < 270°
Dec:(−) for Lat. contrary name

Z_1: same sign as B
Z_2: (−) for F > 90°

Lat. / A		84°			85°			86°			87°			88°			89°			Lat. / A	
LHA/F		A/H	B/P	Z_1/Z_2	A/H	B/P	Z_1/Z_2	A/H	B/P	Z_1/Z_2	A/H	B/P	Z_1/Z_2	A/H	B/P	Z_1/Z_2	A/H	B/P	Z_1/Z_2	LHA	
°	°	° ′	° ′	°	° ′	° ′	°	° ′	° ′	°	° ′	° ′	°	° ′	° ′	°	° ′	° ′	°	°	°
0	180	0 00	6 00	90·0	0 00	5 00	90·0	0 00	4 00	90·0	0 00	3 00	90·0	0 00	2 00	90·0	0 00	1 00	90·0	180	360
1	179	0 06	6 00	89·0	0 05	5 00	89·0	0 04	4 00	89·0	0 03	3 00	89·0	0 02	2 00	89·0	0 01	1 00	89·0	181	359
2	178	0 13	6 00	88·0	0 10	5 00	88·0	0 08	4 00	88·0	0 06	3 00	88·0	0 04	2 00	88·0	0 02	1 00	88·0	182	358
3	177	0 19	6 00	87·0	0 16	5 00	87·0	0 13	4 00	87·0	0 09	3 00	87·0	0 06	2 00	87·0	0 03	1 00	87·0	183	357
4	176	0 25	5 59	86·0	0 21	4 59	86·0	0 17	3 59	86·0	0 13	3 00	86·0	0 08	2 00	86·0	0 04	1 00	86·0	184	356
5	175	0 31	5 59	85·0	0 26	4 59	85·0	0 21	3 59	85·0	0 16	2 59	85·0	0 10	2 00	85·0	0 05	1 00	85·0	185	355
6	174	0 38	5 58	84·0	0 31	4 58	84·0	0 25	3 59	84·0	0 19	2 59	84·0	0 13	1 59	84·0	0 06	1 00	84·0	186	354
7	173	0 44	5 57	83·0	0 37	4 58	83·0	0 29	3 58	83·0	0 22	2 59	83·0	0 15	1 59	83·0	0 07	1 00	83·0	187	353
8	172	0 50	5 57	82·0	0 42	4 57	82·0	0 33	3 58	82·0	0 25	2 58	82·0	0 17	1 59	82·0	0 08	0 59	82·0	188	352
9	171	0 56	5 56	81·0	0 47	4 56	81·0	0 38	3 57	81·0	0 28	2 58	81·0	0 19	1 59	81·0	0 09	0 59	81·0	189	351
10	170	1 02	5 55	80·1	0 52	4 55	80·0	0 42	3 56	80·0	0 31	2 57	80·0	0 21	1 58	80·0	0 10	0 59	80·0	190	350
11	169	1 09	5 53	79·1	0 57	4 55	79·0	0 46	3 56	79·0	0 34	2 57	79·0	0 23	1 58	79·0	0 11	0 59	79·0	191	349
12	168	1 15	5 52	78·1	1 02	4 53	78·0	0 50	3 55	78·0	0 37	2 56	78·0	0 25	1 57	78·0	0 12	0 59	78·0	192	348
13	167	1 21	5 51	77·1	1 07	4 52	77·0	0 54	3 54	77·0	0 40	2 55	77·0	0 27	1 57	77·0	0 13	0 58	77·0	193	347
14	166	1 27	5 49	76·1	1 12	4 51	76·1	0 58	3 53	76·0	0 44	2 55	76·0	0 29	1 56	76·0	0 15	0 58	76·0	194	346
15	165	1 33	5 48	75·1	1 18	4 50	75·1	1 02	3 52	75·0	0 47	2 54	75·0	0 31	1 56	75·0	0 16	0 58	75·0	195	345
16	164	1 39	5 46	74·1	1 23	4 48	74·1	1 06	3 51	74·0	0 50	2 53	74·0	0 33	1 55	74·0	0 17	0 58	74·0	196	344
17	163	1 45	5 44	73·1	1 28	4 47	73·1	1 10	3 50	73·0	0 53	2 52	73·0	0 35	1 55	73·0	0 18	0 57	73·0	197	343
18	162	1 51	5 42	72·1	1 33	4 45	72·1	1 14	3 48	72·0	0 56	2 51	72·0	0 37	1 54	72·0	0 19	0 57	72·0	198	342
19	161	1 57	5 41	71·1	1 38	4 44	71·1	1 18	3 47	71·0	0 59	2 50	71·0	0 39	1 53	71·0	0 20	0 57	71·0	199	341
20	160	2 03	5 38	70·1	1 42	4 42	70·1	1 22	3 46	70·0	1 02	2 49	70·0	0 41	1 53	70·0	0 21	0 56	70·0	200	340
21	159	2 09	5 36	69·1	1 47	4 40	69·1	1 26	3 44	69·0	1 04	2 48	69·0	0 43	1 52	69·0	0 22	0 56	69·0	201	339
22	158	2 15	5 34	68·1	1 52	4 38	68·1	1 30	3 43	68·0	1 07	2 47	68·0	0 45	1 51	68·0	0 22	0 56	68·0	202	338
23	157	2 20	5 32	67·1	1 57	4 36	67·1	1 34	3 41	67·1	1 10	2 46	67·0	0 47	1 50	67·0	0 23	0 55	67·0	203	337
24	156	2 26	5 29	66·1	2 02	4 34	66·1	1 38	3 39	66·1	1 13	2 44	66·0	0 49	1 50	66·0	0 24	0 55	66·0	204	336
25	155	2 32	5 26	65·1	2 07	4 32	65·1	1 41	3 38	65·1	1 16	2 43	65·0	0 51	1 49	65·0	0 25	0 54	65·0	205	335
26	154	2 38	5 24	64·1	2 11	4 30	64·1	1 45	3 36	64·1	1 19	2 42	64·0	0 53	1 48	64·0	0 26	0 54	64·0	206	334
27	153	2 43	5 21	63·1	2 16	4 27	63·1	1 49	3 34	63·1	1 22	2 40	63·0	0 54	1 47	63·0	0 27	0 53	63·0	207	333
28	152	2 49	5 18	62·1	2 21	4 25	62·1	1 53	3 32	62·1	1 24	2 39	62·0	0 56	1 46	62·0	0 28	0 53	62·0	208	332
29	151	2 54	5 15	61·1	2 25	4 23	61·1	1 56	3 30	61·1	1 27	2 37	61·0	0 58	1 45	61·0	0 29	0 52	61·0	209	331
30	150	3 00	5 12	60·1	2 30	4 20	60·1	2 00	3 28	60·1	1 30	2 36	60·0	1 00	1 44	60·0	0 30	0 52	60·0	210	330
31	149	3 05	5 09	59·1	2 34	4 17	59·1	2 04	3 26	59·1	1 33	2 34	59·0	1 02	1 43	59·0	0 31	0 51	59·0	211	329
32	148	3 11	5 06	58·1	2 39	4 15	58·1	2 07	3 24	58·1	1 35	2 33	58·0	1 04	1 42	58·0	0 32	0 51	58·0	212	328
33	147	3 16	5 02	57·1	2 43	4 12	57·1	2 11	3 21	57·1	1 38	2 31	57·0	1 05	1 41	57·0	0 33	0 50	57·0	213	327
34	146	3 21	4 59	56·1	2 48	4 09	56·1	2 14	3 19	56·1	1 41	2 29	56·0	1 07	1 39	56·0	0 34	0 50	56·0	214	326
35	145	3 26	4 55	55·1	2 52	4 06	55·1	2 18	3 17	55·1	1 43	2 27	55·0	1 09	1 38	55·0	0 34	0 49	55·0	215	325
36	144	3 31	4 52	54·1	2 56	4 03	54·1	2 21	3 14	54·1	1 46	2 26	54·0	1 11	1 37	54·0	0 35	0 49	54·0	216	324
37	143	3 36	4 48	53·2	3 00	4 00	53·1	2 24	3 12	53·1	1 48	2 24	53·0	1 12	1 36	53·0	0 36	0 48	53·0	217	323
38	142	3 41	4 44	52·2	3 05	3 57	52·1	2 28	3 09	52·1	1 51	2 22	52·0	1 14	1 35	52·0	0 37	0 47	52·0	218	322
39	141	3 46	4 40	51·2	3 09	3 53	51·1	2 31	3 07	51·1	1 53	2 20	51·0	1 16	1 33	51·0	0 38	0 47	51·0	219	321
40	140	3 51	4 36	50·2	3 13	3 50	50·1	2 34	3 04	50·1	1 56	2 18	50·0	1 17	1 32	50·0	0 39	0 46	50·0	220	320
41	139	3 56	4 32	49·2	3 17	3 47	49·1	2 37	3 01	49·1	1 58	2 16	49·0	1 19	1 31	49·0	0 39	0 45	49·0	221	319
42	138	4 01	4 28	48·2	3 21	3 43	48·1	2 41	2 58	48·1	2 00	2 14	48·0	1 20	1 29	48·0	0 40	0 45	48·0	222	318
43	137	4 05	4 24	47·2	3 24	3 40	47·1	2 44	2 56	47·1	2 03	2 12	47·0	1 22	1 28	47·0	0 41	0 44	47·0	223	317
44	136	4 10	4 19	46·2	3 28	3 36	46·1	2 47	2 53	46·1	2 05	2 10	46·0	1 23	1 26	46·0	0 42	0 43	46·0	224	316
45	135	4 14	4 15	45·2	3 32	3 32	45·1	2 50	2 50	45·1	2 07	2 07	45·0	1 25	1 25	45·0	0 42	0 42	45·0	225	315

Lat. / A		84°			85°			86°			87°			88°			89°			Lat. / A	
LHA/F		A/H	B/P	Z_1/Z_2	A/H	B/P	Z_1/Z_2	A/H	B/P	Z_1/Z_2	A/H	B/P	Z_1/Z_2	A/H	B/P	Z_1/Z_2	A/H	B/P	Z_1/Z_2	LHA	
°	°	° ′	° ′	°	° ′	° ′	°	° ′	° ′	°	° ′	° ′	°	° ′	° ′	°	° ′	° ′	°	°	°
45	135	4 14	4 15	45·2	3 32	3 32	45·1	2 50	2 50	45·1	2 07	2 07	45·0	1 25	1 25	45·0	0 42	0 42	45·0	225	315
46	134	4 19	4 11	44·2	3 36	3 29	44·1	2 53	2 47	44·1	2 09	2 05	44·0	1 26	1 23	44·0	0 43	0 42	44·0	226	314
47	133	4 23	4 06	43·2	3 39	3 25	43·1	2 55	2 44	43·1	2 12	2 03	43·0	1 28	1 22	43·0	0 44	0 41	43·0	227	313
48	132	4 27	4 01	42·2	3 43	3 21	42·1	2 58	2 41	42·1	2 14	2 01	42·0	1 29	1 20	42·0	0 45	0 40	42·0	228	312
49	131	4 31	3 57	41·2	3 46	3 17	41·1	3 01	2 38	41·1	2 16	1 58	41·0	1 31	1 19	41·0	0 45	0 39	41·0	229	311
50	130	4 36	3 52	40·2	3 50	3 13	40·1	3 04	2 34	40·1	2 18	1 56	40·0	1 32	1 17	40·0	0 46	0 39	40·0	230	310
51	129	4 40	3 47	39·2	3 53	3 09	39·1	3 06	2 31	39·1	2 20	1 53	39·0	1 33	1 16	39·0	0 47	0 38	39·0	231	309
52	128	4 43	3 42	38·2	3 56	3 05	38·1	3 09	2 28	38·1	2 22	1 51	38·0	1 35	1 14	38·0	0 47	0 37	38·0	232	308
53	127	4 47	3 37	37·2	3 59	3 01	37·1	3 12	2 25	37·1	2 24	1 48	37·0	1 36	1 12	37·0	0 48	0 36	37·0	233	307
54	126	4 51	3 32	36·1	4 03	2 57	36·1	3 14	2 21	36·1	2 26	1 46	36·0	1 37	1 11	36·0	0 49	0 35	36·0	234	306
55	125	4 55	3 27	35·1	4 06	2 52	35·1	3 17	2 18	35·1	2 27	1 43	35·0	1 38	1 09	35·0	0 49	0 34	35·0	235	305
56	124	4 58	3 22	34·1	4 09	2 48	34·1	3 19	2 14	34·1	2 29	1 41	34·0	1 39	1 07	34·0	0 50	0 34	34·0	236	304
57	123	5 02	3 17	33·1	4 12	2 44	33·1	3 21	2 11	33·1	2 31	1 38	33·0	1 41	1 05	33·0	0 50	0 33	33·0	237	303
58	122	5 05	3 11	32·1	4 14	2 39	32·1	3 23	2 07	32·1	2 33	1 35	32·0	1 42	1 04	32·0	0 51	0 32	32·0	238	302
59	121	5 08	3 06	31·1	4 17	2 35	31·1	3 26	2 04	31·1	2 34	1 33	31·0	1 43	1 02	31·0	0 51	0 31	31·0	239	301
60	120	5 12	3 00	30·1	4 20	2 30	30·1	3 28	2 00	30·1	2 36	1 30	30·0	1 44	1 00	30·0	0 52	0 30	30·0	240	300
61	119	5 15	2 55	29·1	4 22	2 26	29·1	3 30	1 56	29·1	2 37	1 27	29·0	1 45	0 58	29·0	0 52	0 29	29·0	241	299
62	118	5 18	2 49	28·1	4 25	2 21	28·1	3 32	1 53	28·1	2 39	1 25	28·0	1 46	0 56	28·0	0 53	0 28	28·0	242	298
63	117	5 21	2 44	27·1	4 27	2 16	27·1	3 34	1 49	27·1	2 40	1 22	27·0	1 47	0 54	27·0	0 53	0 27	27·0	243	297
64	116	5 23	2 38	26·1	4 30	2 12	26·1	3 36	1 45	26·1	2 42	1 19	26·0	1 48	0 53	26·0	0 54	0 26	26·0	244	296
65	115	5 26	2 33	25·1	4 32	2 07	25·1	3 37	1 42	25·1	2 43	1 16	25·0	1 49	0 51	25·0	0 54	0 25	25·0	245	295
66	114	5 29	2 27	24·1	4 34	2 02	24·1	3 39	1 38	24·1	2 44	1 13	24·0	1 50	0 49	24·0	0 55	0 24	24·0	246	294
67	113	5 31	2 21	23·1	4 36	1 57	23·1	3 41	1 34	23·1	2 46	1 10	23·0	1 50	0 47	23·0	0 55	0 23	23·0	247	293
68	112	5 34	2 15	22·1	4 38	1 53	22·1	3 42	1 30	22·0	2 47	1 07	22·0	1 51	0 45	22·0	0 56	0 22	22·0	248	292
69	111	5 36	2 09	21·1	4 40	1 48	21·1	3 44	1 26	21·0	2 48	1 05	21·0	1 52	0 43	21·0	0 56	0 22	21·0	249	291
70	110	5 38	2 04	20·1	4 42	1 43	20·1	3 46	1 22	20·0	2 49	1 02	20·0	1 53	0 41	20·0	0 56	0 21	20·0	250	290
71	109	5 40	1 58	19·1	4 44	1 38	19·1	3 47	1 18	19·0	2 50	0 59	19·0	1 53	0 39	19·0	0 57	0 20	19·0	251	289
72	108	5 42	1 52	18·1	4 45	1 33	18·1	3 48	1 14	18·0	2 51	0 56	18·0	1 54	0 37	18·0	0 57	0 19	18·0	252	288
73	107	5 44	1 46	17·1	4 47	1 28	17·1	3 49	1 10	17·0	2 52	0 53	17·0	1 55	0 35	17·0	0 57	0 18	17·0	253	287
74	106	5 46	1 40	16·1	4 48	1 23	16·1	3 51	1 06	16·0	2 53	0 50	16·0	1 55	0 33	16·0	0 58	0 17	16·0	254	286
75	105	5 48	1 33	15·1	4 50	1 18	15·1	3 52	1 02	15·0	2 54	0 47	15·0	1 56	0 31	15·0	0 58	0 16	15·0	255	285
76	104	5 49	1 27	14·1	4 51	1 13	14·1	3 53	0 58	14·0	2 55	0 44	14·0	1 56	0 29	14·0	0 58	0 15	14·0	256	284
77	103	5 51	1 21	13·1	4 52	1 08	13·0	3 54	0 54	13·0	2 55	0 41	13·0	1 57	0 27	13·0	0 58	0 13	13·0	257	283
78	102	5 52	1 15	12·1	4 53	1 03	12·0	3 55	0 50	12·0	2 56	0 37	12·0	1 57	0 25	12·0	0 59	0 12	12·0	258	282
79	101	5 53	1 09	11·1	4 54	0 57	11·0	3 56	0 46	11·0	2 57	0 34	11·0	1 58	0 23	11·0	0 59	0 11	11·0	259	281
80	100	5 55	1 03	10·1	4 55	0 52	10·0	3 56	0 42	10·0	2 57	0 31	10·0	1 58	0 21	10·0	0 59	0 10	10·0	260	280
81	99	5 56	0 57	9·0	4 56	0 47	9·0	3 57	0 38	9·0	2 58	0 28	9·0	1 59	0 19	9·0	0 59	0 09	9·0	261	279
82	98	5 56	0 50	8·0	4 57	0 42	8·0	3 58	0 33	8·0	2 58	0 25	8·0	1 59	0 17	8·0	0 59	0 08	8·0	262	278
83	97	5 57	0 44	7·0	4 58	0 37	7·0	3 58	0 29	7·0	2 59	0 22	7·0	1 59	0 15	7·0	1 00	0 07	7·0	263	277
84	96	5 58	0 38	6·0	4 58	0 31	6·0	3 59	0 25	6·0	2 59	0 19	6·0	1 59	0 13	6·0	1 00	0 06	6·0	264	276
85	95	5 59	0 31	5·0	4 59	0 26	5·0	3 59	0 21	5·0	2 59	0 16	5·0	2 00	0 10	5·0	1 00	0 05	5·0	265	275
86	94	5 59	0 25	4·0	4 59	0 21	4·0	3 59	0 17	4·0	3 00	0 13	4·0	2 00	0 08	4·0	1 00	0 04	4·0	266	274
87	93	6 00	0 19	3·0	5 00	0 16	3·0	4 00	0 13	3·0	3 00	0 09	3·0	2 00	0 06	3·0	1 00	0 03	3·0	267	273
88	92	6 00	0 13	2·0	5 00	0 10	2·0	4 00	0 08	2·0	3 00	0 06	2·0	2 00	0 04	2·0	1 00	0 02	2·0	268	272
89	91	6 00	0 06	1·0	5 00	0 05	1·0	4 00	0 04	1·0	3 00	0 03	1·0	2 00	0 02	1·0	1 00	0 01	1·0	269	271
90	90	6 00	0 00	0·0	5 00	0 00	0·0	4 00	0 00	0·0	3 00	0 00	0·0	2 00	0 00	0·0	1 00	0 00	0·0	270	270

N. Lat: for LHA > 180° … $Z_n = Z$
for LHA < 180° … $Z_n = 360° - Z$

S. Lat.: for LHA > 180° … $Z_n = 180° - Z$
for LHA < 180° … $Z_n = 180° + Z$

AUXILIARY TABLE

Sign of $corr_1$ for F′. *Reverse* sign if $F > 90°$. ↓

Sign for $corr_2$ for A′. ↓

F′ +	1	2	3	4	5	6	7	8	9	10	11	12	13	14	15	16	17	18	19	20	21	22	23	24	25	26	27	28	29	□	− A′
F′ −	59	58	57	56	55	54	53	52	51	50	49	48	47	46	45	44	43	42	41	40	39	38	37	36	35	34	33	32	31	30	+ A′
P°	′	′	′	′	′	′	′	′	′	′	′	′	′	′	′	′	′	′	′	′	′	′	′	′	′	′	′	′	′	′	$Z_2°$
1	0	0	0	0	0	0	0	0	0	0	0	0	0	0	0	0	0	0	0	0	0	0	0	0	0	0	0	0	1	1	89
2	0	0	0	0	0	0	0	0	0	0	0	0	0	0	1	1	1	1	1	1	1	1	1	1	1	1	1	1	1	1	88
3	0	0	0	0	0	0	0	0	0	1	1	1	1	1	1	1	1	1	1	1	1	1	1	1	1	1	1	1	2	2	87
4	0	0	0	0	0	0	0	1	1	1	1	1	1	1	1	1	1	1	1	1	1	2	2	2	2	2	2	2	2	2	86
5	0	0	0	0	0	1	1	1	1	1	1	1	1	1	1	1	1	2	2	2	2	2	2	2	2	2	2	2	3	3	85
6	0	0	0	0	1	1	1	1	1	1	1	1	1	1	2	2	2	2	2	2	2	2	2	3	3	3	3	3	3	3	84
7	0	0	0	0	1	1	1	1	1	1	1	1	2	2	2	2	2	2	2	2	3	3	3	3	3	3	3	3	4	4	83
8	0	0	0	1	1	1	1	1	1	1	2	2	2	2	2	2	2	3	3	3	3	3	3	3	3	4	4	4	4	4	82
9	0	0	0	1	1	1	1	1	1	2	2	2	2	2	2	3	3	3	3	3	3	3	4	4	4	4	4	4	5	5	81
10	0	0	1	1	1	1	1	1	2	2	2	2	2	2	3	3	3	3	3	3	4	4	4	4	4	5	5	5	5	5	80
11	0	0	1	1	1	1	1	2	2	2	2	2	2	3	3	3	3	3	4	4	4	4	4	5	5	5	5	5	6	6	79
12	0	0	1	1	1	1	1	2	2	2	2	2	3	3	3	3	4	4	4	4	4	5	5	5	5	5	6	6	6	6	78
13	0	0	1	1	1	1	2	2	2	2	2	3	3	3	3	4	4	4	4	4	5	5	5	5	6	6	6	6	7	7	77
14	0	0	1	1	1	1	2	2	2	2	3	3	3	3	4	4	4	4	5	5	5	5	6	6	6	6	7	7	7	7	76
15	0	1	1	1	1	2	2	2	2	3	3	3	3	4	4	4	4	5	5	5	5	6	6	6	6	7	7	7	8	8	75
16	0	1	1	1	1	2	2	2	2	3	3	3	4	4	4	4	5	5	5	6	6	6	6	7	7	7	7	8	8	8	74
17	0	1	1	1	1	2	2	2	3	3	3	4	4	4	4	5	5	5	6	6	6	6	7	7	7	8	8	8	8	9	73
18	0	1	1	1	2	2	2	2	3	3	3	4	4	4	5	5	5	6	6	6	6	7	7	7	8	8	8	9	9	9	72
19	0	1	1	1	2	2	2	3	3	3	4	4	4	5	5	5	6	6	6	7	7	7	7	8	8	8	9	9	9	10	71
20	0	1	1	1	2	2	2	3	3	3	4	4	4	5	5	5	6	6	6	7	7	8	8	8	9	9	9	10	10	10	70
21	0	1	1	1	2	2	3	3	3	4	4	4	5	5	5	6	6	6	7	7	8	8	8	9	9	9	10	10	10	11	69
22	0	1	1	1	2	2	3	3	3	4	4	4	5	5	6	6	6	7	7	7	8	8	9	9	9	10	10	10	11	11	68
23	0	1	1	2	2	2	3	3	4	4	4	5	5	5	6	6	7	7	7	8	8	9	9	9	10	10	11	11	11	12	67
24	0	1	1	2	2	2	3	3	4	4	4	5	5	6	6	7	7	7	8	8	9	9	9	10	10	11	11	11	12	12	66
25	0	1	1	2	2	3	3	3	4	4	5	5	5	6	6	7	7	8	8	8	9	9	10	10	11	11	11	12	12	13	65
26	0	1	1	2	2	3	3	4	4	4	5	5	6	6	7	7	7	8	8	9	9	10	10	11	11	11	12	12	13	13	64
27	0	1	1	2	2	3	3	4	4	5	5	5	6	6	7	7	8	8	9	9	10	10	10	11	11	12	12	13	13	14	63
28	0	1	1	2	2	3	3	4	4	5	5	6	6	7	7	8	8	8	9	9	10	10	11	11	12	12	13	13	14	14	62
29	0	1	1	2	2	3	3	4	4	5	5	6	6	7	7	8	8	9	9	10	10	11	11	12	12	13	13	14	14	15	61
30	0	1	1	2	2	3	3	4	4	5	5	6	6	7	7	8	8	9	9	10	10	11	11	12	12	13	13	14	14	15	60
31	1	1	2	2	3	3	4	4	5	5	6	6	7	7	8	8	9	9	10	10	11	11	12	12	13	13	14	14	15	15	59
32	1	1	2	2	3	3	4	4	5	5	6	6	7	7	8	8	9	10	10	11	11	12	12	13	13	14	14	15	15	16	58
33	1	1	2	2	3	3	4	4	5	5	6	7	7	8	8	9	9	10	10	11	11	12	13	13	14	14	15	15	16	16	57
34	1	1	2	2	3	3	4	4	5	6	6	7	7	8	8	9	10	10	11	11	12	12	13	13	14	14	15	15	16	16	56
35	1	1	2	2	3	3	4	5	5	6	6	7	7	8	9	9	10	10	11	11	12	13	13	14	14	15	15	16	17	17	55
36	1	1	2	2	3	4	4	5	5	6	6	7	8	8	9	9	10	11	11	12	12	13	14	14	15	15	16	16	17	18	54
37	1	1	2	2	3	4	4	5	5	6	7	7	8	8	9	10	10	11	11	12	13	13	14	14	15	16	16	17	17	18	53
38	1	1	2	2	3	4	4	5	6	6	7	7	8	9	9	10	10	11	12	12	13	14	14	15	15	16	17	17	18	18	52
39	1	1	2	3	3	4	4	5	6	6	7	8	8	9	9	10	11	11	12	13	13	14	14	15	16	16	17	18	18	19	51
40	1	1	2	3	3	4	4	5	6	6	7	8	8	9	10	10	11	12	12	13	13	14	15	15	16	17	17	18	19	19	50

F′ +/−	1 / 59	2 / 58	3 / 57	4 / 56	5 / 55	6 / 54	7 / 53	8 / 52	9 / 51	10 / 50	11 / 49	12 / 48	13 / 47	14 / 46	15 / 45	16 / 44	17 / 43	18 / 42	19 / 41	20 / 40	21 / 39	22 / 38	23 / 37	24 / 36	25 / 35	26 / 34	27 / 33	28 / 32	29 / 31	□ / 30	−/+ A′
P°	′	′	′	′	′	′	′	′	′	′	′	′	′	′	′	′	′	′	′	′	′	′	′	′	′	′	′	′	′	′	Z_2°
41	1	1	2	3	3	4	5	5	6	7	7	8	9	9	10	10	11	12	12	13	14	14	15	16	16	17	18	18	19	20	49
42	1	1	2	3	3	4	5	5	6	7	7	8	9	9	10	11	11	12	13	13	14	15	15	16	17	17	18	19	19	20	48
43	1	1	2	3	3	4	5	5	6	7	8	8	9	10	10	11	12	12	13	14	14	15	16	16	17	18	18	19	20	20	47
44	1	1	2	3	3	4	5	6	6	7	8	8	9	10	10	11	12	13	13	14	15	15	16	17	17	18	19	19	20	21	46
45	1	1	2	3	4	4	5	6	6	7	8	8	9	10	11	11	12	13	13	14	15	16	16	17	18	18	19	20	21	21	45
46	1	1	2	3	4	4	5	6	6	7	8	9	9	10	11	12	12	13	14	14	15	16	17	17	18	19	19	20	21	22	44
47	1	1	2	3	4	4	5	6	7	7	8	9	10	10	11	12	12	13	14	15	15	16	17	18	18	19	20	20	21	22	43
48	1	1	2	3	4	4	5	6	7	7	8	9	10	10	11	12	13	13	14	15	16	16	17	18	19	19	20	21	22	22	42
49	1	2	2	3	4	5	5	6	7	8	8	9	10	11	11	12	13	14	14	15	16	17	17	18	19	20	20	21	22	23	41
50	1	2	2	3	4	5	5	6	7	8	8	9	10	11	11	12	13	14	15	15	16	17	18	18	19	20	21	21	22	23	40
51	1	2	2	3	4	5	5	6	7	8	9	9	10	11	12	12	13	14	15	16	16	17	18	19	19	20	21	22	23	23	39
52	1	2	2	3	4	5	6	6	7	8	9	9	10	11	12	13	13	14	15	16	17	17	18	19	20	20	21	22	23	24	38
53	1	2	2	3	4	5	6	6	7	8	9	10	10	11	12	13	14	14	15	16	17	18	18	19	20	21	22	22	23	24	37
54	1	2	2	3	4	5	6	6	7	8	9	10	11	11	12	13	14	15	15	16	17	18	19	19	20	21	22	23	23	24	36
55	1	2	2	3	4	5	6	7	7	8	9	10	11	11	12	13	14	15	16	16	17	18	19	20	20	21	22	23	24	25	35
56	1	2	2	3	4	5	6	7	7	8	9	10	11	12	12	13	14	15	16	17	17	18	19	20	21	22	22	23	24	25	34
57	1	2	3	3	4	5	6	7	8	8	9	10	11	12	13	13	14	15	16	17	18	18	19	20	21	22	23	23	24	25	33
58	1	2	3	3	4	5	6	7	8	8	9	10	11	12	13	14	14	15	16	17	18	19	20	20	21	22	23	24	25	25	32
59	1	2	3	3	4	5	6	7	8	9	9	10	11	12	13	14	15	15	16	17	18	19	20	21	21	22	23	24	25	26	31
60	1	2	3	3	4	5	6	7	8	9	10	10	11	12	13	14	15	16	16	17	18	19	20	21	22	23	23	24	25	26	30
61	1	2	3	3	4	5	6	7	8	9	10	10	11	12	13	14	15	16	17	17	18	19	20	21	22	23	24	24	25	26	29
62	1	2	3	4	4	5	6	7	8	9	10	11	11	12	13	14	15	16	17	18	19	19	20	21	22	23	24	25	26	26	28
63	1	2	3	4	4	5	6	7	8	9	10	11	12	12	13	14	15	16	17	18	19	20	20	21	22	23	24	25	26	27	27
64	1	2	3	4	4	5	6	7	8	9	10	11	12	13	13	14	15	16	17	18	19	20	21	22	22	23	24	25	26	27	26
65	1	2	3	4	5	5	6	7	8	9	10	11	12	13	14	15	15	16	17	18	19	20	21	22	23	24	24	25	26	27	25
66	1	2	3	4	5	5	6	7	8	9	10	11	12	13	14	15	16	16	17	18	19	20	21	22	23	24	25	26	26	27	24
67	1	2	3	4	5	6	6	7	8	9	10	11	12	13	14	15	16	17	17	18	19	20	21	22	23	24	25	26	27	28	23
68	1	2	3	4	5	6	6	7	8	9	10	11	12	13	14	15	16	17	18	19	19	20	21	22	23	24	25	26	27	28	22
69	1	2	3	4	5	6	7	7	8	9	10	11	12	13	14	15	16	17	18	19	20	21	21	22	23	24	25	26	27	28	21
70	1	2	3	4	5	6	7	8	8	9	10	11	12	13	14	15	16	17	18	19	20	21	22	23	23	24	25	26	27	28	20
71	1	2	3	4	5	6	7	8	9	9	10	11	12	13	14	15	16	17	18	19	20	21	22	23	24	25	26	26	27	28	19
72	1	2	3	4	5	6	7	8	9	10	10	11	12	13	14	15	16	17	18	19	20	21	22	23	24	25	26	27	28	29	18
73	1	2	3	4	5	6	7	8	9	10	11	11	12	13	14	15	16	17	18	19	20	21	22	23	24	25	26	27	28	29	17
74	1	2	3	4	5	6	7	8	9	10	11	12	12	13	14	15	16	17	18	19	20	21	22	23	24	25	26	27	28	29	16
75	1	2	3	4	5	6	7	8	9	10	11	12	13	14	14	15	16	17	18	19	20	21	22	23	24	25	26	27	28	29	15
76	1	2	3	4	5	6	7	8	9	10	11	12	13	14	15	16	16	17	18	19	20	21	22	23	24	25	26	27	28	29	14
77	1	2	3	4	5	6	7	8	9	10	11	12	13	14	15	16	17	18	19	19	20	21	22	23	24	25	26	27	28	29	13
78	1	2	3	4	5	6	7	8	9	10	11	12	13	14	15	16	17	18	19	20	21	22	22	23	24	25	26	27	28	29	12
79	1	2	3	4	5	6	7	8	9	10	11	12	13	14	15	16	17	18	19	20	21	22	23	24	25	26	27	27	28	29	11
80	1	2	3	4	5	6	7	8	9	10	11	12	13	14	15	16	17	18	19	20	21	22	23	24	25	26	27	28	29	30	10

For P > 80°, use 80°

For $Z_2 < 10°$, use 10°

USE OF CONCISE SIGHT REDUCTION TABLES (continued)

4. *Example.* (b) Required the altitude and azimuth of *Vega* on 2009 July 29 at UT $04^h\ 48^m$ from the estimated position 152° west, 15° south.

1.	Assumed latitude	$Lat =$ 15° S		
	From the almanac	$GHA =$ 99° 39′		
	Assumed longitude	151° 39′ W		
	Local hour angle	$LHA =$ 308		
2.	Reduction table, 1st entry			
	$(Lat, LHA) = (15, 308)$	$A =$ 49 34	$A° = 50$, $A' = 34$	
		$B = +66$ 29	$Z_1 = +71{\cdot}7$,	$LHA > 270°$
3.	From the almanac	$Dec = -38$ 48		*Lat* and *Dec* contrary
	Sum $= B + Dec$	$F = +27$ 41	$F° = 28$, $F' = 41$	
4.	Reduction table, 2nd entry			
	$(A°, F°) = (50, 28)$	$H =$ 17 34	$P° = 37$	
			$Z_2 = 67{\cdot}8, Z_2° = 68$	
5.	Auxiliary table, 1st entry			
	$(F', P°) = (41, 37)$	$corr_1 =$ −11		$F < 90°$, $F' > 29'$
	Sum	17 23		
6.	Auxiliary table, 2nd entry			
	$(A', Z_2°) = (34, 68)$	$corr_2 =$ +10		$A' > 30'$
7.	Sum = computed altitude	$Hc = +17°$ 33′		$F > 0°$
8.	Azimuth, first component	$Z_1 = +71{\cdot}7$		same sign as *B*
	second component	$Z_2 = +67{\cdot}8$		$F < 90°$, $F > 0°$
	Sum = azimuth angle	$Z =$ 139·5		
	True azimuth	$Zn =$ 040°		S *Lat*, $LHA > 180°$

5. *Form for use with the Concise Sight Reduction Tables.* The form on the following page lays out the procedure explained on pages 284-285. Each step is shown, with notes and rules to ensure accuracy, rather than speed, throughout the calculation. The form is mainly intended for the calculation of star positions. It therefore includes the formation of the Greenwich hour of Aries (*GHA* Aries), and thus the Greenwich hour angle of the star (*GHA*) from its tabular sidereal hour angle (*SHA*). These calculations, included in step 1 of the form, can easily be replaced by the interpolation of *GHA* and *Dec* for the Sun, Moon or planets.

The form may be freely copied, however, acknowledgement of the source is requested.

Date & UT of observation h m s	Body	Estimated Latitude & Longitude ° ′ ° ′

Step	Calculate Altitude & Azimuth	Summary of Rules & Notes
Assumed latitude Assumed longitude	$Lat =$ ° $Long =$ ° ′	Nearest estimated latitude, integral number of degrees. Choose $Long$ so that LHA has integral number of degrees.
1. From the almanac:	$Dec =$ ° ′	Record the Dec for use in Step 3.
GHA Aries h Increment m s SHA	$=$ ° ′ $=$ ° ′ $SHA =$ ° ′	Needed if using SHA. Tabular value. for minutes and seconds of time.
$GHA = GHA\ Aries + SHA$	$GHA =$ ° ′	Remove multiples of 360°.
Assumed longitude	$Long =$ ° ′	West longitudes are negative.
$LHA = GHA + Long$	$LHA =$ °	Remove multiples of 360°.
2. Reduction table, 1st entry $(Lat, LHA) = ($ °, °) record A, B and Z_1.	$A =$ ° ′ $A° =$ ° $A' =$ ′	nearest whole degree of A. minutes part of A.
	$B =$ ° ′ $Z_1 =$ °	B is minus if $90° < LHA < 270°$. Z_1 has the same sign as B.
3. From step 1 $F = B + Dec$	$Dec =$ ° ′ $F =$ ° ′ $F° =$ ° $F' =$ ′	Dec is minus if contrary to Lat. Regard F as positive until step 7. nearest whole degree of F. minutes part of F.
4. Reduction table, 2nd entry $(A°, F°) = ($ °, °) record H, P and Z_2.	$H =$ ° ′ $P° =$ ° $Z_2 =$ °	nearest whole degree of P.
5. Auxiliary table, 1st entry $(F', P°) = ($ ′, °) record $corr_1$	$corr_1 =$ ′	$corr_1$ is minus if $F < 90°$ & $F' > 29'$, or if $F > 90°$ & $F' < 30'$.
6. Auxiliary table, 2nd entry $(A', Z_2°) = ($ ′, °) record $corr_2$	$corr_2 =$ ′	$Z_2°$ nearest whole degree of Z_2. $corr_2$ is minus if $A' < 30'$.
7. Calculated altitude = $H_C = H + corr_1 + corr_2$	$H_C =$ ° ′	H_C is minus if F is negative, and object is below the horizon.
8. Azimuth, 1st component 2nd component	$Z_1 =$ ° $Z_2 =$ °	Z_1 has the same sign as B. Z_2 is minus if $F > 90°$. If F is negative, $Z_2 = 180° - Z_2$
$Z = Z_1 + Z_2$	$Z =$ °	Ignore the sign of Z. N Lat: If $LHA > 180°$, $Z_n = Z$, or if $LHA < 180°$, $Z_n = 360° - Z$, S Lat: If $LHA > 180°$, $Z_n = 180° - Z$, or if $LHA < 180°$, $Z_n = 180° + Z$.
True azimuth	$Z_n =$ °	

For use with *The Nautical Almanac's* Concise Sight Reduction Tables pages 284-318.

INDEX TO SELECTED STARS, 2009

Name	No	Mag	SHA	Dec
			°	°
Acamar	**7**	3·2	315	S 40
Achernar	**5**	0·5	335	S 57
Acrux	**30**	1·3	173	S 63
Adhara	**19**	1·5	255	S 29
Aldebaran	**10**	0·9	291	N 17
Alioth	**32**	1·8	166	N 56
Alkaid	**34**	1·9	153	N 49
Al Na'ir	**55**	1·7	28	S 47
Alnilam	**15**	1·7	276	S 1
Alphard	**25**	2·0	218	S 9
Alphecca	**41**	2·2	126	N 27
Alpheratz	**1**	2·1	358	N 29
Altair	**51**	0·8	62	N 9
Ankaa	**2**	2·4	353	S 42
Antares	**42**	1·0	112	S 26
Arcturus	**37**	0·0	146	N 19
Atria	**43**	1·9	108	S 69
Avior	**22**	1·9	234	S 60
Bellatrix	**13**	1·6	279	N 6
Betelgeuse	**16**	Var.*	271	N 7
Canopus	**17**	−0·7	264	S 53
Capella	**12**	0·1	281	N 46
Deneb	**53**	1·3	50	N 45
Denebola	**28**	2·1	183	N 15
Diphda	**4**	2·0	349	S 18
Dubhe	**27**	1·8	194	N 62
Elnath	**14**	1·7	278	N 29
Eltanin	**47**	2·2	91	N 51
Enif	**54**	2·4	34	N 10
Fomalhaut	**56**	1·2	15	S 30
Gacrux	**31**	1·6	172	S 57
Gienah	**29**	2·6	176	S 18
Hadar	**35**	0·6	149	S 60
Hamal	**6**	2·0	328	N 24
Kaus Australis	**48**	1·9	84	S 34
Kochab	**40**	2·1	137	N 74
Markab	**57**	2·5	14	N 15
Menkar	**8**	2·5	314	N 4
Menkent	**36**	2·1	148	S 36
Miaplacidus	**24**	1·7	222	S 70
Mirfak	**9**	1·8	309	N 50
Nunki	**50**	2·0	76	S 26
Peacock	**52**	1·9	53	S 57
Pollux	**21**	1·1	244	N 28
Procyon	**20**	0·4	245	N 5
Rasalhague	**46**	2·1	96	N 13
Regulus	**26**	1·4	208	N 12
Rigel	**11**	0·1	281	S 8
Rigil Kentaurus	**38**	−0·3	140	S 61
Sabik	**44**	2·4	102	S 16
Schedar	**3**	2·2	350	N 57
Shaula	**45**	1·6	96	S 37
Sirius	**18**	−1·5	259	S 17
Spica	**33**	1·0	159	S 11
Suhail	**23**	2·2	223	S 43
Vega	**49**	0·0	81	N 39
Zubenelgenubi	**39**	2·8	137	S 16

No	Name	Mag	SHA	Dec
			°	°
1	*Alpheratz*	2·1	358	N 29
2	*Ankaa*	2·4	353	S 42
3	*Schedar*	2·2	350	N 57
4	*Diphda*	2·0	349	S 18
5	*Achernar*	0·5	335	S 57
6	*Hamal*	2·0	328	N 24
7	*Acamar*	3·2	315	S 40
8	*Menkar*	2·5	314	N 4
9	*Mirfak*	1·8	309	N 50
10	*Aldebaran*	0·9	291	N 17
11	*Rigel*	0·1	281	S 8
12	*Capella*	0·1	281	N 46
13	*Bellatrix*	1·6	279	N 6
14	*Elnath*	1·7	278	N 29
15	*Alnilam*	1·7	276	S 1
16	*Betelgeuse*	Var.*	271	N 7
17	*Canopus*	−0·7	264	S 53
18	*Sirius*	−1·5	259	S 17
19	*Adhara*	1·5	255	S 29
20	*Procyon*	0·4	245	N 5
21	*Pollux*	1·1	244	N 28
22	*Avior*	1·9	234	S 60
23	*Suhail*	2·2	223	S 43
24	*Miaplacidus*	1·7	222	S 70
25	*Alphard*	2·0	218	S 9
26	*Regulus*	1·4	208	N 12
27	*Dubhe*	1·8	194	N 62
28	*Denebola*	2·1	183	N 15
29	*Gienah*	2·6	176	S 18
30	*Acrux*	1·3	173	S 63
31	*Gacrux*	1·6	172	S 57
32	*Alioth*	1·8	166	N 56
33	*Spica*	1·0	159	S 11
34	*Alkaid*	1·9	153	N 49
35	*Hadar*	0·6	149	S 60
36	*Menkent*	2·1	148	S 36
37	*Arcturus*	0·0	146	N 19
38	*Rigil Kentaurus*	−0·3	140	S 61
39	*Zubenelgenubi*	2·8	137	S 16
40	*Kochab*	2·1	137	N 74
41	*Alphecca*	2·2	126	N 27
42	*Antares*	1·0	112	S 26
43	*Atria*	1·9	108	S 69
44	*Sabik*	2·4	102	S 16
45	*Shaula*	1·6	96	S 37
46	*Rasalhague*	2·1	96	N 13
47	*Eltanin*	2·2	91	N 51
48	*Kaus Australis*	1·9	84	S 34
49	*Vega*	0·0	81	N 39
50	*Nunki*	2·0	76	S 26
51	*Altair*	0·8	62	N 9
52	*Peacock*	1·9	53	S 57
53	*Deneb*	1·3	50	N 45
54	*Enif*	2·4	34	N 10
55	*Al Na'ir*	1·7	28	S 47
56	*Fomalhaut*	1·2	15	S 30
57	*Markab*	2·5	14	N 15

*0·1 — 1·2

ALTITUDE CORRECTION TABLES 10°-90°—SUN, STARS, PLANETS

SUN

OCT.—MAR. App. Alt.	Lower Limb	Upper Limb	APR.—SEPT. App. Alt.	Lower Limb	Upper Limb
° ′	′	′	° ′	′	′
9 33	+10·8	−21·5	9 39	+10·6	−21·2
9 45	+10·9	−21·4	9 50	+10·7	−21·1
9 56	+11·0	−21·3	10 02	+10·8	−21·0
10 08	+11·1	−21·2	10 14	+10·9	−20·9
10 20	+11·2	−21·1	10 27	+11·0	−20·8
10 33	+11·3	−21·0	10 40	+11·1	−20·7
10 46	+11·4	−20·9	10 53	+11·2	−20·6
11 00	+11·5	−20·8	11 07	+11·3	−20·5
11 15	+11·6	−20·7	11 22	+11·4	−20·4
11 30	+11·7	−20·6	11 37	+11·5	−20·3
11 45	+11·8	−20·5	11 53	+11·6	−20·2
12 01	+11·9	−20·4	12 10	+11·7	−20·1
12 18	+12·0	−20·3	12 27	+11·8	−20·0
12 36	+12·1	−20·2	12 45	+11·9	−19·9
12 54	+12·2	−20·1	13 04	+12·0	−19·8
13 14	+12·3	−20·0	13 24	+12·1	−19·7
13 34	+12·4	−19·9	13 44	+12·2	−19·6
13 55	+12·5	−19·8	14 06	+12·3	−19·5
14 17	+12·6	−19·7	14 29	+12·4	−19·4
14 41	+12·7	−19·6	14 53	+12·5	−19·3
15 05	+12·8	−19·5	15 18	+12·6	−19·2
15 31	+12·9	−19·4	15 45	+12·7	−19·1
15 59	+13·0	−19·3	16 13	+12·8	−19·0
16 27	+13·1	−19·2	16 43	+12·9	−18·9
16 58	+13·2	−19·1	17 14	+13·0	−18·8
17 30	+13·3	−19·0	17 47	+13·1	−18·7
18 05	+13·4	−18·9	18 23	+13·2	−18·6
18 41	+13·5	−18·8	19 00	+13·3	−18·5
19 20	+13·6	−18·7	19 41	+13·4	−18·4
20 02	+13·7	−18·6	20 24	+13·5	−18·3
20 46	+13·8	−18·5	21 10	+13·6	−18·2
21 34	+13·9	−18·4	21 59	+13·7	−18·1
22 25	+14·0	−18·3	22 52	+13·8	−18·0
23 20	+14·1	−18·2	23 49	+13·9	−17·9
24 20	+14·2	−18·1	24 51	+14·0	−17·8
25 24	+14·3	−18·0	25 58	+14·1	−17·7
26 34	+14·4	−17·9	27 11	+14·2	−17·6
27 50	+14·5	−17·8	28 31	+14·3	−17·5
29 13	+14·6	−17·7	29 58	+14·4	−17·4
30 44	+14·7	−17·6	31 33	+14·5	−17·3
32 24	+14·8	−17·5	33 18	+14·6	−17·2
34 15	+14·9	−17·4	35 15	+14·7	−17·1
36 17	+15·0	−17·3	37 24	+14·8	−17·0
38 34	+15·1	−17·2	39 48	+14·9	−16·9
41 06	+15·2	−17·1	42 28	+15·0	−16·8
43 56	+15·3	−17·0	45 29	+15·1	−16·7
47 07	+15·4	−16·9	48 52	+15·2	−16·6
50 43	+15·5	−16·8	52 41	+15·3	−16·5
54 46	+15·6	−16·7	56 59	+15·4	−16·4
59 21	+15·7	−16·6	61 50	+15·5	−16·3
64 28	+15·8	−16·5	67 15	+15·6	−16·2
70 10	+15·9	−16·4	73 14	+15·7	−16·1
76 24	+16·0	−16·3	79 42	+15·8	−16·0
83 05	+16·1	−16·2	86 31	+15·9	−15·9
90 00			90 00		

STARS AND PLANETS

App Alt.	Corr^{n}
° ′	′
9 55	−5·3
10 07	−5·2
10 20	−5·1
10 32	−5·0
10 46	−4·9
10 59	−4·8
11 14	−4·7
11 29	−4·6
11 44	−4·5
12 00	−4·4
12 17	−4·3
12 35	−4·2
12 53	−4·1
13 12	−4·0
13 32	−3·9
13 53	−3·8
14 16	−3·7
14 39	−3·6
15 03	−3·5
15 29	−3·4
15 56	−3·3
16 25	−3·2
16 55	−3·1
17 27	−3·0
18 01	−2·9
18 37	−2·8
19 16	−2·7
19 56	−2·6
20 40	−2·5
21 27	−2·4
22 17	−2·3
23 11	−2·2
24 09	−2·1
25 12	−2·0
26 20	−1·9
27 34	−1·8
28 54	−1·7
30 22	−1·6
31 58	−1·5
33 43	−1·4
35 38	−1·3
37 45	−1·2
40 06	−1·1
42 42	−1·0
45 34	−0·9
48 45	−0·8
52 16	−0·7
56 09	−0·6
60 26	−0·5
65 06	−0·4
70 09	−0·3
75 32	−0·2
81 12	−0·1
87 03	0·0
90 00	

App. Alt. / Additional Corr^{n}

2009

VENUS

Jan. 1–Jan. 28
May 23–July 10

°	′
0	+0·2
41	+0·1
76	

Jan. 29–Feb. 21
May 1–May 22

°	′
0	+0·3
34	+0·2
60	+0·1
80	

Feb. 22–Mar. 9
Apr. 15–Apr. 30

°	′
0	+0·4
29	+0·3
51	+0·2
68	+0·1
83	

Mar. 10–Apr. 14

°	′
0	+0·5
26	+0·4
46	+0·3
60	+0·2
73	+0·1
84	

July 11–Dec. 31

°	′
0	+0·1
60	

MARS

Jan. 1–Nov. 27

°	′
0	+0·1
60	

Nov. 28–Dec. 31

°	′
0	+0·2
41	+0·1
76	

DIP

Ht. of Eye (m)	Corr^{n} (′)	Ht. of Eye (ft.)
2·4	−2·8	8·0
2·6	−2·9	8·6
2·8	−3·0	9·2
3·0	−3·1	9·8
3·2	−3·2	10·5
3·4	−3·3	11·2
3·6	−3·4	11·9
3·8	−3·5	12·6
4·0	−3·6	13·3
4·3	−3·7	14·1
4·5	−3·8	14·9
4·7	−3·9	15·7
5·0	−4·0	16·5
5·2	−4·1	17·4
5·5	−4·2	18·3
5·8	−4·3	19·1
6·1	−4·4	20·1
6·3	−4·5	21·0
6·6	−4·6	22·0
6·9	−4·7	22·9
7·2	−4·8	23·9
7·5	−4·9	24·9
7·9	−5·0	26·0
8·2	−5·1	27·1
8·5	−5·2	28·1
8·8	−5·3	29·2
9·2	−5·4	30·4
9·5	−5·5	31·5
9·9	−5·6	32·7
10·3	−5·7	33·9
10·6	−5·8	35·1
11·0	−5·9	36·3
11·4	−6·0	37·6
11·8	−6·1	38·9
12·2	−6·2	40·1
12·6	−6·3	41·5
13·0	−6·4	42·8
13·4	−6·5	44·2
13·8	−6·6	45·5
14·2	−6·7	46·9
14·7	−6·8	48·4
15·1	−6·9	49·8
15·5	−7·0	51·3
16·0	−7·1	52·8
16·5	−7·2	54·3
16·9	−7·3	55·8
17·4	−7·4	57·4
17·9	−7·5	58·9
18·4	−7·6	60·5
18·8	−7·7	62·1
19·3	−7·8	63·8
19·8	−7·9	65·4
20·4	−8·0	67·1
20·9	−8·1	68·8
21·4		70·5

Ht. of Eye	Corr^{n}
m	′
1·0	−1·8
1·5	−2·2
2·0	−2·5
2·5	−2·8
3·0	−3·0
See table ←	
m	′
20	−7·9
22	−8·3
24	−8·6
26	−9·0
28	−9·3
30	−9·6
32	−10·0
34	−10·3
36	−10·6
38	−10·8
40	−11·1
42	−11·4
44	−11·7
46	−11·9
48	−12·2
ft.	′
2	−1·4
4	−1·9
6	−2·4
8	−2·7
10	−3·1
See table ←	
ft.	′
70	−8·1
75	−8·4
80	−8·7
85	−8·9
90	−9·2
95	−9·5
100	−9·7
105	−9·9
110	−10·2
115	−10·4
120	−10·6
125	−10·8
130	−11·1
135	−11·3
140	−11·5
145	−11·7
150	−11·9
155	−12·1

App. Alt. = Apparent altitude = Sextant altitude corrected for index error and dip.

CONVERSION OF ARC TO TIME

0°–59°		60°–119°		120°–179°		180°–239°		240°–299°		300°–359°			0'.00	0'.25	0'.50	0'.75
°	h m	°	h m	°	h m	°	h m	°	h m	°	h m	'	m s	m s	m s	m s
0	0 00	**60**	4 00	**120**	8 00	**180**	12 00	**240**	16 00	**300**	20 00	**0**	0 00	0 01	0 02	0 03
1	0 04	**61**	4 04	**121**	8 04	**181**	12 04	**241**	16 04	**301**	20 04	**1**	0 04	0 05	0 06	0 07
2	0 08	**62**	4 08	**122**	8 08	**182**	12 08	**242**	16 08	**302**	20 08	**2**	0 08	0 09	0 10	0 11
3	0 12	**63**	4 12	**123**	8 12	**183**	12 12	**243**	16 12	**303**	20 12	**3**	0 12	0 13	0 14	0 15
4	0 16	**64**	4 16	**124**	8 16	**184**	12 16	**244**	16 16	**304**	20 16	**4**	0 16	0 17	0 18	0 19
5	0 20	**65**	4 20	**125**	8 20	**185**	12 20	**245**	16 20	**305**	20 20	**5**	0 20	0 21	0 22	0 23
6	0 24	**66**	4 24	**126**	8 24	**186**	12 24	**246**	16 24	**306**	20 24	**6**	0 24	0 25	0 26	0 27
7	0 28	**67**	4 28	**127**	8 28	**187**	12 28	**247**	16 28	**307**	20 28	**7**	0 28	0 29	0 30	0 31
8	0 32	**68**	4 32	**128**	8 32	**188**	12 32	**248**	16 32	**308**	20 32	**8**	0 32	0 33	0 34	0 35
9	0 36	**69**	4 36	**129**	8 36	**189**	12 36	**249**	16 36	**309**	20 36	**9**	0 36	0 37	0 38	0 39
10	0 40	**70**	4 40	**130**	8 40	**190**	12 40	**250**	16 40	**310**	20 40	**10**	0 40	0 41	0 42	0 43
11	0 44	**71**	4 44	**131**	8 44	**191**	12 44	**251**	16 44	**311**	20 44	**11**	0 44	0 45	0 46	0 47
12	0 48	**72**	4 48	**132**	8 48	**192**	12 48	**252**	16 48	**312**	20 48	**12**	0 48	0 49	0 50	0 51
13	0 52	**73**	4 52	**133**	8 52	**193**	12 52	**253**	16 52	**313**	20 52	**13**	0 52	0 53	0 54	0 55
14	0 56	**74**	4 56	**134**	8 56	**194**	12 56	**254**	16 56	**314**	20 56	**14**	0 56	0 57	0 58	0 59
15	1 00	**75**	5 00	**135**	9 00	**195**	13 00	**255**	17 00	**315**	21 00	**15**	1 00	1 01	1 02	1 03
16	1 04	**76**	5 04	**136**	9 04	**196**	13 04	**256**	17 04	**316**	21 04	**16**	1 04	1 05	1 06	1 07
17	1 08	**77**	5 08	**137**	9 08	**197**	13 08	**257**	17 08	**317**	21 08	**17**	1 08	1 09	1 10	1 11
18	1 12	**78**	5 12	**138**	9 12	**198**	13 12	**258**	17 12	**318**	21 12	**18**	1 12	1 13	1 14	1 15
19	1 16	**79**	5 16	**139**	9 16	**199**	13 16	**259**	17 16	**319**	21 16	**19**	1 16	1 17	1 18	1 19
20	1 20	**80**	5 20	**140**	9 20	**200**	13 20	**260**	17 20	**320**	21 20	**20**	1 20	1 21	1 22	1 23
21	1 24	**81**	5 24	**141**	9 24	**201**	13 24	**261**	17 24	**321**	21 24	**21**	1 24	1 25	1 26	1 27
22	1 28	**82**	5 28	**142**	9 28	**202**	13 28	**262**	17 28	**322**	21 28	**22**	1 28	1 29	1 30	1 31
23	1 32	**83**	5 32	**143**	9 32	**203**	13 32	**263**	17 32	**323**	21 32	**23**	1 32	1 33	1 34	1 35
24	1 36	**84**	5 36	**144**	9 36	**204**	13 36	**264**	17 36	**324**	21 36	**24**	1 36	1 37	1 38	1 39
25	1 40	**85**	5 40	**145**	9 40	**205**	13 40	**265**	17 40	**325**	21 40	**25**	1 40	1 41	1 42	1 43
26	1 44	**86**	5 44	**146**	9 44	**206**	13 44	**266**	17 44	**326**	21 44	**26**	1 44	1 45	1 46	1 47
27	1 48	**87**	5 48	**147**	9 48	**207**	13 48	**267**	17 48	**327**	21 48	**27**	1 48	1 49	1 50	1 51
28	1 52	**88**	5 52	**148**	9 52	**208**	13 52	**268**	17 52	**328**	21 52	**28**	1 52	1 53	1 54	1 55
29	1 56	**89**	5 56	**149**	9 56	**209**	13 56	**269**	17 56	**329**	21 56	**29**	1 56	1 57	1 58	1 59
30	2 00	**90**	6 00	**150**	10 00	**210**	14 00	**270**	18 00	**330**	22 00	**30**	2 00	2 01	2 02	2 03
31	2 04	**91**	6 04	**151**	10 04	**211**	14 04	**271**	18 04	**331**	22 04	**31**	2 04	2 05	2 06	2 07
32	2 08	**92**	6 08	**152**	10 08	**212**	14 08	**272**	18 08	**332**	22 08	**32**	2 08	2 09	2 10	2 11
33	2 12	**93**	6 12	**153**	10 12	**213**	14 12	**273**	18 12	**333**	22 12	**33**	2 12	2 13	2 14	2 15
34	2 16	**94**	6 16	**154**	10 16	**214**	14 16	**274**	18 16	**334**	22 16	**34**	2 16	2 17	2 18	2 19
35	2 20	**95**	6 20	**155**	10 20	**215**	14 20	**275**	18 20	**335**	22 20	**35**	2 20	2 21	2 22	2 23
36	2 24	**96**	6 24	**156**	10 24	**216**	14 24	**276**	18 24	**336**	22 24	**36**	2 24	2 25	2 26	2 27
37	2 28	**97**	6 28	**157**	10 28	**217**	14 28	**277**	18 28	**337**	22 28	**37**	2 28	2 29	2 30	2 31
38	2 32	**98**	6 32	**158**	10 32	**218**	14 32	**278**	18 32	**338**	22 32	**38**	2 32	2 33	2 34	2 35
39	2 36	**99**	6 36	**159**	10 36	**219**	14 36	**279**	18 36	**339**	22 36	**39**	2 36	2 37	2 38	2 39
40	2 40	**100**	6 40	**160**	10 40	**220**	14 40	**280**	18 40	**340**	22 40	**40**	2 40	2 41	2 42	2 43
41	2 44	**101**	6 44	**161**	10 44	**221**	14 44	**281**	18 44	**341**	22 44	**41**	2 44	2 45	2 46	2 47
42	2 48	**102**	6 48	**162**	10 48	**222**	14 48	**282**	18 48	**342**	22 48	**42**	2 48	2 49	2 50	2 51
43	2 52	**103**	6 52	**163**	10 52	**223**	14 52	**283**	18 52	**343**	22 52	**43**	2 52	2 53	2 54	2 55
44	2 56	**104**	6 56	**164**	10 56	**224**	14 56	**284**	18 56	**344**	22 56	**44**	2 56	2 57	2 58	2 59
45	3 00	**105**	7 00	**165**	11 00	**225**	15 00	**285**	19 00	**345**	23 00	**45**	3 00	3 01	3 02	3 03
46	3 04	**106**	7 04	**166**	11 04	**226**	15 04	**286**	19 04	**346**	23 04	**46**	3 04	3 05	3 06	3 07
47	3 08	**107**	7 08	**167**	11 08	**227**	15 08	**287**	19 08	**347**	23 08	**47**	3 08	3 09	3 10	3 11
48	3 12	**108**	7 12	**168**	11 12	**228**	15 12	**288**	19 12	**348**	23 12	**48**	3 12	3 13	3 14	3 15
49	3 16	**109**	7 16	**169**	11 16	**229**	15 16	**289**	19 16	**349**	23 16	**49**	3 16	3 17	3 18	3 19
50	3 20	**110**	7 20	**170**	11 20	**230**	15 20	**290**	19 20	**350**	23 20	**50**	3 20	3 21	3 22	3 23
51	3 24	**111**	7 24	**171**	11 24	**231**	15 24	**291**	19 24	**351**	23 24	**51**	3 24	3 25	3 26	3 27
52	3 28	**112**	7 28	**172**	11 28	**232**	15 28	**292**	19 28	**352**	23 28	**52**	3 28	3 29	3 30	3 31
53	3 32	**113**	7 32	**173**	11 32	**233**	15 32	**293**	19 32	**353**	23 32	**53**	3 32	3 33	3 34	3 35
54	3 36	**114**	7 36	**174**	11 36	**234**	15 36	**294**	19 36	**354**	23 36	**54**	3 36	3 37	3 38	3 39
55	3 40	**115**	7 40	**175**	11 40	**235**	15 40	**295**	19 40	**355**	23 40	**55**	3 40	3 41	3 42	3 43
56	3 44	**116**	7 44	**176**	11 44	**236**	15 44	**296**	19 44	**356**	23 44	**56**	3 44	3 45	3 46	3 47
57	3 48	**117**	7 48	**177**	11 48	**237**	15 48	**297**	19 48	**357**	23 48	**57**	3 48	3 49	3 50	3 51
58	3 52	**118**	7 52	**178**	11 52	**238**	15 52	**298**	19 52	**358**	23 52	**58**	3 52	3 53	3 54	3 55
59	3 56	**119**	7 56	**179**	11 56	**239**	15 56	**299**	19 56	**359**	23 56	**59**	3 56	3 57	3 58	3 59

The above table is for converting expressions in arc to their equivalent in time; its main use in this Almanac is for the conversion of longitude for application to LMT (*added* if *west*, *subtracted* if *east*) to give UT or vice versa, particularly in the case of sunrise, sunset, etc.

m 0 s	SUN PLANETS ° ′	ARIES ° ′	MOON ° ′	v or d ′	Corr^n ′	v or d ′	Corr^n ′	v or d ′	Corr^n ′
00	0 00·0	0 00·0	0 00·0	0·0	0·0	6·0	0·1	12·0	0·1
01	0 00·3	0 00·3	0 00·2	0·1	0·0	6·1	0·1	12·1	0·1
02	0 00·5	0 00·5	0 00·5	0·2	0·0	6·2	0·1	12·2	0·1
03	0 00·8	0 00·8	0 00·7	0·3	0·0	6·3	0·1	12·3	0·1
04	0 01·0	0 01·0	0 01·0	0·4	0·0	6·4	0·1	12·4	0·1
05	0 01·3	0 01·3	0 01·2	0·5	0·0	6·5	0·1	12·5	0·1
06	0 01·5	0 01·5	0 01·4	0·6	0·0	6·6	0·1	12·6	0·1
07	0 01·8	0 01·8	0 01·7	0·7	0·0	6·7	0·1	12·7	0·1
08	0 02·0	0 02·0	0 01·9	0·8	0·0	6·8	0·1	12·8	0·1
09	0 02·3	0 02·3	0 02·1	0·9	0·0	6·9	0·1	12·9	0·1
10	0 02·5	0 02·5	0 02·4	1·0	0·0	7·0	0·1	13·0	0·1
11	0 02·8	0 02·8	0 02·6	1·1	0·0	7·1	0·1	13·1	0·1
12	0 03·0	0 03·0	0 02·9	1·2	0·0	7·2	0·1	13·2	0·1
13	0 03·3	0 03·3	0 03·1	1·3	0·0	7·3	0·1	13·3	0·1
14	0 03·5	0 03·5	0 03·3	1·4	0·0	7·4	0·1	13·4	0·1
15	0 03·8	0 03·8	0 03·6	1·5	0·0	7·5	0·1	13·5	0·1
16	0 04·0	0 04·0	0 03·8	1·6	0·0	7·6	0·1	13·6	0·1
17	0 04·3	0 04·3	0 04·1	1·7	0·0	7·7	0·1	13·7	0·1
18	0 04·5	0 04·5	0 04·3	1·8	0·0	7·8	0·1	13·8	0·1
19	0 04·8	0 04·8	0 04·5	1·9	0·0	7·9	0·1	13·9	0·1
20	0 05·0	0 05·0	0 04·8	2·0	0·0	8·0	0·1	14·0	0·1
21	0 05·3	0 05·3	0 05·0	2·1	0·0	8·1	0·1	14·1	0·1
22	0 05·5	0 05·5	0 05·2	2·2	0·0	8·2	0·1	14·2	0·1
23	0 05·8	0 05·8	0 05·5	2·3	0·0	8·3	0·1	14·3	0·1
24	0 06·0	0 06·0	0 05·7	2·4	0·0	8·4	0·1	14·4	0·1
25	0 06·3	0 06·3	0 06·0	2·5	0·0	8·5	0·1	14·5	0·1
26	0 06·5	0 06·5	0 06·2	2·6	0·0	8·6	0·1	14·6	0·1
27	0 06·8	0 06·8	0 06·4	2·7	0·0	8·7	0·1	14·7	0·1
28	0 07·0	0 07·0	0 06·7	2·8	0·0	8·8	0·1	14·8	0·1
29	0 07·3	0 07·3	0 06·9	2·9	0·0	8·9	0·1	14·9	0·1
30	0 07·5	0 07·5	0 07·2	3·0	0·0	9·0	0·1	15·0	0·1
31	0 07·8	0 07·8	0 07·4	3·1	0·0	9·1	0·1	15·1	0·1
32	0 08·0	0 08·0	0 07·6	3·2	0·0	9·2	0·1	15·2	0·1
33	0 08·3	0 08·3	0 07·9	3·3	0·0	9·3	0·1	15·3	0·1
34	0 08·5	0 08·5	0 08·1	3·4	0·0	9·4	0·1	15·4	0·1
35	0 08·8	0 08·8	0 08·4	3·5	0·0	9·5	0·1	15·5	0·1
36	0 09·0	0 09·0	0 08·6	3·6	0·0	9·6	0·1	15·6	0·1
37	0 09·3	0 09·3	0 08·8	3·7	0·0	9·7	0·1	15·7	0·1
38	0 09·5	0 09·5	0 09·1	3·8	0·0	9·8	0·1	15·8	0·1
39	0 09·8	0 09·8	0 09·3	3·9	0·0	9·9	0·1	15·9	0·1
40	0 10·0	0 10·0	0 09·5	4·0	0·0	10·0	0·1	16·0	0·1
41	0 10·3	0 10·3	0 09·8	4·1	0·0	10·1	0·1	16·1	0·1
42	0 10·5	0 10·5	0 10·0	4·2	0·0	10·2	0·1	16·2	0·1
43	0 10·8	0 10·8	0 10·3	4·3	0·0	10·3	0·1	16·3	0·1
44	0 11·0	0 11·0	0 10·5	4·4	0·0	10·4	0·1	16·4	0·1
45	0 11·3	0 11·3	0 10·7	4·5	0·0	10·5	0·1	16·5	0·1
46	0 11·5	0 11·5	0 11·0	4·6	0·0	10·6	0·1	16·6	0·1
47	0 11·8	0 11·8	0 11·2	4·7	0·0	10·7	0·1	16·7	0·1
48	0 12·0	0 12·0	0 11·5	4·8	0·0	10·8	0·1	16·8	0·1
49	0 12·3	0 12·3	0 11·7	4·9	0·0	10·9	0·1	16·9	0·1
50	0 12·5	0 12·5	0 11·9	5·0	0·0	11·0	0·1	17·0	0·1
51	0 12·8	0 12·8	0 12·2	5·1	0·0	11·1	0·1	17·1	0·1
52	0 13·0	0 13·0	0 12·4	5·2	0·0	11·2	0·1	17·2	0·1
53	0 13·3	0 13·3	0 12·6	5·3	0·0	11·3	0·1	17·3	0·1
54	0 13·5	0 13·5	0 12·9	5·4	0·0	11·4	0·1	17·4	0·1
55	0 13·8	0 13·8	0 13·1	5·5	0·0	11·5	0·1	17·5	0·1
56	0 14·0	0 14·0	0 13·4	5·6	0·0	11·6	0·1	17·6	0·1
57	0 14·3	0 14·3	0 13·6	5·7	0·0	11·7	0·1	17·7	0·1
58	0 14·5	0 14·5	0 13·8	5·8	0·0	11·8	0·1	17·8	0·1
59	0 14·8	0 14·8	0 14·1	5·9	0·0	11·9	0·1	17·9	0·1
60	0 15·0	0 15·0	0 14·3	6·0	0·1	12·0	0·1	18·0	0·2

m 1 s	SUN PLANETS ° ′	ARIES ° ′	MOON ° ′	v or d ′	Corr^n ′	v or d ′	Corr^n ′	v or d ′	Corr^n ′
00	0 15·0	0 15·0	0 14·3	0·0	0·0	6·0	0·2	12·0	0·3
01	0 15·3	0 15·3	0 14·6	0·1	0·0	6·1	0·2	12·1	0·3
02	0 15·5	0 15·5	0 14·8	0·2	0·0	6·2	0·2	12·2	0·3
03	0 15·8	0 15·8	0 15·0	0·3	0·0	6·3	0·2	12·3	0·3
04	0 16·0	0 16·0	0 15·3	0·4	0·0	6·4	0·2	12·4	0·3
05	0 16·3	0 16·3	0 15·5	0·5	0·0	6·5	0·2	12·5	0·3
06	0 16·5	0 16·5	0 15·7	0·6	0·0	6·6	0·2	12·6	0·3
07	0 16·8	0 16·8	0 16·0	0·7	0·0	6·7	0·2	12·7	0·3
08	0 17·0	0 17·0	0 16·2	0·8	0·0	6·8	0·2	12·8	0·3
09	0 17·3	0 17·3	0 16·5	0·9	0·0	6·9	0·2	12·9	0·3
10	0 17·5	0 17·5	0 16·7	1·0	0·0	7·0	0·2	13·0	0·3
11	0 17·8	0 17·8	0 16·9	1·1	0·0	7·1	0·2	13·1	0·3
12	0 18·0	0 18·0	0 17·2	1·2	0·0	7·2	0·2	13·2	0·3
13	0 18·3	0 18·3	0 17·4	1·3	0·0	7·3	0·2	13·3	0·3
14	0 18·5	0 18·6	0 17·7	1·4	0·0	7·4	0·2	13·4	0·3
15	0 18·8	0 18·8	0 17·9	1·5	0·0	7·5	0·2	13·5	0·3
16	0 19·0	0 19·1	0 18·1	1·6	0·0	7·6	0·2	13·6	0·3
17	0 19·3	0 19·3	0 18·4	1·7	0·0	7·7	0·2	13·7	0·3
18	0 19·5	0 19·6	0 18·6	1·8	0·0	7·8	0·2	13·8	0·3
19	0 19·8	0 19·8	0 18·9	1·9	0·0	7·9	0·2	13·9	0·3
20	0 20·0	0 20·1	0 19·1	2·0	0·1	8·0	0·2	14·0	0·4
21	0 20·3	0 20·3	0 19·3	2·1	0·1	8·1	0·2	14·1	0·4
22	0 20·5	0 20·6	0 19·6	2·2	0·1	8·2	0·2	14·2	0·4
23	0 20·8	0 20·8	0 19·8	2·3	0·1	8·3	0·2	14·3	0·4
24	0 21·0	0 21·1	0 20·0	2·4	0·1	8·4	0·2	14·4	0·4
25	0 21·3	0 21·3	0 20·3	2·5	0·1	8·5	0·2	14·5	0·4
26	0 21·5	0 21·6	0 20·5	2·6	0·1	8·6	0·2	14·6	0·4
27	0 21·8	0 21·8	0 20·8	2·7	0·1	8·7	0·2	14·7	0·4
28	0 22·0	0 22·1	0 21·0	2·8	0·1	8·8	0·2	14·8	0·4
29	0 22·3	0 22·3	0 21·2	2·9	0·1	8·9	0·2	14·9	0·4
30	0 22·5	0 22·6	0 21·5	3·0	0·1	9·0	0·2	15·0	0·4
31	0 22·8	0 22·8	0 21·7	3·1	0·1	9·1	0·2	15·1	0·4
32	0 23·0	0 23·1	0 22·0	3·2	0·1	9·2	0·2	15·2	0·4
33	0 23·3	0 23·3	0 22·2	3·3	0·1	9·3	0·2	15·3	0·4
34	0 23·5	0 23·6	0 22·4	3·4	0·1	9·4	0·2	15·4	0·4
35	0 23·8	0 23·8	0 22·7	3·5	0·1	9·5	0·2	15·5	0·4
36	0 24·0	0 24·1	0 22·9	3·6	0·1	9·6	0·2	15·6	0·4
37	0 24·3	0 24·3	0 23·1	3·7	0·1	9·7	0·2	15·7	0·4
38	0 24·5	0 24·6	0 23·4	3·8	0·1	9·8	0·2	15·8	0·4
39	0 24·8	0 24·8	0 23·6	3·9	0·1	9·9	0·2	15·9	0·4
40	0 25·0	0 25·1	0 23·9	4·0	0·1	10·0	0·3	16·0	0·4
41	0 25·3	0 25·3	0 24·1	4·1	0·1	10·1	0·3	16·1	0·4
42	0 25·5	0 25·6	0 24·3	4·2	0·1	10·2	0·3	16·2	0·4
43	0 25·8	0 25·8	0 24·6	4·3	0·1	10·3	0·3	16·3	0·4
44	0 26·0	0 26·1	0 24·8	4·4	0·1	10·4	0·3	16·4	0·4
45	0 26·3	0 26·3	0 25·1	4·5	0·1	10·5	0·3	16·5	0·4
46	0 26·5	0 26·6	0 25·3	4·6	0·1	10·6	0·3	16·6	0·4
47	0 26·8	0 26·8	0 25·5	4·7	0·1	10·7	0·3	16·7	0·4
48	0 27·0	0 27·1	0 25·8	4·8	0·1	10·8	0·3	16·8	0·4
49	0 27·3	0 27·3	0 26·0	4·9	0·1	10·9	0·3	16·9	0·4
50	0 27·5	0 27·6	0 26·2	5·0	0·1	11·0	0·3	17·0	0·4
51	0 27·8	0 27·8	0 26·5	5·1	0·1	11·1	0·3	17·1	0·4
52	0 28·0	0 28·1	0 26·7	5·2	0·1	11·2	0·3	17·2	0·4
53	0 28·3	0 28·3	0 27·0	5·3	0·1	11·3	0·3	17·3	0·4
54	0 28·5	0 28·6	0 27·2	5·4	0·1	11·4	0·3	17·4	0·4
55	0 28·8	0 28·8	0 27·4	5·5	0·1	11·5	0·3	17·5	0·4
56	0 29·0	0 29·1	0 27·7	5·6	0·1	11·6	0·3	17·6	0·4
57	0 29·3	0 29·3	0 27·9	5·7	0·1	11·7	0·3	17·7	0·4
58	0 29·5	0 29·6	0 28·2	5·8	0·1	11·8	0·3	17·8	0·4
59	0 29·8	0 29·8	0 28·4	5·9	0·1	11·9	0·3	17·9	0·4
60	0 30·0	0 30·1	0 28·6	6·0	0·2	12·0	0·3	18·0	0·5

m 2 s	SUN PLANETS ° ′	ARIES ° ′	MOON ° ′	v or d ′	Corrⁿ ′	v or d ′	Corrⁿ ′	v or d ′	Corrⁿ ′
00	0 30·0	0 30·1	0 28·6	0·0	0·0	6·0	0·3	12·0	0·5
01	0 30·3	0 30·3	0 28·9	0·1	0·0	6·1	0·3	12·1	0·5
02	0 30·5	0 30·6	0 29·1	0·2	0·0	6·2	0·3	12·2	0·5
03	0 30·8	0 30·8	0 29·3	0·3	0·0	6·3	0·3	12·3	0·5
04	0 31·0	0 31·1	0 29·6	0·4	0·0	6·4	0·3	12·4	0·5
05	0 31·3	0 31·3	0 29·8	0·5	0·0	6·5	0·3	12·5	0·5
06	0 31·5	0 31·6	0 30·1	0·6	0·0	6·6	0·3	12·6	0·5
07	0 31·8	0 31·8	0 30·3	0·7	0·0	6·7	0·3	12·7	0·5
08	0 32·0	0 32·1	0 30·5	0·8	0·0	6·8	0·3	12·8	0·5
09	0 32·3	0 32·3	0 30·8	0·9	0·0	6·9	0·3	12·9	0·5
10	0 32·5	0 32·6	0 31·0	1·0	0·0	7·0	0·3	13·0	0·5
11	0 32·8	0 32·8	0 31·3	1·1	0·0	7·1	0·3	13·1	0·5
12	0 33·0	0 33·1	0 31·5	1·2	0·1	7·2	0·3	13·2	0·6
13	0 33·3	0 33·3	0 31·7	1·3	0·1	7·3	0·3	13·3	0·6
14	0 33·5	0 33·6	0 32·0	1·4	0·1	7·4	0·3	13·4	0·6
15	0 33·8	0 33·8	0 32·2	1·5	0·1	7·5	0·3	13·5	0·6
16	0 34·0	0 34·1	0 32·5	1·6	0·1	7·6	0·3	13·6	0·6
17	0 34·3	0 34·3	0 32·7	1·7	0·1	7·7	0·3	13·7	0·6
18	0 34·5	0 34·6	0 32·9	1·8	0·1	7·8	0·3	13·8	0·6
19	0 34·8	0 34·8	0 33·2	1·9	0·1	7·9	0·3	13·9	0·6
20	0 35·0	0 35·1	0 33·4	2·0	0·1	8·0	0·3	14·0	0·6
21	0 35·3	0 35·3	0 33·6	2·1	0·1	8·1	0·3	14·1	0·6
22	0 35·5	0 35·6	0 33·9	2·2	0·1	8·2	0·3	14·2	0·6
23	0 35·8	0 35·8	0 34·1	2·3	0·1	8·3	0·3	14·3	0·6
24	0 36·0	0 36·1	0 34·4	2·4	0·1	8·4	0·4	14·4	0·6
25	0 36·3	0 36·3	0 34·6	2·5	0·1	8·5	0·4	14·5	0·6
26	0 36·5	0 36·6	0 34·8	2·6	0·1	8·6	0·4	14·6	0·6
27	0 36·8	0 36·9	0 35·1	2·7	0·1	8·7	0·4	14·7	0·6
28	0 37·0	0 37·1	0 35·3	2·8	0·1	8·8	0·4	14·8	0·6
29	0 37·3	0 37·4	0 35·6	2·9	0·1	8·9	0·4	14·9	0·6
30	0 37·5	0 37·6	0 35·8	3·0	0·1	9·0	0·4	15·0	0·6
31	0 37·8	0 37·9	0 36·0	3·1	0·1	9·1	0·4	15·1	0·6
32	0 38·0	0 38·1	0 36·3	3·2	0·1	9·2	0·4	15·2	0·6
33	0 38·3	0 38·4	0 36·5	3·3	0·1	9·3	0·4	15·3	0·6
34	0 38·5	0 38·6	0 36·7	3·4	0·1	9·4	0·4	15·4	0·6
35	0 38·8	0 38·9	0 37·0	3·5	0·1	9·5	0·4	15·5	0·6
36	0 39·0	0 39·1	0 37·2	3·6	0·2	9·6	0·4	15·6	0·7
37	0 39·3	0 39·4	0 37·5	3·7	0·2	9·7	0·4	15·7	0·7
38	0 39·5	0 39·6	0 37·7	3·8	0·2	9·8	0·4	15·8	0·7
39	0 39·8	0 39·9	0 37·9	3·9	0·2	9·9	0·4	15·9	0·7
40	0 40·0	0 40·1	0 38·2	4·0	0·2	10·0	0·4	16·0	0·7
41	0 40·3	0 40·4	0 38·4	4·1	0·2	10·1	0·4	16·1	0·7
42	0 40·5	0 40·6	0 38·7	4·2	0·2	10·2	0·4	16·2	0·7
43	0 40·8	0 40·9	0 38·9	4·3	0·2	10·3	0·4	16·3	0·7
44	0 41·0	0 41·1	0 39·1	4·4	0·2	10·4	0·4	16·4	0·7
45	0 41·3	0 41·4	0 39·4	4·5	0·2	10·5	0·4	16·5	0·7
46	0 41·5	0 41·6	0 39·6	4·6	0·2	10·6	0·4	16·6	0·7
47	0 41·8	0 41·9	0 39·8	4·7	0·2	10·7	0·4	16·7	0·7
48	0 42·0	0 42·1	0 40·1	4·8	0·2	10·8	0·5	16·8	0·7
49	0 42·3	0 42·4	0 40·3	4·9	0·2	10·9	0·5	16·9	0·7
50	0 42·5	0 42·6	0 40·6	5·0	0·2	11·0	0·5	17·0	0·7
51	0 42·8	0 42·9	0 40·8	5·1	0·2	11·1	0·5	17·1	0·7
52	0 43·0	0 43·1	0 41·0	5·2	0·2	11·2	0·5	17·2	0·7
53	0 43·3	0 43·4	0 41·3	5·3	0·2	11·3	0·5	17·3	0·7
54	0 43·5	0 43·6	0 41·5	5·4	0·2	11·4	0·5	17·4	0·7
55	0 43·8	0 43·9	0 41·8	5·5	0·2	11·5	0·5	17·5	0·7
56	0 44·0	0 44·1	0 42·0	5·6	0·2	11·6	0·5	17·6	0·7
57	0 44·3	0 44·4	0 42·2	5·7	0·2	11·7	0·5	17·7	0·7
58	0 44·5	0 44·6	0 42·5	5·8	0·2	11·8	0·5	17·8	0·7
59	0 44·8	0 44·9	0 42·7	5·9	0·2	11·9	0·5	17·9	0·7
60	0 45·0	0 45·1	0 43·0	6·0	0·3	12·0	0·5	18·0	0·8

m 3 s	SUN PLANETS ° ′	ARIES ° ′	MOON ° ′	v or d ′	Corrⁿ ′	v or d ′	Corrⁿ ′	v or d ′	Corrⁿ ′
00	0 45·0	0 45·1	0 43·0	0·0	0·0	6·0	0·4	12·0	0·7
01	0 45·3	0 45·4	0 43·2	0·1	0·0	6·1	0·4	12·1	0·7
02	0 45·5	0 45·6	0 43·4	0·2	0·0	6·2	0·4	12·2	0·7
03	0 45·8	0 45·9	0 43·7	0·3	0·0	6·3	0·4	12·3	0·7
04	0 46·0	0 46·1	0 43·9	0·4	0·0	6·4	0·4	12·4	0·7
05	0 46·3	0 46·4	0 44·1	0·5	0·0	6·5	0·4	12·5	0·7
06	0 46·5	0 46·6	0 44·4	0·6	0·0	6·6	0·4	12·6	0·7
07	0 46·8	0 46·9	0 44·6	0·7	0·0	6·7	0·4	12·7	0·7
08	0 47·0	0 47·1	0 44·9	0·8	0·0	6·8	0·4	12·8	0·7
09	0 47·3	0 47·4	0 45·1	0·9	0·1	6·9	0·4	12·9	0·8
10	0 47·5	0 47·6	0 45·3	1·0	0·1	7·0	0·4	13·0	0·8
11	0 47·8	0 47·9	0 45·6	1·1	0·1	7·1	0·4	13·1	0·8
12	0 48·0	0 48·1	0 45·8	1·2	0·1	7·2	0·4	13·2	0·8
13	0 48·3	0 48·4	0 46·1	1·3	0·1	7·3	0·4	13·3	0·8
14	0 48·5	0 48·6	0 46·3	1·4	0·1	7·4	0·4	13·4	0·8
15	0 48·8	0 48·9	0 46·5	1·5	0·1	7·5	0·4	13·5	0·8
16	0 49·0	0 49·1	0 46·8	1·6	0·1	7·6	0·4	13·6	0·8
17	0 49·3	0 49·4	0 47·0	1·7	0·1	7·7	0·4	13·7	0·8
18	0 49·5	0 49·6	0 47·2	1·8	0·1	7·8	0·5	13·8	0·8
19	0 49·8	0 49·9	0 47·5	1·9	0·1	7·9	0·5	13·9	0·8
20	0 50·0	0 50·1	0 47·7	2·0	0·1	8·0	0·5	14·0	0·8
21	0 50·3	0 50·4	0 48·0	2·1	0·1	8·1	0·5	14·1	0·8
22	0 50·5	0 50·6	0 48·2	2·2	0·1	8·2	0·5	14·2	0·8
23	0 50·8	0 50·9	0 48·4	2·3	0·1	8·3	0·5	14·3	0·8
24	0 51·0	0 51·1	0 48·7	2·4	0·1	8·4	0·5	14·4	0·8
25	0 51·3	0 51·4	0 48·9	2·5	0·1	8·5	0·5	14·5	0·8
26	0 51·5	0 51·6	0 49·2	2·6	0·2	8·6	0·5	14·6	0·9
27	0 51·8	0 51·9	0 49·4	2·7	0·2	8·7	0·5	14·7	0·9
28	0 52·0	0 52·1	0 49·6	2·8	0·2	8·8	0·5	14·8	0·9
29	0 52·3	0 52·4	0 49·9	2·9	0·2	8·9	0·5	14·9	0·9
30	0 52·5	0 52·6	0 50·1	3·0	0·2	9·0	0·5	15·0	0·9
31	0 52·8	0 52·9	0 50·3	3·1	0·2	9·1	0·5	15·1	0·9
32	0 53·0	0 53·1	0 50·6	3·2	0·2	9·2	0·5	15·2	0·9
33	0 53·3	0 53·4	0 50·8	3·3	0·2	9·3	0·5	15·3	0·9
34	0 53·5	0 53·6	0 51·1	3·4	0·2	9·4	0·5	15·4	0·9
35	0 53·8	0 53·9	0 51·3	3·5	0·2	9·5	0·6	15·5	0·9
36	0 54·0	0 54·1	0 51·5	3·6	0·2	9·6	0·6	15·6	0·9
37	0 54·3	0 54·4	0 51·8	3·7	0·2	9·7	0·6	15·7	0·9
38	0 54·5	0 54·6	0 52·0	3·8	0·2	9·8	0·6	15·8	0·9
39	0 54·8	0 54·9	0 52·3	3·9	0·2	9·9	0·6	15·9	0·9
40	0 55·0	0 55·2	0 52·5	4·0	0·2	10·0	0·6	16·0	0·9
41	0 55·3	0 55·4	0 52·7	4·1	0·2	10·1	0·6	16·1	0·9
42	0 55·5	0 55·7	0 53·0	4·2	0·2	10·2	0·6	16·2	0·9
43	0 55·8	0 55·9	0 53·2	4·3	0·3	10·3	0·6	16·3	1·0
44	0 56·0	0 56·2	0 53·4	4·4	0·3	10·4	0·6	16·4	1·0
45	0 56·3	0 56·4	0 53·7	4·5	0·3	10·5	0·6	16·5	1·0
46	0 56·5	0 56·7	0 53·9	4·6	0·3	10·6	0·6	16·6	1·0
47	0 56·8	0 56·9	0 54·2	4·7	0·3	10·7	0·6	16·7	1·0
48	0 57·0	0 57·2	0 54·4	4·8	0·3	10·8	0·6	16·8	1·0
49	0 57·3	0 57·4	0 54·6	4·9	0·3	10·9	0·6	16·9	1·0
50	0 57·5	0 57·7	0 54·9	5·0	0·3	11·0	0·6	17·0	1·0
51	0 57·8	0 57·9	0 55·1	5·1	0·3	11·1	0·6	17·1	1·0
52	0 58·0	0 58·2	0 55·4	5·2	0·3	11·2	0·7	17·2	1·0
53	0 58·3	0 58·4	0 55·6	5·3	0·3	11·3	0·7	17·3	1·0
54	0 58·5	0 58·7	0 55·8	5·4	0·3	11·4	0·7	17·4	1·0
55	0 58·8	0 58·9	0 56·1	5·5	0·3	11·5	0·7	17·5	1·0
56	0 59·0	0 59·2	0 56·3	5·6	0·3	11·6	0·7	17·6	1·0
57	0 59·3	0 59·4	0 56·6	5·7	0·3	11·7	0·7	17·7	1·0
58	0 59·5	0 59·7	0 56·8	5·8	0·3	11·8	0·7	17·8	1·0
59	0 59·8	0 59·9	0 57·0	5·9	0·3	11·9	0·7	17·9	1·0
60	1 00·0	1 00·2	0 57·3	6·0	0·4	12·0	0·7	18·0	1·1

m 4	SUN PLANETS	ARIES	MOON	v or d	Corrⁿ	v or d	Corrⁿ	v or d	Corrⁿ
s	° ′	° ′	° ′	′	′	′	′	′	′
00	1 00·0	1 00·2	0 57·3	0·0	0·0	6·0	0·5	12·0	0·9
01	1 00·3	1 00·4	0 57·5	0·1	0·0	6·1	0·5	12·1	0·9
02	1 00·5	1 00·7	0 57·7	0·2	0·0	6·2	0·5	12·2	0·9
03	1 00·8	1 00·9	0 58·0	0·3	0·0	6·3	0·5	12·3	0·9
04	1 01·0	1 01·2	0 58·2	0·4	0·0	6·4	0·5	12·4	0·9
05	1 01·3	1 01·4	0 58·5	0·5	0·0	6·5	0·5	12·5	0·9
06	1 01·5	1 01·7	0 58·7	0·6	0·0	6·6	0·5	12·6	0·9
07	1 01·8	1 01·9	0 58·9	0·7	0·1	6·7	0·5	12·7	1·0
08	1 02·0	1 02·2	0 59·2	0·8	0·1	6·8	0·5	12·8	1·0
09	1 02·3	1 02·4	0 59·4	0·9	0·1	6·9	0·5	12·9	1·0
10	1 02·5	1 02·7	0 59·7	1·0	0·1	7·0	0·5	13·0	1·0
11	1 02·8	1 02·9	0 59·9	1·1	0·1	7·1	0·5	13·1	1·0
12	1 03·0	1 03·2	1 00·1	1·2	0·1	7·2	0·5	13·2	1·0
13	1 03·3	1 03·4	1 00·4	1·3	0·1	7·3	0·5	13·3	1·0
14	1 03·5	1 03·7	1 00·6	1·4	0·1	7·4	0·6	13·4	1·0
15	1 03·8	1 03·9	1 00·8	1·5	0·1	7·5	0·6	13·5	1·0
16	1 04·0	1 04·2	1 01·1	1·6	0·1	7·6	0·6	13·6	1·0
17	1 04·3	1 04·4	1 01·3	1·7	0·1	7·7	0·6	13·7	1·0
18	1 04·5	1 04·7	1 01·6	1·8	0·1	7·8	0·6	13·8	1·0
19	1 04·8	1 04·9	1 01·8	1·9	0·1	7·9	0·6	13·9	1·0
20	1 05·0	1 05·2	1 02·0	2·0	0·2	8·0	0·6	14·0	1·1
21	1 05·3	1 05·4	1 02·3	2·1	0·2	8·1	0·6	14·1	1·1
22	1 05·5	1 05·7	1 02·5	2·2	0·2	8·2	0·6	14·2	1·1
23	1 05·8	1 05·9	1 02·8	2·3	0·2	8·3	0·6	14·3	1·1
24	1 06·0	1 06·2	1 03·0	2·4	0·2	8·4	0·6	14·4	1·1
25	1 06·3	1 06·4	1 03·2	2·5	0·2	8·5	0·6	14·5	1·1
26	1 06·5	1 06·7	1 03·5	2·6	0·2	8·6	0·6	14·6	1·1
27	1 06·8	1 06·9	1 03·7	2·7	0·2	8·7	0·7	14·7	1·1
28	1 07·0	1 07·2	1 03·9	2·8	0·2	8·8	0·7	14·8	1·1
29	1 07·3	1 07·4	1 04·2	2·9	0·2	8·9	0·7	14·9	1·1
30	1 07·5	1 07·7	1 04·4	3·0	0·2	9·0	0·7	15·0	1·1
31	1 07·8	1 07·9	1 04·7	3·1	0·2	9·1	0·7	15·1	1·1
32	1 08·0	1 08·2	1 04·9	3·2	0·2	9·2	0·7	15·2	1·1
33	1 08·3	1 08·4	1 05·1	3·3	0·2	9·3	0·7	15·3	1·1
34	1 08·5	1 08·7	1 05·4	3·4	0·3	9·4	0·7	15·4	1·2
35	1 08·8	1 08·9	1 05·6	3·5	0·3	9·5	0·7	15·5	1·2
36	1 09·0	1 09·2	1 05·9	3·6	0·3	9·6	0·7	15·6	1·2
37	1 09·3	1 09·4	1 06·1	3·7	0·3	9·7	0·7	15·7	1·2
38	1 09·5	1 09·7	1 06·3	3·8	0·3	9·8	0·7	15·8	1·2
39	1 09·8	1 09·9	1 06·6	3·9	0·3	9·9	0·7	15·9	1·2
40	1 10·0	1 10·2	1 06·8	4·0	0·3	10·0	0·8	16·0	1·2
41	1 10·3	1 10·4	1 07·0	4·1	0·3	10·1	0·8	16·1	1·2
42	1 10·5	1 10·7	1 07·3	4·2	0·3	10·2	0·8	16·2	1·2
43	1 10·8	1 10·9	1 07·5	4·3	0·3	10·3	0·8	16·3	1·2
44	1 11·0	1 11·2	1 07·8	4·4	0·3	10·4	0·8	16·4	1·2
45	1 11·3	1 11·4	1 08·0	4·5	0·3	10·5	0·8	16·5	1·2
46	1 11·5	1 11·7	1 08·2	4·6	0·3	10·6	0·8	16·6	1·2
47	1 11·8	1 11·9	1 08·5	4·7	0·4	10·7	0·8	16·7	1·3
48	1 12·0	1 12·2	1 08·7	4·8	0·4	10·8	0·8	16·8	1·3
49	1 12·3	1 12·4	1 09·0	4·9	0·4	10·9	0·8	16·9	1·3
50	1 12·5	1 12·7	1 09·2	5·0	0·4	11·0	0·8	17·0	1·3
51	1 12·8	1 12·9	1 09·4	5·1	0·4	11·1	0·8	17·1	1·3
52	1 13·0	1 13·2	1 09·7	5·2	0·4	11·2	0·8	17·2	1·3
53	1 13·3	1 13·5	1 09·9	5·3	0·4	11·3	0·8	17·3	1·3
54	1 13·5	1 13·7	1 10·2	5·4	0·4	11·4	0·9	17·4	1·3
55	1 13·8	1 14·0	1 10·4	5·5	0·4	11·5	0·9	17·5	1·3
56	1 14·0	1 14·2	1 10·6	5·6	0·4	11·6	0·9	17·6	1·3
57	1 14·3	1 14·5	1 10·9	5·7	0·4	11·7	0·9	17·7	1·3
58	1 14·5	1 14·7	1 11·1	5·8	0·4	11·8	0·9	17·8	1·3
59	1 14·8	1 15·0	1 11·3	5·9	0·4	11·9	0·9	17·9	1·3
60	1 15·0	1 15·2	1 11·6	6·0	0·5	12·0	0·9	18·0	1·4

m 5	SUN PLANETS	ARIES	MOON	v or d	Corrⁿ	v or d	Corrⁿ	v or d	Corrⁿ
s	° ′	° ′	° ′	′	′	′	′	′	′
00	1 15·0	1 15·2	1 11·6	0·0	0·0	6·0	0·6	12·0	1·1
01	1 15·3	1 15·5	1 11·8	0·1	0·0	6·1	0·6	12·1	1·1
02	1 15·5	1 15·7	1 12·1	0·2	0·0	6·2	0·6	12·2	1·1
03	1 15·8	1 16·0	1 12·3	0·3	0·0	6·3	0·6	12·3	1·1
04	1 16·0	1 16·2	1 12·5	0·4	0·0	6·4	0·6	12·4	1·1
05	1 16·3	1 16·5	1 12·8	0·5	0·0	6·5	0·6	12·5	1·1
06	1 16·5	1 16·7	1 13·0	0·6	0·1	6·6	0·6	12·6	1·2
07	1 16·8	1 17·0	1 13·3	0·7	0·1	6·7	0·6	12·7	1·2
08	1 17·0	1 17·2	1 13·5	0·8	0·1	6·8	0·6	12·8	1·2
09	1 17·3	1 17·5	1 13·7	0·9	0·1	6·9	0·6	12·9	1·2
10	1 17·5	1 17·7	1 14·0	1·0	0·1	7·0	0·6	13·0	1·2
11	1 17·8	1 18·0	1 14·2	1·1	0·1	7·1	0·7	13·1	1·2
12	1 18·0	1 18·2	1 14·4	1·2	0·1	7·2	0·7	13·2	1·2
13	1 18·3	1 18·5	1 14·7	1·3	0·1	7·3	0·7	13·3	1·2
14	1 18·5	1 18·7	1 14·9	1·4	0·1	7·4	0·7	13·4	1·2
15	1 18·8	1 19·0	1 15·2	1·5	0·1	7·5	0·7	13·5	1·2
16	1 19·0	1 19·2	1 15·4	1·6	0·1	7·6	0·7	13·6	1·2
17	1 19·3	1 19·5	1 15·6	1·7	0·2	7·7	0·7	13·7	1·3
18	1 19·5	1 19·7	1 15·9	1·8	0·2	7·8	0·7	13·8	1·3
19	1 19·8	1 20·0	1 16·1	1·9	0·2	7·9	0·7	13·9	1·3
20	1 20·0	1 20·2	1 16·4	2·0	0·2	8·0	0·7	14·0	1·3
21	1 20·3	1 20·5	1 16·6	2·1	0·2	8·1	0·7	14·1	1·3
22	1 20·5	1 20·7	1 16·8	2·2	0·2	8·2	0·8	14·2	1·3
23	1 20·8	1 21·0	1 17·1	2·3	0·2	8·3	0·8	14·3	1·3
24	1 21·0	1 21·2	1 17·3	2·4	0·2	8·4	0·8	14·4	1·3
25	1 21·3	1 21·5	1 17·5	2·5	0·2	8·5	0·8	14·5	1·3
26	1 21·5	1 21·7	1 17·8	2·6	0·2	8·6	0·8	14·6	1·3
27	1 21·8	1 22·0	1 18·0	2·7	0·2	8·7	0·8	14·7	1·3
28	1 22·0	1 22·2	1 18·3	2·8	0·3	8·8	0·8	14·8	1·4
29	1 22·3	1 22·5	1 18·5	2·9	0·3	8·9	0·8	14·9	1·4
30	1 22·5	1 22·7	1 18·7	3·0	0·3	9·0	0·8	15·0	1·4
31	1 22·8	1 23·0	1 19·0	3·1	0·3	9·1	0·8	15·1	1·4
32	1 23·0	1 23·2	1 19·2	3·2	0·3	9·2	0·8	15·2	1·4
33	1 23·3	1 23·5	1 19·5	3·3	0·3	9·3	0·9	15·3	1·4
34	1 23·5	1 23·7	1 19·7	3·4	0·3	9·4	0·9	15·4	1·4
35	1 23·8	1 24·0	1 19·9	3·5	0·3	9·5	0·9	15·5	1·4
36	1 24·0	1 24·2	1 20·2	3·6	0·3	9·6	0·9	15·6	1·4
37	1 24·3	1 24·5	1 20·4	3·7	0·3	9·7	0·9	15·7	1·4
38	1 24·5	1 24·7	1 20·7	3·8	0·3	9·8	0·9	15·8	1·4
39	1 24·8	1 25·0	1 20·9	3·9	0·4	9·9	0·9	15·9	1·5
40	1 25·0	1 25·2	1 21·1	4·0	0·4	10·0	0·9	16·0	1·5
41	1 25·3	1 25·5	1 21·4	4·1	0·4	10·1	0·9	16·1	1·5
42	1 25·5	1 25·7	1 21·6	4·2	0·4	10·2	0·9	16·2	1·5
43	1 25·8	1 26·0	1 21·8	4·3	0·4	10·3	0·9	16·3	1·5
44	1 26·0	1 26·2	1 22·1	4·4	0·4	10·4	1·0	16·4	1·5
45	1 26·3	1 26·5	1 22·3	4·5	0·4	10·5	1·0	16·5	1·5
46	1 26·5	1 26·7	1 22·6	4·6	0·4	10·6	1·0	16·6	1·5
47	1 26·8	1 27·0	1 22·8	4·7	0·4	10·7	1·0	16·7	1·5
48	1 27·0	1 27·2	1 23·0	4·8	0·4	10·8	1·0	16·8	1·5
49	1 27·3	1 27·5	1 23·3	4·9	0·4	10·9	1·0	16·9	1·5
50	1 27·5	1 27·7	1 23·5	5·0	0·5	11·0	1·0	17·0	1·6
51	1 27·8	1 28·0	1 23·8	5·1	0·5	11·1	1·0	17·1	1·6
52	1 28·0	1 28·2	1 24·0	5·2	0·5	11·2	1·0	17·2	1·6
53	1 28·3	1 28·5	1 24·2	5·3	0·5	11·3	1·0	17·3	1·6
54	1 28·5	1 28·7	1 24·5	5·4	0·5	11·4	1·0	17·4	1·6
55	1 28·8	1 29·0	1 24·7	5·5	0·5	11·5	1·1	17·5	1·6
56	1 29·0	1 29·2	1 24·9	5·6	0·5	11·6	1·1	17·6	1·6
57	1 29·3	1 29·5	1 25·2	5·7	0·5	11·7	1·1	17·7	1·6
58	1 29·5	1 29·7	1 25·4	5·8	0·5	11·8	1·1	17·8	1·6
59	1 29·8	1 30·0	1 25·7	5·9	0·5	11·9	1·1	17·9	1·6
60	1 30·0	1 30·2	1 25·9	6·0	0·6	12·0	1·1	18·0	1·7

6m s	SUN PLANETS ° ′	ARIES ° ′	MOON ° ′	v or d ′	Corrn ′	v or d ′	Corrn ′	v or d ′	Corrn ′
00	1 30·0	1 30·2	1 25·9	0·0	0·0	6·0	0·7	12·0	1·3
01	1 30·3	1 30·5	1 26·1	0·1	0·0	6·1	0·7	12·1	1·3
02	1 30·5	1 30·7	1 26·4	0·2	0·0	6·2	0·7	12·2	1·3
03	1 30·8	1 31·0	1 26·6	0·3	0·0	6·3	0·7	12·3	1·3
04	1 31·0	1 31·2	1 26·9	0·4	0·0	6·4	0·7	12·4	1·3
05	1 31·3	1 31·5	1 27·1	0·5	0·1	6·5	0·7	12·5	1·4
06	1 31·5	1 31·8	1 27·3	0·6	0·1	6·6	0·7	12·6	1·4
07	1 31·8	1 32·0	1 27·6	0·7	0·1	6·7	0·7	12·7	1·4
08	1 32·0	1 32·3	1 27·8	0·8	0·1	6·8	0·7	12·8	1·4
09	1 32·3	1 32·5	1 28·0	0·9	0·1	6·9	0·7	12·9	1·4
10	1 32·5	1 32·8	1 28·3	1·0	0·1	7·0	0·8	13·0	1·4
11	1 32·8	1 33·0	1 28·5	1·1	0·1	7·1	0·8	13·1	1·4
12	1 33·0	1 33·3	1 28·8	1·2	0·1	7·2	0·8	13·2	1·4
13	1 33·3	1 33·5	1 29·0	1·3	0·1	7·3	0·8	13·3	1·4
14	1 33·5	1 33·8	1 29·2	1·4	0·2	7·4	0·8	13·4	1·5
15	1 33·8	1 34·0	1 29·5	1·5	0·2	7·5	0·8	13·5	1·5
16	1 34·0	1 34·3	1 29·7	1·6	0·2	7·6	0·8	13·6	1·5
17	1 34·3	1 34·5	1 30·0	1·7	0·2	7·7	0·8	13·7	1·5
18	1 34·5	1 34·8	1 30·2	1·8	0·2	7·8	0·8	13·8	1·5
19	1 34·8	1 35·0	1 30·4	1·9	0·2	7·9	0·9	13·9	1·5
20	1 35·0	1 35·3	1 30·7	2·0	0·2	8·0	0·9	14·0	1·5
21	1 35·3	1 35·5	1 30·9	2·1	0·2	8·1	0·9	14·1	1·5
22	1 35·5	1 35·8	1 31·1	2·2	0·2	8·2	0·9	14·2	1·5
23	1 35·8	1 36·0	1 31·4	2·3	0·2	8·3	0·9	14·3	1·5
24	1 36·0	1 36·3	1 31·6	2·4	0·3	8·4	0·9	14·4	1·6
25	1 36·3	1 36·5	1 31·9	2·5	0·3	8·5	0·9	14·5	1·6
26	1 36·5	1 36·8	1 32·1	2·6	0·3	8·6	0·9	14·6	1·6
27	1 36·8	1 37·0	1 32·3	2·7	0·3	8·7	0·9	14·7	1·6
28	1 37·0	1 37·3	1 32·6	2·8	0·3	8·8	1·0	14·8	1·6
29	1 37·3	1 37·5	1 32·8	2·9	0·3	8·9	1·0	14·9	1·6
30	1 37·5	1 37·8	1 33·1	3·0	0·3	9·0	1·0	15·0	1·6
31	1 37·8	1 38·0	1 33·3	3·1	0·3	9·1	1·0	15·1	1·6
32	1 38·0	1 38·3	1 33·5	3·2	0·3	9·2	1·0	15·2	1·6
33	1 38·3	1 38·5	1 33·8	3·3	0·4	9·3	1·0	15·3	1·7
34	1 38·5	1 38·8	1 34·0	3·4	0·4	9·4	1·0	15·4	1·7
35	1 38·8	1 39·0	1 34·3	3·5	0·4	9·5	1·0	15·5	1·7
36	1 39·0	1 39·3	1 34·5	3·6	0·4	9·6	1·0	15·6	1·7
37	1 39·3	1 39·5	1 34·7	3·7	0·4	9·7	1·1	15·7	1·7
38	1 39·5	1 39·8	1 35·0	3·8	0·4	9·8	1·1	15·8	1·7
39	1 39·8	1 40·0	1 35·2	3·9	0·4	9·9	1·1	15·9	1·7
40	1 40·0	1 40·3	1 35·4	4·0	0·4	10·0	1·1	16·0	1·7
41	1 40·3	1 40·5	1 35·7	4·1	0·4	10·1	1·1	16·1	1·7
42	1 40·5	1 40·8	1 35·9	4·2	0·5	10·2	1·1	16·2	1·8
43	1 40·8	1 41·0	1 36·2	4·3	0·5	10·3	1·1	16·3	1·8
44	1 41·0	1 41·3	1 36·4	4·4	0·5	10·4	1·1	16·4	1·8
45	1 41·3	1 41·5	1 36·6	4·5	0·5	10·5	1·1	16·5	1·8
46	1 41·5	1 41·8	1 36·9	4·6	0·5	10·6	1·1	16·6	1·8
47	1 41·8	1 42·0	1 37·1	4·7	0·5	10·7	1·2	16·7	1·8
48	1 42·0	1 42·3	1 37·4	4·8	0·5	10·8	1·2	16·8	1·8
49	1 42·3	1 42·5	1 37·6	4·9	0·5	10·9	1·2	16·9	1·8
50	1 42·5	1 42·8	1 37·8	5·0	0·5	11·0	1·2	17·0	1·8
51	1 42·8	1 43·0	1 38·1	5·1	0·6	11·1	1·2	17·1	1·9
52	1 43·0	1 43·3	1 38·3	5·2	0·6	11·2	1·2	17·2	1·9
53	1 43·3	1 43·5	1 38·5	5·3	0·6	11·3	1·2	17·3	1·9
54	1 43·5	1 43·8	1 38·8	5·4	0·6	11·4	1·2	17·4	1·9
55	1 43·8	1 44·0	1 39·0	5·5	0·6	11·5	1·2	17·5	1·9
56	1 44·0	1 44·3	1 39·3	5·6	0·6	11·6	1·3	17·6	1·9
57	1 44·3	1 44·5	1 39·5	5·7	0·6	11·7	1·3	17·7	1·9
58	1 44·5	1 44·8	1 39·7	5·8	0·6	11·8	1·3	17·8	1·9
59	1 44·8	1 45·0	1 40·0	5·9	0·6	11·9	1·3	17·9	1·9
60	1 45·0	1 45·3	1 40·2	6·0	0·7	12·0	1·3	18·0	2·0

7m s	SUN PLANETS ° ′	ARIES ° ′	MOON ° ′	v or d ′	Corrn ′	v or d ′	Corrn ′	v or d ′	Corrn ′
00	1 45·0	1 45·3	1 40·2	0·0	0·0	6·0	0·8	12·0	1·5
01	1 45·3	1 45·5	1 40·5	0·1	0·0	6·1	0·8	12·1	1·5
02	1 45·5	1 45·8	1 40·7	0·2	0·0	6·2	0·8	12·2	1·5
03	1 45·8	1 46·0	1 40·9	0·3	0·0	6·3	0·8	12·3	1·5
04	1 46·0	1 46·3	1 41·2	0·4	0·1	6·4	0·8	12·4	1·6
05	1 46·3	1 46·5	1 41·4	0·5	0·1	6·5	0·8	12·5	1·6
06	1 46·5	1 46·8	1 41·6	0·6	0·1	6·6	0·8	12·6	1·6
07	1 46·8	1 47·0	1 41·9	0·7	0·1	6·7	0·8	12·7	1·6
08	1 47·0	1 47·3	1 42·1	0·8	0·1	6·8	0·9	12·8	1·6
09	1 47·3	1 47·5	1 42·4	0·9	0·1	6·9	0·9	12·9	1·6
10	1 47·5	1 47·8	1 42·6	1·0	0·1	7·0	0·9	13·0	1·6
11	1 47·8	1 48·0	1 42·8	1·1	0·1	7·1	0·9	13·1	1·6
12	1 48·0	1 48·3	1 43·1	1·2	0·2	7·2	0·9	13·2	1·7
13	1 48·3	1 48·5	1 43·3	1·3	0·2	7·3	0·9	13·3	1·7
14	1 48·5	1 48·8	1 43·6	1·4	0·2	7·4	0·9	13·4	1·7
15	1 48·8	1 49·0	1 43·8	1·5	0·2	7·5	0·9	13·5	1·7
16	1 49·0	1 49·3	1 44·0	1·6	0·2	7·6	1·0	13·6	1·7
17	1 49·3	1 49·5	1 44·3	1·7	0·2	7·7	1·0	13·7	1·7
18	1 49·5	1 49·8	1 44·5	1·8	0·2	7·8	1·0	13·8	1·7
19	1 49·8	1 50·1	1 44·8	1·9	0·2	7·9	1·0	13·9	1·7
20	1 50·0	1 50·3	1 45·0	2·0	0·3	8·0	1·0	14·0	1·8
21	1 50·3	1 50·6	1 45·2	2·1	0·3	8·1	1·0	14·1	1·8
22	1 50·5	1 50·8	1 45·5	2·2	0·3	8·2	1·0	14·2	1·8
23	1 50·8	1 51·1	1 45·7	2·3	0·3	8·3	1·0	14·3	1·8
24	1 51·0	1 51·3	1 45·9	2·4	0·3	8·4	1·1	14·4	1·8
25	1 51·3	1 51·6	1 46·2	2·5	0·3	8·5	1·1	14·5	1·8
26	1 51·5	1 51·8	1 46·4	2·6	0·3	8·6	1·1	14·6	1·8
27	1 51·8	1 52·1	1 46·7	2·7	0·3	8·7	1·1	14·7	1·8
28	1 52·0	1 52·3	1 46·9	2·8	0·4	8·8	1·1	14·8	1·9
29	1 52·3	1 52·6	1 47·1	2·9	0·4	8·9	1·1	14·9	1·9
30	1 52·5	1 52·8	1 47·4	3·0	0·4	9·0	1·1	15·0	1·9
31	1 52·8	1 53·1	1 47·6	3·1	0·4	9·1	1·1	15·1	1·9
32	1 53·0	1 53·3	1 47·9	3·2	0·4	9·2	1·2	15·2	1·9
33	1 53·3	1 53·6	1 48·1	3·3	0·4	9·3	1·2	15·3	1·9
34	1 53·5	1 53·8	1 48·3	3·4	0·4	9·4	1·2	15·4	1·9
35	1 53·8	1 54·1	1 48·6	3·5	0·4	9·5	1·2	15·5	1·9
36	1 54·0	1 54·3	1 48·8	3·6	0·5	9·6	1·2	15·6	2·0
37	1 54·3	1 54·6	1 49·0	3·7	0·5	9·7	1·2	15·7	2·0
38	1 54·5	1 54·8	1 49·3	3·8	0·5	9·8	1·2	15·8	2·0
39	1 54·8	1 55·1	1 49·5	3·9	0·5	9·9	1·2	15·9	2·0
40	1 55·0	1 55·3	1 49·8	4·0	0·5	10·0	1·3	16·0	2·0
41	1 55·3	1 55·6	1 50·0	4·1	0·5	10·1	1·3	16·1	2·0
42	1 55·5	1 55·8	1 50·2	4·2	0·5	10·2	1·3	16·2	2·0
43	1 55·8	1 56·1	1 50·5	4·3	0·5	10·3	1·3	16·3	2·0
44	1 56·0	1 56·3	1 50·7	4·4	0·6	10·4	1·3	16·4	2·1
45	1 56·3	1 56·6	1 51·0	4·5	0·6	10·5	1·3	16·5	2·1
46	1 56·5	1 56·8	1 51·2	4·6	0·6	10·6	1·3	16·6	2·1
47	1 56·8	1 57·1	1 51·4	4·7	0·6	10·7	1·3	16·7	2·1
48	1 57·0	1 57·3	1 51·7	4·8	0·6	10·8	1·4	16·8	2·1
49	1 57·3	1 57·6	1 51·9	4·9	0·6	10·9	1·4	16·9	2·1
50	1 57·5	1 57·8	1 52·1	5·0	0·6	11·0	1·4	17·0	2·1
51	1 57·8	1 58·1	1 52·4	5·1	0·6	11·1	1·4	17·1	2·1
52	1 58·0	1 58·3	1 52·6	5·2	0·7	11·2	1·4	17·2	2·2
53	1 58·3	1 58·6	1 52·9	5·3	0·7	11·3	1·4	17·3	2·2
54	1 58·5	1 58·8	1 53·1	5·4	0·7	11·4	1·4	17·4	2·2
55	1 58·8	1 59·1	1 53·3	5·5	0·7	11·5	1·4	17·5	2·2
56	1 59·0	1 59·3	1 53·6	5·6	0·7	11·6	1·5	17·6	2·2
57	1 59·3	1 59·6	1 53·8	5·7	0·7	11·7	1·5	17·7	2·2
58	1 59·5	1 59·8	1 54·1	5·8	0·7	11·8	1·5	17·8	2·2
59	1 59·8	2 00·1	1 54·3	5·9	0·7	11·9	1·5	17·9	2·2
60	2 00·0	2 00·3	1 54·5	6·0	0·8	12·0	1·5	18·0	2·3

m 8 s	SUN PLANETS ° ′	ARIES ° ′	MOON ° ′	v or d ′	Corrⁿ ′	v or d ′	Corrⁿ ′	v or d ′	Corrⁿ ′
00	2 00·0	2 00·3	1 54·5	0·0	0·0	6·0	0·9	12·0	1·7
01	2 00·3	2 00·6	1 54·8	0·1	0·0	6·1	0·9	12·1	1·7
02	2 00·5	2 00·8	1 55·0	0·2	0·0	6·2	0·9	12·2	1·7
03	2 00·8	2 01·1	1 55·2	0·3	0·0	6·3	0·9	12·3	1·7
04	2 01·0	2 01·3	1 55·5	0·4	0·1	6·4	0·9	12·4	1·8
05	2 01·3	2 01·6	1 55·7	0·5	0·1	6·5	0·9	12·5	1·8
06	2 01·5	2 01·8	1 56·0	0·6	0·1	6·6	0·9	12·6	1·8
07	2 01·8	2 02·1	1 56·2	0·7	0·1	6·7	0·9	12·7	1·8
08	2 02·0	2 02·3	1 56·4	0·8	0·1	6·8	1·0	12·8	1·8
09	2 02·3	2 02·6	1 56·7	0·9	0·1	6·9	1·0	12·9	1·8
10	2 02·5	2 02·8	1 56·9	1·0	0·1	7·0	1·0	13·0	1·8
11	2 02·8	2 03·1	1 57·2	1·1	0·2	7·1	1·0	13·1	1·9
12	2 03·0	2 03·3	1 57·4	1·2	0·2	7·2	1·0	13·2	1·9
13	2 03·3	2 03·6	1 57·6	1·3	0·2	7·3	1·0	13·3	1·9
14	2 03·5	2 03·8	1 57·9	1·4	0·2	7·4	1·0	13·4	1·9
15	2 03·8	2 04·1	1 58·1	1·5	0·2	7·5	1·1	13·5	1·9
16	2 04·0	2 04·3	1 58·4	1·6	0·2	7·6	1·1	13·6	1·9
17	2 04·3	2 04·6	1 58·6	1·7	0·2	7·7	1·1	13·7	1·9
18	2 04·5	2 04·8	1 58·8	1·8	0·3	7·8	1·1	13·8	2·0
19	2 04·8	2 05·1	1 59·1	1·9	0·3	7·9	1·1	13·9	2·0
20	2 05·0	2 05·3	1 59·3	2·0	0·3	8·0	1·1	14·0	2·0
21	2 05·3	2 05·6	1 59·5	2·1	0·3	8·1	1·1	14·1	2·0
22	2 05·5	2 05·8	1 59·8	2·2	0·3	8·2	1·2	14·2	2·0
23	2 05·8	2 06·1	2 00·0	2·3	0·3	8·3	1·2	14·3	2·0
24	2 06·0	2 06·3	2 00·3	2·4	0·3	8·4	1·2	14·4	2·0
25	2 06·3	2 06·6	2 00·5	2·5	0·4	8·5	1·2	14·5	2·1
26	2 06·5	2 06·8	2 00·7	2·6	0·4	8·6	1·2	14·6	2·1
27	2 06·8	2 07·1	2 01·0	2·7	0·4	8·7	1·2	14·7	2·1
28	2 07·0	2 07·3	2 01·2	2·8	0·4	8·8	1·2	14·8	2·1
29	2 07·3	2 07·6	2 01·5	2·9	0·4	8·9	1·3	14·9	2·1
30	2 07·5	2 07·8	2 01·7	3·0	0·4	9·0	1·3	15·0	2·1
31	2 07·8	2 08·1	2 01·9	3·1	0·4	9·1	1·3	15·1	2·1
32	2 08·0	2 08·4	2 02·2	3·2	0·5	9·2	1·3	15·2	2·2
33	2 08·3	2 08·6	2 02·4	3·3	0·5	9·3	1·3	15·3	2·2
34	2 08·5	2 08·9	2 02·6	3·4	0·5	9·4	1·3	15·4	2·2
35	2 08·8	2 09·1	2 02·9	3·5	0·5	9·5	1·3	15·5	2·2
36	2 09·0	2 09·4	2 03·1	3·6	0·5	9·6	1·4	15·6	2·2
37	2 09·3	2 09·6	2 03·4	3·7	0·5	9·7	1·4	15·7	2·2
38	2 09·5	2 09·9	2 03·6	3·8	0·5	9·8	1·4	15·8	2·2
39	2 09·8	2 10·1	2 03·8	3·9	0·6	9·9	1·4	15·9	2·3
40	2 10·0	2 10·4	2 04·1	4·0	0·6	10·0	1·4	16·0	2·3
41	2 10·3	2 10·6	2 04·3	4·1	0·6	10·1	1·4	16·1	2·3
42	2 10·5	2 10·9	2 04·6	4·2	0·6	10·2	1·4	16·2	2·3
43	2 10·8	2 11·1	2 04·8	4·3	0·6	10·3	1·5	16·3	2·3
44	2 11·0	2 11·4	2 05·0	4·4	0·6	10·4	1·5	16·4	2·3
45	2 11·3	2 11·6	2 05·3	4·5	0·6	10·5	1·5	16·5	2·3
46	2 11·5	2 11·9	2 05·5	4·6	0·7	10·6	1·5	16·6	2·4
47	2 11·8	2 12·1	2 05·7	4·7	0·7	10·7	1·5	16·7	2·4
48	2 12·0	2 12·4	2 06·0	4·8	0·7	10·8	1·5	16·8	2·4
49	2 12·3	2 12·6	2 06·2	4·9	0·7	10·9	1·5	16·9	2·4
50	2 12·5	2 12·9	2 06·5	5·0	0·7	11·0	1·6	17·0	2·4
51	2 12·8	2 13·1	2 06·7	5·1	0·7	11·1	1·6	17·1	2·4
52	2 13·0	2 13·4	2 06·9	5·2	0·7	11·2	1·6	17·2	2·4
53	2 13·3	2 13·6	2 07·2	5·3	0·8	11·3	1·6	17·3	2·5
54	2 13·5	2 13·9	2 07·4	5·4	0·8	11·4	1·6	17·4	2·5
55	2 13·8	2 14·1	2 07·7	5·5	0·8	11·5	1·6	17·5	2·5
56	2 14·0	2 14·4	2 07·9	5·6	0·8	11·6	1·6	17·6	2·5
57	2 14·3	2 14·6	2 08·1	5·7	0·8	11·7	1·7	17·7	2·5
58	2 14·5	2 14·9	2 08·4	5·8	0·8	11·8	1·7	17·8	2·5
59	2 14·8	2 15·1	2 08·6	5·9	0·8	11·9	1·7	17·9	2·5
60	2 15·0	2 15·4	2 08·9	6·0	0·9	12·0	1·7	18·0	2·6

m 9 s	SUN PLANETS ° ′	ARIES ° ′	MOON ° ′	v or d ′	Corrⁿ ′	v or d ′	Corrⁿ ′	v or d ′	Corrⁿ ′
00	2 15·0	2 15·4	2 08·9	0·0	0·0	6·0	1·0	12·0	1·9
01	2 15·3	2 15·6	2 09·1	0·1	0·0	6·1	1·0	12·1	1·9
02	2 15·5	2 15·9	2 09·3	0·2	0·0	6·2	1·0	12·2	1·9
03	2 15·8	2 16·1	2 09·6	0·3	0·0	6·3	1·0	12·3	1·9
04	2 16·0	2 16·4	2 09·8	0·4	0·1	6·4	1·0	12·4	2·0
05	2 16·3	2 16·6	2 10·0	0·5	0·1	6·5	1·0	12·5	2·0
06	2 16·5	2 16·9	2 10·3	0·6	0·1	6·6	1·0	12·6	2·0
07	2 16·8	2 17·1	2 10·5	0·7	0·1	6·7	1·1	12·7	2·0
08	2 17·0	2 17·4	2 10·8	0·8	0·1	6·8	1·1	12·8	2·0
09	2 17·3	2 17·6	2 11·0	0·9	0·1	6·9	1·1	12·9	2·0
10	2 17·5	2 17·9	2 11·2	1·0	0·2	7·0	1·1	13·0	2·1
11	2 17·8	2 18·1	2 11·5	1·1	0·2	7·1	1·1	13·1	2·1
12	2 18·0	2 18·4	2 11·7	1·2	0·2	7·2	1·1	13·2	2·1
13	2 18·3	2 18·6	2 12·0	1·3	0·2	7·3	1·2	13·3	2·1
14	2 18·5	2 18·9	2 12·2	1·4	0·2	7·4	1·2	13·4	2·1
15	2 18·8	2 19·1	2 12·4	1·5	0·2	7·5	1·2	13·5	2·1
16	2 19·0	2 19·4	2 12·7	1·6	0·3	7·6	1·2	13·6	2·2
17	2 19·3	2 19·6	2 12·9	1·7	0·3	7·7	1·2	13·7	2·2
18	2 19·5	2 19·9	2 13·1	1·8	0·3	7·8	1·2	13·8	2·2
19	2 19·8	2 20·1	2 13·4	1·9	0·3	7·9	1·3	13·9	2·2
20	2 20·0	2 20·4	2 13·6	2·0	0·3	8·0	1·3	14·0	2·2
21	2 20·3	2 20·6	2 13·9	2·1	0·3	8·1	1·3	14·1	2·2
22	2 20·5	2 20·9	2 14·1	2·2	0·3	8·2	1·3	14·2	2·2
23	2 20·8	2 21·1	2 14·3	2·3	0·4	8·3	1·3	14·3	2·3
24	2 21·0	2 21·4	2 14·6	2·4	0·4	8·4	1·3	14·4	2·3
25	2 21·3	2 21·6	2 14·8	2·5	0·4	8·5	1·3	14·5	2·3
26	2 21·5	2 21·9	2 15·1	2·6	0·4	8·6	1·4	14·6	2·3
27	2 21·8	2 22·1	2 15·3	2·7	0·4	8·7	1·4	14·7	2·3
28	2 22·0	2 22·4	2 15·5	2·8	0·4	8·8	1·4	14·8	2·3
29	2 22·3	2 22·6	2 15·8	2·9	0·5	8·9	1·4	14·9	2·4
30	2 22·5	2 22·9	2 16·0	3·0	0·5	9·0	1·4	15·0	2·4
31	2 22·8	2 23·1	2 16·2	3·1	0·5	9·1	1·4	15·1	2·4
32	2 23·0	2 23·4	2 16·5	3·2	0·5	9·2	1·5	15·2	2·4
33	2 23·3	2 23·6	2 16·7	3·3	0·5	9·3	1·5	15·3	2·4
34	2 23·5	2 23·9	2 17·0	3·4	0·5	9·4	1·5	15·4	2·4
35	2 23·8	2 24·1	2 17·2	3·5	0·6	9·5	1·5	15·5	2·5
36	2 24·0	2 24·4	2 17·4	3·6	0·6	9·6	1·5	15·6	2·5
37	2 24·3	2 24·6	2 17·7	3·7	0·6	9·7	1·5	15·7	2·5
38	2 24·5	2 24·9	2 17·9	3·8	0·6	9·8	1·6	15·8	2·5
39	2 24·8	2 25·1	2 18·2	3·9	0·6	9·9	1·6	15·9	2·5
40	2 25·0	2 25·4	2 18·4	4·0	0·6	10·0	1·6	16·0	2·5
41	2 25·3	2 25·6	2 18·6	4·1	0·6	10·1	1·6	16·1	2·5
42	2 25·5	2 25·9	2 18·9	4·2	0·7	10·2	1·6	16·2	2·6
43	2 25·8	2 26·1	2 19·1	4·3	0·7	10·3	1·6	16·3	2·6
44	2 26·0	2 26·4	2 19·3	4·4	0·7	10·4	1·6	16·4	2·6
45	2 26·3	2 26·7	2 19·6	4·5	0·7	10·5	1·7	16·5	2·6
46	2 26·5	2 26·9	2 19·8	4·6	0·7	10·6	1·7	16·6	2·6
47	2 26·8	2 27·2	2 20·1	4·7	0·7	10·7	1·7	16·7	2·6
48	2 27·0	2 27·4	2 20·3	4·8	0·8	10·8	1·7	16·8	2·7
49	2 27·3	2 27·7	2 20·5	4·9	0·8	10·9	1·7	16·9	2·7
50	2 27·5	2 27·9	2 20·8	5·0	0·8	11·0	1·7	17·0	2·7
51	2 27·8	2 28·2	2 21·0	5·1	0·8	11·1	1·8	17·1	2·7
52	2 28·0	2 28·4	2 21·3	5·2	0·8	11·2	1·8	17·2	2·7
53	2 28·3	2 28·7	2 21·5	5·3	0·8	11·3	1·8	17·3	2·7
54	2 28·5	2 28·9	2 21·7	5·4	0·9	11·4	1·8	17·4	2·8
55	2 28·8	2 29·2	2 22·0	5·5	0·9	11·5	1·8	17·5	2·8
56	2 29·0	2 29·4	2 22·2	5·6	0·9	11·6	1·8	17·6	2·8
57	2 29·3	2 29·7	2 22·5	5·7	0·9	11·7	1·9	17·7	2·8
58	2 29·5	2 29·9	2 22·7	5·8	0·9	11·8	1·9	17·8	2·8
59	2 29·8	2 30·2	2 22·9	5·9	0·9	11·9	1·9	17·9	2·8
60	2 30·0	2 30·4	2 23·2	6·0	1·0	12·0	1·9	18·0	2·9

m 10	SUN PLANETS	ARIES	MOON	v or d	Corr^n	v or d	Corr^n	v or d	Corr^n
s	° ′	° ′	° ′	′	′	′	′	′	′
00	2 30·0	2 30·4	2 23·2	0·0	0·0	6·0	1·1	12·0	2·1
01	2 30·3	2 30·7	2 23·4	0·1	0·0	6·1	1·1	12·1	2·1
02	2 30·5	2 30·9	2 23·6	0·2	0·0	6·2	1·1	12·2	2·1
03	2 30·8	2 31·2	2 23·9	0·3	0·1	6·3	1·1	12·3	2·2
04	2 31·0	2 31·4	2 24·1	0·4	0·1	6·4	1·1	12·4	2·2
05	2 31·3	2 31·7	2 24·4	0·5	0·1	6·5	1·1	12·5	2·2
06	2 31·5	2 31·9	2 24·6	0·6	0·1	6·6	1·2	12·6	2·2
07	2 31·8	2 32·2	2 24·8	0·7	0·1	6·7	1·2	12·7	2·2
08	2 32·0	2 32·4	2 25·1	0·8	0·1	6·8	1·2	12·8	2·2
09	2 32·3	2 32·7	2 25·3	0·9	0·2	6·9	1·2	12·9	2·3
10	2 32·5	2 32·9	2 25·6	1·0	0·2	7·0	1·2	13·0	2·3
11	2 32·8	2 33·2	2 25·8	1·1	0·2	7·1	1·2	13·1	2·3
12	2 33·0	2 33·4	2 26·0	1·2	0·2	7·2	1·3	13·2	2·3
13	2 33·3	2 33·7	2 26·3	1·3	0·2	7·3	1·3	13·3	2·3
14	2 33·5	2 33·9	2 26·5	1·4	0·2	7·4	1·3	13·4	2·3
15	2 33·8	2 34·2	2 26·7	1·5	0·3	7·5	1·3	13·5	2·4
16	2 34·0	2 34·4	2 27·0	1·6	0·3	7·6	1·3	13·6	2·4
17	2 34·3	2 34·7	2 27·2	1·7	0·3	7·7	1·3	13·7	2·4
18	2 34·5	2 34·9	2 27·5	1·8	0·3	7·8	1·4	13·8	2·4
19	2 34·8	2 35·2	2 27·7	1·9	0·3	7·9	1·4	13·9	2·4
20	2 35·0	2 35·4	2 27·9	2·0	0·4	8·0	1·4	14·0	2·5
21	2 35·3	2 35·7	2 28·2	2·1	0·4	8·1	1·4	14·1	2·5
22	2 35·5	2 35·9	2 28·4	2·2	0·4	8·2	1·4	14·2	2·5
23	2 35·8	2 36·2	2 28·7	2·3	0·4	8·3	1·5	14·3	2·5
24	2 36·0	2 36·4	2 28·9	2·4	0·4	8·4	1·5	14·4	2·5
25	2 36·3	2 36·7	2 29·1	2·5	0·4	8·5	1·5	14·5	2·5
26	2 36·5	2 36·9	2 29·4	2·6	0·5	8·6	1·5	14·6	2·6
27	2 36·8	2 37·2	2 29·6	2·7	0·5	8·7	1·5	14·7	2·6
28	2 37·0	2 37·4	2 29·8	2·8	0·5	8·8	1·5	14·8	2·6
29	2 37·3	2 37·7	2 30·1	2·9	0·5	8·9	1·6	14·9	2·6
30	2 37·5	2 37·9	2 30·3	3·0	0·5	9·0	1·6	15·0	2·6
31	2 37·8	2 38·2	2 30·6	3·1	0·5	9·1	1·6	15·1	2·6
32	2 38·0	2 38·4	2 30·8	3·2	0·6	9·2	1·6	15·2	2·7
33	2 38·3	2 38·7	2 31·0	3·3	0·6	9·3	1·6	15·3	2·7
34	2 38·5	2 38·9	2 31·3	3·4	0·6	9·4	1·6	15·4	2·7
35	2 38·8	2 39·2	2 31·5	3·5	0·6	9·5	1·7	15·5	2·7
36	2 39·0	2 39·4	2 31·8	3·6	0·6	9·6	1·7	15·6	2·7
37	2 39·3	2 39·7	2 32·0	3·7	0·6	9·7	1·7	15·7	2·7
38	2 39·5	2 39·9	2 32·2	3·8	0·7	9·8	1·7	15·8	2·8
39	2 39·8	2 40·2	2 32·5	3·9	0·7	9·9	1·7	15·9	2·8
40	2 40·0	2 40·4	2 32·7	4·0	0·7	10·0	1·8	16·0	2·8
41	2 40·3	2 40·7	2 32·9	4·1	0·7	10·1	1·8	16·1	2·8
42	2 40·5	2 40·9	2 33·2	4·2	0·7	10·2	1·8	16·2	2·8
43	2 40·8	2 41·2	2 33·4	4·3	0·8	10·3	1·8	16·3	2·9
44	2 41·0	2 41·4	2 33·7	4·4	0·8	10·4	1·8	16·4	2·9
45	2 41·3	2 41·7	2 33·9	4·5	0·8	10·5	1·8	16·5	2·9
46	2 41·5	2 41·9	2 34·1	4·6	0·8	10·6	1·9	16·6	2·9
47	2 41·8	2 42·2	2 34·4	4·7	0·8	10·7	1·9	16·7	2·9
48	2 42·0	2 42·4	2 34·6	4·8	0·8	10·8	1·9	16·8	2·9
49	2 42·3	2 42·7	2 34·9	4·9	0·9	10·9	1·9	16·9	3·0
50	2 42·5	2 42·9	2 35·1	5·0	0·9	11·0	1·9	17·0	3·0
51	2 42·8	2 43·2	2 35·3	5·1	0·9	11·1	1·9	17·1	3·0
52	2 43·0	2 43·4	2 35·6	5·2	0·9	11·2	2·0	17·2	3·0
53	2 43·3	2 43·7	2 35·8	5·3	0·9	11·3	2·0	17·3	3·0
54	2 43·5	2 43·9	2 36·1	5·4	0·9	11·4	2·0	17·4	3·0
55	2 43·8	2 44·2	2 36·3	5·5	1·0	11·5	2·0	17·5	3·1
56	2 44·0	2 44·4	2 36·5	5·6	1·0	11·6	2·0	17·6	3·1
57	2 44·3	2 44·7	2 36·8	5·7	1·0	11·7	2·0	17·7	3·1
58	2 44·5	2 45·0	2 37·0	5·8	1·0	11·8	2·1	17·8	3·1
59	2 44·8	2 45·2	2 37·2	5·9	1·0	11·9	2·1	17·9	3·1
60	2 45·0	2 45·5	2 37·5	6·0	1·1	12·0	2·1	18·0	3·2

m 11	SUN PLANETS	ARIES	MOON	v or d	Corr^n	v or d	Corr^n	v or d	Corr^n
s	° ′	° ′	° ′	′	′	′	′	′	′
00	2 45·0	2 45·5	2 37·5	0·0	0·0	6·0	1·2	12·0	2·3
01	2 45·3	2 45·7	2 37·7	0·1	0·0	6·1	1·2	12·1	2·3
02	2 45·5	2 46·0	2 38·0	0·2	0·0	6·2	1·2	12·2	2·3
03	2 45·8	2 46·2	2 38·2	0·3	0·1	6·3	1·2	12·3	2·4
04	2 46·0	2 46·5	2 38·4	0·4	0·1	6·4	1·2	12·4	2·4
05	2 46·3	2 46·7	2 38·7	0·5	0·1	6·5	1·2	12·5	2·4
06	2 46·5	2 47·0	2 38·9	0·6	0·1	6·6	1·3	12·6	2·4
07	2 46·8	2 47·2	2 39·2	0·7	0·1	6·7	1·3	12·7	2·4
08	2 47·0	2 47·5	2 39·4	0·8	0·2	6·8	1·3	12·8	2·5
09	2 47·3	2 47·7	2 39·6	0·9	0·2	6·9	1·3	12·9	2·5
10	2 47·5	2 48·0	2 39·9	1·0	0·2	7·0	1·3	13·0	2·5
11	2 47·8	2 48·2	2 40·1	1·1	0·2	7·1	1·4	13·1	2·5
12	2 48·0	2 48·5	2 40·3	1·2	0·2	7·2	1·4	13·2	2·5
13	2 48·3	2 48·7	2 40·6	1·3	0·2	7·3	1·4	13·3	2·5
14	2 48·5	2 49·0	2 40·8	1·4	0·3	7·4	1·4	13·4	2·6
15	2 48·8	2 49·2	2 41·1	1·5	0·3	7·5	1·4	13·5	2·6
16	2 49·0	2 49·5	2 41·3	1·6	0·3	7·6	1·5	13·6	2·6
17	2 49·3	2 49·7	2 41·5	1·7	0·3	7·7	1·5	13·7	2·6
18	2 49·5	2 50·0	2 41·8	1·8	0·3	7·8	1·5	13·8	2·6
19	2 49·8	2 50·2	2 42·0	1·9	0·4	7·9	1·5	13·9	2·7
20	2 50·0	2 50·5	2 42·3	2·0	0·4	8·0	1·5	14·0	2·7
21	2 50·3	2 50·7	2 42·5	2·1	0·4	8·1	1·6	14·1	2·7
22	2 50·5	2 51·0	2 42·7	2·2	0·4	8·2	1·6	14·2	2·7
23	2 50·8	2 51·2	2 43·0	2·3	0·4	8·3	1·6	14·3	2·7
24	2 51·0	2 51·5	2 43·2	2·4	0·5	8·4	1·6	14·4	2·8
25	2 51·3	2 51·7	2 43·4	2·5	0·5	8·5	1·6	14·5	2·8
26	2 51·5	2 52·0	2 43·7	2·6	0·5	8·6	1·6	14·6	2·8
27	2 51·8	2 52·2	2 43·9	2·7	0·5	8·7	1·7	14·7	2·8
28	2 52·0	2 52·5	2 44·2	2·8	0·5	8·8	1·7	14·8	2·8
29	2 52·3	2 52·7	2 44·4	2·9	0·6	8·9	1·7	14·9	2·9
30	2 52·5	2 53·0	2 44·6	3·0	0·6	9·0	1·7	15·0	2·9
31	2 52·8	2 53·2	2 44·9	3·1	0·6	9·1	1·7	15·1	2·9
32	2 53·0	2 53·5	2 45·1	3·2	0·6	9·2	1·8	15·2	2·9
33	2 53·3	2 53·7	2 45·4	3·3	0·6	9·3	1·8	15·3	2·9
34	2 53·5	2 54·0	2 45·6	3·4	0·7	9·4	1·8	15·4	3·0
35	2 53·8	2 54·2	2 45·8	3·5	0·7	9·5	1·8	15·5	3·0
36	2 54·0	2 54·5	2 46·1	3·6	0·7	9·6	1·8	15·6	3·0
37	2 54·3	2 54·7	2 46·3	3·7	0·7	9·7	1·9	15·7	3·0
38	2 54·5	2 55·0	2 46·6	3·8	0·7	9·8	1·9	15·8	3·0
39	2 54·8	2 55·2	2 46·8	3·9	0·7	9·9	1·9	15·9	3·0
40	2 55·0	2 55·5	2 47·0	4·0	0·8	10·0	1·9	16·0	3·1
41	2 55·3	2 55·7	2 47·3	4·1	0·8	10·1	1·9	16·1	3·1
42	2 55·5	2 56·0	2 47·5	4·2	0·8	10·2	2·0	16·2	3·1
43	2 55·8	2 56·2	2 47·7	4·3	0·8	10·3	2·0	16·3	3·1
44	2 56·0	2 56·5	2 48·0	4·4	0·8	10·4	2·0	16·4	3·1
45	2 56·3	2 56·7	2 48·2	4·5	0·9	10·5	2·0	16·5	3·2
46	2 56·5	2 57·0	2 48·5	4·6	0·9	10·6	2·0	16·6	3·2
47	2 56·8	2 57·2	2 48·7	4·7	0·9	10·7	2·1	16·7	3·2
48	2 57·0	2 57·5	2 48·9	4·8	0·9	10·8	2·1	16·8	3·2
49	2 57·3	2 57·7	2 49·2	4·9	0·9	10·9	2·1	16·9	3·2
50	2 57·5	2 58·0	2 49·4	5·0	1·0	11·0	2·1	17·0	3·3
51	2 57·8	2 58·2	2 49·7	5·1	1·0	11·1	2·1	17·1	3·3
52	2 58·0	2 58·5	2 49·9	5·2	1·0	11·2	2·1	17·2	3·3
53	2 58·3	2 58·7	2 50·1	5·3	1·0	11·3	2·2	17·3	3·3
54	2 58·5	2 59·0	2 50·4	5·4	1·0	11·4	2·2	17·4	3·3
55	2 58·8	2 59·2	2 50·6	5·5	1·1	11·5	2·2	17·5	3·4
56	2 59·0	2 59·5	2 50·8	5·6	1·1	11·6	2·2	17·6	3·4
57	2 59·3	2 59·7	2 51·1	5·7	1·1	11·7	2·2	17·7	3·4
58	2 59·5	3 00·0	2 51·3	5·8	1·1	11·8	2·3	17·8	3·4
59	2 59·8	3 00·2	2 51·6	5·9	1·1	11·9	2·3	17·9	3·4
60	3 00·0	3 00·5	2 51·8	6·0	1·2	12·0	2·3	18·0	3·5

12m s	SUN PLANETS ° ′	ARIES ° ′	MOON ° ′	v or d ′	Corrn ′	v or d ′	Corrn ′	v or d ′	Corrn ′
00	3 00·0	3 00·5	2 51·8	0·0	0·0	6·0	1·3	12·0	2·5
01	3 00·3	3 00·7	2 52·0	0·1	0·0	6·1	1·3	12·1	2·5
02	3 00·5	3 01·0	2 52·3	0·2	0·0	6·2	1·3	12·2	2·5
03	3 00·8	3 01·2	2 52·5	0·3	0·1	6·3	1·3	12·3	2·6
04	3 01·0	3 01·5	2 52·8	0·4	0·1	6·4	1·3	12·4	2·6
05	3 01·3	3 01·7	2 53·0	0·5	0·1	6·5	1·4	12·5	2·6
06	3 01·5	3 02·0	2 53·2	0·6	0·1	6·6	1·4	12·6	2·6
07	3 01·8	3 02·2	2 53·5	0·7	0·1	6·7	1·4	12·7	2·6
08	3 02·0	3 02·5	2 53·7	0·8	0·2	6·8	1·4	12·8	2·7
09	3 02·3	3 02·7	2 53·9	0·9	0·2	6·9	1·4	12·9	2·7
10	3 02·5	3 03·0	2 54·2	1·0	0·2	7·0	1·5	13·0	2·7
11	3 02·8	3 03·3	2 54·4	1·1	0·2	7·1	1·5	13·1	2·7
12	3 03·0	3 03·5	2 54·7	1·2	0·3	7·2	1·5	13·2	2·8
13	3 03·3	3 03·8	2 54·9	1·3	0·3	7·3	1·5	13·3	2·8
14	3 03·5	3 04·0	2 55·1	1·4	0·3	7·4	1·5	13·4	2·8
15	3 03·8	3 04·3	2 55·4	1·5	0·3	7·5	1·6	13·5	2·8
16	3 04·0	3 04·5	2 55·6	1·6	0·3	7·6	1·6	13·6	2·8
17	3 04·3	3 04·8	2 55·9	1·7	0·4	7·7	1·6	13·7	2·9
18	3 04·5	3 05·0	2 56·1	1·8	0·4	7·8	1·6	13·8	2·9
19	3 04·8	3 05·3	2 56·3	1·9	0·4	7·9	1·6	13·9	2·9
20	3 05·0	3 05·5	2 56·6	2·0	0·4	8·0	1·7	14·0	2·9
21	3 05·3	3 05·8	2 56·8	2·1	0·4	8·1	1·7	14·1	2·9
22	3 05·5	3 06·0	2 57·0	2·2	0·5	8·2	1·7	14·2	3·0
23	3 05·8	3 06·3	2 57·3	2·3	0·5	8·3	1·7	14·3	3·0
24	3 06·0	3 06·5	2 57·5	2·4	0·5	8·4	1·8	14·4	3·0
25	3 06·3	3 06·8	2 57·8	2·5	0·5	8·5	1·8	14·5	3·0
26	3 06·5	3 07·0	2 58·0	2·6	0·5	8·6	1·8	14·6	3·0
27	3 06·8	3 07·3	2 58·2	2·7	0·6	8·7	1·8	14·7	3·1
28	3 07·0	3 07·5	2 58·5	2·8	0·6	8·8	1·8	14·8	3·1
29	3 07·3	3 07·8	2 58·7	2·9	0·6	8·9	1·9	14·9	3·1
30	3 07·5	3 08·0	2 59·0	3·0	0·6	9·0	1·9	15·0	3·1
31	3 07·8	3 08·3	2 59·2	3·1	0·6	9·1	1·9	15·1	3·1
32	3 08·0	3 08·5	2 59·4	3·2	0·7	9·2	1·9	15·2	3·2
33	3 08·3	3 08·8	2 59·7	3·3	0·7	9·3	1·9	15·3	3·2
34	3 08·5	3 09·0	2 59·9	3·4	0·7	9·4	2·0	15·4	3·2
35	3 08·8	3 09·3	3 00·2	3·5	0·7	9·5	2·0	15·5	3·2
36	3 09·0	3 09·5	3 00·4	3·6	0·8	9·6	2·0	15·6	3·3
37	3 09·3	3 09·8	3 00·6	3·7	0·8	9·7	2·0	15·7	3·3
38	3 09·5	3 10·0	3 00·9	3·8	0·8	9·8	2·0	15·8	3·3
39	3 09·8	3 10·3	3 01·1	3·9	0·8	9·9	2·1	15·9	3·3
40	3 10·0	3 10·5	3 01·3	4·0	0·8	10·0	2·1	16·0	3·3
41	3 10·3	3 10·8	3 01·6	4·1	0·9	10·1	2·1	16·1	3·4
42	3 10·5	3 11·0	3 01·8	4·2	0·9	10·2	2·1	16·2	3·4
43	3 10·8	3 11·3	3 02·1	4·3	0·9	10·3	2·1	16·3	3·4
44	3 11·0	3 11·5	3 02·3	4·4	0·9	10·4	2·2	16·4	3·4
45	3 11·3	3 11·8	3 02·5	4·5	0·9	10·5	2·2	16·5	3·4
46	3 11·5	3 12·0	3 02·8	4·6	1·0	10·6	2·2	16·6	3·5
47	3 11·8	3 12·3	3 03·0	4·7	1·0	10·7	2·2	16·7	3·5
48	3 12·0	3 12·5	3 03·3	4·8	1·0	10·8	2·3	16·8	3·5
49	3 12·3	3 12·8	3 03·5	4·9	1·0	10·9	2·3	16·9	3·5
50	3 12·5	3 13·0	3 03·7	5·0	1·0	11·0	2·3	17·0	3·5
51	3 12·8	3 13·3	3 04·0	5·1	1·1	11·1	2·3	17·1	3·6
52	3 13·0	3 13·5	3 04·2	5·2	1·1	11·2	2·3	17·2	3·6
53	3 13·3	3 13·8	3 04·4	5·3	1·1	11·3	2·4	17·3	3·6
54	3 13·5	3 14·0	3 04·7	5·4	1·1	11·4	2·4	17·4	3·6
55	3 13·8	3 14·3	3 04·9	5·5	1·1	11·5	2·4	17·5	3·6
56	3 14·0	3 14·5	3 05·2	5·6	1·2	11·6	2·4	17·6	3·7
57	3 14·3	3 14·8	3 05·4	5·7	1·2	11·7	2·4	17·7	3·7
58	3 14·5	3 15·0	3 05·6	5·8	1·2	11·8	2·5	17·8	3·7
59	3 14·8	3 15·3	3 05·9	5·9	1·2	11·9	2·5	17·9	3·7
60	3 15·0	3 15·5	3 06·1	6·0	1·3	12·0	2·5	18·0	3·8

13m s	SUN PLANETS ° ′	ARIES ° ′	MOON ° ′	v or d ′	Corrn ′	v or d ′	Corrn ′	v or d ′	Corrn ′
00	3 15·0	3 15·5	3 06·1	0·0	0·0	6·0	1·4	12·0	2·7
01	3 15·3	3 15·8	3 06·4	0·1	0·0	6·1	1·4	12·1	2·7
02	3 15·5	3 16·0	3 06·6	0·2	0·0	6·2	1·4	12·2	2·7
03	3 15·8	3 16·3	3 06·8	0·3	0·1	6·3	1·4	12·3	2·8
04	3 16·0	3 16·5	3 07·1	0·4	0·1	6·4	1·4	12·4	2·8
05	3 16·3	3 16·8	3 07·3	0·5	0·1	6·5	1·5	12·5	2·8
06	3 16·5	3 17·0	3 07·5	0·6	0·1	6·6	1·5	12·6	2·8
07	3 16·8	3 17·3	3 07·8	0·7	0·2	6·7	1·5	12·7	2·9
08	3 17·0	3 17·5	3 08·0	0·8	0·2	6·8	1·5	12·8	2·9
09	3 17·3	3 17·8	3 08·3	0·9	0·2	6·9	1·6	12·9	2·9
10	3 17·5	3 18·0	3 08·5	1·0	0·2	7·0	1·6	13·0	2·9
11	3 17·8	3 18·3	3 08·7	1·1	0·2	7·1	1·6	13·1	2·9
12	3 18·0	3 18·5	3 09·0	1·2	0·3	7·2	1·6	13·2	3·0
13	3 18·3	3 18·8	3 09·2	1·3	0·3	7·3	1·6	13·3	3·0
14	3 18·5	3 19·0	3 09·5	1·4	0·3	7·4	1·7	13·4	3·0
15	3 18·8	3 19·3	3 09·7	1·5	0·3	7·5	1·7	13·5	3·0
16	3 19·0	3 19·5	3 09·9	1·6	0·4	7·6	1·7	13·6	3·1
17	3 19·3	3 19·8	3 10·2	1·7	0·4	7·7	1·7	13·7	3·1
18	3 19·5	3 20·0	3 10·4	1·8	0·4	7·8	1·8	13·8	3·1
19	3 19·8	3 20·3	3 10·7	1·9	0·4	7·9	1·8	13·9	3·1
20	3 20·0	3 20·5	3 10·9	2·0	0·5	8·0	1·8	14·0	3·2
21	3 20·3	3 20·8	3 11·1	2·1	0·5	8·1	1·8	14·1	3·2
22	3 20·5	3 21·0	3 11·4	2·2	0·5	8·2	1·8	14·2	3·2
23	3 20·8	3 21·3	3 11·6	2·3	0·5	8·3	1·9	14·3	3·2
24	3 21·0	3 21·6	3 11·8	2·4	0·5	8·4	1·9	14·4	3·2
25	3 21·3	3 21·8	3 12·1	2·5	0·6	8·5	1·9	14·5	3·3
26	3 21·5	3 22·1	3 12·3	2·6	0·6	8·6	1·9	14·6	3·3
27	3 21·8	3 22·3	3 12·6	2·7	0·6	8·7	2·0	14·7	3·3
28	3 22·0	3 22·6	3 12·8	2·8	0·6	8·8	2·0	14·8	3·3
29	3 22·3	3 22·8	3 13·0	2·9	0·7	8·9	2·0	14·9	3·4
30	3 22·5	3 23·1	3 13·3	3·0	0·7	9·0	2·0	15·0	3·4
31	3 22·8	3 23·3	3 13·5	3·1	0·7	9·1	2·0	15·1	3·4
32	3 23·0	3 23·6	3 13·8	3·2	0·7	9·2	2·1	15·2	3·4
33	3 23·3	3 23·8	3 14·0	3·3	0·7	9·3	2·1	15·3	3·4
34	3 23·5	3 24·1	3 14·2	3·4	0·8	9·4	2·1	15·4	3·5
35	3 23·8	3 24·3	3 14·5	3·5	0·8	9·5	2·1	15·5	3·5
36	3 24·0	3 24·6	3 14·7	3·6	0·8	9·6	2·2	15·6	3·5
37	3 24·3	3 24·8	3 14·9	3·7	0·8	9·7	2·2	15·7	3·5
38	3 24·5	3 25·1	3 15·2	3·8	0·9	9·8	2·2	15·8	3·6
39	3 24·8	3 25·3	3 15·4	3·9	0·9	9·9	2·2	15·9	3·6
40	3 25·0	3 25·6	3 15·7	4·0	0·9	10·0	2·3	16·0	3·6
41	3 25·3	3 25·8	3 15·9	4·1	0·9	10·1	2·3	16·1	3·6
42	3 25·5	3 26·1	3 16·1	4·2	0·9	10·2	2·3	16·2	3·6
43	3 25·8	3 26·3	3 16·4	4·3	1·0	10·3	2·3	16·3	3·7
44	3 26·0	3 26·6	3 16·6	4·4	1·0	10·4	2·3	16·4	3·7
45	3 26·3	3 26·8	3 16·9	4·5	1·0	10·5	2·4	16·5	3·7
46	3 26·5	3 27·1	3 17·1	4·6	1·0	10·6	2·4	16·6	3·7
47	3 26·8	3 27·3	3 17·3	4·7	1·1	10·7	2·4	16·7	3·8
48	3 27·0	3 27·6	3 17·6	4·8	1·1	10·8	2·4	16·8	3·8
49	3 27·3	3 27·8	3 17·8	4·9	1·1	10·9	2·5	16·9	3·8
50	3 27·5	3 28·1	3 18·0	5·0	1·1	11·0	2·5	17·0	3·8
51	3 27·8	3 28·3	3 18·3	5·1	1·1	11·1	2·5	17·1	3·8
52	3 28·0	3 28·6	3 18·5	5·2	1·2	11·2	2·5	17·2	3·9
53	3 28·3	3 28·8	3 18·8	5·3	1·2	11·3	2·5	17·3	3·9
54	3 28·5	3 29·1	3 19·0	5·4	1·2	11·4	2·6	17·4	3·9
55	3 28·8	3 29·3	3 19·2	5·5	1·2	11·5	2·6	17·5	3·9
56	3 29·0	3 29·6	3 19·5	5·6	1·3	11·6	2·6	17·6	4·0
57	3 29·3	3 29·8	3 19·7	5·7	1·3	11·7	2·6	17·7	4·0
58	3 29·5	3 30·1	3 20·0	5·8	1·3	11·8	2·7	17·8	4·0
59	3 29·8	3 30·3	3 20·2	5·9	1·3	11·9	2·7	17·9	4·0
60	3 30·0	3 30·6	3 20·4	6·0	1·4	12·0	2·7	18·0	4·1

m 14 s	SUN PLANETS ° ′	ARIES ° ′	MOON ° ′	v or d ′	Corrn ′	v or d ′	Corrn ′	v or d ′	Corrn ′
00	3 30·0	3 30·6	3 20·4	0·0	0·0	6·0	1·5	12·0	2·9
01	3 30·3	3 30·8	3 20·7	0·1	0·0	6·1	1·5	12·1	2·9
02	3 30·5	3 31·1	3 20·9	0·2	0·0	6·2	1·5	12·2	2·9
03	3 30·8	3 31·3	3 21·1	0·3	0·1	6·3	1·5	12·3	3·0
04	3 31·0	3 31·6	3 21·4	0·4	0·1	6·4	1·5	12·4	3·0
05	3 31·3	3 31·8	3 21·6	0·5	0·1	6·5	1·6	12·5	3·0
06	3 31·5	3 32·1	3 21·9	0·6	0·1	6·6	1·6	12·6	3·0
07	3 31·8	3 32·3	3 22·1	0·7	0·2	6·7	1·6	12·7	3·1
08	3 32·0	3 32·6	3 22·3	0·8	0·2	6·8	1·6	12·8	3·1
09	3 32·3	3 32·8	3 22·6	0·9	0·2	6·9	1·7	12·9	3·1
10	3 32·5	3 33·1	3 22·8	1·0	0·2	7·0	1·7	13·0	3·1
11	3 32·8	3 33·3	3 23·1	1·1	0·3	7·1	1·7	13·1	3·2
12	3 33·0	3 33·6	3 23·3	1·2	0·3	7·2	1·7	13·2	3·2
13	3 33·3	3 33·8	3 23·5	1·3	0·3	7·3	1·8	13·3	3·2
14	3 33·5	3 34·1	3 23·8	1·4	0·3	7·4	1·8	13·4	3·2
15	3 33·8	3 34·3	3 24·0	1·5	0·4	7·5	1·8	13·5	3·3
16	3 34·0	3 34·6	3 24·3	1·6	0·4	7·6	1·8	13·6	3·3
17	3 34·3	3 34·8	3 24·5	1·7	0·4	7·7	1·9	13·7	3·3
18	3 34·5	3 35·1	3 24·7	1·8	0·4	7·8	1·9	13·8	3·3
19	3 34·8	3 35·3	3 25·0	1·9	0·5	7·9	1·9	13·9	3·4
20	3 35·0	3 35·6	3 25·2	2·0	0·5	8·0	1·9	14·0	3·4
21	3 35·3	3 35·8	3 25·4	2·1	0·5	8·1	2·0	14·1	3·4
22	3 35·5	3 36·1	3 25·7	2·2	0·5	8·2	2·0	14·2	3·4
23	3 35·8	3 36·3	3 25·9	2·3	0·6	8·3	2·0	14·3	3·5
24	3 36·0	3 36·6	3 26·2	2·4	0·6	8·4	2·0	14·4	3·5
25	3 36·3	3 36·8	3 26·4	2·5	0·6	8·5	2·1	14·5	3·5
26	3 36·5	3 37·1	3 26·6	2·6	0·6	8·6	2·1	14·6	3·5
27	3 36·8	3 37·3	3 26·9	2·7	0·7	8·7	2·1	14·7	3·6
28	3 37·0	3 37·6	3 27·1	2·8	0·7	8·8	2·1	14·8	3·6
29	3 37·3	3 37·8	3 27·4	2·9	0·7	8·9	2·2	14·9	3·6
30	3 37·5	3 38·1	3 27·6	3·0	0·7	9·0	2·2	15·0	3·6
31	3 37·8	3 38·3	3 27·8	3·1	0·7	9·1	2·2	15·1	3·6
32	3 38·0	3 38·6	3 28·1	3·2	0·8	9·2	2·2	15·2	3·7
33	3 38·3	3 38·8	3 28·3	3·3	0·8	9·3	2·2	15·3	3·7
34	3 38·5	3 39·1	3 28·5	3·4	0·8	9·4	2·3	15·4	3·7
35	3 38·8	3 39·3	3 28·8	3·5	0·8	9·5	2·3	15·5	3·7
36	3 39·0	3 39·6	3 29·0	3·6	0·9	9·6	2·3	15·6	3·8
37	3 39·3	3 39·9	3 29·3	3·7	0·9	9·7	2·3	15·7	3·8
38	3 39·5	3 40·1	3 29·5	3·8	0·9	9·8	2·4	15·8	3·8
39	3 39·8	3 40·4	3 29·7	3·9	0·9	9·9	2·4	15·9	3·8
40	3 40·0	3 40·6	3 30·0	4·0	1·0	10·0	2·4	16·0	3·9
41	3 40·3	3 40·9	3 30·2	4·1	1·0	10·1	2·4	16·1	3·9
42	3 40·5	3 41·1	3 30·5	4·2	1·0	10·2	2·5	16·2	3·9
43	3 40·8	3 41·4	3 30·7	4·3	1·0	10·3	2·5	16·3	3·9
44	3 41·0	3 41·6	3 30·9	4·4	1·1	10·4	2·5	16·4	4·0
45	3 41·3	3 41·9	3 31·2	4·5	1·1	10·5	2·5	16·5	4·0
46	3 41·5	3 42·1	3 31·4	4·6	1·1	10·6	2·6	16·6	4·0
47	3 41·8	3 42·4	3 31·6	4·7	1·1	10·7	2·6	16·7	4·0
48	3 42·0	3 42·6	3 31·9	4·8	1·2	10·8	2·6	16·8	4·1
49	3 42·3	3 42·9	3 32·1	4·9	1·2	10·9	2·6	16·9	4·1
50	3 42·5	3 43·1	3 32·4	5·0	1·2	11·0	2·7	17·0	4·1
51	3 42·8	3 43·4	3 32·6	5·1	1·2	11·1	2·7	17·1	4·1
52	3 43·0	3 43·6	3 32·8	5·2	1·3	11·2	2·7	17·2	4·2
53	3 43·3	3 43·9	3 33·1	5·3	1·3	11·3	2·7	17·3	4·2
54	3 43·5	3 44·1	3 33·3	5·4	1·3	11·4	2·8	17·4	4·2
55	3 43·8	3 44·4	3 33·6	5·5	1·3	11·5	2·8	17·5	4·2
56	3 44·0	3 44·6	3 33·8	5·6	1·4	11·6	2·8	17·6	4·3
57	3 44·3	3 44·9	3 34·0	5·7	1·4	11·7	2·8	17·7	4·3
58	3 44·5	3 45·1	3 34·3	5·8	1·4	11·8	2·9	17·8	4·3
59	3 44·8	3 45·4	3 34·5	5·9	1·4	11·9	2·9	17·9	4·3
60	3 45·0	3 45·6	3 34·8	6·0	1·5	12·0	2·9	18·0	4·4

m 15 s	SUN PLANETS ° ′	ARIES ° ′	MOON ° ′	v or d ′	Corrn ′	v or d ′	Corrn ′	v or d ′	Corrn ′
00	3 45·0	3 45·6	3 34·8	0·0	0·0	6·0	1·6	12·0	3·1
01	3 45·3	3 45·9	3 35·0	0·1	0·0	6·1	1·6	12·1	3·1
02	3 45·5	3 46·1	3 35·2	0·2	0·1	6·2	1·6	12·2	3·2
03	3 45·8	3 46·4	3 35·5	0·3	0·1	6·3	1·6	12·3	3·2
04	3 46·0	3 46·6	3 35·7	0·4	0·1	6·4	1·7	12·4	3·2
05	3 46·3	3 46·9	3 35·9	0·5	0·1	6·5	1·7	12·5	3·2
06	3 46·5	3 47·1	3 36·2	0·6	0·2	6·6	1·7	12·6	3·3
07	3 46·8	3 47·4	3 36·4	0·7	0·2	6·7	1·7	12·7	3·3
08	3 47·0	3 47·6	3 36·7	0·8	0·2	6·8	1·8	12·8	3·3
09	3 47·3	3 47·9	3 36·9	0·9	0·2	6·9	1·8	12·9	3·3
10	3 47·5	3 48·1	3 37·1	1·0	0·3	7·0	1·8	13·0	3·4
11	3 47·8	3 48·4	3 37·4	1·1	0·3	7·1	1·8	13·1	3·4
12	3 48·0	3 48·6	3 37·6	1·2	0·3	7·2	1·9	13·2	3·4
13	3 48·3	3 48·9	3 37·9	1·3	0·3	7·3	1·9	13·3	3·4
14	3 48·5	3 49·1	3 38·1	1·4	0·4	7·4	1·9	13·4	3·5
15	3 48·8	3 49·4	3 38·3	1·5	0·4	7·5	1·9	13·5	3·5
16	3 49·0	3 49·6	3 38·6	1·6	0·4	7·6	2·0	13·6	3·5
17	3 49·3	3 49·9	3 38·8	1·7	0·4	7·7	2·0	13·7	3·5
18	3 49·5	3 50·1	3 39·0	1·8	0·5	7·8	2·0	13·8	3·6
19	3 49·8	3 50·4	3 39·3	1·9	0·5	7·9	2·0	13·9	3·6
20	3 50·0	3 50·6	3 39·5	2·0	0·5	8·0	2·1	14·0	3·6
21	3 50·3	3 50·9	3 39·8	2·1	0·5	8·1	2·1	14·1	3·6
22	3 50·5	3 51·1	3 40·0	2·2	0·6	8·2	2·1	14·2	3·7
23	3 50·8	3 51·4	3 40·2	2·3	0·6	8·3	2·1	14·3	3·7
24	3 51·0	3 51·6	3 40·5	2·4	0·6	8·4	2·2	14·4	3·7
25	3 51·3	3 51·9	3 40·7	2·5	0·6	8·5	2·2	14·5	3·7
26	3 51·5	3 52·1	3 41·0	2·6	0·7	8·6	2·2	14·6	3·8
27	3 51·8	3 52·4	3 41·2	2·7	0·7	8·7	2·2	14·7	3·8
28	3 52·0	3 52·6	3 41·4	2·8	0·7	8·8	2·3	14·8	3·8
29	3 52·3	3 52·9	3 41·7	2·9	0·7	8·9	2·3	14·9	3·8
30	3 52·5	3 53·1	3 41·9	3·0	0·8	9·0	2·3	15·0	3·9
31	3 52·8	3 53·4	3 42·1	3·1	0·8	9·1	2·4	15·1	3·9
32	3 53·0	3 53·6	3 42·4	3·2	0·8	9·2	2·4	15·2	3·9
33	3 53·3	3 53·9	3 42·6	3·3	0·9	9·3	2·4	15·3	4·0
34	3 53·5	3 54·1	3 42·9	3·4	0·9	9·4	2·4	15·4	4·0
35	3 53·8	3 54·4	3 43·1	3·5	0·9	9·5	2·5	15·5	4·0
36	3 54·0	3 54·6	3 43·3	3·6	0·9	9·6	2·5	15·6	4·0
37	3 54·3	3 54·9	3 43·6	3·7	1·0	9·7	2·5	15·7	4·1
38	3 54·5	3 55·1	3 43·8	3·8	1·0	9·8	2·5	15·8	4·1
39	3 54·8	3 55·4	3 44·1	3·9	1·0	9·9	2·6	15·9	4·1
40	3 55·0	3 55·6	3 44·3	4·0	1·0	10·0	2·6	16·0	4·1
41	3 55·3	3 55·9	3 44·5	4·1	1·1	10·1	2·6	16·1	4·2
42	3 55·5	3 56·1	3 44·8	4·2	1·1	10·2	2·6	16·2	4·2
43	3 55·8	3 56·4	3 45·0	4·3	1·1	10·3	2·7	16·3	4·2
44	3 56·0	3 56·6	3 45·2	4·4	1·1	10·4	2·7	16·4	4·2
45	3 56·3	3 56·9	3 45·5	4·5	1·2	10·5	2·7	16·5	4·3
46	3 56·5	3 57·1	3 45·7	4·6	1·2	10·6	2·7	16·6	4·3
47	3 56·8	3 57·4	3 46·0	4·7	1·2	10·7	2·8	16·7	4·3
48	3 57·0	3 57·6	3 46·2	4·8	1·2	10·8	2·8	16·8	4·3
49	3 57·3	3 57·9	3 46·4	4·9	1·3	10·9	2·8	16·9	4·4
50	3 57·5	3 58·2	3 46·7	5·0	1·3	11·0	2·8	17·0	4·4
51	3 57·8	3 58·4	3 46·9	5·1	1·3	11·1	2·9	17·1	4·4
52	3 58·0	3 58·7	3 47·2	5·2	1·3	11·2	2·9	17·2	4·4
53	3 58·3	3 58·9	3 47·4	5·3	1·4	11·3	2·9	17·3	4·5
54	3 58·5	3 59·2	3 47·6	5·4	1·4	11·4	2·9	17·4	4·5
55	3 58·8	3 59·4	3 47·9	5·5	1·4	11·5	3·0	17·5	4·5
56	3 59·0	3 59·7	3 48·1	5·6	1·4	11·6	3·0	17·6	4·5
57	3 59·3	3 59·9	3 48·4	5·7	1·5	11·7	3·0	17·7	4·6
58	3 59·5	4 00·2	3 48·6	5·8	1·5	11·8	3·0	17·8	4·6
59	3 59·8	4 00·4	3 48·8	5·9	1·5	11·9	3·1	17·9	4·6
60	4 00·0	4 00·7	3 49·1	6·0	1·6	12·0	3·1	18·0	4·7

16^m	SUN PLANETS	ARIES	MOON	v or d	Corr^n	v or d	Corr^n	v or d	Corr^n
s	° ′	° ′	° ′	′	′	′	′	′	′
00	4 00·0	4 00·7	3 49·1	0·0	0·0	6·0	1·7	12·0	3·3
01	4 00·3	4 00·9	3 49·3	0·1	0·0	6·1	1·7	12·1	3·3
02	4 00·5	4 01·2	3 49·5	0·2	0·1	6·2	1·7	12·2	3·4
03	4 00·8	4 01·4	3 49·8	0·3	0·1	6·3	1·7	12·3	3·4
04	4 01·0	4 01·7	3 50·0	0·4	0·1	6·4	1·8	12·4	3·4
05	4 01·3	4 01·9	3 50·3	0·5	0·1	6·5	1·8	12·5	3·4
06	4 01·5	4 02·2	3 50·5	0·6	0·2	6·6	1·8	12·6	3·5
07	4 01·8	4 02·4	3 50·7	0·7	0·2	6·7	1·8	12·7	3·5
08	4 02·0	4 02·7	3 51·0	0·8	0·2	6·8	1·9	12·8	3·5
09	4 02·3	4 02·9	3 51·2	0·9	0·2	6·9	1·9	12·9	3·5
10	4 02·5	4 03·2	3 51·5	1·0	0·3	7·0	1·9	13·0	3·6
11	4 02·8	4 03·4	3 51·7	1·1	0·3	7·1	2·0	13·1	3·6
12	4 03·0	4 03·7	3 51·9	1·2	0·3	7·2	2·0	13·2	3·6
13	4 03·3	4 03·9	3 52·2	1·3	0·4	7·3	2·0	13·3	3·7
14	4 03·5	4 04·2	3 52·4	1·4	0·4	7·4	2·0	13·4	3·7
15	4 03·8	4 04·4	3 52·6	1·5	0·4	7·5	2·1	13·5	3·7
16	4 04·0	4 04·7	3 52·9	1·6	0·4	7·6	2·1	13·6	3·7
17	4 04·3	4 04·9	3 53·1	1·7	0·5	7·7	2·1	13·7	3·8
18	4 04·5	4 05·2	3 53·4	1·8	0·5	7·8	2·1	13·8	3·8
19	4 04·8	4 05·4	3 53·6	1·9	0·5	7·9	2·2	13·9	3·8
20	4 05·0	4 05·7	3 53·8	2·0	0·6	8·0	2·2	14·0	3·9
21	4 05·3	4 05·9	3 54·1	2·1	0·6	8·1	2·2	14·1	3·9
22	4 05·5	4 06·2	3 54·3	2·2	0·6	8·2	2·3	14·2	3·9
23	4 05·8	4 06·4	3 54·6	2·3	0·6	8·3	2·3	14·3	3·9
24	4 06·0	4 06·7	3 54·8	2·4	0·7	8·4	2·3	14·4	4·0
25	4 06·3	4 06·9	3 55·0	2·5	0·7	8·5	2·3	14·5	4·0
26	4 06·5	4 07·2	3 55·3	2·6	0·7	8·6	2·4	14·6	4·0
27	4 06·8	4 07·4	3 55·5	2·7	0·7	8·7	2·4	14·7	4·0
28	4 07·0	4 07·7	3 55·7	2·8	0·8	8·8	2·4	14·8	4·1
29	4 07·3	4 07·9	3 56·0	2·9	0·8	8·9	2·4	14·9	4·1
30	4 07·5	4 08·2	3 56·2	3·0	0·8	9·0	2·5	15·0	4·1
31	4 07·8	4 08·4	3 56·5	3·1	0·9	9·1	2·5	15·1	4·2
32	4 08·0	4 08·7	3 56·7	3·2	0·9	9·2	2·5	15·2	4·2
33	4 08·3	4 08·9	3 56·9	3·3	0·9	9·3	2·6	15·3	4·2
34	4 08·5	4 09·2	3 57·2	3·4	0·9	9·4	2·6	15·4	4·2
35	4 08·8	4 09·4	3 57·4	3·5	1·0	9·5	2·6	15·5	4·3
36	4 09·0	4 09·7	3 57·7	3·6	1·0	9·6	2·6	15·6	4·3
37	4 09·3	4 09·9	3 57·9	3·7	1·0	9·7	2·7	15·7	4·3
38	4 09·5	4 10·2	3 58·1	3·8	1·0	9·8	2·7	15·8	4·3
39	4 09·8	4 10·4	3 58·4	3·9	1·1	9·9	2·7	15·9	4·4
40	4 10·0	4 10·7	3 58·6	4·0	1·1	10·0	2·8	16·0	4·4
41	4 10·3	4 10·9	3 58·8	4·1	1·1	10·1	2·8	16·1	4·4
42	4 10·5	4 11·2	3 59·1	4·2	1·2	10·2	2·8	16·2	4·5
43	4 10·8	4 11·4	3 59·3	4·3	1·2	10·3	2·8	16·3	4·5
44	4 11·0	4 11·7	3 59·6	4·4	1·2	10·4	2·9	16·4	4·5
45	4 11·3	4 11·9	3 59·8	4·5	1·2	10·5	2·9	16·5	4·5
46	4 11·5	4 12·2	4 00·0	4·6	1·3	10·6	2·9	16·6	4·6
47	4 11·8	4 12·4	4 00·3	4·7	1·3	10·7	2·9	16·7	4·6
48	4 12·0	4 12·7	4 00·5	4·8	1·3	10·8	3·0	16·8	4·6
49	4 12·3	4 12·9	4 00·8	4·9	1·3	10·9	3·0	16·9	4·6
50	4 12·5	4 13·2	4 01·0	5·0	1·4	11·0	3·0	17·0	4·7
51	4 12·8	4 13·4	4 01·2	5·1	1·4	11·1	3·1	17·1	4·7
52	4 13·0	4 13·7	4 01·5	5·2	1·4	11·2	3·1	17·2	4·7
53	4 13·3	4 13·9	4 01·7	5·3	1·5	11·3	3·1	17·3	4·8
54	4 13·5	4 14·2	4 02·0	5·4	1·5	11·4	3·1	17·4	4·8
55	4 13·8	4 14·4	4 02·2	5·5	1·5	11·5	3·2	17·5	4·8
56	4 14·0	4 14·7	4 02·4	5·6	1·5	11·6	3·2	17·6	4·8
57	4 14·3	4 14·9	4 02·7	5·7	1·6	11·7	3·2	17·7	4·9
58	4 14·5	4 15·2	4 02·9	5·8	1·6	11·8	3·2	17·8	4·9
59	4 14·8	4 15·4	4 03·1	5·9	1·6	11·9	3·3	17·9	4·9
60	4 15·0	4 15·7	4 03·4	6·0	1·7	12·0	3·3	18·0	5·0

17^m	SUN PLANETS	ARIES	MOON	v or d	Corr^n	v or d	Corr^n	v or d	Corr^n
s	° ′	° ′	° ′	′	′	′	′	′	′
00	4 15·0	4 15·7	4 03·4	0·0	0·0	6·0	1·8	12·0	3·5
01	4 15·3	4 15·9	4 03·6	0·1	0·0	6·1	1·8	12·1	3·5
02	4 15·5	4 16·2	4 03·9	0·2	0·1	6·2	1·8	12·2	3·6
03	4 15·8	4 16·5	4 04·1	0·3	0·1	6·3	1·8	12·3	3·6
04	4 16·0	4 16·7	4 04·3	0·4	0·1	6·4	1·9	12·4	3·6
05	4 16·3	4 17·0	4 04·6	0·5	0·1	6·5	1·9	12·5	3·6
06	4 16·5	4 17·2	4 04·8	0·6	0·2	6·6	1·9	12·6	3·7
07	4 16·8	4 17·5	4 05·1	0·7	0·2	6·7	2·0	12·7	3·7
08	4 17·0	4 17·7	4 05·3	0·8	0·2	6·8	2·0	12·8	3·7
09	4 17·3	4 18·0	4 05·5	0·9	0·3	6·9	2·0	12·9	3·8
10	4 17·5	4 18·2	4 05·8	1·0	0·3	7·0	2·0	13·0	3·8
11	4 17·8	4 18·5	4 06·0	1·1	0·3	7·1	2·1	13·1	3·8
12	4 18·0	4 18·7	4 06·2	1·2	0·4	7·2	2·1	13·2	3·9
13	4 18·3	4 19·0	4 06·5	1·3	0·4	7·3	2·1	13·3	3·9
14	4 18·5	4 19·2	4 06·7	1·4	0·4	7·4	2·2	13·4	3·9
15	4 18·8	4 19·5	4 07·0	1·5	0·4	7·5	2·2	13·5	3·9
16	4 19·0	4 19·7	4 07·2	1·6	0·5	7·6	2·2	13·6	4·0
17	4 19·3	4 20·0	4 07·4	1·7	0·5	7·7	2·2	13·7	4·0
18	4 19·5	4 20·2	4 07·7	1·8	0·5	7·8	2·3	13·8	4·0
19	4 19·8	4 20·5	4 07·9	1·9	0·6	7·9	2·3	13·9	4·1
20	4 20·0	4 20·7	4 08·2	2·0	0·6	8·0	2·3	14·0	4·1
21	4 20·3	4 21·0	4 08·4	2·1	0·6	8·1	2·4	14·1	4·1
22	4 20·5	4 21·2	4 08·6	2·2	0·6	8·2	2·4	14·2	4·1
23	4 20·8	4 21·5	4 08·9	2·3	0·7	8·3	2·4	14·3	4·2
24	4 21·0	4 21·7	4 09·1	2·4	0·7	8·4	2·5	14·4	4·2
25	4 21·3	4 22·0	4 09·3	2·5	0·7	8·5	2·5	14·5	4·2
26	4 21·5	4 22·2	4 09·6	2·6	0·8	8·6	2·5	14·6	4·3
27	4 21·8	4 22·5	4 09·8	2·7	0·8	8·7	2·5	14·7	4·3
28	4 22·0	4 22·7	4 10·1	2·8	0·8	8·8	2·6	14·8	4·3
29	4 22·3	4 23·0	4 10·3	2·9	0·8	8·9	2·6	14·9	4·3
30	4 22·5	4 23·2	4 10·5	3·0	0·9	9·0	2·6	15·0	4·4
31	4 22·8	4 23·5	4 10·8	3·1	0·9	9·1	2·7	15·1	4·4
32	4 23·0	4 23·7	4 11·0	3·2	0·9	9·2	2·7	15·2	4·4
33	4 23·3	4 24·0	4 11·3	3·3	1·0	9·3	2·7	15·3	4·5
34	4 23·5	4 24·2	4 11·5	3·4	1·0	9·4	2·7	15·4	4·5
35	4 23·8	4 24·5	4 11·7	3·5	1·0	9·5	2·8	15·5	4·5
36	4 24·0	4 24·7	4 12·0	3·6	1·1	9·6	2·8	15·6	4·6
37	4 24·3	4 25·0	4 12·2	3·7	1·1	9·7	2·8	15·7	4·6
38	4 24·5	4 25·2	4 12·5	3·8	1·1	9·8	2·9	15·8	4·6
39	4 24·8	4 25·5	4 12·7	3·9	1·1	9·9	2·9	15·9	4·6
40	4 25·0	4 25·7	4 12·9	4·0	1·2	10·0	2·9	16·0	4·7
41	4 25·3	4 26·0	4 13·2	4·1	1·2	10·1	2·9	16·1	4·7
42	4 25·5	4 26·2	4 13·4	4·2	1·2	10·2	3·0	16·2	4·7
43	4 25·8	4 26·5	4 13·6	4·3	1·3	10·3	3·0	16·3	4·8
44	4 26·0	4 26·7	4 13·9	4·4	1·3	10·4	3·0	16·4	4·8
45	4 26·3	4 27·0	4 14·1	4·5	1·3	10·5	3·1	16·5	4·8
46	4 26·5	4 27·2	4 14·4	4·6	1·3	10·6	3·1	16·6	4·8
47	4 26·8	4 27·5	4 14·6	4·7	1·4	10·7	3·1	16·7	4·9
48	4 27·0	4 27·7	4 14·8	4·8	1·4	10·8	3·2	16·8	4·9
49	4 27·3	4 28·0	4 15·1	4·9	1·4	10·9	3·2	16·9	4·9
50	4 27·5	4 28·2	4 15·3	5·0	1·5	11·0	3·2	17·0	5·0
51	4 27·8	4 28·5	4 15·6	5·1	1·5	11·1	3·2	17·1	5·0
52	4 28·0	4 28·7	4 15·8	5·2	1·5	11·2	3·3	17·2	5·0
53	4 28·3	4 29·0	4 16·0	5·3	1·5	11·3	3·3	17·3	5·0
54	4 28·5	4 29·2	4 16·3	5·4	1·6	11·4	3·3	17·4	5·1
55	4 28·8	4 29·5	4 16·5	5·5	1·6	11·5	3·4	17·5	5·1
56	4 29·0	4 29·7	4 16·7	5·6	1·6	11·6	3·4	17·6	5·1
57	4 29·3	4 30·0	4 17·0	5·7	1·7	11·7	3·4	17·7	5·2
58	4 29·5	4 30·2	4 17·2	5·8	1·7	11·8	3·4	17·8	5·2
59	4 29·8	4 30·5	4 17·5	5·9	1·7	11·9	3·5	17·9	5·2
60	4 30·0	4 30·7	4 17·7	6·0	1·8	12·0	3·5	18·0	5·3

18^m	SUN PLANETS	ARIES	MOON	v or d	Corr^n	v or d	Corr^n	v or d	Corr^n
s	° ′	° ′	° ′	′	′	′	′	′	′
00	4 30·0	4 30·7	4 17·7	0·0	0·0	6·0	1·9	12·0	3·7
01	4 30·3	4 31·0	4 17·9	0·1	0·0	6·1	1·9	12·1	3·7
02	4 30·5	4 31·2	4 18·2	0·2	0·1	6·2	1·9	12·2	3·8
03	4 30·8	4 31·5	4 18·4	0·3	0·1	6·3	1·9	12·3	3·8
04	4 31·0	4 31·7	4 18·7	0·4	0·1	6·4	2·0	12·4	3·8
05	4 31·3	4 32·0	4 18·9	0·5	0·2	6·5	2·0	12·5	3·9
06	4 31·5	4 32·2	4 19·1	0·6	0·2	6·6	2·0	12·6	3·9
07	4 31·8	4 32·5	4 19·4	0·7	0·2	6·7	2·1	12·7	3·9
08	4 32·0	4 32·7	4 19·6	0·8	0·2	6·8	2·1	12·8	3·9
09	4 32·3	4 33·0	4 19·8	0·9	0·3	6·9	2·1	12·9	4·0
10	4 32·5	4 33·2	4 20·1	1·0	0·3	7·0	2·2	13·0	4·0
11	4 32·8	4 33·5	4 20·3	1·1	0·3	7·1	2·2	13·1	4·0
12	4 33·0	4 33·7	4 20·6	1·2	0·4	7·2	2·2	13·2	4·1
13	4 33·3	4 34·0	4 20·8	1·3	0·4	7·3	2·3	13·3	4·1
14	4 33·5	4 34·2	4 21·0	1·4	0·4	7·4	2·3	13·4	4·1
15	4 33·8	4 34·5	4 21·3	1·5	0·5	7·5	2·3	13·5	4·2
16	4 34·0	4 34·8	4 21·5	1·6	0·5	7·6	2·3	13·6	4·2
17	4 34·3	4 35·0	4 21·8	1·7	0·5	7·7	2·4	13·7	4·2
18	4 34·5	4 35·3	4 22·0	1·8	0·6	7·8	2·4	13·8	4·3
19	4 34·8	4 35·5	4 22·2	1·9	0·6	7·9	2·4	13·9	4·3
20	4 35·0	4 35·8	4 22·5	2·0	0·6	8·0	2·5	14·0	4·3
21	4 35·3	4 36·0	4 22·7	2·1	0·6	8·1	2·5	14·1	4·3
22	4 35·5	4 36·3	4 22·9	2·2	0·7	8·2	2·5	14·2	4·4
23	4 35·8	4 36·5	4 23·2	2·3	0·7	8·3	2·6	14·3	4·4
24	4 36·0	4 36·8	4 23·4	2·4	0·7	8·4	2·6	14·4	4·4
25	4 36·3	4 37·0	4 23·7	2·5	0·8	8·5	2·6	14·5	4·5
26	4 36·5	4 37·3	4 23·9	2·6	0·8	8·6	2·7	14·6	4·5
27	4 36·8	4 37·5	4 24·1	2·7	0·8	8·7	2·7	14·7	4·5
28	4 37·0	4 37·8	4 24·4	2·8	0·9	8·8	2·7	14·8	4·6
29	4 37·3	4 38·0	4 24·6	2·9	0·9	8·9	2·7	14·9	4·6
30	4 37·5	4 38·3	4 24·9	3·0	0·9	9·0	2·8	15·0	4·6
31	4 37·8	4 38·5	4 25·1	3·1	1·0	9·1	2·8	15·1	4·7
32	4 38·0	4 38·8	4 25·3	3·2	1·0	9·2	2·8	15·2	4·7
33	4 38·3	4 39·0	4 25·6	3·3	1·0	9·3	2·9	15·3	4·7
34	4 38·5	4 39·3	4 25·8	3·4	1·0	9·4	2·9	15·4	4·7
35	4 38·8	4 39·5	4 26·1	3·5	1·1	9·5	2·9	15·5	4·8
36	4 39·0	4 39·8	4 26·3	3·6	1·1	9·6	3·0	15·6	4·8
37	4 39·3	4 40·0	4 26·5	3·7	1·1	9·7	3·0	15·7	4·8
38	4 39·5	4 40·3	4 26·8	3·8	1·2	9·8	3·0	15·8	4·9
39	4 39·8	4 40·5	4 27·0	3·9	1·2	9·9	3·1	15·9	4·9
40	4 40·0	4 40·8	4 27·2	4·0	1·2	10·0	3·1	16·0	4·9
41	4 40·3	4 41·0	4 27·5	4·1	1·3	10·1	3·1	16·1	5·0
42	4 40·5	4 41·3	4 27·7	4·2	1·3	10·2	3·1	16·2	5·0
43	4 40·8	4 41·5	4 28·0	4·3	1·3	10·3	3·2	16·3	5·0
44	4 41·0	4 41·8	4 28·2	4·4	1·4	10·4	3·2	16·4	5·1
45	4 41·3	4 42·0	4 28·4	4·5	1·4	10·5	3·2	16·5	5·1
46	4 41·5	4 42·3	4 28·7	4·6	1·4	10·6	3·3	16·6	5·1
47	4 41·8	4 42·5	4 28·9	4·7	1·4	10·7	3·3	16·7	5·1
48	4 42·0	4 42·8	4 29·2	4·8	1·5	10·8	3·3	16·8	5·2
49	4 42·3	4 43·0	4 29·4	4·9	1·5	10·9	3·4	16·9	5·2
50	4 42·5	4 43·3	4 29·6	5·0	1·5	11·0	3·4	17·0	5·2
51	4 42·8	4 43·5	4 29·9	5·1	1·6	11·1	3·4	17·1	5·3
52	4 43·0	4 43·8	4 30·1	5·2	1·6	11·2	3·5	17·2	5·3
53	4 43·3	4 44·0	4 30·3	5·3	1·6	11·3	3·5	17·3	5·3
54	4 43·5	4 44·3	4 30·6	5·4	1·7	11·4	3·5	17·4	5·4
55	4 43·8	4 44·5	4 30·8	5·5	1·7	11·5	3·5	17·5	5·4
56	4 44·0	4 44·8	4 31·1	5·6	1·7	11·6	3·6	17·6	5·4
57	4 44·3	4 45·0	4 31·3	5·7	1·8	11·7	3·6	17·7	5·5
58	4 44·5	4 45·3	4 31·5	5·8	1·8	11·8	3·6	17·8	5·5
59	4 44·8	4 45·5	4 31·8	5·9	1·8	11·9	3·7	17·9	5·5
60	4 45·0	4 45·8	4 32·0	6·0	1·9	12·0	3·7	18·0	5·6

19^m	SUN PLANETS	ARIES	MOON	v or d	Corr^n	v or d	Corr^n	v or d	Corr^n
s	° ′	° ′	° ′	′	′	′	′	′	′
00	4 45·0	4 45·8	4 32·0	0·0	0·0	6·0	2·0	12·0	3·9
01	4 45·3	4 46·0	4 32·3	0·1	0·0	6·1	2·0	12·1	3·9
02	4 45·5	4 46·3	4 32·5	0·2	0·1	6·2	2·0	12·2	4·0
03	4 45·8	4 46·5	4 32·7	0·3	0·1	6·3	2·0	12·3	4·0
04	4 46·0	4 46·8	4 33·0	0·4	0·1	6·4	2·1	12·4	4·0
05	4 46·3	4 47·0	4 33·2	0·5	0·2	6·5	2·1	12·5	4·1
06	4 46·5	4 47·3	4 33·4	0·6	0·2	6·6	2·1	12·6	4·1
07	4 46·8	4 47·5	4 33·7	0·7	0·2	6·7	2·2	12·7	4·1
08	4 47·0	4 47·8	4 33·9	0·8	0·3	6·8	2·2	12·8	4·2
09	4 47·3	4 48·0	4 34·2	0·9	0·3	6·9	2·2	12·9	4·2
10	4 47·5	4 48·3	4 34·4	1·0	0·3	7·0	2·3	13·0	4·2
11	4 47·8	4 48·5	4 34·6	1·1	0·4	7·1	2·3	13·1	4·3
12	4 48·0	4 48·8	4 34·9	1·2	0·4	7·2	2·3	13·2	4·3
13	4 48·3	4 49·0	4 35·1	1·3	0·4	7·3	2·4	13·3	4·3
14	4 48·5	4 49·3	4 35·4	1·4	0·5	7·4	2·4	13·4	4·4
15	4 48·8	4 49·5	4 35·6	1·5	0·5	7·5	2·4	13·5	4·4
16	4 49·0	4 49·8	4 35·8	1·6	0·5	7·6	2·5	13·6	4·4
17	4 49·3	4 50·0	4 36·1	1·7	0·6	7·7	2·5	13·7	4·5
18	4 49·5	4 50·3	4 36·3	1·8	0·6	7·8	2·5	13·8	4·5
19	4 49·8	4 50·5	4 36·6	1·9	0·6	7·9	2·6	13·9	4·5
20	4 50·0	4 50·8	4 36·8	2·0	0·7	8·0	2·6	14·0	4·6
21	4 50·3	4 51·0	4 37·0	2·1	0·7	8·1	2·6	14·1	4·6
22	4 50·5	4 51·3	4 37·3	2·2	0·7	8·2	2·7	14·2	4·6
23	4 50·8	4 51·5	4 37·5	2·3	0·7	8·3	2·7	14·3	4·6
24	4 51·0	4 51·8	4 37·7	2·4	0·8	8·4	2·7	14·4	4·7
25	4 51·3	4 52·0	4 38·0	2·5	0·8	8·5	2·8	14·5	4·7
26	4 51·5	4 52·3	4 38·2	2·6	0·8	8·6	2·8	14·6	4·7
27	4 51·8	4 52·5	4 38·5	2·7	0·9	8·7	2·8	14·7	4·8
28	4 52·0	4 52·8	4 38·7	2·8	0·9	8·8	2·9	14·8	4·8
29	4 52·3	4 53·1	4 38·9	2·9	0·9	8·9	2·9	14·9	4·8
30	4 52·5	4 53·3	4 39·2	3·0	1·0	9·0	2·9	15·0	4·9
31	4 52·8	4 53·6	4 39·4	3·1	1·0	9·1	3·0	15·1	4·9
32	4 53·0	4 53·8	4 39·7	3·2	1·0	9·2	3·0	15·2	4·9
33	4 53·3	4 54·1	4 39·9	3·3	1·1	9·3	3·0	15·3	5·0
34	4 53·5	4 54·3	4 40·1	3·4	1·1	9·4	3·1	15·4	5·0
35	4 53·8	4 54·6	4 40·4	3·5	1·1	9·5	3·1	15·5	5·0
36	4 54·0	4 54·8	4 40·6	3·6	1·2	9·6	3·1	15·6	5·1
37	4 54·3	4 55·1	4 40·8	3·7	1·2	9·7	3·2	15·7	5·1
38	4 54·5	4 55·3	4 41·1	3·8	1·2	9·8	3·2	15·8	5·1
39	4 54·8	4 55·6	4 41·3	3·9	1·3	9·9	3·2	15·9	5·2
40	4 55·0	4 55·8	4 41·6	4·0	1·3	10·0	3·3	16·0	5·2
41	4 55·3	4 56·1	4 41·8	4·1	1·3	10·1	3·3	16·1	5·2
42	4 55·5	4 56·3	4 42·0	4·2	1·4	10·2	3·3	16·2	5·3
43	4 55·8	4 56·6	4 42·3	4·3	1·4	10·3	3·3	16·3	5·3
44	4 56·0	4 56·8	4 42·5	4·4	1·4	10·4	3·4	16·4	5·3
45	4 56·3	4 57·1	4 42·8	4·5	1·5	10·5	3·4	16·5	5·4
46	4 56·5	4 57·3	4 43·0	4·6	1·5	10·6	3·4	16·6	5·4
47	4 56·8	4 57·6	4 43·2	4·7	1·5	10·7	3·5	16·7	5·4
48	4 57·0	4 57·8	4 43·5	4·8	1·6	10·8	3·5	16·8	5·5
49	4 57·3	4 58·1	4 43·7	4·9	1·6	10·9	3·5	16·9	5·5
50	4 57·5	4 58·3	4 43·9	5·0	1·6	11·0	3·6	17·0	5·5
51	4 57·8	4 58·6	4 44·2	5·1	1·7	11·1	3·6	17·1	5·6
52	4 58·0	4 58·8	4 44·4	5·2	1·7	11·2	3·6	17·2	5·6
53	4 58·3	4 59·1	4 44·7	5·3	1·7	11·3	3·7	17·3	5·6
54	4 58·5	4 59·3	4 44·9	5·4	1·8	11·4	3·7	17·4	5·7
55	4 58·8	4 59·6	4 45·1	5·5	1·8	11·5	3·7	17·5	5·7
56	4 59·0	4 59·8	4 45·4	5·6	1·8	11·6	3·8	17·6	5·7
57	4 59·3	5 00·1	4 45·6	5·7	1·9	11·7	3·8	17·7	5·8
58	4 59·5	5 00·3	4 45·9	5·8	1·9	11·8	3·8	17·8	5·8
59	4 59·8	5 00·6	4 46·1	5·9	1·9	11·9	3·9	17·9	5·8
60	5 00·0	5 00·8	4 46·3	6·0	2·0	12·0	3·9	18·0	5·9

20m s	SUN PLANETS ° ′	ARIES ° ′	MOON ° ′	v or d ′	Corrn ′	v or d ′	Corrn ′	v or d ′	Corrn ′
00	5 00·0	5 00·8	4 46·3	0·0	0·0	6·0	2·1	12·0	4·1
01	5 00·3	5 01·1	4 46·6	0·1	0·0	6·1	2·1	12·1	4·1
02	5 00·5	5 01·3	4 46·8	0·2	0·1	6·2	2·1	12·2	4·2
03	5 00·8	5 01·6	4 47·0	0·3	0·1	6·3	2·2	12·3	4·2
04	5 01·0	5 01·8	4 47·3	0·4	0·1	6·4	2·2	12·4	4·2
05	5 01·3	5 02·1	4 47·5	0·5	0·2	6·5	2·2	12·5	4·3
06	5 01·5	5 02·3	4 47·8	0·6	0·2	6·6	2·3	12·6	4·3
07	5 01·8	5 02·6	4 48·0	0·7	0·2	6·7	2·3	12·7	4·3
08	5 02·0	5 02·8	4 48·2	0·8	0·3	6·8	2·3	12·8	4·4
09	5 02·3	5 03·1	4 48·5	0·9	0·3	6·9	2·4	12·9	4·4
10	5 02·5	5 03·3	4 48·7	1·0	0·3	7·0	2·4	13·0	4·4
11	5 02·8	5 03·6	4 49·0	1·1	0·4	7·1	2·4	13·1	4·5
12	5 03·0	5 03·8	4 49·2	1·2	0·4	7·2	2·5	13·2	4·5
13	5 03·3	5 04·1	4 49·4	1·3	0·4	7·3	2·5	13·3	4·5
14	5 03·5	5 04·3	4 49·7	1·4	0·5	7·4	2·5	13·4	4·6
15	5 03·8	5 04·6	4 49·9	1·5	0·5	7·5	2·6	13·5	4·6
16	5 04·0	5 04·8	4 50·2	1·6	0·5	7·6	2·6	13·6	4·6
17	5 04·3	5 05·1	4 50·4	1·7	0·6	7·7	2·6	13·7	4·7
18	5 04·5	5 05·3	4 50·6	1·8	0·6	7·8	2·7	13·8	4·7
19	5 04·8	5 05·6	4 50·9	1·9	0·6	7·9	2·7	13·9	4·7
20	5 05·0	5 05·8	4 51·1	2·0	0·7	8·0	2·7	14·0	4·8
21	5 05·3	5 06·1	4 51·3	2·1	0·7	8·1	2·8	14·1	4·8
22	5 05·5	5 06·3	4 51·6	2·2	0·8	8·2	2·8	14·2	4·9
23	5 05·8	5 06·6	4 51·8	2·3	0·8	8·3	2·8	14·3	4·9
24	5 06·0	5 06·8	4 52·1	2·4	0·8	8·4	2·9	14·4	4·9
25	5 06·3	5 07·1	4 52·3	2·5	0·9	8·5	2·9	14·5	5·0
26	5 06·5	5 07·3	4 52·5	2·6	0·9	8·6	2·9	14·6	5·0
27	5 06·8	5 07·6	4 52·8	2·7	0·9	8·7	3·0	14·7	5·0
28	5 07·0	5 07·8	4 53·0	2·8	1·0	8·8	3·0	14·8	5·1
29	5 07·3	5 08·1	4 53·3	2·9	1·0	8·9	3·0	14·9	5·1
30	5 07·5	5 08·3	4 53·5	3·0	1·0	9·0	3·1	15·0	5·1
31	5 07·8	5 08·6	4 53·7	3·1	1·1	9·1	3·1	15·1	5·2
32	5 08·0	5 08·8	4 54·0	3·2	1·1	9·2	3·1	15·2	5·2
33	5 08·3	5 09·1	4 54·2	3·3	1·1	9·3	3·2	15·3	5·2
34	5 08·5	5 09·3	4 54·4	3·4	1·2	9·4	3·2	15·4	5·3
35	5 08·8	5 09·6	4 54·7	3·5	1·2	9·5	3·2	15·5	5·3
36	5 09·0	5 09·8	4 54·9	3·6	1·2	9·6	3·3	15·6	5·3
37	5 09·3	5 10·1	4 55·2	3·7	1·3	9·7	3·3	15·7	5·4
38	5 09·5	5 10·3	4 55·4	3·8	1·3	9·8	3·3	15·8	5·4
39	5 09·8	5 10·6	4 55·6	3·9	1·3	9·9	3·4	15·9	5·4
40	5 10·0	5 10·8	4 55·9	4·0	1·4	10·0	3·4	16·0	5·5
41	5 10·3	5 11·1	4 56·1	4·1	1·4	10·1	3·5	16·1	5·5
42	5 10·5	5 11·4	4 56·4	4·2	1·4	10·2	3·5	16·2	5·5
43	5 10·8	5 11·6	4 56·6	4·3	1·5	10·3	3·5	16·3	5·6
44	5 11·0	5 11·9	4 56·8	4·4	1·5	10·4	3·6	16·4	5·6
45	5 11·3	5 12·1	4 57·1	4·5	1·5	10·5	3·6	16·5	5·6
46	5 11·5	5 12·4	4 57·3	4·6	1·6	10·6	3·6	16·6	5·7
47	5 11·8	5 12·6	4 57·5	4·7	1·6	10·7	3·7	16·7	5·7
48	5 12·0	5 12·9	4 57·8	4·8	1·6	10·8	3·7	16·8	5·7
49	5 12·3	5 13·1	4 58·0	4·9	1·7	10·9	3·7	16·9	5·8
50	5 12·5	5 13·4	4 58·3	5·0	1·7	11·0	3·8	17·0	5·8
51	5 12·8	5 13·6	4 58·5	5·1	1·7	11·1	3·8	17·1	5·8
52	5 13·0	5 13·9	4 58·7	5·2	1·8	11·2	3·8	17·2	5·9
53	5 13·3	5 14·1	4 59·0	5·3	1·8	11·3	3·9	17·3	5·9
54	5 13·5	5 14·4	4 59·2	5·4	1·8	11·4	3·9	17·4	5·9
55	5 13·8	5 14·6	4 59·5	5·5	1·9	11·5	3·9	17·5	6·0
56	5 14·0	5 14·9	4 59·7	5·6	1·9	11·6	4·0	17·6	6·0
57	5 14·3	5 15·1	4 59·9	5·7	1·9	11·7	4·0	17·7	6·0
58	5 14·5	5 15·4	5 00·2	5·8	2·0	11·8	4·0	17·8	6·1
59	5 14·8	5 15·6	5 00·4	5·9	2·0	11·9	4·1	17·9	6·1
60	5 15·0	5 15·9	5 00·7	6·0	2·1	12·0	4·1	18·0	6·2

21m s	SUN PLANETS ° ′	ARIES ° ′	MOON ° ′	v or d ′	Corrn ′	v or d ′	Corrn ′	v or d ′	Corrn ′
00	5 15·0	5 15·9	5 00·7	0·0	0·0	6·0	2·2	12·0	4·3
01	5 15·3	5 16·1	5 00·9	0·1	0·0	6·1	2·2	12·1	4·3
02	5 15·5	5 16·4	5 01·1	0·2	0·1	6·2	2·2	12·2	4·4
03	5 15·8	5 16·6	5 01·4	0·3	0·1	6·3	2·3	12·3	4·4
04	5 16·0	5 16·9	5 01·6	0·4	0·1	6·4	2·3	12·4	4·4
05	5 16·3	5 17·1	5 01·8	0·5	0·2	6·5	2·3	12·5	4·5
06	5 16·5	5 17·4	5 02·1	0·6	0·2	6·6	2·4	12·6	4·5
07	5 16·8	5 17·6	5 02·3	0·7	0·3	6·7	2·4	12·7	4·6
08	5 17·0	5 17·9	5 02·6	0·8	0·3	6·8	2·4	12·8	4·6
09	5 17·3	5 18·1	5 02·8	0·9	0·3	6·9	2·5	12·9	4·6
10	5 17·5	5 18·4	5 03·0	1·0	0·4	7·0	2·5	13·0	4·7
11	5 17·8	5 18·6	5 03·3	1·1	0·4	7·1	2·5	13·1	4·7
12	5 18·0	5 18·9	5 03·5	1·2	0·4	7·2	2·6	13·2	4·7
13	5 18·3	5 19·1	5 03·8	1·3	0·5	7·3	2·6	13·3	4·8
14	5 18·5	5 19·4	5 04·0	1·4	0·5	7·4	2·7	13·4	4·8
15	5 18·8	5 19·6	5 04·2	1·5	0·5	7·5	2·7	13·5	4·8
16	5 19·0	5 19·9	5 04·5	1·6	0·6	7·6	2·7	13·6	4·9
17	5 19·3	5 20·1	5 04·7	1·7	0·6	7·7	2·8	13·7	4·9
18	5 19·5	5 20·4	5 04·9	1·8	0·6	7·8	2·8	13·8	4·9
19	5 19·8	5 20·6	5 05·2	1·9	0·7	7·9	2·8	13·9	5·0
20	5 20·0	5 20·9	5 05·4	2·0	0·7	8·0	2·9	14·0	5·0
21	5 20·3	5 21·1	5 05·7	2·1	0·8	8·1	2·9	14·1	5·1
22	5 20·5	5 21·4	5 05·9	2·2	0·8	8·2	2·9	14·2	5·1
23	5 20·8	5 21·6	5 06·1	2·3	0·8	8·3	3·0	14·3	5·1
24	5 21·0	5 21·9	5 06·4	2·4	0·9	8·4	3·0	14·4	5·2
25	5 21·3	5 22·1	5 06·6	2·5	0·9	8·5	3·0	14·5	5·2
26	5 21·5	5 22·4	5 06·9	2·6	0·9	8·6	3·1	14·6	5·2
27	5 21·8	5 22·6	5 07·1	2·7	1·0	8·7	3·1	14·7	5·3
28	5 22·0	5 22·9	5 07·3	2·8	1·0	8·8	3·2	14·8	5·3
29	5 22·3	5 23·1	5 07·6	2·9	1·0	8·9	3·2	14·9	5·3
30	5 22·5	5 23·4	5 07·8	3·0	1·1	9·0	3·2	15·0	5·4
31	5 22·8	5 23·6	5 08·0	3·1	1·1	9·1	3·3	15·1	5·4
32	5 23·0	5 23·9	5 08·3	3·2	1·1	9·2	3·3	15·2	5·4
33	5 23·3	5 24·1	5 08·5	3·3	1·2	9·3	3·3	15·3	5·5
34	5 23·5	5 24·4	5 08·8	3·4	1·2	9·4	3·4	15·4	5·5
35	5 23·8	5 24·6	5 09·0	3·5	1·3	9·5	3·4	15·5	5·6
36	5 24·0	5 24·9	5 09·2	3·6	1·3	9·6	3·4	15·6	5·6
37	5 24·3	5 25·1	5 09·5	3·7	1·3	9·7	3·5	15·7	5·6
38	5 24·5	5 25·4	5 09·7	3·8	1·4	9·8	3·5	15·8	5·7
39	5 24·8	5 25·6	5 10·0	3·9	1·4	9·9	3·5	15·9	5·7
40	5 25·0	5 25·9	5 10·2	4·0	1·4	10·0	3·6	16·0	5·7
41	5 25·3	5 26·1	5 10·4	4·1	1·5	10·1	3·6	16·1	5·8
42	5 25·5	5 26·4	5 10·7	4·2	1·5	10·2	3·7	16·2	5·8
43	5 25·8	5 26·6	5 10·9	4·3	1·5	10·3	3·7	16·3	5·8
44	5 26·0	5 26·9	5 11·1	4·4	1·6	10·4	3·7	16·4	5·9
45	5 26·3	5 27·1	5 11·4	4·5	1·6	10·5	3·8	16·5	5·9
46	5 26·5	5 27·4	5 11·6	4·6	1·6	10·6	3·8	16·6	5·9
47	5 26·8	5 27·6	5 11·9	4·7	1·7	10·7	3·8	16·7	6·0
48	5 27·0	5 27·9	5 12·1	4·8	1·7	10·8	3·9	16·8	6·0
49	5 27·3	5 28·1	5 12·3	4·9	1·8	10·9	3·9	16·9	6·1
50	5 27·5	5 28·4	5 12·6	5·0	1·8	11·0	3·9	17·0	6·1
51	5 27·8	5 28·6	5 12·8	5·1	1·8	11·1	4·0	17·1	6·1
52	5 28·0	5 28·9	5 13·1	5·2	1·9	11·2	4·0	17·2	6·2
53	5 28·3	5 29·1	5 13·3	5·3	1·9	11·3	4·0	17·3	6·2
54	5 28·5	5 29·4	5 13·5	5·4	1·9	11·4	4·1	17·4	6·2
55	5 28·8	5 29·7	5 13·8	5·5	2·0	11·5	4·1	17·5	6·3
56	5 29·0	5 29·9	5 14·0	5·6	2·0	11·6	4·2	17·6	6·3
57	5 29·3	5 30·2	5 14·3	5·7	2·0	11·7	4·2	17·7	6·3
58	5 29·5	5 30·4	5 14·5	5·8	2·1	11·8	4·2	17·8	6·4
59	5 29·8	5 30·7	5 14·7	5·9	2·1	11·9	4·3	17·9	6·4
60	5 30·0	5 30·9	5 15·0	6·0	2·2	12·0	4·3	18·0	6·5

22m s	SUN PLANETS ° ′	ARIES ° ′	MOON ° ′	v or d ′	Corrn ′	v or d ′	Corrn ′	v or d ′	Corrn ′
00	5 30·0	5 30·9	5 15·0	0·0	0·0	6·0	2·3	12·0	4·5
01	5 30·3	5 31·2	5 15·2	0·1	0·0	6·1	2·3	12·1	4·5
02	5 30·5	5 31·4	5 15·4	0·2	0·1	6·2	2·3	12·2	4·6
03	5 30·8	5 31·7	5 15·7	0·3	0·1	6·3	2·4	12·3	4·6
04	5 31·0	5 31·9	5 15·9	0·4	0·2	6·4	2·4	12·4	4·7
05	5 31·3	5 32·2	5 16·2	0·5	0·2	6·5	2·4	12·5	4·7
06	5 31·5	5 32·4	5 16·4	0·6	0·2	6·6	2·5	12·6	4·7
07	5 31·8	5 32·7	5 16·6	0·7	0·3	6·7	2·5	12·7	4·8
08	5 32·0	5 32·9	5 16·9	0·8	0·3	6·8	2·6	12·8	4·8
09	5 32·3	5 33·2	5 17·1	0·9	0·3	6·9	2·6	12·9	4·8
10	5 32·5	5 33·4	5 17·4	1·0	0·4	7·0	2·6	13·0	4·9
11	5 32·8	5 33·7	5 17·6	1·1	0·4	7·1	2·7	13·1	4·9
12	5 33·0	5 33·9	5 17·8	1·2	0·5	7·2	2·7	13·2	5·0
13	5 33·3	5 34·2	5 18·1	1·3	0·5	7·3	2·7	13·3	5·0
14	5 33·5	5 34·4	5 18·3	1·4	0·5	7·4	2·8	13·4	5·0
15	5 33·8	5 34·7	5 18·5	1·5	0·6	7·5	2·8	13·5	5·1
16	5 34·0	5 34·9	5 18·8	1·6	0·6	7·6	2·9	13·6	5·1
17	5 34·3	5 35·2	5 19·0	1·7	0·6	7·7	2·9	13·7	5·1
18	5 34·5	5 35·4	5 19·3	1·8	0·7	7·8	2·9	13·8	5·2
19	5 34·8	5 35·7	5 19·5	1·9	0·7	7·9	3·0	13·9	5·2
20	5 35·0	5 35·9	5 19·7	2·0	0·8	8·0	3·0	14·0	5·3
21	5 35·3	5 36·2	5 20·0	2·1	0·8	8·1	3·0	14·1	5·3
22	5 35·5	5 36·4	5 20·2	2·2	0·8	8·2	3·1	14·2	5·3
23	5 35·8	5 36·7	5 20·5	2·3	0·9	8·3	3·1	14·3	5·4
24	5 36·0	5 36·9	5 20·7	2·4	0·9	8·4	3·2	14·4	5·4
25	5 36·3	5 37·2	5 20·9	2·5	0·9	8·5	3·2	14·5	5·4
26	5 36·5	5 37·4	5 21·2	2·6	1·0	8·6	3·2	14·6	5·5
27	5 36·8	5 37·7	5 21·4	2·7	1·0	8·7	3·3	14·7	5·5
28	5 37·0	5 37·9	5 21·6	2·8	1·0	8·8	3·3	14·8	5·6
29	5 37·3	5 38·2	5 21·9	2·9	1·1	8·9	3·3	14·9	5·6
30	5 37·5	5 38·4	5 22·1	3·0	1·1	9·0	3·4	15·0	5·6
31	5 37·8	5 38·7	5 22·4	3·1	1·2	9·1	3·4	15·1	5·7
32	5 38·0	5 38·9	5 22·6	3·2	1·2	9·2	3·5	15·2	5·7
33	5 38·3	5 39·2	5 22·8	3·3	1·2	9·3	3·5	15·3	5·7
34	5 38·5	5 39·4	5 23·1	3·4	1·3	9·4	3·5	15·4	5·8
35	5 38·8	5 39·7	5 23·3	3·5	1·3	9·5	3·6	15·5	5·8
36	5 39·0	5 39·9	5 23·6	3·6	1·4	9·6	3·6	15·6	5·9
37	5 39·3	5 40·2	5 23·8	3·7	1·4	9·7	3·6	15·7	5·9
38	5 39·5	5 40·4	5 24·0	3·8	1·4	9·8	3·7	15·8	5·9
39	5 39·8	5 40·7	5 24·3	3·9	1·5	9·9	3·7	15·9	6·0
40	5 40·0	5 40·9	5 24·5	4·0	1·5	10·0	3·8	16·0	6·0
41	5 40·3	5 41·2	5 24·7	4·1	1·5	10·1	3·8	16·1	6·0
42	5 40·5	5 41·4	5 25·0	4·2	1·6	10·2	3·8	16·2	6·1
43	5 40·8	5 41·7	5 25·2	4·3	1·6	10·3	3·9	16·3	6·1
44	5 41·0	5 41·9	5 25·5	4·4	1·7	10·4	3·9	16·4	6·1
45	5 41·3	5 42·2	5 25·7	4·5	1·7	10·5	3·9	16·5	6·2
46	5 41·5	5 42·4	5 25·9	4·6	1·7	10·6	4·0	16·6	6·2
47	5 41·8	5 42·7	5 26·2	4·7	1·8	10·7	4·0	16·7	6·3
48	5 42·0	5 42·9	5 26·4	4·8	1·8	10·8	4·1	16·8	6·3
49	5 42·3	5 43·2	5 26·7	4·9	1·8	10·9	4·1	16·9	6·3
50	5 42·5	5 43·4	5 26·9	5·0	1·9	11·0	4·1	17·0	6·4
51	5 42·8	5 43·7	5 27·1	5·1	1·9	11·1	4·2	17·1	6·4
52	5 43·0	5 43·9	5 27·4	5·2	2·0	11·2	4·2	17·2	6·5
53	5 43·3	5 44·2	5 27·6	5·3	2·0	11·3	4·2	17·3	6·5
54	5 43·5	5 44·4	5 27·9	5·4	2·0	11·4	4·3	17·4	6·5
55	5 43·8	5 44·7	5 28·1	5·5	2·1	11·5	4·3	17·5	6·6
56	5 44·0	5 44·9	5 28·3	5·6	2·1	11·6	4·4	17·6	6·6
57	5 44·3	5 45·2	5 28·6	5·7	2·1	11·7	4·4	17·7	6·6
58	5 44·5	5 45·4	5 28·8	5·8	2·2	11·8	4·4	17·8	6·7
59	5 44·8	5 45·7	5 29·0	5·9	2·2	11·9	4·5	17·9	6·7
60	5 45·0	5 45·9	5 29·3	6·0	2·3	12·0	4·5	18·0	6·8

23m s	SUN PLANETS ° ′	ARIES ° ′	MOON ° ′	v or d ′	Corrn ′	v or d ′	Corrn ′	v or d ′	Corrn ′
00	5 45·0	5 45·9	5 29·3	0·0	0·0	6·0	2·4	12·0	4·7
01	5 45·3	5 46·2	5 29·5	0·1	0·0	6·1	2·4	12·1	4·7
02	5 45·5	5 46·4	5 29·8	0·2	0·1	6·2	2·4	12·2	4·8
03	5 45·8	5 46·7	5 30·0	0·3	0·1	6·3	2·5	12·3	4·8
04	5 46·0	5 46·9	5 30·2	0·4	0·2	6·4	2·5	12·4	4·9
05	5 46·3	5 47·2	5 30·5	0·5	0·2	6·5	2·5	12·5	4·9
06	5 46·5	5 47·4	5 30·7	0·6	0·2	6·6	2·6	12·6	4·9
07	5 46·8	5 47·7	5 31·0	0·7	0·3	6·7	2·6	12·7	5·0
08	5 47·0	5 48·0	5 31·2	0·8	0·3	6·8	2·7	12·8	5·0
09	5 47·3	5 48·2	5 31·4	0·9	0·4	6·9	2·7	12·9	5·1
10	5 47·5	5 48·5	5 31·7	1·0	0·4	7·0	2·7	13·0	5·1
11	5 47·8	5 48·7	5 31·9	1·1	0·4	7·1	2·8	13·1	5·1
12	5 48·0	5 49·0	5 32·1	1·2	0·5	7·2	2·8	13·2	5·2
13	5 48·3	5 49·2	5 32·4	1·3	0·5	7·3	2·9	13·3	5·2
14	5 48·5	5 49·5	5 32·6	1·4	0·5	7·4	2·9	13·4	5·2
15	5 48·8	5 49·7	5 32·9	1·5	0·6	7·5	2·9	13·5	5·3
16	5 49·0	5 50·0	5 33·1	1·6	0·6	7·6	3·0	13·6	5·3
17	5 49·3	5 50·2	5 33·3	1·7	0·7	7·7	3·0	13·7	5·4
18	5 49·5	5 50·5	5 33·6	1·8	0·7	7·8	3·1	13·8	5·4
19	5 49·8	5 50·7	5 33·8	1·9	0·7	7·9	3·1	13·9	5·4
20	5 50·0	5 51·0	5 34·1	2·0	0·8	8·0	3·1	14·0	5·5
21	5 50·3	5 51·2	5 34·3	2·1	0·8	8·1	3·2	14·1	5·5
22	5 50·5	5 51·5	5 34·5	2·2	0·9	8·2	3·2	14·2	5·6
23	5 50·8	5 51·7	5 34·8	2·3	0·9	8·3	3·3	14·3	5·6
24	5 51·0	5 52·0	5 35·0	2·4	0·9	8·4	3·3	14·4	5·6
25	5 51·3	5 52·2	5 35·2	2·5	1·0	8·5	3·3	14·5	5·7
26	5 51·5	5 52·5	5 35·5	2·6	1·0	8·6	3·4	14·6	5·7
27	5 51·8	5 52·7	5 35·7	2·7	1·1	8·7	3·4	14·7	5·8
28	5 52·0	5 53·0	5 36·0	2·8	1·1	8·8	3·4	14·8	5·8
29	5 52·3	5 53·2	5 36·2	2·9	1·1	8·9	3·5	14·9	5·8
30	5 52·5	5 53·5	5 36·4	3·0	1·2	9·0	3·5	15·0	5·9
31	5 52·8	5 53·7	5 36·7	3·1	1·2	9·1	3·6	15·1	5·9
32	5 53·0	5 54·0	5 36·9	3·2	1·3	9·2	3·6	15·2	6·0
33	5 53·3	5 54·2	5 37·2	3·3	1·3	9·3	3·6	15·3	6·0
34	5 53·5	5 54·5	5 37·4	3·4	1·3	9·4	3·7	15·4	6·0
35	5 53·8	5 54·7	5 37·6	3·5	1·4	9·5	3·7	15·5	6·1
36	5 54·0	5 55·0	5 37·9	3·6	1·4	9·6	3·8	15·6	6·1
37	5 54·3	5 55·2	5 38·1	3·7	1·4	9·7	3·8	15·7	6·1
38	5 54·5	5 55·5	5 38·4	3·8	1·5	9·8	3·8	15·8	6·2
39	5 54·8	5 55·7	5 38·6	3·9	1·5	9·9	3·9	15·9	6·2
40	5 55·0	5 56·0	5 38·8	4·0	1·6	10·0	3·9	16·0	6·3
41	5 55·3	5 56·2	5 39·1	4·1	1·6	10·1	4·0	16·1	6·3
42	5 55·5	5 56·5	5 39·3	4·2	1·6	10·2	4·0	16·2	6·3
43	5 55·8	5 56·7	5 39·5	4·3	1·7	10·3	4·0	16·3	6·4
44	5 56·0	5 57·0	5 39·8	4·4	1·7	10·4	4·1	16·4	6·4
45	5 56·3	5 57·2	5 40·0	4·5	1·8	10·5	4·1	16·5	6·5
46	5 56·5	5 57·5	5 40·3	4·6	1·8	10·6	4·2	16·6	6·5
47	5 56·8	5 57·7	5 40·5	4·7	1·8	10·7	4·2	16·7	6·5
48	5 57·0	5 58·0	5 40·7	4·8	1·9	10·8	4·2	16·8	6·6
49	5 57·3	5 58·2	5 41·0	4·9	1·9	10·9	4·3	16·9	6·6
50	5 57·5	5 58·5	5 41·2	5·0	2·0	11·0	4·3	17·0	6·7
51	5 57·8	5 58·7	5 41·5	5·1	2·0	11·1	4·3	17·1	6·7
52	5 58·0	5 59·0	5 41·7	5·2	2·0	11·2	4·4	17·2	6·7
53	5 58·3	5 59·2	5 41·9	5·3	2·1	11·3	4·4	17·3	6·8
54	5 58·5	5 59·5	5 42·2	5·4	2·1	11·4	4·5	17·4	6·8
55	5 58·8	5 59·7	5 42·4	5·5	2·2	11·5	4·5	17·5	6·9
56	5 59·0	6 00·0	5 42·6	5·6	2·2	11·6	4·5	17·6	6·9
57	5 59·3	6 00·2	5 42·9	5·7	2·2	11·7	4·6	17·7	6·9
58	5 59·5	6 00·5	5 43·1	5·8	2·3	11·8	4·6	17·8	7·0
59	5 59·8	6 00·7	5 43·4	5·9	2·3	11·9	4·7	17·9	7·0
60	6 00·0	6 01·0	5 43·6	6·0	2·4	12·0	4·7	18·0	7·1

m 24 s	SUN PLANETS ° ′	ARIES ° ′	MOON ° ′	v or d ′	Corrⁿ ′	v or d ′	Corrⁿ ′	v or d ′	Corrⁿ ′
00	6 00·0	6 01·0	5 43·6	0·0	0·0	6·0	2·5	12·0	4·9
01	6 00·3	6 01·2	5 43·8	0·1	0·0	6·1	2·5	12·1	4·9
02	6 00·5	6 01·5	5 44·1	0·2	0·1	6·2	2·5	12·2	5·0
03	6 00·8	6 01·7	5 44·3	0·3	0·1	6·3	2·6	12·3	5·0
04	6 01·0	6 02·0	5 44·6	0·4	0·2	6·4	2·6	12·4	5·1
05	6 01·3	6 02·2	5 44·8	0·5	0·2	6·5	2·7	12·5	5·1
06	6 01·5	6 02·5	5 45·0	0·6	0·2	6·6	2·7	12·6	5·1
07	6 01·8	6 02·7	5 45·3	0·7	0·3	6·7	2·7	12·7	5·2
08	6 02·0	6 03·0	5 45·5	0·8	0·3	6·8	2·8	12·8	5·2
09	6 02·3	6 03·2	5 45·7	0·9	0·4	6·9	2·8	12·9	5·3
10	6 02·5	6 03·5	5 46·0	1·0	0·4	7·0	2·9	13·0	5·3
11	6 02·8	6 03·7	5 46·2	1·1	0·4	7·1	2·9	13·1	5·3
12	6 03·0	6 04·0	5 46·5	1·2	0·5	7·2	2·9	13·2	5·4
13	6 03·3	6 04·2	5 46·7	1·3	0·5	7·3	3·0	13·3	5·4
14	6 03·5	6 04·5	5 46·9	1·4	0·6	7·4	3·0	13·4	5·5
15	6 03·8	6 04·7	5 47·2	1·5	0·6	7·5	3·1	13·5	5·5
16	6 04·0	6 05·0	5 47·4	1·6	0·7	7·6	3·1	13·6	5·6
17	6 04·3	6 05·2	5 47·7	1·7	0·7	7·7	3·1	13·7	5·6
18	6 04·5	6 05·5	5 47·9	1·8	0·7	7·8	3·2	13·8	5·6
19	6 04·8	6 05·7	5 48·1	1·9	0·8	7·9	3·2	13·9	5·7
20	6 05·0	6 06·0	5 48·4	2·0	0·8	8·0	3·3	14·0	5·7
21	6 05·3	6 06·3	5 48·6	2·1	0·9	8·1	3·3	14·1	5·8
22	6 05·5	6 06·5	5 48·8	2·2	0·9	8·2	3·3	14·2	5·8
23	6 05·8	6 06·8	5 49·1	2·3	0·9	8·3	3·4	14·3	5·8
24	6 06·0	6 07·0	5 49·3	2·4	1·0	8·4	3·4	14·4	5·9
25	6 06·3	6 07·3	5 49·6	2·5	1·0	8·5	3·5	14·5	5·9
26	6 06·5	6 07·5	5 49·8	2·6	1·1	8·6	3·5	14·6	6·0
27	6 06·8	6 07·8	5 50·0	2·7	1·1	8·7	3·6	14·7	6·0
28	6 07·0	6 08·0	5 50·3	2·8	1·1	8·8	3·6	14·8	6·0
29	6 07·3	6 08·3	5 50·5	2·9	1·2	8·9	3·6	14·9	6·1
30	6 07·5	6 08·5	5 50·8	3·0	1·2	9·0	3·7	15·0	6·1
31	6 07·8	6 08·8	5 51·0	3·1	1·3	9·1	3·7	15·1	6·2
32	6 08·0	6 09·0	5 51·2	3·2	1·3	9·2	3·8	15·2	6·2
33	6 08·3	6 09·3	5 51·5	3·3	1·3	9·3	3·8	15·3	6·2
34	6 08·5	6 09·5	5 51·7	3·4	1·4	9·4	3·8	15·4	6·3
35	6 08·8	6 09·8	5 52·0	3·5	1·4	9·5	3·9	15·5	6·3
36	6 09·0	6 10·0	5 52·2	3·6	1·5	9·6	3·9	15·6	6·4
37	6 09·3	6 10·3	5 52·4	3·7	1·5	9·7	4·0	15·7	6·4
38	6 09·5	6 10·5	5 52·7	3·8	1·6	9·8	4·0	15·8	6·5
39	6 09·8	6 10·8	5 52·9	3·9	1·6	9·9	4·0	15·9	6·5
40	6 10·0	6 11·0	5 53·1	4·0	1·6	10·0	4·1	16·0	6·5
41	6 10·3	6 11·3	5 53·4	4·1	1·7	10·1	4·1	16·1	6·6
42	6 10·5	6 11·5	5 53·6	4·2	1·7	10·2	4·2	16·2	6·6
43	6 10·8	6 11·8	5 53·9	4·3	1·8	10·3	4·2	16·3	6·7
44	6 11·0	6 12·0	5 54·1	4·4	1·8	10·4	4·2	16·4	6·7
45	6 11·3	6 12·3	5 54·3	4·5	1·8	10·5	4·3	16·5	6·7
46	6 11·5	6 12·5	5 54·6	4·6	1·9	10·6	4·3	16·6	6·8
47	6 11·8	6 12·8	5 54·8	4·7	1·9	10·7	4·4	16·7	6·8
48	6 12·0	6 13·0	5 55·1	4·8	2·0	10·8	4·4	16·8	6·9
49	6 12·3	6 13·3	5 55·3	4·9	2·0	10·9	4·5	16·9	6·9
50	6 12·5	6 13·5	5 55·5	5·0	2·0	11·0	4·5	17·0	6·9
51	6 12·8	6 13·8	5 55·8	5·1	2·1	11·1	4·5	17·1	7·0
52	6 13·0	6 14·0	5 56·0	5·2	2·1	11·2	4·6	17·2	7·0
53	6 13·3	6 14·3	5 56·2	5·3	2·2	11·3	4·6	17·3	7·1
54	6 13·5	6 14·5	5 56·5	5·4	2·2	11·4	4·7	17·4	7·1
55	6 13·8	6 14·8	5 56·7	5·5	2·2	11·5	4·7	17·5	7·1
56	6 14·0	6 15·0	5 57·0	5·6	2·3	11·6	4·7	17·6	7·2
57	6 14·3	6 15·3	5 57·2	5·7	2·3	11·7	4·8	17·7	7·2
58	6 14·5	6 15·5	5 57·4	5·8	2·4	11·8	4·8	17·8	7·3
59	6 14·8	6 15·8	5 57·7	5·9	2·4	11·9	4·9	17·9	7·3
60	6 15·0	6 16·0	5 57·9	6·0	2·5	12·0	4·9	18·0	7·4

m 25 s	SUN PLANETS ° ′	ARIES ° ′	MOON ° ′	v or d ′	Corrⁿ ′	v or d ′	Corrⁿ ′	v or d ′	Corrⁿ ′
00	6 15·0	6 16·0	5 57·9	0·0	0·0	6·0	2·6	12·0	5·1
01	6 15·3	6 16·3	5 58·2	0·1	0·0	6·1	2·6	12·1	5·1
02	6 15·5	6 16·5	5 58·4	0·2	0·1	6·2	2·6	12·2	5·2
03	6 15·8	6 16·8	5 58·6	0·3	0·1	6·3	2·7	12·3	5·2
04	6 16·0	6 17·0	5 58·9	0·4	0·2	6·4	2·7	12·4	5·3
05	6 16·3	6 17·3	5 59·1	0·5	0·2	6·5	2·8	12·5	5·3
06	6 16·5	6 17·5	5 59·3	0·6	0·3	6·6	2·8	12·6	5·4
07	6 16·8	6 17·8	5 59·6	0·7	0·3	6·7	2·8	12·7	5·4
08	6 17·0	6 18·0	5 59·8	0·8	0·3	6·8	2·9	12·8	5·4
09	6 17·3	6 18·3	6 00·1	0·9	0·4	6·9	2·9	12·9	5·5
10	6 17·5	6 18·5	6 00·3	1·0	0·4	7·0	3·0	13·0	5·5
11	6 17·8	6 18·8	6 00·5	1·1	0·5	7·1	3·0	13·1	5·6
12	6 18·0	6 19·0	6 00·8	1·2	0·5	7·2	3·1	13·2	5·6
13	6 18·3	6 19·3	6 01·0	1·3	0·6	7·3	3·1	13·3	5·7
14	6 18·5	6 19·5	6 01·3	1·4	0·6	7·4	3·1	13·4	5·7
15	6 18·8	6 19·8	6 01·5	1·5	0·6	7·5	3·2	13·5	5·7
16	6 19·0	6 20·0	6 01·7	1·6	0·7	7·6	3·2	13·6	5·8
17	6 19·3	6 20·3	6 02·0	1·7	0·7	7·7	3·3	13·7	5·8
18	6 19·5	6 20·5	6 02·2	1·8	0·8	7·8	3·3	13·8	5·9
19	6 19·8	6 20·8	6 02·5	1·9	0·8	7·9	3·4	13·9	5·9
20	6 20·0	6 21·0	6 02·7	2·0	0·9	8·0	3·4	14·0	6·0
21	6 20·3	6 21·3	6 02·9	2·1	0·9	8·1	3·4	14·1	6·0
22	6 20·5	6 21·5	6 03·2	2·2	0·9	8·2	3·5	14·2	6·0
23	6 20·8	6 21·8	6 03·4	2·3	1·0	8·3	3·5	14·3	6·1
24	6 21·0	6 22·0	6 03·6	2·4	1·0	8·4	3·6	14·4	6·1
25	6 21·3	6 22·3	6 03·9	2·5	1·1	8·5	3·6	14·5	6·2
26	6 21·5	6 22·5	6 04·1	2·6	1·1	8·6	3·7	14·6	6·2
27	6 21·8	6 22·8	6 04·4	2·7	1·1	8·7	3·7	14·7	6·2
28	6 22·0	6 23·0	6 04·6	2·8	1·2	8·8	3·7	14·8	6·3
29	6 22·3	6 23·3	6 04·8	2·9	1·2	8·9	3·8	14·9	6·3
30	6 22·5	6 23·5	6 05·1	3·0	1·3	9·0	3·8	15·0	6·4
31	6 22·8	6 23·8	6 05·3	3·1	1·3	9·1	3·9	15·1	6·4
32	6 23·0	6 24·0	6 05·6	3·2	1·4	9·2	3·9	15·2	6·5
33	6 23·3	6 24·3	6 05·8	3·3	1·4	9·3	4·0	15·3	6·5
34	6 23·5	6 24·5	6 06·0	3·4	1·4	9·4	4·0	15·4	6·5
35	6 23·8	6 24·8	6 06·3	3·5	1·5	9·5	4·0	15·5	6·6
36	6 24·0	6 25·1	6 06·5	3·6	1·5	9·6	4·1	15·6	6·6
37	6 24·3	6 25·3	6 06·7	3·7	1·6	9·7	4·1	15·7	6·7
38	6 24·5	6 25·6	6 07·0	3·8	1·6	9·8	4·2	15·8	6·7
39	6 24·8	6 25·8	6 07·2	3·9	1·7	9·9	4·2	15·9	6·8
40	6 25·0	6 26·1	6 07·5	4·0	1·7	10·0	4·3	16·0	6·8
41	6 25·3	6 26·3	6 07·7	4·1	1·7	10·1	4·3	16·1	6·8
42	6 25·5	6 26·6	6 07·9	4·2	1·8	10·2	4·3	16·2	6·9
43	6 25·8	6 26·8	6 08·2	4·3	1·8	10·3	4·4	16·3	6·9
44	6 26·0	6 27·1	6 08·4	4·4	1·9	10·4	4·4	16·4	7·0
45	6 26·3	6 27·3	6 08·7	4·5	1·9	10·5	4·5	16·5	7·0
46	6 26·5	6 27·6	6 08·9	4·6	2·0	10·6	4·5	16·6	7·1
47	6 26·8	6 27·8	6 09·1	4·7	2·0	10·7	4·5	16·7	7·1
48	6 27·0	6 28·1	6 09·4	4·8	2·0	10·8	4·6	16·8	7·1
49	6 27·3	6 28·3	6 09·6	4·9	2·1	10·9	4·6	16·9	7·2
50	6 27·5	6 28·6	6 09·8	5·0	2·1	11·0	4·7	17·0	7·2
51	6 27·8	6 28·8	6 10·1	5·1	2·2	11·1	4·7	17·1	7·3
52	6 28·0	6 29·1	6 10·3	5·2	2·2	11·2	4·8	17·2	7·3
53	6 28·3	6 29·3	6 10·6	5·3	2·3	11·3	4·8	17·3	7·4
54	6 28·5	6 29·6	6 10·8	5·4	2·3	11·4	4·8	17·4	7·4
55	6 28·8	6 29·8	6 11·0	5·5	2·3	11·5	4·9	17·5	7·4
56	6 29·0	6 30·1	6 11·3	5·6	2·4	11·6	4·9	17·6	7·5
57	6 29·3	6 30·3	6 11·5	5·7	2·4	11·7	5·0	17·7	7·5
58	6 29·5	6 30·6	6 11·8	5·8	2·5	11·8	5·0	17·8	7·6
59	6 29·8	6 30·8	6 12·0	5·9	2·5	11·9	5·1	17·9	7·6
60	6 30·0	6 31·1	6 12·2	6·0	2·6	12·0	5·1	18·0	7·7

26m s	SUN PLANETS ° ′	ARIES ° ′	MOON ° ′	v or d ′	Corrn ′	v or d ′	Corrn ′	v or d ′	Corrn ′
00	6 30·0	6 31·1	6 12·2	0·0	0·0	6·0	2·7	12·0	5·3
01	6 30·3	6 31·3	6 12·5	0·1	0·0	6·1	2·7	12·1	5·3
02	6 30·5	6 31·6	6 12·7	0·2	0·1	6·2	2·7	12·2	5·4
03	6 30·8	6 31·8	6 12·9	0·3	0·1	6·3	2·8	12·3	5·4
04	6 31·0	6 32·1	6 13·2	0·4	0·2	6·4	2·8	12·4	5·5
05	6 31·3	6 32·3	6 13·4	0·5	0·2	6·5	2·9	12·5	5·5
06	6 31·5	6 32·6	6 13·7	0·6	0·3	6·6	2·9	12·6	5·6
07	6 31·8	6 32·8	6 13·9	0·7	0·3	6·7	3·0	12·7	5·6
08	6 32·0	6 33·1	6 14·1	0·8	0·4	6·8	3·0	12·8	5·7
09	6 32·3	6 33·3	6 14·4	0·9	0·4	6·9	3·0	12·9	5·7
10	6 32·5	6 33·6	6 14·6	1·0	0·4	7·0	3·1	13·0	5·7
11	6 32·8	6 33·8	6 14·9	1·1	0·5	7·1	3·1	13·1	5·8
12	6 33·0	6 34·1	6 15·1	1·2	0·5	7·2	3·2	13·2	5·8
13	6 33·3	6 34·3	6 15·3	1·3	0·6	7·3	3·2	13·3	5·9
14	6 33·5	6 34·6	6 15·6	1·4	0·6	7·4	3·3	13·4	5·9
15	6 33·8	6 34·8	6 15·8	1·5	0·7	7·5	3·3	13·5	6·0
16	6 34·0	6 35·1	6 16·1	1·6	0·7	7·6	3·4	13·6	6·0
17	6 34·3	6 35·3	6 16·3	1·7	0·8	7·7	3·4	13·7	6·1
18	6 34·5	6 35·6	6 16·5	1·8	0·8	7·8	3·4	13·8	6·1
19	6 34·8	6 35·8	6 16·8	1·9	0·8	7·9	3·5	13·9	6·1
20	6 35·0	6 36·1	6 17·0	2·0	0·9	8·0	3·5	14·0	6·2
21	6 35·3	6 36·3	6 17·2	2·1	0·9	8·1	3·6	14·1	6·2
22	6 35·5	6 36·6	6 17·5	2·2	1·0	8·2	3·6	14·2	6·3
23	6 35·8	6 36·8	6 17·7	2·3	1·0	8·3	3·7	14·3	6·3
24	6 36·0	6 37·1	6 18·0	2·4	1·1	8·4	3·7	14·4	6·4
25	6 36·3	6 37·3	6 18·2	2·5	1·1	8·5	3·8	14·5	6·4
26	6 36·5	6 37·6	6 18·4	2·6	1·1	8·6	3·8	14·6	6·4
27	6 36·8	6 37·8	6 18·7	2·7	1·2	8·7	3·8	14·7	6·5
28	6 37·0	6 38·1	6 18·9	2·8	1·2	8·8	3·9	14·8	6·5
29	6 37·3	6 38·3	6 19·2	2·9	1·3	8·9	3·9	14·9	6·6
30	6 37·5	6 38·6	6 19·4	3·0	1·3	9·0	4·0	15·0	6·6
31	6 37·8	6 38·8	6 19·6	3·1	1·4	9·1	4·0	15·1	6·7
32	6 38·0	6 39·1	6 19·9	3·2	1·4	9·2	4·1	15·2	6·7
33	6 38·3	6 39·3	6 20·1	3·3	1·5	9·3	4·1	15·3	6·8
34	6 38·5	6 39·6	6 20·3	3·4	1·5	9·4	4·2	15·4	6·8
35	6 38·8	6 39·8	6 20·6	3·5	1·5	9·5	4·2	15·5	6·8
36	6 39·0	6 40·1	6 20·8	3·6	1·6	9·6	4·2	15·6	6·9
37	6 39·3	6 40·3	6 21·1	3·7	1·6	9·7	4·3	15·7	6·9
38	6 39·5	6 40·6	6 21·3	3·8	1·7	9·8	4·3	15·8	7·0
39	6 39·8	6 40·8	6 21·5	3·9	1·7	9·9	4·4	15·9	7·0
40	6 40·0	6 41·1	6 21·8	4·0	1·8	10·0	4·4	16·0	7·1
41	6 40·3	6 41·3	6 22·0	4·1	1·8	10·1	4·5	16·1	7·1
42	6 40·5	6 41·6	6 22·3	4·2	1·9	10·2	4·5	16·2	7·2
43	6 40·8	6 41·8	6 22·5	4·3	1·9	10·3	4·5	16·3	7·2
44	6 41·0	6 42·1	6 22·7	4·4	1·9	10·4	4·6	16·4	7·2
45	6 41·3	6 42·3	6 23·0	4·5	2·0	10·5	4·6	16·5	7·3
46	6 41·5	6 42·6	6 23·2	4·6	2·0	10·6	4·7	16·6	7·3
47	6 41·8	6 42·8	6 23·4	4·7	2·1	10·7	4·7	16·7	7·4
48	6 42·0	6 43·1	6 23·7	4·8	2·1	10·8	4·8	16·8	7·4
49	6 42·3	6 43·4	6 23·9	4·9	2·2	10·9	4·8	16·9	7·5
50	6 42·5	6 43·6	6 24·2	5·0	2·2	11·0	4·9	17·0	7·5
51	6 42·8	6 43·9	6 24·4	5·1	2·3	11·1	4·9	17·1	7·6
52	6 43·0	6 44·1	6 24·6	5·2	2·3	11·2	4·9	17·2	7·6
53	6 43·3	6 44·4	6 24·9	5·3	2·3	11·3	5·0	17·3	7·6
54	6 43·5	6 44·6	6 25·1	5·4	2·4	11·4	5·0	17·4	7·7
55	6 43·8	6 44·9	6 25·4	5·5	2·4	11·5	5·1	17·5	7·7
56	6 44·0	6 45·1	6 25·6	5·6	2·5	11·6	5·1	17·6	7·8
57	6 44·3	6 45·4	6 25·8	5·7	2·5	11·7	5·2	17·7	7·8
58	6 44·5	6 45·6	6 26·1	5·8	2·6	11·8	5·2	17·8	7·9
59	6 44·8	6 45·9	6 26·3	5·9	2·6	11·9	5·3	17·9	7·9
60	6 45·0	6 46·1	6 26·6	6·0	2·7	12·0	5·3	18·0	8·0

27m s	SUN PLANETS ° ′	ARIES ° ′	MOON ° ′	v or d ′	Corrn ′	v or d ′	Corrn ′	v or d ′	Corrn ′
00	6 45·0	6 46·1	6 26·6	0·0	0·0	6·0	2·8	12·0	5·5
01	6 45·3	6 46·4	6 26·8	0·1	0·0	6·1	2·8	12·1	5·5
02	6 45·5	6 46·6	6 27·0	0·2	0·1	6·2	2·8	12·2	5·6
03	6 45·8	6 46·9	6 27·3	0·3	0·1	6·3	2·9	12·3	5·6
04	6 46·0	6 47·1	6 27·5	0·4	0·2	6·4	2·9	12·4	5·7
05	6 46·3	6 47·4	6 27·7	0·5	0·2	6·5	3·0	12·5	5·7
06	6 46·5	6 47·6	6 28·0	0·6	0·3	6·6	3·0	12·6	5·8
07	6 46·8	6 47·9	6 28·2	0·7	0·3	6·7	3·1	12·7	5·8
08	6 47·0	6 48·1	6 28·5	0·8	0·4	6·8	3·1	12·8	5·9
09	6 47·3	6 48·4	6 28·7	0·9	0·4	6·9	3·2	12·9	5·9
10	6 47·5	6 48·6	6 28·9	1·0	0·5	7·0	3·2	13·0	6·0
11	6 47·8	6 48·9	6 29·2	1·1	0·5	7·1	3·3	13·1	6·0
12	6 48·0	6 49·1	6 29·4	1·2	0·6	7·2	3·3	13·2	6·1
13	6 48·3	6 49·4	6 29·7	1·3	0·6	7·3	3·3	13·3	6·1
14	6 48·5	6 49·6	6 29·9	1·4	0·6	7·4	3·4	13·4	6·1
15	6 48·8	6 49·9	6 30·1	1·5	0·7	7·5	3·4	13·5	6·2
16	6 49·0	6 50·1	6 30·4	1·6	0·7	7·6	3·5	13·6	6·2
17	6 49·3	6 50·4	6 30·6	1·7	0·8	7·7	3·5	13·7	6·3
18	6 49·5	6 50·6	6 30·8	1·8	0·8	7·8	3·6	13·8	6·3
19	6 49·8	6 50·9	6 31·1	1·9	0·9	7·9	3·6	13·9	6·4
20	6 50·0	6 51·1	6 31·3	2·0	0·9	8·0	3·7	14·0	6·4
21	6 50·3	6 51·4	6 31·6	2·1	1·0	8·1	3·7	14·1	6·5
22	6 50·5	6 51·6	6 31·8	2·2	1·0	8·2	3·8	14·2	6·5
23	6 50·8	6 51·9	6 32·0	2·3	1·1	8·3	3·8	14·3	6·6
24	6 51·0	6 52·1	6 32·3	2·4	1·1	8·4	3·9	14·4	6·6
25	6 51·3	6 52·4	6 32·5	2·5	1·1	8·5	3·9	14·5	6·6
26	6 51·5	6 52·6	6 32·8	2·6	1·2	8·6	3·9	14·6	6·7
27	6 51·8	6 52·9	6 33·0	2·7	1·2	8·7	4·0	14·7	6·7
28	6 52·0	6 53·1	6 33·2	2·8	1·3	8·8	4·0	14·8	6·8
29	6 52·3	6 53·4	6 33·5	2·9	1·3	8·9	4·1	14·9	6·8
30	6 52·5	6 53·6	6 33·7	3·0	1·4	9·0	4·1	15·0	6·9
31	6 52·8	6 53·9	6 33·9	3·1	1·4	9·1	4·2	15·1	6·9
32	6 53·0	6 54·1	6 34·2	3·2	1·5	9·2	4·2	15·2	7·0
33	6 53·3	6 54·4	6 34·4	3·3	1·5	9·3	4·3	15·3	7·0
34	6 53·5	6 54·6	6 34·7	3·4	1·6	9·4	4·3	15·4	7·1
35	6 53·8	6 54·9	6 34·9	3·5	1·6	9·5	4·4	15·5	7·1
36	6 54·0	6 55·1	6 35·1	3·6	1·7	9·6	4·4	15·6	7·2
37	6 54·3	6 55·4	6 35·4	3·7	1·7	9·7	4·4	15·7	7·2
38	6 54·5	6 55·6	6 35·6	3·8	1·7	9·8	4·5	15·8	7·2
39	6 54·8	6 55·9	6 35·9	3·9	1·8	9·9	4·5	15·9	7·3
40	6 55·0	6 56·1	6 36·1	4·0	1·8	10·0	4·6	16·0	7·3
41	6 55·3	6 56·4	6 36·3	4·1	1·9	10·1	4·6	16·1	7·4
42	6 55·5	6 56·6	6 36·6	4·2	1·9	10·2	4·7	16·2	7·4
43	6 55·8	6 56·9	6 36·8	4·3	2·0	10·3	4·7	16·3	7·5
44	6 56·0	6 57·1	6 37·0	4·4	2·0	10·4	4·8	16·4	7·5
45	6 56·3	6 57·4	6 37·3	4·5	2·1	10·5	4·8	16·5	7·6
46	6 56·5	6 57·6	6 37·5	4·6	2·1	10·6	4·9	16·6	7·6
47	6 56·8	6 57·9	6 37·8	4·7	2·2	10·7	4·9	16·7	7·7
48	6 57·0	6 58·1	6 38·0	4·8	2·2	10·8	5·0	16·8	7·7
49	6 57·3	6 58·4	6 38·2	4·9	2·2	10·9	5·0	16·9	7·7
50	6 57·5	6 58·6	6 38·5	5·0	2·3	11·0	5·0	17·0	7·8
51	6 57·8	6 58·9	6 38·7	5·1	2·3	11·1	5·1	17·1	7·8
52	6 58·0	6 59·1	6 39·0	5·2	2·4	11·2	5·1	17·2	7·9
53	6 58·3	6 59·4	6 39·2	5·3	2·4	11·3	5·2	17·3	7·9
54	6 58·5	6 59·6	6 39·4	5·4	2·5	11·4	5·2	17·4	8·0
55	6 58·8	6 59·9	6 39·7	5·5	2·5	11·5	5·3	17·5	8·0
56	6 59·0	7 00·1	6 39·9	5·6	2·6	11·6	5·3	17·6	8·1
57	6 59·3	7 00·4	6 40·2	5·7	2·6	11·7	5·4	17·7	8·1
58	6 59·5	7 00·6	6 40·4	5·8	2·7	11·8	5·4	17·8	8·2
59	6 59·8	7 00·9	6 40·6	5·9	2·7	11·9	5·5	17·9	8·2
60	7 00·0	7 01·1	6 40·9	6·0	2·8	12·0	5·5	18·0	8·3

28m	SUN PLANETS	ARIES	MOON	v or d	Corrn	v or d	Corrn	v or d	Corrn
s	° ′	° ′	° ′	′	′	′	′	′	′
00	7 00·0	7 01·1	6 40·9	0·0	0·0	6·0	2·9	12·0	5·7
01	7 00·3	7 01·4	6 41·1	0·1	0·0	6·1	2·9	12·1	5·7
02	7 00·5	7 01·7	6 41·3	0·2	0·1	6·2	2·9	12·2	5·8
03	7 00·8	7 01·9	6 41·6	0·3	0·1	6·3	3·0	12·3	5·8
04	7 01·0	7 02·2	6 41·8	0·4	0·2	6·4	3·0	12·4	5·9
05	7 01·3	7 02·4	6 42·1	0·5	0·2	6·5	3·1	12·5	5·9
06	7 01·5	7 02·7	6 42·3	0·6	0·3	6·6	3·1	12·6	6·0
07	7 01·8	7 02·9	6 42·5	0·7	0·3	6·7	3·2	12·7	6·0
08	7 02·0	7 03·2	6 42·8	0·8	0·4	6·8	3·2	12·8	6·1
09	7 02·3	7 03·4	6 43·0	0·9	0·4	6·9	3·3	12·9	6·1
10	7 02·5	7 03·7	6 43·3	1·0	0·5	7·0	3·3	13·0	6·2
11	7 02·8	7 03·9	6 43·5	1·1	0·5	7·1	3·4	13·1	6·2
12	7 03·0	7 04·2	6 43·7	1·2	0·6	7·2	3·4	13·2	6·3
13	7 03·3	7 04·4	6 44·0	1·3	0·6	7·3	3·5	13·3	6·3
14	7 03·5	7 04·7	6 44·2	1·4	0·7	7·4	3·5	13·4	6·4
15	7 03·8	7 04·9	6 44·4	1·5	0·7	7·5	3·6	13·5	6·4
16	7 04·0	7 05·2	6 44·7	1·6	0·8	7·6	3·6	13·6	6·5
17	7 04·3	7 05·4	6 44·9	1·7	0·8	7·7	3·7	13·7	6·5
18	7 04·5	7 05·7	6 45·2	1·8	0·9	7·8	3·7	13·8	6·6
19	7 04·8	7 05·9	6 45·4	1·9	0·9	7·9	3·8	13·9	6·6
20	7 05·0	7 06·2	6 45·6	2·0	1·0	8·0	3·8	14·0	6·7
21	7 05·3	7 06·4	6 45·9	2·1	1·0	8·1	3·8	14·1	6·7
22	7 05·5	7 06·7	6 46·1	2·2	1·0	8·2	3·9	14·2	6·7
23	7 05·8	7 06·9	6 46·4	2·3	1·1	8·3	3·9	14·3	6·8
24	7 06·0	7 07·2	6 46·6	2·4	1·1	8·4	4·0	14·4	6·8
25	7 06·3	7 07·4	6 46·8	2·5	1·2	8·5	4·0	14·5	6·9
26	7 06·5	7 07·7	6 47·1	2·6	1·2	8·6	4·1	14·6	6·9
27	7 06·8	7 07·9	6 47·3	2·7	1·3	8·7	4·1	14·7	7·0
28	7 07·0	7 08·2	6 47·5	2·8	1·3	8·8	4·2	14·8	7·0
29	7 07·3	7 08·4	6 47·8	2·9	1·4	8·9	4·2	14·9	7·1
30	7 07·5	7 08·7	6 48·0	3·0	1·4	9·0	4·3	15·0	7·1
31	7 07·8	7 08·9	6 48·3	3·1	1·5	9·1	4·3	15·1	7·2
32	7 08·0	7 09·2	6 48·5	3·2	1·5	9·2	4·4	15·2	7·2
33	7 08·3	7 09·4	6 48·7	3·3	1·6	9·3	4·4	15·3	7·3
34	7 08·5	7 09·7	6 49·0	3·4	1·6	9·4	4·5	15·4	7·3
35	7 08·8	7 09·9	6 49·2	3·5	1·7	9·5	4·5	15·5	7·4
36	7 09·0	7 10·2	6 49·5	3·6	1·7	9·6	4·6	15·6	7·4
37	7 09·3	7 10·4	6 49·7	3·7	1·8	9·7	4·6	15·7	7·5
38	7 09·5	7 10·7	6 49·9	3·8	1·8	9·8	4·7	15·8	7·5
39	7 09·8	7 10·9	6 50·2	3·9	1·9	9·9	4·7	15·9	7·6
40	7 10·0	7 11·2	6 50·4	4·0	1·9	10·0	4·8	16·0	7·6
41	7 10·3	7 11·4	6 50·6	4·1	1·9	10·1	4·8	16·1	7·6
42	7 10·5	7 11·7	6 50·9	4·2	2·0	10·2	4·8	16·2	7·7
43	7 10·8	7 11·9	6 51·1	4·3	2·0	10·3	4·9	16·3	7·7
44	7 11·0	7 12·2	6 51·4	4·4	2·1	10·4	4·9	16·4	7·8
45	7 11·3	7 12·4	6 51·6	4·5	2·1	10·5	5·0	16·5	7·8
46	7 11·5	7 12·7	6 51·8	4·6	2·2	10·6	5·0	16·6	7·9
47	7 11·8	7 12·9	6 52·1	4·7	2·2	10·7	5·1	16·7	7·9
48	7 12·0	7 13·2	6 52·3	4·8	2·3	10·8	5·1	16·8	8·0
49	7 12·3	7 13·4	6 52·6	4·9	2·3	10·9	5·2	16·9	8·0
50	7 12·5	7 13·7	6 52·8	5·0	2·4	11·0	5·2	17·0	8·1
51	7 12·8	7 13·9	6 53·0	5·1	2·4	11·1	5·3	17·1	8·1
52	7 13·0	7 14·2	6 53·3	5·2	2·5	11·2	5·3	17·2	8·2
53	7 13·3	7 14·4	6 53·5	5·3	2·5	11·3	5·4	17·3	8·2
54	7 13·5	7 14·7	6 53·8	5·4	2·6	11·4	5·4	17·4	8·3
55	7 13·8	7 14·9	6 54·0	5·5	2·6	11·5	5·5	17·5	8·3
56	7 14·0	7 15·2	6 54·2	5·6	2·7	11·6	5·5	17·6	8·4
57	7 14·3	7 15·4	6 54·5	5·7	2·7	11·7	5·6	17·7	8·4
58	7 14·5	7 15·7	6 54·7	5·8	2·8	11·8	5·6	17·8	8·5
59	7 14·8	7 15·9	6 54·9	5·9	2·8	11·9	5·7	17·9	8·5
60	7 15·0	7 16·2	6 55·2	6·0	2·9	12·0	5·7	18·0	8·6

29m	SUN PLANETS	ARIES	MOON	v or d	Corrn	v or d	Corrn	v or d	Corrn
s	° ′	° ′	° ′	′	′	′	′	′	′
00	7 15·0	7 16·2	6 55·2	0·0	0·0	6·0	3·0	12·0	5·9
01	7 15·3	7 16·4	6 55·4	0·1	0·0	6·1	3·0	12·1	5·9
02	7 15·5	7 16·7	6 55·7	0·2	0·1	6·2	3·0	12·2	6·0
03	7 15·8	7 16·9	6 55·9	0·3	0·1	6·3	3·1	12·3	6·0
04	7 16·0	7 17·2	6 56·1	0·4	0·2	6·4	3·1	12·4	6·1
05	7 16·3	7 17·4	6 56·4	0·5	0·2	6·5	3·2	12·5	6·1
06	7 16·5	7 17·7	6 56·6	0·6	0·3	6·6	3·2	12·6	6·2
07	7 16·8	7 17·9	6 56·9	0·7	0·3	6·7	3·3	12·7	6·2
08	7 17·0	7 18·2	6 57·1	0·8	0·4	6·8	3·3	12·8	6·3
09	7 17·3	7 18·4	6 57·3	0·9	0·4	6·9	3·4	12·9	6·3
10	7 17·5	7 18·7	6 57·6	1·0	0·5	7·0	3·4	13·0	6·4
11	7 17·8	7 18·9	6 57·8	1·1	0·5	7·1	3·5	13·1	6·4
12	7 18·0	7 19·2	6 58·0	1·2	0·6	7·2	3·5	13·2	6·5
13	7 18·3	7 19·4	6 58·3	1·3	0·6	7·3	3·6	13·3	6·5
14	7 18·5	7 19·7	6 58·5	1·4	0·7	7·4	3·6	13·4	6·6
15	7 18·8	7 20·0	6 58·8	1·5	0·7	7·5	3·7	13·5	6·6
16	7 19·0	7 20·2	6 59·0	1·6	0·8	7·6	3·7	13·6	6·7
17	7 19·3	7 20·5	6 59·2	1·7	0·8	7·7	3·8	13·7	6·7
18	7 19·5	7 20·7	6 59·5	1·8	0·9	7·8	3·8	13·8	6·8
19	7 19·8	7 21·0	6 59·7	1·9	0·9	7·9	3·9	13·9	6·8
20	7 20·0	7 21·2	7 00·0	2·0	1·0	8·0	3·9	14·0	6·9
21	7 20·3	7 21·5	7 00·2	2·1	1·0	8·1	4·0	14·1	6·9
22	7 20·5	7 21·7	7 00·4	2·2	1·1	8·2	4·0	14·2	7·0
23	7 20·8	7 22·0	7 00·7	2·3	1·1	8·3	4·1	14·3	7·0
24	7 21·0	7 22·2	7 00·9	2·4	1·2	8·4	4·1	14·4	7·1
25	7 21·3	7 22·5	7 01·1	2·5	1·2	8·5	4·2	14·5	7·1
26	7 21·5	7 22·7	7 01·4	2·6	1·3	8·6	4·2	14·6	7·2
27	7 21·8	7 23·0	7 01·6	2·7	1·3	8·7	4·3	14·7	7·2
28	7 22·0	7 23·2	7 01·9	2·8	1·4	8·8	4·3	14·8	7·3
29	7 22·3	7 23·5	7 02·1	2·9	1·4	8·9	4·4	14·9	7·3
30	7 22·5	7 23·7	7 02·3	3·0	1·5	9·0	4·4	15·0	7·4
31	7 22·8	7 24·0	7 02·6	3·1	1·5	9·1	4·5	15·1	7·4
32	7 23·0	7 24·2	7 02·8	3·2	1·6	9·2	4·5	15·2	7·5
33	7 23·3	7 24·5	7 03·1	3·3	1·6	9·3	4·6	15·3	7·5
34	7 23·5	7 24·7	7 03·3	3·4	1·7	9·4	4·6	15·4	7·6
35	7 23·8	7 25·0	7 03·5	3·5	1·7	9·5	4·7	15·5	7·6
36	7 24·0	7 25·2	7 03·8	3·6	1·8	9·6	4·7	15·6	7·7
37	7 24·3	7 25·5	7 04·0	3·7	1·8	9·7	4·8	15·7	7·7
38	7 24·5	7 25·7	7 04·3	3·8	1·9	9·8	4·8	15·8	7·8
39	7 24·8	7 26·0	7 04·5	3·9	1·9	9·9	4·9	15·9	7·8
40	7 25·0	7 26·2	7 04·7	4·0	2·0	10·0	4·9	16·0	7·9
41	7 25·3	7 26·5	7 05·0	4·1	2·0	10·1	5·0	16·1	7·9
42	7 25·5	7 26·7	7 05·2	4·2	2·1	10·2	5·0	16·2	8·0
43	7 25·8	7 27·0	7 05·4	4·3	2·1	10·3	5·1	16·3	8·0
44	7 26·0	7 27·2	7 05·7	4·4	2·2	10·4	5·1	16·4	8·1
45	7 26·3	7 27·5	7 05·9	4·5	2·2	10·5	5·2	16·5	8·1
46	7 26·5	7 27·7	7 06·2	4·6	2·3	10·6	5·2	16·6	8·2
47	7 26·8	7 28·0	7 06·4	4·7	2·3	10·7	5·3	16·7	8·2
48	7 27·0	7 28·2	7 06·6	4·8	2·4	10·8	5·3	16·8	8·3
49	7 27·3	7 28·5	7 06·9	4·9	2·4	10·9	5·4	16·9	8·3
50	7 27·5	7 28·7	7 07·1	5·0	2·5	11·0	5·4	17·0	8·4
51	7 27·8	7 29·0	7 07·4	5·1	2·5	11·1	5·5	17·1	8·4
52	7 28·0	7 29·2	7 07·6	5·2	2·6	11·2	5·5	17·2	8·5
53	7 28·3	7 29·5	7 07·8	5·3	2·6	11·3	5·6	17·3	8·5
54	7 28·5	7 29·7	7 08·1	5·4	2·7	11·4	5·6	17·4	8·6
55	7 28·8	7 30·0	7 08·3	5·5	2·7	11·5	5·7	17·5	8·6
56	7 29·0	7 30·2	7 08·5	5·6	2·8	11·6	5·7	17·6	8·7
57	7 29·3	7 30·5	7 08·8	5·7	2·8	11·7	5·8	17·7	8·7
58	7 29·5	7 30·7	7 09·0	5·8	2·9	11·8	5·8	17·8	8·8
59	7 29·8	7 31·0	7 09·3	5·9	2·9	11·9	5·9	17·9	8·8
60	7 30·0	7 31·2	7 09·5	6·0	3·0	12·0	5·9	18·0	8·9

30^{m} (s)	SUN PLANETS	ARIES	MOON	v or d	Corrⁿ	v or d	Corrⁿ	v or d	Corrⁿ
s	° ′	° ′	° ′	′	′	′	′	′	′
00	7 30·0	7 31·2	7 09·5	0·0	0·0	6·0	3·1	12·0	6·1
01	7 30·3	7 31·5	7 09·7	0·1	0·1	6·1	3·1	12·1	6·2
02	7 30·5	7 31·7	7 10·0	0·2	0·1	6·2	3·2	12·2	6·2
03	7 30·8	7 32·0	7 10·2	0·3	0·2	6·3	3·2	12·3	6·3
04	7 31·0	7 32·2	7 10·5	0·4	0·2	6·4	3·3	12·4	6·3
05	7 31·3	7 32·5	7 10·7	0·5	0·3	6·5	3·3	12·5	6·4
06	7 31·5	7 32·7	7 10·9	0·6	0·3	6·6	3·4	12·6	6·4
07	7 31·8	7 33·0	7 11·2	0·7	0·4	6·7	3·4	12·7	6·5
08	7 32·0	7 33·2	7 11·4	0·8	0·4	6·8	3·5	12·8	6·5
09	7 32·3	7 33·5	7 11·6	0·9	0·5	6·9	3·5	12·9	6·6
10	7 32·5	7 33·7	7 11·9	1·0	0·5	7·0	3·6	13·0	6·6
11	7 32·8	7 34·0	7 12·1	1·1	0·6	7·1	3·6	13·1	6·7
12	7 33·0	7 34·2	7 12·4	1·2	0·6	7·2	3·7	13·2	6·7
13	7 33·3	7 34·5	7 12·6	1·3	0·7	7·3	3·7	13·3	6·8
14	7 33·5	7 34·7	7 12·8	1·4	0·7	7·4	3·8	13·4	6·8
15	7 33·8	7 35·0	7 13·1	1·5	0·8	7·5	3·8	13·5	6·9
16	7 34·0	7 35·2	7 13·3	1·6	0·8	7·6	3·9	13·6	6·9
17	7 34·3	7 35·5	7 13·6	1·7	0·9	7·7	3·9	13·7	7·0
18	7 34·5	7 35·7	7 13·8	1·8	0·9	7·8	4·0	13·8	7·0
19	7 34·8	7 36·0	7 14·0	1·9	1·0	7·9	4·0	13·9	7·1
20	7 35·0	7 36·2	7 14·3	2·0	1·0	8·0	4·1	14·0	7·1
21	7 35·3	7 36·5	7 14·5	2·1	1·1	8·1	4·1	14·1	7·2
22	7 35·5	7 36·7	7 14·7	2·2	1·1	8·2	4·2	14·2	7·2
23	7 35·8	7 37·0	7 15·0	2·3	1·2	8·3	4·2	14·3	7·3
24	7 36·0	7 37·2	7 15·2	2·4	1·2	8·4	4·3	14·4	7·3
25	7 36·3	7 37·5	7 15·5	2·5	1·3	8·5	4·3	14·5	7·4
26	7 36·5	7 37·7	7 15·7	2·6	1·3	8·6	4·4	14·6	7·4
27	7 36·8	7 38·0	7 15·9	2·7	1·4	8·7	4·4	14·7	7·5
28	7 37·0	7 38·3	7 16·2	2·8	1·4	8·8	4·5	14·8	7·5
29	7 37·3	7 38·5	7 16·4	2·9	1·5	8·9	4·5	14·9	7·6
30	7 37·5	7 38·8	7 16·7	3·0	1·5	9·0	4·6	15·0	7·6
31	7 37·8	7 39·0	7 16·9	3·1	1·6	9·1	4·6	15·1	7·7
32	7 38·0	7 39·3	7 17·1	3·2	1·6	9·2	4·7	15·2	7·7
33	7 38·3	7 39·5	7 17·4	3·3	1·7	9·3	4·7	15·3	7·8
34	7 38·5	7 39·8	7 17·6	3·4	1·7	9·4	4·8	15·4	7·8
35	7 38·8	7 40·0	7 17·9	3·5	1·8	9·5	4·8	15·5	7·9
36	7 39·0	7 40·3	7 18·1	3·6	1·8	9·6	4·9	15·6	7·9
37	7 39·3	7 40·5	7 18·3	3·7	1·9	9·7	4·9	15·7	8·0
38	7 39·5	7 40·8	7 18·6	3·8	1·9	9·8	5·0	15·8	8·0
39	7 39·8	7 41·0	7 18·8	3·9	2·0	9·9	5·0	15·9	8·1
40	7 40·0	7 41·3	7 19·0	4·0	2·0	10·0	5·1	16·0	8·1
41	7 40·3	7 41·5	7 19·3	4·1	2·1	10·1	5·1	16·1	8·2
42	7 40·5	7 41·8	7 19·5	4·2	2·1	10·2	5·2	16·2	8·2
43	7 40·8	7 42·0	7 19·8	4·3	2·2	10·3	5·2	16·3	8·3
44	7 41·0	7 42·3	7 20·0	4·4	2·2	10·4	5·3	16·4	8·3
45	7 41·3	7 42·5	7 20·2	4·5	2·3	10·5	5·3	16·5	8·4
46	7 41·5	7 42·8	7 20·5	4·6	2·3	10·6	5·4	16·6	8·4
47	7 41·8	7 43·0	7 20·7	4·7	2·4	10·7	5·4	16·7	8·5
48	7 42·0	7 43·3	7 21·0	4·8	2·4	10·8	5·5	16·8	8·5
49	7 42·3	7 43·5	7 21·2	4·9	2·5	10·9	5·5	16·9	8·6
50	7 42·5	7 43·8	7 21·4	5·0	2·5	11·0	5·6	17·0	8·6
51	7 42·8	7 44·0	7 21·7	5·1	2·6	11·1	5·6	17·1	8·7
52	7 43·0	7 44·3	7 21·9	5·2	2·6	11·2	5·7	17·2	8·7
53	7 43·3	7 44·5	7 22·1	5·3	2·7	11·3	5·7	17·3	8·8
54	7 43·5	7 44·8	7 22·4	5·4	2·7	11·4	5·8	17·4	8·8
55	7 43·8	7 45·0	7 22·6	5·5	2·8	11·5	5·8	17·5	8·9
56	7 44·0	7 45·3	7 22·9	5·6	2·8	11·6	5·9	17·6	8·9
57	7 44·3	7 45·5	7 23·1	5·7	2·9	11·7	5·9	17·7	9·0
58	7 44·5	7 45·8	7 23·3	5·8	2·9	11·8	6·0	17·8	9·0
59	7 44·8	7 46·0	7 23·6	5·9	3·0	11·9	6·0	17·9	9·1
60	7 45·0	7 46·3	7 23·8	6·0	3·1	12·0	6·1	18·0	9·2

31^{m} (s)	SUN PLANETS	ARIES	MOON	v or d	Corrⁿ	v or d	Corrⁿ	v or d	Corrⁿ
s	° ′	° ′	° ′	′	′	′	′	′	′
00	7 45·0	7 46·3	7 23·8	0·0	0·0	6·0	3·2	12·0	6·3
01	7 45·3	7 46·5	7 24·1	0·1	0·1	6·1	3·2	12·1	6·4
02	7 45·5	7 46·8	7 24·3	0·2	0·1	6·2	3·3	12·2	6·4
03	7 45·8	7 47·0	7 24·5	0·3	0·2	6·3	3·3	12·3	6·5
04	7 46·0	7 47·3	7 24·8	0·4	0·2	6·4	3·4	12·4	6·5
05	7 46·3	7 47·5	7 25·0	0·5	0·3	6·5	3·4	12·5	6·6
06	7 46·5	7 47·8	7 25·2	0·6	0·3	6·6	3·5	12·6	6·6
07	7 46·8	7 48·0	7 25·5	0·7	0·4	6·7	3·5	12·7	6·7
08	7 47·0	7 48·3	7 25·7	0·8	0·4	6·8	3·6	12·8	6·7
09	7 47·3	7 48·5	7 26·0	0·9	0·5	6·9	3·6	12·9	6·8
10	7 47·5	7 48·8	7 26·2	1·0	0·5	7·0	3·7	13·0	6·8
11	7 47·8	7 49·0	7 26·4	1·1	0·6	7·1	3·7	13·1	6·9
12	7 48·0	7 49·3	7 26·7	1·2	0·6	7·2	3·8	13·2	6·9
13	7 48·3	7 49·5	7 26·9	1·3	0·7	7·3	3·8	13·3	7·0
14	7 48·5	7 49·8	7 27·2	1·4	0·7	7·4	3·9	13·4	7·0
15	7 48·8	7 50·0	7 27·4	1·5	0·8	7·5	3·9	13·5	7·1
16	7 49·0	7 50·3	7 27·6	1·6	0·8	7·6	4·0	13·6	7·1
17	7 49·3	7 50·5	7 27·9	1·7	0·9	7·7	4·0	13·7	7·2
18	7 49·5	7 50·8	7 28·1	1·8	0·9	7·8	4·1	13·8	7·2
19	7 49·8	7 51·0	7 28·4	1·9	1·0	7·9	4·1	13·9	7·3
20	7 50·0	7 51·3	7 28·6	2·0	1·1	8·0	4·2	14·0	7·4
21	7 50·3	7 51·5	7 28·8	2·1	1·1	8·1	4·3	14·1	7·4
22	7 50·5	7 51·8	7 29·1	2·2	1·2	8·2	4·3	14·2	7·5
23	7 50·8	7 52·0	7 29·3	2·3	1·2	8·3	4·4	14·3	7·5
24	7 51·0	7 52·3	7 29·5	2·4	1·3	8·4	4·4	14·4	7·6
25	7 51·3	7 52·5	7 29·8	2·5	1·3	8·5	4·5	14·5	7·6
26	7 51·5	7 52·8	7 30·0	2·6	1·4	8·6	4·5	14·6	7·7
27	7 51·8	7 53·0	7 30·3	2·7	1·4	8·7	4·6	14·7	7·7
28	7 52·0	7 53·3	7 30·5	2·8	1·5	8·8	4·6	14·8	7·8
29	7 52·3	7 53·5	7 30·7	2·9	1·5	8·9	4·7	14·9	7·8
30	7 52·5	7 53·8	7 31·0	3·0	1·6	9·0	4·7	15·0	7·9
31	7 52·8	7 54·0	7 31·2	3·1	1·6	9·1	4·8	15·1	7·9
32	7 53·0	7 54·3	7 31·5	3·2	1·7	9·2	4·8	15·2	8·0
33	7 53·3	7 54·5	7 31·7	3·3	1·7	9·3	4·9	15·3	8·0
34	7 53·5	7 54·8	7 31·9	3·4	1·8	9·4	4·9	15·4	8·1
35	7 53·8	7 55·0	7 32·2	3·5	1·8	9·5	5·0	15·5	8·1
36	7 54·0	7 55·3	7 32·4	3·6	1·9	9·6	5·0	15·6	8·2
37	7 54·3	7 55·5	7 32·6	3·7	1·9	9·7	5·1	15·7	8·2
38	7 54·5	7 55·8	7 32·9	3·8	2·0	9·8	5·1	15·8	8·3
39	7 54·8	7 56·0	7 33·1	3·9	2·0	9·9	5·2	15·9	8·3
40	7 55·0	7 56·3	7 33·4	4·0	2·1	10·0	5·3	16·0	8·4
41	7 55·3	7 56·6	7 33·6	4·1	2·2	10·1	5·3	16·1	8·5
42	7 55·5	7 56·8	7 33·8	4·2	2·2	10·2	5·4	16·2	8·5
43	7 55·8	7 57·1	7 34·1	4·3	2·3	10·3	5·4	16·3	8·6
44	7 56·0	7 57·3	7 34·3	4·4	2·3	10·4	5·5	16·4	8·6
45	7 56·3	7 57·6	7 34·6	4·5	2·4	10·5	5·5	16·5	8·7
46	7 56·5	7 57·8	7 34·8	4·6	2·4	10·6	5·6	16·6	8·7
47	7 56·8	7 58·1	7 35·0	4·7	2·5	10·7	5·6	16·7	8·8
48	7 57·0	7 58·3	7 35·3	4·8	2·5	10·8	5·7	16·8	8·8
49	7 57·3	7 58·6	7 35·5	4·9	2·6	10·9	5·7	16·9	8·9
50	7 57·5	7 58·8	7 35·7	5·0	2·6	11·0	5·8	17·0	8·9
51	7 57·8	7 59·1	7 36·0	5·1	2·7	11·1	5·8	17·1	9·0
52	7 58·0	7 59·3	7 36·2	5·2	2·7	11·2	5·9	17·2	9·0
53	7 58·3	7 59·6	7 36·5	5·3	2·8	11·3	5·9	17·3	9·1
54	7 58·5	7 59·8	7 36·7	5·4	2·8	11·4	6·0	17·4	9·1
55	7 58·8	8 00·1	7 36·9	5·5	2·9	11·5	6·0	17·5	9·2
56	7 59·0	8 00·3	7 37·2	5·6	2·9	11·6	6·1	17·6	9·2
57	7 59·3	8 00·6	7 37·4	5·7	3·0	11·7	6·1	17·7	9·3
58	7 59·5	8 00·8	7 37·7	5·8	3·0	11·8	6·2	17·8	9·3
59	7 59·8	8 01·1	7 37·9	5·9	3·1	11·9	6·2	17·9	9·4
60	8 00·0	8 01·3	7 38·1	6·0	3·2	12·0	6·3	18·0	9·5

32^{m}	SUN PLANETS	ARIES	MOON	v or d	Corr^{n}	v or d	Corr^{n}	v or d	Corr^{n}
s	° ′	° ′	° ′	′	′	′	′	′	′
00	8 00·0	8 01·3	7 38·1	0·0	0·0	6·0	3·3	12·0	6·5
01	8 00·3	8 01·6	7 38·4	0·1	0·1	6·1	3·3	12·1	6·6
02	8 00·5	8 01·8	7 38·6	0·2	0·1	6·2	3·4	12·2	6·6
03	8 00·8	8 02·1	7 38·8	0·3	0·2	6·3	3·4	12·3	6·7
04	8 01·0	8 02·3	7 39·1	0·4	0·2	6·4	3·5	12·4	6·7
05	8 01·3	8 02·6	7 39·3	0·5	0·3	6·5	3·5	12·5	6·8
06	8 01·5	8 02·8	7 39·6	0·6	0·3	6·6	3·6	12·6	6·8
07	8 01·8	8 03·1	7 39·8	0·7	0·4	6·7	3·6	12·7	6·9
08	8 02·0	8 03·3	7 40·0	0·8	0·4	6·8	3·7	12·8	6·9
09	8 02·3	8 03·6	7 40·3	0·9	0·5	6·9	3·7	12·9	7·0
10	8 02·5	8 03·8	7 40·5	1·0	0·5	7·0	3·8	13·0	7·0
11	8 02·8	8 04·1	7 40·8	1·1	0·6	7·1	3·8	13·1	7·1
12	8 03·0	8 04·3	7 41·0	1·2	0·7	7·2	3·9	13·2	7·2
13	8 03·3	8 04·6	7 41·2	1·3	0·7	7·3	4·0	13·3	7·2
14	8 03·5	8 04·8	7 41·5	1·4	0·8	7·4	4·0	13·4	7·3
15	8 03·8	8 05·1	7 41·7	1·5	0·8	7·5	4·1	13·5	7·3
16	8 04·0	8 05·3	7 42·0	1·6	0·9	7·6	4·1	13·6	7·4
17	8 04·3	8 05·6	7 42·2	1·7	0·9	7·7	4·2	13·7	7·4
18	8 04·5	8 05·8	7 42·4	1·8	1·0	7·8	4·2	13·8	7·5
19	8 04·8	8 06·1	7 42·7	1·9	1·0	7·9	4·3	13·9	7·5
20	8 05·0	8 06·3	7 42·9	2·0	1·1	8·0	4·3	14·0	7·6
21	8 05·3	8 06·6	7 43·1	2·1	1·1	8·1	4·4	14·1	7·6
22	8 05·5	8 06·8	7 43·4	2·2	1·2	8·2	4·4	14·2	7·7
23	8 05·8	8 07·1	7 43·6	2·3	1·2	8·3	4·5	14·3	7·7
24	8 06·0	8 07·3	7 43·9	2·4	1·3	8·4	4·6	14·4	7·8
25	8 06·3	8 07·6	7 44·1	2·5	1·4	8·5	4·6	14·5	7·9
26	8 06·5	8 07·8	7 44·3	2·6	1·4	8·6	4·7	14·6	7·9
27	8 06·8	8 08·1	7 44·6	2·7	1·5	8·7	4·7	14·7	8·0
28	8 07·0	8 08·3	7 44·8	2·8	1·5	8·8	4·8	14·8	8·0
29	8 07·3	8 08·6	7 45·1	2·9	1·6	8·9	4·8	14·9	8·1
30	8 07·5	8 08·8	7 45·3	3·0	1·6	9·0	4·9	15·0	8·1
31	8 07·8	8 09·1	7 45·5	3·1	1·7	9·1	4·9	15·1	8·2
32	8 08·0	8 09·3	7 45·8	3·2	1·7	9·2	5·0	15·2	8·2
33	8 08·3	8 09·6	7 46·0	3·3	1·8	9·3	5·0	15·3	8·3
34	8 08·5	8 09·8	7 46·2	3·4	1·8	9·4	5·1	15·4	8·3
35	8 08·8	8 10·1	7 46·5	3·5	1·9	9·5	5·1	15·5	8·4
36	8 09·0	8 10·3	7 46·7	3·6	2·0	9·6	5·2	15·6	8·5
37	8 09·3	8 10·6	7 47·0	3·7	2·0	9·7	5·3	15·7	8·5
38	8 09·5	8 10·8	7 47·2	3·8	2·1	9·8	5·3	15·8	8·6
39	8 09·8	8 11·1	7 47·4	3·9	2·1	9·9	5·4	15·9	8·6
40	8 10·0	8 11·3	7 47·7	4·0	2·2	10·0	5·4	16·0	8·7
41	8 10·3	8 11·6	7 47·9	4·1	2·2	10·1	5·5	16·1	8·7
42	8 10·5	8 11·8	7 48·2	4·2	2·3	10·2	5·5	16·2	8·8
43	8 10·8	8 12·1	7 48·4	4·3	2·3	10·3	5·6	16·3	8·8
44	8 11·0	8 12·3	7 48·6	4·4	2·4	10·4	5·6	16·4	8·9
45	8 11·3	8 12·6	7 48·9	4·5	2·4	10·5	5·7	16·5	8·9
46	8 11·5	8 12·8	7 49·1	4·6	2·5	10·6	5·7	16·6	9·0
47	8 11·8	8 13·1	7 49·3	4·7	2·5	10·7	5·8	16·7	9·0
48	8 12·0	8 13·3	7 49·6	4·8	2·6	10·8	5·9	16·8	9·1
49	8 12·3	8 13·6	7 49·8	4·9	2·7	10·9	5·9	16·9	9·2
50	8 12·5	8 13·8	7 50·1	5·0	2·7	11·0	6·0	17·0	9·2
51	8 12·8	8 14·1	7 50·3	5·1	2·8	11·1	6·0	17·1	9·3
52	8 13·0	8 14·3	7 50·5	5·2	2·8	11·2	6·1	17·2	9·3
53	8 13·3	8 14·6	7 50·8	5·3	2·9	11·3	6·1	17·3	9·4
54	8 13·5	8 14·9	7 51·0	5·4	2·9	11·4	6·2	17·4	9·4
55	8 13·8	8 15·1	7 51·3	5·5	3·0	11·5	6·2	17·5	9·5
56	8 14·0	8 15·4	7 51·5	5·6	3·0	11·6	6·3	17·6	9·5
57	8 14·3	8 15·6	7 51·7	5·7	3·1	11·7	6·3	17·7	9·6
58	8 14·5	8 15·9	7 52·0	5·8	3·1	11·8	6·4	17·8	9·6
59	8 14·8	8 16·1	7 52·2	5·9	3·2	11·9	6·4	17·9	9·7
60	8 15·0	8 16·4	7 52·5	6·0	3·3	12·0	6·5	18·0	9·8

33^{m}	SUN PLANETS	ARIES	MOON	v or d	Corr^{n}	v or d	Corr^{n}	v or d	Corr^{n}
s	° ′	° ′	° ′	′	′	′	′	′	′
00	8 15·0	8 16·4	7 52·5	0·0	0·0	6·0	3·4	12·0	6·7
01	8 15·3	8 16·6	7 52·7	0·1	0·1	6·1	3·4	12·1	6·8
02	8 15·5	8 16·9	7 52·9	0·2	0·1	6·2	3·5	12·2	6·8
03	8 15·8	8 17·1	7 53·2	0·3	0·2	6·3	3·5	12·3	6·9
04	8 16·0	8 17·4	7 53·4	0·4	0·2	6·4	3·6	12·4	6·9
05	8 16·3	8 17·6	7 53·6	0·5	0·3	6·5	3·6	12·5	7·0
06	8 16·5	8 17·9	7 53·9	0·6	0·3	6·6	3·7	12·6	7·0
07	8 16·8	8 18·1	7 54·1	0·7	0·4	6·7	3·7	12·7	7·1
08	8 17·0	8 18·4	7 54·4	0·8	0·4	6·8	3·8	12·8	7·1
09	8 17·3	8 18·6	7 54·6	0·9	0·5	6·9	3·9	12·9	7·2
10	8 17·5	8 18·9	7 54·8	1·0	0·6	7·0	3·9	13·0	7·3
11	8 17·8	8 19·1	7 55·1	1·1	0·6	7·1	4·0	13·1	7·3
12	8 18·0	8 19·4	7 55·3	1·2	0·7	7·2	4·0	13·2	7·4
13	8 18·3	8 19·6	7 55·6	1·3	0·7	7·3	4·1	13·3	7·4
14	8 18·5	8 19·9	7 55·8	1·4	0·8	7·4	4·1	13·4	7·5
15	8 18·8	8 20·1	7 56·0	1·5	0·8	7·5	4·2	13·5	7·5
16	8 19·0	8 20·4	7 56·3	1·6	0·9	7·6	4·2	13·6	7·6
17	8 19·3	8 20·6	7 56·5	1·7	0·9	7·7	4·3	13·7	7·6
18	8 19·5	8 20·9	7 56·7	1·8	1·0	7·8	4·4	13·8	7·7
19	8 19·8	8 21·1	7 57·0	1·9	1·1	7·9	4·4	13·9	7·8
20	8 20·0	8 21·4	7 57·2	2·0	1·1	8·0	4·5	14·0	7·8
21	8 20·3	8 21·6	7 57·5	2·1	1·2	8·1	4·5	14·1	7·9
22	8 20·5	8 21·9	7 57·7	2·2	1·2	8·2	4·6	14·2	7·9
23	8 20·8	8 22·1	7 57·9	2·3	1·3	8·3	4·6	14·3	8·0
24	8 21·0	8 22·4	7 58·2	2·4	1·3	8·4	4·7	14·4	8·0
25	8 21·3	8 22·6	7 58·4	2·5	1·4	8·5	4·7	14·5	8·1
26	8 21·5	8 22·9	7 58·7	2·6	1·5	8·6	4·8	14·6	8·2
27	8 21·8	8 23·1	7 58·9	2·7	1·5	8·7	4·9	14·7	8·2
28	8 22·0	8 23·4	7 59·1	2·8	1·6	8·8	4·9	14·8	8·3
29	8 22·3	8 23·6	7 59·4	2·9	1·6	8·9	5·0	14·9	8·3
30	8 22·5	8 23·9	7 59·6	3·0	1·7	9·0	5·0	15·0	8·4
31	8 22·8	8 24·1	7 59·8	3·1	1·7	9·1	5·1	15·1	8·4
32	8 23·0	8 24·4	8 00·1	3·2	1·8	9·2	5·1	15·2	8·5
33	8 23·3	8 24·6	8 00·3	3·3	1·8	9·3	5·2	15·3	8·5
34	8 23·5	8 24·9	8 00·6	3·4	1·9	9·4	5·2	15·4	8·6
35	8 23·8	8 25·1	8 00·8	3·5	2·0	9·5	5·3	15·5	8·7
36	8 24·0	8 25·4	8 01·0	3·6	2·0	9·6	5·4	15·6	8·7
37	8 24·3	8 25·6	8 01·3	3·7	2·1	9·7	5·4	15·7	8·8
38	8 24·5	8 25·9	8 01·5	3·8	2·1	9·8	5·5	15·8	8·8
39	8 24·8	8 26·1	8 01·8	3·9	2·2	9·9	5·5	15·9	8·9
40	8 25·0	8 26·4	8 02·0	4·0	2·2	10·0	5·6	16·0	8·9
41	8 25·3	8 26·6	8 02·2	4·1	2·3	10·1	5·6	16·1	9·0
42	8 25·5	8 26·9	8 02·5	4·2	2·3	10·2	5·7	16·2	9·0
43	8 25·8	8 27·1	8 02·7	4·3	2·4	10·3	5·8	16·3	9·1
44	8 26·0	8 27·4	8 02·9	4·4	2·5	10·4	5·8	16·4	9·2
45	8 26·3	8 27·6	8 03·2	4·5	2·5	10·5	5·9	16·5	9·2
46	8 26·5	8 27·9	8 03·4	4·6	2·6	10·6	5·9	16·6	9·3
47	8 26·8	8 28·1	8 03·7	4·7	2·6	10·7	6·0	16·7	9·3
48	8 27·0	8 28·4	8 03·9	4·8	2·7	10·8	6·0	16·8	9·4
49	8 27·3	8 28·6	8 04·1	4·9	2·7	10·9	6·1	16·9	9·4
50	8 27·5	8 28·9	8 04·4	5·0	2·8	11·0	6·1	17·0	9·5
51	8 27·8	8 29·1	8 04·6	5·1	2·8	11·1	6·2	17·1	9·5
52	8 28·0	8 29·4	8 04·9	5·2	2·9	11·2	6·3	17·2	9·6
53	8 28·3	8 29·6	8 05·1	5·3	3·0	11·3	6·3	17·3	9·7
54	8 28·5	8 29·9	8 05·3	5·4	3·0	11·4	6·4	17·4	9·7
55	8 28·8	8 30·1	8 05·6	5·5	3·1	11·5	6·4	17·5	9·8
56	8 29·0	8 30·4	8 05·8	5·6	3·1	11·6	6·5	17·6	9·8
57	8 29·3	8 30·6	8 06·1	5·7	3·2	11·7	6·5	17·7	9·9
58	8 29·5	8 30·9	8 06·3	5·8	3·2	11·8	6·6	17·8	9·9
59	8 29·8	8 31·1	8 06·5	5·9	3·3	11·9	6·6	17·9	10·0
60	8 30·0	8 31·4	8 06·8	6·0	3·4	12·0	6·7	18·0	10·1

34m	SUN PLANETS	ARIES	MOON	v or d	Corrn	v or d	Corrn	v or d	Corrn
s	° ′	° ′	° ′	′	′	′	′	′	′
00	8 30·0	8 31·4	8 06·8	0·0	0·0	6·0	3·5	12·0	6·9
01	8 30·3	8 31·6	8 07·0	0·1	0·1	6·1	3·5	12·1	7·0
02	8 30·5	8 31·9	8 07·2	0·2	0·1	6·2	3·6	12·2	7·0
03	8 30·8	8 32·1	8 07·5	0·3	0·2	6·3	3·6	12·3	7·1
04	8 31·0	8 32·4	8 07·7	0·4	0·2	6·4	3·7	12·4	7·1
05	8 31·3	8 32·6	8 08·0	0·5	0·3	6·5	3·7	12·5	7·2
06	8 31·5	8 32·9	8 08·2	0·6	0·3	6·6	3·8	12·6	7·2
07	8 31·8	8 33·2	8 08·4	0·7	0·4	6·7	3·9	12·7	7·3
08	8 32·0	8 33·4	8 08·7	0·8	0·5	6·8	3·9	12·8	7·4
09	8 32·3	8 33·7	8 08·9	0·9	0·5	6·9	4·0	12·9	7·4
10	8 32·5	8 33·9	8 09·2	1·0	0·6	7·0	4·0	13·0	7·5
11	8 32·8	8 34·2	8 09·4	1·1	0·6	7·1	4·1	13·1	7·5
12	8 33·0	8 34·4	8 09·6	1·2	0·7	7·2	4·1	13·2	7·6
13	8 33·3	8 34·7	8 09·9	1·3	0·7	7·3	4·2	13·3	7·6
14	8 33·5	8 34·9	8 10·1	1·4	0·8	7·4	4·3	13·4	7·7
15	8 33·8	8 35·2	8 10·3	1·5	0·9	7·5	4·3	13·5	7·8
16	8 34·0	8 35·4	8 10·6	1·6	0·9	7·6	4·4	13·6	7·8
17	8 34·3	8 35·7	8 10·8	1·7	1·0	7·7	4·4	13·7	7·9
18	8 34·5	8 35·9	8 11·1	1·8	1·0	7·8	4·5	13·8	7·9
19	8 34·8	8 36·2	8 11·3	1·9	1·1	7·9	4·5	13·9	8·0
20	8 35·0	8 36·4	8 11·5	2·0	1·2	8·0	4·6	14·0	8·1
21	8 35·3	8 36·7	8 11·8	2·1	1·2	8·1	4·7	14·1	8·1
22	8 35·5	8 36·9	8 12·0	2·2	1·3	8·2	4·7	14·2	8·2
23	8 35·8	8 37·2	8 12·3	2·3	1·3	8·3	4·8	14·3	8·2
24	8 36·0	8 37·4	8 12·5	2·4	1·4	8·4	4·8	14·4	8·3
25	8 36·3	8 37·7	8 12·7	2·5	1·4	8·5	4·9	14·5	8·3
26	8 36·5	8 37·9	8 13·0	2·6	1·5	8·6	4·9	14·6	8·4
27	8 36·8	8 38·2	8 13·2	2·7	1·6	8·7	5·0	14·7	8·5
28	8 37·0	8 38·4	8 13·4	2·8	1·6	8·8	5·1	14·8	8·5
29	8 37·3	8 38·7	8 13·7	2·9	1·7	8·9	5·1	14·9	8·6
30	8 37·5	8 38·9	8 13·9	3·0	1·7	9·0	5·2	15·0	8·6
31	8 37·8	8 39·2	8 14·2	3·1	1·8	9·1	5·2	15·1	8·7
32	8 38·0	8 39·4	8 14·4	3·2	1·8	9·2	5·3	15·2	8·7
33	8 38·3	8 39·7	8 14·6	3·3	1·9	9·3	5·3	15·3	8·8
34	8 38·5	8 39·9	8 14·9	3·4	2·0	9·4	5·4	15·4	8·9
35	8 38·8	8 40·2	8 15·1	3·5	2·0	9·5	5·5	15·5	8·9
36	8 39·0	8 40·4	8 15·4	3·6	2·1	9·6	5·5	15·6	9·0
37	8 39·3	8 40·7	8 15·6	3·7	2·1	9·7	5·6	15·7	9·0
38	8 39·5	8 40·9	8 15·8	3·8	2·2	9·8	5·6	15·8	9·1
39	8 39·8	8 41·2	8 16·1	3·9	2·2	9·9	5·7	15·9	9·1
40	8 40·0	8 41·4	8 16·3	4·0	2·3	10·0	5·8	16·0	9·2
41	8 40·3	8 41·7	8 16·5	4·1	2·4	10·1	5·8	16·1	9·3
42	8 40·5	8 41·9	8 16·8	4·2	2·4	10·2	5·9	16·2	9·3
43	8 40·8	8 42·2	8 17·0	4·3	2·5	10·3	5·9	16·3	9·4
44	8 41·0	8 42·4	8 17·3	4·4	2·5	10·4	6·0	16·4	9·4
45	8 41·3	8 42·7	8 17·5	4·5	2·6	10·5	6·0	16·5	9·5
46	8 41·5	8 42·9	8 17·7	4·6	2·6	10·6	6·1	16·6	9·5
47	8 41·8	8 43·2	8 18·0	4·7	2·7	10·7	6·2	16·7	9·6
48	8 42·0	8 43·4	8 18·2	4·8	2·8	10·8	6·2	16·8	9·7
49	8 42·3	8 43·7	8 18·5	4·9	2·8	10·9	6·3	16·9	9·7
50	8 42·5	8 43·9	8 18·7	5·0	2·9	11·0	6·3	17·0	9·8
51	8 42·8	8 44·2	8 18·9	5·1	2·9	11·1	6·4	17·1	9·8
52	8 43·0	8 44·4	8 19·2	5·2	3·0	11·2	6·4	17·2	9·9
53	8 43·3	8 44·7	8 19·4	5·3	3·0	11·3	6·5	17·3	9·9
54	8 43·5	8 44·9	8 19·7	5·4	3·1	11·4	6·6	17·4	10·0
55	8 43·8	8 45·2	8 19·9	5·5	3·2	11·5	6·6	17·5	10·1
56	8 44·0	8 45·4	8 20·1	5·6	3·2	11·6	6·7	17·6	10·1
57	8 44·3	8 45·7	8 20·4	5·7	3·3	11·7	6·7	17·7	10·2
58	8 44·5	8 45·9	8 20·6	5·8	3·3	11·8	6·8	17·8	10·2
59	8 44·8	8 46·2	8 20·8	5·9	3·4	11·9	6·8	17·9	10·3
60	8 45·0	8 46·4	8 21·1	6·0	3·5	12·0	6·9	18·0	10·4

35m	SUN PLANETS	ARIES	MOON	v or d	Corrn	v or d	Corrn	v or d	Corrn
s	° ′	° ′	° ′	′	′	′	′	′	′
00	8 45·0	8 46·4	8 21·1	0·0	0·0	6·0	3·6	12·0	7·1
01	8 45·3	8 46·7	8 21·3	0·1	0·1	6·1	3·6	12·1	7·2
02	8 45·5	8 46·9	8 21·6	0·2	0·1	6·2	3·7	12·2	7·2
03	8 45·8	8 47·2	8 21·8	0·3	0·2	6·3	3·7	12·3	7·3
04	8 46·0	8 47·4	8 22·0	0·4	0·2	6·4	3·8	12·4	7·3
05	8 46·3	8 47·7	8 22·3	0·5	0·3	6·5	3·8	12·5	7·4
06	8 46·5	8 47·9	8 22·5	0·6	0·4	6·6	3·9	12·6	7·5
07	8 46·8	8 48·2	8 22·8	0·7	0·4	6·7	4·0	12·7	7·5
08	8 47·0	8 48·4	8 23·0	0·8	0·5	6·8	4·0	12·8	7·6
09	8 47·3	8 48·7	8 23·2	0·9	0·5	6·9	4·1	12·9	7·6
10	8 47·5	8 48·9	8 23·5	1·0	0·6	7·0	4·1	13·0	7·7
11	8 47·8	8 49·2	8 23·7	1·1	0·7	7·1	4·2	13·1	7·8
12	8 48·0	8 49·4	8 23·9	1·2	0·7	7·2	4·3	13·2	7·8
13	8 48·3	8 49·7	8 24·2	1·3	0·8	7·3	4·3	13·3	7·9
14	8 48·5	8 49·9	8 24·4	1·4	0·8	7·4	4·4	13·4	7·9
15	8 48·8	8 50·2	8 24·7	1·5	0·9	7·5	4·4	13·5	8·0
16	8 49·0	8 50·4	8 24·9	1·6	0·9	7·6	4·5	13·6	8·0
17	8 49·3	8 50·7	8 25·1	1·7	1·0	7·7	4·6	13·7	8·1
18	8 49·5	8 50·9	8 25·4	1·8	1·1	7·8	4·6	13·8	8·2
19	8 49·8	8 51·2	8 25·6	1·9	1·1	7·9	4·7	13·9	8·2
20	8 50·0	8 51·5	8 25·9	2·0	1·2	8·0	4·7	14·0	8·3
21	8 50·3	8 51·7	8 26·1	2·1	1·2	8·1	4·8	14·1	8·3
22	8 50·5	8 52·0	8 26·3	2·2	1·3	8·2	4·9	14·2	8·4
23	8 50·8	8 52·2	8 26·6	2·3	1·4	8·3	4·9	14·3	8·5
24	8 51·0	8 52·5	8 26·8	2·4	1·4	8·4	5·0	14·4	8·5
25	8 51·3	8 52·7	8 27·0	2·5	1·5	8·5	5·0	14·5	8·6
26	8 51·5	8 53·0	8 27·3	2·6	1·5	8·6	5·1	14·6	8·6
27	8 51·8	8 53·2	8 27·5	2·7	1·6	8·7	5·1	14·7	8·7
28	8 52·0	8 53·5	8 27·8	2·8	1·7	8·8	5·2	14·8	8·8
29	8 52·3	8 53·7	8 28·0	2·9	1·7	8·9	5·3	14·9	8·8
30	8 52·5	8 54·0	8 28·2	3·0	1·8	9·0	5·3	15·0	8·9
31	8 52·8	8 54·2	8 28·5	3·1	1·8	9·1	5·4	15·1	8·9
32	8 53·0	8 54·5	8 28·7	3·2	1·9	9·2	5·4	15·2	9·0
33	8 53·3	8 54·7	8 29·0	3·3	2·0	9·3	5·5	15·3	9·1
34	8 53·5	8 55·0	8 29·2	3·4	2·0	9·4	5·6	15·4	9·1
35	8 53·8	8 55·2	8 29·4	3·5	2·1	9·5	5·6	15·5	9·2
36	8 54·0	8 55·5	8 29·7	3·6	2·1	9·6	5·7	15·6	9·2
37	8 54·3	8 55·7	8 29·9	3·7	2·2	9·7	5·7	15·7	9·3
38	8 54·5	8 56·0	8 30·2	3·8	2·2	9·8	5·8	15·8	9·3
39	8 54·8	8 56·2	8 30·4	3·9	2·3	9·9	5·9	15·9	9·4
40	8 55·0	8 56·5	8 30·6	4·0	2·4	10·0	5·9	16·0	9·5
41	8 55·3	8 56·7	8 30·9	4·1	2·4	10·1	6·0	16·1	9·5
42	8 55·5	8 57·0	8 31·1	4·2	2·5	10·2	6·0	16·2	9·6
43	8 55·8	8 57·2	8 31·3	4·3	2·5	10·3	6·1	16·3	9·6
44	8 56·0	8 57·5	8 31·6	4·4	2·6	10·4	6·2	16·4	9·7
45	8 56·3	8 57·7	8 31·8	4·5	2·7	10·5	6·2	16·5	9·8
46	8 56·5	8 58·0	8 32·1	4·6	2·7	10·6	6·3	16·6	9·8
47	8 56·8	8 58·2	8 32·3	4·7	2·8	10·7	6·3	16·7	9·9
48	8 57·0	8 58·5	8 32·5	4·8	2·8	10·8	6·4	16·8	9·9
49	8 57·3	8 58·7	8 32·8	4·9	2·9	10·9	6·4	16·9	10·0
50	8 57·5	8 59·0	8 33·0	5·0	3·0	11·0	6·5	17·0	10·1
51	8 57·8	8 59·2	8 33·3	5·1	3·0	11·1	6·6	17·1	10·1
52	8 58·0	8 59·5	8 33·5	5·2	3·1	11·2	6·6	17·2	10·2
53	8 58·3	8 59·7	8 33·7	5·3	3·1	11·3	6·7	17·3	10·2
54	8 58·5	9 00·0	8 34·0	5·4	3·2	11·4	6·7	17·4	10·3
55	8 58·8	9 00·2	8 34·2	5·5	3·3	11·5	6·8	17·5	10·4
56	8 59·0	9 00·5	8 34·4	5·6	3·3	11·6	6·9	17·6	10·4
57	8 59·3	9 00·7	8 34·7	5·7	3·4	11·7	6·9	17·7	10·5
58	8 59·5	9 01·0	8 34·9	5·8	3·4	11·8	7·0	17·8	10·5
59	8 59·8	9 01·2	8 35·2	5·9	3·5	11·9	7·0	17·9	10·6
60	9 00·0	9 01·5	8 35·4	6·0	3·6	12·0	7·1	18·0	10·7

36^m	SUN PLANETS	ARIES	MOON	v or d	Corr^n	v or d	Corr^n	v or d	Corr^n
s	° ′	° ′	° ′	′	′	′	′	′	′
00	9 00·0	9 01·5	8 35·4	0·0	0·0	6·0	3·7	12·0	7·3
01	9 00·3	9 01·7	8 35·6	0·1	0·1	6·1	3·7	12·1	7·4
02	9 00·5	9 02·0	8 35·9	0·2	0·1	6·2	3·8	12·2	7·4
03	9 00·8	9 02·2	8 36·1	0·3	0·2	6·3	3·8	12·3	7·5
04	9 01·0	9 02·5	8 36·4	0·4	0·2	6·4	3·9	12·4	7·5
05	9 01·3	9 02·7	8 36·6	0·5	0·3	6·5	4·0	12·5	7·6
06	9 01·5	9 03·0	8 36·8	0·6	0·4	6·6	4·0	12·6	7·7
07	9 01·8	9 03·2	8 37·1	0·7	0·4	6·7	4·1	12·7	7·7
08	9 02·0	9 03·5	8 37·3	0·8	0·5	6·8	4·1	12·8	7·8
09	9 02·3	9 03·7	8 37·5	0·9	0·5	6·9	4·2	12·9	7·8
10	9 02·5	9 04·0	8 37·8	1·0	0·6	7·0	4·3	13·0	7·9
11	9 02·8	9 04·2	8 38·0	1·1	0·7	7·1	4·3	13·1	8·0
12	9 03·0	9 04·5	8 38·3	1·2	0·7	7·2	4·4	13·2	8·0
13	9 03·3	9 04·7	8 38·5	1·3	0·8	7·3	4·4	13·3	8·1
14	9 03·5	9 05·0	8 38·7	1·4	0·9	7·4	4·5	13·4	8·2
15	9 03·8	9 05·2	8 39·0	1·5	0·9	7·5	4·6	13·5	8·2
16	9 04·0	9 05·5	8 39·2	1·6	1·0	7·6	4·6	13·6	8·3
17	9 04·3	9 05·7	8 39·5	1·7	1·0	7·7	4·7	13·7	8·3
18	9 04·5	9 06·0	8 39·7	1·8	1·1	7·8	4·7	13·8	8·4
19	9 04·8	9 06·2	8 39·9	1·9	1·2	7·9	4·8	13·9	8·5
20	9 05·0	9 06·5	8 40·2	2·0	1·2	8·0	4·9	14·0	8·5
21	9 05·3	9 06·7	8 40·4	2·1	1·3	8·1	4·9	14·1	8·6
22	9 05·5	9 07·0	8 40·6	2·2	1·3	8·2	5·0	14·2	8·6
23	9 05·8	9 07·2	8 40·9	2·3	1·4	8·3	5·0	14·3	8·7
24	9 06·0	9 07·5	8 41·1	2·4	1·5	8·4	5·1	14·4	8·8
25	9 06·3	9 07·7	8 41·4	2·5	1·5	8·5	5·2	14·5	8·8
26	9 06·5	9 08·0	8 41·6	2·6	1·6	8·6	5·2	14·6	8·9
27	9 06·8	9 08·2	8 41·8	2·7	1·6	8·7	5·3	14·7	8·9
28	9 07·0	9 08·5	8 42·1	2·8	1·7	8·8	5·4	14·8	9·0
29	9 07·3	9 08·7	8 42·3	2·9	1·8	8·9	5·4	14·9	9·1
30	9 07·5	9 09·0	8 42·6	3·0	1·8	9·0	5·5	15·0	9·1
31	9 07·8	9 09·2	8 42·8	3·1	1·9	9·1	5·5	15·1	9·2
32	9 08·0	9 09·5	8 43·0	3·2	1·9	9·2	5·6	15·2	9·2
33	9 08·3	9 09·8	8 43·3	3·3	2·0	9·3	5·7	15·3	9·3
34	9 08·5	9 10·0	8 43·5	3·4	2·1	9·4	5·7	15·4	9·4
35	9 08·8	9 10·3	8 43·8	3·5	2·1	9·5	5·8	15·5	9·4
36	9 09·0	9 10·5	8 44·0	3·6	2·2	9·6	5·8	15·6	9·5
37	9 09·3	9 10·8	8 44·2	3·7	2·3	9·7	5·9	15·7	9·6
38	9 09·5	9 11·0	8 44·5	3·8	2·3	9·8	6·0	15·8	9·6
39	9 09·8	9 11·3	8 44·7	3·9	2·4	9·9	6·0	15·9	9·7
40	9 10·0	9 11·5	8 44·9	4·0	2·4	10·0	6·1	16·0	9·7
41	9 10·3	9 11·8	8 45·2	4·1	2·5	10·1	6·1	16·1	9·8
42	9 10·5	9 12·0	8 45·4	4·2	2·6	10·2	6·2	16·2	9·9
43	9 10·8	9 12·3	8 45·7	4·3	2·6	10·3	6·3	16·3	9·9
44	9 11·0	9 12·5	8 45·9	4·4	2·7	10·4	6·3	16·4	10·0
45	9 11·3	9 12·8	8 46·1	4·5	2·7	10·5	6·4	16·5	10·0
46	9 11·5	9 13·0	8 46·4	4·6	2·8	10·6	6·4	16·6	10·1
47	9 11·8	9 13·3	8 46·6	4·7	2·9	10·7	6·5	16·7	10·2
48	9 12·0	9 13·5	8 46·9	4·8	2·9	10·8	6·6	16·8	10·2
49	9 12·3	9 13·8	8 47·1	4·9	3·0	10·9	6·6	16·9	10·3
50	9 12·5	9 14·0	8 47·3	5·0	3·0	11·0	6·7	17·0	10·3
51	9 12·8	9 14·3	8 47·6	5·1	3·1	11·1	6·8	17·1	10·4
52	9 13·0	9 14·5	8 47·8	5·2	3·2	11·2	6·8	17·2	10·5
53	9 13·3	9 14·8	8 48·0	5·3	3·2	11·3	6·9	17·3	10·5
54	9 13·5	9 15·0	8 48·3	5·4	3·3	11·4	6·9	17·4	10·6
55	9 13·8	9 15·3	8 48·5	5·5	3·3	11·5	7·0	17·5	10·6
56	9 14·0	9 15·5	8 48·8	5·6	3·4	11·6	7·1	17·6	10·7
57	9 14·3	9 15·8	8 49·0	5·7	3·5	11·7	7·1	17·7	10·8
58	9 14·5	9 16·0	8 49·2	5·8	3·5	11·8	7·2	17·8	10·8
59	9 14·8	9 16·3	8 49·5	5·9	3·6	11·9	7·2	17·9	10·9
60	9 15·0	9 16·5	8 49·7	6·0	3·7	12·0	7·3	18·0	11·0

37^m	SUN PLANETS	ARIES	MOON	v or d	Corr^n	v or d	Corr^n	v or d	Corr^n
s	° ′	° ′	° ′	′	′	′	′	′	′
00	9 15·0	9 16·5	8 49·7	0·0	0·0	6·0	3·8	12·0	7·5
01	9 15·3	9 16·8	8 50·0	0·1	0·1	6·1	3·8	12·1	7·6
02	9 15·5	9 17·0	8 50·2	0·2	0·1	6·2	3·9	12·2	7·6
03	9 15·8	9 17·3	8 50·4	0·3	0·2	6·3	3·9	12·3	7·7
04	9 16·0	9 17·5	8 50·7	0·4	0·3	6·4	4·0	12·4	7·8
05	9 16·3	9 17·8	8 50·9	0·5	0·3	6·5	4·1	12·5	7·8
06	9 16·5	9 18·0	8 51·1	0·6	0·4	6·6	4·1	12·6	7·9
07	9 16·8	9 18·3	8 51·4	0·7	0·4	6·7	4·2	12·7	7·9
08	9 17·0	9 18·5	8 51·6	0·8	0·5	6·8	4·3	12·8	8·0
09	9 17·3	9 18·8	8 51·9	0·9	0·6	6·9	4·3	12·9	8·1
10	9 17·5	9 19·0	8 52·1	1·0	0·6	7·0	4·4	13·0	8·1
11	9 17·8	9 19·3	8 52·3	1·1	0·7	7·1	4·4	13·1	8·2
12	9 18·0	9 19·5	8 52·6	1·2	0·8	7·2	4·5	13·2	8·3
13	9 18·3	9 19·8	8 52·8	1·3	0·8	7·3	4·6	13·3	8·3
14	9 18·5	9 20·0	8 53·1	1·4	0·9	7·4	4·6	13·4	8·4
15	9 18·8	9 20·3	8 53·3	1·5	0·9	7·5	4·7	13·5	8·4
16	9 19·0	9 20·5	8 53·5	1·6	1·0	7·6	4·8	13·6	8·5
17	9 19·3	9 20·8	8 53·8	1·7	1·1	7·7	4·8	13·7	8·6
18	9 19·5	9 21·0	8 54·0	1·8	1·1	7·8	4·9	13·8	8·6
19	9 19·8	9 21·3	8 54·3	1·9	1·2	7·9	4·9	13·9	8·7
20	9 20·0	9 21·5	8 54·5	2·0	1·3	8·0	5·0	14·0	8·8
21	9 20·3	9 21·8	8 54·7	2·1	1·3	8·1	5·1	14·1	8·8
22	9 20·5	9 22·0	8 55·0	2·2	1·4	8·2	5·1	14·2	8·9
23	9 20·8	9 22·3	8 55·2	2·3	1·4	8·3	5·2	14·3	8·9
24	9 21·0	9 22·5	8 55·4	2·4	1·5	8·4	5·3	14·4	9·0
25	9 21·3	9 22·8	8 55·7	2·5	1·6	8·5	5·3	14·5	9·1
26	9 21·5	9 23·0	8 55·9	2·6	1·6	8·6	5·4	14·6	9·1
27	9 21·8	9 23·3	8 56·2	2·7	1·7	8·7	5·4	14·7	9·2
28	9 22·0	9 23·5	8 56·4	2·8	1·8	8·8	5·5	14·8	9·3
29	9 22·3	9 23·8	8 56·6	2·9	1·8	8·9	5·6	14·9	9·3
30	9 22·5	9 24·0	8 56·9	3·0	1·9	9·0	5·6	15·0	9·4
31	9 22·8	9 24·3	8 57·1	3·1	1·9	9·1	5·7	15·1	9·4
32	9 23·0	9 24·5	8 57·4	3·2	2·0	9·2	5·8	15·2	9·5
33	9 23·3	9 24·8	8 57·6	3·3	2·1	9·3	5·8	15·3	9·6
34	9 23·5	9 25·0	8 57·8	3·4	2·1	9·4	5·9	15·4	9·6
35	9 23·8	9 25·3	8 58·1	3·5	2·2	9·5	5·9	15·5	9·7
36	9 24·0	9 25·5	8 58·3	3·6	2·3	9·6	6·0	15·6	9·8
37	9 24·3	9 25·8	8 58·5	3·7	2·3	9·7	6·1	15·7	9·8
38	9 24·5	9 26·0	8 58·8	3·8	2·4	9·8	6·1	15·8	9·9
39	9 24·8	9 26·3	8 59·0	3·9	2·4	9·9	6·2	15·9	9·9
40	9 25·0	9 26·5	8 59·3	4·0	2·5	10·0	6·3	16·0	10·0
41	9 25·3	9 26·8	8 59·5	4·1	2·6	10·1	6·3	16·1	10·1
42	9 25·5	9 27·0	8 59·7	4·2	2·6	10·2	6·4	16·2	10·1
43	9 25·8	9 27·3	9 00·0	4·3	2·7	10·3	6·4	16·3	10·2
44	9 26·0	9 27·5	9 00·2	4·4	2·8	10·4	6·5	16·4	10·3
45	9 26·3	9 27·8	9 00·5	4·5	2·8	10·5	6·6	16·5	10·3
46	9 26·5	9 28·1	9 00·7	4·6	2·9	10·6	6·6	16·6	10·4
47	9 26·8	9 28·3	9 00·9	4·7	2·9	10·7	6·7	16·7	10·4
48	9 27·0	9 28·6	9 01·2	4·8	3·0	10·8	6·8	16·8	10·5
49	9 27·3	9 28·8	9 01·4	4·9	3·1	10·9	6·8	16·9	10·6
50	9 27·5	9 29·1	9 01·6	5·0	3·1	11·0	6·9	17·0	10·6
51	9 27·8	9 29·3	9 01·9	5·1	3·2	11·1	6·9	17·1	10·7
52	9 28·0	9 29·6	9 02·1	5·2	3·3	11·2	7·0	17·2	10·8
53	9 28·3	9 29·8	9 02·4	5·3	3·3	11·3	7·1	17·3	10·8
54	9 28·5	9 30·1	9 02·6	5·4	3·4	11·4	7·1	17·4	10·9
55	9 28·8	9 30·3	9 02·8	5·5	3·4	11·5	7·2	17·5	10·9
56	9 29·0	9 30·6	9 03·1	5·6	3·5	11·6	7·3	17·6	11·0
57	9 29·3	9 30·8	9 03·3	5·7	3·6	11·7	7·3	17·7	11·1
58	9 29·5	9 31·1	9 03·6	5·8	3·6	11·8	7·4	17·8	11·1
59	9 29·8	9 31·3	9 03·8	5·9	3·7	11·9	7·4	17·9	11·2
60	9 30·0	9 31·6	9 04·0	6·0	3·8	12·0	7·5	18·0	11·3

38^m	SUN PLANETS	ARIES	MOON	v or d	Corr^n	v or d	Corr^n	v or d	Corr^n
s	° ′	° ′	° ′	′	′	′	′	′	′
00	9 30·0	9 31·6	9 04·0	0·0	0·0	6·0	3·9	12·0	7·7
01	9 30·3	9 31·8	9 04·3	0·1	0·1	6·1	3·9	12·1	7·8
02	9 30·5	9 32·1	9 04·5	0·2	0·1	6·2	4·0	12·2	7·8
03	9 30·8	9 32·3	9 04·7	0·3	0·2	6·3	4·0	12·3	7·9
04	9 31·0	9 32·6	9 05·0	0·4	0·3	6·4	4·1	12·4	8·0
05	9 31·3	9 32·8	9 05·2	0·5	0·3	6·5	4·2	12·5	8·0
06	9 31·5	9 33·1	9 05·5	0·6	0·4	6·6	4·2	12·6	8·1
07	9 31·8	9 33·3	9 05·7	0·7	0·4	6·7	4·3	12·7	8·1
08	9 32·0	9 33·6	9 05·9	0·8	0·5	6·8	4·4	12·8	8·2
09	9 32·3	9 33·8	9 06·2	0·9	0·6	6·9	4·4	12·9	8·3
10	9 32·5	9 34·1	9 06·4	1·0	0·6	7·0	4·5	13·0	8·3
11	9 32·8	9 34·3	9 06·7	1·1	0·7	7·1	4·6	13·1	8·4
12	9 33·0	9 34·6	9 06·9	1·2	0·8	7·2	4·6	13·2	8·5
13	9 33·3	9 34·8	9 07·1	1·3	0·8	7·3	4·7	13·3	8·5
14	9 33·5	9 35·1	9 07·4	1·4	0·9	7·4	4·7	13·4	8·6
15	9 33·8	9 35·3	9 07·6	1·5	1·0	7·5	4·8	13·5	8·7
16	9 34·0	9 35·6	9 07·9	1·6	1·0	7·6	4·9	13·6	8·7
17	9 34·3	9 35·8	9 08·1	1·7	1·1	7·7	4·9	13·7	8·8
18	9 34·5	9 36·1	9 08·3	1·8	1·2	7·8	5·0	13·8	8·9
19	9 34·8	9 36·3	9 08·6	1·9	1·2	7·9	5·1	13·9	8·9
20	9 35·0	9 36·6	9 08·8	2·0	1·3	8·0	5·1	14·0	9·0
21	9 35·3	9 36·8	9 09·0	2·1	1·3	8·1	5·2	14·1	9·0
22	9 35·5	9 37·1	9 09·3	2·2	1·4	8·2	5·3	14·2	9·1
23	9 35·8	9 37·3	9 09·5	2·3	1·5	8·3	5·3	14·3	9·2
24	9 36·0	9 37·6	9 09·8	2·4	1·5	8·4	5·4	14·4	9·2
25	9 36·3	9 37·8	9 10·0	2·5	1·6	8·5	5·5	14·5	9·3
26	9 36·5	9 38·1	9 10·2	2·6	1·7	8·6	5·5	14·6	9·4
27	9 36·8	9 38·3	9 10·5	2·7	1·7	8·7	5·6	14·7	9·4
28	9 37·0	9 38·6	9 10·7	2·8	1·8	8·8	5·6	14·8	9·5
29	9 37·3	9 38·8	9 11·0	2·9	1·9	8·9	5·7	14·9	9·6
30	9 37·5	9 39·1	9 11·2	3·0	1·9	9·0	5·8	15·0	9·6
31	9 37·8	9 39·3	9 11·4	3·1	2·0	9·1	5·8	15·1	9·7
32	9 38·0	9 39·6	9 11·7	3·2	2·1	9·2	5·9	15·2	9·8
33	9 38·3	9 39·8	9 11·9	3·3	2·1	9·3	6·0	15·3	9·8
34	9 38·5	9 40·1	9 12·1	3·4	2·2	9·4	6·0	15·4	9·9
35	9 38·8	9 40·3	9 12·4	3·5	2·2	9·5	6·1	15·5	9·9
36	9 39·0	9 40·6	9 12·6	3·6	2·3	9·6	6·2	15·6	10·0
37	9 39·3	9 40·8	9 12·9	3·7	2·4	9·7	6·2	15·7	10·1
38	9 39·5	9 41·1	9 13·1	3·8	2·4	9·8	6·3	15·8	10·1
39	9 39·8	9 41·3	9 13·3	3·9	2·5	9·9	6·4	15·9	10·2
40	9 40·0	9 41·6	9 13·6	4·0	2·6	10·0	6·4	16·0	10·3
41	9 40·3	9 41·8	9 13·8	4·1	2·6	10·1	6·5	16·1	10·3
42	9 40·5	9 42·1	9 14·1	4·2	2·7	10·2	6·5	16·2	10·4
43	9 40·8	9 42·3	9 14·3	4·3	2·8	10·3	6·6	16·3	10·5
44	9 41·0	9 42·6	9 14·5	4·4	2·8	10·4	6·7	16·4	10·5
45	9 41·3	9 42·8	9 14·8	4·5	2·9	10·5	6·7	16·5	10·6
46	9 41·5	9 43·1	9 15·0	4·6	3·0	10·6	6·8	16·6	10·7
47	9 41·8	9 43·3	9 15·2	4·7	3·0	10·7	6·9	16·7	10·7
48	9 42·0	9 43·6	9 15·5	4·8	3·1	10·8	6·9	16·8	10·8
49	9 42·3	9 43·8	9 15·7	4·9	3·1	10·9	7·0	16·9	10·8
50	9 42·5	9 44·1	9 16·0	5·0	3·2	11·0	7·1	17·0	10·9
51	9 42·8	9 44·3	9 16·2	5·1	3·3	11·1	7·1	17·1	11·0
52	9 43·0	9 44·6	9 16·4	5·2	3·3	11·2	7·2	17·2	11·0
53	9 43·3	9 44·8	9 16·7	5·3	3·4	11·3	7·3	17·3	11·1
54	9 43·5	9 45·1	9 16·9	5·4	3·5	11·4	7·3	17·4	11·2
55	9 43·8	9 45·3	9 17·2	5·5	3·5	11·5	7·4	17·5	11·2
56	9 44·0	9 45·6	9 17·4	5·6	3·6	11·6	7·4	17·6	11·3
57	9 44·3	9 45·8	9 17·6	5·7	3·7	11·7	7·5	17·7	11·4
58	9 44·5	9 46·1	9 17·9	5·8	3·7	11·8	7·6	17·8	11·4
59	9 44·8	9 46·4	9 18·1	5·9	3·8	11·9	7·6	17·9	11·5
60	9 45·0	9 46·6	9 18·4	6·0	3·9	12·0	7·7	18·0	11·6

39^m	SUN PLANETS	ARIES	MOON	v or d	Corr^n	v or d	Corr^n	v or d	Corr^n
s	° ′	° ′	° ′	′	′	′	′	′	′
00	9 45·0	9 46·6	9 18·4	0·0	0·0	6·0	4·0	12·0	7·9
01	9 45·3	9 46·9	9 18·6	0·1	0·1	6·1	4·0	12·1	8·0
02	9 45·5	9 47·1	9 18·8	0·2	0·1	6·2	4·1	12·2	8·0
03	9 45·8	9 47·4	9 19·1	0·3	0·2	6·3	4·1	12·3	8·1
04	9 46·0	9 47·6	9 19·3	0·4	0·3	6·4	4·2	12·4	8·2
05	9 46·3	9 47·9	9 19·5	0·5	0·3	6·5	4·3	12·5	8·2
06	9 46·5	9 48·1	9 19·8	0·6	0·4	6·6	4·3	12·6	8·3
07	9 46·8	9 48·4	9 20·0	0·7	0·5	6·7	4·4	12·7	8·4
08	9 47·0	9 48·6	9 20·3	0·8	0·5	6·8	4·5	12·8	8·4
09	9 47·3	9 48·9	9 20·5	0·9	0·6	6·9	4·5	12·9	8·5
10	9 47·5	9 49·1	9 20·7	1·0	0·7	7·0	4·6	13·0	8·6
11	9 47·8	9 49·4	9 21·0	1·1	0·7	7·1	4·7	13·1	8·6
12	9 48·0	9 49·6	9 21·2	1·2	0·8	7·2	4·7	13·2	8·7
13	9 48·3	9 49·9	9 21·5	1·3	0·9	7·3	4·8	13·3	8·8
14	9 48·5	9 50·1	9 21·7	1·4	0·9	7·4	4·9	13·4	8·8
15	9 48·8	9 50·4	9 21·9	1·5	1·0	7·5	4·9	13·5	8·9
16	9 49·0	9 50·6	9 22·2	1·6	1·1	7·6	5·0	13·6	9·0
17	9 49·3	9 50·9	9 22·4	1·7	1·1	7·7	5·1	13·7	9·0
18	9 49·5	9 51·1	9 22·6	1·8	1·2	7·8	5·1	13·8	9·1
19	9 49·8	9 51·4	9 22·9	1·9	1·3	7·9	5·2	13·9	9·2
20	9 50·0	9 51·6	9 23·1	2·0	1·3	8·0	5·3	14·0	9·2
21	9 50·3	9 51·9	9 23·4	2·1	1·4	8·1	5·3	14·1	9·3
22	9 50·5	9 52·1	9 23·6	2·2	1·4	8·2	5·4	14·2	9·3
23	9 50·8	9 52·4	9 23·8	2·3	1·5	8·3	5·5	14·3	9·4
24	9 51·0	9 52·6	9 24·1	2·4	1·6	8·4	5·5	14·4	9·5
25	9 51·3	9 52·9	9 24·3	2·5	1·6	8·5	5·6	14·5	9·5
26	9 51·5	9 53·1	9 24·6	2·6	1·7	8·6	5·7	14·6	9·6
27	9 51·8	9 53·4	9 24·8	2·7	1·8	8·7	5·7	14·7	9·7
28	9 52·0	9 53·6	9 25·0	2·8	1·8	8·8	5·8	14·8	9·7
29	9 52·3	9 53·9	9 25·3	2·9	1·9	8·9	5·9	14·9	9·8
30	9 52·5	9 54·1	9 25·5	3·0	2·0	9·0	5·9	15·0	9·9
31	9 52·8	9 54·4	9 25·7	3·1	2·0	9·1	6·0	15·1	9·9
32	9 53·0	9 54·6	9 26·0	3·2	2·1	9·2	6·1	15·2	10·0
33	9 53·3	9 54·9	9 26·2	3·3	2·2	9·3	6·1	15·3	10·1
34	9 53·5	9 55·1	9 26·5	3·4	2·2	9·4	6·2	15·4	10·1
35	9 53·8	9 55·4	9 26·7	3·5	2·3	9·5	6·3	15·5	10·2
36	9 54·0	9 55·6	9 26·9	3·6	2·4	9·6	6·3	15·6	10·3
37	9 54·3	9 55·9	9 27·2	3·7	2·4	9·7	6·4	15·7	10·3
38	9 54·5	9 56·1	9 27·4	3·8	2·5	9·8	6·5	15·8	10·4
39	9 54·8	9 56·4	9 27·7	3·9	2·6	9·9	6·5	15·9	10·5
40	9 55·0	9 56·6	9 27·9	4·0	2·6	10·0	6·6	16·0	10·5
41	9 55·3	9 56·9	9 28·1	4·1	2·7	10·1	6·6	16·1	10·6
42	9 55·5	9 57·1	9 28·4	4·2	2·8	10·2	6·7	16·2	10·7
43	9 55·8	9 57·4	9 28·6	4·3	2·8	10·3	6·8	16·3	10·7
44	9 56·0	9 57·6	9 28·8	4·4	2·9	10·4	6·8	16·4	10·8
45	9 56·3	9 57·9	9 29·1	4·5	3·0	10·5	6·9	16·5	10·9
46	9 56·5	9 58·1	9 29·3	4·6	3·0	10·6	7·0	16·6	10·9
47	9 56·8	9 58·4	9 29·6	4·7	3·1	10·7	7·0	16·7	11·0
48	9 57·0	9 58·6	9 29·8	4·8	3·2	10·8	7·1	16·8	11·1
49	9 57·3	9 58·9	9 30·0	4·9	3·2	10·9	7·2	16·9	11·1
50	9 57·5	9 59·1	9 30·3	5·0	3·3	11·0	7·2	17·0	11·2
51	9 57·8	9 59·4	9 30·5	5·1	3·4	11·1	7·3	17·1	11·3
52	9 58·0	9 59·6	9 30·8	5·2	3·4	11·2	7·4	17·2	11·3
53	9 58·3	9 59·9	9 31·0	5·3	3·5	11·3	7·4	17·3	11·4
54	9 58·5	10 00·1	9 31·2	5·4	3·6	11·4	7·5	17·4	11·5
55	9 58·8	10 00·4	9 31·5	5·5	3·6	11·5	7·6	17·5	11·5
56	9 59·0	10 00·6	9 31·7	5·6	3·7	11·6	7·6	17·6	11·6
57	9 59·3	10 00·9	9 32·0	5·7	3·8	11·7	7·7	17·7	11·7
58	9 59·5	10 01·1	9 32·2	5·8	3·8	11·8	7·8	17·8	11·7
59	9 59·8	10 01·4	9 32·4	5·9	3·9	11·9	7·8	17·9	11·8
60	10 00·0	10 01·6	9 32·7	6·0	4·0	12·0	7·9	18·0	11·9

40m s	SUN PLANETS ° ′	ARIES ° ′	MOON ° ′	v or d ′	Corrn ′	v or d ′	Corrn ′	v or d ′	Corrn ′
00	10 00·0	10 01·6	9 32·7	0·0	0·0	6·0	4·1	12·0	8·1
01	10 00·3	10 01·9	9 32·9	0·1	0·1	6·1	4·1	12·1	8·2
02	10 00·5	10 02·1	9 33·1	0·2	0·1	6·2	4·2	12·2	8·2
03	10 00·8	10 02·4	9 33·4	0·3	0·2	6·3	4·3	12·3	8·3
04	10 01·0	10 02·6	9 33·6	0·4	0·3	6·4	4·3	12·4	8·4
05	10 01·3	10 02·9	9 33·9	0·5	0·3	6·5	4·4	12·5	8·4
06	10 01·5	10 03·1	9 34·1	0·6	0·4	6·6	4·5	12·6	8·5
07	10 01·8	10 03·4	9 34·3	0·7	0·5	6·7	4·5	12·7	8·6
08	10 02·0	10 03·6	9 34·6	0·8	0·5	6·8	4·6	12·8	8·6
09	10 02·3	10 03·9	9 34·8	0·9	0·6	6·9	4·7	12·9	8·7
10	10 02·5	10 04·1	9 35·1	1·0	0·7	7·0	4·7	13·0	8·8
11	10 02·8	10 04·4	9 35·3	1·1	0·7	7·1	4·8	13·1	8·8
12	10 03·0	10 04·7	9 35·5	1·2	0·8	7·2	4·9	13·2	8·9
13	10 03·3	10 04·9	9 35·8	1·3	0·9	7·3	4·9	13·3	9·0
14	10 03·5	10 05·2	9 36·0	1·4	0·9	7·4	5·0	13·4	9·0
15	10 03·8	10 05·4	9 36·2	1·5	1·0	7·5	5·1	13·5	9·1
16	10 04·0	10 05·7	9 36·5	1·6	1·1	7·6	5·1	13·6	9·2
17	10 04·3	10 05·9	9 36·7	1·7	1·1	7·7	5·2	13·7	9·2
18	10 04·5	10 06·2	9 37·0	1·8	1·2	7·8	5·3	13·8	9·3
19	10 04·8	10 06·4	9 37·2	1·9	1·3	7·9	5·3	13·9	9·4
20	10 05·0	10 06·7	9 37·4	2·0	1·4	8·0	5·4	14·0	9·5
21	10 05·3	10 06·9	9 37·7	2·1	1·4	8·1	5·5	14·1	9·5
22	10 05·5	10 07·2	9 37·9	2·2	1·5	8·2	5·5	14·2	9·6
23	10 05·8	10 07·4	9 38·2	2·3	1·6	8·3	5·6	14·3	9·7
24	10 06·0	10 07·7	9 38·4	2·4	1·6	8·4	5·7	14·4	9·7
25	10 06·3	10 07·9	9 38·6	2·5	1·7	8·5	5·7	14·5	9·8
26	10 06·5	10 08·2	9 38·9	2·6	1·8	8·6	5·8	14·6	9·9
27	10 06·8	10 08·4	9 39·1	2·7	1·8	8·7	5·9	14·7	9·9
28	10 07·0	10 08·7	9 39·3	2·8	1·9	8·8	5·9	14·8	10·0
29	10 07·3	10 08·9	9 39·6	2·9	2·0	8·9	6·0	14·9	10·1
30	10 07·5	10 09·2	9 39·8	3·0	2·0	9·0	6·1	15·0	10·1
31	10 07·8	10 09·4	9 40·1	3·1	2·1	9·1	6·1	15·1	10·2
32	10 08·0	10 09·7	9 40·3	3·2	2·2	9·2	6·2	15·2	10·3
33	10 08·3	10 09·9	9 40·5	3·3	2·2	9·3	6·3	15·3	10·3
34	10 08·5	10 10·2	9 40·8	3·4	2·3	9·4	6·3	15·4	10·4
35	10 08·8	10 10·4	9 41·0	3·5	2·4	9·5	6·4	15·5	10·5
36	10 09·0	10 10·7	9 41·3	3·6	2·4	9·6	6·5	15·6	10·5
37	10 09·3	10 10·9	9 41·5	3·7	2·5	9·7	6·5	15·7	10·6
38	10 09·5	10 11·2	9 41·7	3·8	2·6	9·8	6·6	15·8	10·7
39	10 09·8	10 11·4	9 42·0	3·9	2·6	9·9	6·7	15·9	10·7
40	10 10·0	10 11·7	9 42·2	4·0	2·7	10·0	6·8	16·0	10·8
41	10 10·3	10 11·9	9 42·4	4·1	2·8	10·1	6·8	16·1	10·9
42	10 10·5	10 12·2	9 42·7	4·2	2·8	10·2	6·9	16·2	10·9
43	10 10·8	10 12·4	9 42·9	4·3	2·9	10·3	7·0	16·3	11·0
44	10 11·0	10 12·7	9 43·2	4·4	3·0	10·4	7·0	16·4	11·1
45	10 11·3	10 12·9	9 43·4	4·5	3·0	10·5	7·1	16·5	11·1
46	10 11·5	10 13·2	9 43·6	4·6	3·1	10·6	7·2	16·6	11·2
47	10 11·8	10 13·4	9 43·9	4·7	3·2	10·7	7·2	16·7	11·3
48	10 12·0	10 13·7	9 44·1	4·8	3·2	10·8	7·3	16·8	11·3
49	10 12·3	10 13·9	9 44·4	4·9	3·3	10·9	7·4	16·9	11·4
50	10 12·5	10 14·2	9 44·6	5·0	3·4	11·0	7·4	17·0	11·5
51	10 12·8	10 14·4	9 44·8	5·1	3·4	11·1	7·5	17·1	11·5
52	10 13·0	10 14·7	9 45·1	5·2	3·5	11·2	7·6	17·2	11·6
53	10 13·3	10 14·9	9 45·3	5·3	3·6	11·3	7·6	17·3	11·7
54	10 13·5	10 15·2	9 45·6	5·4	3·6	11·4	7·7	17·4	11·7
55	10 13·8	10 15·4	9 45·8	5·5	3·7	11·5	7·8	17·5	11·8
56	10 14·0	10 15·7	9 46·0	5·6	3·8	11·6	7·8	17·6	11·9
57	10 14·3	10 15·9	9 46·3	5·7	3·8	11·7	7·9	17·7	11·9
58	10 14·5	10 16·2	9 46·5	5·8	3·9	11·8	8·0	17·8	12·0
59	10 14·8	10 16·4	9 46·7	5·9	4·0	11·9	8·0	17·9	12·1
60	10 15·0	10 16·7	9 47·0	6·0	4·1	12·0	8·1	18·0	12·2

41m s	SUN PLANETS ° ′	ARIES ° ′	MOON ° ′	v or d ′	Corrn ′	v or d ′	Corrn ′	v or d ′	Corrn ′
00	10 15·0	10 16·7	9 47·0	0·0	0·0	6·0	4·2	12·0	8·3
01	10 15·3	10 16·9	9 47·2	0·1	0·1	6·1	4·2	12·1	8·4
02	10 15·5	10 17·2	9 47·5	0·2	0·1	6·2	4·3	12·2	8·4
03	10 15·8	10 17·4	9 47·7	0·3	0·2	6·3	4·4	12·3	8·5
04	10 16·0	10 17·7	9 47·9	0·4	0·3	6·4	4·4	12·4	8·6
05	10 16·3	10 17·9	9 48·2	0·5	0·3	6·5	4·5	12·5	8·6
06	10 16·5	10 18·2	9 48·4	0·6	0·4	6·6	4·6	12·6	8·7
07	10 16·8	10 18·4	9 48·7	0·7	0·5	6·7	4·6	12·7	8·8
08	10 17·0	10 18·7	9 48·9	0·8	0·6	6·8	4·7	12·8	8·9
09	10 17·3	10 18·9	9 49·1	0·9	0·6	6·9	4·8	12·9	8·9
10	10 17·5	10 19·2	9 49·4	1·0	0·7	7·0	4·8	13·0	9·0
11	10 17·8	10 19·4	9 49·6	1·1	0·8	7·1	4·9	13·1	9·1
12	10 18·0	10 19·7	9 49·8	1·2	0·8	7·2	5·0	13·2	9·1
13	10 18·3	10 19·9	9 50·1	1·3	0·9	7·3	5·0	13·3	9·2
14	10 18·5	10 20·2	9 50·3	1·4	1·0	7·4	5·1	13·4	9·3
15	10 18·8	10 20·4	9 50·6	1·5	1·0	7·5	5·2	13·5	9·3
16	10 19·0	10 20·7	9 50·8	1·6	1·1	7·6	5·3	13·6	9·4
17	10 19·3	10 20·9	9 51·0	1·7	1·2	7·7	5·3	13·7	9·5
18	10 19·5	10 21·2	9 51·3	1·8	1·2	7·8	5·4	13·8	9·5
19	10 19·8	10 21·4	9 51·5	1·9	1·3	7·9	5·5	13·9	9·6
20	10 20·0	10 21·7	9 51·8	2·0	1·4	8·0	5·5	14·0	9·7
21	10 20·3	10 21·9	9 52·0	2·1	1·5	8·1	5·6	14·1	9·8
22	10 20·5	10 22·2	9 52·2	2·2	1·5	8·2	5·7	14·2	9·8
23	10 20·8	10 22·4	9 52·5	2·3	1·6	8·3	5·7	14·3	9·9
24	10 21·0	10 22·7	9 52·7	2·4	1·7	8·4	5·8	14·4	10·0
25	10 21·3	10 23·0	9 52·9	2·5	1·7	8·5	5·9	14·5	10·0
26	10 21·5	10 23·2	9 53·2	2·6	1·8	8·6	5·9	14·6	10·1
27	10 21·8	10 23·5	9 53·4	2·7	1·9	8·7	6·0	14·7	10·2
28	10 22·0	10 23·7	9 53·7	2·8	1·9	8·8	6·1	14·8	10·2
29	10 22·3	10 24·0	9 53·9	2·9	2·0	8·9	6·2	14·9	10·3
30	10 22·5	10 24·2	9 54·1	3·0	2·1	9·0	6·2	15·0	10·4
31	10 22·8	10 24·5	9 54·4	3·1	2·1	9·1	6·3	15·1	10·4
32	10 23·0	10 24·7	9 54·6	3·2	2·2	9·2	6·4	15·2	10·5
33	10 23·3	10 25·0	9 54·9	3·3	2·3	9·3	6·4	15·3	10·6
34	10 23·5	10 25·2	9 55·1	3·4	2·4	9·4	6·5	15·4	10·7
35	10 23·8	10 25·5	9 55·3	3·5	2·4	9·5	6·6	15·5	10·7
36	10 24·0	10 25·7	9 55·6	3·6	2·5	9·6	6·6	15·6	10·8
37	10 24·3	10 26·0	9 55·8	3·7	2·6	9·7	6·7	15·7	10·9
38	10 24·5	10 26·2	9 56·1	3·8	2·6	9·8	6·8	15·8	10·9
39	10 24·8	10 26·5	9 56·3	3·9	2·7	9·9	6·8	15·9	11·0
40	10 25·0	10 26·7	9 56·5	4·0	2·8	10·0	6·9	16·0	11·1
41	10 25·3	10 27·0	9 56·8	4·1	2·8	10·1	7·0	16·1	11·1
42	10 25·5	10 27·2	9 57·0	4·2	2·9	10·2	7·1	16·2	11·2
43	10 25·8	10 27·5	9 57·2	4·3	3·0	10·3	7·1	16·3	11·3
44	10 26·0	10 27·7	9 57·5	4·4	3·0	10·4	7·2	16·4	11·3
45	10 26·3	10 28·0	9 57·7	4·5	3·1	10·5	7·3	16·5	11·4
46	10 26·5	10 28·2	9 58·0	4·6	3·2	10·6	7·3	16·6	11·5
47	10 26·8	10 28·5	9 58·2	4·7	3·3	10·7	7·4	16·7	11·6
48	10 27·0	10 28·7	9 58·4	4·8	3·3	10·8	7·5	16·8	11·6
49	10 27·3	10 29·0	9 58·7	4·9	3·4	10·9	7·5	16·9	11·7
50	10 27·5	10 29·2	9 58·9	5·0	3·5	11·0	7·6	17·0	11·8
51	10 27·8	10 29·5	9 59·2	5·1	3·5	11·1	7·7	17·1	11·8
52	10 28·0	10 29·7	9 59·4	5·2	3·6	11·2	7·7	17·2	11·9
53	10 28·3	10 30·0	9 59·6	5·3	3·7	11·3	7·8	17·3	12·0
54	10 28·5	10 30·2	9 59·9	5·4	3·7	11·4	7·9	17·4	12·0
55	10 28·8	10 30·5	10 00·1	5·5	3·8	11·5	8·0	17·5	12·1
56	10 29·0	10 30·7	10 00·3	5·6	3·9	11·6	8·0	17·6	12·2
57	10 29·3	10 31·0	10 00·6	5·7	3·9	11·7	8·1	17·7	12·2
58	10 29·5	10 31·2	10 00·8	5·8	4·0	11·8	8·2	17·8	12·3
59	10 29·8	10 31·5	10 01·1	5·9	4·1	11·9	8·2	17·9	12·4
60	10 30·0	10 31·7	10 01·3	6·0	4·2	12·0	8·3	18·0	12·5

42^{m}	SUN PLANETS	ARIES	MOON	v or d	Corr^{n}	v or d	Corr^{n}	v or d	Corr^{n}
s	° ′	° ′	° ′	′	′	′	′	′	′
00	10 30·0	10 31·7	10 01·3	0·0	0·0	6·0	4·3	12·0	8·5
01	10 30·3	10 32·0	10 01·5	0·1	0·1	6·1	4·3	12·1	8·6
02	10 30·5	10 32·2	10 01·8	0·2	0·1	6·2	4·4	12·2	8·6
03	10 30·8	10 32·5	10 02·0	0·3	0·2	6·3	4·5	12·3	8·7
04	10 31·0	10 32·7	10 02·3	0·4	0·3	6·4	4·5	12·4	8·8
05	10 31·3	10 33·0	10 02·5	0·5	0·4	6·5	4·6	12·5	8·9
06	10 31·5	10 33·2	10 02·7	0·6	0·4	6·6	4·7	12·6	8·9
07	10 31·8	10 33·5	10 03·0	0·7	0·5	6·7	4·7	12·7	9·0
08	10 32·0	10 33·7	10 03·2	0·8	0·6	6·8	4·8	12·8	9·1
09	10 32·3	10 34·0	10 03·4	0·9	0·6	6·9	4·9	12·9	9·1
10	10 32·5	10 34·2	10 03·7	1·0	0·7	7·0	5·0	13·0	9·2
11	10 32·8	10 34·5	10 03·9	1·1	0·8	7·1	5·0	13·1	9·3
12	10 33·0	10 34·7	10 04·2	1·2	0·9	7·2	5·1	13·2	9·4
13	10 33·3	10 35·0	10 04·4	1·3	0·9	7·3	5·2	13·3	9·4
14	10 33·5	10 35·2	10 04·6	1·4	1·0	7·4	5·2	13·4	9·5
15	10 33·8	10 35·5	10 04·9	1·5	1·1	7·5	5·3	13·5	9·6
16	10 34·0	10 35·7	10 05·1	1·6	1·1	7·6	5·4	13·6	9·6
17	10 34·3	10 36·0	10 05·4	1·7	1·2	7·7	5·5	13·7	9·7
18	10 34·5	10 36·2	10 05·6	1·8	1·3	7·8	5·5	13·8	9·8
19	10 34·8	10 36·5	10 05·8	1·9	1·3	7·9	5·6	13·9	9·8
20	10 35·0	10 36·7	10 06·1	2·0	1·4	8·0	5·7	14·0	9·9
21	10 35·3	10 37·0	10 06·3	2·1	1·5	8·1	5·7	14·1	10·0
22	10 35·5	10 37·2	10 06·5	2·2	1·6	8·2	5·8	14·2	10·1
23	10 35·8	10 37·5	10 06·8	2·3	1·6	8·3	5·9	14·3	10·1
24	10 36·0	10 37·7	10 07·0	2·4	1·7	8·4	6·0	14·4	10·2
25	10 36·3	10 38·0	10 07·3	2·5	1·8	8·5	6·0	14·5	10·3
26	10 36·5	10 38·2	10 07·5	2·6	1·8	8·6	6·1	14·6	10·3
27	10 36·8	10 38·5	10 07·7	2·7	1·9	8·7	6·2	14·7	10·4
28	10 37·0	10 38·7	10 08·0	2·8	2·0	8·8	6·2	14·8	10·5
29	10 37·3	10 39·0	10 08·2	2·9	2·1	8·9	6·3	14·9	10·6
30	10 37·5	10 39·2	10 08·5	3·0	2·1	9·0	6·4	15·0	10·6
31	10 37·8	10 39·5	10 08·7	3·1	2·2	9·1	6·4	15·1	10·7
32	10 38·0	10 39·7	10 08·9	3·2	2·3	9·2	6·5	15·2	10·8
33	10 38·3	10 40·0	10 09·2	3·3	2·3	9·3	6·6	15·3	10·8
34	10 38·5	10 40·2	10 09·4	3·4	2·4	9·4	6·7	15·4	10·9
35	10 38·8	10 40·5	10 09·7	3·5	2·5	9·5	6·7	15·5	11·0
36	10 39·0	10 40·7	10 09·9	3·6	2·6	9·6	6·8	15·6	11·1
37	10 39·3	10 41·0	10 10·1	3·7	2·6	9·7	6·9	15·7	11·1
38	10 39·5	10 41·3	10 10·4	3·8	2·7	9·8	6·9	15·8	11·2
39	10 39·8	10 41·5	10 10·6	3·9	2·8	9·9	7·0	15·9	11·3
40	10 40·0	10 41·8	10 10·8	4·0	2·8	10·0	7·1	16·0	11·3
41	10 40·3	10 42·0	10 11·1	4·1	2·9	10·1	7·2	16·1	11·4
42	10 40·5	10 42·3	10 11·3	4·2	3·0	10·2	7·2	16·2	11·5
43	10 40·8	10 42·5	10 11·6	4·3	3·0	10·3	7·3	16·3	11·5
44	10 41·0	10 42·8	10 11·8	4·4	3·1	10·4	7·4	16·4	11·6
45	10 41·3	10 43·0	10 12·0	4·5	3·2	10·5	7·4	16·5	11·7
46	10 41·5	10 43·3	10 12·3	4·6	3·3	10·6	7·5	16·6	11·8
47	10 41·8	10 43·5	10 12·5	4·7	3·3	10·7	7·6	16·7	11·8
48	10 42·0	10 43·8	10 12·8	4·8	3·4	10·8	7·7	16·8	11·9
49	10 42·3	10 44·0	10 13·0	4·9	3·5	10·9	7·7	16·9	12·0
50	10 42·5	10 44·3	10 13·2	5·0	3·5	11·0	7·8	17·0	12·0
51	10 42·8	10 44·5	10 13·5	5·1	3·6	11·1	7·9	17·1	12·1
52	10 43·0	10 44·8	10 13·7	5·2	3·7	11·2	7·9	17·2	12·2
53	10 43·3	10 45·0	10 13·9	5·3	3·8	11·3	8·0	17·3	12·3
54	10 43·5	10 45·3	10 14·2	5·4	3·8	11·4	8·1	17·4	12·3
55	10 43·8	10 45·5	10 14·4	5·5	3·9	11·5	8·1	17·5	12·4
56	10 44·0	10 45·8	10 14·7	5·6	4·0	11·6	8·2	17·6	12·5
57	10 44·3	10 46·0	10 14·9	5·7	4·0	11·7	8·3	17·7	12·5
58	10 44·5	10 46·3	10 15·1	5·8	4·1	11·8	8·4	17·8	12·6
59	10 44·8	10 46·5	10 15·4	5·9	4·2	11·9	8·4	17·9	12·7
60	10 45·0	10 46·8	10 15·6	6·0	4·3	12·0	8·5	18·0	12·8

43^{m}	SUN PLANETS	ARIES	MOON	v or d	Corr^{n}	v or d	Corr^{n}	v or d	Corr^{n}
s	° ′	° ′	° ′	′	′	′	′	′	′
00	10 45·0	10 46·8	10 15·6	0·0	0·0	6·0	4·4	12·0	8·7
01	10 45·3	10 47·0	10 15·9	0·1	0·1	6·1	4·4	12·1	8·8
02	10 45·5	10 47·3	10 16·1	0·2	0·1	6·2	4·5	12·2	8·8
03	10 45·8	10 47·5	10 16·3	0·3	0·2	6·3	4·6	12·3	8·9
04	10 46·0	10 47·8	10 16·6	0·4	0·3	6·4	4·6	12·4	9·0
05	10 46·3	10 48·0	10 16·8	0·5	0·4	6·5	4·7	12·5	9·1
06	10 46·5	10 48·3	10 17·0	0·6	0·4	6·6	4·8	12·6	9·1
07	10 46·8	10 48·5	10 17·3	0·7	0·5	6·7	4·9	12·7	9·2
08	10 47·0	10 48·8	10 17·5	0·8	0·6	6·8	4·9	12·8	9·3
09	10 47·3	10 49·0	10 17·8	0·9	0·7	6·9	5·0	12·9	9·4
10	10 47·5	10 49·3	10 18·0	1·0	0·7	7·0	5·1	13·0	9·4
11	10 47·8	10 49·5	10 18·2	1·1	0·8	7·1	5·1	13·1	9·5
12	10 48·0	10 49·8	10 18·5	1·2	0·9	7·2	5·2	13·2	9·6
13	10 48·3	10 50·0	10 18·7	1·3	0·9	7·3	5·3	13·3	9·6
14	10 48·5	10 50·3	10 19·0	1·4	1·0	7·4	5·4	13·4	9·7
15	10 48·8	10 50·5	10 19·2	1·5	1·1	7·5	5·4	13·5	9·8
16	10 49·0	10 50·8	10 19·4	1·6	1·2	7·6	5·5	13·6	9·9
17	10 49·3	10 51·0	10 19·7	1·7	1·2	7·7	5·6	13·7	9·9
18	10 49·5	10 51·3	10 19·9	1·8	1·3	7·8	5·7	13·8	10·0
19	10 49·8	10 51·5	10 20·2	1·9	1·4	7·9	5·7	13·9	10·1
20	10 50·0	10 51·8	10 20·4	2·0	1·5	8·0	5·8	14·0	10·2
21	10 50·3	10 52·0	10 20·6	2·1	1·5	8·1	5·9	14·1	10·2
22	10 50·5	10 52·3	10 20·9	2·2	1·6	8·2	5·9	14·2	10·3
23	10 50·8	10 52·5	10 21·1	2·3	1·7	8·3	6·0	14·3	10·4
24	10 51·0	10 52·8	10 21·3	2·4	1·7	8·4	6·1	14·4	10·4
25	10 51·3	10 53·0	10 21·6	2·5	1·8	8·5	6·2	14·5	10·5
26	10 51·5	10 53·3	10 21·8	2·6	1·9	8·6	6·2	14·6	10·6
27	10 51·8	10 53·5	10 22·1	2·7	2·0	8·7	6·3	14·7	10·7
28	10 52·0	10 53·8	10 22·3	2·8	2·0	8·8	6·4	14·8	10·7
29	10 52·3	10 54·0	10 22·5	2·9	2·1	8·9	6·5	14·9	10·8
30	10 52·5	10 54·3	10 22·8	3·0	2·2	9·0	6·5	15·0	10·9
31	10 52·8	10 54·5	10 23·0	3·1	2·2	9·1	6·6	15·1	10·9
32	10 53·0	10 54·8	10 23·3	3·2	2·3	9·2	6·7	15·2	11·0
33	10 53·3	10 55·0	10 23·5	3·3	2·4	9·3	6·7	15·3	11·1
34	10 53·5	10 55·3	10 23·7	3·4	2·5	9·4	6·8	15·4	11·2
35	10 53·8	10 55·5	10 24·0	3·5	2·5	9·5	6·9	15·5	11·2
36	10 54·0	10 55·8	10 24·2	3·6	2·6	9·6	7·0	15·6	11·3
37	10 54·3	10 56·0	10 24·4	3·7	2·7	9·7	7·0	15·7	11·4
38	10 54·5	10 56·3	10 24·7	3·8	2·8	9·8	7·1	15·8	11·5
39	10 54·8	10 56·5	10 24·9	3·9	2·8	9·9	7·2	15·9	11·5
40	10 55·0	10 56·8	10 25·2	4·0	2·9	10·0	7·3	16·0	11·6
41	10 55·3	10 57·0	10 25·4	4·1	3·0	10·1	7·3	16·1	11·7
42	10 55·5	10 57·3	10 25·6	4·2	3·0	10·2	7·4	16·2	11·7
43	10 55·8	10 57·5	10 25·9	4·3	3·1	10·3	7·5	16·3	11·8
44	10 56·0	10 57·8	10 26·1	4·4	3·2	10·4	7·5	16·4	11·9
45	10 56·3	10 58·0	10 26·4	4·5	3·3	10·5	7·6	16·5	12·0
46	10 56·5	10 58·3	10 26·6	4·6	3·3	10·6	7·7	16·6	12·0
47	10 56·8	10 58·5	10 26·8	4·7	3·4	10·7	7·8	16·7	12·1
48	10 57·0	10 58·8	10 27·1	4·8	3·5	10·8	7·8	16·8	12·2
49	10 57·3	10 59·0	10 27·3	4·9	3·6	10·9	7·9	16·9	12·3
50	10 57·5	10 59·3	10 27·5	5·0	3·6	11·0	8·0	17·0	12·3
51	10 57·8	10 59·6	10 27·8	5·1	3·7	11·1	8·0	17·1	12·4
52	10 58·0	10 59·8	10 28·0	5·2	3·8	11·2	8·1	17·2	12·5
53	10 58·3	11 00·1	10 28·3	5·3	3·8	11·3	8·2	17·3	12·5
54	10 58·5	11 00·3	10 28·5	5·4	3·9	11·4	8·3	17·4	12·6
55	10 58·8	11 00·6	10 28·7	5·5	4·0	11·5	8·3	17·5	12·7
56	10 59·0	11 00·8	10 29·0	5·6	4·1	11·6	8·4	17·6	12·8
57	10 59·3	11 01·1	10 29·2	5·7	4·1	11·7	8·5	17·7	12·8
58	10 59·5	11 01·3	10 29·5	5·8	4·2	11·8	8·6	17·8	12·9
59	10 59·8	11 01·6	10 29·7	5·9	4·3	11·9	8·6	17·9	13·0
60	11 00·0	11 01·8	10 29·9	6·0	4·4	12·0	8·7	18·0	13·1

44m	SUN PLANETS	ARIES	MOON	v or d	Corrⁿ	v or d	Corrⁿ	v or d	Corrⁿ
s	° ′	° ′	° ′	′	′	′	′	′	′
00	11 00·0	11 01·8	10 29·9	0·0	0·0	6·0	4·5	12·0	8·9
01	11 00·3	11 02·1	10 30·2	0·1	0·1	6·1	4·5	12·1	9·0
02	11 00·5	11 02·3	10 30·4	0·2	0·1	6·2	4·6	12·2	9·0
03	11 00·8	11 02·6	10 30·6	0·3	0·2	6·3	4·7	12·3	9·1
04	11 01·0	11 02·8	10 30·9	0·4	0·3	6·4	4·7	12·4	9·2
05	11 01·3	11 03·1	10 31·1	0·5	0·4	6·5	4·8	12·5	9·3
06	11 01·5	11 03·3	10 31·4	0·6	0·4	6·6	4·9	12·6	9·3
07	11 01·8	11 03·6	10 31·6	0·7	0·5	6·7	5·0	12·7	9·4
08	11 02·0	11 03·8	10 31·8	0·8	0·6	6·8	5·0	12·8	9·5
09	11 02·3	11 04·1	10 32·1	0·9	0·7	6·9	5·1	12·9	9·6
10	11 02·5	11 04·3	10 32·3	1·0	0·7	7·0	5·2	13·0	9·6
11	11 02·8	11 04·6	10 32·6	1·1	0·8	7·1	5·3	13·1	9·7
12	11 03·0	11 04·8	10 32·8	1·2	0·9	7·2	5·3	13·2	9·8
13	11 03·3	11 05·1	10 33·0	1·3	1·0	7·3	5·4	13·3	9·9
14	11 03·5	11 05·3	10 33·3	1·4	1·0	7·4	5·5	13·4	9·9
15	11 03·8	11 05·6	10 33·5	1·5	1·1	7·5	5·6	13·5	10·0
16	11 04·0	11 05·8	10 33·8	1·6	1·2	7·6	5·6	13·6	10·1
17	11 04·3	11 06·1	10 34·0	1·7	1·3	7·7	5·7	13·7	10·2
18	11 04·5	11 06·3	10 34·2	1·8	1·3	7·8	5·8	13·8	10·2
19	11 04·8	11 06·6	10 34·5	1·9	1·4	7·9	5·9	13·9	10·3
20	11 05·0	11 06·8	10 34·7	2·0	1·5	8·0	5·9	14·0	10·4
21	11 05·3	11 07·1	10 34·9	2·1	1·6	8·1	6·0	14·1	10·5
22	11 05·5	11 07·3	10 35·2	2·2	1·6	8·2	6·1	14·2	10·5
23	11 05·8	11 07·6	10 35·4	2·3	1·7	8·3	6·2	14·3	10·6
24	11 06·0	11 07·8	10 35·7	2·4	1·8	8·4	6·2	14·4	10·7
25	11 06·3	11 08·1	10 35·9	2·5	1·9	8·5	6·3	14·5	10·8
26	11 06·5	11 08·3	10 36·1	2·6	1·9	8·6	6·4	14·6	10·8
27	11 06·8	11 08·6	10 36·4	2·7	2·0	8·7	6·5	14·7	10·9
28	11 07·0	11 08·8	10 36·6	2·8	2·1	8·8	6·5	14·8	11·0
29	11 07·3	11 09·1	10 36·9	2·9	2·2	8·9	6·6	14·9	11·1
30	11 07·5	11 09·3	10 37·1	3·0	2·2	9·0	6·7	15·0	11·1
31	11 07·8	11 09·6	10 37·3	3·1	2·3	9·1	6·7	15·1	11·2
32	11 08·0	11 09·8	10 37·6	3·2	2·4	9·2	6·8	15·2	11·3
33	11 08·3	11 10·1	10 37·8	3·3	2·4	9·3	6·9	15·3	11·3
34	11 08·5	11 10·3	10 38·0	3·4	2·5	9·4	7·0	15·4	11·4
35	11 08·8	11 10·6	10 38·3	3·5	2·6	9·5	7·0	15·5	11·5
36	11 09·0	11 10·8	10 38·5	3·6	2·7	9·6	7·1	15·6	11·6
37	11 09·3	11 11·1	10 38·8	3·7	2·7	9·7	7·2	15·7	11·6
38	11 09·5	11 11·3	10 39·0	3·8	2·8	9·8	7·3	15·8	11·7
39	11 09·8	11 11·6	10 39·2	3·9	2·9	9·9	7·3	15·9	11·8
40	11 10·0	11 11·8	10 39·5	4·0	3·0	10·0	7·4	16·0	11·9
41	11 10·3	11 12·1	10 39·7	4·1	3·0	10·1	7·5	16·1	11·9
42	11 10·5	11 12·3	10 40·0	4·2	3·1	10·2	7·6	16·2	12·0
43	11 10·8	11 12·6	10 40·2	4·3	3·2	10·3	7·6	16·3	12·1
44	11 11·0	11 12·8	10 40·4	4·4	3·3	10·4	7·7	16·4	12·2
45	11 11·3	11 13·1	10 40·7	4·5	3·3	10·5	7·8	16·5	12·2
46	11 11·5	11 13·3	10 40·9	4·6	3·4	10·6	7·9	16·6	12·3
47	11 11·8	11 13·6	10 41·1	4·7	3·5	10·7	7·9	16·7	12·4
48	11 12·0	11 13·8	10 41·4	4·8	3·6	10·8	8·0	16·8	12·5
49	11 12·3	11 14·1	10 41·6	4·9	3·6	10·9	8·1	16·9	12·5
50	11 12·5	11 14·3	10 41·9	5·0	3·7	11·0	8·2	17·0	12·6
51	11 12·8	11 14·6	10 42·1	5·1	3·8	11·1	8·2	17·1	12·7
52	11 13·0	11 14·8	10 42·3	5·2	3·9	11·2	8·3	17·2	12·8
53	11 13·3	11 15·1	10 42·6	5·3	3·9	11·3	8·4	17·3	12·8
54	11 13·5	11 15·3	10 42·8	5·4	4·0	11·4	8·5	17·4	12·9
55	11 13·8	11 15·6	10 43·1	5·5	4·1	11·5	8·5	17·5	13·0
56	11 14·0	11 15·8	10 43·3	5·6	4·2	11·6	8·6	17·6	13·1
57	11 14·3	11 16·1	10 43·5	5·7	4·2	11·7	8·7	17·7	13·1
58	11 14·5	11 16·3	10 43·8	5·8	4·3	11·8	8·8	17·8	13·2
59	11 14·8	11 16·6	10 44·0	5·9	4·4	11·9	8·8	17·9	13·3
60	11 15·0	11 16·8	10 44·3	6·0	4·5	12·0	8·9	18·0	13·4

45m	SUN PLANETS	ARIES	MOON	v or d	Corrⁿ	v or d	Corrⁿ	v or d	Corrⁿ
s	° ′	° ′	° ′	′	′	′	′	′	′
00	11 15·0	11 16·8	10 44·3	0·0	0·0	6·0	4·6	12·0	9·1
01	11 15·3	11 17·1	10 44·5	0·1	0·1	6·1	4·6	12·1	9·2
02	11 15·5	11 17·3	10 44·7	0·2	0·2	6·2	4·7	12·2	9·3
03	11 15·8	11 17·6	10 45·0	0·3	0·2	6·3	4·8	12·3	9·3
04	11 16·0	11 17·9	10 45·2	0·4	0·3	6·4	4·9	12·4	9·4
05	11 16·3	11 18·1	10 45·4	0·5	0·4	6·5	4·9	12·5	9·5
06	11 16·5	11 18·4	10 45·7	0·6	0·5	6·6	5·0	12·6	9·6
07	11 16·8	11 18·6	10 45·9	0·7	0·5	6·7	5·1	12·7	9·6
08	11 17·0	11 18·9	10 46·2	0·8	0·6	6·8	5·2	12·8	9·7
09	11 17·3	11 19·1	10 46·4	0·9	0·7	6·9	5·2	12·9	9·8
10	11 17·5	11 19·4	10 46·6	1·0	0·8	7·0	5·3	13·0	9·9
11	11 17·8	11 19·6	10 46·9	1·1	0·8	7·1	5·4	13·1	9·9
12	11 18·0	11 19·9	10 47·1	1·2	0·9	7·2	5·5	13·2	10·0
13	11 18·3	11 20·1	10 47·4	1·3	1·0	7·3	5·5	13·3	10·1
14	11 18·5	11 20·4	10 47·6	1·4	1·1	7·4	5·6	13·4	10·2
15	11 18·8	11 20·6	10 47·8	1·5	1·1	7·5	5·7	13·5	10·2
16	11 19·0	11 20·9	10 48·1	1·6	1·2	7·6	5·8	13·6	10·3
17	11 19·3	11 21·1	10 48·3	1·7	1·3	7·7	5·8	13·7	10·4
18	11 19·5	11 21·4	10 48·5	1·8	1·4	7·8	5·9	13·8	10·5
19	11 19·8	11 21·6	10 48·8	1·9	1·4	7·9	6·0	13·9	10·5
20	11 20·0	11 21·9	10 49·0	2·0	1·5	8·0	6·1	14·0	10·6
21	11 20·3	11 22·1	10 49·3	2·1	1·6	8·1	6·1	14·1	10·7
22	11 20·5	11 22·4	10 49·5	2·2	1·7	8·2	6·2	14·2	10·8
23	11 20·8	11 22·6	10 49·7	2·3	1·7	8·3	6·3	14·3	10·8
24	11 21·0	11 22·9	10 50·0	2·4	1·8	8·4	6·4	14·4	10·9
25	11 21·3	11 23·1	10 50·2	2·5	1·9	8·5	6·4	14·5	11·0
26	11 21·5	11 23·4	10 50·5	2·6	2·0	8·6	6·5	14·6	11·1
27	11 21·8	11 23·6	10 50·7	2·7	2·0	8·7	6·6	14·7	11·1
28	11 22·0	11 23·9	10 50·9	2·8	2·1	8·8	6·7	14·8	11·2
29	11 22·3	11 24·1	10 51·2	2·9	2·2	8·9	6·7	14·9	11·3
30	11 22·5	11 24·4	10 51·4	3·0	2·3	9·0	6·8	15·0	11·4
31	11 22·8	11 24·6	10 51·6	3·1	2·4	9·1	6·9	15·1	11·5
32	11 23·0	11 24·9	10 51·9	3·2	2·4	9·2	7·0	15·2	11·5
33	11 23·3	11 25·1	10 52·1	3·3	2·5	9·3	7·1	15·3	11·6
34	11 23·5	11 25·4	10 52·4	3·4	2·6	9·4	7·1	15·4	11·7
35	11 23·8	11 25·6	10 52·6	3·5	2·7	9·5	7·2	15·5	11·8
36	11 24·0	11 25·9	10 52·8	3·6	2·7	9·6	7·3	15·6	11·8
37	11 24·3	11 26·1	10 53·1	3·7	2·8	9·7	7·4	15·7	11·9
38	11 24·5	11 26·4	10 53·3	3·8	2·9	9·8	7·4	15·8	12·0
39	11 24·8	11 26·6	10 53·6	3·9	3·0	9·9	7·5	15·9	12·1
40	11 25·0	11 26·9	10 53·8	4·0	3·0	10·0	7·6	16·0	12·1
41	11 25·3	11 27·1	10 54·0	4·1	3·1	10·1	7·7	16·1	12·2
42	11 25·5	11 27·4	10 54·3	4·2	3·2	10·2	7·7	16·2	12·3
43	11 25·8	11 27·6	10 54·5	4·3	3·3	10·3	7·8	16·3	12·4
44	11 26·0	11 27·9	10 54·7	4·4	3·3	10·4	7·9	16·4	12·4
45	11 26·3	11 28·1	10 55·0	4·5	3·4	10·5	8·0	16·5	12·5
46	11 26·5	11 28·4	10 55·2	4·6	3·5	10·6	8·0	16·6	12·6
47	11 26·8	11 28·6	10 55·5	4·7	3·6	10·7	8·1	16·7	12·7
48	11 27·0	11 28·9	10 55·7	4·8	3·6	10·8	8·2	16·8	12·7
49	11 27·3	11 29·1	10 55·9	4·9	3·7	10·9	8·3	16·9	12·8
50	11 27·5	11 29·4	10 56·2	5·0	3·8	11·0	8·3	17·0	12·9
51	11 27·8	11 29·6	10 56·4	5·1	3·9	11·1	8·4	17·1	13·0
52	11 28·0	11 29·9	10 56·7	5·2	3·9	11·2	8·5	17·2	13·0
53	11 28·3	11 30·1	10 56·9	5·3	4·0	11·3	8·6	17·3	13·1
54	11 28·5	11 30·4	10 57·1	5·4	4·1	11·4	8·6	17·4	13·2
55	11 28·8	11 30·6	10 57·4	5·5	4·2	11·5	8·7	17·5	13·3
56	11 29·0	11 30·9	10 57·6	5·6	4·2	11·6	8·8	17·6	13·3
57	11 29·3	11 31·1	10 57·9	5·7	4·3	11·7	8·9	17·7	13·4
58	11 29·5	11 31·4	10 58·1	5·8	4·4	11·8	8·9	17·8	13·5
59	11 29·8	11 31·6	10 58·3	5·9	4·5	11·9	9·0	17·9	13·6
60	11 30·0	11 31·9	10 58·6	6·0	4·6	12·0	9·1	18·0	13·7

46m	SUN PLANETS	ARIES	MOON	v or d	Corrⁿ	v or d	Corrⁿ	v or d	Corrⁿ
s	° ′	° ′	° ′	′	′	′	′	′	′
00	11 30·0	11 31·9	10 58·6	0·0	0·0	6·0	4·7	12·0	9·3
01	11 30·3	11 32·1	10 58·8	0·1	0·1	6·1	4·7	12·1	9·4
02	11 30·5	11 32·4	10 59·0	0·2	0·2	6·2	4·8	12·2	9·5
03	11 30·8	11 32·6	10 59·3	0·3	0·2	6·3	4·9	12·3	9·5
04	11 31·0	11 32·9	10 59·5	0·4	0·3	6·4	5·0	12·4	9·6
05	11 31·3	11 33·1	10 59·8	0·5	0·4	6·5	5·0	12·5	9·7
06	11 31·5	11 33·4	11 00·0	0·6	0·5	6·6	5·1	12·6	9·8
07	11 31·8	11 33·6	11 00·2	0·7	0·5	6·7	5·2	12·7	9·8
08	11 32·0	11 33·9	11 00·5	0·8	0·6	6·8	5·3	12·8	9·9
09	11 32·3	11 34·1	11 00·7	0·9	0·7	6·9	5·3	12·9	10·0
10	11 32·5	11 34·4	11 01·0	1·0	0·8	7·0	5·4	13·0	10·1
11	11 32·8	11 34·6	11 01·2	1·1	0·9	7·1	5·5	13·1	10·2
12	11 33·0	11 34·9	11 01·4	1·2	0·9	7·2	5·6	13·2	10·2
13	11 33·3	11 35·1	11 01·7	1·3	1·0	7·3	5·7	13·3	10·3
14	11 33·5	11 35·4	11 01·9	1·4	1·1	7·4	5·7	13·4	10·4
15	11 33·8	11 35·6	11 02·1	1·5	1·2	7·5	5·8	13·5	10·5
16	11 34·0	11 35·9	11 02·4	1·6	1·2	7·6	5·9	13·6	10·5
17	11 34·3	11 36·2	11 02·6	1·7	1·3	7·7	6·0	13·7	10·6
18	11 34·5	11 36·4	11 02·9	1·8	1·4	7·8	6·0	13·8	10·7
19	11 34·8	11 36·7	11 03·1	1·9	1·5	7·9	6·1	13·9	10·8
20	11 35·0	11 36·9	11 03·3	2·0	1·6	8·0	6·2	14·0	10·9
21	11 35·3	11 37·2	11 03·6	2·1	1·6	8·1	6·3	14·1	10·9
22	11 35·5	11 37·4	11 03·8	2·2	1·7	8·2	6·4	14·2	11·0
23	11 35·8	11 37·7	11 04·1	2·3	1·8	8·3	6·4	14·3	11·1
24	11 36·0	11 37·9	11 04·3	2·4	1·9	8·4	6·5	14·4	11·2
25	11 36·3	11 38·2	11 04·5	2·5	1·9	8·5	6·6	14·5	11·2
26	11 36·5	11 38·4	11 04·8	2·6	2·0	8·6	6·7	14·6	11·3
27	11 36·8	11 38·7	11 05·0	2·7	2·1	8·7	6·7	14·7	11·4
28	11 37·0	11 38·9	11 05·2	2·8	2·2	8·8	6·8	14·8	11·5
29	11 37·3	11 39·2	11 05·5	2·9	2·2	8·9	6·9	14·9	11·5
30	11 37·5	11 39·4	11 05·7	3·0	2·3	9·0	7·0	15·0	11·6
31	11 37·8	11 39·7	11 06·0	3·1	2·4	9·1	7·1	15·1	11·7
32	11 38·0	11 39·9	11 06·2	3·2	2·5	9·2	7·1	15·2	11·8
33	11 38·3	11 40·2	11 06·4	3·3	2·6	9·3	7·2	15·3	11·9
34	11 38·5	11 40·4	11 06·7	3·4	2·6	9·4	7·3	15·4	11·9
35	11 38·8	11 40·7	11 06·9	3·5	2·7	9·5	7·4	15·5	12·0
36	11 39·0	11 40·9	11 07·2	3·6	2·8	9·6	7·4	15·6	12·1
37	11 39·3	11 41·2	11 07·4	3·7	2·9	9·7	7·5	15·7	12·2
38	11 39·5	11 41·4	11 07·6	3·8	2·9	9·8	7·6	15·8	12·2
39	11 39·8	11 41·7	11 07·9	3·9	3·0	9·9	7·7	15·9	12·3
40	11 40·0	11 41·9	11 08·1	4·0	3·1	10·0	7·8	16·0	12·4
41	11 40·3	11 42·2	11 08·3	4·1	3·2	10·1	7·8	16·1	12·5
42	11 40·5	11 42·4	11 08·6	4·2	3·3	10·2	7·9	16·2	12·6
43	11 40·8	11 42·7	11 08·8	4·3	3·3	10·3	8·0	16·3	12·6
44	11 41·0	11 42·9	11 09·1	4·4	3·4	10·4	8·1	16·4	12·7
45	11 41·3	11 43·2	11 09·3	4·5	3·5	10·5	8·1	16·5	12·8
46	11 41·5	11 43·4	11 09·5	4·6	3·6	10·6	8·2	16·6	12·9
47	11 41·8	11 43·7	11 09·8	4·7	3·6	10·7	8·3	16·7	12·9
48	11 42·0	11 43·9	11 10·0	4·8	3·7	10·8	8·4	16·8	13·0
49	11 42·3	11 44·2	11 10·3	4·9	3·8	10·9	8·4	16·9	13·1
50	11 42·5	11 44·4	11 10·5	5·0	3·9	11·0	8·5	17·0	13·2
51	11 42·8	11 44·7	11 10·7	5·1	4·0	11·1	8·6	17·1	13·3
52	11 43·0	11 44·9	11 11·0	5·2	4·0	11·2	8·7	17·2	13·3
53	11 43·3	11 45·2	11 11·2	5·3	4·1	11·3	8·8	17·3	13·4
54	11 43·5	11 45·4	11 11·5	5·4	4·2	11·4	8·8	17·4	13·5
55	11 43·8	11 45·7	11 11·7	5·5	4·3	11·5	8·9	17·5	13·6
56	11 44·0	11 45·9	11 11·9	5·6	4·3	11·6	9·0	17·6	13·6
57	11 44·3	11 46·2	11 12·2	5·7	4·4	11·7	9·1	17·7	13·7
58	11 44·5	11 46·4	11 12·4	5·8	4·5	11·8	9·1	17·8	13·8
59	11 44·8	11 46·7	11 12·6	5·9	4·6	11·9	9·2	17·9	13·9
60	11 45·0	11 46·9	11 12·9	6·0	4·7	12·0	9·3	18·0	14·0

47m	SUN PLANETS	ARIES	MOON	v or d	Corrⁿ	v or d	Corrⁿ	v or d	Corrⁿ
s	° ′	° ′	° ′	′	′	′	′	′	′
00	11 45·0	11 46·9	11 12·9	0·0	0·0	6·0	4·8	12·0	9·5
01	11 45·3	11 47·2	11 13·1	0·1	0·1	6·1	4·8	12·1	9·6
02	11 45·5	11 47·4	11 13·4	0·2	0·2	6·2	4·9	12·2	9·7
03	11 45·8	11 47·7	11 13·6	0·3	0·2	6·3	5·0	12·3	9·7
04	11 46·0	11 47·9	11 13·8	0·4	0·3	6·4	5·1	12·4	9·8
05	11 46·3	11 48·2	11 14·1	0·5	0·4	6·5	5·1	12·5	9·9
06	11 46·5	11 48·4	11 14·3	0·6	0·5	6·6	5·2	12·6	10·0
07	11 46·8	11 48·7	11 14·6	0·7	0·6	6·7	5·3	12·7	10·1
08	11 47·0	11 48·9	11 14·8	0·8	0·6	6·8	5·4	12·8	10·1
09	11 47·3	11 49·2	11 15·0	0·9	0·7	6·9	5·5	12·9	10·2
10	11 47·5	11 49·4	11 15·3	1·0	0·8	7·0	5·5	13·0	10·3
11	11 47·8	11 49·7	11 15·5	1·1	0·9	7·1	5·6	13·1	10·4
12	11 48·0	11 49·9	11 15·7	1·2	1·0	7·2	5·7	13·2	10·5
13	11 48·3	11 50·2	11 16·0	1·3	1·0	7·3	5·8	13·3	10·5
14	11 48·5	11 50·4	11 16·2	1·4	1·1	7·4	5·9	13·4	10·6
15	11 48·8	11 50·7	11 16·5	1·5	1·2	7·5	5·9	13·5	10·7
16	11 49·0	11 50·9	11 16·7	1·6	1·3	7·6	6·0	13·6	10·8
17	11 49·3	11 51·2	11 16·9	1·7	1·3	7·7	6·1	13·7	10·8
18	11 49·5	11 51·4	11 17·2	1·8	1·4	7·8	6·2	13·8	10·9
19	11 49·8	11 51·7	11 17·4	1·9	1·5	7·9	6·3	13·9	11·0
20	11 50·0	11 51·9	11 17·7	2·0	1·6	8·0	6·3	14·0	11·1
21	11 50·3	11 52·2	11 17·9	2·1	1·7	8·1	6·4	14·1	11·2
22	11 50·5	11 52·4	11 18·1	2·2	1·7	8·2	6·5	14·2	11·2
23	11 50·8	11 52·7	11 18·4	2·3	1·8	8·3	6·6	14·3	11·3
24	11 51·0	11 52·9	11 18·6	2·4	1·9	8·4	6·7	14·4	11·4
25	11 51·3	11 53·2	11 18·8	2·5	2·0	8·5	6·7	14·5	11·5
26	11 51·5	11 53·4	11 19·1	2·6	2·1	8·6	6·8	14·6	11·6
27	11 51·8	11 53·7	11 19·3	2·7	2·1	8·7	6·9	14·7	11·6
28	11 52·0	11 53·9	11 19·6	2·8	2·2	8·8	7·0	14·8	11·7
29	11 52·3	11 54·2	11 19·8	2·9	2·3	8·9	7·0	14·9	11·8
30	11 52·5	11 54·5	11 20·0	3·0	2·4	9·0	7·1	15·0	11·9
31	11 52·8	11 54·7	11 20·3	3·1	2·5	9·1	7·2	15·1	12·0
32	11 53·0	11 55·0	11 20·5	3·2	2·5	9·2	7·3	15·2	12·0
33	11 53·3	11 55·2	11 20·8	3·3	2·6	9·3	7·4	15·3	12·1
34	11 53·5	11 55·5	11 21·0	3·4	2·7	9·4	7·4	15·4	12·2
35	11 53·8	11 55·7	11 21·2	3·5	2·8	9·5	7·5	15·5	12·3
36	11 54·0	11 56·0	11 21·5	3·6	2·9	9·6	7·6	15·6	12·4
37	11 54·3	11 56·2	11 21·7	3·7	2·9	9·7	7·7	15·7	12·4
38	11 54·5	11 56·5	11 22·0	3·8	3·0	9·8	7·8	15·8	12·5
39	11 54·8	11 56·7	11 22·2	3·9	3·1	9·9	7·8	15·9	12·6
40	11 55·0	11 57·0	11 22·4	4·0	3·2	10·0	7·9	16·0	12·7
41	11 55·3	11 57·2	11 22·7	4·1	3·2	10·1	8·0	16·1	12·7
42	11 55·5	11 57·5	11 22·9	4·2	3·3	10·2	8·1	16·2	12·8
43	11 55·8	11 57·7	11 23·1	4·3	3·4	10·3	8·2	16·3	12·9
44	11 56·0	11 58·0	11 23·4	4·4	3·5	10·4	8·2	16·4	13·0
45	11 56·3	11 58·2	11 23·6	4·5	3·6	10·5	8·3	16·5	13·1
46	11 56·5	11 58·5	11 23·9	4·6	3·6	10·6	8·4	16·6	13·1
47	11 56·8	11 58·7	11 24·1	4·7	3·7	10·7	8·5	16·7	13·2
48	11 57·0	11 59·0	11 24·3	4·8	3·8	10·8	8·6	16·8	13·3
49	11 57·3	11 59·2	11 24·6	4·9	3·9	10·9	8·6	16·9	13·4
50	11 57·5	11 59·5	11 24·8	5·0	4·0	11·0	8·7	17·0	13·5
51	11 57·8	11 59·7	11 25·1	5·1	4·0	11·1	8·8	17·1	13·5
52	11 58·0	12 00·0	11 25·3	5·2	4·1	11·2	8·9	17·2	13·6
53	11 58·3	12 00·2	11 25·5	5·3	4·2	11·3	8·9	17·3	13·7
54	11 58·5	12 00·5	11 25·8	5·4	4·3	11·4	9·0	17·4	13·8
55	11 58·8	12 00·7	11 26·0	5·5	4·4	11·5	9·1	17·5	13·9
56	11 59·0	12 01·0	11 26·2	5·6	4·4	11·6	9·2	17·6	13·9
57	11 59·3	12 01·2	11 26·5	5·7	4·5	11·7	9·3	17·7	14·0
58	11 59·5	12 01·5	11 26·7	5·8	4·6	11·8	9·3	17·8	14·1
59	11 59·8	12 01·7	11 27·0	5·9	4·7	11·9	9·4	17·9	14·2
60	12 00·0	12 02·0	11 27·2	6·0	4·8	12·0	9·5	18·0	14·3

48m	SUN PLANETS	ARIES	MOON	v or d	Corrn	v or d	Corrn	v or d	Corrn
s	° ′	° ′	° ′	′	′	′	′	′	′
00	12 00·0	12 02·0	11 27·2	0·0	0·0	6·0	4·9	12·0	9·7
01	12 00·3	12 02·2	11 27·4	0·1	0·1	6·1	4·9	12·1	9·8
02	12 00·5	12 02·5	11 27·7	0·2	0·2	6·2	5·0	12·2	9·9
03	12 00·8	12 02·7	11 27·9	0·3	0·2	6·3	5·1	12·3	9·9
04	12 01·0	12 03·0	11 28·2	0·4	0·3	6·4	5·2	12·4	10·0
05	12 01·3	12 03·2	11 28·4	0·5	0·4	6·5	5·3	12·5	10·1
06	12 01·5	12 03·5	11 28·6	0·6	0·5	6·6	5·3	12·6	10·2
07	12 01·8	12 03·7	11 28·9	0·7	0·6	6·7	5·4	12·7	10·3
08	12 02·0	12 04·0	11 29·1	0·8	0·6	6·8	5·5	12·8	10·3
09	12 02·3	12 04·2	11 29·3	0·9	0·7	6·9	5·6	12·9	10·4
10	12 02·5	12 04·5	11 29·6	1·0	0·8	7·0	5·7	13·0	10·5
11	12 02·8	12 04·7	11 29·8	1·1	0·9	7·1	5·7	13·1	10·6
12	12 03·0	12 05·0	11 30·1	1·2	1·0	7·2	5·8	13·2	10·7
13	12 03·3	12 05·2	11 30·3	1·3	1·1	7·3	5·9	13·3	10·8
14	12 03·5	12 05·5	11 30·5	1·4	1·1	7·4	6·0	13·4	10·8
15	12 03·8	12 05·7	11 30·8	1·5	1·2	7·5	6·1	13·5	10·9
16	12 04·0	12 06·0	11 31·0	1·6	1·3	7·6	6·1	13·6	11·0
17	12 04·3	12 06·2	11 31·3	1·7	1·4	7·7	6·2	13·7	11·1
18	12 04·5	12 06·5	11 31·5	1·8	1·5	7·8	6·3	13·8	11·2
19	12 04·8	12 06·7	11 31·7	1·9	1·5	7·9	6·4	13·9	11·2
20	12 05·0	12 07·0	11 32·0	2·0	1·6	8·0	6·5	14·0	11·3
21	12 05·3	12 07·2	11 32·2	2·1	1·7	8·1	6·5	14·1	11·4
22	12 05·5	12 07·5	11 32·4	2·2	1·8	8·2	6·6	14·2	11·5
23	12 05·8	12 07·7	11 32·7	2·3	1·9	8·3	6·7	14·3	11·6
24	12 06·0	12 08·0	11 32·9	2·4	1·9	8·4	6·8	14·4	11·6
25	12 06·3	12 08·2	11 33·2	2·5	2·0	8·5	6·9	14·5	11·7
26	12 06·5	12 08·5	11 33·4	2·6	2·1	8·6	7·0	14·6	11·8
27	12 06·8	12 08·7	11 33·6	2·7	2·2	8·7	7·0	14·7	11·9
28	12 07·0	12 09·0	11 33·9	2·8	2·3	8·8	7·1	14·8	12·0
29	12 07·3	12 09·2	11 34·1	2·9	2·3	8·9	7·2	14·9	12·0
30	12 07·5	12 09·5	11 34·4	3·0	2·4	9·0	7·3	15·0	12·1
31	12 07·8	12 09·7	11 34·6	3·1	2·5	9·1	7·4	15·1	12·2
32	12 08·0	12 10·0	11 34·8	3·2	2·6	9·2	7·4	15·2	12·3
33	12 08·3	12 10·2	11 35·1	3·3	2·7	9·3	7·5	15·3	12·4
34	12 08·5	12 10·5	11 35·3	3·4	2·7	9·4	7·6	15·4	12·4
35	12 08·8	12 10·7	11 35·6	3·5	2·8	9·5	7·7	15·5	12·5
36	12 09·0	12 11·0	11 35·8	3·6	2·9	9·6	7·8	15·6	12·6
37	12 09·3	12 11·2	11 36·0	3·7	3·0	9·7	7·8	15·7	12·7
38	12 09·5	12 11·5	11 36·3	3·8	3·1	9·8	7·9	15·8	12·8
39	12 09·8	12 11·7	11 36·5	3·9	3·2	9·9	8·0	15·9	12·9
40	12 10·0	12 12·0	11 36·7	4·0	3·2	10·0	8·1	16·0	12·9
41	12 10·3	12 12·2	11 37·0	4·1	3·3	10·1	8·2	16·1	13·0
42	12 10·5	12 12·5	11 37·2	4·2	3·4	10·2	8·2	16·2	13·1
43	12 10·8	12 12·8	11 37·5	4·3	3·5	10·3	8·3	16·3	13·2
44	12 11·0	12 13·0	11 37·7	4·4	3·6	10·4	8·4	16·4	13·3
45	12 11·3	12 13·3	11 37·9	4·5	3·6	10·5	8·5	16·5	13·3
46	12 11·5	12 13·5	11 38·2	4·6	3·7	10·6	8·6	16·6	13·4
47	12 11·8	12 13·8	11 38·4	4·7	3·8	10·7	8·6	16·7	13·5
48	12 12·0	12 14·0	11 38·7	4·8	3·9	10·8	8·7	16·8	13·6
49	12 12·3	12 14·3	11 38·9	4·9	4·0	10·9	8·8	16·9	13·7
50	12 12·5	12 14·5	11 39·1	5·0	4·0	11·0	8·9	17·0	13·7
51	12 12·8	12 14·8	11 39·4	5·1	4·1	11·1	9·0	17·1	13·8
52	12 13·0	12 15·0	11 39·6	5·2	4·2	11·2	9·1	17·2	13·9
53	12 13·3	12 15·3	11 39·8	5·3	4·3	11·3	9·1	17·3	14·0
54	12 13·5	12 15·5	11 40·1	5·4	4·4	11·4	9·2	17·4	14·1
55	12 13·8	12 15·8	11 40·3	5·5	4·4	11·5	9·3	17·5	14·1
56	12 14·0	12 16·0	11 40·6	5·6	4·5	11·6	9·4	17·6	14·2
57	12 14·3	12 16·3	11 40·8	5·7	4·6	11·7	9·5	17·7	14·3
58	12 14·5	12 16·5	11 41·0	5·8	4·7	11·8	9·5	17·8	14·4
59	12 14·8	12 16·8	11 41·3	5·9	4·8	11·9	9·6	17·9	14·5
60	12 15·0	12 17·0	11 41·5	6·0	4·9	12·0	9·7	18·0	14·6

49m	SUN PLANETS	ARIES	MOON	v or d	Corrn	v or d	Corrn	v or d	Corrn
s	° ′	° ′	° ′	′	′	′	′	′	′
00	12 15·0	12 17·0	11 41·5	0·0	0·0	6·0	5·0	12·0	9·9
01	12 15·3	12 17·3	11 41·8	0·1	0·1	6·1	5·0	12·1	10·0
02	12 15·5	12 17·5	11 42·0	0·2	0·2	6·2	5·1	12·2	10·1
03	12 15·8	12 17·8	11 42·2	0·3	0·2	6·3	5·2	12·3	10·1
04	12 16·0	12 18·0	11 42·5	0·4	0·3	6·4	5·3	12·4	10·2
05	12 16·3	12 18·3	11 42·7	0·5	0·4	6·5	5·4	12·5	10·3
06	12 16·5	12 18·5	11 42·9	0·6	0·5	6·6	5·4	12·6	10·4
07	12 16·8	12 18·8	11 43·2	0·7	0·6	6·7	5·5	12·7	10·5
08	12 17·0	12 19·0	11 43·4	0·8	0·7	6·8	5·6	12·8	10·6
09	12 17·3	12 19·3	11 43·7	0·9	0·7	6·9	5·7	12·9	10·6
10	12 17·5	12 19·5	11 43·9	1·0	0·8	7·0	5·8	13·0	10·7
11	12 17·8	12 19·8	11 44·1	1·1	0·9	7·1	5·9	13·1	10·8
12	12 18·0	12 20·0	11 44·4	1·2	1·0	7·2	5·9	13·2	10·9
13	12 18·3	12 20·3	11 44·6	1·3	1·1	7·3	6·0	13·3	11·0
14	12 18·5	12 20·5	11 44·9	1·4	1·2	7·4	6·1	13·4	11·1
15	12 18·8	12 20·8	11 45·1	1·5	1·2	7·5	6·2	13·5	11·1
16	12 19·0	12 21·0	11 45·3	1·6	1·3	7·6	6·3	13·6	11·2
17	12 19·3	12 21·3	11 45·6	1·7	1·4	7·7	6·4	13·7	11·3
18	12 19·5	12 21·5	11 45·8	1·8	1·5	7·8	6·4	13·8	11·4
19	12 19·8	12 21·8	11 46·1	1·9	1·6	7·9	6·5	13·9	11·5
20	12 20·0	12 22·0	11 46·3	2·0	1·7	8·0	6·6	14·0	11·6
21	12 20·3	12 22·3	11 46·5	2·1	1·7	8·1	6·7	14·1	11·6
22	12 20·5	12 22·5	11 46·8	2·2	1·8	8·2	6·8	14·2	11·7
23	12 20·8	12 22·8	11 47·0	2·3	1·9	8·3	6·8	14·3	11·8
24	12 21·0	12 23·0	11 47·2	2·4	2·0	8·4	6·9	14·4	11·9
25	12 21·3	12 23·3	11 47·5	2·5	2·1	8·5	7·0	14·5	12·0
26	12 21·5	12 23·5	11 47·7	2·6	2·1	8·6	7·1	14·6	12·0
27	12 21·8	12 23·8	11 48·0	2·7	2·2	8·7	7·2	14·7	12·1
28	12 22·0	12 24·0	11 48·2	2·8	2·3	8·8	7·3	14·8	12·2
29	12 22·3	12 24·3	11 48·4	2·9	2·4	8·9	7·3	14·9	12·3
30	12 22·5	12 24·5	11 48·7	3·0	2·5	9·0	7·4	15·0	12·4
31	12 22·8	12 24·8	11 48·9	3·1	2·6	9·1	7·5	15·1	12·5
32	12 23·0	12 25·0	11 49·2	3·2	2·6	9·2	7·6	15·2	12·5
33	12 23·3	12 25·3	11 49·4	3·3	2·7	9·3	7·7	15·3	12·6
34	12 23·5	12 25·5	11 49·6	3·4	2·8	9·4	7·8	15·4	12·7
35	12 23·8	12 25·8	11 49·9	3·5	2·9	9·5	7·8	15·5	12·8
36	12 24·0	12 26·0	11 50·1	3·6	3·0	9·6	7·9	15·6	12·9
37	12 24·3	12 26·3	11 50·3	3·7	3·1	9·7	8·0	15·7	13·0
38	12 24·5	12 26·5	11 50·6	3·8	3·1	9·8	8·1	15·8	13·0
39	12 24·8	12 26·8	11 50·8	3·9	3·2	9·9	8·2	15·9	13·1
40	12 25·0	12 27·0	11 51·1	4·0	3·3	10·0	8·3	16·0	13·2
41	12 25·3	12 27·3	11 51·3	4·1	3·4	10·1	8·3	16·1	13·3
42	12 25·5	12 27·5	11 51·5	4·2	3·5	10·2	8·4	16·2	13·4
43	12 25·8	12 27·8	11 51·8	4·3	3·5	10·3	8·5	16·3	13·4
44	12 26·0	12 28·0	11 52·0	4·4	3·6	10·4	8·6	16·4	13·5
45	12 26·3	12 28·3	11 52·3	4·5	3·7	10·5	8·7	16·5	13·6
46	12 26·5	12 28·5	11 52·5	4·6	3·8	10·6	8·7	16·6	13·7
47	12 26·8	12 28·8	11 52·7	4·7	3·9	10·7	8·8	16·7	13·8
48	12 27·0	12 29·0	11 53·0	4·8	4·0	10·8	8·9	16·8	13·9
49	12 27·3	12 29·3	11 53·2	4·9	4·0	10·9	9·0	16·9	13·9
50	12 27·5	12 29·5	11 53·4	5·0	4·1	11·0	9·1	17·0	14·0
51	12 27·8	12 29·8	11 53·7	5·1	4·2	11·1	9·2	17·1	14·1
52	12 28·0	12 30·0	11 53·9	5·2	4·3	11·2	9·2	17·2	14·2
53	12 28·3	12 30·3	11 54·2	5·3	4·4	11·3	9·3	17·3	14·3
54	12 28·5	12 30·5	11 54·4	5·4	4·5	11·4	9·4	17·4	14·4
55	12 28·8	12 30·8	11 54·6	5·5	4·5	11·5	9·5	17·5	14·4
56	12 29·0	12 31·1	11 54·9	5·6	4·6	11·6	9·6	17·6	14·5
57	12 29·3	12 31·3	11 55·1	5·7	4·7	11·7	9·7	17·7	14·6
58	12 29·5	12 31·6	11 55·4	5·8	4·8	11·8	9·7	17·8	14·7
59	12 29·8	12 31·8	11 55·6	5·9	4·9	11·9	9·8	17·9	14·8
60	12 30·0	12 32·1	11 55·8	6·0	5·0	12·0	9·9	18·0	14·9

50 m	SUN PLANETS	ARIES	MOON	v or d	Corrⁿ	v or d	Corrⁿ	v or d	Corrⁿ
s	° ′	° ′	° ′	′	′	′	′	′	′
00	12 30·0	12 32·1	11 55·8	0·0	0·0	6·0	5·1	12·0	10·1
01	12 30·3	12 32·3	11 56·1	0·1	0·1	6·1	5·1	12·1	10·2
02	12 30·5	12 32·6	11 56·3	0·2	0·2	6·2	5·2	12·2	10·3
03	12 30·8	12 32·8	11 56·5	0·3	0·3	6·3	5·3	12·3	10·4
04	12 31·0	12 33·1	11 56·8	0·4	0·3	6·4	5·4	12·4	10·4
05	12 31·3	12 33·3	11 57·0	0·5	0·4	6·5	5·5	12·5	10·5
06	12 31·5	12 33·6	11 57·3	0·6	0·5	6·6	5·6	12·6	10·6
07	12 31·8	12 33·8	11 57·5	0·7	0·6	6·7	5·6	12·7	10·7
08	12 32·0	12 34·1	11 57·7	0·8	0·7	6·8	5·7	12·8	10·8
09	12 32·3	12 34·3	11 58·0	0·9	0·8	6·9	5·8	12·9	10·9
10	12 32·5	12 34·6	11 58·2	1·0	0·8	7·0	5·9	13·0	10·9
11	12 32·8	12 34·8	11 58·5	1·1	0·9	7·1	6·0	13·1	11·0
12	12 33·0	12 35·1	11 58·7	1·2	1·0	7·2	6·1	13·2	11·1
13	12 33·3	12 35·3	11 58·9	1·3	1·1	7·3	6·1	13·3	11·2
14	12 33·5	12 35·6	11 59·2	1·4	1·2	7·4	6·2	13·4	11·3
15	12 33·8	12 35·8	11 59·4	1·5	1·3	7·5	6·3	13·5	11·4
16	12 34·0	12 36·1	11 59·7	1·6	1·3	7·6	6·4	13·6	11·4
17	12 34·3	12 36·3	11 59·9	1·7	1·4	7·7	6·5	13·7	11·5
18	12 34·5	12 36·6	12 00·1	1·8	1·5	7·8	6·6	13·8	11·6
19	12 34·8	12 36·8	12 00·4	1·9	1·6	7·9	6·6	13·9	11·7
20	12 35·0	12 37·1	12 00·6	2·0	1·7	8·0	6·7	14·0	11·8
21	12 35·3	12 37·3	12 00·8	2·1	1·8	8·1	6·8	14·1	11·9
22	12 35·5	12 37·6	12 01·1	2·2	1·9	8·2	6·9	14·2	12·0
23	12 35·8	12 37·8	12 01·3	2·3	1·9	8·3	7·0	14·3	12·0
24	12 36·0	12 38·1	12 01·6	2·4	2·0	8·4	7·1	14·4	12·1
25	12 36·3	12 38·3	12 01·8	2·5	2·1	8·5	7·2	14·5	12·2
26	12 36·5	12 38·6	12 02·0	2·6	2·2	8·6	7·2	14·6	12·3
27	12 36·8	12 38·8	12 02·3	2·7	2·3	8·7	7·3	14·7	12·4
28	12 37·0	12 39·1	12 02·5	2·8	2·4	8·8	7·4	14·8	12·5
29	12 37·3	12 39·3	12 02·8	2·9	2·4	8·9	7·5	14·9	12·5
30	12 37·5	12 39·6	12 03·0	3·0	2·5	9·0	7·6	15·0	12·6
31	12 37·8	12 39·8	12 03·2	3·1	2·6	9·1	7·7	15·1	12·7
32	12 38·0	12 40·1	12 03·5	3·2	2·7	9·2	7·7	15·2	12·8
33	12 38·3	12 40·3	12 03·7	3·3	2·8	9·3	7·8	15·3	12·9
34	12 38·5	12 40·6	12 03·9	3·4	2·9	9·4	7·9	15·4	13·0
35	12 38·8	12 40·8	12 04·2	3·5	2·9	9·5	8·0	15·5	13·0
36	12 39·0	12 41·1	12 04·4	3·6	3·0	9·6	8·1	15·6	13·1
37	12 39·3	12 41·3	12 04·7	3·7	3·1	9·7	8·2	15·7	13·2
38	12 39·5	12 41·6	12 04·9	3·8	3·2	9·8	8·2	15·8	13·3
39	12 39·8	12 41·8	12 05·1	3·9	3·3	9·9	8·3	15·9	13·4
40	12 40·0	12 42·1	12 05·4	4·0	3·4	10·0	8·4	16·0	13·5
41	12 40·3	12 42·3	12 05·6	4·1	3·5	10·1	8·5	16·1	13·6
42	12 40·5	12 42·6	12 05·9	4·2	3·5	10·2	8·6	16·2	13·6
43	12 40·8	12 42·8	12 06·1	4·3	3·6	10·3	8·7	16·3	13·7
44	12 41·0	12 43·1	12 06·3	4·4	3·7	10·4	8·8	16·4	13·8
45	12 41·3	12 43·3	12 06·6	4·5	3·8	10·5	8·8	16·5	13·9
46	12 41·5	12 43·6	12 06·8	4·6	3·9	10·6	8·9	16·6	14·0
47	12 41·8	12 43·8	12 07·0	4·7	4·0	10·7	9·0	16·7	14·1
48	12 42·0	12 44·1	12 07·3	4·8	4·0	10·8	9·1	16·8	14·1
49	12 42·3	12 44·3	12 07·5	4·9	4·1	10·9	9·2	16·9	14·2
50	12 42·5	12 44·6	12 07·8	5·0	4·2	11·0	9·3	17·0	14·3
51	12 42·8	12 44·8	12 08·0	5·1	4·3	11·1	9·3	17·1	14·4
52	12 43·0	12 45·1	12 08·2	5·2	4·4	11·2	9·4	17·2	14·5
53	12 43·3	12 45·3	12 08·5	5·3	4·5	11·3	9·5	17·3	14·6
54	12 43·5	12 45·6	12 08·7	5·4	4·5	11·4	9·6	17·4	14·6
55	12 43·8	12 45·8	12 09·0	5·5	4·6	11·5	9·7	17·5	14·7
56	12 44·0	12 46·1	12 09·2	5·6	4·7	11·6	9·8	17·6	14·8
57	12 44·3	12 46·3	12 09·4	5·7	4·8	11·7	9·8	17·7	14·9
58	12 44·5	12 46·6	12 09·7	5·8	4·9	11·8	9·9	17·8	15·0
59	12 44·8	12 46·8	12 09·9	5·9	5·0	11·9	10·0	17·9	15·1
60	12 45·0	12 47·1	12 10·2	6·0	5·1	12·0	10·1	18·0	15·2

51 m	SUN PLANETS	ARIES	MOON	v or d	Corrⁿ	v or d	Corrⁿ	v or d	Corrⁿ
s	° ′	° ′	° ′	′	′	′	′	′	′
00	12 45·0	12 47·1	12 10·2	0·0	0·0	6·0	5·2	12·0	10·3
01	12 45·3	12 47·3	12 10·4	0·1	0·1	6·1	5·2	12·1	10·4
02	12 45·5	12 47·6	12 10·6	0·2	0·2	6·2	5·3	12·2	10·5
03	12 45·8	12 47·8	12 10·9	0·3	0·3	6·3	5·4	12·3	10·6
04	12 46·0	12 48·1	12 11·1	0·4	0·3	6·4	5·5	12·4	10·6
05	12 46·3	12 48·3	12 11·3	0·5	0·4	6·5	5·6	12·5	10·7
06	12 46·5	12 48·6	12 11·6	0·6	0·5	6·6	5·7	12·6	10·8
07	12 46·8	12 48·8	12 11·8	0·7	0·6	6·7	5·8	12·7	10·9
08	12 47·0	12 49·1	12 12·1	0·8	0·7	6·8	5·8	12·8	11·0
09	12 47·3	12 49·4	12 12·3	0·9	0·8	6·9	5·9	12·9	11·1
10	12 47·5	12 49·6	12 12·5	1·0	0·9	7·0	6·0	13·0	11·2
11	12 47·8	12 49·9	12 12·8	1·1	0·9	7·1	6·1	13·1	11·2
12	12 48·0	12 50·1	12 13·0	1·2	1·0	7·2	6·2	13·2	11·3
13	12 48·3	12 50·4	12 13·3	1·3	1·1	7·3	6·3	13·3	11·4
14	12 48·5	12 50·6	12 13·5	1·4	1·2	7·4	6·4	13·4	11·5
15	12 48·8	12 50·9	12 13·7	1·5	1·3	7·5	6·4	13·5	11·6
16	12 49·0	12 51·1	12 14·0	1·6	1·4	7·6	6·5	13·6	11·7
17	12 49·3	12 51·4	12 14·2	1·7	1·5	7·7	6·6	13·7	11·8
18	12 49·5	12 51·6	12 14·4	1·8	1·5	7·8	6·7	13·8	11·8
19	12 49·8	12 51·9	12 14·7	1·9	1·6	7·9	6·8	13·9	11·9
20	12 50·0	12 52·1	12 14·9	2·0	1·7	8·0	6·9	14·0	12·0
21	12 50·3	12 52·4	12 15·2	2·1	1·8	8·1	7·0	14·1	12·1
22	12 50·5	12 52·6	12 15·4	2·2	1·9	8·2	7·0	14·2	12·2
23	12 50·8	12 52·9	12 15·6	2·3	2·0	8·3	7·1	14·3	12·3
24	12 51·0	12 53·1	12 15·9	2·4	2·1	8·4	7·2	14·4	12·4
25	12 51·3	12 53·4	12 16·1	2·5	2·1	8·5	7·3	14·5	12·4
26	12 51·5	12 53·6	12 16·4	2·6	2·2	8·6	7·4	14·6	12·5
27	12 51·8	12 53·9	12 16·6	2·7	2·3	8·7	7·5	14·7	12·6
28	12 52·0	12 54·1	12 16·8	2·8	2·4	8·8	7·6	14·8	12·7
29	12 52·3	12 54·4	12 17·1	2·9	2·5	8·9	7·6	14·9	12·8
30	12 52·5	12 54·6	12 17·3	3·0	2·6	9·0	7·7	15·0	12·9
31	12 52·8	12 54·9	12 17·5	3·1	2·7	9·1	7·8	15·1	13·0
32	12 53·0	12 55·1	12 17·8	3·2	2·7	9·2	7·9	15·2	13·0
33	12 53·3	12 55·4	12 18·0	3·3	2·8	9·3	8·0	15·3	13·1
34	12 53·5	12 55·6	12 18·3	3·4	2·9	9·4	8·1	15·4	13·2
35	12 53·8	12 55·9	12 18·5	3·5	3·0	9·5	8·2	15·5	13·3
36	12 54·0	12 56·1	12 18·7	3·6	3·1	9·6	8·2	15·6	13·4
37	12 54·3	12 56·4	12 19·0	3·7	3·2	9·7	8·3	15·7	13·5
38	12 54·5	12 56·6	12 19·2	3·8	3·3	9·8	8·4	15·8	13·6
39	12 54·8	12 56·9	12 19·5	3·9	3·3	9·9	8·5	15·9	13·6
40	12 55·0	12 57·1	12 19·7	4·0	3·4	10·0	8·6	16·0	13·7
41	12 55·3	12 57·4	12 19·9	4·1	3·5	10·1	8·7	16·1	13·8
42	12 55·5	12 57·6	12 20·2	4·2	3·6	10·2	8·8	16·2	13·9
43	12 55·8	12 57·9	12 20·4	4·3	3·7	10·3	8·8	16·3	14·0
44	12 56·0	12 58·1	12 20·6	4·4	3·8	10·4	8·9	16·4	14·1
45	12 56·3	12 58·4	12 20·9	4·5	3·9	10·5	9·0	16·5	14·2
46	12 56·5	12 58·6	12 21·1	4·6	3·9	10·6	9·1	16·6	14·2
47	12 56·8	12 58·9	12 21·4	4·7	4·0	10·7	9·2	16·7	14·3
48	12 57·0	12 59·1	12 21·6	4·8	4·1	10·8	9·3	16·8	14·4
49	12 57·3	12 59·4	12 21·8	4·9	4·2	10·9	9·4	16·9	14·5
50	12 57·5	12 59·6	12 22·1	5·0	4·3	11·0	9·4	17·0	14·6
51	12 57·8	12 59·9	12 22·3	5·1	4·4	11·1	9·5	17·1	14·7
52	12 58·0	13 00·1	12 22·6	5·2	4·5	11·2	9·6	17·2	14·8
53	12 58·3	13 00·4	12 22·8	5·3	4·5	11·3	9·7	17·3	14·8
54	12 58·5	13 00·6	12 23·0	5·4	4·6	11·4	9·8	17·4	14·9
55	12 58·8	13 00·9	12 23·3	5·5	4·7	11·5	9·9	17·5	15·0
56	12 59·0	13 01·1	12 23·5	5·6	4·8	11·6	10·0	17·6	15·1
57	12 59·3	13 01·4	12 23·8	5·7	4·9	11·7	10·0	17·7	15·2
58	12 59·5	13 01·6	12 24·0	5·8	5·0	11·8	10·1	17·8	15·3
59	12 59·8	13 01·9	12 24·2	5·9	5·1	11·9	10·2	17·9	15·4
60	13 00·0	13 02·1	12 24·5	6·0	5·2	12·0	10·3	18·0	15·5

m 52	SUN PLANETS	ARIES	MOON	v or d	Corrⁿ	v or d	Corrⁿ	v or d	Corrⁿ
s	° ′	° ′	° ′	′	′	′	′	′	′
00	13 00·0	13 02·1	12 24·5	0·0	0·0	6·0	5·3	12·0	10·5
01	13 00·3	13 02·4	12 24·7	0·1	0·1	6·1	5·3	12·1	10·6
02	13 00·5	13 02·6	12 24·9	0·2	0·2	6·2	5·4	12·2	10·7
03	13 00·8	13 02·9	12 25·2	0·3	0·3	6·3	5·5	12·3	10·8
04	13 01·0	13 03·1	12 25·4	0·4	0·4	6·4	5·6	12·4	10·9
05	13 01·3	13 03·4	12 25·7	0·5	0·4	6·5	5·7	12·5	10·9
06	13 01·5	13 03·6	12 25·9	0·6	0·5	6·6	5·8	12·6	11·0
07	13 01·8	13 03·9	12 26·1	0·7	0·6	6·7	5·9	12·7	11·1
08	13 02·0	13 04·1	12 26·4	0·8	0·7	6·8	6·0	12·8	11·2
09	13 02·3	13 04·4	12 26·6	0·9	0·8	6·9	6·0	12·9	11·3
10	13 02·5	13 04·6	12 26·9	1·0	0·9	7·0	6·1	13·0	11·4
11	13 02·8	13 04·9	12 27·1	1·1	1·0	7·1	6·2	13·1	11·5
12	13 03·0	13 05·1	12 27·3	1·2	1·1	7·2	6·3	13·2	11·6
13	13 03·3	13 05·4	12 27·6	1·3	1·1	7·3	6·4	13·3	11·6
14	13 03·5	13 05·6	12 27·8	1·4	1·2	7·4	6·5	13·4	11·7
15	13 03·8	13 05·9	12 28·0	1·5	1·3	7·5	6·6	13·5	11·8
16	13 04·0	13 06·1	12 28·3	1·6	1·4	7·6	6·7	13·6	11·9
17	13 04·3	13 06·4	12 28·5	1·7	1·5	7·7	6·7	13·7	12·0
18	13 04·5	13 06·6	12 28·8	1·8	1·6	7·8	6·8	13·8	12·1
19	13 04·8	13 06·9	12 29·0	1·9	1·7	7·9	6·9	13·9	12·2
20	13 05·0	13 07·1	12 29·2	2·0	1·8	8·0	7·0	14·0	12·3
21	13 05·3	13 07·4	12 29·5	2·1	1·8	8·1	7·1	14·1	12·3
22	13 05·5	13 07·7	12 29·7	2·2	1·9	8·2	7·2	14·2	12·4
23	13 05·8	13 07·9	12 30·0	2·3	2·0	8·3	7·3	14·3	12·5
24	13 06·0	13 08·2	12 30·2	2·4	2·1	8·4	7·4	14·4	12·6
25	13 06·3	13 08·4	12 30·4	2·5	2·2	8·5	7·4	14·5	12·7
26	13 06·5	13 08·7	12 30·7	2·6	2·3	8·6	7·5	14·6	12·8
27	13 06·8	13 08·9	12 30·9	2·7	2·4	8·7	7·6	14·7	12·9
28	13 07·0	13 09·2	12 31·1	2·8	2·5	8·8	7·7	14·8	13·0
29	13 07·3	13 09·4	12 31·4	2·9	2·5	8·9	7·8	14·9	13·0
30	13 07·5	13 09·7	12 31·6	3·0	2·6	9·0	7·9	15·0	13·1
31	13 07·8	13 09·9	12 31·9	3·1	2·7	9·1	8·0	15·1	13·2
32	13 08·0	13 10·2	12 32·1	3·2	2·8	9·2	8·0	15·2	13·3
33	13 08·3	13 10·4	12 32·3	3·3	2·9	9·3	8·1	15·3	13·4
34	13 08·5	13 10·7	12 32·6	3·4	3·0	9·4	8·2	15·4	13·5
35	13 08·8	13 10·9	12 32·8	3·5	3·1	9·5	8·3	15·5	13·6
36	13 09·0	13 11·2	12 33·1	3·6	3·2	9·6	8·4	15·6	13·7
37	13 09·3	13 11·4	12 33·3	3·7	3·2	9·7	8·5	15·7	13·7
38	13 09·5	13 11·7	12 33·5	3·8	3·3	9·8	8·6	15·8	13·8
39	13 09·8	13 11·9	12 33·8	3·9	3·4	9·9	8·7	15·9	13·9
40	13 10·0	13 12·2	12 34·0	4·0	3·5	10·0	8·8	16·0	14·0
41	13 10·3	13 12·4	12 34·2	4·1	3·6	10·1	8·8	16·1	14·1
42	13 10·5	13 12·7	12 34·5	4·2	3·7	10·2	8·9	16·2	14·2
43	13 10·8	13 12·9	12 34·7	4·3	3·8	10·3	9·0	16·3	14·3
44	13 11·0	13 13·2	12 35·0	4·4	3·9	10·4	9·1	16·4	14·3
45	13 11·3	13 13·4	12 35·2	4·5	3·9	10·5	9·2	16·5	14·4
46	13 11·5	13 13·7	12 35·4	4·6	4·0	10·6	9·3	16·6	14·5
47	13 11·8	13 13·9	12 35·7	4·7	4·1	10·7	9·4	16·7	14·6
48	13 12·0	13 14·2	12 35·9	4·8	4·2	10·8	9·5	16·8	14·7
49	13 12·3	13 14·4	12 36·2	4·9	4·3	10·9	9·5	16·9	14·8
50	13 12·5	13 14·7	12 36·4	5·0	4·4	11·0	9·6	17·0	14·9
51	13 12·8	13 14·9	12 36·6	5·1	4·5	11·1	9·7	17·1	15·0
52	13 13·0	13 15·2	12 36·9	5·2	4·6	11·2	9·8	17·2	15·1
53	13 13·3	13 15·4	12 37·1	5·3	4·6	11·3	9·9	17·3	15·1
54	13 13·5	13 15·7	12 37·4	5·4	4·7	11·4	10·0	17·4	15·2
55	13 13·8	13 15·9	12 37·6	5·5	4·8	11·5	10·1	17·5	15·3
56	13 14·0	13 16·2	12 37·8	5·6	4·9	11·6	10·2	17·6	15·4
57	13 14·3	13 16·4	12 38·1	5·7	5·0	11·7	10·2	17·7	15·5
58	13 14·5	13 16·7	12 38·3	5·8	5·1	11·8	10·3	17·8	15·6
59	13 14·8	13 16·9	12 38·5	5·9	5·2	11·9	10·4	17·9	15·7
60	13 15·0	13 17·2	12 38·8	6·0	5·3	12·0	10·5	18·0	15·8

m 53	SUN PLANETS	ARIES	MOON	v or d	Corrⁿ	v or d	Corrⁿ	v or d	Corrⁿ
s	° ′	° ′	° ′	′	′	′	′	′	′
00	13 15·0	13 17·2	12 38·8	0·0	0·0	6·0	5·4	12·0	10·7
01	13 15·3	13 17·4	12 39·0	0·1	0·1	6·1	5·4	12·1	10·8
02	13 15·5	13 17·7	12 39·3	0·2	0·2	6·2	5·5	12·2	10·9
03	13 15·8	13 17·9	12 39·5	0·3	0·3	6·3	5·6	12·3	11·0
04	13 16·0	13 18·2	12 39·7	0·4	0·4	6·4	5·7	12·4	11·1
05	13 16·3	13 18·4	12 40·0	0·5	0·4	6·5	5·8	12·5	11·1
06	13 16·5	13 18·7	12 40·2	0·6	0·5	6·6	5·9	12·6	11·2
07	13 16·8	13 18·9	12 40·5	0·7	0·6	6·7	6·0	12·7	11·3
08	13 17·0	13 19·2	12 40·7	0·8	0·7	6·8	6·1	12·8	11·4
09	13 17·3	13 19·4	12 40·9	0·9	0·8	6·9	6·2	12·9	11·5
10	13 17·5	13 19·7	12 41·2	1·0	0·9	7·0	6·2	13·0	11·6
11	13 17·8	13 19·9	12 41·4	1·1	1·0	7·1	6·3	13·1	11·7
12	13 18·0	13 20·2	12 41·6	1·2	1·1	7·2	6·4	13·2	11·8
13	13 18·3	13 20·4	12 41·9	1·3	1·2	7·3	6·5	13·3	11·9
14	13 18·5	13 20·7	12 42·1	1·4	1·2	7·4	6·6	13·4	11·9
15	13 18·8	13 20·9	12 42·4	1·5	1·3	7·5	6·7	13·5	12·0
16	13 19·0	13 21·2	12 42·6	1·6	1·4	7·6	6·8	13·6	12·1
17	13 19·3	13 21·4	12 42·8	1·7	1·5	7·7	6·9	13·7	12·2
18	13 19·5	13 21·7	12 43·1	1·8	1·6	7·8	7·0	13·8	12·3
19	13 19·8	13 21·9	12 43·3	1·9	1·7	7·9	7·0	13·9	12·4
20	13 20·0	13 22·2	12 43·6	2·0	1·8	8·0	7·1	14·0	12·5
21	13 20·3	13 22·4	12 43·8	2·1	1·9	8·1	7·2	14·1	12·6
22	13 20·5	13 22·7	12 44·0	2·2	2·0	8·2	7·3	14·2	12·7
23	13 20·8	13 22·9	12 44·3	2·3	2·1	8·3	7·4	14·3	12·8
24	13 21·0	13 23·2	12 44·5	2·4	2·1	8·4	7·5	14·4	12·8
25	13 21·3	13 23·4	12 44·7	2·5	2·2	8·5	7·6	14·5	12·9
26	13 21·5	13 23·7	12 45·0	2·6	2·3	8·6	7·7	14·6	13·0
27	13 21·8	13 23·9	12 45·2	2·7	2·4	8·7	7·8	14·7	13·1
28	13 22·0	13 24·2	12 45·5	2·8	2·5	8·8	7·8	14·8	13·2
29	13 22·3	13 24·4	12 45·7	2·9	2·6	8·9	7·9	14·9	13·3
30	13 22·5	13 24·7	12 45·9	3·0	2·7	9·0	8·0	15·0	13·4
31	13 22·8	13 24·9	12 46·2	3·1	2·8	9·1	8·1	15·1	13·5
32	13 23·0	13 25·2	12 46·4	3·2	2·9	9·2	8·2	15·2	13·6
33	13 23·3	13 25·4	12 46·7	3·3	2·9	9·3	8·3	15·3	13·6
34	13 23·5	13 25·7	12 46·9	3·4	3·0	9·4	8·4	15·4	13·7
35	13 23·8	13 26·0	12 47·1	3·5	3·1	9·5	8·5	15·5	13·8
36	13 24·0	13 26·2	12 47·4	3·6	3·2	9·6	8·6	15·6	13·9
37	13 24·3	13 26·5	12 47·6	3·7	3·3	9·7	8·6	15·7	14·0
38	13 24·5	13 26·7	12 47·9	3·8	3·4	9·8	8·7	15·8	14·1
39	13 24·8	13 27·0	12 48·1	3·9	3·5	9·9	8·8	15·9	14·2
40	13 25·0	13 27·2	12 48·3	4·0	3·6	10·0	8·9	16·0	14·3
41	13 25·3	13 27·5	12 48·6	4·1	3·7	10·1	9·0	16·1	14·4
42	13 25·5	13 27·7	12 48·8	4·2	3·7	10·2	9·1	16·2	14·4
43	13 25·8	13 28·0	12 49·0	4·3	3·8	10·3	9·2	16·3	14·5
44	13 26·0	13 28·2	12 49·3	4·4	3·9	10·4	9·3	16·4	14·6
45	13 26·3	13 28·5	12 49·5	4·5	4·0	10·5	9·4	16·5	14·7
46	13 26·5	13 28·7	12 49·8	4·6	4·1	10·6	9·5	16·6	14·8
47	13 26·8	13 29·0	12 50·0	4·7	4·2	10·7	9·5	16·7	14·9
48	13 27·0	13 29·2	12 50·2	4·8	4·3	10·8	9·6	16·8	15·0
49	13 27·3	13 29·5	12 50·5	4·9	4·4	10·9	9·7	16·9	15·1
50	13 27·5	13 29·7	12 50·7	5·0	4·5	11·0	9·8	17·0	15·2
51	13 27·8	13 30·0	12 51·0	5·1	4·5	11·1	9·9	17·1	15·2
52	13 28·0	13 30·2	12 51·2	5·2	4·6	11·2	10·0	17·2	15·3
53	13 28·3	13 30·5	12 51·4	5·3	4·7	11·3	10·1	17·3	15·4
54	13 28·5	13 30·7	12 51·7	5·4	4·8	11·4	10·2	17·4	15·5
55	13 28·8	13 31·0	12 51·9	5·5	4·9	11·5	10·3	17·5	15·6
56	13 29·0	13 31·2	12 52·1	5·6	5·0	11·6	10·3	17·6	15·7
57	13 29·3	13 31·5	12 52·4	5·7	5·1	11·7	10·4	17·7	15·8
58	13 29·5	13 31·7	12 52·6	5·8	5·2	11·8	10·5	17·8	15·9
59	13 29·8	13 32·0	12 52·9	5·9	5·3	11·9	10·6	17·9	16·0
60	13 30·0	13 32·2	12 53·1	6·0	5·4	12·0	10·7	18·0	16·1

54m	SUN PLANETS	ARIES	MOON	v or d	Corrn	v or d	Corrn	v or d	Corrn
s	° ′	° ′	° ′	′	′	′	′	′	′
00	13 30·0	13 32·2	12 53·1	0·0	0·0	6·0	5·5	12·0	10·9
01	13 30·3	13 32·5	12 53·3	0·1	0·1	6·1	5·5	12·1	11·0
02	13 30·5	13 32·7	12 53·6	0·2	0·2	6·2	5·6	12·2	11·1
03	13 30·8	13 33·0	12 53·8	0·3	0·3	6·3	5·7	12·3	11·2
04	13 31·0	13 33·2	12 54·1	0·4	0·4	6·4	5·8	12·4	11·3
05	13 31·3	13 33·5	12 54·3	0·5	0·5	6·5	5·9	12·5	11·4
06	13 31·5	13 33·7	12 54·5	0·6	0·5	6·6	6·0	12·6	11·4
07	13 31·8	13 34·0	12 54·8	0·7	0·6	6·7	6·1	12·7	11·5
08	13 32·0	13 34·2	12 55·0	0·8	0·7	6·8	6·2	12·8	11·6
09	13 32·3	13 34·5	12 55·2	0·9	0·8	6·9	6·3	12·9	11·7
10	13 32·5	13 34·7	12 55·5	1·0	0·9	7·0	6·4	13·0	11·8
11	13 32·8	13 35·0	12 55·7	1·1	1·0	7·1	6·4	13·1	11·9
12	13 33·0	13 35·2	12 56·0	1·2	1·1	7·2	6·5	13·2	12·0
13	13 33·3	13 35·5	12 56·2	1·3	1·2	7·3	6·6	13·3	12·1
14	13 33·5	13 35·7	12 56·4	1·4	1·3	7·4	6·7	13·4	12·2
15	13 33·8	13 36·0	12 56·7	1·5	1·4	7·5	6·8	13·5	12·3
16	13 34·0	13 36·2	12 56·9	1·6	1·5	7·6	6·9	13·6	12·4
17	13 34·3	13 36·5	12 57·2	1·7	1·5	7·7	7·0	13·7	12·4
18	13 34·5	13 36·7	12 57·4	1·8	1·6	7·8	7·1	13·8	12·5
19	13 34·8	13 37·0	12 57·6	1·9	1·7	7·9	7·2	13·9	12·6
20	13 35·0	13 37·2	12 57·9	2·0	1·8	8·0	7·3	14·0	12·7
21	13 35·3	13 37·5	12 58·1	2·1	1·9	8·1	7·4	14·1	12·8
22	13 35·5	13 37·7	12 58·3	2·2	2·0	8·2	7·4	14·2	12·9
23	13 35·8	13 38·0	12 58·6	2·3	2·1	8·3	7·5	14·3	13·0
24	13 36·0	13 38·2	12 58·8	2·4	2·2	8·4	7·6	14·4	13·1
25	13 36·3	13 38·5	12 59·1	2·5	2·3	8·5	7·7	14·5	13·2
26	13 36·5	13 38·7	12 59·3	2·6	2·4	8·6	7·8	14·6	13·3
27	13 36·8	13 39·0	12 59·5	2·7	2·5	8·7	7·9	14·7	13·4
28	13 37·0	13 39·2	12 59·8	2·8	2·5	8·8	8·0	14·8	13·4
29	13 37·3	13 39·5	13 00·0	2·9	2·6	8·9	8·1	14·9	13·5
30	13 37·5	13 39·7	13 00·3	3·0	2·7	9·0	8·2	15·0	13·6
31	13 37·8	13 40·0	13 00·5	3·1	2·8	9·1	8·3	15·1	13·7
32	13 38·0	13 40·2	13 00·7	3·2	2·9	9·2	8·4	15·2	13·8
33	13 38·3	13 40·5	13 01·0	3·3	3·0	9·3	8·4	15·3	13·9
34	13 38·5	13 40·7	13 01·2	3·4	3·1	9·4	8·5	15·4	14·0
35	13 38·8	13 41·0	13 01·5	3·5	3·2	9·5	8·6	15·5	14·1
36	13 39·0	13 41·2	13 01·7	3·6	3·3	9·6	8·7	15·6	14·2
37	13 39·3	13 41·5	13 01·9	3·7	3·4	9·7	8·8	15·7	14·3
38	13 39·5	13 41·7	13 02·2	3·8	3·5	9·8	8·9	15·8	14·4
39	13 39·8	13 42·0	13 02·4	3·9	3·5	9·9	9·0	15·9	14·4
40	13 40·0	13 42·2	13 02·6	4·0	3·6	10·0	9·1	16·0	14·5
41	13 40·3	13 42·5	13 02·9	4·1	3·7	10·1	9·2	16·1	14·6
42	13 40·5	13 42·7	13 03·1	4·2	3·8	10·2	9·3	16·2	14·7
43	13 40·8	13 43·0	13 03·4	4·3	3·9	10·3	9·4	16·3	14·8
44	13 41·0	13 43·2	13 03·6	4·4	4·0	10·4	9·4	16·4	14·9
45	13 41·3	13 43·5	13 03·8	4·5	4·1	10·5	9·5	16·5	15·0
46	13 41·5	13 43·7	13 04·1	4·6	4·2	10·6	9·6	16·6	15·1
47	13 41·8	13 44·0	13 04·3	4·7	4·3	10·7	9·7	16·7	15·2
48	13 42·0	13 44·3	13 04·6	4·8	4·4	10·8	9·8	16·8	15·3
49	13 42·3	13 44·5	13 04·8	4·9	4·5	10·9	9·9	16·9	15·4
50	13 42·5	13 44·8	13 05·0	5·0	4·5	11·0	10·0	17·0	15·4
51	13 42·8	13 45·0	13 05·3	5·1	4·6	11·1	10·1	17·1	15·5
52	13 43·0	13 45·3	13 05·5	5·2	4·7	11·2	10·2	17·2	15·6
53	13 43·3	13 45·5	13 05·7	5·3	4·8	11·3	10·3	17·3	15·7
54	13 43·5	13 45·8	13 06·0	5·4	4·9	11·4	10·4	17·4	15·8
55	13 43·8	13 46·0	13 06·2	5·5	5·0	11·5	10·4	17·5	15·9
56	13 44·0	13 46·3	13 06·5	5·6	5·1	11·6	10·5	17·6	16·0
57	13 44·3	13 46·5	13 06·7	5·7	5·2	11·7	10·6	17·7	16·1
58	13 44·5	13 46·8	13 06·9	5·8	5·3	11·8	10·7	17·8	16·2
59	13 44·8	13 47·0	13 07·2	5·9	5·4	11·9	10·8	17·9	16·3
60	13 45·0	13 47·3	13 07·4	6·0	5·5	12·0	10·9	18·0	16·4

55m	SUN PLANETS	ARIES	MOON	v or d	Corrn	v or d	Corrn	v or d	Corrn
s	° ′	° ′	° ′	′	′	′	′	′	′
00	13 45·0	13 47·3	13 07·4	0·0	0·0	6·0	5·6	12·0	11·1
01	13 45·3	13 47·5	13 07·7	0·1	0·1	6·1	5·6	12·1	11·2
02	13 45·5	13 47·8	13 07·9	0·2	0·2	6·2	5·7	12·2	11·3
03	13 45·8	13 48·0	13 08·1	0·3	0·3	6·3	5·8	12·3	11·4
04	13 46·0	13 48·3	13 08·4	0·4	0·4	6·4	5·9	12·4	11·5
05	13 46·3	13 48·5	13 08·6	0·5	0·5	6·5	6·0	12·5	11·6
06	13 46·5	13 48·8	13 08·8	0·6	0·6	6·6	6·1	12·6	11·7
07	13 46·8	13 49·0	13 09·1	0·7	0·6	6·7	6·2	12·7	11·7
08	13 47·0	13 49·3	13 09·3	0·8	0·7	6·8	6·3	12·8	11·8
09	13 47·3	13 49·5	13 09·6	0·9	0·8	6·9	6·4	12·9	11·9
10	13 47·5	13 49·8	13 09·8	1·0	0·9	7·0	6·5	13·0	12·0
11	13 47·8	13 50·0	13 10·0	1·1	1·0	7·1	6·6	13·1	12·1
12	13 48·0	13 50·3	13 10·3	1·2	1·1	7·2	6·7	13·2	12·2
13	13 48·3	13 50·5	13 10·5	1·3	1·2	7·3	6·8	13·3	12·3
14	13 48·5	13 50·8	13 10·8	1·4	1·3	7·4	6·8	13·4	12·4
15	13 48·8	13 51·0	13 11·0	1·5	1·4	7·5	6·9	13·5	12·5
16	13 49·0	13 51·3	13 11·2	1·6	1·5	7·6	7·0	13·6	12·6
17	13 49·3	13 51·5	13 11·5	1·7	1·6	7·7	7·1	13·7	12·7
18	13 49·5	13 51·8	13 11·7	1·8	1·7	7·8	7·2	13·8	12·8
19	13 49·8	13 52·0	13 12·0	1·9	1·8	7·9	7·3	13·9	12·9
20	13 50·0	13 52·3	13 12·2	2·0	1·9	8·0	7·4	14·0	13·0
21	13 50·3	13 52·5	13 12·4	2·1	1·9	8·1	7·5	14·1	13·0
22	13 50·5	13 52·8	13 12·7	2·2	2·0	8·2	7·6	14·2	13·1
23	13 50·8	13 53·0	13 12·9	2·3	2·1	8·3	7·7	14·3	13·2
24	13 51·0	13 53·3	13 13·1	2·4	2·2	8·4	7·8	14·4	13·3
25	13 51·3	13 53·5	13 13·4	2·5	2·3	8·5	7·9	14·5	13·4
26	13 51·5	13 53·8	13 13·6	2·6	2·4	8·6	8·0	14·6	13·5
27	13 51·8	13 54·0	13 13·9	2·7	2·5	8·7	8·0	14·7	13·6
28	13 52·0	13 54·3	13 14·1	2·8	2·6	8·8	8·1	14·8	13·7
29	13 52·3	13 54·5	13 14·3	2·9	2·7	8·9	8·2	14·9	13·8
30	13 52·5	13 54·8	13 14·6	3·0	2·8	9·0	8·3	15·0	13·9
31	13 52·8	13 55·0	13 14·8	3·1	2·9	9·1	8·4	15·1	14·0
32	13 53·0	13 55·3	13 15·1	3·2	3·0	9·2	8·5	15·2	14·1
33	13 53·3	13 55·5	13 15·3	3·3	3·1	9·3	8·6	15·3	14·2
34	13 53·5	13 55·8	13 15·5	3·4	3·1	9·4	8·7	15·4	14·2
35	13 53·8	13 56·0	13 15·8	3·5	3·2	9·5	8·8	15·5	14·3
36	13 54·0	13 56·3	13 16·0	3·6	3·3	9·6	8·9	15·6	14·4
37	13 54·3	13 56·5	13 16·2	3·7	3·4	9·7	9·0	15·7	14·5
38	13 54·5	13 56·8	13 16·5	3·8	3·5	9·8	9·1	15·8	14·6
39	13 54·8	13 57·0	13 16·7	3·9	3·6	9·9	9·2	15·9	14·7
40	13 55·0	13 57·3	13 17·0	4·0	3·7	10·0	9·3	16·0	14·8
41	13 55·3	13 57·5	13 17·2	4·1	3·8	10·1	9·3	16·1	14·9
42	13 55·5	13 57·8	13 17·4	4·2	3·9	10·2	9·4	16·2	15·0
43	13 55·8	13 58·0	13 17·7	4·3	4·0	10·3	9·5	16·3	15·1
44	13 56·0	13 58·3	13 17·9	4·4	4·1	10·4	9·6	16·4	15·2
45	13 56·3	13 58·5	13 18·2	4·5	4·2	10·5	9·7	16·5	15·3
46	13 56·5	13 58·8	13 18·4	4·6	4·3	10·6	9·8	16·6	15·4
47	13 56·8	13 59·0	13 18·6	4·7	4·3	10·7	9·9	16·7	15·4
48	13 57·0	13 59·3	13 18·9	4·8	4·4	10·8	10·0	16·8	15·5
49	13 57·3	13 59·5	13 19·1	4·9	4·5	10·9	10·1	16·9	15·6
50	13 57·5	13 59·8	13 19·3	5·0	4·6	11·0	10·2	17·0	15·7
51	13 57·8	14 00·0	13 19·6	5·1	4·7	11·1	10·3	17·1	15·8
52	13 58·0	14 00·3	13 19·8	5·2	4·8	11·2	10·4	17·2	15·9
53	13 58·3	14 00·5	13 20·1	5·3	4·9	11·3	10·5	17·3	16·0
54	13 58·5	14 00·8	13 20·3	5·4	5·0	11·4	10·5	17·4	16·1
55	13 58·8	14 01·0	13 20·5	5·5	5·1	11·5	10·6	17·5	16·2
56	13 59·0	14 01·3	13 20·8	5·6	5·2	11·6	10·7	17·6	16·3
57	13 59·3	14 01·5	13 21·0	5·7	5·3	11·7	10·8	17·7	16·4
58	13 59·5	14 01·8	13 21·3	5·8	5·4	11·8	10·9	17·8	16·5
59	13 59·8	14 02·0	13 21·5	5·9	5·5	11·9	11·0	17·9	16·6
60	14 00·0	14 02·3	13 21·7	6·0	5·6	12·0	11·1	18·0	16·7

56m s	SUN PLANETS ° ′	ARIES ° ′	MOON ° ′	v or d ′	Corrn ′	v or d ′	Corrn ′	v or d ′	Corrn ′
00	14 00·0	14 02·3	13 21·7	0·0	0·0	6·0	5·7	12·0	11·3
01	14 00·3	14 02·6	13 22·0	0·1	0·1	6·1	5·7	12·1	11·4
02	14 00·5	14 02·8	13 22·2	0·2	0·2	6·2	5·8	12·2	11·5
03	14 00·8	14 03·1	13 22·4	0·3	0·3	6·3	5·9	12·3	11·6
04	14 01·0	14 03·3	13 22·7	0·4	0·4	6·4	6·0	12·4	11·7
05	14 01·3	14 03·6	13 22·9	0·5	0·5	6·5	6·1	12·5	11·8
06	14 01·5	14 03·8	13 23·2	0·6	0·6	6·6	6·2	12·6	11·9
07	14 01·8	14 04·1	13 23·4	0·7	0·7	6·7	6·3	12·7	12·0
08	14 02·0	14 04·3	13 23·6	0·8	0·8	6·8	6·4	12·8	12·1
09	14 02·3	14 04·6	13 23·9	0·9	0·8	6·9	6·5	12·9	12·1
10	14 02·5	14 04·8	13 24·1	1·0	0·9	7·0	6·6	13·0	12·2
11	14 02·8	14 05·1	13 24·4	1·1	1·0	7·1	6·7	13·1	12·3
12	14 03·0	14 05·3	13 24·6	1·2	1·1	7·2	6·8	13·2	12·4
13	14 03·3	14 05·6	13 24·8	1·3	1·2	7·3	6·9	13·3	12·5
14	14 03·5	14 05·8	13 25·1	1·4	1·3	7·4	7·0	13·4	12·6
15	14 03·8	14 06·1	13 25·3	1·5	1·4	7·5	7·1	13·5	12·7
16	14 04·0	14 06·3	13 25·6	1·6	1·5	7·6	7·2	13·6	12·8
17	14 04·3	14 06·6	13 25·8	1·7	1·6	7·7	7·3	13·7	12·9
18	14 04·5	14 06·8	13 26·0	1·8	1·7	7·8	7·3	13·8	13·0
19	14 04·8	14 07·1	13 26·3	1·9	1·8	7·9	7·4	13·9	13·1
20	14 05·0	14 07·3	13 26·5	2·0	1·9	8·0	7·5	14·0	13·2
21	14 05·3	14 07·6	13 26·7	2·1	2·0	8·1	7·6	14·1	13·3
22	14 05·5	14 07·8	13 27·0	2·2	2·1	8·2	7·7	14·2	13·4
23	14 05·8	14 08·1	13 27·2	2·3	2·2	8·3	7·8	14·3	13·5
24	14 06·0	14 08·3	13 27·5	2·4	2·3	8·4	7·9	14·4	13·6
25	14 06·3	14 08·6	13 27·7	2·5	2·4	8·5	8·0	14·5	13·7
26	14 06·5	14 08·8	13 27·9	2·6	2·4	8·6	8·1	14·6	13·7
27	14 06·8	14 09·1	13 28·2	2·7	2·5	8·7	8·2	14·7	13·8
28	14 07·0	14 09·3	13 28·4	2·8	2·6	8·8	8·3	14·8	13·9
29	14 07·3	14 09·6	13 28·7	2·9	2·7	8·9	8·4	14·9	14·0
30	14 07·5	14 09·8	13 28·9	3·0	2·8	9·0	8·5	15·0	14·1
31	14 07·8	14 10·1	13 29·1	3·1	2·9	9·1	8·6	15·1	14·2
32	14 08·0	14 10·3	13 29·4	3·2	3·0	9·2	8·7	15·2	14·3
33	14 08·3	14 10·6	13 29·6	3·3	3·1	9·3	8·8	15·3	14·4
34	14 08·5	14 10·8	13 29·8	3·4	3·2	9·4	8·9	15·4	14·5
35	14 08·8	14 11·1	13 30·1	3·5	3·3	9·5	8·9	15·5	14·6
36	14 09·0	14 11·3	13 30·3	3·6	3·4	9·6	9·0	15·6	14·7
37	14 09·3	14 11·6	13 30·6	3·7	3·5	9·7	9·1	15·7	14·8
38	14 09·5	14 11·8	13 30·8	3·8	3·6	9·8	9·2	15·8	14·9
39	14 09·8	14 12·1	13 31·0	3·9	3·7	9·9	9·3	15·9	15·0
40	14 10·0	14 12·3	13 31·3	4·0	3·8	10·0	9·4	16·0	15·1
41	14 10·3	14 12·6	13 31·5	4·1	3·9	10·1	9·5	16·1	15·2
42	14 10·5	14 12·8	13 31·8	4·2	4·0	10·2	9·6	16·2	15·3
43	14 10·8	14 13·1	13 32·0	4·3	4·0	10·3	9·7	16·3	15·3
44	14 11·0	14 13·3	13 32·2	4·4	4·1	10·4	9·8	16·4	15·4
45	14 11·3	14 13·6	13 32·5	4·5	4·2	10·5	9·9	16·5	15·5
46	14 11·5	14 13·8	13 32·7	4·6	4·3	10·6	10·0	16·6	15·6
47	14 11·8	14 14·1	13 32·9	4·7	4·4	10·7	10·1	16·7	15·7
48	14 12·0	14 14·3	13 33·2	4·8	4·5	10·8	10·2	16·8	15·8
49	14 12·3	14 14·6	13 33·4	4·9	4·6	10·9	10·3	16·9	15·9
50	14 12·5	14 14·8	13 33·7	5·0	4·7	11·0	10·4	17·0	16·0
51	14 12·8	14 15·1	13 33·9	5·1	4·8	11·1	10·5	17·1	16·1
52	14 13·0	14 15·3	13 34·1	5·2	4·9	11·2	10·5	17·2	16·2
53	14 13·3	14 15·6	13 34·4	5·3	5·0	11·3	10·6	17·3	16·3
54	14 13·5	14 15·8	13 34·6	5·4	5·1	11·4	10·7	17·4	16·4
55	14 13·8	14 16·1	13 34·9	5·5	5·2	11·5	10·8	17·5	16·5
56	14 14·0	14 16·3	13 35·1	5·6	5·3	11·6	10·9	17·6	16·6
57	14 14·3	14 16·6	13 35·3	5·7	5·4	11·7	11·0	17·7	16·7
58	14 14·5	14 16·8	13 35·6	5·8	5·5	11·8	11·1	17·8	16·8
59	14 14·8	14 17·1	13 35·8	5·9	5·6	11·9	11·2	17·9	16·9
60	14 15·0	14 17·3	13 36·1	6·0	5·7	12·0	11·3	18·0	17·0

57m s	SUN PLANETS ° ′	ARIES ° ′	MOON ° ′	v or d ′	Corrn ′	v or d ′	Corrn ′	v or d ′	Corrn ′
00	14 15·0	14 17·3	13 36·1	0·0	0·0	6·0	5·8	12·0	11·5
01	14 15·3	14 17·6	13 36·3	0·1	0·1	6·1	5·8	12·1	11·6
02	14 15·5	14 17·8	13 36·5	0·2	0·2	6·2	5·9	12·2	11·7
03	14 15·8	14 18·1	13 36·8	0·3	0·3	6·3	6·0	12·3	11·8
04	14 16·0	14 18·3	13 37·0	0·4	0·4	6·4	6·1	12·4	11·9
05	14 16·3	14 18·6	13 37·2	0·5	0·5	6·5	6·2	12·5	12·0
06	14 16·5	14 18·8	13 37·5	0·6	0·6	6·6	6·3	12·6	12·1
07	14 16·8	14 19·1	13 37·7	0·7	0·7	6·7	6·4	12·7	12·2
08	14 17·0	14 19·3	13 38·0	0·8	0·8	6·8	6·5	12·8	12·3
09	14 17·3	14 19·6	13 38·2	0·9	0·9	6·9	6·6	12·9	12·4
10	14 17·5	14 19·8	13 38·4	1·0	1·0	7·0	6·7	13·0	12·5
11	14 17·8	14 20·1	13 38·7	1·1	1·1	7·1	6·8	13·1	12·6
12	14 18·0	14 20·3	13 38·9	1·2	1·2	7·2	6·9	13·2	12·7
13	14 18·3	14 20·6	13 39·2	1·3	1·2	7·3	7·0	13·3	12·7
14	14 18·5	14 20·9	13 39·4	1·4	1·3	7·4	7·1	13·4	12·8
15	14 18·8	14 21·1	13 39·6	1·5	1·4	7·5	7·2	13·5	12·9
16	14 19·0	14 21·4	13 39·9	1·6	1·5	7·6	7·3	13·6	13·0
17	14 19·3	14 21·6	13 40·1	1·7	1·6	7·7	7·4	13·7	13·1
18	14 19·5	14 21·9	13 40·3	1·8	1·7	7·8	7·5	13·8	13·2
19	14 19·8	14 22·1	13 40·6	1·9	1·8	7·9	7·6	13·9	13·3
20	14 20·0	14 22·4	13 40·8	2·0	1·9	8·0	7·7	14·0	13·4
21	14 20·3	14 22·6	13 41·1	2·1	2·0	8·1	7·8	14·1	13·5
22	14 20·5	14 22·9	13 41·3	2·2	2·1	8·2	7·9	14·2	13·6
23	14 20·8	14 23·1	13 41·5	2·3	2·2	8·3	8·0	14·3	13·7
24	14 21·0	14 23·4	13 41·8	2·4	2·3	8·4	8·1	14·4	13·8
25	14 21·3	14 23·6	13 42·0	2·5	2·4	8·5	8·1	14·5	13·9
26	14 21·5	14 23·9	13 42·3	2·6	2·5	8·6	8·2	14·6	14·0
27	14 21·8	14 24·1	13 42·5	2·7	2·6	8·7	8·3	14·7	14·1
28	14 22·0	14 24·4	13 42·7	2·8	2·7	8·8	8·4	14·8	14·2
29	14 22·3	14 24·6	13 43·0	2·9	2·8	8·9	8·5	14·9	14·3
30	14 22·5	14 24·9	13 43·2	3·0	2·9	9·0	8·6	15·0	14·4
31	14 22·8	14 25·1	13 43·4	3·1	3·0	9·1	8·7	15·1	14·5
32	14 23·0	14 25·4	13 43·7	3·2	3·1	9·2	8·8	15·2	14·6
33	14 23·3	14 25·6	13 43·9	3·3	3·2	9·3	8·9	15·3	14·7
34	14 23·5	14 25·9	13 44·2	3·4	3·3	9·4	9·0	15·4	14·8
35	14 23·8	14 26·1	13 44·4	3·5	3·4	9·5	9·1	15·5	14·9
36	14 24·0	14 26·4	13 44·6	3·6	3·5	9·6	9·2	15·6	15·0
37	14 24·3	14 26·6	13 44·9	3·7	3·5	9·7	9·3	15·7	15·0
38	14 24·5	14 26·9	13 45·1	3·8	3·6	9·8	9·4	15·8	15·1
39	14 24·8	14 27·1	13 45·4	3·9	3·7	9·9	9·5	15·9	15·2
40	14 25·0	14 27·4	13 45·6	4·0	3·8	10·0	9·6	16·0	15·3
41	14 25·3	14 27·6	13 45·8	4·1	3·9	10·1	9·7	16·1	15·4
42	14 25·5	14 27·9	13 46·1	4·2	4·0	10·2	9·8	16·2	15·5
43	14 25·8	14 28·1	13 46·3	4·3	4·1	10·3	9·9	16·3	15·6
44	14 26·0	14 28·4	13 46·5	4·4	4·2	10·4	10·0	16·4	15·7
45	14 26·3	14 28·6	13 46·8	4·5	4·3	10·5	10·1	16·5	15·8
46	14 26·5	14 28·9	13 47·0	4·6	4·4	10·6	10·2	16·6	15·9
47	14 26·8	14 29·1	13 47·3	4·7	4·5	10·7	10·3	16·7	16·0
48	14 27·0	14 29·4	13 47·5	4·8	4·6	10·8	10·4	16·8	16·1
49	14 27·3	14 29·6	13 47·7	4·9	4·7	10·9	10·4	16·9	16·2
50	14 27·5	14 29·9	13 48·0	5·0	4·8	11·0	10·5	17·0	16·3
51	14 27·8	14 30·1	13 48·2	5·1	4·9	11·1	10·6	17·1	16·4
52	14 28·0	14 30·4	13 48·5	5·2	5·0	11·2	10·7	17·2	16·5
53	14 28·3	14 30·6	13 48·7	5·3	5·1	11·3	10·8	17·3	16·6
54	14 28·5	14 30·9	13 48·9	5·4	5·2	11·4	10·9	17·4	16·7
55	14 28·8	14 31·1	13 49·2	5·5	5·3	11·5	11·0	17·5	16·8
56	14 29·0	14 31·4	13 49·4	5·6	5·4	11·6	11·1	17·6	16·9
57	14 29·3	14 31·6	13 49·7	5·7	5·5	11·7	11·2	17·7	17·0
58	14 29·5	14 31·9	13 49·9	5·8	5·6	11·8	11·3	17·8	17·1
59	14 29·8	14 32·1	13 50·1	5·9	5·7	11·9	11·4	17·9	17·2
60	14 30·0	14 32·4	13 50·4	6·0	5·8	12·0	11·5	18·0	17·3

58ᵐ s	SUN PLANETS ° ′	ARIES ° ′	MOON ° ′	v or d ′	Corrⁿ ′	v or d ′	Corrⁿ ′	v or d ′	Corrⁿ ′
00	14 30·0	14 32·4	13 50·4	0·0	0·0	6·0	5·9	12·0	11·7
01	14 30·3	14 32·6	13 50·6	0·1	0·1	6·1	5·9	12·1	11·8
02	14 30·5	14 32·9	13 50·8	0·2	0·2	6·2	6·0	12·2	11·9
03	14 30·8	14 33·1	13 51·1	0·3	0·3	6·3	6·1	12·3	12·0
04	14 31·0	14 33·4	13 51·3	0·4	0·4	6·4	6·2	12·4	12·1
05	14 31·3	14 33·6	13 51·6	0·5	0·5	6·5	6·3	12·5	12·2
06	14 31·5	14 33·9	13 51·8	0·6	0·6	6·6	6·4	12·6	12·3
07	14 31·8	14 34·1	13 52·0	0·7	0·7	6·7	6·5	12·7	12·4
08	14 32·0	14 34·4	13 52·3	0·8	0·8	6·8	6·6	12·8	12·5
09	14 32·3	14 34·6	13 52·5	0·9	0·9	6·9	6·7	12·9	12·6
10	14 32·5	14 34·9	13 52·8	1·0	1·0	7·0	6·8	13·0	12·7
11	14 32·8	14 35·1	13 53·0	1·1	1·1	7·1	6·9	13·1	12·8
12	14 33·0	14 35·4	13 53·2	1·2	1·2	7·2	7·0	13·2	12·9
13	14 33·3	14 35·6	13 53·5	1·3	1·3	7·3	7·1	13·3	13·0
14	14 33·5	14 35·9	13 53·7	1·4	1·4	7·4	7·2	13·4	13·1
15	14 33·8	14 36·1	13 53·9	1·5	1·5	7·5	7·3	13·5	13·2
16	14 34·0	14 36·4	13 54·2	1·6	1·6	7·6	7·4	13·6	13·3
17	14 34·3	14 36·6	13 54·4	1·7	1·7	7·7	7·5	13·7	13·4
18	14 34·5	14 36·9	13 54·7	1·8	1·8	7·8	7·6	13·8	13·5
19	14 34·8	14 37·1	13 54·9	1·9	1·9	7·9	7·7	13·9	13·6
20	14 35·0	14 37·4	13 55·1	2·0	2·0	8·0	7·8	14·0	13·7
21	14 35·3	14 37·6	13 55·4	2·1	2·0	8·1	7·9	14·1	13·7
22	14 35·5	14 37·9	13 55·6	2·2	2·1	8·2	8·0	14·2	13·8
23	14 35·8	14 38·1	13 55·9	2·3	2·2	8·3	8·1	14·3	13·9
24	14 36·0	14 38·4	13 56·1	2·4	2·3	8·4	8·2	14·4	14·0
25	14 36·3	14 38·6	13 56·3	2·5	2·4	8·5	8·3	14·5	14·1
26	14 36·5	14 38·9	13 56·6	2·6	2·5	8·6	8·4	14·6	14·2
27	14 36·8	14 39·2	13 56·8	2·7	2·6	8·7	8·5	14·7	14·3
28	14 37·0	14 39·4	13 57·0	2·8	2·7	8·8	8·6	14·8	14·4
29	14 37·3	14 39·7	13 57·3	2·9	2·8	8·9	8·7	14·9	14·5
30	14 37·5	14 39·9	13 57·5	3·0	2·9	9·0	8·8	15·0	14·6
31	14 37·8	14 40·2	13 57·8	3·1	3·0	9·1	8·9	15·1	14·7
32	14 38·0	14 40·4	13 58·0	3·2	3·1	9·2	9·0	15·2	14·8
33	14 38·3	14 40·7	13 58·2	3·3	3·2	9·3	9·1	15·3	14·9
34	14 38·5	14 40·9	13 58·5	3·4	3·3	9·4	9·2	15·4	15·0
35	14 38·8	14 41·2	13 58·7	3·5	3·4	9·5	9·3	15·5	15·1
36	14 39·0	14 41·4	13 59·0	3·6	3·5	9·6	9·4	15·6	15·2
37	14 39·3	14 41·7	13 59·2	3·7	3·6	9·7	9·5	15·7	15·3
38	14 39·5	14 41·9	13 59·4	3·8	3·7	9·8	9·6	15·8	15·4
39	14 39·8	14 42·2	13 59·7	3·9	3·8	9·9	9·7	15·9	15·5
40	14 40·0	14 42·4	13 59·9	4·0	3·9	10·0	9·8	16·0	15·6
41	14 40·3	14 42·7	14 00·1	4·1	4·0	10·1	9·8	16·1	15·7
42	14 40·5	14 42·9	14 00·4	4·2	4·1	10·2	9·9	16·2	15·8
43	14 40·8	14 43·2	14 00·6	4·3	4·2	10·3	10·0	16·3	15·9
44	14 41·0	14 43·4	14 00·9	4·4	4·3	10·4	10·1	16·4	16·0
45	14 41·3	14 43·7	14 01·1	4·5	4·4	10·5	10·2	16·5	16·1
46	14 41·5	14 43·9	14 01·3	4·6	4·5	10·6	10·3	16·6	16·2
47	14 41·8	14 44·2	14 01·6	4·7	4·6	10·7	10·4	16·7	16·3
48	14 42·0	14 44·4	14 01·8	4·8	4·7	10·8	10·5	16·8	16·4
49	14 42·3	14 44·7	14 02·1	4·9	4·8	10·9	10·6	16·9	16·5
50	14 42·5	14 44·9	14 02·3	5·0	4·9	11·0	10·7	17·0	16·6
51	14 42·8	14 45·2	14 02·5	5·1	5·0	11·1	10·8	17·1	16·7
52	14 43·0	14 45·4	14 02·8	5·2	5·1	11·2	10·9	17·2	16·8
53	14 43·3	14 45·7	14 03·0	5·3	5·2	11·3	11·0	17·3	16·9
54	14 43·5	14 45·9	14 03·3	5·4	5·3	11·4	11·1	17·4	17·0
55	14 43·8	14 46·2	14 03·5	5·5	5·4	11·5	11·2	17·5	17·1
56	14 44·0	14 46·4	14 03·7	5·6	5·5	11·6	11·3	17·6	17·2
57	14 44·3	14 46·7	14 04·0	5·7	5·6	11·7	11·4	17·7	17·3
58	14 44·5	14 46·9	14 04·2	5·8	5·7	11·8	11·5	17·8	17·4
59	14 44·8	14 47·2	14 04·4	5·9	5·8	11·9	11·6	17·9	17·5
60	14 45·0	14 47·4	14 04·7	6·0	5·9	12·0	11·7	18·0	17·6

59ᵐ s	SUN PLANETS ° ′	ARIES ° ′	MOON ° ′	v or d ′	Corrⁿ ′	v or d ′	Corrⁿ ′	v or d ′	Corrⁿ ′
00	14 45·0	14 47·4	14 04·7	0·0	0·0	6·0	6·0	12·0	11·9
01	14 45·3	14 47·7	14 04·9	0·1	0·1	6·1	6·0	12·1	12·0
02	14 45·5	14 47·9	14 05·2	0·2	0·2	6·2	6·1	12·2	12·1
03	14 45·8	14 48·2	14 05·4	0·3	0·3	6·3	6·2	12·3	12·2
04	14 46·0	14 48·4	14 05·6	0·4	0·4	6·4	6·3	12·4	12·3
05	14 46·3	14 48·7	14 05·9	0·5	0·5	6·5	6·4	12·5	12·4
06	14 46·5	14 48·9	14 06·1	0·6	0·6	6·6	6·5	12·6	12·5
07	14 46·8	14 49·2	14 06·4	0·7	0·7	6·7	6·6	12·7	12·6
08	14 47·0	14 49·4	14 06·6	0·8	0·8	6·8	6·7	12·8	12·7
09	14 47·3	14 49·7	14 06·8	0·9	0·9	6·9	6·8	12·9	12·8
10	14 47·5	14 49·9	14 07·1	1·0	1·0	7·0	6·9	13·0	12·9
11	14 47·8	14 50·2	14 07·3	1·1	1·1	7·1	7·0	13·1	13·0
12	14 48·0	14 50·4	14 07·5	1·2	1·2	7·2	7·1	13·2	13·1
13	14 48·3	14 50·7	14 07·8	1·3	1·3	7·3	7·2	13·3	13·2
14	14 48·5	14 50·9	14 08·0	1·4	1·4	7·4	7·3	13·4	13·3
15	14 48·8	14 51·2	14 08·3	1·5	1·5	7·5	7·4	13·5	13·4
16	14 49·0	14 51·4	14 08·5	1·6	1·6	7·6	7·5	13·6	13·5
17	14 49·3	14 51·7	14 08·7	1·7	1·7	7·7	7·6	13·7	13·6
18	14 49·5	14 51·9	14 09·0	1·8	1·8	7·8	7·7	13·8	13·7
19	14 49·8	14 52·2	14 09·2	1·9	1·9	7·9	7·8	13·9	13·8
20	14 50·0	14 52·4	14 09·5	2·0	2·0	8·0	7·9	14·0	13·9
21	14 50·3	14 52·7	14 09·7	2·1	2·1	8·1	8·0	14·1	14·0
22	14 50·5	14 52·9	14 09·9	2·2	2·2	8·2	8·1	14·2	14·1
23	14 50·8	14 53·2	14 10·2	2·3	2·3	8·3	8·2	14·3	14·2
24	14 51·0	14 53·4	14 10·4	2·4	2·4	8·4	8·3	14·4	14·3
25	14 51·3	14 53·7	14 10·6	2·5	2·5	8·5	8·4	14·5	14·4
26	14 51·5	14 53·9	14 10·9	2·6	2·6	8·6	8·5	14·6	14·5
27	14 51·8	14 54·2	14 11·1	2·7	2·7	8·7	8·6	14·7	14·6
28	14 52·0	14 54·4	14 11·4	2·8	2·8	8·8	8·7	14·8	14·7
29	14 52·3	14 54·7	14 11·6	2·9	2·9	8·9	8·8	14·9	14·8
30	14 52·5	14 54·9	14 11·8	3·0	3·0	9·0	8·9	15·0	14·9
31	14 52·8	14 55·2	14 12·1	3·1	3·1	9·1	9·0	15·1	15·0
32	14 53·0	14 55·4	14 12·3	3·2	3·2	9·2	9·1	15·2	15·1
33	14 53·3	14 55·7	14 12·6	3·3	3·3	9·3	9·2	15·3	15·2
34	14 53·5	14 55·9	14 12·8	3·4	3·4	9·4	9·3	15·4	15·3
35	14 53·8	14 56·2	14 13·0	3·5	3·5	9·5	9·4	15·5	15·4
36	14 54·0	14 56·4	14 13·3	3·6	3·6	9·6	9·5	15·6	15·5
37	14 54·3	14 56·7	14 13·5	3·7	3·7	9·7	9·6	15·7	15·6
38	14 54·5	14 56·9	14 13·8	3·8	3·8	9·8	9·7	15·8	15·7
39	14 54·8	14 57·2	14 14·0	3·9	3·9	9·9	9·8	15·9	15·8
40	14 55·0	14 57·5	14 14·2	4·0	4·0	10·0	9·9	16·0	15·9
41	14 55·3	14 57·7	14 14·5	4·1	4·1	10·1	10·0	16·1	16·0
42	14 55·5	14 58·0	14 14·7	4·2	4·2	10·2	10·1	16·2	16·1
43	14 55·8	14 58·2	14 14·9	4·3	4·3	10·3	10·2	16·3	16·2
44	14 56·0	14 58·5	14 15·2	4·4	4·4	10·4	10·3	16·4	16·3
45	14 56·3	14 58·7	14 15·4	4·5	4·5	10·5	10·4	16·5	16·4
46	14 56·5	14 59·0	14 15·7	4·6	4·6	10·6	10·5	16·6	16·5
47	14 56·8	14 59·2	14 15·9	4·7	4·7	10·7	10·6	16·7	16·6
48	14 57·0	14 59·5	14 16·1	4·8	4·8	10·8	10·7	16·8	16·7
49	14 57·3	14 59·7	14 16·4	4·9	4·9	10·9	10·8	16·9	16·8
50	14 57·5	15 00·0	14 16·6	5·0	5·0	11·0	10·9	17·0	16·9
51	14 57·8	15 00·2	14 16·9	5·1	5·1	11·1	11·0	17·1	17·0
52	14 58·0	15 00·5	14 17·1	5·2	5·2	11·2	11·1	17·2	17·1
53	14 58·3	15 00·7	14 17·3	5·3	5·3	11·3	11·2	17·3	17·2
54	14 58·5	15 01·0	14 17·6	5·4	5·4	11·4	11·3	17·4	17·3
55	14 58·8	15 01·2	14 17·8	5·5	5·5	11·5	11·4	17·5	17·4
56	14 59·0	15 01·5	14 18·0	5·6	5·6	11·6	11·5	17·6	17·5
57	14 59·3	15 01·7	14 18·3	5·7	5·7	11·7	11·6	17·7	17·6
58	14 59·5	15 02·0	14 18·5	5·8	5·8	11·8	11·7	17·8	17·7
59	14 59·8	15 02·2	14 18·8	5·9	5·9	11·9	11·8	17·9	17·8
60	15 00·0	15 02·5	14 19·0	6·0	6·0	12·0	11·9	18·0	17·9

TABLES FOR INTERPOLATING SUNRISE, MOONRISE, ETC.

TABLE I—FOR LATITUDE

Tabular Interval 10°	5°	2°	Difference between the times for consecutive latitudes: 5m	10m	15m	20m	25m	30m	35m	40m	45m	50m	55m	60m	1h 05m	1h 10m	1h 15m	1h 20m
° ′	° ′	° ′	m	m	m	m	m	m	m	m	m	m	m	m	h m	h m	h m	h m
0 30	0 15	0 06	0	0	1	1	1	1	1	2	2	2	2	2	0 02	0 02	0 02	0 02
1 00	0 30	0 12	0	1	1	2	2	3	3	3	4	4	4	5	05	05	05	05
1 30	0 45	0 18	1	1	2	3	3	4	4	5	5	6	7	7	07	07	07	07
2 00	1 00	0 24	1	2	3	4	5	5	6	7	7	8	9	10	10	10	10	10
2 30	1 15	0 30	1	2	4	5	6	7	8	9	9	10	11	12	12	13	13	13
3 00	1 30	0 36	1	3	4	6	7	8	9	10	11	12	13	14	0 15	0 15	0 16	0 16
3 30	1 45	0 42	2	3	5	7	8	10	11	12	13	14	16	17	18	18	19	19
4 00	2 00	0 48	2	4	6	8	9	11	13	14	15	16	18	19	20	21	22	22
4 30	2 15	0 54	2	4	7	9	11	13	15	16	18	19	21	22	23	24	25	26
5 00	2 30	1 00	2	5	7	10	12	14	16	18	20	22	23	25	26	27	28	29
5 30	2 45	1 06	3	5	8	11	13	16	18	20	22	24	26	28	0 29	0 30	0 31	0 32
6 00	3 00	1 12	3	6	9	12	14	17	20	22	24	26	29	31	32	33	34	36
6 30	3 15	1 18	3	6	10	13	16	19	22	24	26	29	31	34	36	37	38	40
7 00	3 30	1 24	3	7	10	14	17	20	23	26	29	31	34	37	39	41	42	44
7 30	3 45	1 30	4	7	11	15	18	22	25	28	31	34	37	40	43	44	46	48
8 00	4 00	1 36	4	8	12	16	20	23	27	30	34	37	41	44	0 47	0 48	0 51	0 53
8 30	4 15	1 42	4	8	13	17	21	25	29	33	36	40	44	48	0 51	0 53	0 56	0 58
9 00	4 30	1 48	4	9	13	18	22	27	31	35	39	43	47	52	0 55	0 58	1 01	1 04
9 30	4 45	1 54	5	9	14	19	24	28	33	38	42	47	51	56	1 00	1 04	1 08	1 12
10 00	5 00	2 00	5	10	15	20	25	30	35	40	45	50	55	60	1 05	1 10	1 15	1 20

Table I is for interpolating the LMT of sunrise, twilight, moonrise, etc., for latitude. It is to be entered, in the appropriate column on the left, with the difference between true latitude and the nearest tabular latitude which is *less* than the true latitude; and with the argument at the top which is the nearest value of the difference between the times for the tabular latitude and the next higher one; the correction so obtained is applied to the time for the tabular latitude; the sign of the correction can be seen by inspection. It is to be noted that the interpolation is not linear, so that when using this table it is essential to take out the tabular phenomenon for the latitude *less* than the true latitude.

TABLE II—FOR LONGITUDE

Long. East or West	Difference between the times for given date and preceding date (for east longitude) or for given date and following date (for west longitude): 10m	20m	30m	40m	50m	60m	1h + 10m	1h + 20m	1h + 30m	1h + 40m	1h + 50m	1h + 60m	2h 10m	2h 20m	2h 30m	2h 40m	2h 50m	3h 00m
°	m	m	m	m	m	m	m	m	m	m	m	m	h m	h m	h m	h m	h m	h m
0	0	0	0	0	0	0	0	0	0	0	0	0	0 00	0 00	0 00	0 00	0 00	0 00
10	0	1	1	1	1	2	2	2	2	3	3	3	04	04	04	04	05	05
20	1	1	2	2	3	3	4	4	5	6	6	7	07	08	08	09	09	10
30	1	2	2	3	4	5	6	7	7	8	9	10	11	12	12	13	14	15
40	1	2	3	4	6	7	8	9	10	11	12	13	14	16	17	18	19	20
50	1	3	4	6	7	8	10	11	12	14	15	17	0 18	0 19	0 21	0 22	0 24	0 25
60	2	3	5	7	8	10	12	13	15	17	18	20	22	23	25	27	28	30
70	2	4	6	8	10	12	14	16	17	19	21	23	25	27	29	31	33	35
80	2	4	7	9	11	13	16	18	20	22	24	27	29	31	33	36	38	40
90	2	5	7	10	12	15	17	20	22	25	27	30	32	35	37	40	42	45
100	3	6	8	11	14	17	19	22	25	28	31	33	0 36	0 39	0 42	0 44	0 47	0 50
110	3	6	9	12	15	18	21	24	27	31	34	37	40	43	46	49	0 52	0 55
120	3	7	10	13	17	20	23	27	30	33	37	40	43	47	50	53	0 57	1 00
130	4	7	11	14	18	22	25	29	32	36	40	43	47	51	54	0 58	1 01	1 05
140	4	8	12	16	19	23	27	31	35	39	43	47	51	54	0 58	1 02	1 06	1 10
150	4	8	13	17	21	25	29	33	38	42	46	50	0 54	0 58	1 03	1 07	1 11	1 15
160	4	9	13	18	22	27	31	36	40	44	49	53	0 58	1 02	1 07	1 11	1 16	1 20
170	5	9	14	19	24	28	33	38	42	47	52	57	1 01	1 06	1 11	1 16	1 20	1 25
180	5	10	15	20	25	30	35	40	45	50	55	60	1 05	1 10	1 15	1 20	1 25	1 30

Table II is for interpolating the LMT of moonrise, moonset and the Moon's meridian passage for longitude. It is entered with longitude and with the difference between the times for the given date and for the preceding date (in east longitudes) or following date (in west longitudes). The correction is normally *added* for west longitudes and *subtracted* for east longitudes, but if, as occasionally happens, the times become earlier each day instead of later, the signs of the corrections must be reversed.

GPO U.S. GOVERNMENT PRINTING OFFICE: 2008—343-392

INDEX TO SELECTED STARS, 2009

Name	No	Mag	SHA	Dec
			°	°
Acamar	**7**	3·2	315	S 40
Achernar	**5**	0·5	335	S 57
Acrux	**30**	1·3	173	S 63
Adhara	**19**	1·5	255	S 29
Aldebaran	**10**	0·9	291	N 17
Alioth	**32**	1·8	166	N 56
Alkaid	**34**	1·9	153	N 49
Al Na'ir	**55**	1·7	28	S 47
Alnilam	**15**	1·7	276	S 1
Alphard	**25**	2·0	218	S 9
Alphecca	**41**	2·2	126	N 27
Alpheratz	**1**	2·1	358	N 29
Altair	**51**	0·8	62	N 9
Ankaa	**2**	2·4	353	S 42
Antares	**42**	1·0	112	S 26
Arcturus	**37**	0·0	146	N 19
Atria	**43**	1·9	108	S 69
Avior	**22**	1·9	234	S 60
Bellatrix	**13**	1·6	279	N 6
Betelgeuse	**16**	Var.*	271	N 7
Canopus	**17**	−0·7	264	S 53
Capella	**12**	0·1	281	N 46
Deneb	**53**	1·3	50	N 45
Denebola	**28**	2·1	183	N 15
Diphda	**4**	2·0	349	S 18
Dubhe	**27**	1·8	194	N 62
Elnath	**14**	1·7	278	N 29
Eltanin	**47**	2·2	91	N 51
Enif	**54**	2·4	34	N 10
Fomalhaut	**56**	1·2	15	S 30
Gacrux	**31**	1·6	172	S 57
Gienah	**29**	2·6	176	S 18
Hadar	**35**	0·6	149	S 60
Hamal	**6**	2·0	328	N 24
Kaus Australis	**48**	1·9	84	S 34
Kochab	**40**	2·1	137	N 74
Markab	**57**	2·5	14	N 15
Menkar	**8**	2·5	314	N 4
Menkent	**36**	2·1	148	S 36
Miaplacidus	**24**	1·7	222	S 70
Mirfak	**9**	1·8	309	N 50
Nunki	**50**	2·0	76	S 26
Peacock	**52**	1·9	53	S 57
Pollux	**21**	1·1	244	N 28
Procyon	**20**	0·4	245	N 5
Rasalhague	**46**	2·1	96	N 13
Regulus	**26**	1·4	208	N 12
Rigel	**11**	0·1	281	S 8
Rigil Kentaurus	**38**	−0·3	140	S 61
Sabik	**44**	2·4	102	S 16
Schedar	**3**	2·2	350	N 57
Shaula	**45**	1·6	96	S 37
Sirius	**18**	−1·5	259	S 17
Spica	**33**	1·0	159	S 11
Suhail	**23**	2·2	223	S 43
Vega	**49**	0·0	81	N 39
Zubenelgenubi	**39**	2·8	137	S 16

No	Name	Mag	SHA	Dec
			°	°
1	*Alpheratz*	2·1	358	N 29
2	*Ankaa*	2·4	353	S 42
3	*Schedar*	2·2	350	N 57
4	*Diphda*	2·0	349	S 18
5	*Achernar*	0·5	335	S 57
6	*Hamal*	2·0	328	N 24
7	*Acamar*	3·2	315	S 40
8	*Menkar*	2·5	314	N 4
9	*Mirfak*	1·8	309	N 50
10	*Aldebaran*	0·9	291	N 17
11	*Rigel*	0·1	281	S 8
12	*Capella*	0·1	281	N 46
13	*Bellatrix*	1·6	279	N 6
14	*Elnath*	1·7	278	N 29
15	*Alnilam*	1·7	276	S 1
16	*Betelgeuse*	Var.*	271	N 7
17	*Canopus*	−0·7	264	S 53
18	*Sirius*	−1·5	259	S 17
19	*Adhara*	1·5	255	S 29
20	*Procyon*	0·4	245	N 5
21	*Pollux*	1·1	244	N 28
22	*Avior*	1·9	234	S 60
23	*Suhail*	2·2	223	S 43
24	*Miaplacidus*	1·7	222	S 70
25	*Alphard*	2·0	218	S 9
26	*Regulus*	1·4	208	N 12
27	*Dubhe*	1·8	194	N 62
28	*Denebola*	2·1	183	N 15
29	*Gienah*	2·6	176	S 18
30	*Acrux*	1·3	173	S 63
31	*Gacrux*	1·6	172	S 57
32	*Alioth*	1·8	166	N 56
33	*Spica*	1·0	159	S 11
34	*Alkaid*	1·9	153	N 49
35	*Hadar*	0·6	149	S 60
36	*Menkent*	2·1	148	S 36
37	*Arcturus*	0·0	146	N 19
38	*Rigil Kentaurus*	−0·3	140	S 61
39	*Zubenelgenubi*	2·8	137	S 16
40	*Kochab*	2·1	137	N 74
41	*Alphecca*	2·2	126	N 27
42	*Antares*	1·0	112	S 26
43	*Atria*	1·9	108	S 69
44	*Sabik*	2·4	102	S 16
45	*Shaula*	1·6	96	S 37
46	*Rasalhague*	2·1	96	N 13
47	*Eltanin*	2·2	91	N 51
48	*Kaus Australis*	1·9	84	S 34
49	*Vega*	0·0	81	N 39
50	*Nunki*	2·0	76	S 26
51	*Altair*	0·8	62	N 9
52	*Peacock*	1·9	53	S 57
53	*Deneb*	1·3	50	N 45
54	*Enif*	2·4	34	N 10
55	*Al Na'ir*	1·7	28	S 47
56	*Fomalhaut*	1·2	15	S 30
57	*Markab*	2·5	14	N 15

*0·1 — 1·2

ALTITUDE CORRECTION TABLES 0°–35°— MOON

App. Alt.	0°–4° Corrⁿ	5°–9° Corrⁿ	10°–14° Corrⁿ	15°–19° Corrⁿ	20°–24° Corrⁿ	25°–29° Corrⁿ	30°–34° Corrⁿ	App. Alt.
′	° ′	° ′	° ′	° ′	° ′	° ′	° ′	′
00	**0** 34·5	**5** 58·2	**10** 62·1	**15** 62·8	**20** 62·2	**25** 60·8	**30** 58·9	**00**
10	36·5	58·5	62·2	62·8	62·2	60·8	58·8	**10**
20	38·3	58·7	62·2	62·8	62·1	60·7	58·8	**20**
30	40·0	58·9	62·3	62·8	62·1	60·7	58·7	**30**
40	41·5	59·1	62·3	62·8	62·0	60·6	58·6	**40**
50	42·9	59·3	62·4	62·7	62·0	60·6	58·5	**50**
00	**1** 44·2	**6** 59·5	**11** 62·4	**16** 62·7	**21** 62·0	**26** 60·5	**31** 58·5	**00**
10	45·4	59·7	62·4	62·7	61·9	60·4	58·4	**10**
20	46·5	59·9	62·5	62·7	61·9	60·4	58·3	**20**
30	47·5	60·0	62·5	62·7	61·9	60·3	58·2	**30**
40	48·4	60·2	62·5	62·7	61·8	60·3	58·2	**40**
50	49·3	60·3	62·6	62·7	61·8	60·2	58·1	**50**
00	**2** 50·1	**7** 60·5	**12** 62·6	**17** 62·7	**22** 61·7	**27** 60·1	**32** 58·0	**00**
10	50·8	60·6	62·6	62·6	61·7	60·1	57·9	**10**
20	51·5	60·7	62·6	62·6	61·6	60·0	57·8	**20**
30	52·2	60·9	62·7	62·6	61·6	59·9	57·8	**30**
40	52·8	61·0	62·7	62·6	61·6	59·9	57·7	**40**
50	53·4	61·1	62·7	62·6	61·5	59·8	57·6	**50**
00	**3** 53·9	**8** 61·2	**13** 62·7	**18** 62·5	**23** 61·5	**28** 59·7	**33** 57·5	**00**
10	54·4	61·3	62·7	62·5	61·4	59·7	57·4	**10**
20	54·9	61·4	62·7	62·5	61·4	59·6	57·4	**20**
30	55·3	61·5	62·8	62·5	61·3	59·5	57·3	**30**
40	55·7	61·6	62·8	62·4	61·3	59·5	57·2	**40**
50	56·1	61·6	62·8	62·4	61·2	59·4	57·1	**50**
00	**4** 56·4	**9** 61·7	**14** 62·8	**19** 62·4	**24** 61·2	**29** 59·3	**34** 57·0	**00**
10	56·8	61·8	62·8	62·4	61·1	59·3	56·9	**10**
20	57·1	61·9	62·8	62·3	61·1	59·2	56·9	**20**
30	57·4	61·9	62·8	62·3	61·0	59·1	56·8	**30**
40	57·7	62·0	62·8	62·3	61·0	59·1	56·7	**40**
50	58·0	62·1	62·8	62·2	60·9	59·0	56·6	**50**

HP	L	U	L	U	L	U	L	U	L	U	L	U	L	U	HP
′	′	′	′	′	′	′	′	′	′	′	′	′	′	′	′
54·0	0·3	0·9	0·3	0·9	0·4	1·0	0·5	1·1	0·6	1·2	0·7	1·3	0·9	1·5	**54·0**
54·3	0·7	1·1	0·7	1·2	0·8	1·2	0·8	1·3	0·9	1·4	1·1	1·5	1·2	1·7	**54·3**
54·6	1·1	1·4	1·1	1·4	1·1	1·4	1·2	1·5	1·3	1·6	1·4	1·7	1·5	1·8	**54·6**
54·9	1·4	1·6	1·5	1·6	1·5	1·6	1·6	1·7	1·6	1·8	1·8	1·9	1·9	2·0	**54·9**
55·2	1·8	1·8	1·8	1·8	1·9	1·8	1·9	1·9	2·0	2·0	2·1	2·1	2·2	2·2	**55·2**
55·5	2·2	2·0	2·2	2·0	2·3	2·1	2·3	2·1	2·4	2·2	2·4	2·3	2·5	2·4	**55·5**
55·8	2·6	2·2	2·6	2·2	2·6	2·3	2·7	2·3	2·7	2·4	2·8	2·4	2·9	2·5	**55·8**
56·1	3·0	2·4	3·0	2·5	3·0	2·5	3·0	2·5	3·1	2·6	3·1	2·6	3·2	2·7	**56·1**
56·4	3·3	2·7	3·4	2·7	3·4	2·7	3·4	2·7	3·4	2·8	3·5	2·8	3·5	2·9	**56·4**
56·7	3·7	2·9	3·7	2·9	3·8	2·9	3·8	2·9	3·8	3·0	3·8	3·0	3·9	3·0	**56·7**
57·0	4·1	3·1	4·1	3·1	4·1	3·1	4·1	3·1	4·2	3·2	4·2	3·2	4·2	3·2	**57·0**
57·3	4·5	3·3	4·5	3·3	4·5	3·3	4·5	3·3	4·5	3·3	4·5	3·4	4·6	3·4	**57·3**
57·6	4·9	3·5	4·9	3·5	4·9	3·5	4·9	3·5	4·9	3·5	4·9	3·5	4·9	3·6	**57·6**
57·9	5·3	3·8	5·3	3·8	5·2	3·8	5·2	3·7	5·2	3·7	5·2	3·7	5·2	3·7	**57·9**
58·2	5·6	4·0	5·6	4·0	5·6	4·0	5·6	4·0	5·6	3·9	5·6	3·9	5·6	3·9	**58·2**
58·5	6·0	4·2	6·0	4·2	6·0	4·2	6·0	4·2	6·0	4·1	5·9	4·1	5·9	4·1	**58·5**
58·8	6·4	4·4	6·4	4·4	6·4	4·4	6·3	4·4	6·3	4·3	6·3	4·3	6·2	4·2	**58·8**
59·1	6·8	4·6	6·8	4·6	6·7	4·6	6·7	4·6	6·7	4·5	6·6	4·5	6·6	4·4	**59·1**
59·4	7·2	4·8	7·1	4·8	7·1	4·8	7·1	4·8	7·0	4·7	7·0	4·7	6·9	4·6	**59·4**
59·7	7·5	5·1	7·5	5·0	7·5	5·0	7·5	5·0	7·4	4·9	7·3	4·8	7·2	4·8	**59·7**
60·0	7·9	5·3	7·9	5·3	7·9	5·2	7·8	5·2	7·8	5·1	7·7	5·0	7·6	4·9	**60·0**
60·3	8·3	5·5	8·3	5·5	8·2	5·4	8·2	5·4	8·1	5·3	8·0	5·2	7·9	5·1	**60·3**
60·6	8·7	5·7	8·7	5·7	8·6	5·7	8·6	5·6	8·5	5·5	8·4	5·4	8·2	5·3	**60·6**
60·9	9·1	5·9	9·0	5·9	9·0	5·9	8·9	5·8	8·8	5·7	8·7	5·6	8·6	5·4	**60·9**
61·2	9·5	6·2	9·4	6·1	9·4	6·1	9·3	6·0	9·2	5·9	9·1	5·8	8·9	5·6	**61·2**
61·5	9·8	6·4	9·8	6·3	9·7	6·3	9·7	6·2	9·5	6·1	9·4	5·9	9·2	5·8	**61·5**

DIP

Ht. of Eye	Corrⁿ	Ht. of Eye	Ht. of Eye	Corrⁿ	Ht. of Eye
m		ft.	m		ft.
2·4		8·0	9·5		31·5
	−2·8			−5·5	
2·6		8·6	9·9		32·7
	−2·9			−5·6	
2·8		9·2	10·3		33·9
	−3·0			−5·7	
3·0		9·8	10·6		35·1
	−3·1			−5·8	
3·2		10·5	11·0		36·3
	−3·2			−5·9	
3·4		11·2	11·4		37·6
	−3·3			−6·0	
3·6		11·9	11·8		38·9
	−3·4			−6·1	
3·8		12·6	12·2		40·1
	−3·5			−6·2	
4·0		13·3	12·6		41·5
	−3·6			−6·3	
4·3		14·1	13·0		42·8
	−3·7			−6·4	
4·5		14·9	13·4		44·2
	−3·8			−6·5	
4·7		15·7	13·8		45·5
	−3·9			−6·6	
5·0		16·5	14·2		46·9
	−4·0			−6·7	
5·2		17·4	14·7		48·4
	−4·1			−6·8	
5·5		18·3	15·1		49·8
	−4·2			−6·9	
5·8		19·1	15·5		51·3
	−4·3			−7·0	
6·1		20·1	16·0		52·8
	−4·4			−7·1	
6·3		21·0	16·5		54·3
	−4·5			−7·2	
6·6		22·0	16·9		55·8
	−4·6			−7·3	
6·9		22·9	17·4		57·4
	−4·7			−7·4	
7·2		23·9	17·9		58·9
	−4·8			−7·5	
7·5		24·9	18·4		60·5
	−4·9			−7·6	
7·9		26·0	18·8		62·1
	−5·0			−7·7	
8·2		27·1	19·3		63·8
	−5·1			−7·8	
8·5		28·1	19·8		65·4
	−5·2			−7·9	
8·8		29·2	20·4		67·1
	−5·3			−8·0	
9·2		30·4	20·9		68·8
	−5·4			−8·1	
9·5		31·5	21·4		70·5

MOON CORRECTION TABLE

The correction is in two parts; the first correction is taken from the upper part of the table with argument apparent altitude, and the second from the lower part, with argument HP, in the same column as that from which the first correction was taken. Separate corrections are given in the lower part for lower (L) and upper (U) limbs. All corrections are to be **added** to apparent altitude, *but 30′ is to be subtracted from the altitude of the upper limb.*

For corrections for pressure and temperature see page A4.

For bubble sextant observations ignore dip, take the mean of upper and lower limb corrections and subtract 15′ from the altitude.

App. Alt. = Apparent altitude = Sextant altitude corrected for index error and dip.